MW01485138

HANDBOOK OF
HUMAN FACTORS

HANDBOOK OF HUMAN FACTORS

Edited by
GAVRIEL SALVENDY
Purdue University

A Wiley-Interscience Publication
JOHN WILEY & SONS
New York · Chichester · Brisbane · Toronto · Singapore

Library of Congress Cataloging-in-Publication Data:

Handbook of human factors.

  "A Wiley-Interscience publication."
  1. Human engineering—Handbooks, manuals, etc.
I. Salvendy, Gavriel, 1938–
TA166.H275   1986   620.8'2   86-9083
ISBN 0-471-88015-9

Printed in the United States of America

10 9 8 7 6 5 4

ADVISORY BOARD

CONTRIBUTORS

Earl A. Alluisi
Chief Scientist
Air Force Human Resources
    Laboratory
Brooks Air Force Base, Texas

Lynn Y. Arnaut
Doctoral Candidate
Department of Industrial Engineering
    and Operations Research
Virginia Polytechnic Institute and State
    University
Blacksburg, Virginia

M. M. Ayoub
Horn Professor of Industrial
    Engineering and Biomedical
    Engineering
Texas Technological University
Lubbock, Texas

Woodrow Barfield
Doctoral Candidate
School of Industrial Engineering
Purdue University
West Lafayette, Indiana

Dennis B. Beringer
Assistant Professor
University of Wisconsin-Madison
Madison, Wisconsin

Nicholas A. Bond, Jr.
Director
Office of Naval Research
Science Liaison Office Far East
Tokyo Akasaka Center
APO San Francisco, California

Donald M. Broadbent
External Staff Scientist
Department of Experimental
    Psychology
University of Oxford
Oxford, England

John F. Brock
Manager, Man-Machine Science
Honeywell Systems and Research
    Center
Minneapolis, Minnesota

Hans-Jörg Bullinger
Professor Dr.-Ing. habil and Director
Fraunhofer-Institut für
    Arbeitswirtschaft und Organization
    (IAO)
Universität Stuttgart
Stuttgart, West Germany

Don B. Chaffin
Director, Center for Ergonomics and
Professor of Industrial and Operations
    Engineering
University of Michigan
Ergonomics Center
Ann Arbor, Michigan

Tien Chien-Chang
Assistant Professor
School of Industrial Engineering
Purdue University
West Lafayette, Indiana

Julien M. Christensen
Chief Scientist: Human Factors and
    Logistics
Universal Energy Systems
Dayton, Ohio

Gerald P. Chubb
Principal Consultant
SofTech, Inc.
Fairborn, Ohio

E. Nigel Corlett
Professor and Head
Department of Production Engineering
  and Production Management
University of Nottingham
Nottingham, England

Brian Crist
Ergonomic Consultant
Human Factors Group
Rochester, New York

William H. Cushman
Ergonomic Consultant
Human Factors Group
Rochester, New York

Sara J. Czaja
Visiting Associate Professor
Department of Industrial Engineering
State University of New York at Buffalo
Amherst, New York

Louis E. Davis
Professor of Organizational Sciences
Graduate School of Management and
  Chairman of the Center for Quality
  of Working Life
Institute of Industrial Relations
University of California
Los Angeles, California

Colin G. Drury
Professor
Department of Industrial Engineering
State University of New York at Buffalo
Amherst, New York

Ray E. Eberts
Assistant Professor
School of Industrial Engineering
Purdue University
West Lafayette, Indiana

Ward Edwards
Professor and Director
Social Science Research Institute
University of Southern California
University Park
Los Angeles, California

Jay Elkerton
Doctoral Candidate
Virginia Polytechnic Institute and State
  University
Blacksburg, Virginia

John R. Etherton
Safety Engineer
National Institute for Occupational
  Safety and Health
Division of Safety Research
Morgantown, West Virginia

Donald L. Fisher
Associate Professor
Department of Industrial Engineering
  and Operations Research
University of Massachusetts
Amherst, Massachusetts

Ralph E. Flexman
Professor Emeritus of Aviation
University of Illinois
Champaign-Urbana, Illinois

Patrick J. Foley
Professor
Department of Industrial Engineering
University of Toronto
Toronto, Ontario, Canada

King-Sun Fu
Goss Distinguished Professor of
  Engineering
School of Electrical Engineering
Purdue University
West Lafayette, Indiana

Chaya Garg-Janardan
Doctoral Candidate
School of Industrial Engineering
Purdue University
West Lafayette, Indiana

Irwin L. Goldstein
Professor and Chair
Department of Psychology
University of Maryland
College Park, Maryland

Etienne Grandjean
Professor Emeritus and Former
  Director of the Department of
  Ergonomics
Swiss Federal Institute of Technology
Zurich, Switzerland

Joel S. Greenstein
Professor
Department of Industrial Engineering
Clemson University
Clemson, South Carolina

Susan M. Grey
Research Associate
Department of Production Engineering
  and Production Management
University of Nottingham
Nottingham, England

Lewis F. Hanes
Manager, Human Sciences
Westinghouse R & D Center
Pittsburgh, Pennsylvania

John E. Harrigan
Professor
Architecture and Environmental Design
California Polytechnic State University
San Luis Obispo, California

Martin G. Helander
Associate Professor
Department of Industrial Engineering
State University of New York at Buffalo
Amherst, New York

Hal W. Hendrick
Professor and Chairman
Department of Human Factors
Institute of Safety and Systems
  Management
University of Southern California
Los Angeles, California

Ronald A. Hess
Professor
Department of Mechanical Engineering
University of California
Davis, California

Dennis H. Holding
Professor
Department of Psychology
University of Louisville
Louisville, Kentucky

Carl Graf Hoyos
Professor and Head,
Department of Psychology
Technical University of Munich
Munich, West Germany

Stacy R. Hunt
Manager, Human Factors Department
General Electric Corporation
King of Prussia, Pennsylvania

Bernard C. Jiang
Research Assistant
Department of Industrial Engineering
Texas Technological University
Lubbock, Texas

Barry L. Johnson
Director
Division of Biomedical and Behavioral
  Science
National Institute for Occupational
  Safety and Health
Cincinnati, Ohio

Dylan M. Jones
Senior Lecturer
Department of Applied Psychology
University of Wales Institute of Science
  and Technology
Cardiff, Wales

Barry H. Kantowitz
Professor
Department of Psychological Sciences
Purdue University
West Lafayette, Indiana

Peter Kern
Head of Department of Ergonomics
Fraunhofer-Institut für
    Arbeitswirtschaft und Organization
    (IAO)
Universität Stuttgart
Stuttgart, Germany

James L. Knight, Jr.
Senior Research Psychologist
AT&T Consumer Products Laboratory
Neptune, New Jersey

Koji Kobayashi
Chairman of the Board and Chief
    Executive Officer
NEC Corporation
Tokyo, Japan

Stephan A. Konz
Professor
Department of Industrial Engineering
Kansas State University
Manhattan, Kansas

Volker Korndörfer
Head, Robotics Department
Fraunhofer-Institut für
    Arbeitswirtschaft und Organization
    (IAO)
Universität Stuttgart
Stuttgart, West Germany

Karl H. E. Kroemer
Professor and Director
Ergonomics Laboratory
Industrial Engineering and Operations
    Research Department
Virginia Polytechnic Institute and State
    University
Blacksburg, Virginia

Patrick C. Kyllonen
Personnel Research Psychologist
Air Force Human Resources
    Laboratory
Brooks Air Force Base, Texas

K. Ronald Laughery, Jr.
President
Mirco Analysis and Design
Boulder, Colorado

Kenneth R. Laughery, Sr.
Luce Professor
Psychology Department
Rice University
Houston, Texas

David R. Lenorovitz
Senior Staff Human Factors Engineer
Computer Technology Associates, Inc.
Englewood, Colorado

Michael E. McCauley
Vice President
Monterey Technologies, Inc.
Carmel, California

Ann Majchrzak
Assistant Professor
Organizational Behavior and Human
    Resources
Krannert Graduate School of
    Management
Purdue University
West Lafayette, Indiana

Robert W. Mason
Technical Advisor
Division of Biomedical and Behavioral
    Sciences
National Institute for Occupational
    Safety and Health
Cincinnati, Ohio

David Meister
Senior Scientist
Human Factors Program
U.S. Navy Personnel Research and
    Development Center
San Diego, California

Dwight P. Miller
Member of Technical Staff
Statistics, Computing, and Human
    Factors Division 7223
Sandia National Laboratories
Albuquerque, New Mexico

Timothy H. Monk
Assistant Professor
Institute of Chronobiology
Cornell Medical Center-Westchester
White Plains, New York

John B. Moran
Director
Division of Safety Research
National Institute for Occupational
  Safety and Health
Morgantown, West Virginia

Neville D. Moray
Professor
Departments of Industrial Engineering
  and Psychology
University of Toronto
Toronto, Ontario
Canada

Werner F. Muntzinger
Senior Scientist
Fraunhofer-Institut für
  Arbeitswirtschaft und Organization
  (IAO)
Universität Stuttgart
Stuttgart, Germany

Shimon Y. Nof
Associate Professor
School of Industrial Engineering
Purdue University
West Lafayette, Indiana

John F. O'Brien
Electric Power Research Institute
Palo Alto, California

Hobart G. Osburn
Professor
Department of Psychology
University of Houston
Houston, Texas

Masamitsu Oshima
President and Chairman of the Board
The Medical Information System
  Development Center
Tokyo, Japan

Barbara Paramore
Senior Scientist
Essex Corporation
Alexandria, Virginia

Mark D. Phillips
Senior Staff Engineer
Computer Technologies Associates
Englewood, Colorado

A. Alan B. Pritsker, President
Pritsker and Associates
West Lafayette, Indiana

Teresa L. Roberts
Senior Scientist
Xerox Office Systems Division
Palo Alto, California

Frederick H. Rohles, Director
Institute for Environmental Research
Kansas State University
Manhattan, Kansas

Walter Rohmert
Professor and Head
Institut für Arbeitswissenschaft der
  Technischen
Hochscule Darmstadt
Darmstadt, West Germany

Ellen F. Roland
Senior Associate
Roland and Associates Corporation
Monterey, California

Richard M. Ronk
Senior Researcher
Division of Safety Research
National Institute for Occupational
  Safety and Health
Morgantown, West Virginia

John C. Ruth
Vice President, Marketing
McDonnell-Douglas Electronics
  Company
St. Charles, Missouri

Gavriel Salvendy
NEC Professor of Industrial
  Engineering
Purdue University
West Lafayette, Indiana

Richard Schweickert
Associate Professor
Department of Psychological Sciences
Purdue University
West Lafayette, Indiana

Stanley E. Seashore
Program Director Emeritus
Institute of Social Research
University of Michigan
Ann Arbor, Michigan

Joseph L. Selen
Doctoral Candidate
Department of Industrial Engineering
Texas Technological University
Lubbock, Texas

Tapas K. Sen
Division Manager
Work Relations Research
AT & T
New York, New York

Joseph Sharit
Assistant Professor
Department of Industrial Engineering
State University of New York at Buffalo
Amherst, New York

Sylvia B. Sheppard
Principal Scientist
Computer Technology Associates
Lanham, Maryland

Thomas B. Sheridan
Professor
Department of Mechanical Engineering
Massachusetts Institute of Technology
Cambridge, Massachusetts

Ben Shneiderman
Associate Professor
Department of Computer Science
University of Maryland
College Park, Maryland

Carol A. Simpson
Co-Owner
Psycho-Linguistic Research Associates
Menlo Park, California

Arnold M. Small
Professor Emeritus
Department of Human Factors
Institute of Safety
University of Southern California
Fullerton, California

Karl U. Smith
Professor Emeritus
Department of Psychology
University of Wisconsin
Madison, Wisconsin

Michael J. Smith
Associate Professor
Department of Industrial Engineering
University of Wisconsin
Madison, Wisconsin

Thomas J. Smith
Assistant Professor
Occupational Science Program
Department of Kinesiology
Simon Fraser University
Burnaby, British Columbia, Canada

Robert D. Sorkin
Professor
Department of Psychological Sciences
Purdue University
West Lafayette, Indiana

Edward A. Stark
Senior Staff Scientist
Link Flight Simulation Division
The Singer Company
Binghamton, New York

Alan D. Swain
Distinguished Member of the Technical
    Staff
Statistics, Computing and Human
    Factors Division
Sandia National Laboratories
Albuquerque, New Mexico

Robert W. Swezey
Director
Behavioral Sciences Research Center
Science Applications, Inc.
McLean, Virginia

Donald I. Tepas
Professor
Department of Psychology
The University of Connecticut
Storrs, Connecticut

Harold P. Van Cott
Vice President and Director
Systems Operability and Design Group
Essex Corporation
Alexandria, Virginia

Gerald J. Wacker
Manager of Office Systems
Xerox Corporation
El Segundo, California

Donald E. Wasserman
Director of Engineering and Operations
National Center for Rehabilitation
   Engineering
Wright State University
Dayton, Ohio

Mark Weiser
Associate Professor
Department of Computer Science
University of Maryland
College Park, Maryland

Christopher D. Wickens
Professor
Institute of Aviation
Aviation Research Laboratory
University of Illinois-Willard Airport
Savoy, Illinois

Beverly H. Williges
Senior Research Associate
Department of Industrial Engineering
   and Operations Research
Virginia Polytechnic Institute and State
   University
Blacksburg, Virginia

Robert C. Williges
Professor
Department of Industrial Engineering
   and Operations Research
Virginia Polytechnic Institute and State
   University
Blacksburg, Virginia

David D. Woods
Senior Engineer
Human Sciences Department
Westinghouse R & D Center
Pittsburgh, Pennsylvania

Bernhard Zimolong
Professor
Psychologisches Institut
Bochum, West Germany

FOREWORD

The *Handbook of Human Factors* is being published at a time when all phases of the U.S. economy are undergoing great technological change. All sections of the economy—manufacturing, service industries, agriculture, education, communications, and so on—are faced today with the toughest competition in our history. This competition is not only internal but worldwide. The technological edge that the United States had in most all phases of the economy in the 1950s and 1960s has been reduced dramatically in the 1970s and early 1980s.

We are now, along with our worldwide competitors, in a race to take advantage of the technological breakthroughs that have occurred, mainly in computer science and communications. These changes are occurring particularly in manufacturing, service industries, and communication, and are requiring major changes in the role of people.

This revolution has increased the necessity for more emphasis on the relationship between all workers from the top executives down to the manufacturing, service, and sales operators. Although the difference between success and failure in any endeavor has always been the people involved, today the human factor is even more important.

The publication of the *Handbook of Human Factors* is therefore particularly timely. Regardless of what phase of the economy a person is involved in, this handbook is a very useful tool. Every area of human factors from environmental conditions and motivation to the use of new communications systems, robotics, and business systems is well covered in the handbook by experts in every field.

E. M. ESTES
Retired President
General Motors Corporation

PREFACE

This handbook is concerned with the role of humans in complex systems, the design of equipment and facilities for human use, and the development of environments for comfort and safety. Thus this handbook provides vital information about the effective design and use of systems requiring the interaction among human, machine (computer), and environment.

In a literal sense, human factors is as old as the machine and the environment, for it was aimed at designing them for human use. However, it was not until World War II that human factors has emerged as a separate discipline.

Although I have referred to this discipline as human factors it is appropriate to point out that there is a large body of practitioners and researchers here and abroad who refer to this discipline as "ergonomics." For example, the U.S. national organization to which members of this discipline belong to is called the Human Factors Society but the international society to which it belongs to is called the International Ergonomics Association.

The field of human factors has developed and broadened considerably since its formal inception more than 40 years ago and has generated a body of knowledge in the following broad areas of specialization:

The human factors function
Human factors fundamentals
Functional analysis
Job and organizational design
Environmental equipment and workplace design
Environmental design
Design for health and safety
Design of selection and training systems
Performance modeling
Human factors in the design and use of computing systems
Applications of human factors in computer systems

The foregoing list shows how broad the field has become. It shows, too, that human factors has broadened its concern to include many highly relevant management issues and questions such as design and utilization of artificial intelligence and expert systems and the optimal utilization of human resources. As such, the handbook should be of value to all human factors specialists, engineers, applied and experimental psychologists, and computer scientists.

Such a breadth of subject matter presents a serious challenge to represent successfully the entire field of human factors in a single handbook. I did not believe in

1983, when this all began, that any one person could properly select the subjects to be included in the handbook without serious distortions to fit his or her own particular area of knowledge and bias. Accordingly, an advisory board composed of experts in the more important areas of human factors was invited to advise the editor in planning the contents of the handbook. The advisory board members are listed on page v. I sincerely appreciate their excellent counsel and advice during the preparation of the handbook. Nevertheless, any sampling deficiencies that remain are of course my own responsibilities.

The 5403 manuscript pages of the 66 chapters constituting the handbook were written by 103 people. In creating this handbook the authors gathered information from 3850 references and presented 316 tables and 569 figures to provide theoretically based and practically oriented material for use by both practitioners and researchers.

As should be apparent from the perusal of the chapters of the handbook, we have made every effort to obtain authors with diverse training and professional affiliations from the United States and other countries throughout the world. Each author contributing to this handbook was given the following objectives:

1. The handbook should serve the following people:
   a. Practicing human factors engineers
   b. Nonhuman factors type engineers.
2. The handbook should serve the following types of organizations:
   a. The very small, medium, and large corporations.
   b. Continuous process and discrete part manufacturing industries.
   c. Service industries including hospitals; banking; insurance; post offices; hotels, motels, and restaurants; armed services; local, state, and federal government; universities; distribution; marketing; labor relations; and the legal profession.
3. It is planned that six notions will run, as a theme, through the handbook:
   a. Techniques and methods that will assist human factors specialist in the design and operation of a work environment requiring increased interaction between humans and "intelligent" machines.
   b. "Human use of human beings" can lead to greater productivity and is in fact compatible with profit motive.
   c. The past half century has seen the effective application of techniques that subdivide activities to improve operations. The next decade will see a sharp emphasis on the study of "total systems" in order to optimize operations through the integration of subsystems or parallel systems. The handbook will describe this expansion of analytical capability. The need for maintaining both approaches will be indicated because they are complementary and not mutually exclusive concepts.
   d. Because the purpose of the handbook is the application of knowledge to solving real-world problems, it will present all available useful tables, graphs, nomographs, and formulas pertaining to the application and use of human factors methodologies. The scope and limitation of each methodology will be reviewed and a step-by-step description of its use will be illustrated.
   e. Examples will be utilized to demonstrate the use of various methodolo-

gies. From these examples the reader may draw inferences on the use of the presented methodology to specific work situations.

   f. Because human factors and its methodologies are practiced and utilized in both manufacturing and service industries, it is imperative to recognize both these areas during the writing of the various chapters.

The many contributing authors came through magnificently. I thank them all most sincerely for agreeing so willingly to create this handbook with me.

Each chapter submitted was peer reviewed. The following individuals kindly contributed to the review process:

John Boose	Karl H. E. Kroemer
Donald E. Broadbent	Kenneth R. Laughery, Sr.
E. Nigel Corlett	K. Ronald Laughery, Jr.
Colin G. Drury	Ernest J. McCormick
Hubert E. Dunsmore	David Meister
Cindelyn Eberts	Neville Mcray
Ray Eberts	Robert Murray
King-Sun Fu	Benjamin W. Niebel
Mikell Groover	Shimon Y. Nof
Peter A. Hancock	Herschel C. Self
Lewis F. Hanes	Dennis R. Short
Martin G. Helander	Michael J. Smith
Hal W. Hendrick	Robert D. Sorkin
Richard S. Hirsch	Robert C. Williges
Carl G. Hoyos	Neil J. Zimmerman
Stephan A. Konz	

The index for the Handbook was prepared by the most able Cindelyn Eberts of West Lafayette, Indiana, in cooperation with all of the chapter authors.

I had the privilege of working first with Thurman R. Poston, and subsequently with James L. Smith, our Wiley editors, who significantly facilitated my editorial work. I was fortunate to have during the preparation of this handbook the most able assistance of my secretaries, Loretta Bowman and Diana Huffer, in carrying out a diversified set of administrative and secretarial work.

Finally, I was most pleased to have the opportunity of working with my daughter, Laura, on selected aspects of the handbook—it was a most enjoyable experience. My wife, Catherine, my parents, Paul and Katarina, and my children, Laura and Kevin, have made the preparation of this handbook possible. My love and sincere thanks go to them.

GAVRIEL SALVENDY

West Lafayette, Indiana
August 1986

CONTENTS

HANDBOOK OF
HUMAN FACTORS

PART 1
THE HUMAN FACTORS FUNCTION

CHAPTER 1.1

THE HUMAN FACTORS PROFESSION

JULIEN M. CHRISTENSEN

Universal Energy System, Inc., Dayton, Ohio

1.1.1 ORIGINS AND DEVELOPMENT

The concept of human factors is as old as mankind; in fact, it may have marked the beginning of mankind (Dart, 1959; Clark, 1960; Oakley, 1952). Dart (1959) tells us that ". . . long before he knew how to fashion weapons and tools from stone, man had discovered another and livelier material for his primitive skill." The first tools may well have been fashioned from osteodontokeratic materials and not from wood and stone. The point is that the pebble tools that *Australopithecus* Prometheus selected and the scoops that he fashioned from the cannon bones of antelope constitute clear evidence of specific, intelligent reactions to an environment. In their interactions with environmental features people had recognized that their effectiveness could be increased significantly by even slight modifications in those features. Improved controls and displays—or "knobs and dials," as they are often termed—are modern examples of the profound effects that relatively minor changes in the environment can sometimes have on performance.

1.1.1.1 The Age of Tools

In retrospect, progress during the Age of Tools was agonizingly slow. This may have been due to early man's residence in a benevolent climate that required only simple shelters and provided him with relatively ample supplies of food. Be that as it may, let us skip forward a few log units in time and examine the period that is often termed the Industrial Revolution. Christensen (1962), following the lead of an anonymous author (1960), uses the term "revolutions" because there appear to have been three fairly distinct phases to this period. The next few sections rely heavily on my previous work (Christensen, 1962).

1.1.1.2 The Age of Machines

Phase 1 of the Industrial Revolution, the Age of Machines, covered a period of approximately 120 years (1750 to 1870). La Mettrie had published his controversial book, *L'homme Machine,* in 1748. At least two lessons are to be learned from La Mettrie's writings. First, the people of today are as sensitive about being compared to machines as they were in La Mettrie's time. This automatically closes the minds of many who should listen when, for example, someone describes the human as an element in a closed-loop control system. Second, we can infer from La Mettrie's position that much can be learned about human behavior by considering how machines operate under similar circumstances. Cannot much of the renewed interest in cognitive behavior be attributed at least partly to the digital computer? Humans are "in and of" their systems; they are not apart from them.

The Age of Machines also witnessed brilliant invention in the textile industry; in 1801, for example, Jacquard employed punched cards to aid in the programming of weaving machines. The hazards that had been associated with steam engines were significantly reduced when Watt designed the self-regulating governor. (This invention marked the beginning of automation and cybernetics.)

Fitts (1963), drawing upon the 1828 charter of the Institution of Civil Engineering, reports that engineers defined their profession as "the art of directing the great sources of power in nature to the use and convenience of man." Thus even in 1828 engineering was recognized as having a significant artistic component; however, we firmly oppose their position that nature was meant to be exploited simply for the "use and convenience of man."

One hundred and fifty years later, the Honorable Brockway McMillan, then the Assistant Secretary of the Air Force for Research and Development, reemphasized the significance of the creative process in engineering. He stated, "Engineering is a creative process, one whose characteristic elements scarcely have names; they are certainly not the parts of the process usually packaged into courses and taught in the engineering curriculum. Yet until the engineer masters these characteristically creative steps he no more deserves the title 'engineer' than a plasterer deserves the title 'architect' " (McMillan, 1962).

1.1.1.3 The Age of Power

Phase 2, the Power Revolution, covered roughly the next 75 years (1870 to 1945), and was characterized by broad and varied applications of powered machinery in transportation, industry, and agriculture. During this period, a tenfold increase in the horsepower that was available to each person in the United States was realized approximately every 30 years. Pioneers from engineering and psychology made contributions that today would be included under the human factors rubric. Names such as

Frederick W. Taylor, the Gilbreths, Muensterberg, and Binet come to mind. However, the emphasis by and large was on adapting people to their work, resulting in extensive investigations in the areas of selection, classification, training, adjustment of work schedules, and so on. Taylor, however, made contributions that were clearly "human engineering" in nature. Consider, for example, his experimentation with different sized shovels.

It was also during this period that the concept of "one best way" assumed prominence. This unfortunate concept persists even today. We say "unfortunate" because of the overwhelming evidence supporting individual differences. Although there may be *one* way that is better than any other *one* way, because of the individual differences that exist among people, even this way will not be optimal for most of those engaged in a particular task. Norbert Wiener undoubtedly had this concept in mind when he declared the following:

> *It is a degradation to a human being to chain him to an oar and use him as a source of power; but it is an almost equal degradation to assign him purely repetitive tasks in a factory, which demand less than a millionth of his brainpower. But it is simpler to organize a factory or galley which uses individual human beings for a trivial fraction of their worth than it is to provide a world in which they can grow to their full stature.*
>
> WIENER, 1950.

Research that today would be considered within the province of human factors was being conducted in the late 1930s at the University of Cambridge in England and in the early 1940s at the U.S. Army Air Force's School of Aviation Medicine. However, during World War II, as in World War I, the emphasis was almost entirely on selection, classification, and training, although near the end of that conflict human factors engineering (then generally termed "engineering psychology") began to emerge as a distinct discipline. People such as Bartlett, Drew, and Craig in Great Britain and Fitts, McFarland, Williams, and Taylor in the United States pioneered the movement. In the Soviet Union the name of Lomov assumed, and retains, international prominence.

Engineering was beginning formally to recognize the broadening of its responsibilities. The official definition of industrial engineering, as adopted by the American Institute for Industrial Engineers in 1961, reads:

> *Industrial engineering is concerned with the design, improvement, and installation of integrated systems of* people, *material, equipment and energy. It draws upon specialized knowledge and skills in the mathematical, physical and* social sciences *together with the principles and methods of engineering analysis and design to specify, predict and evaluate the results to be obtained from such systems. Weston, 1961. (emphasis added).*

This definition reveals a sensitivity to the need to consider the social sciences as well as the physical sciences in the development of modern systems. In fact, the above is not a bad definition of *systems* engineering.

However, it would be misleading even to imply that human factors was immediately recognized as a profession having a significant contribution to make in the development of systems. Early practitioners can testify to the perplexity, doubt, skepticism, and even ridicule with which they were sometimes met when they first attended systems development meetings. To their credit, it was the authorities in the Department of Defense (DOD) who first actively supported the human factors profession. Even today many industries are not yet at the stage of development with respect to human factors that DOD was a generation ago.

Some of the delay in the acceptance of human factors can be laid directly at the door of the profession itself. When human factors specialists were invited to participate in the early stages of systems development, they usually found that their tools and data were not adequate. Data usually were not expressed as functions that were related or relatable to important systems criteria. For some time, human factors specialists had to be content to examine drawings and mock-ups of equipment and systems—a stage usually too late to allow fundamental changes in concept or assignment of activity. In fact, human factors specialists became known as "knobs and dials" people, probably because that area was where most of their knowledge resided and because, being in the development process so late, they could change little except a few controls and displays. This, unfortunately, is still true in much of non-DOD industry, even though the tools for handling early involvement by human factors specialists is probably now on a par with those of other disciplines (with *all* needing improvement). As stated elsewhere (Section 1.1.1.5), successful product liability suits are unquestionably hastening at least the recognition, if not wholehearted acceptance, of human factors by a significant proportion of industry. The systems model (see Section 1.1.3) provides a framework that enables those who adhere to its principles and procedures to recognize, treat, and integrate the contributions of all those who would contribute to systems design, development, and acquisition. Finally, participation in a development program that follows the systems model discloses points of weakness in the data, principles, and methods of the fields of specialization that need to be addressed.

Why was early support of human factors in the United States afforded by the Department of Defense? It may have been due to recognition of the fact that military systems were becoming so complex that conventional techniques of selection, classification, and training could not be relied upon to assure acceptable operation and maintenance of systems. It became evident to many that the capabilities and limitations of people had to be considered early in design and throughout development, acquisition, and utilization—selection and training, on the one side, and human engineering on the other were beginning to be recognized as opposite sides of the same coin. The possibilities for meaningful *trade-offs* between the two areas seem limitless, although means for treating these trade-offs quantitatively have been agonizingly slow in their development.

1.1.1.4 A New Age: Machines for Minds

The end of World War II ushered in Phase 3 of the Industrial Revolution—the Machines for Minds phase. This era has already witnessed significant progress in the application of fission, space exploration, automation, and, of course, the high-speed digital computer. In human factors, exciting advancements or promise of advancements have been made in such areas as artificial intelligence, expert systems, evoked potentials, predictor displays, and many others. In the DOD and its supporting industries, human factors has expanded well beyond the displays and controls era and entered the cognitive area, including contributions to the design of complex command, control, communications, and information centers.

Unquestionably, the greatest of all of these is the computer, for it invades the province of the human mind wherein reside such profound attributes as creativity, judgment, emotion, motivation, and so on. Those who design computers will derive considerable benefit from increased understanding of the central nervous system, just as those scientists who are exploring the central nervous system will derive benefit from familiarity with computer developments.

1.1.1.5 A Significant Current Development: Products Safety

With some notable exceptions, the response of non-DOD industries to human factors has been slow and often something less than enthusiastic. This condition probably is attributable to a number of factors, one of which is the fact that colleges of engineering, with the notable exception of their departments of industrial engineering, have been unconscionably slow to prescribe even minimum training in human factors for their students. Another reason could be the lack of data that directly and convincingly demonstrate the impact of human factors on the "bottom line."

Current developments, however, suggest that society is attaching increased importance to *protection* (safety). The products liability movement has changed the dictum, "Let the buyer beware" to "Let the designer–manufacturer–vendor beware," and is modifying the ennuyé attitude that some had toward human factors. Nothing gets the attention of an industry regarding the safe design of its products as effectively as the payment of a large amount of money due to legal action taken against them for personal injury or death. As I have said elsewhere, "Those who have devoted their professional careers to the human factors/ergonomics movement watch with wonder and awe as the legal profession does through the courts what they have been unable to do" (Christensen, 1977).

The products safety/products liability movement, then, has tended to focus the attention of the courts on the human factors issues that relate to the danger associated with the operation and maintenance of equipment and systems. We define *danger* as a combination of hazard and risk (i.e., $D = H \times R$). *Hazard* is defined as a set of circumstances that could cause injury or death. *Risk* is the probability of occurrence of a hazardous event. It should be apparent that danger can be reduced to zero in any situation by the complete elimination of all hazards or by the reduction of risk to zero.

The elimination of hazards is very much a matter of concern for human factors specialists as well as for safety engineers, attorneys, and others. Risk, on the other hand, is usually somewhat more under the control of the individual (or an authority figure) who (presumably) after making an assessment of the hazards associated with an event decides whether or not engagement is worth the risk.

After identification of hazards, we recommend the employment of a model consisting of only five words when performing a safety analysis of a product or system. These five words are DESIGN–REMOVE–GUARD–WARN–TRAIN; except for "train," they are listed in order of priority. That is, once a hazard is identified the designer has an obligation to try to *design* the item so as to eliminate the hazard. When feasible and affordable, "design" is generally the preferred way to reduce hazards.

The *remove* criterion is illustrated by the use of remote manipulators or robots to reduce exposure of humans to irradiated materials, toxic substances, and so forth. The construction of highway bridges over railroads is another example of what can be done to remove people from dangerous interfaces.

If analysis discloses that hazards still exist after the investigation of the "design" and "remove" factors, the investigator should consider the possibility of *guarding* the hazardous interface. Requirements for guards can be found in Hammer (1976).

If residual hazard exists after attention to "design," "remove," and "guard," the designer then has a duty to *warn*. In a very real sense, a warning is an admission by the designer that, despite his best efforts, some hazards still remain in the product. Guidelines for warning devices and labels can be found in Christensen (1983).

The fifth term in the model is *train*. No one expects an individual without appropriate training to operate or maintain a complex system. However, the nature and level of skills, knowledge, and abilities that are required for safe and effective operation and maintenance must be specified by those in the best position to define them—usually someone on the staff of the design or systems engineer. The vendor should then assure that the purchaser or user has met or will meet these competency requirements before operating the equipment/system.

Finally, a quotation that I have used elsewhere seems appropriate with respect to the present status of products safety/products liability. In his *Ominiana* in 1812, S. T. Coleridge said, "To all new truths, or renovations of old truths, it must be as in the ark between the destroyed and the about-to-be-renovated world. The raven must be sent out before the dove, and ominous controversy must precede peace and the olive wreath."

1.1.1.6 Future Developments

Greater attention to product safety is only one trend that we see being established as we rush toward the twenty-first century. Others include more extensive use of instructional systems that are more responsive to individual differences in acquisition rates, retention, and interests.

There will be greater attention to individual differences, based on increased respect for the individual. Elsewhere I have declared that a careful analysis of the skills, knowledge, and abilities required in many DOD jobs, for example, would almost certainly reveal that many of the jobs could be done by people with physical handicaps (e.g., operation of computers and many record-keeping jobs). Certainly these people are being deprived of opportunity and privilege when they cannot serve in the armed services of our country. The human factors profession has a responsibility to see that as many different jobs as possible are made available to everyone who is qualified, regardless of race, sex, creed, age, or physical condition. The country is gradually becoming aware of the enormous human resources that have been wasted because of prejudice against women, minority races, the handicapped, and older citizens. The opportunity to work and serve in some useful capacity must be made available to the broadest possible spectrum of individuals. Proper and careful application of what is known about equipment/systems design, working conditions, training, and so on can contribute toward this worthwhile goal.

As more and more traditional industrial jobs succumb to automation, opportunities will increase for the application of human factors to the service industries and to process control industries. Often the things manipulated in the latter are ideas and concepts; the profession must press forward in its attempts better to understand such concepts as cognition, creativity, and motivation.

Settlements in space will require not only additional knowledge regarding the effects of such environmental factors as weightlessness but also improved understanding of inter- and intragroup relationships and of the maintenance of satisfactory performance in the absence of the accolades of fellow citizens that even now are beginning to diminish. Human factors, properly applied, can help assure the success of future space programs. Traditional concerns with the individual and his workplace must be expanded to include greater consideration of the performance of groups and direct and indirect effects on society. The challenges of new frontiers such as space should be viewed as unparalleled opportunities to discover more about people and mankind. It was Alexander Pope who reminded us that the proper study of mankind is man. We cannot imagine what the exploration of space will reveal about people and mankind.

Finally, the potential impact of the computer on technological societies defies the imagination. We are barely over the threshold of the Age of Computers and yet we can already envisage these devices completely dominating the activities that are routine, uninteresting, and perhaps hazardous to physical and mental health. But this is just the beginning. Computers will disclose new relationships; they will create (at least according to any reasonable definition of the term). Most exciting of all, however, advancements in computer technology will inspire insight and will stimulate formulation and test of hypotheses and development of theories regarding the great mysteries of the universe that might never otherwise have occurred. Prudent men will not consider the computer their enemy but will accept its development and availability as an incredibly competent ally that will immeasurably further our understanding not only of the physical universe but also of that most mysterious of the universes within the universe—the human mind.

This larger responsibility of the primary profession that is supported by human factors—engineering—has been addressed recently by Pletta (1984). I agree with Pletta that if engineering is to achieve the degree of professionalism that it seeks and deserves, then the curriculum must be revised and expanded to five or even more years. Engineering is clearly one of the most important professions in a technical society. The time that engineers currently spend in formal education is not commensurate with the responsibilities that they have been asked to assume.

1.1.1.7   A Word on Definitions

Considerable confusion exists with respect to the terms used to describe the various areas and subareas of human factors We prefer to use the term human factors or ergonomics to cover both *protection and performance* and *research and applications.* "Human factors engineering" (HFE) and "applied ergonomics" are concerned with the *application* of the data and principles of human factors and ergonomics to the design of equipment, subsystems, and systems. HFE is an engineering enterprise.

The research basis for the field of human factors is found in virtually all the so-called life sciences—biology, anthropology, physiology, neurology, psychology, and so on. Research aimed directly at the solution of a design problem is appropriately termed *engineering psychology, engineering anthropology,* and so on. A piece of research in engineering psychology, for example, would be distinguished from a piece of research in experimental psychology by the nature of the abscissa. The abscissa for an investigation in engineering psychology should be a dimension or factor that can be manipulated directly by the designer—size of dial, amount of illumination, resistance in a control, and so on.

As stated above, selection refers to the process of identifying those people who possess appropriate aptitudes and capabilities, whereas training refers to the process of changing the response characteristics of people so that they can interact more effectively with equipment and systems (and, of course, with ideas, concepts, other people, and even natural features of the environment). Selection and training must be included under the broad umbrella of human factors because the development of an effective, symbiotic relationship between man and machine depends upon a carefully orchestrated consideration of the capabilities and limitations of the humans on the one hand and of the design of the hardware *and software* on the other. Unfortunately, designers often make assumptions about what people can do (or can be taught to do) that are as erroneous as was the "one best way" assumption discussed earlier. (We are reminded of a statement made by Lt. General Willard W. Scott, Jr., The Commander of West Point, at a recent meeting at that great institution. During a discussion regarding the capability of various army tanks, one officer observed, "Gentlemen, the best tank in the field is the tank with the best crew!") Superiority in systems can never be achieved without excellence of human components, of hardware components, and of optimization of the interactions among them.

1.1.2   THE ATTAINMENT AND MAINTENANCE OF COMPETENCE IN HUMAN FACTORS

Industrial engineers have long admonished their practitioners to "go to the floor" when they wish to determine the activities that are associated with a job. We restate this as "go to the field" (as well as to the floor). Irrespective of terminology, learning what people actually do is fundamental to the improvement of operations, retrofit of equipment, design of systems, revision of procedures, and specification of selection and training requirements.

Progress in human factors, as in any profession, depends on the availability and proper employment of its methods and tools. Let us briefly examine one that is currently receiving considerable attention and one that should receive more attention.

1.1.2.1   Tools and Methods

It is axiomatic that any field of science or technology makes progress as a function not only of the capabilities of its members but also of the development and application of improved tools and methods. Human factors is no exception. Although "tools and methods" are treated expertly elsewhere in this handbook, we presume to mention two here. The first is, in our opinion, fundamental. We must be able to identify, define, and analyze the activities of individuals *and teams of individuals* before we can hope to achieve the symbiosis that Licklider describes (personal communication). Task analysis, a product of the seminal mind of Robert Miller (1963), represents a significant improvement over past methods of activity analysis in that, properly gathered, the data can be useful to all the human factors specialties. However, adequate treatment of the interlocking, interactive activities of *teams* of individuals, could, in our opinion, benefit from further development of task analysis.

As our second example, we view the development of improved computer-based simulation as particularly significant with respect to the development of improved human factors methodology. To be able to investigate alternative dynamic man–machine interrelationships well before any "tin-bending" occurs is a resource that should provide a viable method for effective early participation during the systems development and acquisition process—an answer to the "too little, too late" criticism that has frequently been levied at the profession. Trade-off studies, which are fundamental to all systems development, can be conducted while there is still time to implement promising alternatives. This is yet another example, then, of how the contributions of the human factors profession can be enhanced because of the availability of the digital computer.

For a description of other tools and methods available to human factors practitioners, see Chapter 1.1 of this handbook.

Table 1.1.1.   Percentage Distribution of Educational
Background of the Membership of the Human Factors Society

Academic Specialty	% 1980	% 1984
Psychology	52.9	50.7
Engineering	17.7	16.3
Human factors/ergonomics	6.3	5.8
Industrial design	4.3	3.7
Medicine/physiology/life sciences	2.1	3.7
Education	2.6	1.1
Business administration	2.2	2.2
Computer science	(<1.0)	1.3
Other[a]	6.6	6.5
Students and not specified	5.3	8.7
Total	100.00	100.00

Source: M. Knowles, Ed., *Directory and Yearbook,* The Human
Factors Society, 1980 and 1984.

[a] "Other" includes physics, anthropology, sociology, architecture,
industrial management, and operations research.

1.1.2.2   The Nature of the Human Factors Society Membership

The 1984 Directory of the Human Factors Society contains the names of more than 3500 members (Knowles, 1984). A summary of the membership by educational specialty is shown in Table 1.1.1 for the years 1980 and 1984. There are no startling changes in percentage distribution between 1980 and 1984. The percentage of psychologists decreased slightly, as did the percentage of engineers. The percentage of medical/life scientists showed the greatest proportional increase. Unfortunately, the percentage of educators decreased. Two-thirds of the members list psychology or engineering as their primary profession. Although this fraction is reducing, it is happening very slowly.

Well over 90% of the members had at least a bachelor's degree, with nearly 40% holding the Ph.D. degree. Most of the bachelor's, master's, and Ph.D. degrees were granted in psychology, with engineering second in number. This difference in favor of psychology is probably a reflection of the fact that engineers are considered professionals (e.g., eligible to take the professional engineer's examination) upon being awarded the bachelor's degree, whereas in many states, psychologists are required to have a Ph.D. in order to hold a professional license. Other members of the Human Factors Society have degrees in human factors/ergonomics, industrial design, medicine, physiology, life sciences, education, business administration, physics, anthropology, sociology, architecture, industrial management, and operations research. Such diversity is certain to enliven and energize a society, while complicating issues associated with certification, accreditation, and professional conduct. (The Executive Council of Human Factors Society has established committees to examine and consider the knotty problems associated with certification and accreditation.)

1.1.2.3   Opportunities in Formal Education

As of 1982, there were 52 colleges and/or universities that offered advanced degrees in "human factors" (Sanders and Strother, 1982). Twenty-six of these were offered in departments of industrial engineering or some variant thereof, such as industrial engineering and operations research. Eighteen were located in departments of psychology. Four of the remaining eight were offered in engineering departments other than industrial, and the final four were divided equally among human factors, occupational health and safety, operations research, and management science.

Many members have degrees in both one of the sciences and in engineering (e.g., a B.S. in engineering and a Ph.D. in experimental psychology). Such an educational background should prepare an individual exceptionally well for work in the human factors field.

1.1.2.4   Continuing Education and Experience

Graduation from a sound academic program is, of course, only the first step in the development of a qualified practitioner in any profession. True competence is developed only after years of experience. Attendance at short courses, workshops, and symposia; attendance at professional meetings; and self-study all help the professional in his quest for new degrees and varieties of competence and in his attempts to maintain the skills, knowledge, and abilities that he already possesses.

Numerous short courses are currently available to the human factors professional. The oldest continuous short course in human factors was started by Paul M. Fitts, who is generally credited as having been the father of human factors in the United States. This two-week course has been offered annually in the College of Engineering at the University of Michigan since 1958. After the untimely death of Professor Fitts, the course has been carried on by an outstanding Fitts student, Richard W. Pew.

1.1.2.5 Professional Societies of Relevance to Human Factors

This section and the next three sections are concerned with the professional societies and publications that have relevance to the human factors profession. We are indebted to Dr. Pew for the use of these materials, which have been taken with very slight modification from materials that he prepared for the aforementioned short course (Pew, 1984).

1. Human Factors Society
    P.O. Box 1369
    Santa Monica, CA 90406
    (213)394-1811

2. Division 21 of the American
    Psychological Association (APA)
    Society of Engineering Psychologists
    1200 17th Street, NW
    Washington, DC 20036

3. Systems, Man and Cybernetics (formerly Man-Machine Systems Group)
    of the Institute of Electrical and Electronics Engineers (IEEE)
    345 East 47th Street
    New York, NY 10017

4. Ergonomics Society of Great Britain, continental affiliates, Canada and Australia

5. Society of Automotive Engineers
    400 Commonwealth Drive
    Warrendale, PA 15096

6. American Society of Mechanical Engineers
    345 East 47th Street
    New York, NY 10017

7. Institute of Industrial Engineers (formerly American Institute of Industrial Engineers)
    25 Technology Park/Atlanta
    Norcross, GA 30092

8. Operations Research Society of America
    428 East Preston Street
    Baltimore, MD 21202

9. Acoustical Society of America
    428 East 45th Street
    New York, NY 10017

10. Optical Society of America
    2000 L Street, NW
    Washington, DC 20036

11. Society for Information Display
    654 N Sepulveda Blvd
    Los Angeles, CA 90049

12. Illuminating Engineering Society
    c/o United Engineering Center
    345 East 47th Street
    New York, NY 10017

13. American Society of Agricultural Engineers
    2950 Niles Road
    St. Joseph, MI 49085

14. Industrial Designers Society of America
    1750 Old Meadow Road
    McLean, VA 22101

15. Association for Computing Machinery
    Special Interest Group in Computer-Human Interaction
    1133 Avenue of the Americas
    New York, NY 10036

16. American Industrial Hygiene Association
    475 Wolf Ledges Parkway
    Akron, OH 44311
17. Human Factors Association of Canada
    Box 200
    Downsview, Ontario M3M, 3BN
    Canada
18. Systems Safety Society
    Box A
    Newport Beach, CA 92663
19. American Society of Safety Engineers
    850 Busse Highway
    Park Ridge, IL 60068

Societies exist also in the Soviet Union, the countries of the Eastern bloc, and Japan.

1.1.2.6 Journal Publications of Relevance to Human Factors

Pew (1984) lists 24 regularly published journals that contain materials related to one or more areas of human factors. This list does not include the numerous professional journals and trade journals that occasionally publish a human factors article.

Journals of the Societies

A. Primary
   1. *Human Factors* (HFS)
   2. *Ergonomics* (ES)
   3. *IEEE Transactions on Systems, Man and Cybernetics* (a merger of the *IEEE Transactions on Man-machine Systems* and the *IEEE Transactions on Systems Sciences & Cybernetics*) (IEEE)

B. Secondary
   4. *Information Display* (SID)
   5. *Journal of Experimental Psychology, Human Perception and Performance* (APA)
   6. *Journal of Applied Psychology* (APA)
   7. *Quarterly Journal of Experimental Psychology* (British Experimental Psychology Society)
   8. *Journal of the Acoustical Society of America* (ASA)
   9. *Journal of the Optical Society of America* (OSA)
   10. *Industrial Engineering* (AIIE)
   11. *Operations Research* (ORSA)
   12. *Vision Research* (ARVO)
   13. *Perception and Psychophysics*
   14. *Professional Safety* (American Society of Safety Engin.)
   15. *Hazard Prevention* (Systems Safety Society)
   16. *PsycSCAN: Applied Psychology* (American Psychological Assoc.)

Independent Journals

   1. *Perceptual and Motor Skills*
      Box 1141
      Missoula, MT 59801
   2. *Journal of Motor Behavior*
      726 State Street
      Santa Barbara, CA 93101
   3. *Behavior Research Methods and Instrumentation*
      Psychonomic Press
      295 Pine Avenue
      Goleta, CA 93017

4. *Le Travail Humain*
   Presses Universitaires de France,
   Departement des Periodigues
   12, rue Jean-de-Beauvais, Paris V.
   Tel. 033-48-03
   C.C.P. Paris 1-302-69

5. *The Journal of Human Ergology*
   Human Ergology Research Association
   Jinrui Dotaigaku Kenkyukai,
   Business Center for Academic Societies Japan,
   2-4-16 Yayoi, Tokyo 113, Japan

6. *Applied Ergonomics*
   IPC Business Press Ltd.
   205 East 42nd Street
   New York, NY 10017

7. *International Journal of Man-Machine Studies*
   Academic Press
   111 Fifth Avenue
   New York, NY 10003

8. *Behavior and Information Technology*
   Taylor & Francis, Inc.
   242 Cherry Street
   Philadelphia, PA 19106-1906

1.1.2.7 Human Factors References of Historical Interest [from Pew, (1984)]

Chapanis, A., Garner, W. R., and Morgan, C. T. (1947). *Applied experimental psychology.* New York: Wiley.

McFarland, R. C. (1946). *Human Factors in air transport design.* New York: McGraw-Hill.

Sinaiko, H. W. (1961). *Selected papers on human factors in the design and use of control systems.* New York: Dover.

Woodson, W. E., and Conover, D. W. (1970). *Human-engineering guide for equipment designers* (2nd ed.). Berkeley, CA: University of California Press.

1.1.2.8 General Human Factors References [from Pew, (1984)]

Bailey, R. W. (1982). *Human performance engineering: A guide for system designers.* Englewood Cliffs, NJ: Prentice Hall.

Chapanis, A., Ed. (1975). *Ethnic variables in human factors engineering.* Baltimore: Johns Hopkins University Press.

Chapanis, A. (1959). *Research techniques in human engineering.* Baltimore: Johns Hopkins University Press.

Damon, A., Stoudt, H. W., and McFarland, R. A. (1966). *The human body in equipment design.* Cambridge, MA: Harvard University Press.

DeGreen, K., Ed. (1970). *Systems psychology.* New York: McGraw-Hill.

Eastman Kodak Company (1983). *Ergonomic design for people at work.* Belmont, CA: Lifetime Learning Publications.

Grandjean, E. (1971). *Fitting the task to the man,* 2nd ed. London: Taylor & Francis.

Howell, W. C., and Goldstein, I. L. (1971). *Engineering psychology.* New York: Appleton-Century Crofts.

Huchingson, R. Dale (1981). *New horizons for human factors in design.* New York: McGraw-Hill.

Kantowitz, B. H., and Sorkin, R. D. (1983). *Human Factors: Understanding people–system relationships.* New York: Wiley.

Konz S. (1979). *Work Design.* Columbus, OH: Grid Publishing.

Kraiss, K. F. and Moraal, J. (1976). *Introduction to human engineering.* Bonn, West Germany: Verlag TUV Rheinland.

Lindsay, P. H., and Norman, D. A. (1977). *Human information processing: An introduction to psychology,* 2nd ed. New York: Academic.

McCormick, E. J., and Sanders, M. S. (1982). *Human factors in engineering and design,* 5th ed. New York: McGraw-Hill.

Meister, D. (1971). *Human factors: Theory and practice.* New York: Wiley-Interscience.

Meister, D. (1976). *Behavioral foundations of system development.* New York: Wiley.

Meister, D., and Rabideau, G. F. (1965). *Human factors evaluation in system development.* New York: Wiley.

Muckler, F. A., Ed. (1984). *Human factors review: 1984,* Santa Monica, CA: The Human Factors Society.

National Aeronautics and Space Administration (1978). *Anthropometric Source Book,* Vols. 1–3 (NASA Reference Publication 1024). Washington, DC.

National Aeronautics and Space Administration (1973). *Bioastronautics Data Book* (NASA SP 3006). Washington, DC: US Government Printing Office, 1973.

Parsons, H. M. (1972). *Man–machine system experiments.* Baltimore: The Johns Hopkins University Press.

Poulton, E. C. (1972). *Environment and human efficiency,* Springfield, IL: Charles C Thomas.

Sanders, M. S. and McCormick, E. J. (1976). *Workbook for human factors in engineering and design.* Dubuque, IA: Kendall/Hunt.

Sheridan, T. B., and Ferrell, W. R. (1974). *Man–machine systems,* Cambridge, MA: MIT Press.

United States Air Force (1977). *Human factors engineering,* 3rd ed. (Design Handbook 1–3). Wright-Patterson Air Force Base, Ohio.

United States Department of Defense (1981). Military Standard MIL-STD 1472C, Human Factors Design Criteria for Military Systems Equipment and Facilities, May 1981.

United States Department of Defense (1981). Military Handbook MIL-HDBK 759A, Human Factors Engineering Design for Army Material, June 1981.

Van Cott, H. P., and Kinkade, R. G., Eds. (1972). *Human engineering guide to equipment design.* Washington, DC: U.S. Government Printing Office.

Wickens, C. D. (1984). *Engineering psychology and human performance,* Columbus, OH: Charles E. Murrell.

Woodson, W. E. (1981). *Human factors design handbook.* New York: McGraw-Hill.

1.1.3 THE SYSTEMS MODEL—A VALUABLE AND NECESSARY INTEGRATOR

Cultural systems have several common characteristics. Included among these are the following:

1. Common purpose of their components
2. Common information network
3. Interaction among their components
4. Constraints imposed on their components

Examination of the complexity of modern systems suggests that effective design requires that a well-defined plan be followed that will assure timely, systematic consideration of relevant data, principles, and practices from all the contributing disciplines. The final design is invariably the result of numerous compromises ("trade-offs") made with respect to the criteria established for the system (cost, performance, maintainability, etc.)

1.1.3.1 Example of a Systems Model

The complexity of modern systems demands that a systematic, well-defined set of procedures be applied initially and throughout the various phases or stages of systems development and acquisitions. We believe that development of even the simplest products can benefit from adherence to the "systems approach"; we seriously doubt that complex systems can be developed successfully without such an approach.

A possible systems model is shown in Figure 1.1.1 (Christensen, 1984). It is important to remember that this is a *process,* continuous over time, and not a succession of discrete steps. The entire enterprise is generally under the direction of a systems engineer, who by training and/or experience understands the importance of developing, integrating, evaluating, and applying in a timely, coordinated manner relevant data and principles from all contributing disciplines. Very large systems may require both a systems engineer and a systems manager; the latter generally would be concerned with financial matters, personnel, administration, public relations, and so on.

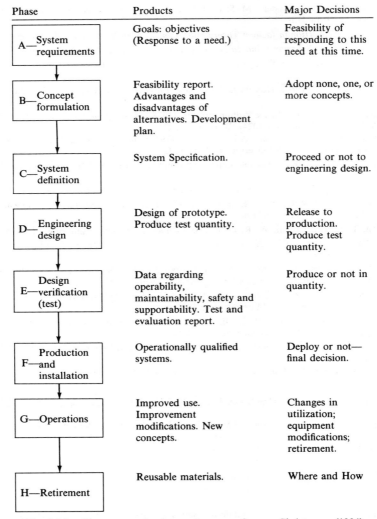

Phase	Products	Major Decisions
A—System requirements	Goals: objectives (Response to a need.)	Feasibility of responding to this need at this time.
B—Concept formulation	Feasibility report. Advantages and disadvantages of alternatives. Development plan.	Adopt none, one, or more concepts.
C—System definition	System Specification.	Proceed or not to engineering design.
D—Engineering design	Design of prototype. Produce test quantity.	Release to production. Produce test quantity.
E—Design verification (test)	Data regarding operability, maintainability, safety and supportability. Test and evaluation report.	Produce or not in quantity.
F—Production and installation	Operationally qualified systems.	Deploy or not— final decision.
G—Operations	Improved use. Improvement modifications. New concepts.	Changes in utilization; equipment modifications; retirement.
H—Retirement	Reusable materials.	Where and How

Fig. 1.1.1. The systems development process. Source: Christensen (1984).

Note the last block in the diagram, "Retirement." The shortages of various materials, protection of the environment, safety hazards, and elimination of public eyesores suggest that attention be given to the final disposition of a system before it is made available to the public.

The systems model is essentially an elaboration of a traditional problem-solving model that virtually everyone has encountered at sometime during his or her educational career. The general model consists of four steps:

Formulation of the problem (definition)

Development of possible alternative solutions

Test and selection of one (or more) alternatives

Implementation

The "formulation" phase includes essentially the requirements and conceptual phases of the detailed model shown in Figure 1.1.1. The "development and consideration of alternatives" phase includes essentially system definition and engineering design. The "test and selection" phase corresponds roughly to design verification, and the "implementation" phase corresponds to production, installation, operations, and retirement.

1.1.3.2   Evaluation of the Systems Approach

The systems model, at least as employed in practice, has received both accolades and criticisms. Advocates suggest that among the advantages is the fact that its proper employment virtually demands an integrated approach to development and acquisition. Criteria by which to judge effectiveness must be defined. Constraints become evident. (An interesting and rather complete description of a system can be developed by specifying the constraints that are placed on the components of the system, including people, as they become elements or subsystems in a larger system.) The need to conduct appropriate trade studies becomes evident, usually at the first meeting of the systems team. Control and display interfaces are considered earlier and, it is hoped, more thoroughly. The nature of the contributions of the so-called "minor" disciplines becomes evident very early in the process. The need to develop predictions regarding manpower, training, and training program requirements (including simulator requirements) is faced early in development, as it should be. Timely delivery of the inputs of all participants forces early estimates of milestones (and the ensuing embarrassment if they are not met). Finally, adherence to the requirements of the model should have a unifying effect not only among the members of the design team but also between the design team and the management–administrative team. It should encourage early and continuous interaction with the customer.

The systems approach has been criticized for leading to too many tiers of management, production of unnecessary paper, and enchantment with procedures. Overstudy of the problem before taking timely, appropriate action has occurred. Unless the systems engineer is a wise manager, some of the team members are apt to feel neglected and that their inputs are considered relatively unimportant. Others claim that the procedures are so stereotyped that the expression of initiative and creativity by individual participants is virtually impossible. Cost overruns are common and projected delivery schedules nearly always prove to be optimistic. Finally, the systems engineer must constantly be alert for interaction among components. The assumption that components and subsystems can be extracted, studied in isolation, and even redesigned and then reinserted in the system without consideration of possible interactive effects (the additivity principle) is a most dangerous one. It should be assumed that interactions exist until proven otherwise.

Training in systems development methods, good communications within the team and with significant people (e.g., customers) outside the team, a sensitivity to interaction, and a generous measure of understanding of the views of others generally will assure that the system will be acquired on time and within budget, that the system will meet its requirements, and that its employment will be an enduring source of pride and satisfaction for all members of the systems team.

The information in this handbook provides the methods, the data, and the principles that are needed by human factors specialists in the achievement of their lofty objectives—safe, effective, satisfying contributions to the systems of our technological society.

REFERENCES

Anon. (1960). *Colliers' Encyclopedia*, Vol. 10.

Christensen, J. M. (1962). The evaluation of the systems approach in human factors engineering. *Human Factors, 4*(1).

Christensen, J. M. (1977). Implications of product liability for engineering design and education. *Engineering Education, 68*(3).

Christensen, J. M. (1983). Human factors in hazard/risk evaluation. In Forsberg, H. G., and Belfrage, B. Eds., *Human reliability in complex technical systems.* Stockholm: INGENJORSVETENSKAP-SAKADEMIEN.

Christensen, J. M. (1984). The nature of systems development. In R. W. Pew and P. Green, Eds., *Human factors engineering.* Ann Arbor, MI: University of Michigan College of Engineering.

Clark, J. D. (1960, July). Human ecology during pleistocene and later times in Africa south of the Sahara. *Current Anthropology.*

Dart, R. A. (1959). *Adventures with the missing link.* New York: Viking.

Fitts, P. M. (1963). *Human factors engineering—concepts and theory.* Ann Arbor, MI: The University of Michigan Engineering Summer Conferences.

Hammer, W. (1976). *Occupational safety management and engineering.* Englewood Cliffs, NJ: Prentice-Hall.

Knowles, M. G., Ed. (1980, 1984). *Human Factors Society Directory and Yearbook,* Santa Monica, CA: The Human Factors Society.

McMillan, B. (1962). Opening Address, First Congress on the Information Sciences, Hot Springs, VA.

Miller, R. B. (1963). Task description and analysis. In R. M. Gagne, Ed., *Psychological principles in system development.* New York: Holt, Rinehart & Winston.

Oakley, K. P. (1952). *Man, the toolmaker.* London: British Museum of Natural History.

Pew, R. W. (1984). Introduction to human factors engineering, In R. W. Pew and P. Green, Eds., *Human factors engineering.* Ann Arbor, MI: The University of Michigan College of Engineering.

Pletta, D. H. (1984). Engineering's emerging public purpose. *The Best of Tau Beta Pi, 75*(4).

Sanders, M. S., and Strother, L. (1982). *Directory of graduate human factors programs in the U.S.A.,* Santa Monica, CA: The Human Factors Society.

Weston, A., Ed. (1961). The emerging role of industrial engineering. *J. Industrial Engineering, 12*(2).

Wiener, N. (1950). *The human use of human beings; Cybernetics and society.* Boston: Houghton Mifflin Company.

CHAPTER 1.2

SYSTEMS DESIGN, DEVELOPMENT, AND TESTING

DAVID MEISTER

U.S. Navy Personnel Research and Development Center, San Diego, California

1.2.1 INTRODUCTION AND OVERVIEW

No description of human factors (the discipline, henceforth abbreviated as HF) would be complete or indeed truthful if it did not describe the application of HF principles and data to systems design, development, and testing. Indeed, the discipline began during World War II as an attempt to apply the principles of applied experimental psychology to the solution of wartime problems.

We start with the definition of basic terms. A *human–machine system* (also known in the past as a *man–machine system*) is an organization of men and women and the machines they operate and maintain in order to perform assigned jobs that implement the purpose for which the system was developed. This is the mere skeleton of a definition; the reader should refer to Meister, 1971 and 1981, for more detail. Aircraft, ships, trains, stock exchanges, hospitals, schools, fire departments, and welfare and corporate institutions are all systems, some making use of many sophisticated machines, others using few machines. All of them, however, satisfy the essentials of a system: people organized by means of procedures and processes to fulfill a superordinate purpose, in which the output of their activity is some end result different from the initial inputs these people made to the system operation. The major elements of a system that must be addressed during system development are hardware and software, personnel, operating procedures, and technical data.

Development is the process of transforming the system requirement (what the system is supposed to do, as described in words and numbers on paper) into the actual functioning system. Development encompasses (1) *design,* which is everything other than fabrication that is required to produce a functioning system, and (2) *testing,* which is the evaluation of design and of the system to ensure that these satisfy specifications, standards, and requirements.

HF principles and data (the outputs of research) are applied to the design, development, and testing of systems. The individual who applies HF to the system development process is the HF engineer (henceforth called the HFE).

This chapter discusses what the HFE does in applying HF to system development. In this process the HFE becomes one of a group of engineers of various specialities functioning interactively as a design team. In this role the HFE acts more as an engineer than a researcher; in fact, most of the time the HFE is not a researcher at all, but rather a practitioner (in the sense in which a physician is a practitioner).

Subsequent chapters deal with specialized topics such as methodology, robotics, or computer software. All the HF research performed on these, if relevant, will eventually be applied in system development. Hence it is impossible to understand the meaning or relevance of that research unless one understands what system development is and how the HFE functions in it.

Because the subject matter of system development encompasses so many subtopics, it is impossible in a single chapter to describe these as completely as one might wish. The reader is urged to supplement this chapter by reading the referenced books and papers. This chapter discusses the following topics which stem from the variables that affect the success of HF in system development: (1) general characteristics of system development; (2) the governmental background of development; (3) the various development phases and their functions; (4) the questions that arise in each phase; (5) what the HFE does in these phases; (6) the methods and the data available; (7) the constraints that affect the HFE; (8) the relationship of the HFE and the engineer, and the latter's attitudes toward HF; (9) the adequacy of research support for system development HF.

Because system development is highly pragmatic, our description of it must be highly realistic, containing problems and difficulties, "warts and all."

Note also that this chapter is largely a description of the development of military or other governmental systems, because development of civilian systems (such as automobiles) has not been adequately described in the literature. However, to the extent that development processes are logical, stemming from development *needs,* what is described here should apply also to civilian system development, although with fewer of the government's formal documentation requirements.

1.2.2 GENERAL CHARACTERISTICS OF SYSTEM DEVELOPMENT

Before prodceeding to a detailed analysis of the individual development phases it is useful to "get the broad picture" by examining their overall salient characteristics.

1.2.2.1 Molecularization

The development process as a whole is one of working from broad, molar functions to progressively more molecular tasks and subtasks. When the HFE first encounters system requirements, he or she deals with the total system and major subsystems; as development proceeds, the way leads down to subtasks at the level of switch activation and the placement of controls and displays. Actions taken at an earlier, more molar (system, subsystem) level have profound consequences for molecular (task, subtask, component) levels. For example, if the developer decides in Preliminary Design that information will be presented via a cathode-ray tube (CRT), in Detail Design all the problems involved in using a CRT, including display brightness, resolution, and ambient lighting, must be faced. This is why the HFE should insist on being part of the design team from the beginning of development.

1.2.2.2 Requirements as Forcing Functions

System requirements drive the design tasks that are performed during development. The criterion of design adequacy is always the system requirement, and design options are developed to satisfy that requirement. That is why the absence of formal behavioral requirements in the design specifications or development contract makes it much more difficult for the HFE to secure adequate consideration of behavioral inputs.

1.2.2.3 System Development as Discovery

Initially there are many unknowns about the system. In some systems (fortunately only a few) development may begin even before system requirements have been thoroughly specified. Often system requirements are changed during development; the military has a habit of complicating original design by adding new requirements. In addition, system requirements can be implemented in at least several different ways; early on, the designer may not know the best way or indeed the range of choices available. Similarly, the behavioral implications of system requirements are unknowns to the HFE. For example, what kind of workload will personnel be exposed to? Progressively, as choices among design options are made, these unknowns are clarified until, when the system is produced, installed, and operationally tested, almost everything about the system is known. Even then there are unknowns, because between the first prototype and the last production model the system can still produce nasty surprises that require problem investigation and resolution.

1.2.2.4 System Development as Transformation

System development for the designer is the transformation of system requirements into physical mechanisms to perform system operations, such as hardware, software, and procedures. For the HFE system development is also a transformation but somewhat more complex: from the physical requirement to the behavioral implications of that requirement to the physical mechanisms for implementing these behavioral implications. Almost without exception the major system requirements are physical, for example, speed, range, endurance, power consumption, and reliability. Almost never is there an explicit behavioral requirement. Behavioral requirements are inferred from a concept of how the system should function ideally; for example, the system should not impose too heavy a burden on its personnel, or the pilot of an aircraft should have adequate visibility outside his cockpit. The HFE examines the physical requirement, determines its behavioral implications, and suggests a physical mechanism for encompassing these implications. In an absurdly simplistic, and therefore more obviously illustrative example, the HFE asks himself, what does it mean that the system must be operated outdoors in arctic conditions? Well, operators will probably have to wear gloves; in consequence it would be advisable to make controls large to accommodate for gloved clumsiness. The operators will respond more slowly to stimuli; hence system events requiring fast responses should be slowed if possible. This example is relatively obvious, but others are by no means so simple. For example, the operator may be required to correlate several channels of information presented in overlapping fashion. What is the significance of this for the operator? What physical implementing mechanisms should be recommended for reducing errors resulting from this requirement? The HFE is engaged in this process of implication/transformation throughout system development but it is most apparent early in development, for example, during a mission/function analysis.

1.2.2.5 Time

Because of all the unknowns that must be resolved, system development is time-driven. There is never enough time for the HFE to do the analyses, studies, tests, and so on that would be done given one's "druthers." In this respect system development is a degraded process. The implication for the HFE is that whatever is done must be timely or it is wasted effort. Nor can the HFE rage against this as being uniquely irrational; everyone is affected by the frenetic pace, including, and most prominently, the engineer with primary design responsibility.

Another aspect of time is that, depending on the size and complexity of the system, development may last over several years. For example, Detail Design and Testing for the Atlas (first intercontinental ballistic missile) system lasted from 1956 to 1962. Long development time permits the HFE to get to know the major "players" in the development game; but long development is usually associated with large projects and it is difficult for the HFE to keep "tabs" on all the relevant phases of such large projects.

1.2.2.6   Cost

Cost is another forcing function. First of all, if support money is tight, there may not be a HF program at all or it may be severely curtailed (fewer analyses and evaluations than are desirable) or it may be aborted early in order to reallocate the HF money to another (supposedly more important) development effort that is suffering financially. Second, a HFE recommendation cannot be too costly or it will be automatically rejected. That is why recommending redesign after hardware has been, as they say, "bent," is like spitting into the wind.

1.2.2.7   Iteration

System development is iterative. There are two reasons for iteration. (1) As we shall see, the same questions and activities arise and must be performed at different development phases; (2) where a required analysis has not been performed earlier or because there is new information which was not available earlier, or even because the design team cannot make up its mind, design becomes a process of "cut and try" in which a design solution is proposed, examined, and rejected, the designer tries again, the solution is rejected again, until progressively the design begins to approximate the final configuration. For example, my colleagues, responsible for the console design of the propulsion control system in the new DDG-51 class of Navy destroyers, have made and sent to the primary designer at least three different sets of control panel drawings.

1.2.2.8   Design Competition

Where the system is large, its design is performed by a team of specialists such as electrical, hydraulic, controls, reliability, weights engineers, and even HFEs. One engineering group is prime and one designer in that group is the prime coordinator of the team. The dominant group/engineer exercises veto power over the others, and the less influential groups, among whom one finds the HF group, must function under constraints established by the dominant one. Inputs from supporting team members are funneled to the prime designer and somehow integrated (by processes that are covert and hence still unclear). System development may be viewed as a process of choosing among competing design options. The inputs made to the designer by the various specialists describe various aspects of those options and competing interests. Competition here is a matter of how well each design option satisfies system requirements. Each specialist group has its own special interest: the reliability group is concerned for equipment reliability; the weights group is concerned about reducing system weight as much as possible, and so forth. Interests clash and may have to be reconciled. For example, to improve reliability in a fighter fire control system the reliability engineer desires redundancy, but the weights engineer fears the additional weight of that redundancy. The HFE has his or her own constituency: the human, whose interests must be compromised with those of the others. This does not mean that the design team functions discordantly; it is merely that each point of view is pushed vigorously and must be reconciled.

Because of the competitive aspect of system development and the parochial attitude most engineers have, it is important that the HF group be located *within* the contractor's engineering organization and not as a customer service or other nonengineering specialty unit. Experience has shown that HF outside of engineering has a much more difficult position to maintain than if it is part of engineering.

1.2.2.9   Priorities

These are assigned on the basis of criteria, the most important being the anticipated effect of the input on performance capability, because that is what the designer is most concerned about. Another criterion is absence of side effects. For example, a design option may promise superior performance capability but its resulting reliability requirements are too great to accept. A design input should not require redesign of any other subsystem or component because this will cause delays. If at all possible the design recommendation should fit within traditional guidelines, or be similar to the manner in which a design problem has characteristically been solved; this is because the average designer is highly conservative. And, all other things being equal, the input should not be too difficult to implement.

1.2.2.10   Relevance

The implication of these unwritten "rules" for the HFE is that to be considered meaningful his or her design inputs must be relevant to performance capability. All other things being equal, a HF input directly relevant to a hardware or software or even a procedural aspect of the system will be more gracefully received by the designer than an input dealing with an aspect related solely to personnel, such as selection or training or technical documentation. HF activities and methods related directly to system design are therefore viewed as more critical than ones not so directly related. Design relevance as perceived by engineers is critical for the acceptance and judged value of behavioral inputs during development.

1.2.2.11 Design Reviews

The designer with primary responsibility is first among equals. Obviously the designer works as part of a design team but his or her decisions, made after consultation with others on the team, are extremely influential, even if subject to reversal by higher authority. Design decisions are subject to periodic formal design reviews, which are examinations about the adequacy of the design to date, and it is possible through this mechanism for the HFE who feels his or her inputs have received short shrift to influence the designer's decisions.

What has been said so far suggests that much as one would prefer to believe that system development is a logical, rigorous, rational process of intellectual discovery and creation, that ideal is distorted by nonintellective influences such as unrealistic cost and schedule estimates; organizational confusion; lack of communication among designers; and, not least for the HFE, prejudice against HF.

1.2.3 BEHAVIORAL INPUTS

During development certain behavioral inputs are made by the HFE. Behavioral inputs during development may consist of four types (the term "may" indicates that the HFE may not be required or may not be able to provide all of these inputs):

1. Those relating to the *selection* and *acquisition* of system personnel, that is, determination of the number of personnel required by the system, description of the jobs to be performed in the new system, and description of the skills and skill levels required of operating and maintenance personnel.

2. Those relating to personnel *training,* that is, specification of length (in time) of course, student throughput (number) during that period, and equipment facilities (e.g., trainers, simulators, plant) and instructors needed for training.

3. Those affecting the *design* of the system hardware, software, and procedures, the intent being to ensure that these development products satisfy at least minimal HF standards, that is, design and/or review/evaluation of man–machine interfaces (usually displays and control panels); design and/or review/evaluation of computer software; design and/or review/evaluation of job procedures; review of technical data; specification of the required characteristics of the working environment (e.g., lighting, workplace layout); and prediction of personnel performance.

4. Those related to *testing* the personnel elements of the system and evaluating their operational effectiveness, that is, writing of test plans, including specification of personnel performance criteria and measures; statistical and experimental designs; design, review, and evaluation of test scenarios; conduct of personnel performance tests; and analysis of test data, development of conclusions, and writing of test reports.

1.2.4 GOVERNMENT MANAGEMENT PROCESSES

1.2.4.1 Introduction

System development flows in two parallel channels that overlap. Assuming that the system is being developed for the government, there are a set of background governmental processes, the major activities of which are *planning* the system, *monitoring* its development after a contract has been let to industry, and *evaluating* the product after it has been fabricated. Except in rare instances, government agencies do not design systems themselves (or rather it would be more correct to say that they do not carry design through the complete development cycle including production, although some agencies may perform very early design). Sometime during Preliminary Design government contracts out the development process to industry, which completes development, performs initial system testing, and turns the system over to government for final evaluation testing.

This section describes the governmental activities involved in managing the system development process. The reason for giving the reader this (largely) Department of Defense (DOD) background is because most HFEs working on development of new systems do so, directly or indirectly, for the military. There are exceptions, of course, such as HFEs working for other than DOD agencies and HFEs developing commercial systems such as airliners, but this market for HFEs is rather small. Because this is so, the HFE should be aware of the governmental procedures influencing the systems he works on.

The government's management plan for system development proceeds and overlaps (in parallel) with actual hardware/software development. The paperwork in these management exercises suggests only slightly what the HFE actually does during development; that is discussed in later sections. The governmental processes described below essentially create an adversary situation (like a trial) in which the system sponsor (customer) or project manager must continuously justify and defend the worth of his system against the implied presumption that the government would be better off *without* the system.

For systems developed for completely commercial use, these governmental activities do not apply. However, if the system is being developed at least partially with a view to government (as well as civilian) purchase, the developer must be aware that the system will have to satisfy certain requirements implicit in these activities.

The system development phases are logical, sequential, and overlapping. They include *System Planning, Preliminary Design, Detail Design, Testing and Evaluation, Production,* and, once the system has been turned over to the customer or sponsor, and is in routine usage, *Operations.*

DOD, which is the major governmental procurer of new systems, formalizes these phases by breaking them down into a series of milestones at which the system under development is critically reviewed before permission is given to continue development. Among the elements to be reviewed are those dealing with personnel. (For descriptions of this cycle and the relationship of behavioral factors to the various milestones, see Holshouser, 1975; Meister, 1983; Price, Fiorello, Lowry, Smith, and Kidd, 1980; Sawyer, Fiorello, Kidd, and Price, 1981). The four decision points and the new names given by the government to the phases are as follows:

1. *Milestone I—Program Initiation.*   Requires that a mission need be demonstrated in a document called the Mission Element Need Statement (MENS).
2. *Milestone II—Demonstration and Validation.*   Depends on recommendations made in the Decision Coordinating Paper (DCP).
3. *Milestone III–Full-Scale Engineering Development.*   Based on updated versions of the DCP.
4. *Milestone IIIA–Production and Deployment.*   Same as (3).

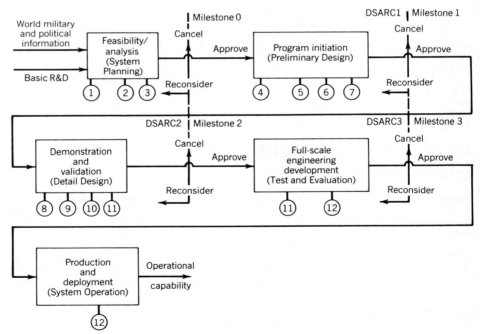

Fig. 1.2.1.   Military system acquisition (modified from Sawyer, Fiorello, Kidd and Price, 1981). (1) Mission profile/scenario analysis, (2) Function analysis—Function flow diagrams, (3) Decision/Action diagrams, (4) Function allocation trade-offs, (5) Time lines, (6) Operational sequence diagrams, (7) Task descriptions, (8) Workload analysis, (9) Task analysis, (10) Mock-ups, (11) Evaluation of man–machine interfaces, (12) Operational testing.

DOD Directive 5000.1 (Department of Defense, 1977a) specifies that human engineering factors are to be included as *constraints* on system design (note the negative connotation) and as a system element, starting with initial concept studies and continuing throughout development.

DOD Directive 5000.2 (Department of Defense, 1977b) establishes the Defense System Acquisition Review Council (DSARC) and the individual service DSARCs responsible for reviewing DCPs and for making recommendations concerning program direction. Holshouser (1975) and Meister (1983) list the questions that must be answered at each DSARC milestone. Figure 1.2.1 graphically depicts the various development phases, the DSARC milestones, and the HFE inputs made in each phase.

DOD Directive 5000.3 (Department of Defense, 1977c) requires that all systems be subject to test and evaluation (T & E), that T & E begin as early as possible and continue throughout development, and that acquisition schedules be based on T & E milestones.

Individual government agencies may also specify their own requirements. For example, the Naval Material Command, which is responsible for acquisition of all Navy systems, has just issued NAVMAT Instruction 3900.9A, which mandates a HF program throughout system development (Naval Material Command, 1984). These directives and instructions are presumably mandatory. However, in practice, because of engineering attitudes toward HF (to be described later), they are from a HF standpoint often evaded and fulfilled only halfheartedly.

The DOD management process described below is a very formal one, befitting the amount of taxpayer money involved. It is unclear whether industry engages in similar planning for commercial systems, although it would seem reasonable that commercial analogues of the military activities would be performed also.

The studies and analyses in the various DOD phases are iterative; documents are continually updated, amplified, and reviewed in each phase. Hence the behavioral inputs made to these documents must also be repeatedly updated. See Meister (1983) for a listing of the many organizational entities involved in the Navy in reviewing the various documents.

1.2.4.2 Feasibility/Analysis Phase

The major activity in this phase is the identification and definition of a mission need. Such a need is either a technological development that counters a known threat or the recognition of a strategic or tactical threat that requires the development of a new weapon system. The mission need is analyzed either in MENS or an operational requirement (OR).

MENS describes the mission area and need in terms of the tasks to be performed; projected threat assessment; existing capabilities to accomplish the mission; need in terms of deficiency in that capability; known constraints to solutions, such as cost and time; effect of lack of capability; and a plan for identifying and exploring alternative systems.

Although it would appear from the preceding list that MENS does not require behavioral inputs, one of the known constraints to solutions may be lack of personnel and training resources (although this would hardly restrain government planners). MENS must also be reviewed to ensure that at least one of the alternatives identified in the document keeps personnel requirements within current levels.

1.2.4.3 Program Initiation Phase

The major activities in this phase are as follows: (1) reestablishment of the mission need (is it still valid?); (2) a survey of available technology to identify areas of technical inadequacy of proposed systems; (3) the beginning of a definition of an acquisition strategy; and (4) preparation and issuing of documentation required for the milestone I decision. Preliminary Design is initiated. There are three major documents:

1. The development proposal (DP) is prepared by the command responsible for the system acquisition. It presents alternatives and trade-offs to achieve a range of capabilities. It forms the basis for the decision coordinating paper (DCP), which is the most important document for reaching a decision.
2. The DCP summarizes the program, the acquisition strategy, the alternatives considered, direction needed by decision authority, and additional requirements issued only by the acquisition executive.
3. The test and evaluation master plan (TEMP) describes how the new system will be evaluated.

Note that these and all other documents are developed for the entire system; HF has only a minor role in them. From a behavioral standpoint, the DP is reviewed to ensure that manpower estimates included in the alternatives presented are accurate and that there is consideration of training, support, and human resources factors that would affect the introduction of the new system. The DCP is reviewed to ensure that estimated manning levels are included; manpower requirements are compared with those of a base-line (predecessor) system, if one exists; potential trade-offs among manpower, design, and logistical elements are analyzed; and the training concepts that will be analyzed during the Demonstration and Validation phase are identified.

1.2.4.4 Demonstration and Validation Phase

At this point alternative system concepts have been identified, the subsystems targeted for advanced development have been determined, the mission need has been defined, and acquisition plans have been developed. In this phase Detail Design of the system is initiated, the management plan is developed further, TEMP is established, the integrated logistics support (ILS) plan is established, requests for proposals for system/subsystem development are written, prototypes of systems under development are constructed, and preparations are made for the Milestone II decision.

The ILS plan is particularly important from the standpoint of behavioral inputs because two of its principal elements are personnel and training. It is reviewed to ensure that personnel implications, including life cycle costs, are adequately addressed. This includes estimates of personnel in terms of numbers and skills, unique personnel resource constraints (e.g., introduction of new or critical skills), life cycle cost estimates for personnel and training, the training concept, and the scheduling of manpower, training, and equipment so that all three coincide.

TEMP is also important for HF because tests must be developed to determine whether personnel can perform required tasks adequately in the new system and to ensure that system characteristics do not have a negative impact on the ability of personnel to perform their jobs. TEMP, which is written about the time of DSARC I, is used both as an input to the DSARC decision process and as a T & E management plan.

A special section of TEMP must be reserved for personnel/HF. An outline of what is supplied in that section is provided later.

Another major input into DSARC II is the training plan, which is developed to describe the training resources required to support the manpower requirements.

After the Demonstration and Validation phase is over and the service secretary is prepared to recommend the preferred system for full-scale engineering development, this recommendation is documented in an updated DCP and reviewed by DSARC prior to decision by the Secretary of Defense. DSARC II reaffirms the mission element need and updating of the threat and asks whether the system in development meets mission element needs; system trade-offs have produced the optimum balance in cost, performance, and schedule; risks are acceptable; planning for selection of major subsystems is underway; the T & E results support the recommendation; and TEMP identifies the T & E to be accomplished prior to DSARC II and III.

With regard to behavioral issues, DSARC II personnel documentation should (1) provide the rationale for personnel estimates, (2) identify any unique skills required, (3) estimate personnel requirements for maintenance to be performed below depot level, and (4) discuss T & E plans.

1.2.4.5 Full-Scale Engineering Development Phase

During this phase, (1) detailed ILS specifications are developed, (2) requests for proposals (RFPs) from contractors for full-scale system development are written, (3) full-scale engineering development of the system is completed and production planning/preparation begins, (4) T & E (developmental) continues, and (5) preparations are made for the Milestone III decision. The various document inputs that were developed previously (e.g., ILS, TEMP) are updated.

1.2.4.6 Production and Deployment Phase

The Milestone III decision indicates whether the system under development will go into full-scale production and be operationally deployed.

This frenetic paper chase is pursued only at governmental levels and has little impact on actual system development at the engineering facility unless the sponsor fails to answer milestone questions adequately and the system is canceled (this last occurs more frequently than one would expect, especially for systems having no political protection).

The government directives setting up the processes we have described make little mention of HF and do not serve to advance HF interests because purely governmental functions, with the exception of monitoring, do not much influence the contractor, and the government's monitoring of contractor engineering development is quite weak, particularly with regard to HF.

1.2.5 HUMAN FACTORS DEVELOPMENT QUESTIONS

The DOD also mandates that certain behavioral activities be performed during the system development process. MIL-H-46855B (Department of Defense, 1979) requires that *as appropriate* for a particular project the following analyses and evaluations be performed: definition and allocation of system functions; information flow and processing analysis; estimates of potential operator/maintainer processing capabilities; identification of human roles; task analysis; analysis of critical tasks; workload analysis; studies, experiments, laboratory tests; development of mock-ups and models; dynamic simulation; application of HF to Preliminary and Detail Design; and test and evaluation of design.

The term "as appropriate" in the preceding paragraph was italicized because it describes a major qualification to these activities and a severe migraine to HFEs. Manifestly, the type and number of HF activities in any single development project vary depending on the complexity of the project, whether or not it has a predecessor system, the degree of similarity between the new and the prior system, and which behavioral activities were performed on that predecessor system. For minor developments that represent only modest changes in design, many if not most HF activities might be inappropriate; a very new, complex system might well demand all of them.

Consequently program management must be allowed to adjust the scope of its HF activities to system needs. For those program managers indifferent or hostile to HF, this qualifying phrase offers an easy rationale for reducing the HF scope on their projects, whether the reduction is warranted or not. Because of this qualification, the governmental customer who wishes any or all of these activities to be performed must specify them and provide money to support them, for most program managers regard HF as a requirement above and beyond routine engineering design, however much they claim that "good" human engineering is a routine part of that design. It is a fair bet that commercial system development includes only a few most important of these HF activities.

The activities specified in MIL-H-46855B (Department of Defense, 1979) are required because each system development phase raises behavioral questions that must be answered. These (taken from Meister, 1982) are listed in Table 1.2.1.

The HF activities performed to answer these questions are listed in abbreviated form in Table 1.2.2 (which is based on Parks and Springer, 1975). Despite categorization by developmental phases, behavioral activities tend to overlap these phases.

Table 1.2.1 Behavioral Questions Arising in System Development

Feasibility Analysis and Program Initiation

1. Assuming a predecessor system, what changes in the new system require changes in numbers and types of personnel?
2. What changes in tasks to be performed will require changes in personnel, selection, training, and system operation?

Demonstration/Validation

3. Of the various design options available at this time, which is the most effective from the standpoint of HF?
4. Will system personnel be able to perform all required functions effectively?
5. Will personnel encounter excessive workload?
6. What factors are responsible for potential error and can these be eliminated or compensated for in the design?

Full-Scale Engineering Development

7. Which is the better of two or more proposed subsystem/equipment configurations?
8. What level of personnel performance can one achieve with each configuration and does this level satisfy system requirements?
9. Will personnel encounter excessive workload and what can be done about it?
10. What training should be provided to personnel?
11. Are the following adequately human engineered: hardware/software; procedures; technical data; total job design?

Production/Deployment

12. Have all system dimensions affected by behavioral variables been properly human engineered?
13. Are system personnel able to do their jobs effectively?
14. Does the system satisfy its personnel requirements?
15. What design inadequacies exist that must be rectified?

Table 1.2.2   HF Activities in System Development

Developmental Phase	Behavioral Activity
Feasibility analysis and program initiation	Review engineering planning documents to ensure that behavioral factors have been considered.
	Review predecessor system behavioral analyses and other documentation.
	Compare predecessor and proposed system in terms of number and type of personnel required, new training needed, and other behavioral factors.
	Predict areas of significant behavioral impact and difficulty.
	Perform the following analyses (optional): Mission analysis Determination of functions
Demonstration and validation	Continue performance of mission and function analysis.
	As required, perform the following analyses: Function allocation Behavioral trade studies Task description/identification Task analysis Special analyses (e.g., information flow, workload) Develop HF section of TEMP
	Review and critique proposed designs.
Full- scale engineering development	Review and evaluate man–machine interface design drawings and procedures.
	Perform detailed behavioral trade studies.
	Recommend equipment features and control/display hardware to designers.
	Participate in design reviews.
	Make quantitative predictions of personnel performance.
	Update HF section of TEMP.
	Develop training requirements/program.
	Construct mock-up and conduct mock-up tests.
	Participate in developmental tests.
Production and deployment	Conduct HF part of operational tests.
	Evaluate and recommend system improvement modifications.

1.2.6   HFE ACTIVITIES DURING SYSTEM DEVELOPMENT

This section expands upon the list of HF activities in Table 1.2.2.

1.2.6.1   System Planning

The planning activities described in the preceding section are almost always conducted by personnel working for the individual government agencies sponsoring system development, for example, the Naval Sea Systems Command for a new class of warships. DOD employs a large number of HFEs and those located at headquarter facilities, together with a few at other government laboratories, may be involved in system planning. The "user community" (i.e., military services) are also involved in system planning. Those HFEs working for potential contractors such as McDonnell/Douglas, Boeing, Litton, and IBM normally have no part in this very early planning. When the system is contracted out for further development to a company in Preliminary or Detail Design, the HFEs working for that company become involved in development but have no role in government planning. The government's HFEs may continue to pursue the management planning activities described previously, and a few may be required to monitor the contractor's efforts.

At this point it might be wise to talk about industry development of products (e.g., appliances, office furniture) rather than systems, because HFEs may be involved in this also. Generally, although not invariably, most products are smaller and simpler than systems, but for the ergonomist the major difference between them is that the system incorporates its operators, whereas a product does not. In a system its operators form one of a number of subsystems and the ergonomist/designer attempts to program their behavior by specifying, for example, their selection criteria and their training to achieve maximum output and efficiency. The user of a product such as a toaster or washer is not part of a toaster/washer system; the product can be used as he or she wishes, although the developer attempts to program the user's behavior for safety reasons by, for example, providing use instructions. Because of this difference between the role of personnel in products and systems, the commercial product developer need not perform as many ergonomic analyses and evaluations as does the military system developer (Meister, 1973, 1984).

For a commercial system the developer will perform only those behavioral analyses and evaluations considered most important. In general the developer is more concerned about aesthetics and packaging, for which the industrial designer is responsible. If the operator is considered to play a critical role in the system, the HFE will be involved to some extent in system planning. Generally, even in military system development, unless the operator is considered highly important to system functioning, relatively little attention will be paid to operator concerns throughout system development.

Ideally the HFE will be involved in development from the start of system planning, but this rarely occurs, primarily because system planners view the system almost completely in terms of physical parameters. Efforts have been made in recent years to develop procedures such as the Navy's HARD-MAN Program (HARDMAN, 1979), which will involve HF more completely in the government's management processes, but it is unclear just how successful these have been.

At the start of new development two situations may exist. Often there is a predecessor system and the new system is an updated version of a previous one, cr the new system is a new class of a well-known general type of system. For example, the new system may be an updated command control system with greatly increased surveillance capability, which requires a new team organization to process information. The HFE may be asked to analyze the new system in terms of its similarities to and differences from the previous system to determine the impact of the new system on HF; what additional personnel, if any, will be required? What will they be asked to do? What will be the effect of this on required training? What performance difficulties, if any, might be expected from the new team organization? Will these have an impact upon system reliability?

Where the new system is a new class of a general system-type, for example, a highly advanced air superiority fighter, it may not be possible for the HFE to make a comparison with a single predecessor but if he or she is familiar with the general state of aircraft system development, the HFE may be asked to point out any special behavioral problems that might arise.

Occasionally there is no predecessor system and here, if the operator is important to system functioning, the HFE's contribution will become more significant. Certainly this must have been the case (or one would hope so) when the first space vehicles were being developed. This is because when no predecessor system exists there are many more unknowns about human performance that require examination.

The HFE's major role in a truly new system is to raise behavioral questions that need answering and to attempt to find answers or at least see that some resolution of the concerns noted is attempted. These concerns are communicated to the primary system planner in the form of memoranda and meeting discussions.

The HFE in a truly new system makes use of *mission* and *function* and (sometimes) *task analysis* (to the extent that the HFE has information). A primary requirement for this analysis is creative imagination and the facility for asking questions. The HFE imagines what the system might be like and how it might function under various design options; deduces from this the functions (and tasks) that must be performed; defines the behavioral unknowns in the system; examines alternative options (where these exist); points out potential problems, and suggests means of studying these problems.

System planning should conclude with a specification of system requirements (what the system must do to a particular level, e.g., range and speed, including alternative strategies for accomplishing these requirements). Included in the specification should be the anticipated effect of these alternatives on human performance.

System planning for a very new system is limited primarily by lack of information about its parameters. However, this very lack of definition gives the HFE an opportunity to influence development because it permits behavioral factors to be emphasized.

1.2.6.2 Preliminary Design

Despite tenuous demarcation lines between System Planning and Preliminary Design, in Preliminary Design the first outlines of the system configuration begin to take shape. If they are not already known, the following should be developed:

1. Specification of the individual subsystems and their functions (performed by system engineers).
2. Determination of number of system personnel required, their functions, and tasks (performed by HFEs).
3. At a gross level, description of the proposed means of implementing system functions (performed primarily by system engineers with inputs from HFEs and other members of the design team).

Preliminary Design is a complete design of the system at a system/subsystem level; ordinarily there is only limited developmental testing. In addition to the specification of functions and tasks, alternative design concepts are examined, and preliminary equipment drawings and procedures are developed and reviewed. Questions of technological feasibility are examined and resolved (however, such questions are rarely of interest to the HFE, except where the technology involves extensive or important operator interaction). In Preliminary Design those drawings that are of special interest to the HFE are of the man–machine interface (first level only). In Detail Design the Preliminary Design drawings are refined, revised, and brought down to a component level.

The mission analysis begun in system planning is now continued and expanded by the HFE. This is because the HFE may not have been involved in previous system planning; or the mission analysis performed may have been deficient; but, most important, because with the passage of time and continuing analysis by system engineers much more information about the system is/should be available to the HFE than previously. This additional information permits a more refined behavioral analysis.

The goal of HF mission analysis is twofold: determination of functions to be performed by system personnel; and specification of the behavioral implications of mission requirements.

For the HFE mission analysis requires gathering all available information about the system mission (which may mean applying to system engineers and reading documents), developing a sequential-event scenario (narrative description) of how the system is assumed to perform its missions, and analyzing system information in terms of goals, required inputs and outputs, performance requirements, environmental factors, and performance constraints (see Meister, 1971). The HFE looks for any of these that may have significant negative effects on human performance or will require special HF provisions.

More detailed functions are determined by progressively examining the inputs and outputs of more molar functions. If the overall function is, for example, to monitor engine status and some of the components of that status are engine temperature and turbine speed, the higher-order function is partitioned into more molecular functions of "monitor steam temperature," "monitor turbine speed," and so forth. If there is a predecessor system some of the mission and function analysis outputs of that system may be transferrable to the new one.

Function allocation is the process by which alternative design options are considered and one option is selected for implementation. It is *not* the simplistic determination of whether the man or the machine alone will do the job, as in the case of the Fitts list or some more advanced versions of it (Geer, 1981). Function allocation is in fact the essence of design, because it involves conceptualizing alternative ways in which the system/equipment/personnel could function. Prior to this, subsystem functions have been neutral in the sense that the way in which they would be implemented has not been specified. Each design option should now be described and analyzed in terms of inputs, constraints, performance requirements, and criteria, in much the same way as was done in mission analysis.

The major problems the HFE faces in function allocation are that many designers do not consider all possible design alternatives, do not analyze them systematically or communicate their details to others, and fail to make a systemic comparison of the alternatives.

In an ideal system development world the HFE would develop alternative system configurations, but the HFE (1) almost never has primary responsibility for design of a subsystem or equipment, and (2) rarely possesses the engineering knowledge that is needed, because the design alternatives must first be conceptualized and accepted or rejected in terms of physical mechanisms before one can consider how the operator will function in the configuration.

This usually places the HFE in the position of having to wait until the designer develops design alternatives or recommends a single option before the operator can be examined in that option. In either case the HFE should examine each option from the standpoint of its effect on personnel and indicate to the designer the behavioral strong and weak points of each.

In the process of developing design alternatives the engineer may find certain behavioral uncertainties that require what is called a "behavioral trade study." For example, such a study might be performed if there was a choice between using a manned aircraft or a Remote Piloted Vehicle (RPV) to perform surveillance and the critical question was the comparative accuracy of the pilot or RPV controller to maintain a designated track. A behavioral trade study is the systematic analysis of a design problem and the application of information from the literature and data from test results, and so on, to the problem. If the problem is sufficiently important, it is conceivable that the HFE might perform what is called a "quick and dirty" study (not elaborate, not necessarily fully controlled), but in most cases because there is usually no time for a formal experiment the trade study depends on extrapolation from already existent data.

Once the design alternatives are analyzed there are very formal procedures for making quanitative comparisons among them. These may involve paired comparisons based on criteria weights (Meister, 1971). There are also computerized methods for selecting an optimal design (see Meister, 1985, for a description of these). Although such methods are quite sophisticated, they are not often utilized (Williges and Topmiller, 1980). The comparison of design alternatives and selection of one by the designer is a somewhat intuitive process, although it may be performed on a consensus basis (within the design team).

The selection of the one final design configuration depends upon a set of criteria in which HF adequacy is only one and among the least valued of that set. In one study (Meister and Farr, 1966) designers were asked to rank the importance of criteria such as performance capability, reliability, and producibility. HF was ranked last. HF in the design choice plays an "exclusionary" role; just as in DOD Directive 5000.1 (Department of Defense, 1977a) it is considered as a constraint on system design; that is, any configuration in which the operator cannot function is of course completely excluded (e.g., a configuration that requires a single operator to lift 500 pounds manually); any configuration that presents great difficulty to the operator would be eliminated if all the configurations were otherwise equal. In no case, however, is a configuration *accepted* because it is best from a HF standpoint.

Task description/identification is described in Chapter 3.4; because of this our treatment of this activity is curtailed, despite its great importance (also see Meister, 1985). Task description/identification requires the listing for each function in each design alternative of all the sequential actions (tasks) that must be performed to implement that funciton. It is essentially taxonomic classification. Because it is taxonomic (and hence largely a matter of choice), task identification is fairly simple. The primary difficulty for the HFE arises in deciding on the level of detail a task should describe, for example, at the level of the individual control or display manipulation—"reads meter"—or at a somewhat more molar level—"determines that temperature is within acceptable range." By itself, task identification has no real effect upon design; it is only when the HFE analyzes the task that there is potentially an impact on design.

Task analysis asks questions about the task description. The HFE begins by selecting those tasks that should be analyzed (those that are critical, apparently complex, demanding of the operator, etc.), for the analytic process is too time consuming to waste on simple, routine tasks. Tasks are analyzed, as are mission and functions, in terms of their behavioral implications for equipment design, manning, selection, training, and test and evaluation. A complete list of the questions that can be asked would be much too long for this chapter but are included in Meister, 1985. Despite its importance to HF, task analysis has not, in my experience, had much specific influence upon equipment design, because it does not suggest design characteristics. However, task analysis enables the HFE to anticipate operator performance difficulties that may hamper system operations, including situations that may cause excessive workload, high probability of error, and so on. If such situations are found, the HFE explores them further with the primary designer to secure resolution of the problem or at least to alert the designer to the problem. Task analysis is also necessary for manning, selection, and training, but these do not have direct impact on the primary design.

The culmination of the analytic process is the application of *man–machine design analysis* (human engineering design principles) to the man–machine interface (MMI), which is usually a control panel, console, cockpit, or work station. These design principles have been described in detail by Van Cott and Kinkade (1972), among others.

1.2.6.3 Detail Design

Detail Design from a process standpoint is merely a continuation of Preliminary Design; Preliminary Design analyses are continued but at a more detailed level of the equipment rather than the system/ subsystem. During Preliminary and Detail Design one of the most important things the HFE does is to review equipment drawings of the MMI for adherence to principles of "good" human engineering (e.g., MIL-STD 1472C, Department of Defense, 1981). This is not the same as analysis and evaluation of alternative design configurations performed during function allocation in Preliminary Design. Those configurations describe how the *subsystem* will function; they are at a significantly more molar level than that of the MMI, which describes how the individual equipment will be operated, and with which we deal in Detail Design.

The design drawings that describe the MMI are what are called "top level" drawings, meaning that they represent the surface of the equipment, what the operators sees in terms of controls and displays arranged in a particular manner. This is the MMI for the operator. There is another MMI for the maintenance man which describes the *internal* arrangement of components, circuit relationships, accessibility to components, location of test points, and so on. The relationship between operator performance and control/display layout is relatively obvious (although not yet quantified); the relationship between internal equipment architecture and maintenance (troubleshooting) performance is quite complex and obscure, filled with uncertainty, because there is no preset step-by-step procedure for analyzing the architecture in order to determine the cause of failures. Most often the HFE evaluates the external MMI through drawings while ignoring the technician's internal MMI. The HFE should not be faulted for this; it is simply that the internal MMI is a problem infinitely more complex than that of the external MMI and available research (which is quite sparse) has provided little or no aid.

Drawing evaluation is attribute evaluation. An attribute is a characteristic that the MMI *should*, or, in the case of negative characteristics, should *not* have because of its effect on operator performance. The problem in attribute evaluation is to determine which MMI dimensions are important to personnel performance. The MMI has many dimensions, for example, size, shape, weight, color, number of controls and displays, and control-display arrangement, some of which are important, others less so. As far as I know, there has been no attempt to secure agreement among HFEs on a priority scale of importance, probably because most of those characteristics have not been tested against operator performance.

Beyond importance, MMI characteristics must be recognizable. Some equipment characteristics that can be conceptualized are very difficult to discern, even when the product being evaluated is three-dimensional. For example, the dynamic characteristics of a display, for example, frequency of display presentation or amount of information transmitted, are not readily observable because they are largely perceptual/cognitive. If MMI characteristics are described in a checklist, they must be expressed in such a way that the checklist user is quite definite about what they describe. How well HFEs can in fact recognize various attributes has never really been tested.

Much of the information the HFE wishes to derive from the drawing must be inferred. For example, the fact that meters are shown on a drawing suggests a requirement for a monitoring function, but may not describe the nature of what is being monitored.

Other information sources such as task analyses and specifications must supplement the drawing.

The development of equipment drawings begins very early in design and continues through various informal and formal design revisions until the drawing is released to production. Revisions are made as alternative design configurations are considered, revised, and refined. This is the iteration referred to previously as a general characteristic of system development. If the HFE wishes to affect the design he is evaluating, he cannot wait until the final drawing, because by that time design details have "hardened into concrete" and the evaluation must be timely.

In evaluating the MMI drawing the HFE begins either with a formal written procedure describing how the MMI will be utilized (although this may not be available if the development process is just beginning) and/or with a rough mental model of its operation. That is because the evaluation is based on the anticipated performance of the operator. In reviewing a control panel layout, the HFE asks questions such as, are controls and displays laid out in accordance with (1) sequence of operation (which is the most desirable condition because it is easier to learn and perform); or, if this is not possible because the control/operating procedure is nonsequential, is the layout in accordance with (2) the frequency with which controls must be operated or displays monitored; the relative importance of controls and displays; or in terms of similar functions (all controls mediating the same function located together, etc.). Above all, the HFE asks, what will be the effect of the design on the operator's performance?

A complete listing of all the questions that can be asked about a MMI would be lengthy and tedious at best; a MMI checklist based on MIL-STD-1472C (Department of Defense, 1981) for example, would run to hundreds of questions. That is because for each relevant equipment characteristic an evaluation question can be generated. For practical reasons, therefore, most HFEs do not evaluate to that degree of detail. Instead they scan the drawings to focus on a few critical aspects, which may vary from equipment to equipment, depending on the equipment's function.

Meister (1985) presents a list of 20 inputs, outputs, and physical panel characteristics that may lead to operator error. These include such features as rapidly changing displays, requirements for coordination with other operators, a large number of controls, and displays crowded together. The HFE need only think of all the job aspects that are undesirable, although their precise effect on operator performance is unknown.

The HFE must determine that not only does a design inadequacy exist, but that its probable effect on operator performance will be seriously negative. The HFE should not be concerned with trivialities because the designer will reject these automatically. This requires the HFE to determine (1) the effect of the deviant MMI characteristic on operator performance (e.g., will it frequently cause an error?); (2) the effect of that operator performance on the functioning of the subsystem of which the operator is a part. The HFE has difficulty making the first determination because he lacks a quantitative data base relating MMI characteristics to performance. Because of this the HFE's evaluation of drawings is highly intuitive and therefore dissatisfying. The determination of system effects can be made from information provided by systems engineers.

The output of the HFE's evaluation is a series of qualitative judgments indicating that such and such characteristics are unsatisfactory. (It is much easier to determine the negative attributes of the MMI than its positive ones.) If a formal checklist is applied, it is possible by counting positive and negative attributes to derive a quantitative "figure of merit" for the equipment represented by the drawing, but this is of little value and is almost never done.

Because in the early stages of system design a detailed operating procedure or task analysis is often lacking, and a data bank of error probabilities associated with equipment characteristics is unavailable, the evaluation of design drawings places a premium on the HFE's imagination and capability, especially when the MMI is complex. Empirical data on the consistency with which the HFE evaluates drawings during system development do not exist and there is no way of assessing the validity of the evaluations made except in a controlled (and hence artificial) experiment. In a study by Meister and Farr (1966), when five "experts" were presented with nine control panel drawings and evaluated them using specified checklist items and rankings of layout adequacy, consistency among the experts was poor.

As was pointed out at the start of this chapter, a significant problem in design evaluation is timeliness. When an HFE depends on sufficient notice by designers that a new MMI drawing has been produced, one may find that one has been "overlooked" and the drawing has been formally approved before one has had a chance to review it. For this reason HFEs working in system development have agitated for many years to secure "sign-off" authority, which means that the drawing cannot be formally approved unless their signature is appended. Designers have fought equally hard to prevent this dilution (as they see it) of their authority, claiming that the additional signature will unduly delay issuance of the drawing. So far the designers have been more successful in their arguments than the HFE has been. To ensure being "on top" of design activities, the HFE will have to make continuing periodic checks of the engineers designing the MMI.

Design evaluation is most often informal, a matter of the HFE standing over the designer's desk and commenting verbally on the drawing. This has the disadvantage that no record is kept of the comments made by the HFE; a verbal informal design review should always be followed by a formal memorandum to the designer. A *design review*, on the other hand, is a very formal meeting of all members of the design team for the purpose of formally reviewing their work and making binding decisions. The design review is held at periodic intervals, a week, two weeks, or monthly, for example. There may also be a design inspection at less frequent intervals at which customer representatives are present.

The importance of the design review is that it provides a formal podium for evaluation and criticism of the design to date. Notes of comments made and decisions rendered are recorded as part of the design archive. If for one reason or another the HFE feels that he or she has been "shortchanged," the design review provides a setting at which one must be heard. The HFE should be aggressive in asking questions, raising objections, and making recommendations. (Not too aggressive, however; one does not want to alienate the other members of the design team.)

As has been pointed out, the demarcation between Preliminary Design and Detail Design is quite tenuous, with most HF functions continuing to be performed at a more molecular level. The new element in Detail Design, at least as regards the HFE, is involvement in testing. Although testing in the sense of physical performance and data recording may have gone on in a very limited manner during Preliminary Design, it is only with Detail Design that it becomes really important. Testing is performed in three forms: mock-up testing; developmental tests; and planning for operational testing (OT). Our treatment of these topics is not complete, because the material is treated at greater length in Chapter 9.1.

Mock-ups

The most common types of mock-up are full-scale static and functioning mock-ups. The static mock-up for the HFE is usually a representation of a control panel, console, or aircraft cockpit. Occasionally a complete work station consisting of one or more consoles and other equipments is mocked up.

The static mock-up may be built of plywood, styrofoam, and other simple materials; the controls and displays it contains may be paper drawings or actual components. It may be possible to throw switches on and off, but the static mock-up is very limited: it cannot even be programmed to demonstrate equipment functioning.

The functioning mock-up can perform functions varying in sophistication from a demonstration of a preprogrammed sequence of equipment operations to routine operations, making use of microcomputers; the latter make the mock-up almost indistinguishable from what is ordinarily called a simulator.

A reduced-scale model is often used as part of the visual simulation of an airborne platform (e.g., a video camera scanning a 1/10 scale model of a terrain) but it can also be used for some less sophisticated applications, such as facility layouts (Pope, Williams, and Meister, 1978).

For the HFE the mock-up, unlike other system products, is not a routine development product. It is essentially a tool to assist the HFE by permitting limited testing of the MMI. Buchaca (1979) has provided a detailed listing of mock-up applications. These include aiding in the presentation of design concepts to management and the customer, as part of design reviews and inspections; aiding in review of the MMI; studying maintainability characteristics in terms of accessibility, component location, and so on; aiding in the detailed design of packaging and mounting, installation, room arrangement, and cable routing; developing and refining operating, maintenance, and installation procedures; and functioning as a training aid for user personnel.

The static mock-up is popular with HFEs because of its relative cheapness. It can be readily modified to agree with design changes. Indeed, if these changes are sufficiently great, as in proceeding from one prototype to another, another mock-up may be built quite easily. The more functional a mock-up is (the closer its characteristics approach those of an operating equipment), the more expensive it is. That cost does not usually come out of the engineering development budget unless the mock-up is required as part of engineering development (e.g., aircraft development; if the HF program has to absorb its cost, it should have an excellent reason for developing it). That reason can only be to answer questions that can be answered in no other way.

One can use a mock-up as a three-dimensional drawing and evaluate it with a human engineering checklist or a subject can be placed in or at the mock-up and asked to perform relevant tasks, such as throwing switches in accordance with an operating procedure. If the mock-up contains equipment that would be routinely checked, the HFE can have the subject find components, trace harnesses, connect or disconnect cables, remove and replace cover plates, and so on.

Further examples of mock-up use can be secured from Hawkins (1974); McLane, Weingartner, and Townsend (1966); Janousek (1970); Seminara and Gerrie (1966); and Gravely and Hitchcock (1980).

The mock-up as it has been described is a representation of hardware. The mock-up as it applies to computer software is quite different. To compare alternative software configurations it is necessary to develop a program for each alternative, but once this is done, these programs can be tried out on an already available computer. Even if one did not wish to develop and try out alternative programs, the one program selected as the design choice can (indeed, must be) tried out to discover and remedy its "bugs." From that standpoint mock-up testing of software programs is inherent in the development of the programs and can therefore be implemented as a routine part of computerized system development rather than as something peculiar to HF.

Development Tests

The Department of Defense (see Holshouser, 1977) specifies that certain tests be conducted during development and denotes them as developmental tests (DT & E). DT & E is required for all acquisition programs and is conducted in four major phases:

1. DT-I is conducted during the planning phase to support the program initiation decision. It consists primarily of analysis and studies to derive the human factors/system requirements.

2. DT-II is conducted during the demonstration/validation phase to support the full-scale development decision. It demonstrates that design risks have been identified and minimized. It is normally conducted at the subsystem/component level, up to and including employment of engineering models for final evaluation.

3. DT-III is conducted during the full-scale engineering development phase to support the first major production decision. It demonstrates that the design meets its specifications in performance, reliability, maintainability, supportability, survivability, system safety, and electromagnetic vulnerability.

4. DT-IV is conducted after the first major production decision to verify that product improvements, or correction of design deficiencies discovered during operational evaluation, follow-on test and evaluation, or operational employment, are effective.

Developmental tests are usually engineering equipment tests designed to study equipment design characteristics. They include tests of breadboards and prototypes, first production articles, qualification tests, and engineering mock-up inspections. Because developmental tests are specifically oriented to engineering purposes, for example, tests of materials, nondestructive reliability tests, their objectives often fail to encompass human factors. Nevertheless, it is sometimes possible for the HFE to gather information of behavioral value from these tests; the HFE ought to be on the lookout for those that have some value for HF. Those include any test that involves the MMI and particularly a test in which personnel take part by "playing" the role of operators.

The methods used to secure behavioral data from developmental tests do not differ substantially from those used with mock-ups, for example, recording of performance time and errors made, use of checklists, observations and interviews. (If there is an advantage of the developmental test over the mock-up, it is that the stimulus situation—the equipment—is more realistic, although actual test operations may not be.)

The great disadvantage of gathering data from developmental tests is that they *are* developmental. This means that unless design is more or less stable at the time of the test, a changing system configuration may render the collected information useless for evaluation purposes.

If a developmental test indicates a system inadequacy, the system is likely to be changed, which is exactly the reason for the test. The HFE may wish to consider just how stable design is before endeavoring to collect data from such a test.

The Operational Test

Department of Defense directives require that before a system is accepted by the military, it must be tested operationally. The military divides OT & E (Operational Test and Evaluation) into two major categories: initial OT & E, which is all testing accomplished prior to the first major production decision; and follow-on OT & E, which is all testing thereafter. Ordinarily, OT & E is performed by customer personnel (i.e., the military or civilian HFE representatives of the military); only rarely in initial OT & E will contractor personnel be involved in this test. Follow-on OT & E will be conducted by the government agency, with the prototype system under its control, its test planning performed by service

HFEs and others, utilizing operational personnel as subjects. One should not, however, assume a complete dichotomy between DT and OT; these tests overlap, with later DT and earlier OT shading into each other.

In the Army the final acceptance tests of the systems are planned and conducted by the Operational Test and Evaluation Agency, Ft. Belvoir, Virginia; in the Navy, by the Operational Test and Evaluation Force, Norfolk, Virginia; and in the Air Force, by the Flight Test Center, Edwards Air Force Base, California. Depending on the characteristics of the individual system, its financing and scheduling, the sequence of testing may be compressed and combined into fewer but more complex tests. Do nonmilitary systems proceed through such detailed tests? Data on this point are lacking. However, the *logic* of HF testing is independent of the military or civilian contexts; what is described here is how the tests *should* be performed; but often there are many deviations (General Accounting Office, 1984).

Certain requirements exist if a prototype system is to be tested in a truly operational mode. The system must have been fabricated as a complete entity; personnel to operate the test system must be representative of (similar to) those who will eventually operate the system and must have been trained to do so; procedures to operate the system as it is designed to function operationally must have been written; an environment representative of the one(s) in which the system will routinely function (e.g., Alaska in winter for cold-weather operations) must be prepared for testing.

There are many constraints on OT, not the least those involving financial, political, and schedule factors. In consequence OT as performed for the military is much less sophisticated than most controlled laboratory studies. For example, the military has a preference for relatively simple, subjective measures of performance, without any elaborate HF instrumentation.

At the same time test management expects much less of HF in these tests than HFEs perhaps expect of themselves. One can view OT theoretically as an attempt to determine the relationship between a particular subsystem output (personnel performance) and total system output. HFEs want to know how the personnel subsystem affects system output and how the nonpersonnel elements of the system affect personnel performance. What test management wants to know is much simpler: can personnel do their jobs, and what serious behavioral problems, if any, exist.

In the course of conducting the HF part of an OT the HFE makes use of many techniques (e.g., objective measures, observations, interviews, and ratings). There is no single methodology that one can point to as being distinctively OT. What makes OT distinctive is that it provides a framework in which these other techniques can be applied; it organizes and systematizes them. OT adapts general methods to a specific system and problem. Another distinction of OT is that it attempts to measure under conditions of *maximum fidelity* to the operational environment. In this OT differs fundamentally from laboratory-oriented psychological studies, which are only minimally concerned, if at all, with fidelity to a particular setting. The HFE must decide whether the system will be routinely exercised in a special environment such as the desert; if not, there is no need to simulate that environment. The development of the OT plan (the TEMP) is a major planning task for evaluators because of the many facets (only one of which, the behavioral aspect, we are concerned with) that large, complex systems display. Depending on the size of the system, the planning task may require the continuing services of half a dozen personnel for the better part of a year or more because of the many reviews and revisions the TEMP undergoes.

The HFE is usually only one among a number of specialists developing the test plan, and HF requirements must be compromised with those of others. The test plan may have a separate section dealing with personnel performance, but HF requirements do not take precedence over those of the engineering specialties (rather the reverse is true).

The HF part of the TEMP describes the purpose of the test generally and specifically; indicates any experimental comparisons that must be made; specifies performance standards, criteria, and measures; indicates the test instruments and how data are to be collected; and reviews subject characteristics, the test schedule, and data analysis methodology (see Meister, 1985).

The amount of control the HFE will have over the test situation and the constraints HF must face may determine the kinds of measures that can be applied and the data collection methodology. If, for example, the HFE cannot interview system personnel because system operations occupy them fully, or if the workplace is so cramped that a video camera cannot be used, the HFE must take this factor into account in planning.

Of all the activities the HFE performs during development, planning and conduct of OT are perhaps the most effective. To a large extent this is because the HFE's formal university training (if his or her background is psychology) has emphasized performance measurement. It would be inappropriate to describe OT methodology because Chapter 10.1 deals with this, but certain points should be noted.

In planning the HF OT the HFE will be faced with the critical questions of *standards* and *measures*. A standard is some performance output that personnel are expected to achieve. A measure is a personnel performance or system output that can be used to describe the *quality* of that performance.

The purpose of OT as it applies to personnel is to verify that personnel can perform to explicit or implicit standards imposed by the system and its mission. To determine whether personnel are performing adequately one must compare their performance with some standard. The problem for the HFE is that standards exist for the overall system—for example, an aircraft is supposed to achieve a designated speed, range, or fuel consumption—but the system standard almost never describes what personnel must do.

Theoretically the standard for any task can be derived by function/task analysis from system/mission requirements. However, the thread relating an individual task to a system requirement may be extremely long and convoluted and hence difficult to disentangle.

Lacking human performance standards, it is impossible for OT to verify human performance except in a very limited way. If the system completes its missions adequately (a system engineering judgment because it deals with terminal outputs), and no outstanding operator performance deficiencies are noted, this *suggests* that personnel can do their jobs.

From the standpoint of the HFE whose training has been scientific this is a very unsatisfactory situation even though it will satisfy most test managers. If one has no standards with which to evaluate performance, OT serves as merely a context in which to discover human performance inadequacies, although from a purely developmental standpoint the uncovering of such deficiencies and their subsequent remediation are very important.

Because a standard is ultimately a value judgment, it is reasonable to utilize subject matter experts, for example, experienced mechanics, to attempt to derive a specific quantitative standard for a task.

In any system the number of performance outputs that *could* be measured might bewilder the HFE because there is literally an embarrassment of riches. It is therefore necessary to select a subset of these performance outputs as measures.

All measures are not equally useful in describing performance because some of them may be only indirectly related to performance. Moreover, measures may be more or less detailed, depending on the task level they describe. To measure rifle firing proficiency, for example, one could record the tremor of the finger in squeezing the trigger or, preferably, the number of hits on a target. Measurement adequacy is determined by the immediacy of its relationship to the system output. The level at which one measures may determine in part *how* the data are collected. For example, instrumentation would probably be required if one wished to measure trigger pressure but not if one measured error in hitting the target.

The HFE would prefer, all things considered, to make use of objective, observable measures that are almost always performance time and errors. The problem is that although it is not hard to measure these it is more difficult to make sense of them. Unless objective measures are *critical* for a task they mean very little. For example, if there is no maximum time in which a task must be performed, who cares how long it takes? Because in the absence of a performance standard objective measures may not be very meaningful, the HFE may have to rely on subjective evaluation of performance by supervisors or experts or on the opinions of subject test personnel. For an instructor pilot, for example, to say that a trainee's performance is adequate involves a performance standard, even if that standard can only be inferred.

1.2.6.4 Production

This phase overlaps Detail Design and Test and Evaluation. Because it is largely a fabrication activity and hence the province of industrial engineers, only a few HFEs involve themselves in factory problems, which from a behavioral standpoint center on workmanship difficulties.

1.2.6.5 Test and Evaluation

The major HF activity in this phase is the conduct of the operational test. It is important to remember that the HFE is not in charge of the OT; the test is performed primarily for engineering purposes.

If all goes well, OT is performed as it was planned in the TEMP. Data collectors and data collection forms and instrumentation are placed in readiness at the test site; the system is energized; subjects (test personnel) proceed to perform in accordance with operational procedures. However, even in follow-on OT events rarely move precisely as desired: equipment breaks down or performs erratically; scheduled events are delayed; weather (when it is important) does not cooperate; test management may decide to deviate from the test plan. The HFE in charge of HF testing must be alert and quick to modify the plans, if necessary.

One of the potentially serious limitations on all forms of OT is that there may be an insufficient number of data points because of these potential disruptions. The number of task "runs" may be insufficient for practical, much less statistical, confidence. Fortunately test management is much less impressed with statistical tests of significance than with practical answers to questions such as was the mission completed successfully; were all required tasks performed correctly; could the operating procedures be carried out as written? Mission/task accomplishment, procedures, human engineering of equipment, technical documentation—these are the things the test manager is most concerned about with regard to HF, not number of errors made or elaborate statistical analyses, which have little meaning for the manager. For test management the test is performed primarily to uncover and resolve serious difficulties. The HFE should include statistics that are relevant to the test purposes, but statistics are useful primarily as supporting, not primary, evidence.

The HF part of the final test report must of course contain the usual research study headings: test purpose; method; results; conclusions; but what test management looks for are conclusions and recommendations. There are two types of HF conclusions: how well did personnel perform and *problem areas*.

All problems described should be significant ones, significant, that is, with regard to system goals and outputs; test management is not much interested in problems that do not have significant impact on the system mission, unless these can be easily remedied by a change in written procedure or labeling, for example. Urgent problems (which must be documented in terms of effect on system performance) should also be remediable by means other than by redesign of hardware or software. Only if a problem is highly significant for system success, a criterion that does not include personnel unless it is a safety problem, will a design change be considered and even then every other possible means of solving the problem will be explored first. This may involve the HFE in a certain amount of problem investigation, even detective work, resembling criminal detection, if the cause of the problem is not immediately obvious from the test results.

OT is not scientific research and the HFE should not expect it to be, although it is possible to gather data that, when combined with other data from system tests, aid in building up a corpus of data for personnel performance prediction purposes.

Human Factors in Selection and Training

The preceding sections have emphasized the role of the HFE in design of the primary system, hardware, software, procedures, and all other job aspects. The HFE may also be involved in establishing personnel requirements for selection and the development of training. Both these areas are discussed in detail in Chapters 8.1 and 8.2, but a few words should be said here to set the stage.

Selection requirements should be derived from the task analysis performed in Preliminary and Detail Design. Specific procedures for performing a skills requirements analysis are given in Chapter 8.1 but if the system has a predecessor the HFE will probably take advantage of the analysis performed in the development of that system, modifying selection criteria only to satisfy demands imposed by equipment modifications.

In System Planning one of the activities performed by the HFE was to examine the changes that had to be made in the new system (assuming it had a predecessor) and assess the expected impact on training requirements. At a later stage, probably some time in Detail Design when the system configuration has "settled down," work begins on the development of a formal training course. If the system is a military one (and even if it is not; ISD is becoming popular in commercial training), the training course is developed in accordance with principles of Instructional System Design (ISD), which are described in Chapter 8.4 and in Meister (1985). The starting point for the ISD process is a detailed job and task analysis that builds on the primary system task analysis. Whether ISD is implemented by the contractor's training staff or by a special training group within the customer's facility depends on the individual arrangements made for the specific system. The contractor will almost certainly be responsible for initial factory training of system personnel.

For a system of any magnitude training is almost always developed by training specialists whose background and expertise vary widely from that of the HFE, whose responsibility is primary system design. Training specialists have a background in education, HFEs, more one in psychology. Even their job titles may differ. For smaller systems and companies that do not employ many specialists, the primary design HFE may also be given the training responsibility.

1.2.7 THE HFE AND THE DESIGNER

1.2.7.1 Introduction

In this discussion the term "designer" represents everyone working in system development other than the HFE (e.g., project manager, system engineer, design engineer, draftsman.) The relationship between the HFE and the designer is one of the two most important factors determining the HFE's effectiveness. The other important factors are the data, principles, and design guidelines that support and direct what the HFE does (to be discussed below). The importance of the relationship between the designer and the HFE is that the former controls the decisions made during system development.

Comparatively little research has been done on the relationship between the HFE and designer. As a consequence, whatever we do know is based mostly on HFE experience and is largely anecdotal. However, these anecdotes have been common currency long enough so that we must take them seriously.

What we do know can be summarized in a series of statements: (1) Most designers are at best neutral, at worst actively hostile, toward HF. They do not read HF material and they accord HF a low priority in the design process. (2) Their attitudes toward HF are bolstered by a series of beliefs or assumptions that are largely erroneous or simplistic. (3) We know very little about the basis on which the designer designs. He has difficulty communicating his rationale for his design decisions but, from what we can observe, his design process is extremely conservative, not completely systematic, somewhat primitive, and based on the principle of least effort. (4) Many HFEs tend to be somewhat reactive toward the designer and the design process because there must first be a design option before it can be examined for behavioral aspects. Consequently many designers tend to view the HFE merely as a design critic. (5) Because of the designer's somewhat negative attitude toward HF, the HFE spends a good deal of time in efforts to indoctrinate and persuade the designer to assume a more positive viewpoint, with somewhat dubious success.

1.2.7.2   Designer Attitudes Toward HF

A number of studies have attempted to assess designer attitudes toward HF. As has been pointed out, designers rank HF last in terms of the criteria they would employ in judging the worth of a design. HFEs feel that the attitude of designers to HF is generally negative, although some HFEs feel that it has improved slightly over the years (Meister, 1979). Anecdotal evidence suggests that HF is generally ignored by designers so that HFEs must make a determined effort to secure access to design drawings and meetings. Efforts to provide designers with HF information tend to founder because of their reluctance to use HF handbooks such as Van Cott and Kinkade (1972) and Woodson (1981). Meister and Farr (1967) observed in their sample of engineers that only one or two had a copy of a standard HF reference available to them. Despite this, HF researchers expend much effort in writing HF handbooks which their authors believe will influence designer attitudes positively. The problem of how to communicate effectively with designers has tended to preoccupy HFEs over the years.

1.2.7.3   Designer Misconceptions

These are not formal assumptions but rather inarticulate belief structures that can be deduced from designer actions and arguments. If asked to agree or disagree with statements describing these beliefs, designers might well disagree, because these beliefs are relatively unconscious. Yet these misconceptions are extremely powerful barriers to the acceptance of HF activities and inputs. Designers feel that:

1.   Whatever the characteristics of the system, the human is sufficiently flexible that he can compensate for design inefficiencies. This is a reflection of our historical and cultural self-confidence that Americans are resourceful, innovative, and inventive. There is partial support for this: minor inadequacies can be overcome; but HFEs feel that this is not true of major ones. Moreover, the ability to overcome design inadequacies (the bailing wire and string syndrome) is probably true only as long as system personnel are unstressed. Stress in combat tends to rigidify personnel and magnify system weaknesses.

2.   The system can "buffer" or compensate for the effects of personnel inadequacies (e.g., errors, increased response time, reduced performance quality) so that inefficient human performance does not really have a significant negative effect on the system. If this were true there would be no need to consider personnel in design of a system except for selection and training. There is, however, considerable empirical evidence related to accidents and safety, for example, that this is not true (see, for example, Meister, 1971). The counterargument is that few systems can be made completely "personnel-proof," and if they were completely automatized at tremendous expense and complexity, our concern for inadequate operator performance would then transfer to the maintenance technician, whose performance would then degrade because of added equipment complexity.

3.   Another misconception is that including HF inputs in design will not solve the human–system mismatch problem; in other words, that HF as a discipline is ineffective. Great efforts have been made to counter this argument by compilations of case histories documenting HF contributions to design and system efficiency (e.g., Price, Fiorello, Lowry, Smith, and Kidd, 1980; Sawyer, Fiorello, Kidd, and Price, 1981). However, when inputs of various disciplines enter into the design process, it is very difficult to extract any single contribution that was significantly more important than another. As a result, convincing empirical evidence testifying to the worth of HF does not yet exist.

4.   Another designer objection to HF is that the cost of including HF in design is excessive. I have heard that the maximum cost of a full-scale HF design program (primarily salaries) ranges from 1 to 3 % of the total system cost, which if true, seems a small enough price to pay for enhancing operator performance. When designers object to HF cost, however, they refer to the cost of modifying already designed configurations or hardware. They assume that HFE recommendations will always involve design modifications. The counterargument is that if the HFE is involved in the design process at the same time that other engineering disciplines are involved, his inputs are made before design is "cast in concrete"—and so there is no added cost.

5.   Designers commonly assert that "good" engineering automatically includes consideration of system personnel and HF. Hence special efforts such as those of the HFE are not needed. This is the way it should be, but in practice this is usually not the case. The designer concerned about equipment problems has no inclination to think about personnel factors. Moreover, HF is a specialized discipline;

it is not true that anyone (without training) can apply it just because he or she is human and can empathize with a human operator. Even that empathy is misplaced: the designer identifies with an idealized operator who has the same background and training as the designer's own, but most operators are much less skilled.

1.2.7.4 How the Engineer Designs

We are concerned about how the engineer designs because ultimately all HF principles applied to design must be incorporated by the engineer. We assume that how engineers analyze design problems determines how they use HF information and how they include operator considerations in design. Information has value to the designer only to the extent that it can be related to the design task. If, then, the HFE knows how the engineer designs, he or she can perhaps do a better job of supplying usable HF data.

Studies examining how designers make use of HF inputs have used three methods: First, in the structured design exercise, or simulation, a design problem developed by the researcher was presented to the individual engineer for solution. The major variables in studies using this technique (e.g., Meister and Sullivan, 1967; Meister, Sullivan and Askren, 1968; Eastman, 1968; Lintz, Askren, and Lott, 1971) were type and the presence (or absence) of HF data made available to the engineer subjects.

The results of these studies suggest that engineers do not perform much deliberate systematic design analysis but exhibit what Meister (1971) calls "an obsession with hardware." They rely heavily on design solutions that had worked well for them in the past. This prevents them from considering novel approaches to design. The speed with which the engineer develops his design concept is matched by his relative inflexibility in modifying that concept later, except in minor details. The engineer's use of HF information is determined almost exclusively by design requirements. The engineer prefers to design with a minimum of advice or constraint from others (hence the notion of the HFE as a "critic") and of course lacks patience to consider operator factors in making design decisions. In general, the kind of HF inputs the engineer understands best (even though he or she may not like them) are those that constrain the design (e.g., maximum number of personnel for which equipment is to be designed) or that deal with concrete system operations.

In the second method as utilized by Meister and Farr (1967), Rogers and Armstrong (1977), and Rogers and Pegden (1977), engineer subjects are presented with selected types of HF data. Typically this information has been extracted from existing HF data documentation such as Van Cott and Kinkade (1972) and developed in alternative formats (Meister and Sullivan, 1967). The variables in studies using this technique have been types of information and formats. Measurement consisted of assessing the users' preferences for alternative formats, and their ability to extract the "correct" information from the various types of data presentations.

The results of these studies suggest that designers are more receptive to data phrased in quantitative, graphic, or tabular terms than to data that are qualitative and verbal; and that they resist complex verbalisms, probably because they are not verbally fluent.

They are relatively indifferent to abstract, general HF inputs. They prefer that all inputs relate very specifically to their design problem. This effectively eliminates most general written HF materials.

In the third method groups of "experts" are asked about the importance of certain HF factors on system design or operation (see Whalen and Askren, 1974; Potempa, Lintz, and Luckew, 1975). Skill is one of the most important of these factors because a high skill requirement in the operator of equipment drives up the personnel cost associated with the system. These studies have demonstrated that operational personnel can estimate the impact of certain types of system/equipment characteristics on the numbers and types of personnel required (Potter, Korkan, and Dieterly, 1975; Potempa et al., 1975; Whalen and Askren, 1974).

Does HF data impact on designers? Meister et al. (1968) concluded that manpower quantity and personnel skill constraint data do have some small impact on the equipment configuration, a finding based on detailed case studies of the design processes of six system engineers, who designed to a specification under controlled conditions. In a follow-on study Meister, Sullivan, Finley, and Askren (1969a) found that the amount and timing of human resources data inputs do exercise some influence on design. Although system requirements are paramount, the results of these studies indicate the importance of emphasizing HF in system documentation; this helps keep the concept in front of the designer's mind and leads to greater consideration of HF. Meister, Sullivan, Finley, and Askren (1969b) found that the engineer relates a number of design concepts and characteristics, such as test points, internal components, checkout and troubleshooting procedures, and type of test equipment required, to the skill level of the maintenance technician.

Askren (1976) hypothesized that engineers and managers resist considering the operator, with his various attributes and costs, as a hardware design constraint, but that manpower-related factors would be acceptable to engineers and management as a "tie breaker" when all other engineering factors were equal. (This relates to the "exclusionary" principle mentioned previously.)

Lintz, et al. (1971) found a negative correlation $(-.32)$ between utilization of HF data by design engineers and experience; HF data related to costs and numbers were considered almost three times more valuable than data related to skill type or personnel availability. However, Lintz et al. (1971) consider that although designers will include HF data in engineering design trade studies, the trade-off process very much depends on the personal style of the engineer.

Askren (1976) theorized that before HF data could be effective, it would be necessary to provide data to the engineer regarding the effect on the operator of "choice point alternatives" in the design process. Unfortunately, engineer subjects did not agree on the value of any trade-off study parameter, including HF data, nor did they ever request HF data for use in solving the problems presented (Askren and Lintz, 1975).

Hornick, Robinson, Rogers, and Sullivan (1981) attempted to determine the kinds of skill concepts engineers apply to their designs and whether the sophistication of these skill concepts could be increased by presenting engineers with a structured framework based on behavioral research.

Engineers were asked to estimate skill levels required by operation and maintenance tasks, first in an *unstructured* format by listing the most important tasks to be performed in the operation and maintenance of equipment on which they had recent design experience. The engineers had to indicate the skill and the degree of skill required by these tasks. Then in a *structured* format they sorted cards describing specified tasks into five levels of a specified cognitive or psychomotor skill.

The results of this study suggest that engineers have relatively few and nondifferentiated concepts of skill, although they consider that equipment maintenance requires a higher level of skill, oriented primarily on cognitive capabilities, whereas operating tasks require fewer psychomotor abilities. The effect of the difference between the two formats indicates that it may be possible to increase the sophistication of engineers' skill concepts by providing them with a structured situation procedure that leads them through the skill-analysis process.

What can one say in summary of the research on the engineer as designer? The paucity of data on the designer and design situations is frustrating; most of the studies cited are from the 1960s and early 1970s. Since then, nothing has been published on the subject. We still have only a very unclear idea of how the engineer designs; engineers are not very introspective or verbal and it is quite possible that they are unaware of their own processes so that they cannot transmit them to others. Because so much depends on system requirements, it is clear that only *authority*, represented by the customer for the system, will force the designer to pay attention. This can be done only by including HF considerations as design requirements in contractual documents.

1.2.8   DATA SUPPORT FOR HUMAN FACTORS

In addition to designer attitudes an essential factor determining HF effectiveness (if we define effectiveness as the incorporation of HF inputs in design) is the availability of data and design guidelines. The term "data" as we use it is all-inclusive; the data may be either "raw" (quantitative values in individual studies) or "processed" (compilations of data in a taxonomized format such as that found in human reliability data bases (e.g., Munger, Smith, and Payne, 1962). The term also includes principles and conclusions (usually qualitative) and procedures for performing analyses and evaluations (also usually qualitative).

Data that can be applied to HF efforts come in several forms: military standards, the most important of which in design is MIL-STD 1472C (Department of Defense, 1981); procedural reports (e.g., Berson and Crooks, 1976); reference books such as Van Cott and Kinkade (1972) and Woodson (1981); teaching texts such as McCormick and Sanders (1982); human factors journals such as *Human Factors* and *Ergonomics;* and proceedings of symposia such as those of the Human Factors Society, the International Ergonomics Association, and special conferences. General literature such as psychological journals and books on learning theory and industrial psychology are of limited value to the HFE.

Before considering how useful these sources are for HF during system development, it is necessary to determine just what data the HFE needs. We assume that the more relevant a datum is to the analyses/evaluations the HFE performs, and to the design questions listed in Table 1.2.1, the more useful the data are; and data in quantitative form should be more useful than qualitative data.

These are generalities with which few would quarrel. However, I assume more specifically that underlying HF analyses and evaluations is an attempt to draw a quantitative relationship between some characteristic(s) of the system or the job as a whole and its effect on anticipated operator performance. A characteristic is desirable (from a HF standpoint) if it produces enhanced operator performance; it is poor if it results in lowered performance, for example, greater likelihood of error, slower response time, or poorer response quality. The essence of mission analysis, for example, is the drawing of behavioral implications from system requirements; and what are those implications if they do not describe the relationships between requirements and performance, preferably in quantitative form?

These implied relationships suggest the necessity for a data base consisting of error probabilities (because all performance is probabilistic) and response times as a function of individual (and combinations of) system characteristics. No such adequate data base exists today, although a very primitive prototype was published in 1962 (Munger et al., 1962) and is still used. There is a tremendous need to update and expand that base.

HF in design rests on certain principles that are most clearly enunciated for military systems in

MIL-STD 1472C (Department of Defense, 1981) and for software design in Smith and Aucella (1983). For the most part, these contain relatively simple and obvious behavioral requirements, the implementation of which presents not much difficulty. There are, however, many *inferred* behavioral requirements (perhaps more important than the obvious ones) for which the implementing mechanisms are not so apparent. It can be argued that there is a behavioral requirement to reduce the operator's workload to the point that he learns the operating procedure easily, feels no pressure during operation, and makes no inordinate number of errors. How does the HFE implement this requirement? By reducing the number of controls and displays to a minimum, by reducing the flow of information? As systems become progressively more complex, the relatively simple, overt requirements one finds in MIL-STD 1472C are likely to become less relevant and the inferred behavioral requirements more important. For the complex, inferred requirements what are needed are design guidelines the HFE can follow. These are procedures for incorporating behavioral requirements in design. Ideally, they should contain the following:

1. Description of the behavioral requirement to be incorporated in design, for example, reduce operator workload.
2. Alternative ways of incorporating the requirement into design, that is, alternative design solutions of the problem, for example, reducing the number of controls and displays, reducing the amount of information transmitted by displays, or providing "prompting" displays.
3. The effect of each design alternative on the operator's performance.
4. Advantages of incorporating the alternative design features described in (2).

This last would permit the HFE to perform a cost–benefit analysis of the alternative design solutions. The advantage of a design alternative is the enhanced performance that will result from the feature compared with performance without that alternative. The reason for item (4) is that some designers manifest tremendous resistance to the concept of behavioral inputs in design, based in part on skepticism that the design recommendation the HFE wishes to incorporate will have any discernible effect on operator performance. The only way the HFE can counter this argument is to demonstrate by empirical data that the design feature is worth implementing.

The two types of data needed (quantitative performance data bases and design guidelines) interact because the effects and advantages of a design guideline are ideally expressed in quantitative performance terms such as successful task accomplishment or error probability.

Needless to say, such HF design guidelines do not exist, primarily because research leading to their development has not been performed. One wonders *when* that research will be performed.

Presently available HF research studies job design in operator performance terms but does not specifically deal with the design (physical equipment or procedural) variables that determine that performance. For example, we have many studies of workload measurement and prediction, but few if any that investigate workload as a function of specific design characteristics. From the standpoint of the HFE, the research is tangential to his real concerns; it just misses the mark. In addition, each study is idiosyncratic, each utilizing a different methodology, a different metric, and so on. There are few if any sustained, systematic research efforts seeking to answer design questions. Material in individual reports is quantitative; however, individual items not combined with material from other reports, not formatted in data base terms, have very limited utility unless by chance they are specifically relevant to the HFE's immediate problem. Data from a single report represent only a single instance and hence one can have only limited confidence in their validity. The HFE working in system development has no time, however, to perform the scholarly task of integrating data from many individual studies.

The effect of this lack of applicable data is to make the HFE rely largely on common sense and intuition, which are admirable in themselves, but they are not science, for science must go beyond common sense and intuition.

The need for additional research is almost a cliché in the conclusions of research reports; and for HF the great need is for research to support HF in system development. It is not that too little research is being performed (although more research is always desirable) but that so little of it is meaningful to the HFE. The research interests of the government (which sponsors the vast majority of HF research) are directed to the exotic, to the sophisticated, to "advancing the cutting edge of technology" as it is expressed in Washington. HF research is driven in large part by the newest technology, rather than by the apparently mundane design questions of Table 1.2.1. So, for example, the most attractive topics at present to those who fund HF research may be artificial intelligence, microprocessor applications to command, control and communications systems, decision-making mechanisms, and so on. It is necessary to say "at present," for governmental research interests change frequently because sponsors expect results rapidly and often prefer to switch their money to a new topic that now appears to have greater success potential.

For example, efforts over the years to gain support for the development of quantitative performance data bases have been rejected on the ground that their development is not "research." For anyone who truly understands the data needs of system development further comment is unnecessary.

1.2.9 IS HUMAN FACTORS WORTHWHILE?

This is perhaps not a question that one should ask of a scientific discipline; it would never be asked of HF in its research mode; and certainly it is not one that is asked of engineering. It is necessary, however, to raise the question because many HFEs wonder about the utility of what they do. They recognize that the sociocultural milieu in which they function (see Perrow, 1983, for an excellent discussion of HF in system development from a sociological point of view) is very negative and they wonder how much they achieve for their efforts.

Another way of asking the question is, does HF achieve its goal of developing more effective systems? There is no hard evidence on this point. Many HFEs reason in reverse; what they see of systems with serious human engineering deficiencies because these systems had little or no HF consideration suggests that, if there were no HF involvement at all, our systems would be much less effective than they are. This reasoning has no empirical confirmation.

One might ask, given the most positive circumstances in which to provide HF inputs, can HF do what it purports to do? Again, there is no empirical evidence, but it is necessary to recognize the severe disciplinary deficiencies under which we labor, deficiencies not so much in methodology as in supporting data. It is, however, in the power of HFEs to make some small efforts to improve *on their own* (i.e., without the aid of formal experimental studies). A small step forward might be to follow my suggestion (Meister, 1983) that HFEs voluntarily keep detailed records of what they do and how they do it, and to publish these, so that a more accurate and complete picture of HFE activity and system development circumstances can be built up.

For most HFEs working in system development most of the topics discussed in the remainder of this book have significance only if they provide or will ultimately provide data or principles useful to HF in system development. This is probably unfair to HF research, which has its own rationale. Originally (World War II) the HF researcher and the practitioner were one and the same, but increasingly both have become individual (and, to a great extent, noninteractive) specialities. This is unfortunate because each has much to offer the other.

REFERENCES

Askren, W. B. (1976, March). Human resources as engineering design criteria, (Report AFHRL 62703F 11240103). Wright-Patterson Air Force Base, OH: Air Force Human Resources Laboratory.

Askren, W. B., and Lintz, L. M. (1975). Human resources data in system design trade studies. *Human Factors, 17,* 4–12.

Berson, B. L., and Crooks, W. H. (1976, September). Guide for obtaining and analyzing human performance data in a material development project (Report PTR 1026-3). Woodland Hills, CA: Perceptronics.

Buchaca, N. J. (1979, July 1). Models and mockups as design aids (Technical Document 266, Revision A). San Diego, CA: Naval Ocean Systems Center (AD A109 511).

Department of Defense (1977a, January). Major system acquisitions (DOD Directive 5000.1). Washington, DC.

Department of Defense (1977b, January). Major system acquisition process (DOD Directive 5000.2). Washington, DC.

Department of Defense (1977c, January). Test and evaluation (DOD Directive 5000.3), Washington, DC.

Department of Defense (1979, January 31). Human engineering requirements for military systems (MIL-H-46855B). Washington, DC.

Department of Defense (1981, May 2). Human engineering design criteria for military systems, equipment and facilities (MIL-STD 1472C). Washington, DC.

Eastman, C. M. (1968, February). Exploration of the cognitive process in design (Report AFOSR-68-1374). Madison, WI: University of Wisconsin, Environmental Design Center.

Geer, C. W. (1981, September). Human Engineering Procedures Guide (Report AFAMRL-TR-81-35). Wright-Patterson Air Force Base, OH: Aerospace Medical Research Laboratories (AD-A108 643).

Gravely, M. L, and Hitchcock L. (1980). The use of dynamic mock-ups in the design of advanced systems. *Proceedings,* Human Factors Society Annual Meeting, 5–8.

General Accounting Office (1984, February 24). The Army needs more comprehensive evaluations to make more effective use of its weapon system testing (Report GAO/NSIAD-84-40). Washington, DC: General Accounting Office.

Hardman Project Office, Chief of Naval Operations (OP-112) (1979, July). Documented description of the weapon system acquisition process (Report HR 79-01). Washington, DC.

Hawkins, E. D. (1974, March). Application of helicopter mockups to maintainability and other related disciplines (Report AD 786500). Texarkana, TX: Red River Army Depot.

Holshouser, E. L. (1975, September 30). Translation of DSARC milestones into human factors engineering requirements, (Technical Publication TP-75-58). Point Mugu, CA: Pacific Missile Test Center.

Holshouser, E. L. (1977, June 30). Guide to human factors engineering general purpose test planning (GPTP) (Technical Publication TP-77-14). Point Mugu, CA: Pacific Missile Test Center.

Hornick, R. J., Robinson, J. E., Rogers, J. G., and Sullivan, D. (1981, July). Design engineers' concepts of skills for system operation and maintenance (Technical Note 81-20). San Diego, CA: Navy Personnel Research and Development Center.

Janousek, J. A. (1970). The use of mockups in the design of a deep submergence rescue vehicle. *Human Factors, 12,* 63–68.

Lintz, L. M., Askren, W. B., and Lott, W. J. (1971, June). System design trade studies: The engineering process and use of human resources data (Report AFHRL TR-71-24). Wright-Patterson Air Force Base, OH: Air Force Human Resources Laboratory (AD-732 201).

McCormick, E. J., and Sanders, M. S. (1982). *Human factors in engineering and design,* (5th ed.) New York: McGraw-Hill.

McLane, J. F., Weingartner, W. J., Townsend, J. C. (1966, May). Evaluation of functional performance of an integrated ship control conning council by operator personnel (Report MEL-RND-333-65). Annapolis, MD: Marine Ergonomic Laboratory (AD-482211).

Meister, D. (1971). *Human factors: Theory and practice.* New York: Wiley.

Meister, D. (1973). The future of ergonomics as a system discipline. *Ergonomics, 16,* 267–280.

Meister, D. (1979). The influence of government on human factors research and development. *Proceedings,* Human Factors Society Annual Meeting, Boston, MA, 5–13.

Meister, D. (1981). *Behavioral research and government policy.* Elmwood, NY: Pergamon.

Meister, D. (1982). The role of human factors in system development. *Applied Ergonomics, 13*(2), 119–124.

Meister, D. (1983, March). Behavioral inputs to the weapon system acquisition process (Report SR 83-21). San Diego, CA: Navy Personnel Research and Development Center.

Meister, D. (1983). Are our methods any good? A way to find out. *Proceedings,* Human Factors Society Annual Meeting, Norfolk, VA, 75–79.

Meister, D. (1984, May). A catalogue of erogomic design methods. *Proceedings,* International Conference on Occupational Ergonomics, Toronto, Canada, 17–25.

Meister, D. (1985). *Behavioral analysis and measurement methods.* New York: Wiley.

Meister, D., and Farr, D. E. (1966, September). The methodology of control panel design (Report AMRL-TR-66-28). Wright-Patterson Air Force Base, OH: Aerospace Medical Research Laboratories.

Meister, D., and Farr, D. E. (1967). The utilization of human factors information by designers. *Human Factors, 9,* 71–87.

Meister, D., and Sullivan, D. J. (1967, March). A further study of the use of human factors information by designers (Final Report). Canoga Park, CA: Bunker-Ramo Corporation (Contract 5-4974-00).

Meister, D., Sullivan, D. J., and Askren, W. N. (1968, September). The impact of manpower requirements and personnel resources data on system design (Report AFHRL TR-68-44). Wright-Patterson Air Force Base, OH: Air Force Human Resources Laboratory (AD-678 864).

Meister, D., Sullivan, D. J., Finley, D. L., and Askren, W. B. (1969a, October). The effect of amount and timing of human resources data on subsystem design (Report AFHRL TR-69-22). Wright-Patterson Air Force Base, OH: Air Force Human Resources Laboratory (AD-699 577).

Meister, D., Sullivan, D. J., Finley, D. L., and Askren, W. B. (1969b, October). The design engineer's concept of the relationship between system design characteristics and technician skill level (Report AFHRL TR-69-23). Wright-Patterson Air Force Base, OH: Air Force Human Resources Laboratory (AD-699 578).

Munger, S. J., Smith, R. W., and Payne, D. (1962, January). An index of electronic equipment operability: Data store (Report AIR-C43-1/62-RP(1)). Pittsburgh, PA: American Institute for Research.

Naval Material Command, (1984, April 9). Human factors in naval material (NAVMATINST 3900.9A). Chief of Naval Material, Washington, DC.

Parks, D. L., and Springer, W. E. (1975, June 1). Human factors engineering analytic process definition and criterion development for CAFES (Report D180-18750-1). Seattle, WA: Boeing Aerospace Company (AD-A040 478).

Perrow, C. (1983). The organizational context of human factors engineering. *Administrative Science Quarterly, 28,* 521–541.

Pope, L. T., Williams, H. L., and Meister, D. (1978, December). Analysis of concepts for maintenance of MCM ship vehicle systems (Special Report 79-7). San Diego, CA: Navy Personnel Research and Development Center (CONFIDENTIAL).

Potempa, K. W., Lintz, L. M., and Luckew, R. S. (1975). Impact of avionic design characteristics on technical training requirements and job performance. *Human Factors, 17,* 13–24.

Potter, N. R., Korkan, K. D., and Dieterly, D. L. (1975, June). A procedure for quantification of technological changes on human resources (Report AFHRL TR-75-33). Wright-Patterson Air Force Base, OH: Air Force Human Resources Laboratory (AD-A014 335).

Price, H. E., Fiorello, N., Lowry, J. C., Smith, M. G., and Kidd, J. S. (1980, July). The contribution of human factors in military system development: Methodological considerations (Report TR 476). Alexandria, VA: Army Research Institute.

Rogers, J. G., and Armstrong, R. (1977). Use of human engineering standards in design. *Human Factors, 19,* 15–23.

Rogers, J. G., and Pegden, C. D. (1977). Formatting and organization of a human engineering standard. *Human Factors, 19,* 55–61.

Sawyer, C. R., Fiorello, M., Kidd, J. S., and Price, H. E. (1981, July). Measuring and enhancing the contributions of human factors in military system development: Case studies of the application of impact assessment methodologies (Technical Report 519). Alexandria, VA: Army Research Institute.

Seminara, J., and Gerrie, J. K. (1966). Effective mockup utilization by the industrial design–human factors team. *Human Factors, 4,* 347–359.

Smith, S. L., and Aucella, A. (1983, March). Design guidelines for the user interface to computer-based information systems (Report ESD-TR-83-122, MTR-8857). Bedford, MA: MITRE Corporation (AD-A127 345).

Van Cott, H. P., and Kinkade, R. G. Eds., (1972). *Human engineering guide to equipment design,* rev. ed. Washington, DC: U.S. Government Printing Office.

Whalen, G. V., and Askren, W. B. (1974, December). Impact of design trade studies on system human resources (Report AFHRL TR-74-89). Wright-Patterson Air Force Base, OH: Air Force Human Resources Laboratory (AD-A009 639).

Williges, R. C., and Topmiller, D. A. (1980, April 30). Task III final report. Technology assessment of human factors engineering in the Air Force, Wright-Patterson AFB, OH: Aerospace Medical Research Laboratories.

Woodson, W. E. (1981). *Human factors design handbook.* New York: McGraw-Hill.

PART 2
HUMAN FACTORS
FUNDAMENTALS

Chapter 2.1

SENSATION, PERCEPTION, AND SYSTEMS DESIGN

PATRICK FOLEY
NEVILLE MORAY

University of Toronto

2.1.1 INTRODUCTION

It is of more than theoretical interest that Henri Pieron gave the subtitle *Guide de la Vie* to his book *Sensation* published in 1944. As he says in the introduction to the English translation (Pieron, 1952),

> Sensation comes in to guide the living organism in its global behavior towards the external world. Knowledge of this world is built up in order to facilitate the necessary relations with the objects and forces constituting it. The value and scope of our knowledge cannot therefore be independent of the fundamental processes which condition it. The study of sensation is accordingly of the foremost importance in practice as well as theory.

"Guide de la vie," "pathways to knowledge," call them what you will, the study of sensory processes is complex and challenging.

Even more challenging is the task of attempting to distill, within the few pages alloted in this handbook, the essence of the thousands and thousands of studies carried out to further our understanding of the structural elements and the functioning of these sensory systems. Tabulations of specific sensitivities and resolution limits already exist in many forms and are readily accessible. The more specific chapters in this handbook relating to display systems, lighting, noise, and so on, include such data as they relate to the problem at hand. Therefore, a collation of such data here would be redundant, and, what is much more important, would do little to resolve the main problem, which is that to be meaningful they have to be exhaustive. A comprehensive listing is provided by Mowbray and Gebhard (1958), and shown in Tables 2.1.1 to 2.1.3. Table 2.1.1 demonstrates clearly the inadequacy of the common concept of "the five senses" by specifying various possible sensations, the sense organ involved, and the nature of the stimulating energy. Table 2.1.2 compares the operating ranges for each modality and the number of discrimination steps within each range. The distinction between relative and absolute discriminations is interesting and important, even though our knowledge here is quite limited. Table 2.1.3 is a more detailed examination of ranges and discriminations for vision and audition in particular. The important point here is that such tables specify only *boundaries,* in that, for example, the lower limits can be achieved only under very special conditions. What can be achieved under "normal" operating conditions is something else again.

It is common to find such tabulations in textbooks of perception and ergonomics, but it is our contention that such tables are seriously misleading. It is true that the data reported in them are well established. But it is most important to realize that *in most cases the facts cannot be used as a simple look-up table to predict what will happen when observer is confronted by a particular display.*

Because many ergonomists and designers have a strong wish that such data could be so used, it is worth discussing at some length why we believe that they cannot.

For historical reasons, there is a tendency to think about sensation and perception in two ways that are misleading. Information reaches the brain as a result of signals from the observer's (external or internal) environment. The signals are encoded or transduced by the receptors and their associated neural mechanisms. The result is a stream of information that is passed to the higher parts of the brain such as the appropriate sensory cortex. Because of this sequence of events, it is natural to think that if we know the anatomy and physiology of the sense organs and the related psychological facts about the perceptual correlates of physical variables we will be able to define the properties of sensation and perception, and to predict from the physical properties of the input what will be perceived. Such a claim is closely related on the one hand to the classical associationist view of psychology. From that point of view, wholes that we perceive are the sum of the parts of sensation. Sensation supplies the "bricks" of perception, which, if attended, become the "buildings" we perceive. This view finds a more modern justification in certain computational approaches to pattern analysis and in certain neurophysiological data, the details of which we cannot go into here. But in essence, the assumption is that what an observer perceives is made up of components of sensory information; and by understanding the laws governing the latter, we will understand the former. If light of a certain intensity and wavelength is shone on the retina, or if sound of a certain frequency and intensity falls on the ear, then on successive occasions identical values of the physical variables will cause identical values of perceptual variables. Perception is a "bottom-up" process.

In fact, that is not the case.

The second mistake is to think of perception as analogous to photography or sound recording. In much psychological literature there is an implicit assumption that conscious awareness of an incoming stimulus is the end and purpose of perception. If there is a cat in the environment, the purpose of perception is to provide one with an accurate picture (in one's consciousness) of the cat. If a bird

Table 2.1.1 A Survey of Man's Senses and the Physical Energies that Stimulate Them

Sensation	Sense Organ	Stimulated By	Originating
Sight	Eye	Some electromagnetic waves	Externally
		Mechanical pressure	Externally or internally
Hearing	Ear	Some amplitude and frequency variations of the pressure of surrounding media	Externally
Rotation	Semicircular canals	Change of fluid pressures in inner ear	Internally
	Muscle receptors	Muscle stretching	Internally
Falling and rectilinear movement	Semicircular canals	Position changes of small, bony bodies in the inner ear	Internally
Taste	Specialized cells in tongue and mouth	Chemical substances dissolvable in saliva	Externally (contact)
Smell	Specialized cells in mucous membrane at the top of the nasal cavity	Vaporized chemical substances	Externally
Touch	Skin mainly	Surface deformation	On contact
Vibration	None specific	Amplitude and frequency variations of mechanical pressure	On contact
Pressure	Skin and underlying tissue	Deformation	On contact
Temperature	Skin and underlying tissue	Temperature changes of surrounding media or of objects contacted	Externally and on contact
		Mechanical movement	
		Some chemicals	
Cutaneous pain	Unknown, but thought to be free nerve endings in the skin	Intense pressure, heat, cold, shock, and some chemicals	Externally and on contact
Subcutaneous pain	Thought to be free nerve endings	Extreme pressure and heat	Externally and on contact
Position and movement	Muscle nerve endings	Muscle stretching	Internally
	Tendon nerve endings	Muscle contraction	Internally
(Kinesthesis)	Joints	Unknown	Internally

Source: G. H. Mowbray and J. W. Gebhard, "Man's Senses as Information Channels," (Report CM-936), The Johns Hopkins University Applied Physics Laboratory (May 1958).

sings, the purpose of perception is to hear accurately the song. If one is a control room operator, the purpose of perception is to see the level of the mercury in the thermometer or the position of the pointer on a gauge, or to hear the alarm buzzer when the process goes off limits. In discussions of perception, it is often assumed that pattern analysis, pattern recognition, and perception are synonymous, and that their purpose is to provide the observer with an accurate representation to consciousness of the state of the environment.

In fact that is not the case.

The purpose of the perceptual system is not to produce pictures in the mind's eye or sounds in the mind's ear. *The purpose of the perceptual system is to provide the sufficient conditions for adaptive behavior.* This is particularly clear when we examine perception in relation to an applied field such as human factors. The pilot, the control room operator, and the person working with a remote manipula-

Table 2.1.2 Man's Senses as Informational Channels: A Comparison of the Intensity Ranges and Intensity Discrimination Abilities of the Senses

Sense	Intensity Range		Intensity Discrimination	
	Smallest Detectable	Largest Practical	Relative	Absolute
Vision	2.2 to 5.7 $\times$ 10^{-10} ergs	Roughly, the brightness of snow in the mid-day sun, or about 10^9 times the threshold intensity	With white light, there are about 570 discriminable intensity differences in a practical range	With white light, 3 to 5 absolutely identifiable intensities in a range of 0.1 to 50 ml
Audition	1 $\times$ 10^{-9} ergs/cm²	Roughly, the intensity of the sound produced by a jet plane with afterburner or about 10^{14} times the threshold intensity	At a frequency of 2000 Hz, there are approximately 325 discriminable intensity differences	With pure tones about 3 to 5 identifiable steps
Mechanical vibration	For a small stimulator on the fingertip, average amplitudes of 0.00025 mm can be detected	Varies with size of stimulator, portion of body stimulated, and individual. Pain is usually encountered about 40 db above threshold	In the chest region a broad contact vibrator with amplitude limits between 0.05 and 0.5 mm provides 15 discriminable amplitudes	3 to 5 steps
Touch pressure	Varies considerably with body areas stimulated and the type of stimulator. Some representative values: Ball of thumb—0.026 erg Fingertips—0.036 to 1.090 ergs Arm—0.032 to 0.113 erg	Pain threshold	Varies enormously for area measured, duration of stimulus contact, and interval between presentation of standard and comparison stimuli	
Smell	Widely variant with type of odorous substances. Some representative values: Vanillin—2 $\times$ 10^{-7} mg/m³. Mercaptan (C_2H_5SH)—4 $\times$ 10^{-5} mg/m³. Diethyl ether—($C_2H_5OC_2H_5$) 1.0 mg/m³	Largely unknown Largely unknown	No data available No data available	No data available No data available
Taste	Widely variant with type and temperature of taste substance. Some representative values: Sugar—0.02 Mo Quinine sulfate—4 $\times$ 10^{-7} Mo	Not known	No data available	No data available

Table 2.1.2 (*Continued*)

Sense	Intensity Range		Intensity Discrimination	
	Smallest Detectable	Largest Practical	Relative	Absolute
Temperature	Sensation of heat results from a 3 sec exposure of 200 cm² of skin at rate of 1.5×10^{-4} g-cal/cm²/sec	Pain results from a 3-sec exposure of 200 cm² of skin at a rate of 0.218 g-cal/cm²/sec	No data available	No data available
Kinesthesis	Joint movements of 0.2 to 0.7° at a rate of 10°/min can be detected. Generally, the larger joints are the most sensitive	Unknown	Nc data available	No data available
Angular acceleration	Dependent on the type of indicator used 1. Skin and muscle senses 1°/sec² 2. Nystagmic eye movements 1°/sec² 3. Oculogyral illusion 0.12°/sec²	Unconsciousness or "blackout" occurs for positive "G" forces of 5 to 8 G lasting 1 sec or more Negative forces of 3 to 4.5 G cause mental confusion, "red-vision," and extreme headaches lasting sometimes for hours following stimulation	No data available	No data available
Linear acceleration	In aircraft—0.02 G for accelerative forces and 0.08 G for decelerative forces	For forces acting in the direction of the long axis of the body, the same limitations as for angular acceleration apply	No data available	No data available

Source: G. H. Mowbray and J. W. Gebhard, "Man's Senses as Information Channels," (Report CM-936), The Johns Hopkins University Applied Physics Laboratory (May 1958).

tor need to perceive the state of the environment in order to *act* on the environment. The woman driving an automobile, the man ironing his shirt, and the cashier in a supermarket need perceptual information in order to *act* appropriately. It follows that the properties of perception can be understood only as a component in a closed-loop system, which is shown in Figure 2.1.1.

Several things follow from the closed-loop iterative nature of this process. One of the most important is that accuracy is not the most important characteristic of perception. Very often, a very rough estimate of the state of the environment will suffice for action. Moreover, if the action causes unexpected results, the next iteration can be used either to obtain more accurate information or to initiate more appropriate action, or both. This amounts to saying that actions can be directed either to the task or process with which the operator is concerned (what has traditionally been called the operator's "response" or "output") *or to changing the state of the operator.*

The action taken may be to look more closely at a gauge, or to listen more intently to a sound. It may be to decide to spend more time in looking at a strip-chart record (rather than merely taking a quick glance at it) so as to ensure that a more accurate judgment about its meaning can be made.

Table 2.1.3   A Comparison of the Frequency Ranges and Frequency Discrimination Abilities of Some of the Senses

Sense	Wavelength or Frequency Range		Wavelength or Frequency Discrimination	
	Lowest	Highest	Relative	Absolute
Vision Hue	300 mμ	1500 mμ	At medium intensities there are about 128 discriminable hues in the spectrum	12 or 13 hues
Interrupted white light	Unlimited	At moderate intensities and with a duty cycle of 0.5, white light fuses at about 50 interruptions per second	At moderate intensities and with a duty cycle of 0.5, it is possible to distinguish 375 separate rates of interruption in the range of 1 to 45 interruptions per second	No greater than 5 or 6 interruption rates can be positively identified on an absolute basis
Audition Pure tones	20 Hz	20,000 Hz	Between 20 and 20,000 Hz at 60 db loudness, there are approximately 1800 discriminable steps	4 or 5 tones
Interrupted white noise	Unlimited	At moderate intensities and with a duty cycle of 0.5, interrupted white noise fuses at about 2000 interruptions per second	At moderate intensities and with a duty cycle of 0.5, it is possible to distinguish 460 separate interruption rates in the range of 1 to 45 interruptions/sec	Unknown
Mechanical vibration	Unlimited	Unknown, but reported to be as high as 10,000 Hz with high intensity stimulation	Between 1 and 320 Hz, there are 180 discriminable frequency steps	Unknown

Source: G. H. Mowbray and J. W. Gebhard, "Man's Senses as Information Channels," (Report CM-936), The Johns Hopkins University Applied Physics Laboratory (May 1958).

It may be to walk to a different part of the control room, to call up a different display on a CRT, or to alter the brightness or contrast to improve the legibility of the display. In short, a major part of perception is not pattern analysis but tactical decisions about what information to process, and how to process it.

Perception is a dynamic process to a large extent under the control of the perceiver. It is as much, if not more, a "top-down" as a "bottom-up" process, and it is for that reason that the first

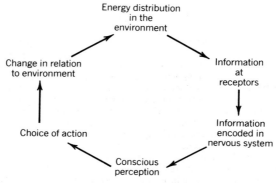

Fig. 2.1.1. Cybernetics of perception.

attitude discussed above is mistaken. The use to which an observer puts a given physical input is dynamic, changing from moment to moment. The description of a stimulus in physical terms is therefore inadequate to predict what an observer will perceive. It is for this reason that engineer designers who lack training in human factors and ergonomics cannot use published tables of perceptual properties in the way in which they can use tables of properties of materials, steam tables, and so on.

We define "perception" in this chapter to involve consciousness. To perceive something is to be aware of it and, as William James stated a hundred years ago, to pay attention to it. Perception is not, therefore, the same as pattern analysis. Both common experience and experimental research make the difference clear. Most readers will have had the experience, while driving at night, of suddenly realizing that they have traveled some distance without being aware of it. On such occasions, one ceases consciously to process information for a significant length of time. But the highly practiced skill of driving can be controlled by the output of the brain's pattern analyzing mechanisms without conscious perception. Experimental work on attention supports this view of the dissociation of pattern analysis and conscious attention (Moray, 1970), and in particular the experiments by Oswald, Taylor, and Treisman (1960) on pattern recognition by sleeping "observers" leaves no doubt on the point. The *contents of conscious perception are the result of operations by the observer on the output of the brain's pattern-analyzing mechanisms* and are not the operation of those mechanisms themselves. Perception is constructed by the observer, not determined by the parameters of the physical signals that fall on the receptors.

This again underlines the essentially unstable and dynamic nature of perception. Even when we consider relatively simple perception judgments different answers will be obtained on different occasions.

2.1.2 LIMITATIONS ON HANDBOOK DATA

Tables 2.1.1 to 2.1.3 summarize classical findings on the psychophysics of perception. In view of our contention about the instability of psychological measures, how should such tables be used?

The absolute limits of sensitivity are set primarily by the physiological and anatomical characteristics of the sense organs. Those sensitivities are of concern only to designers in very special, marginal situations. Even then, it is important to note the difference between physiological and psychological limits. It is conceivable that a designer might wish to know the absolute detectability of a light if the problem of interest was, say, the detection of radar targets, or of navigation lights at night. The physiological data are clear: one quantum of light energy falling on a dark-adapted rod is sufficient to change its state. In that sense the eye is as sensitive as any detector can be. Similarly, the absolute sensitivity of the hair cells in the inner ear is within an order of magnitude of the level of energy corresponding to the random motion of air molecules striking the eardrum. But stimulation at those energies result in a sensation of neither light nor sound. At least 6 to 10 quanta are required if an observer is to detect a flash with a probability of about .5. Moreover, as we shall see in a later section of this chapter, the number of quanta required vary with the subjective criterion adopted by the observer. Furthermore, those values are determined in highly artificial laboratory conditions, using very highly trained, highly motivated, dark-adapted observers who are concentrating wholly on the task of detecting a light, undistracted by any other concerns. Similar comments could be made *pari passu* about auditory signals near the physiological threshold.

The *designer's task is to ensure that under the conditions in which users will employ the equipment being designed the levels of physical energy falling on the receptors will be at least several orders of magnitude greater than the absolute physiological sensitivities.*

Consider the highly significant experiment carried out by Hecht et al. (1942) to answer the straightforward question, "What is the least amount of energy required to give a sensation of light?" The full answer is equally straightforward; approximately 10 to 15 quanta of light energy, at a wavelength of 510 nm, presented for a duration of 1 msec, over a circular retinal area subtending 10 min of arc, centered at 20° from the fovea in the temporal retina, to an eye dark adapted for at least 40 min immediately prior to testing. These conditionins are by no means arbitrary: each was selected on the basis of known data, as necessary to answer the specific question asked. The important point here is that the answer is equally specific. It is context dependent.

Consider the further question, "What is the resolving power of the human eye?," on the surface an equally straightforward question. But a logical extension, "resolution for what?," adds a new dimension—points of light, a fine line, a vernier displacement, a Landoldt C, a Snellen E, a checkerboard, a grating? And even when we attempt to apply Fourier methods of optical image evaluation to avoid these problems by deriving the spatial modulation transfer function of the eye, we are still left with further questions as to the antecedent and concomitant conditions—adaptation level, luminance, contrast and other surround conditions, area of the retina stimulated, static or dynamic presentation, and so on. These are psychophysical contextual variables. We can complicate the question still more by bringing in variables associated with the psychological context, expectancy, motivation, and training, to name a few of the more important.

For example, Foley, Lavery, and Abbey (1964), carried out an experiment to test the claim of Shvarts (1957) that scotopic visual acuity could be improved by manipulating purely motivational variables. Scotopic visual acuity refers to the ability to resolve fine details at very low light levels. Shvarts asserted that human visual sensitivity can be raised to an almost unlimited degree. Interpreting his experimental evidence purely on the basis of a motivational variable, he argued that were increased motivation the level of excitation in certain cerebral centers is raised, thus facilitating "conditioned connections between the visual stimulus and its adequate response." Contrary to Shvarts's hypothesis, improvement in scotopic acuity with practice is shown to be the result of learning, and specific to the stimulus used. No improvement takes place until specific knowledge of results is introduced, and the improvement does not transfer to another target configuration. The results are shown in Figure 2.1.2 and are quite clear. However, it should be emphasized that they do show a considerable improvement in scotopic acuity as a result of training for a specific target, and this can be important in the human factors context.

Finally, the answer to the question "At what frequency will an intermittent light be perceived as just flickering?" has been demonstrated to depend upon such factors as age, intelligence, race, sex, stimulus area, stimulus luminance, stimulus wave form, surround area, surround luminance, surround wave form, adaptation, circadian rhythm, stress and anxiety, ambient temperature, anoxia, and caffeine and other drugs—the list goes on and on.

These three questions have not been chosen simply to make a point. They are in fact typical of the kinds of questions that are asked, about all the sensory modalities, of human factors specialists, and sometimes *by* human factors specialists. They are not easy to answer, but they do underline the incontrovertible fact that there are times when it may be more efficient to consult a specialist than an handbook.

As we said earlier, tabulations have to be exhaustive to be useful. It is not enough simply to specify a threshold without, at the same time, specifying the conditions under which that threshold was obtained. The threshold acuity for a fine line, under optimal conditions, is given as 0.5 sec of visual angle, approximately equivalent to a wire, for example, $\frac{1}{16}$ inch in width, viewed at a distance of $\frac{1}{4}$ mile. Yet low flying pilots have considerable difficulty seeing power lines. Even a cursory glance at the paper by Blackwell (1946), on contrast thresholds of the human eye, should make clear not only the amount of careful work required to answer an elementary question about the detection characteristics of the human eye, but also how difficult it is to give an uncomplicated answer to the question as to whether pilots should be able to see power lines at a reasonable distance, a not unreasonable question.

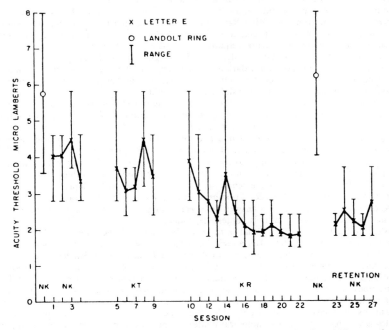

Fig. 2.1.2. Change in acuity threshold as a result of learning and motivation. NK = no knowledge of result. KT = knowledge of average performance. KR = knowledge of results on each trial. Source: Foley et al. (1964).

The values listed in Tables 2.1.1 to 2.1.3 and used in the above examples were obtained in very refined laboratory investigations into what are called *psychophysical functions,* or the laws of psychophysics, in which properties of the energy stimulating the receptors are related to their effects on the organizer, experienced as sensation or perception. (The distinction between sensation and perception generally is a matter of complexity—we sense simple dimensions such as warmth and brightness; we perceive objects, pictures, speech, and so forth).

We have seen how the application of psychophysical data to the solution of ergonomic problems is context dependent. We must now discuss the extent to which the psychophysical values are themselves context dependent.

2.1.3 PSYCHOPHYSICAL LAWS

The well-known psychophysical laws such as the Weber–Fechner Law or Stevens' Power Law should be regarded as limiting cases. They describe boundaries at which perceptual difficulties occur, and it is the task of the designer to ensure that the situations in which equipment is used do not push users close to those boundaries.

Consider, for example, the most general form of the Weber–Fechner Law. As a result of the extensive experimental work which began in the nineteenth century, and which was part of the foundations of modern psychology, the concept of the just noticeable difference (jnd) was developed as a way of describing relative sensitivities. The early form of the psychophysical function can be summarized by saying that the magnitude of change in a physical stimulus that will just be noticed by an observer is a constant proportion of the stimulus:

$$\frac{\Delta I}{I} = \text{constant} \overset{\Delta}{=} \text{jnd}$$

This law holds, more or less, for all sensory modalities provided that the magnitude, *I,* is neither close to the absolute threshold nor extremely large.

The experimental paradigm used to construct jnds consists in providing a standard stimulus of magnitude *I,* and a range of other stimuli of slightly greater or less magnitude. The observer compares each stimulus with the standard, and judges it to be "the same" or "different," or "greater," "less," "equal," and so on, the exact task depending on the particular paradigm. The results of such an experiment produce curves of the type shown in Figure 2.1.3.

The jnd is usually defined as the magnitude of ΔI that will cause a judgment of "greater," or "different" with a probability 50% greater than chance. For many dimensions on which stimuli can vary—brightness, color, loudness, weight, length, and so—the jnd is of the order of magnitude of 1%.

It is that kind of investigation which leads to the conclusion that there are many thousands of discriminably different lights. Because the sensitivity of the visual system has a range of about 10 log units of brightness, half a log unit of wavelength, and in addition the dimension of saturation (purity of color), one can add up the number of "1%" jnds on each dimension, and multiply the numbers by the number of dimensions. But once again these data are not useful as such except in very special

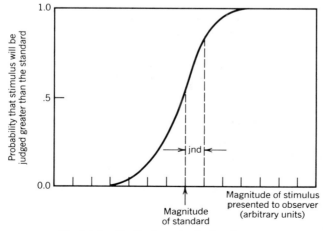

Fig. 2.1.3. Idealized psychophysical function.

cases. Pollack (1952), in a classic series of papers on auditory psychophysics, noted that such psychophysical arithmetic would lead one to conclude that an observer should be able to discriminate between many thousands of stimuli. But if such an observer is asked to *identify* stimuli, as distinct from *comparing* them, less than a dozen can be reliably identified, regardless of the magnitude of the difference between them.

The research paradigm is again far removed from everyday tasks. One can imagine special cases—for example, paint manufacturers or hi-fi system designers—to whom single jnds could be of importance. Even then it must be noted that the magnitude of a jnd depends on a number of factors, such as motivation, adaptation, practice, and fatigue, that are held constant in the laboratory investigation but that vary in most situations in daily life. All these factors can change the slope of the psychophysical functions shown in Figure 2.1.3, and hence can change the magnitude of the jnd, although the latter will almost always be of the order of a few percent in the mid range of a sensory dimension, and certainly less than 10%.

A designer could then approach the question of choosing discriminable stimuli (signals) simply by taking an arbitrarily large multiple of the jnd of the dimension in question. As a rule of thumb, it is probably safe to say that a change in the magnitude of a stimulus of $\Delta I = 0.1I$ will "never" be missed. It is actually possible to go further and provide some theoretical justification for such an heuristic. The slope of the psychophysical function in Figure 2.1.3 can be modeled using the normal (Gaussian) curve. If there were no variability in the observer, there would be some level of ΔI at which, for example, the observer's judgment would suddenly change from "less than" to "greater than" as the magnitude of the comparison stimulus was increased from below to above that of the standard with which it was being compared. The fact that the psychophysical function is not a vertical straight line but an ogive can be explained if we assume that the "differential threshold" varies in a random manner from moment to moment and that the position is distributed normally. In fact, the ogival form of the curve in Figure 2.1.3 is the integral of the normal curve. In choosing the define the jnd as a ΔI which gives rise to a "different" or "greater" judgment on a certain proportion of trials we are effectively defining it as a cutoff point at some standard deviation of the normal distribution.

The magnitude of the standard deviation increases as the slope of the psychophysical curve becomes flatter. And if we want to be sure that the observer will "never" make an incorrect judgment when shown as ΔI of some magnitude, we can choose a ΔI that is so far out, say, three or four deviations, that the observer will always reject it as coming from the same distribution. In fact, for most dimensions on which stimuli vary, four standard deviations would produce a "very noticeable difference" of the order of 10% or greater, thus justifying the heuristic.

For most users, it is not important to know the smallest jnd. What is required in design is to choose stimulus differences that are very large compared with the relevant jnd.

There is a further limit on "absolute identifiability" which is due to a separate mechanism, and which was mentioned in connection with Pollack's work on hearing. As a rule of thumb, and regardless of the dimension on which the stimulus varies, people can identify in an absolute sense only less than about 10 stimuli. That is, if a listener has no comparison standard, but just hears a number of sounds, the latter will be sorted into about eight classes, regardless of how many there really are. If someone identifies a number of colors, not more than about 8 to 10 color names can be used reliably in the absense of standards for comparison. There are of course, exceptions, such as people with "absolute pitch." But for most purposes no more than 8 to 10 categories can be used in absolute identification, regardless of their separation. That is, even if 20 colors are separated each from the other by, say, 20 jnds of wavelength, they will be identified no more accurately than if they were separated by, say, three jnds, and will in each case by sorted into about eight categories.

More modern work on the nature of the psychophysical function by Stevens has led to the general acceptance of the Power Law as a description of the way in which the magnitude of sensation varies with the magnitude of the physical stimulus: $S \propto P^k$ where S is the magnitude of sensation, P is the magnitude of the physical stimulus, and k is an exponent characteristic of the sensory modality. If $k = 1$, then sensation changes directly with the physical magnitude: the perceived length of a line is very close to its physical length. If $k > 1$, then the sensation grows more rapidly than the physical stimulus; an example is electrical shock. If $k < 1$ the sensation grows less rapidly than the physical magnitude; an example is the brightness of white light.

The form of the psychophysical function implies that people should be quite good at "cross model matching," so that they would, for example, be asked to adjust a tone until its loudness was as loud as a particular light was bright. Surprisingly, this turns out to be possible, and Stevens' power functions can predict the matches quite accurately. However, although it is accepted that Stevens' Law is an improvement on the Weber–Fechner Law, the same kind of comments on its use will be relevant.

2.1.4 APPLICATION OF PSYCHOPHYSICS: THE EXAMPLE OF COLOR SPECIFICATION

Some of the most precise specifications of psychophysical values are in the area of color vision, where international standards have been adopted for many years. It is therefore particularly instructive to consider how a designer might use psychophysical tables of the highest quality.

As an example, let us consider the problem of specifying a color code for particular application. Here we are concerned with general principles, and the necessity for an understanding of color vision at the various levels involved, starting with the problems of specification itself and ending with the problem of selection.

How can we specify a particular color? We could, as in the Munsell system, produce an atlas containing material standards, selected and organized according to some criteria. In the Munsell system, colors are assigned a position in a three-dimensional space, the coordinates being hue, value, and chroma. One selects a color from the atlas, reads off the coordinates, for example, 5R.6/10, and then whoever has to reproduce that color simply looks up the color corresponding to these coordinates in the atlas, and works to a visual match. There are obvious drawbacks to this technique, the most important being that it can only work for colors that are capable of production as material standards, thus excluding, as the most obvious example, signal lights. The subjective nature of the final matching procedure is another obvious drawback, as is the difficulty of specifying tolerances. The system is described in detail in many publications, but it is recommended that Munsell's own description of the system be consulted (Munsell, 1941).

A more general system, capable of being utilized to specify any color, including signal lights, was developed by the CIE (Commission Internationale de L'Eclairage), based on a physical analysis of the electromagnetic energy entering the eye, and the psychophysical response of the eye of this energy. The system is worth describing in detail, for an understanding of the principles and results of the procedures enables the human factors specialist to deal with most, if not quite all, of the problems likely to arise.

Let us start at the beginning. If an observer with normal color vision is asked to classify a random assortment of color samples, he or she will usually separate them into heaps each of a different *hue,* that is, the greens, the blues, the reds, and so on. These heaps can then be arranged in a *hue circle.* Starting with the blues, for example, next to them can be placed the blue-greens, then the green-blues, the greens, the yellow-greens, the yellows, the orange, the reds, the red-blues or purples, and thus back to the blues again. These heaps will not be homogeneous, however; some reds, for example, will be much "redder" than others, such as maroons and pale pinks. These differences are differences in *saturation,* and the reds can be ordered according to their degree of saturation, with the most saturated on the circumference of the hue circle, the least saturated at the center, and the others along what might be termed the "red radius." The classification will still present problems, there will be some reds that have the same redness, that is, saturation, but that differ according to their *brightness,* and these would have to be ordered along an orthogonal dimension, ranging from very dark to very bright. All the samples will now be accounted for. The relation of these dimensions to the Munsell dimensions of *hue, chroma,* and *value,* should be obvious.

One might expect these psychophysical dimensions to be correlated with physical dimensions, and this is to some extent the case: hue with wavelength, saturation with the admixture of wavelengths, that is, purity, and brightness with intensity. However, these relations are neither exclusive nor one to one. Our perception of color for a given combination of these physical variables depends on many factors: the colors to which we have just been exposed, the other colors in the immediate vicinity, the part of the retina stimulated, and the retinal size of the object, to name a few. The important thing to remember is that color is not simply a property of objects. It is the result of an interaction between radiant energy and the visual system. The component wavelengths of radiant energy from a light source are modified by the selective spectral characteristics of the objects from which they are reflected, or through which they are transmitted, to the visual system. On the other hand, the characteristics of the visual system in turn determine the response to this input. Both are important and both must be considered.

The colors we see in everyday life can be thought of as mixtures of their component wavelengths. We see the resultant, the eye being essentially an integrative mechanism. This leads logically to a consideration of color mixture, of which there are two kinds, subtractive and additive.

1. *Subtractive.* Mixing pigments is a subtractive process. If yellow and blue pigments are mixed the result is green. The yellow pigment reflects light in the red, yellow, and green regions of the spectrum, and absorbs light from all other regions. The blue pigment reflects in the green, blue, and violet regions, absorbing all others. In the mixture of the two, the only region not absorbed is the green, and this then is the color we see. Note that the actual mixture takes place in the material.

2. *Additive.* In additive color mixing, the mixing is done in the observer's eye. Examples of additive mixing are given by the simultaneous projection of colored lights on the same spot on a screen—that is, the presentation of small colored dots or thin lines close together so that the eye is unable to resolve them—and by the rotation of a disc made up of colored sectors at a speed such that the individual sectors cannot be resolved—the Maxwell disc. In these examples of additive mixtures, lights of different colors enter the eye either simultaneously or in a periodic sequence of high frequency, and stimulate either the same spot on the retina or very closely related elements.

Because we are interested here in the specification of color, and because subtractive mixtures do not provide a unique or simple method of specification, we deal only with additive mixtures. These mixtures obey three simple laws, formulated by Grassman in 1853.

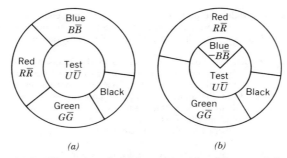

(a)   (b)

Fig. 2.1.4.   Illustration of additive color mixing (Grassman's Laws).

1. Any color can be matched by a linear combination of three primary colors.
2. Lights of the same color produce identical effects in the mixture, regardless of their spectral composition.
3. If, of a mixture of two or more components, one or more components are steadily changed while the other(s) remain constant, the color of the mixture steadily changes.

A simple experimental arrangement is show in Figure 2.1.4a, a schematic diagram of a Maxwell disk, with the color to be matched, $U\bar{U}$, in the center, and the three primaries, $\bar{R}$, $\bar{G}$, and $\bar{B}$, in varying amounts R, G, and B, surrounding it. The black sector serves to lighten or darken the color produced. The disk is rotated at high speed, and the two colors matched by varying the proportions of $\bar{R}$, $\bar{G}$, and $\bar{B}$. In some cases it may not be possible to make this match unless one of the primaries is added to the test color, as shown in Figure 2.1.4b. Because we take this primary away from the others it becomes a negative amount of that primary. Grassman's First Law then enables us to express the relationship between the test color and the primaries by means of a simple equation:

$$U\bar{U} = R\bar{R} + G\bar{G} + B\bar{B} \qquad (1)$$

That is, U parts of the test color $\bar{U}$ are matched by an additive mixture R parts of the primary color $\bar{R}$, G parts of the primary color $\bar{G}$, and B parts of the primary color $\bar{B}$.

When one of the primaries is added to the test color, the equation becomes, in the example shown,

$$U\bar{U} = R\bar{R} + G\bar{G} - B\bar{B} \qquad (1a)$$

Any color may be thought of then as a vector in three-dimensional space, hence, the bar over the letter symbol. The amounts (sector openings) of the primaries necessary to obtain a match with the test color are called the tristimulus values of the test color with respect to the given set of primaries. The tristimulus values R, G, and B of a test color $U\bar{U}$ will therefore be the coordinates of the color vector $U\bar{U}$ in the color space that is defined by the three primaries.

Grassman's Laws, then, enable us to specify any color in terms of three numbers, the tristimulus values. To make this the basis for a specification system, however, certain conditions must be standardized, that is, the primaries, the light source, and a standard observer representing the average color matching responses of a large number of individuals selected.

The CIE in 1931 adopted a system that specified in detail all the requirements listed above.

Any color can be thought of as an additive mixture of all the monochromatic components in its spectrum. Each monochromatic component in its turn can now be specified in terms of the tristimulus values of the given primaries for which it is a match. Therefore, the tristimulus values of the color can be determined by simply adding the corresponding tristimulus values of each monochromatic component. In accordance with the notation used previously, Equation (1), given that the tristimulus values of each monochromatic component, λ, are R, G, and B, then the tristimulus values of a color $U\bar{U}$ are as follows:

$$R = \sum_{\lambda} R_\lambda, \qquad G = \sum_{\lambda} G_\lambda, \qquad B = \sum_{\lambda} B_\lambda$$

The tristimulus values of the monochromatic components depend on the primaries chosen, and the amount of radiant energy at each wavelength in the spectrum, that is, the spectral composition of the test color. Assuming this to be N_λ, determined by measurement, and the primaries given, then the tristimulus values of the color $U\bar{U}$ can now be written more specifically as

$$R = \sum_\lambda N_\lambda \bar{r}_\lambda, \qquad G = \sum_\lambda N_\lambda \bar{g}_\lambda, \qquad B = \sum_\lambda N_\lambda \bar{b}_\lambda$$

Here $\bar{r}_\lambda$, $\bar{g}_\lambda$, and $\bar{b}_\lambda$ are still tristimulus values, but refer to the monochromatic components of an equal energy spectrum. They are known as the color mixture functions with respect to the primaries $\bar{R}$, $\bar{G}$, and $\bar{B}$. If the primaries $\bar{R}$, $\bar{G}$, and $\bar{B}$ are given in the form of monochromatic lights of wavelengths 700, 546, and 436 nm, respectively, the average color mixture functions for a large group of observers result in the curves shown in Figure 2.1.5. This figure indicates that we can match, for example, light of wavelength = 500 nm, if the primaries $\bar{R}$, $\bar{G}$, $\bar{B}$ are set at amounts $\bar{r}(500) = -0.08$, $\bar{g}(500) = 0.1$, and $\bar{b}(500) = 0.05$, respectively.

The negative component of the color mixture functions in Figure 2.1.3 makes them inconvenient, and hence the CIE selected another primary system that provides color mixture functions that are positive throughout the spectrum. The CIE primaries are called the $\bar{X}$, $\bar{Y}$, and $\bar{Z}$ primaries, and the color mixture functions related to them are labled x_λ, y_λ, and z_λ. The tristimulus values that may be computed from them are known as the X, Y, and Z tristimulus values. The x_λ, y_λ, and z_λ color mixture functions are shown in Figure 2.1.6.

At the same time the CIE defined three light sources, Illuminants A, B, and C, the relative energy distributions of which are defined over the visible spectrum. We can then be more specific regarding the spectral composition, N_λ, of the test color, because it is determined by the spectral energy distribution of the light source, P_λ, and the spectral reflectance, B_λ, or the spectral transmittance, T_λ, of the test object. That is, for a reflecting object, $N_\lambda = p_\lambda B_\lambda$, and the equations now become, in the CIE primary system

$$X = k \sum_\lambda B_\lambda P_\lambda \bar{x}_\lambda, \qquad Y = k \sum_\lambda B_\lambda P_\lambda \bar{y}_\lambda, \qquad Z = k \sum_\lambda B_\lambda P_\lambda \bar{z}_\lambda$$

where k is a normalization factor $= 100/\Sigma P_\lambda \bar{y}_\lambda$, by virtue of which the luminous reflectance of a perfectly diffusing, perfectly reflecting surface equals 100. An important feature of this system is that any computed Y value gives directly the luminance of a light source, or the luminous reflectance of an object, or the luminous transmittance of a filter. This is because the color mixture function $\bar{y}$ was made indentical to the luminous efficiency function of the standard observer.

Given the tristimulus values, the chromaticity coordinates, x and y, are computed as follows:

$$x = \frac{X}{X + Y + Z}, \qquad y = \frac{Y}{X + Y + Z}$$

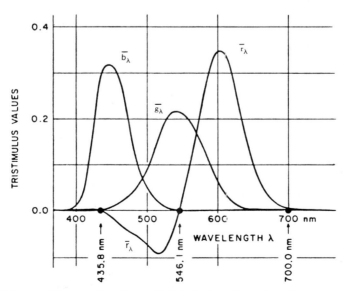

Fig. 2.1.5. Color-matching functions, $\bar{r}_\lambda$, $\bar{g}_\lambda$, $\bar{b}_\lambda$ (tristimulus values of equal-energy spectrum) in the primary system $R = 700.0$ nm, $G = 546.1$ nm, $B = 435.8$ nm. Source: Wysecki and Stiles, 1967.

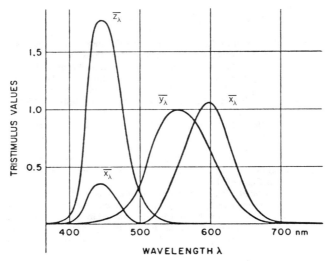

Fig. 2.1.6. 1931 CIE color-matching functions, $\bar{x}_\lambda$, $\bar{y}_\lambda$, $\bar{z}_\lambda$. Source: Wysecki and Stiles, 1967.

The chromaticity coordinates specify a color in a plane cross section of the color space. This chromaticity diagram is shown in Figure 2.1.7.

The spectrum colors, from 380 to 760 nm, form the spectrum locus. Also shown in the diagram are the coordinates of the CIE illuminants A, B, and C. The straight line joining the two ends of the spectrum represent the purples, that is, additive mixtures of red (760 nm) and blue (380 nm) in varying proportions.

One of the many useful features of this diagram is that, given the x, y coordinates of a color, and the light source used in determining these coordinates, the color may be specified in terms of its dominant wavelength and purity, which are correlates of the psychological attributes hue and saturation, referred to above. The brightness is related to the luminous reflectance, or transmittance, given by the tristimulus value Y. Figure 2.1.8 gives two examples of this approach.

Let Z be a color with coordinates xa, ya and C be the reference light source. Then the dominant wavelength of A is found by producing CA until it intersects the spectrum locus, and reading the wavelength at the point of intersection. (P is the intersection on the strength line joining the two ends of the spectral locus, of the projection of CB.) The purity of A is simply the ratio CA/CS, which may be measured on the diagram, but can simply be computed by the following formula:

$$\text{Purity A} = \frac{xa - xc}{xs - xc}$$

If, as in the case of color B, the projection of CB does not intersect with the spectrum, the line is produced backward until it does, at T, which then specifies the complementary dominant wavelength of B. The purity is still given by the ratio CB/CP, which may be measured or calculated as before.

This, then, is the CIE system. Given standard sources and the color mixture functions of the standard observer, we need only measure the spectral reflectance or transmittance of the pigment or dye, for example, to arrive at a specification of the color in terms of two numbers, the chromaticity coordinates, x, y. The utility of the system is obvious: we can specify signal lights as easily as we can surface colors; matching a specification does not require the acquisition of color atlases; and tolerance specification is straightforward and graphic. Figure 2.1.9 is a standard example, (CIE, 1959). More details can be found in standard sources such as the book *Color Science* (Wysecki and Stiles, 1967).

We are now in a position to examine the second part of the problem we initially set out to tackle, the selection of colors for our color code. Here the diagram also proves to be extremely valuable. The problem of selection depends critically on discriminability. Figure 2.1.10 shows MacAdam's discrimination ellipses for normal observers, (MacAdam, 1942). Note that each ellipse has been magnified by a factor of 10, for clarity. Note also that these confusion zones are, in fact, elliptical, and that the major axes of the ellipses vary in orientation as one proceeds from the lower left to the upper left. It is also apparent that the color space, as represented in the diagram, is not uniform. However, the diagram does allow us to specify colors for a color code, for normal observers. Given that, we still

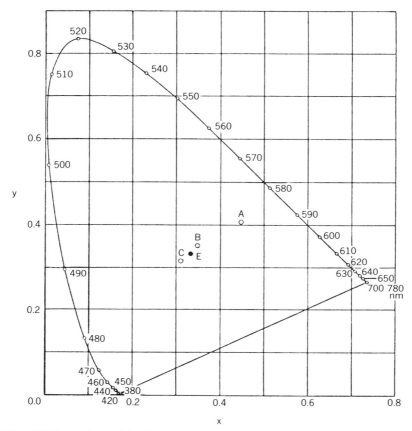

Fig. 2.1.7. CIE (x, y)–chromaticity diagram with spectrum locus, purple line, the chromaticity points of CIE standard sources A, B, C, and the equal-energy stimulus E. Source: Wysecki and Stiles, 1967.

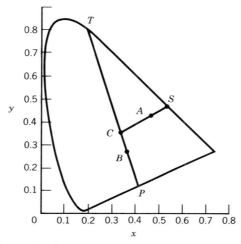

Fig. 2.1.8. Example for specification of a color in terms of dominant wavelength and periodicity from the CIE diagram. Source: Wysecki and Stiles, 1967.

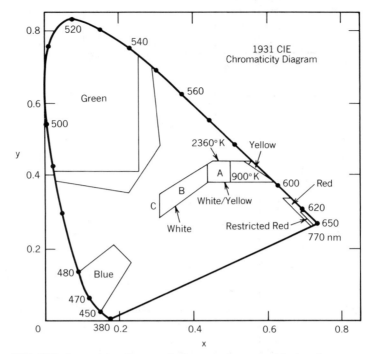

Fig. 2.1.9. 1931 CIE chromaticity diagram showing certain acceptable domains recommended for signal lights. Source: Wisecki and Stiles, 1967.

have the problem of specifying discriminability for nonnormal observers such as observers with defective color vision. What are their characteristics?

We stated earlier that any color could be matched by an additive mixture of three primary colors, by Grassman's First Law. This is true only for normal observers, for the male population, approximately 92%. There exist also other classes of observers, who behave in different but predictable ways. Briefly, a further 6% of the male population, although still requiring three primary colors, use different proportions of these primaries to match particular colors. Normal observers, requiring three primaries, are known as normal trichromats. Observers who require three primaries but in different proportions from the normal are known as anomalous trichromats. This category can be subdivided according to the particular colors affected. If, for example, more red is required in a red/green mixture to match a given yellow, the observer is defined as protanomalous; if more green, then as deuteranomalous. One percent of the male population is protanomalous, and 5% deuteranomalous. A third category, tritanomalous, is so rare that it need not concern us here. In addition there are other observers, 2% of the male population, who can match any color by an additive mixture of only two primary colors. They are known as dichromats, and can also be subdivided into two main categories, protanopes and deuteranopes, each accounting for 1% of the male population. Again, a third category, tritanopes, is so rare that they need not concern us here. Finally, there are also observers who need only one primary to match any other color. They are known as monochromats, account for <0.001% of the male population, and, incidentally, are the only ones who can properly be called "color blind."

Because they usually also suffer from very poor visual acuity, photophobia and so on, they also need not concern us in this context. If we are to take account of these individuals for coding purposes, then we need the same information about their confusions, and in the same form, as we have for the normal. The CIE diagram allows us to do this, and Figure 2.1.9 shows the results of investigations undertaken by Pitt (1935). Figure 2.1.11 shows the confusion loci for a protanope and for a deuteranope. Note that they are distinct and predictable. We now and only now have all the information we need to specify a color code.

However, the problems raised at the beginning of this chapter still apply. We must not be misled by the rigor with which this information was obtained. In Table 2.1.3, the statement is made that although there are some 128 discriminable hues in the spectrum on a relative basis, there are only some 12 or 13 hues discriminable on an absolute basis. That is, only a very small number can be absolutely identified as "that" hue.

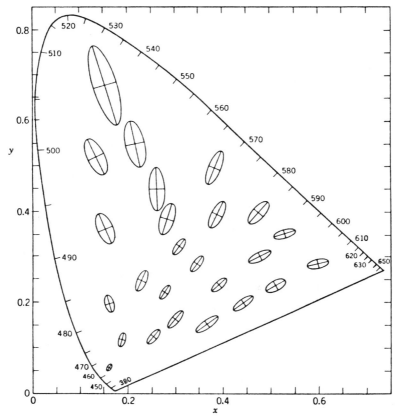

Fig. 2.1.10. 1931 CIE chromaticity diagram showing MacAdam's ellipses (10 times enlarged). Source Wisecki and Stiles, 1967.

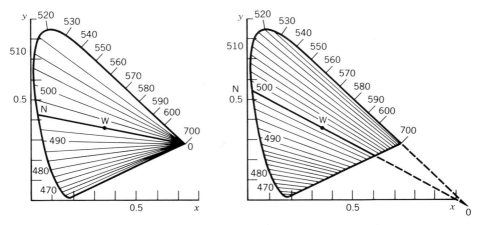

Fig. 2.1.11. Chromaticity confusions of protanopes (left) and deuteranopes (right). N refers to the neutral point, W to white light, and O the copunctal point. Source: Graham, 1965.

Furthermore, even when we have taken into account the situational context of the color sensitivity of the observer, we must remember that in general both lights and colored objects or surfaces are viewed in the presence of other colors and under varying conditions of illumination. As lights age, whether fluorescent or incandescent, the composition of their spectra alters, and so the composition of the light falling on and being reflected by surfaces in the environment changes, and may alter the relative discriminability of the color.

Again, we are dealing with context-dependent sensory phenomena.

2.1.5 MECHANISMS OF PERCEPTION

It was mentioned above that research on psychophysics is always conducted under laboratory conditions in which full attention is directed to the stimulus, and that only under such conditions are quantitative results such as those listed in Tables 2.1.1 to 2.1.3 obtained. Considering that fact more closely brings us to two psychological mechanisms that are of absolutely fundamental importance in translating laboratory results into useful principles of ergonomic design. They are the speed–accuracy trade-off (SATO) and attention.

Let us assume that we have some way of measuring the perceived magnitude of a stimulus. In particular, let us assume a situation in which we measure that magnitude while we place the observer under great time pressure to make his judgment. For example, in an experiment by one of the authors, listeners heard groups of three sounds, each 100 msec in duration, separated by 400 msec. The task was to judge whether the second sound was louder than the first, and to make a response before the onset of the third sound. Thus the observer had to make a response in less than half a second. If we examine the perceived magnitude of the second sound, and relate it to how long it took for the observer to make his response (all the observers were male), we find data such as those in Figure 2.1.12.

Although the magnitude of the physical event was constant, the observer behaved as if it were phenomenologically weaker when judgments were made rapidly. The shape of the curve suggests that the longer the observer waited before making a response, the more accurate was that response. This is a form of the SATO phenomenon. If we ask an observer to respond rapidly, then the probability of error increases. If we ask an observer to be accurate, then the observer requires more time. Observers behave as if they "collect" information about which to make judgments and from which to construct perceptions. The time over which information can be usefully integrated in this way may extend over several hundreds of milliseconds at least. It depends on the sense modality, the nature of the signal being observed, and subjective factors within the observer.

In the world of human–machine interactions SATO is of the greatest importance. For example, if an operator is examining circuit boards or components as a quality-control inspector and the production line is moving at a fixed rate, then the operator is under time pressure and SATO operates, even though attention is entirely directed to the one task. Thus paced work tends to promote rapid perceptual judgments rather than accurate ones, and self-paced work accuracy at the expense of speed. Management policy and incentive schemes can emphasize one or another aspect of the SATO.

Even when there is no overt time pressure from the task, there may be implicit pressure owing

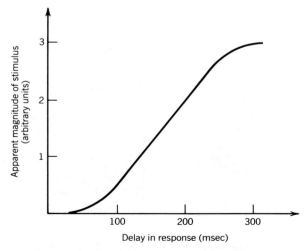

Fig. 2.1.12. Speed–accuracy trade-off in perception.

to limits of attention. In real-world tasks, operators seldom make more than two to three eye movements per second, and accurate perception is possible only within an area of about 2° of arc around the point of fixation. Hence in a large control room with perhaps several hundred displays, an operator does not have enough time to spend reading the information from each display. Some displays go unnoticed for long periods, and even those that are important and that are frequently examined are continuously subject to SATO effects. Hence there is an absolute necessity for clear design according to good ergonomic principles. Good design maximizes the rate at which information can be acquired, and minimizes the sampling time (fixation time) required to acquire information. It steepens the slope of curves such as that in Figure 2.1.12.

It follows that there is an intimate relation between the tactics of attention adopted by an observer, the way in which SATO affects perception, and the physical attributes of displays. To achieve swift and accurate perception requires (1) that appropriate physical signals be displayed, (2) that the observer pay attention in an appropriate way, and (3) that appropriate settings of speed–accuracy criteria be chosen. For example, consider a warning light so bright that the probability of not noticing it if it is looked at directly is .001. It would be easy to describe color and brightness characteristics to ensure such a level of detectability. But now suppose that it is one of a large array of warning lights in a control room with several hundred such lights—perhaps the annunciator board of a power station control room. What is the probability that the operator will be looking in the correct direction the moment the light becomes illuminated? If that probability is as low as .01, then the probability of seeing the warning as soon as it comes on is not .999, but .009. The psychophysical description of the properties of the light is not, without the whole context of the industrial ecology of the operator's task, sufficient to guarantee that the light will be perceived. On the other hand, if such psychophysical data are *not* used, then the probability of accurate perception-given ideal attention and SATO settings will be very low.

Despite our warnings about the dynamic nature of perception, we still believe that there is value in tables of psychophysical data, such as those in Tables 2.1.1 to 2.1.3.

To understand their correct role, we must first consider in more detail the relation between "bottom-up" and "top-down" mechanisms in perception.

2.1.5.1 "Bottom-Up" and "Top-Down" Processing in Perception

The emphasis of this chapter has been, up to a point, on the components into which the perceived experience may be broken down, and the underlying anatomical and physiological mechanism. This in turn implies the kind of atomistic or "bottom-up" approach to perception that has been emphasized so far. We concentrate on the components of sensation from which perceptions are constructed. But the history of research into perception has always had a second strand and this, the "top-down" approach, has become increasingly recognized in the last 20 years as the "information processing" attitude to perception has become increasingly prominent. (Historically, this second strand has always had a prominent place in European work, although largely neglected for many decades in North America.) This "top-down" approach emphasizes the active strategies of information processing brought to perception by the observer. What is perceived on any given occasion is determined by the interaction of these two sorts of information processing. The definition of a display in terms of physics, anatomy, physiology, and psychophysics ensures that information is available for the brain to use, but whether that information is used accurately by the observer depends also on "top-down" processing.

Particularly when perception is considered in the context of ergonomics, it is necessary to emphasize that the purpose of perception is to produce adaptive action by the observer—to enable the latter to *do* something as a result of what is perceived. As stated earlier, perception is part of a cybernetic closed-loop information processing system. Stimulation of the receptors leads to perception of the state of the environment. That leads to actions and to expectations about the future state of the environment. In turn, the latter changes the way in which future input is processed and here alters the way in which information from the receptions is interpreted, which in turn leads to a change in attitudes, action and expectations, and so on. *The bottom-up attitude to perception can provide design guidelines that are necessary for accurate perception but are not sufficient for accurate perception. The top-down determinants of perception make the outcome of perception, given a particular input, dynamic and unstable.*

Only by taking into account both bottom-up and top-down processing can a satisfactory "design for perception" be achieved in such a way that perception will guide action adaptively.

2.1.5.2 Principles of Top-Down Processing

Four main determinants of top-down perceptual processing will be discussed:

1. Innate mechanisms producing "Gestalten"
2. Long-term expectations and utilities including perceptual "stereotypes"

3. Short-term expectations and utilities

4. Attention

Gestalten

Koffka (1935) and others established the importance of "Gestalten" in perception in the 1930s. Human observers behave as if there are certain "hard-wired" properties of the nervous system which impose structure on incoming information and force perception into certain modes unless the observer makes a strong effort to disregard them. For example, in Figure 2.1.13*a*, the values of three variables are plotted on a graph. Without any effort on the part of the observer, it is natural to see the dotted and dashed lines each as a single, continuous entity. Indeed it is almost impossible not to see the lines as continuous.

Even in Figure 2.1.13*b*, we have no difficulty in following the lines. This tendency was called "common fate" by Gestalt psychologists. Another such feature is "good figure." Nearly circular patterns tend to be seen as circles, slightly broken lines to be seen as complete, and so on, and items close together to be seen as a unit (Figure 2.1.14). In a field full of moving elements, those elements whose movements are correlated are automatically seen as a unified set, standing out from elements that are stationary or that are moving in a different way and are mutually correlated. In the case of sound, Gestalten are formed as a result of rhythm, pitch, and so on. For example, a soprano voice stands out as a unit from a background of deeper voices even in a monaural message, and a series of high-pitched tones is heard as a single "message" against a background of voices without any effort by the observer.

The importance of Gestalt principles is in that they seem to represent a basic, natural mode of perception. A display that matches Gestalt principles is easy to read, and is correctly perceived with little or no effort and minimal attention by the observer. A display that departs from the principles of good Gestalten is hard to interpret, and preoccupies attention. A display that violates Gestalt principles is also likely to cause perceptual errors, for the same mechanism that helps to interpret good Gestalten correctly tends to force inputs into correspondence with good Gestalten even when they are different. Gestalten are preferred modes of perception.

An example of the use of such Gestalt principles in a display is the well-known recommendation as to the layout of a set of several gauges (Figure 2.1.15). If the gauges are arranged so that under normal operating conditions all the set points are aligned, the display is perceived as a series of straight lines. A gauge with an abnormal reading stands out clearly as a figure from the background of straight

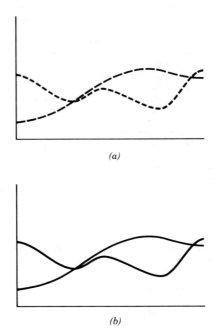

(a)

(b)

Fig. 2.1.13. Gestalt principles in perception: At crossing points lines are "automatically" seen as units continuing in the same direction, not as lines that touch and change direction or symbol sharply.

Fig. 2.1.14. Gestalt principles in perception: Identical "units" become spontaneously organized into thin columns or large squares depending on spatial context.

lines. If such an alignment is not used, each gauge must be individually examined to determine the presence of abnormal reading.

Stereotypes

There are some grounds for believing that the tendency to perceive Gestalten is hard-wired into the neural circuits of the brain. If so, they represent truly universal principles of perceptual organization common to all humans. In addition there are principles that have almost as great generality within quite large groups of people who have in common a culture, education, or professional training and standards. These may best be called perceptual "stereotypes," and represent expectations that are constant over long periods in an individual. They are not as universal as Gestalten, and because of this, they are principles of perceptual organization that can aid accurate perception and at the same time may cause confusion. They are frequently associated with patterns of adaptive action, in what are then called "stimulus–response stereotypes."

An obvious example is the tendency to associate the color red with "hot" and blue with "cold," red with "danger" and green with "safe," red with "stop" and green with "go." Similarly, many people expect that in a circular display, a clockwise movement represents an increasing value, and an anticlockwise movement represents a decreasing value. In each case there is at least one group of people to whom the opposite interpretation is the "natural" one. Thus, to a physicist, blue is hotter than red (because of the relation of temperature to the emission spectrum of black-body radiation); and to anyone thinking in terms of a valve or tap, a clockwise motion closes the valve and so decreases the flow. In the case of color coding, a particularly interesting problem arises in the electrical power industry. It is common to recommend a "green board" philosophy in control room panel design. When the plant is normal, any lights that are lit are green and they turn red when the abnormal situation develops. But in electrical engineering, it has been common for many years to use red for a "live" circuit in which current is flowing, and green for an open circuit. In the latter case, the mapping is "unsafe—red; safe—green," (which is similar to "stop—red; go—green"). But there is a conflict in that when normal generation and transmission of power are taking place, the transmission control board has red lights, whereas the generating control board is green.

The point here is that in general operators do not examine each component of a display in turn, but view a complex collection of displays as a whole, seeing an overall pattern that means "normal" or "abnormal." Only when an abnormality is detected is a detailed examination made of each instrument in turn. A particularly clear example is the case of switch positions. If a North American is asked to glance quickly at a collection of equipment to make sure that all power has been switched off, he or

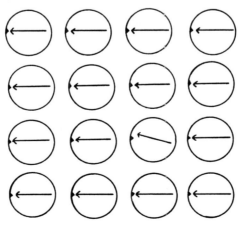

Fig. 2.1.15. Use of Gestalt principles to enable rapid detection of an abnormal instrument setting without the need to read each instrument.

she expects to see all the switches in the "down" position if all the equipment is turned off. The entire array of switches is seen as a single pattern, rather than each switch being individually scanned. But equipment manufactured in Great Britain is made in accordance with the British stereotype— that switches in the "up" position are "off." So, if the equipment includes British components, it is not possible to perceive the "safety state" of the equipment at a glance, but memory must be involved to decide which component is "on" and which "off."

Stereotypical expectations cause a physical stimulus to be perceived not merely as the physical stimulus but, automatically, *as what the physical stimulus stands for.* That is, the meaning is normally perceived automatically, not just the pattern at the receptors. In this sense, the stereotype forces perception to take certain values, much in the same way that neural mechanisms force Gestalten to be perceived. But because stereotypes are learned, and are not universal, what is seen or heard by different observers is different for a given input (consider the difference in the perception of a gunner or a fireman on hearing someone shout "fire!").

It is important to emphasize that we are claiming not merely that a different interpretation will be put on a given perception by people having different stereotypes, but that in a very real sense, what is perceived is different before the observer has time to reflect on its meaning.

It follows that it is important for a designer to establish what stereotypes shape the perception of those who will use a display in order to map the physical stimulus onto perception correctly. Under stress, time pressure, or fatigue, an observer will tend to perceive what the stereotype expects, even if no trouble arises during normal situations.

Short-Term Expectations

The stereotypes referred to above are stable and difficult to alter. In addition, there is a constant dynamic interaction between physical stimulation and the expectations of the observer. Fluctuation in what is perceived can occur on an extremely short time scale—in some cases in fractions of a second.

The best way to emphasize the interaction between the effect of the physical properties of inputs and the effect of subjective factors is by introducing the Theory of Signal Detection (TSD) in the context of fault detection during quality control.

Consider the case in which a quality-control inspector is watching a series of printed circuit boards passing by on a production line. The task is to detect the presence of faulty solder joints. Intuitively, we feel that the inspector is more likely to report a large fault than a small one, that appropriate lighting will improve the detectability of a fault, and so forth. That is, the physical properties of the signal to be detected (here a bad solder joint) determine the probability that a fault will be perceived.

Consider now the case in which a board comes by that has a solder joint that is in some small degree different in appearance from the standard, normal joint which the inspector has been trained to accept. Will the inspector perceive it as a fault, or not? There is good empirical evidence that at least two factors influence the way in which he decides what has been seen. If the inspector knows that there are many faulty components coming through, this particular ambiguous case will more probably be judged faulty than if the inspector believes that few faulty components are being produced. Secondly, the inspector will have some mental model of what his judgments are worth. To reject a good board is costly to the company; to let by a faulty board is also costly if the completed unit is to be returned under warranty. Moreover, it may be that the inspector's pay scale is tied in some way to the relative number of false alarms, correct detections, missed faults, and so on, that are made. If missed faults are costly, then when in doubt the inspector should judge the component faulty; if the rejection of boards that are actually satisfactory is costly, then the inspector should judge the component to be normal. In other words, *the judgment as to what is seen is affected by the expectations of the observer and the payoffs associated with the outcome of the judgments to be made, given a situation in which the physical input to the receptors is constant.*

Note that the word "judgment," rather than "perception," has been made in the above passage. That is, when the observer considers what is perceived, the *response* made is affected by expectation and payoff, (expected utility). There is considerable evidence that conscious perception itself can be regarded as an observer's response to the output of the pattern-analyzing mechanism of the brain. As a result, under many circumstances what is perceived is also affected by expectation and payoff, not merely what is done about what is perceived.

TSD offers a formal way to represent these phenomena, and because they are so important, and so ubiquitous, it is worth discussing TSD in some detail. For a particularly clear account of the application of TSD to practical problems, see Swets and Pickett (1983), who also provide computer programs for the appropriate data analysis.

TSD proposes that all perceptual judgments occur under uncertainty, owing to "noise" in the perceptual system. "Noise" is here defined as any events that are not caused by the signal that the observer is trying to detect. Noise may be due to glare in a visual task, the hum due to 60 Hz mains frequency of the electrical supply in an auditory task. It may be due to random activity in the nervous system.

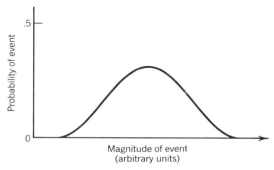

Fig. 2.1.16. Probability distribution of events observed in the absence of an input.

We assume that there is some place in the nervous system where activity associated with messages from the receptors is measured and related to the likelihood that it was caused by real signal. Because of random fluctuations in the environment and in the nervous system, the level of activity fluctuates even in the absence of a real input. The distribution of this activity is shown in Figure 2.1.16. (Think of what one sees if one shuts one's eyes while sitting in the dark—not black, but a rapidly varying pattern of visual noise.)

We now assume that when a real signal arrives, the activity due to that signal is added to the activity due to the background noise, with the result that the mean of the distribution is shifted along the likelihood axis (See Figure 2.1.17).

At any moment when the observer is asked to say whether or not a signal (e.g., poor solder joint) is present the observer examines the evidence presented, and must decide from which distribution it comes. For very large signals (S_L) or very low levels of mere noise (N_L) there is no problem. But for intermediate levels, a statistical decision is called for (Event E_i). The event could come from either distribution. What should the observer do?

The observer should choose some level on the likelihood axis at which to place a decision cutoff. Call this C. Now, if the observed activity is greater than C, the observer should say "yes, this was a signal"; if less than C, the observer should say "no, what is present is merely noise." But where should the cutoff be placed?

Notice that the cutoff gives rise to four areas under the two distributions, labeled TP, FP, TN, and FN in Figure 2.1.17. (These areas are respectively often called Hit, False Alarm, Correct Rejection, and Miss.) These areas can also be conveniently represented on a matrix:

		What is actually present	
		Signal S	No Signal N
What is said to be present by the observer	"Yes" Signal Present	TP	FP
	"No" Signal Absent	FN	TN

It is not appropriate in this chapter to develop the formal mathematics of TSD. But the following intuitive development will show the importance of the concepts. We can now make a rational choice for the position of C_0. If the observer knows that signals occur in many observations, C_0 should move to the left along the likelihood axis, so that when in doubt, a judgment of "signal present" will be made. If FPs are relatively cheap, or TPs very valuable, then again C_0 should move to the left. Conversely, if signals are rare, FNs costly, or TNs valuable, the position of C_0 should move to the right. It is possible to define precisely where C_0 should be placed when the signal probabilities and the payoffs associated with the cells in the decision matrix are specified.

The consequence of moving C_0 is that the proportion of TFs can be moved from 0 to 1 independently of any change in the strength of the signal. That is, for faults of a given detectability, either none, all, or some intermediate proportion of them will be detected, depending on the expectations of the inspector and the perceived values of TPs, FPs, TNs, and FNs. This result is due to the contribution of "top-down" subjective factors to perception.

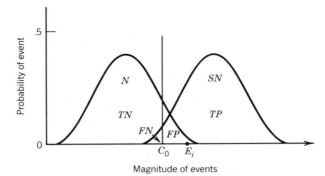

Fig. 2.1.17. Theory of signal detection. N is the distribution of events in the absence of an input. SN is the distribution of events given real inputs. The area TP is the probability of a true positive judgment (a "hit"). The area FP is the probability of a false positive judgment. The area TN is the probability of a true negative judgment. The area FN is the probability of a false negative judgment (a "miss"). C_0 is the observer's cutoff criterion. E_i is the magnitude of an ambiguous stimulus. For discussion see text.

On the other hand, suppose we leave C_0 in a constant position and increase the magnitude of the signal. As the signal magnitude increases, so the mean of the SN distribution moves to the right. As the magnitude of the signal decreases, so the mean of the SN distribution moves to the left. In each case, the proportion of signals detected (the probability of TP) changes. Signal strength is thus represented, in TSD, by the separation of the mean of the distributions (usually called d' in the literature). This is the contribution of the physical properties of the task and bottom-up processing to the proportion of signals detected.

Because it is an axiom of this approach to perceptual judgment that *all* judgments are made under statistical uncertainty, it follows that all perceptual decisions must be described in terms of two parameters, one (signal magnitude) to do with the physical properties of the signals to be processed, the other (C_0) to do with the decision criterion adopted by the operator. Elsewhere in this volume, more detailed treatment of TSD is related to watch-keeping and inspection tasks, and evidence is presented to show that when TSD is used as a normative model, humans can approach very close to optimal performance. For now the following observation at a qualitative level is sufficient: if an observer is showing perceptual behavior that is inadequate, there may be two kinds of cause. If the inefficiency can be shown to be related to the properties of the "signals," then attention should first be paid to the design of the work station, equipment, lighting displays, and so on. But if the inefficiency is due to subjective factors, then the appropriate way to improve perceptual judgments is by training or motivation. The observer's expectations about the probability of "signals," and his or her understanding of the payoff matrix associated with different judgments should be altered. Only when the latter has been optimized should attention be paid to hardware redesign. For more details of TSD applications see Swets and Pickett (1983).

There are a number of assumptions made in TSD that may often not be satisfied in practical situations. But more important than the details of the method which may or may not allow its quantitative applications to a particular task is the general conceptual framework, and the following principle. *Perception is determined by the interaction of the physical parameters of the stimulus with the subjective control of the perceptual mechanisms by the observer.*

Attention

A major determinant both of the context and quality of perception is attention, which operates both bottom-up and top-down. Certain aspects of a stimulus are able to attract attention, with the result that the observer switches to the new stimulus. Characteristics of inputs that attract attention include high intensity, sudden changes, movement, and emotional importance (such as the occurrence of a person's own name). The first of these are stimulus characteristics; they cause perception to be interrupted by this salience. Emotional importance, on the other hand, must be a top-down effect because there is no physical property that involves emotional importance as such.

Little use can be made of emotional characteristics in the design of human machine systems. But the salience of signals is of considerable importance if it is necessary to try to attract an observer's attention. Flashing lights and trains of sounds are for this reason used as warning signals. Such highly salient signals will attract attention, although they cannot guarantee that the observer will not redirect attention voluntarily away from salient signal once the distraction has been noted.

The mechanisms of attention have been intensively studied during the last 30 years. It is well established that attention can operate both by changing the observer's sensitivity and by changing the decision criterion as discussed in the preceding section. Attention can change the observer's sensitivity to an incoming signal by means of an internal "gain control." It can also change the willingness of an observer to judge that a particular signal is present, by means of a criterion shift. Moreover, at least in hearing, changes in sensitivity and criterion can occur several times a second. Attention can be highly labile.

A major topic of research has been the extent to which an observer can process more than one message at a time, that is, whether "parallel processing" is possible. The bulk of the evidence suggests that most incoming inputs are always processed in parallel to the point of pattern analysis. Attention seems to modulate the contents of consciousness, operating on the output of pattern analysis mechanisms. Indeed very highly practiced skills can be performed without conscious attention in an "automatic" mode of processing: the incoming information is used to produce a response, but it is not perceived.

It is possible to set up laboratory experiments that suggest that parallel processing can occur in a variety of paradigms. But although such results are of great theoretical interest they are of little practical importance. For the purpose of system design it should always be assumed that an observer pays attention to only one source of information at a time. A few cases exist where parallel processing appears to play a role in applied settings. For example, Fisk and Schneider (1981) found that the vigilance decrement in a detection task could be reduced if observers were given sufficient training for their behavior to be "automatic." And if the displays in control rooms are appropriately designed it is sometimes possible for operators to view the display "as a whole" rather than fixating each instrument in turn (L. Bainbridge, 1985, personal communication).

Attention is limited particularly with regard to vision. Because of the structure of the eye a source of information must be fixated if it is to be perceived accurately. With the exception of advanced "integrated" displays that have been designed to display multiple information in a single window, an observer pays attention to what is fixated. Moreover, because in "real life" tasks not more than two or three eye movements per second are made, attention is severely limited.

There are several dimensions of auditory signals that allow an observer to select one message and ignore another. Of these the most effective is a difference in the position of the source of sound, but others that can be used are differences in pitch, intensity, rhythm, or quality of voice (Broadbent, 1958; Moray, 1970). It is often thought that because auditory signals can be perceived without the need to orient toward the source of information, hearing is a good way to present multiple alarms. But there is good evidence that if more than about six messages arrive, confusion results, and if they happen simultaneously, masking and interference occur unless stereophonic separation is used. In particular, the tendency to rely on intensity to guarantee the reception of an auditory signal should be resisted, because operators often switch off warnings if they are so loud as to be annoying or distracting.

In applied settings visual attention dominates information processing. In a large conventional control room, for example, operators walk from display to display, or allow their eyes to move over the displays, fixating first one and then another. The amount of information to be assimilated is often far greater than can be handled in the available time, and as a result tactics of attention develop.

The first of these is the SATO discussed earlier. In general, the longer the display is fixated, the more accurately its value is perceived. If an observation has to be cut short in order to transfer to another display, then perception is less accurate.

The second tactic, which has not been investigated extensively, is the intelligent allocation of attention. In real systems, it is often the case that several variables are correlated ("coupled"). There is some evidence that in such cases, observers tend to look at one or two displays from each subsystem, and rely on the correlation to know that the subsystem as a whole is operating normally (Iosif, 1968). When an abnormal state is detected, attention tends to lock onto that subsystem and others are ignored ("cognitive tunnel vision," or "cognitive lockup").

Underlying both those tactics, however, are more general patterns of visual attention. At least 10 mathematical models have been proposed for the dynamics of visual attention, and most of them have been successfully tested on at least one occasion. Although observers have the impression that they are choosing where they look, they are usually extremely inaccurate if asked to describe the way in which they attend to displays (such as the cockpit of an aircraft). In real life, operators characteristically are highly practiced, performing the same task for many days or months. During this time, they develop mental models (usually not available to consciousness) of the statistical properties of the displays. Dynamic visual attention is driven by those models to keep the observer's uncertainty about the state of the system to a minimum. It is known that at least the following variables play a role in controlling dynamic visual attention. First is the rate at which the display varies: the greater the bandwidth the more frequently is the display sampled. Second is the value of the information displayed: the more the information is worth, the more frequently it is sampled. Third is the cost of observation: the more costly an observation, the less frequently is the display sampled. Fourth is forgetting: as time passes since the last observation, the observer becomes less certain of its value even if it varies only slightly or not at all. Fifth is the coupling between displays. For an extensive review of dynamic visual attention see Moray (1983, 1986).

2.1.5.3   Advanced Technology

At this time little is known about the impact of new technology on patterns of attention and perception. Computer-generated displays, in which different areas of a single screen may be altered, windowed, overlaid, and so on, may produce new patterns of attention. Basically the variables mentioned in the preceding paragraph are certainly the basic determinants, but new patterns may emerge. Anyone involved in designing such displays should be aware that both the psychophysics of computer-generated images and the tactics of attention may be different from the descriptions currently to be found in the literature. Moray and Rotenberg in our laboratory have recently found that in a process control task there were more than 20 times as many fixtures on instruments as there were actual actions to change the system state variables, and almost no fixations were made elsewhere than the display. It follows that understanding attention is far more important than noting control actions in understanding the information processing load on an operator. Foley and Harwood in our laboratory have found conclusive evidence that the visual system is affected differently by a CRT display from the way it is affected by a conventional display even when color, brightness, text font, contrast, and material displayed are all equivalent.

2.1.6   CONCLUSIONS

An understanding of basic anatomy and physiology, of psychophysics, and of the principles of perception and attention is essential for appreciating the problem faced by an operator in a human–machine system. All actions are based on information, past and present, received from the environment. The information is encoded through the psychophysical functions. Past knowledge is stored as a more-or-less veridical mental models of the task in human–machine systems, and new information is interpreted through that model. That interpretation is carried out in the light of the operators' beliefs about the probability and value of events.

The final outcome is intelligent and adaptive behavior. Tables in handbooks can provide base lines and boundary conditions. People provide usable perception.

REFERENCES

Blackwell, H. R. (1946). Contrast thresholds of the human eye. *Journal of the Optical Society of America, 50,* 624–643.

Broadbent, D. E. (1958). *Perception and communication.* London: Pergamon.

CIE (1959). *Colors of light signals.* (Publication CIE #2, W.1.3.3). Paris: CIE. Berean Center.

Fisk, A. D. and Schneider, W. (1981). Control and automatic processing during tasks requiring sustained attention: a new approach to vigilance. *Human Factors, 23,* 737–750.

Foley, P. J., Lavery, J., and Abbey, D. (1964). Scotopic acuity and knowledge of results. *Perceptual and Motor Skills, 18,* 505–508.

Graham, C. H. (Ed.) (1965). *Vision and visual perception.* New York: Wiley.

Hecht, S., Schlaer, S., and Pirenne, M. H. (1942). Energy, quanta, and vision. *Journal of General Physiology, 25,* 819–840.

Iosif, G. (1968). La strategie dans la surveillance des tableaux de commande. I. Quelques facteurs determinants de caractere objectif. *Revue Roumanien des Sciences Sociales-Psychologiques, 12,* 147–161.

Koffka, K. (1935). *Principles of gestalt psychology.* New York: Harcourt, Brace.

MacAdam, D. L. (1942). Visual sensitivities to color differences in daylight. *Journal of the Optical Society of America, 32,* 247–274.

Moray, N. P. (1970). *Attention: Selective processes in vision and hearing.* London: Hutchinson.

Moray, N. P. (1983). Attention is dynamic visual display in man-machine systems. In R. Parasurman, Ed., *Varieties of attention.* New York: Academic, pp. 485–513.

Moray, N. (1986). Monitoring behaviour and supervisory control. In K. Boff, L. Kaufmann, and J. Beattie, Eds., *Handbook of human perception and performance.* New York: Wiley.

Mowbray, G. H., and Gebhard, J. W. (1958). Man's senses as information channels (Report CM-936). Baltimore: Applied Physics Laboratory, The John Hopkins University.

Munsell, A. H. (1941). *A color notation.* Baltimore: Munsell Color.

Oswald, I., Taylor, A., and Treisman, M. (1960). Discriminative responses to stimulation during human sleep. *Brain, 83,* 440–448.

Pieron, H. (1952). *The sensations.* New Haven, CT: Yale University Press.

Pitt, F. H. G. (1935). Characteristics of dichromatic vision (Special Report 200). London: Medical Research Council.

Pollack, I. (1952). The information of elementary auditory displays. I. *Journal of the Acoustical Society of America, 24,* 745–749.

Shvarts, L. A. (1957). Raising the sensitivity of the visual analyser. *Psychology in the Soviet Union.* Palo Alto, CA: Stanford University Press.

Swets, J., and Pickett, R. (1983). *Evaluation of diagnostic systems.* New York: Academic.

Tobias, J. V. (Ed.) (1970). *Foundations of modern auditory theory,* Vol. 1. New York: Academic.

Tobias, J. V. (Ed.) (1972). *Foundations of modern auditory theory,* Vol. 2. New York: Academic.

Wysecki, G., and Stiles, W. S. (1967). *Color science.* New York: Wiley.

Major Reference Sources for Psychophysical Data

Graham, C. H. (1965). *Vision and visual perceptions.* New York: Wiley.

Tobias, J. V. (1970, 1972). *Foundations of modern auditory theory,* Vols. 1, 2. New York: Academic.

Van Cott, H. P., and Kinkade, R. G. (1972). *Human engineering guide to equipment design* (Rev. ed.) Washington, D.C.: U.S. Government Printing Office.

CHAPTER 2.2

INFORMATION PROCESSING, DECISION-MAKING, AND COGNITION

CHRISTOPHER D. WICKENS

**Institute of Aviation and Department of Psychology,
University of Illinois at Urbana-Champaign**

The author would like to thank Dr. Martha Weller for her helpful editing and comments on the text of this chapter.

In any task to which the human is assigned, information must be processed. Events and objects in the world must be perceived and interpreted, and then either responded to immediately or stored in memory for later action.

Figure 2.2.1 provides a representation of human information processing that explicitly labels each of these mental activities. Information relayed through the senses is first perceived. This process of perceptual recognition involves some match between the sensory information and a "template" or representation of the recognized object stored in permanent long-term memory (Chapter 2.1). Once a stimulus is identified, a decision must be made as to what action to take. In this case, a response may be selected immediately, or the information may be maintained for some period of time in working memory. If the latter course of action is chosen, the stored information may either be given a more permanent status in long-term memory (learned), forgotten altogether, or used to generate a response. Once a response is selected (either immediately or after memory storage), it must be executed, normally through a process of coordinated muscular control, operating somewhat independently of the selection that preceded it (Fitts and Peterson, 1964; see also Chapter 2.9).

Finally, as indicated in the figure, the consequences of a response normally become available again to perception as *feedback*. This feedback may either be *intrinsic*—such as the feeling of the fingers, the sound of a key press, or the sound of one's voice; or *extrinsic*—such as a light that appears on a video display to acknowledge that a command was received. Feedback of both forms is generally helpful to performance, particularly for the novice, and when it is immediate.

Most of the activities described in Figure 2.2.1 can take place rapidly, but under constraints reflecting the capacity of the various mental operations involved. These capacities are of two generic forms: (1) Each operation has limits in the *speed* of its functioning and in the amount of information that can be processed in a given unit of time. (2) There are limits on the total attention, "mental energy," or resources available to the information processing system. These limits are represented by the "pool" of attentional resources shown at the top of Figure 2.2.1.

This chapter discusses some of the fundamental memory, decisional, response, and attentional limitations of human performance. The approach then focuses on limits of particular mental operations, rather than on characteristics of the entire system. Although it is true that this approach cannot be used exclusively to model performance in more complex settings it is equally true that these more complex modeling efforts must account for the basic limitations and characteristics of the subprocesses. These subprocesses are the concern of this chapter and are examined in five major subsections: (1) the relation of displayed and perceived information to its storage in working memory, (2) the limitations

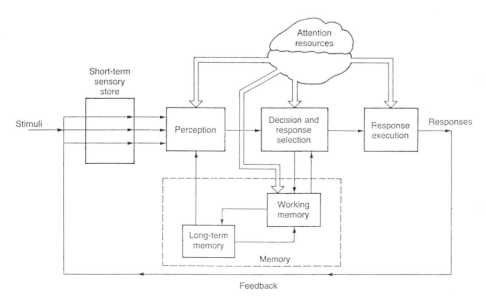

Fig. 2.2.1. A model of human information processing. Source: Wickens, 1984a.

of working memory itself, (3) the limitations of cognitive processes related to decision-making and diagnosis, (4) the limitations of response processes as they are manifested in discrete responding tasks, and (5) the limitations and characteristics of attention as they influence the human's ability to carry out two tasks concurrently.

2.2.1 PERCEPTION AND MEMORY

The basic limits of perception, related to absolute and relative sensitivity and the processing of color, are addressed in Chapters 2.1 and 5.1, and Chapter 2.4 addresses issues of learning and forgetting. This chapter focuses on how characteristics of memory and higher mental functioning bear upon perception. These aspects of perception are becoming increasingly important in highly automated systems in which machines replace the human's response capabilities (Wiener and Curry, 1980; Wickens, 1984a), because the humans who must monitor such systems often need to gather large amounts of perceptual information, storing this in working memory before infrequent responses are initiated.

2.2.1.1 Expectancy

As shown in Figure 2.2.1, perceptual recognition can be described as the association between an incoming stimulus event and a recognized "template" that is stored in long-term memory (Neisser, 1967; Rummelhart, 1977). For example, perception of the letter A results both because the features of the A are analyzed by the visual system and because A is a familiar concept frequently experienced, and therefore has a strong representation in memory. Although sensory evidence is thereby necessary for perception to take place, that perception may be facilitated by *expectancies* of which stimuli are likely to occur. The role of expectancy in perception is quite pronounced and can trade off with the sensory quality of a stimulus, such that if expectancy is high, perception will still be possible even if the stimulus is degraded. Data provided by Tulving, Mandler, and Baumal (1964) illustrate this trade off explicitly. They presented subjects with sentences of the form, "I'll complete my studies at the ————," and

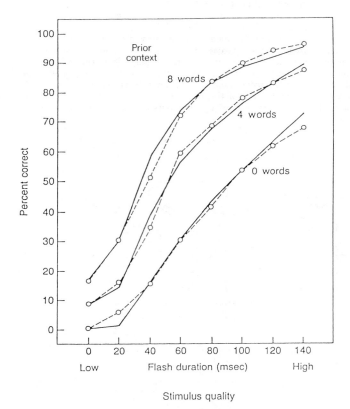

Fig. 2.2.2.   The effects of prior context (eight words, four words, and no words), and stimulus quality (exposure duration) on word recognition. After Tulving, Mandler, and Baumal, 1964.

displayed the final word for very brief durations, thereby producing a degraded sensory stimulus. The experimenters could adjust both the duration (and therefore the degraded quality) of the stimulus, and the amount of prior word context between eight, four, and no letters. The results, shown in Figure 2.2.2, illustrate the almost perfect trade-off in recognition accuracy between stimulus quality and redundancy. As one of these variables decreases, the other may be increased to maintain a constant level of recognition performance. Similar trade-offs may be observed when listening to degraded speech with varying amounts of context-determined expectancy, under varying levels of stimulus degradation (see Chapter 6.1.).

Unfortunately, precise parameters of the trade-off are difficult to specify for two reasons: (1) The absolute level of stimulus degradation is determined by a large number of different factors (e.g., poor viewing, small text, visual glare, vibration), and their combined effect is difficult to predict. (2) Expectancy depends to a great extent upon the redundancy of the language in the communications systems, along with the operator's knowledge of that redundancy. The latter is a subjective factor that is hard to quantify. Nevertheless, a firm design guideline can be imposed such that any anticipated decrease in perceptual quality can be compensated to some degree by increasing stimulus expectancy. The latter in turn can be accomplished by (1) restricting vocabulary size of the potential message set, (2) increasing the sequential constraints between pairs of items, or (3) using familiar overlearned material (Wickens, 1984a).

2.2.1.2   Object Integrality

Our perceptual system is limited in its ability to process information from different stimulus objects at one time (see Chapter 2.1). In contrast, we can process in parallel the several dimensions of a single object (Lappin, 1967; Kahneman and Henik, 1981). Therefore, if several variables must be coded on a display, and their values must be interrelated, these variables will be more rapidly and efficiently interpreted when they are represented in an integrated object format. For example, Jacob, Egeth, and Bevon (1976) required subjects to classify nine-dimensional stimulus patterns. These patterns could be represented by any of the four formats shown in Figure 2.2.3. The nine dimensions of the

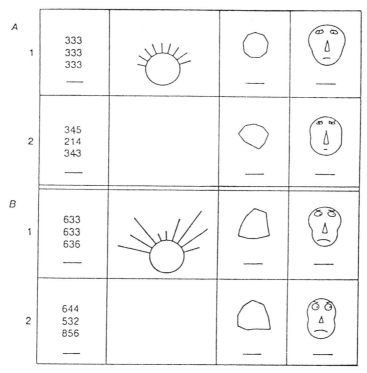

Fig. 2.2.3.   Four categories of stimuli for the display of multivariate data. Within each column (except column 2) are two examples of each of two assigned categories (A and B). Source: Jacob, Egeth, and Bevon, 1976.

two formats on the left (the digital and the "glyph" displays) are clearly separate from each other, whereas on the right, these dimenisons (the radii of a polygon surrounded by contours, and the features of a face) are more closely integrated into a single object. The face display has the added feature of increased familiarity. Jacob et al. found that performance was both more rapid and accurate with the two object-like displays on the right, with the greatest advantage shown by the face display. In a dynamic system monitoring task, Carswell and Wickens (1984) found a corresponding advantage of a polygon display over a separated bargraph display more like that shown in the second column of Figure 2.2.3. The polygon object display has been employed as a display of critical safety parameters in nuclear power plants (Woods, Wise, and Hanes, 1981).

It should be noted, however, that separate displays may be effectively employed in those instances when multiple variables can be processed independently and need not be compared or related, as is the case in determining whether any particular system parameter exceeds a critical value. In such instances, the physical appearance (location, orientation, color) of the critical value should be identical across all displays. For example, if all rotating meters on a given panel have an out of tolerance level, that level should be at the same angle for all.

2.2.1.3 Stimulus–Central Processing Compatibility

Displayed information can often be arranged along a continuum defining the extent to which that information is inherently spatial-analog (e.g., information concerning relative locations, analog transformations, or continuous motion) or linguistic-symbolic and verbal (e.g., a set of instructions, alphanumerics codes, directions, or logical operations). The two end points on this "verbal-spatial" continuum also appear to define two different working memory systems that may be used to perform the mental transformations or rehearsal necessary to carry out the tasks (Baddeley and Hitch, 1974). These memory systems may be labeled as "spatial" and "verbal." Furthermore, any one of four "formats" of information display may be used to present task-related information as shown in Figure 2.2.4. These display formats are defined in terms of sensory modality (auditory versus visual) and processing code (verbal versus spatial/analog).

Often, a system designer may need to make a choice of which format to use for displaying a given piece of information. To aid this choice there exist guidelines, represented in Figure 2.2.4, that specify high-compatibility linkages between stimulus formats and central processing operations (Wickens, Sandry, and Vidulich, 1983; Wickens, Vidulich, and Sandry-Garcia, 1984). This is known as stimulus–central processing (S-C) compatibility. Specifically, tasks with spatial-analog requirements (e.g., navigation, tracking, or assessing relative locations) are found to be best served by visual, particularly visual-spatial, formats. In contrast, tasks that make extensive use of verbal or phonetic working memory, when digits and words must be rehearsed, are found to be better served by auditory–verbal (i.e.,

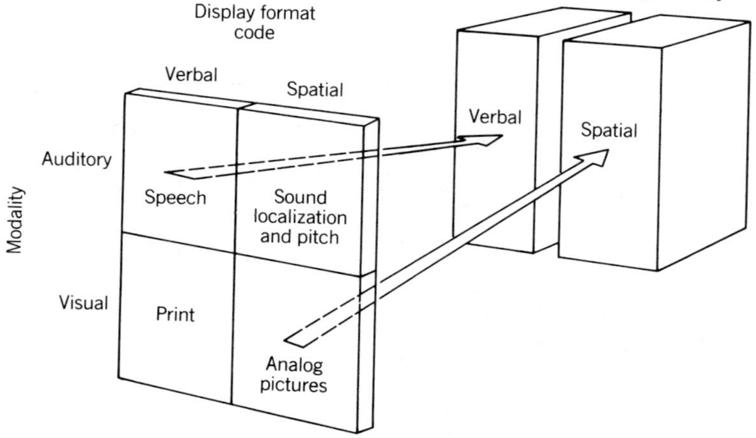

Fig. 2.2.4. Guidelines for display compatibility. Four formats of information display defined by two modalities (auditory and visual) and two codes (spatial and verbal) of perception. The arrows indicate the optimal assignment of formats to working memory.

speech) input. Because the speech channel is necessarily serial and transient, it may be important to "echo" speech messages with redundant visual print, particularly if the message is long. The auditory-spatial format (i.e., tone pitch or intensity, apparent spatial location) does not appear to be optimal for processing any sort of information, except *simple* warning alerts, redundant cues, or when other formats are heavily overloaded (Thompson, 1981; Vinge and Pitkin, 1972; Wickens et al., 1984).

The importance of redundant coding in different formats should be reemphasized. Whether information is presented for immediate use (Colquhoun and Baddeley, 1969; Garner and Fefoldy, 1970) or procedural instructions (Booher, 1975), redundant use of different formats has provided benefits. For example, redundant presentation of information in the auditory and visual modality, such as that used to present sonar signals by Colquhoun and Baddeley (1969), will accommodate transient shifts in noise within the processing environment (e.g., visual glare, background noise, verbal distraction) which may influence one format more than another. The redundant use of spatial (pictorial) and printed instructions, as examined by Booher (1975), will address the strengths of different ability groups in the population (e.g., users of high spatial versus high verbal ability). The cost of building in redundant formats in terms of extra display space or weight will usually be well compensated by the increased reliability that redundancy brings.

A second aspect of S-C compatibility pertains directly to ordered, analog systems and quantities. People have "internal models" that describe their conceptions of such quantities. These internal models often (but not always) correspond to the way those systems behave in the real world (Gentner and Stevens, 1983; McClosky, 1983). Efforts should be made to display analog information in a configuration that is consistent with those models (Roscoe, 1968). For example, altitude and temperature are both commodities with which people associate "high" and "low" end points. Therefore, the S-C compatible display of these quantities should be oriented *vertically*, with high levels at the top, low levels at the bottom, and with upward movement of the quantity indicated by upward movement on the display (Figure 2.2.5a). The latter two criteria can be perfectly satisfied by the use of fixed-scale, moving-pointer displays (Figure 2.2.5a), whereas only one or the other of these criteria can be satisfied by moving-tape displays (Figure 2.2.5b and c) (Wickens, 1984). A second disadvantage of both moving-scale displays is that scale values become difficult to read when the variable is changing rapidly.

The general issue of display compatibility as it pertains to the design of computer systems is discussed in more detail in Chapter 11.2.

2.2.1.4 Predictive Displays

Humans are not good at predicting future states on the basis of present information. Whether the task is one of predicting bottlenecks in industrial task assignment, future pressure or temperature gradients in a chemical or energy process, future trajectories of aircraft or ships, or future collision potentials in air traffic control, or extrapolating statistical growth curves, such predictions are poorly made, and usually load heavily on the operator's limited cognitive resources. As a consequence, it is

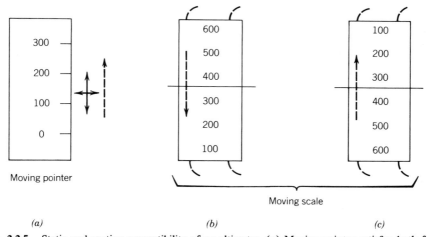

| (a) | (b) | (c) |

Fig. 2.2.5. Static and motion compatibility of an altimeter. (*a*) Moving pointer satisfies both forms of compatibility. (*b*) Moving tape with high values at the top satisfies static compatibility, because high readings are at the top. However, an increase in the displayed quantity causes a downward movement of the display, thus violating dynamic compatibility. (*c*) Moving tape with low values at the top satisfies dynamic compatibility, but violates static compatibility.

not surprising that predictive displays have almost uniformly been found to improve control and scheduling performance (e.g., West and Clark, 1974; Whitefield, Ball, and Ord, 1980; Smith and Crabtree, 1975; Jensen, 1981). These displays unburden working memory from the computational complexities of figuring out the future state of the process or system on the basis of its present state. They also relieve memory from the need to apply algorithms for predicting how the system responds to its present and expected inputs.

Predictive displays typically work according to one of three principles. If it is possible for the system actually to know the inputs to which the human operator must respond before they arrive, then these inputs can be directly displayed. This principle, known as *input preview,* is exemplified in a manufacturing or scheduling process by a display indicating (i.e., predicting) those units about to arrive at the operator's work station. In vehicle or chemical process control, however, it is often impossible to know the future state of the system because of environmental uncertainties that can be known only statistically (e.g., future turbulence). Nevertheless, future state may be inferred by a fast computer simulation of the system dynamics based upon the current system state and its inputs (Sheridan, 1981). Such a simulation may be demanding of computer resources, however, and it is possible to produce approximations of future system state by simple linear regression in which future values are predicted from linear combinations of present values. This third technique was employed effectively by Jensen (1981) in predictive aircraft displays.

Humans can and will use predictive information as long as it is accurate. However, in most systems the accuracy of prediction is a diminishing function of time into the future. This function will fall off more rapidly as (1) the environment has greater uncertainty (e.g., increased air turbulence in an aircraft), (2) the system has lower inertia and is more responsive to inputs (i.e., systems with shorter time constants), and (3) the operator has greater opportunity to exercise control. All three factors should be taken into account when the system designer chooses the *span of prediction,* that is, how far into the future, predictive information will be given.

2.2.1.5 Spatial Cognition

The limits of spatial working memory are often evident when people must navigate and deal with geographical information. Principles concerning the optimal relation between perception and memory in this domain can be specified with regard to *frames of reference,* and the use of maps versus route lists.

Frames of Reference

Displays that depict movement normally portray some set of moving elements against the frame of a stable background. These displays, however, often differ with respect to which element is moving and which is stable, that is, to the *frame of reference* of movement. For example, the conventional aircraft attitude indicator portrays the horizon as moving and the aircraft symbol as stable. This is an "inside-out" frame of reference, showing the same relative motion on the display that the pilot would view out of the cockpit. In contrast, the air traffic controller sees a stable map with a moving aircraft symbol: an "outside-in" frame of reference. In different systems, there may be a number of reasons for selecting one frame of reference over the other; consequently, it is impossible to assert that one is necessarily superior (Roscoe, 1980; Wickens, 1984a). It *is* possible, nevertheless, to specify three principles that can and should influence the choice of reference frames under particular circumstances.

Constancy of Reference Frames. People have a difficult time rapidly reorienting between frames of reference. This kind of situation easily produces control blunders, because the direction of movement required for the controller to compensate for a given display movement may well be compatible within one frame of reference and wholly incompatible within another. Where frames must be switched, it seems appropriate to ensure that the two displays are quite different from each other so that confusion is minimized. For example, the displays should be located at different places, have different colors, or use some other distinct differentiating attribute.

The Principle of the Moving Part. This principle (Roscoe, 1968) assumes that people have certain expectations—an internal model—of what actually moves in a system. The element that moves on the display should be the same and move in the same direction as the operator's expectation of motion, which reflects the internal model. The stable element in the model should be stable on the display. Thus, for example, in the case of a remotely operated robot, control should be easier if the image is an outside-in picture of the robot than an inside-out picture from a camera mounted on the robot itself. This is because the operator assumes the robot to be the moving element in a stationary world. Such an outside-in depiction also leads to less total movement of visual elements on the display, often a source of distraction. Unfortunately, in navigational "electronic map" displays, the outside-in framework with a fixed north-up map can lead to difficulties whenever the controlled vehicle is heading south. In this case, the principle of the moving part is partly violated because a rightward control movement produces a leftward movement of the vehicle symbol on the display.

Compatibility with Operator's Viewpoint. The operator seated within a controlled vehicle has, by definition, an inside-out view. Thus the aircraft pilot who views a moving horizon indicator of aircraft attitude (an inside-out view) has a perspective that is identical to the view outside the cockpit (Roscoe, 1980). In this case the static compatibility between the display and the outside world is preserved, even though the principle of the moving part is violated by the use of an inside-out display.

Fortunately, both the problems related to outside-in displays (reversed frame of reference in map displays when heading south, and incompatibility with the operator's viewpoint when seated in a vehicle), as well as the violation of motion expectancies of the inside-out display, can be addressed by a compromise display that uses the algorithm of *frequency separation* (Fogel, 1959; Roscoe and Williges, 1975; Roscoe, 1980). Using the frequency-separated algorithm, rapid changes in display properties are driven by the outside-in principle, whereas relatively low-frequency changes are driven according to inside-out principles. To see how this principle operates, consider a navigational display in which a vehicle is heading "up" the display. If the vehicle turns to the right, a high-frequency change, the vehicle symbol also turns to the right. If it maintains this new course, a lower-frequency behavior, the entire map now slowly rotates counterclockwise until the vehicle's new heading is again upward. Such a system preserves the principle of the moving part (important for rapid inputs), while generally adhering to a compatible static view of the outside world (for one who is inside the vehicle), and maintains a heading-up rather than "north-up" orientation (for navigational displays).

Maps versus Route Lists

Providing travelers with information on how to navigate from point A to point B represents one important issue in the human factors of spatial cognition. Two major navigational aids are typically used, maps and route lists. Maps are spatial and, unless they are rotated to a heading-up framework, tend to be outside-in, world-referenced displays. Route lists (statements of "turn left," "turn right," or "go straight"), in contrast, tend to be verbal, inside-out, ego-referenced displays (note, however, that if the directions "east" and "west" are used instead of "right" and "left," then these become world-referenced). Which of these aids is best seems to depend on the task at hand and the likelihood of navigational errors.

Navigation. For actual nagivation of a vehicle there is some evidence that route lists (or ego-referenced commands) are superior (Wetherell, 1979). This is because, no matter what the momentary orientation of the vehicle, the commanded turn (left–right) is always compatible with the operator's frame of reference.

Errors. It is clear, however, that the route list is good only so long as one is on-route. If a mistake in navigation is made, the route list becomes worthless, whereas the map allows one to regain the original route, or to find a new way to the goal.

Planning. It is also evident that any planning that requires adoption of a world frame of reference—for example, arranging a meeting place, judging the relative location of two landmarks, or determining alternative courses of navigation—will be best served by world-referenced maps (Bartram, 1980; Baty, 1976; Wickens, 1984a).

Thorndyke and Hayes-Roth (1982) have found that people can possess different forms of knowledge about an environment. These forms, *route knowledge* and *survey knowledge,* correspond very closely to route lists and maps, respectively. Route and survey knowledge may be achieved by different sorts of training (navigation and map study, respectively) and show the same relative advantages and disadvantages as do route lists and maps, respectively.

Distortions of Spatial Recognition

Whether our knowledge of an area is represented by route or survey knowledge, this knowledge seems to possess a systematic bias or distortion, that of structuring the world within the framework of a N–S–E–W "grid" (Tversky, 1981; Howard and Kerst, 1981; Wickens, 1984a). For example, people's memory reconstructions of geographical areas that have features running in diagonal directions, corners that are not right angles, or curves and twists, tend to force or "distort" these features into much more of a N–S–E–W framework.

2.2.1.6 Comprehension

One goal of good display design is to enable rapid and accurate comprehension or understanding of the displayed material. Such a goal is of equal importance in written procedures. Here cognitive psychologists have identified certain important principles of human information processing that make procedural information difficult or easy to comprehend. Three important factors are considered below.

Logical Reversals

Whenever a reader or listener is required to reverse logically the meaning of a statement to translate from a physical sequence of words to an understanding of what is intended, comprehension seems to be hindered. One example of this is provided by the use of negatives. We comprehend more rapidly that a particular light should be "on" than that it should be "not off." A second example of logical reversals is falsification. It is faster to understand that a proposition should be true than that it should be untrue or false. Experiments by Clark and Chase (1972) suggest that these differences are not simply the result of the greater number of words or letters that normally occur in reversed statements, but result from the cognitive difficulties in processing such statements as well.

These investigators used the experimental analog of an operator who reads a verbal instruction (i.e., "Check to see that valve X is closed") and verifies the statement either against the physical state of the valve or against his own memory of the state. The statement may be either affirmative ("valve is closed") or negative ("valve is not open"). Furthermore, it may be either true, agreeing with the state of the valve, or false. From their research, Clark and Chase drew two important conclusions with direct design implications.

1.   Statements that contain negatives invariably take longer to verify than those that do not. Therefore, where possible, instructions should contain only positive assertions (i.e., "Check to see that the switch is on") rather than negative ones ("Check to ensure that the switch is not off").

2.   Whether a statement is verified as true or false influences verification time in a more complex way. If the statement contains no negatives (is positive), then true statements are verified more rapidly than false. However, if statements contain negatives, then false statements are verified more rapidly than true ones.

When these conclusions are used to help the designer phrase proper instructions or to predict the time that will be required for operators to respond to instructions, the meaning of "true" and "false" must be carefully defined. In application, the actual state of a system may take on different values with different probabilities. "True" must therefore be defined as the most likely state of a system. For example, if a switch is normally in an up position, the instruction should read: "Check to ensure that the switch is up," or "Is the switch up?" Because this position has the greatest frequency, such a statement is normally verified as a true positive. Furthermore, as long as negatives in wording are avoided, then the principle will always hold that affirmations are processed faster than falsifications.

The results of the basic laboratory work on the superiority of positives over negatives have been confirmed in at least one applied environment, the formatting of highway traffic-regulation signs. Experiments have suggested that prohibitive signs, whether veral ("no left turn") or symbolic (◯) are more difficult to comprehend than permissive signs such as "right turn only" (Dewar, 1976; Whitaker and Stacey, 1981).

Absence of Cues

People have difficulty extracting information from the absence of cues. Fowler (1980) articulates this point in his analysis of an aircraft crash near the airport at Palm Springs, California. He notes that the *absence* of an R symbol on the pilot's airport chart in the cockpit was the only indication of the critical information that the airport did not have radar. Because terminal radar is something pilots come to depend upon and the lack of radar is highly significant, Fowler argues that it is far more logical to call attention to the absence of this information by the *presence* of a visible symbol, than it is to indicate the presence of this information with a symbol. In general, the presence of a symbol should be associated with information that an operator needs to know rather than with certain expected environmental conditions.

Order Reversals

Many times instructions are intended to convey a sense of ordered events. This order is often in the time domain (procedure X is followed by procedure Y). When instructions are to convey a sense of order, it is important that the elements in the instructions be congruent with the order of events. For example, if subjects are to learn to verify that the order of elements is $A > B > C$, it is better to say, "A is greater than B, and B is greater than C," than "B is greater than C, and A is greater than B," or "B is less than A, and C is less than B" (DeSoto, London, and Hendel, 1965). In the first case, the physical ordering of information, A B B C, conforms with the intended "true" ordering (A B C). In the last two cases, it does not (B C A B or B A C B). Furthermore, in the third case the word "less" is used to verify an ordering that is specified in terms of "greater." This represents an additional form of cognitive reversal. The problems with order reversals would dictate that procedural instructions should read, "Do A, then do B," rather than "Prior to B, do A," because the former preserves the actual sequencing of events in the ordering of statements on the page.

The notion of congruence of ordering appears to be a specific case of the more general finding that people comprehend active sentences more easily than passive ones. Active sentences (e.g., "The malfunction caused the symptoms") preserve an ordering (first the malfunction, then the symptoms) that is more congruent with a mental causal model of the process than is that of a passive sentence (e.g., "The symptoms were caused by the malfunction"). Research has found that passive sentences require both a longer time to comprehend and a greater capacity to hold in working memory (Savin and Perchonock, 1965).

2.2.2 WORKING MEMORY LIMITATION

Human memory has often been dichotomized into two components—long-term memory, which is our permanent store of information and facts about the world, and short-term or "working" memory, our limited store of "conscious" information, which is temporary, fragile, and limited. There is ample evidence from both physiological and behavioral data to support the distinction between these two memory systems. The organization, storage, and properties of long-term memory relate very directly to issues of learning and training (see Chapter 2.4). These are beyond the scope of this chapter, which focuses instead on the limiting characteristics of working memory: a system that can hold but a few items of information for but a short period of time, in a specific code or format that relies heavily upon the human's limited processing resources. These bottlenecks and some of the potential ways around them are detailed in this section.

2.2.2.1 Time and Attention

Without rehearsal, information held in working memory appears to be forgotten entirely within 10 to 20 sec and considerable loss of information occurs over even shorter durations, leading, for example, to the transposition of digits in a phone number (Peterson and Peterson, 1959). Of course, material may be maintained for longer periods of time by rehearsal, but such rehearsal competes for attention with other concurrent perceptual and cognitive activities (Klatzky, 1980; Underwood, 1976).

2.2.2.2 Space

Working memory is also limited by the number of unrelated items that it can hold even when full attention is devoted to rehearsal, this number ranging somewhere between 5 and 9. Miller (1956) has associated the capacity of working memory with "the magic number 7 ± 2." Larger groups than this, such as a phone number with an area code (10 digits) or a nine-digit zip code, exceed the capacity and will likely have one or more items forgotten or transposed before recall takes place. Time, space, and attention interact as shown in Figure 2.2.6. A single item not exceeding capacity may, in fact,

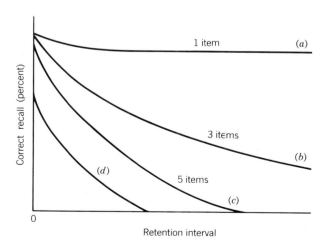

Fig. 2.2.6. Hypothetical relation between retention interval, number of items held in working memory, and recall performance. Curve *d* represents a number of items that exceeds working memory capacity because recall is not perfect even with no delay.

be retained for a long period in working memory without being rehearsed. As more items are added, the prevention of rehearsal exerts a systematically greater effect on the decay curve.

These limits are dangerously restrictive in many complex systems and can often lead humans to enter perceived data into machines incorrectly. The system designer should carefully consider any task demands that will impose loads on working memory and ensure that the information that must be stored is not represented in a transient way. As noted in Section 2.1.3, visual echoes of synthesized speech information will accomplish this purpose. However, even where material is visually presented, there will be some difficulties encountered if more than 7 ± 2 values must be simultaneously compared and contrasted. This is the case, for example, in computerized menus or procedures manuals that offer sets of options. Within any given list, these options should be constrained to no more than the working memory capacity. If more options than this must be presented then they should be configured hierarchically, so that each cluster in which the elements belong together is presented separately. In addition to these systems design guidelines, there are a number of other properties of working memory which, if properly understood and exploited, can circumvent some of its most restrictive limitations. These are now considered.

2.2.2.3 Chunking

Miller (1956) noted that the capacity limits of 7 ± 2 hold only for items that are not related to each other in the operator's experience. For example, 11 unrelated letters exceed capacity. If, however, 11 letters constitute two words, LOW PRESSURE, there are no longer 11 spaces demanded in working memory, but only two. That is, there are two "chunks." The same chunking principles hold true with other strings of items such as phone prefixes (472), area codes (303), or abbreviations (FBI). The individual items within these groups have been experienced together and are "glued" together as a single coherent unit in long-term memory.

In any system in which strings of items may need to be memorized, it is important for the system designer to try to capitalize upon chunking principles by using item strings that are either familiar or are presented in such a way that chunking can be more readily carried out by the user (Wickens, 1984a). For example, the use of letters rather than digit strings in creating codes is often an advantage, because the greater richness and meaningfulness of letter strings will more readily allow them to be associated in memory (i.e., to "build" chunks; Conrad, 1964). Also, long strings of unrelated digits or letters should have gaps between every third or fourth letter, for these create perceptual units that help induce chunking behavior. This is what is done with the social security number. Chunks with three to four items between gaps seem to be the optimum size that leads to best memory storage (Wickelgren, 1964). When certain items within the strings are in fact related to each other, it is important to group these related items close together and to separate them from unrelated groups. This is, of course, typically done in phone numbers, which are written as 333 6195, not 3336195, or 3336 195. In the former case the digits related in the prefix are separated from the rest of the number. In the latter cases they are not.

It should be noted that the principles of chunking extend well beyond the simple memory of digit and letter strings and have been employed to describe such phenomena as memory of chessboard piece position (Chase and Simon, 1973), computer programs (Norcio, 1981), and cues in decision-making (Payne, 1980). For example, Norcio has found that statements in a computer program are better retained if the physical gaps or indentations on the page separate different chunks of the program statement. Thus where printed information must be read, stored, and used (as in instructions, procedures, or recipes), the retention process can be aided by determining the logical chunks that are grouped together in long-term memory because of their common associations, and by physically separating these chunks from others.

The close association that exists between perception and memory is reinforced by the finding that a single object acts as a sort of chunk that supports the memory of its various attributes (Kahneman and Henik, 1981; Wickens, 1984a). As an example, Yntema (1963) found that subjects showed considerably better memory for a small number of objects that varied on a greater number of attributes than for many objects which varied on a few attributes. For example, in an air traffic control problem the altitude, air speed, heading, and size of two aircraft would be better retained than the altitude and airspeed of four aircraft, even though in each case eight items are to be held in working memory. Where possible, then, from a memory standpoint, operators should be given the responsibility for supervising many attributes of a smaller number of objects, rather than the converse, whether these "objects" are aircraft, units under production, or units such as manufacturing systems, engines, or chemical processes.

2.2.2.4 Similarity and Forgetting

A second variable that influences the loss of information from working memory is the *similarity* both between the items in a group that are to be remembered and between those items and other competing activities. The first of these, intragroup similarity, describes the greater forgetting shown of a string of similar looking letters (QOGDUQDUQ) than of a string of different ones (XPOLNTYNK). Because much rehearsal uses an acoustic "loop," acoustic similarity between items also leads to increased forgetting. Thus, the string EGBZCDT is more likely to be recalled incorrectly than an acoustically dissimilar string, XFEYWMU (Conrad, 1964), even when these are displayed visually.

Yntema (1963) also found that when several attributes of an object must be retained in working memory, memory is best if each attribute has its own, distinct, identifiable code that is dissimilar from the code of other attributes. For example, in an air traffic control problem, altitude might be coded in terms of raw feet (4100), heading in terms of compass direction (315), and air speed in 100 knot-units (2.8). Furthermore, the codes could be made physically different in other respects, that is, different colors or sizes. This distinction would maintain each code's unique appearance, maximize the intra-item differences, and so reduce the likelihood of interference.

A second way to maximize differences between items that must be held in working memory is to eliminate completely redundant elements that are identical across items. For example, when an air traffic controller must deal with several aircraft from one fleet, all possessing similar identifier codes (AI3404, AI3402, AI3401), the interference due to similarity across items makes it difficult for the controller to maintain their separate identity in working memory (Fowler, 1980).

2.2.2.5 Processing Code Similarity

As noted in Section 2.1.3, working memory may be defined in terms of two distinct subsystems (Baddeley and Hitch, 1974). *Verbal working memory* is used to store verbal linguistic material (letters, digits, words) and is phonetic and acoustic in its properties, whereas *spatial working memory* maintains analog, spatial, and pictorial properties of the environment and is more visual in its characteristics. Both memory systems show the same general properties of rapid loss, limited capacity, and chunking. Their distinction is important because it defines another dimension along which similarity can vary. A memory load that contains a mixture of spatial and verbal items will be better retained than one containing the same number of exclusively verbal or spatial items. Of equal importance, the verbal-spatial dichotomy helps to predict the kind of forgetting that will be caused by *retroactive interference,* that is, disruption of memory caused by activity performed between the time the material is encoded and recalled. Memory for verbal material is heavily disrupted by perceptual, cognitive, and motor activities that are also verbal and vocal, but less so by concurrent spatial activities. The converse relation holds for memory for spatial material (Baddeley and Hitch, 1975; Baddeley and Lieberman, 1980; Wickens, 1984a).

2.2.2.6 Strategies

The human's working memory operations can be modified to some extent by employing various performance strategies. Two of these in particular appear to offer some benefit in overcoming its fragile limitations. The first is related to chunking, where Chase and Ericsson (1981) have demonstrated that it is possible to train subjects to achieve apparently phenomenal feats of memory—hearing and recalling strings of more than 60 unrelated digits. This is done by emphasizing chunking strategies as described in Section 2.2.3. One subject, for example, was an avid runner and chunked strings of three to four digits as "typical" running times for a given distance. For example, the string 3542 might be chunked as one cognitive unit, a 3 min 54.2 sec mile run. Although the extremely long digit strings memorized by these subjects were obtained only with extensive training, it does appear quite feasible to train operators to look for and create wherever possible any association of subunits that are physically presented close together, whenever memory for unrelated strings is required.

A second strategy, one of "purging memory," has been successfully employed by Bjork (1972) to reduce the interference effects that result when a series of items must be held in memory, used, and then can be forgotten. Under these circumstances one often finds that older items, no longer needed, continue to interfere with the newer relevant items—a phenomenon known as *proactive interference.* Bjork found that the simple procedure of instructing subjects consciously to purge their memory of the now unwanted information appears to be sufficient to eliminate the undesirable effects of proactive interference.

A final mnemonic strategy that can be performed to aid retention in working memory, as well as to improve the storage of material into long-term memory (learning), is to capitalize on humans' rich memory for visual images. This is a special case of chunking whereby a set of items is associated with distinct visual images (material already experienced and stored in long-term memory). Then these images may be tied together by creating a new and sometimes bizarre image. Experiments by Bower and Reitman (1972) have demonstrated the considerable value of this mnemonic strategy in improving memory (see also Yates, 1966, and Luria, 1968 for a further discussion of mnemonic devices).

2.2.3 DECISION-MAKING AND DIAGNOSIS

The preceding section documented the restricting limits of working memory. Complex decision-making must often rely upon working memory to entertain, weigh, and compare the various alternatives;

consequently, it is no surprise that decision-making suffers from a number of memory-related drawbacks. Most decisions that humans perform can be placed into one of three general categories: choice, decision under uncertainty, and diagnosis. Each of these can be represented in simplified form by the two-outcome examples shown in Figure 2.2.7. Each of these cases, in turn, may generalize to far more complex problems.

2.2.3.1 Choice

In choice (Figure 2.2.7a) two objects are offered, each differing on their values along two attributes, A_1 and A_2. The choice of the object with the highest value is derived by multiplying importance weights on each attribute by the attribute value for each object, and then summing these products for each object across attributes. The object with the highest value is the optimal choice. The same representation is applicable as well to the choice between two courses of action, for example, two treatments that a physician may prescribe, or two alternative actions that a plant supervisor may take when confronted with a system failure. In this case the attribute values now reflect the values of different anticipated outcomes, which may each result from different courses of action. This extension from object choice to action choice makes the issue of uncertainty relevant; uncertainty is dealt with in Section 2.2.3.2.

The major difficulties or limitations in human choice performance result from the cognitive working memory loads that are placed on people as they must perform the mental multiplications necessary to consider and integrate all attributes of all objects or actions (Slovic, Fischoff, and Lichtenstein, 1977). Instead of loading memory in this fashion, people often resort to simplifying strategies in which major components of the information are ignored. For example, they may attend only to the most important attribute, and eliminate from consideration all objects that do not score highly on that attribute (Tversky, 1972).

A second human limitation in choice tasks is the tendency of decision-makers to seek and try to

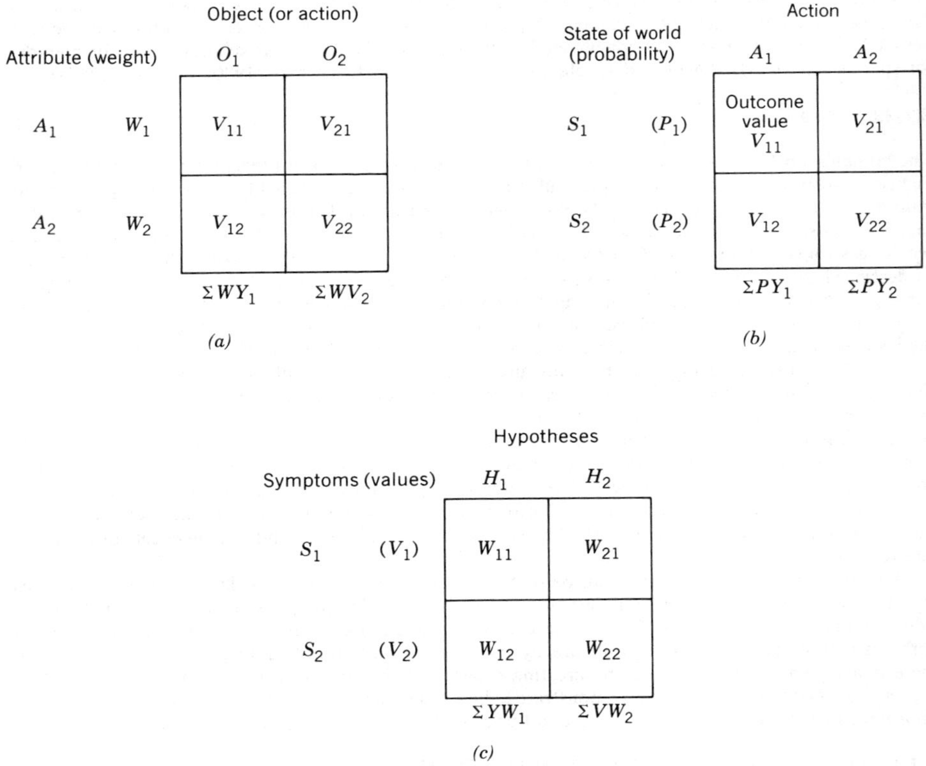

Fig. 2.2.7. Schematic representations of the operations involved in (a) choice, (b) decision-making under risk, (c) diagnosis. In each example the object, action, or hypothesis that has the highest sum of products at the bottom of the matrix is the optimal choice.

use far more information (i.e., attribute values) than the limited attention and memory system can readily incorporate. In fact, it is sometimes found that decision-making quality not only fails to improve as more information sources become available, but actually deteriorates (Samet, Weltman, and Davis, 1976; Oskamp, 1965; Schroeder and BenBassat, 1975).

Multiattribute utility theory has been offered as a successful analytical tool or decision aid to assist people in dealing with the complex choices between several objects or courses of action, differing on several attributes of varying importance (Edwards, 1977). The theory is based on a structural mechanism for analyzing the decision problem into its component attributes and values, and has been found to offer "better" (more optimal and satisfying) choices than those that are arrived at through intuitive decision making (Keeney, 1977; see Chapter 9.1 for a more detailed description of multiattribute utility theory).

2.2.3.2 Uncertainty

Decision-making under uncertainty (Figure 2.2.7b) is formally identical to the choice task between courses of action. However, in this case, the attributes that were listed in the left column of Figure 2.2.7a are replaced by possible states of the world in 2.2.7b, and the attribute weights replaced by the associated probabilities of those states. Each action combined with a state of the world generates an outcome and an associated value depicted in the cells of the matrix. As an example of decision-making under uncertainty, consider the power plant supervisor facing a potential malfunction. The two courses of action in Figure 2.2.7b are to shut down a turbine or to keep it running. The two states of the world are that the turbine either is or is not malfunctioning. The optimal choice of action here may be calculated by computing an *expected value* for each of the four outcomes, that is, the probability of the state of the world, multiplied by the value associated with the outcome of the cell. These expected value products are then summed across each action and the action with the highest score (total expected value) is the optimal one to be chosen.

In addition to the problems that were documented above for the choice paradigm, the added element of uncertainty or risk in Figure 2.2.7b, when considered in conjunction with value, leads to further departures from optimal behavior. This is because people's intuitive estimates of probability and value do not always correspond to the actual objective values of these quantities. In the first place, ample research has documented that subjective probability tends to overestimate the frequency of very rare events and to underestimate the frequency of very frequent ones (Sheridan and Ferrell, 1974). Overestimation of the likelihood of rare events will, for example, lead one to inflate the expected value of rare positive outcomes (such as winning a lottery). This bias accounts in part for people's tendency to gamble at an expected loss.

Secondly, people do not appear to treat subjective value, known as *utility*, as a linear function of objective value. That is, $10.00 added to $1000 is perceived as a smaller gain in utility than is $10.00 added to $100. Instead, the function appears to be more like that shown in Figure 2.2.8, a function

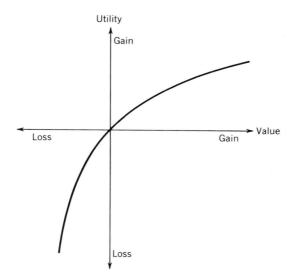

Fig. 2.2.8. The hypothetical relation between value and utility.

whose shape has two important implications: (1) the gain of a fixed amount of value is perceived as having a smaller utility, the greater the value to which the gain is added; and (2) a disproportionately greater perceived change (i.e., loss or gain, respectively) in utility is perceived as the result of the loss of a fixed amount of value than of an equivalent gain. The manner in which these perceptual distortions from the objective levels of probability and value can bias risky decision-making is sometimes complex, but their overall influence on decision behavior has been well documented (Tversky and Kahneman, 1974; Edwards, Lindman, and Phillips, 1965).

2.2.3.3 Diagnosis

The third form of decision-making—diagnosis—is quite similar in many respects to decisions made under risk and is represented in Figure 2.2.7c. In human factors applications this representation describes the task of troubleshooting. Here, for example, the troubleshooter has two possible hypotheses about the state of a malfunction, a broken pump and a leaky pipe. He also observes two symptoms (s)— perhaps low water pressure and high temperature. To perform an initial diagnosis, the value or state of these symptoms (v) must be multiplied by diagnostic weights (W) one for each symptom—hypothesis combination. Then the products are summed across hypotheses, and that hypothesis with the highest sum is assumed to be the most likely hypothesis. At this point, the hypothesis may be further tested for confirmation or refutation.

Of course, the task of diagnosis is susceptible to the same limits of working memory and mental arithmetic as are decision and choice. Mehle (1982), for example, has argued that it is difficult for people to hold in working memory or entertain more than two or three hypotheses at once. In addition, the tasks of diagnosis and hypothesis testing are hampered by five further characteristics that make the particular human biases even more pronounced.

Representativeness

For optimal performance the proper diagnosis should be based only in part on the diagnostic weights, representing the degree of "match" between the particular symptoms shown and the hypothesized state. For example, a broken pump might be characterized by a particular set of high and low pressure readings at points leading into and away from the pump, respectively. In addition to this *symptomatic* information, however, a second important factor is the *prior probability* that one hypothesis or the other is true. In the foregoing example, it may be that a leaky pipe is 100 times more likely to occur than a malfunctioning pump. In this case, the optimal diagnosis (that with the greatest probability of being correct) should be biased toward the pipe failure even if the pattern of symptoms appears more similar to that produced by the pump failure. Bayes' Theorem (Edwards, Lindman, and Phillips, 1965; see Chapter 9.1) provides an optimal prescription for how to weight prior probability and symptomatic evidence in reaching a diagnosis. Yet a good deal of experimental evidence summarized by Tversky and Kahneman (1974) and Kahneman, Slovic, and Tversky (1982) suggests that people often tend to ignore prior probabilities and focus exclusively on how similar the pattern of symptoms are to the hypothesized state. That is, they are too much influenced by the *representatives* of the pattern. Tversky and Kahneman refer to this bias as the representativeness heuristic.

Cue Reliability: The "as if" Heuristic

Symptoms or cues may vary both in their reliability (i.e., how likely an observed cue value is to be accurate) and in their diagnostic weighting (i.e., how readily a symptom can discriminate between two hypotheses). In fact, the absolute value of the correlation between a cue value and the probability that a hypothesis is true may vary anywhere from 0 (an unreliable or undiagnostic cue) to 1.0 (completely reliable and diagnostic). Yet when faced with a number of such cues or symptoms of varying degrees of reliability, people tend to treat them all "as if" they were of equal (and often perfect) reliability (Wickens, 1984a). This "equalizing" of a continuous variable (degree of correlation) seems to represent another example of the cognitive simplification that is carried out in order to reduce the demands on working memory.

Cue Salience

If there is one variable that leads people to weight certain cues more than others, it is not the information value of the cue, but its salience—brightness, loudness, or centrality in the visual field (Wallsten, 1980). Hence if an operator is presented several symptom values on a physical display, all of which *should* be weighted equally, more weight tends to be given to those that are at the top or center of the display, are the largest, or are the most distinctive. The tendency to focus on salient cues has been found to be enhanced under time stress (Wallsten, 1980).

Anchoring and the Confirmation Bias

A hypothesis that is initially chosen may be further tested by seeking more data to either confirm or refute it. However, in this hypothesis testing process, people manifest what is sometimes called the *anchoring* heuristic (Tversky and Kahneman, 1974). They lay their cognitive anchor firmly on the initially chosen hypothesis and do not readily accept new data in favor of a competing hypothesis. This tendency occurs, at least partly, as a result of a systematic *confirmation bias* in hypothesis testing. Cues are sought and found that support the already chosen hypothesis. Those cues that are ambivalent (i.e., consistent with both the chosen and the competing hypothesis) are perceived only as supporting the chosen hypothesis, whereas cues that refute the hypothesis are not sought and therefore not found (Mynatt Doherty, and Tweney, 1977; Wason and Johnson-Laird, 1972). Furthermore, even when those cues are noticed, they are less likely to be remembered (Levine, 1965).

Use of Negative Evidence

A given hypothesis may be directly supported by the presence of a particular noticeable symptom. However, it may also be indirectly supported by the *absence* of a symptom which, had it been present, would have supported a competing hypothesis. Yet people seem to have a difficult job using such "negative information" to narrow down hypotheses sets in troubleshooting sorts of tasks (Hunt and Rouse, 1981), a point noted in the context of comprehension in Section 2.2.1.6.

2.2.3.4   Corrective Solutions

A global consideration of these limits and biases suggests that diagnosis is a task for which humans are ill-suited, a conclusion that has been reached by others as well (see Rasmussen and Rouse, 1981, for an exhaustive treatment of fault diagnosis and a discussion of much of the research in this area). There have been basically three approaches that have been taken to improving the human component of decision-making and diagnostic systems. The first of these has involved efforts to develop *decision aids* that might unburden working memory and unbias certain tendencies in order to allow human diagnosis to proceed more optimally. These range from working memory "unburdening" procedures that keep track of diagnostic evidence as it is observed, to more complex procedures that help perform Bayesian calculations on the likelihood of competing hypotheses in the face of new evidence (Edwards, Phillips, Hays, and Goodman, 1968), or that assist the operator in selecting relevant cues for hypothesis testing (Samet, Weltman, and Davis, 1976). Several investigators have discussed the relative merits and limits of machine aid or computer diagnosis in such areas as medicine (Lusted, 1976), nuclear power (Kiguchi and Sheridan, 1979), and aviation (Wiener and Curry, 1980).

The second effort, referred to as "debiasing," involves training and making the decision-maker aware of the sorts of biases discussed above. To learn, for example, that one has a tendency to ignore evidence that favors a competing hypothesis may make one more sensitive to consider other alternatives in the future. In this regard a number of efforts have been successfully demonstrated. For example, Koriat, Lichtenstein, and Fischoff (1980) found that forcing forecasters to entertain reasons why their forecasts might not be correct reduced their biases toward overconfidence in the accuracy of the forecast. Lopes (1982) achieved some success at training subjects away from nonoptimal anchoring biases when processing multiple information sources. She called subjects' attention to their tendency to anchor on initial stimuli that may not be informative, and had them anchor instead on the most informative sources. When this was done, the biases were reduced. Tversky and Kahneman (1974) have suggested that decision-makers should be taught to encode events in terms of probability rather than frequency, because probabilities intrinsically account for events that did not occur (negative evidence) as well as those that did.

The third effort, similar in many respects to debiasing, simply involves repeated training in the decision-making process, and may be of two forms. It may involve training in the subject matter, which allows operators to appreciate the co-occurrence of and correlation between symptom values. This enables the skilled decision-maker to recognize "syndromes" and make fairly rapid, effortless diagnoses of commonly occurring failures (Rasmussen, 1981; Phelps and Shanteau, 1978). However, such familiarity does not necessarily assist in dealing with novel, unexpected failures, or combinations of failures. At a different level, training may focus more on *generic* sorts of troubleshooting skills. These include such skills as using "split half" strategies—tests that eliminate half of the alternatives—or instruction on the use of negative evidence (Rouse and Hunt, 1984).

2.2.4   RESPONSE PROCESSES

We perceive and we act. Even the monitor of highly automated systems must be prepared at some time to intervene with manual or voice control in case of a failure, or when it is necessary to change automated modes of operation. Under such circumstances it is essential that the action be selected rapidly and correctly. The following section describes some of the main variables that influence the speed and accuracy with which actions are produced. The first part of this section specifically addresses factors that influence the speed of correct responses, the second considers the nature of mistakes and errors in responding, and the final part relates measures of the speed of information processing to the specific *stages* of processing described in Figure 2.2.1.

2.2.4.1   Choice Reaction Time

Degree of Choice

When a response is made to an environmental event, there is usually some uncertainty in the nature of the event and in the choice of an appropriate action. Only rarely is but a single stimulus expected and a single response appropriate. The sprinter at the starting line represents an example of this absence of alternative actions. However, even the sprinter must continuously make a choice of whether to respond or not, and an incorrect choice may lead to a false start. Research over the last several decades has indicated that the minimum time to respond to stimulus events—the reaction time—is heavily influenced by the amount of uncertainty of choice: making one of two possible responses to one of two possible events is faster than making one of three, which in turn is faster than one of four, and so forth. In fact, the relation between reaction time and the degree of choice follows a fairly predictable mathematical law known as the *Hick–Hyman Law* (Hick, 1952; Hyman, 1953). The formal expression of the law is

$$RT = a + b \log_2 N$$

where N is the number of possible equi-likely stimulus–response pairings that could occur in a given context. Because $\log_2 N$ is formally equivalent to the information content of a stimulus in *bits*, the Hick–Hyman law may be rewritten as

$$RT = a + b H_S$$

where H_S is the information content of the stimulus. This function is shown in Figure 2.2.9.

Probability and Expectancy

Stimuli that are expected are responded to rapidly whereas those that are not yield slow and sometimes erroneous responses (Fitts, Peterson, and Wolpe, 1963). A major source of expectancy results from the *probability* or *frequency* with which events occur. Probability can also be represented in an informational context because the information conveyed by an event whose probability is P is equal to $\log_2(1/P)$. The average information conveyed by a *series* of events with differing probability is simply the weighted average of the individual events' information values. That is,

$$H_{AV} = p_i \log_2 \left(\frac{1}{P_i}\right)$$

These event probabilities may be treated either as absolute probabilities or conditional probabilities given a particular context (i.e., $p(X/Y)$, which may be read "the probability of X given the context

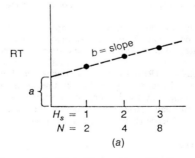

(a)

Fig. 2.2.9. The figure illustrates the Hick–Hyman Law: The linear relation between stimulus information ($H_S = \text{Log}_2 N$) and reaction time. Reaction time can then be expressed as $RT = a + bH_S$. The actual values of reaction time cannot be precisely predicted without knowing the value of other variables that influence reaction time (e.g., compatibility, stimulus intensity).

of Y"). The distinction between absolute and conditional probably is important. For example, in the English alphabet, $p(u)$ is fairly low, but $P(u/q)$ is nearly 1.0.

The usefulness of information theory in describing human information processing in reaction time is demonstrated because the Hick–Hyman Law is found to apply just as well when information is manipulated by probability and context as when it is manipulated by the number of possible stimulus–response pairs (Hyman, 1953; Fitts, Peterson, and Wolpe, 1963).

In summary then, the Hick–Hyman Law provides a powerful tool for predicting information processing latency of operators confronted with a set of possible events that may occur. The Hick–Hyman Law has important implications for system design because the information content of a set of events (warning lights, messages, and so forth) may be defined as important variables that increase the uncertainty of a message set to be responded to. Anything that increases this information content (i.e., increasing the number of possible messages, or varying their relative frequency) can be expected to increase mental workload, increase the chance of errors, and slow the processing time.

Stimulus Discriminability

The speed of response is slowed as the set of possible stimuli that may occur become more similar to each other (Miller and Pachella, 1973). Sometimes systems designers can do nothing about the similarity between stimuli to which operators must respond, if these stimuli occur naturally in the environment. However, whenever artificial symbols or codes are used to present information to the operator it is of paramount importance to try to maintain dissimilarity. This can be done by trying to keep the ratio of features that distinguish between a set of stimuli to shared features as low as possible (Tversky, 1977). Suppose two engines are coded as ENG00141A and ENG00142A. It will take longer to discriminate between these two codes, and they will be more likely to be confused, than if they were simply coded ENG1A and ENG2A, or more optimally E1 and E2. Of course, sometimes the additional material is necessary to distinguish these items from still others (.e.g, from ENG00142B). But where this extra material is not necessary, processing is facilitated if it is deleted.

Stimulus–Response Compatibility

The physical relationship or *compatibility* between a set of stimuli and a set of responses can have a profound influence on the speed of response. Certain compatibility relations are spatially defined. For example, stimuli that are to the right should be responded to with response devices that are also located to the right, and with a rightward movement or clockwise rotation of those devices (Simon, 1969) Furthermore, physical arrays of stimuli in a certain orientation should preserve the same orientation for their corresponding responses (Fitts and Seeger, 1953). Where possible, the response made to a stimulus should be physically close to the stimulus itself.

As noted in Section 2.2.1.3, display compatibility is a special case of stimulus–response compatibility, concerning the importance of orienting a displayed quantity so that it is congruent with, and changes in the same physical direction as, the operator's mental model of the quantity (Roscoe, 1968; Wickens, 1984a). This should be true whether or not an immediate response is required to the displayed quantity.

Compatibility can also define the relation between the movement of an adjustment control and the feedback perceived on a dial or indicator. Movement of an indicator in an "expected" direction will be rapidly confirmed. One that moves in the opposite direction from expectations (i.e., is incompatible), may cause the operator inadvertently to reverse the control input, as if believing that his initial action was incorrect. Figure 2.2.10 shows a series of control–display relationships that represent "population stereotypes," that is, expected relations between control and display movement (Loveless, 1963; Wickens, 1984). Sometimes, as in the example of Figure 2.2.10c, these relations are obvious, because all principles are in concordance. However, in other instances the panel is laid out in such a way that two principles may be in conflict. Then population stereotypes indicate which principle is the strongest and will dominate. The following are five general principles of control–display stereotypes.

1. *Clockwise Stereotype.* There is a basic tendency to rotate a dial clockwise to change a variable.

2. *Clockwise to Increase Stereotype.* There is a stereotype to rotate a dial clockwise to increase the value on a display or to move a linear display upwards or to the right (see Figure 2.2.10c and d).

3. *Proximity of Movement Stereotype.* With any rotary control the arc of the rotating element that is closest to the moving display is assumed to move in the same direction as the display. The panel in Figure 2.2.10c conforms to this principle. That in Figure 2.2.10d is neutral, because the movement of the closest arc is at right angles to the display movement. However, the panel in Figure 2.2.10e violates the principle. Note here that the two arrows of motion move in opposite directions. Figure 2.2.10e presents an example of a conflict of principles because this relation conforms to Principle 2 (clockwise increase) but must violate Principle 3. Under such circumstances Loveless concludes that Principle 3 dominates, so that the arrangement in Figure 2.2.10e would be better if the displayed quantity moved downward instead of upward. Figure 2.2.10c of course conforms to both principles and so would be best.

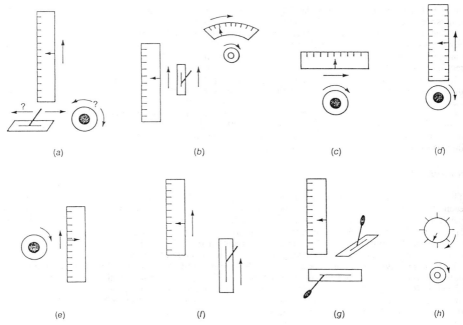

Fig. 2.2.10. Control–display relations and their population stereotypes. See discussion in the text for description of each figure.

4. *Congruence.* Where possible, linear motions of control and display should be along the same axis (Figure 2.2.10f, not g), and rotational motions of control and display should be in the same direction (Figure 2.2.10b and h). However, note that in Figure 2.2.10h the principle of congruence (4) is pitted against that of proximity of movement (3). Here Loveless concludes that congruence will dominate. However, the design shown in Figure 2.2.10b in which only the top half of the scale is read eliminates the problems of Figure 2.2.10h.

5. *Congruence of Location.* When there are several controls, each controlling one of several displays, and each control cannot be placed close to its display, then the configuration of displays should be congruent with the configuration of controls. A square array of four displays should be associated with a square array of controls. Where this is not achieved, population stereotypes break down. The classic example of this breakdown is in the association of controls to burners on the stove. Chapanis and Lindenbaum (1959) and Shinar and Acton (1978) report that there is a wide diversity of population stereotypes concerning which control is associated with which burner.

Probably the most important message conveyed by the research on spatial compatibility and direction of motion stereotypes is that where possible every effort should be made to use configurations in which all principles are congruent. Where some principles are opposed, even though one may be stronger than another, the net stereotype of movement is still weakened, and the likelihood of an inadvertent error is increased. This might be particularly true in times of crisis or stress, when the consequences of an error are greater (Loveless, 1963).

Compatibility can also be defined in terms of input and output *modality*. Wickens et al. (1983, 1984), for example, have found that tasks that are basically spatial in terms of their central processing demands are better served by inputs that are visual and responses that are manual, whereas tasks that demand verbal central processing activities, such as storing alphanumeric information, are better served by auditory input and speech responses. This principle can serve as a useful guideline for deciding which tasks are best served by voice recognition and synthesis technology. In particular, it apears that any task where responses are naturally represented as linguistic units (words, letters, or digits), benefit from using voice response, rather than keyboard entry. In contrast, tasks that require intrinsically continuous spatial-analogue responses such as continuous vehicle control are better served by spatially guided manual responses.

2.2.4.2   Stimulus Pacing

Sometimes the human operator must respond rapidly to a series of uncertain stimuli that occur in sequence. This is the task of the postal mail sorter, typist, or message transcriber. Under these circumstances the timing between successive stimuli and responses contributes four further influences to processing efficiency as described below.

Information Processing Rate—The Decision Complexity Advantage

Research has well documented that when two stimuli, each requiring a choice response, are presented sufficiently close in time, the reaction time to the second is delayed (Kantowitz, 1974). This phenomenon, known as the *psychological refractory period,* clearly generalizes to the case when several stimuli must be processed in rapid succession (Welford, 1967), and indicates that there are definite limits to the number of choice responses that can be made per unit time. Debecker and Desmedt (1970) have found that with the simplest of all possible choices (1 bit), this limit is approximately 2 decisions/sec. At the same time, there are clearly circumstances when humans must (and can) transmit information at a greater rate than this. The skilled typist when transcribing text may transmit up to 8 to 10 bits/sec. One reason for this greater rate is that each key stroke on the typewriter transmits more than the minimum 1 bit/response.

The contrast between the choice reaction time experiment and the skilled typist illustrates a fundamental and well documented principle of information processing described by Wickens (1984a) as the *decision complexity advantage* (Alluisi, Muller, and Fitts, 1957; Broadbent, 1971). That is, more information can be transmitted by the human operator per unit time when this information is represented by a smaller number of more complex decisions, than by a greater number of simple decisions. In other words, increasing decision complexity slows the speed of each individual decision somewhat, as shown in Figure 2.2.9, but this slowing is more than compensated for by the increase in information per decision. Thus the product of (decisions/sec) × (information/decision) is greater when the second term of this equation is large and the first is small than when the reverse is true.

The decision complexity advantage has implications for response devices. For maximum information transmission these devices should be designed to yield high amounts of information per response. Voice transmission clearly does this, and the typewriter yields more information/response than does the Morse code key. However, typewriter keyboards are inherently limited by their size and the speed of finger movement from one key to the next (Norman and Fisher, 1982). This limit would suggest the advantages that might be gained from using *chording* devices in which different combinations of keys are depressed simultaneously. It has been shown that these devices used in courtroom stenotypers (Seibel, 1964), postal sorting equipment (Conrad and Longman, 1965), and redesign of commerical typewriters (Lockhead and Klemmer, 1959; Gopher, 1983) can indeed use the decision complexity advantage to allow more rapid transmission rate.

The decision complexity advantage also has implications for human interaction with computerized data bases, or hierarchical menu selection systems. Any such system can be described by a "breadth" of choice alternatives available to the operator at one point in time, and a "depth," describing the number of sequential choices that must be made to reach the bottom of the hierarchical menu. The implications of the decision complexity advantage, confirmed in an experiment by Miller (1981) is that menu search can be carried out more rapidly with a few "broad" levels (a shallow structure) than with many narrow levels (a "deep" structure). Of course, the breadth at a given level should be somewhat constrained by the working memory capacity of the novice user, as described in Section 2.2.2.

Speed Stress versus Load Stress

Closely related to the decision complexity advantage is the general finding that operators can process high rates of information emanating from a single source more rapidly than lower rates emanating from several sources (Conrad, 1951; Goldstein and Dorfman, 1978). Thus information transmission (information /sec/source) × (number of sources) is higher when the first term is higher than the second. The two terms define speed stress and load stress, respectively. For example, suppose that the quality-control inspector must monitor certain features of a product on three different channels, viewed on a video display. The implications of Goldstein's and Dorfman's conclusions are that performance is more efficient if all products are presented at one rather than three separate locations.

Preview

As discussed in Section 2.2.1.4, information preview is beneficial. This is particularly true in serial response tasks such as typing or transcription (Hershon and Hillix, 1965; Shaffer and Hardwick, 1970). Shaffer (1973) in a careful study of the effects of preview on performance of a skilled typist, found that increasing the amount of preview from no letters (no preview) progressively upward increases typing speed up to a preview of approximately eight letters, beyond which no further benefits are realized. This appears to be an amount that is sufficient usually to leave one untyped word in the preview window. As noted in Section 2.2.1.4, the advantages of preview are not restricted to typing tasks, but are evident in tasks such as industrial scheduling, driving, and flying as well.

Pacing

Pacing refers to the scheduling under which successive stimuli follow one another, and is a relevant concept in such industrial tasks as mail sorting, assembly line work, or inspection. The pacing schedule may be described by two dimensions. One of these is the extent to which the scheduling is *force-paced* or *self-paced*. In a force-paced schedule, successive stimuli occur at a regular, externally controlled *interstimulus interval*. In a self-paced schedule, the scheduling is determined by the speed of the operator's reaction. The self-paced timing may be carried out in a number of ways, for example, by imposing a fixed delay between the subjects' response and the subsequent stimulus (the *response–stimulus* interval), or allowing the operator to adjust the interstimulus interval. Across either the forced- or self-paced schedule, the second dimension refers to the time separation between successive stimuli (or successive responses). In either a force-paced or a self-paced schedule, the interstimulus or response–stimulus interval can be set to a wide range of values.

The research that has been done concerning the effect of pacing schedules on performance does not conclusively favor one or the other under all circumstances (e.g., Salvendy and Smith, 1981; Conrad, 1960; Wickens, 1984a). Across a fairly wide range of timing intervals there appears to be little difference between them. At short intervals with laboratory reaction time tasks (high speed stress), there is some evidence that the force-paced schedule provides more accurate performance for an equal speed of output (Waganaar and Stakenburg, 1975), although this conclusion appears to be reversed at the slightly slower speed of simulated industrial tasks (Basila and Salvendy, 1979).

It is important to realize that the relative merits of the self-paced schedule will grow to the extent that processing complexity becomes more *variable* across stimuli; variability in turn is governed by variability in stimulus quality or any of the factors discussed in Section 2.2.4.1. With extensive variability in processing latency the self-paced schedule allows the operator to compensate for long responses at one time by short responses at another. With the force-paced schedule, on the other hand, the designer must make a choice: leave the interval long (sufficient to accommodate the longest expected processing latencies) or make the interval short. The former choice leads to an accumulation of waiting "dead time." The latter causes some stimuli to have insufficient time for processing, with errors as a consequence.

2.2.4.3 Errors

The preceding discussion has focused on task variables such as compatibility, probability, and pacing as these influence processing latency. Most of the variables that increase latency can also be expected to decrease the expected accuracy of a response, given that latency is maintained at a constant level. Ample data, for example, suggest that error rates are higher when stimulus–response (S–R) compatibility is reduced, when the probability of a stimulus is lower, or when the number of possible S–R pairs is increased (Broadbent, 1971). Thus when human performance is compared between two systems, it is generally safe to say that the system to which the human responds fastest is best in terms of accuracy as well. This positive relationship between latency and error implies that no matter how small latency differences may seem, these differences should be considered when evaluating performance, because they may sometimes be expressed as a difference in accuracy or reliability. (A more detailed treatment of the nature of human error and reliability may be found in Chapter 2.8.)

There is one notable exception to the expected positive correlation between latency and error rate, and this is the *speed–accuracy trade-off* (SATO). When operators are instructed to respond rapidly they typically increase their error rate, and when they are requested to respond accurately, response latency increases (Pachella, 1974; Wickelgren, 1977). Thus whenever two systems, conditions, or subjects differ from each other in terms of their *speed–accuracy set*, it is no longer safe to assume that the faster system (or subject) is necessarily best. Instead, to assess the relative merits of the two systems, one must have available an implicit weighting scheme to indicate how much of a change in latency corresponds to how much of a change in accuracy.

These relations are illustrated graphically in Figure 2.2.11. The reaction time and accuracy data are cross-plotted for the processing of information from two different hypothetical displays, under each of two different speed–accuracy sets. It is apparent that performance with display A is in all respects better than with display B. But it is also clear that performance under set I is not necessarily superior to that under set II. If accuracy is all that counts, then I is better than II. But if latency is the variable of interest, then set II is superior. If a 10% change in accuracy is viewed as functionally equivalent to a 500 msec change in latency, then the two conditions are in some sense "equivalent."

There appear to be two major factors that cause changes in the SATO. First, and most intuitively,

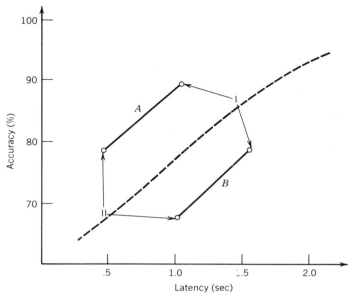

Fig. 2.2.11. The speed–accuracy trade-off. Conditions A and B differ in their overall merits of performance. A is superior to B. Points I and II, however, differ in their speed–accuracy set. The dashed line traces out the diminishing gains in accuracy that occur at extremely long latencies.

any change in *instructions* emphasizing one variable over the other will have the expected effect. In this regard, it is important to realize that the function plotting latency versus accuracy seems to run in a manner shown by the dashed line in Figure 2.2.11. The asymptotic nature of this line as it approaches 100% accuracy indicates that instructions to maintain *perfect* accuracy (i.e., maintain zero error rates) may pass a point of diminishing returns such that processing latency becomes intolerably long (Pachella, 1974). Humans appear to perform more optimally, in terms of maximizing the rate of information transmitted per unit time, if allowed to make a small percentage of errors (Seibel, 1972; Rabbitt, 1981).

Arousal is the second variable that influences the SATO. Any variable such as stress or drugs that increases the level of arousal tends to shift performance in the direction from points I to points II in Figure 2.2.11, yielding faster but less accurate performance. It should be noted that these arousal shifts may result from both environmental and drug-realted stressors (Hockey, 1984), from factors that increase the amount of preparation or anticipation of a stimulus (Posner, 1978), and from certain intrinsic task-related characteristics. In the latter category, for example, stimuli in the auditory modality tend to be more arousing (and therefore more likely to be responded to erroneously) than visual stimuli.

2.2.4.4 Stages of Information Processing

There is by now a fair consensus among cognitive psychologists that information passes through a number of separate processing stages as stimulus is transformed to response (Smith, 1968; Sternberg, 1969; McClelland, 1979; Wickens, 1984a). From the perspective of human factors, the importance of the distinction between processing stages results because knowing that a particular environmental stressor, chemical toxicant, or system characteristic influences one processing stage and not another has potential implications for system redesign or reconfiguration. For example, knowing that a given stressor influences response processes and not encoding should lead the designer to focus on improvement of the control, rather than the display interface.

Sternberg's (1969, 1975) memory search task and the *additive factors* logic presented are at once cornerstones to the identification of separate processing stages and methodological diagnostic tools for pinpointing the locus of influence of system variables on these stages. Theoretical treatment of both the task and the additive factors logic is presented elsewhere (Sternberg, 1969, 1975; Wickens, 1984a); this section focuses only upon its application.

In the basic Sternberg task, the subject is given a set of two to five unrelated letters to retain in working memory. These may be presented auditorily or visually. The number of items is known as the *memory set size* (*M*). After the memory set is presented, then a series of *probe* stimuli are presented. These probe stimuli are letters that have a roughly equal chance of being drawn from the memory set or not. The subjects' task is to respond as rapidly as possible if the probe is (positive response) or is not (negative response) contained in the memory set. The probe may be presented either auditorily or visually, and the two alternative responses may be given either via key presses or speech. On different trials *M* may be set to different values, although often two and four is convenient.

When positive and negative response times are plotted as a function of *M*, the data typically appear as shown in Figure 2.2.12*a*, in which reaction time (*RT*) is a linearly increasing function of *M*. Sternberg has argued that these data reveal two basic components of processing latency: (1) The linear *slope* of the function indicates that *RT* increases by a fixed amount each time *M* is incremented. Sternberg argues that this slope indexes the speed with which working memory can be searched— that is, it is a measure of central processing efficiency. The slopes for positive and negative responses are typically identical. (2) The intercept of the function is a value that is independent of M. This intercept provides an estimate of the time to process information during stimulus encoding and response, that is, the time unrelated to central processing. The intercept is typically slightly higher for negative than for positive responses.

When total processing latency is neatly partitioned in this fashion, using the formula $RT = a + bM$, it is then possible to examine what happens to the function shown in Figure 2.2.12*a*, as the subject performing the Sternberg task is influenced by some agent (e.g., increased stress, arousal, work-load, toxic poisoning, aging). Generally, three patterns of effects can be anticipated: (1) a change in *intercept*, shown in Figure 2.2.12*b*, which indicates that the agent influences stimulus encoding or

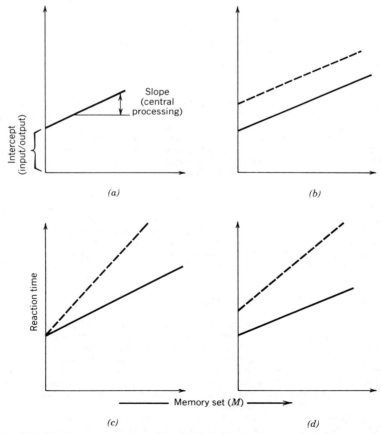

Fig. 2.2.12. The Sternberg memory search task. Each figure plots reaction time as a function of memory set size, *M*. (*a*) The basic slope and intercept. (*b*) The effect of an agent (dashed line) on the intercept, reflecting a slowing of input or output processing. (*c*) The effect of the agent on the slope (central processing latency). (*d*) The joint effect of the agent on both slope and intercept.

response processes, but leaves central processing unaffected; (2) a change in *slope*, shown in Figure 2.2.12c, which indicates that central processing operations in working memory are retarded by the agent; (3) a combination of effects, shown in Figure 2.2.12d, which diagnoses the change to result from both "peripheral" (perception and response) and central processes. Such a technique using the Sternberg task has been employed, for example, to assess the influence of mercury poisoning (the agent) on working memory functioning (Smith and Langolf, 1981), the influence of workload manipulations (the agent) on processing efficiency (Crawford, Pearson, and Hoffman, 1978), or the differential effects of aging (the agent) on processing stages (Salthouse and Sonberg, 1982; Braune and Wickens, 1985).

2.2.5 ATTENTION

The preceding sections have described the limitations and characteristics of component processes such as perception, memory, response, and decision. The concept of *attention* refers to a more global limitation on the resources available to time-share activities and to process information in parallel. In considering guidelines for system design, it is useful to think of attention in terms of three characteristics, each of which represent its different manifestations. (1) *Selective attention* refers to the ability to distribute this limited resource to the necessary and appropriate channels of information at the optimal times. For example, if an aircraft pilot disregards a critical diminishing altimeter reading while concentrating on a potential cockpit malfunction, this represents a failure of selective attention (Wiener, 1977). (2) *Divided attention* refers to the ability to process information for two tasks, or along two channels in parallel. One's ability to control a vehicle while listening to aural communications offers an example of successful divided attention. (3) *Attention as a resource* refers to the quantitative effortful dimension of attention. The task of maintaining seven digits in working memory demands more attention or resources than maintaining five digits.

2.2.5.1 Selective Attention

Under many circumstances it is reasonable to approximate the human operator as a single-channel processor, who is capable of dealing with only one source of information at a time (Moray, 1984; Broadbent, 1971, see also Chapter 2.1). This is particularly true in the industrial control room in which operators must monitor banks of instruments, each providing essentially one channel of information, and each requiring sufficient visual acuity to read so that little information can be gained from peripheral viewing.

In situations such as this, Moray (1984) (see Chapter 2.1), has argued that the selective allocation of attention is guided by the operator's internal model of the statistical properties of the environment. Information sources are sampled with a frequency that is inversely proportional to their information content (Senders, 1964), and pairs of sources are sampled according to their sequential associations. If the operator's mental mode is correct, as demonstrated by the instrument scan patterns of a well-trained aircraft pilot, then the pattern of samples can be used as a guideline for positioning instruments. Those instruments that are sampled frequently should be located centrally and relatively close together. Pairs of instruments that must or should be sampled in sequence should also be positioned close together. For example, McRuer, Jex, Clement, and Graham (1967) used a model of fixation transitions to derive the optimal layout for a Boeing 707 Instrument Landing System display, and obtained a close correspondence between the model-derived display and that which was actually certified by the FAA.

Finally, Harris and Christhilf (1980) have examined the visual fixation dwell time associated with aircraft instrument scanning. They note that fixations may be dichotomized into two qualitatively different types. (1) Check readings are extremely short, averaging around $\frac{1}{2}$ sec dwell time, and are essentially for monitoring whether an instrument needle is at its expected level. (2) Information extraction fixations are considerably longer (usually greater than one second) and are used when new information regarding position or velocity is to be derived.

2.2.5.2 Divided Attention

Although the single-channel model provides a useful approximation for examining selection, it is clear that humans are often capable of carrying on two or more activities simultaneously. For example, the automobile driver's vision may be fixated on a region in the middle of the road while peripheral vision processes velocity information imparted by the "streaming" of the adjacent background. Removal of these latter cues impairs driving performance. To say, however, that parallel processing is possible is not to say that it is perfect, that is, that the two channels of information may be processed in parallel as well as either is processed alone.

In environments that impose high information processing demands on the human operator, such as that of the industrial process controller during a system failure, or the fire chief coordinating units at a major fire, parallel processing is a tremendous asset. Thus, it is important to consider four factors that can influence the extent of parallel processing.

Spatial Proximity

Because the eye has only limited resolution beyond the fovea, it is often nearly impossible to parallel-process detailed visual stimuli that are spatially separated. Hence, moving stimuli closer together at least increases the possibility of parallel processing. However, a narrow spatial separation may lead to other costs associated with confusion and visual clutter. Dynamic stimuli in close spatial proximity to other stimuli that are relevant slow the processing of the relevant stimuli (Eriksen and Eriksen, 1976). Similarly, in the auditory domain messages that emanate from a common spatial source (such as two adjacent speakers) and share other features in common (such as common voice characteristics, or semantic content) are more readily confused, and semantic information in one channel interferes with understanding of the other (Egan, Carterette, and Thwing, 1954; Treisman, 1964). Thus, in terms of auditory message systems, it is probably best to avoid high degrees of physical similarity between separate information sources.

Finally, in the visual world there is evidence that similar spatial locations may help, but does not guarantee parallel processing. In both laboratory tasks (Neisser and Becklan, 1976), and a simulated aircraft head-up display superposed on a visual runway, (Fisher, Haines, and Price, 1982), some investigators have found that critical events in one spatially overlapping channel may be entirely ignored as the other channel is processed. In summary, moving stimuli together in the visual world may foster parallel processing to some extent, but it does not guarantee that such processing is obtained. In the auditory domain such as configuration is probably ill-advised.

Common Object

The concept of object integrality was discussed in Section 2.1.2 and is not treated in detail here. The point is reiterated, however, that the various attributes of a single perceptual object (in contrast to several objects in close proximity) *do* benefit from a good deal of parallel processing. Hence integrated "object displays" (Figure 2.2.4) or the use of various attributes of a single object (i.e., color, size, and shape) to convey multiple information channels is likely to improve the degree of parallel information processing (Lappin, 1967; Wickens, 1984a).

Separated Resources

Figure 2.2.1 portrays attention as a resource that can be distributed to different stages of processing depending upon task demands. Tasks that are more difficult demand more resources, allowing fewer to be available for other concurrently performed tasks. Several investigations, summarized by Wickens (1980; 1984b) and Navon and Gopher (1979) have suggested that there are multiple processing resources. All tasks do not compete for the same "undifferentiated pool" of resources as indicated in Figure 2.2.1, just as all heating systems do not rely on the same fuel source: some use coal, some gas, some oil. The implications of the multiple resources viewpoint are that two tasks that demand separate resources are time-shared effectively, whereas two tasks with common resource demands interfere with each other.

One proposal identifying the structural dimensions of the human processing system that defines these resources was presented by Wickens (1980, 1984a,b), and is shown in Figure 2.2.13. According to his view, separate resources may be defined in terms of three dichotomous dimensions. These are as follows. (1) Stages of information processing: Perception and central processing (i.e., working memory) operations use different resources from those underlying response processes. (2) Modalities of input: Visual processing employs different resources from auditory processing. (3) Codes of information processing: Handling spatial information uses different resources from those involved with handling verbal information. This third dichotomy refers to perception (speech and print versus graphics and pictures), central processing (spatial working memory versus memory for linguistic information), and response processes (speech output versus spatially guided manual responses). The multiple resources model presented in Figure 2.2.13 does not predict that tasks demanding separate resources are perfectly time-shared. It does predict that time-sharing efficiency is improved (the decrement from single task to dual task performance is reduced) to the extent that a pair of time-shared tasks use different levels along the three dichotomous dimensions shown in Figure 2.2.13. Some examples of tests of the theory along the three dimensions in applied contexts are as follows.

Wickens et al (1983) examined differences in both input modality (auditory versus visual) and output code (speech versus manual control) for interfacing a verbal task in a flight simulator environment. Both the auditory input and voice response were found to produce better time-sharing performance than their visual manual counterparts. General findings also suggest that the benefits to speech control in dual task settings are best realized not in the speech-controlled task itself, but in the concurrent task (Aretz, 1983; Wickens et al., 1983). However, in adhering to multiple resources theory as a

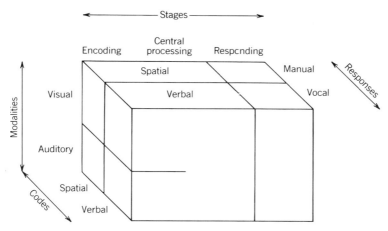

Fig. 2.2.13. A model of the multiple resources within the human processing system (after Wickens, 1984a). To the extent that tasks share common cells in the three-dimensional representation, they will be more likely to interfere.

guideline for implementing voice recognition and synthesis (VRAS) technology, it is important not to violate the principles of S-C-R compatibility as described in Section 2.2.1.3 and 2.2.4.1. That is, VRAS technology is most beneficial in both single- and dual-task settings when verbal rather than spatial tasks are interfaced (Wickens et al., 1983, 1984).

The role of central processing codes (spatial and verbal) in influencing task interference was discussed in Section 2.2.2.5. Tasks that demand spatial central processing activities interfere with concurrent tasks that are spatial and visual and are time-shared fairly efficiently with tasks that are verbal and phonetic, whereas the converse relation holds for verbal tasks. For example, Wickens and Weingartner (1985) found that the spatial task of process control monitoring could be carried out quite efficiently with a concurrent auditory linguistic task, but not with a concurrent task that required the use of spatial working memory. As a design guideline for predicting when tasks will and will not interfere, it can be assumed that almost any task requiring moving, positioning, or orienting objects in space or performing other analog transformations is predominantly spatial. Tasks for which words, language, or logical operations are natural mediators are verbal.

The dichotomy of processing resources by a stage of processing dimensions is used to account for the fact that tasks demanding perception and memory often compete with each other, but not with tasks whose primary demands are related to responding. Hence requiring an operator to respond to events on a monitored display will not disrupt performance of the monitoring itself (particularly if speech responses are used), but adding an additional perceptual channel to monitor, or imposing working memory demands, will be likely to do so.

Task Difficulty

Some tasks are more difficult than others. This may be either because they impose greater processing requirements (remembering seven rather than four digits, for example), or because a given person is less practiced or skilled (remembering seven digits without and with chunking). It is important to recognize that difficulty will affect not only the level of performance but also the resources demanded in performing a task. In fact, difficulty may affect the latter without influencing the former. Difficulty and resource demands are represented schematically in the *performance resource function,* shown in Figure 2.2.14 (Norman and Bobrow, 1975). Each function plots the level of performance on a given task as a function of the resources invested into the task. The two functions might represent the performance of two versions of a single task that differs in its level of difficulty (or the level of practice of an operator performing the task). Note that at full investment of resources there is no observable difference in performance. Only as resources are diverted to a concurrent task are the differences in performance efficiency revealed.

2.2.5.3 Attention as a Resource: Mental Workload

The performance resource function shown in Figure 2.2.14 reflects the resource demand or *mental workload* imposed upon the operator in performing a task. The concept of workload is discussed in

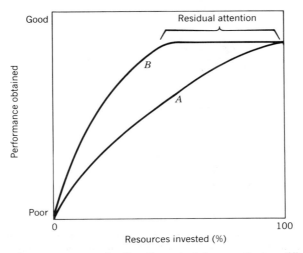

Fig. 2.2.14. The performance resource function Curve A: A less practiced or difficult task in which there is no residual attention. Curve B: A more practiced or easy task, in which the amount of residual attention is shown.

Chapter 3.5 from a perspective that emphasizes its physiological aspects. In this section we describe the concept from a performance perspective. Human factors interest in workload has arisen from the realization that good performance is not all that is important for a good system, but that performance must be obtainable without an excessive cost to human processing resources. That is, preferably some degree of "residual attention" should be available before performance starts to decrease, as shown by curve B in Figure 2.2.14. Thus the operator should be able to handle unexpected demands without consuming the resources necessary to maintain a certain criterion level of primary task performance (Rolfe, 1973; Wickens, 1984a). The issue of how to assess this margin of residual capacity has been discussed in a number of volumes (Moray, 1979; *Human Factors Journal,* 1979) but these discussions offer no clear consensus as to what is the "best" technique. Wierwille and Williges (1978) offer an exhaustive tabulation of techniques available, and Wierwille and Williges (1980) present an annotated bibliography of workload research. Described below are three categories of techniques, with a brief description of some of the strengths and weaknesses of each.

Secondary Tasks

A *secondary task* imposed while the operator is performing the primary task provides the most direct estimate of the amount of available resources. The easier the primary task, the more available resources there are, and the better the secondary task is performed (Rolfe, 1973; Ogden, Levine, and Eisner 1979). For example, Damos (1978) has documented how secondary tasks can reveal differences in flight proficiency between flight instructors and flight students, and Dornic (1980) has shown that the differences in secondary task performance reveal the differences in cognitive demands between speaking a first and second language, even as language performance is similar. Garvey and Taylor (1959) observed differences in the performance of secondary tasks time-shared with two manual control systems that were performed equally well under single-task conditions.

There are, however, two primary limitations to the secondary task measures of mental workload. First, they are often *obtrusive,* disrupting the performance of the very primary task whose performance they are intended to assess. In critical real-time environments, such as an aircraft approach for landing, this disruption may be dangerous. Secondly, they are sometimes insensitive to real demand changes in the primary task (Wickens, 1984a,b). The reason for this insensitivity is evident from the multiple resources model shown in Figure 2.2.13. If a primary task demands different resources from the secondary, the latter does not offer a valid measure of the resource demands of the former and will underestimate its resource demands. Hence secondary tasks should have resource demands that are qualitatively similar to those of the primary tasks they are intended to measure. As examples, the critical instability tracking task (Jex and Clement, 1979) or measure of tapping regularity (Michon, and Van Doorne 1967 provides a good workload measure for any manual control task; the task of time estimation (Hart, 1975), or the Sternberg memory search task (see Section 2.2.4.4) offers good estimates of perceptual/cognitive workload. Primary tasks that are predominantly visual should employ visual secondary tasks for accurate workload estimates (Wickens et al., 1983). A more exhaustive listing of types of secondary tasks and criteria for their application may be found in Ogden et al. (1979).

Subjective Measures

Subjective ratings of task difficulty represent perhaps the most acceptable measure of workload from the standpoint of the actual system user, who feels quite comfortable in simply stating, or ranking, the subjective feelings of "effort" or attention demands encountered in performing a given task or set of tasks (Eggemeier, 1981; Moray, 1982; Reid, Shingledecker, and Eggemeier, 1981). Some have argued (Sheridan, 1980) that these measures come nearest to tapping the essence of mental workload. Indeed, there may be circumstances in which a system designer would feel more comfortable using data concerning how an operator feels about a task than concerning how the task is performed. It is important, however, to know how accurately an operator can assess the demands imposed upon his or her limited resources, what the dimensions underlying that assessment might be, and how these are scaled.

Cooper–Harper Scale. Perhaps the oldest and best validated subjective measure of workload is the Cooper–Harper rating scale of aircraft handling qualities, a decision-tree procedure that rates handling qualities on a 10 point scale (Cooper and Harper, 1969). Within the relatively restricted domain of the tracking or manual control task (and the system-dynamics dimension of task difficulty; see Chapter 2.7), the Cooper–Harper ratings provide a reliable and acceptable measure. In fact, Jex and Clement (1979) found that this measure of manual control workload correlated quite highly ($r = .96$) with a measure of residual capacity as assessed by a secondary tracking task. Wierwille and Casali (1983) have recently adopted the Cooper–Harper rating procedure to the measurement of mental workload, and demonstrated its applicability to perceptual and cognitive tasks. The decision-tree procedures used to elicit the workload ratings are shown in Figure 2.2.15.

Multidimensional Workload Scales. Sheridan (1980) has proposed that three dimensions define the subjective experience of mental workload. These are related to the proportion of time busy or information processing load, the mental effort invested in a task or its complexity, and the "emotional stress" of the task. Reid, Shingledecker, and Eggemeier (1981) have formalized a rating procedure along these three dimensions known as subjective workload assessment technique (SWAT). They have found that subjects are readily able to rate tasks along the three dimensions, and there appears to be a fair degree of agreement between raters on the relative ratings of different tasks along the three dimensions. Reid et al. also propose a method for combining the estimated values on the dimensions to produce a single workload rating. It is less clear, however, whether these dimensions are truly independent, whether they capture all the variability there is in subjective workload, or how they relate to the processing resource dimensions that underlie task performance.

An extensive series of investigations by Wickens and Yeh (1983, Yeh and Wickens 1984) in which a series of diverse tasks were rated on a series of bipolar rating scales (Childress, Hart, and Bortalussi, 1982) reveals that the terms "perceptual-cognitive effort" and "response load" accurately captured most of the variance in the subjective experience of those tasks. Because their tasks were not performed under conditions of high stress or anxiety, it is reasonable to propose that the *stress* term should be added when that variable appears to be relevant.

Subjective-Performance Dissociations. The primary difficulties observed with subjective ratings relate to the fact that they do not always agree with performance. Yeh and Wickens (1984) have documented cases in which tasks that are performed better than others are rated as having a higher subjective workload. In these cases, the system designer is left with uncertainty as to which is the "better" system. Wickens and Yeh have offered guidelines in this regard by identifying two generic factors that inflate subjective ratings of difficulty relative to performance decrements. These are (1) motivational or incentive factors that induce subjects to invest more resources into the task (e.g., display variables that provide more precise input information that induces more precise control fall into this category), and (2) task configurations that have multiple task elements, or multiple objects to be encoded. They also indentified two variables that underestimate subjective ratings relative to performance decrements: (1) tasks that compete for common resources as shown in Figure 2.2.13 (e.g., two visual tasks as opposed to a visual and an auditory task); (2) single tasks that are increased in their resource demand (e.g., increasing the disturbance frequency of an input that is to be tracked).

Physiological Measures

Physiological measures of task workload are discussed extensively by Romhert (Chapter 3.5) and are not covered here except to note that these measures, as do secondary tasks, appear to be differentially sensitive to different demands within the multiple resource space. For example, evoked brain potentials tend to provide better measures of perceptual/cognitive demands than of response-related processing

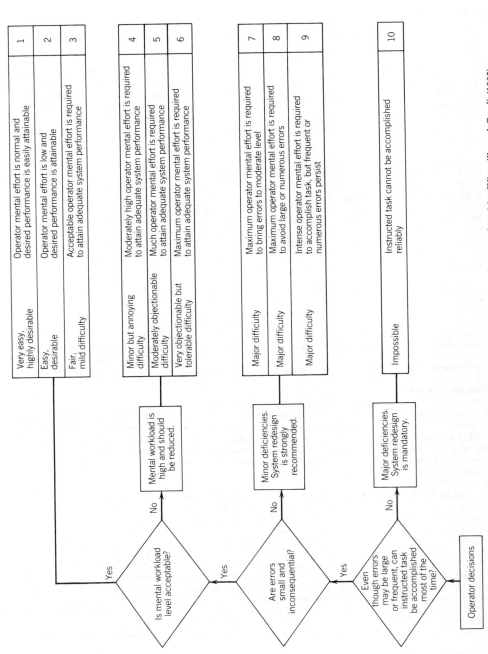

Very easy, highly desirable	Operator mental effort is normal and desired performance is easily attainable	1
Easy, desirable	Operator mental effort is low and desired performance is attainable	2
Fair, mild difficulty	Acceptable operator mental effort is required to attain adequate system performance	3
Minor but annoying difficulty	Moderately high operator mental effort is required to attain adequate system performance	4
Moderately objectionable difficulty	Much operator mental effort is required to attain adequate system performance	5
Very objectionable but tolerable difficulty	Maximum operator mental effort is required to attain adequate system performance	6
Major difficulty	Maximum operator mental effort is required to bring errors to moderate level	7
Major difficulty	Maximum operator mental effort is required to avoid large or numerous errors	8
Major difficulty	Intense operator mental effort is required to accomplish task, but frequent or numerous errors persist	9
Impossible	Instructed task cannot be accomplished reliably	10

Mental workload is high and should be reduced.

Is mental workload level acceptable?

Minor deficiencies. System redesign is strongly recommended.

Are errors small and inconsequential?

Major deficiencies. System redesign is mandatory.

Even though errors may be large or frequent, can instructed task be accomplished most of the time?

Operator decisions

Fig. 2.2.15. Decision tree procedures for eliciting subjective workload ratings, employed by Wierwille and Casali (1983).

100

(Isreal, Wickens, and Chesney, 1980). Heart-related measures offer more sensitive measures of response load (Kalsbeek and Sykes, 1967; Derrick and Wickens, 1984), and pupil diameter measures (Beatty, 1981) appear to offer measures that are sensitive to demands across all processing resources. A series of investigations by Wierwille and his colleagues have contrasted heart rate with other subjective and secondary task measures of pilot workload, involving both flying and cognitive tasks (Wierwille and Connor, 1983; Casali and Wierwille, 1984; Wierwille, Rahimi, and Casali, 1985).

2.2.6 SUMMARY

This chapter has described several components and constraints on human information processing. Knowledge of and adherence to the guidelines presented here will help, but will not guarantee, the design of an effectively human factored system. This is because the principles presented represent only one component of the important human factors issues. In the first place, critical factors related to motor control (Chapter 2.7) sensory processing (Chapter 2.1), and anthropometry, (Chapter 2.5), for example, are treated elsewhere. Without adherence to these principles, an otherwise well-designed system may be totally unusable.

Secondly, complex task performance often reflects more than the sum of the component processing capabilities, because these capabilities may combine and interact in unpredictable ways. Hence equal attention must be paid to models of complex performance such as those described in Part 9 of this handbook. Finally, the principles described here have purposefully been presented at a fairly generic, rather than a system-specific level (i.e., computer, aircraft, industrial assembly). In order to translate these generic guidelines into specific design recommendations, further scrutiny is required from the system's perspective of the kind presented in Part 10 of the handbook.

Nevertheless, the processing characteristics described in this chapter are generally based upon decades of well controlled empirical research, and it is safe to say that an understanding of those characteristics and adherence to the guidelines presented will lead to an improved system design.

REFERENCES

Alluisi, E., Muller, P. I., and Fitts, P. M. (1957). An information analysis of verbal and motor response in a force-paced serial task. *Journal of Experimental Psychology, 53,* 153–158.

Aretz, A. J. (1983). A comparison of manual and voice response modes for the control of aircraft systems. In L. Haugh and A. Pope, Eds., *Proceedings,* 27th Annual Meeting of the Human Factors Society. Santa Monica, CA: Human Factors Society.

Baddeley, A. D. and Hitch, G. (1974). Working memory. In G. Bower, Ed., *Recent advances in learning and motivation,* (Vol. 8). New York: Academic.

Baddeley, A. D., and Lieberman, K. (1980). Spatial working memory. In R. S. Nickerson, Ed., *Attention and Performance VIII.* Hillsdale, NJ: Erlbaum.

Bartram, D. J. (1980). Comprehending spatial information: The relative efficiency of different methods of presenting information about bus routes. *Journal of Applied Psychology, 65,* 103–110.

Basila, B. and Salvendy, G. (1979). Non-work related movements in machine-paced and self-paced work. In C. Bensel, Ed., *Proceedings,* 23rd Annual Meeting of the Human Factors Society. Santa Monica, CA: Human Factors Society.

Baty, D. L. (1976). Evaluating a CRT map predictor for airborne use. *IEEE Transactions on Systems, Man, and Cybernetics, 6,* 209–215.

Beatty, J. (1982). Task-evoked pupillary responses, processing load, and the structure of processing resources. *Psychological Bulletin, 91,* 276–292.

Bjork, R. A. (1972). Theoretical implications of directed forgetting. In A. W. Melton and E. Martin Eds., *Coding processes in human memory.* Washington, DC: Winston.

Booher, H. R. (1975). Relative comprehensibility of pictorial information and printed words in proceduralized instructions. *Human Factors, 17,* 266–277.

Bower, G. H. and Reitman, J. S. (1972). Mnemonic elaboration in multilist learning. *Journal of Verbal Learning and Verbal Behavior, 253,* 478–485.

Braune, R. and Wickens, C. D. (1985). The functional age profile: An objective decision criterion for the assessment of pilot performance capacities and capabilities. *Human Factors, 27,* in press.

Broadbent, D. (1971). *Decision and stress.* New York: Academic.

Carswell, C. M. and Wickens, C. D. (1984). Stimulus integrality in displays of system input-output relationships: A failure detection study. *Proceedings,* 28th Annual Meeting of the Human Factors Society. Santa Monica, CA: Human Factors Society.

Casali, J. G., and Wierwille, W. W. (1984). On the measurement of pilot perceptual workload: A comparison of assessment techniques addressing sentivity and intrusion issues. *Ergonomics, 27,* 1033–1050.

Chapanis, A. and Lindenbaum, L. E. (1959). A reaction time study of four control-display linkages. *Human Factors, 1,* 1–14.

Chase, W. and Chi, M. (1979). Cognitive skill: Implications for spatial skill in large-scale environments (Technical Report No. 1). Pittsburgh, PA: University of Pittsburgh Learning and Development Center.

Chase, W. G. and Ericsson, A. (1981). Skilled memory. In S. A. Anderson, Ed., *Cognitive skills and their acquisition.* Hillsdale, NJ: Erlbaum.

Chase, W. G. and Simon, H. A. (1973). The mind's eye in chess. In W. G. Chase, Ed., *Visual information processing.* New York: Academic.

Clark, H. H. and Chase, W. G. (1972). On the process of comparing sentences against pictures. *Cognitive Psychology, 3,* 472–517.

Colquhoun, W. P. and Baddeley, A. D. (1969). Evaluation of auditory, visual, and dual-mode displays for prolonged sonar monitoring in repeated sessions. *Human Factors, 17,* 425–437.

Conrad, R. (1951). Speed and load stress in sensori-motor skill. *British Journal of Industrial Medicine, 8,* 1–7.

Conrad, R. (1960). Letter sorting machines—paced, "lagged," or unpaced. *Ergonomics, 3,* 149–157.

Conrad, R. (1964). Acoustic comparisons in immediate memory. *British Journal of Psychology, 55,* 75–84.

Conrad, R. and Longman, D. S. A. (1965). Standard typewriter vs. chord keyboard—an experimental comparison. *Ergonomics, 8,* 77–88.

Cooper, G. E., and Harper, R. P. (1969). The use of pilot ratings in the evaluation of aircraft handling qualities (NASA Ames Technical Report NASA TN-D-5153). Moffett Field, CA: NASA Ames Research Center.

Crawford, B. M., Pearson, W. H., and Hoffman, M. S. (1978). Multipurpose switching and flight control workload. *Proceedings,* Sixth Symposium on Psychology in the Department of Defense, U.S. Airforce Academy, pp. 219–221.

Damos, D. (1978). Residual attention as a predictor of pilot performance. *Human Factors, 20,* 435–440.

Debecker, J. and Desmedt, R. (1970). Maximum capacity for sequential one-bit auditory decisions. *Journal of Experimental Psychology, 83,* 366–373.

Derrick, W. L. and Wickens, C. D. (1984). A multiple processing resource explanation of the subjective dimensions of operator workload (University of Illinois Technical Report EPL-84-2/ONR-84-2). Champaign, IL: Engineering Psychology Laboratory.

Desoto, C. B., London, M., and Handel, S. (1965). Social reasoning and spatial paralogic. *Journal of Personal and Social Psychology, 2,* 513–521.

Dewar, R. E. (1976). The slash obscures the symbol on prohibitive traffic signs. *Human Factors, 18,* 253–258.

Dornic, S. S. (1980). Language dominance, spare capacity, and perceived effort in bilinguals. *Ergonomics, 23,* 369–378.

Edwards, w. (1977). How to use multiattribute utility measurement for social decision making. *IEEE Transactions on Systems, Man, and Cybernetics, 7,* 326–340.

Edwards, W., Lindman, H., and Phillips, L. D. (1965). Emerging technologies for making decisions. In T. M. Newcomb, Ed., *New directions in psychology II.* New York: Holt, Rinehart, & Winston.

Edwards, W., Phillips, L. D., Hays, W. L. and Goodman, B. C. (1968). Probabilistic information processing systems: design and evaluation. *IEEE Transactions on Systems, Science, and Cybernetics, 4,* 249–265.

Egan, J., Carterette, E., and Thwing, E. (1954). Some factors affecting multichannel listening. *Journal of the Acoustical Society of America, 26,* 774–782.

Eggemeier, T. F. (1981). Current issues in subjective assessment of workload. In R. Sugarman, Ed., *Proceedings,* 25th Annual Meeting of the Human Factors Society. Santa Monica, CA: Human Factors Society.

Eriksen, B. A., and Eriksen, C. W. (1979). Effects of noise letters upon the identification of a target letter in a non-search task. *Perception and Psychophysics, 16,* 143–149.

Fisher, E., Haines, R., and Price, T. (1980, December). Cognitive issues in head-up displays (NASA Technical Paper 711). Washington, D.C.: NASA.

Fitts, P. M. and Peterson, J. R. (1964). Information capacity of discrete motor responses. *Journal of Experimental Psychology, 67,* 103–112.

Fitts, P. M., Peterson, J. R., and Wolpe, G. (1963). Cognitive aspects of information processing II: Adjustments to stimulus redundancy. *Journal of Experimental Psychology, 65,* 423–432.

Fitts, P. M. and Seeger, C. M. (1953). S-R compatibility: Spatial characteristics of stimulus and response codes. *Journal of Experimental Psychology, 46,* 199–210.

Fogel, L. J. (1959). A new concept: The kinalog display system. *Human Factors, 1,* 30–37.

Fowler, F. D. (1980). Air traffic control problem: A pilot's view. *Human Factors, 22,* 645–654.

Garner, W. R. and Fefoldy, G. L. (1970). Integrality of stimulus dimensions in various types of information processing. *Cognitive Psychology, 1,* 225–241.

Garvey, W. D. and Taylor, F. V. (1959). Interactions among operator variables, system dynamics, and task-induced stress, *Journal of Applied Psychology, 43,* 79–85.

Gentner, D. and Stevens, A. L. (1983). *Mental Models.* Hillsdale, NJ: Erlbaum.

Goldstein, I. L. and Dorfman, P. W. (1978). Speed stress and load stress as determinants of performance in a time-sharing task. *Human Factors, 20,* 603–610.

Gopher, D. (1983). On the contribution of vision-based imagery to the acquisition and operation of a transcription skill. In W. Printz and A. Sanders, Eds., *Cognition and Motor Processes.* New York: Springer.

Harris, R. L. and Christhilf, D. M. (1980). What do pilots see in displays? *Proceedings,* 24th Annual Meeting of the Human Factors Society. Santa Monica, CA: Human Factors Society.

Hart, S. G. (1975). Time estimation as a secondary task to measure workload. *Proceedings,* 11th Annual Conference on Manual Control (NASA TMX-62, N75-33679, 53). Washington, DC: U.S. Government Printing Office, pp. 64–77.

Hershon, R. L. and Hillix, W. A. (1965). Data processing in typing: Typing rate as a function of kind of material and amount exposed. *Human Factors, 7,* 483–492.

Hick, W. E. (1952). On the rate of gain of information. *Quarterly Journal of Experimental Psychology, 4,* 11–26.

Hockey, R. (1984). Varieties of attentional state: The effects of environment. In R. Parasuraman and R. Davies, Eds., *Varieties of attention.* New York: Academic, pp. 449–479.

Howard, J. H. and Kerst, R. C. (1981). Memory and perception of cartographic information for familiar and unfamiliar environments. *Human Factors, 23,* 495–504.

Hunt, R. and Rouse, W. (1981). Problem-solving skills of maintenance trainees in diagnosing faults in simulated power plants. *Human Factors, 23,* 317–328.

Hyman, R. (1953). Stimulus information as a determinant of reaction time. *Journal of Experimental Psychology, 45,* 423–432.

Isreal, J. B., Wickens, C. D., Chesney, G. L., and Donchin, E. (1980). The event-related brain potential as an index of display-monitoring workload. *Human Factors, 22,* 211–224.

Jacob, R. J. K., Egeth, H. E., and Bevon, W. (1976). The face as a data display. *Human Factors, 18,* 189–200.

Jensen, R. J. (1981). Prediction and quickening in prospective flight displays for curved landing and approaches. *Human Factors, 23,* 333–364.

Jex, H. R., and Clement, W. F. (1979). Defining and measuring perceptual-motor workload in manual control tasks. In N. Moray, Ed., *Mental workload, its theory and measurement.* New York: Plenum.

Kahneman, D., and Henik, A. (1981). Perceptual organization and attention. In M. Kubovy and J. R. Pomerantz, Eds., *Perceptual Organization.* Hillsdale, NJ: Erlbaum, pp. 181–209.

Kahneman, D., Slovic, P., and Tversky, A., Eds. (1982). *Judgment under uncertainty: Heuristics and biases.* New York: Cambridge University Press.

Kalsbeek, J. W., and Sykes, R. W. (1967). Objective measurement of mental load, *Acta Psychologica, 27,* 253–261.

Kantowitz, B. H. (1974). Double stimulation. In B. H. Kantowitz, Ed., *Human information processing.* Hillsdale, NJ: Erlbaum.

Keeney, R. L. (1977). The art of assessing multiattribute utility functions. *Organization Behavior in Human Performance 19,* 267–310.

Kiguchi, T., and Sheridan, T. B. (1979). Criteria for selecting measures of plant information with applications to nuclear reactors, *IEEE Transactions on Systems, Man, and Cyberneties, 9,* 165–175.

Klatzky, R. L. (1980). *Human memory: Structures and processes.* San Francisco: Freeman.

Koriat, A., Lichtenstein, S., and Fischoff, B. (1980). Reasons for confidence. *Journal of Experimental Psychology: Human Learning & Memory, 6,* 107–118.

Lappin, J. S. (1967). Attention in the identification of stimuli in complex visual displays. *Journal of Experimental Psychology, 75,* 321–328.

Levine, M. (1965). Hypothesis behavior by humans during discrimination learning. *Journal of Experimental Psychology, 71,* 331–338.

Lockhead, G. R. and Klemmer, E. T. (1959). An evaluation of an 8-key word-writing typewriter (IBM Research Report RC-180). Yorktown Heights, NY: IBM Research Center.

Lopes, L. (1982). Toward a procedural theory of judgment. Wisconsin Human Information Processing Program WHIP Report #17. University of Wisconsin.

Loveless, N. E. (1963). Direction of motion stereotypes: A review. *Ergonomics, 5,* 357–383.

Luria, A. R. (1968). *The mind of a mnemonist.* New York: Basic.

Lusted, L. B. (1976). Clinical decision making. In D. Dombal and J. Grevy, Eds., *Decision making and medical care.* Amsterdam: North-Holland.

McClelland, J. L. (1979). On the time-relations of mental processes: An examination of processes in cascade. *Psychological Review, 86,* 287–330.

McClosky, J. (1983). Intuitive physics. In D. Gentner and A. L. Stevens, Eds., *Mental models.* Hillsdale, NJ: Erlbaum.

McRuer, D., Jex, H., Clement, W., and Graham, D. (1967). Development of a system's analysis theory of manual control displays (TR-163-1). Systems Technology, Inc. Hawthorne, California.

Mehle, T. (1982). Hypothesis generation in an automobile malfunction inference task. *Acta Psychologica, 52,* 87–116.

Michon, J. A., and Van Doorne, H. (1967). A semi-portable apparatus for measuring perceptual motor load. *Ergonomics, 10,* 67–72.

Miller, D. P. (1981). The depth/breadth tradeoff in heirarchical computer menus. In R. Sugarman (Ed.) *Proceeding 25th Annual Meeting of the Human Factors Society.* Santa Monica, CA: Human Factors.

Miller, G. A. (1956). The magical number seven plus or minus two: Some limits on our capacity for processing information. *Psychological Review, 63,* 81–97.

Miller, G. A., and R. Pachella (1973). On the locus of the stimulus probability effect. *Journal of Experimental Psychology, 101,* 501–506.

Moray, N., Ed. (1979). *Mental workload: Its theory and measurement,* New York: Plenum.

Moray, N. (1982). Subjective mental load, *Human Factors, 23,* 25–40.

Moray, N. (1984). Attention to dynamic visual displays in man–machine systems. In R. Parasuraman and R. Davies, Eds., *Varieties of attention.* New York: Academic, pp. 485–512.

Mynatt, C. R., Doherty, M. E., and Tweney, R. D. (1977). Confirmation bias in a simulated research environment: an experimental study of scientific inference. *Quarterly Journal of Experimental Psychology, 29,* 85–95.

Navon, D., and D. Gopher (1979). On the economy of the human processing system. *Psychological Review, 86,* 254–255.

Neisser, U. (1967). *Cognitive psychology.* New York: Appleton-Century–Crofts.

Neisser, U. and Becklan, R. (1975). Selective looking: attention to pure tones. *Cognitive Psychology, 7,* 480–494.

Norcio, A. F. (1981). Human memory processes for comprehending computing programs (Technical Report AS-2-81). Annapolis, MD: U.S. Naval Academy, Applied Sciences Department.

Norman, D., and Bobrow, D. (1975). On data-limited and resource-limited processing. *Journal of Cognitive Psychology, 7,* 44–60.

Norman, D. A., and Fisher, D. (1982). Why alphabetic keyboards are not easy to use: Keyboard layout doesn't matter much. *Human Factors, 24,* 509–520.

Ogden, G. D., Levine, J. M., and Eisner, E. J. (1979). Measurement of workload by secondary tasks. *Human Factors, 21,* 529–548.

Oskamp, S. (1965). Overconfidence in case-study judgments. *Journal of Consulting Psychology, 29,* 261–265.

Pachella, R. (1974). The use of reaction time measures in information processing research. In B. H. Kantowitz, Ed., *Human information processing.* Hillsdale, NJ: Erlbaum.

Payne, J. W. (1980). Information processing theory: Some concepts and methods applied to decision research. In T. S. Wallsten, Ed., *Cognitive processes in choice and decision behavior.* Hillsdale, NJ: Erlbaum.

Peterson, L. R., and Peterson, M. J. (1959). Short term retention of individual verbal items. *Journal of Experimental Psychology, 58,* 193–198.

Phelps, R. H. and Shanteau, J. (1978). Livestock judges: How much information can an expert use? *Organization Behavior in Human Performance, 21,* 209–219.

Posner, M. I. (1978). *Chronometric explorations of the mind.* Hillsdale, NJ: Erlbaum.

Rabbitt, P. M. A. (1981). Sequential reactions. In D. H. Holding, Ed., *Human skills.* New York: Wiley.

Rasmussen, J. (1981). Models of mental strategies in process control. In J. Rasmussen & W. Rouse, Eds., *Human detection and diagnosis of system failures.* New York: Plenum.

Rasmussen, J., and W. B. Rouse (1981). *Human detection and diagnosis of system failures.* New York: Plenum.

Reid, G. B., Shingledecker, C., and Eggemeier, T. (1981). Application of conjoint measurement to workload scale development. In R. Sugarman, Ed., *Proceedings,* 25th Annual Meeting of the Human Factors Society. Santa Monica, CA: Human Factors Society.

Rolfe, J. M. (1973). The secondary task as a measure of mental load. In W. T. Singleton, J. G. Fox, and D. Whitefield, Eds., *Measurement of man at work.* London: Taylor & Francis, pp. 135–148.

Roscoe, S. N. (1968). Airborne displays for flight and navigation. *Human Factors, 10,* 321–332.

Roscoe, S. N. (1980). *Aviation psychology.* Ames, IA: Iowa State University Press.

Roscoe, S. N., and Williges, R. C. (1975). Motion relationships and aircraft attitude guidance displays: A flight experiment. *Human Factors, 17,* 374–387.

Rouse, W. B. and Hunt, R. M. (1984). Human problem-solving in fault and diagnosis tasks. In R. Rouse, Ed., *Advances in man–machine system research.* Greenwich, CT: JAI.

Rummelhart, D. (1977). *Human information processing.* New York: Wiley.

Samet, M. G., Weltman, G., and Davis, K. B. (1976, December). Application of adaptive models to information selection in C3 systems (Technical Report PTR-1033-76-12). Woodland Hills, CA: Perceptronics.

Salthouse, T. A., and Somberg, B. L. (1982). Isolating the age deficit of speeded performance. *Journal of Gerontology, 37,* 59–63.

Salvendy, G., and Smith, M. J., Eds. (1981). *Machine Pacing and Occupational Stress.* London: Taylor & Francis.

Savin, H. B., and Perchonock, E. (1965). Grammatical structure and the immediate recall of English sentences. *Journal of Verbal Learning and Verbal Behavior, 4,* 348–353.

Schroeder, R. G., and Benbassat, D. (1975). An experimental evaluation of the relationship of uncertainty to information used by decision makers. *Decision Sciences, 6,* 556–567.

Seibel, R. (1964). Data entry through chord, parallel entry devices. *Human Factors, 6,* 189–192.

Seibel, R. (1972). Data entry devices. In H. S. VanCott and R. G. Kinkade, Eds., *Human engineering guide to equipment design.* Washington, DC: U.S. Government Printing Office.

Senders, J. (1964). The human operator as a monitor and controller of multidegree of freedom systems. *IEEE Transactions on Human Factors in Electronics, 5,* 2–6.

Shaffer, L. H. (1973). Latency mechanisms in transcription. In S. Kornblum, Ed., *Attention and performance IV.* New York: Academic.

Shaffer, L. H., and Hardwick, J. (1970). The basis of transcription skill. *Journal of Experimental Psychology, 84,* 424–440.

Sheridan, T. (1980). Mental workload: What is it? Why bother with it? *Human Factors Society Bulletin, 23,* 1–2.

Sheridan, T. B. (1981). Understanding human error and aiding human diagnostic behavior in nuclear power plants. In J. Rasmussen, and W. B. Rouse (Eds.), *Human Detection and Diagnosis of Systems Failures.* New York: Plenum.

Sheridan, T. B., and Ferrell, L. (1974). *Man–machine system.* Cambridge, MA: MIT Press.

Shinar, D., and Acton, M. B. (1978). Control-display relationships on the four burner range: Population stereotypes versus standards. *Human Factors, 20,* 13–17.

Simon, J. R. (1969). Reaction toward the source of stimulus. *Journal of Experimental Psychology, 81,* 174–176.

Slovic, P., Fischoff, B., and Lichtenstein, S. (1977). Behavioral decision theory. *Annual Review of Psychology, 28,* 1–39.

Smith, E., (1968). Choice reaction time: Analysis of the major theoretical positions. *Psychological Bulletin, 69,* 77–110.

Smith, H. T., and Crabtree, R., (1975). Interactive planning. *International Journal of Man–Machine Studies, 7,* 213.

Smith, P., and Langolf, G. D. (1981). The use of Sternberg' memory-scanning paradigm in assessing effects of chemical exposure. *Human Factors, 23,* 701–708.

Sternberg, S. (1969). The discovery of processing stages: Extensions of Donders' method. *Journal of Experimental Psychology, 27,* 276–315.

Sternberg, S. (1975). Memory scanning: New findings and current controversies, *Quarterly Journal of Experimental Psychology, 27,* 1–32.

Thompson, D. A. (1981). Commercial aircrew detection of system failures: State of the art and future trends. In J. Rasmussen and W. B. Rouse, Eds., *Human detection and diagnosis of system failures.* New York: Plenum.

Thorndyke, P., and Haynes-Roth, B. (1982). Differences in spatial knowledge acquired through maps and navigation. *Cognitive Psychology, 14,* 560–589.

Treisman, A. (1964). The effect of irrelevant material on the efficiency of selection listening. *American Journal of Psychology, 72,* 533–546.

Tulving, E., Mandler, G., and Baumal, R. (1964). Interaction of two sources of information in tachistoscopic word recognition. *Canadian Journal of Psychology, 18,* 62–71.

Tversky, A. (1977). Features of similarity, *Psychological Review, 84,* 327–352.

Tversky, A., and Kahneman, D. (1974). Judgment under uncertainty: Heuristics and biases. *Science, 185,* 1124–1131.

Tversky, B. (1981). Distortion of memory for maps. *Cognitive Psychology, 13,* 407–433.

Underwood, G. (1976). *Attention and memory.* Oxford: Permagon.

Vinge, E., and Pitkin, E. (1972). Human operator for aural compensatory tracking. *IEEE Transactions on Systems, Man, and Cybernetics, 2,* 504–512.

Waganaar, W. A., and Stakenburg, H. (1975). Paced and self-paced continuous reaction time. *Quarterly Journal of Experimental Psychology, 27,* 559–563.

Wallsten, T. S., (1980). *Cognitive processes in choice and decision behavior.* Hillsdale, NJ: Erlbaum.

Wason, P. C. and Johnson-Laird, P. M. (1972). *Psychology of reasoning: Structure and content.* London: Batsford.

Weiner, E. L. (1977). Controlled flight into terrain accidents: System induced errors. *Human Factors, 19,* 171–185.

Weiner, E. L., and Curry, R. E. (1980). Flight deck automation: Promises and problems. *Ergonomics, 23,* 995–1012.

Welford, A. T. (1967). Single channel operation in the brain. *Acta Psychologica, 27,* 5–21.

West, B., and Clark, J. A. (1974). Operator interaction with a computer controlled distillation column. In E. Edwards and F. P. Lees, Eds., *The human operator in process control.* London: Taylor & Francis.

Wetherell, A. (1979). Short-term memory for verbal and graphic route information. *Proceedings,* Meeting of the Human Factors Society. Santa Monica, CA: Human Factors Society.

Whitaker, L. A. and Stacey, S. (1981). Response times to left and right directional signals. *Human Factors, 23,* 447–452.

Whitefield, D., Ball, R., and Ord, G. (1980). Some human factors aspects of computer-aiding concepts for air traffic controllers. *Human Factors, 22,* 569–580.

Wickelgren, W. A., (1964). Size of rehearsal group in short-term memory. *Journal of Experimental Psychology, 68,* 413–419.

Wickelgren, W. (1977). Speed accuracy tradeoff and information processing dynamics. *Acta Psychologica, 41,* 67–85.

Wickens, C. D. (1980). The structure of attentional resources. In R. Nickerson and R. Pew, Eds., *Attention and performance VIII,* Hillsdale, NJ: Erlbaum.

Wickens, C. D. (1984a). *Engineering psychology and human performance.* Columbus, OH: Charles Merrill.

Wickens, C. D. (1984b). Processing resources in attention. In R. Parasuraman & R. Davies, Eds., *Varieties of attention.* New York: Academic, pp. 63–98.

Wickens, C. D., Sandry, D., and Vidulich, M. (1983). Compatibility and resource competition between modalities of input, central processing, and output: Testing a model of complex task performance. *Human Factors, 25,* 227–248.

Wickens, C. D., Vidulich, M., and Sandry-Garza, D. (1984). Principles of S-C-R compatibility with spatial and verbal tasks: The role of display–control location and voice-interactive display–control interfacing. *Human Factors, 26,* 533–543.

Wickens, C. D. and Weingartner, A. (1985). Process control monitoring: The effects of spatial and verbal ability and concurrent task demand. *Proceedings of Second Mid-Central Ergonomic/Human Factors Conference—Trends in Ergonomics/Human Factors,* Vol. II. Lafayette, IN: Purdue University. (1983).

Wickens, C. D. and Yeh, Y.-Y. (1983). The dissociation of subjective ratings and performance: A multiple resources approach, In L. Haugh and A. Pope, Eds., *Proceedings,* 27th Annual Conference of the Human Factors Society. Santa Monica, CA: Human Factors Society.

Wierwille, W. and Casali, J. (1983). A validated rating scale for global mental workload measurement applications. In L. Haugh and A. Pope, Eds., *Proceedings of the Human Factors Society.* Santa Monica, CA: Human Factors Society.

Wierwille, W. W., and Connor, S. A. (1983) Evaluation of 20 workload measures using a psychomotor task in a moving-base aircraft simulator. *Human Factors, 25,* 1–16.

Wierwille, W. W., Rahimi, M., and Casali, J. G. (1985). Evaluation of 16 measures of mental workload using a simulated flight task emphasizing mediational activity. *Human Factors, 27,* 489–502.

Wierwille, W. W. and Williges, R. C. (1978, September). *Survey and analysis of operator workload assessment techniques.* (Report No. S-78-101). Blacksburg, VA: Systemetrics.

Wierwille, W. W. and Williges, B. H. (1980, March). *An annotated bibliography on operator mental workload assessment.* (Report SY-27R-80). Patuxent River, MD: Naval Air Test Center.

Woods, D., Wise, J., and Hanes, L. (1981). An evaluation of nuclear power plant safety parameter display systems, In R. C. Sugarman, Ed., *Proceedings,* 25th Annual meeting of the Human Factors Society. Santa Monica, CA: Human Factors Society.

Yates, F. H. (1966). *The art of memory.* Chicago: University of Chicago Press.

Yeh, Y.-Y., and Wickens, C. D. (1984). Why do performance and subjective workload measures dissociate? *Proceedings,* 28th Annual Meeting of the Human Factors Society. Santa Monica, CA: Human Factors Society.

Yntema, D. (1963). Keeping track of several things at once. *Human Factors, 6,* 7–17.

CHAPTER 2.3
MOTIVATION

CARL GRAF HOYOS

Technical University of Munich

2.3.1 WHAT IS MOTIVATION?

Attempts to explain human behavior in terms of experience, knowledge, and abilities (in short, in terms of a set of conditions prerequisite to achievement as well as the effective regulations that accompany them) have generally left open the question of "why" a person tackles a certain task and is occupied with it for a shorter or longer period of time. Many interpreters of human behavior, scientists as well as laymen, have repeatedly attempted to find a determining force in human behavior by means of posing the question *why,* that is, with which goals, with which inner motives, has someone just initiated this activity rather than another one. To answer these questions, a special discipline has been established within the field of psychology. This discipline, which is known as motivation research, has developed as a result of the inadequacy of traditional explanations of human behavior. In the attempt to fill this vacuum, numerous constructs have been devised and introduced into this field of research: drive, impulse, goal, need, motive, and others. These clearly serve an important purpose in the everyday observation and evaluation of human beings: they comply with the human need to correlate behavior as observed from the outside to conditions within a person, that is, to attribute these with qualities that become effective in certain situations and influence behavior. This tendency also applies to the professional and economic behavior of persons, which managers, supervisors, economists, and engineers attempt to understand and explain. Thus, behavior can be seen as a function of ability and motivation and can be formulated as follows. Because motivational influences, as will be shown, manifest themselves in the decisions of individuals, the following can be stated as well: achievement = (abilities × decisions) (Poulton, 1970; Rockwell, 1972).

What kinds of decisions are involved? The literature in this field regularly refers to decisions based on questions such as the following:

1. Which alternatives for action are chosen?
2. With how much persistence does someone pursue a goal?
3. How intensively does one strive toward one's goal?
4. What degree of difficulty does one prefer if one can choose among several demanding tasks?

Although such questions can be asked with respect to all kinds of situations, their relationship to behavior in work situations and/or in man–machine systems is especially clear. The term behavior in work situations, or work behavior, generally refers to the accomplishment of work tasks (for a definition of work task, see Section 2.3.2.1) that are either assigned to an employee, as is generally the case, or are at times determined by the employee. According to Hacker (1978, p. 54), "the work task is the basic starting point for every psychological analysis of the work process." Human work activity, according to Hacker, is characterized by the following:

1. It is a conscious, goal-oriented activity.
2. It is set on the realization of a goal in the form of an anticipated product.
3. The anticipated product existed in the form of an idea prior to the commencement of the activity itself.
4. It is willfully directed toward the established goal.
5. Personality development runs parallel to the production process.

The first characteristic is particularly important to motivation psychology, especially in view of the above-mentioned questions and the explicit goal setting expressed in Locke's theory (see Section 2.3.2.1). Steers and Porter (1979) have differentiated between three types of goals that can influence the effort and performance of an employee:

1. "Organization-wide" or "department-wide goals" are generally goals that affect whole groups of individuals. They may, for example, determine their future in the organization, may supply them with information concerning expectations the organization has of them, or may inspire a feeling of self-worth in those employees who identify with their organization.

2. "Task-goals" are more specific goals, which define tasks for one or more individuals in such a way that each employee knows what he or she is responsible for.

3. "Personal goals" or "levels of aspiration" are goals set by the individual. A "task-goal" can, for example, call for a 10% sales increase, whereas an individual finds only 5% realistic. Thus 5% is the "personal goal," the "level of aspiration."

The goals that one actually sets for oneself are a result of effort on the part of the management, the objective conditions influencing perception, the redefinition of the tasks (Hackman, 1970), and the aspirations of the employee. Thus it is quite difficult to formulate a clear hypothesis on the relationship between motivation and achievement. The goals of the organization and of the employees concentrate not only on the accomplishment of work tasks and on the resulting performance but also on a further series of factors such as entering into the organization; within-group cooperation; problem solving, for example, in quality circles; readiness to train and further qualify oneself; staying with the organization (i.e., low absentee rate, little fluctuation); and safety and accident prevention. In studying work motivation it is important to consider the fact that some of these goals are incompatible and that conflicts can therefore arise for members of the organization. Measures supporting safety and those advocating achievement, for example, often contradict each other.

How does the concept of work fit into the system of personal goals and motives? It is quite possible that an occupation or a place of work might have its own defined class of demands and incentives. This applies especially to achievement-oriented individuals insofar as the execution of work tasks supplies them with the best measurement and acknowledgment of their own proficiency. This is of course less true for persons with a strong need for group identity, which can be satisfied in a work team as well as in a sports club. These persons have classes of situations of motivational equivalency that are not limited to their occupations. Various researchers have tackled this problem by investigating the relationship between work and leisure (Kabanoff, 1980; Kabanoff and O'Brien, 1980). Results of studies on work restructuring by Oldham, Hackman, and Pearce (1976), Steers and Spencer (1977), and Kleinbeck (1981) agree with these considerations. Persons with a strong need for growth or achievement are more likely to profit from measures that have the goal of increasing the incentive values of certain work tasks: they show better work results and greater satisfaction. Persons less motivated with respect to achievement should rather experience a feeling of satisfaction resulting from good work relations.

According to the "expanded model for motivation" (Heckhausen, 1980), which is described in the next section, this is a question of the varying consequences that different groups of employees expect with regard to the work process. They regard their behavior at work as an effective instrument by means of which their goals can be met. Using this approach, the dilemma of having to assume various motives for the same behavior can be avoided; instead, it becomes necessary to differentiate more clearly between the persons in question. Motivation theory attempts to explain why these goals are set and accepted and which goals are set, as well as how intensively they are pursued. Motivation, as interpreted today, is a complex process with a number of components and their interrelations. All in all, it is a hypothetical construct that cannot be observed, and that can be regarded as a link between demands and incentives of a situation, acting on or making a decision concerning that situation. "Thus the manner in which the concept of motivation is expounded on remains quite abstract. This concept attempts to find a common denominator for that which regularly occurs between the conditions prior to and the activity which follows a situation" (Heckhausen, 1980, p. 30). From this it follows that an adequate measurement of motives and motivation is difficult. Products of human fantasy, as are developed, for example, when a person is shown pictures such as those contained in the Thematic Apperception Test (TAT) and creates stories based on these can provide insights into human motivation. Also, attempts have been made to construct questionnaires with the help of which motives can me measured. Kleinbeck (1980) and Kleinbeck, Schmidt, and Ottmann (1982) have shown that indicators of the strength of motivation can be found by application of the secondary task technique.

One of the main problems in the application of motivation theories to work/operator behavior in the man–machine system is making these variables operational and obtaining adequate data. It is only in experimental situations that it has been possible to specify and measure human decisions with precision, for example, in tests of dexterity (Atkinson, 1958; Schneider, Gallitz, and Meise, 1973). Behavior in organizations can generally not be measured in the same manner. Evaluations of effort and persistence as observed by a superior or as judged by the employee, such as are common in field research, have proved to be quite problematic. The dilemma that has to date been inherent to field research on motivation therefore continues to exist today: what needs explaining, that is, what the dependent variable is, is perfectly clear, yet is extremely difficult to operationalize and measure. Projects dealing with the restructuring of work and, in connection with these, the development of new methods (e.g., Job Diagnostic Survey, JDS) have made the most progress in this area (Hackman and Oldham, 1975; Hackman, 1976; Ernst and Kleinbeck, 1980). In this chapter we want to consider several theories and results of motivation research, but must limit ourselves to those approaches that have proved or can prove meaningful to the fields of organizational and work psychology.

2.3.2   PROCESS THEORIES OF MOTIVATION

A good general picture of the development and status of motivation psychology can be obtained by means of differentiation between two groups of theories which appear in the work of Campbell, Dunnette, Lawler, and Weick (1970) and Campbell and Pritchard (1976), that is, between (1) process theories and (2) theories defined by content. The first group attempts to explain the causes and goals of behavior and to determine which variables or components are primarily involved in this process as well as

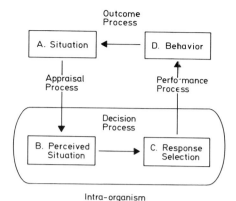

Fig. 2.3.1.   A paradigm for analysis of stress cycle (after McGrath, 1976).

how these interact. Theories defined by content tend to give information on the question of the type of needs that motivate an individual or on which class of goals he prefers to strive for. Thus theories of this type provide us with explanations concerning both possible alternatives for action and the type of choice that can be made between alternatives. This division is, however, not considered exclusive and exact. In fact, process theories seldom neglect to refer to specific goals and needs, whereas content-oriented theories rarely dispense with a discussion of the manner in which variables function. Process theories of motivation attempt to explain the way in which various components combine to evoke a *decision* and action in a demanding situation. An appropriate model can serve to demonstrate where the "site" of the influence of the motivation process in the execution of an action can be found. The cycle of perceptions, decisions, and achievement outputs in performing an action is shown in a simple yet convincing form in a behavior model developed by McGrath (1976) for the purpose of understanding psychological stress. In this cycle, four components are combined with one another: two can be classified as object, and two as subject components (Figure 2.3.1).

The object components in a work or man–machine system are the work situation (work task, equipment, work environment) and the observable behavior of the operator in the work situation, the result of which is work output or, in more general terms, activity output. The perception of a situation (as a result of information processing) and choice of action are specified as the subject components. These are connected by means of decision-making processes. Motivation affects both these components; its primary concern is to explain the decision that is ultimately made.

2.3.2.1   Motivation by Expectation and Incentives

A group of motivation theories known as "expectancy–valence theories," or instrumentality theories, and as valence–instrumentality–expectancy (VIE) theories (a combination of the first two) can be applied here. In recent years these theories have assumed a leading position, especially in the field of organizational and work psychology (Locke, 1976). We therefore concern ourselves primarily with these theories. VIE theories, which are all more or less derived from Kurt Lewin's work, have their roots in studies done by McClelland, Atkinson, and others on achievement motivation (Atkinson, 1964; Atkinson and Feather, 1966), in Vroom's (1964) work on motivation in the work process; and in studies on decision-making processes which have led to the development of the SEU model (subjective expected utility). Heckhausen has combined these approaches in his "expanded model of motivation" (Figure 2.3.2).

Figure 2.3.2 shows a sequence of events that are entirely compatible with McGrath's circular model. However, the fact crucial to the motivation process is the following: The person involved in an activity anticipates this sequence and converts the parameters into a decision, which theoreticians summarily call an "effort." This word refers to decisions such as those already mentioned, for example, choice between certain tasks, or a decision to continue or terminate work on a given task. Once a decision has been made, tasks can be accomplished and results and consequences can be seen. Using the model developed by Heckhausen, we discuss the individual components of the process of motivation.

A subject within a situation has the opportunity to strive for a certain goal. Intensity of effort to reach the goal depends on the valence of the expected outcome and on the expectancy to reach the goal. "The outcome of an action acquires its valence from the sum of all the valences of the ensuing consequences (. . .). Instrumentality denotes the expected causal connection between an outcome and a consequent event" (Heckhausen, 1977, p. 287). There are expectancies to reach a goal, that is, to

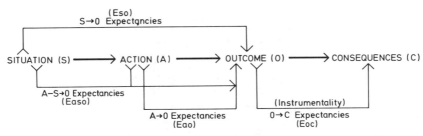

Fig. 2.3.2. Expectancies relating different stages in a sequence of events (after Heckhausen, 1977).

produce an outcome. A subject may perceive a probability to reach an outcome by one's own activity (E_{ao}) or expect that external circumstances will produce the outcome (E_{so}) or build up an expectancy that combines both forms to E_{aso}.

Situation and Task

Every action takes place within the context of a situation—in the case of work-related actions, in a work situation. One of the chief characteristics of the work situation is the existence of work tasks, which are determined by the employee (rather an exception than the rule) or which are assigned. According to Hackman's (1970) widely used definition, a task consists of a certain stimulus as well as instructions explaining how this stimulus shall be used and which goals are to be attained. This information can be very exact and detailed, as is the case for processing instructions, but it can also be formulated in a more general way, leaving a considerable amount of freedom, as is often the case for work safety regulations. These task components—stimuli, instructions on procedures, and instructions concerning goals—constitute demands for the employee; meeting these demands can be regarded as conforming to a role [compare Graen's (1969) version of the valence–expectancy model]. In perceiving a situation an operator recognizes a task along with its components and processes them, a procedure that Hackman (1970) has called redefinition of the task. Several aspects of motivation can be identified in this process. One must accept the goals that have been assigned and must set these for oneself. The perception of the situation activates certain expectations, for example, the expectation that one can accomplish the task using the abilities and knowledge one has, that is, that one is equal to the task. In some cases the perception of the situation also contains some indications as to the incentives connected with the work. As early as 1959 studies conducted by Herzberg et al. brought to the attention of researchers the importance of work content as a control condition relevant to motivation. Hackman and Oldham (1976) have reviewed a series of studies on the effect work content has in encouraging motivation. In their analysis they have determined five so-called core dimensions to which they attribute a considerable degree of influence over the control of motivation (Figure 2.3.3). In accordance with the extent to which they manifest themselves in the work activity of the employee, these five core dimensions lead to the specific modes of experience that Hackman and Oldham call "critical psychological states." The dimensions "skill variety," "task identity," and "task significance" give the employee a more or less strong feeling that his work has meaning. Autonomy at work gives a person the feeling that he is responsible for the results of his work, and the feedback he receives allows him to evaluate the level of performance he has attained in terms of the actual result of his work. The core dimensions can be mathematically combined in such a way that a motivating potential score (MPS) results, according to the following rule:

$$\text{(MPS)} = \frac{\left(\dfrac{\text{Skill}}{\text{variety}} + \dfrac{\text{Task}}{\text{identity}} + \dfrac{\text{Task}}{\text{significance}}\right)}{3} \times \text{Autonomy} \times \text{Job feedback}$$

To obtain data for these core dimensions, Hackman and Oldham (1975) have developed the above-mentioned JDS. The ways in which individuals experience work situations and the practical knowledge they gain through confrontation with their work can have various consequences, even measurable quantities such as high-quality performance, absentee rates, and fluctuation. Hackman and Oldham's model greatly enriched motivation research and stimulated a number of studies that have, at least in part, verified their hypothesis.

Expectations

The operator carefully considers the prospects of achieving the task goal, that is, develops expectations with regard to the results of his or her actions, in short, "action-outcome expectations," also known

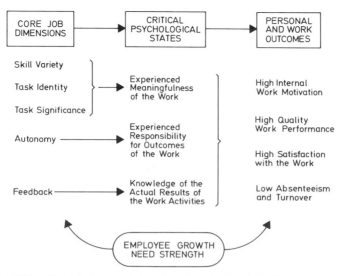

Fig. 2.3.3.   The job characteristics of work motivation (after Hackman, 1979).

as "subjective success probability." These expectations deal with the question of whether an appropriate amount of effort will lead to the accomplishment of the task goal. "Subjective probability" means, what are my changes of achieving this immediate goal, for example, of tackling a certain step in the processing of a work piece or of solving a construction problem? As Campbell and Pritchard (1976) have emphasized, subjective probabilities are conditional probabilities (the achievement of a goal depends, for example, on the abilities of an individual).

An outcome can sometimes be arrived at without the assistance of the operator. The operator must know how probable this is (Heckhausen calls this a "situation-outcome expectation"); this knowledge stems from one's experience. Heckhausen has postulated a "situation-action-outcome expectation" to express the sum of the various types of expectations mentioned above. "It refers to the subjective degree of probability with which external and variable circumstances can increase or decrease the action-outcome-expectation and thus lead to a resulting action-outcome-expectation" (Heckhausen, 1980, p. 621). An operator must consider how present occurrences come to pass before expectations concerning future occurrences can be developed, asking, for example, whether an occurrence is the result of his or her behavior or whether it is exclusively the result of external circumstances. Thus causal attribution is made, in that the operator assumes the person or environment to be the cause of the occurrence. According to Rotter, individual tendencies exist which place the primary emphasis on either the internal or the external causes (Rotter, 1966; Phares, 1976).

Within the realm of possible causes for success and failure certain internal and external causes have proved decisive for causal attribution; the most important research in this connection has been conducted by Weiner (1980, 1982), who continued the work of Heider, Kelley, and others.

Internal Causes.   An operator can, for the start, attribute the accomplishment of a task to his or her ability, that is, capacity and skill; the operator must then consider how much effort has been exerted in trying to attain the goal. This attribution is to be understood literally: the operator "recognizes" these factors as causes of success or failure. The term *external causes* first brings to mind concepts such as luck and chance, and then external determining factors in general. In the case of task execution, the assignment of a task with a certain degree of difficulty can be regarded as the manifestation of the external determining factors. Obviously more stable and variable dimensions are present in both pairs of causes: competence is enduring, effort is variable; task difficulty is also regarded as stable, chance (luck) is variable. Weiner (1980) has classified these causes in a two-dimensional classification scheme (Table 2.3.1). He has reported on numerous studies that have provided indications of the symbols and data persons use to determine the causes of success and failure. Experienced task difficulty is, for example, a function of the observed performance of persons other than oneself in the execution of these and similar tasks. If one completes a task successfully, and this is also the case for the majority of other observable persons, then this success tends to be attributed to the simplicity of the task, that is, to a stable external dimension. If a majority of persons are unsuccessful in their attempts to accomplish a task and one is successful, then this success tends to be attributed to one's own ability. The basis on which further decisions can be made depends on whether success is attributed to internal or external, to stable or variable causes.

Table 2.3.1   A Two-Dimensional Classification Scheme for the Perceived Determinants of Achievement Behavior (After Weiner, 1980)

	Locus of Control	
Stability	Internal	External
Stable	Ability	Task difficulty
Unstable	Effort, mood, fatigue, illness	Luck

The above-mentioned expectations result from the concept of internal and external control: if the operator believes that the desired result will be attained even without his assistance, then he has a situation-outcome expectation. If an expectation of this type prevails, the operator will find it unnecessary and impossible to intervene in the flow of events. If the operator, from a subjective and objective point of view, has only a small amount of control over the flow of events and the way in which an activity result is achieved, then he or she is said to experience *lack of controllability*. According to Seligman (1974) this can lead to a state of helplessness. However, a clearer explanation of what control refers to in work situations is necessary when making such considerations. One must, for example, differentiate between influencing sources of noise and establishing a relationship between the result of an action and its consequences (Folkman, 1984).

In contrast, frequently internal causal attributions constitute an expectation that one can attain a result that is relevant to a motive by means of personally intervening in the flow of events. The internal causes are talent and effort. According to Meyer (1973), the operator must evaluate the adequacy of his or her abilities for the accomplishment of the task that has been set. Of course this is possible only if one has accumulated experience with similar tasks and has thus developed a subjective concept according to which one feels one can recognize one's abilities. People in fact tend to over- or underestimate their own abilities. If individuals are asked to evaluate their own performance in cases where they have received little feedback, those who have a high estimate of their own talent tend to give realistic evaluations whereas those with a low estimate of their talent tend to underrate their achievements (Meyer, 1984).

Consequences

The immediate results of an activity generally have broader consequences. In the case of behavior in work organizations rewards associated with task achievement come to mind. As a consequence of one's activity one can also learn something about one's own proficiency; that is, one can make a positive assessment of one's own work. In general, consequences of activity can consist of self-assessment and assessments made by others. These assessments can be made at various times. A reward can immediately follow the result of a work activity. If, however, one aspires toward success in one's field of work as a consequence of one's activity, this goal can first be realized a considerable amount of time after the completion of a certain activity. A certain kind of expectation, referred to in Figure 2.3.2 as "instrumentality," is geared toward consequences. It specifies the probability with which fulfilling a task will lead to certain consequences. Vroom (1964) introduced the aspect of the consequences of activities and the instrumentality of work results to their realization to the field of work motivation, thus extending the valence–expectancy theory to include the future, an aspect which had until then been missing. Instrumentality can assume values between $+1.0$ and -1.0; that is, it can be definite that the result of an activity will give rise to certain consequences, yet the same result can bring about the nonoccurrence of other consequences or be neutral with respect to still other consequences, whereas the expectations cited earlier can assume values between 0 and 1.0. Many individuals actually experience no feedback, for example, in the form of recognition, from their supervisors, regardless of whether they have made an effort or not (instrumentality $= 0$).

Hackman and Oldham's model (see above), in which experience plays an important role in determining how effectively a worker is performing a job, should be taken into consideration here. Researchers who feel committed to neobehaviorism have made abundant use of the possibility of systematically supplying the worker with feedback on his activity (Sulzer-Azroff, 1982). Systematic reinforcements have proved highly successful when applied to the sensitive area of work safety (Komaki, 1983). Latham (1982) has formulated demands for the way in which feedback should be given: feedback must be accurate, must be given at set, if possible at intermittent, intervals and must be interpretable by the addressee.

Valence

The individual attaches a certain type of meaning to the results and consequences of an activity; that is, these have a valence that lets them appear to be attractive goals. The concept of valence was introduced by Lewin and has since been taken up by numerous researchers; it is sometimes referred to by other names. Atkinson and Vroom's reference to affective valence, according to which occurrences that help stimulate pleasure and avoid displeasure have a higher valence, reveals that the origin of this concept is to be found in hedonistic approaches. Yet valence must and should not be interpreted only in this manner. It refers, above all, to "prospective values" (Heckhausen) in general, that is, to certain incentives or, according to Atkinson, to the attractivity of goals. Atkinson sees valence as a combination of motives and rewards and thus shows the origin of valence in personal experience and external feedback. According to Vroom (1964), the result of an activity generally has no value in itself, but takes on value in its capacity to bring about consequences to which the operator attaches meaning. Consequences are therefore said to have incentive character or simply to be incentives. The theories discussed here can then also be called incentive theories. As already stated, the goals in question can be imminent or distant goals. Various authors have attempted to place the classical needs on this level (Campbell and Pritchard, 1976); as seen from the point of view of the organization, these are the incentives and rewards that can be offered in return for performance (Rosenstiel, 1975).

The valence of an activity outcome is thus a product of the valence of consequences and the instrumentality of the primary outcome to the achievement of further-reaching goals. According to Figure 2.3.2 as well as to Campbell and Pritchard (1976), activity itself has a valence. Herzberg, Mausner, and Snyderman (1959) in their well-known "motivator-hygiene theory" believe that "work itself" has value that is worth striving for.

Interaction of Components

Decisions concerning alternatives for action, levels of effort, and so on develop in the interaction of the components of the motivation process. In recent literature this process has been attributed with a calculatory character. According to Meyer (1973), the operator calculates the amount of effort needed for the accomplishment of a task. The task difficulty and one's own talent serve as points of orientation. The manner in which the components of motivation are incorporated in this "calculation" has been the theme of much deliberation. Most authors, for instance, Atkinson and Vroom, combine expectations and valence by means of multiplication; in other words, one sees the effect of an activity goal on behavior if one places valence in relation to the prospects of achieving that goal. This is plausible, because neither attractive but unattainable goals nor unattractive, easily attainable goals are worth an exertion of effort. In empirical studies, relating components by means of multiplication has proved only slightly superior to a simple addition of the same components. In any case, the following holds true: one's decision to occupy oneself with a certain task and to exert a certain amount of effort in connection with this task is a function of (1) one's personal estimation of the probability of achieving the task goal and (2) the valence of the outcome. The latter derives from the valence of the consequences, which are a product of the outcome and the personal estimation of the probability that the activity result will lead to these consequences.

Set Goals and Reference Norms

As has already been shown, activity outcomes are evaluated by the operator and/or by others. These evaluation processes assume the existence of reference norms or set goals. In the accomplishment of an achievement, the level of demands made by the operator assumes the role of a reference norm: the extent to which an activity outcome is experienced as being successful depends on the level or the quality of the performance being strived for. In the work process, reference norms are sometimes supplied by the task: instructions concerning goals are components of well-defined tasks (see above). Their role in the motivation process has, however, not yet been expounded on as far as we know. On the other hand, an area of research established by Locke has concerned itself with the function of explicit goal setting. As Locke and his various colleagues have been able to confirm in the course of laboratory and field studies, achievement is influenced by the way in which goals are set: performance is always better when concrete goals are set than when a nonspecific intensive effort is demanded (Do your best!); that is, if goals are attainable, goals and achievements are related in a linear fashion. In addition, performance is better when the goals are demanding than when they are modest. As a result of these studies, the motivating effect of feedback becomes more evident. As the psychology of learning recognized long ago, the learning of skills and the production of work results to a large extent depend on the information an operator receives concerning his or her performance. As Locke and his colleagues have been able to demonstrate, information on results improves performance, especially in those cases where these are connected to concrete goals.

Locke obtained his first results from laboratory studies in which test persons were asked to execute very simple tasks, for example, to add pairs of numbers. Hinrichs (1978) and Locke et al. (1981) summarized the ensuing field studies, in which the basic tenets of this approach were verified. The assumption that more difficult goals led to better task performance could be verified for a wide variety of tasks: typing, coding, prose learning, perceiving speed, card sorting, and other tasks, most of which were laboratory tasks. However, field studies on transportation activities, logging, and office and research activities showed a clear increase in productivity when employees were presented with well-defined goals and were informed about the progress of their achievements. But field studies also revealed the limitations of and the specific conditions that apply to the measures that were derived from motivation theory. Concrete goal setting is not very meaningful for activities that depend strongly on machine processing. In connection with democratization attempts, the question of whether employees should set their own goals or whether these goals should be set for them has gained significance. No evidence for a clear answer to this question has yet been found; Hinrichs (1978) recommends a participative approach. In a field experiment involving logging crews, Latham and Yukl (1975) found that for uneducated loggers, participative goal setting resulted in better performance than assigned goal setting. Therefore, participation influences goal setting only when there is a free choice of goals. When goal difficulty is held constant, participation does not seem to affect performance. As can be expected, performance motivation alone is seldom effective in concrete work situations; other motives are, as a rule, also activated. The test persons in Latham's and Saari's (1979) experiment set higher goals for themselves—task difficulty remaining constant—when the experimenter showed them encouragement and support. According to the studies that do exist, goal setting clearly influences the interactions with external evaluations, that is, with certain consequences of performance results, as is postulated in the VIE theories. The studies mentioned here primarily refer to the amount of work done. The aspect of quality has thus far clearly been neglected. It has often been shown that, as goals become more difficult, it becomes more and more important to differentiate between quantity and quality. These begin to assume increasingly divergent values or develop a reciprocal relationship to each other, which expresses itself in a negative correlation of both variables (Kleinbeck, 1980).

Evaluation and Criticism

The variations on the VIE theories of motivation show a clear advance over earlier and one-sided approaches, such as those that took only incentives or only motives into consideration. They avoid the difficulties that regularly arise when one tries to explain individual actions such as taking up work or accomplishing a work task in terms of specific motives without taking the instrumentality for further-reaching goals into account. Various authors have reported on and given critical appraisals of these theories and studies which attempt to validate them (Heneman and Schwab, 1972; Lawler, 1973; Mitchell, 1974; Neuberger, 1974; Locke, 1975; Rosenstiel, 1975; Campbell and Pritchard, 1976; Heckhausen, 1980). Many assumptions were empirically verified; many proved invalid. All in all, the balance is not as favorable as might have been expected in view of the elegance of the theory. It is obvious that various methodical problems have not yet been solved, for example, the measurement of expectation and the evaluations of performance. The studies conducted so far have probably been too general and have tried to test the model as a whole. Campbell and Pritchard (1976) suggest first concentrating on the individual components and studying their variability, dependency on situations, and measurability.

2.3.2.2　Motivation According to the Principle of Equilibrium

Numerous observations have shown us how the initiation of activities and the control of behavior are affected by conditions and changes in the environment. Behaviorist theories, which offer an abstract model according to which behavior is controlled by incentives and assigned contingencies, immediately come to mind here. Effects such as reduction of drive, optimal states of stimulation, and solutions to conflicts, that is, conditions of inner equilibrium in general, also referred to as homeostasis of the inner milieu, function here as guiding theoretical values. These approaches, which are in part already historical, need not be elaborated on here. Several theories that place stronger emphasis on the interaction between the individual and the environment are being discussed more intensively today. In principle they are concerned with trying to determine whether a lack of equilibrium and incongruencies between a person and the environment constitute a motivating force. It may appear somewhat arbitrary to subsume the following approaches under the concept of lack of equilibrium; on the other hand, it would be equally arbitrary to regard equilibrium or congruence as incentives.

Cognitive Dissonance

Probably the oldest and most thoroughly researched approach is the theory of cognitive dissonance (Festinger, 1957, 1964). It assumes that a person strives toward congruence between the cognitive representation of the environment and information presently available. One experiences differences between one's own perception of things and the actual situation, here referred to as cognitive dissonances, as unpleasant and is therefore motivated to restore a state of congruence. This can be attempted by means of various strategies, for example, resisting new information that might lead to an increase in the amount of dissonance, or reevaluating dissonant elements in order to reduce the amount of dissonance.

Control Motivation

A major discrepancy between a person and the environment exists when a person is only minimally able to influence the flow of events. One is then at the mercy of external influences and a victim of uncontrollable processes. One experiences this situation as being unpleasant, perhaps even threatening, and is motivated to take measures that will help one regain control. This is a case of "control motivation," the origins and applications of which have already been referred to above insofar as they relate to the work situation. In the so-called reactance theory (Brehm, 1966, 1972), reduced control over a situation was explained in terms of motivation theory: a person who believes that it is possible to engage in activities A, B, and C, but then realizes that some of these possibilities are not open, then experiences reactance. The limitation of one's options for activity motivates a person to make an effort to win back one's state of "freedom," that is, to regain a greater amount of control over the flow of events in the environment. This approach has found support in studies on causal attribution and on the experience of lack of controllability. Research on so-called risk acceptance has used a similar approach (Otway and Winterfeldt, 1980): risks tend to be more readily accepted if one feels one can influence the risk (natural disasters versus engaging in dangerous sport activities).

Equity Theory

An "equity theory" pertaining directly to the realities inherent to organizations was formulated some time ago (Adams, 1965). This theory postulated the search of the individual for a certain state of equilibrium with his or her social environment—especially from the point of view of monetary remuneration. Individuals compare the relationship between effort and returns that characterizes their own activities with the corresponding relationship that they perceive as applying to others. A similarity between these relationships produces a feeling of equity, and dissimilarity, of inequity. The latter experience generally produces tension, which in turn motivates the implementation of measures that reduce tension. Appropriate measures can be of a purely cognitive nature, for example, reevaluation of the compared relationships, or can consist of actions that change the effort–return relationship for the person evaluating it or for others. This theory has definitely proved valuable as an explanatory approach for the analysis of certain remuneration systems. Because the desire for equity influences the evaluation of activity outcomes, Campbell and Pritchard (1976), as well as Lawler (1973) before them, want to subsume this theory under the VIE theories.

2.3.2.3 Evaluation

In general, theories that attempt to explain behavior by means of homeostatic tendencies (more congruence between the cognitive representation of the environment and the current situation one experiences, more control, less uncertainty) have not yet sufficiently proved themselves in the area of organizational psychology; this is, however, true for most other theories as well. Yet their potential contribution to an understanding of work behavior should not be underestimated, especially because, as Neuberger (1974) has demonstrated, more recent approaches, even if not directly in connection with the theories mentioned here, no longer show the weaknesses of their—for the most part behaviorist—predecessors. The interrelationship with the environment is, for example, more strongly emphasized (theories that are concerned only with the stabilization of the inner milieu have not been mentioned here); in addition, it became evident that not only short-term changes in behavior but also long-term perspectives in which equilibrium was striven for can be explained. Finally, concepts of system and control theory can serve as explanatory aids and thus be profited from.

2.3.3 CONTENT-ORIENTED THEORIES

Content-oriented theories of motivation attempt to explain which classes of goals individuals strive toward, which alternatives for action they perceive, and which needs they have. Laypersons as well as experts assume that everyone has certain needs that one strives to fulfill or certain motives according to which one acts. If one considers the manifold possibilities for activity that persons can choose from and pursue, then conjectures and needs or motives should explain the choice of a certain alternative, whereby a motive is surrounded by a complex motivation process that process theories of motivation shall elucidate.

The Motive

What does one tend to associate with the term motive, a construct that is used as a general designation for many other terms such as need and drive? Very diverse opinions have been expressed regarding the nature of a motive: a motive as an innate need, as a state of deprivation, or as a disturbance of inner equilibrium or of equilibrium with one's environment. Atkinson (1964) described a motive as a learned disposition to strive for a certain type of satisfaction; consequently he sees a motive as an affective component of the motivation process. According to McClelland's (1951) two-factor theory, a motive can have two directions, approach toward and avoidance of something, that is, a tendency to search for pleasure and a tendency to avoid displeasure. In the area of performance, for example, these motives are called "hope for success" and "fear of failure." In recent years cognitive interpretations of the term motive have become more common. Meyer (1976) and Kleinbeck and Schmidt (1976), for example, describe the achievement motive as the striving to attain new information concerning one's own talent or proficiency. A situation in which it is possible to obtain this information can activate the performance motive.

Heckhausen (1980) largely disregards assumptions of the kind just mentioned and describes motives as highly generalized content classes of events, in the realization of which the operator has a personal interest. Here a motive is seen from the point of view of the consequences of activities, and is thus independent of speculations concerning needs, drives, and so on. In this case a motive has anticipatory impetus, because meaningful events are seen as lying in the future. Events that have taken place are changes in the environment and therefore new stimulus configurations. These supply the operator with feedback if he sees them as consequences of his own activity or at least in relation to his anticipation. According to this intrepretation, motives are more than wishes and needs in the traditional sense. Whereas older traditional definitions see motives, in the sense of needs, as being located within a person, more recent cognitive approaches tend to overemphasize the role of the environment; that is, they interpret motives largely from a situational point of view. In contrast, Nuttin (1984) sees needs as being firmly implanted in the unity of the individual with his environment. "Thus, behavioral needs can be conceived and defined as 'required relationships' between the individual and the world. More precisely, needs are behavioral relationships in as far as they are 'required' to insure the individual's optimal functioning." (Nuttin, 1984, p. 60).

Classes of Motives

It may be more interesting to ask whether acceptable lists of possible and typical human motives exist than to ask about their nature. To answer these questions it would be necessary to report on an immense number of very diverse approaches and theories. That is not the intention of this chapter. Both in the past and at present, research has concerned itself with relatively few motives (needs, drives) in greater detail: with social motives, especially affiliation, power, helping, aggression, and the achievement motive. For the classification and organization of motives Maslow's (1954) attractive concept is given preference, especially in the field of organizational psychology, even though very little empirical proof of its validity is available. Maslow works with a hierarchy of needs: physiological needs, the need for safety/security, social needs, needs of the ego, and the need for self-actualization as the "peak" need. Alderfer (1969, 1972) has presented a newer version of this concept, which has clearly proved superior to the work of his predecessors. He introduces the following groups of needs:

1. *"Existence" Needs.*   These are needs that must be satisfied in order to secure the survival of the individual (basic needs).
2. *"Relatedness" Needs.*   These are motives that have the goal of establishing contact, feeling the proximity of others, and being cared for—in short, contact needs.
3. *"Growth Needs.*   These constitute one's desire to have a formative and innovative influence on the situation in which one finds oneself (desire for self-actualization).

In this connection a number of assumptions have been formulated, such as the following:

1. The less the basic needs are satisfied, the more their goals are striven toward.
2. The less the contact need is satisfied, the stronger the basic needs become.
3. The less the desire for self-actualization is satisfied, the stronger the contact need becomes (according to Rüttinger, Rosenstiel, and Molt, 1974).

Growth needs have received special attention in connection with job design and job redesign. Hackman and Oldham (1976) have postulated a connection between the objective and subjective core dimensions of the work situation and employee satisfaction—moderated by factors such as the strength of growth needs. The concept of growth needs or higher-order needs has proved to be too imprecise. Steers (1975) as well as Steers and Spencer (1977) have shown, however that achievement motivation influences work behavior.

Nuttin (1984) has adopted a different approach for differentiating between human motives. He has established three levels of relational functioning: biological, psychosocial, and ideational. On these levels individuals seek to develop themselves and to establish specific forms of functional contacts with objects. These contacts can be considered as being the basis of human needs or motives.

Organizational Incentives

The needs referred to in the classifications of motives (e.g., Maslow, 1954; Alderfer, 1969) can be satisfied both by self-evaluations and by evaluations made by others. Thus, information concerning one's own proficiency can be obtained by experiencing success and by receiving feedback from others. The differentiation between self-evaluation and evaluation by others has led to the introduction of the terms intrinsic and extrinsic motivation. In work organizations, external evaluations can be communicated to an employee directly by supervisors, by means of the organization of the remuneration and bonus system as well as of the work itself. One can gain insights into the substance of motivation by considering the incentives offered by the organization, which have a verifiable effect on the experience and behavior of the employees.

Studies on work satisfaction supply a key to the understanding of the motivating effect of organizational conditions. The concept of work satisfaction has been interpreted in many diverse ways. Neuberger and Allerbeck (1978, p. 14) have compared very different definitions of this concept and have worked out the following interpretation, in which they express what they consider to be the main theme common to these definitions: Satisfaction can be seen ". . . as an attitude towards work, which is structured and judgmental and which possibly influences work activity." They continue, "The existing conditions in the environment are evaluated according to the amount of positive feeling they evoke" (compare Locke, 1976). In dealing with the question of incentives of work behavior, it is interesting to note a tendency toward the interpretation of work satisfaction in relation to specific aspects rather than only in general terms. The well-known and frequently tested Job Description Index (JDI) (Smith, Kendall, and Hulin, 1969) is organized according to the following aspects: work, supervisors, colleagues, pay, and promotion.

By differentiating between these aspects of the work environment in numerous studies, it has been possible to determine which classes of events are meaningful to an employee, that is, contain incentives that can contribute to motivation. However, these aspects of work organizations do not necessarily correspond to the generally acknowledged classes of motives. Pay is, without doubt, a multivalent category; supervisors influence both an employee's aspirations toward achievement and his or her contact motive. The studies conducted by Herzberg et al. (1959) have been the most persistent attempts to direct attention toward the incentive character of aspects of work and the work environment. In the first study, in the spirit of the critical incidents method, the researchers asked technicians and other employees to describe situations and events in which they had experienced a high degree of satisfaction. Episodes in which the persons interviewed were especially dissatisfied were likewise collected. These episodes were analyzed in detail in an effort to reveal factors that had contributed to an experience of satisfaction or dissatisfaction.

In the process of this analysis, a thesis, which has since become the subject of animated discussions, was formulated, stating that experiences within an organization could not be represented on a continuous scale ranging from "satisfied" to "dissatisfied," but should rather be evaluated using two scales: satisfaction versus lack of dissatisfaction. In those episodes that had led to a high degree of satisfaction, the following themes or factors were most frequently referred to: performance, recognition, work itself, responsibility, promotion, and pay.

The second group consisted of organizational policies, relations with subordinates, colleagues, and supervisors, work conditions, pay (which has a multivalent position), and technical aspects of leadership.

Herzberg and his colleagues initially referred to the first group of factors as "satisfiers" and finally called them "motivators," because they have a positive influence on the person performing an activity and contribute to his or her work-related development. This is primarily a question of factors that have to do with work content. The factors of the second group were called "dissatisfiers" or "hygiene factors," a term derived from the concept of medical hygiene: Its failure to exist or negative characterization is detrimental to the motivation of the employee; its existence, however, does not necessarily inspire positive motivation. This is to be seen primarily in connection with aspects of the work environment.

Results and interpretations of the work of Herzberg et al. have led to numerous controversies and, above all, to methodical and theoretical contradictions. More recent studies in various countries have led to further confirmations of the theory of Herzberg and coworkers.

Hackman and Oldham (1976), with the help of the JDS, have tried to comprehend the way in which motivation can be encouraged by work content. Recently the orientation toward general values, as an aspect of work motivation, has been more strongly emphasized than the consideration of needs. This tendency is currently of interest because—as numerous studies have shown—a change in values, especially as regards work, industry, and technology, is taking place in industrial nations. According to Inglehart (1977), who uses Maslows's motivation theory as his point of departure, the values of large population groups are shifting from material to postmaterial values. Work is still seen as something very important; it is, however, no longer regarded as a value per se, but is instead seen as competing with other values, such as those connected with leisure time, environmental protection, and health (Rosenstiel, 1984; Vaassen, 1984). The results presented here on value research are difficult to interpret, because little consensus has yet been found concerning the definition of the term value and concerning the way in which a person's judgments should be measured.

Evaluation

All in all, the attempts to classify motives present a somewhat confusing picture. The desire to find "the" classification will quite definitely remain frustrated. Research concerns itself with motives such as achievement, contact, power, and others; these are, however, only a few of the goals that operators can potentially set for themselves. Which motives are stimulated or which classes of meaningful events an operator is capable of perceiving depend to a large extent on the incitement and reinforcement mechanisms that exist in an organization or, in some cases, in society in general. Thus, a careful analysis of a situation must always precede an analysis of motivation. What is especially necessary at present is to view the individual in his or her interaction with the environment. The question of whether there are, in fact, motives for which no incentives exist or (conjectured) incentives that are not related to needs has as yet received no satisfactory answer. Demands stating that social, economic, or cultural affairs must be organized in accordance with human needs offer no immediate concrete suggestions as to the way in which this organization might be accomplished.

2.3.4 SUMMARY

Interpretations and explanations of human behavior by the persons who perform activities, by naive observers, and by scientists suggest the usefulness of introducing constructs with the help of which the choice of goals and their difficulty and intensity as well as the persistence in pursuing these goals can be explained. These constructs are known as incentives, needs, and so on, or more generally, as motives; the process during which these are activated and become effective is called motivation. Motivation is a hypothetical construct which is, as such, not observable and can therefore only be measured indirectly, for example, with the help of products of human fantasy or questionnaires. When we try to explain certain sectors of behavior, for example, work behavior, a problem arises because it is not clear whether a type of motivation exists that is specific to work. The effect of motivation, that is, the dependent variables such as choice of task difficulty, can be made operational readily in experimental studies, but only with considerable difficulties in field studies.

The development and present status of motivation psychology can be outlined by differentiating between process theories and theories determined by content. Process theories attempt to explain how behavior develops, which components are involved, and how these interact. The first process theory designated as such was the valence–expectancy theory, sometimes known as the valence–instrumentality–expectancy (VIE) theory; this theory is the most widely recognized process theory today. In this theory (owing to the existence of certain variations it would be more accurate to speak of a group of theories), the decision one makes to occupy oneself with a certain task and to exert a certain amount of effort in attempting to accomplish that task is a function of (1) one's subjective evaluation of the probability of accomplishing the task goal and (2) the valence of the activity outcome. The latter results from the valence of the consequences of the outcome and from the subjective evaluation of the probability with which the outcome will lead to these consequences. The situation and the task must always be carefully examined, especially in the case of work systems, in order for the causative and regulative environmental conditions to be recognized. As part of the process of expectation, causal attribution, that is, the opinion of the person performing an activity concerning the way in which internal and external conditions influence the course of events, plays an important role, insofar as it determines the extent to which one sees a connection between one's own activities and results or expects certain occurrences to take place independent of one's efforts.

Activity as goal-oriented behavior depends on set goals and reference norms. Experimental and field studies with exactly defined performance goals have supplied researchers with insights concerning the way in which these goals and norms function. These defined goals have almost always been accompanied by increased productivity, but they also depend on social and other influences. Approaches in which disturbances of the equilibrium between a person and the environment are seen as constituting a motivating force were regarded as a group of process theories. The theory of cognitive dissonance is probably the best known theory belonging to this group. Equilibrium is also disturbed in the case of loss of control. The theory of equity, which attributes a motivating force to the divergence between one's own effort–returns behavior and that observed in the activities of others, is an approach that has more in common with the methods of organizational psychology. This force works toward balancing the equilibrium by affecting changes in attitude and behavior.

Theories of motivation that are determined by content contain more information about the type of needs that motivate an individual and the class of goals that individual prefers to strive for. A discussion of well-known motive groups and classifications follows some considerations as to the "nature" of a motive (a number of widely differing interpretations are increasingly giving way to a cognitive interpretation). The former include social and achievement motives, to which researchers have paid a great deal of attention, and also the well-known classifications made by Maslow and Alderfer, which have over the years received considerable recognition in the field of organizational psychology. It was expected that studies on work satisfaction would offer some insights into the classes of motives that are involved in work situations and have manifested themselves as incentives.

It is not easy to say how to apply motivation theory to human factors issues and how human factors specialists can utilize respective knowledge, insofar as ergonomic activities of all kinds have their motivational implications. Therefore, the main aim of a chapter on motivation theory is to make human factors specialists sensitive to the motivational implications of their work. Their considerations may be guided by the following principle.

Human factors specialists (ergonomists) should see as their main task the optimization of decisions of operators in person–machine systems with respect to human needs as well as to the functioning of the total system in question (see Chapter 1.2). Decisions can be made only if the system provides degrees of freedom for decision-making. Degrees of freedom are related, as discussed in the beginning of the chapter, to selecting between different tasks, to show more or less commitment to the task, and so on. One of the main results of the recent debate on work restructuring was that more autonomy at work increases work satisfaction. Human factors specialists (ergonomists) can serve this principle in two ways. (1) They should take care that operations for running a system are not too detailed and not put into a fixed and rigid order. The possibility must exist for the operator controlling a system to select procedures and treatments according to his or her knowledge of and experience with goals, components, and processes of the system. (2) The teaching of decision-making must have a high rank among goals of staff training.

The kind of motivational implications to be considered in ergonomic work can be demonstrated by some examples.

Most operations in person–machine systems are based on a certain amount of processing information; the willingness of the operator to seek sufficient information depends on rewards and feedback (see Chapter 2.4).

Many tasks that human operators have to accomplish in modern person–machine systems, particularly monitoring automated processes, are boring. For motivational reasons monitoring jobs should be enriched with ambitious supplementary tasks.

Decisions must frequently be made under conditions of uncertainty and/or risk. This burdens the operator with high responsibility. The HF specialist should be aware of this and recommend a design that allows responsibility to be shared among operators. Team principles can be encouraging and rewarding (see Chapter 4.2).

One of the main activities of a HF specialist is to allocate functions to a human operator or to a machine. Such decisions are made mostly according to functional properties of the subsystem in question. However, this decision is also a matter of motivation. A human operator should not only perform a certain task better than a machine, but the task must be attractive and present an opportunity to develop his or her abilities (see Chapter 1.3).

REFERENCES

Adams, J. S. (1965). Inequity in social exchange. In L. Berkowitz, Ed., *Advances in experimental social psychology,* Vol. 2. New York: Academic.

Alderfer, C. P. (1969). An empirical test of a new theory of human needs. *Organizational Behavior and Human Performance, 4,* 142–175.

Alderfer, C. P. (1972). *Existence, relatedness, and growth. Human needs in organizational settings.* New York: Free Press.

Atkinson, J. W. (1957). Motivational determinants of risk-taking behavior. *Psychological Review, 62,* 359–372. Also in J. W. Atkinson, Ed., *Motives in fantasy, action, and society.* Princeton, NJ: Van Nostrand, 1958.

Atkinson, J. W. (1969). *An introduction to motivation.* Princeton, NJ: Van Nostrand.

Atkinson, J. W., and Feather, N. T., Eds. (1966). *A theory of achievement motivation.* New York: Wiley.

Brehm, J. W. (1966). *A theory of psychological reactance.* New York: Academic.

Brehm, J. W. (1972). *Responses to losses of freedom: a theory of psychological reactance.* Morristown, NJ: General Learning.

Campbell, J. R., Dunnette, M. D., Lawler, E. E., and Weick, K. E. (1970). *Managerial behavior, performance, and effectiveness.* New York: McGraw-Hill.

Campbell, J. P., and Pritchard, R. D. (1976). Motivation theory in industrial and organizational psychology. In M. D. Dunnette, Ed., *Handbook of industrial and organizational psychology.* (pp. 63–130) Chicago: Rand McNally.

Ernst, G., and Kleinbeck, U. (1980). Die Wirkung leistungsthematischer Anregungsgehalte von Arbeits-situationen auf die Einstellung zur Arbeit. *Zeitschrift für Arbeitswissenschaft, 34* (6 NF), 97–102.

Festinger, L. (1957). *A theory of cognitive dissonance.* Evanston, IL: Row, Peterson.

Festinger, L. (1964). *Conflict, decision, and dissonance.* Palto Alto, CA: Stanford University Press.

Folkman, S. (1984). Personal control and stress and coping processes: a theoretical analysis. *Journal of Personality and Social Psychology, 46,* 839–852.

Graen, G. (1969). Instrumentality theory of work motivation: Some experimental results and suggested modifications. *Journal of Applied Psychological Monograph, 53.*

Hacker, W. (1978). *Allgemeine Arbeits- und Ingenieurpsychologie.* Bern: Huber.

Hackman, J. R. (1970). Tasks and Task Performance in Research on Stress. In J. E. McGrath, Ed., *Social and psychological factors in stress.* (pp. 202–237) New York: Holt, Rinehart & Winston.

Hackman, J. R. (1976). Work design. In J. R. Hackman and J. L. Suttle, Eds., *Improving life at work.* (pp. 96–162) Santa Monica, CA: Goodyear.

Hackman, J. R., and Oldham, G. R. (1975). Development of the job diagnostic survey. *Journal of Applied Psychology, 60,* 159–170.

Hackman, J. R., and Oldham, G. R. (1974). Motivation through the design of work: test of a theory (Technical Report No. 6). Department of Administrative Sciences, Yale University.

Hackman, J. R., and Oldham, G. R. (1976). Motivation through the design of work: Test of a theory. *Organizational Behavior and Human Performance, 16,* 250–279.

Heckhausen, H. (1977). Achievement motivation and its constructs: A cognitive model. *Motivation and Emotion, 1,* 283–329.

Heckhausen, H. (1980). *Motivation und Handeln.* Berlin: Springer.

Heneman, H. G., and Schwab, D. P. (1972). Evaluation of research on expectancy theory predictions of employee performance. *Psychological Bulletin, 78,* 1–90.

Herzberg, F., Mausner, B., and Snyderman, B. (1959). *The motivation to work.* New York: Wiley.

Hinrichs, J. R. (1978): *Practical management for productivity.* New York: Van Nostrand.

Inglehart, R. (1977). *The silent revolution. Changing values and political styles among western publics.* Princeton, NJ: Princeton University Press.

Kabanoff, B. (1980). Work and nonwork: A review of models, methods, and findings, *Psychological Bulletin, 88,* 60–77.

Kabanoff, B., and O'Brien, G. E. (1980). Work and leisure: A task attributes analysis. Journal of Applied Psychology, 65, 596–609.

Kleinbeck, U. (1980). Arbeitsmotivation. in C. Graf Hoyos et al., Eds., *Grundbegriffe der Wirtschaftspsy-chologie.* (pp. 434–445) Munich: Kösel.

Kleinbeck, U. (1981). Eignet sich die Theorie der Leistungsmotivation für die Begründung von Arbeitsstrukturierungsmassnahmen? In U. Kleinbeck and G. Ernst, Eds., *Zur Psychologie der Arbeitsstrukturierung.* (pp. 32–41) Frankfurt: Campus.

Kleinbeck, U., and Schmidt, K.-H. (1976). Motivationsanalyse von Leistungen: der Instrumentalitätsas-pekt. In H.-D. Schmalt and W.-U. Meyer, Eds., *Leistungsmotivation und Verhalten.* Stuttgart: Klett.

Kleinbeck, U., Schmidt, K.-H., and Ottmann, W. (1982). Motivationspsychologische Aspekte der Bewe-gungssteuerung. *Archiv für die gesamte Psychologie, 134,* 181–195.

Komaki, J. L. (1983). A behavioral approach to work motivation. In Conference Papers *International Seminar on Occupational Accident Research* (pp. 67–97), Stockholm, September 5–9

Latham, G. P. (1982). Behavior-based Assessment of Organizations. In L. W. Frederiksen, Ed., *Hand-book of organizational behavior management.* (pp. 95–115) New York: Wiley.

Latham, G. P., and Saari, L. M. (1979). Importance of supportive behavior relationships in goal setting. *Journal of Applied Psychology, 64, 151–156.*

Latham, G. P., and Yukl, G. A. (1975). Assigned versus participative goal setting with educated and uneducated wood workers. *Journal of Applied Psychology, 60,* 299–302.

Latham, G. P., and Yukl, G. A. (1977). Effects of assigned and participative goal setting on performance and job satisfaction. *Journal of Applied Psychology, 61,* 161–171.

Lawler, E. E. (1973). *Motivation in work organizations.* Monterey, CA: Brooks/Cole.

Locke, E. A. (1975). Personnel attitudes and motivation, *Annual Review of Psychology, 26,* 457–480.

Locke, E. A. (1976). The nature and causes of job satisfaction. In M. D. Dunnette, Ed., *Handbook of industrial and organizational psychology.* (pp. 1297–1349) Chicago: Rand McNally.

Locke, E. A., Shaw, K. N., Saari, L. M., and Latham, G. P. (1981). Goal setting and task performance: 1969–1980. *Psychological Bulletin, 90,* 125–152.

Maslow, A. H. (1959). *Motivation and personality.* New York Harper.

McClelland, D. C. (1951). *Personality.* New York: Dryden.

McGrath, J. W. (1976). Stress and behavior in Organizations. In M. D. Dunnette, Ed., *Handbook of industrial and organizational psychology.* (pp. 1351–1395) Chicago: Rand McNally.

Meyer, W.-U. (1973). *Leistungsmotiv und Ursachenerklärung von Erfolg und Misserfolg,* Stuttgart: Klett.

Meyer, W.-U. (1976). Leistungsorientiertes Verhalten als Funktion von wahrgenommener eigener Begabung und wahrgenommener Aufgabenschwierigkeit. In H.-D. Schmalt and W.-U. Meyer, Eds., *Leistungsmotivation und Verhalten.* (pp. 101–135) Stuttgart: Klett.

Meyer, W.-U. (1984). Das Konzept von der eigenen Begabung: Auswirkungen, Stabilität und vorauslaufende Bedingungen. *Psychologische Rundschau, 35,* 136–150.

Mitchell, T. R. (1974). Expectancy models of job satisfaction, occupational preference and effort: A theoretical, methodological and empirical appraisal. *Psychological Bulletin, 81,* 1053–1077.

Neuberger, O. (1974). *Messung der Arbeitszufriedenheit.* Stuttgart: Kohlhammer.

Neuberger, O., and Allerbeck, N. (1978). *Messung und Analyse von Arbeitszufriedenheit* (Schriften zur Arbeitspsychologie No. 26). Bern: Huber.

Nuttin, J. (1984). *Motivation, planning, and action.* Hillsdale, NJ: Erlbaum.

Oldham, G. R., Hackman, J. R., and Pearce, J. L. (1976). Conditions under which employees respond positively to enriched work. *Journal of Applied Psychology, 61,* 395–403.

Otway, H. J., and Winterfeldt, D. V. (1980). Risikoverhalten in der Industriegesellschaft. In C. Graf Hoyos et al., Eds., *Grundbegriffe der Wirtschaftspsychologie.* (pp. 512–522) Munich: Kösel.

Phares, E. J. (1976). *Locus of control in personality.* Morristown, NJ: General Learning.

Poulton, E. C. (1970). *Environment and human efficiency.* Springfield, IL: Charles C Thomas.

Rockwell, T. (1972). Skills, judgment and information acquisition in driving. In T. W. Forbes, Ed., *Human factors in highway traffic safety research,* (pp. 133–164) New York: Wiley.

Rosenstiel, L. V. (1975). *Die motivationalen Grundlagen des Verhaltens in Organisationen: Leistung und Zufriedenheit,* Berlin: Duncker und Humblot.

Rosenstiel, L. V. (1984). Wandel der Wertezielkonflikte bei Führungskräften. In R. Blum and M. Steiner, Eds., *Aktuelle Probleme der Betriebswirtschaftslehre* (Festschrift für Louis Perridon). Berlin: Duncker und Humbolt.

Rotter, J. B. (1966). Generalized expectancies for internal versus external control of reinforcement. *Psychological Monographs, 80.*

Rüttinger, B., Rosenstiel, L. V., and Molt, W. (1974). *Motivation des wirtschaftlichen Verhaltens.* Stuttgart: Kohlhammer.

Schneider, K., Gallitz, H., and Meise, C. (1973). *Motivation unter Erfolgsrisiko.* Göttingen: Hogrefe.

Seligman, M. E. P. (1974). Depression and learned helplessness. In R. J. Friedman and M. M. Katz, Eds., *The psychology of depression.* New York: Wiley.

Smith, P. C., Kendall, L. M., and Hulin, C. L. (1969). *The measurement of satisfaction in work and retirement: A strategy for the study of attitudes.* Chicago: Rand McNally.

Steers, R. M. (1975). Effects of need for achievement on the job performance—job attitude relationship. *Journal of Applied Psychology, 60,* 678–682.

Steers, R. M., and Porter, L. W. (1979). *Motivation and work behavior.* New York: McGraw-Hill.

Steers, R. M., and Spencer, D. G. (1977). The role of achievement motivation in job design. *Journal of Applied Psychology, 62,* 472–479.

Sulzer-Azaroff, B. (1982). Behavioral approaches to occupational health and safety. In L. W. Frederiksen, Ed., *Handbook of organizational behavior management.* New York: Wiley.

Vaassen, B. (1984). Die Bedeutung der Arbeit—widersprüchliche Ergebnisse der empirischen Wertefortschung. *Psychologie und Praxis—Zeitschrift für Arbeits- und Organisationspsychologie, 28*(2), 98–108.

Vroom, V. H. (1964). *Work and motivation.* New York: Wiley.

Weiner, B. (1972). *Theories of motivation.* Chicago: Rand McNally.

Weiner, B. (1976). *Theorien der Motivation.* Stuttgart: Klett.

Weiner, B. (1980). *Human motivation.* New York: Holt, Rinehart & Winston.

Weiner, B. (1982). An attribution theory of motivation and emotion. In H. W. Krohne and L. Laux, Eds., *Achievement, stress, and anxiety.* (pp. 223–245) Washington, D.C.: Hemisphere.

CHAPTER 2.4
LEARNING AND FORGETTING FACTS AND SKILLS

PATRICK C. KYLLONEN*
EARL A. ALLUISI

Air Force Human Resources Laboratory
Brooks Air Force Base, Texas

* Now at the Department of Educational Psychology, University of Georgia, Athens, Georgia 30602.

124

2.4.1 OVERVIEW

"Learning," wrote the theorist McGeoch back in 1942, "is a change in performance as a result of practice." It seemed a reasonable definition at the time in light of then-current theories fashioned to accommodate animal- and human-psychomotor learning. The problem with McGeoch's definition is that both its assertions are wrong in light of today's understanding of learning.

First, learning can occur without changing performance (Zimmerman and Rosenthal, 1972). Second, practice is not necessary for learning to occur. Social learning theory (Bandura, 1977; Rosenthal and Zimmerman, 1978) has provided numerous demonstrations that we can learn solely through observation or simply by being told. Though perhaps less useful, a more nearly correct statement about learning today might simply relate it to some kind of "change."

Despite the fact that we do not have a better definition of learning, we do know some things about when, how, and under what conditions learning takes place. Documenting the state of what we do know about learning is the purpose of this chapter.

This chapter is divided into four sections. The first provides an account of the human as an information processing system that serves as the basis for our discussion of learning phenomena. Although many of the major principles of learning can be discussed without reference to an information processing perspective, we believe that such a perspective can serve as an organizing framework in which apparently diverse phenomena can be seen to be related. Thus, the adoption of such a framework is not merely an indulgence in abstract theory, but rather, it can be viewed as a facilitator for the understanding of learning principles themselves.

The second section derives principles of learning and implications for training from the information processing account. In this section, we review a variety of learning situations and discuss the facilitating and debilitating conditions that affect different kinds of learning.

The third section deals with noncognitive correlates and determinants of learning and performance, with special emphasis on mood and motivation.

The final section focuses on implications for practice and explores specific applications of learning principles. There we suggest techniques for increasing the effectiveness of learning programs through changes in (1) individual learner strategies, (2) the training regime, and (3) the system that the learner will ultimately be operating. Throughout the chapter, the focus is on the kind of *complex* learning that takes place when humans acquire new knowledge and skills in training programs or operate in worklike settings.

2.4.2 THE HUMAN INFORMATION PROCESSING SYSTEM

The topic of the human as an information processing system has been explored in depth elsewhere in this handbook, and that conceptualization is adopted here in discussing how people acquire facts and skills. The information processing conceptualization is a general approach to studying human thought processes. It is not a specific theory, although researchers have formulated numerous theories that might be called instances of information processing theories.

In general, the information processing approach (like any of the specific information processing theories) views the human as an active participant in the learning process. An information processing analysis traces the flow of information typically starting with some initial perception, through a series of cognitive processes, to either an observable response by the human or merely a change in the contents of memory. For example, after viewing a picture, we might first engage in the process of recognizing the picture, then perhaps associate it with something we have experienced before; then we might attempt to store the image of the picture in an act of remembering it.

Compared to the more traditional behaviorist theories, information processing theories are more concerned with the way in which incoming information is interpreted. Because of this emphasis on what might be called *cognitive mediation,* information processing theories tend to be far more powerful and therefore better suited than behaviorist theories to explaining complex human thought processes such as those that might be invoked while learning.

Perhaps the most distinguishing feature of the information processing approach is to be found in its general characterization of the human in analogy with a computer system. Like a computer, the human can be thought of as a symbol-manipulating device consisting of (1) a more or less permanent memory containing all the data available to the system, (2) a temporary main memory holding the data currently being operated upon or processed, (3) software or programs that are essentially lists of instructions to be executed, and (4) a central processing unit that actually cycles through the instruction list. This section is organized along the lines of this metaphor, that is, by the four listed components of the human information processing system that we have relabeled (1) declarative memory, (2) working

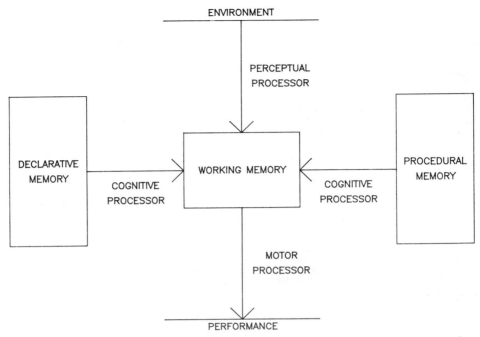

Figure 2.4.1 The flow of information through the human cognitive system. Boxes depict memories; links depict information transformation actions accomplished during the various types of processing cycles.

memory, (3) procedural memory, and (4) the processing system, respectively. Figure 2.4.1 provides a sketch of the flow of information through the cognitive system, and thus it can be used as a framework within which to organize the discussion of the system components that follows.

2.4.2.1 Declarative Memory

Declarative memory is also called long-term, propositional, conceptual, secondary, or semantic memory, depending on the context and the writer. It is the system component that contains one's factual knowledge. For many purposes, this kind of knowledge is assumed to be stored permanently, or at least to decay very slowly over time. Table 2.4.1 summarizes the key characteristics of this memory component.

Factual knowledge includes knowledge of the identity of a machine part, knowledge of what the capital of Argentina is, knowledge that 3 plus 4 equals 7, knowledge of what you did yesterday, and so forth. One of the key theoretical issues in memory research over the past decade and a half has been the question of how this knowledge is organized.

It is convenient to think of declarative memory as a network in which the nodes represent concepts and the node-connecting links represent relations among the concepts. For example, suppose you are told that "John is a janitor" and "John likes baseball." A network representation of these ideas could consist of three nodes, labeled John, janitor, and baseball, with links labeled "occupation" and "hobby" connecting the nodes.

Another feature of the organization of declarative memory, which can be represented quite naturally in the network formulation, is that information is stored hierarchically. That is, we are able to think about a concept at various levels of detail or grain sizes. We can know that it is possible to write a computer program to accomplish some data transformation feat without working out the details of how the program might really be coded.

The hierarchical characteristic of memory organization is a feature that can be exploited in designing learning materials. It has long been known that humans find it considerably easier to remember materials that have a built-in organizational structure (Bousfield, 1953; Tulving and Pearlstone, 1966), and easier yet if that structure is made quite apparent (Bower, Clark, Lesgold, and Winzenz, 1969). In learning lists, for example, if we notice that some of the to-be-remembered items are components of system A and the rest are components of system B, we find it much easier to remember the list than if we

Table 2.4.1 Characteristics of Declarative Memory

Primary function	Stores meaning of inputs
Capacity	Unlimited
Contents	Semantic codes (primarily)
	Spatial codes
	Acoustic codes
	Motor codes
	Temporal codes
Information units	Concepts
	Propositions
	Schemata/frames/scripts
Organization	Node-link hierarchy—allows default values, property inheritance, multiple levels of complexity
Learn/forget processes	Learning by being told (advice-taking)
	Encoding (encoding specificity principle)
	Storage (very slow decay)
	Retrieval

thought of the items as a series of independent entities. Textbook writers and authors of technical manuals alike take advantage of this learning principle by organizing the material by topic and subtopic.

To understand the nature of learning as it occurs in everyday settings, it is important to consider what the basic unit of information is that is represented in this network. Learning can be thought of as a three-stage process of encoding, storage, and retrieval. *Encoding* refers to the process of extracting the essential information from some input. In information processing terms, we talk about transforming an external stimulus into a set of symbols. *Storage* refers to the process of forming a copy of the symbol set in declarative memory, so that it can be *retrieved* at a later time.

In these terms, then, what is it that we encode, store, and retrieve? What is the nature of the basic unit of information? What do the symbols extracted during encoding depict? Consider the kind of learning that has to do with verbal *concepts*. At one level, we can speak of words as the basic units, and words or the concepts to which they refer can be represented symbolically as a set of basic features. The word's features correspond roughly to elements in the dictionary entry, such as its part of speech, its various definitions and uses, and examples of its use. Certainly we remember individual words and their features. Much of verbal learning research has concerned itself with identifying the conditions under which apparently simple memory acts, such as remembering a particular word and the context within which it was learned, are performed. Sophisticated models of concepts have been proposed to account for our knowledge of the meaning of various words (e.g., Smith and Medin, 1981; Tversky, 1977).

Considerable attention has been given in recent years not only to how we remember and represent words and concepts in isolation, but also to how we remember and represent the associations among various concepts. One popular method for representing relations among individual units, employed in a number of memory theories (Anderson, 1976; 1983; Anderson and Bower, 1973; Kintch, 1974; Norman and Rumelhart, 1975), has treated the *proposition* as the main building block of declarative memory. The proposition, in these theories, encodes the meaning of various utterances and written sentential units. One of the key ideas to emerge from this literature is that humans are quite adept at remembering the meaning of some input string (such as a sentence), even though some of the surface features (such as its syntax) are not retained very long. This would appear to be a fairly reliable and distinctive feature of memory: It is "wired for meaning." We retain the meaning of the pictures we view and the utterances we hear long after the surface features of those incoming experiences are lost (e.g., Bransford and Franks, 1971).

Besides concepts and propositions, a still higher order of representation of what Rumelhart and Norman (1983) have called "supra-sentential knowledge" is needed. For example, each of us has a great deal of knowledge of various mundane events such as going to a restaurant, visiting a doctor, waking up in the morning (Bower, Black, and Turner, 1979), and writing something—a letter, a paper, even a chapter or an instruction manual. This mundane knowledge facilitates communication because it allows the listener or reader to infer many of the events that occurred even though they are not explicitly stated.

Thus, if we are told that John wrote a machine assembly manual, we can infer that John described a machine to be assembled, wrote a step-by-step list of procedures to follow in the assembly, drew or had drawn some pictures corresponding to the various stages of the assembling activity, had his secretary type a draft of the manual, and perhaps even had the draft typeset. We could also infer that he wrote it either in response to his boss's wish or because he had created a machine for others to use. We are quite capable of making these inferences even though such events were not part of the original message. The memory units that enable inferences of this type have in various places been called *scripts* (Schank and Abelson, 1977), *schemata* (Rumelhart and Ortony, 1977), and *frames* (Minsky, 1975).

Two of the important properties of such memory structures are that they allow *property inheritance* and include *default values*. *Property inheritance* refers to the fact that a memory unit can inherit a property from a higher-order memory unit. For example, if we are told that John wrote an assembly manual for a widge machine, we can readily infer what John did even if we do not know what a widge machine is. We assume that a widge machine is a kind of machine, and that the properties of machines in general hold for the widge machine.

Default values refer to the fact that we readily assume certain aspects of the situation without being told them directly. For example, we can assume that the widge machine has moving parts because most machines do. We also might assume (though with less confidence) that the widge machine is made of metal, is colored gray or beige, and has an electric power supply. Although we know that these attributes of the widge machine are not necessarily true, we often find communication and learning facilitated by making the inference that they are true, unless we are told otherwise. For example, if we are told that the widge machine in question is the European model, we can draw the important implication about its power supply without being told directly that it is designed to run on 220 V, 50 Hz circuits. We do that because we assume that the machine has a power supply that runs on electricity, again without having been told that directly. The obvious advantage of such complex, well-organized representations is that they serve as shorthand expressions for a complex set of characteristics. The disadvantage is that we frequently find ourselves misled by making assumptions about items and events that are not valid.

Orthogonal to the issue of the breadth and complexity (grain size) of memory units is what might be called the content or format issue. In the preceding unit-size discussion, we conveniently referred to only verbal-like concepts and expressions. But certainly we remember information that is not so neatly put into words. We remember spatial arrangements, we remember motor/kinesthetic patterns, and we remember inputs of an acoustic nature. Although certain theorists have pushed the claim that all information can be represented propositionally, it is convenient in many applications to consider alternative representational formats.

At one time it was widely assumed that working memory (to be discussed) held information primarily in an acoustic format, whereas long-term memory held information in a semantic format. The claim was made on the basis of studies showing that short-term memory confusion errors were most likely to result from similar-sounding inputs (e.g., *d* is likely to be confused with *b;* Conrad, 1964), and long-term memory errors resulted from inputs of similar meanings. However, it has been shown that semantic errors can be made in short-term memory (Shulman, 1970), and we are introspectively aware of having long-term representations of sound and spatial patterns. Thus, it is useful to consider that both long- and short-term memories are capable of representing information in a variety of formats.

The apparent robust patterns of confusability among stimuli is probably due to the statistical relations between code types and memories. As a species, most communication, which reflects working memory processes, has occurred acoustically (widespread literacy in the population is a relatively recent trend). On the other hand, the number of ideas of which we can be aware, which reflects long-term memory, far exceeds the number of phonemes we have available. We have more to say elsewhere in our discussion of working memory about the coding issue and related phenomena such as the modality-specific interference effect.

A final feature of the organization of memory that is useful in our discussion of learning has to do with its context-sensitive nature. Numerous research studies have shown that learners' recall is best when the test conditions most closely mimic those under which the materials were orginally studied (see discussion in Tulving, 1983, pp. 223–251). In one especially picturesque depiction of this phenomenon, Godden and Baddeley (1975) had subjects study word lists while either under water or on land. Subjects recalled the words much better if they were tested in the same conditions in which they studied (about a 12% effect).

This phenomenon has been related to what Tulving has called the *encoding specificity principle*. In its most recent formation this principle asserts that ". . . recollection of an event, or a certain aspect of it, occurs if and only if properties of the trace of the event are sufficiently similar to the properties of the retrieval information" (Tulving, 1983, p. 223). Tulving has documented a wide variety of situations in which this principle has been shown to operate. Among these are learning under the influence of marijuana, learning messages spoken with either a male or female voice, and learning under conditions of imaginative processing or not. In reach case the recall of items learned under one set of conditions (e.g., male voice) was better if the test occurred under similar conditions (e.g., being cued with a male voice). Although there are documented exceptions to this principle, the effect has occurred in a sufficiently wide variety of contexts to be regarded as fairly robust.

2.4.2.2 Working Memory

Information that is the subject of current cognitive activity is assumed to be stored in *working memory*, the key characteristics of which are listed in Table 2.4.2. Current cognitive activity includes not only the focused objects of conscious attention, but also other related information that affects processing. For example, the focus of conscious attention while reading might be on the individual word or word clusters that are taken in on a single eye fixation. But it can be shown that stimuli at the periphery of the current fixation can affect processing, despite the fact that attention is not centered on the peripheral stimuli (McConkie and Rayner, 1975).

Another demonstration that peripheral working-memory elements may affect processing is the set of studies related to what had at one time been called *subliminal perception*. These studies have shown that the phenomenon of memory priming (i.e., preparing memory to process an input more easily) can be created with a stimulus about which a subject has no conscious awareness. For example, Marcel (1983) has shown that word-recognition time (i.e., the time it takes to determine whether a string of letters forms a word or a nonword) can be affected by a priming stimulus (a word shown seconds prior to the test word) that is shown so briefly that it cannot be identified (subjects show no conscious awareness of it).

These and other phenomena can perhaps be explained most easily if working memory is assumed not to be a separate storehouse of information, but rather, to be that portion of declarative memory that is currently active. Because activation is a finite quantity, and because it is assumed to spread with time, working memory is in a constant state of flux, shifting with the focus of attention and with the peripheral elements related to those foci.

At one time, working memory was thought of as simply short-term contrasted with long-term storage. Two features were believed to distinguish the two storage systems. The first was rate of decay. An item in long-term storage was assumed to decay in strength only very slowly, if at all, whereas an item in short-term storage was assumed to decay rapidly. Numerous studies (reviewed by Card, Moran, and Newell, 1983, pp. 35–39) have shown that short-term recall data can be reasonably well fit by an exponential decay function. The rate parameter of that function is a function of the number of items held in working memory: holding more items leads to faster decay.

Because the rate of decay is exponential, it is convenient to speak of the half-life of an item in working memory (i.e., the time after which the probability of successful retrieval is less than or equal to .5). The constant, 73 sec, has been used as the half-life for one item, and 7 sec for the half-life of three times (which comes close to the maximum momentary capacity; Card et al., 1983). These constants must be viewed as rough or working approximations because numerous factors determine how long we can retain short-term information. For example, if we rehearse items (as we do after the typically one-sided conversation with the directory-assistance recording), we can retain them indefinitely. On the other hand, if we attend to other items immediately after presentation of the target items, the target items decay more rapidly.

The second feature that was thought to distinguish the two storage systems was the type of code that was represented in the store. It was assumed that the short-term store recorded primarily (if not exclusively) the phonological code, whereas the long-term store recorded the meaning of inputs. These conclusions were based on studies that showed that confusion errors on memory-span tests, in which letters were presented visually, were based on the sound of the letters (e.g., Conrad, 1964).

A nice elaboration of this view was provided in a series of studies that demonstrated a modality-specific interference effect (Baddeley, 1966; Brooks, 1968; Santa, 1977). These studies appeared to demonstrate the necessity of positing two short-term memory codes: spatial and acoustic. Items of a particular code type interfere with same-code-type items more than with the other code-type items. That is, spatial items interfere with other spatial items, and acoustic items interfere with other acoustic

Table 2.4.2 Characteristics of Working Memory

Primary function	Center of all thought/learning
	Subset of declarative memory, temporary storage, in flux
Capacity	Highly limited (rapid decay)
Contents	Primarily, but not exclusively acoustic codes
Information units	Same as declarative memory
Organization	Same as declarative memory
Learn/forget processes	Interference (similar stimuli)
	Displacement (3 to 7 slots)
	Temporal decay

items, but spatial and acoustic items do not interfere as much with each other. In the Brooks (1968) study, subjects were asked to make judgments about a mental image of a block letter or about a sentence. In one condition, they made their yes–no response by pointing to *y*'s and *n*'s displayed on a piece of paper (the spatial-response condition), and in the other condition, they merely said "yes" or "no" to respond (the verbal/acoustic-response condition). It was shown that subjects were far better at making judgments about the mental image when they responded in the verbal/acoustic condition, and far better at making linguistic judgments when they responded in the spatial condition.

There have been numerous demonstrations in recent years that have complicated the concept of separability of short- and long-term stores, but the general concept still seems useful, at least for heuristic purposes. Modality-specific interference is now seen to be more likely a result of interference due to general similarity. Same-modality stimuli are more similar than other-modality stimuli and thus more likely to be interfering. This general-similarity argument applies to a wide range of phenomena, including release from proactive interference (D. Wickens, 1972). C. Wickens and colleagues (1984; Wickens and Boles, 1983; Wickens, Sandry, and Vidulich, 1983; Wickens, Vidulich, and Sandry, 1983) have recently summarized the evidence for cross-modality interference phenomena. But probably the important point for any specific application is not whether an exact featural distinction can be made for the two kinds of memory stores, but rather, whether an identified factor is likely to enhance or retard short-term decay. Decay is greater with increased time, displacement, and interference; and similar stimuli create interference.

2.4.2.3 Procedural Memory

A well-known epistemological distinction contrasts two kinds of knowledge: *knowledge that* and *knowledge how* (Ryle, 1949). *Knowledge that* is knowledge of facts, principles, relations, or what we described earlier as declarative knowledge. This can be contrasted with *knowledge how*—how to do something, such as solve a mathematical problem, answer a question, program a computer, or ride a bicycle. Artificial intelligence researchers have found the distinction useful in representing complex knowledge bases (Winograd, 1975) and researchers in human thought processes have found it useful in modeling learning and performance phenomena (Anderson, 1976).

Thus *procedural memory,* also called *production memory,* is memory for production knowledge, or *knowledge how,* just as *declarative memory* is memory for declarative knowledge or *knowledge that.* Procedural and declarative memories share many features, one of which is their relative permanence. That is, each is a kind of long-term memory characterized by slow decay over time. The main features of this memory store are listed in Table 2.4.3.

A recent study that nicely illustrates the difference between declarative and procedural memory was conducted by Schmalhofer (1982). In the study, two groups of students learned some principles of programming in the LISP computer language widely used in artificial intelligence circles, by reading through the first chapter of a popular textbook. One group consisted of computer science students who knew at least three programming languages; the other group consisted of psychology students who knew no programming languages. Neither group had had any exposure to LISP, which differs in important respects from other common computer languages. After both groups had an equal time

Table 2.4.3 Characteristics of Procedural Memory

Primary function	Permanent store of how-to knowledge
Capacity	Unlimited
Contents	Same as declarative memory
Information units	Productions (if–then rules) specific to general
Organization	Flat
Learn/forget processes	Learning by doing (practice)
	General operators (problem solving)
	Analogy
	Generalization
	Discrimination
	Strengthening

to study, their declarative memory for the text was tested—that is, both groups of students were tested on how much of the textual material they could remember.

Somewhat surprisingly, the novice psychology students remembered just about as much as the computer science students. The students then were tested for how well they could solve simple problems drawn from the material presented in the first chapter. Here, in constructing simple programs, large differences were found between the two groups: The computer science students were much better than the psychology students at applying the principles they had learned.

We have already noted that declarative memory can be represented as a node–link network structure. Procedural memory is differently represented; specifically, cognitive scientists have found it convenient to represent procedural or production knowledge in terms of *if–then* or *condition action* rules called *productions*. The entire set of such rules for a given procedure or production is referred to as a *production system*.

The *if* part of the rule contains one or more clauses that serve as the signal for whether the production can be applied or "fired" on any given cycle of the cognitive processor. If the clauses match active items in working memory, then the production fires. The *then* side of the rule contains a set of one or more "actions" taken when the production fires. An action can be either a change in the contents of working memory (i.e., the activation of a new item or items) or an actual physical action by the system such as a movement of the eyes to a different stimulus or the pressing of a response button. An example of a production rule that some of us might employ in a frustrating situation (such as our not knowing how to fix some device) is as follows:

If the goal is to fix the device, and there are no more techniques available to try,
Then hit the side of the device with a sharp blow.

This production would fire if the two conditional clauses in the *if* part agreed with items in working memory. That is, if working memory contained the two items (1) "the goal is to fix the device" and (2) "there are no more techniques available to try," the production would fire. The result would be that working memory would now contain the new assertion "hit the side of the device with a sharp blow." Then a second production that had as its condition "*if* the goal is to hit the side of some device with a sharp blow" would fire. The result of such a firing would presumably be the actual physical action of hitting the device.

The example suggests that production rules are somewhat similar to the stimulus–response bonds of traditional behavioral psychology. The stimulus is analogous to the condition part of a production rule, and the response is analogous to the action part. But there are key differences between the two formulations. First, productions need not be specific. (Indeed, in the example, the device represented in the rule did not have to be a specific device, but could be any device.) Secondly, the clauses on neither the condition nor the action side of the production have to be actual stimuli or concrete actions; either or both can be symbolic. (Thus the goal in the example production was not an actual stimulus that could be observed, but rather, an entity inferred from an understanding that human behavior is generally goal driven.) Anderson (1976) has provided an in-depth discussion of these and other characteristics of production systems versus stimulus–response systems.

Compared to learning in declarative memory, relatively little research has been devoted to understanding how learning new productions occurs. But it is likely that there are some differences in the two kinds of learning. We can imagine learning new facts simply by being told them. On the other hand, learning new procedures or new rules would seem to depend largely on practice. The more practice we have, the better we will be in exercising various kinds of cognitive and motor skills.

One of the intriguing ideas to come out of recent research on production learning suggests that what comes about as a result of increased practice is the capability for more carefully determined selections of which specific productions to apply in a given instance. Thus early stages of skill development might be characterized by frequent errors in applying productions, and later stages by few errors—that is, by knowing which situation calls for which "technique."

2.4.2.4 The Processing Cycle

Thought, as an information processing activity, takes place in real time, and so it is useful to consider the mechanisms that give thought its real-time character. Several current memory theories (e.g., J. R. Anderson, 1983; Card et al., 1983; Newell, 1973; Newell and Simon, 1972) assume that the basic unit of processing activity is the recognition-act, or production, cycle (the action of a production firing in response to satisfaction of the conditions of the production). Card and his colleagues have shown that many fairly simple cognitive activities can be modeled nicely by assuming that the time for a production cycle is about 100 msec for perceptual processing (the initial translation of an external stimulus into memory) and 70 msec each for cognitive processing (simple translations, decisions, or choices) and motor processing (actual execution of a simple motor response). These and other characteristics of the processing cycle are listed in Table 2.4.4.

Table 2.4.4   Characteristics of the Processing Cycle

Definition	Production rule—pattern matching and execution
Kinds of cycles	Perceptual Cognitive Motor
Duration	Perceptual (100 msec) Cognitive (70 msec) Motor (70 msec)
Mediating variables	Stimulus intensity/data quality Amount of effort Practice (decrease by a power law)

For example, consider the simple reaction time test in which a person is required to press a button as soon as a specific displayed signal is seen. In Card et al.'s (1983) analysis, this activity is presumed to require the execution of one perceptual processing cycle (to perceive the stimulus), one execution of the cognitive processor (to initiate a command for moving the finger), and one execution of the motor processor (to actually move the finger to the button). Summing the three cycle times gives a total of 240 msec. This is approximately what is observed in simple reaction time tasks.

There are, of course, many conditions that can cause these cycle time values to deviate from the averages. First, cycle time varies with stimulus intensity or the quality of incoming data (people react faster to visual stimuli of greater brightness or contrast, or to auditory stimuli of greater loudness or distinctiveness). Second, cycle time can be shortened by greater effort (e.g., in response to increased task demands or information loading). Third, cycle time can be shortened as a function of practice.

A wide variety of cognitive tasks has been shown (Lewis, 1978; Newell and Rosenbloom, 1981) to be well fit by a power law of the form

$$RT = aT^b \tag{2.4.1}$$

where RT is total response time for the specific task, a is the time for executing the task on the first trial, T is trial number, and b is the rate constant (a negative constant to represent the fact that time decreases with practice). Although other mechanisms can influence the response-time decreases associated with practice (e.g., different productions may apply; productions may change their nature with practice), the power law seems to hold regardless of the actual cause for the decrease in response time.

2.4.2.5   Summary: Human Information Processing System

Four components of the human information processing system have been described: declarative memory, working memory, production memory, and the processing cycle. These constitute the foundation for understanding the various learning phenomena and principles of learning presented in the sections that follow.

Declarative memory is assumed to store the more or less permanent body of factual knowledge possessed by the individual. Knowledge in this system is assumed to be organized hierarchically. It represents the meaning of inputs and is context-sensitive, and learning materials are optimized to the extent that they reflect these characteristics of the system. Likewise, individual learners tend to optimize their learning by taking into account the characteristics of this memory. For example, we benefit by being shown how the materials to be remembered are organized, and it is useful as a study technique to attempt to make meaningful even apparently meaningless material. Finally, we remember best in situations that most closely resemble the situation under which the learning took place.

Procedural memory stores the procedural knowledge of the cognitive system—that is, the knowledge of how to do things. This type of knowledge can be represented in the form of *if–then* rules called *productions*. One of the key ways in which procedural memory differs from declarative memory is in how learning occurs. In contrast with the learning of facts, the learning of procedures occurs slowly and is facilitated by lots of practice.

Working memory is that portion of declarative memory that is currently the subject of attention, or closely related to the subject of attention. Two features that characterize information in working memory status are that (1) such information decays fairly rapidly, and (2) the code is primarily acoustic, although there is some semantic and spatial coding involved in certain contexts. Some strategies under the learner's control, such as rehearsal, can slow down the rate of decay in working memory. On the other hand, interference caused by same-modality stimuli, or stimuli similar in other respects, can increase the rate of decay and thereby interfere with learning.

The *recognition-act, or production, cycle* is the mechanism that governs the rate of information processing. For many purposes it can be assumed that the most elementary cycles take approximately 70 to 100 msec, but cycle times vary with such factors as stimulus intensity, processing effort, and practice. Elementary cycles are additive, with three involved even in a simple reaction task (perceptual, cognitive, and motor processing cycles).

2.4.3 VARIETIES OF LEARNING

For much of the history of psychology, learning research focused on how people memorize lists of words, nonsense syllables, and associated pairs of such stimuli. But more recently, learning phenomena have been reexamined with a new emphasis on how people acquire complex cognitive skills such as the ability to type, to program computers, or to understand complex domains such as social science and radiology. Several recent books (e.g., Anderson, 1981; Chipman, Segal, and Glaser, 1985; Segal, Chipman, and Glaser, 1985), as well as a new journal (*Cognition and Instruction*), exemplify these trends. New studies have examined complex learning topics such as learning to solve geometry and physics problems, and learning to use a text editor. Moreover, new proposals have surfaced for characterizing the variety of learning acts that an individual may perform.

With this newly expanded focus on complex (as well as simple) learning processes, it is useful to consider the broad classes or varieties of learning. A preliminary taxonomy that the organization of this section reflects is the trimodal proposal of Rumelhart and Norman (1978).

Specifically, the trimodal taxonomy recognizes three kinds of learning—*accretion, restructuring,* and *tuning. Accretion* refers to the acquisition of facts in declarative memory. *Restructuring* refers to the initial acquisition of procedures in procedural memory. *Tuning* refers to the process of modifying existing procedures in procedural memory to make such procedures quicker and more reliable. Characteristics of these three types of learning are listed in Table 2.4.5. It should be noted that the trimodal

Table 2.4.5 Characteristics of the Three Types of Learning

Accretion

Memory affected	Declarative
Type of learning	Concept/relation acquisition through encoding (learning by observing, being told, taking advice)
Learning processes	Stimulus discrimination Response learning Association
Learning strategies	Elaboration, interactive imagery, recoding, method of loci, pegword method, PQ4R, networking
Other variables that affect the process	Stimulus concreteness, distinctiveness, and pattern Study-test similarity

Restructuring

Memory affected	Procedural
Type of learning	Production acquisition through practice
Learning processes	Compilation Proceduralization Composition
Learning strategies	Use general operators (problem-solving methods); use analogy
Other variables that affect the process	Working memory load Feedback immediacy and reliability Number of positive experiences Number of negative experiences (failures)

Tuning

Memory affected	Procedural
Types	Production modification through practice
Learning processes	Generalization Discrimination Production strengthening
Learning strategies	Feedback guidance from external source (tutor) Feedback guidance from internal model
Other variables that affect the process	Mapping consistency Same as restructuring learning

taxonomy is used here as a beginning concept about which the section is organized, but that does not imply that all of Rumelhart and Norman's (1978) delineations have been followed. Rather, although the three kinds of learning are considered separately, particular attention is paid to the conditions that enhance and retard each.

2.4.3.1 Accretion

Accretion refers to the process of acquiring new facts or relationships, or augmenting memory with a record of new events. In terms of the information processing framework we outlined in the preceding section, accretion amounts to the process of transferring (or copying) a simple memory structure (a small set of related nodes and links) from temporary storage in working memory to permanent storage in declarative memory. Many everyday memory acts fall into this category: learning a telephone number, learning to associate a name with a face, learning (or committing to memory) one's scheduled activities for the day, or learning a list of groceries to be purchased. In general, most of the learning activities that may be described as memorizing can be classified as learning by accretion.

Much of traditional learning research in psychology has dealt primarily with accretion, and so it is this kind of learning about which psychologists know most. The classic accretion-learning situation is that in which subjects study word or nonsense syllable paired associates. A three-stage model that distinguishes the stages of stimulus discrimination, response learning, and association has been used to characterize such associative learning. Each of these stages has been found to affect the probability that a particular pair can be memorized.

The Three-Stage Model

Stimulus discrimination refers to the process of differentiating between the various stimuli or classes of stimuli to which responses are required. Stimuli that are more difficult to differentiate from one another are generally harder to learn. The previously cited Conrad (1964) study of acoustic confusions in short-term memory demonstrated this general phenomenon. Specifically, Conrad found that a memory-span test for letters was harder if the letters shared phonemes (e.g., d, b, c, t) than if they did not (e.g., a, m, o, k). Sperling, Parish, Pavel, and Desaulniers (1984) have used a similar interpretation in explaining the fact that we can remember longer strings of digits than of letters; that is, digits tend to share fewer phonemes than letters.

Response learning refers to that part of the associative-learning process that has to do with unitizing the response to the stimulus. Because the response itself must be learned, it can be considered as a serial list to be memorized. Thus if the response pattern is complex (e.g., if it is a multipart response), learning is more difficult.

Association is the process of finding a link to connect two stimuli (or a stimulus with a response). Earlier memory models assumed that the probability of forming an association between two items varied directly with the amount of time the two items were together in working memory (e.g., Anderson and Bower, 1973). Some theorists included the notion of intentionality. However, many studies have shown that both factors may be superfluous. For example, there is evidence that the memorability of stimuli may be related not to intention to learn, but rather, to the amount of processing that is done on inputs (Craik and Lockhart, 1972). Other studies have shown that duration in working memory predicts recall only if the item is actively processed (Bradley and Glenberg, 1983). These highly reliable empirical findings defy our intuitive theories of memory: Many of us believe that if we concentrate hard and rehearse at length ("burn it in"), we will remember better. But the empirical evidence collected over the past decade or so tells us a different story: We learn better when we actively attempt to relate arbitrary stimuli into a meaningful pattern.

Encoding Strategies (Mnemonic Devices)

The fact that human memory is not merely a repository of uninterpreted incoming stimuli, but rather, an active processing system, has given rise to our use of countless memory tricks or what are called *mnemonic devices.* These are procedures that a person can use to increase the probability that a studied pair of stimuli or list of items will be recalled. Books have been written to teach such techniques (e.g., Belleza, 1982; Furst, 1958; Higbee, 1977; Lorayne and Lucas, 1974; Yates, 1966).

Several well-documented mnemonic techniques have passed the scrutiny of empirical validation. One class of techniques might be called *encoding strategies.* The common thread uniting these apparently diverse methods is that they require the user to translate arbitrary, unconnected, low-meaningful material into highly meaningful, interactive information.

Perhaps the simplest encoding strategy is the *method of elaboration* using interactive sentences. Suppose the task of a novice LISP programmer is to remember that the LISP function CAR returns the first element in a list. This can be treated essentially as a paired-associates learning task in which CAR is the stimulus term and *returns the first element in a list* is the response term. One way to commit the pair to memory would be to create a sentence using the two parts: "The import CAR salesman *returned the very first* Fiat he received in the shipment because it was the wrong color."

A closely related technique employs *interactive imagery* to connect two items. It has been shown that lists of concrete nouns are much easier to remember than lists of abstract nouns because the former are easier to image. Indeed, the concreteness of a noun's referent is one of the best predictors of its memorability (Paivio, 1971). It is likely that one of the reasons why imagery works so well as a mnemonic device is that it makes memories more distinctive. Our memory for pictures is much better than our memory for words because pictures stand out more distinctively from other pictures than words do (Shepard, 1967).

It is not just imagery per se, however, that makes items more memorable. Bower (1970a) has shown that a pair of items can best be remembered if the imagery is interactive rather than separate. Learners instructed to image referents to noun pairs vividly did not recall the nouns any better than did a control group instructed merely to rehearse the nouns. However, subjects instructed to image vividly the paired nouns interactively, recalled substantially better (by almost a factor of two: 30 to 53% recall). This effect is explained by the network model of memory discussed earlier. That is, when treated as simply a task to remember the names for two concrete objects, subjects have to encode and store two separate memory traces. But when given interactive instructions, subjects integrate the two objects into a sentence (by instruction) of the form "actor–action–object." Thus the memory task is reduced to the encoding and storage of a single memory trace, rather than two. And that is essentially a smaller-scale memory task.

Recoding Strategies (Mnemonic Devices)

A second type of mnemonic device is one in which the to-be-remembered item is recoded to a form that makes it more accessible. This *recoding strategy* can best be illustrated and is most useful with inherently meaningless memory items such as digits (e.g., memorizing telephone numbers). The strategy is to set up a recoding table in long-term memory. One application of this strategy is to associate certain consonants with each of the nine digits ($1 = d$ or t; $2 = n$; $3 = m$; $4 = r$, etc.) and commit them to memory so well that they can be applied almost effortlessly. When presented with a digit string, the user translates each of the digits into its associated letter and then attempts to make words out of the resultant consonants, adding vowels where necessary. Anderson (1980, p. 124) has described using this strategy to memorize the telephone number of the White House (202-456-1414) by translating it into the consonant string, "n-z-n-r-l-sh-t-r-t-r." In turn, this letter string can be padded to produce the words, "nose on a relish traitor." Then, by imagining some interactive image involving the President punching the nose of someone caught illegally exporting relish secrets, the user of this strategy can easily recall the image and construct the digit sequence.

Techniques like this require the user to commit to memory a usable translation scheme such as the digit-letter association used in the example. But there are variations, and individuals have been known to employ already established idiosyncratic coding schemes. For example, Chase and Ericsson (1981) described one individual who increased his memory span to 81 digits by using the sophisticated translation scheme of translating digits into running times (e.g., 3–5–9 might be translated to 3 min 59 sec). Because the individual knew a great deal about running times (he was a practicing runner), this translation produced a more meaningful code for him. Similarly, Hunter (1977) documented the technique of the mathematician Aiken, who, when presented with a digit string to be memorized, would translate the digits into number facts (a great many of which he knew).

Why Do Recoding Strategies Work?

There are several nonconflicting explanations of the way in which such techniques work. First, they tend to increase the meaningfulness of the materials to be remembered, and meaningfulness has long been known to aid memory and retrieval. Secondly, they provide more entry points for retrieval. That is, when an elaborated connection is established between two items to be associated, entering the network structure at any of the nodes in the elaboration provides a pathway back to the item nodes that constitute the memory task. For example, if you now try to recall what the example telephone number was, and whose it was, you will probably be able to recall only some of the features of the elaboration, but not the number: It was the White House telephone number, and had something to do with the president and a thief (aha, the "nose on a relish thief," no, the "nose on a relish traitor"). Had the associated digit-letter translation table been committed to memory, it would take little or no effort to reproduce the desired digit string, 202-456-1414.

The Method of the Loci

Unfortunately, the recoding techniques do not always lead the user reliably to converge back to the original to-be-remembered inputs. An alternative and very old scheme makes use of the human's excellent abilities to perceive (and image) visual space. It is widely known as the *method of the loci*, and was employed by the ancient Romans for memorizing lists of items. To use this method, one imagines a familiar walk, such as a walk through one's own house or around the block. Each landmark in the walk serves as a focal point around which to imagine the successive items in the list to be memorized.

For example, if the task were to remember a grocery list with items such as bread, milk, and potatoes, then one might imagine bread sitting at the front door (the first stop on the walk), milk spilled on the floor in front of the hall closet (the second stop), and potatoes being mashed in the front hall (the third stop). When it is time to recall the items, the user merely takes the mental walk, recalling the items as they are spotted along the way.

One can see that this method works by providing definite and well-learned memory nodes from which to search for the target items (cf. Bower, 1970b). Chase and Ericsson (1981) have shown how such a *retrieval structure* can facilitate recall in numerous situations through the use of a variety of nmemonic strategies.

For example, the *pegword method* is a similar (isomorphic) technique. With it, the user first commits to memory an ordered list of pegwords (a bun could be first, a shoe second, a tree third, etc.). Then the pegwords are imaged with the items to be remembered: a bun beside a loaf of bread, a shoe with a milk bottle in it, and a tree with hanging potatoes as fruit. In order to recall the list of groceries, one retrieves the pegwords (bun, shoe, and tree), the associated images, and the list of grocery items with the pegword images—bread, milk, and potatoes.

Other Factors Affecting Accretion

Thus far we have concentrated on mnemonic strategies for the user, but the items to be remembered or other aspects of the memory situation itself can also affect the probability of recall. Stimuli (words, sentences, stories) that are more concrete (Paivio, 1971), more interesting (R. C. Anderson, 1983), more distinctive (Von Restorff, 1933), or arranged in a nonlinear spatial pattern (Bellezza, 1983) are more likely to be remembered than are stimuli that do not have these characteristics.

As was discussed earlier, all these factors are moderated by the degree of similarity in the study and test situations. The classic study here was conducted by showing that subjects performed much better on a final examination if they were tested in the same room where they studied (Abernathy, 1940). This has practical implications for how to study for the Graduate Record Examination (GRE) or any kind of licensing examination: namely, one should try to study under the same or as-similar-as-possible conditions as those under which the test will be administered (and even in the same room, if possible).

Learning to Forget

Forgetting is often viewed as merely the flip side of learning, and in many ways this is true. For example, to the extent that incompletely learned materials are more easily and quickly *forgotten* (i.e., not recallable), one way to forget is to employ a study method at odds with those recommended, thereby minimizing the effective learning. However, long-term forgetting also occurs in other ways. One is simply through decay over time. Without reviving a memory trace, the strength of that trace (i.e., the ease and probability of its being recalled) will decrease approximately as a power function of time (see Equation 2.4.1). If you have not seen or thought about your friend Fubre lately, you might very well have forgotten Fubre's middle name, even if you did know it at one time. Another way in which forgetting occurs is through interference. Indeed, many of the conditions that enhance learning can be seen as means for overcoming interference. Making a to-be-remembered item more meaningful, more distinctive, or more vivid are ways of separating the item—of making it stand apart from the millions of facts that we routinely have stored away in memory.

Textbook and Lecture Learning

To complete this discussion of learning by accretion, two additional techniques must be reviewed. These are methods for learning meaningful text and lecture materials. One is the *PQ4R method* for studying chapters in textbooks (Thomas and Robinson, 1972; similar to the *SQ3R method,* for *Survey, Question, Read, Recite, Review*). This technique prescribes that the learner first *preview* and generate *questions* about the material; then *read* it carefully, *reflect* on what it is saying while it is being read, and *recite* as much of the material as possible; and finally, *review* the material, paying particular attention to the questions that were developed before beginning the reading of it.

The PQ4R method incorporates many of the principles developed here, especially insofar as it encourages the active participation of the learner in extracting the meaning of the material. It may be that the process of generating and then answering questions actually promotes deep meaningful processing of the material, and is responsible for the method's success (Anderson, 1980, p. 219). The technique of continually generating hypotheses about what the meaning or outcome of a textual unit will be has been shown to promote dramatically enhanced understanding and recall of the material, despite the fact that the reader may generate erroneous hypotheses along the way (Rumelhart, 1981).

The general principle (of engaging the student in the process of questioning and hypothesizing about the information to be learned) is now believed to be quite an important tutorial device. It is showing up in suggestions for teaching (e.g., Collins and Smith, 1982) and in the design of a number of intelligent tutoring systems (for examples, see Sleeman and Brown, 1982).

The second method—not quite as complete or as well specified as the PQ4R—is the *method of networking*. It is especially useful for organizing information presented in the form of a lecture. A technique many students employ when listening to lectures is that of attempting to construct an outline of the material presented. Networking is essentially an elaboration of outlining—removing the inherent hierarchical constraints of the outline and encouraging the depiction of relations between the ideas presented. Thus the listener attempts to draw a network that represents the ideas and their associations to each other and to other knowledge the listener has. Although the technique has not yet been widely validated, some studies have shown it to be a sound memory enhancer (Bethell-Fox, 1983). It appears that the technique works by virtue of its forcing the listener to organize the material, and to process it actively and relate it to things that are already known. Thus it forces the learner to become an active participant in bringing meaning to the material to be studied.

2.4.3.2 Restructuring

Restructuring refers to the act of acquiring a new procedure or set of procedures for dealing with novel situations. In our information processing terminology, restructuring is the process of creating a new procedure or production (or set of productions) and adding it to production memory. Acquiring a production is tantamount to learning a new rule for handling some situation.

Although Rumelhart and Norman (1978) restricted their definition of restructuring to the acquisition of complex skill over the long term, we find it useful to define it here more generally as *skill learning through rule acquisition* or, therefore, as the *augmentation of production memory*.

An example of restructuring was described earlier in the context of learning to program in the LISP computer language. Other examples are learning to operate a machine (such as a word processor or an electric blender), learning to respond to a changing display (such as the monitor in an air traffic control environment), or initial learning of a motor skill (such as typing).

Compared to accretion, restructuring leads to a much more gradual change in memory, and in the subsequent use of that memory. We simply do not instantly acquire procedural skills the same way we (almost) instantly acquire facts.

We have all had the experience of acquiring a new psychomotor skill, such as properly swinging a golf club. At first, after initially being instructed on how to do it, exercising the newly acquired skill feels awkward. We execute it slowly and erratically. We check our every move. We try to make sure that what we are doing matches what we were instructed to do. With much practice we begin to execute the skill more smoothly. We become less conscious of the individual steps in the procedure. We may even lose the ability to verbalize exactly what it is that we are doing.

Then, during the final stages, after we have acquired considerable proficiency, the skill becomes "second nature." We can execute the desired movements accurately, smoothly, and effortlessly, perhaps even without being fully aware of what it is we are doing. Fitts (1964) characterized this process of skill acquisition with a three-stage model in which skill progresses from a *cognitive stage,* through an *associative stage,* to an *autonomous stage.*

The principal characteristic of the cognitive stage is that the learner is aware of the verbal description representation of the skill. At the associative stage, the learner applies the skill more reliably and efficiently, ironing out the bugs in the procedure; but each skill execution step is consciously tied to a verbal statement about what it is supposed to be done. At the final, autonomous stage, the learner begins to lose awareness of the individual steps that make up the skill and continues to refine the skill to the point where active attention is no longer required for smooth performance.

A similar three-stage model has been proposed to explain the acquisition of cognitive skills, such as learning to troubleshoot electronic equipment, to solve geometry problems, or to program computers (J. R. Anderson, 1983; Neves and Anderson, 1981). This model proposes that cognitive skills progress from a *declarative stage,* through a *knowledge compilation stage,* to a final *production-tuning stage.*

At the declarative stage, the skill components are represented as declarative facts that are processed by general productions. At this stage, the skill is executed step by step. At the knowledge compilation stage, the declarative step-by-step representation begins to be replaced by a more specialized procedural representation. The skill facts (the step-by-step instructions) are essentially transformed into procedures at this point. Then, during the tuning phase, as the skill (represented as a smooth production set) is tailored to be applied more and more reliably, its declarative representation may completely disappear.

The Declarative Stage

J. R. Anderson (1983) has proposed two general methods for skills to be initially acquired during the declarative stage. One is the *use of general operators* applied to a declarative representation. Here, the learner typically studies a text and commits various facts to memory. For example, in the previously cited Schmalhofer (1982) study, subjects studied Chapter 1 of the LISP programming text. They committed to memory various facts about the data structures in the language, along with the workings of a few elementary functions. However, the students' knowledge of the factual material did not guarantee that they could solve problems using that material.

J. R. Anderson (1983) has proposed that to solve such problems, it is necessary that the learner apply a set of general-purpose productions that operate on the newly acquired declarative information. General-purpose productions are those that can be applied across a wide variety of problem-solving situations, such as mathematical puzzles, language comprehension, and spatial-relations problems. Examples of such general-purpose rules are (1) break the problem down into smaller problems, (2) determine what happens in the limiting case, (3) use a method that has generally been successful in the past, and (4) try, one at a time, each of the set of methods presented in the text for finding such relations. We all have storehouses of handy heuristics such as these for dealing with novel problems.

One of the problems associated with the acquisition of a cognitive skill in this fashion is that it severely strains working memory, thereby causing frequent errors and slow performance. The working-memory burden is essentially a function of the learner's having to keep active a lot of newly acquired declarative information. One way of reducing some of the burden is by making the application of the general productions more automatic and effortless.

There are numerous courses offered at universities (typically in engineering departments), as well as books written for the layperson, on the general procedures for solving problems. Especially useful in this regard is Wickelgren's (1974) book, *How to Solve Problems,* in which many of these general methods are described. Although firm evidence on the utility of these methods is hard to come by, Wickelgren at least provides anecdotal evidence that practice with his general methods improves problem-solving ability. This suggests that such kinds of practice should facilitate skill acquisition as the first step in the acquisition of a new cognitive skill.

The second of J. R. Anderson's (1983) general methods for initial skill acquisition involves the *use of analogy.* Here, after having read the initial textual material, the learner attempts to solve test problems by finding similar problems, the solutions of which are known. Thus by appropriate substitution (or analogy), the student is able to solve the test problem.

The utility of analogy as a means for acquiring skills is currently disputed. On the one hand, using analogies to solve problems can be a trivial expedient and one that might not carry the learner any closer to true understanding of the skill. Indeed, Anderson, Farrell, and Sauers (1984) argue that students ought to be discouraged from using analogy, and instead encouraged to use the general-operator method.

On the other hand, there may be many situations for which general-operator solutions are simply too difficult to come by. In these cases, it may be that analogy is the most effective teaching tool available. In this regard, Rumelhart and Norman's (1981) discussion of the characteristics of a good analogy can be summarized in four principles. The first is that the analogy should be made to a domain in which the student already has a sufficient degree of knowledge. For example, Gentner (1983) attempted to instruct students on the properties of electrical current by relating these analogously to water flow. Unfortunately, she found out later (after the teaching method met with substantially less-than-expected success) that her students really did not come to the instructional situation with a very good understanding of the principles of water flow.

The second principle is that the target and source domains should differ by a minimum number of dimensions. What often seems like a brilliant analogy to a teacher may be completely opaque to the student. The two systems involved in the analogy may differ in so many respects that the student does not see any relations that will help in understanding their similar nature.

The third principle is that operations that are natural in the target domain should also be natural in the source domain. Rumelhart and Norman discussed the problems of teaching fractions to school children using the analogy of a sliced (i.e., fractionated) pie. With such an analogy, the concepts of adding and subtracting fractions fall out quite naturally. However, the analogy is of no use in trying to teach fractional multiplication and division. (What is a half a slice times a third of a slice?)

Finally, the fourth principle is that inappropriate operations should be inappropriate in both target and source domains. Adherence to this principle reduces the probability that the learner will adopt an inappropriate operation or a wrong feature from the analogy.

The Knowledge Compilation Stage

After the initial problem-solving experience with newly acquired declarative knowledge, knowledge pertinent to the development of the cognitive skill is said to become compiled. During this stage of cognitive skill development, skill execution becomes smoother primarily because more and more of the information that previously had to be maintained in working memory is now permanently stored in the form of more special-purpose productions. Problem solving in the specific domain becomes less a matter of applying factual knowledge, and more a matter of calling upon already learned procedures. This represents the very beginning of the development of expertise. The learner begins to recognize situations for which routine assemblies of the actions that make up the total situation response have already been stored. That is, the learner begins to recognize characteristic response requirements for common situations, rather than having to search memory for possible ways to respond to each specific situation that arises.

J. R. Anderson (1983) has discussed two steps in the compilation stage: composition and proceduralization. Both can be understood as processes that operate on the production rules in production memory.

Composition is the process of combining two temporally contiguous productions with a common goal into a single production. To illustrate, consider the following example, based loosely on a production system developed by Polson and Kieras (1984) for deleting a word in a manuscript editing task. Suppose that after performing the delete task a few times, the learner now has stored the following two productions that perform the task:

Production 1.   *If* the goal is to delete <a specific word>, and the word begins at <x, y>, and the cursor is not at <x, y>,
Then move the cursor to <x, y>.

Production 2.   *If* the goal is to delete <a specific word>, and the word begins at <x, y>, and the cursor is at <x, y>,
Then press the delete button.

After applying these two productions sequentially, composition may cause a third production to be formed as follows:

Production 3.   *If* the goal is to delete <a specific word>, and the word begins at <x, y>, and the cursor is not at <x, y>,
Then move the cursor to <x, y> and press the delete button.

In this case, composition has essentially replaced a two-step operation with a one-step operation. One of the advantages of the one-step operation, of course, is speed. If we assume that productions are executed at a more or less constant rate, then the composed production solution will be twice as fast as the two-step solution (recall our earlier argument that the production cycle is the basic metric of information processing speed). However, the ability to introspect about the nature of the individual steps in the procedure is lost with composition. That is, once production 3 is acquired, the learner begins to lose access to the individual steps and therefore is unable to say precisely what is being done. At this point, the learner might not be a very effective teacher for a novice. At different times, many of us have been on both sides of an interchange in which the teacher is trying to tutor a composed skill. The teacher says, "If you see this, then just do x, then y, then z." The student replies, "But how do you do x?"

The example also reveals a key ingredient for the promotion of composition learning. That is, for productions to be composed, they must be united by a common goal (in this case the goal of deleting a word). Consider what would happen if the two productions were not executed in sequence, but rather, were interrupted by the application of a fourth production that did not have this goal, such as the following (based on a similar example by J. R. Anderson, 1983):

Production 4.   *If* the boss taps you on the back and asks you a question,
Then answer the question.

We would not want composition to combine productions 1 and 4, because that would produce the useless production:

Production 5.   *If* the goal is to delete <a specific word>, and the word begins at <x, y>, and the cursor is not at <x, y>, and the boss taps you on the back, and asks you a question,
Then move the cursor to <x, y> and answer the question.

This would slow down learning. Thus the principle is that to promote learning it is useful to allow the learner to work with a consistent goal set. One way to accomplish this is by avoiding the application of data-driven productions (such as the one described)—for example, by setting conditions so that interruptions are less likely. Many mathematicians and statisticians seem to have acquired this principle implicitly, as demonstrated by their having set up their work areas in ways and places that minimize the probability of distractions. A favorite technique is to use low lighting in a quiet study, and to rely nearly exclusively on a narrow-beam desk lamp for light. That way, one tends to minimize the opportunities to be interrupted by distracting data (such as the sight of a monograph on the bookshelf that one has been dying to read).

In the laboratory, this principle has been shown to be the most important determinant of whether a skill becomes automatized (i.e., able to be applied with little effort and low cost on the processing system). Schneider and Fisk (1982) found that skills exercised in a consistent mapping condition (i.e., a condition in which specific patterns are consistently related to specific responses) become automatized after very few trials. On the other hand, skills exercised in a varied mapping condition (i.e., a condition in which stimuli and response pairings are changed frequently) may lead to the prevention of process automaticity.

Finally, mention should be made about the time course of learning by composition. If composition were the only learning mechanism, one would expect that learning should follow an exponential rather than a power law. Also, if two productions were composed after every cycle, the rate parameter should equal one-half. The power function essentially represents the fact that the rate of change slows over time (Newell and Rosenbloom, 1981); that is, composition is less efficient after more and more practice. Neves and Anderson (1981) provide a tentative explanation for why this should be so, and they also show how composition can produce power-law learning.

Proceduralization works hand in hand with composition to make productions more powerful and therefore to accelerate the development of cognitive skill. The essence of proceduralization is to replace variables with constants in production clauses. This is a means for putting declarative knowledge directly into the production. Consider the earlier example for deleting a word. The positions of the cursor and the target (the x, y coordinates) are two variables. In order for production 1 to be selected, the subject must retain in working memory both the x, y location of the cursor and the x, y location of the target word. However, suppose that a common delete operation occurred when the word was located at a specific x, y location such as 1, 1. Further suppose that a special function key was available for getting to 1, 1. Then, through compilation, the subject might acquire a production that replaced the two variables $<x, y>$ with the constants $<1, 1>$ and $<$the 1, 1 key$>$, respectively.

The learning-enhancement principle from this example is much the same as that from composition: Consistent practice promotes learning. If the editing task only infrequently required the learner to perform the editing operation at location $<1, 1>$, then this production might never be proceduralized. Practice on specific tasks is most likely to lead to proceduralized productions.

2.4.3.3 Tuning

Tuning is the process of making existing procedures more efficient by changing (restricting or enlarging) the number of situations in which they apply, or just by making them apply more reliably in appropriate circumstances. Thus, tuning is related to and logically follows from restructuring. However, the characteristics of tuning and restructuring are sufficiently distinct to permit us to discuss them separately.

Examples of *learning through tuning* are typically found in activities characterized by the maxim, *practice makes perfect.* Improvements in motor skills (such as typing) or cognitive skills (such as pattern detection or aircraft recognition) that result from hours or even years of practice qualify as tuning. Such improvements continue on long after the skill is initially composed and proceduralized.

Because procedures can be represented as production sets, it is useful to consider tuning as the modification of individual productions. Productions can be made more broad through the *process of generalization.* They can be made to apply in more restricted situations through the *process of discrimination.* They can be made to apply more reliably simply through practice or exercise and the concomitant *process of production strengthening.*

Although these three tuning mechanisms (generalization, discrimination, and strengthening) are familiar to the traditional behaviorist, they can also be shown to be quite easily adaptable and subsequently much more powerful in an information processing account. J. R. Anderson (1983) has described how productions can be modified in these three ways.

Generalization

By means of generalization, the conditions under which a production applies can be extended in two ways. One is by deleting a test in the *if* side of the production. For example, the previously cited delete-word production set could be generalized to a delete-anything (word, string, number) production if the test for whether the target is a word were removed.

The second means for accomplishing a generalization is a reverse of composition: Change a constant to a variable. For example, if the student were provided with production 5 instead of first learning the general production 1, then the student would tend with practice to generalize production 5 to form production 1.

Discrimination

Discrimination is essentially the reverse of generalization. The learner modifies existing productions by attaching an additional test clause on the *if* side. The classic and most transparent case of discrimination occurs in early language acquisition. Children learn general rules such as "attach *ed* to represent a past action." However, with practice and some failures (e.g., "goed," "hitted," "runned"), the child learns to be more discriminating in his or her application of the *ed* rule. The discrimination process produces an extra test to ensure that the verb in question is not an exception to the general rule.

Strengthening

Production strengthening is the means by which individual productions come to be called upon more reliably in target situations. With more and more practice in executing a specific production, the learner becomes more and more reliable in applying that production in specific circumstances.

However, there can be debilitating effects of production strengthening as well. Strong productions, like old habits, may be hard to extinguish. J. R. Anderson (1983) has proposed that each time a production is successfully applied, its strength is increased by a unit amount. If the production application results in a faulty response (or some other form of negative feedback), the strength of the production is decreased. In computer simulations of problem solving, Anderson used a decrease of 25% in the production's strength.

Practice and Feedback

In all three forms of tuning, the most important determinants of learning success are *practice* and *feedback*. The learner must practice for tuning to occur, and feedback must follow the exercise of productions for the learner to determine whether newly formed productions are useful to retain.

In Anderson's system, generalization and discrimination occur more or less indiscriminately. Strengthening is the mechanism that provides us the useful productions we desire, and strengthening is controlled principally by feedback. Feedback can be produced by an external source, such as a tutor, but it can also be produced internally. That is, a learner can be guided by an internalized model of what skilled performance is like. The model can serve to provide intrinsic feedback. Indeed, this kind of feedback is an important contribution to the effectiveness of learning by doing (Anzai and Simon, 1979).

2.4.3.4 Summary: Varieties of Learning

This section on the *varieties of learning* has presented a taxonomy of learning types, with particular attention paid to the conditions (of both the learner and the learning situation) that affect learning and subsequent performance. Following a proposal by Rumelhart and Norman (1978), three kinds of learning were discussed: *accretion, restructuring,* and *tuning.*

Learning by accretion is facilitated by conditions that make use of the general findings that meaningful material is easier than arbitrary material to learn. Thus this kind of learning is facilitated when the learner imposes or extracts meaning from the material studied, and several mnemonic devices and strategies for doing this were described. Likewise, the teacher, trainer, or tutor can take advantage of this principle by arranging the learning situation so that the meaningfulness and organization of the material is obvious to the learner. Remember, meaningful material that is organized and evokes concrete imagery is fastest learned and longest remembered.

Learning by restructuring was broken down for convenience into two sequential stages as suggested by J. R. Anderson (1983). In the first stage, the learner solves problems after committing relevant facts to memory. Problem solving is accomplished either by the use of *general-purpose procedures* or by analogy. A learner can improve problem-solving abilities by studying standard general-purpose procedures and problem-solving methods, practicing them, and applying them in a variety of situations. For *learning by analogy,* different strategies were described for the tutor to use to create more useful analogies. In the second restructuring stage, the learner compiles factual knowledge into procedural knowledge through the operations of *composition* and *proceduralization.* In the case of both these learning mechanisms, consistent environments without interruptions tend to facilitate learning.

Learning by tuning is the final stage of the learning process. It continues long after skills are initially acquired. Two mechanisms, *generalization* and *discrimination,* were described in the context of *tuning* or *tailoring productions.* Additionally, the mechanism of *strengthening* was shown to interact with the other tuning mechanisms. The most important determinants of tuning success are *practice* by the learner in the presence of *feedback,* which can be provided either intrinsically to the learning task or by an external agent such as the teacher, trainer, or tutor.

2.4.4 NONCOGNITIVE LEARNING VARIABLES

The foregoing sections of this chapter have concentrated on the cognitive correlates and determinants of learning. Recent findings regarding the noncognitive variables that affect the learning process are reviewed in this section.

Although recent books have examined the role played by affective and conative variables in learning and cognition (e.g., Clark and Fiske, 1982; Snow and Farr, in press), systematic research programs in this area have only recently been launched. Relatively little is known as yet about the noncognitive side of learning, and that which is known appears to fall into two categories of research: (1) that on the conative and motivational determinants of learning, and (2) that on affective aspects, such as the effects of mood states on learning and memory. Thus, *motivation* and *mood* are the major headings that follow.

2.4.4.1 Motivation

To the average person on the street (or the average manager on the line), the most important determinant of learning success would seem to be *motivation*. Yet, surprisingly, numerous laboratory studies have shown the intention to learn (one form of motivation) does not directly affect learning success. Rather, the amount of processing of the input stimulus is more directly related to the probability of the stimulus is being remembered.

For example, Craik and Lockhart (1972) found that if subjects were instructed to make a judgment about the category of a word, such words would be remembered on a surprise test given at the end of the session much better than if subjects were told to make a more superficial judgment (e.g., of whether the word was upper- or lowercase). Likewise, Woodward, Bjork, and Jongeward (1973) found that the amount of time an item is simply rehearsed (maintenance rehearsal) does not affect the probability that it will be recalled. Thus, the data suggest that to learn, it is not enough to merely *want* to learn something; rather, it is equally (or more) important that learners know how to study, and study that way.

Yet, intention or motivation to learn can affect whether a stimulus is even attended to in the first place, and whether the subject will persist in trying alternative methods for studying (or to figure out) the stimulus. Gitomer and Glaser (in press) found that one difference between skilled (i.e., successful) and less-skilled (i.e., unsuccessful) problem solvers is that when confronted with an unusual task (or a standard task presented in an unusual format), the skilled performer tries different approaches until finding one that works. The less skilled individual tends to give up more easily and more quickly.

However, the most important motivational consideration for many applications has to do *not* with determining who is motivated, but rather with determining *how the learning task can be made more motivating*. We know that more interesting learning materials motivate, and thus, it is not surprising that interest is a powerful predictor of what will be remembered. R. C. Anderson (1983) has estimated that in normal text, *interest* (defined by subjective ratings) is 30 times more important than readability in predicting recall.

Another way of determining what is motivating is to look at the characteristics of computer arcade games, which, at least until very recently, have proved to be extremely motivating. It has been suggested that instructional programs can be made more interesting (even fun to do) by using the same characteristics that contribute to the compellingness of video arcade games (Malone and Lepper, 1983).

What are these characteristics? Malone (1981) found that the characteristics of video games children like most (1) have explicit goals, (2) use sound effects, (3) have an element of randomness, (4) employ graphics rather than words to convey instructions and feedback, and (5) keep score. Instructional programs can be designed to include many of these features.

2.4.4.2 Mood

Mood has been shown to play an important role in learning. Not only do chronically depressed patients have learning difficulties (Beck, 1967), but learning in an experimentally induced depressed mood state has been shown to be 30% less efficient than learning in a neutral or elated state (Ellis, Thomas, and Rodriguez, 1984; Leight and Ellis, 1981).

An interesting finding of these studies is that mood can be changed so easily. The typical procedure for inducing a depressed-mood state, the so-called *Velton technique* (Velton, 1968), is simply to have subjects read a set of statements such as "I just don't care about anything—life just isn't any fun." The reverse procedure, to put one in an elated mood (or to get one out of the depressed mood), is to have the person read statements such as "I feel so good I almost feel like laughing." More elaborate and systematic methods for overcoming the debilitating effects of depressive and anxious mood states have been described in the literature on test anxiety. Chapters in the volume *Test Anxiety: Theory, Research, and Applications* (Sarason, 1980) describe various modeling techniques, relaxation exercises, and desensitization methods for overcoming anxiety in learning and testing situations.

Mood also has been shown to operate as a context variable in producing study–test interaction effects (Bower, Monteiro, and Gilligan, 1978; Leight and Ellis, 1981; Isen, Shalker, Clark, and Karp, 1978). That is, we recall better those items that were learned in a mood state similar to the testing state. If we are sad, we will more likely recall those items that were learned while we were sad than those learned while we were happy. This has been called the *mood–congruity effect*. It can be likened to other state-dependent retrieval effects documented by Tulving (1983) and discussed elsewhere in this chapter.

Thus, to recall something that was learned in a particular mood state, it is useful to reinstate that mood. Why this is so is less than well understood, but theoretical advances have tended to incorporate the findings into information-processing terms similar to the framework presented earlier in this chapter (Bower and Cohen, 1982; Clark, 1982).

2.4.5   FORTY-THREE PRACTICAL APPLICATIONS OF LEARNING PRINCIPLES

The principles of learning developed earlier in this chapter lend themselves to practical applications within three broad classes that correspond to the three participants in the learning activity—the *student,* the *teacher,* and the *system designer* who creates a system that a student will have to learn to operate.

The first class of applications is to the student's learning strategies. If students have a better understanding of how learning occurs, how thought proceeds, and how memory is organized, they should be better able to learn more effectively by applying this knowledge. One who reads this chapter carefully, for example, and who learns all the material presented, should be a much more effective learner.

The second class of applications is in teaching, or the use of instructional systems. This includes not only conventional classroom instruction, but also personal tutoring systems, human or computerized, and advice giving, data-base management, and expert systems designed to instruct students.

The third class of applications has to do with systems design. The extent to which the system is easy to learn should be an important concern of the designer. This is not to say that ease of learning is the only, or even the most important consideration, but the training requirements should certainly be of substantial concern to the system designer. Of what benefit would it be to design a system that humans could not learn to operate?

Thus, in this final section, 43 practical applications of learning principles are presented under three headings that correspond to the three classes of principles; namely, *learner strategies principles* (14 applications), *instructional systems design principles* (20 applications), and *principles for designing learnable systems* (11 applications). All 43 applications are listed in Table 2.4.6. The focus in this section is on the application—not on why these principles work, but rather on how to use them. The rationale or justification of the practical applications can be inferred by the serious student from the earlier sections of the chapter.

Table 2.4.6   Forty-Three Practical Applications of Learning Principles

Learner Strategies Principles

1. Search for meaning while studying.
2. Use strategies for minimizing memory load.
3. Study, study, study.
4. Avoid detention with extraneous details.
5. Permit liberal recall time.
6. Employ advance organizer methods.
7. Employ the recitation method (self-imposed study test).
8. Space practice.
9. Learn parts first.
10. Overlearn skill components.
11. Seek good models of skilled performance.
12. Seek knowledge of results.
13. Concentrate on study material.
14. Relax and adopt a good mood.

Instructional Systems Design Principles

Domain General Principles
15. Task-analyze the learning domain.
16. Organize instructional goals around behavioral objectives.

Learning from Text
17. Provide previews, reviews, and tests.
18. Provide study strategies.

Problem Solving and High-Performance Skills
19. Show positive and negative instances of concept.
20. Identify the goal structure of the problem.
21. Shape successive approximations to target performance.
22. Minimize working-memory load.
23. Keep the learning task interesting.

Table 2.4.6 (*Continued*)
_____

24. Provide immediate feedback on errors.
25. Maintain active learner participation.
26. Provide instruction in context.
27. Encourage use of general problem-solving procedures.
28. Consider the learning task a rules acquisition task.
29. Avoid changing requirements in midstream.
30. Maximize critical skills trials.
31. Achieve operational fidelity.
32. Mix component training.
33. Train under mild speed stress.
34. Train time-sharing skills.

Principles for Designing Learnable Systems

35. Consider learning ultimate-performance trade-offs.
36. Consider what tasks will be performed by the system.
37. Specify alternative methods for performing tasks.
38. Allow error recovery.
39. Use color/graphics to highlight changing information.
40. Clean display to eliminate irrelevancies.
41. Avoid abstract information; use concrete information.
42. Avoid elaborate rationales for procedural instructions.
43. Create conditions for practice and testing.
_____

2.4.5.1 Learner Strategies Principles

These *learner strategy applications* of learning principles are meant to be tips for the learner to use to become a more effective student. They have been adapted from a variety of sources including both cognitive and educational psychology textbooks, as well as how-to-study guides. Anderson (1980), Gage and Berliner (1979), and Bower and Hilgard (1981) provide similar recommendations, along with a rationale based on cognitive theory.

    1. *Search for Meaning while Studying.* A central theme of the learning principles in this chapter has been that memory is organized to process the meaning of inputs (Anderson and Bower, 1973; Kintch, 1974; Norman and Rumelhart, 1975). It follows that to make the most efficient use of memory, the learner should actively seek out the meaning of the materials being studied.

    When study material is inherently arbitrary, as many lists of procedures or lists of items in a taxonomy seem to be at first, the learner can use some of the *mnemonic techniques* described earlier to impose meaning. When meaning is already present in the material, as in the case of functioning systems or texts, the learner should attempt to *understand the overall system,* where the components fit into the picture, and why.

    These strategies may seem cumbersome and even counterintuitive because they require that the learner not merely passively encode the incoming information, but also actively elaborate upon the material, relating the new material to that which has already been learned. Despite the "up-front" investment intrinsic to active elaboration strategies, their use will have tremendous payoff in ultimate learning efficiency.

    2. *Use Strategies for Minimizing Memory Load.* It has long been known that there is a real and severe limitation to the amount of information that can be processed at any one time (Miller, 1956), and working memory quickly becomes overloaded when it is processing incoming data streams. Several strategies for reducing memory burden were described earlier in this chapter. From the standpoint of the learner, two of the most important are *rehearsal* and *chunking*.

    Most adults are well aware of the benefit of rehearsal (e.g., repeating a telephone number to remember it), but it works best when only the material to be remembered is rehearsed (e.g., hang up the receiver after hearing the telephone number so that no more data will be received).

    Chunking is an effective means for reducing the amount of information active in working memory. Seek out familiar patterns within the incoming data (e.g., familiar number patterns such as 1985). Remembering such "chunks" of information makes it easier to recall the elements that are subsumed by them.

3. *Study, Study, Study.* Many learners naively believe that they have learned a body of material once they have been exposed to it. But countless experiments have shown that the more the material is studied, the stronger will be the memory trace for that material (Wickelgren, 1981). Mental expertise has been shown to develop even after long and extensive practice (Lewis, 1978; Newell and Rosenbloom, 1981)—there is no limit to how well something can be known! By using the techniques described, one's study efforts will be more effective and efficient. However, mere study by itself is not necessarily efficient; it pays to study wisely, using the study techniques described here.

4. *Avoid Detention with Extraneous Details.* Textbooks and instruction manuals sometimes clutter their presentation of the primary material with irrelevant details. Although it is sometimes difficult for the learner to determine what is central and what peripheral (this is something that makes learning difficult), maximum learning efficiency is obtained by focusing on the material that is directly related to the learning goals. Some studies have shown, for example, that students may remember a particular topic better by studying the outline/summary for the topic than by studying the text that is outlined (Reder and Anderson, 1980). Writers who include irrelevancies in their text may believe that doing so can contribute to learner elaborations (compare Number 1 above). However, the learner's elaborations are more effective than those provided by the text or the instructor.

5. *Permit Liberal Recall Time.* Studies have shown that subjects can continue to recall little-known facts hours after the recall test begins (Nickerson, 1980; Williams, 1978). The key to such feats of recall is that the subject must be patient in tracking associations. (Some forms of psychotherapy are grounded in this principle.) If a recall failure is experienced, and if it is important to remember some item or event, one should follow associative leads by thinking about things related to the target memory trace. (Where was I? What day of the week was it? Who told me it?)

6. *Employ Advance Organizer Methods.* One of the more-powerful techniques for studying textual materials is the use of advance organizers (Ausabel, 1960; Frase, 1975). Many articles, manuals, and texts provide outlines or abstracts of the material they cover. It pays to study such preview sources carefully before embarking on the study of the main body of material. While studying advance organizers, reflect on what is being asserted and what is being left out. Develop questions about the to-be-studied material based on the organizer material.

7. *Employ the Recitation Method (Self-Imposed Study Test).* An effective technique for learning material presented either in a lecture or in writing is to attempt to recall (or recite) as many as possible of the points from the material after studying it (Anderson, 1980; Smith and Collins, 1981). The poor learner moves on to the next chapter, whereas the good learner reviews the chapter just studied.

8. *Space Practice.* One of the hallmarks of the skillful learner is the ability to apply learning skills in a variety of contexts. Repetitive practice of a skill for hours at a time provides a more limited context for the skill than intermittent practice spaced over a longer period of time. Thus, related to the notion of encoding specificity (Tulving, 1983), a skill practiced in a variety of temporal settings will probably develop into a more useful, generalizable skill. Fourteen hours of study spread over a week, in two-hour blocks per day, should produce more effective learning than 14 hrs in a row. This pertains to learning both text and skills. Thus an effective learner strategy is to mix the materials to be studied—two hours on one topic, then two hours on the next, and so forth.

9. *Learn Parts First.* Learners often attempt to accomplish too much, too fast, in attempting to attain quick mastery of a complex skill. It is more efficient to attempt to master the components of the skill one at a time (Anderson, 1980; Schneider, 1982). For example, in learning integration methods, an efficient technique is to study one or two, and practice them extensively, then study one or two more and practice them extensively. This is better than studying all the methods before practicing any, and then turning to the mixed sets typically provided at the ends of sections and chapters in texts.

10. *Overlearn Skill Components.* The acquisition of complex skills such as computer programming or mathematical problem solving puts severe strain on working-memory resources. One way to overcome the processing overload is to automatize lower-order skills or skill components by overlearning them (Schneider, 1982). For example, in learning a new programming language, it is useful to practice thoroughly the fundamentals in simple function writing before attempting to write a complex multifunction program. It is also useful to review the fundamental skills occasionally even at advanced stages of skill development. (Even professional baseball players or golfers find it useful to practice the elementary components of their skills.)

11. *Seek Good Models of Skilled Performance.* It is often a good technique to observe expert performance of a skill (Bandura, 1977; Keele, 1973). For example, studying a computer program written by an expert can be a useful way of learning programming technique and style (Anderson, 1980).

12. *Seek Knowledge of Results.* Although feedback on problem-solving efforts is not always available, when it is available (e.g., from a tutor, or worked-out examples in a text), it should be used as much as possible. It is also important that such feedback be studied immediately while the context and approach to the problem solving are fresh in memory (Anderson, Boyle, Farrell, and Reiser, 1984a).

13. *Concentrate on Study Material.*   Learning occurs as a result of transformations of information in working memory (e.g., Anderson, 1983) and thus for successful learning to occur, it is necessary to get the proper information into working memory by focusing attention on the immediate study problem at hand. Although this point should be obvious, it is surprising how often learners ignore it. Sometimes would-be learners even create distractions (e.g., by playing radio or television while studying). The serious learner minimizes the probability of interruptions and thus creates an efficient study environment that enhances a focusing of attention on the task at hand which facilitates learning.

14. *Relax and Adopt a Good Mood.*   Relaxation exercises designed to reduce test anxiety have been described by Meichenbaum (1977). Such general stress reducers can be effective in certain situations. Also, a positive, nondepressed mood can facilitate learning and mood can be enhanced through various exercises designed to promote positive thinking (e.g., the Velton, 1968, technique previously cited).

2.4.5.2 Instructional System Design Principles

The 20 applications of learning principles that follow (i.e., Numbers 15 to 34) pertain to instructional systems. For the most part, they are equally applicable to the subject-matter expert who designs such systems and to the instructor who uses them. The applications are stated at a fairly general level. They may be used in a variety of instructional systems, ranging from traditional classroom instruction to sophisticated computerized intelligent tutorial systems, and covering a wide variety of subject-matter domains.

Domain General Principles

Most complex skills are appropriately viewed as combinations of fact and skill knowledge. Thus, the following two applications are general to both fact and skill instruction. These two techniques form the nucleus of the traditional instructional system design (ISD) approach.

15. *Task-Analyze the Learning Domain.*   Elaborate descriptions of methods for conducting detailed analyses of learning tasks exist (e.g., Gagné and Briggs, 1977). These methods should be reviewed and used as a guide to the design and use of good instructional systems.

16. *Organize Instructional Goals around Behavioral Objectives.*   Instruction should be goal oriented, especially in the workplace, where instruction time is subtracted directly from employee productivity. Gagné (1985) has provided numerous examples of how complex instructional goals can be broken down into hierarchies of simpler and more tractable behavioral objectives.

Learning from Text

The next two applications pertain primarily to learning from texts (i.e., fact learning).

17. *Provide Previews, Reviews, and Tests.*   Text materials are more quickly and better learned when they are associated with outline previews, chapter reviews, and tests as some texts provide (Reder and Anderson, 1980).

18. *Provide Study Strategies.*   Numerous techniques for improving memory for textlike material are reviewed in the earlier sections of this chapter—for example, various mnemonic strategies, the PQ4R method, and networking techniques (also reviewed in introductory cognitive psychology texts such as Anderson, 1980). The initial time required to learn these strategies can be considerable. It would take at least a day, and perhaps even as long as a week, to learn all of them. The instructor or instructional system designer must judge the extent to which the benefit to be gained would justify the "up-front" investment of time. Obviously, each paragraph of text cannot be preceded by an hour-long course on study methods, but it might be quite profitable to spend two days on the topic at the start of a three-month course.

Problem Solving and High-Performance Skills

The remaining 16 applications to instructional systems (Numbers 19 to 34) pertain principally to programs for teaching problem solving and high-performance cognitive skills. They are based primarily on suggestions by Schneider (1982, 1984) and by Anderson et al. (1984a), who developed them in the context of creating instructional systems to teach computer programming and geometry problem solving (Anderson et al.), and air traffic control (Schneider). They are especially important in the context of developing intelligent tutoring systems, which has become a popular forum for applications of cognitive psychology in recent years (see Sleeman and Brown, 1982).

19. *Show Positive and Negative Instances of Concepts.*   When first presenting some concept (such as the definition of a system or a component), instructors and instructional designers often overlook the benefits of presenting nonexamples of the concept. The availability of both examples and nonexamples can help the learner to establish the boundaries of the concept to be learned. Humans are notoriously

unreliable in seeking and accepting negative evidence against a working hypothesis (Snyder and Swann, 1978; Wason and Johnson-Laird, 1972), and thus explicit provision of negative instances may serve to counteract this characteristic.

20. *Identify the Goal Structure of the Problem.* Nearly every reasonably complex problem-solving task can be represented as a goal tree in which the top-level goal state is broken down into subgoal (i.e., intermediate) states, which may be broken down still further until the level of starting data (the "givens" in the problem) is reached. Several introductory texts or artificial intelligence provide examples (e.g., Nilsson, 1981; Winston, 1984). Given this representation, the student's task is to find a path that leads from the starting point to the top-level goal. To traverse the tree successfully, the student must search the space of paths and intermediate states, some of which are fruitful, some of which are off track, looking for those paths connecting subgoal states that ultimately lead to the goal state.

Anderson et al. (1984a) claim that one of the difficulties encountered by students who are trying to learn a new domain, such as geometry or computer programming, is that the search space concept is never explicitly presented. Thus they have developed computerized tutors that show students the structure of the goal tree, and from this students learn the importance of searching through the problem space represented by the tree. It is likely that students would develop more realistic expectations of the requirements for reasoning in a new domain if such a technique were more widely employed by instructors. Note that this suggests performing a kind of cognitive task analysis of the learning task, substituting intermediate cognitive subgoal states for the more conventional behavioral objective states of traditional task analysis.

21. *Shape Successive Approximations to Target Performance.* Shaping is the traditional behaviorist term to describe a training system in which successively better (closer-to-goal) performances are rewarded. As an instructional design strategy, the application of this principle has two components. One is that the tutor should be able to recognize and accept partially correct solutions. Unfortunately, relatively few of the more traditional frame-based computer-assisted instruction programs bother to build in this capability. The second component is that extensive advice should be provided during the early stages of problem solving or skill learning, with less and less advice as the learner becomes more skilled (Anderson et al., 1984a; Smith and Collins, 1981).

22. *Minimize Working-Memory Load.* One of the most important reasons why new cognitive skills are so difficult to acquire is that working-memory capacity is often exceeded, and that makes it difficult to process information (Anderson, Farrell, and Sauers, 1984b). There are three major ways in which the instructor or instructional system designer can reduce memory load for the learner.

One is by training those problem-solving strategies that minimize load. Even if the load-minimization strategy is not the best ultimately, in many cases it can serve as a good starter strategy. Of course, this approach is appropriate only when multiple strategies are available for learning the task.

The second is to provide external memory aids (e.g., on the computer screen in computerized instruction), so that the learner does not have to keep track of all the data that have to be processed. Windows or notes on the screen can be used to keep track of the goals, subgoals, and intermediate results, the history of actions attempted, intermediate calculations and states, and so on. Also, diagrams can be used to display complex relations that otherwise might be difficult to grasp.

The third way to reduce working-memory load is to introduce new tasks slowly. If the learner attempts to employ too many new skills at once, learning will suffer. By introducing only one skill at a time, the teacher or designer is reducing the load.

23. *Keep the Learning Task Interesting.* Malone (1981) has discussed at length the variables that seem to contribute to what makes computer arcade games fun. Among these are the availability of short-term goals, the use of immediate feedback, the adaptability of the task to the user's skill level, and various other motivation tricks such as exciting graphics and animation. It may well be that skill-learning tasks in general can benefit from use of these attributes in producing learner motivation. Schneider (1982) claims that the most effective motivators he has used have been an inexpensive sound chip for making arcade-like sounds, and a graphic facility for animating objects such as letters or words (e.g., the word "blows up" if it is correctly detected).

24. *Provide Immediate Feedback on Errors.* Feedback can play a role not only in motivation, but also in providing information. If the student commits an error, but does not know it is an error, the procedure leading to the error will become strengthened in memory. Thus, it is important to provide immediate feedback in order to reduce the strength of the erroneous procedure (Anderson et al., 1984a).

25. *Maintain Active Learner Participation.* Although humans are perfectly capable of learning solely through observation of a model (Bandura, 1977; Rosenthal and Zimmerman, 1978), overly passive observation can increase the probability of boredom, inattention, and poor learning (Schneider, 1982, 1984). Good teachers ask questions of students in the middle of lectures to keep them on their toes. With computerized instruction, opportunities for inviting learner participation are even easier. In general, the more actively the learner is involved, the more learning will take place. Teachers and designers should insert frequent tests even in situations where the learner is attempting to acquire a skill primarily by observing.

26. *Provide Instruction in Context.* Many studies in human problem solving have demonstrated

an enormous context-specific effect for problem-solving skills (e.g., Wason and Johnson-Laird, 1972). The teacher or designer cannot assume that a skill acquired in an artificial context will automatically transfer to the operational context. Thus, it is useful to train component skills in a situation similar to the target-performance context where this is feasible.

27. *Encourage Use of General Problem-Solving Procedures.* Anderson et al. (1984a) structure their tutors so that students are not allowed to rely on sample solutions for solving proof and programming problems. Thus, their tutors force the student to apply general procedures for solving problems. Although students prefer to use analogy, its indiscriminate use promotes shallower processing and there is always the possibility that the student will acquire a wrong mapping feature from the analogy. The use of general problem-solving procedures produces better learning generally.

28. *Consider the Learning Task a Rules Acquisition Task.* In cognitive-skill domains such as mathematics, programming, electronics, and word processing, it is useful to consider the student's task to be that of acquiring a set of production (if–then) rules for learning how to respond in various situations. Numerous examples representing complex skill domains have been developed in the last few years (e.g., Anderson et al., 1984a; Brown and VanLehn, 1980; Card et al., 1983; Polson and Kieras, 1984).

The advantage of such an approach is that it makes explicit the learning requirements for various tasks. Polson and Kieras, for example, developed a system for determining how well some skills acquired on one task transfer to different problem-solving tasks. Brown and VanLehn have developed a theory of children's "bugs" experienced during performance of subtraction problems. By representing the student as a set of productions, Anderson et al. claimed that it should be possible to set a more appropriate level of discourse for instruction at any particular point in the student's progression through the stages of skill development. That is, at early stages of skill, it is more appropriate to provide finer-grained specific instruction. As skill progresses, instruction is more appropriately scaled to higher and more general levels of specificity.

29. *Avoid Changing Requirements in Midstream.* Schneider and Fisk (1982) have shown that the use of a consistent rule for mapping stimuli to responses leads to rapid development of automatic processing, whereas inconsistent rules prevent it. Thus, consistency between what the learner sees and how she or he is to respond is a key factor in determining how quickly the skill will be acquired (Schneider, 1982).

30. *Maximize Critical Skills Trials.* In some operational tasks of a monitoring or vigilance nature, there is typically such a long lag between the events that require the learner to respond, that the learner gets very few trials for practicing the response. Because learning is a function of practice, it will pay to alter the training situation to allow the exercise of such skills. As an example of how this can be accomplished, consider a training system developed by Schneider (1982) which gave trainees more practice on an air traffic controller task by increasing the rate at which events were displayed (by as much as 100 to 1). When using this feature, care should be taken that the more frequent responses in the training situation do not establish unrealistic expectancies that will degrade the learner's performance on the job.

31. *Achieve Operational Fidelity.* To promote the generalizability of skills, it is useful to allow the exercising of such skills under various conditions, some (many, most, or even all) of which duplicate those of the operational situation (Schneider, 1982, 1984).

32. *Mix Component Training.* It is better to avoid strictly sequential component-skill training, opting instead for iterative component practice (Schneider, 1984).

33. *Train under Mild Speed Stress.* Learners are more likely under mild speed pressure to develop process automaticity; with more liberal performance-time allowances, students tend to rely more on slow-moving controlled processes (Schneider, 1982). Where timing is critical, for example, in many piloting and air traffic control tasks, the ability to respond quickly is critical, and it is influenced by training.

34. *Train Time-Sharing Skills.* When the operational context of the skill requires it to be performed in tandem with other skills (i.e., time-sharing situations), then it is useful to train the skills together so that the time-sharing aspects of the total task are learned (Schneider, 1984).

2.4.5.3 Principles for Designing Learnable Systems

Learnability is another of the "ilities" that needs to be considered in designing systems that humans have to learn to operate and maintain. The prudent designer should strive to design systems that are easy to learn—that is, that have few training requirements.

The nine practical applications of learning principles that follow (Numbers 35 to 43) are intended as general guides to be used in designing systems that will be easier to learn to use. Included are applications to both the design of the system and the instructional manual that explains it. The availability and use of computers in the design process make it quite feasible to integrate these two system components that were nearly always treated separately in the past.

35. *Consider Learning Ultimate-Performance Trade-Offs.* System learnability and expected ultimate-performance levels are not necessarily perfectly correlated (Card et al., 1983; Roberts, 1979).

Systems that are easy to learn to use might prove limited in performance capabilities after the user has acquired considerable expertise. On the other hand, systems that are extremely difficult to learn might never reach their high-performance design capability because the manning and training requirements are too great to be met in practice. Thus, the designer must evaluate the trade-offs and take them into account in reaching a decision regarding the final design.

36. *Consider what Tasks Will Be Performed By the System.* The criterion for the designer's decisions should be based on knowledge of how the system will be used in the operational setting—by whom, when, how, and to accomplish what tasks (Card et al., 1983). A system designed to be used by first-term military personnel should not take several years to learn to operate or maintain.

37. *Specify Alternative Methods for Performing Tasks.* Tasks should be do-able by more than one method (Card et al., 1983). A method for novices might be designed to provide intermediate checks, displays of the results of intermediate steps, extensive documentation for each step, and perhaps even a menu that lists possible subsequent steps. The presentation of such supplementary information should reduce the working-memory load for the novice, and improve both learning and performance. On the other hand, an expert user might be slowed down by this level of detail. Thus, by creating novice-to-expert alternative methods, the designer could seek to maximize both the learnability of the system and the system's ultimate-performance levels.

38. *Allow Error Recovery.* Novices can be slowed down tremendously in their learning of how to operate a system when they spend a considerable amount of their time trying to recover from errors. This relates to our earlier discussion of what fosters knowledge compilation (J. R. Anderson; 1983, R. C. Anderson, 1983). The designer can eliminate some of this otherwise wasted time and effort by establishing a procedure for dropping back to a previous system state when one or more errors are committed (Card et al., 1983).

39. *Use Colors/Graphics to Highlight Changing Information.* Another method for reducing the working-memory burden on the learner is to cue relevant aspects of displayed information that might otherwise have to be kept in mind. For example, new information on the display can be highlighted with color or motion, but in ways that are not confused with the highlighting that is typical (and beneficial) in displaying critical information such as warnings and danger signals. Current intelligent tutoring system development work (Anderson et al., 1984a; Sleeman and Brown, 1982) devotes considerable attention to this issue.

40. *Clean Display to Eliminate Irrelevancies.* Although a display can serve as an important aid to memory, it should not be used as a general-purpose scratch pad. The unnecessary display of information that is irrelevant to the task can add to the operator's memory load. Nonpertinent information draws attention, usurping valuable and limited attentional resources that might otherwise be devoted to processing task-relevant information. As is the case with normal text processing (e.g., Reder and Anderson, 1980), the designer can improve comprehensibility by limiting the presentation of information to only that information critical to the learning or performance task.

41. *Avoid Abstract Information; Use Concrete Information.* Pictures and graphic symbols can be more powerful than words in communicating large amounts of information, and they carry the additional advantage of greater memorability (Shepard, 1967) as a result of distinctiveness, as was discussed in a foregoing section. The designer should consider ways for expressing information pictorially wherever possible.

42. *Avoid Elaborate Rationales for Procedural Instructions.* Procedure lists should specify in a straightforward fashion what it is the operator should do. In the same way that irrelevant text can impede learning the main points of a passage (Reder and Anderson, 1980), elaborate rationales may impede learning the steps in procedural instructions.

43. *Create Conditions for Practice and Testing.* The system designer should keep in mind that operators learn by practicing. Therefore, it is important to enable practicing of operator skills. As was discussed in an earlier section of this chapter, acquired knowledge and cognitive skills decay, albeit slowly, over time (Wickelgen, 1981). Thus even the best-learned skills must be practiced to maintain high levels of performance (Schneider, 1984).

2.4.5.4 Summary: Forty-Three Practical Applications

This final section of the chapter discussed 43 principles of learning that we believe should be of fairly direct practical utility to the student, the teacher, and the systems designer concerned with learnability issues.

Fourteen applications of learning principles were outlined for the student interested in more effective ways to study. The general themes characterizing these applications are that the serious student attempts actively to process new material, relating it to what he or she already knows, uses certain proven learning strategies, and persists in concentrated study and self-testing.

Twenty applications of learning principles were provided for those concerned with designing instructional systems. It is important that the instructional designer first consider the nature of the learning that is required—that is, learning from text, learning to solve problems, or learning high-performance cognitive skills—for each suggests different points of emphasis. It is also important for the instructional designer to consider the duration of the instructional treatment. Short-duration treatments do not often warrant substantial time investment in developing general study skills, for example.

The final nine practical applications were tailored for the system designer concerned with the issue of how to make a particular system easier to learn to use by the operators that will have to learn to use it. With the availability of automated, and in many cases computer-based systems where manual-based systems were once the norm, it is now possible and even feasible for the designer to begin considering learnability issues as an increasingly integral part of the job of systems design. We suggest that designers approach their task in a fashion similar to that used by instructional systems designers: that they consider exactly what tasks will ultimately be performed by the system being designed, and that the variety of possible ways for performing those tasks be considered. The system designer must, of course, always keep in mind the relation between how easy the system is to learn, and how powerful the system must be. Given such considerations, the designer can take advantage of the power and flexibility of the computer for presenting data and choices simply, for allowing the operator recovery procedures, for enabling practice of operator skills, and for using informative graphics for a variety of purposes.

REFERENCES

Abernathy, E. M. (1940). The effect of changed environmental conditions upon the results of college examinations. *Journal of Psychology, 10,* 293–301.

Anderson, J. R. (1976). *Language, memory, and thought.* Hillsdale, NJ: Erlbaum.

Anderson, J. R. (1980). *Cognitive psychology and its implications.* San Francisco, CA: Freeman.

Anderson, J. R., Ed. (1981). *Cognitive skills and their acquisition.* Hillsdale, NJ: Erlbaum.

Anderson, J. R. (1983). *Architecture of cognition.* Cambridge, MA: Harvard University Press.

Anderson, J. R. and Bower, G. H. (1973). *Human associative memory.* Hillsdale, NJ: Erlbaum.

Anderson, J. R., Boyle, C. F., Farrell, R., and Reiser, B. J. (1984a). *Cognitive principles in the design of computer tutors* (Technical Report No. ONR 84-1). Pittsburgh, PA: Carnegie-Mellon University.

Anderson, J. R., Farrell, R., and Sauers, R. (1984b). Learning to program in LISP. *Cognitive Science, 8,* 87–129.

Anderson, R. C. (1983, May). *Cognition, instruction, and interest.* Paper presented at the Office of Naval Research Conference on Aptitude, Learning, and Instruction: Conative and affective process analysis, Stanford, CA.

Anzai, Y., and Simon, H. A. (1979). The theory of learning by doing. *Psychological Review, 86,* 124–140.

Ausabel, D. P. (1960). The use of advance organizers in the learning and retention of meaningful verbal learning. *Journal of Educational Psychology, 51,* 267–272.

Baddeley, A. D. (1966). Short-term memory for word sequences as a function of acoustic, semantic, and formal similarity. *Quarterly Journal of Experimental Psychology, 18,* 362–365.

Bandura, A. (1977). *Social learning theory.* Englewood Cliffs, NJ: Prentice-Hall.

Beck, A. (1967). *Depression; Clinical, experimental, and theoretical aspects.* New York: Harper & Row.

Bellezza, F. S. (1982). *Improve your memory skills.* Englewood Cliffs, NJ: Prentice-Hall.

Bellezza, F. S. (1983). The spatial-arrangement mnemonic. *Journal of Educational Psychology, 75,* 830–837.

Bethell-Fox, C. E. (1983). Visual metuorlsure as a learning strategy. Unpublished dissertation, Stanford University, Palo Alto, CA.

Bousfield, W. A. (1953). The occurrence of clustering in the recall of randomly arranged associates. *Journal of General Psychology, 49,* 229–240.

Bower, G. H. (1970a). Imagery as a relational organizer in associative learning. *Journal of Verbal Learning and Verbal Behavior, 9,* 529–533.

Bower, G. H. (1970b). Analysis of a mnemonic device. *American Scientist, 58,* 496–510.

Bower, G. H., Black, J. B., and Turner, T. J. (1979). Scripts in memory for text. *Cognitive Psychology, 11,* 177–220.

Bower, G. H., Clark, M. C., Lesgold, A. M., and Winzenz, D. (1969). Hierarchical retrieval schemes in recall of categorized word lists. *Journal of Verbal Learning and Verbal Behavior, 8,* 323–343.

Bower, G. H., and Cohen, P. R. (1982). Emotional influences in memory and thinking: Data and theory. In M. S. Clarke and S. T. Fiske, Eds. *Affect and cognition: The seventeenth annual Carnegie symposium on cognition.* Hillsdale: NJ: Erlbaum, pp. 291–331.

Bower, G. H., and Hilgard, E. R. (1981). *Theories of learning,* 5th ed. Englewood Cliffs, NJ: Prentice-Hall.

Bower, G. H., Monteiro, K. P., and Gilligan, S. G. (1978). Emotional mood as a context for learning and recall. *Journal of Verbal Learning and Verbal Behavior, 17,* 573–585.

Bradley, M. M., and Glenberg, A. M. (1983). Strengthening associations: Duration, attention, or relation. *Journal of Verbal Learning and Verbal Behavior, 22,* 650–666.

Bransford, J. D., and Franks, J. J. (1971). The abstraction of linguistic ideas. *Cognitive Psychology, 2,* 331–350.

Brooks, L. R. (1968). Spatial and verbal components of the act of recall. *Canadian Journal of Psychology, 22,* 349–368.

Brown, J. S., and VanLehn, K. (1980). Repair theory: A generative theory of bugs in procedural skills. *Cognitive Science, 4,* 379–426.

Card, S. K., Moran, T. P., and Newell, A. (1983). *The psychology of human–computer interaction.* Hillsdale, NJ: Erlbaum.

Chase, W. G., and Ericsson, K. A. (1981). Skilled memory. In J. R. Anderson, Ed. *Cognitive skills and their acquisition.* Hillsdale, NJ: Erlbaum.

Chipman, S. F., Segal, J. W., and Glaser, R., Eds. (1985). *Thinking and learning skills,* Vol. 2, *Research and open questions.* Hillsdale, NJ: Erlbaum.

Clark, M. S. (1982). A role for arousal in the link between feeling states, judgments, and behavior. In M. S. Clark and S. T. Fiske, Eds. *Affect and cognition: The seventeenth annual Carnegie symposium on cognition.* Hillsdale, NJ: Erlbaum, pp. 263–290.

Clark, M. S. and Fiske, T. S., Eds. (1982). *Affect and cognition: The seventeenth annual Carnegie symposium on cognition.* Hillsdale, NJ: Erlbaum.

Collins, A., and Smith, E. E. (1982). Teaching the process of reading comprehension. In D. Detterman and R. Sternberg, Eds. *How and how much can intelligence be increased?* Norwood, NJ: Ablex.

Conrad, R. (1964). Acoustic confusions in immediate memory. *British Journal of Psychology, 55,* 75–84.

Craik, F. I. M., and Lockhart, R. S. (1972). Levels of processing: A framework for memory research. *Journal of Verbal Learning and Verbal Behavior, 11,* 671–684.

Ellis, H. C., Thomas, R. L., and Rodriguez, I. A. (1984). Emotional mood states and memory: Elaborative encoding, semantic processing, and cognitive effort. *Journal of Experimental Psychology: Learning, Memory, and Cognition, 10,* 470–482.

Fitts, P. (1964). Perceptual-motor skill learning. In A. W. Melton, Ed. *Categories of human learning.* New York: Academic, pp. 243–285.

Frase, L. T. (1975). Prose processing. In G. H. Bower, Ed. *The psychology of learning and motivation,* Vol. 9. New York: Academic.

Furst, B. (1958). *Stop forgetting: How to develop your memory and put it to productive use.* Garden City, NY: Doubleday.

Gage, N., and Berliner, D. (1979). *Educational psychology.* Chicago: Rand-McNally.

Gagné, R. M. (1985). *The conditions of learning and theory of instruction,* 4th ed. New York: Holt, Rinehart, & Winston.

Gagné, R. M., and Briggs, L. J. (1977). *Principles of instructional design,* 2nd ed. New York: Holt, Rinehart, & Winston.

Gentner, D. (1983, May). Paper presented at the Office of Naval Research Contractors Meeting on Problem-Solving and Mental Models, Berkeley, CA.

Gitomer, D. H., and Glaser, R. (in press). If you don't know it, work on it: Knowledge, self-regulation and instruction. In R. E. Snow & M. J. Farr, Eds., *Aptitude, learning, and instruction,* Vol. 3, *Conative and affective process analysis.* Hillsdale, NJ: Erlbaum.

Godden, D. R., and Baddeley, A. D. (1975). Context-dependent memory in two natural environment: on land and under water. *British Journal of Psychology, 66,* 325–331.

Higbee, K. L. (1977). *Your memory: How it works and how to improve it.* Englewood Cliffs, NJ: Prentice-Hall.

Hunter, I. M. L. (1977). Mental calculation. In P. N. Johnson-Laird and P. C. Wason, Eds., *Thinking.* New York: Cambridge University Press, pp. 35–45.

Isen, A. M., Shalker, T. E., Clark, M., and Karp, L. (1978). Positive affect, accessibility of material in memory, and behavior: A cognitive loop? *Journal of Personality and Social Psychology, 36,* 1–12.

Keele, S. W. (1973). *Attention and human performance.* Pacific Palisades, CA: Goodyear.

Kintch, E. (1974). *The representation of meaning in memory.* Hillsdale, NJ: Erlbaum.

Leight, K., and Ellis, H. (1981). Emotional mood states, strategies, and state dependency in memory. *Journal of Verbal Learning and Verbal Behavior, 20,* 251–266.

Lewis, C. H. (1978). *Speed and practice.* Unpublished manuscript, Pittsburgh, PA: Carnegie Mellon University, Department of Psychology.

Lorayne, H., and Lucas, J. (1974). *The memory book.* New York: Ballantine Books or Stein & Day.

Malone, T. W. (1981). Toward a theory of intrinsically motivating instruction. *Cognitive Science, 4,* 333–369.

Malone, T. and Lepper, M. (1983, May). *Motivation, cognition, and computerized instruction.* Paper presented at the Office of Naval Research Conference on Aptitude, Learning, and Instruction: Conative and Affective Process Analysis, Stanford, CA.

Marcel, T. (1983). Conscious and unconscious perception: Experiments on visual masking and word recognition. *Cognitive Psychology, 15,* 197–237.

McConkie, G. W., and Rayner, K. (1975). The span of the effective stimulus dervive a fixation in reading. *Perception and psychysics 17,* 578–596. Washington, DC: U.S. Office of Education.

McGeoch, J. A. (1942). *The psychology of human learning.* New York: Longmans.

Meichenbaum, D. (1977). *Cognitive behavior modification: An integrative approach.* New York: Plenum.

Miller, G. A. (1956). The magical number seven, plus or minus two: Some limits on our capacity for processing information. *Psychological Review, 63,* 81–97.

Minsky, M. (1975). A framework for representing knowledge. In P. H. Winston, Ed., *The psychology of computer vision.* New York: McGraw-Hill.

Neves, D. M., and Anderson, J. R. (1981). Knowledge compilation: Mechanisms for the automatization of cognitive skills. In J. R. Anderson, Ed., *Cognitive skills and their acquisition.* Hillsdale, NJ: Erlbaum.

Newell, A. (1973). Production systems: Models of control structures. In W. G. Chase, Ed., *Visual information processing.* New York: Academic.

Newell, A., and Rosenbloom, P. (1981). Mechanisms of skill acquisition and the law of practice. In J. R. Anderson, Ed., *Cognitive skills and their acquisition.* Hillsdale, NJ: Erlbaum.

Newell, A., and Simon, H. A. (1972). *Human problem solving.* Englewood Cliffs, NJ: Prentice-Hall.

Nickerson, R. (1980). Motivated retrieval from archival memory. *Nebraska Symposium on Motivation, 28,* 73–120.

Nilsson, N. J. (1981). *Principles of artificial intelligence.* Palo Alto, CA: Tioga.

Norman, D. A. and Rumelhart, D. E. (1975). *Explorations in cognition.* San Francisco, CA: Freeman.

Paivio, A. (1971). *Imagery and verbal processes.* New York: Holt, Rinehart, & Winston.

Polson, P. G., and Kieras, D. E. (1984). A formal description of users' knowledge of how to operate a device and user complexity. *Behavior Research Methods, Instrumentation, & Computers, 16,* 249–255.

Reder, L. M., and Anderson, J. R. (1980). A comparison of texts and their summaries: Memorial consequences. *Journal of Verbal Learning and Verbal Behavior, 19,* 121–134.

Rosenthal, T. L., and Zimmerman, B. J. (1978). *Social learning and cognition.* New York: Academic.

Rumelhart, D. E. (1981, January). *Understanding understanding.* (Technical Report No. CHIP 100). San Diego, CA: University of California, Center for Human Information Processing.

Rumelhart, D. E., and Norman, D. A. (1978). Accretion, tuning, and restructuring: Three modes of learning. In J. W. Cotton & R. Klatzky, Eds., *Semantic factors in cognition.* Hillsdale, NJ: Erlbaum.

Rumelhart, D. E., and Norman, D. A. (1981). Analogical processes in learning. In J. R. Anderson, Ed., *Cognitive skills and their acquisition.* Hillsdale, NJ: Erlbaum.

Rumelhart, D. E., and Norman, D. A. (1983, June). *Representation in memory* (Technical Report No. CHIP 116). San Diego, CA: University of California, Center for Human Information Processing.

Rumelhart, D. E., and Ortony, A. (1977). The representation of knowledge in memory. In R. C. Anderson, R. J. Spiro, and W. E. Montague, Eds., *Schooling and the acquisition of knowledge.* Hillsdale, NJ: Erlbaum.

Ryle, G. (1949). *The concept of mind.* London: Hutchinson.

Santa, J. L. (1977). Spatial transformations of words and pictures. *Journal of Experimental Psychology: Human Learning and Memory, 3,* 418–427.

Sarason, I. E., Ed. (1980). *Test anxiety: Theory, research, and applications.* Hillsdale, NJ: Erlbaum.

Schank, R. C., and Abelson, R. (1977). *Scripts, plans, goals and understanding.* Hillsdale, NJ: Erlbaum.

Schmalhofer, F. J. (1982). *Comprehension of a technical text as a function of expertise.* Unpublished doctoral dissertation, University of Colorado, Boulder.

Schneider, W. (1982, April). *Automatic/control processing concepts and their implications for the training of skills—final report* (Technical Report No. 8101). Champaign, IL: University of Illinois, Human Attention Research Laboratory.

Schneider, W. (1984, August). *Training high performance skills: Fallacies and guidelines* (Technical Report No. HARL-ONR-8301). Champaign, IL: University of Illinois, Human Attention Research Laboratory.

Schneider, W., and Fisk, A. D. (1982). Degree of consistent training: Improvements in search performance and automatic process development. *Perception and Psychophysics, 31*, 160–168.

Segal, J. W., Chipman, S. F., and Glaser, R., Eds. (1985). *Thinking and learning skills:* Vol. 1, *Relating instruction to research.* Hillsdale, NJ: Erlbaum.

Shepard, R. N. (1967). Recognition memory for words, sentences, and pictures. *Journal of Verbal Learning and Verbal Behavior, 6*, 156–163.

Shulman, H. G. (1970). Encoding and retention of semantic and phonemic information in short-term memory. *Journal of Verbal Learning and Verbal Behavior, 9*, 499–508.

Sleeman, D., & Brown, J. S., Eds. (1982). *Intelligent tutoring systems.* New York: Academic.

Smith, E. E., and Medin, D. L. (1981). *Categories and concepts.* Cambridge, MA: Harvard University Press.

Snow, R. E., and Farr, M., Eds. (in press). *Aptitude, learning, and instruction,* Vol. 3: *Conative and affective process analysis.* Hillsdale, NJ: Erlbaum.

Snyder, M., and Swann, W. B. (1978). Hypothesis-testing processes in social interaction. *Journal of Personality and Social Psychology, 36*, 1202–1212.

Sperling, G., Parish, D. H., Pavel, M., and Desaulniers, D. H. (1984, November). *Auditory list recall: Phonemic structure, acoustic confusability, and familiarity.* Paper presented at the Meeting of the Psychonomic Society, San Antonio, TX.

Thomas, E. L., and Robinson, H. A. (1972). *Improving reading in every class: A sourcebook for teachers.* Boston: Allyn and Bacon.

Tulving, E. (1983). *Elements of episodic memory.* New York: Oxford University Press.

Tulving, E., and Pearlstone, Z. (1966). Availability versus accessibility of information in memory for words. *Journal of Verbal Learning and Verbal Behavior, 5*, 381–391.

Tversky, A. (1977). Features of similarity. *Psychological Review, 84*, 327–352.

Velton, E. (1968). A laboratory task for induction of mood states. *Behavior Research and Therapy, 6*, 473–482.

Von Restorff, H. (1933). Über die Wirkung von bereichsbildungen im Spurenfeld. *Psychologie Forschung, 18*, 299–342.

Wason, P. C., and Johnson-Laird, P. N. (1972). *Psychology of reasoning: Structure and content.* Cambridge, MA: Harvard University Press.

Wickelgren, W. A. (1974). *How to solve problems.* San Francisco: Freeman.

Wickelgren, W. A. (1981). Human learning and memory. *Annual Review of Psychology, 32*, 21–52.

Wickens, C. D. (1984). *Engineering psychology and human performance.* Columbus, OH: Merrill.

Wickens, C., and Boles, D. B. (1983, December). *The limits of multiple resource theory: The role of task correlation/integration in optimal display formatting* (Technical Report No. EPL-83-5/ONR-83-5). Champaign-Urbana, IL: University of Illinois, Engineering Psychology Program.

Wickens, C. D., Sandry, D. L., and Vidulich, M. (1983). Compatibility and resource competition between modalities of input, central processing, and output. *Human Factors, 25*, 227–248.

Wickens, C. D., Vidulich, M., and Sandry, D. L. (1983). S-C-R compatibility: Paul Fitts in the 1980's. In R. Jenson, Ed., *Second symposium on aviation psychology.* Columbus: Ohio State University, Institute of Aviation.

Wickens, C. D. (1972). Characteristics of word encoding. In A. W. Melton and E. Martin, Eds., *Coding processes in human memory.* Washington, DC: Winston.

Williams, M. D. (1978). *The process of retrieval from very long term memory* (Technical Report No. CHIP-75). San Diego, CA: University of California, Center for Human Information Processing.

Winograd, T. (1975). Frame representations and the declarative-procedural controversy. In D. G. Bobrow and A. Collins, Eds., *Representation and understanding.* New York: Academic.

Winston, P. H. (1984). *Artificial intelligence.* 2nd ed. Reading, MA: Addison-Wesley.

Woodward, A. E., Bjork, R. A., and Jongeward, R. H. (1973). Recall and recognition as a function of primary rehearsal. *Journal of Verbal Learning and Verbal Behavior, 12*, 608–617.

Yates, F. A. (1966). *The art of memory.* Chicago: University of Chicago Press.

Zimmerman, B. J., and Rosenthal, T. L. (1972). Concept attainment, transfer, and retention through observational learning and rule provision. *Journal of Experimental Child Psychology, 14*, 139–150.

CHAPTER 2.5
ENGINEERING ANTHROPOMETRY

KARL H. E. KROEMER

Virginia Polytechnic Institute and State University
Blacksburg, Virginia

2.5.1 INTRODUCTION: THE HUMAN AS A SYSTEM ELEMENT

Data describing the dimensions of the human body, that is, anthropometric* information, are available to the engineer just as is information on other components of the operator–task–equipment–environment system. Engineers must become familiar with these data and with their applications so that they can design work systems, equipment, and tools for proper fit to the human, to achieve safe and efficient operation.

For the planned design of manned systems, or for the evaluation of existing ones, the following logic sequence should be applied:

1. **Task Analysis.** Allocate duties to either person or machine, considering the operational environment. For this:
   a. Review operator dimensions and capabilities essential for the design task.
   b. Review machine specifications and capabilities needed.
2. **Operator–Machine Interface Design.** Use computer models, templates, and drawings. Check the designed person–equipment interface by mock-ups.
3. **Test of Designed System.** Evaluate the ease and efficiency of task performance.

Figure 2.5.1 indicates this sequence. There is no reason to treat information on the human in a less systematic and conscious manner than information on machine characteristics; in fact, the human is

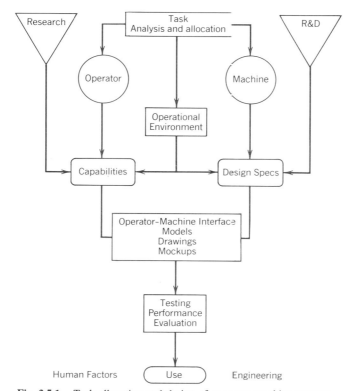

Fig. 2.5.1. Task allocation and design of operator–machine systems.

---

* The word anthropometry is derived from the greek *anthropos,* man, and *metrein,* to measure; it does not mean the same as "anthropomorphic."

the most important part in a manned system and should therefore be considered first. Hence, information on human dimensions and capabilities is a very important component of engineering knowledge and engineering data.

2.5.2 DEVELOPMENT AND SCOPE

At the end of the thirteenth century Marco Polo described the different body sizes and body builds of the tribes and races that he saw on his world travels. Hence, physical anthropology as a recording and comparing science is often traced to him. Blumenbach's *On the Natural Differences in Mankind* contained complete anthropometric data around 1800. The Belgian statistician Quetelet applied statistics to anthropometric and somatographic surveys in the middle 1800s. At about the same time, a new offspring now called "biomechanics" (see Chapter 2.6 in this book) indicated that engineers were becoming aware of data on the human body and its mechanical properties for design purposes.

This rapidly increasing interest in anthropometric information made standarization of measuring methods desirable. For this, the body was posed into certain repeatable postures, and bony landmarks on the body identified from which to take measurements. Conventions of anthropologists in Monaco in 1906 and six years later in Geneva provided standardization in anthropometric methods. Martin's *Lehrbuch der Anthropologie* (first edition in 1914) became the authoritative textbook and handbook for many decades. New engineering applications, newly developed measuring techniques, and advanced statistical procedures gave reason to update standardization in the 1960s (by Hertzberg) and 1980s (by ISO Technical Committee 159). The armed forces and NASA have been in the forefront of applying anthropometric data, often with computerized design models, an approach followed by some industries.

Data on anthropometry (and biomechanics) are needed for general design standards and specific requirements, used both in the design of new systems and in the evaluation of existing ones so that systems, products, machines, and tools "fit" the human operator and user, with the purpose of achieving "ease and efficiency."

2.5.3 APPLICATIONS: HUMAN–EQUIPMENT INTERFACES

The primary application areas for anthropometric data are in design of

Work space
Equipment, tools, controls
Protective clothing

Examples for these applications are the "cockpits" in airplanes, automobiles, earthmoving equipment, tanks, submarines, and surface ships. Other typical applications are hand-operated controls, helmets, face masks, and survival clothing for mine rescues. The design of space ships as developed in the United States and in the Soviet Union depends much on human data.

Design "contours" are established by the outline of the human body or its segments. These contours establish, for example, the minimal sizes of openings through which the body must pass (doors, hatches, etc.) or through which body segments must fit for maintenance purposes, or of openings in machinery so *small* that human body parts (e.g., fingers) cannot pass so that they may not be injured. Similarly, contours establish design dimensions for protective clothing.

Another set of interfaces is established by points or landmarks on the human body. For example, the eye establishes a design specification for the "line" or "field" of sight, such as used in aircraft design; or the "eye ellipse" (all possible locations of the drivers' eyes) used in automobile design. Another major interface is established by "butt and back" for design of seats in vehicles, or offices. The foot design landmark establishes the heel rest or "package origin" in automobile design, determining the location of pedals. Finally, the hand establishes a very important design variable in terms of its outer, inner, or preferred reach envelope.

2.5.4 TERMINOLOGY AND STANDARDIZATION

Figure 2.5.2 shows the measuring planes and the most often used terms in describing the human body. Figures 2.5.3 and 2.5.4 indicate anatomical landmarks used in anthropometry.

To avoid ambiguity, the following terms are used in anthropometry:

Height. A straight-line, point-to-point vertical measurement.
Breadth. A straight-line, point-to-point horizontal measurement running across the body or segment.
Depth. A straight-line, point-to-point horizontal measurement running fore–aft.
Distance. A straight-line, point-to-point measurement between landmarks.
Circumference. A closed measurement following a body contour, hence usually not circular.
Curvature. A point-to-point measurement following the contour, usually not circular.

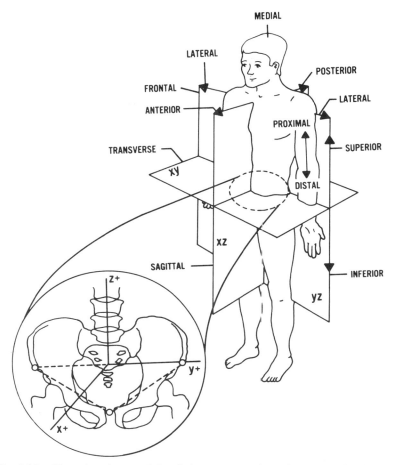

Fig. 2.5.2. Measuring planes and descriptive terms used in anthropometry (NASA, 1978).

Traditionally, these and other measures (such as weight and volume) are taken in the metric system.

For the measurements, the body is placed in an upright straight posture, with body segments at 0, 90, or 180° to each other, for example, "stand erect, heels together; butt, shoulder blades, and back of head touching a wall; arms extended straight forward, fingers straight. . . ." If needed, the head is positioned in the "Frankfurt plane: pupils on the same horizontal level; right tragion (approximately earhole) and the lowest point of the right orbit (eye socket) also in a horizontal plane." Figure 2.5.5 shows typical measuring postures.

Although these standard procedures allow uniform and repeatable measurements, there remain several problems. First, the measurement postures employed are not the ones found at work. Hence, they must be corrected and adjusted to reflect actual working postures. Second, many of the measurements are not spatially related to each other. For example, in the lateral view, stature, eye height, and shoulder height are not in the same plane. Third, although most measurements are taken to bony landmarks, some are taken along soft tissues. This may cause problems, for example, in the fitting of face masks. Although bony landmarks establish more reliable measuring points than soft tissue contours, their relations to the skeletal links and articulations of the human body must be established.

2.5.5 ANTHROPOMETRIC SURVEYS AND MEASUREMENT TECHNIQUES

Traditionally, anthropometric surveys were performed for other than engineering application purposes; only during the last few decades were surveys specifically aimed at providing design data, particularly for the armed forces. Until recently, therefore, information on civilian populations was scarce whereas soldiers were well described.

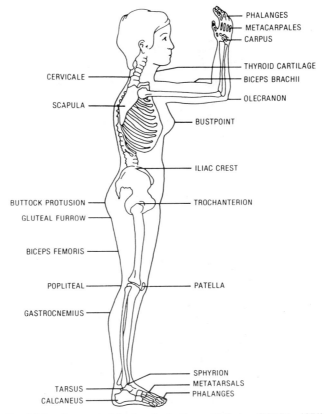

Fig. 2.5.3. Anatomical landmarks in the sagittal view (NASA, 1978).

Large-scale anthropometric surveys are very expensive, time-consuming, and difficult to perform. However, on the basis of existing anthropometric information one can do small though highly directed surveys in which only key dimensions are measured. Then, other dimensions are derived statistically from those key dimensions. For this, regression models and matching procedures have proved to be effective.

Measurement techniques usually follow the traditional procedure of using instruments that require physical contact with the subject. Their accuracy (reflecting the true value) and precision (repeatability) are sufficient for practical purposes, and their application is rather simple and straightforward, but they have two disadvantages: they are slow and they cannot describe the body in motion. Hence newer techniques (mostly photographic procedures) have been explored and used on several occasions.

There is an interplay between the aspects of "sampling" (who, when, where, how many), the measuring procedures (traditional tools, photography, holography, space-spotting), and the purposes of the data (nutritional survey, orthopedic survey, clothing survey, equipment design survey). Hence organization and procedures of such surveys may be very complicated. However, new approaches to anthropometric information gathering are likely to change the assessment procedures very much from those employed just a few years ago.

2.5.6 MEASURING INSTRUMENTS

Early measurements of body sizes used standard human dimensions as comparison units, for example, span of hand, thumb, or foot, often derived from the body dimensions of the rulers and kings. Today's anthropometry still uses a small set of simple traditional instruments:

Anthropometers, basically straight rods with one fixed and one movable arm, with the distance between these two arms indicated on the ruler.

Spreading and sliding calipers, used to measure shorter distances than with the anthropometer.

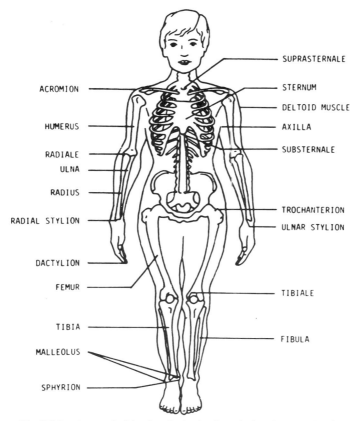

Fig. 2.5.4. Anatomical landmarks in the frontal view (NASA, 1978).

Tapes to measure along contours.

Scales to measure weight (mass).

In addition, several specialized instruments are used, such as cones to measure grip circumference or boards with holes to measure finger size. Some of these instruments are incorporated into measuring systems, such as "boxes" to insert foot or head, or vertical walls with scale markings to position the body and to take measurements. The dimensions measured are usually directly read from the scales on the instruments; however, one can use analog or digital readout devices that allow automatic recording of the measured results in computer-compatible means.

Photography is another way to take human body dimensions. Photographic measures may rely on one camera with mirrors or on several cameras to take pictures from usually orthogonal directions. More complex systems use stereographic images, including holography, or ultrasonics or thermography. Of course, one may take either still pictures or movies. Photography or videotaping can store the image of the whole body or of segments thereof, thus allowing different dimensions to be taken at any time from the picture. However, it is difficult to identify landmarks, particularly bony landmarks, on the image of contours.

Obviously, the traditional way of using physical measurements needs experienced measurers and is slow, cumbersome, and prone to error. However, it is simple and easily understood. Further, there is the practical point of view that the current data base relies on measurements taken in this way; one cannot simply discard all this information. Still, advances in measurement technology quickly might lead to different techniques, for instance, of some type of body imaging.

2.5.7 TYPES OF BODY BUILD

People come in different sizes, and with different proportions. Hence, there has been always interest in associating certain traits with the appearance of the body. Categorization of body builds into different

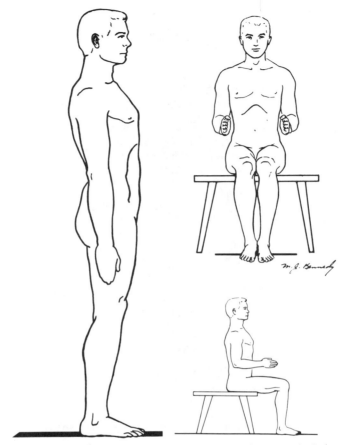

Fig. 2.5.5. Typical postures used in anthropometric data gathering.

types is called somatotyping.* Early somatotyping schemes were developed by Hippocrates (about 400 BC), who thought that body types, and particularly temperaments, were determined by body fluids. He distinguished between the "moist" type determined by "black gall," the "dry" type governed by "yellow gall," the "cold" type characterized by "slime," and the "warm" type governed by "blood."

The psychiatrist Ernst Kretschmer developed in the 1920s a three-type system intended to describe psychological traits. His basic body types are somewhat similar to the ones used in the 1940s by W. H. Sheldon, whose ratings were meant to be purely descriptive of the body proportions. Sheldon's typology was originally based on intuitive assessment, not on actual measurements, which were included later by his disciples. Table 2.5.1 describes the body types, and the terms used to describe them.

Unfortunately, these somatotypes have not proved to be reliable predictors of performance or capabilities, and hence are of little use for engineers.

2.5.8 VARIABILITY OF ANTHROPOMETRIC DATA

There are many causes for and symptoms of variability of anthropometric data. They can be divided into three groups: (1) intraindividual variations, (2) interindividual variations, and (3) secular variations.

DNA combinations are a major cause of size variability: about 2.4×10^9 possible chromosome combinations exist. The individual genetic endorsement comprises the genotype (cellular differentiation) and the phenotype, which determines the biologically measurable characteristics.

The environment may influence body size by altitude, temperature, sunlight, and topography, includ-

---

* From the greek *soma,* body.

Table 2.5.1 Body Typologies

Descriptor	Typology	
	Kretschmer	Sheldon
Lean, slender, fragile	Asthenic (leptosomic)	Ectomorphic
Stocky, stout, soft, round	Pyknic	Endomorphic
Muscular	Athletic	Mesomorphic

ing soil type. Obviously, nutrition has indirect and direct effects: overeating increases body sizes in the direction of obesity, whereas lack of nutrition leads to slenderness and to a reduction in height.

The effects of aging are obvious. During the growing years stature, weight, and accompanying dimensions increase. During early adulthood, in the twenties, many dimensions become reasonably stable. With increasing age, certain dimensions begin to change again, heights are reduced, and circumferences and weight increase. Table 2.5.2 approximates changes in key body dimensions with age.

Because the U.S. population is a composite of many different races, it is difficult and in many cases pointless to explore the differences associated with ethnic origin in body sizes. In general terms, the differences between U.S. white and black ethnic groups are relatively small, but there may be rather distinct differences between these two groups and people of oriental origin. Figures 2.5.6 and 2.5.7 indicate schematically some of these differences, which, however, are of no consequence for the design of equipment for the total U.S. population. One should remember that U.S. population statistics include *all* ethnic groups.

There are also some differences in body sizes among different professions. However, the often postulated and occasionally demonstrated differences between white and blue collar groups are not very clear and again of no consequence for design that aims to fit the whole population. The same holds true for left–right asymmetry, referring to differently developed body segments depending on one's preference to perform activities with the left or right hand or foot. Again, those differences are too small to warrant consideration in design for the total population.

There are, of course, differences in body dimensions of males and females. Figures 2.5.8 and 2.5.9 indicate that many of the heights and circumferences are distinctly different between males and females. However, there is also significant overlap in many dimensions, for example, in lower leg length and in buttock circumference. Thus although there are anthropometric disparities that may be of importance for special design purposes, other dimensions are sufficiently similar to allow "unisex" designs. A

Table 2.5.2 Approximate Changes in Stature with Age

Age (years)	Change (cm)	
	Females	Males
1 to 5[a]	+36	+36
5 to 10	+28	+27
10 to 15	+22	+30
15 to 20	+ 1	+ 6
20 to 35[b]	0	0
35 to 40	− 1	0
40 to 50	− 1	− 1
50 to 60	− 1	− 1
60 to 70	− 1	− 1
70 to 80	− 1	− 1
80 to 90	− 1	− 1

Source: Based on data by VanCott and Kinkade (1972).

[a] Average stature at age 1: females 74 cm, males 75 cm.

[b] Average maximal stature: females 161 cm, males 174 cm.

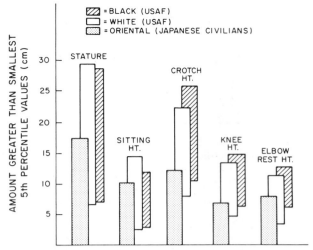

Fig. 2.5.6. Comparison of 5th to 95th percentile male heights (NASA, 1978).

typical example for this is the adjustment range of arm rests on chairs, as reflected by the anthropometric dimension "elbow rest height." Figure 2.5.10 shows the differences and overlaps in a bivariate graph of stature and weight for female and male U.S. Air Force soldiers.

There are also transient diurnal body size changes. For example, a person's body weight varies by up to 1 kg/day owing to changes in body water content. Stature may be reduced at the end of the day by up to 5 cm, mostly because of changes in postures and in thickness of spinal disks. Leaning erect against a wall during measurement as opposed to free standing may increase stature by up to 2 cm. Measuring stature in the prone position increases stature up to 3 cm as compared to the standing position. Circumferences change with different conditions: for example, hip and buttock circumferences are quite different while sitting as compared to standing, chest circumference changes with breathing, biceps circumference reflects muscle flexion, and so on. Garments can change body dimensions very much: for example, "foundations" worn by females change certain circumferences; "short sleeve" clothing is quite different from outdoor winter clothes.

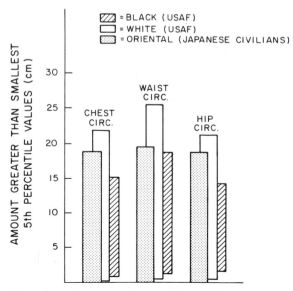

Fig. 2.5.7. Comparison of 5th to 95th percentile male circumferences (NASA, 1978).

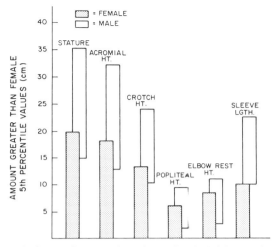

Fig. 2.5.8. Comparison of 5th to 95th percentile male and female heights and lengths (NASA, 1978).

Often discussed are "secular" changes in adult body sizes, particularly observed during the last several decades. Current estimates are that in the United States stature recently has increased by about 1 cm per decade, and during the same time weight has increased by about 2 kg. There is much speculation as to whether these secular changes will continue at the same speed. The observation periods with reliable body measures are too short to be certain; however, there is reason to believe that the observed changes depend on interactions between genetic and environmental factors such as nutrition, health, care, and sanitation. NASA and the U.S. Air Force had to deal with this problem in the 1960s, when design guidelines for the space shuttle to be used in the 1980s had to be developed. At that time it was determined that the body size increases would not significantly affect design standards, because within that period of about two decades no dramatic changes should be expected. This statement appears to be true also for industrial applications, where very few pieces of equipment, or tools, have to be fitted so exactly that the rather minute changes in anthropometry would require design standard changes within relatively short periods of time.

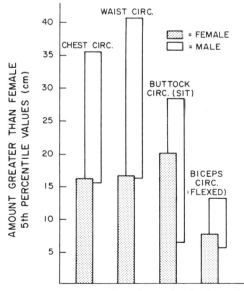

Fig. 2.5.9. Comparison of 5th to 95th percentile male and female circumferences (NASA, 1978).

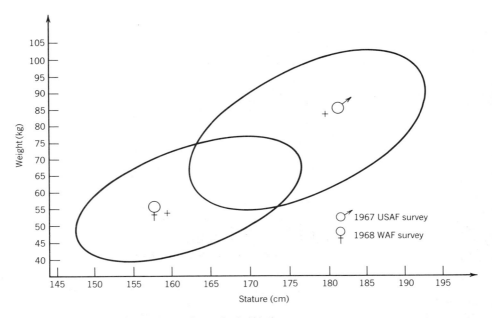

*Approximately 95% of each group is contained within the
appropriate ellipse

"NASA RP-1024, 1978"

Fig. 2.5.10. Stature and weight distributions for female and male soldiers (NASA, 1978).

2.5.9   ANTHROPOMETRIC DESIGN DATA

The most complete set of information on the body size of the adult U.S. population is contained in
NASA's *Anthropometric Source Book* (NASA, 1978). The data mostly refer to military populations.
Hence various attempts have been made to use this information in combination with the rather few
sources of reliable data on civilians to estimate the current body dimensions of the U.S. civilian population. Table 2.5.3 presents such estimates for U.S. civilians, female and male, for the age group of
approximately 20 to 60 years. This table indicates that the 50th percentile values of stature are 160.5
cm for females and 173.6 cm for males. (This corresponds to about 5'3" and 5'7", respectively.)

Table 2.5.3   U.S. Civilian Body Dimensions, Female/Male, in Centimeters for Ages 20 to 60 Years[a]

| | Percentiles | | | |
	5th	50th	95th	Std. Dev.
Heights				
Stature (height)	149.5/161.8	160.5/173.6	171.3/184.4	6.6/6.9
Eye height	138.3/151.1	148.9/162.4	159.3/172.7	6.4/6.6[b]
Shoulder (acromion) height	121.1/132.3	131.1/142.8	141.9/152.4	6.1/6.1[b]
Elbow height	93.6/100.0	101.2/109.9	108.8/119.0	4.6/5.8
Knuckle height	64.3/69.8	70.2/75.4	75.9/80.4	3.5/3.2
Height, sitting	78.6/84.2	85.0/90.6	90.7/96.7	3.5/3.7
Eye height, sitting	67.5/72.6	73.3/78.6	78.5/84.4	3.3/3.6[b]
Shoulder height, sitting	49.2/52.7	55.7/59.4	61.7/65.8	3.8/4.0[b]
Elbow rest height, sitting	18.1/19.0	23.3/24.3	28.1/29.4	2.9/3.0
Knee height, sitting	45.2/49.3	49.8/54.3	54.5/59.3	2.7/2.9
Popliteal height, sitting	35.5/39.2	39.8/44.2	44.3/48.8	2.6/2.8
Thigh clearance height	10.6/11.4	13.7/14.4	17.5/17.7	1.8/1.7

Table 2.5.3   (*Continued*)

	Percentiles			
	5th	50th	95th	Std. Dev.
Depths				
Chest depth	21.4/21.4	24.2/24.2	29.7/27.6	2.5/1.9[b]
Elbow–fingertip distance	38.5/44.1	42.1/47.9	46.0/51.4	2.2/2.2[b]
Buttock–knee distance, sitting	51.8/54.0	56.9/59.4	62.5/64.2	3.1/3.0
Buttock-popliteal distance, sitting	43.0/44.2	48.1/49.5	53.5/54.8	3.1/3.0
Forward reach, functional	64.0/76.3	71.0/82.5	79.0/88.3	4.5/5.0
Breadths				
Elbow-to-elbow breadth	31.5/35.0	38.4/41.7	49.1/50.6	5.4/4.6
Hip breadth, sitting	31.2/30.8	36.4/35.4	43.7/40.6	3.7/2.8
Head dimensions				
Head breadth	13.6/14.4	14.54/15.42	15.5/16.4	0.57/0.59
Head circumference	52.3/53.8	54.9/56.8	57.7/59.3	1.63/1.68
Interpupillary distance	5.1/5.5	5.83/6.20	6.5/6.8	0.44/0.39
Hand Dimensions				
Hand length	16.4/17.6	17.95/19.05	19.8/20.6	1.04/0.93
Breadth, metacarpal	7.0/8.2	7.66/8.88	8.4/9.8	0.41/0.47
Circumference, metacarpal	16.9/19.9	18.36/21.55	19.9/23.5	0.89/1.09
Thickness, metacarpal III	2.5/2.4	2.77/2.76	3.1/3.1	0.18/0.21
Digit 1				
Breadth, interphalangeal	1.7/2.1	1.98/2.29	2.1/2.5	0.12/0.13
Crotch–tip length	4.7/5.1	5.36/5.88	6.1/6.6	0.44/0.45
Digit 2				
Breadth, distal joint	1.4/1.7	1.55/1.85	1.7/2.0	0.10/0.12
Crotch–tip length	6.1/6.8	6.88/7.52	7.8/8.2	0.52/0.46
Digit 3				
Breadth, distal joint	1.4/1.7	1.53/1.85	1.7/2.0	0.09/0.12
Crotch–tip length	7.0/7.8	7.77/8.53	8.7/9.5	0.51/0.51
Digit 4				
Breadth, distal joint	1.3/1.6	1.42/1.70	1.6/1.9	0.09/0.11
Crotch–tip length	6.5/7.4	7.29/7.99	8.2/8.9	0.53/0.47
Digit 5				
Breadth, distal joint	1.2/1.4	1.32/1.57	1.5/1.8	0.09/0.12
Crotch–tip length	4.8/5.4	5.44/6.08	6.2/6.99	0.44/0.47
Foot Dimensions				
Foot length	22.3/24.8	24.1/26.9	26.2/29.0	1.19/1.28
Foot breadth	8.1/9.0	8.84/9.79	9.7/10.7	0.50/0.53
Lateral malleolus height	5.8/6.2	6.78/7.03	7.8/8.0	0.59/0.54
Weight (kg)	46.2/56.2	61.1/74.0	89.9/97.1	13.8/12.6

[a] Courtesy of Dr. J. T. McConville, Anthropology Research Project, Yellow Springs OH 45387 and Dr. K. W. Kennedy, USAF-AMRL-HEG, OH 45433.

[b] Estimated by the author.

This is somewhat less (about 1 inch) than has been published for U.S. military populations, indicating that indeed the military is an anthropometrically select group.

Altogether, most anthropometric data follow reasonably well a normal or Gaussian distribution. Hence, one can equate the 50th percentile with the average, or mean, value. In this case, one is allowed to use parametric statistics which mostly rely on using the average $\bar{x}$, the standard deviation S, and the sample size N. The most important formulas are given in Table 2.5.4. Of course, these apply only if the variable x is normally distributed.

A percentile is the value of a variable (e.g., eye height) below which is a known percentage of all values (say, 5%) and above which is the rest (in this case, 95%). If one designs, for example, to fit persons between the 5th and 95th percentile, one knows that this design will fit the central 90%, but it is too large for 5% and too small for another 5% of the intended user population. The use of percentiles is a major step away from the simple but false assumption that one could design for the mythical "average person." No such phantom has ever existed, nor will. It has been shown repeatedly that a person average in one dimension (say, stature) is not very likely to be average in other dimensions (for example, weight, leg length, arm circumference). A person "average" in many or all dimensions simply does not exist; using this ghost as a design principle is inexcusable. Of course, one cannot make the assumption either that there would be persons who have all 5th percentile dimensions, or 80th, for example. A similar faulty assumption underlies the idea that one could use constant proportions, such as alleging that leg length is a given percentage of stature, or that there is a constant ratio between weight and stature. Proportioning assumes again that one can simply use average dimensions for both numerator and denominator of that ratio—this obviously can lead to a multiplication of errors.

A better procedure is to select carefully the most critical dimension and to determine the appropriate percentile cutoff points at the lower and higher ends of the distribution. Though one would normally use a symmetrical distribution (such as designing for the 5th through 95th percentiles) there may be reasons to select a nonsymmetrical user population. Table 2.5.5 presents the multiplication factors and procedures to calculate the most important percentile points.

An example for the use of anthropometric data for the design of equipment is presented in Table 2.5.6. Here, the heights of the seat, of a support surface, and of a footrest were calculated for computer work stations, assuming that the 1st percentile female through 99th percentile male user had to be accommodated. Starting point for the calculations was "popliteal height, sitting" to determine the necessary height of the seat. To this was added the thigh clearance height to determine the needed opening under the support surface height. (For anthropometric data used, see Table 2.5.3.) The data show that different adjustment ranges and design dimensions result from the three design strategies used.

Of course, there are usually some assumptions that one must make, and some data that are not so appropriate or reliable, but are the "best available," or one's "best estimate." Two examples follow. In the assessments of adjust heights for computer work stations, the assumption was made that one must consider the possibility that a person adjusting the seat to the highest position in order to fit long legs also may have very thick thighs (which is not necessarily true). The computer work stations were to fit very small females as well as very big males (which may be an excessively wide design range). However, if such assumptions are rationally made and clearly stated ("p_1 females through p_{99} males") one can discuss the appropriateness in a reasoned manner and hence either continue to use these assumptions, or readjust them. The point is that exact anthropometric data are available and should be used in the same exacting way as is done usually in engineering design.

Among the decisions to be made by the design engineer is the population to be fitted. The data in Table 2.5.3 present information separately for females and for males. This information was used

Table 2.5.4 Statistics in Anthropometry ("Normal" Distribution Assumed)

Mean	$\bar{x} = \dfrac{\Sigma x}{N}$
	coincides with median (50th percentile) and mode
Range	$x_{max} - x_{min}$
Standard deviation	$S = (\text{variance})^{1/2} = \left[\dfrac{\Sigma(x - \bar{x})^2}{N} \right]^{1/2}$
Skewness	$\dfrac{\Sigma(x - \bar{x})^3}{N}$
Coefficient of correlation	$R = \dfrac{S_{xy}}{(S_x S_y)^{1/2}}$

Table 2.5.5 Calculation of Percentiles

Percentile p Associated With			
Below Mean $x_b = \bar{x} - kS$	Above Mean $x_a = \bar{x} + kS$	Central Percent Included in the Range x_b to x_a	k
0.5	99.5	99	2.576
1	99	98	2.326
2	98	96	2.06
2.5	97.5	95	1.96
3	97	94	1.88
5	95	90	1.65
10	90	80	1.28
15	85	70	1.04
16.5	83.5	67	1.00
20	80	60	0.84
25	75	50	0.67
37.5	62.5	25	0.32
50	50	0	0

in the preceding example to fit the small females and the large males. However, one might also decide to develop a composite population, for example, one that consists of $a\%$ females and $b\%$ males. In this case, one needs certain computational procedures to develop such a composite population. Table 2.5.7 presents the formulas needed for this procedure.

The U.S. worker population has changed dramatically from just a few decades ago. For example, the life expectancy has increased by approximately 25 years from 1900 to 1980, now being about 73 years. With the current birth rate, approximately 18% of all Americans will be over 65 years of age in 2020. Furthermore, it is estimated that by 1995 two-thirds of all U.S. workers may be female. These increases in females and older persons among the working population may have significant effects on design and selection of equipment, tools, and clothing.

Table 2.5.6 Effects of Three Design Strategies on Heights of Equipment (in Centimeters) for Computer Work Stations (Based on 1st Percentile Female and 99th Percentile Male Anthropometry)

	Strategy		
Equipment Height	Adjust Heights of Seat and of Support Surface	Adjust Seat Height but Keep Support Surface Fixed	Keep Height of Seat Fixed but Adjust Support Surface
Support surface[a]			
Maximum	72.7	72.7 (fixed)	72.7
Minimum	47.3		62.2
Adjustment range	25.4		10.5
Seat[b]			
Maximum	52.7	61.2	52.7 (fixed)
Minimum	35.8	52.7	
Adjustment range	16.9	8.5	
Footrest			
Maximum	Not needed	27.4	16.9
Minimum		0.0	0.0
Adjustment range		27.4	16.9

Source: Adapted from K. H. E. Kroemer, Design parameters for video display terminal stations. *Journal of Safety Research, 14:* 131–136 (1983).

[a] Assuming 2 cm table thickness.

[b] Including 2 cm for heels.

Table 2.5.7 Percentile Values of Composite Populations

Given: Two samples a and b; $a\% + b\% = 100\%$ of the composite population.

To determine at what percentile of the composite population is a specific value of x, one proceeds stepwise (for this, one needs to know mean $\bar{x}$ and standard deviation S of the distribution of the variable x for both samples a and b).

Step 1: Determine factors k associated with x in the samples a and b. For sample a:

$$x_a = \bar{x}_a - k_a S_a \quad \text{if } x_a < \bar{x}_a$$
$$x_a = \bar{x}_a + k_a S_a \quad \text{if } x_a > \bar{x}_a$$
$$k_a = \frac{|x_a - \bar{x}_a|}{S_a} \tag{1a}$$

Similarly, for sample b:

$$k_b = \frac{|x_b - \bar{x}_b|}{S_b} \tag{1b}$$

Step 2: Obtain factor k associated with x in the combined population:

$$k = ak_a + bk_b \tag{2}$$

Step 3: Determine percentile p associated with k; use Table 2.5.5. If percentiles for each x are known in each group, one may simply add the proportioned percentiles:

$$p = ap_a + bp_b \tag{3}$$

Source: Modified from K. H. E. Kroemer, Engineering anthropometry. In D. J. Oborne and M. M. Gruneberg, Eds., *The physical environment at work*. London: Wiley, 1983, pp. 39–68.

2.5.10 (INSTEAD OF A) SUMMARY

The following excerpt appeared in the *Washington Post* on May 25, 1984.

Lt. J. G. Chimpanzee

The Navy has adopted new flight training standards that will require its aviators, as a whole, to have longer arms and shorter legs.

The standards will exclude 73 percent of all college-age women and 13 percent of the college-age men, according to a military spokesman.

Capt. Frank Dully, commanding officer of the Naval Aerospace Medical Research Laboratory, said the new standards were devised because some aviation candidates could not reach rudder pedals or see over instrument panels. Some taller pilots were so tightly wedged that their helmets bumped the aircrafts' canopies.

"We found out that manufacturers are still building airplanes the way they want, but God is not making people to fit them," Dully said.

Previously, 39 percent of the female applicants and 7 percent of the men were ineligible to become aviation candidates because of their size.

REFERENCES AND BIBLIOGRAPHY

Chapanis, A., Ed. (1975). *Ethnic variables in human factors engineering.* Baltimore: The John Hopkins University Press.

Department of Defense Military Handbook (1980). Anthropometry of U.S. military personnel (metric) (DOD-HDBK-743). Washington, DC: U.S. Government Printing Office.

Easterby, R., Kroemer, K. H. E., and Chaffin, D. B., Eds. (1982). *Anthropometry and biomechanics.* New York: Plenum.

NASA (1978). *Anthropometric source book,* 3 vols. (NASA Reference Publication 1024). Houston, TX: NASA.

Roebuck, J. A., Kroemer, K. H. E., and Thomson, W. G. (1975). *Engineering anthropometry methods.* New York: Wiley.

Van Cott, H. P. and Kinkade, R. G., Eds. (1972). *Human engineering guide to equipment design.* Washington, DC: U.S. Government Printing Office.

CHAPTER 2.6

BIOMECHANICS OF THE HUMAN BODY

KARL H. E. KROEMER

**Virginia Polytechnic Institute and State University
Blacksburg, Virginia**

2.6.1 SCOPE, DEVELOPMENT, AND STATUS

Biomechanics explains the characteristics of a biological system, here the human body, in mechanical terms. This definition makes the usefulness and applicability of this scientific approach clear, as well as its limitations. Biomechanics relies heavily on basic information from anthropology, particularly anthropometry and anatomy, and from physiology and orthopedics, whereas basic research procedures and techniques obviously stem from physics (mechanics), mathematics, and computer sciences. Biomechanical research and applied work are of particular interest to engineers.

The origins of biomechanics reach back to the middle 1600s: Alfonso Giovani Borelli's "biomechanical model" of the human body represented the long bones as straight links connected at articulations representing body joints; muscles were considered to be "engines" spanning one or two body articulations and moving the system components. The link–joint–muscle model still underlies current "stickman" approaches. About 100 years ago, a series of research papers was published in Germany that discussed mechanical attributes of the human body, such as mass, properties, and the effects of external and internal forces. In more recent years, high-performance aircraft, space travel, automobile crash experiments, and the sports sciences have benefited much from the mechanical consideration of human body behavior. To characterize body components one often uses analogies such as the following:

Bones	Structural members, central axes, lever arms
Contours	Surfaces of geometric bodies
Flesh	Volumes, masses
Joints	Bearing surfaces and articulations
Joint linings	Lubricants
Muscles	Motors, dampers, or locks
Nerves	Control and feedback circuits
Organs	Generators, consumers
Tendons	Cables transmitting pull forces
Tissue	Elastic load-bearing surfaces, springs

These analogies indicate again the simplifications often imposed by biomechanical modeling.

Biomechanics is still a developing scientific and engineering field, as compared to the more mature area of engineering anthropometry. Research approaches, measurement procedures, and data bases are still being established. However, the scarcity of engineering data in some areas does not excuse disregarding the biomechanical knowledge already at hand. A rather substantial data and knowledge base exists, for example, in (static) muscle strength for the assessment of lift capabilities, and regarding responses to vibrations and impacts. In other areas the scientific and applied progress is fast; hence the engineer is well advised to follow the progress reported in the scientific and engineering literature, for example, in the journals *Biomechanical Engineering, Biomechanics, Ergonomics, Human Factors,* and *Spine.*

2.6.2 MECHANICAL BASICS

By definition, "mechanics is the study of forces and their effects on masses." Within this field one distinguishes between statics and dynamics. In statics, one studies masses at rest, in equilibrium as a result of acting forces. In dynamics, the object of studies are motions of masses. Within the subset of dynamics, in kinematics movement is studied without consideration of the forces, whereas in kinetics the forces bringing about the movement are the major study objects.

Biomechanics is based on Newton's laws. The first states that a mass remains at uniform motion (which includes at rest) until acted upon by unbalanced forces; the second is derived from the first, indicating that force is the result of acceleration times mass; the third indicates that action must be opposed by reaction.

"Force" is one of the important aspects in biomechanics. Unfortunately, it is not a basic unit of the physical system, but a derived unit (according to Newton's second law). This has the interesting result that no device exists that measures force directly. All force measuring devices use other physical phenomena (often displacements) which then are transformed and calibrated in force units. In this context it should be remembered that the pound (lb), gram (g), and ounce (oz) are not force but mass units. The more appropriate terms poundal, slug, gram-force, and pond are no longer in general use. The correct unit is the Newton: 1 lb_f is approximately 4.45 N; 1 kg_f = 1 kp $\simeq$ 9.81 N.)

Force has vector qualities, described not only by its magnitude but also by its direction. Unfortunately, the directional aspect is occasionally forgotten in human "strength" assessment in sports and physical education.

Torque (moment) is the product of force and its lever arm (perpendicular to the vector direction) to the articulation about which it acts. This lever arm is often called the "mechanical advantage."

2.6.3 ANTHROPOMETRIC BASICS

Current biomechanics relies much on data gathered in traditional anthropometry, adapted and often simplified to fit the mechanical framework and terminology. Conversely, systematic biomechanical approaches have influenced definitions and procedures in anthropometry as well. In the preceding chapter Figure 2.5.2 indicates the location of the basic reference planes: the midsagittal plane divides the body into a left and right half; the frontal plane divides the body into anterior and posterior sections; and the transverse plane cuts the body into superior and inferior parts. The location of the midsagittal plane is rather well defined, but the locations of the frontal or transverse planes are not. However, it is usually assumed that the three planes meet in the pubic area, so that their common intersection, the origin of an XYZ axis system, is close to the center of mass of the upright standing body.

In the tradition of Borelli, the human skeletal system is usually simplified into links connecting at joints. The long bones between major articulations are thus reduced to straight lines between the joints, with the number of links (and joints) dependent on the needs for accuracy and reality of the model. Figure 2.6.1 shows a typical link–joint system used in biomechanical modeling. This is obviously a great simplification from more realistic descriptions such as shown in Figures 2.5.3 and 2.5.4. Note, for example, that the model of the spinal column has only two articulations between the cervical and sacral junctures. Clearly, this cannot be a realistic model of the spinal column with its many degrees of freedom in many articulations. However, for a given purpose, such a simplification may be acceptable and practical.

An example for such purposeful simplification is the modeling of the wrist joint. With respect to the forearm, the hand may be flexed and extended, abducted and adducted, and pronated and supinated.

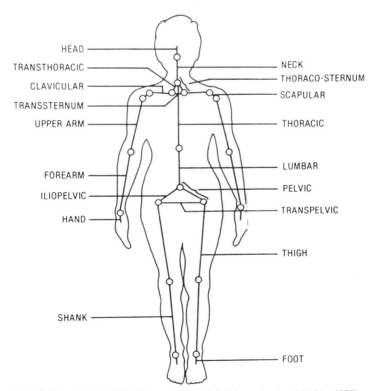

Fig. 2.6.1.   Typical link–joint system used in biomechanics (NASA, 1978).

Although the flexions and ductions can be correctly represented by a condyloid joint with two degrees of freedom, it is certainly anatomically false to place the rotatory motions also into the wrist joint (as is often done in modeling) because the twisting pronation and supination in fact take place within the forearm. Still, this procedure may be sufficiently realistic for certain modeling purposes, provided the limitations of this simplification are clearly understood and respected.

There are no unambiguous and easily understood terms to describe the motions of the body and of its segments. Obviously, just a few lines above it was tedious to describe the several motion capabilities of the hand with respect to the forearm. The traditional medical–anatomical terminology is often unclear and does not distinguish between displacements of a body segment and the motions occurring in the articulations. Furthermore, a null position is often implied but not defined. Attempts to design better notation systems have not yet met with general acceptance. Several ballet notations exist that describe positions and motions of body segments to the satisfaction of choreographers and dancers, but they are cumbersome in engineering use. We still need to develop a notation system that is exact and easily applied.

2.6.4 MUSCLE STRENGTH

The assessment of human muscle strength is a task that has been approached by anthropometrists, biomechanicists, physiologists, and orthopedists. Strength data are of particular interest to engineers for the design of equipment and controls that must be operated by muscular strength exertion. "Strength" is understood here as the force, or torque, that can be transmitted (usually by hand or foot) to an outside object in a single voluntary effort.

2.6.4.1 Generation of Strength

With current technology, no means exist to measure reliably the amount of contractile force developed within a muscle in vivo: no "force cell" exists that could be inserted into the muscle or its tendon attachments to the bone, to measure the actually developed force. Hence, one must measure strength indirectly, that is, as the result of the (unknown) muscular force, M, pulling at its lever arm, l, in the attempt to rotate a body segment around its nearest joint against an external resistance. Figure 2.6.2 shows this using the example of the biceps muscle trying to rotate the forearm around the elbow joint. As this sketch indicates, even under the (untrue) assumption of a constant muscular force the amount of torque developed depends on the changing lever arm, acting at right angles between the force vector and the elbow joint. Figure 2.6.3 details the biomechanical conditions further. Here, it is recognized that the actual direction of the contraction force vector, M, is at an angle β to its vector component, S, which is perpendicular to the lever arm, l. Of course the angle and the amount of muscular contraction force, M, depend on the elbow angle α. S, M, and β are unknown and not directly measurable. Even the lever arm, l, is not easy to discern. However, all these are being counteracted by an external measurable force, H, in this case applied by the hand against an external measuring device. This hand force, H, acts at a known lever arm, f, around the elbow joint. For equilibrium, the following is true:

$$T = Sf \tag{1}$$

and

$$T = Pl \tag{2}$$

with

$$P = M \cos \beta \tag{3}$$

Hence

$$T = Sf = Pl \tag{4}$$

Solving for the hand force,

$$S = M \cos \beta \frac{l}{f} \tag{4a}$$

or for the muscular contraction force,

$$M = \frac{S}{\cos \beta} \frac{f}{l} \tag{4b}$$

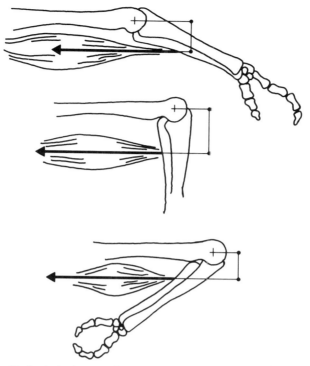

Fig. 2.6.2. Mechanical advantage (lever arm) of the biceps as a function of elbow angle.

Of course the angle β depends on the anatomical conditions existing and on the elbow angle α.

If the resultant muscle torque exceeds the external resisting torque, the muscles involved shorten while contracting and hence rotate the body segment involved by decreasing the angle spanned by the muscle. This is called a *concentric* motion. Obviously, the dynamics of acceleration, constant motion, and decelerations of muscle and of segment and external masses involved can become quite complex. On the contrary, if the muscular torque is weaker than the externally acting torque, the muscles involved stretch while contracting. This is called an *eccentric* motion. Again, the dynamics involved may be rather complex. However, if there is equilibrium between the muscular and the external torque, muscle length does not change. This is called an *isometric** muscular contraction, and given the condition of equilibrium between forces and torques, this constitutes the rather simple case of statics.

The strength exerted is a function of the skill of the person to position the body so that maximal strength capabilities are used at the best leverages, by providing body support in such a way that

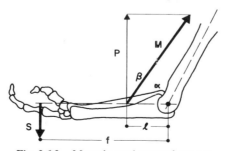

Fig. 2.6.3. Muscular and external torques.

---

* Isometric, from the Greek *isos,* equal, and *metrein,* measure.

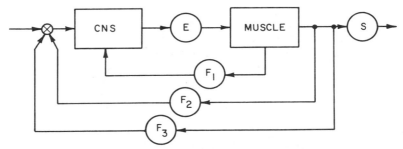

Fig. 2.6.4. Model of muscle strength exertion (modified from Kroemer, 1979, with permission of the Human Factors Society).

highest active forces are facilitated by suitable reaction forces at the support surfaces, and by careful timing of the use of muscular kinetic and by positional static energy. Figure 2.6.4 depicts a simple model of the regulation of muscle strength exertion. To generate strength, S, (such as applied to a control lever, or a dynamometer) a learned or innate "executive program" is called up in the central nervous system (CNS). "Subroutines" are used to modify the general program to adjust it to the given condition. Excitation impulses are then sent along the efferent pathways, E, to the muscles involved. (These impulses can be monitored via an electromyogram, EMG.) At the muscle, the innervated muscle fibers are triggered to contract. Depending on muscle capacity, mechanical advantages (as just discussed), and other variables, an according amount of strength, S, is generated for exertion to an external object.

The results of the programmed excitations are monitored through several afferent feedback systems, F, of which three loops are depicted. The shortest and most direct is a reflex-type loop, which utilizes primarily Ruffini organs in the body joints and Golgi and spindle organs at the muscle, which report on location, length, and tension generated through the muscular effort. The second feedback loop uses exteroceptors. These are primarily kinesthetic, reporting the touch and pressure sensations at the interfaces between the body and the external objects. The third feedback loop also relies on exteroceptors, reporting vision and sound signals. Any deviations of the afferent signals coming from the receptors are compared with prestored expected values and, if different, lead to corrections in the efferent signals generated in the CNS.

Even such a simple model of muscle strength control provides information about how to affect and how to monitor strength exertion. It shows that monitoring (via EMG) the signals sent forward along the efferent pathways to the muscle does not predict the exact amount of strength actually generated because muscle force output varies with training, fatigue, length, and mechanical advantages. The model also shows that one can influence the strength exertion by manipulating the exteroceptorial feedback. For example, a person not allowed visual and auditory information about the actual results would find it rather difficult to generate the same submaximal force repeatedly.

All measurements of muscle strength include the effects of motivation. The subject decides, consciously or not, what percentage of actually available strength will be exhibited in the given situation. This fact is reflected in the technical term "maximum voluntary contraction" (MVC). Extreme motivation may lead to extraordinary feats of strength, such as shown by the proverbial mother lifting a car from her child. Personal danger or competition may lead to very high motivation, with the possibility of damage to muscles or to tendons and their attachments to the bones. In contrast, other testing situations may motivate a subject to exhibit only a portion of the actually available strength; in an extreme situation a person might like the experimenter to believe that only very little strength is actually available to obtain fraudulent compensation for a faked loss of strength due to an accident. (Testing without feedback may unveil this—see above.) The main point is that, at present, all strength measurements depend on the voluntary participation of the subject.

2.6.4.2 Measurement Techniques

Because the static case with no change in muscle length and hence no motion is so relatively simple, most muscle strength measurements have been conducted with muscles at constant length while exerting their strength. In the real world, however, many muscle strength exertions do not take place under isometric (static) conditions, but under dynamic circumstances where motion occurs while force or torque is exerted. Hence displacement must be considered together with its time derivatives, speed, acceleration, jerk, and so on. Because all these are likely to change during the motion, over time, complex physical and physiological conditions exist that are difficult to control and to assess. One simple way to measure muscular capability is to determine the maximal amount of mass (weight) that a person can lift. This was formerly called an *isotonic* test; however, this is a false term because no "constant tonus" of the muscle is generated by the constant external weight. Therefore, this misleading term is no longer used; it has been replaced by the better descriptive term *isoinertial*, which indicates that a constant external mass is the recipient of muscular energy. The advantage of the inertial procedure is its simplicity and its realism, for example, in tests of lifting ability (Kroemer, 1983). However, the isoinertial test does not yield a direct measurement (in Newton or Newton meter) of actually generated force or torque.

Another way of controlling the measurement conditions is to keep the first derivative of displacement, velocity, constant. This condition is called *isokinetic*. In the last decade or so, measurement devices have been developed that allow angular displacements to be kept at constant speed, even if large amounts of torque are applied to the device. This actually exerted torque is measured while the device and its attached body segment are at constant angular motion; problems arise, however, at the beginning and the end of the motion, when the body segment must be accelerated to achieve the required speed, and be decelerated at the end. So far, few isokinetic strength data have been published (Kamon, Kiser, and Pytel, 1982).

Muscle strength is a complex function of inherent muscular capabilities, mechanical advantages, skill, and motivation. Muscles can actively contract only but may be passively elongated by external pull, which they resist by developing tension. Hence, muscle tension depends also on its length. Figure 2.6.5 shows the strength–length relationship schematically. Obviously, the muscle cannot contract actively below its shortest possible length. Its largest active contractile force is developed at about resting length. If further elongated, the active contractile force diminishes, reaching zero at about double resting length; however, the muscle resists the stretching by passive tension, being largest near double its resting length. (If stretched farther, the muscle or its tendon attachments may be damaged.) Hence, the total force that can be generated by the muscle is a summation of active contraction and passive resistance to external stretch. The largest muscular force as a result of that summation is at about 120% of resting length.

The force actually available at the foot, or hand, depends on the strength of the muscles involved (as just discussed) and on the relative locations of the limbs involved. These link positions determine the mechanical advantages by changing lever arms with the relative displacements of the body segments. Thus, there is an interaction between geometry and location of the body segments, and strength capabilities of the muscles involved, as sketched in Figures 2.6.3, 2.6.4, and 2.6.5. A combination of these variables usually leads to force or torque output curves as schematically shown in Figure 2.6.6. The strength available for control operation, or lifting of loads, is usually rather low near the extreme positions of the body segments, and larger in intermediate positions. However, one cannot necessarily assume that the maximum strength is available somewhere in a middle position; for example, in extending the knee joint and pushing forward with the foot (such as an automobile driver pushing hard on a pedal) the maximal forward force is available with the knee almost fully extended.

Given the knowledge about human muscle strength capabilities, data on force or torque capabilities are not as simply applicable as many engineers might hope they are. The two primary data sources, the design handbook by VanCott and Kinkade (1972) and the NASA sourcebook (1978), present almost exclusively static strength data measured under a variety of experimental conditions. These conditions include body posture (standing or sitting), body support (reaction forces), interaction with the measuring device (configuration and location of pedal or handle), gender (most data were collected on males, few on females), musculoskeletal components involved, and so forth. Hence these data represent isolated conditions and the engineer trying to use them must ascertain that the circumstances for which strength data are sought are indeed sufficiently similar to those under which the data were measured and for which they are represented.

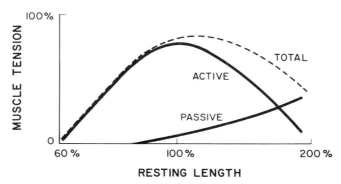

Fig. 2.6.5. Muscle force as a function of muscle length.

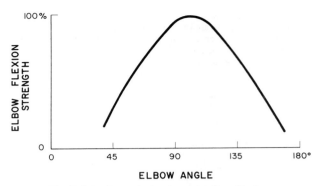

Fig. 2.6.6.   Strength developed in elbow flexion.

Because no comprehensive set of data is available, the designer often has to resort to either actually measuring strength capabilities under the conditions existent in the given case or contacting ergonomic laboratories in which such data are customarily measured. Aside from industry or government (mostly military) installations, several universities host large ergonomics laboratories, such as the following:

SUNYAB	University of Michigan
Professor C. G. Drury	Professor D. B. Chaffin
IE Department	IOE Department
Amherst, NY 14260	Ann Arbor, MI 48109
Texas Tech University	Virginia Tech (VPI)
Professor M. M. Ayoub	Professor K. H. E. Kroemer
IEOR Department	IEOR Department
Lubbock, TX 79409	Blacksburg, VA 24061

Other sources can be obtained from the Human Factors Society (1124 Montana Avenue, Santa Monica, CA 90406) or the American Industrial Hygiene Association (475 Wolf Ledges Parkway, Akron, OH 44311-1087).

2.6.5   KINEMATIC CHAINS

Two examples of working positions were briefly mentioned above: the driver of a vehicle, pressing a foot on a pedal while stabilizing the body (i.e., providing reaction force) by thrusting the back against the backrest of the seat; and the person trying to push hard with the hands, positioning the body to provide the best reaction, including the placement of the feet on the floor, bracing if possible against a solid object. In each case one recognizes that the force exerted with the foot, or hand, must be transmitted through all body segments involved until proper reaction force is provided by an external solid object. Newton's Third Law indicates that only so much force can be generated as counterforce exists. The earlier mentioned "stickman" models reflect a series of connected body links which transmit all generated forces from the point of application to the point of body support.

For the case of a person trying to lift or push with the hands while standing on an inclined surface, the kinematic chain is sketched in Figure 2.6.7. Hand forces may be exerted in horizontal or vertical directions. They generate torques around the wrist joint. These torques must be transmitted around the elbow (E), the shoulder (S), and along the cervical (C), thoracic (T), and lumbar (L), sections of the spine. Then the torque is transmitted around the hip (H), knee (K), and ankle (A), joints and finally finds its counterparts at the foot (F), where it may again be separated into horizontal and vertical forces and their lever arms.

In this chain, the weakest link determines the amount of force or torque that can be transmitted to the outside. If, for example, mechanical advantages at the shoulder were inappropriate, the torque that can be generated there would be limiting for the whole system. Or, if only limited strain could be tolerated at the lumbar spine, then this area would constitute the weak link in the total chain. Finally, the inclination α of the plane on which the foot rests and the coefficient of friction available there determine the capability of the whole system to transmit force, or torque, at the hands.

A set of basic equations provides the basis for a computational analysis for this system. These equations are as follows:

Basic Equations

$$H_x + F_x = 0$$
$$H_y + F_y = 0$$
$$H_z + F_z = 0$$
$$T_{Hxz} + T_{Fxz} = 0$$
$$T_{Hyz} + T_{Fyz} = 0$$
$$T_{Hxy} + T_{Fxy} = 0$$

The coefficient of friction μ and the angle α at which the floor is inclined determine the equations at the foot. Thus

$$F_x = \mu F_z \qquad \text{if } \mu \leqslant 1$$

With the slope angle α

$$F_x = \mu F_z \cos \alpha$$

Of course, when motion must be considered, dynamic conditions come into play according to Newton's Second Law.

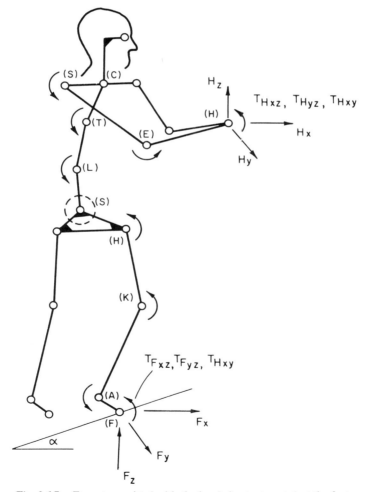

Fig. 2.6.7. Force transmitted with the hands is counteracted at the foot.

2.6.6 LOW BACK BIOMECHANICS

Low back overexertion injuries constitute, at present, the most expensive single cause of compensatory payments in U.S. industry. Many of these injuries are associated with manual material handling, particularly lifting, often attributed to excessive strain at the lower lumbar region of the spinal column. It is thought that in many cases the compression-bearing capacity of the lumbar disks is exceeded. Furthermore, relative displacements of the vertebrae may lead to a variety of problems. The resulting complex strains may result in traumatic or cumulative injuries.

Biomechanically it is rather interesting to consider the action of muscles and of load-bearing surfaces at the lower back. As Figure 2.6.8 shows schematically, all muscles involved exert coordinated pulling actions to balance the vertebral column. This results in a compression force that must be transmitted along the spinal column, aggravated by the weight of upper body mass and by an external load. The resulting force vector, **S**, must be transmitted primarily by the spinal disks. For simplicity, it is often assumed that this vector is perpendicular to the vertebral surface, generating simple compression loading in the disk. This is, of course, not true if the surfaces are at angles to each other, either sagittally or frontally, caused by bending or twisting the back. In the lumbar spine, the inferior surface of the lowest lumbar vertebra and the superior surface of the sacrum are even in the erect standing condition at a angle inclined about 45° in the sagittal view. Obviously, actual geometries and angular displacements of the load-bearing surfaces and the mechanical strain characteristics of the disk cannot truly be approximated by the simple assumption of compression loading. Research in this area is active (Chaffin and Andersson, 1984).

While attempting to lift, for example, one tightens the muscles of the stomach, particularly the rectus abdominis and external and internal oblique muscles, and generates a column of compressed air in the abdominal cavity. This column can carry some of the downward-directed load, but the largest portion of the load by far is transmitted through the disks of the spinal column, with the highest load in the lower part of the lumbar section. This is where many of the back injuries associated with heavy material handling occur.

In fact, even when handling no external load one may overstrain the lower section of the spinal column by simply transmitting the weight of the upper body. It is rather self-evident, though not yet quantitatively well researched, that bending and twisting of the body (of the spinal column) can significantly increase that load. For the rather simple case of leaning straight forward, in the midsagittal plane, one realizes from Figure 2.6.8 that the majority of the counteraction to the forward-displaced load of the body masses must be generated by the erector spinae and latissimus dorsi muscles. These

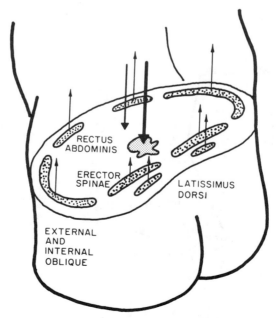

Fig. 2.6.8. Cut through the trunk at the lumbar section to show primary muscles and the spinal column (schematic, modified from a manuscript sketch by A. Schultz, about 1980).

have very short lever arms with respect to the spinal column. Hence they must generate very large pull forces to keep the forward-bent body in balance. Such very large pull forces accordingly generate very high compression forces along the spinal column. Hence forward bending of the body can overload a spinal column. This condition is clearly much aggravated if forces must be exerted to a load in front of the body, particularly when bending over it, in attempted lifting or lowering. Much biomechanical research has been focused on this problem, and although some of the phenomena are reasonably well understood, others still need research efforts.

The spinal column is one of the most vulnerable and vital parts of the human body. Injuries to the spinal column in the course of sports activities, in automobile accidents, and among military fliers and parachutists, for example, have triggered much interest in this field. A survey of the state of knowledge in spinal biomechanics is provided by Kazarian (1981). Understanding injury mechanisms allows the avoidance of hazards by suitable design of equipment and procedures.

2.6.7 BODY POSTURES

Standardized upright and stretched body postures help anatomists, anthropometrists, and biomechanicists to measure body characteristics. However, these are not the postures assumed at work, nor are they in fact convenient, comfortable, or voluntarily assumed. What, then, is a healthy and appropriate body posture at work? Many unexplained traditions and theories exist, most of which do not seem to be very relevant or well based to apply in today's conditions. For example, the evolution from four-legged to upright walking has been blamed for many spinal problems. Orthopedists and mothers used to exhort us to "sit straight," but relaxed learning against a backrest is so much more comfortable. Although many medical, physiological, orthopedic, and anatomical questions still must be answered, some biomechanical observations and experiences exist.

If at all possible, a person should be allowed to change body postures often and freely. Forcing an operator into a given posture that must be maintained over considerable times always becomes uncomfortable, if not unbearable, rather quickly.

Sitting is preferable to standing, provided the work to be performed does not require much hand force or extensive body movements. Large forward foot forces can be exerted while sitting with a solid backrest. Standing, on the other hand, allows exertion of more hand force and permits larger body motions, but the total weight of the body including the legs must be transmitted through the feet, whereas in sitting only the weight of the upper body must be transmitted to the seat while the feet rest on the floor.

When sitting, a comfortable chair must be provided. Though the concept of comfort is a difficult one, certain attributes are obvious: to allow change of sitting postures, transmit force through accommodating but not spongy surfaces, provide a large and adjustable back support with suitable upholstery, and permit easy adjustments in sitting height and sitting angles

Lying, mostly supine, is occasionally used in high-performance aircraft, and in high-acceleration maneuvers in spacecraft, because blood flow to the brain is less impeded when the acceleration vector is transversing the body. Furthermore, a larger load-bearing surface is provided. Lying or half-lying positions are also customarily assumed for relaxation, and may be suitable even for work, such as in computer operation. Interestingly, the so-called "neutral body position in weightlessness" shown in Figure 2.6.9 is (if turned on its back) quite similar to the one assumed by persons floating relaxed in water, as observed half a century ago. In fact, this is also a posture that has guided the contours of easy chairs. This "neutral" posture may be one in which the combined resulting pulls of muscles at the segments are in comfortable balance.

Under normal conditions, for healthy persons, a body position is suitable if it fulfills two requirements: The first is that the weight of body segments must be transmitted, in terms of a kinematic chain, in the least strainful way. This means, for example, for a seated person that headrest, full-size backrest, armrests, and the seatpan should be receiving the weight of head, arms, trunk, and thighs, while the weight of the lower leg and feet is transmitted to the floor or a suitable footrest. An example is the workplace equipment, particularly the seat, at a computer work station. The second requirement is that forces to be exerted toward outside objects, mostly with hands and/or feet, be counteracted at the shortest possible distance by the reaction forces provided by support surfaces. The typical example for this is the automobile driver whose foot forces are primarily taken up by the lower portion of the backrest.

For manual material handling, particularly lifting of loads from the floor, the most adequate posture is to bend the knees and to lean only slightly forward, and to straighten legs (and back) while lifting the load in front of, and close to the body (NIOSH, 1981). This reduces the biomechanical strain of body components generally to a minimum. Dangerous postures in manual material handling are particularly those in which the back is severely bent, and in which sideward twisting of the spinal column occurs. However, to avoid overexertion injuries, it is much preferable to provide ergonomically designed work tasks and equipment that make strainful lifting, lowering, pushing, pulling, and carrying unnecessary (Snook 1978; Snook, Campanelli, and Hart 1978). Training people how to "lift correctly" is much less effective than ergonomically correct job design.

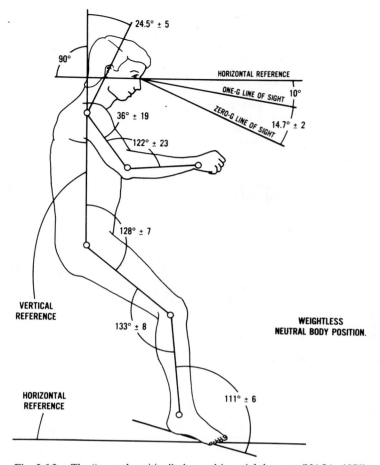

Fig. 2.6.9. The "neutral position" observed in weightlessness (NASA, 1978).

2.6.8 ERGONOMIC MODELS: A SUMMARY

A model is a simplification of an actual complex system, developed and used for understanding the behavior of the original system. Because the biomechanics of the human body are so complex and the field of biomechanics is so relatively young, no single biomechanical model of the human body exists at this time, but many models explain details. Based on theoretical "models," systematic "procedures" can be developed that provide valid biomechanical measurement "techniques" (Kroemer, 1984).

Several recent publications provide overviews of the state of the art in certain biomechanical subdisciplines. For example, the biomechanics of impacts were described by King and Chou in 1976. In 1984, King reviewed models of the musculoskeletal system, including human gait. Also in 1984, Chaffin and Andersson discussed static and dynamic modeling of the biomechanics of lifting activities, which had been reviewed in detail by Ayoub, Mital, Asfour, and Bethea in 1980. In fact, Chaffin and Andersson's book is the first to compile the existing knowledge in one comprehensive text. In 1982, Kroemer provided a short review of the status and future direction in the development of ergonomic person–equipment interface models. On the more practical side, the September 1982 issue of the *Journal of IEEE Computer Graphics* contains a number of articles discussing animation procedures useful in depicting human body motions.

Many other biomechanical models exist, for example, of the soft tissues, of the cardiovascular system, and of cellular and microscopic biological elements. However, no comprehensive model of the biomechanics of the human is at hand; in fact, even major biomechanical variables (e.g., muscle strength) have not been well modeled yet. This indicates the state of flux in this rather young but quickly developing scientific and engineering discipline. The field of biomechanics still relies much on procedures and data inputs from its parent disciplines, for example, anatomy, orthopedics, and physiology, in combination with mechanical aspects, but it is rapidly developing into a discipline of its own standing. Even in its current state, it can provide the engineer and designer with a wealth of information and with much guidance regarding the proper design of systems within which people must perform safely and well.

REFERENCES

Ayoub, M. M., Mital, A., Asfour, S. S., and Bethea, N. J. (1980). Review, evaluation and comparison of models for predicting lifting capacity. *Human Factors, 22,* 257–269.

Chaffin, D. B., and Andersson, G. B. J. (1984). *Occupational biomechanics.* New York: Wiley.

Kamon, E., Kiser, D., and Pytel, J. (1982). Dynamic and static lifting capacity and muscular strength of steelmill workers. *American Industrial Hygiene Association Journal, 43,* 853–857.

Kazarian, L. (1981). Injuries to the human spinal column: Biomechanics and injury classification. *Exercise and Sports Science Reviews, 9,* 297–352.

King, A. I. (1984, May). A Review of Biomechanical Models. *The Journal of Biomechanical Engineering, 106,* 97–104.

King, A. I., and Chou, C. C. (1976). Mathematical modeling, simulation and experimental testing of biomechanical system crash response. *Journal of Biomechanics, 9,* 301–317.

Kroemer, K. H. E. (1979). A new model of muscle strength regulation. *Proceedings, Annual Conference, Human Factors Society, Boston, MA, 29 October–1 November 1979.* Santa Monica, CA: Human Factors Society, pp. 19–20.

Kroemer, K. H. E. (1982). Ergonomics model of the human operator. *Proceedings, Workstation Space Human Factors," Leesburg, VA 24–26 August 1982.* Washington, DC: NASA, pp. A23–A28.

Kroemer, K. H. E. (1983). An isoinertial technique to assess individual lifting capability. *Human Factors, 25*(5), 493–506.

Kroemer, K. H. E. (1984). Ergonomics of manual material handling: Review of models, methods, and techniques. *Proceedings, International Conference on Occupational Ergonomics, 7–9 May 1984,* Vol. 2. Rexdale, Ontario: HFAC-IEA, pp. 56–60.

NASA (1978). *Anthropometric Source Book,* 3 Vols. (NASA Reference Publication 1024). Houston, TX: NASA.

NIOSH (1981). *Work practices guide for manual lifting* (Technical Report 81-122). Cincinnati, OH: NIOSH.

Snook, S. H. (1978). The design of manual handling tasks. *Ergonomics, 21,* 963–985.

Snook, S. H., Campanelli, R. A., and Hart, J. W. (1978). A study of three preventive approaches to low back injury. *Journal of Occupational Medicine, 20*(7), 478–481.

Van Cott, H. P. and Kinkade, R. G., Eds. (1972). *Human equipment guide to equipment design.* Washington, DC: U.S. Government Printing Office.

BIBLIOGRAPHY

Astrand, P. O. and Rodahl, K. (1977). *Textbook of work physiology,* 2nd ed. New York: McGraw-Hill.

Clauser, C. E., McConville, J. T., and Young, J. W. (1969). *Weight, volume, and center of mass of segments of the human body* (AMRL-TR-69-70). Wright-Patterson Air Force Base, OH: Aerospace Medical Research Laboratory.

Easterby, R., Kroemer, K. H. E., and Chaffin, D. B., Eds. (1982). *Anthropometry and biomechanics.* New York: Plenum.

Hay, J. G. (1973). The center of gravity of the human body. *Proceedings Kinesiology III.* Washington, D.C.: American Association for Health, Physical Education, and Recreation, pp. 20–44.

Kaleps, I., Clauser, C. E. C., Young, J. W., Chandler, R. I., Zehner, G. F., and McConville, J. T. (1984). Investigation into the mass properties of the human body and its segments. *Ergonomics, 27,* No. 12, 1225–1237.

Kroemer, K. H. E. (1983). *Ergonomics manual for manual material handling,* 2nd rev. ed. Blacksburg, VA: Polytechnic Institute and State University.

Kroemer, K. H. E. (1984). Ergonomics of video display terminal stations. *American Industrial Hygiene Association ergonomic guide,* Akron, OH.

Roebuck, J. A., Kroemer, K. H. E., and Thomson, W. G. (1975). *Engineering anthropometry methods.* New York: Wiley.

CHAPTER 2.7

MANUAL CONTROL
AND TRACKING

JAMES L. KNIGHT, JR.

AT&T Consumer Products Laboratory
Neptune, New Jersey

I wish to acknowledge the contribution of Helen Fairbrother, who prepared the section of this chapter on performance assessment methodology and dependent variables, and who offered many useful comments and much help concerning other portions as well. I also gratefully acknowledge a variety of useful recommendations and improvements suggested by several anonymous reviewers of this chapter. Finally, I would like to thank Mary L. Knight, who provided careful and thorough editorial assistance. The views and opinions expressed in this chapter are those of the author and do not necessarily reflect the opinions of AT&T Consumer Products Laboratory.

2.7.1 INTRODUCTION

It is justifiable to describe the area of manual control and tracking as fundamental to human factors in general and to human engineering in particular. The history of this area is long, probably traceable to Donders' "reaction time" laboratory; the "B-type" (choice) reaction task can be viewed as an elemental pursuit-mode step tracking task. Study in this area blossomed with wartime research into human control of complex electromechanical systems including aircraft, ships, and weapons. During this period, very robust engineering models of the human operator, applicable to a wide variety of tasks, were developed. More recently, information processing approaches have increasingly focused on underlying cognitive aspects of tracking and control performance. The advent of new, sophisticated controls made possible by computers, and of many innovative response–entry devices and control tasks arising from human interactions with computer, is providing a new domain for application of manual tracking and control technology.

In this chapter the basic elements and concepts of this theory and technology are presented. The extant data base is vast, so the content of this chapter is broad rather than deep. References are provided to lead the reader to more detailed sources. This chapter is intended to present models and guidelines at an "engineering" level of accuracy: They should facilitate the correct ordering of design alternatives in terms of performance impact and should allow some realistic assessment of potential cost/benefit trade-offs. The theory and technology of manual control and tracking have wide applicability. Manual control and/or tracking are involved in quickly reaching toward and pressing a pushbutton; in driving a car, flying aircraft, or monitoring and controlling chemical processes; in the elemental motions of production workers on an assembly line; and in quickly turning a radio dial accurately to tune in a station.

Adams (1961) defined the basic attributes of the tracking task as follows:

1. A paced, externally programmed input or command signal defines the operator's motor response, which is performed by manipulating a control mechanism.
2. The control mechanism generates an output signal.
3. The input signal minus the output signal is the tracking error quantity, and the operator's requirement is the null of this error.

It should be noted that the pacing element mentioned in Adams' first attribute may be "internal" (arising from a person's desire to finish a tracking task as quickly as possible) or "external" (arising from the independent movement of an object being tracked, for example). Most tracking tasks outside the laboratory are largely self-paced or internally paced. Most laboratory tracking tasks, from which the majority of tracking data is drawn, are externally paced. With self-paced tasks the operator has an important dimension of control lacking in tasks with rigid external pacing. The flexibility of self-pacing usually makes the tracking task easier: tracking and control task difficulty usually increase as pacing becomes more external and rigid. With self-pacing, the operator can control the rate of movement of the "target" being followed. During periods of low variability (e.g., driving along a nearly straight road) the operator can speed up and during periods of high variability (driving through a section of winding road) the operator can slow down the task. By this adjustment the operator can keep the demands of the tracking or control task near his or her limits of ability, thereby maximizing performance.

This chapter is structured around the conceptual models shown in Figures 2.7.1a and b. Figure 2.7.1a represents the control task and Figure 2.7.1b represents the tracking task. These two paradigms are closely related. In the control task, the human operator observes the status of some time-changing system. The operator's responses have a direct effect upon this system. The system can exhibit complex behavior and the form of this behavior (e.g., movement) affects the operator's ability to achieve adequate control. So an analysis of the way in which common types of system behavior affect performance is one important section in this chapter.

Before the operator can respond to the system's time-varying behavior, the operator must perceive its status. The way in which information about the system's status is presented to the operator is the second major determinant of tracking performance and is covered in another major section of this chapter.

Finally, the operator must make physical responses to change the system's status in some desired way. But to affect the system, the operator's responses usually pass through a complex control mechanism beginning with, perhaps, a joy stick, then through electronics and servomotors, hydraulic systems, and so forth before actually interacting with the controlled system. The characteristics of the path between the operator's responses and the controlled system is the third major determinant of control and tracking performance discussed in this chapter.

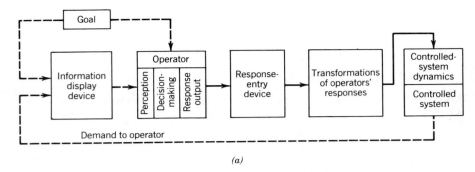

(a)

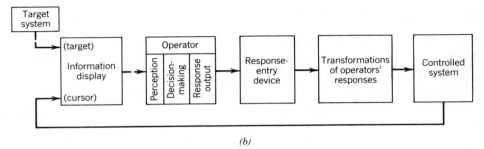

(b)

Fig. 2.7.1. Structural models of (a) control and (b) tracking tasks.

The human operator closes the control loop and his or her psychomotor, information-processing, and cognitive characteristics and skills determine the essential nature of the overall control process. Therefore, an information-processing model of the human operator is presented first to provide a framework for understanding the behavioral data of the three major sections noted above.

In the control task the operator's responses act directly upon the "target" system to keep it within tolerance or to move it to a desired state or position. The operator's goal may be shown as a zero point or fixed reference line on a display screen or meter. In the tracking task, the operator's responses do not affect the time-varying target system. Rather they affect the operator's "controlled system." Traditionally, the target system's representation on the operator's display is referred to as *the target*. The representation of the controlled system on the operator's display is referred to as *the cursor*. The operator's responses that affect the cursor's movement may be subject to complicated dynamics just as in the control task. In the tracking task the operator tries to make the controlled system follow the changes in the "target" system. From the operator's vantage, the task is to make the cursor closely follow the target.

The theory and parameters underlying performance in both tasks are closely related and are now reviewed.

2.7.2   AN INFORMATION-PROCESSING MODEL OF THE HUMAN OPERATOR

Even apparently simple manual tasks depend upon a complex series of mental activities. Therefore performance can be understood best within the context of human information-processing abilities. This section contains a general information-processing model of the human operator focusing on features that are important determinants of manual control and tracking performance.

Crossman and Seymour (1957) identified five predominantly cognitive elemental activities or steps involved in most psychomotor tasks including manual control and tracking. These are:

Plan
Initiate
Control
End
Check

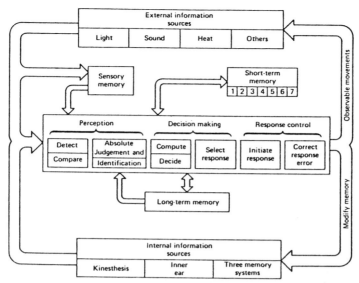

Fig. 2.7.2. Information-processing model of the human operator.

These five mental activities can be used to describe an operator's global task (i.e., controlling a system to keep it within tolerance) as well as the minute, individual control movements used to correct momentary system deviations from a desired state.

The operator's ability to perform these critical cognitive activities, and therefore the ability to perform manual control tasks effectively, rests upon fundamental cognitive processes and functions. These basic mental functions and processes, often referred to as "information-processing stages" are shown in Figure 2.7.2, which represents an information-processing model of the human operator.

In this model, which will provide a framework for considering manual control performance, the operator is continually presented with (on the left side of Figure 2.7.2) information about the task being performed. Some of this information concerns the states of the target and controlled systems, and some concerns the responses the operator is currently making to keep the system within tolerance. The operator uses (i.e., processes) this information and generates a control action that emerges on the right side of Figure 2.7.2. In this model, the operator is viewed as a channel through which information flows.

Three major information processing stages are shown in Figure 2.7.2: perception, decision-making, and response control. Also shown are three memory systems (sensory, short-term, and long-term memory) for storing information.

In the perceptual processing stage, current information about the states of the target and controlled system is detected and encoded into an appropriate physiological format. This information may arrive from a visual display or other symbolic system representation. In the decision-making stage, the newly entered information is used by the operator to plan or select a control response to keep the system within tolerance or bring it back into tolerance. Finally, once an appropriate response has been selected, the response control stage generates neural signals that cause the selected response to be performed by the operator's muscles. The emitted response has an effect upon the controlled system and information about this effect is fed back into the operator via the perceptual stage. Information about the response is also stored in the operator's memory systems.

2.7.2.1 Levels of Control Sophistication

Although the model in Figure 2.7.2 and the associated description suggest a simplistic operator, driven only by momentary error signals, this is not an accurate view. The model incorporates control processes on at least three levels. The lowest level is indeed very closely bound to the momentary error signal presented to the operator. The operator makes ad hoc corrections to unanticipated error conditions. At this level, the operator's performance can be modeled in terms of mechanistic feedback devices whose performance depends almost exclusively on inherent information transmission lag times and internal noise levels. Performance on this level relies relatively little upon available short- and long-term memory systems. What use is made primarily involves selecting appropriate responses based upon learned characteristics of the response-entry device (e.g., spring tension in a joystick).

On the next higher level, the operator makes use of inherent predictability in the time-varying changes of the target system. This predictability becomes most useful to an operator only after experience and practice have allowed development of an internal model of the target system's behavior. This internal model is maintained in the operator's memory. Initially this model may be only rudimentary, relying on the fact that all practical system behaviors are statistically predictable: for some value x, the state of the system at time T is predictable on the basis of its state at time $T - x$.

For inexperienced operators, x is small. But with practice and the development of a more effective internal model of the target system, x becomes larger. The operator is no longer bound to respond only to momentary, unpredicted errors, but makes responses that anticipate target system behavior. This level of performance depends upon short- and long-term memory systems and much more sophisticated activity in the decision-making stage. However, even though this stage is making more complex decisions, the overall control task may become easier because the decision-making stage is not continually occupied by an excessive number of simple, but time-consuming, low-level control decisions.

Finally, at an even higher level, the human operator can develop or make use of previously developed, general-purpose strategies to make the control task easier while simultaneously improving performance. For example, an operator may switch from a "position-matching" strategy (in which many repositioning movements are generated) to a "speed-matching" strategy in which relatively infrequent speed adjustments are made. This additional level of response integration eases the operator's task by allowing more time between control responses. The operator must be aware of the system dynamics as well as his or her own capabilities and limitations. A central focus of manual control and tracking theory has become the usually complex strategies that the operator devises to overcome, or at least minimize, the impact of both human and hardware limitations. A variety of typical strategies and their impact upon performance are described in this chapter.

The levels of control sophistication outlined above can be related to the inner and outer control loops discussed in Chapter 9.5. The outer control loop is associated with global plans and strategies designed to achieve a control objective within the constraints of the operator's abilities and control-system characterisitics. Typically, the operator is aware of the strategies formulated at this level and is conscious of individual response selections and strategy adjustments. The inner control loop is most closely associated with the two lower levels of control sophistication. Especially after practice the operator's responses are likely to be automatic and demand relatively little conscious attention.

It is important to note that the two higher levels of control activity outlined above require a level of cognitive sophistication far beyond that associated with simple negative-feedback servo-correction systems. These higher abilities are made possible by the complexity of the information-processing stages shown in Figure 2.7.2 and by the availability of memory systems. The information transmission capacity (bits per unit time) of each stage is a critical limiting factor in overall tracking performance.

2.7.2.2   The Structure of the Manual Control Task

The human operator, as represented in the foregoing model, is only one component in the closed-loop information path that characterizes most manual control tasks. Figure 2.7.3 is a more complete schematic representation of a typical manual tracking task showing the human operator as well as other components and information transformations. In this diagram, which includes a second-order control system* the main elements that determine tracking performance can be seen. These are as follows:

1. The human operator, represented in more detail by the information-processing model of Figure 2.7.2.
2. The intrinsic behavior of the target system. The changing state of the target system is the primary source of "demand" signal that motivates the operator to generate control responses.
3. The manner in which the target system's state is displayed to the operator.
4. The controls by which the operator's responses are entered into the system and the transformations imposed on these responses.

Each of these components, beginning with the operator, is considered in detail in one of the four main sections of this chapter.

---

* It is a *second-order* control system because two integrations of the operator's response are applied to the controlled system's behavior.

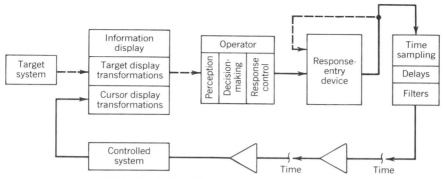

Fig. 2.7.3. Typical tracking task with second-order control.

2.7.2.3 Information-Processing Limits on Manual Control Performance

Control and tracking performance depend upon, and is limited by, the information-processing capacities of the three major stages and by the storage characteristics of the three memory systems.

Limits on manual control performance arise from three important characteristics of the major information processing stages:

They have limits as to the amount of information they can process per unit time.

They require a minimum time in which to perform their functions.

They have "background noise levels." For example, the response control stage always operates in the context of muscle tremor that limits the precision of fine manual control performance.

If information arrives too rapidly, a stage may become overloaded and unable to operate effectively. This limit to the rate at which a stage can handle (i.e., transmit) information is its channel capacity.

Psychological research on human information-processing capacities has often utilized small sets of discrete stimuli and responses, such as rows of lights and buttons. Manipulating parameters such as stimulus and response set sizes and probability distributions over these sets has been a frequent method for determining the channel capacities of individual processing stages as well as the overall human "channel." In manual control tasks, the operator must often respond to a continuously changing stimulus (e.g., the movements of the system being controlled). Therefore, it is appropriate to visualize the operator not only as responding to a limited number of discrete system states (although this may be precisely the operator's task in some step-tracking situations) but as responding to a continuously changing demand signal. In the continuous case, it is typical to describe information processing limits in terms of *bandwidth* (measured in Hertz) rather than channel capacity (measured in bits per second).

Estimates of human channel capacity based upon responses to discrete stimuli yield maximum values on the order of 10 bits/sec. Studies of tracking performance in which the operator must follow a continuously varying input signal yield bandwidth estimates of about 1 Hz: Humans are not very good at following signal components above 1 Hz.

Channel capacity estimates based upon continuous-response situations typically are smaller than those based upon discrete-response tasks. Characteristics of the operator such as the psychological refractory period limit the rate at which successive independent responses can be made to about two per second, regardless of information content.

Some Sources of Information Overload

A stage can transmit information at its maximum rate only if the information is "coded" in the proper format. If the stage must perform complex transformations on the information it receives to produce an appropriate output response, then the rate at which information is handled is below maximum.

In common manual control tasks the operator may have to compute derivatives of presented information to determine the effects of previous responses, and perform multiple integrations to select an appropriate control response. This situation occurs in the case of systems with high "order of control."

The need to perform complex data transformations imposes particular demands upon the decision-making stage. In general, reducing the degree of processing required to map input information into a desired response improves information transmission rate and control performance.

The channel capacity of a stage can be reached in at least three ways. First, a control task may be inherently difficult and present information to a particular stage at an excessive rate. Manual-control performance can improve if the processing capacity of the affected stage(s) increases.

Inexperienced operators are prone to a second source of stage overload. Typically, much of the information available to an operator is either irrelevant or redundant. A novice operator fails to recognize this and attempts to process more information than necessary. This results in overload and consequently low performance.

An operator following a continuously moving target on a display must keep track of constantly changing target positions. This becomes more difficult as the target path becomes more random and as movement speed increases. In such difficult conditions, an experienced operator can take advantage of redundancy in the target's motion: for some value, x, the position of the target at time T is predictable by its position at time $T - x$. The experienced operator need only observe the moving target every x time units. Actually, even experienced operators observe the location of moving targets about twice as often as necessary. However, less skilled operators sample this information far more frequently, imposing an unnecessary load on various information-processing stages.

The process whereby an operator comes to attend to only essential information in a task is a critical mechanism in the development of manual control proficiency. To take advantage of redundancy in the track of the target whose path is being followed, an internal model must be developed by the operator. The internal model allows the operator to make predictions about the future state (e.g., position) of the target, thus eliminating the need to sample excessively and process information from external sources.

The availability of an internal model whereby an operator can predict the behavior of a target or controlled system is one of the most significant factors in achieving proficient manual control. In some cases, because of the complexity or rate of movement of the target or cursor, an *external* model may be explicitly provided to the operator. This external model can often serve as a partial substitute for an internalized representation, thus enhancing performance. Essentially, the external model allows the operator to predict future system states, just as does an internal model, thus reducing the need for low-level control responses based upon unexpected moment-to-moment error signals.

The third way in which information-processing overload may occur is when two tasks compete for an operator's attention and simultaneously present information to the same limited-capacity stage. In this case the operator may choose to process the information from only one task, thus drastically reducing performance on the neglected task. Or, the operator may choose to process some information from each task, thus producing milder degradation in both cases.

Although the issues of time-sharing and attention distribution are beyond the scope of this chapter, it is particularly relevant to multidimensional tracking performance. An operator may be required to track a visual target that is moving only horizontally on a display. Alternatively, the operator may be required to track a target moving in both the horizontal and vertical dimensions. This case represents multidimensional tracking. Even when the operator attempts to disregard movement in one dimension completely (e.g., tries to follow *only* the horizontal movements of the target) performance on the chosen dimension (i.e., in this case, horizontal) will generally be inferior to that obtained in one-dimensional tracking.

Two important factors influencing multidimensional tracking performance are as follows:

1. *Degree of Dimensional Integration.*   Tracking one visual target that is moving both horizontally and vertically is easier than tracking targets varying in two *different* sensory dimensions: It is easier than tracking two *separate* visual targets. Similarly, moving a single joystick to follow vertical and horizontal target movements is usually easier than manipulating two separate joysticks (one for horizontal and one for vertical control responses). Dimensional integration influences the "structural interference" among tracking performances on each dimension.

2. *Degree of Conceptual Integration.*   To the extent target system behavior and controlled-system responses to the operator's actions are similar in the various tracking dimensions, the overall multidimensional tracking task will be easier. For one thing, only one, rather than several, internal models need to be developed to predict system behavior. Also, there is less likelihood of "conflict" among responses selected closely in time. For example, if leftward target drift is counteracted with rightward control stick movement, then two-dimensional tracking will be best if upward target movement is counteracted with downward control movement. In this case a single response-selection rule (move control opposite to target drift) applies to each tracking dimension. Essentially, conceptual integration refers to the extent the same "rules" can be applied to tracking in each dimension.

Psychological Refractory Period Limitations

Another important source of performance limitation is a fixed delay of about 300 msec that must separate successive outputs from the decision-making stage. This is the so-called psychological refractory period (PRP). If information is presented to the decision-making stage within 300 msec of a previous decision, decision-making is delayed until the PRP has elapsed. This refractory delay does not decline with practice and is one reason why operators are sometimes modeled as *intermittent* servomechanisms in the analysis of closed-loop tracking performance. Such mechanisms sample and respond to system changes at discrete intervals, rather than continuously.

Although the PRP cannot be eliminated, it is minimized when stimulus–response (S–R) compatibility is high. S–R compatibility refers to the "naturalness" of the relationship between a stimulus and the response with which it is associated. For example, if an operator is using a joystick to control a cursor and if the cursor drifts away from a target, then a compatible response would be to move the joystick *toward* the target to bring the cursor closer. Even with high S–R compatibility the PRP delay may decline only to about 200 msec.

Summary Perspective of the Operator

The foregoing model of the human operator as a series of stages sensitive to information flow rate (i.e., transmission rate) emphasizes two important aspects of the human operator:

1. The need for information selection mechanisms and efficient information transformations (i.e., coding schemes) to avoid overload.
2. The need for internal (or, occasionally, external) models of the target and controlled systems that allow the operator to utilize inherent redundancy in system behavior and predict future system states.

The proficient operator usually possesses effective internal models of the target and controlled systems. These models allow the operator to predict accurately the future states of the systems and to select and process only essential information.

2.7.3 MEASURING CONTROL AND TRACKING PERFORMANCE

This part of the chapter concerns certain aspects of the methodology of typical tracking and control experiments. Familiarity with the dependent variables commonly used in control and tracking experiments is important to system designers for four reasons:

1. Display and control manufacturers may provide data about the performance aspects of their products. Engineers involved in designing systems that use these products need guidelines for evaluating those data.
2. Some factors affecting results of an *experiment* about a particular type of tracking system may be important in the environment in which a related control system is actually implemented. This may apply particularly to asymmetric transfer effects or to aspects of control isomorphism.
3. System development usually requires some form of testing during the design stages. To avoid surprises about the quality of performance once the system is complete, designers should focus on the appropriate variables during the design-stage testing.
4. Knowledge of experimental variables and procedures may also provide insight to the complexity and sophistication required for valid control-system simulations.

Issues related to design of evaluation experiments and dependent variables for assessing performance are now discussed.

2.7.3.1 Two Common Problems in Control and Tracking Experiments

To evaluate the validity of experimental data, it is critical to consider the conditions under which the results were obtained. Two problems, transfer of training and range effects, frequently produce misleading data in tracking and control studies.

Transfer of Training

Transfer of training occurs when one task has an effect on another task performed later. Positive transfer occurs when practicing Task 1 improves performance of Task 2. When performance of Task 2 suffers after performance of Task 1, the transfer is negative. As long as Task 1 and Task 2 affect each other the same way, the effects of transfer upon the validity of experimental results can be controlled by proper statistical balancing of testing orders.

However, when Task 1 and Task 2 do not have the same effects on each other, transfer is asymmetric, and only results from the first-completed task should be evaluated. Using average performance over all trials may bias results in favor of the conditions helped by positive transfer. The effects of asymmetric transfer can be avoided with "between-groups" experimental designs: in these designs each operator experiences only one experimental condition. Therefore, practice in one condition cannot affect performance in another.

Range Effects

A second problem occurs when an operator experiences a range of values for a particular experimental variable. In this case transfer biases tend to favor unequally the variable values in the middle of the range. Such range effects may lead the experimenter to conclude, mistakenly, that particular values should be used. In reality however, shifting the range of tested values would likewise shift the apparently "optimum" value to the middle of the new range.

2.7.3.2   Measures of Control and Tracking Performance

A variety of measures are available to quantify control and tracking performance. Tracking performance scores are usually based on error measurements; common scores are derived from errors in position, errors in time, or errors in phase. Some of these scores are more useful than others.

Errors in Position

Position errors reflect distance between cursors and targets. Position errors at the time of cursor-movement direction reversals are particularly useful. An operator overshoots when he or she does not respond quickly enough to a reversal of target-movement direction. Similarly, undershooting occurs when the operator anticipates the change in direction, and reverses the control movement before the track actually turns.

Overshooting and undershooting patterns may reveal performance strategies and response biases. Overshooting and undershooting at reversals may also indicate how well the operator has extracted the average time between reversals for a particular track. If the operator undershoots when the time between reversals is long for this particular track and overshoots when the time between reversals is short, then he or she may be predicting when the average reversal should occur, and changing direction based on this estimation.

Frequency Analysis of Responses

The operator's response track can be analyzed into its component frequencies. From this analysis, it can be determined what proportion of the response is composed of frequencies that were present in the target track, and what proportion is uncorrelated with the target. The latter portion is "remnant." Large remnants indicate the operator's use of nonlinear strategies such as dither.

Errors in Time and Phase

Errors in time can be measured at reversals, by comparing the time at which the track reverses direction with the time the operator reverses. The average timing error (or time lag) can be determined by filtering the response and tracking functions, then cross-correlating the amplitudes of the filtered tracks. The average time lag at the frequency of the filter is the time the response curve has to be moved forward in time to maximize the cross-correlation. This average time lag can be converted to a phase lag. Phase lags can be determined only for those frequencies actually present in the target's track.

Overall Measures of Error

Overall measures of error reflect the average distance between the cursor and the target. There are several such measures. Some use the sign of the momentary error whereas others do not.

Constant Position Error.   This is the arithmetic mean of the distance error. It indicates the extent to which average position is above or below the track. Because this measure is computed from signed errors, integrated over time, positive and negative errors tend to cancel one another.

Standard Deviation of the Error Distribution.   This measure is sometimes called "variable error" and corresponds to the standard deviation in statistics. It is computed by summing the squared differences between the constant position error and each individual error. This is one of the best overall indications of how well the operator is performing.

Average Absolute Error.   This is also called "modulus mean error." It is the arithmetic average of the unsigned momentary position errors. It reflects *both* variable error and the constant position error. The distribution of this metric is J-shaped, so it is less well suited for parametric statistical tests than constant position error.

Root-Mean-Squared Error. The root-mean-squared (RMS) error is the square root of the sum of the momentary errors, each squared. Like standard deviation of the error, RMS error is compatible with parametric statistics.

The constant position error and the standard deviation of the error are statistically independent, though they may be correlated in a given sample. Average absolute error and RMS error are correlated with each other and with the other two measures. The amount of correlation between average absolute error and RMS error varies by population and task. When the correlation is high, RMS error is preferable because it can be calculated more precisely from a record of a given length. RMS errors have another advantage in that they can be combined easily. Thus the total error in a system containing several operators can be determined by squaring and summing the component RMS errors, then taking the square root of the total. Unless the constant position error is high and in one direction, it provides a fairly accurate estimate of the total system error. When constant position error is high, combining the RMS errors will underestimate the system error.

2.7.3.3 Other Scoring Methods

Most other scoring methods are not as useful or reliable as those described above. Occasionally these measures may provide rough estimates to more useful measures, but they should not be substituted because they may also be very misleading. One measure, time on target, warrants particular comment because it is so frequently encountered and is prone to misinterpretation. Time on target is simply the total duration of time that the cursor is aligned with the target. This measure is easily measured with simple equipment but it is quite misleading because it does not differentiate large and small errors. Also, the proportion of time on target depends upon the *size* of the target as well as the operator's path-following performance. If time on target is used, nonparametric statistics are appropriate. Time on target can be converted to RMS error; however, at least 50% more tracking time is needed to provide an accurate conversion.

2.7.4 SYSTEM BEHAVIOR—EFFECTS ON OPERATOR'S PERFORMANCE

Whatever demand signal an operator attempts to follow, whether a moving cursor on a display or the edges of a roadway, the movement of the signal from moment to moment can be represented in a common mathematical representation. This consists of a summation of pure sine waves:

$$F(t) = C_0 + \sum_{n=1}^{\infty} C_n \cos(n \omega_0 t - \theta_n) \tag{1}$$

The highest-frequency sine wave with nonzero amplitude in the sum determines the fastest movement that can be exhibited by the signal.

However, even though any system behavior an operator might be required to follow can be described in this mixture-of-sine-waves representation, the exact form of the signal path has important consequences for tracking performance. In following different types of tracks, a range of qualitatively distinct strategies is frequently seen and different elemental components of skill are important. In this section, effects upon performance of four commonly encountered demand-signal patterns are reviewed. These patterns (Figures 2.7.4a–d) are:

Tracks with a single step (Figure 2.7.4a)
Tracks with many steps (Figure 2.7.4b)
Ramps (Figure 2.7.4c)
Smoothly varying combinations of sine waves (Figure 2.7.4d)

2.7.4.1 Tracks with a Single Step

Single-step tracks contain an isolated, sudden change of position. Often the time of the jump is unpredictable. The operator must make a quick, corresponding control response to bring the system back into tolerance or to move the cursor to a new position. The quick change in position corresponds to the presence of very-high-frequency sine wave components in the Fourier (mixture-of-sine-waves) representation of a single-step track. Because the human operator has a bandwidth of approximately 1 Hz, these high-frequency components will not be accurately represented in his or her responses, and so tracking performance will not be perfect.

Typical examples of single-step tracking include making a quick highway lane change (example

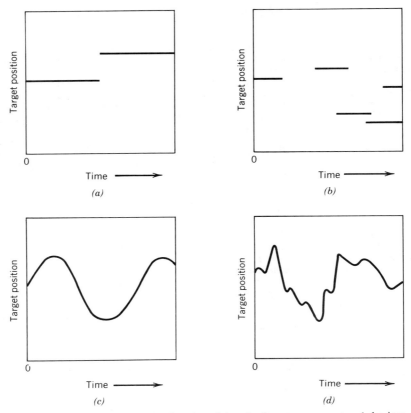

Fig. 2.7.4.   Target position as a function of time for four common system behaviors.

from Poulton, 1974), moving the finger quickly to a push button, or moving to a new station by turning the dial on a radio.

Single-step tracking performance depends upon the following principal factors:

The Operator's reaction time, applicable in the case of temporally unexpected movements or when the particular required control response is not known to the operator in advance

The amplitude (i.e., distance) of the required control movement

The accuracy (i.e., size of the target area to which the operator intends to move) of the response

Response pacing demands (i.e., whether responses must meet a time deadline)

In addition, performance also depends upon the characteristics of the display that presents information to the operator and upon characteristics of the response-entry device by which the operator's response enters the control loop.

Reaction Time

If the target's position jump is unexpected, or if the operator does not know which of several jumps will be made until one occurs, then the control response does not begin until at least one reaction time after the jump. The extent of the reaction-time lag depends on numerous factors that can be related to the information-processing characteristics of the Operator.

The change in target system must be first be detected. We may assume the operator's attention is directed toward the information display device and that the jump is represented by a large (>>JND)*

---

* The just noticeable difference (JND) is the minimum stimulus change in a particular sensory dimension that is detectable.

Table 2.7.1 Minimum Reaction Time Delays for Various Sensory Modalities

Modality	Delay (msec)
Vision	180
Audition	140
Touch	155
Proprioception (bodily rotation)	520

Source: Table 1.2, Brebner and Welford, 1980.

display signal. In this case the minimum possible reaction time depends primarily upon the particular sensory system that is being stimulated. It also depends upon time needed by the response control stage to activate appropriate muscles. Minimum reaction times for signals presented in several sensory modalities are given in Table 2.7.1.

If the Operator does not know in advance which control response will be required, then the decision-making stage introduces additional increments to reaction time delay.

Extent of Control Movement

Once initiated, large control movements accelerate and attain faster speeds than short movements. However, this additional speed is usually not enough to compensate the added distance and so long responses require more time than short responses. Of somewhat more interest is the relation between movement distance and accuracy. Craik and Vince (1963) showed that movement accuracy is approximately proportional to movement distance for quick movements longer than about 2 cm:

$$\frac{E}{D} \approx 0.05 \tag{2}$$

where E is average absolute error and D is intended movement distance. This relationship is Craik's Ratio Rule (Poulton, 1974). For control movements less than 2 cm, the value of Craik's ratio is no longer approximately constant but rises very rapidly. At 0.25 cm it rises to about 0.20, and for intended movements less than 0.1 cm error rises to about 0.80. This reflects the impact of physiological noise (i.e., muscle tremor) from the response-control stage.

Speed/Accuracy Trade-offs in Control Movements

The speed of a control movement depends on the difficulty of the response. Fitts (1954) defined an Index of Movement Difficulty (ID) by analogy to information theory:

$$ID(\text{bits}) = \log\left(\frac{2A}{W}\right) \tag{3}$$

where A is the amplitude (i.e., extent) of the intended movement and W is the width of the "target" area in which the operator wishes the movement to terminate. Thus W is a measure of the intended precision or accuracy of the control movement.

Fitts found that the time needed to complete a movement once started (MT) could be predicted from movement ID:[*]

$$MT = k_m + C_m \times ID \tag{4}$$

where k_m is a delay constant that depends upon the body member being used for responding (i.e., the foot and hand yield different values for k_m; for hand movements, a typical value is 0.177 sec); and C_m (sec/bit) is a measure of information handling capacity. Typically, C_m is 0.1 sec/bit or greater.

---

[*] Alternative versions of "Fitts's Law" have been derived from a variety of underlying theoretical models of movement control. Welford (1968) proposed the following version: $MT = C \log(\{W/A\} + 0.5)$, where $C \approx 100$ msec/bit. These alternative forms generally agree closely with Fitts's, original version but offer somewhat improved matches to data obtained with particularly wide or narrow targets.

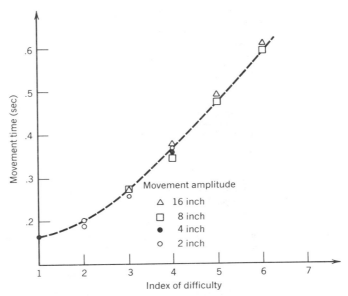

Fig. 2.7.5. Demonstration of Fitts's Law.

For repetitive responses such as moving back and forth between two positions, C_m is slightly lower than for single, isolated movements.

It can be seen [from Equation (3)] that each doubling of the required accuracy (i.e., *halving* the target width) increases ID by one unit. This adds [Equation (4)] one time increment (the exact value of the increment depends on the channel capacity of the response-control stage) to the movement time.

The relationship between ID (movement accuracy *if* movement distance is constant) and MT is Fitts's Law. Fitts's Law implies that constant ratios of movement accuracy and movement distance produce constant movement times. This is illustrated in Figure 2.7.5. However, in some situations Fitts's Law does not always hold, especially for very narrow (<0.25 inch) or very wide (>2 inch) targets. In these cases operators do not distribute their errors appropriately over the specified targets: For small targets too many responses miss, whereas for large targets responses may not take advantage of all available target width, confining their distribution to a rather narrow central portion of the target. Thus the operator may be moving to an "effective" target whose width does not correspond to the actual target. To calculate the effective target width, a normal distribution of movement extents may be assumed. Then the variance of the movement distribution is estimated by calculating the percentage of movements falling outside the target boundaries (i.e., the percentage of misses). Then the "effective" target width is taken to be ±1 standard deviation of the inferred movement distribution.

Visual Guidance in Single-Step Tracking

Control movements that take less than 300 msec cannot benefit from visual guidance during their execution (Keele and Posner, 1968). For such movements there is insufficient time to process visual information, select the proper response correction, and send it to the appropriate muscles.

Although visual guidance of very quick control movements is not possible, they may still be subject to in-course correction based on other sources of information (for example, from the kinesthetic sensory system, which responds more quickly than the visual system). Also, response selection processes may continue even after an initial command has been sent to the response-control stage. Occasionally, an incorrect control response may begin and then be almost immediately amended. Corrections to movement direction can sometimes occur less then 0.1 sec after a response movement has started. Gibbs (1965) found 10% of incorrect responses in step-tracking experiments to be corrected in 50 msec or less. This suggests that a second response command can be issued immediately after the first command and without processing new visual information.

Two Response Strategies for Single-Step Tracking

For relatively large, quick control movements the operator may make a series of separate, progressively smaller responses. Each response reduces the remaining distance to the target by 95% on average.

However, when the operator has closely approached the target, the proportional error reduction will be much less than 95%, because Craik's Ratio Rule does not apply to movements less than about 2 cm. Therefore this "successive approximation" response-control strategy is not efficient for small, fine movements where a cursor must be precisely aligned with a target.

An alternative response control strategy is to monitor the movement continuously. This "closed-loop monitoring" strategy requires the operator to process a large amount of information and so movement speed is slow. With the successive approximation strategy, the end point of each movement segment is programmed before it begins. With the slower, closed-loop monitoring strategy, the stopping point of a movement is determined one reaction time before the end of the movement.

In many cases, movements in single-step tracking may be two-phase responses (Welford, 1968). In the first phase the operator makes a quick, open-loop ballistic movement, characteristic of the successive approximation control strategy. In the second, final phase, the operator makes a slow, final positioning movement characteristic of the closed-loop monitoring strategy. These two phases correspond to the MOVE and POSITION elements of Crossman and Seymour's (1957) taxonomy, given at the start of this chapter. Because the POSITION element involves continuous response monitoring, it is the source of high information-processing demands upon the operator. Hence, two POSITION responses cannot usually be carried out at the same time. However, the operator can often make two MOVE responses simultaneously because the combined information processing demand is low.

2.7.4.2 Tracks with Many Steps

Tracks with many steps contain discrete jumps spaced closely together in time. Usually, the important time interval is that between the end of the response to step i and the presentation of step $i + 1$. As this interval becomes greater than 3 or 4 sec, the interaction between successive response movements decreases (except as related to timing associated with response anticipation).

A common example of multistep tracking is following the position of a target on a radar screen where the target jumps to a new position every time the sweep line revolves. Another example occurs in video arcade games where the player must quickly move a cursor to dodge obstacles that continually appear in its path. Common laboratory models of multistep tracking tasks include the double-stimulation paradigm (Kantowitz, 1974) and sequential reaction time tasks, as well as traditional multistep tracking paradigms.

The two most noteworthy features of multistep tracking are the following:

PRP effects

Statistical aspects of the step sequence (which affect its predictability)

Effects of the Psychological Refractory Period

Because of the PRP (Section 2.7.2) the response to a step that follows another by less than about 300 msec is delayed by approximately $300 - I$ msec (where I is the interval separating the first and second steps). This is the case if the operator attempts to make separate, independent response choices for each control movement. This delay represents the fixed interval that must separate independent response selections by the decision-making stage. Processing of information about a second step is delayed until the 300 msec PRP has elapsed. Although the magnitude of the PRP can be modified slightly, it cannot be completely eliminated. Therefore operators typically adopt control strategies to minimize the impact of PRP delays upon overall tracking performance.

Response Grouping Strategy. The strategy in which the operator makes independent responses to each successive step almost inevitably leads to PRP delay of responses that follow others. This delay is *added* to the normal reaction time delays that would characterize a response to the same step presented in a single-step tracking situation. An alternative strategy involves delaying the selection of a response to the first step and selecting a "compound" response to *both* the first *and* second steps, after the second step arrives (Halliday, Kerr, and Elithorne, 1960). By selecting a compound, "grouped" response, the decision-making stage is required to make only one response selection, thus minimizing the PRP effect. However, this strategy does have some costs. First, selection of a response to the first step must be delayed, thus reducing first-step tracking performance. However, if the interval I between the two steps is small, then the resulting first-response delay (which will approximately equal I) will be small compared to the second-response PRP delay, $300 - I$ msec, that would result from the independent-response strategy.

The grouped-response strategy also affects first-response accuracy. Usually the response to the first step is affected by the characteristics of the second response. For example, if the second step requires a movement in the opposite direction to the first step, then the first response is usually too small— the operator begins moving in the direction of the second step before completing the first-step response. Also, if the operator waits for the second step before beginning to respond, the first step may no longer be visible, and so the operator will have to move on the basis of his or her memory of the first step's position.

Effects of Statistical Aspects of Step Sequence

As the operator gains experience in responding to a sequence of steps in a track, he or she learns some of the statistical aspects of the sequence. For example, the operator learns the range of step sizes, the average step size, and the distribution of interstep times. The operator also learns statistical dependencies among successive stimuli.

This knowledge can improve tracking performance but it also leads to range effects and can disrupt performance when the operator must respond to situations that violate expectations based upon the internalized model of the step sequence.

Generally, range effects will cause the operator to make the most accurate control responses to steps of average size. Responses to large steps will usually be too small and responses to small steps will usually be too large. Similarly, responses will be made most rapidly to steps that follow a preceding step by an interval that is average for the step sequence. These types of range effects produce performance advantages for steps with values in the middle of the range of possible values.

Cross (cited by Noble and Trumbo, 1967) compared effects of step size, interstep interval, and step direction. Cross found that size and interval produce comparable range-effect benefits when responses to steps that were average in these dimensions were made. However, Cross also found that the range effect does *not* occur if the operator is uncertain about more than one dimension of successive steps. For example, if the size of successive steps is not fixed, a range effect for step size emerges only if step direction *and* interstep interval are constant.

Two Tracking Strategies Based on Step-Size Distribution. In addition to the "average step size" strategy just mentioned, operators frequently exhibit two other strategies based upon the probability distribution of step sizes. In the "maximizing" strategy, the operator always prepares for the most frequent (i.e., modal) step and therefore responses to this step, when it occurs, are fastest and most accurate. This is a very effective strategy when the probability distribution is very uneven because the benefit from being prepared for one step outweighs the penalty associated with being unprepared for all other possible responses. Trumbo (1970) found that every one of 16 operators adopted the maximizing strategy when one particular step occurred 92% of the time.

When the probability distribution is more even, the operator may adopt a "probability matching" strategy in which the likelihood of preparing for one or another step is matched to its probability of occurrence. For example, Trumbo (1970) found that when the probability of the modal step size was 0.7, most operators prepared for that step about 70% of the time.

The maximizing and probability-matching strategies are most likely to appear when

Response speed is emphasized.

Operators are experienced enough to have developed an internal model of the step sequence statistics.

Operators are encouraged to anticipate the appearance of steps.

Control Strategies for Repetitive Step Sequences

Even when the operator knows which two steps will closely follow one another, if a separate-response strategy is used PRP delays emerge. When the sequence of steps is repetitive (and therefore perfectly predictable) the operator can adopt another strategy known as "locking" (Slack, 1953) to minimize or eliminate PRP delays. In this strategy the operator develops a sequence of responses to match the sequence of steps. This response sequence is started and the operator makes only timing (i.e., phase) adjustments to keep the responses locked to the appearance of corresponding steps. This is a highly efficient strategy that can also be used in following continuous sine wave tracks. However, the disadvantage of this strategy is that unexpected changes in the step sequence pattern may not be detected, resulting in gross performance disruptions for a period of time.

Even when operators do not adopt a locking strategy, they may be using the predictability of the step sequence to guide their responses. For example, when Slack (1953) inserted an out-of-place step into the regular sequence, operators often made a response that would have been appropriate for the expected step.

2.7.4.3 Ramp Tracks

In a ramp track, the target being followed moves continuously in a uniform direction. Characteristics of the movement rate differentiate various types of ramp tracking tasks and lead to different effects on tracking performance. In rate ramps, the target moves with constant velocity; in acceleration ramps, constant acceleration occurs; in delta-acceleration ramps, a constant rate of change of acceleration is used.

Few real-life tasks involve ramp tracking because, from the vantage of an observer, objects moving past appear to change velocity even when they actually have constant speed. An exception, described by Poulton (1974) is a person filming, from the center of a circular track, a horse racing around the track: here the velocity of the tracked object (the horse) appears constant from the perspective of the observer.

Nevertheless, ramp tracking is important because it allows experimental investigation of steady-state tracking performance (unlike step-tracking tasks, which involve short-term discontinuities in the track of the followed target). As Poulton (1974) points out:

> Changes in his [the operator's] responses must therefore [in the steady-state ramp tracking task] be determined by his limitations, and by the strategies which he uses to overcome his limitations. Experiments with ramp tracks can thus indicate the nature of man's limitations and strategies.
>
> POULTON, 1974, p. 92.

Two Tracking Strategies for Velocity Ramps

In following a target moving with constant velocity, the operator may continuously monitor the difference between the target position and his or her cursor position. At regular, closely spaced time intervals, quick correction responses are made to move the cursor to the current target position or slightly ahead of it. This tracking strategy is position matching. The operator will typically make corrections about two times per second. This time interval between corrections remains about the same regardless of ramp velocity. Thus doubling the velocity approximately doubles the average absolute error. Thus the trigger for correction responses is not the amount of momentary error (the distance between the cursor and the target) but the passage of about one reaction time delay. Essentially, the operator must make numerous response-selection decisions, and this imposes a high information-processing load. Psychological refractory period delays also limit the rate at which corrective responses can be made.

During the (approximately) 0.5 sec between successive responses, the operator must select and begin a response and then wait while error accumulates enough to determine the selection of a particular corrective response.

This strategy is different from the way a typical mechanical tracking mechanism might work. Usually, such mechanisms continuously lag behind the tracked system, nulling the error that existed one time-lag unit previously. Craik (1947) examined the velocity ramp-tracking performance of a servo-mechanism with a lag of 0.25 sec. The device produced an oscillating error around the target path with a frequency of 0.5 sec, similar to that found with human operators. Unlike the human, however, the servomechanism, which is always responding to the error that existed 0.25 sec before, continues to oscillate even after the ramp stops. Damping must be added to the servomechanism to cause this oscillation gradually to fade. In contrast, the human operator, after making his or her last response, sees that error no longer builds up and so immediately stops making corrective responses.

Rate Matching. In contrast to the position-matching strategy in which the operator makes numerous position-changing responses, the rate-matching strategy allows the operator to make relatively infrequent rate-changing responses. The operator observes the target track until the velocity of the target can be estimated. Then the operator selects a response that causes the cursor to move at a constant rate similar to the target's. In order to catch up initially with the target, the selected cursor velocity is slightly higher than the target's. When the cursor eventually passes the target, a new constant cursor velocity must be selected so the target will catch up to the cursor. Craik (1947) found the operators made about one-quarter as many corrective responses per unit time (about 0.5 per sec) using a rate-matching strategy as when using a position-matching strategy (about 2 per sec).

Although two pure alternative strategies were just described, in actual tracking tasks humans typically use mixtures of these different approaches. For example, Craik (1947) describes a strategy like rate matching but with intermittent position corrections imposed at frequent intervals.

Failure of Fitts's Law

Fitts's Law is inaccurate when the target at which the operator is aiming is not stationary (as it is in step-tracking tasks). Nonstationary targets occur in ramp tracking when the operator detects a position error and attempts to move the cursor into alignment with the (continuously moving) target. The degree of inaccuracy depends upon the size of the target (greater for small targets), target velocity (greater for higher velocities), and the order of control (Section 2.7.6; greater for *position-control* than for rate-control systems) (Jagacinski, Repperger, Ward, and Moran, 1980). Jagacinski et al. have proposed an alternate index of movement difficulty that more accurately predicts target capture times in the case of moving targets:

$$ \text{MT} = c + dA + e(V + 1)\left(\left(\frac{1}{W}\right) - 1\right) \tag{5} $$

where *c, d,* and *e* are regression coefficients dependent upon order of control; *A* is intended movement amplitude; *V* is target velocity; and *W* is target width. *A, V,* and *W* are measured in degrees of visual angle.

2.7.4.4   Sine Wave Tracks

Although all the track types discussed so far can be described as combinations of sine waves, we now consider combinations that produce smooth, continuously changing paths for the operator to follow. This type of path may be described as continuous with varying velocities and accelerations. Tracking such target paths represents a very common real-life task. From the standpoint of an observer on the side of the road, the rate at which a car approaches from one side, passes in front, and then recedes on the other side follows a sine wave function. Also the driver of a car must counteract disturbances that would otherwise cause the car to deviate from a straight path. The form of these disturbances can be represented as a combination of sine waves requiring continuous smooth tracking by the driver.

In this section we first consider tracks represented by single sine waves, then tracks composed of multiple sine waves in combination.

Sine Wave Frequency Effects

Humans have a tracking bandwidth of approximately 1 to 2 Hz. Therefore, tracking performance usually declines when the operator must follow sine wave tracks with components above 1 Hz. However, if the track is easily predictable, as it is with single sine waves, the operator may adopt strategies such as "locking" to overcome this limitation partially. Figure 2.7.6 shows a typical example of single sine wave tracking performance as a function of frequency (data after Noble, Fitts, and Warren, 1955). The "knee" in these data is at about 1.75 Hz, corresponding to the operator's bandwidth.

Locking Strategies as a Function of Track Frequency

With single sine wave tracks, operators quickly adopt locking strategies in which they attempt to generate a sine wave response matched to the apparent frequency and amplitude of the signal being followed. Phase correction responses are made as required to keep the response matched to the followed track. This is a very efficient strategy because it greatly reduces the number of response selection decisions that the operator must make.

For low-frequency signals (0.25 Hz) Poulton (1950) found that 10 of 11 operators in his study adopted a locking strategy within one cycle of the start of a single sine wave track. With a 1 Hz track Poulton reports that operators made phase and/or amplitude correction responses about once every 4 sec. Phase errors may be as large as 90° before the operator institutes a corrective adjustment to the sine wave response he or she is producing.

The locking strategy, by reducing the number of response selection decisions that must be made, allows the operator to follow predictable sine wave signals with frequencies about 1 Hz. Ellson and Gray (1948) found that untrained operators could lock to tracks up to 2 Hz. However, they were usually unable to do so with 3 Hz signals. For these high frequencies, the responses usually had the

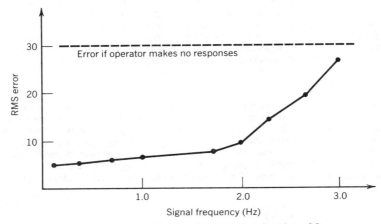

Fig. 2.7.6.   Sine wave tracking performance as a function of frequency.

proper amplitude but were out of phase. It is difficult for the operator to correct phase errors because he or she cannot determine whether the sine wave response being generated is too high or too low in frequency. Similar limits to locking-based tracking performance have been found for highly practiced operators as well (Noble, Fitts, and Warren, 1955).

Effects Seen in Closed-Loop Transfer Function

The closed-loop transfer function shows the proportion of each frequency in the input signal that appears in the output response of the operator. It is useful to consider this function specifically because it reveals several important effects upon the operator's performance and strategies of input signal characteristics. Increasing input signal frequencies has three effects that can be seen in the closed-loop transfer function.

1. The average amplitude of the transfer function decreases as the maximum frequency of the input signal increases (Elkind, 1956). The operator tends to make smaller-sized responses with increasing input frequencies. This may reflect a deliberate strategy: If the operator makes very small responses when the input signal begins to go beyond his or her bandwidth, this at least ensures that tracking error will not be larger than if the operator stopped responding at all. If the operator were to respond wildly, then his or her responses might easily increase tracking error above the no-response base line.

2. The time lag in the operator's responses increases as the input frequency increases. Essentially, for very low frequencies, the signal is highly predictable and so the operator is responding to expected changes in the signal. At high frequencies the signal is less predictable, imposing a higher information processing load on the operator. Hence response time lags increase. Elkind's (1956) closed-loop transfer functions show time lags of about 100 msec with 0.5 Hz signals and about 180 msec with 2.4 Hz signals.

3. The remnant is that portion of the operator's output response spectrum that is not present in the input signal. The size of the remnant increases with increasing input signal frequency. Elkind's (1956) data show an increase in remnant power from 10 to 60% of the track power as the track frequency increases from 0.5 to 2.4 Hz.

Effects of Sine Wave Amplitude

The velocity and acceleration in a sine wave path are directly proportional to the amplitude of the signal. Therefore, effects similar to those found for changes in velocity and acceleration with ramp tracks should be seen. With sine waves, it is found that doubling the amplitude approximately doubles the average absolute error in tracking. (This is the case if no changes in control gain or display magnification are made.) Actually, the error does not quite double in step with amplitude, but is slightly less. Thus, there is a slight advantage to larger sine wave amplitudes. This is partly attributable to the fact that these signals generally make more effective use of the display screen when they are presented to the operator: Tracking performance with sine wave signals is best when the maximum variation of the signal covers the entire width of the display.

Combinations of Sine Waves

When sine waves are combined, the path appears to become more irregular and less predictable to the operator. Thus the difficulty of the tracking task increases. In particular, the usefulness of the locking strategy is greatly reduced. Even with combinations of three sine waves, operators still make use of some predictability in the signal path. For one thing, operators can learn some recurring patterns in tracks composed of small numbers of sine waves. Also, they are able to take advantage of short-term redundancy in the signal. For tracks containing a top sine wave frequency of 0.33 Hz Poulton (1952b) found operators were able to predict the track's movement up to about 0.5 cycles (of the top frequency) ahead.

Finally, even with very complex signals, such as those produced by low-pass-filtered white noise, the operator can still learn some features of the track, such as:

Average and maximum amplitude

Average position of the target on the display

Average frequency of reversals

Average velocities and accelerations

Frequency spectrum

With combinations of sine waves, characteristics of the component with greatest amplitude will largely control performance. For example, the highest frequency component controls the number of response direction reversals if it has the highest amplitude (Elkind, 1956).

2.7.5 DISPLAYING SYSTEM BEHAVIOR—EFFECTS ON OPERATOR'S PERFORMANCE*

To perform a tracking or control task the operator *must* have one item of information: an indication of the difference between the current system state and the desired state. Without this "error" signal the operator has no idea when a corrective response is needed. However, providing additional information usually enhances an operator's performance. The types of additional information that are provided and the manner in which information is displayed to the operator have great influence upon performance.

In realistic tracking/control tasks, four types of information may be displayed:

1. "Error" information
2. Target position information. This shows to "where" the operator should move the controlled-system or cursor
3. Position of the controlled-system or cursor
4. "Performance aids" that provide specialized guidance to the operator in particularly complex tracking situations

2.7.5.1 Pursuit and Compensatory Displays

Two basically different information display modes are Pursuit and Compensatory. Pursuit, or "true motion," mode presents both target position and cursor position, and the operator can immediately perceive the error signal as the difference between these two positions.

Compensatory displays present only the direction and magnitude of the error signal—the difference between the target's position and that of the controlled system or cursor. In contrast to "true motion" pursuit displays, compensatory displays present "relative motion" information.

Most "real-life" tracking tasks performed without electronic displays involve pursuit mode. The simplest case is moving the finger to a target push button: The position of the target (the push button) and the controlled system (the finger) are both always visible. Another example is driving a car. The position of the goal (the edge or center of the road) and the controlled system (the car) are visible and the operator can judge the independent changes in each (e.g., the curvature of the road and the movement of the car).

"Real-life" tracking tasks using compensatory mode usually involve aiming. Imagine an aircraft flying in a cloudless blue sky. A person might have the task of keeping the aircraft centered in a camera sight. This task involves compensatory-mode tracking: As the target (the aircraft) moves away from the center of the sight, the cameraperson knows there is an aiming error but cannot tell from the display alone whether the error is due to the aircraft's motion or to movement of the camera. With compensatory mode, target movement and controlled-system movement are completely confounded.

Because target and controlled-system motion cannot be distinguished in compensatory mode, compensatory displays are sometimes more difficult to interpret. Poulton (1974) describes the example of "radar-assisted collisions." Ships use Plan Position Indicators (PPIs) to show the position of other ships *relative* to the position of the "controlled ship." The position of the "controlled ship" is always the center of the PPI display. The bearing and distance of other ships are easy to see from the PPI display but the independent movements of the controlled ship and those appearing on the PPI cannot be distinguished. This makes it easy for the controlled-ship's captain to tell that the ship is in danger of collision but does not provide clear information on what action to take to avoid the collision. Furthermore, the PPI display may be misleading in that it can appear to show that two ships are moving *directly toward* each other when they are actually moving perpendicularly. Changes in the error vector can be misinterpreted as changes in position of one or the other ship. This is illustrated in Figure 2.7.7.

Pursuit displays almost always produce better tracking performance than compensatory displays. However, this enhanced performance may be at the cost of a more expensive or complex display system (e.g., a meter versus a CRT). Pursuit displays must be able to show two separate position indicators (i.e., that of the controlled system or cursor and that of the target or goal to which the operator tries to move the cursor). Because position variation may be much greater than the error signal (i.e., the difference between the two positions), large display sizes may be needed when pursuit, rather than compensatory, displays are used, in order to achieve adequate display resolution of the error signal. The designer must make a trade-off between resolution of the error signal, display size

---

* The initial portion of this section concerns visual displays on which target status and controlled-system status are coded by position on the display. Some of the recommendations and generalizations apply only to this case. (For example, pursuit mode is not always better than compensatory mode with auditory presentations.) Some exceptions to the generalizations of the first part of this chapter can be found in later sections dealing with nonvisual displays.

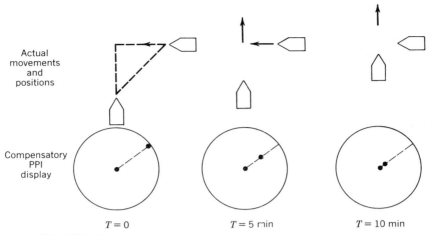

Fig. 2.7.7. Real versus apparent motion on a relative motion PPI display.

(which may be at a premium), the benefits of pursuit mode, and expected performance changes in particular tracking task situations. Various options are available to the display designer, in addition to changing display size, that may make pursuit mode possible even in limited-space environments.

Intermediate Display Modes

In pursuit mode the operator has information on target and cursor positions. In compensatory mode only error information is provided. Intermediate conditions are possible with error + target position information, error + cursor position, and cursor + target position only. Poulton (1952b) compared conditions in which explicit target position, cursor position, and error signals were given in five combinations. Results showed that a pure compensatory (error only) condition was worst. Providing cursor position as well did not produce statistically significant improvement. However, providing target position information in addition to error information did significantly improve performance. Poulton found performance to be best with pure pursuit mode (target and cursor position only).

Interestingly, adding explicit error information to the pure pursuit display actually degraded performance. Providing the operator with additional information is not always beneficial, particularly if it diverts attention from necessary information, causes the operator to adopt inefficient control strategies, or affects the operator's level of stress or motivation.

In addition to qualitative display changes, quantitative changes may produce "intermediate" display modes. Senders (1953) used a display in which target position, cursor position, and error were all explicitly presented. However, position changes were multiplied by a constant P ($P = 1.0, 0.75, 0.50, 0.25$), and the error signal was multiplied by $1 - P$. The pure pursuit condition corresponded to $P = 1$ and the pure compensatory condition corresponded to $P = 0$. Senders found pure pursuit to produce the best performance and pure compensatory to produce the worst performance. But adding only 25% (i.e., $P = 0.25$) position information to the compensatory display produced performance almost equal to pure pursuit. Thus highly attenuated target and cursor position information greatly improves performance. This is a design option for dealing with the display size requirements of pursuit mode.

A "real-life" intermediate mode situation might arise in driving. With full vision, the driver can see how the curvature of the road contributes to changing error (deviation of car from center of lane) magnitude. In limited vision conditions, such as fog, the driver may be able to see only a few yards ahead. It may not be immediately apparent when sideways drift of the car is due to movement of the steering wheel or changing curvature of the road.

Poulton (1974) describes this situation of driving with restricted vision as "pure pursuit" mode and driving with visibility ahead as "pursuit with preview." It is not always clear whether a situation is a pursuit or compensatory task. Here, the "limited-vision" case is described as compensatory because the driver is unable to identify the source of error (i.e., road curvature or steering error). This is the critical element that distinguishes compensatory from pursuit mode. The attribution of error components is much easier when both the controlled system and the target move against a stationary reference background. Then the operator can see how the target is moving, how the controlled system is moving, and how each movement contributes to momentary error

A general recommendation, therefore, is to provide structured backgrounds upon which targets and cursors move. This structure will facilitate movement detection and distance judgment (Hammerton and Tickner, 1970a).

Sources of Pursuit-Mode Superiority

There are several reasons why a properly designed pursuit-mode display is usually superior to (and always at least as good as) a compensatory-mode display. First, performance improves as the operator develops accurate internal models of the path of the target and of the dynamics of the controlled system. The error signal available in compensatory mode confounds the contributions of the operator's responses and the target's movements. Thus it becomes more difficult for the operator to develop accurate, predictive models of either. In many situations the dynamics of the controlled system is so complicated that the operator is unable to learn them unless they can be "studied" apart from the variability introduced by target movement.

Also, pursuit mode often yields a more S–R compatible situation than does compensatory mode: with pursuit mode, the operator moves the joy stick (for example) toward the target whereas in compensatory mode the operator moves the joy stick *away* from the "target" (i.e., the error marker).

Finally, the ambiguity of compensatory-mode displays can impose additional loads upon the operator's decision-making processes. Misinterpretation of compensatory-mode displays is relatively easy. (The example of "radar-assisted collisions" shows how a changing error vector can be misinterpreted as a changing target position vector.) Such misinterpretations become more likely in conditions of stress.

Display Magnification

One of the presumed benefits of compensatory-mode display is that the size of the error display is not limited by the scaling needed to show all positions of the target and cursor. Based on this reasoning it might be beneficial to magnify the error signal in a compensatory display to the largest extent permitted by the display's physical size. Poulton (1974, p. 166) suggests, however, that increasing the magnitude of the error signal beyond that which would be inferred from the corresponding pursuit display does not improve performance. Indeed, performance may be degraded with magnified displays early in practice: the error introduced by an unskilled operator may *add* to the error caused by the target's movements. With magnification, the resulting combined error signal may go beyond the limits of the display. Furthermore, the higher speed of error vector movement may degrade the operator's ability to follow it.

Experimental results have almost always shown advantages for magnified displays. However, Poulton (1974) suggests that these results most likely reflect range or transfer effects that produce bias in favor of magnified conditions. The general recommendations that can be made are as follows:

For pursuit displays, magnification should be such that target and cursor movement utilize the full available display area without exceeding it.

For compensatory displays, the error display should be at least as large as it would be in a corresponding pursuit mode implementation, but larger magnification probably is not helpful.

Effects of Display Mode on Step and Ramp Tracking

When the target's motion is very simple, the fact that its display is confounded with the operator's actions is not likely to degrade performance significantly. This situation obtains with step and ramp tracks and to a lesser degree with single sine waves.

With step tracks the change in the error signal happens so quickly that the display appears to move just as does the target track in pursuit mode. The only significant difference between the pursuit and compensatory mode displays is S–R compatiblity: with pursuit mode the operator moves the control toward the target whereas the compensatory mode requires control movements *away* from the error indicator. This incompatible situation probably reduces performance, especially early in training.

Display-Mode Effects on Sine Wave Tracks

Even tracking single sine waves (at least early in training) performance with compensatory mode is about half that with pursuit mode (Poulton, 1952a). However, Pew, Duffendack, and Frensch (1967) report that after extended practice with both pursuit and compensatory displays, single sine wave tracking performance with the two modes can become comparable.

The task is more complicated when the operator must track combinations of several sine waves. Again the compensatory mode produces about twice the tracking error of pursuit mode [four-sine wave track with top frequency of about 1 Hz (Poulton, 1952a)]. The data show less practice improvement with compensatory mode than was found in the case of single sine wave tracking. It is less likely that the performance with compensatory mode would equal pursuit mode performance even after extended practice. It is more difficult for the operator to develop an adequate internal model of the complex track's statistical attributes (e.g., mean position) with compensatory-mode information display.

The data cited above were obtained with relatively high-frequency tracks and a zero-order (see Section 2.7.6) control system. When very-low-frequency signals are followed, the task becomes more like ramp tracking and, as discussed previously, the advantage of pursuit mode may be small. Also, with high order of control the relative advantage of pursuit mode may diminish. However, for visual, position-based displays, compensatory mode will not be superior to pursuit mode when performance is properly measured.

This general advantage for pursuit mode is valid only for visual, position-based (analog) information displays. Tracking performance with other coding and sensory systems is discussed below.

2.7.5.2 Augmented Displays

So far the presentation of the first three basic, tracking information items (target position, cursor position, and error), has been discussed. In *augmented* displays, additional performance aids are presented as well. These additional information items may be combined with the three basic items (as is done with "quickened" displays) or may be presented separately (as with "switching lines" in phase-plane displays). Two important kinds of display augmentation will be discussed: quickened displays and predictor displays.

Quickened Displays

Quickened displays add low "order-of-control" components to the *display* of the controlled system's movements. (Order of control is discussed in Section 2.7.6. Essentially it refers to the number of integrations (over time) that occur between the operator's control movements and the movements of the controlled system.)

A display-quickened, second-order system is shown in Figure 2.7.8. Figure 2.7.8 shows that a "position" component has been added to the *display* of the controlled system's actual movement. It is critical to note that the resulting display does *not* accurately show the controlled system's true movement or position. The added "position" component changes only the display, not the controlled system. This is the difference between display quickening and control quickening. In the former, the low order-of-control components are added to the display; in the latter the low order-of-control components are added to the controlled system's movements.

Display quickening can be beneficial because it improves the "apparent" responsiveness of the operator's control. This enhances S–R compatibility. But more importantly, display quickening can provide additional information showing the operator when to make certain control movements. This is an advantage if the operator would otherwise have to judge when to make these responses.

For example, to track a "step" with a second-order (acceleration) control system, an optimal strategy is to accelerate maximally toward the target. Then, when the cursor reaches halfway, maximum deceleration is applied. Without display quickening, the operator must judge when the cursor is midway to the target. If position and rate components are added to the display, then the controlled system appears to reach the target (i.e., the cursor *does* actually reach the target but the *controlled system* does *not*) at just the moment when the operator should apply maximum deceleration. This provides a much better signal to the operator than a judgment whether the cursor is halfway to the target.

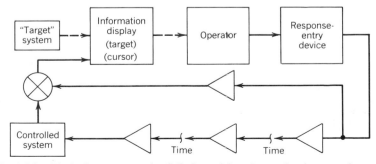

Fig. 2.7.8. Block diagram example of display-quickened second-order control system.

Problems with Display Quickening.　Although display quickening can aid performance with high order-of-control systems, the advantages are not universal. (Control quickening, in contrast, almost always produces performance benefits.) Some of the limitations and problems of display quickening are the following:

1.　Display quickening masks the true effects of the operator's control movements. Thus it is more difficult for the operator to develop an accurate model of the controlled system's dynamics. With compensatory mode, display quickening also masks true target movement, making it more difficult for the operator to develop an accurate internal predictive model of target motion.

2.　The operator has an inaccurate and sometimes misleading indication of controlled-system movement and position. The operator does not know the true error in his or her performance. This may give the operator a false sense of security, causing him or her no longer to devote adequate attention to the display or to stop responding when the controlled system is actually in a dangerous situation.

This problem is particularly apparent with simple ramp and single sine wave tracks. The added rate and position components in the display of the cursor's movement can make it appear that the target is being tracked perfectly. (For example, the added apparent rate makes it appear that the controlled system has acquired the proper velocity to follow the steadily moving target. In fact, because the extra rate component is virtual rather than real, the controlled system's rate is too small and it continuously falls behind. A similar lag condition occurs when following simple sine wave tracks.)

3.　Quickened displays may cause the operator to adopt suboptimal control strategies. For example, the optimal control strategy for tracking a "step" with a second-order control is full acceleration followed by full deceleration. The quickened display shows the operator when to apply full deceleration but, by its appearance, induces the operator to adopt a "graded" deceleration strategy.

Phase-Plane Displays

Phase-plane displays are typically used in compensatory mode applications. Error magnitude (position discrepancy between cursor and target) is displayed as usual. However, rate information is also presented. Unlike quickened displays, the rate information is not combined with position information but is presented on an independent display dimension perpendicular to the position error component. On a typical phase-plane display, a sine wave target path causes the cursor (which now has a vertical coordinate dependent on target rate and a horizontal coordinate dependent on target position) to move in a clockwise direction around the periphery of the display. The operator's objective is to move the cursor to the center of the display (where position and rate error components are both zero).

Aside from adding rate error to the standard compensatory mode position-error information, phase-plane displays may add a "switching line." Whenever the cursor crosses the switching line the operator should move the control fully in the opposite direction. Thus the switching line provides the same type of timing information that was available with the quickened display described above. The shape of the switching line depends upon the order of control of the controlled system. For a first-order system the switching line is straight; for a second-order system it is S-shaped; for high orders of control the shape of the line is time dependent.

Predictor Displays

Predictor displays utilize a computer model of the controlled system to predict its position at some future time, based upon its current position, its control dynamics, and expected operator actions. Usually, the operator action built into the model is slowly bringing the control back to the center or neutral position. The predictive information provided by this type of display is usually in addition to the momentary position information normally provided by pursuit or compensatory displays.

Predictor displays differ from quickened displays in that they (predictor displays) give an accurate picture of the current situation. Because "virtual" position and rate components are added with quickened displays, these displays provide an inaccurate presentation of the current control situation. For this reason predictor displays may be more helpful in complex situations than quickened displays.

Some predictor displays show only the current situation and the situation at one future moment. Others show the "path" that the controlled system will follow for some following time. Pew (1966) describes a rudimentary predictor display of the first type. This display, also known as a "velocity vector" display, shows the current controlled system's position as a dot. A vector starting at this dot has a length and direction such that it predicts the controlled system's position 0.5 sec later (assuming the operator makes no control movements to change the current rate). Thus the length of the vector represents the controlled system's velocity.

The submarine depth display described by Kelly (1962) is an example of the more complex type of predictor display that shows the controlled system's path over a segment of future time. McLane and Wolf (1966) compared a predictor display for submarine steering with three alternative quickened displays and found the predictor display to be superior to all.

Preview

In tracking, the operator must keep the cursor matched to the momentary position of the target, so the basic pursuit display shows only the *current* positions of target and cursor. In this condition the operator's responses to target movement will always be delayed by at least one reaction time (unless the operator can accurately predict the target's movements).

Response delays can be minimized by providing a preview of the target's path for some time into the future. For example, driving performance is improved when the driver can see the road further ahead, compared to conditions such as fog, where the driver must respond to unexpected road curves. Preview allows the operator to prepare responses in advance of the time they will be needed. Also, longer response sequences can be preprogrammed by the decision-making stage, thus reducing the rate at which individual response decisions must occur.

How much preview does the operator need to improve performance maximally? The answer depends upon the predictability of the target's movement, the frequencies in the target's movement spectrum, and the dynamics (particularly time lags) of the controlled system. For unpredictable, high-frequency (>0.5 Hz) tracks, the preview should at least be 0.5 sec + control time lag. The 0.5 sec component represents the operator's response generation rate (reaction time + time to observe effect of response).

In the case of zero control lag, extending the preview beyond 0.5 sec may continue to improve performance but to a far lesser degree than the benefit afforded by the 0.5 sec preview. Poulton (1974) recommends that the preview extend at least to the next reversal in track direction. However, this suggests that the amount of needed preview increases for low-frequency tracks. Probably a more consistent recommendation is to provide at least 0.5 sec of preview in addition to constant system time lags.

Effects of Visual "Noise"

The information concerning the position of the target and/or controlled system may be masked by visual noise. This noise degrades performance, but the effects are different for masking of target versus masking of controlled-system information. When the visual representation of the controlled system is noisy, performance is higher than when the target representation is degraded. This is because the operator has other, somewhat redundant information regarding the controlled system: the operator knows where his or her control is set and how it has been moved. This reduces the impact of visual noise. However, the operator is unable to distinguish visual noise in the target's display from actual target movement. Thus it is more important to maintain a noise-free display of target information than of controlled-system information.

In compensatory mode, both target and controlled-system movement contribute to the information that the operator sees. Thus effects of visual noise, whether arising from the controlled system or the target, degrade performance equally.

Intermittent Displays

The issue of intermittent displays is important for several reasons. First, in complex tasks the operator may have to monitor several displays. This requires distribution of attention among several signal sources and naturally produces intermittent perception of each display. In addition, operators blink or otherwise momentarily do not see a display. Some common tracking situations frequently require tracking with an intermittently displayed cursor. Computer display terminals generally use blinking cursors and the operator must move the cursor over the display screen. Finally, in some situations it might be necessary to display several displays at the same location, for example, on a CRT. The operator may select among several displays in sequence and this situation demands intermittent display presentation.

When the operator controls the sampling process, performance is likely to be better than when the sampling follows a rigid sequence. The operator can vary the sampling sequence according to momentary information processing demands and can devise sampling strategies for unusual periods of high information load, particularly emanating from one or a few displays.

Two parameters of the sampling process are particularly critical:

Sampling period (the time between samples)

The duration of each sample

In addition, for complex situations the sequence of sampling among displays may be important.

Results of Poulton (1974, p. 229) show that even when tracking completely predictable, single sine waves, sampling as often as once per second with a sample duration of 0.7 sec (70% duty cycle) may reliably reduce performance.

Knight (1980) showed that cursor blinking at 4 blinks/sec with a 50% duty cycle could slow cursor positioning speed compared to a 100% duty cycle condition. Thus even fairly rapid sampling can produce performance deficits. Just as visual noise has a greater impact when it affects the target's representation, so sampling of the cursor display is likely to have greater impact on performance than sampling the setting of the operator's cursor control device.

Although a track can be exactly reconstructed from momentary samples of its position taken at twice the highest frequency component, it is important to remember that the human operator may not be able effectively to apply a Fourier transform to the sampled values and thereby obtain an effective substitute for continuous view of the track. In addition to requiring higher than twice the maximum frequency sampling, the human's performance is degraded by very short samples. When tracking very-low-frequency (0.17 Hz) but complex sine wave paths, Poulton (1952b) reports performance decrements for view durations shorter than 0.2 sec.

With an intermittent display, tracking performance during short (2 sec) blackouts can decline sharply but the operator is often able to reacquire the target very quickly after its reappearance (Hammerton and Tickner, 1970b). Tracking performance during blackouts may be poor even if the target is following a simple acceleration or delta-acceleration ramp path. In this case the operator may incorrectly attribute a constant velocity to the target during the blackout interval, thus accumulating position error.

2.7.5.3 Alternative Visual Coding Methods

So far, only the situation in which target and controlled-system "movement" are represented as changes in spatial position on a display screen has been considered (all cited data have been based on this situation). However, visual codes other than position can be used to present target and controlled-system status information. Alternative coding systems may be useful for conserving display space, presenting multiple tracking problems on the same display (thus avoiding the need for the operator to redirect his vision), and avoiding confusion among different tracking problems.

In pursuit mode the operator must make comparisons between the magnitudes of two signals, computing the (error) difference between them. In compensatory mode, the operator need only compare the error signal to a constant reference. This makes some coding dimensions better suited for compensatory mode than for pursuit mode. If these coding systems are used for pursuit mode, performance is likely to be *worse* than that obtainable with compensatory mode. The general recommendations and effects noted in the preceding sections remain valid unless otherwise noted in Table 2.7.2.

2.7.5.4 Alternative Sensory Modalities

Just as with alternative visual coding dimensions, it may be useful to employ alternative sensory dimensions. This may avoid "structural" interference among simultaneous tracking tasks: the operator cannot look in two different locations at once but *can* look and listen at the same time. Potential alternative sensory dimensions are listed in Table 2.7.3.

2.7.6 CONTROL RESPONSIVENESS—EFFECTS ON OPERATOR'S PERFORMANCE

After the operator has determined the present state of the system being controlled, an appropriate response must be selected to keep it in tolerance or bring it back into tolerance. The operator's response

Table 2.7.2   Alternate Visual Coding Methods for Status Information

Coding Method		
Magnitude	Direction	Considerations
Flash rate	Side (left/right)	Good; compensatory mode only
Flash contrast (modulation depth)	Flash rate	Flash rate only fair as direction code; compensatory mode only
Speed		Pursuit and compensatory performance is low (>60% modulus mean error) even with high S-R compatibility
Brightness	Brightness difference	For compensatory mode, worse than position code with foveal vision but *better* for peripheral vision (Moss, 1964)

Table 2.7.3   Alternative Sensory Dimensions for Information Presentation

Dimension		
Magnitude	Direction	Considerations
Sound level	Direction	In simple step tracking (3 state) or very-low-frequency sine waves (0.03 Hz) may be almost as good as visual display
Tone-pulse rate	Direction	Much worse than vision
"Distance" of stereo "image"	Direction of stereo "image"	Audio stereo image is a much worse "target" than visual image. Much worse pursuit and compensatory performance than vision
		Compensatory conditions usually *better* than pursuit when auditory loudness and direction codes are used
Pitch	Relative pitch	Performance about half that of comparable visual display
		Pitch coding better for pursuit mode
Tone-pulse rate	Pitch	In a multi-task situation, a compensatory display using these dimensions + a visual display for the other task produced *better* performance than two visual displays (Ellis et al., 1953)
Rate of cutaneous vibration	Direction on cutaneous vibration	For a simple step task compensatory performance was equal to visual rudimentary visual display
Amplitude of cutaneous vibration	Direction of cutaneous vibration	Poor; average absolute error about 2.5 times greater than visual condition

must be entered into the controlled system. The manner in which this is accomplished has a critical influence upon performance. At least three important areas can be identified:

1. The muscle-systems and limbs used to make the response
2. Physical characteristics of the control device (e.g., joy stick, trackball, stylus) used for response entry
3. Transformations, time delays, and noise that are applied to the operator's responses on the control device

Because the first two items are dealt with in detail elsewhere in this book (Chapter 9.5) they are mentioned only briefly here. Consideration is limited to factors specifically affecting tracking performance. The remainder of this section concentrates on the third item—the effects of transformations made upon the output of the response-entry device.

2.7.6.1   Muscles and Limbs Used for Control Activation

Generally, all the muscle groups and limbs that might be used for control activation produce qualitatively similar tracking performance. This is because tracking and control performance is mediated by central information-processing activities. However, quantitative differences are found among alternative muscle groups and limbs. Some of these are due to different masses, centers of gravity, and so on. The joy stick is a response-entry device that has been designed in versions suitable for thumb, hand, and forearm activation and so the effects of different muscle groups upon tracking performance can be compared. In easy tracking conditions Hammerton and Tickner (1966) found all three types of joy stick control to produce the same performance. But in more difficult conditions the order of performance (best to worst) was hand, thumb, forearm. In the most difficult condition, the forearm target acquisition time was 188% that of the hand target acquisition time.

Control Characteristics

The effects of numerous control parameters upon tracking performance have been investigated and most of these characteristics are discussed in Chapter 9.5. The characteristics most relevant to tracking and control performance include the following:

Friction and damping (viscous damping is preferred)

Inertia (useful when rate matching strategy is employed)

Backlash (consistently degrades performance)

Dead space (consistently degrades performance)

Control centering (spring-centering generally improves performance)

Pressure- versus movement-responsive controls (task dependent)

In many systems, the operator's responses on a control activate servomotors or hydraulic mechanisms, for example. Usually in these cases, the operator does not have an accurate "feel" of the system being controlled. In particular, backlash and dead space (hysteresis) may still be present in the system but the operator may no longer be aware of them. If a control system has these actuation defects, the response-activation device should provide appropriate sensations to the operator rather than masking the defects.

The relative performance of pressure- versus movement-responsive controls is task dependent. Pressure-response controls do not require movement over a distance, but produce an output signal related to the pressure that the operator applies to them. Movement-response controls produce output signals related to their position. However, with spring centering, position and pressure can be related directly. This is one reason spring centering is beneficial: The operator has two, redundant sources of kinesthetic information regarding the position to which the control has been moved. The joint effect of these two information sources is to permit more accurate positioning.

Generally, pressure-response controls produce better performance in tracking situations where response time lags are the principal sources of performance deficit. These situations include tracking high-frequency signals and using velocity and higher-order control systems. Movement-responsive controls are better for ramp tracking and other situations involving position-setting (e.g., step tracking).

2.7.6.2 Transformations of the Control Device Output

Usually the effect of the operator's response upon the controlled system is not direct: a 1 inch movement of a joy stick often does not produce an immediate, 1 inch movement of the controlled system. Often the effect of the response is delayed, or "noise" is added to it, or it is integrated, multiplied, or otherwise mathematically transformed. Some of the important consequences of such manipulations are now reviewed.

S–R Compatibility

As discussed previously, S–R compatibility refers to the "naturalness" or "directness" of the relationship between the operator's control movement and the effect on the system as seen by the operator (i.e., if the operator is looking at a system representation on a display, the S–R compatibility is related to the effect on the displayed representation, rather than the system itself). High S–R compatibility is beneficial because it reduces the need for complex information transformations by the operator.

Zero-order control system, control quickening, short system time lags, and proper physical alignment of the control device and the display screen all improve performance partly because they enhance S–R compatibility.

Control Gain, Display Gain, and CD Ratio

Control gain refers to the sensitivity of the system to the operator's movement of the response-entry device. When gain is low, a *large* movement of the control is needed to produce a moderate effect on the controlled system. When gain is high, small control movements produce large effects on the system. When the operator is controlling a mechanical system, gain can be thought of in terms of gearing ratios. A high gain is equivalent to a high gear ratio and the operator generally has to exert large forces on the control to make it move; however, the effect of the small control movements upon the system is large. Low gain is equivalent to low mechanical gear ratio and the control forces required of the operator are small. However, they will be exerted over longer distances to produce substantial effects on the system.

The gain of a control can be altered in two ways:

1. The output at the extremes of the control movement can be changed. For example, if a joy stick moves ± 3 inches (maximum extent) to produce ± 3 inch movements in the controlled system, then changing the arrangement so that ± 6 inch system movements are produced for maximum joy stick movements doubles the control gain. The effect on the system of each unit of joy stick movement has been doubled.

2. The control gain may be increased *without* increasing the *maximum* control effect. If the control gain is doubled, as just described, but the range of joy stick movement is restricted to ± 1.5 inches, then the maximum system effect remains unchanged (± 3 inches).

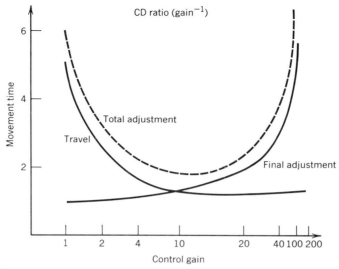

Fig. 2.7.9. Effect of control gain and CD ratio on cursor positioning time (after Jenkins and Connor, 1949).

Control gain is distinct from *display* gain. Display gain represents the relationship between controlled-system movement and movement of the system's *representation* on the operator's display. This relation can be varied independently of control gain.

The control/display (CD) ratio is the ratio of control gain to display gain. A CD ratio of 1 means that the controlled cursor on the operator's display moves the same distance as the control device. CD ratio is the inverse of sensitivity. A low CD ratio means that small movements of the control result in large movements of the controlled cursor.

A CD Ratio of 1 represents the most compatible situation and thus usually has an inherent advantage.

For controls that the operator cannot quickly move from one extreme position to the other (e.g., multiturn dials, trackballs, cranks), target acquisition time is a very broad U-shaped function of CD ratio (Figure 2.7.9, after data of Jenkins and Connor, 1949).

With low CD ratios (i.e., high control gain or sensitivity) the operator can quickly move to the general vicinity of the target. However, precise, final positioning movements are difficult and slow. With high CD ratios (i.e., low gain/sensitivity) the final positioning phase is easier but the initial, gross movement phase is slow. The range of best performance depends on characteristics of the control device but is quite broad. The data in Figure 2.7.9 show that CD ratios from 4 to 20 produce near-optimal performance.

Although dials generally produce U-shaped performance/CD ratio functions, joy stick controls typically produce J-shaped curves. This is because, with joy sticks, the operator can move the control quickly from one extreme position to the other, and so control gain does not make the initial quick positioning phase of target acquisition much easier. For joy sticks and other controls in which the full range of movement can be covered very quickly, CD ratios near 1 are usually optimal. Generally, target acquisition time can be predicted from Fitts's Law, which applies over the portion of the J-shaped function to the left of the "knee" where acquisition time begins to rise rapidly.

When joy sticks or other rapid-positioning controls are used, the maximum movement of the control should utilize the entire extent of the display: moving the joy stick from one side to the other should move the controlled cursor from one side of the display to the other.

Order of Control

Order of control refers to the number of mathematical integrations over time that are applied to the operator's response.* Thus in a zero-order system, a joy stick movement by the operator might cause

---

* For simplicity the operator's response is assumed to be a movement, say, in inches, and the controlled system to be one that is varying in position. In general, neither of these situations need apply: the operator's response might be a pressure on an isometric joy stick and the system's behavior might involve change in temperature, for example.

a change in position of the controlled system. In a first-order control system, a position change on the joy stick would cause a change in velocity of the controlled system. In a second-order system, joy stick position corresponds to system acceleration, and in a third-order control system, joy stick position corresponds to change in accleration (a so-called delta-acceleration system).

Generally, low order of control enhances performance. As mentioned above, the zero-order (position) control system has the greatest S–R compatibility. Also, because the control-order integrations occur over time, the zero-order system has the shortest time lag between the operator's response and the effect of that response.

Low order of control also reduces the number of response movements the operator must make. For example, in a step-tracking task, the operator need make only one response to cause a cursor to move from position X and stop at position Y. With a first-order (rate control) system, the operator first makes a control movement to impart a velocity to the cursor. Then, just before the cursor reaches position Y, the operator must make a second movement, returning the joy stick to the center (zero-velocity point). Finally, high-order control systems insert phase lags into the operator's response patterns and reduce the amplitude of the operator's responses inversely with frequency. For example, in a first-order (rate control) system, the effective amplitude of the Operator's responses is halved each time frequency is doubled: if the joy stick was moved ±1 inch to follow a 0.5 Hz sine wave path, then at 1 Hz a ±2 inch movement would be required. With a second-order control system, the effective amplitude of the operator's responses is reduced by a factor of 4 each time frequency doubles.

Thus the operator must modulate the amplitude of his or her control responses as a function of signal frequency as well as signal amplitude. This interaction of frequency and amplitude in higher order-of-control systems makes the operator's task much more complex.

Dither.   Although the effect of frequency upon response amplitude with nonzero orders of control usually makes tracking more difficult, it also makes a control strategy known as "dither" useful. Dither refers to rapid, small oscillatory movements of the response-entry device that the operator may make. Because effective response amplitude is attenuated owing to the high frequency of the operator's responses, the controlled system remains steady. If the operator tried to keep the control joy stick steady then any small positional drift would not have an attenuated effect and could cause a relatively large system movement.

Control Aiding and Quickening

Control aiding and control quickening both refer to the same process of combining low and high order-of-control components. The manner in which this addition is accomplished is the distinction between control aiding and quickening. These two alternatives are schematically represented in Figure 2.7.10. With control aiding, higher order-of-control components are added to the movement of a system

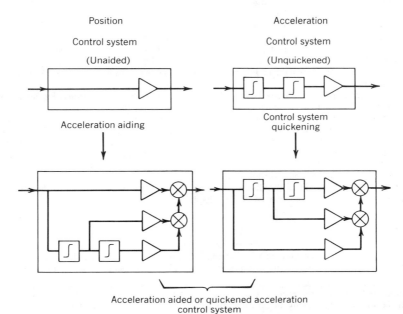

Fig. 2.7.10.   Aided and quickened control architectures.

otherwise governed by low order of control. This may be beneficial because it reduces the complexity of the operator's information processing task. The classic example concerns gunnery tracking systems. The operator's task is to follow a target moving with approximately constant velocity. By adding a rate (first-order) component to the normal position (zero-order) component, the tracking system "automatically" maintains a movement similar to that of the target In this case, the aiding function can be viewed as removing the burden from the operator of computing an integral of the control-position changes over time.

Although aiding can be beneficial, it should be used cautiously, for performance is usually best when low order of control is used: Adding higher order-of-control components often is found to degrade performance, or at least not significantly enhance performance, above that achievable with the pure low order-of-control system.

Control quickening differs from *display* quickening. With control quickening the lower-order-of-control component is added to the actual movement of the controlled system. With display quickening the lower-order-of-control component is added only to signal that is *displayed* to the operator. Thus display quickening can introduce a discrepancy between the actual system state and the information presented to the operator. This discrepancy can sometimes degrade performance. However, actual control quickening is almost always beneficial. The performance benefits of control quickening/aiding are illustrated by the data of Birmingham, Kahn, and Taylor (1954). In this experiment the operator had to keep a moving dot positioned within a square target area. With a third-order control system (delta acceleration) this task proved very difficult; when a zero-order (position) component was added, operators were able to perform almost perfectly.

Effects of Order of Control in Specific Situations

In almost all situations, low order of control produces better performance than higher order of control. Therefore the following sections really concern the types of performance deficits that are introduced by higher orders of control and some of the methods available to minimize these deficits.

Step Tracks. For step tracks a zero-order control system has the highest S–R compatibility and produces the best performance. For example, in a comparison of zero- and first-order control, Gibbs (1962) found target acquisition times of 0.55 sec for the position control and 0.9 sec for the rate control. Hammerton (1963) found 1.2 sec for a rate-control condition and also 8.5 sec for a second-order (acceleration) control. Adding a position component to a rate-control system can substantially reduce time to move to the target position. However, the final positioning task (getting the cursor to stay at the target position) is still difficult because of the velocity component that the cursor has when it reaches the target. The operator must make a reverse movement to counteract this rate. The accurate selection of the proper reverse movement is difficult (the operator cannot just return to the center control position, as with a pure rate-control system, to stop the cursor on target) because of the position component that has been added to the control system.

In general, however, step tracking improves when a position component is added to a rate-control system and the degree of benefit increases as the amplitude of the position component becomes larger with respect to the rate component.

There is at least one situation when a quickened rate-control system may produce superior performance. This is multistep tracking, where target jumps are of constant distance and at regular intervals. This situation would occur if an operator was following a target on a PPI display and the target was moving with constant velocity: With each radar sweep, the target jumps to a new position.

If the rate set into the cursor matches the target's stepping movements, then the cursor is near the spot at which the target appears after each radar sweep. Minor position corrections are done quickly, taking advantage of the position component in the quickened control system.

Velocity Ramp Tracks. When tracking single targets, even first-order (velocity) control systems have not been empirically shown to be better than zero-order (position) control systems. However, it is reasonable to expect that this could be demonstrated in situations where the velocity of the target remains constant for substantial intervals. In this situation the operator could simply hold the control in a fixed position, making occasional correction responses as needed.

Although this expected, potential advantage has not been clearly demonstrated with single-target tracking, it has been demonstrated in tracking *multiple* targets that are each moving with approximately constant velocity. In this case, when the operator diverts his or her attention from one target to make tracking corrections for another target, the cursor (which has a rate component) continues approximately to follow the nonattended targets.

Sine Wave Tracks. The advantage of zero-order control is maintained when tracking sine waves. With *pursuit* tracking, even first-order control greatly reduces S–R compatibility.* With zero-order

---

* This disadvantage is minimized when compensatory displays are used.

control, the operator always moves the joy stick (for example) toward the target to reduce tracking error. First-order control introduces a phase lag of 90°, so the operator must move the joy stick in the direction of the target for one-quarter cycle and then away from the target in the next quarter cycle. Because the first-order system introduces a 90° phase lag, the operator must compensate by trying to respond *ahead* of the current target position. This becomes increasingly difficult as the frequency of the track increases or as more and more sine wave components are added to make the track more complex.

The situation is even worse with second-order (acceleration) control systems in which 180° phase lags are introduced. Thus the very incompatible situation of having always to move the joy stick *away* from the momentary target position is introduced. The need to predict at least one-half cycle ahead makes second-order tracking of complex or high-frequency sine wave signals very difficult.

With *compensatory* displays the relative disadvantage of first-order control systems is reduced. This is because S–R compatibility is increased: With a compensatory display and zero-order control system the operator must always move the joy stick *away* from the momentary target position in order to recenter the target. With a rate control, the operator moves the joy stick *toward* the target and then returns the control to the center position when the target reaches the center.

Although performance improves when a compensatory, rather than a pursuit, display is used with a rate-control system, the performance is still generally inferior to that which could be obtained with a zero-order control system. Furthermore, the relative improvement to first-order performance associated with a compensatory display is likely to occur only with low-frequency, high predictable tracks, because the operator must still compensate for the phase lag of the rate-control system by predicting (90°) ahead. This becomes very difficult when the track is complex and the time available to calculate the predictions is short (i.e., with tracks containing high frequencies).

Adding High Order-of-Control Components. In some situations the addition of high order-of-control components, such as acceleration, would appear to be potentially beneficial. For example, in following an airplane moving overhead, the speed of the aircraft appears to increase until the plane is overhead—that is, the aircraft appears to accelerate. Therefore, adding an acceleration component to a tracking system might help an operator follow this track. Speight and Bickerdike (1968) found a beneficial effect of adding a small acceleration component to a zero-order control system. However, this result may be attributable to asymmetric transfer and range effects, because it was not obtained with a between-groups experimental design.

Furthermore, even if the addition of an acceleration component aids tracking while the aircraft is approaching, the advantage may disappear when the aircraft passes overhead and then appears to decelerate. In this case the operator must move the control *past* its center position to introduce a corresponding deceleration of the controlled cursor. This is likely to be complicated by the zero-order control component.

Control System Time Lags

The effect of the operator's response upon the controlled system may be delayed. This control lag should be distinguished from display lag, in which the effect upon the system is immediate but the change in the system is not immediately shown to the operator. Control time lags are generally detrimental when applied to zero-order control systems. The detriment to performance may occur even when the lags are not immediately perceptible to the operator. The form of the performance detriment depends upon the nature of the time delay function.

The simplest case is transmission lag: a constant delay intervenes between the operator's action and its effect on the system. Exponential lag represents the effect of a single-pole, low-pass filter (Figure 2.7.11a): When the operator makes a response there is an immediate, but small, effect on the system. But the full effect of the response is not achieved till later.

The "time constant" of the equivalent filter is the delay between the operator's response and the moment the effect on the system has reached 67% of its final value. Because an exponential time lag represents the effect of a low-pass filter between the operator's control and the controlled system, it is apparent that exponentially lagged systems remove high-frequency components from the operator's responses. Sigmoid time lags represent the equivalent effect of a two-pole, low-pass filter. Effectively such a filter has a sharper high-frequency roll-off than a one-pole filter. Hence high-frequency components in the operator's responses are more completely eliminated when sigmoid control lags are introduced.

Transmission lags disrupt performance because they require the operator to predict further ahead. The difficulty of doing this and therefore the performance deficit associated with transmission delays increases with the complexity and highest frequency component of the tracked signal. With unpredictable, high-frequency signals, even very small transmission lags degrade performance.

With completely predictable signals the operator may be able to compensate for control system transmission lags once they are known. The effectiveness of the compensation will decrease as the time lag grows.

Exponential and sigmoid delays usually (but, as discussed below, not always) degrade performance.

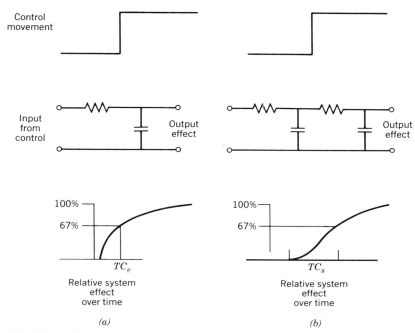

Fig. 2.7.11. Time lags introduced by control filtering (*a*) exponential and (*b*) sigmoid.

This occurs for several reasons. First, both these lag types effectively reduce control gain: A large control movement produces only a small effect on the system, at least initially. Second, to compensate for the slow growth of the effect upon the system (i.e., the lowering of control gain) the operator must compensate by making initially larger-than-normal control movements in order to acquire a target quickly. With an exponentially lagged zero-order control in step tracking, the operator must move the control *past* the position that would correspond to the correct placement of the cursor. Then, when the cursor nears the desired target position the operator must move the joy stick in the opposite direction, back toward the target. The operator stops the movement with the joy stick at a position corresponding to the target's location. This is like the movements required with a rate-control system except the operator would move the joy stick to the center position with first-order control. Just as with the rate-control system, the low S–R compatibility of the lagged position-control system degrades performance.

However, there is one situation in which exponential lags are beneficial. Performance with a second-order (acceleration) control system can be improved by adding an exponential lag. This makes the second-order system behave somewhat like a first-order (rate-control) system. The benefit can be seen in the data of Tickner and Poulton (1972), which show a decrease in target acquisition time from 11.9 sec with an acceleration control to 5.2 sec with a "lagged rate" control, with an exponential time constant of 0.5 sec. Even with a time constant of 4 sec, the addition of an exponential lag to the acceleration system appreciably reduced target acquisition time (to 7.9 sec).

Control Effects in Multidimensional Tracking

Just as multidimensional tracking presented special display issues, a number of particular control effects also emerge, including

    Cross-coupling
    Differing orders of control

Cross-coupling means that control movement on one dimension (say, the horizontal) produces a response component on the other (say, the vertical). Therefore, to produce a horizontal cursor movement the operator might need to move a joy stick with a compensatory vertical sloping movement. Symmetrical cross-coupling is like rotating the entire display about the center: the operator must offset all joy stick (for example) movements by the same angular amount (e.g., 10°).

Symmetrical cross-coupling reduces S–R compatibility. However, Bernotat (1970) reports negligable performance deficits for cross-couplings up to 45°: 90° cross-coupling does degrade performance significantly.

Asymmetric cross-coupling occurs when control movement along one of two dimensions produces a component in the other; but control movement in the other dimension does not produce a corresponding cross-coupling. Operators can compensate for constant asymmetric cross-coupling well, just as they can for constant symmetrical cross-coupling.

With variable cross-coupling, tracking becomes much more difficult. Even though operators can compensate for constant symmetrical cross-coupling of 45°, *variable* cross-coupling up to this amount increases average absolute error by a factor of 3 (compared to a no cross-coupling condition) even after 6 days of practice (McLeod, 1973). Variable cross-coupling of only 23° almost doubles average absolute error. Operators apparently do not learn to compensate for variable cross-coupling.

In some cases different orders of control may be applied to response components in different movement dimensions. This generally degrades performance. For one thing, transfer of training from one order of control to another is usually asymmetric and often negative (because the movements and phase lags appropriate for one order of control are usually inappropriate for another). Furthermore, distributing the operator's experience between two control systems makes it more difficult to develop an effective internal model of either control system. In two-dimensional tracking environments, the same control dynamics should be applied to each movement dimension.

2.7.6.3 Effects of Control System Nonlinearities

All the transformations of control output discussed so far have been linear in that they involve summations, delays, or multiplications. The "pattern" of the operator's responses remains the same after these transformations. With nonlinear transformations, the pattern of the operator's response is changed. The relationship between control output level and system effect varies as a function of the control output level. For example, control gain may be nonlinear so that gain is low when output is low (i.e., the operator is making small responses) and high when the control output is larger. Dead space is an example of this type of nonlinearity: the operator must make a response that exceeds some threshold size before *any* effect on the system occurs.

Another common nonlinearity involves step functions: the system effect suddenly jumps from one value to another when the operator's (continuous) response magnitude reaches some critical level. Or the operator's response may be sampled at discrete time intervals so the system effect jumps each time a new sample is made. Finally, the operator may have a response-entry device (such as a set of pushbuttons) that permits only a set of discrete responses to be made, thus causing jumps (i.e., nonlinearities) in system effects.

Generally, nonlinearities degrade performance, particularly when they are combined with *low* order of control. With high order of control, nonlinearities may be beneficial or may induce operators to adopt efficient control strategies.

Control nonlinearities should be distinghished from *display* nonlinearities. Control nonlinearity occurs between the operator's responses and their effect upon the controlled system. Even when control nonlinearities are present, the operator may continue to see an accurate representation of the current system state. Thus the operator always has a valid representation of the error that his or her performance is producing.

Display nonlinearity occurs between the controlled system and its representation on the operator's display. For example, a 1 m movement of the system may be represented as a 2 cm movement on the operator's display near the display center (e.g., when the controlled system is almost in the proper position) but as a 1 cm movement near the edges of the display. This gives the operator misleading information concerning the actual error that his or her performance is generating. Although both display and control nonlinearities generally degrade performance, special caution is warranted with display nonlinearity. Control nonlinearity is sometimes helpful for high order-of-control systems but display nonlinearity is not.

Continuous Nonlinearities

Figure 2.7.12 illustrates three common types of nonlinearities in control gain which have the following interpretations:

Type 1.   Control gain is lower for small responses so the operator must make proportionately larger control movements to effect fine adjustments in the controlled system.

Type 2.   Control gain is higher for small responses than for large responses. This type of nonlinearity is commonly found on engine controls such as the gas pedal of cars: Once the control nears its extreme position, the engine just cannot produce much additional power so the effect of control movement is small.

Type 3.   Control gain is highest for response sizes in the middle of the response-size range.

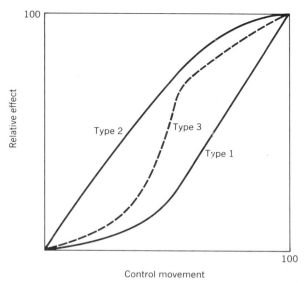

Fig. 2.7.12.   Three types of control system nonlinearities.

Wortz, McTee, Swartz, Rhinelander, and Dalhamer (1965) report that the effects of these nonlinearity types are task dependent. Type 2 nonlinearity (and linear) produced best performance for tracking ramps with multiple constant-velocity sections; type 1 nonlinearity was worst. However, for multiple-step tracking, type 3 nonlinearity was the best of the three and type 1 was again the worst.

The steering mechanism in cars contains nonlinearities. It is a second-order (acceleration-control) system in which the position of the steering wheel determines an angular acceleration.

Type 1 nonlinearity is introduced so that small adjustments of the steering wheel have small effects on angular acceleration when the car is moving in a straight line: When the steering wheel is turned very sharply the gain of the control system increases. It is likely that this arrangement does not improve the normal road-following performance of the driver at roughly constant speed. However, this nonlinearity does accommodate the *qualitatively* different situations in which the driver adjusts the steering wheel: the high gain for large wheel excursions probably improves the ability of the driver to make very sharp maneuvers (such as parking and turning 90° corners) at very slow speeds. It probably does not improve the driver's ability to maintain a straight path over a gradually curving road and may degrade performance when the driver begins to lose control and makes large control movements that produce unexpectedly large system effects.

Stepped Nonlinearities

Step nonlinearities may result from either time or amplitude sampling of the operator's responses. With time sampling, the magnitude of the operator's response is observed at regular or irregular time intervals and the sample is applied as an input to the system until the next sample is taken. As the sampling rate becomes higher the detrimental effects of this type of nonlinearity decrease. Sampling at less than twice the highest frequency in the operator's response spectrum is likely to produce performance deficits.

These deficits arise both because the operator's intended responses are not faithfully passed to the controlled system and because new "alias" frequencies are added to the operator's response spectrum. Alias frequencies are generated when frequencies with periods shorter than one-half the sampling interval are present. These can be filtered but this can potentially introduce new sources of performance deficits associated with time lags.

Amplitude sampling can be viewed as providing the operator with only a limited number of control positions (i.e., a discrete set of responses). For example, an operator may have only a single switch with which he or she can turn an engine on or off. By controlling the duty cycle of the engine with this binary control, acceleration and rate may be regulated. For systems with low order of control, amplitude sampling degrades performance: Either the operator's positioning accuracy is limited by the number of available responses, or the number of individual responses that must be made to achieve fine positioning becomes very large.

For systems with higher orders of control, amplitude sampling may be beneficial or at least not detrimental. Because each added order of control involves an integration over time, the effect is to "smooth out" jumps between sampled control values. This smoothing is not sufficient with first-order systems to prevent performance detriment (Bennet, 1956). However, with second-order (acceleration control) systems, on–off–on control is an effective control strategy.

With this method the operator causes the controlled cursor to accelerate toward a target at maximum rate. Then, when the cursor reaches the halfway point, the operator applies maximum acceleration (i.e., deceleration) in the opposite direction. By providing only two push buttons (one for maximum acceleration in each direction) the operator controls only response timing, not response magnitude. This reduces the information-processing load on the operator. Also, the time delay in going from one maximum acceleration to another is reduced compared to a continuous-control implementation, where time would be needed to move the control from one position to the other.

The general recommendation is that sampled control should not be used with lower than second order-of-control systems.

2.7.7 CONCLUSIONS

Some of the important elements of manual control and tracking have been reviewed. The most important of these can be summarized as follows:

1. The operator's performance limitations arise from his or her limited information—processing capacity, inherent time delays such as the PRP, and structural, biomechanical characteristics. The operator's effective bandwidth is about 1 Hz.
2. Skilled control and tracking performance rests upon effective predictive models of the target system and controlled-system dynamics.
3. Task characteristics that enhance the operator's ability to predict the future and thus minimize his or her inherent limitations improve performance.
4. Task characteristics that promote rapid development of accurate internal predictive models enhance performance.
5. Normally beneficial task characteristics therefore include pursuit mode, track preview, low order of control, noise-free displays, low-frequency tracks, continuous rather than intermittent information display, minimal time lags, accurate status displays (i.e., very conservative use of display quickening, display filtering, etc.), predictive displays, and control (not display) quickening.

REFERENCES

Adams, J. A. (1961). Human tracking behavior. *Psychological Bulletin, 58,* 55–79.

Craik, K. J. W. (1947). Theory of the human operator in control systems. 1. The operator as an engineering system. *British Journal of Psychology, 38,* 56–61.

Bennet, C. A. (1956). Sampled-data tracking: Sampling of the operator's output. *Journal of Experimental Psychology, 51,* 429–438.

Bernotat, R. K. (1970). Rotation of visual reference systems and its influence on control quality. *IEEE Transactions on Man–Machine Systems, MMS-II,* 129–131.

Birmingham, H. P., Kahn, A., and Taylor, F. V. (1954). A demonstration of the effects of quickening in multiple-coordinate control tasks (Report 4333). Washington, DC: U.S. Naval Research Laboratory.

Brebner, J. M. T. and Welford, A. T. (1980). Introduction: An historical background sketch. In A. T. Welford, Ed., *Reaction times.* London: Academic.

Craik, K. J. W. and Vince, M. A. (1963). Psychological and physiological aspects of control mechanisms. *Ergonomics, 6,* 419–440.

Crossman, E. R. F. W. and Seymour, W. D. (1957). The nature and acquisition of industrial skills. London: Department of Scientific and Industrial Research.

Elkind, J. I. (1956). Characteristics of simple manual control systems (Technical Report 111). Lexington, MA: MIT Lincoln Laboratory.

Ellis, W. H. B., Burrows, A., and Jackson, K. F. (1953). Presentation of air speed shile deck-landing: Comparison of visual and auditory methods (UK RAF Flying Personnel Research Committee Report 841). London: Air Ministry.

Ellson, D. G. and Gray, F. E. (1948). Frequency responses of human operators following a sine wave input (Memorandum Report MCREXD-694-2N). Wright-Patterson Air Force Base, OH: USAF Air Material Command.

Fitts, P. M. (1954). The information capacity of the human motor system in controlling the amplitude of movement. *Journal of Experimental Psychology, 47,* 381–391.

Gibbs, C. B. (1962). Controller design: Interactions of controlling limbs, time-lags and gain in positional and velocity systems. *Ergonomics, 5,* 385–402.

Gibbs, C. B. (1965). Probability learning in step-input tracking. *British Journal of Psychology, 56,* 233–242.

Halliday, A. M., Kerr, M. and Elithorne, A. (1960). Grouping of stimuli and apparent exceptions to the psychological refractory period. *Quarterly Journal of Experimental Psychology, 12,* 72–89.

Hammerton, M. (1963). An investigation into the optimal gain of a velocity control system. *Ergonomics, 66,* 450–543.

Hammerton, M. and Tickner, A. H. (1970a). Structured and blank backgrounds in a pursuit tracking task. *Ergonomics, 13,* 719–722.

Hammerton, M. and Tickner, A. H. (1970b). The effect of temporary obscuration of the target on a pursuit tracking task. *Ergonomics, 13,* 723–725.

Jagacinski, R. J., Repperger, D. W., Ward, S. L., and Moran, M. S. (1980). A test of Fitts' Law with moving targets. *Human Factors, 22,* 225–233.

Jenkins, W. L. and Connor, M. B. (1949). Some design factors in marking settings a linear scale. *Journal of Applied Psychology, 33,* 395–409.

Kantowitz, B. H. (1974). Double stimulation. In B. H. Kantowitz, Ed., *Human information processing: Tutorials in performance and attention.* Hillsdale, NJ: Erlbaum.

Keele, S. W. and Posner, M. I. (1968). Processing of visual feedback in rapid movements. *Journal of Experimental Psychology, 77,* 155–158.

Kelly, C. R. (1962). Predictor instruments look into the future *Control Engineering, 9,* 86–90.

Knight, J. L. (1980). Some human factors aspects of cursor positioning. *Processings of the 24th Annual Convention, Human Factors Society,* 63.

McLane, R. C. and Wolf, J. D. (1966). Symbolic and pictorial displays for submarine control. *Proceedings of the Second Annual NASA-University Conference on Manual Control* (Report SP-128). Washington, DC: U.S. National Aeronautics and Space Administration, pp. 213–228.

McLeod, P. D. (1973). Response strategies with a cross-coupled control system. *Journal of Experimental Psychology, 97,* 64–69.

Moss, S. M. (1964). Tracking with a differential brightness display: II. Peripheral tracking. *Journal of Applied Psychology, 48,* 249–254.

Noble, M., Fitts, P. M. and Warren, C. E. (1955). The frequency response of skilled subjects in a pursuit tracking task. *Journal of Experimental Psychology 49,* 249–256.

Noble, M. and Trumbo, D. (1967). The organization of skilled response. *Organizational Behavior and Human Performance, 2,* 1–25.

Pew, R. W. (1966). Performance of human operators in a three-state relay control system with velocity-augmented displays. *IEEE Transactions on Human Factors in Electronics, HFE-7,* 77–83.

Pew, R. W., Duffendack, J. C., and Frensch, L. K. (1967). Sine-wave tracking revisited. *IEEE Transactions on Human Factors in Electronics, HFE-8,* 130–134.

Poulton, E. C. (1950). Perceptual anticipation in tracking (Report 118). Cambrdige, England: UK MRC Applied Psychology Unit.

Poulton, E. C. (1952a). Perceptual anticipation in tracking with two-pointer and one-pointer displays. *British Journal of Psychology, 43,* 222–229.

Poulton, E. C. (1952b). The basis of perceptual anticipation in tracking. *British Journal of Psychology, 43,* 295–302.

Poulton, E. C. (1974). *Tracking skill and manual control.* New York: Academic Press.

Senders, J. W. (1953). The influence of surround on tracking performance. Part 1. Tracking on combined pursuit and compensatory one-dimensional tasks with and without a structured surround (Technical Report 52-229, Part 1). Wright-Patterson Air Force Base, OH: USAF Wright Air Development Center.

Slack, C. W. (1953). Learning in simple one-dimensional tracking. *American Journal of Psychology, 66,* 33–44.

Speight, L. R. and Bickerdike, J. S. (1968). Control law optimization in simulated visual tracking of aircraft. *Ergonomics, 11,* 231–247.

Tickner, A. H. and Poulton, E. C. (1972). Acquiring a target with an unlagged acceleration control system and with 3 lagged-rate control systems. *Ergonomics, 15,* 49–56.

Trumbo, D. (1970). Acquisition and performance as a function of uncertainty and structure in serial tracking tasks. *Acta Psychologica, 33,* 252–266.

Welford, A. T. (1968). *Fundamentals of skill.* London: Methuen.

Wortz, E. C., McTee, A. C., Swartz, W. F., Rhinelander, T. W. and Dalhamer, W. A. (1965). Effects of control-display displacement functions on pursuit and compensatory tracking. *Aerospace Medicine, 36,* 1042–1047.

CHAPTER 2.8

HUMAN ERROR AND
HUMAN RELIABILITY

DWIGHT P. MILLER
ALAN D. SWAIN

Sandia National Laboratories
Albuquerque, New Mexico

Special thanks go to Dr. Robert G. Easterling of Sandia National Laboratories and to Dr. Michael G. Miller of the U.S. Library of Congress, who edited the manuscript, and to Janis A. Purcell, without whose word processing skills this submission would not have been possible.

2.8.1 OVERVIEW

The importance of a human error depends on its social and economic consequences. A forgotten postage stamp might delay some correspondence, but a forgotten emergency procedure may cause a jumbo jet to crash. The psychological roots of these two errors may be very similar, but it is the latter kind of mistake that receives careful investigation and for which means of prevention are sought. Of all the aircraft accidents investigated by the National Transportation Safety Board in recent years, more than 70% were cited as attributable to "pilot error." Similar statistics dominate the incident reports made to the Aviation Safety Reporting System. This broad category, which includes everything from poor manual control to inadequate flight planning, identifies events not attributable to mechanical/ electrical breakdowns or sudden severe weather (acts of God). Why is this statistic so high? Because the system has evolved over the years to become extremely safe. That is, the relentless effect of gravity has encouraged the production of highly reliable machines, frequent inspections, and rigid operating rules. Because people still have major responsibilities in communications, aircraft control, decision-making, and traffic management, their errors dominate the statistics. As a system improves through equipment reliability, the occurrence rate of system failures decreases, but of the remaining failures the proportion resulting from human error increases. Therefore if we are to make further improvements in these relatively safe systems, we must concentrate on the reduction of human error.

This chapter is intended to help the reader understand how human error can degrade system performance, how the work situation and tasks can be analyzed for human error potential, and how the equipment or task characteristics might be modified to reduce the likelihood of human error. Reviewing the psychological theories and processes that lead to errors is beyond the scope of this chapter. The reader may refer to Norman (1981) for a more thorough discussion of this topic. Nor will this chapter provide an extensive, thorough review of the recent literature concerning human error and human reliability. Meister has examined the field from this perspective in a review article (Meister, 1984).

The following sections examine the nature of human error on the psychological as well as the systems level. Various classifications of errors are defined and the impact of performance-shaping factors on error likelihood is addressed. Human reliability analysis is explained and several techniques are examined. A final section deals with strategies for reducing human error.

2.8.2 WHAT IS HUMAN ERROR?

2.8.2.1 System Context

Before one can consider the relationship of human error and system performance, one must know what human error is. As defined by Rigby (1970), human error is any member of a set of human actions that exceeds some limit of acceptability. It is an out-of-tolerance action, where the limits of acceptable performance are defined by the system. Most errors are unintentional, inadvertent actions that are inappropriate in the given situation. There are some errors that are intentional. They occur when someone intends to perform an act that is incorrect, but believes it to be correct or to represent a better method. Malevolent behavior, which is deliberate behavior calculated to damage the system, is not considered to be part of human error in this treatment.

Not all human errors result in system degradation. An error can be recovered or corrected before it results in undesirable consequences to the system. This recovery factor may be the result of human redundancy, for example, someone checking another's work, or the system itself may detect an error before any consequences accrue. Human errors can be studied independently of their consequences on the system, and this independence of contexts helps in the analysis of human error. On the personnel-task level, factors that increase or decrease the likelihood of successful completion of the task can be studied. We also may look at similar tasks under similar conditions and generalize about the probability of error. On the other hand, the task in its relationship to other tasks in executing system functions determines the importance of that error in terms of its impact on system performance.

Reliability is the antithesis of error likelihood. It is the probability that no errors occur. Reliability is conventionally defined (Evans, 1976) as the probability of successful performance of a mission. (Throughout the chapter, the term "reliability" is used in the sense of the probability of being correct or performing satisfactorily. It is not used in the manner of psychologists to denote consistency or repeatability of some measure of human performance.) Meister (1966) defined human reliability as

The probability that a job or task will successfully be completed by personnel at any required stage in system operation within a required minimum of time (if the time requirement exists).

Swain and Guttmann (1983) defined human reliability as

The probability that a person (1) correctly performs some system-required activity in a required time period (if time is a limiting factor) and (2) performs no extraneous activity that can degrade the system.

Human reliability analysis (HRA) is a method by which human reliability is estimated. The closely associated system attribute of availability refers to the probability that the system or component is available for use when it is needed.

The basic expression of error likelihood used throughout this chapter is the *human error probability* (HEP). The HEP is the probability that when a given task is performed, an error will occur. An HEP is calculated as the ratio of errors committed to the number of opportunities for that error, or an estimate of that ratio.

$$\text{HEP} = \frac{\text{number of errors}}{\text{number of opportunities for error}} \tag{1}$$

The denominator represents the exposure to the task or task element of interest. A *task* is a level of job behavior that describes the performance of a meaningful job function or any unit of behavior that contributes to the accomplishment of system goal or function. A *task element* refers to any identifiable subtask or division of a task. An example of a task would be machinist milling a part to a given design specification. Task elements contributing to the task would include turning on the machine, adjusting the machine setting, retrieving the piece to be machined, and so on. Unfortunately, the denominator of Equation (1) is often difficult to determine because the opportunities may be covert, unrecorded, or part of a procedure whose steps appear to be continuous.

The assumed frequency distribution of the HEP variable over people and conditions has not been consistent in the literature. For instance, Askren and Regulinski (1969) found that the Weibull distribution gave the best fit to their data. Swain and Guttmann (1983) have assumed a log-normal distribution. When the performance of skilled people is considered, it is reasonable to assume that most HEPs fall near the low end of the error distribution. Although systematic collections of actuarial data do not exist to support this presumption, the preponderance of available human performance data (e.g., reaction times) indicates that a log-normal or similar distribution prevails. Mills and Hatfield (1974) found that, within limits, assumptions as to the distribution of task times are not very meaningful.

Any estimate of an HEP is associated with uncertainty. The uncertainty discussed in this chapter refers to the combination of imperfect knowledge and stochastic variability. Because estimated HEPs are usually estimates of performance based on some relevant data and judgment, the uncertainty expressed reflects the anticipated spread of HEPs across the assumed distribution, not uncertainty in the sense of confidence limits. When HEPs are assigned to tasks or task elements as estimates of performance, they are expressed as point estimates, HEPs, and uncertainty bounds (UCBs). The range between the lower and upper UCBs is assumed to include at least 90% of the true HEPs for a given task or activity (Swain and Guttmann, 1983). The large uncertainty associated with estimated HEPs and hard data renders the assumed underlying distribution less critical.

2.8.2.2 Types of Human Error

There are several ways of classifying human error. In the system context, primary interest is on those errors that constitute erroneous inputs to the system. Stated differently, incorrect inputs to the system stem from incorrect human outputs. If human output is examined without regard to internal processes, a classification scheme emerges that relates human output to system requirements. Such a scheme has been introduced by Swain (1963) and used by many of the HRA methodologists over the past 20 years. This scheme is illustrated in the following paragraph.

A person can make an error if he does something incorrectly, fails to do something he should, or fails to do something in time. An *error of omission* occurs when an operator omits a step in a task, or the entire task. Thus when a mechanic changes engine oil and puts the new oil in without replacing the oil-pan drain plug, his omission error results in wasted oil and a messy garage floor. An *error of commission* occurs when an operator does the task, but does it incorrectly. This is a broad category, encompassing *selection errors, sequence errors, time errors,* and *qualitative errors.* In the oil-change example, if the mechanic chooses the wrong-sized socket for tightening the drain plug, he has made a selection error. If he adds new oil before draining the old oil, he has made an error in sequence. If he cannot finish the job in the allotted time, he has made a time error. If he puts too little torque on the oil-pan drain plug, causing it to leak slowly, he has made a qualitative error.

context. Of course, to obtain estimates of the HEPs for these human outputs, it is often necessary for the HRA analyst to consider the underlying behavioral processes.

Norman (1981) suggests a different means of classifying human error. Being more concerned with the psychological bases for errors, Norman categorizes his "action slips" by their presumed sources. The groups of action slips are based on a model called the activation-trigger-schema system or ATS. The ATS assumes that action sequences are controlled by schemas that are sensorimotor knowledge structures. The model's operation is based on the activation and selection of schemas and uses a triggering mechanism that requires appropriate conditions. Any given task is modeled as a hierarchy of schemas. The highest-level parent schema is equated with the concept of intention. The action sequence is made up of a parent schema that orchestrates activity by initiating subschemas that control component parts of the action sequence, called child schemas. Modeling action in this manner permits numerous opportunities for errors in the action sequence. Errors can take place in the selection of intention and in component specification. Schemas can be triggered out of order, omitted, or substituted for by inappropriate schemas. Thus Norman's three major classification of slips based on their presumed sources are as follows:

1. *Slips that result from errors in the formation of the intention.*   These include decision-making and problem-solving errors.

2. *Slips that result from faulty activation of schemas.*   These include forgetting the intention, misordering an action sequence, and skipping or repeating steps in an action sequence.

3. *Slips that result from faulty triggering of active schemas.*   These include triggering an activated schema at an inappropriate time, blending components from competing schemas, and failure to trigger owing to an insufficient match in triggering conditions.

Norman's method of classification may work well in the theoretical discussion of the causation of human error. However, it becomes unwieldy in the context of human reliability analysis. Norman's method is based on the ATS model, without which it loses structure. In order for human reliability analysts to use a classification method such as this, they would have to agree on the precepts of the ATS model. Because most psychological theories of behavior are controversial, this is highly unlikely. Furthermore, the analyst would be forced to analyze human error at its psychological source, and this depth of analysis is unreasonable for analysts untrained in cognitive psychology. Perhaps the most significant problem with this classification is that a single human error that degrades system performance may be consistently classified across people and systems as a commission error (e.g., keyboard digit entry) in Swain's classification system, but may be identified as different types of action slips depending on what the operator was thinking at the time. This obviously would cause havoc in an empirically based data system. In studying the relationships of human errors to system performance, we want to know what kinds of errors are made and how frequently they are made. For this purpose, Swain's classification is appropriate. Once errors are identified as being too frequent in a given work situation, the psychological modeling may lead to ways to reduce them.

High levels of cognitive functioning such as strategizing, problem-solving, and decision-making present significant problems for HRA. In order to identify human error, one must identify the criterion of performance. The criterion is usually well defined, discrete as opposed to continuous, and known by the operator or the system prior to execution of the task. Unfortunately, these properties are foreign to high-level cognitive tasks. Consider the decision of selecting the best job offer. There is no correct decision. The candidate must optimize the selection on several criteria of interest such as salary, location, job security, and so on, but even these characteristics are anticipated, not known. A great-looking job may turn out to be a poor career move three years after the decision. Sometimes an investment is made based on all the latest market information. A week later the market may change significantly and make the investment a losing proposition. It is difficult to consider human error in tasks where information is uncertain, or where the criteria change with time.

Complex, cognitive tasks such as these typically require optimization on several criteria, result in trade-offs among criteria, and involve time-variant characteristics. Therefore, there can be no right or wrong decision. There may be a best decision for any given set of characteristics and for any given point in time, but that may not be known until much later. Human reliability analysis cannot adequately handle or predict human behavior in these types of tasks. Swain and Guttmann (1983) attempt to address cognitive tasks, but only to a limited degree. They feel comfortable in analyzing diagnostic behavior that leads to simple, rule-directed decisions. Diagnosis of system faults based on limited information can be studied in terms of human error. The fault is due to either cause A or cause B and the operator or maintenance person must infer which is correct by acquiring information and testing his hypothesis. His conclusion is either correct or incorrect. His corrective action is usually dictated by standard maintenance or repair practice or operating rule, but not by informed judgment.

Although this error classification scheme receives some criticism from experimental psychologists because it is not based on behavioral disfunctions, it adequately addresses the events of interest in an HRA. Similar to the analysis of equipment reliability, attention is paid to human output in the system

2.8.3 WHY PEOPLE MAKE ERRORS

Analyzing human errors, estimating their probabilities, and finding ways to reduce them require an understanding of why people make errors. This section explores the aspects of people and systems that precipitate human errors.

2.8.3.1 Work-Situation Approach

The *work-situation approach* (Swain, 1969b) to improving system reliability finds its roots in the basic human engineering philosophy, which advocates that the system should be designed for the user, not vice versa. Whereas traditional industrial thinking often has put the burden of human error on the worker and his or her persumed lack of competence, the work-situation approach examines the task demands, equipment, and work environment for characteristics that predispose the worker to make errors. Although traditional industrial strategy has relied heavily on personnel selection, matching workers to jobs, training programs, and motivational schemes promoting mental alertness to reduce error and increase productivity, the work-situation approach emphasizes the identification of error-predisposing conditions and their elimination or modification. This approach assumes that errors usually occur for reasons other than poor worker attitudes and that few workers consciously and deliberately make errors.

Error-likely situations (Swain and Guttmann, 1983) are identified as work situations where the ergonomics are so poor that errors are likely to occur. These situations make demands on workers that are not compatible with their capabilities, limitations, experience, and expectations. Any design that violates a strong population stereotype could be considered error likely. An *accident-prone situation* is an error-likely situation that fosters human errors likely to result in personal injury or equipment damage. The term accident-prone is often applied to specific people. An accident-prone person is one who has a proportionately greater share of accidents when compared to one's peers. This concept lost credibility when statistical analysis demonstrated that people labeled accident-prone had not experienced more mishaps than the number expected due to chance alone (Mintz and Blum, 1961). Seeking to weed out accident-prone individuals is not a useful approach to error reduction. Most modern industrial-safety specialists conclude that it is more cost-effective to look for work situations, not people, that are accident-prone.

2.8.3.2 Performance-Shaping Factors

Swain (1967) introduced the term *performance-shaping factor* (PSF) to describe any factor that influences human performance. In this chapter, the discussion is limited to PSFs that influence human reliability. External PSFs are those outside the individual, brought to bear by the environment or task. Situational task and equipment characteristics (PSFs) that predispose industrial workers to increased errors include the following (after Meister, 1971):

1. *Inadequate Work Space and Work Layout.* Precision tasks require adequate work space and proper layout. If parts bins are not arranged in accordance with assembly procedures, the probability of selecting an incorrect part increases. Poor seating or work station design can lead to fatigue, decreased productivity, and increased errors.

2. *Poor Environmental Conditions.* Inadequate lighting increases the difficulty of fine motor tasks such as wiring small components to circuit boards. High temperature and noise level decrease motivation and effort levels and increase error rates.

3. *Inadequate Human Engineering Design.* Poorly designed control panels, machines, and test equipment can lead to sequence and selection errors. Most automobile drivers can identify with turning on the windshield wipers when attempting to turn on the headlights owing to the similarity of the controls.

4. *Inadequate Training and Job Aids Procedures.* Workers who do not receive adequate job information or practice make more errors when starting a new job. Poorly written operating manuals and procedures lead to uncertainty and errors on the part of machine or system operators.

5. *Poor Supervision.* If new workers don't receive informative feedback on their work from their supervisors, they may be unaware of their errors or how to correct their incorrect methods. Poor supervision can also lead to poorly motivated workers or overworked personnel owing to demanding work schedules.

Internal PSFs are those that operate within the individual. They are human attributes such as skills, abilities, and attitudes that the worker brings to the job. If training has been adequate, however, internal PSFs generally have less impact than external PSFs on human reliability. This is fortunate because the external PSFs are nearly all under the control of industrial management and, if identified

Table 2.8.1 Example of Internal PSFs

Training/experience	Task knowledge
Skill level	Social factors
Intelligence	Physical condition
Motivation/attitude	Sex differences
Emotional state	Strength/endurance
Perceptual abilities	Stress level

as contributing to errors, can be modified. Some examples of internal PSFs are listed in Table 2.8.1.

Stress, a very important internal PSF, is the body's physiological or psychological response to an external or internal stressor. When task demands cannot be adequately met by the worker's abilities, performance usually suffers owing to excessive task loading, a psychological stressor. Stress usually has a nonmonotonic effect on performance. At very low levels of stress, there is not enough arousal to keep a person sufficiently alert to do a good job. Similarly, at high stress levels, performance usually deteriorates as the stressor increases or persists for long periods of time. Somewhere between low and high levels of stress, there is a level that does not hamper performance, called the optimal level of stress, which varies with task complexity and other factors (Fitts and Posner, 1967; Welford, 1973). Disruptive stress can increase the possibility of error by a factor of 2 to 5 according to Swain and Guttman (1983) and extremely high levels of stress can result in even higher degrees of performance degradation.

Another internal PSF that can have a significant influence on task performance is experience. Practice and skills acquired in training and on the job increase proficiency and decrease the number of errors. Different jobs and tasks require differing amounts of experience to reach equivalent levels of proficiency. Consider the skills necessary in handling baggage as a skycap versus those required to become a commercial airline pilot. As a temporary expedient for HRA, Swain addresses only two levels of skill to simplify the human performance model. A *novice* is someone who has not yet acquired all the skills or knowledge necessary to perform a job's tasks effectively and efficiently. A *skilled* worker is someone who has.

Stress and inexperience are such influential internal PSFs that their combination can increase a worker's probability for error by a factor of as much as 10. Swain uses the model shown in Table 2.8.2 to account for the effects of stress and experience on human error probability in doing routine tasks.

Table 2.8.2 Model Accounting for Stress and Experience in Performing Routine Tasks

	Increase in Error Probability	
Stress Level	Skilled	Novice
Very low	×2	×2
Optimum	×1	×1
Moderately High	×2	×4
Extremely High	×5	×10

2.8.3.3 Task Complexity

Tasks differ in the amount of mental processing necessary for successful completion. Simple tasks such as sweeping the floor require very little cognitive effort, whereas complex tasks such as brain surgery demand much more. People can develop very specialized skills with experience and show great variability when assessed on internal PSFs such as intelligence and emotional makeup. However, as a population, humans generally possess similar performance limitations, and process information in the same manner. Experimental psychologists have been demonstrating this phenomenon and recording their findings since the late 1800s. These universal limitations in capability cause people to make more errors while performing tasks that are more complex.

Memory limitations are known to anyone who has ever taken a history exam or tried to remember a new 10-digit telephone number. Although long-term memory limitations hamper remembering historical facts for days and weeks after exposure, short-term or temporary memory's volatility often makes us forget a phone number before dialing is completed. (See Chapter 2.4 for a more detailed discussion of memory characteristics.)

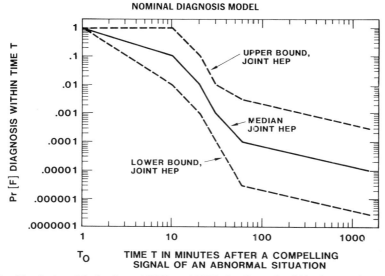

Fig. 2.8.1. Nominal model of estimated HEPs and UCBs for diagnosis within time T of one abnormal event by control room personnel.

The knowledge or skills required to perform skilled work can be retained for long periods of time after learning is complete. Workers usually do not have difficulty maintaining skill level through vacation periods or during job rotations. However, industrial task sequences in which several task elements have to be performed in a specific order are strongly affected by memory limitations. Written procedures and detailed checklists were developed to assist people in remembering all the task elements and their correct sequential order. Swain and Guttmann (1983) estimate that for nuclear power plant operators, the probability of omitting one task out of five decreases by a factor of more than 100 when written procedures are used, compared with responding to detailed oral instructions alone. Procedures with checkoff provisions reduce error potential by an additional factor of three.

Diagnosing a system's status and acquiring an understanding of a problem within a system is a complex task. Usually, the more complex the system is, the more complex a fault diagnosis becomes. A trained system engineer, operator, or technician uses his knowledge of the system along with status indications and logic to arrive at a diagnosis. Often a systematic elimination of potential faults leads to the correct diagnosis. Successful diagnosis of problems in complex systems takes considerable cognitive effort and time. If time is severely limited, the potential for erroneous diagnosis increases greatly. Swain has developed a model that demonstrates the relationship between diagnosis error likelihood and time after the initial detection of an abnormal event in a nuclear power plant control room. The probability of failure to correctly diagnose drops from certainty at zero minutes, to .1 in 10 min, and to .0001 in 1 hr (see Figure 2.8.1).

Variations in task complexity place different demands upon human capabilities and, as a result, cause different rates of errors. Rasmussen (1981) has grouped industrial performance into three behavior categories based on task complexity and experience level. The term *skill-based behavior* has been used to refer to performance of overlearned routine tasks by virtually autonomous patterns of behavior. Assembly-line workers who do the same, simple assembly tasks every day operate in the skill-based behavior mode. Rasmussen uses the term *rule-based behavior* to address behavior that requires more conscious effort in following general rules where modifications in behavior patterns are made in response to changing situational characteristics. The behavior mode associated with the most complex tasks is called *knowledge-based behavior*. This term is applied to unfamiliar situations, where original behavior is required. Most industrial tasks involve skill- and rule-based behavior. Although this behavioral categorization scheme has proved to be useful in a qualitative sense, no attempt has been made by Rasmussen to relate the terms to quantitative estimates of error likelihood.

2.8.4 HUMAN RELIABILITY ANALYSIS

2.8.4.1 Background

The basic theoretical underpinnings of human reliability come from conventional equipment reliability technology. Individual system-component reliabilities are combined according to the configuration of the system and its processing sequences. The various functional sequences are modeled by use of functional event trees or system fault trees. Component reliabilities are assigned based upon the manufacturer's predictions or failure-rate data collected from the field. Each potential functional sequence can be represented by a different path through the system fault tree. Component reliabilities in each path are combined by mathematical rules (depending on the series or parallel system configuration) to produce a reliability figure for that path through the fault tree and its event sequence corrolate.

Human reliability analysis (HRA) models task sequences in a similar manner. Human reliability figures for component tasks or task elements are combined mathematically to synthesize the error probability for the entire task or task sequence. Systematic HRAs permit logical examination of man–machine relationships, potential errors, and estimates of relative frequencies of particular task events. Not only do HRAs provide the human reliability analyst with estimates of human error probability (and reliability), but when combined with system reliability analyses, they assess the detrimental effects human errors have on the system.

Although the combinative techniques are very similar and straightforward for the equipment and human reliability analyses, developing the component and task element reliability estimates are worlds apart. Manufacturers and users of machines, components, equipment, and instrumentation record maintenance and breakdown events, thereby creating a data base. Because particular equipment models remain in service over many years, the data base increases, and the reliability predictions based upon past performance improve in accuracy and confidence. Unfortunately, the same is not true for human reliability data. Very few industries collect and record data on the errors made by their workers. Those that do don't always publish their findings. Human performance is also extremely variable with a multitude of PSFs influencing different tasks in different ways. With the absence of a large, high-quality data base, many human reliability analysts are often forced to resort to using laboratory data or their own subjective estimates. Section 2.8.6 of this chapter discusses the data problem in more detail.

2.8.4.2 Analysis and Prediction

The purpose of HRA is to analyze the man–machine system and predict the potential for human error. This can be done qualitatively or quantiatively. A qualitative HRA, usually performed in system design or modification, is used to reduce the effects of human error in the system to some tolerable level. In its simplest form, a qualitative HRA might involve an informal task analysis, an appraisal of how many instances the same error would occur in a given period (one month, or one year), an evaluation of the consequences or costs incurred, and a decision as to whether it should be corrected. Where several potential errors are being analyzed, they might even be evaluated on a five- or seven-point scale for relative likelihood. This approach could be called semiquantitative. In contrast, a quantitative HRA estimates the effects of human error in a system by assigning HEP's and UCBs to the individual task elements and it predicts the overall probability of failure for a task on an absolute scale. Quantitative HRAs can be used in *probabilistic risk assessments* (PRAs) where the resulting probabilities of failure for task sequences are combined with probabilities of failure for the equipment in a system fault tree or a system event tree. The results of quantitative HRAs have recently been incorporated into PRAs of critical nuclear power plant operations in an effort by the Nuclear Regulatory Commission (NCR) to evaluate which event scenarios present the greatest risk to the public. Refer to Kolb et al. (1982) for an example of a nuclear plant PRA done for the NRC's Interim Reliability Evaluation Program.

The first step in any HRA, regardless of technique, is to understand completely the tasks to be performed by the operator and their relationships to system performance. Thus the HRA becomes part of a *man–machine systems analysis* (MMSA) (Swain and Guttmann, 1983), which is a general method used to identify and evaluate existing or potential human performance problems within the system context. The major elements in an MMSA are outlined below:

1. *Understand and Describe System Goals and Functions of Interest.* The analyst must gain an understanding of the functions, purposes, and goals of the system, as well as the functions people perform to attain system goals. The points of interaction among people and between people and the system must be identified. Flowcharts may aid in the understanding or description of the system on the part of the analyst.

2. *Understand and Describe Task Responsibilities and Performance Criteria.* After having identified a task or task sequence of importance or interest, the analyst must perform a task analysis. A task analysis identifies the specific behaviors required of the people in a man–machine system. As stated in Section 2.8.2, a task is a level of job behavior that describes the performance of a meaningful job function and contributes to the accomplishment of a system function or goal. There are many forms of task analysis that support various levels of analytic detail. All forms, however, involve describing the sequence of task elements in terms of perceptual, cognitive, and motor behaviors. These behaviors are then analyzed in terms of initiating cues, time requirements, feedback, precision requirements,

communication, manipulation, controls used, and coordination with other tasks. Tasks are analyzed in relation to their criteria for success and difficulties are assessed.

3. *Assess Situational Personnel Characterisitcs.* In this step, external PSFs such as air quality, temperature, noise level, lighting, and supervision are assessed. Sources of information include observations at the plant, and interviews with management, supervision, and people working on the line. Particular interest is on those situational characteristics likely to have adverse effects on human performance. Likewise, the skills, experience, training, and motivation of the workers who operate and maintain the plant system are identified.

4. *Analyze Tasks for Potential Human Errors.* The capabilities and limitations of the personnel should be compared with the demands of the tasks to be performed. Any mismatch between these two sets of factors will increase the potential for human error. This comparison can be accomplished by someone knowledgeable in human factors, preferably observing the actual work situation. The best way for the analyst to determine task complexity is to perform the tasks himself using the available written procedures, if appropriate. Other effective methods include interviewing operators who perform the tasks and observing talk-throughs or walk-throughs given by the workers. Additional information can be acquired by asking the operators to talk through hypothetical, yet realistic, work-situation scenarios.

5. *Estimate the Likelihood of the Potential Human Errors.* At this point in the MMSA, either a quantitative or qualitative HRA is performed. The objective of this step and the next one is to provide an estimate of the importance of each potential error identified in Step 4. An error's importance is a function of its frequency, the chances for recovery, the severity of potential consequences, and the cost of making changes to improve the situation. The HRA predicts the frequency of the error and its potential for recovery prior to system consequences. Section 2.8.5 describes several HRA techniques.

6. *Estimate the Consequences of the Human Errors.* One of the more influential determinants of an error's importance, the consequences or system impact, must be considered. In production industries, an error in fabrication or inspection may lead to a part's rejection or to a substandard part being shipped. In the nuclear power industry, an unrecovered error may lead to radiation exposure, and several errors in succession, as in the Three Mile Island accident, could lead to core meltdown. Errors that lead to trivial consequences are not of interest to management and often go uncorrected. Some systems demonstrate great insensitivity to substantial variations in estimated HEPs for particular tasks. A *sensitivity analysis* can be performed wherein a task's HEP is manipulated and its effect on the system is ascertained.

7. *Recommend Changes to Improve the System.* In this optional step, remedial suggestions are made to improve system reliability. The analyst has identified "important" errors, and in this step the analyst and a systems expert can formulate solutions for the problem areas. Obviously, solutions that are inexpensive and easy to implement will be looked upon most favorably by management. The solutions may involve improving ambient conditions, altering tasks to reduce task complexity, adding inspections to improve recovery potential, and improving the quality of training. A follow-up analysis after the recommendations have been implemented can analyze system improvement by comparing modified system performance with original system performance. Thus the MMSA technique can become part of an iterative process of system improvement.

2.8.4.3 Historical Perspective

One of the earliest, systematic applications of HRA to a system reliability study was done at Sandia National Laboratories in a 1952 classified study of an aircraft nuclear weapon system. The approach taken was to estimate human error rates for the relevant human tasks in ground and air ordnance handling and incorporate the estimates into reliability equations in the same manner as equipment and system events. Because there were no failure-rate data available, the analysts speculated that ground tasks deserved a .01 error probability and tasks performed in the air were about one-half as reliable, for an error probability of .02. Through the 1950s, HRAs were performed on critical tasks involved with nuclear weapon manufacturing and field handling at Sandia. Although the quantitative aspects of the analyses were questionable in their validity owing to lack of data, qualitatively they were sound, and they could be used to compare alternative operating methods, thereby increasing system safety and reliability.

It wasn't until 1961 that a major research effort was made to develop a data base of human reliability figures (Munger, Smith, and Payne, 1962; Payne and Altman, 1962) known as the AIR Data Store. In 1962, Rook presented an HRA technique (developed by Swain and Rook) and its application to production tests at the Human Factors Society Annual Meeting (Rook, 1962). At that same meeting, Swain proposed a large-scale human error data bank to facilitate the use of the newly developed quantitative HRA techniques. By 1964, there was enough concern in the scientific community for human reliability that a symposium held in Albuquerque drew more than 70 people. Four different human reliability studies were presented at the symposium:

1. A description of the AIR Data Store and its application using a simple error product rule.
2. The Technique for Human Error Rate Prediction (THERP), the Sandia method initiated by Swain and Rook.
3. The generation of reliability estimates through a computerized Monte Carlo program at Aeronutronics.
4. The use of Aerojet-General Corporation's expert ratings of operator performance to calibrate the AIR Data Store for a specific application.

Meister (1964) evaluated the four studies and pointed out that there was a lack of available data on human error rates and other performance attributes, and that some of the four methods relied on the product rule combination of task-element failure probabilities to estimate complete task failure probabilities. The product rule assumes that task elements are completely independent. This assumption lacks validity in that task elements performed by one individual or by several individuals in a given situation have certain degrees of interdependence. The danger in ignoring dependence among task elements is that the resulting products of several small numbers are extremely small, and represent overly optimistic (nonconservative) estimates of human reliability. In analyzing human reliability, it is always desirable to predict reliability conservatively, so that any major discrepancies with reality will be on the side of fewer errors and better system performance.

The human reliability field grew out of its infancy in the late 1960s with several notable developments. Altman (in Askren, 1967) described a molar classification scheme for human error that embodied situational and other factors known to influence performance. Four months later, Meister (Hornyak, 1967) introduced the Bunker-Ramo data bank, which, unlike the molecular AIR Data Store, listed task reliabilities rather than task-element reliabilities.

During this same period, progress in the field was reported at several scientific symposia. Swain (1967b) introduced a method to include event dependencies in HRA techniques. Williams (Blanchard and Harris, 1967) also developed a method at the Martin Marietta Corporation to take event dependencies into account. His approach was similar to THERP in that it corrected estimated task error rates to account for the influence of other event contingencies.

Owing to the lack of objective human error data, HRA researchers investigated the use of subject-matter experts' estimates of human error probabilities. Problems were realized in the experts' lack of skill in reporting in the language of probabilities. Using relative comparisons simplified the task but scaling was needed to correct rankings to absolute probabilistic values. Swain (1967a) attempted to use expert judgments and conventional scaling techniques to develop an interval scale of task error likeliness and to calibrate this scale with some known error rates to convert it to a ratio scale. Another technique grew out of an experimental study conducted by the U.S. Air Force (Askren and Regulinski, 1969) to determine the feasibility of modeling the probability of error as a function of the length of time a task requires. They concluded that modeling reliability in the time domain was practicable. Refer to Swain (1969) for a more complete account of the early developments in human reliability.

During the early 1970s two methods emerged as the foremost HRA methodologies. Swain and his colleagues had developed THERP to the point where it could be successfully applied to industrial real-world prediction and subsequent design decisions. Siegel and his co-workers (Siegel and Federman, 1971; Siegel and Wolf, 1969) performed numerous experiments in modifying variables (PSFs) in their computer models, running the simulations, and examining the results.

During the late 1970s and early 1980s, several alternative approaches were investigated, both in the United States and Europe, in response to the international concern over the safety and reliability of the nuclear power industry. Some of these are examined in the following section. Reference is made to Meister's review (1984), which covers published HRA methodologies from the 1970s to the present, and which compares the Swain and Siegel techniques.

2.8.5 HUMAN RELIABILITY ANALYSIS TECHNIQUES

Unfortunately, space does not allow the description of all HRA techniques that have been developed. This section does describe the review several techniques that have been published and applied to real-world problems, as well as some recently developed techniques. Refer to Meister (1984), Embrey (1976), and Pew, Feehrer, Baron, and Miller (1977) for recent descriptions and reviews of the following HRA methodologies that are not reviewed here.

1. The Siegel and Wolf technique.
2. Systems Analysis of Integrated Networks of Tasks (SAINT).
3. The Lockheed Georgia technique.
4. The AIR Data Store.
5. The Establishment of Personnel Performance Standards (TEPPS).
6. The Askren–Regulinski approach.

2.8.5.1 Technique for Human Error Rate Prediction (THERP)

One of the oldest and most widely used HRA techniques is the Technique for Human Error Rate Prediction (THERP) developed by Swain and Rook in the early 1960s. Swain and Guttmann (1983) have defined the quantitative technique and define THERP as:

> *A method to predict human error probabilities and to evaluate the degradation of a man–machine system likely to be caused by human errors alone or in connection with equipment functioning, operational procedures and practices, or other system and human characteristics that influence system behavior.*

The THERP approach uses conventional reliability technology modified to account for greater variability and interdependence of human performance as compared with that of equipment performance. The procedures of THERP are similar to those employed in conventional reliability analysis, except that human task activities are substituted for equipment outputs. The steps involved follow the general format of an MMSA:

1. Define the system failures that may be influenced by human errors and for which probabilities are to be estimated.
2. Identify, list, and analyze human operations performed and their relationships to system tasks and functions of interest.
3. Estimate the relevant human error probabilities.
4. Determine the effects of the human errors on the system failure events of interest. (This usually involves integration of the HRA with a system reliability analysis.)
5. Recommend changes to the system in order to reduce system failure rate to an acceptable level. Steps 2 through 4 can be repeated to evaluate the changes.

THERP can be used to generate quantitative estimates of task reliability estimates, interdependence of human activities, the effect of event PSFs, equipment performance, and other system influences. Although not really a hypothetical model, as some perceive it to be, THERP was developed as a practical, applied technique—a fast and relatively simple method of providing recommendations to system designers and analysts who need quantitative estimates of the effects of human errors on system performance. Figure 2.8.2 shows the four phases of the THERP approach.

The basic tool used to model tasks and task sequences is the HRA event tree. The HRA event tree starts with any convenient point in an activity sequence and works forward in time. Based on an underlying task analysis, it follows the same sequence in time, showing graphically the parts of the task analysis relevant to the HRA. It should not be confused with the fault tree approach, which starts with a fault and works backward in time. THERP models events as a sequence of binary decision nodes. At each node, the task is done either correctly or incorrectly. At every binary branching, the probabilities of the events must sum to 1.0. The probabilities assigned to all human activities depicted by tree limbs are conditional probabilities, except for those in the first branching. If the first branching represents a carry-over from some other tree, or a task based on some previous event likelihood, it too will be a conditional probability. Capital letters are used to denote failures and their probabilities. Lowercase letters are used to indicate successes (or desirable states) and their probabilities. Figure 2.8.3 illustrates a very simple example of an event tree representing the performance of two tasks, "A" and "B." Note that because task "A" is always performed first, the probabilities associated with task "B" are all conditional on the outcome of "A." The interdependence of tasks "A" and "B" are represented by the symbols $b|a$, $B|a$, $b|A$, and $B|A$. If the conditional relationships are understood the notation can be shortened to b and B.

Once the HRA event tree is constructed and the estimates of the conditional probabilities of success or failure are assigned to each limb, the probability of each path through the tree is calculated. A series system has total success only if both tasks "A" and "B" are successful, represented by one success path whose probability of success (Pr[S]) is equal to the product of a and $b|a$. Conversely, a parallel system succeeds if either task is performed successfully, and fails only when both tasks fail. To calculate the total success probability for more than one success path, the individual path Pr[S] values are summed. Because Pr[S] $= 1 -$ Pr[F] and vice versa, only success or failure need be calculated for the system, for its complement can be derived easily.

Although the modeling of tasks with event trees and using conventional reliability mathematics to derive task success probabilities is rather straightforward, the assignment of HEPs to the individual task elements' failure branches of the event tree requires substantial judgment on the part of the analyst. Swain and Guttmann (1983) recommend that someone knowledgeable in human performance technology perform this portion of the analysis. Obviously, data acquired from the specific industrial situation under analysis would constitute the best estimate of error for the task. However, because such human error data are seldom available, other sources must be used. Swain and Guttmann have

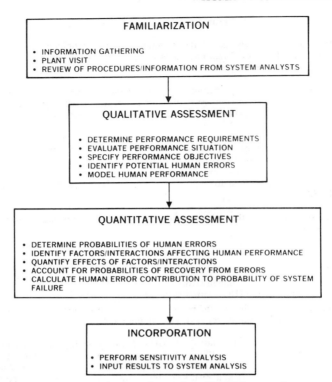

Fig. 2.8.2. The four phases of the THERP approach.

searched through all the available literature for human error data and found few that were applicable to nuclear power operation tasks. The information found was combined with their own expert judgments and human performance models to create an interim HEP data bank. Much of their *Handbook of Human Reliability Analysis* presents the models and predictive HEP estimates that are not only useful in nuclear power operations but also applicable to many other industrial settings. When applicable HEP data or estimates are located, they are referred to as *nominal HEPs* in THERP. These represent the probability of human error without considering the influence of plant-specific and task-specific PSFs. Once these are taken into account, as discussed in Section 2.8.3, a *basic HEP* is formed, which represents the probability of human error for that task in isolation. A *conditional HEP* is a modification of the basic HEP to account for influences of other tasks or events that may include the preceding task elements, tasks, and the number of people involved in performing the task.

One of the major problems in modeling tasks as sequences of behaviors, each with its own probability of failure, is the determination of how the failure or success of one task may be related to the failure or success of another task. One of the strengths of THERP is its ability to account for the interactions with a model of dependence. Two events are independent if the conditional probability of one event is the same whether or not the other event has occurred. That is, the probability of success on task "B" is the same regardless of success or failure on task "A." Or, in terms of Figure 2.8.3, $b = b|a = b|A$ and $B = B|A = B|a$. If events have any influence on one another they are not independent, but dependent. Failure to take dependence into account can lead to overoptimistic estimates of task error probabilities, which is an undesirable outcome.

The dependence model presented in NUREG/CR-1278 deals with the continuum of zero dependence (or complete independence) to complete positive dependence. Positive dependence assumes a positive relationship between events such that failure on the first task increases the probability of failure on the second task. The same relationship holds for probability of success. Dependence can occur between people, as when several technicians perform a task together or when one worker checks the accuracy of another in the form of an inspection. Consider the case in which an assembly-line mechanic must torque a bolt on a unit and, in a subsequent inspection, a checker verifies the torque. If the checker observes the torquing operation and then checks the torque against the criterion used by the mechanic (instead of looking up the torque specification in the assembly manual), he has assumed that the mechanic has used the correct value. This assumption makes the checker's task dependent upon the

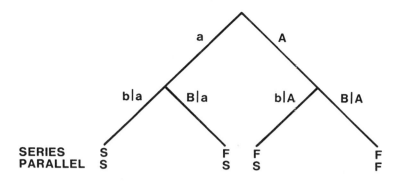

SERIES	S			F	F		F
PARALLEL	S			S	S		F

TASK "A" = THE FIRST TASK

TASK "B" = THE SECOND TASK

a = PROBABILITY OF SUCCESSFUL PERFORMANCE OF TASK "A"

A = PROBABILITY OF UNSUCCESSFUL PERFORMANCE OF TASK "A"

b|a = PROBABILITY OF SUCCESSFUL PERFORMANCE OF TASK "B" GIVEN a

B|a = PROBABILITY OF UNSUCCESSFUL PERFORMANCE OF TASK "B" GIVEN a

b|A = PROBABILITY OF SUCCESSFUL PERFORMANCE OF TASK "B" GIVEN A

B|A = PROBABILITY OF UNSUCCESSFUL PERFORMANCE OF TASK "B" GIVEN A

FOR THE SERIES SYSTEM:

$Pr[S] = a(b|a)$

$Pr[F] = 1 - a(b|a) = a(B|a) + A(b|A) + A(B|A)$

FOR THE PARALLEL SYSTEM:

$Pr[S] = 1 - A(B|A) = a(b|a) + a(B|a) + A(b|A)$

$Pr[F] = A(B|A)$

Fig. 2.8.3. Example of an HRA event tree.

torquing task. If the mechanic used the wrong value, he would most likely fail to torque the bolt to specification, and the checker would not catch the error. However, if the mechanic used the correct value and the checker verified the torque using the correct criterion, the inspection task would be valid. Thus the probability of error in inspection (B) changes as the result of the success (a) or failure (A) of the torquing task, or $B = B|a < B|A$. This increase in $B|A$ can be reflected in the event-tree model to avoid an overly optimistic estimation of the failure probability for the torquing/inspection process.

Dependence can also occur within an individual as several related tasks are performed. In errors of omission, the failure to perform the first task in a closely related group of tasks increases the likelihood that the second will also be omitted. Sometimes the tasks are so highly related that the entire sequence is represented by one HEP for omission. These tasks would be considered completely dependent. In errors of commission, the fact that the first task is done incorrectly and without recovery increases the likelihood that the remaining similar tasks are. If a mechanic torques a bolt to the wrong value, chances are high that the other identical bolts remaining in the sequence will also be incorrectly torqued. Swain and Guttmann (1983) have developed a five-level scale of dependence and guidelines on how it should be applied to task situations. For each level (zero, low, medium, high, complete), a formula is provided to calculate the conditional probability of failure (or success) on task "N" given the failure or success on the previous task "N − 1." The formulas alter the conditional probability from its original value at zero dependence to 1.0 at complete dependence.

The THERP approach can be difficult to comprehend without seeing the entire process unfold. Following is a short example that uses THERP to illustrate the HRA event tree, the use of collective expert judgment by qualified HRA specialists to assign conditional HEPs to the failure limbs in the tree, and a check on the accuracy of the predictions made in the study.

2.8.5.2 An Example Using THERP

The example is based on a major defect problem that occurred in the assembly area at a Department of Energy contractor's facility. As in most industrial applications of HRA, major cognitive aspects of behavior (e.g., diagnosis and decision-making) contribute negligibly to the problem at hand and can be ignored. Furthermore, the stress level of the assembler-technician (AT) for the operation in question could be considered optimal, and stress effects could be ignored. With two exceptions, dependence effects were estimated directly, one of the methods described in Chapter 10 of NUREG/CR-1278 (Swain and Guttmann, 1983). The two exceptions employed the positive dependence model described in the same chapter.

The major defect was the result of an AT failing to remove one of two shorting devices from plugs in an electronic programmer before connecting the two plugs to a junction box in a major assembly. The shorting devices are black, carbon-filled, foam, static pads inserted directly into the plugs prior to handling. The purpose of the conductive static pads is to short out the pins in the two identical-appearing connectors as a protection against static electricity during movement and handling of the programmer. Each static pad is cut to a size to fit inside the connector. Once a pad has been inserted, a yellow, plastic dust cover is placed over the plug and pad and is secured with a rubber band. This helps compress the static pad onto the pins in the plug. An uncompressed pad in use at the time of the AT's error was about $\frac{3}{8}$ inch thick. After compression, it may be around $\frac{1}{4}$ inch thick and will not protrude from the connector.

The programmer is transported to another assembly area where a different AT inserts it into a rigid foam support that is part of a major assembly. At this point, the AT must remove the static pads and connect the two programmer plugs to a junction box. As noted, this AT failed to remove the static pad from one of the plugs, and it was not possible to perform later electrical tests to detect this defect. The error was discovered when the major assembly was disassembled in a random-sample, quality-assurance procedure. In discussions and talk-throughs with the person who made the error, it was apparent that the task of removing a pad that happened to be stuck in a plug defined an error-likely situation. At the time that the defect was detected, 284 major assemblies had been fabricated and delivered to the customer. Because the defect resulted in a significant loss in reliability for the major assembly, it was necessary to estimate the probability that there might be other undetected defective units with either one or two static pads left in. Fortunately for the analysis effort, only one AT had been involved, so that any other similar errors would have been made by him. Therefore, individual differences in ATs could be ignored.

A detailed task analysis was performed at the industrial facility and from it an HRA event tree was prepared (Figure 2.8.4). This tree (without the assigned probabilities) and the results of the task analysis were shown to two other HRA specialists who concurred with the structure of the tree. The three analysts independently assigned HEPs to each failure limb of the tree. In making these assignments each analyst had to take into account dependencies from one human action to another. Although the estimated HEPs assigned to each limb are subjective, in no case did the three analysts' estimates differ by more than a factor of 3. In most cases the differences were less than a factor of 2, and a consensus was reached for all estimates. The nature of the judgments made and how they are assigned to event tree limbs are described below.

Independence of AT activities was assumed between different programmers because they occur at intervals of 1 hr or longer. However, because the AT removes the yellow dust covers and static pads from the two plugs on a given unit within a minute or less, there can be different levels of dependence between actions performed on the same unit. It is even possible to forget to remove the pad from the first plug, but the pad's falling out from the second plug may remind an AT that he or she forgot to remove the first pad. Thus, most of the limbs in the tree represent conditional probabilities involving some level of dependence between different actions. It was assumed that if a static pad falls out (or dangles from the plug), the pad has been (or will be) removed. Thus, limb A in the event tree (Figure 2.8.4) states, "P1 falls; pad removed." If the static pad sticks up in the connector when the yellow dust cover is removed, the AT either remembers or forgets to reach up and remove the pad. It should be noted that although an outline of the procedure was available, this was such a familiar job that the AT did not consult it. Furthermore, if he had, he would have found no mention of the step to remove the pads. This action was considered as skill-of-the-craft, not requiring a written step. Thus limb A states, "P1 sticks," which is shorthand for the undesirable situation of the pad sticking in the connector and not being visible to the operator. Limb D states, "P1 pad missed," which means that the AT failed to remember to remove the pad. Limb h represents the pad from P2 falling out (or dangling), in which case a successful removal of that pad is assumed. Limb i represents the recovery of the human error of forgetting the P1 pad; this recovery occurs because of the assumption that the presence of the fallen or dangling P2 pad reminds the AT to check for the presence of the P1 pad, and he recovers from his previous oversight.

The AT was used as a subject-matter expert to estimate probabilities of .8 and .2 for, respectively, the pad falling out (or dangling) and the pad sticking up in the connector. All the other estimated conditional probabilities were made by the HRA specialists. The basic probability of .05 (limb D) of an AT's not removing a stuck pad is based on a downward adjustment of a .15 nominal inspection error probability for the detection of defects that occur at a rate of about .01 or less (McCornack, 1961). The downward adjustment reflects the much higher "defect rate" in the assembly operation and the resulting AT's higher expectancy for "defects." This estimated adjustment to an HEP of .05

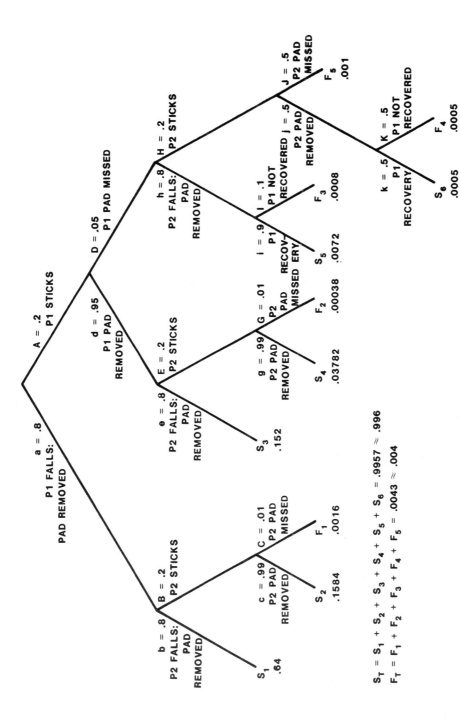

Fig. 2.8.4. HRA event tree for static pad problem.

$S_T = S_1 + S_2 + S_3 + S_4 + S_5 + S_6 = .9957 \approx .996$

$F_T = F_1 + F_2 + F_3 + F_4 + F_5 = .0043 \approx .004$

concurs with the relevant studies of Fox and Haslegrave (1969) and Wilkerson (1964). Note that the HEP for limb C is a factor of 5 lower than that for limb D. The difference here is that, in the case of error C, there is a conditioning effect of the AT's having noted that the pad fell out of the first plug. Therefore, even though the pad sticks in P2, the immediately preceding experience of having witnessed the pad in P1 (limb a) should help remind him to take out the stuck pad in P2. The factor of 5 difference was based on collective judgment.

Error J represents the same type of error (i.e., failing to extract a stuck pad), but here the conditioning effects are quite different. The pad in P1 stuck (limb A), but the AT failed to extract it (limb D). This error also predisposes the AT to fail to extract the stuck pad in P2. In the judgment of the analysts, this represented a "high dependence level" between the two human actions. Using the dependence model from NUREG/CR-1278 (Swain and Guttmann, 1983), this is equivalent to assigning an HEP of about .5. On the other hand, if the AT did remove the second pad (event j), it was estimated (event k) that this removal would remind him to check for the pad he had previously failed to remove from P1. This retroactive dependence was also defined as high dependence. Similarly, the recovery of the P1 error (limb A) represented by limb i is based on a cue provided by limb h. That is, the falling (or dangling) of the pad from P2 would likely remind the operator that this event did not occur for P1 and that he had failed to extract the pad from the P1 connector. This visual cue is more compelling than the cue afforded by reaching up and removing the pad from P2 (limb j). Consequently, the estimated recovery factor of .9 for limb i is higher than the .5 estimated for limb k. Again, these estimates are based on the collective judgment of the analysts, and estimates of retroactive influence are especially speculative. Event g has a fairly high success probability because the operator has already removed the pad from P1 (limb d). This should provide a good cue to remove the stuck pad from P2.

Using the HEP estimates and equations shown in Figure 2.8.4, the joint probability that a programmer will have one or two stuck pads and that the operator will fail to remove the stuck pad or pads is estimated as .004. A conservative estimate of this probability was calculated as .01 by increasing the values used for the HEPs by a factor of 2 and assigning complete rather than high dependence. An optimistic estimate of the defect probability was calculated as .0005 by reducing the values used for the HEPs by a factor of 5. These recalculations illustrate the use of sensitivity analysis, a technique employed when dealing with large uncertainties in estimating HEPs. In addition to estimating the defect probability for a single, randomly selected unit, the expected number of defective units out of the 284 in question, and the probabilities of one or more, two or more, three or more, and four or more defective units were calculated, as shown in Table 2.8.3.

All the estimates in Table 2.8.3 were based on the assumption that the 80/20 split estimated by the AT was approximately correct. However, the HEP estimates would not change materially unless the true percentage of pads sticking were as low as about 1%, in which case the AT would have a very strong expectancy of not experiencing a stuck pad, and the analysts would then estimate a basic

Table 2.8.3   Expected Number of Defective Programmers and Probability of x or More Defective Units in a Population of 284

	Estimated Probabilities		
	Best	Pessimistic	Optimistic
Defect probability for a single randomly selected unit	0.004	0.01	0.0005
Probability of no defective units	0.32	0.06	0.87
Probability of x or more defective units[a]			
1	0.68	0.94	0.13
2	0.32	0.77	0.01
3	0.11	0.53	0.001
Expected number (Pr × N)	1.14	2.84	0.14

[a] These calculations are based on the following equations:

$$\Pr[x \text{ defective units}] = \binom{284}{x} p^x (1 - p)^{284-x}$$

where p = the best, pessimistic, or optimistic probability of 0.004, 0.01, or 0.0005.

HEP of about 0.15 instead of the 0.05 used in the example analysis. Based on the AT's experience with the assembly operation in question, it was doubted that his estimate of 20% would be off by anything approaching the factor of 20 required to get the percentage of stuck pads down to 1%. On the other hand, if the AT underestimated the frequency of stuck pads by a factor of 2, the analysts would retain the same HEP estimates as above. Such an underestimation is not very likely in view of the AT's statement that the pad usually falls out or dangles. However, if the AT were this much in error, one would substitute 0.4 for 0.2 in the event tree, and the total failure estimate of 0.004 becomes $0.009 \approx 0.01$. These recalculations illustrate another use of sensitivity analysis.

To sum up the HRA, reference to Table 2.8.3 indicates that the analysts' best estimate of the probability of there being one or two stuck pads in any of the 284 programmers selected at random was 0.004, and it was further estimated that there was about a $0.68 \approx 0.7$ probability of there being at least one programmer in these 284 units with one or two static pads still in place. Because the consequences of an optimistic assessment of the probability of another defect occurring could not be tolerated, it was decided to check the accuracy of the 0.004 estimate. At considerable cost, 64 additional major assemblies were disassembled to check for similar errors. No more pads were found. Thus the fraction of defective components found in the total sample was $1/65 = 0.015$. To check the agreement of the data (i.e., one failure in 65 tests) with the assignment of $p = 0.004$, a statistical test of significance was performed. The significance level (SL) was computed by

$$SL = \Pr[x \geq 1 | p = 0.004, n = 65]. \tag{2}$$

For purposes of calculating SL, the sample of 65 was treated as if it were a random sample, although it was not, because it was defined to include the one defective programmer. Note that the SL computed by including the previously discovered failure will be low and will err on the side of indicating disagreement with the assessment. Because every programmer is assumed to have a constant failure probability of $p = 0.004$, SL can be determined by evaluating terms of the binominal distribution,

$$SL = \sum_{x=1}^{n} \binom{n}{x} p^x (1-p)^{n-x} = 1 - (1-p)^n \tag{3}$$

where n = sample size = 65, and p = assessed failure probability = 0.004. Then SL = $1 - (1 - 0.004)^{65} = 0.23$. Thus, given that the true probability of one or two retained pads per programmer is 0.004, the probability that a sample of 65 will have at least one bad component is 0.23. Therefore, there is no reason to reject the above assessment of 0.004. Calculating the SL using the pessimistic estimate of 0.01 yields SL = 0.48, so this estimate also cannot be rejected.

If only the verification of the 0.004 estimate is considered, it would have been preferable that all 284 programmers had been checked for the defect, but this was not done because of the very substantial costs involved. Had such a complete check been done and no further defects found, the true probability of defective units would be $1/284 = 0.00352$, which happens to round to the same 0.004 as the assessment based on HRA methods only.

The overall results, including the statistical analysis, indicate that even though subjective judgment is involved, properly trained analysts can collectively perform HRAs using THERP with a degree of accuracy adequate for practical industrial problems.

2.8.5.3   Maintenance Personnel Performance Simulation (MAPPS)

An outgrowth of earlier simulation modeling work by Siegel and his colleagues (Siegel and Federman, 1971; Siegel and Wolf, 1969), MAPPS is a digital simulation model that provides reliability estimates for maintenance performance (Siegel, Bartter, Wolf, Knee, and Haas, 1984). In order to perform the simulation, the analysts must first perform task analyses to determine the tasks required and identify the associated subtasks for the individual(s) involved. For each subtask, input data based on ratings (e.g., various PSFs) or actual measurements (e.g., performance times) are entered along with selected parameter values (PSFs). A computer then simulates the performance of each subtask using the model algorithms and a Monte Carlo technique of simulation. The output of the simulation includes probability of success, time to completion, areas of operator overload, idle time, and level of stress. Changing parameter values and reiterating the simulation can demonstrate the effects of a particular parameter or subtask performance. Using this iterative method, the analyst can forecast the results of a potential design improvement prior to implementation.

The input parameters are not treated independently, but interactively, to determine their collective effects on subtask performance. The following PSFs are quantified by algorithms internal to the simulation and the input variables specified by the analyst:

1.   Maintainer's and work crew's abilities in terms of intellective capacity and perceptual-motor abilities.

2. Fatigue effects, expressed as performance decrement due to number of hours of performance. Recovery in the form of rest is considered to lessen the fatigue level.

3. Heat effects, considered as having a moderating effect on intellective and perceptual-motor abilities.

4. Subtask ability requirements by type of maintainer (e.g., maintenance mechanics, electricians) are identified including assembly, disassembly, and communication.

5. Accessibility values for tasks such as removing and replacing components.

6. Clothing impediment to perceptual-motor ability based on the interaction between accessibility and subtask difficulty.

7. Quality of maintenance procedures.

8. Stress effect, based on four stressors:
   a. *Time stress,* the ratio of needed time to available time.
   b. *Communication stress,* the percent of message comprehension as a function of ambient noise and message length.
   c. *Radiation stress,* the stress as a linear function of radiation dosage beyond 800 mrems.
   d. *Ability difference stress,* the maintainers' ability differences within the work crew.

9. Aspiration level of the individual based on the ratio of successfully completed subtasks to the actual number of subtasks attempted.

10. Organizational climate, policies, administrative structure, and values affect the detection of errors and their recovery.

The simulation output of task success probability is based on the difference between the subtasks' difficulty and the maintainer's ability. The following equation is based on a model presented by Rasch (Lord and Novick, 1968) for item analysis in test construction. It is used to calculate p, the probability of success, completing a test item where $x =$ the difference between task difficulty and the maintainer's ability (ranges from -5 to 5), and e is the base of the system of natural logarithms (2.71828):

$$p = \frac{e^x}{1 + e^x} \tag{4}$$

This probability of success is computed for every person involved in the subtask. The probabilities are then summed and weighted to yield the work crew's average probability of success.

The MAPPS technique represents an attempt to model human performance for maintenance tasks. The model has the advantage of using a fairly comprehensive set of PSFs that interact with each other, and incorporates recovery factors such as the effects of rest and checks enforced by quality-control policies. The model is computer-based, which allows the analyst to simulate the same task repeatedly and assess how the inputs affect the outputs. Although numerous inputs, many of which involve subjective ratings, are required of the analyst, default values are provided for unknown PSFs. The opacity of the simulation algorithms weakens face validity and could lead to a lack of confidence on the part of the analyst. The MAPPS technique is similar to THERP in that both require a task analysis, account for individual and task PSFs, yield estimates of probability of successful task/subtask completion, and permit the analyst to perform sensitivity analyses. The techniques differ in the algorithms used to model the effects of PSFs on performance and in the method by which the analyst arrives at the final estimate of task reliability. The authors of MAPPS are currently attempting to validate the model.

2.8.5.4 Operator-Action Tree Method (OAT)

Another quantitative HRA technique developed through the efforts of Wreathall, Fragola, and Hall (Hall et al., 1982) concentrates on diagnosis and decision errors of nuclear power plant operators after the initiation of an accident. These authors stated that the THERP approach with its detailed task analysis emphasizes molecular elements, and that a more holistic approach is required for analyzing diagnosis and decision errors to derive a time-reliability curve. Instead of accounting for an individual's performance under highly situational PSFs, the OAT model describes the statistical performance of a group of operators responding to generalized situations. Performance estimates in the form of HEPs are then incorporated into an event tree or fault tree analysis.

The OAT approach is based on the assumption that human response to an environmental event consists of three activities: (1) perception, (2) diagnosis, and (3) response. The basic operator action tree shown in Figure 2.8.5 is based on the potential for error in each of the three activities.

The second major assumption is that time available for diagnosis is the dominant factor determining the probability of failure. That is, given a short period of time, people will fail to diagnose a situation correctly more often than when given a longer period. The thinking interval or time interval available for diagnosis is delimited by the operator's first awareness of an abnormal situation and the initiation

BASIC OPERATOR ACTION TREE

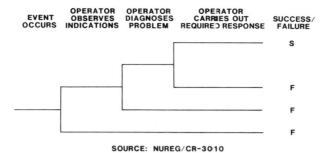

SOURCE: NUREG/CR-3010

Fig. 2.8.5. Example of a basic operator-action tree.

of the selected response. A model, referred to as the time–reliability correlation, was developed to demonstrate the relationship between probability of failure and time available. A log–log plot of this relationship is shown in Figure 2.8.6. Because no data were available at the time of the OAT method's development, an interim functional relationship was developed by the consensus of an engineering psychologist and two systems analysts. Since its inception, some data have been gathered relating response time and error probability that substantiate the authors' claim that available time is an important factor in correctly performing cognitive tasks in an accident situation.

The application of the OAT method and time–reliability correlation to a risk or reliability analysis consists of the following steps:

1. Identify relevant plant safety functions from the event trees. Discussions with operators provide insight into what their concerns are at the time of interest in the accident scenario (e.g., removal of decay heat from the core of a pressurized water reactor).
2. Identify the generational actions required to achieve the plant safety functions.
3. Identify the relevant displays that present relevant alarm indications and the time available for the operators to take the appropriate mitigating actions.
4. Represent the errors in the fault trees or event trees of the PRA.
5. Estimate the probabilities of the errors. Once the thinking interval has been established, the nominal error probability is calculated from the time–reliability relationship, such as that in Figure 2.8.6. A "reluctance factor" may be used to modify the nominal value to account for reluctance on the part of the operators to initiate corrective action that in itself may compromise plant safety.

TIME-RELIABILITY CURVE

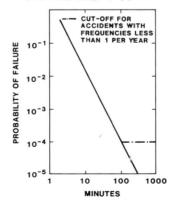

SOURCE: NUREG/CR-3010

Fig. 2.8.6. The time–reliability curve.

The OAT method applies a relatively general time-response model to a relatively specific set of performance circumstances. Perhaps on the average, the model can predict probabilities of cognitive errors. However, for any single prediction, the uncertainty will be rather high. The present model does not account for the potentially large effects of PSFs such as experience or stress on diagnosis and decision-making. In fact, the authors limit its use to errors of omission and to screening analysis only. Evidently errors of commission confound the criterion of performance by altering the state of the plant, and redirect the safety function of importance. The OAT method provides little in the way of confident estimates of cognitive error, and makes one wonder if the concept of speed–accuracy trade-off (a phenomenon found in choice reaction-time studies, e.g., Swanson and Briggs, 1969) has been taken too far. The OAT method's interim model can be used for making "ballpark" estimates of cognitive omission errors in the absence of useful cognitive data.

Time–reliability curves different from Figure 2.8.6 appear in NUREG/CR-2815 (Papazoglou, Bari, Buslik, Hall, Ilberg, Samanta, Teichmann, Youngblood, El-Bassioni, Fragola, Lofgren, and Vesely, 1984) and NUREG/CR-1278. In both documents, a "screening analysis diagnosis" curve appears that is a modification of a curve developed by Fragola (curve 17 in Figure 6–1 of NUREG/CR-3010). The modification was made to fit the stress and dependence models in NUREG/CR-1278. A less pessimistic set of time–reliability estimates called the "nominal diagnosis model" appears in NUREG/CR-1278 with some general rules on how to modify the values for differences in PSFs. This nominal diagnosis model is based on a modification of the screening diagnosis curve (from NUREG/CR-3010) using more realistic assumptions about control-room manning and dependence levels (see Figure 2.8.1). In summary, however, it should be said that all these curves representing diagnosis errors are highly speculative, and considerably more work in this area is needed.

2.8.5.5 Expert Judgment Techniques

One approach to quantitative HRA is the use of subject-matter experts to make judgments on the likelihood of human error in task performance. Expert judgments can be used to generate HEP estimates for use in task-synthesis/task-simulation techniques or to estimate the reliability of a particular task in full context.

Recently, Seaver and Stillwell (1983) have taken two psychological-scaling techniques and developed a set of procedures for implementing them to generate HEP estimates for nuclear power operation tasks. One method, *paired comparisons,* has the expert judge whether a human error is more likely in task "A" or task "B." This is relatively easy because numerical estimates are not required of the expert. However, paired-comparison judgments are required between all pairs of a set of tasks from a number of experts, and this can be a laborious and time-consuming exercise. Moreover, to convert the interval scale of error likeliness to a ratio scale of HEPs, at least two (preferably more) tasks from the paired comparison exercise must have some known HEPs, which are used to calibrate the interval scale. Another method, *direct numerical estimation,* requires experts to provide HEP estimates for each task. Although direct numerical estimation is a more difficult task than paired comparison, the expert does not have to make nearly as many judgments. Another advantage of direct numerical estimation is that it can be used to obtain estimates of uncertainty bounds. Figure 2.8.7 illustrates a response scale and task statement developed for data collection using this technique.

These two expert judgment procedural techniques were recently evaluated in the field by Comer, Kozinsky, Eckel, and Miller (1983). Nineteen boiling water reactor instructors served as subject-matter experts. The tasks evaluated ranged in scope from choosing a wrong switch from a set of similar-looking switches that are functionally arranged, to restoring off-site power in the event of a station blackout. Here, the former is a task element that is generalizable to many tasks, whereas the latter relates to an entire scenario or task and is context-specific.

The results of the evaluation provided encouraging support for the use of expert judgment. For the task-level estimates, direct estimates obtained by averaging across judges and paired comparisons showed extremely high correlation ($r = 0.94$) when the same judges were used in each technique. The task element-level direct estimates demonstrated a similarly high correlation ($r = 0.89$). When compared to the appropriate HEP estimates in NUREG/CR-1278, correlation coefficients of 0.68 and 0.40 were achieved ($p \le 0.05$), respectively, for the direct estimates and paired comparisons. Additionally, interjudge consistency was good and yielded highly significant coefficients of concordance for both techniques. The authors could not conclude that the expert judgment techniques had predictive validity because no "true" HEPs are available for comparison. The two techniques are, however, easy to implement, with complete processing instructions obviating the participation of a psychological-scaling expert.

2.8.5.6 Success Likelihood Index Methodology (SLIM)

The *Success Likelihood Index Methodology* (SLIM) developed by Embrey (1983) and his co-workers (Embrey, Humphreys, Rosa, Kirwan, and Rea, 1984) is another HRA technique that uses expert judgment to develop HEP estimates. The SLIM technique is a systematic method for scaling task

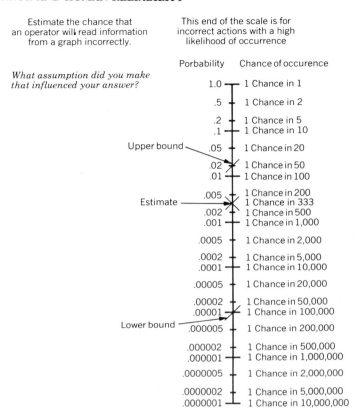

Fig. 2.8.7. Example of a task statement and response scale for direct numerical estimation.

success likelihood as a function of the conditions (PSFs) influencing successful completion of the task. An absolute measure of success probability for scaled tasks can be calculated after calibrating the scale with reference tasks of known reliability.

Multiple expert judges are used, either working together or alone, in order to benefit from a broad range of experience and to reduce biases inherent within individual judgments. The underlying assumption of SLIM is that the HEP for a given situation depends on the combination of effects from a small set of PSFs. In the procedure, judges are asked to assess the relative importance (weight of each PSF) with regard to its impact on the task of interest. A second, independent assessment is then made regarding the state or level of the PSFs in the task situation. This second numerical rating is a statement concerning how good or how bad the PSF is. Having identified the small set of PSFs, weighted their respective importance, and rated their relative goodness, one then multiplies the respective weights and ratings for each PSF. These products are then summed to produce the Success Likelihood Index (SLI), varying from 0 to 100. The SLI represents the judge's belief regarding the positive or negative effects of the PSFs on task success.

The SLIM approach assumes that the SLI has a functional relationship to the expected long-term probability of success. This relationship is considered to be logarithmic where

$$\log \Pr[S] = a(\text{SLI}) + b \tag{5}$$

and a and b are empirically derived constants. To calibrate the relationship empirically, at least two tasks of known reliability must be included in the task set evaluated. Solving simultaneous equations for two unknowns yields coefficient a and constant b. Substituting SLI values and taking antilogs yield $\Pr[S]$ for each task. If two tasks of known reliability cannot be obtained, the author suggests having the judges use direct estimation for the two reference tasks. Direct estimation is also recommended for estimating uncertainty bounds for the SLI-derived HEPs, either individually or by group consensus.

Several evaluations of SLIM have been performed by its originator. A pilot experiment was first carried out by Embrey (1983) to evaluate the feasibility of SLIM. In that study, six tasks were used for which known failure probabilities were a variable. A high correlation ($r = 0.98$) was found between the HEPs calculated using SLIM and the actual values. However, interjudge agreement was not significant, and the range of success probabilities studied was 0.32 to 0.95. Typically, industrial human errors of interest have much larger success probabilities (i.e., very small HEPs) that are characteristically much more difficult to estimate.

A more comprehensive study assessed 21 tasks of known reliability, which formed three groups of seven tasks each, roughly corresponding to Rasmussen's skill-, rule-, and knowledge-based behaviors. The tasks ranged in error probabilities from 10^{-1} to 5×10^{-5}. The eight judges included four reliability analysts, two nuclear operations people, and two human factors specialists. Only six PSFs were considered in deriving the SLIs for the 21 tasks, and even this small number presented problems for the judges. When common PSF weights were used for the seven tasks in each category, an insignificant correlation was found to exist between log HEPs and the SLIs (-0.47). When equal weights for all 67 PSFs were used, the correlation went up to a significant level (-0.60). When three tasks were removed from the analysis, the correlation rose to -0.71. No values of interjudge agreement were reported with these results.

A second SLIM study used 13 judges evaluating eight critical actions embedded in several nuclear power operations scenarios. Seven PSFs were suggested by the experimenters, and consensus definitions were developed by the judges prior to task evaluation. The PSF weights and ratings assessed by the judges were considered independently in the estimation of SLI values. Reference HEP estimates were obtained by direct estimation of two "boundary conditions" by the judges. The boundary conditions represented the best and worst possible task situations in terms of the PSFs. Interjudge consistency approached significance when calculated for three judges who participated in all the assessments.

A third study is currently being run to evaluate the implementation of SLIM through the use of an interactive computer program called MAUD (Multi-Attribute Utility Decomposition). The SLIM–MAUD approach incorporates procedural changes made as a result of the previous studies (Embrey et al., 1984). The SLIM technique was tested in a truly interactive fashion, with each evaluation using the latest variant of the SLIM procedures.

SLIM has been demonstrated to be a method for scaling a task's likelihood of success as a function of the contextual, influential PSFs. In order to determine quantitative HEPs for the tasks, the scale must be calibrated using reference tasks. The face validity of SLIM, though less than that of the conventional psychological scaling techniques noted above, may be adequate for the analyst who must rely on expert judgment techniques. The method's content validity, interjudge consistency, and utility have yet to be conclusively demonstrated. The technique is still in the developmental stage.

2.8.5.7 Sociotechnical Approach to Assessing Human Reliability (STAHR)

A method has recently been developed by Phillips and co-workers (Phillips, Humphreys, and Embrey, in press, a; Phillips, Humphreys, Embrey, and Selby, in press, b), that arrives at expert judgments using an iteratively derived group consensus to asses human reliability of nuclear power plant tasks. Heavily influenced by the SLIM–MAUD approach, the *Socio-Technical Approach to Assessing Human Reliability* (STAHR) consists of a "technical component," an influence diagram, and a "social component." The technical component involves developing a description of the "target event" (task) as well as the general setting and conditions leading up to the event. The modeling technique of the influence diagram, borrowed from decision theory, is potentially easier to use than event trees or fault trees. Modeling only dependencies between events, the influence diagram organizes them as a system of conditional probabilities. The following example taken from Phillips et al. (in press, a), may help to illustrate the notion of the influence diagram.

The top node on Figure 2.8.8 indicates the target event. For example, if an alarm in the control room signals that some malfunction has occurred and the operator attempts to correct the malfunction by following established procedures, one target event might be that the operator correctly performs a specified step in the procedure. The influence diagram shows three major influences on the target event. One is the quality of the information available to the operator, a second is the extent to which the organization of the nuclear power stations contributes to getting the work done effectively, and the third influence is the impact of personal and psychological factors pertaining to the operators themselves. Another way of saying this is that the effective performance of the target event depends on the physical environment, the social environment, and personal factors. Each of these factors is itself influenced by other factors. The quality of information is largely a matter of good design of the control room and of the presence of meaningful procedures. The organization is requisite, that is, it facilitates getting the required work done effectively, if the operations department has a primary role at the power station, and if the organization at the power station allows the effective formation of teams. Personal factors will contribute to effective

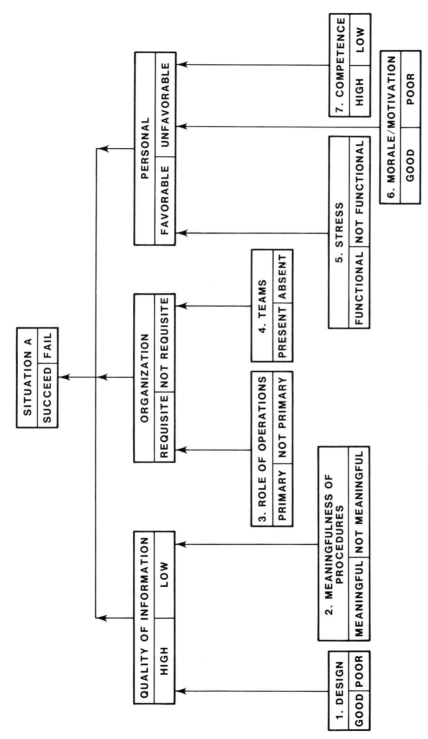

Fig. 2.8.8. Example of an influence diagram.

performance of the target event if the level of stress experienced by operators is helpful, if morale and motivation of the operators are good, and if the operators are highly competent.

As in the activities described previously, the remainder of the method relies on the knowledge and judgment of several subject-matter experts. The authors do not stipulate the backgrounds necessary for this group, but in a field test the following roles were represented:

1. Group consultant and facilitator
2. Technical moderator
3. Reactor operators trainer
4. Thermohydraulic engineer and procedures specialist
5. Probabilistic risk analyst
6. Reliability and systems analyst
7. Human reliability specialist
8. Reactor operator

The group of experts participates in assessing the "weights of evidence" for various levels of influences on adjacent levels and on the target event (the mechanics of this process go beyond the scope of this chapter). The target event is then assessed, conditional on the middle-level influences. The unconditional target event probability and the unconditional weight of evidence of the middle influences are then calculated with a computer. These results are compared with holistic judgments made by the assessors. Any discrepancies are discussed and revised. The process outlined above is then repeated as necessary until the assessors have finished refining their judgments.

The STAHR technique depends on the coalescing of many experts' diverse opinions into a consensus judgment. Although this process at times can be difficult and frustrating, the authors contend that "diversity of viewpoint" is a key criterion in composing the groups. The STAHR approach is still in the early developmental stage, so it is not possible to judge its utility fairly. However, as with any expert judgment method, the inconvenience of using a large, diverse group of subject-matter experts will eliminate the consideration of using this technique for many human reliability analysts.

In summation, the various techniques available to the analyst have different strengths and weaknesses that should be considered prior to use. The THERP technique is the only one in general use; the others are being used on an experimental basis. Although the techniques' differences are emphasized above, there are some wide commonalities among them, such as the use of task analysis and estimating the influences of PSFs on performance. Although each technique has been tested by its developers, no program has yet comparatively evaluated the various techniques on a given set of practical problems. However, an NRC research program (*Risk Methods Integration and Evaluation Program,* or RMIEP) is underway to evaluate the convergent validity of selected techniques for specific classes of tasks.

2.8.5.8 Incorporation of HRAs Into PRAs

As previously mentioned, a prime motivation for conducting an HRA might be to support a larger-scale PRA. Several PRAs have been performed in the nuclear power industry (Arkansas Nuclear One, Kolb et al., 1982; Zion PRA, 1981; Indian Point PRA, 1982; Seabrook PRA, Pickard, Lowe, and Garrick, Inc., 1983), and a few have been applied to the chemical processing and petroleum industries (proprietary reports). The process of integrating human-system interactions into the PRA framework should be systematic and preferably standardized across an industry. The process has been studied, and the results of these investigations have led to several recommended structural guidelines. These studies are discussed below.

The Electric Power Research Institute (EPRI) sponsored a project called *Systematic Human Action Reliability Procedure* (SHARP) (Hannaman and Spurgin, 1983). SHARP's objectives were to provide a structured, level-by-level approach to the incorporation of human interactions into PRAs, and to enhance the documentation and reproducibility of the study. There was no intention to develop new HRA models or techniques, or to rely on any given HRA approach. Instead, a framework was developed to accommodate the models and analyses chosen by the analyst.

The framework that evolved came from studying previous PRA efforts, abstracting the incorporation process, identifying the difficulties, and exploring the characteristics of an idealized framework. As a result, a seven-step procedure was developed to provide guidance for the system analyst. Each step has defined objectives, inputs, outputs, activities, and rules. The basic goals for each step are as follows:

1. *Definition.*   Ensure that many different types of human interactions are adequately considered in the study.
2. *Screening.*   Identify the human interactions that are significant to the operation and safety of the plant.

3. *Breakdown.*   Develop a detailed description of important human interactions by defining the key influence factors necessary to complete the modeling. The human interaction modeling consists of representation, impact assessment, and quantification.

4. *Representation.*   Select and apply techniques for modeling important human interactions in logic structures. Such methods help identify additional significant human actions that might impact the system logic trees.

5. *Impact Assessment.*   Explore the impact of significant human actions identified in the preceding step on the system logic trees.

6. *Quantification.*   Apply appropriate data or other quantification methods to assign probabilities for the various interactions examined, determine sensitivities, and establish uncertainty ranges.

7. *Documentation.*   Include all necessary information for the assesment to be traceable, under-standable, and reproducible.

SHARP is not considered a "cookbook," but a menu of steps, activities, and rules that can be selectively applied by the analyst in a PRA study. The SHARP framework appears to be a useful set of procedures, which some industry analysts consider could form the basis of an industry standard.

Another document that discusses the integration of HRA into PRA is the *Interim Reliability Evaluation Program Procedures Guide* (Carlson, Gallup, Kolaczkowski, Kolb, Stack, Lofgren, Horton, and Lobner, 1983). Although the report reviews analysis procedures predominantly associated with plant systems availability and accident sequences, a major section deals with human reliability and procedural analysis methods and the role they play in the PRA.

Other major contributions in this area include the *Probabilistic Safety Analysis Procedures Guide* (Papazoglou et al., 1984), the *IEEE/ANS PRA Procedures Guide* (Nureg/CR-2300, 1983), and *A Procedure for Conducting a Human Reliability Analysis for Nuclear Power Plants* (Bell and Swain, 1983).

2.8.6   THE DATA PROBLEM

The most serious problem in the rather young field of human reliability is the lack of data. As seen in the preceding section, all HRA techniques require human error data at some point in their analysis-prediction processes. The synthesis and simulation techniques require nominal HEPs as basic elements that can be combined or manipulated to reflect the specific task sequences and contextual PSFs. Data are needed not only to perform HRAs but also to validate the assumptions, models, and procedures that comprise the HRA techniques. This section discusses the inherent problems of human error data collection, some of the more ambitious attempts that have been made, and the current thinking behind the formation of a future data base.

2.8.6.1   Generalizability of Data

The common thread running through the problems associated with human error data is that of generaliza-bility. In a very literal sense, every error made by a person is unique. It has its own causes, motivations, influences (PSFs), action sequences, and system impacts or consequences. However, in the spirit of scientific progress, we assume that there are enough similarities in error characteristics that we might combine similar errors into groups. We also assume that errors that are similar from a behavioral and situational perspective have similar probabilities of occurrence. It is this premise that allows analysts quantitatively to predict HEPs for a given set of conditions. If error data cannot be assimilated or combined, and error data cannot be generalized to similar sets of conditions, then the science of human reliability analysis cannot thrive.

There are many impediments hindering the combination of error data. A prerequisite for a data base is a taxonomic structure for organizing the data. Ideally, the most efficient use of existing data would prescribe a single, taxonomic structure that everyone could use. Different approaches to HRA don't even necessarily agree on the scope of behaviors to be considered for a data point. Some HRA developers wish to find holistic data that would apply toward a worker's successfully completing a functional task (e.g., landing an airplane). Others desire data covering task elements (e.g., selecting a radio frequency). Still others need data elsewhere on the spectrum of task scope. Unfortunately, owing to the multitude of theoretical approaches to HRA, the level of detail required, and the many different applications areas (e.g., nuclear power operations, aviation, and automobile driving), universal consensus on a generalizable taxonomy is a virtual impossibility.

Another impediment to the mass aggregation of human error data is the variation in the type of sources from which the data are collected. There are four basic sources of human error data; the field, simulator activities, laboratory experiments, and expert judgment. All things being equal, the quality (or content validity) of the four data types follows the same order. Field data, collected in the industrial setting, will be the most applicable to the tasks studied in the HRA. Unfortunately, field data are the most difficult to collect. In order to document human errors during active industrial shifts, either an observer has to observe and record the incident, or the worker who commits the error must report it. There are obvious difficulties in recording someone else's errors, and an understandable reluctance on the part of the worker to confess to making an error. The National Aeronautics and Space Administration's Aviation Safety Reporting System (ASRS) and the NRC's Licensee Event Reports program are two of the more successful self-reporting, data-collection mechanisms. It is unfortunate that neither is adequately equipped to produce HEP data.

Simulator exercises that may involve initial or refresher training or a qualification for certification provide excellent opportunities to record the participant's action and to detect errors (Beare, Dorris, Bovell, Crowe, and Kozinsky, 1983). Opportunities for error (HEP denominator information) are as readily observed and recorded as the error events themselves (HEP numerator information). The drawbacks to simulator data include the high cost and the fact that the participants are performing with unusually high motivations and expectancies for nonroutine events. Hence these data are usually accepted by the analyst only if some calibration factor can be used in conjunction with the raw error data.

Experimental studies provide human error data. The literature of the general behavioral sciences provides some data, but because most experiments test hypotheses and collect performance variables such as reaction time and trials to criterion, the error data are usually controlled (i.e., held constant while other variables fluctuate), not collected, or not reported. The lack of substantial theoretical work on human error has led to a dearth of laboratory experiments designed to study its characteristics and mechanisms. Studies can be designed to collect human error data for a specific application, but this would be very costly, especially if low probability events were to be studied.

Expert judgment, discussed in the preceding section, is the least valid of the data sources. Basically, subject-matter experts judge the relative or absolute likelihood of error for several task descriptions, and these responses are converted to usable HEPs by mathematical scaling techniques. Obvious disadvantages include the expert not having full task information and lacking a basic familiarity (or "feel" for) the quantitative medium (e.g., probabilities, odds) in which he must report. On the other hand, the method can be practical in that it takes less time and money than the other data collection methods. Expert judgment does not really produce HEP data, but estimates, and therefore is not considered to be a reliable "data"-gathering technique.

A data base containing diverse data types from many different types of sources must tag data items with source information so that the data user can judge its quality and adopt an appropriate level of confidence. The data should be well documented in general, with personal and situational PSFs, system information, and dates when the data were collected. One of the problems with error data collected in the field and in simulators is that they pertain to the system as defined at that time. As changes and improvements are made within it, the system no longer operates in the manner in which it once did. The effect of time variance may, in some situations, invalidate the generalizability of data that are extremely similar in all other respects. For a more comprehensive discussion of data base criteria and potential problems refer to Blanchard (1975), and Miller (1982).

2.8.6.2 Data-Base Review

Material attempts at constructing formal, human error data bases have been few. The first realistic attempt was the AIR Data Store (Munger, Smith, and Payne, 1962). The behavioral literature was searched for the time and error data pertaining to molecular task elements (e.g., turning a rotary knob). These data were incorporated into the Data Store, which was organized by task element and ergonomic design features. In order to obtain task reliability, the user simply multiplied all the appropriate task element reliabilities. Task time could be obtained by adding up the task element times. Although the AIR Data Store did not receive much use, it was the first general set of error data that could be applied to various systems.

Irwin, Levitz, and Freed (1964) expanded on the AIR Data Store to create a data base specific to maintenance and inspection duties on the Titan II missile system. Because the methodology and data were expanded and modified to meet the needs of that particular system for the Aerojet General Company, the data base was of limited use to the rest of the human-factors community.

Meister also attempted to expand the Data Store in the Bunker-Ramo data bank (Hornyak, 1967). Subjective expert judgment was used to estimate the effects of PSFs on performance. Meister demonstrated the format for a detailed data bank, although it was never developed.

Topmiller, Eckel, and Kozinsky (1982) performed a detailed review of existing human reliability data banks in an effort to evaluate available human error data and categorization schemes for nuclear power plant PRA applications. The three previously described data banks are reviewed and reprinted in the report. Two additional existing data bases and four currently operating data systems were also reviewed. These were the following:

1. Technique for Establishing Personnel Performance Standards (TEPPS) (Blanchard, Mitchell, and Smith, 1966)
2. Operational Recording and Data System (OPREDS) (Osga, 1981)

3. Air Force Inspection and Safety Center (AFISC) (U.S. Air Force, 1971)
4. Aviation Safety Reporting System (ASRS) (Federal Aviation Administration, 1979)
5. Nuclear Plant Reliability Data System (NPRDS) (Southwest Research Institute, 1980)
6. Safety-Related Operator Action (SROA) Program (Barks, Kozinsky, and Eckel, 1982)

All nine data bases were evaluated against three categories: those for users (e.g., ease of use, specificity), those for data processing and evaluation (e.g., quantification, recoverability), and those for collection (e.g., relevance, cost). Among the past data bases reviewed, the AIR Data Store received the highest mean rating on the equally weighted criteria. The SROA data base topped the list of active data bases. It should be noted that the criteria were tuned to the application of nuclear power operations and that different conclusions could be reached for different applications. The report concluded that a human reliability data bank be established that would be specifically tailored for nuclear power plant PRA applications.

2.8.6.3 Human Reliability Data Bank

Following the Topmiller et al. (1982) review, the NRC funded a three-year research program to develop the conceptual framework and implementation procedures for a human reliability data bank that would support HRAs for incorporation into nuclear power plant PRAs (Comer, Kozinsky, Eckel, and Miller, 1983). The concepts, data processing methods, and data retrieval techniques were evaluated in a feasibility study (Miller and Comer, 1985; Comer, Seaver, Stillwell, and Gaddy, 1985) and were modified based on the findings of the study (Comer and Donovan, 1985). Some of the features that make this data bank attractive and feasible are as follows:

1. Incoming data are screened to ensure they meet the minimum criteria (e.g., quantitative HEPs and error rates).
2. Administrative reviews and procedures are set up to minimize erroneous categorization of incoming data.
3. The data bank is set up to receive field, simulator, laboratory, and analytic data.
4. PSF information is "tagged" to the data.
5. Rules and procedures are established for data combination and aggregation.
6. Three different specificity levels offer flexibility in accommodating data of differing levels of task analysis detail.
7. The data bank user has a set of procedures to help him in finding the appropriate data of interest.
8. A data clearinghouse is established to provide additional information and assistance to the data bank users.

The human reliability data bank, if implemented, will have the primary purpose of supporting nuclear power PRA activities. If it is successful, the scope of its uses may be widened to support human reliability activities applicable to other industries as well.

2.8.7 STRATEGIES FOR DEALING WITH HUMAN ERROR

2.8.7.1 Change the Worker

The first step in reducing human error is to identify its causes correctly. Usually the work situation can be improved through ergonomic redesign to reduce errors and increase productivity. However, in some cases, the work situation is well designed and the tasks are reasonable, but the worker still commits an unacceptably high number of errors. The poor performance may be due to personnel factors such as deficient dexterity, poor vision, partial deafness, or inadequate skills. These deficiencies are usually screened out in personnel selection or improved through providing perceptual aids (e.g., prescription lenses) and training. On-the-job training programs and probationary work periods can give new workers special consideration in the form of relaxed work standards. After a short period, however, the recently hired personnel are expected to perform up to the normal standards.

Industries that require highly skilled performance or decision-making in positions of considerable responsibility (and risk) usually require certification (e.g., airline pilots and nuclear power station control-room operators). In order to maintain certification, refresher training and requalification are usually necessary.

Physical and mental attributes often determine whether there will be a good match between the worker and job. Tall people tend to be better basketball players than short people. Soviet tank crews are selected from the shortest of the qualifying troops owing to the tank's tactically ideal, low profile and the concomitant cramped quarters. Often industrial workers are shifted from job to job until a good match is found. Some people can never adjust to rotating shift work. If individual differences are recognized and accounted for, errors can be reduced.

Motivational and emotional problems can lead to unintentional errors in the workplace. Poorly motivated workers or operators can commit unacceptably high errors, compromise plant safety, and cause poor morale among fellow workers. Sometimes a high error rate is the first overt sign that a worker is having emotional problems. The sooner this is recognized by the supervisor, the sooner the problem can be addressed.

2.8.7.2 Change the Work Situation

Too often in American industry the worker is blamed for making errors, producing defects, and initiating accidents, when, in fact, the work situation itself is poorly designed and predisposes the worker to error. The work-situation approach, introduced in Section 2.8.3, assumes that the work situation can be improved to take into account human limitations, thereby reducing the likelihood of worker error. Typically, a systems safety engineer, ergonomist, or similarly trained expert examines an industrial process to identify error-likely situations that may lead to accidents or unacceptably high defect rates. Several of the HRA techniques discussed in Section 2.8.5 are based on this method. When ergonomic deficiencies are identified, the specialist can assess the impact on errors and recommend design changes. The changes may involve modifications of environmental conditions, supervision methods, performance feedback, equipment design, processing techniques, and job aids. Some of the standard ergonomic design guides include those of Woodson (1981) and VanCott and Kincade (1972) and MIL-STD-1472C (1981).

Practical considerations make this approach susceptible to missing key factors, which may contribute to error-likely situations. Without interviewing the regular operators or getting detailed walk-throughs or talk-throughs, the specialist may not gain sufficient understanding of the tasks involved to identify the factors contributing to errors. Chances are extremely high that the specialist will never observe an instance of the error being studied, especially if it is thought to be a low-probability event.

An alternative approach to identifying situation-caused errors involves worker participation. One version is called the error-cause removal (ECR) program and consists of six basic elements (Swain, 1973):

1. Management supervision, engineering, and production personnel are educated in the value of an ECR program.
2. Production workers and ECR team coordinators are trained in the techniques to be employed.
3. Production workers (the ECR team) learn to report errors and error-likely situations, analyze these reports to determine causes, and develop candidate-design situations to remove or appropriately modify these error causes.
4. Ergonomists and other specialists evaluate the candidate-design solutions in terms of worth and cost, and select the best of these situations or develop alternate solutions.
5. Management implements the best design solutions and provides recognition for efforts of production personnel in the ECR program.
6. Aided by continuing inputs from the ECR program, ergonomists and other specialists evaluate the changes in the production process.

Perhaps the best empirical support for the ECR program and the concept of ECR teams is found in the results of a study by Chaney (1969). Although this study was not directed specifically at reducing production errors (the primary goal was to increase production rate), the approach was similar to that advocated for the ECR program. The Chaney study demonstrated three important findings relevant to the ECR approach:

1. Worker participation in a program to improve production is feasible given the use of appropriate participative techniques by the team leader.
2. The greater the level of worker participation, the better the results.
3. Valuable design suggestions to improve performance in industrial operations can come out of such a program.

The ultimate in worker participation in programs to reduce errors and defective products is probably found in Japan's quality-control (QC) circles (Juran, 1967). The QC circle is a small group of departmental work leaders and line operators who help solve quality problems. The QC circle movement originated in Japan about 1963 and has some similarities to an ECR team. In Japan there are certain aspects of the industrial culture that contribute to the willingness of workers to participate in such programs, for example,

1. The existence of a blame-free atmosphere.
2. The acceptance of workers as part of management's quality team.
3. The recognition of the worker as, indeed, being a subject-matter expert.
4. The management's action on recommendations made by the worker team.

In addition, management is not shamed by imperfection, but sincerely motivated to discover inefficiencies and correct them.

2.8.7.3   Reduce the System Impact

In most cases, human errors cannot be eliminated completely but can be reduced to a tolerable level. In systems where human error has been reduced to a low level using previously described techniques, but remains above the tolerable level, the impact of the error on the system can be reduced. In other words, the error occurs but its system impact is reduced or eliminated through early detection and remedial action.

Systems can be designed to experience graceful degradation instead of total failure in response to a critical human error. NASA has used this approach for decades. When equipment failures or human errors occur, they are detected quickly and recovered before mission success is compromised. Often this involves resorting to backup equipment or "hot spares" (equipment that is kept running in the event that it is needed for backup purposes). Redundancy is the key to a forgiving system. Human redundancy can be designed into critical operations. Checkers or inspectors can be used to verify that a human task is performed correctly or on time. Machines can also be used to monitor human performance. Radar altimeters in airplanes were developed to provide redundancy for barometric altimeters, but can also alert the pilot when the plane has flown below an acceptable preset altitude.

Systems tolerant of human error can be made extremely reliable and safe. The airline industry has evolved to the point where it is rare that any single human error will precipitate a fatal crash. In fact, the system has so many checks, redundancies, safety rules, and quality standards, that several (a minimum of three or four) serious errors must be made in succession to cause a severe accident. Few industries can afford to become as sophisticated as the airline industry, but it does provide a model toward which other industries can strive.

REFERENCES

Askren, W. B., Ed. (1967, May). *Symposium on reliability of human performance in work* (AMRL-TR-67-88). Wright-Patterson Air Force Base, OH: Aerospace Medical Research Laboratories.

Askren, W. B., and Regulinski, T. L. (1969). Quantifying human performance for reliability analysis of systems. *Human Factors, 11*(4), 393–396.

Barks, D. B., Kozinsky, E. J., and Eckel, J. S. (1982, May). *Nuclear power plant control room task analysis: Pilot study for pressurized water reactors* (General Physics Corporation and the Oak Ridge National Laboratory, NUREG/CR-2598). Washington, DC: U.S. Nuclear Regulatory Commission.

Beare, A. N., Dorris, R. E., Bovell, C. R., Crowe, D. S., and Kozinsky, E. J. (1984). *A simulator-based study of human errors in nuclear power plant control room tasks* (General Physics Corporation and Sandia National Laboratories, NUREG/CR-3309). Washington, DC: U.S. Nuclear Regulatory Commission.

Bell, B. J., and Swain, A. D. (1983, May). *A procedure for conducting a human reliability analysis for nuclear power plants* (Sandia National Laboratories, NUREG/CR-2254). Washington, DC: U.S. Nuclear Regulatory Commission.

Blanchard, R. E. (1975). Human performance and personnel resource data store design guidelines. *Human Factors, 17*(1), 25–34.

Blanchard, R. E. and Harris, D. H., Eds. (1967, June). *Man–machine effectiveness analysis, a symposium.* Los Angeles, CA: Los Angeles Chapter, Human Factors Society.

Blanchard, R. E., Mitchell, M. B., and Smith, R. L. (1966, June). *Likelihood of accomplishment scale for a sample of man–machine activities.* Santa Monica, CA: Dunlap and Associates.

Carlson, D. D., Gallup, D. R., Kolaczkowski, A. M., Kolb, D. J., Stack, D. W., Lofgren, E., Horton, W. H., and Lobner, P. R. (1983, January). *Interim reliability evaluation program procedures guide* (Sandia National Laboratories, NUREG/CR-2728). Washington, DC: U.S. Nuclear Regulatory Commission.

Chaney, F. B. (1969). Employee participation in manufacturing job design, *Human Factors, 11*(2), 101–106.

Comer, M. K., Kozinsky, E. J., Eckel, J. S., and Miller, D. P. (1983, February). *Human reliability data bank for nuclear power plant operations,* Vol. 2, *A data bank conception and system description*

(General Physics Corporation and Sandia National Laboratories, NUREG/CR-2744). Washington, DC: U.S. Nuclear Regulatory Commission.

Comer, M. K., Seaver, D. A., Stillwell, W. G., and Gaddy, C. D. (1984). *Generating human reliability estimates using expert judgment: Paired comparisons and direct numerical estimation,* Vol. 1, Main Report, Vol. 2, Appendices, NUREG/CR-3688). Washington, DC: U.S. Nuclear Regulatory Commission.

Comer, M. K., Donovan, M. D., and Gaddy, C. D. (1985). *Human reliability data bank: Evaluation results* (General Physics Corporation and Sandia National Laboratories, NUREG/CR-4009). Washington, DC: U.S. Nuclear Regulatory Commission.

Comer, M. K., and Donovan, M. D. (1985). *Specification of a human reliability data bank for conducting HRA segments of PRAs for nuclear power plants* (General Physics Corporation and Sandia National Laboratories, NUREG/CR-4010). Washington, DC: U.S. Nuclear Regulatory Commission.

Embrey, D. E. (1976, July). *Human reliability in complex systems: An overview* (NCSR.R10). Warrington, England: National Centre of Systems Reliability, United Kingdom Atomic Energy Authority.

Embrey, D. E. (1983, May). *The use of performance shaping factors and quantified expert judgment in the evaluation of human reliability: An initial appraisal* (Brookhaven National Laboratory, NUREG/CR-2986). Washington, DC: U.S. Nuclear Regulatory Commission.

Embrey, D. E., Humphreys, P., Rosa, E. A., Kirwan, B., and Rea, K. (1984). *SLIM–MAUD: An approach to assessing human error probabilities using structural expert judgment,* Vol. 1, *Overview of SLIM–MAUD,* March; Vol. 2, *Detailed analysis of the technical issues* (draft), May (NUREG/CR-3518). Washington, DC: U.S. Nuclear Regulatory Commission.

Evans, R. A. (1976). Reliability optimization. In E. J. Henley and J. W. Lynn, Eds., *Generic techniques in systems reliability assessment,* Leyden, The Netherlands: Noodhoff International Publishing, pp. 117–131.

Federal Aviation Administration (1979, June 15). Aviation safety reporting program (FAA Advisory Circular 00–46B). Washington, DC.

Fitts, P. M., and Posner, M. I. (1967). *Human performance.* Belmont, CA: Brooks/Cole Publishing Co.

Fox, J. C., and Haslegrave, C. M. (1969). Industrial inspection efficiency and the probability of a defect occurring. *Ergonomics, 12*(5), 713–721.

Hall, R. E., Fragola, J., and Wreathall, J. (1982, November). *Post event human decision errors: Operator action tree/time reliability correlation* (Brookhaven National Laboratory, NUREG/CR-3010). Washington, DC: U.S. Nuclear Regulatory Commission.

Hannaman, G. W., and Spurgin, A. J. (1983, December). *Systematic human action reliability procedure (SHARP)* (Report No. NUS-4486), San Diego, CA: NUS Corporation.

Hornyak, S. J. (1967). *Effectiveness of display subsystem measurement and prediction techniques* (Report No. TR-67-292). Griffiss Air Force Base, NY: Rome Air Development Center.

Indian point probabilistic safety study (1982). Vol. 1, Section 0—Methodology and Section 1—Plant analysis. New York: Power Authority of the State of New York and Consolidated Edison Company of New York, Inc.

Irwin, I. A., Levitz, J. J., and Freed, A. M. (1964). *Human reliability in the performance of maintenance* (Report No. LRP 317/TDR-63-218). Sacramento, CA: Aerojet General Corp.

Juran, J. M., (1967). The QC circle phenomenon. *Industrial Quality Control, 23,* 329–336.

Kolb, G. J., Kunsman, D. M., Bell, B. J., Brisbin, N. L., Carlson, D. D., Hatch, S. W., Miller, D. P., Roscoe, B. J., Stack, D. W., Worrell, R. B., Robertson, J., Wooton, R. O., McAhren, S. H., Ferrell, W. L., Galyean, W. J., and Murphy, J. A. (1982, June). *Interim reliability evaluation program: Analysis of the Arkansas Nuclear One—Unit 1 Nuclear Power Plant,* Vols. 1 and 2 (Sandia National Laboratories, NUREG/CR-2787). Washington, DC: U.S. Nuclear Regulatory Commission.

Lord, F. M., and Novick, M. R. (1968). *Statistical theories of mental scores,* Reading, MA: Addison-Wesley.

McCornack, R. (1961). *Inspector accuracy: A study of the literature* (SCTM-53-61(14)). Albuquerque, NM: Sandia National Laboratories.

Meister, D. (1964). Methods of predicting human reliability in man–machine systems, *Human Factors, 6*(6), 621–646.

Meister, D. (1966). Human factors in reliability. Section 12 in W. G. Ireson, Ed., *Reliability handbook.* New York: McGraw-Hill.

Meister, D. (1971, November). *Comparative analysis of human reliability models* (Report L0074-107). Westlake Village, CA: Bunker-Ramo Electronics Systems Division.

Meister, D. (1984). Human reliability. In F. A. Muckler, Ed., *Human Factors Review: 1984.* Santa Monica, CA: Human Factors Society.

Miller, D. P. (1982). Human performance data bank. In R. E. Hall, J. E. Fragola, and W. J. Luckas, Jr., Eds. *Conference record of the 1981 IEEE standards workshop in human factors and nuclear safety, August 30–September 4, 1981, Myrtle Beach, SC.* New York: Institute of Electrical and Electronic Engineers, Inc.

Miller, D. P., and Comer, K. (1985). Process evaluation of the human reliability data bank. *Proceedings of the Twelfth Water Reactor Safety Research Information Meeting* (NUREG/CP-0048), Vol. 6. Washington, DC: U.S. Nuclear Regulatory Commission

MIL-STD-1472C (1981, May). Military Standard, *Human engineering design criteria for military systems, equipment, and facilities.* Washington, DC: U.S. Department of Defense.

Mills, R. G., and Hatfield, S. A. (1974). Sequential task performance: Task module relationships, reliabilities, and times, *Human Factors, 16*(2), 117–128.

Mintz, A., and Blum, M. L. (1961). A re-examination of nuclear accident proneness concept. Chap. 57 in E. A. Fleishman, Ed., *Studies in personnel and Industrial Psychology.* Homewood, IL: Dorsey Press.

Munger, S. J., Smith, R. W., and Payne, D. (1962, January). *An index of electronic equipment operability: Data store* (AIR-C43-1/62-RP(1)). Pittsburgh, PA: American Institutes for Research.

Norman, D. A., Categorization of action slips, *Psychological Review, 88*(1), 1–15.

NUREG/CR-2300 (1983, January). *PRA procedures guide,* Vols. 1 and 2, American Nuclear Society/ Institute for Electrical and Electronic Engineers. Washington, DC: U.S. Nuclear Regulatory Commission.

Osga, G. A. (1981, March). *Guidelines for development, use, and validation of human performance data bank for NTDS combat operations* (Systems Exploration, Inc., Naval Ocean Systems Center, Contract No. NOV-00123-80-D0263); San Diego, DA: OSGA.

Papazoglou, I. A., Bari, R. A., Buslik, A. J., Hall, R. E., Ilberg, D., Samanta, P. K., Teichmann, T., Youngblood, R. W., El-Bassioni, A., Fragola, J., Lofgren, E., and Vesely, W. (1984, January). *Probabilistic Safety Analysis Procedures Guide* (Brookhaven National Laboratory NUREG/CR-2815). Washington, DC: U.S. Nuclear Regulatory Commission.

Payne, D., and Altman, J. W. (1962, January). *An index of electronic equipment operability: Report of development* (AIR-C43-1/62-FR). Pittsburgh, PA: American Institutes for Research.

Pew, R. W., Feehrer, C. E., Baron, S., and Miller, D. C. (1977, March). *Critical review and analysis of performance models applicable to man–machine evaluation* (AFOSR-TR-77-0520). Washington, DC: Air Force Office of Scientific Research, Bolling Air Force Base.

Phillips, L. D., Humphreys, P., and Embrey, D. (in press, a). Appendix D: A socio-technical approach to assessing human reliability (STAHR). In *A pressurized thermal shock evaluation of the Calvert Cliffs unit 1 nuclear power plant.* Oak Ridge, TN: Oak Ridge National Laboratory.

Phillips, L. D., Humphreys, P., Embrey, D., and Selby, D. L. (in press, b). Appendix E: Quantification of operator actions by STAHR Methodology. In *A pressurized thermal shock evaluation of the Calvert Cliffs unit 1 nuclear power plant.* Oak Ridge, TN: Oak Ridge National Laboratory.

Pickard, Lowe, and Garrick, Inc. (1983, December). *Seabrook Station Probabilistic Safety Assessment* (PLG-0300). Newport Beach, CA.

Rasmussen, J. (1981, August). *Human errors. A taxonomy for describing human malfunctions in industrial installations* (Report No. RISØ-M-2304). Roskilde, Denmark: RISØ National Laboratory.

Rigby, L. V. (1970, May). The nature of human error. In *Annual technical conference transactions of the ASQC.* Milwaukee, WI: American Society for Quality Control, pp. 457–466.

Rook, L. W. (1962, June). *Reduction of human error in industrial production* (SCTM 93–62(14)). Albuquerque, NM: Sandia National Laboratories.

Seaver, D. A., and W. G. Stillwell, (1983, March). *Procedures for using expert judgment to estimate human error probabilities in nuclear power plant operations* (Decision Science Consortium and Sandia National Laboratories, NUREG/CR-2743). Washington, DC: U.S. Nuclear Regulatory Commission.

Siegel, A. I., and Wolf, J. A. (1969). *Man–machine simulation models.* New York: Wiley.

Siegel, A. I., and Federman, P. J. (1971). *Prediction of human reliability. Part I: Development and test of a human reliability predictive technique for application in electronic maintainability.* Warminster, PA: Naval Air Development Center.

Siegel, A. I., Bartter, W. D., Wolf, J. J., Knee, H. E., and Haas, P. M. (1984). *Maintenance personnel performance simulation (MAPPS) model: Summary description* (Applied Psychological Services and Oak Ridge National Laboratory, NUREG/CR-3626), Vol. 1. Washington, DC: U.S. Nuclear Regulatory Commission.

Southwest Research Institute (1980, December). *Reporting procedures manual for the nuclear plant reliability data system.* San Antonio, TX: Southwest Research Institute.

Swain, A. D. (1963, August). *A method for performing a human factors reliability analysis* (Monograph SCR-685). Albuquerque, NM: Sandia National Laboratories.

Swain, A. D. (1967, January). Field calibrated simulation. IV-A-1 to IV-A-21 in *Proceedings of the Symposium on Human Performance Quantification in Systems Effectiveness.* Washington, DC: Naval Material Command and the National Academy of Engineering.

Swain, A. D., (1967, May). Some limitations in using the simple multiplicative model in behavior quantification. In W. B. Askren, Ed. *Symposium on reliability of human performance in work* (AMRL-TR-67-88). Wright-Patterson Air Force Base, OH: Aerospace Medical Research Laboratory, pp. 17–32.

Swain, A. D. (1969, July). Overview and status of human factors reliability analysis. In *Proceedings of 8th annual reliability and maintainability conference.* Denver, CO: American Institute of Aeronautics and Astronautics, pp. 251–254.

Swain, A. D. (1969, August). A work situation approach to improving job safety. In *Proceedings, 1969 Professional Conference, American Society of Safety Engineers.* Chicago, IL: pp. 233–257.

Swain, A. D. (1973). Design of industrial jobs a worker can and will do. *Human Factors, 15*(2), 129–136.

Swain, A. D., and Guttmann, H. E. (1983, August). *Handbook of human reliability analysis with emphasis on nuclear power plant applications* (Sandia National Laboratories, NUREG/CR-1278). Washington, DC: U.S. Nuclear Regulatory Commission.

Swanson, J. M., and Briggs, G. E. (1969). Information processing as a function of speed versus accuracy. *Journal of Experimental Psychology, 81*(2), 223–229.

Topmiller, D. A., Eckel, J. S., and Kozinsky, E. J. (1982, December). *Human reliability data bank for nuclear power plant operations,* Vol. 1: *A review of existing human reliability data banks* (General Physics Corporation and Sandia National Laboratories, NUREG/CR-2744). Washington, DC: U.S. Nuclear Regulatory Commission.

U.S. Air Force (1971, December). Life sciences accident and incident classification elements and factors (AFISC Operating Instruction AFISCM 127-6).

Van Cott, H. P., and Kinkade, R. G. Eds. (1972). *Human engineering guide to equipment design,* rev. ed. Washington, DC: U.S. Nuclear Regulatory Commission.

Welford, A. T. (1973). Stress and performance, *Ergonomics, 16*(5), 567–580.

Wilkerson, R. G. (1964). Artificial signals as an aid to an inspection task, *Ergonomics, 7*(1), 63–72.

Woodson, W. E. (1981). *Human factors design handbook.* New York: McGraw-Hill.

Zion Probabilistic Safety Study (1981). Chicago, IL: Commonwealth Edison Co.

CHAPTER 2.9

FEEDBACK-CONTROL MECHANISMS OF HUMAN BEHAVIOR

THOMAS J. SMITH

Simon Fraser University
Burnaby, British Columbia

KARL U. SMITH

University of Wisconsin
Madison, Wisconsin

The area of theory and research known as behavioral–physiological cybernetics studies feedback control or self-regulation of behavioral and physiological function. Of primary interest are those reciprocal regulatory interactions between a motor mechanism and a receptor, as mediated by the brain, in which feedback from the receptor both regulates and is controlled by motor response. This interaction has the fundamental properties of being dynamic, closed-loop, self-regulatory, and continuous in operation. In typical instances of behavior, muscle action generates self-stimulation of receptors, as suggested in Figure 2.9.1a. Muscle action also serves to control patterns and sources of environmental stimulation, as indicated in Figure 2.9.1b.

Muscle groups in action not only control the environment by means of feedback, but also generate various kinds of physiological feedback that affect all aspects of vital organization, that is, bioenergy production, organic metabolism, organic function, and brain function. Control of bioenergy production and internal visceral operations by means of physiological feedback of muscular activity is illustrated in Figure 2.9.2.

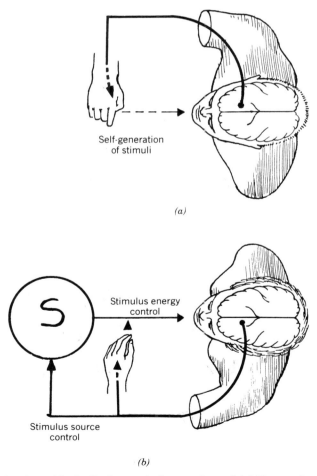

Self-generation
of stimuli

(a)

Stimulus energy
control

Stimulus source
control

(b)

Fig. 2.9.1. Modes of muscular feedback control of sensory input. (a) Self-generation of sensory input. (b) Control of environmental sensory input.

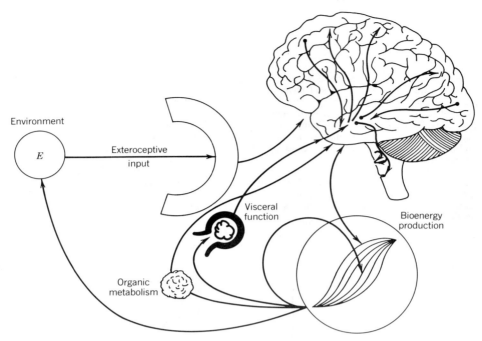

Fig. 2.9.2. Modes of muscular feedback control of bioenergy production, organic metabolism, and visceral function.

2.9.1 ORIGINS OF FEEDBACK CONCEPTS

Appreciation of the fact that living systems somehow have the ability to self-regulate their own activities dates back about 100 years. Pflüger (1875) provided an explanation for maintenance of steady-state or static conditions in living organisms by asserting, "The cause of every need of a living being is also the cause of the satisfaction of the need." Fredericq (1887) also called attention to the importance of biological self-regulation:

> The living being is an agency of such sort that each disturbing influence induces by itself the calling forth of compensatory activity to neutralize or repair the disturbance.

These early speculations regarding biological self-regulation were based on the work of Bernard (1865), who emphasized the importance of the internal environment in the establishment and maintenance of stable conditions in the body:

> It is the fixity of the milieu interieur which is the condition of free and independent life, and all the vital mechanisms, however varied they may be, have only one object, that of preserving constant the conditions of life in the internal environment.

2.9.1.1 Homeostasis versus Homeokinesis

To describe internal constancy, Cannon (1932) coined the term *homeostasis* to designate the coordinated physiological processes that maintain most of the steady states of the body. According to Cannon, "If a state remains steady, it does so because any tendency towards change is automatically met by increased effectiveness of the factor or factors which resist the change."

The term homeostasis, which suggests equilibrium, steady states, and stability, presents an oversimplified picture of physiological integration. Although factors such as temperature, pH, and electrolyte concentrations appear to exhibit relative constancy, most if not all such conditions vary, often over a broad range, and the mechanisms that regulate these conditions are dynamic and highly variable. We suggest that the term *homeokinesis* more accurately describes the variable balance, kinetic characteristics, and integrative control processes of most biological systems, conditions, and mechanisms. With respect to the mechanisms of behavior, the responding organism is designed to be dynamic rather than static, to behave constantly and integratively during its lifetime. The organization of behavior depends on constant motion generated by the motion systems themselves. This is true of physiological mechanisms as well as behavior. All living organisms self-regulate their own activities by dynamically and continuously controlling both the sources of stimulus input and their responses to that input.

2.9.1.2   Behavioral–Physiological Cybernetics

Self-regulation in living systems is mediated by feedback-control mechanisms. The term feedback control was used first by electrical engineers to describe automatic control of machine operations. The term referred to the return of a feedback signal of the effects of a machine's output to its input to correct its operation. Recognizing the similarity between such ongoing control in mechanical and living systems, Wiener (1948) coined the term *cybernetics*, from the Greek word "kybernetes" or "helmsman," to refer to the study of feedback-controlled guidance or steering in both living and nonliving systems. In the cybernetic analogy adopted by Wiener, behavior and physiological mechanisms are seen as control systems which, like self-controlled automatic machines utilizing feedback, generate action in following environmental changes and correct or modulate this action in terms of sensed feedback information.

Wiener's cybernetic analogy provided the basis for the *servomechanism model* of behavior and physiological function. A servomechanism has three main components: an action (response) mechanism, a feedback (stimulus) detector, and a controller or computer which translates the feedback signal into corrective action. Thus any system functioning as a servomechanism has three main properties: (1) it generates movement toward a target or defined path; (2) it detects error by comparing the effect of this action with the true path; and (3) it uses the error signal to redirect the action. This is essentially a negative-feedback model of compensatory action, and it is the model emphasized in most human engineering texts (Kantowitz and Sorkin, 1983).

Negative-feedback control, as in the servomechanism model, represents only one mechanism of behavioral and physiological regulation. Closed-loop control of behavioral and physiological functioning includes also *positive-feedback control* and *positive-* and *negative-feedforward control*. In positive-feedback control, an initial disturbance in a system actuates a series of events that tend to amplify the magnitude of the disturbance. For example, there are a number of metabolic and endocrine pathways in which production of a particular product enhances the subsequent rate of production of that product. It can be shown theoretically and experimentally that, left unchecked, positive-feedback control systems are inherently unstable and may exhibit wild oscillations. In behavioral and physiological systems, however, positive and negative control mechanisms are customarily coupled so that instabilities are ultimately brought under control.

In feedforward control, the detection of an error or disturbance in a system is fed forward to control, either positively or negatively, the state of the system at some future point in time. A behavioral example is visual-manual tracking, such as a hunter shooting a moving target. If one treats the target as static (a negative-feedback situation), one would always miss. Instead one uses the present trajectory of the target (the initial disturbance) to project its position in space some 1 or 2 sec into the future and aims for that point instead. The most dramatic example of feedforward control is the ability of humans to control time itself by planning, projecting, scheduling, and predicting.

The field of behavioral/physiological cybernetics is based on experimental study of the actual feedback and feedforward mechanisms of behavior and physiological functioning, rather than on abstract conceptual or mathematical models. The negative-feedback model is inadequate to deal with the many kinds of feedback-controlled interactions that are involved in both primary and integrated levels of systems control in the human body, upon which the significant phenomena of biological adaptation are based. Feedback control takes many systems forms and can be based on many quantitative modes of positive adjustment beyond detection of differences or errors. It may be involved in regulation of order, sequence, summing, separation, elaboration and multiplication, integration, and serial interaction. It may interrelate movements, movements and receptor functions, and movements and physiological functioning, all on a positive active basis, as contrasted with a negative-feedback homeostatic basis.

2.9.1.3   Research on Tracking Systems

One of the significant results of extending the cybernetic analogy to describe behavioral and physiological functions as servomechanisms was to point up the significance of the study of tracking and steering behavior. Limited studies of eye and head tracking of visual targets and displays were made before World War II, mainly with animals as subjects. The onset of war instigated numerous research and training investigations of target tracking to solve problems of radar operation, gunnery, aircraft piloting, and other military operations, studies that fostered a growing understanding of human-factors design of machine operations. In this context, the term human-factors design refers to the compliance or correspondence between the properties of an operating machine and the properties of behavior and physiological function involved or needed in the operation.

The first major human-factors problem of design of military machines arose in connection with different types or modes of control of cursors and indicators in tracking systems, as in radar and gunnery. Three types of tracking control were developed, direct, velocity, and aided, the last two of which were power-aided. Aided control combines direct and motor control of a tracking cursor. Heated arguments arose during and after World War II regarding the relative accuracy of the three types of tracking control. The problem was resolved by Lincoln and Smith (1952) in one of the first comprehensive cybernetic experiments on human-factors design subsequent to the war. Electromechanical apparatus was designed to produce controlled conditions of the three types of tracking comparable to tracking of aircraft targets, and to shift rapidly from one type of control to another. The results indicated clearly that direct tracking is superior to power-aided tracking.

2.9.1.4 Effects of Delaying the Feedback Signal

Study of aided-tracking systems led to explicit recognition of the problem of delayed feedback (Simon and Smith, 1956). In aided tracking the rate and extent of the operator's movement of the control affect the rate and extent of movement of the cursor. The sensitivity of the system can be changed by adjusting the relation between the two rates—a relation that is specified in terms of the aided-tracking time constant. For tracking aircraft targets, engineers determined that a time constant of 0.5 sec produced maximum accuracy in this type of system. Lincoln and Smith identified this time constant as a feedback-delay factor in tracking control and concluded that the delay accounted for the inferiority of aided tracking to direct tracking.

In earlier wartime research on the B-29 gun system, it had been realized that a particular type of machine delay influenced accuracy in tracking. This gun system was controlled by analog computers. As a result of an engineering error in designing the system, servomechanism control of the guns by gunsight and computer was unstable. To correct this instability, the sight input to the computer was filtered mechanically by slotting the azimuth, elevation, and range input gears, thereby introducing significant delays between movement of the gunner and movement of the guns, especially when he tried to slew on target. Study of this gun system in operation was one of the first direct studies of the effects of feedback delay on machine performance (Smith, 1945), and constituted an early demonstration of the closed-loop nature of tracking behavior.

2.9.1.5 Theories of Tracking Behavior

Extensive interest in tracking as a human-factors problem generated the first systematic theories of machine behavior. Most of these theories focused on the attempt to explain the oscillatory nature of tracking errors and involved different interpretations of the time characteristics of those oscillations. A theory based on the servomechanism analogy viewed the tracker as operating much like an electromechanical control system (Fitts, 1951). Craik (1947) and Elkind (1953) described the relations between tracking error and target speed in terms of a linear mathematical model. Licklider (1960) maintained that the relationship was quasi-linear. A reaction-time theory of the error was based on the observed magnitude of the aided-tracking time constant (Mechler, Russell, and Preston, 1949). Searle (1951) and Vince (1948) thought that this reaction time or delay was produced by intermittency in perception. Others (Fitts, 1951; Birmingham and Taylor, 1954) assumed that the response time in error control was produced by machine delay.

Feedback analyses of the effects of intermittency of visual feedback on reproduction of learned visual patterns indicated that there are many types of such interruption of sensory feedback signals in tracking. These include periodic blanking, averaging, displacing, extraneously interrupting, differentiating, and transforming such signals (Smith and Sussman, 1969). There is no general theory of intermittency or sampled data that covers all these patterns of sensory input.

These theoretical analyses of tracking have served to direct interest to many of the time-defined aspects of tracking behavior, especially those related to feedback delay. However, they have had limited relevance to research designed to study other characteristics and conditions of tracking. They are inapplicable to some of the problems we discuss in later sections, for example, questions regarding similarities and differences between vigilant tracking of continuously moving targets and steering behavior, and questions regarding the integration of different movements, such as postural, travel, and manipulative movements, in tracking functions. Experimental analyses of the parametric variations in the specialization, integration, and organization of movements and their physiological substrates, and feedback control by the motor system, have advanced far beyond early theoretical formulations of machine behavior and its human-factors requirements.

2.9.1.6 Assessing Control-System Response

One major contribution of engineering and mathematical approaches to control-system analysis has been the demonstration that dynamic evaluation of the response characteristics of behavioral and physiological control systems has marked advantages over steady-state analysis, in that it offers improved signal-to-noise benefits, and it also reveals kinetic differences in the response characteristics of various elements of the control system. These advantages have spawned the successful application of various methods of dynamic analysis to study of physiological control processes. These methods appear to have general applicability to characterizing control-system response dynamics, although they have limited usefulness for deciphering the actual control mechanisms themselves.

Dynamic analysis of control-system response is based on applying various stimulus patterns, or forcing functions, to the system and assessing the response kinetics. Forcing functions commonly used include the step, the sinusoid, the ramp, the impulse, and the pseudo-random binary sequence (PRBS). Typical applications include assessing: (1) ventilatory response to step, ramp, impulse, and PRBS exercise work loads (Bennett, Reischl, Grodins, Yamashiro, and Fordyce, 1981; Whipp and Ward, 1980); (2) ventilatory response to step and sinusoid patterns of inhaled CO_2 (Bellville, Fleischli, and Defares, 1969; Grodins, 1963); (3) blood gas and pH responses to step changes in inhaled carbon dioxide (T. J. Smith, 1977); and swimming performance responses to sinusoidally varying training impulses (Calvert, Banister, Savage, and Bach, 1976). Response-time constants established with these studies range from a few seconds for the first phase of the ventilatory response to exercise to 15 days for the fatigue response and 50 days for the second phase of the swimming performance response to training. The PRBS study of Bennett and colleagues (1981) was particularly effective in resolving the rapid component of the ventilatory response to exercise, with the established time constant of 7 sec implicating a neural pathway in exercise hyperpnea. Generally, the established success of these dynamic analysis methods argues for their routine incorporation into behavioral and physiological control-system research.

2.9.2 BEHAVIORAL CYBERNETICS AND MACHINE DESIGN

By about 1960, the various analyses and studies of feedback control in living systems had set the stage for a broad-based research program in behavioral cybernetics, especially in relation to tool using and operation of machines. The problems that gave direction to this research were defined in terms of accepted knowledge in several areas: (1) the distinctions made by engineers concerning the three component sectors of machine function, that is, control components such as levers, push buttons, or wheels; actuator or power components for magnification, reduction, integration, or transmission; and slave or operational components that cut, smash, grasp, move, turn, track, and so on; (2) the need to consider systems factors of motor coordination and integration, especially of postural, travel, and manipulative movements, in specifying the characteristics of machine behavior; (3) the necessity for spatial and temporal compliance between machine components and motor-sensory feedback interactions involved in machine operation; and (4) the recognition that most machine and tool operations, with minor exceptions, involve either vigilant tracking or projective steering behavior.

2.9.2.1 Multidimensionality in Body Motion

A program of development of several types of manlike anthropomorphous machines encountered a problem that could not be resolved by applying limited servomechanism concepts (Mosher and Murphy, 1965). The problem arose particularly in the design of a walking machine consisting of a cab to hold the operator, fitted with 12 ft articulated steel legs with feet attached (Figure 2.9.3). It was intended that when the operator moved his legs, the legs of the machine would move compliantly. The original design of the machine provided for force and position compliances between the sensed action of the operator's legs and feedback from the action of the machine legs, but with this design the operator could not balance the machine or walk up and down on an incline.

A complex motion pattern such as walking involves several movement systems of the body. The basic system provides *postural support* by means of stabilizing movements of the large muscles of the body. Movements of the legs in walking—as of the arms in throwing—are described as *travel* movements, because they transport the limbs freely through space. The third main type of movement upon which skills depend is described as *manipulative* or *articulative* movements. These are the finely controlled movements of the hands and fingers, movements of the mouth and face in speech, and—in walking—the adjustments of the feet and toes on the walking surface. According to cybernetic theory, bodily motion patterns are integrated by means of intrinsic feedback circuits relating the various movement systems, not by chemical states of inhibition and excitation at central synapses, as postulated by Sherrington (1906).

Recognizing the need for integrating postural, travel, and manipulative movements, and for spatial compliance of movement and visual feedback in performance, Smith (1966c) advised that the walking machine would have to sense the relation between leg movement and torso position on a continuous basis, and that the operator would have to have spatially compliant feedback of this torso–leg relationship in two dimensions. When spatially compliant integrative feedback was incorporated into the design of a prototype of the walking machine, the operator could articulate the legs to walk and maintain balance.

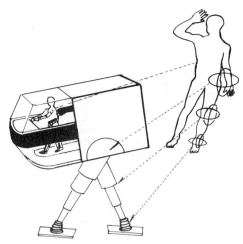

Fig. 2.9.3. Manlike (anthropomorphous) walking machine or pedipulator of Mosher (Mosher and Murphy, 1965). The design and operation of this machine was a first concrete test of servomechanism versus multiple-feedback control theory applied to coordination of movements. The machine is a prototype of all anthropomorphous robots.

2.9.2.2 Multidimensional Feedback in Tool Using

In tool and machine behavior, the operator integrates his movements by utilizing various kinds of feedback from a number of sources, both intrinsic and extrinsic to the body. From the operation itself comes *reactive feedback* from control operations, that is, the self-generated visual, auditory, tactual, kinesthetic, and proprioceptive information concerning the operator's own movements in adjusting controls; *instrumental feedback,* that is, the information sensed by the operator concerning the actions of the machine; and *operational feedback,* that is, the feedback that the operator gets from the persisting effects of machine operations on the environment. For example, in the familiar operation of handwriting, the writer gets reactive feedback from his own movements, instrumental feedback from movements of the pen or pencil, and operational feedback from the written words (Figure 2.9.4).

For optimally efficient machine operations, properties of the three main sectors of machine function must comply spatially and temporally with the distinctive spatial and temporal properties of reactive, instrumental, and operational feedback. Further, the machine must be designed and operated to permit optimal integration of these compound motor-sensory feedback interactions which are related to the three sectors of machine action. This integration typically involves coordination of postural (related to overall machine action), travel, and manipulative movements, and of bilateral arm, leg, head, and eye movements.

The multidimensional feedback concept also can be applied to design of hand tools. In tools such as a pliers or screw driver, the control element is the handle, the operational element is the working end of the tool, and the actuator is the mechanical leverage provided in the linkage between the handle and the operational component. The principles of spatial and temporal compliance of reactive, instrumental, and operational feedback determine the optimal human-factors design and characteristics of such tools.

These principles were tested in limited ways by devising television techniques, illustrated in Figure 2.9.5, to control reactive, instrumental, and operational feedback separately in tracing and writing tasks (Smith and Smith, 1966). As the subject traced a line between points arranged in a circle, he could not see his hand or the tracing tool directly, but viewed them on a closed-circuit television monitor. The critical technique was to use special lighting, markers that were either black or white, and black or white gloves for the subject's hand. The lighting was such that the subject could not see a white marker or the subject's own white-gloved hand. Thus it was possible to eliminate visual feedback from the tracing tool or from the hand of the subject. (Kinesthetic and tactual reactive feedback could not be eliminated.) Operational feedback could be eliminated by using a marker or stylus that left no trace. In an experiment comparing accuracy of tracing under various feedback conditions, it was found that performance was best when subjects received all three types of visual feedback, reactive, instrumental, and operational. Limiting visual feedback to one of these decreased accuracy to similar levels for the three conditions.

A second study used the same lighting and masking techniques to vary the amount and source of

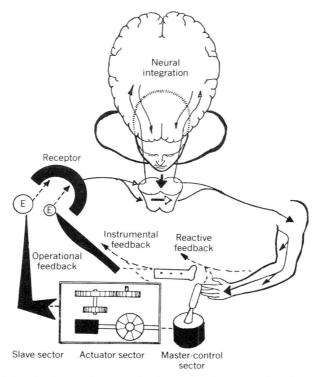

Fig. 2.9.4. Spatial feedback compliances needed in reactive, instrumental, and operational feedback in the operation of tools and machines.

Fig. 2.9.5. Laboratory closed-circuit television setup used to vary the reactive, instrumental, and operational feedback of target-tracing and location motions. By use of special lighting, black and white gloves, and black and white tracing pencils, any component of the performance—hand of the subject, tracing tool (as shown), or trace—could be eliminated from the subject's vision.

instrumental feedback in using a marker to trace through successive gates of a visual maze. Instrumental feedback was varied so that the subject could see either the tip of the marker, the cone including the tip, the shank, cone, and tip of the marker, or a narrow band on the shank some 3 inches above the tip. Error-time scores were lowest with the most complete instrumental feedback, the shank–cone–tip condition. They were markedly and significantly highest for the condition in which instrumental feedback was limited to sight of the narrow band on the shank of the marker. Scores were significantly higher for the tip-only condition than for the cone-tip condition. In other words, performance was best with the most complete instrumental feedback, but with limited feedback, it was better when the source was the operational tip of the tool than when it was a band displaced from the tip.

A series of studies of handwriting explored the relationship of its timing and legibility to the nature of feedback from the writing operation. It was found that optimal legibility depends on an instrument designed to provide optimal instrumental and operational feedback (Smith, Kao, and Knutson, 1969).

2.9.2.3   Effects of Feedback Displacement

Many tool and machine operations are performed under conditions that displace or distort visual feedback in various ways. For example, using a mirror to guide one's movements produces a condition of spatially displaced feedback. The experiment comparing four conditions of instrumental feedback was repeated under conditions of reversed and inverted visual feedback—conditions that were produced readily by manipulating the televised image of performance. It was found that error-time scores were significantly higher with the displaced visual feedback than with normally oriented feedback for all four conditions. Further, error-time scores were significantly higher with inverted feedback than with reversed feedback for all four conditions. These results along with many others show that the efficiency of tool-using performances depends not only on the type and amount of feedback, but also on the spatial relationships between feedback and movement. When a condition of spatial displacement introduces a noncompliance between perception and motion, performance is affected adversely.

The studies described so far have analyzed performances guided and controlled by continuous dynamic visual feedback from the actions involved. Because in some machine behavior visual feedback may be restricted to the static aftereffects of the operation, an experiment was designed to compare the effectiveness of dynamic visual feedback of motion with that of static visual feedback from a completed operation (Smith and Smith, 1962). The task was to mark a pattern of dots in two concentrically arranged circles. In the dynamic condition, the subject saw the action of his hand, of the marker, and of the dots made. In the static condition, he saw only the completed pattern of dots at the end of performance. Both performance accuracy and learning speed were markedly and significantly superior with dynamic feedback. Static information about aftereffects of performance—what has long been known in learning psychology as knowledge of results—is a type of delayed feedback that may or may not serve to guide movement and to produce learning.

The results of these various experiments indicate that tool and machine operations are more efficient when adequate reactive, instrumental, and operational feedback are available, when the various sources of feedback are spatially and temporally compliant, and when they provide continuous dynamic information concerning the operation in progress.

2.9.2.4   Steering Behavior

Early tracking theories made no distinction between vigilant tracking of a moving target, which requires continuous detection of target action, and steering of machines and vehicles within the confines of a stationary path or a specified direction of movement. In the belief that tracking and steering are distinctively different performances, a research program was launched to study some of the characteristics and problems of driving a car.

One experimental setup closely simulated an actual driving situation to investigate effects of spatially distorting visual feedback to the driver (Kao and Smith, 1969). Many visors or windshields distort the driver's view of the road, especially in the case of a short driver in a large car, who views the road from a low position. To assess the effects of such spatially distorted feedback, closed-circuit television methods were devised to displace angularly the visual image of the road ahead.

Figure 2.9.6 illustrates how a television camera fixed to the roof of an experimental station wagon could be varied in position from a center point on the front edge of the roof to a far left and a far right position. In each case the camera gave an image of the roadway in front of the car. The driver's direct vision of the road was blocked; instead, the driver viewed a television monitor mounted on the windshield directly in front of him or her. Subjects, who were experienced drivers, drove this vehicle on an S-shaped roadway 55 ft long, marked with red pylons every 5 ft. Touching a pylon was scored as an error. Results showed that steering performances with the left and center camera positions, which represented only limited space displacement of visual feedback to a driver seated in a car with the steering wheel on the left, did not differ significantly in accuracy. By contrast, the right camera position produced a highly significant increase in driving errors. The findings indicated

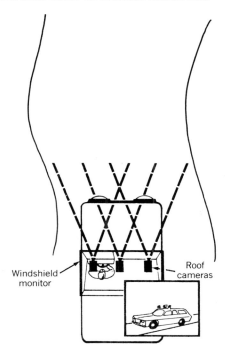

Fig. 2.9.6. How closed-circuit television was used to displace angularly visual feedback of the road ahead in steering an automobile (based on Kao and Smith, 1969).

that although drivers can tolerate limited angular displacement of feedback, their performance deteriorates as the displacement progresses beyond a critical angle.

A second setup that simulated driving a car was devised by combining a real-time, analog–digital–analog computer control system with optical, closed-circuit television, and videotape instrumentation (Smith, 1971b). The subject-driver was located in a Volkswagon bucket seat and watched a moving road display on a television monitor (Fig. 2.9.7). Horizontal position of the road display could be controlled by aligning it with a car-hood marker centered at the bottom edge of the television screen. The computer system was programmed to produce changes in the road display simulating progressive changes that might occur as a car moves along a road. Also, the computer measured driver error in aligning the car-hood marker with a center line in the moving road display.

The computer system also could be programmed to introduce feedback delays between the driver's adjustment of steering wheel and resulting change in position of the road display. Delays used were equivalent to the steering-feedback delays that can occur in actual cars and trucks due variously to powered steering, momentum resulting from high speeds, skidding, and loose mechanical linkages between the steering column and the front wheels. Although it is nearly impossible to measure the effects of such delays on accuracy of steering under actual driving conditions, the computerized steering setup made such measurement possible.

In a first experiment, the effects of three steering delays were measured (Smith, Kao, and Kaplan, 1970). Nine male licensed drivers were divided into three groups, and each group practiced the steering task for 4 days with a 0.0, 0.2, or 0.4 sec feedback delay. In 10-trial tests on the fifth day, the group that had practiced with zero feedback delay showed a 50% reduction in steering error. The groups that had practiced with delayed feedback, however, showed little or no learning. Steering error increased progressively with delay magnitude, with the 0.4 sec delay producing roughly twice the level of error as that shown in the zero-delay condition. On the other hand, steering behavior was not affected as severely as vigilant target tracking by comparable feedback delays.

The computerized steering setup was used to study coordination of manual steering movements and eye movements under variable conditions of feedback delay (Putz and Smith, 1970a, 1970b, 1970c; Smith, 1971b). Driver eye movements were registered with a photoelectric transducer that did not interfere with vision of the road display. The computer system processed signals from both manual and eye movements and displayed them on an oscillograph. One study revealed that with zero feedback delay, manual steering movements more frequently than not led or anticipated the direction of movement

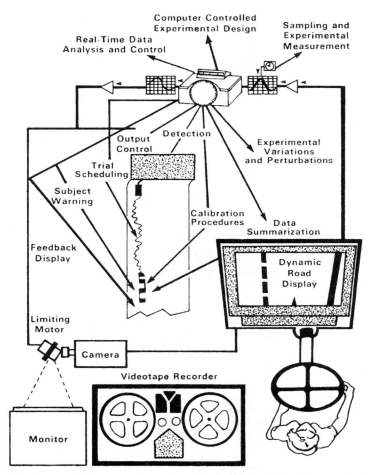

Fig. 2.9.7. Laboratory real-time-computer, videotape, closed-circuit-television, mechanical-optical setup used to produce visual feedback delay of steering movements and perceptual feedback delay in simulated vehicle steering (based on Smith, Kao, and Kaplan, 1970).

of the road display, whereas the eye-tracking movements were correlated in time with the road-display movements. With steering-feedback delays, this anticipative or feedforward control by the manual steering movements was lost; with feedback delay, the manual steering action consistently lagged movement of the road display and corresponding eye-tracking movements. This finding confirmed the view that predictive steering is different from vigilant tracking of unpredictable targets. The two behaviors show distinctively different patterns of movement integration and feedback control.

2.9.2.5 Feedforward Control

To investigate the differences between projective steering and vigilant tracking and steering, the driving simulator was modified to permit variation in the extent of exposure of the road display. The section of the display immediately above the car-hood marker on the bottom edge of the monitor screen was hidden to produce a *perceptual delay,* as contrasted to a steering-movement feedback delay, when a delay is introduced between steering action and visual feedback of the effects of that action. In the condition of perceptual delay, the driver could not see the road immediately in front of the car but had to view the road farther ahead. Thus feedback concerning the accuracy of steering movements in following the road was delayed, for the position of the car always lagged behind the part of the road that could be seen (Smith, 1971b).

An experiment compared the effects of perceptual delays with vigilant steering-movement delays of the same magnitudes. Vigilant steering was produced by giving subject drivers only a small extent of the road display immediately in front of the marker that had to be aligned with movement of the display. Perceptual delays in projective steering were varied by hiding more or less of the road immediately in front of the driver. Steering-movement delays in vigilant steering were varied by introducing specified delay intervals between the driver's movements and the effects of those movements on the televised display. Feedback delays of 0.0, 0.2, 0.4, 0.6, 0.8, and 1.0 sec were used for both projective and vigilant steering. The results showed that small perceptual delays of 0.2 and 0.4 sec did not degrade steering half as much as vigilant steering delays of the same magnitudes. With larger delays, however, projective steering was degraded to the same degree as vigilant steering. Perceptual delays exceeding 0.2 sec produced a significant increase in saccadic (fast corrective) movements of the eyes, which occurred only infrequently with zero delay.

The findings show clearly that the servomechanism model of vigilant tracking does not apply to projective steering. Steering of vehicles in a projected path or roadway differs in movement integration and control from vigilant steering and tracking, in which predictive control of movement can occur in only a limited way. The result most significant for human-factors practices in machine and driver-training design is that coordination of manual steering and eye movements is affected differently under the two types of guidance, and with different feedback delays in each type of guidance.

Many simpler instrumental behaviors are controlled in a projective or predictive way. Handwriting, for example, is understood best as a projective steering process involving integration of postural, travel, and manipulative movements.

The integration of complex motion patterns involves predictive or feedforward control based on tracking–feedback relations between movements in *body-movement tracking*. As suggested in Figure 2.9.8, the sensory (visual, auditory, tactual, kinesthetic, or proprioceptive) change produced by a first movement is controlled as feedback by a second movement that in turn generates a compliant sensory input controlled as feedback by the first movement, and so on. The timing between such intermuscle tracking movements is on a split-second basis (Sperry, 1939; Stetson, 1905, 1951; Stetson and Bauman, 1933; Stetson and McDill, 1923).

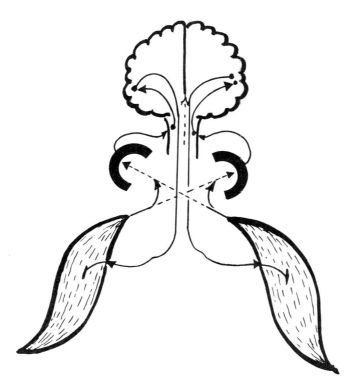

Fig. 2.9.8. Principle of crossed motor-sensory feedback control in body-movement tracking. The immediate sensory or sensory-neural feedback of one body movement is detected and tracked by a second movement as feedback. Because the sensory neuromotor mechanisms of the second movement can predict or project the space pattern of the first movement, the feedback timing of the coordination can be on a split-second basis.

Using the laboratory computer system, experiments were designed to measure the systems interactions of body-movement tracking (Smith and Arndt, 1969). One experiment found that in simulated swimming movements performed by expert swimmers, respiratory movements track and anticipate arm movements (A. Luetke, cited by Smith and Smith, 1970). Studies of tracking and vehicle steering showed that eye and hand movements are coordinated by eye-tracking of targets and road displays, changes in which are controlled by arm movements (Putz and Smith, 1970a, 1970b, 1970c; Smith, 1971b). In binocular eye-movement tracking, it appears that the subordinate eye tracks movements of the dominant eye; the response of one eye lags the other by some hundredths of a second (Schmidt, Putz, and Smith, 1970; Smith, Schremser, and Huang, 1970; Smith, Schremser, and Putz, 1971).

2.9.2.6 Safety in Vehicle Operation

Laboratory analyses of steering indicate first that safety considerations in vehicle design should take into account the differential effects of both perceptual and steering-movement delays in projective steering. Delayed steering-movement feedback may represent the deadly split second responsible for loss of vehicle control in emergency stituations. This human factor is especially significant for the inexperienced or aging driver. Very long feedback delays in the steering systems of heavy industrial mobile machines, aircraft, and ships make it necessary to require very specialized training of operators on the specific machines.

A second safety consideration in vehicle operation is how machine feedback from sudden jerks, skids, turns, and so on affects the stability of the operator. An analysis of the safety problems of mobile industrial machines and vehicles, such as cranes, lift trucks, diggers, and cars, showed that a major safety and hazard-control feature in the design of the vehicles is the provision made for controlling the posture and related steering movements of the operator (Smith, 1979). Seat belts on cranes, trucks, and cars, in addition to protecting driver and passengers from collision injury even more importantly help maintain the posture of the driver and thus aid steering movements in emergency situations. Machine feedback that affects the stability of the driver combined with delay in steering feedback can transform an emergency into a disaster.

2.9.3 DELAYED SENSORY FEEDBACK

We have described a number of experiments and observations demonstrating the deleterious effects of feedback delay on instrumental or machine behavior. Many other studies have shown that other, unaided, behaviors are similarly affected. Lee (1950a, 1950b, 1951) opened this challenging new area of research by rigging up a dual audiotape device that both recorded the speech of a subject and played it back to his ears via electrical transmission. The subject wore headphones which, in addition to conveying the recorded speech sounds to his ears, also prevented him from hearing his speech via air transmission of sound. With this setup, slight delays were introduced between the subject's speech movements and his auditory feedback from those movements. The effects were dramatic. In almost all subjects, delays of about 0.2 sec caused drastic speech disturbances—stammering, pauses, errors, and sometimes complete blocking.

Lee's research initiated many diverse research studies of the effects of delayed feedback on both instrumental and unaided behavior. As described graphically in Figure 2.9.9, several methods were devised for such research and applied to tracking, steering, handwriting and graphic behavior, posture, head movements, and other patterns of behavior (Smith, 1962). The most versatile laboratory technique was a computer system comparable to that represented in Figure 2.9.7. This type of system permitted diversified research on the split-second time relations characteristic of the feedback-control circuits coordinating body movements (Smith, Myziewski, Mergen, and Koehler, 1963).

When the adverse effects of machine-induced delays first were noted in tracking operations, the phenomenon was thought of only in relation to machine behavior. The growing body of research showing that delayed sensory feedback degrades the organization and effectiveness of all kinds of body movements brought about the realization that behavior of all kinds is feedback controlled. Research on delayed feedback provides the most convincing evidence for this central concept of behavioral cybernetics: that behavior is controlled throughout by dynamic feedback relations between movement and the feedback it generates (Smith, 1972b, 1973; Smith and Smith, 1973; Smith and Smith, 1970). Behavior systems are closed-loop systems. They do not respond to environmental stimuli in an open-loop manner, but control their own actions by means of self-generated feedback.

2.9.3.1 Effects of Feedback Delay

The main research findings on the effects of delayed sensory feedback are summarized in the following paragraphs. First, it has been found that all motor-sensory mechanisms are degraded to some extent in accuracy, timing, and integration by the introduction of feedback delay.

Feedback delays adversely affect the integration of more complex movement coordinations, such

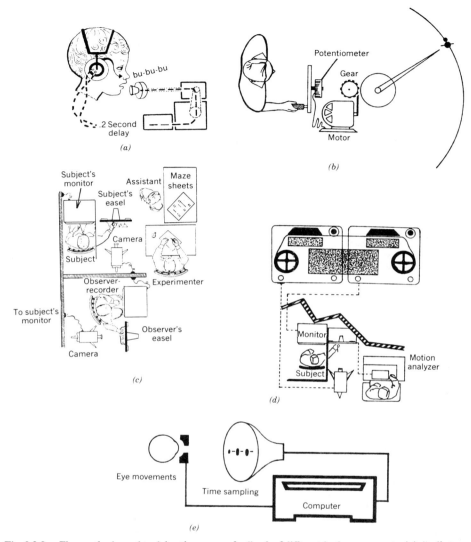

Fig. 2.9.9. Five methods used to delay the sensory feedback of different body movements. (*a*) Audiotape recorder equipped with a delay loop. (*b*) Powered (aided) and damped tracking systems. (*c*) Cross-linked dual closed-circuit television setup. (*d*) Videotape recorder equipped with a delay loop. (*e*) Analog–digital–analog computer setup.

as bilateral movements, just as they degrade specific motor-sensory interactions, indicating that different motor-sensory mechanisms are coordinated by crossed body-movement tracking interactions.

Some motor-sensory mechanisms, including speech, singing, and bilateral movement coordinations, show peak disturbances with delays of specific magnitudes. By contrast, other motor-sensory mechanisms are degraded progressively as a function of magnitude of delay interval. The speech of young adults is disturbed most severely with auditory feedback delays of about 0.2 sec, and less severely with shorter or longer delays. However, the delay interval causing maximal disturbance to speech is usually longer than 0.2 sec for both children and older adults, indicating that feedback timing is an intrinsic aspect of motor-sensory interactions that changes with age. As a result of an intrinsic increase in closed-loop timing, older adults show maximal disturbances of speech with auditory delays somewhat longer than the peak delays found for young adults.

As shown in studies of projective steering, movements involving predictive control are affected less severely than vigilant activities by delays of small magnitudes. Discrete or repetitive learned activities may be affected very little if it is possible to span movement-to-movement intervals by memory, as coded in predictive body-tracking relationships.

In tasks requiring vigilant tracking or steering, little or no learning occurs with practice under conditions of feedback delay. Projective steering with short delay intervals can be improved with practice, but learning tends to be unstable.

Complex, precise movements are affected more by feedback delay than less complex and precise tasks.

When feedback delays occur in combination with other perturbations, such as spatial displacement of feedback, all major aspects of motor control, perception, and integration are affected more severely than by delay alone. This finding is especially important to human-factors design of machines which may introduce feedback delays along with spatial noncompliances (Putz and Smith, 1971).

All machine and tool performances that have been studied are affected adversely by feedback delays, including those in which delay is introduced by powered steering or tracking. Machine-produced feedback delays combine with intrinsic response times in controlling environmental situations to define the oscillatory nature of tracking and steering errors.

2.9.3.2 Delayed Feedback as a Human-Factors Issue

The significance of motor-sensory feedback delay is still a central human-factors issue because of the influence of operant-learning theory in psychology and human-factors engineering. Operant-learning psychologists always have equated sensory feedback with reinforcement—the type of aftereffect of a response that provides reward or punishment in a learning situation. Thus, from the first disclosures about delayed feedback, most learning psychologists have considered it comparable to the reinforcement interval in operant learning, that is, the time between a response to be learned and presentation of the reinforcement. This misconception still exists even though it is well established that, although learning may occur with delayed reinforcements, significant intervals of feedback delay in motor-sensory operations impair learning or make it impossible. General summaries of the critical differences between reinforcement and feedback have described distinctive differences between the effects of feedback delay and reinforcement delay (Ammons, 1956; Smith and Smith, 1966). As K. Smith (1954) first pointed out, the alleged learning effects of delayed operant rewards and punishments are artifacts of direct physiological feedback effects of the motor activity, either of the primary response to be learned or of the so-called reinforcing responses.

Basing human-factors design of machines, performances, and complex operating systems on the specifications of operant-learning theory is a simplistic approach and doomed to failure. For optimal efficiency and safety, it is absolutely essential to design such systems according to cybernetic principles of maximizing effective feedback and minimizing feedback delay.

2.9.3.3 Control Theories of Fast Movement Coordinations

Research on delayed feedback is especially significant in assessing the various theories proposing that motor skills are not controlled by sensory feedback but are programmed by the brain (Arbib, 1981; Keele, 1981, 1973; Kelso, Holt, Kugler, and Turvey, 1980; Kelso and Stelmach, 1976; Landers and Christina, 1978; McCloskey, 1981; Poulton, 1981; Schmidt, 1976; Stelmach, 1978; Stelmach and Requin, 1980). In view of the speed of some skill patterns, such as serial movements of the fingers in piano playing and typing, coupled with other evidence showing that skilled movements persist despite sensory deafferentiation of the moving limb, some theorists hold that sensory feedback is not needed for their timing and control (Keele, 1981, 1973).

Although it is next to impossible to test the concept of central brain programming experimentally, the feedback-control concept has been tested repeatedly, with many different experimental disigns, and for many different movement coordinations, including piano playing. Specific experiments have demonstrated that piano playing and other even more refined musical instrumentation, such as flute and violin skills, are degraded sharply by computer-controlled auditory feedback delay (Ansell and Smith, 1966; Smith and Smith, 1966). After a brief initial exposure to auditory feedback delay, highly talented violin, piano, flute, and cello players absolutely refused to participate further in the experiments. Once they discovered the effects of such delay on their performances, they felt that the experience would perturb their skills indefinitely. Regarding eye–hand coordinations, such as typing, it should be noted that any task involving visual monitoring depends specifically on eye-movement control of visual feedback. It has been demonstrated in a variety of computerized experiments that delaying the visual feedback of eye movements impairs all forms of eye-movement tracking (Coleman, Ruff, and Smith, 1970; Smith, Putz, and Molitor, 1969; Putz and Smith, 1970b; Smith, 1970).

The basic premise of the motor program concept is that movements may be initiated and guided under efferent control of a prestructured motor program without reference to afferent input from the periphery. Proof that complex tasks such as piano playing and typing are programmed by the brain independently of sensory feedback would require experiments showing that they are unaffected by delayed motor-sensory feedback, and also are unaffected by displaced visual feedback. We have shown experimentally, however, that all graphic tasks—drawing, writing letters, writing words—are radically affected in performance (accuracy or legibility) and learning, not only by delayed visual feedback, but also by inversion, reversal, combined inversion and reversal, and angular displacement of visual feedback (Smith, 1962; Smith and Smith, 1966; W. M. Smith, McCrary, and Smith, 1960), and that the fine manipulative movements are affected most. Research on displaced vision is discussed in more detail in the next section, as is research showing that reversal of the visual feedback of eye movements makes visual fixation impossible.

In our opinion, this research on the adverse effects of temporally delayed or spatially perturbed visual feedback on motor performance and motor learning poses a fundamental difficulty to the advocates of motor programming, most of whom have ignored its implications (the research is not cited by Arbib (1981), Keele (1981), McCloskey (1981) or Poulton (1981), nor in earlier work by Keele (1973), Kelso and Stelmach (1976), or Schmidt (1976)). Whatever neuronal modifications may take place in the learning of a skilled motor act, the fact that the performance of this act, as well as the learning process itself, are severely impaired by even slight perturbations in normal temporal and/or spatial feedback relationships suggests that peripheral motor-sensory and central brain mechansims subserving motor control are tightly and reciprocally coupled or yoked in guidance and execution of dynamic movements. Conceptual approaches that assume that, under some conditions, the motor-sensory substrates of motor behavior may be subsidiary or derivative to central mechanisms thus disregard the tight, reciprocal integration of muscle, receptor, and brain center activity, which represents the key to both feedback and projective features of dynamic motor control.

As has been shown in the experiments on automobile steering discussed earlier, the nature of feedback control is specialized in tasks that are to some extent projective or predictive, in which case the perceptual-motor system is able to span a limited interval of time in projecting its control processes (Rack, 1981; Roland, 1978). This is not to say the predictive patterns of movement are programmed in the brain; rather, we would say that predictive control depends largely on body-tracking relationships that are mediated by intrinsic proprioceptive, kinesthetic, and tactual receptor systems.

Body-movement tracking undoubtedly occurs in the cross-linking of finger movements in manual skills and the coordination of lip, tongue, cheek, and jaw movements in speech. Finger movements of each hand can be controlled separately in split-second timing both by direct tactual control and by body-movement tracking of one finger by another. The thumb probably leads in most finger tracking, and the other fingers track the thumb motion tactually and proprioceptively. Such linked tracking coordinations probably can reduce serial finger-movement timing to the tremor interval of about 0.1 sec or less. If the fingers are used serially with a rocking movement of the hand, as in a fast run on the keyboard, projective proprioceptive control by the hand produces successive finger actions at intervals of perhaps a few hundredths of a second. Much the same type of linked tracking probably occurs between the big toe and other toes in taking off in the stride in running and walking. The fact that the thumb, big toe, and lip are projected so prominently on the somesthetic cortex suggests that these mechanisms have evolved as lead body-movement tracking components for other finer digital and face movements.

As noted above, the underlying assumption of the brain programming models of movement coordination is that movements may involve their own closed feedback control before learning, but that learning establishes a brain program which permits feedforward control of a coordinated pattern without the necessity of movement-specific feedback. Conclusive evidence supporting the validity of this model is not available. However, Sussman and Smith (Smith and Sussman, 1969; Sussman and Smith, 1969, 1970b) tested the role of feedback factors in learning and short-term memory by varying the persistence of a visual pattern to be learned. A curving line was produced on a moving oscillograph record by a moving spot, and the persistence of the pattern was varied from simple exposure of the moving spot to exposure of the complete line. The accuracy of subjects asked to reproduce the line pattern varied as a function of the persistence of the perceptual pattern. Similar results were found whether the dynamic line pattern was learned by observation or by manually tracking the line as it appeared on the record, although accuracy of reproduction was somewhat superior with the observation group. Learning and short-term memory thus were shown to be a function of the persistence of sensory feedback controlled either by eye tracking or manual tracking movements.

In these experiments, subjects reproduced the visual pattern by operating a hand control to generate a moving line. When visual feedback of these reproductive movements was delayed, accuracy of reproduction was degraded for both the observation and manual tracking groups. The degree of impairment varied as a function of the magnitude of the feedback delay. Neither perceptually learned nor manually learned space-related movements lost their dependence on feedback regulation.

The studies discussed above are based on dynamic systems measurement of intermotor and motor-sensory coordinations, which defines a new area of research that holds the promise of being directly relevant to human factors and ergonomic designs for skilled motor performance. The foregoing experiments and exploratory observations point up the wide variety of feedback-control relationships that can be analyzed in controlled studies. Two or more body movements can be matched in parallel tracking relationships. The same movements can be coupled serially in a linked relationship, in which one movement provides a tracking signal to the second and vice versa. One movement can body-track a second in a compensatory or negative-feedback relationship. Groups of movements can be combined to track the sensory effects of a single movement or the combined effect of another group of movements. One movement can be made to track the differential relationship between two other movements. Or one movement can be made to add to, multiply, or elaborate temporally or spatially the sensory effects of a first movement. In substance, the systems integration of body movements by means of intermuscular feedback can be studied and understood without recourse to open-ended, abstract concepts of cognitive programming by central brain states, but based instead on theory and experimental research on body-movement tracking, which puts analysis of muscular integration on a concrete objective basis subject to well-defined experimental designs.

2.9.4 NEUROGEOMETRIC ORGANIZATION OF BEHAVIOR

More than a century ago, Helmholtz (1856–1866) wrote a massive tract, *Physiological Optics,* on human vision and space perception that exerted a major influence on psychological theory and practice. A key part of the book consisted of an empirical theory of space perception, the view that space perception is learned through the association of visual and movement experience. Helmholtz's main evidence concerned the effects of and adjustment to space-displaced vision, as related to corrected strabismus and other visual defects. Numerous psychologists beginning with Stratton (1896, 1897, 1899) have investigated space-displaced vision experimentally, usually by having subjects wear telescopic spectacles, prisms, mirrors, or other optical devices to displace the subject's view of the environment and his or her own movements. The main interest was to determine to what extent space perception is learned and thus independent of other developmental factors. If space perception and movements in space are entirely learned, then a person should be able to learn to perform tasks under any spatial arrangements imposed by machines or other environmental conditions.

A detailed analysis of the methods and results of all research on space-displaced vision up to 1961 (Smith and Smith, 1962) found that subjects made some adjustment but not full adjustment to displacements, even after long periods of exposure. Different types of displacements (angular displace-ments, inversion, reversal, combined inversion and reversal) produced movement disturbances of varying severity, followed by varying degrees of adjustment.

One prominent example of such research, carefully evaluated by Smith and Smith (1962), is the reversed, inverted, and angularly displaced vision experiments of Kohler (1955), which often are cited as having resolved the basic question of learning versus maturation in the development of space percep-tion. In one study, goggle-mounted prisms were used to reverse the vision of two subjects (one of whom was Kohler himself) for 24 and 37 day periods. In a second study, a mirror affixed to a headband and mounted horizontally above the eyes was used to invert the visual field of three subjects, for 6, 9, and 10 day periods. Finally, goggle-mounted prisms also were used to displace angularly the vision of a series of subjects by 5 to 30° for periods of time ranging from 5 to 124 days. From these studies, it was concluded that human subjects can adapt completely to spatial perturbation of the visual field, that such adaptation occurs naturally through learning as a matter of course, and that perceptual reorientation is a generalized phenomenon encompassing all motor-sensory modalities.

The validity of these conclusions, and the importance attached to the work by subsequent investiga-tors, can be criticized from a number of standpoints. In making his observations, Kohler (1955) relied almost exclusively on introspective—subjective—reports by subjects of the perceived effects of distorted vision. Few controlled, objective studies of motor performance effects were conducted. Considering the limited number of subjects used, and the limited number of critical observations made under some experimental conditions, one cannot be sure to what extent subjective bias influenced the conclu-sions. Secondly, the equipment used to distort vision apparently introduced confounding visual perturba-tions that were disregarded in the analysis. For example, the reversing prisms also reversed depth cues, and neither of the two reversed-vision subjects apparently ever corrected for depth disorientation. The inverting mirror could be viewed only through a slit which limited the field of view. Because the mirror was head-mounted, the environment moved with every movement of the head, but not when the eyes alone moved without head movement. The combined effects of the mirror in inverting vision and in causing the environment to move with head movement created nausea in some subjects. Generally speaking, Kohler's analysis (1955) reveals a disturbing lack of awareness of the spatial variables involved in various modes of visual displacement.

Finally, the reports by Kohler (1955) of his visual disorientation experiments contain a number of apparent discrepancies, nonspecific generalities, and examples of incomplete reporting. The main factual question raised by the research is to what extent subjects established relatively normal and efficient visuomotor coordination and perception under conditions of inverted, reversed, or angularly displaced vision. A careful reading of the reports reveals that complete adaptation was not demonstrated clearly in any of the experiments on inversion and reversal. Neither of the two reversed-vision subjects was reported unequivocally to have achieved complete adaptation. Despite lack of specific comment in the paper, it can be inferred that two of the three mirror-inversion subjects did not achieve complete perceptual reorientation. For the angular-displacement subjects, the specific experimental observations suggest that none completely adapted to displacements exceeding 15°, although the results are interpreted otherwise. The specific observations also suggest that the perceptual reorientation achieved with any of the conditions was not generalized but highly specific to particular movements.

From their detailed review of the displaced vision experiments of Kohler (1955) and others, Smith and Smith (1962) conclude that the idea that adjustment to displaced vision involves general reorientation of the perceptual world is untenable, in that adjustment appears to be specific to relearned movements, just as learning of movements and skill patterns under normal visual conditions likewise is very specific. Also, adaptation to displaced vision depends on activity-controlled sensory feedback. Subjects adapt more effectively through active movement than by being moved passively (Held and Hein, 1958). Learning with displaced vision is highly unstable; any stress or other disturbance degrades the level of adjustment already achieved. Readjustment to normal orientation of the visual field is rapid and requires no practice, indicating again that adjustment to displacement is task- and movement-specific. Finally, most of the experimental methods and techniques were judged to be poorly designed, so that the results were subject to interpretations other than the conclusions reached by particular investigators.

The great variation of performance and learning with different kinds of visual displacements indicates that the Helmholtz theory is inadequate. Our cybernetic restatement of the theory, incorporating the principle of neurogeometric organization of behavior, holds that although certain aspects of space perception and visually controlled movement are learned, the nature and degree of learning are determined by the nature and degree of spatial compliance between muscular control and sensory input, and that the brain is organized genetically as a result of evolution, and by growth, maturation, and learning, as a space-structured and space-organized integrative system (Smith and Smith, 1962). The neurogeometric principle is a main postulate of cybernetic theory of learning and perception, and also is a central principle of human-factors design of tool using and machine operations as well as of unaided performance.

2.9.4.1 Experiments on Displaced Vision

A research program spanning more than two decades developed new laboratory techniques to investigate the characteristics of visually controlled performance and the extent to which it can be modified by learning. The most useful innovations were the adaptation of closed-circuit television and computer methods to measure the effects of displacing the visual field in all possible dimensions (Smith and Smith, 1962; Smith, 1963; W. M. Smith, Smith, Stanley, and Harley 1956). These two methods are illustrated in Figure 2.9.10, along with prisms, telescopic spectacles, and mirrors as used in many other studies of visual displacement. In laboratory television methodology (Figure 2.9.5), closed-circuit setups were used so that a subject could observe his or her movements only indirectly by watching the televised image of those movements on a monitor. The scanning circuits of the camera were fitted with switches to change the direction of monitor scanning, making possible inversion, reversal, and combined inversion–reversal of the visual field viewed by the subject. Controlled angular displacements were obtained by angularly displacing the television camera in different planes. To use the laboratory computer system to study spatially displaced visual feedback, the computer was programmed to reverse or alter the graphic coordinates of cathode-ray memory displays or graphic displays constituting visual feedback of the subject's performance.

A critical feature of this research was the demonstration that the television and computer methods used for displacing vision are superior to the optical techniques used in prior studies. One advantage is that the television and computer systems permitted better experimental control and balanced design, and more rigorous statistical analysis. A further crucial advantage is that because a video camera was used to transform the visual field, no artificial constraints were placed on vision or on head–eye movement. Consequently, the performance tasks studied could be carried out in a more natural and realistic setting, with minimal confounding effects of the displacement apparatus itself.

The results of numerous studies (Smith, 1961a, 1961b, 1963, 1964, 1970, 1971a; Smith and Smith, 1962) showed that the performance and learning of spatially distinct motor-sensory interactions were affected quite differently by feedback displacements of different kinds. Generally, limited angular displacements caused the least disturbance, with reversal, inversion–reversal, and inversion causing progressively more, in that order. Because inversion–reversal does not alter the intrinsic directional relationships within the feedback pattern itself, it causes less disturbance as a rule than inversion without reversal. In general, directional compliance between a displaced visual pattern and specific movements determined the degree of disturbance of performance and the extent of learning that would occur in compensating the displacement. The principle of spatial feedback compliance applies across the board to controlled experimental findings on the effects of displacing visual feedback of body motions.

Feedback patterns displaced angularly to progressively larger angles caused little disturbance to movements until a *critical angle* of displacement was reached, at which point performance deteriorated rapidly (Gould and Smith, 1962). The size of the critical angle varied with a number of factors, including the precision and complexity of the motion pattern being performed. More precise and more complex movements are less tolerant of angular feedback displacements than less precise or

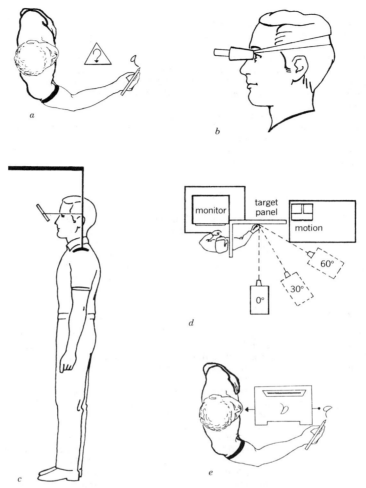

Fig. 2.9.10. Five methods used to space-displace and distort the sensory feedback of body movements. (*a*) Prism. (*b*) Spectacles. (*c*) Mirror. (*d*) Closed-circuit television. (*e*) Analog–digital–analog computer system.

simpler movements; that is, simpler movements have a larger *normal range* of displacement before the breakdown point is reached. In learning a new skill, a machine skill in which feedback is displaced for example, the type of movement and the critical displacement angle determine the extent of learning required. Also, relative spatial displacement and compliances between two or more distinct movements and their sensory feedback patterns determine the degree of integration of these movements and the amount of learning needed to coordinate them effectively.

A question never answered in conventional learning psychology is how the myriad response patterns that make up a person's behavior repertoire are kept specialized and distinct in memory. Experimental findings from this research program indicate that all movements and their sensory feedback control are specialized by spatial feedback relationships—by relative spatial displacement between a movement and its controlling feedback. Thus different motor-sensory interactions are spatially distinct and owe their specialization to their spatial organization.

In a comprehensive experiment, four groups of eight subjects each practiced a task demanding eight distinct directions of movement, each group with a different condition of visual feedback—normal, reversed, inverted, or inverted-reversed (Smith and Wargo, 1963). Well over three-fourths of the 32 possible combinations of movement direction and displacement condition were significantly different from each other in terms of error scores, supporting our view that movements are specialized in terms of spatial compliances in motor-sensory feedback. Additional studies have shown that travel and manipulative movements and direct and compensatory movements are specialized in both performance and learning in relation to the specific conditions and spatial compliances of motor-sensory feedback (Smith, 1963, 1964; Smith and McDermid, 1964).

Although research has shown that the space-structuring of behavior is subject to learned change, other research shows that maturational factors also are involved. Because maturation of the brain apparently is more or less complete by age 10, an experiment was designed to test whether a change in the spatial organization of motor-sensory functions could be detected at about this age. Boys 9, 10, and 11 years of age were tested for their ability to perform compensatory movements that corrected for reversal, inversion, or combined inversion and reversal of the visual field of performance (Greene and Smith, 1963). All the 9-year-old subjects could learn to some extent to perform a direct response with displaced vision, but only one or two of these showed any ability to learn the compensatory or corrective type of movement. The ability to make compensatory adjustments under conditions of displaced feedback emerged at age 10, and was complete by age 11.

2.9.4.2 Feedback Control of Neurogeometric Organization

On the assumption that both maturational and learning flexibility of spatially coordinated behavior and the brain mechanisms that integrate it evolved gradually in the animal kingdom, we would anticipate that the deficiencies produced by displacement of visual feedback would be greater for lower animals than for higher. This is in fact the case. After anatomical rotation of the eyes of flies (von Holst, 1954; von Holst and Mittlestädt, 1950), and of larval frogs and salamanders (Sperry, 1942, 1951; Stone, 1944), these animals displayed persisting inversions and reversals of visually guided movements that were not readjusted in adulthood. Hess (1956) and Pfister (1955) reported similar results for chickens fitted with goggles that reversed and angularly displaced the animals' visual fields. Foley (1940) found, however, that a monkey wearing telescopic goggles made some readjustment to the resulting inversion and reversal of the visual field. The device initially caused more drastic disturbance in the monkey's behavior than any reported for human subjects; for 6 days it appeared comatose and nauseous.

Jacobsen (1967) showed that even amphibians can recover from eye rotation if the operation is carried out early in larval life—at a period of development earlier than that of the animals used by Sperry and Stone. The fact that larval amphibians recovered their ability to relate movements to visual feedback after very early anatomical displacement, but not after a later operation, indicates that feedback-controlled activity determines neurogeometric organization of the brain very early in life but not after the spatial relations between motor and sensory systems have been established.

Studies of higher animals have shown even more clearly that development of directional specificity in the visuomotor system depends on feedback-controlled activities and thus can be impaired by early restriction of vision. Several studies of birds and mammals kept in the dark during infancy showed that a limited period of dark rearing does not affect the animal's ability to turn toward a light or track a moving light, but it typically retards development of visually controlled responses to spatial patterns and specific objects (Walk, 1965).

Several studies of kittens illustrate the need for feedback-controlled activity in mammalian development. Kittens are born with their eyes closed. A developmental study showed that they did not react to moving stripes in the visual field when their eyes first opened, but gradually developed optokinetic-following movements (Warkentin and Smith, 1937). Their visual acuity in distinguishing and following moving lines reached nearly adult level at 28 days, at which time the kittens displayed accurate visual control of the forepaws in jumping and preparatory to landing. These spatially controlled predictive movements appeared suddenly as a result of normal maturation. However, the development of such behavior can be retarded by restricting normal visually controlled activities. For example, kittens that had been deprived of normal visual feedback of body movements by restraining them in a body holder did not discriminate moving and nonmoving objects when released, and did not discriminate depth from a surface edge (Held and Hein, 1963; Riesen and Aarons, 1959). Kittens raised in the dark for several weeks and then fitted with a collar to restrict vision of their forepaws and surfaces below their line of sight were retarded in developing visual placing reactions (Hein and Held, 1967). Kittens raised in darkness for a year before exposure to light were deficient in all visually organized movements but gradually recovered, whether kept in the laboratory or placed in a home (Baxter, 1966). Recovery was about three times as fast in the home as in the laboratory, indicating that the speed of recovery depended on the amount and nature of visuomotor activity.

Research on primates is too limited to determine whether or not their space-structured behavior is retarded more than that of lower mammals by visual deprivation. Infant monkeys confined for 5 weeks in an apparatus that blocked vision of arm and body movements developed reaching and grasping movements only to a limited extent during the last half of the restriction period (Held and Bauer, 1967). Of five infant chimpanzees kept in darkness for 16 months, four were given momentary exposure to light and patterns each day and a fifth was given $1\frac{1}{2}$ hr of normal vision each day (Riesen, 1950). After 16 months, the fifth animal was the only one that displayed normal visual behavior. The other animals were sensitive to light but not to visual patterns, nor did they respond to objects moved toward them. After 6 months, their recovery was still incomplete. It was determined later that these severely deprived animals had suffered degenerative effects in the retina.

Persons blind from infancy whose sight is restored by surgery cannot immediately use vision to control behavior in space (von Senden, 1932). They may show limited figure-ground perception but cannot identify common figures. Blind children may show more precise tactual discrimination at age six than sighted children (Pick, Pick, and Klein, 1967). However, organized performance of tactual and auditory tasks is more limited in blind than in sighted children (Axelrod, 1959). The latter finding suggests that visual deprivation in human infancy may restrict overall maturation of integrative motor-sensory functions. The adverse effects of early deafness on maturation of integrative (cognitive) behavior can be interpreted in the same way.

2.9.4.3 Neurogeometric Organization of the Brain

The spatial organization of behavior is controlled and integrated by spatially organized mechanisms of the central nervous system. Direction- and movement-specific neurons detect spatial differences in stimulation within a specific receptor, between different receptors, and between activity in afferent and efferent paths (Smith and Smith, 1962). Studies indicating the existence of such neurons in the brain, retina, and tactual systems (Barlow and Hill, 1963; Barlow, Hill, and Levick, 1964; Lettvin, Maturana, McCulloch, and Pitts, 1959) soon were followed by research showing that the directional specificity of brain neurons, in higher animals especially, depends on feedback-controlled activities during development. One study showed that visual deprivation in mammalian infancy led to specific deficiencies in brain neurons (Wiesel and Hubel, 1963). When kittens were deprived of vision in one eye for 2 to 3 months after birth, neurons of their visual cortex responded normally to stimulation of the nondeprived eye, but very few neurons responded to stimulation of the deprived eye. A related study indicated that the specific nature of motor-sensory activities during early development may lead to overdevelopment as well as underdevelopment of different brain centers (Rosenzweig and Leiman, 1968). In kittens, deprivation of vision led to hyperdevelopment of the somesthetic (tactual-proprioceptive) areas of the cerebral cortex, apparently as a result of the animals' forced use of nonvisual motor-sensory feedback during the deprivation period. This finding is consistent with the reports noted earlier that blind children may show more precise tactual discrimination than sighted children even though blindness leads to deficiencies in performing organized tasks.

An overview of more recent research results finds them consistent with results of the earlier studies in showing that maturation of spatially organized behavior and of brain neurons underlying it depends on space-structured activity in infancy and is impaired by deprivation of such activity (Barlow, 1975; Blakemore, 1977; Boothe, 1981; Hein and Diamond, 1972; Hein, Vital-Durand, Salinger, and Diamond 1979; Hubel, 1967, 1978; Norton, 1981; Stein and Gordon, 1981). Maturation depends both on self-generated activity, as perceived by the animal, and on movement-related environmental visual stimulation (Hein and Diamond, 1972). Deprivation of visually guided movements in infancy affects maturation of neurons in the primary visual areas of the cortex and in other cortical areas and brain centers as well (Norton, 1981; Stein and Gordon, 1981). The functions of the visual areas of the cortex and the effects of deprivation of space-guided movements on brain neuron maturation are nonspecific; specificity of neuron development appears to be a function of extended maturation (Barlow, 1975; Barlow and Pettigrew, 1971; Berkley, 1981). Maturation of refined spatial vision in primates requires similarly refined patterns of stimulation in infancy (Boothe, 1981).

Especially significant are the studies of Jacobson (1973) and Jacobson and Hunt (1973) showing that neurogeometric connections linking the sensory and motor systems of larval frogs develop at the embryonic level as a two-coordinate system, in two phases. First, one series of neurons matures defining one coordinate of visuomotor control, and somewhat later a second series of neurons matures to define a different coordinate of control. Our interpretation of these results is that these coordinates probably underlie visuomotor control in the vertical and horizontal dimensions.

Research on maturation of the brain provides a number of clues about the relative plasticity of development and function of brain centers and neurons involved in spatially guided behavior. There apparently is a primary period of maturation of visual function that is genetically encoded as a process of within-body activity and internal environmental stimulation. It may proceed in terms of movement-generated self-stimulation within the normal range of visual environments (Boothe, 1981). Subsequent to this period, there is a long-term period of plasticity in visual maturation in which various types of activity-induced and guided maturation occur (Aslin, Alberts, and Peterson, 1981; Barlow, 1975; Berkley, 1981; Blakemore, 1977; Eccles, 1973). Demonstrations of plasticity of brain neuron maturation dependent on peripheral motor-sensory feedback in space-guided and self-generated activity are consistent with our view that maturation and learning both represent dynamic feedback-controlled processes of plastic gene expression.

Analysis of the conditions of neuron degeneration (Cowan, 1973) suggests that decline in neuron connections and death of brain neurons may be accelerated by movement limitations and restrictions.

Just as infant restriction has been found to impair brain-neuron maturation, the differential survival and elaboration of neuron functioning in adulthood is probably related to organized motor-sensory activities. The picture of brain maturation and development that appears to be emerging is that a high level of diversified space-organized activity is essential for optimal development throughout life. It follows that the persistence of organized activities in older adults would retard the degenerative effects of aging. Research has shown that skilled behaviors that are used regularly in a job or profession decline very little with age if they continue to be used daily (Greene and Smith, 1962; Smith, 1965). The pace of walking persists in persons who continue to walk. Writing and speaking skills if used regularly persist with little change far beyond the so-called retirement age of 65.

When Helmholtz formulated his empirical theory over a century ago, he did not have our current understanding of the interacting processes of maturation and learning. His distinction between innate and learning factors—a distinction that has influenced much of psychology to the present day—does not conform to what we know about developmental interactions. Yet the very type of evidence assembled by Helmholtz (1850, 1856–1866) to support empiricism—evidence concerning the effects of space-displaced vision and recovery from space-distorted vision and visual deprivation early in life—has proliferated into the broadly based body of scientific evidence showing the necessity of space-organized activity for the maturation of space-organized perception and movement. Maturation no longer is understood as a genetic unfolding independent of activity and stimulation, but rather as an activity-guided course of behavioral and neuronal development. Helmholtz's (1924–1925) theory stands as a preliminary statement of the idea that space perception develops through feedback-controlled activity and experience. Our present understanding is that the processes of activity-based maturation of brain and behavior are not completely flexible, but reflect species-specific limitations of behavior organization. The specific characteristics and limitations of the human species manifested in its development are the human factors that dictate human-factors design.

2.9.4.4 Reversing Visual Feedback of Eye Movements

A critical test of Helmholtz's empiricism was carried out by devising techniques for reversing visual feedback of eye movements, that is, reversing the spatial relationships between movement of the eyes and the resulting displacement of the visual image on the retina. With other experimental procedures of reversing or inverting visual feedback, such as wearing telescopic spectacles or other devices, the normal relationship between eye movements and image displacement of the retina is not disturbed. By contrast, reversing retinal feedback of eye movements not only gave the retina and the brain a reversed visual pattern of the environment, but also reversed visual feedback of the dynamic oculomotor control of this pattern.

The two methods used to reverse visual feedback of eye movements are shown in Figure 2.9.11. The first consisted of wearing in one eye a scleral lens that was fitted on its front surface with a miniature reversing dove prism. The device could be rotated to produce inversion as well. The second method was to yoke an eye-tracking cursor spot on a cathode-ray tube to the eye movements of a subject by means of the feedback-control operations of a computer system. The subject was given two tasks: to use this spot yoked to his eye movements as a cursor in tracking a second target spot on the CRT display, and to adjust the position of the eye-movement-controlled spot when it was displaced by computer control. Computer programming then was used to reverse the movement of the cursor spot in relation to the eye movements controlling it.

Both methods of reversing retinal feedback of eye movements caused marked perturbation of eye-movement control and nearly complete inability to adjust to the displacement (Schmidt, Gottlieb, Coleman, and Smith, 1974; Smith and Molitor, 1969; Smith, Putz, and Molitor, 1969). The findings were consistent with results of animal experiments and neurophysiological studies which suggest that the primary oculomotor control components of space-organized vision are of maturational origin and not subject to alteration by learning.

2.9.5 BEHAVIORAL–PHYSIOLOGICAL INTEGRATION

Living systems rely upon an impressive array of behavioral, physiological, cellular, and molecular control mechanisms to achieve integrated patterns of activity and function (Adolph, 1982). We attribute integration of behavioral and physiological control processes primarily to motor activity. Our approach is described by the principle of homeokinesis, which, as we have said, recognizes that motor activity is the normal and constant condition of the organism.

From a purely physical and bioengineering point of view, muscular behavior dominates all body functions because it is associated with by far the largest part of the variable bioenergy exchanges in the body. Many lines of research converge to suggest the dynamic nature of this domination. The energy state of the body related to cellular energy metabolism, as well as to higher levels of organ-systems function (cardiovascular, pulmonary, etc.), is reciprocally linked to and depends on the level and pattern of motor behavior.

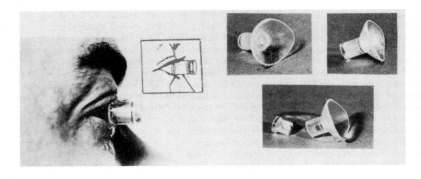

a

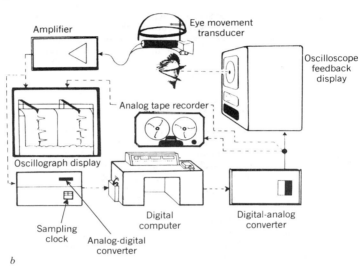

b

Fig. 2.9.11. (*a*) Optical method of reversing retinal feedback of eye movements. (*b*) Computer technique of reversing retinal feedback of eye movements by yoking a calibrated cursor spot to eye movement (based on Smith and Molitor, 1969).

2.9.5.1 Motor-System Control Mechanisms

Closed-loop mechanisms of the compound muscle systems of the body operate dynamically to control relative orientation, length, velocity, and force of muscle shortening and lengthening during motor behavior. Systematic analyses of motor interactions indicate some nine major ways in which multijoint muscle arrangements interact during skilled motor performance (Smith and Smith, 1970). These modes or parameters of multijoint muscular coordination in skill include: (1) reciprocal interactions of opposed extensor and flexor muscles; (2) interaction of postural, travel, and manipulative movements; (3) bilateral interaction of muscle groups; (4) bilateral dominance and subordination of muscle groups (laterality); (5) interaction of body and receptor movements; (6) interaction of ventilatory with other body movements; (7) interaction of ballistic (thrown) and tense (continuously controlled) movements; (8) integration of movements in relation to tools or machines; and (9) social integration of movements. This last includes those specialized social activities that depend for their organization on interactions among the motor systems of two or more persons. In all these integrations, the interactions of concentric and eccentric muscle contractions operate to govern relative stretch and shortening of muscle fibers. In regulating muscle stretch, the various movement coordinations also regulate production of muscle energy through a complex system of feedback circuits.

The closed circuits of muscle activity and its kinesthetic and proprioceptive mechanisms have been of research interest for nearly a century, but the feedback-control nature of the circuits was pointed out first by Wiener (1948). The most significant aspect of the control process is that contraction of a particular muscle actuates its associated muscle-spindle and tendon-organ stretch receptors, thus generating sensory feedback to the brain that affects most strongly the motoneurons attached to the particular muscle and affects its antagonists and other synergically related muscles in diminishing strength (Henneman, 1980). These integrative motor relationships, involving spinal-cord and brain connections, apparently constitute a hierarchical feedback network controlling groups of muscles on both sides of the body (Binder and Stuart, 1980; Nashner, 1979). Motor units forming a pool are recruited by this network according to a size principle (Henneman, 1981). Small motoneurons are excited more readily than large, so that a given level of activity impinging on a motoneuron pool will activate those neurons at and below the size responsive to the activity level.

The role of muscle stretch receptors, and the gamma efferent system, in integrative feedback control of a muscle pool and the coordination of muscle groups still is not understood clearly. Several significant facts stand out. The Golgi tendon organs signal muscle tension (Houk, Crago, and Rymer, 1980). The two main types of receptors associated with muscle spindles provide different information to the brain regarding muscle activity (Matthews, 1980, 1981). The primary afferents of the spindle indicate both muscle-fiber length and velocity of stretch, whereas the secondary afferents indicate only length. P. B. C. Matthews (personal communication) assumes that the primary afferents control the sensitivity of stretch-controlled feedback circuits. It has been demonstrated that gamma efferent activity is changed in parallel with alpha-fiber contraction, showing coordinate control of muscle-spindle sensitivity within a motor pool. In this interaction, the spindle remains primed to provide length and movement-velocity information at whatever length the muscle happens to be (Houk and Rymer, 1981).

The main idea derived from probing the nature of motor control via the gamma efferent system and related muscle-spindle, tendon, joint, and tactual receptors is that muscular contraction at the level of the muscle pool is a compound behavioral–physiological integrative process. Bundles of contracting fibers do not operate in simple reflexes, but in interlinked feedback circuits in which muscles track each other in mutual interrelationships for specific patterns of motor action and integration.

The integrative process whereby a motor unit or pool can regulate its own sensory input sensitivity under variable conditions of operation, as in the case of the coordinate action of the alpha and gamma motor mechanisms and their receptors, carries with it the assumption of dynamic or homeokinetic regulation. By coordinate action, a primary motor system can guide its activity so as to generate a signal that will prepare the action of its coordinate to provide unperturbed ongoing control. This is the essence of behavioral steering mechanisms, as described earlier. Quite clearly, the servomechanism negative-feedback model describes neither predictive steering nor projective control that can occur between primary motor-unit activity and its gamma efferent muscle-spindle coordinates. The graded smooth movements made possible by kinesthetic-proprioceptive-tactual control would seem to involve a modulating mechanism to provide predictive adjustments of individual muscles and muscle groups (Harris and Henneman, 1980; Rack, 1981). We have demonstrated repeatedly that the nature of such movements can be altered by perturbations of exteroceptive (including tactual) feedback.

2.9.5.2 Motor Control of Energy Metabolism

The mechanisms of motor activity and skilled motor integration have their most fundamental adaptive role in governing the efficiency of energy production for contraction and for other metabolic and organic processes linked to energy production. The sliding filament model of muscle contraction (Gergely, 1978; Huxley, 1969) describes contraction in terms of four major components—calcium ion, two principal contractile proteins, actin and myosin, and adenosine triphosphate (ATP), the energy source. Actin and myosin are rodlike molecules (filaments) arranged in parallel patterns within the muscle cell, and connected by cross bridges formed in the presence of calcium by ends of the myosin filaments (myosin heads) binding with specialized sites on adjacent actin filaments. During contraction, cross bridges form, change conformation, and release in a repetitive contractile cycle, causing the actin and myosin filaments to be pulled past one another and resulting in overall shortening of the muscle.

Energy for the conformational change is provided by ATP, which binds to the myosin head and loses its terminal high-energy phosphate during the contractile cycle to form the derivative breakdown products, adenosine diphosphate (ADP) and inorganic phosphate (P_i). The energy made available by this reaction changes the molecular conformation of the cross bridge and pulls the two contractile filaments past one another. Once a cross bridge is released, ADP and P_i remain bound to the myosin head until another ATP is bound by the head, at which point the two breakdown products are released and a new cross bridge is formed, initiating a new contractile cycle.

The integration of these contractile events with the metabolic events surrounding ATP production is quite precise and is most obvious during exercise or muscular work. In particular, the rates of both ATP production and oxygen consumption by the cell during work are closely geared to the rate of ATP utilization (hydrolysis), which in turn is determined primarily by the work rate (Holloszy et al., 1978; Newsholme, 1978). Although the metabolic rate as measured by oxygen consumption of a muscle cell may vary over a 300-fold range between rest and maximal work, it is observed that the levels of ATP and ADP remain constant and very low throughout this range. These observations suggest that both the precursor (ADP) and the product (ATP) of energy metabolism constitute regulated variables, and that mechanisms must exist in the cell for tightly coupling energy-consuming with energy-generating reactions.

Atkinson (1965, 1966) was one of the first to describe the feedback-control mechanisms underlying energy metabolic regulation. He noted that ADP has a positive feedback effect on those metabolic reactions that enhance ATP synthesis, namely, the reactions involved in converting glycogen to glucose, in converting glucose to pyruvate through the glycolytic pathway, and in providing electrons for mito-chondrial oxidative phosphorylation through pyruvate to carbon dioxide conversion in the citric acid cycle. Conversely, ATP has a negative feedback effect on the glycolytic rate and on the reaction feeding pyruvate into the citric acid cycle. Collectively, these feedback influences have the effect of accelerating glycolysis and oxidative phosphorylation to enhance ATP synthesis when ATP utilization increases, but decelerating these same pathways when ATP utilization decreases. Because muscle contrac-tile activity is the major determinant of ATP utilization, particularly during exercise, feedback control by ATP and its breakdown products over the rate of ATP synthesis enables precise coupling between work rate and energy metabolism.

Recently the feedback mechanisms responsible for this coupling have been further clarified (Holloszy et al., 1978; Newsholme, 1978). It appears that it is the ATP/ADP ratio in the cell, also regulated at a remarkably constant level, rather than the ADP or ATP concentrations alone, that determines the rate of oxygen consumption. Because the chemically similar ADP and ATP each competitively inhibit at least eight enzymes that use the other adenine nucleotide as substrate, a constant ATP/ADP ratio may be essential for optimal kinetic efficiency of energy transfer during muscular work.

Evidence that the rate of fuel oxidation is tied to the requirements of the contractile process raises two questions: how mobilization of fuel from body stores is linked to the rate of fuel utilization by muscle, and how the fuel and energy requirements of muscle tissue are integrated with those of other tissues. Of the blood-borne fuels, only fatty acids and glucose represent significant energy sources for sustained muscular activity. The fact that the blood glucose level stays relatively constant across a broad range of work and fuel utilization rates suggests that blood glucose is a key regulatory target in energy metabolism. Undoubtedly, a major reason for this circumstance is that some tissues (e.g., brain, kidney) rely exclusively on glucose as a fuel source, so blood glucose levels must be maintained across a broad range of activity. The degradation of liver cell glycogen to glucose is the critical step that determines the rate at which glucose is provided to the blood. This step is stimulated by circulating epinephrine, which is elevated during physical activity, and is under precise feedback control by small changes in blood glucose (Newsholme, 1978). In turn, blood glucose levels are tied to the rate of glycolysis and oxidative phosphorylation in muscle cells, and thereby to the rate of ATP utilization and the rate of muscle contraction. Thus these linkages integrate the energy metabolic pathways of two different tissues, liver and muscle.

Fat also is metabolized via a series of regulatory linkages which integrate the metabolism of more than one tissue. The release of free fatty acids into the blood from adipose tissue is stimulated by circulating epinephrine, and feedback-regulated by the level of blood-borne fatty acids. In turn, blood fatty acid levels are influenced in large part by the rate of fatty acid oxidation in muscle for energy production, and thereby by the level of energy utilization and contractile activity. It also is known that fatty acid oxidation decreases glucose utilization by muscle. These relationships, then, represent a regulatory integration of metabolism of two fuel sources, fat and carbohydrate, spanning three tissues, muscle, liver, and adipose.

The relationships just described suggest that exercise or muscular activity can regulate efficiency of muscle energy production and its molecular processes. For example, it has been shown that physical training causes an increase in the size and number of skeletal muscle cell mitochondria, with a concomi-tant rise in the capacities of the cell to oxidize fat and carbohydrates (Åstrand and Rodahl, 1977; Holloszy et al., 1978). Furthermore, trained individuals derive a higher percentage of energy from fat oxidation and less from carbohydrate metabolism than do the untrained, at submaximal work loads. These metabolic adaptations are accompanied by a characteristic training increase in the aerobic capacity of muscle tissue.

These research findings indicate overall a direct regulatory linkage between multijoint motor activity and primary processes of energy metabolism at the cellular level. Physical activity actuates neurohor-monal sympathetic pathways which promote release of both glucose and fatty acid as fuel into the bloodstream. Levels of these blood-borne fuels are precisely regulated, along with the ATP/ADP ratio in muscle cells, through a network of positive and negative feedback-control mechanisms spanning multiple tissues. The relationships betwen contractile activity, energy consumption, energy production, fuel utilization, and fuel mobilization are illustrated in Figure 2.9.12, which suggests how coordinate interactions of compound muscle groups control the stretch of particular muscles or motor units, and hence regulate internal molecular muscle-energy production and its related patterns of organic-physiological function and integration. The parameters of coordinate muscle interaction involved in skilled motor adaptations are listed to the left in the figure. The dimensions of molecular and physiological feedback related to the muscle group undergoing contraction and shortening are indicated by the arrows relating this muscle group to different vital levels of activity.

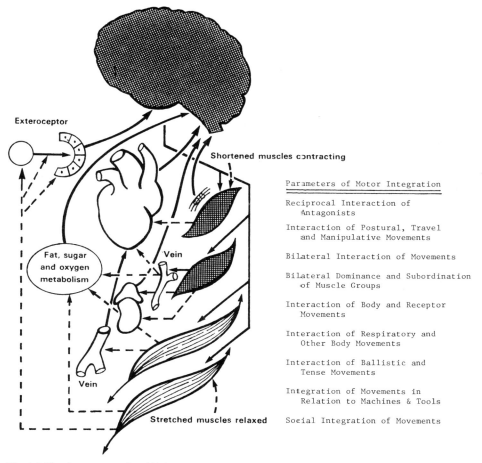

Fig. 2.9.12. Parameters of multijoint muscular coordination that act dynamically to control the relative stretch and shortening of interacting muscles in work so as to regulate the efficiency of bioenergy production within muscle and the efficiency of organic metabolism and visceral function that backstop bioenergy production.

2.9.5.3 Integration of Motor and Organic Function

We have described evidence for the feedback integration of motor behavior and energy metabolism at the cellular and tissue level. Equally compelling evidence points to the central role of motor behavior in regulating organ systems involved in thermoregulation, salt and water balance, renal function, neuroendocrine activity, gastrointestinal function, and, of primary interest here, cardiovascular and pulmonary functions. All cells and tissues rely upon adequate fuel and energy for normal function, and the mobilization and utilization of both fuel and energy for organ-systems function appear to be precisely feedback-regulated by contractile activity of muscle.

The regulation of organ systems by motor activity is most apparent during exercise or physical exertion, and physiologists long have recognized the exercise response as a valuable tool for delineating the nature and limits of motor-organic control processes (Mitchell, 1980). During the transition from rest to maximal exertion, many physiological variables change dramatically whereas others change hardly at all (Åstrand and Rodahl, 1977). Cardiac output may increase five- to sixfold, with a dramatic redistribution of blood flow to working muscle, skin, and heart and away from gut and kidney. Heart rate may show a three- to fourfold increase, and stroke volume, one- to twofold. Pulmonary ventilation may show a 20- to 30-fold increase. Oxygen uptake increases 10- to 20-fold, with a 30 to 90% increase in the portion of oxygen allotted to working muscle. Heat production shows a six- to sevenfold increase.

By contrast, a number of variables appear to be highly regulated, showing little variation during exercise. Blood acidity (pH) and carbon dioxide levels remain constant, body temperature relatively so. Mean arterial blood pressure remains constant or drops slightly, even though systolic pressure is elevated markedly. These constancies protect the brain and other organ systems that operate comfortably only within a relatively narrow range of temperature and pH levels.

A major function of the cardiorespiratory system is to insure adequate provision of blood-borne oxygen to meet the metabolic demands of working muscle. The tight coupling that exists between a number of cardiorespiratory variables, such as cardiac output and heart rate, and the level of oxygen consumption involves both metabolic and neurogenic feedback mechanisms. A rise in blood pressure activates parasympathetic neurons, which release acetylcholine to cause slowing of the heart, a drop in cardiac output, and dilation of peripheral vasculature, with a consequent reduction in blood pressure. Increased oxygen demand during exercise activates sympathetic neurons, which release norepinephrine and epinephrine to cause acceleration of the heart, increased cardiac output, and a complex vascular response involving vasodilation in working muscle and skin and vasoconstriction in nonessential tissues. As noted, glucose and fatty acids also are mobilized by these sympathetic hormones to fuel the increased energy demand. Increases in tissue carbon dioxide or hydrogen ion concentrations, or reduction in oxygen level, can cause vasodilation independently of neurogenic influences. Adenosine, a product of ATP utilization, also appears to be a vasodilatory metabolite in heart, brain, and skeletal muscle tissue (Berne Knabb, Ely, and Rubio, 1983; Sparks, 1980). These various effects ensure that cardiac output and muscle tissue perfusion are enhanced in accordance with the increased oxygen demand of working muscle.

The most likely neurogenic mechanism for integrating heart and vasomotor responses with skeletal muscle contractile activity is a circuit involving feedback from chemosensitive neurons and/or from mechanoreceptors in skeletal muscle, geared respectively to the level of muscle metabolism and to the degree of motor unit recruitment and both dependent on work rate (Mitchell and Schmidt, 1983; Perez-Gonzales, 1981; Rowell, 1980; Stone and Liang, 1984). It also appears likely that independent thermoregulatory pathways exist for feedback-coupling the level of heat production in working muscle to vasomotor and sweat gland responses in the skin (Nadel, 1980).

The other major system responsible for providing oxygen to working muscle is the ventilatory system. Low oxygen, high carbon dioxide, and/or high hydrogen ion levels in the blood activate peripheral chemoreceptors in the aortic arch and carotid sinus, as well as chemosensitive neurons in medullary respiratory centers. Input from stretch receptors in the lung and aortic arch also affects ventilatory control. The exact mechanism regulating the large increase in ventilation from rest to moderate exercise is not known. To regulate levels of carbon dioxide and hydrogen ion in the blood, which remain virtually constant between rest and moderate exercise, it appears likely that chemical and/or mechanical stimulation from working muscle is the primary means of balancing carbon dioxide excretion with production (Dempsey, Vidruk, and Mastenbrook, 1984; Wasserman, 1984; Whipp, Ward, and Wasserman, 1984).

In summary, regulatory integration of motor behavior with cardiorespiratory control during exercise involves some combination of metabolic and neural feedback from working muscle. Overall, we can identify six major feedback effects of coordinate muscular activity that play a regulatory role in metabolic and organic functioning. As suggested in Figure 2.9.13, these are (1) metabolic feedback—the relationships of energy and fuel exchanges with activity; (2) organic-mechanical feedback—the control of cardiorespiratory function by muscular contraction; (3) organic-functional feedback—the physiological effects of activity on level of specific organic functions; (4) neurohormonal feedback—primarily the release of epinephrine by the adrenal gland; (5) interoceptive feedback—stimulation of interoceptors by somatic muscular activity; and (6) kinesthetic-proprioceptive feedback—stimulation of muscle, tendon, and joint receptors and activation of the gamma-efferent system by muscular activity. Our homeokinetic principle holds that the efficiency of integrated control of the various parameters of metabolic and organic function in sleep, rest, moderate activity, sustained exertion, and maximal exertion is regulated specifically by coordinated patterns of motor skill. Such dynamic motor-physiological integrations appear to be basic to the overall systems organization of the organism in both adaptation and development.

The integrated motor–physiological relationships outlined in Figure 2.9.13 have a number of human-factors implications. For example, it is not appropriate to consider behavioral and physiological function separately in the assessment of the performance and health implications of work demands. Evidence for this point is provided by a recent study of pulp workers and a matched group of townspeople, which found a significant correlation between the level of skill in psychomotor performance and the adequacy of cardiorespiratory function, in both groups (T. J. Smith et al., 1982). Work that makes inappropriate or excessive demands of the motor system may also perturb organic and metabolic functioning and thus induce generalized health and safety problems. Finally, training should be considered a systems process with anticipated systems benefits. Behavioral performance training on a given task very likely enhances task-related physiological efficiency; conversely, generalized cardiorespiratory fitness training should be expected to enhance task-specific performance.

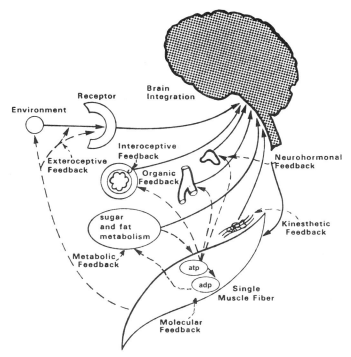

Fig. 2.9.13.   Parameters of physiological feedback in voluntary muscular contraction and skilled coordinate muscular activity.

2.9.5.4   Motor-System Control of Physiological Efficiency

The analysis of behavioral–physiological integration is of central importance to human-factors science in that it establishes bridges between design features of performance situations and processes, and the factors and mechanisms determining efficiency of vital integration at all levels—molecular, cellular, organic, muscular, and behavioral. These bridges are established and controlled by the motor system, through motor-pool integration and the parameters of skilled integration of multijoint motor mechanisms.

Our current understanding of behavioral–physiological integration also refines our understanding of performance efficiency. In conventional terms, work efficiency has been defined as the external work output of the skeletal motor system in moving the body and external loads through space, relative to energy consumed and heat produced in the process. However, external work performance is always accompanied by forms of muscular work not customarily included in the equation (e.g., work of ventilatory muscles in moving gas, work of heart muscle in moving blood), plus underlying energy-consuming processes at the cellular and organic level required to maintain systems integrity during work performance. Thus physiological efficiency, or more particularly bioenergetic and internal efficiency of vital organization, reflects a complex set of variable behavioral–physiological integrations that can be adjusted to variable conditions of tension and stretch imposed on muscle while doing work. It is our belief that the primary determinants of physiological efficiency are the human-factors design features of both the performance situations, and the processes of growth, maturation, learning, training, and physical conditioning essential to skilled motor coordination.

Human-factors/ergonomics science has revolutionized industrial engineering, athletics, sports, exercise practices, and rehabilitation by finding that the variance in human performance that can be achieved by adjusting conditions of human-factors design can exceed the variance attributable to practice or to individual differences in a normal population. The basis for the central role of human-factors design in determining performance efficiency can be found in the theories and facts concerning behavioral–physiological integration.

The human organism is confronted daily with highly variable conditions and demands for muscular activity. The motor system and its bioenergetic mechanisms have evolved to be adjusted to sleep, rest, active relaxation, moderate activity, sustained exertion, maximal exertion, fatigue, stress, emotion, and emergency performance. The diversified bioenergetic and physiological feedback circuits that are specifically locked to skilled coordinate motor activities clearly provide a variable means of adjusting every organic and cellular system of the body to the demands imposed on the motor system. It follows that human-factors principles of design for all types of performance and adaptation may be critical in determining not only the degree of skill that can be executed or developed, but also the physiological efficiency of the system in various types of activity. By directing attention to the objectively discernible linkages between behavior and physiological organization at all levels, the analysis developed here offers guidelines for orienting human-factors design and operation of performance situations to the organism as a whole.

2.9.5.5 Voluntary Control of Organic Processes

A special field of research and practice deals with what is known as biofeedback, or physiological feedback of body movements. Some of the early research in behavioral cybernetics investigated subjects' abilities to control their own physiological functions by using enhanced or substitute sensory feedback linked to an organic function or process. Computerized studies were carried out on self-regulation of brain waves (Smith and Ansell, 1965), heart rate and blood pressure (Ansell, Waisbrot, and Smith, 1967), breathing (Henry, Junas, Smith, and Ansell, 1965; Henry, Junas, and Smith, 1967; Smith and Henry, 1967), and muscular contractions (Rubow and Smith, 1971). Subjects trying to effect changes in the observed indicators of such physiological functions used a number of body-movement techniques, including relaxation, changed breathing patterns, postural changes, attentional alertness, and choking up (to control pressure on the carotid arteries). Some of the techniques used were similar to those sometimes used by experienced offenders to manipulate lie-detector indicators.

Subjects differed markedly in their ability to control changes in brain waves and heart rate, with some persons demonstrating quite marked control, especially over heart rate. However, trying to control heart functions or brain waves voluntarily is a different matter altogether from controlling visual or auditory signals generated by extrinsic muscular contractions, including breathing movements. An analysis of respiratory movements revealed that external somatic respiration represents one of the most tractable and refined tracking systems of the body. The ability of some subjects to control heart rhythm apparently was based on the fact that changes in respiratory pressure can induce related changes in the heartbeat.

The first therapeutic application of biofeedback techniques in the behavioral cybernetics laboratory was a program designed to train patients by means of direct sensory-feedback conditioning for self-control of petit-mal attacks of visually elicited epilepsy (Ansell, Booker, and Forster, 1966). Although patients achieved some success in learning to control the onset of attacks, their improvement was on the whole temporary and related to the clinical setting.

A second therapeutic study used electromyographic techniques to help a woman regain motility of a face partially paralyzed by accidental injury to the left side (Booker, Rubow, and Coleman, 1969). After a damaged section of the facial nerve had been replaced with a branch of the hypoglossal nerve, feedback training was initiated to help her regain control of the sagging facial muscles. Transduced signals from electrodes placed on each cheek were processed through the computer system to control cursor spots on a cathode-ray tube. Using the fully functional muscles on the right side of her face, the patient could move the right cursor spot up and down. At the start she could produce minimal movement of the left cursor spot, but only with great effort and exaggerated facial distortions. After practice in moving the spots compliantly, she was asked to use them to track a centrally located target spot as it moved up and down. She regained considerable self-control over her left facial muscles in 3 weeks of intermittent training sessions, but maintaining the improvement required retraining sessions after intervals of 3 to 6 months. Although the procedures used for muscle-feedback training were comparable to those used generally for skill training, difficulties arose because of isolation of the muscle groups to be trained (Rubow and Smith, 1971).

Biofeedback training has been interpreted by many learning and clinical psychologists as a demonstration of operant learning. An analysis of the learning and physiological factors involved in heart-rate control indicated that anything that could be achieved in self-regulation of organic functions by strictly operant reward-and-punishment techniques could be achieved more directly and effectively by well-designed cybernetic procedures that give a split-second sensory feedback signal of organic functioning (Ansell et al., 1967). At present, however, biofeedback training is being approached more from the operant than the cybernetic point of view. It also has been popularized as a quick-fix or do-it-yourself therapeutic technique for controlling stress, pain, and other organic effects. Evaluating the various techniques objectively is difficult, because they may be used in combination with hypnosis, yoga, transcendental meditation, relaxation therapy, drugs, and questionable methods of operant conditioning (DiCara et al., 1974; Barber et al., 1975/76; Kamiya, Barber, Miller, Shapiro, and Stoyva, 1977).

The inherent weaknesses and limitations of biofeedback methods noted in the initial computerized studies appear not to have been overcome. Furthermore, analysis of the extensive literature up to 1982 suggests that, with only rare exceptions, research in the area is relatively uncontrolled with respect to experimenter error, social participation by patients, placebo effects, direct and indirect suggestion, unrecognized motor adjustment strategies used by patients, operant manipulation of subjects and patients, determining the extent of transfer of therapeutic results outside the laboratory or clinic, and determining the extent to which the externalized organic signals actually represented significant organic changes that would occur only with the methods used. In a word, the techniques thus far developed have limited and generally unreliable significance for human-factors design of medical therapy, rehabilitation, and movement training.

2.9.5.6 Motor-System Control of Maturation and Learning

The homeokinetic principle of motor control applies not only to performance and physiological functioning, but also to maturation and learning, those processes of development by means of which movements are coordinated for increased efficiency and accuracy. Our comprehensive cybernetic theory of self-regulation holds that the direct sensory and physiological feedback effects of movement define the course of maturation and learning and their interrelationships and continuity (Smith, 1966a, 1966b, 1968, 1971c, 1972b, 1977; Smith and Smith, 1966; Smith and Smith, 1969). Evidence concerning brain maturation already reviewed supports this view. The maturational processes that define the primary species-specific detection mechanisms of the central nervous system very likely involve dendritic growth of cortical neurons, whereas learning very likely involves specialization of neuronic detection mechanisms at the molecular level (Lynch and Baudry, 1984). Both kinds of processes establish the neurogeometry of the brain.

From the earliest days of learning research, it has been recognized that learning requires activity as well as giving rise to changes in activity. Research in behavioral cybernetics has shown that learning to track self-generated targets in visual-manual tracking proceeds more rapidly and reaches a higher level of accuracy than learning to track a target generated externally in the environment (Sussman and Smith, 1969, 1970a). These and other findings, some of which were described earlier, suggest that the efficiency of learning under many specialized conditions is determined by the learner's activity and motor-sensory feedback. The implication for human-factors research is to investigate in many performance areas the relative value of self-generated and environmentally related movements in learning.

Most learning research has been based on the S-R (stimulus–response) model of behavior control, or as modified to include a reinforcement after the response. Cybernetic theory holds that the processes of relating behavior to the external environment are not controlled by external stimuli but rather are self-regulated processes wherein coordinated movements and integrated organic functioning are used to control the environment and the nature of external stimuli. Five modes of integrated self-control of the external environment typically are involved in moment-to-moment adjustments. As illustrated in Figure 2.9.14, these include: (1) self-stimulation of receptors by movement and organic functioning; (2) movement control of the physical properties of environmental stimuli; (3) movement control of the sources of stimulation; (4) control of the transmission of environmental stimuli; and (5) movement control of orientation, interaction, and sensitivity of receptors. Cybernetic feedback control of the environment is clearly indicated in the case of the eyes, which combine some eight distinctive patterns of accommodative, orienting, and integrative movements to determine visual perception of space and to guide movements.

A basic question regarding the homeokinetic model of learning is what kinds of organic metabolic and organic functional changes occur as a consequence of the learner's activity—that is, practice—during learning. Numerous lines of research suggest that every distinctive movement of the body carries with it distinctive changes in breathing, heart function, circulation, and neurohormonal secretion. Even verbal social behavior is now known to involve such changes. Thus the course of learning is affected by the feedback effects on brain function of these organic and physiological changes as well as of movement-controlled exteroceptive inputs.

Although exercise physiologists (Åstrand and Rodahl, 1977; Poortmans and Niset, 1981) have not identified the diverse activity-related responses of the organic-metabolic and organic-functional system as learning changes, the fact is established that exercise activity at every level of exertion produces persisting changes in fat, sugar, oxygen, endorphin, adrenalin, and other metabolic processes as well as in heart function, circulation, liver (glycogen) function, and respiration (Terjung, 1984). We believe that these direct physiological feedback effects of movements, consisting of either functional or both functional and structural changes in certain organs, supersede the effects of any delayed artificial reward or punishment that may be manipulated to accompany a specific pattern of movement, although movements linked directly to rewards or punishments have their own physiological feedback effects affecting the learning of an integrated behavior pattern (K. Smith, 1954). Physiological and organic changes have the status of primary movement-controlled learning effects because they can affect the brain in split-second timing in relation to movements and molecular energy exchanges in muscles that control them.

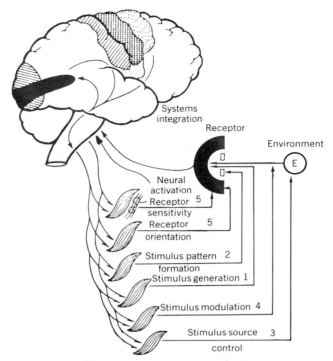

Fig. 2.9.14. Specialized levels of neural activation, receptor control, stimulus pattern formation, stimulus modulation, stimulus selection, and environmental control involved in visually guided behavior.

2.9.6   FEEDBACK CONTROL OF SOCIAL BEHAVIOR

Social interdependence begins with the fertilization of the germ cell and continues through all phases of life-span development. A cybernetic approach to understanding this interdependence describes social interactions in feedback terms that conform to tracking and steering models. The many diverse forms of social behavior—imitation, following, caretaking, speech, social coordination in work, play, and athletic or artistic performance—all can be analyzed as tracking activities controlled by social feedback.

Social tracking is a closed-loop linking of the motor and sensory processes of two or more persons in a process comparable to body-movement tracking in individuals. In social tracking, the movements of a first person generate sensory imputs to the receptors of a second person, who in controlling this input by movements as feedback produces a compliant sensory input to the receptors of the first person, and so on. The effectiveness of social tracking and its attendant physiological efficiency are based on neurogeometric compliance in motor-sensory feedback control between interacting individuals (Smith, 1971a, 1972a).

Variable feedback modes determine different patterns of motor-sensory coordination and interaction in social tracking. As suggested in Figure 2.9.15a and b, social tracking may involve parallel or matched activities, as in imitation or marching, or series-linked, as in speech communication. It may consist of positive feedback control, in which two persons complement each other, or negative, in which they negate or oppose each other. It may be compensatory, in which one person corrects for limitations of a second, or elaborative, in which one person elaborates and extends the actions of another. It may be integrative, in which the two persons adjust their movements to produce a special combined effect, or differential, in which they integrate movements to produce a differential effect.

Similar variations in patterns of social tracking occur in interactive social control of the environment. As suggested by Figure 2.9.16a, one person may track events in the external environment while a second tracks the movements of the first, both interacting variously in the feedback relationships described above. In group social tracking (Figure 2.9.16b), one person may track the relationships (similarities and differences) between two or more other people, using any one or a combination of the variable interactive feedback patterns described above.

Within groups, social tracking takes many forms, depending on the number of persons and their relationships, and their functions and bases of organization. Figure 2.9.17 illustrates individual–group

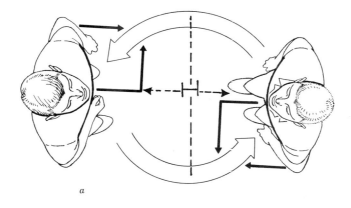

a

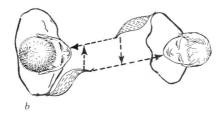

b

Fig. 2.9.15. (*a*) Matched and (*b*) series-linked social tracking. In the series-linked mode, movements of one person generate sensory input to the second that is controlled as feedback and thus instigates a serial movement by the second person, and so on.

(*a*), intragroup (*b*), intergroup (*c*), and institutionalized or cultural (*d*) social tracking. Again, all the variable forms of feedback control described above may characterize particular instances of these general patterns.

The effectiveness and efficiency of social interactions is based on specialization of the parameters of compound movement coordination (see Figure 2.9.12) (Smith, 1973). Variations in social interplay of integrated body movements occur in social interaction between two or more persons with the result that various kinds of social skills in family life, play, education, sexual interaction, work, sports, and artistic performance achieve patterns and levels of coordination not to be found in individual behavior.

2.9.6.1 Physiological Compliance in Social Tracking

Social tracking, like skilled motor coordinations in the individual, involves feedback control of the efficiency of bioenergy production and of organic metabolism and function. This control is mediated through all the parameters of kinesiological feedback: muscular control of molecular, organic-metabolic, organic-functional, organic-mechanical, kinesthetic-proprioceptive, and interoceptive feedback. These control patterns take on a special dynamic characteristic in social tracking, for interacting persons may develop compliances in their physiological or organic processes, as compliant heart rates or movements of respiration.

Compliant variations in organic or physiological functions have been reported in personal interactions (Malmo, Boag, and Smith, 1957), in active and passive participation in two-person groups (Nowlin, Endorfer, Bogdonoff, and Nichols, 1968), in relation to changes in social behavior (Boyd and Di Mascio, 1954), in relation to conformity pressure (Bogdonoff, Back, Klein, Estes, and Nichols, 1962), and during psychotherapeutic interviews (Coleman, Greenblatt, and Solomon, 1956). When therapist and client in counseling reached some correlation or agreement, moderation occurred in both participants in the galvanic skin reflex, heart rate, and skin temperature (Bales, 1954).

Using computerized methods, Sauter (1971) measured the relative accuracy of respiratory and manual tracking in two interacting subjects when the visual effects of manual movements of one perturbed the visual feedback signal of respiratory tracking of the other, and vice versa; and when these two types of control were combined to give a systems feedback signal of the combined tracking. The first or crossed interaction was most accurate. Respiratory movements were as precise in adjusting

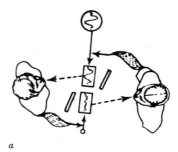

a

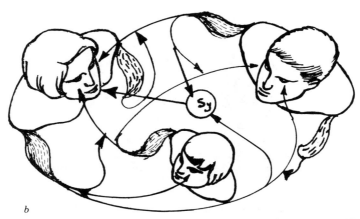

b

Fig. 2.9.16. (*a*) Environmentally related social tracking and (*b*) group tracking.

to the sensory effects of socially generated movements as they are in tracking other movements in individual body-movement coordinations.

Research on physiological compliance and its feedback effects in social interactions is presently very limited, but is important in demonstrating that these interactions occur and can be studied experimentally. Compliance of organic and physiological function may be important in social coordination and attachments, and in maturation and learning in social settings (Smith, 1972b, 1973).

2.9.6.2 Feedback Principles of Joint Performances

Early research on social coordination in discrete performances and in tracking tasks with terminal feedback (Burnstein and Wolff, 1964); Glanzer, 1962; Hall, 1957; Johnston, 1966; Johnston and Nawrocki, 1966; Pryer and Bass, 1957; Rosenberg and Hall, 1958; Zajonc, 1962) indicated that team interactions have the characteristics of feedback control and can be described meaningfully in terms of the social-tracking model. Results showed generally that when previously unrelated persons try to coordinate their movements in joint tracking tasks for which they have had no training, they may show little or no improvement.

In an attempt to clarify some of the characteristics of joint performances, computerized techniques were devised to measure the systems effects of the movements of two persons engaged in joint tracking tasks, in which each person tried to adjust the single tracking cursor to follow a target. The differences between each person's movements were measured continuously and integrated. Because an output of the integrated measure was used to control the tracking cursor directly, neither person got feedback information concerning the accuracy of his or her own movements. After many specific experiments, it was concluded that learning and performance with such an integrated systems feedback of socially coordinated movements was not greatly superior to the team performance and learning that has been found with so-called terminal feedback (Kao and Smith, 1971). A positive result of this and other

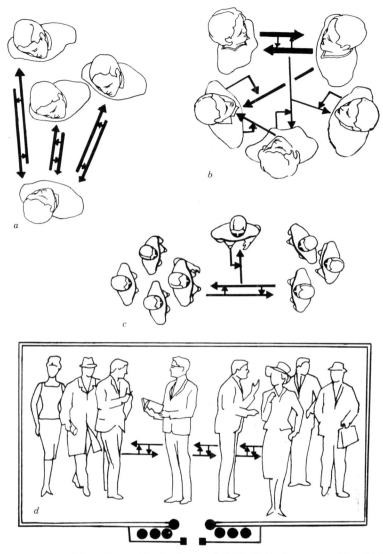

Fig. 2.9.17. Group social tracking and feedback control. (*a*) Individual–group tracking. (*b*) Intragroup tracking. (*c*) Intergroup tracking. (*d*) Intrainstitutional (cultural) social tracking.

research (Sauter and Smith, 1971; Ting, Smith, and Smith, 1972) was finding that behavioral matching of the dynamic movement characteristics of interacting individuals and human-factors control of the modes of feedback received by each person were more significant in determining social-tracking accuracy than repetitive practice.

Accuracy and efficiency of social performance and learning depend on the specific type of feedback information that interacting persons get of each other's movements. Integrated systems feedback of the joint feedback effects of two or more interacting persons (as obtained by combining the errors or correct sensory indicators of the joint movements) is by itself generally not effective as a determinant of accuracy of social performance and learning. Direct sensory feedback of movements involved in the interactions is required for each person. Terminal feedback (so-called learning feedback, knowledge of results, or reinforcement) is ineffective in learning joint interacting movements in social tracking and team learning.

Cybernetic research on joint and team performances indicated that when coordinate movements are to be learned, learning is optimized by pairing an experienced, skilled person with a novice learner.

Also, when movements are learned by imitation, as in child training, teaching, or coaching, learning is enhanced if the teacher or imitatee adjusts his or her movements to conform to the level of ability of the imitator.

REFERENCES

Adolph, E. F. (1982). Physiological integrations in action. *The Physiologist, 25*(2) Suppl., 1–67.

Ammons, R. B. (1956). Effects of knowledge of performance: A survey and tentative theoretical formulation. *Journal of General Psychology, 54,* 279–299.

Ansell, S., and Smith, K. U. (1966). Application of a hybrid computer system to human factors research. *American Society of Mechanical Engineers Human Factors, 69,* 1–8.

Ansell, S., Booker, H. E., Forster, H., and Smith, K. U. (1966). Application of real-time computer system to the clinical treatment of photosensitive epilepsy. *Proceedings of the 19th annual conference on engineering in medicine and biology, 18* (4), 126.

Ansell, S., Waisbrot, A., and Smith, K. U. (1967). Hybrid computer analysis of self-regulated heart-rate control. In S. A. Yefsky, Ed., *Proceedings of law enforcement, science, and technology,* Vol. 1 (pp. 403–418). New York: Thompson.

Arbib, M. A. (1981). Perceptual structures and distributed motor control. In V. B. Brooks, Ed., *Handbook of physiology,* Section 1: *The nervous system,* Vol. 2 *Motor control,* Part 2 (pp. 1449–1480). Bethesda, Md: American Physiology Society.

Aslin, R. N., Alberts, J. R., and Peterson, M. R., Eds. (1981). *Development of perception: Psychobiological perspectives,* Vol. 2, *The visual system.* New York: Academic.

Åstrand, P-O, and Rodahl, K. (1977). *Textbook of work physiology: Physiological bases of exercise,* 2nd ed. New York: McGraw-Hill.

Atkinson, D. E. (1965). Biological feedback control at the molecular level. *Science, 150,* 851–862.

Atkinson, D. E. (1966). Regulation of enzyme activity. *Annual Review of Biochemistry, 35,* 85–124.

Axelrod, S. (1959) *Effects of early blindness.* New York: American Foundation for the Blind.

Bales, R. F. (1951). *Interaction process analysis,* Cambridge: Addison-Wesley.

Barber, T. X., DiCara, L. V., Kamiya, J., Miller, N. E., Shapiro, D., and Stoyva, J., Eds. (1975/76). *Biofeedback and self-control.* Chicago: Aldine.

Barlow, H. G. (1975) Visual experience and cortical development. *Nature, 258,* 199–204.

Barlow, H. B., and Hill, R. M. (1963). Selective sensitivity to direction of movement in ganglion cells of the rabbit retina. *Science, 139,* 412–414.

Barlow, H. B., and Pettigrew, J. D. (1971). Loci of specificity of neurons of the visual cortex in young kittens. *Journal of Physiology, 218,* 98P–100P.

Barlow, H. B., Hill, R. M., and Levick, W. R. (1964). Retinal ganglion cells responding selectively to direction and speed of image motion in the rabbit. *Journal of Physiology, 173,* 377–407.

Baxter, B. L. (1966). Effect of visual deprivation during postnatal maturation on the electroencephalogram of the cat. *Experimental Neurology, 14,* 224–237.

Bellville, J. W., Fleischli, G., and Defares, J. G. (1969). A new method of studying regulation of respiration—the response to sinusoidally varying CO_2 inhalation. *Computers and Biomedical Research, 2,* 329–349.

Bennett, F. M., Reischl, P., Grodins, F. S., Yamashiro, S. M., and Fordyce, W. E. (1981). Dynamics of ventilatory response to exercise in humans. *Journal of Applied Physiology, 51*(1), 194–203.

Berkley, M. A. (1981). Animal models of visual development: Behavioral evaluation of some physiological findings in cat visual development. In R. N. Aslin, J. R. Alberts, and M. R. Peterson, Eds., *Development of perception: Psychobiological perspectives,* Vol. 2, *The visual system* (pp. 197–215). New York: Academic.

Bernard, C. (1865). *Introduction à l'étude de la médecine expérimentale.* Paris: Baillière. Trans. by H. C. Greene (1929). New York: Macmillan.

Berne, R. M., Knabb, R. M., Ely, S. W., and Rubio, R. (1983). Adenosine in the local regulation of blood flow: A brief overview. *Federation Proceedings, 42*(15), 3136–3142.

Binder, N. D., and Stuart, D. G. (1980). Motor unit-muscle receptors interactions: Design features of the neuromuscular control system. In J. E. Desmedt, Ed., *Spinal and supraspinal mechanisms of voluntary motor control and locomotion,* Vol. 8, *Progress in Clinical Neurophysiology* (pp 72–98). Basel: S. Karger.

Birmingham, H. P., and Taylor, F. W. (1954). *A human engineering approach to the design of man-operated continuous control systems* (Report No. 4333). Washington, DC: U. S. Navy Research Laboratory.

Blakemore, C. (1977). Genetic instructions and developmental plasticity in the kitten's visual cortex. *Philosophical Transactions of the Royal Society, London, B, 278,* 425–434.

Bogdonoff, M. D., Back, K. W., Klein, R. F., Estes, E. H., Jr., and Nichols, C. (1962). The physiologic response to conformity pressure in man. *Annals of Internal Medicine, 57,* 389–397.

Booker, H. E., Rubow, R. T., and Coleman, P. J. (1969). Simplified feedback in neuromuscular retraining: An automated approach using electromyographic signals. *Archives of Physical Medicine and Rehabilitation, 50,* 621–625.

Boothe, R. G. (1981). Development of spatial vision in infant macaque monkeys under conditions of normal and abnormal visual experience. In R. N. Aslin, J. R. Alberts, and M. R. Peterson, Eds., *Development of perception: Psychobiological perspectives* Vol. 2, *The visual system* (pp. 217–242). New York: Academic.

Boyd, R. W., and Di Mascio, A. (1954). Social behavior and autonomic physiology (A socio-physiologic study). *Journal of Nervous and Mental Disease, 120,* 207–218.

Burnstein, W. W., and Wolff, P. C. (1964). Shaping of three-man teams on a multiple DRL-DRH schedule using collective reinforcement. *Journal of the Analysis of Behavior, 7, 379–409.*

Calvert, T. W., Banister, E. W., Savage, M. V., and Bach, T. (1976) A systems model of the effects of training on physical performance. *IEEE Transactions on Systems, Man, and Cybernetics, 6*(2), 94–102.

Cannon, W. B. (1932). *The wisdom of the body.* New York: Norton.

Coleman, P., Ruff, C., and Smith, K. U., (1970). Effects of feedback delay on eye-hand synchronization in steering behavior. *Journal of Applied Psychology, 54,* 271–277.

Coleman, R., Greenblatt, M., and Solomon, H. C. (1956). Physiological evidence of rapport during psychotherapeutic interviews. *Diseases of the Nervous System, 17*(3), 71–77.

Cowan, W. M. (1973). Neuronal death as a regulative mechanism in the control of cell number in the nervous system. In M. Rockstein, Ed., *Development and aging in the nervous system* (pp. 19–41). New York: Academic.

Craik, K. J. W. (1947). Theory of the human operator in control systems. I. The operator as an engineering system. *British Journal of Psychology, 38,* 56–61.

Dempsey, J. A., Vidruk, E. H., and Mastenbrook, S. M. (1980). Pulmonary control systems in exercise. *Federations Proceedings, 39*(5), 1498–1505.

DiCara, L. V., Barber, T. X., Kamiya, J., Miller, N. E., Shapiro, D., and Stoyva, J., Eds. (1974). *Biofeedback and self-control.* Chicago: Aldine.

Eccles, J. C. (1973). Trophic influences in the mammalian central nervous system. In M. Rockstein, Ed., *Development and aging in the nervous system* (pp. 89–103). New York: Academic.

Elkind, J. I. (1953). *Tracking response characteristics of the human operator* (Air Research Defense Command Memorandum No. 40). Dayton, OH: U. S. Air Force Human Factors Operational Research Laboratory.

Fitts, P. M. (1951). Engineering psychology and equipment design. In S. S. Stevens, Ed., *Handbook of experimental psychology* (pp. 1287–1340). New York: Wiley.

Foley, J. P., Jr. (1940). An experimental investigation of the effect of prolonged inversion of the visual field in the rhesus monkey. *Journal of Genetic Psychology, 56,* 21–51.

Fredericq, L. (1887). "Methode des Gekreutzen Kreislaufs. *Bulletin de l'Académie r. de Belgique. Classe des Sciences, 13*(4), 417.

Gergely, J. (1978). Some molecular aspects of muscle contraction and relaxation. In F. Landry and W. A. R. Orban, Eds., *Third international symposium on biochemistry and exercise. Regulatory mechanisms in metabolism during exercise* (pp. 43–59). Miami: Symposia Specialists.

Glanzer, M. (1962). Experimental study of team training and team functioning. In R. Glaser, Ed., *Training research and education* (pp. 379–409). Pittsburgh: University of Pittsburgh Press.

Gould, J., and Smith, K. U. (1962). Angular displacement of the visual feedback of motion. *Science, 137,* 619–620.

Greene, D., and Smith, K. U. (1962). Scientific motion study and aging processes in performance. *Ergonomics, 5,* 155–164.

Greene, P., and Smith, K. U. (1963). Maturation of performance with space-displaced vision. *Science, 141,* 727–728.

Grodins, F. S. (1963). *Control theory and biological systems,* New York: Columbia University Press.

Hall, R. L. (1957). Group performance under feedback that confounds responses of group members. *Sociometry, 20,* 297–305.

Harris, D. A., and Henneman, E. (1980). Feedback signals from muscle and their efferent control.

In V. B. Mountcastle, Ed., *Medical physiology,* 14th ed., Vol. 2 (pp. 703–717). St. Louis: C. V. Mosby.

Hein, A., and Diamond, R. M. (1972). Locomotory space as a prerequisite for acquiring visually guided reaching in kittens. *Journal of Comparative and Physiological Psychology, 81,* 394–398.

Hein, A., and Held, R. (1967). Dissociation of the visual placing response into elicited and guided components. *Science, 158,* 390–392.

Hein, A., Vital–Durand, F., Salinger, W., and Diamond, R. (1979). Eye movements initiate visuomotor development in the cat. *Science, 204,* 1321–1322.

Held, R., and Bauer, J. A., Jr. (1967). Visually guided reaching in infant monkeys after restricted rearing. *Science, 155,* 718–720.

Held, R., and Hein, A. V. (1958). Adaptation of disarranged hand-eye coordination contingent upon reafferent stimulation. *Perceptual and Motor Skills, 8,* 87–90.

Held, R., and Hein, A. V. (1963). Movement-produced stimulation in the development of visually guided behavior. *Journal of Comparative and Physiological Psychology, 56,* 872–876.

Helmholtz, H. L. F. von (1850). *Mueller's Archive von Anatomie und Physiologie,* 276–364.

Helmholtz, H. L. F. von (1856–1866). *Handbuch der physiologischen optik,* 3 vols., Hamburg and Leipzig: Voss.

Helmholtz, H. L. F. von (1924–1925). *Physiological optics,* Vol. 1. Trans. from 3rd German ed., J. P. C. Southall, Ed. Rochester, NY: Optical Society of America.

Henneman, E. (1980). Organization of the motor systems. A preview. In V. B. Mountcastle, Ed., *Medical physiology,* 14th ed., Vol. 2. (pp. 669–673). St. Louis: C. V. Mosby.

Henneman, E. (1981). Recruitment of motoneurons: The size principle. In J. E. Desmedt, Ed., *Motor unit types, recruitment and plasticity in health and disease. Progress in clinical neurophysiology,* Vol. 9 (pp. 26–60). Basel: S. Karger.

Henry, J., Junas, R., Smith, K. U., and Ansell, S. (1965). Remote cybernetic analysis of delayed feedback of breath pressure control: Applications of space medicine. *Proceedings AIAA fourth manned space flight meetings, St. Louis, 1965* (pp. 125–137). New York: Association of Aeronautics and Astronautics.

Henry, J., Junas, R., and Smith, K. U. (1967). Experimental cybernetic analysis of delayed feedback of breath pressure control. *American Journal of Physical Medicine, 46,* 1317–1331.

Hess, E. H. (1956). Space perception in the chick. *Scientific American, 195*(1), 71–80.

Holloszy, J. O., Winder, W. W., Fitts, R. H., Rennie, M. J., Hickson, R. C., and Conlee, R. K. (1978). Energy production during exercise. In F. Landry and W. A. R. Orban, Eds., *Third international symposium on biochemistry of exercise: Regulatory mechanisms in metabolism during exercise* (pp. 61–74). Miami: Symposia Specialists.

Holst, E. von (1954). Relations between the central nervous system and the peripheral organs. *British Journal of Animal Behaviour, 2,* 89–94.

Holst, E. von, and Mittlestädt, H. (1950). Das reafferenz-prinzip. *Die Naturwissenshaften, 20,* 464–476.

Houk, J. C., and Rymer, W. Z. (1981). Neural control of muscle length and tension. In V. B. Brooks, Ed., *Handbook of physiology,* Section 1: *The nervous system,* Vol. 2, *Motor control,* Part 1 (pp. 257–323). Bethesda, MD: American Physiological Society.

Houk, J. C., Crago, P. E., and Rymer, W. Z. (1980). Functional properties of the golgi tendon organs. In J. E. Desmedt, Ed., *Spinal and supraspinal mechanisms of voluntary motor control and locomotion. Progress in clinical neurophysiology,* Vol. 8 (pp. 33–43). Basel: S. Karger.

Hubel, D. H. (1967). Effects of distortion of sensory input on the visual system of kittens. *The Physiologist, 10*(1), 17–45.

Hubel, D. H. (1978). Effects of deprivation on the visual cortex of cat and monkey. *Harvey Lecture, 72,* 1–51.

Huxley, H. E. (1969). The mechanisms of muscular contraction. *Science, 164,* 1356–1366.

Jacobsen, M. (1967). Retinal ganglion cells: Specification of central connections in larval xenopus laevis. *Science, 155,* 1106–1108.

Jacobson, M. (1973). Genesis of neuronal locus specificity. In M. Rockstein, Ed., *Development and aging in the nervous system* (pp. 105–119). New York: Academic.

Jacobson, M. E., and Hunt, R. K. (1973). The origins of nerve cell specificity. *Scientific American, 228*(2), 26–35.

Johnston, W. A. (1966). Self-evaluation in a simulated team. *Psychonomic Science, 6,* 261–262.

Johnston, W. A., and Nawrocki, L. H. (1966). *The effect of simulated social feedback on the performance of good and poor trackers* (OSP Technical Report). Arlington, VA: U.S. Air Force.

Kamiya, J., Barber, T. X., Miller, N. E., Shapiro, D., and Stoyva, J., Eds. (1983). *Biofeedback and self-control.* Chicago: Aldine.

Kantowitz, B. H., and Sorkin, R. D. (1983). *Human factors: Understanding people-system relationships.* New York: Wiley.

Kao, H., and Smith, K. U. (1969). Cybernetic television methods applied to feedback analysis of automobile safety. *Nature, 222,* 299–300.

Kao, H., and Smith, K. U. (1971). Social feedback: Determination of social learning. *Journal of Nervous and Mental Disease, 152,* 289–297.

Keele, S. W. (1973). *Attention and human performance.* Palisades, CA: Goodyear.

Keele, S. W. (1981). Behavioral analysis of movement. In V. B. Brooks, Ed., *Handbook of physiology,* Section 1: *The nervous system.* Vol. 2. *Motor control,* Part 2 (pp. 1391–1414). Bethesda, MD: American Physiology Society.

Kelso, J. A. S., Holt, K. G., Kugler, P. N., and Turvey, M. T. (1980). On the concept of coordinate structures as dissipative structures. II. Empirical lines of convergence. In G. E. Stelmach and J. Requin, Eds., *Tutorials in motor behavior* (pp. 49–70). Amsterdam: North Holland.

Kelso, J. A. S., and Stelmach, G. E. (1976). Central and peripheral mechanisms in motor control. In G. E. Stelmach, Ed., *Motor Control. Issues and trends* (pp. 1–40). New York: Academic.

Kohler, I. (1955). Experiments with prolonged optical distortion. *Acta Psychologica, 11,* 176–178.

Landers, D., and Christina, R., Eds., (1978). *Psychology of motor behavior and sports.* Urbana, IL: Human Kinetics.

Lee, B. S. (1950a). Some effects of side-tone delay. *Journal of the Acoustical Society of America, 22,* 639–640.

Lee, B. S. (1950b). Effects of delayed speech feedback. *Journal of the Acoustical Society of America. 22,* 824–826.

Lee, B. S. (1951). Artificial stutter. *Journal of Speech and Hearing Disorders, 16,* 53–55.

Lettvin, J. Y., Maturana, H. R., McCulloch, W. S., and Pitts, W. H. (1959). What the frog's eye tells the frog's brain. *Proceedings.* Institution of Radio Engineers of Australia, 47, 1940–1951.

Licklider, J. C. R. (1960). Quasi-linear operator models in the study of manual tracking. In R. D. Luce, Ed., *Developments in mathematical psychology* (pp. 167–279). Glencoe, IL: Free Press.

Lincoln, R. S., and Smith, K. U. (1952). Systematic analysis of the factors determining accuracy in visual tracking. *Science, 116,* 183–187.

Lynch, G., and Baudry, M. (1984). The biochemistry of memory: A new and specific hypothesis. *Science, 224,* 1057–1063.

Malmo, R. B., Boag, T. J., and Smith, A. A. (1957). Physiological study of personal interaction. *Psychosomatic Medicine, 19*(2), 105–119.

Matthews, P. B. C. (1980). Developing views on the muscle spindle. In J. E. Desmedt, Ed., *Spinal and supraspinal mechanisms of voluntary motor control and locomotion. Progress in clinical neurophysiology,* Vol. 8 (pp. 12–27). Basel, Switzerland: S. Karger.

Matthews, P. B. C. (1981). Muscle spindles: Their messages and their fusimotor supply. In V. B. Brooks, Ed., *Handbook of physiology,* Section 1: *The nervous system,* Vol. 2, *Motor control,* Part 1 (pp. 189–228). Bethesda, MD: American Physiological Society.

McCloskey, D. I. (1981). Corollary discharges: Motor commands and perception. In V. B. Brooks, Ed., *Handbook of physiology,* Section 1: *The nervous system,* Vol. 2. *Motor control,* Part 2 (pp. 1415–1447). Bethesda, MD: American Physiology Society.

Mechler, E. A., Russell, J. B., and Preston, M. G. (1949). The basis of the aided-tracking time constant. *Journal of the Franklin Institute, 248,* 327–334.

Mitchell, J. H. (1980). Regulation in physiological systems during exercise: Introduction. *Federation Proceedings, 39*(5), 1479–1480.

Mitchell, J. H., and Schmidt, R. F. (1983). Cardiovascular reflex control by afferent fibers from skeletal muscle receptors. In J. T. Shepherd and F. M. Abboud, Eds., *Handbook of physiology,* Section 2: *The cardiovascular system,* Vol. 3, *Peripheral circulation and organ blood flow,* Part 2 (pp. 623–658). Bethesda, MD: American Physiological Society.

Mosher, R. S., and Murphy, W. (1965, November). Human control factors in walking machines. *American Society of Mechanical Engineers Human Factors.* Chicago.

Nadel, E. R. (1980). Circulatory and thermal regulations during exercise. *Federation Proceedings, 39*(5), 1491–1497.

Nashner, L. M. (1979). Organization and programming of motor activity during posture control. In E. Granit and O. Pompeiano, Eds., *Reflex control of posture and movement. Progress in brain research,* Vol. 50 (pp. 177–184). Amsterdam: Elsevier/North-Holland Biomedical.

Newsholme, E. A. (1978). Control of energy provision and utilization in muscle in relation to sustained exercise. In F. Landry and W. A. R. Orban, Eds., *Third international symposium on biochemistry of exercise: regulatory mechanisms in metabolism during exercise* (pp. 3–27). Miami, FL: Symposia Specialists.

Norton, T. T. (1981). Development of the visual system and visually guided behavior. In R. N. Aslin, J. R. Alberts, and M. R. Peterson, Eds., *Development of perception: psychobiological perspectives,* Vol. 2, *The visual system* (pp. 113–154). New York: Academic.

Nowlin, J., Endorfer, C., Bogdonoff, M. D., and Nichols, C. R. (1968). Physiologic response to active and passive participation in a two-person interaction. *Psychosomatic Medicine, 30*(1), 87–94.

Perez-Gonzales, J. F. (1981). Factors determining the blood pressure responses to isometric exercise. *Circulation Research, 48*(6), Suppl. I, I-76–I-86.

Pfister, H. (1955). *Über das verhalten der hühner beim tragen von prismen* (Doctoral Dissertation). Innsbruck, Austria: University of Innsbruck.

Pflüger, E. (1875). Beiträge zur lehre von der respiration. I. Über die physiologische verbrennung in den lebendigen organismen. *Archiv Für die Gesamte Physiologie des Menschen und der Tiere, 10,* 251–367.

Pick, H. L., Jr., Pick, A. D., and Klein, R. E. (1967). Perceptual integration in children. In L. P. Lipsitt and C. C. Spiker, Eds., *Advances in child development and behavior,* Vol. 3 (pp. 191–223). New York: Academic.

Poortmans, J., and Niset, G., Eds. (1981). *Biochemistry of exercise,* 2 vols. Baltimore: University Park Press.

Poulton, E. C. (1981). Human manual control. In V. B. Brooks, Ed., *Handbook of physiology,* Section 1: *The nervous system,* Vol. 2, *Motor control,* Part 2 (pp. 1337–1389). Bethesda, MD: American Physiology Society.

Pryer, M. W., and Bass, B. M., (1957). Some effects of feedback on behavior in groups. *Sociometry, 22,* 56–63.

Putz, V., and Smith, K. U. (1970a). Feedback analysis of learning and performance in steering and tracking behavior. *Journal of Applied Psychology, 54,* 239–247.

Putz, V., and Smith, K. U. (1970b). Steering feedback delay: Effects on binocular coordination and hand-eye synchronization. *American Journal of Optometry and Physiological Optics, 47,* 234–238.

Putz, V., and Smith, K. U. (1970c). Dynamic motor factors in determining effects of retinal feedback reversal and delay. *American Journal of Optometry and Physiological Optics, 47,* 372–383.

Putz, V., and Smith, K. U. (1971). Human factors in operating systems defined by delay and displacement of retinal feedback. *Journal of Applied Psychology, 55,* 9–21.

Rack, P. M. H. (1981). Limitations of somatosensory feedback in control of posture and movement. In V. B. Brooks, Ed., *Handbook of physiology,* Section 1: *The nervous system,* Vol. 2, *Motor control,* Part 1 (pp. 229–256). Bethesda, MD: American Physiological Society.

Riesen, A. H. (1950). Arrested vision. *Scientific American, 183*(1), 16–19.

Riesen, A. H., and Aarons, L. (1959). Visual movement and intensity discrimination in cats after early deprivation of pattern vision. *Journal of Comparative and Physiological Psychology, 52,* 142–149.

Roland, P. E. (1978). Sensory feedback to the central cortex during voluntary movement in man. *The Behavioral and Brain Sciences, 1,* 129–171.

Rosenberg, S., and Hall, R. L. (1958). The effect of different social feedback conditions upon performance in dyadic teams. *Journal of Abnormal and Social Psychology, 57,* 271–277.

Rosenzweig, M. R., and Leiman, A. L. (1968). Brain functions. *Annual Review of Psychology, 19,* 55–98.

Rowell, L. B. (1980). What signals govern the cardiovascular response to exercise? *Medicine and Science in Sports and Exercise, 12*(5), 307–315.

Rubow, R., and Smith, K. U. (1971). Feedback parameters of electromyographic learning. *American Journal of Physical Medicine, 50,* 115–131.

Sauter, S. (1971). *Psychophysiological feedback components of social tracking* (Master's Thesis). Madison, WI: University of Wisconsin.

Sauter, S., and Smith, K. U. (1971). Social feedback: Quantitative division of labor in social interactions. *Journal of Cybernetics, 1,* 80–93.

Schmidt, J., Putz, V., and Smith, K. U., (1970). Binocular coordination: Feedback of synchronization of eye movements for space perception. *American Journal of Optometry and Physiological Optics, 47,* 679–689.

Schmidt, J., Gottlieb, M., Coleman, P., and Smith, K. U. (1974). Reversal of the retinal feedback of binocular eye motions in depth vision. *American Journal of Optometry and Physiological Optics, 51,* 382–399.

Schmidt, R. A. (1976). The schema as a solution to some persistent problems in motor learning theory. In G. E. Stelmach, Ed., *Motor control. Issues and trends* (pp. 41–65). New York: Academic.

Searle, L. V. (1951). *Psychological studies of tracking behavior. IV. The intermittency hypothesis as a basis for predicting optimum aided-tracking time constants* (Report No. 3872). Washington, DC: U.S. Navy Research Laboratory.

Senden, M. von (1932). *Raum- und gestaltauffassung bei operierten blindgeborenen vor und nach der Operation.* Leipzig, Germany: Barth.

Sherrington, C. S. (1906). *The integrative action of the nervous system.* New York: Yale University Press.

Simon, J., and Smith, K. U. (1956). Theory and analysis of aided pursuit tracking in relation to target speed and aided-tracking time constant. *Journal of Applied Psychology, 40,* 302–306.

Smith, K. (1954). Conditioning as an artifact. *Psychological Review, 61,* 217–225.

Smith, K. U. (1945). *Behavioral systems analysis of aircraft gun systems* (Special Report). Washington, DC: U.S. Air Force Air Materials Command.

Smith, K. U. (1961a). The geometry of human motion and its neural foundations: I. Perceptual and motor adaptation to displaced vision. *American Journal of Physical Medicine, 40,* 71–87.

Smith, K. U. (1961b). The geometry of human motion and its neural foundations: II. Neurogeometric theory and its experimental basis. *American Journal of Physical Medicine, 40,* 109–129.

Smith, K. U. (1962). *Delayed sensory feedback and behavior.* Philadelphia: Saunders.

Smith, K. U. (1963). Sensory feedback analysis in visual science: A new theoretical-experimental foundation for physiological optics. *American Journal of Optometry and Physiological Optics, 40,* 365–417.

Smith, K. U. (1964). Sensory feedback analysis in medical research: II. Spatial organization of neurobehavioral systems. *American Journal of Physical Medicine, 43,* 49–84.

Smith, K. U. (1965). *Behavior organization and work.* Madison, WI: College Printing.

Smith, K. U. (1966a). Cybernetic theory and analysis of learning. In E. Bilodeau, Ed., *Acquisition of skill* (pp. 425–482). New York: Academic.

Smith, K. U. (1966b). Cybernetic foundations of learning science. In *Proceedings of the XXth meeting of the association of computing machinery* (pp. 1–16). New York: Association of Computing Machinery.

Smith, K. U. (1966c). *Review of the principles of human factors in design of the exoskeleton and four-legged pedipulator* (Report to General Electric Company Special Products Division). Madison, WI: University of Wisconsin Behavioral Cybernetics Laboratory.

Smith, K. U. (1968). Outlook for development and application of behavioral cybernetic and educational feedback designs in instructional technology (pp. 1–21). Washington, DC: Federal Commission on Instructional Technology.

Smith, K. U. (1970). Inversion and delay of the retinal image: Feedback systems analysis of a classical problem of space perception. *American Journal of Optometry and Physiological Optics, 47,* 175–204.

Smith, K. U. (1971a) *Real-time computer analysis of body motions: Systems feedback analysis and techniques in rehabilitation.* Washington, DC: Social and Rehabilitation Administration.

Smith, K. U. (1971b). Experimental systems analysis of delayed steering feedback. In I. E. Asmussen, Ed., *Psychological aspects of driver behavior,* Vol. 1, *Driver behavior,* Section 1.2. Voorburg, Netherlands: Institute of Road Safety Research.

Smith, K. U. (1971c). Behavioral cybernetic foundations of the systems factors in educational technology. In *Second Prague symposium on educational technology* (pp. 1–18). Prague, Czechoslovakia: Czechoslovak Cybernetic Association.

Smith, K. U. (1972a). Social tracking in the development of educational skills, *American Journal of Optometry and Physiological Optics, 49,* 50–60.

Smith, K. U. (1972b). Cybernetic psychology. In R. N. Singer, Ed., *The Psychomotor domain* (pp. 285–348). New York: Lea and Febiger.

Smith, K. U. (1973): Physiological and sensory feedback of the motor system: Neural-metabolic integration for energy regulation in behavior. In J. Maser, Ed., *Efferent organization and the integration of behavior* (pp. 19–66). New York: Academic.

Smith, K. U. (1977). Feedback learning theory as the foundation of educational, training, and rehabilitative technology. *Educational Technology Journal, 17*(10), 18.

Smith, K. U. (1979). *Human factors and systems principles for occupational safety and health.* Cincinnati: National Institute of Occupational Safety and Health, Training and Manpower Division.

Smith, K. U., and Ansell, S. (1965). Closed-loop computer system for the study of the sensory feedback effects of brain rhythms. *American Journal of Physical Medicine, 44,* 125–137.

Smith, K. U., and Arndt, R. (1969). Experimental computer automation in perceptual-motor and social behavioral research. *Journal of Motor Behavior, 1,* 11–28.

Smith, K. U., and Henry, J. (1969). Cybernetic foundations for rehabilitation. *American Journal of Physical Medicine, 46,* 379–467.

Smith, K. U., and McDermid, C. (1964). Sensory feedback analysis of compensatory response to displaced vision. *Journal of Applied Psychology, 48,* 63–68.

Smith, K. U., and Molitor, K. (1969). Adaptation to reversal of retinal feedback of eye-movements. *Journal of Motor Behavior, 1,* 69–87.

Smith, K. U., and Smith, M. F. (1966). *Cybernetic principles of learning and educational design.* New York: Holt, Rinehart & Winston.

Smith, K. U., and Smith, M. F. (1973). *Psychology: Introduction to behavior science.* Boston: Little Brown.

Smith, K. U., and Smith, T. J. (1969). Systems theory of therapeutic and rehabilitative learning with television. *Journal of Nervous and Mental Disease, 148,* 386–429.

Smith, K. U., and Smith, T. J. (1970). Feedback mechanisms of athletic skill and learning. In L. Smith, Ed., *Motor skill and learning* (pp. 83–195). Chicago: Athletic Institute.

Smith, K. U., and Smith, W. M. (1962). *Perception and motion: An analysis of space-structured behavior.* Philadelphia: Saunders.

Smith, K. U., and Sussman, H. (1969). Cybernetic theory and analysis of motor learning. In E. Bilodeau and I. Bilodeau, Eds., *Principles of skill acquisition* (pp. 103–139). New York: Academic.

Smith, K. U., and Wargo, L. (1963). Specialization of space-organized sensory feedback mechanisms and learning. *Perceptual and Motor Skills, 16,* 749–756.

Smith, K. U., Myziewski, M., Mergen, J., and Koehler, J. (1963). Computer systems control of delayed auditory feedback. *Perceptual and Motor Skills, 17,* 343–354.

Smith, K. U., Kao, H., and Knutson, R. (1969). An experimental cybernetic analysis of handwriting and penpoint design. *Ergonomics, 12,* 453–458.

Smith, K. U., Putz, V., and Molitor, K. (1969). Eye-movement retina delayed feedback, *Science, 166,* 1542–1544.

Smith, K. U., Kao, H., and Kaplan, R. (1970). Human factors analysis of driver behavior by experimental systems methods. *Journal of Accident Analysis and Prevention, 2,* 11–20.

Smith, K. U., Schremser, R., and Huang, A. (1970). Binocular synchronization. *Nature, 228*(13), 40–41.

Smith, K. U., Schremser, R., and Putz, V. (1971). Binocular saccadic time differences in reading. *Journal of Applied Psychology, 55,* 251–258.

Smith, T. J. (1977). *Ventilatory and chemical buffering of hypercapnic acidosis, following abrupt administration of inspiratory CO_2* (Doctoral Dissertation). Madison, WI: University of Wisconsin.

Smith, T. J., Banister, E. B., Brown, S., Fadl, S. M., Berry, J., Oliver, C., and Sprince, N. (1982). A behavioral/physiologic health study of kraft pulp mill workers. *American industrial hygiene conference abstracts* (p. 141). Cincinnati: American Industrial Hygiene Association.

Smith, W. M., Smith, K. U., Stanley, R., and Harley, W. (1956). Analysis of performance in televised visual fields: Preliminary report. *Perceptual and Motor Skills, 6,* 195–198.

Smith, W. M., McCrary, J. W., and Smith, K. U. (1960). Delayed visual feedback and behavior. *Science, 132,* 1013–1014.

Sparks, H. V. (1980). Mechanism of vasodilation during and after ischemic exercise. *Federation Proceedings, 39*(5), 1487–1490.

Sperry, R. W. (1939). Action current study in movement coordination. *Journal of General Psychology, 20,* 295–313.

Sperry, R. W. (1942). Reestablishment of visuomotor coordinations by optic nerve regeneration. *Anatomy Record, 84,* 470.

Sperry, R. W. (1951). Mechanisms of neural maturation. In S. S. Stevens, Ed., *Handbook of experimental psychology* (pp. 236–280). New York: Wiley.

Stein, B. E., and Gordon, B. (1981). Maturation of the superior colliculus. In R. N. Aslin, J. R. Alberts, and M. R. Peterson, Eds., *Development of perception: Psychobiological perspectives,* Vol. 2, *The visual system* (pp. 157–196). New York: Academic.

Stelmach, G. E., Ed. (1978). *Information processing in motor learning and control.* New York: Academic.

Stelmach, G. E., and Requin, J. (1980). *Tutorials in motor behavior.* Amsterdam: North Holland.

Stetson, R. H. (1905). A motor theory of rhythm and discrete succession. *Psychological Review, 12,* 250–270; 293–350.

Stetson, R. H. (1951). *Motor phonetics: A study of speech movements in action.* Amsterdam: North Holland.

Stetson, R. H., and Bauman, H. D. (1933). The action current as measure of muscle contraction. *Science, 77,* 219–221.

Stetson, R. H., and McDill, J. A. (1923). Mechanism of different types of movement. *Psychological Monographs, 32*(3), 18–40.

Stone, H. L., and Liang, I. Y. S. (1984). Cardiovascular response and control during exercise. *American Review of Respiratory Disease, 129*(2), Suppl., S13–S16.

Stone, L. S. (1944). Functional polarization in retinal development and its reestablishment in regenerating retinas of rotated grafted eyes. *Proceedings of the Society of Experimental Biology and Medicine, 57,* 13–14.

Stratton, G. M. (1896). Some preliminary experiments in vision without inversion of the retinal image. *Psychological Review, 3,* 611–617.

Stratton, G. M. (1897). Vision without inversion of the retinal image. *Psychological Review, 4,* 341–360.

Stratton, G. M. (1899). The spatial harmony of touch and sight *Mind, 8,* 492–505.

Sussman, H., and Smith, K. U. (1969). Analysis of memory as a feedforward control mechanism. *Journal of Motor Behavior, 2,* 101–117.

Sussman, H., and Smith, K. U. (1970a). Delayed feedback in steering during learning and transfer of learning. *Journal of Applied Psychology, 54,* 334–342.

Sussman, H., and Smith, K. U. (1970b). Sensory feedback modes as determinants of learning and memory. *Journal of Educational Research, 64,* 64–66.

Terjung, R. L., Ed. (1984). *Exercise and sports science reviews,* Vol. 12. Lexington, MA: Collamore Press.

Ting, T., Smith, M., and Smith, K. U. (1972). Social feedback factors in rehabilitative processes and learning. *American Journal of Physical Medicine, 51,* 86–101.

Vince, M. A. (1948). Corrective movements in a pursuit task. *Quarterly Journal of Experimental Psychology, 1,* 85–103.

Walk, R. D. (1965). The study of visual depth and distance perception in animals. In D. S. Lehrman, R. A. Hinde, and E. Shaw, Eds., *Advances in the study of behavior* (pp. 99–154). New York: Academic.

Warkentin, J., and Smith, K. U. (1937). The development of visual acuity in the cat. *Journal of Genetic Psychology, 50* 371–399.

Wasserman, K. (1984). Coupling of external to internal respiration. *American Review of Respiratory Disease, 129*(2), Suppl., S21–S24.

Whipp, B. J., and Ward, S. A. (1980). Ventilatory control dynamics during muscular exercise in man. *International Journal of Sports Medicine, 1,* 146–159.

Whipp, B. J., Ward, S. A., and Wasserman, K. (1984). Ventilatory responses to exercise and their control in man. *American Review of Respiratory Disease, 129*(2), Suppl. S17–S20.

Wiener, N. (1948). *Cybernetics, or control and communication in the animal and the machine.* New York: Wiley.

Wiesel, T. N., and Hubel, D. H. (1963). Single cell responses in striate cortex of kittens deprived of vision in one eye. *Journal of Neurophysiology, 26,* 1003–1017.

Zajonc, R. B. (1962). The effects of feedback and percentage of group success on individual and group performance. *Human Relations, 15,* 149–161.

CHAPTER 2.10

SPEECH COMMUNICATION

R. D. SORKIN AND B. H. KANTOWITZ

Purdue University
West Lafayette, Indiana

Substantial portions of this chapter are reprinted from Chapter 9 of B. H. Kantowitz and R. D. Sorkin, *Human Factors: Understanding People–System Relationships,* New York: Wiley, 1983; with permission from John Wiley & Sons, Inc.

2.10.1 INTRODUCTION

The human-factors aspects of speech communication have charged dramatically in the past few years. The widespread use of microprocessors has brought about an entirely new world of speech technology, including the computer generation of speech and the computer recognition of speech and speakers. This technology has already spread to many consumer, industrial, and military applications. With it has come a variety of human factors problems that cannot be solved by the application of existing tables and formulas.

Probably the best tool that one can have in this area is a good understanding of the basic principles of speech production and perception. We try to achieve that goal in the first part of this chapter; then we consider the problem of assessing the characteristics of a communication channel. In the last section of the chapter we consider the new area of speech technology, with discussion of the principles of wave-form digitization and speech bandwidth compression.

2.10.2 SPEECH PRODUCTION AND PERCEPTION

One strategy for studying the speech signal is based on speech production: How do we use the different parts of our vocal system (our chest, tongue, vocal cords, lips etc.) to encode the information that we want to convey in speech? That is, what particular movements and positions of these organs correspond to the consonant and vowel sounds that we intend to generate? A completely different strategy is to study the acoustic properties of the transmitted speech signal. How can we decode a complex speech signal to tell what speech message (e.g. consonant or vowel) was transmitted? This approach studies the perceptual side of the speech system. Both these approaches have been valuable in acquiring information about the speech system, but a good many questions still remain.

One can appreciate how complex the problem is by considering that even though a specific place for the tongue usually accompanies a particular vowel sound, the exact location of the tongue depends on what sounds were made immediately before and after the one that we are studying. Furthermore, when we make that vowel sound in rapidly spoken speech, we may never actually reach that tongue position. That is, the "normal" tongue position that we associate with producing a particular vowel might be only a target position for the tongue, rather than an absolute requirement. Trying to pin down the speech sounds from the perceptual side is at least as difficult. In a given period of spoken speech, one doesn't know precisely where the sound for one syllable ends and another begins. Furthermore, the exact acoustic properties of a given syllable may be different each time they are repeated or when they appear as parts of different words. So we face large problems of complexity and variability in describing speech stimuli.

2.10.2.1 Phonemes

To the nonspeech scientist, the basic units of speech are the vowel and consonant sounds. Speech scientists have defined a smaller unit: the phoneme. The phoneme is defined as the smallest speech sound that can change the meaning of a word. Phonemes form syllables and words; in order to define or produce a word, we need only to generate a specific sequence of phonemes. Unfortunately, the phoneme is really more of a theoretical definition than it is a precise description of the spoken segments of sound that form our speech alphabet. Most phonemes can't be characterized by a specific sound that would allow us to perfectly identify it, if we could hear it all by itself. Table 2.10.1 illustrates the phonemes of English and the phonetic symbols used to indicate them.

We can gain some understanding of speech production by examining the way phonemes are produced by our speech mechanism. Figure 2.10.1 shows the major parts of our vocal tract. Figure 2.10.2 is a schematic diagram of this system, showing the energy source for pushing air through it, and the important air cavities. A vowel is generated by vibration of the vocal cords in the larynx by air pressure from the lungs. The vibratory motion of the vocal cords is something like one's lip vibration when one plays a trumpet or trombone. The vocal folds vibrate at 80 to 400 vibrations per second. These sound vibrations are then acted on by the various vocal cavities such as the pharynx, the oral cavity, and the nasal cavity, and by their respective openings. When the slide on your trombone is moved, the dimensions of the internal cavity of the instrument are changed—the same thing happens when one places one's tongue in different positions inside one's oral cavity. As we change the size and shape of these cavities and openings, we accentuate or minimize different frequencies that were present in the original vibrations of the vocal cords. For the vowel sounds, the most important control of these structures is in the tongue and lips.

The generation of the consonant sounds differs in several respects from the vowels. When we

Table 2.10.1   The Phonemes of English and Their Phonetic Symbols

	Consonants				Vowels		Diphthongs	
p	pill	θ	thigh	i	beet		ay	bite
b	bill	ð	thy	I	bit		æw	about
m	mill	š	shallow	e	bait		ɔy	boy
t	till	ž	measure	ε	bet			
d	dill	č	chip	æ	bat			
n	nil	ǰ	gyp	u	boot			
k	kill	l	lip	ʊ	put			
g	gill	r	rip	ʌ	but			
ŋ	sing	y	yet	o	boat			
f	fill	w	wet	ɔ	bought			
v	vat	ʌ	whet	a	pot			
s	sip	h	hat	ə	sofa			
z	zip			i	marry			

Source: From H. Clark and E. Clark, *Psychology and Language: An introduction to psycholinguistics,* copyright 1977 by Harcourt Brace Jovanovich, Inc., and reproduced by permission.

generate consonant sounds, we constrict part of the mouth to stop or restrict the passage of air. The major differences among the consonants have to do with where and how in the oral cavity we make that constriction, and whether there is also a vibration of the vocal cords. Table 2.10.2 summarizes these aspects of consonant generation.

These dimensions of phoneme production can provide a way to classify any speech segment into which phoneme was transmitted; it is only necessary to specify where the element fell on the place, manner, and voicing dimensions. However, we have pointed out that the production of such a sound segment depends on what sounds precede and follow it, so the classification task is actually more difficult. Furthermore, for natural-sounding speech, there are also aspects of speech production that extend in time over more than one sound segment. Stress and intonation pattern are examples. Stress has to do with the different emphasis given to syllables in a spoken sentence, and usually involves a change in the intensity, pitch, or duration of the vowel part of a syllable. Intonation generally involves a change in the pitch of the voice, as in the rising intonation at the end of a question. The speed of

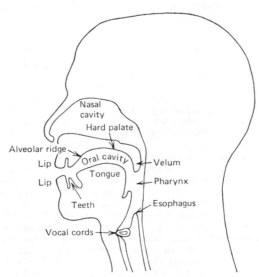

Fig. 2.10.1.   Important parts of the speech production system. (From *Psychology and Language: An Introduction to Psycholinguistics* by H. H. Clark and E. V. Clark, p. 181. Copyright 1977 by Harcourt Brace Jovanovich, Inc., and reproduced by permission.)

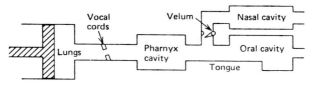

Fig. 2.10.2. Schematic diagram of the speech production system.

transmitting phonemes in spoken speech is normally about 12 phonemes per second. We can understand speech at rates as high as 30 or more phonemes per second (Liberman, Cooper, Shankweiler, and Studdert-Kennedy, 1967).

2.10.2.2 Speech Spectrograms

We can examine the acoustic nature of the speech signal, and try to determine which phoneme was generated from the acoustic information, rather than from the production information. A useful tool for this purpose is the speech spectrogram, shown in Figure 2.10.3. The horizontal axis of this plot is the time from the beginning to the end of the utterance, and the vertical dimension is frequency from 0 to 8000 Hz. The darkness of the lines indicates the intensity of the sound at each frequency.

The second word spoken in the spectrogram of Figure 2.10.3 is "factors." It can be seen that certain frequencies are emphasized during the vowel segments of the sound. The enhanced frequencies are at approximately 800, 2200, 3000, and 4200 Hz. These bands of frequencies are known as formants, and correspond to the mechanical resonances of the cavities of the vocal system. When we change the position of the tongue, for example, we change the size of the oral cavity and pharynx, and thus the frequencies at which these cavities enhance the sound frequencies generated by the vocal cords. The first formant is defined as the lowest resonant frequency, and the second formant is the next-highest resonant frequency.

The situation is somewhat different for consonants. For consonants, there are quite rapid changes in the formants, and sometimes the energy spread is very broad and abrupt. In some cases the sound source is very much like noise. The consonants of English probably carry most of the information conveyed in a word. The problem is that there is no simple one-to-one relationship between the consonant phonemes and the spectrographic patterns associated with those phonemes (Liberman et al., 1967).

2.10.2.3 Effects of Phoneme Context

The spectrographic patterns for many of the consonant sounds are identical; in order for us to determine which consonant phonemes were spoken, we have to look at cues in the vowel sounds that immediately follow the consonant. For example, to discriminate among the consonants /b/, /g/, or /d/, we would need to know the change in the frequency of the second formant of the vowel phoneme that came after the consonant. The important cues for identifying most of the consonant phonemes are not fixed acoustical properties, but depend on the speech sounds that follow them in time. Some speech sounds are relatively independent of the immediate speech context; the vowels and the consonants /s/, /z/, /š/, /ž/, /č/, /ǰ/ are examples (Cole and Scott, 1974).

Another speech cue that depends on the immediate context is voice-onset time (vot), the time between the release of air pressure from the lips and the beginning of vocal cord vibration, as in the syllables /ba/, /pa/, /da/, and /ta/. This time between pressure release and voicing is usually about zero for the /ba/ sound and about 60 msec for the /pa/ sound. Suppose we made a tape of the /ba/ sound. We could make that /ba/ sound into a /pa/ sound by adding a silent gap of 60 msec at the right place in the tape. But suppose that gap were less than 60 msec but more than zero; what phoneme would we hear? Playing with such an artificial stimulus illustrates an important speech phenomenon known as *categorical perception*. Suppose you listened to a list of such sounds that has its vot vary continuously from 0 to 60 msec. You would be impressed by the fact that this list sounds like the sequence: ba, ba, ba, ba, ba, pa, pa, pa, pa. There is a *sudden* transition in the phoneme that you hear, even though you know that the actual physical cue varies slowly from one extreme to the other. Apparently our speech perception system forces us to put these inputs into discrete categories, and this process depends on our analysis of the immediate speech context as well as the specific speech sound.

The dependence of our speech perception on the stimulus context extends quite a bit deeper than the preceding or following phoneme sounds. The characteristics of a speaker's voice affects our perception of individual phoneme sounds, as does the grammar of the spoken sentences, and the listener's expectations and familiarity with the spoken material. All these contextual factors are critical in determining the intelligibility of a speech channel.

Table 2.10.2 **Place and Manner of Articulation of the English Consonants**

Place of Articulation

Manner of Articulation (Unvoiced / Voiced)	Bilabial — The lips are together.	Labiodental — The bottom lip is against upper front teeth.	Dental — The tongue is against the teeth.	Alveolar — The tongue is against the alveolar ridge.	Palatal — The tongue is against the hard palate.	Velar — The tongue is against the velum.	Glottal — The glottis is constricted.
STOPS Complete closure and release of pressure	p / b			t / d		k / g	
FRICATIVES Constriction that produces turbulence		f / v	θ / ð	s / z	š / ž		h
AFFRICATES Complete closure followed by constriction					č / ǰ		
NASALS Velum is lowered to allow air through nasal cavity	m			n		ŋ	
LATERALS Shape tongue to enable opening at sides				l			
SEMI-VOWELS Shape tongue to enable opening in middle	w			r	y		

Source: From H. Clark and E. Clark, *Psychology and Language: An Introduction to Psycholinguistics,* copyright 1977 by Harcourt Brace Jovanovich, Inc., and reported by permission.

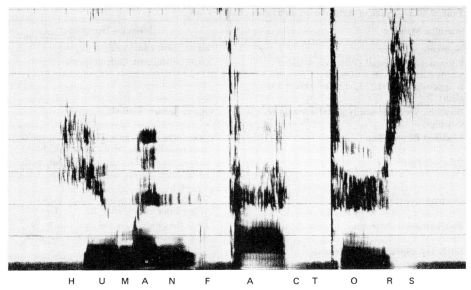

H   U   M   A   N   F   A   C   T   O   R   S

Fig. 2.10.3. Speech spectrogram of "human factors," female speaker. (Courtesy of Robert Sorkin.)

2.10.3   CHANNEL INTELLIGIBILITY AND QUALITY

In this section we discuss the evaluation of speech communication channels such as telephone and radio communication. Evaluating a speech channel depends on what aspects of communication are considered to be important. The intelligibility of the speech may be a major concern. It may be necessary to specify how sensitive the channel will be to disruption by noise, or how rapidly messages may be transmitted over the channel. In some cases where speech is easily understandable, the sound "quality" may be of major interest: Will the channel's users perceive the speech as being of acceptable quality? Solving problems of intelligibility or channel quality may involve some compromises in either the training or the instrumentation for a system. So an early human-factors analysis of the channel can be important in arriving at an effective system design.

2.10.3.1   Articulation and Intelligibility

Roughly speaking, the terms intelligibility and articulation refer to the same general question: How good is speech transmission over the channel? Articulation testing usually means a check on the recognition of syllables or phonemes, whereas intelligibility testing usually refers to the comprehension of words, sentences, or the total message. Some of the many articulation and intelligibility tests that have been devised are illustrated in Table 2.10.3. The stimuli in these tests include nonsense syllables, words, and sentences. In some cases the correct answer can be predicted with some accuracy from the preceding speech context, as in the high-predictability item in the SPIN test.

2.10.3.2   Message Familiarity and Predictability

Two important factors affecting the intelligibility of a channel are the listener's familiarity and experience with the message set and his or her ability to predict the message given only partial or degraded information. The predictability of the message is a function of the size of the message set and the relative probability of a message being chosen from the message set. The least predictable case is when the messages are all equally probable. Communication is optimal when the message set is small and when the listener is highly familiar with the set of possible messages.

These familiarity and predictability factors interact with the particular conditions of reception such as the signal-to-noise ratio and whether (and how) the listener uses both ear channels. In general, highly predictable message sets tolerate much poorer signal-to-noise ratios than do more unpredictable sets. These principles are evident in the results from intelligibility and articulation tests with different kinds of test materials, summarized in Table 2.10.3 and Figure 2.10.4. Figure 2.10.4 illustrates how intelligibility increases as a function of the speech signal-to-noise ratio. Notice that the intelligibility is poorer for the nonsense syllables than for the phonetically balanced (PB) word lists. In a phonetically balanced word list, phonemes occur with approximately the same frequency as they occur in English.

Table 2.10.3 Types of Articulation and Intelligibility Tests

Stimulus Material	Stimulus (Response)
Nonsense syllables	monz, nihf, nan, zeef, . . .
Phonetically balanced (PB) monosyllabic words (the occurrence of different phonemes is approximately the same as in spoken English)	smile, strife, pest, end, heap, . . .
Bisyllabic spondaic words (equal stress on both syllables)	again, farmer, football, . . .
Modified Rhyme Test	rang (rang, fang, gang, hang, bank, sank) bark (mark, bark, park, dark, lark, hark)
Triword Test	badge, bayed, mat (batch, base, bat, bash, bathe, base, bayed, bays, beige, mat, fat, that, rat, vat) . . .
Interrogative sentences	What do you saw wood with? (saw) What letter comes after B? (C)
Five key word sentences	*Deal* the *cards from* the *top,* you *bully Jerk* the *cord,* and *out tumbles* the *gold.*
SPIN test (Speech in babble noise)	
High predictability	The watchdog gave a warning growl. (growl)
Low predictability	The old man discussed the dive. (dive)

Source: After Egan (1948), House, Williams, Hecker, and Kryter (1965), Kalikaw, Stevens, and Elliot (1977), and Sergeant, Atkinson, and Lacroix (1979).

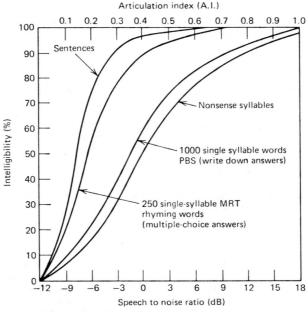

Fig. 2.10.4. Intelligibility of different test materials as a function of speech-to-noise ratio. (From J. C. Webster, "Speech Interference Aspects of Noise," p. 197, in D. M. Lipscomb, Ed., *Noise and Audiology,* copyright 1978 by University Park Press, and reproduced by permission.)

A study of the effects of message set size and intelligibility was performed by Miller, Heise, and Lichten (1951), who showed that words chosen from specified lists of 2, 4, 8, 32, 256, or 1000 words were identified better, the shorter the size of the list.

Message sets where the response alternatives are not defined or constrained to the listener are more difficult and more susceptible to the degrading effects of noise. These factors of listener familiarity and specification of the message set are involved in the use of defined vocabularies in certain communication environments. An example is the international word-spelling alphabet: Alpha, Bravo, Charlie, and so on. Using this vocabulary allows very high intelligibility for alphabetic information under noisy conditions and with speakers and listeners of different nationalities.

2.10.3.3 Effects of Sentence Context

The sentence context can provide important information for speech intelligibility, as shown in the superior results with sentence material in Figure 2.10.4. The size of the message set is effectively reduced, and the predictability of the possible messages is increased by the sentence context. An experiment by Miller et al. (1951) showed that words presented in five-word sentences are identified more accurately than words presented in isolation under noisy conditions. These effects are shown in finer detail by an experiment by Miller and Isard (1963). They presented people with lists of three different types of sentences:

1. *Grammatical sentences.*   Trains carry passengers across the country.
2. *Anomalous sentences.*   Trains steal elephants around the highways.
3. *Ungrammatical strings of words.*   On trains hire elephants the simplify.

The lists of sentences were designed so that they contained the same words equally often, despite the differences in word sequences. The grammatical sentences conformed to the grammatical rules and meaningfulness of ordinary English, whereas the anomalous sentences preserved the grammatical rules but not the meaning. As Figure 2.10.5 shows, accuracy was greatest on the grammatical, a little less on the anomalous, and worst on the ungrammatical sentences. The experiment showed that listeners employ information at higher levels than the segments of phonemes, to help identify which words were transmitted.

There are a variety of speech phenomena that illustrate this point, that information from sentence structure and meaning is used to recognize particular phoneme segments. Under conditions of noise or other degradation of the speech signal, a listener fills in perceptually what is missing, and "hears" the speech as normal. Warren (1970) produced an interesting demonstration of this with an illusion known as the *phonemic restoration effect*. He presented subjects with a recorded normal sentence in which a short portion of the sentence had been removed and replaced with the sound of a cough. Most people who hear the recording don't report that there are any sounds missing from the sentence. Those that do usually select the wrong syllable as having been replaced. A tone in place of the cough has a similar effect. Even when we know how the recording is made, we still hear the added sound as being present along with a normal and complete sentence. We're not sure where in the sentence the extra sound occurred. This phenomenon is significant to us because the obliteration of parts of normal speech is a common occurrence in ordinary communication, particularly in noisy situations. We fill in these wiped out segments automatically; add to these disruptions the sloppiness and speed of normal speech and it is easy to see that our perception of speech must involve high levels of

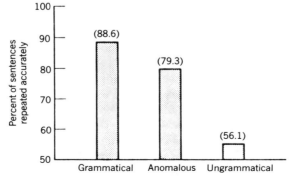

Fig. 2.10.5.   Percent of sentences received correctly as a function of the sentence context. (After Miller and Isard, 1963.)

information processing. Sometimes when we hear such degraded speech, we automatically misperceive a reasonable substitute (from a phonemic, syntactic, and semantic viewpoint).

2.10.3.4 Speech Intensity

An important channel characteristic is the speech signal-to-noise ratio, shown as the horizontal dimension of Figure 2.10.4. The intensity of speech depends on the speaker and the amount of vocal effort the speaker is expending. Speech levels at different amounts of vocal effort range from peak pressures of 70 dB at a whisper, to 110 dB at a shout. At a given level of vocal effort the range of levels spans about 40 dB. The range across different speech sounds is about 30 dB. A system designed to reproduce the extreme range, from the minimum levels of soft speech sounds during a whisper to the peak levels of loud speech sounds during a shout, would have a dynamic range of 80 dB. But it is possible for trained speakers and listeners to communicate effectively over systems with dynamic ranges of 20 dB or less.

Under ordinary conditions, a speaker automatically increases vocal effort to compensate for an added background noise. As the noise increases, the increase in effort soon becomes insufficient to compensate for the increasing noise (Kryter, 1970). Apparently if one speaks in noisy conditions, one's speech is more intelligible than it would be if it was recorded in quiet and the noise were added later. However, at some point of vocal effort such as shouting, intelligibility drops and the advantage is lost. Thus in testing a channel it is a good idea to use previously recorded materials in order to avoid the complex interaction between the channel characteristics and the speaker's way of speaking.

2.10.3.5 Binaural Listening

Another important aspect of the speech channel is whether the speech (and noise) is presented to one or both of the listener's ears. A 15 dB advantage in detection can result from the binaural presentation of a tonal signal. Similar advantages can be gained from the presentation of speech signals over the two auditory channels. The effect depends on the ability of the auditory system to separate the speech and noise signals, using an analysis of the differences in sound phase and intensity at the two ears. This analysis is also involved in the so-called cocktail party effect, which describes our ability to attend selectively to different conversations in a crowded, noisy room. This process is made possible by our having two ears and our being able to localize the direction of different sound sources. If one ear were to be plugged, our understanding of speech in such conditions would be severely degraded. Briefly, we are able to use information residing in physical cues about the desired and undesired signals to screen out irrelevant signals. The cues in the cocktail party situation involve the spatial location of the speaker—but we can employ other cues such as the frequency range of the speaker. For example, it is much easier to attend selectively to a woman speaker against a babble of men's voices than to a man's voice in the same background. Signals that are physically similar to the desired input produce the greatest interference with it, because we are unable to do any filtering on the basis of physical differences between the desired and undesired signal.

2.10.3.6 Distortion of Speech

Sometimes the communication channel distorts the speech signal in some way. We can tolerate a variety of different types of distortion before intelligibility is severely degraded, but certain types are much worse than others. Some types of deliberate distortion of the speech signal have been used to reduce the normal range of frequencies of speech or to code more effectively the intensity range of speech for radio transmission. Another type of manipulation removes time segments of the speech signal. In one technique the speech is turned on and off at a rapid rate; during the off periods the channel can be shared by another speech message. In addition to distorting the original speech wave form in some way, all these techniques remove some information from the speech signal. In the last section of this chapter we discuss the general problem of coding the speech signal for efficient transmission.

Peak Clipping

One technique for modifying the intensity characteristics of the speech signal is to compress the range of intensities that the channel will pass. Peak clipping is the process by which positive and negative peaks of the speech wave form are clipped off; only the remaining wave form is transmitted. The remaining wave form can be amplified so that its new peaks are equal to the old ones. Figure 2.10.6 illustrates the effects of peak clipping a speech wave form. Suppose that regardless of the moment-to-moment intensity of your speech, your microphone always transmitted either a full positive or a full negative voltage. This is an example of infinite peak clipping, and is approximated by the 20 dB clipping case shown in Figure 2.10.6. Peak clipping and then reamplifying the wave form to the previous

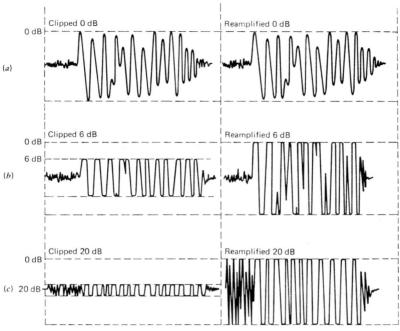

Fig. 2.10.6. Speech samples before and after 6 and 20 dB of peak clipping. The right-hand samples are reamplified until their peak amplitudes are the same as the original sample. (From J. C. R. Licklider, D. Bindra, and I. Pollack, "The Intelligibility of Rectangular Speech Waves," *American Journal of Psychology, 61,* 3. Copyright 1948 by Karl M. Dallenbach, and reproduced by permission of University of Illinois Press.)

peak amplitude has the effect of increasing the power of the consonant sounds relative to the vowels. It also increases the average speech power, making the speech more intelligible in certain kinds of noise than unclipped speech.

Filtering the Speech Spectrum

The ordinary speech spectrum spans a frequency range from below 100 to over 8000 Hz. In fact, the contribution of different frequencies to the intelligibility of speech varies somewhat over this spectrum. The importance of different regions of the speech spectrum can be determined by feeding the speech signal through a band-pass filter, a device that passes a defined band of frequencies and rejects any frequencies outside that band. For example, all the frequencies above 1700 Hz could be rejected without much effect on intelligibility. Conversely, all the frequencies below 1700 Hz could be filtered out with much the same effect. Different speech sounds are affected by these different manipulations; the consonants are hurt much more by rejecting the high frequencies, and vowels are hurt more by cutting out the lows. These results are related to the general effects of noise on speech. In high noise levels, the vowels remain intelligible longer as the noise level increases, because more of the speech energy is in the vowels. If we filter out and discard a 3000 Hz band of frequencies centered at 1700 Hz, intelligibility drastically decreases. On the other hand, if we pass only those middle frequencies and reject everything else, there is almost no effect on intelligibility. How much of a bandwidth is needed for essentially no decrease in intelligibility? Perfectly satisfactory communication is obtainable with a band of frequencies from 800 to 2500 Hz. Highly trained speakers and listeners can communicate effectively with defined message sets over channels having much narrower bandwidths. Remember that eliminating these "extraneous" speech frequencies may remove nonspeech information such as cues about the speaker's sex, age, or emotional state, and may have drastic effects on the perceived quality of the communication channel.

2.10.3.7 Articulation Index

If some frequency regions are more important than others to speech intelligibility, then some regions should be more sensitive to interference from noise. If we could specify those frequency regions, we

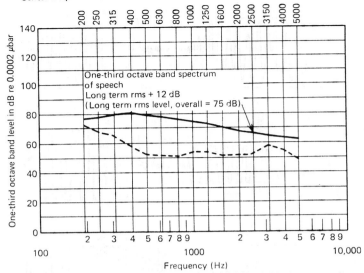

Fig. 2.10.7. Example of the calculation of an AI by the one-third octave band method. (This material is reproduced with permission from American National Standard, *Methods for the Calculation of the Articulation Index,* ANSI S3.5-1969, p. 17, figure 6. Copyright 1969 by the American National Standards Institute. Copies of this standard may be purchased from the American National Standards Institute, 1430 Broadway, New York, NY 10018.)

would have a way to predict the disruptiveness of a given spectral distribution of noise. This idea is the basis for the Articulation Index (AI) (Beranek, 1947; French and Steinberg, 1947).

The AI is a technique for predicting the effects of noise on speech by considering the speech signal-to-noise ratios in a set of frequency bands spaced across the speech spectrum. Originally, a set of 20 different frequency bands were defined that were assumed to each contribute an equal amount to speech intelligibility. Under ideal circumstances, each band contributes $\frac{1}{20}$ or 0.05 to the maximum AI score of 1.0. Under noisy conditions some of these bands would contribute proportionally less than that 0.05 maximum, depending on the signal-to-noise ratio within that band. A simpler procedure for computing the AI is outlined in Figure 2.10.7 and Table 2.10.4.

The more that the speech spectrum of the speaker in a particular application differs from that of the noise, the larger the resulting AI. In Figure 2.10.4 the AI is plotted across the top of the graph, showing how intelligibility varied as a function of the AI. Figure 2.10.8 shows how AI varies with the level of a fairly representative noise spectrum that might be found in a typical "open plan" office layout (Pirn, 1971).

The AI measure seems to be a good predictor of speech intelligibility over a variety of noise conditions and speech manipulations (Kryter, 1962) but is inappropriate for systems where there is much speech processing for compressing the speech band. In those cases, you may have to evaluate directly the intelligibility of the system under consideration. An AI value below 0.3 will usually be unsatisfactory for most applications. An AI of 0.5 could be satisfactory for a limited vocabulary situation; an AI of 0.7 would be required for sentence comprehension better than 99%. Automatic techniques for assessing the AI of a given channel have been developed. One of these involves the calculation of a Speech Transmission Index (STI) that allows consideration of the effects of a variety of channel manipulations, such as distortions in the temporal aspects of the signal. The interested reader is referred to an article by Steeneken and Houtgast (1980) for further information about this index.

2.10.3.8   Channel Quality

The particular communication channel of interest may have AI values that are larger than 0.8 or more. How does one evaluate channels with such high indexes of intelligibility? Webster (1978) suggests that at high AI values, practically all intelligibility tests yield such high scores as to be very inefficient at discriminating among different systems. In such cases one may have to employ tests involving reaction time measurements, quality judgments, or tasks involving the presence of competing messages or of a speech babble as the noise.

Table 2.10.4   Computation of the AI

The ANSI procedure involves comparing the noise levels and the peak speech levels in 15 one-third octave bands.

1. Plot the one-third octave spectrum of the speech signal to be evaluated; this is shown in Figure 2.10.7 as the idealized male speaker spectrum.
2. Plot the one-third octave noise spectrum; a sample spectrum is plotted in the figure.
3. Determine, for each band, the difference in decibels between the peak speech level and the noise level. (If the noise is bigger, assign a zero difference; if the signal is more than 30 dB bigger, assign it a 30 dB difference.)
4. Multiply these decibel differences by the appropriate weighting factor given in column 3, and enter these numbers in the fourth column.
5. Add up the entries in the fourth column and you have the AI. (The reason for the 30 dB limitation is that the speech signal is assumed to have a 30 dB dynamic range within each band. The calculation is actually an estimate of the proportion of the 30 dB "used up" by the noise in each band.)

Band	Speech Peaks Minus Noise (dB)	Weight	Columns 2 × 3
200	4	0.0004	0.0016
250	10	0.0010	0.0100
315	13	0.0010	0.0130
400	24	0.0014	0.0336
500	26	0.0014	0.0364
630	26	0.0020	0.0520
800	24	0.0020	0.0480
1000	21	0.0024	0.0504
1250	18	0.0030	0.0540
1600	18	0.0037	0.0666
2000	15	0.0037	0.0555
2500	15	0.0034	0.0510
3150	6	0.0034	0.0204
4000	8	0.0024	0.0192
5000	12	0.0020	0.0240

$$AI = 0.5357$$

Source: This material is reproduced with permission from American National Standard, *Methods for the Calculation of the Articulation Index,* ANSI S 3.5-1969, page 17, Figure 6. Copyright by the American National Standards Institute. Copies of this standard may be purchased from the American National Standards Institute, 1430 Broadway, New York: N.Y. 10018.

One technique asks people to compare the quality of speech on a test channel and a standard channel. In an experiment by Munson and Karlin (1962), listeners compared speech on several communication channels to speech on a high-fidelity monaural system. Munson and Karlin measured how much noise had to be added to the high fidelity channel to make it appear equal in quality to the channel under test. The Transmission Preference Level (TPL) measures this noise. Figure 2.10.9 illustrates this measure for some practical types of speech circuits. Webster (1978) has calculated the AI values that correspond to these values and they are included in the figure. It is evident that in some applications we might need to specify the quality of a given channel at AI's larger than 1.0. The reason for this is not simply the quality as perceived by a prospective user. The speed and ease with which speech can be understood and information processed from the channel may be an important function of the channel characteristics at AI's greater than 0.8. Other approaches to the assessment of speech quality are discussed by Nickerson and Huggins (1977). Additional aspects of the problem of perceiving speech in noise are discussed in Chapter 6.1.

2.10.4   SPEECH TECHNOLOGY

Communication engineers have long been interested in compressing the speech channel to as narrow a band of frequencies as possible, while still preserving good intelligibility and speech quality. The reason for this interest is that a given radio or wire channel has a practical limit on the band of

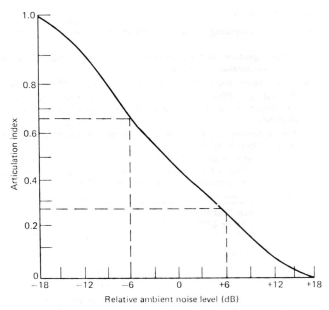

Fig. 2.10.8. AI as a function of background noise. Each 6 dB increment in the ambient-noise level generally results in an AI decrease of 0.2. (From R. Pirn, "Acoustic Variables in Open Planning," *Journal of the Acoustical Society of America*, 1971, 49, 1342, and reproduced by permission.)

frequencies that it can pass. A given channel can carry a limited number of voice or other (video, etc.) messages. Advances in satellite and fiber-optics telecommunication have somewhat reduced the need for more efficient use of bandwidth, but there are still many applications where it is desirable.

There is an important new reason for learning how to encode the speech signal efficiently. The availability of inexpensive and efficient digital electronics has made it practical to perform various computations on speech signals. Applications of this new technology include the computer processing of speech for encryption or privacy, low data rate communications, the storage and generation of

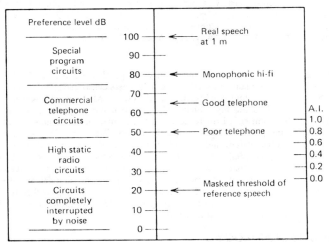

Fig. 2.10.9. Transmission preference levels and AI values for different types of communication circuits. (From J. C. Webster, "Speech Interference Aspects of Noise," 211, in D. M. Lipscomb, Ed., *Noise and Audiology,* copyright 1978 by University Park Press, and W. A. Munson and J. E. Karlin. "Isopreference Method for Evaluating Speech-Transmission Circuits," *Journal of the Acoustical Society of America,* 1962, *34,* 773, and reproduced by permission.)

synthetic speech by computer from internal coded information or from written text, and the recognition by computer of spoken material or of particular speakers. Today, we can see widespread use of speech encryption devices, consumer products that speak, computer instruction devices that generate spoken instructions to workers, and the spoken entry of limited speech vocabularies to computers. In this section, we discuss briefly basic aspects of the new technology and some of the potential human-factors problems.

2.10.4.1 Wave-Form Sampling and Quantization

The first operation that must be performed on a wave form, if we are going to process it by computer, is to convert it from analog to digital form. Figure 2.10.10 illustrates a typical analog signal (a portion of a vowel sound) that we have digitized for computer processing. Notice that the sampling operation selects a number of particular moments in time to be measured; the rest of the wave form is ignored. In fact, this sampling operation does not discard as much information as one might think. Suppose the signal of interest to us contains frequency components in a band, W, that is 3000 Hz wide. (Perhaps it includes frequencies from 500 to 3500 Hz.) We can sample the signal at a rate of at least $2W$, or 6000 times per second, without losing any information about these different frequency components. The reason for this is that it takes only two pieces of information to specify the amplitude and phase of any particular frequency component; to reconstruct one second of our signal we would need a number of samples equal to two times the number of frequency components. This principle is known as the *sampling theorem*.

What kind of precision is needed to specify the amplitude of each of those $2W$ samples? This question is called the *quantization* problem. We need to divide the amplitude scale (as we did the time dimension) into a number of small but measurable increments. But how small must each of these increments be? In Figure 2.10.10 we quantized each sample to one of 16 values (e.g., 4 bits). You can see that the result is a bit crude. A quantization precision of 12 bits would specify each amplitude sample to a precision of one in 4096, or about 0.02%, a value that would be unnecessarily precise for speech purposes. Both the sampling rate and the precision of quantization will affect how well the final wave form matches the original wave form.

The quantization and sampling operations have another effect, in addition to discarding information that is in the original signal. These operations add noise to the signal. Imagine a sinusoidal signal that is quantized to a precision of 4 bits, at a sampling rate of 20 samples per cycle. The resultant wave form will still look approximately sinusoidal, but it will have various square edges at different points along the signal. If the frequency components present in this new signal were analyzed, there would be many more frequencies present than in the original sinusoid, which had only one component. Of course that is to be expected whenever we perform some nonlinear operation such as sampling and quantizing. But these additional components would be heard as noise—and they might even interfere with our ability to hear the information in the original signal that we wanted to preserve. Schroeder, Atal, and Hall (1979) suggested an interesting way to make the noise components less evident in highly processed speech by adjusting the intensities of the speech frequencies so that they mask some of the noise frequencies.

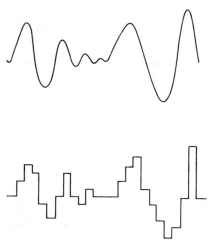

Fig. 2.10.10.   Analog signal (top) and sampled and quantized version of analog signal (bottom). Each sample can take one of 16 values (4 bits).

2.10.4.2 Bit-Rate Reduction

Choosing the sampling rate and the precision of quantization will obviously depend on the specific application. For the high-fidelity digital reproduction of music, we might want a sampling rate of 50,000 samples per second and a quantization of 16 bits per sample. Transmitting such signals would take a transfer rate of $50,000 \times 16$ or 800,000 bits per second. A personal computer might have 262,000, 8-bit bytes or about 2×10^6 bits in its memory. Thus it could be used to store only $2\frac{1}{2}$ sec of that high-fidelity digitized music. Fortunately, the speech channel can be a lot narrower in bandwidth than the high-fidelity music channel. An 8 bit quantization and a 5000 Hz sampling rate would result in quite acceptable speech. The bit rate would then be $5000 \times 8 = 40,000$ bits/sec. The computer could hold almost 1 min of that speech.

Actually we can hedge that last figure in a number of ways. We can cut down the quantization accuracy; remember that amplitude clipping of the speech wave form does not have a large effect on intelligibility—so we could use much less than the 8 bits (in fact, we could compress the range of speech intensity by using a log transformation prior to digital conversion). A more elegant approach is to restrict the intensity range (in bits) of each subsequent sample to some increment or decrement that depends on the intensity of the prior sample value. This approach takes advantage of the fact that there is a correlation between the intensity of successive speech samples. The speech mechanism is a physical system that has mass and inertia, and so there are physical limitations on the sounds that can be made right after other sounds. This fact has led to the development of special pulse code modulation techniques that take advantage of this predictable aspect of speech. A very elegant technique carries the procedure even further; essentially it works by quantizing only those parts of the signal that cannot be predicted from the already coded signal (Atal and Hanauer, 1971). Such systems have produced speech with rates of 2400 bits/sec and lower.

Even lower bit rates are possible if additional computation can be performed on the input signal. For example, it is possible to extract information about the particular phoneme sequence, transmit that data, and then reconstruct the speech signal from the phoneme sequence at the destination. We mentioned at the outset of this chapter that phoneme rates in normal speech are normally about 12 per second; if it takes, say, 6 bits to encode each phoneme, it is easy to see that a bit rate of 72 bits/ sec would suffice to transmit speech. At such a rate, the 262,000 byte memory we mentioned earlier would hold about 8 hr of speech.

It is quite likely that evaluations of the effectiveness of proposed and existing systems for encoding and generating speech at low bit rates will be needed. It is difficult at this point to present any rigid formula, such as the AI, that would allow performance predictions to be made. Two approaches to the problem of developing objective measures of channel quality for such systems are proposed by Barnwell (1979) and Mermelstein (1979); the reader may refer to these studies or to the *IEEE Transactions on Acoustics, Speech, and Signal Processing* for further information about this problem. There may be important perceptual and cognitive limitations when low bit-rate speech is employed. For example, the processing time required for certain tasks is greater when synthetic rather than natural speech is used (Pisoni, 1982). It may be necessary to evaluate directly prospective systems with suitable articulation or intelligibility tests of the sort that we have described, or with performance tests in the actual proposed application. This type of evaluation, prior to fixing the system design, may be easier to perform than in the past, because it is likely that the prospective synthesis system will first be implemented on a computer.

It is possible to generate intelligible speech from text by specifying the operations that will cause a phoneme sequence to be generated by a computer speech synthesizer. These systems were developed out of the pioneering work of speech scientists and engineers during the past 30 years. Better speech generation involved the use of rules for making the acoustic transitions between phonemes and for modifying the pitch and timing of each speech segment as a function of the phoneme context. The logical next step was to construct a program for generating speech from printed text. Several systems have been developed for accomplishing this function. The reverse side of the speech synthesis problem is the automatic recognition of spoken words by machine. Both these applications of speech technology are discussed at length in Chapter 11.5; some problems in the use of synthetic speech signals are also discussed in Chapter 3.3.

2.10.5 SUMMARY

Human-factors questions will be involved in many systems that have the speech channel as a vital component. The production and perception of speech are highly complex processes; for example, speech perception depends on many factors including the acoustic, syntactic, and semantic context of the input signal. Techniques are available for the evaluation of the intelligibility of speech communication channels. These techniques require specification of the nature of the speech messages (particularly their predictability and familiarity) and the noise background (particularly its level in different frequency bands relative to the speech signal). There are several reasons why the designer should be aware of new techniques for evaluating speech communication systems. First, measures of intelligibility may not accurately assess the perceived quality of a channel, as judged by prospective users. Second, intelligibility measures may be imprecise measures of performance (such as processing time) in tasks that make high-level cognitive and memory demands on the user. Third, existing measures may be inappropriate under conditions of extensive reduction in bandwidth or bit rate, such as with digitally encoded or processed speech or with synthetic speech. Under these conditions, the potential impact of system or signal characteristics on information processing or user acceptability may be significant.

REFERENCES

American National Standards Institute (1969). Methods for the calculation of the Articulation Index (ANSI S3.5-1969).

Atal, B. S., and Hanauer, S. (1971). Speech analysis and synthesis by linear prediction of the speech wave. *Journal of the Acoustical Society of America, 50,* 637–655.

Barnwell, T. P., III (1979). Objective measures for speech quality testing. *Journal of the Acoustical Society of America, 66,* 1656–1663.

Beranek, L. L. (1947). The design of speech communication systems. *Institute of Radio Engineers, Proceedings, 35,* 880–890.

Clark, H. and Clark, E. (1977). *Psychology and language.* New York: Harcourt Brace Jovanovich.

Cole, R. A. and Scott, B. (1974). Toward a theory of speech perception. *Psychological Review, 81,* 348–374.

Egan, J. P. (1948). Articulation testing methods. *Laryngoscope, 58,* 955–991.

French, N. R., and Steinberg, J. C. (1947). Factors governing the intelligibility of speech sounds. *Journal of Acoustical Society of America, 19,* 90–119.

House, A. S., Williams, C. E., Hecker, M. H. L., and Kryter, K. D. (1965). Articulation-testing methods: Consonantal differentiation with a closed-response set. *Journal of the Acoustical Society of America, 37,* 158–166.

Kalikow, D. N., Stevens, K. N., and Elliot, L. L. (1977). Development of a test of speech intelligibility in noise using sentence materials with controlled word predictability. *Journal of the Acoustical Society of America, 61,* 1337–1351.

Kryter, K. D. (1962). Validation of the Articulation Index. *Journal of the Acoustical Society of America, 34,* 1698–1702.

Kryter, K. D. (1970). *The effects of noise on man.* New York: Academic.

Liberman, A. M., Cooper, F. S., Shankweiler, D. P., and Studdert-Kennedy, M. (1967). Perception of the speech code. *Psychological Review, 74,* 431–461.

Licklider, J. C. R., Bindra, D., and Pollack, I. (1948). The intelligibility of rectangular speech-waves. *American Journal of Psychology, 61,* 1–20.

Mermelstein, P. (1979). Evaluation of a segmental SNR measure as an indicator of the quality of ADPCM coded speech. *Journal of the Acoustical Society of America, 66,* 1664–1667.

Miller, G. A., Heise, G. A., and Lichten, W. (1951). The intelligibility of speech as a function of the context of the test materials. *Journal of Experimental Psychology, 41,* 329–335.

Miller, G. A., and Isard, S. (1963). Some perceptual consequences of linguistic rules. *Journal of Verbal Learning and Verbal Behavior, 2,* 217–228.

Munson, W. A., and Karlin, J. E. (1962). Isopreference method for evaluating speech-transmission circuits. *Journal of the Acoustical Society of America, 34,* 762–774.

Nickerson, R. S., and Huggins, A. W. F. (1977, February). The assessment of speech quality (Technical Report No. 3486, DARPA Project MDA093-75-C-0180). Bolt Beranek & Newman.

Pirn, R. (1971). Acoustical variables in open planning. *Journal of the Acoustical Society of America, 49,* 1339–1345.

Pisoni, D. B. (1982). Perceptual evaluation of voice response systems: Intelligibility, recognition & understanding. Workshop on Standardization for Speech I/O Technology, National Bureau of Standards (ICST), Gaithersburg, Maryland, March 18–19.

Sergeant, L., Atkinson, J. E., and Lacroix, P. G. (1979). The NSMRL tri-word test of intelligibility (TTI). *Journal of the Acoustical Society of America, 65,* 218–222.

Schroeder, M. R., Atal, B. S., and Hall, J. L. (1979). Optimizing digital speech coders by exploiting masking properties of the human ear. *Journal of the Acoustical Society of America, 66,* 1647–1652.

Steeneken, H. J. M. and Houtgast, T. (1980). A physical method for measuring speech-transmission quality. *Journal of the Acoustical Society of America, 67,* 318–326.

Warren, R. M. (1970). Perceptual restoration of missing speech sounds. *Science, 167,* 392–393.

Webster, J. C. (1978). Speech interference aspects of noise. In D. M. Lipscomb, Ed., *Noise and audiology.* Baltimore: University Park, pp. 193–228.

PART 3
FUNCTIONAL ANALYSIS

CHAPTER 3.1
SURVEYS IN ORGANIZATIONS

STANLEY E. SEASHORE

University of Michigan
Ann Arbor, Michigan

This chapter, with some amendments, is reprinted with permission from *Handbook of Organizational Behavior,* edited by Jay W. Lorsch and issued by Prentice-Hall, 1986.

This chapter aims to raise some issues of strategic choice in the design, conduct, and interpretation of organizational inquiries using survey methods. These choices include, first, the question whether a survey is to be preferred over alternative methods. There follow issues relating to the design and conduct of a survey plan and choice among alternative technical features. Some associated issues concern the formulation of the topics to be covered, ethical aspects of survey practice, and the linkage of survey results to organizational purposes, policies, and plans of action. The reference throughout this chapter is to interview or questionnaire surveys with organization members.

3.1.1 THE NATURE OF SURVEYS

A manager (or anyone else, for that matter) wanting to learn something about an organization and its members can use any or all of several methods (Seashore, Lawler, Mirvis, and Cammann, 1983; Stone, 1978). The manager can examine records (Is turnover rising?), talk directly with some members who are in a good position to observe and report their views on matters of interest (Are your people satisfied with the new wage agreement?), have a consultant inquire by interview into the adequacy of coordination between two departments and report on suggested methods for improving it, or monitor the organization's suggestion to note the recurrent themes of dissatisfaction, preference, and problem solution. These plus other methods are all familiar and commonly used ways to collect information about the attitudes, opinions, beliefs, experiences, observations, and intentions of members of the organization. The substance may relate to policies, programs, facilities, persons, procedures, work environments, external environments, or anything else of interest. These are all "surveys" in the limited sense that they seek to obtain information directly from or about members of the organization. They all have their respective merits and limitations. The more thorough and systematic survey methods addressed here aim to strengthen the merits and moderate the limitations.

The employee survey approach to the assessment of organizations is not a new thing, but has a history of development spanning four decades. The application expanded gradually during the decades of the 1950s and 1960s as to both the number and the variety of employing organizations. Reliable current information is not available, but it is apparent that a majority of the larger work establishments, both public and private, have made some use of employee surveys, that many now have provisions for continuous or recurrent surveys, and that smaller establishments increasingly find it feasible and useful to conduct employee surveys. It has long ceased to be a novelty, or primarily a "research" rather than a "management" activity, and it has an established place among the roster of practices associated with the information getting, performance surveillance, and policy review functions of management. Many firms have in-house staff competence for such work and numerous consulting and research firms offer professional services to clients (Backstrom and Hursh-Cesar, 1981).

It can not be said that there is a typical practice, for the variations in program, purpose, method, conceptual basis, and application strategies are many. However, the following brief description treats one of the several common varieties of practice.

This chapter is written with primary reference to surveys of entire organizations, of subunits, or of categories of employees targeted because of the kinds of jobs they occupy. The procedures and issues to be described, however, can also be adapted to the kinds of problems commonly encountered in the application of ergonomic, human-factor, and work-system design tasks which usually encompass specific, unique, and local limitations. Adaptive suggestions appear at the end of this chapter.

3.1.2 AN EXAMPLE OF AN ORGANIZATIONAL SURVEY*

The establishment in question is a machine manufacturing firm employing over 3000 people in four plants, all located in the same city. The firm is prosperous and technologically advanced, with an established place in its industry. There are three labor unions with jurisdictions, respectively, over blue collar, white collar, and certain professional employees. The firm has a history of good employee relations, few occasions of organized labor dispute, and liberal employee benefit programs and services. Employee surveys had been conducted before, for special limited purposes but not on a regular basis. The survey was not stimulated by any crisis or unusual issues, but by a desire on the part of the management to have a comprehensive assessment of the current concerns of employees with respect

---

* This description is taken with permission and with minor alterations from *Labour and Society, 1,* (2), 71–73, (1976).

to their employment, sources of satisfaction and dissatisfaction, and opinions about various programs, policies, and working conditions then prevailing or planned.

Although the firm had the staff capability for such a survey, they preferred in this instance to engage an outside group, partly to ensure neutrality and anonymity for the survey, and partly to avoid staff overload.

A general conception of the purposes, content, and methods of the survey was worked out jointly with the consultants and then discussed with each of the three labor unions. Consent to the survey was given by the officers of each of the unions, with the provision that each would have opportunity to review the specific questions to be asked of their members, and each would have equal access with management to the statistical summaries resulting from the survey, but not necessarily to the consultant's interpretive analyses and recommendations. Guarantees of anonymity were provided for all individuals participating in the survey. Participation would be voluntary. All categories of jobs and levels would be included.

The survey process included: (1) preliminary interviews with a variety of employees to compose a roster of issues and topics of concern to employees, managers, and union officers; (2) a 50 min questionnaire completed at the workplace on paid time by a stratified random sample of employees and managers; (3) lengthy structured interviews with supervisors of major work departments and sections to obtain background, work organization, and work characteristics; (4) information from firm records concerning absences, turnover, productivity, and product quality for selected departments and sections (not for individuals). The content of the survey focused upon the perceptions, beliefs, and reactions of individual respondents regarding their own specific jobs and work environment. Subsidiary information used as analytic and interpretive aids included demographic information, reports of family and community factors that might be associated with the individual's work life, and the "work group effectiveness" data indicated by (4) above. With respect to the respondents' questionnaire reports, about 250 specific questionnaire responses covering about 50 topical areas were produced and recombined into about 40 generic conceptual variables representing environmental attributes (e.g., "physical discomfort," "job challenge," "pay equity," "work load") and individual responses to these attributes (e.g., "satisfaction," "preference," "motivation," "expectation").

The consultant's reports, planned jointly with the management and unions, provided statistical summaries for each question, each topical area, and each derived indicator. These data were shown in ways to allow comparison among work departments and sections, and among demographic classes of employees (e.g., by sex, age, union membership, length of service, type and level of job). The reports were provided to the management and to each of the unions. The consultants provided technical explanations of the data (not interpretations) and suggested procedures for the further review, interpretation, and application of the results by various special-interest groups within the firm. The interpretive process included special attention to issues suggested by (1) relatively high rates of employee dissatisfaction or preference; (2) evidence of "importance" derived from association between attributes of the working environment and such consequences as relatively high individual satisfaction or relatively high work group effectiveness; and (3) evidence of inequitable advantage to some group or beliefs about such inequity. Some limited attention was given to comparisons between the firm's own data and similar data from other unidentified firms provided by the consultant from previous studies.

The reader is reminded that this is but one example of several survey approaches to the assessment of a specific establishment, and is not to be regarded as "typical" or "prevailing." The varieties of practice vary along several dimensions, including the source of initiative, the definition of objectives or purposes to be served, the arrangements for control over the survey processes and outcomes, and the conceptual and methodological choices. It will serve, however, to illustrate issues to be discussed.

3.1.3 ADVANTAGES OF THE SURVEY METHOD

The main advantages of the survey method, compared to the alternatives of similar purpose, do not need much explication. They are as follows:

1. Surveys can be conducted with assured anonymity or confidentiality for individuals, thus allowing (although not guaranteeing) candor in reporting private opinions, treatment of "hot" topics that are not normally in management information systems and records, and access to information that does not exist except in the respondent's heads.

2. With sufficiently large samples of respondents, it is possible to apply statistical analysis techniques. The quality of the data can be assessed, subgroups can be compared, variations as well as consensus uniformities can be displayed, correlations can be determined, and the like.

3. Surveys with standardized questions and formats allow replication to detect changes over time, and allow extension to other parts or subgroups of the organization.

4. Cost effectiveness: a questionnaire survey can elicit, at relatively low cost, information from many people covering a wide range of topics. Interview surveys are much more costly per respondent, but may be cost-effective in other respects.

5. Sampling: respondents may be selected in ways that ensure representativeness, or absence of serious sample bias.

6. In some situations (with consent of respondents and tight confidentiality provisions) the survey data can be linked with "open" records—absences, productivity, subsequent turnover, pay, and the like—to allow the assessment of associations between what people say and what they do.

3.1.4 SOME LIMITATIONS AND RISKS

There are a number of limitations and risks that quite often proscribe the conduct of an employee survey or reduce its potential scope and utility. Among them are the following:

1. *Ambiguity of Purpose.* The initiators of a survey, usually an upper management or staff group, may find it impossible to reach sufficient consensus about the purposes of the survey, the priorities of content, and the procedures for post-survey review and interpretation of the results (Sirota, 1974). Managements (and unions) may already be so committed to the policies and programs at issue that the results of a survey are unlikely to have any practical utility in forward planning.

2. *Distrust.* Some initial level of trust in management and union leadership is needed for employees to volunteer freely, accept assurances of anonymity or confidentiality, and have confidence that sensible and considered interpretations and actions will follow.

3. *Unacceptable Topics.* In any organization, certain topics of great interest may be disallowed, on grounds that the employees have little information, context, or experience as a basis for forming opinions, or on grounds that a topic is too controversial. For example, management may prefer not to solicit employee views on practices that, for legal reasons, can not be altered; unions may not want open information about members' views on issues under negotiation.

4. *Organizational Disturbance.* The effective conduct of a survey requires consultation and information exchange with employees, scheduling of activities, and (usually) absence from workplace for interviews or questionnaires. Also, the survey is unavoidably a public event that may call attention to quiescent issues or induce unrealistic expectations about subsequent actions.

Such considerations, along with some others not mentioned, are not trivial. They often lead the knowledgeable professional staff member or an external consultant to dissuade the organization from undertaking a survey.

3.1.5 THE UTILITY OF SURVEYS

The preceding example of an organizational survey illustrates one common and relatively simple type of application. The aim was primarily to get a description of the current state of the organization on a broad spectrum of factors of interest to the management. The poll-like descriptive information referred to the *employees* as to their opinions, intentions and preferences, to the *organization* as to certain policies and programs, to the *physical and social work environment* as seen by the employees, *management–union* relationships, and the like. The information was used in a rather simple form: tables and graphs showing the mean responses and distributions (variances) on each variable or index, set up to allow comparison of demographic subgroups. Although some supplemental analysis did occur—for example, a search for patterns of work group composition and attitudes associated with high group performance—the utility was primarily to allow scanning for information that might suggest the existence of potential problems or provide reassurance that things were in a satisfactory state.

Organizational surveys, however, have a wide range of potential uses and a single survey can support multiple purposes. These multiple uses arise in part, of course, from the scope of information content in the survey. They arise mainly from the interpretational strategy that is invoked and from the statistical treatment of the results.

Prediction

Managers often want a basis for estimating the future behavior of their employees, or for understanding the conditions under which employees would do one thing or another. Some examples follow: (1) How many employees, and what kinds, expect to be looking for a job elsewhere during the coming year? (2) Under new retirement policies, just introduced or being considered, how many think they would opt for early retirement, for deferred decision, or for continued employment beyond the traditional age limit? How many more would choose early (or deferred) retirement if attractive bonuses were offered? (3) A firm plans to move its headquarters from an urban center to a suburb; if the move were made, about how many (and what kinds of) employees would expect to commute from their present homes, would move to new home locations, or would choose or be compelled to quit the firm? (4) How many of the eligible employees plan to sign up for a stock purchase plan to be initiated next year?

Such questions can be asked directly, are persuasively related to sensible forward planning, and can be answered directly by the employees for themselves. Supplemental questions can be used to amplify the considerations used in making their own forward predictions. Of course, some will be uncertain what they will do later, and some will change their minds. Changing external conditions (inflation, job market, etc.) might intervene to modify the future behaviors. Still, the information may be reassuring, or it may be arresting, as in the case of a firm with a history of moderate voluntary turnover where it was learned that over half of the work force definitely planned to quit and felt confident of finding acceptable jobs elsewhere.

The prediction of future behavior need not always rest upon poll-like self-reported intentions. If one has a confidently held theory or model about the correlates of the behavior of interest then predictions can be made from these correlates. Such a model can often be discovered empirically by analyses of survey data in conjunction with firm records, and the model can then be applied to estimating behavioral changes or comparative rates in the future. For example, the conditions associated with differential absenteeism rates in different plants or departments may be determined in this manner, and the model used to estimate the rates to be expected in the future under constant or changed conditions. The analytic procedures are analogous to those of operations research but applied to personnel rather than to a work system.

Explanation

Quite often the aim to be served by a survey is not description or prediction, but rather the investigation of causes—a search for an understanding of the factors leading to favorable outcomes (to be fostered) or unfavorable outcomes (to be moderated). Following are two examples: (1) An engineering design firm questioned their historical rates and high costs of rework—"finished" designs that had to be redone. Designers and draftsmen were surveyed to get a detailed description of the conditions and events associated with their most recent instances of rework. There resulted a short roster of frequently recurring "causes," some of which were inherent in the work and not controllable by the firm and some of which suggested minor modifications of work flow, information exchange practices, and location of responsibilities. Rework costs dropped substantially within a few months. (2) A firm with many branch offices wanted to understand better why some offices performed much more effectively than others. A survey of management practices reported by sales-persons in each of the offices suggested, by comparison among offices, a strong influence arising from the local support systems for new sales staff and variations in local approaches to coordination and decision-making. With this added understanding, programs for coaching and training could be designed in a way targeted to these kinds and locations of deficiency, previously unrecognized. In both these cases, other approaches might well have led to similar advantage, but the survey method was relatively quick and inexpensive, and the people affected by the changes "owned" and responded to the new information because it was their own, not imposed. To conduct such inquiries involving complex analyses of survey data it is necessary to have information about some "outcome" or "result" that is of concern—rework costs in one example, and sales volume in the other—to which the survey data can be linked.

Monitoring Change

Firms undergoing evolutionary changes (e.g., growth) or significant transitions (reorganization, new technologies) are put at some risk from unintended side effects upon staff that may accompany the main and planned changes. It may be judged worthwhile to monitor potential areas of impact to get early signals of gains to be preserved or losses to be moderated. Two examples follow: (1) A firm composed of relatively autonomous units, attracted by the potential utility of new information management technologies, set out to centralize the production planning, purchasing, and distribution scheduling for all units. Some feared that side effects might be a loss of local management initiatives, diminished sense of responsibility, "distance" from customers, delays in problem solving, and the like. To complement their established information systems for monitoring fiscal and production indicators, they introduced periodic questionnaire surveys of management people at all levels, giving the capability of detecting changes that might warrant some early corrective attention. (2) At least three large U.S. firms—probably more—maintain a service unit in the corporate office to conduct employee surveys upon request of the local manager in any unit of the firm. A standard questionnaire with some local amendments is used, derived from some years of experience as to the kinds of variables that may change and that are most useful to managers. Many units, but not all, elect to take such a "reading" of their organizational health at intervals of 2 or 3 years.

Monitoring programs such as these may be targeted to a particular one-time set of issues and subgroups in the firm, or may be a more generalized and broadly based screening procedure to detect trends of change that may require action or justify self-congratulation.

Evaluating Programs

Most firms introduce, from time to time, some programs of change intended to accomplish specific purposes. The purposes may be, for example, to reduce sexual harrassment, to accelerate the promotion of outstanding junior staff, to improve understanding of policy changes, or to improve the generation and flow of ideas about work procedures. Some information is available from supervisory and managerial reports about the progress and success of such programs, but often the crucial information is available earlier and more realistically through direct reports—that is, a survey—from the people who are expected to be affected by a program. Early surveys can provide guidance in altering, fine tuning, or abandoning a program; later surveys can reveal the degree of success in attaining the intended purposes. Unanticipated problems in the conduct of the program may be revealed. Frequently, the needed information is not of a kind that can be observed and reliably reported by supervisors and managers. Often brief, small-sample surveys using interviews or questionnaires can serve to track the progress of the program. Often there are valuable by-products: the organization learns something about how to change itself effectively, and the employees gain some influence in helping to guide the programs.

Deciding

It occasionally happens that a firm is confronted with a choice between alternative and incompatible lines of action or policy when the choice rests mainly upon the preferences of the employees. These preferences may be assumed, or roughly estimated, at some risk, or they may be measured through an advisory poll. Shall we introduce flexi-time? What additional fringe benefit is most preferred? How about the 4-day week?

Basic Research

All managers run their organizations on the basis of some set of values and some set of implicit or explicit theories about people and organizations. These are all subject to examination and change. Many issues can be illuminated by basic research procedures that may include the use of surveys as one feature—perhaps a central feature—of the inquiry. Most of the progress made in recent decades in rethinking issues of organizational design, organizational policies, personnel practices, and the like has come from such basic research conducted by, or with the support and collaboration of, managers willing to question the contemporary folklore and inherited wisdom of managers. The necessary posture to serve this purpose has two aspects: (1) a view that any organization is a kind of "natural experiment" in which variations in practice can be compared against some criterion of organizational effectiveness, and an organization may be purposely changed in an experimental way with provisions for evaluating the outcome. (2) Such basic research is likely to be undertaken without any short-term commitment or even an expectation that a particular course of action will be demanded by the results; instead, the aim is to gain some insight into organizational functioning that may be applicable to future problems and decisions.

   The number of such unresolved issues is still enormous. What is the optimal rate of turnover (renewal) for an organization? How do superior managers get identified and advanced through a career line? What are the conditions under which "participative" practices improve organizational performance? What are the main problems that arise when organizations are merged, or divided?

3.1.6 PLANNING A SURVEY: DESIGN ISSUES

The technology of organizational surveys has become quite sophisticated, and there exist books and journal articles treating the design issues in operational detail (Dunham and Smith, 1979). The intent in the following pages is only to highlight the main areas of choice in design.

3.1.6.1 Interviews versus Questionnaires

The factors involved in this choice may be economic, conceptual, political, or technical, in some weighted mix. If a large population of respondents is intended, then the economic considerations carry weight, for the questionnaire survey can be administered at a much smaller cost per respondent. Hiring or training of skilled interviewers is costly, interviews usually take more time (and involve two people) than a questionnaire of similar content, some travel or scheduling costs may be involved, and the costs of converting an interview record into quantified form for statistical analyses can be formidable. If the population of respondents is relatively small, the cost consideration may become minor, for the savings from a questionnaire administration tend to be offset by substantially added costs of preparing and pretesting a sound questionnaire, compared with a workable interview guide.

   On the conceptual side, a deficiency or risk of a "paper and pencil" questionnaire is that one must know in advance, or make good judgments about, what one wants to ask, how to ask it, and what constraints in interpretation one is willing to accept. The questionnaire is highly efficient in getting standardized responses to prescribed questions. The interview is highly efficient in accepting unanticipated responses, clarifying ambiguous meanings, and adapting the interview somewhat to the particular case.

"Political" considerations may involve issues of credibility, candor, and respect for respondents. For example, if the survey treats controversial topics and the results are to be subject to debate and challenge, then one wants the data to be from a representative and large sample of respondents and one wants the data to flow from response to statistical summary with minimal intervention for coding, screening, and interpretation of unique responses. A questionnaire or a questionnaire-like poll is preferred. In some situations where privacy and confidentiality are especially important, the questionnaire procedure offers added assurance to skeptical respondents. In some situations, the relatively higher (or lower) status of the interviewer can inhibit candor. Matters of language, gender, culture, and ethnicity may demand the neutrality of a uniform questionnaire or, in some situations, the sympathy of an interview by "one of us."

Among the technical factors that bear upon a choice between questionnaires and interviews, most arise from either constraints imposed by the intended analysis plans or constraints of comparison. If trend data are required, or comparisons among organizational units, for example, the procedures must be identical in all cases. If previously standardized scales and indexes are to be employed, the prior methods for getting the data must be replicated.

Most organizations with experience in organizational surveys use *both* interview and questionnaire procedures, in selective ways. They are often used in tandem—for example, an initial, small-scale interview survey to aid the formulation of topics, questions, and response formats to be then used in a larger-scale questionnaire survey.

3.1.6.2 Populations and Samples

Many organizational surveys include all members of the organization, or all members of a given class (all supervisors) or a given organizational unit (engineering department). No issues of sampling arise, for such a "total sample" can not be improved upon. However, when the base population is large, or its size and boundaries not known, some considered sampling procedure must be employed.

The elements to be sampled are usually persons, but quite often the interest is not in persons, as such, but in some other kinds of "object"—for example, jobs, events, decisions, groups, places, times, or products. In addition, the sampling need not seek representativeness, but may be purposefully non-representative. Thus, an organizational survey may be designed to include all those people involved in a sample of product redesign decisions, or those who are members of a representative sample of work groups. In some situations a purposive sample is most efficient even though not representative of the base population. Some examples of purposive sampling follow: (1) A firm wanted to compare high-productivity and low-productivity branches, and surveyed members of the top 10 and bottom 10 branches, ignoring the middle mass of "representative" branches. (2) A firm concerned about misinterpretations of their annual leave policy statement interviewed some employees, chosen mainly on grounds of variety and convenience, until the last 20 interviews produced no new information; the "sample" was small and unrepresentative but sufficient to identify the only five ways the existing policy statement could be misinterpreted.

When representative samples are needed, there are two main issues to be considered. *First,* how to ensure against a biased sample? Randomness is the key, but "random" does not mean merely "casual." Methods of varying convenience and precision for ensuring randomness exist. *Second,* how large a sample is needed? The answer to that question depends upon the degree of sampling error that is judged tolerable in the survey results and, more importantly, upon the adequate representation of any subgroups that will be singled out for comparisons or for special analyses. For example, if one wants to make a special analysis of middle managers, then it is necessary to consider whether a simple random sample will produce the requisite number of this class of respondents.

Those with some acquaintance with survey sampling have a useful bag of tricks that can be employed to obtain adequate samples without inflating the scale of a survey. An example is the purposeful oversampling of certain population categories that would otherwise produce too few respondents; this can be done without loss of the means for having random representation of the whole of the base population. Also, it is possible to make a complex sample design that provides interlocking subsamples, thus achieving simultaneously and efficiently a good representation of the base population, of departments or work groups, and of jobs.

3.1.6.3 Instrument Development and Pretesting

Whether a questionnaire or an interview procedure is chosen, a crucial step in design is the preparation and pretesting of the instrument to be used (Belson, 1981; Garden, 1969; Sudman and Bradburn, 1982).

Whatever the survey is to be about, it is likely that some other organization with a similar purpose has already prepared and used a questionnaire or interview schedule. As a rule, it is better to adopt all or parts of such "pretested" instruments if they fit the case. The fit is seldom perfect, however, and it is likely that some modifications will be needed to conform to the local terminology practices and organizational structure, or to include some topics unique to the new situation. There exist some general questionnaires and interview schedules in the public domain, available to all. Some are proprietary and can be had only on permission or in connection with the owner's services. A number of topic-specific scales and indexes in the public domain can be incorporated in an instrument that is otherwise locally developed (for example, scales for measuring job satisfaction, motivational properties of jobs, job stress, organizational structure, and supervisory style). Such instruments or components are likely to have gone through several revisions and to have documented properties as to reliability and validity (Cammann, Fichman, Jenkins, and Klesh, 1983; Dunham, Smith, and Blackburn, 1977; Smith, Kendall, and Hulin, 1969).

Whether one opts for a standard instrument or a tailor-made instrument, or some combination of the two, it is wise to provide a pretest with a small number of diverse employees to ensure that the questions are unambiguous, the response categories clear, and none of the questions are unnecessarily offensive or threatening.

3.1.7 PLANNING A SURVEY: ORGANIZATIONAL ISSUES

There are a number of issues of organizational policy and practice that should be considered as part of the survey planning process. Failure to do so is likely to result in such issues coming up later in circumstances that make their resolution troublesome.

3.1.7.1 Confidentiality

Some surveys involve topics and respondent populations for whom confidentiality, privacy of views, and anonymity are not feasible or not a matter of great concern. The usual case, however, puts a high premium upon assurance to the respondents that their views can not be used personally either for individual advantage or adversely. Some surveys contain hazardous information—about the behavior of one's supervisor, evaluations of top management, intentions to quit, drug usage at work, areas of dissatisfaction, and the like. Employees would be naive or foolish to respond with candor unless they feel assured against harm and see no advantage in misrepresenting their views.

Anonymity usually can be assured with credibility if employees have a modicum of trust in the organization and if various symbolic and practical steps are taken. The usual minimum steps are to state clearly the firm's intention to respect anonymity, to omit respondents' names and identifying numbers from the questionnaires, to designate openly the persons responsible for maintaining anonymity, and to ask few questions of kinds that could conceivably be employed to identify a particular respondent. A further step might be to separate access to the data from the organization by using an outside agency to collect, store, and analyze the data. In some few instances, organizations have seen fit to form a committee of employees to oversee the arrangements for handling and protection of the data.

A potential cost of total respondent anonymity is a constraint upon the certain legitimate and valuable uses of the survey data. The main point here is that without identification of the respondents there can be no later resurvey of the same population and no linkage of the survey data with organizational records. Ordinarily, this is not a matter of concern. For some purposes, however, identification of respondents is desired. Two examples follow: (1) To understand the delayed effects of different kinds of job stress upon individuals it is desirable to resurvey the same individuals and analyze the two surveys jointly. (2) To understand the conditions associated with absenteeism, high productivity, career progress, and the like, it might be necessary to tap the organization's personnel and operating records. It is possible, or course, to maintain confidentiality even when anonymity is forfeited, and employees will ordinarily assent to this provided their cooperation is voluntary and the procedures for protection are plausibly explained.

3.1.7.2 Participation in Planning

The planning of a survey inevitably raises a fundamental question regarding the location of control over its purposes, design, and the use of the results. Aside from any sense of "rights" to a share in such control on the part of top managers, staff specialists, supervisors, union officers, or other groups, there is the further question whether the quality and utility of the survey might be enhanced by their participation in the planning. The question arises with particular force in an organization with little past experience with surveys, for then the risk is greater that people will misunderstand features of the plan and be less confident that their concerns and interests will be taken into account.

Some considerations weigh against broadly extended participation in the planning. Participation is likely to extend the planning time and to require diversion of some payroll hours from other work. Some ideas may be advocated that then have to be rejected; rejection may be on technical grounds, on grounds of policy or legality, or simply because there is no way to accommodate an overabundance of suggestions.

On the positive side, early discussions with various interest groups or their representatives are likely to produce some good ideas, will certainly forestall some misunderstandings that otherwise would arise, and will very likely help to ensure cooperation in the conduct of the survey.

3.1.7.3 Voluntarism

It is a rare situation in which an organization can, with impunity, require employees to be interviewed, or to fill out a questionnaire, on matters that seem to intrude upon their sphere of privacy. Studies have shown that in most organizations employees (including managers) have rather clear ideas about the kinds of information the management is entitled to demand (Can you work overtime next Saturday?) and the kinds employees may legitimately withhold (Why can't you work overtime on Saturday?). Accordingly, surveys are usually conducted in a way that makes it possible for an employee, without risk, to decline to participate.

The principle of voluntarism arises out of respect for the prevailing norms of individual privacy, but also from the practical fact that anyone so coerced will be free to, and probably stimulated to, respond with less than total candor. The burden falls upon the management to make sure that the prospective respondents are informed about the nature and purposes of the survey, as well as about their safe option to decline. Some will do so. The proportion of decliners may be as low as 1 or 2%, under optimal conditions, or as high as 30% or even more if practical problems prevent some participation (illness, vacations, illiteracy) or if the levels of disinterest or distrust are high. Participation rates of 85 to 95% are commonly attained; if they are much lower, and if representativeness is important, there is a need to evaluate the risk of distortion from selective, biased declination.

The steps taken to ensure cooperation are usually those customary for the organization, and may include some or all of the following: (1) a letter of explanation to each prospective respondent; (2) meetings to present the plan and respond to questions; (3) supporting communications from union officials or the respondent's own supervisor; (4) articles in the house organ; (5) a "hot line" for anonymous queries about the survey plan; (6) assurance that interviews or questionnaire administration will be on company time, not on personal time.

3.1.7.4 Reporting the Results

With any organizational survey, questions will arise about the reporting of results. Who will get to see the results? In what detail? When? For what purposes? These questions arise for some people out of mere curiosity, interest in the organization, or wonderment about how one's own views compare with those of others. There is usually some sense of reciprocity: those who volunteer information may feel they are entitled to get some in return. For others, the concern about the reporting and use of the results is much more intense and more closely linked with their role, status, and responsibilities in the organization. The results may reflect well or badly upon their own performance, or they may be highly pertinent to current organizational programs, problems, and tasks for which they are responsible. The data may contain some potential both for risk and for help.

Prudence requires that questions of reporting be anticipated and, as part of the early planning, that some guiding principles be worked out and tested for acceptability. The primary factors to be considered are the intended uses of the data by the organization, and matters of respondent privacy and data confidentiality. The results must get to those who are to use it, in a form and detail matched to the intended use. The information must be in a form that protects respondents against undue disclosure. Confidentiality must be weighed against the costs of secrecy and the possible gains from openness.

The resolution of these questions of reporting can not be generally prescribed for each survey has some unique features of content and context. An example can be offered for the common case of a questionnaire survey that is broad in topical scope and diverse respondent populations.

Such a survey probably should not be undertaken at all unless the plans include review of the information by all or many of the people whose organizational performance and future actions are implicated. One form of reporting is to undertake a "data feedback" program in a "waterfall" fashion designed to encourage understanding of the information, diagnosis of potential areas of action, and planning of feasible actions. For this, the survey data are summarized in statistical form aggregated separately for each of the organizational units and delivered to each unit along with some assistance in its interpretation. Thus a top management group might get the data summarized for the organization as a whole and for a few major subparts of the organization. At the next organizational level, each management group would get its own unit's data as well as the pooled data for the organization as a whole for comparison. Thus in a "waterfall" manner, the pertinent information is provided at some or all levels of the organization in a form and detail appropriate in each case, with great flexibility as to open or limited exposure to information about others' units and great flexibility as to the topics to be reported. Some selectivity as to topics covered is usually needed, either to highlight the more pertinent ones or simply to avoid information overload. Receiving groups may be invited to request any supplemental information from the survey that they need, and may be invited to report their interpretations, problem definitions, or proposed lines of response (Rickards and Bersant, 1980; Dodd and Pesci, 1977).

In addition to directing the survey information to user groups in some fashion that compels attention, such as that described, an organization may also wish to make public within the organization the results that are of general interest to employees. The objectives and methods would be more in the style of public relations than of problem solving and action. Members of the organization are likely to be interested in knowing of some of the general findings from the survey and particularly about what is being done to use the survey for organizational benefit. Failure to report may feed the rumor mill or leave an impression that no one is paying attention to the views of those surveyed. Conducting a survey creates the expectation that something will be done with it, and no news may be considered bad news. As one manager put it, although with excessive drama, "A survey is like pulling the pin of a hand grenade . . . . You can't just stand there holding it."

In most organizational surveys, the main problems of survey reporting arise because of the sheer volume of information created, some of it inconsequential, and from the fact that some issues will require extensive analysis before the meaning becomes clear. An effective strategy often is a sequential one, with early screening of the results by users for priority topics, and later stages of statistical analysis in collaboration with the user groups. For example, a given department may be notably lower than the others with respect to employees' confidence that their career aspirations will be realized; the review group may then ask the data base what kinds of people in what kinds of jobs feel singularly disadvantaged, and this may help to decide what can be done about it, if anything.

There exist some effective conventions for reporting survey data in ways that protect privacy. Many adopt the rule that no data will be aggregated and released for subgroups of small size, or the rule that remarks in interviews, or write-ins on questionnaires, will not be cited verbatim if they contain any potentially identifying features.

3.1.7.5 Professional Help

Although organizational surveys can be done by persons lacking professional qualifications and experience, and sometimes with a satisfactory result, an organization is well advised to get qualified advice. At a minimum, such a person should review the plans as to purpose, instrumentation, administration arrangements, analysis and reporting, and the intended interpretive and use strategies. Such a person is likely to have good ideas that will moderate risks and add to effectiveness and efficiency. Most organizations, however, will want to have the survey managed—not merely inspected—by a professionally competent person, either an inhouse staff specialist or one obtained from the outside. Even firms that maintain an inside staff with survey qualifications often seek outside help; they may do so to balance staff work load, or to have the help of people with special qualifications for a particular kind of survey, or to gain the added assurance of respondent protection and cooperation by locating the survey data elsewhere.

The inside professional has the advantages of familiarity with the organization, knowledge of past and concurrent events that might bear on effective planning, and continuity and presence throughout the whole period of planning, conduct of the survey, and applications of the results. The outside professional is likely to have the advantages of having done a wide variety of surveys under different conditions, an acquaintance with resources and instruments that are available, full attention to the survey, and dropping off the payroll when no longer needed.

Competent professional help is readily available. Some individuals and small firms specialize in such service. Larger general consulting firms are likely to have an appropriate capability. Many universities have staff and support facilities that can be made available to organizations; they are likely to be found in the department of psychology or the school of business administration. Organizations that wish further specialization and pertinent experience might find a suitable person with expertise, for example, in hospital organizations, social agencies, or public administrative organizations.

3.1.8 PLANNING A SURVEY: ANALYSIS AND INTERPRETATION

The interpretation of organizational survey data can be approached at several different levels of complexity, each with distinctive features of convenience, cost, and analytic power. When planning the survey it is essential to look ahead to the intended uses of the data, and to build into the design the features necessary to allow use in that way (Sonquist and Dunkelberg, 1977).

Much can be learned about an organization (or an issue) by simply scanning the "raw" aggregated data displayed graphically or as simple tables of means and distributions. For this use one need only bring to bear one's own judgment whether a given result seems good or bad, alarming or reassuring, or whatever. It helps to compare observations among several scanners, and it helps to have had some prior experience in "reading" survey data.

However, the raw data can be deceptive, the meaning and action implication may not be apparent, the necessary collateral information may not be clearly in mind or easily found, and the joint review of several related measures may become complex enough to require simplification through multivariate statistical analyses. Thus, it usually happens that a group of managers examining the aggregated data from their survey may have some difficulty arriving at a common view of what the data mean, and therefore what they might do about it. They may observe that a substantial proportion of their people report general satisfaction with their jobs, say, 85%. To some this seems to mean that satisfaction prevails, and all is well. Someone else will inevitably ask, "Shouldn't they *all* be satisfied? What are we doing wrong?" Someone else will ask, "What difference does it make to our year-end bottom line? None, I'd guess." Another will ask, "I wonder *what* they are dissatisfied about—maybe we need their ideas about some things that should be changed." In short, the interpretations may require assumptions and speculations. Questions get asked that can be clarified by a more sophisticated look at the data. An experienced "reader" might help by observing that job satisfaction rates generally run a little higher than 85%, that job satisfaction—at least most of the time—is only weakly associated with productivity but is usually a factor in absences and turnover, that some dissatisfaction is inevitable and probably not a bad thing provided it is a transient condition, with few people being persistently and chronically dissatisfied.

Beyond such direct reading and interpretation of the survey news about one's organization there are several further steps that can be taken if the survey is well designed: (1) comparison among unlike or contrasting groups within the organization, (2) comparison with like groups—the same organization at an earlier time, or other organizations; and (3) testing the meaning of the data against some criteria. The criteria may be "outcomes" that represent the organization's values and goals, or other correlates of some organizational significance.

3.1.8.1 Comparing Unlike Groups

It is usually the case that an organizational survey covers enough respondents of different kinds and in different parts of the organization to allow useful analytic comparisons. To pursue the job satisfaction illustration, one can ask the data set whether dissatisfaction is concentrated among women compared with men, among younger employees, among midlevel managers, compared with those at higher or lower levels, or among those long in their present jobs compared with those that have moved, or whether the dissatisfaction rates are similar in all departments. The possibilities are many. The patterns of association of dissatisfaction with demographic, organizational, and personal characteristics will often clarify the issue of what dissatisfaction "means" in the organization and will often suggest whether corrective steps are needed and might reasonably be taken.

The same interpretive strategy can be applied to matters that may be of more importance to the organization than employee satisfaction. Suppose that a rather large number of people report that they have important skills and abilities that are not being used in their present jobs. No doubt some are overestimating the value of their unused skills, but the question arises whether such underutilization (certainly very costly) is distributed generally, is concentrated in certain organizational units, or is concentrated in certain categories of employees.

For example, one large organization found that nearly half of their people in technical positions felt that they had little or no chance to use their best skills. Comparison of this group with others showed that this view was reported unduly by employees who were relatively young, had been recruited from universities during the last few years, and were employed in units other than R & D and general business management functions. Further, a comparatively large proportion of them felt their promotional opportunities were limited, and were considering a move to some other employer. The managers knew that there was some disaffection among these employees, but the strength of it surprised them. They wanted some turnover (the leavers often went with customer firms) but felt they were at risk of losing too many and perhaps the best ones. There followed a reconsideration of their recruitment program, and the introduction of a fast promotion program and more challenging jobs for a limited number of the more promising young employees.

Some users of survey data routinely scan with a preselected, limited array of demographic, positional, and attitudinal variables to have a condensed report of instances of "off standard" patterns of association when tested against the standards set by the prevailing conditions within the organization. Such a procedure is unguided by theory or purpose, and risks turning up some information that proves trivial upon examination, but it can uncover unanticipated problems (or the absence of "problems").

3.1.8.2 Comparing Like Groups

Nearly all users of organizational survey data wish for some external norms or standards that will help them evaluate their own survey results. How do we compare with other organizations? How do we compare with other firms similar to ours? This is a sensible idea and sometimes a practical one as well. Although there is little reason to assume that the prevailing norm is optimal, or a suitable standard for one's own organization, still it can be useful to know whether one's own results are typical, or unusually high or low.

Two conditions are needed to make such comparisons. First, one has to use an interview schedule or questionnaire that is at least in part a "standard" one—that is, one that has been used in other organizations. Second, one has to have access to the survey results from other organizations along with enough information to allow judgment about their comparability.

Access to appropriate data from other organizations is not always easy and not always useful, but should be considered as part of the early planning. In large multi-unit organizations with a continuing survey program there is likely to be a central data bank allowing comparison of the results for any unit with the norms from other similar units in the firm. Some service firms and agencies maintain data banks from surveys they had conducted in the past in diverse organizations; these may allow comparison of one's own data with those of other (unidentified) organizations of similar character. There are also a few public data banks from national sample surveys, cutting across all types of employment, which potentially can allow comparisons based upon specific categories of employees— for example, secretaries, factory production workers, managers, teachers, and the like. There are a few industry associations whose members agree to use a common survey procedure and to pool their results, without member identification, to provide shared industry norms.

To plan upon such comparisons outside one's own organization it is essential to locate the source and arrange access in advance. It is also necessary to accept all or much of the "standard" instrument at a loss of flexibility in designing an instrument tailored to local issues and conditions. As usual, costs and benefits must be weighed. The anticipated benefits are usually simple: to get a fix on whether one's own organization is typical, or is, instead, off-standard in significant ways. The interpretations are often not straightforward, for one has to assume that the compared organizations are really similar, without unique conditions that reasonably "explain" and justify the differences that are found.

The most informative comparison, when feasible, is between the survey results obtained and those from the same survey of the same organization at a prior time. This allows detection of trends of change. Managers are often more concerned about where their organization is going than in where it is now. This accounts for the increasing use by organizations of periodic surveys—say, at intervals of 2 or 3 years—to monitor progress toward, or away from, their preferred organizational conditions.

3.1.8.3 Impact Analysis and Diagnosis

The foregoing approaches to the interpretation of survey results rest upon normative judgments and group comparisons. They invite and inform decisions about what issues to attend to and they can identify some of the conditions that pertain to choosing a course of action. In many cases, this is sufficient. However, the survey method of inquiry allows analyses that are targeted to selected issues and designed to guide actions and policy changes. The procedures involve a multi-factor analysis of the conditions associated with some criterion that is valued.

As an example, a firm took pride in their salary program for middle level managers, which emphasized "individualized" salary increases over across-the-board scaled increases, large differentials in salary increases based on merit, and a policy of confidentiality among peers about salaries. They wanted (their criterion) their people to expect pay keyed to performance and also to be satisfied with their own pay. Their survey asked, among other things, how satisfied respondents were with the amount of their salary and with the fairness of their salaries compared with others in the firm. Analysis showed that dissatisfaction was not, as expected, limited to those whose salaries were objectively lower than those of others of similar age, service, and rank but occurred also among well-salaried "high flyers." A dozen possible explanatory variables were checked out singly and in various combinations; one stood out as a significant "cause" of dissatisfaction which seemed to have no offsetting gain or constraint. This factor was the confidentiality policy, which apparently was being honored in some departments but not in others. Respondents who reported "knowing the salaries of most of the people in your group" tended strongly to be more satisfied with their own pay and with the equity of the salary determination system; this was true equally for those relatively over- or underpaid, and was unrelated to the size, function, or average salary in the many work groups. It appeared that the confidentiality policy was having an effect in this firm opposite to that intended. This could not have been discovered by direct inquiry nor by debate among policy-makers. The firm declared an "open information" policy and instructed the managers accordingly.

To undertake such a targeted impact analysis, one must anticipate such a use for the survey data and ensure that the survey instrument contains a set of questions that allows a reasonably inclusive analytic resolution of the issue. There is a catch-22, however, for to ask the optimal questions one should know in advance what variables are likely to be pertinent. In most situations, this dilemma solves itself to some degree—the issue is probably recognized as an issue because people on the scene have different ideas about it. These competing ideas suggest many of the facets of the issue that should be incorporated in the survey instrument. Many of the issues that are likely to arise have been researched by others or debated in print, and these sources can be used to help formulate the set of questions that is needed.

The criterion to be targeted in an impact analysis can be almost any organizational state or outcome thought to be important, so long as it is variable (not a constant) within the organization and so long as it can be measured with some degree of accuracy. The issue may refer to individual employees, provided that the criterion can be measured for individuals and linked with their survey data, or it can refer to work groups, organizational units, or critical events. Such analyses have been done, for example, to clarify such questions as the following: Why do some employees fail to attain the work effort and added pay that are available under an incentive pay system? What are the characteristics of work groups with low (or high) absence rates? What are the trade-offs among employees of different categories as to a choice between more pay or an improved fringe benefit? Are the new CRT work stations more, or less, stressful than those not yet brought under the new production technology?

Such analyses are likely to require some fairly sophisticated statistical and theoretical competence. They typically involve imposing controls for personal and situational factors that might otherwise leave the results ambiguous, and may require testing for data quality. Often they require simultaneous treatment of interacting factors in multivariate analysis strategies. This work is not for the novice. Professional help is needed.

3.1.8.4 Interpreting Subjective Measures

Users of organizational surveys, especially if not experienced, are often skeptical about the accuracy and validity of survey data. Are not attitudes unstable and likely to change? What is the use of measuring opinions if the employees are not well informed about the complexities of the issue? Do not people often behave in ways incompatible with their expressed views? People seem to have an overall positive or negative mental set that colors their response to all questions. Many of the questions provide for vague answers, such as "many" or "very little," with no reference to hard numbers.

These are grounds for skepticism. The accuracy and validity of survey data should be challenged by managers at least as rigorously as they challenge their other sources of information. The confidence with which managers use their familiar operating and fiscal data arises not so much from their intrinsic superior quality as from awareness of their limitations and experience in their use. They know that an index of quarterly profit may be as much a construction of art and policy as a report of "true" profit, and that a tally of in-process reject rates is, at best, a soft approximation of current quality problems. Both, however, have familiar limitations, and both serve reasonably well for detecting trends or unsatisfactory conditions.

In some respects, the quality of survey data can be assessed statistically. Professional and frequent users of such information routinely conduct checks for reliability, stability, concurrent validity, predictive validity, and internal consistency of response. The occasional user is not likely to undertake such tests, but relies on the counsel of someone who is familiar with the pitfalls, the ways of maximizing reliability, and the like.

Aside from the technology of data quality (which always shows the measures to be less than perfect and occasionally shows them to be unusable) there are considerations of a conceptual sort that are of help. One is statistical: Although any individual respondent's report of opinion may be erroneous or distorted in some individualistic way (i.e., unreliable) the aggregation of reports from similar respondents can "average out" some of the error to produce a measure for the population that is highly reliable. Another is interpretational: Employees may hold attitudes and opinions that are uninformed, irrational, or naive (or in other ways unlike those of the interpreter), but still these attitudes are the basis upon which people act; the attitude is a useful fact in itself, quite apart from its objective "correctness." The stability of attitudes and opinions is highly variable; when conditions change, when there is little initial basis for taking a view, and new information is received, then attitudes do change—as anyone should know from following the presidential election polls; many other attitudes and opinions are highly resistant to change. The absence of anchored equal-interval scales (as in the case of dollars, lineal feet, or pounds) can be distressing at first, but the experienced user of survey data soon learns that scales of variation (i.e., deviations from a norm or mean) work well when the interpretations rest upon relationships among measures more than upon their absolute scale values. There can be no "zero point" on a scale of employees' trust of top management, although there may be a neutral point meaning "no opinion."

Not all components of an organizational survey are "subjective" and, for some purposes, a survey may emphasize a preference for data of objective reference. This range of possibilities arises because the employee respondents may answer queries in different roles: as subjects of inquiry (What is your opinion?), as part of a panel of expert observers (How frequently is the work in your area interrupted by parts delivery failure?), or as sole sources of the objective facts (How do you usually get to work—on foot, by personal auto, car pool, or public transportation?). The preference for objective or subjective information should arise entirely from the requirements of the issue addressed, not from any belief that objective data are inherently more reliable, more valid or more useful. For example, in an inquiry about pay it may be useful to know how much, or at what rate, the respondent is objectively paid, but for some purposes it is more important to know whether the respondent regards the pay to be "sufficient for your normal personal and family expenses." The two measures will have similar degrees of accuracy, but they convey information that is different in meaning and possibly uncorrelated.

3.1.9 PLANNING A SURVEY: ACHIEVING AN INTEGRATED PLAN

It should be evident from the foregoing pages that a plan for an organizational survey has many component parts and that each part offers some choice among alternatives. Also, each organization is unique to some extent as to the constraints to be recognized and opportunities that might be exploited. It would appear, in principle, that the variations are endless and the choice points very numerous.

In practice, matters are more simple than that. First, the components of a workable plan as to the design features tend to be interdependent in such a way that a few key choices will preclude many of the related options that exist in principle and will make many of the remaining consequent choices virtually self-determining. Second, an organization does not realistically have all of the options, for there are some constraints arising from its own history, resources of time and money, readiness to handle information of an unfamiliar kind, and the like.

The starting point, of course, is to formulate some conception of purposes that might be served by an organizational survey. These purposes may intrude themselves in the form of specific difficulties or problems that press for resolution. They may arise from speculations about trends and prospective future problems, or from an interest in an overview appraisal of the health of the organization. Thus the defined purposes may press toward a survey narrowly focused upon selected current issues, one that allows treatment in depth of some aspects of organizational life that are of longer range concern, or a survey that asks broadly, "How are we doing?" A survey can be tailored to the purpose, but the purpose must be known and must be acknowledged by those organization members who are to try to use the results. This self-evident truth is often overlooked, with the consequence of having a survey that fails to match the needs, expectations, and capabilities of the users.

3.1.9.1 Design Elements

The main elements in the technical design of a survey have been mentioned in prior pages: topical coverage, form of instrumentation, selection of samples or targeted categories of respondents, advanced planning for an analysis and reporting procedure, and preparation of the prospective users for their work of interpretation and action.

The crucial idea is that these elements are all so interdependent that they have to be considered as a set, not as separate choices. For example, if the purpose is to get a broad "reading" of the conditions of the organization compared with external standards or norms, then it follows that the instrumentation will be a standard one, that participative design is precluded, that the coverage will be broad with some forfeiting of depth of topical coverage, that the respondents will be a representative or total sample of organization members, that the analysis and reporting plan will be relatively simple and must conform to that established by the external source of norms, that sophisticated analysis of specific issues is probably precluded, and so on.

It is evident that there are trade-offs among design features, with each feature establishing some requirements as to others and forfeiting of advantage in others. There is no objectively ideal or optimal design that fits all organizations. A design can be optimized to fit the priorities imposed by the purposes of the survey or to fit the values and preferences of the deciders. It is usual for two or three (perhaps more) alternative overall designs to be reviewed and weighed before a choice is made.

3.1.9.2 Fit to Context

Quite independently of the technical design features that have been partially enumerated, there is the question of fit to the particular organization. There usually exist both some constraints that have to be accommodated and some opportunities that invite attention. For example, if there is some doubt whether employees have full confidence in the protection of their anonymity or in the considered use of the survey results, then it is likely that some risky topics will be left untreated and that no attempt will be made to link the survey returns to individual data of record. Similarly, if there has been an earlier survey in the same organization, attention goes to the option of replicating the earlier design in whole or in part to gain the advantage of assessing trends of change. If there exist many work groups of similar function but highly variable performance (a kind of "natural" experiment) then attention may turn to designing the technical features of the survey to allow examination in depth of the causes of higher and lower performance. All organizations offer some constraints and opportunities of such kinds.

3.1.9.3 Organizational Linkages

One of the early strategic choices deserves special mention in this terminal paragraph, namely, the choice whether, at one extreme, to regard the survey as a top management activity primarily for informing the upper levels of management and selected staff groups or, at the other extreme, to embed the survey in the whole of the organization's review and decision-making system (Passmore and Friedlander, 1982). Of course, there are intermediate positions as well. The posture taken on this matter has much to do with the approach to planning and with the extent of utilization of the survey results. Consider this example: A firm had a history of being relatively open with information and committed to engagement of employees in matters that concerned them. A "steering committee" was formed to plan the survey—a rather large group, with half the members nominated by the three unions and half by the management. External consultants worked with the steering committee to formulate and adopt the plans. A "standard" questionnaire was used, with numerous amendments and additions developed by the committee. It was agreed in advance that following the survey there would be formed a number of smaller issue-oriented task groups to review the survey results (along with any other available and relevant information) and to recommend any actions that seemed appropriate and feasible. The management reserved the right to appraise the recommendations for possible acceptance, but plainly was predisposed to be accepting unless there were strong contrary factors. The survey analysis and reporting was planned to be responsive to the special requirements of each of the task groups, each of which could request additional analytic steps as their understanding of their task developed. Of the dozen task groups, a few reported that they had no recommendations to offer. The others, over a time span of 18 months, produced a number of acceptable proposals for changes in the firm's policies and practices, including a successful major reorganization of one work unit.

Such an approach is not feasible or not desired in many organizations. It is one way to get good value from an organizational survey.

3.1.10   WORKPLACE AND WORK-SYSTEM APPLICATIONS

Survey methods are potentially applicable to the solution of human factor, ergonomic, and other work-system problems when the information required is not readily obtainable in other ways. There are common situations of this kind. They include the following: (1) The needed information may be historic, available only through the recollections of people who have observed past events (e.g. operation delays, equipment breakdown, inadequate response to off-standard situations). (2) The informants are numerous, or dispersed in time or space, and thus not readily interviewed informally (e.g., operators of mobile equipment, observers of rare but crucial events, persons in long-linked sequential operations). (3) The information has not been or can not be collected easily through some reporting and record accumulation system. (4) The needed information concerns private, individual mental states, or activities (e.g., their attitudes, intentions, knowledge, preferences, and suggestions concerning work performance, workplace equipment, or work environment).

This reasoning suggests that many human-factor and ergonomic inquiries may be aided by the application of survey methods. However, others may be exempted: there may be so few informants that direct conference with them is preferred; the informants may be inaccessible; the ergonomic solutions may be so obvious, urgent, and compelling as to preclude benefit from a survey; the problem may be solvable only through experimentation under laboratory conditions or through computerized simulation.

Many readers will readily think of situations, past or future, that invite the use of survey methods in their own work. Some concrete examples are provided to stimulate thought.

Tellers and other customer service representatives of a large bank were surveyed weekly, for several months, to collect an inventory in detail about otherwise unreported customer complaints that had some bearing upon the bank's facilities, policies, and operating procedures.

Design engineers were surveyed regarding their most recent experience of work delay or rework, and their opinions as to the causal factors.

Assembly line workers were surveyed regarding the incidence and nature of observed "near accidents," that is, potentially serious accidents that would have occurred but for luck or extraordinary unprescribed actions.

In connection with the design of a new chemical plant, operators in a similar existing plant were surveyed regarding their main problems with existing work station layout and instrumentation, and their suggestions for improvements for the new plant.

Secretaries were surveyed for information about their time allocation to various tasks, as a basis for modifications of their work stations and the reallocation of some tasks.

Vehicle maintenance mechanics and solid waste collection operators in a major city were surveyed by questionnaire and interview to identify desired improvements in the design of the compactor vehicles. There resulted several modifications, overlooked or previously rejected by the vehicle specification staff, which gained maintenance economy, fewer vehicles out of operation, added comfort, and reduced pickup time per ton.

The foregoing instances illustrate the use of numerous survey respondents, not as subjects of study, but as informants about attributes of their work and work facilities. The following instances concern the employees' own attitudes, suggestions, intentions, and the like.

A firm had reason to believe that it had "overdesigned" many of its jobs as to simplification, routinization, required training time, and latitude for employee judgment. A survey of employees, using a standard instrument and some interviews, led to the redesign of many of the jobs to get a better balance between the technical ideals and the tolerable psychological demands of the work. Costs, employee turnover, uncorrected errors, and employee dissatisfaction were reduced.

A multistation assembly operation was provided with operating supplies and hand tools (e.g. adhesives, power screw drivers) chosen mainly on the basis of slightly lower cost and known technical adequacy. Some operator suggestions led to the systematic trial of alternative supplies and tools, using a survey method to assess employee preferences. There appeared a clear majority consensus and preference on several items; their introduction improved productivity and employee satisfaction at trivial increased supply cost.

A firm enlarging its work area and capacity surveyed present employees as to the properties of their work areas (space, atmospheric conditions, supply sources, traffic, ease of communication with interdependent others, and the like), and used this "preference inventory" to achieve an improved plan for the new facilities.

A reader wishing to inquire about such applications would do well to begin by reading an introductory text on survey design and a few examples of their application. *Organizational Surveys,* by R. Dunham and F. Smith (Glenview, IL: Scott, Foresman, 1979) is a brief operational guide to survey methods, enlarging considerably on some of the issues discussed in this chapter and providing references to supplementary sources that may be needed. Most instances of application do not get into print in a detailed way, but some do. An application to a problem of rework delays and costs in an engineering design organization is described by W. Hancock, B. Macy, and S. Peterson in "Assessment of Technologies and Their Utilization," a chapter in *Assessing Organizational Change* (S. Seashore, E. Lawler, P. Mirvis, and C. Cammann, Eds.; New York: Wiley, 1983). For an application in the optimization of job design, see *Work Redesign,* by R. Hackman and C. Oldham (Reading, MA: Addison-Wesley, 1980). An application to skilled maintenance workers in a production facility is described by W. Hancock in the *Journal of Contemporary Business* (1982, *11*(2), 107–114), and an elaboration of the theory and method is provided by J. Liker and W. Hancock in "A Method for Detecting Systems Barriers to Engineering Productivity" (Technical Report 84-25, 1984, I&OE Department, School of Engineering, University of Michigan).

REFERENCES

Backstrom, C. H. and Hursh-Cesar, G. (1981). *Survey research.* New York: Wiley.

Belson, W. A. (1981). *The design and understanding of survey questions.* Aldershot, England: Gower Press.

Berdie, D. and Anderson, J. (1974). *Questionnaires: Design and use.* Metuchen, NJ: The Scarecrow Press. 1974.

Cammann, C., Fichman, M., Jenkins, G., Jr., and Klesh, J. (1983). Assessing the attitudes and perceptions of organizational members. In S. Seashore et al., Eds., *Assessing organizational change.* New York: Wiley, pp. 71–138.

Dodd, W. E. and Pesci, M. L. (1977, June). Managing morale through survey feedback. *Business Horizons.*

Dunham, R. B. and Smith, F. J. (1979). *Organizational surveys.* Glenview, IL: Scott, Foresman & Company.

Dunham, R. B., Smith, F. J. and Blackburn, R. S. (1977). Validation of the index of organizational reactions with the IDT, MSQ and focus scales. *Academy of Management Journal, 20,* 420–432.

Gorden, R. (1969). *Interviewing: Strategy, techniques, and tactics.* Homewood, IL: Dorsey Press.

Passmore, W. and Friedlander, C. (1982, September). An action-research program for increasing employee involvement in problem solving. *Administrative Science Quarterly, 27*(3), 343–342.

Rickards, T. and Bessant, J. (1980). A mirror for change: Survey feedback experiences. *Leadership and Organizational Development Journal, 1,* 10–14.

Seashore, S. E., Lawler, E. E., Mirvis, P. H., and Cammann, C. (1983). *Assessing Organizational Change: A Guide to Methods, Measures, and Practice.* New York: Wiley.

Sirota, D. (1974). Why managers don't use attitude survey results. In S. W. Gellerman, *Behavioral Science in Management.* Baltimore: Penguin Books.

Smith, P. C., Kendall, L. M. and Hulin, C. L. (1969). *The Measurement of Satisfaction in Work and Retirement.* Chicago: Rand McNally.

Sonquist, J. and Dunkelberg, W. (1977). *Survey and opinion research: Procedures for processing and analysis.* Englewood Cliffs, NJ: Prentice-Hall.

Stone, E. (1978). *Research Methods in Organizational Behavior.* Santa Monica, CA: Goodyear.

Sudman, S. and Bradburn, N. M. (1982). *Asking questions: A practical guide to questionnaire design.* San Francisco: Jossey-Bass.

CHAPTER 3.2

ANALYTIC TECHNIQUES FOR FUNCTION ANALYSIS

KENNETH R. LAUGHERY, SR.

Rice University
Houston, Texas

K. RONALD LAUGHERY, JR.

Micro Analysis and Design
Boulder, Colorado

3.2.1	**INTRODUCTION**	**330**
3.2.1.1	Terminology	330
3.2.1.2	Uses of Functional Analysis	330
3.2.1.3	The Hierarchical Nature of Functional Analysis	331
3.2.1.4	Methods of Representation	331
3.2.2	**FLOW ANALYSIS**	**331**
3.2.2.1	Process Charts	331
3.2.2.2	Flow Diagrams	336
3.2.2.3	Operational Sequence Diagrams	336
3.2.2.4	Other Chart Techniques	337
3.2.2.5	Utility and Limitations of Flow Analysis Techniques	338
3.2.3	**TIME-LINE ANALYSIS**	**339**
3.2.3.1	Time-Line Charts	341
3.2.3.2	Gantt Charts	341
3.2.3.3	Other Techniques and Challenges	342
3.2.4	**NETWORK ANALYSIS**	**342**
3.2.4.1	Network Diagrams	342
3.2.4.2	Mathematical Procedures	343
3.2.4.3	Applications	343
3.2.5	**CONCLUSIONS**	**353**
	REFERENCES	**353**

3.2.1 INTRODUCTION

The design of complex systems typically involves the use of a variety of analytic techniques. These techniques include procedures for analyzing relationships between the various components, or subsystems, that make up the overall system. The components can vary widely, from many different types of equipment to humans with differing experiences and competencies. Likewise, relationships between components can take many forms, such as communications or physical interactions. Hence it is important to have available analytic tools for exploring, understanding, and modifying how a complex system functions.

In this chapter some of the analytic techniques that have been used for such purposes are presented. Other chapters also present techniques that are employed for similar purposes. Chapters 1.2 (system design, development, and testing), 2.8 (human error and reliability), and 9.4 (stochastics, network models) are examples.

The focus of this chapter is on the interactions among system components, including people, and the flow of material or information. The techniques described have been developed and used for designing new systems and for analyzing those already in existence. It is an understatement to say that there has been great variety in the techniques that have been employed as well as in the terminology used to describe them. Yet an analysis of the techniques reveals that there are a large number of variations on a limited set of basic approaches. These basic approaches include the following:

The analysis of the sequence of events in a system

The analysis of events with respect to time

The analysis of structure or relationships among system components, including those other than human

In this chapter these approaches are referred to as flow analysis, time-line analysis, and network analysis. Aspects of all three are covered.

3.2.1.1 Terminology

Much of the work on the analysis of event sequences in systems originated in industrial engineering (Barnes, 1968; Niebel, 1976; Turner, Mize, and Case, 1978). The term *process analysis* has been used to refer to a group of techniques for analyzing the various steps involved in a procedure. Usually this has been in the context of some manufacturing process. The concept of *functional analysis* is more recent and is primarily associated with more formal systems analysis procedures. A function can be viewed as a logical unit of behavior of a human or machine component that is necessary to accomplish the mission of the system.

In this chapter we use the term functional analysis and regard it as essentially synonymous with process analysis. It is concerned with exploring and understanding the dynamic as well as static (structural) aspects of systems. As already noted, it encompasses a variety of more specific techniques that can be grouped into three basic approaches: flow analysis, time-line analysis, and network analysis. There are other terms that are frequently used in ways more or less synonymous with these, such as sequential analysis and link analysis. We do not attempt to sort out the definitions of these and other terms, or the nuances in their use. Rather, we attempt to be consistent in using the terms and focus on the purposes for which the techniques are used and the methods themselves.

The three techniques for carrying out functional analyses may be viewed as means for modeling systems. That is, the various types of charts, diagrams, tables, matrices, equations, and computer programs that make up the techniques to be discussed permit the analysis of dynamic and structural properties. These models make it possible, for example, to trace the flow of materials through a manufacturing process before it is constructed in order to determine problems or flaws in the design. Similarly, such models provide a means for analyzing existing display-control panel designs on the basis of operator behavior patterns in order to optimize better the arrangement of display and control elements.

3.2.1.2 Uses of Functional Analysis

The range of purposes or uses of functional analysis is quite broad. As the above examples indicate, the techniques may be applied during the design phase or they may be applied to an operating system. During design such analyses aid in determining the logical flow of events, the allocation of functions to machines and to people, the distribution of work load across machines and/or people, the coordination of events in time, and the arrangement of equipment and people in space. Each of these purposes represents a major decision category in design where errors or suboptimal strategies may result in serious cost, performance, or safety consequences. The availability of system representations or models such as these techniques can be extremely useful for exploring alternatives in each of these decision categories.

Functional analysis is also frequently used for solving problems in existing systems. Delays, excessive inventories, equipment breakdowns, accidents, defective products, dissatisfied customers, employee dissatisfaction, absenteeism, human errors, and of course, excessive costs are among the many reasons why one may choose to take a new look at an existing system. When applied to existing systems, some of the techniques of functional analysis may be even more useful because some types of data may be available for the analysis. For example, network analyses are often used in considering the design or redesign of display-control panel layouts. In analyzing an existing panel, data on sequence of use or frequency of use of components may be available and can serve as the basis for applying the technique.

3.2.1.3 The Hierarchical Nature of Functional Analysis

Frequently a functional analysis consists of analyzing the system at more than one level. In this sense the analysis may be a hierarchical effort and almost always proceeds in a top-down fashion. The components or subsystems making up the level at which one starts in turn define the analyses at the next level. In other words, each phase in the analysis is the basis for the analyses in the subsequent phase. This approach provides an increasingly detailed view of the system where each decomposition cycle provides more, but better defined, functions. The ultimate desired level of detail depends, of course, on the purposes for which the analysis is being carried out.

3.2.1.4 Methods of Representation

The specific methods used in functional analysis range from the formal and rigorously quantitative to the informal and qualitative. Some of the procedures have their roots in the early work of industrial engineers whereas others have been developed by people in human factors, operations research, and other fields. The models can take a variety of forms. Among the more qualitative methods, a number of different charting, diagraming, and tabular techniques have been developed and used for representing aspects of system functioning. Some of the more formal, quantitative methods have been associated with operations research. Linear programming, network analysis, and computer simulation are examples of such techniques. These techniques have been particularly useful for solving certain classes of design problems such as sequencing or routing. In recent years the techniques for carrying out functional analyses have been moving in the direction of more formal, quantitative methods. It is likely this trend will continue. The following sections discuss the three general approaches: flow analysis, timeline analysis, and network analysis.

3.2.2 FLOW ANALYSIS

Flow analysis has been defined as the detailed examination of the progressive travel, of either personnel or material, from place to place and/or from operation to operation (ANSI, 1982). We would add "information" to this definition as another type of entity that may travel through the system. The basic idea of flow analysis is to represent the sequence of events to take place in a proposed or existing system. This representation, in turn, is useful for understanding and analyzing the functions involved in accomplishing the objectives or mission of the system.

One important characteristic of the techniques of flow analysis is that they do not quantitatively represent time. Dynamic events are represented in sequence and in this sense their flow across time is accounted for, but the time requirements of individual or aggregate events are not specified. An implication of this lack of time specification is that the flow analysis techniques are essentially used for representing order information about serial events. More specifically, some entity or class of entities is traced through the various processes or operations involved in the system. Examples would be materials as they flow through a manufacturing plant, people as they carry out a sequence of activities in a job, and information as it flows through a series of operations in a computer program.

Many of the flow analysis techniques presented here originated in industrial engineering, particularly with regard to methods analysis. They have been used widely in the design and analysis of manufacturing processes. This use includes both planning new systems and analyzing existing systems. Generally, the techniques can be characterized as qualitative in that they involve diagramming and charting procedures as opposed to mathematical, statistical, or computerized methods.

3.2.2.1 Process Charts

The process chart is a graphic representation of the separate steps or events involved in a process. It may show the steps that some material or information goes through, or it may indicate the sequence

◯	Operation—An operation occurs when an object, person, or information is intentionally changed.
⬦	Transportation—A transport occurs when an object, person, or information moves or is moved from one location to another.
☐	Inspection—An inspection occurs when an object, person, or information is examined or tested for identification, quality, or quantity.
D	Delay—A delay occurs when an object, person, or information waits for the next planned action.
▽	Storage—A storage occurs when an object, person, or information is kept under control and authorization is required for removal.
▣	Combined operation—Inspection is performed within an operation.
◉	Combined operation—An operation is performed while a product is in motion.

Fig. 3.2.1 Basic symbols used in process charts.

of activities carried out by a person. There exists today a widely accepted symbol set for making up process charts. Originally, the Gilbreths proposed a set of 40 symbols that they used for constructing such charts (Gilbreth and Gilbreth, 1921). Later the American Society of Mechanical Engineers established a set of five symbols as a standard (ASME, 1947). This latter set is made up of the first five symbols in Figure 3.2.1. Combined symbols are also used as illustrated by the last two symbols in Figure 3.2.1.

There are many types or formats of process charts that are used in flow analysis. Several are shown and discussed here. A more complete presentation of the various formats as well as procedures for constructing the charts may be found in other sources (Barnes, 1968; Niebel, 1976; Turner, et al., 1978; Kadota, 1982). The overall objective of the techniques has already been stated, namely, to represent the flow of events and activities (functions) involved in system performance. More specifically, the purpose is to ask questions about each of the functions such as the following:

1. Is this function necessary? Can it be eliminated?
2. Can this function be combined with another?
3. Are the functions properly sequenced, or should the sequence be changed?
4. Can this function be improved?

Operation Process Charts

The operation process chart is one of the simplest charting formats. As in virtually all the charting techniques the sequence of activities or events proceeds from the top of the chart downward. Only operations and inspections are shown. Brief verbal notes or explanations are used to define each of the operations and inspections. Transports, delays, and storage are not represented in this technique. Such charts are usually intended to provide a rather quick overall view of the essential functions in a process.

Flow Process Charts

A flow process chart is an expansion of the operation process chart. In addition to operations and inspections, these charts depict transports, delays, and storage.

Flow process charts are used to represent the flow of various entities. One type of use, known as a product-type or material-type chart, is to show the flow of some entity as it proceeds through some set of operations. An example of this type of chart is shown in Figure 3.2.2. It represents the flow of buffing wheels in a recoating process.

A second type of flow process chart is the person type. Such charts are used to represent the sequence of activities carried out by a person in performing some task. The steps a pilot goes through in preparing an airplane for takeoff would be a task or procedure that lends itself to such a charting technique. These person-type charts are frequently employed in job and task analysis.

A third type of flow process chart is known as the equipment type. Here the procedures that take place on or are carried out with a piece of equipment are represented. Examples might be a forklift truck in a warehouse or a copying machine in an office setting.

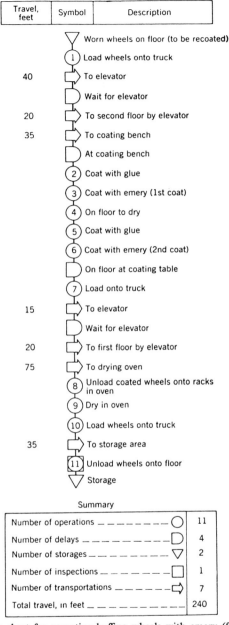

Travel, feet	Symbol	Description

Summary

Number of operations — — — — — — — ○	11
Number of delays — — — — — — — — D	4
Number of storages — — — — — — — ▽	2
Number of inspections — — — — — — □	1
Number of transportations — — — — — ⇨	7
Total travel, in feet — — — — — — — —	240

Fig. 3.2.2. Flow process chart for recoating buffing wheels with emery (from Barnes, 1968, Figure 29, by permission).

In describing the above three types or uses of flow process charts, the examples were fairly simple. These charts are frequently used to model much more complex situations such as the operation of an entire manufacturing plant. An example of a somewhat more complex chart, for making, filling, and sealing a can used in shipping instruments, is shown in Figure 3.2.3. In such cases the charts obviously consist of many more events or activities. Furthermore, when dealing with large, complex systems the charting technique typically involves a number of charts representing different levels of system definition. As noted in the introduction, this hierarchical analysis should proceed in a top-

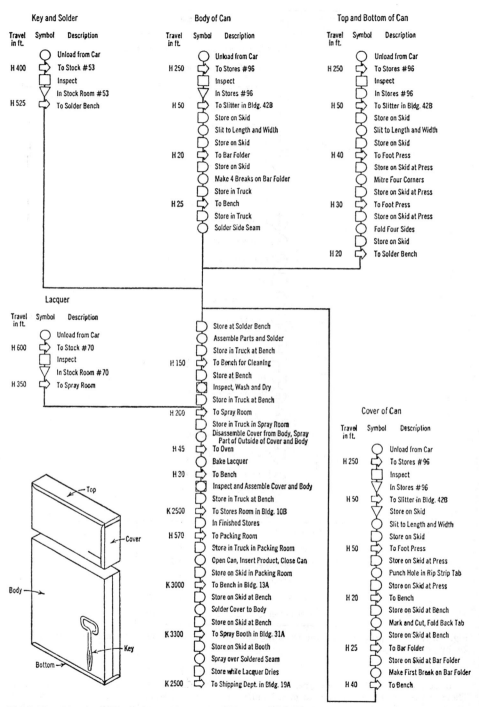

Fig. 3.2.3. Flow process chart for making, filling, and sealing a rectangular can for shipping instruments (from Barnes, 1968, Figure 41, by permission).

down fashion. A chart is first prepared for the system at its most general level of functioning. Many of the functions specified at this level are then the focus of an individual flow process chart depicting the flow of events within that function. In Figure 3.2.3, for example, the operation "assemble parts and solder" may be the subject of a separate flow process chart. These "second-level" functions may in turn be further broken down and charted, and so on. This progressive breakdown of the system functions is continued until a level of specificity is reached that provides answers to the design and analysis questions being posed.

As noted, there are many different formats for flow process charts. Figures 3.2.2 and 3.2.3 are examples. Figure 3.2.4 is another format, and it represents the flow of a requisition document in an office. The form used has the charting symbols preprinted. A line is simply drawn connecting appropriate symbols to depict the sequence of events. These are a few examples; there are others. Obviously the most useful format depends on the events and activities being charted.

Present Method ☒ Proposed Method ☐	PROCESS CHART	
SUBJECT CHARTED Requisition for small tools	DATE _____	
Chart begins at supervisor's desk and ends at typist's desk in purchasing department	CHART BY J. C. H.	
	CHART NO. R 136	
DEPARTMENT Research laboratory	SHEET NO. 1 OF 1	

DIST. IN FEET	TIME IN MINS.	CHART SYMBOLS	PROCESS DESCRIPTION
		●⇨□D▽	Requisition written by supervisor (one copy)
		○⇨□D▽	On supervisor's desk (awaiting messenger)
65		○⇨□D▽	By messenger to superintendent's secretary
		○⇨□D▽	On secretary's desk (awaiting typing)
		●⇨□D▽	Requisition typed (original requisition copied)
15		○⇨□D▽	By secretary to superintendent
		○⇨□D▽	On superintendent's desk (awaiting approval)
		○⇨■D▽	Examined and approved by superintendent
		○⇨□D▽	On superintendent's desk (awaiting messenger)
20		○⇨□D▽	To purchasing department
		○⇨□D▽	On purchasing agent's desk (awaiting approval)
		○⇨■D▽	Examined and approved
		○⇨□D▽	On purchasing agent's desk (awaiting messenger)
5		○⇨□D▽	To typist's desk
		○⇨□D▽	On typist's desk (awaiting typing of purchase order)
		●⇨□D▽	Purchase order typed
		○⇨□D▽	On typist's desk (awaiting transfer to main office)
		○⇨□D▽	
		○⇨□D▽	
		○⇨□D▽	
		○⇨□D▽	
		○⇨□D▽	
		○⇨□D▽	
		○⇨□D▽	
105		3 4 2 8	Total

Fig. 3.2.4. Flow process chart of office requisition procedure (from Barnes, 1968, Figure 37, by permission).

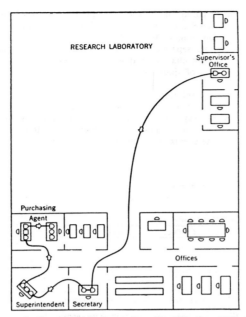

Fig. 3.2.5. Flow diagram of office requisition procedure (from Barnes, 1968, Figure 36, by permission).

3.2.2.2 Flow Diagrams

The process charts represent symbolically the various functions involved in some process. Flow diagrams are drawings or pictorial representations of the setting in which the functions are carried out. Their purpose is to show *where* the various steps in a procedure take place.

Figure 3.2.5 presents a flow diagram of the office requisition procedure charted in Figure 3.2.4. The transport functions are better visualized in the diagram. It should be noted, however, that this particular diagram does not communicate so well as the flow process chart information about one of the critical problems associated with the requisition procedure, the eight delays. The chart in Figure 3.2.4 permits the analyst to note these delays conveniently by simply scanning the function columns. Thus both charts and diagrams may be useful.

3.2.2.3 Operational Sequence Diagrams

Operational sequence diagrams (OSDs) were developed by Dunlap and Associates (Kurke, 1961). The purpose of the OSD is to depict graphically the information-decision sequences a system must undergo to complete a mission. OSDs are used to represent the flow of these sequences through a system. Kurke notes three uses:

1. To establish sequence-of-operations requirements between subsystem interfaces at various levels of system analysis.
2. To depict the logical result of each of several decision-action sequences.
3. To evaluate panel layout and work-space designs.

The central concept of an OSD is the decision. Information comes to the component (frequently, but not always, a human), a decision is made, and an action is taken. A standard set of symbols, shown in Figure 3.2.6, is employed in constructing an OSD. Note that in addition to the shape of the symbols, they may be coded in other ways. Single-lined symbols represent manual operations, and double-lined symbols are automatic operations. Solidly shaded symbols are used to depict no action or information, and half-shaded symbols indicate partial information or incorrect operations owing to noise or error sources in the system.

Bateman (1979) used the OSD technique to analyze problems associated with passengers changing planes from one airline to another at the Dallas–Fort Worth International Airport. The airport uses an automated ground transportation system consisting of 21 km of guideway connecting 14 stations

⬡	Operator decision.
☐	Action, for example, control operation.
▽	Transmitted information.
○	Received information, e.g., indicator display.
∪	Previously stored information, e.g., knowledge.

The above single-lined symbols represent manual operations. The following symbols show special codes.

▭	Double-lined symbols represent automatic operations.
▨	Solid symbols indicate no action or no information.
◪	Half-filled symbols indicate partial information or incorrect operations owing to noise or error sources in the system.

Fig. 3.2.6. Symbols used in operational sequence diagrams.

at four terminals, a hotel, and two remote parking areas. There are 51 powered vehicles operating on five separate color-coded routes. The typical passenger user is changing from one airline to another, is not familiar with the system, and is likely to be in a hurry. The passenger knows the airline and flight number he/she is going to, and is trying to determine which vehicle to get on and at which station to get off. The information about which vehicle to get on is divided among five sources: a bulletin board, a television monitor, a panel showing gate to station relations, a sign over the door of the vehicle, and a display inside the vehicle. While the first three of these sources are being consulted, at least one vehicle is likely to arrive, and the passenger in a hurry is likely to be tempted to board it.

Figure 3.2.7 shows the OSD constructed by Bateman (1979) representing the information-decision-action sequences in the system. The half-shaded received-information symbols indicate that there are four opportunities for an error, and the five triangles show the five different displays that are required to obtain two pieces of information (which vehicle to get on and at which station to get off).

This particular OSD proved quite useful in identifying the problems associated with an existing system in which costly errors were being made. Had such a chart been constructed and analyzed during the design phase of the system, the frequent problem of passengers boarding the wrong vehicle (and the problems that accompany such mistakes) might have been avoided.

3.2.2.4 Other Chart Techniques

There are other charting techniques for representing the flow of material, activities, or information in a system. One is known as FAST, the Functional Analysis Systems Technique (Bytheway, 1971). FAST is a charting procedure that uses two-word (verb, noun) statements to define functions. The purpose and procedure are similar to constructing the operation process chart in that they focus on defining the functions and their sequencing. The standard charting symbols are not used; rather, the operations are specified as verb–noun statements, and the chart is made up of boxes containing these statements.

Another example is the traditional flow-charting technique employed in developing computer programs. Here again, the standard symbol set is different from the set developed by industrial engineers. The basic purpose and approach is similar, however, in that the sequence of operations, storage, decisions, and so forth is being represented.

There are a number of charting techniques that have been developed for simultaneously representing more than one sequence of events or activities. These charts are referred to as multiprocess or multiple activity charts. Three of these techniques are used in industrial engineering and are known as left-hand/right-hand charts, gang charts, and triple resource charts. In a left-hand/right-hand chart the activities of the two hands are listed separately. Figure 3.2.8 shows a gang chart simultaneously representing the activities of 10 different people involved in unloading canned goods from a freight car. A triple resource chart is typically used for simultaneously representing the activities of the person, the operations performed by the equipment, and the operations performed on the material.

An important point about these three multiprocess charting examples is that although time is not quantitatively specified, it is represented in that the activities of the two hands, the 10 workers, or the human–material–equipment shown on any given line in the chart are occurring during the same time period. Indeed, in some applications of these charts, time is quantitatively specified. These applications are covered in Section 3.2.3 on time-line analysis.

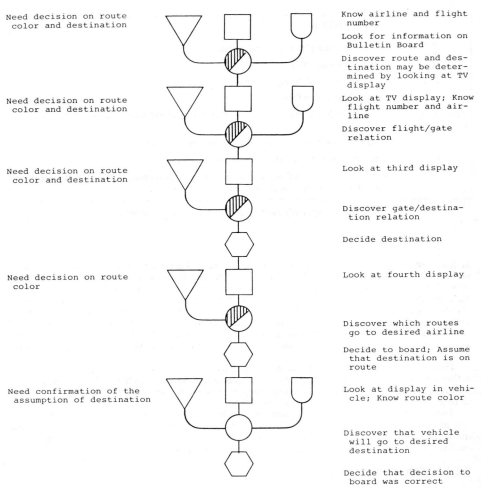

Need decision on route
color and destination

Need decision on route
color and destination

Need decision on route
color and destination

Need decision on route
color

Need confirmation of the
assumption of destination

Know airline and flight
number

Look for information on
Bulletin Board

Discover route and des-
tination may be deter-
mined by looking at TV
display

Look at TV display; Know
flight number and air-
line

Discover flight/gate
relation

Look at third display

Discover gate/destina-
tion relation

Decide destination

Look at fourth display

Discover which routes
go to desired airline

Decide to board; Assume
that destination is on
route

Look at display in vehi-
cle; Know route color

Discover that vehicle
will go to desired
destination

Decide that decision to
board was correct

Fig. 3.2.7. Operational sequence diagram for airport ground transportation information system (from R. P. Bateman, "Design Evaluation of an Interface between Passengers and an Automated Ground Transportation System," *Proceedings of the Human Factors Society 23rd Annual Meeting,* 1979, p. 120. Copyright 1979, by the Human Factors Society, Inc. and reproduced by permission).

3.2.2.5   Utility and Limitations of Flow Analysis Techniques

As with all of the techniques presented in this chapter, the primary purpose of flow analysis is to assist the designer and analyst of a system. The *process* of constructing a chart or diagram of the functions involved can be exceptionally useful. Information must be gathered, functions must be defined, and sequences must be arranged. In short, the work, the thinking, and the organization that go into the chart construction process can contribute significantly to the identification of problems and the definition of possible solutions.

The display of information in the charts and diagrams can, of course, lead to the identification of problems and alternative solutions. One example of such an outcome concerns the requisition procedure shown in Figure 3.2.4. The chart shows eight different delays in processing the requisition. Changes in typing and approval procedures reduced the number of delays from eight to three.

A third example of the utility of a chart concerns the operational sequence diagram for the airport ground transportation system shown in Figure 3.2.7. The construction of the OSD and the chart itself clearly identified the problems faced by the passengers. It also suggested some solutions having to do with the consolidation of the different information sources as well as one more far-reaching

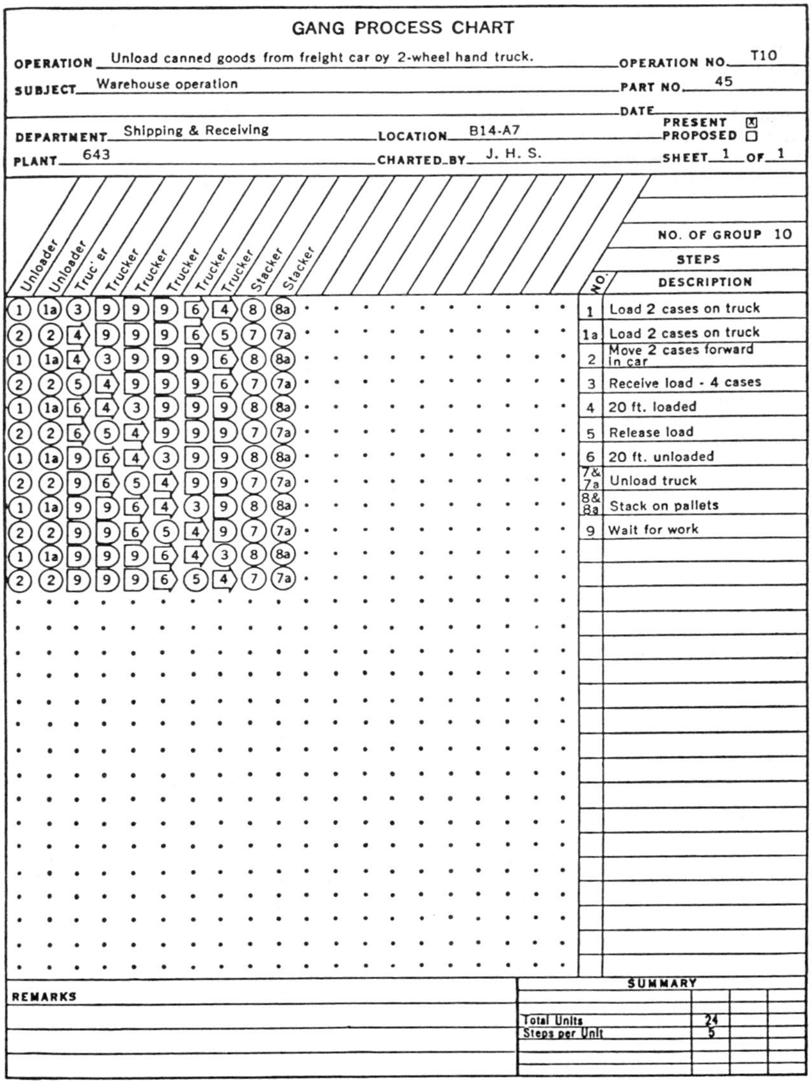

GANG PROCESS CHART

OPERATION ___Unload canned goods from freight car oy 2-wheel hand truck.___ OPERATION NO.___T10___

SUBJECT ___Warehouse operation___ PART NO.___45___

DATE

DEPARTMENT ___Shipping & Receiving___ LOCATION ___B14-A7___ PRESENT ☒ PROPOSED ☐

PLANT ___643___ CHARTED BY ___J. H. S.___ SHEET ___1___ OF ___1___

NO. OF GROUP 10

STEPS

												NO.	DESCRIPTION
Unloader	Unloader	Trucker	Trucker	Trucker	Trucker	Trucker	Trucker	Stacker	Stacker				
①	⑴ₐ	③	9	9	9	⑥	⑷	⑻	⑻ₐ			1	Load 2 cases on truck
②	②	⑷	9	9	9	⑥	⑸	⑺	⑺ₐ			1a	Load 2 cases on truck
①	⑴ₐ	⑷	③	9	9	9	⑥	⑻	⑻ₐ			2	Move 2 cases forward in car
②	②	⑸	⑷	9	9	9	⑥	⑺	⑺ₐ			3	Receive load - 4 cases
①	⑴ₐ	⑹	⑷	③	9	9	9	⑻	⑻ₐ			4	20 ft. loaded
②	②	⑹	⑸	⑷	9	9	9	⑺	⑺ₐ			5	Release load
①	⑴ₐ	9	⑹	⑷	③	9	9	⑻	⑻ₐ			6	20 ft. unloaded
②	②	9	⑹	⑸	⑷	9	9	⑺	⑺ₐ			7&7a	Unload truck
①	⑴ₐ	9	9	⑹	⑷	③	9	⑻	⑻ₐ			8&8a	Stack on pallets
②	②	9	9	⑹	⑸	⑷	9	⑺	⑺ₐ			9	Wait for work
①	⑴ₐ	9	9	9	⑹	⑷	③	⑻	⑻ₐ				
②	②	9	9	9	⑹	⑸	⑷	⑺	⑺ₐ				

REMARKS

SUMMARY

Total Units	24	
Steps per Unit	5	

Fig. 3.2.8. Gang chart (from Barnes, 1968, Figure 48, by permission).

solution in which all the information was stored on a computer and the passenger simply interacted with the computer from a terminal.

The primary limitation of process analysis is that it does not include specific time information such as mean time to perform functions. Consequently, detailed analyses of system dynamics are not possible. Time-line analysis adds this dimension.

3.2.3 TIME-LINE ANALYSIS

Time-line analysis has been defined as an analytical technique for the derivation of human performance requirements which attends to both the functional and temporal loading for any given combination of tasks (ANSI, 1982). This definition can be expanded to include in addition to human performance requirements the requirements of any system component such as a vehicle, a piece of factory equipment, or a computer. The technique is used primarily for work-load prediction, analysis, and scheduling.

The allocation of work load may include the scheduling of assignments for an individual human or machine as well as the distribution and scheduling of assignments across people and machines.

Generally time-line analysis is used to help anticipate or solve resource allocation problems. The optimal use of resources (people or equipment) requires that they neither be underutilized nor that their capacity be exceeded. On occasion the resources may be fixed, and the problem becomes one of predicting when a task or function may be completed (the turnaround time for running a computer program in a batch operation). Some of the benefits of time-line analysis can be the completion of work on time, the minimization of errors, overtime, delays, and inventories, and the high utilization of facilities and people.

In the analysis of relationships between components in a system, time has some properties that make it a uniquely valuable base line. It never changes direction (it does not back up), and its velocity is always constant.

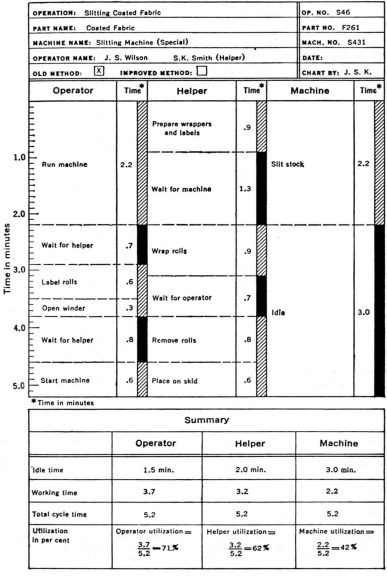

Fig. 3.2.9. Human–machine chart (from Barnes, 1968, Figure 64, by permission).

3.2.3.1 Time-Line Charts

When time information is added to some of the flow analysis charting techniques, the result is a time-line analysis. An example drawn from traditional industrial engineering methods is the human–machine chart shown in Figure 3.2.9. The chart is used to analyze the interaction between operators and equipment. As indicated in the summary portion of this chart, there is a high proportion of idle time which would probably cause one to consider alternative procedures in the execution of this function. The chart enables one to see the relationships between the activities of the different components in planning such alternatives.

Another form of time-line chart is listing tasks or functions in a column on the left side of a chart and then simply using lines to portray the events and times horizontally. An example of this technique is shown at the top of Figure 3.2.10. Working from this time-line display an experienced analyst knowing which tasks are assigned to each operator can then construct a work-load profile for each person involved in carrying out the tasks. Such a profile is shown at the bottom of Figure 3.2.10. An analysis of the profile indicates that problems may well exist in the workload of operator A, because there are times (such as at the 15 min mark) when the sum of the requirements of the individual tasks would appear to exceed the operator's capacity.

3.2.3.2 Gantt Charts

One of the time-line techniques that originated in the 1940s in response to the need for better management and control of complex projects was the Gantt chart. A horizontal bar chart is used with functions plotted vertically and time plotted horizontally. An example of a Gantt chart is shown in Figure 3.2.11. The chart shows both completed and planned work by shading the bars to represent the portion of the task that has been completed. At a given point in time, say, at the end of week 2, one can determine which functions are ahead of schedule (task 3) and which are behind (task 4). There are many variations on this type of chart. For example, special symbols or color codes can be used to represent conditions such as materials shortages or machine breakdowns.

Although the Gantt chart provides an excellent graphic display for showing project progress, its utility in complex systems is limited. It does not, for example, describe the dependencies or interactions among the various functions. In general, it has been replaced with more formal computerized analytic procedures for analyzing complex systems. Nevertheless, in the right circumstances it may still be useful for sequencing and scheduling.

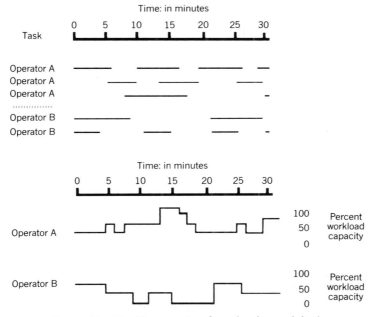

Fig. 3.2.10. Time-line procedure for estimating work loads.

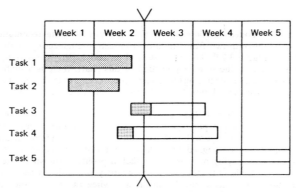

Fig. 3.2.11. Gantt chart (from R. P. Sadowski and D. J. Medeiros, "Planning and Control for Manufacturing Systems and Projects," in G. Salvendy, Ed., *Handbook of Industrial Engineering.* Copyright 1982, Figure 11.2.5(b), by permission).

3.2.3.3 Other Techniques and Challenges

There are numerous computer programs available today that can take as input time-line data and produce work-load profiles for the system entities. Such programs not only remove some of the drudgery from such analyses, but they also make it feasible to analyze time-line information in more complex situations. But the challenging issue in time-line analysis today is not the plotting or computational work involved. Rather, it is estimating the proportion of processing capacity taken up by some task or series of tasks. This issue is simple in dealing with most machines, it is interesting in dealing with computers, but it achieves its most challenging form in dealing with the human.

The capacity issue in understanding and scheduling work loads in humans is challenging for two reasons. First, as the proportion of intellectual, as opposed to physical, tasks in jobs continues to increase, there is an increasing need to be able to determine when the work load assigned to a person is too little or too much. This question arises more and more frequently, and optimal design requires reasonable answers. The second reason is that while there is an increasing amount of human-factors research in the area of mental work load, the answers are not yet available. Parks (1979) and Wickens (1984) present useful discussions of this issue.

A final comment is in order with regard to time-line analysis as presented here. This discussion has been limited. We have presented some of the charting techniques and noted some of their applications in dealing with relatively simple systems involving a few machines, a few people, or a few tasks. Much more sophisticated techniques are available for time-based functional analysis in complex systems. We have limited our comments because many of the more formal techniques are presented in other chapters in this volume. Chapter 9.4 (stochastics network models) is particularly relevant. Some of the materials presented in the next section of this chapter are also relevant.

3.2.4 NETWORK ANALYSIS

Network analysis has been defined as a technique used in planning a project consisting of a sequence of activities and their interrelationships within a network of activities making up a project (ANSI, 1982). This definition focuses on one of the major application areas of the technique; namely, its use in project planning and control. Another application area of network analysis that has been common in industrial engineering and human factors concerns layout or arrangement problems. Plant layout, work environment layout, and the arrangement of display-control panels are examples. In human factors, *link analysis* is a term frequently applied to such efforts.

3.2.4.1 Network Diagrams

A network can be viewed as a graphic representation of the entities that make up a system and their relationships to each other. There are two basic elements to a network diagram, nodes (representing the functions and/or entities) and arcs (representing the relationships). The terminology used in network analysis is not always consistent. Nodes may be referred to as points or vertexes and the arcs may be labeled links, edges, branches, or relations. Figure 3.2.12 shows an example of a network diagram. In most diagrams nodes are represented by circles and arcs by lines.

Networks can be used to represent a great variety of entities and relationships. Entities may be people, buildings, equipment, work stations, displays, controls, some state in a development process,

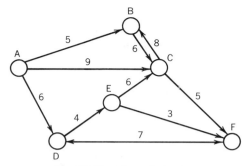

Fig. 3.2.12. Network diagram.

and so on. Similarly, arcs may represent many types of relationships. Some examples of relationships follow:

1. *Distance.* The distance (centimeters, feet, miles, etc.) between nodes.
2. *Time.* The time required for people, materials, information, and so on to move from one node to another.
3. *Frequency.* A count of the number of times some relationship occurs between two nodes. Examples would be the number of times an employee walks from one location to another in a work environment and the number of eye movements between two displays on a console.
4. *Importance.* Usually a rating of the importance between nodes. For example, communications between A and B may be more important than communications between C and D. Importance ratings are often used in planning layouts or arrangements where it is not yet possible to obtain actual distance, time, or frequency measures.

The magnitude of relationship measures may be noted on a network graph as in Figure 3.2.12. In addition to magnitude, network relationships may also have the property of direction. Indeed, in the application of network analysis to project planning and control, the directional property is fundamental. Often one does not proceed with task C until tasks A and B are completed. The directional property can represent asymmetry, such as the frequency or time for going from A to B being different from B to A. Directional properties of arcs are usually indicated on diagrams by arrows on the lines as in Figure 3.2.12. Where the magnitude values between two nodes are asymmetric (different for the two directions), separate lines may be drawn as between B and C.

3.2.4.2 Mathematical Procedures

Recent years have witnessed the development of sophisticated mathematical procedures for carrying out network analyses. These formal procedures are particularly useful for dealing with large, complex systems and generally require computers for computation. Numerous publications present descriptions of the mathematics of network analysis (Elmaghraby, 1970; Ford and Fulkerson, 1962; and Whitehouse, 1973).

3.2.4.3 Applications

Network analysis has been applied to a variety of problems. Some of the problem categories are planning, scheduling, distribution, and layout or arrangement.

Project Planning and Control

In Section 3.2.3 Gantt charts were briefly described as a technique for project planning, scheduling, and monitoring. It was pointed out that they are limited to relatively simple systems. Dependencies and interactions among functions are not represented, adjustments and rescheduling are not easy to represent, and it is difficult to determine the effects that delays may have.

Two network procedures emerged in the late 1950s that overcame the Gantt chart limitations and that have proved especially useful for planning and monitoring large projects. These procedures are the critical path method (CPM) and the program evaluation and review technique (PERT). Both CPM and PERT employ network structures to represent the sequence of activities on a project. For analysis with either of these techniques a project must have four properties:

1. The project consists of well defined functions or tasks with identifiable beginning and end points.

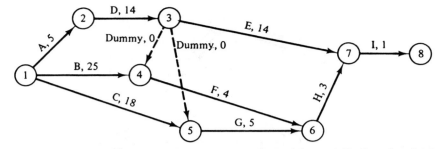

Fig. 3.2.13. Critical path network (from W. C. Turner, J. H. Mize, and K. Case, *Introduction to Industrial and Systems Engineering*, copyright 1978, p, 324. Reprinted by permission of Prentice-Hall, Inc., Englewood Cliffs, NJ).

2. The functions or tasks may be started or stopped independently of each other within any given sequence.

3. The time required for each function or task can be determined or estimated.

4. The functions or tasks have precedence relationships and must be performed in order.

Figure 3.2.13 shows an example of a critical path network for one of CPM's most common uses, a construction job. The diagram contains a good deal of information. The characteristics and meaning of the diagram are as follows:

1. A circle represents a node which in turn represents some state in the completion of the project.

2. A line with an arrow represents one of the functions that must be completed as part of the project. Its length has no meaning.

3. Arrows (functions) leaving a node cannot be started until all functions coming into a node are completed.

4. A dashed line represents a dummy function. These functions are used to show precedence relationships and are not real—they have a time duration of zero.

It is common to use node numbers as labels, although they do not usually define precedence. Also, it is common to label each line with the function it represents. Because such labeling can create excessive clutter, codes are frequently used. In Figure 3.2.13 the letters A through I might represent such construction functions as roofing, plumbing, and so on. The numbers on the lines represent time, and in the case of a construction project might refer to days.

From the diagram it is then possible to see the necessary sequence of the various functions. It is also possible to compute time requirements of various paths through the network including the longest path, the "critical path." In Figure 3.2.13 the critical path is through nodes 1, 2, 3, 7, and 8, and it includes functions A, D, E, and I. The time requirement is 34 days. This critical path is important because it defines the minimum time for completion of the project. It also defines the functions for which there is no "slack time"; that is, any slippage in the time to perform those functions will delay project completion. Note, however, that other tasks will have slack time in the sense that it is not necessary for them to start at the earliest possible time in order for the project to be completed on time.

PERT is much like CPM except that it allows for uncertainty in the estimates of time to perform a function. Whereas CPM requires a single time estimate for each function, three times are estimated in PERT. These estimates are referred to as a most likely time, an optimistic time, and a pessimistic time. In PERT it is possible to compute statistical estimates of completion time, or, turning it around, it is possible to compute the probability of completion by a certain date. In a sense PERT is also more realistic, because in many projects it is not possible to predict accurately time requirements for all functions.

In large projects the networks and the computational procedures can become quite complex. Computer programs are available to enable one to use the techniques in such situations. Although PERT and CPM were designed primarily as a tool for planners of large projects, they can also be useful tools for human factors specialists. If the functions in a PERT analysis represent operator tasks, then one can analyze best, average, and worst time to complete the operator's job. Additionally, by using these techniques to analyze a mutli-operator system, individual slack time and operator work load may be evaluated.

Clearly there are some advantages to the techniques such as the following:

1. They encourage a formal, logical approach.

2. They encourage detailed, long-range planning.

3. They focus attention on critical functions.
4. They permit an analysis of the effects of technical and procedural changes.
5. They permit an analysis of the effects of various resource allocation alternatives.

Numerous sources exist for a more detailed description of the CPM and PERT techniques (Moder and Phillips, 1970).

Layout and Arrangement

Industrial engineers and human-factors specialists have used network analysis procedures for dealing with layout and arrangement problems. In planning the layout of a business office facility, for example, the issue may be where to locate the different departments such as sales, the president, the controller, and the production supervisor. In the layout of a specific manufacturing plant the concern might be to locate different equipment, storage areas, loading docks, and so on. In laying out a control-display panel, the problem may be where to put the various controls and displays.

Such problems lend themselves to network analysis. In terms of diagrams, the nodes in the first example would be the departments, and the relationships might be interactions between personnel in the departments. In the manufacturing plant example, the nodes would be the equipment, storage areas, and docks, and the relationships could be the movement of materials between them. In the control-display panel layout, the nodes would be controls and displays and the relationships could be frequency of movement between the various nodes.

The objective of network analyses in such situations is to optimize the layout with respect to some criterion. In these three examples the criterion might be to minimize the total distance that people must travel, the distance through which materials must be moved, or the hand/eye travel time. In such problems a first question concerns the appropriate choice of a measure of the relationship. The frequency of interpersonal interactions, the amount of material moved, and the frequency and distance of eye movements would be likely measures. In planning new facilities these measures would have to be estimates based on ratings of "experts" or on actual measures taken from similar existing facilities. For example, if one were designing new facilities to be occupied by existing departments, it would be possible to obtain the interaction frequencies in the current environment and apply them to the design of the new one.

Often in designing new facilities the relationships between entities cannot be quantified and one works with qualitative measures such as ratings of importance of being close together. Flow process charts can be quite useful in the development of such ratings. A widely used nomenclature has evolved for coding such ratings as follows:

A	Absolutely necessary
E	Especially important
I	Important
O	Ordinary closeness OKay
U	Unimportant
X	Not desirable

The values of relationships, qualitative or quantitative, are usually organized into what industrial engineers call a from–to chart or an activity relationship diagram, which are simply matrices containing the values. Figure 3.2.14 shows a from–to chart for interactions between staff members in an office.

To From	Office Manager	Secretary	Typist	Clerk	⋯	Totals
Office Manager	—	10	5	1	⋯	
Secretary	20	—	10	10	⋯	
Typist	10	30	—	20	⋯	
Clerk	5	25	10	0	⋯	
·	·	·	·	·		
·	·	·	·	·		
Totals						

Fig. 3.2.14. From–to chart for office staff interactions (from W. C. Turner, J. H. Mize and K. Case, *Introduction to Industrial and Systems Engineering*, copyright 1978, p. 101. Reprinted by permission of Prentice-Hall, Inc., Englewood Cliffs, NJ).

Notice that the relationships are quantitative (number of trips per day) and that the values between nodes (people) is asymmetric; that is, the office manager goes to the secretary 10 times but the secretary goes to the office manager 20 times. Figure 3.2.15 presents an activity relationship diagram for components in a dairy. The letters in the cells are "rated importance" of being close.

Based on the kind of relationship information contained in Figure 3.2.15, the next step in the planning or analysis of layouts is to construct a link diagram (also called a string diagram). The link diagram is a schematic representation of the relationships between the entities being arranged. Its purpose is to display graphically possible layouts of the facility. Figure 3.2.16 shows a string diagram for the dairy whose relationships were defined in Figure 3.2.15. In this diagram the nodes are the various activities or functions that go on. They are shape coded on the basis of the type of function. The importance of the relationship between nodes in this diagram is represented by the number of connecting lines.

Link diagrams are essentially possible solutions to layout or arrangement problems. Typically, when working from relationship data one draws many diagrams, or arrangements. The central issue, of course, is how one goes about constructing link diagrams from the data. Both qualitative heuristic approaches have been suggested and formal computerized procedures have been developed.

As noted earlier, the objective in this type of analysis is to optimize the layout with respect to some criterion. Hence the manner in which one constructs link diagrams depends on the criterion. We assume here that the objective in the layout problem is to arrange the elements so that the greater the frequency or importance of the relationships between them, the closer they are together. A heuristic that is frequently employed is to start with the element that is most important or has the highest frequency relationships with other elements and locate this entity at the center of the diagram. Next, the element that has the most important or most frequent relationship to the first is selected and located close by. These elements are then connected by a line. In order to indicate the magnitude of the relationship, the line may be labeled with a number reflecting a quantitative value. Magnitude

Activity or Function		Activity or Function Number													
		15	14	13	12	11	10	9	8	7	6	5	4	3	2	1
1. Carton Storage		U	U	U	U	U	U	E	E	E	U	U	U	U	U	
2. Unloading Area		U	U	U	U	U	U	U	U	U	U	U	U	I		
3. Raw Milk Storage		U	U	U	U	U	U	U	U	U	U	U	I			
4. Homogenizer		U	U	U	U	U	U	U	U	U	U	I				
5. Pasteurizer		U	U	U	U	U	U	U	U	U	I					
6. Pasteurization Storage Tanks		O	U	U	U	U	E	I	I	I						
7. Filling Machine	1	O	E	U	I	O	A	U	U							
8. Filling Machine	2	O	E	U	I	O	A	U								
9. Filling Machine	3	O	E	U	I	O	A									
10. Cooler		U	U	A	U	I										
11. Byproduct Room		O	U	U	X											
12. Restroom		U	U	O												
13. Truck Loading Area		U	U													
14. Metal Crate Loading		U														
15. Laboratory																

Fig. 3.2.15. Activity relationship diagram for dairy (from W. C. Turner, J. H. Mize, and K. Case, *Introduction to Industrial and Systems Engineering*, copyright 1978, p. 102. Reprinted by permission of Prentice-Hall, Inc., Englewood Cliffs, NJ).

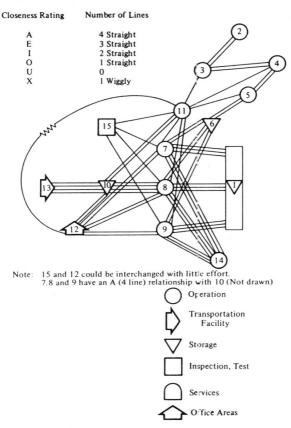

Closeness Rating · Number of Lines

A · 4 Straight
E · 3 Straight
I · 2 Straight
O · 1 Straight
U · 0
X · 1 Wiggly

Note: 15 and 12 could be interchanged with little effort.
7, 8 and 9 have an A (4 line) relationship with 10 (Not drawn)

○ Operation

Transportation
Facility

▽ Storage

□ Inspection, Test

⬭ Services

⌂ Office Areas

Fig. 3.2.16. Link (or string) diagram for dairy (from W. C. Turner, J. H. Mize, and K. Case, *Introduction to Industrial and Systems Engineering,* copyright 1978, p. 103. Reprinted by permission of Prentice-Hall, Inc., Englewood Cliffs, NJ)

may also be indicated by multiple lines as in a string diagram, which reflects a qualitative importance rating (see Figures 3.2.15 and 3.2.16). The next element to be added to the diagram is then selected on the basis of its relationships to the first two. The general heuristic is to sequence the selection of elements in a greatest-to-least order of relationships. The placement of elements in the diagram is guided by the magnitudes of the relationships. This procedure continues until all elements and their relationships are represented. The criteria for placement are usually to minimize distances and to avoid crossing lines.

In carrying out this diagram construction procedure with more than a few elements, problems are soon encountered. Minimizing one relationship conflicts with minimizing another. In short, one has an overall optimization problem. The technique usually involves making revisions or rearrangements during the construction procedure. Often several diagrams are drawn before a satisfactory layout is achieved.

The two major tasks in using a network or link analysis approach to solving layout and arrangement problems are obtaining good relationship data and constructing the optimal layouts. Getting good relationship data of course depends heavily on the situation. In planning and developing new systems it is frequently necessary to rely on ratings from experts or analyses of similar situations. In some cases it is possible to construct simulations such as mock-ups to obtain the type of frequency, time, or distance data discussed earlier. Similarly, where the problem involves developing alternatives to existing situations such data may be obtained. One of the first applications of link analysis to human-factors problems was the work of Fitts, Jones, and Milton (1950) in which eye movements between aircraft instruments were obtained and used to design alternative instrument panel layouts.

Often acquiring relationship data can be automated. Haygood, Teel, and Greening (1964) reported one of the first efforts to use a computer to record such information. In the context of analyzing a complex flight control system console they automated counting the number of times each component was used and the pairings of component use. Today's computer technology readily permits such data acquisition in many situations. Display and control systems where computers are central components lend themselves to such applications. Indeed, problems associated with human–computer interaction, such as keyboard layout, can be addressed in this manner.

Many formal computerized procedures have been developed for constructing layouts from relationship data. Such techniques are especially useful for dealing with large systems in which the number and variety of entities and relationships can overwhelm less formal graphic procedures. It is beyond the scope of this chapter to present these techniques in detail, but a few will be noted briefly and references cited to more extensive descriptions.

Four layout and arrangement packages are known as CORELAP—Computerized Relationship Layout Planning (Lee and Moore, 1967), ALDEP—Automated Layout Design Program (Seehof and Evans, 1967), CAPABLE—Controls and Panel Arrangement by Logical Evaluation (Bonney and Williams, 1977) and CRAFT—Computerized Relative Allocation and Facilities Technique (Armour and Buffa, 1963). CORELAP, ALDEP, and CAPABLE are basically construction procedures in that initial layouts are generated. CRAFT is an improvement technique in that it requires an initial layout and makes changes in it. Francis and White (1974) discuss the use of the CORELAP, ALDEP, and CRAFT techniques for dealing with plant layout problems. The techniques have also been applied more directly to human factors problems.

Bartlett and Smith (1973) used the CRAFT program to design an aircraft instrument panel. The input to CRAFT consists of four types of information:

1. An initial layout.
2. The area required by each element.
3. A "volume array," which consists of relationships (frequencies) between elements.
4. A "cost array," which consists of some sort of cost coefficients, such as travel times for each operator eye or hand movement between elements. These values could also represent factors such as importance ratings.

CRAFT begins by determining centroids for the elements in the initial layout and then computing interelement distances which are stored in a distance matrix. The cost of the layout is then computed by calculating the product of the volume (frequencies), cost (travel times), and distances. The procedure next considers all pairwise or three-way (user's option) interchanges between elements that have equal areas or that have a common border. The interchange offering the greatest reduction in total cost is then made. CRAFT next considers interchanges in the new layout in the same way, making the interchange that results in the greatest total cost reduction. This procedure continues until no interchange can be found that reduces total cost or until the program is terminated by the user.

There are a couple of additional characteristics of CRAFT worth noting. First, the final layout is developed by making improvements from an initial layout. Because the final layout thus depends on the initial arrangement, several initial layouts should be tried. It is also possible to specify "dummy" elements in CRAFT. Dummy elements are used to represent fixed areas where elements may not be located. Figure 3.2.17 shows the final layout of the aircraft instrument panel in the Bartlett and Smith study.

CORELAP is a computer program that works from a relationship matrix to construct a layout. Cullinane (1977) applied the program to the design of a layout for a minicomputer laboratory. The input to the program consists of the relationship matrix for the elements and the area required for each element. Figure 3.2.18 presents the relationship matrix for this particular design problem. CORELAP begins by computing a "total closeness rating" for each element, which is the sum of all closeness relationships involving that element. The numerical values used are A = 6, E = 5, I = 4, O = 3, U = 2, and X = 1. The element having the highest rating is assigned to the center of the layout. In the event of a tie, the element requiring the largest area is selected. The element with the highest relationship to the first is then selected to add to the layout. In the event of a tie, the element with the highest total rating is selected. The third element is selected on the basis of its relationships to the first two. This procedure continues until all elements have been selected. Once an element has been selected a placement decision must be made. The decision is made on the basis of a placement rating which is the sum of the weighted closeness ratings of the element and its neighbors. A tie-breaking rule is based on boundary lengths.

The final layout for the minicomputer facility is shown in Figure 3.2.19. It can be seen that CORELAP develops layouts by growing out from the center. The program does not have the capability of assigning elements to certain locations, nor can it be constrained to work within some overall shape, such as a building space or panel. Consequently, the layouts generated may have unrealistic shapes. Its primary use, therefore, is to generate initial layouts. One of the significant values of the technique is that it may be used to handle large problems. For layouts involving 12 elements only 66 pairs of relationships exist, and an initial link diagram can be accomplished manually in a reasonable time period. For a

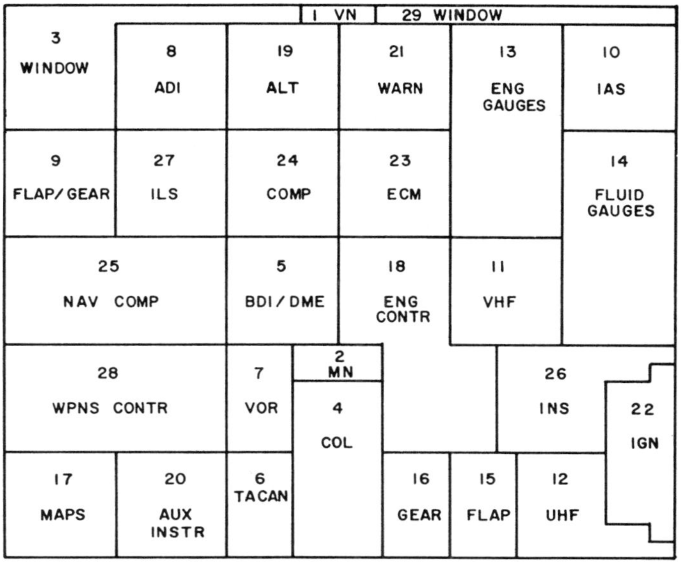

3 WINDOW	8 ADI	19 ALT	21 WARN	13 ENG GAUGES	10 IAS
9 FLAP/GEAR	27 ILS	24 COMP	23 ECM		14 FLUID GAUGES

(Top header of panel: I VN · 29 WINDOW)

Fig. 3.2.17. CRAFT layout of aircraft instrument panel (from N. W. Barlett and L. A. Smith, "Design of Control and Display Panels using Computer Algorithms," *Human Factors*, 1973, *15*, 6. Copyright 1973, by The Human Factors Society, Inc. and reproduced by permission).

problem involving 45 elements, however, approximately 1000 relationships exist. The CORELAP program readily handles such problems.

CAPABLE is another layout design program that was developed by Bonney and Williams (1977). Details of the procedure are not presented here. One example of its application was to design the layout of control devices in a bar and rod mill. Twenty-seven controls and related elements located

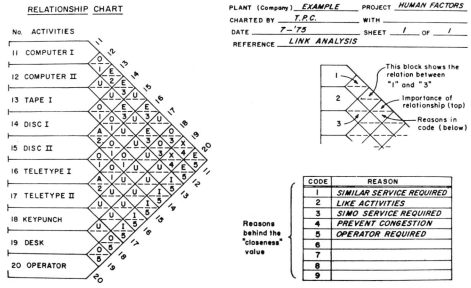

Fig. 3.2.18. Relationship matrix for minicomputer laboratory (from T. P. Cullinane, "Minimizing Cost and Effort in Performing a Link Analysis," *Human Factors*, 1977, *19*, 153. Copyright 1977, by The Human Factors Society, Inc. and reproduced by permission).

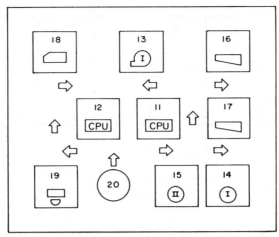

Fig. 3.2.19. CORELAP layout of minicomputer laboratory (from T. P. Cullinane, "Minimizing Cost and Effort in Performing a Link Analysis," *Human Factors,* 1977, *19,* 155. Copyright 1977, by The Human Factors Society, Inc. and reproduced by permission).

on floor, wall, and table surfaces were arranged. The results were estimated to lower significantly the number of required movements and to improve the comfort rating of the operators' task.

Task Network Modeling

Task network modeling is a technique that is an outgrowth of some of the task description technologies such as flow diagrams and operational sequence diagrams. Computer languages exist, such as SAINT (Systems Analysis of Integrated Networks of Tasks), which are specifically designed for the study of task network models. These languages facilitate the study of the dynamic effects on human performance of a system architecture as defined by a static system description such as an operational sequence diagram.

In a task network model, human performance is separated into a series of subtasks, and relationships among the subtasks are represented by a network. Each node of the network represents a discrete subtask performed by the human. The network structure defines the order in which the human performs the subtasks. Therefore, the subtask that the human is performing at any point in the simulation is represented by the node of the network which is currently being executed. Associated with each node are the following attributes:

1. Time to complete the subtask.
2. If the subtask is always to be followed by the same subtask, the subtask to be performed after the current subtask is completed.
3. If the subtask can be followed by several subtasks, the list of possible subtasks to be performed after the current subtask is completed as well as the way that each possible next task is selected.
4. Any user-defined relationships that define the effect on the system of the human performing a subtask, or vice versa.
5. Subtask number (arbitrarily assigned).
6. Subtask name.

Once these attributes are determined for each subtask, a computer model of the task network can be readily developed via any of the computer modeling languages designed for task network simulation. Once the model is computerized, simulation experiments can be conducted by varying the subtask attributes or the structure of the network itself. For example, if one wants to explore the overall effect on the system of the slower performance of one or several of the subtasks, the "time to perform" attribute of those subtasks could be changed and the simulation run. Because performance times are probabilistic (i.e., they are randomly sampled from a known distribution), we can obtain estimates of the probability distribution representing time to perform the overall task network.

To provide an example of a task network, consider the human dialing a telephone. Figure 3.2.20 presents the network. As one follows through the network, the human picks up the receiver, determines

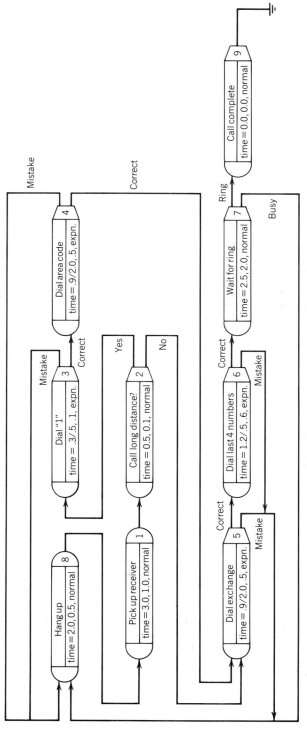

Fig. 3.2.20. Network model of human dialing a telephone.

whether the call is long distance, dials the appropriate numbers, and then redials if the number is busy or a mistake is made. This model was used to study the time savings that would be realized if a touch-tone system were installed to replace a dial system as a function of the number of long-distance calls.

Although task network models are best suited for human tasks that are largely procedural in nature, the modeling languages being used for studying task networks easily facilitate the formation of hybrid models. For example, subtasks of a task network model that involve decision-making behavior can be modeled via artificial intelligence; those tasks that involve psychomotor skills can be modeled via control theory. Because many user–system tasks are procedural and involve some level of cognitive behavior (e.g., decision-making), it is proposed that task network models are well suited for modeling many types of user–system interaction processes. By the nature of most human–computer systems, the interaction between the human and computer can be described as a set of steps that are performed sequentially with the human sometimes having to select from several courses of action available. Associated with each of these steps is a set of human performance parameters including time to perform, the probability of making an error, and other aspects of the behavior. Human tasks that can be characterized in this manner are ideally suited for task network modeling. This is particularly true because task network models can be supplemented by other modeling techniques when required. Additionally, computer languages designed for task network simulation (e.g., SAINT) are usually capable of simulating other parts of the system (e.g., hardware, the operator environment).

Recently, a microcomputer version of SAINT has been developed by Laughery (1984). This microcomputer version is designed to allow the simulation of fairly substantial human operator tasks, consisting of perhaps several hundred subtasks. Also, model construction software has been developed for Micro SAINT so that developing human operator models is a relatively straightforward effort which does not require computer programming skills. Work currently underway on Micro SAINT is aimed at developing a human operator simulation tool that can be used by any human-factors specialist with a minimum of training.

A Final Example

A final example of the application of network analysis in human factors is the innovative work of Harper and Harris (1975). The purpose of this project was to take police intelligence data and use it to portray relationships among suspected criminals, to determine the structure of criminal organizations, and to identify the nature of suspected criminal activities. In this analysis the entities or nodes were people and the relationships or links were contacts between them.

The procedure began with assembling available information in order to prepare an association matrix of relationships between a number of suspected people. Three categories of relationships were

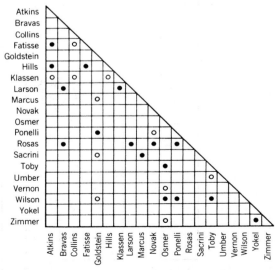

Fig. 3.2.21. Association matrix between people in police intelligence study (from W. R. Harper and D. H. Harris, "The Application of Link Analysis to Police Intelligence," *Human Factors,* 1975, *17,* 161. Copyright 1975, by The Human Factors Society, Inc. and reproduced by permission).

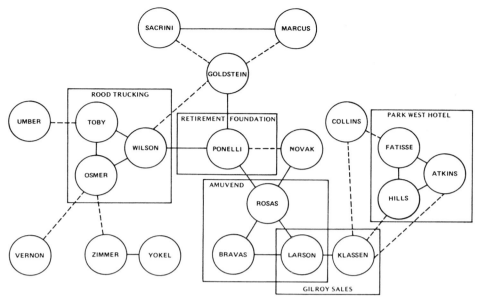

Fig. 3.2.22. Link diagram in police intelligence study (from W. R. Harper and D. H. Harris, "The Application of Link Analysis to Police Intelligence," *Human Factors,* 1975, *17,* 161. Copyright 1975, by The Human Factors Society, Inc. and reproduced by permission).

defined: strong, weak, and none. A strong link might be two relatives who are seen together frequently. A weak link would consist of frequent telephone calls from one individual to a place of business owned by another. Figure 3.2.21 shows an association matrix from one of their studies. The solid circles represent strong links, the open circles represent weak links, and the empty cells reflect no links. The association matrix was then used to construct the link diagram shown in Figure 3.2.22. Solid lines represent strong links and dashed lines are weak links. The boxes represent possible organizations suggested by the structure of the diagram.

This type of analysis offers two important contributions to law enforcement efforts. First, large amounts of intelligence information can be summarized and integrated into an understandable format. Second, new hypotheses about individuals and organizations can be formulated from the diagram. Indeed, such an analysis was useful in identifying a criminal organization engaged in robbery, prostitution, and gambling in northern California. Similarly, the technique was useful in southern California for identifying an organization engaged in narcotics, real estate fraud, and pornography.

3.2.5   CONCLUSIONS

In this chapter we have described a number of analytic techniques for dealing with systems. The techniques have ranged from informal qualitative procedures such as flow analysis, to more formal quantitative techniques, such as task network simulation and modeling. Some of the qualitative procedures are often useful for "getting a feel" for potential designs in new systems or for zeroing in on problems in existing situations. The more formal quantitative techniques can also be applied to such design and evaluation problems when relevant data is available.

In general, progress over the past few decades has made increasingly powerful analytic tools available to the human-factors specialist. A central factor in these developments has been the computer. The computational power of current computers and the plummeting costs of such power make it possible to apply formal analytic procedures to many large complex systems involving people.

There are many analytic techniques besides those covered in this chapter that are useful to human factors. As noted, a number of these techniques are presented in other chapters in this volume. Undoubtedly progress in related methodologies can be expected to continue at a fast pace in the future.

REFERENCES

ANSI (1982). *Industrial engineering terminology* (ANSI Standard Z94.0). New York: American National Standards Institute.

Armour, G. C., and Buffa, E. S. (1963). A Heuristic algorithm and simulation approach to relative location of facilities. *Mathematical Science, 9.*

ASME (1947). *Operation and flow process charts* (ASME Standard 101). New York: American Society of Mechanical Engineers.

Barnes, R. M. (1968). *Motion and time study.* New York: Wiley.

Bartlett, M. W., and Smith, L. A. (1973). Design of control and display panels using computer algorithms. *Human Factors, 15*(1).

Bateman, R. P. (1979). Design evaluation of an interface between passengers and an automated ground transportation system. *Proceedings of the Human Factors Society 23rd Annual Meeting.*

Bonney, M. C., and Williams, R. W. (1977). CAPABLE, a computer program to layout controls and panels. *Ergonomics, 20*(3).

Bytheway, C. W. (1971). FAST diagrams for creative function analysis. *Journal of Value Engineering,* pp. 6–10.

Cullinane, T. P. (1977). Minimizing cost and effort in performing a link analysis. *Human Factors, 19*(2).

Elmaghraby, S. E. (1970). *Network models in management science.* New York: Springer.

Fitts, P. M., Jones, R. E., and Milton, J. L. (1950). Eye movements of aircraft pilots during instrument landing approaches. *Aeronautical Engineering Review, 9.*

Ford, L. R., and Fulkerson, D. R. (1962). *Flows in networks,* Princeton, NJ: Princeton University Press.

Francis, R. L., and White, J. A. (1974). *Facility layout and location: An analytical approach,* Englewood Cliffs, NJ: Prentice-Hall.

Gilbreth, F. B., and Gilbreth, L. M. (1921). Process charts. *Transactions of the ASME, 43.*

Harper, W. R., and Harris, D. H. (1975). The application of link analysis to police intelligence. *Human Factors, 17*(2).

Haygood, R. C., Teel, K. S., and Greening, C. P. (1964). Link analysis by computer. *Human Factors, 6*(1).

Kadota, T. (1982). Charting techniques. In G. Salvendy, Ed., *Handbook of industrial engineering.* New York: Wiley.

Kurke, M. I. (1961). Operational sequence diagrams in system design. *Human Factors, 3*(1), 66–78.

Laughery, K. R., Jr. (1975). Instructions for the use of Micro SAINT: FORTRAN 77 Version 1.0. (Technical Report). Boulder, CO: Micro Analysis and Design.

Lee, R. C., and Moore, J. M. (1967). CORELAP—COmputerized RElationship LAyout Planning. *Journal of Industrial Engineering, 18.*

Moder, J. J., and Phillips, C. R. (1970). *Project management with CPM and PERT.* New York: Van Nostrand Reinhold.

Niebel, B. W. (1976). *Motion and time study.* Homewood, IL: Richard D. Irwin.

Parks, D. L. (1979). Current workload methods. In N. Moray, Ed., *Mental workload: Its theory and measurement.* New York: Plenum Press.

Seehof, G. M., and Evans, W. O. (1967). Automated layout design program. *Journal of Industrial Engineering. 18.*

Turner, W. C., Mize, J. H., and Case, K. (1978). *Introduction to Industrial and Systems Engineering,* Englewood Cliffs, NJ: Prentice-Hall.

Whitehouse, G. E. (1973). *Systems analysis and design using network techniques.* Englewood Cliffs, NJ: Prentice-Hall.

Wickens, C. (1984). *Engineering psychology and human performance.* Columbus, OH: Charles E. Merrill.

CHAPTER **3.3**

ALLOCATION OF FUNCTIONS

BARRY H. KANTOWITZ
ROBERT D. SORKIN

Purdue University
West Lafayette, Indiana

3.3.1 INTRODUCTION

The human-factors specialist who divides work between people and machines is performing allocation of functions. Certain activities (or functions) are assigned to the human and other activities are automated. Allocation of functions determines not only how well a person–machine system will operate but also the quality of working life for the personnel who must toil within that system. Any human-factors specialist who designs work that other people must perform has a grave responsibility. Thus allocating functions is one of the most important tasks given to system designers (Bailey, 1982, p. 221).

This chapter reviews and evaluates techniques developed over the past 30 years to enhance allocation of functions. There are many tools that the designer can, and indeed should, use but not all are equally effective. This chapter starts with the oldest tools and shows how newer tools have been proposed to improve allocation of functions. We also look to the future and briefly examine some very new tools that are being developed now but are not yet suitable for practical use. Because allocation of functions determines human work load, we discuss the nature of human work load and especially mental work load. Finally, the chapter ends with a practical discussion of allocation of functions in the context of robotics and CAD/CAM.

Allocation of functions occurs only in the context of a person–machine system designed to fulfill specified system goals. Thus before examining the details of allocation, we must first define a person–machine system and explore the evaluation of system goals. We can then progress to our main goal of reviewing the actual techniques used to allocate functions between people and machines.

3.3.2 PERSON–MACHINE SYSTEMS

Although it is possible to define a system formally with mathematical symbols and set theory (Hall and Fagen, 1956), we use an informal verbal definition. A person–machine system is an arrangement of people and machines interacting within an environment in order to achieve a set of system goals. The human factors specialist tries to optimize the interaction between people and machine elements of the system, while taking the environment into account. A representative schematic of a person–machine system is shown in Figure 3.3.1. The right half of the diagram represents the machine subsystem as it appears to the human-factors specialist. Visual and other displays represent the internal equipment status in a form the human can understand. Controls allow the human operator to make changes in the internal equipment. These two boxes represent some of the most important human-factors aspects of the machine and everything else is bunched into a single box labeled internal equipment status. The engineers who designed this internal equipment spent months arranging detailed subsystems that are not represented directly in Figure 3.3.1. From their viewpoint this is a gross oversimplification of the machine subsystem and it is almost insulting to represent all their effort by a single box. This oversimplification does not mean that the human-factors specialist fails to appreciate the importance of this engineering accomplishment. It does mean that to a large extent the detailed inner workings of the machine—the gears and cogs, integrated circuits, and digital logic—are irrelevant. The human-factors specialist must design the specifications for displays and controls and also help the engineering team to ensure that system dynamics are compatible with human skills. Other specialists are responsible for implementing these specifications and the human factors specialist does not particularly care whether a necessary time delay is achieved by a bucket brigade circuit or by a delay line.

The human subsystem is represented by the left side of Figure 3.3.1. Information from the displays is perceived. This information is processed and decisions are made. Motor responses are made to alter control settings. Figure 3.3.1 is not a complete representation of the human. There is nothing to show the workings of the brain and central nervous system directly. Although the brain is important, the human-factors specialist does not need to know what is happening in individual neurons in order to optimize the system. Figure 3.3.1 highlights those portions of the person–machine system that are of greatest importance to human factors. It neglects other important aspects of machines and people that are less crucial to human factors.

Perhaps the most important part of Figure 3.3.1 is the vertical line separating the machine subsystem from the human subsystem. This line represents the interface between human and machine. Information flows across this interface in both directions: from the machine to the person and from the person to the machine. Thus Figure 3.3.1 represents a closed-loop system because one can start at any point in the diagram and trace one's way around the complete system, eventually returning to the starting point.

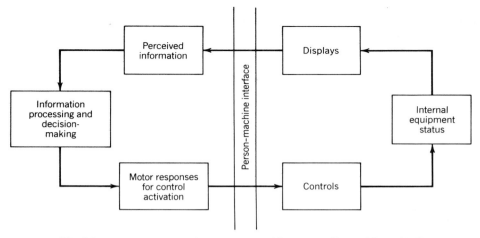

Fig. 3.3.1. Representation of the person–machine system (from Meister, 1971).

3.3.2.1 Defining System Goals

Before anyone can start to design a person–machine system, one must know what the system is supposed to do. This seems obvious at first, but we shall soon see that systems have several goals and that some of these goals compete with other goals. It is often impossible to accomplish all the desirable system goals. When this happens, it is better to discover it at the start of the design process rather than after the system goes out the door to customers. In many firms, systems goals are specified by management (or by a customer) and the human-factors team is simply given a set of goals to accomplish. Trade-offs among goals are seldom clearly specified. The team may produce a system that realizes the ambiguous goals they were given, only to be castigated later on when management realizes those goals were not what it really wanted. Thus a prudent human-factors team will spend some time analyzing a list of system goals and even going back to management for clarification when goals conflict.

An example will make this conflict more visible. Let us suppose we have been hired to create a list of system goals for the postal service. We can start by writing down several desirable goals.

1. Speedy service
2. Frequent mail delivery
3. Home delivery to all U.S. residents
4. Low cost
5. Guaranteed transit time

Already we can see some problems with this very short list. The first problem is ambiguity. Precisely how speedy is "speedy service?" Does delivery to "all U.S. residents" include hermits living in caves on the slopes of Mount Hood? These goals must be refined to be more specific. For example, frequent mail delivery could be defined as twice a day. In Europe at one time housewives mailed in their daily grocery orders that were delivered the same afternoon. Or frequent mail delivery could mean once per day seven days a week. Is 22 cents a letter "low cost"?

Even after these ambiguities are cleared up, the problem of conflict among goals remains. A system that delivers mail to 95% of U.S. residents will incur very high costs in rural or other low-density areas. This conflicts with the goal of low cost. Should residents in these areas pay a surcharge for daily delivery, should they have less frequent delivery than urban areas, or should these rural delivery costs be absorbed by the entire system? Management may lack the information required to decide. Thus instead of immediately setting off to design a system, the human-factors team may first have a specify the trade-offs among conflicting system goals. For example, home delivery to 80% of residents may cost 18.5 cents per letter while delivery to 95% might raise the cost to 35 cents. Without knowledge of these trade-offs a rational decision cannot be made.

3.3.2.2 Levels of Technology

There are two schemes for classifying levels of technology. First, we can organize technology by its general level of complexity and detail. One such scheme (Carson, 1970) has three levels of complexity.

The first and lowest level concerns relatively simple responses to components of the system such as its knobs and dials. The focus of interest is the individual element rather than general system properties. The human-factors specialist working at this level must keep track of many small details. At the second level the flow of behavior is more important than individual components. This would include tasks such as tracking and driving a car along a winding road. Here the specialist is concerned with more general human capabilities, such as avoiding an overload of information. The third and highest level involves the total system and decisions as to what tasks should be performed by people and what other tasks by machines. Machines are broadly defined, and this third level also includes relationships between people and their environments. For example, the famous architect Le Corbusier defined a house as "a machine to live in."

A second classification scheme creates levels of technology by focusing on the contributions of people and machines to the total system. At the lowest level, the person supplies both power and control of the system. A human wielding a shovel is functioning at this lowest level of technology. The first improvement has the machine supplying power, but the human still exercises control. A human operating a punch press is one example of this level of technology. At the next level the machine supplies power and information and the human still controls. Any factory where people read displays and turn controls functions at this level. Finally, the highest level of technology has the machine supplying power, information, and control while the human monitors the operation. An automated plant run by computers is an example of this level.

Human factors is needed at all levels of technology. Indeed, one might even argue that it is needed more at the lowest levels because there are many more people interacting with lower-level systems; for example, there are more hammers than there are nuclear power plants. However, allocation of functions is required most when the system is complex. It is only at the highest levels of technology that the machine subsystem has the sophistication to accomplish functions requiring the symbolic manipulation of information. Therefore, the remainder of this chapter ignores lower levels of technology. Instead, we focus upon the highest levels where great system complexity implies more choices for system designers. As technology becomes more sophisticated, the human tends to be more a monitor of system behavior than an active controller, and function allocation must take this into account.

3.3.2.3   A Philosophical Design Dichotomy

But before the human-factors specialist can get down to work, a more general philosophical question must be raised. One approach to the human as monitor is to realize that most of the time the system works quite well without human intervention. Indeed, the system often works better if the human keeps hands off. For example, in the Three Mile Island accident the automatic safety systems worked as designed and turned on the emergency pumps. The human operators erred and manually turned off the pumps. Therefore, one possible conclusion is that the human should be eliminated entirely from systems, or if that is not possible–for example, there are legal requirements that operators be present in nuclear power plants–then the person's role should be minimized. This will, on the average, decrease the opportunity for human error and so increase system reliability. Furthermore, it will lower operating costs by substituting machines for people.

Another philosophy is to keep the human operator involved as much as possible in the system, even if this means creating artificial tasks, such as logging in display readings by hand, to keep the person busy. Then, should the machine part of the system fail, the human will be able to leap in and fix the problem. (The minimization philosophy assumes that the human probably won't be able to fix the problem anyway.) Thus this philosophy argues that in landing a probe on the moon, it is best to have human astronauts along in case something unexpected goes wrong. The human is clever and adaptive and can often solve unanticipated problems.

We can't tell the reader which philosophy to believe or which a company should adopt. However, our own personal beliefs bias us against including a human in a system when there is little for the person to contribute. In fact, such an undemanding job is quite stressful and unpleasant. We believe that the human-factors expert has an obligation to provide meaningful work for people. This leads us to the next topic, dividing work between people and machines.

3.3.3   TECHNIQUES FOR FUNCTION ALLOCATION

Having discussed systems in general, we are now prepared to get down to the nuts and bolts of the chapter. This section is divided into four parts. First we cover the historical foundations of allocation of functions reaching back to its ancient roots in 1951. Then we look at current methods for function allocation that are related to these ancient roots. After a brief discussion of the perils of automation we examine a relatively new notion in allocation, that of dynamic allocation including biocybernetic methods.

3.3.3.1 Historical Antecedents

The ground rules for allocation of functions were laid down by the late Paul Fitts (1951). His ideas were logical and clearly articulated. His fundamental axiom was to render unto the human that at which the human excels and to render unto the machine that at which the machine excels. The machine subsystem and the human subsystem could be viewed as competitors vying for the various and sundry functions performed within the system. Each subsystem was rewarded with the functions that it could perform better than the other subsystem. If a function required rapid arithmetic calculation, or lifting great weights, it was awarded to the machine. If a function required pattern detection in noise, or handling rare overloads of information, it was assigned to the human. Functions were evaluated, at least during early design stages, in isolation, and allocation was based upon the relative efficiency of human versus machine for each particular function. Later, of course, these allocations were summed to check that one subsystem did not receive an undue share of the work. Such deficiencies could be remedied by shifting marginal allocations back to the other subsystem to lighten the burden of the overloaded subsystem, especially if that subsystem happened to be human.

This approach fostered the creation of wonderful tables of functions that contrasted human and machine performance on a wide variety of functions. These tables were staples of human factors and no self-respecting text or handbook dared be without them. Accordingly, we have included Table 3.3.1 as a representative effort. It is reproduced in its entirety without any editorial changes or additions. Reading (and writing) such tables is great fun, especially as technology creeps up and gains on the human. For example, at one time machines could not interpret fingerprints so that pattern recognition functions were better performed by people. Now, however, artificial intelligence systems can read fingerprints so this function can be relocated into the machine column. We leave as a pastime for the reader, identification of other such technological advances related to speech perception and generation, and so forth. Alas, because the human is evolving at a far slower rate than machines, we have been unable to locate any examples of functions moving from the machine to the human column.

As the reader may suspect, this traditional approach, although still practiced today, encountered several difficulties. There was no problem in building tables of functions and updating them as machines became more capable. But there were substantial problems in using these tables. Although they were a convenient framework for the first steps in function allocation, the final design seldom looked much like the initial attempt. Indeed, efforts by designers at rebuilding the tables based upon the actual allocations were soon abandoned because in most cases the lack of fit was obvious. Designers of real systems complained about the lack of good allocation algorithms and consigned the lengthy tables to the dustbins of academia where they were considered suitable for the instruction of introductory students in human factors. Indeed, the first author of this chapter recalls more than one occasion on which he was confronted by frustrated system designers who complained that tables they had duly memorized as students were of so little help in accomplishing function allocation for real systems.

Anyone who has not attempted allocating functions may find this astonishing. Certainly, assigning functions to subsystems based upon their ability to perform the function appears to be reasonable and logical. What went wrong? One of the most trenchant analyses of this paradox was put forth by Jordan (1963) in an article that should be required reading for all human-factors specialists. Jordan articulated the fundamental flaw in the clean engineering approach laid down by Fitts. Any table that can compare human and machine, especially if numerical indexes of relative performance or equations can be listed, as any good engineer would attempt, is bound to favor the machine. Machines and people are not really comparable subsystems, even though human-factors specialists go to great length to relate the two. This fundamental truth is rediscovered by each generation of human-factors specialists. For example, now there is a debate about equating the reliability of people and machines and much effort has been devoted to establishing estimated probabilities of human error in various

Table 3.3.1. Functional Advantages and Disadvantages of Men and Machines

Functional Area	Man	Machine
Data sensing	Can monitor low-probability events not feasible for automatic systems because of number of events possible	Limited program complexity and alternatives; unexpected events cannot be handled adequately
	Absolute thresholds of sensitivity are very low under favorable conditions	Generally not as low as human thresholds
	Can detect masked signals effectively in overlapping noise spectra	Poor signal detection when noise spectra overlap

Table 3.3.1 (*Continued*)

Functional Area	Man	Machine
	Able to acquire and report information incidental in primary activity	Discovery and selection of incidental intelligence not feasible in present designs
	Not subject to jamming by ordinary methods	Subject to disruption by interference and noise
Data processing	Able to recognize and use information, redundancy (pattern) of real world to simplify complex situations	Little or no perceptual constancy or ability to recognize similarity of pattern in spatial or temporal domain
	Reasonable reliability in which the same purpose can be accomplished by different approach (corollary of reprogramming ability)	High reliability may increase cost and complexity; particularly reliable for routine repetitive functioning
	Can make inductive decisions in new situations; can generalize from few data	Virtually no capacity for creative or inductive functions
	Computation weak and relatively inaccurate; optimal game theory strategy cannot be routinely expected	Can be programmed to use optimum strategy for high-probability situations
	Channel capacity limited to relatively small information throughput rates	Channel capacity can be enlarged as necessary for task
	Can handle variety of transient and some permanent overloads without disruption	Transient and permanent overloads may lead to disruption of system
	Short-term memory relatively poor	Short-term memory and access times excellent
Data transmitting	Can tolerate only relatively low imposed forces and generate relatively low forces for short periods	Can withstand very large forces and generate them for prolonged periods
	Generally poor at tracking though satisfactory where frequent reprogramming required; can change to meet situation. Is best at position tracking where changes are under 3 radians per second	Good tracking characteristics over limited requirements
	Performance may deteriorate with time; because of boredom, fatigue, or distraction; usually recovers with rest	Behavior decrement relatively small with time; wear maintenance and product quality control necessary
	Relatively high response latency	Arbitrarily low response latencies possible
Economic properties	Relatively inexpensive for available complexity and in good supply; must be trained	Complexity and supply limited by cost and time; performance built in
	Light in weight, small in size for function achieved; power requirement less than 100 watts	Equivalent complexity and function would require radically heavier elements, enormous power and cooling resources
	Maintenance may require life support system	Maintenance problem increases disproportionately with complexity
	Nonexpendable; interested in personal survival; emotional	Expendable; non-personal; will perform without distraction

Source: From Bekey, G. A. (1970). The human operator in control systems. In K. B. DeGreen Ed., *Systems Psychology.* New York: McGraw-Hill. Used with permission.

tasks (e.g., Swain and Guttmann, 1980). However, this entire approach has been severely criticized on the grounds that people and machines have fundamentally different metrics for reliability (Adams, 1982). Reliability in machines is measured by mean hours between failures and percentage of machines that fail in a given time. Because human reliability cannot be easily characterized in these terms it is awkward to say the least, and unlikely to say the most, that human and machine subsystems can be combined to produce meaningful reliability values for the total system.

The fallacy, noted by Jordan, is that machines and people are not comparable. Side-by-side comparisons in the long run will fail because they inevitably produce the following conclusion: People are flexible but inconsistent, whereas machines are consistent but inflexible. Therefore, Jordan suggests the concept of comparability be replaced by the more accurate concept of complementarity. Instead of thinking about whether a task should be performed by a person or by a machine, we should instead realize that functions are performed by people and machines together. Activities must be shared between people and machines and not just allocated to one or to the other.

In many automated systems the human is allocated only those functions that are too expensive or too hard to assign to the machine. The human is regarded as a link between subsystems and is provided only that information and control needed by a link. When the system fails, the human cannot take over manually. Links cannot suddenly become supervisors. We return to the issue of manual backup raised by Jordan later in the chapter. For now, we reiterate Jordan's main point (page 165):

Man is not a machine, at least not a machine like the machines men make.

Allocation of functions between people and machines must go beyond the clear engineering rules we would follow when allocating functions between two machine subsystems.

3.3.3.2 Current Procedures

At first glance, one might conclude that Jordan's analysis dealt a fatal blow to allocation of functions based upon tables of relative merit for people versus machines. However, human factors is, if anything, a pragmatic discipline. An imperfect tool is still better than no tool at all. So many designers continue to use tables of relative merit either because they do not find Jordan's criticism convincing, or more likely because they are not familiar with anything better. This section examines alternatives and extensions to the familiar tables of relative merit.

Let the Machine do It

One simple approach that requires no tables at all is to delegate as many functions as possible to the machine (Chapanis, 1970). This approach appears to be considerably less elegant than the rational deliberative process where the designer carefully decides if the human or the machine is better suited to some specific task. But it is quite sensible. The reliability of machines can be increased less expensively than can the reliability of people. There are many systems where overall system performance can be enhanced, or at least maintained, by assigning most functions to the machine subsystem.

However, there is one outstanding risk associated with this design approach. The human-factors specialist must check to determine that the leftover functions leave the human subsystem with a reasonable set of tasks to perform. Unreasonable sets of tasks arise in two main ways. First, the load created by the tasks may not match human capabilities. Although overload is the first defect that designers look for, underload can be equally fatal to the human. A design strategy of delegating the maximum number of functions to the machine is likely to produce underload for the human operator.

Stress is created when task demands do not match human capabilities (see Kantowitz and Sorkin, 1983, Chapter 19). Although it is obvious that overload can create stress, it is equally true that stress is also created by severe underload. Perhaps an example will make this point clearer. Imagine an assembly line with filled bottles of Coca Cola passing by on a conveyor belt. A strong light is behind the bottles. The job is to inspect the bottles for sediment, and so on as they move past the light. This task is so boring that occasional 7-UP bottles must be inserted to keep the human inspector awake and functioning at all. Few humans would be happy to perform this job 8 hours a day. (Indeed, it is possible to train a pigeon to do this task!)

Unreasonable sets of tasks can arise even though the amount of work load is acceptable. This second way of producing dissatisfaction in the human subsystem occurs when the functions left over for the human do not form a coherent set. If the operator cannot fathom the cognitive structure behind a set of functions—perhaps because the designer has failed to provide such a structure—morale and performance are impaired. This is a more subtle error in function allocation than simple underload or overload. The human-factors specialist must also consider the mental representation that the human subsystem will construct to link his or her assigned functions.

Function allocation whereby the human is left with whatever the machine can't do often is accompanied by a design philosophy that regards the human as a nuisance. Such designers would prefer not to have humans in the system at all. In some cases the human is there merely to satisfy legal requirements or to meet union stipulations in the face of automation. The presence of the human is rationalized as the subsystem of last resort. If all else fails, at least the human can pull the plug and prevent further damage. This, of course, is the extreme case of an allocation policy that gives the human operator responsibility for only the abnormal system conditions. When the system is operating properly, the designer has intended that the human keep hands off. We dislike such a design philosophy. It makes human incompetence a self-fulfilling prophecy. If the human has not been allocated functions when the system is operating properly, it is most unlikely that he or she can operate effectively in any manual backup mode.

A Balanced Approach

The system designer, faced with the realities of a messy world, does not perform allocation of functions in the tidy manner portrayed in most textbooks. The spirit of the working designer has been captured by Bailey (1982, Chapter 11) and this section draws heavily upon his remarks. Bailey divides allocation into four categories: allocated to machine by management, allocated to human or machine by requirements, allocated by a systematic allocation procedure, and unable to allocate.

In an ideal world, a system designer would be given a set of system goals and could then derive system functions needed to realize the goals. The designer would have complete freedom to assign the functions. Usually some of the designer's decisions have been preempted by the time the problem reaches his or her desk. Management will have decided a priori that certain functions are to be assigned to machines. Although these a priori decisions may not be the result of a careful analysis, the designer is stuck with these constraints, which often derive from management's belief that automation is an innate good. (This belief is discussed in the following section.)

System requirements dictate other function allocations. Some functions cannot be assigned to machines for obvious reasons. For example, we would not want a machine to be able to launch a nuclear attack based upon its belief that incoming missiles have been detected. On a less dramatic level, in the United States a licensed operator is required to throw the final switch that powers up a nuclear power plant, although this is well within the technical capability of automated equipment. Similarly, some requirements cannot reasonably be accomplished by people. A requirement to calculate the cube root of a five-digit number in less than 1 sec must be met by a machine. The designer is not free to alter the system requirements and so cannot alter the function allocation.

In a real system, the designer is at first unable to allocate some functions. Sometimes additional analyses allow the designer to break these functions down into smaller functions that can be allocated. However, some functions call for intimate cooperation between people and machine, as noted by Jordan, and these cannot be allocated to one subsystem alone. Instead, the designer must think in terms of complementarity rather than comparability. Each subsystem must have the ability to hold up its share of a dialogue. For example, a computer might present a list of alternatives. The human would select one alternative and the human–computer interaction would continue recursively until a system objective was attained.

Finally, we come to those functions that the designer can allocate to people or to machines. Here the traditional functional allocation system taught in classrooms may be useful. We do not reject using standard tables of relative ability; we reject only the idea that these tables should automatically be applied to all system functions. It should be clear that the three categories discussed above have greatly diminished the set of functions that can be easily allocated to man or to machine.

Once the designer has completed function allocations, all system functions are assigned to a person, a machine, or to a person/machine dialogue. At this point the designer should document the reasons behind the particular allocation that has been decided. Most designers would now feel that their job has been completed. However, if time permits, the designer (or another designer) should consider alternative allocation schemes. Although the original scheme may be satisfactory, it probably is not optimal. If a system will be in use for several years, the extra cost of considering alternative allocations during system development will be repaid many times over during the operational life of the system. The temptation to assign functions to people if the function cannot be quickly implemented in hardware and software should be resisted.

A Formal Approach

The balanced approach just described captures the flavor of a designer at work but does not offer any quantitative comparisons to augment general guidelines. A more formal approach suggested by Meister (1985) provides a prescriptive method that documents how allocation of functions should be performed in order to create an optimal design solution to a design problem. Meister's formal procedure has five steps.

1. Consider all possible ways a function can be implemented. Usually, the typical engineer restricts the set of ways to those that have worked well in the past. The behavioral human-factors specialist must prod the engineer also to consider manual modes of operation that involve an operator.

2. Write a narrative description of design alternatives as shown in Table 3.3.2. This allows a qualitative comparison of alternatives. The human-factors specialist must determine that the operator is indeed capable of performing the assigned functions for various alternatives.

3. Establish criteria, and their relative weights, for the entire system. Standard criteria for system evaluation include cost, performance, reliability, maintainability, personnel requirements, safety, and so on. In order to quantify the relative importance of these criteria, Meister suggests making all possible pairwise comparisons among criteria. In each pairwise comparison, the more important criterion gets a score of one; the other gets a score of zero. These scores are summed and divided by the total score to obtain weights on a scale of zero to one. Although we approve of this procedure, we must also point out that the actual numerical weight values are not that important. The outcome of each binary comparison does not take into account how much more important is the preferred criterion. For example, a comparison of performance requirements versus cost may favor performance as the more important criterion. Similarly, a comparison of performance versus maintainability may also favor performance. Both these choices count equally in determining the final weights. However, in the first comparison (performance versus cost) performance may be twice as important as cost, whereas in the second comparison performance may be 10 times more important than maintainability.

It is easy to create more complex scaling procedures that generate more accurate relative weights. However, the major benefit of this procedure may be more that it forces the designer to list systematically and compare all criteria together, rather than the actual numbers such as comparison generates.

Table 3.3.2. Design Alternatives for Allocation of Functions

Alternative 1 (Operator Primarily)	Alternative 2 (Human–Machine Mix)	Alternative 3 (Machine Primarily)
Sonarman detects target signal on scope, examines brightness, shape, recurrence, movement, etc., and reports "probable submarine" or "nonsubmarine target"	Sonarman detects target signal on scope. Associated computer also detects signal, records it, and searches library of standard signals. Computer displays to sonarman original signal and comparison signal on sonar gear, together with the probability of its being a submarine. Sonarman decides on basis of own analysis and computer information whether target signal is submarine or nonsubmarine and reports accordingly	When a signal having a strength above a specified threshold is received by the sonar array, a computer associated with the detection apparatus automatically records the signal, analyzes its strength, brightness, recurrence, etc. according to preprogrammed algorithms, compares it with a library of standard sonar signals, and displays an indicator reading "probable submarine"
Operator Functions		
Detection of signal	Detection of signal	Take action on receipt of "probable submarine" signal
Analysis of signal	Analysis of signal	
Decision-making	Decision-making	
Reporting of decision	Reporting of decision	
Machine Functions		
Display of signal	Detection of signal	Detection of signal
	Recording of signal	Analysis of signal
	Searching of comparison signals	Decision-making
	Analysis of signal	Display of conclusion
	Display of information	
Advantages/Disadvantages		
No machine backup for operator inadequacies	Operator/machine each back each other up	No operator backup for machine inadequacies

Source: From Meister, 1985.

4. A similar procedure is used to create weights for each design alternative. Each criterion is paired with each alternative and scores of zero and one are assigned.

5. The final quantitative step combines the weights achieved in steps 3 and 4. The cross-products (criteria × alternatives) are summed over each alternative. The alternative with the highest score is preferred.

As a psychometric procedure, this scaling technique could be refined considerably. Nevertheless, it is useful as a convenient alternative to the primarily qualitative methods of allocation currently being used. At the very least it forces the designer to lay out the options (and biases) systematically. This is an important step in the right direction.

3.3.3.3 Perils of Automation

There is an unfortunate tendency, especially among system designers with little training in human factors, to believe that automation is an innate good. In this section, we attempt to refute this stereotype by examples from aviation where flight-deck automation is an important topic. Our examples are drawn from a NASA-industry workshop on automation reported by Boehm-Davis, Curry, Wiener, and Harrison (1983).

The workshop identified five problems associated with automation. First, newly automated systems seldom provide all of the anticipated benefits. For example, the first version of the ground proximity warning system produced many false alarms. This could result in decreased safety and increased crew work load.

Second, failure of automated equipment leads to problems of credibility. If users have an option, they will not rely upon equipment they do not trust.

Third, training requirements are often increased by automation. The user must be able to operate the equipment in two modes, automatic and manual. Furthermore, it is usually the case that automatic operation deprives the crew of practice in the manual mode. This loss of proficiency may cause problems when the manual mode need be invoked. Finally, the increased complexity of new automated cockpit systems that often perform more functions than the equipment they replace can make the new systems more difficult to learn.

Fourth, designers often fail to anticipate new problems that the automated systems will create because they have instead focused upon the benefits of the new system. For example, incidents have already occurred when pilots entered incorrect way-point coordinates in inertial navigation systems. Indeed, it seems that this was a likely cause for the demise of a Korean airliner that wandered into Russian airspace and was shot down.

Fifth, automation makes pilots into system monitors rather than active controllers. Some pilots dislike this new role. Furthermore, a pilot might not be mentally prepared suddenly to take over control from the automatic systems in an emergency.

This last point is related to unthinking reliance upon automatic systems. Pilots may not notice when an automatic system is not performing properly, as is illustrated by the controlled flight into terrain of Eastern flight 401.

The following story of Eastern Airlines flight 401 is drawn from Danaher (1980). Flight 401 was a L-1011 wide-body jet approaching Miami International Airport for a night landing. However, it was diverted because the instrument panel did not show that the nose landing gear was properly locked in the down position. The pilot turned on the autopilot, setting it to cruise at 2000 ft. This reduced the cockpit work load so that the crew could devote their attention to checking of the apparent landing gear malfunction. The preoccupied crew failed to notice that the autopilot had been accidentally switched off, sending the plane into a gradual decline.

Miami approach control, although not technically responsible for flight 401 once it had left the area, noticed that 401 was down to an altitude of 400 ft. The controller radioed: "Eastern," ah, 401 how are things comin along out there?" The crew replied immediately that they were making progress and intended to return for a landing at Miami. Less than 30 second later flight 401 hit the ground. The impact destroyed the plane and killed 99 people out of the 176 aboard.

It is ironic that the automated equipment functioned perfectly yet did not prevent this accident. The crew relied on the autopilot and failed to notice it had been switched off. The air traffic controller knew that flight 401 was at a dangerously low altitude, yet then-current procedures did not require him to notify the flight of its altitude, because it is the pilot's responsibility to keep track of altitude. (The controller thought his altitude reading might have been wrong because it was possible for the equipment to display incorrect information for a very short period of time between updates.) There is now an automatic minimum altitude warning that alerts controllers to possibly dangerous altitude deviations by planes under their control. The rules have also been changed so that the controller now notifies the pilot. However, maintaining proper altitude is still the responsibility of the pilot.

The major lesson of this accident is that human error cannot always be prevented merely by adding more automatic machinery. In fact, unthinking reliance on automatic systems such as the autopilot can at times contribute to accidents instead of eliminating them.

3.3.3.4 Dynamic Allocation

A new technique may eventually remove the need for a system designer to perform any function allocation by allowing the user to decide allocation on-line. This is the best approach (if feasible) because it frees the user from the Procrustean bed created by the system designer. Because the first commandment of human factors is "Honor thy user," it is quite desirable to let the user decide upon allocation of function.

A simple example of such dynamic allocation is the cruise control on a car. When the driver wishes to allocate the function of maintaining constant vehicle speed to the machine subsystem, he or she engages the cruise control. When such an allocation is not desirable, as in heavy urban traffic, the driver reserves this function. The various automatic systems now available for flight-deck automation are more sophisticated examples of dynamic allocation. The pilot decides whether or not to engage these automatic functions.

Dynamic allocation is merely a logical progression in technological progress that offers greater freedom for allocation of functions. At one time, allocation was fixed in hardware. The allocation of the system designer could not be changed without rebuilding the entire system. As technology advanced, elements of systems were incorporated into software. Software is relatively easy to change. This trend will continue and soon software will be more important than hardware. This means that allocation of functions is no longer written in stone and silicon. This implies that design emphasis will eventually shift from correctly anticipating user needs in allocation to evaluating system behavior as well as the performance and satisfaction of the human subsystem.

The ultimate in dynamic allocation requires no conscious effort from the user. When work load gets too large, the system automatically picks up more of the burden to relieve the human operator. Current research in biocybernetics (Donchin, 1984) offers an interesting prospect. It has been well established in the laboratory, and field tests are now underway, that operator work load can be measured by such biological signals as heart rate and brain waves. We can anticipate a day when biological sensors are routinely included in the equipment of airplane pilots and other workers who are subject to excessive information loads. How this information will be used to improve system performance while avoiding some of the perils of automation discussed previously will be a fascinating challenge for human factors in the future.

3.3.4 ALLOCATION AND WORK LOAD

A successful allocation of functions achieves two major aims. First, all system goals can be readily accomplished. Second, the human subsystem is presented with a coherent set of functions that offer a reasonable work load: not too high and not too low. It is important to realize that these two major aims are not always entirely compatible. The system designer, or management, must decide which aim is foremost. For example, avoiding underload might dictate assigning some functions to the human that the machine could do better, and this may decrease overall system performance. Some years ago, general design philosophy favored achieving all system goals even if the jobs left for the human were less than ideal. Now the pendulum is swinging the other way. Indeed, in Norway it is legal for workers to sue their employers if the workers believe that their job is too boring! It is to be hoped that a clever designer can steer a path between Scylla and Charybdis to attain both major aims. Achieving system goals has been discussed in a preceding chapter in this handbook. Here we focus upon measuring operator work load.

A complete review of work load exceeds the scope of this chapter. The interested reader is directed to discussions of work load in general (Hart, 1985) and mental work load in particular (Kantowitz, 1985a; 1985b). As systems become more flexible owing to the dominance of software (see Section 3.3.3.4), it will become more important to measure work load so that possible changes in allocation can be evaluated within the operating system instead of within the mind of the designer.

As our society becomes more and more technological, more and more workers are engaged in mental work rather than physical work. Of course, even such physical work as digging ditches requires some mental effort to guide the shovel, but most would agree that the mental requirements of most modern jobs exceed the physical requirements. An office worker or executive seated at a video display terminal does some physical work pressing keys but this part of the job is less taxing than the cognitive effort that determines which keys should be pressed. Physical work load and energy expenditure can be measured fairly accurately by such means as oxygen consumption and heat generated by the body. Because more and more work is based primarily on mental, rather than physical, work, considerable effort has gone into establishing methods for measuring mental workload. The ultimate goal would be a single number that describes or predicts how much mental work load is associated with any given task. Unfortunately, it is not likely that this goal will ever be accomplished (Johanssen et al., 1979). Mental work load is too complex a concept to be summarized by a single number.

A recent NATO conference on mental work load (Moray, 1979) concluded that there is no single definition of it. Many operational definitions were offered depending on the background of the individual researcher. Good arguments can be made for relating mental work load to information processing and attention (experimental psychologists), time available to perform a task (system engineers), control engineering (electrical and industrial engineers), and stress and arousal (physiological psychologists and physiologists). Behavioral measures of mental work load have been classified into 14 tasks or methods divided into three categories (Williges and Wierwille, 1979): subjective opinions, spare mental capacity, and primary task.

Subjective opinions can be collected either by rating scales or by questionnaires and/or interviews. A rating scale provides a psychometric technique for ordering opinions in a mathematically consistent manner whereas an interview or questionnaire obtains data that are not quantified. The most popular rating scale is the Cooper–Harper, 1969). More recent efforts at subjective assessment of mental work load (Eggemeier, 1981) have used a mathematical procedure called conjoint measurement to obtain a work load scale (Reid, Shingledecker, and Eggemeier, 1981). Although this work is at an early stage of development so that not all the psychometric assumptions have been validated empirically, it does appear to be promising. Multidimensional scaling is another approach that has been used (Derrick, 1981). Perhaps this concentration of effort will soon produce a reliable and valid subjective scale of mental work load (Moray, 1982).

Spare mental capacity is the largest category. It is based on the assumption that attention is limited and that the human channel has an upper bound (Rolfe, 1971). Measures of spare capacity have been derived from information theory (Senders, 1970) and assorted computer simulations. But by far the most data have been collected using the secondary task paradigm (Ogden, Levine, and Eisner, 1979).

In this paradigm an extra task, called the secondary task, must be performed together with the task of interest, called the primary task. Operators are instructed to devote their attention to the primary task and to perform the secondary task only when this will not affect their ability to do the primary task. Decrements in secondary-task performance are interpreted as evidence of primary-task work load. Thus if a given secondary task can be performed at 80% efficiency with Primary Task A, but only with 50% efficiency with Primary Task B, then we conclude that Task B imposes a greater work load. Although this logic is correct, there are many technical problems in selecting appropriate secondary tasks (Kantowitz, 1985c). Because human-factors specialists have only recently become aware of the limitations of the limited channel model (Sanders, 1979) that imply there is no universal secondary task (Pew, 1979), many of these secondary-task results are difficult to interpret. Although a few researchers have tried to use theoretical models to guide selection of secondary tasks (Wickens, 1980), most have used arbitrary combinations of primary and secondary tasks that prevent generalization of results beyond the particular combination of tasks selected. Because it is not practical to perform experiments with all possible combinations of the wide variety of tasks used to assess spare mental capacity, it is clear that progress in this area will depend more on new theoretical developments than on acquisition of more data based upon the incorrect limited- or single-channel model.

The primary-task method is based upon the assumption that increasing mental work load will cause a decrement in primary-task performance. Indeed, one researcher goes so far as to state that if the mission was successfully completed, there was no overload (Albanese, 1977)! This approach may be successful if work load is indeed quite high. But analysis of a primary task that is associated with low to medium mental work loads is often too insensitive to reveal effects of additional tasks or procedural alterations.

The practicing engineer who lacks the equipment and experience needed to use the secondary-task method of measuring workload is probably best served by subjective ratings. The easiest way to obtain a subjective rating is simply to ask an operator to give a number, usually from 1 to 7 or from 1 to 100, on an arbitrary scale. Verbal descriptions are attached to the ends of the scale to serve as anchor points (see Table 3.3.3). Although the exact meaning of the particular response, say 4.6, is unclear, one still can compare sets of ratings. Thus we can say that allocation scheme A yields the lowest work-load rating when compared to alternative allocation schemes B, C, and so on. Ratings are best obtained from experienced workers as they function on-line, but if this is too inconvenient, acceptable ratings can be obtained based upon memory.

A more sophisticated technique uses several subscales in addition to an overall work-load scale. These subscales can be combined statistically to yield a calculated work-load rating. Table 3.3.3 shows a set of subscales developed by the Human Performance Group, NASA-Ames Research Center. Overall work load is calculated by assigning a weight to each subscale. Weights are different for different individuals. The weights are arrived at by asking each person to compare all possible pairs of subscales and to decide which member of the pair is more important in determining his or her own work load. The number of times a particular subscale is chosen represents its weight. (Weights can be normalized by dividing by the total number of paired comparisons to make each weight less than one, but this does not alter the logic of this method.) Finally, the calculated overall work load rating is produced by multiplying the weight of each subscale by the rating (e.g., from 1 to 100) on that subscale and summing all these products. Note that this is a time-consuming process because each person must not only give a numerical rating for each subscale but also must compare all possible pairs of subscales for relative importance. Because there is usually a high correlation between calculated work-load scores and scores obtained by asking operators simply to rate directly overall work load,

Table 3.3.3.   Rating Scale Descriptions

Title	End Points	Descriptions
Overall work load	Low, high	The total work load associated with the task, considering all sources and components
Task difficulty	Low, high	Whether the task was easy or demanding, simple or complex, exacting or forgiving
Time pressure	None, rushed	The amount of pressure you felt due to the rate at which the task elements occurred. Was the task slow and leisurely or rapid and frantic?
Performance	Failure, perfect	How successful you think you were in doing what we asked you to do and how satisfied you were with what you accomplished
Mental sensory effort	None, impossible	The amount of mental and/or perceptual activity that was required (e.g., thinking, deciding, calculating, remembering, looking, searching)
Physical effort	None, impossible	The amount of physical activity that was required (e.g., pushing, pulling, turning controlling, activating)
Frustration level	Fulfilled, exasperated	How insecure, discouraged, irritated, and annoyed versus secure, gratified, content, and complacent you felt
Stress level	Relaxed, tense	How anxious, worried, uptight, and harassed or calm, tranquil, placid, and relaxed you felt
Fatigue	Exhausted, Alert	How tired, weary, worn out, and exhausted or fresh, vigorous, and energetic you felt
Activity type	Skill based, rule based, knowledge based	The degree to which the task required mindless reaction to well-learned routines or required the application of known rules or required problem solving and decision-making

the extra cost of using subscales may not always be justified. However, calculated work-load ratings usually have less variability than directly estimated ratings.

3.3.5   FUNCTION ALLOCATION IN MANUFACTURING

Automation, robotics, and flexible manufacturing systems (FMS) have drastically altered the working relationships between people and machines in manufacturing. There is an urgent need for more knowledge of allocation of functions in this new production environment. However, the vast majority of publications on robotics and FMS neglect the role of the human in these systems. Even the few publications that mention people tend to be concerned with worker motivation and potential union problems. This section reviews those rare human-factors studies that have dealt directly with function allocation in manufacturing and makes some educated guesses about the future development of this key area.

Some authors are unduly pessimistic about the problems robotization creates for allocation of functions. Noro and Okada (1983) state that the table of relative merit approach used traditionally is but a "fortunate result of the past" and will not work in modern manufacturing. They believe that rapid technological advances in this area will keep human-factors experts from creating effective algorithms for function allocation. We disagree and offer the research discussed in this section as a counterargument.

Detailed tables of relative merit have been generated by Nof, Knight, and Salvendy (1980), Kamali, Moodie, and Salvendy (1982), and Hwang, Barfield, Chang, and Salvendy (1984); this same approach has also been extended to computer-aided design by Barfield and Salvendy (1984). A later chapter in this handbook presents these tables, so they are not duplicated here. Several case studies show how these tables can effectively improve allocation of functions and so enhance system performance. For example, Kamali et al. (1982) use a relative-merit table to allocate functions in a Transaxle assembly operation to automation, robot, or human. One function the system must perform is gauging the height of a retainer. This could be performed by any of the three subsystems. However, based upon the table, this task was assigned to automation because it was too simple and boring for a human and a robot would be too expensive. Similar analyses were used to allocate other functions.

Because modern manufacturing represents a new issue for human-factors efforts, it is hardly surprising that the pioneering research described above is based upon the traditional table of relative merit.

We would expect that as research in this area becomes more sophisticated, some of the more recent techniques described in Section 3.3.3 will begin to be applied. One example of a later approach that stresses the symbiotic relationship between system components is called the SIMBIOSIS model (Parsons and Kearsley, 1982). It has nine types of activities that are often required in robot systems: surveillance, intervention, maintenance, backup, input, output, supervision, inspection, and synergy. However, because the description provided by Parsons and Kearsley was general and lacked any detailed examples, it is impossible to determine precisely how functions are allocated in the SIMBIOSIS model.

Another aspect of complementarity, as opposed to comparability, was noted by Domas and Helander (1984). In the manufacturing industry, component parts have been redesigned to better fit robot capabilities. This is called Design for Automation (DFA), and for it several new principles have been developed. They include such empirical guidelines as design for unidirectional assembly, eliminate parts requiring extremely tight tolerances, eliminate parts that are too bulky or too small, and so forth. The big surprise for manufacturing engineers was that once DFA principles have been applied to make automatic assembly easier, the parts are also easier to assemble for humans. Indeed, after the product has been redesigned it is often the case that the simplified assembly is allocated to people because the capital expense required for automation is no longer justified!

Our conclusion is that allocation of functions in manufacturing will proceed in much the same way as in more traditional areas of human factors. Tables of relative merit already exist, and will be used by designers. The next stage, moving from comparability to complementarity, is already underway. We expect more studies stressing complementarity in the near future. We have been unable to locate published studies of dynamic allocation in manufacturing, but would not be surprised to discover that dynamic allocation is already being considered in industry. We fully expect to have some examples of dynamic allocation for the next edition of this handbook.

REFERENCES

Adams, J. A. (1982). Issues in human reliability. *Human Factors, 24,* 1–10.

Albanese, R. A. (1977). Mathematical analysis and computer simulation in military mission workload. *Proceedings of the AGARD conference on methods to assess workload.* AGARD-CPP-216.

Bailey, R. W. (1982). Human performance engineering. Englewood Cliffs, NJ: Prentice Hall.

Barfield, W., and Salvendy, G. (1984). Computer aided design: Human factors considerations. *Proceedings of the Human Factors Society, 28,* 654–658.

Bekey, G. A. (1970). The human operator in control systems. In K. B. DeGreen, Ed., *Systems psychology.* New York: McGraw Hill.

Boehm-Davis, D. A., Curry, R. E., Wiener, E. L., and Harrison, R. L. (1983). Human factors of flight-deck automation: Report on a NASA-industry workshop. *Ergonomics, 26,* 953–961.

Carson, D. H. (1970). Human factors and elements of urban housing. In T. K. Sen, Ed., *Human factors applications in urban development.* New York: Riverside Research Institute.

Chapanis, A. (1970). Human factors in system engineering. In K. B. DeGreene, Ed., *Systems psychology.* New York: McGraw Hill.

Cooper, G. E. and Harper, R. P. (1969). The use of pilot rating in the evaluation of aircraft handling qualities (TN-D-5153). Moffett Field, CA: NASA, Ames Research Center.

Danaher, J. W. (1980). Human error in ATC system operations. *Human Factors, 22,* 535–545.

Derrick, W. L. (1981). The relationship between processing resource and subjective dimensions of operator workload. *Proceedings of the Human Factors Society, 25,* 532–536.

Domas, K. and Helander, M. (1984). Manual versus robotic assembly: Some implications of product design. *Proceedings of the Human Factors Society, 28,* 659–663.

Donchin, M. (1984). Cognitive psychophysiology. Hillsdale, NJ: Erlbaum.

Eggemeier, F. T. (1981). Current issues in subjective assessment of workload. *Proceedings of the Human Factors Society, 25,* 513–517.

Fitts, P. M. (1951). *Human engineering for an effective air navigation and traffic control system.* Washington, DC: National Research Council.

Hall, D. A., and Fagen, R. E. (1968). Definition of system. In W. Buckley, Ed., *Modern systems research for the behavioral scientist.* Chicago: Aldine.

Hart, S. G. (1985). Theory and measurement of human workload. NASA working paper, Ames Research Center.

Hwang, S., Barfield, W., Chang, T., and Salvendy, G. (1984). Integration of humans and computers in the operation and control of flexible manufacturing systems. *International Journal of Production Research, 22,* 841–856.

Johanssen, G., Moray, N., Pew, R., Rasmussen, J., Sanders, A., and Wickens, C. (1979). Final report of experimental psychology group. In N. Moray, Ed., *Mental workload.* New York: Plenum.

Jordan, N. (1963). Allocation of functions between man and machines in automated systems. *Journal of Applied Psychology, 47,* 161–165.

Kamali, J., Moodie, C. L., and Salvendy, G. (1982). A framework for integrated assembly systems: humans, automation and robots. *International Journal of Production Research, 20,* 431–448.

Kantowitz, B. H. (1985a). Mental work. In B. M. Pulat and A. Alexander, Eds., *Industrial ergonomics.* Institute of Industrial Engineering.

Kantowitz, B. H. (1985b). Mental workload. In P. A. Hancock, Ed., *Human factors psychology.* Amsterdam: Elsevier.

Kantowitz, B. H., (1985c). Stages and channels in human information processing: A limited review and analysis of theory and methodology. *Journal of Mathematical Psychology, 29,* 135–174.

Kantowitz, B. H., and Sorkin, R. D. (1983). *Human factors: Understanding people–system relationships.* New York: Wiley.

Meister, D. (1971). *Human factors: Theory and practice.* New York: Wiley.

Meister, D. (1985). Behavioral analysis and measurement methods. New York: Wiley.

Moray, N. (1979). *Mental workload.* New York: Plenum.

Nof, S. Y., Knight, J. L., Jr., and Salvendy, G. (1980). Effective utilization of industrial robots—A job and skills analysis approach. *AIIE Transaction, 12,* 216–225.

Noro, K. and Okada, Y. (1983). Robotization and human factors. *Ergonomics, 26,* 985–1000.

Ogden, G. D., Levine, J. M., and Eisner, E. J. (1979). Measurement of workload by secondary task. Human Factors, 21, 529–548.

Parsons, H. M. and Kearsley, G. P. (1982). Robotics and human factors: Current status and future prospects. *Human Factors, 24,* 535–552.

Pew, R. (1979). Secondary tasks and workload measurement. In N. Moray, Ed., *Mental workload.* New York: Plenum.

Moray, N. (1982). Subjective mental workload. *Human Factors, 24,* 25–40.

Reid, G. B., Shingledecker, C. A., and Eggemeier, F. T. (1981). Application of conjoint measurement to workload scale development. *Proceedings of the Human Factors Society, 25,* 522–526.

Rolfe, J. M. (1971). The secondary task as a measure of mental load. In W. T. Singleton, Ed., *Measurement of man at work.* London: Taylor & Francis.

Sanders, A. F. (1979). Some remarks on mental load. In N. Moray, Ed., *Mental workload.* New York: Plenum.

Senders, J. (1970). The estimation of operator workload in complex systems. In K. B. DeGreen, Ed., *Systems Psychology.* New York: McGraw-Hill.

Swain, A., and Guttmann, H. E. (1980). *Handbook of human reliability analysis with emphasis on nuclear power plant applications.* Washington, DC: U.S. Nuclear Regulatory Commission.

Wickens, C. D. (1980). The structure of attentional resources. In R. S. Nickerson, Ed., *Attention and performance VIII.* Hillsdale, NJ: Erlbaum.

Williges, R. C. and Wierwille, W. W. (1979). Behavioral measures of aircrew mental workload. *Human Factors, 21,* 549–574.

CHAPTER 3.4
TASK ANALYSIS

COLIN G. DRURY

State University of New York at Buffalo
Amherst, New York

BARBARA PARAMORE
HAROLD P. VAN COTT

Essex Corporation
Alexandria, Virginia

SUSAN M. GREY
E. NIGEL CORLETT

University of Nottingham

3.4.1 INTRODUCTION

Task analysis is a formal methodology, derived from systems analysis, which describes and analyzes the performance demands made on the human elements of a system. By concentrating on the human element in systems analysis, it can compare these task demands with known human capabilities.

The goal of task analysis is to provide the basis for integrating human and machine into a total human–machine system. A task analysis defines the performance required of humans, just as an engineering analysis defines the performance required of hardware and a program flow chart defines the performance of software. All three are necessary.

In this chapter, the orgins and antecedents of task analysis are explored for the understanding they can provide of modern developments in both military and industrial systems. These modern developments are described in some detail to show that there is no single method of task analysis applicable to all jobs. Finally, an example of a large, complex system is presented in detail to illustrate both the techniques of task analysis and the methods of collecting the information required.

3.4.2 EARLY HISTORY

Numerous instruments have been developed to record, measure, and analyze the physical movements of the human body at work and in sports. These attempts to capture the overt or observable behavior of men and women have been used for numerous purposes: to identify the performance demands of jobs; to identify the skills and knowledge needed to perform them; to establish the number and types of people needed for a system requiring more than a single person; and to design or to evaluate the interface between personnel and the equipment and tool systems with which they must work.

Although many techniques used for describing and analyzing human performance have focused on capturing the overt or observable physical behavior of individual performance, more recently new methods have been employed to describe and analyze the covert, nonobservable cognitive activities or processes that underlie overt behavior. These new methods have been used to develop integrated theories of cognitive performance, to define the knowledge requirements of jobs and job tasks, and to develop "expert systems" that use computers to perform the cognitive tasks once performed by people.

The methods used to analyze observable and covert human behavior have been given various names; activity analysis, job analysis, time study, decision analysis, ergonomic analysis, and task analysis are among the most common. Each of these terms is generally associated with somewhat different approaches and objectives, and not all of them are forms of task analysis. All, however, have the common goal of identifying the performance demands of jobs and job tasks, so the term used in this chapter is task analysis.

One of the best known early methods of task analysis was developed by Frederick W. Taylor, an industrial engineer. This method, called time study, was used by Taylor to describe, analyze, and improve the efficiency of workers in various jobs such as factory assembly line operations. Taylor's method helped to pinpoint ways in which a workplace could be redesigned to economize on the number of movements, the energy expenditure, and the time required to assemble components or to perform other human tasks. Taylor's method was extended by the Gilbreths, and subsequently by others. Time and motion study remains a standard technique in the repertory of industrial engineering (Mundel, 1970).

Another antecedent of task analysis was the method of job analysis developed in the 1930s by the U.S. Department of Labor. This method was devised to establish a factual and consistent basis for identifying personnel qualification requirements and for recruiting, selecting, placing, and promoting personnel. Other applications have included job redesign, vocational counseling, and the development of training.

The impetus for a still different type of task analysis came about with the development of electromechanical systems in the 1940s and 1950s–in particular, military aircraft and weapon control systems. These systems involved people in new kinds of relationships with equipment and sharply altered the nature of human tasks. Associated with the introduction of these new and sophisticated systems was the need for written operating and maintenance procedures to be followed by personnel in performing tasks in complex systems. Finally, as systems and tasks became more complex, a need arose for some valid basis for identifying the training content of jobs.

In 1953, the Air Force published *A Method for Man–Machine Task Analysis,* developed by Robert B. Miller (1953a). This report described a method for analyzing the operator's job in any human–machine system as part of a system's linkages from input to output functions. Miller's method defined operator tasks with reference to those system functions. In another report Miller (1953b) addressed the use of task analysis to define the maintenance task requirements for new systems. In both cases the purpose was to assist the process of planning for training and training equipment.

Since Miller's initial development of human–machine task analysis as a method, his basic approach has been modified in the procedures used, and extended in application, both by himself (Miller, 1956, 1963, 1974) and by others. The form of a task analysis, as is demonstrated frequently in this chapter, depends upon the purpose for which it is performed. Hence a wide variety of different forms have been developed, (e.g., Vallerie, 1978), to meet specific purposes. An early major extension was by Van Cott and Altman (1956), who developed a method for using task analysis information in system design and development. Shortly thereafter the preparation of task analysis information became a requirement in the procurement of major military systems. Task analysis was used as a way to generate a common reference data base for integrated development of selection and training procedures including the determination of the Qualitative and Quantitative Personnel Requirements Information (QQPRI) for new systems (Demaree et al., 1982; Swain, 1962).

Another application of task analysis is to determine whether the activity or task demands of a job fall within the capabilities of workers. The following example of such an analysis by Drury (1983) gives the flavor of task analysis and one particular application of the data developed by using it.

The system was an automated kitting line that would preassemble kits of parts using computer and robotic technology and send the kits to a manual assembly line. At the conceptual stages ergonomics advice was sought about the place where parts from suppliers first encountered the system—the port of entry. The operator here would have to enter shipping information from each shipment into a computer, which would then print a bar code that later stages of the computer-controlled system could read. The task analysis (shown in Table 3.4.1) is merely one column of task description and one column of task analysis. From this analysis ergonomics advice was given for each task. Note that this information was made available to the designer very early in system design, while there was still time to rectify computer dialogues, conveyor heights, and equipment positions. As actual hardware is specified more exactly, new ergonomics problems are likely to arise but by then more detailed task analyses are possible.

3.4.3 HUMAN PERFORMANCE IN TECHNOLOGICAL SYSTEMS

In order to understand and use task analysis it is necessary to first understand human performance in technological systems.

Advanced human–machine systems consist of hardware, software, and personnel. These components act together to accomplish some mission or output goal. The accomplishment of an output goal depends on a number of variables that represent system functions. In other words, system functions are activities that control the variables that influence the goal output of a system.

System functions may be executed by personnel or by hardware–software components of the system, but often involve both. Personnel performance requirements in function execution depend on the degree of automation of the system. At the lowest level there is no automation, only mechanization: personnel directly control the hardware and monitor its performance parameters and results by means of displays of sensor outputs, by direct perception, or by a combination of these. As automation increases, the "machine" is increasingly able to control its own performance (e.g., to maintain a functional parameter at a given value without human involvement; or at a higher level, to maintain several parameters in proper relationship; or at a still higher level, to alter control schemes to optimize parameter relationships across functions for different modes and conditions of operation).

With higher levels of automation, human performance requirements become increasingly supervisory in nature. Personnel verify, monitor, and evaluate the execution of system functions by the hardware and software constituents, and adjust and coordinate their performance as required to safeguard the system and maintain productivity.

The human constituents of a system bear the ultimate responsibility for recognizing, interpreting, compensating for, and correcting or mitigating the consequences of deficiencies, failures, and malfunctions in the hardware and software and in their own performance. Thus "human error" or "judgment error" are terms found frequently in reports on system failures. In many cases, however, system failures can be attributed to the poor design of tasks for human capabilities, to defective interfaces between task performers and equipment subsystems, to inadequate training, to poorly conceived operating or maintenance procedures, or to other situational factors. In-depth investigation and analysis of accidents and injuries provide proof that often inadequate attention is given to describing, evaluating, and facilitating the human performance required by the system.

The broad objective of task analysis is to meet these needs. Human error cannot be completely eliminated by systematic forecasting, description, and analysis of human tasks during system development, nor by the evaluation of actual task performance during system test and evaluation, but such an effort can go a long way to achieving system designs that make the best use of human capabilities, provide adequate margins of safety, and are more effective in achieving system goals economically.

Table 3.4.1 Preliminary Task Analysis of Port of Entry Operator

Task	Ergonomics Problem
1. Move pallet to side of conveyor.	1.1. Pushing or pulling heavy "vehicle."
2. Enter shipping information from invoice.	2.1. Invoice and computer dialogue should be in same format. Dialogue should be written to be "user friendly" to materials handling operator.
3. Computer checks shipping information.	3.1. What happens if discrepancies are found—e.g., wrong codes on invoice, unordered parts?
4. Operator places carton on conveyor.	4.1. Height, reach, weight of carton.
	4.2. Operator has to turn carton right way up to read information and ensure carton enters system correctly oriented.
5. Operator enters code from carton into computer.	5.1. Is product code and/or meaningful name on carton in a position with contrast & size to make it readable?
	5.2. Can operator remember 7-digit code long enough to enter it (2–5 sec)?
	5.3. How does operator handle obviously damaged or wrong cartons?
6. Computer prints bar code for carton.	6.1. Is printer convenient to operator?
	6.2. Does meaningful name appear on bar code sticker?
7. Operator puts bar code sticker on carton and checks code against box code.	7.1. Height, location, and orientation of bar code are acceptable within wide limits—no problem.
	7.2. Can operator compare two 7-digit numbers reliably? Meaningful names will help.
8. Operator places carton to conveyor.	8.1. Is release button convenient or does next computer input release carton?
9. Go to next carton (task 4).	9.1. Does operator have to put all cartons of one product code onto conveyor together? If so, he must hunt through pallet each time.
	9.2. Does operator have to keep count of boxes?
	9.3. If the last box has been coded does computer keep all of order "in bond" until it checks boxes against invoice for discrepancies? If so, how are mistakes rectified?

3.4.4 THE NATURE OF TASKS IN TECHNOLOGICAL SYSTEMS

Definition of a Task

Although there are numerous definitions of "task," there is general agreement that a "task" is a set of human actions that contributes to a specific functional objective and ultimately to the output goal of a system.

Thus, although the definition of what constitutes a task is always somewhat arbitrary, when the objective of a task is defined the basic content of the task becomes evident. Control over the scope or size of tasks is accomplished in large part by the definition of their objectives. An effort must always be made to keep the chunks of work action that are designated as tasks about the same in scope or size. This ensures a consistent level of description of all of the human performance requirements of a system.

Task Characteristics

Characteristics that are useful in defining tasks and controlling their size can be stated as follows:

1. Task actions are related to each other not only be their objective but also by their occurrence in time. One of the concerns of task analysis is to establish and evaluate the time distribution of actions within and across tasks. Task actions include perceptions, discriminations, decisions, control actions, and communications. Every task involves some combination of these different types of cognitive and physical action.

2. Each task has a starting point that can be identified as a stimulus or cue for task initiation. A cue is often not a single item of data or information. It may consist of several data points, received closely in time or dispersed over a longer time, which together have significance as a cue that an action is to be taken.

3. Each task has a stopping point that occurs when information or feedback is received that the objective of the task has been accomplished.

4. Task cues and feedback may be provided by instrumentation or direct sensory perception, or they may be generated administratively, say, by a supervisor or co-worker.

5. A task is usually, but not necessarily, defined as a unit of action performed by one individual.

Distinction Among Discrete, Continuous Control, and Branching Tasks

In the analysis of system operations a distinction should be made among three types of tasks—discrete or procedural tasks, continuous or tracking tasks, and branching tasks. A discrete or procedural task requires an individual to execute a series of separate action elements (not necessarily in an invariant sequence), in response to specific stimuli and/or instructions given in a procedure document.

A continuous task is one that requires an operator to operate a control continuously, such as a joystick or steering wheel, and to maintain an output within given limits, while sampling deviations from those limits that are viewed directly or displayed. There are also continuous monitoring tasks, which consist of sampling deviations while system performance is controlled by an automated subsystem.

Continuous control tasks extend over a relatively long period of time, for example, throughout a phase of operations or for the duration of a mission. Throughout this period the cycle of control adjustment and/or deviation sampling is repeated more or less continuously. In describing such a task it is necessary to find a way of indicating that the cycle repeats and the time interval over which it repeats. While a continuous task is being sustained, various discrete tasks may have to be executed. Thus the distinction between the two types of tasks affects assessment of work load and staffing requirements. Also, the two types of tasks require different skills and different strategies for training and performance assessment.

A branching task is a variant of a discrete task where task sequence is determined largely by the outcome of particular "choice" tasks in the operation. Examples are found in inspection or troubleshooting operations where tests are made for fault conditions which, if found, necessitate some new action. The typical task description for branching operation is done at a cruder level than for sequential operations, often almost at a function level. Analysis is often limited to annotations around a block diagram or omitted entirely until each block in the diagram is further subdivided into a linear sequence. It is usual to use the standard flow chart symbols defined as in Figure 3.4.1.

The task description then resembles a flow process chart except that what is being charted is what happens to an operator rather than to a product. It is conventional to have the chart start at the top and proceed downward and/or to the right as time progresses.

3.4.5 DEFINITIONS IN TASK ANALYSIS

Definition

As McCormick (1976) has stated, there are many variants of task analysis, "But there is also . . . an underlying theme. The theme relates to the dissection of human work into 'tasks,' and the further analysis thereof." Another definition that seems to have general applicability is as follows:

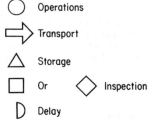

Fig. 3.4.1. Standard flow charting symbols.

Task analysis is a process of identifying and describing units of work, and analyzing the resources necessary for successful work performance. Resources in this context are both those brought by the worker (e.g., skills, knowledge, physical capabilities) and those which may be provided in the work environment (e.g., controls, displays, tools, procedures/aids).

Distinction Between Task and Job Analysis

The terms task analysis and job analysis are often used interchangeably, as if they were equivalent. They are not. A task analysis addresses in detail the specific exchanges between personnel and the equipment components of a particular system (or class of system if the designs are sufficiently alike). It does so without necessarily specifying by which person or in which job given tasks or parts of tasks are performed. A job analysis addresses the activities of a given occupational category or position. Viewed in an organization or system context, this set of activities may be seen as a subset of the total activities that go on in the operations, maintenance, or other aspects of the organization or system. On the other hand, system tasks may be a subset of all the tasks performed by an individual in his or her job.

Some methods of job analysis use description and analysis, but the approach to defining tasks and the level of detail are not the same as in a human–machine system task analysis, especially one that addresses system operation. The Department of Labor method, mentioned earlier, is an example of a job analysis method that employs a form of task analysis, as is the method prescribed in the Department of Defense Interservice procedures for Instructional System Development (Bronson et al., 1975). Functional job analysis (Fine and Wiley, 1971) is another example.

Distinction Between Task Description and Task Analysis

Methods for task analysis all entail three different kinds of activity: (1) system description and analysis; (2) specification of the human task requirements of the system; (3) analysis, resynthesis, interpretation, evaluation, and transformation of the task requirements in light of knowledge and theory about human characteristics. The last of these activities is properly called "task analysis." The second of these activities is task description. The first activity provides the basis for task description, whereas the task description provides the information needed for task analysis. This section addresses all three activities, which together have come to be called "task analysis." It is an imprecise term but one that no one is likely to be able to change at this late date.

3.4.6 RECENT DEVELOPMENTS IN TASK ANALYSIS

The ideas on task analysis generated and refined in the United States in connection with military and aerospace applications have led to further lines of developments in manufacturing industry. Although there were some early attempts to use task analysis in industry, (e.g., Verdier, 1960), the major thrust has been more recent and more European in flavor. Singleton's widely used book, *Man–Machine Systems* (1974), contains examples of using task analysis techniques in industry. He continued the psychology-based tradition of his predecessors and even emphasized the closed-loop nature of control in most tasks by having each task use a "trigger" to initiate an action toward a desired goal and a "check" to provide feedback on task completion. Tasks are listed in sequential order, numbered by rows, and the task description is by column.

Continuing Singleton's methods, Drury (1983) presented a task analysis of the task of aligning a lamp in the lamp holder used in a photocopying machine. The "trigger" column was omitted because in practice it was almost always the successful completion of the preceding action in such a repetitive task. The usual task number column was followed by "purpose," "action," and "check" columns to complete the task description. For task analysis, columns headed "control problems" and "display problems" were used, with "postural problems" summarized at the end to save space on the form. It was found that a good step between task analysis and redesign of the workplace was to enumerate formally the requirements for a good design, again using the three headings of controls, displays, and posture. This step forced the analyst to state clearly what would be necessary to solve the current problems, without necessarily solving the problems. In our experience, too many ergonomists leap straight from the analysis to solutions, which can be easily rejected by others on the design team. Stating the design requirements allows all team members to join in the design phase, helping to ensure team ownership of the resulting designs.

The resulting completed task analysis form (Table 3.4.2) gives a good idea of the information that can be gained from the discipline of using a formal methodology. In this example, it appeared obvious to all concerned (before the analysis) that the knee obstruction under the bench was the only real problem. The task analysis showed that the visual display was an equal cause of awkward postures. A mock-up was built to test an alternative design using video cameras to replace the current optical system and a deformable lamp holder to replace the X and Y joysticks.

Well before the Singleton and Drury papers, methodologies were being developed for even more

Table 3.4.2 Completed Industrial Task Analysis

Task No.	Purpose	Action	Check	Control Problems	Display Problems
010–040	Insert lamp and holder	Not considered for analysis	Kinesthetic	—	—
050	Adjust height	Estimate distances from top and bottom of coil to edges of rectangle. Use thumb wheel to set equal gaps.	Visual	Thumb wheel uses R hand which is also used as tilt joystick as 050, 060 are combined.	Contrast and brightness fall off at top and bottom of screen.
060	X tilt adjust	Estimate vertical from any line on screen (not dotted). Adjust RH joystick in X direction only. Compensate for position changes with LH joystick.	Visual	Must coordinate LH and RH joystick movements to keep image central. High "stiction" in tilt mechanism gives jerky action.	Too many vertical lines, including max tilt lines, cause confusion.
070	X tilt/height clamp	Tighten lamp clamp screws using torque driver No 1. Check still in alignment after tightening.	Kin (torque) visual	Must hold holder by LH while tightening clamp screws, no place to hold. Lamp house in way of screwdriver. Must now tighten X, Y position clamp screws to provide firm base for operation 080.	—
080	Y tilt adjust and clamp	With X lamp off and Y lamp on, check Y tilt against verticals on screen. Adjust using Y direction only movement of RH joystick. Compensate for position changes with LH joystick.	Visual	As for 060, both points.	As for 060
		Tighten tilt adjust screws with socket wrench. Check alignment after tightening.	Kin (torque) visual	Must hold holder by LH while tightening screws. No place to hold. Torque reaction moves chair on wheels—must grip bench between knees to compensate. Must reach over lamp house.	Image obscured by socket wrench.
090	X, Y position adjust and clamp	With both lamps on, check X, Y position against verticals on screen. Adjust X position with LH joystick and clamp RH position screw with screwdriver 2.	Visual	Extreme arm abduction with R arm to tighten screws. Must loosen X, Y position screws before adjustment. Must hold position with LH joystick while tightening screws.	Too many images on screen, both hard to see and hard to tell which is which. Image obscured by screwdriver. Images very low contrast. Two images here have two different standards on same screen.
		Adjust Y position and clamp LH position screw with screwdriver 2. Check alignment after tightening.	Kin Visual		
100	Unclamp holder, remove	Not analyzed			

376

General Problems	Design Requirements
Displays	1. *Displays*
"Exit pupil," of display is small and low down, forcing operator to bend head downward for maximum contrast and brightness.	a. Brighter and more contrasty.
	b. Separate X, Y to prevent confusion
	c. Raise display to reduce bending
Dark work area in midst of lighter surroundings—should be reversed.	d. Less restricted display viewing area
	2. *Controls*
	a. Less static friction on tilt adjust
	b. Less inertia on position adjust
	c. Increase compatibility of controls/displays to make adjustments more natural
	d. Provide for direct holding of lamp while tightening various screws
Posture	3. *Posture*
Knee position obstruction forces operator to sit too far back—hence extreme reaching and bending.	a. Remove knee obstruction to get closer to task
	b. Have smallest possible lamp housing to allow freedom of arm positioning
	c. Remove need for extreme should abduction in task 090

Source: From Drury, 1983.

detailed analyses of skills. The skills-analysis-based training movement in the United Kingdom (e.g., Crossman, 1956; Seymour, 1954, 1966, 1968) was developing task-analytic techniques for probing the perceptual and motor difficulties encountered in industrial skill learning. For each task in the job, the knowledge and skill required are determined. To do this, the analyst must work with the operator to consider, for instance, what knowledge the operator requires to set up, operate, or inspect the work, how this will be obtained, and how and when it will be used. Sources of information, whether they be handbooks, written instructions, or memory, must be considered, as must the limits of the operator's discretion with regard to these instructions. When considering skill content, the analyst must decide what the operator does and how he does it. As Seymour (1954) points out:

> We must start with the observable and proceed to the invisible. In other words we commence with the effector processes and go on to study the receptor processes.

Thus the skilled performance is observed as closely as possible over several cycles, with each movement being recorded in detail. The analyst then gains additional information by questioning, attempting the task himself, or various other methods appropriate to the case under study. It should be noted at this point that attempting the task oneself is a technique to be used with caution because the analyst is not the skilled operator and may easily misinterpret between observed and experienced behavior. However, it is often critical to see and hear through the eyes and ears of the operator. At the end of this stage, the analyst should have a detailed statement of the performance of the task, stating the eye fixations, muscular control, sensory checks, points that tell the operator whether the task is proceeding satisfactorily, and the times required by the operator to perform the various parts of his task. The recording of this analysis can be done on the form presented by Seymour (1954, p. 42).

One outgrowth of this work was what has come to be known as hierarchical task analysis (HTA). Developed by Annett and Duncan (1967) and refined and demonstrated by Duncan (1974), Shepherd (1976), and Piso (1981), HTA was used to develop training procedures for complex, process industries. As with any other task analysis method, HTA starts with the overall objectives of the system and describes an operation that fulfills these objectives. The operation is then redefined as a series of suboperations plus a plan that defines how these suboperations are linked. Each suboperation is examined to determine whether it is adequately defined to the satisfaction of both the analyst and the potential operator of the system. If it is not adequately defined, it is progressively redefined.

The question of stopping at an approrpriate level of description is an important feature of hierarchical task analysis. Annett and Duncan propose the "$P \times C$ rule" to help the analyst make rational decisions regarding where stopping the analysis is appropriate. The rule states that further redescription is unnecessary where the product of probability of inadequate performance (P) and the cost of inadequate performance (C) is acceptable. This rule is used "because if either the value of P or C is at or near zero, then their products also will be at or near zero. This therefore gives a quick indication of operations that are not critical or that present little learning difficulty. The $P \times C$ rule is applied to each operation in turn. Again if the value obtained is unacceptable, then that operation is further broken down, and the rule applied to each subordinate operation. This will most likely yield a large irregular block diagram, but one in which only the essential operations will be represented" (Stammers and Patrick, 1975, p. 55). When these operations have been described, they must be linked together with a plan. This may be as simple as a fixed sequence (e.g., start-up procedure) or may be more like Bainbridge's "goal-directed program," which varies the sequence and even the operations depending on the data received and decision made.

The National Council for System Reliability (1982) points out a number of distinct advantages of hierarchical task analysis:

1. It provides the analyst with a powerful means of coming to grips with some very complex tasks. If only one level of redescription is chosen to describe a task, the analyst needs only one plan to state how operations are selected and sequenced, but if this one plan governs many operations, it may prove impossible for the analyst to state. On the other hand, to use a hierarchy of operations, several plans are needed but each of these governs fewer operations and is far easier for the analyst to decipher.

2. A hierarchy of simpler plans make it easier for the trainee to learn, concentrating practice on rationally indentified parts of the task, rather than trying to master the complexities of the whole task from the outset. It is also more apparent to the trainee how the various goals and subgoals of the task interrelate.

3. It enables the analyst to examine some parts of the task in detail while leaving other parts in far less detail as required.

The original format for recording a hierarchical task analysis was tabular, with tasks ordered by hierarchy of task, not plan; hence tasks in a single plan may be widely separated. Shepherd (1976) devised an alternative format which, while remaining tabular, emphasizes the sequences of the plans, with task subordinate to plans.

Piso (1981) accepted Shepherd's (1976) approach but noted that, as with Annett and Duncan's method, it is designed to be used for training purposes. When extending the method for other purposes, he proposed that the following questions must be answered:

1. What task information should be gathered and described?
2. How should this information be collected?
3. Supposing the specific task aspects are defined, how can we set up a hierarchical diagram of the tasks?
4. What level of description is necessary? (Is the $P \times C$ criterion applicable?)

Piso applied these questions successfully to the analysis of the process-control tasks of glass-furnace operators. He discussed the first two under the heading of "information collection." The purpose of their analysis was to reveal how the operators controlled the different process variables, and so the analysis was restricted to control tasks. Based on the control-loop model (i.e., perception → decision → operation/action → evaluation/feedback), the following questions were formulated:

1. What is the goal of the control task?
2. What information does the operator use for his decision to act?
3. When and under what conditions does the operator decide to take action?
4. How is this action carried out?
5. What are the consequences of this action? How does the operator receive feedback and how does he react afterwards?
6. How often will the task be carried out?
7. Who will carry out the task?
8. What kind of problems can occur?

An initial process analysis identified the different control tasks. The above questions were used as the basis for semi-structured interviews with experienced operators.

With regard to the third question, of the design of the hierarchical diagram, the control loop sequence was used to divide the task into three phases—a prephase, action phase, and evaluation phase. The prephase is split into perception and decision and covers questions 2 and 3 above. The action phase is split into a sequence of operations, with reference to question 4. Information for the evaluation phase can be obtained from question 5. An illustration of how this was done for one of the control tasks is shown in Figure 3.4.2. Information collected from questions 1, 6, 7, and 8 is represented using Shepherd's tabular format as in Figure 3.4.3.

Although nonmanual skills have always been part of supervisory and managerial jobs, in recent years there has been a rapid growth of shop-floor jobs that largely involve nonmanual skills, particularly in the process industries. Here the skill is essentially a mental one, with operators being concerned with decision-making and the handling of information in circumstances that rarely occur in a predetermined order. Typically, operators exercise control over continuous processes, the most predominant skill being what Beishon (1967) terms the "control skill." Beishon notes that the operator controls a complex process to achieve goals of quality and quantity of output by using process knowledge of the relationships between input and output.

The method of *signal-flow graph analysis* can be usefully applied to the description of a control task. It was designed for application to the human operator by Bainbridge, Beishon, and Crossman from a technique developed by electrical engineers to facilitate the analysis of electrical networks. This technique links the variables in the system together so that a diagram is formed to indicate the control complexity of the system. The variables are drawn as circles or "nodes," and are linked together by lines, or branches, which represent the system functions, or the causal dependency relating two variables. In electrical engineering, the gain of that function of the system interspersed between the points of action of the variables would be written alongside it and various rules used to combine different parts of the system. In the ergonomic use of this technique, it has been found useful to put the time relationships between variables alongside these function lines.

A signal-flow graph is not so much a way of obtaining additional information about a system as a way of presenting this information so that it is simple to analyze. The following types of information are necessary to draw a signal-flow graph and to make the preliminary job analysis.

1. *The Aims of the System.* These can be specified as the tolerance within which the system variables must be maintained. These variables may be either process variables or engineering limits of the structure.

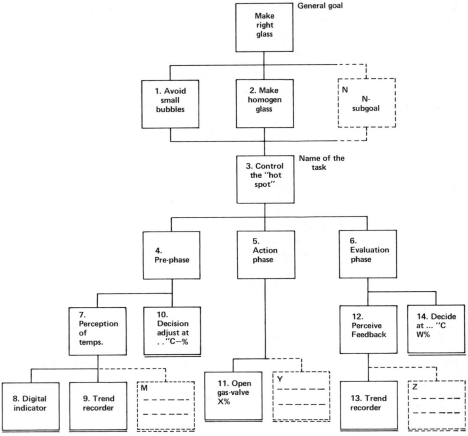

Fig. 3.4.2. Hierarchical diagram of a process control task: controlling a hot spot in glassmaking (from Shepherd, 1976).

2. *How the System Works.* The relations between variables in the system and how these affect the characteristics of the product must be determined. The inputs, outputs, and internal variables of the system must be identified. The inputs and outputs are of two types:

   a. physical inputs and outputs of the process, for example, materials and energy; and
   b. control points, that is, inputs, points at which system variables can be controlled, and outputs, displays of relevant variables.

3. The part the operator plays in the system, in terms of

   a. the person–machine relationships, the system functions for which the person is responsible; and
   b. the operating method. This can be described under three headings: routine actions, for example switching sequences; control rules, their interactions, and relative importance; any time sequence of operation that may be used, because of regular or irregular cycles in the process.

Actually drawing the graph is a heuristic process, although standard symbols have been defined (e.g., Beishon, 1967). A first step in drawing the signal-flow graph is to choose the boundaries of the system; this may be difficult. Suitable boundary criteria are operator's responsibilities, engineering units or subsystems, that is, sections each with a small number of inputs and outputs from other parts of the system.

In principle, the graph is drawn by starting with any variable, usually the principal output, the product, enumerating the variables affecting it, enumerating the variables affecting each of these, and so on, drawing each variable as a circle, connecting it to the variables that it affects by branches,

No.	Task, goal, operation description	Remarks	IPC	MI
1	Avoid small bubbles 1. When the temperature in the furnace is too low, the chemical reaction is delayed and small bubbles remain in the glass			
2	etc. . . .			
3	Control the "hot spot" 3. The operator must keep the temperature of the air in the middle of the furnace at a certain value – – – – – – – – – – – 4. Pre-phase 5. Action phase: regulate total gas flow 6. Evaluation phase	The operator has no assistance for this control task		
4	5,6 . . .			
7	Perception of temperatures 7. Two information sources were mentioned in the interviews: 8,9 – – – – – – – – – – – 8. Digital temperature indication 9. Trend recorder	–both sources are located on a control panel –operator looks at these temperatures every 10 minutes		

Fig. 3.4.3.   Data from Figure 1 in tabular format (from Shepherd, 1976).

and indicating the direction and general nature of the transmittance of the branch. A network of branching chains is obtained. Each chain should be continued until it ends in an input, or output, or returns to a variable earlier in the chain, forming a feedback loop. Whenever there is a path through the process from an independent variable to a dependent variable, the dependent variable can be connected with the independent variable by a path through a control function to form a closed control loop. This control function may be carried out either by the person or automatically. By describing the system in this way it is possible to study the control responsibilities of the person and the machine and the effect of reallocating these functions.

This process of drawing the graph is usually not simple or immediately possible, but consists of trial and error and further reference to the sources of information. The completeness of the signal-flow graph is self-checking (except for functions that act in parallel) because, to be a logical structure, all the functions must be specified.

Mathematical techniques are available for combining functions to eliminate uninformative variables and leave ones showing important structural elements.

With some modifications, the signal-flow graph method can be applied to systems that include human operators. In the usual signal-flow graph, the structure that carries out a function is not specified, but when people carry out one of the functions, it is useful to be able to specify their position in the system. The line of the signal-flow graph represents a function; when this is carried out by people, they can be represented by the function diagram symbol for a structure, a box. This box, the person, cannot be connected to the variable circle by continuous lines, because these represent functions, so dotted lines are used.

The boxes representing operator functions can be collected at one side of the graph, and can be grouped together. Such groups might represent functions all controlling the same dependent variable, or affected by the same control variable. In a large system, a group might represent those functions carried out by the same operator. This provides a means of studying the interactions of operator responsibilities. It should also be possible to draw a signal-flow graph for the communication structure.

Although the signal-flow graph is primarily a task description tool, it can aid in task analysis by making explicit the decisions involved. The signal-flow graph provides two methods that might be useful in the measurement of these decisions:

1. The number of common members of the above-described groups of control functions reflects the amount of interaction between system variables, and therefore the numbers of factors the person must take into account when making a control action.

2. Combination of the machine functions of a loop the person controls shows the order of control the person must apply.

While these methods were being developed in Europe, in the United States the Office of Naval Research sponsored the development of a job analysis methodology that was to be the culmination of many years of effort by E. J. McCormick and his co-workers. As a means of cataloging job demands, they developed a Check List of Worker Activities, which was used to analyze 250 jobs in the steel industry. Fourteen factors emerged from a factor analysis of these data (Palmer and McCormick, 1961). These 14 factors were later reduced to four and used as the basis for the Worker Activity Profile (Gordon, 1963). Tests of this methodology, and further factor analyses (McCormick, Cunningham, and Gordon, 1967), led eventually to the ONR-sponsored Position Analysis Questionnaire or PAQ (McCormick, Jeanneret, and Mecham, 1969).

The PAQ itself underwent numerous tests and revisions, being tested on over 800 jobs. Inter-rater reliability was found to be high when measured on 60 jobs. In its final form (Form B), the PAQ has the structure shown in Table 3.4.3. The PAQ consists of 187 jobs elements (items) that characterize or imply various types of basic human behaviors that are involved in jobs in general. The elements are organized into six major divisions. These divisions and their subdivisions are given in Table 3.4.3, along with the number of job elements in each. Some examples of job elements are as follows:

Use of written materials
Use of touch
Coding/decoding
Use of precision tools
Use of keyboard devices
Advising
Specified work pace
Number of personnel supervised

In the analysis of jobs with the PAQ, the analyst makes a response for each job element, using a specfic scale for each job element. There are six types of scales, as follows: extent of use; amount of time; importance to the job; possibility of occurrence (used only with hazards); applicability; and special scales. All are five-point scales except the applicability scale (which requires merely a presence or absence decision). Examples of three of the scales are given below.

Amount of Time	Extent of Use	Importance
Does not apply	Does not apply	Does not apply
1. Infrequent/rarely	1. Nominal/very infrequent	1. Very minor
2. Under $\frac{1}{3}$ of the time	2. Occasional	2. Low
3. Between $\frac{1}{3}$ and $\frac{2}{3}$ of the time	3. Moderate	3. Average
4. Over $\frac{2}{3}$ of the time	4. Considerable	4. High
5. Almost continually	5. Very substantial	5. Extreme

Examples of these scales would be question 92 (section 3.3) asking for the amount of time spent walking, question 17 (section 1.1.2) asking for the extent of use of tactual information sources, and question 175 (section 6.2) asking for the importance of keeping a specified work pace. Special scales are used only rarely, for example, a six-point scale for number of personnel supervised in question 134 (section 4.5.1). In general, a great effort has been made to simplify the collection of data, resulting in a most workable instrument. The inter-rater reliability in one study was found to be reasonably high, with a mean coefficient of .79.

A factor analysis of data for 2200 jobs resulted in the identification of 45 factors (called job dimensions) that represent the "structure" of jobs in the labor force. Certain examples of these job dimensions are as follows:

Interpreting what is sensed
Watching devices/materials for information

Processing information
Performing handling/related manual activities
Exchanging job-related information
Being in a stressful/unpleasant environment
Being alert to changing conditions

The PAQ is clearly a different type of job analysis instrument from typical task analysis instruments in that it provides for the analysis of jobs in terms of basic human behaviors that cut across various types of jobs, rather than in terms of specific tasks that characterize the work activities of particular jobs or small groups of jobs. Thus the PAQ job elements cannot be considered as "tasks" as defined in this chapter; for example, many of the PAQ job elements do not have observable ". . . starting points that can be identified as a stimulus or cue. . . ." However, the distinction between task analysis and task description made in this chapter has a parallel to the PAQ, in that the PAQ provides for

Table 3.4.3   Structure of Position Analysis Questionnaire (PAQ)

Area of Coverage	Number of Questions		
1. Information Input	Total 35		
1.1 Sources of job information		20	
1.1.1 Visual sources of job information			14
1.1.2 Nonvisual sources of job information			6
1.2 Discrimination and perceptual activities		15	
1.2.1 Discrimination activities			8
1.2.2 Estimation activities			7
2. Mediation Processes	Total 14		
2.1 Decision-making and reasoning		2	
2.2 Information processing activities		6	
2.3 Use of stored information		6	
3. Work Output	Total 50		
3.1 Use of physical devices		29	
3.1.1 Hand tools			6
3.1.2 Other hand devices			5
3.1.3 Stationary devices			1
3.1.4 Control devices			9
3.1.5 Mobile and transportation equipment			8
3.2 Integrative manual activities		8	
3.3 General body activities		7	
3.4 Manipulation/coordination abilities		6	
4. Interpersonal Activities	Total 36		
4.1 Communications		10	
4.2 Miscellaneous interpersonal relationships		3	
4.3 Amount of personal contact		1	
4.4 Types of personal contact		14	
4.5 Supervision and coordination		8	
4.5.1 Supervision given			7
4.5.2 Supervision received			1
5. Work Situation and Job Context	Total 18		
5.1 Physical working conditions		12	
5.2 Psychological and sociological aspects		6	
6. Miscellaneous aspects	Total 36		
6.1 Work schedule, method of pay, and apparel		21	
6.2 Job demands		12	
6.3 Responsibility		3	

the "analysis" of jobs in terms of preestablished units of work behavior, rather than for the "description" of jobs (such as in an essay, verbal manner).

Although the PAQ does not serve the same job-specific uses in human factors programs as conventional task analysis techniques, it can serve various purposes in connection with related personnel management functions (such as those involved in what is sometimes referred to as the personnel subsystem). Its primary uses in organizations have been in such areas as personnel selection (it provides the basis for deriving estimates of aptitudes, thus eliminating the need for conventional test validation procedures); personnel development and training; the development of career ladders; performance appraisal (the job dimension scores provide the basis for such appraisal); and job evaluation (it serves as the basis for establishing rates of pay without the need for conventional job evaluation procedures). Over 400 organizations have used the PAQ for such personnel-related purposes or for research, involving the analysis of over 100,000 positions representing about 3000 different jobs.* A few organizations have used the PAQ in combination with conventional task analysis techniques in connection with their personnel management programs.

A second technique to have arisen rather directly from this work and approach is the AET method, an acronym for Arbeitswissenschaftliches Erhebungsverfahren zur Tatigkeitsanalyse, or "ergonomic job analysis." Originally developed by Landau and Rohmert at Darmstadt (Landau, 1978; Rohmert and Landau, 1979, 1981), the AET technique first describes the work system (objects, equipment, and environment) and the task. It then analyzes this task in terms of job demands for perception, decision, and response. The full system with the number or characteristics to be measured is shown in Table 3.4.4.

Again, the distinction between description (A and B) and analysis (C) is clearly seen. A coding scheme exists for each of the characteristics, covering a diversity from use of footrests, through illumination intensity, to shift work conditions. In practice, part B, that part concerned with the details of the task sequence, is not covered as deeply as it would be with other task-analytic techniques. However, the integration of task with equipment and environment in the same analysis procedure gives a unique tool for detecting potential mismatches between job demands and human capabilities.

The AET has seen rather widespread use in European industry with many thousands of jobs currently in the data base held by its authors. The publication of the English language version (Rohmert and Landau, 1983) should lead to adoption of the AET in North America.

Recent examples of task analysis in U.S. industry were provided by Drury (1983). His example of a branching analysis at a stock exchange shows how at a higher level the task is branching, but each subtask can be considered using a sequential task analysis. Table 3.4.5 and Figure 3.4.4 show a proposed direct data entry keyboard for use in a major stock exchange. A keyboard is used to record trades between brokers, bid prices, and offer prices. It is held by the reporter standing in the trading crowd and is designed to replace mark-sense cards which the reporter currently marks and hands to another person to feed into a card reader. A key issue in the study was whether or not the keyboard would be as rapid as the mark-sense card system.

A task description of the reporter's job was obtained by interviewing reporters and observation of their activities on the trading floor. This task description was then used as a task analysis of the new keyboard, for example, in specifying push-button sizes and the use of a mixture of dedicated and general-purpose keys. A sequence of dedicated keys was used for the major functions (VOLUME, BID, TRANSMIT, NEXT, etc.) and laid out according to task sequences defined by the task description of Figure 3.4.4. A numeric key pad, in telephone format rather than calculator format to accord with frequent telephone use stereotypes, was provided to key in numerical information. Finally, an alphabetic keyboard was provided for setting up which option (an option is an agreement to buy or sell stock for a specific price before a specific date) or stock was associated with which dedicated key and for entering rarely traded stocks and options that did not merit dedicated keys. The alphabetic keyboard was provided with a hinged cover to prevent accidental operation and to reduce visual choices in most operation modes.

Task analysis did not stop when the design was finished. All the tasks in Figure 3.4.4 were subjected to a sequential task analysis to determine whether all procedures were feasible on the keyboard and to estimate the speed of operation. Figure 3.4.4 shows how the keyboard is set up at the start of trading and then used in three modes for three different types of stocks or options traded in that crowd. After any transaction is recorded, new transactions of the same type can be recorded in an abbreviated format as shown.

The detailed analysis, shown in Table 3.4.5 for "Transact main option" was part of the final pencil-and-paper analysis of feasibility. After the remarks column are "number of key strokes" and "time" columns, estimated from Neal's (1977) data on rate of entry of strings of numerical information. Comparing the predicted time with the observed time of the current system demonstrated that the keyboard was always at least as rapid for data entry.

---

* The PAQ is copyrighted by the Purdue Research Foundation and is available from the University Bookstore, W. State Street, W. Lafayette, IN 47906. Processing of PAQ data is carried out by PAQ Services, Inc., 1625 North, 1000 East, Logan, UT 84321.

Table 3.4.4 The AET Methodology: Contents

Part A—Work System Analysis	Total 143

1. Work objects   33
    1.1 Material work objects (physical condition, special properties of the material, quality of surfaces, manipulation delicacy, form, size, weight, dangerousness)
    1.2 Energy as work object
    1.3 Information as work object
2. Equipment
    2.1 Working equipment   36
       2.1.1 Equipment, tools, machinery to change the properties of work objects
       2.1.2 Means of transport
       2.1.3 Controls
    2.2. Other equipment
       2.2.1 Displays, measuring instruments
       2.2.2 Technical aids to support human sense organs
       2.2.3 Work chair, table, room
3. Work environment   74
    3.1 Physical environment
       3.1.1 Environmental influences
       3.1.2 Dangerousness of work and risk of occupational diseases
    3.2 Organizational and social environment
       3.2.1 Temporal organization of work
       3.2.2 Position in the organization of work sequence
       3.2.3 Hierarchical position in the organization
       3.2.4 Position in the communication system
    3.3 Principles and methods of remuneration
       3.3.1 Principles of remuneration
       3.3.2 Methods of remuneration

Part B—Task Analysis	Total 31

1. Tasks relating to material work objects
2. Tasks relating to abstract work objects
3. Person related tasks
4. Number and repetitiveness of tasks

Part C—Job Demand Analysis	Total 42

1. Demands on perception   17
    1.1 Mode of perception
       1.1.1 Visual
       1.1.2 Auditory
       1.1.3 Tactile
       1.1.4 Olfactory
       1.1.5 Proprioceptive
    1.2 Absolute/relative evaluation of perceived information
    1.3 Accuracy of perception
2. Demands for decision   8
    2.1 Complexity of decision
    2.2 Pressure of time
    2.3 Required knowledge
3. Demands for response/activity   17
    3.1 Body posture
    3.2 Static work
    3.3 Heavy muscular work
    3.4 Light muscular work, active light work
    3.5 Strenuousness and frequency of movements

Table 3.4.5 Task Analysis of Keyboard Entry for "Transact Main Option" from Figure 3.4.4

	Task Description		Task Analysis		
Purpose	Action (Keys Pressed)	Check (Display)	Remarks	Number of Key Strokes	Time (sec)
Enter option and month	MONTH 1	BLY in "stock" JAN in "month"	BLY is main option	2	1.1
Enter strike price	STRIKE 70	70 in "strike"		3	0.8
Initiate volume	VOLUME	T in "b/o/t"	Neither bid nor offer pressed; must be trade	4	1.0
Enter volume	250	250 in "volume"		5	1.3
Enter price	PRICE 15·0	15·0 in "price"			
Identify buyer	BUY 694	694 in "broker" B in "buy/sell"		4	1.0
Transmit data	TX	No change	TX button will have no effect if pressed again	1	0.3

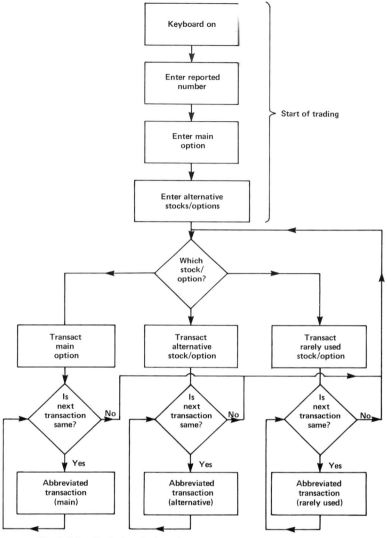

Fig. 3.4.4. Task descriptions for stock market keyboard entry task.

A final example of a task analysis devised for a specific industrial task is shown in Figure 3.4.5. Drury and Wick (1984) were faced with problems of musculoskeletal injuries in a shoe manufacturing company. The task description form thus concentrates on a detailed recording of the angles of each limb at each step in the task. Note that in highly repetitive operations, a fixed sequence is followed and hence a sequential task description format is the logical choice. Task analysis in this study was the comparison of angles and forces recorded with human abilities to produce these angles and forces without injury. The simplest case was to count the daily incidence of "damaging wrist motions," which were defined as the wrist being away from a neutral position when a grip force was being applied. Much medical evidence (see e.g., Armstrong et al., 1984, for review) shows that these conditions figure prominently in the etiology of cumulative trauma disorders of the upper limbs.

An important use of this "postural task analysis" was to perform the same analysis twice, once on the original job and again on a prototype workplace reconfigured to reduce or eliminate the problems found in the first analysis. The same operator was used, usually after a week's continuous production use of the prototype. Counting the number of "damaging wrist motions" before and after in this way clearly showed where improvement had taken place. It was also a salutary discipline for the analyst and ergonomist: were all the problems found on the first analysis and did the redesign actually

JOB ANALYSIS

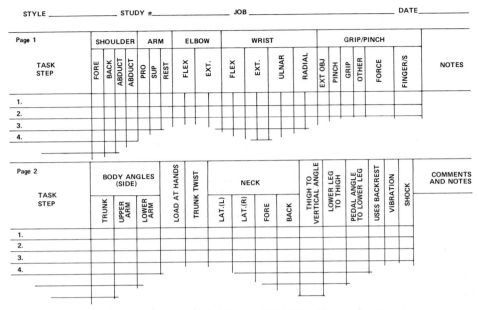

Fig. 3.4.5. Task analysis of musculoskeletal system demands.

produce the expected improvements? Such a before-and-after analysis, reported by Drury and Wick for three jobs, showed very large improvements in "damaging wrist motions" in all cases. These same improvements were incidentally accompanied by perceived comfort and productivity improvements.

Further use of the same basic methodology was made in order to produce a methodology for matching job demands to the reduced capabilities of already injured workers. A data base was constructed, using headings similar to those of Figure 3.4.4, and a program written to interrogate all the 300 jobs on the data base to determine which jobs do not exceed the capabilities of the returnees. If no suitable matches are found, then near-matches are examined to see where low-cost job changes could result in the successful return to work of an injured employee. (The system has currently been in operation for almost 2 years with useful results.) However, the point to be emphasized from the standpoint of task analysis is that a completed task analysis is just as useful for selection/placement decisions as it is for equipment design or training.

Again it should be noted that the variation in task analysis format is the result of the different requirements placed on the task analysis. There is no general-purpose task analysis, only specific classes of application.

3.4.7 PLANNING FOR TASK ANALYSIS

Formal planning is important in understanding task description and analysis. For a complex system a number of task analysts may have to work simultaneously to complete the effort on schedule. The task analysis schedule may have to be phased with other system development and evaluation activities. In addition, a thorough and useful task analysis usually requires participation by a number of different types of personnel in addition to task analysis specialists: design engineers, system safety analysts, instrumentation and control engineers, operations managers and staff, training specialists, procedures writers, and others. A multidisciplinary team with broad, multidisciplinary composition helps ensure that the thorough knowledge of a system and its operation necessary for complete and technically accurate task identification and description is available. For these reasons, the following planning steps are recommended:

Establish the objectives and scope of the task analysis.
Establish a task data collection model that reflects the needs of the task analysis (information requirements, data formats).
Identify the personnel requirements of individuals on the task analysis team.

Prepare a schedule for the task analysis effort.

Obtain management support.

Develop a task analysis program plan and uniform procedures for the team members to follow

Set up a quality control method for reviewing the task descriptive data for completeness and technical accuracy.

3.4.8 PHASES OF TASK ANALYSIS

3.4.8.1 System Description and Analysis

Man–machine system task analysis is a top-down undertaking. Each phase of analysis provides an increasingly detailed view of the human–machine interaction requirements. Each phase provides the basis for a subsequent phase of analysis although the process may unfortunately stop short of detailed task analysis because of cost or time constraints.

The principal objective of the system description and analysis phase is task identification. It also provides task analysts with the overall understanding of the system and its context that is important if the resulting product is to be useful.

System Functional Analysis

This effort generally begins with a review of the organizational context of which the system is a part. In this phase the goal outputs of the system are specified, including any specific constraints and criteria associated with their achievement. The analysis then proceeds to specify the system functions necessary to accomplish the goal output. A system function is an activity that maintains control over a variable that influences the goal output.

It is sometimes necessary, before defining system functions, to define potential modes of operation, or major system states, because they involve different functions or different ways of executing functions.

Examples of system functions include the following:

Target detection and tracking

Vehicle guidance

Reactivity control

Heat removal

Pressure control

Inventory control

Functions may have to be subdivided or decomposed in subfunctions if they are complex.

The next step is to examine the equipment configuration of the system. One or more separate subsystems of electrical, electronic, and/or mechanical components generally are designed to serve each function. In some cases subsystems may serve more than one function. In any case, the engineering design is studied to determine how its subsystems and components accomplish various system functions.

The analysis continues by describing the general process of function execution, including human as well as equipment roles. Specific criteria for function performance should be identified.

Operational Sequence Analysis

In this process the operational flow and the relationships of functions and human and equipment actions in time are described. In the analysis of military systems this process is called "mission analysis."

Because a military system may have more than one mission, each unique mission must be analyzed. Likewise, an industrial system may have several modes of operation or a number of distinctly different states in which operations are conducted. These should also be analyzed separately.

A preliminary analysis of system functions provides a basis for selecting operating sequences to ensure that all functional requirements are addressed.

The steps in performing an operational sequence analysis are listed below.

1. *Develop an Operating Sequence Profile.* This is a description of the sequence of major events or phases of operations.

2. *Develop a Narrative Description of the Sequence.* This defines the starting and ending conditions, the major events or phases, the functions invoked by the events or phases, and the major human and equipment actions. Figure 3.4.6 is a sample description of an aircraft system mission. Figure 3.4.7 is a description of an operating sequence in an industrial setting (a nuclear power plant).

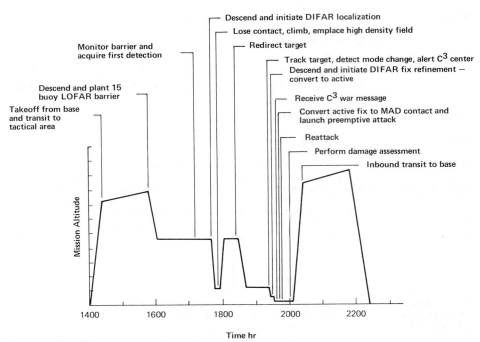

Fig. 3.4.6. Mission description for an aircraft.

Operating Sequence Overview

Plant Name:

Operator Function/Subfunction: Supervise and control plant operations./Restore plant to a safe condition.

NSSS Type: C-E, PWR

Operating Sequence ID: 07

CR Type: Single

Operating Sequence: Loss of Condenser Vacuum

Initial Conditions: The plant is operating at full power. Four circulating water pumps are running. Three of four air removal pumps are running. The plant is in a normal lineup with a boron dilution in progress. CEDMCS is operating in manual sequential.

Initiating Event: One circulating water pump trips, resulting in drop in condenser vacuum.

Expected Progression of Action: Annunciator alarm alerts the crew to malfunction. The crew verifies pump trip and initiates investigation of cause. Condenser vacuum starts to decrease; an additional air removal pump starts automatically to prevent further decrease in vacuum. The crew stops the dilution and reduces power manually to restore vacuum. The cause of the circulating water pump failure is diagnosed as a spurious breaker relay trip. The relay is reset; the pump is restarted and runs normally. The crew restores the plant to full power and normal lineup.

Final Conditions: The plant is operating at full power. Four circulating water pumps are running; three of four air removal pumps are running.

Major Systems: 4.16 kV power, Circulating Water System (CWS), Condenser Air Removal (CAR), Reactor Coolant System (RCS), Feedwater System (FWS), Control Element Drive Mechanism Control System (CEDMCS).

Fig. 3.4.7. Operating sequence description for a nuclear power plant.

3. *Develop Functional Flow Diagrams.* Functional flow diagrams depict the interrelationships between functions and performance requirements and their sequence. These diagrams are developed in increasing detail until a level suitable for allocation or identification of behavioral functions and identification of tasks is reached. Figure 3.4.8 is a sample diagram from a functional flow analysis of an aircraft system mission.

4. *Develop Action–Decision Diagrams.* Diagrams of this kind may be prepared instead of or in addition to functional flows. They may be prepared at various levels of detail. Like functional flows they may not, in early stages, distinguish between what is done by people and what is done by equipment or software. Figure 3.4.9 illustrates the nature of a decision–action diagram. The example shown is an emergency response sequence in a nuclear power plant. In this case, the functional allocation had been made so that the diagram shows only human operator performance requirements.

The final step in operational sequence description is the preparation of operational sequence diagrams (OSDs). An OSD is a detailed description of operator, equipment, and software actions in which each of these elements is sequentially integrated with the other in a time sequence. Flow chart symbology of the type shown in Figure 3.4.1 is used to identify the nature of the required actions. OSDs are a tool for human factors engineering of design. They are not necessary for task identification and description.

Alternative for Analysis of Established Systems

The preceding techniques should be used when there is no operating example of the system being analyzed. When a system does exist operating procedures and other system descriptive documentation may be used as for operating sequence definition and task identification. In this case the analysis may proceed directly to the identification of the behavioral functions and tasks that can be determined from the interface design, from safety analyses, procedures, technical specifications, and other system documents. A detailed procedure for task analysis using this approach as the point of departure is contained in a report on *Task Analysis of Nuclear Power Plant Control Room Crews,* recently performed for the U.S. Nuclear Regulatory Commission (Burgy et al., 1983).

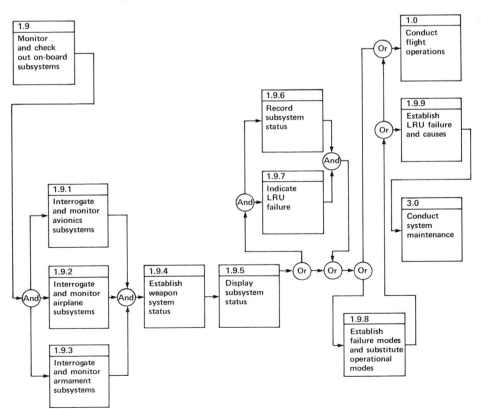

Fig. 3.4.8. Example of a functional flow analysis for an aircraft mission.

A Note About Task Analysis of Maintenance and Other Support Functions

So far we have discussed task analysis from the perspective of system operations in which personnel interact extensively with equipment and software to execute system functions. In maintenance and other support functions the human–machine dynamics characteristic of operations are not present. The functions belong to personnel who use tools and equipment to execute them. Time considerations are also different for maintenance and support functions. For example, stimuli for task action do not appear in a continuous stream as they do in most operations. Although there is an order for performance of steps within each task and some normal ordering of tasks, each task can be considered separately. Task overlaps in time usually are unimportant and do not have to be considered.

In maintenance task analyses a system review is performed to identify and categorize the equipment to be maintained and the types of equipment and tools to be used. Behavioral functions are defined,

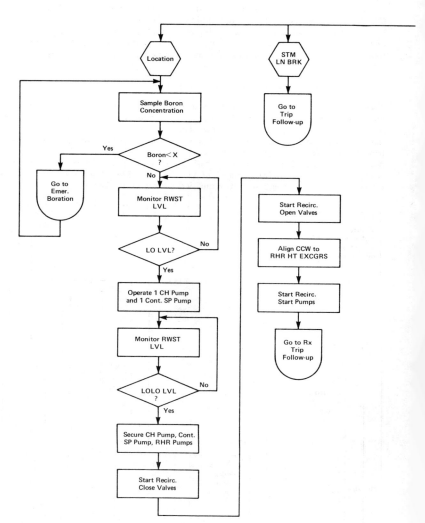

Fig. 3.4.9. The decision-making algorithm.

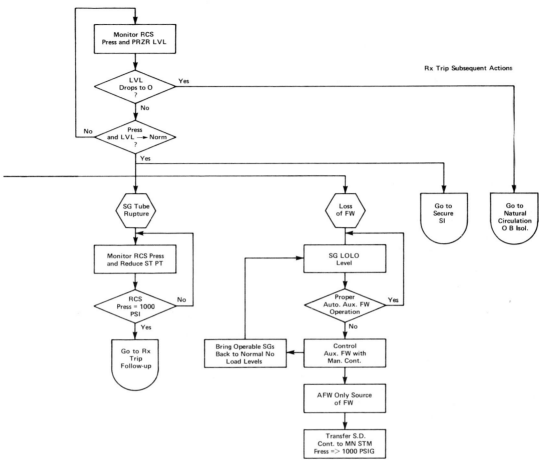

Fig. 3.4.9. *(Continued)*

which represent categories of human actions performed on the various types of equipment (e.g., checking, servicing, calibration, troubleshooting). A task list is generated by integrating and further analyzing equipment types and behavioral functions. Miller (1956) offers guidance for maintenance task analysis in his *Suggested Guide to Position-Task Description.*

3.4.8.2 Task Description Task List

The first step in task description is to prepare a task list for each mission or operating sequence from the results of the system description/analysis. The task list indicates the scope and sequencing of the total array of human performance requirements in the sequence. Verification of the completeness and accuracy of the initial list is desirable, although changes may be in order as the descriptive analysis process. Verification may involve system design and engineering experts and operations experts depending on the stage of development of the system. The following information items are useful to include in a task list:

Mission/operating sequence identification: an alphanumeric code and/or title.

System function identification: an alphanumeric code and/or title.

Behavioral function identification: an alphanumeric code and/or title.

Task sequence number, based on time of task initiation relative to initiation of other tasks. Task sequencing may not be absolute; an acceptable sequence should be established.

Task title: a summary statement of the human performance requirement.

Task purpose/objective: the immediate reason for performing the task, or the output expected from the task.

Detailed Task Description

Each task is described in terms of a set of information requirements established according to the intended use of the task-descriptive data. The information items suggested for a task description are essential parts of the description, regardless of its application. Items commonly included are the following:

Work-area location: where a task is accomplished.

Task description: includes equipment used, displays used, controls used, actions taken (detailed, sequential description of each task step), support equipment and aids used, feedback regarding individuals's actions.

Frequency of performance: once, twice, and so forth per shift, daily, weekly, monthly, and so on.

Performance time: estimated hours and minutes task will take or the maximum permissible time.

Criticality: effect of failure to perform on the success or failure of mission/operating sequence. Criticality is usually expressed as a relative ranking and is based on expert judgment.

Performance criteria and standards: criteria for successful task execution.

The preceding list was adapted from Woodson (1981). In addition, the following items may be included:

Error potential: most likely significant human errors based on system operating experience, observations of performance, or expert judgment.

Severity of error consequences: a relative ranking or defined scale. Value developed by expert judgment of the severity of consequences of an error. This item of information may be used in place of a "criticality" assessment.

Hazards/safety factors: any known or suspected hazards or requirements for avoiding or preventing hazards.

Task difficulty: usually a relative ranking developed by expert judgment.

Tools and equipment: tools, test equipment, protective gear, reference manuals, and so on.

Skills and knowledge: types and levels of skill or knowledge pertinent to the selection and training of individuals for the job or tasks.

Physical characteristics: special physical requirements necessary for effective task performance (e.g., body size, strength, dexterity, coordination, physiological tolerance to specific and anticipated environmental stresses).

3.4.8.3   Descriptive Data Collection

Any of four different basic techniques for task-descriptive data collection may be used. The choice of technique depends on the status of system development and implementation, the expertise of the analysis team, and the availability of other personnel to contribute to the effort. It is preferable to use a combination of the following:

Documentation Review. It is sometimes possible to gather substantial information from system descriptions and schematics, panel front drawings, technical specifications, existing procedures, and training materials for similar systems if they do not exist for the system under study. These materials may be used to prepare an initial task description.

Questionnaire Survey. For systems in operation or with an already operational counterpart, questionnaires may be used to gather task-descriptive information from subject matter experts (SMEs). Respondents may include trainers, design/operations engineers, and operations personnel.

Interviewing. Personal interviews may also be used to gather data from SMEs. A formal questionnaire may be used to structure the interviews or the task data form may be used. When analysts have ready access to SMEs, interviewing may be a fairly informal matter of asking questions as needs arise during data development.

Observation. Observation of simulated or actual performance exercises may be used initially to collect task-descriptive data or to verify, correct, and fill in gaps in a preliminary analysis developed by any of the above means. Observation requires that there be at least a mock-up, simulation, or prototype of the system. It may also be possible to perform observations on a similar system, provided system differences can be specified and their effects on performance requirements identified with confidence.

In making observations for the purpose of describing tasks, a slow, task-by-task performance of the sequence during which personnel actions can be discussed is useful. This may be followed by a real-time performance of the sequence or an approximation to it, allowing task and personnel interactions and time and movement factors to be examined.

3.4.8.4 Applications Analyses

System- and task-descriptive data provide an essential reference and input source for a variety of further types of analysis. It is these further analyses, here referred to as "applications analyses," that yield products for use in personnel performance development.

Applications analyses require principles, theory, and techniques from different fields of the behavioral sciences. Those who will perform applications analyses should, ideally, participate in formulating the descriptive data collection model, but may not necessarily conduct the descriptive parts of the analysis.

Some of the applications of system- and task-descriptive data are identified below. Also identified are some of the analysis techniques that may be used for each application. References are given to further sources of information concerning these techniques.

1. For human engineering of design
   a. Development of operational sequences diagrams, to analyze the allocation and time sequencing of information decision and action requirements between personnel hardware and software (e.g., as a basis for assessing and optimizing function allocation) (Geer, 1976).
   b. Comparison of information and control needs specified through task descriptive analysis to characteristics of actual or proposed displays and controls, in light of human engineering criteria for design of controls and displays (to optimize instrumentation and control design) (U.S. Nuclear Regulatory Commission, 1981).
   c. Link analysis to define and represent graphically the frequency and criticality of interfaces between operators and work station or work area components and between different members of an operating crew (to optimize work station or work area design) (Woodson, 1981; Chapanis, 1962).
2. For performance prediction, personnel selection criteria and measures, and training
   a. Development of behavioral taxonomies and classifications of task attributes or dimensions of performance (to identify the classes of skill necessary to be trained, for example) as a basis for task selection for training (Finley et al., 1970).
   b. Critical incident analysis (Flanagan, 1954; American Institutes for Research, 1973) to establish critical tasks and personnel skills/attributes.
   c. Development of task skill and knowledge requirements and learning hierarchies, (Bronson et al., 1975).
3. For studies of human/system reliability see fault tree techniques (such as Swain's Technique for Human Error Rate Prediction (Swain, 1964; Swain and Guttman, 1980) and computer simulations of operator performance (Whitmore and Parks, 1974; Parks and Springer, 1976; Kozinski et al., 1984).
4. For analysis and modeling of mental activity see empirical observation, interviewing, and experimentation in real or simulated task situations—to explore the knowledge bases, recognition mechanisms, and strategies employed for search, problem diagnosis, and evaluation of solutions to operational perturbations (Bainbridge et al., 1968; Bainbridge, 1974).

3.4.9 AN EXAMPLE OF TASK DESCRIPTION AND ANALYSIS

A summary is provided here of the approach to task analysis used in a recent study of control room operations in nuclear power plants (Burgy et al., 1983). This is an example of analyzing an established system, but the techniques are also applicable to systems that are still at an early stage of development.

Scope of the Project

The project was designed to provide task data for evaluation of six areas:

1. Human engineering of new control rooms and of retrofits in existing control rooms
2. Staffing requirements (number, skill mix)
3. Operator qualification and training requirements
4. Normal, off-normal, and emergency procedures
5. Job performance aids
6. Communications

Another objective of the project was to format the data and create an on-line task data-base management system to facilitate analyses for applications in the areas identified above.

Methdology

The unit of analysis chosen was an operating sequence, which had been explicitly defined in the system description/analysis phase, in terms of initial plant conditions, sequence-initiating event, plant changes, major clusters of control room crew activity in response to those changes, and subsystems and major components involved. Twenty-four operating sequences were defined. They were selected to encompass a representative range of system functions, operating modes, and operator roles or behavioral functions.

In the project a sharp distinction was made between task-descriptive and task-analytic data. Descriptive data were defined as those items that characterize the task and can be collected by direct observation, and/or by consulting plant documentation and operators to verify factual information that may be difficult to discern from observation. Task-analytic data were defined as those items that represent conclusions and inferences derived from task-descriptive data and other information and analyses necessary for particular applications. A third category was defined to consist of static contextual data (e.g., panel layouts), which also may be needed for certain applications of task data but are readily available and need not be established through observation/analysis of performance. These distinctions were judged necessary for the design of an efficient multipurpose data base. The task-descriptive data form a standard reference set that is distinct from additional data needed for particular applications.

A defined-field task-descriptive data format was used. The format was based on a "model sentence" to promote consistency in the content and construction of each task description. Strict control was exerted over vocabulary used in defined fields. Emphasis was placed on performance observation for data collection. Demonstration analyses were performed to show the use of the data for various human factors analysis purposes (as identified in the project objectives).

Steps in Task Description

The data were developed in four iterative steps: tabletop analysis; on-site data collection; analysis of audiovisual records obtained on site to complete data development; and quality control review and data base entry.

    A.   Tabletop analysis
        1.   Development of operating sequence
        2.   Task identification, development of preliminary task list
        3.   Preliminary task description
    B.   On-site data collection
        1.   Review of tabletop analysis results with plant system/operations experts
        2.   Correction of preliminary results to obtain a closer approximation of actual performance requirements
        3.   Performance of operating sequences involving rehearsal; real-time exercise to obtain an audiovisual record; crew review of videotape and recorded explanation of their actions; debriefing question/interview sessions
    C.   Follow-up review of audiovisual records of performance
        1.   Extraction of performance time data at the task element level
        2.   Reorganization of tasks, revisions, incorporation of additional data based on performance explanations of operators
    D.   Quality control and data entry
        1.   Data check for clarity and conformance to vocabulary and format controls
        2.   Key entry and key verification; automated edit checks
        3.   Final review of printout

Task Description Format and Vocabulary Controls

Figure 3.4.10 is an example of a task-descriptive format for an on-line task data file. The upper part of the form provides plant, sequence, system function, and task identification. Task identification includes the task name (a summary description of the required action), objective, and cues.

The detailed task description is recorded on the lower part of the form. Each action, or "element," in the task is written across the form on a separate line. (In this analysis, the scope of the tasks was made small enough that they could be analyzed into elements without intermediate division into subtasks.)

ACTION-INFORMATION REQUIREMENTS DETAILS (AIRD)

PLANT: _____ UNIT: _____ REVIEWER: _____ Sheet _____ of _____

DATE: _____

SEQUENCE: ____Alternate Shutdown Cooling____ NO.: _____ INITIATING CUES: Cooldown requirements not met;
FUNCTION: ____Heat Removal____ NO.: _____ all rods in past 06; RPV pressure LT 300 psig
TASK NAME: ____Initiate flow thru RHR heat exchgr____ NO.: _____ TERMINATING CUES:
TASK OBJECTIVE: ____To establish Suppression Chamber water cooling____ Flow through Hx in band (_____ – _____ gpm)

REMARKS:

NATURE OF TASK: Continuous _____ Discrete X

BEHAVIORAL ELEMENTS

Line No.	Verb	System/ Subsystem	Component	Parameter	State/ Direction	Value/ Range	Units/ Rate	Precision	Accuracy	Resp Time	Trend Req?	If Yes . . .	If No . . .	Comments	on AIRS Sh. No.	Seq Prbm
	positions	RHR	valve (Hx bypass)		close											
	observes	RHR	valve (Hx bypass)		closed											
	positions	RHR	valves (flow contr)		open (proper lineup to SC)											
	observes	RHR	valves (flow contr)		open											
	positions	RHR	pump		start											
	observes	RHR	pump		on											
	observes	RHR	pump	amps	in band											
	observes	RHR	pump	discharge pressure	in band											
	observes	RHR	Hx	flow (shell out)	in band											

Fig. 3.4.10. Task description format.

397

PROCESSES	ACTIVITIES	SPECIFIC BEHAVIORS	DEFINITIONS
1. Perceptual	1.1 Searching for and Receiving Information	1.1.1 Inspects 1.1.2 Observes 1.1.3 Reads 1.1.4 Monitors 1.1.5 Scans 1.1.6 Detects	To examine carefully, or to view closely with critical appraisal. To attend visually to the presence or current status of an object, indication, or event. To examine visually information which is presented symbolically. To keep track of over time. To quickly examine displays or other information sources to obtain a general impression. To become aware of the presence or absence of a physical stimulus.
	1.2 Identifying Objects, Actions Events	1.2.1 Identifies 1.2.2 Locates	To recognize the nature of an object or indication according to implicit or predetermined characteristics. To seek out and determine the site or place of an object.
2. Cognitive	2.1 Information Processing	2.1.1 Interpolates 2.1.2 Verifies 2.1.3 Remembers	To determine or estimate intermediate values from two given values. To confirm. To retain information (short-term memory) or to recall information (long-term memory) for consideration.
	2.2 Problem Solving and Decision Making	2.2.1 Calculates 2.2.2 Chooses 2.2.3 Compares 2.2.4 Plans 2.2.5 Decides 2.2.6 Diagnoses	To determine by mathematical processes. To select after consideration of alternatives. To examine the characteristics or qualities of two or more objects or concepts for the purpose of discovering similarities or differences. To devise or formulate a program of future or contingency activity. To come to a conclusion based on available information. To recognize or determine the nature or cause of a condition by consideration of signs or symptoms or by the execution of appropriate tests.
3. Motor	3.1 Simple/Discrete	3.1.1 Moves 3.1.2 Holds 3.1.3 Pushs/Pulls	To change the location of an object. To apply continuous pressure to a control. To exert force away from/toward the actor's body.
	3.2 Complex/Continuous	3.2.1 Positions 3.2.2 Adjusts 3.2.3 Types	To operate a control which has discrete states. To operate a continuous control. To operate a keyboard.
4. Communication		4.0.1 Answers 4.0.2 Informs 4.0.3 Requests 4.0.4 Records 4.0.5 Directs 4.0.6 Receives	To respond to a request for information. To impart information. To ask for information. To document something, as in writing. To ask for action. To be given written or verbal information.

Fig. 3.4.11. Classification of tasks, example.

The arrangement of fields or columns for element description is such that the element can be read as a sentence. The model for the sentence is as follows:

1. Subject (assumed to be an operator)
2. Performs the following action (a verb describing the behavior, e.g., observes, adjusts)
3. Addressing the following plant component and/or parameter (the object of the action)
4. Using the following control room component (the means of action—a control or display or other means).

In Figure 3.4.10, the first element may be read as follows: "(Operator) positions RHR heat exchanger bypass valve closed." The means of action is described on the second page of the form (not illustrated). The empty fields in Figure 3.4.10 are used to record quantitative data (such as the specific value or operating band of a parameter, units of measure, rate of change of a parameter) that further define the operator's information and control capability requirements A menu of allowable terms is used to control entries in each of the verbal fields for element description.

In most task-descriptive analyses, control is exerted over at least the terms used for task actions or behaviors. This facilitates grouping of tasks for subsequent analysis, such as analysis to identify essential types of performance. Figure 3.4.11 shows one classification frequently used for this purpose.

3.4.10 CONCLUSIONS

Task analyses from a wide variety of disciplines and application areas have been shown to have a number of common elements. First, they focus on a comparison between the task demands placed on the human by the system and the capabilities which the human brings to the system. The distinction between task description, which defines the task demands, and task analysis, which compares these with human capabilities, is fundamental to all the methods presented.

Second, task analysis is an on-going activity in systems design. A crude analysis early in the design process can help define appropriate human machine function allocation and avoid gross human/system mismatches. Such an analysis can be progressively refined to form the basis of hardware and software changes, training requirements, and job aids. In this sense, a task analysis is unavoidable because training is always required in any system complex enough to need a human-factors specialist. If a task analysis will eventually be needed for training, then logic suggests that it be performed early enough in systems design to give hardware and software designs the opportunity to benefit from insights into the human component.

Finally, procedures manuals that are used both as a training aid and as a hard-copy job aid (e.g., Polimo and Braby, 1980) require detailed task descriptions and the warnings of difficulties in human-performed tasks that only the analysis part of task analysis can provide. In a very real sense, task analysis should be the basis of most human-factors design effort. This has long been recognized in the military, where equipment design, training, and job-aid preparation occur in parallel as a matter of policy. Increasingly, industry is recognizing the need to apply such techniques to its own operations, whether they be nuclear power plants or automobile production lines.

REFERENCES

American Institutes for Research (1973). *The critical incident technique: A bibliography.* Palo Alto, CA.

Annett, J., and Duncan, K. D. (1967). Task analysis and training design. *Occupational Psychology, 41,* 211–221.

Armstrong, T. J., et al. (1984). Analysis of jobs for control of upper extremity cumulative trauma disorders. *Proceedings of 1984 International Conference on Occupational Ergonomics, Toronto, Ontario,* pp. 416–420.

Bainbridge, L. (1974). Analysis of verbal protocols from a process control task. In E. Edwards and F. P. Lees, (Eds.), *The human operator in process control.* London: Taylor and Francis.

Bainbridge, L., et al. (1968). A study of real-time human decision-making using a plant simulator. *Operations Research Quarterly, 19* (Special Conference Issue), 91–106.

Beishon, R. J. (1967). Problems of task description in process control. *Ergonomics, 10*(2), 177–186.

Bronson, R. K., et al. (1975). *Interservice procedures for instructional systems development: Executive summary and model.* Fort Benning, GA: U.S. Army Combat Arms Training Board. (Report available from the National Technical Information Service, Springfield, VA., 22161. NTIS Accession Number AD-A19 486.)

Bronson, R. K., et. al. (1975). Interservice procedures for instructional systems development: Phase I—analysis, and Phase II—design. Fort Benning, GA.: Naval Training Device Center, Army Combat Arms Training Board. (Reports available from the National Technical Information Service,

Springfield, VA 22161. NTIS Accession Number AD-A019 487, Phase I; Ad-A019 488, Phase II.)

Burgy, D., et al. (1983). *Task analysis of nuclear power plant control room crews.* Volume 1: *Project approach and methodology* (NUREG/CR-3371) Washington, DC: U.S. Nuclear Regulatory Commission.

Chapanis, A. (1962). *Research techniques in human engineering.* Baltimore, MD: The Johns Hopkins University Press.

Crossman, E. R. F. W. (1956). Perceptual activity in manual work. *Research, 9,* 42–49.

Demaree, R. G., et al. (1982). *Development of qualitative and quantitative personnel requirements information* (Report MRL-TDR-62-4). Wright-Patterson Air Force Base, OH: Behavioral Sciences Laboratory, Aerospace Medical Division.

Drury, C. G. (1983). "Task analysis methods in industry. *Applied Ergonomics, 14*(1), 19–28.

Drury, C. G., and Wick, J. (1984). Ergonomics applications in the shoe industry. *Proceeding of International Conference on Occupational Ergonomics, Toronto, Ontario,* pp. 489–493.

Duncan, K. D. (1974). Analytical techniques in training design. In E. Edwards and F. P. Lees, (Eds.), *The Human Operator in Process Control.* London: Taylor and Francis.

Fine, S. A., and Wiley, W. W. (1971). *An introduction to functional job analysis.* Washington, DC: The W. E. Upjohn Institute for Employment Research.

Finley, D. L., et al. (1970). *Human performance prediction in man–machine systems.* Volume I: *A technical review* (NASA CR-1614). Washington, DC: National Aeronautics and Space Administration.

Flanagan, J. C. (1954). The critical incident technique. *Psychological Bulletin, 51*(28), 28–35.

Geer, C. W. (1976). *Analysts's guide for the analysis sections of MIL-H-46855.* Warminster, PA: Naval Air Development Center.

Gordon, G. G. (1963). An investigation of the dimensionality of worker-oriented job variables. Unpublished Ph.D. Thesis. Lafayette, IN: Purdue University.

Kozinski, E. J., et al. (1984). *Safety-related operator actions: Methodology for developing criteria.* Oak Ridge, TN: Oak Ridge National Laboratory, ORNL/TM8942, NUREG/CR-3515.

Landau, K. (1978). Das Arbeitswissenschaftliche Erheburgsverfahren zur Tatigkeitsanalyse-AET. Dissertation am Fachbereich Maschinenbau der Technische Hohschule, Dramstadt.

Landua, K., and Rohment, W. (1981). Fallbeispiele zur Arbeitsanalysi. Bern-Suttgart-Wien: Hans Huber.

McCormick, E. J. (1976). Job and task analysis. In M. D. Dunette, (Ed.), *Handbook of organizational and industrial psychology.* Chicago: Rand McNally.

McCormick, E. J., Cunningham, J. W., and Gordon, G. G. (1967). Job dimensions based on factorial analyses of worker-oriented job variables. *Personnel Psychology, 20,* 417–430.

McCormick, E. J., Jeanneret, P. R., and Mecham, R. C. (1969). The development and background of the Position Analysis Questionnaire (Report #5). Lafayette, IN: Occupational Research Center, Purdue University.

McCormick, E. J., Mecham, R. C., and Jeanneret, P. R. (1977). *PAQ technical manual.* Logan, UT: PAQ Services, Inc. (distributed by University Book Store, West State Street, West Lafayette, IN 47906).

Miller, R. B. (1953, June). *A method for man–machine task analysis.* Wright-Patterson Air Force Base, OH: Wright Air Development Center, Air Research and Development Command. (Report available from the Defense Technical Information Center, Alexandria, Va. 22314. DTIC Accession Number AD-15721.)

Miller, R. B. (1953, March). *Anticipating tomorrow's maintenance job.* Chanute Air Force Base, Il: Human Resources Research Center, Air Training Command. (Report available from the Defense Technical Information Center, Alexandria, VA 22314. DTIC Accession Number AD-13060.)

Miller, R. B. (1956, April). *A suggested guide to position-task description.* Lowry Air Force Base, Co: Armament Systems Personnel Research Laboratory. (Report available from the National Technical Information Service, Springfield, Va 22161. NTIS Accession Number AD-606010.)

Miller, R. B. (1963). "Task description and analysis." In R. M. Gagne, (Ed.), *Psychological principles in system development.* New York: Holt, Rinehart and Winston.

Miller, R. B. (1974, May). *A method for determining task strategies.* Silver Spring, MD: American Institutes for Research. (Report available from the National Technical Information Service, Springfield, VA 22161. NTIS Accession Number AD-783847.)

Mundel, M. E. (1970). *Motion and time study,* 4th ed. Englewood Cliffs, NJ: Prentice-Hall.

National Council for Systems Reliability (1982).

Neal, A. S. (1977). Time intervals between keystrokes, records and fields in data entry with skilled operators. *Human Factors, 19,* 163–173.

Palmer, G. J., Jr. and McCormick, E. J. (1961). A factor analysis of job activities. *Journal of Applied Psychology, 45,* pp. 289–294.

Parks, D. L., and Springer, W. E. (1976). *Human factors engineering analytic process definition and criterion development for CAFES* (Boeing Report D180-18750-1). Seattle, WA: Boeing Aerospace Co., Research and Engineering Division.

Piso, E. (1981). Task analysis for process-control tasks: The method of Annett, et al., applied. *Journal of Occupational Psychology, 54,* 247–254.

Polimo, A. M., and Braby, R. (1980). *Learning of procedures in Navy technical training.* An evaluation of strategies and formats (TAEG Technical Report 84). Orlando, FL: TAEG.

Rohmert, W., and Landau, K. (1979). *Das Arbeitswissenschaftliche Erheburgsverfahren zur Tatigkeitsanalyse (AET).* Bern-Stuttgart-Wien: Hans Huber.

Rohmert, W., and Landau, K. (1983). *A new technique for job analysis.* London: Taylor and Francis.

Seymour, W. D. (1954). *Industrial training for manual operations.* London: Isaac Pitman.

Seymour, W. D. (1966). *Industrial skills.* London: Isaac Pitman.

Seymour, W. D. (1968). *Skills analysis training.* London: Isaac Pitman.

Shepherd, A. (1976). An improved tabular format for task analysis. *Journal of Occupational Psychology, 49,* 93–104.

Singleton, W. T. (1974). *Man–machine systems.* London: Penguin.

Swain, A. D. (1962). *System and task analysis, a major tool for designing the personnel subsystem.* (Report SCR-457). Albuquerque, NM: Sandia Corp.

Swain, A. D. (1964). *THERP* (Reprint SC-R-64-1338) Albuquerque, NM: Sandia Corp.

Swain, A. D., and Guttman, H. E. (1980). *Handbook of human reliability analysis with emphasis on nuclear power plant applications* (NUREG/CR-1278). Washington, DC: U.S. Nuclear Regulatory Commission.

U.S. Nuclear Regulatory Commission (1981). *Guidelines for control room design reviews* (NUREG-0700, Section 3.7). Washington, DC: U.S. Nuclear Regulatory Commission.

Vallerie, A. (1978, February). *Survey of task analysis methods* (RN-80-17). Washington, DC: Army Research Institute (AD A096 868).

Van Cott, H. P., and Altman, J. W. (1956). *Procedures for incorporating human engineering considerations in the design of weapon systems* (WADC Technical Report 56-488). Wright-Patterson Air Force Base, OH: Wright Air Development Center, Air Research and Development Command.

Verdier, P. A. (1960). *Basic human factors for engineers.* New York: Exposition Press.

Whitmore, C. D. and Parks, D. L. (1974). *Computer Aided Function-Allocation Evaluation System (CAFES), Phase IV, Volume 1* (Boeing Report No. D180-18433-1). (See in particular "Human Operator Simulation.") Seattle, WA: Boeing Aerospace Co., Research and Engineering Division.

Woodson, W. E. (1981). *Human factors design handbook.* New York: McGraw-Hill.

CHAPTER 3.5

PHYSIOLOGICAL AND PSYCHOLOGICAL WORK LOAD MEASUREMENT AND ANALYSIS

WALTER ROHMERT

**Institut für Arbeitswissenschaft
der Technischen Hochschule Darmstadt
Darmstadt, West Germany**

3.5.1 WHY IS WORK LOAD ANALYSIS NEEDED?

This question would not be asked if the subject of human work load were not mainly human. But because everybody is human, each person tends to think that he or she automatically knows most of what needs to be known about a person when matching up a working situation to him or her. On the other hand, the amount of specialist knowledge already available about humans is such that there is probably no one person, even a specialist, who can learn and know it all.

If a problem arises that is outside the limits of our general knowledge, we usually call upon a specialist for assistance. But if we have a complex problem to deal with involving humans, we still tend to use our own subjective opinion or consult the subjective opinions of others instead of calling in a specialist, perhaps because the existence of human-factors specialists is not yet widely known.

Low output, poor quality, errors, and accidents are the usual symptoms of a problem that, among other things, may need work load measurement attention. The more the worker/operator is stressed, the greater the need to ensure a good match between the person and the task to minimize the risk of error/overload and maximize accuracy and output. This is a fundamental reason why the specialist's knowledge about work analysis is needed in modern industry and particularly during the process of designing human-at-work systems.

Work physiology and occupational psychology attempt to define humans in their relation to the workplace, working process, and means of work, and to study the structure and the functions of the human body with the objective of matching labor conditions to human requirements and of adapting humans to the conditions of work intensity, working environment, and working process. The dynamic behavior of any human-at-work system depends on the subsystems "human" and "work," human being described by one's special qualifications and one's work load, and *work* by the degree of rationalization and/or mechanization.

The prime purpose of any human-at-work system is the production of goods or services, which in general requires a flow of energy, material, and information. Certain of the systems, owing to the relationship between the two elements, impose a load on the "human" component which requires that we know something about human capabilities in the particular situation. In addition to discussing the effects of work load it is particularly appropriate to comment on suitable or imputable work load intensities valid for individuals as well as for groups of different types of workers. Beginning with the latter aspect a suitable work load contains a prognosis out of data from studies of actual working activities, but this ignores the possibility of development, and it is sometimes more correct to evaluate inherent capabilities, as is done, for example, when the possibility of improvement by means of muscular training is taken into account in the evaluation of muscular strength.

There is of course continuous development in what constitutes a socially acceptable level of work load. Such changes may have a *time* base, such as in the requirements expected daily, weekly, monthly, or even over a lifetime, or an *environmental* base, such as the expectations under adverse or dirty conditions. Our concept of what constitutes a suitable work load is also influenced by technical development, which has led to the mechanization of heavy muscular work and the creation of new activities. Owing to these changes it is impossible to specify suitable work load numerically. This chapter deals only with the physiological and psychological aspects of the whole problem; clearly other human, social, and technological aspects must be taken into account.

3.5.2 PRACTICABLE APPROACH FOR DETERMINING WORK LOAD

Starting from the problem of how to objectify work load we note that the conception of assessing work load can be understood axiomatically. Work load can be defined as the total of all determinable influences for the working person (Edholm and Weiner, 1981). Using this pragmatic definition of work load, one can derive plausible descriptions of work load for manual operations, for example. On the other hand, there seems to be another type of work load, meaning mental work load. What does one have in mind when speaking of mental work load? As Rasmussen (1979) points out, there is an overtone of physical effort in the phrase. A load is something that imposes a burden on a structure, or makes it approach the limit of its performance along some dimension. Go far enough along that dimension, and the system fails in some way. In the case of mental work load, the central concept is the rate at which information is processed by the human operator, basically the rate at which decisions are made and the difficulty of making the decisions.

The overtone of physical effort in the phrase of work load is acceptable, if manual tasks are considered. Thus in a first and very simple attempt it would be sufficient, for assembly tasks in industry, for example, to classify and to count the basic motions within one manual assembly cycle (Laurig, 1982). Such a simple approach permits surprising insight into differences between various types of manual operations, especially when utilizing the analytical differentiation in the predetermined-motion-time systems (PMT systems) between the left and the right hands. Furthermore, the lengths of the motions given in PMT analyses can be utilized, too. The sum of these lengths during one cycle time can be understood as some sort of an index. Table 3.5.1 shows such simple indexes pointing out differences between work load of examples of assembly operations chosen at random. If the assembly task consists of driving screws into a rectifier, one will find 14 "standard elements" or "basic motions" for the left and 124 for the right hand within a total assembly cycle. The sum of "distances moved" or "lengths of motions" is 64 cm for the left and 310 cm for the right hand within the total assembly cycle. When a crank is assembled the differences between the left and right hands show only a factor of 1 in this example. When contact springs are assembled no differences are found at least in the work load of the right or the left hand.

For comparing and discussing the indexes in Table 3.5.1, the duration of work load must be comparable. This can be achieved by relating the proposed indexes to the total cycle time (see Table 3.5.2).

Using those indexes one benefits from the axiomatic concept of the PMT systems. Although there is a lack of scientific foundation for these PMT systems, the application of these systems according to well-defined rules leads to the same results at all times, even for different users. This is a consequence of the axiomatic objectivity of those systems. Therefore, it should be allowed for practical purposes to derive pragmatically defined indexes of work load from these PMT systems and to use them in an axiomatic way.

But what does it mean to use such an index? Does it mean a higher work load, if the number of basic motion elements of both hands is great? Does it mean a low work load, if the total length of motions of the hands is small? Is it possible to recognize a critical work load with regard to man's capabilities?

In terms of physics, load might be higher when the duration of moving the arms is longer. But this is not true if we consider the human effort in making longer or shorter arm movements. Not only geometric dimensions but also functional factors may be considered, such as optimal movement directions (see Figure 3.5.1), optimal points of application of muscular strength and its optimal direction (see Figure 3.5.2), and areas of most effective or accurate muscular strength (see Figure 3.5.3).

From Figure 3.5.1 one can learn what directions of movements are favorable when starting the movement at a certain point of the normal working height and working area. The broken lines are circles around the center of the shoulder joint; they describe the manual working area. If, for example, one wishes to start a movement from a certain point of this area one must bring the circle with its center to this point. If the circle is turned in such a way that the black zone is touching one of the thick lines, then the movement of the hand can be fulfilled faster in that direction, the darker the sector of the circle. The lines and the various bright and dark colored circle sectors in Figure 3.5.1 are the results of many experiments with the aim of showing the best conditions for quick hand movements.

Figure 3.5.2 shows optimal points of application of muscular strength in a certain direction. In this figure all points are connected by isodynes, that is, curves for which the muscular strengths exercised horizontally in the direction of the back of the hand show the same value. One can see from the four diagrams that the highest strength values are given in the plane of symmetry of the human body ($\beta = 0°$). In this plane we also find the greatest differences in strength when the position of the hand is altered only several centimeters.

Figure 3.5.3 shows the maximum strength in vertical pull dependent on the grip height. As can be seen from the diagram strength is maximal in carrying a load with both arms extended vertically; a relative maximum is found if the hands are located in the horizontal plane of the shoulder joint.

Table 3.5.1   Comparison of Three Assembly Tasks Using Simple Indexes of Work Load

| | | Number of "Standard Elements" or "Basic Motions" Related to the Total Cycle | | Sum of "Distances Moved" or "Lengths of Motions" Related to the Total Cycle (cm) | |
	Assembly Task	Left	Right	Left	Right
Example 1	Drive screws into a rectifier	14	124	64	310
Example 2	Assemble contact springs	47	45	614	535
Example 3	Assemble a crank	54	91	633	1057

Source: Laurig (1982).

Table 3.5.2 Comparison of the Examples Given in Table 3.5.1 with Regard to Different Cycle Times

	Number of "Standard Elements" or "Basic Motions" Related to the Predetermined Total Cycle Time		Sum of "Distances Moved" or "Lengths of the Motions" (cm) Related to the Predetermined Total Cycle Time	
	Left	Right	Left	Right
Example 1	0.72	6.38	3.29	15.90
Example 2	2.80	2.68	36.60	31.80
Example 3	2.32	3.92	28.60	45.60

Source: Laurig (1982).

There are minima in muscular strength in the horizontal plane through the knee joint and elbow joint.

The simple examples given in Figures 3.5.1–3.5.3 show that even physical work load must be evaluated in relation to human capacities and abilities. The utility of such arbitrarily constructed indexes as shown in Tables 3.5.1 and 3.5.2 still seems to be limited. Therefore, there might be a certain interest in basing the derivation of such indexes on more scientific physiological and psychological criteria. Following the concept of the degree of control, the basic motions can be classified into mainly proprioceptively (Åstrand and Rohdal, 1977) or mainly associatively controlled. With regard to the terminology of work physiology or ergonomics, the physiological terms "proprioceptive" and "associative" can be related to the terms "mainly muscular work" and "mainly sensorimotor work" (Rohmert, 1983). Consequently, it is possible to classify in terms of PMT systems, for example, the basic motion "reach" as "muscular work load" and "grasp" as "sensorimotor work load."

There are similar human-related restrictions if we do consider not physical work load but mental work load (Moray, 1979). To describe mental work load, Moray (1979) differentiates load measures as normative measures, performance measures as empirical measures, and effort measures as physiological measures. Moray shows that there would seem to be two empirical starting points to the definition of mental load. One might regard error and latency scores as indexes of work load. If errors rise or latency increases, the task is more difficult for the worker/operator. In some respect this would be expected from a priori theories. Both information theory and signal detection theory, for example, would lead one to expect such a relation. Empirically the picture becomes more complicated owing to the human operator's ability to choose a speed–accuracy trade-off point in the light of one's under-

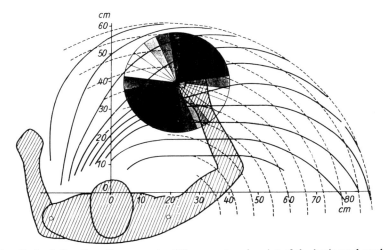

Fig. 3.5.1. Optimal direction of movements of the arm at each point of the horizontal working area.

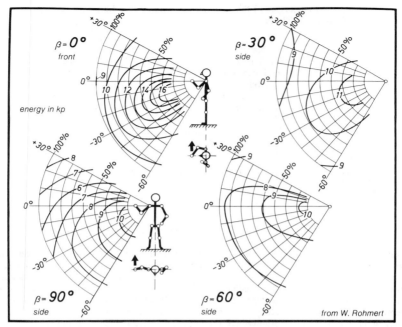

Fig. 3.5.2. Isodynes, that is, lines of constant maximal strength, of the right arm exercised horizontally in the direction of the back of the hand (strengths expressed in kilograms).

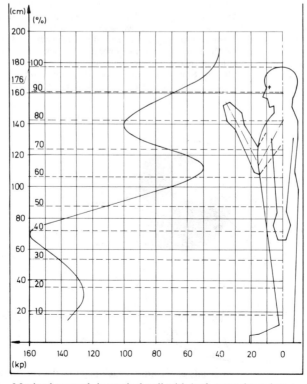

Fig. 3.5.3. Maximal strength in vertical pull with both arms dependent on grip height.

standing of the situation, that is to say, how one imagines causality to be directed, what one sees as a payoff structure, and so on. But at least in situations where errors and latency are positively correlated, it is certain that the operator's load is increasing. The payoff structure might be regarded as part of the boundary conditions, and in order to predict performance accurately it must somehow be stabilized.

Moray (1979) explains that the second starting point might be to take seriously the concept of effort. This term seems to mean human capacity-related concepts or subjective-feelings concepts, experienced by the worker/operator. But we cannot be sure that in real-life tasks low-load monitoring tasks do feel stressful, too, unless they are associated with danger. However, one question that Moray (1979) poses that must be answered is the direction of causality between, for example, physiological state, effort, and performance. Here we must ask, does load change the physiological state, or can the state alter the load? If there is an optimal level of arousal, what exactly does that mean? Is the net average information transmission rate hoop-shaped with respective arousal? Or is it merely the feeling of effort that is hoop-shaped? If a person changes his position on the arousal continuum, does that person feel more or less stress, feel more or less efficient, or in fact become more or less efficient?

In defining work load, Jahns (1973) divides the broad area of work load into three functionally relatable attributes: input load, operator effort, and performance or work result. Input load concerns factors or events external to the human operator whereas operating effort is internal to the human operator. In my own concept (Rohmert, 1983) I use the term "stress." Performance is traditionally defined as purposeful data outputs generated by the human operator which serve as inputs to other components of the human–machine–environment system and may provide feedback on effort adequacy. Input load may be task-induced or situation-induced. Situation-induced or environmental variables are noise, vibration, climate, illumination, social contacts, and so on. As far as effort is concerned, we speak of "strain" in our concept. Also performance or work result measures can be regarded as strain.

3.5.3 HISTORICAL REVIEW OF PHYSIOLOGICAL RESEARCH ON WORK LOAD

Work load cannot be measured in strictly physical terms, as is immediately evident when the following types are listed:

Heavy dynamic muscular work

One-sided dynamic muscular work (work of small muscle groups)

Static muscular work

Mental work (with strain on concentration and attention)

Uniform work with monotonous surroundings

Additional effects of atmospheric conditions (temperature, humidity, ventilation, radiation, clothing, acclimatization)

Strain due to other environmental influences (e.g., *physical*—noise, glare, lack of light, vibration; and *sociological*—person–person problems or influences, individual and group problems)

To evaluate work load completely it would be necessary to take into account how such factors combine to affect humans at work. This implies that evaluating work intensity does not mean measuring "work" but rather the "load" in relation to objective data relating to the activity of humans in industrial performances, which can be measured in physical terms or in terms of work study or which can only be described by words, reports, sketches, drawings, or pictures. In addition to the heaviness of work load and the nature of work, work load is also influenced by its duration. The determination of all demands of the work according to its nature, heaviness, and duration and its assessment for the purpose of better physiologically adapting humans to work and/or work to humans can be effected by various methods. As with job evaluation it is essentially a two-stage process. Initially, in each case the nature of the assessment of the demands of the worker's operations must be determined—a qualitative process. Secondly, a quantitative ranking is necessary for the demands, which then enables the particular operation to be contrasted with other work at other work places or other processes.

Assessment of the demands that the work makes on the worker can be effected summarily or analytically, the two methods differing by the degree to which the individual types of demands are subdivided.

For the quantitative ranking of the demands and for determining work load numerically, the principles of ranking and stepwise classification are used. In ranking, operations are arranged in one or more sequences in such a way that those requiring the highest evaluation are at the peak of the sequence or series and those with the least evaluation are at the end. The second principle, that of stepwise classification, is characterized in that first of all stages with different evaluations are fixed, and into these work intensity is classified.

By combining the summary and analytical methods of consideration on the one hand with the principles of ranking and stepwise classification on the other, four basic methods are obtained (see

Table 3.5.3   Methods of Arranging Industrial Operations According to Work Load Demands Made on the Worker

Method of Quantitative Classification of Work Load	Method of Qualitative Comprehension of the Demands that the Activity, Workplace, and Working Process make on the Worker	
	Summary Consideration	Analytical Consideration
Ranking	Drawing up a work load sequence (one single load series for all activities, workplaces, and working processes)	Drawing up a number of demand sequences (one sequence for each type or nature of demand)
Stepwise classification	Drawing up work load class (with one single load class scale for all activities, workplaces, and working processes)	Drawing up load class (with a demand class scale for each type of demand)

Table 3.5.3), to which, or to combinations of which, almost all usual methods of assessment of work load can be attributed. In each case, groups of tasks with the same or similar work load requirements are obtained.

The French chemist Lavoisier discovered as long ago as 1789 that energy expenditure increases during human work and proposed to compare, by means of oxygen consumption measurement, the physical and material amount of the activity of a philosopher, author, or composer with that of a heavy muscular worker. This was possibly the first proposed scientific job evaluation system, but 100 years later it was admitted by physiologists that any assessment of mental work load based on oxygen consumption was futile, because neither the limits of mental capacity nor mental fatigue could be determined. Research workers all over the world agreed to apply Lavoisier's principle only to muscular work. Energy expenditure in more than 100 different activities and more than 100 professions has been measured and today we have tables of results in several languages (Passmore and Durnin, 1955; Spitzer and Hettinger, 1964; Katsuki, 1960). In these excellent reviews are given more than 200 references to the experimental work of research groups or schools of work physiology all over the world.

Table 3.5.4   Estimating Table of Energy Expenditure

Position Body or Movement (A)[a]	Net Energy Expended (kcal/min)	Type of Work (B)[b]		Net Energy Expended (kcal/min)
Sitting	0.3	Hand work	Light	0.3– 0.6
			Medium	0.6– 0.9
Kneeling	0.5		Heavy	0.9– 1.2
Crouching	0.5	One-arm work	Light	0.7– 1.2
			Medium	1.2– 1.7
Standing	0.6		Heavy	1.7– 2.2
Stooping	0.8	Both-arms work	Light	1.5– 2.0
			Medium	2.0– 2.5
Walking[c]	1.7–3.5		Heavy	2.5– 3.0
Climbing (without load, inclination > 10°)	0.75 per meter height	Whole-body work	Light	2.5– 4.0
			Medium	4.0– 6.0
			Heavy	6.0– 8.5
			Very heavy	8.5–11.5

[a] In most cases estimates of A are very easy.

[b] When starting with estimate B (type of work), one should use the average value; later, with improved skill, estimates can be made in finer steps.

[c] In walking, the first value is for a speed of 2 km/hr, and the second value for 4.5 km/hr.

Estimated energy expenditure = A + B.

Table 3.5.5   Examples of Estimating Energy Expenditure of Work

Activity	Net Energy Expenditure (kcal/min)			Duration of Activity (min)	Energy per Activity (kcal)
	A	B	Total		
Raking leaves on a turf (slowly walking, light work with both arms)	1.7	1.5	3.2	25.0	80.0
Planing planks (stooping, light whole-body work)	0.8	3.5	4.3	3.6	15.5

Although the number of determined activities and professions is high, nevertheless there is no complete survey. Therefore we resort to using estimating tables similar to that shown in Table 3.5.4, which is valid for an average man 175 cm body height, 70 kg body weight, and 30 years of age, with two examples of the use of this table being shown in Table 3.5.5.

Work physiology can now classify work load, but to what end? It is suggested that by undertaking experiments the following three questions might be solved.

3.5.3.1   How Much of a Worker's Food Consumption Is Due to His Work?

In several experiments the measurement of energy expenditure by means of a gas meter (Figure 3.5.4) has proved to be a simple and reliable method for determining the nutritional requirements for specific activities in a profession, in the realm of sports, or in the military area. In some nonvoluntary large-scale experiments it was possible to show to what extent a lack of food is responsible for a lack in production of a whole nation or a group within the population (e.g., mining, iron, and steel industry;

Fig. 3.5.4.   Gas meter of the Max-Planck-Institut für Arbeitsphysiologie Dortmund to determine energy expenditure in the field.

Kraut and Keller, 1961). In a similar vein Kerkhoven (1962) showed, after contrasting the food intake and energy expenditure of Dutch and Nigerian workers, that it would be desirable not only on humanitarian but also on economic grounds to increase the food intake of the latter. It is in very heavy work requiring an energy expenditure of 4000 kcal/day that a reduction in food effects the greatest decrease of production, and if food intake is only 3400 kcal/day, production decreases to about 70%; this figure is reduced to 55% when the food intake is only 3000 kcal/day. There is also some influence in light work; the energy required for rest or leisure can be reduced only a little.

By applying these considerations to a whole nation—including all heavy muscular workers, children, and elderly people—one can calculate that an average amount of food intake of 3000 kcal/day per person is required, whereas an amount of 1800 kcal excludes any physical work worth mentioning. The efficiency of food, that is, the ratio between the economically profitable and the total calories, reaches a maximum at 3000 kcal/day. Higher values lead to a decrement in efficiency owing to the additional food intake not being realized into work but rather into body weight increment. Too low nutrition of a nation is of course also inefficient owing to the decreasing ratio between production and food intake. Our existing knowledge of the balance between energy inputs and outputs means that it is now possible to give advice about nutrition to the governments of developing countries.

3.5.3.2 For What Type of Work or Type of Tool Is the Energy Requirement a Minimum?

Research work in this area has brought a good deal of knowledge about such basic activities as walking, climbing, carrying a load, weight lifting, cranking, and cycling, and a lot of practical rules have been learned from this, such as the following:

Work with optimal speed.
Lift as little body weight as possible.
Use body weight or inertia as counter-force.
Never drop on the floor work pieces such as bricks that have to be piled up again.
Deliver work pieces to a height of at least 0.5 m! And so on.

Experiments on the transport of loads by muscle power have shown that it is possible for the daily output of a human to be doubled by halving the energy consumption for the same pile.

From those general rules a number of suggestions for mechanization in industry emerge. Knowledge of energy requirements and efficiency is furthermore an important basis for all climatic–physiological calculations.

3.5.3.3 What Is the Suitable Amount of Daily Energy Expenditure?

Although it would appear useful to apply engineering methods of calculating energy requirements or degrees of efficiency to the problems of humans at work there are certain basic limitations. It seems reasonable to believe that suitable energy expenditure would have a certain maximum value for a similar group of people and therefore it would be worthwhile discovering by how much the work in industry or forestry or agriculture exceeds this value. However, it is an error to think that each increment of the degree of efficiency will facilitate work; the almost unlimited endurance of machine capacity has been applied also to humans and their work. By evaluating many results and experiences with high energy requirements a maximum value of 2000 kcal expended per 8 hr of work has been deduced (Rohmert and Rutenfranz, 1983).

Engineers became very interested in this figure, thinking that the necessity for interruptions in work for energy reasons arises from the fact that the human body has over fairly long periods of time only about 2000 kcal at its disposal to expend on heavy muscular work when using large groups of muscles. An increased output could be accomplished only with difficulty by the absorption of food and is scarcely found to be exceeded in practical industrial investigations. On the basis of this daily limit of energy transformation, which is acceptable from energy considerations, for those operations whose energy transformation exceeds this limit, the work-free time to be allotted during the work time in order that 2000 kcal/shift shall not be exceeded can be calculated (REFA, 1958).

From more recent investigations (Gupta and Rohmert, 1964; Rohmert, 1963), however, we know that keeping to this limit of the daily transformation of energy does not in fact ensure that each prolonged work period is free from fatigue. Accordingly the allowances calculated for limiting the daily energy transformation should not be described as relaxation allowances. The German Work Study Federation REFA has excluded the formula of calculation of work-free time in the current edition of the "REFA-Buch."

The decisive factor for the acceptability of muscular work is the adequate provision of the oxygen consumption of the muscles throughout the working period. It comes as no surprise, therefore, to find that at no time has static muscular work fit into the concept of a maximum suitable limit of energy expenditure. In any professional activity static muscular contractions are involved, which can cause muscular fatigue with very low levels of energy expenditure. To cope with this difficulty in the past, work physiology did with static muscular load what was done with mental load, that is, excluded it from considerations. It took an additional 50 years to remedy this situation (Monod, 1956; Rohmert, 1960c; Scherrer and Monod, 1967).

With regard to suitable work intensity, work physiologist Müller and his colleagues (Karrasch and Müller, 1951; Müller, 1953) evolved the concept of endurance limit (EL). EL is a physiological limit that characterizes the ability of the human body during muscular work to bring various inner balances into the highest possible "steady-state" of muscular work. EL can be expressed in terms of endurance time, energy expenditure, heart rate, or other physiological terms. After numerous investigations, the behavior of the heart frequency in the course of the working and resting periods of the day has proved to be the safest measure for fulfilling the requirement of EL (see also Monod, 1967). From investigations of heart rate we know that fatigue and suitable work intensity are not linked unconditionally to the limit of energy expenditure of 2000 kcal/day. This is shown, based on a more detailed conception of acceptability of work intensity, in Table 3.5.6.

Efficiency is almost the same in cycling and pedaling, but the mass of the working muscles is three times higher in cycling, and consequently we find it to be three times higher when expressed in either mechanical work or energy expenditure terms. A similar muscle mass is used in cycling and shoveling with a light shovel and, although efficiency is different, the energy expenditure is the same when using a heavy shovel with a higher load and a smaller number of throws; however, there is a high static load of muscles involved and the consequent decrease of blood flow through these muscles leads to a reduction of work in physics or energy expenditure terms.

These examples serve to illustrate three main influences on what constitutes acceptable levels of work intensity:

1. Amount of working muscle mass
2. Type of work (static or dynamic muscular work)
3. Speed of movements

In practical working situations one cannot analyze the extent to which heavy muscles are working statically or dynamically, but knowledge of this is necessary in order to evaluate whether blood flow is sufficient or not. Neither can muscle mass nor blood flow be determined in practical work situations; hence the idea of Lavoisier is not suitable for evaluation of work intensity. Only by analyzing the physiological conditions of muscular fatigue and recovery can we deduce suitable work load.

Of all physiological criteria used for measuring fatigue the behavior of heart rate stands out. Christensen (1931), Asmussen, Christensen, and Nielsen (1939), Karrasch and Müller (1951), Brouha and Harrington (1957), Monod (1967), Rohmert (1960a, b) and others have studied the correlation between muscular fatigue and heart rate. Relating heart rate and energy expenditure, Figure 3.5.5 shows the main differences that are important in any physiological evaluation. The figure shows the nonlinear relation between energy expenditure and heart frequency caused by the type of work, speed of movements, and amount of working muscle mass. There is a further nonlinear influence of the duration of work that is not shown in the diagram. If working time were included each curve of Figure 3.5.5 would have several parameters, which would show that the longer the period of work, the steeper the increase in heart rate.

Table 3.5.6 Suitable Daily Work Intensity in Various Activities of Men

Type of Work	Suitable Work in 8 Hr		
	Degree of Efficiency (%)	Total Amount of Work (mkp)	Energy (kcal)
Cycling (60 revolutions/min)	23.5	200,000	2000
Cranking (both arms, 45 revolutions/min)	19.5	100,000	1200
Pedaling (one foot, 60 treads/min)	23.0	57,000	600
Shoveling (22 tosses of 4 kp/min with a light shovel)	5.0	43,000	2000
Shoveling (4 tosses of 11 kp/min with a heavy shovel)	3.0	22,000	1700

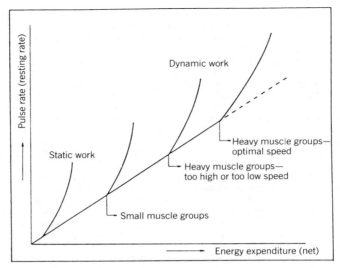

Fig. 3.5.5.   Heart rate and energy expenditure in muscular work.

The historical review of physiological research on work load shows the necessity of considering humans as the measure of work. Under these aspects, physiological and psychological work load measurement and analysis could be described in terms of ergonomics. In ergonomics physiological and psychological work load measurement and analysis are based on the concepts of work, stress, and strain.

3.5.4   ERGONOMICS CONCEPT OF ANALYZING WORK, STRESS AND STRAIN

3.5.4.1   The Concept of Ergonomics

Ergonomics (as a science and as a technique) is related to the analysis of problems of people in their various working conditions within their real-life situations. Ergonomists try to analyze these relations, conditions, and real-life situations with the aim of harmonizing demands and capacities, pretensions and actualities, longings and constraints. The aim of this section is to explain the needs of ergonomics research within the four main tasks of analyzing, measuring, evaluating, and designing. Ergonomics is based on the theory of the human-at-work system and the stress and strain concept.

It may often be assumed that research in ergonomics is difficult because the very concept of ergonomics seems to consist of a conglomerate of various sciences. Despite this complexity there is a common aspect that can be represented in the study of human needs, characteristics, abilities, and skills as applied to the design, production, and management of high quality and humane products or services, within a given work environment and social surroundings. In order to achieve a central goal, ergonomics requires investigations into the human being, the existing human living and working situations, as well as a certain understanding of the functional necessities of future human work conditions.

Time budget studies have been described by Graf and Rutenfranz (1958). Such a study determines the amount of time spent working, traveling to and from work, and sleeping, as well as that time classified as leisure time in a social context and that time that cannot be exactly classified. This is done for every half hour of a 24 hr day. The results are expressed as a percentage of the total time.

Figure 3.5.6 shows the results of an analysis of 141 observed midday shifts at 10 different computer centers. It is quite obvious that time, duration, and location of work determine how the other time elements are distributed within a working day. It is also apparent that work during the midday shift in this example leaves the worker less free time, sometimes even split up.

If one questions the general value of such an elementary diagram, one should remember that purposeful work means more than simply the performance of daily activities. The idea includes the worker understanding and accepting the work as that which it really is, "tilling the soil"—that work means both happiness and toil, satisfaction and sweat on the brow, pleasure and the expenditure of energy. The reality of work would be lost if one of these preceptions of work were suppressed. Free time and sleeping time would be void of their very meaning. The repose between working hours means more to the worker than simply time not spent working. The outcome of work is of daily significance throughout a lifetime. When the working day is viewed in this way, the worker is not

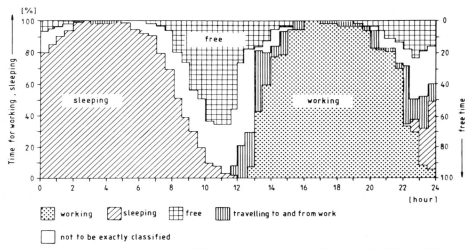

Fig. 3.5.6. Time budget study (10 different computer centers, 141 observed midday shifts).

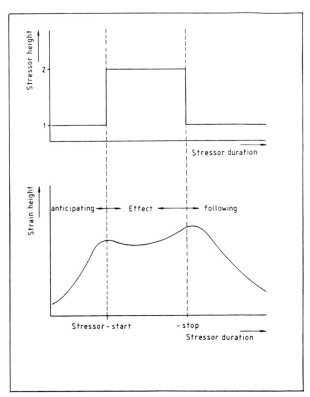

Fig. 3.5.7. Example of possible strain reactions caused by stress.

413

just seen as locked up in a—however pleasantly designed—production system, but rather as an individual. The ergonomist, who strives for the exploration and analysis of human work, must not fit this rude image of a social engineer.

The main point that should be obtained from Figure 3.5.6 is that isolation of the elements should be avoided because every element exercises influence on the work day. This infers that

1.   Every elementary work analysis should be repeated in order to determine the type and extent of reactive and anticipated effects on the worker; this infers

2.   That analytical methods should be well managed by the researcher in order to avoid artifacts. Ergonomic laboratory experiments cannot satisfactorily simulate work and life realities. Test subjects in laboratory and field experiments are considerably different from a worker in his living and working environment, and this infers

3.   That the results obtained through research are falsely interpreted if the independent variables are related to the dependent variables in various ways as seen by the researcher and the test subject.

A solid scientific knowledge of ergonomics is required in order to apply adequately the methods and interpret meaningfully the results given in Figure 3.5.6, keeping in mind the bearing they have on the real objects of these results, that is, the worker, the employer, and the entire social surroundings.

If we pick out one element of the worker's daily life, this element can show a plurality of effects. Fig. 3.5.7 gives an example: if there is a time series of one stressor with two distinct heights, the

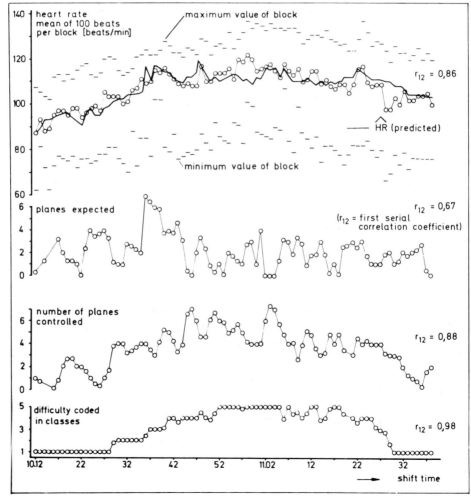

Fig. 3.5.8.   Variation of heart rate and stress factors.

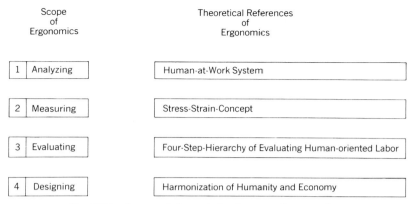

Fig. 3.5.9.   Scope and theoretical references of ergonomics.

reaction to this stressor—called strain—must not follow the transient response; this is because other stressors could be neglected. This we found in a study of air-traffic controllers (approach control—pickup position at Frankfurt airport) that a maximum reaction of heart rate (upper time series in Fig. 3.5.8) did not occur at the highest degree of mental load (the two lowest time series in Fig. 3.5.8) but at the top of the time series for the number of aircraft expected.

The general value of this result means that all conclusions from experimental findings are wrong, if they ignore important stressors or a distinct feedback from another distinct stressing element of the working day. This means that using only ergonomics analytical procedures does not follow the workers' real-life complexity; many more epidemiologically oriented procedures must be regarded.

3.5.4.2   Scope and Theoretical References of Ergonomics

Obtaining benefits from ergonomics means analyzing, measuring, evaluating, and designing of human-at-work systems (Figure 3.5.9). The theory of the stress–strain concept ensures the ergonomic tasks of measuring and evaluating human labor that kind of dignity which respects man's individuality. If a four-step hierarchy is considered, human labor is evaluated with respect to the criteria of ability, tolerability, acceptability, and satisfaction with the aim of combining the human and economic share contributes to the human-at-work system.

The problems in ergonomic research can be compared to those of the social sciences, for example, psychology (Schneewind, 1975, see Figure 3.5.10). The psychological exercises of description, explanation, prediction, and alteration are comparable to the ergonomic processes of analysis, measurement, evaluation, and design. Even without describing this comparison completely it is obvious that ergonomics is strictly oriented toward a quantitative goal.

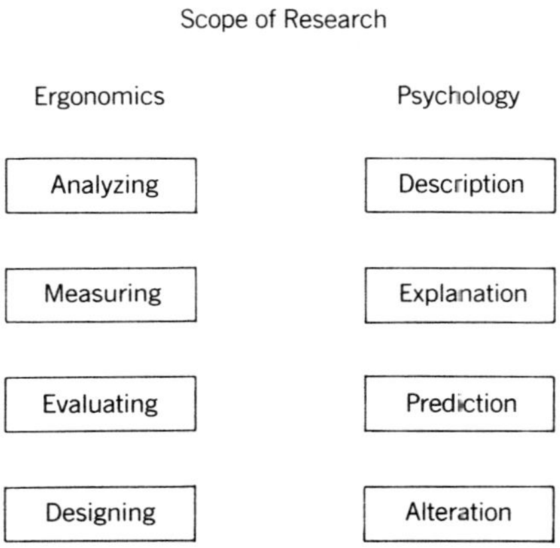

Scope of Research

Ergonomics   Psychology

Analyzing   Description

Measuring   Explanation

Evaluating   Prediction

Designing   Alteration

Fig. 3.5.10.   Scope of research in ergonomics and psychology.

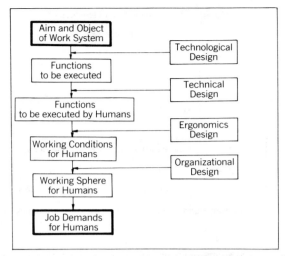

Fig. 3.5.11. Influences to job demands for a human in a working system by different design areas.

Ergonomic analysis and measurement is based upon the theory of the human-at-work system and the relationship between the requirements placed on the worker and the resulting strain. The requirements are not only influenced by the technical sophistication of the work system. Four different design areas are important for the analysis of work (Figure 3.5.11). Objective demands on people result from the aim of the work system and its design in a four-step process. *Technological design* selects the type of technique and fixes the basis for production capacity and general working conditions. *Technical design* is related to the utilization of technical equipment based on applied technology. The result of this design is the functional work partition between human and machine. Separating humans from the work process makes them dependent on the influences of technological and technical design. *Ergonomic design* covers all aspects of adapting work to humans, in particular to their abilities, capacities, and needs. The main objectives of ergonomic design are improved working conditions and increased output by humans. *Organizational design* means the division of labor, and problems of specialization and job enlargement have to be solved. In addition, consideration must be given to problems of the dependence of the different tasks on each other, on problems of working time and rest pauses, the daily position of work and rest, and problems of shift work, and so on.

Job demands can be described in a very general human-at-work system. This is defined as a model of the relations between humans and their tasks (see Figure 3.5.12). The model is illustrated for a very complex human task, namely, the task of air-traffic controlling. This task may be considered as a special type of a human-at-work system with many different physiological and psychological work

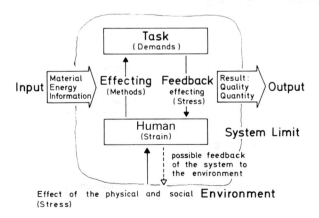

Model of the Working System

Fig. 3.5.12. Model of the human-at-work system.

load components. The task of the controller is to give external instructions to a pilot in an aircraft and to maintain internal coordination with other controllers. Besides input measures such as presented information about the aircraft to be guided and energy to handle miscellaneous material and technical equipment, the controller needs methods for solving the task gained by training and experience. With these methods he or she is effective in the work system. By solving the task, the controller is subject to demands that present certain resistances (work load or difficulties) and that must be overcome, thereby obtaining feedback. The sources of this feedback, which influences a person performing a task, may be called stressors. Stressors depend not only on the work load and the difficulties of the task and its duration, but also on the work environment with its physical and chemical components (e.g., climate, noise, illumination) and its social components (e.g., leadership, management relations, communication problems with other controllers or staff) which are effective as stress components both within and outside the working system limits. Within the person, stress leads to a distinguishable strain, depending not only on stressors but also on different individual controllers' capacities, abilities, or skills. The result for the controller's task is shown in components of both quality and quantity of the control performance. By the system limits (i.e., the controller's workplace or his functional area), the relations between humans and work are separated from the environment.

Such a general and rough description of the working system "controller's performance" provides a possible basis for detailed work analyses.

With regard to the controller and the control function there are three starting points from which to analyze the controller's input and his or her share in the air-traffic control system productivity (see Figure 3.5.12):

1. The demands of the task (and all concrete elements of the working system), which means a special kind of task analysis or job evaluation.
2. The qualifications of the air-traffic control officers, which determine differences between the methods of working by experienced controllers and trainees, respectively.
3. The capacity of the air-traffic control officers (as a human-related part of the working system, with regard to the quality and quantity of the system's output).

To each of these starting points for analyzing human performance, there is attached the disadvantage that the feedback that influences a person doing work is neglected. All these starting points neglect the analysis of strain. If, however, an air-traffic control system contains human input and if future developments are going to need the controller's input too, it seems really necessary to consider the strain put on humans and its contribution to the system's reliability. This is one reason for analyzing strain in more detail.

The theory of the stress–strain concept indicates that strain cannot be determined wholly by a consideration of the specific work load. Strain also depends on the individual characteristics, abilities, skills, and needs of the working person. The cause–effect diagram given in Figure 3.5.13 is valid only for analysis of work processes that are specifically determined for the worker. This concept has been expanded for the past 10 years for the purpose of ergonomic field research and scientific instruction whenever human activities at occupational workplaces can possibly be executed in different manners. In this view the external stress, which is found in a purely objective analysis and which the individual person is confronted with and required to cope with, is accompanied by the working person's individual possibilities of mastering the objective stress in a specifically subjective way. Thus the effects of stress cannot only be found objectively and on a bottleneck basis, that is, from physiological and biochemical data or from personal experience, but also they must include data of performances and efforts.

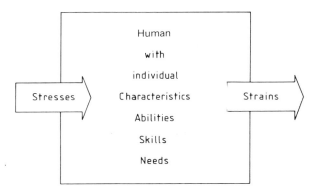

Fig. 3.5.13. Relationship between stress and strain.

3.5.4.3 The Ergonomic Stress and Strain Concept

The scheme in Figure 3.5.14 shows these considerations during the first step of a job relevant work analysis.

The work tasks of either producing things or rendering services are done under certain conditions of performance and environment, for example, the dimensions and factors inherent to specific tasks and situations involve objective work demands that can be described by a profile of the necessary muscular and mainly nonmuscular demands. The difficulty, the duration, and the composition (e.g., the simultaneous and successive distribution of these specific demands) result in the combined stresses which all the exogenous effects of a working system exert on the working person. Stress is found in a performance-specific work analysis to be merely relevant to working objects and situations.

The discussion of the stress–strain concept begun with Figure 3.5.14 is continued in Figure 3.5.15. The performance-relevant work analyses are followed by a performance regulation. The combined stresses are classified along ergonomically defined kinds of stresses, which are actively coped with or passively put up with, depending on the specific subjective behavior of the working person. The active case involves activities directed toward the efficiency of the working system, whereas the passive case induces reactions (voluntary or involuntary reactions) that are mainly concerned with minimizing stress. The relation between stress and activity is decisively influenced by the individual characteristics and needs of the working person. The main factors of influence are those determining performance from the field of motivation and concentration, and those from the field of disposition, which are mostly referred to as abilities and skills.

It is only in the special case of static muscular work of a single group of muscles that task, demand, stress, and performance are identical. But as soon as another group of muscles is involved in static work, there is a physiological scope of performance that is called into action. For this case, we are able to prove that a parallel use of groups of muscles prolongates the maximum endurance as the stress is shifted among the groups of muscles more frequently. We also could demonstrate the same effect for dynamic ergometer work. For constant ergometer work we found both variations in the conditions of performing ergometer work (variations of pedaling speed and strength) and subject-relevant variations in the behavior of the ergometer driver (e.g., different use of body weight while pedaling). Because self-regulation is already employed in strictly dynamometric and ergometric muscular work, we may rightly expect it to be involved in mainly nonphysical work with higher cognitive shares. Hacker (1976) has proved that self-regulation is employed in these cases as well.

When these findings are transferred onto the enlarged stress–strain concept, feedbacks turn up: these feedbacks influence the conditions under which a task is performed, as the specific dimensions and factors that are involved in the accomplishment of a task are altered through self-regulation; in addition these feedbacks alter the working person's subject-relevant behavior both in the active way of coping with stress and in the passive toleration of stress; finally, we may also expect the individual characteristics and needs of the working person to be influenced by feedbacks. The corresponding directions of feedback are outlined in Figure 3.5.15, whereas they were omitted from Figure 3.5.14 for the sake of clarity.

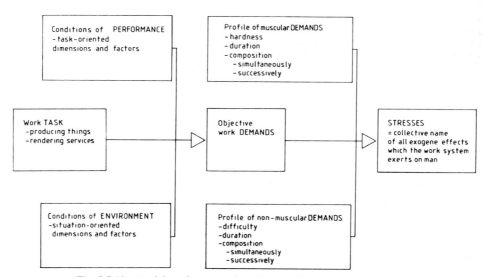

Fig. 3.5.14. Activity-relevant work analyses in the stress–strain concept.

In the stress–strain concept, the possibly regulated or reactive activities supply measurable performances.

Figure 3.5.16 completes the stress–strain concept for the strain exerted on the working person within the working system. The stresses relevant to behavior, which are manifest in certain activities, cause individually different strains and are defined by the reaction of physiologic or biochemical indicators—functional dimensions of the employed physiological organic systems. Because these indicators can be perceived they are susceptible to psychophysical scaling, which represents the strain as experienced by the working person. In a behavioristic approach, the existence of strain can also be derived from an activity analysis. This approach is mainly employed when activities are predominantly determined by the working person and not by strain. The advantage of the physiological–biochemical approach versus the behavioristic and psychophysical approaches may be seen in the reduced possibility of voluntary influence in the working person. The intensity by which indicators of strain react upon stresses or activities and performances is determined by characteristics and characteristic curves that are derived from the functions of the corresponding organic systems and comply with the properties and needs of the working person.

Despite constant stresses the indicators derived from the fields of activity, performance, and strain may vary in their temporal extent. Such temporal variations are to be interpreted as processes of adaptation of organic systems, whose positive effects cause a reduction of strain/improvement of activity or performance (e.g., through training); in the negative case, however, they result in increased strain/ reduced activity or performance (e.g., fatigue, monotony). The negative effects are likely to appear, if so-called endurance limits are exceeded in the course of the working process. The positive effects may come into action, if the available abilities and skills are improved in the working process itself, for example, when the threshold of training stimulation is slightly exceeded. Thus adaptation proves to feed back upon the individual characteristics and needs of the working person as a result of the strain situation. For reasons of clarity these directions of feedback are not outlined in Figure 3.5.16.

When the process of adaptation is carried beyond defined thresholds, the employed organic system may be damaged so as to cause a partial or total deficiency of its functions. Typical examples of such damages are noise-induced hearing loss and occupational hazards, such as a multitude of chemical

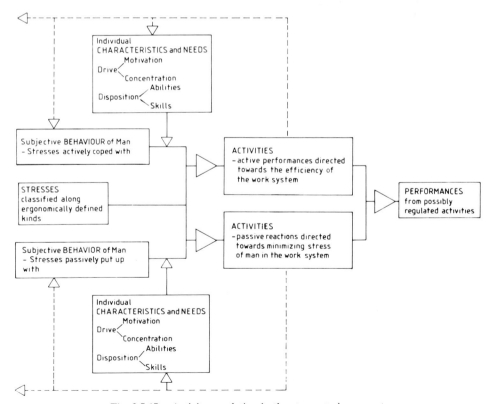

Fig. 3.5.15. Activity regulation in the stress–strain concept.

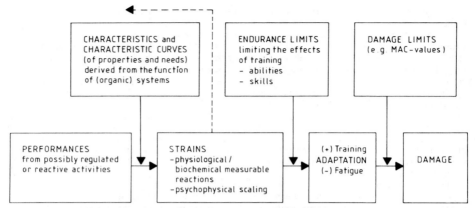

Fig. 3.5.16. Strains exerted on man in the working system.

agents, which endanger the working person's health in the sense of risk factors, as the threshold limits of maximum acceptable concentrations—MAC values—are exceeded.

The differentiated stress–strain concept, as outlined in Figures 3.5.14 to 3.5.16, permits the segmentation of the difficult general analysis of working systems into aspects that are more easily dealt with and can be related to measuring and rating approaches. A comprehensive analysis of human activities is composed of the partial analyses given in Figure 3.5.17.

The scope of ergonomics is not only related to analyzing and measuring human work, stress, and strain. It is also necessary to evaluate the result of research or the application of experience to human-at-work systems. Designing human work shows close links between economic and human aspects of the design, and it is clearly quite illogical to pursue human aims without considering economic aims of design. Remember that, in a wider sociological context, economic design aims must always be of service to humanity design aims; otherwise, in the long term they become self-destructive and pointless.

3.5.4.4 A Four-Level Hierarchy for Assessment of Working Conditions

To assess the results of the analysis of human work in the context of an overall human-at-work system, apart from the primary aims of humanity and profitability, there are four cardinal criteria, the fulfilment of which contributes to the primary aims. Therefore, the evaluation of any concrete analyzed work system must satisfy a four-level hierarchy: ability; tolerability; acceptability; and satisfaction.

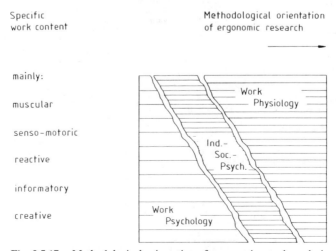

Fig. 3.5.17. Methodological orientation of ergonomics work analysis.

The ability of a human to do the job is a basic prerequisite for profitability. Human capabilities and characteristics must enable the human to do the work before the work can be carried out at all. This limit of human ability means a *level of practicability* of using human input. If the result of work analysis does not consider these maximal capacities, even if only required for a limited time by people at their work, then the ergonomist is obliged to give up all claim to human power, but instead must design a completely automatic work system. Therefore the level of practicability of human work makes it imperative that human functions are catered to (e.g., body dimensions, muscular strength, speed and accuracy of movements, sensory-organ functions). In most cases, we are not only interested in the "on off" completion of a job (as in sports), but we require a daily repetition in the work during a normal shift's duration and over the total working life's period. Under these longer lasting conditions the demands placed on humans must be lower. The highest level of demand that can be endured to the end of the normal working life without any work-related damage to health or normal human functional abilities is called the *level of tolerability*. The qualitative and quantitative output of a person affects profitability. Then there is effectiveness in the sense of general efficiency. Not all working conditions are equally conducive to quality and quantity of output or to incurring minimum costs; this, in turn, affects profitability.

The cardinal criteria for assessing the humanity of work and working conditions are health, well-being, and job satisfaction. However, it is important to see that well-being and satisfaction always include long-term tolerability of work in terms of health. Therefore the second level of tolerability in assessing work systems also makes imperative claims on the designer as well as the worker and the trade unions, or the employers and the employers' federations.

The third criterion level means that working conditions should be accepted by those who are social partners on the shop floor or at the labor agreement level. Unfortunately conditions that are not tolerable over the total lifetime can be accepted by local agreement as well as imposed by labor laws. Therefore the third *level of acceptability* of human work must respect the hierarchical level of tolerability first of all. Understandably, social and perhaps individual values have a large part to play here. Also, there are likely to be changes in attitudes in the course of time. This is shown clearly by the development of the modern slogan of "improving quality of life." There are, of course, human working conditions that were acceptable in the past or that will be accepted in times of high unemployment, which in normal conditions today are assessed to be inhumane. However, the more important postulate for humanity is to respect the conditions of long-term tolerability.

The hierarchy of assessment of analyzed human working conditions that starts with the individual level of man's ability ends at an individual level, the highest *level of satisfaction*. Both the levels between are mainly oriented to more or less homogeneous groups and not necessarily to the individual. Striving for conditions of highest job satisfaction does not necessarily ensure tolerable conditions. For example, if someone is very content with his or her working conditions it might nevertheless be that the work load is intolerably high owing to one stress factor being too high, not necessarily in intensity but in duration. This can occur if a person works too much overtime or omits rest pauses,

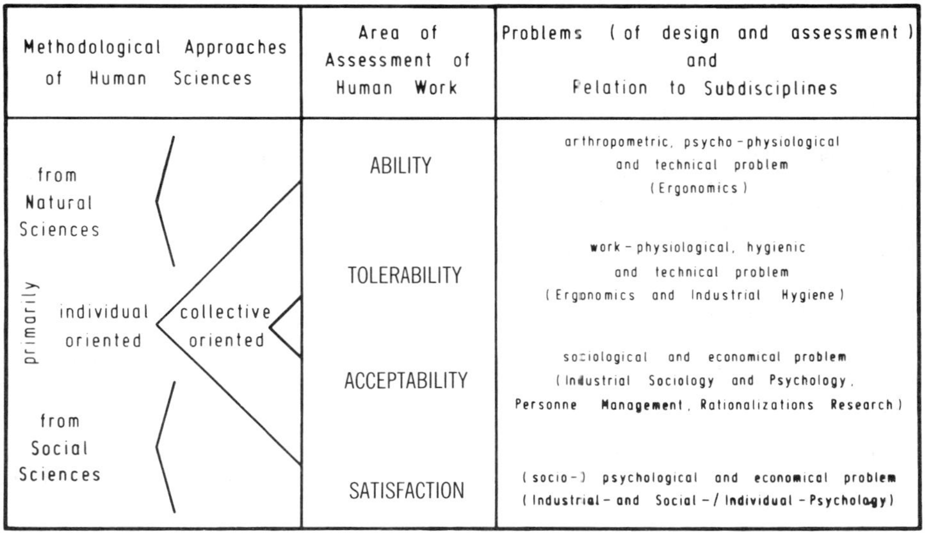

Methodological Approaches of Human Sciences		Area of Assessment of Human Work	Problems (of design and assessment) and Relation to Subdisciplines
from Natural Sciences		ABILITY	arthropometric, psycho-physiological and technical problem (Ergonomics)
individual oriented	collective oriented	TOLERABILITY	work-physiological, hygienic and technical problem (Ergonomics and Industrial Hygiene)
		ACCEPTABILITY	sociological and economical problem (Industrial Sociology and Psychology, Personne Management, Rationalizations Research)
from Social Sciences		SATISFACTION	(socio-) psychological and economical problem (Industrial- and Social-/ Individual-Psychology)

(primarily)

Fig. 3.5.18. A four-level hierarchy for assessment of human work.

recovery periods, and holidays under the illusion that he or she is doing a very satisfactory job. This example shows the importance of the basic levels in the hierarchy, as well as the necessity of setting and controlling design aims merely at these basic levels. A person does not know what is right and useful with regard to the defined tolerable conditions of health and well-being. This is shown by the examples of a person deciding when to take meals or working time schedules and rest periods. There are very critical limits to the self-regulation of human working conditions. There is a real need in disciplines, methods, techniques, and knowledge to adapt work to humans and humans to work to ensure "tolerable" working conditions. Here is a real challenge for the human-factor sciences.

Figure 3.5.18 summarizes the general principles of the four-level hierarchy of evaluating either the state of working condition or the effects of work on humans. For each of these four levels there are individual as well as collective standards that permit a rank order to occur in each case. Allied to each level we find methods and techniques from different disciplines that are involved in the problems of analyzing and assessing human working conditions. In this way we find important subdisciplines of the human sciences. This is not intended as a claim for a hierarchy of scientific disciplines; nevertheless, there is an important role for disciplines that use the methods and techniques of the natural sciences, the results of which have two important advantages in the context of human-at-work problems. Firstly, the results of applying methods of natural sciences to work analysis problems are reproducible; one can fix standards that can be measured. Secondly, these results (standards) have to stand the burden of being accepted and agreed upon, in accordance with human laws, as in mathematics. These two facts are, of course, extremely important in the application of knowledge to the field of work analysis where policies based on sociopolitical interests exist. In this context, ergonomics is classed with the natural sciences and plays an important role in fixing standards of human work based on the evaluation of the first two levels of the hierarchy, namely, the levels of ability and tolerability.

3.5.5   AN APPROACH TO ASSESS WORK LOAD IN ASSEMBLING TASKS USING ERGONOMIC METHODOLOGY

With the ergonomics concept of work, stress, and strain in mind, an example in assessing work load in assembling tasks is presented here (Rohmert and Martin, 1984). Ergonomics methodology used in this example is given in Table 3.5.7. In this table the first column shows the conditions of work (task-related, human-related, related to activity, performance, and strain, and finally with regard to the result of the assembling system and the consequences for the assembling operator). The succeeding columns of the matrix contain the methods of assessment of work load (r, rough; f, fine), the items of description, and the criteria of evaluation of the working conditions. The last column of the matrix contains references to the literature, which we used in our field studies.

Table 3.5.7 is based on the experience of more than 20 years of field studies. The purpose of these studies is to avoid high strain and low capacity through work design oriented towards future human-at-work systems. Also, these methods are used to redesign existing human-at-work systems.

The references cited in the table are related to many field studies that were carried out in my department. I thank the managers in the field

Who support ergonomic research

Who are willing to apply the results derived from the field studies

Who take the initiative to train staff members, engineers, doctors, and work study people in the broad field of ergonomics

Unfortunately the results have not been published in English, but only in German. By reading the references one could get an impression of the sociopolitical situation in West German industries. Many studies mentioned were financially supported by the Ministry for Research and Technology of the Federal German Republic. By the results of the studies it will be shown to what extent ergonomics helps in increasing capacity and in limiting strain. It is the real aim of ergonomics to harmonize profitability and humanity.

3.5.6   SUMMARY

Work load measurement as well as work load analysis is the task of work study. The ergonomic approach to these problems is oriented toward the stress and strain concept. An historical review of physiological research on work load was given. It was shown that not only physiological and experimental methods are useful. Different work load measures can also be ranked.

The concept of ergonomics means that work load can be expressed by stress measures. However, it is very important to know that these stress measures are evaluated through strain measures. The difference between stress and strain is that strain depends not only on objective data, but is also influenced by the characteristics and needs of the individual worker.

As a result of roughly 20 years of field studies a matrix is presented that shows the dependence

Table 3.5.7 Matrix of Conditions and Criteria of Evaluation of Work and Assessment of Work Load Used In Ergonomics Field Studies Related To Assembling Tasks

Conditions of Work	Methods of Assessment of Work Load (r, rough; f, fine)	Items of Description	Criteria of Evaluation of the Working Conditions	References
Task-related				
Task	AET (r) (task analysis) Analysis of job plan, priority graph analysis (f)	(see AET)	(see AET)	Rohmert and Landau, 1979 Fuchs, 1971
Object	AET (r) Measuring ergonomics dimensional parameters (f)	Weight, shape, and size of object	(see AET) Relation to maximal strengths	Rohmert and Landau, 1979 Rohmert and Jenik, 1972
Workplace layout	AET (r) Somatographic studies (f)	Inner and outer measures of workplace, geometry of sight	(see AET) Body dimensions of workers, body position, posture, position of articulations, etc.	Rohmert and Landau, 1979 Jenik, 1972; Martin, 1981; Jenik, 1975
	Biomechanical studies (f)	Handling forces, movement distances, and speeds		
Linkage of work	AET (r) Somatographic studies (f)	(see AET) Reaching area	(see AET) Body dimensions of workers	Rohmert and Landau, 1979 Jenik, 1972; Martin, 1981
Time standards	AET (r) (task analysis) PMT studies (f)	(see AET) Number (movement elements)	(see AE1) Relative frequency of motor and sensory elements	Rohmert and Landau, 1979 Antis, 1969; BMFT, 1976a, b
Input/output of information	AET (r)	(see AET)	(see AET)	Rohmert and Landau, 1979
Environment	AET (r)	(see AET)	(see AET)	Rohmert and Landau, 1979
Physical	Measuring of noise, illumination, and climate (f)	Physical and psychophysiological measures (dB(A), etc.)	Limits of tolerability and comfort	UVV, 1974
Organizational	AET (r) Linkage studies of the working system (f)	(see AET) Buffer size, waiting times	(see AET) Pressure of timing, time pressure of decisions	Rohmert and Landau, 1979 Wucherpfennig, 1978
Social	AET (r) Layout study for assembling system (f)	(see AET) Arrangement of work places	(see AET) Possibility of communication	Rohmert and Landau, 1979

423

Table 3.5.7 (*Continued*)

Conditions of Work	Methods of Assessment of Work Load (r, rough; f, fine)	Items of Description	Criteria of Evaluation of the Working Conditions	References
Human-related				
Capacity conditions oriented to task				
Energetic	Measuring individual capacity related to the work content (f)	Performance rate, actual cycle time, frequency of errors	Relation to averages of greater collective of co-workers	REFA, 1971
Informational	Tests for measuring informatory work capacity (calculating tests, etc.) (f)	Speed of perception and processing of information	Endurance limit, endurance limit of strain	Luczak, 1975
Capacity conditions oriented to workplace				
Anthropometric	Dimensional description of body proportions of the workers (f)	Body dimensions of workers	Reaching area of arms, etc.	Jenik, 1962; Martin, 1981
Physiological	Measuring maximal endurance and strength (dynamometric and ergometric measuring) (f)	Efficiency of physiological subsystems of the human body, age, sex, etc.	Maximal endurance, endurance limit, endurance limit of strain	Rohmert, 1962; Martin, 1982
Biomechanical	Measuring maximal strength (f)	Maximal strength within the movement range	Stress level in relation to maximal reactions and stress duration	Rohmert and Jenik, 1972; Rohmert, 1960c; Jenik, 1973
Activities and actual Effects (behavior of performance)				
Stresses	AET (demands) (r)	(see AET)	(see AET)	Rohmert and Landau, 1979
	Measuring of average capacities (f)	Averages of cycle time, times of cyclic elements	Stress duration, stress level	Rohmert and Rutenfranz, 1975
	Measuring biomechanical stresses (f)	Force, distance, speed		
	Reductive stress assessment (f)	Force, distance, speed	Stress level and duration	Laurig, 1974

Activities	Measuring momentary activity and tendency of changes in activity (f)		
1. Motor activity	Mechanical activity, emergy expenditure	Efficiency	BMFT, 1980; BMFT, 1976a, b
2. Sensory, informatory activity	Oculographic activity	Frequency (relatively and absolutely)	Haider, 1977
Performances	Measuring momentary cycle time and of tendency of changes in cycle time (f)	Single characteristic and distribution of cycle time	Cycle time in relation to standard time	Rohmert and Rutenfranz, 1975
Strains	Measuring the height of strain, momentary and trends (f)	Central cardiac strain (heart rate), peripheral strain (muscle activity)	Endurance limit of strain, limit of tolerability	Laurig, 1974; Martin, 1982; Rohmert, 1962; Luczak, 1975
Output of the system (energetic and informatory changes)			
Quantitative	Measuring quantitative output (f)	Number of pieces per shift or parts of shift	Relation to standard or average quantitative output	REFA, 1971
Qualitative	Measuring qualitative output (f)	Quality of product, error frequency	Relation to quality aim (related to working or assembly system)	REFA, 1971
Consequences			
Fatigue	Self-rating of the subjective state (r)	Relative characteristic at different interview times	Level of rated fatigue	Nitsch and Udris, 1976
	Measuring of trends in strain, degree of fatigue on the basis of stress and strain analysis (f)	Speed of change in fatigue-conditioned strain	Endurance limit, endurance limit of strain	Luczak, 1975; Laurig, 1974; Rohmert and Rutenfranz, 1975; BMFT 1980; BMFT, 1976a, b
Adaptation	Measuring of strain, performance, and activity trends after phases of work interruption (f)	Change of work strategy, change of parameters in relation to each other	Duration of adaptation, level of strain in adaptation phases	Martin, 1982
Recovery	Measuring of strain before and after phases of work interruption (f)	Degrees of recovery, speed of recovery-conditioned change in strain	Endurance limit, endurance limit of strain	Rohmert, 1962; Martin, 1982; Rohmert, 1960c; Luczak, 1979
	Deductive assessment of rest allowances (f)			

Table 3.5.7 (*Continued*)

Conditions of Work	Methods of Assessment of Work Load (r, rough; f, fine)	Items of Description	Criteria of Evaluation of the Working Conditions	References
Changes in labour conditions				
Damage	Measuring the state of illness, absenteeism (r) Studies of work medicine (f)	Degree of damage	Relation of illness and absenteeism to work	Tichauer, 1976
Training	Survey of physiological and psychological capacity conditions (f)	Dynamometric and ergometric experimental results and information processing tasks	Change of physiological and psychological efficiency	Schmidtke, 1973
Learning	Survey of job-related skills	Performance rate	Relation to results of earlier tests and results of a greater number of workers	Schmidtke, 1973

Source: Rohmert and Martin (1984).

of the conditions of work and the methods of assessment of work load. References are listed to the most important field studies which used different types of work load measurements and analyses as a prerequisite for designing human-at-work systems with high capacity and acceptable strain.

REFERENCES

Antis, W. (1969). *Die MTM-Grundbewegung,* Pittsburgh, PA: Maynard.

Asmussen, I., Christensen, E. H., and Nielsen, M. (1939). Die Bedeutung der Körperstellung für die Pulsfrequenz bei Arbeit. *Scandinavisches Archiv Fuer Physiologie, 81,* 225–233.

Åstrand, B.-O., and Rohdal, K. (1977). *Textbook of work physiology.* New York: McGraw-Hill.

BMFT—Bundesministerium für Forschung und Technologie (1976a). *Entwicklung und Einführung verbesserter Arbeitsstrukturen in der elektrotechnischen Industrie—Robert Bosch GmbH.* Bonn.

BMFT—Bundesministerium für Forschung und Technologie (1976b). *Flexible Verknüpfung von automatisierten Kurztaktarbeitsblöcken mit manuellen Langtaktarbeitsblöcken im Bereich der Erzeugnis-Montage und -Prüfung.* Bonn.

BMFT—Bundesministerium für Forschung und Technologie (1980). *Ergonomische Untersuchung der Belastung und Beanspruchung in bestehenden und neuen Arbeitsstrukturen im Bereich der Aggregatefertigung der Volkswagenwerk AG.* Bonn.

Brouha, L., and Harrington, M. E. (1957). Heart rate and blood pressure reactions of men and women during and after muscular exercises. *Lancet,* 79–80.

Christensen, E. H. (1931). Beiträge zur Physiologie schwerer körperlicher Arbeit. IV. Mitt.: Die Pulsfrequenz während und unmittelbar nach schwerer körperlicher Arbeit. *Arbeitsphysiologie, 4,* 453–463.

Edholm, O. G., and Weiner, J. S., Eds. (1981). *The principle and practice of human physiology.* London: Academic Press.

Fuchs, W. (1971). *Methodik der Erstellung von Zeit-Modellen zur Ablaufplanung in Arbeitssystemen.* Berlin: Beuth-Vertrieb.

Graf, O., and Rutenfranz, J. (1958). *Zur Frage der Belastung von Jugendlichen* (Forschungsbericht Wirtschafts- und Verkehrsministerium Nordrhein-Westfalen No. 619). Köln: Westdeutscher Verlag.

Gupta, M. N., and Rohmert, W. (1964). *Muscular fatigue during transport of load in the horizontal plane* (Report No. 5). New Delhi: Ministry of Labour and Employment, Industrial Physiology Division.

Hacker, W. (1976). Psychische Regulation von Arbeitstätigkeiten: Innere Modelle, Strategien in Mensch-Maschine-Systemen, Belastungswirkungen. In W. Hacker, Ed., *Psychische Regulation von Arbeitstätigkeiten.* Berlin: Deutscher Verlag der Wissenschaften.

Haider, E. (1977). *Beurteilung von Belastung und zeitvarianter Beanspruchung des Menschen bei kompensatorischen Regeltätigkeiten: Simulation—Feldstudien—Modelle,* VDI-Fortschr.-Ber. Reihe 17, No. 5. Düsseldorf: VDI-Verlag.

Jahns, D. W. (1973). *A concept of operator workload in manual vehicle operations* (Report No. 14). Meckenheim: Forschungsinstitut für Anthropotechnik.

Jenik, P. (1972). Maschinen menschlich konstruiert. *Maschinenmarkt, 78,* 87–90.

Jenik, P. (1975). *Biomechanische Analyse ausgewählter Arbeitsbewegungen des Armes.* Berlin: Beuth-Vertrieb.

Karrasch, K., and Müller, E. A. (1951). Das Verhalten der Pulsfrequenz in der Erholungsperiode nach körperlicher Arbeit. *Arbeitsphysiologie, 14,* 369–382.

Katsuki, S., (1960). Relative metabolic rate of industrial work in Japan. Tokyo: Institute of Labor Sciences.

Kerkhoven, C. L. (1962). Kennelly's law. *Work Study and Industrial Engineering,* 48–60.

Kraut, H., and Keller, W. (1961). Arbeit und Ernährung. In *Handbuch der gesamten Arbeitsmedizin, Band 1: Arbeitsphysiologie.* Berlin: Urban und Schwarzenberg, pp. 471–511.

Laurig, W. (1974). *Beurteilung einseitig dynamischer Muskelarbeit.* Schriftenreihe Arbeitswissenschaft und Praxis. Berlin: Beuth-Vertrieb.

Laurig, W. (1982). *Grundzüge der Ergonomie—Einführung,* Berlin: Beuth-Vertrieb.

Luczak, H. (1975). *Untersuchungen informatorischer Belastung und Beanspruchung des Menschen,* VDI-Fortschr.-Ber. Reihe 10, No. 2. Düsseldorf: VDI-Verlag.

Luczak, H. (1979). *Arbeitswissenschaftliche Untersuchungen von maximaler Arbeitsdauer und Erholzeiten bei informatorisch-mentaler Arbeit nach dem Kanal- und Regler-Mensch-Modell sowie superponierten Belastungen am Beispiel Hitzearbeit,* VDI-Fortschr.-Ber. Reihe 10, No. 6. Düsseldorf: VDI-Verlag.

Martin, K. (1981). "Videosomatographie—Ein neues Hilfsmittel bei der Arbeitsgestaltung." *FB/JE, 30,* 21–26.

Martin, K. (1982). *Untersuchungen von Ermüdung und Erholungszeiten bei einseitig dynamischer Muskelarbeit,* VDI-Fortschr.-Ber. Reihe 17, No. 10. Düsseldorf: VDI-Verlag.

Monod, H. (1956). *Contribution à l'étude du travail statique.* Thèse Doct. Med., Foulon, Paris.

Monod, H. (1967). La validité des mesures de frequence cadiaque en ergonomie. *Ergonomics, 10,* 485–537.

Moray, N. (1979). *Mental work load—its theory and measurement,* New York: Plenum.

Müller, E. A. (1953). The Physiological basis of rest pauses in heavy work. *Quarterly Journal of Experimental Physiology, 38,* 205–215.

Nitsch, J., and Udris, I. (1976). *Beanspruchung im Sport.* Bad Homburg: Limpert.

Passmore, R., and Durnin, J. V. G. A. (1955). Human energy expenditure. *Physiological Reviews, 35,* 801–840.

Rasmusen, J. (1979). Reflections on the concept of operator workload. In N. Moray, Ed., *Mental work load—its theory and measurement.* New York: Plenum.

REFA (1958). *REFA-Buch,* Vol. 2, Zeitvorgabe, Munich: Hanser Verlag.

REFA (1971). *REFA-Methodenlehre des Arbeitsstudiums,* Vol. 1. Munich: Hanser Verlag.

Rohmert, W. (1960a). Ermittlung von Erholungspausen für statische Arbeit des Menschen, *Internationale feitschrift Fuer Angewandie Physiologie Einschliesslich Arbeits Physiologie, 18,* 123–164.

Rohmert, W. (1960b). Zur Theorie der Erholungspausen bei dynamischer Arbeit. *Internationale Zeitschrift Fuer Angewandie Physiologie Einschliesslich Arbeits Physiologie, 18,* 191–212.

Rohmert, W. (1960c). *Statische Arbeit des Menschen.* Berlin: Beuth-Vertrieb.

Rohmert, W. (1962). *Untersuchungen über Muskelermüdung und Arbeitsgestaltung.* Berlin: Beuth-Vertrieb.

Rohmert, W. (1963). Bestimmung der Erholungszeit bei körperlicher Schwerarbeit aufgrund physiologischer Arbeitsstudien. In *Arbeitsstudien heute und morgen,* Berlin: Beuth-Vertrieb, pp. 103–116.

Rohmert, W. (1983). Formen menschlicher Arbeit. In W. Rohmert and J. Rutenfranz, Eds., *Praktische Arbeitsphysiologie.* Stuttgart: Thieme.

Rohmert, W., and Jenik, P. (1972). *Maximalkräfte von Frauen im Bewegungsraum der Arme und Beine.* Berlin: Beuth-Vertrieb.

Rohmert, W., and Landau, K. (1979). *Das Arbeitswissenschaftliche Erhebungsverfahren zur Tätikgkeitsanalyse AET. Handbuch mit Merkmalheft.* Bern: Hans Huber.

Rohmert, W., and Martin, K. (1984). *Arbeitswissenschaftliche Begleitforschung für das Projekt AEG: Entwicklung von überbetrieblich anwendbaren Entscheidungs- und Handlungshilfen für die Planung, die Einführung und den Einsatz neuer Arbeitsstrukturen in der Montage,* Forschungsbericht HdA, No. HA 84–004. Bonn: Bundesministerium für Forschung und Technologie.

Rohmert, W., and Rutenfranz, J., Eds. (1975). *Arbeitswissenschaftliche Beurteilung der Belastung und Beanspruchung an unterschiedlichen industriellen Arbeitsplätzen.* Bonn: Bundesminister für Arbeit und Sozialordnung, Referat Öffentlichkeitsarbeit.

Rohmert, W., and Rutenfranz, J., Eds. (1983). *Praktische Arbeitsphysiologie.* Stuttgart: Thieme.

Scherrer, J., and Monod, H. (1960). Le travail musculaire locale et la fatigue che l'homme, *Journale de Physiologie,* (Paris), *52,* 419–501.

Schmidtke, H., Ed. (1973). *Ergonomie 1—Grundlagen menschlicher Arbeit und Leistung.* Munich: Carl Hanser.

Schneewind, K. (1975). Psychologie—Was ist das?" *Trierer Universitätsreden,* Vol. 5. Trier: NCO-Verlag.

Spitzer, H., and Hettinger, Th. (1964). *Tafeln für den Kalorienumsatz bei körperlicher Arbeit.* Berlin: Beuth-Vertrieb.

Tichauer, E. R. (1976). Occupational biomechanics and development of work tolerance. In P. V. Komi, *Biomechanics V-A.* Baltimore: University Park Press, pp. 493–505.

UVV (1974). *Unfallverhütungsvorschrift Lärm,* UBG 121.

Wucherpfennig, D. (1978). *Zeitliche Bindung bei manueller Fließarbeit. Ergebnisse von Arbeitsplatzuntersuchungen und eine Zusammenstellung von Gestaltungsregeln.* Berlin: Beuth-Vertrieb.

PART 4
JOB AND ORGANIZATION DESIGN

CHAPTER 4.1
JOB DESIGN

LOUIS E. DAVIS

University of California
Los Angeles

GERALD J. WACKER

Xerox Corporation
El Segundo, California

4.1.1 INTRODUCTION

Every organization is composed of roles. Through their roles, people in an organization have systematic relationships with each other and with their work. When people occupy the roles, then we have a functioning organization.

The relationships among roles and people in organizations, however, are quite complex. With people come personal needs, aspirations, social attitudes, creativity, and so on. Because of this, organizations are more than just work systems; they also become small societies.

To design jobs in this context, we must consider not only the formal work contents of jobs, but also the occupational and social characteristics of roles, with sensitivity to individual needs and expectations. If jobs are well designed, the organization can address its needs for productivity and adaptiveness, and jobholders can address their needs for a meaningful working life. This chapter examines the factors and decision processes that lead to the design of jobs.

4.1.2 DEFINITION OF JOB

There is often confusion about what "job" means. In its narrowest meaning, a job is *a one-time task*, as in the "job shop" sense. In another usage, a job is *a specific set of ongoing tasks to be performed by an individual*, as in the "job description" sense. In the broadest sense, one's job is *the totality of one's role in an organization*, including one's career path.

The distinction between the second and third meanings proved to be important in the design of jobs in a new food-processing plant (Davis and Wacker, in preparation). The organization design team, examining the problems in the company's existing plants, discovered that no set of detailed job descriptions could really convey what was required of workers—to respond as a team to unforeseen problems—nor could detailed job descriptions address what the workers desired from the company—recognition, challenge, and a secure career path. As a result of this discovery, the jobs in the new plant were designed such that each person's job is to be a member of one of five teams. Each team has a set of responsibilities, such as operating a packaging line for a product, maintaining certain machines, inspecting materials, and so on. These responsibilities are specified in detail. However, the tasks performed by any individual vary from day to day depending on several unpredictable factors. Therefore, an individual's task assignments are necessarily temporary; a person may operate a machine one period and do another task the next. In a narrow "job-description" sense, one's job is a particular task assignment that may change daily; in a broad "role" sense, one's job is to help carry out the responsibilities assigned to the team, to participate in team decisions, to cross-train, and to use one's judgment to contribute to the team's productivity, quality, and development.

More elaborate descriptions of various types of job designs, including teams, are given later in this chapter. We now turn to a more basic issue introduced by the above example—the need to design jobs so that jobholders can deal with instability, interruptions, and uncertainty.

4.1.3 JOB DESIGN UNDER CONDITIONS OF UNCERTAINTY

One cannot predict with certainty how people will respond in their jobs. People differ from one another in skill, experience, creativity, and aspirations. Moreover, a person's attendance and responses vary from day to day, because of variations in health, mood, outside commitments, and so on. The first challenge of job design is to create roles that enable jobholders to deal with human uncertainty.

Designers can generally predict the actions and capabilities of a machine with much greater certainty than those of a person. However, this fact can be quite misleading in job design, as several authors have noted (Davis and Taylor, 1979; Jordan, 1963; Waddel, 1956). Tasks that are so routine as to be reduced to a formula are almost always more efficiently done by machine. *The most important role of people in a work setting is to deal with technological uncertainty*, such as mechanical breakdowns, unusual production requirements, disruptive events, and processes that involve value judgments. The second challenge of job design is to create roles that enable jobholders to deal with technological uncertainty.

Finally, people have a perennial tendency to form social bonds and to help each other. Some of the human and technological uncertainty can be counteracted by mutually supportive relationships among employees. The formation of these relationships, however, is also uncertain. The third challenge of job design is to create roles that enable jobholders to deal with social uncertainty.

A short case will illustrate how jobholders must cope with uncertainty. A computerized information system for room assignment and accounting was installed in a hospital. Soon afterward, the chief of surgery asked the night receptionist to assign a particular patient to a particular room for the next day. Unfortunately, the computer system had already reserved that room for someone else, and there was no provision for the receptionist to reexamine that reservation and exercise her own judgment.

Two scenarios can be envisioned. First, the receptionist could shrug her shoulders and tell the surgeon, "I only work here; if you want something, then *you* can try talking to the computer." The surgeon might have to wait until morning and then spend time searching for the proper forms and approvals. The second scenario is what actually occurred. The receptionist had previously discovered that, after five successive invalid entries into the computer, it would display the message "revert to manual." The old manual procedure involved calling a floor nursing supervisor and arranging room assignments with her. So the receptionist "subverted" the computer system in order to respond to an unprogrammed event—the surgeon's particular need. Had the receptionist not accidentally discovered how to control her technology, had she not been familiar with the alternate procedure, had she not been able to communicate with the floor nursing supervisor, and had she not been motivated, then an ineffective outcome for the surgeon would have occurred. In this case the technology for room assignment was not designed to respond to the unforeseen event. In order for the organization to adapt to this event, spontaneous coordination was required of the receptionist and the floor supervisor.

4.1.4   DESIGNING JOBS TO MEET THREE KINDS OF ORGANIZATIONAL NEEDS

To design jobs is to create an organizational setting by which people, through their roles, relate systematically with each other and with some product or service. This involves looking at the needs of the organization using three frameworks.

4.1.4.1   Production-System Framework

From this perspective the organization is viewed as a transforming agency or flow process, in which certain inputs are processed into certain outputs (a product or service). Jobs must be designed to meet needs for efficient operations, high-quality outputs, well maintained equipment, properly recorded and processed information, and proper care of materials. Of particular importance in job design is the utilization of people to meet these needs in the face of technological uncertainty. Several factors affect this: the reliability of equipment, the predictability of the production process, the complexity of the technology, the level of automation, the scale of production, and the rate of technological innovation.

4.1.4.2   Miniature-Society Framework

From this perspective the organization is viewed as a social institution in which members behave according to shared understandings, role expectations, loyalties, career paths, rewards, conflicts, traditions, and so on. Jobs must be designed to meet needs for hiring, training, and coordination of people, for maintaining the social order and resolving conflicts, for discovery and exchange of knowledge, and for adaptation to changing conditions in the organization's environment. Several factors affect how these needs are addressed in job design: the labor turnover rate, absentee rate, union-management requirements, the stability of organizational goals, financial requirements, and legal and regulatory constraints. The first two of these factors have both mediating and outcome effects for job design. On the one hand, poorly designed jobs are likely to exacerbate labor turnover, absenteeism, and grievances. On the other hand, high turnover and absence—regardless of cause—give rise to greater needs for hiring, training, coordination, and evaluation.

4.1.4.3   Individual Framework

From this perspective the organization is viewed as a collection of particular individuals, each with his or her own goals, commitments, and life style that only partially intersect with the organization. Work organizations, along with other institutions in society, provide individuals with opportunities to meet their personal needs and aspirations. Indeed, the organization's ability to meet its product or service goals depends in part on its ability to offer its work force these opportunities. The following are of particular importance to job design (Davis, 1977; Engelstad, 1979):

1. The need for work to be challenging and meaningful.
2. The need to be able to learn on the job, which requires that the job provide specific performance criteria and feedback.
3. The need to have some scope of decision-making involving discretion and judgment, and be evaluated on the basis of objective outcomes.
4. The need for social support in the workplace—to have others to rely on for help and understanding.

Table 4.1.1   Quality of Working Life Criteria Checklist

Physical Environment

Safety
Health
Attractiveness
Comfort

Compensation

Pay
Benefits

Institutional Rights and Privileges

Employment security
Justice and due process
Fair and respectful treatment
Participation in decision-making

Job Content

Variety of tasks
Feedback
Challenge
Task identity
Individual autonomy and self-regulation
Opportunity to use skills and capacities
Perceived contribution to product or service

Internal Social Relations

Opportunity for social contact
Recognition for achievements
Provision of interlocking and mutually supportive roles
Opportunity to lead or help others
Team morale and spirit
Small-group autonomy and self-regulation

External Social Relations

Job-related status in the community
Few work restrictions on outside life style
Multiple options for engaging in work (e.g., flexible work hours, part-time options, shared jobs. and subcontracting)

Career Path

Learning and personal development
Opportunities for advancement
Multiple career path possibilities

---

5. The need for recognition of one's contribution.
6. The need to relate one's work role with one's outside life.
7. The need for a desirable future—career paths, not dead-end jobs.
8. The need for options in one's job to accommodate individual differences and circumstances.

Individual needs are presented in a somewhat different format, in the Quality of Working Life Criteria Checklist, given in Table 4.1.1.

4.1.5 DECISIONS INVOLVED IN DESIGNING JOBS

Job design involves four sets of decisions: (1) deciding what tasks will be performed by the work force; (2) deciding how these tasks will be grouped together and assigned to individuals; (3) deciding

Table 4.1.2 Needs and Decisions Involved in Job Design

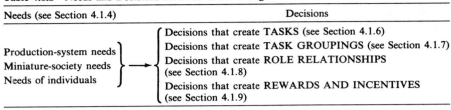

Needs (see Section 4.1.4)	Decisions
Production-system needs Miniature-society needs Needs of individuals	Decisions that create TASKS (see Section 4.1.6) Decisions that create TASK GROUPINGS (see Section 4.1.7) Decisions that create ROLE RELATIONSHIPS (see Section 4.1.8) Decisions that create REWARDS AND INCENTIVES (see Section 4.1.9)

how individuals will relate to each other so that their work can be coordinated; (4) deciding how individuals will be rewarded for their performance as members of the organization. *All these decisions should take into account the needs of the organization as a production system, as a miniature society, and as a collection of individuals* (see Table 4.1.2).

4.1.6 DECIDING WHAT TASKS WILL BE PERFORMED BY THE WORK FORCE

There are two types of tasks that are performed by the work force, technical tasks and organizational tasks.

4.1.6.1 Technical Tasks

Technical tasks contribute directly to the production system. The work place technology—equipment, materials, configuration, controls, technical manuals, information system, procedure manuals, and so on—determines what technical tasks will be performed by the work force.

Unfortunately, those who design workplace technology often fail to consider the organization as a miniature society and as a collection of individuals when making these decisions. This can lead unnecessarily to poor jobs. In one case (Davis and Taylor, 1979) a plant was designed to produce an aerosol spray product. In that design several machines for filling and capping the spray cans were set up about 50 ft apart from each other. The machines automatically uprighted, filled, capped, and packaged the spray cans. The system was designed so that workers performed only two technical tasks. The first was to insert the spray valve with its attached small plastic tube into the large hole in the top of each upright can, brought to the worker on a circular conveyor belt. The second technical task, only ocassionally required, was to press a stop switch on the machine in the event of trouble. The second task appears to have been the main reason for having workers stationed at each machine; the first task was simply left unautomated to give workers something to keep them busy because they were stationed at the machine anyway. This design left workers socially isolated, performing meaningless work, with no opportunity to learn new skills or progress on a career path. Although the design would have been acceptable if people were simply machine-like elements in a production system, it is unacceptable when one considers the consequences of ignoring the social and individual needs of people.

Probably the best-known example of these consequences is the Chevrolet Vega plant, hailed in 1970 as the world's most technologically advanced assembly line. By 1972 it was gripped by labor discontent, sabotage, absenteeism, and poor quality. This culminated in a strike in 1972, which attracted national attention when local union leaders appeared on television and decried the dehumanized jobs.

Contrast the above with the decisions made in a recently designed food-processing plant (Davis and Wacker, in preparation). The existing technology involved a series of tasks—manually unloading 100 lb sacks of raw material from boxcars and stacking the sacks on pallets, to be stored until their contents was manually dumped into hoppers. From that point, semi-automated equipment sorted, processed, and packaged the product. In older plants the unloading and dumping tasks were performed by men, whereas the operation of the semi-automatic equipment was done mostly by women. This was unacceptable for the new plant, because it would have created social barriers in the plant and blocked people's opportunity to learn all phases of the operation. With the backing of top management, the designers initiated an industry-wide conversion to semi-bulk, mechanical handling of the raw material. Additional automated machinery was developed in the plant so that all the tasks to be performed by the work force could be done by either men or women.

4.1.6.2 Organizational Tasks

Organizational tasks (sometimes called social-system tasks) involve coordination, planning, hiring, job assignment, training, performance appraisal, problem solving, discipline, and so on. The organizational rules and procedures, authority structure, compensation system, personnel system, and production control system determine what organizational tasks will be performed by the work force.

In the past, organizational tasks were performed exclusively by supervisors and/or specialists. The trend over the past decade or so has been to design the organization so that workers can participate in organizational tasks that affect their own roles and responsibilities. "Quality circles," for example, are a mechanism toward this end, although they often fail when not part of an overall effort to redesign jobs to enable workers to perform necessary organizational tasks.

4.1.7   DECIDING HOW TASKS WILL BE GROUPED TOGETHER AND ASSIGNED

Grouping tasks together is a matter of creating boundaries within the organization—partitioning the work force's tasks into segments so that individuals know what work they are responsible for. These groupings occur in two realms, technical and social.

4.1.7.1   Technical-System Realm

Decisions about technical systems can create or eliminate options in grouping tasks together. Even such an apparently simple decision as the location of a switch or meter, or the mode of input and output for a computer, allocates technological resources to particular work stations and thus constrains the grouping of tasks. For example, railroad spurs and loading docks are typically designed so that shipping and receiving tasks are performed in one physical location. In one recently designed paper mill, however, the job designers rejected this arrangement, because it would not have permitted those employees responsible for the first step of the production process to control their input, a major source of variability for them. They built separate sets of spurs and docks for shipping and receiving. The extra cost was more than repaid by high performance of work teams whose job designs focused on workers' responsibility to control productivity and quality.

Other examples of technological designs used to group tasks include U-shaped assembly configurations, so that jobholders are not restricted to a one-machine/one-job mode of organization; buffer areas, so that a jobholder can interrupt a production-line task to perform another needed task and later catch up with the production-line task; and self-propelled assembly carriers (Gyllenhammar, 1977), so that the assembly path can be changed easily by work teams themselves.

4.1.7.2   Organizational (Social-System) Realm

Decisions about organization structure, supervision, and so forth can create or eliminate options in grouping and assigning tasks. For example, the designers of a new food-processing plant (Davis and Wacker, in preparation) integrated quality-control tasks with production tasks in order to maximize feedback. They created roles in which teams were responsible for both production and quality-control objectives.

4.1.8   DECIDING HOW INDIVIDUALS WILL RELATE TO EACH OTHER

Technical-system decisions establish the infrastructure by which materials and information connect jobholders. The conveyor belt, for example, links people by virtue of transporting materials between work locations, although it can also isolate people if it imposes a fixed pace and configuration of work. Information is transferred by physical proximity as well as by equipment such as telephones and computers. The use of glass rather than opaque walls is an example of how architectural decisions can impact on job design, enabling people to visually relate to each other.

Organizational resources also link individuals. Communication procedures, reporting relationships, promotion paths, meetings, team identities, and so on provide linkages.

Personal and interpersonal resources are frequently overlooked linkages between jobs. Acquaintanceships, personal commitments to organizational goals, social skills, and familiarity with tasks outside one's own job all help to integrate the many jobs into a single system. Often it is the individual's own decisions, within the overall organizational and community culture, that enable these personal and interpersonal resources to develop.

4.1.9   DECIDING HOW INDIVIDUALS WILL BE REWARDED

Reward systems consist of (1) the rewards given to individuals, including pay, promotions, perquisites, opportunities for further training, visibility, and responsibility; (2) the criteria for earning rewards, including supervisory evaluations, peer evaluations, achievement of objectives, piecework, commission plans, formal educational achievements, in-house tests, seniority, and so on. By creating incentives for appropriate behaviors, reward systems help tie the needs of individuals to the needs of the organization. The behaviors that are rewarded must be congruent with other aspects of the job design. If, for example, jobs are designed for dealing with uncertainty, then seniority or piecework-based rewards will be less effective than skill-based rewards.

Designers of a food-processing plant (Davis and Wacker, in preparation) wanted certain organizational tasks—training and work assigning—to be grouped together with technical tasks so that workers would cross-train each other. However, workers would have had no incentive to cross-train under the company's previously existing reward system. In the new design, a person is paid according to the number of task-related skills and knowledge he or she has mastered, regardless of the person's particular work assignment at a particular time. Under this new reward structure, workers can address their own needs for pay and advancement, while simultaneously addressing the organization's needs for flexibility, when they cross-train each other. Cross-training then becomes a secondary reward, and it is used as an incentive for attendance and cooperation: teams reward a member who has contributed to the team by giving him or her opportunities to cross-train, which then lead to advancement.

During the plant's start-up period, management decided to suspend cross-training until the plant had reached a certain level of efficiency, reasoning that it would be premature to train a person at a new task when he or she was not yet proficient at a first task. That decision, however, removed a major linkage between production-system needs and individual and miniature-society needs, by dislodging an incentive that teams could use to control their members equitably. This weakened the job design. Other means had to be established to give teams the facility to control their members until cross-training was reopened.

4.1.10   TYPES OF JOB DESIGNS

4.1.10.1   Historically Evolved or Undesigned Jobs

One way to design jobs is to let them "grow, like Topsy." This laissez-faire approach to job design characterized production systems before the Industrial Revolution, about (1780). Jobs were called "trades," and changed little from generation to generation. Questions of who would perform what tasks, and how, were based on traditions and rules of thumb, and enforced by guilds.

Some tradition-determined jobs remain today, mostly among the professions. For example, the roles of physician and nurse are not designed simply as an extension of the technology of medicine. These job designs rest on social traditions that are now protected by legislation and guilds of doctors and nurses.

Probably the best-known critic of undesigned jobs was Frederick W. Taylor (1911). He urged, at the turn of this century, that traditions give way to more deliberate ways of making job-design decisions.

4.1.10.2   Machine Model of Jobs

The machines that emerged during the Industrial Revolution not only meant that new jobs would have to be designed around the new technology, but also gave rise to a belief that jobs and organizations could be designed with the same certainty as machines. This machine metaphor for job design originated with Adam Smith (1970/1776) and Charles Babbage (1965) and was further developed by F. W. Taylor (1911).

Smith, Babbage, and Taylor urged managers to design each job as narrowly and precisely as possible, as if all events could be anticipated and specified. They advised that each task be made a job unto itself. Complex tasks such as maintenance were to be assigned to specialized maintenance jobs, decision tasks such as inspection to specialized inspector roles, and so on. It was further advocated that when tasks were grouped together for assignment to individuals, an individual's assigned tasks should all be at the same skill level, so that as few individuals as possible would be responsible for high-skill tasks.

Problems have arisen with jobs designed on these principles. One such problem is "job myopia," or the "It's not my job" syndrome. Important needs go unmet because those who are first aware of a problem shrug it off as not part of their jobs. (And indeed, the problem may not fall within their narrow job scopes.) Those who are formally responsible for the problem do not learn of it until it becomes serious enough to warrant formal communication. By then problem solving is awkward, inefficient, and too late. Further hampering effectiveness, the specialists ignore and/or create problems according to their own work loads rather than according to the broader needs of the organization. Many workers see the ineffectiveness of this system, but they do not take initiative because to do so is seen as encroaching on another's territory. Much energy that would be available for cross-training and spontaneous problem solving is instead diverted to protecting and aggrandizing individuals' narrow specialties. Ultimately, workers become alienated and lose their sense of organizational community when they see that they are stuck in a system of machine-like, dead-end jobs which neither contributes to organizational effectiveness nor addresses their needs for challenging, meaningful work.

In the machine model of jobs it was assumed that, because every eventuality could be specified in the job design, there would be no necessity for people to exercise any individual discretion. Indeed, it was assumed that to protect the system people should be prohibited from discretionary activity. Incidents of "working to rule" or "malicious obedience" show that assumption to be false. For example, British railway workers, instead of striking 35 years ago, virtually halted train service simply by coming to work and doing exactly what their job descriptions and standard operating procedures called for, no more and no less. They stopped the railroad cold by leaving their discretion at home. Similarly, in New Jersey, police enforced all traffic laws without any exercise of discretion, and thus managed to cripple traffic flow on key arteries. These cases remind us that rules and job specifications must be tempered by individual discretion, and that the success of the organization rests partly on each employee's willingness to exercise that discretion. Today's sophisticated automated technology requires very substantial use of discretion, because the most important work is intervention to minimize downtime.

4.1.10.3 Job Enlargement and Job Enrichment

Growing recognition of the organizational and personal costs of the machine model led to a search for remedies. By 1940 *job enlargement* (Conant and Kilbridge, 1965) was introduced as a palliative. Its assumption was that the deficiency in the machine model lay in the narrowness of jobs and skills, and that the cure was to string more of these together. Job enlargement did not, however, address the other weaknesses of the machine model.

A different remedy was *job enrichment* (Herzberg, 1966), introduced in the 1960s. Like job enlarge-ment is retained the one-person/one-job unit of analysis derived from the machine model. It made a significant contribution, however, by distinguishing "vertical" from "horizontal" grouping of tasks—a vertical grouping including the more discretionary tasks such as planning, set-up, measurement, and control—and by urging managers to create "whole" jobs by grouping together into each single job the tasks needed to achieve some measurable result.

A plastic bag plant tripled its output of automatic bag-making machines while enriching the jobs in its bag-making department. In the unenriched jobs (Figure 4.1.1) the tasks associated with the front end of the machines were grouped into one set of jobs, and back-end tasks into another set of jobs. Because the front-end jobs required more skill, the positions, entitled "operators," were filled by men at higher pay. The back-end jobs, entitled "inspector-packers" were filled by women at lower pay. Much of the work on the machines involved dealing with breakdowns and errors. It was felt that these could be greatly reduced if each operator were assigned to perform all the tasks associated with one particular machine (Figure 4.1.2). In addition to this a new compensation plan was introduced in which all workers are paid a monthly salary rather than an hourly wage. Both men and women hold the newly enriched operator jobs, and the company reported an increase in both productivity and job satisfaction.

Job enrichment alerted managers to the needs for broader skills, for the use of discretion, for feedback of results, and for the inclusion in a job of interrelated tasks associated with the product or service. It did not, however, address issues of relationships between jobs, organizational tasks, or career paths.

4.1.10.4 Self-Maintaining Work Teams

Beginning in the late 1940s, British and American researchers (Emery, 1969; Herbst, 1974; Rice, 1958; Trist, Higgin, Murray, and Pollack, 1963) created a body of literature called sociotechnical

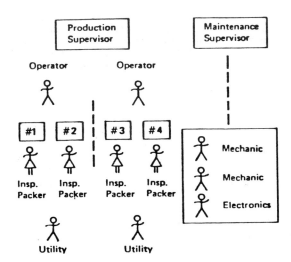

Total: 11 (plus 2 supervisors)

Fig. 4.1.1. Staffing of a bag-making shift with initial fragmented jobs. From L. E. Davis and A. B. Cherns, Eds., *The Quality of Working Life,* Volume 2, The Free Press, New York, 1975, p. 277.

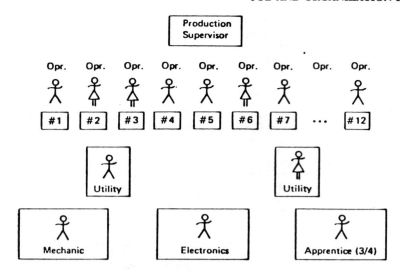

Fig. 4.1.2. Staffing of bag-making shift with current enriched jobs. From L. E. Davis and A. B. Cherns, (Eds.), *The Quality of Working Life,* Volume 2, The Free Press, New York, 1975, p. 277.

systems theory, and a model of job design that uses the small group, rather than the individual, as the main unit of analysis and basic building block of the organization. The organization is divided into segments, each of which reflects a distinct state change in the product or service process, with its own measurable outcomes. The responsibility for carrying out the activities associated with each segment, and for achieving the measurable outcomes, is assigned to a team. Teams are constituted to be *self-maintaining,* that is, to contain all the technical and organizational tasks, and resources, necessary to achieve their goals. Each team can be thought of as a "mini-business." Because these units are composed of a *group* of individuals who must use teamwork to achieve their outcomes, they are less vulnerable to human uncertainty, and more likely to draw on the forces of social cohesion in the service of work-related goals.

In contrast to the deterministic, complete specification of the machine model of job design, self-maintaining work teams are designed according to a principle called "minimal critical specification." The team is given only those few specifications necessary for interface with the whole organization. The team can then exercise discretionary options according to its learnings from experience (see Table 4.1.3). In some of the latest job designs, work teams take responsibility for organizational learning, cross-training, peer counseling, internal coordination, and adaptation to uncertainty.

Work teams require compatible decisions about technical systems. At Volvo's Kalmar plant in Sweden, a new technology for automobile assembly was developed to support the team form of job design (Gyllenhammar, 1977). Computer-monitored carriers enable each work team to lay out its own work and to control the pace of its work. Volvo estimated that the new technology cost 10%

Table 4.1.3   Job-Design Decisions for Self-Maintaining Work Teams

Typically Specified by Job Designers, with Participation by Workers when Available	Typically Left for Team Decision-Making, with Manager Consultation when Needed
Measurable input specifications	Task assignment scheduling
Measurable output specifications—quantity and quality	Work methods
	Work pace
Equipment and resources	Work hours
Work-station layout options	Counseling and discipline
Compensation and advancement plan	Internal information flows
External information flows	Internal leadership
	Team membership

more than a conventional assembly line, but expected to recover the costs in increased flexibility and higher worker motivation. At Philips in the Netherlands, televisions are assembled in "production islands," which are U-shaped assembly lines that enable assembly workers to interact and cooperate (den Hertog, 1977). In a new Canadian polypropylene plant, a process control computer displays decision-aiding information rather than making decisions (Davis and Sullivan, 1980). The responsibility for final decisions rests with work teams. This design was chosen because of the technological uncertainty contained in the large number of uncontrollable variables flowing in the system. It was expected that by involving work teams in day-to-day technical decisions, they would develop ways to control those variables better.

Similarly, work teams require compatible decisions about compensation systems. One approach is to provide work teams with group bonuses. Another approach, which satisfies a number of team requirements, is "pay by knowledge and skill." Knowledge and skills needed to perform tasks are identified and grouped into modules. These modules are graded to form levels. Standards for training and testing are developed to provide a means for advancing through the levels, which are associated with pay increases. Individuals are paid not according to what task they perform, but by what tasks they have demonstrated themselves capable of performing. An individual's pay level does not, however, bestow formal authority over other team members because teams operate as a group of peers and make decisions by consensus. This compensation and progression system encourages teamwork, cross-training, and acquisition of skills.

Work teams also require an appropriate organizational culture, which must be internalized by extensive training and reinforced until team members have gained experience at teamwork and can socialize new members. One of the most critical areas for a team is to learn how to deal internally with members who do not meet team membership norms and standards (excessive absence, unwillingness to help others, etc.). For teams to develop a style of consensual decision-making, managers must deal with teams in a manner that is neither neglectful nor interfering. Thus managers themselves must undergo appropriate training and socialization and practice similar behaviors by structuring themselves as teams.

In one case, a plant was organized into five teams consisting of 14 to 28 members per team (Davis and Wacker, in preparation). Each team is given responsibility not only for production goals and equipment maintenance, but also for quality control, interteam coordination, safety, discipline, and so on. For each of these organizational responsibilities, the teams designate individuals to act as "quality control coordinator," "work assignment coordinator," "vacation schedule coordinator," or "training coordinator," for example. These organizational tasks are rotated periodically, and along with the technical tasks such as those associated with production, quality control, and maintenance, make up individuals' work assignments. Communication coordinators from each team meet with the operations manager every day; quality-control coordinators meet with the quality-control manager as needed; training coordinators with the personnel manager; and so forth.

4.1.11   ANALYTIC METHODS IN SUPPORT OF JOB DESIGN

This section presents some methods for collecting and analyzing data to help make job-design decisions—to help identify tasks to be performed by the work force, task groupings, and relationships among individuals. Although these methods are useful for focusing attention on certain aspects of job design, they are really quite limited as design tools. Actual job-design decisions should be made from an overall concept of the needs of the organization as a production system, as a miniature society, and as a collection of individuals. Job design is most effective when it is part of organization design.

4.1.11.1   Transformation Flow Chart

To conceptualize the organization as a production system, it is useful to draft a flow chart showing how the product or service progresses from its input state to its output state. This flow chart is helpful for enabling job designers to conceptualize production requirements without being locked into specific technological or social decisions. The transformation flow chart aids designers to transcend the existing equipment, layout, and task structure, for example.

The transformation flow chart, an example of which is given in Figure 4.1.3, shows (1) the progressive states of the product or service, and (2) the "unit operations" by which the product or service is transformed from one state into another. Product or service states are of three types: inputs, throughputs, and outputs. An output is a finished product, final state, or waste product that leaves the system.

Unit operations are expressed in terms of the state changes in a process of transformation or service. If particular machines connote particular changes in product state, then it may be useful to express the unit operations in terms of these machines. Inspection is not typically shown as a separate unit operation if it serves merely to verify that a transformation occurred. On the other hand, creating or updating permanent records, such as a medical history, may be considered separate unit operations. Storage is usually not considered a unit operation unless some change of product state—whether desired or undesired—could occur during the storage. Decision-making, calculation, setup, positioning, and

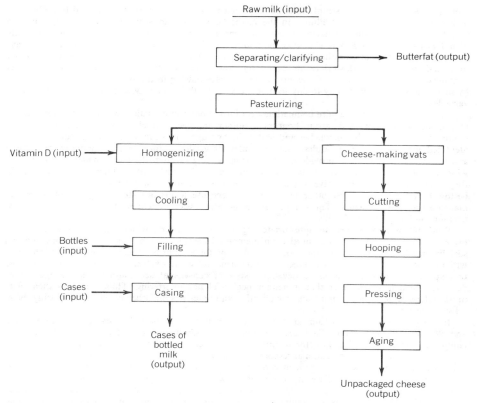

Fig. 4.1.3. Transformation flow chart for a dairy.

so on, are not expressed as unit operations by themselves, but rather are considered tasks within a unit operation.

The selection of equipment, layout, and job structure should enable employees to visualize the transformation flow-chart as they work. Job or team boundaries should encompass one or more unit operations, so that each job or team is allocated a distinct piece of the transformation process, and can measure its progress in terms of changes in the state of the product.

4.1.11.2   Variance Analysis

Modern job design seeks to address performance requirements under all conditions, not just under stable, steady-state conditions as assumed under the machine model. With complete technological certainty and environmental stability, there would be no deviation or disturbance in the transformation process. Most production systems, however, are faced with pockets of technological uncertainty, where human judgment and intervention are critical for controlling the process. Variance analysis is a technique of identifying those areas of disturbance and variability, so that jobs can be designed to control variability in the production process.

A "variance" is an unwanted discrepancy between a desired state and an actual state. Sometimes a desired state is specified as a range of tolerance within which deviation is permissible. A variance occurs when some variable in the production system is outside its range of tolerance.

There are two types of variances: *product state (or service state) variances,* which pertain to input, throughput, and output specifications, and *system variances,* which pertain to equipment and the ambient physical environment (e.g., dust, humidity, heat).

Although the concept of variance could be extended to include discrepancies in skill and motivation, this would confound too many variables together. Variance analysis simply gives the designer a detailed understanding of the technological uncertainty and environmental instability in the production system, in effect alerting the designer to identify all contingencies likely to interrupt the flow of work.

The first step in variance analysis is to list all variances that could impede the production or service process. Often it is useful to organize the variance list in the format shown in Figure 4.1.4.

Unit operation	Product state variances	System variances

Fig. 4.1.4.   Variance list format.

Some guidelines for identifying variances follow:

1. A variance should be listed with reference to the unit operation in which it would occur, regardless of where the variance is ultimately detected.
2. Variances may be stated as deviations—"water too cold"—or as variables—"water temperature."
3. Variances are better expressed in state terms rather than in process terms—"dirty dishes" rather than "unwashed dishes."
4. Variances should be expressed so as to reflect the degree of precision and objectivity of the product or process specifications—"water too cold," when the specification is imprecise or subjective; "water below 92°C," when more precise.

The second step in variance analysis is to identify the dependency or causal relationship that may exist among the variances. An aid for this is a variance interrelatedness matrix, an example of which is shown in Figure 4.1.5, used in the redesign of jobs in a paper mill. A cell in the matrix represents the interaction of the variance in its row with the variance in its column, as in a mileage chart on a road map. A blank cell indicates no relatedness, whereas a "3" indicates relatedness of great importance, such as between the variances numbered 22 and 42 in Figure 4.1.5. In designing jobs, it is necessary to group tasks together and create relationships between individuals so as to provide the basis for coping with the interrelatedness of variances.

The third step is the identification of "key variances," those whose control is most critical to the successful outcome of the production or service system. Because a large percentage of problems are usually caused by a small percentage of variances, key-variance identification helps job designers see the critical role of people in the production system. A key variance (1) can seriously impair quantity, quality, cost, people, or resources; (2) interacts with or causes disturbances in other variables; (3) cannot be predicted with certainty as to time, place, frequency, or intensity of occurrence; (4) can be detected, prevented, corrected, or otherwise controlled by timely, appropriate human action.

The fourth step in variance analysis is to draft a table of key variance control, the format of which is shown in Table 4.1.4. This should contain brief descriptions of how, where, and by whom each key variance can be detected, corrected, and prevented. It should also describe how information about each key variance can be transmitted. "Feedforward" is the information that is transmitted from the point where a variance is detected to the point where it can be corrected or kept from getting worse. "Feedback" is the information that is transmitted from the point of detection to the point where further occurrences can be prevented.

The fifth step is to construct a table of the skills, knowledge, information, and authority needed for jobholders to be able to control key variances. The format for this table is shown in Table 4.1.5. This table indicates the significant content that should be designed into jobs to enable the work force to deal with the technological and environmental uncertainty in the production system.

4.1.11.3   Technological Assessment

Although variance analysis helps identify specific areas of technological and environmental uncertainty, technological assessment provides more general information. Each unit operation can be assessed in several dimensions:

Automation

A process is automated when it does not require human assistance to perform under stable conditions. Human assistance is required, however, to control variances that exceed the system's automatic capabil-

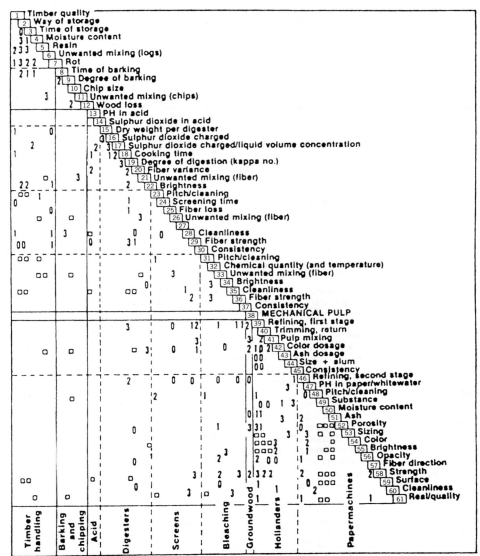

Fig. 4.1.5. Variance interrelatedness matrix. From P. H. Englestad, "Sociotechnical Approach to Problems of Process Control," in L. E. Davis and J. C. Taylor, Eds., *Design of Jobs,* Goodyear, Santa Monica, CA, 1979, p. 205.

ity. The greater the degree of automation, the greater the proportion of employee time spent on maintenance, regulation, diagnosis, planning, and control, and the less the proportion spend on routine operational tasks (Davis and Taylor, 1979; Hazlehurst, Bradbury, and Corlett, 1969). In manual systems the work force is more likely to be idle when the equipment is not functioning, yet in automated systems people would be busiest when the system goes down. In systems that are partially automated, jobs should be designed to be as flexible as possible in order to make full use of human resources both when the equipment is functioning normally and when system variances occur.

Table 4.1.4 Format of Key Variance Control Tabl

Key Variance	Occurrence	Detection	Feedforward	Correction	Feedback	Prevention	Suggestions for Better Control	
							Technical Modifications	Changes in Jobs and Organizations
Unit operation:	Where:	Where:	Channels:	Where:	Channels:	Where:		
Variance:	How:	How:	Lead time:	How:	Lag time:	How:		
	By whom:	By whom:		By whom:		By whom:		
Unit operation:	Where:	Where:	Channels:	Where:	Channels:	Where:		
Variance:	How:	How:	Lead time:	How:	Lag time:	How:		
	By whom:	By whom:		By whom:		By whom:		

Table 4.1.5 Format of Table of Skills, Knowledge, Information, and Authority Needed to Control Key Variances

Key Variance to be Controlled	Skills Needed	Knowledge Needed	Information Needed	Local Authority Needed

Programmability

Some technologies require only that people follow a program or recipe. Other technologies, based on less exact sciences, cannot specify programmed action—people must develop some intuition or heuristics based on experience (trial and error). In the latter technologies, jobs should be designed to encourage the use of discretion, learning, and sharing of information.

Subjectivity

With some technologies, output quality can be determined objectively. With others, however, specifications are vague and variance detection is subjective. A food-processing manager once remarked casually, "We could almost train monkeys to operate the equipment; what really takes skill is knowing how to recognize the color and taste of a good product, and what to do to get it consistently good." If people must make subjective judgments, they need appropriate feedback to develop and maintain a sensitivity to product quality.

Stability

Although most variances occur during the performance of some unit operation, some can occur "spontaneously" during storage, transport, waiting, and so on. Variance control is more critical when product states or equipment are unstable.

Equivalency

In a redisign of a paper mill, Engelstad (1979) reported that four pulp digesters were treated as equivalent or interchangeable pieces of equipment. In fact, however, one of the digesters was particularly efficient with certain kinds of wood fiber. A few operators discovered this, but their jobs were designed so that they had no incentive for sharing this information with others. On the other hand, jobs in an aluminum smelter (Wacker, 1979), were redesigned to take account of the fact that each smelting furnace tended to develop its own "personality." Workers performed a wide range of tasks for a small group of furnaces in order to learn the peculiarities of each.

Variability

Some technologies, such as computers, are in either an "up" (operating) or "down" (not operating) state, whereas others can function in intermediate states, such as an automobile in need of a tune-up. Jobs should be designed to recognize the different roles of people required in different states of the technology.

4.1.11.4 Task Ratings

The principal tasks to be performed by the work force can be rated according to the quality of working life checklist shown in Table 4.5.1 (see also Hackman and Oldham, 1975). These ratings can be used to guide the grouping of tasks so that each job or team has a reasonable balance of both undesirable and desirable tasks, isolated and collaborative work, simple and complex tasks, and so on.

Task ratings can be misleading in job design. The quality of a job is more than the sum of its tasks, because the interrelationship among the tasks should form a meaningful whole job. The contents of a job should further career opportunities, and the job should connect the individual with the functioning of the organization.

4.1.11.5 Mobility Analysis

An important source of job-design information resides in the actual capabilities of employees. Mobility analysis addresses a simple question: What jobholders can move to other jobs should the need arise? A matrix type of chart (see Table 4.1.6) can show mobility among a group of employees.

4.1.11.6 Interaction Analysis

As with mobility analysis, a chart can be made to show the relative frequency by which people interact with each other on work-related matters. This analysis helps reveal interdependencies among tasks and jobs, as well as social relationships that have evolved among jobholders.

4.1.11.7 Responsibility Analysis

Part of job design is the allocation of responsibility for making decisions, for use of resources, and for achieving results. Table 4.1.7 gives a format for compiling such information.

Table 4.1.6 Mobility Analysis Format

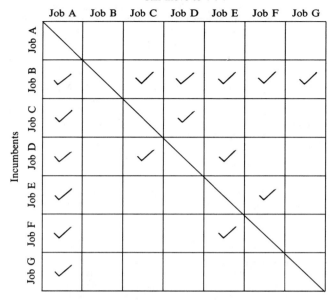

Can move to . . .

Incumbents	Job A	Job B	Job C	Job D	Job E	Job F	Job G
Job A							
Job B	✓		✓	✓	✓	✓	✓
Job C	✓			✓			
Job D	✓		✓		✓		
Job E	✓					✓	
Job F	✓				✓		
Job G	✓						

In a redesign of supervisory jobs (Davis and Valfer, 1965), two modifications in supervisors' responsibilities were introduced separately into a number of aircraft instrument repair shops. One of the modifications was to shift from a functional division of labor to a product-based division of labor, thus giving each shop responsibility for all tasks required to produce a product. The other modification added a responsibility—inspection and quality control—to the shops involved. The supervisors involved in the second modification initially performed the inspection themselves, but later delegated it to subordinates. Compared with other shops where no changes took place, attitudes, productivity, and quality improved in the modified shops. As supervisors concerned themselves with planning and controlling external events impinging on their shops, their subordinates took on some of the characteristics of work teams.

4.1.11.8 Loci of Organizational Stability and Instability

Organizational stability indicators, such as labor turnover rate, absenteeism rate, grievance rate, incidents of antisocial behavior, and attitude measures, can point to special needs for job design. The increasing heterogeneity of the work force (e.g., age, sex, ethnicity, education), together with the faster rates of social change and shorter lead times of market shifts and technological innovations, has focused attention on internal stability. There are both the desire to preserve employee skills and motivation within the organization and the desire to adapt to changing conditions.

Organizational stability presents a dual challenge to job design: first, to design jobs that abate the causes of instability, and second, to design jobs that enable the organization to cope with instability.

A low turnover rate can indicate those jobs that are attractive to jobholders, but it can also indicate dead-end jobs. Tracking turnover can also show the existing paths of advancement (career ladders).

4.1.12 CRITERIA FOR GROUPING TASKS TO CREATE JOB AND TEAM BOUNDARIES

The following criteria are useful for grouping tasks and structuring jobs and teams:

Each task group is a meaningful unit of the organization—a mini-business or identifiable craft.

Task groups are separated by stable buffer areas.

Each task group has definite, identifiable inputs and outputs.

Each task group has associated with it definite criteria for performance evaluation, such as productivity, quality, efficiency, yield, maintenance, cost, waste, errors, cross-training, skill acquisition, customer complaints, machine utilization, and downtime.

Each task group can get timely feedback about the results of its actions, and feedforward about what to expect in the future.

Table 4.1.7 Responsibility Analysis Format

Job or Team	Tasks	Discretionary Aspects of Task Performance	Latitude for Decisions	Available Resources	Accountability for Results	Sources of Feedback

Each task group has resources to measure and control variances that occur within its area of responsibility.

Highly interdependent tasks are grouped together.

Tasks are grouped around mutual cause–effect relationships.

Tasks are grouped around common skills, knowledge, or data bases.

Jobs are linked to facilitate cross-training.

Jobs incorporate opportunities for skill acquisition relevant to career advancement.

Jobs or teams each contain a balance of group and individual tasks, desirable tasks, and simple and complex tasks.

Jobs or teams each contain a balance of comfortable work environments and if necessary stressful ones.

Jobs or teams have the skills, knowledge, feedback, and authority for self-regulation and of adaptation to variations in their environments.

Within a team, all the tasks required to achieve its goals are identified and can be allocated to team members as work assignments.

The work stations of a team are physically and/or temporally proximate to each other to permit face-to-face relationships.

Teams perform tasks that maintain relationships both within and among teams.

4.1.13 STRATEGIES FOR DESIGNING JOBS

Although it is beyond the scope of this chapter to describe the overall process by which job designers should manage their function, this may be found elsewhere (Davis, 1982; Davis and Wacker, 1982). In general, job design is not effective if it is simply delegated to specialists. The most effective and readily accepted jobs are those that are designed or redesigned participatively. Line managers, employees, engineers, union officials, and others must all be involved in the process in order for the most desirable outcomes to be achieved.

Although existing organizations might do well to begin job redesign with a reexamination of managerial strategy, technology, and organizational structure, there are usually a number of constraints that cannot be removed. Job redesign is often just a part of a continuous stream of changes the organization makes to meet requirements coming from its environment. One of the objectives of job redesign is to provide an experience that sets a precedent for future redesign needs.

Usually there is a complex system of forces that underlies the job redesign endeavor. In an aluminum smelter (Wacker, 1979), rising labor turnover rates, increasing difficulty in recruiting workers, technological obsolescence, and more aggressive government regulations of occupational safety and health conditions and equal employment opportunities for minorities and women led to efforts to automate some of the most physically demanding and undesirable tasks. The new technology required less manpower and new task groupings. But the new technology alone did not solve the organization's problems of turnover, recruitment, and efficiency. Thus the job redesign was directed not only at suiting the new technology, but also at creating more attractive and challenging jobs, at attaining work force flexibility so that the organization could function with a broad range of turnover rates, and at training and advancement geared to individual readiness, regardless of vacancies created by turnover.

Employee suggestions for, and their reactions to, redesigned jobs are influenced by their past experiences and their expectations for the future. In one plant employees saw their needs as being met by petitioning for a second time clock, so that they would not have to lose time in clocking out for lunch. It was beyond their experience to know that job redesign could address more fundamental aspects to an improved quality of working life. In the aluminum smelter cited previously, the job redesign resulted in fewer status differences between smelter jobs. Although most of the employees welcomed the new opportunities and new technology, some complained about the loss of their former perquisites. Attitudes improved among the employees whose jobs were changed as well as those whose jobs had not been changed, because the latter group looked forward to an imminent changeover. Their jobs now had a future.

Because of the influence of past experiences and future expectations, implementation is more critical in job redesign than in new job design. It is important to foster *ownership of change:* those responsible for carrying out and living with proposed redesigns should feel that the changes have come out of their own interests and efforts. A widespread commitment to a proposal is a vital factor to its ultimate success, for two reasons: (1) emotional identification with the redesign and the desire to put forth the effort to make it work, and (2) understanding the rationale of the design sufficiently to anticipate and minimize the effects, that is, to reduce the uncertainty over the consequences for the individual.

In both new design and redesign, the job-design consultant has a very different role from that of a conventional consultant. Because the design process is multiphasic and includes many parties, the internal and/or external consultant performs five basic functions:

Training and Guidance. The principles of job design and the methods of analysis are taught to members of design task forces and steering committees, who, in turn, make the design decisions. The consultant does not design the jobs, but rather guides the principals through the job-design endeavor.

Process Facilitation. The consultant helps plan and facilitate the process in all phases of the design, helping participants clarify their roles and identify emerging issues.

Mediating. When several distinct groups are party to a redesign project, for example, company and union, the consultant may act as a neutral broker and go-between in order to ensure consensual agreement for design and implementation.

Research. The assessment of job-design results may require research methods for which the consultant has special training and experience.

Change Agent. As innovative job designs are developed, the consultant acts as a carrier of these innovations to other parts of the institution of which the organization is a part. The role here is not one of "expert" so much as "idea broker."

4.1.14 ROLE OF THE HUMAN-FACTORS/ERGONOMICS SPECIALIST

Because job designs must meet organizational and individual as well as production needs, the human-factors specialist is advised to share the organizational task of designing jobs with other parties including line managers and the jobholders themselves. A full-fledged job-design effort includes (1) negotiating the scope of the project, (2) developing a design team and perhaps also a steering committee to oversee the project, (3) formulating a design philosophy and design criteria, (4) analyzing the technical and organizational tasks, (5) formulating several alternative design plans, (6) evaluating the potential costs and benefits of each plan, (7) gaining consensus for the final design plan, and (8) managing the processes of implementation and evolution. The human-factors specialist—in a classical "staff" or analyst role—participates in steps 3 through 6, using the analytic tools described in this chapter and elsewhere in this handbook, taking care that social and individual needs as well as production needs are accounted for. The practitioner could cast off the "specialist" role and take on a "leadership" role, participating in all steps, often with the aid of a job-design consultant.

4.1.15 SUMMARY AND CONCLUSION

Job design is a process of creating roles through which people in an organization can relate systematically with each other and with some product or service. Job design involves deciding what tasks will be performed by the work force, deciding how these tasks will be grouped together and assigned to individuals, deciding how individuals will relate to each other so that their work can be coordinated, and deciding how they will be rewarded. These decisions must take into account the needs of the organization as a production system, as a miniature society, and as a collection of individuals, embracing technical and economic criteria and also the needs and expectations that jobholders carry into the organization from the larger society. Although older models of job design rested on assumptions of certainty, modern practice focuses on enabling the organization to adapt to uncertainty. Because job design rests partially on technical-system and social-system factors, which may place unnecessary constraints on roles and relationships, the process of job design is inseparable from issues of organization design.

The high rate of change surrounding most organizations requires dynamic, multifaceted organizational structures and jobs. Job contents and relationships are under continual need to adapt to change. In modern organizations a well-managed evolutionary process is needed, in which the job contents and relationships are continually examined and modified as needed. This evolutionary process is most successful when it is participative, involving jobholders in the redesign of their own jobs and work relationships.

REFERENCES

Babbage, C. (1965). *On the economy of machinery and manufacturers* (4th ed. enlarged; originally published 1835). New York: Augustus M. Kelley.

Conant, E. H., and Kilbridge, M. D. (1965). An interdisciplinary analysis of job enlargement: technology, costs, and behavioral implications. *Industrial and Labor Relations Review, 18,* 377–390.

Davis, L. E. (1977). Evolving alternative organization designs, *Human Relations, 30,* 261–273.

Davis, L. E. (1982). Organization design. In G. Salvendy, Ed., *Handbook of industrial engineering.* New York: Wiley.

Davis, L. E., and Sullivan C. S. (1980). A Labour-Management Contract and Quality of Working Life. *Journal of Occupational Behavior, 1,* 29–41.

Davis, L. E., and Taylor, J. C. (1979). Technology and job design. In L. E. Davis and J. C. Taylor, Eds., *Design of jobs,* 2nd ed. Santa Monica, CA: Goodyear.

Davis, L. E., and Valfer, E. S. (1965). Supervisory job design. *Ergonomics, 8,* 1.

Davis, L. E., and Wacker, G. J. (1982). Job design. In G. Salvendy, Ed., *Handbook of Industrial Engineering.* New York: Wiley.

Davis, L. E., and Wacker, G. J. (in preparation). *Designing the adaptive organization.*

Emery, F. E., Ed. (1969). *Systems Thinking,* London: Penguin.

Engelstad, P. H. (1979). Sociotechnical apporach to problems of process control. In L. E. Davis and J. C. Taylor, Eds., *Design of jobs,* 2nd. ed. Santa Monica, CA: Goodyear.

Gyllenhammar P. G. (1977). *People at Work,* Reading, MA: Addison-Wesley.

Hackman, J. R., and Oldham, G. R. (1975). Development of the Job Diagnostic Survey. *Journal of Applied Psychology, 60,* 159–170.

Hazlehurst, R. J., Bradbury, R. J., and Corlett, E. N. (1969). A comparison of the skills of machinists on numerically controlled and conventional machines. *Occupational Psychology, 43* 3, 169–182.

Herbst, P. G. (1974). *Socio-Technical Design.* London: Tavistock.

den Hertog, J. F. (1977). The search for new leads in job design. *Journal of contemporary business, 6* 2, 49–66.

Herzberg, F. (1966). *Work and the nature of man.* Cleveland: World.

Jordan, N. (1963). Allocation of functions between man and machine in automated systems. *Journal of Applied Psychology, 47,* 161–165.

Rice, A. K. (1958). *Productivity and social organization: The ahmedabad experiment.* London: Tavistock.

Smith, A. (1970). *The wealth of nations.* London: Penguin (originally published 1776).

Taylor, F. W. (1911). *The principles of scientific management.* New York: Harper & Row.

Trist, E. L., Higgin, G. W., Murray, H., and Pollack, A. B. (1963). *Organizational choice.* London: Tavistock.

Wacker, G. J. (1979). Evolutionary job design: A case study (working paper). Madison, WI: Department of Industrial Engineering, University of Wisconsin.

Waddel, H. L. (1956). The fundamentals of automation. In H. B. Maynard, Ed., *Industrial engineering handbook.* pp. 325–331. New York: McGraw-Hill.

BIBLIOGRAPHY

Csikszentmihalyi, M. (1976). *Beyond boredom and anxiety.* San Francisco: Jossey-Bass.

Ford, R. N. (1979). *Why jobs die and what to do about it.* New York: AMACOM.

Herzberg, F. (1976). *The managerial choice: To be efficient and human.* New York: Dow Jones.

Hill, P. (1976). *Towards a new philosophy of management.* London: Gower.

Mumford, E., and Sackman, H., Eds. (1975). *Human choice and computers.* Amsterdam: North-Holland.

Nadler, D. A., Hackman, J. R., and Lawler, E. E. (1979). *Managing organizational behavior.* Boston: Little, Brown.

Schon, D. A. (1971). *Beyond the stable state.* New York: Random House.

Sheppard, H. L., and Herrick, N. Q. (1972). *Where have all the robots gone?* New York: Free Press.

CHAPTER **4.2**

PARTICIPATIVE GROUP TECHNIQUES

TAPAS K. SEN

AT&T
New York, New York

I am grateful to Drs. Richard J. Campbell, J. Douglas Carroll, Mary Carol Day, Kenneth Rifkin, Gavriel Salvendy, Ms. Maureen Tierney, and Mr. D. L. Lemasters for their valuable comments and suggestions in the preparation of this chapter.

4.2.1 INTRODUCTION

The past few decades have witnessed an impressive evolution of human factors from its early concentration on human–machine design. Today, human-factors specialists are involved in a broad spectrum of activities such as criminology, human–computer communication, environmental design, training technology, consumer products, safety systems, and several other areas. The fundamental objective of all these activities is to enhance or optimize the performance of the individual, the system, or the work organization under study. This chapter is devoted to the last topic, that is, how to improve organizational performance or effectiveness.

Human-factors scientists have developed various methods to contribute to the improvement of organizational performance. Job design (Chapter 4.1) and Organizational design (Chapter 4.3) are two such approaches discussed elsewhere in this handbook. This chapter describes yet another approach called participative group techniques, which require the involvement of an entire work group in setting its goals and finding ways of improving its effectiveness.

The concept of using a small work group as the basic unit for work design and work innovation has existed for a number of years. However, industries have started using this approach only recently. Depending on its specific characteristics, this approach has been given different names by its sponsors and practitioners. In the early 1950s, the concept of sociotechnical system was developed at the Tavistock Work Research Institute in England (Trist, Higgin, Murray, and Pollack, 1963). A strategy studied there was the reorganization of work by Welsh coal miners. One of the concepts that emerged was that of autonomous work teams in which the workers were given the freedom to design the work process without any rigid procedure being imposed upon them by higher authorities. This concept was later developed further by Einar Thorsrud and his coworkers at the Work Research Institute in Norway. It was called Democracy at Work (Emery and Thorsrud, 1976). More recently, a similar approach called The Quality of Work Life has been used by General Motors (Guest, 1979) and AT&T (Kofke et al., 1983). A similar effort at Ford has been called "Employee Involvement" (UAW–Ford, 1983).

The following three forces of change have influenced the recent adoption of the worker involvement approach by industries.

1. *Worldwide Competition.* Competition in major businesses such as the automobile, textile, electronics, and communicating industries has put pressure on management to improve quality and productivity, cut costs, and increase customer satisfaction. Success in these efforts requires a business strategy that is dynamic and flexible and an organization culture that values minimum supervisory controls and maximum worker commitment.

2. *The Advent of the So-Called Computer-Mediated Technology* This technology requires new worker skills and fundamental changes in the manufacturing process, in work design, and in management methods.

3. *The Emergence of New Work Values* These values are expressed by today's better-trained and better-educated workers whose work expectations are more varied than the basic pay, benefits, and safety concerns of the traditional workers (Skelly, 1981). In addition to these concerns, the new workers expect to participate in meaningful decisions that directly affect their day-to-day work. They value self-expression through work and look for opportunities to make contributions and be recognized.

Whether one speaks of Quality of Work Life (QWL), Employee Involvement (EI) or Industrial Democracy, a major element of this process is the concept of union–management collaboration and cooperation. This has been the single most important ingredient in the success of worker involvement efforts at AT&T, Ford, and General Motors. Although traditionally the unions and management have maintained an adversarial relationship, both parties have come to realize that there are situations when a joint effort in addressing some issues could benefit both parties. In the case of AT&T, for example, this realization led both parties to agree, during their 1980 bargaining, to establish joint union–management committees to explore ways of improving quality of work life as well as service to the customer.

For human-factors specialists, it is essential to understand the types of issues that are addressed in QWL or EI groups and how these issues eventually get resolved with union–management collaboration. In particular, where technical issues are addressed by a work group, the human-factors specialist can make a meaningful contribution by serving as a subject-matter expert to the work group. This is a relatively unfamiliar territory for the human-factors specialists, so experience and involvement are necessary before significant contributions can be made in solving contemporary industrial problems.

4.2.2   A Human Resource Strategy Model

Although group techniques such as QWL and EI encourage local innovation, the specific model or the strategies used may vary from group to group. One such model developed by the author is discussed here. Figure 4.2.1 describes the fundamental components of the model. Figure 4.2.1a describes the basic principles and Figure 4.2.1b the basic strategies. First, let us look at the principles in Figure 4.2.1a.

The base line of the triangle in Figure 4.2.1a is *profit*. It is indeed known as the "bottom line" in the business world. One must make profit in order to survive. Thus, it is vitally important that the union and all workers share this conviction. Every employee must believe that he or she has a responsibility to contribute to the company's profit.

However, profit can be made only if a business has something to sell—a product or a service. Productivity, then, becomes an essential medium. It is the vehicle that delivers the profit. Again, all employees must understand this principle and fully participate in the process that produces the goods or services.

The third principle in Figure 4.2.1a is a very important one for management to understand and follow. It says that the most important element of the productivity process is *people*. Machines are only aids to people. It's people who operate the machines and provide maintenance. It's people who manage work and make an organization successful. To be successful, a company needs the support of all of its people.

The major challenge faced by modern management is how to fully utilize and manage this vital resource— the people. How can management create an innovative and effective organization? What strategy should they follow in managing people? Let us look at Figure 4.2.1b, which describes three basic elements of such a strategy.

1. *The Individual Employee.*   Once the proper employee selection has been made, the supervisor or the manager must ensure that proper training and development opportunities are provided to the employee and that an environment is created for the employee to make relevant contributions. Among other things, such an environment must include effective communication and an effective reward/recognition system.

2. *Interaction among Employees.*   In a group process, all the members of a group must interact with one another. This interaction enables the group to establish group objectives, explore innovative work design, enhance productivity, and improve quality. The supervisor or the manager must ensure that proper interactions take place among group members and provide the necessary training and education on how to be effective in a group process.

3. *Integration of all Subgroup Activities.*   This integration is important because it enables the group to deliver its objectives. The supervisor or the manager should play the role of the integrator. In fact, in a participative group process, the supervisor becomes the integrator, as opposed to continuing the traditional role of controlling, monitoring, and regulating.

In the model described above, the ideal situation requires that the principles of people, productivity, and profit be appropriately coupled with the strategies of individual development, effective interaction, and productive integration. A successful manager is the one who understands the important relationship between the two base lines in Figure 4.2.1, profit on the one hand and individual development on the other.

The primary role of a human-factors specialist in such a model is that of a systems specialist. He or she can be extremely valuable in developing methods for optimum interaction and integration, and in providing the group with techniques for evaluating its success. If the group work involves

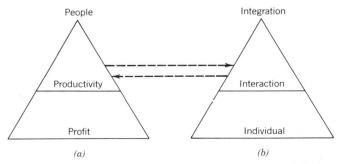

Fig. 4.2.1.   A human resource strategy model. (*a*) Basic principles. (*b*) Basic strategies.

any design activity that requires the implementation of human factors or ergonomics considerations, the specialist can serve as a teacher or a technical advisor. In any case, it is very important that the human-factors specialists are aware of such group activities going on within their company and make their services available.

4.2.3   MODELS OF PARTICIPATIVE GROUP TECHNIQUES

Group techniques, or processes that use participation as the central theme, can vary in their details depending on the parties involved. However, they all share some common elements. Three processes are discussed here, one involving nonexempt workers, one involving middle management, and one involving top management. In discussing these, an attempt is made to relate each process to a Leadership Model for Managing Change, developed by the author. The model assumes that managing change is a dynamic process and that it requires the top leadership of any organization to be continually involved in three levels of activities, or roles, as described below.

1. *Create, Catalyze, and Care.*   Any change brings along with it a certain amount of ambiguity or uncertainty. This ambiguity usually produces anxieties among employees and causes a subsequent loss of work efficiency. To avoid this loss of work efficiency, top leadership must *create* a vision of the future. This vision has to be articulated in terms of new corporate goals and must be communicated to all employees. Next, the top leadership must serve as a *catalyst* to generate enthusiasm among employees about the new vision. How this enthusiasm gets generated depends on individual management style, but one essential ingredient is *care*. Top management must demonstrate that they sincerely care about and value their employees. Care is essential to develop trust in top management, to generate true enthusiasm for the new vision, and to encourage employees to work toward the attainment of new corporate goals.

2. *Design, Develop, and Demonstrate.*   A method, an approach, or a system is needed if a new product or a new service is to be offered. Often this would require new business strategies or tactics. All these must be *designed* and *developed*. The extent to which top leadership gets directly involved in the design or development phase depends on the nature of the product, service, or strategies, but it is essential that top leadership allocate proper resources and provide guidance as required. Above all, top leadership has to *demonstrate* that the new product or service is feasible and that it has a good potential for success at the marketplace. This is needed to maintain high morale and enthusiasm among the employees and to project a good image of top management outside the company.

3. *Implement, Involve, and Improve.*   Once a prototype has been designed and developed or the basic demonstration project has been completed, the new process or the system must be implemented throughout the organization. Here top management must make sure that *all* employees are *involved* in this implementation phase. This strategy of total employee involvement in the implementation phase has made Japanese management highly successful in their business in recent years (Ouchi, 1981; Pascale and Athos, 1981).

We mentioned earlier that managing change is a dynamic process. Thus once a new system is in place for a new service or product, a continuous process of evaluation must continue. The marketplace can change abruptly, and therefore *improvement* of the process is needed, when required, to remain competitive. Perhaps most of this improvement or enhancement comes from the workers directly involved in the work, but top leadership must be alert, remain in touch with employees at all levels, and provide resources and encouragement whenever necessary.

We now turn to the three group techniques mentioned earlier.

4.2.3.1   A Process for Worker Involvement

Earlier we mentioned that a major element of the worker involvement process is union–management collaboration. The experience of companies with active unions strongly suggest that if QWL or similar group techniques are to be successful, a joint union–management support is essential. Usually, a worker involvement process has three components, although the specifics may vary from situation to situation.

The first component is a *top-level* union–management committee. Usually, it is made up of a few members of top management and an equal number of top leaders of the union. The CWA/AT&T model also includes an outside consultant as a member of the top-level committee. This person serves as a joint consultant to both the union and the company on QWL matters. All expenses are equally shared by the union and the company.

The main purpose served by this committee is providing leadership. Simply put, this committee makes QWL happen. It provides advice and counsel to local worker involvement or QWL committees and develops strategies to broaden the effort within the company. Such strategies may include developing educational modules such as seminars or workshops for top leadership in both the union and the company.

The second element is a *local* union–management committee. Again, like the top-level committee, this one is made up of both union and management representatives, usually equal in numbers. Participation is voluntary. Unlike the top-level committee, however, this committee gets involved with the QWL teams in their respective areas and provides help as needed by the teams.

The third component is the *worker involvement teams.* They are usually small in size and are called by various names such as Quality Circles, Quality Control Circles, QWL teams, problem-solving teams, and others (Kofke et al., 1983; Simmons and Mares, 1983; Thompson, 1982). Although the specific nature of these teams may differ from group to group, there are certain aspects fairly common to all. Five of these common characteristics are listed below.

1. Members of a team usually belong to the same work group or they work on the same project. They include nonexempt workers, union representatives, and the group supervisor.
2. Members sign up voluntarily.
3. The teams meet at frequent intervals, usually once a week or once every 2 weeks, during working hours.
4. They discuss work-related issues and recommend solutions to their management.
5. A union and/or management person called the facilitator frequently helps the team with group problem-solving techniques or related topics. Other training, if required, is provided by the company.

Further details about the structure and operation of these teams, their strengths as well as their shortcomings, are given below.

Quality circles (QC) is the most commonly used term for these groups. The idea was conceived by the Japanese Union of Scientists and Engineers in the 1960s. The basic objective of these groups was to ensure high quality of their products through the practice of statistical quality-control principles. The supporters of QC in Japan strongly believed that the *correct* solution to a problem can be arrived at only through a cooperative problem-solving process that involves both the manager and the workers.

Quality circles arrived on the American scene in the mid-1970s and spread quickly. During the first 8 to 9 years, most large American businesses embraced some form of a QC program. There are both similarities and differences between the Japanese and American approaches to QC. We limit our discussion, however, to the QC-type worker involvement programs as practiced in American business.

The Team Structure

Usually, the group or the team consists of a small number of employees, 10 or 12 at the most. It includes the workers and representatives of both management and the union. The workers are either co-members of a work unit reporting to the same supervisor or members of different work groups involved in a similar task. The former is more common.

The Team Objectives

Quality circles or worker involvement teams are brought together primarily to solve or recommend solutions to problems related to productivity and/or quality. However, once the teams start working together, other issues related to health, safety, and work practices may come up and be addressed by the team members. Often, some of these issues are familiar to human factors, and the teams can benefit from advice and counsel of human-factors professionals on these issues.

The Team Operation

Usually the teams meet once every week for an hour, sometimes once every 2 weeks. the important thing is that they should meet regularly, once the meeting interval has been jointly decided by the union and management. If a meeting has to be canceled owing to some emergency, management should do it in consultation with the union and the workers. Although participation is voluntary, all meetings take place on company time; that is, workers receive their normal pay while they attend team meetings.

The following is a list of items that have been found to contribute to successful team operation.

Each team should elect a leader.

Someone should be responsible for taking minutes and circulating them to team members.

An agenda should be set for each meeting and members should come prepared to discuss agenda items.

Each group should have a facilitator.

The Role of a Facilitator

The success of QC or EI teams depends heavily on the team's ability to identify the right issues, to collect relevant data, to perform proper analysis, to devise appropriate solutions, and finally to develop a group consensus. In all these activities, a facilitator can play a very significant role.

A facilitator, sometimes called a resource person, should be someone other than the group leader or the recorder of minutes. He or she helps the group by reviewing what's going on, filtering out unimportant issues, ensuring that the group maintains its focus, and helping them with group problem-solving techniques. Depending on the task and the team, the facilitator may also serve as the communication link between the team and upper management, and a resource for technical information, where applicable. Because of the importance of the facilitator, many organizations with a good QC or EI track record have developed effective facilitator training programs.

The Role for Human Factors

As mentioned elsewhere in this chapter, QC or EI is a new area for human factors. However, a significant contribution can be made in this area by the human-factors professionals. Currently, many human-factors practitioners work in isolation as specialists. Their involvement in a participative problem-solving process would certainly strengthen their role and enhance their chances for broader recognition within their organizations. In the participative mode, four new roles are suggested here for the human factors professional.

Facilitation.   The human-factors specialist can serve as the facilitator.

Group Consultation.   The human-factors specialist can play a major role as a group consultant to a QC or worker involvement team. In this role, he or she can help the group by providing various ideas for the group to address.

Technical Advice.   In this role, the human-factors person is called upon to suggest technical solutions to problems being considered by the group, or to help the group structure a technical problem in a way such that solutions can be discussed.

Problem-Solving Assistance.   The human-factors person is a well-trained problem solver. He or she could use this training to help the teams learn good problem-solving techniques.

Quality circles have certainly gone through a faddish stage in the United States. A recent article by Lawler and Mohrman (1985) provides some excellent suggestions about effective use of QC. However, based on my own experience with worker involvement teams in the area of QWL, the following "do's and don't's" are recommended for those who would like to try worker involvement for the first time.

1. The union must be involved from the beginning. The fundamental QWL or QC goals should be jointly formulated by the union and the company management.

2. Every effort should be made to demonstrate top management commitment for participative/worker involvement programs.

3. Resources should be made available to the worker involvement teams. This should include meeting facilities, various training required, and technical advice and support where needed.

4. The teams should get quick feedback on recommendations made by them to upper management.

5. A reward system should be in place to recognize teams making significant contributions.

6. Most often teams get isolated from their major work group. A process should be developed to facilitate communication between a QC/worker involvement team and the larger work group/groups of the team members.

7. The teams often start with easy problems and soon reach a plateau. Encouragement is needed from management to explore and address new issues.

8. Middle management resistance is a big hurdle. Every effort should be made to keep the communication channel open between the teams and their middle management. The primary objective of QC/worker involvement is to enhance employee satisfaction and work effectiveness. Middle management neither does nor should lose any control because of worker participation.

9. Every team activity must be regarded as a line activity, never a staff or personnel program.

10. To the human-factors specialist, this gives an opportunity to provide technical assistance and interpret human factors to a broader audience. However, he or she should never try to dominate the group. All decisions must be group decisions, rather than one expert's decisions.

11. For new starters, success is a slow process. Patience and encouragement are needed from upper management to sustain team momentum and motivation. The human-factors professional can help by providing the teams with new ideas.

Table 4.2.1 Leadership Roles for QWL Committees or Teams

| | Degree of Involvement | | |
Roles	Top-Level Committee	Local Committee	Team
Create, catalyze, and care	High	—	—
Design, develop, and demonstrate	High to medium	High to medium	Medium
Implement, involve, and improve	Medium to low	Medium	High to medium

Let us now examine how the three committee activities described above relate to the three leadership roles of the managing change model discussed earlier. Table 4.2.1 shows this relationship.

The first role, that is, create, catalyze, and care, is primarily the task of the top-level committee. Thus no involvement by the local committee or the team is indicated. The role of designing and developing the prototype belongs mainly to the team or the local committee. However, when basic strategies are concerned, the top-level committee is heavily involved. That is why, for both the top-level and the local committee, high to medium involvement is shown. In implementing or improving the process, the team is heavily involved, although both the top-level committee and the local committee play an important role in providing guidelines to the team, evaluating progress, and suggesting improvements when outside market forces require changes for survival.

Let us now look at some of the benefits that may accrue from such participative activities. Published documents show that participation and teamwork can produce significant improvements in profit and productivity. Quality improvements can also be equally impressive. Cost savings or quality improvement of over 50% have been reported (Serrin, 1984).

Union–management collaboration improves business climate and grievances go down significantly, leaving union leaders and management more time to devote to other productive issues (Heckscher, Kofke, Sen, Straw, and Tierney, 1985).

Despite this, employee involvement of participative management still remains fairly unknown to most managers. Broad training and education of management at all levels must be developed. Moreover, good research data on how to design effective QWL groups are not readily available. Often, design errors in the QWL committee structure cause failures (Lawler and Ledford, 1982). Human-factors specialists can certainly find these areas offering them a new challenge to make significant contributions.

4.2.3.2 A Participative Model for Middle Management

Much of what has been written recently on employee participation has dealt with nonexempt workers and the kind of union–management cooperation discussed in Section 3.1. However, if participation and involvement improve both work efficiency and effectiveness, they certainly should be useful for management-level employees as well. This awareness is slowly gaining momentum in the business world as it strives to improve white-collar productivity (Williams, 1976). Ford, for example, is encouraging participative approaches among its salaried employees. Its program is called "Participative Management and Salaried Employee Involvement" (PM/EI).

The middle management model discussed here can be used by any organizational unit involving middle and lower management. The basic strategy of this model requires that the process of employee participation be used as an intervention. It also requires the head of the unit to be personally involved as the change agent. There is no union involvement because participants in this model are not members of any union. Also, unlike the worker involvement process discussed in Section 3.1, participation is not voluntary. The unit head as the change agent has the challenging task of energizing all the members of the unit to participate fully in the process and enhance their personal contributions to the organization's goals.

The model uses two types of groups, a core group and subunit groups. The core group again can be of two types. In the first type, the core group consists of the top two or three levels of management within the unit. In the second case, it consists of representatives from all levels of employees. It is recommended that the size of the core group be kept under 20 Larger groups do not work out well. If the unit size is large, members of the core group should have a fixed term, such as 1 year, so that others can participate in the group as well. The major functions of the core group are as follows:

1. Identify key organizational issues
2. Set organizational goals
3. Devise new strategies
4. Provide support to the subunit groups

The subunit groups contain all members of the respective subunits. The major functions of these groups are as follows:

1. Identify work-related problems
2. Develop group consensus on how to solve them
3. Submit proposals for solutions to the core group for approval
4. Communicate final results to the core group

The Role of Opinion Surveys

The process of key issues identification can be greatly facilitated by the use of employee opinion surveys. A properly designed survey can produce valuable data for the employee participation groups. The survey can bring to the surface major organizational issues and identify both areas of strength and areas of weakness. Feedback of results to the group members can help them sort out the important issues and develop effective solutions. The combination of survey and feedback is also helpful in building trust between employees and their management.

The AT&T Survey Process

Employee attitude or opinion surveys are used by many companies. Some of these surveys are designed specifically for QWL type work. The AT&T approach is described here and provides an example of how these surveys are used to make employee participation more efficient.

The AT&T survey has been used as an adjunct to its QWL and other participative efforts. The underlying assumption of this survey—called the Work Relationships Survey—is that work involves four fundamental areas of relationships or interactions. They are as follows:

1. *Relationships with People.* This includes people such as subordinates, co-workers, supervisor, upper management, customers, suppliers, and others.
2. *Relationships with Physical Work Environment.* This includes human factors or ergonomics issues such as the design of tools and machines, the work space, the job itself, and the overall work system.
3. *Relationships with Policies and Practices.* This includes work rules and various corporate policies, including personnel policies.
4. *Relationships with Corporate Goals.* More specifically, this deals with the gap between employee and corporate goals and values. This is the central theme of today's interest in corporate culture.

Given these four basic areas of work relationships, the AT&T model assumes that for each one of these four areas, one must maintain a minimum level of satisfaction for work to be meaningful but should strive for an optimum level of satisfaction which is associated with a high level of work effectiveness. At this optimum level, individual work effectiveness almost reaches a plateau.

The Work Relationships Survey has been used in AT&T by both middle-management participative teams and union–management QWL teams. Some have used it as a tool to trigger interest in QWL or employee participation. In addition, the survey also has been used as an evaluative instrument to measure the effectiveness of a QWL-type intervention. In a classic experiment, Lemasters and his colleagues at AT&T used a unique combination of participative management and employee opinion survey to transform an organization from a noncompetitive to a successful market-oriented enterprise (Lemasters et al., in preparation).

Strengths versus Concerns

In addition to using "favorable," "unfavorable," and "neutral" results, as is commonly done with survey data, the Work Relationships Survey has also used two other measures that have been found useful by QWL practitioners. The first is the measure of "strengths and concerns." Figure 4.2.2 defines this measure in a graphic form.

As Figure 4.2.2 shows, in order for a relationships area to be recognized as an area of strength, the corresponding "% favorable" result from the survey must be *above 50%* and the "% unfavorable" result must be *under 30%*. The reverse, that is, "% favorable" under 50% and "% unfavorable" above 30%, is considered an area of concern or weakness. It should be pointed out that the actual cutoff points are chosen arbitrarily here and different organizations can choose their own criteria.

Importance versus Satisfaction

The second concept used by the Work Relationships Survey is the gap between "importance" and "satisfaction." For each major subject category included in the survey, the respondent was asked to judge how "important" it was to him or her and how 'satisfied' he or she was with the same. A gap exists when the importance score is higher than the corresponding satisfaction score. This gap suggests

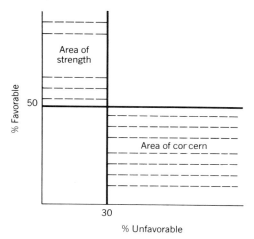

Fig. 4.2.2. Definition of strength and concern areas.

room for improvement. The matrix in Figure 4.2.3 shows how these data can be conveniently presented. The third quadrant (lower left), with asterisks, highlights those issues that have high importance but low satisfaction. The high and low cutoff points could be set by each group.

The two kinds of results described above, strengths versus concerns and importance versus satisfaction, are convenient to use and usually are very helpful to the participative teams in sorting out the top priority issues for action planning.

Let us now examine how the activities of the core and subunit groups relate to the three leadership roles in our managing change model discussed earlier (Table 4.2.2).

The role of create, catalyze, and care belongs to the core group. However, as the leader and the change agent, the unit head has the primary responsibility to help the group create the vision and teach them how to catalyze the rest of the organization. This is somewhat different from the union–management committee discussed earlier, in which there is no single leader. The core group also has the responsibility of designing and developing the new strategies, and demonstrating their usefulness. If a survey is used, it is the core group that should decide the issues to be addressed and take part in the design. The subunit groups get heavily involved in the implementation of strategies and in the process of improving their group outputs. The core group makes certain that the employees remain involved and provides them guidance and assistance as required.

4.2.3.3 Impact Scaling: A Participative Process for Top Management

The management of corporations may be centralized or decentralized. In the decentralized mode, independent units usually run themselves as profit centers, with minimal guidance from the top. In the case of centralized management, a top executive group controls the policies and strategies that guide the various units or departments. In either case, however, the unit heads seldom interact with one another to optimize the overall company objectives. From a corporate point of view, these units suboptimize their own performance.

	IMPORTANCE	
	High	Low
High	2	1
Low	******* 3	4

SATISFACTION

Fig. 4.2.3. Importance versus satisfaction.

Table 4.2.2   Leadership Roles for Middle-Management Participative Team

Roles	Degree of Involvement	
	Core Group	Subunit Group
Create, catalyze, and care	High	—
Design, develop, and demonstrate	High	Low
Implement, involve, and improve	Low	High

Suboptimization works as long as the business environment does not change unpredictably and historical trend lines predict the business future fairly accurately. However, in today's world, social, political, economic, or technological changes take place rapidly, sometimes unpredictably. Therefore, competitive survival depends on a well-coordinated management team that develops integrated strategies for optimum corporate performance. Seldom can the success of an isolated unit ensure a healthy survival of the total corporation.

The participative group technique to be described here was developed at AT&T (Carroll, 1977; Sen and Bozzomo, 1977). Its primary objective is to help the chief executive of any organization develop a strategic "corporate convergence," jointly with his or her immediate subordinates (this team is called the core group in this process). By "corporate convergence," we mean that the core group, after a thorough discussion of the issues, *converges* on a common understanding of the issues and then develops a strategic plan to address those issues. The process is called impact scaling for two reasons:

1. Its main focus is on the possible *impact* of environmental changes on the future health of the corporation.

2. It uses a statistical technique called "multidimensional scaling" which aids in producing future scenarios of the organization under study. These scenarios are examined by the core group before appropriate strategic decisions are made.

The impact scaling process requires the involvement of two groups, the core group as defined above and a staff group that provides research and administrative support to the core group. Once the strategies are formulated, however, all other subunits within the organization get involved in developing tactics to respond to the strategies. The process consists of six steps:

Step 1. Identify Strategic Health Indicators

This step requires the core group to identify, select, and agree on the key health indicators in the strategic areas of the business. These indicators should be no more than 10 or 12 in number. The indicators selected must be measurable, objectively or subjectively, so that their movement in the right direction can be evaluated.

This indicator selection process requires a few meetings (two to four) of the core group, during which an extensive discussion takes place among the group members. Usually, a "consensor" or a similar machine is used which allows anonymous rating of each issue by core group members. A graphic display of the rating distribution stimulates group discussion. Such discussions clarify the interpretations of the issues and then another rating exercise follows. The iterative process continues until group convergence is reached. It should be noted that this process is somewhat different from the well-known Delphi technique which has been used in similar tasks (Sackman, 1974).

The staff group prepares a background paper on a number of possible indicators. Each core group member is required to be familiar with the contents of this paper before coming to the meeting.

Step 2. Select Key Events

In this step, the core group members identify events that have either happened or are likely to happen within the planning horizon. Usually this horizon is taken to be 5 years, but may vary depending on the situation. The events could be either internal (important business decisions) or external (social, political, economic, or technological) to the organization. In any case, the events to be discussed must be those that would have some impact on the key health indicators selected in Step 1. The key event window in Figure 4.2.4 is used as a guide by the core group in selecting the events. First, the events are placed in the four quadrants according to their overall importance in terms of probability of happening and impact. Then no more than 20 or 25 most important events are chosen, based on

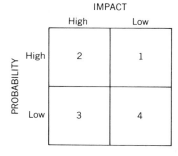

Fig. 4.2.4.   Key event window.

their overall importance. Quadrant 2 events are evaluated first, then quadrant 3 events, and finally quadrant 1 events. Quadrant 4 events, that is, low-probability, low-impacts cases, are discarded.

The staff group prepares background papers on a number of events for the core group. These papers provide the basic points for discussion by the core group. The event selection process follows the same rating and discussion exercise as described in Step 1 for the key health indicators. It is recommended that a trained facilitator be used for the group discussions on events and indicators. However, the chief executive must be actively involved and serve as a tie-breaker should a tie occur.

Step 3. Make Event Assessments

This step is optional. If event probability is not the subject of interest, this step is not required. However, it should be used if the estimate of event probabilities would make planning more realistic. The core group members are required, in this step, to provide two kinds of probability estimates. First, each one provides an estimate of the probability of each event occurring within the planning horizon. We call it the *unconditional probability*. Then for each event, the members provide other estimates of the probability of its happening, given that each other event happens. We call these extimates the *conditional probabilities.*

The matrix in Figure 4.2.5 shows an example of the kind of data this rating exercise would generate. We call it the interaction matrix data. It is similar to the cross-impact matrix (Gordon and Hayward, 1968) and the related technique of cross-impact analysis, on which a lot of research has been done (Blackman, 1973; Jackson and Lawton, 1976; Mitroff and Turoff, 1976). The top row in Figure 4.2.5

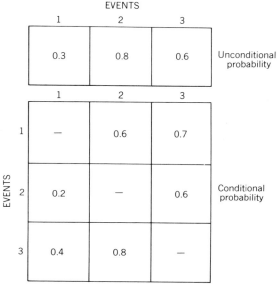

Fig. 4.2.5.   Interaction matrix.

provides the unconditional probabilities of three events used in this example. The lower matrix provides the conditional probabilities. Each number within a cell denotes the probability of the row event, given that the column event happens. Note that the matrix is not symmetrical, which happens in most cases. This means that unless two events are orthogonal (unrelated), in which case the conditional probability is the same as the unconditional probability, the cross-impacts of events may be different depending upon which event happens first (i.e., the impact of event A on B is not necessarily the same as that of B on A).

Step 4. Make Impact Estimates

This step is *not* optional. It requires each core group member to provide, for each event, its impact on each key health indicator. The impact matrix shown in Figure 4.2.6 is an example of the kind of data this step would generate. In this example, three key health indicators are used. They are income, dividend, and market share. For each of the three events shown in Figure 4.2.5, we show what their impact would be on the three indicators, in terms of percentage increase or decrease, assuming that the event happens. Usually this increase or decrease is defined relative to the trend forecasts, which are made assuming no environmental changes. Thus, the top left means that if event 1 happens, it increases income by 5% over the trend forecast of income. Similarly, if event 2 happens, it decreases income by 4% from trend forecast, and so on.

Step 5. Perform Statistical Analyses

In this step, two types of analyses are performed. The *first* one takes the interaction matrix data obtained in step 3 and performs a multidimensional scaling analysis as explained by Carroll (1977) in his discussion of the impact scaling model. The output of this analysis summarizes the major areas of concern inherent in the interaction matrix data. In the mathematical model of Impact Scaling, these areas of concern are called *dimensions*.

Figure 4.2.7 shows an example of a multidimensional scaling graphical output. The output shown is called an *environmental scenario*. The data are from an impact scaling study (Study A) in which 13 events were used by the core group. Four types of events were used—political, employee-related, pricing-related, and competition-related. Two dimensions are plotted here, one against the other. Each one is bipolar; that is, the extreme ends have different meanings. The horizontal dimension reflects the concern of what kind of competition the company would encounter within the planning horizon and what kind of competitive responses the company could provide. The second dimension reflects the concern of employee needs on the one hand and the flexibility of pricing products and services on the other hand.

The events are plotted as data points against these two dimensions (x–y plot). They are indicated by their classification codes. The projection of each event on each dimension reflects how important that dimension is to that event. The vector called Prob Environ indicates the most probable business environment within the framework of these areas of concern. In this example, it means that the most probable business environment within the planning horizon is the one in which the company would keep facing new competition and new pricing challenges.

Although senior executives find the interaction matrix exercise time-consuming, the outputs are quite valuable in helping them focus their attention in the right strategic direction. Sometimes it also

| | INDICATORS | | |
	Income	Dividend	Market share
EVENTS 1	+5%	+2%	+3%
EVENTS 2	−4%	0	0
EVENTS 3	+2%	0	0

Fig. 4.2.6. Impact matrix.

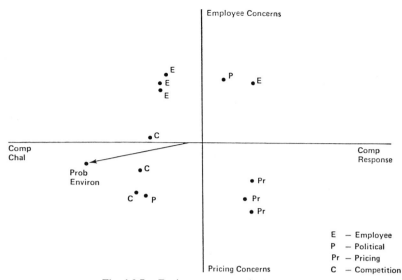

Fig. 4.2.7. Environment scenario (study A).

can identify planning gaps. One such example is shown in Figure 4.2.8 from another impact scaling study (Study B). Sixteen events were used in this study. Two dimensions are shown here and only those events are plotted that were significantly related to these two dimensions. Three types of events, related to the employees (EMP), marketing (MARK), and outside competing companies (OCC), were used.

The horizontal axis (dimension) reflects the strategic and tactical concerns, and the vertical axis represents the dimension of competing forces on the one hand and company responses on the other.

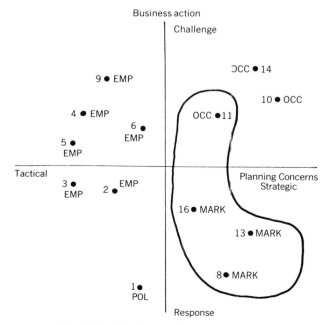

Fig. 4.2.8. Environmental scenario (study B).

Three significant competing forces OCC 10, OCC 11, and OCC 14 were identified but it was discovered that all the marketing responses, MARK 8, MARK 13, and MARK 16 would successfully respond to only one competing force, OCC 11. Thus there was a planning gap, and the core group went to work on the appropriate responses to forces OCC 10 and OCC 14.

The *second* type of statistical analysis performed in this step uses the impact matrix data of step 4 and produces what is called the *base scenario* and the *pessimistic scenario*. The base scenario shows the *expected impact* on the business indicators of all the events. In producing the base scenario, no change in the probability estimates of the interaction matrix is assumed. The pessimistic scenario, on the other hand, assumes that all the undesirable events would happen (probability of 1.0). Figure 4.2.9 shows the base and pessimistic scenario results obtained in Study B. Ten business indicators were used.

Indicators 1 and 2 were income-related indicators. When the core group saw the results of the pessimistic scenario, it was quite obvious to them that proactive action was necessary to ensure healthy income in the future.

Step 6. Produce Enhanced Scenario

To help the core group make appropriate proactive decisions, a chart called an impact map is used as shown in Figure 4.2.10. Only fictitious data are used in this figure for the purpose of demonstration. The horizontal axis is the probability of events (0 to 1.0) and the vertical axis is the degree of impact (positive and negative). In this example, events 2, 13, 18, 27, 68, and 108 have negative impacts, with high probability of occurrence. Events 19, 23, 61, 71, and 72, on the other hand, have positive impacts, also with high probability of occurrence. In situations like this, two kinds of proactive actions can be taken: (1) For those events that are *controllable,* actions should be taken to make the desirable events happen and undesirable events not happen. That will make the corresponding probability values 1.0, and 0, respectively. (2) For *uncontrollable* events, actions should be taken to enhance the positive impacts and mitigate the negative impacts.

The core group in Study B performed the exercise described above and produced new probability values (0 or 1.0) for those events they could control. Where impact changes were also possible, the group agreed on new impact estimates they thought they could accomplish. The impact scaling model was run again with these new data. The result was the enhanced scenario as depicted in Figure 4.2.9. The enhanced scenario, it should be noted, produces a much better income expectation (indicators 1 and 2) than the base scenario.

After such proactive strategic decisions are made, the strategies are communicated down through

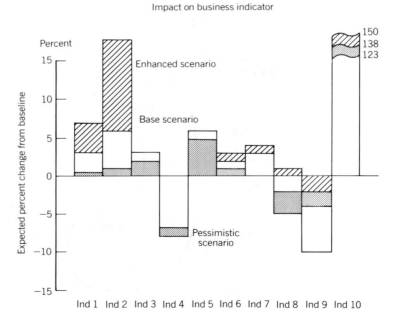

Fig. 4.2.9. Future impact scenarios (study B).

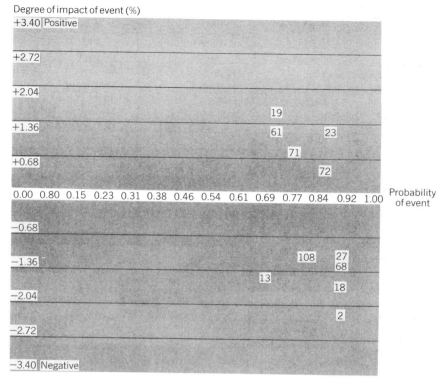

Fig. 4.2.10. Impact map.

the management levels, and appropriate subunits are chosen to design tactical plans for the success of the strategies. Figure 4.2.11 summarizes the process and describes the various steps involved.

1. Proper environmental analysis is performed; interaction and impact Matrix data are collected.
2. The impact scaling model fitting process is run.
3. Strategic plans are formulated by the core group.
4. Integrated tactical plans are developed by subunits to respond to the strategies. A transition plan is required if current tactical plans are designed to produce different objectives.

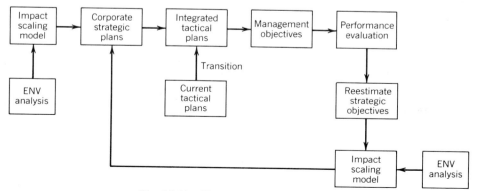

Fig. 4.2.11. The integrated impact scaling process.

Table 4.2.3 Leadership Roles for Top Management Participative Team

Roles	Degree of Involvement		
	Core Group	Staff Group	Subunit Group
Create, catalyze, and care	High	—	—
Design, Develop, and demonstrate	Medium to high	Medium to high	Low
Implement, involve, and improve	Low	Medium to High	High

5. The integrated tactical plans are translated into management objectives for individual employees at all levels.

6. At certain predetermined intervals, the performance of the tactical plans is evaluated. Two conclusions may be derived if the performance results are not satisfactory: either the tactical plans are not working efficiently, or the strategies developed are unrealistic. In the former case, corrective actions are taken for the tactical plans. In the latter case, new objectives are formulated and the impact scaling model fitting process is run again. If new environmental data are available, they also are included as input to the model. The new model outputs are now used to develop new strategies and the process is repeated.

Let us now go back to our managing change model and examine the roles played by the various groups involved in the impact scaling process, directly or indirectly. As Table 4.2.3 shows, the core group has the primary responsibility for the create, catalyze, and care role. The chief executive plays the most important role. However, his or her immediate subordinates play equally important roles in their respective units. The staff group or the subunits do not have primary responsibility in this area.

For the next role, design, develop, and demonstrate, both the core group and the staff group have medium to high involvement depending on whether they are working on strategic or tactical plans. The subunits get somewhat involved in this role when tactical plans are being worked on.

For the next role, implement, involve, and improve, mostly the staff group and the subunits get heavily involved. However, the core group must continually show their interest in the tactical plans and provide the necessary resources.

Finally, let us see how human-factors practitioners could be involved in such a process. Three major human-factors roles can be identified. *First,* the human-factors specialists can use such a process themselves to identify and recommend to top management the kind of technology that would be most beneficial to their company. It should include critical considerations like cost, labor versus capital supply, productivity enhancement, new worker skills/supply, job design and work satisfaction. *Second,* human-factors specialists should be included in the staff group. They could contribute to the background papers related to new technology and assist the core group with any questions they may have on human factors or ergonomics related matters. *Third,* human-factors specialists could make significant contributions to the development of tactical plans and the evaluation of their effectiveness.

REFERENCES

Blackman, A. W., Jr. (1973). A cross-impact model applicable to forecasts for long-range planning. *Technological Forecasting and Social Change, 5,* 233–242.

Carroll, J. D., (1977). Impact scaling: Theory, mathematical model, and estimation procedures. *Proceedings of the Human Factors Society—21st Annual Meeting,* pp. 513–517.

Emery, F., and Thorsrud, E. (1976). *Democracy at work: The report of the Norwegian Industrial Democracy Program.* Leiden: Martinius Nijhoff.

Gordon, T. J., and Hayward, H. (1968). Initial experiments with the cross impact matrix method of forecasting. *Futures, 1,* 100–117.

Guest, R. (1979, July/August). Quality of work life—learning from Tarrytown. *Harvard Business Review,* 76–87.

Heckscher, C., Kofke, G. E., Sen, T. K., Straw, R., and Tierney, M. (1985). Quality of work life: AT&T and the Communication Workers of America examine process after three years. Washington, DC: *U.S. Department of Labor.*

Jackson, J. E., and Lawton, W. H. (1976). Some probability problems associated with cross-impact analysis. *Technological Forecasting and Social Change, 8,* 263–273.

Kofke, G., Donohue, J., Heckscher, C., Jensen, B., Ketchum, W., Maccoby, M., Straw, R., and Williamson, L. (1983, February). *Roadmap for successfully managing quality of work life.* CWA/Bell System Joint National Committee on Working Conditions and Service Quality Improvement. Basking Ridge, NJ: AT&T.

Lawler, E. E., III, and Ledford, G. E., Jr. (1982). Productivity and the quality of work life. *Human Resources Productivity.* New York: Executive Enterprises Publications Co.

Lawler, E. E., III, and Mohrman, S. A. (1985, January-February). Quality circles after the fad. *Harvard Business Review,* 64–71.

Lemasters, D. L., et al. (In preparation). Participative management and organizational change. AT&T.

Mitroff, I. I., and Turoff, M. (1976). On the distance between cross-impact models: A set of metric measures for cross-impact analysis. *Technological Forecasting and Social Change, 8,* 275–283.

Ouchi, W. (1981). *Theory Z: How American business can meet the Japanese challenge.* Reading, MA: Addison-Wesley.

Pascale, R. T., and Athos, A. G. (1981). *The art of Japanese management.* New York: Simon & Schuster.

Sackman, H. (1974, April). *Delphi Assessment: Expert opinion, forecasting, and group process.* U.S. Air Force Project Rand, R-1283-PR.

Sen, T. K., and Bozzomo, R. E. An application of impact scaling for corporate strategic planning. *Proceedings of the Human Factors Society—21st Annual Meeting,* pp. 518–519.

Serrin, W. (1984, December 2). Giving workers a voice of their own. *The New York Times.*

Simmons, J. and Mares, W. (1983). *Working together,* Chapter 6. New York: Alfred A. Knopf.

Skelly, F. (1981). "How to manage new values vs. old values: Performance is the key. In L. K. O'Leary, Ed., *Quality of work life, the human dimensions.* New York: AT&T.

Thompson, P. C. (1982). Quality circles at Martin Marietta Corporation. In R. Zager and M. P. Rosow, Eds., *The innovative organization.* New York: Pergamon.

Trist, E., Higgin, G. W., Murray H., and Pollack, A. B. (1963). *Organizational choice.* London: Tavistock.

Williams, E., (1976). *Participative management: Concepts, theory and implementation.* Atlanta, GA: College of Business Administration, Georgia State University.

The UAW-Ford National Joint Committee on Employee Involvement (1983). *EI handbook II.* Ford Motor Company.

CHAPTER 4.3

ORGANIZATIONAL DESIGN

HAL W. HENDRICK

University of Southern California
Los Angeles

Traditionally, human factors/ergonomics has tended to focus on the design of specific jobs, work groups, and related human–machine interfaces. Typically, in system design, the operations to be required of the system to accomplish its purposes are identified. These operations, in turn, are analyzed to identify the specific functions that constitute them. The human-factors specialist often comes into the design process at this point, and based on his or her professional knowledge of human performance capabilities and limitations, assists in allocating these functions to humans or machines. From this point in the process, the human-factors specialist's knowledge of human–machine technology is applied to designing specific jobs, integrating jobs into work groups, and then designing specific human–machine interfaces including controls, displays, work-space arrangements, and work environments.

Although applied within a systems analysis framework, most of the above described activities actually are at the individual, team, or at best, subsystem level. In short, they represent what herein shall be referred to as human-factors applications at the *microergonomic* level. The focus of this section is on the application of human factors in system design at the *macroergonomic* or overall organizational level. Conceptually, it is entirely possible to do an outstanding job of microergonomically designing a system's components, modules, and subsystems, yet fail to reach relevant system effectiveness goals or criteria because of inattention to the macroergonomic design of the system.

The classic example of this problem was the introduction of the *longwall* method of mining in a British deep-seam coal mine. The traditional mining system was largely manual in nature. It utilized teams of small, fairly autonomous groups of miners. Control over work was exercised by the group itself. Each miner performed a variety of tasks, and most jobs were interchangeable among workers. The workers derived considerable satisfaction out of being able to complete the entire "task." Sociotechnically, the psychosocial characteristics, the characteristics of the external culture, the task requirements, and the system's organizational design were congruent. The more automated, technologically advanced longwall method replaced this costly manual method of mining. No longer restricted to working a short face of coal, miners now could extract coal from a long wall. However, this new and more technically efficient system resulted in an organizational design that was not congruent with the psychosocial and cultural characteristics of the work force. Rather than small groups, shifts of 10 to 20 men were required. Workers were restricted to narrowly defined tasks. Job rotation was not possible. There now was a high degree of interrelationship among the tasks of the three shifts, and problems from one shift carried over to the next, thus holding up labor stages in the extraction process. This complex and highly rigid design was very sensitive to both productivity and social disruptions. Instead of improved productivity, low production, absenteeism, and intergroup rivalry became common (Degreene, 1973). Later, in studies of other coal mines by the Tavistock Institute of Human Relations in London (Trist, Higgin, Murray, and Pollock, 1963), this conventional longwall method was compared with a *composite* longwall method in which the system's organizational design utilized a combination of the new technology and features of the old psychosocial work structure of the manual system. In comparison with the conventional longwall method, the composite longwall system reduced the interdependence of the shifts, increased the variety of skills utilized by each worker, and permitted self-selection by the workers of their team members. Production was significantly higher than for either the conventional longwall or the old manual system. Absenteeism and other measures of poor morale dropped dramatically.

Based on the Tavistock Institute studies, Emory and Trist (1960) concluded that different organizational designs can utilize the same technology. The key is to select the organizational design, or subset of designs, that is most effective in terms of (1) the people who will constitute the human portion of the system, and (2) relevant external environments, and then employ the available technology in a manner that achieves congruence.

Although different technologies can utilize different organizational designs, technology, *once employed in the design of a system*, does constrain the subset of possible designs. With the introduction of microelectronics and related automation into managerial, administrative, production, logistical, marketing, and other facets of our modern complex systems, the systems' organizational designs become far more constrained then in more traditional, labor-intensive human–machine systems. Because of this progressively increasing automation, it increasingly has become important to consider first what is the optimal macroergonomic design for the system before proceeding to the microergonomic design of human–machine modules, subsystems, and interfaces. In short, "a top-down ergonomic approach is essential to insure that the dog wags the tail, and not visa versa" (Hendrick, 1984).

4.3.1 SOME BASIC CONCEPTS

In this section the major dimensions and major generalizable models of organizational design and their use in the macroergonomic design of systems are reviewed. In order to provide a common framework, a few basic concepts are clarified first.

4.3.1.1 The Meaning of Organization

An organization may be defined as *"the planned coordination of two or more people who, functioning on a relatively continuous basis and through division of labor and a hierarchy of authority, seek to achieve a common goal or set of goals"* (Robbins, 1983, p. 5). Breaking this definition down, we can note the following.

1. Planned coordination of collective activities implies *management*. Activities do not just emerge, but are premeditated. Implicit in planned coordination in complex, high-technology organizations is a need for information and decision support systems to facilitate management of the system. The design of these support systems must be compatible with the organization's design.

2. Because organizations are made up of more than one person, individual activities must be designed and functionally allocated so as to be complementary, balanced, harmonized, and integrated to ensure effective, efficient, and relatively economical functioning. This design activity first requires a macroconception of the overall organizational design in order to ensure that the result will be an effectively functioning *system*.

3. Organizations accomplish their activities and functions through a division of labor and a hierarchy of authority. Thus organizations have *structure*. How this structure is designed is likely to be critical to an organization's functioning.

4. The collective activities and functions of an organization are oriented toward achieving a common goal or set of goals. From a system design standpoint, this implies that criteria for assessing the effectiveness of an organization's design exist, and should be explicitly identified, weighted, and utilized in evaluating feasible alternative designs for the overall organization.

4.3.1.2 Organizational Structure

We noted in paragraph 3 above that the concept of organization, with its division of labor and hierarchy of authority, implies *structure*. An organization's structure may be conceptualized as having three components: *complexity, formalization,* and *centralization* (Robbins, 1983).

1. *Complexity* refers to the degree to which organizational activities are differentiated, and the extent to which integrating mechanisms are utilized to coordinate and facilitate the functioning of the differentiated components.

2. *Formalization* is the degree to which an organization relies on rules and procedures to direct the behavior of the people in a human–machine system.

3. *Centralization* refers to the extent to which the locus of decision-making authority is either centralized or dispersed downward within the hierarchy.

The nature of each of these components, including operational measures and design guidelines for each, is discussed below.

4.3.1.3 Organizational Design

Organizational design refers to the design of an organization's structure to achieve the organization's goals. From a macroergonomics standpoint, as part of the system design process, organizational design involves (1) identifying the system's purpose or goals; (2) making explicit the relevant measures of organizational effectiveness (OE); weighting them, and subsequently utilizing these OE measures as criteria for evaluating feasible alternative structures; (3) systematically developing the design of the three major components of organizational structure; (4) systematically considering the system's *technology, psychosocial,* and relevant *external environment* variables as moderators of organizational structure; and (5) deciding on the general *type* of organizational structure for the system. These five "steps" of the organizational design process are considered in turn and constitute the remainder of this section.

4.3.2 ORGANIZATIONAL GOALS

Goals may be defined as the *objectives* of the organization. They are the desired states of affairs that organizations are designed to achieve. Goals may be classified by *criteria, focus,* and *time frame* (Szilagyi and Wallace, 1983).

4.3.2.1 Classification by Criteria

Raia (1974) has identified the following as among the most frequently used criteria.

1. *Productivity.* Productivity goals usually are expressed as levels of output per unit or per worker across the organization. Units produced per employee per day, costs per unit of production, or income generated per employee are commonly used operational measures of productivity.

2. *Market.* Market goals can be operationally defined in a variety of ways. Examples might be to increase the market share for a given product by 10% (market share goal), or to sell a specified number of units next year (output oriented goal).

3. *Resources.* Organizations sometimes establish goals concerning changes in their resource base. Examples might include reducing the company's long-term debt by 20 million dollars in 5 years (financial base goal), increase the production capacity by 30% in 3 years (physical resource goal), or decrease turnover by 5% next year (human resource goal).

4. *Profitability.* Profitability usually is expressed as a ratio, such as net income, earnings per share, or return on investment.

5. *Innovation.* Because of rapid technology change, increasingly important to many systems is the development of new products to maintain their competitive position. An innovation goal might be to develop a new, more efficient manufacturing process within 3 years, or to develop a new computer having a specified increased data processing capacity by a given date.

6. *Social Responsibility.* In part, because of culturally based psychosocial changes in the work force, social responsibility goals are becoming increasingly important to an organization's effectiveness. These goals might center around such factors as improving the quality of work life or reducing pollution.

Achieving the goals within each of these classes may be facilitated or inhibited by the organization's design.

4.3.2.2 Classification by Focus

Szilagyi and Wallace (1983) note that classifying goals by focus entails describing the *nature of the action* that will be taken. They identify three frequently used categories.

1. *Maintenance Goals.* Maintenance goals usually are stated as the specific level of activity or action that is to be sustained over time. An example for an airline would be to have at least 80% of its aircraft in service at one time.

2. *Improvement Goals.* Any goal that includes an *action* verb is likely to be an improvement goal, as it indicates a specific change that is wanted. Examples might be "increasing" market share, "decreasing" accidents, or "improving" return on investment.

3. *Development Goals.* Development goals are similar to improvement goals, but refer to some form of growth, expansion, learning, or advancement. Examples might be increasing the number of new products introduced, increasing the educational level of managers, or increasing plant capacity.

An advantage of expressing goals in terms of focus is that they are easily understood by persons having diverse backgrounds, such as managers, production workers, and system designers, and serve to orient their thinking and activities.

4.3.2.3 Classification by Time Frame

As we shall see below, when one considers environmental influences on complexity, classification of goals by time frame can be very useful in the organizational design process. Using this approach, goals usually are classified as long, intermediate, or short term.

1. *Short-Term Goals.* Short-term goals usually concern those that cover a period of 12 months or less. Frequently, production goals take this form.

2. *Intermediate-Term Goals.* Intermediate goals usually span a period of from 1 to 3 years. Often, sales organizations have goals with an intermediate time orientation.

3. *Long-Term Goals.* These are goals that typically will cover a period of 3 years or longer. Research and development goals most often fall into this category.

Although the goals for any given organizational activity can be a combination of all three time frames, as suggested by the illustrative examples above, system activities tend to have a predominant time orientation. Often it is the nature of the functional environmental demands on a given activity that determines its particular time orientation. For example, research and development of new products take 3 to 5 years or more in most cases. Thus this function must be oriented at least this far into the future in order to be aware of product requirements to use as research and development design criteria. Failure for the research and development function to have a long-term goal orientation likely would result in the development of products that are obsolete or unwanted by the time they are produced and marketed.

4.3.2.4 Hierarchical Nature of Goals

It should be noted that when goals are translated into objectives, they become *ends.* In analyzing these goals, system designers must evaluate alternatives as to *how* they will be achieved, or the *means.* The means selected at one hierarchical level become the goals or ends for the next lower level. The hierarchical flow of means–ends for a system has strong design implications for the structural differentiation of the organization. To a significant extent, the division of labor within the organization will be a direct outcome of the means-ends analysis (Szilagyi and Wallace, 1983). For example, if the overall goal for a new system is to improve transportation in a large metropolitan area, a number of alternatives could be considered, such as new highways or a subway, streetcar, bus, light rail, or monorail system. The choice of system, in turn, serves as the ends at the next hierarchical level, for it strongly affects selection of the types of organizational units that will be involved in operating and maintaining the system (i.e., from a system design standpoint, selection of units would be based on answering the question, "What types of functions are required to best accomplish the type of system selected?"). The approaches (means) chosen by the system designers for meeting the operational and maintenance goals within the units in turn serve as the goals for the next, or subunit, level of organization. The approaches (means) chosen for meeting these ends at the subunit level in turn affect selection of the subunit functions to be designed into the system and hence the grouping of activities into subunits. The system design choices for enabling these subunits to meet their goals in turn affect their division of labor.

4.3.3 ORGANIZATIONAL EFFECTIVENESS CRITERIA

If, as part of the system design team, human-factors specialists are to participate in the macroergonomic design of organizations, they must have a means of evaluating the relative *effectiveness* of various structural arrangements that appear feasible. During the 1960s and early 1970s numerous studies were carried out to identify criterion measures of organizational effectiveness (OE). A review of these studies by Campbell (1977) identified 30 different criteria of OE. These are presented in Table 4.3.1.

One thing that is reflected by these 30 criteria is that no single measure of OE is sufficient. Different organizational functions are likely to require evaluation based on different sets of characteristics. The task of the human-factors specialist, working with other members of the system design team, is to establish which combination of OE criteria are relevant for evaluating each aspect of a proposed organizational design, and to weight them in terms of their importance to system functioning. Those OE criteria that are selected then must be operationally defined in terms that are meaningful for the particular system (e.g., aircraft manufacturer, public utility, oil refinery) and system subfunctions (e.g., sales, marketing, production).

It should be noted that often the selection of specific OE criteria, and their relative weighting, are value judgments. The stated goals for a system (themselves often a reflection of value judgments) can help considerably in the selection of OE criteria, because some criteria are direct reflections of these goals. Others, however, are less tied to a specific, explicitly stated goal. For example, OE criterion 15 from Table 4.3.1, *flexibility/adaptation*, may be very important to the effectiveness of a particular organization or subfunction in responding to its external environment, but it may not be explicitly stated. Rather, it may be implicit across several goals as possibly important to their attainment.

A second factor to note is that a number of criteria represent *competing values*. For example,

Table 4.3.1 Criteria and Measures of Organizational Effectiveness

1. *Overall effectiveness.*   The general evaluation that takes into account as many criteria facets as possible. It is measured usually by combining archival performance records or by obtaining overall ratings or judgments from persons thought to be knowledgable about the organization.

2. *Productivity.*   Usually defined as the quantity or volume of the major product or service that the organization provides. It can be measured at three levels: individual, group, and total organization via archival records or ratings or both.

3. *Efficiency.*   A ratio that reflects a comparison of some aspect of unit performance to the costs incurred for that performance.

4. *Profit.*   The amount of revenue from sales left after all costs and obligations are met. Percentage return on investment or percentage return on total sales are sometimes used as alternative definitions.

5. *Quality.*   The quality of the primary service or product provided by the organization that may take many operational forms, which are determined largely by the kind of product or service provided by the organization.

6. *Accidents.*   The frequency of on-the-job accidents resulting in lost time.

7. *Growth.*   Represented by an increase in such variables as total work force, plant capacity, assets, sales, profits, market share, and number of innovations. It implies a comparison of an organization's present state with its own past state.

8. *Absenteeism.*   The usual definition stipulates unexcused absences, but even within this constraint there are a number of alternative definitions (e.g., total time absence versus frequency of occurrence).

Table 4.3.1 (*Continued*)

9. *Turnover.*   Some measure of the relative number of voluntary terminations, which is almost always assessed via archival records.

10. *Job satisfaction.*   Has been conceptualized in many ways but the modal view might define it as the individual's satisfaction with the amount of various job outcomes that he or she is receiving.

11. *Motivation.*   In general, the strength of the predisposition of an individual to engage in goal-directed action or activity on the job. It is not a feeling of relative satisfaction with various job outcomes but is more akin to a readiness or willingness to work at accomplishing the job's goals. As an organizational index, it must be summed across people.

12. *Morale.*   The model definition seems to view morale as a group phenomenon involving extra effort, goal communality, commitment, and feelings of belonging. Groups have some degree of morale, whereas individuals have some degree of motivation (and satisfaction).

13. *Control.*   The degree, and distribution, of management control that exists within an organization for influencing and directing the behavior of organization members.

14. *Conflict/cohesion.*   Defined at the cohesion end by an organization in which the members like one another, work well together, communicate fully and openly, and coordinate their work efforts. At the other end lies the organization with verbal and physical clashes, poor coordination, and ineffective communication.

15. *Flexibility/adaptation.*   Refers to the ability of an organization to change its standard operating procedures in response to environmental changes.

16. *Planning and goal setting.*   The degree to which an organization systematically plans its future steps and engages in explicit goal-setting behavior.

17. *Goal consensus.*   Distinct from actual commitment to the organization's goals, consensus refers to the degree to which all individuals perceive the same goals for the organization.

18. *Internalization of organizational goals.*   Refers to the acceptance of the organization's goals. It includes the belief that the organization's goals are right and proper.

19. *Role and norm congruence.*   The degree to which the members of an organization agree on such things as desirable supervisory attitudes, performance expectations, morale, role requirements, and so on.

20. *Managerial interpersonal skills.*   The level of skill with which managers deal with supervisors, subordinates, and peers in terms of giving support, facilitating constructive interaction, and generating enthusiasm for meeting goals and achieving excellent performance.

21. *Managerial task skills..*   The overall level of skills with which the organization's managers, commanding officers, or group leaders perform work-centered tasks and tasks centered on work to be done and not the skills employed when interacting with other organizational members.

22. *Information management and communication.*   Completeness, efficiency, and accuracy in analysis and distribution of information critical to organizational effectiveness.

23. *Readiness.*   An overall judgment concerning the probability that the organization could successfully perform some specified task if asked to do so.

24. *Utilization of environment.*   The extent to which the organization interacts successfully with its environment and acquires scarce and valued resources necessary to its effective operation.

25. *Evaluations by external entities.*   Evaluations of the organization, or unit, by the individuals and organizations in its environment with which it interacts. Loyalty to, confidence in, and support given the organization by such groups as suppliers, customers, stockholders, enforcement agencies, and the general public would fall under this label.

26. *Stability.*   The maintenance of structure, function, and resources through time and, more particularly, through periods of stress.

27. *Value of human resources.*   A composite criterion that refers to the total value or total worth of the individual members, in an accounting or balance sheet sense, to the organization.

28. *Participation and shared influence.*   The degree to which individuals in the organization participate in making the decisions that affect them directly.

29. *Training and development emphasis.*   The amount of effort that the organization devotes to developing its human resources.

30. *Achievement emphasis.*   An analog to the individual need for achievement referring to the degree to which the organization appears to place a high value on achieving major new goals.

Source: Adapted with permission from John P. Campbell, "On the Nature of Organizational Effectiveness," in P. S. Goodman, J. M. Pennings, and Associates, Eds., *New Perspectives on Organizational Effectiveness.* San Francisco: Jossey-Bass, 1977, pp. 36–41.

flexibility/adaptation versus *stability,* and *participation and shared influence* versus *control* represent several of the more obvious conflicting pairs. Striking the right balance between these competing values in the design of a particular system's organizational structure may be critical to its success. Factors that can aid making these value judgments are presented in the discussion of the next two "steps" in the organizational design process.

4.3.4 DESIGNING THE DIMENSIONS OF ORGANIZATIONAL STRUCTURE

As was previously noted in defining "organizational structure," the structure of a human–machine system can be conceptualized as having three core dimensions. These are *complexity, formalization,* and *centralization* (Robbins, 1983).

4.3.4.1 Complexity

Complexity refers to the degree of *differentiation* and *integration* that exists within an organization. Organizational structures embody three major types of differentiation. These are *horizontal differentiation* or the degree of horizontal separation between units, *vertical differentiation* or the depth of organizational hierarchy, and *spatial dispersion* or degree to which the location of an organization's facilities and personnel are dispersed geographically. An increase in any one of these forms of differentiation increases an organization's complexity.

As the differentiation of an organization increases, the need for integrating devices also increases. These mechanisms are needed because with greater differentiation of the organization's activities, the difficulty of communication, coordination, and control also increases. Some of the more common integrating mechanisms that can be designed into the system are formal rules and procedures, liaison positions, committees, system integration offices, and computerized information and decision support systems. It also should be noted that vertical differentiation in itself is a form of integration mechanism, because a manager at one level serves to coordinate and control the activities of several lower-level managerial or worker positions.

4.3.4.2 Horizontal Differentiation

Horizontal differentiation refers to the degree of job specialization and departmentalization that is designed into the organization's structure. Job specialization leads to greater complexity because it requires more sophisticated and expensive methods of control. Yet specialization is common to virtually all high-technology systems because of the inherent efficiencies in the division of labor. Adam Smith (1970/1776) demonstrated this point more than 100 years ago. He noted that 10 men, each doing particular tasks (job specialization), could produce about 48,000 straight pins per day. However, if each man worked separately and independently, performing all the production tasks, those 10 workers combined would be lucky to make 200.

Division of labor creates groups of specialists. The way in which human-factors personnel group these specialists is known as *departmentalization.* Departments can be designed into a system on the basis of (1) function, (2) simple numbers, (3) product or services, (4) client, (5) geography, and (6) process. Most large corporations will use all six (Robbins, 1983).

1. *Function.* This is the most popular basis for grouping activities. Thus in a typical manufacturing firm there may be departments devoted to manufacturing, sales, research and development, and accounting.
2. *Simple Numbers.* This is the simplest way of grouping activities. For example, if we have 30 employees and three supervisors, we could simply divide the employees into groups of 10, and assign a supervisor to each. This method is most likely to be utilized in small organizations, at the lowest level of more complex organizations, and with unskilled or semiskilled jobs.
3. *Product or Service.* This approach is effectively utilized in the design of large corporations. Each product division, such as Chevrolet, Pontiac, Oldsmobile, Buick, and Cadillac in the case of General Motors, operates with considerable autonomy, supported by its own functionally grouped departments. This form of departmentalization is to be preferred when there are diverse or rapidly changing product or service lines.
4. *Client.* The type of client or customer serviced can be the most effective basis for ergonomically organizing activities. For example, an aerospace manufacturing corporation might have separate sales organizations for military, commercial, and private aviation customers. This method of departmentalizing activities should be considered whenever the system's clients have distinctly different needs or service requirements that require consolidation for more effective, comprehensive service.
5. *Geography.* Geography or territory can be a particularly useful basis of departmentalization for facilitating product or service distribution, or for any situation where localized knowledge enhances decision-making or responsiveness. For these reasons, sales organizations frequently are organized by territory.

6. *Process.* This form of organization is useful where a customer or product must go through a series of units because of specialized equipment or personnel needs. In this approach, different activities and skills are grouped in terms of the process that must be followed. For example, on U.S. Air Force bases personnel service activities, such as the issuance of identification cards and motor vehicle registration, are grouped in a customer service office. These offices utilize military police, personnel administrators, and other occupational specialties.

Horizontal differentiation requirements can be *quantitatively* assessed by using two basic measures. These are (1) the number of occupational specialties that will be required, and (2) the level of training to be designed into the jobs. The greater the number of occupations and the longer the period of training to be required by the system's design, the greater the need for differentiation in the design of the system's organizational structure.

4.3.4.3 Vertical Differentiation

The measure of vertical differentiation is simply the number of levels separating the chief executive position from the jobs directly involved with the system's output. In general, as the size of an organization increases, the need for vertical differentiation also increases (Mileti, Gillespie, and Haas, 1977). In one study, size alone was found to account for 50 to 59% of the variance (Montanari, 1976). A key factor underlying this size–vertical differentiation relationship is *span of control*. Any one supervisor is limited in the number of subordinates that he can direct effectively (Robbins, 1983). Thus as the number of positions increase, the number of first-level supervisory jobs that must be designed into the system also increases. This increase, in turn, increases the number of second-, third-, and higher-level managerial positions required. For example, if the span of control is eight, and the organization has 512 worker positions, the number of first-level supervisory jobs is 64, of second-level managerial positions, 8, and of third-level, 1 (e.g., the chief executive officer or plant manager). There are thus three levels of supervision. If the organization has 4096 employees, the number of supervisory levels increases to four (i.e., 512 first-level supervisors, 64 second-level managers, 8 third-level, and 1 fourth-level).

Although span of control limitations underlie the size–vertical differentiation relationship, these limitations can be quite varied, depending on a number of factors. Thus although a span of control of eight requires four hierarchical levels of supervision with 4096 employees, a span of control of 16 would require only three supervisory levels. Thus as span of control can be increased, organizations tend to become flatter and broader in shape. A major factor affecting the span of control that is desirable is the degree of *professionalization* of employee positions. In general, as the degree of professionalization (education and skill requirements) designed into employee jobs increases, the employees are better able to function autonomously with only a minimum of supervision. Thus the span of control for a given supervisor can be increased. Related to professionalization, certain *types* of jobs require less direction and control than others. A number of other variables also can affect vertical differentiation requirements. These include the type of technology, environmental factors, and psychosocial variables. These are discussed separately below.

4.3.4.4 Spatial Dispersion

Spatial dispersion refers to the performance of a system's activities in multiple locations. In a sense, it is an extended dimension to horizontal and vertical dispersion in that it is possible to separate both power and task centers geographically. Spatial dispersion measures include (1) the number of geographic locations within the organization, (2) the average distance of the separated places from the organization's headquarters, and (3) the number of employees in these separated locations in relation to the number in the headquarters location (Hall, Haas, and Johnson, 1967). In general, complexity increases as the number of geographically separated units from the headquarters, the proportion of personnel located at these separated units, and the average distance of these units from the headquarters increase.

The Horizontal, Vertical, Spatial Differentiation Relation

When one looks at many very large or very small organizations, it would seem that there is a high intercorrelation among the three types of differentiation. Large high-technology corporations, such as ITT, General Motors, and Exxon, are characterized by a high level of all three kinds of differentiation. The corner grocery store, local service station, and neighborhood dry cleaner, on the other hand, have little of any kind of differentiation. However, beyond these two extremes there is little systematic relationship among the types of differentiation. Some fairly large organizations, such as an army battalion, are characterized by a high degree of vertical differentiation, but relatively little horizontal differentiation or geographical dispersion; universities typically have little vertical differentiation or geographical dispersion, but a high degree of horizontal differentiation (Hage and Aiken, 1967); some small retail chains have little horizontal or vertical differentiation, but geographically dispersed units. In short, the optimal level of each type of differentiation for an organization has to be individually assessed in terms of the variables that affect it. Some of these factors we already have noted in describing the measurement and nature of each type of differentiation. Other key variables are discussed in Section 4.3.5.

4.3.4.5 Integration

Once the nature and degree of differentiation have been determined, human-factors specialists must pay particular attention to the resultant integration needs of the system. In part, the nature of the integration devices that will be utilized will be determined by the degree of formalization and centralization utilized in the system's organizational design, because formalization devices and centralization are themselves integrating mechanisms. The extent to which computer-based information and decision support systems form part of the systems technology also will be a factor, for these too are integrating mechanisms. They thus form part of the system's organizational design, and care must be taken to ensure that they are ergonomically designed to be an integrated, compatible part of the system's organizational structure.

The type of technology to be utilized in the system, external environment variables, and psychosocial factors also help to determine the optimal integration mechanisms that should be designed into the system's structure. These are considered below.

4.3.4.6 Formalization

For ergonomic design purposes, formalization can be defined as the degree to which jobs within the organization are standardized. In highly formalized designs, jobs allow for little employee discretion over what is to be done, when it is to be done, and how it is to be accomplished (Robbins, 1983). There are explicit job descriptions, extensive rules, and clearly defined procedures covering work processes. Often the design of the system hardware and human–machine interfaces in themselves prohibit employee discretion. Where formalization is low, jobs are designed to allow employees considerable freedom to exercise discretion. Employee behavior thus is relatively unprogrammed, and usually allows for considerably greater use of one's mental capacities. Low formalization also usually necessitates greater training or experience as a part of the individual job requirements.

In general, the simpler and/or more repetitive the jobs to be designed into the system, the greater the value of formalization for effective system integration and functioning. The greater the professionalism designed into the jobs, the less the need for, or utility of, high formalization. In fact, it is likely to inhibit both employee motivation and effective functioning. Related to this is the kind of work to be performed. Thus production jobs with stable repetitive characteristics lend themselves to relatively high formalization; whereas research and development or sales, which may require considerable flexibility, innovation, or responsiveness to change are likely to be stifled by a high degree of formalization. As is discussed below, the degree of predictability and stability or uncertainty and change in the relevant external environments of the organization and its constituent units also affect the degree of formalization that should be ergonomically designed into the system. Thus from a macroergonomics perspective, in considering the classic trade-offs between selection, training, and hardware skill requirements to be designed into the system, the human-factors specialist also must consider (1) the relative stability of the external environments with which the organization and its constituent units interact, and (2) the related degree of formalization that is optimal for effective system functioning. With the widespread introduction of computer-based information and decision support systems, these last two considerations are equally important to the design of system software.

4.3.4.7 Centralization

Centralization refers to the degree to which formal decision-making is concentrated in an individual, unit, or level (usually high in the organization), thus permitting employees (usually low in the organization) minimal input into the decisions affecting their jobs (Robbins, 1983). It should be noted that centralization essentially is concerned with *decision discretion*. Where decisions are delegated downward, but policies or other formalization mechanisms exist in the system to constrain the discretion of lower-level positions, there is in reality increased centralization. Conversely, the transference of information in systems requires filtering. Decisions often are made at intermediate hierarchical levels as to what information gets passed upward to higher management. The greater the extent to which the system is designed to permit information to be reduced, summarized, selectively omitted, or embellished, the less the extent to which the actual decision is concentrated and controlled by the centralized decision maker. In short, the less is the actual degree of centralization.

The filtering of information to the decision maker, discussed above, illustrates that from a systems standpoint the actual making of the decision in itself does not determine the degree of centralization. Rather, it is the degree of control that one holds over the decision-making *process* that is the true measure of centralization. With this in mind, it should be noted that the actual decision only indicates the *intended* action. As the decision passes down through the hierarchical levels to those who will actually execute it, it too may undergo filtering. The system's design may even formally permit modification, delay, or rejection of the decision. This kind of formal filtering is a desirable ergonomic design feature where conditions require safeguards to prevent implementation of a decision in the light of conditions, perhaps unknown to the decision maker, that could be detrimental to the system, or that could pose a safety hazard. From an ergonomic design standpoint, however, it is equally important to *prevent* filtering of the decision where it is likely to inhibit effective system functioning (i.e., where it is likely to dilute the desired level of true centralization).

In general, centralization is desirable (1) when a comprehensive perspective is required, such as in strategic decision-making, (2) when it provides significant economies, (3) for financial, legal, and other decisions where they clearly can be done more efficiently when centralized, or (4) when operating in highly stable and predictable relevant external environments. Decentralization is desirable (1) when the design of a given manager's job will result in taxing or exceeding human information processing and decision making capacity; (2) in order to enable the organization to respond rapidly to changing or unpredictable conditions at the point where change is occurring; (3) in order to provide more detailed "grass roots" input into decisions; (4) to provide greater motivation to employees by allowing them to participate in decisions that affect their jobs, (5) more fully utilize their mental capacities, and (6) increase their sense of psychological significance to the organization and to themselves; (7) to gain greater employee commitment to, and support for, decisions by involving them in the process; and (8) to provide greater training opportunity for lower-level managers.

More will be said about psychosocial, environmental, and technological factors as determinants of centralization below.

4.3.4.8 Relationship of Centralization, Complexity, and Formalization

Centralization and Complexity

In general, research indicates an inverse relationship between centralization and complexity (Robbins, 1983; Hage and Aiken, 1967; Child, 1972). For example, as the number of occupational specialties and related training requirements designed into a system increase, the expertise required to make sound decisions also increases, thus forcing decentralization for effective system functioning. Conversely, the simpler and more repetitive the jobs, the greater the utility of centralized decision making.

Centralization and Formalization

No clear, simple relationship has been found between centralization and formalization. Other factors tend to moderate these relationships. For example, with a high degree of job professionalization, both low formalization and decentralization tend to be optimal. On the other hand, with unskilled repetitive jobs, both formalization and centralization tend to be needed (Hage and Aiken, 1967). However, the *type* of decision moderates this relationship. Professionals expect decentralization of decisions that affect their work directly, together with a low level of formalization, but *not* (1) decisions concerning personnel issues (e.g., salary and performance appraisal procedures) where the predictability that comes with standardization is desired, and (2) strategic decision-making, which is enhanced by the more comprehensive perspective of centralization and which has little direct impact on their daily activities (Robbins, 1983).

Formalization and Complexity

The relationship between formalization and complexity tends to be a function of (1) the *direction* of differentiation, and (2) the degree of professionalization (Robbins, 1983). High horizontal differentiation, if achieved by increasing the number and kinds of routine repetitive tasks, results in the need for a high degree of formalization (Pugh, Hickson, Hinings, and Turner, 1968). If it is achieved by increasing the number and kinds of highly skilled, complex positions (professionalization) low formalization should be optimal, along with decentralized decision-making (Hage, 1965). High vertical differentiation usually involves designing in an increased number of managerial and technical specialists (professionalization) and nonroutine tasks. Thus for optimal functioning, low formalization should be incorporated into the organization's design for these positions (Robbins, 1983; Hage, 1965).

4.3.5 SOCIOTECHNICAL SYSTEM COMPONENTS AS MODERATORS OF ORGANIZATIONAL DESIGN

Based on the Tavistock Institute Studies, mentioned earlier, Emory and Trist (1960) coined the term *sociotechnical system* to convey more adequately the nature of complex human–machine systems. The sociotechnical system concept views organizations as open systems engaged in transforming inputs into desired outcomes. *Open* systems means that sociotechnical organizations have permeable boundaries exposed to the environments in which they exist. These environments thus permeate or enter the organization along with the inputs to be transformed. The primary ways in which environmental changes enter the organization are through the people who work in it, through its marketing or sales function, and through its materials or other input functions (Davis, 1982).

As transformation agencies, organizations are in constant interaction with their environment. They receive inputs from their environment, transform these into desired outputs, and export outputs to their environment. Organizations bring two critical factors to bear on the transformation process, technology in the form of a *technological subsystem,* and people in the form of what, in human-factors parlance, is known as the *personnel subsystem.* The design of the technical subsystem defines the tasks to be performed. The design of the personnel subsystem prescribes the ways in which the tasks are performed. Each interacts with the other at every human–machine interface. The technical and personnel subsystems are thus mutually interdependent. Both subsystems operate under *joint causation,* meaning that they both are affected by causal events in the environment. The technological subsystem, once designed, is relatively stable and fixed. It thus falls to the personnel subsystem to adapt further to environmental change. Joint causation leads to the related sociotechnical concept of *joint optimization.* Joint optimization means that because the technological and personnel subsystems respond jointly to causal events, optimizing one subsystem and fitting the second to it results in suboptimization of the joint *system.* Thus maximizing overall system effectiveness requires jointly optimizing the technological and personnel subsystems. The need for joint optimization requires the joint design of the technical and personnel subsystems in order to develop the best possible fit between the two, given the objectives and requirements of each, and of the overall system (Davis, 1982).

As is inferred above, the design of an organization's structure involves consideration of three major sociotechnical system components that interact and affect optimal organizational design. These are (1) the *technological subsystem,* (2) the occupational roles and relationships or *personnel subsystem,* and (3) selected characteristics of the relevant *external environments* that permeate the organization. Each of these components has been studied in relation to its effects on the three organizational design dimensions, reviewed earlier, and empirical models have emerged.

4.3.5.1 Technology

Technology, as a determinant of organizational design, has been operationally classified in three distinctively different ways: (1) by the mode of production, or *production technology;* (2) by the action individuals perform upon an object to change it, or *knowledge-based technology;* and (3) by the strategy selected for reducing uncertainty, which is determined by technology, or *technological uncertainty.* From each of these classification schemes, themselves empirically derived, a major generalizable model of the technology–organizational design relationship has been empirically developed.

Woodward: Production Technology

The classic series of studies of technology as a determinant of organizational structure was conducted by Joan Woodward and her associates (1965). She and her colleagues studied 100 manufacturing firms in South Essex, England having at least 100 employees. The organizations varied greatly in terms of size, type of industry, managerial levels (2 to 12), span of control (2 to 12 at the top; 20 to 90 at the first-line supervisory level), and ratio of line employees to staff personnel (less than $1:1$ to more than $10:1$). Through interviews, observations, and review of company records, the following, among other factors, were noted for each firm: (1) the organization's mission and significant historical events, (2) the manufacturing processes and methods utilized, and (3) the organization's success, including changes in share of the market, relative growth or stagnation within its industrial field, and fluctuation of its stock prices.

As part of her analysis, Woodward identified three modes of technology: (1) *unit,* (2) mass, and (3) *process* production. These modes were seen as representing categories on a scale of increasing *technological complexity.* At the least complex end were the unit and small batch producers. These firms produced custom-made products. Next were the large batch or mass production firms, such as those that produce automobiles and other more or less standardized products using predictable, repetitive production steps. The organizations highest in production complexity were long-run, heavily automated process production firms, such as oil and chemical refiners. Three important organizational structure variables were found to increase as technological complexity increased. First, as *technological complexity increased, the degree of vertical differentiation also increased.* For each class of technology, the successful firms tended to have the median number of hierarchical levels for that category. This optimum for unit producers was three, for mass it was four, and for process, six. The less successful firms in each category had a noticeably greater or lesser number of levels. Second, as *technological complexity increased, the optimal ratios of administrative support staff to industrial line personnel increased.* Third, as *technological complexity increased, the span of control of the top-line managers increased.* Woodward's findings for the successful firms in each technology mode are summarized in Table 4.3.2 for the three organizational design dimensions, and were as follows.

1. *Unit production* firms had low complexity. First-line supervisors had relatively narrow spans of control and there was little line and staff differentiation. Jobs were widely rather than narrowly defined. Formalization and centralization both were low.

2. *Mass production* organizations had high complexity. First-line supervisors had wide spans of control, there was clear line and staff differentiation, and jobs were narrowly defined. Formalization and centralization both were high.

3. *Process production* units had high vertical differentiation with wide spans of control, including the lower supervisory levels. There was little line and staff differentiation. Formalization and centralization both were low.

Several follow-up studies have lent support to Woodward's findings (Harvey, 1968; Zwerman, 1970). It should be noted, however, that she implies causation when, in fact, her methodology really establishes only correlation. In applying Woodward's findings in organizational design, another caution should be observed. Woodward's data were collected from within a single culture and at a particular time. In a different culture, or at some other time, the sociocultural and other environmental factors might be very different, and thus result in somewhat different interactions with production mode in terms of their influence on organizational design.

Perrow: Knowledge-Based Technology

A major shortcoming of Woodward's model is that it applies only to manufacturing firms, which constitute less than half of all organizations. Perrow (1967) has empirically developed a more generalizable model of the technology–organizational design relationship using a *knowledge-based* rather than a *production* classification scheme. He began by defining technology as the action one performs upon an object in order to change that object. This action requires some form of technical knowledge. Using this approach, he identified two underlying dimensions of knowledge-based technology. The first of these concerns the number of exceptions encountered in one's work, or *task variability*. The second has to do with the type of search procedures one has available for responding to task exceptions, or *task analyzability*. These procedures can range from "well defined" to "ill defined." At the "well defined" end of the continuum, solving problems involving task exceptions can be accomplished using logical and analytical reasoning. At the "ill defined" end there are no readily available formal search procedures, and one must rely on experience, judgment, and intuition. The combination of these two dimensions, when dichotomized, yields a two-by-two matrix having four cells. As shown in Table 4.3.3, each cell represents a different knowledge-based technology.

1. *Routine* technologies have few exceptions and well defined problems. Mass production units fall within this category. Routine technologies are best accomplished through standardized coordination and control procedures, and are associated with high formalization and centralization.

2. *Nonroutine* technologies have many exceptions and difficult to analyze problems. Aerospace operations would be an example. These technologies require flexibility, and thus should be decentralized and have low formalization.

3. *Engineering* technologies have many exceptions, but can be handled using well defined rational-logical processes. They therefore lend themselves to centralization, but require the flexibility that is achievable through low formalization.

4. *Craft* technologies typically involve relatively routine tasks, but problems rely heavily on experience, judgment, and intuition for decision. Thus problem solving must be done by those with the particular expertise. This requires decentralization and low formalization.

Perrow's model has been supported by considerable empirical research, both in the private and public sectors (e.g., Van deVen and Delbecq, 1974; Hage and Aiken, 1969; Magnusen, 1970).

Table 4.3.2 Summary of Woodward's Findings on the Design Features of Effective Organizations

Organizational Structure	Mode of Production		
	Unit	Mass	Process
Complexity			
Vertical differentiation	Low	Moderate	High
Horizontal differentiation	Low	High	Moderate
Formalization	Low	High	Low
Centralization	Low	High	Low

Table 4.3.3   Perrow's Technology Classification

		Task Variability	
		Routine with Few Exceptions	High Variety with Many Exceptions
Problem Analyzability	Well Defined and Analyzable	Routine	Engineering
	Ill Defined and Unanalyzable	Craft	Nonroutine

Thompson: Technological Uncertainty

Thompson (1967) has demonstrated that the type of technology determines a *strategy* for reducing uncertainty, and that specific structural arrangements facilitate uncertainty reduction. He identifies three types of technologies, based on the tasks that an organization performs, (1) long-linked, (2) mediating, and (3) intensive.

1. A *long-linked* technology accomplishes its tasks by a sequential interdependence of its units, as might be illustrated by an automobile assembly line. Because it involves a fixed sequence of repetitive steps, the major uncertainties in this type of organization are at the input and output sides. Thus management responds to uncertainty by controlling inputs and outputs. This is accomplished primarily through planning and scheduling, and suggests a moderately complex and formalized structure.

2. A *mediating* technology is one that links clients on both the input and output sides, thus performing a mediating or interchange function. In short, mediators, such as banks, utility companies, and the post office, link units that otherwise are independent. A mediating technology is characterized by a pooled or parallel interdependence of the different units. These otherwise independent units are bound together by rules, regulations, and standard operating procedures. They thus function best with low complexity and high formalization.

3. An *intensive* technology is one that represents a customized response to a diverse set of contingencies. It involves a variety of techniques that are drawn upon to transform an object from one state to another. The particular techniques employed are, at least in part, based upon feedback from the object itself. The classic example is a hospital, where the object being transformed is the patient. The techniques that can be employed are varied, and are selected largely on the patient's condition and response to previously used techniques. The major uncertainty is the object itself. The flexibility of response, such as many alternatives, is a must for effective system functioning. An intensive technology thus operates best with a structure characterized by high complexity and low formalization.

Thompson's model, unfortunately, has not been fully tested empirically. The lack of data therefore makes it impossible to draw a definitive conclusion regarding the model's validity (Robbins, 1983). The one study of consequence analyzed 297 subunits for 17 business and industrial firms (Mahoney and Frost, 1974). This study provided partial support for Thompson's model by demonstrating that long-linked and mediating technologies were associated closely with formalization and advanced planning, whereas intensive technologies were characterized by mutual adjustments to other units.

4.3.5.2   Environment

Critical not only to the functioning of organizations but to their very survival is their ability to adapt to their environments. In terms of open systems, organizations require monitoring and feedback mechanisms to follow and sense changes in their relevant task environments, and the capacity to make responsive adjustments. "Relevant task environment" refers to that part of the organization's external environment that is made up of the firm's critical constituencies that can positively or negatively influence the organization's effectiveness. Neghandi (1977), based on field studies of 92 industrial firms in five different countries, has identified five external environments that significantly impact on organizational functioning.

1. *Socioeconomic.*   Particularly the degree of stability, nature of the competition, and availability of materials and workers.

2. *Educational.*   The availability of facilities and programs, and the educational level and aspirations of workers.

3. *Political.*   The degree of stability, and the governmental attitudes toward (1) business (friendliness versus hostility), (2) control of prices, and (3) "pampering" of industrial workers.

4. *Cultural.*   Social status and caste system, values and attitudes toward work, management, etc., and the nature of trade unions and union–management relationships.

The particular task environments that are relevant are different for each organization in type, qualitative nature, and importance. The particular weighted combination of relevant task environments for a given organization or subunit can be thought of as its *specific task environment.* A major determinant of an organization's specific task environment is its *domain,* or the range of products or services offered and markets served (Robbins, 1983). Domain is important because it determines the points at which the organization depends on its specific environment (Thompson, 1967).

A second major determinant of an organization's environment is its *stakeholders.* The stakeholders in the firm's environment include stockholders, lenders, members of the organization, customers, users, governmental agencies, and the local community, among others. Each of these has interests in the organization and a potential for action that could significantly affect the organization's future. The design of the organization must be capable of responding to the objectives and related actions of its various stakeholders. For example, as a transformation agency, the organization's technological subsystem has to respond to technical, economic, market, and other aspects of its environmental domain; has to meet governmental regulations for pollution, safety, and so on; has to meet the needs of its stockholders; and has to satisfy the needs of its workers for interesting work, careers, and quality social relationships on the job, among others (Davis, 1982).

Of particular importance to organizational design is the fact that all relevant task environments vary along two critical dimensions. These are (1) the degree of *change* and (2) the degree of *complexity* (Duncan, 1972). The degree of change refers to the extent to which a given task environment is dynamic or remains stable over time. The degree of complexity refers to whether the components of an organization's specific environment are few or many in number. For example, does the firm interact with many or few customers, suppliers, competitors, governmental agencies, and so on. These two environmental dimensions in combination determine the *environmental uncertainty* of an organization. Figure 4.3.1 illustrates this relationship, and gives the environmental characteristics and a representative industry for four different levels of uncertainty.

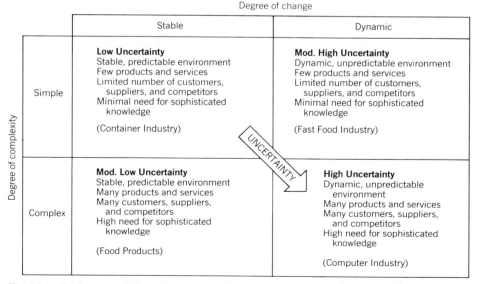

Fig. 4.3.1   Environmental dimensions. Adapted from Robert Duncan, "What Is the Right Organization Structure? Decision Tree Analysis Provides the Answer," *Organizational Dynamics,* Winter 1979, p. 63.

Two major generalizable models have been empirically derived for assessing environmental dimensions as determinants of organizational structure. One model focuses on environmental uncertainty. The other treats environmental uncertainty as one of several key environmental dimensions affecting structure, albeit the most important.

Burns and Stalker: Environmental Uncertainty

From their studies of 20 English and Scottish industrial firms, Burns and Stalker found that the type of organizational structure that worked effectively in a relatively stable and simple organizational environment was very different from that required for a more dynamic and complex environment. For stable environments, *mechanistic structures,* characterized by high complexity, formalization, and centralization, were the most efficient. These structures typically had routine tasks, programmed behaviors, and were slow in their capacity to respond to change. For unstable, more complex environments, *organic structures,* characterized by flexibility and adaptability, were the more successful. In organic structures, emphasis is placed on lateral rather than vertical communication, influence based on expertise and knowledge rather than authority and position, information exchange rather than directives from above, loosely defined responsibilities, and conflict resolution by interaction rather than by superiors. Thus a structure with low vertical differentiation, decentralization, and low formality is optimal (Burns and Stalker, 1961). Similar findings were implicit in Emory and Trist's (1965) analysis of the effects of environmental instability on sociotechnical systems. A comparison of the organizational design features of mechanistic versus organic systems is shown in Table 4.3.4.

Lawrence and Lorsh: Subunit Environment and Design Complexity

In a complex organizational environment, firms typically develop specialized subunits to deal with specific parts of the environment, or what was referred to earlier as the different relevant task environments. Lawrence and Lorsh (1969) conducted field studies to determine which type of organizational design was best for coping with different economic and market environments. They studied many companies in various industries including plastics, food, and containers, which varied considerably in their degree of environmental uncertainty. Based on their study of these organizations, Lawrence and Lorsh identified five major variables that can be assessed regarding subunit environments to determine the optimal degree of horizontal differentiation. These are (1) *uncertainty of information* (low, moderate,

Table 4.3.4 Design Features of Effective Mechanistic and Organic Organizations

Mechanistic	Design Feature	Organic
	Complexity	
High: small spans of control	Vertical differentiation	Low: large spans of control
High: highly specialized tasks	Horizontal differentiation	Low: low task specialization
High: many required because of high differentiation and low autonomy	Integrating mechanisms	Low: few required because of low differentiation and high autonomy
	Centralization	
High: decision-making reserved for management		Low: decision-making relegated to lowest level at which competency and skill exists
	Formalization	
High: low autonomy and high differentiation require formal rules and procedures		Low: high autonomy and low differentiation require few formal rules and procedures

or high), (2) *time span of feedback* (short, medium, or long), (3) pattern of *goal orientation* (focus of tasks), (4) pattern of *time orientation* (short, medium, or long), and (5) pattern of *interpersonal relationships* (task or social). In general, the more dissimilar the functions on one or more of these dimensions, the stronger the likelihood that they should be differentiated into separate subunits for effective functioning (Lawrence and Lorsh, 1969).

Lawrence and Lorsh also found that the greater the differentiation, the greater the need for integrating mechanisms. They noted that differentiation tends to encourage different viewpoints and thus greater conflict. Integrating mechanisms are needed to resolve these conflicts to the organization's benefit. Lawrence and Lorsh further noted that the more interdependent the tasks of the major subunits, the more information processing is required for effective integration.

Their research also found level of uncertainty of the external environment to be of foremost importance in selecting the structure appropriate for effective functioning. Subunits with more stable environments (e.g., production) tended to have high formalization, whereas those operating in less predictable environments (e.g., research and development) had low formalization.

In summary, based on Lawrence and Lorsh's research, whenever an organization's design does not fit its mission, environment, and resources, its functioning is likely to suffer. A mismatch between the organization's task and degree of differentiation results in a loss of relevant information. Differentiation, shifts in mission or resources, and environmental changes each can create integration problems unless adequate integration mechanisms are designed into the organization's structure. High levels of interdependence between subunits require particularly careful attention to information processing mechanisms to ensure effective integration.

4.3.5.3 Personnel Subsystem

Critical to an organization's design are at least three major aspects of the personnel subsystem. These are (1) the degree of professionalism, or the skills and training requirements of the organization and its constituent subunits, (2) demographic characteristics of the work force to be employed in the organization, and (3) psychosocial aspects of the work force.

Degree of Professionalism

Robbins (1983) notes that formalization can take place on the job or off. When it is done on the job, formalization is *external* to the employee. The rules, procedures, and human–machine interfaces are designed to limit employee *discretion*. This characterizes unskilled and semiskilled positions, and is what is meant by the term "formalization." Professionalism, on the other hand, creates *internalized* formalization of behavior through a socialization process that is an integral part of training or education. That is, persons learn the values, norms, and expected behavior patterns of the job *before* entering the organization. Thus there is a trade-off, from an ergonomic design standpoint, between formalization of the organizational design and professionalization of the jobs in the system design process. Because jobs in the overall organization or in selected subunits are designed to require persons with professional training (i.e., with considerable formal education or skills training), they also can be designed and integrated to allow for low formalization, including considerable employee decision discretion. To the extent that the individual jobs are designed to be performed with little education or skills training, formalization should be designed into the organization's structure.

Demographic Factors

Many demographic characteristics of the work force that will form the organization's personnel subsystem could potentially interact with the organization's design. Those that are most striking within the United States and, to a greater or lesser extent, in other industrialized countries are (1) the rapidly increasing number of women entering the work force, including entry into what traditionally were male-dominated occupations; (2) the "graying" of the work force, and (3) demographic shifts in psychosocial characteristics.

1.   *Women.*   Although women are entering the work force in progressively increasing numbers, and entering professional, highly skilled, or "professionalized" blue collar jobs, as well as managerial positions, there is as yet no clear indication as to how these demographic changes will or should affect organizational design.

2.   *"Graying" of the Work Force.*   Presently, within the United States and elsewhere, as the post-World War II baby boom bulge moves through the work force, the average age of the force is "graying" at the rate of about 6 months per year. This trend began in the late 1970s and will continue well into the 1990s, resulting in an older, more experienced, more mature, and better trained work force. As a result of this shift, the work force thus is becoming more "professionalized." It is likely that organizational designs will need to accommodate to this change by becoming somewhat less formalized. It also is likely to be necessary to decentralize more of the decision-making process if these employees

are to feel fully utilized and remain motivated towards their work. Designers of future systems and modifiers of old ones will need to consider these factors carefully, particularly in designing those functions and jobs where high formalization and centralization traditionally have characterized the organization's structure.

3. *Psychosocial Shifts.*   Yankelovich (1979), based on extensive longitudinal studies of work force attitudes and values, notes that those workers born after World War II have very different views and feelings about work, and that these conceptions and values will affect profoundly work systems in America in the next two decades. Yankelovich refers to this group of workers as the "new breed." This "new breed" of workers has three principal values that distinguish them from those of the mainstream of older workers: (1) the increasing importance of leisure, (2) the symbolic significance of the paid job, and (3) the insistence that jobs become less depersonalized and more meaningful. According to Yankelovich, when asked what aspects of work are more important, the "new breed" person stresses "being recognized as an individual," and "the opportunity to be with pleasant people with whom I like to work." From an organizational design standpoint, these values translate into a need for more decentralized and less formalized organizational structures, and attendant greater professionalism designed into individual jobs and human–machine interfaces than has characterized traditional bureaucracies. Because decentralization and lack of formalization permit greater participation in the decision-making process by employees, these design characteristics both allow for greater individual recognition and respect of an employee's worth and enhance meaningful social relationships on the job.

Psychosocial Factors

Harvey, Hunt, and Schroder (1961) have identified a higher-order structural personality dimension, concreteness–abstractness of thinking or *cognitive complexity,* as underlying different conceptual systems for perceiving reality. In general, the degree to which a given culture or subculture (1) encourages by its child-rearing and educational practices active exposure to new experiences or diversity, and (2) provides, through affluence, education, communications media, and transportation systems, opportunities for exposure to diversity, the more cognitively complex the persons of that culture will become. Active exposure to diversity increases one's opportunity to develop new conceptual categories, and shades of gray within categories in which to store experiential data, or *differentiation.* With an active exposure to diversity one also learns new rules and combinations of rules for *integrating* information and deriving more insightful conceptions of problems and solutions. Thus relatively closed-minded approaches to new experiences and/or a lack of exposure to diversity results in the development of relatively limited differentiation and integration in one's conceptualizing of reality, or concrete functioning. Conversely, an active or open-minded exposure to considerable diversity leads to the development of a high degree of differentiation and integration in one's conceptualizing, or abstract functioning. Relatively concrete adult functioning consistently has been found to be characterized by a high need for structure and order and for stability and consistency, closeness of beliefs, absolutism, authoritarianism, paternalism, and ethnocentrism. Concrete persons tend to see their views, values, norms, and institutional structures as relatively static and unchanging. Abstract adult functioning is characterized by a low need for structure and order, openness of beliefs, relativistic thinking, empathy, and a strong people orientation. Abstract persons also have a dynamic conception of their world and *expect* their views, values, norms, and institutional structures to change (Harvey, 1963; Harvey et al., 1961).

Hendrick (1979, 1981, 1984) has found evidence to suggest that relatively concrete work groups and managers function best under high centralization, vertical differentiation and formalization, or mechanistic organizational designs. In contrast, abstract or more cognitively complex work groups and managers function best in organic structures.

Because of (1) the shifts to more permissive child-rearing practices that tend to encourage relativistic rather than absolutist thinking, (2) a higher level of general education, (3) greater affluence, and (4) the development of far superior communications and transportation systems since World War II, the majority of the work force born after World War II are functioning at a more cognitively complex level than the mainstream of their older colleagues. Whereas approximately 80% of those born prior to World War II are relatively concrete in their functioning, this is true of less than half of those born after World War II in the United States. As the pre-World War II work force moves out of our organizations, and more of the "new breed" move in, the trend toward a progressively more cognitively complex work force will continue (Hendrick, 1981). The result likely will be a progressively increasing demand for more organic organizational structures than typify present bureaucratic organizations.

Personnel Subsystem Implications for Organizational Design

Many of the available data on personnel subsystem determinants of organizational design are in the form of either attitude survey results or projections of psychosocial and demographic studies. In spite of their somewhat tenuous nature, there is a convergence of these data dealing with different personnel subsystem dimensions that lends credence to the conclusion. They all indicate a need for organizations of the future to be designed to be as vertically undifferentiated, decentralized, and lacking in formalization as their environments and technology will permit.

4.3.6 CHOOSING THE RIGHT STRUCTURAL FORM

Thus far we have considered the basic dimensions of organizational structure and the sociotechnical system components as modifiers of these dimensions. Ultimately, these factors must be integrated into an overall structural form. In this regard, a variety of *types* of overall structural form are available to the systems designers. Just as the design of the individual dimensions of organizational structure can enhance or inhibit organizational functioning, and just as the particular design chosen for each of these dimensions has particular ergonomic design implications, the same also is true for the type of overall structural form chosen for the organization. This section (1) reviews the four types of organization most commonly found, (2) discusses the advantages and disadvantages of each, and (3) provides guidelines for determining when each type is and is not likely to be an appropriate choice. The four general types of organizational form discussed herein are the classical or *machine bureaucracy,* the *professional bureaucracy,* the *matrix organization,* and the *free-form* design. It should be noted at this point that large, complex organizations may, and often do, have relatively autonomous units that will have different overall forms. The smaller the organization, however, the greater the likelihood of a single overall type of organizational design.

4.3.6.1 Classical or Machine Bureaucracy

This form of organization had its roots in two streams of thought, (1) *scientific management* and (2) the *ideal bureaucracy.*

1. *Scientific Management.* The last few years of the nineteenth century were characterized by the accumulation of resources and a rapidly developing technology in American and European industry. During this period labor became highly specialized, and industrial engineers were called upon to help design organizations and to optimize efficiency. These engineers were involved in the designing of the equipment, installation layout, and procedures, and even made suggestions for managing the work force (Lawrence, Kolodny, and Davis, 1977). One of these engineers, Fredrick W. Taylor (1911), stood out among his colleagues and has had a major impact on the shaping of classical organizational theory through his concepts of "scientific management." The essence of his concepts of organization are implicit in his four basic principles of managing (Szilagyi and Wallace, 1983, p. 458).

First. Develop a science for each element of man's work that replaces the old rule-of-thumb method.

Second. Scientifically select and train, teach, and develop the workman. In the past he chose his own work and trained himself as best he could.

Third. Heartily cooperate with the men in order to ensure all the work is being done in accordance with the principles of the science that has been developed.

Fourth. Provide equal division of work and responsibility between the management and the workmen. The management takes over all work for which they are more qualified than the workmen. In the past, almost all the work and the greater part of the responsibility were thrown upon the men.

As may be seen, Taylor advocated scientific analysis rather than pure common sense and intuition to job and organizational design. In addition, he emphasized the importance of cooperation and developed scientific principles to achieve this in the design of organizations and their components. Finally, he advocated clear job definition through specialization (Szilagyi and Wallace, 1983).

2. *"Ideal Bureaucracy."* Although heavily influenced by Taylor's concepts of scientific management, the classical bureaucratic design was conceptualized by Max Weber at the beginning of the twentieth century. Weber recommended that organizations adhere to the following design principles (1946, p. 214).

1. All tasks necessary to accomplish organization goals must be divided into highly specialized jobs. A worker must master his trade, and this expertise can be more readily achieved by concentrating on a limited number of tasks.

2. Each task must be performed according to a "consistent system of abstract rules." This practice allows the manager to eliminate uncertainty due to individual differences in task performance.

3. Offices or roles must be organized into hierarchical structure in which the scope of authority of superordinates over subordinates is defined. This system offers the subordinates the possibility of appealing a decision to a higher level of authority.

4. Superiors must assume an impersonal attitude in dealing with each other and subordinates. This psychological and social distance enables the superior to make decisions without being influenced by prejudices and preferences.

5. Employment in a bureaucracy must be based on qualifications, and promotion is to be decided on the basis of merit. Because of this careful and firm system of employment and promotion, it is assumed that employment will involve a lifelong career and loyalty from employees.

Weber assumed that strict adherence to these organizational design principles was the "one best way" to achieve organizational goals. By implementing a structure that emphasized efficiency, stability and control, Weber believed organizations could obtain optimal efficiency (Szilagyi and Wallace, 1983).

These theoretical principles of Taylor's scientific management and Weber's ideal bureaucracy have resulted in what today is referred to as the *machine bureaucracy* type of organizational design. Its basic structural characteristics are as follows.

1. *Division of Labor.*   Each person's job is narrowly defined, and consists of relatively simple, routine, and well-defined tasks.

2. *A Well-Defined Hierarchy.*   A relatively tall, clearly defined, formal hierarchical structure of positions and offices in which each lower office is under the supervision and control of a higher one. Tasks and positions tend to be grouped by function. Line and staff functions are clearly distinguished and are kept separate.

3. *High Formalization.*   A dependence on formal rules and procedures to ensure uniformity and to regulate employee behavior.

4. *High Centralization.*   Decision-making is reserved for management. There is relatively little employee decision discretion.

5. *Career Tracks for Employees.*   Members are expected to pursue a career within the organization, and career tracks form part of the organizational design for all but the most unskilled positions.

Advantages

The primary advantages of the machine bureaucracy are efficiency, stability, and control over the organizations functioning. Narrowly defined jobs with a clear set of routinized tasks minimize the likelihood of error, better enable individuals to know their own function and the roles of others, require comparatively few prerequisite skills, and minimize training time and costs. Formalization ensures stability and a smooth, integrated pattern of functioning. Centralization ensures control and further enhances stability.

Disadvantages

There are at least two major disadvantages to the machine bureaucracy type of design. First, the machine bureaucracy tends to result in jobs that are lacking in intrinsic motivation, and that fail to utilize fully the mental and psychological capacities of the workers. Secondly, machine bureaucracies tend to be inefficient in responding to environmental change and nonroutine situations.

When to Use

When the education and skill levels of the available labor pool are relatively low, system operations can be largely routinized, and the relevant external environments are comparatively simple and stable, a machine bureaucracy design is likely to be the most effective for the overall organization. To the extent that the above stated conditions do not exist, one or another of the other three forms of overall organization is likely to be preferable.

4.3.6.2 Professional Bureaucracy

Professionalism was previously defined as the degree of training and education, and related internalized formalization of behavior required by the design of specific jobs. The professional bureaucracy design is one that relies on a relatively high degree of professionalism in the jobs that make up the organization. Its major differences from the machine bureaucracy design lie in the areas of (1) job scope and decision discretion, (2) centralization, and (3) formalization. In professional bureaucracies jobs are more broadly defined, somewhat less routinized, and allow for greater employee decision discretion. Thus there is less need for formalization and *tactical* decision-making is decentralized. Like machine bureaucracies, positions are grouped functionally, they are hierarchical, and *strategic* decision-making usually remains centralized.

Advantages

There are at least three major advantages to the professional bureaucracy type of design as compared with a machine bureaucracy. First, professional bureaucracies can more effectively cope with complex environments and nonroutine tasks. Secondly, jobs tend to be more intrinsically motivating and to make greater use of the mental and psychological capabilities of the employees. Thirdly, less managerial control and tactical decision-making is required, thus freeing management to give greater attention to long-range planning and strategic decision-making.

Disadvantages

Professional bureaucracies are not as efficient as machine bureaucracies for coping with simple environments. They require a more highly skilled labor force and attendant greater training time and expense. Control is not as tight, and both the line and staff functions and procedures are likely to be less clear. The management skills required also tend to be more sophisticated (e.g., a greater reliance on tolerance for ambiguity, and on pursuasive and facilitation skills rather than on a simple and direct authoritarian style).

When to Use

A professional bureaucratic design is to be preferred when the relevant external environments are generally complex and at least moderately stable, and there is an available applicant pool of professionalized workers to staff the organization. This form of overall design is less likely to be effective if the relevant external environments are either simple or highly dynamic and if the available applicant pool is not professionalized. This form of design also is less desirable if the available management pool is highly authoritarian and concrete, rather than cognitively complex in its functioning.

4.3.6.3 The Concept of Adhocracy

A major disadvantage of both the machine and professional bureaucratic forms of organizational design is that they tend to be inefficient in responding to highly dynamic relevant external environments. It is primarily for this reason that the two more recent forms of organization, the *matrix* and *free-form* designs, have emerged. Collectively, these two newer forms have been referred to as *adhocracy* designs, as distinct from *bureaucracy* designs. An adhocracy can be described as a "rapidly changing, adaptive, temporary system organized around problems to be solved by groups of relative strangers with diverse professional skills" (Bennis, 1969). In terms of their structural dimensions, adhocracies are characterized by moderate to low complexity, low formalization, and decentralization (Robbins, 1983).

Low Complexity

Because adhocracies are staffed predominantly by professionals, horizontal differentiation tends to be high. Vertical differentiation, however, tends to be low. The low vertical differentiation reflects the low formalization and decentralized decision-making, and the need for flexibility in responsiveness that would be inhibited by many layers of administration. Additionally, because of the professional staffing, the need for supervision is minimized. With low vertical differentiation the need for integrating mechanisms tends to be moderate to low. In particular, there is an absence of formalization as an integrating mechanism.

Low Formalization

As noted previously, the greater the *internal* formalization provided by professionalism, the less the requirement for, or desirability of, external organizational formalization. The rules and procedures that exist in adhocracies often tend to be informal and unwritten. Flexibility of response is more essential than adherence to formalized procedure. This lack of formalization is a key distinction between professional bureaucracies and adhocracy designs. In a professional bureaucracy problems still can be classified into some category and treated in a largely routinized manner. In adhocracies, the nature of many of the problems dealt with tends to be unique and not subject to a routinized decision process. Adhocracies thus tend to depend on decentralized "teams" of professionals for decision-making (Robbins, 1983).

Adhocracies, then, tend to be characterized by flexible, adaptive structures in which multidisciplinary teams are formed around specific problems or objectives. Whereas bureaucratic forms tend to have relatively stable departments, adhocracy organizations have constantly changing units. New units form to deal with new problems or objectives whereas old ones either are (1) dissolving as problems are solved or objectives are met, or (2) changing in makeup as different stages of a project are reached, resulting in a need for different skills or levels of effort.

Advantages

Adhocracy designs had their origin in the task forces of World War II. The military created ad hoc teams to accomplish specific missions, and then disbanded them when the mission was completed.

The roles played by the team were flexible and often interchangeable. The time span for the existence of these teams varied greatly and was determined by mission requirements. As required, subunits could be added, created from within, or severed from the unit. The result of these features was the ability to react efficiently, creatively, and in a timely manner to a dynamic environment, and thus be highly effective. These same characteristics and resultant advantages characterize the adhocracies of today. When the ability to be adaptive and innovative and to respond rapidly to changing situations and objectives is required, when these responses require the collaboration of persons possessing different specialties, and/or when the tasks are nonprogrammed, the adhocracy forms of organization tend to be more effective than the bureaucratic forms (Robbins, 1983).

Disadvantages

All adhocracies have at least there major disadvantages. These are (1) conflict, (2) social and psychological stress, and (3) inherently inefficent structure (Robbins, 1983).

1. *Conflict.*   Because there are no clear boss–subordinate relationships there is ambiguity over lines of authority and responsibility. Conflict thus is an integral part of adhocracy.
2. *Psychological and Sociological Stress.*   Because the structure of teams or units is temporary, work role interfaces also are not stable. However, the establishment or dismantling of human relationships is a slower psychosocial process, and is stressed any time there is significant organizational change. Thus some employees, particularly those who tend to be concrete in their conceptual functioning, are likely to be strained by these stresses and find it difficult to cope.
3. *Inherent Inefficiency.*   By comparison with bureaucratic forms of organization, adhocracies lack both in precision and in the expediency that comes with routinization of function and structural stability. It is only where these inefficiencies are more than offset by the gains in efficiency in terms of responsiveness and/or innovation that an adhocratic design should be considered.

4.3.6.4   The Matrix Design

Of the two major forms of adhocracy, the matrix design has been the more widely used. Basically, the matrix is a combination of departmentalization by function and by product or project. Functional departments, typical of bureaucracies, exist and tend to be lasting. Unlike bureaucracies, however, members of the functional departments are farmed out to project or product teams as new projects or product lines develop and the combined technical expertise of the individual departments is required. As the need for a given department's professional input to the interdisciplinary team is no longer required, or the level of effort reduces, individuals return to the "home" department or transfer to another team. The project or product team manager supervises the team's interdisciplinary effort, but each team member also has a functional department supervisor. The matrix design thus breaks a fundamental design concept of bureaucracy, *unity of command.*

Temporary Versus Permanent Matrix Structures

The type of matrix thus far discussed, depicted in Figure 4.3.2, is referred to as a *temporary matrix* in that the project or product teams are not lasting. In contrast, *permanent matrix* designs have relatively lasting interdisciplinary units superposed over the functional structure. Anderson notes that large business colleges often have this form of organization (Robbins, 1983). Here, superposed on the functional adacemic departments, such as accounting and management, are relatively lasting product structures such as undergraduate programs, graduate programs, and executive development programs. Directors of the product structures utilize faculty from the various functional departments to achieve their goals. This organizational arrangement enables the goals of the interdisciplinary programs to be met via centralized program direction, integration, and direct accountability. The functional departments enable the maintenance of scholarly and professional activities within the specific disciplines.

Advantages

The primary advantage of matrix designs is that they afford the best of two worlds, the stability and professional support of depth of functional departmentalization, and the interdisciplinary response capability of ad hoc teams, such as characterize free-form designs.

Disadvantages

Over and above those characteristics of all adhocracy designs that cause problems, the major disadvantage of matrix organizations is that employees must serve two bosses. One is their functional department head, who tends to be relatively long term and somewhat remote from the team member's immediate tasks; the other is the project or product team director who tends to be comparatively short term

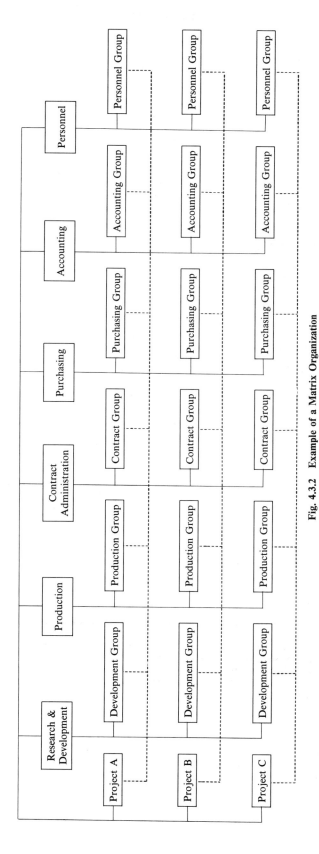

Fig. 4.3.2 Example of a Matrix Organization

but immediate to the employees' present tasks. Serving two masters with overlapping supervisory responsibility and different goals, responsibilities, and orientations frequently creates conflict and can disrupt organizational functioning. A second major problem for the employee is that when assigned too long to a project or projects, the employee may have difficulty keeping technically current and may lose contact with his or her functional department. Both these consequences can adversely affect the employee's career.

When to Use

The matrix organization is particularly well suited for responding to dynamic and complex relevant external environments where both interdisciplinary responsiveness and providing for functional depth in the individual disciplines are considered essential.

4.3.6.5 Free-Form Designs

The last and newest of our general types of organizational structures is the *free-form design* (Pascucci, 1968). The free-form organization resembles an amoeba in that it continually changes its shape in order to survive (Szilagyi and Wallace, 1983). the focus of free-form designs is responsiveness to change in highly dynamic, complex, and competitive environments.

In free-form organizations functional departmentalization is replaced by a *profit center* arrangement. Profit centers are results oriented and are managed as teams. Collectively, they constitute the organizational team. Free-form designs thus are characterized by very low formalization and hierarchical differentiation, and decision-making is highly decentralized. A very heavy reliance is placed on professionalism as expressed through participation, internalized formalization, and autonomy. As with matrix organizations, project teams are created, changed, and disbanded as required to meet organizational goals and problems. Unlike matrix organizations, there is no underlying functional departmentalization or "home" structure. Managers and employees alike require a great deal of flexibility, tolerance for ambiguity, and ability to handle change in free-form organizations. They thus need to be at least somewhat cognitively complex in their conceptual functioning.

Advantages

The primary advantage and basic purpose of free-form designs is the ability to respond to highly competitive, complex, and dynamic environments with speed and innovation.

Disadvantages

The free-form organization has essentially the same disadvantages as matrix adhocracies, only to a greater extent. Conflict, sociopsychological stress, and inherent administrative inefficiency are an integral part of a continuously changing and amorphous organizational structure. It thus requires a highly professionalized work force to succeed.

When to Use

Free-form designs should be considered (1) when the organization's success or survival critically depends on speed of response and innovation, and (2) the available work force and management pool are highly professionalized. These features tend to characterize high-technology organizations operating in highly dynamic, complex, and competetive environments.

4.3.6.6 Ergonomic Implications of Different Organizational Forms

Earlier, we noted the ergonomics implications of decisions regarding the specific dimensions of organizational design and related sociotechnical system factors. Over and above these considerations, the choice of *type* of overall organizational structure brings its own unique set of ergonomic challenges. Typically, these have to do with how to design human–machine interfaces to overcome or minimize the disadvantages of a given organizational form and to enhance its advantages. Illustrative of these human-factors challenges are the following.

Machine Bureaucracies

How can human-factors specialists help in broadening the structure of jobs and in increasing employee decision discretion so as to (1) make jobs more intrinsically interesting and to utilize more fully employee psychosocial and mental capacities, yet (2) maintain the stability, control, and relatively low personnel and training costs that have made this form of organization so successful? As the work force continues to age and become better educated and less cognitively concrete in its conceptual functioning, these issues will have to be addressed.

Professional Bureaucracies

How can human-factors specialists help in designing human–machine interfaces on a systems level so as to take full advantage of the greater skills and internal formalism of a professional work force; in particular, how can human-factors specialists help professional bureaucracies to become more responsive to dynamic environments and nonroutinized problems?

Matrix Organizations

How can human-factors specialists help facilitate the dual objectives of interdisciplinary project team membership and functional departmental membership; in particular, what kinds of inputs to human–machine interfaces at the systems level can human-factors specialists make to minimize the problems of dual membership and attendant disruptions of organizational functioning?

Free-Form Designs

How can human-factors specialists contribute to human–machine interface design at the systems level to provide the flexibility and fluidity of task assignment and human–machine, human–human, and machine–machine interfaces required by these organizations? Can human-factors specialists also contribute to making these organizations more inherently efficient in the light of their design fluidity?

The answers to these challenges will vary with the technology, external environments, personnel subsystem composition, and perhaps other factors. By focusing both the science and practice of human factors on these issues at the macroergonomics level, human-factors specialists should be capable of making significant contributions to both productivity and the quality of work life within organizations. Ultimately, the field of human factors should be able to contribute directly to organizational theory and to advancing the state of the art of organizational design.

REFERENCES

Bennis, W. G. (1969, July–August). Post bureaucratic leadership. *Transaction*, p. 45.

Burns, T., and Stalker, G. M. (1961). *The management of innovation*. London: Tavistock.

Campbell, J. P. (1977). On the nature of organizational effectiveness in P. S. Goodman, J. M. Pennings, et al., Eds., *New perspectives on organizational effectiveness*. San Francisco: Jossey-Bass.

Child, J. (1972, June). Organization structure and strategies for control: A replication of the Aston study. *Administrative Science Quarterly*, 163–177.

Davis, L. E. (1982). Organization design. In G. Salvendy, Ed., *Handbook of industrial engineering*. New York: Wiley.

Degreene, K. (1973). *Sociotechnical systems*. Englewood Cliffs, NJ: Prentice-Hall.

Duncan, R. B. (1972, September). Characteristics of organizational environments and perceived environmental uncertainty. *Administrative Science Quarterly*, 315.

Emory, F. E., and Trist, E. L. (1965). The causal texture of organization environments *Human Relations, 18*, 21–32.

Emory, F. E., and Trist, E. L. (1960). Sociotechnical Systems. In C. W. Churchman and M. Verhulst, Eds., *Management science: Models and techniques*, Vol. 2. Oxford: Pergamon.

Hage, J. (1965, December). An axiomatic theory of organizations. *Administrative Science Quarterly*, 303.

Hage, J., and Aiken, M. (1967, June). Relationship of centralization to other structural properties. *Administrative Science Quarterly*, 72–91.

Hage, J., and Aiken, M. (1969, September). Routine technology, social structure, and organizational goals. *Administrative Science Quarterly*, 366–377.

Hall, R. H., Haas, J. E., and Johnson, N. J. (1967, December). Organizational size, complexity, and formalization. *American Sociological Review*, 905–912.

Harvey, E. (1968, April). Technology and the structure of organizations. *American Sociological Review*, 247–259.

Harvey, O. J. (1963). System structure, flexibility and creativity. In O. J. Harvey, Ed., *Experience, structure and adaptability*, New York: Springer.

Harvey, O. J., Hunt, D. E., and Schroder, H. M. (1961). *Conceptual systems and personality organization*. New York: Wiley.

Hendrick, H. W. (1979). Differences in group problem solving as a function of cognitive complexity. *Journal of Applied Psychology, 64,* 518–525.

Hendrick, H. W. (1981). Abstractness, conceptual systems, and the functioning of complex organizations In G. W. England, A. R. Negandhi, and B. Wilpert, Eds., *The functioning of complex organizations.* (pp. 25–50) Combridge, MA: Oelgeschiager, Gunn & Hain.

Hendrick, H. W. (1984a). Cognitive complexity, conceptual systems and organizational design and management: Review and ergonomic implications. In H. W. Hendrick and O. Brown, Jr., Eds., *Human factors in organizational design and management.* (pp. 15–26) Amsterdam: North-Holland.

Hendrick, H. W. (1984b). Wagging the tail with the dog: Organizational design considerations in ergonomics. *Proceedings of the Human Factors Society 28th Annual Meeting.*

Kerr, C., and Rosow, J. M., Eds. (1979). *Work in America: The next decade.* New York: Van Nostrand Reinhold.

Lawrence, P. R., and Lorsh, J. W. (1969). *Organization and environment.* Homewood, IL: Irwin.

Lawrence, P. R., Kolodny, H., and Davis, S. The human side of the matrix. *Organizational dynamics,* pp. 43–61.

Magnusen, K. (1970). *Technology and organizational differentiation: A field study of manufacturing corporations,* unpublished doctoral dissertation. Madison, WI: University of Wisconsin.

Mahoney, T. A., and Frost, P. J. (1974). The pole of technology in models of organizational effectiveness. *Organizational Behavior and Human Performance,* (pp. 122–138).

Mileti, D. S., Gillespie, D. F., and Haas, J. E. (1977, September). Size and structure in complex organizations. *Social Forces,* pp. 208–217.

Montanari, J. R. (1976). *An expanded theory of structural determination: An empirical investigation of the impact of managerial discretion on organizational structure,* unpublished doctoral dissertation. Boulder, CO: University of Colorado.

Negandhi, A. R. (1977). A model for analysing organizations in cross-cultural settings: A conceptual scheme and some research findings. In A. R. Negandhi, G. W. England, and B. Wilpert, Eds., *Modern Organizational Theory,* Kent, OH: Kent State University Press.

Pascucci, J. J. (1968, September–October). The emergence of free-form management. *Personnel Administration,* pp. 33–41.

Perrow, C. (1967 April). A framework for the comparative analysis of organizations *American Sociological Review,* 194–208.

Pugh, D. S., Hickson, D. J., Hinings, C. R., and Turner, C. (1968, June). Dimensions of organization structure. *Administrative Science Quarterly,* 75.

Raia, A. (1974). *Managing by objectives.* Glenview, IL: Scott, Foresman.

Robbins, S. R. (1983). *Organizational theory: The structure and design of organizations.* Englewood Cliffs, NJ: Prentice-Hall.

Smith, A. (1970). *The wealth of nations.* London: Penguin (originally published 1776).

Szilagyi, A. D., Jr., and Wallace, M. J., Jr. (1983). *Organizational behavior and performance,* 3rd ed., Glenview, IL: Scott, Foresman.

Taylor, F. W. (1911). *Principles of scientific management,* New York: Harper.

Thompson, J. D. (1967). *Organizations in action,* New York: McGraw-Hill.

Trist, E. L., Higgin, G. W., Murray, H., and Pollock, A. B. (1963). *Organizational choice.* London: Tavistock.

Van de Ven, A. H., and Delbecq, A. L. (1979, June). A task contingent model of work-unit structure. *Administrative Science Quarterly,* pp. 183–197.

Weber, M. (1946). *Essays in sociology* (trans. H. H. Gerth and C. W. Mills). New York: Oxford.

Woodward, J. (1965). *Industrial organization: Theory and practice.* London: Oxford University Press.

Yankelovich, D. (1979). Work, values and the new breed. In C. Kerr and J. M. Rosow, Eds., *Work in America: The decade ahead.* New York: Van Nostrand Reinhold, pp. 3–26.

Zwerman, W. L. (1970). *New perspectives on organization theory.* Westport, CT: Greenwood.

CHAPTER **4.4**
DESIGN FOR OLDER PEOPLE

ARNOLD M. SMALL

University of Southern California
Los Angeles, California

4.4.1 INTRODUCTION

The first purpose of this chapter on aging is to broaden one's perspective and knowledge concerning those of our population who live beyond middle age. The second purpose is to encourage recognition of, and attention to, those features of older persons that are germane to successful and safe product and environmental design. The third purpose is to promote that depth of understanding of the characteristics and needs of older adults that will result in raising the appropriate questions about design for this population and lead to seeking the answers from such sources as may be available, including competent consultants in the human-factors profession.

The group of persons considered herein is essentially that whose members may be described as upper middle age and beyond. No sharp line of demarcation is justified but age 55 may serve as a general indicator, especially since it includes the latter portion of many people's working life.

In view of the inclusion in this handbook of a section on design for the handicapped, that portion of the older population so classified is not dealt with here.

4.4.2 OLDER ADULTS IN THE UNITED STATES

The inclusion of this chapter in the handbook implies there must be special or additional considerations involved when designing for "older" people as compared to "younger" people. This is in fact true. Changes do occur over the life-span of individuals which need to be understood, considered, and made a part of design accountability when a human interface is involved.

Actually, when design accommodates the relevant characteristics of older people, it sometimes results in having positive value for younger generations as well. A case in point is the human factor of strength. Leg and arm strengths reach their maximum by about 25 years of age. During the age span of 30 to 65 years a usually gradual decline then occurs, resulting in a reduction to about 50% of that at age 25. Clearly if design entails a strength requirement and it is made compatible with the 65 year old person's capability, those younger may experience even greater ease, accuracy, and reliability of performance, less energy expenditure, and possibly greater speed of performance, especially if a repetitive task is involved. Such salutory effects do not always result but should be included when weighing the efficacy of meeting requirements related to characteristics of older persons.

Before we consider the characteristics associated with aging, we note that added importance for doing so exists in view of demographic changes taking place in the United States and other countries as well. The total population of the United States, 113,530,000 in 1900, doubled in 1980, then increased 30% to 233,267,000 at the end of 1980. The progression of median age with these growths is shown in Figure 4.4.1 from 1810 through 1982 (the latest reported by the U.S. Census Bureau) and beyond as projected. (The median is the point at which half the total lies above and half below.)

As significant as the median age progression are the number and proportion of elderly in a population (Borgatta and McClusky, 1980). At the same time that the total population doubled (1900–1980), the proportion of those 65 years of age and over increased nearly seven times. At present the nation has approximately 23.5 million persons over 65, 13.0 million of whom are women, and 10.5 million men. For those persons *60 years* of age and older, the total is about 32.9 million or 15.2% of the resident population. This 60-plus group has increased twice as fast as the total population since 1970. Furthermore, the baby boom of the 1950s will produce a related gray boom of the 2010s.

Further analysis of growth *within* the over-65 age group reveals that the 75-plus segment has increased numerically 10 times and the 85-plus group about 17 times since 1900. This over-85 group now numbers 2.5 million persons and is the fastest growing group. By the year 2030, the 75-plus group may be more than 33% of the past-65 population, and the past-85 group 10% of it.

Put in another way, 1 of every 25 persons (4%) in 1900 was over 65; now it is about 1 of every 9 (11+%). By 2020 it is expected to be 1 of every 6 (16+%), or possibly 5. Put in terms of numbers and assuming the total population as currently forecast for 2020 to be 300.0 million, the number of persons over 65 would be about 50.0 million. Those over 75 years in that group would number about 16.3 million. Those over 85 years would number 4.9 million. Figure 4.4.2 displays the comparison between the present and forecast figures.

The makeup of our nation's population is clearly in the process of change, resulting in a significant increase in the total and in the proportion of older adults.

Several generalizations concerning older adults should be pointed out at this point inasmuch as many misconceptions are held in our country about them. Several negative stereotypes of old people as a whole group are among them. These include regarding older persons as poor, incompetent, unteachable, sick, unhappy, and lonely.

Of the 23.5 million persons over 65, the able-bodied, competent, and active constitute roughly 80

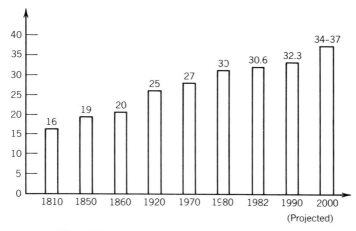

Figure 4.4.1. Median age of the U.S. population.

to 85%. People in this group may have some chronic indispositions but this does not constitute sickness usually. Surveys show the majority own their homes, have an income sufficient to maintain an acceptable way of life, and are active in personal, group, and civic activities. Quite a number travel in groups, continue to work, participate in physical fitness classes, teach, do volunteer work, and participate in special educational courses. They report as high levels of life satisfaction as do younger people.

The remainder of the 65 and older (approximately 2 million) include those who suffer major physical, mental, or social losses requiring a range of supportive and restorative health and social care and services. They receive more public attention, which probably serves as the basis for some of the overall negative stereotypes. This group includes those whom one would identify as handicapped.

It is important to recognize that people grow old in different ways, progress at different rates, and are affected materially by it at different times in the course of their lives. This creates a very heterogeneous older population and makes classifications very difficult. No two people are alike, but the aged are more dissimilar than young generations. As Koucelik (1982) points out, the adults between 18 and 35 years of age are actually the most alike of all the population segments, contrary to the stereotype of sameness applied to the older generations by the uninformed.

This heterogeneity plus the nature and amount of gerontological investigations and research relevant to design for older persons make it nearly impossible at this time to announce much *specific* design-oriented material for assuring designs compatible with the needs of older persons. The characteristics of older people to be presented herein identify and describe the several categories that must be considered in designing for the older populations. The snares existing in unwarranted classifications within this large older population, however, require that the specific segment of people for which the design is

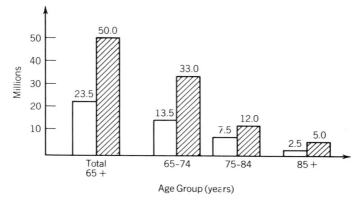

Figure 4.4.2. Number and distribution of persons 65 years and older in the United States. Unfilled bar, 1982; filled bar, 2020.

intended be defined and studied further, providing bench marks and criteria for assuring product design effectiveness and acceptance. Also necessary is the subsequent determination of feasibility and evaluation.

The problem of effective design standards is an area of concern in meeting the needs of such a varied population. There are some basic codes, general population standards, and design criteria that provide broad guidance for appropriate design, and progress is being made toward more definitive specific developments. A good example of this is Koucelik's *Aging and the Product Environment* (1982).

The importance of an adequate design evaluation, based upon the design and acceptance criteria carefully established for the product and before production go-ahead, is vital for all concerned. The reader is referred to Chapanis' *Research Techniques in Human Engineering* (1959) as one source of guidance in formulating and executing such evaluations.

4.4.3 CHARACTERISTICS OF OLDER ADULTS: THE LAST HALF OF LIFE

In this section information is provided concerning the following characteristics:

Physical (anthropometric) and biomechanical

Psychomotor performance

Sensory-perceptual

Cognitive, attention, information processing, decision-making

Attitudinal

Stress

Reference has already been made to the variability existing within the older population, and this applies to findings on the above characteristics. Variability increases as a cohort group advances in age as well as in cross-sectional studies at a given time. An example of the presence of this variability is the increasing number of people who live to age 75 without any significant changes. This number is increasing. In the second half of life, age is a poor predictor of physical, intellectual, or social performance.

The data and information available relating to the areas dealt with herein come from both cross-sectional investigations and longitudinal ones, the majority probably from the former. A cross-sectional study measures the differences in characteristics among an age-heterogeneous group's members at a given point in time. The longitudinal study measures changes over several years, starting with groups (cohorts) whose member's ages cluster within a narrow range such as 5 to 10 years; the clusters may cover both young and old adults at the time the study begins. Thus the change in characteristics, if any, and its rate may be ascertained for both the group and the individual.

The results of many studies on changes with age are expressed in terms of proportions of values for young populations as the base line. Space does not permit incorporation of much of this base-line quantitative data involved, but references are provided for this purpose. The most readily available source, however, may be found in relevant chapters of this handbook.

The results of attributive-type research on human characteristics therefore are not often expressed in terms directly susceptible for use in design guidance. In any specific instance, one must first judge whether design guidance interpretations are warranted and, second, exercise great care in fashioning those believes warranted, which should result in some indication of their merit, that is, confidence in their use for design. This applies, of course, to research on human characteristics at any age.

The material following is mostly illustrative in nature. For fuller treatment the reader is encouraged to investigate the references provided here, both those of general interest and those specific to topics in this chapter.

4.4.4 PHYSICAL (ANTHROPOMETRIC) AND BIOMECHANIC CHARACTERISTICS

The changes in body measurements, movement, reach, and muscle strength present a varied picture.

Weight tends to increase up to age 50, then stabilize to age 60, with a small increase thereafter.

Hip measurements show increases throughout adult life.

Head circumference, length, and breadth increase slightly.

Height and sitting height decrease from age 50 on.

Upper arm and calf circumference and abdomen measurements, after increasing in midlife, decline in later years.

Table 4.4.1 illustrates the change in height and weight for men and women from age 50. Copious measurements of structural body dimensions may be found in Van Cott and Kinkade (1972).

Table 4.4.1 Height and Weight of Male and Female Americans at Different Ages

	Male				Female			
	Height (in)		Weight (lb)		Height (in)		Weight (lb)	
Age (yr)	Mean	S.D.	Mean	S.D.	Mean	S.D.	Mean	S.D.
50–59	67.3	2.6	165	25	62.8	2.4	148	28
60–69	66.8	2.4	162	24	62.2	2.4	146	28
70–79	66.5	2.2	157	24	61.8	2.2	144	27
80–89	66.1	2.2	151	24				

Functional body dimensions include many items such as reach and range of movement in various body positions and at various angles and radii of gyration of body units. Copious data are included by Van Cott and Kinkade (1972) and by Woodson and Conover (1964). Probably the most significant changes in these regards for older adults are in the reduction in the viable extremes of body positions taken, in range of reaches, and flexibility, especially of the limbs. These come about owing to changes in muscle, joint, and spinal column characteristics in later life. In designing, therefore, the best approach appears to be the use of the lower values in overall population spread, thus meeting the needs of most people of all ages.

Muscle strength and power decrease with age substantially by age 75–80 and are affected by numerous factors such as age, sex, body build and position, handedness, fatigue, environmental conditions, and motivation. Leg strength for men in the younger brackets varies from 400 to 600 lb (depending on position), arm extensor muscles 48 lb, arm biceps 60 lb, and grip force 95 lb. These values and many more (see Van Cott and Kinkade, 1972; Woodson and Conover, 1964; and Chaffin and Anderson, 1984) are usually less for women and their decline is steeper. At age 50, women can exert approximately one-half as much force as men. For men, strength reaches a maximum about age 25, of which about 95% is present at age 40, 80% at 50 to 60, and as little as 50% progressively beyond 65. Hand strength declines about 16%. Muscle strength decrement proceeds at somewhat different rates in parts of the body and tends to decrease in rate based on differential use. Hand and arm are less affected by aging than trunk or leg. For further information relating to strength, forces exerted and dynamic applications, see the references noted above.

4.4.5 PSYCHOMOTOR PERFORMANCE

Probably the most prevalent and significant change with age is the slowing of behavior. This applies not only to relatively simple sensory and motor processes but to more complex processes involving higher mediating mental operations. (See Birren, Woods, and Williams, 1980; and Welford, 1977.)

Most studies of reaction time have taken place in highly controlled laboratory situations. As might be expected, they result in the shortest reaction times for human beings both for simple (single-choice) and complex (multiple-choice) conditions. Reaction times in Table 4.4.2 are from laboratory tests of young adults in their twenties typically.

Studies involving older persons generally indicate a 20% increase in simple reaction time by age 60 compared to the 20 year olds. In the case of multiple-choice reaction time, with equiprobable signals, reaction time increases with the number of signals involved, ranging from 300 msec for 2 signals to 600 msec for 10 for both young and old adults, with the 20% differential tending to increase with the number of signals.

Table 4.4.2 Range of Simple and Complex Reaction Times (Laboratory)

Sense Modality	Range (msec)
Touch	100–160
Audition	130–170
Vision	150–200
Temperature	150–200
Smell	200–500
Taste	200–1200
Pain	700–1000

Response times also increase for all age groups, but progressively more for the older when (1) signal probabilities vary, (2) their discriminability from one another is made more difficult, (3) the compatibility between the required response with the presented signal is reduced, and (4) there is stress, for example, increased work load. In general, the greater the degree of the above, the greater the increase in response time, and it may be nonlinear. Quantitative research dealing with this complexity is meager at present.

4.4.6 SENSATION AND PERCEPTION

Vision

The variations with age in vision and audition are probably the most commonly recognized changes and have been measured extensively over the years. Inasmuch as glasses are available to correct visual deficiencies, tests of vision for older persons from a functional point of view involve wearing their glasses. A visual acuity (Snellen) of 20/50 or worse represents a real impairment. The U.S. National Health Survey of April, 1968 showed corrected best vision (static acuity) in the better eye to be 20/50 or worse for about 9% of those 55 to 64 years of age, 18% for ages 65 to 74, and 30% for ages 75 to 79. In Figure 4.4.3 it is noted that unless a substantial deterioration existed in both eyes, activity was not impaired in general (see Welford, 1981).

Static visual acuity decreases steadily over the years from age 15 to 90 years. It increases with level of illumination for all age levels (see Anderson and Palmore, 1974).

Loss in visual accommodation is essentially linear with age with depth of focus dropping from 10 diopters at age 10 to about 1 at age 60 to 70.

Dark adaptation varies with age progressively and rather markedly in rate and final level achieved, the largest effect being for ages 60 to 89 (see Corso, 1981). Once adapted, the average person in the older group requires 240 times as much light to see a target as one in the youngest group.

Differential threshold (discriminable differences) in vision is the keenest for young adults (20 to 25) and results in 570 discriminable intensities with white lights, the largest for any sense modality. As visual acuity decreases with age, this number is reduced somewhat proportionally.

The effects of glare are most pronounced for persons 75 to 85 years old, who require a 50- to-70-fold increase in target luminance compared to persons 5 to 15 years old. Large individual variations exist in the older group.

Reduction in the size of the visual field appears minimal up to age 55. From then on a gradual shrinkage takes place, especially by age 75.

Color vision tests reveal a loss in color sensitivity clearly evident at age 70, progressing steadily to age 90, when one-sixth of those have more than 50% incorrect judgments.

Audition

As with vision, changes in hearing are among the most commonly occurring and recognized as related to aging. Hearing is, in fact, one of the major problems for many older adults. The major changes in the hearing threshold increase (hearing loss) occur progressively with age and frequency above 1000 cycles per second (Hz) and to an increasing extent. Figure 4.4.4 shows this higher loss with frequency and age for men and women combined.

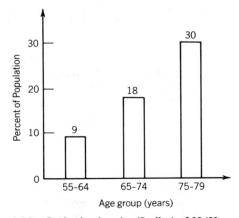

Figure 4.4.3. Static visual acuity (Snellen) of 20/50 or worse.

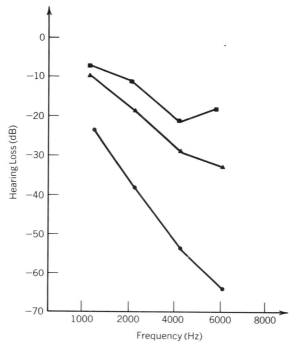

Figure 4.4.4. Hearing loss as a function of frequency and age, men and women combined (reference: hearing threshold for age 20). ■ Age 50; ▲ age 65; ● age 82.

Differences in hearing loss between men and women exist with women having more loss at 500 and 1000 Hz whereas men have significantly more loss at 3000, 4000, and 6000 Hz. At the high frequencies, men tend to experience loss 6 years younger than women (Eisdorfer and Wilke, 1974).

Even though hearing loss tends to increase as a function of age, it has been found in longitudinal studies over a 24 year period that the overall pattern of individual audiograms for healthy ears remains quite stable.

The ability to discriminate between sounds of very small differences in pitch (differential threshold expressed in hertz) as a function of age is less well understood than other issues in audition. Corso (1981) cites literature (König) which reports that the ability to discriminate frequency seems to deteriorate in approximately linear manner between ages 25 and 55 years, after which the difference in frequency required for detection increases more markedly, especially for higher frequencies.

The number of discriminable differences in loudness is best for younger ears and leads to a total of 325 at 2000 Hz, about the most sensitive point in the hearing threshold. With hearing thresholds at higher frequencies being progressively higher, this number is reduced relatively. The effect of the hearing losses that occur with older adults may cause further change but this matter may be complicated by other considerations such as recruitment (see Van Cott and Kinkade, 1972).

Corso (1981) reports a progressive decrease in percent correct for speech intelligibility with age. Compared to an age group of 20 to 29 years, the decrease for age 40 to 49 is 5%, at 60 is 10%, at 65 is 17%, and at 80 to 89 is 35%.

For information on the changes with age for the other sensory systems (vestibular, taste, smell, and skin senses) the reader is referred to Corso (1981).

4.4.7 COGNITION

Schaie (1983), in summarizing many studies involving age changes in cognitive abilities such as intellectual ability, reasoning, word fluency, verbal meaning, and educational aptitude, concludes that those changes before age 60 are trivial in magnitude but that by age 74, reliable decrements do occur for all such abilities. In terms of magnitude of these changes, he notes that "prior to age 60, no decrement in excess of .2 SD can be observed, whereas by age 81, the magnitude of decrement is approximately one population standard deviation (SD) for most variables. In other words, it is typically the late 60s and the 70s during which most individuals seem to experience significant ability decrement. Even so it is typically only by age 81 that one can show that the average older person will fall below the middle range of performance for young adults."

Memory retention of older adults for familiar faces, places, and other information learned one to three decades earlier is good. In addition, as Fozard points out (1980) "older persons remember information known directly to both young and elderly adults as well as do young persons. They recall information unique to their cohort better than do young adults who could only have known the same information second-hand. Such findings dispel the widespread belief that good memory by older adults is restricted to material of idiosyncratic importance."

Total knowledge is known to increase with age and older people retrieve information from their larger information base as efficiently as younger people do from their smaller base.

The speed of retrieval from long-term memory depends upon the degree of familiarity with what is to be recalled and the compatibility existing in the relationship between stimuli involved, responses involved, or a combination thereof. These relationships derive from one's physiological nature, learning, and culture, especially in early life. They give rise to expectancies in our experience from which population stereotypes frequently emerge. Compatibility conceptually is a continuum and may vary from 100% to zero (complete reversal) (See McCormick, 1976).

Compatibility is regarded as one of the mediating factors in human information processing providing the basis for decisions about actions to be taken. A. M. Small, Sr. and Fitts and Seeger have characterized stimulus–response (S–R) compatibility as follows: "A task involves compatible S–R relations to the extent that the ensemble of stimulus and response combinations comprising a task (or a problem or situation) results in a high rate of information transfer." (See McCormick, 1976.) In this context, compatibility implies a process of information transformation or recoding in that activity and assumes that the degree of compatibility is at a maximum when the recoding processes required are at a minimum. This applies also to stimulus–stimulus (S–S) and response–response (R–R) compatibilities.

Studies show that lack of compatibility increases task latency as a function of age. Fozard (1980) reports latency of 525 msec at age 33 with 100% compatibility compared with 680 msec for 72 year olds, a difference of 155 msec. For the same task with total incompatibility (relationship reversal), the results for the same age groups were 650 and 900 msec, a difference of 250. Some evidence indicates also a tendency for errors to occur more frequently with low compatibility and increasingly with age.

Compatibility is thus an important design factor with respect to age effects, especially for information displays and controls as related both to physical features and modes of operation. Likewise important are compatible grouping and placement of displays and controls.

When the most compatible relationships are not obvious, it is necessary to identify them on the basis of empirical experiments. (See Chapanis, 1959).

Corso (1976) indicates some evidence exists concerning visual perception for older people being due to either a loss in visual acuity or some aspect of "cognitive rigidity," with or without age held constant. If a lack of perceptual flexibility (reorganizational resistance) is involved in a given visual feature, such as figural aftereffects, central information processing stages are directly involved in which age effects do occur.

Older persons are able to learn despite popular opinions to the contrary. They appear to be more selective in what they undertake to learn and then proceed to learn it, often reaching the proficiency level of younger persons. The rate of their learning is slower, however, than that of young adults, becoming more marked with complexity of the material to be learned. As a result, the total time span required is longer.

There are limits, though, on what may be successfully attempted, set by the changes with age in strength, mobility of body members, viable postures, and so forth in regard to the nature and limits of motor or mental skill demands. Once learned, the speed of performance tends to be slower for complex operations than for young adults, but often with fewer errors committed.

An example of the foregoing is the increasing number of older adults who are showing interest and activity in computer matters from an intellectual as well as from a user point of view.

4.4.8   ATTITUDES

In this section a number of factors are identified—which have been observed or inferred—that seem to relate to older people's behavior. At present no substantive proof exists that validates them, but some appear to offer at least partial explanation of a number of findings related to slowing with age, selective learning, product design, acceptance, and safety.

For the older adult, the accumulation of many years of experience provides the opportunity for development in numerous ways. Among these are dealing with uncertainty, risk taking, making value and significance judgments, making errors, interacting with people, modes of personal expression, and analysis of one's needs related to products and environment.

In turn, and for some, this experience appears to result in the following:

1. Taking a cautious stance in responding to tests, surveys, and so on
2. Deliberation in decision-making
3. Seeking reduction in uncertainty
4. A management approach in risk taking
5. Increased concern for safety
6. Selectivity in learning new skills and retraining
7. A more balanced view of the future with technological developments and applications
8. Greater awareness of product deficiencies and need for more explicit human-factors requirements in product and environment design, congruent with characteristics and needs of older persons
9. Strategies and tactics for accomplishing the above, including purchasing

It seems clear that these developments have a meaningful relation to and provide emphasis on the timely and effective application of those human factors in design related to aging.

Other factors concerning productivity, retraining, and adaptability are discussed in Coates and Kirbey (1982).

4.4.9 STRESS

Stress is mentioned briefly here primarily as related to the older worker. Older workers tend to be quite responsive to management's control of stress in relation to work accomplishment, productivity, and quality. Thus their perception and awareness of effective human engineering design features of the equipment, tools, and environment provided for their use tend to lower stress, improve safety, and increase output and worker satisfaction to a greater degree than in the case of younger workers.

If stress reaches an overload condition with a task involving sequential elements, total interruption is likely to occur.

4.4.10 EPILOGUE

The reader is reminded that the foregoing sections deal with older people in the aggregate. As mentioned at the outset, those reliable and valid data available represent only a minor portion of what are needed. Those available do not allow a high degree of accuracy in defining deficits occurring as a function of aging.

The number of facets and factors involved and the increasing variability with age make gerontology a particularly difficult but important field for investigation. Eisdorfer and Cohen have remarked that "Aging should not be considered a sufficient explanation for change without supporting data." Research on aging is expanding in this country and abroad.

Finally, the large and growing number of older persons in our country and the increasing variability among them give rise to the expectation that there will be those whose last half of life will involve very few deficits of note or that they will adapt to those few that may exist. This in fact is happening!

REFERENCES

Birren, J. E., and Schaie, K. W. (1977). *Handbook of the psychology of aging.* New York: Van Nostrand Reinhold.*

Chapanis, A. (1959). *Research techniques in human engineering.* Baltimore: Johns Hopkins Press.*

Eisdorfer, C., and Lawton, M. P., Eds. (1973). *The psychology of adult development and aging.* Washington, DC: American Psychological Association.*

Koucelik, J. A. (1979). *Human factors and environmental design.* In T. O. Byerts, S. C. Howell, and L. A. Pastalan, Eds., *Environmental context of aging.* New York: Garland.*

Koucelik, J. A. (1982). *Aging and the product environment.* New York: Van Nostrand Reinhold.

Palmore, E., Ed. (1974). *Normal aging II.* Durham, NC: Duke University Press.*

Poon, L. W., Ed. (1980). *Aging in the 1980s.* Washington, DC: American Psychological Association.*

Schaie, K. W., Ed. (1983). *Longitudinal studies of adult psychological development.* New York: The Guilford Press.*

Van Cott, H. P., and Kinkade, R. G., Eds. (1972). *Human engineering guide to equipment design.* Washington, DC: U.S. Government Printing Office.

* Selected reading.

Physical (Anthropometric) and Biomechanical Characteristics

Chaffin, D. B., and Anderson, G. (1984). *Occupational biomechanics.* New York: Wiley.

Damon, A., and Stoudt, H. W. (1963). The functional anthropometry of old men. In *Human Factors, 6,* 485–491.*

Friedlander, J. S., Costa, P. T., Bosse, R., Ellis, E., Rhoads, J. G., and Stoudt, H. W. (1977). Longitudinal physique changes among healthy white veterans at Boston. *Human Biology, 49,* 54–558.*

Van Cott, H. P., and Kinkade, R. G., Eds. (1972). *Human engineering guide to equipment design.* Washington, DC: U.S. Government Printing Office.

Welford, A. T. (1958). *Aging and human skill.* Westport, CT: Greenwood Press.*

Woodson, W. E., and Conover, D. W. (1964). *Human engineering guide for equipment designers,* 2nd ed. Berkeley, CA: University of California Press.

Psychomotor Performance

Birren, J. E., Woods, A. M., and Williams, M. V. (1980). "Behavioral slowing with age: Causes, organization and consequences." In L. W. Poon, Ed., *Aging in the 1980s.* Washington, DC: American Psychological Association.

Welford, A. T. (1977). Motor performance. In J. E. Birren and K. W. Schaie, Eds., *Handbook of the psychology of aging.* New York: Van Nostrand Reinhold.

Welford, A. T. (1981). Signal, noise, performance and noise. *Human Factors, 23*(1), 97–109.

Sensation–Perception

Anderson, J., Jr., and Palmore, E. (1974). Longitudinal evaluation of ocular function. In E. Palmore, Ed., *Normal aging II.* Durham, NC: Duke University Press.

Corso, J. F. (1981). *Aging sensory systems and perception.* New York: Praeger.

Eisdorfer, C., and Wilke, F. (1974). Auditory changes. In E. Palmore, Ed., *Normal aging II.* Durham, NC: Duke University Press.

Cognition

Fozard J. L. (1980). The time for remembering. In L. W. Poon, Ed., *Aging in the 1980s.* Washington, DC: American Psychological Association.

McCormick, E. J. (1976). *Human factors engineering and design.* New York: McGraw-Hill.

Schaie, K. W. (1983). The Seattle longitudinal study: A 21-year exploration of psychometric intelligence in adulthood. In K. W. Schaie, Ed., *Longitudinal studies of adult psychological development.* New York: Guilford Press.

Attitudes and Epilogue

Borgatta, E. F., and McCluskey, N. G., Eds. (1980). *Aging and society: Current research and policy perspectives.* Beverly Hills, CA: Sage.

Coates, G. D., and Kirbey, R. H. (1982). Organismic factors and individual human performance and productivity. In E. A. Alluisi and E. A. Fleishman, Eds., *Human performance and productivity,* Vol. III. Hillsdale, NJ: Erlbaum.

PART 5
EQUIPMENT AND WORKPLACE DESIGN

CHAPTER 5.1
DESIGN OF VISUAL DISPLAYS

MARTIN G. HELANDER

State University of New York at Buffalo
Amherst, New York

An operator needs displays to present or summarize information. Speedometers, altimeters, compasses, and radarscopes are all examples of displays that were created to supplement the human senses. The design of military display panels was one of the first applications of human-factors engineering. More recently, civilian applications, for example, in automobiles, processing, and nuclear power plants, have become equally important.

This chapter gives an overview of traditional as well as recent human-factors research on display parameters. Types of mechanical displays and legibility and readability are first described. The optimum location of display in the visual field is discussed. Measurement of display illumination is then briefly reviewed.

Some technical characteristics of CRT terminals are discussed including display resolution, video bandwidth, choice of type of display, and measures to reduce reflections on the screen. Finally, visual aspects of color displays and choice of colors are covered, and some principles for graphics design are reviewed.

5.1.1 MECHANICAL DISPLAY INDICATORS

Mechanical display indicators present information on either a fixed or stationary scale, using a pointer or reference marker. There are three basic types of mechanical display indicators: direct reading counters, moving pointers on a fixed scale, and moving scales with fixed pointers; see Figure 5.1.1.

A fixed scale requires the most panel space and constant illumination across the entire scale. The scale length is limited. A moving scale saves panel space because the entire scale does not have to be displayed. Only a small section has to be illuminated and the use of a tape allows long scales. A counter is the most economical in terms of space and illumination. This type of display is limited only by the number of counter drums.

Depending on the type of information displayed, each of these indicators has advantages and disadvantages; see Table 5.1.1. For displaying quantitative information, counters generally minimize errors and reading time. However, counters are often difficult to read when the values change rapidly. Scales are less efficient for the display of quantitative information (Grether and Baker, 1972). Qualitative information is best displayed on a fixed scale with a moving pointer, because a change is easily detected without reading individual numbers. Moving scales and counters require individual numbers to be read and are not optimum for qualitative information.

For setting specific values, a counter is preferred, followed by a fixed scale with moving pointer.

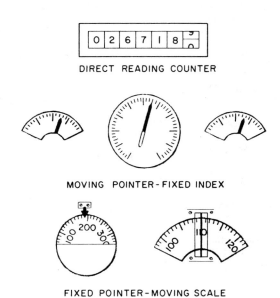

DIRECT READING COUNTER

MOVING POINTER-FIXED INDEX

FIXED POINTER-MOVING SCALE

Fig. 5.1.1. Basic types of mechanical displays.

Table 5.1.1   Choice of Mechanical Display Indicator as a Function of Task

Use of Display	Type of Task	Display Typically Used for	Type of Display Preferred
Quantitative reading	Exact numerical value	Time from a clock; rpm from tachometer	Counter
Qualitative reading	Trend; rate of change	Rising temperature; ship off course	Moving pointer
Check reading	Verifying numeral value	Process control	Moving pointer
Setting to desired value	Setting target bearing; setting course	Compass	Counter or Moving pointer
Tracking	Continuous adjustment of desired value	Following moving target with cross hair	Moving pointer
Spatial orientation	Judging position and movement	Navigation aids	Moving pointer or moving scale

A moving scale is more ambiguous with respect to the position of the setting. For tracking, a moving pointer is preferred. This provides the simplest relationship to manual control motion.

5.1.2   LEGIBILITY AND READABILITY

There are several important human-factors principles that enhance the readability of a display. Table 5.1.2 summarizes some of the important considerations that apply to the design of warnings, signals, labels, and other displays (Woodson, 1981).

Table 5.1.2   Principles that Enhance the Effectiveness of Visual Displays

Conspicuity

The sign should attract attention and be located where people will be looking. Three main fctors determine the amount of attention people devote to a sign: prominence, novelty, and relevance.

Emphasis

The most important words should be emphasized. For example, a sign might emphasize the word "danger" by using larger characters and borderlines.

Legibility

Legibility may be optimized by enhancing the contrast ratio of the characters against the background, and by using type fonts that are easy to read.

Intelligibility

Make clear what the hazard is and what may happen if the warning is ignored. Use as few words as possible, avoiding acronyms and abbreviations. Tell the operator exactly what to do.

Visibility

The sign or the label should be visible under all expected viewing conditions, including day and night viewing, bright sunlight, and so forth.

Maintainability

Materials must be chosen that resist the aging and wear due to sunlight, rain, cleaning detergents, soil, vandalism, and so forth.

Standardization

Use standard words and symbols whenever they exist. Although many existing standards may not follow these recommendations, they are usually well established and it might be confusing to introduce new symbols.

5.1.2.1 Print Style

Capital letters are generally recommended for all displays. For black letters on a white background, the stroke width-to-height ratio should be approximately $1:6$ to $1:8$. When white letters are presented on a black background, the recommended width-to-height ratio is $1:8$ to $1:10$ (Berger, 1944a, b). This difference in optimum stroke widths of white and black characters depends upon a phenomena referred to as irradiation; the white character or white background bleeds into its darker surroundings. An example of irradiation is found on visual display terminals where dark characters on a light background are less visible than light characters on a dark background, if the width-to-height ratio remains the same.

5.1.2.2 Size of Print

The preferred height of letters and numerals for labels depends upon viewing distance, ambient illumination, and the importance of the text. Peters and Adams (1959) proposed the following formula to calculate appropriate character height:

$$H = 0.0022D + K_1 + K_2 \tag{1}$$

where D = viewing distance
$\quad K_1$ = correction factor for viewing distance and illumination
$\quad K_2$ = correction factor for importance of the material

Table 5.1.3 presents the recommended height of letters and numerals for several values of K_1 and K_2. As an example, for a reading distance of 14 inch and a K_1 value of 0.06, a letter size of 0.09 inch is recommended for nonimportant information and 0.17 inch for important information. As viewing distance increases or reading conditions get worse, the size of characters should be increased.

5.1.3 LOCATION OF DISPLAYS IN THE VISUAL FIELD

The optimal arrangement of displays should consider both the capacity and limitations of the human operator and task requirements. These factors are summarized below under the following subheadings: limitations of the visual field, frequency and sequence of display use, and logical and perceptual factors for grouping of display information.

5.1.3.1 Limitations of the Visual Field

The location and arrangement of displays should be adapted to the operator's "normal" viewing angle, the field of view, and the characteristics of eye movements. Once a normal viewing angle has been established, it is possible to assign priorities to locations in the visual field. Obviously, the central visual field should be reserved for the most important displays, with the peripheral visual field for displays that are less important.

Many human-factors manuals cite a normal viewing angle for a seated operator of about 15°

Table 5.1.3 Table of Heights of Letters and Numerals Recommended for Labels and Markings on Panels, for Varying Distance and Conditions, Derived from Formula Height (in.) = 0.0022D + $K_1 + K_2$[a]

Viewing Distance, D (in.)	0.0022D Value	Nonimportant Markings $K_2 = 0.0$			Important Markings $K_2 = 0.075$		
		$K_1 = 0.06$	$K_1 = 0.16$	$K_1 = 0.26$	$K_1 = 0.06$	$K_1 = 0.16$	$K_1 = 0.26$
14	0.0308	0.09	0.19	0.29	0.17	0.27	0.37
28	0.0616	0.12	0.22	0.32	0.20	0.30	0.40
42	0.0926	0.15	0.25	0.35	0.23	0.33	0.43
56	0.1232	0.18	0.28	0.38	0.25	0.35	0.45

[a] Applicability of K_1 values:
$K_1 = 0.06$ (above 1.0 fc, favorable reading conditions)
$K_1 = 0.16$ (above 1.0 fc, unfavorable reading conditions)
$K_1 = 0.16$ (below 1.0 fc, favorable reading conditions)
$K_1 = 0.26$ (below 1.0 fc, unfavorable reading conditions)

below the horizontal. Figure 5.1.2 illustrates the visual field with eye rotations and head rotations, and the normal viewing angle (U.S. Department of Defense, 1981).

The use of a normal viewing angle may be an oversimplification, because the preferred line of sight and its dynamic range vary with task requirements. For some tasks, such as flying an airplane, in which the operator is strapped to the seat and fairly immobile, it is reasonable to assume low variability of viewing angles. However, for a VDT operator, viewing angle often changes as the operator shifts his/her weight on a chair. Therefore, it may be appropriate for designers to use measures of angle variability ranging from the 5th to the 95th percentile, just as is done with anthropometric measures; see Figure 5.1.3.

Grandjean (1984) noted great variability in the angles of various joints, such as the elbow and shoulder, among VDT operators. This variability should be considered in the design of workplaces to the same extent as anthropometric variability. Research is needed to document the variability in viewing angles for different tasks and work postures. Most experts agree that the "normal viewing angle" should be slightly below the horizontal, at least for sitting operators. The 15° assumption may, however, be misleading; most individuals working at a desk assume viewing angles of up to 60°, and recent data for VDT operators indicate that the average preferred viewing angle is about 35° (Kroemer, 1984). For most applications, the normal viewing angle can assume just about any value between 0 and 50°. It may therefore be advisable to obtain measures that are relevant to the particular task.

As mentioned above, once the normal viewing angle has been established, the visual field can be divided into areas with different priorities. Although the visual field covers 180° horizontally and 150° vertically (with eye movements, but stationary head), attention cannot be equally allocated over the entire visual field. Designers must therefore determine which areas are the most important.

Sanders (1970) analyzed head and eye movements in a target detection task. He distinguished three attentional areas in the visual field: *the stationary field*, where peripheral viewing is sufficient; *the eye field*, where the supplementary use of eye movements is required; and *the head field*, where head movements are also necessary. For visual targets presented within a visual angle less than 30°

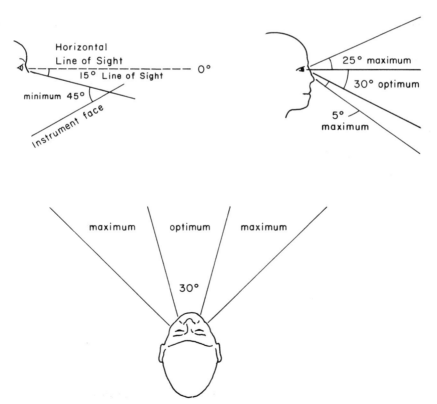

Fig. 5.1.2. Normal line of sight and preferred viewing angles. (*Source:* U.S. Department of Defense, 1981.)

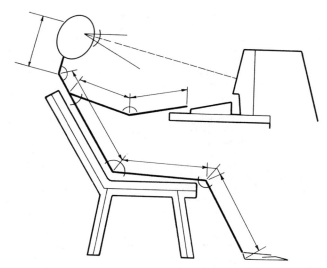

Fig. 5.1.3. To establish typical work postures and viewing angles, it is necessary to obtain values of both anthropometric measures and joint angles.

from the viewing angle, there were rarely eye movements. For targets presented between 30° and 80° beyond the viewing angle, there were eye movements; and for targets beyond 80° there were also complementary head movements; see Figure 5.1.4.

In the stationary field, two inspections may be performed simultaneously by a single glance without altering the eye fixation point. This means that one fixation is sufficient to perceive the status of two displays, an important consideration for design. In the eye field, eye movements are necessary. Therefore, one display may be located in the field of direct vision, whereas other displays are placed in the periphery. Displays located in the peripheral field demand attention through "preattentive" processes (Neisser, 1966). Such processes assume reflexive eye movements to the part of the display where motion is detected. These eye movements are not under voluntary control, because they occur before the operator has consciously processed the information.

For displays in the head field, head movements are required. These displays are outside the peripheral range, so that pre-attentive processes and reflex eye movements are disabled (Kraiss, 1976).

5.1.3.2 Frequency and Sequence of Use of Displays

To make optimum use of the visual field, it is essential to know the importance of various displays and how the operator will monitor one or more displays. Measures of frequency, sequence, and criticality

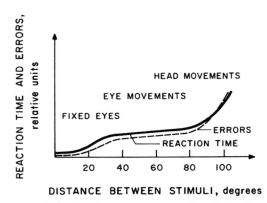

Fig. 5.1.4. Reaction time and identification errors as a function of stimulus separation in the visual field. (*Source:* Sanders, 1970.)

of use form the basis for the determination of the importance of displays. The most frequently used displays should be positioned in convenient locations. Large groups of important but infrequently used displays can be positioned in less convenient locations. Displays normally used in sequence should be arranged left-to-right or top-to-bottom.

Similar considerations hold true for the locations of controls and the interaction between controls and their corresponding displays. The position of each control must therefore be selected so that it is consistent with the display; see Figure 5.1.5.

Task analysis is an important tool for analyzing the frequency of use and the criticality of displays, and assigning displays to one of three categories: primary displays, secondary displays, and emergency displays. Depending on their classification, displays may be given different priorities in the visual field. For example, to use Sanders' (1977) concepts, primary displays could be located in the stationary field, and secondary and emergency displays in the eye field.

Displays that are related to a particular function should be grouped together, and not intermixed with displays for other functions. Displays for emergency situations should be isolated from others and their information easily detectable and accessible.

5.1.3.3 Logical and Perceptual Factors for Grouping of Display Information

When there are several displays in the work environment that are consistently associated with different subtasks of the operation, it is helpful to use these associations for making the physical layout and grouping of displays. In general, displays should be located so that they structure the task and make it more meaningful to the operator. This may include geometric arrangements of the displays to mimic the physical arrangement of functions being monitored. Many such arrangements rely on the Gestalt theory for grouping information. This theory postulates that, by using concepts such as proximity and similarity, the observer will group the displays in a more meaningful and useful way. He or she will organize the displays in the visual field according to perceptual rules that subconsciously operate within an observer; to most operators the grouping is immediate and self-evident. Figure 5.1.6 shows an example of the Gestalt theory (Kahneman, 1973).

Similar principles can now be used for the design of displays and associated labels and controls (Kraiss, 1976). Figure 5.1.7 shows several examples of how these Gestalt principles can be incorporated into display designs.

In Figure 5.1.7a, displays have been divided into subgroups. Following the Gestalt law of proximity, the operator automatically associates similar functions for the displays in each group. Figure 5.1.7b shows that the design and arrangement of dials can facilitate the identification of normal conditions.

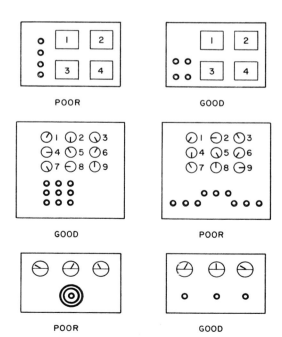

Fig. 5.1.5. Controls must be positioned to enhance the interaction with the corresponding displays.

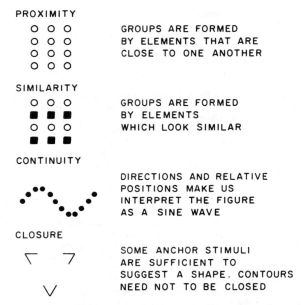

Fig. 5.1.6 The Gestalt theory may be used in design to emphasize groups or interrelationships between different display units.

For example, if normal conditions are represented by nine o'clock positions, deviation from this position is rapidly identified. The absence of such normal positions increases monitoring time by requiring each indicator to be read individually (Dashevsky, 1964). This example utilizes the Gestalt law of continuity. The observer only has to look for exceptions, or deviations, not read all dials.

5.1.4 CODING OF DISPLAY INFORMATION

In order to make information on a display easy to detect and understand, information may be coded in various ways. By coding, artificial features are added that make it easier to distinguish the information and reduce the information processing time. There are several different ways of coding information: by alphanumeric notations, by color, by geometric shapes, by angle of inclination, by size or form, by luminance, or by flash rate. An overview of advantages and disadvantages of several visual coding methods is given in Table 5.1.4 (Grether and Baker, 1972). An extensive presentation of coding methods is given in a human engineering data base by Meister (1984).

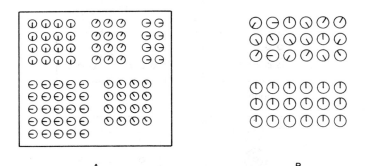

A   B

Fig. 5.1.7. Principles of Gestalt laws used in design of displays. (*A*) The Gestalt law of proximity used for grouping of displays. (*b*) The Gestalt laws of continuation and closure facilitate check reading of the dials. "Normal" at 12 o'clock for all dials.

Table 5.1.4 Comparison of Coding Methods

Code	Number of Code Steps[a]		Evaluation	Comment
	Maximum	Recommended		
Color				
Lights	10	3	Good	Location time short. Little space required. Good for qualitative coding. Larger alphabets obtainable by combining saturation and brightness with the color code. Ambient illumination not critical factor
Surfaces	50	9	Good	Same as above, except ambient illumination must be controlled. Has broad application
Shapes				
Numerals and letters	Unlimited		Good	Especially for identification. Uses little space
Geometric	15	5	Fair	Particularity useful for symbolic representations. Some shapes may be more difficult to dissiminate
Pictorial	30	10	Good	Allows direct association. Requires high display resolution
Magnitude				
Area	6	3	Fair	Requires large space. Easy to find information
Length	6	3	Fair	Requires large space. Easy to find information
Luminance	4	2	Poor	May interfere with other signals. Ambient
Stereo-depth	4	2	Poor	Limited population of users. Difficult to instrument
Angle of inclination	24	12	Good	Particularity good for quantitative evaluations, but applications are limited to round instruments, such as dials and clocks
Flash rate	5	2	Fair	Difficult to differentiate between flash rates. Good for attracting attention

[a] The maximum number assumes a high training and use level of the code. Also, a 5% error in decoding must be expected. The recommended number assumes operational conditions and a need for high accuracy.

Depending on the task, visual codes differ in applicability. It may therefore be misleading to provide general guidelines for the number of levels of each coding method. The information in Table 5.1.4 should, therefore, only be considered a rough guideline. Ideally, the requirement for coding should be investigated for each specific case using analytical methods, such as task analysis, link analysis, or simulation. Different design alternatives can then be tested while simultaneously evaluating the subject's performance, work load, or eye movement behavior. The combination of analyses and simulation represents the most powerful technique for approaching the problem of display arrangement or panel layout (Kelley, 1968).

5.1.5   COGNITIVE FACTORS IN DISPLAY DESIGN

Not only are the legibility and the visibility of displays important, but also the message itself should be designed so that it is easy to understand and minimizes reading time and errors. Broadbent (1977) pointed out that there are many different levels of language that can be used to portray intended meaning. A label or a message can be expressed in an active, passive, or negative manner, as illustrated by the following:

Active   The large lever controls the depth of the cut.

Passive   The depth of the cut is controlled by the large lever.

Negative The small lever does not control the depth of the cut.

Active statements are generally easier to understand than passive or negative statements (McCormick and Sanders, 1982).

Symbols may also be effective in transmitting information to the user. They have long been used for traffic signs, public information signs, and as a replacement for labels to denote product use. Recently, symbols have also become popular in computer applications, where "icons" are sometimes used as alternatives to command names.

Depending upon the type of information conveyed, symbols may be more or less appropriate. In general, symbols are appropriate for concrete information, such as indicating the location of a restroom. For more abstract concepts, however, symbols are more difficult to use. The National Bureau of Standards (Lerner, 1981), for example, failed to devise a comprehensible symbol for an exit sign.

Many machine manufacturers prefer to use symbols rather than labels, because symbols, unlike labels, do not have to be translated when the machine is exported to non-English speaking countries. Unfortunately, the understandability of the proposed symbols is often low. Cultural differences and lack of standardized precise meaning are problems. As an example, Figure 5.1.8 gives examples of symbols proposed by SAE for use on construction machines (Helander, 1983).

In an attempt to regulate the design of public symbols, Zwaga and Easterby (1982) developed a standard for evaluating symbols that are submitted to the International Standards Organization (ISO) for acceptance. Every symbol to be approved by ISO as an international symbol must be tested in six countries and the average comprehension must be at least 66%.

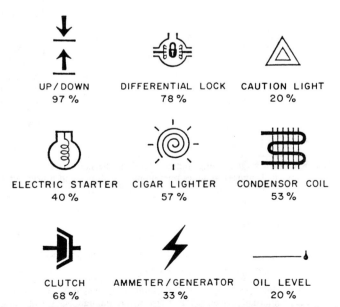

PERCENT CORRECT RESPONSES

UP/DOWN 97%	DIFFERENTIAL LOCK 78%	CAUTION LIGHT 20%
ELECTRIC STARTER 40%	CIGAR LIGHTER 57%	CONDENSOR COIL 53%
CLUTCH 68%	AMMETER/GENERATOR 33%	OIL LEVEL 20%

Fig. 5.1.8.   Examples of symbols for construction machines considered for adoption by the Society of Automotive Engineers. Percentages of untrained observers who recognized the meaning of the symbols. (*Source:* Helander and Schurick, 1983.)

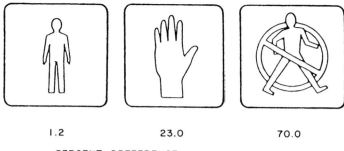

        1.2   23.0   70.0

              PERCENT PREFERENCE

Fig. 5.1.9. To portray "DON'T WALK," a prohibitive sign with a slash is the best alternative in this case, although the slash is generally counter to good design practice. (*Source:* Robertson, 1977.)

Dewar (1976) found that prohibitive traffic signs require longer processing times because, unlike permissive signs, they do not explicitly state the desired action. An example is a "NO LEFT TURN" symbolic traffic sign. The driver must first understand what not to do in order to conclude that only driving straight and right turns are allowed. A permissive sign, with arrows to the right and straight, contains less information and reduces driver reaction time. Dewar also pointed out that a slash indicating "DO NOT" in a prohibitive sign obscures legibility. He therefore recommended a partial slash that does not cover the entire symbol. Despite the advantage of permissive symbolic signs over prohibitive signs, it is sometimes difficult to design appropriate symbols. Figure 5.1.9 shows that to portray the meaning of "DON'T WALK," a prohibitive sign is much more effective than other alternatives (Robertson, 1977).

There are also motivational aspects of display design. A study by Johansson and Backlund (1970) investigated the recall of various road signs. Drivers were stopped by police at a road barrier and asked to describe the most recent road sign that they had passed. Signs containing information relevant to the immediate driving task had a higher likelihood of being recalled than did signs that contained more general information, such as "general danger"; see Figure 5.1.10. These results suggest that the relevance of signs affect the amount of attention given to them.

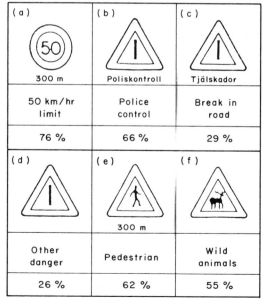

Fig. 5.1.10. Percent of drivers recalling road signs 1 min after passing them. (*Source:* Johansson and Backlund, 1970.)

5.1.5.1 Pictorial Displays

It is often helpful to supplement displays with pictorial features, such as a pointer shaped to resemble the object to which it refers. Pictorial features can be useful for position orientation and movements in three-dimensional space. An aircraft orientation display, for example, is aided by a pictorial representation of the attitude of the aircraft. As shown in Figure 5.1.11, two alternatives are available. The first is a picture of the horizon as the pilot sees it. The aircraft remains stationary and the horizon tilts. The second alternative provides a perspective as seen from behind the aircraft, where the aircraft is tilting against a fixed horizon. The tilting horizon is often referred to as an "inside-out" display and the tilting airplane as an "outside-in" display.

Kelley (1968) pointed out that the inside-out model may present problems with motion incompatibility and lead to control reversals. By using a tilting aircraft (outside-in model), this motion incompatibility can be eliminated. Perhaps the outside-in model is more appropriate with respect to the internal model that a pilot develops, assuming that his/her internal model is not a fixed vehicle in a moving environment but a moving vehicle in a fixed environment. Johnson and Roscoe (1972), however, found that pilot visibility aspects are important. If the pilot has full visibility of the horizon, and is able to switch from the instrument to the outside view, it is best to use the "inside-out" model. This is more compatible with what the pilot is observing.

5.1.6 MEASUREMENT OF ILLUMINATION

This section is limited to factors that are particularly relevant to the photometric measurement of displays. Techniques for measuring and describing light are given elsewhere in this handbook.

Light can be described as radiant energy traveling in the form of electromagnetic waves and having a wavelength proper for evoking a visual sensation. Measurement of light is a special case of measurement of radiant energy. The term *radiometry* applies to the measurement of all radiant energy, whereas the term *photometry* applies to the measurement of light, where the spectral energy has been weighted at each wavelength according to the wavelength sensitivity of the eye. Figure 5.1.12 shows the weighting function of a standard observer as adopted by CIE in 1931 (Judd and Wyszecki, 1967). Although this weighting curve provides a reasonable estimate of the average observer, there is actually enormous variability among individuals, even among those having normal color vision.

The most important radiometric and photometric terms for the measurement of visual displays are listed below (Hardesty and Projector, 1973; Snyder, 1980; Farrell and Booth, 1984).

Radiant power, also called radiant flux, is the time-rate of flow of radiant energy. It is measured in joules per second or watts.

Luminous power, also called luminous flux, is radiant power (energy per time) weighted according to the spectral sensitivity of the eye; see Figure 5.1.12. This is the visible part of radiant power, and is measured in lumens (lm).

Illuminance, formerly called illumination, is the flux incident on a unit surface area. It is measured in lux in the SI system and footcandles (fc) in the English system.

For quantifying photometric qualities of displays, luminance is the most frequently used measure. The term brightness was formerly used as a synonym, but today the term brightness is restricted to subjective sensations associated with a given luminance. The definition of luminance can be understood by reference to Figure 5.1.13, which illustrates a light source with an intensity of 1 candela radiating flux uniformly in all directions.

A *candela* is defined as $\frac{1}{60}$ of the intensity of 1 cm² of a black-body radiator at the freezing point of platinum (2047°K).

Fig. 5.1.11. Two alternatives for displaying aircraft altitude. To the left as the pilot sees it (inside-out), and to the right as an outside onlooker would see it (outside-in).

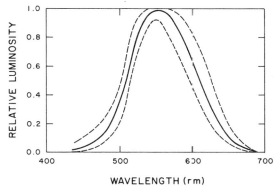

Fig. 5.1.12. Luminosity function for 1931 CIE Standard Observer compared to the range of values obtained for 37 subjects. (*Source:* Judd and Wyszecki, 1967.)

A spherical surface subtends a solid angle (angle in space) of 4π or 12.57 steradians. Stated differently, a sphere of 1 ft (unit) radius has a surface area of 12.57 ft². Thus the total flux emitted by a 1 candela uniform light source is 12.57 lm, and the illuminance on the surface of a 1 ft sphere is 1 lm/ft², or 1 fc. Correspondingly, the illuminance on a 1 m sphere is 1 lm/m² or 1 lux. There is a simple conversion between lux and footcandle. Because there are 10.76 ft² in 1 m², 1 fc equals 10.76 lux.

Figure 5.1.13 also illustrates the inverse square law of illuminance, which says that illuminance is inversely proportional to distance squared from the source of illumination. The illuminance on the surface located 1 m from the light source is therefore 1/10.76 fc. The inverse square law does not hold true for light sources other than point sources and therefore is not applicable for displays (or luminaires) with extended luminous surfaces.

Luminance is the luminous power reflected or emitted from a surface in a specified direction. The surface is defined in terms of the apparent area as viewed from a specified direction. Measurement units are candelas per square meter (cd/m²) in the SI system and footlamberts (fL) in the English system.

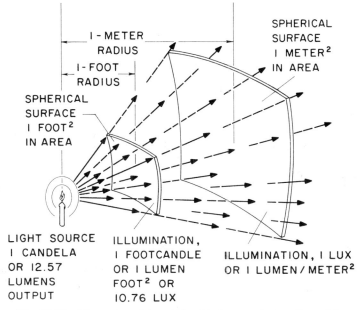

Fig. 5.1.13. Illustration of the relationship between footcandles and lux.

For surfaces that are free of specular (mirrorlike) reflections, with the incoming light evenly reflected (diffused) in all directions, the luminance of the surface may be determined by multiplying the incident illuminance by the reflectance of the particular surface. In the international system (SI), reflectance is given by the following formula:

$$\text{reflectance (\%)} = \frac{\text{luminance (cd/m}^2) \times \pi}{\text{illuminance (lux)}} \qquad (2)$$

In English units, reflectance is determined by the following formula:

$$\text{reflectance (\%)} = \frac{\text{luminance (fL)}}{\text{illuminance (fc)}} \qquad (3)$$

The reflectance of a surface is determined by comparing it to a surface with known reflectivity. A gray standard with 18% reflectance or a white standard with 80% reflectance may be used for this comparison. The standard is superimposed on the surface with unknown reflectance and, by measuring the luminance from both surfaces, the value of the unknown reflectance may be derived.

Figure 5.1.14 illustrates the conversion from illuminance to luminance for a CRT display with and without a neutral density filter. Filters reduce the amount of light that passes through them; they are called neutral density filters if they do so without changing the color of the light. The purpose of the neutral density filter is to enhance the contrast between the characters and the background. Ambient light passes through the filter once before it is reflected by the screen surface. The reflected

Effect of Neutral Density Filters on Display Contrast Ratio

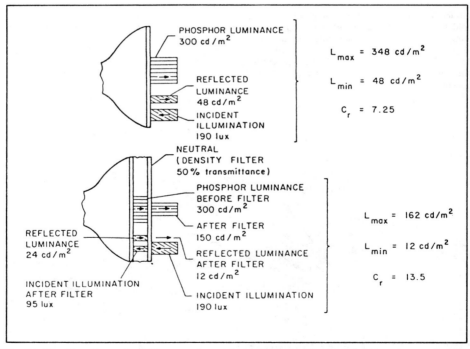

Fig. 5.1.14. The effects of a 50% transmittance neutral density filter on display contrast ratio (C_r) and character luminance L_{max}. If it is assumed that the phosphor has a reflectance of 0.8, the reflected luminance for the no-filter condition is $(190 \times 0.8)/\pi = 48$ cd/m². The character luminance is then $300 + 48 = 348$ cd/m², and the contrast ratio is $348/48 = 7.25$. In the bottom figure, the incident illuminance is reduced twice by the filter, but the character luminance only once. This increases the contrast ratio to $162/12 = 13.9$, which enhances visibility. Although the character luminance is reduced (162 vs. the former 348 cd/m²), the gain in contrast appreciably increases legibility. (*Source:* Farrell and Booth, 1984.)

light passes through the filter a second time on its way to the viewer. Some light is lost each time that it passes through the filter. However, the light emitted by the characters passes through the filters only once. As a result, the luminance of the characters is reduced less than that of the reflected background, thereby enhancing the character's contrast with its background (Farrell and Booth, 1984). Figure 5.1.14 indicates a contrast ratio of 7.25 without the filter and 13.5 with the filter. At the same time the character luminance is reduced from 348 to 162 cd/m². The increased contrast enhances legibility, more than making up for the loss in luminance. The reduced character luminance may, however, detract from visibility if the luminance is close to threshold (e.g., character luminance less than 30 cd/m²).

5.1.7 CONTRAST RATIO

To be visible, information displayed on a CRT must have either higher or lower luminance than the surrounding areas. The difference between the target and its background is referred to as contrast. There are several alternative expressions for quantifying contrast. All of them form a ratio between the luminance of the target and that of the background. The simplest expression of contrast is

$$\text{contrast ratio, } C_R = \frac{L_{max}}{L_{min}} \tag{4}$$

where L_{max} = luminance emitted by the area (either target or background) of the greatest intensity (cd/m²)

L_{min} = luminance emitted by the area of least intensity (cd/m²)

Modulation contrast, often preferred by engineers and optical scientists, is an alternative expression of contrast ratio:

$$\text{modulation contrast, } C_M = \frac{L_{max} - L_{min}}{L_{max} + L_{min}} \tag{5}$$

Modulation contrast is often used as a basis for measuring display image quality as well as quantifying the characteristics of the human visual system. Below, we explore the concepts of modulation transfer function (MTF) to express display image quality, and contrast threshold function to express the characteristics of the visual system. These concepts are used in conjunction with Fourier analysis in order to analyze mathematical expressions of the degradation of images as they are transferred through optical systems, such as camera lenses, or reproduced on visual displays (Snyder, 1980; Cornsweet, 1970). Similar techniques have been applied for many years to the analysis of temporal properties of systems, such as telephones and high fidelity photographs. The technique, also referred to as linear systems analysis, decomposes any repetitive wave form into its component frequencies, each with a specific amplitude and phase relationship. If all the frequencies are appropriately combined, the resulting summation is the original wave form. Figure 5.1.15 illustrates how a nearly square wave luminance distribution can be reproduced by summing a series of sine waves of different frequencies and amplitudes. The procedure by which a given wave form is decomposed into a set of sine waves is called Fourier analysis. Conversely, the procedure by which a wave form is composed by adding sine wave functions is called Fourier synthesis.

The human visual system may be described in terms of its sensitivity to stimuli of different spatial frequencies. The concept of spatial frequency is fundamental to the measurement of the contrast threshold function and MTF (Cornsweet, 1970). In Figure 5.1.16, the contrast threshold is plotted as a function of spatial frequency (Snyder, 1980). The human eye is most sensitive to spatial frequencies around 3 to 5 cycles per degree, whereas higher frequencies (i.e., little space between gratings) and lower frequencies (more space between gratings), are more difficult to perceive. At higher luminance levels, the most detectable frequency is a little higher, about 6 cycles per degree.

Any optical system, whether a lens or a display, will degrade the image of the original figure. This degradation, which varies with the size of the image, can be quantified using a display modulation transfer function. Although large features, corresponding to low spatial frequencies, are reproduced without much degradation, smaller features, corresponding to high spatial frequencies, are more difficult to reproduce because they are relatively more affected by noise and blur in the system, resulting in greater degradation. Figure 5.1.17 shows an MTF curve for a display system. Note that for the low spatial frequencies, the reproduction is perfect, with contrast $C = 1$, whereas MTF and contrast roll off at higher spatial frequencies.

Blur, resolution, and MTF are all related to the sharpness of display characters. There are, however, no standard procedures for measuring these parameters. For characters formed on a CRT, a photometer with a microscanning slit is typically used to measure luminance across the profile of a single dot in a character. Figure 5.1.18 shows examples of luminance distributions for a dot matrix character in which one pixel (dot) is turned off (Murch, 1984). The contrast between the written and unwritten

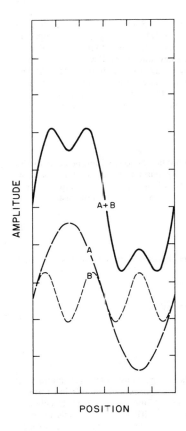

AMPLITUDE

A + B

A

B

POSITION

Fig. 5.1.15. Additions of sine waves to synthesize a square wave.

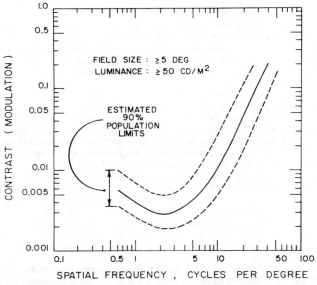

Fig. 5.1.16. Contrast threshold as a function of spatial frequency. Note that the visual system is most sensitive to spatial frequencies around 3 to 5 cycles per degree. (*Source:* Snyder, 1980.)

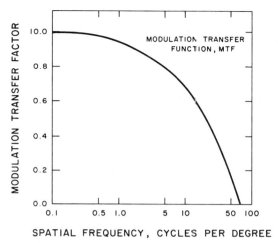

Fig. 5.1.17. Modulation transfer function (MTF) curve for a display system. (*Source:* Snyder, 1980.)

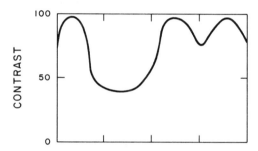

Fig. 5.1.18. Luminance distribution for three display dots with one intermediate dot turned off.

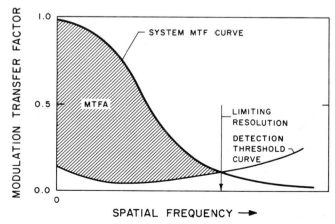

Fig. 5.1.19. Modulation transfer function area (MTFA) is the area between the system MTF curve and the human contrast threshold function. (*Source:* Snyder, 1980.)

areas is 45% when one pixel is missing, shown by the dip in the curve. This is due to the Gaussian distribution of dot luminances which causes individual dots to overlap. As the spatial frequency (spacing) of dots increases, the contrast between dots decreases, and the MTF gradually decreases.

It is now possible to compare the MTF of a display with the contrast threshold function of the eye; see Figure 5.1.19. As the spatial frequency increases beyond a certain value, the MTF curve crosses the contrast threshold function. This crossing point corresponds to the limit of resolution, beyond which the human observer cannot perceive finer details on the particular display. Of course the limit of resolution varies, depending upon the luminance of the display and the modulation contrast.

There have been several different models for quantifying image quality in displays. The most recent of these have utilized the MTF or derivatives thereof. In particular, the area enclosed by the MTF and the contrast threshold function, called modulation transfer function area (MTFA), has proved to be a good summary measure of image quality (Snyder, 1980). The main difference between an MTF and the MTFA is that the MTFA takes into account the sensitivity of the human eye. The MTFA measure has been validated against subjective estimates of image quality, using a variety of different images ranging from face recognition to military reconnaissance. MTFA correlated well with performance measures.

In order to get a measure of MTFA for a CRT display, it is necessary to use a microphotometer to obtain an accurate description of the luminance profile of a dot or pixel. This procedure is very time-consuming, and requires specialized equipment. As a short cut, Beaton (1984) suggested that the following formula may be used for approximating MTFA for displays with spots that have a luminance profile that is approximately normally distributed:

MTFA = 10^A

where A = $1.48 + 0.6 * V_D - 1.07 * W_D - 1.62 * A_B - 0.17 * V_D * A_B$
$+ 0.59 * W_D * A_B + 0.48 * L_M * A_B + 0.06 * V_d * L_M * A_B$

where V_D = viewing distance in meters; 38 cm $< V_D <$ 102 cm   (6)

L_M = $\log_{10}$ of the peak display luminance in cd/m²; 1.3 $< L_M <$ 2.5

W_D = the full width of the gaussian spot at the half amplitude point in mm; .15 mm $< W_D$ $<$ .76 mm

A_B = $\log_{10}$ of the reflected luminance from the display screen; 0 $< A_B <$ 1.7

The required value of MTFA depends on whether the display is used for graphics or for alphanumerics. Generally, a high-resolution display has an MTFA value of 10 or greater, whereas a moderately high resolution display falls in the range of 7 to 10. For most office applications where the equipment is used for displaying alphanumeric information, the MTFA should be greater than 5.

5.1.8   SCREEN RESOLUTION AND CHARACTER DESIGN

Characters on a CRT display are usually formed by dot matrices. The dots are generated by turning the electron beam on and off as it sweeps over the phosphor. In order to retain character symmetry for oddly shaped characters such as "M," an odd number of dots in each direction are used. The most common dot matrix sizes use 5 × 7 and 7 × 9 dots. An uppercase letter may, for example, be five dots wide and seven dots high. A descender is the leg of a character that drops below the rest of the character, such as in g or y. Two extra vertical dots are required for true descenders. Lowercase characters with true descenders are easier to read than characters without them. The minimum acceptable size of a dot matrix is 5 × 7, and 7 × 9 or 9 × 11 are generally preferred. Because there is usually at least one dot space between row of characters, 10 vertical dots are required for a 5 × 7 matrix, including 2 for descenders and 1 for spacing. Likewise, 12 or 14 lines are required for a 7 × 9 and 9 × 11 dot matrix, respectively.

The legibility of characters also depends on the type of font used. For 5 × 7 characters, the Huddleston font is generally considered best and for 5 × 9 or 9 × 11 matrix sizes, the Huddleston and the Lincoln/Mitre fonts are equally good (see Figure 5.1.20). In essence, the closer the characters resemble regular stroke characters, the easier they are to read. Character legibility is therefore improved when the dots fill most of the space in a character. Because of this, high resolution dot matrices are preferred. In addition, because square dots fill more empty space, they are preferred over round dots and oblique dots (Snyder and Maddox, 1978).

It is sometimes difficult to avoid confusion between certain characters in a dot matrix alphabet. For example, for the Huddleston font, operators commonly confuse "Y" and "V," "4" and "1," and "Z" and "2." For other matrices, some common confusions occur with "5" and "S," and "2" and "7." Before equipment is ordered, the confusion of characters should be investigated and, because there is no standardization of fonts among different manufacturers, care should be taken by the designer and user to select fonts that maximize legibility.

The optimum character size depends on the task. It appears that smaller characters and more densely packed text may be best for quick scanning. Larger characters should be employed in visual

Fig. 5.1.20. Example of Huddleston font in a 9 × 11 matrix. (*Source:* Snyder and Maddox, 1978.)

search tasks and in tasks involving typing from material on the screen, where recognition of each individual character is important.

The requirements for character legibility for a VDT are approximately the same as for other forms of written material. Shurtleff (1980) observed that, although the resolution of the CRT display is usually less than that of printed material, contrast may be higher. Character sizes as small as 10 min of arc ($\frac{10}{60}$ of a degree) are acceptable, providing contrast is sufficiently high; see Figure 5.1.21. However, most researchers recommend a character height of at least 14 min of arc, which would also make the character legible at lower contrast. Characters larger than about 25 min of arc are not appropriate for reading tasks, because they disrupt normal eye movement patterns by forcing the eye to move more than for smaller size characters. Shurtleff (1980) suggested that a maximum character size should be about 22 min of arc. This gives a preferred range of 14 to 22 min of arc. Most researchers agree that an optimum character size for reading is about 18 min of arc (Shurtleff, 1980). For a viewing distance of 24 inches, this corresponds to a character size of 0.12 inches, and for a viewing distance of 20 inches, about 0.10 inch. The latter character height is, for most applications, considered a practical minimum. In general, character height in inches for a viewing distance of D in. and an angular character height of A arc min is given by $H = AD/3484$. For visual search tasks, it is advantageous to use larger characters; 22 to 24 min of arc is generally considered adequate.

On a standard line, 12 inch VDT display with 240 horizontal scanning lines and an active area of 8.5 by 6.4 inches, the distance between horizontal lines is 6.4/240, or 0.027 inch. For a character with five scanning lines, the height is about 0.13 inch, which corresponds to (3484)(0.13)/24 min, or to approximately 20 min of arc at or about 24 inches, a normal viewing distance.

Current word processing and data processing displays range from a few lines to a full page of approximately 60 lines. Because displays vary in size, the characters necessarily vary in size. It is clear that full-page displays are desirable for formatting pages, but are difficult to read because of the resulting small characters. Similarly, it is obvious the larger characters used in partial page displays increase legibility, but that formatting of the page is more difficult. There is no useful research to suggest optimal levels of display information density; such research is yet required. At this stage, one can only point out that the choice of display size and text density depends on task requirements. Some tasks typically need less information than others, and might use a smaller display with fewer characters.

5.1.8.1 Video Bandwidth

To reduce costs, most VDT screens use commercial television screens with 525 scanning lines (Farrell and Booth, 1984). A commercial TV has a limited bandwidth of about 4 million Hz (4 MHz). This

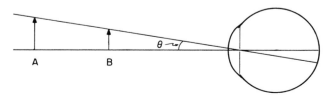

Fig. 5.1.21. From the point of the observer it is appropriate to define the size of objects in terms of visual angle. The two objects A and B subtend the same visual angle and produce the same size retinal image.

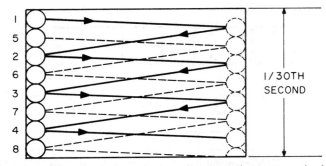

Fig. 5.1.22. The electronic beam is swept horizontally across the screen, activating the phosphor when it is turned on. This screen exemplifies 2 : 1 interlace.

means that during one second, there may be 4 million "changes" on the screen. For a commercial TV, each frame is refreshed at a 30 Hz rate. In order to reduce flicker sensitivity for normal TV viewing distances, every other line is refreshed every other time the beam sweeps across the screen; see Figure 5.1.22. This is referred to as a 2 : 1 interlace. For VDT workplaces the operator is sitting much closer to the screen and is thus appreciably more sensitive to the flicker on the interlace screen than a person at a greater distance. Therefore, adjacent lines on VDT screens are usually superposed, so that the resulting lines are refreshed at 60 Hz.

Of the original 525 lines on a display screen, about 480 are available for writing; the rest are lost for various reasons. Because adjacent lines are superposed on a VDT, there are only 240 lines available for writing. Alternatively, the 2 : 1 interlaced 480 line display would allow 480 lines, but many users would perceive the flickering. Using a 5 × 7 dot matrix, which requires 10 horizontal lines, it is then possible to fit a maximum of 240/10, or 24 rows of characters. For a 7 × 9 dot matrix, with 12 vertical lines, the maximum is 20 rows of characters.

The standard screen with 4 MHz video bandwidth has a maximum of about 470 addressable dots in the horizontal direction. Using a 5 × 7 dot matrix, this corresponds to about 80 characters. However, if the resolution of the dot matrix is increased to 7 × 9, then only 60 characters can be displayed. There is of course a simple solution to this dilemma, namely, to increase the bandwidth. If, at an increase in dollar cost, the video bandwidth is increased from 4 to, say, 10 MHz, then the resolution of the dot matrix can be increased to 9 × 11. There are obviously several cost trade-offs, with important implications for the ergonomic design of the screen. Full-page word processing displays use a higher (20 MHz) video bandwidth. Although such displays are considerably more expensive, it seems likely that their use will increase in the future.

5.1.8.2   Choice of Type of Display

In the past, CRT displays have dominated the market. Recently, however, several alternative display technologies, such as light-emitting diodes (LED), electroluminescence (EL) displays, plasma displays, liquid crystal displays, electrochromatic displays, and electrophoretic displays, have become available. For a review of these technologies, see Snyder (1980). There are several differences among the display technologies, including the type of power source, use of dynamic gray scale, use of colors, resolution, and the effect of ambient illumination. These factors are summarized in Figure 5.1.23 (Snyder, 1980). Because flat panels have much smaller space requirements than CRT displays, it may be predicted that flat panel displays are going to become much more common.

5.1.9   CONTROL OF SCREEN REFLECTIONS

Display screen reflection of ambient light is a significant problem in the workplace. This section will explain different types of screen reflections and measures available to reduce them.

There are two types of screen reflections: specular reflections and veiling reflections; see Figure 5.1.24. Specular reflections produce mirrorlike images on the screen, for example, reflections of the operator, luminaires, and other objects in the room.

A veiling reflection is a diffuse reflection produced by light falling on the screen surface. Most of the veiling reflection is caused by the phosphor, which has an irregular surface, similar to the surface of a piece of paper. The irregularity of the surface causes reflections to be spread in all directions. Veiling reflections increase the luminance of both the screen background and the characters, thereby reducing the contrast ratio of displayed characters.

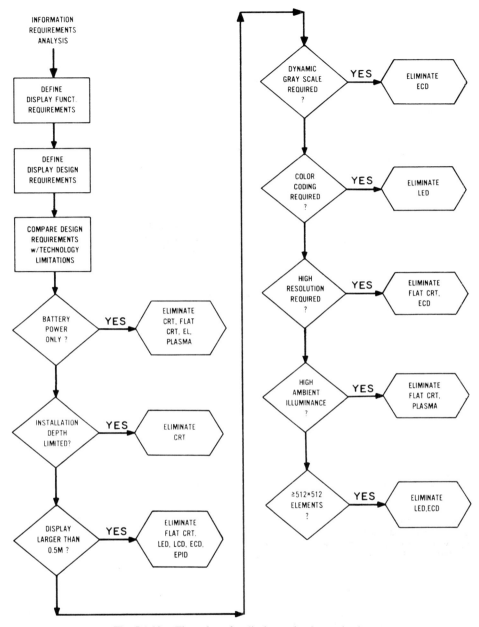

Fig. 5.1.23. Flow chart for display technology selection.

Screen reflections reduce the legibility of characters and often cause operator discomfort. Reflections can be minimized through measures taken at the source of the reflection and at the work station; see Table 5.1.5.

The ideal combination of measures to reduce screen reflection depends on the characteristics of each workplace, including considerations of office and work-station layout, required illumination levels, location of windows, and type of VDT screen. It is difficult to give definitive recommendations without analyzing all these factors. The measures described in Table 5.1.5 are therefore applicable to varying extents, depending upon the circumstances.

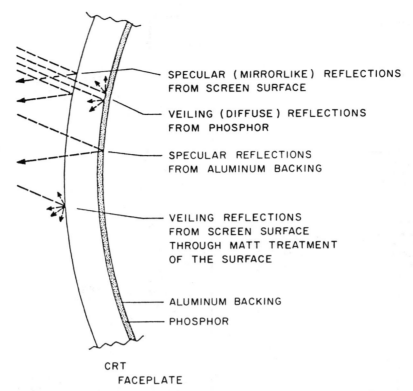

SPECULAR (MIRRORLIKE) REFLECTIONS
FROM SCREEN SURFACE

VEILING (DIFFUSE) REFLECTIONS
FROM PHOSPHOR

SPECULAR REFLECTIONS
FROM ALUMINUM BACKING

VEILING REFLECTIONS
FROM SCREEN SURFACE
THROUGH MATT TREATMENT
OF THE SURFACE

ALUMINUM BACKING

PHOSPHOR

CRT
FACEPLATE

Fig. 5.1.24. Illustration of different kinds of reflections from a screen surface.

5.1.9.1 Covering of Windows

Covering the window permanently with building materials has negative esthetic and motivational consequences, and should be used only as a last resort. Three different types of window covering may be
used: dark film, louvers or mini-blinds, and curtains.

A dark film attached to the window reduces the general level of illumination in the office. Vertical
louvers shield the sun, while allowing workers to see outside. Horizontal louvers, on the other hand,
do not allow outside viewing. In addition to shielding sunlight, louvers block window reflections during
night work, when the darkness outside turns windows into reflecting mirrorlike surfaces.

White curtains have high reflectance and create a highly luminous surface. Darker colored curtains
may be preferred, therefore.

5.1.9.2 Lighting Control

Luminaires should be positioned so that screen reflections are minimized. There are three types of
measures that can be used to improve office illumination: control of location and direction of illumination,
use of indirect lighting, and use of task illumination. The optimum design solution depends on the
office layout and can be perfected only in the real office environment. The following gives some general
advice with respect to possible locations and types of luminaires used at a work station.

VDT office environments have somewhat different lighting requirements than traditional offices.
Nonshielded luminaires mounted in the ceiling should not be used. Rather, luminaires that give a
confined, downward distribution of light are recommended. Figure 5.1.25a shows that, by using a
light cone confied to 100° on a luminaire, the operator is not subject to direct glare. However, if
luminaire C had been positioned closer, the operator would have been exposed to glare. Likewise,
luminaire A does not produce any screen reflections, because the screen is outside the illumination
zone. If luminaire A had been closer to the operator, there could have been specular reflections on
the screen. Luminaire B, which is necessary for illuminating the workplace, is the most prominent

Table 5.1.5 Measures for Reducing Screen Reflections[a]

Measure	Advantage	Disadvantage
	At the Source	
Cover windows		
Dark film	Reduces veiling and specular reflections	Difficult to see out
Louvers or mini-blinds	Excludes direct sunlight, reduces veiling and specular reflections	Must be readjusted in order to see out
Curtains	Reduces veiling and specular reflections	Difficult to see out
Lighting control		
Control of location and direction of illumination	Reduces veiling reflections, may eliminate specular reflections	None
Indirect lighting	Reduces specular reflections, economy of office space by moving work stations closer	None
Task illumination	Reduces veiling reflection, increases visibility of source document	None
	At the Work Station	
Move workstation	Reduces veiling and specular reflection	None
Tiltable screen	Reduces specular reflection	Readjustment necessary
Tilted screen filter	Eliminates specular reflection	Bulky arrangement for large screens
Screen filters and treatments		
Neutral density (gray) filter	Reduces veiling reflection, increases character contrast and visibility	Less character luminance
Color filter (same color as phosphor)	Reduces veiling reflection, increases character contrast and visibility	Less character luminance
Micro mesh, micro louver	Reduces veiling reflection, increases contrast	Limited angle of visibility, nonembedded filters get dirty
Polaroid filter	Reduces veiling reflection, increases contrast and visibility	Decreased character luminance
Quarter wavelength anti-reflection coating	Eliminates specular reflection	Expensive, difficult to maintain
Matte (frosted) finish of screen surface	Decreases specular reflections	Increases character edge spread (fuzziness, increases veiling reflections)
CRT screen hold	Reduces veiling and specular reflection	Difficult to avoid shadow on screen
Sunglasses (gray, brown)	None—contrast unchanged	Less character luminance and visibility
Reversed video	Reduces specular reflections	Increased flicker sensitivity
Screening of luminaires and windows	Reduces specular reflections	Might create isolated workplaces

[a] Note that the optimal combination of measures depends on the particular office, equipment, office layout, and types of VDT.

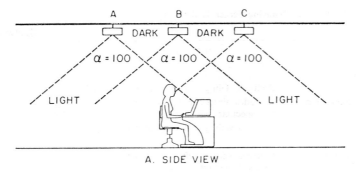

A. SIDE VIEW

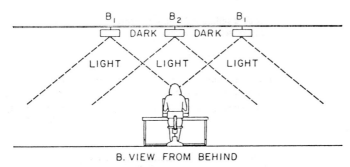

B. VIEW FROM BEHIND

Fig. 5.1.25. Luminaire A may generate specular reflections, B washes out screen contrast, and C may cause direct glare. Luminaire B_2, positioned right above the operator, may cause reflections from the work-station table. B_1 is a better location.

source for causing veiling luminance on the screen. In addition, it may cause reflected glare from the keytops and the table top.

Figure 5.1.25*b* shows the seated operator as seen from behind. The luminaire previously referred to as B may actually be positioned above the operator (location B_2) or to the side of the operation (location B_1). Luminaire B_1 does not cause direct glare, because the operator's eyes are outside the illuminated zone. Luminaire B_2 is closer, and produces more veiling luminance and reflected glare from objects on the table, and is therefore less suitable than B_1.

The light distribution of a luminaire can be specified by the use of polar curves which represent a vertical section through the light. These curves are classified in terms of British zonal (BZ) numbers ranging from BZ 1 to BZ 10. For BZ 1 the maximum intensity is downward and for BZ 10, most of the illumination is in the horizontal direction. Figure 5.1.26 shows the BZ curves 1 to 4, all of which represent light distributions suitable for VDT office environments.

Indirect Lighting

A common suggestion for reducing specular reflections is to use indirect illumination (Carlsson, 1979; Kokoschka and Bodmann, 1980). This can be achieved by positioning luminaires low and directing some of the illumination upward; see Figure 5.1.27. The work station is then illuminated both directly from the luminaire and indirectly by light reflected from the ceiling. This has two advantages:

1. The luminaires can be moved closer to the work station without changing the angle for the shielded dark zone (compare Figure 5.1.25). Work stations (and associated lighting systems) can then be moved closer without creating glare problems, thereby making better use of office space.
2. The illumination reflected from the ceiling does not create specular reflections on the screen.

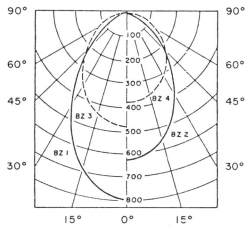

Fig. 5.1.26. BZ (British zonal) curves 1 to 4 represent light distribution that is suitable for a VDT office.

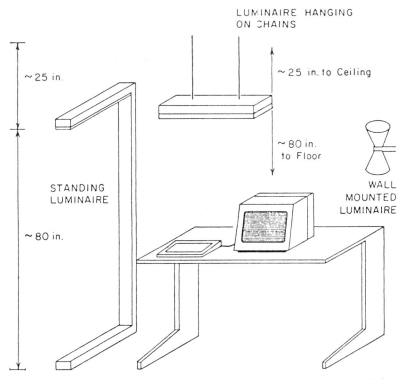

Fig. 5.1.27. Lower height luminaires with restricted light cone reduce glare. The luminaires may be open on the top so that about 60% of the illumination is directed upward. It is then reflected by the ceiling so that the workplace is illuminated partly by diffuse indirect illumination.

For indirect lighting systems, it has been recommended that about 40% of the illumination should be directed downward and 60% upward (Carlsson, 1979). In order to avoid patterns on the ceiling owing to differences in reflectance, both the ceiling and the luminaires should be the same color, preferably white (Carlsson, 1979). Figure 5.1.27 shows three designs of indirect, lower-height lighting systems: free standing, hanging from the ceiling, or mounted on the wall.

Task illumination may be used to enhance the visibility of the source document to an appropriate level (about 500 to 600 lux) without increasing the illumination on the screen surface (above 200 lux). There are several alternatives for task illumination, such as a spotlight mounted on the ceiling and directed at the source document, or it might be a lamp mounted on the tabletop.

5.1.9.3   Moving the Work Station

Screen reflections can be reduced by moving the VDT work stations away from windows, or by lowering the ambient illumination level; see Figure 5.1.28. To minimize both direct glare from windows and screen reflections of them, the screen surface should be positioned perpendicular to windows.

5.1.9.4   Tiltable Screen

Tilting the screen helps minimize screen reflections. Just as with a rearview mirror in a car, the tilt mechanism makes it possible to compensate for differences in eye positions. A range of adjustability from approximately 15 to 20° backward to 5° forward is usually adequate.

5.1.9.5   Screen Filters and Treatments

The main purpose of filters is to improve character-screen contrast; see Figure 5.1.14. The light from the ambient illumination passes through the filter before it reaches the screen surface. It is then reflected and passes through the filter again on the way back to the operator. Because the light from the characters passes through the filter only once, character luminance is less reduced than screen luminance and the contrast ratio increases. Lowering the character luminance level unfortunately decreases visibility. Filters should therefore not be used for screens with low character illuminance.

Specular reflections from the front surface of filters are not reduced unless the surface of the filter is treated. Such treatments may involve an anti-reflective coating or a matte finish that blurs specular reflection.

There are two types of *color filters:* neutral density filters and color filters. A neutral density filter is grey and reduces the energy of wavelengths by an equal percentage. Color filters allow only specified

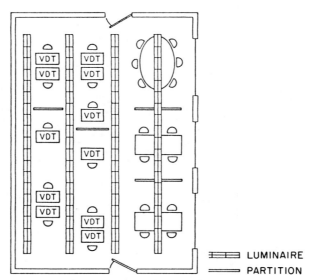

═══ LUMINAIRE

═══ PARTITION

Fig. 5.1.28. Several designs are illustrated: (1) The VDT work stations are located away from the windows. (2) The VDT screens are positioned perpendicular to the windows, thereby avoiding direct glare and screen reflections. (3) Luminaires are positioned to the side of, rather than behind, work stations. (4) Partitions are used to screen off illumination from the windows and the luminaires.

color bands to pass and absorb other wavelengths. Colored filters should match the color of the phosphor. For a phosphor with a distinct green color, such as P31, a green filter will pass about 70% of the green light, but much less of the other colors. This enhances character contrast, and does not significantly reduce the luminance of the characters. Ambient illumination, which is typically white, is absorbed by the filter, except for the green portion. The veiling reflection from the screen background is thereby minimized, and character contrast is improved.

Micro-louvers or micro-mesh filters allow only light parallel to the walls of the small openings of the filter to reach the screen surface, thereby reducing the amount of illumination falling on the screen. At the same time, the viewing angle is restricted. To avoid specular reflections from the filter surface, micro-mesh filters are usually not embedded in plastic. As a result, they tend to collect dust and dirt which reduce character luminance. Although the filter can be cleaned, in practice this is rarely done, and several manufacturers have stopped promoting micro-mesh filters.

Circular polarizers may also be used to enhance the character-background contrast. The incoming light is polarized vertically after passing through the first polarizing filter. A quarter-wave filter transforms the transmitted polarized light into two components polarized at right angles, with one component retarded relative to the other by $\frac{1}{4}$ of a wavelength. This is called circular polarized light. In Figure 5.1.29 the light is right polarized, but changes to left polarization after being reflected from the screen. After passing back through the quarter-wave filter, it is polarized horizontally, and is therefore extinguished by the polarizer. It should be noted that the front surface of a circular polarizer itself may cause specular reflections. In some cases these may be controlled by the other measures mentioned below.

On *quarter-wavelength filters*, the outside surface is coated with an anti-reflection coating that is optically a quarter wavelength ($\lambda/4$) in thickness, the same kind of single-layer coating that is used on the surface of coated camera lenses. Light rays reflected from the front and back surfaces are therefore in counterphase (half-wavelength difference), canceling each other, which extinguishes the reflections. The anti-reflection coating significantly reduces specular reflection. The main disadvantage is that antireflection coatings are difficult to maintain. For example, a fingerprint adds thickness to the coating so that it loses its nonreflective characteristics. Anti-reflection coating is fairly expensive and when oily, fingerprinted, or dirty, loses its effectiveness; thus it might not be appropriate to use in some industrial environments.

Specular reflections may also be reduced by *matte treatment* (frosting) of the screen (or filter) surface. This eliminates the shiny, mirrorlike surface of the screen and makes the image of a reflected object more diffuse. Specular reflections are thereby transformed to veiling reflections, which reduces the contrast between the characters and the background and spreads the character luminance over a wider surface. Obviously this is a trade-off; there should not be an excess of either specular or veiling reflections.

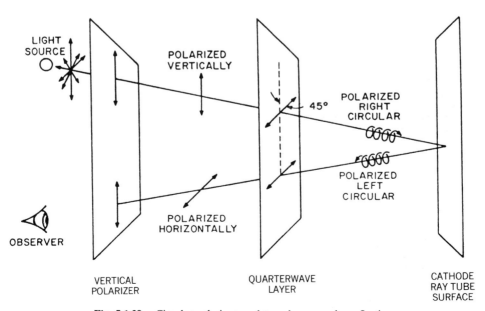

Fig. 5.1.29. Circular polarizer used to reduce specular reflection.

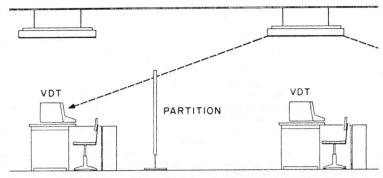

Fig. 5.1.30. A partition may be used to reduce screen reflections.

A *hood,* positioned on top of the CRT screen, may be used to screen out the ambient illumination. It may be difficult to position the hood without casting a shadow on the screen from the edge of the hood. Such shadows are annoying and hoods should be used only when they can effectively screen off a specific illumination source.

Many VDT operators wear *sunglasses.* The use of sunglasses with a neutral gray or brown tint does not increase contrast, because they reduce the luminance level of the characters and the background equally, and visibility is therefore impoverished. Also, reflections in the sunglasses reduce display contrast. Sunglasses of the same color as the phosphor (e.g., green for P39 or P31) would enhance character contrast, but might be rejected by the operators for aesthetic reasons.

Reversed video, or black characters on a light background, reduces visible specular reflections on the screen. The light background washes out the reflections and the effect is similar to putting milk in coffee, which makes it difficult to see reflections in the cup. Reversed video has a disadvantage in that the luminous surface is larger, which increases the sensitivity to flicker from the screen. Whether these or other characteristics warrant the use of reversed video has not yet been proved (Helander, Billingsley, and Schurick, 1984).

Sometimes it is difficult to position luminaires without inducing screen reflections. If so, *partitions* may be useful; see Figure 5.1.30. The use of partitions produces an illumination environment that is similar to a small office where there are less disturbing specular reflections from light sources than in open-plan offices.

5.1.10 PRESBYOPIA AND ACCOMMODATIVE RANGE

The ability of the eye to accommodate to different visual distances decreases with ages. Figure 5.1.31 shows that from the age of 5 to 70 the normal accommodative range decreases from 14 to 1 diopter. This old-age phenomenon is called presbyopia, and is due to loss of elasticity of the lens in the eye. It causes hyperopes to wear glasses with a positive or magnifying corrective lens in order to see clearly at near distances, or they may wear reading glasses, or bifocals with the lower part of the glasses corrected to fit the close viewing distance. Myopes may have less difficulty with presbyopia because, depending on their refractive error, they may be able to see nearby objects simply by removing their spectacles. Although the results of presbyopia may differ, the same dioptic ranges apply regardless of type of correction required. As an example, assume that a 40 year old individual has a 5 diopter accommodative amplitude. Such an individual, if emmetropic (good vision for objects at infinity), will be able to focus on objects located at any distance from 0 to 5 diopters, corresponding to a range of infinity to $\frac{1}{5}$, or 0.2 m. An individual who is 3 diopters myopic will be able to focus on objects at any distance from 3 to 8 diopters (0.33 to 0.13 m). A person 3 diopters hyperopic will be able to focus on objects at any distance from −3 diopters (optically beyond infinity) to 2 diopters (0.5 m).

Presbyopia may cause considerable problems with CRT terminals. Owing to the large size of the terminals, they are often located at a height above normal viewing angles. As a result, the borderlines for the bifocals are too low. In addition, a typical viewing distance for a CRT is approximately 25 in. However, the lower part of the bifocal is usually ground for clear vision at 12 inches and the upper part for clear vision at 200 inches. To increase the legibility, the operator is forced to turn his or her head back and move closer to the screen. This is a very uncomfortable posture and often causes neck pain and headaches. Two measures can be taken to overcome this problem: (1) The screen can be lowered and moved closer to the operator. This is appropriate for a small screen, but

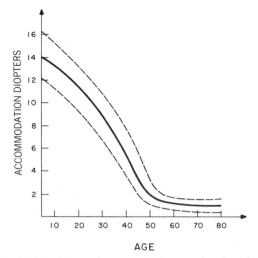

Fig. 5.1.31. Range of accommodation as a function of age.

may not be convenient for large screens. (2) Special screen viewing glasses (terminal glasses) that have been designed for the actual viewing distance can be used.

5.1.11 VISUAL ASPECTS OF COLOR DISPLAYS

During the past few years, color displays have become increasingly common. In this section we review some of the capabilities and limitations of color perception and some of the physical properties of color displays.

5.1.11.1 Specification of Chromaticity

In the human eye there are three types of color receptors. The visible energy at a particular wavelength stimulates not just one of these receptors, but all of them to differing degrees. As a result the human perceives two color attributes, hue and brightness. Contrary to popular belief, the three types of receptors, known as red, green, and blue cells, do not tell the brain anything about redness, greenness, or blueness (Thorell, 1983). Actually, each of these three receptors is sensitive to a range of wavelengths; see Figure 5.1.32 (Cornsweet, 1970).

The signals from the receptors on the retina are processed by higher-order brain cells. There are complicated correlations in the retina and brain that make it possible to add information from one

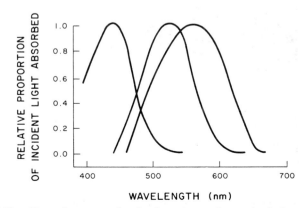

Fig. 5.1.32. Absorption spectra for the three classes of cones in the human retina.

cell while subtracting information from other cells; thereby some of the neural signals are excitatory and others are inhibitory. In the monkey brain, three classes of so-called color opponent cells have been identified (DeValois, Abramov, and Jacobs, 1966). There are cells that excite to red and inhibit to green (or vice versa), cells that excite to yellow and inhibit to blue (or vice versa), and cells that excite to white and inhibit to black (or vice versa). It is evident that, by having not only additive but subtractive color elements, the range of combinations from three primaries increases considerably.

Similar considerations were adopted in 1931 by the International Commission on Illumination (CIE) in devising a method for specifying colors. Figure 5.1.33 shows an apparatus that makes it possible to calculate a resulting hue as the subjective equivalent of a mixture of three primaries, red, green, and blue (Hardesty and Projector, 1973).

If the primaries are well chosen, a substantial number of all possible real colors can be reproduced by adding different mixes of the three primaries. However, for some colors, one of the primaries must be added to the color being matched, rather than to the other two primaries. To obtain a color match with all existing colors it is thus necessary sometimes to use negative quantities of one of the primaries. Because this would be inconvenient, CIE eliminated the negative values by a mathematical transformation. The resulting three 1931 CIE primaries, X, Y, and Z, are purely imaginary; they do not exist physically and therefore can be used only for analytical purposes; no color matching like that described in Figure 5.1.33 can be accomplished.

Figure 5.1.34 shows the wavelength distributions of the three primaries, X, Y, and Z. The distribution of X is bimodal and the distribution of Y is identical to the luminosity function of the 1931 CIE standard observer; see Figure 5.1.12. Values of the distribution curves are also given in Table 5.1.6.

To specify a color on a display, the distribution of the spectral radiant energy of the color is multiplied, wavelength by wavelength, by the three distributions ($\bar{x}$, $\bar{y}$, $\bar{z}$) of the spectral tristimulus values; see Figure 5.1.35. These distributions are hence used to measure the radiant energy of the image. The areas of the resulting curves are calculated by adding the contributions for each individual wavelength. The summation is usually carried out at wavelength intervals of 5 or 10 nm. As a result, the tristimulus values X, Y, and Z are obtained. The tristimulus values X, Y, and Z are then normalized, using the equations at the bottom of the figure, so that the resulting chromaticity coordinates x, y, and z sum to unity:

$$x + y + z = 1 \qquad\qquad\qquad (7)$$

To specify a color, only two of the three coordinates have to be given, because the value of the remaining coordinate is always implied. Most commonly, only x and y are given and plotted in the CIE chromaticity diagram; see Figure 5.1.36. The redundant z axis is sometimes thought of as projecting out from the page.

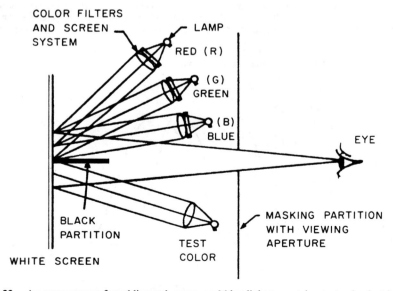

Fig. 5.1.33. An arrangement for adding red, green, and blue light to match a test color in tristimulus colorimetry.

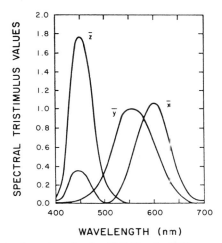

WAVELENGTH (nm)

Fig. 5.1.34. Spectral tristimular values for the 1931 Standard Observer. Because CIE primaries X, Y, and Z do not exist physically, the easiest way to illustrate them is in terms of the amount of each required to provide a visual match to the spectral colors. These quantities are referred to as spectral tristimulus values. (*Source:* Farrell and Booth, 1984.)

By definition, the coordinates for white are $x = y = z = 0.33$. This is referred to as the illuminant or achromatic reference point, and is denoted by C.

The chromaticity coordinates of a color define the dominant wavelength and purity. (The corresponding subjective sensations are called hue and saturation.) The dominant wavelength and purity of a color may be calculated by drawing a line through the achromatic reference point C and through the coordinates representing the color. The dominant wavelength is the intersection of the line with the borderline of the CIE color space. Purity of the color may be calculated by using the formula:

$$\text{Purity} = \frac{A}{A + B} \tag{8}$$

Table 5.1.6   Distributions of the Spectral Tristimulus Values for the 1931 Standard Observer

Wave-length (λ) (nm)	$\bar{x}_\lambda$	$\bar{y}_\lambda$	$\bar{z}_\lambda$	Wave-length (λ) (nm)	$\bar{x}_\lambda$	$\bar{y}_\lambda$	$\bar{z}_\lambda$
380	0.0014	0.0000	0.0065	550	0.4334	0.9950	0.0087
390	0.0042	0.0001	0.0201	560	0.5945	0.9950	0.0039
400	0.0143	0.0004	0.0579	570	0.7621	0.9520	0.0021
410	0.0435	0.0012	0.2074	580	0.9163	0.8700	0.0017
420	0.1344	0.0040	0.6456	590	1.0263	0.7570	0.0011
430	0.2839	0.0116	1.3856	600	1.0622	0.6310	0.0008
440	0.3483	0.0230	1.7471	610	1.0026	0.5030	0.0003
450	0.3362	0.0380	1.7721	620	0.8544	0.3810	0.0002
460	0.2908	0.0600	1.6692	630	0.6424	0.2650	0.0000
470	0.1954	0.0910	1.2876	640	0.4479	0.1750	0.0000
480	0.0956	0.1390	0.8130	650	0.2835	0.1070	0.0000
490	0.0320	0.2080	0.4652	660	0.1649	0.0610	0.0000
500	0.0049	0.3230	0.2720	670	0.0874	0.0320	0.0000
510	0.0093	0.5030	0.1582	680	0.0468	0.0170	0.0000
520	0.0633	0.7100	0.0782	690	0.0227	0.0082	0.0000
530	0.1655	0.8620	0.0422	700	0.0114	0.0041	0.0000
540	0.2904	0.9540	0.0203	710	0.0058	0.0021	0.0000
				720	0.0029	0.0010	0.0000

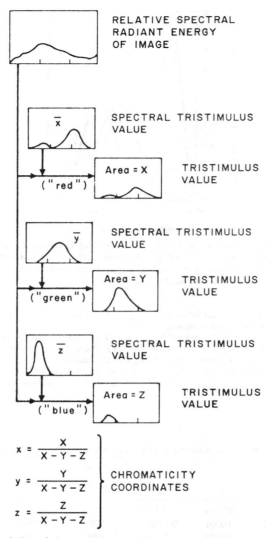

Fig. 5.1.35. Calculation of chromaticity coordinates. (Adapted from Farrell and Booth, 1984.)

It should be recognized that luminance is not specified in the CIE space. This has considerable negative consequence, and Figure 5.1.36 illustrates that misleading results may be obtained if colors are described only in terms of chromaticity without reference to luminance or reflectance. For example, note the closeness of the CIE coordinates corresponding to a Hershey chocolate bar and red lipstick (Hendley and Hecht, 1949). A chocolate bar primarily illuminated by a light that is not seen by an observer does look red, but this requires a special setup.

5.1.11.2 Color Discrimination

For choosing screen colors, it is important to understand how many different colors can be used and how different they must be in terms of chromaticity coordinates. Based on research on just-noticeable-differences (JNDs), MacAdam (1948) plotted areas representing 10 JNDs in the CIE chromaticity space; see Figure 5.1.37. The areas, referred to as MacAdam ellipses, are much smaller in the blue and the red parts of the diagram than in the green. This is an obvious drawback because, ideally, the CIE space should be designed so that the distance between any two points in the diagram could

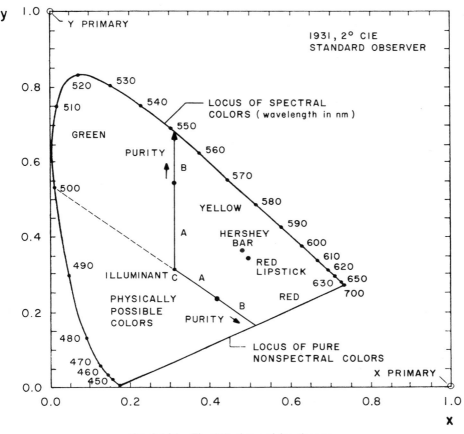

Fig. 5.1.36. The CIE chromaticity diagram.

be used in order to assess the magnitude of perceived color differences. In order to make the ellipses uniform in size, the 1931 CIE color was transformed using the following formulas:

$$u' = \frac{2x}{6y + x + 1.5} \tag{9}$$

$$v' = \frac{4.5y}{6y - x + 1.5} \tag{10}$$

This color space, known as the 1976 CIE uniform color space (UCS), produces MacAdam ellipses much more uniform in size. The 1976 color space can be used to calculate differences in color contrast between two colors.

Lippert (1984), in some not-yet published work, derived the following formula to calculate the color difference, ΔE, between pairs of colors with different luminance:

$$\Delta E = \sqrt{\left(\frac{230}{Y_T} \Delta Y\right)^2 + (200\Delta u')^2 + (200\Delta v')^2} \tag{11}$$

where Y_T = target luminance (cd/m²)

ΔY = difference in luminance between target and background (cd/m²)

$\Delta u'$ = difference in u' according to Equation (9)

$\Delta v'$ = difference in v' according to Equation (10)

For adequate discrimination, any pair of two colors should differ by a minimum of 20 ΔE units. This formula, which was derived for characters of one color contrasted against a different color background, generated superior results compared to results of other methods of calculating color contrast.

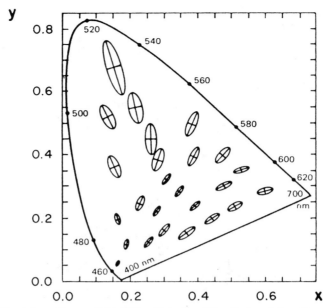

Fig. 5.1.37. MacAdam ellipses of 10 JNDs. Note the different size of the ellipses, indicating the nonuniformity of the CIE space.

Research is presently underway to investigate the generalizability of this formula to targets other than alphanumeric characters.

5.1.12 USE OF COLOR DISPLAYS

Color displays are often preferred because many designers and users feel that color is more pleasing and possibly more realistic than black and white. There are, however, several technological as well as perceptual limitations to the use of color. This section briefly reviews these problems, which include the effect of ambient illumination on CRT displays, limitations in contrast sensitivity, and chromostereopsis owing to the lack of correction for color aberration in the optical system of the human eye.

CRT displays are not capable of reproducing highly saturated colors; see Figure 5.1.38. They can produce saturations adequate for most tasks. Lack of extreme saturations does not necessarily present a problem to users, because the nervous system is capable of adapting to various conditions, which makes it possible to perceive colors appropriately in different lighting. In fact, the visual system adapts so well that it can hardly appreciate the differences between fluorescent whites and incandescent whites. Such color differences are, of course, obvious in color photographs in which objects appear greenish blue if photographed under fluorescent light and yellow under incandescent illumination. With appropriate illumination, color photographs, even if off-color in daylight, do not appear off-color.

Ambient illumination produces veiling luminance on the screen, washing out screen contrast and reducing the number of colors that can be perceived. The primary effect of this veiling luminance is a desaturation of colors. Thus colors that are relatively far from the CIE white point, for example, blue-green and yellow, become desaturated, causing them to be virtually indiscriminable from white. The ambient light mixes with the emitted light to create an additive mixture based on the proportions of each. Depending on the color of the light, the desaturation may be different. Because most ambient light sources are broad band, there is a desaturation of all three phosphors. However, fluorescent light, which contains a greater amount of short wavelength energy, desaturates blue more than green or red (Murch, 1984). Figure 5.1.38 shows the influence of sunlight on the range of display color (Laycock and Viveach, 1982). With these limitations in mind, it is obvious that color coding that may be used under low levels of illumination is less effective when the ambient level of illumination is raised.

5.1.12.1 Task Considerations in Choice of Color Coding

The number of colors appropriate for use on displays depends on the application. For imaging and solids modeling in computer-aided design, hundreds of shades are used effectively, but for simple

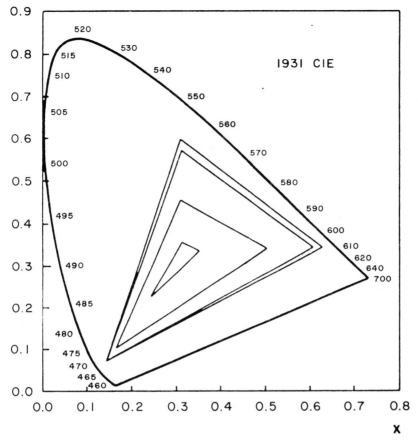

Fig. 5.1.38. The outer triangle depicts the maximum color rendering capability of a color CRT display. As the ambient illumination increases, colors are diluted and the size of the triangle shrinks.

coding of information, more than 10 would cause problems. Care should be taken not to use more colors than are needed by the particular application. In general, color coding may be helpful if:

The display is unformatted.

Symbol density is high.

The operator must search for specified information.

Task analysis can be used to analyze the number of colors necessary for a given display. A task is broken down into subtasks and task elements. The information requirements for each subtask are analyzed and may be classified in terms of importance and frequency of use. Based on such considerations, information is classified as primary or secondary, which helps to determine the optimum location on the screen. Once information has been classified, it is possible to code it by color, size, luminance, and so forth.

Several studies have investigated colors that can be used for color coding. The selection of color codes should take into consideration established meanings that are associated with the colors. Myers (1967) recommended that red, yellow, and green should be reserved for "Danger," "Caution," and "Safe," respectively. Military standards (U.S. Department of Defense, 1981) specify that displays shall conform to a color coding scheme whereby:

Red is used to alert an operator that the system or any part of the system is inoperative. Examples of indicators are "NO-GO," "ERROR," "FAILURE," and "MALFUNCTION."

Flashing red is used to denote emergency conditions that require immediate action.

Yellow is used to advise an operator of marginal situations in which caution is necessary or unexpected delay may be encountered.

Green is used to indicate a fully operational system, that all conditions are satisfactory.

White is used to indicate system conditions that do not have a right or wrong, such as expressing alternatives or transient conditions.

Blue is used to indicate an advisory light, but preferential use of blue should be avoided.

These symbolic meanings of colors have been adopted by the U.S. Department of Defense with one exception, that they do not necessarily apply to design of cockpits, where special consideration must be given due to problems of dark adaptation.

5.1.12.2 Perceptual Limitations in Color Coding

There are several perceptual limitations in the choice of colors. The optics of the human eye, unlike the optics of a camera, is not corrected for color aberration. The focal length of the eye is shorter for shorter wavelengths; that is, the refractive power of the eye is different for short and long wavelengths: about 3 diopters greater for saturated blue than for saturated red. For a CRT that has less saturated colors, the difference is only 1 diopter. This causes a three-dimensional or depth effect called chromostereopsis, in which the red and blue parts are perceived at different distances. Because it is difficult to focus on both at the same time, the eye's accommodation mechanism may drive the focusing of the eye back and forth. Murch (1984) investigated the amount of accommodation necessary for a CRT display. He concluded that reds, oranges, yellows, and greens can be viewed together without refocusing, but cyan or blue can not be viewed with red. It is generally agreed that chromostereopsis should be avoided, because it may be annoying and may cause visual fatigue.

Reaction time studies of different colors have revealed that, for low levels of illumination, response time is shortest to the blue part of the spectrum and longest to the red. The dimmer the light, the more critical the wavelength, so that for mesopic vision (low level illumination) corresponding to 0.0004 to 0.04 cd/m², the eye is most sensitive to the blue part of the spectrum. However, for higher illumination levels, i.e. 0.2 cd/m², reaction time is fastest for red, followed by green and yellow (Tyte, Wharf, and Ellis, 1975). These results are consistent with the differential sensitivity of the cones and the rods in the eye, and may partly be due to the loss of sensitivity of the eye to the red part of the spectrum in low illuminance environments, see Figure 5.1.39.

There have been several comparisons of the relative legibility of different colors in alphanumeric recognition studies. Table 5.1.7 (reproduced from Snyder, 1980) compares results from studies by Meister and Sullivan (1969) and Rizy (1967). There was good agreement between the studies, with red and yellow being the best colors, and blue and green the poorest. However, these studies were limited in scope because they investigated the contrast between color against a black or achromatic background without any measurements of the saturation or purity of the stimulus.

With very extensive practice and under ideal conditions, human observers can in an absolute sense (without comparison to a color chart, etc.) identify up to 50 colors (Hanes and Rhodes, 1959). This number, however, far exceeds any reasonable number for operational conditions outside of the laboratory. With less practice, but under laboratory conditions, it has been found that, as the number of colors

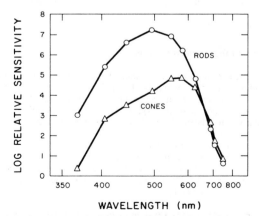

Fig. 5.1.39. Photopic (cone) and scotopic (rod) spectral sensitivity curves. Cone sensitivity peaks at 555 nm (yellowish green), while rods peak at 510 nm (green).

Table 5.1.7 Relative Ranking of Different Colors in
Alphanumeric Recognition Studies (Lower Ranking Indicates
Better Performance)

Color	Meister and Sullivan (1969)	Rizy (1967)
Red	2	1
Yellow	1	2
Magenta	4	3
White	3	4
Cyan	5	5
Blue	7	6
Green	6	7

increases, the number of identification errors also increases; see Table 5.1.8. If the operator's task requires absolute identification of a color (in the absence of other colors for comparison), five appears to be the maximum number of colors for high accuracy, but 10 is acceptable if minor errors are permissible.

Individuals suffering from common forms of color deficiency experience difficulty in distinguishing colors that differ only in their respective amounts of red and green. Colors of constant luminance should therefore not differ solely in the amount of red and green they contain, but also in the amount of blue.

The sensitivity to variations in spatial frequency is different for colors than for black and white. Figure 5.1.40 illustrates that, for black and white vision, there is a peak in spatial frequency sensitivity at about 5 Hz. This corresponds to the size of a single character on an alphanumeric display at a normal viewing distance. For color perception, the maximum sensitivity is shifted to lower spatial frequencies, with yellow/blue attenuating more than red/green variations. For color, larger objects, such as the entire screen surface, are therefore more easily perceived than smaller objects, such as characters.

In general, the smaller the object, the poorer the color will appear to the eye. Even for fine gratings, where spatial patterns can be detected, color may not be recognizable. For example, a low spatial frequency grating of yellow and blue stripes is indistinguishable from a black/white grating. The black/white channel conveys high spatial frequency information, such as edges, sharp contours, and fine detail, whereas the color conveys low spatial frequency information, such as the global aspects of shapes. It is therefore easier to appreciate color in large objects and black and white in small objects.

According to Thorell (1983), this may explain why color works so well in search tasks. When searching for an item, the first things to look for are global characteristics, such as the overall shape or color. For example, a color-coded word on a screen subtending 2° across the retina would correspond to 0.25 cycles per degree, a spatial frequency to which the color channels are quite sensitive. Thus color coding of whole words should stimulate visual search mechanisms quite well, but is not particularly effective for smaller items, such as individual characters on the screen, for which the spatial frequency is about 2 cycles per degree, or higher. This may suggest that colors could be used for coding whole words or the background, rather than the symbols or single characters. It also suggests that larger display screens will enable higher color quality, because larger areas of color can be produced (Thorell, 1983). For alphanumeric displays, perhaps the best compromise is to use large background areas for color coding and to specify the characters in high-contrast black or white.

Owing to the insensitivity of the retinal periphery to saturated red and green, these colors should not be used in the periphery of the display. Yellow and blue are good peripheral colors, although

Table 5.1.8 Percent Incorrect Responses in Indicating
Colors as a Function of Number of colors

Number of Colors	Percent Incorrect Responses
10	2.5
12	4.5
15	5.4
17	28.6

Source: Hanes and Rhodes, 1959.

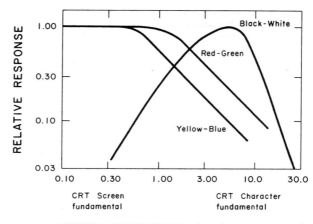

Fig. 5.1.40. Schematized spatial contrast sensitivity functions for brightness-varying (black-white) and color-varying (red-green and yellow-blue) gratings.

the latter should not be used for text and thin lines. Complementary color pairs, for example, red/green and yellow/blue, make good combinations for color displays.

The sensitivity of the eye decreases with increasing age. This is due to the gradual decline of transparency of the lens and the vitreous humor in the eye. It is therefore advisable to avoid colors with low luminance on displays used by older individuals.

Saturated blue should not be used for the presentation of fine detail, because the central part of the fovea is relatively insensitive to that color. For similar reasons, blue is excellent for background color. In addition, blue is perceived clearly in the periphery of the visual field, thus making it a good background for large fields (Murch, 1984).

5.1.13 DESIGN OF GRAPHICS

During the past few years, sophisticated software has resulted in the increasing availability of graphics. Unfortunately, there has not yet been much research on graphics. In general, a graph is considered superior to tables or scales if the shape of the depicted function is important for decision-making or

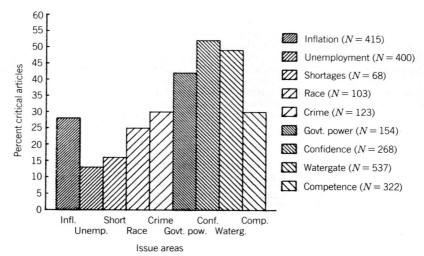

Fig. 5.1.41. Example of bad graphics design. (Graph from Miller, Goldenberg, and Erbring, 1979.) Reprinted with permission.

when interpolation is necessary (VanCott and Kinkade, 1972). Engel and Granda (1975) recommended the use of graphics to supplement explanations in text and for visualization of spatial information. Tufte (1983) suggested several principles of graphic design:

Graphical excellence is that which gives the viewer the greatest number of ideas in the shortest time with the least ink in the smallest space. Graphical excellence is nearly always multivariate. Graphical excellence requires telling the truth about the data.

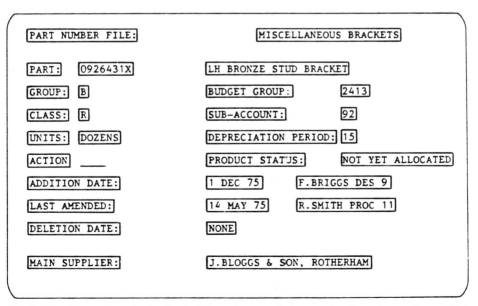

Fig. 5.1.42. Illustration of reduction in information complexity as a function of screen layout. This figure was taken from a paper by T. Stewart and appeared in *Applied Ergonomics,* vol. 7, number 3, pp. 137–146, and was published by Butterworth Scientific Limited, Guildford, Surrey, UK. Reprinted with permission of the publisher.

In order to present graphical information effectively, Tufte claimed that it is important to maximize data information and minimize redundant information. Figure 5.1.41 provides an example of what not to do.

This graph exemplifies several principles of how graphics should not be designed (Tufte, 1983):

Inadequate substance of information.

Crinkly lettering all in uppercase sans serif.

Pointlessly ordered cross-hatching.

Labels written in computer abbreviations.

Optical vibration caused by moire patterns.

Overly busy vertical scaling with more percentage markers than actual data points.

Observed values of percentages should be printed instead.

Tufte (1983) concluded that, because the information consists of a few numbers and many words, it would be better presented in a table.

Related to graphics design is the issue of information density and layout complexity. Bonsiepe (1968) proposed a method for quantifying the complexity of a page in bits. The following formula is used to quantify complexity:

$$C = -N \sum_{n=1}^{m} p_n \log_2 p_n \tag{12}$$

where C = complexity (bits)

        N = number of events (i.e., widths or heights)

        m = number of event classes (i.e., number of unique widths or heights)

        p_n = probability of concurrence of the nth event (based on the frequency of events within that class)

Tullis (1983) showed how the technique can be used for redesigning information presented on a CRT; see Figure 5.1.42. For Figure 5.1.42a, the original version, the following calculations are obtained:

> 32 horizontal distances in 22 unique classes = 140 bits
> 36 vertical distances in 11 unique classes   = 122 bits
> Overall complexity = 262 bits

For Figure 5.1.42b, the redesigned version,

> 30 horizontal distances in 6 unique classes = 67 bits
> 30 vertical distances in 10 unique classes   = 98 bits
> Overall complexity = 165 bits

REFERENCES*

Baker, C. H., and Grether, W. F. (1954). Visual presentation of information (Report WADL-TR-54160). Wright-Patterson Air Force Base, OH: Wright Air Development Center.

Beaton, R. (1984). Personal communication. Beaverton, OR: Tektronix.

Berger, C. (1944a). I, Stroke-width, form and horizontal spacing of numerals as determinants of the threshold of recognition. *Journal of Applied Psychology, 28,* 208–231.

Berger, C. (1944b). II, Stroke-width, form and horizontal spacing of numerals as determinants of the threshold of recognition. *Journal of Applied Psychology, 28,* 336–340.

Bonsiepe, G. A. (1968). A method of quantifying order in typographic design. *Journal of Typographic Research, 2,* 203–220.

Broadbent, D. E. (1977). Language and ergonomics. *Ergonomics, 8*(1), 15–18.

Carlsson, L. (1979). *Ljus och belysningskrav vid arbete med bildskärmar på tidningsföretag* (in Swedish). Stockholm: Tidningarnas Arbetsmiljö-Kommitte.

Cornsweet, T. A. (1970). *Visual perception.* New York: Academic.

* NOTE: To order documents published by the U.S. Government, give AD number, where listed (in parentheses), and order from National Technical Information Service, U.S. Dept. of Commerce, Springfield, VA 22161.

Dashevsky, S. G. (1964). Check-reading accuracy as a function of pointer alignment, patterning, and viewing angle. *Journal of Applied Psychology, 48,* 344–347.

De Valois, R. L., Abramov, I., and Jacobs, G. H. (1966). Analysis of response patterns of LGH cells. *Journal of the Optical Society of America, 56,* 966–977.

Dewar, R. E. (1976). The slash obscures the symbol on prohibitive traffic signs. *Human Factors, 18,* 253–258.

Engel, S. E., and Granda, R. E. (1975). *Guidelines for man/display interfaces* (Tech. Report TR 00.2720). Poughkeepsie, NY: IBM.

Farrell, R. J., and Booth, J. M. (1984). *Design handbook for imagery interpretation equipment.* Seattle, WA: Boeing Airspace Company.

Grandjean, E. (1984). Postural problems at office machine workstations. In: E. Grandjean, Ed., *Ergonomics and health in modern offices.* Philadelphia: Taylor & Francis.

Grether, W. F., and Baker, C. A. (1972). Visual presentation of information. In H. A. VanCott and R. G. Kinkade, Eds., *Human engineering guide to equipment design.* Washington, DC: U.S. Government Printing Office.

Hanes, R. M., and Rhodes, M. V. (1959). Color identification as a function of extended practice. *Journal of the Optical Society of America, 49,* 1060–1064.

Hardesty, G. K. C., and Projector, T. H. (1973). *Navships display illumination design guide.* Washington, DC: U.S. Government Printing Office.

Helander, M. G., Billingsley, P. A., and Schurick, J. M. (1984). An evaluation of human factors research on visual display terminals in the workplace. *Human Factors Review,* 55–130.

Helander, M. G., and Schurick, J. M. (1983). Evaluation of symbols for construction machines. Presented at Human Factors Society 26th Annual Meeting. Santa Monica, CA: Human Factors Society.

Hendley, C. D., and Hecht, S. (1949). The colors of natural objects and terrains and their relation to visual color deficiency. *Journal of the Optical Society of America, 39,* 870–873.

Johansson, G., and Backlund, F. (1970). Drivers and road signs. *Ergonomics, 13,* 741–759.

Johnson, S. L., and Roscoe, S. N. (1972). What moves, the airplane or the world? *Human Factors, 14*(2), 107–129.

Judd, D. B., and Wyszeki, G. (1967). *Color in business, science and industry.* New York: Wiley.

Kahneman, D. (1973). *Attention and Effort.* Englewood Cliffs, NJ: Prentice-Hall.

Kelley, C. R. (1968). *Manual and automatic control.* New York: Wiley.

Kraiss, K. F. (1976). Vision and visual displays. In K. F. Kraiss and J. Moraal, Eds. *Introduction to human engineering.* Köln, West Germany: Verlag TÜV Rheinland.

Kroemer, K. H. E. (1984). Personal communication. Blacksburg, VA: Virginia Polytechnic Institute.

Kokoschka, S., and Bodmann, H. W. (1980). Kontrast und Beleuchtungsniveau am Bildschirmarbeitsplats (in German). *Proceedings of the 19th Session of the International Commission on Lighting,* pp. 305–310.

Laycock, J., and Viveach, J. P. (1982, April). Calculating the perceptibility of monochrome and colour displays viewed under various illumination conditions. *Displays.*

Lerner, H. D. (1981). *Evaluation of exit directional symbols.* NBSIR 81-2268. Washington, DC: National Bureau of Standards.

Lippert, T. M. (1984). *A standardized color-difference metric of legibility.* Blacksburg, VA: Human Factors Laboratory, Virginia Polytechnic Institute and State University.

MacAdam, D. L. (1948). Color discrimination and the influence of color contrast on visual acuity. *Optique Physiologigue Coleurs, 28,* 161–173.

MacAdam, D. L. (1966). Color science and color photography. *Journal of Photographic Sciences, 14,* 229–250.

Meister, D., and Sullivan, D. J. (1969). *Guide to human engineering design for visual displays* (Report N0014-68-C-0278). Washington, DC: Office of Naval Research (AD 693 237).

Meister, D. (1984). *Human engineering data base for design and selection of cathode ray tube and other display systems* (Report NPRDC TR 84-51). San Diego, CA: Navy Personnel Research and Development Center (AD 145 704).

Miller, A. H., Goldenberg, Z. N., and Erbring, L. (1979). Type-set politics: Impact of newspapers on public confidence. *American Political Review, 73,* 67–84.

Murch, G. M. (1984). Human factors of displays. *Seminar Lecture Notes.* Los Angeles, CA: Society for Information Display.

Myers, W. S. (1967). *Accommodation effects in multicolor displays* (Tech. Rep. AFFDL-TR-67-161). Wright-Patterson Air Force Base, OH.

National Research Council (1983). *Video displays, work, and vision.* Washington, DC: National Academy Press.

Neisser, U. (1966). *Cognitive psychology.* New York: Appleton-Century-Crofts.

Peters, G. A., and Adams, B. B. (1959). Three criteria for readable panel markings. *Product Engineering, 30*(21), 123–131.

Rizy, E. F. (1967). *Dichroic filter specifications for color additive displays. II. Further exploration of tolerance areas and influence of other display variables* (Report RADC-TR-67-513). Griffith Air Force Base, NY: Rome Air Development Center (AD 659 346).

Robertson, H. D. (1977). Pedestrian preferences for symbolic signal displays. *Traffic Engineering, 47*(6), 38–42.

Sanders, A. F. (1970). Some aspects of the selective process in the functional visual field. *Ergonomics, 13*(1), 101–117.

Shurtleff, D. A. (1980). *How to make displays legible.* LaMirada, CA: Human Interface Design.

Snyder, H. L., and Maddox, M. E. (1978). *Information transfer from computer-generated dot-matrix displays.* Report No. HFL-78-3, Blacksburg, VA: Department of IEOR, Virginia Polytechnic Institute and State University.

Snyder, H. L. (1980). *Human visual performance and flat panel display image quality.* Blacksburg, VA: Virginia Polytechnic Institute and State University. Human Factors Laboratory.

Thorell, L. G. (1983). Using color on displays: A biological perspective. *1983 SID Seminar Lecture Notes.* Los Angeles, CA: Society for Information Display.

Tufte, E. R. (1983). *The visual display of quantitative information.* Cheshire, CT: Graphics Press.

Tullis, T. S. (1983). The formatting of alphanumeric displays: A review and analysis. *Human Factors, 25*(6), 657–682.

Tyte, R., Wharf, J., and Ellis, B. (1975). Visual response times in high ambient illumination. *SID Digest,* 98–99.

U.S. Department of Defense (1981). *Military Standard 1472 C.* Human engineering design criteria for military systems, equipment and facilities. Washington, DC.

VanCott, H. P. and Kinkade, R. C., Eds. (1972). *Human engineering guide in equipment design.* Washington, DC: U.S. Government Printing Office.

Woodson, W. E. (1981). *Human factors design handbook.* New York: McGraw-Hill.

Zwaga, H., and Easterby, R. (1982). Developing effective symbols for public information: The ISO testing procedure. *Proceedings of the International Ergonomics Association Conference.* Tokyo.

CHAPTER 5.2

DESIGN OF AUDITORY AND TACTILE DISPLAYS

ROBERT D. SORKIN

Purdue University
West Lafayette, Indiana

5.2.1. CHOOSING A NONVISUAL DISPLAY CHANNEL

A display, whether utilizing a visual or nonvisual channel, is a device designed to convey information as quickly and errorlessly as possible. In certain systems and task environments the most appropriate design choice for a display channel will use a nonvisual mode, either an auditory or tactile channel. This chapter discusses the characteristics of the auditory and tactile display channels as well as specific guidelines for the design of auditory and tactile displays.

It is evident that the visual display channel is the channel of choice for conveying information at high rates to a human operator in a complex system. However, certain task and situational factors may indicate that a nonvisual channel is to be preferred. Table 5.2.1 summarizes some of the aspects of a choice between visual and auditory display modalities. In some cases the use of a nonvisual channel may be mandatory, as in underwater work where visibility is very poor. Another factor is cost; a nonvisual channel may be less costly to implement in terms of money, weight, or power usage. An example of this is a portable marine radar system which utilizes an auditory rather than a visual display.

The most frequent use of the auditory channel is in situations where the displayed information occurs randomly and must immediately capture the attention of the operator. Thus alerting, warning, and alarm displays are usually auditory displays or augment visual displays (such as annunciators). Obviously the near omnidirectional aspect of the auditory display is an important reason for this usage. A major part of the section on auditory displays concerns the design of auditory warning systems.

5.2.1.1 Information Rates and Codes

In general, the information rates available via the auditory or tactile channel are lower than those for the visual channel. Very high information rates (exceeding 50 to 75 bits/sec) have been obtained with these channels, but the practical limits are much lower. The maximum information rate for the auditory channel is estimated from the perception of spoken speech (75 bits/sec at a normal rate); high rates for the tactile channel are estimates obtained from highly experienced Braille and Optacom users (25 to 50 bits/sec). These high rates are relevant only in situations involving very highly skilled

---

Table 5.2.1   When to Use the Auditory or Visual Form of Presentation

Use auditory presentation if:

1. The message is simple.
2. The message is short.
3. The message will not be referred to later.
4. The message deals with events in time.
5. The message calls for immediate action.
6. The visual system of the person is overburdened.
7. The receiving location is too bright or dark-adaptation integrity is necessary.
8. The person's job requires him to move about continually.

Use visual presentation if:

1. The message is complex.
2. The message is long.
3. The message will be referred to later.
4. The message deals with location in space.
5. The message does not call for immediate action.
6. The auditory system of the person is overburdened.
7. The receiving location is too noisy.
8. The person's job allows him to remain in one position.

Source: Deathridge, 1972.

persons. A more realistic limit on the information rate for either the auditory or tactile display channel is probably below 20 to 25 bits/sec. Most applications of these channels involve much lower rates, perhaps 5 bits/sec or less. High rates generally are required in applications where the auditory or tactile channel is employed as a sensory replacement channel for the deaf or blind, or in applications involving very highly trained operators.

The potential use of the auditory or tactile modality in moderate to high information rate applications invariably raises the important design question of how to code the information so that it is communicated effectively over the display channel. For example, in some auditory displays the code employed is "natural," in that there is a direct relationship between the spatial and temporal components of the original information and the displayed components. In such cases special training may not be required for the operator to "decode" the display output. A simple example is the sound of an automobile engine starting up. This uncoded sound provides an important source of information in knowing how to coordinate operation of the starter switch and gas pedal. An example of a high rate "natural" tactile channel is found in a technique, known as the Tadoma method, used by some deaf-blind persons for speech perception. The user of this technique places his or her hands on the speaker's face so that the vibrations of the speaker's larynx, the opening and closing of the jaw, and the air flow at the mouth can be felt. These tactile cues are directly related to the speech-generating movements of the speaker and enable good speech recognition by experienced Tadoma users.

In a display employing an "artificial" display code, there is no natural relationship between the spatial and temporal elements of the original information and the display output. For example, a future speech reception device for the deaf might first perform a (computer) recognition of the speech phonemes, followed by display of these phonemes. The phoneme sequence could be displayed via a tactile display system stimulating the skin of the user's hands or body. It is possible to design a display device (see Section 5.23) that, given sufficient user training, will allow very high information rates. In general, the particular choice of a "natural" or "artificial" display system code may involve significant design trade-offs between the amount of user training required, the potential user population, and the ultimate performance possible with the system.

5.2.2 THE AUDITORY DISPLAY CHANNEL

The auditory channel is effective for communicating different types of information to a human operator. In some applications, loading of the auditory channel has nearly approached the limits possible for safe and efficient system operation. For example, on the flight decks of some passenger aircraft, there are 30 to 40 auditory alarm and warning signals to be discriminated, identified, and responded to. This is in addition to the heavy use of the auditory channel for speech communication. In subsequent sections of this chapter we discuss the discriminative capacities of the auditory channel and principles for specifying the characteristics of auditory and speech message signals.

5.2.2.1 Sensitivity

The auditory system is sensitive to a wide range of sound intensities, from about 0.0002 microbar (0.0002 dynes/cm²) of pressure change at the ear to 200 microbar, a dynamic range of 6 log units. Sound pressure level, L_p, in decibels (dB) is the conventional way to describe sound intensity:

$$L_p = 20 \log_{10}\left(\frac{p}{p_r}\right) \tag{1}$$

where p and p_r are rms measures of sound pressure and p_r is a reference pressure of 0.0002 microbar. Table 5.2.2 gives sound pressure levels and sound pressures for some typical acoustic environments. Sound pressure level is normally measured with a sound pressure level meter, which weights the component frequencies present in a sound by one of several filter characteristics designated as A, B, or C, and integrates and averages the input sound for about 200 msec (on the fast setting) to more than 500 msec (on the slow setting). The A-filter setting approximates the human response to sound at different frequencies. A measurement made at these settings is indicated as 72 dB(A). The slow time setting and the A-filter weighting are standard for most applications.

In some cases it is important to specify the sound stimulus more precisely than possible with an ordinary sound pressure level meter. For example, sounds of very brief duration may be employed in auditory display devices or may be present as potential interfering noise. Such brief sounds cannot be measured accurately on the ordinary SPL meter owing to the comparatively long integration time. Another limitation is that in many situations the precise spectral nature of the sound is of concern. One may need to specify the level of each frequency component of a sound. In these cases a frequency analyzer is required; typical instruments are octave band or $\frac{1}{3}$ octave band analyzers, which specify the sound level in each of a set of adjacent filter bands. Analog and digital wave analyzers are also available for more precise measurement at much better frequency resolution.

Table 5.2.2   Sound Pressures and Levels for Typical Environment Conditions

Sound Pressure (microbar)	Sound Pressure Level[b]	Environmental Conditions
0.0002	0	Threshold of hearing
0.00063	10	Rustle of leaves
0.002	20	Broadcasting studio
0.0063	30	Bedroom at night
0.02	40	Library
0.063	50	Quiet office
0.2	60	Conversational speech (at 1 m)
0.63	70	Average radio
1.0[a]	74	Light traffic noise
2.0	80	Typical factory
6.3	90	Subway train
20	100	Symphony orchestra (fortissimo)
63	110	Rock band
200	120	Aircraft takeoff
2000	140	Threshold of pain

Source: From F. A. White, *Our Acoustic Environment,* p. 39, copyright 1975 by John Wiley & Sons, Inc., and reprinted by permission.

[a] 1 microbar = 1 dyne/cm^2 = 74 dB is a common reference pressure for instrument calibration.

[b] Reference is 2×10^{-4} microbar.

The auditory system is sensitive over a range of frequencies from about 20 to 20,000 Hz, with best sensitivity around 2000 to 4000 Hz. The bottom curve of Figure 5.2.1 shows how the *absolute threshold* varies with frequency. The absolute threshold is defined as the lowest level of a stimulus needed for the listener to detect the stimulus. Absolute threshold levels at the best sound frequencies are around 0 dB (or 0.0002 microbar). The sound levels needed for detecting signals in practical situations are higher, depending on the level of background noise, the observer's hearing, and the specific observing conditions.

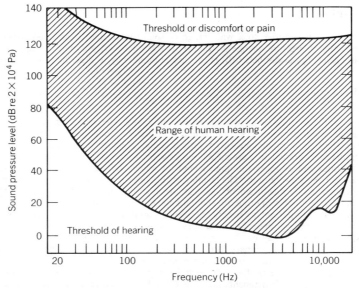

Fig. 5.2.1.   The range of human hearing. (From F. A. White, *Our Acoustic Environment,* p. 114. Copyright 1975 by John Wiley & Sons, Inc., and reproduced by permission.)

For frequencies below 500 Hz, there is almost no loss of hearing (<10 dB) with age, but that is not the case for high frequencies. Above 10,000 Hz, sizable portions of the population will have losses of 20 dB or more by age 45 (Green, 1976). In addition, long time exposure to noise will produce severe losses in the 4000 to 6000 Hz region. These aging and noise-induced loss factors may result in a high degree of variability in the high-frequency hearing ability of a user population.

The human listener can discriminate small changes in the level of a sound; the smallest detectable change in a stimulus is defined as the *difference threshold*. In general, difference thresholds increase with stimulus level; the larger the stimulus, the more difficult it is to detect a small change in level. The difference threshold level for pure tone sound signals (sinusoidal signals) is as small as 0.5 dB, over a wide range of starting levels and frequencies. However, information-carrying changes in the level of a practical auditory display would have to be many times greater than 0.5 dB. Such a level-dependent code would have to be limited to use with pure tone displays and under conditions consisting of relatively noninterfering noise backgrounds.

Very small changes in tone frequency can be discriminated under some conditions, depending on the initial frequency. The general relationship is given by

$$\frac{\Delta f}{f} = 0.005 \tag{2}$$

where Δf is the frequency difference threshold and f is the initial frequency. For frequencies below 500 Hz, Δf is a constant value of about 4 Hz. In practical display systems a much larger Δf would be required for an information-carrying frequency code, and critical concern would have to be given to the complete ensemble of display signals and to the possible presence of interfering signals in the display environment.

The *pitch* that a listener experiences when hearing a sound does not necessarily correspond to the frequency region having the heaviest concentration of sound energy; it is affected by other features of the complex. Pitch is a more practical dimension for use as a display code than frequency. The pitch of auditory display signals and the use of pitch as a display code are considered in Section 5.2.2.5.

5.2.2.2 Loudness

The loudness of sound stimuli depends on both sound level and frequency. An increase in level of 10 dB corresponds to approximately a doubling in the perceived loudness. This relationship can be seen in the following equation:

$$S = 10^{0.03(L_p - 40)} \tag{3}$$

where S is the loudness of a tone in units of *sones,* and L_p is the sound level of the tone. The constant 40 in the equation corresponds to the reference level for loudness; for example, a 40 dB tone at 1000 Hz is defined to have a loudness of 1 sone. The loudness of tones at frequencies other than 1000 Hz is given by the set of equal loudness curves shown in Figure 5.2.2. The 40 dB curve shows the level of different frequency tones required to match a 40 dB, 1000 Hz, tone in loudness.

Increasing the background noise conditions (or using persons with partial hearing loss) can result in a change in the relationship given by Equation (3). In such cases loudness increases much more rapidly than doubling, as the level of sound is increased 10 dB. This can result in very uncomfortable and annoying loudness levels.

In general, levels above 90 dB(A) are probably annoying to most people and may be disruptive of communications and the performance of other tasks. The next two sections discuss how to calculate display levels that are appropriate for a particular noise environment.

5.2.2.3 Display Level

At low auditory signal levels the human detection of a signal is essentially a probabilistic process, subject to fluctuations in the external noise background and in the auditory system of the observer. The probability that a signal will be detected increases from zero to 100% as the level of the signal is increased over a range of perhaps 20 dB or more.

The *masked threshold* is defined as the level required for 75% correct detection of the signal when presented to the observer in a two-interval task. In a two-interval task the observer reports which one of two defined observation intervals randomly contains the signal; both contain noise. A signal that is 6 dB above the masked threshold will allow essentially perfect detection performance in a controlled test situation. That is, the signal will be detectable on almost all presentations with a negligible error rate.

Fidell (1978) studied the effectiveness of audible warning signals in eliciting accurate and rapid responses in a simulated driving task. He found that the effectiveness of the signals was a function of

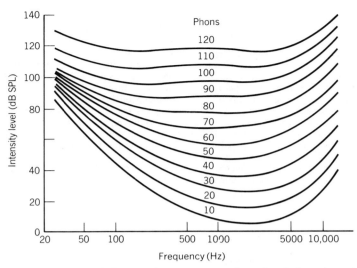

Fig. 5.2.2. Equal loudness contours. From each curve one may determine the intensity and frequency of tones that are equal in loudness to the referenced, 1000 Hz tone. (After Fletcher and Munson, 1933.)

their detectability, and was relatively independent of their spectral or temporal characteristics. Signals about 15 to 16 dB above the masked threshold were sufficiently high in level to ensure effective and consistent performance by his subjects.

Fidell and Teffeteller (1981) studied the relationship between the detectability of sounds and their noticeability. Their study approached the question of display level from the opposite direction: how high in level must a signal be to ensure notice by people engaged in an absorbing, unrelated task? They found that the noticeability of sounds was related to the degree of cognitive involvement in the task. The effectiveness and intrusiveness of the sound was objectively related to the sound's detectability; virtually everyone will notice an intruding sound that is 22 dB above the masked threshold.

Once a sound is 15 dB above the masked threshold, most of the influence of the background is gone and increasing the level further has little beneficial effect. Furthermore, overly high signal levels should be avoided because they can produce a number of negative effects ranging in seriousness from annoyance and interference with other tasks to permanent hearing loss. For example, the problem of overly loud cockpit warning signals has been discussed by Patterson and Milroy (1979) and Patterson (1982). They pointed out that some of the auditory warning signals on present aircraft are far too loud, given the background noise level. These signals probably interfere with communication, are aversive and annoying to the flight crew, and tend to hold crew attention beyond the point where the particular problem has been identified. Patterson (1982) recommends a level of 30 dB above masked threshold as the upper limit for auditory warning signals on aircraft flight decks. If absolute levels greater than 115 dB are required, channels other than the auditory channel should be considered.

Table 5.2.3 summarizes recommendations with respect to the level of auditory display signals. Procedures for computing the masked threshold level in particular background noise environments are discussed in the next section.

Table 5.2.3 Summary of Recommendations for the Level of Auditory Signals

1. Signal levels 6 to 10 dB above the masked threshold will be sufficient for 100% detectability in controlled test situations.

2. Signal levels 15 to 16 dB above the masked threshold will be sufficient for situations requiring a rapid response to a signal (such as a warning signal).

3. The level of an auditory warning signal should be less than 30 dB above the masked threshold, in order to minimize operator annoyance and the disruption of communication.

4. Nonauditory channels should be considered for environments that require very high absolute sound levels ($\geq$ 115 dB).

5.2.2.4 Computation of Masked Threshold

Two important questions in determining the prediction of the masked threshold level are (1) whether the interfering sound(s) is concentrated at a few individual frequencies and (2) whether the signal to be detected is shorter in duration than about 100 msec. In general, sound stimuli interfere with the detection of a signal only when frequencies of the interfering sound are near frequency components of the signal. However, very intense interfering tones give rise to harmonics within the auditory system, and these distortion products may interfere with the detection of signal frequencies remote from the original noise components. Table 5.2.4 summarizes some of the effects of interference from tones on signal detection. By intense tones we mean levels exceeding 80 to 90 dB.

Most interfering sounds are not composed of only one or two very intense frequencies but are a complex mixture of many frequencies. For short-duration signals, the whole noise spectrum is relevant, although for long ones we need only consider a narrow band of frequencies around the signal frequency. For example, most engine and factory noise sources have essentially continuous noise spectra. The important parameter of such noise sources is the *noise spectrum level*, $L_{spectrum}$, the noise level per unit bandwidth. This measure is useful because the auditory system is relatively insensitive to the phase of individual noise components and to noise components far away in frequency from the ones to be detected. $L_{spectrum}$ is defined in terms of the noise power density, N_o, the noise power per cycle. Then

$$L_{spectrum} = 10 \log N_o \tag{4}$$

The *overall* level of the noise, L_{noise}, is

$$L_{noise} = 10 \log(N_o \times \text{BW}) = 10 \log N_o + 10 \log \text{BW} \tag{5}$$

where BW is the band of noise frequencies contributing to the total noise power. The important quantity in specifying the masked threshold for short-duration tones (<100 msec) is the ratio of signal energy, E_s, to noise power density. The signal energy, E_s, is equal to the product of the signal power, P_s, times the signal duration, t. Thus

$$10 \log E_s = 10 \log P_s t = 10 \log P_s + 10 \log t = L_s + 10 \log t \tag{6}$$

where P_s is the signal power, t is the signal duration, and L_s is the signal level in decibels.

The masked threshold level depends on a number of listener and task factors such as the listener's uncertainty about the time of signal occurrence and the requirement for precise listener knowledge of the signal characteristics. In laboratory detection tasks the masked threshold for brief (<100 msec) duration tones is approximately an E_s/N_o ratio of 40 or less. This corresponds to a level difference between the signal and noise of 16 dB, for example:

$$10 \log \left(\frac{E_s}{N_o}\right) = 10 \log 40 = 16 \text{ dB} \tag{7}$$

For example, consider the case of a 100 msec, 78 dB tone signal in noise with $L_{spectrum}$ of 52 dB. Then

$$10 \log \left(\frac{E_s}{N_o}\right) = 78 + 10 \log(0.1) - 52 = 16 \text{ dB} \tag{8}$$

This signal would be just at the masked threshold.

When the signal is longer in duration than approximately 100 msec, the masked threshold depends

---

Table 5.2.4 Summary of the Effects of Interfering Tones on Signal Detection

1. Most interference occurs when the signal tone is near to the frequency of the interfering tone.

2. As the interfering tone increases in intensity, the interference spreads to additional signal frequencies.

3. The effect of the interfering tone is increased when it is below the signal frequency, rather than above.

4. The interaction of intense tones results in the generation of additional tones, including harmonics of the original tones and tones at summation and difference frequencies.

Source: From B. H. Kantowitz and R. D. Sorkin, *Human Factors: Understanding People-System Relationships,* p. 523, copyright 1983 by John Wiley & Sons, Inc., and reproduced by permission.

on signal *power* rather than signal energy. This is because at that point the signal duration begins to exceed the auditory system's ability to integrate energy. At signal durations exceeding 100 msec, the important masking is produced by the noise power in a band of frequencies around the signal frequency. This band of noise frequencies can be thought of as arising from a hypothetical listening filter that the listener centers on the signal's frequency. If the power of the signal exceeds the power of the noise coming through that filter, the signal is audible. The characteristics of this effective listening filter determines how effective the noise is in interfering with the signal components.

If the noise is essentially continuous and level over frequency, the filter can be approximated by an equivalent rectangular filter of bandwidth BW. The bandwidth of this filter is approximately proportional to the filter center frequency, and varies with the particular signal detection condition and the listener age; minimal bandwidth is approximately 0.06 times f_c, where f_c is the filter center frequency. A practical bandwidth for a wider range of conditions and middle-aged listeners is $0.15f_c$ (Patterson, 1982). For example, if the signal of interest were at 1000 Hz, the equivalent rectangular filter would have a bandwidth of 150 Hz. Assuming a level, continuous noise of 48 dB spectrum level, the total noise power level effective in making the signal would be, in decibels,

$$L_{noise} = 10 \log N_o + 10 \log (150)$$
$$= 48 + 21.8 = 69.8 \text{ dB} \tag{9}$$

Thus the signal would have to be at least 69.8 dB to be audible in this noise. This method can be used if the noise is moderately constant over frequency, and does not change more than 6 dB over the bandwidth of the filter.

If the noise spectrum is not smooth, and cannot be approximated by a constant spectrum level, a more complex procedure must be employed to compute the masked threshold level. For an arbitrary noise spectrum, $N(f)$, the required signal power at threshold is given by

$$p_s = \int_{-\infty}^{\infty} N(f)W(f)\,df \tag{10}$$

That is, the threshold signal power is equal to the integral of the product of the noise spectrum, $N(f)$, times the filter function, $W(f)$.

An approximation for the auditory filter has been developed by Roy Patterson and his colleagues (Patterson, Nimmo-Smith, Weber, and Milroy, 1982), called the Roex filter, for *ro*unded *ex*ponential filter. This filter is given by the equation

$$W(g) = (1 - r)(1 + pg)e^{-pg} + r \tag{11}$$

where g is the normalized separation from the center frequency, f_c, to the evaluation point, f; for example,

$$g = \frac{|f - f_c|}{f_c} \tag{12}$$

In Equation (11), p determines the pass band of the filter and r is a constant restricting its dynamic range. Then for an arbitrary noise spectrum, $N(g)$, the threshold signal power is given by

$$P_s = f_c \int_0^{0.8} N(g)[(1 - r)(1 + pg)e^{-pg} + r]dg \tag{13}$$

where f_c converts the integral variable from relative to absolute frequency, and the integration range is restricted to a relative frequency of 0.8. This equation yields a practical threshold prediction provided that the signal is of moderate duration and the noise has more than four components in the filter range.

Patterson et al. (1982) have estimated values of p for data obtained from 16 listeners varying in age from 23 to 75 years. Figure 5.2.3 illustrates how p depends on listener age. A p value of 25 would be appropriate to employ for middle-aged listeners. The r parameter is largely independent of age below age 60 (Patterson et al., 1982), and if the noise spectrum does not vary by more than 30 dB over the width of the filter band, r can be ignored.

If the noise spectrum is not dominated by any pure tone components, Patterson (1982) suggests a step function approximation to compute P_s. One starts from well below the signal frequency and considers successive steps along the frequency scale until one is well above the signal frequency. The step width employed should be chosen so that there is less than a $+3$ or -6 dB deviation in level for

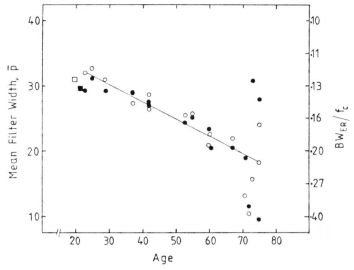

Fig. 5.2.3. Average filter-width values, p, plotted as a function of age. The left and right ears are designated by filled and open circles, respectively. The right-hand ordinate shows the relative bandwidth of the filter ($BW_{ER}/f_o \simeq 4 \ p^{-1}$). (From R. D. Patterson, I. Nimmo-Smith, D. L. Weber, and R. Milroy, The deterioration of hearing with age: Frequency selectivity, the critical ratio, the audiogram, and speech threshold, *Journal of the Acoustical Society of America*, 1982, 72, 1793. Copyright 1982 by the American Institute of Physics and reproduced by permission.)

any step. Then for each noise step within the filter range (centered at f_c), compute and sum the contribution of each step, where each step is given by

$$N_i \ [p^{-1}(2 + pC)e^{-pC} - p^{-1}(2 + pF)e^{pF}]$$

where N_i is the noise power in step i and C and F are the closer and further edges of the step (in relative frequency). Then,

$$P_s = \sum_{all \ i} N_i [p^{-1}(2 + pC_i)e^{-pC_i} - p^{-1}(2 + pF_i e^{-pF_i}] \tag{14}$$

Note that $N_i = 10^{(L_n/10)}$ where L_n is the total noise level in each step.

5.2.2.5 Display Pitch

Pitch is an important characteristic of sound signals useful for encoding information in auditory displays. The pitch of a sound signal is normally associated with the prominent frequency component of a sound's spectrum. However, there are at least two important exceptions to this statement. The first exception is that pitch generally corresponds to the period, or repetition rate, of a sound wave form. For example, if a trombone, a bassoon, or a guitar play the note C3, the C below middle C, they produce very different wave forms. But these wave forms all have exactly the same repetition rate.

A second exception concerns sounds with harmonically related frequency components. The pitch of such a signal generally is at a much lower frequency than is present in the signal, that of the "missing fundamental" frequency. A sound composed of the harmonically related components at 1200, 1400, 1600, and 1800 Hz has the same low pitch as a sound with a single component at 200 Hz. This low-frequency pitch is perceived even though no signal energy is present at low frequencies and even in the presence of interfering noise at low frequencies. This phenomenon accounts for why we perceive low-pitched sounds on the ordinary telephone or on small portable radios. Neither of these systems is capable of reproducing the low-frequency energy associated with low-pitched sounds; they are heard because of the sensitivity of the auditory pitch mechanism to the harmonic structure of musical or vocal sounds.

Pitch provides a useful code for carrying auditory display information. The pitch of signals that contain harmonically related components is stable and relatively insensitive to the relative level of the components or to masking of some components, provided that there are a sufficient number of components. The more frequency components, the easier it is to generate a set of different display sounds that have discriminable and recognizable pitch and sound quality (timbre). In general, signals with most of their energy in the first five harmonics sound fuller or more sonorous than signals with significant energy in higher harmonics. The latter sound "sharper." The presence of additional, but inharmonic, components can also add a shrill characteristic to a sound. Table 5.2.5 provides specific design criteria related to the spectral dimension for auditory warning and alarm signals.

5.2.2.6 Display Duration, Shape, and Temporal Pattern

Several aspects of the temporal form and extent of auditory display signals are important factors in the detectability, coding, and listener reaction to these signals. In this section we refer to the duration or temporal shape of the signal wave form *envelope,* rather than to the individual rapid pressure changes that constitute the signal wave form.

Duration

The minimum duration signal burst should be at least 100 msec to ensure reliable detection. Patterson (1982) recommends pulse durations of 100 msec (plus 25 msec rise and fall times and quarter-sine shaping) as near optimum envelope parameters for flight-deck warning applications. The use of signals shorter than 150 msec allows for a low signal on/off ratio and manipulation of the temporal pattern of pulses for coding purposes. Another advantage of short signals is that the off periods can be long enough to allow minimal disruption of ongoing speech communication. The redundancy of normal speech enables listeners to fill in parts of words and sentences obliterated by (sufficiently) brief auditory signals.

Pulse Shape

The rise and decay times of each signal pulse and the particular shaping of the pulse during these onset and offset periods are other important factors. Fast rates of change in a sound act to draw a listener's attention. Existing flight-deck warning signals have extremely rapid rise times (off to 100 dB in 10 msec) but Patterson (1982) finds such abrupt onsets inappropriate for the flight deck situation. Very rapid sound onsets are associated with catastrophic events in a listener's environment and may produce an involuntary startle reflex. This startle reflex may produce an operator reaction that is inconsistent with the desired response.

Sound onset rates of 10 dB/msec or faster appear to a listener as instantaneous and produce a startle response if the final level is sufficiently high. In addition, a rapid change in the power of any sound produces a spectral spread of energy that may appear as a loud bang. Onset rates of slower than 1 dB/msec can be perceived as rising and produce little startle if the final level is below 90 dB.

In many practical situations, a display signal has to be moderately high in level in order to be heart above the background. Thus the 1 dB/msec limit is probably a reasonable limit on the onset rate. Generally, the offset rate should match the onset characteristic. The onset pulse shape is not

Table 5.2.5 Spectral Considerations for Auditory Warning and Alarm Signals

1. The pitch of warning sounds should be between 150 and 1000 Hz.
2. Signals should have at least four prominent frequency components. As a consequence:
   a. The chance of masking by other sounds is minimized.
   b. There is a minimum of changes to pitch and sound quality under masking conditions.
   c. The number of different distinctive signal codes that may be generated is maximized.
3. Signals with harmonically regular frequency components should be used rather than inharmonic spectra.
   a. Lower-priority warning signals should have most of their energy in the first five harmonics.
   b. Higher-priority, immediate action signals should have relatively more energy in harmonics 6 through 10.
   c. High-priority signals can be made very distinctive by incorporating a small number of additional inharmonic components.
4. The prominent frequency components for signals should be in the range from 1000 to 4000 Hz. At least four of the first 10 harmonics should be prominent.
5. Rapid glides (100 msec) in the fundamental frequency of a signal (providing that the above constraints are not violated) can be an effective way to signal urgency and command listener attention.

Source: After Patterson, 1982.

Fig. 5.2.4. Temporal code for national standard fire signal.

critical, but should be concave down or linear. No wave form overshoot or ringing should be evident in the acoustic signal. For signals 20 to 30 dB above the masked threshold, an onset and offset of 20 to 30 msec duration meet the above criteria.

Temporal Pattern

Patterson and Milroy (1980) studied the learning and retention of flight-deck warning signals. They found that confusion often occurred between warning signals that shared the same pulse repetition rate, even though the signals had quite different spectra. Using signals with a variety of temporal patterns would minimize these confusions and greatly improve performance. Patterson (1982) suggested that pulse rate is a preferred way to code the urgency or priority of a warning signal. Coding high urgency by high pulse rate avoids the traditional practice of using a disruptively high level signal with very rapid onset time.

A temporally coded signal is the proposed solution to the problem of choosing a standard national fire-alarm signal. The National Fire Prevention Association asked the Committee on Hearing and Bioacoustics and Biomechanics (CHABA) of the National Research Council to recommend an auditory display signal for use as a standard national fire alarm (Swets et al., 1975). The signal was to be universally recognizable by all people and to be easily distinguished from other alarms. A further consideration was to be able to adapt existing fire alarms to the new signal.

Because fire alarms exist in a variety of different acoustic environments with different environmental noise spectra, the CHABA committee felt that recommending a signal having a specific spectral configuration would be impractical; a signal 15 dB above threshold in one environment might be near threshold in another. The solution was a temporal code consisting of a repeating sequence of two short signals followed by a long signal and a time-out, as shown in Figure 5.2.4. The durations are 400 to 600 msec for an on period, and 300 to 600 msec for an off period; onset and offsets should be at least 10 dB in 100 msec.

A prototype of an advanced, immediate action warning proposed by Patterson (1982) is shown in Figures 5.2.5 and 5.2.6. This signal is composed of a burst of six pulses. The spectral nature of each pulse is fixed. The grouping pattern of four pulses followed by two pulses at longer spacings enables generation of a characteristic rhythm. The spectral nature of the basic pulse combines with the temporal rhythm to make the signal distinctive and easily identified. Figure 5.2.5 shows how the level contour of the pulses and the pulse separations can be changed to yield signals with a greater impression of urgency.

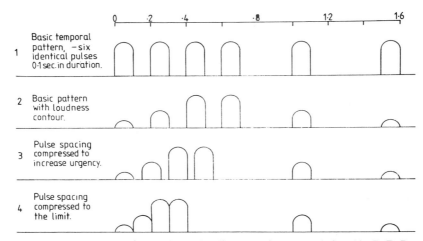

Fig. 5.2.5. Component patterns for an advanced auditory warning system designed by R. D. Patterson. (From R. D. Patterson, Guidelines for Auditory Warning Systems on Civil Aircraft, CAA Paper 82017, Civil Aviation Authority, London, November 1982; a patent application has been filed for an integrated warning system of this type.)

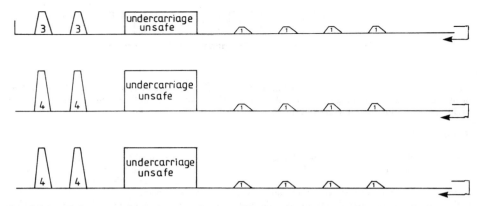

Fig. 5.2.6. Time course of complete warning in auditory warning system designed by R. D. Patterson. (From R. D. Patterson, Guidelines for Auditory Warning Systems on Civil Aircraft, CAA Paper 82017, Civil Aviation Authority, London, November 1982; a patent application has been filed for an integrated warning system of this type.)

Patterson discussed how this scheme can be developed to include informational changes in the overall level of the signals and the inclusion of a voice warning signal, to create a warning system integrated in the flight-deck environment (see Figure 5.2.6 and Section 5.2.9).

5.2.2.7 Binaural Displays

As a two-input system, the normal human auditory system possesses binaural signal processing abilities that are useful in certain auditory display situations. The most obvious processing ability is the localization of the source of a sound in the head's azimuthal plane. Sensitivity to the time and intensity differences between the sound waves arriving at the ears (interaural time and intensity differences) provides the basis for this ability.

The interaural time difference in a sound is a crucial cue to sound location for sound frequencies below 1000 Hz. Under laboratory testing conditions, a listener can detect interaural time differences as short as 10 μsec. At high frequencies, the interaural intensity difference is a more significant cue to localization, because the shadowing effect of the head is much greater for higher frequencies. A listener can detect interaural intensity differences as small as 0.7 dB.

The use of ongoing interaural time differences as a localization cue actually extends to high- as well as low-frequency signals. In such cases the high-frequency signal must be sufficiently complex that the time difference appears as a delay in the envelope of the wave form at one ear (McFadden and Pasanen, 1978). If signals are to be localized easily, they should include components that are spread across the entire frequency spectrum. However, even a broad-spectrum signal may be quite difficult to localize if presented in a small, hard walled environment.

This sensitivity to interaural time and intensity also enables the human listener to detect signals that would otherwise be buried in the background noise. One can demonstrate up to a 15 dB signal level advantage for detection conditions in which there are interaural differences in the signal or noise input to each ear. That is, one can drop the signal level 15 dB below the level that could be detected with identical inputs to both ears (or listening with one ear alone).

In addition to improving detectability, these binaural cues also enable the improved processing of speech messages in noise. Licklider (1948) described a simple technique to improve the intelligibility of headphone-presented speech in a free field noise environment. The problem situation was a pilot (wearing headphones) in a noisy cockpit. Licklider found that reversing the wires to one headphone, for example, inverting the speech wave form (or shifting its phase by 180°), greatly improved the intelligibility of the speech. Localization cues also enable us to attend selectively to spatially separated conversations in a crowded, noisy room: this is the so-called "cocktail party" effect. In such an environment, covering one ear tightly will cause a sudden decrease in the ability to hear separate conversations.

Stereophonic Displays

The use of both ear inputs in an auditory display system is fundamental to stereophonic sound techniques. Two important aspects of sound localization related to sound localization and stereophony should be noted. The first aspect is known as the *precedence* effect (Gardner, 1968). The second concerns the cues that are critical for reproducing a full stereophonic impression over headphone-presented displays.

The precedence effect is a characteristic of the auditory system that enables much better sound localization than might be expected in actual physical environments. In a normal room situation there may be many surfaces from which a sound will echo and reverberate. Sound reaching the listener's head from these multiple sources presents a multitude of complex interaural time and intensity difference cues. The best information about the true source of the sound is the time and intensity cues associated with the very *first* sound wave that strikes the listener's ears. Later arriving wave forms will include cues relevant to the multiple echo signals, in addition to the original signal. The precedence effect relates to the way the auditory system normally suppresses all localization cues but those associated with the first wave front.

This phenomenon can be demonstrated easily by the reader with a stereo system hooked to matched speakers. While seated midway between the speakers, switch the system to the monaural position so that identical signals are heard by each ear. The sound will appear to be localized in the space midway between the two speakers. Now shift your body position a small amount to the left; the sound will now appear to be coming *only* from the left speaker. The first-to-arrive wave fronts from the left speaker override the slightly delayed cues from the right speaker, creating the effect. This is a problem for "stereophonic" systems that employ loudspeaker reproduction; the sounds reaching the listener's ears (except in very special systems) carry time and intensity cues relating to the loudspeakers' position and unrelated to the sound position of the recorded source.

The second aspect of stereophonic displays concerns headphone presentation. If loudspeaker reproduction always involves problems involving the listener's head position, perhaps headphone presentation could offer a solution. The problem with headphone-presented material is that it is usually perceived by the listener to be inside the head rather than in the space around the listener. This phenomenon can be minimized by incorporating additional interaural cues employed by the normal auditory system. The sound spectra reaching each ear from an actual external source are separately shaped or filtered by the attenuation characteristic of each external ear and ear canal. This spectral shaping depends highly on the direction of the sound and provides important cues about the sound location. Reconstructing these cues in headphone-presented material produces the illusion that the sound originates in the external space around the listener's body (Wightman, Kistler, and Walczar, 1983).

Except for stereophonic sound reproduction systems, the use of localizing auditory displays is not very common. However, sound localization is obviously an important aspect of how emergency vehicles such as ambulances, police cars, and fire trucks may signal their position and direction. Caelli and Porter (1980) have studied the problems involved in the use of emergency vehicle sirens. They found very poor judgment accuracy of emergency vehicle location and distance, particularly when the emergency vehicle was approaching the rear of the observer's vehicle. Driving with only the driver's window open increased localization errors owing to the acoustic effects of a single opening in the car; closing all the windows improved localization but greatly reduced the detectability of the ambulance siren.

These experiments confirmed the results of interviews with experienced ambulance drivers concerning the inadequacy of siren function on the road. Under normal driving conditions the ambulance siren is very poorly heard and localized. Caelli and Porter suggested that the use of flashing lights on emergency vehicles was probably a superior strategy to siren use. They also suggested the use of a warning signal *inside* the car that could be triggered by radio by approaching emergency vehicles.

We can imagine a similar system in which the actual siren sounds could activate a localizing auditory display system inside the car. External microphones on the auto's roof would pick up external sounds and feed them into a microcomputer. The computer would continuously analyze such sound for acoustic signatures typical of sirens and other emergency vehicles. It would also compute the direction and estimated distance of the siren. The computer would then generate a localizable signal on a set of acoustic speakers located around the periphery of the vehicle interior, simultaneously suppressing the output of the vehicle's radio/stereo system. This naturally coded and localizable signal would then enable the driver to take appropriate evasive action. The cost effectiveness of such a system would be a significant bar to its use.

5.2.2.8 Coding and Decision Factors

As in the design of any complex display system, systems utilizing the auditory channel will involve questions about several nonsensory parameters. These questions may relate to the complexity of the signal codes to be employed, to requirements for operator training and retraining, and to other task factors. In this section we focus on two questions that are often involved in the implementation of practical auditory display systems: the complexity of the display code and alarm decision factors.

Code Complexity

Most applications of the auditory display channel involve the use of multiple tone and speech message announcement signals. A prime example is warning signals on aircraft flight decks. Federal Aviation Administration (FAA, 1977 a, b) reports have pointed out that:

1. There is no consistent utilization philosophy for auditory alerting signals.
2. There is only a partial standard for the acoustic codes employed.
3. The number of signals employed is increasing; newer aircraft have 14 to 17 alerting signals (including some voice signals).

A NASA report by Cooper (1977) suggested a limit of four or five signals even though some current aircraft now have as many as 30 or more. Patterson and Milroy (1980) studied the learning and retention of aircraft warning signals. They found that although large sets of warnings can be learned, it can require considerable learning time and regular retraining. Patterson's (1982) recommendation is that up to six immediate action warning signals plus two *attensons* may be used provided that:

1. The warning sounds have distinctive temporal and spectral patterns.
2. The perceived urgency of the warnings matches their priority.
3. The warning sounds are reinforced by key-word voice warnings with good speech quality.

An *attenson* is a special warning sound that signals the priority level of a particular warning to follow. For example, in addition to the six immediate action signals suggested above, Patterson recommends a single attenson signal that would indicate a lower-priority class of "abnormal condition" alerts. This signal would precede the occurrence of specific voice warnings detailing the abnormal condition. A third priority attenson, at the "advisory level," could signal the need to check a visual display array. Patterson's recommendation is discussed further in Section 5.2.2.9 and shown in Table 5.2.6.

Alarm Decision Factors

The nature of the auditory channel makes it ideal for use with alarm and alerting signals. In a variety of large system environments utilizing such signals, however, a number of decision-related problems have arisen. These problems have been observed on aircraft flight decks, in automated factories, and in the control rooms of nuclear power plants. Usually these alarm displays are driven by an automated subsystem which processes incoming data about system status and compares those data to expected normal/abnormal criteria. When the process data exceed (or fall below) certain preset threshold values, an alarm is generated, and operator intervention may occur. Yet the literature indicates that operators frequently have trouble appropriately identifying, assigning priorities, and responding to conditions with these displays (Cooper, 1977; Banks and Boone, 1981).

One critical display parameter that has been identified in these problems is the false alarm rate, the probability that an alarm will be signaled when the underlying system state is not a true alarm state. Sorkin and Woods (1985) performed a signal detection analysis of the general problem of automated monitor-human operator systems. They point out that in many systems with automated alarm subsystems, the designer is tempted to set the operating point of the automated subsystem so as to minimize the subsystem's miss rate (the probability that a true alarm state exists and no alarm is signaled). Such a design strategy maximizes the false alarm rate of the subsystem (as well as its hit rate, the probability that it will correctly signal a true alarm condition). This strategy may make the automated subsystem appear to be very effective but it may have serious consequences on the performance of the overall system. Given high subsystem alarm rates, busy human operators may not be able to attend to and process all the signaled events. The net result may be overall human-automated system performance that is quite poor, depending upon the precise nature of the interactions between the human and automated subsystem and the operator monitoring strategies.

Sorkin and Woods recommended that the operating parameters of the automated alarm subsystem be set with appropriate consideration of the overall system performance and operator task requirements. They pointed out that a multilevel warning signal may be preferable under busy operator conditions. A two- or three-priority level alarm, for example, could indicate the degree of confidence that the automated subsystem has in its alert; a definite problem report would have one code, a potential problem another. Warnings usually induced by particular operator actions or by test procedures could fall in the latter category.

5.2.2.9 Speech Message Displays

Speech is often considered as a means to relieve information overload on a primary visual task, because it enables a very high rate of information transmission over the auditory channel. In addition, the technology of speech generation has made automatic speech message systems practical in a variety of system environments. However, a number of questions exist regarding optimal ways to generate speech message signals and how to integrate these signals with other auditory signals and other operator tasks. Research in this area is developing rapidly; in this section some general principles and problems of speech message systems are discussed and some general design guidelines given.

Speech Level and Spectrum

The energy of a speech signal varies about 30 dB across the various speech sounds, with the vowel sounds containing most speech energy and the consonants the least. The vowel sounds have concentrations of acoustic energy in harmonically related bands of frequencies called formants. These frequencies correspond to mechanical resonances of the vocal tract. The first formant is between 200 and 800 Hz, depending on the vowel and the speaker; the second is in the region around 1500 Hz; the third is around 2400 Hz; and the fourth is around 3500 Hz. Most of the acoustic energy is in the first two formants and very little energy is above 3500 Hz. Because of the way the speech system generates consonants, much more high frequency than low frequency energy is involved, some above 5000 Hz. However, the redundancy of natural speech allows one to filter out the very high and very low frequencies of a speech signal without much effect on intelligibility.

The appropriate level and spectrum for speech signal displays follow the same overall considerations addressed in Sections 2.1 to 2.5; that is, the frequency range should be restricted to 500 to 5000 Hz. Because considerable speech information is carried by consonant sounds having lower speech power, shorter durations, and higher frequencies than vowel sounds, reproducing the necessary dynamic range of the signal may be a problem. For example, in a background of intense noise, some high-level speech sounds would have to be very high above the noise level in order for the weaker speech sounds to be detectable. It may be desirable to process the speech message prior to using it as a display signal, such as by boosting speech components that are above 500 Hz or by compressing (or restricting) the amplitude of the speech waveform. For speech messages presented in the flight deck environment, Patterson (1982) recommends a boost of 3 dB/octave in the region of 0.5 to 5 kHz.

Artificially Generated Speech

In many speech message systems, speech is generated by artificial means, usually by some digital to analog conversion process. In some environments it is thought desirable to have the speech appear to be computer generated, in order to preserve the distinctiveness of that particular speech message channel. Computer-generated speech has been of two general types, (1) speech derived from originally spoken speech that has been digitized and processed for more efficient, lower bit-rate transmission, and (2) speech synthesized directly from text or programmed strings using algorithms for the acoustic generation of phoneme sequences (including phonome transition rules, stress rules, etc.). Speech generated by either technique may, depending on the techniques used, appear very natural sounding or may have a distinctive accent or machine-like sound.

Artificial speech is generally not as redundant as natural speech and therefore is much more sensitive to the effects of the linguistic and task context, the operator training, the background noise, and other manipulation of spectrum or level. Artificial speech may incur deficits in the information processing of certain tasks and may demand greater attentional resources than natural speech. For example, Pisoni (1982) and Luce, Feustal, and Pisoni (1983) have shown that in certain tasks there can be perceptual and cognitive consequences of using synthetic rather than natural speech. As techniques for speech encoding and speech synthesis-by-rule improve, these particular problems may decrease somewhat.

The use of artificial speech allows the system designer much greater control of such speech parameters as fundamental frequency (pitch) and speech rate, as well as the sex, accent, and other distinctive characteristics of the speaker. Thus to some extent the designer can tailor the voice characteristics of the particular speech message system to suit the task environment and noise background. Simpson and Marchionda-Frost (1984) studied the effects of different speech message rates and fundamental frequencies on the accuracy and speed of responses to speech messages. The messages were presented to pilots simultaneously engaged in a simulated (video game) flying task. They found that for the synthesized messages in their study (1) there were no significant effects of voice pitch or speech rate on accuracy, but that (2) response time decreased as speech rate increased. The fundamental speech frequencies employed were 70, 90, and 120 Hz; speech rates were 123, 156, and 178 words/min. Some pilots reported that they believed they might occasionally miss a message spoken at the fastest rate; others reported that messages presented at the slowest rate took time and attention away from the primary flying task. Thus there may be an attentional trade-off in the best speech rate to use in a specific application.

Message Format

One ongoing concern in the design of speech message displays has been the speech message format and the use of auditory alerting signals as prefixes to the speech message. The major questions relate to the relative effectiveness of messages with (1) monosyllabic versus polysyllabic words, (2) key-word messages (such as two words in isolation) versus full sentence format (semantic context) messages, and (3) speech messages with or without preceding alerting tones.

As is the case for natural speech, synthetic speech polysyllabic words are more intelligible than monosyllabic words. Similarly, words in sentences are more intelligible than words in isolation. The syllable and sentence context increases the redundancy of the messages and results in improved intelligibility. Response time to messages in sentence format has also been found to be shorter than to keyword messages (Simpson, 1976; Hart and Simpson, 1976). In summarizing the results of several studies, Simpson and Navarro (1984) found that cockpit warning messages composed of two monosyllabic words are generally inferior to sentence format messages. They noted that messages composed of two polysyllabic words fall in between those two extremes, in some cases performing as effectively as messages in sentence format.

Simpson and Williams (1980) studied synthesized voice warnings under two different linguistic formats (key word or key word plus additional word semantic context) with and without an alerting tone prefix. They found that the tone prefix increased rather than decreased the response time for either linguistic format (measured from the onset of the tone prefix). There were no differences in response time to the different linguistic context conditions even though the semantic context messages were longer in duration than the key word messages. The failure of the beneficial effect for the tone prefix may have been due to the distinctive nature of the synthesized speech employed, which allowed the speech messages also to serve an alerting function.

In a related study, Hakkinen and Williges (1984) investigated the hypothesis that it is the distinctive nature of synthetic speech that eliminates the benefit of a tone prefix. They measured response times to synthesized speech warnings with and without warning tone prefixes, under two conditions: (1) a condition in which synthesized speech was used only for warning messages, and (2) a condition in which synthetic speech was used for both warning messages and other nonurgent, task-related information. The alerting tone prefix was used only for warning messages. In condition 1, their results replicated those of Simpson and Williams, indicating an increase in response time for the tone prefix case. However, in condition 2, the alerting tone did not increase response time but improved performance over the no-prefix case. These results suggest that the distinctiveness of synthesized speech can, in certain circumstances, provide an alerting as well as an informational function in a speech warning message.

Both the Simpson and Williams and Hakkinen and Williges studies used only one type of tone prefix; that is, the tone could perform only an alerting function. One can imagine that the use of more than one tone prefix could result in further advantage for the tone prefix conditions. Moreover, it is evident that advanced systems such as aircraft flight decks will include additional applications of synthesized speech, thereby reducing the distinctiveness and alerting capacity of a no-prefix speech warning message. Patterson (1982) has advocated limiting the use of speech warnings for signaling immediate-action emergency conditions in aircraft because of the ongoing use of the speech channel for other purposes (including synthesized speech) and because of the difficulty of communicating during a long-duration speech warning.

Patterson suggests that the speech message channel should be employed primarily for abnormal condition displays rather than emergency warnings. The requirements for abnormal condition warnings are such that, (1) time and disruption are less critical, and (2) the number of alternatives to be signaled is larger, and the versatility and reliability of full format speech messages can be realized. Patterson suggests an integrated approach to the flight-deck warning problem that would include a heirarchy of warnings. His suggestion for an advanced auditory-speech warning system is illustrated in Table 5.2.6 and Figures 5.2.5 and 5.2.6. An alternative philosophy for the design of speech display systems is presented in the chapter in this volume by Simpson *et al.,* on speech controls and displays.

Table 5.2.6   Integrated Tone and Speech Warning System

Priority	Purpose	Result	Description
Highest	Emergency condition	Immediate	Auditory pulse sequences integrated with brief, key word warning (see Figure 5.2.6)
Second	Abnormal condition alert	Immediate crew awareness	Specific auditory prefix followed by 1 of 10 full-format voice messages, repeating after suitable pause
Third	Advisory alert	Crew check of visual display panel	Specific auditory signal

Source: After Patterson, 1982.

5.2.3 THE TACTILE DISPLAY CHANNEL

The tactile display modality is rarely employed as a primary display channel except in unusual situations. It is employed more frequently as an adjunct channel, for example, in the coding of the shape and size of knobs and of similar controls. This section discusses the design of tactile display systems for particular applications, including tactile channel devices designed to replace the visual or speech channel of sensory impaired persons.

5.2.3.1 Tactile Channel Sensitivity

The tactile channel can be used with either electrical or mechanical skin transducers. Electrical stimulation of the skin is termed electrocutaneous or electrotactile stimulation. The absolute threshold for electrotactile stimulation is very low, about 10^{-7} W-sec. Above-threshold stimuli require a display device to produce 0.17 to 2.9 mA of current (about 290 μW to 80 mW (Sherrick, 1984).

There are two potential problems associated with use of the electrotactile system as a display channel. The first problem is the small range of intensity from the absolute threshold to the pain threshold; that is, the function relating perceived intensity (sensory magnitude) to electric current intensity is quite steep. This limits the usable dynamic range for encoding electrotactile channel information; an operator may not be able to discriminate many levels of intensity. However, an advantage of this steep function is the relatively low amount of energy needed for stimulation at intense sensation levels. The second problem is high variability in the effect of a given stimulus depending upon the location of the electrode and the nature of the electrode–skin contact. In spite of these problems the electrotactile channel has not been ruled out as a practical display channel, for example, as a tactile aid for the deaf (Sherrick, 1984). More serious practical constraints exist in the use of thermal or chemical tactile transducer systems. For these reasons, this section emphasizes consideration of display transducers that apply mechanical pressure or vibration stimulation to the skin.

The major types of skin are hairy and glabrous skin. The glabrous skin is on the palm and the sole; most other skin surfaces contain hair follicles. A number of different mechanoreceptor organs are found in the skin, including free nerve endings (including those that surround the hair follicles), the Merkel "disks" near the hairs, and the Ruffini endings. Glabrous skin such as the fingertips include a high density of Meissner corpuscles; somewhat deeper in the skin and in the joints are the Pacinian corpuscles.

Verrillo (1966) has studied the vibrotactile sensitivity of the skin to different vibration frequencies (10 to 3000 Hz). Figure 5.2.7 illustrates how threshold skin displacement depends on vibrotactile transducer frequency, for different size contractor areas. The mid- to high-frequency regions of the curves are probably mediated by the Pacinian system; the receptor system or systems responsible for the low-frequency part of the curve have not been clearly identified (Gescheider, Verrillo, and Van Doren, 1982). Perceptible skin displacements in the mid-frequency region require about 0.1 μW of

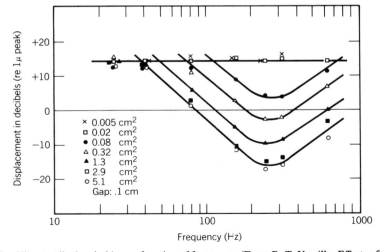

Fig. 5.2.7. Vibrotactile thresholds as a function of frequency. (From R. T. Verrillo, Effects of contactor area on the vibrotactile threshold, *Journal of the Acoustical Society of America*, 1966, *35*, 1965, and reproduced by permission.)

mechanical power applied over an area of 0.6 cm² at the fingertip (Sherrick, 1984). To stimulate the skin at levels from 10 to 40dB above threshold takes from 1 μW to 10 mW of mechanical power.

Sensitivity to cutaneous stimulation depends greatly on the part of the body stimulated. Weinstein (1968) has assessed aspects of our ability to detect, discriminate, and localize simple pressure stimuli applied to different regions of the male and female body. Figures 5.2.8 through 5.2.11 illustrate the results of his studies. Figures 5.2.8 and 5.2.9 illustrate absolute pressure thresholds for males and females. As might be expected, lowest thresholds are in the face area, followed by the fingers and upper body. Figure 5.2.10 illustrates two-point discrimination thresholds for males. These thresholds indicate the ability to distinguish a stimulus composed of two separated pressure points from a single pressure stimulus. Again, thresholds are lowest in the face and hand. Figure 5.2.11 shows point localization thresholds for males. These thresholds indicate how well a subject can localize the position of a point stimulus relative to a reference stimulus.

Cholewiak and Craig (1984) have examined the recognition, discrimination, and masking of vibrotactile spatial patterns presented to the finger, palm, and thigh. They reported that the type of display and the kinds of patterns affect the level of performance from these three sites. For example, if good response to single changes in long-duration patterns is desired, the thigh would be a good candidate for a display site; however, pattern recognition or discrimination would best be presented to the finger. In some applications, of course, the hands must be reserved for normal uses in sensing and manipulating the worker's environment.

Active Touch

The preceding discussion of tactile ability rests mainly on the results of methods that assess the sense of "passive touch," that is, stimulation presented passively to the skin of the hand or other body area. However, we usually employ the tactile sense modality in a quite different, "active" manner. People normally investigate and explore the surfaces of objects via pressing, rubbing, squeezing, and other actions. The result of these complex behaviors is a correlated stimulation of the internal senses that are sensitive to tendon and joint position and muscle state, as well as the skin receptor systems and the interaction between these various systems. For example, active touch is the mode of operation employed in the discrimination of different control or knob shapes. Gibson (1962) has argued that the ability to identify object shape is much better if an active, rather than a passive, touching mode

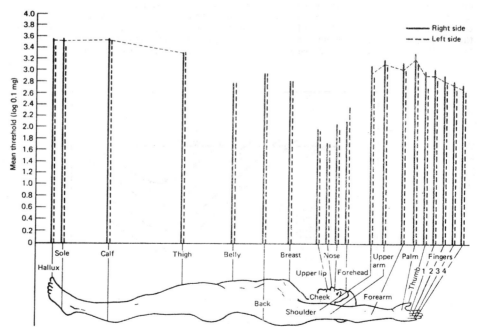

Fig. 5.2.8. Pressure sensitivity thresholds for males. (From S. Weinstein, Intensive and extensive aspects of tactile sensitivity as a function of body part, sex, and laterality, in D. R. Kenshalo, Ed., *The Skin Senses,* 1968, pp. 195–218. Courtesy of Charles C Thomas, Publisher, Springfield, IL.)

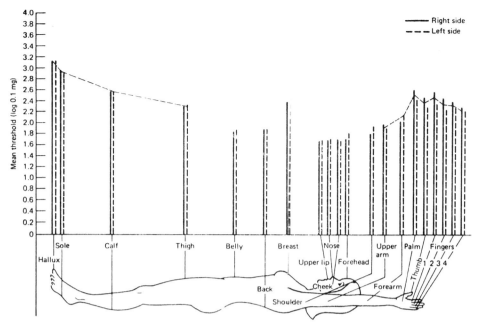

Fig. 5.2.9. Pressure sensitivity thresholds for females. (From S. Weinstein, Intensive and extensive aspects of tactile sensitivity as a function of body part, sex, and laterality, in D. R. Kenshalo, Ed., *The Skin Senses*, 1968, pp. 195–218. Courtesy of Charles C Thomas, Publisher, Springfield, IL.)

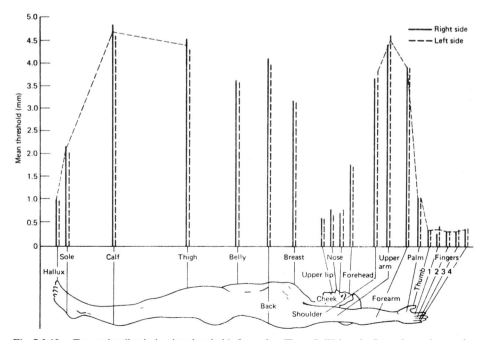

Fig. 5.2.10. Two-point discrimination thresholds for males. (From S. Weinstein, Intensive and extensive aspects of tactile sensitivity as a function of body part, sex, and laterality, in D. R. Kenshalo, Ed., *The Skin Senses*, 1968, pp. 195–218. Courtesy of Charles C Thomas, Publisher, Springfield, IL.)

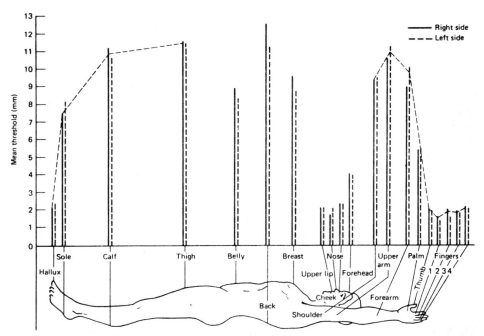

Fig. 5.2.11. Point localization thresholds for males. (From S. Weinstein, Intensive and extensive aspects of tactile sensitivity as a function of body part, sex, and laterality, in D. R. Kenshalo, Ed., *The Skin Senses,* 1968, pp. 195–218. Courtesy of Charles C Thomas, Publisher, Springfield, IL.)

is employed. Other investigators have disputed the idea that there is a true disparity in tactile system processing capacity in these two modes (Schwartz, Perey, and Azulay, 1975).

The active touch mode is also employed to obtain information about the properties of object surfaces. For example, it is often necessary for craftsmen to gauge the smoothness of work surfaces prior to application of a coating material. A technique employed by some craftspeople enables increases in the accuracy with which they may detect surface roughness. The technique is to place an intermediate sheet of paper between the fingers and the surface to be evaluated. Judgments of roughness are higher than if the fingers touch the surface directly. The reason for this phenomenon seems to be in the sensory masking or interference produced by lateral (shear) forces acting on the skin in the direct contact case. These shear forces interfere with the sensory signals conveying roughness and orientation. When a thin membrane is placed between the fingers and the surface, these shear forces are greatly reduced, and performance improves. Figures 5.2.12 and 5.2.13 illustrate results of experiments by Lederman (1978) on this phenomenon. Lederman showed that the size of the effect of the intermediate membrane depends on the magnitude of the shear force. The design of tactile displays requiring active touching movement by the user might be improved by use of a thin covering membrane between the display device and the hand.

When two tactile stimuli are offset in time, there may be complex effects on the resultant sensation. Two offset stimuli that are perceptually resolved when presented simultaneously may be fused when offset by a time duration of less than 2 msec. Greater time offsets may yield the sensation that the stimuli are closer spatially than when presented simultaneously. Successive stimulation of spatially separated sites can also produce very compelling movement (including gouging or hopping) sensations. These temporal and spatial phenomena may be significant causes of interference between the components of a dynamic, complex tactile display (Sherrick, 1984).

Another example of the type of interference that may exist between the components of a tactile display is in the use of tactile displays of graphic information for the blind. Area patterns or textures can act to interfere with the efficient extraction of information from such displays. For example, with a display consisting of a raised line graph plotted against a raised background grid there is an interfering effect of the reference grid on tracking the raised line. Barth (1984) found that depressing the grid ("incising" the grid lines) below the level of the surface improved graph reading performance significantly. This finding was in spite of the general result that raised line material generally leads to superior

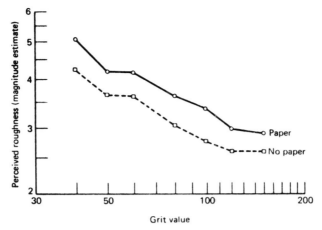

Fig. 5.2.12. Perceived roughness (magnitude estimates) as a function of sandpaper grit in paper and no-paper conditions. (From S. J. Lederman, "Improving one's touch" . . . and more. *Perception and Psychophysics,* 1978, *24,* 155. Copyright 1978 by the Psychonomic Society, Inc., and reproduced by permission.)

performance than incised line material. Isolating the principal and reference data on a tactile display via this technique minimized the negative effects of interference between these display components.

Shape and texture coding has been employed in many systems that employ knobs and controls. The military, for example, has standardized the control shapes used for different aircraft control functions such as landing gear and throttle. Figure 5.2.14 illustrates such controls. It is possible, by trial and error, to design a particular set of shape or texture-coded controls such that confusion between any two controls will be minimal. Figure 5.2.15 illustrates such a set of controls.

5.2.3.2   The Tactile Channel as a Supplementary Display Channel

The ability to use the tactile channel may be crucial to the performance of some tasks. For example, salvage divers are frequently required to "work in the dark" and rely heavily on their tactile sense.

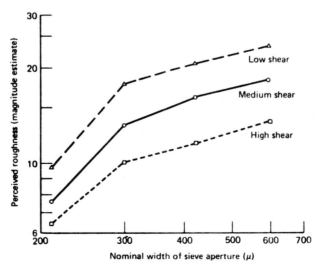

Fig. 5.2.13. Perceived roughness (magnitude estimates) as a function of nominal width of seive aperture and shear. (From S. J. Lederman, "Improving one's touch" . . . and more. *Perception and Psychophysics,* 1978, *24,* 158. Copyright 1978 by the Psychonomic Society, Inc., and reproduced by permission.)

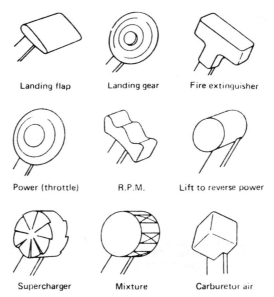

Fig. 5.2.14.   Standard aircraft controls using shape coding.

Banks and Goehring (1979) conducted a study of military salvage divers performing a standardized set of underwater tasks such as disassembling shackles and hatches. They investigated diver performance under four conditions of decreasing visibility and two different tactile conditions, wearing gloves and not wearing gloves. Wearing gloves had a clearly deleterious effect on performance; the effect increased as vision was degraded. Other studies of divers, including that of Curley and Bachrach (1981), have also demonstrated the importance of the tactile channel in underwater conditions of reduced visibility.

Fig. 5.2.15.   A set of shapes that can be discriminated by touch alone.

When the ordinary visual channel may be expected to be degraded, the system designer should make provisions for adequate employment of the tactile modality.

In systems placing very high information loads on the visual channel, the tactile (or auditory channel) may be employed as a supplementary display source. An example is in high-performance aircraft where very high visual and auditory work loads may occur. Several different types of tactile and electrotactile tracking displays were studied by Triggs, Lewison, and Sanneman (1974). They presented aircraft pitch and roll information via arrays of tactile and electrotactile stimulators. They found little difference between the electrotactile and vibrotactile two-dimensional tracking displays, but they did find that some forms of coding the pitch and roll information on the display were more effective than others. They also compared performance to that with a continuous visual display. Although performance was better on the visual display, the difference could probably be attributed to the time-sampled and amplitude-quantized nature of the tactile displays used, rather than to any inherent band-width differences between human visual and tactile tracking.

Hirsch (1974) studied a single-axis visual tracking task that simulated aircraft attitude control. Error rate information was supplied via two vibrotactile stimulators on the thumb and index fingers. He found that providing rate information improved performance over the visual display alone. Hirsch suggested that there are many practical control situations where a tactile display can be used to provide feedback otherwise unavailable to the operator.

Another example of providing a tactile display as a supplement to a high visual load control task is that employed by Jagacinski, Miller, and Gilson (1979). Figure 5.2.16 illustrates their tactual display and the control-display relationships. The tactile display is a variable height slide in the control handle; error is displayed as a proportional displacement of the slide either forward or backward on the stick. Movement of the stick in the appropriate direction eliminates the displayed protrusion as it reduces the control stick position error.

A novel type of tactile display consists of an air or fluid stream directed against the skin. Loomis and Collins (1978) studied the properties of a system consisting of a water jet directed against a thin watertight membrane between the water jet and the skin. This system allows the generation of a point stimulus moving over the skin with almost zero frictional force. They found extremely high sensitivities to small and rapid changes in the position of the point stimulus. Devices of this type may be a practical way to implement highly quantized and high-speed tactile display systems.

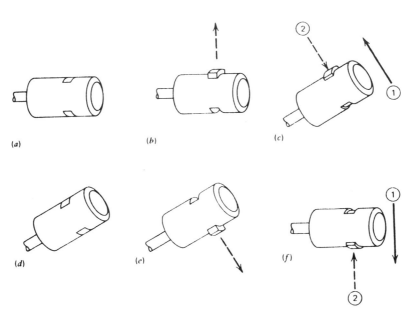

Fig. 5.2.16. Control–display relationships for tactual display studied by Jagacinski, Miller & Gibson (1979). The displayed error is indicated in (b) and (e). The control movement shown as (1) results in a decrease in the displayed error, (2). (From *Human Factors,* 1979, *21,* 80. Copyright 1979 by *The Human Factors Society,* Inc., and reproduced by permission.)

5.2.3.3 The Tactile Channel as a Sensory Replacement Channel

Perhaps the most interesting problems of tactile display design arise in the use of the tactile channel as a replacement or supplementary channel for the deaf or blind. Many systems have been developed for these purposes; this section summarizes some of the major design aspects of these display systems.

A variety of tactile and electrotactile systems have been developed for use by the profoundly hearing impaired but there has not yet been widespread production or use of such aids. This is in contrast to the relatively large number of patients who have had surgery for the implantation of a cochlear prosthetic device to stimulate (electrically) their auditory nerve directly. However, given the large population of people for whom implant surgery is not desirable, and given the developmental nature of cochlear implantation, an effective tactile aid is considered to be highly desirable (Sherrick, 1984). Indeed, a tactile aid for use by blind persons in reading ordinary printed text (the Optacon) is in widespread use.

Design Requirements

Sherrick (1984) has considered many of the major design requirements for tactile or electrotactile aids; his discussion provides useful information for the design of tactile displays for other specialized purposes. One requirement is for the aid to be small in size and weight. Tactile aids for the sensory impaired person will probably have to be worn on the body (or carried) and therefore must be rugged, relatively unobtrusive, and comfortable for long periods of usage. The probability of use by small children means that the display must be effective over a small skin area, as well as small in size.

In addition to the normal design requirements of low battery drain, low distortion, and appropriate frequency response, the system must be limited in its emission of acoustic energy. By their very nature, vibration displays generate acoustic as well as mechanical energy; this unwanted acoustic output may be a source of interference to persons or equipment near the display user. In the case of devices that directly code acoustic energy to tactile energy, acoustic feedback could cause the system to self-oscillate.

The display system should also have a wide dynamic range, perhaps 40 dB, to take advantage of the range of sensitivity of different parts of the body. Another important consideration is that the system be relatively insensitive to the contact pressure between the skin and the mechanical transducer. That is, the amount of static force on the transducer produced by the skin should not influence the output force of the transducer. In many applications the coupling between the display transducer and the skin will be highly variable; this should not affect the display's output characteristics.

A number of additional design questions involve the choice of the number of sites on the skin to be stimulated, the number (and range) of vibratory frequencies or channels, and the number (and range) of intensities to be discriminated at each vibration frequency. The other major question concerns the nature of the code that relates the original information to these intensive, spectral, and spatial dimensions. Various types of single and multiple channel tactile display systems and codes have been developed for the sensory handicapped. Some information is available on the effective information rates that have been achieved with these systems, but it is not yet possible to outline a general set of display specifications for the tactile display designer.

Tactile Systems for the Deaf

Tactile systems developed as aids to the deaf have included both single-site, single-frequency and multiple-site devices. Multiple-channel devices have coded speech frequency by the linear spatial location on the skin and speech power by vibration amplitude. That is, the amplitude of stimulation on different parts of the skin corresponds to the speech power in different frequency bands. Other multiple-channel devices have coded speech frequency and power as two orthogonal spatial axes on the skin. Brooks and Frost (1984) discuss an example of the former system, a tactile vocoder system using $16\frac{1}{3}$-octave filter channels each driving one of a 16 solenoid array placed on the subject's forearm. They found that subjects could learn reliably to identify many words of a small set of words after 40 to 55 hr of intensive training (70 words for one subject and 150 for another). However, the comparative value of different systems and design parameters (number of frequency channels, display size, etc.) has not been established.

More complex codes specific to the speech signal could be based on articulatory or phonetic aspects of the speech input. An articulatory code would encode and display characteristics of the input based on speech production. For example, one technique being employed to study use of the Tadoma method is the construction of an artificial speaker (head) that exhibits the mouth, jaw, throat movement, and air flow characteristics of a live speaker. "Speech" from a properly implemented artificial speaker of this type would be decipherable by a skilled Tadoma user. Of course, such an articulatory-based tactile display system would be useful only for speech sounds. A system based on the phonetic or information carrying speech units might encode and display the English phonemes and other aspects of speech stress, intonation, and timing. This type of system would probably be much more computation-intensive.

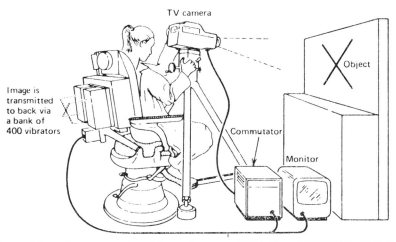

Fig. 5.2.17. Schematic representation of the tactile system, TVSS. (From B. W. White, F. A. Saunders, L. Scadden, P. Bach-Y-Rita, and C. C. Collins. Seeing with the skin, *Perception and Psychophysics,* 1970, 7, 23. Copyright 1970 by Psychonomic Journals, Inc., and reproduced by permission.)

Tactile Systems for the Blind

Some systems developed for presenting visual information to the blind convert an optical image of a scene into a tactile display presented to the subject's skin. In one system, the tactile vision substitution system (TVSS) shown in Figure 5.2.17, the tactile image is derived from a TV camera viewing the scene in front of the subject (White, Saunders, Scadden, Bach-Y-Rita, and Collins, 1970). The tactile display was an array of 20 X 20 vibrators on the subject's back. Subjects with this system were able

Fig. 5.2.18. Optacon tactual array. (Courtesy James C. Craig.)

Pisoni, P. B. (1982, March). Perceptual evaluation of voice response systems: Intelligibility, recognition, and understanding. *Workshop of Standardization for Speech I/O Technology.* Gaithersburg, MD: National Bureau of Standards, pp. 183–192.

Schwartz, A. S., Perey, A. J., and Azulay, A. (1975). Further analysis of active and passive touch in pattern discrimination. *Bulletin of the Psychonomic Society, 6,* 7–9.

Sherrick, C. E. (1984). Basic and applied research on tactile aids for deaf people: Progress and prospects. *Journal of the Acoustical Society of America, 75,* 1325–1342.

Sherrick, C. E., and Cholewiak, R. W. (in press). Cutaneous sensitivity. In K. Boff, L. Kaufman, and J. Thomas, Eds. *Handbook of perception and human performance.*

Simpson, C. A. (1976, May). Effects of linguistic redundancy on pilot's comprehension of synthesized speech. *Proceedings of the Twelfth Annual Conference on Manual Control,* Moffett Field, CA. National Aeronautics and Space Administration, Ames Research Center, NASA, TMX-73170.

Simpson, C. A., and Marchionda-Front, K. (1984). Synthesized speech rate and pitch effects on intelligibility of warning messages for pilots. *Human Factors, 26,* 509–518.

Simpson, C. A., and Navarro, T. (1984, May). Intelligibility of computer generated speech as a function of multiple factors. *Proceedings of the National Aerospace and Electronics Conference,* Dayton, OH.

Simpson, C. A., and Williams, D. H. (1980). Response time effects of alerting tone and sematic context for synthesized voice cockpit warnings. *Human Factors, 22,* 319–330.

Sorkin, R. D., and Woods, D. D. (1985). Systems with human monitors: A signal detection analysis. *Human–Computer Interaction, 1,* 49–75.

Swets, S. A., Green, D. M., Fay, T. H., Kryter, K. D., Nixon, C. M., Riney, J. S., Schultz, T. J., Tanner, W. P., Jr., and Whitcomb, M. A. (1975). A proposed standard fire alarm signal. *Journal of the Acoustical Society of America, 57,* 756–757.

Triggs, T. J., Lewison, W. H., and Sanneman, R. (1974). Some experience with flight-related electrocutaneous and vibrotactile displays. In F. Geldard, Ed. *Cutaneous communication systems and devices.* Austin, TX: Psychonomic Society, pp. 57–64.

Verrillo, R. T. (1966). Effect of contactor area on the vibrotactile threshold. *Journal of the Acoustical Society of America, 35,* 1962–1966.

Weinstein, S. (1968). Intensive and extensive aspects of tactile sensitivity as a function of body part, sex, and laterality. In D. R. Kenshalo, Ed. *The skin senses.* Springfield, IL: Charles C Thomas.

White, B. W., Saunders, F. A., Scadden, L., Bach-Y-Rita, P., and Collins, C. C. (1970). Seeing with the skin. *Perception and Psychophysics, 7,* 23–27.

White, F. A. (1975). *Our acoustic environment.* New York: Wiley.

Wightman, F. L., Kistler, D., and Walczar, N. (1983). Localization in man: Importance of outer-ear structures. *Journal of the Acoustical Society of America, 73,* 51.

CHAPTER 5.3
DESIGN OF CONTROLS*

HANS-JÖRG BULLINGER
PETER KERN
WERNER F. MUNTZINGER

Fraunhofer-Institut für Arbeitswirtschaft und Organisation, Stuttgart

---

* See also Chapter 11.4 on human factors in computer hardware, Chapter 11.5 on speech controls and displays, and Chapter 5.2 on design of auditory and tactile displays.

5.3.1 INTRODUCTION

Controls constitute interface elements in the human–machine system through which a human transfers mechanical energy or information to the technical system for performing automatic control functions. The human receives the haptic and proprioceptive information required to perform a task from the control. Design, arrangement, and task of the controls have a considerable influence on the strain to which humans are subjected as well as on the effectiveness and safety of the system. The influencing factors of the "operating effectiveness" are represented in Figure 5.3.1. Owing to the interaction between human, control, and technical system, it is obvious that controls cannot be regarded as machine elements, as is done in many standards and regulations.

The design, selection, and arrangement of the controls must be made with special consideration of the criteria of ergonomics. The design dimensions of the control such as shape, size, material, and surface as well as the control task must be compatible with the anatomical, anthropometric, and physiological marginal conditions of man. Anthropometric parameters must be taken into account in dimensioning the control (size, shape) and for positioning of the control [grasp area of the hand–arm system (HAS) and/or step area of the foot–leg system (FLS)]. Anatomical marginal conditions, such as the motion range of the joints, are of importance to the maximum actuating range, as are physiological data to the transmission of forces and moments. Furthermore, the human parameters that vary from one individual to the next must also be accounted for in the ergonomic design and arrangement of the controls. It is the intention of this chapter to give the design engineer of technical systems such as work equipment and vehicles, etc.—in which energy and/or information must be transferred from human to the technical system via controls—guidance in the selection and arrangement of controls. The ergonomically suitable control can be selected only after the design engineer has defined the control task. General requirements for the definition of the design dimensions—shape, size, material, and surface—that are of importance to the manufacturer of controls, are discussed in Section 5.3.3.

5.3.2 CONTROL CLASSIFICATION

As guidance for the design engineer in the selection of controls, a classification and evaluation of the wide variety of controls can only be made in connection with the control task and the coupling conditions that exist between the contact element of the control and the extremity involved.

5.3.2.1 Control Task

Resistance, accuracy, and *speed*—these are regarded as the most important parameters of the control task. These performance parameters are particularly influenced by the type of control and by the dimensioning and positioning of the control. The maximum isometric muscle forces of humans in the HAS and FLS motion range determined experimentally by Rohmert (1966), Caldwell (1959), and others constitute the basis for the transmission of forces to controls. The most important influencing variables regarding the maximum isometric actuating moments—besides human muscular capacity— are technical parameters such as force application point, direction of force, and the type of coupling. Actuation of a control should take place within the maximum force output range. In hand- and foot-operated controls, the actuating forces or moments can be increased if the direction of force transmission is chosen so that the body can find a support (backrest) or if it is possible to translate the body's weight into actuating forces. The maximum isometric actuating moments for spoked handwheels (Mainzer, 1982) are shown in Figure 5.3.2 by way of example. As can be seen from Figure 5.3.2, the body's weight can be applied if the handwheel is arranged at low positions, resulting in high actuating moments. However, it is possible to translate the body's weight effectively only if the handwheels have a large diameter; this is not so for small diameters, which can be seen from the rather flat curve in Figure 5.3.2.

The actuating moment must not be, however, the exclusive criterion for the arrangement of controls, because—as shown in the above example—the high actuating moment is produced only under an unfavorable posture, thus requiring additional static work. In the case of frequent actuation of the control, the actuation resistance, on account of the continuous performance limit of man, should not exceed 15% of the maximum force. If the actuation times are very short (≤ 5 sec), the resistance can account for approximately 50% of the maximum force. In designing safety equipment, it must be ensured that, regarding the force to be developed, even the weakest worker (fifth percentile) will still be able to actuate the control. When electric, pneumatic, and hydraulic servo-amplifiers are used, provision will have to be made for a resistance, to obtain a tactile and proprioceptive feedback (see Section 5.3.5) and to dampen disturbing influences caused by the operator and by the environment and imparted to the control.

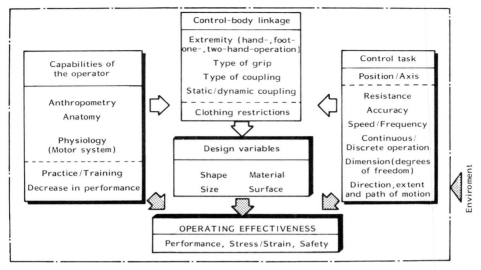

Fig. 5.3.1. Factors influencing operating effectiveness.

The actuating accuracy depends on the following factors: type of control, type of coupling, control dynamic, forcing function (type of track), motion range, environmental influences, and so on. The actuating speed depends primarily on the resistance and the type of control used. The highest speeds can be obtained over a large motion area by using a hand or finger crank, because no regrasping of the control will be necessary due to the static coupling. The dimensions, that is, the crank radius in the above example, and the arrangement of the control element are further parameters. The motion speed of the HAS, which depends on the direction of motion on a horizontal plane and is a function of the motion amplitude, approximately corresponds to an ellipse, owing to physiological and biomechanical marginal conditions (Schmidtke and Stier, 1960). These findings have shown that motions with the right arm can be made more quickly from bottom left to top right than motions from bottom right to top left, which can be attributed to the biomechanical properties of the elbow joint. Considering

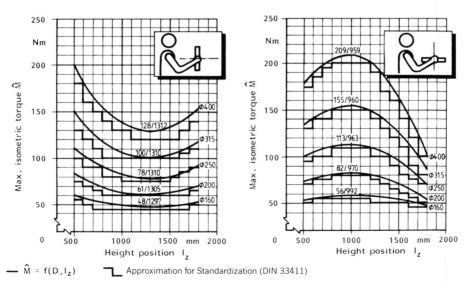

— $\hat{M} = f(D, l_z)$ ⌐ Approximation for Standardization (DIN 33411)

Fig. 5.3.2. Isometric torques of handwheels [$T = f$(control diameter, height, direction)]. (*Source:* Mainzer, 1982.)

the motion amplitude, this speed-to-accuracy relationship can be best described by the empirical relation of Fitts (1964) (Fitts's Law: $MT = a + b^2 \log(2A/W)$, where MT is movement time, A amplitude, W target size, and a and b constants). This relation is also applicable to the motions of controls.

5.3.2.2   Geometrics of Control Movements

The degrees of freedom of a control planned in the design are defined by the *dimensionality*. On the one hand, a multidimensional control task can be performed by several one-dimensional controls or, on the other hand, by one multidimensional control. Thus, for example, the longitudinal and transverse dynamics of a manually shifted motor vehicle is controlled by four one-dimensional controls and one two-dimensional gear-shift lever. However, it would also be possible to perform this task with one two-dimensional control lever, but this would entail greater technical complexity. In multidimensional control elements, problems are posed by the disturbance in the various directions.

According to the *type of movement,* a distinction is made between translationally actuated and rotationally actuated controls. Depending on the scope of movement, rotational (handwheel) and quasirotational control movements (lever) may occur. The performance parameters are essentially determined by the *direction of movement,* which results from the orientation of the control axis and the position of the control to the worker (front-, side-, height-position).

Information input by controls can be either *continuous* or *discontinuous (discrete)*, that is, in steps or stages. In case of digital input, it must be differentiated whether the control can assume two positions, using mechanical stops, or several positions. To set the intermediate positions, provision will have to be made for mechanical ratchets or for appropriate displacement-resistance characteristics, so that the position can be safely identified. For rotational one-hand control, the control positions should be located between 15 and 45°.

5.3.2.3   Control/Body Linkage

The worker is able to translate the information gained from mental processing into the FLS or HAS. On the one hand, the extremity to be used is governed by the specified performance parameters; on the other hand, in complex control tasks it must be ensured for ergonomic reasons that the stresses and strains to which the upper and the lower extremities are subjected are as balanced as possible. Generally speaking, foot-operated controls should be used only when actuating speed and accuracy do not play a major role. The *type of coupling* can be determined according to whether coarse-motor or sensory-motor control tasks are to be performed (for typology of types of coupling between hand and control, see Bullinger and Solf, 1979).

Clasping by the hand or foot should be aimed at for coarse-motor control tasks, whereas gripping by the fingers should be aimed at for sensory-motor control tasks. Gripping by the fingers with the fingertips as coupling on the control is especially advantageous in sensory-motor tasks, because it is here that the modal fields of the skin have the greatest tactile information capability, owing to the distribution of the peripheral receptors (Pacini's Meissner's, and Merkel's corpuscles) and the central neurones supplying this field. The scope of motion of the HAS is steadily decreased in the types of coupling—contact (touch) to grasp—because the joints of the hands and fingers cannot become active owing to the coupling. The movements are further reduced in the event of a two-hand coupling on the control. Analogous to hand-operated controls are different types of coupling between foot and control. Actuation is possible with the forefoot, with the heel, or with the whole foot. Here, the type of coupling is also governed by the control task to be handled.

The transmission of energy to the control element may be *positive* or *frictional*. In frictional coupling, the power is transmitted through the friction forces that exist between the control and the hand or foot, with the necessary normal forces produced by the coupling forces. In frictional coupling, the coefficient of friction that exists between control element and hand is important (see Section 5.3.3); this coefficient is determined by the material and the surface structure of the control element. A positive coupling permits greater actuating forces to be applied, but at the same time it also confines the possible movements of the HAS.

Frictional coupling is favorable for dynamic coupling on the control (regrasping). If the scope of movement of the HAS is not sufficient for the required actuating distance, *static coupling* changes to *dynamic coupling*. In dynamic coupling, regrasping is necessary on the control, so that the required movement can be performed. If the possibilities of movement are exhausted in static coupling, a larger scope of movement can be achieved only by movements of the trunk, resulting in additional stress and strain of the worker. The range of static movement of a rotationally actuated control by gripping it with five fingers is determined with approximately 3 rads, by the scope of pronation and supination movements of the proximal and distal radio-ulnar joint, with the axis of rotation arranged in the sagittal-horizontal axis. For dynamic coupling, special requirements must be made with respect to the design parameters shape, size, and surface (Kern, Muntzinger, and Solf, 1984).

On the basis of the variables discussed above, the most important hand- and foot-operated controls are represented in Figure 5.3.3. The controls have been arranged into three main groups, based on

Fig. 5.3.3. Hand- and foot-operated control devices and their operational characteristics and control functions. ○ Not suitable; ○ acceptable; ● recommended.

Path of C. motion	Control	Dimension [mm]	Force F [N] / Moment M [Nm] (D / R)	Force F [N] / Moment M [Nm] (M)	2 positions	>2 positions	Continuous adjustment	Precise adjustment	Quick adjustment	Large force application	Tactile feedback	Setting visible	Accidental actuation
Turning movement	Handwheel 	D : 160 – 800 d : 30 – 40	D 160 – 200 mm 200 – 250 mm	M 2 – 40 Nm 4 – 60 Nm	◐	◐	●	●	◐	●	○	○	◐
	Crank 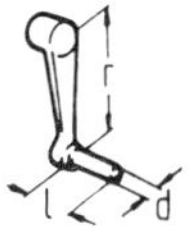	Hand (Finger) r : <250 (<100) l : 100 (30) d : 32 (16)	R <100 mm 100 – 250 mm	M 0,6 – 3 Nm 5 – 14 Nm	◐	◐	●	◐	●	◐	◐	◐	○
	Rotary knob 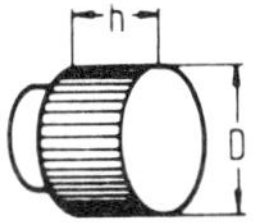	Hand (Finger) D : 25–100 (15–25) h : >20 (>15)	D 15 – 25 mm 25 – 100 mm	M 0,02 – 0,05 Nm 0,3 – 0,7 Nm	◔	◐	●	●	●	○	○	○	◐
	Rotary selector switch 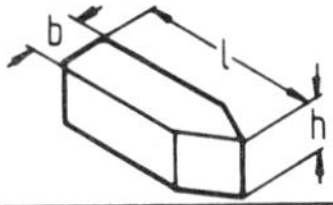	l : 30 – 70 h : >20 b : 10 – 25	D 30 mm 30 – 70 mm	M 0,1 – 0,3 Nm 0,3 – 0,6 Nm	●	●	◔	◔	◔	◐	◐	●	●
	Thumbwheel 	b : >8	0,4 – 5 N		◐	◐	●	●	●	○	○	○	●
	Rollball 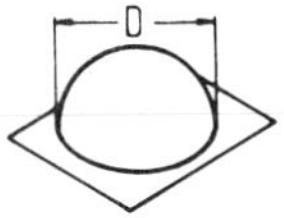	D : 60 – 120	0,4 – 5 N		○	○	●	●	◔	○	○	○	●

Path of C. motion: **Swivelling movement**

Control	Dimension (mm)	Force F [N] Moment M [Nm]	2 positions	> 2 positions	Continuous adjustment	Precise adjustment	Quick adjustment	Large force application	Tactile feedback	Setting visible	Accidental actuation
Lever 	d : 30 – 40 l : 100 – 120	10 – 200 N	●	●	●	◕	◐	●	◐	◐	○
Joystick 	s : 20 – 150 d : 10 – 20	5 – 50 N	●	●	●	●	◐	◔	◐	◐	○
Toggle switch 	b : >10 l : >15	2 – 10 N	●	◐	○	○	●	○	●	●	○
Rocker switch 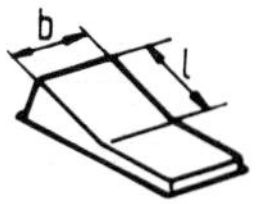	b : >10 l : >15	2 – 8 N	●	○	○	○	●	○	●	●	◕
Rotary disk 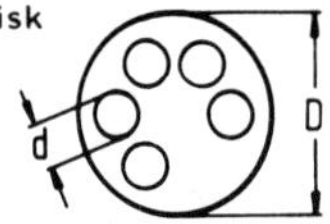	d : 12 – 15 D : 50 – 80	1 – 7 N	●	◕	◔	○	◔	○	○	○	◕
Pedal 	b : 50 – 100 l : 200 – 300 l : 50 – 100 (Forefoot)	Sitting : 16 – 100 N Standing : 80 – 250 N	◔	◔	◐	◐	●	●	◐	○	○

Fig. 5.3.3. (*Continued*)

582

Path of C. motion	Control	Dimension (mm)	Force F [N] / Moment M [Nm]	2 positions	>2 positions	Continuous adjustment	Precise adjustment	Quick adjustment	Large force application	Tactile feedback	Setting visible	Accidental actuation
Linear movement	Handle (Slide) 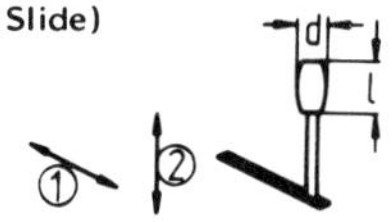	d : 30 – 40 l : 100 – 120	F_1 : 10 – 200 N F_2 : 7 – 140 N	●	●	●	◕	◐	●	◐	◐	○
	D-Handle 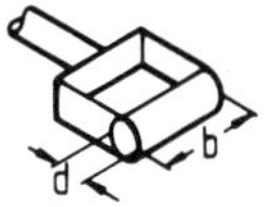	d : 30 – 40 b : 110 – 130	10 – 200 N	●	●	●	◔	◔	●	◔	◔	○
	Push button 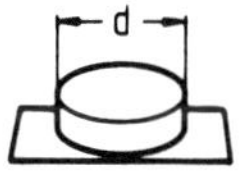	Finger : d >15 Hand : d >50 Foot : d >50	Finger : F = 1 – 8 N Hand : F = 4 – 16 N Foot : F = 15 – 90 N	●	○	○	○	●	○ ◐ ●	○	○	●[b] ○
	Slide 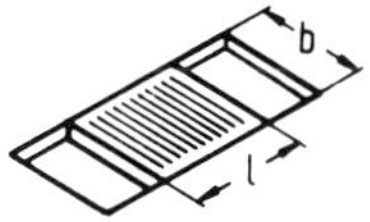	l : >15 b : >15	1 – 5 N (Touch grip)	●	◕	◐	◐	◐	○	○	◕	●
	Slide 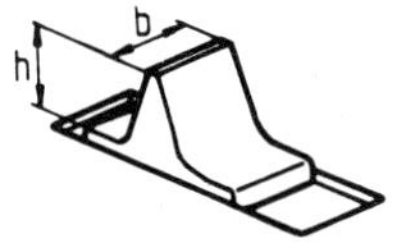	b : >10 h : >15	1 – 10 N (Thumb-finger grip)	●	◕	◔	◕	◐	◐	○	◕	◐
	Sensor key 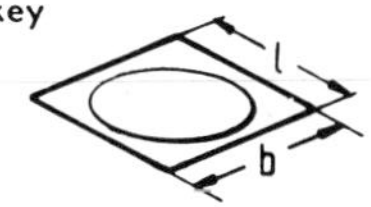	l : >14 b : >14		●	○	○	○	●	○	○	○	◐

[b] Recessed installation of the control

Fig. 5.3.3. (Continued)

583

whether a rotational, quasirotational (swiveling), or translational movement of the control is involved. Besides the characteristic dimensions, the figure also includes information regarding the permissible actuating forces. The evaluation of the controls with respect to the criteria—actuating movement (continuous, discrete), accuracy, speed, and so on—is made by means of qualitative statements.

The large number of ergonomic findings available in various forms dealing with the properties of hand and foot controls often make it difficult for the design engineer to find the optimum control for specific requirements. Figure 5.3.3 is based on the most important ergonomic findings relevant to controls. The selection is aggravated by the interactions that exist among the different variables. The assistance of a computer in the selection of controls permits an easier, better, and faster selection of the control most favorable to a given control task. All properties and dimensions of the various controls are stored in a data base. An algorithm developed by Pulat (1980) for the selection of controls divides the total benefit of a control into a task-specific benefit (force, accuracy, speed, etc.) and into a non-task-specific benefit (space required, coding, etc.). If the benefit of a characteristic is below the benefit requested by the design engineer, the total benefit will become zero. The control with the greatest total benefit is regarded as optimal for the requirements made by the design engineer. The *computer-assisted selection* will be extended by a *computer-assisted arrangement of controls and displays* in the future.

5.3.3   DESIGN PARAMETERS OF CONTROLS

5.3.3.1   Conventional Control Elements

The general design principles and the marginal conditions to be taken into account are discussed now for some controls, by way of example (Figure 5.3.3). The following information is particularly important to the manufacturer of controls. As shown in Figure 5.3.1, the design variables—*shape, size, material,* and *surface*—are important factors influencing the operating effectiveness, with the characteristics performance (force, speed, accuracy), stress and strain of the worker, and safety criteria. For defining the design variables, the control task, the coupling conditions, and the capabilities and traits of humans are important. Important influencing factors for the characteristics of the longitudinal and cross-sectional form of the control are mainly the anatomy and the type of coupling, and the anthropometry of fingers and hand for the dimensions of the control.

The positioning of the control relative to the worker (side, top, front, orientation) is regarded as a further influencing variable with respect to the shape of the control. Because controls are manufactured as standardized elements without accounting for the position to be determined by the design engineer, it is necessary that the shape of the control should satisfy the ergonomic criteria for all prevailing positions. For these reasons, anthropomorphic shapes of controls with recessed grips for the phalanges are not suitable as standard controls. Anthropomorphic shapes of controls are also unfavorable for dynamic coupling on the control, because access to the control is possible only at quite specific points, thus impairing quick regrasping on the control. Further design hints regarding shape and size are now given for one-hand-operated controls—for both coarse- and sensory-motor control tasks—and for one-hand- and two-hand-operated disk handwheels. Details on material and surface are also given. These details are of general validity and can also be applied to other controls illustrated in Figure 5.3.3.

For one-hand operation, an approximate ellipsoid of revolution has proved to be the optimum shape for the transmission of large moments (Kern et al., 1984). The shape of the control illustrated in Figure 5.3.4 permits the fingers and palm to couple on the rear and front sides of the control and thus great actuating moments can be produced. This shape of the control element applies to a diameter range of between 70 and 100 mm. Control diameters above 100 mm do not permit a further increase in the actuating moment, because the end phalanges can no longer couple on the rear side of the

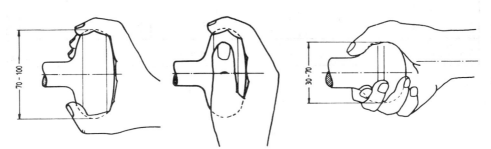

Fig. 5.3.4.   Shapes for one-hand-operated controls.

control element. As shown in Figure 5.3.4, this shape is very favorable for the arrangement of the control element in both the sagittal-horizontal axis and in the horizontal-frontal axis. For the diameter range of 30 to 70 mm, the shape of the control changes, because the gripping by five fingers (with the fingers equally spaced) changes to a four-finger grip (thumb extended) (Figure 5.3.4, right).

Extra-fine profiles have proved to be advantageous with respect to the transmission of power under unfavorable surrounding conditions such as when the hand or the control is very dirty. Figure 5.3.5 shows the maximum actuating moments of one-hand-operated controls in relation to the shape of the control under favorable and unfavorable surrounding conditions (oil-contaminated). As the figure shows, the greatest actuating moment can be transmitted with the positively actuated dihedral part C_2 and the profiled control C_5, when the controls and/or the hands are dirty. In dynamic coupling on the control, the unfavorable regrasping in C_2 is regarded as a disadvantage. When controls are used in which heavy dirt can be expected, it must be ensured that the shape of the control is so defined that the deposit of dirt and fluids will not be possible in the coupling zone. Under favorable surrounding conditions, the control C_6 coated with a pressure-anthropomorphic material will be advantageous (coat thickness 6 mm, 55 Shore). Highly elastic material surfaces are used so that the shape of the control can be constantly adapted to factors specific to a given situation, such as variation of the anthropometric dimensions of the hand or different coupling conditions, depending on position and arrangement. As a result of the partial adaptation of the shape of the control by the coupling forces, a large coupling area is ensured, permitting great static and dynamic actuating moments.

Sensory-motor control tasks with low resistance should be performed by gripping with five fingers. Coupling of the controls on the fingertips is necessary in fine-motor control tasks, because this is where the skin has the highest tactile sensitivity. The front and rear sides of the control are not used as coupling area. In a diameter range of from 50 to 125 mm, a cylindrical disk with a thickness of 20 to 25 mm is a suitable shape. Smaller controls are primarily actuated by the fingers.

Designing the shape of one-hand- and two-hand-operated disk handwheels is regarded as problematic. For safety aspects, disk handwheels must have a closed stay. The relationship between maximum isometric actuating moment and the rim diameter has been approximated on the basis of experimental tests by a regression function of the second degree for different control arrangements (A_x sagittal-horizontal axis; A_y horizontal-frontal axis; $l_{z\,K}$, $l_{z\,E}$, $l_{z\,S}$, positioning at knee, elbow, and parting level) (Kern et al., 1984). The regression function permits a rim diameter of approximately 45 mm to be derived ($dM/dd = 0 \rightarrow d_{\text{opt}}$). The results are shown in Figure 5.3.6. For reasons of economy and weight, the calculated rim diameter cannot be realized. Considerable savings of material and weight can be achieved by taking suitable design measures in which the stay of the control can be used as coupling area (control shape; see Figure 5.3.7). In the design, it will also have to be taken into account that the shape must satisfy criteria of ergonomics and safety for all relevant positions and arrangements. As shown in Figure 5.3.7, the shape of the control satisfies ergonomic criteria for both two-hand operation in the sagittal-horizontal axis and for one-hand operation in the horizontal-frontal axis; that is, a large coupling area and thus minimum strain of the hand are guaranteed in all positions.

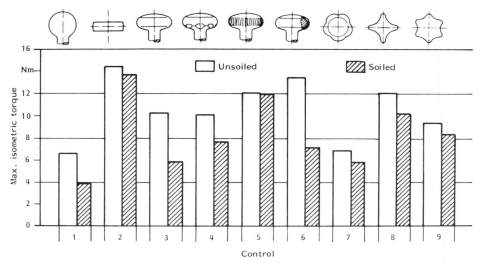

Fig. 5.3.5. Torques exerted at various control alternatives ($D = 80$ mm) under soiled and unsoiled conditions. (*Source:* Kern, Muntzinger, and Solf, 1984.)

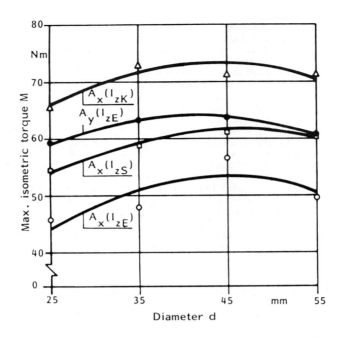

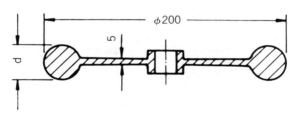

Axis	Posi-tion	Constant $(\hat{M} = a + b \cdot d + c \cdot d^2)$			$\hat{M}_{max}$ (Nm)	d_{opt} (mm)
		a	b	c		
A_x	I_{zK}	36,85092	1,64992	-0,01890	72,85	43,6
A_x	I_{zE}	9,89322	1,92908	-0,02153	53,10	44,8
A_x	I_{zS}	31,02859	1,25693	-0,01299	61,43	48,4
A_y	I_{zE}	34,66092	1,41146	-0,01709	63,80	41,3

Fig. 5.3.6. Torques exerted at handwheels (disk type) with different rim diameters. (*Source:* Kern, Muntzinger, and Solf, 1984.)

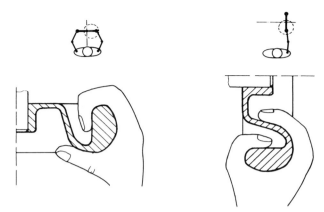

Fig. 5.3.7. Shape for handwheels (disk type) and coupling conditions for different handwheel arrangements. (*Source:* Kern, Muntzinger, and Solf, 1984.)

In the case of frictional coupling of the extremity with the control, the power is transmitted by friction forces. The magnitude of the friction force F_F can be determined by the law of friction ($F_F = \mu \cdot F_N$, where μ is the coefficient of friction and F_N the normal force). In a frictional coupling, the normal force will be developed by the coupling force of the hand or fingers. The coefficient of friction is determined by anatomical-physiological parameters such as size of the coupling area, surface structure, and the skin's degree of moisture on the one hand, and by the material's properties such as surface roughness and profiles on the other hand. In frictional coupling—contrary to positive power transmission to the control—the material must also be selected under the aspect of the frictional behavior that exists between hand/fingers and the control. The reason why special importance is attached to the selection of the materials is that unsuitable materials and surface structures of controls very quickly lead to heavy strain and destruction of the upper skin layers. Figure 5.3.8 represents the coefficients of friction—standardized to the material of plexiglass—of 29 common materials for hand contact (Bullinger, Kern, and Solf, 1979). To determine the coefficient of friction of the materials, the samples were moved over the stretched hand at a defined normal force of 40 N. Figure 5.3.8 shows at the top the coefficients of static and sliding friction (μ_1, μ_2). The deformation of the skin until the setting in of sliding friction is shown at the bottom. For smooth materials, a high correlation could be seen between the surface roughness and the coefficient of friction. When the control elements are actuated with the fingers, the friction force, among other things, depends on the direction. If the friction force is transmitted along the fingers, there will be smaller coefficients of friction than in force transmission at right angles to the fingers. In translationally actuated controls, the force will be transmitted longitudinally and transversely, and predominantly at right angles to the fingers in rotationally actuated controls.

If the surface of the control is profiled, the size and form of the profiles and the profile spacing will be important. Moreover, the direction of the profiles relative to the hand or the fingers is also relevant for power transmission. Because the strain on, and the danger of injury to, the skin becomes higher with increasing profile size, only fine profiles (profile spacing < 3 mm) are permissible for control-element surfaces if the control must be operated with wet, oily, or dirty hands. In this case the profiles must be vertically orientated to the direction of force. Figure 5.3.9 shows the relative coefficients of friction versus the form and direction of profiles. As shown by the results, the effectiveness of the profiles over smooth surfaces depends heavily on the normal force. Under smaller loads per unit area (normal forces), profiled surfaces show smaller coefficients of friction than smooth surfaces. The effective coupling area, and thus the adhesive force between hand and material, is considerably reduced by the profiles. Under higher loads per unit area, greater coefficients of friction result in the case of profiled surfaces, owing to a certain interlocking between skin and profiles.

During experimental tests in which the effectiveness of the different profiles was checked on real control operations, the subjects had to perform a pursuit tracking task (Kern et al., 1984). The control element (diameter = 70 mm, length = 80 mm) was alternately operated by two hands over a movement range of five revolutions (see Figure 5.3.10). The tracking error was used as an evaluation variable for the control quality, and the rate error as a measure for the continuity of the control movement. Figure 5.3.11 shows the tracking error relative to the shape of the control, its profile, and the resistance. A significant influence could be evidenced regarding the profile and shape of the control. The largest tracking error was noted for the hexahedral part C_2 over the total resistance range. The nonprofiled cylinder C_1 is favorable only if the resistance is very low. Minor errors arise in the profiled controls

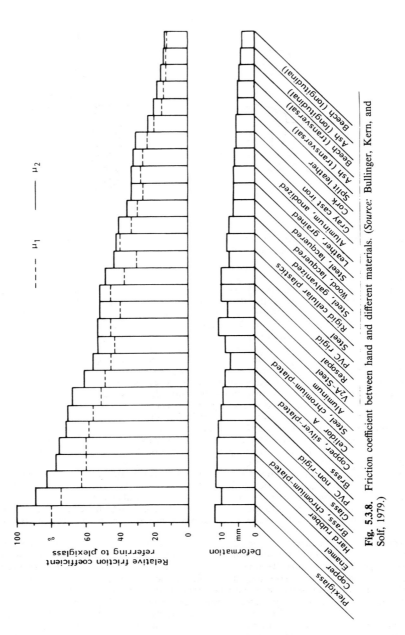

Fig. 5.3.8. Friction coefficient between hand and different materials. (*Source:* Bullinger, Kern, and Solf, 1979.)

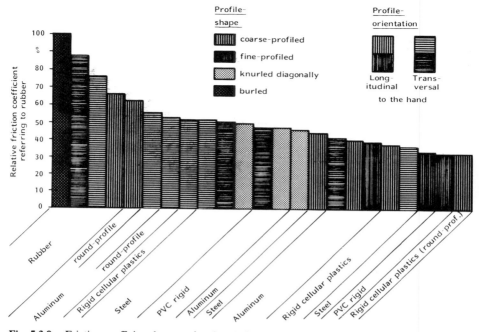

Fig. 5.3.9. Friction coefficient between hand and different surface profiles. (*Source:* Bullinger, Kern, and Solf, 1979.)

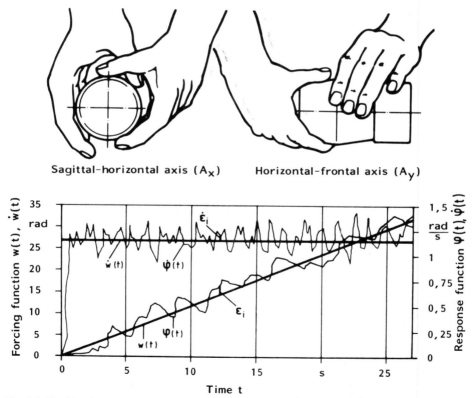

Sagittal-horizontal axis (A$_x$)   Horizontal-frontal axis (A$_y$)

Fig. 5.3.10. Two-hand-operated control and the forcing/response function of the pursuit tracking task. (*Source:* Kern, Muntzinger, and Solf, 1984.)

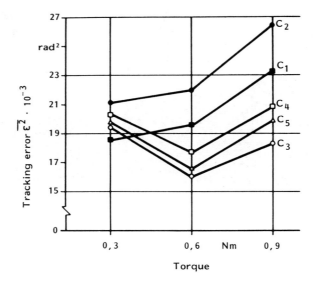

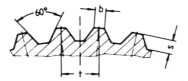

Control	Parameters of the surface		
	t (mm)	b (mm)	s (mm)
C_1	Cylinder	($R_Z = 1,27\,\mu m$)	
C_2	Hexahedral control	($R_Z = 1,51\,\mu m$)	
C_3	1	0,4	0,6
C_4	3	0,6	1,2
C_5	6	0,8	2,2

Fig. 5.3.11. Effects of control surface and resistance on tracking error. (*Source:* Kern, Muntzinger, and Solf, 1984.)

at higher resistances (0.4 to 0.9 Nm). As can be seen in Figure 5.3.11, there is a characteristic pattern for the profiled controls over the resistance range. The smallest tracking error can be seen in the medium resistance range. At high resistances, the abrupt movements due to static friction can be seen as a cause for the low control efficiency and, at low resistances, the missing proprioceptive information and the transfer of inadvertent movements (also compare Seibt, 1971).

If relative movements occur between control and hand during operation, the *thermal conductivity* of the control material will also have an influence on the operating effectiveness. The thermal conductivity of relevant control materials may vary between 0.15 W/Km (PVC rigid) and 70 W/Km (steel). At high actuating speeds, the frictional heat developed—for example, by operating a crank with a fixed handle—can be dissipated only via the hand, which may result in an inadmissibly high temperature rise in the hand. In static coupling of the hand/fingers on the control, materials with low thermal conductivity are required—particularly at low ambient temperatures—so that the hand's temperature will not be imparted to the control too rapidly, resulting in hypothermia of the hand.

If *gloves* have to be worn for performing control tasks, this must be accounted for in the dimensioning of the controls. Provision will have to be made for a plus or minus allowance of approximately 10% for inner and outer dimensions. With respect to the degree of fulfilling the control task, a reduced tactile sensitivity and lower mobility of the hand must be expected in sensory-motor control tasks when gloves are worn (Taylor and Berman, 1982). Similar restrictions regarding the actuating performance must be expected if foot controls are operated by a person wearing *heavy boots*.

5.3.3.2 Specialized Input Devices

Various special controls have been developed along with the increasing number of computer-aided work systems. The selection of these information input elements such as keyboard, isometric and isotonic joystick, roll ball, mouse, light pen or gun, touch-sensitive display, and graphic tablet (digitizer) must be oriented toward the kind of information input, for example, data input, drafting, or cursor positioning.

The keyboard is exclusively suitable for the input of alphanumeric data. For cursor positioning, the direct input with a joystick, for example, is much faster than positioning by keys. For positioning tasks, the mouse has slight advantages over the joystick (Card, English, and Burr, 1978). The relatively large movement area is regarded as a disadvantage of the mouse, because this requires an additional work area. In a large number of inputs by a light pen, an increase in the stress and strain of the HAS of the operator can be expected, owing to static hold energy. For the input of information via touch-sensitive displays, provision will have to be made for fields of 12 × 12 mm with a spacing of 14 mm for static systems (control rooms, control panels, etc.), and fields of 17 × 17 mm with a spacing of 19 mm for dynamic systems (vehicles), owing to the anthropometry of the fingers (Rühmann, 1984). As in the case of light pens, heavy fatigue in the arm can be expected over longer input times. In the operation of touch-sensitive displays, Rühmann (1984) has found only a negligible reduction in the operating accuracy under the influence of mechanical vibrations.

Because isometric joysticks are especially suitable for rate control systems and higher-order control systems, and isotonic joysticks for position control systems, an isotonic or a spring-centered control with a low spring rate should be used for cursor control, which constitutes a discrete positioning task. The joystick should be operated by grasping with two or three fingers. The roll ball can also be used for two-dimensional cursor control. The advantage of the roll ball is that the control sensitivity can be adjusted over a wide range. Experimental tests conducted with an isotonic joystick, a roll ball, and a finger-actuated electronic tablet have shown that the electronic tablet is more favorable for graphic input than the joystick and the roll ball (Pitrella and Holzhausen, 1982). No significant difference could be seen between the joystick and the roll ball. For cursor positioning tasks, there were also no significant differences among the three alternative input systems. For further input systems for the human–computer interface, also see Chapter 11.4. Provision must also be made for further special controls for persons whose myodynamic capacity or anatomical mobility is restricted (compare Chapter 11.5).

5.3.4 ARRANGEMENT OF CONTROLS

The arrangement of controls is governed by human capabilities and traits such as anatomy, anthropometry, and physiology, on the one hand, and by the characteristics of the technical system to be manipulated, on the other hand. The most important criteria to be taken into account in the arrangement of controls are the following:

Movement range of the HAS or FLS

Movement-physiological marginal conditions (requirements made on actuating accuracy, speed, force, torque)

Coupling conditions

Frequency and importance of information input

Possibilities of visual feedback

Sequence of the process to be controlled (sequence of activities)

Spatial compatibility regarding the technical system or the displays

Safeguards against inadvertent operation

Operation while sitting or standing

As to the first two aspects, it must be considered whether the controls will have to be actuated by women or men, or by both, and what percentile range will be relevant to the collective (grasping reach capability of the HAS and FLS for the different percentiles of men and women; see Chapter 5.4. For the transmission of greater forces and moments, for example, by means of handwheels, provision will have to be made for operation while standing, and for informative control tasks, provision will have to be made for operation while sitting. This largely eliminates the need for static posture energy. Depending on the type of control used, its preferred vertical position will be between the elbow level and the shoulder level. Rotational controls actuated by both hands should be positioned in the median plane; one-hand actuated controls should be positioned in the sagittal plane (shoulder joint). Controls that must be actuated very accurately should be arranged in the grasping reach of the forearms. An armrest will increase the accuracy of control movements.

In the design of control panels and control rooms having a large number of controls and displays, further criteria will have to be taken into account to minimize unnecessary mental processes and to avoid time-consuming successive movements. When arranging/grouping several controls, the interaction that exists between these controls (functions) will have to be considered in addition. A distinction can be made between two principles. In a grouping according to the sequence of the process to be controlled, the controls will have to be arranged in the order of their operation. The arrangement in this *sequential grouping* should be made for an operating sequence from left to right or from top to bottom. In a *functional grouping*, the controls with the same functions will have to be clustered, with the spatial compatibility having to be considered in addition (e.g., left switch for left turbine). Further improvements regarding the operating effectiveness can be expected if the *operating frequency* is also used as an arrangement criterion. Thus frequently used controls should be arranged in the central movement or actuating range and infrequently operated controls in the peripheral range. Vital controls, such as emergency stop switches, must also be arranged in the central area.

To permit a perceptive and cognitive organization of the visual field, controls should be combined to make up a matrix-like control field (Neumann and Timpe, 1976). The grouping effect can be achieved by suitable spacings and/or by color marking and boundary lines. If two controls have to be operated simultaneously, the controls will have to be grouped in such a way that operation with different extremities (right hand–left hand, foot–hand) is possible. The spatial correlation between controls and displays should also be ensured. The direct spatial allocation of displays and controls is optimal. The control is located directly below or adjacent to the associated display. If this allocation cannot be implemented for reasons of accessibility of the controls, it is recommended that a separate display and control field be used. The spatial organization of the controls within the grasping reach of the operator must be identical with the spatial organization of the displays within the visual field. For the design of control panels, vehicle cabins, and so on, also see specialized literature, such as Schmidtke and Rühmann, (1978) and Chapter 5.4. For the arrangement of flight control and navigation instruments in aircraft cockpits, there are various standards, such as the "basic-t grouping," that have to be observed.

To permit errorfree operation of the controls without inadvertently actuating any neighboring controls, certain distances between the controls must be observed, depending on the type of control used. In Table 5.3.1, the minimum and, at the same time, optimum distances between two neighboring controls are listed for the most important controls (Applied Ergonomics Handbook, 1974; Grandjean, 1979). If the operator wears gloves, provision will have to be made for a corresponding allowance. If the rear side has to be used as coupling area in one-hand- and two-hand-operated handwheels, a free space of between 20 and 35 mm will have to be provided between the rear side of the control and the technical system, depending on the control element involved.

Integrated controls can be of advantage for different reasons, such as for reducing the number of controls, of simultaneous actuation of different functions. For simultaneous actuation of two functions (continuous and discrete), a lever will be suitable, for example, that is operated by grasp, and that has an integrated rocker switch or push button to be operated with the thumb. If different control sensitivities are required for coarse and fine adjustment, knobs of different diameters can be arranged on the same axis of rotation. However, provision should not be made for more than three concentric knobs. For dimensioning of three concentric knobs, see McCormick (1976). To perform satisfactorily

Table 5.3.1. Space Between Adjacent Controls

Control	Extremity	Distance between Controls (mm)	
		Minimum	Optimum
Push button	Finger	20	50
Toggle switch	Finger	25	50
Lever	Hand	50	100
	Both hands	75	125
Handwheel	Both hands	75	125
Knob and rotary selector switch	Hand	25	50
Pedal	One foot	50	100

Source: Grandjean, 1979.

the control tasks "quick turning" over a wide actuating range and "precise actuation," a handwheel will be suitable that is provided with a crank handle, which may be collapsible.

It is also possible to combine controls with scales and displays. For control tasks with several discrete positions, the control often assumes the shape of a hand (rotary selector switch). The positions are coded on the panel with digits, with a reasonable contrast being required between panel, hand (control), and digits. If the number of control positions ≤12, the coding should correspond to that of a watch. This permits the operator to perceive the control position quickly, especially if there are several controls. Another variant is the use of a disk with a scale fixed to the rear side of the control and a marking on the panel. The dimensions must be so defined that the view of the scale or display cannot be obstructed by the fingers during operation.

5.3.5 CONTROL DYNAMICS

There are many technical systems in which a direct linkage of control and system is not possible, because the actuating forces required for controlling the system do not lie within human capabilities. Therefore, the system is often controlled indirectly by means of suitable servo-amplifier devices. The decoupling of the control and the system permits the displacement-resistance characteristic to be optimized and it thus permits an improved utilization of the proprioceptive and haptic information. Proprioceptive perception is made through the sensors (muscular spindles, Golgi's corpuscles, etc.) of the extrafusal muscles, the tendons, and the joints. By an existing actuation resistance, the displacement feedback is supplemented by an additional force feedback. Resistance is also necessary to reduce control errors resulting from tremor of the hands and mechanical vibrations. The amount of resistance can be derived from the measurable status variables of the system or from simulated data of the system model. Owing to the immense technical complexity of such active controls, the generation of a control "feel" can often be realized only by displacement, rate, and acceleration-proportional actuating forces. When movement-proportional actuating forces (inertia, damping, elastic force) and a control sensitivity V are used, the control element transfer function $H_C(s)$ illustrated in Figure 5.3.12 is obtained.

A nonlinear transfer characteristic is also possible if, for example, static and sliding friction or backlash exists in the control. Generally speaking, the transfer characteristics of the system $H_S(s)$ will have to be taken into account for achieving an optimum adaptation of the control element dynamics.

5.3.5.1 Linear Mechanical Transfer Characteristics

A high degree of *inertia* of the controls causes a great time constant and impairs quick directional changes and accurate control movements. Owing to the existing inertia of the arm, Poulton (1974) does not think it necessary to apply additional inertial forces in the operation of joysticks. In connection with other dynamic resistances, particularly with a spring-centered control, the inertia also impairs the manual tracking performance. In rotationally actuated controls, however, inertia may improve the continuity of the control process at the expense of a reduced adjusting speed.

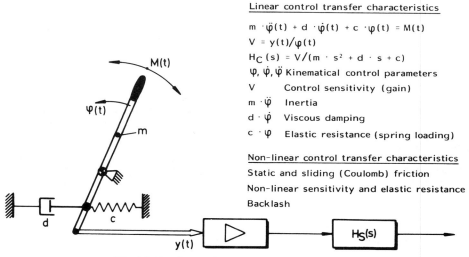

Linear control transfer characteristics

$$m \cdot \ddot{\varphi}(t) + d \cdot \dot{\varphi}(t) + c \cdot \varphi(t) = M(t)$$
$$V = y(t)/\varphi(t)$$
$$H_C(s) = V/(m \cdot s^2 + d \cdot s + c)$$

$\varphi, \dot{\varphi}, \ddot{\varphi}$ Kinematical control parameters

V Control sensitivity (gain)

$m \cdot \ddot{\varphi}$ Inertia

$d \cdot \dot{\varphi}$ Viscous damping

$c \cdot \varphi$ Elastic resistance (spring loading)

Non-linear control transfer characteristics

Static and sliding (Coulomb) friction

Non-linear sensitivity and elastic resistance

Backlash

Fig. 5.3.12. Types of control transfer characteristics.

Rate-proportional resistance (*viscous damping*) generally increases the control performance by smoothing the actuating movement and, like all dynamic resistances, reduces human disturbances (tremor, inadvertent movements of the hand) and environmental disturbances (mechanical vibrations, gravitational forces). Viscous damping has proved to be most effective in connection with elastic force, and most unfavorable in combination with inertia. A favorable effect of viscous damping is particularly achieved in connection with controlled systems of zero order ($H_S(s) = K$) (Rühmann, 1978). In a two-dimensional step tracking (position control system), Kraiss (1970) has been able to evidence that the damping of a joystick, in contrast to elastic forces, has an insignificant influence on the fine movement structure. In controlled systems of higher order, a control damped by viscous means will result in very poor control performance.

Elastic resistance (*spring loading*) supports the positional proprioception, supplies information regarding the zero point, and thus is of special advantage to the optimum utilization of the secondary control loop of the human motor system. The zero-point information is of great importance to rate control systems ($H_S(s) = K/s$) and acceleration control systems ($H_S(s) = K/s^2$). For reasons of safety, spring loading has further advantages, because the control automatically returns to its zero point as soon as it is released. To achieve a favorable control action, the spring rate should not be dimensioned too low. Very small restoring forces result in overriding, whereas very great restoring forces lead to underriding. By contrast, a certain amount of initial stress in the control in its zero position causes a reduction in the control performance (Bahrick, Fitts, and Schneider, 1955). A combination of elastic resistance and heavy inertia will also have an adverse effect on the control action. To reduce overriding, 9 to 20 N is specified as the minimum value for the restoring resistance in joysticks and hand levers, at maximum displacement. 130 N is recommended as the upper limit value that should not be exceeded, so that excessive muscular tension of the operator and underriding effects will be avoided. In pedals, the actuation time increases with rising restoring force. When the design of the restoring force remains within reasonable limits and the accuracy of maintaining a specific pedal position is the evaluation criterion instead of actual time, then there are no influences on the performance.

Isotonic (free of restoring forces; $m, d, c \rightarrow 0$) and *isometric* (free of displacement; $c \rightarrow \infty$) controls are regarded as special cases. The controlled variable $y(t)$ is proportional to the displacement on the one hand, and to the force on the other hand. The advantages of the isometric control are the shorter deceleration times of the neuromuscular system, because no movements have to be made, the low degree of cross talk between the operating directions, and the reduced fault susceptibility to mechanical vibrations. Nevertheless, the advantage becomes evident only in controlled systems of higher order in the high-frequency range. Preference should be given to isotonic and spring-loaded elements in position control systems and low-frequency systems. The absence of the zero-point information in an isotonic control is the main influencing factor for the loss of performance in the control of systems of higher order. McRuer and Magdaleno (1966) determined the transfer function of a human–machine system for an isometric control, an isotonic control, and a spring-loaded control in different controlled systems ($H_S(s) = K$, $H_S(s) = K/s$, $H_S(s) = K/s^2$). Figure 5.3.13 shows the open-loop transfer characteristics of an acceleration control system. In the higher-frequency range, there was a large amplification of the amplitude response and a reduced phase angle in the case of the isometric control. In the low-frequency range, amplitude response and phase characteristic were nearly identical.

5.3.5.2   Nonlinear Mechanical Transfer Characteristics

Nonlinear transfer characteristics are due to *static* and *sliding friction, mechanical backlash,* and *nonlinear spring characteristics*. Friction is composed of a static part and a sliding part (Coulomb's friction). There is no interrelation between Coulomb's friction and the kinematic parameters of the control; that is, the proprioceptors do not receive any information. Static friction has an especially adverse effect on the control performance at the reversal points of the actuating movements, owing to the large static part. Abrupt movements are induced on account of the high resistance that is temporarily effective at $\dot{\phi}(t) = 0$. If the scope of movements is very small, precise actuation is thus not possible. This effect is particularly applicable in the case of low moments of inertia and great control gain. Adequate static friction has the advantage that the hand or foot can rest on the control without actuating it. The results obtained in experimental tests made by various authors with respect to the influence of Coulomb's friction on the tracking performance or adjusting time are not uniform. For knobs, it is true that the coarse adjusting time considerably increases with growing friction, whereas the fine adjusting time does not depend on the existing friction. Consequently, friction in particular has a negative effect, when speed is of greater importance in a control task and accuracy of less importance.

Backlash at any rate impairs the control accuracy. The control error is approximately linear to the existing backlash and is even magnified by additional system friction. If backlash exists, the negative effect is intensified with the increasing magnification factor of the controlled system. Among others, Gibbs (1962) has provided evidence of this in adjusting tests made with a joystick. This negative influence is retained in varying the control–display ratio (C/D ratio), the order of the technical system, and the actuating organ (thumb, hand, forearm). In all cases, the highest control performance is

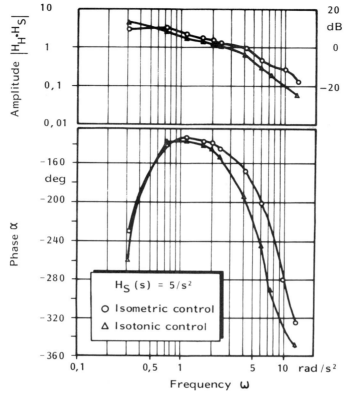

Fig. 5.3.13. Open-loop transfer characteristics for an isometric and isotonic control (pressure and free-moving stick). (*Source:* McRuer and Magdaleno, 1966.)

obtained for the backlash-free joystick. The above findings are applicable to all types of controls; for knobs, the influence of backlash is smaller, however. The adverse effects of backlash are worse, the higher the requirements made as to control accuracy. Besides nonlinearities regarding the restoring force (nonlinear elastic resistance), such as backlash and friction, nonlinearities may occur regarding the controlled variable (*dead zone* or *dead space*). It is basically true that a dead zone has an adverse effect in controlled systems of low order, whereas a minor dead zone may be advantageous in systems of higher order (Rühmann, 1981). The dead zone of a control may extend over the total range of movement (on/off controls).

For *nonlinear degressive spring characteristics,* more favorable values are obtained only for the fine adjustment time. For a system of zero order and first order, the coarse adjustment time remains invariant between linear and degressive spring characteristic (Rothbauer, 1978). A degressive spring characteristic supplies better zero-point information to the operator, and this is important in rate and acceleration control systems. In discrete controls such as keyboards and push buttons, the resistance must increase and finally drop sharply, to indicate that actuation has been performed. This proprioceptive feedback is frequently supplemented by an acoustic feedback.

5.3.5.3 Control Sensitivity

On account of their adaptation capability, humans can largely adapt their actions to the control sensitivity (gain) V. The functional relation between tracking error and gain represents a U-shaped pattern with a wide area of nearly constant control quality. An increase in the tracking error occurs only with a relatively high or low gain. The selection of gain depends primarily on whether emphasis is laid on criteria of speed or on criteria of accuracy. The system dynamics $H_S(s)$ must be known for the adaptation of the control sensitivity. To be able to compensate large deviations, a higher gain will be necessary in systems of higher order. The amount of gain is determined by the stability of the total system. In nonlinear gains—that is, a small displacement of the control will receive little gain and a high displacement of the control will receive large gain—improvements become evident regarding the control performance and the operating activity. Results show that use of such a nonlinear control gain can improve tracking performance by 10% and reduce the operator's input frequency by 25% (Krüger, 1982).

5.3.5.4 Active Controls

Active controls supply information regarding the system status via the control to the operator. The increase of the restoring moment in an automobile's steering, dependent on the road speed, for example, constitutes an active control. By the feedback of suitable status variables (compare Figure 5.3.14) and the control element dynamics deduced therefrom, a higher degree of performance and reduced stress and strain of the operator can be achieved in comparison with passive controls. Active controls are only advantageous in high-frequency and complex systems.

The functioning of active controls is based on the fact that the operator transfers a force to the control, whereas the displacement of the control is controlled by the technical system. An active control can also be realized by force zero-point shift (Röger, 1978). The displacement (force zero shift) of the control is effected by an actuator and is proportional to the tracking error (Figure 5.3.14). This principle induces the operator to take control action in a quite specific direction. In addition to the transfer of quantitative information, an active control can also be used to transfer qualitative information, for example, mechanical vibrations as warning information.

5.3.5.5 Effect of Mechanical Vibrations

Under the influence of mechanical vibrations, the effectiveness of a human–machine system can be improved by defining the dynamics of a control in a suitable manner, taking the system action into account. For position and rate control systems, Rühmann (1978) suggests viscously damped controls for damping of the movements around the wrist joint induced by the vibrations, and isometric controls for acceleration control systems. In lateral frequency excitation of higher frequency (> 4 Hz), the hand–arm vibrations are damped to a greater extent by spring-centered controls than by isometric controls; in the low-frequency range (≤ 4 Hz), the conditions are reversed (Allen, Jex, and Magdaleno, 1973).

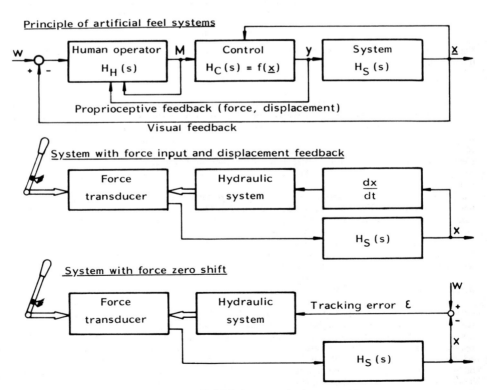

Fig. 5.3.14. Artificial feel systems (active controls).

5.3.6 CONTROL CODING

The objective of coding controls is to guarantee their quick and safe identification. The coding is based on the capability of visual and tactile discrimination of controls with different control functions. Coding can be realized by *color, shape, surface structure, dimension, position, text,* and *symbols.* For reasons of esthesiophysiology, color coding is not suitable as a primary type of coding, because the perception of colors depends heavily on the lighting conditions at a workplace and because disturbances of the color sense of humans are relatively frequent. By using a coding combination in which, for example, symbol coding and color coding are superposed, the disadvantages of one type of coding can mostly be compensated. Colors have an inherent element of meaning that must be taken into account in the coding (e.g., red—danger; yellow—caution, attention). Shape and surface coding of the controls permits easy visual and tactile identification (for shape and surface coding, see Woodson and Conover, 1966; and Damon, Stoudt, and McFarland, 1966). Tactile information is retained, even without visual contact and under unfavorable lighting conditions.

If great transfer forces are involved, the shape must satisfy the anatomical and anthropometrical marginal conditions, despite the coding, so as not to cause overstrain of the hand and fingers. For these reasons, very complex shapes—for example, to give the control a symbolic meaning—can be used only with very low resistances. In size coding, the controls used for the same control tasks are given identical dimensions. Here, a maximum of three sizes should be used, with a minimum difference of 20% from one size to the next. When gloves have to be worn, the tactile perception is impaired in shape and surface coding. In identical technical systems, an equivalent position coding is required (e.g., accelerator pedal and brake pedal in an automobile) to avoid transfer effects when the operator changes from one system to another. Coding by text and symbols is based on the functional identification of controls and is effective only under good lighting and visual checking conditions. Brief signs and common abbreviations, which are easily comprehensible without any learning, are suitable for this kind of coding. Only such symbols are permissible that, if partly covered, do not allow another meaning to be implied. Alphanumeric characters have a high degree of unambiguity, but they are linked with the language involved and call for comparatively much space. The unique allocation of the coding to the control must be ensured; it is recommended that the marking appear above the control. Text and symbol coding results in a wide variety of coding possibilities. For the optimum size and line thickness of alphanumeric characters relative to the visual distance, see, for example, Chapanis and Kinkade (1972). The minimum symbol size for the normal-vision operator is a function of contrast, surrounding brightness, distance of observation, the geometric structure of the symbols, and the manner of symbol production (dot screen technique or line screen technique). For design information regarding this subject, see Geiser (1977).

5.3.7 CONTROL–SYSTEM, CONTROL–DISPLAY, AND DISPLAY–SYSTEM COMPATIBILITY

To ensure high effectiveness and safety of the system and a reduction of the response time and learning phase, a high degree of *compatibility* is required between control, technical system, and display. Compatibility exists if the direction of movement of the control coincides with the direction of movement of the technical system or the observable system variables (stimulus–response compatibility). Thus on clockwise turning of the steering wheel of an automobile, we expect a change of direction to the right. The objective of compatible system design is to take into account the generally existing or acquired stereotypes of humans. If movement and perception stereotypes are taken into account—which under certain circumstances may vary in the population, as, for example, in the case of left-handers—decoding steps and thus mental processing work will not be necessary for the operator. For the direction of movement of the control and the function or response of the technical system, recommendations are given in Table 5.3.2. The operation of a valve is an exception to these movement-effect stereotypes. To open the valve, it must be turned anticlockwise; to close it, it must be turned clockwise. The specifications contained in Table 5.3.2 may differ in different national standards and regulations.

In complex technical systems, the system action can often be manipulated only through displays. Between control and display and between display and system action, design principles will also have to be observed, so that the required compatibility will be guaranteed. The movement of the controls and that of the associated display element should be identical in their directions. For example, if a control is turned to the right, the hand on a dial must move clockwise; on a horizontal scale, it must move to the right, and on a vertical scale, it must move upward. A rising value should bring about a displacement of the hand in clockwise direction on a dial, a displacement to the right on a horizontal scale, and a displacement upward on a vertical scale. In the case of a combination of a rotational control and a vertical scale, it is also important whether the rotational control is arranged to the right or to the left of the scale. If operated in clockwise direction and with the control arranged on the right side, the hand or pointer will move upward and, if arranged on the left side, it will

Table 5.3.2. Recommended Control Movements

Function	Control Action
On	Up, right, forward, pull (switch knobs)
Off	Down, left, rearward, push (switch knobs)
Right	Clockwise, right
Left	Counterclockwise, left
Up	Up, rearward
Down	Down, forward
Retract	Rearward, pull, counterclockwise, up
Extend	Forward, push, clockwise, down
Increase	Right, up, forward
Decrease	Left, down, rearward

move downward. Displays with fixed pointer and movable scale, in connection with controls, inevitably lead to incompatibilities. Therefore, such combinations should not be used. Further design principles must be observed if display and controls are arranged on different levels. In this connection, see Bradley (1954) and Chapter 5.1. Besides the stimulus–response compatibility, the position compatibility is also of importance. Position compatibility exists if a meaningful spatial allocation has been established between controls, displays, and the associated sensors and effectors (see section 5.3.4).

Control–display ratio (*C/D ratio*) as a further important design variable is the ratio between the deflection of the display (pointer) and the displacement of the control. If pointer and control element are performing a translational movement, the following will apply to the C/D ratio: $R = C/D$, where C = displacement of control part (mm), and D = deflection of display marker (mm). With an optimum determination of the C/D ratio, it will be possible to reduce the adjusting time drastically. Generally speaking, a very large C/D ratio will be required for exacting requirements of accuracy, and a small C/D ratio for quick actuating activities over a large area. In fine adjustment, the actuating time will exponentially rise with decreasing C/D ratio. The large display deflection in less control movement will cause multiple overshooting in the target area, resulting in an increase in the adjusting time. Because adjusting tasks mostly constitute coarse and fine adjustment, a reasonable compromise will have to be found. A C/D ratio of 0.1 to 0.4 is stated for rotationally actuated controls, and of 2.5 to 4 for levers (Chapanis and Kinkade, 1972). Although the diameter of a control does not have any influence, display size and time delays of the technical system (system dynamics) and tolerance band of the reference value are important influencing factors that have to be taken into account in determining the C/D ratio.

5.3.8 SAFETY REQUIREMENTS

Spurious actuation of a control may be caused by the operator, by unauthorized persons, or by surrounding influences such as mechanical vibrations or the falling down of objects. Inadvertent actuation of a control by the operator may be caused by slipping off a control and actuating a neighboring control, because of inadequate distance between one control and the next, getting one's clothing caught, and/or supporting oneself by holding onto a control. Inadvertent actuation of a control may also result from wrong operation by the operator, for example, due to unfavorable coding. The design principles discussed above should therefore also be seen in the light of safety aspects.

The marginal conditions of safety must be taken into consideration as early as in the definition of the design parameters (Section 5.3.3). Shape and surface of the control must be designed so that slipping off a control is prevented, in order to avoid injury to the worker and spurious actuation. A push button may, for example, be provided with a concave top or with extra-fine profiles. If controls have to be mounted on rotating shafts, provision will have to be made for clutches, so that the transmission of power can be interrupted. For actuation, it will first be necessary to apply an axial force. If this cannot be realized, controls will have to be used, in which getting clothing caught is not possible; that is, disk handwheels with a closed stay must be used instead of cranks and spoked handwheels.

In the case of important controls, in which an inadvertent actuation would endanger persons and the system, provision will have to be made for additional safeguards. Such undesirable actuation can be prevented or minimized by the following design measures, which are based on various cause-and-effect principles such as the interruption or complication of the flux of force.

Covering of the control

Recessing or shielding of the control

Controls not to be arranged within the supporting and main movement area of arms and legs

Direction of control movement is not identical to the direction of movement of the body or its extremities

Making provision for large tripping forces

Making provision for frictionally actuated instead of positively actuated controls (e.g., finger slide instead of push button)

Detachable controls (controls must not be interchangeable)

Two-dimensional control (one degree of freedom for unlocking), lockable control, multifunctional control (one control for unlocking or for simultaneous operation (two-hand control, dead man's control)

Moreover, keys are to be regarded as special forms of controls that can be used for rotational actuating movements for two or several steps. By removal of the key, a high degree of safety is guaranteed against unauthorized and inadvertent operation. The design measures mentioned above will have to be supplemented by notices in the form of the various coding possibilities. It is obvious that safety measures in part do not satisfy the requirements of ergonomics. Covers and recessed controls, for example, prevent quick actuation of controls, large tripping forces place greater stresses and strains on the operator, and the movements of controls in directions other than the preferred anatomical directions do not permit any favorable transfer of forces. Thus it may be necessary to compromise, with due consideration of the priorities. Neudörfer (1982) has compiled a systematic catalog for avoiding the undesirable actuation of controls, including examples of application.

REFERENCES

Allen, R. W., Jex, H. R. and Magdaleno, R. E. (1973). Manual control performance and dynamic response during sinusoidal vibration (AMRL-TR-73-78). Wright Patterson Air Force Base, Ohio.

Applied Ergonomics Handbook (1974). Guilford, Great Britain: IPC Science and Technology Press.

Bahrick, H. P., Fitts, P. M., and Schneider, R. (1955). Reproduction of simple movements as a function of factors influencing proprioceptive feedback. *Journal of Experimental Psychology, 49,* 445–454.

Bradley, J. V. (1954). Desirable Control–Display Relationships for Moving-Scale Instrument Report No. WADC-TR-54-423, Wright Air Development Center, WPAFB, Ohio.

Bullinger, H.-J., and Solf, J. J. (1979). *Ergonomische Arbeitsmittelgestaltung I–III.* Bremerhaven: Wirtschaftsverlag NW.

Bullinger, H.-J., Kern, P., and Solf, J. J. (1979). *Reibung zwischen Hand und Griff.* Bremerhaven: Wirtschaftsverlag NW.

Caldwell, L. S. (1959). The effect of the special position of a control on the strength of six linear hand movements (Report 411). Fort Knox, Ky: U.S. Army Medical Research Laboratory.

Card S. K., English, W. K., and Burr, B. J. (1978). Evaluation of mouse, rate-controlled isometric joystick, step keys, and text keys for text selection on a CRT. *Ergonomics, 21*(8), 601–613.

Chapanis, A., and Kinkade, R. G. (1972). Design of Controls. In H. P. van Cott and R. G. Kinkade, Eds., *Human Engineering Guide to Equipment Design.* Washington, DC.

Damon, A., Stoudt, H. W., and McFarland, R. A. (1966). The human body in equipment design. Cambridge, MA: Harvard University Press.

Fitts, P. M. (1964). The information capacity of the human motor system in controlling the amplitude of movement. *Journal of Experimental Psychology, 47,* 381–391.

Geiser, G. (1977). Mensch–Maschine–Kommunikation in Leitständen. PDV-Bericht 131.

Gibbs, C. B. (1962). Controller design: Interactions of controlling limbs, time-lags and gains in positional and velocity systems, *Ergonomics, 5,* 385–402.

Grandjean, E. (1979). Physiologische Arbeitsgestaltung, Thun: Ott-Verlag.

Kern, P., Muntzinger, W. F., and Solf, J. J. (1984). Entwicklung von normungsfähigen, ergonomisch richtig gestalteten Bedienteilen, BMFT-HdA.

Kraiss K.-F. (1970). Beitrag zur Optimierung des Steuerkraftverlaufs von Bedienelementen (Forschungsbericht No. 4). Forschungsinstitut für Anthropotechnik. Wachtberg-Werthhoven.

Krüger, W. (1982). Untersuchung der nichtlinearen Bediensignalverstärkung bei einer kontinuierlichen Trackingaufgabe (Bericht No. 54). Forschungsinstitut für Anthropotechnik. Wachtberg-Werthhoven.

Mainzer, J. (1982). Ermittlung und Normung von Körperkräften-dargestellt am Beispiel der statischen Betätigung von Handrädrern. *Fortschritt-Berichte,* Reihe 17, Nr. 12, VDI-Verlag GmbH, Dusseldorf.

McCormick, E. J. (1976). *Human factors in engineering and design.* McGraw-Hill.

McRuer, D. T., and Magdaleno, R. E. (1966). Human pilot dynamics with various manipulators (AFFDL-TR-66-138). Wright-Patterson Air Force Base, OH.

Neudörfer, A. (1982). Systematischer Katalog zum Vermeiden unerwünschter Betätigungen von Bedienteilen. *Werkstatt und Betrieb, 115*(12), 225–236.

Neumann, J., and Timpe, K. P. (1976). Psychologische Arbeitsgestaltung. Berlin: VEB Deutscher Verlag der Wissenschaften.

Pitrella, F. D., and Holzhausen, K. P. (1982). Selection and experimental comparison of computer input devices (Forschungsbericht 57). Forschungsinstitut für Anthropotechnik, Wachtberg-Werthhoven.

Poulton, E. C. (1974). *Tracking skill and manual control.* New York: Academic.

Pulat, B. M. (1980). A Computer Aided Panel Design and Evaluation System, Ph.D. Dissertation, North Carolina State University.

Röger, W. (1978). Das Bedienelement als Informationsträger bei Bahnführungsaufgaben. Dissertation an der TU Braunschweig.

Rohmert, W. (1966). Maximalkräfte von Männern im Bewegungsraum der Arme und Beine (Forschungsbericht No. 1616 des Landes Nordrhein-Westfalen). Köln, West Germany: Westdeutscher Verlag.

Rothbauer, G. (1978). Zum Problem des Bewegungswiderstandes bei einfachen und komplexen Stellbewegungen des Armes (Bericht No. 40). Forschungsinstitut für Anthropotechnik, Wachtberg-Werthhoven.

Rühmann, H. (1978). Untersuchung über den Einfluss der mechanischen Eigenschaften von Bedienelementen auf die Steuerleistung des Menschen bei stochastischen Rollschwingungen. Forschungsbericht aus der Wehrtechnik, BMV/g-FBWT 78–11.

Rühmann, H. (1981). Schnittstellen in Mensch–Maschine–Systemen. In H. Schmidtke, Ed., *Lehrbuch der Ergonomie.* München: Carl-Hanser-Verlag.

Rühmann, H. (1984). Die Schwingungsbelastung in Mensch-Maschine-Systemen, *Fortschritt-Berichte,* Reihe 17, Nr. 22, VDI-Verlag GmbH, Dusseldorf.

Schmidtke, H., and Rühmann, H.-P. (1978). Ergonomische Gestaltung von Steuerständen. Bremerhaven: Wirtschaftsverlag NW.

Schmidtke, H., and Stier, F. (1960). Der Aufbau komplexer Bewegungsabläufe aus Elementarbewegungen (Forschungsberichte des Landes Nordrhein-Westfalen, No. 822). Köln, West Germany: Westdeutscher Verlag.

Seibt, F. (1971). Steuerleistung in Abhängigkeit vom Übersetzungsverhältnis und von Coulombscher Reibung im Bedienelement. Dissertation an der TU München.

Taylor, R. M., and Berman, J. V. F. (1982). Ergonomic aspects of aircraft keyboard design: The effects of gloves and sensory feedback on keying performance. *Ergonomics, 25*(11), 1109–1123.

Woodson, W. E. and Conover, D. W. (1966). Human engineering guide for equipment designers. Berkeley, CA: University of California Press.

CHAPTER 5.4

BIOMECHANICAL ASPECTS OF WORKPLACE DESIGN*

DON B. CHAFFIN

The University of Michigan, Ann Arbor,

---

* See also Chapter 2.6 on biomechanics of the human body and Chapter 7.2 on manual material handling.

5.4.1 OBJECTIVES OF WORK-SPACE DESIGN

Work-space design objectives vary in emphasis with the orientation of the designer and intended user. In general, the architect/interior designer strives for pleasing aesthetics through the use of colors, materials, illumination, and shapes of objects. The industrial engineer emphasizes economy of motion, that is, arranging objects to avoid redundant movements, large motions, and waiting times. More recently, the ergonomist seeks to design a work space that minimizes biomechanical stresses by assuring that the worker remains erect, keeps work objects close to the body, is properly braced, and has good foot traction.

Because the design objective of the industrial engineer for motion economy is so closely related to that of the ergonomist for minimizing postural stress, the work-space guidelines and limitations related to both are discussed in this chapter. Of course it is hoped that aesthetic qualities will also be considered in all work-space design, but specific recommendations to do so are beyond the scope of this presentation.

5.4.2 TRADITIONAL WORK-SPACE DESIGN CRITERIA—MOTION ECONOMY

At the turn of the century economy of motions in manual work was accomplished by time studies of skilled workers. This continues today to be a reasonable approach, given that there exists an operating workplace/worker. The steps involved in such an analysis are as follows:

1. *Development of a Preferred Method of Performing a Job.* This assumes that the task or job can be performed using different methods. To develop a preferred method an analyst must (1) state the objective of the operation, (2) identify available methods to accomplish that objective, (3) use feasible methods on a trial basis, and (4) select the best method to meet the objective.

2. *Preparation of a Standard Practice.* This requires that the preferred method be stated *formally.* This includes (1) sketching of the workplace, with the machines or tools used and with relevant dimensions of objects handled at the workplace, (2) listing of abnormal working conditions that could affect work performance (e.g., dust, heat, illumination), and (3) tabulating the sequence of motions required of a worker to complete the operation.

3. *Determination of a Time Standard.* This is accomplished by having a skilled worker perform the operation while being timed; that is, a *time study* is performed.

In a methods laboratory, jobs are mocked up and performed while being time studied. The arrangement of objects to be manually handled and the general work space is systematically varied. By such an iterative approach, with diligence and common sense developed from years of experience in an industry, work spaces can be designed to maximize productivity.

Many professionals have contributed to these early motion and time analysis methods, as discussed by Barnes (1968) and Niebel (1972). One of the early major contributors to this field was Frederick W. Taylor, sometimes referred to as the father of modern time analysis. Taylor's study of shoveling rice, coal, and iron ore in 1898 was a rigorous, comprehensive investigation to determine the optimal shovel size, shoveling frequency, and rest allocation necessary to maximize the total output of a group of workers (Copley, 1923). Taylor's evaluation showed, for instance, that a shovel that would allow about 200 N of iron to be lifted each time resulted in maximizing the total daily output for a group of physically strong workers. By a combination of (1) careful selection and training of workers for such work, (2) provision of special shovels for different weight materials, and (3) paying bonuses for above average output, Taylor was able to demonstrate that 140 men could perform the same amount of work as was previously done by 400 to 600 workers (Copley, 1923).

Though the shovel example is more of a tool change than a work-space change, it amply demonstrates the approach and possible benefits derived by such an empirical method of job design.

About the same time that Taylor's studies were being published, Frank and Lillian Gilbreth were advocating that manual activities in industry ought to be carefully categorized to minimize fatigue and monotony so as to best maximize productivity and employment of the handicapped (Barnes, 1968). One of the earliest and most quoted studies performed by Frank Gilbreth was of bricklaying (Gilbreth, 1911). By using photographs of the bricklayers in his construction business he was able to demonstrate that fatigue and wasted motions were not necessary for high output if (1) the bricks were first sorted for good and bad bricks, (2) they were oriented with the best side facing out, and (3) they were presented to the bricklayer at a *comfortable working height by means of an adjustable scaffold*. His work methods resulted in well over double the output per man-hour than had previously been reported in the industry.

One of the major contributions of the Gilbreths was that of *micromotion analysis,* a term coined by them in a paper to the American Society of Mechanical Engineers (Gilbreth, 1912). In this procedure, great attention is given to categorizing individual motions, referred to as *elemental motions.* These were carefully timed and became the basis later for one of the first predetermined time systems. Once the necessary elements of a job were known the Gilbreths advocated trying out different work-space arrangements in simulations of the job.

In 1924, Segur developed time prediction equations for various standardized classes of motions from film analysis of industrial operations during World War I (Niebel, 1972). Similar data were published by Holmes, Engstrom, Barnes, Quick and others during the 1930s. What these authors disclosed is that the time required of people to perform certain basic or *elemental motions* is about the same for different people. Thus the time to perform a complex manual job can be predicted by describing the job as a sequence of elemental motions that have known time requirements, and then summing the normative times required for each elemental motion. Most of the normative elemental motion times have been derived by performing frame-by-frame analysis of films to determine the time taken in various industrial operations. The resulting elemental time values have become known as *predetermined elemental motion times.* Because different groups of experts performed the analyses, slightly different motion classifications were used, and hence different motion–time prediction systems developed, many of which exist today.

Use of Predetermined Time Systems for Work-space Design

Because it is not often possible physically to design a work space by mocking it up and empirically studying the effects on a worker's performance time, predetermined motion–time systems are often preferred. Such systems provide the ability to stimulate the effects of varied work spaces "on paper."

What follows is a brief description of the Methods–Time Measurement (MTM) predetermined motion–time system. It is not meant to act as a thorough discourse on the use of this particular system, but rather should be viewed as an example of one popular system used by thousands of practitioners in industries throughout the world. For a thorough understanding of the use of such systems the reader should consult textbooks by Barnes (1968), Niebel (1972), Karger and Bayha (1966), and Konz (1979).

The MTM system developed by Maynard and others in the late 1940s is based on film and time studies of various industrial jobs. It relies on a description of manual activities with reference to a well-defined set of basic elemental motions. It should be noted that to use this particular motion classification system with accuracy and consistency requires training in a special course of 24 to 80 classroom hours offered by the MTM Association.

The resulting time values for each elemental motion in the MTM procedure are given in units of one hundred-thousandths of an hour (0.00001 hr), and are referred to as one time-measurement unit (TMU). Use of such a fraction of an hour was chosen as the basic unit of measure because production rates in industry are often quoted in units produced per hour. For general reference, one TMU equals 0.0006 min or 0.036 sec. It should also be noted that the resulting time values that are published in tables are proprietary, and thus written permission to duplicate them is required from the MTM Association, Fairlawn, New Jersey.

As an example of the time required to move various objects about in the workplace, MTM predicted time values are presented in Table 5.4.1 for different types of human *moves* common in industry. The time values presented in Table 5.4.1 vary depending on the distance of a particular move, as well as (1) three types of terminal conditions and (2) the load being transported. The effect of the terminal conditions and length of motion are self-explanatory. The load effect is more complicated in that it assumes that the weight held in one hand adds an initial time delay (referred to as a "constant" in Table 5.4.1) plus a proportional delay (referred to as a "factor"). Thus if one moves a load of 4 kg mass with one hand a distance of 10 cm to an indefinite location, then the move time would be 6.8(1.07) + 2.8 = 10.1 TMU, or 0.36 sec. If two hands are used, then the load is considered to be halved when choosing specific table values.

It should be clear from inspection of Table 5.4.1 that in the interest of maximum worker performance objects to be moved should be arranged to be (1) close together, (2) of light weight, and (3) easily positioned at the end of the motion.

Similar types of time values are available in tabular look-up form for the following motions:

Reach—a motion of the unloaded hands or fingers.

Position—small motions necessary when aligning object to be released at end of motion.

Release—either a distinct motion of fingers or the release of an object at the end of a motion without such an overt motion.

Table 5.4.1 Predicted Move Time Data Wherein a Move Is Defined as a Motion of the Hand Required to Transport an Object[a]

Maximum Distance Moved (cm)	Time (TMU) A	Time (TMU) B	Time (TMU) C	Hand in Motion B
Less than 4	2.0	2.0	2.0	1.7
4	3.1	4.0	4.5	2.8
6	4.1	5.0	5.8	3.1
8	5.1	5.9	6.9	3.7
10	6.0	6.8	7.9	4.3
12	6.9	7.7	8.8	4.9
14	7.7	8.5	9.8	5.4
16	8.3	9.2	10.5	6.0
18	9.0	9.8	11.1	6.5
20	9.6	10.5	11.7	7.1
22	10.2	11.2	12.4	7.6
24	10.8	11.8	13.0	8.2
26	11.5	12.3	13.7	8.7
28	12.1	12.8	14.4	9.3
30	12.7	13.3	15.1	9.8
35	14.3	14.5	16.8	11.2
40	15.8	15.6	18.5	12.6
45	17.4	16.8	20.1	14.0
50	19.0	18.0	21.8	15.4
55	20.5	19.2	23.5	16.8
60	22.1	20.4	25.2	18.2
65	23.6	21.6	26.9	19.5
70	25.2	22.8	28.6	20.9
75	26.7	24.0	30.3	22.3
80	28.3	25.2	32.0	23.7

Weight Allowance

Wt (kg) up to	Constant TMU	Factor	Case and Description
1	0	1.00	A
2	1.6	1.04	Move object to other hand or against stop
4	2.8	1.07	
6	4.3	1.12	B
8	5.8	1.17	Move object to approximate or indefinite location
10	7.3	1.22	
12	8.8	1.27	
14	10.4	1.32	
16	11.9	1.36	C
18	13.4	1.41	Move object to exact location
20	14.9	1.46	
22	16.4	1.51	

[a] Values are not to be duplicated without permission from MTM Association for Standards and Research, Fairlawn, NJ.

Disengage—an involuntary (rebound) motion often required when two objects suddenly come apart under exertion.

Grasp—an overt motion necessary to gain control of an object.

Eye focus/travel—the time required for the eyes to move and accommodate to provide visualization of an object.

Turn/apply pressure—the manipulation of controls, tools, and objects necessary to turn an object by rotation of the hand about the long axis of the forearm.

Body, leg/foot motion—the motion of transporting the body with values given per step for varied conditions.

Simultaneous motions—rules are given so that some motions can be performed together. For example, both the right and left hand reach to an object; therefore only the greatest of the two time values are used in the standard time prediction.

A sample MTM job analysis is given in Table 5.4.2 using standard coding notations to specify the elemental motion conditions expected of a worker performing the job; see also Figure 5.4.1. Note: The brackets indicate simultaneous motions and thus only the greatest time is entered in the center column to be used in predicting the total time of the task. In the example a tote box weighing 200 N (22 kg) is lifted to a workbench and one part weighing 10 N is removed to the bench from those other parts jumbled in the tote box. By raising the pallet to the bench height using a mechanical, hydraulic, electric, or pneumatic pallet lift device, considerable time could be saved (plus reducing biomechanical stresses on the worker's back).

Limitations on the Use of Predetermined Time Systems in Work-space Design

It should be clear that minimizing worker performance times must be a major consideration in planning a work space. The ability to simulate physical working conditions in a full-scale but adjustable work

Table 5.4.2   Sample MTM Analysis of Lifting a Tote Box from a Pallet to a Workbench and Removing a Part from the Tote Box; see Figure 5.4.1

Activity Description	Left-Hand Elements	Motion Time (TMU)	Right-Hand Elements or Body Motions
Sidestep to pallet.		17.0	SS12C1
Stoop to tote box, and		29.0	S
Reach to tote box during stoop.	~~R10B~~		~~R10B~~
Grasp handles on tote box,	GIA	2.0	GIA
Arise, and			AS
Lift tote box during arise.	~~M10B~~$\frac{22}{2}$	31.9	~~W10B~~$\frac{22}{2}$
Sidestep toward bench with tote box, and			
Move tote box during sidestep.	M12B$\frac{22}{2}$	23.7	SS12C
			M12B$\frac{22}{2}$
Release tote box on bench.	RL1	2.0	RL1
Reach into tote box.		12.9	R10C
Grasp part in box.		7.3	G4A
Move part to bench.		10.6	M18B
Release part on bench.		2.0	RL1

Total Time Required: 138.4 TMU or 4.98 sec

Source: Courtesy J. Foulke.

space along with the ability to time study specific tasks carefully is essential. Unfortunately, few companies have maintained adequate facilities and staffs to perform such physical simulations, although with increased costs of labor, combined with health and safety requirements, more companies are rebuilding this capability.

If a full-scale simulation is not feasible, then the use of predetermined motion–time (PMT) systems is at least warranted for good work-space planning. Unfortunately, even the most sophisticated PMT system does not reflect the effects of a workplace that has the worker in an awkward biomechanical posture. For instance, motion–time values as shown in Table 5.4.1 are not altered by a worker performing the motions overhead, or with arms horizontally extended in front of the body. Such awkward postures degrade an operator's performance time, perhaps by 20% or more, depending on the specific posture and load required.

Furthermore, PMT systems assume a "nonfatigued" person. Moving parts from a pallet on the floor to a table is much more energy-consuming than moving the parts from a pallet at knuckle height to a table. Depending on the frequency and weight of the parts, considerably more fatigue and performance degradation would be expected in the former case than in the latter.

In the same context, if a workbench is not adjusted to an individual's anthropometry, as discussed in Chapter 2.6, then localized muscle fatigue and performance decrements develop in the back, shoulder, and arm muscles. This aspect is discussed further in the following section.

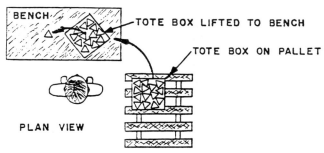

Fig. 5.4.1.   Sample MTM analysis. *Tote box* weighs 200N and measures $30 \times 45 \times 20$ cm.; *bench* is 76 cm high; *pallet* is 10 cm high; *parts* weigh 10N and are jumbled in tote box; see Table 5.4.2 (courtesy J. Foulke).

5.4.3   CONTEMPORARY WORK-SPACE DESIGN TO REDUCE BIOCHEMICAL STRESS

The use of a detailed motion and time analysis procedure should be an integral part of any good work-space design method. But so also should a detailed biomechanical analysis, which would allow comparison of work-space conditions that could produce increased muscle fatigue and other more serious chronic musculoskeletal disease, as discussed by Chaffin and Andersson (1984).

Background on Muscle Fatigue and Tendon and Ligament Inflammation

The skeletal muscles responsible for motion and postural maintenance are not well adapted to prolonged static loading, even at relatively light loads of about 30% of a person's maximum strength. This is depicted for back muscles, which are probably the best adapted muscles for static work, in Figure 5.4.2. If a person simply inclines the torso 30° forward from an erect posture, due for instance to a workbench being about 4 inches below the individual's standing erect knuckle height, the back muscles would probably be contracted to a level above 25% of an average person's strength. Thus, as depicted in Figure 5.4.3, muscle fatigue (pain and loss of function) would develop in a matter of minutes in such a posture. Designing a workbench height to minimize such stooped postures is also critical to avoid shoulder pain, as discussed by Cailliet (1981). In this case, in an attempt to work on an object below standing erect knuckle height the worker "droops" the shoulder, causing static stretch of a small muscle in the upper back, which fatigues rapidly, resulting in a very painful "trigger point." If prolonged, the muscle and its tendon become inflamed, and the pain can persist for weeks after the insult.

The acute pain of an exhaustive muscle exertion can be thought of as a precursor or warning sign. If the pain causing exertion is continued on a daily basis adaptation may occur; that is, the muscle becomes stronger and gains in endurance with a concomitant decrease in pain levels. If the exertion is so heavy that pain does not decrease or even increases, not only is the muscle tissue unable to adapt, but an inflammatory process can be initiated in tendon and joint tissues involved in the exercise. Such inflammation can develop as a reaction to mechanical strain of the tissues when there is not adequate rest to allow physiological adaptation. In particular, inflammation of tendons, or *tendonitis,* can occur when a tendon has been repeatedly tensed. Swelling from inflammation of the tendon may itself contribute to pain. With further exertion the tendon collagen fibers become separated. At the point of greatest stress, often where the tendon passes around an adjacent bone or ligament structure, the collagen fibers can be pulverized, leaving debris containing calcium salts. These calcium salts and circulatory fluids within the injured tendons produce further swelling and pain (Cailliet, 1981). If the harmful work activity is continued, the degeneration will involve the surrounding tendon synovia and possibly the bursa membrane of the adjacent joint. Because the synovia and bursa provide a low-friction sliding mechanism for tendons, muscles, and bones, any inflammation and resulting degener-

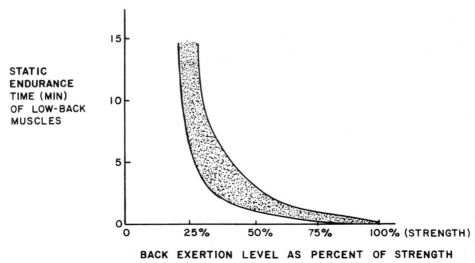

Fig. 5.4.2.   Range of back muscle endurance values as obtained using various torso forward inclination postures and loads to vary exertion levels (Jorgensen, 1970).

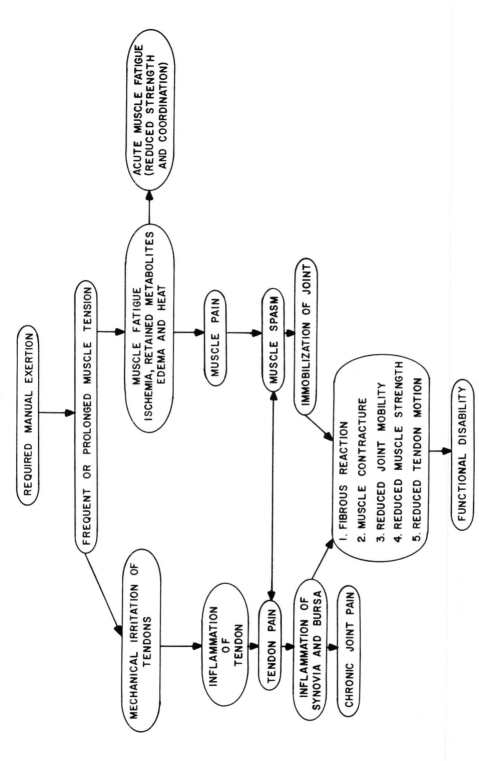

Fig. 5.4.3. Sequence of events producing musculoskeletal pain and functional disability because of frequent or prolonged muscle tension.

ation of these tissues can produce limitation of motion as well as pain. If restricted mobility remains, muscles and fascia shorten, producing a muscle contracture. A permanent functional disability is often the final result.

The sequence of events leading to a permanent dysfunction is diagrammed in Figure 5.4.3. The figure also illustrates that natural somatic reactions occur because of motion-related pain. Thus the joint can be immobilized by antagonistic muscle actions. These muscle contractions "splint" the joint, and often lead to spastic and painful muscles. They also promote fibrous reactions or adhesions between adjoining sliding tissues in the synovia, bursa, and joint capsular articulating surfaces (Cailliet, 1981).

The avoidance of the conditions described requires the recognition that the musculoskeletal system is well suited to produce repeated motion at low force loads. When the loads become high, are prolonged or frequent in nature, or involve awkward postures, however, the muscles, tendons, synovia, bursa, and bone of the articulating structures become painful and inflamed, and begin to degenerate. What follows is a discussion of various workplace and machine control layout considerations that are meant to be helpful in reducing the incidence and severity of these types of problems.

5.4.4 GUIDELINES FOR WORKPLACE AND MACHINE CONTROL LAYOUT

It should be apparent from the preceding discussion that if frequent or sustained muscle exertions are required, care must be taken to minimize the external load required and avoid unnatural postures. This section attempts to develop some practical guidelines regarding the placement of machine controls and general workplace layout. Because seated operator activities are so prevalent in industry today, shoulder, arm, and neck problems are first discussed in that context. Following that is a general discussion of machine control locations and work surface heights.

Shoulder-Dependent Overhead Reach Limitations

The shoulder joint represents one of the most complex biomechanical structures one can imagine. The ball-in-socket arrangement of the glenohumeral joint (which is really a ball on a comparatively flat socket) provides a large amount of mobility for the arm, and is the true *shoulder joint* per se. Mobility for the arm is further enhanced by six additional joints which constitute the trunk–arm complex sometimes referred to as the *shoulder girdle*. These allow the glenohumeral joint to translate and rotate, providing additional reach capability. In providing such extreme mobility, intrinsic stability has been sacrificed. In fact, the stability of the shoulder joint during normal motions largely depends upon the ligaments and joint capsules, and on the extensive musculature of the upper torso. This is described in detail by Chaffin and Andersson (1984).

Of particular concern in the specification of arm work requirements is that the hands *not* have to reach frequently or be held for sustained periods above the shoulder. Jobs that require such elevated arm activities have been shown to create "degenerative tendonitis" in the biceps and supraspinatus muscles (Bjelle, Hagberg, and Michaelsson, 1979; and Herberts, Kadefors, Andersson, and Peterson, 1981). If the arm is held in an elevated posture (e.g., when overhead welding) shoulder muscle fatigue and biceps tendonitis has been identified as a major concern in the workplace, especially for older workers who have reduced joint mobility (Herberts, Kadefors, and Broman, 1980). It has also been shown by Hagberg (1981a), using an EMG analysis technique, that the upper part of the trapezius muscle rapidly fatigues and becomes painful when the arm is flexed and held at 90° (shoulder height). These same studies showed that the rate of muscle fatigue development in several shoulder muscles was 10 to 20 times slower when the shoulder load moment was below 20% of the shoulder flexion strength moment, compared to when the load moment was approximately 40% of the strength moment. This latter result should indicate a major concern to work-space designers when coupled with the fact that strength moment for the shoulder in flexion and abduction is much lower when the arm is elevated than when in lower positions.

The effect of holding the arm in various elevated positions has been measured by Chaffin (1973) using EMG frequency spectra shifts and a subjective pain rating to measure endurance times of young healthy males. Figures 5.4.4 and 5.4.5 depict these findings. From these studies it is concluded that sustained, elevated arm work, especially if supporting a load, must be minimized to avoid shoulder muscle fatigue and associated tendonitis problems (Hagberg, 1981b). It should also be clear that with repeated exertions of short duration (perhaps only a few seconds), similar muscle fatigue will develop if the relative load on the muscles is over approximately 40% of the expected strength and rest periods between contractions are less than about 10 times the contractile period.

Further, even without a hand load any elevation of the arm in abduction or forward flexion above about 90° greatly increases the stress on various tendon-ligament-capsular tissues, as indicated by greatly increased passive resistive moments toward the end of the volitional range-of-motion of the shoulder (Engin, 1980). In this regard, it is known that acute tendonitis of the shoulder muscles can be induced by high velocity arm motions, as in tossing materials (Nolan, 1979). If materials are to be tossed, the receiving container should be placed low to minimize arm elevations, above 50° from vertical.

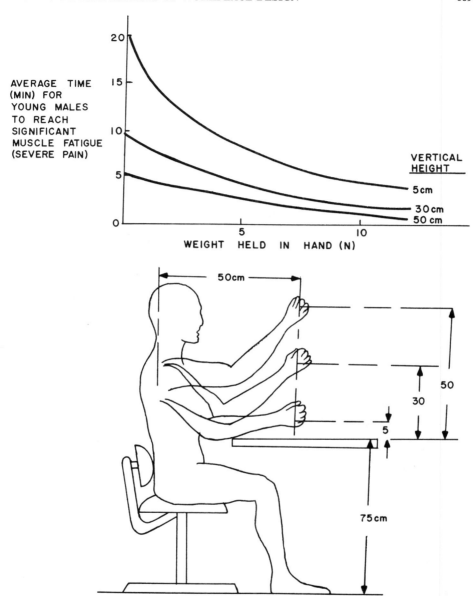

Fig. 5.4.4. Expected time to reach significant shoulder muscle fatigue for varied arm flexion postures. The larger the flexion angle, the earlier fatigue will develop (Chaffin, 1973).

Shoulder- and Arm-Dependent Forward Reach Limits

When reaching forward the shoulder joint is flexed and the elbow becomes extended. If a load is held in the hands, the load moments at the elbow and shoulder can become large relative to the flexor strength moments required at both joints. If the arm and forearm are elevated to nearly horizontal when reaching forward, a load of only 56 N held in the hands will create a load moment at the shoulder equivalent to the flexor strength moment predicted for the average female, and a 115 N load would equal the shoulder lifting strength of an average male. Clearly, from the preceding discussions, such loads could be held in these positions only for a few seconds before shoulder fatigue would develop. With weaker or older subjects these endurance limits would be even lower.

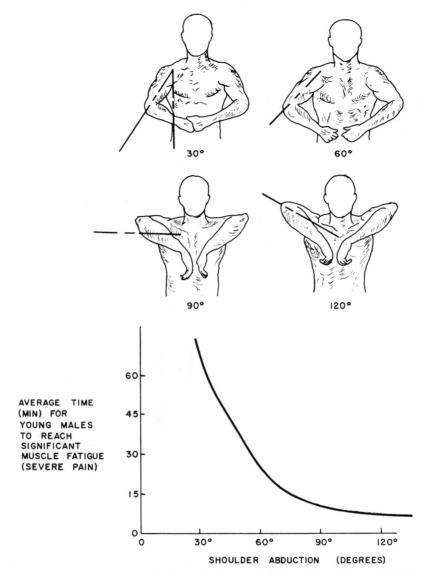

Fig. 5.4.5. Expected time to reach significant shoulder muscle fatigue for varied arm abduction angles (Chaffin, 1973).

Varied forward reach arm postures and hand load conditions were used in a study of shoulder muscle fatigue (Chaffin, 1973). The results are shown in Figure 5.4.6 as the mean endurance times of five healthy young males. The endurance times were based on EMG frequency shifts and subjective pain levels.

In the same study by Chaffin (1973) an adjustable elbow support was also used. This external body support reduced the load moment on the shoulders. The resulting muscle fatigue then developed in the elbow flexor muscles instead of the shoulder muscles, with the mean endurance times as shown in Figure 5.4.7.

It should be clear from all these data that loads of even nominal amounts cannot be supported for sustained periods, especially if the arm or forearm is elevated in a forward reach posture. This means that the use of a padded forearm support is important to relieve the load moments at the shoulder and elbow, as advocated by Chaffin (1973), Grandjean (1982), and Konz (1979). Powered

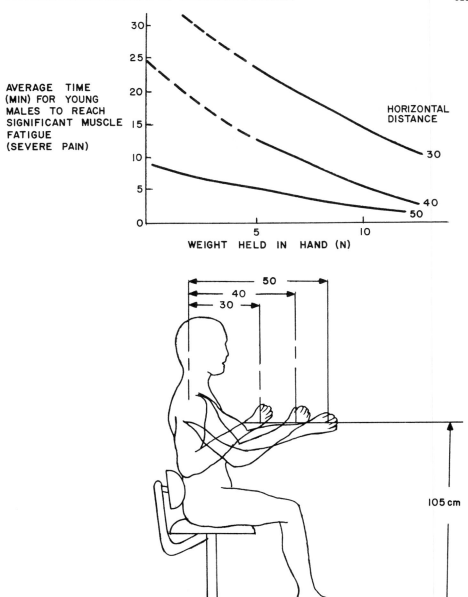

Fig. 5.4.6. Expected time to reach significant shoulder muscle fatigue for different forward arm reach postures. The elbow is unsupported. The greater the reach, the shorter the endurance time (Chaffin, 1973).

hand tools used frequently during a workday (often used when driving screws repeatedly in an assembly) or for sustained periods (when welding, spraying, or grinding) should be suspended from a tool balancer device to minimize the weight effect (Konz, 1979). Also, workpieces or assemblies should not have to be supported by one hand while the other performs a required operation. Good workplace design provides adjustable fixtures that support the workpiece in proper orientations for the operator, taking account of both visual and manual task requirements (Chaffin, 1973).

In this latter regard, it should be recognized that arm/forearm postures are often dictated by

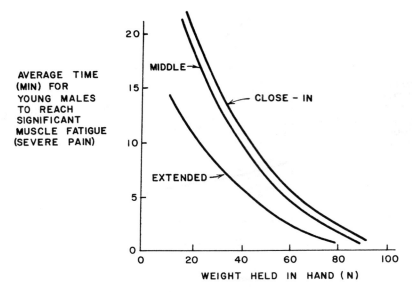

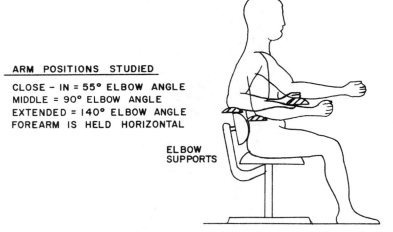

Fig. 5.4.7. Expected time to reach significant shoulder and arm muscle fatigue for different arm postures and hand loads with the elbow supported. The greater the reach, the shorter the endurance time (Chaffin, 1973).

hand orientation around the long axis of the forearm. If the hand is supinated (palm up), then the arm normally is adducted and close to the torso. If the task requires the hand to be prone (palm down) then the arm normally is more abducted and elevated. If the hand is located in a position that already requires the arm to be elevated, then using a prone hand posture may further the arm elevation. Realization of this interdependence between hand orientation and arm postures is important, particularly in designing handles on tote boxes, hand-tool configurations, and controls on machines.

In reference to this latter point, if a control or a screw must be turned in a clockwise direction requiring outward rotation or supination of the hand about the long axis of the forearm, it is also important that the elbow be flexed to 90° or less (Tichauer, 1978). This provides good mechanical advantage for the biceps brachii, which is a major outward rotator of the forearm as well as an

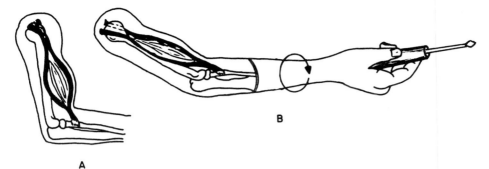

Fig. 5.4.8. Forceful outward rotation (supination) of the forearm and hand is best achieved with the elbow flexed (A). When extended (B), the biceps brachii is not positioned to assist in rotation of the ulna and it produces high joint compression forces (Tichauer, 1978).

elbow flexor muscle. According to Tichauer (1978), by having the elbow flexed, biceps contraction rotates the proximal head at the ulna about the radius without producing a large compression stress within the elbow joint, as is the case when the elbow is extended. This is illustrated in Figure 5.4.8. Clearly the flexed elbow reduces the tendon force requirement on the biceps tendon as well as assisting the supinator and brachio-radialis muscles which are also involved in forearm supination. When studying the effects of elbow angle in operations requiring frequent supination exertions, Tichauer (1978) compared injuries of two groups of workers. Twenty-three subjects with elbow pain were found among workers who performed repeated forearm supination activities with the elbow extended to 130° or more, whereas no cases were reported for workers using an elbow angle of 85° or less.

Rotational strength performance of the forearm is also enhanced by having the elbow flexed, as reported by Rohmert (1966). In his studies five healthy males attempted to rotate a vertical handle placed at various locations relative to their reach and shoulder locations. Table 5.4.3 and Figure 5.4.9 present the means of the static strength values obtained. What is clearly evident in these values is a trend toward lower strengths when the elbow is extended to provide maximum reach. This result was consistent at all four shoulder angles. Thus from biomechanical, strength, and epidemiological points of view, it appears that work activities involving forearm rotation should be performed with the elbow flexed.

When a person attempts to perform a maximal push or pull exertion with one hand in a seated posture, the resulting strength performance is very dependent upon shoulder and elbow angles. Because of the interaction of strength performance and posture, biomechanical strength models are important in understanding the complexities of such three-dimensional exertions. In this regard, the three-dimensional strength model described by Chaffin and Andersson (1984) was used to predict one-handed exertions of the human operator performing various seated static exertions (Garg and Chaffin, 1975). Some resulting one-handed strength predictions for an average male in a seated posture are presented in Figure 5.4.10. Inspection of these predictions once again discloses how complex asymmetric strengths

Table 5.4.3 Forearm Rotation Strengths in Various Arm Postures

Direction of Exertion	Shoulder Angle	Location of Handgrip as Percentage of Maximal Grip Distance		
		50%	75%	100%
Rotate clockwise	30°	17.5	16.6	7.9
(supination)	0	18.4	15.9	5.8
	−30°	15.6	13.3	4.9
	−60°	11.6	9.1	6.7
Rotate counterclockwise	30°	16.9	15.0	10.9
(pronation)	0	17.9	15.5	10 .9
	−30°	20.8	18.2	11.8
	−60°	22.8	20.1	13.7

Source: Rohmert, 1966.

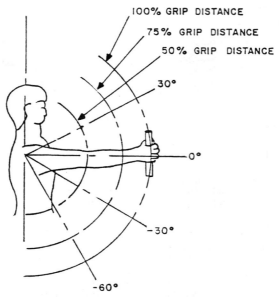

Fig. 5.4.9. Forearm rotation strengths; see Table 5.4.3 (Rohmert, 1966).

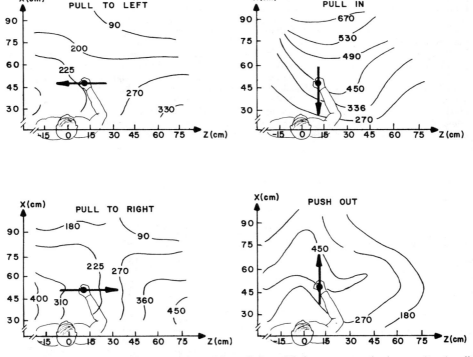

Fig. 5.4.10. One-handed static strength model predictions (N) for average males in seated and well braced posture with hand located in horizontal plane 25 cm above hip joints (Chaffin, 1973).

are to rationalize. In general, exertions in a lateral direction (to the left or right) with the hand close to the body require arm humeral rotation at the shoulder, and result in much lower values than exertions in the sagittal plane (in and out). These latter types of exertions use arm flexion and extension, involving the stronger biceps, brachialis, and triceps muscles of the arm.

It is also clear in Figure 5.4.10 that when the arm is close to its reach limit (elbow is straighter or locked in extension) push and pull actions can become quite high, being limited only by torso strenghts and balance. In this regard, if the seat and foot support do not brace the person, then the reactive load moments and forces created at the torso and legs will become limiting; that is, the person will fall off the seat or the seat will tip over. For further information on seated work, see Chapter 11.1.

The consideration of bracing a seated person to allow one-arm maximum push and pull exertions is experimentally addressed by Laubach (1978). In his experiment, aircraft seats were used with full shoulder and lap belts to brace the torso. The seat back was tilted at various angles and the seat was moved fore and aft to alter leg postures. The resulting static strengths of Air Force recruits were found to be greatly affected by each of these changes in a complex fashion.

It must be concluded from all these results that specification of an optimal machine control location or the best position to locate stock is not simply determined. The direction of the exertion must first be determined (e.g., lift, push-inward, pull to left, rotate). Then the normative muscle endurance and strength curves presented earlier in Figures 5.4.4 through 5.4.10 should be consulted for general guidance. A biomechanical strength model determination, as described by Chaffin and Andersson (1984) may also be warranted. Finally, because of the various complexities described, if the manual function is strenuous (i.e., it involves a frequent or forceful exertion) then a mock-up should be constructed and representative workers should perform the required exertions to determine if further corrections are necessary to control potential musculoskeletal problems.

Neck/Head Posture Work Limitations

Often when performing precise motions a worker is required to lean forward to obtain good vision of the work area. Some common examples of such postural requirements occur when inspecting parts for small visual defects, assembling equipment with small or intricate parts, or reading text material written on paper or on a visual display unit.

The requirement of work postures with the nect flexed forward can occur because of a combination of (1) the seat height being too high, (2) the seat placement being too far back from work area, (3) a workbench or table being too low, or (4) the visual demands of the task requiring a specific eye location (e.g., to look into a near-vertical microscope, or into a part assembly lying flat on a workbench, or to view a meter or visual display poorly located relative to the relaxed seated eye location of an individual).

It appears that the cervical extensor muscles may produce muscle fatigue symptoms and pain quite quickly when head inclination angles become significant. Kumar and Scaife (1979) noted subjective reports of neck discomfort during the workday, and found these to be related to neck inclination angle. Using electromyograph frequency shifts in the cervical muscles and a subjective pain rating, Chaffin (1973) found that the endurance time of five young healthy females was greatly decreased when the neck inclination angle exceeded 30°, as shown in Figure 5.4.11.

As has been discussed earlier in this chapter, acute muscle fatigue due to prolonged static exertions, if recurrent for months or years, can be a precursor to other more serious and chronic musculoskeletal disorders. The cervical spine is no exception. For this reason, it is imperative that neck pain symptoms be considered as having a potential occupational biomechanics etiology. The worker complaining of such should be evaluated to determine if the head is inclined more than a normal 20 to 30° for any prolonged period of time. If so, adjustments should be made in the chair, workbench, and workpiece location to provide vision with a more vertical neck and head orientation, or frequent mini-rest periods should be scheduled to allow the worker to move the head into a relaxed, upright posture. The latter is particularly necessary if the work being done is highly scheduled or in other ways mentally demanding. Such mental concentration has a tendency to result in postural muscles being "over-tensed," thus further contributing to the muscle fatigue process.

Torso Postural Considerations in Workbench Height Limitations

As indicated in the preceding section, the lumbar extensor muscles are required to stabilize the torso when standing in a stooped posture. As the torso is inclined from vertical the load moment at the lumbar disks increases as the sine of the inclination angle. Thus at approximately 30° of torso inclination the load moment is 50% of its maximum value achieved at 90° (horizontal).

Prolonged, forward stooped postures are often necessary for workers performing work in a standing or seated posture when the workbench is too low. Electromyographs of the lumbar erector spinae muscles have been obtained on workers, and have corroborated muscle fatigue of the low back when the workers were required to be in stooped postures for prolonged periods of the day (Habes, 1980).

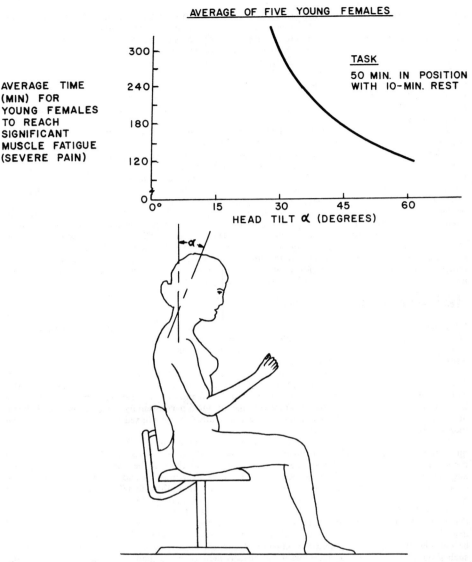

AVERAGE OF FIVE YOUNG FEMALES

AVERAGE TIME (MIN) FOR YOUNG FEMALES TO REACH SIGNIFICANT MUSCLE FATIGUE (SEVERE PAIN)

TASK
50 MIN. IN POSITION WITH 10-MIN. REST

HEAD TILT α (DEGREES)

Fig. 5.4.11. Neck extensor fatigue versus head tilt angle. The more the head is tilted, the earlier fatigue develops (Chaffin, 1973).

Subjective discomfort levels in the lower back also have been directly correlated with the load moment at the hip joint by Boussenna, Corlett, and Pheasant (1982). A study by Jorgensen (1970) revealed that back muscle endurance times in various stooped postures decreased appreciably when the postures required more than about 30% of isometric back strength (as shown earlier in Figure 5.4.2). From these latter results, the author speculates that most men and 85 to 90% of women would be capable of maintaining a 20° stooped posture during the day. Because the muscle endurance-tension curve shown earlier in Figure 5.4.2 is so asymptotic at about 30% of strength, and because the load-moment increases rapidly for each degree of torso inclination above 20°, any additional flexion would become increasingly difficult for people to tolerate.

Conversely, concern must be shown for a workbench that is too high. Tichauer (1968) found that the metabolic cost of packing groceries increased while performance decreased if the work height was such that the arms had to be abducted to an angle of 20° or more.

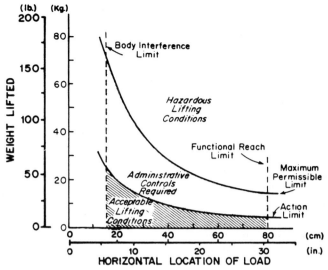

Fig. 5.4.12. Maximum weight versus horizontal location for infrequent lifts from floor to knuckle height (NIOSH, 1981).

In short, adjustment of the height of a workbench when standing and the seat–workbench combination when seated is necessary to accommodate a large proportion of the population. Konz (1979) reviewed various recommendations in this regard, and concluded that having the work area about 5 cm below the elbow when standing or seated in an erect posture was optimal. This provides adequate clearance for necessary elbow and forearm motions and requires little stooping or shoulder rotation if a person needs to rest or stabilize the forearm on the workbench. Based on 5th and 95th percentile male and female workers, this would require a workbench height adjustable between 94 and 115 cm.

If heavy objects are to be lifted and carried, as opposed to prolonged light work on a bench top, the objects should be presented at a lower height, approximately the center of grip of the hands when standing erect with the arms at the sides. The NIOSH recommendation in this regard is for heavy objects to be stored at 75 cm (NIOSH, 1981). In this same regard, the most important aspect of storing heavy objects is to provide a work space wherein the worker can hold or lift an object in a fashion that maintains the load as close to the body as possible. The need to keep a load close to the body is exemplified in Figure 5.4.12. The "horizontal location of load" in the figure is measured from a worker's foot location (at the ankles) to the center of mass of the load. The two limits shown (i.e., the *maximum permissible limit,* which denotes hazardous lifting for most workers, and the *action limit,* which describes hazardous lifting for some workers) are very sensitive to the horizontal location of the load. Simply providing clearance for a person's feet when next to a conveyor will increase the permissible load by 30% or more. In addition to the vertical and horizontal location of loads to be lifted, it is also necessary to assure a good working surface, that is, one that provides high foot traction, is free of any tripping hazards, and is level (or less than about a 4° grade).

5.4.5 SUMMARY

It has been shown in this chapter that localized musculoskeletal injury can be caused by sustained or frequent exertions. At first there may be simple, acute muscle fatigue. If the causal conditions persist, however, inflammation of tendons, synovia, bursa, and adjacent joint structures result in more severe and chronic symptoms and functional limitations.

Several job design considerations are necessary to prevent such cumulative trauma. First there must be the recognition that muscular strength is quite limited in certain postures, and any exertion in these "weaker" postures will require the muscles to work close to or at their maximum capacity, producing high fatigue rates. Secondly, muscles cannot sustain static contractions over about 15 to 30% of their strength without fatigue. This is particularly relevant to the specification of work conditions requiring prolonged unusual postures, for example, stooped-over, reaching above the head, leaning to one side, or standing on one foot. Also, when handling a heavy tool or object, it can create high load moments at the shoulder and back if not held close to the body with the torso in an erect posture. These high load moments result in rapid muscle fatigue and associated shoulder and low back tissue cumulative trauma.

The length of time provided for muscles to rest between contractions is important in prolonging the endurance time. Unfortunately, if the load on the muscle is high relative to its strength, and the contraction must be sustained for more than 5 to 7 sec, rest allowances of 20 to 50 times the contraction period may be necessary to allow recovery from the resulting fatigue process. Once again, any static exertion of even a few seconds should be of limited strength demand, that is, less than 20% of maximum.

Lastly, it has been shown that fatiguing muscle loads are often associated with job postural requirements, which in turn depend on the degree of matching between the workplace/machine control and display layout dimensions and individual worker anthropometry. This leads to the recommendation that future workplace and machine designs must have a large range of adjustability to accommodate the normal anthropometric variations in worker populations. This is true of both sitting and standing workplace designs.

REFERENCES

Barnes, R. M. (1968). *Motion and time study.* (pp. 10–20 and 487–510). New York: Wiley.

Bjelle, A., Hagberg, M., and Michaelsson, G. (1973). Occupational and individual factors in acute shoulder-neck disorders among industrial workers. *British Journal of Industrial Medicine, 15*(4), 346–354.

Boussenna, M., Corlett, E. N., and Pheasant, S. T. (1981). The relationship between discomfort and postural loading at the joints. *Ergonomics, 25*(4), 315–322.

Cailliet, R. (1981). *Shoulder pain,* 2nd ed., Philadelphia: F. A. Davis, pp. 38–53.

Chaffin, D. B. (1973). Localized muscle fatigue—definition and measurement. *Journal of Occupational Medicine, 15*(4), 346–354.

Chaffin, D. B. and Andersson, G. (1984). *Occupational biomechanics,* New York: Wiley.

Copley, F. B. (1923). *Frederick W. Taylor,* Vol. I. New York: Harper and Bros., pp. 10, 56.

Engin, A. E. (1980). On the biomechanics of the shoulder complex. *Journal of Biomechanics, 13,* 575–590.

Garg, A., and Chaffin, D. B. (1975, March). A biomechanical computerized simulation of human strengths. *AIIE, 7,* 1–15.

Gilbreth, F. B. (1911). *Motion study.* Princeton, NJ: D. Van Nostrand. p. 88.

Gilbreth, F. B. (1972). The present state of the art of industrial management. *Transactions of the ASME, 34,* 1224–1226.

Grandjean, E. (1982). *Fitting the task to the man—An ergonomic approach.* London: Taylor & Francis.

Habes, D. (1980). Low-back EMG and pain in stooped posture. Ann Arbor: University of Michigan, *Center for Ergonomics.*

Hagberg, M. (1981a). Workload and fatigue in repetitive arm elevations. *Ergonomics, 24*(7), 543–555.

Hagberg, M. (1981b). Electromyographic signs of shoulder muscular fatigue in two elevated arm positions. *American Journal of Physiological Medicine, 60*(3), 111–121.

Herberts, P., Kadefors, R., Andersson, G., and Peterson, I. (1981). Shoulder pain in industry: An epidemiological study on welders. *Acta Orthopedica Scandinevika, 52,* 229–306.

Herberts, P., Kadefors, R., and Broman, H. (1980). Arm positioning in manual tasks—an electromyographic study of localized muscle fatigue, *Ergonomics, 23*(7), 655–665.

Jorgensen, K. (1970). Back muscle strength and body weight as limiting factors for work in standing slightly-stooped position, *Scandinavian Journal of Rehabilitation Medicine, 2,* 149–153.

Karger, D. W., and Bayha, F. H. (1966). *Engineered work measurement.* New York: Industrial Press, 89–402.

Konz, S. (1979). *Work design.* Columbus, OH: Grid Publishing Co., pp. 103–144.

Kumar, S., and Scaife, W. G. S. (1979). A precision task, posture, and strain. *Journal of Safety Research, 11*(1), 28–36.

Laubach, L. L. (1978). Human muscular strength. In Webb Associates, Ed., *Anthropometric Source Book* (NASA No. 1024). Washington, DC: U.S. National Aeronautics and Space Administration.

National Institute for Occupational Safety and Health (1981). *Work practices guide for practices guide for manual lifting* (Technical Report 81–122). Cincinnati, OH: NIOSH, pp. 129–144.

Neibel, B. W. (1972). *Motion and time study.* (pp. 417–462) Homewood, IL: Irwin.

Nolan, M. F. (1979). Internal rotator-adductor tendonitis. *Physical Medicine, 56,* 254–278.

Rohmert, W. (1966). *Maximalkraefte von Maennern im Bewegungsraum der Arme und Beine.* Köln, West Germany: Westdeutscher Verlag.

Taylor, F. W. (1929). *The principles of scientific management,* New York, Harper and Bros., p. 36.

Tichauer, R. (1968). Potential of biomechanics for solving specific hazard problems. *Proceedings of ASSE 1968 Conference,* Park Ridge, IL: American Society of Safety Engineers, pp. 149–187.

Tichauer, R. (1978). *The biomechanical basis of ergonomics.* New York, Wiley-Interscience, p. 38.

PART 6
ENVIRONMENTAL DESIGN

CHAPTER 6.1
NOISE

D. M. JONES

University of Wales Institute of Science and Technology, Cardiff

D. E. BROADBENT

University of Oxford

Noise in the form of unwanted sound is an inevitable side effect of human and mechanical activity. It impinges on the most delicate and sensitive of receptors, consisting of mechanical, hydraulic, and neurological components, fashioned for a range of purposes that are central to human well-being. The ear is not simply a passive device for recording sound; it has intimate connections with the mechanisms of alertness, it plays a central role in transmitting language, and it also serves as a medium for rich and diverse aesthetic experience. Noise may pose a threat to all these functions of human hearing.

This chapter is concerned with outlining the effects of noise on four main areas: loss of hearing, communication, efficiency at work, and general well-being. Before embarking on an account of these effects of noise we examine the ways in which sound may be described and measured.

6.1.1 THE MEASUREMENT OF SOUND

6.1.1.1 Frequency and Intensity

The portion of the ear that is responsible for hearing is primarily sensitive to variations in pressure. These variations are usually transmitted through the air, although any elastic medium will serve the purpose of relaying sound. Oscillations in pressure set up in an object are transmitted through the air by successive displacement of molecules from their stable positions.

The pattern of displacement of molecules is shown in Figure 6.1.1 for what is known as a plane wave, illustrated by the reciprocating motion of a piston in an air-filled tube. Regular motion of the piston produces alternate areas of compression and rarefaction within the tube. In the area adjacent to the face of the piston, as the piston moves to the right (Figure 6.1.1a), the molecules are compressed (producing an area of relatively high pressure); a move to the left (Figure 6.1.1b) produces an area of rarefaction (an area of relatively low pressure). With several cycles of such activity, the pattern of rarefaction and compression appears to move down the tube (Figure 6.1.1c). This pattern may be charted by observing the level of pressure at a single point in the tube over some interval of the piston's activity. Figure 6.1.1d charts the pattern of pressure variation produced by the simple harmonic motion of the piston. This pattern of change, known as a sine wave, corresponds to the pattern observed in recording a single tone, such as that produced by a properly struck tuning fork.

Two features of the way in which the action of the piston may vary are important to our discussion. First, as the size of the excursions of the piston increase, so does the amplitude of the resulting sine wave. This brings about a change in the intensity of the sound (measured in physical terms) which may be accompanied by the perception of an increase in loudness (the subjective attribute of intensity). Second, an increase in the number of excursions of the piston will result in an increased frequency of peaks and troughs of the wave. The frequency of the sound is the number of peaks per second and is expressed in hertz (Hz). Changes in frequency may result in changes in the sensation of pitch (the subjective attribute of frequency). The range to which the human ear is sensitive stretches from 20 Hz to 20 KHz and is referred to as the audio-frequency range. It is important to distinguish parameters of sound that are the result of physical measurement and the psychological effects associated with such changes: there is no simple one-to-one correspondence between them (for further details of these phenomena, see Moore, 1984).

6.1.1.2 The Decibel

The unit of measurement of intensity is the decibel. So large is the range of pressures to which the ear is sensitive (1 : 100,000,000,000,000) that a logarithmic scale of magnitude is used to encompass it. In the decibel scale the pressure produced by a source (p) is expressed in relation to a standard reference value (p_{ref}), which corresponds roughly to the minimum pressure detectable by the human ear (taken as 20 micropascals). Levels expressed in this way are referred to as sound pressure levels (SPL) thus:

$$\text{Number of decibels (SPL)} = 20 \log \left(\frac{p}{p_{ref}} \right)$$

Decibels may also be used to express ratios between two magnitudes, and in this case the standard reference value is not employed. If in doubt, the reader should consult the appropriate sources for further details of the manipulation of decibels (see, for example, Kohler, 1984). Caution should be exercised over the use of the decibel scale. For example, doubling of sound pressure corresponds to an increase in SPL of 6 dB and multiplication of sound pressure by a factor of 10 increases the SPL

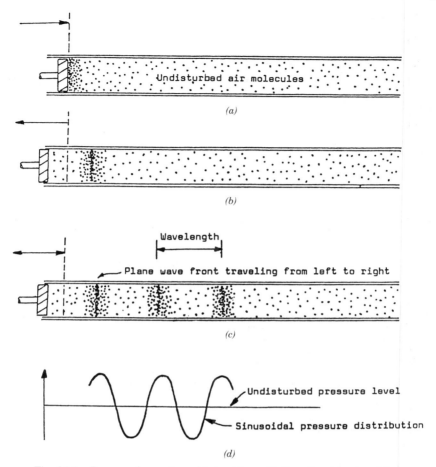

Fig. 6.1.1.   Pressure changes brought about by oscillation of a piston in a tube.

by 20 dB. Doubling sound energy, on the other hand, corresponds to an increase of only 3 dB. Notice that subjective loudness increases less rapidly than energy or pressure. Roughly, a 10 dB increase doubles the loudness as judged by a listener.

6.1.1.3   Sound Level Meters

The two parameters of primary importance in the measurement of noise are its SPL and its frequency characteristics or spectrum. Details of the latter are rarely specified, because of measurement difficulties. SPL is usually measured by a sound level meter. A number of different models of sound level meter are available commercially, but several features are common to all. First, a microphone is used to transform pressure variations into analog electrical signals. Second, electronic networks amplify the voltages to a level that can be displayed via a meter or digital display. At the same time, currently available devices usually offer the facility for differential attenuation of the frequency range. The reason for this is that the ear is not equally sensitive to all frequencies and the selection of one of a range of weighting networks can be used to simulate the action of the ear. Three such networks are usually provided—A, B, and C—and their characteristics are shown in Figure 6.1.2. Historically, these weightings were intended to simulate the sensitivity of the ear at low (A), medium (B), and high (C) intensities. As Figure 6.1.2 shows, a higher SPL is required in order to produce the same numerical reading on an A-weighted network as on a C-weighted network. Generally, the A and C weightings have found greatest favor. The latter gives essentially equal weight to all frequencies. The A weighting gives greater weight to all frequencies, which contribute more to effects on people, and this is the rationale for using it.

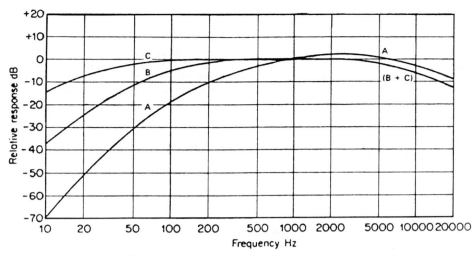

Fig. 6.1.2. Weighting networks typically employed in sound level meters.

More sophisticated forms of meter will allow switching between electronically set "slow" and "fast" rates of response of the display. Setting the meter to the slow rate of response allows the observer to discern the overall level of sound even though the variations in pressure would not be steady enough to allow a reading. Another concession made to rapid variations in pressure is the provision of a "hold" facility, which displays the maximum (or sometimes minimum) of a set of readings. This is particularly useful when the noise is of the impulse type, where the rise in intensity is extremely rapid, and the "hold" facility allows the observer to read the maximal value at leisure. A detailed analysis of very fast changes is best accomplished using an oscilloscope.

Another facility that may be provided on a meter (or provided as an accessory) is a series of filters that pass on for analysis and display sounds only from a limited range of frequencies (known as band-pass filters). In this way, only the energy in a band of sound covering an octave (an interval on the frequency continuum whose upper bound is twice that of its lower bound), or some other convenient interval (such as half- or third-octave bands), may be shown. By taking readings from each filter, set in turn to cover the whole of the audio-frequency range, a convenient, if somewhat crude, picture emerges of the distribution of constituent frequencies in the overall sound.

The advent of microelectronics has allowed the range of facilities to be increased without a corresponding penalty of weight, convenience, or cost. As a result one or both of the following measures are now usually available.

L10, L50, L90, Ln. Each case gives the SPL exceeded for, respectively, 10, 50, 90, or $N\%$ of the total duration of the period of observation. Values may be weighted or unweighted.

Leq. Usually, sound varies in intensity over time. The Leq value for such a signal is the dB(A) level of a constant sound which, if continued over the same interval, would represent the same total energy as the variable sound.

Dose meters are small, simple, and inexpensive versions of the sound level meter which assess total noise exposure at the workplace. They may be worn for long periods without inconvenience. The dose may be expressed as a proportion of the maximum permitted 8 hr dose. By means of relatively simple calculations, the Leq value may also be derived from the reading. Dose meters also signify, by means of a simple visual alarm, whether a specified maximum peak level has been exceeded. Such devices should be used with very great care to ensure that a reading is representative of the exposure. Artifacts may occur, for example, by intentional or unintentional shouting into the microphone. Wearers should be advised of the danger of false readings and if possible be supervised during the period of recording.

In the next section, we examine the effects of noise and begin by discussing what is arguably the most serious effect of exposure to noise, that of hearing loss.

6.1.2 NOISE-INDUCED HEARING LOSS

Exposure to intense sound results in a loss of hearing. This simple assertion conceals a plethora of complications. The main question at issue is, how much noise produces how much hearing loss?

Typically, the main parameter of interest is the threshold: the level at which sounds are detected. Ideally the threshold should be measured before and after exposure to loud noise and the degree of loss ascertained from the difference between the two readings. Three types of hearing loss resulting from exposure to loud noise are usually distinguished: first, temporary threshold shift (TTS) is a short-lived and reversible elevation of the level at which sounds are heard; second, noise-induced permanent threshold shift (NIPTS) is a long-term effect of noise exposure where the hearing loss is not reversible; third, acoustic trauma, which is the result of a single, usually short, exposure to extremely intense noise such as may arise from gunfire or explosions.

Before discussing TTS and NIPTS it should be noted that the notion of hearing loss is not without ambiguity. It may at different times be used to refer to the degree of physiological damage to the tissues in the inner ear, to the ability to detect sound, or to the ability to discriminate one sound from another. Moreover, the degree of loss, as measured by the ability to listen to tones, for example, may bear a tenuous relationship to the degree of handicap. This is in part because handicap is itself difficult to define and may encompass social, domestic, and occupational aspects of life and the criterion for handicap will be different for each of these activities.

6.1.2.1   Temporary Threshold Shift

Although several features can be discerned in the recovery of hearing, that value corresponding to the threshold 2 min after exposure is taken to be representative of the temporary loss (Ward, 1976). The degree of threshold shift is a function of the frequency, intensity, and duration of the noise. Both for pure tones and narrow bands of noise, the greatest elevation of threshold is found at points relatively remote from the stimulating frequency, typically one-half to one octave, though occasionally as much as two octaves, above the upper bound of the stimulating sound (Ward, 1983). TTS shows a linear increase with increasing SPL, at least for moderate intensities of sound (80 to 105 dB SPL) and for exposures of less than 8 hr (see Figure 6.1.3). Among the factors known to diminish the impact of noise on TTS are the range over which the acoustic energy is spread and the degree of change in the spectral composition of the sound during the period of exposure. The growth of TTS to a constant noise level reaches an asymptote in 8 to 12 hr. As a general rule, if TTS at 2 min does not exceed 25 dB, recovery occurs in 16 hr. This means that although the worker at the end of an 8 hr shift may have significant impairment during the hours of rest that usually follow work, hearing is relatively good during sleep and is normal when work commences the following day.

In addition to hearing loss, TTS may be accompanied by the experience of subjective tinnitus, which is often a ringing sound heard in one or both ears (Terry, Jones, Davis, and Slater, 1983).

6.1.2.2   Noise-Induced Permanent Threshold Shift

The primary concern with the study of NIPTS is the safe level of noise to which individuals should be exposed during their working lives. The specification of guidelines for such an exposure should

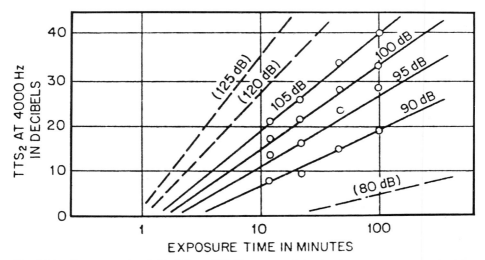

Fig. 6.1.3.   Temporary threshold shift at 4000 Hz measured 2 min after exposure to a noise band centered at 1700 Hz for a range of SPL and durations. (From Ward, Glorig, and Sklar, 1959, with the permission of W. D. Ward and the American Institute of Physics.)

include (1) the stipulation of the limit of exposure for continuous noise in the typical working day, along with (2) some rule for trading exposure time and intensity, so that increased intensity is compensated for by shorter exposure in some lawful way, and (3) some safe upper limit in both intensity and number of short bursts of noise.

NIPTS may arise from a variety of sources, and part of the difficulty of understanding the effects of occupational exposure is in assessing the amount of hearing loss contributed by these other factors. Three types of factors have been identified as contributing to the overall hearing loss: effects due to age (presbycusis); effects due to nonacoustic agents such as industrial chemicals, ototoxic, drugs or illness (nosoacusis); and effects of noise exposure outside the work setting (socioacusis). Each of these factors may contribute in different degrees to the overall loss of hearing. Moreover, the influence of these factors may be as great as that arising from occupational sources. For example, it has been estimated that the typical socioacoustic exposure may be as high as 80 dB(A) for 8 hr (Schori and McGatha, 1978), whereas the recommended maximum for occupational exposure may be 90 dB(A) over a similar interval.

A fairly clear picture has emerged of the way in which NIPTS varies with exposure to noise of occupational origin. An analysis of survey findings suggests that exposure to noise below 80 dB(A) over the working day is innocuous (Passchier-Vermeer, 1974). Exposure to 85 dB(A) produces a loss of the order of 15 dB at high frequencies but many people are unaffected. It may be argued that this is unlikely to produce handicap. With noise at a level of 90 dB(A) over the working day there will be an appreciable number of people whose loss of hearing is well in excess of 20 dB. At intensities higher than 90 dB(A) the degree and scope of loss increase markedly: severe losses at high frequencies together with modest losses at low frequencies (see Figure 6.1.4). These findings serve as the basis for legislation restricting the exposure to high levels of noise.

For NIPTS, the changes in sensitivity with frequency appear to be less dependent on the type of noise to which the individual is exposed than is that for TTS. NIPTS first appears in the region of 4000 Hz and, with increasing exposure to noise, spreads to adjacent frequencies (see Figure 6.1.5). One reason for susceptibility to hearing loss in this region is that the ear is particularly sensitive to frequencies in this region. Insofar as there is a dependence of hearing loss on the spectrum of noise, high frequencies are rather more serious. This means that the A-weighted noise level is a better measure of risk to hearing than the unweighted measure.

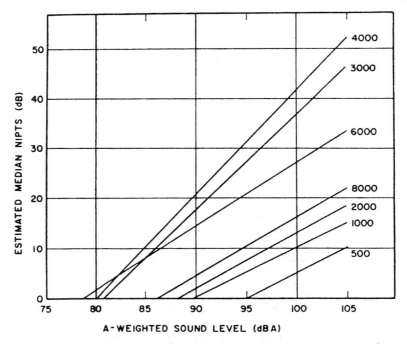

Fig. 6.1.4. Estimated NIPTS at various frequencies produced by 10 years or more of exposure to industrial noise. The noise level is the A-weighted level for 8 hr/day, 200 days/year. (From Ward, 1983, with the permission of W. D. Ward.)

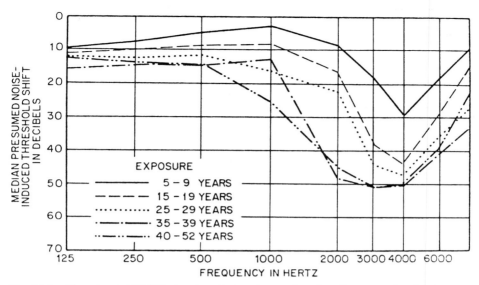

Fig. 6.1.5. The estimated NIPTS at various frequencies as a function of exposure duration in years. (From Taylor, Pearson, Mair, and Burns, 1965, with the permission of the American Institute of Physics.)

The trading relation between time and intensity is based on "the equal energy rule" so that if 90 dB(A) for 8 hr is acceptable, then for every halving of exposure time, the overall intensity can be increased by 3 dB. Some investigators have taken a less cautious approach by advocating 5 or 6 dB as an appropriate halving factor (e.g., Kraak, 1982). Some sounds extend the physical response of the ear beyond its "elastic limits." In recognition of this danger, most countries place an upper limit of 140 dB peak SPL to any single exposure. Additionally, some limit to the number of impulses may also be specified or caution may be counseled (see Passchier-Vermeer, 1983).

6.1.2.3 Hearing Conservation

Three methods may be adopted to reduce the risk of hearing loss from noise. The first is that the level of noise be reduced by the design or modification of machines. The transmission of sound from such machines may also be reduced by the use of baffles or screens. Such solutions are best implemented by specialist engineers. A second approach is to test the susceptibility of individuals to hearing loss. Unfortunately, tests or indexes of susceptibility are poor at predicting the degree of risk of hearing loss. A third approach, dealt with in some detail here, is to attenuate the sounds at the worker's ear by means of plugs or muffs. This is a convenient means of hearing conservation that has the virtue of being inexpensive but is not without its drawbacks.

Broadly speaking, hearing protectors are of two types: earplugs and earmuffs. Earplugs are either inserted in or held against the entrance to the external ear canal. They are made from a variety of materials, such as silicone, clay, or rubber, and may be either premolded or shaped to fit by the user. This latter type is usually disposable. Earmuffs are used to enclose the external ear completely by means of a pair of earcups held against the head by a sprung headband. The earmuffs fit closely to the side of the head by means of a soft cushion which forms an effective acoustic seal around the edge of the earcup. Earmuffs may be incorporated within one of a wide range of protective devices such as helmets and welding masks, or may have earphones inserted within the earcup.

The degree of attenuation offered by hearing protectors varies from one version to another. Maximum protection is offered when earplugs and earmuffs are combined, the combination of the two devices giving an advantage of between 5 and 15 dB (depending on frequency) over earmuffs alone. Figure 6.1.6 gives some indication of the likely attenuation with various devices, although it should be emphasized that, in practice, the degree of attenuation depends on factors influencing the fit of the protector. Once the degree of attenuation provided by the hearing protector is known, the noise exposure of the wearer may be adjusted accordingly (the reader is advised to consult the legislation and codes of practice in the country of interest). In addition to the considerations of attenuation, factors influencing the choice of device include its comfort, safety, and durability.

TYPE OF PROTECTOR	FREQUENCY RANGE IN HERTZ				
	1- 20	20 -100	100 - 800	800 - 8000	> 8000
 EARPLUGS	5–10	5–20	20–35	30–40	30–40
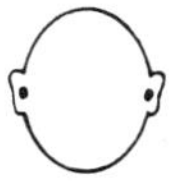 SEMIINSERT EARPLUGS	5–10	5–20	15–20	25–40	30–40
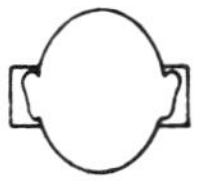 EARMUFFS	0–2	2–15	15–35	30–45	35–45
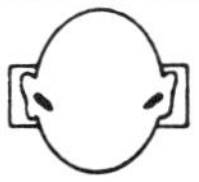 EARPLUGS AND EARMUFFS COMBINED	10–15	15–25	25–45	30–60	40–60
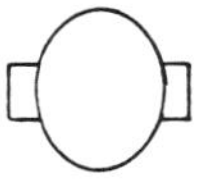 COMMUNICATION HEADSETS	0–2	2–10	10–30	25–40	30–40
 HELMET	0–2	2–7	7–20	20–55	30–55
 SPACE HELMET (TOTAL HEAD ENCLOSURE)	3–8	5–10	10–25	30–60	30–60

ENTRIES SHOW APPROXIMATE MINIMA AND MAXIMA OF PROTECTION
AVAILABLE FROM VARIOUS TYPES OF DEVICES, IN DECIBELS.

Fig. 6.1.6. The expected effects, in decibels, of a range of hearing protectors as a function of frequency. (From C. M. Harris, *Handbook of Noise Control.* Copyright © 1979. Reprinted with permission of McGraw-Hill Book Company.)

6.1.2.4 Effects of Hearing Protectors on Talking and Listening

In continuous noise at levels between 85 and 105 dB SPL, hearing protectors improve the reception of speech although this advantage does not extend to those who have hearing loss at high frequencies (Howell and Martin, 1975). This effect occurs despite the fact that noise attenuates the speech and the noise in equal measure (the signal-to-noise ratio is the same with and without ear defenders). In order to overcome the effects of noise the level of speech would have to be increased. The most likely reason for the advantage given by hearing protectors for the reception of speech is that speech of over 85 dB is known to produce significant degrees of aural distortion, and this effect is reduced by attenuating the overall level of sound, thus reducing speech as well as noise. At levels of continuous noise below 80 dB SPL (with speech below 75 dB SPL), hearing protectors may reduce the intelligibility of speech for the listener. This is because the quieter portions of speech are attenuated by the hearing protector to levels below those of the threshold of audibility.

The intelligibility of speech produced by the talker may be diminished when hearing protectors are worn. Wearing hearing protectors in noise causes the speaker to lower the level of the voice by 3 to 4 dB. A speaker regulates the loudness of the voice on the basis of the information about its loudness reaching the ear via bones of the skull and through the air. When hearing protectors are worn, the level of airborne speech sounds reaching the ear is reduced and speech is transmitted to the ear primarily by bone conduction. This gives the speaker the impression that there is less external noise to counteract while talking to others and thus the level of speech is adjusted to a lower level than it should be for the noisy setting.

Unfortunately, in continuous loud noise the beneficial effect of wearing hearing protectors for the listener is more than outweighed by the detrimental effect of hearing protectors on the speaker. In intermittent noise, conventional hearing protectors severely disrupt the reception of speech when the noise is temporarily in abeyance. However, this shortcoming can be overcome partly by the use of amplitude-sensitive earplugs, which increase the degree of attenuation as the level of noise breaches a certain threshold of intensity (Mosko and Fletcher, 1971).

Although the detectability of sounds against a noise background is therefore little affected by hearing protection, there is the possibility that unexpected sounds such as approaching vehicles or other hazards may be less "attention-getting." Reduction of noise at the source is the preferred method of hearing protection, for this and other reasons.

In the next section we discuss the way noise impairs, among other things, the perception of simple signals, alarms, and speech.

6.1.3   EFFECTS OF NOISE ON COMMUNICATION

Noise may interfere with the detection of auditory signals; this effect is known as masking. Masking will have an effect on the detection of relatively simple auditory signals, such as warnings and alarms, but by far the most pervasive effect of masking is that on speech.

6.1.3.1   Masking of Nonspeech Sounds

Figure 6.1.7 shows that as a tone is brought progressively closer to the center of a masker (a narrow band of noise), increasingly high levels of sound must be used for the tone to be heard. A striking feature of the family of curves shown is that at progressively higher levels of masker the curve describing the degree of masking becomes more asymmetric. The effect of the masker is spread upward along the frequency continuum. It follows that a signal below the lower bound of a masker of this sort will be a more effective alarm than one above the upper bound of the masker. Only that portion of a masker close in frequency to the signal will contribute to the masking of the signal. The range of sounds that contributes to the masking of, say, a tone is known as the critical band. The width of this band is roughly 50 Hz at frequencies below 1000 Hz and some 10% of the center frequency above 100 Hz. It follows that noise falling outside these bounds will not contribute to any great degree to the masking of tones within it (see Webster, 1984, for further details).

The most usual case is of the masking of contemporaneous sounds; however, masking can occur when a masker either precedes or follows a sound. Masking may also be "central" rather than peripheral when the perception of a signal at one ear is impaired by the presence of a masker in the other ear. In certain circumstances, the power of the ear to localize sound may be used to militate against the effects of masking by capitalizing on the phenomenon known as "masking level difference." The most striking effects of this phenomenon are to be found at frequencies below 1500 Hz. For example, if a signal in the form of a sine wave from a single source is presented simultaneously to each ear, accompanied by noise to both ears (again from the same source), then, as expected, the tone will be masked in proportion to the signal-to-noise ratio. However, a lower estimate of masking will be obtained by simply inverting the signal (making it out of phase) at one ear or by presenting noise to both ears and presenting the signal to one ear only (Gebhardt and Goldstein, 1972). These techniques can be used to improve speech intelligibilty over earphones in noisy surroundings; this effect also occurs with locating speech sources at an angle to sources of masking noise.

Sound may be used to signify various states of a machine, for example, as in the use of alarms. In more complex systems, several different sorts of alarms may be required. In this case, in addition to observing rules about the intensity of the alarm in relation to the background sound, care must be exercised over the selection of spectral and temporal features of the alarms in order that they may be distinguished from one another.

One set of guidelines developed for use on civilian aircraft embodies a number of interesting suggestions that may be adapted to other circumstances (Patterson, 1983), as follows: (1) The alarm should

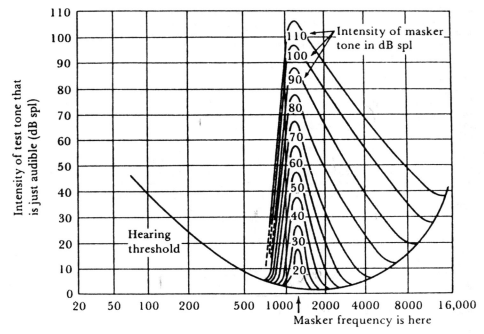

Frequency of test tone (Hz)

Fig. 6.1.7. Masking curves for a narrow band of noise centred at 1200 Hz. Each curve represents the intensity of a tone that is just audible in masker at the given intensity. (Adapted from E. Zwicker and B. Scharf (1965). A model of loudness summation. *Psychological Review, 72*, 3–26. Copyright © 1965 by the American Psychological Association. Reprinted/Adapted by permission of the author.

be at least 15 dB above the level at which it can just be detected in the background noise. An upper limit of 25 dB above this threshold should be observed when the background levels are already high to avoid the danger of overloading the ear if the level of the alarm is excessively high. (2) If pulses of sound are used in the alarm the length of pulse should ideally be 100 to 150 msec and should rise (and fall) in 20 to 30 msec. Pulses of shorter duration should be reserved to signify more urgent events. (3) The use of a distinctive temporal pattern (of five or more pulses) should aid identification of warnings when more than one warning is possible. (4) A sound with several harmonically related components (that is, containing component sounds that are multiples of each other in frequency) helps to give the impression of a single fused sound which again helps identification. (5) Synthesized speech may be presented during an alarm, but its prominence should depend on the purpose of the alarm. If the alarm should be responded to immediately, the spoken component should be brief and should not be repeated when the alarm is repeated. If the alarm is less urgent, a longer spoken portion may be used, and this can be presented at each repetition of the signal. For examples of the way in which warning sounds can be designed for optimum detection, see Chapter 5.2 by Sorkin.

6.1.3.2 Masking of Speech

Predicting the effects of noise on speech presents a host of complexities simply because speech is a broad-band complex acoustical signal containing periodic and aperiodic components which include impulse sound. Effects on the talker and on the listener may be distinguished.

Effects on the Talker

Loud noise causes the overall level of speech to be raised spontaneously. In most cases this increased speech output does not match the increase in intensity of the noise. Values as low as a 3 dB increase have been recorded in response to a 10 dB elevation in noise (van Heusden, Plomp, and Pols, 1979), although in other cases more modest values between 5 and 10 dB in response to similar elevations have been recorded (Pearsons, Bennett, and Fidell, 1977).

In the absence of masking sounds, the overall level of speech that is adopted spontaneously varies widely. When face to face with a listener 1 m away, the range of speech can be as great as 35 to 90 dB SPL (Pickett, 1956), although there is evidence that this range is less in settings outside the laboratory (with a typical range of 48 to 70 dB(A); Pearsons et al., 1977). When speaking into a microphone people affect a spontaneous increase in the level of speech, with the range of intensities being typically 54 to 72 dB SPL. Idealized data for a range of distances and background noises are shown in Figure 6.1.8. As a general rule, voice level will be raised as the background noise level increases, as the task of communicating a message is made more demanding or when talking in front of an audience or into a microphone (Webster, 1984).

One response to the difficulties of communicating in noise is for the person to shout. Increasing vocal effort has side effects on the intelligibility of the speech. The intelligibility of speech is relatively constant at levels of speech in the range from 50 to 78 dB but intelligibility diminishes above and below this range (Pickett, 1956). The average spectral composition of speech tends to change when the level exceeds this range, which may account in part for the drop in intelligibility. As the output increases frequencies around 1000 Hz predominate with frequencies below this region being less in evidence (Webster, 1979). In addition, there are residual effects of hoarseness from shouting for any length of time.

Effects on the Listener

When speech is presented through electronic circuitry such as that on a television, the effect of ambient noise may be overcome by the simple expedient of turning up the volume control. For example, as the intensity of noise is increased from 35 to 57 dB(A) the volume control on a television will be set in the range 55 to 70 dB(A), a trading relation of 0.7 dB (Pearsons et al., 1977). Although these data, which were drawn from field studies, call for a 20 dB difference between speech and noise for purposes of listening in the domestic environment, other studies suggest that this difference may be too large and that values of the order of 5 dB meet the criterion of "just comfortable" and a difference of 15 dB meets the "most comfortable" criterion (Pols, van Heusden, and Plomp, 1980).

When face to face, individuals may regulate the distance from one another to overcome the effect of noise. However, it is very difficult to predict these changes because the distances adopted have as much to do with the social proprieties or the physical constraints as with the acoustic properties of a setting. Another useful way of overcoming the effect of noise is speech (lip) reading. Even when speech reception is already fairly good (with half the sentences being heard correctly, the so-called speech reception threshold), improvements equivalent to a reduction in noise level by 5 to 8 dB can

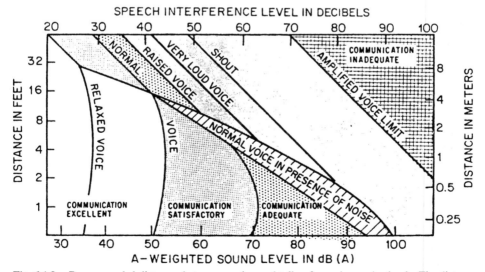

Fig. 6.1.8. Recommended distances between speaker and talker for various voice levels. The distance is plotted as a function of A-weighted sound level along the lower abscissa and in terms of speech interference level along the upper abscissa. The distances given are for speaker and listener not facing one another. For face-to-face communication, the value of noise level given in the figure may be increased by 5 dB. (From C. M. Harris (1979). *Handbook of Noise Control.* New York: McGraw-Hill. Copyright © 1979. Reprinted with the permission of McGraw-Hill Book Company.)

be brought about by this means even in those untrained in speech-reading skills. In less favorable conditions for reception of speech, the advantage gained by lip reading is much greater and may be equivalent to an improvement of 15 to 25 dB in terms of signal-to-noise ratio (Waltzman and Levitt, 1978).

Levels of the human voice may range from 35 to 90 dB SPL, the usable range being between 50 and 80 dB(A). The range between just audible and just tolerable levels of speech is roughly 90 dB, but the preferred level for listening corresponds to that for production, that is, between 50 and 80 dB(A). A number of indexes of the likely effects of masking have been developed; each has shortcomings to some degree. Part of the difficulty with the development of such indexes is that the effect of what is being said is often as great as the effect of the ratio between signal and noise. In addition, there are effects such as that of the degree of practice at listening for sounds and familiarity with the voice (both of which influence intelligibility), which are difficult to incorporate into such indexes.

6.1.3.3 Measures of Interference with Speech by Noise

Despite its several shortcomings, one of the best ways of predicting masking properties of noise on speech intelligibility is to use the integrated A-weighted measure. The main advantage of this type of measure (Leq with A weighting) is that it is readily available on sound level meters, with the accessibility and simplicity of measurement that this implies (Klumpp and Leonard, 1963). One can therefore use the meter to take a reading, and then consult Figure 6.1.8 with measurement and the distance of communication in feet, to determine if communication will be adequate. Averaging methods, such as the articulation index and the speech interference level are more complex but in general give better predictions of interference.

The Articulation Index (AI)

This measure is based on the idea that each of a range of narrow bands of frequencies makes a contribution to the total degree of interference. In its original form, 20 bands were used and the general procedure was one of calculating the difference between the speech level and the noise levels in each of the bands. In the original method there were more bands in the 2000 Hz region than in the 500 Hz region to give due recognition to the fact that some frequencies contribute more than others to speech intelligibility. Variants of the method involve the use of bands of a fixed size (such as third-octave or octave intervals) and the use of weighting factors to achieve the same end. For a detailed analysis of the procedure consult Kryter (1970).

The accuracy of AI depends in large part upon a range of factors, including the type of speech material, the size of the vocabulary, and the context, such as that provided by a sentence. There is a wide variety of tests incorporating these factors to different degrees; some test the intelligibility of single words and others test whole sentences. One type, the PB (phonetically balanced) test, uses lists of monosyllabic words constructed to represent the distribution of phonemes in the English language. This kind of material may be presented with or without a range of alternative responses. Where response alternatives are given they are usually similar to the word of interest and are called "rhyme tests." Tests of sentence comprehension take various diverse forms, with variations in the type of sentence delivered and the type of response required (for an overview, see Webster, 1984).

Speech Interference Level (SIL)

This measure, unlike AI, relies on the measurement of noise alone. SIL is calculated from the arithmetic mean of the sound pressure levels for octave bands centered at 500, 1000, 2000, and 4000 Hz. The resulting index gives the level of the speech signal that should yield satisfactory reception in a given noise. Using Figure 6.1.8, one may assess the combination of distance and vocal effort that will give rise to the SIL in question. For example, an SIL of 60 may be achieved in "a normal voice in the presence of noise" with a speaker-to-listener distance of a little over 1 m.

6.1.3.4 Recommendations: Effects of Noise on Speech

The following tactics may be employed to improve the intelligibility of speech: (1) standardized phraseology and procedures should be developed by limiting the vocabulary to a relatively small number of polysyllabic words. (2) Only a limited number of talkers should use a system such as a paging system or public address system. Speakers and listeners improve as a result of experience. (3) Hearing protectors should be employed but in difficult circumstances they should be used in conjunction with a headphone/microphone system. (4) Speech should be amplified (but not to levels that overload the ear), using noise-canceling microphones if necessary.

In the next section we extend our discussion of the effects of noise on communication to include recent work on the way that noise may interfere with material that is presented visually and on some of the cognitive processes involved in reading and writing.

6.1.4   EFFECTS ON READING

Until recently, it was thought that the effect of background speech on reading and language was rather small. A series of studies has by now shown that speech at relatively low levels can disrupt memory for items presented visually, which in practical terms means that other people's conversations are likely to disrupt the activities of reading and writing. This concern is in part due to the increasing prevalence of office work, the trend toward open-plan offices, and the likely implications for efficiency in such settings. In addition to the concern over efficiency at work, there is some suggestion that the development of reading performance may be impaired as a result of exposure to noise at home (Cohen, Glass, and Singer, 1973) and in the classroom (Crook and Langdon, 1974).

Two further features of the effect of irrelevant speech on material presented visually make the topic an important one. The first is that the effect seems to be independent of the overall intensity of background speech. For example, the disruptive effect of speech on memory for visually presented material remains roughly the same over a range of intensities from 40 to 76 dB(A) (Colle, 1980; Salame and Baddeley, 1983). The second important feature is that the effect is one that depends on the type of material being heard. Noise, be it continuous, intermittent, white, or pink, produces little or no effect in this kind of task (Murray, 1965). However, even speech in a language unfamiliar to the listener disrupts this type of performance. The effect in part depends on the degree to which the visual material and the distracting speech share similar speech sounds or phonemes (see Figure 6.1.9).

One view of these effects is that the visual image is translated by the observer into a kind of acoustic image and that this image is very difficult to discriminate in the presence of other acoustic images introduced from material that is heard. Thus the interference of one type of material with another occurs in some part of memory in which materials are stored in their acoustic (phonological) form (Salame and Baddeley, 1983). Another possibility is that the interference occurs at an earlier stage when the two types of material are competing for entry into the nervous system; in other words, the effect is at the level of perception (Broadbent, 1983). This view is based on the observation made in other contexts that if, when an individual is trying to decide on a course of action, erroneous information irrelevant to the judgment is presented, the judgment is made more difficult. Regardless of which of these views is correct, the important practical point about this area of research is that types of noise that were hitherto regarded as safe can now be regarded as having the potential to disrupt performance.

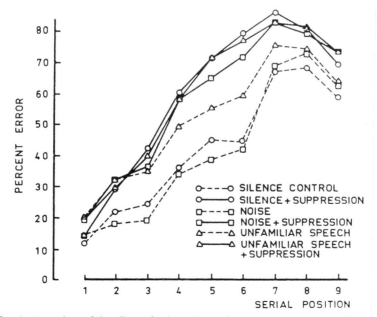

Fig. 6.1.9.   A comparison of the effects of noise and speech upon memory for a list of items presented visually. The presentation position of the items to be remembered is shown on the abscissa. On the ordinate, the percentage error for each of the serial positions is shown. The suppression condition shown in the legend is one in which rehearsal is prevented. (From Salame and Baddeley, 1983. Reprinted by permission of P. Salame.)

Comprehension of the material being read suffers in loud noise. Not only does it take longer to read a passage, but it seems that memory for the contents of the passage is poor. Systematic studies of the type of errors produced in proofreading tasks indicate that although superficial features of the text are understood, part of the deeper meaning of the passage is lost in the presence of noise. Noise as low as 68 dB(A) impairs the detection of contextual errors (grammatical mistakes, omissions, and the presence of incorrect and inappropriate words) but both the detection of noncontextual errors (misspellings and typographical errors) and the average rate of work are unimpaired (Weinstein, 1977).

When writing in noise, efficiency may also be diminished because of the difficulty of retrieving material from that part of long-term memory concerned with the meaning of words, the so-called "semantic memory." In laboratory versions of such tasks, one may be asked to produce the appropriate word when given the name of the category of meaning from which the word is drawn (e.g., "four-footed animal") and a clue to the first letter of the word (e.g., "D"). The main variable of interest is the time in which the correct spoken response ("dog") is produced. By the simple expedient of changing the first letter clue (e.g., to "Y"), the instance may be made more difficult to retrieve (the correct response being "yak"). Noise at levels as low as 80 dB appears to facilitate the process of retrieving easy instances but the retrieval of more difficult instances is impaired (Eysenck, 1975). This type of effect may be produced by a shift in the confidence in which each type of material is recalled. After reading a story in loud noise, people become more cautious about recalling rare (or difficult) words and less cautious about recalling common words, which is a finding in line with the ones just discussed for the speed of response (Jones, Thomas, and Harding, 1982).

In the next section the discussion of performance is extended to include tasks whose primary component is not verbal.

6.1.5 EFFECTS ON TASK PERFORMANCE

The effects of noise on performance may be divided conveniently into four categories: (1) effects of arousal, (2) effects of lack of control, (3) strategic effects, and (4) effects on attention.

6.1.5.1 Effects of Arousal

This effect takes three main forms. The first is an effect of short, often unpredictable, noise bursts that may startle the individual. The second type of arousing effect is one due to the variability of acoustic input, which may be produced by occasional bursts of sound and reduced by continuous unvarying noise. The third is that loud sounds raise the general excitability or responsiveness of the person; this may be good or bad for working efficiency, depending on the general state of the person.

Bursts of Noise

Unexpected bursts of noise produce marked but transient changes in the physiological response. Three categories of response to unpredicted noises may be distinguished: startle response, orienting reflex, and defense reflex (Burns, 1979). The startle response is a potentially protective muscular response (including eye closure, facial muscle contraction, and head jerk), whereas the orienting reflex is a general alerting response (a "what is it?" reaction), and the defense reflex is a response to intense sound stimuli that become interpreted as harmful. The magnitude of the orienting response diminishes with repeated presentation of the noise burst (a phenomenon known as habituation) as the novelty of the stimulus diminishes. The startle response also habituates, although there is some evidence that it never completely disappears even after many repetitions of the stimulus (Landis and Hunt, 1939). The defense response may increase with repeated presentation of the stimulus in conditions where the significance of the noise burst becomes regarded as malignant.

The effects of bursts of noise on performance tend to be localized in the period up to 30 sec following the burst. Some attempts have been made to establish the type of task that would be most susceptible to such effects with generally equivocal results. Bursts of noise slow the speed of a simple motor response only when the burst arrives during the execution of the response rather than, for example, during the presentation of the stimulus (Fisher, 1973). In another type of task involving mental arithmetic, the effect of noise appears to be on the intake of information, whereas in the same task the period of mental calculation was relatively immune to the effect of bursts (Woodhead, 1964). Results of yet another study suggest that if bursts (at 96 dB(A)) are timed to coincide with each of a string of visual digits there are more errors when the digits are subsequently recalled than if the bursts arrive between the digits (Salame and Wittersheim, 1978). This last study once again suggests that the intake of information is disrupted by noise.

Based on these results, one general rule might be that the elements of the task that are susceptible to disruption by noise bursts are those that are "data limited." Activities of this kind include short perceptual or motor events in which the deployment of compensatory effort on the part of the subject is not possible. Such perceptual and motor effects may be a reflection of the startle effect described above, although no study has yet sought to confirm this view. Thus higher mental activity such as that involved in calculation is immune to the specific effects of infrequent bursts because of the way in which short-term memory may compensate for such brief disruptions.

One important consideration is the extent to which these short-term responses diminish with repeated presentation of the burst. Although it is clear that elements of the physiological response such as the orienting reflex are eliminated by habituation, the startle response may resist complete elimination. Even after long periods of exposure such as that found in competition marksmen, residual effects of the eye-blink and head-jerk response are found (Landis and Hunt, 1939). Laboratory studies have confirmed that hearing blank pistol shots (with a peak level of 124 dB SPL) is still capable of disrupting performance of a skilled tracking task after many shots have been fired (May and Rice, 1971). Performance in complex tasks is disrupted if uninterrupted vision and steady posture are necessary for the successful execution of the task.

Effects that extend to the order of 30 sec beyond the burst are likely to be ones associated with strategic effects rather than arousing effects (see Woodhead, 1966). As the effect of startle becomes less pronounced, the subject may gain tactical advantage by anticipating the appearance of a noise burst and overcome its effects by compensatory effort.

Variability of the Acoustic Background

A second type of arousing effect may be found because noise, of a variable or intermittent type, may improve performance in vigilance tasks after a period of long-continued work. These tasks involve the detection of small changes in the flow of information arising from one or more sources over long unbroken periods of time. Typically, ability to detect such signals diminishes as the time at work increases. Varied and irrelevant auditory stimulation stems the decline usually found in quiet, at least when events within the task are presented at a low rate (McGrath, 1963). This suggests that in long, monotonous tasks, variable noise may serve to raise the level of arousal and hence improve efficiency (see Figure 6.1.10).

Similarly, even continuous noise, if loud, can reduce the effect of other conditions that produce drowsiness. For example, the effect of loud noise may counteract that of the loss of sleep. In isolation, noise and sleep loss have a deleterious effect at the end of a 40 min period of continuous work. Yet when both stressors are present together, the net effect is one of an improvement in performance (Wilkinson, 1963). Because it is generally regarded that both very high and very low levels of arousal may give rise to inefficiency, it may be argued that the de-arousing effect of sleep loss is being counteracted by the arousing effect of noise (see Broadbent, 1971, for an overview).

Judgments of confidence about the presence of a signal in a task involving sustained attention also show this kind of effect. Noise on its own reduces the incidence of judgments of intermediate confidence, but the addition of sleep loss cancels this tendency (Hartley and Shirley, 1977). Evidence of an interaction between noise and environmental warmth is inconsistent as is the effect with incentive. Some drugs, including alcohol (Colquhoun and Edwards, 1975) and chlorpromazine (Hartley, Couper-

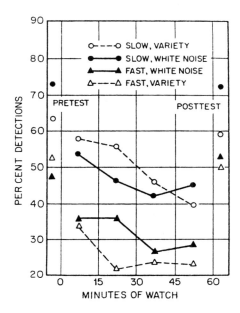

Fig. 6.1.10. Percentage of signals detected as a function of time spent at a visual vigilance task under two different rates of stimulus presentation (slow vs. fast) and two different conditions of auditory stimulation (variable sound vs. continuous white noise). (From Buckner, D. N. and McGrath, J. J. (1963) *Vigilance: A Symposium.* New York: McGraw-Hill. Copyright © 1963. Reprinted with permission of McGraw-Hill Book Company.)

Smartt, and Henry, 1977), cancel the effects of noise, which once again points to the effects of noise on the general state of the individual.

Individuals may also differ in their general level of arousal. Neurotic individuals, long thought to be particularly susceptible to annoyance by noise (see Thomas and Jones, 1982, for an overview), are also more susceptible to the damaging effect of noise on performance (von Wright and Vauras, 1980). Exposure to noise itself increases anxiety (Standing and Stace, 1980). The addition of noise-induced anxiety to that already present from chronic anxiety associated with neuroticism leads to decreased efficiency.

The picture of the way in which noise interacts with other agents is very complex and details of the mechanism responsible for such effects are not well understood. However, sufficient is known to lend weight to the generalization that the effects of noise are at least in part due to the general state of the organism. In addition, these findings suggest that in industrial settings, where noise is in the company of many stresses, the combined effect of these stresses could be much more than is suggested from their study in isolation.

6.1.5.2   Effects of Lack of Control

Noise, in addition to being a physical entity, also occurs "in a context of cognition and social circumstances" (Glass and Singer, 1972, p. 157). Typical of the type of studies showing this kind of effect is a series by Glass and Singer (1972). Subjects were exposed to 109 dB(A) bursts of random conglomerate noise for 25 min and their performance compared to a group who undertook the experiment in quiet. Subjects were given simple verbal, numerical, and motor tasks to perform in noise. At the end of the exposure, one or a combination of the following tasks was given: a proofreading task, a task measuring tolerance to the frustration of attempting to solve problems that were in fact insoluble, and the Stroop color-word test.

Those tasks undertaken during the period of exposure seemed immune to the effects of noise. However, performance of the tasks in the period following exposure was impaired by noise. This impairment could be reduced in a variety of ways. One way was to suggest that the subjects could switch the noise off if they wished, but the experimenter preferred that they should not do so: subjects were in this case imbued with "perceived control." Few subjects exercised control and only those not to do so were included in the final analysis. The result of introducing perceived control was to improve the level of performance following noise to that of the control group who received no noise. Other means of reducing the aftereffects of noise included making the noise bursts more predictable (by making them regular or by signaling their onset). Noise need not be loud in order to produce aftereffects; in one case, 65 dB(A) was sufficient, as long as the bursts were not predictable.

Since the original work of Glass and Singer aftereffects of noise have been demonstrated in a wide range of circumstances including continuous loud noise (e.g., Jones, Auburn, and Chapman, 1982). Subsequent research has confirmed that these effects could be observed with unpredictable and uncontrollable stress of any kind and were not produced by noise alone. Furthermore, anticipation of a loud noise stressor appears to be sufficient to impair performance and the expectation of control counteracts this effect. In one study, performance was impaired in subjects who were expecting noise but were told that they would not after all receive the noise. Another group who were told that they would receive the noise with perceived control showed no such deficit after being told that noise would not be presented (Cohen and Spacapan, 1983).

The issue of control over noise has a bearing on response to noise within the community. Annoyance may in part be governed by the feeling of being able to control the source of the noise as well as the acoustic properties of the noise (Graeven, 1975). Some features of performance associated with a lack of control, namely, the failure to persist at problem solving tasks, have been observed in children at schools beneath a busy approach path to an airport (Cohen, Evans, Krantz, and Stokols, 1980). These effects will be touched upon again when we discuss noise annoyance.

6.1.5.3   Strategic Effects

The effects of noise on performance change when details of the task, the experimental setting, or the subject population are altered. This has led some researchers to argue that these are effects of the particular strategy adopted. The idea of a strategy suggests that the same task may on different occasions be performed by using slightly different mental operations within the person. According to this view we should not think of a straightforward depression of efficiency in noise, but rather of some activities being favored and others not.

Fine-grained analysis of performance may reveal that effects of noise that were originally thought to be mechanical and involuntary are in fact of the strategic type. One study involved a task in which subjects were required to observe a stream of items which stopped unexpectedly and to recall as many of the items as they could remember (Hamilton, Hockey and Rejman, 1977). Noise (at 85 dB(C)) impaired memory for the items remote from the end of the list and tended to improve the performance of the last few in the list. However, subsequent experiments (Smith, 1983) were able to show that this effect was reversed if people were trying to remember only a few items. In this case, they were able to go back to the earliest of the items they were supposed to recall. When they were asked to remember more material, on the other hand, they started at the very last items (these being produced intact), followed by those further in the past (relatively less well remembered). Thus it is not a deficiency in the way the material is stored that is changed in noise; rather, it is the choice of a way to recall the material.

This effect of noise appears only if the individual has a variety of means at his or her disposal to perform the task. For instance, noise disrupts the tendency normally found in quiet for words of similar meaning to be recalled together (called "clustering"), even though when the words were originally presented, they were dispersed throughout a list (Hormann and Osterkamp, 1966). The extent of the effect depends on the nature of the list to be remembered, and the effect does not appear if the list is very easy to cluster or very difficult to cluster, these two instances being ones where there is one very dominant strategy, albeit a different one in each case (Smith, Jones, and Broadbent, 1981).

Once a strategy is adopted, noise tends to increase the likelihood of its continued adoption even when circumstances might suggest otherwise. Moreover, rapid alternation between different types of tasks is particularly damaging to performance in noise (Dornic and Fernaeus, 1981). The reluctance to abandon a strategy in noise is shown by studies examining the effects of noise on the speed of response to subtle changes in the features of a task. When signals in a task are not equiprobable, noise produces faster reaction to probable signals and slower reactions to rare signals. If, without warning, the signals become equiprobable, in noise the pattern of responding previously established tends to be carried over (Smith, 1985).

Another line of attack is to try to identify the most prevalent types of strategy. A good deal has been learned about the effects of noise from the study of memory and the way that items are rehearsed while a person is trying to remember something. A whole range of stresses increases the tendency to "parrot back" a list of items, in other words, to remember the items verbatim (Dornic, 1975). Such an effect has been observed in a variety of ways. For instance, noise improves performance if items are to be recalled in the same order as they were presented, but impairs performance if the items are required in different order (Hamilton, Hockey, and Quinn (1972). The effect of an improvement of recall in order is present even when the words in the list are known to the subject in advance or when the words are given to the subject to be placed in the order in which they were originally presented (Wilding and Mohindra, 1980). An interesting feature of these findings is that noise appears to improve at least one aspect of performance.

One reason for the improved memory for order is that the tendency for words in memory to be repeated, as if in an attempt to remember something by saying it over and over internally, becomes more emphatic in loud noise. When this covert type of articulation is suppressed by getting the individual to say something irrelevant during the period of memorization, then the effect of noise no longer appears (Wilding and Mohindra, 1980). Moreover, conditions that encourage rehearsal or the expectation that recall will be required (and hence favor rehearsal) are the ones that are sensitive to the effects of noise (Breen-Lewis and Wilding, 1983).

Noise also appears to slow the rate of rehearsal but the outcome of such slowing depends upon the nature of the material to be remembered. Recall of letters that sound the same is improved by this slowing of rehearsal, but recall of less confusable items might be impaired by it. Slower rehearsal means that there are fewer items in memory, and this will be an advantage when those items are potentially confusable; recall of more distinct sets of items would not necessarily gain advantage from the same reduction in numbers of items (Wilding and Mohindra, 1983).

6.1.5.4   Effects on Attention

Attention During Prolonged Work

Early in the history of noise research, effects of very intense noise were noted on vigilance tasks. Our understanding of the way in which vigilance performance was influenced by noise had to await the development of sophisticated theories of signal detection (see Davies and Parasurman, 1982). These theories distinguish between effects on the efficiency of detection of response bias (which is roughly equated with the observer's readiness to make a response for a given amount of evidence) and perceptual sensitivity (the efficiency with which signal and nonsignal events can be discriminated).

Loud Noise Influences Response Bias

If people are required to state the confidence of their judgment that an event is a signal, noise tends to increase the tendency to use extreme categories of judgment at the expense of intermediate categories: they are more prepared to assert that they are sure that a signal is or is not there (Broadbent and Gregory, 1965). From what is known about vigilance performance in quiet, we may predict what will happen in noise to the number of signals detected. When signals are very unlikely, people report the presence of a signal only when they have high confidence, and doubtful judgments that something is present do not produce a report. The increased certainty that results from noise then gives more correct reports. If signals are probable, however, people report them unless they are certain that no signal was present. Doubtful judgments of the absence of a signal tend to get reported as positive detections. In that case noise reduces the numbers of reports of signals.

The prevailing level of confidence, in addition to its effects on the reporting of signals, will have effects on the way in which observers check on the state of a display. If several sources are involved, it is possible to chart the process of interrogation by offering the observer brief glimpses of a state of each display. Typically, some displays receive more attention than others, and this tendency is exacerbated in loud noise (Hockey, 1973). The action of noise in this case seems to be one of exaggerating those biases that already prevail about where significant events requiring action are most likely to occur.

Effects have also been noted on another class of vigilance tasks that are sensitive to levels of noise as low 80 dB(C) (Jones, Smith, and Broadbent, 1979). In this case the detection of signals places a heavy reliance on memory, and the periods over which performance was assessed were relatively short. Because of these differences, the results may arise out of changes of the strategic type.

Some 11 experiments have used the serial reaction task to study sustained attention. In its most frequently used form, the task consists of five lights and five response disks, each arranged in a pentagon. The lighting of a bulb is contingent on a disk, not necessarily the correct one, being tapped. The bulbs are lit in an unpredictable sequence and because a response has to be made before the light comes on, the rate of work is determined by the person performing the task. Typically, the task lasts some 40 min. Seven of these experiments have found errors reliably increased at levels in excess of 90 dB(A). Three studies have found that the incidence of exceptionally long responses was greater in the same levels of noise (see Davies and Jones, 1984, for a detailed discussion).

Some of these results depend on the details of the procedure involved, and there is debate about the role in which noise serves to mask acoustic cues in the performance of a task (see Broadbent, 1978; Poulton, 1979). However, one of the experiments showing the usual result of an increase in errors over time (see Figure 6.1.11) employed a number of precautions to eliminate the effects of acoustic cues including the use of a silent keyboard and ear defender headphones (Jones, 1983). Note

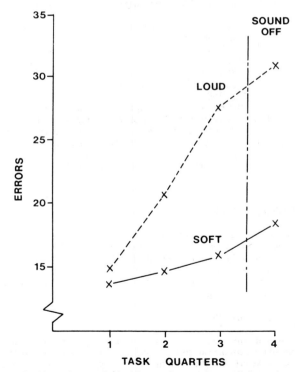

Fig. 6.1.11. Effects of white noise on the incidence of errors in a serial reaction task lasting 40 min. In both loud (90 dB(C)) and soft (60 dB(C)) conditions the noise is switched off at the end of the third quarter.

that in this case when the noise is switched off (shown by the dotted vertical line) performance continues to deteriorate.

Time Sharing

Everyday tasks are made up of a range of different activities each with different priorities. The effect of noise on tasks of this sort is to swing resources away from elements of low priority and toward those that are seen to be subjectively more important (Hockey, 1970). Some variation will be observed in the patterning of response in tasks involving several different elements; this is to be expected in view of the different demands made by the tasks and by the type of instructions given for their execution (see, for example, Forster, 1978, for a detailed discussion).

Elements within a memory task can also be assigned high or low priority. The effect of noise bears a striking resemblance to that found with tasks involving motor skills just described. The usual procedure is to present a list of words with each word arranged in one of a number of positions. Instructions emphasize recall of the words in correct order without mention of recalling locations. Once the recall of words is complete (high-priority element) the recall of locations is required (low-priority element). The presence of noise at 95 dB usually improves the recall of words but the recall of locations deteriorates (Davies and Jones, 1975).

The effect of noise depends on priority assigned to each element. The effect does not depend on whether locations or words are to be recalled because the same effect of priority appears if the recall of words is made secondary to the recall of locations. Neither does the effect depend on the order in which the elements are recalled because it is possible to arrange the task so that the low-priority element is recalled first; again noise improves the recall of the high-priority element at the expense of the low-priority element (Smith, 1982). Noise produces improvements in efficiency on some components of the setting, usually the one emphasized by instruction, with the component of minor importance showing significant deterioration.

6.1.5.5 Effects of Noise on Productivity

Before leaving this section on performance, it is important to note the results of the relatively few studies that have dealt with the issue of the effects of noise on efficiency at the workplace.

Noise tends to be associated with accidents. Mean noise level at the place of work correlates very strongly with the frequency (but not the severity) of accidents (Kerr, 1950). Younger and less experienced workers seem to be more susceptible to the effects of noise on accidents, and the effects seem to be greater in levels of noise above 95 dB(A) (Cohen, 1974). It is of course possible that the effect of noise on accidents arises because noisy jobs are also difficult and dangerous ones. However, the incidence of accidents is reduced when ear defenders are introduced, which suggests that the effect of noise is one of intensity rather than of exposure to risk (Cohen, 1976). Other studies have shown that productivity can be increased when the level of noise is reduced either by introducing ear defenders (Weston and Adams, 1935) or by acoustical treatment of the workplace (Broadbent and Little, 1960). This effect is not due to the speed of action, but to greater accuracy.

6.1.5.6 Summary

The results of laboratory experiments on noise and performance have shown a range of effects. Changes in efficiency may arise at levels of noise as low as 80 dB(C) and these may take the form of either an improvement or deterioration in task performance. The following activities may be at risk: (1) those that have to be conducted over long intervals especially if the noise is continuous, (2) those that involve steady posture and uninterrupted vision, which are at risk from sudden bursts of noise, (3) those that are thought to be relatively unimportant or occur infrequently, (4) those requiring an effort on the part of the person to initiate action, especially if the noise appears to be beyond the person's control, (5) those that demand the comprehension of the meaning of material rather than rote learning, and (6) those that demand some flexibility of response, such as a response to a sudden change in circumstances.

6.1.6 EFFECTS OF NOISE ON WELL-BEING

In this section we deal with the rather more general effects of noise that are found in the community. The problem is important to human-factors engineers and ergonomists because the noise of operations may produce complaints from local residents or indeed from the work force.

6.1.6.1 Noise Annoyance

The idea of "noise annoyance" carries with it manifold suggestions of discomfort, disturbance, displeasure, unease, irritation, and complaint. From our own experience of noise we would not expect the degree of annoyance to be some simple function of the intensity of the noise. Certainly, it is the case

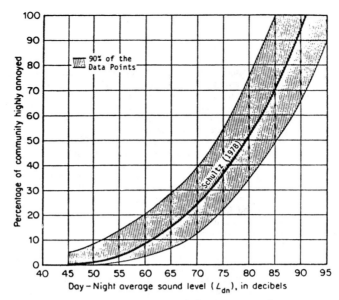

Fig. 6.1.12. The relation between noise exposure and the percentage of survey respondents claiming to be highly annoyed. (From Schultz, 1978, with the permission of the author and the American Institute of Physics.)

that the proportion of people expressing annoyance increases as the intensity of sound increases, but we know too that the circumstances in which a given sound arises also account in part for the annoyance.

The intensity of noise needed to produce a given level of annoyance is less when the noise is heard indoors than when it is heard outdoors (Robinson, Bowsher, and Copeland, 1963). Expectations about the noise source cloud the judgment of annoyance, as is shown by the tendency for aircraft to be judged more noisy than an automobile, even when they are at the same intensity (Robinson, Copeland, and Rennie, 1961). In addition, the difficulty of performing a task in noise may also shape attitudes to subsequent encounters with that noise (Moser and Jones, 1983).

As we would expect, the level of noise plays a major role in the generation of annoyance. Two lines of evidence lend substantial weight to this view. First, if an analysis is made of social surveys that measure levels of annoyance with changes in intensity, carefully separating out the effects of response bias (the willingness to use certain categories of annoyance) from those of the true sensitivity of noise annoyance to intensity, the effects on annoyance are clearly shown in the latter measure (Fidell, Horonjeff, and Green, 1981). The other line of evidence is the degree to which annoyance may be predicted from physical measurements alone. A synthesis of the results of several large-scale studies is shown in Figure 6.1.12 (after Schultz, 1978). On the abscissa is shown the metric Ldn, a measure based on a 24 hr Leq (see above) with a 10 dB weighting to levels measured between 22.00 and 06.00 hr. On the ordinate is given the proportion who report themselves to be highly annoyed by the noise. Although the general rule that Figure 6.1.12 represents is not without its critics, it serves as a useful approximation to the likely effects of noise within the community and points to a lawful association between intensity and annoyance (see Fidell, 1984, for a discussion).

Spectrum as well as level is important, high frequencies normally being more annoying than low at the same intensity, and this again is a reason for using A-weighted measurements. For more sophisticated corrections for spectral characteristics, see Kryter (1970).

There are also some sounds that appear to be annoying to a extent that is out of proportion to their physical intensity. Perhaps the best example is the sound produced by drawing fingernails across a blackboard. One analysis suggests that these less intense sounds are simply more distinct or detectable (Fidell and Teffeteller, 1981). Thus far, however, we have no means of knowing in advance whether a sound will show exceptional detectability (see Broadbent, 1975, for a discussion of various possibilities).

Reducing the level of the noise will reduce annoyance but to a rather variable degree because of social and individual factors whose influence is difficult to predict.

6.1.6.2 Noise and Health

So great and pervasive is the effect of noise on our everyday lives that the idea of it influencing our mental and physical health is intuitively appealing. In view of the strong and stereotyped manner in which the human physiological system responds to noise (see Section 6.1.5 above), it is perhaps reasonable to suppose that this contributes to stress. However, a feature common to all such physiological responses is that they habituate (become reduced in magnitude) on successive exposure to the same noise stimulus (see for example, Burns, 1979). It may be argued, however, that those persons who habituate slowly are most at risk from the likely systemic damage caused by the repeated elicitation of these reflexes (Jansen, 1961).

A very wide range of physiological function has been shown to be sensitive to noise (see Burns, 1979; Rehm, 1983) and the implications of such changes for health are still not understood. However, evidence is accumulating of a higher incidence of symptoms in noise-exposed groups. Studies of the effects of industrial noise and health have noted increase in the incidence of nausea, headache, anxiety, sexual impotence, and heart disease (for example, Jansen, 1961). In each case, however, it is difficult to distinguish the effects of the type of job and the effects of noise. Noisy jobs are usually the most physically demanding, dirty, and dangerous. The influence of these factors alone might account for the higher incidence of symptoms.

Studies of the effects of noise in the community, arising from aircraft or traffic, though not suffering the shortcomings of industrial studies, are undermined by a different class of problems. The particular context in which questions about health are asked seems to be an important variable: if no reference is made to aircraft noise while questioning respondents about their health, the overall statistical relation between aircraft noise and physical symptoms is diminished (Grandjean, 1974). One way of overcoming the effect of bias in verbal responses is to use some physiological measure instead. In these cases changes in physiological activity are found, such as an increase in blood pressure of children living near airports (Cohen et al., 1980), but again these effects may arise owing to a number of confounding factors, such as a bias due to socioeconomic class or bias in the way in which blood pressure is recorded.

Another device for assessing the effects of noise on health, this time in an indirect way, is to compare the use of primary health care facilities in areas representing different levels of noise exposure from aircraft. In these cases the use of drugs (particularly sleeping tablets, antacids, sedatives, and drugs for the treatment of hypertension) was higher in the noisy areas and the use of primary health care more frequent (Grandjean, 1974). Such studies run the risk of confounding some other factor, not associated with noise exposure, such as social class. At least one study has managed to show that the use of primary health care facilities is, in addition to being higher in areas of more intense noise, related to the degree of annoyance expressed about health. However, a causal connection between noise and health is still not clearly demonstrated.

One indirect way in which noise might influence mental health is through its action on hearing loss. The incidence of hearing loss is higher in psychiatric populations than would be expected for the population as a whole (Kay and Roth, 1961). Moreover, the incidence of psychiatric symptoms is slightly higher for those attending outpatient clinics for hearing loss than is present in the general population (Mahapatra, 1974). The misperception of speech and lack of auditory stimulation that accompany hearing loss have been implicated in the development of paraphrenia, a state that resembles schizophrenia and is often associated with delusions and hallucinations (Clark, 1984). Despite some suggestive evidence, the association with hearing loss and mental health can be regarded only as tentative; many more studies must be undertaken before any kind of definitive picture emerges.

Despite the methodological difficulties, it seems safe to conclude that noise does not cause changes in mental health. It is the case that people with psychological problems tend to be annoyed by noise. It is also the case that noise increases annoyance, but there is no evidence that noise increases the incidence of people with problems (Broadbent, 1980).

6.1.6.3 Social Effects of Noise

It is clear from our discussion of noise annoyance that the meaning of sound and the context in which it appears are important. In this section, the effects of noise on specific aspects of social conduct are discussed. Although these social effects are indirect, they should be borne in mind by human-factors engineers as possibilities in noisy workplaces.

A range of effects associated with living in noisy neighborhoods has been noted, including high crime rate and truancy. However, such effects may arise from factors such as family size, density of dwellings, and age of family members rather than from noise per se. Evidence of a reduction of casual social interaction associated with noisy areas (e.g., Appleyard and Lintell, 1972), although usually attributed to the stressful effects of noise, is more likely to arise as a side effect of noise on the masking of speech.

The most widely studied of the social behaviors is that of helping. Typically, a confederate of the experimenter feigns some distressing incident and the behavior of those who are passing by is observed for signs of helping. Comparisons are made between the amount of help offered in the presence and absence of noise. A range of studies shows that individuals exposed to noise above 80 dB(A) in such circumstances are less likely to grant interviews (Boles and Hayward, 1978), less likely to help pick up a dropped package or deck of index cards, and less likely to give help in response to a request for change (Page, 1977).

There are several possible interpretations of these findings. The first is that noise increases the chance of failing to notice the victim's plight. This may arise in one or a combination of three ways: (1) attention is narrowed by noise and the incident is not noticed (Cohen and Spacapan, 1984), (2) the sounds associated with the incident are masked by the noise and it goes unnoticed, and (3) people walk more quickly through noisy settings and thus are less likely to notice incidental details (Jones, Chapman, and Auburn, 1981). The other possibility is that the presence of noise does not influence the perception of the incident but may make helping appear to be more difficult. According to this view individuals avoid rather than fail to perceive the victim's plight. Because of noise, potential helpers become passersby.

Noise appears to potentiate the expression of anger. In a situation where a person is made angry (by, for example, being given electric shocks and told of failure in a laboratory task) and is given the opportunity to return the shocks, the angry person behaves more aggressively by giving more shocks than one who has not been angered. This effect is even more pronounced in noise. If loud noise is played when the person is being angered, the subsequent behavior is even more aggressive (Donnerstein and Wilson, 1976).

There is some evidence that judgment of others also appears to change in noise. The findings show rather mixed effects. Some studies show the effect of noise is to judge others more severely (Sauser, Arauz, and Chambers, 1978) whereas others show that (for females at least) noise leads to reduced interpersonal distance (reflecting increasing attraction) even when conversation was prohibited (Mathews, Canon, and Alexander, 1974). Part of the trend in these and other studies is a systematic change in the way that evidence is weighed. When asked to judge the general characteristics of others from a sketchy outline there is a tendency for more extreme judgments to be made in loud noise (70 to 90 dB) irrespective of the type of attribute being judged (Siegel and Steele, 1980). This trend for noise to produce extreme judgments is one that has been observed in tasks involving memory and sustained attention and may be a general feature of the response to noise (see Section 6.1.5).

In summary, the effects of noise on our social life seem to be varied. In some ways we may regard them as extensions of the effects of noise on communication and performance that have already been discussed. These include the way in which noise influences our perception of speech, the tendency for people to avoid noise, and the way in which noise tends to bias decisions toward using categories of judgment that are certain.

6.1.7 CONCLUSIONS AND RECOMMENDATIONS

1. For an 8 hr working day the overall level of noise should ideally be no greater than 85 dB(A). Exposure should be halved for every 3 dB increase in level above this.

2. The best way of reducing noise, if possible, is at the source. This can sometimes be done by modification of machinery, introduction of silencers, and similar means (see, for example, Chapters 25 through 30 in Harris, 1979). If this cannot be done, the next most desirable possibility is to place shielding between the source and the person (see, for example, Chapters 21 through 24 in Harris, 1979).

3. An economical way in which to reduce the risk to hearing is by wearing ear defenders. In continuous noise this is likely to improve the efficiency of the listener and may make the quality of speech produced by the talker slightly worse. In intermittent noise the efficiency of the listener may be severely disrupted in the periods of quiet.

4. Masking of simple sounds is reduced as the signal-to-noise ratio is increased. Those masking sounds close in frequency to the masker contribute most to its masking. If masking sounds are concentrated in a narrow band of frequencies, signals should be at frequencies remote from it (preferably below the lower bound of the masker). Masking effects may be partly overcome by capitalizing on the capacity of the ear to localize sound.

5. Alarms should be more than just detectable in background noise and preferably 15 dB above it. The timing and spectral composition of the sound should be chosen to make the alarm distinctive from others that might be sounded at the same time, as well as to convey urgency.

6. The bounds for the satisfactory reception of speech extend from 50 to 80 dB(A). Sustaining speech above 70 dB(A) usually results in fatigue. There may be subtle psychological side effects of listening or speaking in noise in the form of a reduction in spare capacity to deal with other tasks.

7. The precise effects of noise on speech should be assessed empirically. Two well-researched techniques are available. The articulation index requires measurement of the speech and noise signals; and an estimate of interference based on knowledge of the noise alone is employed in the speech interference level measure. As a first approximation, the A-weighted sound level of the noise may be used with Figure 6.1.8 to assess the level of speech required to meet certain criteria of acceptability.

8. The effects of noise on speech may be reduced by (1) amplifying speech (although there will be an upper limit), (2) excluding noise by using noise-canceling microphones in conjunction with headphones, (3) restricting the range of vocabulary to agreed, polysyllabic sets of words, and (4) using experienced speakers and listeners.

9. Performance will be temporarily disrupted by sudden bursts of noise. Those activities involving

the intake of information and ones requiring a steady posture are most at risk from this kind of effect.

10. Even at very low levels, the presence of background speech will have damaging effects on the way verbal materials, even in written form, are remembered. Noise also has an effect on the retrieval of words. These two findings suggest that reading and writing are particularly susceptible to disruption by background speech.

11. Continuous unvarying noise will have damaging effects on performance at the end of a long period of work, these effects typically appearing at levels of noise in excess of 90 dB SPL.

12. The strategy usually favored in quiet is more emphatically adopted in noise. Whether this will improve performance depends on the particular circumstances. If the characteristics of the task change without notice then the effect of noise will be to slow the process of adaptation to new circumstances. These effects tend to occur at levels of noise between 80 and 90 dB SPL.

13. With verbal materials, noise tends to increase the chance that they will be remembered by rote rehearsal, which means that features of the material associated with meaning are less well remembered. Beneficial or deleterious effects will be found at levels in excess of 80 dB(C).

14. Noise diverts attention away from less important elements of a task to those that are regarded as being more important.

15. Judgments become more extreme in loud noise: more confidence is expressed in the adequacy of a decision even though, on the basis of evidence, this might be unwarranted. This effect is found in cases as diverse as sustained attention, recognition memory, and social judgments.

16. Noise reduces helpfulness and increases the chance of aggressive behavior.

17. Annoyance is determined by both the intensity of the sound and the attitude of the individual. Feelings of control over the noise influence both the degree of annoyance and effect on performance.

18. Evidence of the effect of noise on physical health is equivocal. Noise does not affect mental health. People in poor mental health have a tendency to be more annoyed by noise.

REFERENCES

Appleyard, D., and Lintell, M. (1972). Environmental quality of city streets: The residents' viewpoint. *Journal of the American Institute of Planners, 38,* 84–101.

Boles, W. E., and Hayward, S. C. (1978). Effects of urban noise and sidewalk density upon pedestrian co-operation and tempo. *Journal of Social Psychology, 104,* 29–35.

Breen-Lewis, K., and Wilding, J. (1983). Noise, time of day and test expectations in recall and recognition. *British Journal of Psychology, 75,* 51–63.

Broadbent, D. E. (1971). *Decision and stress.* New York: Academic.

Broadbent, D. E. (1975). Waves in the eye and ear. *Journal of Sound and Vibration, 41,* 113–125.

Broadbent, D. E. (1978). The current state of noise research: Reply to Poulton. *Psychological Bulletin, 85,* 1052–1067.

Broadbent, D. E. (1980). Noise in relation to annoyance. Performance and mental health. *Journal of the Acoustical Society of America, 68,* 15–17.

Broadbent, D. E. (1983). Recent advances in understanding performance in noise. In G. Rossi, Ed., *Noise as a public health problem.* Milan: Centro Ricerche e Studi Amplifon.

Broadbent, D. E., and Gregory, M. (1965). Effects of noise and of signal rate upon vigilance analysed by means of decision theory. *Human Factors, 7,* 155–162.

Broadbent, D. E., and Little, F. A. J. (1960). Effects of noise reduction in a work situation. *Occupational Psychology, 34,* 133–140.

Burns, W. (1979). Physiological effects of noise. In C. M. Harris, Ed., *Handbook of noise control.* New York: McGraw-Hill.

Clark, C. R. (1984). The effects of noise on health. In D. M. Jones and A. J. Chapman, Eds., *Noise and society.* Chichester, Great Britain: Wiley.

Cohen, A. (1974). Industrial noise and medical, absence and accident record data on exposed workers. In W. D. Ward, Ed., *Proceedings of the International Congress on Noise as a Public Health Problem.* Washington, DC: U.S. Environmental Protection Agency.

Cohen, A. (1976). The influence of a company hearing conservation program on extra-auditory problems in workers. *Journal of Safety Research, 8,* 146–162.

Cohen, S., and Spacapan, S. (1983). The after effects of anticipating noise exposure. In G. Rossi, Ed., *Noise as a Public health problem.* Milan: Centro Ricerche e Studi Amplifon.

Cohen, S., and Spacapan, S. (1984). The social psychology of noise. In D. M. Jones and A. J. Chapman, Eds., *Noise and Society.* Chichester, Great Britain: Wiley.

Cohen, S., Glass, D. C., and Singer, J. E. (1973). Apartment noise, auditory discrimination and reading ability in children. *Journal of Experimental Social Psychology, 9,* 407–422.

Cohen, S., Evans, G. W., Krantz, D. S., and Stokols, D. (1980). Physiological, motivational, and cognitive effects of aircraft noise on children: Moving from the laboratory to the field. *American Psychologist, 35,* 231–243.

Colle, H. A. (1980). Auditory encoding in visual short-term recall: effects of noise intensity and spatial location. *Journal of Verbal Learning and Verbal Behavior, 19,* 722–735.

Colquhoun, W. P., and Edwards, R. S. (1975). Interaction of noise with alcohol on a task of sustained attention. *Ergonomics, 18,* 81–89.

Crook, M. A., and Langdon, F. J. (1974). The effects of aircraft noise in schools around London Airport. *Journal of Sound and Vibration, 34,* 221–232.

Davies, D. R., and Parasuraman, R. (1981). *The psychology of vigilance.* New York: Academic.

Davies, D. R., and Jones, D. M. (1975). The effects of noise and incentives upon retention in short-term memory. *British Journal of Psychology, 66,* 61–68.

Davies, D. R., and Jones, D. M. (1984). Effects of noise on performance. In W. Tempest, Ed., *The Noise Handbook.* London: Academic.

Donnerstein, E., and Wilson, D. W. (1976). Effects of noise and perceived control on ongoing and subsequent aggressive behavior. *Journal of Personality and Social Psychology, 34,* 774–781.

Dornic, S. (1975). Some studies on the retention of order information. In P. M. A. Rabbitt and S. Dornic, Eds., *Attention and performance,* Vol. 5. New York: Academic.

Dornic, S., and Fernaeus, S. E. (1981). Type of processing in high-load tasks: the differential effect of noise (Report No. 576). Stockholm: Department of Psychology, University of Stockholm.

Egan, J. P., and Hake, H. W. (1950). On the masking of a simple auditory stimulus. *Journal of the Acoustical Society of America, 22,* 622–630.

Eysenck, M. W. (1975). Effects of noise, activation level, and response dominance on retrieval from semantic memory. *Journal of Experimental Psychology, 104,* 143–148.

Fidell, S. (1984). Community response to noise. In D. M. Jones and A. J. Chapman, Eds., *Noise and society.* Chichester, Great Britain: Wiley.

Fidell, S. and Teffeteller, S. (1981). Scaling the annoyance of intrusive sounds. *Journal of Sound and Vibration, 78,* 291–298.

Fidell, S., Horonjeff, R., and Green, D. (1981). Statistical analyses of urban noise. *Noise Control Engineering, 16,* 75–80.

Fisher, S. (1973). The "distraction effect" and information processing complexity. *Perception, 2,* 78–89.

Forster, P. M. (1978). Attentional selectivity: a rejoinder to Hockey, *British Journal of Psychology, 69,* 505–506.

Gebhardt, C. J., and Goldstein, D. P. (1972). Frequency discrimination and the MLD. *Journal of the Acoustical Society of America, 51,* 1228–1232.

Glass, D. C., and Singer, J. E. (1972). Urban stress: Experiments on noise and social stressors. New York: Academic.

Graeven, D. B. (1975). Necessity control and predictability of noise annoyance. *Journal of Social Psychology, 95,* 85–90.

Grandjean, E. (1974). Sozio-psychologische Untersuchungen vor der Fluglarms. Berne: Eidgenossisches Lustant, Bundeshaus.

Hamilton, P., Hockey, G. R. J., and Quinn, J. G. (1972). Information selection, arousal and memory. *British Journal of Psychology, 63,* 181–189.

Hamilton, P., Hockey, G. R. J., and Rejman, R. (1977). The place of the concept of activation in human information processing theory: An integrative approach. In S. Dornic, Ed., *Attention and performance,* Vol. 6. New York: Lawrence Erlbaum.

Harris, C. M., Ed. (1979). *Handbook of Noise Control,* 2nd ed. New York: McGraw-Hill.

Hartley, L. R., and Shirley, E. (1977). Sleep loss, noise, and decisions. *Ergonomics, 20,* 481–482.

Hartley, L., Couper-Smartt, J., and Henry, T. (1977). Behavioural antagonism between chlorpromazine and noise in man. *Psychopharmacology, 55,* 97–101.

Hockey, G. R. J. (1970). Effect of loud noise on attentional selectivity. *Quarterly Journal of Experimental Psychology, 22,* 28–36.

Hockey, G. R. J. (1973). Changes in information selection patterns in multisource monitoring as a function of induced arousal shifts. *Journal of Experimental Psychology, 101,* 35–42.

Hormann, H., and Osterkamp, J. (1966). Uber den Einfluss von Kontinvierlichem Larm auf die Organisation von Gedachtrisinhalten. *Zeitschrift fur Experimentelle und Angewandte Psychologie, 13,* 31–38.

Howell, K., and Martin, A. M. (1975). An investigation of the effects of hearing protectors on vocal communication in noise. *Journal of Sound and Vibration, 41* (2), 181–196.

Jansen, G. (1961). Adverse effects of noise on iron and steel workers. *Stahl und Eisen, 81,* 217–220.

Jones, D. M. (1983). Loud noise and levels of control. In G. Rossi, Ed. *Noise as a public health problem.* Milan: Centro Ricerche e Studi Amplifon.

Jones, D. M., Auburn, T. C., and Chapman, A. J. (1982). Perceived control in continuous loud noise. *Current Psychological Research, 2,* 111–122.

Jones, D. M., Chapman, A. J., and Auburn, J. C. (1981). Noise in the environment: A social perspective. *Journal of Environmental Psychology, 1,* 43–59.

Jones, D. M., Smith, A. P., and Broadbent, D. E. (1979). Effects of moderate intensity noise on the Bakan vigilance task. *Journal of Applied Psychology, 64,* 627–634.

Jones, D. M., Thomas, J. R., and Harding, A. (1982). Recognition memory for prose items in noise. *Current Psychological Research, 2,* 33–44.

Kay, D. W. K., and Roth, M. (1961). Environmental and hereditary factors in the schizophrenias of old age. *Journal of Mental Science, 107,* 649–686.

Kerr, W. A. (1950). Accident proneness and factory departments. *Journal of Applied Psychology, 34,* 167–170.

Klumpp, R. G., and Leonard, J. L. (1963). Observer variability in reading noise level with meters. *Sound, 2,* 25–29.

Kryter, K. D. (1970). *Effects of noise on man.* New York: Academic.

Kohler, H. K. (1984). The measurement of noise. In D. M. Jones and A. J. Chapman, Eds., *Noise and society.* Chichester, Great Britain: Wiley.

Kraak, W. (1982). Investigations on criterial for the risk of hearing loss due to noise. In J. V. Tobias and E. E. Schubert, Eds., *Hearing research and theory,* Vol. 1. New York: Academic.

Kryter, K. D. (1970). *The effects of noise on man.* New York: Academic.

Landis, C., and Hunt, W. A. (1939). The Startle Pattern. New York: Farrar and Rinehart.

Mahapatra, S. B. (1974). Psychiatric and psychosomatic illness in the deaf. *British Journal of Psychiatry, 125,* 450–451.

Mathews, K., Canon, L., and Alexander, K. (1974). The influence of level of empathy and ambient noise on the body buffer zone. *Personality and Social Psychology Bulletin, 1,* 367–359.

May, D. N., and Rice, C. G. (1971). Effects of startle due to pistol shot on control precision performance. *Journal of Sound and Vibration, 15,* 197–202.

McGrath, J. J. (1963). Irrelevant stimulation, and vigilance performance. In D. N. Buckner and J. J. McGrath, Eds., *Vigilance: A symposium.* London: McGraw-Hill.

Moore, B. C. J. (1984). The perception of sound. In D. M. Jones and A. J. Chapman, Eds., *Noise and society.* Chichester, Great Britain: Wiley.

Moser, G., and Jones, D. M. (1983). Annoyance and performance. In G. Rossi, Ed., *Noise as a public health problem.* Milan: Centro Ricerche e Studi Amplifon.

Mosko, J. F., and Fletcher, J. L. (1971). Evaluation of the gundefender earplug: temporary threshold shift and speech intelligibility. *Journal of the Acoustical Society of America, 49,* 1732–1734.

Murray, D. J. (1965). The effects of white noise on the recall of vocalized lists. *Canadian Journal of Psychology, 19,* 333–345.

Nixon, C. W. (1979). Hearing protective devices: Ear protectors. In C. M. Harris, Ed., *Handbook of noise control.* New York: McGraw-Hill.

Page, R. A. (1977). Noise and helping behavior. *Environment and Behavior, 9,* 311–334.

Passchier-Vermeer, W. (1974). Hearing loss due to continuous exposure to steady state broad-band noise. *Journal of the Acoustical Society of America, 56,* 1585–1593.

Passchier-Vermeer, W. (1983). Measurement and rating of impulse noise in relation to noise-induced hearing loss. In G. Rossi, Ed., *Noise as a public health problem.* Milan: Centro Ricerche e Studi Amplifon.

Patterson, R. D. (1983). Guidelines for auditory warning systems on civil aircraft: A summary and a prototype. In G. Rossi, Ed., *Noise as a public health problem.* Milan: Centro Ricerche e Studi Amplifon.

Pearsons, K. S., Bennett, R. L., and Fidell, S. (1977). Speech levels in various noise environments (EPA-600/1-77-025). Washington, DC: U.S. Environmental Protection Agency.

Pickett, J. M. (1956). Effects of vocal force on the intelligibility of speech sounds. *Journal of the Acoustical Society of America, 28,* 902–905.

Pols, L. C. W., van Heusden, E. and Plomp, R. (1980). Preferred listening level for speech disturbed by fluctuating noise. *Applied Acoustics, 13,* 267–279.

Poulton, E. C. (1979). Composite model for human performance in continuous noise. *Psychological Review, 86,* 361–375.

Rehm, S. (1983). Research on extra-aural effects of noise since 1978. In G. Rossi, Ed., *Noise as a public health problem.* Milan: Centro Ricerche e Studi Amplifon.

Robinson, D. W., Bowsher, J. M., and Copeland, W. C. (1963). On judging the noise from aircraft in flight. *Acustica, 13,* 324–336.

Robinson, D. W., Copeland, W. C., and Rennie, A. J. (1961). Motor vehicle noise measurement. *Engineer, 211,* 493–497.

Salame, P., and Baddeley, A. D. (1983). Differential effects of noise and speech on short-term memory. In G. Rossi, Ed., *Noise as a public health problem.* Milan: Centro Ricerche e Studi Amplifon.

Salame, P., and Wittersheim, G. (1978). Selective noise disturbance of the information input in short-term memory. *Quarterly Journal of Experimental Psychology, 30,* 693–794.

Sauser, W. I., Arauz, C. G., and Chambers, R. M. (1978). Exploring the relationship between level of office noise and salary recommendations: A preliminary research note. *Journal of Management, 4,* 57–63.

Schori, T. R., and McGatha, E. A. (1978). A real-world assessment of noise exposure. *Journal of Sound and Vibration, 12,* 24–30.

Schultz, J. (1978). Synthesis of social surveys on noise annoyance. *Journal of the Acoustical Society of America, 64,* 377–405.

Siegel, J. M., and Steele, C. M. (1980). Environmental distraction, and interpersonal judgements. *British Journal of Social and Clinical Psychology, 19,* 23–32.

Smith, A. P. (1982). The effects of noise and task priority on recall of order and location. *Acta Psychologica, 74,* 245–256.

Smith, A. P. (1983). The effects of noise and memory load on a running memory task. *British Journal of Psychology.*

Smith, A. P. (1985). Noise biased probability and serial reaction. *British Journal of Psychology, 76,* 89–95.

Smith, A. P., Jones, D. M., and Broadbent, D. E. (1981). The effects of noise on recall of categorized lists. *British Journal of Psychology, 72,* 299–316.

Standing, L., and Stace, G. (1980). The effects of environmental noise on anxiety. *Journal of General Psychology, 103,* 263–267.

Taylor, W., Pearson, J., Mair, A., and Burns, W. (1965). Study of noise, and hearing in jute weaving. *Journal of the Acoustical Society of America, 38,* 113–120.

Terry, A. M. P., Jones, D. M., Davis, B. R., and Slater, R. (1983). Parametric studies of tinnitus masking and residual inhibition. *British Journal of Audiology, 17,* 245–256.

Thomas, J. R., and Jones, D. M. (1982). Individual differences in noise annoyance and the uncomfortable loudness level. *Journal of Sound and Vibration, 82,* 189–304.

van Heusden, E., Plomp, R., and Pols, L. C. W. (1979). Effect of ambient noise on the vocal output and the preferred listening level of conversational speech. *Applied Acoustics, 12,* 31–43.

von Wright, J., and Vauras, M. (1980). Interactive effects of noise and neuroticism on recall from semantic memory. *Scandinavian Journal of Psychology, 21,* 97–101.

Waltzman, S. B., and Levitt, H. (1978). SIL as a predictor of face-to-face communication in noise. *Journal of the Acoustical Society of America, 63,* 581–590.

Ward, W. D. (1976). Transient changes in hearing. In G. Rossi and M. Vigone, Eds., *Man and noise.* Milan: Edizioni Minerva Medica.

Ward, W. D. (1983). Noise-induced hearing loss. In D. M. Jones and A. J. Chapman, Eds., *Noise and society.* Chichester, Great Britain: Wiley.

Ward, W. D., Glorig, A., and Sklar, D. L. (1959). Temporary threshold shift from octave-band noise: Applications to damage-risk criteria. *Journal of the Acoustical Society of America, 31,* 522–528.

Webster, J. C. (1979). Effects of noise on speech. In C. M. Harris, Ed., *Handbook of noise control.* London: McGraw-Hill.

Webster, J. C. (1984). Noise and communication. In D. M. Jones and A. J. Chapman, Eds., *Noise and society.* Chichester, Great Britain: Wiley.

Weinstein, N. D. (1977). Noise and intellectual performance: A confirmation and extension. *Journal of Applied Psychology, 62,* 104–107.

Weston, H. C., and Adams, S. (1935). The performance of weavers under varying conditions of noise (Industrial Health Research Board Report No. 65). London: H.M.S.O.

Wilding, J., and Mohindra, N. (1980). Effects of subvocal suppression, articulating aloud and noise on sequence recall. *British Journal of Psychology, 71,* 247–262.

Wilding, J. M., and Mohindra, N. (1983). Noise slows phonological coding and maintenance rehearsal: An explanation for some effects of noise on memory. In G. Rossi, Ed., *Noise as a public health problem.* Milan: Centro Richerche e Studi Amplifon.

Wilkinson, R. T. (1963). Interaction of noise with knowledge of results, and sleep deprivation. *Journal of Experimental Psychology, 66,* 332–337.

Woodhead, M. M. (1964). The effects of bursts of noise on an arithmetic task. *American Journal of Psychology, 77,* 627–633.

Woodhead, M. M. (1966). An effect of noise on the distribution of attention. *Journal of Applied Psychology, 50,* 296–299.

CHAPTER **6.2**
MOTION AND VIBRATION

DONALD E. WASSERMAN

National Center for Rehabilitation Engineering
Wright State University
Dayton, Ohio

Before reading this chapter the reader should be aware that the area of human vibration and motion has many voids in knowledge and is not straightforward. For example, vibration measurement procedures vary widely; there is little at this writing in the way of low-frequency, vibration-absorbing materials that when applied to a given vibrating device will guarantee safety for workers using the device under varied and long-term operating conditions; in the case of whole-body vibration (i.e., head-to-toe) there are very few hard epidemiological health data on vibration effects on workers. The epidemiological data on segmental (hand–arm vibration) are much better. Each vibration problem is unique and limited usually to the tool or machine being analyzed under a particular working condition. Usually, vibration engineers obtain and record measurements (mostly acceleration) of a vibrating source at various locations on the source including points where the human comes into contact with the source. Next, Fourier analysis, modal analysis, transfer function analysis, and mechanical impedance analysis (to name a few) are used on the data in an effort to "fingerprint" effectively the vibrating source under the given work conditions. Next, equations of motion are developed for the given condition, noting system resonance points. Finally, with a knowledge of damping materials and damping techniques, these are applied to the system under test, comparing the theoretical reductions to actual reductions obtained. Thus vibration reduction becomes a repetitive and difficult process until the desired solution is obtained. If damping materials and techniques are insufficient, then various mechanical changes are usually made to the vibrating system (e.g., stiffening the mechanical system, reducing the system mass) in an effort to shift resonance points higher in frequency where damping materials become more effective.

Thus it should be apparent that simple solutions to complex vibration problems do not exist at this writing. In this chapter, I attempt to familiarize the reader with the problems relating to human vibration exposure, how basic measurements are made, and how to interpret these measurements in view of what is known about the effects of vibration on workers. Finally, in an effort to aid the reader in the understanding of vibration a glossary of selected vibration terms is given at the end of the chapter in Section 6.2.13.

6.2.1 INTRODUCTION

Vibration and motion surround us, whether they be in our work environment, at home, or at play. Mechanical vibration impinges on us when we mow our lawns, trim the hedges, perform household repairs, and so on. It impinges on us when we drive our automobiles or operate a motorcycle. At work, vibration appears in truck and bus driving, heavy equipment and farm vehicle operation, chain sawing, metal chipping, and grinding, to name a few. In fact, NIOSH estimates there are some 8 million workers in the United States exposed to occupational vibration (Wasserman, Badger, Doyle, and Margolies, 1974). If we assume that an individual works at the same vibration-susceptible job for 30 years, 50 weeks per year, at a 30 hr work week, the individual can receive up to 45,000 hours of vibration exposure (Wasserman and Badger, 1972).

6.2.2 PHYSICAL VIBRATION

Mechanical vibration is a vector quantity (magnitude and direction) and may be nonperiodic, periodic, or random (Brock, 1973; Randall, 1977; Tse, Morse, and Hinck, 1971). Periodic vibration may best be described as an oscillating motion of a particle, or body, about a reference position, the motion repeating itself (frequency) exactly after a certain period of time. The simplest form of periodic vibration is harmonic motion, which when plotted as a function of time is represented by a sinusoidal curve (Figure 6.2.1a), where T is the period of vibration, that is, the time elapsed between two successive conditions of motion.

The relationship between time (T) (in seconds) and frequency (f) (in hertz, Hz) is

$$f = \frac{1}{T}$$

If the vibration has the form of a pure translational oscillation along a given axis (x), the instantaneous displacement of a body from the reference position can be mathematically described by the equation

$$X = X_{\text{peak}} \sin\left(2\pi \frac{t}{T}\right) = X_{\text{peak}} \sin\left(2\pi f t\right) = X_{\text{peak}} \sin\left(\omega t\right) \tag{1}$$

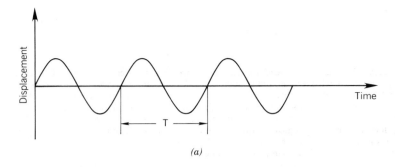

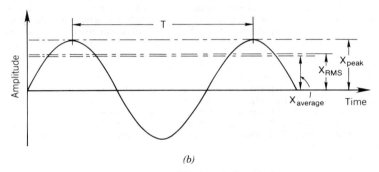

Fig. 6.2.1. (*a*) Pure sinusoidal vibration and (*b*) the relationship between average, rms, and peak amplitude values. (Courtesy B&K Corp.)

where $\omega = 2\pi f$ = angular frequency
$\quad X_{peak}$ = maximum displacement from the referenced position
$\quad\quad t$ = time

The velocity (v) of a moving body is the time rate of change of the displacement, expressed in ft/sec or m/sec:

$$V = \frac{dx}{dt} = \omega X_{peak} \cos (\omega t) = V_{peak} \cos (\omega t) = V_{peak} \sin \left(\omega t + \frac{\pi}{2} \right) \tag{2}$$

The acceleration (a) of the motion is the time rate of change of the velocity expressed in gravitational g units or m/sec² (1 g = 9.81 m/sec²) and is

$$a = \frac{dv}{dt} = \frac{d^2x}{dt^2} = -\omega^2 X_{peak} \sin (\omega t) = -A_{peak} \sin (\omega t) = A_{peak} \sin (\omega t + \pi) \tag{3}$$

From these equations it can be seen that the velocity leads the displacement by a phase angle of 90° and the acceleration leads the velocity by a phase angle of 90°. X_{peak}, V_{peak}, and A_{peak} are characterizing values for the respective magnitudes.

Some practical and useful forms of the above equations are given next for linear acceleration and velocity, where D.A. refers to peak-to-peak or double amplitude:

English Units
acceleration (g) = 0.051D.A. (in.) f^2 (Hz)   (4)
velocity (in./sec) = πD.A. (in.) f (Hz)   (5)

Metric Units

acceleration $(g) = 0.02\text{D.A. (cm)}\, f^2(\text{Hz})$ (6)

velocity $(\text{in./sec}) = \pi \text{D.A. (cm)}\, f(\text{Hz})$ (7)

Magnitude description in terms of peak values is quite useful as long as pure harmonic vibration is considered because it applies directly to the above equations. For more complex vibration other descriptive quantities may be preferred.

The absolute average value is sometimes used when the time history of vibration has to be taken into account:

$$X_{\text{avg}} = \frac{1}{T}\int_0^T x\, dt$$ (8)

A more useful quantity is the root-mean-square (rms) value, which is

$$X_{\text{rms}} = \sqrt{\frac{1}{T}\int_0^T x^2(t)\, dt}$$ (9)

The importance of the rms value as a descriptive quantity is its direct relationship to the energy content of the vibration.

For a pure harmonic motion these relationships are

$$X_{\text{rms}} = \left(\frac{\pi}{2\sqrt{2}}\right) X_{\text{avg}} = \left(\frac{1}{\sqrt{2}}\right) X_{\text{peak}}$$ (10)

General forms of these relationships are

$$X_{\text{rms}} = F_f\, X_{\text{avg}} = \left(\frac{1}{F_c}\right) X_{\text{peak}}$$ (11)

$$F_f = \frac{X_{\text{rms}}}{X_{\text{avg}}}$$ (12)

$$F_c = \frac{X_{\text{peak}}}{X_{\text{rms}}}$$ (13)

F_f and F_c are called "form factor" and "crest factor," respectively, and provide an indication of the wave shape of the vibration being studied (see Figure 6.2.1b).

For pure harmonic motion we obtain

$$F_f = \frac{\pi}{2\sqrt{2}} = 1.11$$ (14)

$$F_c = \sqrt{2} = 1.414$$ (15)

Most mechanical vibration found in daily life is not pure harmonic motion even though much of it may be characterized as periodic. A method for describing this type of vibration is known as Fourier spectrum analysis.

In particular, complex vibration data contain multiple frequencies, all contributing various amplitudes to the total vibration measured by vibration transducers. A Fourier spectrum shows that a complex vibration quantity can be broken down into its constituent parts in both frequency and amplitude (Figures 6.2.2 and 6.2.3). This is usually plotted graphically—the horizontal axis represents frequency, and the vertical axis represents magnitude (e.g., acceleration). The number of vertical lines indicates the total number of vibration frequencies present; the horizontal position of each line gives a single vibration frequency; the height of each line indicates the vibration acceleration amplitude that a given frequency contributes to the total spectrum.

Mathematically,

$$\begin{aligned} F(t) = a_0 &+ a_1 \sin \omega t + a_2 \sin 2\omega t + a_3 \sin 3\omega t + \cdots + a_n \sin (n)\omega t \\ &+ b_1 \cos \omega t + b_2 \cos 2\omega t + b_3 \cos 3\omega t + \cdots + b_n \cos (n)\,\omega t \end{aligned}$$ (16)

Graphically Figure 6.2.2a shows a complex nonharmonic periodic motion waveform. Figure 6.2.2b indicates how this same wave form may be broken down into a sum of harmonically related sine waves. Figure 6.2.3 depicts the Fourier conversion of this wave form from the time to frequency

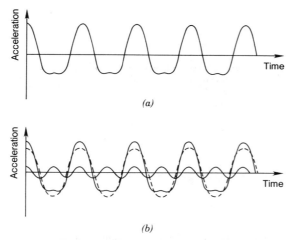

(a)

Fig. 6.2.2. (*a*) An example of complex nonharmonic periodic motion and (*b*) how it may be broken down into a sum of harmonically related sine waves. (Courtesy B&K Corp.)

domains. Figure 6.2.4 depicts the time and frequency domain representations of a pure sinusoidal wave form of Fig. 6.2.3, and a square wave. The interested reader is referred to more detailed texts on this subject for additional information (Morrow, 1963).

Vibration transmissibility (a transfer function) is defined for each vibration frequency present in the spectrum and is the ratio of vibration acceleration appearing at one point to impinging vibration at another point, where both are applied in the same direction. Transmissibility is a method of determining how vibration is altered as it moves through a system: a ratio of unity means there is no alteration of vibration applied at one point and appearing at a second point; a ratio greater than unity indicates an amplification of the original vibration; a ratio of less than unity indicates an attenuation of the original vibration.

Coherence is a measure of the "goodness" of vibration transmissibility data. A coherence of 1 indicates that the measured vibration appearing at a second point, thought to be due to vibration impinging upon the first point, is truly derived from this first point and not from another vibrating source. A coherence of 0 indicates that the measured vibration appearing at a second point is not due to the vibration solely applied to the first point. A value between 0 and 1 is suspect, with possible speculation that the system under examination is nonlinear.

Mathematically, if we assume a linear system, the coherence function is defined as

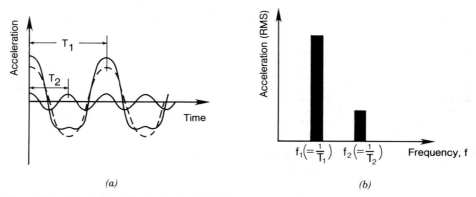

(a)   *(b)*

Fig. 6.2.3. Illustration of how the signal of Figure 6.2.2 can be described in terms of a frequency spectrum. (*a*) Time domain and (*b*) frequency domain. (Courtesy B&K Corp.)

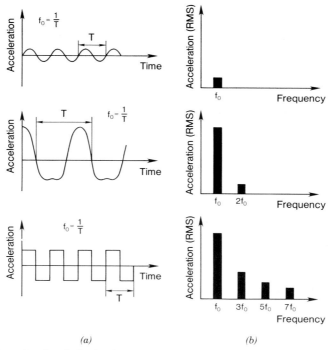

Fig. 6.2.4. Examples of periodic signals and their frequency spectra. (a) Time domain; (b) frequency domain. (Courtesy B&K Corp.)

$$\gamma^2 = \frac{G_{yx}{}^2}{(G_{xx})(G_{yy})} \tag{17}$$

where G_{yx} = cross power spectrum between input and output
G_{xx} = power spectrum of input
G_{yy} = power spectrum of output

6.2.3 HUMAN VIBRATION

It is customary to classify vibration impinging on humans into two general categories: whole-body vibration (vibration transmitted to the entire body through some support such as a vehicular seat or building floor); and segmental (e.g., hand–arm) vibration (vibration locally applied to specific body parts such as the hands and arms from a vibrating hand tool). In the United States alone it is estimated that there are approximately 7 million workers exposed to whole-body vibration and about 1 million workers exposed to hand–arm vibration (1). The term resonance as used in human vibration work refers to the tendency of the human body to act in concert with externally generated vibration (at selected vibration frequencies) and actually to amplify the incoming vibration and exacerbate its effects. In the region of 4 to 8 Hz, more particularly 5 Hz, a human's "whole body" (mostly upper torso) is in resonance (Coermann, 1962) in the vertical direction; in the horizontal and lateral directions resonance is in the 1 to 2 Hz range. Furthermore, various body parts resonate at other frequencies; for example, the head–shoulder system resonates in the 20 to 30 Hz range, and the eyeballs resonate in the 60 to 80 Hz range, to name a few (Grether, 1971). In general, the larger the system mass, the lower the resonate frequency. The concept of resonance is important because it would appear that at resonance frequencies the human is most likely to be susceptible to the effects of vibration exposure. It should be understood by the reader that the concept of resonance is not limited to human exposure to vibration; it appears in virtually all structures. For example, soldiers break cadence when crossing a bridge because the low-frequency vibration produced by marching will set the bridge into resonance and it will most likely collapse. Another example of resonance is that of a crystal goblet held in the hand and subjected to a frequency-varying acoustical field. The goblet will be destroyed at the goblet's natural resonant frequency.

6.2.4 SEGMENTAL (HAND–ARM) VIBRATION

In 1862 a French physician named Maurice Raynaud published a thesis (Raynaud, 1888) in which he described a disease of the hands and fingers of clinical patients. The initial stages of the disease (which now bears his name) are intermittent tingling and/or numbness in the fingers (a neurological component). Later as the disease progresses there appears a peripheral vascular component, namely, intermittent blanching and cyanosis of one or more fingertips (usually not the thumb). The disease progresses to the point where the intermittent blanching becomes more and more frequent with attacks lasting from 15 to 30 min. Eventually primary Raynaud's disease can, in the extreme case, lead to gangrene of one or more of the affected fingers. Cold temperatures tend to trigger attacks and both fever and smoking tend to exacerbate the problem.

It was not until 1913 when Loriga in Italy (Loriga, 1911) first linked Raynaud's disease type symptoms to the use of vibrating hand tools used in the workplace. The most comprehensive study linking hand-tool vibration and cold temperatures to Raynaud's symptomology was first reported in 1918 by Dr. Alice Hamilton (Hamilton, 1918), who investigated the complaints of limestone cutters using vibrating pneumatic tools in Bedford, Indiana.

When Raynaud's disease is related to the work environment it is variously called Raynaud's phenomenon of occupational origin, or vibration white finger (VWF), or "dead hand," to distinguish it from nonoccupational Raynaud's disease. There are various medical conditions that elicit Raynaud type symptomology, such as hand cuts and lacerations, diabetes, and so forth, that a physician must exclude as part of the differential medical diagnosis (see Table 6.2.1) (Taylor and Pelmear, 1975; Wasserman and Taylor, 1977; Wasserman, Taylor, Behrens, Samueloff, and Reynolds, 1982).

Table 6.2.1   Differential Diagnosis—Raynaud's Phenomenon

Medical Condition		
Primary		
Raynaud's disease		Constitutional white finger
Secondary		
Connective tissue disease	a.	Scleroderma
	b.	Systemic lupus erythematosus
	c.	Rheumatoid arthritis
	d.	Dermatomyositis
	e.	Polyarteritis nodosa
	f.	Mixed connective tissue disease
Trauma		
Direct to extremities	a.	Following injury, fracture, or operation
	b.	Of occupational origin (vibration)
	c.	Frostbite and immersion syndrome
To proximal vessels by compression	a.	Thoracic outlet syndrome (cervical rib, scalenus anterior muscle)
	b.	Costoclavicular and hyperabduction syndromes
Occlusive vascular disease	a.	Thromboangiitis obliterans
	b.	Arteriosclerosis
	c.	Embolism
	d.	Thrombosis
Dysglobulinemia	a.	Cold hemagglutination syndrome Cryoglobulinemia Macroglobulinemia
Intoxication	a.	Acro-osteolysis
	b.	Ergot
	c.	Nicotine
Neurogenic	a.	Poliomyelitis
	b.	Syringomyelia
	c.	Hemiplegia

Source: NIOSH Public. No. 82–118.

Table 6.2.2   Stage Assessment of Raynaud's Phenomenon[a] (Taylor–Pelmear Classification System)

Stage	Condition of Digits	Work and Social Interference
00	No tingling, numbness, or blanching of digits	No complaints
0T	Intermittent tingling	No interference with activities
0N	Intermittent numbness	No interference with activities
TN	Intermittent tingling and numbness	No interference with activities
01	Blanching of one or more fingertips with or without tingling and numbness	No interference with activities
02	Blanching of fingers beyond tips. Usually confined to winter	Slight interference with home and social activities. No interference at work
03	Extensive blanching of digits. Frequent episodes in summer as well as winter	Definite interference at work, at home, and with social activities. Restriction of hobbies
04	Extensive blanching. Most fingers. Frequent episodes in summer and winter	Occupational change to avoid further vibration exposure because of severity of signs and symptoms

[a] Complications are not used in this grading.

In the early 1960s it was claimed by Pecora et al. in the United States that Raynaud's phenomenon of occupational origin (VWF) "may have become an uncommon occupational disease approaching extinction in this country" (Pecora, Udel, and Christman, 1960). However, recent studies by me and my associates have shown that this is not the case (14, 16); in fact, our studies recorded a nearly 50% prevalence of VWF in chipper and grinder workers with a mean latency period of 2 years (latency period refers to the time from when a worker began using vibrating hand tools to the appearance of the first white fingertip). This has been determined, after medical examinations and exclusions, using the Taylor–Pelmear classification system (Taylor, Wasserman, Behrens, Samueloff, and Reynolds, 1984) (see Table 6.2.2). The VWF problem is quite high in the United States and has existed, unfortunately, for many years since the Hamilton study (Taylor and Pelmear, 1975; Wasserman and Taylor, 1977; Wasserman, Taylor, Brehens, Samueloff, and Reynolds, 1982; Behrens et al., 1984). Ironically, my colleagues and I repeated the historic Hamilton study in Bedford more than 60 years later, examining some of the offspring of the original participants, using the same pneumatic tools. Results show VWF still exists in high prevalence (Taylor, Wasserman, Behrens, Samueloff, and Reynolds, 1984).

6.2.5   HAND–ARM VIBRATION MEASUREMENT

Because vibration is a vector quantity, hand–arm measurements must be obtained with respect to a standardized biodynamic/biocentric/coordinate system developed by the International Standards Organization (ISO 5349) and shown in Figure 6.2.5. The measuring instruments use a special weighting filter shown in Figure 6.2.6.

Generally vibration measurements are difficult to make. To make matters worse there are few unified methods; however, it is customary to mount three perpendicular, suitable, low-mass (15 grams or less), crystal type accelerometers on a small, light-weight metal cube. This cube in turn is securely fastened to a fixture device that in turn is mounted to the tool under test at the points where the hands grip the tool (Wasserman et al., 1981). When it is not possible to use three accelerometers, a single accelerometer along the major vibration axis may be adequate. Each accelerometer output is

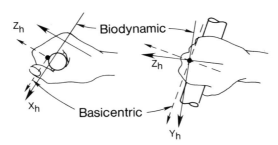

Fig. 6.2.5.   Biodynamic and basicentric coordinate systems for hand–arm measurements (Courtesy ACGIH)

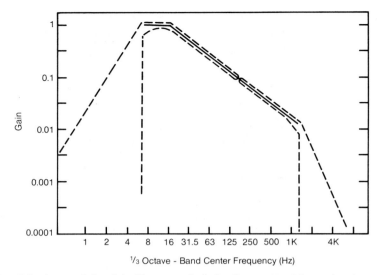

Fig. 6.2.6. Gain characteristics of the filter network used to frequency-weight acceleration components (continuous line). Filter tolerances shown as dashed lines (ISO 5349).

next converted to charge (using a conditioning "charge amplifier") in order to minimize accelerometer signal loss. The amplifier's output is next FM tape recorded. Then spectrum and other analysis can be performed at a later time and the data can be appropriately weighted and compared to ISO 5349. The reader is cautioned that because certain tool vibration acceleration levels are quite high (Brammer and Taylor, 1982) (e.g., for chipper tools, greater than 2000 g, on the chisel for a frequency range of 8 to 1500 Hz), it is entirely possible to destroy the accelerometer and/or obtain erroneous results owing to overloading resulting in DC shifting and saturating of the input conditioning amplifiers (Taylor and Pelmear, 1975; Wasserman and Taylor, 1977). In some instances mechanical filters in series with the measuring accelerometer can help, but one must constantly scrutinize the data to confirm that the mechanical filter is not significantly modifying the actual vibration reaching the accelerometer in the frequency range of interest.

6.2.6 HAND–ARM VIBRATION STANDARDS

As of 1984 there were no mandatory hand–arm standards in the United States. For several years there have been attempts to develop a hand–arm vibration standard. The longest concentrated effort has been with ISO Technical Committee 108, Subcommittee 4. This group has developed ISO/DIS 5349 (Guide for the Measurement and the Evaluation of Human Exposure to Vibration Transmitted to the Hand). The latest version of this document (1983) is the best attempt to bring together much of the worldwide hand–arm epidemiological, clinical, and laboratory studies and is worthy of the reader's consideration. The crux of the document is the time-dependent frequency-weighted (elbow-shaped) curves. The *reverse* of one such curve is depicted in Figure 6.2.6 (solid line).

At this writing the American National Standards Institute S3.39 group is also developing a hand–arm vibration standard.

One of the most progressive groups is the Physical Agents Committee of the American Conference of Government Industrial Hygienists (ACGIH), who with the assistance of several prominent vibration scientists, have developed a series of threshold limit values (TLV) for the hand–arm system,* based principally on VWF epidemiology studies. Recommended TLV values keep the worker only out of Stage 3 and beyond in the Taylor–Pelmear classification system (see Table 6.2.3). These tables values are weighted and measured according to ISO/DIS 5349.

The ACGIH guidelines use the following time-weighted average equation for calculating each of the three perpendicular directions X_h, Y_h, and Z_h. If in *any* direction acceleration is exceeded then the TLV has been exceeded.

$$A_{k\,eq} = \left[\frac{1}{T}\sum_{i=1}^{n}(A_{ki})^2\,T_i\right]^{1/2} = \sqrt{(A_{k1})^2\frac{T_1}{T} + (A_{k2})^2\frac{T_2}{T} + \cdots + (A_{kn})^2\frac{T_n}{T}} \tag{18}$$

---

* Available from ACGIH, 6500 Glenway Ave., Cincinnati, OH 45211.

Table 6.2.3 ACGIH Threshold Limit Values for Exposure of the Hand to Vibration in Either X_h, Y_h, Z_h Directions

Total Daily Exposure Duration[a]	Values of the Dominant,[b] Frequency-Weighted, rms Component Acceleration that Shall not be Exceeded[c] Ak, (Ak_{eq})	
	m/sec²	g
4 hr and less than 8	4	0.40
2 hr and less than 4	6	0.61
1 hr and less than 2	8	0.81
Less than 1 hr	12	1.22

[a] The total time vibration enters the hand per day, whether continuously or intermittently.

[b] Usually one axis of vibration is dominant over the remaining two axes. If one or more vibration axis exceeds the total daily exposure, then the TLV has been exceeded.

[c] Hardly any person exposed at or below the TLVs for vibration contained in this table has progressed to Stage 3 VWF in the Taylor–Pelmear classification (see Table 6.2.2).

where

$$T = \sum_{i=1}^{n} T_i$$

(*Note:* Currently there is commercial instrumentation available to perform ISO 5349 and ACGIH TLV calculations automatically.)

As a result of studies performed by my colleagues and I, NIOSH has publicly issued a current Intelligence Bulletin #38 and a 30 min videotape (#177), both entitled *Vibration Syndrome,* * in an effort to sensitize workers, management, labor, physicians, and so forth, to this problem. The tape and publication are worthy of the reader's consideration.

6.2.7 CURRENT STATUS OF THE VWF PROBLEM

To date there are no mandatory U.S. standards for VWF. Antivibration gasoline chain saws have existed since 1972 and are included in nearly every manufacturer's product line. These saws have helped reduce the prevalence of VWF in those countries when their use has been conscientiously promoted (Wasserman and Taylor, 1977; Brammer and Taylor, 1982). There have been no major antivibration improvements in pneumatic or electrical tools, although it is known that some U.S. firms are working on the problems and one Swedish firm (Atlas Copco Co.) has already begun marketing an antivibration chipping hammer. There are four antivibration glove designs recently introduced in the United States. One glove design uses a special material known as Poron (Rogers Co. Woodstock, CT); the glove is made by Wolverine Corp., Grand Rapids, MI. Another domestic glove uses Sorbothane (Spectrum Sports, Inc., Twinsburg, OH) and is manufactured by the Sager Glove Co., Chicago, IL. The other manufacturers are: Guard-Line Inc., Atlanta, TX; and Steel Grip Safety Apparel Co., Danville, IL.

6.2.8 ELIMINATION OF VWF

Elimination of VWF from the workplace is multifaceted and *all* of the following are recommended: (1) Workers should use both antivibration tools and antivibration gloves. (2) TLV's and standards should be appropriately applied. (3) Vibration exposure times should be determined and work breaks introduced (e.g., 10 min per continuous exposure hour) to avoid constant vibration exposure. (4) If possible, the vibration levels of tools should be measured to obtain design data. (5) A specialized preplacement medical exam should be performed on all workers, especially those with a previous history of peripheral vascular and neurological abnormalities. (6) Workers are advised to have multiple pairs of warm antivibration gloves and warm clothing and not to allow the hands to become chilled. (7) Smoking should be reduced while vibration hand tools are used. (8) The tool should be allowed to do the work by grasping it as lightly as possible, consistent with safe work practice; the tool should

---

* Available from TV Dept., NIOSH, 4676 Columbia Pkwy, Cincinnati, OH 45266.

be rested on a support or workpiece as much as possible. Tools should be operated only when necessary to minimize exposure. (9) If symptoms of tingling, numbness, or signs of white or blue fingers appear, workers should be promptly examined by a physician.

6.2.9   WHOLE-BODY VIBRATION

Unlike hand–arm vibration, the chronic effects of whole-body vibration are not adequately known. Short-term human and animal studies, however, have shown that whole-body vibration may be regarded as a "generalized stressor" and may affect multiple body parts and organs depending on the vibration spectrum and its relationship to various human resonances, particularly 5 Hz. Several whole-body epidemiology studies have been performed; heavy equipment operators (Milby and Spear, 1974; Spear, Keller, Behrens, Hudes, and Tarki, 1924), bus drivers (Gruber and Ziperman, 1974), and truck drivers (Gruber, 1976) all produced mixed results. Generally, they revealed that musculoskeletal difficulties appeared in exposed populations, but not without confounding variables such as lifting, cargo handling, and posture. Examples of laboratory studies of human subjects (Guignard and King, 1972) have shown that during whole-body vibration there are increases in oxygen consumption and pulmonary ventilation. If human subjects are exposed to intense vibration, they may have difficulty in maintaining steady posture.

One study of 78 Russian concrete workers exposed to whole-body vibration showed marked changes in bone structure involving spondylitis deformations, intervertebral osteochondrosis, and calcification of the intervertebral disks and Schmorl's nodes (Rumjanov, 1966). Hypoglycemia, hypocholesterolemia, and low ascorbic acid levels in concrete workers exposed to occupational vibration have also been reported (Puskina, 1967). Gastrointestinal tract changes in gastric secretions and peristaltic motility have been noted in human (Kleiner, 1967) and animal studies. Changes in nerve-conduction velocities owing to vibration have also been reported (Andreeva-Galanina, 1969).

In one Polish study of agricultural and forestry workers a rare clinical description of whole-body vibration sickness is found:

> The first stage is marked by epigastralgia, distension, nausea, loss of weight, drop in visual acuity, insomnia, disorders of the labyrinth, colonic cramps, etc. The second stage is marked by more intense pain concentrated in the muscular and osteoarticular systems. Objective examinations of the workers disclosed muscular atrophy and trophic skin lesions. It is apparent that it is difficult to determine the critical moment at which pathological changes set in, especially due to differences in individual sensitivity to vibration (Jakubowski, 1969).

A study of rats exposed to vibration revealed a drop in lymphocyte count, an increase in granulocyte count, an increased leukocytic alkaline phosphatase activity, faster red cell sedimentation, higher plasma and erythrocyte chloride levels, and a lowering of ascorbic acid and ATP levels of the erythrocytes (Tarnawski, 1969). In a study of liver and kidney function of rats exposed to vibration, ischemia of the liver and kidneys resulted after a single hour of exposure; hyperemia resulted in these organs after 10 days exposure, and after 21 days of exposure, portions of the vascular system ceased functioning (Karmanski, 1969). In a laboratory study on the effects on monkeys of exposure to vibration, gastrointestinal bleeding and lowered hematocrits were noted during the exposure and multiple lesions of the gastric mucosa were seen at necropsy (Sturges, Badger, Slarve, and Wasserman, 1974).

There have been numerous human performance vibration studies throughout the years, mostly in the military sector, showing in general that the lowest subjective discomfort tolerance level occurs around the 5 Hz resonant point. Manual tracking capability is also most seriously affected at this 5 Hz point. Visual acuity is severely impaired in the 1 to 25 Hz range.

On the other hand, performance tasks such as those involving pattern recognition, reaction time, and monitoring appear not to be affected by exposure to vibration (Grether, 1971). Simulated heavy equipment driving tests that compared the effects of a mixture of simultaneous vibratory frequencies (similar to actual occupational vibration) revealed that human subjects performed worse under the mixed conditions, gradually improving as the mixture was replaced by nonresonant single sinusoidal vibratory conditions (Cohen, Wasserman, and Hornung, 1977).

6.2.10   WHOLE-BODY VIBRATION MEASUREMENTS

As with hand–arm measurements, whole-body vibration is measured with respect to standard biodynamic coordinate system (see Figure 6.2.7) according to ISO 2631 (Guide for the Evaluation of Human Exposure to Whole-Body Vibration). The frequency range of interest for the whole-body measurements is generally 1 to 80 Hz, although in certain applications it may be necessary to have a frequency response below 1 Hz down to 0.05 Hz (e.g., some heavy equipment measurements, ship motion studies). Thus for all practical purposes a response from DC to 80 Hz is desirable in measurement accelerometers; the best choices are piezoresistive or strain gauge devices, depending on what must be measured. Both types are configured as Wheatstone bridges and require DC voltage excitation and direct coupled

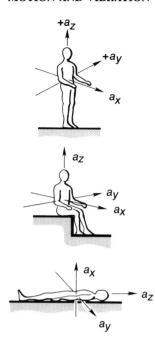

Fig. 6.2.7. Whole-body vibration biodynamic coordinate system (ISO 2631).

operational amplifiers to raise the signal level to the point where it can be accurately recorded on an FM tape recorder. (Crystal accelerometers used for hand–arm measurements are *not* suitable for whole-body measurements because they generally lack the required very-low-frequency response.) The mass of the measuring accelerometers must be very small (1 to 2 grams) to minimize "mass loading errors." Mostly for human measurements, it usually turns out that piezoresistive accelerometers are used because these can be made very light, whereas strain gauge accelerometers are much heavier. Piezoresistive accelerometers are temperature sensitive and must be obtained with internal temperature compensation. Also, these devices tend to "ring" unless they have sufficient internal damping (usually 0.5 to 0.7 damping ratio). Both the strain gauge piezoresistive devices have the distinct advantage of being self-calibrating (whereas crystal devices are not). In practice these two self-calibrating devices are oriented with their sensitive faces pointed toward gravity. The bridge amplifier is zero balanced against the force of gravity; rotating the accelerometer 90° away from gravity causes a voltage deflection unbalance that is exactly equal to 1 g and is linearly incremental. On the other hand, with crystal devices it is necessary to use a small calibration shaker with a built-in reference accelerometer. The device under test is affixed to the calibrator and its output is compared to the reference device under vibration. As an added but not mandatory check, piezoresistive and strain gauge devices can also be calibrated using this second calibration check.

Whole-body measurements must be made at identifiable parts on the body, for example, at the buttocks, while the subject is seated (Wasserman, Asburry, and Doyle, 1979) (see Figure 6.2.8). In this case, a hard rubber disk is used with a partially hollowed out center. In this center a small metal cube is placed with three perpendicular lightweight piezoresistive accelerometers. Simultaneously, a metal bite bar (shown in Figure 6.2.9) can be used with from one to three piezoresistive accelerometers to measure vibration at the subject's jaw. In this case a simple dental impression is made by using warm dental material on one end of the bite bar. The subject bites once into the warm material, it hardens, and the device can be used for measurements. Once the dental material is removed, the bite bar can be cleaned, sterilized, and used for another subject.

Generally, it is desirable to mount accelerometers only on bony protrusions (e.g., knee, shoulder blade) using stiff double-sided carpet tape. Bony sites are preferable because (1) they are easily located and anatomically definable; and (2) there are a minimum number of tissue layers sliding under the measurement site.

Once the whole-body measurements have been made and tape recorded on a multitrack FM tape recorder a spectrum analysis should be performed on each channel with results compared to ISO 2631. In addition, it is useful to examine crest factor and coherence among the various recorded channels. At least one company (B&K) has developed an instrument for comparing vibration data directly to ISO 2631.

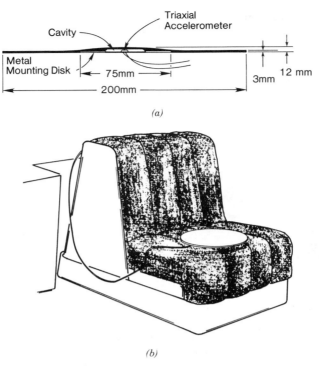

(a)

(b)

Fig. 6.2.8. (a) Triaxial accelerometer embedded in molded hard rubber disk. (b) Seat accelerometer disk shown taped and in use during typical field study.

Fig. 6.2.9a. Top: Metal bite bar (left) with accelerometer and dental impression (right). Bottom: Bite bar in use during typical field study.

6.2.11 WHOLE-BODY VIBRATION STANDARDS

Since 1964 the ISO TC/08/SC4 group of scientists have been developing ISO 2631 (Guide for the Evaluation of Human Exposure to Whole-Body Vibration). The guide originally was developed with a frequency range from 1 to 80 Hz; at this writing there is an addendum to extend the frequency ranges from 0.1 Hz to meet the 1 Hz point in order to take into account motion sickness data from the 0.1 to 0.7 Hz band. The current ISO 2631 document is still under extensive revision, but has been adopted as is by ANSI (document S3.18). Figure 6.2.10 depicts the rms vertical vibration direction (A_z) limits as defined in ISO 2631.

Figure 6.2.11 depicts both the ISO 2631 horizontal and lateral vibration rms direction (A_x,A_y) limits. Each plot graphs acceleration versus frequency as a function of exposure times from 1 min to 8 hr. The A_z plot (Figure 6.2.10) depicts the lowest tolerance level in the human resonance 4 to 8 Hz band. Similarly, the A_x,A_y plot shows the 1 to 2 Hz human resonance band.

Each plot can be thought of in terms of three families of curves (reduced comfort, fatigue decreased proficiency (FDP), and health exposure limits); all curves have the same shape and move up and down in acceleration with respect to FDP. The FDP graphs are shown in Figures 6.2.10 and 6.2.11. FDP refers to the prevention of task performance decrements. To obtain reduced comfort divide the acceleration values by 3.15; to obtain health exposure limits multiply the acceleration values by 2. The reader is cautioned that these curves are to be used *very* carefully; for example, the curves refer to discrete single sinusoidal vibration and not to nonsinusoidal (spectrum conditions usually found in the workplace), the data regarding the health curves are not very strong, and so on. The details of the guide should be directly consulted before using the graphs.

6.2.12 WHOLE-BODY VIBRATION WORK PRACTICES

Because of the paucity of adequate epidemiological/medical/physiological data, ISO 2631 may not be adequate to protect workers from potential *health effects* of whole-body vibration (Wasserman, 1981). Thus in addition to ISO 2631 an interim/supplementary work practice guide should be used (especially if the vibrating source emanates components in the 4 to 8 Hz A_z range or 1 to 2 Hz A_x,A_y range).

1. Limit the time spent by workers on a vibrating surface to no more than is absolutely necessary to perform the job safely.

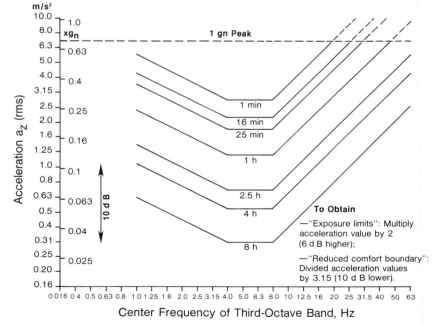

Fig. 6.2.10. Vertical (A_z) vibration exposure limits as a function of exposure time and frequency for fatigue decreased proficiency boundary (ISO 2631).

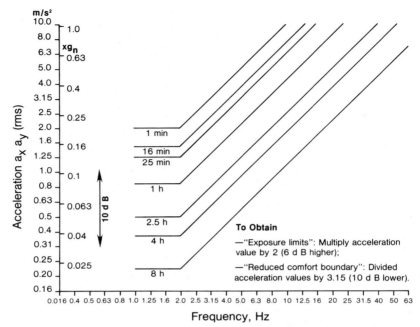

Fig. 6.2.11. Horizontal and lateral (A_x, A_y) vibration exposure limits as a function of exposure time and frequency for fatigue decreased proficiency boundary (ISO 2631).

2. Have machine controls moved, wherever possible, off the vibrating surface.
3. Mechanically isolate the vibrating source, or the surface where the workers are stationed, in order to reduce exposure.
4. Compare measured vibration acceleration to ISO 2631.
5. Carefully maintain vibrating machinery to prevent the development of excess vibration.

6.2.13 DEFINITIONS OF SELECTED VIBRATION TERMINOLOGY*

Displacement; relative displacement: A vector quantity that specifies the change of position of a body, or particle, with respect to a reference frame.

Notes

1. The reference frame is usually a set of axes at a mean position or a position of rest. In general, the displacement can be represented by a rotation vector, a translation vector, or both.
2. A displacement is designated as relative displacement if it is measured with respect to a reference frame other than the primary reference frame designated in the given case. The relative displacement between two points is the vector difference between the displacement of the two points.

Velocity; relative velocity: A vector that specifies the time derivative of displacement.

Notes

1. The reference frame is usually a set of axes at a mean position or a position of rest. In general, the velocity can be represented by a rotation vector, a translation vector, or both.
2. A velocity is designated as relative velocity if it is measured with respect to a reference frame other than the primary reference frame designated in a given case. The relative velocity between two points is the vector difference between the velocities of the two points.

---

* Partly extracted from "Vibration & Shock-Vocabulary" ISO 2041–1975 (E/F).

Acceleration: A vector that specifies the time derivative of velocity.

Notes

1.   The reference frame is usually a set of axes at a mean position or a position of rest. In general, the acceleration can be represented by a rotation vector, a translation vector, or both.
2.   An acceleration is designated as relative acceleration if it is measured with respect to a reference frame other than the primary reference frame designated in a given case. The relative acceleration between two points is the vector difference between the velocities of the two points.

Excitation; stimulus: An external force (or other input) applied to a system that causes the system to respond in some way.

Response of a system: A quantitative expression of the output of the system.

Transmissibility: The nondimensional ratio of the response amplitude of a system in steady-state forced vibration to the excitation amplitude. The ratio may be one of forces, displacements, velocities, or accelerations.

Linear system: A system in which the response is proportional to the magnitude of the excitation.

Note

This definition implies that the dynamic properties of each element in the system can be represented by a set of linear differential equations with constant coefficients, and that the principle of superposition can be applied to the system.

Mechanical system: An aggregate of matter comprising a defined configuration of mass, stiffness, and damping.

Center of gravity: That point through which passes the resultant of the weights of its component particles for all orientations of the body with respect to a gravitational field.

Degrees of freedom: The number of degrees of freedom of a mechanical system is equal to the minimum number of independent generalized coordinates required to define completely the configuration of the system at any instant of time.

Single-degree-of-freedom system: A system requiring but one coordinate to define completely its configuration at any instant.

Multi-degree-of-freedom system: A system for which two or more coordinates are required to define completely the configuration of the system at any instant.

Stiffness: The ratio of change of force (or torque) to the corresponding change in translational (or rotational) displacement of an elastic element.

Compliance: The reciprocal of stiffness.

Transfer function (of a system): A mathematical relation between the output (or response) and the input (or excitation) of the system.

Impedance: The ratio of a harmonic excitation of a system to its response (in consistent units), both of which are complex quantities and both of whose arguments increase linearly with time at the same rate. The term generally applies only to linear systems.

Notes

1.   The concept is extended to nonlinear systems where the term incremental impedance is used to describe a similar quantity.
2.   The terms and definitions relating to impedance apply under sinusoidal conditions only.
3.   The reciprocal of impedance is admittance (the complex ratio of displacement to force) or mobility (the complex ratio of velocity to force).

Mechanical impedance: At a point in a mechanical system, the complex ratio of force to velocity where the force and velocity may be taken at the same or different points in the same system during simple harmonic motion.

Driving-point impedance: In a mechanical sense, the complex ratio of the force to velocity taken at the same point in a mechanical system during simple harmonic motion.

Transfer impedance: In a mechanical sense, the complex ratio of the force taken at one point in a mechanical system to the velocity taken at another point in the same system during simple harmonic motions.

Vibration: The variation with time of the magnitude of a quantity that is descriptive of the motion or position of a mechanical system, when the magnitude is alternately greater and smaller than some average value or reference.

Periodic vibration: A periodic quantity whose values recur for certain equal increments of the independent variable.

Random vibration: A vibration whose magnitude cannot be precisely predicted for any given instant of time.

Note

The probability that the magnitude of a random vibration falls within a given range can be specified by a probability distribution function.

Cycle: The complete range of states or values through which a periodic phenomenon or function passes before repeating itself identically.

Fundamental period; period: The smallest increment of the independent variable of a periodic quantity for which the function repeats itself.

Note

The fundamental period is called the period.

(Cyclic) frequency: The reciprocal of the fundamental period.

Note

The unit of frequency is the hertz (Hz), which corresponds to 1 cycle per second.

Fundamental frequency: (1) Of a periodic quantity, the reciprocal of the fundamental period. (2) Of an oscillating system, the lowest natural frequency. The normal mode of vibration associated with this frequency is known as the fundamental mode.

Harmonic (of a periodic quantity):   A sinusoid whose frequency is an integral multiple of the fundamental frequency.

Peak value; peak magnitude; positive peak value; negative peak value: The maximum value of a quantity during a given interval.

Note

A peak value of an oscillating quantity is usually taken as the maximum deviation of that quantity from the mean value. A positive peak value is the maximum positive deviation and a negative peak value is the maximum negative deviation.

Peak-to-peak value (of an oscillating quantity):   The algebraic difference between the extreme values of the quantity.

Excursion; total excursion (of a vibration):   The peak-to-peak displacement.

Crest factor (of an oscillating quantity); peak-to-rms ratio: The ratio of the peak value to the rms value.

Note

The value of the crest factor of a sine wave is $\sqrt{2}$.

Form factor (of an oscillating quantity):   The ratio of the rms value to the mean value for one-half cycle between two sucessive zero crossings.

Note

The form factor for a sinusoid is $\pi/2\sqrt{2} = 1.111$.

Instantaneous value; value: The value of a variable quantity at a given instant.

Resonance: The vibration frequency at which maximum mechanical energy is transferred from the vibrating source to the human.

Note

At resonance the human is (1) maximally "tuned" to the vibrating source and (2) then is an actual amplification of the incoming vibration by the human.

Vibration isolator: An isolator designed to attenuate the transmission of vibration in a frequency range.
Decibel (dB):   One-tenth of a bel.

Notes

1.   The magnitude of a level in decibels is 10 times the logarithms to the base 10 of the ratio of powerlike quantities; that is,

$$L_p = 10 \log_{10} \frac{x^2}{x_0^2}$$
$$= 20 \log_{10} \frac{x}{x_0}$$

2.   Examples of quantities that qualify as powerlike quantities are sound pressure squared, particle velocity squared, sound intensity, sound energy density, voltage squared. Thus the bel is a unit of sound pressure squared level; it is common practice, however, to shorten this to sound pressure level because ordinarily no ambiguity results from so doing.

Nominal bandwidth (of a filter); bandwidth: The difference between the nominal upper and lower cutoff frequencies. The difference may be expressed (1) in hertz; (2) as a percentage of the pass-band center frequency; or (3) as the interval between the upper and lower nominal cutoffs in octaves.
Constant bandwidth filter: A filter that has a bandwidth of constant value when expressed in hertz. It is independent of the value of the center frequency of the filter.
Proportional bandwidth filter: A filter that as a bandwidth is proportional to the frequency.

Note

Octave bandwidth, one-third octave bandwidth, and so on, are typical bandwidths for proportional bandwidth filters.

Octave: The interval between two frequencies which have a frequency ratio of two.

Note

The interval, in octaves, between any two frequencies is the logarithm to the base 2 (or 3322 times the logarithm to the base 10) of the frequency ratio.

One-half octave; half octave: The interval between two frequencies that have a frequency ratio of $2^{1/2}$, or 1.414.
One-third octave; third octave: The interval between two frequencies that have a frequency ratio of $2^{1/3}$, or 1.2599.
One-tenth decade: The interval between two frequencies that have a frequency ratio of $10^{1/10}$, or 1258.9.

Notes

1.   The difference between $\frac{1}{10}$ decade and $\frac{1}{3}$ octave is less than 0.1%. The two bandwidths can therefore be considered equivalent for practical purposes.
2.   The interval, in decades, between any two frequencies, is the logarithm to the base 10 of the frequency ratio.

REFERENCES

Adreeva-Galanina, E. D. (1969). Towards a solution to the problem of degeneration and regeneration of peripheral nerves following experimental exposure to vibration. *Glg. Tr. Prof. Zabol. 13*, 4–7.

Behrens, V., Taylor, W., Wilcox, T., Wasserman, D., Samueloff, S., and Reynolds, D. (1984). Vibration syndrome in chipping and grinding workers. *Journal for Occupational Medicine, 26,* 765–788.

Brammer, A. J., and Taylor, W., Eds. (1982). *Vibration effects on the hand and arm in industry.* New York: Wiley.

Brock, J. T. (1973). *Mechanical vibration and shock measurements.* Naerum, Denmark: Bruel and Kjaer.

Coermann, R. R. (1962). The mechanical impedance of the human body in sitting and standing positions at low frequencies. *Human Factors, 4,* 225–253.

Cohen, H. H., Wasserman, D. E., and Hornung, R. (1977). Human performance and transmissibility under sinusoidal and mixed vertical vibration. *Ergonomics, 20,* 207–216.

Grether, W. F. (1971). Vibration and human performance. *Human Factors, 13,* 203–205.

Gruber, G. J. (1976). Relationship between whole-body vibration and morbidity patterns among interstate truck drivers (DHEW/NIOSH Pub. No. 77–167).

Gruber, G. J., and Zipperman, H. H. (1974). Relationship between whole-body vibration and morbidity patterns among motor-coach operators (DHEW/NIOSH Pub. No. 75–104).

Guignard, J. C., and King, P. F. (1972). Aeromedical aspects of vibration and noise (AGARDograph No. 17). London: NATO, Technical Editing and Reproduction.

Hamilton, A. (1918). A study of spastic anemia in the hands of stonecutters: An effect of the air hammer on the hands of stonecutters. *Industrial Accidents and Hygiene Series* (Bulletin 236, No. 19) USDOL Bureau of Labor Statistics.

Jakubowski, R. (1969). General characteristics of vibration disease at various workplaces in agriculture and forestry. *Med. Wiej., 4,* 47–51.

Karmanski, J. (1969). Hemodynamic changes in certain parenchymatous organs in albino rats caused by low frequency mechanical vibrations. *Przegl. Lek., 25,* 763–765.

Kleiner, A. I. (1967). Study of the main stomach functions in patients with vibration disease. *Gig. Tr. Prof. Zabol., 11,* 25–27.

Loriga, G. (1911). Pneumatic tools: Occupation and health. *Ball. Inspect. Larboro., 2,* 35–37.

Milby, T. H., and Spear, R. C. (1974). Relationship between whole-body vibration and morbidity patterns among heavy equipment operators (DHEW/NIOSH) Pub. No. 74–131).

Morrow, C. T. (1963). *Shock and vibration engineering.* New York: Wiley.

Pecora, L. J., Udel, M., and Christman, R. P. (1960). Survey of current status of Raynaud's phenomenon of occupational origin. *Journal of the American Industrial Hygiene Association, 21,* 80–83.

Puskina, N. M. (1967). Biochemical blood values in workers exposed to vibration. *Gig. Tr. Prof. Zabol., 11,* 25–27.

Randall, R. B. (1977). *Frequency analysis.* Naerum, Denmark: Bruel and Kjaer.

Raynaud, M. (1888). Local asphyxia and symmetrical gangrene at the extremities (Trans.) In *Selected Monographs.* London: New Sydenham Society. (Original work M.D. Thesis, Paris, 1862.)

Rumjancev, G. I. (1966). Bone structure changes in the spinal column of prefabricated concrete workers exposed to whole-body high frequency (50 Hz) vibration. *Gig. Tr. Prof. Zabol., 11,* 25–27.

Spear, R. C., Keller, C. A., Behrens, V., Hudes, M., and Tarter, D. (1974). Morbidity patterns among heavy equipment operators exposed to whole-body vibrations: Follow-up to a 1974 study (DHEW/NIOSH Pub. No. 77–120).

Sturges, D. V., Badger, D. W., Slarve, R. N., and Wasserman, D. E. (1974). Laboratory studies on chronic effects of vibration exposure. Oslo, Norway: NATO-AGARD Proceedings CPP-145.

Tarnawski, A. (1969). Modification of blood corpuscles by vibration. *Med., 20,* 345–347.

Taylor, W., and Pelmear, P. L., Eds. (1975). *Vibration white finger in industry.* London: Academic.

Taylor, W., Wasserman, D. E., Behrens, V., Samueloff, S., and Reynolds, D. (1984). Effects of the air hammer on the hand of stonecutters: The limestone quarries of Bedford, Indiana, revisited. *British Journal of Industrial Medicine., 41,* 773–776.

Tse, F. S., Morse, I. E., and Hinkle, R. T. (1971). *Mechanical vibrations.* Boston: Allyn and Bacon.

Wasserman, D. E. (1981). Occupational vibration in the foundry. *Procedures of the Symposium on Occupational Health Material Control Technology in the Foundry and Secondary Non-Ferrous Smelting Industries* (DHHS/NIOSH Pub. No. 81–114).

Wasserman, D. E., Asburry, W. A., and Doyle, T. E. (1979). Whole-body vibration of heavy equipment operators. *Shock and Vibration Bulletin, 49,* 47–68.

Wasserman, D. E., and Badger, D. (1972). The NIOSH plan for developing industrial vibration exposure criteria. *Journal of Safety Research, 4,* 146–154.

Wasserman, D. E., Badger, D., Doyle, T. E., and Margolies, L. (1974). Industrial Vibration—An overview. *Journal of American Safety Engineers, 19,* 38–43.

Wasserman, D. E., Reynolds, D., Behrens, V. Taylor, W., Samueloff, S., and Basel, R. (1981). VWF disease in U.S. workers using pneumatic chipping and grinding hand tools. *Engineering,* Vol. 2 (DHHS/NIOSH Pub. No. 82–101).

Wasserman, D. E., and Taylor, W., Eds. (1977). *Proceedings of the International Occupational Hand-Arm Vibration Conference* (DHEW/NIOSH Pub. No. 77–170)

Wasserman, D. E., Taylor, W., Behrens, V., Samueloff, S., and Reynolds, D. (1982). VWF disease in U.S. workers using pneumatic chipping and grinding hand tools. *Epidemiology,* Vol. 1 (DHHS/ NIOSH Pub. No. 82–118).

CHAPTER **6.3**
ILLUMINATION

WILLIAM H. CUSHMAN

Ergonomic Consultant
Rochester, New York

BRIAN CRIST

Ergonomic Consultant
Rochester, New York

6.3.1 OVERVIEW

Light is the stimulus for vision, the primary sensory channel for receiving information about our environment. Our ability to perform most tasks depends, to some extent, on the quantity and quality of the light that illuminates them. Lighting may also affect the way we feel and perceive our environment.

Our discussion begins with a brief overview of photometry (light measurement), including guidelines for measuring luminance and illuminance (Section 6.3.2). The various types of artificial light sources and luminaires are compared in Section 6.3.3. Section 6.3.4 is concerned with the effects of lighting on behavior. The remaining sections are oriented toward applications: guidelines for illumination including task lighting and a comparison of direct versus indirect lighting (Section 6.3.5), measurement and control of glare (Section 6.3.6), task modification as an alternative to increasing illumination (Section 6.3.7), use of models and simulation to evaluate alternate designs (Section 6.3.8), lighting for workplaces with video display terminals (Section 6.3.9), and inspection lighting (Section 6.3.10).

6.3.2 PHOTOMETRY

Photometry is the branch of metrology that is concerned with the measurement of visible light, usually with a sensing device having a spectral sensitivity similar to the average human eye. The fundamental quantity in photometry is luminous flux, which is measured in *lumens*. One lumen (lm) is equal to $\frac{1}{683}$ W for light with a wavelength of 555 nm. In addition to luminous flux, the ergonomist needs to be familiar with two other photometric concepts—illuminance and luminance.

Illuminance, also called illumination, is a measure of the amount of light from ambient and local sources that falls on a surface. In the international or SI system of measurement units, illuminance is measured in *lux*. One lux (lx) is equal to 1 lm/m² of illuminated surface area. Illumination may also be measured in *footcandles* (fc) where 1 fc is equal to one lm/ft². Because 1 m² = 10.76 ft²,

$$1 \text{ fc} = 10.76 \text{ lx}$$

There are two other useful relationships that should be kept in mind, the inverse square law of illumination and the cosine law of illumination. The inverse square law of illumination states that illumination varies inversely with the square of the distance d between the illuminated surface and the source that illuminates it. This relationship may be expressed by the following equation:

$$\text{illuminance} = \frac{I}{d^2}$$

where I is the source intensity. The cosine law of illumination states that the illumination on a surface varies directly with the cosine of the angle between a line perpendicular to the surface and the direction of light. If the angle is represented by θ, then the relationship given above becomes:

$$\text{illuminance} = \cos\theta \left(\frac{I}{d^2}\right)$$

Although these laws apply strictly only for point sources, they may be used to calculate illumination from many nonpoint sources (often called extended sources) without serious error. If an extended source is uniform in brightness and if its maximum dimension is less than one-fifth of the distance to the surface, the error in calculation (owing to the fact that the source is not a point source) is less than 1% (Lum-i-neering Associates, 1979).

Luminance is the photometric correlate of the psychological sensation of brightness. It is related to the amount of light emitted in a given direction by a luminous source (e.g., a tungsten lamp) or by an illuminated surface (e.g., a piece of white paper on a desk); its magnitude does not vary with viewing distance. In the international or SI system of measurement units, luminance is expressed in *candelas per square meter* (cd/m²). (One candela is equal to one lm/steradian in a given direction). Luminance may also be expressed in *footlamberts* (ftL). To convert from footlamberts to candelas per square meter, one uses the following relationship:

$$1 \text{ ftL} = 3.43 \text{ cd/m}^2$$

The luminance of a perfectly diffuse reflecting surface can easily be determined if the illumination on that surface and its reflectance are known. For the SI system the luminance of the surface in candelas per square meter is given by the expression

$$\text{luminance} = \frac{\text{illuminance} \times \text{reflectance}}{\pi}$$

Illuminance is expressed in lux, and reflectance is defined as the ratio of the light reflected from the diffuse surface (reflected luminous flux, to be more precise) to the incident light (incident luminous flux). In other words, reflectance = reflected light/incident light. If the reflecting surface is not a perfect diffuse reflector, then the luminance factor of the surface is used instead of reflectance. The luminance factor for a surface is defined as the ratio of the luminance of that surface when viewed in a given direction to the luminance of a uniform white diffusing surface illuminated in the same way and viewed from the same direction.

The equation for determining the luminance of a surface is slightly different when British units are used [i.e., footcandles (fc) for illuminance and footlamberts (ftL) for luminance]. The equation is

$$\text{luminance} = \text{illuminance} \times \text{reflectance}$$

The difference between this equation and the one given for the SI system is related to differences in the way luminance is defined in the two measurement systems (the SI system and the system based on British units). For additional details, see Boyce (1981, pp. 5–7).

6.3.2.1 Measuring Luminance

A device used to measure luminance is called a photometer. Its sensitivity to light of various wavelengths matches that of the CIE photopic curve. (CIE refers to the Commission Internationale de l'Eclairage, or International Commission on Illumination, an international standards organization.) The most versatile photometers have a small aperture (1° or less) and a direct-reading digital display. Because the luminance of an object does not vary with viewing distance, a photometer may be placed at any convenient distance and pointed toward the luminous surface. An optical device is usually provided to assist the operator with alignment and focusing and to provide an indication of the boundary of the area being measured. If the luminance of the area is not uniform, the reading obtained will be an integrated average. If the luminance of the object varies with the direction from which it is viewed, as is the case with directional light sources and nondiffusing surfaces, then the photometer must view the luminous object from the same angle as the viewer. Zero adjustment and selection of the proper sensitivity range may or may not be necessary. For additional details consult the operating manual provided by the manufacturer of your instrument.

6.3.2.2 Measuring Illuminance

An illumination meter or an illuminance probe that attaches to the system unit of a modular photometer may be used to measure the illumination on a surface. The sensor should have built-in cosine correction (see discussion of cosine law of illumination above.) An illumination meter's sensitivity to light of various wavelengths also matches the CIE photopic curve. To obtain a measurement, place the meter or probe on the surface of interest and observe the reading on the display. Zero adjustment and selection of the proper sensitivity range may or may not be necessary, depending on the unit. Refer to the operating manual of your instrument for additional details.

6.3.3 ARTIFICIAL LIGHT SOURCES

The incandescent lamp is the most commonly used artificial light source. Its popularity stems from a number of factors: low cost; availability in numerous sizes, bulb configurations, and base types; good color rendering qualities; and convenience. However, because light is produced by heating a filament, most of the energy is in the infrared (nonvisible) region of the spectrum. For this reason incandescent lamps are the least efficient (i.e., fewest lumens per watt or lowest *efficacy*) of all commonly used artificial light sources. (Lighting engineers usually use the term "efficacy" rather than "efficiency" because efficiency is usually defined as the ratio of output/input with both quantities expressed in the same units.)

Closely related to standard incandescent lamps are the tungsten–halogen or "quartz" lamps. These lamps contain iodine or bromine in addition to a fill gas and operate at a higher temperature than conventional incandescent lamps. The primary advantages of the "quartz" lamp are higher efficacy, longer lamp life, and a smaller decline in light output over the life of the lamp.

The remaining commonly used artificial light sources are classified as gaseous discharge lamps.

These include fluorescent lamps, mercury lamps, metal halide lamps, and high- and low-pressure sodium lamps. High-pressure lamps are sometimes referred to as "high-intensity discharge lamps" or simply HID lamps.

Gaseous discharge lamps produce light by passing an electric current through a vapor at high or low pressure. The electrons in the arc collide with atoms of the vapor, causing some electrons to shift temporarily into a higher, less stable energy state. As the electrons return to the normal state, photons are released. These photons may be either in the visible or ultraviolet region of the spectrum. If the radiations are in the ultraviolet region, it is necessary to coat the inside surface of the lamp with phosphors. The phosphors produce visible light when irradiated with ultraviolet radiation.

The color characteristics of the light produced by gaseous discharge lamps depend on several factors such as the gas and other materials that fill the envelope, relative pressure (high or low), and phosphor characteristics (if phosphors are used). In general, high-pressure lamps have a more continuous spectral power distribution and often provide better color rendering.

Fluorescent lamps (low-pressure mercury lamps) are the most widely used gaseous discharge lamps, particularly for commercial and industrial applications. Fluorescent tubes are coated with phosphors on the inside surface and contain both mercury and an inert gas. Unlike some discharge lamps, fluorescents provide good color rendering and a high efficacy. The color of the light emitted by a fluorescent lamp depends on the phosphor coating on the tube walls. There are seven standard colors provided by all major manufacturers: White, Cool White, Deluxe Cool White, Warm White, Deluxe Warm White, Natural, and Daylight. The different types of fluorescent lamps also vary significantly in cost and efficacy.

A second type of low-pressure discharge lamp is the low-pressure sodium lamp. These lamps are the most efficient of all artificial light sources. However, the light that they emit is an almost monochromatic yellow. Color rendering is very poor (virtually nonexistent) and, for this reason, low-pressure sodium lamps are suitable only for a few applications (such as roadway lighting) where color discrimination is not important.

Mercury vapor lamps produce light when electrons collide with mercury gas at high pressure. The predominant emissions are in the ultraviolet and blue regions of the spectrum. The efficacy and spectral characteristics of a given lamp depend on the vapor density within the lamp and the phosphor (if any) that is used to coat the inside wall of the bulb or tube. Clear (uncoated) lamps produce a line spectrum rather than a continuous spectrum. As a consequence, color rendering for clear mercury lamps is poor. Lamp life for mercury lamps, however, is usually very long—up to 18,000 hr.

Metal halide lamps are high-pressure mercury lamps with halide salts (such as sodium, thorium, scandium, and indium) added to the arc tube. This modification approximately doubles the lumen output of a typical metal halide lamp (as compared with a standard mercury lamp) with substantial improvements in color rendering owing to a more continuous spectral power distribution. This combination of high efficacy and good color rendering makes the metal halide lamp an attractive choice for many applications.

High-pressure sodium lamps usually have a lower efficacy than their low-pressure counterparts. However, the color rendering and color appearance of the high-pressure lamps are considerably better. Because phosphors are not used to coat the inside walls of the glass envelope, the luminous area of the lamp (the arc tube) may be quite small. Although this simplifies optical control, it increases the potential for glare if shielding is inadequate. An additional disadvantage of high-pressure sodium lamps (and low-pressure sodium lamps as well) is the high initial cost.

6.3.3.1 Comparison of Artificial Light Sources

In selecting an artificial light source for a particular lighting application, the two most important considerations are usually efficacy (inversely related to operating costs) and color rendering. Other important factors include initial costs and lamp life, maintenance requirements including cleaning, and ease of shielding and directional control.

The characteristics of the six most commonly used artificial light sources are listed in Table 6.3.1. The spectral power distribution (SPD) diagrams for four of these sources are given in Fig. 6.3.1. Unfortunately, the SPDs for the more efficient light sources usually have one or more distinct peaks. As a consequence, these sources are not suitable for tasks requiring precise color discrimination.

The CIE Color Rendering Index or CRI (far righthand column in Table 6.3.1) is a measure of how colors would appear when illuminated by a light source of interest (i.e., a "test" source) as compared with how they look when illuminated by a standard or "reference" source. The reference source is usually some form of daylight or an incandescent lamp. The index for the test source is obtained by calculating the changes in the positions of eight test colors in CIE color space that would occur if the source of illumination were changed from the reference source to the test source. The obtained values are appropriately scaled and averaged so that 100 represents perfect color rendering; that is, the test colors would have the same appearance when illuminated by the test source as when illuminated by the reference source.

Table 6.3.1   Characteristics of Artificial Light Sources

Type	Lamp Cost	Lamp Life (yr)	Efficacy (lm/W)	CIE Color Rendering Index
Incandescent	Low	<1	17–23	89–98 (very good)
Fluorescent	Low to moderate	5–8	50–80	52–94 (fair to very good)
Mercury	Moderate	9–12	50–55	15–51 (poor to fair)
Metal halide	Moderate	2–3	80–90	53–78 (fair to good)
High-pressure sodium	High	3–5	85–125	21–29 (poor to fair)
Low-pressure sodium	High	4–5	100–180	Not applicable (very poor)

Several other indexes of color rendering have also been proposed. The Crawford index is based on a comparison of the spectral power distributions of a test source and reference source. The Color Discrimination Index (CDI) attempts to describe the discriminability of a set of test colors when illuminated by a given source. The Color Rendering Capacity (CRC) is an index related to the maximum number of different surface colors that can be discriminated when the test source is the only source of illumination (Xu, 1983).

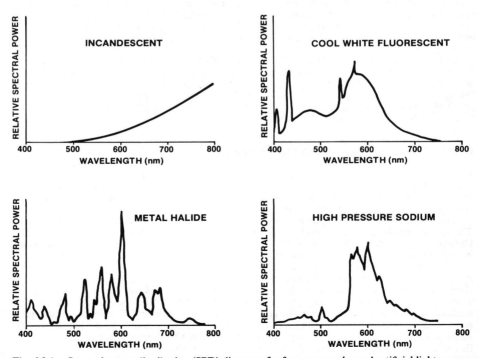

Fig. 6.3.1. Spectral power distribution (SPD) diagrams for four commonly used artificial light sources.

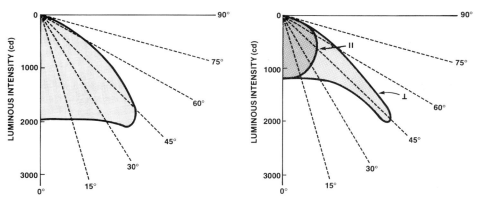

Fig. 6.3.2. Diagrams showing distribution of luminous intensity (candlepower distribution) for two luminaires. Luminous intensity is represented by the radial distance from the origin. Left panel: distribution is symmetrical for all vertical planes passing through the center of the luminaire. Right panel: asymmetric bilateral batwing distribution—inner curve represents vertical plane parallel to the luminaire axis; outer curve represents vertical plane perpendicular to the luminaire axis. For additional details, see text.

For additional reference information on artificial light sources, see Henderson and Marsden (1972) or Kaufman and Haynes (1981a, 1981b).

6.3.3.2 Luminaires and Light Control

Units that house artificial light sources and their related electrical equipment are called *luminaires*. The recessed troffer with fluorescent lamps is probably the most common type of luminaire used for office lighting. A frequently used luminaire for industrial applications is the "high bay" reflector equipped with a HID lamp. Luminaires control and redirect light to where it is needed in a number of ways including reflection, refraction, transmission, and absorption.

6.3.3.3 Luminaire Photometrics

The distribution of luminous intensity (candlepower distribution) for a luminaire is generally represented by one or more polar coordinate graphs. If the distribution is symmetrical for all vertical planes passing through the center, a single graph is sufficient to describe the candlepower distribution for that luminaire. Otherwise, several plots are required. (See Figure 6.3.2. Note: only the lower right quadrant of the candlepower distribution is shown.) The upper left-hand corner of the plot (the origin) represents the center of the luminaire, and nadir is defined as straight down from that point. Luminous intensity is represented by the radial distance from the origin. The left panel in Figure 6.3.2 shows the candlepower diagram for a luminaire that distributes all the usable light downward within a 140° cone. However, the region of highest intensity falls within a 100° cone. The right panel shows the candlepower distribution for a luminaire with a bilateral batwing (not symmetrical for all vertical planes). In this diagram the inner curve represents the vertical plane parallel to the luminaire axis, and the outer curve (i.e., the "batwing") represents the vertical plane perpendicular to the luminaire axis.

A luminaire may be classified by the percentage of light in the upper and lower hemispheres of its candlepower distribution diagram. There are five categories in the CIE classification system: direct, semi-direct, general diffuse, indirect, and semi-indirect. If the percentage of light directed downward is at least 90%, the luminaire is classified as a *direct* lighting unit. If the percentage of light directed upward is at least 90%, the luminaire is classified as an *indirect* lighting unit.

6.3.4 EFFECTS OF LIGHTING ON BEHAVIOR

Numerous laboratory studies and field studies have sought to determine the relationships between lighting and behavior. This section has been prepared to acquaint the reader with some of these fundamental data which are the basis for many current lighting practices. A knowledge of this material is essential in order to understand the rationale for the solutions to lighting problems discussed in later sections. It will also help the reader to find solutions for many other lighting problems that are not discussed here.

6.3.4.1   Illumination and Performance—Task Characteristics

The visual difficulty of a task is closely related to the size of the smallest critical elements and the contrast between the task and background. Generally, performance improves if either size or contrast is increased. In some instances increasing the time for observation or the level of illumination may be beneficial.

The size of the smallest critical task element is usually specified in terms of the visual angle that it subtends. Contrast, however, may be specified in any one of several different ways. Probably the most commonly used definitions for contrast are

$$C = \frac{L_o - L_b}{L_o} \quad \text{and} \quad c = \frac{L_o - L_b}{L_b}$$

Where L_o is the luminance of the object and L_b is the luminance of the background. If the task elements are periodic, contrast (or contrast modulation) is sometimes defined as

$$\frac{L_{max} - L_{min}}{L_{max} + L_{min}}$$

where L_{max} and L_{min} are the maximum and minimum luminances, respectively.

The relationships between size, contrast, illumination, and performance are clearly shown in the results of experiments by Weston (1945, 1962). Subjects examined charts with arrays of Landolt rings differing in size and contrast. The task was to identify all rings having a specified orientation. Some of Weston's findings are shown in Figure 6.3.3. Size is expressed as the visual angle subtended by the gap width (i.e., the smallest critical detail) of the Landolt ring and mean performance is a composite score based on time taken to complete the task and errors.

Figure 6.3.3 shows several fundamental relationships that should be kept in mind when designing or modifying tasks and the lighting that will be used for those tasks. First, performance improves with increases in illumination, but the incremental benefit diminishes with each subsequent increase. Beyond a certain point, further increases in illumination are not beneficial. The level of illumination where the transition occurs depends upon the size and contrast of the target. For example, asymptotic

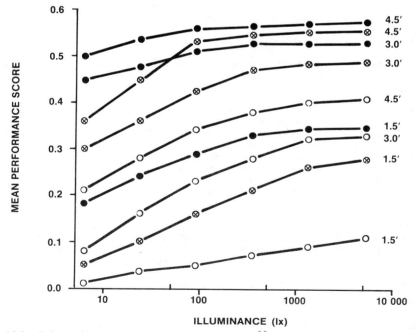

Fig. 6.3.3.   Relationships between task contrast ($\bullet$ = high, $\otimes$ = medium, $\circ$ = low), size of smallest critical task detail (expressed in minutes of visual angle), illumination, and performance (speed and accuracy) for a Landolt ring task. (Adapted from Weston, 1945, Table III.)

performance was obtained at about 100 1x for medium and high contrast targets when the gap subtended 4.5 min. However, when the gap subtended 1.5 min, performance continued to improve as illumination was increased beyond 1000 1x. Second, greater improvements in performance can be obtained by changing the task (i.e., increasing size or contrast) rather than by increasing illumination. Third, performance for the easy visual tasks (large size, high contrast) far exceeds performance for the difficult tasks (small size, low contrast), regardless of the level of illumination. The point to be made here is that simply increasing illumination does not transform a difficult visual task into an easy one.

Office Tasks

Most office lighting studies have shown some improvement in productivity with increases in illumination (e.g., Barnaby, 1980; Hughes and McNellis, 1978; IERI, 1975). It is not clear, however, how much of the change is attributable to better task visibility and how much is related to motivational factors. Nevertheless, the improvements in performance have usually been greater for visually demanding tasks performed by workers over age 45. Improvements for younger workers or for less visually demanding tasks have been much smaller.

The benefits versus cost of increasing illumination in offices have again become a frequently discussed topic. An article in a recent issue of *Office Administration and Automation* (OAA, 1984) argues that the economic benefits of returning to the pre-"energy crisis" illumination levels far outweigh the costs of providing the additional lighting. It describes the ill-fated energy conservation efforts at the Social Security Administration's Baltimore office. Prior to 1974 a lighting level of 1100 1x was provided for office workers at this facility, but in response to the OPEC oil embargo in 1973, the U.S. Government decided to reduce the lighting level by 50% (to 550 1x) in some areas but by as much as 70% (to 325 1x) in others. Productivity in one area where lighting had been reduced declined almost 30%, and employees began complaining of eyestrain and headaches. The original lighting levels were restored, and soon afterward productivity returned to its original level. The author estimated that for every dollar saved by reducing energy consumption, $160 of productivity was lost.

Data such as these must be interpreted with the utmost of caution. The method used to achieve a reduction in overall average illumination was simply to remove lamps. The lighting level at some workplaces must have been far less than the target average of 325 to 550 1x. Furthermore, changes in the distribution of light may have reduced the quality of light at some workplaces as well as the quantity. In addition, the morale of the workers must have been greatly reduced by the sudden decrease in lighting. The point to be made here is that changes other than the decrease in illumination occurred simultaneously, and it is difficult, if not impossible, to sort out the effects attributable to each.

Industrial Tasks

McCormick (1970) has summarized the results of surveys showing an increase in productivity for 15 industrial tasks following an increase in illumination. The tasks included a wide variety of activities such as metal-bearing manufacturing, steel machining, carburetor assembly, inspecting roller bearings, weaving cloth, and letter sorting. In most instances the increase in work output was modest—less than 15% for a 2- to 10-fold increase in illumination. Considering that the original lighting level was generally less than 55 1x and that some of the tasks were visually demanding, the observed increases in productivity do not seem unreasonable. However, the reader is again cautioned that the results of any field study making before and after comparisons without the use of an adequate control group are difficult to interpret because factors other than lighting may be responsible for any observed changes.

6.3.4.2 Illumination, Visual Performance, and Aging

As the eye ages beyond age 40, certain irreversible changes take place. There is an increase in absorption of light by the lens due to an increase in yellow pigmentation. This reduction in the transparency of the lens and an aging-related increase in the scattering of light in the optic media significantly reduce the amount of light that reaches the retina. A third change is a gradual recession in the near point of accommodation and concurrent reduction in the range of accommodation. A fourth change is a reduction in average pupil diameter.

The effects of these changes are very significant. Weale (1963), for example, has estimated that the retinal illuminance for a 60 year old is typically only about one-third of that for a 20 year old. This explains why older persons sometimes benefit from an increase in illumination. The increase in scatter may decrease acuity and contrast sensitivity and aggravate the effects of glare from bright sources of light within the visual field. Hence older persons often have a lower tolerance for glare.

The results of a study by Smith and Rea (1978) show the relationship between illumination and performance for young and old persons doing difficult and easy versions of the same task—proofreading (see Figure 6.3.4). Fig. 6.3.4 shows that increases in illumination beyond 100 1x had only a small effect on performance. Also shown is that the difference in performance between young and old readers

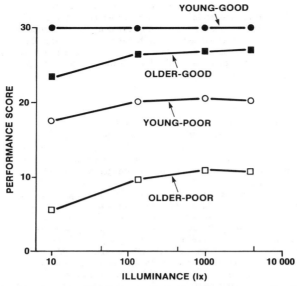

Fig. 6.3.4. Relationships between illuminance and performance on a proofreading task for young and older workers using materials with good and poor print quality. (Adapted from Smith and Rea, 1978.)

was much smaller when print quality was good. These findings suggest that the high levels of illumination frequently recommended for older workers are not always necessary and that the difference between the performance of young and old workers may be reduced by modifying the task.

Color discrimination also diminishes with age. This may be related to the changes in the absorption of light by the lens described earlier, which are not uniform throughout the visible spectrum.

The studies previously cited show that illumination may affect performance, but that the effect depends on the task and worker. Difficult visual tasks require more light than less demanding tasks, and older workers may benefit from an increase in illumination to a greater extent than younger workers. Performance at any given time, however, depends to a large extent on worker motivation. If workers are not motivated, they will not perform well, regardless of the lighting system. For this reason laboratory studies (where motivation is usually very high) may provide better initial estimates of the minimum light levels necessary to perform specific tasks. However, laboratory studies do not provide the final answer because subjects are not required to perform for extended periods of time, something that they might not be able to do if conditions were poor. Hence both laboratory investigations and field studies are essential for determining the answers to lighting questions.

6.3.4.3 Glare and Performance

Glare refers to "the sensation produced by luminance within the visual field that is sufficiently greater than the luminance to which the eyes are adapted to cause annoyance, discomfort, or loss in visual performance and visibility" (Kaufman and Haynes, 1981a). It may be categorized according to its effects—physical discomfort or visual performance decrement (i.e., discomfort glare vs. disability glare)—or its origin (direct glare vs. reflected glare). Direct glare may be caused by one or more bright sources of light in the visual field (i.e., bright sources of light that shine directly into the eyes). Examples of direct glare sources include sunlight and incandescent bulbs. Reflected glare, on the other hand, is caused by light reflected from an object or objects that an observer is viewing. These reflections may be either specular or diffuse. If the reflections are diffuse, they are often referred to as *veiling reflections*. Examples of reflected glare include the reflections often seen on the screens of video display terminals (VDTs) and reflections from the pages of books. Reflected glare usually reduces the apparent contrast of task elements being viewed at the workplace and, unless controlled, may affect task performance.

The effects of direct glare are usually confined to annoyance and discomfort. However, in extreme cases involving very difficult visual tasks, performance may also be affected. The results of a study reported by Luckiesh and Moss (1937) and Luckiesh (1944) illustrate this point. The glare source was a 100 W tungsten-filament incandescent lamp placed at various angular distances from the observer's line of sight. When the glare source was very close to the observer's line of sight (i.e., displaced 5°), visibility of the test object decreased by 84%. However, when the glare source was 40° from the line of sight, visibility of the test object decreased by only 42%.

Reitmaier (1979) studied the effects of veiling reflections on performance of a Landolt ring task. As the veiling reflections superposed on the task increased (i.e., as contrast was reduced), performance declined. Such a performance decrement is readily predictable from the data shown in Figure 6.3.3.

6.3.4.4 Luminance Ratios and Performance

Both visual comfort and visual performance are affected by the relationship between the luminance of an object being viewed and the luminance of the area immediately surrounding that object. If the ratio of the two luminances is sufficiently large, the observer will experience ocular discomfort. In addition, visual performance may be impaired.

For many years lighting authorities have maintained that a task-to-surround luminance ratio of not more than 3 : 1 (or at most 5 : 1) is necessary in order to maintain visual comfort and a high level of visual performance. (If the task is darker than the surround, the maximum recommended luminance ratio is 1 : 3.) The origin of these recommendations may be traced to several studies conducted in the 1920s and 1930s. These studies were concerned with the effect of surround luminance on thresholds such as visual acuity rather than suprathreshold tasks such as those performed by office workers and industrial workers.

Probably the most frequently cited of these studies is Lythgoe's (1932) study concerning the effect of surround luminance on visual acuity. The acuity targets were Landolt rings, and the surround luminance could be adjusted from well below that of the task to well above. Lythgoe's data are usually interpreted as showing that maximum acuity is obtained when the surround luminance is equal to or slightly lower than the task luminance. The data also suggest that a surround that is brighter than the task causes a much greater performance decrement than a surround that is dimmer. However, because there are so few data points, it is possible to make other interpretations. According to Chapanis (1949), Lythgoe himself believed that the best acuity occurs when the luminance of the surround is between $\frac{1}{10}$ and $\frac{1}{100}$ of the visual task's luminance. Nevertheless, several studies could be cited in support of the recommendation for a 3 : 1 maximum ratio (see Boyce, 1981, p. 301 for a review of several recent studies).

The validity of the recommendation for a 3 : 1 maximum task-to-surround luminance ratio still remains in doubt, however. Chapanis (1949) has severely criticized the original work upon which the recommendation is based. In addition, he has shown the recommendation is inconsistent with some of the data that supporters have used as justification.

The results of a recent study by Cushman et al. (1984) suggest that Chapanis's skepticism may be justified. In that study luminance ratios as great as 110:1 did not adversely affect visual comfort or performance on a photo-finishing task.

6.3.4.5 Subjective Factors and Preferences

Different illuminance levels, patterns of light, and colors create different moods and other subjective impressions. Among the more noteworthy contributions in the study of the effects of lighting on subjective impressions are several studies by Flynn and his associates. These investigators have used two powerful methods—factor analysis and multidimensional scaling—to provide a wealth of useful design information.

In one of the earlier studies Flynn, Spencer, Martyniuck, and Hendrick (1973) arranged to have a conference room lighted in six different ways: (1) overhead downlighting, low intensity; (2) peripheral wall lighting, all walls; (3) overhead diffuse lighting, low intensity; (4) a combination of overhead downlighting (low intensity) and peripheral lighting on end walls; (5) overhead diffuse lighting, high intensity; and (6) a combination of overhead downlighting (low intensity), peripheral wall lighting, and overhead diffuse lighting. The subjects rated each room using Semantic Differential rating scales (continuous scales anchored at the ends by opposites). For example, the subjects were asked to rate each room on a "friendly" · · · "hostile" scale, where "friendly" was at one end of the scale and "hostile" was at the other. There were over 20 such scales including the following pairs: pleasant—unpleasant, relaxed—tense, interesting—monotonous, clear—hazy, bright—dim, simple—complex, spacious—cramped, and formal—informal.

Application of factor analysis revealed five underlying dimensions and the scales most strongly associated with each. The dimensions were tentatively identified as evaluative, perceptual clarity, spatial complexity, spaciousness, and formality. As might be expected, there were large differences among the six rooms. The rooms with a combination of lighting on the table and walls had the highest scores on the evaluative dimension (friendly, pleasant, sociable, relaxed, interesting, likeable, etc.) whereas the rooms that provided only overhead diffuse lighting received the lowest. On the perceptual clarity dimension the rooms with higher illuminances received higher scores. A final difference of interest was observed for the spaciousness dimension. Rooms that provided illumination only on the table were perceived as being small and cramped, whereas those with both wall and table lighting or

Table 6.3.2   Summary of Flynn's Work on Subjective Impressions

Subjective Impression	Lighting
Perceptual clarity	Bright, uniform lighting; some wall lighting or high reflectance walls
Spaciousness	Uniform wall lighting
Relaxation	Nonuniform lighting; wall emphasis rather than overhead lighting
Privacy	Nonuniform lighting; low light intensities near user, higher intensities elsewhere
Pleasantness	Nonuniform lighting; some wall lighting or high reflectance walls

Source: Adapted from Flynn, 1977.

wall lighting alone were perceived as being larger and more spacious. A higher level of illumination on the table also helped to enhance the impression of spaciousness.

In a later study Flynn and Spencer (1977) asked people to rate their impressions of a room lit with several combinations of general and wall lighting utilizing four different types of artificial light sources. Illuminance was held fairly constant. Lighting with Cool White fluorescent lamps produced feelings of clarity and distinctness, whereas the use of high-pressure sodium lamps had the opposite effect, producing hazy and vague sensations. When the room was lit with fluorescent lamps, it was perceived as being more pleasant than when lit by mercury lamps. The addition of wall lighting by fluorescent lamps made the disparity even greater.

A brief summary of Flynn's work is given in Table 6.3.2. It provides designers with a set of guidelines, based on scientific research, for creating impressions with lighting. The applicability of the information may be limited, however, to only certain types of interiors. Clearly, additional research would be highly desirable.

The preceding discussion has clearly shown that the types of light sources, luminaires, and luminaire arrangements all contribute to our perceptions of artificially illuminated interior spaces. The remainder of this section considers illuminance preferences for office lighting and a brief discussion of direct user control.

Illuminance on a horizontal work plane is probably the most widely used lighting specification. It is easy to design for and, if applied uniformly across an entire work area, offers the user the greatest flexibility because furnishings can be moved about without much concern for lighting adequacy.

Data concerning illuminance preferences for an office environment have been obtained by Saunders (1969). An illuminance of 200 lx was considered to be inadequate; levels of 600 lx or more were considered to be good. Increases beyond 600 lx, however, had very little effect on the subjects' ratings; that is, the ratings for 1600 lx were about the same as those for 600 lx. Other investigators have also confirmed these results.

Although the study cited above is typical, one can find studies in the literature indicating preferred levels of illumination from 10 to 2000 lx. The preference for an individual performing a given task appears to be a compromise between the effects of visibility, glare, adaptation, custom, and individual eccentricity (Ross and Baruzzini, 1975). This raises two interesting questions. Suppose an individual were allowed to adjust the lighting at his or her workplace. Would that person adjust the illuminance to the optimum level? Would this motivate the individual to perform better? The limited data that are available suggest that performance may sometimes be improved. Lehon (1974) found that self-selection of illumination did improve performance, but the effect was greater for persons with defective vision.

Self-selection of lighting level is a concept that is consistent with current ergonomic thinking that workplaces should be provided with adjustability to accommodate the individual differences of various workers. However, additional research is needed to quantify both the objective and subjective benefits.

6.3.5   ILLUMINATION GUIDELINES

Selection of a satisfactory illuminance for a given workplace involves a balancing of several factors. The visual difficulty (such as the contrast and size of the smallest details), criticalness, frequency, and time available for each task and subtask must be considered. The minimum lighting level for the workplace must be adequate for performing the most difficult and critical tasks. Some consideration for individual differences among workers (such as age) is desirable. Worker comfort, appearance of the work area, psychological factors and impressions (see Section 6.3.4), and worker expectations should be considered. Feasibility and economics are also important factors.

The interrelations among the factors described above are complex, and the designer has to consider the various options that are available if the design objectives are to be met. A lower level of general illumination may be possible if tasks can be modified to make them easier (Section 6.3.7) or if supplementary lighting (also called "task lighting") can be used to illuminate the most difficult tasks (Section 6.3.5). This usually reduces operating costs. The illumination requirements for a workplace may also be reduced by improving lighting quality (Section 6.3.6). Adjustable task lighting is also a good method for accommodating individual differences in seeing ability and personal preferences. The appearance of the work space is important because it may affect worker motivation. Finally, feasibility and economics are determined, to a large degree, by technology and energy costs, which are constantly changing and vary from one location to another. A combination of these factors and user expectations (which vary from culture to culture) may help explain why, until recently, recommendations for illumination at the workplace have been steadily increasing. It also explains why recommendations for illumination have historically been higher in the United States, Holland, and Switzerland than in Germany, England, and France.

The complexities discussed above make precise recommendations for illumination levels very difficult. Flynn (1979) has provided a set of guidelines that correlate with the international recommendations in CIE Report #29 and take into consideration some of the factors discussed above. Recommendations based on this work are given in Table 6.3.3. Note that a range of values, rather than a single value, is suggested for each type of activity. The specific value that is most appropriate within a given range depends on task variables (such as requirements for speed and accuracy and the reflectance of the background) and the visual capabilities of the individual workers. For a more complete discussion, see Kaufman and Haynes (1981a, Section A of the appendix), IES (1982), and IES (1983).

The specification of illuminance ranges for various types of activities has several advantages over earlier approaches that provided specific recommendations for different types of environments (e.g., offices, libraries, shops, factories, roadways, retail businesses). The earlier approach assumed that illumination would be uniform throughout the work area. The recommended value had to be high enough to accommodate older workers doing the most difficult visual tasks that were to be performed. The newer approach is much more flexible. Recommendations are provided for each specific task. Illumination elsewhere in the work space may be lower or higher as needed. The opportunities for energy conservation are greater when lighting is nonuniform, and such a work space is usually aesthetically more appealing as well.

6.3.5.1 Task Lighting

Task lighting (or *local lighting*) is a term that refers to a desk lamp or task illuminator that supplements the illumination provided by general lighting sources such as recessed fluorescent troffers and daylight. When task lighting is used, general lighting may be reduced (to conserve energy) whereas illumination for difficult or critical visual tasks may be increased (to increase productivity or reduce errors). For example, in the recent past an entire large office might have been uniformly lighted at a high level such as 1200 lx. Today, the general illumination for that same area may be as low as 500 lx with supplementary task lighting at specific workplaces. Task lighting is especially effective when a small number of workers occupy a relatively large space.

Task lighting should not be considered as a panacea, however. Objectionable shadows, nonuniformity of illumination, and reflected glare may occur unless special precautions are taken. Task lighting units that are directly in front of the worker (such as those built into office furniture) may pose serious problems if located in the offending zone (primary zone for reflected glare—see Section 6.3.6 for details).

In theory, the problem of specular reflections associated with task lighting may be solved by placing the sources so that the light comes from the left and right sides of the workplace. In practice, however, this approach has not always been acceptable to users (Boyce, 1979). (See also Figure 8 in Shellko and Williams, 1976.) A better approach is to provide an extended source (such as a desk lamp with a linear fluorescent tube) having an articulated arm. Then the task light may be placed directly over the task or moved elsewhere, if necessary, to minimize glare. The articulated arm also permits the user to vary the amount of supplementary illumination by raising or lowering the task light. Alternately, a dimmer control could be provided. The task light should also minimize obstruction of the work surface.

Tasks performed in the traditional office seldom require an illumination of more than 500 to 700 lx. Any additional lighting that is needed may be provided by task lighting units. Lower ambient levels may be appropriate for work areas with VDTs (Section 6.3.9), but if illumination is reduced below 300 lx, the working environment will appear to be too dark, and workers may complain. Similarly, in order to achieve a reasonable balance between ambient and task sources, it is seldom advisable to reduce general lighting in an office to a level below 300 lx.

Table 6.3.3 Recommended Range of Illuminance for Various Types of Tasks

Type of Activity or Area[a]	Range[b] of Illuminance (lx)[c]	
Public areas with dark surroundings	20–50	
Simple orientation for short temporary visits	50–100	
Working spaces where visual tasks are only occasionally performed	100–200	A
Performance of visual tasks of high contrast or large size: reading printed material, typed originals, handwriting in ink, good xerography; rough bench and machine work; ordinary inspection; rough assembly	200–500	
Performance of visual tasks of medium contrast or small size: reading pencil handwriting, poorly printed or reproduced material; medium bench or machine work; difficult inspection; medium assembly	500–1000	B
Performance of visual tasks of low contrast or very small size: reading handwriting in hand pencil on poor quality paper, very poorly reproduced material; very difficult inspection	1000–2000	
Performance of visual tasks of low contrast and very small size over a prolonged period: fine assembly, highly difficult inspection, fine bench and machine work	2000–5000	
Performance of very prolonged and exacting visual tasks: the most difficult inspection, extra fine bench and machine work, extra fine assembly	5000–10,000	C
Performance of very special visual tasks of extremely low contrast and small size: some surgical procedures	10,000–20,000	

Source: Adapted from Flynn, 1979. See also IES (1982) and IES (1983).
[a] See Section 6.3.9 for recommendations for tasks involving video displays.
[b] The choice of a specific value within a given range depends on task variables and the visual capabilities of the worker.
[c] A—General lighting throughout work spaces
  B—Illuminance on task
  C—Illuminance on task obtained by a combination of general and local (supplementary) lighting

6.3.5.2 Direct and Indirect Lighting

The advantages and disadvantages of direct and indirect lighting systems (see Section 6.3.3 for definitions) are summarized in Table 6.3.4. An examination of Table 6.3.4 shows why direct systems are much more popular. Interest in indirect systems, however, has recently increased, particularly for lighting office areas in which VDTs are used. In these areas glare may be the most serious lighting problem, and indirect lighting, in some cases, is an appropriate solution (see Section 6.3.9 for a discussion of lighting for VDT workplaces). Also, satisfactory indirect lighting systems are now easier to design with the improved computer software that is available.

Table 6.3.4   Advantages and Disadvantages of Direct
and Indirect Lighting

Advantages	Disadvantages
Direct Lighting	
Simplicity of design with predictable results	Potential for glare is much greater
High user acceptance	Problems with nonuniformity and shadows
Usually more efficient on lumens/watt basis	
Indirect Lighting	
Potential for glare and shadows is greatly reduced	More difficult to achieve high level of illumination needed for some tasks
	Often less efficient on a lumens/watt basis
	Design of satisfactory indirect systems is more difficult
	Usually requires a high ceiling

Some designers have created luminaires for indirect lighting that combine high-intensity discharge lamps with special optics to increase efficiency. Different types of lamps (e.g., metal halide and high-pressure sodium) may be used in the same lighting unit to achieve better color rendering. These enhanced indirect systems often cost no more than conventional direct fluorescent lighting systems when overall system economics are considered.

6.3.6   MEASUREMENT AND CONTROL OF GLARE

The various types of glare and their effects on performance have been discussed in Section 6.3.4 above. Note: a given source may produce direct glare and reflected glare. Both direct glare and reflected glare may cause discomfort and performance decrements.

Discomfort glare (glare that causes discomfort) is usually a more serious problem than disability glare (glare that adversely affects performance) for offices and industrial environments. In these contexts the contrast of critical task elements is often far above threshold so that performance decrements due to glare are very small and go largely unnoticed. (A notable exception is visual inspection where the critical task elements may be near threshold.) On the other hand, discomfort glare, by definition, never goes unnoticed.

The origins of discomfort glare and disability glare are quite different. Discomfort glare seems to be related to the activity of the muscles of the iris that control pupil diameter. Disability glare, on the other hand, has two known origins: (1) the scattering of light (from the glare source) by the optic media, which reduces the apparent contrast of objects being viewed and the background, and (2) photochemical and neural inhibition in the retina and neural pathways (Hopkinson and Collins, 1970).

Several methods have been developed to assess the potential for discomfort glare. Two widely used methods are the visual comfort probability (VCP) method, used in North America, and the British Glare Index system. Another method is the European Glare Limiting Method. A fourth way to estimate glare potential is to evaluate luminance ratios for objects in the observer's field of view.

The first two methods (VCP and Glare Index) calculate glare sensations from four physical values: the luminance of the glare source, the angular subtense of the glare source at the eye, luminance of the background, and deviation of the glare source from the line of sight. VCP values are expressed in terms of the percentage of persons who would *not* experience discomfort for a given set of viewing conditions. An acceptable VCP is usually considered to be 70% or more. This is inconsistent with the usual ergonomic recommendation to design for 90% acceptability. Limiting values for the Glare Index vary from one application to another.

Methods that are used to assess reflected glare (glare caused by reflections from objects in the field of view) and the potential for disability glare are usually more complicated. One method for evaluating reflected glare is to determine the equivalent sphere illumination (ESI). The ESI for a

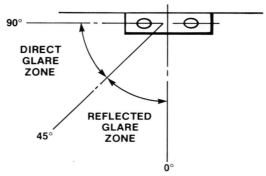

Fig. 6.3.5. Direct glare zone and reflected glare zone for a luminaire. For additional details, see text. (After Lum-i-neering Associates, 1979.)

given task and set of viewing conditions is the amount of illumination the task would require if placed in a photometric sphere to make it just as visible as it is in its real environment. Because ESI is usually lower than the illumination at the workplace, the difference between the two values gives an indication of the amount of reflected glare. Two other useful metrics are the contrast rendering factor (CRF) and the disability glare factor (DGF). The CRF may be used for quantifying the effects of veiling reflections on a task, and the DGF may be used for quantifying the effect of nonuniformity of luminance in the visual field surrounding the task (see Boyce, 1981, for a discussion). See also Hopkinson and Collins (1970) for a review of other methods.

Relative to a luminaire, light radiated in the zone from 45° to 90° is more likely to cause direct glare, whereas light radiated in the zone from 0° to 45° is more likely to fall within the offending zone and cause reflected glare. (See Figure 6.3.5 for an illustration showing direct and reflected glare zones for a luminaire and Figure. 6.3.6 for a diagram showing the reflected glare offending zone. See Section 6.3.4 for definitions of direct and reflected glare.) Hence, by examining the candlepower distribution curve for a given luminaire (e.g., Figure 6.3.2), one can roughly estimate its potential for direct and reflected glare. The designer must decide whether direct glare or reflected glare is likely to pose a more serious problem for a given application and select the luminaires accordingly. Luminaire placement is also critical, particularly when trying to keep light out of the offending zone in order to minimize reflected glare (see Figures 6.3.6 and 6.3.7). For this reason a task lighting source should be equipped with an articulated arm or other means for adjusting its position.

The discussion in the paragraphs above suggests many practical methods for reducing and controlling glare. These methods, as well as others that the authors have found to be helpful, are listed in Table 6.3.5. See also Section 6.3.9 for additional methods that are useful for VDT workplaces.

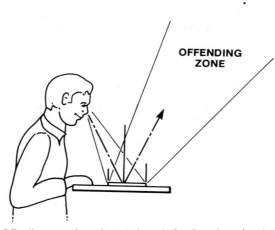

Fig. 6.3.6. Offending zone for reflected glare. (After Lum-i-neering Associates, 1979.)

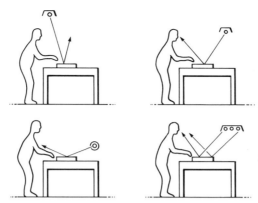

Fig. 6.3.7. Illustration showing relationship between luminaire position and the potential for reflected glare. Left side: low potential for reflected glare. Low-angle lighting (lower left) is useful for emphasizing surface irregularities in some types of visual inspection tasks. Right side: higher potential for reflected glare but useful for some types of inspection. (See also Section 6.3.10 on inspection lighting.) (Adapted from Kaufman and Haynes, 1981b.)

6.3.7 TASK MODIFICATION

Difficult visual tasks can sometimes be modified to make them easier. If the task can be made easier, performance usually improves. Modifications involving an increase in the size or contrast of critical task elements are usually the most effective. Increasing the viewing time and increasing the illumination on the task (previously discussed in Section 6.3.4) may also be beneficial.

Table 6.3.5 Techniques for Controlling Glare

To Control Direct Glare

Position luminaires as far away from the operator's line of sight as practical.

Use several small low-intensity luminaires rather than one large high-intensity luminaire.

Use luminaires with a minimum of luminous intensity in the direct glare zone (see Figures 6.3.5 and 6.3.2), e.g., indirect lighting units or direct lighting units with parabolic louvers or a "batwing" candle-power distribution.

Increase luminance of area around any glare source.

Use task lights with intensity controls and some method for adjusting their position relative to the task.

Reorient workplace, if necessary.

Use light shields, hoods, and visors if other methods are impractical.

See also Section 6.3.9 for methods for reducing glare at VDT workplaces.

To Control Reflected Glare

Position luminaires so no significant amount of reflected light is directed toward the eyes. See Figure 6.3.7.

Use luminaires with diffusing or polarizing lenses.

Use surfaces that diffuse light such as flat paint, nongloss paper, and textured finishes.

Change the orientation of the workplace or task, the viewing angle, or the viewing direction as necessary to improve task visibility.

Keep illuminance level as low as feasible.

Use indirect lighting.

Use combination of top and side lighting.

See also Section 6.3.9 for methods for reducing glare at VDT workplaces.

Source: Adapted from E. J. McCormick's *Human Factors Engineering* (1970) with permission of McGraw-Hill Book Company.

The relationship between size, contrast, illumination, and visual performance has been previously discussed (Section 6.3.4 and Figure 6.3.3). These data show, quite clearly, that increasing the size or contrast of the critical task elements is usually far more effective (in terms of improving performance) than increasing the illumination on the task. For tasks near threshold Ross and Baruzzini (1975) have estimated that increasing the size of the critical task elements may be as much as 60 times more effective than increasing the illumination level. Similarly, increasing the task contrast or increasing the time for observation may be as much as 10 times more effective.

The effective size of the critical task elements (visual angle subtended by the smallest critical detail) may be increased in several ways:

1. Increase the physical size of the task objects (e.g., increase type size, line width, size of critical graphic symbols, size of mechanical parts).
2. Decrease the viewing distance. As the operator moves closer to the task, the visual angle subtended by the smallest critical detail increases.
3. Use a magnifying device such as a magnifying glass or microscope.
4. Orient the task objects so that they are perpendicular to the worker's line of sight.

Several methods are also available for increasing task contrast:

1. Increase contrast between task objects and the background. This may be done by using soft-lead pencils rather than hard-lead pencils, using ink pens rather than pencils, changing typewriter and printer ribbons frequently, using white paper rather than gray paper, adjusting copier exposure controls, adjusting CRT brightness and contrast controls, using surface grazing light, and so on.
2. Decrease reflected glare (see Table 6.3.5).
3. Use contrasting colors for different task objects and the background.

6.3.8 USING MODELS AND SIMULATION

It is sometimes desirable to evaluate several alternate designs before making a final design commitment, particularly if the application is new or if the tasks are very difficult or unusual. Mathematical models, full-size mock-ups, and scale models have been used extensively for this purpose.

Wheelwright (1977) used a $\frac{1}{10}$ scale model to evaluate various lighting concepts for the cargo bay of the U.S. space shuttle. In one experiment two lamp types, three orientation configurations, four reflector designs, and five types of lenses were evaluated. Figure 6.3.8 shows the pattern of illumination for the 90° and 150° reflectors. Notice that the illumination within the cargo bay is much more uniform when the 150° reflectors were used. However, the lamps became annoying glare sources when used with the wider angle reflectors. The glare problem was solved, using the model, by adding baffles on the forward side of the reflectors and by recessing the lamps in the cargo bay liner.

From the results of his experiments using the $\frac{1}{10}$ scale model, Wheelwright was able to determine that the general illumination within the cargo bay would be adequate. The investigation also showed that there was a need to add a second bulkhead light, that a small amount of lamp tilt was beneficial, and that three of the four reflector designs were satisfactory.

The remainder of this chapter is devoted to a discussion of two specific lighting applications: office lighting for workplaces with VDTs and industrial inspection lighting. The section on lighting for VDTs considers differences between the automated office and traditional office and the changes that must be made to accommodate the new technology. The section on inspection lighting is concerned with ways to improve the detectability of blemishes near the limits of visibility. These examples will serve to illustrate ways that the principles discussed above may be applied to solve lighting problems.

6.3.9 LIGHTING FOR WORKPLACES WITH VDTS*

The office has traditionally been designed to accommodate paper-based information systems. A high level of general lighting (e.g., 700 lx or more) has often been recommended because it is suitable for tasks such as typing, filing, reading from paper copy, and handwriting.

The characteristics of the tasks that are performed in an automated office, however, are quite different. The primary displays are VDTs rather than paper documents, and the lighting requirements for VDTs are quite different because information is displayed differently. The difficulties arise because VDTs have the following characteristics:

1. The display surface (i.e., the VDT screen) is vertically oriented. The operator's line-of-sight is at least 20° higher than when viewing paper documents on a horizontal surface. This increases the potential for direct glare from luminaires and windows.
2. Any illumination on the display screen decreases the contrast between the characters (or graphics)

---

* Adapted from Cushman, 1983.

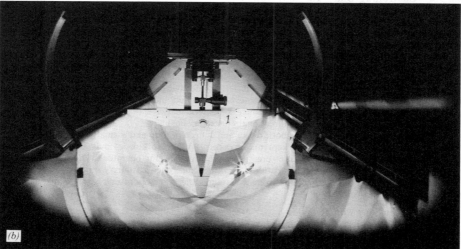

Fig. 6.3.8. Pattern of illumination in cargo bay of $\frac{1}{10}$ scale model of U.S. space shuttle using (*a*) 90° reflectors and (*b*) 150° reflectors. For additional details, see text. (After Wheelwright, 1977; photographs courtesy of NASA.)

and the background because the luminance of the dark areas (usually the background) is increased by a proportionally greater amount than the luminance of the light areas (usually the characters).

3. The screen of a VDT is curved and is often highly reflective. It acts somewhat as a mirror, creating glare as light from bright objects above and behind the operator is reflected from the screen and into the operator's eyes. These reflections reduce display contrast and may partially or totally obscure some of the information on the screen.

6.3.9.1 Illumination Guidelines for VDT Workplaces

There are many differing viewpoints regarding the appropriate level of illumination for VDT workplaces. Helander (1981) has reviewed nine existing and proposed standards for VDTs and found that the

Table 6.3.6   Illumination Guidelines for Workplaces with VDTs

Work NOT involving paper documents	150–400 lx
Work involving paper documents	400–550 lx[a]

Source: After Cushman, 1983.
[a] plus additional task lighting when needed.

recommendations vary considerably—from 200 to over 1000 lx. To add to the confusion, the Illuminating Engineering Society of North America recommends 50 to 100 lx for VDT workplaces, a lower range than suggested by any of the standards or proposals that Helander reviewed (Kaufman and Haynes, 1981b, p. 2–8 and p. 2–5).

A set of illumination guidelines for VDT workplaces based on reasonable compromises and trade-offs is given in Table 6.3.6. The lower range (150 to 400 lx) is ideal for tasks that do not involve transfer of information from paper documents. Illumination is low enough that good display contrast is usually obtainable without contrast enhancing devices. However, as noted earlier, some workers may complain if general illumination is less than 300 lx. The 550 lx maximum is suggested for tasks involving transfer of information from paper documents (e.g., data entry and word-processing applications). Increasing general illumination above 550 lx may reduce display contrast significantly, which makes the task more difficult.

If source documents have poor contrast, 550 lx may be insufficient for reading some of the information on the documents. Supplementary task lighting, directed only at the documents, may be added, but the amount of supplementary lighting should not exceed the amount that is actually needed.

6.3.9.2   Controlling Glare Through Design

Control of glare should be a primary consideration during the design of any work area where VDTs are to be used. Good initial design will eliminate the need for costly retrofitting and modifications at the workplace and VDT. It will also help to minimize worker complaints.

Direct glare from fluorescent lighting units can be eliminated if the units are equipped with specular parabolic louvers. Louver panels with small cells (4 × 4 cm or less) are more effective than panels with larger cells. Retrofitting of existing luminaires with parabolic louver panels having very small cells (e.g., 1.5 × 1.5 cm) is sometimes possible. Other effective methods for controlling direct glare from fluorescent luminaires include use of some pigmented acrylic lenses, black nonparabolic metal louver panels, and prismatic lenses that produce a bilateral batwing distribution (if workplaces can be oriented to take advantage of their directional characteristics).

Indirect lighting systems minimize both direct and reflected glare and virtually eliminate shadows. Provided that glare from daylight is kept under control, VDT workplaces may be oriented in any direction and rearranged without risk of creating glare problems for operators. The luminaires may be free-standing or attached to office partitions, furniture, and other stationary objects.

Glare from windows and reflections from walls and other surfaces must also be controlled. The following methods are suggested:

1. Cover windows with draperies, louver drapes, venetian blinds, or other coverings.
2. Paint walls with paint having a flat (nongloss) finish or cover walls with carpeting.
3. Use office furniture that has a matte or textured finish.

6.3.9.3   Modifications at the Workplace and VDT

Sometimes it is not possible to design or modify a work area so that it is ideal for VDT use. Tenants, for example, may not be able to obtain permission to install new lighting systems or to change window coverings or color schemes. Options are limited, but nevertheless, satisfactory solutions to lighting problems can often be found.

Modifications of the workplace and VDT that are effective methods for reducing glare include the following:

1. Reorient the workplace so that there are no sources of direct glare within the operator's field of view or annoying reflections on the VDT screen. In many instances this may be accomplished by orienting the VDTs so that their screens are perpendicular to windows and overhead luminaires.
2. Tilt the display upward or downward or rotate it a few degrees clockwise or counterclockwise to eliminate specific reflections that appear on the screen.
3. Purchase VDTs that have an antireflection coating or etching on the screen or attach an antiglare filter to reduce reflected glare. Micromesh filters, circular polarizers, neutral filters, and specturally selective filters are commonly used. (Be sure that any glass or plastic filter has an antireflection treatment

on the front surface.) The utility of any given device varies from one environment to another. Therefore, no single type of filter is suitable for all applications. Operator preferences should also be given some consideration. For details concerning the advantages and disadvantages of each type of filter, see Cushman (1983). Note: Some antiglare filters significantly degrade image quality and should be avoided.

4. Consider attaching a hood to shield the screen from ambient light sources.

6.3.10 INSPECTION LIGHTING

Imperfections or defects in manufactured products such as paper, fabrics, mirrors, foils, film, and plastics look different because they transmit or reflect light differently than the defect-free areas around them. The defects may be either nonfunctional appearance blemishes (such as a spot or abraded surface) or more serious flaws (such as cracked solder joints or missing threads in a fabric) that may affect product performance. Some of the defects are not visible, or are just barely visible, under ordinary room lighting. The aim of inspection lighting is to make these defects easier to see.

The five steps in the design of an inspection lighting system and workplace are as follows:

1. Determine the optical characteristics of the defects to be detected and of the defect-free areas of the material.
2. Using the information and data on the optical characteristics of the defects, determine the number and types of light sources that will be needed to make the defects easier to see.
3. Design a work station that places the product or other material at an acceptable viewing distance and that allows the inspector to assume a comfortable posture.
4. Arrange the inspection lights to maximize the visibility of the defects.
5. Reduce or otherwise control glare as much as possible.

These five steps along with general guidelines for designing effective inspection lighting systems and inspection workplaces are discussed in the following sections. Additional information may be found in Faulkner and Murphy (1973), Kaufman and Haynes (1981b), and *Ergonomic Design for People at Work* (Eastman Kodak Company, 1983).

6.3.10.1 Determining the Optical Characteristics of Materials and Defects

Optical characteristics refers to physical properties that affect the transmission and reflection of light. The optical characteristics of the defects themselves that are critical include density, location relative to the surface of the material, the shape and direction of the flaw, and specularity. Characteristics of the materials such as the density, flatness, specularity, texture, thickness, and internal light transmission ("light-piping") should also be determined. Once the optical characteristics of the defects and defect-free areas of the material are known, light sources that enhance the visibility of the defects can be selected.

Product engineers, quality-control personnel, experienced inspectors, and materials handbooks are good sources of information concerning the optical characteristics of materials and defects. A classification scheme that describes defects for a specific product in terms of their physical characteristics (e.g., "scratch," "dig," "wrinkle," or "coating skip") may also be helpful. Microscopic examination may be necessary in order to determine the optical characteristics of very small defects (e.g., irregularities in a shallow embossed pattern). A stereo zoom microscope with a range of 20 to 50 power is recommended for this purpose.

6.3.10.2 Selecting Light Sources for Visual Inspection

Recommended types of lighting units and mounting positions for detecting a wide variety of different defects in various materials are given in Table 6.3.7. To use the table, select the line corresponding to the material and defect combination of interest and read horizontally across to determine the type of light, mounting, and appearance of the defect. For example, to detect irregular surface scratches in metal, use a collimated, round beam source above the surface and perpendicular to it. The defect will appear as shiny lines on a dark background.

A typical diffuse inspection light source might consist of one or more fluorescent lamps in a box or troffer (interior painted white) with a thin white acrylic or opal glass diffuser. To ensure uniform illumination, the lamps should be about 30 cm longer than the width of the area to be inspected.

A collimated inspection lighting system might consist of a row of spotlights mounted just above the far end of an inspection table. Some automobile "driving lights" and certain aircraft lamps are highly recommended because they have a wide, flat beam that is more nearly collimated than a typical spotlight beam. Hence more light is applied where it is needed and less glare is produced.

In practice it is not always necessary to use the optimum light source to detect each type of defect. Searches for several types of defects may be made simultaneously provided that the inspection lighting is adequate for all types of defects that must be detected. Use of a single type of lighting

Table 6.3.7 Guidelines for Selecting and Mounting Inspection Lighting Units for Various Materials and Defects

Material	Defect	Material Properties[a]	Defect Properties[b]	Type of Light[c]	Light Mounting[d]	Defect Appearance[e]
Paper	Voids, contaminants	T: medium to high	D: lower or higher than paper	TR	BP	1
	Bumps or digs (with product flow)	NA	L: above or below surface	C, FB	LA, S	2
	Bumps or digs (across product flow)	NA	L: above or below surface	C, FB	LA, E	2
Coated paper	Discoloration	R: medium to high	L: surface	DI, WA	AP	3
	Irregular matte	R: medium to high	L: part of surface; G: profile differences	C, FB/RB	LA, E/S	1
Fabric	Missing strands	T: almost opaque to high	D: lower than fabric	TR	BP	1
	Loose strands (with product flow)	NA	L: above or below surface	C, FB	LA, S	2
	Loose strands (across product flow)	NA	L: above or below surface	C, FB	LA, E	2
Laminates	Thin coating	T: medium to high	D: lower or higher than material	TR	BP	1
Metals (general)	Bumps or digs (with product flow)	NA	L: above or below surface	C, FB	LA, S	2
	Bumps or digs (across product flow)	NA	L: above or below surface	C, FB	LA, E	2
	Scratches (with product flow)	NA	L: in surface; G: V-grooves	C, FB/RB	LA, E	4
	Scratches (irregular)	NA	L: in surface; G: V-grooves	C, RB	AP, P	4
	Irregular matte in satin finish	R: medium to high	L: part of surface; G: profile differences	C, FB/RB	LA, E/S	1
Glass	Scratches (with product flow)	R: medium to high	L: in surface; G: V—grooves	C, FB/RB	LA, E	4
	Scratches (irregular)	R: medium to high	L: in surface; G: V-grooves	C, RB	AP, P	4

Foil	Scratches (with product flow)	R: medium to high	L: in surface G: V-grooves	C, FB/RB	LA, E	4
	Scratches (irregular)	R: medium to high	L: in surface G: V-grooves	C, RB	AP, P	4
Transparent film	Scratches	T: medium to high	L: in surface G: V-grooves	C, FB/RB	BP, 45	4
Anodized aluminum	Discoloration	R: medium to high	L: on surface	DI, WA	AP	3
Mirrors and chrome parts	Deformations	R: high	L: surface G: flatness deviation	DI, ST	AP	5
Plastic moldings	Voids, bubbles	T: high	L: in, on, or between surfaces	LI	F, L-P	1
	Internal stresses	T: high	D: lower or higher than material L: between surfaces	TR, 2PS	BP, PM	6
Plastic sheet	Voids, bubbles	T: high	L: in, on, or between surfaces	LI	F, L-P	1
Structural models	Internal stresses	T: high	D: Lower or higher than material L: between surfaces	TR, 2PS	BP, PM	6

[a] R—reflectivity; T—transmittance.

[b] D—density; G—geometry; L—location.

[c] C—collimated; DI—diffuse; FB—flat beam; FB/RB—flat beam or round beam; LI—linear; RB—round beam; ST—stripe overlay; TR—transmitted; WA—wide area; 2PS—2 polarizing sheets.

[d] 45–45° angle to surface; AP—above product; BP—below or behind product; E—at end of workplace; E/S—at end or side of workplace; F—focused on edge; LA—low angle (high angle of incidence as measured from the normal); L-P—light piping; P—perpendicular; PM—light and one sheet behind product, one sheet between inspector and product; S—at side of workplace.

[e] 1—lighter or darker than surrounding area; 2—highlights and shadows; 3—enhanced color difference; 4—shiny lines on dark background; 5—bends in stripe reflections; 6—colored contour lines.

also eliminates the need for a sampling strategy for timing and sequencing multiple types of lights. Experimentation and experience are necessary in order to make intelligent trade-off decisions regarding the number of specialized sources that should be used.

6.3.10.3 Designing Inspection Workplaces

Inspection workplaces should be designed so that the product or material being inspected is directly in front of the inspector and, at the same time, is close enough that the smallest anticipated critical defect can be seen without difficulty. The designer should consult an anthropometric data base for guidance in the selection of workplace dimensions (such as inspection surface height) to ensure a comfortable working posture with favorable veiwing angles for all inspectors. Seated workplaces are preferred when the items to be inspected are small, when the inspection is critical, or when viewing is prolonged. Standing workplaces are preferred when the items are large and bulky.

Many problems may arise when attempting to retrofit existing production machinery with an inspection work station. Mirrors may be used to overcome some of the problems related to viewing positions and viewing angles. However, mirrors may reduce the visibility of the defects, and they do require regular cleaning.

6.3.10.4 Mounting the Inspection Lights

The inspection lighting units should be mounted so that they maximize the visibility of all defects of interest. Some experimentation will be necessary to determine the optimum positions. (See Table 6.3.7 for guidelines.)

To ensure a satisfactory inspection lighting system, each collimated light source should be mounted so that its position and angle of incidence may be adjusted. The optimum position of a given lamp may change as conditions change, such as when a manufacturing process is altered. The range of adjustment that is desirable depends on the types of materials to be inspected.

Transilluminators should be mounted as close as feasible to the far side (side away from the inspector) of the material being inspected. For a moving web product the dimension of the transilluminator in the direction of the product movement should be at least 30 cm for each 18 m/min of web travel (Eastman Kodak Company, 1983). The total amount of transilluminated area that is required depends on the width of the product, the number and severity of defects anticipated, and the viewing distance.

The size and placement of simple diffuse sources and those used with stripe overlays depend on the distance between inspector and the product and the viewing angles. Scale drawings should be used to verify that the source will adequately illuminate a sufficiently large area.

The placement of collimated sources for detecting most surface irregularities (bumps, digs, linear scratches, strands, etc.) is very critical. Normally these sources are mounted just above the far end or just above the left or right side of an inspection table so that their beams "graze" the surface of the product. For some applications the angle of incidence (as measured from the normal) may be as great as 80° or more (i.e., the rays are almost parallel to the surface of the product). Experimentation is usually essential in order to determine the proper placement of the sources. A displacement as small as 1 or 2 cm or a very slight change in the angle of incidence may make the difference between a very useful inspection lighting source and one that is barely acceptable. However, positioning of lighting units for the detection of irregular scratches or scratches in a clear web is less critical.

The light-piping technique may be used for detecting internal defects, but its successful application requires careful design. The light admitting edge must be flat and polished in order to let light pass into the material without misdirection or diffusion. Care must also be taken to prevent light from falling onto the surface; if this were to happen, there would be some reduction in the visibility of any internal defects. Showcase lamps with parabolic reflectors are effective sources, although their output may be inadequate for large products. Tubular tungsten-halogen lamps may also be used if the heat that they release can be adequately dissipated before it damages the product.

6.3.10.5 Control of Glare at Inspection Workplaces

Inspectors are sometimes required to detect defects that are near threshold (just barely visible). Hence glare (which decreases contrast and degrades visual performance) must be kept to an absolute minimum for these very difficult inspection tasks. Problems may occur when the inspection workplace is an integral part of a manufacturing system in a production environment where controlling glare has not been a prime concern (e.g., unshielded light sources, highly reflective metal parts, large windows).

Positioning of the inspection light source is critical if glare is to be minimized (see Figure 6.3.7). Reorientation of the workplace to reduce direct and reflected glare from other sources is also advised, where feasible. Otherwise, baffles may be used to shield bright sources. Sections of transilluminators that are not in active use may be covered with masks. Reflective machine parts may sometimes be painted flat black. It is also highly desirable to provide intensity controls for the inspection lights. This will permit the inspector to vary the illumination of the task as needed depending upon his

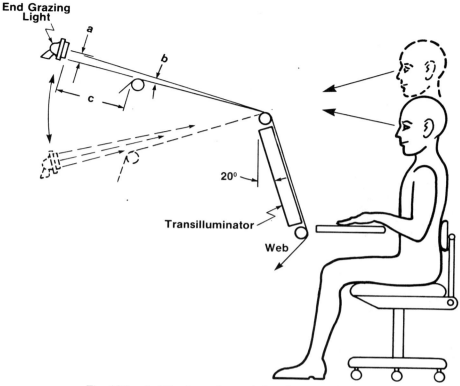

Fig. 6.3.9. A sitting inspection workplace for web products.

individual preference and glare tolerance, and also to adapt to changing characteristics of the defects and the product. For a discussion of other methods for controlling glare, see Table 6.3.5 and Section 6.3.6.

6.3.10.6 Application—A Sitting Inspection Workplace for Web Products

The following example is given to show how the five design steps discussed above may be integrated into a workplace for full-time inspection. The workplace (Figure 6.3.9) is intended to be used for the inspection of web products such as fabric, fabric-based goods, paper, and paper-based products. The upper section is also suitable for opaque products such as laminates and foils. Design tradeoffs (frequently necessary because of conflicting requirements) are also discussed.

The lower viewing section consists of a large transilluminator that may be used for detecting density variations in the base material and coatings. The upper section has an end grazing light which provides a low angle, collimated beam for detecting coating lines, drag lines, and scratches that may occur as webs are manufactured. Reference points for the positions of the viewing sections (including the range of adjustability for the upper section) are determined by the seated eye height of 5th and 95th percentile inspectors. Adjustment of the position and angle of the source relative to the web (dimensions a and b) is provided by the lamp bracket. This latter positioning adjustment provides some flexibility for viewing a variety of defects with different optical characteristics. A flat black paint is recommended for all frame parts.

A suggested source for the grazing light is an automobile driving light with a constant beam height (about 10 cm) and a horizontal spread of about 50°. For this source the distance between the grazing light and the far end of the web (dimension c) should be equal to the product width.

Additional grazing lights may also be mounted along one side of the transilluminator if surface irregularities are anticipated. Some glare shielding might also be desirable.

The work surface, for record forms or a keyboard, is positioned as low as possible to allow for comfortable viewing of the transilluminator section. The maximum length of the transilluminator (which sets a practical upper limit for the speed of web travel) is constrained by the seated eye height and range of tolerable viewing angles for the 5th percentile inspector.

REFERENCES

Barnaby, J. F. (1980). Lighting for productivity gains. *Lighting Design and Application, 10*(2), 20–28.

Boyce, P. R. (1979). Users' attitudes to some types of local lighting. *Lighting Research and Technology, 11,* 158–164.

Boyce, P. R. (1981). *Human factors in lighting.* New York: Macmillan.

Chapanis, A. (1949). How we see: A summary of basic principles. In *Human Factors in Undersea Warfare.* Washington, DC: National Research Council, pp. 3–60.

Cushman, W. H. (1983). Lighting for workplaces with visual display terminals. In *Health and Ergonomic Considerations of Visual Display Units—Symposium Proceedings.* Akron, OH: American Industrial Hygiene Association, pp. 73–85.

Cushman, W. H., Dunn, J. E., and Peters, K. A. (1984). "Workplace luminance ratios: Do they affect subjective fatigue and performance?, *Proceedings of the Human Factors Society 28th Annual Meeting.* Santa Monica, CA: Human Factors Society, 991.

Eastman Kodak Company (1983). *Ergonomic design for people at work,* Vol. 1. Belmont, CA: Lifetime Learning Publications.

Faulkner, T. W., and Murphy, T. J. (1973). Lighting for difficult tasks. *Human Factors, 15,* 149–162.

Flynn, J. E. (1977). A study of subjective responses to low energy and nonuniform lighting systems. *Lighting Design and Application, 7*(2), 6–15.

Flynn, J. E. (1979). The IES approach to recommendations regarding levels of illumination. *Lighting Design and Application, 9*(9), 74–77.

Flynn, J. E., and Spencer, T. J. (1977). The effect of light source colour on user impression and satisfaction. *Journal of the Illuminating Engineering Society, 6,* 167–179.

Flynn, J. E., Spencer, T. J., Martyniuck, O., and Hendrick, C. (1973). Interim study of procedures for investigating the effect of light on impression and behaviour. *Journal of the Illuminating Engineering Society, 3,* 87–94.

Helander, M. (1981). A critical review of human factors standards for visual display units. Paper presented at the American Industrial Hygiene Association seminar Health, Safety, and Ergonomic Considerations of Visual Display Units, Arlington, VA, April, 13–14.

Henderson, S. T., and Marsden, A. M., Eds. (1972). *Lamps and lighting.* New York: Crane, Russak.

Hopkinson, R. G., and Collins, J. B. (1970). *The ergonomics of lighting.* London: Macdonald.

Hughes, P. C., and McNellis, J. F. (1978). Lighting, productivity and the work environment. *Lighting Design and Application, 8*(12), 32–40.

Illuminating Engineering Research Institute (IERI) (1975). Annual Report. New York: Illuminating Engineering Research Institute.

IES (Illuminating Engineering Society of North America) Industrial Lighting Committee (1983). Proposed American national standard practice for industrial lighting. *Lighting Design and Application, 13*(7), 29–68.

IES (Illuminating Engineering Society of North America) Office Lighting Committee (1982). Proposed American national standard practice for office lighting. *Lighting Design and Application, 12*(4) 27–60.

Kaufman, J. E., and Haynes, H., Eds. (1981a). *IES lighting handbook,* reference volume. New York: Illuminating Engineering Society of North America.

Kaufman, J. E., and Haynes, H., Eds. (1981b). *IES lighting handbook,* applications volume. New York: Illuminating Engineering Society of North America.

Lehon, L. H. (1974). The effects of self-selected level of illumination on reading speed and comprehension of visually impaired and normally sighted children. Thesis, Teachers College, Columbia University. Abstract from *Dissertation Abstracts International* (Humanities and Social Sciences), Vol. 35, 4A, 2088-A.

Luckiesh, M. (1944). *Light, vision, and seeing.* New York: Van Nostrand.

Luckiesh, M., and Moss, F. K. (1937). *The science of seeing.* New York: Van Nostrand.

Lum-I-Neering Associates (1979). *Lighting design handbook.* Port Hueneme, CA: U.S. Navy Civil Engineering Laboratory, #AD A074836.

Lythgoe, R. J. (1932). The measurement of visual acuity (Medical Research Council Special Report No. 173). Cited by Chapanis (1949) and Hopkinson and Collins (1970).

McCormick, E. J. (1970). *Human factors engineering.* New York: McGraw-Hill.

Office Administration and Automation (OAA) (1984, May). "Improving the quality of light . . . and work performance." Pp. 38–48.

Reitmaier, J. (1979). Some effects of veiling reflections in papers. *Lighting Research and Technology, 11,* 204–209.

Ross and Baruzzini, Inc. (1975). *Lighting and thermal operations: Energy conservation principles applied to office lighting* (Conservation Paper No. 18). Washington, DC: Federal Energy Administration.

Saunders, J. E. (1969). The role of the level and diversity of horizontal illumination in an appraisal of a simple office task. *Lighting Research and Technology, 1,* 37–46.

Shellko, P. L., and Williams, H. G. (1976). The integration of task and ambient lighting into office furniture. *Lighting Design and Application, 6*(9), 14–23.

Smith, S. W., and Rea, M. S. (1978). Proofreading under different levels of illumination. *Journal of the Illuminating Engineering Society, 8,* 47–52.

Weale, R. A. (1963). *The aging eye.* London: H. K. Lewis.

Weston, H. C. (1962). *Sight, light and work.* London: H. K. Lewis.

Weston, H. C. (1945). The relationship between illuminance and visual efficiency—the effect of brightness contrast (Industrial Health Research Board Report No. 87). London: Great Britain Medical Research Council.

Wheelwright, C. D. (1977). Illumination evaluation using $\frac{1}{10}$ scale model payload bay and payloads (JSC Internal Note 77EW-4). Houston: National Aeronautics and Space Administration.

Xu, H. (1983). A study of the colour rendering capacity of a light source. *Lighting Research and Technology, 15,* 185–189.

CHAPTER 6.4
CLIMATE

FREDERICK H. ROHLES
STEPHAN A. KONZ

Kansas State University, Manhattan

6.4.1 INTRODUCTION

From the standpoint of the thermal environment, the human is classified as a homeotherm (the human must maintain a relatively uniform body temperature that is independent of the environmental temperature). As such, the warm-blooded human contrasts with the cold-blooded reptiles (poikilotherms) whose body temperatures vary with the temperature of the environment.

To maintain a constant body temperature, prehistoric man sought shelter in caves and covered himself with animal skins. Today we estimate people spend about 95% of the time indoors. With the exception of the relatively small segment of the population who work in extreme heat or extreme cold, people work and live in a comfortable and non-thermally stressful indoor climate.

When evaluating the effects of temperature or any other facet of the environment on the individual occupant, we consider three types of criteria. These are *physiological* (health), *performance* (including safety), and *affectivity* (comfort, preference, and acceptability). Physiological criteria (such as body temperature, weight loss or gain, and heart, respiration, and sweat rate) are fairly standardized. Performance criteria are less standardized; they vary from output on an industrial task to laboratory tasks (problem solving, simple arithmetic, finger dexterity, reaction time, and hand–eye coordination). The effects of heat and cold on performance have been the subject of considerable research (an excellent reference for cold is Burton and Edholm, 1970, and for heat is Kerslake, 1972). However, the researcher who does performance research in these areas has difficulty in controlling the motivation of his subjects. Essentially performance is the product of the individual's skill and level of drive (motivation). Thus, if at a given temperature a decrement in performance is observed, it is difficult to decide whether that decrement is due to the temperature or motivation or both. This problem must be considered in evaluating all studies that involve temperature and performance.

Even more elusive is the measurement of feeling or affectivity. The rating scale is the general measuring device. Current research on rating scales has been directed toward quantifying the subjective approach to measuring affectivity.

However, regardless of the criterion used, it is determined by seven factors that constitute the thermal environment. In the discussion that follows these seven factors are addressed; then the features of the thermally comfortable, heat stress, and cold stress conditions are described.

6.4.2 THE SEVEN THERMAL DETERMINANTS AND THEIR MEASUREMENT

The human response to the thermal environment depends on seven variables. (See the glossary, section 6.4.6) for a definition of terms.)

The first of these is the air temperature, or adjusted *dry-bulb temperature.*

The second is relative humidity (rh) or *water vapor pressure;* the latter is the preferred variable to describe the moisture content of the air because it is independent of the air temperature, whereas rh depends on the temperature of the air. Dew-point temperature is another index used to describe the water content in the air; it is the temperature at which moist air becomes saturated (100% rh) with water vapor. (*Measurement:* A sling psychrometer is probably the cheapest and easiest device for measuring temperature and relative humidity. The device shown in Figure 6.4.1 yields the wet-bulb and dry-bulb temperatures and contains a scale for converting these variables into rh.)

The third variable, *mean radiant temperature,* is exemplified by the temperature originating from hot or cold surfaces in the environment such as outside walls or windows. If we are in the center of a large room, radiant temperature is a minor concern because the thermal environment of our immediate space is affected very little by the radiation originating from the walls. In contrast, sitting *beside* a window on a winter day results in an exposure to an asymmetric thermal environment. One side of you is cool and the other is comfortable. To take advantage of the radiant heat provided by the sun, we employ the strategy of opening the draperies during the winter. Closing the draperies in the summer reduces the radiant heat loads. (*Measurement:* Several devices are available for measuring the mean radiant temperature; however, all use a 6 inch diameter copper sphere, painted flat-black. Inside the sphere is either a thermocouple or thermistor that terminates at a digital temperature readout device.)

The fourth variable is *air velocity.* For comfort, the average air movement should not exceed 30 ft/min (0.15 m/sec); this is considered to be "still air" and is relatively imperceptable. Velocities above this level may create discomfort from drafts. (*Measurement:* Air velocity is measured with an anemometer. Several of these are available at various prices. A good-quality anemometer is expensive and sensitive to minor velocity changes.)

Two variables can be combined into an index, *operative temperature;* it is the average of the air and radiant temperatures weighted by the respective heat transfer coefficients. At air velocity of less than 30 ft/min, the operative temperature is the average of the air and radiant temperatures. Another

Fig. 6.4.1. Sling psychrometer.

index that is used in the comfort literature is *new effective temperature.* This physiologically derived index combines the air temperature and rh into a single value.

When the human is placed in a thermal environment, three additional variables must be defined. The first of these is clothing. The human body is continually generating heat through metabolism. If the human is nude, the heat that is generated dissipates into the environment. If the person is clothed, the heat is "trapped" in the garments. The "heat-trapping" quality of the ensemble constitutes its insulating characteristics, measured in "clos."

Table 6.4.1 Clo Units for Individual Items of Clothing[a]

Men		Women	
Item	Clo	Item	Clo
Underwear			
Long underwear upper	0.10	Full slip	0.19
Long underwear lower	0.10	Half slip	0.13
T-shirt	0.09	Long underwear upper	0.10
Sleeveless	0.06	Long underwear lower	0.10
Briefs	0.06	Bra and panties	0.05
Shirt			
Heavy long sleeve	0.29	Heavy blouse	0.29
short sleeve	0.25	Light blouse	0.20
Light long sleeve	0.22	Heavy dress	0.70
short sleeve	0.14	Light dress	0.22
(plus 5% for tie or turtleneck)			
Heavy vest	0.29	Heavy skirt	0.22
Light vest	0.15	Light skirt	0.10
Heavy sweater	0.37	Heavy sweater	0.37
Light sweater	0.20	Light sweater	0.17
Heavy jacket	0.49	Heavy jacket	0.37
Light jacket	0.22	Light jacket	0.17
Heavy trousers	0.32	Heavy slacks	0.44
Light trousers	0.26	Light slacks	0.26
Socks		Stockings	
Knee-high	0.10	Any length	0.01
Ankle length	0.04	Panty hose	0.01
Shoes		Shoes	
Boots	0.08	Boots	0.08
Oxfords	0.04	Pumps	0.04
Sandals	0.02	Sandals	0.02

Source: Sprague and Munson, 1974.
[a] Total intrinsic insulation = 0.82(Σ individual items) clos.

Table 6.4.2   Metabolic Rates of Typical Tasks

Activity	Metabolic Rate (mets)
Reclining	0.8
Seated, quietly	1.0
Sedentary activity (office, dwelling, lab, school)	1.2
Standing, relaxed	1.2
Light activity, standing (shopping, lab, light industry)	1.6
Medium activity, standing (shop assistant, domestic work, machine work)	2.0
High activity (heavy machine work, garage work)	3.0

Table 6.4.1 presents the clo values for individual clothing items. To estimate the clo value of a particular ensemble, the clo values of individual items should be totaled and multiplied by 0.82. As a general rule of thumb, 1 clo will provide about 7.2°C of protection. In other words, an ensemble whose clo value measures 1.0 will afford enough warmth at 20°C to provide comfort equal to nude at 27.2°C. At 65°F (18.3°C) a clothing ensemble that is estimated to keep approximately 90% of the wearers comfortable might consist of the following: shoes, warm socks, briefs, long-sleeve woven shirt, warm trousers, T-shirt, long-sleeve warm sweater, and a warm jacket; the clo value of this ensemble is estimated to be 1.34.

The sixth variable is *physical activity*. Activity can be measured in terms of mets. When we are seated quietly, our metabolic rate is approximately 1 met. Table 6.4.2 presents the metabolic rates of other activities. Figure 6.4.2 shows the relationship between temperature, clo, and activity level.

The seventh variable is *time*. Even though the thermal conditions to which humans are exposed are not constant for long periods, time has received only modest attention as a variable in comfort research. Manual operation of radiators, ventilating fans, thermostats, doors, windows, and window blinds can give rise to large and unsystematic changes in indoor climate. Cyclic variations also occur with automatic controls. There often are abrupt changes in the thermal environment between buildings, or between different areas in the same building. Temporal changes in individual activity alter the metabolic rate; the addition or removal of clothing also can produce variations in heat balance. These situations result in changes in the basic physiological and affective responses of the individual concerned. *In short, the steady-state temperature conditions that characterize most of comfort research are, in practice, the exceptions rather than the rule.*

In general, three types of nonsteady-state conditions can be identified.

The first type involves "step" changes in conditions, such as those experienced in going from an automobile to the supermarket and return.

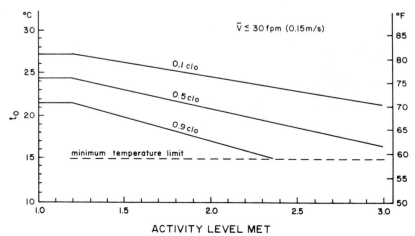

Fig. 6.4.2.   Temperature, clo, and activity level all affect comfort.

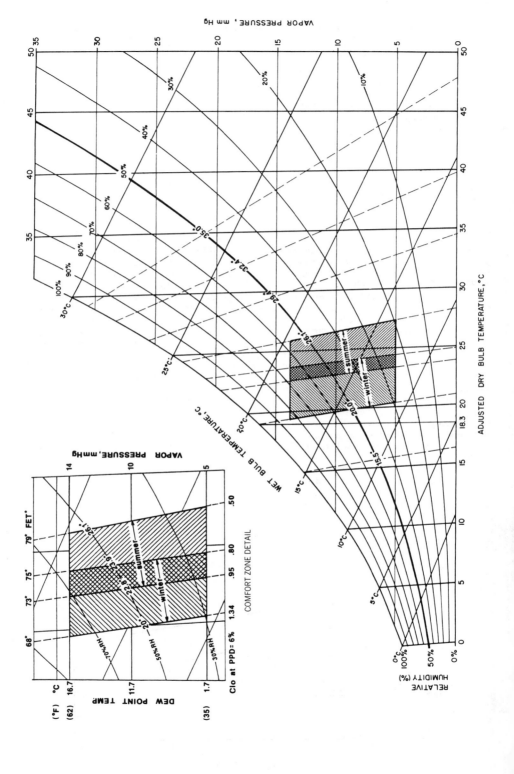

The second nonsteady temperature condition involves "ramps" or drifts; it is similar to those associated with night setback. The effects of these temperature drifts on comfort have been studied by Berglund and Gonzales (1978); they concluded that low temperature drifts (±1°F/hr or 0. 55°/hr) about a thermally neutral temperature are almost indistinguishable from constant temperature conditions. For example, a drift of this magnitude, which causes the temperature to deviate from a neutral temperature by ±3.5°F (1.9°C), will reduce thermal acceptability only from 90 to about 80%.

The third time condition is related to cyclical temperature fluctuations. These are associated with thermostat tolerances, the size and effectiveness of the heating and cooling systems, and the thermal efficiency of the structure (e.g., storm windows, insulation, infiltration). For example, in a well-insulated house with a good heating system, the temperature will rise quickly a small amount above the thermostat setting and diminish slowly. The opposite will occur in a poorly constructed residence with a poor heating system. Here, a long time will be required for the temperature to rise to the thermostat setting, and once reached, the furnace will shut off and the house will cool down quickly.

These seven variables interact to affect our response to the thermal environment. The discussion that follows demonstrates the role of these factors in indoor comfort, heat stress, and cold stress.

6.4.3 INDOOR COMFORT

Standards for thermal comfort have been the product of engineering research over many years. In 1923, the New York State Commission on Ventilation published air quality and thermal comfort criteria for schools, factories, and offices. In the same year, Houghten and Yagloglou addressed the comfort problem and introduced the concept of effective temperature.

Today, the American Society of Heating, Refrigerating, and Air Conditioning Engineers (ASHRAE) represents one of the major sources of thermal comfort research. Four figures from the ASHRAE literature and comfort standard are presented. Figure 6.4.3 shows the comfort zones as projected on a psychrometric chart; Figure 6.4.4 demonstrates how increased air velocity can extend the upper limit of the summer comfort envelope; Figure 6.4.5 shows the relationship between clo and temperature within the comfort zones. Figure 6.4.2 demonstrated the interrelationship between temperature, clo, and activity level.

The percent people dissatisfied (PPD) can be calculated (Rohles, Konz, and Munson, 1980):

PPD = percentage of people dissatisfied (voting other than 3, 4, 5) corresponding to cumulative area from negative infinity for CSIG or HSIG

CSIG = number of standard deviations from 50% for cold conditions (<25.3ET*) for sedentary activity and 0.5 to 0.6 clo
= $10.26 - 0.477(ET^*)$

HSIG = number of standard deviations from 50% for hot conditions (>25.3ET*) for sedentary activity and 0.5 to 0.6 clo
= $-10.53 + 0.34(ET^*)$

ET* = new effective temperature, °C

For example, for an ET* of 18°C, CSIG = +1.67; from a normal table 95% are dissatisfied. For an ET* of 30°C, HSIG = −0.21, and 42% are dissatisfied. At 25.3 ET*, the minimum (6%) are dissatisfied.

Fig. 6.4.3. Psychrometric chart showing the summer and winter comfort envelopes. Six variables are presented on this chart. The adjusted dry-bulb temperature is indicated on the abscissa. The ordinate contains the scales for describing the moisture content in the air; the water vapor pressure is on the right ordinate; the dew-point temperature (on the comfort zone detail only) is on the left ordinate. The diagonally sloping lines are the wet-bulb temperatures. The lines curving horizontally upward across the chart represent relative humidities. New effective temperature (ET*) lines also are drawn. These are the sloping dashed lines that cross the 50% rh line. They are labeled 15.5*, 20.0*, 26.1*, 29.4*, 32.4*, and 35.0* (22.8* and 23.9* are also labeled on the comfort zone detail). Anywhere along any one of these lines an individual will experience the same thermal sensation and will have the same amount of "skin wettedness" due to regulatory sweating. For example, 27°C/30% rh, 26.1°C/50% rh, and 25.5°C/70% rh all are equivalent to a new effective temperature of 26.1°C; as such, all lie on the 26.1°C new effective temperature line. At these conditions we would experience the same thermal sensation. Two comfort envelopes or zones are specified—one for winter and one for summer. It is estimated that the thermal conditions within these envelopes will be acceptable to 94% of the occupants (PPD = 6%) when they wear the clothing ensemble indicated. It is assumed (1) the air velocity within the envelopes is less than 30 ft/min (0.2 m/sec) and (2) that the occupants are engaged in sedentary or near-sedentary activity.

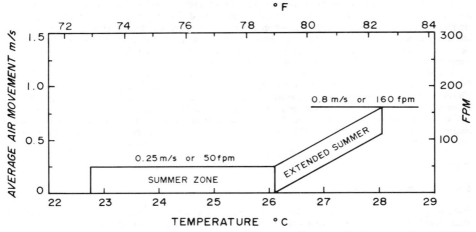

Fig. 6.4.4. Air velocity can be traded off with temperature. At higher velocities, we tolerate higher temperatures.

6.4.4 HEAT STRESS

Table 6.4.3 gives the recommended heat stress Wet-Bulb Globe Temperatures (WBGT). There are two key points about the values: (1) they are recommendations, not legally binding, and (2) they are *not limits;* they are values at which precautions (provision of adequate drinking water, annual physical examinations, training in emergency aid for heat stroke) should *begin* to be taken.

If radiant temperature is close to air temperature:

WBGT = 0.7 NWB + 0.3 GT

If radiant temperature is not close to air temperature:

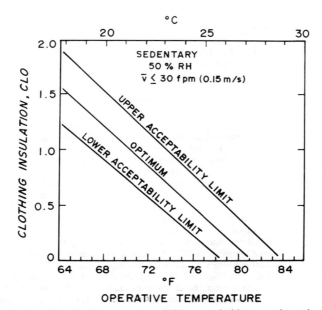

Fig. 6.4.5. Clo can be traded off with temperature. With more clothing, we tolerate lower temperatures.

WBGT $= 0.7$ NWB $+ 0.2$ GT $+ 0.1$ DBT

where WBGT = Wet-Bulb Globe Temperature, °C

NWB = Natural Wet-Bulb Temperature, °C (This is not psychometric wet bulb (WB) but the temperature of a wet wick exposed to *natural* air currents.)

$\qquad$ = WB (for air velocity > 2.5 m/sec)

$\qquad$ = 0.1 DBT $+ 0.9$ WB (for $0.3 < V < 2.5$ m/sec)

GT = Globe temperature, °C

DBT = Dry bulb temperature, °C

WBGT may be calculated from the above equations or by using direct-reading instruments.

Table 6.4.4 gives a brief outline of some considerations for reduction of heat stress. The first possibility is to design away the problem through erogonomic techniques. If the problem remains, there are medical and administrative techniques to reduce problems. If they don't work, then there are some postexposure techniques.

Table 6.4.3 Threshold WBGT Temperatures (°C) as a Function of Air Velocity and Metabolic (Basal + Activity) Rate[a]

Metabolic Rate (W)	Low Air Velocity (<1.5 m/sec)	High Air Velocity (≥ 1.5 m/sec)
Light (<230)	30	32
Moderate (230–350)	27.8	30.5
Heavy (>350)	26	28.9

[a] A velocity of 1.5 m/sec is a "noticeable breeze"; paper blows at 0.8 m/sec.

Table 6.4.4 Checklist for Control of Heat Stress

Ergonomic techniques
$\quad$ Metabolism
$\quad$ Radiation
$\quad$ Convection
$\quad$ Evaporation
$\quad$ Clothing
$\quad\quad$ Conventional
$\quad\quad$ Personal cooling garments
Medical and administrative techniques
$\quad$ Prevention
$\quad\quad$ Fitness
$\quad\quad$ Gender
$\quad\quad$ Age
$\quad\quad$ Acclimatization
$\quad\quad$ Dehydration
$\quad\quad$ Salt
$\quad$ Postexposure
$\quad\quad$ Emergency treatment
$\quad\quad$ Cool-down area
$\quad\quad$ Cool-down temperature and humidity
$\quad\quad$ Rest

6.4.4.1 Engineering (Ergonomic) Techniques

Storage of heat in the body is a function of metabolic rate, radiant heat rate, convection rate, evaporation rate, and conduction rate.

Reducing the metabolic rate should be done by mechanization or job design (e.g., power tool instead of hand tool, hoist or lift truck instead of manual lifting, storing material at hip height instead of floor height) rather than working more slowly (which decreases productivity).

The effect of radiant temperature can be reduced by working "in the shade." Clothing (a mobile shield) is the first line of defense; hats and long-sleeve shirts should be worn. A fixed shield between the person and the heat source is a second line of defense. Roofs and umbrellas protect against the sun; aluminum shields work well against furnace heat.

Increasing convective heat transfer is good if air temperature is less than skin temperature in the heat (about 95°F, or 35°C). Clothing should permit venting and pumping at the neck, waist, wrists, and ankles so air trapped under the clothing can be exchanged. Increasing air velocity (through fans) helps, but note that fan velocity drops off rapidly with distance. Roof ventilators can be used to remove hot air.

Evaporation of sweat is a very effective cooling technique. If the sweat drips off, it does no cooling and is wasted. An example of this is the sweating that occurs under nonpermeable garments. Dehumidification of the air is helpful but tends to be expensive. Evaporation also is improved by increased air velocity; this requires fans that are close to the person and removal of clothing barriers.

In heat, the ability of a garment to transmit moisture is very important. Moisture can be transmitted through the fabric or through garment openings (neck, waist, ankle, arms). Light colors reflect radiant loads (sun, furnaces) but are difficult to keep clean. Special cooling garments (developed though work with astronauts) now are commercially available. There are water-cooling models, ice slab models, and, most popular, air-cooling models. The primary advantage of the air-cooling models is that they supply clean air for breathing in addition to cooling (many hot environments also are dirty).

6.4.4.2 Medical and Administrative Techniques

These techniques basically involve reducing the rise in heart rate, blood pressure, and body temperature.

Physical fitness is desirable; it can be achieved either by training or selection. If workers are selected on the basis of medical exams, an initial physical upon hiring is not sufficient; there must be an annual physical checkup.

Males traditionally have held jobs done during heat stress. There is no evidence that females cannot do work in heat stress. However, small size (generally associated with being female) is a handicap with physical work in heat because the small person uses a greater percent of the available work capacity. Thus big, strong people should be selected, male or female.

Age again is relatively irrelevant; the individual's physical condition is the important factor.

Acclimatization is very important. Acclimatization requires about 10 days of *exercise* in the heat; if there is no exercise, there is no acclimatization. For each day of nonexercise, the equivalent of 0.5 days of acclimatization is lost.

Dehydration of 3% causes physiological and psychological performance changes, 5% gives evidence of heat exhaustion, 7% results in hallucinations, and 10% is very hazardous and leads to heat stroke. Unfortunately, the thirst drive is not sufficient to replace water loss. Thus supervisors need to ensure not only that water is available but that the workers drink it.

Salt deficiency is unusual owing to the large amount of salt stored in the body and the amount of salt in the typical diet. Salt tablets rarely are desirable; some people take them excessively and get stomach and high blood pressure problems. Salt on food usually is sufficient. Maffley (1979) reported a study showing that only if water losses were 8% or more of body weight would salt loss exceed dietary intake. That is, in most situations, salt in commercial salt-replacement drinks just gets excreted in the urine.

6.4.4.3 Postexposure

If the heat stress problem can't be prevented, perhaps the effects can be reduced. Emergency treatment requires training employees (1) to recognize symptoms of heat stroke and heat exhaustion, and (2) to administer emergency treatment. A "cool-down" area can be any convenient location. It should have a telephone and liquids available (perhaps just an insulated container of cool water). Excess clothing (such as protective clothing) should be removed during the cool-down period. A fan is desirable. Artificially cooled areas are most needed if the work environment temperature is over 35°C ET* (Konz, Rohles, and McCullough, 1983).

The amount of rest required in heat stress situations has not been determined. How much rest is needed in heat is a complicated question with many opinions and not much evidence. Extra rest probably is needed for work environments over 36°C ET* (Konz et al., 1983). Millican, Baker, and

Cook (1981) emphasized the importance of self-pacing of workers and the workers deciding the work/rest schedule for each task on each day.

6.4.5 COLD STRESS

What is the limiting cold environmental temperature? It really depends on whether discomfort, dexterity, or death (cell level or organism level) is the criterion. Yet people want a single number that will apply to all situations. Table 6.4.5 gives a simple index, the wind-chill index. The index was kept simple by using the time to freeze a 1 liter bottle of water as the criterion rather than the discomfort, dexterity, or death of a clothed, breathing human. The index is inappropriate in most situations because people exposed to cold change their microenvironment (get out of the wind, exercise, add clothing) as well as adjust internally (shiver, vasoconstrict). However, wind chill is the only index now available.

Reducing cold stress follows two basic strategies: modify the environment or clothing.

6.4.5.1 Environment

Modifying the environment can be as simple as providing a windbreak. Plastic strip doors permit an area to be enclosed and yet permit passage of lift trucks, conveyorized items, or even people during the few seconds of their passage. Relative humidity (vapor pressure) is not very important in the cold because the absolute vapor pressures are so low. From Fig. 6.4.2, note that at 0°C the 100% rh corresponds to 4 mm of water vapor pressure and 50% corresponds to 2 mm; both 2 and 4 mm are very low. However, wet clothing (from rain, melted snow) does present a severe problem.

There are a number of local heating (space heating) alternatives. If the warm air can be retained in the area, electric resistance heat combined with a fan is popular because installation of these modular units is simple, capital cost is low, and there is no pollution problem. Space heating with fuel oil (kerosene) has become more popular with the development of small heaters with various fire-safety features such as automatic shutoff if tipped over. Fuel heaters may present pollutant problems (CO, NO_2, CO_2, SO_2) if there is no air exchange with the air outside the room. Radiant heaters heat by radiation and thus do not heat the intervening air. They are useful in open, windswept areas such as docks and when the person is in a relatively fixed position (such as under-desk radiant heaters for typists).

6.4.5.2 Clothing

Clothing can be added to the head, hands, feet, arms, legs, and torso.

The peripheral vasculature of the head does not constrict; therefore, it is a large source of heat

Table 6.4.5 Wind-chill "Equivalent" Temperatures Predict the Effect of Air Velocity at Various Temperatures[a]

Dry-Bulb Temperature (°C)	Air Velocity (m/sec)							
	2	4	6	8	10	12	14	16
+4	+3.3	+1.8	−5.0	−7.4	−9.1	−10.5	−11.5	−12.3
+2	+1.3	−4.2	−7.6	−10.2	−12.0	−13.5	−14.6	−15.4
0	−0.8	−6.5	−10.3	−12.9	−15.0	−16.5	−17.6	−18.5
−2	−2.8	−8.9	−12.9	−15.7	−17.9	−19.5	−20.7	−21.6
−4	−4.9	−11.3	−15.5	−18.5	−20.8	−22.5	−23.8	−24.8
−6	−6.9	−13.7	−18.1	−21.3	−23.7	−25.5	−26.9	−27.9
−8	−9.0	−16.1	−20.0	−24.1	−26.6	−28.5	−29.9	−31.0
−10	−11.0	−18.5	−23.4	−26.9	−29.5	−31.5	−33.0	−34.1
−12	−13.1	−20.9	−26.0	−29.7	−32.4	−34.5	−36.1	−37.2
−14	−15.1	−23.3	−28.6	−32.4	−35.3	−37.5	−39.1	−40.4
−16	−17.2	−25.7	−31.3	−35.2	−38.2	−40.5	−42.4	−43.5
−18	−19.2	−28.1	−33.9	−38.0	−41.1	−43.5	−45.3	−46.6
−20	−21.2	−30.5	−36.5	−40.8	−44.0	−46.5	−48.3	−49.7

[a] The number in the table gives the temperature at 1.8 m/sec that has the same wind chill as the dry-bulb temperature at a given air velocity. Thus −12° at 2 m/sec, −6° at 4 m/sec, −2° at 6 m/sec, and 0° at 8 m/sec are approximately equivalent.

loss. A stocking cap gives good insulation, can be worn differently to vary its insulation, and is easily stored when not in use. For greater protection, protect the neck and throat, either by a cap that covers the neck or by a hood. Hoods should have drawstrings; a fur-lined front edge reduces the effect of wind.

Cold hands are important because they influence dexterity. Gloves cover each finger individually whereas mittens cover the fingers as a group. Therefore gloves have more surface area to lose heat and one finger can't warm another finger. If finger dexterity is not needed, mittens with liners are better than gloves. If finger dexterity is needed, airtight, close-fitting gloves are satisfactory for moderate cold. For more severe weather, a multilayer approach is desirable, with knit gloves inside and an airtight mitten outside. The mittens should extend past the wrist; important design features are that they are "easy-off" and they are attached to the coat with a cord. Thus the user, when not requiring dexterity, wears both mittens and gloves; when dexterity is needed, the mittens are removed and they hang from the coat, ready to be put on.

Feet also get cold. The best insulator is air. One strategy is two pairs of socks—cotton on the inside and wool on the outside (wool may irritate the skin). Shoes and boots must be oversize. The sole should have minimum conductivity (i.e., no nails). Keep the feet dry with waterproof layers. A simple, general-purpose combination is ordinary work shoes (one size too large, ankle high, with a thick, high-insulation sole) plus rubber galoshes. If wetness is a problem, each person should keep a pair of dry socks (in locker or vehicle).

Legs and torso are not much of a problem. Two, three, or even four layers of thin clothing should be worn; each layer traps air. Moreover, layering permits the removal of some layers when doing exercise. Clothing designed to permit air to escape when exercising (venting) is good.

6.4.6 GLOSSARY

Acceptable thermal environment: An environment that at least 80% of the occupants would find thermally acceptable.

Adjusted dry-bulb temperature (t_{adb}): The average of the air temperature (t_a) and the mean radiant temperature (t_r) at a given location. The adjusted dry-bulb temperature (t_{adb}) is approximately equivalent to operative temperature (t_o) at air velocities less than 0.4 m/sec (80 ft/min) when t_r is less than 50°C (120°F).

Clo value: A numerical representation of a clothing ensemble's thermal resistance. One clo equals the clothing required to keep a resting individual comfortable while seated in an atmosphere of 70°F with rh less than 50% and air movement at 20 ft/min. 1 clo = 0.155 m²K/W (0.88 ft²Fh/BTU).

Dew-point temperature (t_{dp}): The temperature at which moist air becomes saturated (100% relative humidity) with water vapor when cooled at constant pressure.

Mean radiant temperature (t_r): The uniform surface temperature of a radiantly black enclosure in which an occupant would exchange the same amount of radiant heat as in the actual nonuniform space.

Metabolic rate: Rate of energy production of the body. Metabolism is expressed in met units. One met, defined as 58.2 W/m² (18.4 Btu/hr-ft²), is equal to the energy produced per unit surface area of a seated person at rest. The surface area of an average man is about 1.8 m² (19 ft²).

New effective temperature (ET*): The uniform temperature of a radiantly black enclosure at 50% rh, in which an occupant would experience the same comfort, physiological strain, and heat exchange as in the actual environment with the same air velocity. The former index was effective temperature, labeled ET. ET* is the same line but labeled according to where it intersects 50% humidity instead of 100%.

Occupied zone: The region within a space, normally occupied by people, generally considered to be between the floor and 180 cm (6 ft) above the floor and more than 0.6 m (2 ft) from walls or fixed air-conditioning equipment.

Operative temperature (t_o): The uniform temperature of a radiantly black enclosure in which an occupant would exchange the same amount of heat by radiation plus convection as in the actual nonuniform environment. Operative temperature is the average, weighted by respective heat transfer coefficients (h_c and h_r), of the air (t_a) and mean radiant temperatures (t_r):

$$t_o = \frac{h_c t_a + h_r t_r}{h_c + h_r} \tag{1}$$

At air velocities of ≤ 0.4 m/sec (80 ft/min) and $t_r \leq 50$°C (120°F), operative temperature is approximately the simple average of the air and mean radiant temperatures and is equal to the adjusted dry-bulb temperature.

Optimum operative temperature: The temperature that satisfies the greatest possible number of people at a given clothing and activity level.

Relative humidity (rh): The ratio of the mole fraction of water vapor present in the air to the mole fraction of water vapor present in saturated air at the same temperature and barometric pressure; it approximately equals the ratio of the partial pressure or density of the water vapor in the air to the saturation pressure or density.

Skin wettedness: The ratio of actual evaporation to maximum possible evaporation in that environment. Mislabeled as skin wettedness, it actually has nothing to do with percent of the skin that is wet.

REFERENCES

ASHRAE (1981). Physiological principles comfort and health. Chapter 8 in *Handbook of fundamentals.* Atlanta, GA: American Society of Heating, Refrigerating and Air Conditioning Engineers.

ASHRAE (1981). *Thermal environmental conditions for human occupancy* (ASHRAE Standard 55–1981). Atlanta, GA: American Society of Heating, Refrigerating and Air Conditioning Engineers. The standard for indoor thermal comfort.

Berglund, L. and Gonzales, R. (1978). Occupant acceptability of eight-hour long temperature ramps. *ASHRAE, 84,* Pt. 2.

Burton, A. C. and Edholm, O. G. (1970). *Man in a cold environment.* New York: Hafner.

Eastman Kodak Co. (1983). *Ergonomic design for people at work.* Belmont, CA: Lifetime Learning Publications (Wadsworth).

  Chapter V-D contains a general description of the thermal constituents of the environment. Excellent and comprehensive list of references. Written by an industrial ergonomic group for industrial applications.

Kerslake, D. McK. (1972). *The stress of hot environments.* London: Cambridge University Press.

Konz, S., Rohles, F., and McCullough, E. (1983). Male responses to intermittent heat. *ASHRAE Transactions, 89*(1), 79–100.

Maffly, R. (1979). Running out of water. *Emergency medicine.* 57–61.

McIntyre, D. A. (1980). *Indoor climate.* Essex, England: Applied Science Publishers.

  The most comprehensive single source on the human response to the thermal environment. Unmatched in coverage detail; excellent reference list; appendices include Radiation Form Factors for the Human Body, A Summary of Fanger's Comfort Equation, Gagge's Two Node Model of Thermoregulation, and the Formulation of the Standard Effective Temperature.

Millican, R., Baker, R., and Cook, G. (1981). Controlling heat stress. Administrative versus physical control. *American Industrial Hygiene Association Journal, 42*(6), 411–415.

Rohles, F., Konz, S., and Munson, D. (1980). Estimating occupant satisfaction from effective temperature (ET*). *Proceedings of 24th Annual Meeting of the Human Factors Society,* pp. 223–227.

Sprague, C., and Munson, D. (1974). A composite ensemble method for estimating thermal insulating values of clothing, *ASHRAE Transactions, 80*(1), 120–129.

CHAPTER 6.5

HUMAN ENGINEERING FOR SPACE

STACY R. HUNT

**General Electric Corporation
King of Prussia, Pennsylvania**

6.5.1 INTRODUCTION

The focus of this chapter is human engineering as related to traveling, working, and living in space. There is a vast amount of human engineering data on human perceptual, psychomotor, motor, and cognitive functions that is applicable to humans in space. Because these data are very voluminous and well documented in this handbook and other existing literature, they are not replicated here. This chapter concentrates on background information regarding humans in space to provide insight as to where and how to apply existing data and to identify human engineering information that is unique to the space environment. a brief review of the history of manned space travel and the role of humans in space activity is provided to develop a basis for the material that follows.

Manned earth orbital space missions have been successfully undertaken by the United States and the Soviet Union for over two decades. The early manned space programs such as Mercury, Gemini, Apollo, and Soyuz were conducted in spacecraft with small volumes that severely restricted the space for flight crew members and constrained crew mobility. This constraint significantly limited the scope of tasks that could be performed in orbit. These programs showed that humans could successfully perform in the orbital environment many of the tasks that are routinely accomplished on the ground.

The constraints on human mobility in orbit were reduced significantly with extravehicular activity (EVA), which increased in extent and scope as the early space programs matured. During subsequent Apollo missions and the lunar landings, EVA became longer in duration and involved many additional tasks, further demonstrating human ability to perform in both the orbital and lunar environments.

Missions involving flights to the moon and rendezvous and docking with other orbiting vehicles with subsequent return to the earth have also been successfully accomplished. These efforts continued into programs such as the Skylab and Saylut missions where extended stays in space were possible. The Skylab and Saylut missions also provided significant increases in operational and experimental capabilities. Reusable manned space systems capable of carrying payloads into and out of low earth orbit became a reality with the success of the Space Transportation System and the shuttle spacecraft.

Human flexibility, ingenuity, adaptability, and ability to perform in the orbital environment were further demonstrated when the Skylab was saved by EVA and Intravehicular Activity (IVA) repairs. The Soviet space program has had numerous similar examples of space vehicle repair by cosmonauts.

Early results of the shuttle program demonstrated the capability of humans to perform an even wider range of tasks in space. EVA retrieval, manipulation, and capture of relatively large objects have demonstrated the feasibility of human participation in the assembly of large structures in space such as the Space Station. During these space programs, we have also clearly demonstrated the ability to control and manage successfully a wide variety of operational equipment and experimental payloads that have been part of the various missions.

Human-factors contributions and influences have been highly significant in the success of past and current space programs. Very useful data have been obtained on human ability to perform operational activities in space, and a basic understanding of the capabilities and limitations of humans in space has been accomplished. However, it is essential to recognize that in many cases the data on human physiological, psychomotor, sensory, and psychological attributes and capabilities in the space environment have not been obtained under controlled conditions. These data are not sufficiently extensive to provide a quantitative, comprehensive data base on which to provide exacting, broad-based human-factors standards for application to space systems. The human engineer will have to continue to extrapolate from existing ground-based data and use space-based data where available. In addition, the data obtained from flight crew members to date have been collected primarily on a limited population consisting primarily of individuals who were from a highly select population composed of test pilots or other highly trained specialists. Future space missions and Space Stations will include individuals selected from a much broader and more diverse population and will not be trained as extensively or comprehensively. In extrapolating data and establishing design standards, careful consideration of the numerous differences between the ground environment and the environment of space as well as the potential differences between the data of prior flight crew populations and future crew composition is necessary.

The data base of human factors relative to the design of space systems will be dynamic. With the opportunity for further study as exposure to space flight increases, rapid expansion and modification of the human-factors data base can be expected.

6.5.2 THE HUMAN ROLE IN SPACE

Assisting in the definition of the human role in space and the optimization of the effectiveness of our performance in space are critical challenges to the human engineer. The allocation of space tasks to human or machine is a significant and continuing area of study for human engineering.

Table 6.5.1 Sensory, Psychomotor, and Intellectual Functions Critical
to Human Performance in the Space Environment

Sensory and Perceptual Functions		
Vision	*Audition*	*Somesthesis*
Visual acuity	Auditory threshold	Cutaneous
Brightness discrimination	Absolute	Shape
Contrast discrimination	Loudness	Texture
Monocular/binocular	Pitch discrimination	Vibration
depth perception	Sound localization	Temperature
Color/color contrast		Pressure
discrimination		Localization
Motion and rate of		Kinesthesis and
movement detection		proprioception
		Complex (e.g., itch, oily, greasy)

Psychomotor/Motor Functions	Intellectual Functions
Force application	Short-term memory
Eye/hand coordination	Long-term memory
Fine/gross movement	Cognition
Overall body positioning and limb relationships	
Rate of movement	

Studies for allocating tasks to human or machine have usually involved task analyses with subsequent assignment of specific tasks to that approach resulting in best task performance. A number of recent studies and workshops have been completed which in part or in whole focused on the human role or capability in space (Lewis, 1982; Loftus, Bond, and Patton, 1975. McDonnell Douglas, 1984; USAF/NSIA/AIAA, 1982, 1984).

One fundamental method for assessing human capability in space has been to examine our basic sensory, motor, and intellectual capabilities with an emphasis on those functions critical to space activities. Table 6.5.1 shows a number of the human capabilities important to our ability to perform tasks anticipated for the space environment. Table 6.5.2 from Loftus et al. (1975) provides a brief summary of characteristics of human senses and indications for their use. From existing psychological and human engineering data, it is possible to compare fundamental human capability with the required level of performance anticipated for the space task. With the possible exception of the vestibular system, there does not appear to be major alteration in the basic underlying mechanisms of human sensory and perceptual processes as a result of residence in the space environment. There have been a number of reports that visual sensory function may be enhanced in the weightlessness of the space environment. This phenomenon is under study.

Care must be exercised in the extrapolation of task performance data because specific space environment and task elements may have a significant impact on performance levels as compared to earth-based values. Space environment factors that may adversely impact human performance include vibration and acceleration forces, heat, radiation, and weightlessness. These impacts may be task specific. For example, weightlessness may have a negative impact on the ability to apply controlled levels of force if appropriate restraints are not provided for the crewperson, but weightlessness may also permit the crewperson to move large mass objects with relative ease.

Space motion sickness or the space adaptation syndrome (SAS) has been experienced by a significant number of the crew members on past space flights. SAS has been shown to have some impact on task performance especially early in the space mission. Until SAS is more completely understood and the methods for preventing it are fully developed, SAS should be taken into account when developing operations and designing hardware. (SAS is discussed in detail in Section 6.5.4, Space Adaptation Syndrome.)

6.5.3 WEIGHTLESSNESS AND ACCELERATION

6.5.3.1 Weightlessness

Humans have developed in and are continuously exposed to recognized gravitational and inertial forces. In the space environment, astronauts are exposed to significantly different conditions including subgravity

Table 6.5.2 Characteristics of the Senses

Parameter	Vision	Audition	Taste and Smell	Touch	Vestibular
Sufficient stimulus	Light-radiated electromagnetic energy in the visible spectrum. Heavy particles	Sound-vibratory energy, airborne or structural paths	Particles of matter in solution (liquid or aerosol)	Tissue displacement by physical means	Accelerative forces
Spectral range	Wavelengths of 400–700 μm (violet to red)	20–20,000 Hz	Taste: salt, sweet, sour, bitter Smell: fragrant, acid, burnt, and caprylic	> 0 to < pulses/sec	Linear and rotational accelerations
Spectral resolution	120–160 steps in wavelength (hue) varying from 1 to 20 μm	~3 Hz (20–1000 Hz) 0.3% (above 1000 Hz)	—	$\dfrac{\Delta \text{pps}}{\text{pps}} = 0.10$	—
Dynamic range	~90 dB (useful range) for 3×10^{-9} cd/cm^2 (0.00001 mL) to 32 cd/cm^2 (10,000 mL)	~140 dB 0 dB = 0.0002 dyn/cm^2	Taste: = 50 dB 3×10^{-5} to 3% concentration quinine sulfate Smell: 100 dB	~30 dB 0.01–10 mm	Absolute threshold = 0.2°/sec
Amplitude resolution $\Delta I/I$	Contrast = $\Delta I/I$ = 0.015	0.5 dB (1000 Hz at 20 dB or above)	Taste: = 0.20 Smell: 0.10 to 50	$\Delta I/I$ nonlinear and large at low force levels ~0.15	~0.10 change in acceleration
Acuity	1° of visual angle	Temporal acuity (clicks) = 0.001 sec	—	Two-point acuity = 0.1 mm (tongue) to 50 mm (back)	—
Response rate for successive stimuli	~0.1 sec	0.01 sec (tone bursts)	Taste: ~30 sec Smell: ~20–60 sec	Touches sensed as discrete to 20/sec	~1–2 sec nystagmus may persist to 2 min after rapid changes in rotation
Reaction time for simple muscular movement	~0.22 sec	~0.19 sec	—	~0.15 sec (for finger motion, if finger is the one stimulated)	—

Table 6.5.2 (*Continued*)

Parameter	Vision	Audition	Taste and Smell	Touch	Vestibular
Best operating range	500–600 μm (green-yellow) 107.6 lm/m² (10 fc) to 2152 lm/m² (200 fc)	300–6000 Hz 40–80 dB	Taste: 0.1–10% concentration	—	~1 g acceleration directed head to foot
Indications for use	1. Spatial orientation required 2. Spatial scanning or search required 3. Simultaneous comparisons required 4. Multidimensional material presented 5. High ambient noise levels	1. Nondirectional warning or emergency signals 2. Small temporal relations important 3. Poor ambient lighting 4. High vibration or g-forces present	1. Parameter to be sensed has characteristic smell or taste 2. Changes are abrupt	1. Conditions unfavorable for both vision and audition 2. Visual and auditory senses	1. Gross sensing of acceleration information

Source: Loftus, Bond, and Patton, 1975.

and weightlessness. Adaptation to space conditions results in a complex series of physiological and behavioral changes. During longer-duration orbital missions (up to 6 months or more), significant musculoskeletal and cardiovascular system changes can occur. These changes remain areas of major study. Summary discussions regarding human adaptation to subgravity and weightlessness are available (Berry, 1973; Garriott and Doerre, 1977; Johnston and Dietlein, 1977; Pestov and Gerathewohl, 1975).

Humans maintain body posture and position under static and dynamic circumstances by the use of kinesthetic, visual, vestibular, and other sensory stimuli. Use of these sensory inputs with the resulting musculoskeletal system responses is critical to human ability to maintain balance and perform coordinated motor movements. In weightlessness, sensory systems sensitive to gravitational forces are exposed to conditions differing from a normal gravity environment. The resulting effects of weightlessness, gravitational forces, and some types of acceleration forces on human perception are of importance to the human engineer (see the following section).

True weightlessness of extended duration occurs only during orbital flight. However, numerous operational constraints are superposed on the weightlessness of flight as a result of mission activities. To substitute for actual flight studies, a variety of techniques for simulating the conditions of weightlessness are available. These techniques include Keplerian trajectories, where parabolic aircraft flights produce zero-gravity conditions for short durations; neutral buoyancy immersion studies, where subjects are immersed in water and adjusted to neutral buoyancy; and friction-reducing devices such as air bearings and partial or whole body support mechanisms. Each of these approaches has significant limitations. The use of these simulations and the application of data resulting from these studies must be undertaken with consideration of the constraints and impacts of the specific simulation techniques. However, one of the best techniques for the validation of human-factors concepts for space is to conduct tests during orbital flight. With the frequency of shuttle flights projected to rise to 24 flights per year in the near future, NASA will be able to conduct numerous, short-duration evaluations of new human engineering approaches. These studies should provide for rapid, accurate assessment of human engineering concepts for space.

A number of sensory and behavioral phenomena have been reported during space flight. Altered appetite levels have occurred, particularly during episodes of space motion sickness (see Section 6.5.4). Altered color perception and visual acuity effects have also been reported. Visual illusions including the oculogravic and oculogyral illusions have occurred (see the following section).

Readaptation to normal earth gravity following extended weightlessness or other reduced gravity conditions is of concern to the human engineer particularly if high levels of task performance are required immediately after return to earth gravity. Physiological capacity in the normal gravity environment is altered following adaptation to weightlessness. Design of equipment and operational procedures may have to be modified to accommodate altered capability (Berry, 1973; Pestov and Gerathewohl, 1975).

6.5.3.2 Acceleration Forces

During launch and reentry, astronauts are subjected to various levels of acceleration. In future systems flight crews may also be exposed to significant levels of acceleration forces during orbital or other in flight maneuvers. Fraser (1973) provides a summary discussion of performance during sustained acceleration. The general conclusions of Fraser's summary are shown in Table 6.5.3.

Coriolis acceleration is the component acceleration generated by the simultaneous exposure to movement about two axes. This type of acceleration is likely to be encountered in the situation where a space vehicle such as a space station is rotated to produce artificial gravity. Coriolis accelerations produce abnormal vestibular inputs which may result in undesirable, performance-impairing effects. In a rotating space station, these forces would affect the flight crew differently, depending on crew location within the vehicle and the rate and direction of movement of the individual within the vehicle. Woodson (1981) provides a discussion of Coriolis phenomena and recommended design guidelines for equipment design to minimize Coriolis effects.

Visual illusory effects that occur on exposure to acceleration and gravitational forces include the

---

Table 6.5.3 Summary of Performance During Sustained Acceleration

Increased visual reaction time (simple reaction time)

Increased errors on reading task

Possible impairment of higher mental functions

Control and tracking maneuvers impaired above 3–4 g

Higher mental functions may be impaired

Source: Modified from Fraser, 1973.

oculogravic and oculogyral illusions (Graybiel, 1975; Woodson, 1981). The oculogravic illusion is produced in a normal observer when a change in direction of the gravity vertical occurs, such as the conditions occurring in a centrifuge. The oculogravic illusion involves the gravity receptor (otolith) of the vestibular system. The oculogyral illusion is produced when a change in rotation occurs such as when angular rotation is abruptly started or stopped. The oculogyral illusion involves the angular motion sensor (semicircular canals) of the vestibular system. Both these illusions produce perception of movement of the visual field.

6.5.4 SPACE ADAPTATION SYNDROME

Space adaptation syndrome (SAS) is a temporary "space motion sickness" disability that has affected approximately 40 to 50% of the crew members on past U.S. and USSR space missions. SAS varies in intensity from a mild gastrointestinal discomfort to actual nausea and vomiting. In general, the symptoms of SAS are similar to those of motion sickness in a normal earth gravity environment. The symptoms may include some or all of the following: salivation, dizziness, apathy, lethargy, weakness, loss of appetite, anxiety, headache, fatigue, drowsiness, and vomiting. SAS symptoms usually occur early in the flight with the peak at 24 to 36 hr. The symptoms generally do not last longer than 72 hr and usually do not return during a flight (USAF/NSIA/AIAA, 1982, 1983).

Preflight selection of crewpersons to reduce susceptibility to SAS has not been successful on a routine basis. Predictions regarding susceptibility to SAS in prospective crewpersons who have not flown in space have also proved to be inaccurate. Preflight or inflight pharmacological therapy has been used with a moderate degree of success to prevent, reduce in intensity, or fully alleviate SAS. Medication has failed with some crewpersons. Other forms of treatment including biofeedback are under study.

Garriott and Doerre (1977) discuss space motion sickness and crew efficiency on first exposure to zero gravity. They conclude that "a relatively modest amount of crew time may have been lost due to motion sickness on Skylab missions 3 and 4 but that each crew's performance was never substantially impaired for more than 1 day." The Human Role in Space study (McDonnell Douglas, 1984) indicates that performance impairment occurring from SAS varies as a function of the duration of exposure to weightlessness. Human capabilities affected during SAS include discrimination of angular acceleration, cognitive function, visual-motor tracking, manipulative skills, and body positioning.

Until a cure, a fully effective preventive measure, or a reliable method of predicting those persons who are not susceptible to SAS is found, SAS is likely to have an impact on human performance during space flight. The impact will be the greatest on missions of short duration where high levels of performance are likely to be required in the very early portion of the mission.

Task scheduling and time lines for space missions should take into account the possibility of SAS. Task assignments should be developed so that the need for critical performance of specific human functions impaired by SAS is minimized during early portions of the mission. If pharmacological therapy is indicated for the treatment of SAS, the potential effects of the therapy on task performance should be carefully evaluated.

6.5.5 HABITABILITY

6.5.5.1 Definition and Elements of Habitability

Habitability can be thought of as the elements in an environment that influence human comfort, performance, or productivity. Habitability involves the type and quality of the environment and the extent to which elements of the environment sustain overall and personal operational effectiveness, general comfort, desirable attitudes, and morale. There have been many definitions of habitability, all of which eventually deal with the acceptability of the environment to humans. Thus the human becomes the essential element in determining what meets habitability requirements. An extended discussion of the definition and the elements of habitability can be found in Petrov (1975).

Habitability of space vehicles is a major concern for current and future space missions. Figure 6.5.1 shows a number of the many factors that enter into determining habitability of a space vehicle. Spacecraft habitability extends into a wide variety of systems within a spacecraft and, in many cases, involves the interaction of many elements.

A key element in determining the criticality of habitability is mission duration. Generally, as mission duration increases, the importance of habitability also increases. Short-duration space missions such as a 1 to 5 day mission may be successfully accomplished under conditions of habitability that could not be tolerated during 30 day missions. Similarly, the requirements for habitability for a 30 day mission are likely to be markedly different from the requirements for a 180 day or longer mission.

Other key elements determining habitability requirements are crew characteristics including crew size, selection, cultural background, training, educational background, social class, skill levels, and sex. Operational elements and other factors that can significantly impact habitability requirements include work/rest cycles, crew rotation, communications with other crew members and family, recre-

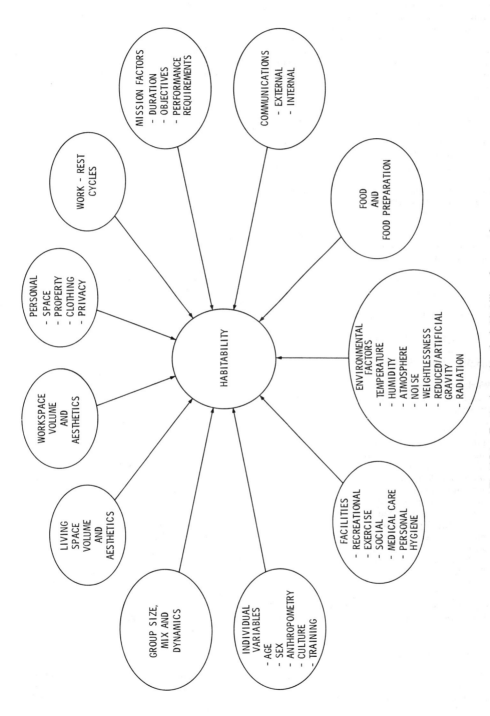

Fig. 6.5.1. Factors impacting the habitability of spacecraft.

ational opportunities and activities, interpersonal relationships with other crew members, including male/female interaction, and other aspects of group dynamics. These factors include not only hardware design issues but also include crew selection, composition, psychological, and sociological factors. (See Section 6.5.8, Psychological/Sociological Aspects of Space Flight.)

Early identification of habitability issues is critical to achieving human-factor goals of providing an environment where flight crew comfort and productivity can be enhanced and maintained through out complete missions. The effect of not rigorously pursuing in a timely fashion the development of appropriate habitability factors can be significant loss of crew performance with possible failure to attain mission objectives, compromise in operational procedures, and/or costly redesign to ensure that habitability needs are met.

6.5.5.2 Designing Spacecraft for Habitability

Designing for habitability includes consideration of spacecraft architecture. Architecture of the vehicle deals with the design of general work spaces, communal areas, and living quarters. This category includes the provision of an appropriate visual environment [e.g., use of color, visual cues to aid in orienting in the weightless environment, windows, noise control (Beranek, 1954, 1960; Peterson and Gross, 1967), and controllable general lighting]. Also in this category are concerns related to the overall operational efficiency and personal acceptability of the work space and living areas, including layout optimization to provide desirable traffic flow patterns, flexibility of layout to accommodate specific mission needs, and provision of private and public or communal areas and other primarily aesthetic elements of interior decorating, especially as applied to living and recreational areas.

Designing spacecraft for habitability frequently results in difficult trade-off situations. For example, the trade-off between the use of vehicle space for work areas or living and recreational areas can become very difficult owing to the limited vehicle space available. In this instance, the habitability requirements for appropriate recreational and living space must be met to ensure the reliability of the human component in what is a complex human–machine system. Unfortunately there are no exact, proved human engineering standards for space requirements in a space vehicle. Petrov (1975) discusses possible classification and allocation of areas for space vehicles.

6.5.6 CREW STATION DESIGN AND OPERATION

6.5.6.1 Crew Station Design

Work-station design has probably received more attention than any other area of human engineering. Numerous extensive sources of work-station information relevant to space vehicle design are available (Brown, 1974; Cakir, Hart, and Stewart, 1982; Calvin and Gazenko, 1975; Dalton, 1974a–f, 1975a, b; General Electric, undated; Gunderson, 1974; Johnson, 1974, 1975; Lewis, 1982; McCormick, 1964; NASA, 1976, 1978; National Research Council, 1983; Petrov, 1975; Sova, 1975; USAF, 1972; USAF/NSIA/AIAA, 1982, 1984; Webb, 1964; Woodson, 1981; Woodson and Conover, 1964). Although these references are of significant value to the human engineer, many require careful interpretation for appropriate direct application to a space system. For example, in using anthropometric data or crew station layout procedures generated for normal gravity work-station design, care must be exercised to ensure that the impact of weightlessness on operator position (e.g., body posture changes) and work-station configurational constraints (e.g., specific work-station configurations resulting from work positions that are related to normal earth gravity) must be taken into account.

Anthropological data and data including the range and characteristics of work/reach envelopes and approaches for methods to obtain needed data of this type are available (General Electric, 1969; Lewis, 1982; McDonnell Douglas, 1984; NASA, 1976; Petrov, 1975).

Principles and methods for work-station layout and arrangement that are appropriate for the design of space systems can be found in McCormick (1964). McCormick's principles include grouping by function, importance, location, sequence, and frequency of use. Other factors discussed by McCormick include preferred location of displays, arrangement of displays, and the location and arrangement of controls. MSFC-STD-512A (NASA, 1976) and MIL-STD-1472C (U.S. Department of Defense, 1981) also provide human engineering guidance for spacecraft control and display requirements and arrangement.

6.5.6.2 Operations

Although a large amount of data exists for work-station configuration and control/display layout, less information is available regarding the layout, configuration, format, and human–machine interactive operational requirements for the complex displays that can be presented on cathode ray tubes (CRT) or similar type displays (Ericsson Information Systems, 1984). With the use of current technology and software, displays with a high degree of human–machine interaction are available on a single hardware interface. This approach permits simplification of hardware configurations for displays in

that many types of displays, differing significantly in the dynamics that can be presented on a common hardware interface. Multiple displays (e.g., windows and panes and also graphics) can be simultaneously displayed creating an "electronic desk." These types of display hardware systems are in use in current spacecraft. Owing to the economy of use of a single display providing many possible displays and the inherent flexibility of this approach, these displays are likely to achieve a very high level of use in complex spacecraft systems of the future.

Guideline human engineering information for the general configuration, format, and interactive features of these types of displays is in the early stage of development. Various types of displays have been identified including such general functional categories as data entry, inquiry, menus, command, and interactive. Galitz (1981) provides human engineering guidance regarding functional categories of screens, color in the use of screen design, and suggested methods for screen design. Additional information dealing with improved productivity and the reduction of human error in the use of computer systems can be found in the works of Bailey (1983) and Galitz (1984).

6.5.7 SPACE SUITS AND EXTRAVEHICULAR ACTIVITY

6.5.7.1 Introduction

A space suit is basically a pressure vessel that surrounds the astronaut and allows the safe performance of useful work in the hostile environment of space. To accomplish this, the suit must provide many functions: supporting atmosphere and atmospheric pressure, temperature control and heat exchange, capability for locomotion, and the services required for normal body functions such as eating, drinking, and handling of waste products. The suit must provide the astronaut with adequate, safe visibility, mobility, and other functions needed for task performance. Requirements of the astronaut for communications with other crew members and sensory inputs such as tactile sensation must be met. The suit must be capable of integrated use with appropriate restraining and locomotion devices to enable effective work to be performed in the weightlessness of space or in the reduced gravity of a lunar surface. The space suit is a highly sophisticated, complex device providing functions very similar to the overall functions of a spacecraft.

The space suit has developed over many years and has varied in characteristics depending on the specific application for which the system was designed. The human engineering problems associated with space suit development are also varied. The same situation remains with current space suits in that a suit designed for use in a lunar environment is likely to differ in significant design characteristics from a suit designed for in-orbit EVA. Jones (1975) discusses the historical development and evolution of space suits relative to their applications.

Projecting possible applications for future space suits suggests that a significant portion of large-scale in-orbit servicing and assembly will be performed by pressure-suited astronauts. The EVA astronaut, using a variety of tools and work aids and with the possible help of teleoperator and robotic devices, will be responsible for the construction and maintenance of future space systems such as Space Station.

The basic elements of human engineering required for future space suit design are present in the approach to and design of current space suits. Smith, as cited by Jones (1975), shows the basic human–machine factors requiring consideration for space suit design as related to the allocation and definition of tasks and the potential limitations of a man and space suit system (Figure 6.5.2). Two major areas of concern for the human engineer relative to space suit design, mobility and sensory modifications, are discussed in the following two sections.

6.5.7.2 Mobility, Range of Motion, and Sizing

Mobility is a major concern in space suit design. Constraints in range of motion at joints in the suit and the loss of flexibility resulting from pressurization of the suit tend to constrain the astronaut's

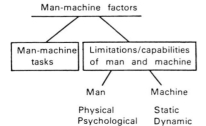

Fig. 6.5.2. Human–machine factors considered in the design of space suit systems. (From W. L. Smith, as cited in Jones, 1975).

ability to move and maintain position. Range of motion limitations at joints have resulted in improved suit designs with increased flexibility through a variety of techniques including articulated joints. Constraints in flexibility due to pressurization have resulted in the use of relatively low-pressure suits. However, the use of low-pressure systems may result in the possibility of the "bends" in some operational circumstances. To avoid the possibility of bends, some low-pressure suit systems such as the Space Shuttle 4 psi suit require considerable prebreathing time (3 hr or more of prebreathing of oxygen) before a flight crew member can effectively use the space suit. This type of constraint not only results in loss of time but may be unacceptable for missions requiring immediate egress after orbit insertion to perform EVA, repeated EVA in a relatively short period, or rapid exit from the spacecraft owing to emergency conditions.

Higher-pressure suits (8 to 14.7 psi) with concomitant reduction in the risk of the bends and with improved flexibility and range of motion are under development.

Most space suits used to date have been essentially custom-fit devices. With the increased heterogeneity of population of the flight crews and other persons who will take part in future space flights, ability to size a suit using a modular approach will become an economic and practical necessity. Sizing of this type can be accommodated by modularization of arm and leg components of the suit. With the provision of space suit features such as long-life mobility joints and the use of dry lubricants at joints, significant improvements in suit life and maintainability can also be accomplished.

6.5.7.3 Sensory Modification

The visual environment of space is different from in a normal earth atmosphere. The intensity of visible radiation is markedly different, and contrast ratios, shadowing, and light scattering are altered. In addition, nonvisible radiation levels are significantly higher in space requiring protective measures. Vision through the space suit helmet is a critical problem.

Table 6.5.4 provides a number of physiological and human-factor areas that influence helmet visor design. The optical properties of the visor are critical. Optical distortions, refractive anomalies, glare, and other undesirable contrast alterations are to be avoided (Graham, 1966).

Pressurized gloves result in decreased tactile feedback to the astronaut. In addition, manual dexterity is significantly reduced. These factors should be considered in allocating tasks and in design of hardware requiring manual manipulation.

6.5.7.4 EVA Tools

The selection or design of tools for use by an astronaut performing EVA tasks is highly dependent on the specific requirements of the tasks and the vehicle systems involved. However, there are a number of requirements and general human engineering guidelines that can be applied to EVA tools. Table 6.5.5 provides a list of these requirements and guidelines (49).

6.5.8 PSYCHOLOGICAL/SOCIOLOGICAL ASPECTS OF SPACE FLIGHT

Information regarding the overall psychological and sociological aspects of space flight is limited. Primary emphasis in this area has been on two categories: crew performance oriented toward mission-related objectives and basic sensory, motor, and perceptual processes as related to physical aspects of the space environment (e.g., weightlessness). Higher-level behavioral phenomena such as interpersonal and group processes, motivational factors, and cognitive function have received little attention.

Table 6.5.4 Physiological Problem Areas That Influence Visor Design

Type of Radiation	Problem Areas
Ultraviolet	Photophthalmia
Visible	Glare
	Visual acuity
	Dark adaptation
	Retinal burns
	Corneal heating
Infrared	Retinal burns
	Corneal heating

Source: Jones, 1975.

Table 6.5.5   Requirements and Human Engineering Guidelines for EVA Tools

Tool manipulation requirement and required force emissions should be compatible with the pressurized EVA glove, suit and 0 g restraints.

Tool grip size should be suitable for use with a pressurized EVA glove.

Tools should not require forces to be exerted in two directions simultaneously.

Tool design should concentrate on gross motor movements.

Tool designs should not require fine motor movements of the hand.

Tool designs should not require two-handed installation or operation.

Reactionless tools have some advantage for the unrestrained astronaut but are not of great value for an adequately restrained astronaut.

Tethering should be provided during tool operations to satisfy 0 g management requirements.

When possible, the EVA tool should be an integral part of the payload.

Storage should be provided for tools for launch, inflight, and reentry.

Power tools should be considered, especially for repetitive, manual tasks.

Power tools should be compact.

Tools should not present any inherent hazard to the crew member.

Regarding crew performance, it is clear that to accomplish manned space flight mission objectives effectively, crew performance must be maintained at high levels throughout many varied mission activities. These mission activities include prelaunch, launch, intravehicular activity (IVA), extravehicular activity (EVA), routine activities, emergency situations, reentry, and landing. The impact of psychological and sociological factors on crew performance has not received major attention or study in the U.S. space program. A limited amount of information, especially from long-duration USSR space flights, indicates that psychological and sociological factors tend to become more significant with increased residence time in the orbiting spacecraft environment.

With an emphasis on mission-related performance, a basic goal related to psychological and sociological factors is to develop a set of conditions in the spacecraft environment that will enhance crew performance by the minimization of human-induced errors and by the optimization of spacecraft design. There are many interdependent factors that affect crew behavioral performance including environmental factors such as weightlessness and radiation, physiological and psychological adaptation to the space environment, work-station design, cultural elements including values and goals, and personality and social factors. To achieve needed levels of crew performance, these interdependent factors must be considered within the context of the space environment.

With respect to basic sensory, motor, and perceptual processes as related to physical aspects of the space environment, a significant amount of effort has been and is being expended on the sensory and motor aspects of the vestibular system. Concern regarding vestibular function is high owing to the actual or potential impact of space adaptation syndrome on crew performance (see Section 6.5.4).

Studies in the area of visual perception changes in the weightlessness of the space environment are underway. At this time data from studies in space are limited, and the existing human engineering data base as extrapolated to the space environment is the best overall source of information for human engineering design guidance (Davson, 1962; General Electric, undated; Parker and West, 1973; USAF, 1958; Webb, 1964; Woodson, 1981).

Other data relevant to the confined conditions of space travel can be found in studies of groups exposed to confinement and isolation, including studies of long-duration submarine confinement, undersea laboratories, expeditions to polar regions, and space simulation studies (USAF/NSIA/AIAA, 1984).

The research and data available regarding adaptation to the space environment has been primarily on relatively short-term physiological and behavioral adaptation phenomena followed by some degree of stability or maintenance of human performance in the weightlessness of the space environment. Additional human engineering considerations should be given to possible effects on behavior, especially as related to longer-duration adaptation and interactive effects between behavioral factors and physiological adaptation.

Crew selection and training will have to be modified for future space missions. Selection and training of most space flight crew members has developed around a test pilot population. Future space flights will involve a much more heterogeneous flight crew with respect to previous selection, training, physical, and psychological characteristics. Unfortunately, recent and current attempts at preselecting crew members specifically on susceptibility to SAS have not been successful (see Section 6.5.4). However, additional application of approaches of this type are warranted.

REFERENCES

Bailey, R. W. (1983). *Human error in computer systems*. Englewood Cliffs, NJ: Prentice-Hall.

Beranek, L. L. (1954). *Acoustics*. New York: McGraw-Hill.

Beranek, L. L. (1960). *Noise reduction*. New York: McGraw-Hill.

Berry, C. A. Weightlessness. In J. F. Parker, Jr., and V. R. West, Eds., *Bioastronautics data book, 2nd ed.* (pp. 349–415) Washington, DC: NASA. (NASA SP-3006)

Brown, J. W. (1974, August). *Skylab Experience Bulletin No. 6: Space Garments for IVA Wear*. Houston, TX: NASA.

Cakir, A., Hart, D. J., and Stewart, T. F. M. (1982). *Visual display terminals*. New York: Wiley.

Calvin, M., and Gazenko, O. G. Eds. (1975). *Foundations of space biology and medicine*, Vols. 1–30 Washington, DC: NASA.

Dalton, M. (1974, June). *Skylab Experience Bulletin No. 1: Translation modes and bump protection*. Houston, TX: NASA. (JSC-09535)

Dalton, M. (1974a, June). *Skylab Experience Bulletin No. 2: Architectural requirements for airlock*. Houston, TX: NASA.

Dalton, M. (1974b, July). *Skylab Experience Bulletin No. 3: Architectural requirements for sleeping quarters*. Houston, TX: NASA.

Dalton, M. (1974c, July). *Skylab Experience Bulletin No. 4: Design characteristics of the sleep restraint*. Houston, TX: NASA. (JSC-09538)

Dalton, M. (1974d, October). *Skylab Experience Bulletin No. 7: An overview of IVA personnel restraint systems*. Houston, TX: NASA.

Dalton, M. (1974e, December). *Skylab Experience Bulletin No. 9: Foot restraint systems*. Houston, TX: NASA. (JSC-09543)

Dalton, M. (1974f, December). *Skylab Experience Bulletin No. 10: Body restraint systems*. Houston, TX: NASA. (JSC-09544)

Dalton, M. (1975a, January). *Skylab Experience Bulletin No. 11: Personal mobility aids*. Houston, TX: NASA. (JSC-09545)

Dalton, M. (1975b, February). *Skylab Experience Bulletin No. 12: Temporary equipment restraints*. Houston, TX: NASA. (JSC-09546)

Davson, H., Ed. (1962). *The eye*, Vol. 4: *Visual optics and the optical space sense*. New York: Academic.

Ericsson Information Systems (1984). *World conference on ergonomics in computer systems proceedings*. September 27–28, 1984. New York.

Fraser, T. M. (1973). Sustained linear acceleration. In J. F. Parker, Jr., and V. R. West, Eds., *Bioastronautics Data Book*, 2nd ed. Washington, DC: NASA, pp. 149–190. (NASA SP-3006)

Galitz, W. O. (1981). *Handbook of screen format design*. Wellesley, MA: QED Information Sciences, Inc.

Galitz, W. O. (1984). *Humanizing office automation*. Wellesley, MA: QED Information Sciences, Inc.

Garriott, O. K., and G. L. Doerre (1977). Crew efficiency on first exposure to zero gravity. In R. S. Johnston and L. F. Dietlein, Eds., *Biomedical results from Skylab*. Washington, DC: NASA, pp. 155–162. (NASA SP-377)

General Electric Co. (undated). *Handbook of human engineering design data for reduced gravity conditions*. Valley Forge, PA. (NASA cont. NAS 8–18117 and NAS 9–8640)

General Electric Co. (1969). *Human engineering criteria for maintenance and repair of advanced space systems*, Vols. 1, 2, and 3. Valley Forge, PA. (GE Doc. No. 69SD4294; NASA cont. NAS 8–21429)

Graham, C. H., Ed. (1966). *Vision and visual perception*. New York: Wiley.

Graybiel, A. (1975). Angular velocities, angular accelerations, and Coriolis accelerations. In M. Calvin and O. G. Gazenko, Eds., *Foundations of space biology and medicine*, Vol. 2. Washington, DC: NASA, pp. 247–304.

Gunderson, R. F. (1974, September). *Skylab Experience Bulletin No. 5: Inflight maintenance as a viable program element*. Houston, TX: NASA. (JSC-09539)

Johnson, M. L. (1974, October). *Skylab Experience Bulletin No. 8: Cleansing provisions within the waste management compartment*. Houston, TX: NASA.

Johnson, M. L. (1975, January). *Skylab Experience Bulletin No. 14: Personal hygiene equipment*. Houston, TX: NASA.

Johnston, R. S., and L. F. Dietlein, Eds. (1977). *Biomedical results from Skylab*, Washington, DC: NASA. (NASA SP-377)

Jones, W. L. (1975) Individual life support systems outside a spacecraft cabin, space suits and capsules. In M. Calvin and O. G. Gazenko, Eds., *Foundations of space biology and medicine*, Vol. 2. (pp. 193–223) Washington, DC: NASA:

Lewis, J. L. (1982). Crew station design. In *Proceedings of the space human factors workshop*. Washington, DC: NASA, pp. IV-3 to IV-22.

Loftus, J. P., Jr., Bond, R. L., and Patton, R. M. (1975). Astronaut activity. In M. Calvin and O. G. Gazenko, Eds., *Foundations of space biology and medicine*, Vol. 2, Washington, DC: NASA, pp. 600–636.

McCormick, E. J. (1964). *Human factors engineering*. New York: McGraw-Hill.

McDonnell Douglas Astronautics Co. (1984). *The human role in space (THURIS)*, Vols. 1, 2, and 3. Huntington Beach, CA. (NASA cont. NAS 8–35611)

NASA (1978, June). *Crew station specifications*. Houston, TX: NASA. (JSC-07387)

NASA (1976). *Man/Systems Requirements for Weightless Environments*. Huntsville, AL: NASA. (MSFC-STD-512A)

NASA (1982). *Space human factors workshop*. August 24–26, 1982. Washington, DC: NASA.

[National Academy of Sciences] National Research Council (1983). *Video displays, work, and vision*. Washington, DC: National Academy Press.

Parker, J. F., Jr., and West, V. R., Eds. (1973). *Bioastronautics data book*, 2nd ed. Washington, DC: NASA. (NASA SP-3006)

Pestov, I. D., and S. J. Gerathewohl. Weightlessness. In M. Calvin and O. G. Gazenko, Eds., *Foundations of space biology and medicine*, Vol. 2. (pp. 305–354) Washington, DC: NASA.

Peterson, A. P. G., and Gross, E. E., Jr. (1967). *Handbook of noise measurement*. West Concord, MA.

Petrov, Y. A. (1975). Habitability of spacecraft. In M. Calvin and O. G. Gazenko, Eds., *Foundations of space biology and medicine*, Vol. 3. (pp. 157–192) Washington, DC: NASA.

Sova, V. (1975, September). *Skylab Experience Bulletin No. 15: Cable management in zero-G*. Houston, TX: NASA. (JSC-09549)

USAF. *Crew stations and passenger accommodations*, 2nd ed. Wright-Patterson Air Force Base, OH. (AFSC DH 2–2)

USAF (1958, November). *Vision in Military Aviation*. Wright-Patterson Air Force Base, OH. (WADC Rep. 58–399; ASTIA No. 207780)

USAF/NSIA/AIAA (1984). *A space operations workshop*. November 29–December 1, 1983. Colorado Springs, CO: USAF/NSIA/AIAA.

USAF/NSIA/AIAA (1982). *Military space systems technology workshop II*. September 23–25, 1982. Albuquerque, NM: USAF/NSIA/AIAA.

U.S. Department of Defense (1981, May). *Human engineering design criteria for military systems, equipment and facilities*. (MIL-STD-1472C)

Webb, P., Ed. (1964). *Bioastronautics data book*. Washington, DC: NASA. (NASA SP-3006)

Woodson, W. E. (1954) *Human engineering guide for equipment designers*. Berkeley, CA: University of California Press.

Woodson, W. E. (1981). *Human factors design handbook*. New York: McGraw-Hill.

Woodson, W. E., and Conover, D. W. (1964). *Human engineering guide for equipment designers*. Berkeley, CA: University of California Press.

CHAPTER 6.6

ERGONOMIC FACTORS IN CHEMICAL HAZARD CONTROL

ROBERT W. MASON

BARRY L. JOHNSON

National Institute for Occupational Safety and Health
Cincinnati, Ohio

6.6.1 INTRODUCTION

The application of ergonomics or human factors to occupational toxicology (Johnson, 1983) and to the control of chemical exposures would appear to have been more incidental than planned. This premise is based on our inability to find a body of information in the literature that would provide those responsible for controlling chemical exposure with human-factors guidelines with which to work. Konz (1979), in his book *Work Design,* includes an excellent review chapter on toxicology, but brings no application of human factors to control of the toxic hazards; instead, control techniques, exclusive of the worker, that are common to the industrial hygiene literature are reiterated. This exclusion of the worker from control strategies seems to have its base half a century ago in the writings of two eminent occupational physicians: Legge, the first Medical Inspector of Factories, Great Britain, and Hamilton of the United States. Legge (1934) wrote:

> (1) *Unless and until the employer has done everything—and everything means a good deal— the workman can do next to nothing to protect himself, although he is naturally willing enough to do his share.*
>
> (2) *If you can bring an influence to bear external to the workman (i.e., one over which he can exercise no control) you will be successful; and if you cannot or do not, you will never be wholly successful. "*

Legge provided some examples of the kinds of external influences he had in mind—substitution of innocuous substances for toxic ones, performance of operations in closed systems, and local ventilation. Examples of controls he enumerated as being not completely effective (because they involve human factors) included respirators, gloves, goggles, and washing conveniences.

Hamilton (1934) wrote that prevention of industrial poisoning is primarily the task of the engineer and that it must never be forgotten that the great majority of the industrial poisons enter the body with the inspired air, and therefore the preventive measures must be planned with a view to keeping the air of the workshop free from poison. She then briefly summarized the preventive measures as follows:

1. Prevent formation or escape of gases, fumes, and dust.
2. If (1) is impossible, remove by means of exhausts.
3. If (2) is impossible, dilute as much as possible by abundant ventilation and by fans blowing air past the face of the worker.
4. If all these are impracticable, protect the worker by a filter mask, absorbent gas mask, or positive pressure air mask.
5. If all the above measures are impracticable, select workers with proved resistance to the poison and put them on short work shifts.
6. For poisons that enter through the skin, prevent escape of fumes and dusts and prevent spilling of liquids.
7. Maintain scrupulous cleanliness of benches, apparatus, and so forth.
8. Provide and launder washable working clothes.
9. For all, provide adequate washing facilities and a clean lunchroom. Workers with poisons that enter through the skin and workers exposed to excessive dust in the air should have shower baths provided, with hot water, soap, and towels.

The control methods advocated by these authors have become the theme in most industrial hygiene literature and the major basis of controlling the hazards of toxic chemicals. Thus Schilling and Hall (1973), Peterson (1977), and Olishifski (1979) discuss control of chemical hazards in terms of substitution, total enclosure, locally applied exhaust ventilation, dust suppression, segregation of processes, general ventilation, general cleanliness, personal hygiene, personal protection, and education, in approximate descending order of preference.

Historically, then, the control of chemical hazards has been placed in the hands of engineers charged primarily with preventing inhalation by enclosure and by ventilation, methods that are commendable because they do "bring influences to bear external to the workman" as Legge advocated.

These methods, which usually involve large capital outlays, are most applicable to routine operations, and thus their application is common to factories, but there remain many chemical exposures that are inadequately controlled because ergonomists and human-factors engineers have not been involved. In this chapter we identify human factors that are important to occupational toxicology, and provide suggestions and examples of how they can be applied to preventing occupational disease.

6.6.2   OCCUPATIONAL TOXICOLOGY AND HUMAN FACTORS

For a toxic substance to have an effect, it must come in contact with the body in sufficient amount and for sufficient time either to affect the tissues it contacts directly or to be absorbed into the body where it can have general, systemic toxic effects.

Because toxic substances first contact the outside of the body we first discuss the effects of contact on the surfaces and then absorption across these surfaces and suggest roles for ergonomists and human-factors engineers in intervention. The outside of the body includes the skin, nails, eyes, teeth, oral passages, nasal passages, upper airways, lungs, and digestive tract. Each of these structures has a history of being affected by exposure to toxic substances in solid, liquid, aerosolized, or gaseous form.

6.6.2.1   Effects of Toxicants on the Skin, Eyes, and Oral Structures

Skin diseases are the most frequently occurring of all occupational diseases. Because the skin is the major area of the body in direct contact with the environment, it is subjected to chemicals in all forms—gases, liquids, and solids present in the air and on tools and working surfaces, in addition to those that workers handle directly. As a result of this substantial body contact, chemical exposures contribute significantly to the totality of occupational skin disorders.

The skin may suffer damage from a variety of substances, including acids, alkalis, halogens, and others. It is damaged as fat solvents remove protective oils; abraded by fibers, metals, and dusts; allergenically sensitized by numerous and diverse substances; and dehydrated. Some substances are keratogenic and carcinogenic to the skin. The hands, including the nails, are particularly vulnerable when they are used as tools, especially by artisans, craftsmen, repairmen, and many others who take advantage of their sense of touch and maneuverability.

Ergonomists and human-factors engineers can play a substantial role in the reduction or elimination of effects of chemicals on the skin through design of jobs, tools, and protective materials (clothing, gloves, and creams) to be used by workers.

Eyes are affected directly by chemical substances that are present in the air as well as by those liquids and solids that get directly into the eye. Substances in the air are often irritating to the eye, and this irritation may serve as a useful indicator or warning to the worker of the presence of substances in the air. Eye irritation has served as the basis for limiting the concentration of many substances in the work environment. Eye irritants, if present at high enough concentration or for sufficient duration, cause tearing and impair vision. The eye is another important tool that the worker uses, and this use places the eye in many vulnerable situations. Human-factors engineers have a large role to play in designing work stations, jobs, and equipment that will reduce the vulnerability of the eye to chemical hazards. Two other areas where human-factors technology can be brought to bear are in the design of protective eyewear and in promoting its use.

The mouth and its parts, including the teeth, have also been used as tools, for example, in pipetting, licking stamps and envelopes, wetting the fingers for turning pages, holding nails, and biting threads, and in this capacity have been exposed to toxic substances. Toxic substances inhaled through the mouth may be dissolved in saliva and mucus, and particulate material may be deposited on the surfaces. Teeth have been directly affected by contact with dusts, acids, and alkalis while breathing through the mouth and while talking, an important occupational activity that has relevance to respirator usage.

6.6.2.2   Effects of Toxicants on Respiratory Passages and Lungs

During normal breathing, most air, 2 to 25 l/min, that enters the body passes through the nasal passages where large particles are deposited, resulting in such effects as nasal cancers in workers exposed to dust in nickel refining, and ulceration and perforation of the nasal septum with exposure to dusts containing mercury, arsenic, and chromium.

Air entering the nasal passages also brings into play the sense of smell, which has been used as a tool to identify chemicals, and can be useful for warning the worker of impending danger for many toxic substances. It is well to note, though, that adaptation, or fatigue, occurs and perception of odors disappears during continuous exposures to certain substances. A consequence of some chronic chemical exposures is permanently diminished or lost sense of smell.

Very reactive chemicals such as ozone, formaldehyde, and chlorine are irritating to the nasal mucosa, and this irritating property can serve as a basis for limiting exposure concentrations as well as warning the worker of impending hazards. Other manifestations of exposure of the nasal passages include tissue congestion, inflammation, and nasal discharge.

The upper airways, trachea, bronchi, and lungs suffer a number of effects from toxic chemicals and dusts, including smooth muscle responses to irritants, sensitization with asthmatic response, chronic bronchitis, emphysema, cancer, and a variety of agent-specific diseases such as silicosis (silica in the lung), asbestosis, and byssinosis (cotton dust in the lung), all of which compromise respiratory function. For example, silicosis and asbestosis induce inflammation, fibrosis, and scarring, thereby reducing elastic properties of the lung and, in many cases, gas exchange.

Under many conditions of work the only way, other than removing the worker from the workplace, to diminish or prevent inhalation of toxic substances is by the use of respirators. Many ergonomic and human-factors issues become involved—comfort, visibility, loss of sense of smell (e.g., an odorless environment), maneuverability (particularly with air-line and self-contained breathing apparatuses), facial characteristics, appearance, perception of need, increased work of breathing, and with self-contained breathing devices, increased work of carrying the device.

6.6.2.3 Effects of Toxicants on Internal Organs

Toxic effects involving the internal organs and tissues occur after absorption of the chemical substance into the body tissue and their magnitude is usually proportional to the absorbed dose. Thus prevention of absorption is the key to eliminating all such effects. Rather than summarizing toxic effects, we present what is known about absorption of toxic substances and suggest ways that this information can be used for ergonomic and human-factors intervention.

6.6.3 Human Factors and Absorption of Toxic Substances

Absorption occurs across all membranes, but at different rates, these rates depending on the anatomy of the membrane, whether active transport is present or not, and upon characteristics, such as solubility and volatility, of the substance being absorbed. In general, the lungs, with about 70 m² of surface area, and the gastrointestinal tract, with about 10 m² of surface area, are anatomically and physiologically adapted to absorption, whereas the skin, with a surface area of about 2 m², is not.

The lungs have a history of being considered as the major site for absorption of toxic chemicals by workers. Consequently more information on occupational toxicology is based on inhalation than on other routes of absorption; for this reason absorption from the lungs must serve as the standard for evaluating absorption from the skin and the digestive tract.

6.6.3.1 Factors Associated with Absorption from the Lungs

The amount of toxic substance that can be absorbed from the lungs is related to the concentration of that substance in the inhaled air and to the volume of air breathed. Concentrations of substances in the air of workplaces are usually expressed in parts per million (ppm) or milligrams per cubic meter (mg/m³); in this chapter, the latter is used. In the United States and many other countries, the major source of information on acceptable concentrations of substances in workroom air is the American Conference of Governmental Industrial Hygienists (ACGIH). For nearly 50 years, that organization has been recommending Threshold Limit Values for substances in workroom air (TLVs®). Currently there are TLVs for over 600 chemical substances (ACGIH, 1984). The TLVs are the 8 hr time-weighted average concentrations that will protect nearly all people exposed. For many substances, there is also a short-term exposure limit (STEL), which is the 15 min time-weighted average exposure that should not be exceeded at any time during a work day; exposures at the STEL should not be repeated more than four times a day, with at least 60 min between exposure periods. For about three dozen substances, a "ceiling limit," which should not be exceeded even instantaneously, is recommended instead of a TLV or STEL. The TLVs range from 0.00006 mg/m³ for proteolytic enzymes to 9000 mg/m³ for carbon dioxide, with the majority in the range of 0.1 to 10 mg/m³ (Table 6.6.1).

Table 6.6.1 summarizes the TLV's (in mg/m³ ranges) and provides some interpretation of these for moderate and heavy work in terms of the amounts that would be available over a workday for absorption by the lungs. Approximate equivalent volumes for liquid materials are presented in metric units which are referred to units that may be more easily visualized by some readers.

The important point is that, for all substances with TLV's less than 1000 mg/m³, the permissible daily dose is not large (less than 2 tablespoons), and for the more toxic substances, those below 1 mg/m³, it is very small (less than 1 drop). It is also well to keep in mind that these amounts represent what might be presented to the lungs for absorption and that the amount absorbed may be substantially less (Table 6.6.2).

Table 6.6.2 presents experimental data obtained from human subjects exposed for 30 min to solvent vapors at various concentrations and at various workloads. The exposure concentration is expressed in milligrams per cubic meter; the amounts of vapor presented to the lung (inhaled) and absorbed are expressed in milligrams. It is worth noting the proportionality of the amount of substance absorbed to the level of work (air breathed) and to the concentration of substance in the air. Thus for xylene, doubling the concentration doubles the amount absorbed; and increasing the air breathed fivefold

Table 6.6.1   Summary of TLVs and Potential for Daily Absorption from the Lung at Light and Heavy Work

| TLV Range (mg/m³) | Number of Substances | Potential for Absorption/Day | | | |
| | | Light Work, 8 Hr | | Heavy Work, 10 Hr | |
		Weight	Volume[a]	Weight	Volume[a]
≤0.1	59	1 mg	1 μl	3 mg	3 μl
≤1	129	10 mg	10 μl	30 mg	30 μl
≤10	162	100 mg	100 μl	300 mg	300 μl
≤100	88	1 g	1 ml	3 g	3 ml
≤1000	82	10 g	10 ml	30 g	30 ml
≤10000	37	100 g	100 ml	300 g	300 ml

[a] Volume at specific gravity of 1.0. The specific gravities of most of the liquid substances in the TLV list are less than 1.5 and range from 0.61 for pentaborane to 3.11 for bromine; for the more dense liquids, the equivalent volumes would be $\frac{1}{2}$ to $\frac{1}{4}$ of those listed, and for the least dense ones about 1.5 times those listed. For comparison, 1 drop is about 50 μl; 1 tsp is about 5 ml; 1 tbls is about 15 ml.

increases the amount absorbed fivefold. An example of how the generalizations made in Table 6.6.1 can be applied when specific data (Table 6.6.2) are not available can be demonstrated with butyl alcohol exposure of 300 mg/m³ at light work and at heavy work.

Because the data in Table 6.6.2 are for 30 min exposures they must be multiplied by 16 for an 8 hr day and by 20 for a 10 hr day. Thus 16 × 220 mg gives about 3.5 g inhaled, compared to the generalization (Table 6.6.1) of 3 g for light work over an 8 hr day; 460 mg × 20 gives about 9.2 g inhaled at heavy work for 10 hr compared to the generalization in Table 6.6.1 of 9 g. About 40% of the inhaled butyl alcohol was absorbed, indicating daily doses of about 1.5 and 3.5 g ($\frac{1}{4}$ to $\frac{1}{2}$ teaspoon) at light and heavy work, respectively.

The information provided in Tables 6.6.1 and 6.6.2 provides a basis for ergonomists and human-factors engineers to enter into the field of controlling toxic occupational hazards. First, these data show that measuring only the concentration in the inhaled air does not provide all the information necessary for evaluating a toxic hazard. Measures of work and of the volume of air breathed may be equally important. Second, the data suggest that there may be many opportunities where evaluating the work situation and redesigning the job could reduce the amount of toxic substance absorbed by as much as would be accomplished by more costly, traditional control techniques that would more than halve the time-weighted average exposure. An example would be installation of conveyer belts to eliminate carrying jobs.

6.6.3.2   Factors Associated with Percutaneous Absorption of Liquids

Absorption of toxic substances through the skin (percutaneous absorption) is recognized by the ACGIH (1984) by attaching a "Skin" notation to the TLV for those substances for which there is "potential contribution to the overall exposure by the cutaneous route, including mucous membranes and eyes, either by airborne, or more particularly, by direct contact with the substance." About 150 substances have the "Skin" notation; most of these substances are liquids, about a third are solids, and a small number are gases (ACGIH, 1980).

The reason for assigning the "Skin" notation to the TLVs is not stated by ACGIH (1980) in about a third of the cases, is based on the LD_{50} (dose that will kill 50% of the animals) for rats or rabbits for another third, and is based on other animal data (sensitization; chronic effects) or human experience (experimental or occupational) for the remainder. Another list, the *Registry of Toxic Effects of Chemical Substances* (Tatken and Lewis, 1983) contains over 1600 substances with an LD_{50} from skin application. Many of these may not be included in the TLV list because they do not have significant industrial use, or the LD_{50} may be so high (several grams per kilogram) as to not pose a significant health hazard from skin absorption. Although quantitative data on rates of absorption of liquids and solids applied to the skin of human subjects would be useful as a basis for assigning the skin notation to TLV's and for evaluating occupational skin exposure, they are sparse (Table 6.6.3).

The area of skin most likely to be exposed in work situations is the hands (about 500 cm² each), although in some operations, and especially in accidents, other and larger areas such as the abdomen and legs become exposed. The absorption rates listed in Table 6.6.4 are mostly for the forearm or

Table 6.6.2 Amounts of Various Chemicals Inhaled (I) and Absorbed (A) During 30 Minutes, at Different Work Levels

Substance	Exposure (mg/m³)	Rest 10 l/min I (mg)	Rest 10 l/min A (mg)	Light 25 l/min I (mg)	Light 25 l/min A (mg)	Medium 38 l/min I (mg)	Medium 38 l/min A (mg)	Heavy 54 l/min I (mg)	Heavy 54 l/min A (mg)	Source
Acetone	740	[261]	115	[495]	228	[737]	339	[1024]	471	Astrand, 1983
Butyl alcohol	300	90	43	220	81	326	130	460	193	Astrand, Ovrum, Lindqvest, and Hultengren, 1976
n-Hexane	600	171	78	406	159					Veulemans, van Vlem, Tanssens, Masschelein, and Leplat, 1982
	360	[105]	[25]	[302]	[47]					
Methyl alcohol	720	[225]	[48]							Sedivec, Mraz, and Flek, 1981
	250	[80]	[45]							
	300	[105]	[58]							
Methylene chloride	870	241	132	676	300					Astrand, Ovrum, and Carlsson, 1975
	1740	479	275	1270	565	1996	574	2781	637	
Styrene	210	57	39	146	95	229	145	355	210	Astrand, 1983
	630	232	140							Astrand, Kilbom, Ovrum, Wahlberg, and Vesterberg, 1974
	1050	374	235							
	1470	503	309							
Toluene	300	93	48	219	107	351	142	517	152	Carlsson, 1982
	375	[100]	72	[300]	174	[450]	228	[650]	277	Carlsson and Lindqvist, 1977
Trichlorethylene	540	155	81	390	179					Astrand and Ovrum, 1976
	1080	322	186	749	378	1164	418	1672	419	
White spirit	1000	293	179							Astrand, Kilbom, and Ovrum, 1975
	1250	321	195	880	357					
	1500	371	182							
	2000	522	252							
	2500	540	344	1690	816					
Xylene	435	126	79	343	219	516	301	762	392	Astrand, Engstrom, and Ovrum, 1978
	870	242	166	618	407					

[a] Brackets indicate calculations from data in the source cited.

Table 6.6.3   Absorption Rates of Liquids through Skin of Human Subjects

Substance	Absorption Rate (μg/cm$^2 \cdot$ min)	Source
Aniline		
In water, 10 g/l	5	Baranowska-Dutkiewicz, 1982
In water, 20 g/l	20	
Aniline	50	Baranowska-Dutkiewicz, 1982
	5	Piotrowski, 1957
Benzene	7	Hanke, Dutkiewicz, and Piotrowski, 1961
	150	Baranowska-Dutkiewicz, 1982
Carbon disulfide	150	Baranowska-Dutkiewicz, 1982
Ethyl benzene	466	Dutkiewicz and Tyras, 1969
Ethylene glycol dinitrate	7	Gross, Kiese, and Resag, 1960
Methyl alcohol	200	Dutkiewicz, Konczalik, and Karwacki, 1980
Methyl n-butyl ketone	6	Divincenzo, Hamilton, Kaplan, Krasavage, and O'Donoghue, 1978
Phenol		
In water; 2.5 g/l	1	Baranowska-Dutkiewicz, 1981
In water; 5.0 g/l	3	
In water; 10.0 g/l	5	
Styrene	200	Dutkiewicz and Tyras, 1969
Toluene	300	Dutkiewicz and Tyras, 1969
Xylene	100	Dutkiewicz and Tyras, 1969
m-Xylene	2	Engstrom, Husman, and Riihimaki, 1977

the entire hand. As reported by Maibach, Feldmann, Milby, and Serat (1971) and Guy and Maibach (1984), these rates are likely to be lower than for other parts of the body, on the average about half. In addition to absorption rates being different for different parts of the body, absorption rates can be tremendously increased (200-fold) for damaged skin (Ilyin, Ivannikov, Parfenov, and Stolyarov, 1975).

Thus estimating the dose of toxicant that may be absorbed from skin exposure requires an evaluation of the rate at which the chemical may be absorbed; the area and type of skin involved, and its integrity; duration of the exposure; and volatility of the material. Duration of exposure and volatility of the compound must be considered together; volatile compounds such as benzene (vapor pressure at 20°C = 75 mm Hg) evaporate from exposed skin in minutes after a splash or removal of the hands from the liquid, whereas less volatile compounds, such as ethylene glycol dinitrate (vapor pressure at 20°C = 0.05 mm Hg), remain on the skin until absorbed or washed off. Thus estimates of the duration of exposure, or time during which absorption can occur, include observed duration of contact with the liquid plus an estimate of the residual contact time after cessation of the observed contact. The amount of residual material remaining on the skin, for example, after removal of the hands from a solution will be quite different for different materials depending on their physical characteristics; immediately after removal of a hand from a beaker of water or ethanol as much as 4 mg/cm^2 may remain on the hand, whereas with light paraffin oil, about twice this much will remain.

As an example of how the data presented in Tables 6.6.1 to 6.6.3 can be utilized to evaluate a

Table 6.6.4   Liquid Contact (cm$^2 \cdot$ min) Required to Absorb Daily Doses Implied by the TLV at Light Work[a]

Absorption Rate (μg/cm$^2 \cdot$ min)	TLV (mg/m^3)				
	0.1	1.0	10	100	1000
1	600	6000	60,000	600,000	
10	60	600	6,000	60,000	600,000
100	6	60	600	6,000	60,000
1000	0.6	6	60	600	6,000

[a] Assumes 50% of amount inhaled is absorbed.

skin exposure problem, let us take the case of methyl alcohol, a solvent likely to be present in preparations for removing paint.

Methyl alcohol is one of the substances with the "Skin" notation, and it has a TLV of 260 mg/m^3 (ACGIH, 1984). According to Table 6.6.1 it would fall in the TLV range ≤ 1000, which indicates the potential for respiratory absorption during 8 hr at light work· at the TLV would be (by linear extrapolation) 2.6 g (or approximately 2.6 ml). Because the data in Table 6.6.2 suggest that, in general, about half of this actually would be absorbed, a reasonable expected daily dose suggested by the TLV would be about 1.3 g; also some data are available in Table 6.6.2 on methyl alcohol absorption at rest to refine this estimate of expected daily dose. If the 45 mg listed for absorption during 30 min is extended linearly to 8 hr, the absorbed dose would be 720 mg which, when multiplied by 2 for correction to light work (which the absorption data in Table 6.6.2 suggest is reasonable), gives 1.4 g compared to the rougher estimate of 1.3 g for the allowable daily dose. The information available in Table 6.6.3 on the rate of absorption of methyl alcohol through the skin is about 200 $\mu g/cm^2 \cdot min$.

If we divide 1.3 g by 200 $\mu g/cm^2 \cdot min$ we find that 6500 $cm^2 \cdot min$ of exposure, or one hand submerged in methyl alcohol for 13 min, would be equivalent to working all day in an atmosphere containing methyl alcohol at its TLV. Generalizing the logic of this example and assuming the allowable daily absorbed dose of a toxicant is half that inhaled at light work at the TLV for 8 hr, Table 6.6.4 was constructed, which gives the $cm^2 \cdot min$ of exposure of the skin to liquids that would result in percutaneous absorption of the daily dose so implied by the TLV. The first entry in Table 6.6.4 was calculated as follows:

If	the TLV is 0.1 mg/m^3, the respiration rate at light work is 25 l/min = 0.025 m^3/min, the workday is 8 hr = 480 min, and the amount absorbed is half that inhaled (0.5)
Then	the absorbed dose is $0.1 \times 0.025 \times 480 \times 0.5 = 0.6$ mg $= 600$ μg
If	the absorption rate is 1 $\mu g/cm^2 \cdot min$
Then	the $cm^2 \cdot min$ of skin exposure required to absorb that dose is $600/1 = 600$ $cm^2 \cdot min$

For order of magnitude interpretations of Table 6.6.4, consider that the area of a thumb is about 15 cm^2, a hand, about 500 cm^2, and the whole body, about 20,000 cm^2; then 60 $cm^2 \cdot min$ is equivalent to holding the thumb in liquid for 4 min; 600 and 6000 $cm^2 \cdot min$ are about equal to holding the hand in liquid for 1 and 10 min, respectively; the highest value tabulated in Table 6.6.4, 600,000 $cm^2 \cdot min$, would be equivalent to whole-body immersion for 30 min, an unlikely situation, but one that might be approached by soaked clothing from a large spill if the clothes were not removed.

Prevention of absorption of liquids through the skin can be accomplished by redesigning equipment and jobs to eliminate the exposure and by providing effective barriers. Coletta, Schwope, Arons, King, and Sivak (1978); Mikatavage, Que Hee, and Ayer (1984); and Stamper, McLeod, Betts, Martinez, and Barardinelli (1984a, 1984b) reported on laboratory evaluations of protective clothing materials for resistance to chemicals. Einert, Adams, Crothers, Moore, and Ottoboni (1963); Guillemin, Murset, Lob, and Riquez (1974); Lauwerys, Dath, Lachapêlle, Buchet, and Roels (1978); Lauwerys et al. (1980); Ishihara, Kanaya, and Ikeda (1976); Hogstedt and Stahl (1980); Fukuchi (1981); and Maxfield, Barnes, Azar, and Trowchimowicz (1975) give information on effectiveness of gloves and barrier creams in laboratory and workplace exposures.

6.6.3.3   Factors Associated with Percutaneous Absorption of Solids

Quantitative data on absorption rates of solid materials applied to the skin were not found, even though about 50 solid substances are included among the TLVs with the "Skin" notation. However, information on amounts of material deposited on the skin of workers is available (Table 6.6.5). In the study of lead arsenate exposure (Wojeck, Nigg, Bramen, Stamper, and Rouseff, 1982), the skin contamination was reported in units of micrograms per square centimeter; when about 20,000 cm^2 of body surface was involved, the total whole-body contamination was up to 1200 mg for lead arsenate. Because the TLV for lead arsenate is 0.15 mg/m^3 (ACGIH, 1984), the maximum acceptable daily respiratory exposure (at heavy work for 10 hr) would be about 4.5 mg (linear extrapolation from Table 6.6.1); thus if only about 0.03% of the whole-body contamination was absorbed (which could be accomplished over 8 hr at an absorption rate of only 0.0005 $\mu g/cm^2 \cdot min$), the maximum acceptable absorption from the respiratory tract at the TLV would likely be exceeded.

In the studies summarized in Table 6.6.5, inhalation exposure was minimal, usually because respirators were worn, and absorption from skin contamination was demonstrated by analysis of body fluids. It is important to note the relatively low concentrations (50 $\mu g/m^3$) found in the air in these studies.

In extremely dusty atmospheres where respirators are commonly worn and concentrations of dusts may be in units of milligrams per cubic meter, whole-body contamination must be several times the

Table 6.6.5   Contamination of the Skin of Workers after Exposure

Substance	Condition of Work	Air Concentration (μg/m³)	Duration (hr)	Skin Contamination (μg/cm²)	Source
Benzidine	Production	0.8	Workday	3–7	Krajewska, Adamiak-Ziemba, Suwalska, and Koskecka, 1980
Azinphosmethyl	Crop spraying	50	2.5–9	0.01–0.05	Franklin et al., 1981
Lead arsenate	Crop spraying	2–5	10	30–60	Wojeck et al., 1982

worst case condition presented in Table 6.6.5. Thus requirements for clean clothes, showers, and protective clothing are obvious when dusts and aerosols are involved.

6.6.3.4   Factors Associated with Percutaneous Absorption of Vapors and Gases

As pointed out by the ACGIH, few quantitative data are available describing absorption of vapors and gases through the skin, but protection of the respiratory tract, while leaving the rest of the body exposed, particularly to a substance with a low TLV, may present a hazardous situation. Absorption of vapors and gases through the skin depends on the concentration in the air, the area of body surface exposed (usually all of it), the temperature and humidity of the air, and characteristics of the substance being absorbed. The little quantitative information available from experimental exposures of human subjects is presented in Table 6.6.6, along with the concentrations of the substances at the TLV and in saturated air (at 20°C) and an absorption factor calculated by dividing the amount absorbed by the product of exposure time and exposure concentration.

This absorption factor [in units of μg/min·mg/m³)] can be used for estimating the doses of gases and vapors absorbed through the skin, if the exposure time and the exposure concentration are known (assuming that the total surface area of the skin is exposed and that, for practical purposes, differences between persons in skin surface area can be ignored). Although the data in Table 6.6.6 for aniline may suggest otherwise, those for nitrobenzene, phenol, and xylene suggest that the absorption factor is independent of substance concentration in the air, which is consistent with the premise that the absorption rate is proportional to the concentration of the substance in air.

If linearity is assumed, then the min·mg/m³ of whole-body exposure to a toxicant while wearing a respirator that might result in absorption of amounts of that toxicant suggested by the TLV can be calculated. Let us take the case of styrene, for which the TLV is 215 mg/m³. At light work, for an 8 hr exposure at that concentration, a worker might inhale 2580 mg of styrene (0.025 m³/min × 480 min × 215 mg/m³) and absorb about half, or about 1300 mg.

If a worker was working in an area where styrene in excess of 215 mg/m³ and wearing a respirator, one might ask "is more protection needed, or how long can the exposure last?" The percutaneous absorption factor for styrene vapor (Table 6.6.6) is 0.1 μg/(min·mg/m³). Dividing 1300 mg = 1,300,000 μg by 0.1 μg, gives 13,000,000 (min × mg/m³), which could be attained by working in a saturated atmosphere (about 14,000 mg/m³) for about 16 hr (960 min). The practical conclusion would be that there would be little need for whole-body protection when working with styrene in a saturated atmosphere.

The logic of this example has been used to construct Table 6.6.7, which gives the min·mg/m³ of exposure of the skin that might result in daily doses suggested by the TLVs.

The first entry in Table 6.6.7 was calculated as follows:

If   the TLV is 0.1 mg/m³, the respiration rate at light work is 25 1/min = 0.025 m³/min, the workday is 8 hr = 480 min, and the amount absorbed is half that inhaled (0.5)

Then   the absorbed dose is 0.1 × 0.025 × 480 × 0.5 = 0.6 mg = 600 μg

If   the whole-body absorption factor is 0.01 μg/(min·mg/m³)

Then   the min·mg/m³ of skin exposure required to absorb that dose is 600/0.01 = 60,000 min·mg/m³

In Table 6.6.6 a tremendous difference in absorption is noted when phenol, nitrobenzene, and aniline are compared with the other compounds; for the former group, at exposure concentrations of 5 to 30 mg/m³, similar amounts are absorbed as with the remaining substances at 1000 to 4000 mg/m³.

The experimentally absorbed doses of aniline at 50 mg/m³ and nitrobenzene at 30 mg/m³ were in excess of the amounts that might be absorbed due to inhalation during light work at the current TLVs of 18 and 5 mg/m³, respectively. Although the experimental exposure concentrations outlined

Table 6.6.6 Vapor Absorption (*A*) Through the Skin of Human Subjects

Substance	Exposure (mg/m³·min)		A (mg)	Absorption Factor μg/(min·mg/m³)	TLV (mg/m³)	Saturated Air (g/m³)	Source
Aniline	5	300	15	10	8	4	Dutkiewicz, 1961
	10	300	20	7			
	20	300	30	5			
	50	300	60	4			
Benzene	1000	420	10	0.02	30		Hanke et al., 1961
Methylchloroform	3300	210	2	0.003	1900	902	Riikimaki and Pfaffli, 1978
Nitrobenzene	5	360	7	4	5	3	Piotrowski, 1967
	10	360	14	4			
	30	360	54	4			
Perchloroethylene	4100	210	48	0.05	335	1000	Riikimaki and Pfaffli, 1978
Phenol	5	360	9	5	19	2	Piotrowski, 1971
	9	360	14	4			
	25	360	47	5			
Styrene	2500	210	60	0.1	215	36	Riikimaki and Pfaffli, 1978
Toluene	2300	210	26	0.05	375	148	Riikimaki and Pfaffli, 1978
Xylene	1300	210	21	0.08	435	555	Rikkimaki and Pfaffli, 1978
	2600	210	44	0.08			

Table 6.6.7   Gas or Vapor Contact (Min·mg/m³) with Whole-Body Skin Surface (2 m²) Required to Absorb Daily Doses Implied by the TLV at Light Work[a]

Absorption Factor [g/(min·mg/m³)]	TLV (mg/m³)				
	0.1	1.0	10	100	1000
0.01	60,000	600,000	6,000,000[b]		
0.1	6,000	60,000	600,000	6,000,000	
1	600	6,000	60,000	600,000	6,000,000
10	60	600	6,000	60,000	600,000

[a] Assumes 50% of amount inhaled is absorbed.

[b] 15,000 mg/m³ for 400 min (about 1 day).

in Table 6.6.6 were usually in excess of the current TLV's, they were substantially less than the respective saturated air concentrations, which are also presented in Table 6.6.6, and which workers might encounter in confined spaces.

Tank cars, storage tanks, silos, sewers, and spaces over open containers are examples of confined spaces that gases may completely occupy, displacing all natural air constituents, or in which the air can be saturated with volatile materials. Consequences of inadvertent entry into such spaces, or of failure of protective equipment, include suffocation (insufficient oxygen), anesthesia (inability to escape), and serious acute poisonings.

The kinds of data needed to determine the amount of protection necessary when entry into confined spaces has to be made have been the subject of this section. When whole-body protection is necessary, ergonomic and human-factors considerations include body-heat production and its removal while wearing the protective clothing; added work associated with wearing the protective clothing and respiratory equipment; visibility; durability of the clothing in regard to the job (e.g., susceptibility to damage from mechanical hazards); and compatibility of the equipment with the job to be done, considering such things as needed maneuverability and restriction of movements imparted to the worker by the protective equipment.

Certainly, insignificant percutaneous absorption of gases and vapors cannot be taken for granted and must be seriously considered, along with appropriate whole-body protection, when respirators are the chosen or necessary control method.

6.6.3.5   Factors Associated with Ingestion of Toxicants

The gastrointestinal tract is involved incidentally in occupational toxicity—substances in the workplace may be accidentally, and sometimes intentionally, ingested in large amounts, but the most common sources of ingestion are from materials entering the mouth from dirty hands or from contaminated food and drink, and from swallowing materials brought up from the lungs by ciliary action and coughing. Common industrial hygiene practices used to prevent ingestion of toxic substances include provision of washing facilities, eating places that are isolated and protected from contamination, and requirements that food and beverages not be taken into the workplace, and indirectly by limiting the inhalation of dusts. Aitio, Jarvisalo, Kilunen, Tossavainen, and Vaittinen (1980) demonstrated that ingestion from deposition on the oral structures of particles too large to be inhaled may also be a problem.

Human factors that are involved include the necessity to eat and drink, habits such as use of tobacco, and the natural tendencies to lick the lips and fingers. At mobile and temporary work stations, such as construction sites, particularly building repair sites, road repair sites, and agricultural work sites, constructing facilities to provide protection is an impossible or impractical consideration. In such cases, involvement of the worker is imperative.

The potential hazards from ingestion can be put into perspective by examining daily doses of systemic poisons that can be absorbed under current occupational exposure limits for substances in the air that workers breathe.

Examination of the TLVs reveals a limit of 1900 mg/m³ for ethyl alcohol and a limit of 0.41 mg/m³ for disulfoton. If we consider that about 10 m³ of air would be inhaled at light work during a workday and that about half of the amount of each toxic substance inhaled would be absorbed, we see that in the first case, it would be necessary to ingest about 10 ml (1900 mg/m³ × 10 m³ × 0.5 = 9500 mg) to equal the daily dose suggested by the TLV, whereas in the latter case ingestion of only 2 μl (0.41 mg/m³ × 10 m³ × 0.5 = 2 mg) would suffice. With few exceptions, all substances for which there are TLVs lie between these extremes, with the majority in the range representative of absorbed doses of less than 5 drops. Pesticides and many metals and their salts are among the substances that have TLVs of 1 mg/m³ or less.

Thus, although early writers on occupational toxicology did not consider ingestion to be a major concern, the substances about which they wrote were generally of lower toxicity than many of the chemicals in use today, the problem of air contamination was substantially greater then, and the criteria of toxicity were different.

6.6.4 SUMMARY OF OCCUPATIONAL TOXICOLOGY

The preceding sections of this chapter have presented information bearing on the effects of substances contacting body surfaces and the absorption of toxicants. We have described the principal routes by which absorption can occur and provided a basis for evaluating resulting doses. Inhalation of toxicants, contact with the skin, and ingestion were discussed. Factors that influence the amount of material absorbed for each of these three routes were detailed. Of particular importance, in a human-factors sense, is absorption of toxicants through the skin, and data were presented that permit human-factors specialists to estimate the amount of toxicants absorbed percutaneously. These calculations will be useful to those designers charged with the design of personal protective equipment, in particular.

In the course of describing how toxic substances can be absorbed, we cited, or in some instances alluded to, human factors that affect why, how, and to what extent persons are occupationally exposed to toxicants. But nothing is as instructive as examples from actual workplace experience. For this reason, the next section presents case studies selected from the Hazard Evaluation program of the National Institute for Occupational Safety and Health. The examples are given in the form of case studies. Each was selected to illustrate opportunities for human-factors practitioners to control workplace exposures to toxic substances.

6.6.5 CASE STUDIES

The opportunities a human-factors specialist has in preventing, or reducing, workers' exposure to toxic substances are considerable, but are sometimes unrecognized. Some examples, presented here as case reports, from the National Institute for Occupational Safety and Health (NIOSH) portray areas where human factors can reduce toxic exposures. The first two case studies were selected from the NIOSH Health Hazard Evaluation program, an activity conducted under the authority of the Occupational Safety and Health Act of 1970. Section 20(a)(6) of the Act authorizes the Secretary of Health and Human Services, following a written request from any employer or authorized representative of employees, to determine whether any substance normally found in the place of employment has potentially toxic effects in such concentrations as used or found. The third case study was selected from a research project sponsored by NIOSH. Each of these case studies is developed here in terms of the problem that was investigated, how the health hazard evaluation (HHE) or research project was conducted, the findings, and areas where a human-factors specialist could contribute to reducing the hazards being investigated.

6.6.5.1 Case Study of a Hazardous Waste Site

The problem of how to dispose of hazardous waste arrived when the United States moved from an agrarian to an industrial economy. A common method has been to store the wastes in hazardous waste sites. In recent years a great concern has been expressed by the public as to the risk posed by these sites. This public concern has been translated into legislative actions that require government and private industry to clean up chemical waste dumps. The first case study is concerned with this problem. In 1980 the Department of Health, State of New Jersey, requested that NIOSH provide on-site industrial hygiene consultations during the removal of hazardous waste from a commercial facility (Costello and Melius, 1981). A 2-acre site had been placed under state jurisdiction in 1978 to remove approximately 50,000 drums of unknown chemicals from the dump. In April 1980, an explosion and fire among the 40,000 drums remaining at the site attracted national attention and led to the HHE request.

The workers engaged in removing the drums of chemicals wore various kinds of protective clothing and equipment. The kind depended on the nature of their work, but also to some degree on the policy of the particular company they worked for. Self-contained breathing apparatus and acid-resistant suits were required where direct skin contact with liquids from ruptured drums was possible. It must be mentioned that the cleanup work was being conducted during the summer months.

The NIOSH HHE focused on the exposure of cleanup workers to chemicals found at the site. An extensive set of area air samples was collected for analysis, and personal air samples were obtained from workers who volunteered to wear sampling pumps. Personal protective equipment was evaluated in both field and laboratory tests for its protective value. The investigators also evaluated the respiratory protection programs of the various companies conducting the waste removal.

Findings from the area industrial hygiene sampling and personal sampling showed that none of the air samples exceeded workplace permissible exposure concentrations established by the U.S. Occupational Safety and Health Administration. What was evident, however, was the presence of a large number, in excess of 20, of different toxic chemicals at the site, with aluminum, calcium, iron, phosphorus, toluene, and xylene being present in largest amounts. In addition the programs in place for respiratory protection, heat stress abatement, and air monitoring were inadequate. These findings merit elaboration.

From a toxicologic perspective, the finding that concentrations of individual toxicants in the air samples were acceptable must be considered in the context of the multiple exposures, and chronic effects. There is no good way to evaluate the health effects produced by chronic low-level exposures to a variety of substances even when they can be measured. The National Research Council (1980) recommended that additivity of chemical exposures be assumed and that the guidelines of the ACGIH (1984) be followed. Most chronic health effects caused by toxicants are dose dependent. That is, the greater the dose, the more likely there will be a toxic response. Single or infrequent exposures to substances, as a general rule, are more likely to be associated with acute injury or intoxication, and are less likely to result in chronic disease than are repeated exposures. The point that must be stressed is that not much is known concerning the chronic toxicologic effects of exposure to combinations of mixtures of toxicants, as is the case in hazardous waste cleanups. The human-factors specialist must assume these kinds of chemical exposure pose significant risk to workers; protective equipment and work methods should be designed and employed to minimize this risk.

The HHE found some shortcomings in the protective equipment provided to workers. In particular, the respiratory protection provided was in need of improvement. Some respirators didn't work, others worked poorly, some were improperly fitted, and some were improperly adjusted. From a toxicologic concern, a properly fitted, functional respirator is necessary to protect against toxicants in vapor, aerosol, or gaseous form. Although engineering controls (e.g., ventilation systems) are to be preferred over personal protective equipment as a means of preventing health risks, there are clearly some jobs—and hazardous waste cleanup is one of them—for which personal protective equipment is the only feasible method of protecting workers. The design of protective equipment, especially respirators, that affords the wearer with comfortable, nonthreatening protection should be a fertile area for human-factors specialists.

The acid-resistant suits ("moon suits") were a problem during hot weather. Three cases of heat illness occurred. The effect of the impermeable clothing is to increase the temperature within the suit. This raises the wearer's body temperature, which in turn increases heart rate and blood flow to the extremities. Loss of body fluid occurs by way of perspiration and pores in the skin open. This changed physiological state will have profound effects on any toxic exposure, irrespective of the route of exposure. Should any chemical penetrate the protective clothing, it will come into contact with a vastly different skin from that found at lower body temperature.

Perspiration on the skin, if not absorbed by clothing, may increase the absorption of the alien chemical, increased blood flow in the skin enhances the transport of chemicals absorbed through the skin, and the increased heart rate more quickly distributes the toxicant throughout the body. The challenge to the human-factors specialist is to design protective clothing and work regimens that do not add to the risk posed by the work itself.

6.6.5.2 Case Study of a Small Business

The control of workers' exposure to toxic substances is particularly acute in medium- to small-scale companies, which together employ as many as 80% of U.S. workers.

Companies that employ only a few employees often do not possess the resources to provide elaborate engineering control systems or costly personal protective equipment, and their access to information on toxic hazards may be limited, but, in total, they employ a large number of workers. The following HHE case report illustrates one of the more common problems, exposure to solvents, found in many medium- to small-scale companies.

NIOSH was asked in 1983 to conduct an HHE of a company that manufactures valves (Wallingford, 1983). The manufacturing was done by production line assembly and required the dipping of parts into solvent baths to remove grease. There were 28 workers engaged in the assembly work per workshift. The specific health concern was with the exposure of workers to the two solvents used in degreasing operations, isopropanol and 1,1,1-trichloroethane (methylchloroform). Workers' exposure occurred when the parts to be degreased were dipped by hand into the solvent baths.

Acute exposure to high concentrations of isopropanol vapor can cause irritation of the eyes, nose, and throat, headache, drowsiness, and incoordination. Ingestion may cause nausea, vomiting, diarrhea, cramps, gastrointestinal pain, unconsciousness, and death. Chronic exposure to isopropanol liquid may cause drying and cracking of the skin. Acute exposure to high concentrations of 1,1,1-trichloroethane vapor may cause irritation of the eyes, central nervous system depression, headache, dizziness, drowsiness, incoordination, nausea, irregular heart beat, and unconsciousness. prolonged exposure to very high concentrations may be fatal. Chronic exposure to the liquid may cause drying, cracking, and inflammation of the skin.

All the degreasing tanks had hinged lids to help suppress solvent vapors when closed. None of the degreasing operations were provided with local exhaust ventilation. Observation by the investigators of work practices indicated that direct skin contact with the degreasing solvents occurred frequently. Many employees never wore gloves.

Interviews with workers indicated that 25 (89%) had experienced, at least once, symptoms consistent with exposure to the degreasing solvents. Twenty employees (71%) reported they experienced dry, cracked, or inflamed skin as a consequence of contact with the degreasing solvents. Twelve (43%) indicated they had experienced symptoms of central nervous system effects. These reports should be understood in the knowledge that area industrial hygiene measurements did not find solvent vapor concentrations in excess of permissible exposure limits. However, it is likely that considerable skin absorption occurred, especially in view of the cracked skin seen on many workers' hands. This problem of skin absorption of the degreasing solvents was further compounded by the observation that the workers used these same two solvents to clean their hands of grease.

This particular HHE exemplifies the problems faced by a human-factors specialist or industrial hygienist in reducing the workers' exposure to solvents. Owing to the modest size of the company, it is unlikely that a mechanized dipping system to effect degreasing could be installed. No, the human-factors specialist would have to intervene by dealing with what is, not what might be. What are some possibilities?

The most promising avenue is to design more effective hand protection, principally by improving the design of protective gloves. Workers in this company didn't wear the gloves provided by their employer, because they were uncomfortable and ill-fitting. Moreover, the designer of impermeable gloves must ensure that the product does indeed block the passage of the targeted chemicals through the glove materials, a goal that has not always been successful (Hogstedt and Stahl, 1980; Einert, Adams, Crothers, Moore, and Ottoboni, 1963; Ishihara et al., 1976).

Two other recommendations made by the HHE team have toxicologic importance. One recommendation was to cease using the degreasing solvents for cleaning the hands. Hand soap works just as well, but without the hazard posed by the solvents. A second recommendation was to consider the use of water-based degreasers in lieu of the organic-based degreasing solvents. The suggestion to change degreasing compounds exemplifies a general principle of control technology. When possible, substitute a less toxic substance for one with greater toxicity. This principle has relevance to the human-factors specialist who designs protective equipment or engineering control systems. For instance, the chemical composition of protective clothing should be carefully selected in order to avoid placing potentially toxic substances against the skin.

6.6.5.3 Case Study of a Reinforced Plastics Operation

The preceding two case studies were gleaned from the NIOSH HHE program. The third case study was drawn from research performed for NIOSH by the University of Kansas (Hopkins, 1981). Whereas the other case studies drew principally upon recommended changes in technology in order to reduce workers' exposures to toxicants, this case study portrays rather nicely the importance of work practices in controlling these exposures.

The reinforced plastics industry supplies a large quantity and variety of consumer products. The fabrication of these products involves the use of many chemicals, some of which are toxic at sufficiently high exposure levels. Much of the work in this industry is accomplished by hand. For this reason, NIOSH was interested in evaluating the efficacy of prescriptive work practices to reduce workers' exposure to chemicals used in the fabrication of products made from reinforced plastics.

Several small-scale to medium-size companies were visited by the investigators. Manufacturing of reinforced, laminated plastic products, such as those manufactured from styrene-containing resins, typically consists of a series of operations. In brief, a mold coated with wax is sprayed with a mixture of pigmented polyester resin and styrene monomer. A catalyst, such as methyl ethyl ketone peroxide, is mixed with the styrene mixture in the spraying operation. After the mold hardens a reinforcing lamination of fibrous glass is added by spraying a mixture of glass fibers, resin–styrene mixture, and catalyst. Additional layers may be added as needed. In the next operation rollers are used to roll out any gas bubbles trapped in the mixture. A finishing process involves removing the mold, trimming it, and polishing the edges of the hull. A chemical of major concern in this industry is styrene. Styrene monomer in liquid or vapor form is irritating to the eyes, nose, throat, and skin. Repeated contact with the skin may produce a dry, scaly, and fissured dermatitis. Acute exposure to high concentrations may produce symptoms of narcosis, cramps, and death due to depression of the central nervous system. The major metabolite of styrene is mandelic acid in urine.

Following selection of the plants for study, an extensive evaluation commenced of the work practices constituting the fabrication of a variety of consumer products. At the same time, an assessment was conducted of education and training materials and programs relevant to the work practices. The third leg of the triad of research methodologies was the development of motivational methods for workers and supervisors that could be used to ingrain desired work practices.

An early discovery of the investigators was the paucity of empirical data pertaining to the effectiveness of work practices to control toxic exposures. This gap must be filled, another opportunity for human-factors specialists. The specific details of the investigation are too lengthy to present here. A summary will suffice to describe the principles of interest to human-factors specialists.

The plant surveys were conducted by a team of psychologists, industrial hygienists, and human-factors engineers. This team identified a large number of work practices that, if modified, would likely result in reduced exposure of workers to styrene. The work practices can be loosely grouped into two categories, those attendant to housekeeping and those specific to production processes. For the former, 28 improved housekeeping practices were developed. Illustrative examples of them are covering floors in work areas with a disposable material (to prevent an accumulation of styrene resin), keeping the spraying booth filters clean, emptying the waste cans on a more frequent basis, covering the resin and gel-coat containers when not in use, and removing empty chemical containers from the work area.

Some representative examples from the list of 11 modified production work practices are the following: sprayers should always work on the upwind side of any source of airborne styrene; spray, lay up, and roll-out work should be done only in prescribed areas (where exhaust systems were installed); and a spray gun operator should not spray within 6 ft of another person. Training materials, principally in the form of videotapes, describing the recommended work practices were developed. Workers were given training by the investigative team in how to employ the recommended work practices. Tests were administered by the investigators to measure how well the workers understood the desired work practices. If needed, further training was provided. As the workers adopted the revised work practices into their everyday work routine, plant supervisors and select members of the investigative team used expressions of social approval to reinforce the workers' use of the new work practices.

The effectiveness of the work practices recommended by the investigators was dramatic. The reduction in ambient air concentrations of styrene ranged from 50 to 80% varying between individual plants. A corresponding reduction in mandelic acid levels in urine was achieved, ranging from 30 to 60%. There was, therefore, clear evidence that the recommended work practices, when coupled with an effective training program, accompanied by social approval from supervisors and co-workers, led to a pronounced reduction of workers' exposure to, and absorption of, styrene.

The implications of this third case study for human-factors specialists are profound. Simple, in some sense, "obvious" changes in work practices led to a marked reduction in exposure to a toxicant. But to arrive at these recommendations required the exercise of considerable knowledge and personal judgment by the investigators, the workers, and company management. Other kinds of work will require different approaches; this is the challenge presented to human-factors specialists.

6.6.6 HUMAN FACTORS EXPERIENCE FROM HHEs

The three case studies described in Section 6.6.5 exemplify the problems faced by human-factors engineers in reducing or preventing workers' exposure to toxic substances. Protective clothing, if available to workers, often doesn't prevent the toxicants in question from reaching the worker. This results from any number of factors, but chief among the reasons are poorly designed equipment, improperly used equipment, inadequate training in its use, and lack of service and maintenance of the equipment. Each of these factors is amenable to improvement by human-factors specialists.

Human factors stated or implied in other NIOSH HHE reports can be loosely grouped into the following categories:

Inadequate personal protective equipment (e.g., clothing that was permeable to chemicals used by the workers)

Poorly designed work processes (e.g., solvent tanks that require workers to reach into them)

Inadequate appreciation of hazards (e.g., not wearing adequate protective clothing)

Poorly arranged work stations (e.g., positioning work processes so that toxic chemicals are carried between stations)

Absent hygiene facilities (e.g., lack of restrooms or hot showers)

Uninformative engineering controls (e.g., ventilation systems that lack controls that inform how well they are operating)

Each of these categories presents challenging problems to the human-factors engineer toward accomplishing the goal of reducing hazardous exposures.

In Table 6.6.8 are summarized those HHE reports in which human-factors considerations were included in the recommendations. For each HHE cited, we state the condition found at the workplace, the effect of this condition on workers, and the solution proposed by NIOSH investigators to abate the condition.

The problems are particularly difficult where the enterprises involved are medium- to small-scale businesses. We believe the problems to be tractable, however, if existing human-factors knowledge is prudently applied and research is undertaken on ways to solve the more difficult problems of preventing human exposure to toxicants.

Table 6.6.8 Human-Factors Issues Found in NIOSH Health Hazard Evaluations

Source	Findings	Effect	Solution Proposed
Rivera (1975)	Poorly designed solvent bath	Solvent inhalation due to poor worker posture	Redesign the tank
Geissert (1977)	Poorly fitted respirators	Respirators not worn; led to exposure to irritating dust	Implement a training program for use and fit of respirators
Price and Thoburn (1977)	Protective clothing not impermeable	Solvents penetrated the skin	Select impervious clothing from available sources
Gilles (1978)	Engineering process deficient	Workers' hands were dipped into oil	Redesign process to eliminate the hazards
Messite and Fannick (1978)	Manual handling of drugs	Skin and oral exposure of workers	Use an instrument instead of hands
Chrostek (1980)	Manual transport of painted materials	Exposure to paint solvents	Rearrange the work area
Tharr, Murphy, and Mortimer (1982)	Assembly process too far from ventilation	Inhalation of solvent vapors	Revise the work practices and rearrange the work area
Hollett and Klemme (1982)	No protective gloves	Skin absorption of solvents	Provide protective gloves
Chrostek and Elesh (1980)	Loading of drums required reaching into them	Inhalation of toxic dusts	Purchase material in smaller drums or tilt the larger drums
Murphy and Lucas (1982)	Worker had to lean over a mold	Inhalation of off-gassing vapors	Place the sand mold upon a platform; revise work practices
Watanabe, Patnode, Singal, and Ferguson (1982)	Loading of hoppers over workers' heads	Spillage led to drug inhalation	Eliminate overhead scooping by adding a mechanical delivery system
Evans (1978)	Bulky gloves provided for fine work	Gloves not worn; skin exposures	Provide thin-gauge gloves; revise work practices
Williams and Hickey (1983)	Inadequate protective clothing	Skin contact with solvents	Provide ventilation and protective clothing
Ruhe, Watanabe, and Stein (1981)	Control panel too far from degreasing tank	Workers had to go to the tank	Relocate a viewing window so that the tank can be seen
Moran and Love (1980)	No restroom	Technicians couldn't clean chemicals from themselves	Provide bathroom facilities
Kominsky (1979)	Manual inspection of a tank's contents	Inhalation of chemicals	Wear a full-face respirator until an alternate procedure is found
Messite and Fannick (1979)	Photographs coated by hand	Oral ingestion and skin absorption of liquids used	Use swabs to apply the dyes; provide protective finger coverings
McManus and Baker (1981)	Barrier creams used inappropriately	Perspiration increased	Discontinue use of creams
Boxer and Mosely (1983)	Lack of gloves, other equipment	Skin contact with adhesives	Provide pins to hold the parts; provide gloves
Almaguer and Kramkowski (1983)	Dip tank solution too low	Workers splashed solution on themselves	Raise the level of the solution in the tank
Lee (1984)	Portable fans led to air turbulence	Toxic fumes were broadcast	Reposition fans; condition air
Roper and Piccirillo (1976)	Improperly worn dust masks	Dust Inhalation	Use half-face respirators; not dust masks
Burroughs (1976)	Manual handling of metal plates	Cutting oils on workers' skin	Use forceps to handle the parts

REFERENCES

ACGIH (1980). *Documentation of the Threshold Limits Values*, 4th ed. Cincinnati, OH: American Conference of Governmental Industrial Hygienists.

ACGIH (1984). *Threshold Limit Values for chemical substances and physical agents in the work environment and biological exposure indices with intended changes for 1984–85*. Cincinnati, OH: American Conference of Governmental Industrial Hygienists.

Aitio, A., Jarvisalo, J., Kiilunen, M., Tossavainen, A., and Vaittinen, P. (1980). Urinary excretion of chromium as an indicator of exposure to trivalent chromium sulphate in leather tanning. *International Archives of Occupational and Environmental Health*, 54, 241–249.

Almaguer, D., and Kramkowski, R. (1983). *NIOSH Health Hazard Evaluation Report HETA 82–116–1319, Modine Manufacturing, McHenry, Illinois*. Springfield, VA: National Technical Information Service No. PB 84–210442.

Astrand, I. (1983). Effect of physical exercise on uptake, distribution and elimination of vapors in man. In V. Fiserova-Bergerova, Ed., *Modeling of inhalation exposure to vapors: Uptake, distribution, and elimination*, Vol. 2, Boca Raton, FL: CRC Press, pp. 107–130.

Astrand, I., and Ovrum, P. (1976). Exposure to trichloroethylene I. Uptake and distribution in man. *Scandinavian Journal of Work, Environment and Health*, 4, 199–211.

Astrand, I., Kilbom, A., Ovrum, P., Wahlberg, I., and Vesterberg, O. (1979). Exposure to styrene I. Concentration in alveolar air and blood at rest and during exercise and metabolism. *Work Environment Health*, 11, 69–85.

Astrand, I., Kilbom, A., and Ovrum, P. (1975). Exposure to white spirit I. Its concentration in alveolar air and blood during rest and exercise. *Scandinavian Journal of Work, Environment and Health*, 1, 15–30.

Astrand, I., Ovrum, P., and Carlsson, A. (1975). Exposure to methylene chloride I. Its concentration in alveolar air and blood during rest and exercise and its metabolism. *Scandinavian Journal of Work, Environment and Health*, 1, 78–94.

Astrand, I., Ovrum, P., Lindqvest, T., and Hultengren, M. (1976). Exposure to butyl alcohol—Uptake and Distribution in man. *Scandinavian Journal of Work, Environment and Health*, 3, 165–175.

Astrand, I., Engstrom, J., and Ovrum, P. (1978). Exposure to xylene and ethylbenzene I. Uptake, distribution, and elimination in man. *Scandinavian Journal of Work, Environment and Health*, 4, 185–194.

Baranowska-Dutkiewicz, B. (1981). Skin absorption of phenol from aqueous solutions in men. *International Archives of Occupational and Environmental Health*, 49, 99–104.

Baranowska-Dutkiewicz, B. (1982). Skin absorption of aniline from aqueous solutions in man. *Toxicology Letters*, 367–372.

Boxer, P. A., and Moseley, C. (1983). *NIOSH Health Hazard Evaluation Report HETA 83-010-1313, Detroit Gasket, Fremont, Ohio*. Springfield, VA: National Technical Information Service No. PB 84-210160.

Burroughs, G. E. (1976). *Health Hazard Evaluation Determination Report No. 75-97-257, C. S. Brainin Company, Mount Vernon, New York*. Cincinnati, OH: National Institute for Occupational Safety and Health.

Carlsson, A. (1982). Exposure to toluene—uptake, distribution and elimination in man. *Scandinavian Journal of Work, Environment and Health*, 8, 43–55.

Carlsson, A., and Lindqvist, T. (1977). Exposure of animals and man to toluene. *Scandinavian Journal of Work, Environment and Health*, 3, 135–143.

Chrostek, W. J. (1980). *Health Hazard Evaluation Determination Report HE 80-108-705, Corporation of Veritas, Philadelphia, Pennsylvania*. Springfield, VA: National Technical Information Service No. PB 81-10644.

Chrostek, W. J., and Elesh, E. (1980). *Health Hazard Evaluation Determination Report HE 79-19-740, Radford Army Ammunition Plant, Hercules Incorporated, Radford, Virginia*. Springfield, VA: National Technical Information Service No. PB 81-170938.

Coletta, G. C., Schwope, A. D., Arons, I. J., King, J. W., and Sivak, A. (1978). *Development of performance criteria for protective clothing used against carcinogenic liquids* (DHEW (NIOSH) Publication No. 79-106). Cincinnati, OH: National Institute for Occupational Safety and Health.

Costello, R., and Melius, J. (1981). *NIOSH Health Hazard Evaluation Report TA 80-77-853, Chemical Control Corporation, Elizabeth, New Jersey*. Springfield, VA: National Technical Information Service No. PB 82-209578.

Divincenzo, G. D., Hamilton, M. L., Kaplan, C. J., Krasavage, W. J., and O'Donoghue, J. L. (1978).

Studies on the respiratory uptake and excretion and the skin absorption of methyl *n*-butyl ketone in humans and dogs. *Toxicology and Applied Pharmacology, 44,* 593–604.

Dutkiewicz, T. (1961). Absorption of aniline vapors in men. In *Proceedings of the 14th International Congress on Occupational Health.* New York: Book Craftsmen Associates, pp. 681–686.

Dutkiewicz, T., and Tyras, H. (1969). Comparative studies on the percutaneous absorption of toluene, ethylbenzene, xylene, and styrene in man (in Polish). *Medycyna Pracy, 20,* (pp. 228–234).

Dutkiewicz, B., Konczalik, J., and Karwacki, W. (1980). Skin absorption and per os administration of methanol in men. *International Archives of Occupational and Environmental Health, 47,* 81–88.

Einert, C., Adams, W., Crothers, R., Moore, H., and Ottoboni, F. (1963). Exposure to mixtures of nitroglycerin and ethylene glycol finitrate. *American Industrial Hygiene Association Journal, 24,* 435–447.

Engstrom, K., Husman, K., and Riihimaki, V. (1977). Percutaneous absorption of *m*-xylene in man. *International Archives of Occupational and Environmental Health, 39,* 181–189.

Evans, W. (1978). *Health Hazard Evaluation Determination Report HE 78-76-548, Western Gear Corporation Flight Structures Division, Jamestown, North Dakota.* Springfield, VA: National Technical Information Service No. PB 81-159443.

Franklin, C. A., Fenske, R. A., Greenhalgh, R., Mathieu, L., Denly, H. V., Leffingwell, J. T., and Spear, R. C. (1981). Correlation of urinary pesticide excretion with estimated dermal contact in the course of occupational exposure to guthion. *Journal of Toxicology and Environmental Health, 7,* 715–731.

Fukuchi, Y., Nitroglycerol concentrations in blood and urine of workers engaged in dynamite production. *International Archives of Occupational Environmental and Health, 48,* 339–346.

Geissert, J. O. (1977). *Health Hazard Evaluation Determination Report No. 75-154-387, Emery Industries, Cincinnati, Ohio.* Springfield, VA: National Technical Information Service No. PB 273821.

Gilles, D. (1978). *Health Hazard Evaluation Determination Report No. 77-88-457, Galion Amco, Inc., Galion, Ohio.* Springfield, VA: National Technical Information Service No. PB 82-194002.

Gross, E., Kiese, M., and Resag, K. (1960). Absorption ethylene glycol dinitrate through the skin (in German). *Archiv fur Toxikologie, 18,* 194–199.

Guillemin, M., Murset, J. C., Lob, M., and Riquez, J. (1974). Simple method to determine the efficiency of a cream used for protection against solvents. *British Journal of Industrial Medicine, 31,* 310–316.

Guy, R. H., and Maiback, H. I. (1984). Correction factors for determining body exposure from forearm percutaneous absorption data. *Journal of Applied Toxicology, 4,* 26–28.

Hamilton, A. (1934). *Industrial toxicology.* New York: Harper Brothers, pp. ix–xix.

Hanke, J., Dutkiewicz, T., and Piotrowski, J. (1961). The absorption of benzene through the skin in man (in Polish). *Medycyna Pracy, 12,* 413–426.

Hogstedt, C., and Stahl, R. (1980). Skin absorption and protective gloves in dynamite work. *American Industrial Hygiene Association Journal, 41,* 367–372.

Hollett, B. A., and Klemme, J. C. (1982). *NIOSH Health Hazard Evaluation Report HETA 81-045C-1217, Uniroyal, Incorporated, Mishawaka, Indiana.* Springfield, VA: National Technical Information Service No. PB 84-183623.

Hopkins, B. L. (1981). *Behavioral procedures for reducing worker exposure to carcinogens (Final Report NIOSH Contract No. 210-77-0040).* Cincinnati, OH: National Institute for Occupational Safety and Health.

Ilyin, L. A., Ivannikov, A. T., Parfenov, Y. D., and Stolyarov, V. P. (1975). Strontium Absorption Through Damaged and Undamaged Human Skin. *Health Physics, 29,* 75–80.

Ishihara, N., Kanaya, A., and Ikeda, M. (1976). *m*-Dinitrobenzene intoxication due to skin absorption. *International Archives of Occupational and Environmental Health, 36,* 161–168.

Johnson, B. L. (1983). Occupational toxicology: NIOSH perspective. *Journal of the American College of Toxicology, 2,* 43–50.

Konz, S. (1979). *Work design.* Columbus, OH: Grid Publishing, pp. 441–428.

Kominsky, J. R. (1979). *Health Hazard Evaluation Determination Report HHE 78-119-637, Texaco, Inc., Bayonne Terminal, Bayonne, New Jersey.* Springfield, VA: National Technical Information Service No. PB 80-163223.

Krajewska, D., Adamiak-Ziemba, J., Suwalska. D., and Kostecka, K. (1980). Evaluation of occupational benzidine exposure of workers in automated benzidine plants (in Polish). *Medycyna Pracy, 31,* 403–410.

Lauwerys, R. R., Dath, T., Lachapelle, J. M., Buchet, J. P., and Roels, H. (1978). The influence of two barrier creams on the percutaneous absorption of *m*-xylene in man. *Journal of Occupational Medicine, 20,* 17–20.

Lauwerys, R. R., Kivits, A., Lhoir, M., Rigolet, P., Houbeau, D., Buchet, J. P., and Roels, H. A. (1980). Biological surveillance of workers exposed to dimethlyformamide and the influence of skin protection on its percutaneous absorption. *International Archives of Occupational and Environmental Health, 45,* 189–203.

Lee, S. A. (1984). *NIOSH Health Hazard Evaluation Report HETA 83-338-1399, Palmer Instruments, Incorporated, Cincinnati, Ohio.* Cincinnati, OH: National Institute for Occupational Safety and Health.

Legge, T. (1934). In S. A. Henry, Ed., *Industrial maladies.* London: Oxford University Press, pp. 2–3.

Maibach, H. I., Feldmann, R. J., Milby, T. H., and Serat, W. F. (1971). Regional variation in percutaneous penetration in man. *Archives of Environmental Health, 23,* 208–211.

Maxfield, M. E., Barnes, J. R., Azar, A., and Trowchimowicz, H. T. (1975). Urinary excretion of metabolite following human exposures to DMF or to DMAC. *Journal of Occupational Medicine, 17,* 506–511.

McManus, K. P., and Baker, E. L. (1981). *NIOSH Health Hazard Evaluation Report HHE 80-084-927, General Electric Company, Lynn, Massachusetts.* Springfield, VA: National Technical Information Service No. PB 83-102848.

Messite, J., and Fannick, N. L. (1978). *Health Hazard Evaluation Determination Report No. 78-6-503, Cumberland Outpatient Department of Beth Israel Hospital, Brooklyn, New York.* Cincinnati, OH: National Institute for Occupational Safety and Health.

Messite, J., and Fannick, N. (1979). *Health Hazard Evaluation Determination Report HE 79-29-566, Giorgi Process, Inc., Yonkers, New York.* Springfield, VA: National Technical Information Service No. PB 81-143778.

Mikatavage, M., Que Hee, S. S., and Ayer, H. E. (1984). Permeation of chlorinated aromatic compounds through Vitron® and nitrile glove materials. *American Industrial Hygiene Association Journal, 45,* 617–621.

Moran, J. M., and Love, J. (1980). *NIOSH Health Hazard Evaluation Report HHE 80-003-785, Steven Janowitz, DDS, Alexandria, Virginia.* Springfield, VA: National Technical Information Service No. PB 82-150152.

Murphy, D. C., and Lucas, C. (1982). *NIOSH Health Hazard Evaluation Report HETA 81-411-1182, Corhart Refractory, Louisville, Kentucky.* Springfield, VA: National Technical Information Service No. PB 84-149885.

National Research Council Panel on Evaluation of Hazards Associated with Maritime Personnel Exposed to Multiple Cargo Vapors (1980). *Principles of Toxicological Interactions Associated with Multiple Chemical Exposures.* Washington, DC: National Acadamy Press.

Olishifski, J. B. (1979). Methods of control. In J. B. Olishifski, Ed., *Fundamentals of industrial hygiene,* 2nd ed. Chicago, IL: National Safety Council, pp. 613–635.

Peterson, J. E. (1977). *Industrial Health.* Englewood Cliffs, NJ: Prentice-Hall, pp. 296–300.

Piotrowski, J. (1957). Quantitative estimation of aniline absorption through the skin in man. *Journal of Hygiene, Epidemiology, Microbiology and Immunology, 1,* 23–32.

Piotrowski, J. (1967). Further investigations on the evaluation of exposure to nitrobenzene. *British Journal of Industrial Medicine, 24,* 60–65.

Piotrowski, J. K. (1971). Evaluation of exposure to phenol: absorption of phenol vapor in the lungs and through the skin and excretion of phenol in urine. *British Journal of Industrial Medicine, 28,* 172–178.

Price, J. H., and Thoburn, T. W. (1977). *Health Hazard Evaluation Determination Report No. 76-60-398, Hayes-Albion Corporation, Wolverine Plastics Division, Milan, Michigan.* Springfield, VA: National Technical Information Service No. PB273749.

Riihimaki, V., and Pfaffli, P. (1978). Percutaneous absorption of solvent vapors in man. *Scandinavian Journal of Work, Environment and Health, 4,* 73–85.

Rivera, R. O. (1975). *Health Hazard Evaluation Determination Report No. 74-135-226 GAF Corporation Equipment Manufacturing Plant, Vestal, New York.* Springfield, VA: National Technical Information Service No. PB 249406.

Roper, C. P., Jr., and Piccirillo, R. E. (1976). *Health Hazard Evaluation Determination Report No. 75-104-325, Olin Corporation, Pisgah Forest, North Carolina.* Springfield, VA: National Technical Information Service No. PB 264806.

Ruhe, R. L., Watanabe, A., and Stein, G. (1981). *NIOSH Health Hazard Evaluation Report HHE*

80-49-808, Superior Tube Company, Collegeville, Pennsylvania. Springfield, VA: National Technical Information Service No. PB 82-232299.

Schilling, R. S. F., and Hall, S. A. (1973). Prevention of occupational disease. In R. S. F. Schilling, Ed., *Occupational health practice.* London: Butterworths, pp. 408–420.

Sedivec, V., Mraz, M., and Flek, J. (1981). Biological monitoring of persons exposed to methanol vapors. *International Archives of Occupational and Environmental Health, 48,* 257–271.

Stampfer, J. F., McLeod, M. J., Betts, M. R., Martinez, A. M., and Barardinelli, S. P. (1984a). Permeation of polychlorinated biphenyls and solutions of these substances through selected protective clothing materials. *American Industrial Hygiene Association Journal, 45,* 634–641.

Stampfer, J. F., McLeod, M. J., Betts, M. R., Martinez, A. M., and Barardinelli, S. P. (1984b). Permeation of eleven protective garment materials by four organic solvents. *American Industrial Hygiene Association Journal, 45,* 642–645.

Tatken, R. L., and Lewis, R. J. (1983). *Registry of toxic effects of chemical substances,* 1981–1982 ed. [DHHS (NIOSH) Publication No. 83-107]. Washington, DC: U.S. National Government Printing Office.

Tharr, D. G., Murphy, D. C., and Mortimer, V. (1982). *NIOSH Health Hazard Evaluation Report HETA 81-455-1229, Red Wing Shoe Company, Red Wing, Minnesota.* Springfield, VA: National Technical Information Service No. PB 84-172592.

Veulemans, H., Van Vlem, E., Janssens, H., Masschelein, R., and Leplat, A. (1982). Experimental human exposures to *n*-hexane—study of the respiratory uptake and elimination, and *n*-hexane concentrations in peripheral venous blood. *International Archives of Occupational and Environmental Health, 49,* 251–263.

Wallingford, K. M. (1983). *NIOSH Health Hazard Evaluation Report HETA 83-170-1346. Xemox Corporation, Cincinnati, Ohio.* Cincinnati, OH: National Institute for Occupational Safety and Health.

Watanabe, A. S., Patnode, R., Singal, M., and Ferguson, R. P. (1982). *NIOSH Health Hazard Evaluation Report HETA 81-322-1228, Mylan Pharmaceuticals, Morgantown, West Virginia.* Springfield, VA: National Technical Information Service No. PB 84-172550.

Williams, T. M., and Hickey, J. L. S. (1983). *NIOSH Health Hazard Evaluation Report HETA 81-311-1250, Uniroyal, Inc., Opelika, Alabama.* Springfield, VA: National Technical Information Service No. PB 84-173038.

Wojeck, G. A., Nigg, H. N., Bramen, R. S., Stamper, J. H., and Rouseff, R. L. (1982). Worker exposure to arsenic in Florida grapefruit spray operations. *Archives of Environmental Contamination and Toxicology, 11,* 661–667.

CHAPTER 6.7

ARCHITECTURE AND INTERIOR DESIGN

JOHN E. HARRIGAN

California Polytechnic State University, San Luis Obispo

6.7.1 INTRODUCTION

We all know that unfavorable environmental conditions may make successful activities and satisfying experiences difficult or impossible (Mather, Kit, Bloch, and Herman, 1970; Pellow, 1981). As we design spaces in which people live and work, we realize that people are frequently influenced, for better or worse, by the extent to which the physical facilities they use are suited to them in terms of the relevant features of the human body, sensory and motor systems, and behavior (McCormick, 1970; Lueder, 1983). Any habitat limits the range and options of its residents, and the prescribed arrangement of the physical environment may increase convenience or complicate the tasks of living (Deasy, 1974; Newman, 1975; Altman and Chemers, 1980). In the area of safety, Neutra and McFarland (1972) state that the features of a facility can prevent accidents, minimize injury, and facilitate rehabilitation, both directly and indirectly. Terms such as "effectiveness," "convenience," "safety," "success," and "satisfaction" stress the need to assure that the immediate environment is formulated as a selective response to individual needs. To this end, the information presented here is designed to help you determine what a space, building, facility, or planned environment means to its users and their activities; to guard against inefficient, inadequate, and unsafe conditions; to establish options so that users can exercise personal preference; and in effect, to achieve the best possible physical and human environment.

The premise defended is that architectural and interior design projects must begin with a critical analysis of the existing situation (Preiser, 1978) and be developed, at least in part, on the basis of human-factors design objectives, criteria, and specifications. How to make these demanding criteria of success possible when our efforts are often limited by available time, funds, resources, and experienced personnel, as well as imposed constraints, guidelines, and policy, is the challenge addressed. The principal response to this challenge is the human-factors program presented here. It is the scope of work that will help you as a member of an architectural or interior design team perfect the environment from the user's point of view. The presentation of the human-factors program is followed by a discussion of recommended research methods and means for reducing and controlling measurement error. A discussion of how to manage and budget your human-factors programming project is included. In the conclusion, a description of what should be contained in your final report is presented.

It should be noted that although the information presented has been directed to human-factors professionals, architects, and interior designers, others may find it of value. Those who employ professional design and planning services, executives who formulate policy for facility development and capital investment, administrators of housing, community service, educational, and recreation programs, developers concerned with community acceptability and marketability, managers of office buildings, industrial plants, and retail stores, and those who plan to design their own environments need a systematic approach to assure full human resources and activities benefits from their capital investments.

A Complex Undertaking

Art, engineering, and science meet when environmental designers work to provide new experiences, places, and ways of doing things. The work of architects, interior designers, and associated professionals combines insight and spontaneity and information and methodology. This is an activity that has yet to be formulated into a comprehensive whole. It probably never will be. The work of environmental designers is characterized by a constant movement between fact and art. They see this not as a problem or shortcoming, but as a requirement for success.

In an imaginative way, the human-factors program may be regarded as a prism which you place between yourself and the situation you are studying to produce a spectrum of justified design specifications. As you develop the required information you should realize that while you are producing information and data during human-factors programming, you are also piling up your own ideas and sharpening your perception of what is a good design or plan. From your own experience you know that design is more often a process of association than a formula-driven process. On the one hand, you consider the needs and wants of the user; on the other hand, you envision physical settings and planned environments that correspond to these expectations and requirements. Inventiveness will not occur without this association. As you are stimulated by your understanding of what is needed, you respond with solutions. Human-factors programming can accelerate this process by directing you to the key considerations of each design feature.

Use of the human-factors program will help you meet your schedule, control costs, and ensure that you don't develop information that is not effectively employable. The phrase "effectively employable" is the key to what is presented. The developed programming scheme does not feed the appetite for information nor does it result in an excess of detail and complication. It is stressed here that good research design is the beginning of a successful programming effort. The environmental designer conducting a program of human-factors research must have understandable objectives. The questions being asked must be clear to everyone Involved and the program of study must effectively assess the space, building, or facility being considered. Behavioral scientists regard research design as the most important step in a program of study. Their feeling is that unless a question or hypothesis can be paired with effective research, speculation will be the only result. We are all familiar with the approach: formulate a theory, specify an hypothesis, and resolve the issue through direct or indirect measurement. Although this approach is difficult to follow in the type of complex, often uncertain situations facing the environmental designer, the intention must be maintained.

The undertaking described in this chapter becomes even more complex when you realize that programming is only a beginning. At the same time you are programming, others are considering other aspects of design development. The design team will be considering site development, structural systems and materials, construction methods, and schedules, developing design concepts, and conducting cost analyses. When all these factors are responded to in final design, expect to see your recommendations modified as conflicts are identified and compromises formulated.

Building Type

The term "building type" refers to the labels, or key words, that order the principal products, professional accomplishments, and archives of architecture, interior design, and associated disciplines. Architectural indexes order the literature cited by building type, such as airports and banks, health centers and hospitals, law buildings and libraries, office buildings and post offices, shopping centers and theaters, and transportation centers and warehouses. The principal assumption of the human-factors program is that you know the history of the building type you are attempting to develop. No one should attempt to develop a human-factors program without knowing, how the subject building type has developed across the years. This is the only way to be sure that the time and resources you devote to the project will produce results with a direct relevance to the professional and technical decisions that must be made during design and planning. You must study your subject building type so you can clarify objectives and identify alternatives. You will work to identify the daily occurrences and activities that are characteristic of existing facilities. You want to identify complex and subtle issues. You must know the existing range of possible options for each design element of your building type. Even if you consider your aims unique, you still must attempt to profit from the design archives of related environments. You must know which design features have worked, which have proved unsuccessful, and for what reasons, under what conditions.

You cannot say you know the history of your subject building type until you have acquainted yourself with the following:

The floor plans, spaces, and areas characteristic of the subject building type

The dimensions, form, and other space attributes that are common

How circulation patterns are used to fulfill activity, safety, security, and efficiency requirements

How space adjacencies relate to user activities and environmental compatibility

The general types of furniture, fixtures, and equipment and the design characteristics that could influence performance, satisfaction, and utility

The reasons why furniture, fixtures, and equipment are placed the way they are

The user groups and activities that require special furnishing, fixtures, placement, signage, and security and safety features and the events that, though of low probability of occurrence, must be considered significant for facility design

The surface design characteristics that could influence user activity and acceptance

The specifications and standards that give full consideration to the effects of temperature, humidity, airflow, sound, noise, illumination, and climatic conditions on performance and comfort

What design features related to surrounding facility characteristics, interfacility flow densities, direction and distances, safety and climatic considerations, site features, and parking and service needs are currently considered important

How new products and evolving technology might change traditional building type characteristics

Though this preparation is exacting, how else can you fulfill the premise that the products of design and planning should be a response to nothing less than a thorough study of what people need and want? The results of your study of the history of a subject building type will be needed as you begin your human-factors program.

6.7.2   HUMAN-FACTORS PROGRAM

The human-factors program shown in Figure 6.7.1 (Harrigan and Harrigan, 1979) identifies the questions you must answer if you intend to plan and design environments from the perspective of user expectations

1.0   PROGRAM CONTEXT

1.1   PROJECT OBJECTIVES

Taking into account the existing situation, anticipated needs, developing events, and image of the future, what are the project objectives?

1.2   ALTERNATIVES

Comparing possible courses of action as the feasibility, consequences, and resource availability, which alternative best corresponds to the stated objectives?

2.0   ORGANIZATION CHARACTERISTICS

2.1   PROGRAMS AND SERVICES

What programs, services, and operational schedules will be in effect during the firt period of facility use?

2.2   ORGANIZATION STRUCTURE

What are the relationships between groups and organizations that use the facility and influence activities within the facility?

2.3   ALTERATION EXPECTANCIES

How soon might it be necessary to modify or expand the facility? What events would most probably lead to the requirement?

5.0   SURFACES

5.1   SURFACE CHARACTERISTICS

What are the specific proposals for surfaces? What are the user effects possibilities of each?

5.2   SIGNAGE

What are the specific proposals for signage?

5.3   DURABILITY AND MAINTAINABILITY

Does the planned facility include areas requiring special attention to durability and maintainability of surfaces?

6.0   CIRCULATION

6.1   INFORMATION FLOW

What individuals and groups must exchange information and what is the nature and frequency of this exchange?

Fig. 6.7.1. Human-factors program. This program identifies the questions you must answer if you intend to design and plan environments from the perspective of user expectations and requirements.

6.2   USER FLOW

How many people will be entering, leaving, and moving about within the facility, for what purpose, and how frequently? What facility spaces are likely to be subject to particularly heavy user flow?

6.3   EQUIPMENT AND MATERIAL FLOW

What is the nature of any equipment and material that must be transported between facility spaces? How will these items be transported, and what is the anticipated frequency of such movements?

6.4   CRITICAL CIRCULATION PATTERN

What are the recommended circulation patterns for facilitating information, user, and equipment and material flow between spaces? In what way is this proposal a response to concerns for efficiency, convenience, safety and security?

3.0   USER CHARACTERISTICS

3.1   USER CATEGORIES

Who will be using the facility? How may these users be grouped into categories? How many individuals would each category include?

3.2   USER ACTIVITY DESCRIPTION

What are the characteristic activities of users? What is known about the extent, time of occurrence, and duration of anticipated activities?

3.3   SOCIOCULTURAL CHARACTERISTICS

What are the customs, styles, norms, and traditions of users? Are these characteristics stable or likely to change?

4.0   SPACE PLANNING

4.1   SPACES

What spaces are needed to support facility users' activities?

4.2   SPACE LAYOUT

What space layouts best correspond to the expectations and requirements of users?

4.3   FURNISHING, FIXTURE, AND EQUIPMENT ALLOCATIONS

What furnishings, fixtures, and equipment, both fixed and mobile, are required for each facility space?

4.4   AMBIENT ENVIRONMENTAL CRITERIA

What provisions should be made for the effect on users of temperature, humidity, airflow, illumination, noise, distractions, annoyances, hazards and climatic conditions?

Fig. 6.7.1.   *(Continued)*

4.5 CONVENIENCE, SAFETY AND SECURITY

Will any user group or activity require special fixtures, furnishings, space layouts, signage, surface treatments, etc.? Even for events which are of low probability of occurrence, what special safety and security measures are necessary?

7.0 SPACE ARRANGEMENT

7.1 PROPOSED SPACE ARRANGEMENT

What is the best way of meeting space adjacency and circulation requirements for this facility? What is the benefit and problem resolution potential of the suggested scheme?

7.2 SPACE REQUIREMENTS

What is the estimated square footage for each facility space?

8.0 SITE CONSIDERATIONS

8.1 SITE REQUIREMENTS

What are the requirements for the site and to what needs do these requirements correspond?

8.2 PLANNING FACTORS

What are the population characteristics of the area surrounding the site? What are the surrounding land use patterns, geographical and historical features and seasonal climatic conditions? What jurisdictional agencies and neighborhood and community groups will be involved in the determination of the allowable uses for the facility site?

8.3 SERVICES IMPACT

What will be the impact of user activities and operations on existing public and private services? Will existing services need to be improved or expanded?

8.4 SITE PLANNING

How should the site be planned in order to achieve maximum compatibility between facility characteristics and site features, user expectations and requirements, neighborhood and community desires and requirements for outdoor space in terms of movement patterns, amenities and landscape development?

Fig. 6.7.1. (*Continued*)

and requirements. If you have already accepted the challenge of user-oriented design and planning, you will recognize that the information system shown is designed to help you during the difficult preliminary tasks of design development. The human-factors program is organized around the theme of programming from the "inside out." It begins—Parts 1.0 Program Context, 2.0 Organization Characteristics, and 3.0 User Characteristics—with an inquiry into the type of activities associated with the planned facility. Parts 4.0 Space Planning, 5.0 Surfaces, and 6.0 Circulation lead you to consider the basic support elements of a facility. Part 7.0 Space Arrangement provides an opportunity to summarize your recommendations. Part 8.0 Site Considerations asks you to step beyond the facility and consider the site and the surrounding areas.

1.0 Program Context, 2.0 Organization Characteristics, and 3.0 User Characteristics

As you develop your answers to the first three sets of questions, you should keep in mind that you are creating an image of what will be going on in the facility and an understanding of the role of the facility within the scope of the client's concerns. As you develop this information, it is difficult to strike a balance between too much and too little. You should limit yourself to information that has a direct application in design development.

Each question item in the human factors program has a distinct application. *1.1 Project Objectives* sets the stage for the entire programming study. All those involved in the effort—the entire design development team, those with financial, operational, and public interests, and those who will use the results of design development, such as specifiers, construction managers, and engineers and contract buyers—must be made aware of the client's goals and objectives and the justification that exists for the proposed project. The second question, *1.2 Alternatives,* provides an opportunity to review the alternatives that the client has considered, to determine the benefits of each, and to establish one alternative as the focus of the programming study. As you develop answers to these questions, you should describe the client's existing situation through answers to such questions as the following: What are the inadequate and satisfactory features of the client's existing facility? How do these inadequacies influence current organization programs and activities? What needs does the client have that are not now being met? What will be the effects of not responding to these needs? What courses of action has the client considered? How do alternatives compare as to feasibility, consequences, and resource conservation? Which considerations will carry the most weight when an alternative is selected?

These questions help achieve mutual understanding between all project participants as to what has to be achieved, setting the basis for cooperation and collaboration. Of course, how much collaboration is desirable is debatable. Any approach to developing design objectives, criteria, and specifications that solicits the full participation of clients and users may make it difficult to proceed expeditiously. Problem closures may be difficult to achieve. Some fear that too much collaboration raises expectations that may not be fulfilled.

Your answers to questions *2.1 Programs and Services* and *2.2 Organization Structure* continue the process of justification, in which every final element of a proposed program or design can be traced to a particular characteristic of the client's situation. These questions direct you to develop a general description of the client's normal way of operating. Question *2.3 Alteration Expectancies* is an attempt to guard against premature obsolescence and to extend the effective life of the facility. Few things should concern you more. If you fail in this undertaking you will soon find that the client will have to correct, with reorganization, additional staffing, or remodelling, problems that could have been avoided.

The questions listed under *3.0 User Characteristics* must be carefully considered, because by creating an image of facility users, their activities, and their characteristic way of behaving, you help maintain a direct concern for the user's situation. In addition, a facility designed to correspond to the expectations and requirements of users has a good chance of being accepted, particularly when individuals see themselves taken into account during design development.

Question *3.1 User Categories* will provide the information that will keep the full spectrum of facility users before the design team at all times. As you work to develop your answer to this question your guideline is: If individuals use the facility, directly influence activities within the facility, or are going to be affected by the presence of the facility, they should be identified. Here you identify organizations, individuals, groups of individuals, those with special needs, and even those "users" who might disrupt or endanger the normal course of daily affairs. Question *3.2 User Activity Description* should be applied to the individuals and groups that you consider particularly important. A complete program could contain a number of such information statements, either for important activities or for critical spaces. As you develop your activity patterns, remember that you are working to keep the design team on target, to identify the activity patterns to which the designer must formulate a specific response. Every facility is required to support a wide range of daily activities. Every square foot, each fixture, furnishing, and equipment item, as well as the funds expended thereon, will eventually have to be justified in terms of activities being supported.

In order to develop an acceptable and effective facility, it is necessary to know as much as possible about the sociocultural character of users. This is the purpose of Question *3.3 Sociocultural Characteristics.* Every facility, large or small, simple or complex, is laden with sociocultural significance. Although you may feel prepared to deal with this complex aspect of the built environment, you must be sure that your experience does not distort the realities of existing social structures and cultural processes. You will be making many references to subjective feelings and preferences. If such feelings and dispositions are taken into consideration, users are likely to appreciate and respond in a positive fashion to the facility design.

4.0 Space Planning, 5.0 Surfaces, and 6.0 Circulation

The character of the program changes at this point. The first three parts of the program are concerned mainly with information development. In the next three parts, you begin to respond to the design implications of your information, working out design objectives, criteria, and specifications. In answer to question *4.1 Spaces,* you will identify the spaces needed. The required features of each are determined as you answer questions *4.2 Space Layout* and *4.3 Furnishing, Fixture, and Equipment Allocations.* In your answer to question *4.4 Ambient Environmental Criteria,* you will include information that you think can help the designer formulate specifications for the facility and justify expenditures directed toward environmental control and energy conservation. Your answer to the last question, *4.5 Convenience, Safety, and Security,* can help the designer eliminate frustration and confusion, prevent accidents, minimize injury, and assure the security of individuals and spaces. The statements developed in response to questions *4.1 Spaces* through *4.5 Convenience, Safety, and Security* should include workable space plans and any information and options that will help the designer produce acceptable and efficient physical settings (see Figures 6.7.4 and 6.7.5). Some considerations may lead to an extended statement: concerns about areas of possible congestion and rapid or distracted movement may lead to a report that emphasizes safety; concerns about stress, fatigue, boredom, and carelessness may lead to a report that emphasizes the development of options for facility users; concern about cost control might lead to a report on options for obtaining programming recommendations; a study of the archives of design may be used to help identify possibilities for environmental features.

As you prepare to answer the questions in the next part of the program, *5.0 Surfaces,* keep in mind that many users base their judgment about whether a facility is "well" or "poorly" designed primarily on their reaction to surfaces. The influences of visual and tactile stimuli on user acceptance can seldom be overestimated. Even the tiniest detail of a physical setting can make a facility more attractive, and at the same time convey symbolic meaning in artistic form (Bettelheim, 1974, p. 108). Bettelheim suggests that a designer should have a deep understanding of what conveys feelings of emotional comfort, stability, and safety. He would not object that surface design features could be the consequence of the designer's personal preferences, experiences, and convictions about the worth of what it is he or she is attempting to achieve. There are many ways to design an attractive setting for human affairs. What matters is the intention with which something is done, the serious thought that goes into design, and wish to please and arrange things to make a strong positive appeal while transmitting pertinent meaning. Providing users with the best possible surrounding to live in—however different the specific forms and details may be—bespeaks of the care that has been taken to give users the best possible physical and human environment, a message the user will seldom fail to receive and appreciate (Bettleheim, 1974, pp. 118–119). You supply the ideas, insights, and suggestions that will help the designer formulate a response to important environmental design features.

5.1 Surface Characteristics is included in the program because surface features and details should not be selected without justification, but should be a response to specific objectives related to user performance and satisfaction. Your specific proposals for signage, in answer to question *5.2 Signage,* should include any information that will make it easier for individuals to use the facility, under both normal and emergency conditions. You can, in your answer to question *5.3 Durability and Maintainability,* alert the designer to spaces that require special attention because of demanding conditions.

In answer to questions *6.1 Information Flow, 6.2 User Flow,* and *6.3 Equipment and Material Flow,* you will develop movement information for each facility space. As you develop your statements concerning circulation, remember that although user flow has traditionally been considered the key to well-integrated facilities, information flow is becoming increasingly more important. Some circulation patterns are related to security or mandated by law, such as the required separation of juvenile offenders from adult offenders in judicial processes and detention. Others may focus on the need to bring employees in a facility into more frequent contact with one another and to create a sense of efficient and thoughtful service to the public. In any institution, housing, or recreational facility, the entire 24 hr routine must be considered. You will have many criteria to meet as you formulate your response to question *6.4 Critical Circulation Pattern* showing the essential and desirable relationships between spaces. The adjacency diagram you produce will be a principal guide for achieving an integrated space arrangement.

7.0 Space Arrangement

By the time you have finished question *6.4 Critical Circulation Pattern,* you are prepared to synthesize your findings and provide the user of the programming report with a summary of the spatial implications of your recommendations. Your answer to question *7.1 Proposed Space Arrangement* might take the form of a single-line plan view drawing that brings together all your space plans in response to your circulation information. Many will consider it premature to become so specific and may wish to formulate only a schematic plan. Nevertheless, a drawing conveys more meaning with greater clarity than any other type of information statement. The more specific you are, the easier it is to develop alternative space arrangements and have a worthwhile critique of your recommendations. In question *7.2 Space Requirements,* all the spaces you have studied should be listed, along with recommended dimensions based on 4.0 Space Planning Information. This will become the basis for an initial estimate of facility cost. If your human-factors program is complete and your findings accurate, you and your client can be confident that the prepared square-foot estimates are trustworthy. When you prepare your human-factors program for publication, you may wish to present these two summary statements first. The rest of your program can be used to justify your recommendations, to show that you have not made any arbitrary decisions and that everything recommended reflects client/user expectations and requirements. If you take this approach, you will need to accompany your space arrangement statement with a brief summary of your reasoning and justification.

8.0 Site Considerations

As you deal with this section keep in mind that the proposed facility and the site selected for it must be compatible in as many ways as possible. For the user, there may be no line of demarcation between the facility and the surrounding area. Nearby residents and the community as a whole will most certainly consider your facility part of their planned environment. You will be expected to take the following points into consideration as you develop your programming information.

Develop and evaluate land use policy.

Allow for side effects by anticipating consequences of facility development, predicting any possible change of events set off by your project.

Maintain an awareness of lines of property and ease of assemblage of land, recognize significant land use trends, avoid fragmenting and wasteful results.

Conserve natural landscape.

Maintain an awareness that attempts at physical change and improvement are connected to the sociocultural structure of a community, that you cannot disregard anyone's social, financial, and cultural investment (State of California, 1977).

All the information that you have prepared up to this point can help you answer question *8.1 Site Requirements*. You should, however, go beyond these requirements and determine, by application of question *8.2 Planning Factors,* what aspects of the selected site or candidate sites might influence facility operations and user activities. In many situations it will be necessary to control the effects of facility-related activities and operations on the surrounding area and community. One aspect of this concern is highlighted in question *8.3 Services Impact*. Question *8.4 Site Planning* serves as a summary statement of final recommendations for site development. Please note that these four questions alone do not deal adequately with the human-factors aspects of environmental planning. When your objective is at a scale beyond that of a space, building, or facility, at the environmental planning scale, you should apply the whole human-factors program at that scale. This means, for instance, that 4.0 Space Planning would be developed in terms of planned areas, that 6.0 Circulation would apply to the entire development, and that 7.0 Space Arrangement would become a proposed plan for the entire area under development.

6.7.3   RESEARCH METHODS

The human-factors program involves more than the research questions presented. It also includes two sets of research methods that represent different ways of answering the questions posed by the human-factors program. The first set of methods presented, direct contact methods, will enable you to determine how clients and users would answer the questions posed in the human factors program. They also provide a characterization of activities and individual behavior, as well as the spaces and facilities in which they occur. The second set, sociocultural methods, are as basic to your research effort as the direct contact methods. These methods provide a great deal of information about the subjective experience of individuals, such as concerns, attitudes, perceptions, and values.

This presentation of research methodologies completes your introduction to the human-factors aspects of environmental design. You should feel confident that these methods can be made to work for you. Although many designers and planners are reluctant to give the impression that they are expert in human-factors research methods, that is exactly the impression you must give. People want assurance that you are skilled in determining their expectations and requirements. The suggested applications, procedures, and techniques are simple. They work because you are guided by the human-factors program. In this instance, you have an advantage over the behavioral scientist, who is always searching for the boundaries of his problem. You have a natural boundary, your project. You already know the key questions, you have an idea of the type of research report you are attempting to produce. The only thing that should concern you is sources of error. In the following pages you will find discussion about sources of error and environmental and observer bias. Here we discuss how thoughtful preparation can help obtain valid and reliable results.

6.7.3.1   Preparation

Before your research methods can be applied, you must prepare a research design (see Section 6.7.5.1). The purpose of a research design is to increase the clarity and precision of the questions asked and the ways you select to answer them. Your research questions originate with the human-factors program and are answered through the methods presented here. As you develop a research design that specifies your informational objectives and selected methods, you should make a preliminary visit to your candidate study site and try to anticipate problems. The critical questions are as follows:

Have we chosen a study site that will help us achieve our informational objectives?

Will the site be available to us?

Under what restrictions will we be operating?

To what degree will our presence affect the situation?

If a facility study, what individuals and activities will best help us understand the daily affairs of the facility?

If a neighborhood study, what family, peer, interest, activity, street, and transitory groups should we study?

If a community study, what public and private, historic and cultural, religious and political, and local and regional organizations should we study?

Will there be uncertainty about what to observe or whom to question?

Will the observers and interviewers be consistent in what they pay attention to and what they record?

How much training should we give observers, interviewers, and study session coordinators?

Should a pilot study be conducted?

Do our human-factors questions and methods match up with the situation, or should more effort go into their development?

When you are seeking to achieve representative data, the extent of your study can be controlled in several ways. While on site, as your observations of actions, movements, spaces, and objects begin to accumulate, you can stop at some arbitrary point and decide whether or not a sufficient basis for representativeness and inference exists. If not, you can weigh the cost of continuing with the expected increase in information. You may also control the extent of your study by establishing a sampling plan prior to entering the field. You might not observe your study site all day, but only during important time periods. You might select from a number of events those of particular importance. You might classify users according to individual characteristics or activities and sample their behavior in proportion to their numbers in the study situation. Your sampling unit may be a portion of a downtown area, a city block, or a section of a housing project. A natural way for a designer to control a study is to arrange the observation program according to the facility units and spaces that are identified with the building type under consideration. The end result of all your preparation is a chart that instructs the study participants where to be, when to be there, and what to observe and record.

6.7.3.2   Direct Contact Methods

The term "direct contact methods" refers to systematic observation, activity analysis, and such common occurences as the interviews you have with clients and users, the conferences and public hearings you attend, and your questionnaire surveys (Webb, Campbell, Schwartz, and Sechrest, 1966; Zeisel, 1981).

Systematic Observation

Observation is the best possible preparation for identifying user needs and activities. In observation programs, your study user activities and their surrounding environment, using notes, photographs, and sketches to record your findings. It is important to remember that many of the design decisions you make will be influenced by observation results that make it essential that you establish a thorough observation program. In observation you cannot simply follow your eye, for you can be overwhelmed by the complexity of the situation to the degree that your approach becomes random and loses its representativeness. Your first impressions will readily indicate what is interesting, but they will seldom help you determine what is significant.

You undertake a program of systematic observation because you believe it is possible to establish justified design objectives for a new facility by observing existing facilities and the activities of users. The time you spend on systematic observation is particularly justified when you are confronted with a situation that is new for you, or one that is complex or highly variable. Even when you think your experience has prepared you well for your design tasks, it is not unusual to find that the results of an observation program have saved you from making a gross error in judgment. Three questions will help guide your program of study.

What Are the General Characteristics of the Situation?   Observation can yield many important insights concerning the general characteristics of your study site. For example, a 1 day excursion through a downtown shopping area can result in a range of interesting and insightful impressions. In public spaces with wide fields of view, you become aware of the varied and rich aspects of the study site. You recognize that the range of sociocultural character is both broad and select. You note which displays are of interest and which goods draw the attention of shoppers moving from place to place. If you are alert, you will recognize which of the actions, movements, and facial expressions are in response to particular events or to spaces with special features. You become aware of which sounds, lights, and spaces appear to attract the attention of shoppers. You might center your observations around a plaza offering space for the stroller, the active child, the socializing youth, and the pensive who may be watching the flow of water in a fountain. You may observe the flow of traffic and the ease or difficulty in obtaining preferred parking. The results of these observations may evolve into a statement of the human and physical characteristics that go together to produce a shopping environment. At the very least, you have guarded against relying too heavily on your own past experience. You may now wish to proceed to a more intensive and systematic study of specific shopper behaviors.

What User Activities Must Be Kept in Mind throughout the Design Program?   Just as you will find it valuable to have depictions of the general characteristics of your study site on hand, you will want to have references that show user activities of particular importance for design deliberations. Activities are often difficult to describe and assess in terms of relative importance. Activity analysis, discussed in the next section, is intended to characterize user activities precisely. At this point, you should attempt to record activities in the selected study area which are generally known to be important and to note how they vary according to time, event, and circumstance. You might find it beneficial to have on hand a photographic sequence that depicts individuals as they interact with one another and with the spaces and objectives around them.

What Do Design Alternatives Look Like?   What has been selected as a final design objective is often best understood in terms of what alternatives have been considered and which are being followed or rejected. Observation data on alternatives can be used to illustrate what has been achieved in similar situations. You need not restrict your alternatives to situations that are similar to yours; examples can come from all types of facilities if they contain significant design achievements. Even an unsuccessful design may contain details that are significant for your deliberation. Your whole presentation may take the form of an occupancy evaluation, demonstrating that various design features make the environment responsive to the expectations and requirements of users or fail to support individuals and their tasks.

Activity Analysis

Sometimes, you can answer your programming questions only when you have detailed information about users' activities. If you study a children's playground, you will readily understand what is being suggested here. You will see children running, throwing, swinging, jumping, climbing, falling, rolling, tumbling, balancing, creeping, hiding, digging, skating, skateboarding, and cycling. To place these activities in order, you may classify them according to user group, game, or event. As you consider these activities, you will form associations between activities and environmental features. Jumping, rolling, tumbling, crawling, and digging will bring to mind landscaping possibilities. Balancing, climbing, swinging, and falling evoke thoughts of bars, ropes, poles, and safety features. Cycling, skating, and skateboarding are associated with hard surfaces, flat and contoured. Pushing and hitting suggest wide fields of view for monitors, so that these activities can be spotted and controlled before they develop into fights. Each game and recreation event brings to mind additional environmental features to support user activities. Activity analysis follows this common-sense approach to identifying important design and planning considerations.

Which Facility Users Must Be Given Special Attention?   The first step in activity analysis is to specify the individuals whose activities should be considered in detail. This would include users under constant and prolonged stress—assembly-line workers, emergency ward staff, police officers, single heads of households, detention facility staff and inmates, and those in understaffed environments. The activities of users prone to fatigue or boredom may be considered in detail—clerk typists, bedridden invalids, students, military personnel, keypunch operators, prisoners, fast-food service employees, draftsmen, and copy and reproduction workers. In your work, you may wish to consider users likely to become disoriented or confused—travelers, first-time users of a facility, those caught in emergency or crisis situations, those who don't speak English, and those with a biochemical dependency. Another category might be people who are emotionally unable to deal with anything other than a perfectly designed environment—senile people, children and adults with learning disabilities, individuals who are easily frustrated, and those with emotional problems.

Some individuals may need special consideration because of physical disabilities that affect vision, speech or motor movement, and coordination. Some individuals will be selected because of a stressful work situation owing to the pressure of the job, detailed and demanding work, or the critical nature of their tasks. Individuals who are particularly important to an organization may be singled out for special attention to emphasize that their concerns and needs are paramount. Listings of individuals to whom activity analysis should be applied may be quite numerous because we are part of a complex culture that employs advanced technology and expanded services and that attempts to create and sustain equal opportunity.

Which of Their Activities Are of Special Importance?   Even the simple movements of disabled individuals (walking, grasping, reaching, sitting) should be given the closest scrutiny. For these people, common daily events such as bathing, cooking, or mailing a letter can assume major importance. On the other hand, the activities that make up the daily routine of ward nurses, clerk typists, laboratory scientists, and technicians may have to be carefully studied. Certain activities may require attention because of the characteristics of the structure, for example, long corridors, many stairs, common spaces, limited space, distracting operations, and hazardous areas. Detailed attention must be given to activities that are likely to cause stress, fatigue, boredom, or confusion, as well as detailed, demanding, or critical functions. Lack of support for such activities will adversely affect convenience, satisfaction, and performance.

What Are the Important Relationships between Users, Their Activities, and the Immediate Environment?   After you have identified the individual and group activities toward which activity analysis will be directed, you must select the best approach to this final step in activity analysis. There are a number of alternatives among the direct contact and sociocultural methods, and it should not be difficult to determine which will work best in a given situation. The methods you select and the way you apply them will depend on the kind of results you want to obtain.

You might wish to understand how activities vary by user and event. Through application of direct contact methods, you would develop a graph that depicts this information. The more specific and quantified the results of activity analysis, the more value they will have. Your activity analysis may require that you characterize occupancy, movement, and communication patterns, seeking to develop activity data in the context of spaces and the circulation between them. Mapping and flow diagrams are excellent formats for occupancy and movement patterns. They rely upon the association between a visual figure and indicated number of users rather than a verbal label and its corresponding numbers, as does a table. Although this information can usually be shown in a table in less space and with less effort, the figures in mapping and flow diagrams convey additional meaning to the reader.

Quite often the goal of activity analysis is to develop a communication matrix. Many facilities must provide communication channels that interconnect individuals and groups in patterns that permit rapid and accurate transmission and collation of information (Kelley and Thibaut, 1968, p. 19). This is particularly true for medical facilities, educational centers, museums, libraries, and government and business offices where information must be routinely developed, stored, retrieved, presented, and exchanged. Even in the area of housing, particularly large housing projects, information may be the key to convenience, safety, security, and the prevention of facility misuse. The first step in developing a communication matrix is to identify each means of communication, such as information displays, signage, communication equipment, face-to-face conversation, telephone conversations, and correspondence. The next is to develop a communication log suitable for your purposes. Measurements of frequency of exchanges and the numbers of individuals involved are achieved by asking participants to maintain a log for a selected period of time.

Activity sequences are frequently used as a guide for space planning details. The sequences of activities occurring in a physical setting should be described as completely as possible. Because this may involve a large quantity of information, the challenge is data reduction. This is the purpose of an activity chart. The activities occurring in a particularly important space are arranged as a sequence of actions and communications, accompanied by a column that shows time of occurrence. Sometimes a sequence of activities can be graphically illustrated in a semipictorial representation.

It is recommended that you begin your research by making a preliminary visit to the study site and try to anticipate and provide for problems with observation and measurement. The way in which you prepare for your study, make and record your observations, and organize your data is critical to the success of your efforts.

Interview

The purpose of the interview is, of course, to ask questions that will motivate people to talk in a manner which is both informative and insightful. You can rely upon the individual and his or her perceptions. Individuals can be very analytical when they discuss their experiences and suggest ways to enhance the settings for their daily affairs. Your biggest possible error would be to ignore an individual as a source of information because he or she doesn't appear to have the right kind of background or preparation to deal with your information needs. You must assume every individual is a potential source of valuable information.

You will first encounter the interview during the selection process. The client has one overriding concern, that is, a facility that fulfills its function and does it well. He or she will seek a designer or firm that demonstrates a willingness to be accountable for a facility design that corresponds to the activities and the personal perferences of its users and enhances their daily experience. In subsequent interviews, clients will attempt to evaluate your rationale and justification for the work proposed. Even at these early meetings, you might provide the client with a copy of the human-factors program to demonstrate your willingness to be accountable for a wide range of considerations, and to structure interviews to your advantage. This type of interview may be characterized as formal. A more suitable form for the many interviews you will conduct yourself is the informal interview, which is essentially conversation—conversation initiated by an explanation of your purpose and structured only by a few key questions or comments. Successful informal interviewing depends less on the exact nature of the questions than on the skill of the interviewer. A well conducted interview is likely to create a high degree of motivation and interest because a person being interviewed feels that one's observations and opinions are really important. One is able to use one's own idiom, to organize one's thoughts in one's own way, and to bring up any aspect of the situation that concerns one personally. The goal is to create the type of interview in which at the end—when the tape is played back—you might find the individual saying, "I never realized I felt that way" (Terkel, 1972, pp. xx–xxi). The importance of the free-flowing, unstructured interview is demonstrated when conversational data drawn from naturally occurring encounters are compared with the results of formal interviews. The bulk of data actually considered significant usually comes from the less structured situations (Van Maanen, 1973, p. 409).

You may wish to structure your interview gently, calling attention to a particular problem while avoiding overdirection. This can be done by showing the individual a photograph of the problem area and asking if he or she is acquainted with this place or has had experience in similar situations. You might ask the person to recall his or her past experience and comment on these recollections. Contrasting situations or different physical settings may also be depicted visually, along with a question asking the individual to express feelings and experiences. This technique is particularly valuable because it brings many issues directly before both the user and the designer and requires few theoretical guidelines. Although the approach may be less than systematic, it helps develop an understanding of the private world of meaning of the user.

Study Sessions

Study sessions may be defined as crucibles for interaction between participants that result in new information or evaluation of existing information. The basic procedure is to assemble a group of six or seven individuals, with additional groups as necessary to achieve a balance of experience, views, interests, and intensity of feeling about the design project. The participants are assigned both individual and shared tasks. A record of progress is maintained, and there are a summary and critique of results. This helps assure that differing opinions about the significance of the information obtained are identified and discussed.

The study sessions are used to acquaint the participants with the scope of concerns identified in the human-factors program, to examine the way they use their present spaces, to help them select the area in the new facility they would like to occupy, to resolve conflicts, and to formulate a final space plan for the facility. In another situation, the study group might be asked to review the result of direct contact and sociocultural methods and to help prepare conclusions. Participants may need to leave the study session room to observe and experience certain aspects of the study environment firsthand.

Questionnaire Survey

It is not always possible to reach everyone whose opinions might be of value. If your design problem indicates the need for broader coverage, you may elect to use a questionnaire survey. Although the questionnaire survey is widely used, it is frequently misused, overextended, and distrusted. In addition, many people dislike or feel little enthusiasm about responding to a questionnaire because it places them in a highly structured and passive situation. In spite of these disadvantages, a skillfully designed questionnaire, such as the *Position Analysis Questionnaire* (McCormick, Jeanneret, and Mecham, 1969), can obtain information from a large number of people with considerably less effort than that needed for some of the other direct contact methods, and may in some cases cause less embarrassment or inconvenience.

The questionnaire depends almost entirely on words. It involves the asking of questions, in one form or another, and the recording of oral or written responses, with the intention of accumulating information about individual experiences, activities, opinions, habits, attitudes, customs, goals, and preferences. In order for this information to be of value, however, great care must be taken both in the formulation of the questions and in the way they are presented. It is also necessary to find people who are interested enough in what you are doing to fill out the questionnaire and who are representative of those affected by your final design and plans.

6.7.3.3   Sociocultural Methods

Sociocultural significance is a fundamental attribute of the built environment. There is no question in the human-factors program that cannot be answered more fully as a result of your assessment of the sociocultural context of your project (Moore and Golledge, 1976; Finsterbusch and Wolf, 1977). Your human-factors programming study must be conducted in such a way that one is ultimately able to see possible relationships between the built environment and the social structures people use to organize their activities and relationships, the preferences for one environmental feature or scheme over another, and the attitudes and customs that permeate daily life. How to respond to such a complex challenge in the limited time allowed us in our projects is what is presented here. Studies based on the model for social science research can often require from 6 months to 2 years to complete. For this reason, it is necessary to follow a rather limited approach in professional situations, in which much less time is normally available.

Here, as with all the methods presented in the handbook, a thorough effort and a systematic progression toward informational objectives are of primary importance. The sociocultural methods are organized into a three-step sequence that should be followed. The goal of the first step, group assessment, is to provide a thorough and complete answer to question 3.1 User Categories. The next step is to study the preferences of the identified groups along the topic lines contained in program sections 4.0 Space Planning, 5.0 Surfaces, 6.0 Circulation, 7.0 Space Arrangement, and selected aspects of 8.0 Site Considerations. Having identified the groups that will use the planned facility and determined their preferences, your study can be extended to assess the attitudes and customs that provide the background for these preferences.

Group Assessment

The first step in group assessment is to identify the people who will use the facility, by role and activity characteristics. If you are following the human-factors program, this process will begin very early in your program. It is seldom ever fully complete, because new individuals and roles will be identified throughout the program. Start your group assessment with a "snowball" survey (Harrigan and Harrigan, 1979). The term "snowball" is suggestive of the application procedure. The initial task is assigned to selected personnel, who are asked to write down as many different facility users as they can think of. This information is edited into an initial listing of facility users. This listing is then sent to another group of facility users, who are asked to review the work of the first group. These people are encouraged to add, delete, and revise the first listing of facility users. The first group is given the same opportunity to revise the listing. In the final step of the "snowball" survey, the general user groups can be asked to review the results and make their contributions. The systematically accumulated results provide a thorough and complete answer to question 3.1 User Categories.

As you work to identify group members, you should be aware that groups have both a formal and an informal dimension. The formal dimension is a reflection of what the organization believes the purpose of the group and the responsibilities of group members to be. The various ways individuals perceive their role and their responsibilities in the group constitute the informal element. Organizational descriptions of groups are found in the programs, policies, regulations, schedules, and planning documents that direct the activities of group members. A content analysis of these documents will reveal the enduring characteristics of groups. As you identify the formal dimension of a group, remember that each member will have developed his own perception of its characteristics. Your goal is to discover the degree of similarity or dissimilarity between organizational descriptions of groups and individual perceptions of the group and personal roles.

The design and planning implications of group dynamics can be understood only through collaboration between designer, client, and user. In the most favorable situation, people will be quite willing to describe and discuss their feelings about group activities and their personal roles. In other cases, it may be necessary to ask one group of individuals to help you define another group, because the prime group will not talk to you or is not available. Harper and Harris (1975, p. 157) developed this approach and used law enforcement personnel to help them determine the structure of criminal organizations. They cite a number of state agencies that have followed this procedure to improve their understanding of organized crime. This technique is especially useful when you need to identify neighborhood and community groups.

Preference Assessment

After you have identified groups of anticipated facility users, their preference can be determined through the use of interviews, questionnaires, and group study sessions. It is at this point that you see where the efficiency of this three-step process is found. By identifying groups first and then going to them directly, you eliminate "blind" preference assessment. Your results are assignable to specific groups, not to an unspecified array of possible facility users.

In preference assessment you present to your identified groups design alternatives and features that have the potential of providing for the expectations and requirements of users. There should be no reluctance to propose design options and possibilities. In order to deal with the question of preference with some efficiency, some kind of boundary is needed. Your presented options will provide this focus and reduce the demands of preference assessment to a minimum. Consider, for example, the case of housing for the elderly. Various individuals and institutions have tried to determine the influence of housing on the life styles of the elderly. After reviewing the literature in this area, it is possible that you would identify two design alternatives as most appropriate for your situation: a high-rise apartment plan and a high-rise hotel plan and all the associated features and services. These alternatives would then become the focus of your preference assessment effort.

The results of your preference assessment study would of course be greatly influenced by such situational factors as region of the country, location within the region, and possible neighborhood peculiarities. Your design alternatives should identify the features of a housing design that are of critical importance to elderly users in a particular situation. At the end of your preference assessment study, you may select one alternative or the other—the high-rise apartment plan or the high-rise hotel plan—or you may decide that some other solution is more appropriate. This final recommendation will now become the focus of your background assessment effort.

Background Assessment

The last part of question 3.3 Sociocultural Characteristics asks: Are these characteristics stable or likely to change? It is because of the possibility for a change in preferences that you must assess these results further. Though definitive studies are best, it is necessary to follow a rather limited approach in professional situations. We must depend on literature research and a limited amount of reassessment of user preferences. In our example, your background investigation would make it clear that the attitudes and customs of the elderly are incredibly diverse and frequently contradictory. A tendency to want to be active is a very positive characteristic of older people. On the other hand, some may exhibit a reduced capacity for adaptation and may be subject to fears of failure if the environmental structure does not clearly indicate what behaviors are possible or appropriate. Some of the elderly wish to be able to manipulate their environment to suit individual needs, and to be able to participate in community, recreational, and social affairs to the extent that they themselves desire. There may be fear of regimentation, and acute dissatisfaction with any housing design that restricts freedom to choose activities or boundaries, or that does not allow free expression of individual life styles. Just as many older people, however, actually prefer a highly structured environment that includes definite boundaries and security precautions, housekeeping services, health care facilities, and readily available companions. Any structure that fails to ensure privacy is likely to be rejected, as will one whose design leads to passivity or isolation. This is the type of information you develop in background assessment. It helps you understand expressed preferences. At the end of this study of the full complexity of the situation facing you, you may wish to revise your design alternatives.

How far must we go in this last step? Wolcott (1975) provides a guideline for background assessment, its goals and limitations. He suggests that a deceptively simple test for judging the adequacy of accounts of attitudes and customs that result from background assessment is to ask whether a person reading them could behave appropriately as a member of the society or social group about which he has been reading, or, more moderately, whether he can anticipate and interpret what occurs in the group as accurately as its own members can. He encourages people who may not be of anthropological persuasion to draw upon the facets of the anthropological approach without feeling that they must make all the commitments and meet all the prerequisites of the professional anthropologist. He assumes that there is a bit of anthropological talent in each of us. He stresses that it is important to look for a connection between little events and big ones. A degree of personal involvement is important for success, yet an advocacy position must be avoided. Finally, Wolcott cautions that background assessment is a high-risk, low-yield venture in terms of the time that must be committed to it and the fact that it is more suited to generating than verifying hunches or hypotheses.

6.7.4 SOURCES OF ERROR

Because every measurement technique produces a degree of error, you must be aware of sources of error and how to guard against them. The following application considerations may help ensure that your results are accurate and make a worthwhile contribution to design development.

6.7.4.1 Basic Considerations

The results of research method applications will assume various forms—photographs, tape recordings, written statements, notes, recollections, and counts and tallies of various individual responses and judgments. Often a combination of photographs and quotations taken from interview and study session material will indicate that a particular viewpoint characterizes users' feelings. If you intend to stress this viewpoint in your deliberations, you should keep a tally of how many individuals expressed it. You also need to state how representative these individuals are of the total group being studied, and indicate the frequency of opposing points of view. Always remember that you cannot base decisions on data alone. You must indicate their source, the methodology by which they were obtained, and how reliable they are. Your best personal safeguard is a healthy skepticism for all the data until you can be confident that they are representative of the situation.

Error is a factor that cannot be overlooked. It can appear inadvertently, as the result of inadequate preparation or haphazard method selection and application. Rosenthal (1976), discussing error, mentions a number of possible sources. Some of these possibilities might not occur to you; some you may be unable to control but should nonetheless keep in mind. Rosenthal would first ask you to be aware that you yourself can become the source of error. As an observer of behavior, you might be inconsistent, make recording or computational errors, or unintentionally slant results. Certain personal attributes such as your sex, age, ethnic background, or religion might influence the way individuals respond to you during interviews, personally administered questionnaires, and study sessions. According to Rosenthal, anxiety, need for approval, hostility, authoritarianism, intelligence, dominance, relative status, and warmth are psychosocial attributes that may influence the behavior of those with whom you are dealing. These same personal and psyshosocial attributes are found in those with whom you have contact, of course, and will have an effect on your diligence and thoroughness. Situational factors such as the physical setting in which you are conducting your research and your prior experiences in a particular situation may restrict your behavior. If you are personally acquainted with those from whom you obtain information, personal regard or disapproval may influence your interpretation of what you see and hear. The relationship between you, your associates, and those who are working for you can also affect performance and results.

There is one additional source of error over which you do have control. Rosenthal stresses that your expectations for the outcome of your study can lead to data and conclusions that are not true representations of what exists. In an interview, you may unintentionally communicate your wishes. You may restrict communication, nonverbally signal approval or disapproval, or expand or reduce the time spent with an interviewee, depending on whether you like or dislike what you are hearing. As you begin to get returns from your study efforts, the initial results may satisfy or dissatisfy you to the extent that you modify your effort to produce more or less of the same. The influence of error on your data can be reduced by self-discipline, careful selection, and training of the individuals who will be responsible for the study.

6.7.4.2 Environmental and Observer Bias

As you identify attitudes and customs and the preferences they produce, you should be fully aware that the effectiveness of these efforts may be limited by environmental and observer bias. A primary source of bias is the environment itself. The attitudes and customs of users are limited, fostered, and often controlled by the characteristics of physical settings they are most familiar with. You must always consider what might change if a different experience existed.

Observer bias may have many sources, as summarized in the following paragraphs (Sparadley, 1972, pp. 21–25).

Cultural Pluralism

In our society, there exist thousands of different roles resulting from specialization, and many distinct life styles exist in every cultural group. As a result, such general distinctions as "elderly," "adolescent," "college student," "handicapped," "lower, middle, and upper class," and "socially ascending and socially fixed" have little significance for guiding our observations.

Ego-, Group-, and Ethno-centrism

The validity and dignity of diverse cultural traditions are too frequently judged in terms of the observer's own culture. Observers must realize that they will always be influenced by their own culture and as a result may misinterpret the behaviors of others. It is helpful to have as a collaborator someone who can represent other points of view, as you proceed with your preliminary fieldwork.

Averaging

Demographic information frequently obscures important cultural differences. Individuals sharing neighborhoods and community facilities, transportation systems, educational institutions, and many other facilities may assign to them very different attributes, values, and meanings that are not apparent in statistical summaries.

Labeling

Once an observer has classified individuals as elderly, youthful, adolescent, or middle aged, or as having particular values or interests, all their observed behaviors and characteristics will be colored by that label (Rosenhan, 1973, p. 253). The label can be so powerful that many of the behaviors of individuals or groups can be overlooked entirely or profoundly misinterpreted. You must recognize the possibility that your perception of the expectations and requirements of users may be shaped entirely by the labels you choose in your initial identification of users.

6.7.4.3 Test and Evaluation Measures

Your attempt to reduce error should be supported by test and evaluation measures. The value of detailed analysis is becoming more apparent as designers deal with increasingly complex environments. All the research measures presented in this handbook can contribute to the assessment of proposed design objectives, criteria, and specifications. You will certainly want to review the presented human-factors information to compare what you are proposing in the context of what is known about human limitations and capabilities and effective environmental features. You will want to model in a suitable scale candidate space plans and will always conclude your test and evaluation steps by collaborating with clients and users to review and critique your programming recommendations.

The Handbook

Human-factors guidelines have something to contribute to every test and evaluation undertaking. As you leaf through this handbook, you will note that most information is highly detailed, often presented in tables and figures. The information presented indicates the range of human limitations and capabilities and tolerable environmental features. If you are to make this information meaningful, it must be placed in the context of your problems. This information will mean very little unless you visualize your facility user unable to move easily through the facility because the space allowed restricts movement; as failing to develop a sense of orientation and direction because the visual surround is confusing and signage inadequate; as experiencing fatigue, stress, and frustration because the environment does not support his or her activities; as attempting to adapt to light and sound levels that are intolerable; or as being exposed to an environment likely to produce accidents because of the failure to meet human requirements and safety specifications. You must give life to your source material by envisioning what the environment means to the user.

Modeling

After you have reviewed the literature that relates to your design options, you will probably decide to construct models of some of your options in order to assess their merit directly. Architects and interior designers traditionally construct models to exhibit design achievements, to help determine the relative acceptability of design alternatives, and to demonstrate the inherent potential of design concepts. Some designers use models throughout the design process to evaluate the conclusions reached at one stage of design before proceeding to the next. As described here, modeling is considered to be a research method that makes certain that design recommendations are characterized and evaluated as realistically as possible. The goal of modeling is to guard against the introduction of undesirable environmental features that interfere with or fail to support user intentions. Without the use of models, opportunities for discovering and evaluating new design concepts and solutions to complex activity–space relation problems may be overlooked. The cost effectiveness of modeling is well established. It is far more economical to work out human-factors considerations and problems with preliminary models than to make changes in the later stages of design or to modify working drawings. Effective use of models during the design process makes it possible to avoid problems that must be corrected later, with organization or people.

A simple and effective preliminary approach to modeling uses two-dimensional paper cutouts of furnishings and equipment at the scale of $\frac{1}{2}$ inch to a foot. In this way, it is possible to experiment with a wide variety of alternative space plans. The scale of $\frac{1}{4}$ inch to a foot may be preferable if working drawings to this scale are already available. A modeling system at the scale of 1 inch to a foot permits the examination of detailed aspects of activity–space relations. If your human-factors program requires a detailed examination of a complex work space, you may find it helpful to prepare models at the scale of 2 inches to a foot, or even move on to full-scale models (Kilbokawa, 1968; Hall, 1976).

Review and Critique

Proposed design objectives, criteria, and specifications should be reviewed and critiqued by your associates, clients, and users. The information you used, the design options developed, the results of your review of human-factors literature, and the models you have constructed should be presented for inspection. This is the safeguard that applies to every step in the human-factors program. Your task should be shared by those who are familiar with the problem area under consideration.

6.7.5   PROGRAM MANAGEMENT

If not properly managed, application of the human-factors program can produce unacceptable demands and confusion about what to do, how to do it, and when to do it, causing inefficient use of resources, project delay, and budget overrun. This consideration should be foremost in your mind as you develop a scope of work and budget for your project.

6.7.5.1   Scope of Work

The first step in programming is to select the questions to be asked and the method of investigation to be employed. It is likely that you will experiment with various types of research design. You will consider the information presented in this chapter in terms of each situation and develop the most economical and efficient research design possible. From project to project your strategy will change. Each question item in the program will be carefully evaluated to determine whether or not it is applicable. You can then select the combination of research methods that best suits your needs. In one type of situation you might pair the three user characteristic questions, 3.1 User Categories, 3.2 User Activity Description, and 3.3 Sociocultural Characteristics with questionnaire survey and systematic observation methods. With this research design you can provide the design team with a ready reference, a human-factors perspective that ensures that future users of a facility are identified, that descriptions of anticipated activities are available, and that the various life styles of users are taken into account. Because the movement of people, information, equipment, and materials is often the key to good facility design, you might develop a scope of work pairing the questions contained in 6.0 Circulation with the activity analysis method. When you are pressed for time or limited by budget, you might consider pairing all eight sections of the human-factors program with one method, study sessions, and work with clients and users to determine what they anticipate in the way of facility features. Certainly for major projects you will want to employ the entire human-factors program, follow the recommended programming sequence, and use as many of the research methods as necessary to establish a precise, thorough, and comprehensive view of user expectations and requirements. Whether you conduct a complete or partial program you should always refer to the rest of this handbook in order to identify additional concerns and methods that will lead to a more complete effort on your part.

Your final scope of work statement will consist of a step-by-step description of your proposed activities and analyses. Figure 6.7.2 provides you with some idea of what might be included. This 1 week intensified period of study, an abbreviated research design, refers to no particular facility. It is a general guide that may be expanded into a more complete scope of work specific to your project and situation.

	MONDAY	TUESDAY	WEDNESDAY	THURSDAY	FRIDAY
MORNING	Meet with client representative who will work with you. – – – – Meet with principal users to explain week's activities.	Systematic Observation	Study Session – – – – Study Session	Interview those identified in study sessions as important to project.	Double-check on selected findings.
	Lunch with those directly responsible for the project.				
AFTERNOON	See all groups to pass out questionnaires and explain purpose of scheduled study sessions.	Study Session – – – – Study Session	Systematic Observation – – – – Collect questionnaires.		Meet with participants to present and critique results.
EVENING	Drive through surrounding neighborhoods and communities.	Begin to organize collected data.	Compile and analyze questionnaire responses.	Summarize questionnaire, study session, and interview results.	

Fig. 6.7.2.   A schedule for an intensive week of study, data collection, and results review.

6.7.5.2   Budget

Upon completion of your scope of work, it is possible to estimate your budget for the project. The following discussion of "person-hours" provides a helpful guide. In Figure 6.7.3 budget is indicated in terms of person-hours. This sample budget is based on the hours required to produce a human-factors programming report for a complex office facility of 20,000 ft² and 86 daily occupants and visitors. The person-hour figures are based on the author's experience in applying human-factors programming. It is an appraisal of the time necessary to complete various phases of a scope of work, which can assist you to make an accurate estimate of person-hours. If you overestimate your needs you might not receive the support you need. On the other hand, if you base your funding on too few person-hours, you will overrun your budget.

The "Milestones" column in Figure 6.7.3 lists all the principal steps in the programming study. Milestones entitled "First Review," "Second Review," "Report Delivery and Presentation," and "Consultation with Design and Planning Group" serve to remind you that you must maintain close liaison with professional associates, clients, and users in order to be sure that the programming information will be used. If you do the programming and work directly with the design team, your programming recommendations will play a significant role in the project. If you present the programming report to the design team and fail to follow through, it will seldom be effectively used. The "Post-occupancy Evaluation" is as important for you as it is for the client, because the human-factors program must always be considered as an experiment, the results of which must be subject to test and evaluation. If the client is unwilling to accept the cost for this effort, you should plan to sponsor it yourself. It is an investment that will assure the fullest profit from the entire programming effort. The client should certainly recognize the value of an assessment of how well the new facility is working. It is possible to identify weaknesses in facility design at this stage, before serious problems develop.

Figure 6.7.3 shows that 160 hr are allocated to 3.0 User Characteristics. Of the eight steps in the scope of work, this requires the most hours for two reasons. First, the taks involved are time consuming in that you will be studing the client/user situation directly and will devote many hours to interviewing, questioning, observing, and study sessions. Second, the information produced (descriptions of user categories, user activities, and user sociocultural characteristics) must be as complete and accurate as possible. Throughout the human-factors programming effort you will refer to this information time and time again. It will, in fact, be your principal reference. Steps 1.0, 2.0, and 3.0 in the program require a total of 216 hours, which amounts to 29% of the total hours for the project (216/750 = 0.29). An additional number of hours, 260, is allocated to steps 4.0 through 8.0, that is, 35% of your total time. The remaining 36% of the allocated time goes into program management and applica-

Milestones	Schedule	Scope of Work	Budget: Man-Hours
Preparation of Programming Proposal	One week effort		40
Contract Negotiation			8
Program Steps	Weeks 1-4	1.0 Program Context	16
		2.0 Organization Characteristics	40
		3.0 User Characteristics	160
First Review	Week 5		8
Program Steps	Weeks 6-9	4.0 Space Planning	60
		5.0 Surfaces	40
		6.0 Circulation	40
		7.0 Space Arrangements	40
		8.0 Site Considerations	80
Second Review	Week 10		8
Report Production	Weeks 11-12	Graphics, typing, production, reproduction	130
Report Delivery and Presentation	As scheduled		8
Consultation with Design and Planning Group	As scheduled		32
Post-occupancy Evaluation	As scheduled		40
			Total 750

Fig. 6.7.3.   A typical human-factors program. This management scheme is applicable to most programming efforts. The total number of person-hours allocated will vary from one project to the next, but the proportion will remain relatively constant.

tions, liaison, and report development. These proportions, 0.29 to 0.35 to 0.36, will remain fairly constant no matter what the nature of the project. If you follow this breakdown, your estimate of person-hours required will be realistic. You should be aware that step 8.0 Site Considerations may require less time if a site has already been selected. If, on the other hand, you have to go through a process of site selection, the number of hours allocated might go far beyond what is shown in Figure 6.7.3.

6.7.6. FINAL REPORT

When we observe in a space, building, or facility that installed furnishings, fixtures, and equipment; ambient environmental conditions; features related to convenience, safety, and security; surface designs and treatments; space and area plans; and structural and site developments provide a perfected environment for users, we wonder how it was accomplished. In design development, who went out of his or her way to see to user needs and wants? Did the client and user represent human-factors concerns in an unequivocal and demanding fashion? Was the design team unusually aware of the user? Did someone else assume the responsibility for human factors—the specifier, the interior designer, the contractor, the craftsmen on the job, or the occupants, through subsequent remodeling? Any or all of these project participants could have been contributors. The goal of this part of the handbook is to identify for such contributors human-factors possibilities from the simple intention to have user-oriented design become an everyday concern to informal and reasoned application to complete human-factors programming efforts.

If you are already experienced in the human-factors aspects of environmental design, you know that the problem is not to find solutions for meeting users' expectations and requirements. Designs that lead to a perfected environment can be developed. There are a great number of possibilities and options. They exist in our archives. We have them in mind as we recall what we did on a similar project. We are inventive and can quite readily produce a perfected environment for the new or unique. What is primarily needed is a good project beginning, an accurate and complete statement of the client's situation and the expectations and requirements of users. A successful human-factors programming report, then, is one where the world of experience to be found in practicing professionals is realized. When you provide designers with a completed human-factors program you make them efficient in realizing their potential for achieving perfected environments.

Consider what you will be doing within the human-factors program. You will be working to identify users, to study their activities, to identify concerns, preferences, and requirements, and then to use this information to establish justified design recommendations. You will use direct contact methods to establish individual points of view concerning the questions posed in the human factors program. You will use sociocultural methods to assess the relationships between the built environment and the group dynamics, preferences, attitudes, and customs that underlie user needs and wants. As a result, you will know a great deal about users, enough to formulate recommendations. Toward this end, your final report should contain specific design objectives, criteria, and specifications, as illustrated in Figure 6.7.4, and the background and justification information similar to what is depicted in Figure 6.7.5. Figure 6.7.5 illustrates a logical approach for converting user expectations and requirements into final programming recommendations. In your first step you will draw together all your research findings. After you organize available information you will be ready to develop and select design elements. In your next activity you will formulate recommendations and then develop the type of final statement you see illustrated in Figure 6.7.4. If you find that more than one recommendation is possible, that is fine. If you supply alternatives, the design team will have more flexibility as it responds to your programming recommendations. Sometimes your study will lead you to conclude that it makes little difference as to how a space is designed, as long as it contains certain essential items. It is also good to note this, providing the designer with the fullest amount of latitude (Chapanis, 1971).

This three-step process—organize available information, select design elements, and formulate recommendations—is not the last step of your programming activity. It is part of the in-process programming effort. From the start, you must be end-product oriented. What is particularly important is that you see all that has been presented as a description of an information management system, and further, that the human-factors program is highly suitable for computer applications and utilization of text-based information system software.

6.7.6.1 An Ongoing Process

The logical approach for converting study results into final programming recommendations depicted in Figure 6.7.5 is a complex undertaking. Each of the three steps noted is demanding. To be successful, you have to rely on your professional experience and your knowledge of the subject building type. As you program, you are required to interject your expertise. What you want to achieve is a balance between insight and analysis. This flexibility in approach is provided for. There is ample opportunity for all your personalized touches. At any time, at any place in the program you can propose candidate design features. Though spontaneity is desirable, you do have a routine to follow.

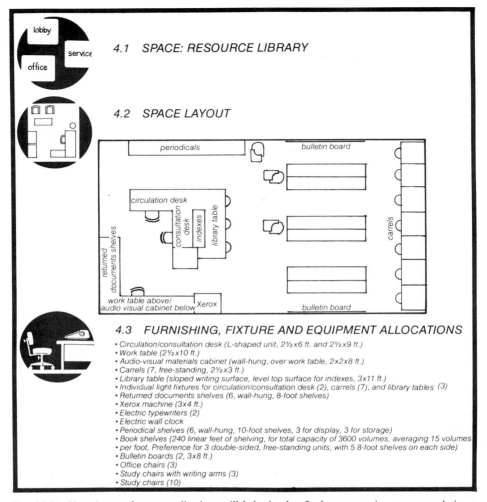

4.1   SPACE: RESOURCE LIBRARY

4.2   SPACE LAYOUT

4.3   FURNISHING, FIXTURE AND EQUIPMENT ALLOCATIONS

- Circulation/consultation desk (L-shaped unit, 2½x6 ft. and 2½x9 ft.)
- Work table (2½x10 ft.)
- Audio-visual materials cabinet (wall-hung, over work table, 2x2x8 ft.)
- Carrels (7, free-standing, 2½x3 ft.)
- Library table (sloped writing surface, level top surface for indexes, 3x11 ft.)
- Individual light fixtures for circulation/consultation desk (2), carrels (7), and library tables (3)
- Returned documents shelves (6, wall-hung, 8-foot shelves)
- Xerox machine (3x4 ft.)
- Electric typewriters (2)
- Electric wall clock
- Periodical shelves (6, wall-hung, 10-foot shelves, 3 for display, 3 for storage)
- Book shelves (240 linear feet of shelving, for total capacity of 3600 volumes, averaging 15 volumes
  per foot, Preference for 3 double-sided, free-standing units, with 5 8-foot shelves on each side)
- Bulletin boards (2, 3x8 ft.)
- Office chairs (3)
- Study chairs with writing arms (3)
- Study chairs (10)

Fig. 6.7.4. Your human-factors applications will help develop final programming recommendations.

For a particular project you will have created a scope of work (see Section 6.7.5.1). As you carry out the scope of work, a data base for your programming deliberations will result. You will be organizing and evaluating information as it develops. A great deal of the information will be readily responded to. You will have little difficulty in arriving at appropriate design objectives, criteria, and specifications. However, in every project you will uncover complex, critical, and subtle issues. Here your design element selection process will be more complex. You will seek additional information, create a range of design feature options, and evaluate their potential for meeting identified expectations and requirements. What you see illustrated in Figure 6.7.4 is an effective format for your final recommendations. This should be accompanied by an information summary that supplies justification, noting what expectations and requirements your recommendation is responding to.

6.7.6.2   Information Management System

To realize the full potential of the human factors program, you need the processing power and utility of a personal computer. As you work with your data to produce design objectives, criteria, and specifications, you need to process information and findings, determine design implications, and arrange blocks of commentary and data to produce a final report with speed and precision. At any time in the human-factors programming process you must be able to present obtained data and promising design

DEVELOPMENT OF DESIGN RECOMMENDATIONS FOR RESOURCE LIBRARY

ORGANIZE AVAILABLE INFORMATION

USERS:	Resource Librarians, Visitors, Telephone/Mail Contacts, Janitor, Handicapped, Repairmen, Vandals
ACTIVITIES:	Library Services, Research, Report Preparation, Studying, Reading, Eating, Listening to Tapes, Answering Telephone
FURNITURE:	Circulation/Consultation Desk, Work Table, Audio-visual Cabinet, Book Shelves, Office Chairs, Study Chairs, Library Table, Card Catalog, Carrels
FIXTURES:	Task-oriented Lighting, General Lighting System, Clock
EQUIPMENT:	Telephones, Typewriters, Xerox Machines, Tape Recorders
ENVIRONMENTAL CONCERNS:	Stuffy Air, Glare, Noise
CONVENIENCE, SAFETY, AND SECURITY:	Braille System, Wheelchairs, Walking-aids, Forced Entry
SURFACE CONCERNS:	Attractive Appearance, Food Stains
CIRCULATION:	Visitors moving throughout the facility, closely associated with classrooms.

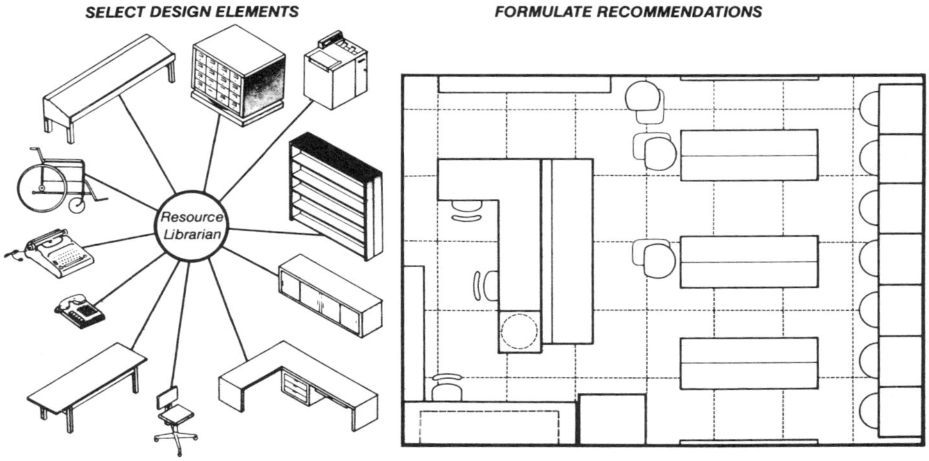

SELECT DESIGN ELEMENTS   **FORMULATE RECOMMENDATIONS**

Fig. 6.7.5.   Research results must be condensed and reformulated for design application.

options to clients and users for their review and critique. You need to, in a timely fashion, brief the design team so they can begin to utilize findings at the earliest time possible. You need to be able to compare candidate design features with archival information and specifications from previous projects. To support these activities the human-factors program was designed as an information management system. Each of the 26 question items in the human-factors program was established as a data file. Data will be stored by data file and indexed by key words identifying project title, data source, subject design element, and building type.

Two text-based information systems are worthy of note. They differ in cost and complication, yet each is an integrated system that can be used to support the human-factors program. The first, Ashton-Tate's *Framework ™*, is the more costly and complicated, yet its application potential is unlimited. The second, Paperback Software International's *Executive Writer ™* and *Executive Filer ™*, is less costly and less complicated, yet it will support all your programming activities. Both will enhance your way of doing things. Both permit flexibility in data sorting and reporting. Each provides for open-ended data file formats. This is a prime requirement because it is difficult to determine length and complexity of data files, which vary from project to project. I am developing a software adjunct, *Design Evaluator-P ™* and *Design Evaluator-DB ™*, for text-based information systems specifically designed for human-factors programming applications. Utilizing *Framework ™* or *Executive Writer ™* and *Executive Filer ™* you will have little difficulty accomplishing this for yourself.

REFERENCES

Altman, I., and Chemers, M. M. (1980). Cultural aspects of environment-behavior. In H. C. Triandis and J. W. Berry, Eds., *Handbook of cross-cultural psychology: Methodology.* Vol. 2. Boston, MA: Allyn and Bacon.

Bettelheim, B. (1974). *A home for the heart.* New York: Knopf.

Chapanis, A. (1971). The search for relevance in applied research. In W. T. Singleton, J. G. Fox, and D. Whitfield, Eds., *Measurement of man at work.* London: Taylor and Francis.

Deasy, C. M. (1974). *Design for human affairs.* New York: Wiley.

Finsterbusch, K., and Wolf, C. P. (1977). *Methodology of social impact assessment.* Stroudsburg, PA: Dowden, Hutchinson and Ross.

Hall, N. B. (1976). *District of Columbia superior court model courtroom evaluation.* Urbana, IL: National Clearinghouse of Criminal Justice Planning and Architecture.

Harper, W. R., and Harris, D. H. (1975). The application of link analysis to police intelligence. *Human Factors, 17,* 157–164.

Harrigan, J., and Harrigan, J. (1979). *Human factors program for architects, interior designers and clients,* rev. ed. San Luis Obispo, CA: Blake Printery.

Kelley, H. H., and Thibaut, J. W. (1968). Group problem solving. In G. Lindzey and E. Aronson, Eds., *The handbook of social psychology,* 2nd ed., Vol. 4. Reading, MA: Addison-Wesley.

Kubokawa, C. C. (1968). Instant modular 3-d mockup for configuring control and display equipment. *IEEE Transactions on Man-machine Systems, 9*(3).

Lueder, R. K. (1983). Seat comfort: A review of the construct in the office environment. *Human Factors, 6,* 701–711.

Mather, W. G., Kit, B., Bloch, G., and Herman, M. (1970). *Man and his job, and the environment: A review and annotated bibliography of selected recent research on human performance.* Washington, DC: National Bureau of Standards Special Publication 319.

McCormick, E. J. (1970). *Human factors engineering.* New York: McGraw-Hill.

McCormick, E. J., Jeanneret, P. R., and Mecham, R. C. (1969). *Position analysis questionnaire.* Lafayette, IN: Purdue Research Foundation.

Moore, G. T., and Golledge, R. G. (1976). *Environmental knowing.* Stroudsburg, PA: Dowden, Hutchinson and Ross.

Neutra, R., and McFarland, R. A. (1972). Accident epidemiology and the design of the residential environment. *Human Factors, 14,* 405–420.

Newman, O. (1975). *Design guidelines for creating defensive space.* Washington, DC: U.S. Government Printing Office.

Pellow, D. (1981). The new urban community: Mutual relevance of the social and physical environments. *Human Organization, 40,* 15–26.

Preiser, W. F. E., Ed. (1978). *Facility programming: Methods and applications.* Stroudsburg, PA: Dowden, Hutchinson and Ross.

Rosenhan, D. L. (1973). On being sane in insane places. *Science, 179,* 250–258.

Rosenthal, R. (1976). *Experimenter effects in behavioral research.* New York: Wiley.

Spradley, J. P. (1972). Adaptive strategies of urban nomads: The ethnoscience of tramp culture. In T. Weaver and D. White, Eds., *The anthropology of urban environments.* Washington, DC: The Society for Applied Anthropology Monograph Series, Monograph Number 11.

State of California (1977). *Urban development strategy for California (review draft),* Sacramento, CA: Office of Planning and Research.

Terkel, S. (1972). *Working.* New York: Pantheon.

Van Maanen, J. (1973). Observations on the making of policemen. *Human Organization, 32,* 407–418.

Webb, E. J., Campbell, D. T., Schwartz, R. D., and Sechrest, L. (1966). *Unobtrusive measures: Nonreactive research in the social sciences.* Chicago, IL: Rand McNally.

Wolcott, H. (1975). Criteria for an ethnographic approach to research in schools. *Human Organizations, 34,* 111–127.

Zeisel, J. (1981). *Inquiry by design: Tools for environment–behavior reseach.* Monterey, CA: Brooks/Cole.

PART 7
DESIGN FOR HEALTH
AND SAFETY

CHAPTER 7.1
HUMAN FACTORS IN OCCUPATIONAL INJURY EVALUATION AND CONTROL

MICHAEL J. SMITH
DENNIS B. BERINGER

University of Wisconsin, Madison

7.1.1 INTRODUCTION

The purpose of this chapter is to examine the role of the "human factor" in the causation and control of occupational injuries. It emphasizes an approach to accident causation that encompasses the human element in terms of the work system. This approach provides the framework for evaluating hazards determining accident potential and for defining the most effective intervention strategies for accident control.

Safety has its roots in many disciplines, so that a system approach to the definition and control of workplace hazards is dictated. One primary discipline is industrial engineering, which deals with factory layout, workplace design, and machinery design. Proper engineering consideration for human limitations, frailties, *and* capabilities is the first step in the elimination of workplace hazards. Thus machinery must be designed that does not tax employee physical strength, producing fatigue and increased accident risk. Likewise, machinery design must also take account of workers' perceptual/motor limitations through controls and displays that are compatible with basic perceptual and movement processes.

Another critical discipline is medicine. The final outcome of an accident is often contingent on the medical treatment that a worker receives immediately after an injury. Prompt treatment by a medical professional is important not only in life-threatening situations, but with less serious injuries as well, for it can affect the course of healing. In terms of prevention, it is often the medical professional that first talks to the injured employee about the accident. This is a time at which valuable information can be collected on the factors involved and the conditions surrounding the accident. In addition, the nature and extent of the injury can often reveal something to the examiner about the physical conditions surrounding the injury and the agent that produced the injury.

Other important disciplines include psychology and sociology, which contribute to understanding the behavioral, emotional, and psychosocial aspects of accident causation and prevention, as well as the rate of recovery from injury. Thus the actions of the injured employee before and during an accident can be of paramount importance in accident causation. The recognition of a hazard and employee behavior around hazardous conditions can also determine whether an accident will occur. In addition, the way in which a group of workers protects its members can influence the accident potential.

Essentially, accident prevention is a multidisciplinary endeavor requiring inputs from many sources to deal with the interacting factors of people, machinery, the work environment, and the organization of work activities. Any successful approach to accident prevention must recognize the need to deal with these diverse factors using the best available tools from various disciplines to organize a systematic and balanced effort.

7.1.2 A MODEL OF SAFETY PERFORMANCE

There are two major viewpoints from which one could examine accidents. First, one could consider which sequential chain of events may lead to unsafe behaviors and accidents and, second, which factors in the work environment influence these sequential events. In this respect one can examine hierarchical models that can be joined to provide this type of information. Ramsey (1978) has proposed an accident sequence model that depicts stages leading to either an accident or avoidance, depending on the result at any given stage (Figure 7.1.1). Each of the stages represents specific behavioral components present within the individual.

It is clear that there are many stages at which an "error" can result in an accident. A particular example may be useful here. Assume that we wish to examine how a worker may become involved in a forklift truck accident, specifically, striking an object with the truck. First, the operator must "see" the object (sensory perception). Second, he must recognize that the truck is on a collision course with the object (information processing). Third, he must decide whether or not to attempt avoiding the object based upon consequences of striking it and, if required, how to avoid it (attitude behavior or decision-making). Finally, the individual must have the ability to avoid a collision with the object through timely action (motor skills, etc.). Failure at any one of these stages, in conjunction with a chance factor, will expose the individual to an accident situation.

All these factors can be referred to as "person" related, dealing with psychological and physical capabilities and functioning. Although these are discussed in further detail below, they interact with influences from a larger sphere, those factors that are related to things or organizations external to the person. These include influences from the physical work environment (machinery and tools, workplace, work tasks) and the organizational structure. This larger reference frame is depicted in Figure

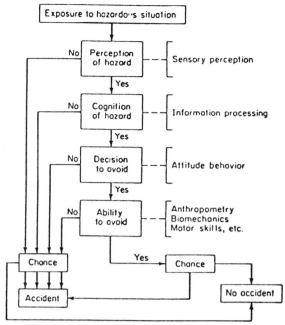

Fig. 7.1.1. Sequential model of accident occurrence. (Adapted from Ramsey, 1978, and E. J. McCormick and M. S. Sanders, *Human Factors in Engineering and Design,* New York: McGraw-Hill, 1982. Used with permission.

7.1.2 with arrows indicating direction in which influence is exerted. Although each set (block) of factors can have effects in isolation from the other factors, they more often than not interact. The following discussion describes the main influences present in each block and some representative interactions.

7.1.2.1 Person

There are, as suggested in Figure 7.1.2, a variety of personal attributes that may affect accident probabilities. These include perceptual/motor capabilities (i.e., vision, hearing, strength, endurance, agility), information processing abilities, and decision-making capabilities (knowledge, personal biases, etc.). The example given earlier for Figure 7.1.1 suggests that the different attributes or "skills" may come into play at different points in the work task. Each will play a greater or lesser role in maintaining safe operation depending on the complexity of the human operator's involvement.

Tasks that require sensory perception beyond "average" levels involve discrimination between very similar states, detection of very weak or faint signals, or in more general terms, detection of subtle differences. In these cases sensory perception capabilities (primarily vision and hearing) may be crucial to accident avoidance. This level of functioning may not be quite as crucial where warning signals

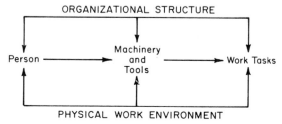

Fig. 7.1.2. Model of Employee Interaction with the workplace.

(changes in process output, overt alarm signals) are very obvious or grossly different from normal states of operation (a loud klaxon horn, for example).

Information processing, however, will be affected by a number of things. The number and types of sources of information will influence how long processing will take; more sources will require more time. Sensory modality (vision, hearing) will also influence processing time. Once the information has been processed, it will be necessary to decide whether or not the situation presents a real hazard. The less radical the departure from normal operations, the longer an accurate evaluation is likely to take.

Having recognized that a hazard exists, the individual must decide how to respond to the threat. This part of the process is exceedingly complex because so many factors influence this stage. The time required to complete this part of the sequence increases both as the number of relevant inputs (information sources) increases and as the number of possible outcomes increases. Although some increase in speed can be accomplished by using "rules of thumb," these may be gross generalizations that are often inappropriate or simply wrong. It should also be noted that when multiple sources of information appear to conflict, this stage will require more time as the individual sorts out the relative importance of each of the cues.

It should also be noted that decision-making will be influenced by personally held beliefs, attitudes, biases, and views transmitted by superiors or "the organizational structure." Given a choice between two possible accident outcomes, the worker may select either that result which will be better received by his organization or, if not congruent, that which is most beneficial personally. Clearly motivation is a substantial factor in the decision making process. (See Wickens, 1984, for more on decision-making.)

Personality characteristics or "biases" have been viewed differently in the past. For many years it was believed that a worker's personality was the most significant factor in accident causation and that certain workers were more "accident prone" than other workers. In a critical review of the literature (Century Research Corporation, 1973) on this topic, it was determined that individual characteristics such as personality, age, sex, and intellectual capabilities are not significant determinants of accident potential or causation. Rather, situational considerations such as the hazard level of the job task and daily personal problems seem to be more important in determining accident risk. Results of several studies suggest that particularly stressful events in the individual's life away from the job may have strong effects on the probability of an accident involvement. Thus there is some evidence that individuals are more or less at risk at different times in their working careers owing to these situational considerations.

Physical capacities such as strength, endurance, and general coordination are also of significance in accident control as they relate to both ability to perform tasks safely and ability to avoid hazards or accidents. There has been a great deal of concern about the entry of women into jobs demanding high energy expenditure and strength requirements owing to a belief that women do not have adequate physical resources to do the work safely. The fact is that there are few jobs in industry that tax the limits of worker strength and endurance, and the majority of women can perform these jobs without an increased risk of an accident. However, there are some jobs that require extensive strength and endurance, these should be limited to workers (male or female) with the necessary physical attributes to do the work safely. Chaffin, Herrin, Keyserling, and Foulke (1977) have defined a test for evaluating worker strength in lifting. Tests such as this allow for appropriate worker selection when job requirements make such selection necessary.

7.1.2.2 Machinery and Tools

As with personal attributes, there are characteristics of the machinery and tools used by the worker that can influence the potential for an accident. A major consideration in this regard is the extent to which machinery and tools limit the worker's use of the most appropriate and effective perceptual/motor skills and energy resources. The relationship between the controls of a machine and the action of that machine dictates the level of perceptual/motor skill necessary to perform a task. In addition, the adequacy of feedback that a worker receives about the action of the machine defines the performance efficiency that can be achieved. The more immediate the feedback is, the better. Of course, the information that is fed back must be of the appropriate quality to allow for efficient performance. (See McCormick and Sanders, 1982, for a detailed discussion of control/display considerations.)

If the action of the controls and the subsequent reaction of the machinery is not compatible with basic perceptual/motor patterns, significant interference with performance can occur. An example will illustrate this influence. An investigation of a fatal accident with a crane used to move scrap iron found that an incompatibility between the actuator lever and the action of the crane boom was a central cause of the accident. When the operator pulled back on the lever the crane boom went forward, and when he pushed forward the crane boom came back toward the cab. This pattern of actuation was contrary to basic perceptual/motor requirements (behavioral stereotypes) that would dictate a compatibility between the movement of the lever and the boom. Thus a compatible design would require that the boom move forward when the lever was pushed forward and back when the lever was pulled back. Over his 20 years of experience with this crane, the operator had learned to compensate for this incompatibility and was able to operate the machine efficiently. Unfortunately, one day while operating the crane a load started to slip. The operator, in his haste to move the slipping load away from a railroad car he was loading, pushed the lever forward to get the load to move forward and out of the way of the train. But this moved the load back into his cab, of course, and killed him. His action under stressful conditions was predictable, for basic perceptual/motor compliance would dictate that a forward motion of the lever should move the crane forward. His 20 years of learning and performing this task in a noncompliant way were forgotten under the stress of the situation, with a fatal result. If the controls and the machine had been compliant in their action, it is likely that no accident would have occurred.

7.1.2.3 Task Factors

The demands of a work activity and the way in which it is conducted can influence the probability of an accident. In addition, the influence of the work activity on employee attention, satisfaction, and motivation can determine behavior patterns that increase accident risk. Thus job task design is an important consideration in accident causation and control. Work task considerations can be separated into the physical requirements and psychological considerations.

The physical requirements influence the amount of energy expenditure necessary to carry out a task. Excessive physical requirements can lead to fatigue, both physiological and mental, which can reduce worker capabilities to recognize and respond to workplace hazards. Typically, relatively high work loads can be tolerated for short periods of time such as 1 hour intervals. However, with longer exposure to heavy work loads and with multiple exposures to shorter-duration heavy work loads, fatigue accumulates and worker capacity is diminished.

There are other task considerations dealing with the content of the task that are related to the physical requirements. These include the pace or rate of work, the amount of repetition in task activities, and work pressure due to production demands. Task activities that are highly repetitive and paced by machinery, as opposed to worker paced, tend to be stressful. Such conditions also diminish a worker's attention to hazards and his or her capability to respond to a hazard. These conditions produce repetitive motion injuries to the musculoskeletal system when the task activity cycle time is short. Thus tasks with relatively low work load and energy expenditure requirements can be very hazardous owing to their high frequency of use of muscles and joints, and their boring nature, which leads to worker inattention to hazards.

There are also psychosocial task content considerations that influence worker attention and motivation. These deal with worker satisfaction with job tasks owing to the amount of control over the work process, the amount of participation in decision-making, the ability to use training, education, and skills in carrying out the task, the amount of esteem associated with the task, and the ability to identify with the end products of the task activity. These considerations are of significance to job stress, which can affect employee ability to recognize and respond to hazards, as well as employee motivation to be concerned with safety considerations. Negative influences can bring about emotional disturbances that limit employee capabilities and motivation. Generally, task considerations are a central aspect in reducing worker fatigue and stress and in enhancing worker motivation and safety behavior.

7.1.2.4 The Work Environment

It does not take a great deal of imagination to picture the major differences between the working environments of an iron foundry and of an insurance office. This dichotomy illustrates a basic consideration in accident causation; that is, the hazard exposure in the work environment can influence the probability of an accident and the seriousness of that accident. In fact, the differences in hazard exposures determine the rates employers pay for workers' compensation insurance. The central concept is one of relative risk; the greater the number of hazards and the more serious their potential to inflict injury, the greater the probability of an accident and the more serious the accident. The hazard potential of various work processes is the most well defined area of safety and can be evaluated using various federal, state, and local standards for worker protection. It is important to note that environmental hazards must be considered in terms of their relationship to worker characteristics rather than to the relative degree of energy release potential. Those situations with the greatest potential to cause harm are typically not the most serious in terms of accident potential because workers take special precautions with such hazards.

Of greater importance than energy release potential is the influence that a hazardous condition has on the basic perceptual/motor processes and motivational considerations of workers. Environmental conditions that hamper the ability of workers to use their senses (poor lighting, excessive noise) and reduce their abilities to respond or react (chemical exposures) increase the potential for accidents. Workplace layout that requires excessive energy expenditure or inconvenient motion patterns or difficult access reduces worker motivation to do a task in the proper way and encourages risk taking. In essence, the environment should be compatible with worker perceptual/motor limitations, energy expenditure capabilities, and motivational needs to encourage recognizing hazards, taking precautions, and desiring to do tasks in the proper way.

7.1.2.5 Organization Structure

The way in which work tasks are organized into plant-wide activities, the style of employee supervision, the motivational climate in the plant, the amount of socialization and interaction between employees, the amount of support employees receive, and management attitude toward safety can all have an influence on plant safety performance. The latter point, management attitude, has often been cited as the most critical element in a successful safety program (Cohen, 1977). If the individuals that run the organization have a disregard for safety considerations, then the management atmosphere will not be one that fosters employee motivation to work safely. Conversely, if the management attitude is one in which safety considerations are paramount (even more important than production goals), then employees show due respect for safety and plant safety performance reflects this respect.

There are other organizational considerations that are important in safety performance that are related to management atmosphere and attitudes. For instance, a management structure that provides for frequent employee interaction with their supervisor and with other employees, as well as frequent social support, will instill an organizational climate that is conducive to cooperative efforts in hazard recognition and control. Such a structure also allows for the motivational climate necessary to encourage appropriate safety behavior.

7.1.2.6 Representative Factors Interactions

All these elements of the safety performance model have to be considered as they operate together and influence each other. The basis of using this model to control accidents is to define how each element contributes to worker behavior individually and as a system. This requires that specific elements cannot be individually maximized but must be optimized in relation to other elements to assure safe worker behavior. Although it is impractical to attempt discussion of all possible factor interactions, some representative examples follow that suggest what one should look for in this area.

Let us choose as our reference example a manual machining operation in a comparatively small manufacturing firm. There are likely to be both "bottom-up" and "top-down" influences, so let us begin with the former and examine how the nature of the tasks and machines can influence other factors. This manufacturing task requires precision boring of cast metal blocks and some grinding of the cast pieces. The two types of activities require different tools and produce different by-products (noise, airborne particulate contaminants, etc.). This may physically separate the two types of activities and produce a similar *organizational* separation: a grinding division and a machining division. This may then produce the need for intermediate supervisory personnel for each division. This separation may affect how safety programs are implemented depending upon the amount of autonomy and authority given to safety personnel (if there are any) and the degree to which the intermediate supervisory personnel support the programs. This is a "bottom-up" effect precipitated by task and machinery requirements.

Effects may also propagate from the top down. The "organizational structure" may decide, at some level of the hierarchy, that production must be increased for the company to remain competitive. This may be communicated to the lower levels as "make more parts" without the necessary qualifying statements concerning quality or safety. The grinder or machinist may then feel tacitly encouraged to take risks in order to increase production (modification of task performance), possibly increasing the probability of accidents.

Assuming this to be the case, several other factors come into play. First, if the individual is working at a comfortable level of output, some small increase in activity may not be detrimental to performance. If, on the other hand, the individual is already working at a strenuous level, any effort to increase output could produce a hazardous situation. Second, increased attention to throughput may decrease attention, as suggested, not only to safe behaviors but also to maintenance of the physical work environment. This could ultimately create additional hazards. Third, machinery and tools may be pressed beyond their normal operating limits or used even when they are not in completely acceptable condition, increasing the probability of producing inferior products or of having accidents.

It should be clear from this brief example that a change in any one aspect of the model may ultimately affect other elements, often in a negative fashion. Lest this section end on a sour note, it should be kept in mind that the converse is true; positive change in one area may propagate positive effects in the others. In order to induce positive change, however, one must be able to measure the potential hazards in the operating environment.

7.1.3 MEASURING HAZARD POTENTIAL AND SAFETY PERFORMANCE

In order to control occupational accidents and related injuries successfully it is necessary to define their nature, and predict when and where they will occur. This requires that some system of hazard detection be developed that can define the frequency of the hazard, its seriousness, and its amenability to control. Traditionally, two parallel systems of information have been used to attain this purpose. One system is hazard identification, such as plant inspections, fault-tree analysis, and employee hazard reporting programs, which have been used to define the nature and frequency of plant hazards. The second system uses employee injury and plant loss control information to define plant problem spots based on the extent of injuries and costs to the organization. When such systems have been integrated, they have been successful in predicting high-risk plant areas or conditions where remedial programs can and should be established for hazard control.

7.1.3.1 Inspection Programs

Hazard identification prior to the occurrence of an occupational injury is a major goal of hazard inspection programs. In the United States such programs have been formalized in terms of federal and state regulations that require employers to monitor and abate recognized hazards. These recognized hazards are defined in federal and state regulations that provide explicit standards of unsafe exposures. The standards can be the basis for establishing an in-plant inspection program, because they determine the specific subject matter to be investigated.

Research has shown that inspections are most effective in identifying permanent fixed physical and environmental hazards that do not vary over time. Inspections are not very effective in identifying transient physical and environmental hazards, or improper workplace behaviors, because these hazards often are not present when the inspection is taking place. A major benefit from inspections, beyond hazard recognition, is the motivational influence on workers gained by demonstrating both management interest in the safety of workers and commitment to a safe working environment. This suggests that an inspection should not be a punitive process where blame laying, confrontation, or punishment is involved. To increase the positive aspects of inspection, indicating the good aspects of a work area as well as hazards is important.

The first step in an inspection program is to develop a checklist that lists all hazards of relevance to the plant based on state and federal standards. Many insurance companies have developed general checklists that can be tailored to a particular plant. These are a good source to start with in drawing up a checklist. A systematic inspection procedure is preferred. This requires that the inspector knows what to look for, where to look for it, and has the proper tools to conduct an effective assessment. In this respect it is important that the checklist be tailored to each work area. To develop a checklist and an inspection procedure for each work area, an analysis of that work area must be undertaken. This analysis should determine (1) the machinery and tools to be inspected, (2) environmental conditions to be inspected, (3) the frequency of inspection necessary to detect and control hazards, (4) the individuals who should conduct the inspection, and (5) the instrumentation needed to make measurements of the hazard(s).

The machinery and tools that require inspection can be determined by (1) the potential for the tool to cause an accident, (2) the seriousness of the resultant effects of an accident, and (3) the number of injuries a specific tool has been identified with. Environmental conditions needing inspection can best be defined by federal, state, and local regulations. The frequency of inspections should be based on the nature of the hazards being evaluated. For instance, once a serious fixed physical hazard has been identified and controlled, it is no longer a hazard and therefore will not have to be regularly inspected. Random spot checking can indicate whether the hazard control remains effective.

Inspections should take place when and where the highest probability of a hazard exists, and reinspection can occur on an incidental basis to assure that hazard control is effectively maintained. The timing of inspections should occur when work processes are operational and be conducted on a recurring basis at regular intervals. According to the National Safety Council (1974), a general inspection of an entire premises should take place at least once a year except for those work areas scheduled for more frequent inspection because of their high hazard level. Because housekeeping is an important aspect of hazard control, inspection of specific work areas should occur at least monthly for cleanliness, clutter, and traffic flow.

The National Safety Council (1974) indicates that a general inspection should cover the following: (1) plant grounds, (2) building and related structures, (3) towers, platforms, or other additions, (4) transportation access equipment and routes, (5) work areas, (6) machinery, (7) tools, (8) materials handling, (9) housekeeping, (10) electrical installations and wiring, (11) floor loading, (12) stairs and stairways, (13) elevators, and (14) roofs and chimneys.

Intermittent inspections are the most common type and are made at irregular intervals, usually on an ad hoc basis. Such inspections are unannounced and are often limited to a specific work area or machine. Their purpose is to keep first-line supervisors and workers alert to safety considerations and hazard detection. Such inspections do not always require a checklist. Systematic informal inspections are made by first-line supervisors on a daily basis in their work area to keep employees aware of safety hazards and also often do not include a checklist. Continuous inspections occur when employees are aware of safety considerations and detect and report hazards as they occur. Maintenance staff can also play a role in defining hazardous conditions while carrying out their duties of machinery repair.

As previously indicated, all employees in an organization can become involved in inspecting for hazards, either formally or informally. Technical, periodic safety inspections should be conducted by the plant safety staff. These persons have special expertise to define hazards. Their expertise can be supplemented by outside experts from insurance companies and government safety agencies. Conducting a formal inspection requires some planning and structure. It first must be determined where to inspect. That decision will be based on hazard and injury potential. A determination must be made whether to give prior warning to the workers in the area to be inspected.

In conducting the inspection, the inspector should identify himself to the department manager and discuss the procedures to be followed, indicating aspects of the inspection that may interfere with the work process. Ways to minimize disruption should be determined with the supervisor. The department manager should be asked to identify special problem areas. At this point a walk-through of the selected work areas is undertaken. The first-line supervisor and worker representatives are asked to participate in the walk-through. As areas are inspected, the first-line supervisor and worker representatives are asked to provide information on specific hazards. The checklist is used to identify each hazard, its nature, exact location, potential to cause serious damage, and possible control measures. During the walk-through, employee input should be solicited. Photographs of hazards can be taken to document their nature and potential seriousness.

Once the inspection is completed a report is prepared that specifies pertinent information about the nature of the hazards, injury potential, and abatement methods. This report must be detailed in a step-by-step fashion for instituting control procedures in a timely manner. Often more than one report is necessary. The person preparing the report must consider the needs and sophistication of the audience in determining the degree of detail that the report contains. A corporate manager who has only a peripheral interest in the inspection is unlikely to wade through a long and detailed report. For this individual's needs, a two-page executive summary or highlights of the results and their implications will suffice. This type of report enables the corporate manager to obtain a quick thumbnail sketch of the major findings and recommendations. A department manager who will be more directly affected by the results of the inspection may require more details, particularly the logic that produced the recommendations and some details about their method of implementation. Essentially, a separate report will have to be written for each audience and be tailored to their individual sophistication levels and needs.

Tailoring reports to meet the needs of a particular audience is only the first step in disseminating the results. It is not sufficient to simply write up the results; they should be shared with concerned parties in a face-to-face meeting. This meeting will give the results greater significance and serve as the basis for further interaction. Such meetings will enhance understanding and allow for in-depth discussion of the findings and recommendations. In fact, it is often wise to have preliminary meetings with interested parties to discuss the results of the inspection before the development of the recommendations. It makes the entire inspection process more relevant to management and worker groups, and facilitates the favorable acceptance of the results and any subsequent recommendations. The parties should be able to identify with and support the inspection conclusions and develop a sense of control regarding what transpires when specific findings are implemented.

7.1.3.2 Injury Statistics

There are four main uses of injury statistics: (1) to identify high risk jobs or work areas, (2) to evaluate company safety performance, (3) to evaluate the effectiveness of hazard abatement approaches, and (4) to identify factors related to injury causation. An injury-reporting recording and analysis system must be able to meet all these uses and requires that detailed information must be collected about the characteristics of injuries, and their frequency and severity. There are two major injury-reporting systems that have been in recent use in the United States. The Occupational Safety and Health Act (OSHA, 1970) established injury reporting and recording requirements that are mandatory for all employers with certain exclusions such as small establishments and government agencies. Regulations have been developed to define how employers are to adhere to these requirements (BLS, 1978). The second system was established by the American National Standards Institute (ANSI) (1969) before the OSHAct of 1970 and also spells out a set of rules by which workplace injuries are reported and recorded. Both systems require detailed reporting of worker injuries to provide information about their nature and the frequency and severity of their occurrence.

The OSHAct requirements specify that any injury to a worker that causes time lost from the job, treatment beyond first aid, transfer to another job, loss of consciousness, or an occupational illness must be recorded on a daily log of injuries (OSHA 200 form; see Figure 7.1.3). This log identifies the person injured, the date and time of the injury, the department or plant location where the injury occurred, and a brief description about the occurrence of the injury highlighting salient facts such as the machinery involved and the nature of the injury. An injury is to be recorded on the day that it occurs. The number of days that the person is absent from the job is also recorded on the employee's return to work. In addition to the daily log, a more detailed form (OSHA 101; see Figure 7.1.4) is filled out for each injury that occurs. This form provides a more detailed description of the nature of the injury, the extent of injury to the employee, the factors that could be related to the cause of the injury, such as the source or agent that produced the injury, and events surrounding the injury occurrence. In many states, workers' compensation forms can be substituted for the OSHA 101 form, as equivalent information is gathered on these forms.

The ANSI reporting and recording system is the forerunner of the OSHA system and the basis of much that is included in the OSHA system. The ANSI Z16.1 (1973) standard specifies what constitutes a reportable injury. This standard is less stringent than the OSHA requirements. For most employers the OSHA requirements take precedence because they are legal requirements that must be met by employers in the United States. In addition, many experts feel that OSHA reporting is more accurate and less open to definitional technicalities. The ANSI standards are of importance, however, because the ANSI Z16.2 (1969) standard establishes a system for describing the events surrounding an injury as well as the nature and extent of the injury. This system is used in almost every state to fill out the OSHA 101 equivalent workers' compensation forms. This recording system provides detailed codes in various categories, such as the agent or source of the injury, the body part injured, and other factors that can be standardized across companies. This allows for the collection of national data on occupational injury causation factors.

Both the OSHAct and ANSI systems specify a procedure for calculating the frequency of occurrence of occupational injuries and an index of their severity. These can be used by companies to monitor their performance. These systems differ but the OSHA system is preferred because there are national data by major industrial categories compiled by the federal government that can serve as a basis of comparison within an industry. Thus a company can determine whether its injury rate is better or worse than other companies in its industry. The OSHA system uses the following formula in determining company annual injury incidence:

$$\text{incidence} = \frac{\text{number of recordable injuries} \times 200,000}{\text{number of hours worked by company employees}}$$

where

1. The number of recordable injuries is taken from the OSHA 200 daily log of injuries.
2. The number of hours worked by employees is taken from payroll records and reports prepared for the Department of Labor or the Social Security Administration.

It is also possible to determine the severity of company injuries. Two methods are typically used. In the first, the total number of days lost due to injuries is compiled from the OSHA 200 daily log and is divided by the total number of injuries recorded on the OSHA 200 daily log. This gives an average number of days lost per injury. In the second, the total number of days lost is multiplied by 1,000,000 and then divided by the number of hours worked by the company employees. This gives a severity index (see ANSI, 1973). Both measures can be compared to an industry average. However, such information is not always available as readily as the injury incidence rate.

Injury incidence and severity information can be used by a company to monitor its injury performance over the years as well as how well it compares with other companies within its industry. Such information provides the basis for making corrections in the company's approach to safety, and can serve as the basis of rewarding managers and workers. However, it must be understood that injury statistics give only a crude indicator of company performance and an even cruder indicator of individual manager or worker performance.

Because injuries are rare events, they do not always reflect the sum total of daily performance of company employees and managers. Thus although they are an accurate measure of safety performance, they are an insensitive measure. Some experts feel that more basic information has to be collected to provide the basis for directing safety efforts. One proposed measure is to use first-aid reports from industrial clinics. These provide information on more frequent events than the injuries required to be reported by the OSHAct. It is thought that these occurrences can provide insights into patterns of hazards and/or behaviors that may lead to the more serious injuries, and that their greater number provides a larger statistical base for determining accident potential.

Yet another proposal is for a company to keep track of all accidents, whether an injury is involved or not. Thus property damage accidents without injury would be recorded, as would "near accidents" and "incidents" that almost produced damage or injury. The proponents of this system feel that a large data base can be established for determining accident causation factors. As with the first-aid reports, the large size of the data base is the most salient feature of this approach. A major flaw in both systems is the lack of uniformity of recording and reporting the events of interest, which means that the programs are useful only within a company and not across companies. Often it is difficult to even get uniformity within a company because the events to be recorded are not as obvious as an injury and may be overlooked. Finally, the method of recording is much more difficult because the nature of the events will differ substantially from injuries, thus making description in a systematic way difficult.

This critique is not aimed at condemning these approaches but at indicating how difficult they

Bureau of Labor Statistics
Log and Summary of Occupational
Injuries and Illnesses

NOTE:	This form is required by Public Law 91-596 and must be kept in the establishment for 5 years. Failure to maintain and post can result in the issuance of citations and assessment of penalties. (See posting requirements on the other side of form.)		RECORDABLE CASES: You are required to record information about every occupational death; every nonfatal occupational illness; and those nonfatal occupational injuries which involve one or more of the following: loss of consciousness, restriction of work or motion, transfer to another job, or medical treatment (other than first aid). (See definitions on the other side of form.)		
Case or File Number	Date of Injury or Onset of Illness	Employee's Name	Occupation	Department	Description of Injury or Illness
Enter a nonduplicating number which will facilitate comparisons with supplementary records.	Enter Mo./day.	Enter first name or initial, middle initial, last name.	Enter regular job title, not activity employee was performing when injured or at onset of illness. In the absence of a formal title, enter a brief description of the employee's duties.	Enter department in which the employee is regularly employed or a description of normal workplace to which employee is assigned, even though temporarily working in another department at the time of injury or illness.	Enter a brief description of the injury or illness and indicate the part or parts of body affected.

Typical entries for this column might be: Amputation of 1st joint right forefinger; Strain of lower back; Contact dermatitis on both hands; Electrocution—body. |
(A)	(B)	(C)	(D)	(E)	(F)
					PREVIOUS PAGE TOTALS ➡
					TOTALS (Instructions on other side of form.) ➡

OSHA No. 200

Fig. 7.1.3. OSHA 200 log of daily injuries. (*Source:* U.S. Department of Labor, Bureau of Labor Statistics, 1978)

are to define and implement. These systems provide a larger base of incidents than that obtained from the limited occurrences in injury recording systems. The main problem is in organizing them into a meaningful pattern. A more fruitful approach to looking at these "after-the-fact" occurrences may be to look at the conditions that can precipitate injuries—hazards. They can provide a large body of information for a sounder statistical base, and can also be organized into meaningful patterns.

7.1.3.3 Hazard Surveys

Inspection and injury analysis systems can be expected to uncover a number of workplace hazards, but they cannot define all the hazards. Many hazards are dynamic and occur only infrequently. Thus they may not be seen during an inspection or may not be reported as a causal factor in an injury. To deal with hazards that involve dynamically changing working conditions or worker behaviors requires a continuously operating hazard identification system. One such system is a hazard survey program. This is a cooperative program between employees and managers to identify and control hazards. Because

U.S. Department of Labor

For Calendar Year 19 _____   Page ____ of____

:Company Name	Form Approved
Establishment Name	O.M.B. No. 1220-0029
Establishment Address	

Extent of and Outcome of INJURY						Type, Extent of, and Outcome of ILLNESS								
Fatalities	Nonfatal Injuries					Type of Illness							Fatalities	Nonfatal Illnesses				
Injury Related	Injuries With Lost Workdays				Injuries Without Lost Workdays	CHECK Only One Column for Each Illness *(See other side of form for terminations or permanent transfers.)*							Illness Related	Illnesses With Lost Workdays				Illnesses Without Lost Workdays
Enter DATE of death. Mo./day/yr. (1)	Enter a CHECK if injury involves days away from work, or days of restricted work activity, or both. (2)	Enter a CHECK if injury involves days away from work. (3)	Enter number of DAYS away from work. (4)	Enter number of DAYS of restricted work activity. (5)	Enter a CHECK if no entry was made in columns 1 or 2 but the injury is recordable as defined above. (6)	Occupational skin diseases or disorders (a)	Dust diseases of the lungs (b)	Respiratory conditions due to toxic agents (c)	Poisoning (systemic effects of toxic materials) (d)	Disorders due to physical agents (e)	Disorders associated with repeated trauma (f)	All other occupational illnesses (g)	Enter DATE of death. Mo./day/yr. (8)	Enter a CHECK if illness involves days away from work, or days of restricted work activity, or both. (9)	Enter a CHECK if illness involves days away from work. (10)	Enter number of DAYS away from work. (11)	Enter number of DAYS of restricted work activity. (12)	Enter a CHECK if no entry was made in columns 8 or 9 (13)

(7) spans columns (a) through (g)

Certification of Annual Summary Totals By _____ Title _____ Date _____

OSHA No. 200

POST ONLY THIS PORTION OF THE LAST PAGE NO LATER THAN FEBRUARY 1.

Fig. 7.1.3. *(Continued)*

the employee is in direct contact with hazards on a daily basis, it is logical to use employee knowledge of hazards in their identification. The information gathered from employees can serve as the basis of a continuous hazard identification system that can be used by management to control dynamic workplace hazards.

A central concept of this approach is that hazards exist in many forms as fixed physical conditions, as changing physical conditions, as worker behaviors, and as an operational interaction that causes a mismatch between worker behavior and physical conditions (Smith, 1973). This concept defines worker behavior as a component critical to the recognition and control of all these hazards. Involving workers in hazard recognition serves to sensitize them to their work environment and acts as a motivator to use safe work behaviors. Such behaviors include using safe work procedures to reduce hazard potential, using compensatory behaviors when exposed to a known hazard, or using avoidance behaviors to keep away from known hazards. The hazard survey program establishes communication between employers and employees about hazards.

The first step in a hazard survey program is to formalize the lines of communication. A primary purpose of this communication network is to get critical hazard information to decision makers as

OSHA No. 101   Form approved
Case or File No. _____   OMB No. 44R 1453

Supplementary Record of Occupational Injuries and Illnesses

EMPLOYER
  1. Name _____
  2. Mail address _____
               (No. and street)   (City or town)   (State)
  3. Location, if different from mail address _____

INJURED OR ILL EMPLOYEE
  4. Name _____ Social Security No. _____
         (First name)   (Middle name)   (Last name)
  5. Home address _____
               (No. and street)   (City or town)   (State)
  6. Age _____   7. Sex: Male_____ Female_____ (Check one)
  8. Occupation _____
            (Enter regular job title, *not* the specific activity he was performing at time of injury.)
  9. Department _____
            (Enter name of department or division in which the injured person is regularly employed, even
            though he may have been temporarily working in another department at the time of injury.)

THE ACCIDENT OR EXPOSURE TO OCCUPATIONAL ILLNESS
  10. Place of accident or exposure _____
               (No. and street)   (City or town)   (State)
      If accident or exposure occurred on employer's premises, give address of plant or establishment in which
      it occurred. Do not indicate department or division within the plant or establishment. If accident oc-
      curred outside employer's premises at an identifiable address, give that address. If it occurred on a pub-
      lic highway or at any other place which cannot be identified by number and street, please provide place
      references locating the place of injury as accurately as possible.
  11. Was place of accident or exposure on employer's premises? _____ (Yes or No)
  12. What was the employee doing when injured? _____
                                          (Be specific. If he was using tools or equipment or handling material,
      _____
                          name them and tell what he was doing with them.)
      _____
  13. How did the accident occur? _____
                              (Describe fully the events which resulted in the injury or occupational illness. Tell **what**
      _____
      happened and how it happened. Name any objects or substances involved and tell how they were involved. Give
      _____
      full details on all factors which led or contributed to the accident. Use separate sheet for additional space.)

OCCUPATIONAL INJURY OR OCCUPATIONAL ILLNESS
  14. Describe the injury or illness in detail and indicate the part of body affected. _____
                                                      (e.g.: amputation of right index finger
      _____
               at second joint; fracture of ribs; lead poisoning; dermatitis of left hand, etc.)
  15. Name the object or substance which directly injured the employee. (For example, the machine or **thing**
      he struck against or which struck him; the vapor or poison he inhaled or swallowed; the chemical or ra-
      diation which irritated his skin; or in cases of strains, hernias, etc., the thing he was lifting, pulling, etc.)
      _____
      _____
  16. Date of injury or initial diagnosis of occupational illness _____.
                                                              (Date)
  17. Did employee die? _____ (Yes or No)
OTHER
  18. Name and address of physician _____
  19. If hospitalized, name and address of hospital _____
      _____
      Date of report _____ Prepared by _____
      Official position _____

Fig. 7.1.4. OSHA 101 form for recording supplemental injury information. (*Source:* U.S. Department of Labor, Bureau of Labor Statistics, 1978)

quickly as possible so that action can be taken to avert an accident. Traditional communication routes in most companies do not allow for quick communication between workers and decision makers and thus serious hazards may not be corrected before an accident occurs. Each company has an established organizational structure that can be used to set up a formalized communication network. For instance, most companies are divided into departments or work units. These can serve as the primary segments to which workers report hazards. These hazards are then communicated to decision makers for action.

Once primary communication units are established, a process to communicate hazard information has to be established. This requires structure and rules. The structure of the program should be simple so that information can flow quickly and accurately. The first step is to designate a decision maker who has the responsibility and authority to respond to serious hazards. This person will act as the chairperson of a hazard committee. The hazard committee is composed of one person from each department or work unit. It is best to have workers elect the committee members from their department, but the chairperson can be appointed by management. The department committee member serves as the primary communication source for the workers in the department. Hazards are reported directly to this person, who then reports them to the committee or, in the case of a serious hazard, immediately to the decision maker.

It is best to have a formal procedure for recording employee-identified hazards. This can be accomplished by a hazard form, which provides a written record of the hazard, its location, and other pertinent information such as the number of employees exposed and possible hazard control measures (see Figure 7.1.5). These forms may be distributed to each employee and may be available from the department committee member. Employees may express their views about the existence of potential hazards anonymously on the forms. Employees should report near-miss accidents, property damage

HAZARD IDENTIFICATION FORM

HAZARD NO. _____

THE STATE OF WISCONSIN
Department of Industry, Labor and Human Relations
Box 2209, Madison, Wisconsin 53701

DATE_____

DIRECTIONS: INCLUDE ONLY ONE HAZARD ON THIS FORM. A HAZARD IS A SET OF CIRCUMSTANCES WHICH CAN RESULT IN BODILY INJURY, IN PROPERTY DAMAGE OR IN A NEAR MISS INCIDENT.

1. DESCRIBE IN DETAIL A WORK HAZARD WHICH HAS THE POTENTIAL TO CAUSE AN ACCIDENT. (PLEASE BE BRIEF BUT COMPLETE.)

2. WHERE IN YOUR WORK PLACE IS THE HAZARD LOCATED?   (BE SPECIFIC) _____

3. HOW MANY WORKERS ARE EXPOSED TO THIS HAZARD? _____

4. HOW OFTEN DOES THIS HAZARD OCCUR?   (IN A DAY, WEEK OR MONTH?) _____

5. WHEN DOES THIS HAZARD OCCUR?   (AT ALL TIMES, RANDOMLY, 9:00 A.M., AT THE END OF THE SHIFT, JUST AFTER BREAK, ON THE NIGHT SHIFT, ON WEEKENDS, ETC.?)

6. WHAT IS THE DURATION OF THIS HAZARD? (LASTS 5 MIN., 1 HR., 1 DAY, ETC.)

CHECK APPROPRIATE BOX

7. IS THIS HAZARD: ☐ ALWAYS PRESENT   ☐ OCCASIONALLY PRESENT?

8. IS IT POSSIBLE TO ANTICIPATE OR PREDICT THE OCCURRENCE OF THIS HAZARD?   ☐ YES   ☐ NO

9. HOW SERIOUS DO YOU FEEL THIS HAZARD IS?

   ☐ EXTREMELY SERIOUS INJURY OR FATALITY AND/OR EXTENSIVE PROPERTY DAMAGE.

   ☐ LOST WORK TIME INJURY AND/OR MODERATE PROPERTY DAMAGE.

   ☐ FIRST AID INJURY ONLY AND/OR MINOR PROPERTY DAMAGE.

10. DO YOU FEEL THAT THIS HAZARD COULD BE CORRECTED BY MODIFYING MACHINE OPERATION OR PRODUCTION LINE PROCEDURES?   ☐ YES   ☐ NO

11. DO YOU FEEL THAT THIS HAZARD COULD BE REDUCED BY INCREASED EMPLOYE TRAINING?   ☐ YES   ☐ NO

12. DO YOU FEEL THAT THIS HAZARD COULD BE CORRECTED BY THE USE OF PERSONAL PROTECTIVE EQUIPMENT OR THE PHYSICAL GUARDING OF MACHINERY?   ☐ YES   ☐ NO

13. DO YOU FEEL THAT THIS HAZARD IS CAUSED BY AN INDIVIDUAL(S) UNSAFE CONDUCT?   ☐ YES   ☐ NO

14. HOW DO YOU FEEL THIS HAZARD COULD BE CORRECTED? _____

DILHR—RES—4253

Fig. 7.1.5. Hazard identification form. (*Source:* Wisconsin Department of Industry, Labor and Human Relations, 1972)

incidents, and potential injury-producing hazards. It is essential in a program such as this that employees be given anonymity and that they be assured that no action will be taken against them for their participation (even if they report silly hazards).

For most hazards, the department committee member will bring employee-reported hazards to the committee's attention at a weekly 1 hr meeting. At this meeting committee members will review the hazards reported by their department, and once all hazards have been reported, they will be rated and ranked. Hazards are rated as to their potential to cause an accident and the extent of injury. The very serious hazards should have been taken care of through direct communication between a committee member and the decision maker and therefore those remaining for the meeting should not be serious. Once rated, hazards can be ranked by their injury potential, their amenability to control, the urgency of control, and their cost. At this point, many nonserious reports or reports that are not really about hazards can be dismissed or discounted. Experience using this type of program indicates that for every 100 hazards reported, 1 is very serious and needs immediate action, 24 require attention quickly to avert a potential accident, 50 require some minor action to improve the quality of working conditions but do not involve a serious hazard, and 25 concern personal complaints of employees that are not related to safety hazards.

An important aspect of having the worker committee evaluate the reported hazards is that employees cannot blame management insensitivity if a low rating is given to a particular hazard. In addition, employees are more likely to agree with a rating given by their peers and adhere to suggested hazard controls. Figure 7.1.6 gives an example of a form used by a hazard evaluation committee to record and rate hazards.

Once hazards have been examined, rated, and ranked by the committee, some plan of action for their control should be developed either by the committee or by company management. The results of the hazards review by the committee and the action to be taken should be fed back to employees after each meeting. Posting the hazard ranking form in work areas after each meeting is a good way to accomplish this.

Employees are expected to fulfill the following duties in this program: (1) examine the workplace to determine if there are hazards, (2) report hazards on the form and return it to the department committee member, (3) make an effort to find out what has happened to the hazard(s) reported, and (4) continue to report hazards as they are observed. This program will provide continuous monitoring of safety hazards using employee input. This participation should stimulate employee awareness of safety and motivate the employee to work more safely. The continued use of the program should encourage the employee to have a vested interest in his or her own safety and the safety of fellow employees. This sense of involvement can carry over into individual work habits.

7.1.4   CONTROLLING WORKPLACE HAZARDS THROUGH HUMAN FACTORS

There are four basic human factors approaches that can be used in concert to control workplace hazards. These are (1) applying methods of workplace and job design to provide working situations that capitalize on worker skills, (2) designing organizational structures that encourage safe working behavior, (3) training workers in the recognition of hazards and proper work behavior(s) for dealing with these hazards, and (4) improving worker safety behavior through work practices improvement. Each of these approaches is based on certain principles that can enhance effective safety performance.

7.1.4.1   Workplace and Job Design

The sensory environment in which job tasks are carried out influences worker perceptual capabilities to detect hazards and respond to them. Being able to see a hazard is an important prerequisite to dealing with it; therefore, workplaces have to provide adequate illumination for visual hazard detection. There is some evidence that appropriate illumination levels can produce significant reductions in accident rates (McCormick, 1976). Not only must the illumination level be appropriate, but illumination should be distributed evenly throughout the plant to eliminate shadows that could hide hazards. Another visual consideration is glare that can come from luminaires or bright or reflective surfaces in the workplace. Such glare can also produce "blind spots" that can hide hazards. Recommendations for specific levels of lighting for particular work operations, as well as means for evaluating glare, are available from ANSI (1973). Perception of warning signs and signals may also be affected by ambient illumination unless care is taken to counter these effects (see McCormick and Sanders, 1982, for specific recommendations).

There are also noise considerations that can influence the successful transmission of hazard information. Work environments that are very loud and do not afford normal conversation can limit the extent of information exchange. In such environments, visual signs are a preferred method for providing safety information. However, in most situations of extreme danger, an auditory warning signal is preferred because it attracts attention more quickly and thus provides for a quicker worker response. In general, warning signals should quickly attract attention, be easy to interpret, and provide information about the nature of the hazard.

TO BE FILLED OUT BY
HAZARD COMMITTEE AND MANAGEMENT

EMPLOYER ______ DIVISION ______
EMPLOYER NO. ______ DEPT. ______

HAZARD STATUS LIST

Hazard Number	Date Hazard Identified	Discription of Hazard	Corr. Cat.*	Hazard Committee Correction Suggestion	Management Comment On Hazard	Management Action	Date Of Action	Hazard Committee Review And Follow Up Of Action

*CORRECTION CATEGORIES

A: IMMEDIATE CORRECTION REQUIRED. WORKERS SHOULD BE PROTECTED FROM HAZARD UNTIL CORRECTION COMPLETED.

B: URGENT. REQUIRES ATTENTION AS SOON AS POSSIBLE.

C: HAZARD SHOULD BE ELIMINATED WITHOUT DELAY, BUT SITUATION IS NOT AN EMERGENCY.

D: NO HAZARD EXISTS BY DETERMINATION OF HAZARD COMMITTEE.

Fig. 7.1.6. Summary sheet of identified hazards. (*Source:* Wisconsin Department of Industry Labor and Human Relations, 1972)

Other environmental conditions such as dust or smoke can also interfere with worker hazard recognition. These and any other conditions that interfere with worker perceptual processes should be eliminated. If this is impossible, then hazard recognition procedures must account for the presence of interference by highlighting hazardous areas to call workers' attention to the hazards, or by providing warnings that can be perceived even under poor perceptual conditions.

In addition to optimizing environmental conditions, machinery layout, use, and design should be considered. Work areas should be designed to allow for traffic flow in a structured manner in terms of the type of traffic, the volume of traffic, and the direction of flow. The traffic flow process should support the natural progression of product assembly. This should eliminate unnecessary traffic and minimize the complexity and volume of traffic. Link analysis can be a particularly useful tool in accomplishing these ends. There should be clearly delineated paths for traffic to use and signs giving directions on appropriate traffic patterns and flow.

Work areas should be designed to provide workers with room to move about in performing tasks without having to assume awkward postures. Task analysis procedures can determine the most economical and safest movement patterns and should serve as the primary basis for determining layout of machinery, work areas, traffic flow, and storage for each work station.

Equipment must conform to principles of proper design so that controls that activate machine action, displays that provide feedback of machine action, and safeguards to protect workers from the action of the machine are compliant with worker skills and expectations. The example of the fatal crane accident demonstrates an important principle of machine design. That is, the action of the machine must be compliant with the action of the controls in spatial characteristics. Other important compliances between controls and machine action are timing and force. The closer in time the action of the machine is to the actuation of the control, the greater the capability of the machine operator to control the machine's action. Likewise, the more compliant the force needed to control the action of the machine is to the results of the action, the greater the operator's control.

The layout of controls on a machine is very important for proper machinery operation, especially in an emergency. In general, controls can be arranged on the basis of (1) their sequence of use, (2) common functions, (3) frequency of use, and (4) relative importance. Any arrangement should take into consideration (1) the ease of access, (2) the ease of discrimination, and (3) safety considerations such as accidental activation. The use of a sequence arrangement of controls is often preferred because it assures smooth, continuous movements throughout the work operation. Generally, to enhance spatial compliance, the pattern of use of controls should sequence from left to right, and from top to bottom. Sometimes controls are more effective when they are grouped by common functions. Often controls are clustered by common functions that can be used in sequence so that a combination of approaches is used.

To control unintentional activation of controls the following steps can be taken: (1) recess the control, (2) isolate the control to an area on the control panel where it will be hard to unintentionally trip, (3) provide protective coverings over the control, (4) provide "lock-out" of the control so that it cannot be tripped unless unlocked, (5) increase the force necessary to trip the control so that extra effort is necessary, and (6) require a specific sequence of control actions such that one unintentional action does not activate the machinery.

A major deficiency in machinery design is the lack of adequate feedback to the machine operator as to the machine action at the point of operation. Such feedback is often difficult to provide because there are few sensors at the point of operation to determine when such action is taking place. However, operators must have some information on the effects of their actuation of controls to be able to perform effectively. Operators may undertake unsafe behaviors to gain some feedback about the machine's performance as the machine is operating, which may put them in contact with the point of operation. This is a very hazardous situation where extreme injury can occur. Thus it is critical to provide some feedback of machine operation to the worker. The more closely this feedback reflects the timing and action of the machinery, the greater the control by the operator. The feedback should be displayed in a convenient location for the operator and is best centered in front of the operator, perpendicular to the line of sight, slightly below the visual horizon (10 to 20°) and at a distance that allows for easy readability.

Work task design is another consideration for controlling safety hazards. Tasks that cause workers to become fatigued or stressed can contribute to accidents. Task design has to be based on considerations that will enhance worker attention and motivation. Thus work tasks must be meaningful, in terms of the content of the work, to eliminate boredom and enhance the worker's mental state. Work tasks should be under the control of the employee and machine-paced operations should be avoided. Tasks should not be repetitive. This latter requirement is often hard to achieve, and when work tasks have to be repeated often, then providing employee control over the pacing of the task reduces stress associated with repetition. Because boredom is also a consideration in repetitive tasks, employee attention can be enhanced by providing frequent breaks from the repetitive activity to do alternate work or take a rest. This brings the employee's attention back into focus.

The question of the most appropriate work schedule is a difficult matter. There is evidence that rotating shift systems produce more occupational injuries than fixed shift schedules (Smith, Colligan, and Tasto, 1982). This implies that fixed schedules are more advantageous for injury control. However, for many younger workers (without seniority and thus often relegated to afternoon and night shifts) this may produce psychosocial problems related to family responsibilities and entertainment needs, leading to stress. Because stress can increase accident potential, the gain from the fixed shift systems owing to improved biological stability may be negated by the psychological instability produced by stress. This suggests that one fruitful approach may be to go to fixed shifts with volunteers working the non-day schedules. Such an approach provides enhanced biological conditions and less psychological distress.

Overtime work should be avoided because of fatigue and stress considerations. It is preferable to have a second shift of workers rather than to overtax the physical and psychological capabilities of employees. Because a second shift is often not economically feasible, some considerations must be established for determining appropriate amounts of overtime. This is a difficult task because there is only limited research on which to base a definitive answer. Therefore, judgment must be used to estimate acceptable amounts of overtime. It is reasonable that job tasks that create high levels of physical fatigue and/or psychological stress should not be performed more than 10 hr in 1 day and 50 hr in 1 week. Jobs that are less fatiguing and stressful can probably be safely performed for up to 12 hr/day. There is some evidence that working more than 50 hr/week can increase the risk of coronary heart disease (Breslow and Buell, 1960; Russek and Zohman, 1958), and therefore working beyond 50 hr/week for extended periods should be avoided.

7.1.4.2 Organizational Design

Organizational policies and practices can have a profound influence on a company's safety record and the safety performance of its employees. To promote safety, organizational policies and practices should demonstrate that occupational safety is important. The first step in this process is to establish a written organizational policy statement on the importance of safety. This should be followed up with written procedures to implement the policy. Such a formalized structure is the foundation on which all safety activities in the company are built. It provides the legitimate basis for undertaking safety-related actions and curtails the frequent arguments among various levels of management about what constitutes acceptable activities. Such a policy statement also alerts employees to the importance of safety.

For employees, the policy statement is the declaration of an intent to achieve a goal. However, employees are skeptical of bureaucratic policies and look for more solid evidence of management commitment. Thus the timing and sequence of safety-related decisions demonstrates how the policy will be implemented and how important safety considerations are. A firm safety policy with no follow-through is worthless, and in fact may be damaging by showing employees the lack of management commitment. This can backfire in terms of poor employee safety attitudes and behaviors that are in keeping with company actions. Thus an employer has to put his "money where his policy is." Otherwise a policy is an empty promise.

Because physical conditions are the most obvious safety hazards, it is important that they be dealt with quickly to demonstrate management commitment. Relations with local, state, and federal safety agencies also reflect on management commitment to safety. Companies with written safety policies and guidelines that have adequate follow-through but that are constantly at odds with government safety officials are sending confusing messages to their employees. It is important to have good public relations with government agencies, even though there may be specific instances of disagreement or even hostility. This positive public image will enchance employee attitudes and send consistent messages to employees about the importance of safety.

In this regard, organizations must ensure an adequate flow of information in the organization. The flow must be both vertical and horizontal within the organizational hierarchy. One approach for dealing with safety communications is to establish communication networks. These are formal structures to ensure that information gets to the people who need to know the message(s). These networks are designed to control the amount of information flow, guarding against information overload, misinforma-tion, or a lack of needed information. Such networks have to be tailored to the specific needs of the organization. They are vital for creating hazard awareness and disseminating general safety information. One example of horizontal information flow is the passing of hazard information from one shift to another in a multi-shift plant. This allows all workers potentially affected by the hazard to be alerted to its presence. Without a communication network, this vital information may not get to all affected employees and an otherwise avoidable accident could occur.

Organizational decision-making is an important motivational tool for enhancing employee safety performance. Decisions about work task organization, methods, and assignments should be delegated to the lowest level in the organization at which they can be logically made; that is, they should be made at the point of action. This approach has a number of benefits. First, this level in the organization has the greatest knowledge of the work processes and operations and of their associated hazards. Such knowledge can lead to better decisions about hazard control. Diverse input to decision making from all organizational levels makes for better decisions because there is more input to work with.

Additionally, this spreading of responsibility through input to decision-making promotes worker and first-line supervisor participation. Such participation has been shown to be a motivator and to enhance job satisfaction (French, 1963). It also gives workers greater control over their work tasks and a greater acceptance of the decisions about hazard control owing to the shared responsibility. All of this leads to decreased stress and increased compliance with safe behavior(s).

Organizations have an obligation to increase company safety by using modern personnel practices. These include appropriate selection and placement approaches, skills training, promotion practices, compensation packages, and employee assistance programs. For safety purposes the matching of worker skills and needs to job task requirements is an important consideration. It is inappropriate to place employees at job tasks for which they lack the proper skills. This will increase injury risk and job stress. Selection procedures must be established to obtain a properly skilled work force. When a skilled worker is not available then training must be undertaken to get skill levels increased before a task is undertaken. This assumes that the employer has carried out a job task analysis and knows the skills required. It also assumes that the employer has devised a way to test for the required skills. Once these two conditions have been met, the employer can optimize the fit between employee skills and job task requirements through selection, placement, and training. Many union contracts require that workers with seniority be given first consideration for promotions. Such consideration is in keeping with this approach as long as the worker has the appropriate skills to do the job task, or the aptitude to be trained to attain the necessary skills.

7.1.4.3 Safety Training

Training workers to improve their skills and to recognize hazardous conditions is a primary means for reducing accidents. Training can be defined as a systematic acquisition of knowledge, concepts, or skills that can lead to improved performance or behavior. Eckstrand (1964) has defined seven basic steps in training. These are (1) defining the training objectives, (2) developing criterion measures for evaluating the training process and outcomes, (3) developing or deriving the content and materials to be learned, (4) designing the techniques to be used to teach the content, (5) integrating the learners and the training program to achieve learning, (6) evaluating the extent of learning, and (7) modifying the training process to improve learner comprehension and retention of the content. These steps provide the foundation for the application of basic guidelines that can be used for designing the training content and for integrating the content and the learner.

In defining training objectives two levels can be established, global and specific. The global objectives are the end goals that are to be met by the training program. For instance, a global objective might be the reduction of eye injuries by 50%. The specific objectives are those that are particular to each segment of the training program, including the achievements to be reached by the completion of each segment. A specific objective might be the ability to recognize eye injury hazards by all employees by the end of the hazard education segment. A basis for defining training objectives is the assessment of company safety problem areas. This can be done using hazard identification methods such as injury statistics, inspections, and hazard surveys. Problems should be identified, ranked in importance, and then used to define objectives.

To determine the success of the training process, criterion for evaluation must be established. Hazard identification measures can be used to determine overall effectiveness. Thus global objectives can be verified by determining a reduction in injury incidence (such as eye injuries) or the elimination of a substantial number of eye hazards. However, it is necessary to have more sensitive measures of evaluation that can be used during the course of training to assess the effectiveness of specific aspects of the training program. This helps to determine the need to redirect specific training segments if they prove to be ineffective. The use of evaluation tools to examine specific objectives is needed. To evaluate the ability of workers to recognize eye hazards, a written or oral examination can be used. Hazards that are not recognized can be emphasized in subsequent training.

The content of the training program should be developed based on the learners' knowledge level, current skills, and aptitudes, and should be flexible enough to allow for individual differences in aptitudes, skills, and knowledge, as well as for individualized rates of learning. The content should allow all learners to achieve a minimum level of safety knowledge and competence by the end of training. The content of the training program has to be directed toward achieving the objectives of the program. The specifics deal with the skills to be learned and the hazards to be recognized and controlled.

There are various techniques that can be used to train workers. Traditionally on-the-job training (OJT) has been emphasized to teach workers job skills and safety considerations. The effectiveness of such training will be based on the skill of the supervisor or lead worker in imparting knowledge and technique as well as their motivation to successfully train the worker. First-line supervisors and lead workers are not educated to be trainers and may lack the skills and motivation to do the best job. Therefore, OJT has not always been successful as the sole safety training method. Because the purpose of a safety training program is to impart knowledge and to teach skills, it is also important to provide classroom experiences to gain knowledge, as well as OJT to attain skills.

Classroom training is used to teach concepts and improve knowledge, and should be carried out in small groups (not to exceed 15 employees). A small group allows for the type of instructor/student interaction needed to monitor class progress, provide proper motivation, and determine each learner's comprehension level. Classroom training should be given in an area free of distractions to allow learners to concentrate on the subject matter. Training sessions should not exceed 30 min, after which workers return to their regular duties. There should be liberal use of visual aids to increase comprehension and make the training more concrete and identifiable to the learners. In addition, written materials should be provided to be taken from the training session for study or reference away from the classroom.

For OJT the major emphasis should be on enhancing skills through observation and practice. Key workers with exceptional skills can be used as role models. Learners can observe these key workers and pick up tips from them. They then can practice what they have learned under the direction of the key workers to increase their skill, obtain feedback on their technique, and be motivated to improve.

Integrating the learner and the training program can best be achieved by attending to some basic principles of safety skills training (see Margolis and Kroes, 1975). These include the following:

1. Making the training situation similar to the actual job conditions. This enhances the transfer of training knowledge and skills to the actual work situation.

2. Explaining the rules to the learners. The learners should be provided with a road map of the training process, what they are expected to learn, and how their performance is expected to improve before they start training.

3. Making sure the basics are well learned. The basic materials provide the foundation on which the rest of the program is built. These materials have to be ingrained in the learner before moving on to the specifics.

4. Providing learners with feedback of their performance on a frequent basis. This will allow learners to modify their responses to improve the learning process. It also can provide a firm foundation for learner motivation.

5. Motivating learners on a specific basis. When a skilled activity is done correctly or a concept learned, the time to provide positive motivation is immediately after the activity is completed. Motivation should always be positive; punishment should not be used. In this regard, when skills are done incorrectly or concepts are not learned, the learners should receive feedback about their performance. This feedback should identify the weakness and how to improve. It should be straightforward and noncritical so as not to demotivate learners.

6. Guiding learners to proper responses. This will speed up skill acquisition. A wrong response or a misunderstanding of a safety concept should be caught quickly before it becomes a bad habit. Thus it is effective to lead learners to proper response modes to develop good habits. However, it must be emphasized that learners also gain valuable information from making mistakes. The key point is to catch the mistakes before they become habits, but not to punish for mistakes, because they can provide valuable learning illustrations.

7. Presenting training materials in stages of difficulty from easiest to most difficult. Learners should be required to study increasingly difficult materials only after they have learned the preceding simpler material that provides the basis for understanding the harder material.

8. Spending the most training time on the most difficult materials. Because training time is limited, the emphasis has to be on those materials that will be harder to learn. This assumes that the hardest materials are also those of most consequence to hazard control. Because the purpose of training is to control hazards, the most emphasis in the training program should be to that end.

9. Using labels to allow learners to identify concepts and skills. This helps learners discriminate between materials being taught, and should make practice, feedback, and transfer to skilled situations easier.

10. Varying the tasks being taught and practiced to increase concept recognition and generalization to different working conditions. This is of special importance for situations in which the learner has to use judgment on the hazard level of new work conditions.

11. Allowing enough time for skills practice. Learners should be given a chance to test their newly acquired knowledge and skills. Frequent practice will allow them to test what they have learned and to keep skills fresh.

12. Encouraging mental practice of skills. Mental practice consists of doing a task in one's head by rehearsing the steps. This establishes sequences of actions in the mind. It is an adjunct to physical practice and should not supplant it.

Once the learner and the training program have been integrated, it will be necessary to evaluate the extent of learning. This can be done by testing learner knowledge and skills. Such testing should be done frequently throughout the training process to provide the learners with performance feedback and to allow for program redirection as needed. Knowledge is best tested by written examinations that test acquisition of facts and concepts. Pictorial examinations (using pictures or slides of working conditions) can be used to determine hazard recognition ability. Oral questioning on a frequent basis can provide the instructor with feedback on the class comprehension of materials being presented, but should not be used for individual learner evaluation because many learners are not highly verbal and could be demotivated by being called on. Skills testing should take place in the work area under conditions that control hazard exposures. Skills can be observed during practice sessions to determine progress under low stress conditions.

The final stage in a training program, once the success of the program has been determined, is to make modifications to improve the learning process. Such modifications should be done on a continuous basis as feedback on learner performance is acquired. In addition, it is necessary to determine if the company objectives have been met by the end of the program. If so, should the objectives be modified? The answers to these questions can lead to modifications in the training program.

7.1.4.4   Hazard Reduction Through Work Practices

A large number of the hazards in the workplace are produced by the interaction between workers and their tools and environment. These hazards cannot be controlled through hazard inspection and machine guarding. They must be controlled by increasing worker recognition of hazards and by proper worker behavior. Such behavior may be evasive action when a hazard occurs or may be the use of safe work procedures to ensure that hazards will not occur. There are few hazard control efforts that do not in some way depend on worker behavior. Making workers aware of hazards is meaningless if the workers do not choose to do something about them. Personal protective equipment is useless if it is not worn, as are inspection systems if hazards are not corrected. It is often true that there is no ideal engineering control method to deal with a hazard. Therefore, it is sometimes necessary to use work practices that will help workers avoid exposure to hazards.

Conard (1983) has defined work practices as employee behaviors, which can be simple or complex, that are related to reducing a hazardous situation in occupational activities. There are a series of steps that can be used in developing and implementing work practices for eliminating occupational hazards. These are (1) the definition of hazardous work procedures, (2) the definition of new work practices to reduce hazards, (3) training employees in the desired work practices, (4) testing the work practices in the job setting, (5) installing the work practices using motivators, (6) monitoring the effectiveness of the work practices, (7) redefining work practices, and (8) maintaining proper worker habits.

In defining hazardous work practices there are a number of sources of information that should be examined. Injury and accident reports such as the OSHA 101 form provide information about the circumstances surrounding an injury. Often worker or management behaviors that contributed to the injury can be identified. Workers are a good source of information about workplace hazards. They can be asked to identify critical behaviors that may be important as hazard sources or as hazard controls. First-line supervisors are also a good source of information because they are constantly observing worker behavior. One should keep in mind, however, that opinions expressed in interviews are influenced by worker adaptation to the task, displaced aggression (from co-workers or supervisors to task), transient mood states, and expectations projected by the interviewer. Questionnaires are somewhat better, but the wording used in the questions must be carefully examined to avoid biasing responses. Given these limitations, the most important source of information may be in directly observing employees at work.

There are a number of considerations when observing employee work behaviors. First, observation must be an organized proposition. Before undertaking observations, it is useful to interview employees and first-line supervisors, and examine injury records to develop a checklist of significant behaviors to be observed. This should include hazardous behaviors as well as those that are used to control hazards. The checklist should identify the job task being observed, the types of behaviors being examined, their frequency of occurrence, and a time frame of their occurrence. The observations should be made at random times so that employees do not change their natural modes of behavior when observed. The time of observation should be long enough for a complete cycle of behaviors associated with a work task(s) of interest to be examined. Two or three repetitions of this cycle should be examined to determine consistency in behavior with an employee and among employees. The recorded behaviors can be analyzed by the frequency and pattern of their occurrence as well as their significance for hazard control. Hot spots can be identified. All behaviors must be grouped into categories in regard to hazard control efforts and then assigned priorities.

The next step is to define the proper work practices that need to be instilled in the worker to control the hazardous procedures observed. Sometimes the observations provide the basis for the good procedures that are to be implemented. Often, however, new procedures must be developed.

There are four classes of work practices that should be considered. These deal with (1) hazard recognition and reporting, (2) housekeeping, (3) doing work tasks safely, and (4) emergency procedures. The recognition of workplace hazards requires that the employee be cognizant of hazardous conditions through training and education, and that the employee actively watch for these conditions. Knowledge is useless unless it is applied. These work practices ensure the application of knowledge and the reporting of observed hazards to fellow workers and supervisors. Housekeeping is a significant consideration for two reasons. A clean working environment makes it easier to observe hazards. It is also a more motivating situation that enhances the use of other work practices. The most critical set of work practices deals with carrying out work tasks safely through correct skill use and hazard avoidance behaviors. This is where the action is between the employee and the environment, and must receive emphasis in instilling proper work practices. There are situations that occur that are extremely hazardous and require the employee to get out of the work area or to stay clear of the work area. These work practices are often life-saving procedures that need special consideration because they are used only under highly stressful conditions (emergencies).

Each of these areas must have work practices explicitly spelled out. These should be statements of the desired behaviors specified in concise, easily understandable language. Statements should typically be one sentence long, and should never exceed three sentences. Details should be excluded unless they are critical to the proper application of the work practice. Once the desired work practices are specified, employees should be given classroom and on-the-job training to teach them the work practices. Training approaches discussed earlier should be applied. These include classroom training as well as an opportunity for employees to test the work practices in the work setting.

To ensure the sustained use of the learned work practices it is important to motivate workers through the use of incentives. There are many types of incentives including (1) money, (2) tokens, (3) privileges, (4) social rewards, (5) recognition, (6) feedback, (7) participation, and (8) any other factors that motivate employees such as enriched job tasks. Positive incentives should be used to develop consistent work practice patterns.

Research had demonstrated that the use of financial rewards in the form of increased hourly wage can have a beneficial effect on employee safety behaviors and reduced hazard exposure. One study (Smith, Anger, Hopkins, and Conard, 1983) evaluated the use of behavioral approaches for promoting employee use of safe work practices to reduce their exposure to styrene. The study was conducted in three plants and had three components: (1) the development and validation of safe work practices for working with styrene in reinforced fiberglass operations, (2) the development and implementation of an employee training program for learning the safe work practices, and (3) the development and testing of a motivational technique for enhancing continued employee use of the safe work practices. Forty-three work practices were extracted from information obtained from a literature search, walk-through plant survey, interviews with employees and plant managers, and input from recognized experts in industrial safety and hygiene. The work practices were pilot tested for their efficacy in reducing styrene exposures. A majority of the work practices were found to be ineffective in reducing styrene exposures, and only those that were effective were incorporated into a worker training program.

The worker training program consisted of classroom instruction followed up with on-the-job application of the material learned in class. Nine 15 min black-and-white videotapes were made to demonstrate the use of safe work practices. Basic information about each work practice and its usefulness was presented, followed by a demonstration of how to perform the work practice. Employees observed one videotape a week for 9 weeks. After each showing, a discussion session was held, followed by on-the-job application of the work practice given by the research training instructor. Once training was completed, each employee was included in the motivational program. This program was based on a financial reward of $10 per week for using the safe work practices. Observations of employee behavior were made by researchers four times daily on a random basis. These observations served as the basis for an employee's receipt of the financial reward.

The effectiveness of the training and motivational programs was measured by examining the changes in employee behavior from before the programs to the end of the study. Employees exhibited approximately 35% of the safe work practices prior to training (41 employees were studied in the three plants). At the end of the study, employees performed approximately 95% of the safe work practices. The real significance of this increased use of safe work practices lies in the effective reduction of employee exposure to styrene. The results indicated a reduction in styrene exposure from before training to the end of the study of 36%, 80%, and 65% for each plant, respectively. The study results demonstrate the effectiveness of behavioral techniques for increasing worker use of safe work practices, as well as the effectiveness of such usage in reducing employee exposures to workplace hazards.

7.1.5   SAFETY PROGRAMS

The preceding materials provide the basis for developing an effective company safety program. There are a number of elements to consider in developing a safety program or upgrading a current program. These include organizational policies, managing various elements of the program, motivational practices, hazard control procedures, dealing with employees, accident investigations, and injury recording. Aspects of each of these have already been discussed and in this section they are integrated into an effective safety program. There has been considerable research into the necessary elements for a successful safety program (see Cohen, 1977) and how these elements should be applied. One primary factor emerges from every study on this subject. A safety program will not be successful unless there is a commitment to the program by top management. This dictates that there be a written organizational policy statement on the importance of safety and the general procedures the company intends to use to meet this policy. Having such a policy is just the first step toward effective management commitment.

Smith, Cohen, Cohen, and Cleveland (1978) have shown that it takes more than a written policy to ensure successful safety performance. It takes involvement on the part of all levels of management in the safety program. For the top managers it means that they must get out onto the shop floor often and talk to employees about plant conditions and safety problems. This can be on a scheduled basis, but seems to be more effective on an informal basis. For middle managers there is a need to participate in safety program activities such as monthly hazard awareness meetings or weekly toolbox meetings. This does not necessitate active presentations by these managers, but it does mean active participation in group discussions and in answering worker questions. These activities expose the upper and middle managers to potential hazard sources and educate them about shop floor problems. It also demonstrates to employees that management cares about their safety and health.

Another aspect of management commitment is the level of resources that are made available for safety programming. Cohen (1977) in reviewing successful programs found that organizational investment in full-time safety staff was a key to good plant safety performance. The effectiveness of such staff was increased, the higher their level in the management structure. The National Safety Council (1974) has suggested that plants with less than 500 employees and a low to moderate hazard level can have an effective program with a part-time safety professional. Larger plants or those with more hazards need more safety staff.

Along with funds for staffing, successful programs also make funds available for hazard abatement in a timely fashion. Thus segregated funds are budgeted to be drawn upon when needed. This gives the safety program flexibility in meeting emergencies when funds may be hard to get quickly from operating departments. An interesting fact about companies with successful safety programs is that they are typically efficient in their resource utilization, planning, budgeting, quality control, and other aspects of general operations, and that they include safety programming and budgeting as just another component of their overall management program. They do not single safety out, or make it special; instead they integrate it into their operations to make it a natural part of daily work activities.

Organizational motivational practices will influence employee safety behavior. Research has demonstrated that organizations that exercise humanistic management approaches have better safety performance (Cohen, 1977; Smith et al., 1978). These approaches are sensitive to employee needs and thus encourage employee involvement. Such involvement leads to greater awareness and higher motivation levels conducive to proper employee behavior. Organizations that use punitive motivational techniques for influencing safety behavior have poorer safety records than those using positive approaches. An important motivational factor is encouraging communication between various levels of the organization (employees, supervisors, managers). Such communication increases participation in safety and builds employee and management commitment to safety goals and objectives. informal communication is often a potent motivator and provides more meaningful information for hazard control.

An interesting research finding is that general promotional programs aimed at enhancing employee awareness and motivation, such as annual safety awards dinners and annual safety contests, are not very effective in influencing worker behavior or company safety performance (Smith et al., 1978). The major reason is that their relationship in time and subject matter to actual plant hazards and safety considerations is so abstract that workers cannot translate the reward to specific actions that need to be taken. It is hard to explain why these programs are so popular in industry given their ineffectiveness. Their major selling point is that they are easy to implement, whereas more meaningful approaches take more effort. They are also highly visible to top management. Thus they are flashy but have very little substance to add to a safety program.

Another important consideration in employee motivation and improved safety behavior is training. There are two general types of safety training that are of central importance. These are (1) skills training, and (2) training in hazard awareness. Training is a key component to any safety program because it is important to employee knowledge of workplace hazards and proper work practices, and it provides the skills necessary to use the knowledge and the work practices. Both formal and informal training seem to be effective in enhancing employee safety performance. Formal training programs provide the knowledge and skills for safe work practices, whereas informal training by first-line supervisors and fellow employees maintains and sharpens learned skills.

All safety programs should have a formalized approach to hazard control. This often includes an inspection system to define workplace hazards, accident investigations, record keeping, a preventive maintenance program, a machine guarding program, review of new purchases to ensure compliance with safety guidelines, and housekeeping requirements. All contribute to a "safety climate" that demonstrates to workers that safety is important. However, the effectiveness of specific aspects of such a formalized hazard control approach has been questioned (Cohen, 1977; Smith et al., 1978). Formalized inspection programs, for instance, have been shown to deal with only a small percentage of workplace hazards (Smith et al., 1971). In fact, Cohen (1977) indicates that more frequent informal inspections may be more effective than more formalized approaches. However, the significance of formalized hazard control programs is that they establish the groundwork for other programs such as work practice improvement and training. In essence, they are the foundation for other safety approaches. They are also a source of positive motivation by demonstrating management interest in employees by providing a clean workplace free of physical hazards. Smith et al. (1978) have demonstrated that sound environmental conditions are a significant contribution to company safety performance and to employee motivation.

REFERENCES

American National Standards Institute (1969). *Method of recording basic facts relating to the nature and occurrence of work injuries* (Z16.2-1969). New York: ANSI.

American National Standards Institute (1973). *Method of recording and measuring work injury experience,* (Z16.1-1973). New York: ANSI.

American National Standards Institute (1973). *American national standard practice for office lighting* (A132.1-1973). New York: ANSI.

Breslow, L. and Buell, P. (1960). Mortality from coronary heart disease and physical activity of work in California. *Journal of Chronic Diseases, 11,* 615–626.

BLS (1978). *Recordkeeping requirements under the Occupational Safety and Health Act of 1970.* Washington, DC: U.S. Department of Labor, Bureau of Labor Statistics.

Century Research Corporation (1973). *Are some people accident prone?,* Arlington, VA: Century Research Corporation.

Chaffin, D., Herrin, G., Keyserling, W. M., and Foulke, J. (1977). *Preemployment strength testing,* Cincinnati, OH: U.S. Department of Health, Education and Welfare, National Institute for Occupational Safety and Health.

Cohen, A. (1977). Factors in successful occupational safety programs. *Journal of Safety Research, 9,* 168–178.

Conard, R. (1983). *Employee work practices,* Cincinnati, OH: U.S. Department of Health and Human Services, National Institute for Occupational Safety and Health.

Eckstrand, G. (1964). *Current status of the technology of training* (AMRL Doc. Tech. Rpt. 64–86). Washington, DC: U.S. Department of Defense.

French, J. (1963). The social environment and mental health. *Journal of Social Issues, 19,* 39–56.

Margolis, B. and Kroes, W. (1975). *The human side of accident prevention.* Springfield, IL: Charles C Thomas.

McCormick, E. (1976). *Human factors guide to engineering,* 4th ed. New York: McGraw-Hill.

McCormick, E. J. and Sanders, M. S. (1982). *Human factors in engineering and design.* New York: McGraw-Hill.

National Safety Council (1974). *Accident prevention manual for industrial operations,* 7th ed. Chicago: NSC.

OSHA (1970). *The Occupational Safety and Health Act of 1970,* Washington, DC: U.S. Congress.

Ramsey, J. D. (1978, May). Ergonomics support of consumer product safety. Paper presented at the American Industrial Hygene Association Conference.

Russek, H. and Zohman, B. (1958). Relative significance of heredity, diet, and occupational stress in CHD of young adults. *American Journal of Medical Science, 235,* 266–275.

Smith, K. U. (1973). Performance safety codes and standards for industry: The cybernetic basis of the systems approach to accident prevention. In J. T. Widner, Ed., *Selected readings in safety* (pp. 356–370). Macon, GA: Academy Press.

Smith, M., Bauman, R., Kaplan, R., Cleveland, R., Derks, S., Sydow, M., and Coleman, P. (1971). *Inspection effectiveness.* Madison, WI: Wisconsin Department of Industry, Labor and Human Relations.

Smith, M. J., Cohen, H., Cohen, A., and Cleveland, R. (1978). Characteristics of successful safety programs. *Journal of Safety Research, 10,* 5–15.

Smith, M., Colligan, M., and Tasto, D. (1982). Shift work health effects for food processing workers. *Ergonomics, 25,* 133–144.

Smith, M., Anger, W. K., Hopkins, B., and Conard, R. (1983). "Behavioral-psychological approaches for controlling employee chemical exposures. In *Proceedings of the Tenth World Congress on the Prevention of Occupational Accidents and Diseases.* Geneva: International Social Security Association.

Wickens, C. D. (1984). *Engineering psychology and human performance.* Columbus, OH: Charles E. Merrill.

CHAPTER 7.2
MANUAL MATERIALS HANDLING

M. M. AYOUB

J. L. SELAN

B. C. JIANG

Texas Tech University
Lubbock, Texas

7.2.1 INTRODUCTION

Manual materials handling (MMH) has been recognized as a major hazard to industrial workers and a major cost to industry. Particular concern has been shown for women and children performing such acts as evidenced by state enacted regulations. All these state regulations have been struck down as unconstitutional because they discriminate against employment of all women without recognition of the large variation in capabilities among women. In addition, these regulations lacked scientific basis.

In 1962 the International labor organization (ILO) published an Information Sheet which stated limits for occasional lifting, as shown in Table 7.2.1. These limits were primarily based on inspection of injury and illness statistics.

Recent statistics indicate the extent and cost of the injuries associated with MMH. In 1979, the National Safety Council estimated that 27% of all industrial injuries were associated with MMH. This amounted to 590,000 injuries and cost approximately 10.4 billion dollars. In 1980, the number of injuries increased to 670,000, despite improved medical care, increased automation in industry, and more extensive use of preemployment examinations.

The economic costs associated with MMH include medical costs, lost work time by both the injured worker and fellow workers, insurance-related costs, loss of material and property damage, lost wages, training cost of a new worker, and administration cost. Moreover, these costs have increased markedly over the years. Snook and Ciriello (1974) reported that, during the period 1938 to 1965, the number of compensable back injuries increased by 11.4% but the average cost per back injury increased by approximately 400%. National Safety Council statistics from 1958 through 1980 depicted in Figure 7.2.1 show the alarming relationship between back injuries and their cost over the years.

The problems associated with MMH are not limited to an isolated group of industries. A diverse number of occupational groups are exposed to MMH stresses, and consequently a diverse number of industries have significant overexertion injury claims (see Tables 7.2.2 and 7.2.3).

Given the frequency, severity, and cost of MMH-related injuries, procedures must be established and followed whereby injuries associated with MMH tasks can be minimized. The purpose of this chapter is to outline such procedures. In hierarchical fashion, these procedures are as follows:

1. *Work practices* based on principles of ergonomics
2. *Job design* such that tasks stay within operator capacities
3. *Employee placement* procedures that match job demands with operator capacities
4. *Training* to reduce back injuries

Several job design and employee placement procedures are described. In addition, psychophysically based lifting capacity prediction equations developed by Ayoub et al. (1983) are presented.

7.2.2 LIFTING CAPACITY DESIGN CRITERIA

The MMH activity most commonly associated with injuries is manual lifting. Therefore, extensive research has been conducted in an effort to determine the lifting capacity of workers. Despite this

Table 7.2.1 Individual Capacity of Lift (Kilograms)

Age (years)	Men	Women
14–16	14.6	9.8
16–18	18.5	11.7
18–20	22.6	13.6
20–35	24.5	14.6
35–50	20.6	12.7
Over 50	15.6	9.8

Source: International Labor Organization, 1962.

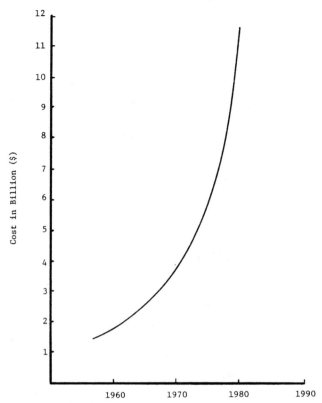

Fig. 7.2.1. Trunk injuries cost over time.

research, it has been noted by Karwowski (1982) that no OSHA regulations exist regarding what constitutes a maximum acceptable or safe weight to lift. This is in part because existing recommendations are based on different methodological approaches assessing different categories of stresses in manual lifting. The three primary criteria used in the determination of lifting capacity are (1) the biomechanical criterion, (2) the physiological criterion, and (3) perceived effort (psychophysical criterion).

7.2.2.1 The Biomechanical Criterion

In general, biomechanics determines what a person can physically do. Biomechanical models attempt to establish the physical stresses imposed on the musculoskeletal system during a lifting action. These stresses serve as the criteria upon which the capacity of lift is based. These physical stresses include reactive forces and torques on various joints of the body, and compressive and shear forces on the lower back (Chaffin, 1975; El-Bassoussi, 1974). The low back, in particular the L4–L5 and L5–S1 disks of the lower back, is especially considered as a basis for load-lifting limits owing to the excessively high forces produced on the low back when lifting (Ayoub and El-Bassoussi, 1976; Park and Chaffin, 1974; and Tichauer, 1971) and the large number of back injuries arising from manual lifting.

The ultimate goal of the biomechanical approach is to set limits on the physical stresses imposed during lifting and then determine the load-lifting capacity based on these limits. Toward this goal, both static and dynamic models of lifting capacity have been developed. Static biomechanical models, such as those developed by Park and Chaffin (1974), assume that the lifting action is performed slowly and smoothly such that forces due to acceleration can be neglected. Dynamic models, such as those developed by Fisher (1967), El-Bassoussi (1974), Ayoub and El-Bassoussi (1976), and Muth, Ayoub, and Gruver (1978), provide data for analyses in the form of time-displacement relationships of the body segments (kinematic analysis) and the forces and torques involved in the motion (kinetic analysis).

Figure 7.2.2, based on the dynamic biomechanical model developed by El-Bassoussi (1974) and Ayoub and El-Bassoussi (1976), shows the effect of weight and horizontal location of load on the

Table 7.2.2   Occurrence of Back Injuries in Various Industries (Selected States, November–December 1980)

Industry	Workers	Percent
Total	906	100
Agriculture, forestry, and fishing	26	3
Mining[a]	1	[b]
Construction	139	15
Manufacturing	378	42
Transportation and public utilities	94	10
Wholesale trade	103	11
Retail trade	90	10
Finance, insurance, and real estate	10	1
Services	53	6
Other industries, not elsewhere classified	12	1

Source: State workers' compensation reports; BLS, 1982.
[a] Limited to oil and gas extraction.
[b] Less than 0.5%.
Note: Owing to rounding, percentages may not add to 100.

Table 7.2.3   Back Injuries in Various Industries/ Occupations Due to Overexertion

Industry Group or Occupation	Overexertion
All industries	25.2[a]
Agriculture, forestry, fishing	18.7
Mining	17.5
Construction	21.4
Manufacturing	25.8
Durable goods	25.6
Nondurable goods	26.2
Transportation, public utilities	26.9
Wholesale trade	29.1
Retail trade	24.6
Finance, insurance, real estate	20.8
Services	28.9
Public administration	19.8
All occupations	25.9
Professional, technical workers	23.3
Managers, administrators	25.9
Sales people	26.1
Clerical workers	26.1
Craftsmen	23.6
Operatives, except transport	26.3
Transport equipment operatives	26.6
Laborers	29.5
Farmers, farm managers	21.6
Farm laborers, foremen	18.7
Service workers	26.1
Private household workers	24.9

Source: BLS, 1979.
[a] Percentage of all types of injuries.

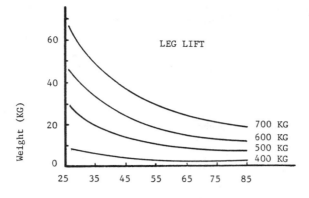

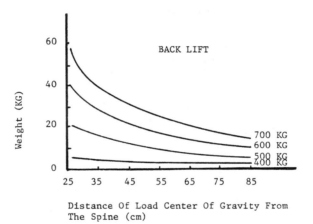

Fig. 7.2.2.   Combination of weight and horizontal location of loads resulting in 400 to 700 kg. Compressive force.

compressive forces on the L5–S1 disk. Compressive forces for both leg (squat) and back (stooped) lifts are shown. The equation

$$H = \left(\frac{W}{2} + 20\right) \text{cm}$$

where W = width of the object away from the body, was used to determine the horizontal location of the load relative to the spine. From data such as these, lifting capacity guidelines can be developed. The NIOSH lifting guidelines (NIOSH, 1981), to be discussed, are an example of such guidelines.

Another approach used in the determination of biomechanical stresses is the measurement of intra-abdominal pressure. The rationale behind the measurement of intra-abdominal pressure is that, when a person lifts a load, pressure in the trunk cavity produces an extensor moment that reduces required contraction forces of the posterior back muscles and consequently minimizes stress on the spinal column. The most common method used in the measurement of intra-abdominal pressure is the ingestion of a pressure transducer that emits FM signals when in the abdominal cavity.

A relationship between intra-abdominal pressure and disk pressure has been shown in several studies, although the results are not without discrepancies. A model developed by Morris, Lucas, and Bleskr (1961) indicates that spinal load is reduced by 30 to 50% owing to peritoneal pressure support. Ortengren et al. (1981) found high correlations between intra-abdominal pressure and disk pressure for symmetric lifting, whereas Shultz, Andersson, Ortengren, Haderspeck, and Nachemson (1982) found weak although significant correlations between the two. More research would appear to be necessary to determine the exact nature of the relationship between intra-abdominal pressure and disk pressure.

7.2.2.2 The Physiological Criterion

The physiological approach may use several criteria, such as oxygen consumption, heart rate, pulmonary ventilation volume, or percent of physical work capacity, as criteria of heaviness of work performed. Generally, the criterion used is the energy expenditure while lifting loads.

Oxygen consumption is generally measured to estimate the energy expenditure required by a lifting task. The measurement of the physiological demands can also be related to an individual's maximum aerobic capacity in order to determine what percent of that capacity is required by a given lifting task.

As with the biomechanical approach, the goal of the physiological approach is to develop limits using metabolic energy expenditure criteria and then determine lifting capacity based on the chosen criterion limits. Several prediction models of metabolic energy expenditure for lifting tasks have been developed (Frederick, 1959; Garg, 1976). Based on physiological criteria, it has been concluded that, for a young male, the 8 hr average metabolic rate should not exceed 5 kcal/min or 33% of the individual's maximum aerobic capacity, and heart rate should not exceed 110 to 115 beats/min (Snook and Irvine, 1967). Figure 7.2.3, based on the model reported by Garg (1976), shows the effect of frequency of lift (lifts/min) and lifting technique on the weight of load that can be lifted to maintain an energy expenditure of 5 kcal/min.

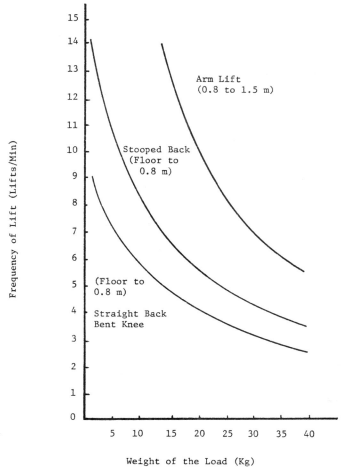

Fig. 7.2.3. Weight of load based on the physiological criterion.

7.2.2.3 The Psychophysical Criterion

The third method employed to determine lifting capacity is the psychophysical approach. Psychophysics deals with the relationship between human sensations and their physical stimuli, this relationship best being described by a power function (Stevens, 1975). The use of psychophysics in lifting tasks requires the subject to adjust the weight of load according to his or her own perception of effort such that the lifting task does not result in overexertion or excessive fatigue. The final weight decided upon by the subject represents the maximum acceptable weight of lift for the given job conditions (frequency of lift, range of lift, container size, etc.).

Several lifting capacity prediction models using the psychophysical approach have been developed (Knipfer, 1974; McConville and Hertzberg, 1966; and McDaniel, 1972). The major limitation with these models has been that they are applicable only to one or two lifting ranges and frequencies of lift. Ayoub et al. (1983) have developed lifting capacity prediction models that are more flexible than the previously developed models in that the models can accommodate several different frequencies and ranges of lift, and different population percentiles and box sizes. These models are presented in detail later.

Because the criterion used in the psychophysical approach is the perceived effort or stress on the part of the subject, it lends itself to the subjective rating of the intensity of the work being performed. One such rating system is the RPE scale (rating of perceived exertion) developed by Borg (1962). The subject is asked to rate work intensity on a scale of 6 to 20 (6 representing very, very light work and 20 representing very, very hard work). The RPE scale values were chosen to be as close as possible to $\frac{1}{10}$ of the corresponding heart rate. Gamberale (1972) and Asfour (1980) found RPE to be linearly related to heart rate and oxygen consumption for both lifting and lowering tasks.

7.2.2.4 Comparison of the Design Criteria

It is the assertion of several investigators that the psychophysical criterion is an appropriate single criterion to use to determine lifting capacity. A problem with the use of the biomechanical or physiological criterion alone is that both biomechanical and physiological stresses are usually present in many lifting tasks. Using the aforementioned physiologically based guidelines proposed by Snook and Irvine (1967), it is readily apparent that an individual could stay within the recommended physiological limits by lifting a very heavy load at a low frequency of lift. However, such a procedure would violate lifting capacity recommendations based on the biomechanical criterion. Conversely, lifting capacity models based solely on the biomechanical criterion are inadequate in dealing with the effects of repetitive lifting on the cumulative physical stresses imposed on the body.

The discrepancies encountered by the use of biomechanical or physiological criteria alone in the determination of lifting capacity become evident when comparing the lifting guidelines presented in Figure 7.2.2 (using the biomechanical criterion) and the lifting guidelines presented in Figure 7.2.3 (using the physiological criterion). Although making comparisons between these two approaches is difficult, the aforementioned problems associated with using only a biomechanical or a physiological criterion can be made more evident. For example, a load of 40 kg is acceptable (i.e., energy expenditure is at or below 5 kcal/min) at low frequencies of lift using the physiological criterion. However, depending on the distance of the load center of gravity from the spine, this weight of load can produce compressive forces in excess of 700 kg.

In general, MMH recommendations based on biomechanical models suggest lifting light loads at higher frequencies of lift, whereas physiological models suggest the lifting of heavier loads at a reduced frequency of lift. Also, it is often assumed by researchers in the area of MMH that only the biomechanical criterion need be considered if the frequency of lift is low, and only physiological criteria need be considered for the lifting of lighter loads at higher frequencies of lift. In many situations this could be a potentially dangerous oversimplification.

Lifting is a task of an extremely complex nature, and because of this it cannot be fully described or explained using only biomechanical or only physiological criteria. Both physiological and biomechanical stresses, among others, are present in nearly every lifting task and, as such, the need exists for a means of determining lifting capacity that can accommodate all these stresses. The virtue of the psychophysical approach is that it attempts to combine the stresses, including the biomechanical and physiological stresses, present in the lifting task under a measure of perceived stress.

7.2.2.5 Manual Materials Handling Capacities

In order to better appreciate the procedures to be discussed in later sections, a brief discussion of the factors affecting MMH is presented. Basically, there are three classes of variables that can affect MMH. They are individual variables, task variables, and environmental variables. These variables, and how they fit into a job design/employee placement paradigm, are shown in Figure 7.2.4.

Individual variables include the age, sex, and body weight of the operator. It has been shown that both maximum oxygen uptake and strength decrease with advancing age (Astrand and Rodahl,

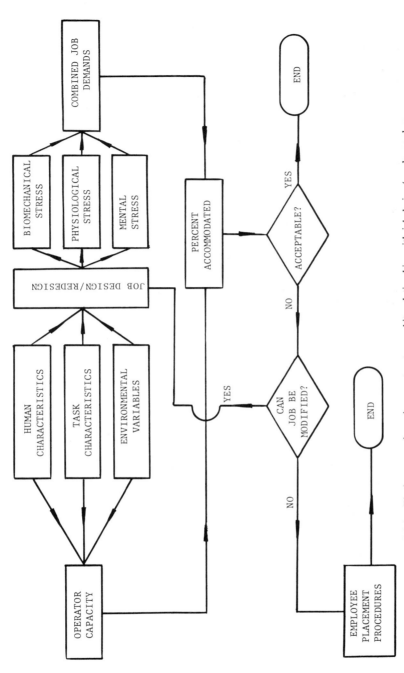

Fig. 7.2.4. The human–task–environment system and its relationships with job design/employee placement.

1970). Females tend to have lower maximum aerobic power and strength than males, with Petrofsky and Lind (1973) reporting that the lifting power of females was approximately 60% of that of males. Kamon and Belding (1971) reported that the relationship between metabolic rate and body weight was linear.

Task variables include the weight of load lifted, frequency of lift, posture, range of lift, and container characteristics. Numerous researchers have demonstrated that an increase in the load lifted results in an increase in metabolic energy rate (e.g., Mital, 1980), and that an exponential relationship exists between work pace and energy expenditure (e.g., Ayoub et al., 1978). Body posture (lifting technique) has been shown to affect metabolic cost of MMH, with free-style lifting being least fatiguing compared to squat or stoop lifting (Brown, 1971). Metabolic energy expenditure increases with an increase in the vertical distance of lift (Aquilano, 1968), whereas the maximum acceptable weight of load is usually highest when lifting from floor to knuckle (e.g., Ayoub et al., 1978). In general, the maximum acceptable weight of load decreases as box size increases in the sagittal plane (Ayoub et al., 1978; Asfour, 1980), although Ciriello and Snook (1978) found no significant difference in weight lifted when box size was increased in the frontal plane.

Manual materials handling capacities and cardiovascular responses can be affected by environmental variables such as thermal stress. Snook and Ciriello (1974) found that weight of load selected by subjects was reduced and heart rate increased as temperature increased. Similar results were obtained by Hafez (1984), who additionally found that oxygen consumption increased as temperature increased.

7.2.2.6 Manual Materials Handling Capacity Data

In this section, MMH capacity data are presented. The data were primarily obtained using the psycho-physical approach. The data to be presented were selected based on their flexibility in terms of factors such as frequency of handling, range of lift/lower, and container size. The lifting capacity models presented here were based on data collected by Ayoub et al. (1978) and Snook (1978). Models were developed to predict both individual and population lifting capacity for six ranges of lift (floor to knuckle, shoulder to reach, etc.). A lifting range assignment based on knuckle height was developed by Ayoub et al., (1978) and is shown in Table 7.2.4.

Lifting capacities for the population were predicted based on frequency of lift, container size, and population percentage. A free-style method of lift was assumed. In using the models, capacity of lift for the six ranges of lift is first predicted based on frequency of lift as shown in Table 7.2.5. These predicted capacities are then adjusted based on box size as shown in Table 7.2.6. Finally, the predicted capacities are adjusted based on the population percentage of interest as shown in Table 7.2.7.

Lifting capacity data based on these equations are presented in Table 7.2.8 for both males and females.

Prediction of capacities of lift for an individual utilizes prediction models developed by Ayoub et al. (1978), presented in Table 7.2.9. These predicted capacities then are adjusted for frequency of lift and box size using the equations given in Tables 7.2.5 and 7.2.6, respectively.

Table 7.2.4   Lifting Range Assignment (KL = Knuckle Level)

Point of Lift Initiation (inches)	Point of Lift Termination (inches)	Range Assignment[a]
0 to KL/2	0 to KL + 10	1
	Kl + 10 to KL + 30	2
	KL + 30 and Above	3
KL/2 to KL	KL/2 to KL	1
	KL to KL + 30	4
	KL + 30 and Above	5
KL to KL + 10	KL to KL + 30	4
	KL + 30 and Above	5
KL + 10 to KL + 20	KL + 10 to KL + 20	4
	KL + 20 and Above	6
KL + 20 and Above	KL + 20 and Above	6

Source: Ayoub et al., 1978.

[a] Range 1 = floor to knuckle, 2 = floor to shoulder, 3 = floor to reach, 4 = knuckle to shoulder, 5 = knuckle to reach, 6 = shoulder to reach.

Table 7.2.5 Capacity Prediction Equations for Lifting Capacity Based on Frequency of Lift

	Frequency of Lift (FY) (lifts/min)			
Range of Lift[a]	$0.1 < FY < 1.0$	(Eq.)	$1.0 \leq FY \leq 12.0$	(Eq.)
	Male Capacity			
F-K	$26.0^{b}FY^{-0.184697}$	1	$26.0 - 0.91(FY - 1)$	7
F-S	$23.3FY^{-0.184697}$	2	$23.3 - 0.91(FY - 1)$	8
F-R	$22.4FY^{-0.184697}$	3	$22.4 - 0.91(FY - 1)$	9
K-S	$23.3FY^{-0.138650}$	4	$23.3 - 0.91(FY - 1)$	10
K-R	$22.7FY^{-0.138650}$	5	$22.7 - 0.91(FY - 1)$	11
S-R	$22.0FY^{-0.138650}$	6	$22.0 - 0.91(FY - 1)$	12
	Female Capacity			
F-K	$17.0^{b}FY^{-0.187818}$	13	$17.0 - 0.5(FY - 1)$	19
F-S	$14.1FY^{-0.187818}$	14	$14.1 - 0.5(FY - 1)$	20
F-R	$12.8FY^{-0.187818}$	15	$12.8 - 0.5(FY - 1)$	21
K-S	$14.0FY^{-0.156150}$	16	$14.0 - 0.5(FY - 1)$	22
K-R	$12.4FY^{-0.156150}$	17	$12.4 - 0.5(FY - 1)$	23
S-R	$12.0FY^{-0.156150}$	18	$12.0 - 0.5(FY - 1)$	24

[a] F = foot; K = knuckle; R = reach; S = shoulder.

[b] Mean capacity for lift based on data from Ayoub et al. (1978) and Snook (1978) for the various ranges of lift for the 50th percentile and 1.0 lift/min. From: Ayoub et al. (1983). *A design guide for manual lifting.* OSHA. Reprinted with permission.

Table 7.2.6 Adjustment of Lifting Capacity Based on Box Size

	Box Size (BX) in Sagittal Plane[a] (CM)			
Range of Lift	$31 \leq BX \leq 46$	(Eq.)	Bx 46	(Eq.)
	Male Capacity			
F-K	$CAP + 0.30(46 - BX)$	1	$CAP + 0.14(46 - BX)$	7
F-S	$CAP + 0.30(46 - BX)$	2	$CAP + 0.14(46 - BX)$	8
F-R	$CAP + 0.30(46 - BX)$	3	$CAP + 0.14(46 - BX)$	9
K-S	$CAP + 0.20(46 - BX)$	4	$CAP + 0.14(46 - BX)$	10
K-R	$CAP + 0.20(46 - BX)$	5	$CAP + 0.14(46 - BX)$	11
S-R	$CAP + 0.20(46 - BX)$	6	$CAP + 0.14(46 - BX)$	12
	Female Capacity			
F-K	$CAP + 0.2(46 - BX)$	13	$CAP + 0.07(46 - BX)$	19
F-S	$CAP + 0.2(46 - BX)$	14	$CAP + 0.07(46 - BX)$	20
F-R	$CAP + 0.2(46 - BX)$	15	$CAP + 0.07(46 - BX)$	21
K-S	$CAP + 0.10(46 - BX)$	16	$CAP + 0.04(46 - BX)$	22
K-R	$CAP + 0.10(46 - BX)$	17	$CAP + 0.04(46 - BX)$	23
S-R	$CAP + 0.10(46 - BX)$	18	$CAP + 0.04(46 - BX)$	24

Source: Ayoub et al. (1983). *A design guide for manual lifting.* OSHA. Reprinted with permission.

[a] CAP = Capacity of lift as determined in Table 7.2.5.

Table 7.2.7   Adjustment of Lifting Capacity Based on Standard Deviation and Frequency

Range of Lift	$0.1 \leq FY < 1.0$	(Eq.)	$1.0 \leq FY \leq 12.0$	(Eq.)
		Frequency (FY)[a]		
		Male Capacity		
F-K	$CAP + Z \cdot 7.66FY^{-0.174197}$	1	$CAP + Z[7.66 - 0.27(FY - 1)]$	7
F-S	$CAP + Z \cdot 6.86FY^{-0.174197}$	2	$CAP + Z[6.86 - 0.24(FY - 1)]$	8
F-R	$CAP + Z \cdot 6.58FY^{-0.174197}$	3	$CAP + Z[6.58 - 0.23(FY - 1)]$	9
K-S	$CAP + Z \cdot 6.67FY^{-0.156762}$	4	$CAP + Z[6.67 - 0.25(FY - 1)]$	10
K-R	$CAP + Z \cdot 6.31FY^{-0.156762}$	5	$CAP + Z[6.31 - 0.24(FY - 1)]$	11
S-R	$CAP + Z \cdot 6.11FY^{-0.156762}$	6	$CAP + Z[6.11 - 0.23(FY - 1)]$	12
		Female Capacity		
F-K	$CAP + Z \cdot 3.12FY^{-0.251605}$	13	$CAP + Z[3.12 - 0.07(FY - 1)]$	19
F-S	$CAP + Z \cdot 2.60FY^{-0.251605}$	14	$CAP + Z[2.60 - 0.06(FY - 1)]$	20
F-R	$CAP + Z \cdot 2.35FY^{-0.251605}$	15	$CAP + Z[2.35 - 0.05(FY - 1)]$	21
K-S	$CAP + Z \cdot 2.57FY^{-0.258700}$	16	$CAP + Z[2.57 - 0.05(FY - 1)]$	22
K-R	$CAP + Z \cdot 2.28FY^{-0.258700}$	17	$CAP + Z[2.28 - 0.05(FY - 1)]$	23
S-R	$CAP + Z \cdot 2.20FY^{-0.258700}$	18	$CAP + Z[2.20 - 0.05(FY - 1)]$	24

Source: Ayoub et al. (1983). *A design guide for manual lifting.* OSHA. Reprinted with permission.

The findings of other researchers provide adjustment factors for other variables affecting MMH capacity. Asfour (1980) had subjects twist in the sagittal plane while lifting, and found that an angle of twist while lifting reduces the maximum acceptable weight of lift by 5%. Garg and Saxena (1980) determined that the absence of handles on the container to be lifted reduces lifting capacity by 7.2%.

Although the emphasis in MMH research has been in the area of lifting, MMH tasks such as lowering, pushing, pulling, and carrying commonly occur in industry. Capacity data for these various MMH activities have been reported by Snook (1978). Samples of these data are presented in Table 7.2.10 for lowering capacity, Table 7.2.11 for pushing capacity, Table 7.2.12 for pulling capacity, and Table 7.2.13 for carrying capacity.

Procedures for Safer Manual Materials Handling Tasks

In this section, procedures whereby the risks associated with MMH tasks can potentially be reduced are summarized. These procedures are

1. Job design/redesign
2. Employee selection
3. Employee training

The three procedures can be viewed as a hierarchy; that is, job design/redesign procedures should be used prior to employee selection procedures, and employee training procedures should be used only as a supplement to job design/redesign or employee placement procedures (refer to Figure 7.2.4). However, prior to the implementation of any of these three procedures, there are several general principles of MMH work design that should be followed. The three procedures listed above can be viewed as means of "fine tuning" a job after coarse adjustments have been made via the general MMH work design principles. The general principles of MMH work design are summarized primarily from Ayoub (1982) and Ayoub et al. (1983).

Manual Materials Handling Task Design

The general principles for MMH work design can be divided into three main areas as follows:

1. Eliminate the need for heavy MMH
2. Decrease MMH demands
3. Minimize stressful body movements

Table 7.2.8 Lifting Capacity Data (Pounds)

Box	Freq.	Pop.	F–K	F–S	F–R	K–S	K–R	S–R
				Male Capacity				
12	1.0	10	88.7	80.4	77.5	78.2	74.4	72.2
12	1.0	50	67.1	61.1	59.0	59.4	56.6	55.0
12	1.0	90	45.4	41.7	40.4	40.5	38.7	37.7
18	1.0	10	78.8	70.5	67.6	71.6	67.8	65.6
18	1.0	50	57.2	51.2	49.1	52.8	50.0	48.4
18	1.0	90	35.5	31.8	30.5	33.9	32.1	31.1
12	5.0	10	77.6	69.7	66.9	67.3	63.7	61.6
12	5.0	50	59.1	53.1	51.0	51.4	48.6	47.0
12	5.0	90	40.5	36.4	35.0	35.4	33.4	32.3
18	5.0	10	67.7	59.8	57.0	60.7	57.1	55.0
18	5.0	50	49.2	43.2	41.1	44.8	42.0	40.4
18	5.0	90	30.6	26.5	25.1	28.8	26.8	25.7
12	9.0	10	66.5	58.9	56.3	56.5	53.0	51.0
12	9.0	50	51.1	45.1	43.0	43.4	40.6	39.0
12	9.0	90	35.6	31.2	29.7	30.2	28.1	26.9
18	9.0	10	56.6	49.0	46.4	49.9	46.4	44.4
18	9.0	50	41.2	35.2	33.1	36.8	34.0	32.4
18	9.0	90	25.7	21.3	19.8	23.6	21.5	20.3
				Female Capacity				
12	1.0	10	52.8	45.0	41.3	41.3	37.0	35.9
12	1.0	50	44.0	37.7	34.7	34.1	30.6	29.7
12	1.0	90	35.1	30.3	28.0	26.8	24.1	23.4
18	1.0	10	46.2	38.4	34.7	38.0	33.7	32.6
18	1.0	50	37.4	31.1	28.1	30.8	27.3	26.4
18	1.0	90	28.5	23.7	21.4	23.5	20.8	20.1
12	5.0	10	47.6	39.9	36.3	36.2	32.0	30.9
12	5.0	50	39.6	33.3	30.3	29.7	26.2	25.3
12	5.0	90	31.5	26.6	24.2	23.1	20.3	19.6
18	5.0	10	41.0	33.3	29.7	32.9	28.7	27.6
18	5.0	50	33.0	26.7	23.7	26.4	22.9	22.0
18	5.0	90	24.9	20.0	17.6	19.8	17.0	16.3
12	9.0	10	42.4	34.8	31.3	31.2	27.0	25.9
12	9.0	50	35.2	28.9	25.9	25.3	21.8	20.9
12	9.0	90	27.9	22.9	20.4	19.3	16.5	15.8
18	9.0	10	35.8	28.2	24.7	27.9	23.7	22.6
18	9.0	50	28.6	22.3	19.3	22.0	18.5	17.6
18	9.0	90	21.3	16.3	13.8	16.0	13.2	12.5

As would be expected, to eliminate the need for heavy MMH is the optimal solution to MMH-related problems. In general, two means exist to accomplish this:

1. The use of mechanical aids such as hoists, lift trucks, lift tables, cranes, elevating conveyors, gravity dumps, and chutes can eliminate (or at least significantly decrease) stresses due to MMH.
2. Changes in the work area layout such that all material is provided at work level can also eliminate heavy MMH. Specifically, the accomplishment of this objective can involve a change in height of either the work level or the worker level.

Table 7.2.9 Regression Coefficients for Maximum Acceptable Weight of Lift Plus Body Weight (kg) for Both Males and Females[a]

Regression Terms	Lifting Ranges					
	F–K	F–S	F–R	K–S	K–R	S–R
Constant term	−101.17	−115.83	−96.62	−83.47	−94.30	−96.59
Arm strength	0.232	0.097	0.014	0.208	0.253	0.011
Shoulder height	0.804	0.863	0.710	0.620	0.730	0.794
Back strength	0.266	0.220	0.268	0.244	0.172	0.221
Abdominal depth	2.707	3.138	3.264	3.077	2.854	2.746
Dynamic endurance	0.640	1.370	0.134	0.164	0.541	0.246

[a] Adjusted to 5 lift/min, box size of 46 cm in sagittal plane.

The methods whereby the need for heavy MMH can be eliminated are summarized in Figure 7.2.5.

If MMH cannot be realistically eliminated, then attempts should be made to decrease the MMH demands of a given job. There are several means by which this second principle of work design can be accomplished:

1. Decrease the weight of the object being handled. The weight of an object can be effectively reduced by assigning the handling to two or more persons, by distributing the load into two or more containers, by reducing the capacity of the container, or by reducing the container weight (e.g., using plastic drums rather than metal drums).

2. Changing the type of MMH activity can decrease the demands of a job. Lifting, lowering, pushing, pulling, carrying, and holding are all types of MMH activities. More importantly, for a given object weight, each type of MMH activity has a relative demand associated with

Table 7.2.10 Lowering Capacity Data

A	B	C	Capacity (kg)						← lowers/min
			K–F			S–K			
			12	4	1	12	4	1	
			Males						
75	76	50	14	21	26	17	20	22	
75	51	50	14	21	27	18	22	24	
49	76	50	16	24	30	17	20	22	
40	51	50	16	25	31	18	22	24	
			Females						
75	76	50	8	13	15	13	13	14	
75	51	50	8	14	15	14	14	15	
49	76	50	9	15	17	13	13	14	
49	51	50	10	16	18	14	14	15	

Source: Snook, 1978.

Note: A = Width of object away from body (cm); B = Vertical distance of lower (cm); C = Percent of industrial population.

Table 7.2.11 Pushing Capacity Data

A	B	Capacity (kg)						← pushes/min
		2.1 m Push			15.2 m Push			
		10	5	1	2	1	0.2	
		Males						
144	50	31, 16[a]	34, 21	38, 25	24, 14	29, 19	37, 25	
95	50	33, 17	37, 22	41, 26	26, 15	33, 18	42, 24	
64	50	30, 17	34, 22	37, 26	22, 15	28, 18	36, 23	
		Females						
135	50	23, 15	26, 18	33, 24	18, 11	19, 13	25, 18	
89	50	24, 16	28, 19	35, 25	21, 12	22, 13	28, 18	
57	50	22, 16	25, 19	32, 25	17, 12	18, 13	24, 17	

Source: Snook, S. H. (1978). "The design of manual handling tasks," *Ergonomics.* London: Taylor & Francis. Reprinted with permission.

Note: A = Vertical distance from floor to hands (cm); B = Percent of industrial population.

[a] Initial forces, sustained forces.

it. As such, it is preferable for a job to require lowering rather than lifting, pulling rather than carrying, and pushing rather than pulling.

3. Changes in the work area layout can decrease MMH-related job demands. Ways in which this can be accomplished are to minimize the horizontal distance between the starting and ending points of a lift (by increasing the height of the lift starting point and/or decreasing the height of the lift termination point), limiting stacking heights to the shoulder height of operators, and keeping heavy objects at the knuckle height of operators. If MMH activities such as carrying, pulling, or pushing are present, work area layout changes that minimize the travel distances for the MMH activities will decrease the job demands.

4. Maximizing the time available to perform the job can decrease job demands. This can be accomplished by reducing the frequency of lift, and/or by incorporating work/rest schedules or job rotation programs into the work design.

Table 7.2.12 Maximum Acceptable Forces of Pull

A	B	Force (kg) with 2.1 m Pull			← pulls/min
		5	1	0.2	
		Males			
144	50	22,16[a]	25, 19	29. 21	
95	50	31, 20	35, 25	40. 28	
64	50	35, 22	39, 27	45. 30	
		Females			
135	50	19, 14	19, 15	22, 17	
89	50	26, 18	27, 19	31, 22	
57	50	29, 18	30, 20	35, 23	

Source: Snook, S. H. (1978). "The design of manual handling tasks," *Ergonomics.* London: Taylor & Francis. Reprinted with permission.

Note: A = Vertical distance from floor to hands (cm); B = Percent of industrial population.

[a] Initial forces, sustained forces.

Table 7.2.13 Maximum Acceptable Weights of Carry

A	B	Weight (kg)						
		2.1 m Carry			8.5 m Carry			← carries/min
		10	1	0.2	3	1	0.2	
				Males				
111	50	18	29	35	17	23	28	
79	50	23	36	45	22	30	38	
				Females				
105	50	14	17	21	31	15	19	
72	50	17	21	26	16	19	23	

Source: Snook, S. H. (1978). "The design of manual handling tasks," *Ergonomics.* London: Taylor & Francis. Reprinted with permission.

Note: A = Vertical distance from floor to hands (cm); B = Percent of industrial population.

The methods whereby job demands can be decreased are summarized in Figure 7.2.6.

The third principle of work design is to minimize stressful body movements required by the job. Specifically, bending and twisting motions imposed on the operator should be reduced.

1. Bending can be reduced by locating objects to be handled within the arm reach envelope of the operator. This can involve, for example, the previously mentioned technique of providing all material at the work level of the operator (by changing the height of the work or worker level) and by avoiding the use of deep shelves that require the operator to bend and reach to obtain objects toward the rear of the shelves.

2. Twisting motions can also be reduced by locating objects within the operator's reach envelope. Work area layouts should be designed such that sufficient space is provided for the entire body to turn and, if the operator is seated, an adjustable swivel chair should be provided.

3. Design considerations should allow the operator to lift objects in a safe manner. Safe lifting incorporates the following: allowing the object to be handled close to the body (by changing the shape of the object or providing better access to the object); using devices such as handles and grips to provide better control of a handled object; balancing the contents of containers; providing rigid containers for increased operator control of the object; and avoiding the lifting of excessively wide objects from the floor level.

The methods whereby stressful body movements can be minimized are summarized in Figure 7.2.7.

7.2.2.7 Job Design/Redesign

In this section, two methods whereby job design/redesign can be conducted to reduce MMH demands are presented. The two methods presented are the job severity index (JSI) developed by Ayoub et al. (1978), and the method reported by National Institute for Occupational Safety and Health (NIOSH,

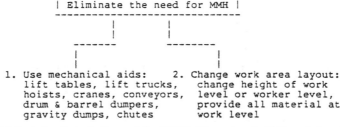

Fig. 7.2.5. Methods to eliminate the need for MMH. (From Ayoub, 1982, and Ayoub et al., 1983.)

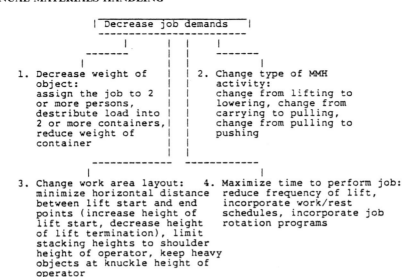

Fig. 7.2.6. Methods to decrease job demands (from Ayoub, 1982, and Ayoub et al., 1983).

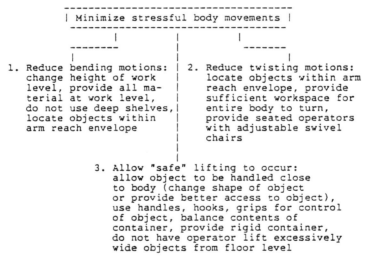

Fig. 7.2.7. Methods to minimize stressful body movements (from Ayoub, 1982; Ayoub et al., 1983; and Chaffin and Baker, 1970).

1981). The JSI is a function of the ratio of job demands to worker capacity. The JSI is stated algebraically as follows:

$$\mathrm{JSI} = \sum_{i=1}^{n} \left(\frac{\mathrm{hours}_i}{\mathrm{hours}_t} \times \frac{\mathrm{days}_i}{\mathrm{days}_t} \right) \sum_{j=1}^{m_i} \left(\frac{F_j}{F_i} \times \frac{\mathrm{WT}_j}{\mathrm{CAP}_j} \right) \tag{1}$$

where n = Number of subtask groups

m_i = Number of tasks in group i

days_i = Exposure days per week for group i

days_t = Total days per week for job

hours_i = Exposure hours per day for group i

hours$_t$ = Number of hours per day that a job is performed

F_j = Lifting frequency for task j

F_i = Total lifting frequency for group i

WT$_j$ = Maximum weight of lift required by task j

CAP$_j$ = The selected applicable maximum acceptable weight of lift adjusted for job factors

The job parameters (exposure hours, frequencies of lift, etc.) are determined through the use of a detailed job analysis. The capacity of the operator is determined based on the psychophysical maximum acceptable weight of lift. If one task is performed, then Equation (1) reduces to

$$\text{JSI} = \frac{\text{job demand}}{\text{operator capacity}} = \frac{\text{WT}}{\text{CAP}} \tag{2}$$

Job design/redesign procedures used in conjunction with the JSI are described in detail elsewhere (Liles, 1985, Ayoub et al., 1983). Briefly, the procedure involves the following:

1. Determine the lifting capacities of the design population and specify an accommodation level. Capacity estimates can be made using the equations given previously or through several other sources. Accommodation in this context refers to the percentage of the population for which the job is to be designed (e.g., 95th percentile male and 50th percentile female).

2. Describe in detail the existing or proposed job design. The first step in the job description is the identification of the distinct tasks that make up the job. Tasks having similar characteristics (slightly different weights of load, container sizes, etc.) are classified as a single lifting task. As a general rule, if the range of item weights exceeds approximately 10 kg, two or more tasks should be defined. The second step is to organize the tasks into task groups, a task group consisting of all tasks performed during the same period of time. The final step is to describe the job and task parameters necessary to compute the JSI. These parameters include (1) percentage of hours per day spent performing each task, (2) number of days per week spent performing each task, (3) lifting range/ranges required for each task (Table 7.2.4 can be used for lifting range classification), (4) dimension of load (i.e., box size) for each task, (5) frequency of lift for each task, and (6) weight of load for each task. The JSI analysis procedure isolates the most stressful task component; therefore the job description needs only to define the task parameter limits from which the worst case component can be taken.

3. Calculate the JSI. An index should be calculated for each lifting task and for the job.

4. Determine acceptability of the job. It has been shown that injury risk is low and relatively constant for all JSI between 0 and 1.5, and significantly higher for JSI values greater than 1.5. If the JSI calculated for the job is less than 1.5, the job can be considered acceptable. JSIs $\geq$ 1.5 indicate a need for design/redesign. The JSIs calculated for each task serve as guides to locate problem areas.

5. If the job is determined to be unacceptable, an iterative process is used to adjust job requirements until an acceptable JSI level (<1.5) is achieved.

An example of the use of the JSI procedure for job design is given below, using the job requirements described in Figures 7.2.8 and 7.2.9 (in this example it is determined if the job must be redesigned based on percentage of the population accommodated).

The 5th percentile male JSI values are calculated in this example.

Given the parameters described in Figures 7.2.8 and 7.2.9, the worker population capacity can be determined for each subtask using the lifting capacity equations presented above. For subtask 1, capacity is first adjusted for frequency of lift using Equation (8) from Table 7.2.5 as follows:

Cap(F–S) = 23.3 − 0.91(FY − 1)

           = 23.30 − 0.91(2 − 1)

           = 22.4 kg

Capacity is then adjusted for box size using Equation (2) from Table 7.2.6 as follows:

Cap(F–S) = CAP + 0.30(46 − BX)

           = 22.4 + 0.30(46 − 40.7)

           = 22.4 + 1.59

           = 23.99 kg

A. Job Title   Stock Room Supplier

B. Job Description (General)   Worker keeps stock room shelves
   filled. Boxes labeled #1 lifted to shelf #1, boxes labeled
   #2 lifted to shelf #2

C. Length of Work Week   C.   5 days

D. Length of Workday (shift)   D.   8 hours

E. Number of shifts   E.   2

F. Number of Lifting Sub-Tasks Defined   F.   2

G. Number of Non-Lifting sub-tasks Defined G.   0

A. Sub-Task #1 Title   Load Box type #1 to shelf #1

B. Hrs/Workday   B.   8

C. Days/Week   C.   5

D. Range of Lift   D. From 0   in. to 57   in.
                                  -----   -----

E. Box Size (sagittal plane)   E.   16   in.

F. Frequency of Lift   F.   2   lifts per min
                             -----   ------

G. Weight   G. to be determined lbs

H. Handles on box (yes or no)   H.   No

I. Does operator twist while lifting (yes or no) I.   No

Fig. 7.2.8. Sample form for job description (adapted from Ayoub et al., 1978).

A. Sub-Task #2 Title   Load Box type #2 to shelf #2

B. Hrs/Workday   B.   8

C. Days/Week   C.   5

D. Range of Lift   D. From 8   in. to 75   in.
                                  -----   -----

E. Box Size (sagittal plane)   E.   24   in.

F. Frequency of Lift   F.   3   lifts per min

G. Weight   G. to be determined lbs

H. Handles on box (yes or no)   H.   Yes

I. Does operator twist while lifting (yes or no) I. yes

Fig. 7.2.9. Sample form for job description (adapted from Ayoub et al., 1978).

The next step is to adjust the capacity such that it reflects the 5th percentile male population (recall that the reference values presented in Table 7.2.5 represent mean capacities). This adjustment is made using Equation (8) from Table 7.2.7 as follows:

$$Cap(F\text{--}S) = CAP + Z[6.86 - 0.24(FY - 1)]$$
$$= 23.99 + (-1.6449)[6.86 - 0.24(2 - 1)]$$
$$= 23.99 + (-1.6449)(6.62)$$
$$= 23.99 + (-10.89)$$
$$= 13.1 \text{ kg}$$

The final adjustment to be made is for the absence of handles on the object to be lifted. As noted, absence of handles decreases capacity by approximately 7.27%; therefore,

$$Cap(F\text{--}S) = (13.1)(0.928) = 12.16 \text{ kg}$$

Similarly, using the appropriate prediction equations and adjustment factors, it can be determined that the worker population capacity of lift associated with subtask 2 is 7.83 kg.

Knowing the predicted population capacities for the subtasks, the JSI of the job and each subtask can be computed as follows:

$$JSI(\text{subtask } 1) = \frac{9.67}{12.16} = 0.7952$$

$$JSI(\text{subtask } 2) = \frac{6.29}{7.83} = 0.8033$$

$$JSI(\text{job}) = \left(\frac{8}{8}\right)\left(\frac{5}{5}\right)\left[\left(\frac{2}{5}\right)\left(\frac{9.67}{12.16}\right) + \left(\frac{3}{5}\right)\left(\frac{6.29}{7.83}\right)\right]$$
$$= 0.8001$$

Based on the calculated JSI values, it can be concluded that the job and its component tasks are acceptable in terms of percentage of the population.

Suppose that the weight of lift required by subtask 1 is 13.6 kg. In order to maintain an acceptable job severity, frequency or job exposure time or both must be reduced. If the frequency of subtask 1 is reduced to 0.1 lift/min, the overall JSI remains at an acceptable level.

$$JSI = \left(\frac{8}{8}\right)\left(\frac{5}{5}\right)\left[\left(\frac{0.1}{3.1}\right)\left(\frac{13.64}{13.95}\right) + \left(\frac{3}{3.1}\right)\left(\frac{6.29}{7.83}\right)\right]$$
$$= 0.8088$$

Another method used for job design is the lifting guidelines established by the National Institute for Occupational Safety and Health (NIOSH) in 1981. NIOSH established load limit recommendations designed to identify hazardous lifting jobs and provide recommendations to alleviate the hazardous elements associated with lifting jobs. The guidelines are based on two limits: an action limit (AL) and a maximum permissible limit (MPL).

The AL is based on the following assumptions:

1. Musculoskeletal injury incidence and severity rates increase moderately in populations exposed to AL lifting conditions.
2. AL lifting conditions would create tolerable compressive forces (350 kg) on the L5–S1 disk of most young, healthy workers.
3. Metabolic rates would not exceed 3.5 kcal/min under AL lifting conditions.
4. More than 75% of women and over 99% of men could lift loads described by the AL.

The MPL attempts to meet the following criteria:

1. Significant increases occur in musculoskeletal injury and severity rates in populations exposed to lifting conditions above the MPL.
2. Lifting conditions above the MPL produce intolerable compressive forces (650 kg) on the L5–S1 in most workers.
3. Metabolic rates exceed 5.0 kcal/min under lifting conditions above the MPL.
4. Only 25% of men and less than 1% of women can perform work above the MPL.

Based on the AL and MPL, three lifting regions are defined. Lifting tasks below the AL represent a nominal risk to workers, tasks above the MPL are considered unacceptable and require engineering

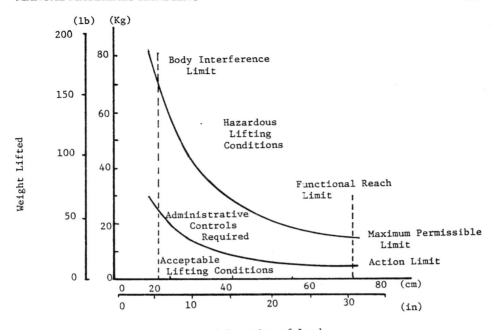

Horizontal Location of Load

Fig. 7.2.10. Lifting regions established by NIOSH guidelines.

controls, and lifting tasks falling between the AL and MPL require administrative (or engineering) controls. These regions are depicted in Figure 7.2.10.

The NIOSH guideline concept is to give an "ideal" weight and adjust it by factors that can be used to improve job design. The modifiable factors are the horizontal location of the object, vertical location of the object, vertical distance lifted, and lifting frequency. These factors are depicted in Figures 7.2.11 through 7.2.14, respectively.

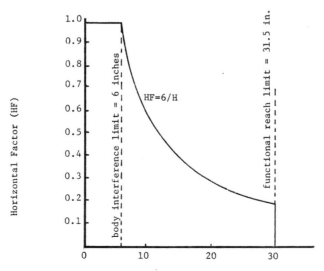

Horizontal Location at Lift Origin Forward of Midpoint Between Ankles (inches)

Fig. 7.2.11. Horizontal factor (from Konz, 1982).

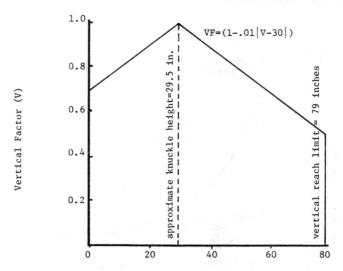

Fig. 7.2.12. Vertical factor (from Konz, 1982).

Based on the above-mentioned factors, the AL and MPL are computed as follows:

$$AL = 40(15/H)(1 - 0.004)|V - 75|(.07 + 7.5/D)(1 - F/F_{max})$$
$$MPL = 3(AL)$$

where H = horizontal location (cm) forward of midpoint between ankles at origin of lift (range 15.0 to 81.0 cm)

V = Vertical location (cm) at origin of lift (range 0 to 178 cm)

D = Vertical travel distance (cm) between origin and destination of lift (range 25 to $(203 - V)$ cm)

F = Average frequency of lift (lifts/min; range 0.2 to 15.0)

F_{max} = Maximum frequency that can be sustained (see Table 7.2.14)

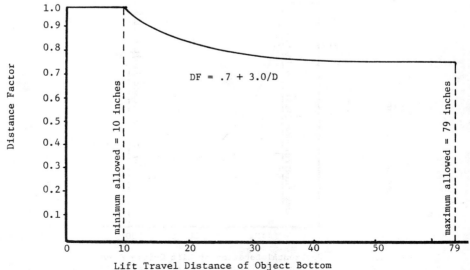

Fig. 7.2.13. Distance factor (from Konz, 1982).

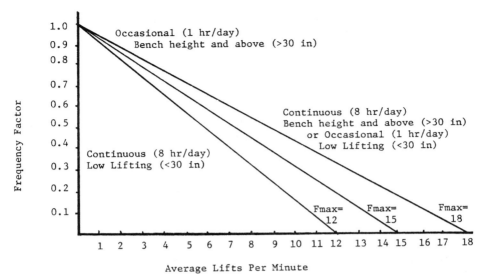

Fig. 7.2.14. Frequency factor (from Konz, 1982).

As an example of the use of the NIOSH guidelines, assume the following lifting conditions:

$$H = 76 \text{ cm}$$
$$V = 102 \text{ cm}$$
$$D = 51 \text{ cm}$$
$$F = 6 \text{ lifts/min}$$
$$F_{max} = 15 \text{ lifts/min}$$

Then

$$AL = 40(76/15)(1 - 0.004|102 - 75|)(0.7 + 51/7.5)(1 - 15/6)$$
$$AL = 40(0.2)(0.892)(0.85)(0.6)$$
$$AL = 3.64 \text{ kg}$$
$$MPL = 10.92 \text{ kg}$$

In other words, given these lifting conditions, a load of 3.64 kg would be acceptable to almost the entire population, whereas a load of 10.92 kg would be hazardous and requires immediate modification.

A comparison of the NIOSH load limit recommendations with the JSI in terms of their ability to evaluate the risk potential of MMH jobs was conducted by Liles et al. (1983). The comparison was based upon two large field studies in which 101 lifting jobs were analyzed using each method. The objectives of the analysis were twofold. First, the analysis determined how well the two methods

Table 7.2.14 Maximum Sustainable Frequency of Life

	Average Vertical Location	
Period (hr)	$V > 75$ cm (30 in.) Standing	$V \leq 75$ cm (30 in.) Stooped
1	18	15
8	15	12

Source: Snook, S. H. (1978). "The design of manual handling tasks," *Ergonomics.* London: Taylor & Francis. Reprinted with permission.

Table 7.2.15 Number of Jobs Observed in Each Risk Category[a] Using JSI or ALR[b]

	JSI			
ALR	L	M	H	Total
L	38	7	0	45
M	13	27	3	43
H	1	5	7	13
Total	52	39	10	101

Source: Liles, Mahagan, and Ayoub, 1983.

[a] L = Low risk category, M = Moderate risk, H = High risk.

[b] ALR = Action limit ratio; ratio of the averaged weight lift to the calculated action limit.

agreed in the assessment of risk potential. Second, the risk assessments were compared to actual injury statistics for the jobs to determine the validity of each method.

Regarding the agreement between the two methods in assessing risk potential, of the 101 jobs, 19 jobs were given higher risk ratings using the NIOSH guidelines than by the JSI method. Ten jobs were given lower risk ratings using the NIOSH guidelines than by the JSI method. Liles et al. (1983) hypothesized that these different evaluations could be due to the following: (1) the JSI method ignores lowering, whereas the NIOSH guidelines assumes that there is no difference between lifting and lowering; (2) the JSI method is more conservative in defining the required weight of lift; and/or (3) each method differs in defining task geometry factors (e.g., load center of gravity). The ratings are shown in Table 7.2.15.

In comparing risk assessments to injury statistics, Liles et al. (1983) found that for number of back injuries, number of lost time back injuries, and number of days lost per lost time injury, the two methods were equally effective at identifying low risk jobs. However, the JSI method appeared to be more effective at differentiating moderate and high risk jobs (Table 7.2.16). For medical expenses, both methods correctly identified low risk jobs, whereas neither differentiated well between moderate and high risk jobs (Table 7.2.17).

Liles, Mahajan, and Ayoub (1983) concluded that, as job analysis and assessment tools, the two methods appeared to have equivalent potential. It was also noted that the JSI method provided procedures for employee placement based on individual capacity, whereas the NIOSH guidelines provided no direct provisions for employee placement.

Table 7.2.16 Back Injury Statistics for Each JSI and ALR[a] Category[b]

	Exposure (hr)	No. Injuries/100 FTE	No. Disabling Injuries/100 FTE	Days Lost/Disabling Injury
JSI				
L	305,333	5.24	2.62	39.75[c]
M	510,485	10.97	7.05	15.11
H	242,063	15.70	11.57	51.07
ALR				
L	308,387	3.89	1.95	52.67[d]
M	406,071	13.79	9.36	41.68[e]
H	343,423	12.23	8.15	14.00

Source: Liles, Mahajan, and Ayoub, 1983.

[a] ALR = Action limit ratio: ratio of the averaged weight of lift to the calculated action limit.

[b] L = Low risk category, M = moderate risk, H = High risk category.

[c] Excluding one serious injury, this statistic is 1.75.

[d] Excluding one serious injury, this statistic is 2.0.

[e] Excluding one serious injury, this statistic is 13.53.

Table 7.2.17 Injury Cost Statistics for Each JSI and ALR[a] Category[b]

	Exposure (hr)	Medical Cost ($/100 FTE)	Total Cost ($/100 FTE)
JSI			
L	192,781	5,014	9,208
M	238,123	16,936	35,092
H	58,727	15,686	36,338
ALR			
L	212,891	4,522	8,320
M	237,667	19,868	41,041
H	39,063	6,159	18,934

Source: Liles, Mahajan, and Ayoub, 1983.

[a] ALR = Action limit ratio; ratio of the averaged weight of lift to the calculated action limit.

[b] L = Low risk category, M = Moderate risk, H = High risk.

7.2.2.8 Employee Placement Procedures

If job design/redesign procedures prove unfeasible or impractical, then an alternative approach to controlling manual lifting hazards is through the use of employee placement procedures. In direct contrast to the job design approach, employee placement procedures attempt to select a worker population whose capacities match or exceed the demands of a given job. Numerous employee screening/placement programs have been and are being used. A critical evaluation of some of these procedures has been presented by Ayoub (1982). A summary of these alternative techniques and their associated shortcomings is presented below:

Procedure	Associated Shortcomings
Back X-ray films	Abnormalities do not necessarily predispose the individual to low back injury
Strength testing	Static strength tests do not take into account repetitive nature of most tasks (preferable are dynamic strength tests); physiological limiting factors are ignored
Medical examinations	Physician cannot match worker capabilities to job demands
Psychological tests	Do not take into account job demands and worker capacities
Job simulators	Expensive, reveal little information regarding level of capacity at which worker must perform to accomplish job
Rating method	Lack of objectivity in determining job demands and worker capacity

Two employee placement procedures will be discussed in greater detail. They are (1) the JSI developed by Ayoub et al. (1978), and (2) the employee strength ratio (ESR) developed by Chaffin, Herrin, Keyserling, and Foulke (1977).

The employee placement procedure used in conjunction with the JSI is similar to the previously described JSI job design procedure. The procedure consists of the following:

1. Determine lifting capacity of the employee. The equations presented in Table 7.2.9 can be used to accomplish this.
2. Describe in detail the job the employee is to be placed into. The job description procedure is identical to that described in the JSI job design procedure.

3. Calculate both an overall JSI and JSIs for the individual tasks.

4. Compare overall JSI and individual task JSIs to critical JSI value of 1.5.

5. Make employee placement decision based on acceptability in terms of JSI. The decision may on occasion require a "judgment call" on the part of the employer. For example, if the overall JSI is acceptable but an individual task JSI is greater than 1.5, consideration may be given to placing the individual in a job involving less risk of injury.

As an example of using the JSI for employee placement, suppose an employee has the following characteristics:

Weight = 81.1 kg
Arm strength = 40.91 kg
Shoulder height = 150 cm
Back strength = 65.77 kg
Abdominal depth = 20 cm
Dynamic endurance = 2 min

If we assume the job requirements presented in Figures 7.2.8 and 7.2.9 and use the appropriate prediction model from Table 7.2.9, the individual's lifting capacity for a floor to shoulder lift (subtask 1) is computed as follows:

$$
\begin{aligned}
\text{Cap(F–S)} &= -115.83 + 0.097(40.91) + 0.863(150) \\
&\quad + 0.220(65.77) + 3.138(20) + 1.370(2) \\
&= -115.83 + 3.97 + 129.45 + 14.47 + 62.76 + 2.74 \\
&= 97.55 \text{ (lifting capacity plus body weight)}
\end{aligned}
$$

If body weight = 81.1 kg, then $97.6 - 81.1 = 16.5$ = predicted lifting capacity F–S. This capacity can then be adjusted for frequency of lift by substituting the obtained capacity for the mean capacity in equation 8 in Table 7.2.5 as follows:

$$
\begin{aligned}
\text{Cap(F–S)} &= 16.5 - 0.91(\text{FY} - 1) \\
&= 16.5 - 0.91(2 - 1) \\
&= 15.59 \text{ kg}
\end{aligned}
$$

Capacity is then adjusted for box size using equation 2 from Table 7.2.6 as follows:

$$
\begin{aligned}
\text{Cap(F–S)} &= 15.59 + 0.3(46.0 - \text{BX}) \\
&= 15.59 + 0.3(46.0 - 40.7) \\
&= 17.18 \text{ kg}
\end{aligned}
$$

The individual's capacity of lift is then adjusted for absence of handles as follows:

$$
\text{Cap(F–S)} = 17.18 \times 0.928 = 15.94 \text{ kg}
$$

Using the same techniques, it can be determined that this individual's capacity given the parameters specified by subtask 2 is 12.59 kg.

Given the predicted lifting capacities, the applicant's JSI if placed on the hypothetical job is determined and found to be less than 1.5. Because the JSI value is less than 1.5, it is concluded that the employee is acceptable for placement on this particular job.

An employee placement procedure similar in concept to the JSI is the employee strength ratio (ESR) developed by Chaffin et al. (1977). The ESR is defined as follows:

$$
\text{ESR} = \frac{\text{object weight lifted on job}}{\substack{\text{Employee strength demonstrated in job} \\ \text{position test}}} \tag{3}
$$

Three static strength tests (arm, legs, and torso) were used in the job position test. The procedure for these tests are described by Chaffin, Herrin, Keyserling, and Foulke (1977). Conceptually, as ESR increases, so does the likelihood of injury. In comparing ESR with medical consequences, Chaffin, Herrin, Keyserling, and Foulke et al. (1977) found only minor positive trends between ESR and back injury days lost and restricted rates. The primary difference between the JSI and the ESR is that the JSI uses capacity to determine the acceptability of an employee for a specific job whereas the ESR makes the placement decision based on the strength of the individual. Capacity here is differentiated from strength in that strength is defined as maximum ability in a single trial whereas capacity is defined as maximum ability in repeated trials. Both ESR and JSI have been tested in the field.

7.2.3 EMPLOYEE TRAINING

A third method whereby MMH-related injuries may be reduced is through the use of an effective training regimen. Although little empirical evidence exists indicating that training does in fact reduce the probability of injuries due to MMH, training used in conjunction with the aforementioned job design and employee placement procedures may prove effective in reducing job severity. The training procedures to be presented here are based largely on recommendations presented in the *Work Practice Guide for Manual Lifting* (NIOSH, 1981), and by Selby (1983) and Ayoub (1982).

The NIOSH guide (1981) suggests three aims for a training program in MMH:

1. Make the trainee aware of the dangers of careless and unskilled MMH.
2. Demonstrate means of avoiding unnecessary stress.
3. Teach employees individually to be aware of what they can handle safely.

It is the position of this chapter that the third aim should be encompassed by proper job design and employee placement procedures rather than letting the worker decide, and therefore emphasis is placed on the first two aims.

The initial issue in any training program regards who is to administer the training. As Selby (1983) notes, all too often this training falls on the shoulders of line supervisors who have little experience in safety and injury prevention. Instructors should have a working knowledge of the scientific principles behind MMH (biomechanics, ergonomics, anatomy, etc.) and have practical experience with the MMH situations and problems specific to the company.

In terms of training program content, an understanding of anatomy (in particular, a working knowledge of the structure and function of the spine) is an essential first ingredient. The basic physics of MMH should also be taught. The NIOSH guide suggests the following in this area:

1. The principle of levers
2. The difference between the force needed to resist gravitational forces on a load and the work needed to raise it
3. The work needed to change the direction of motion
4. Momentum and kinetic energy
5. Newton's Third Law of Motion

This fundamental knowledge serves as the basis for understanding disk pressure and how different body positions affect it. Training in body mechanics should not be limited only to lifting. It should be stressed to workers that body mechanics involves sitting, standing, bending, reaching, stooping, turning, and so on, and that all of these are important in reducing the risk of MMH injuries (Selby, 1983).

The most controversial area of any training program in MMH is instruction on how to lift "properly." The fundamental rule regarding this is that there is no single, correct way to lift. Lifting, like any job, can be done several ways, and because of this on-site training is essential. There are, however, some general rules that can be applied to lifting:

1. Keep the object being handled as close to the body as possible.
2. Utilize the legs as much as possible for leverage rather than the back.
3. Turn the entire body rather than twisting at the trunk.

Leskinen, Stalhammar, Kuorinka, and Troup (1983) employed a dynamic biomechanical model to determine compressive forces imposed on the spine during different lifting techniques. Four lifting techniques were used: a leg lift, a back lift, a load kinetic lift, and a trunk kinetic lift. During the load and trunk kinetic lifts the subjects attempted "to use the kinetic energy of the horizontally moving load or the vertically moving body, respectively, in order to reduce the spinal load".* The leg lift produced the least stress in terms of peak compressive forces. However, the back lift produced the least stress in terms of the compression × time integral, which the authors proposed related the total

---

*From Leskinen, T. P. J., Stalhammar, H. R., Kuorinka, I. A. A., and Troup, J. D. (1983). "A Dynamic Analysis of Spinal Compression with Different Lifting Techniques," *Ergonomics* (p. 595). London: Taylor & Francis. Reprinted with permission.

stress of the lift. The trunk kinetic lift was the most stressful of the lifting techniques. The authors concluded that leg lifts are less stressful on the lower back only to the extent that the load can be lifted between the knees. The advantage of the leg lift was also noted to depend on knee-extensor strength.

The NIOSH guide also suggests some physical factors that can contribute to the prevention of accidents and must to be recognized by the worker:

1. Make sure load is free to move (not stuck)
2. Have lifting aids available
3. Provide proper handles to grasp
4. Supply protective clothing
5. Make sure area for MMH is clean, dry, non-slip, and clear of obstructions

Selby (1983) provides some additional content areas for a MMH training program. She notes the importance for these techniques to be transferable to the home situation (doing yard work, exercising, etc.) so that home injuries are not brought to work by the employee. It is also recommended that employees be given first-aid treatment techniques that they can utilize when back pain occurs. Selby recommends ice massage on the spasmed area, stretching, and aspirin.

Finally, Selby (1983) notes that for every day a person is immobile, 4 days of rehabilitation are necessary for full recovery. Based on this, Selby recommends that light-duty programs be initiated to avoid detriments to muscle tone caused by constant bed rest and to keep the injured worker in touch with fellow workers and management. It should be cautioned, however, that it should be determinable, using an injury risk assessment tool like the JSI, that the light-duty job does in fact present no hazard to the already injured worker.

To close, it should be reiterated that, in order for a MMH training program to be effective, it must be preceeded by application of the work design principles, the job design procedures, and if necessary, the employee placement procedures. Only then should a training program be instituted and relied upon to reduce the likelihood of MMH-related injuries.

REFERENCES

Aquilano, N. J. (1968). A physiological evaluation of time studies for strenuous work as set by stopwatch time studies and two predetermined motion time data systems. *Journal of Industrial Engineering, 19*(9), 425–432.

Asfour, S. S. (1980). Energy cost prediction models for manual lifting and lowering tasks. Ph.D. Dissertation. Lubbock: Texas Tech University.

Astrand, P. O., and Rodahl, K. (1970). *Textbook of work physiology.* New York: McGraw-Hill.

Ayoub, M. A. (1982). Control of manual lifting hazards: I. Training in safe handling. *Journal of Occupational Medicine, 24*(8), 573–577.

Ayoub, M. A. (1982). Control of manual lifting hazards: II. Job Redesign." *Journal of Occupational Medicine, 24*(9), 668–676.

Ayoub, M. A. (1982). Control of manual lifting hazards: III. Preemployment screening. *Journal of Occupational Medicine, 24*(10), 751–761.

Ayoub, M. M., Bethea, N. J., Deivanayagan, S., Asfour, S. S., Bakken, G. M., Liles, D., Mital, A., and Sherif, M. (1978, September). Determination and modeling of lifting capacity (Final Report, HEW (NIOSH) Grant No. 5R010H-00545-02).

Ayoub, M. M., and El-Bassoussi, M. M. (1976). Dynamic biomechanical model for sagittal lifting activities. *Proceedings of the 6th Congress of International Ergonomics Association* (pp. 355–359).

Ayoub, M. M., Gidcumb, C. F., Hafez, H., Intaranont, K., Jiang, B. C., and Selan, J. L. (1983). *A design guide for manual lifting.* OSHA.

Ayoub, M. M., Liles, D., Asfour, S. S., Bakken, G. M., Mital, A., and Selan, J. L. (1982, August). Effects of task variables on lifting capacity (Final Report, HEW (NIOSH) Grant No. 5R010H00798–04).

Borg, G. (1962). *Physical Performance and Perceived Exertion.* Lund, Gleerups.

Brown, J. R. (1971). *Lifting as an industrial hazard.* Ottawa: Canada: Labor Safety Council of Ontario, 1971.

Bureau of Labor Statistics. 1982. *Back injuries associated with lifting* (Bulletin 2144), Washington, D.C.: U.S. Department of Labor.

Chaffin, D. B. (1967). The development of a prediction model for metabolic energy expended during arm activities. Ph.D. Dissertation. Ann Arbor: University of Michigan.

Chaffin, D. B. (1975). Manual materials handling and low back pain. In C. Zenz, Ed., *Occupational medicine, principles and practical applications.* Chicago: Year Book Medical Publishers.

Chaffin, D. B., and Baker, W. H. (1970). A biochemical model for analysis of symmetric sagittal plane lifting. *AIIE Transactions, 2,* 16–27.

Chaffin, D. B., Herrin, G. D., Keyserling, W. M., and Foulke, J. A. (1977). *Pre-employment Strength Testing* (NIOSH Technical Report) DHEW (NIOSH) Publication No. 77–163.

Ciriello, V. M., and Snook, S. H. (1978). The effects of size, distance, height, and frequency of manual handling performance. *Proceedings of the Human Factors Society, 22nd Annual Meeting,* Detroit, MI.

El-Bassoussi, M. M. (1974). A biomechanical dynamic model for lifting in the sagittal plane. Ph.D. Dissertation. Lubbock: Texas Tech University.

Fisher, B. O. (1967). Analysis of spinal stresses during lifting. M.S. Thesis. Ann Arbor: The University of Michigan.

Frederick, W. S. (1959). Human energy in manual lifting. *Modern Materials Handling, 14*(3), 74–76.

Gamberale, F. (1972). Perceived exertion, heart rate, oxygen uptake and blood lactate in different work operations. *Ergonomics, 15*(5), 545–554.

Garg, A. (1976). A metabolic prediction model for manual materials handling jobs. Ph.D. Dissertation. Ann Arbor: University of Michigan.

Garg, A. and Saxena, U. (1980). Container characteristics and maximum weight of lift. *Human Factors, 22*(4), 487–495.

Hafez, H. A. (1984). Manual lifting under hot environmental conditions. Ph.D. Dissertation. Lubbock: Texas Tech University.

International Labor Organization. (1962). Manual lifting and carrying, *International Occupational Safety and Health Information* (Sheet No. 3), Geneva: International Labor Organization.

Kamon, E., and Belding, H. S. (1971). The physiological cost of carrying loads in temperature and hot environment. *Human Factors, 13*(2), 153–161.

Karwowski, W. (1982). A fuzzy sets based model on the interaction between stresses involved in manual lifting tasks. Ph.D. Dissertation. Lubbock: Texas Tech University.

Knipfer, R. E. (1974). Predictive models for the maximum acceptable weight of lift. Ph.D. Dissertation. Lubbock: Texas Tech University.

Konz, S. (1982). NIOSH lifting guidelines. *American Industrial Hygiene Association Journal, 43*(12), 931–933.

Koyl, L., and Hanson, P. (1968). Age, physical ability, and work potential. *National Council on the Aging.* Washington, DC: U.S. Department of Labor, pp. 21–22.

Kroemer, K. H. E. (1983). An isoinertial technique to assess individual lifting capacity. *Human Factors, 25*(5), 493–506.

Leskinen, T. P. J. Stalhammar, H. R., Kuorinka, J. A. A., and Troup, J. D. G. (1983). A dynamic analysis of spinal compression with different lifting techniques. *Ergonomics, 26*(6), 595–604.

Liles, D. H. The application of the job severity index to job design for the control of manual materials handling injury. Accepted to *Ergonomics,* 1984.

Liles, D. H., Mahajan, P., and Ayoub, M. M. (1983). An evaluation of two methods for the injury risk assessment of lifting tasks. *Proceedings of the Human Factors Society,* pp. 279–283.

McConville, J. T., and Hertzberg, H. T. (1966, May). A study of one hand lifting: Final report AMRL-TR-66-17 (Technical Report, Wright-Patterson Air Force Base, OH: Aerospace Medical Research Laboratory.

McDaniel, J. W. (1972). Prediction of acceptable lift capability. Ph.D. Dissertation. Lubbock: Texas Tech University.

Mital, A. (1980). Effect of task variable interactions in lifting and lowering. Ph.D. Dissertation. Lubbock: Texas Tech University.

Morris, J. M., Lucas, D. B., and Bresler, B. (1961, April). Role of the trunk in stability of the spine. *The Journal of Bone and Joint surgery,* 43-A (3), 327–351.

Muth, M. B., Ayoub, M. M., and Gruver, W. A. (1978). A nonlinear programming model for the design and evaluation of lifting tasks. In Colin G. Drury, Ed., *Safety and manual materials handling* (NIOSH Pub. 78–185). NTIS PB-297-660.

NIOSH (1981). *Work practice guide for manual lifting* (Technical Report). Cincinnati, OH: U.S. Department of Health and Human Services.

Ortengren, R., Andersson, G. B. J., and Nachemson, A. L. (1981, January/February). Studies of relationships between lumbar disc pressure, myoelectric back muscle activity, and intra-abdominal (intragastric) pressure. *Spine, 6*(1), 98–103.

Park, K. S., and Chaffin, D. B. (1974). A biomechanical evaluation of two methods of manual load lifting. *Transactions AIIE, 6,* 105–113.

Petrofsky, J. S., and Lind, A. R., (1973). Isometric endurance in men who are overweight. *The Physiologist, 16*(3), 422.

Schultz, A., Andersson, G., Ortengren, R., Haderspeck, K.; and Nachemson, A. (1982). Loads on the lumbar spine: Validation of a biomechanical analysis by measurements of intradiscal pressures and myoelectric signals. *The Journal of Bone and Joint Surgery, 61*(5), 713–720.

Selby, N. C. (1983, August 9). Training procedures to reduce low back injuries. *Proceedings of the U.S. Bureau of Mines Technology Transfer Seminar,* Pittsburgh, PA.

Snook, S. H. (1978). The design of manual handling tasks. *Ergonomics, 21*(12), 963–985.

Snook, S. H., and Irvine, C. H. (1967). "Maximum acceptable weight of lift." *American Industrial Hygiene Association Journal, 28*(4), 322–329.

Snook, S. H., and Ciriello, V. M. (1974). Maximum weights and workloads acceptable to female workers. *Journal of Occupational Medicine, 16*(8), 527–534.

Stevens, S. S. (1975). *Psychophysics: Introduction to its perceptual, neural, and social prospects.* New York: Wiley.

Tichauer, E. R. (1971). A pilot study of the biomechanics of lifting in simulated industrial work situations. *Journal of Safety Research, 3*(3), 98–115.

CHAPTER 7.3
WORK SCHEDULES

DONALD I. TEPAS

**The University of Connecticut
Storrs**

TIMOTHY H. MONK

**Cornell University Medical College, Westchester Division
White Plains, New York**

7.3.1 INTRODUCTION

Through the ages people have developed the notion that regular *day* work of about 8 hr is the normal or natural situation. If a statistical definition of normality is used, this is probably true for the majority of current workers. In fact, a workday longer than 8 hr has been, until very recently, the rule rather than the exception. Nighttime work has been with us since Roman times (Scherrer, 1981). Labor statistics suggest that it is reasonable to conclude that at least a third of urban U.S. households include at least one shift worker. Thus it is also reasonable to propose on a statistical basis that shift work is frequent and common (Presser and Cain, 1983). Clearly, if one adds "moonlighting" and all other alternative work systems (such as flexitime and self-employment) to the shift worker population, the percentage will be even larger.

Despite these statistics, regular 8 hr daytime work has come to be viewed as "normal," and all other work schedules as abnormal, unusual, or unnatural. The negative connotations resulting from these sometimes explicit suggestions may be appropriate in many cases, but they frequently promote an insensitivity to the complex problems that many workers face and suggest erroneously that work schedule selection can be a simple task. In many cases, the work schedules practiced in a given plant or industry have an unknown origin and justification. The effects of a given work schedule on productivity and safety are frequently unknown. Other workplace issues often mask the effects of work schedule or attract more interest owing to traditional labor-management concerns.

Many work schedules can be improved. That is, changes can be made in working hours that will increase productivity and improve the worker's quality of life. Just as one can redesign equipment to improve production as well as worker health and safety, so also can one redesign work schedules to better meet these goals. Managers, unaware of the range of work systems available, may consciously avoid considering the evaluation of current schedules or alternative working hours. This stance is frequently guided by the mistaken notion that it will avoid labor problems. In a similar attitude, organized labor may also avoid consideration of work schedule alternatives under the assumption that they are always a threat to reductions in work hours that have been won in the past. In practice, case studies suggest that failure to attend to sentiments associated with work hours and shift workers may be a more significant contributor to labor strife than either wages or benefits (Imberman, 1983). Despite this and related findings, negotiators continue to concentrate on money rather than work schedule, and the concretion of undesirable and unproductive work schedules continues.

Data indicate that the number of shift workers in the United States has increased (Webb, 1981) and similar trends appear in many other countries. This trend will probably continue, and may very well hold for all forms of alternative work hours. The general objective of this chapter is to provide some of the background and framework needed to guide the introduction of shift work and work schedule changes, and at the same time make the reader more sensitive to the alternatives available, as well as the potential pitfalls associated with poor applications. When considering change, or the development of a new installation, it is very important that those involved consider *all* of the alternatives available, rather than simply duplicating whatever is handy or traditional.

This chapter has four specific objectives. First, a terminology and notation system is presented to describe work schedules in a standard way, and thereby ensure good communication. Second, selected work schedules are described and discussed to demonstrate the variety of systems available. Third, the variables relevant to the selection of a good work schedule are discussed in an effort to increase the reader's sensitivity to the complexity of the human factors involved. Fourth, methodological approaches to the evaluation of existing work schedules, or the selection of new work schedules, are discussed.

7.3.2 TERMINOLOGY AND NOTATION

The initial impression that most people have is that the variety of work schedules is not great. This is false. For example, U.S. fire fighters have been reported to work 150 different work schedules (Tasto and Colligan, 1977). Clearly, there are hundreds (and probably thousands) of different work schedules currently in practice. Undoubtedly, there are many more work schedules that have either been tried and abandoned, or have not as yet been implemented!

Having a great variety of schedules to consider is not a problem per se, in that it simply means there is a greater range of potential options to choose from. However, many schedules are complex, which makes them difficult to compare with other schedules. Some of this complexity exists simply because of the inconsistent use of work schedule terminology. In this section we define work scheduling terms, and demonstrate a notation method that can be used to describe most schedules. These definitions and the notation method are used in the remainder of the chapter. This method models other authors

(Adler and Roll, 1981; Knauth et al., 1983). It also attempts to overcome some conflicts in common U.S. shop-floor usage. Rather than assume that the methodology used is consistent with the literature or immediate industrial experience, the reader should study the terms and notation method presented to assure accurate understanding of subsequent sections of this chapter.

7.3.2.1 Work System Definitions

The work schedule definitions are presented in Table 7.3.1. The basic unit of these definitions is *shift*, the time of day a worker is required to be at the workplace. By definition, all workers who are scheduled to be at a workplace on a regular basis are shift workers, and it is thus proper to refer to those who work during the day as shift workers. Each worker has a work *schedule* for the shifts he or she works. This schedule includes *off time*, hours (usually more than 24 hr) when the worker is not regularly at the workplace. Any workplace that employs more than one individual has a formal or informal way of ensuring that all work requirements are met. When a formal system is used to schedule worker shifts, the term *work system* is used to describe *all* the shift schedules practiced in the given workplace. If such a work system includes schedules for most operations on all 7 days of a week, it is said to have *continuous* work weeks, as opposed to *discontinuous* work week systems, which regularly exclude most Sunday and/or Saturday work.

Some workers have schedules with *permanent hours*, working the same time of day every day. Other workers have schedules with *rotating hours*, working different hours on specified days following a planned schedule. The number of days required for a worker on a given schedule to complete his or her basic and repetitive cycle of shifts and off time is the *basic sequence* of a schedule. By definition, if a worker is employed more than 1 day, the schedule may also vary the day(s) of the week he or she works. In this case, the number of days required for the basic sequence to begin to repeat on the same days of the week is the *major cycle* of a schedule. For many schedules the basic sequence is equal to the length of the major cycle, but for others the basic cycle is longer (never shorter) than the basic sequence.

The variety of work systems that might be used is considerable, because a work system may include many work schedules. A given work system may incorporate work schedules involving both permanent and rotating hours, as well as sequences and cycles of various lengths. Specific examples are presented in subsequent tables.

Table 7.3.1   Work Schedule Definitions

Shift	The time of day on a given day that an individual or a group of individuals are scheduled to be at the workplace.
Off time	The hours a particular individual or group of individuals are *not* normally required to be at the workplace.
Schedule	The sequence of consecutive *shifts* and *off time* assigned to a particular individual or group of individuals as their usual work schedule.
Permanent hours	A *schedule* for an individual or group of individuals that does *not* normally require them to work more than one *shift*. That is, the time of day one works is constant.
Rotating hours	A *schedule* that normally requires an individual or group of individuals to work more than one *shift*. That is, the time of day one works changes.
Discontinuous work weeks	A *schedule* that normally does *not* require an individual or a group of individuals to work on weekends (Saturday and/or Sunday).
Continuous work weeks	A *schedule* that normally requires an individual or group of individuals to work some weekends.
Basic sequence	The minimum number of days required to complete the specific sequence of *shifts* and *off days* constituting a given basic *schedule*. That is, the number of days until a *schedule* begins to repeat.
Major cycle	The minimum number of days required to arrive at a point where the *basic sequence* of a *schedule* begins to repeat on the *same* days of the week. That is, the number of days until the *basic sequence* falls on the same days when it repeats.
Work system	All of the work *schedule*(s) implemented in a given workplace to meet the real or perceived requirements of a given industry, plant, process, or service.

7.3.2.2 Shift Definitions

Table 7.3.2 provides the operational definitions for the shift forms that are presented and discussed in this chapter. The time definition of the *third* shift is a standard one in keeping with the definition of the International Labor Organization for nighttime work. The remaining shift definitions use this as a model and appear to be fairly characteristic of workplace practice in the United States. It is important to note that these definitions limit the use of the terms *day* and *night* to shifts that are significantly *longer* than 8 hr, to avoid confusion and to make notation easier in subsequent tables and discussion.

The specific terms used for shift designation in U.S. workplaces *do* vary from one location to another, and the outside observer, not familiar with local practice, can be easily misled. For example, some plants with discontinuous work weeks refer to the third shift as the first shift because it is the first shift to be worked each week. Although one might argue with some of the terms or definitions found in Table 7.3.2, they should be used to ensure accurate operational discussion of specific work systems with minimal confusion.

In workplaces employing many workers, a shift schedule practiced in common by a number of workers frequently results in these workers being labeled as a *crew*. Using this terminology, a work system is implemented by a number of crews, rather than by work schedules. This terminology can be very misleading, because it leads one to assume that the crew is in fact a group of people working together in an interactive manner. Often this is not the case. The terminology presented in Tables 7.3.1 and 7.3.2 makes *no* such assumptions. A given work shift may be practiced by a single individual or a number of individuals, with or without interactive work duties. Crew designations are not used here, because the value and form of interactive group work systems is not a major focus of this chapter.

7.3.2.3 Work System Notation Method

Given the definitions and terminology presented in the preceding two sections, specific work systems can be schematically presented by a notation system. Table 7.3.3 contains the symbols for the notation system used in this chapter. Definitions for these symbols, when needed, can be found in Tables 7.3.1 and 7.3.2. Table 7.3.4 provides an introduction to the notation. This table shows the work system for a workplace that follows a common practice and limits work time to 8 hr periods during the day on weekdays. The table shows the work schedule for 16 consecutive weeks. This is a discontinuous work week system with permanent hours: only one work shift is used, all workers are assigned to the same schedule, and the schedule is identical every week. In this simple traditional example, the basic sequence and the major cycle are *both* 7 days because the same days of the week are worked every week.

In the following sections, a variety of work system shift combinations are presented and reviewed, using the definitions and notation method outlined above. *Readers who are familiar with the various*

Table 7.3.2 Work Shift Definitions

First shift	A day shift of about 8 hr in duration. In most cases this falls so that at least seven consecutive hours are somewhere between 0600 and 1600 hrs.
Second shift	An afternoon–evening shift of about 8 hr in duration. In most cases this falls so that at least seven consecutive hours are somewhere between 1500 and 0100 hr.
Third shift	A night shift of about 8 hr in duration. In most cases this falls so that at least seven consecutive hours are somewhere between 2200 and 0700 hr.
Day shift	A shift of about 10 or 12 hr in duration that frequently includes eight or more consecutive hours between 1000 and 2200 hr.
Night shift	A shift of about 10 or 12 hours in duration that frequently includes eight or more consecutive hours between 2200 and 1000 hr.
Split shift	Any shift whereby an individual or group of individuals is regularly scheduled to work less than seven consecutive hours, is released from work for *more* than an hour, and then returns to work during the same day for an additional work period.
Shift break	Time within a shift when work is *not* required, usually a time period of *less* than 1 hr in duration. This includes lunch breaks, rest breaks, relief breaks, and all other forms of workplace release from work.
Irregular shift	A variable-length shift where the duration of the shift is not precisely defined in advance and the length of work frequently varies in an unpredictable manner owing to unscheduled events.

Table 7.3.3 Symbols for Work System Notation Tables

Symbol	Meaning
1	First shift
2	Second shift
3	Third shift
D	Day shift
N	Night shift
O	Off time

Days of the week

M	MONDAY
T	TUESDAY
W	WEDNESDAY
R	THURSDAY
F	FRIDAY
S	SATURDAY
K	SUNDAY
S1	A designated schedule for a given work system
S2	Another schedule for the same work system
S3	Another schedule for the same work system
S4	Another schedule for the same work system
M1	The first 28 days of a given designated schedule
M2	Days 29–56 of this schedule (follows M1)
M3	Days 57–84 of this schedule (follows M2)
M4	Days 85–112 of this schedule (follows M3)

Table 7.3.4 Work System with Permanent Hours and Discontinuous Work Weeks Using Only One Schedule[a]

M,T,W,R,F,S,K,M,T,W,R,F,S,K,M,T,W,R,F,S,K,M,T,W,R,F,S,K	
1, 1, 1, 1, 1,O,O, 1, 1, 1, 1, 1,O,O, 1, 1, 1, 1, 1,O,O, 1, 1, 1, 1, 1,O,O	S1M1
1, 1, 1, 1, 1,O,O, 1, 1, 1, 1, 1,O,O, 1, 1, 1, 1, 1,O,O, 1, 1, 1, 1, 1,O,O	S1M2
1, 1, 1, 1, 1,O,O, 1, 1, 1, 1, 1,O,O, 1, 1, 1, 1, 1,O,O, 1, 1, 1, 1, 1,O,O	S1M3
1, 1, 1, 1, 1,O,O, 1, 1, 1, 1, 1,O,O, 1, 1, 1, 1, 1,O,O, 1, 1, 1, 1, 1,O,O	S1M4

[a] Basic sequence = major cycle = 7 days. Notation symbols for this and subsequent work system tables are found in Table 7.3.3.

forms work systems take may wish to skip the remainder of this section and go to section 7.3.3. Readers who are less familiar with shift work practice and terminology are encouraged to read and study all the succeeding sections and the related tables in the order presented.

7.3.2.4 Work Systems Examples—Discontinuous Work Weeks

Table 7.3.5 is another example of a discontinuous work week system with permanent hours. In this case, three work shifts are used, workers are assigned to one of three schedules, the work hours do not vary within a given schedule, and all work shifts are on weekdays. The table shows the work schedule for four consecutive weeks for each of three schedules. Again, the basic sequence and major cycle are both 7 days because the same days of the week are worked every week.

As in the two preceding examples, the system in Table 7.3.6 also is limited to discontinuous work weeks. Unlike the preceding examples, this system has all workers on rotating hours. Three schedules are used, and the table shows the work schedule for four consecutive weeks for each of the three schedules. Each of the three schedules involves work on three different shifts, with five consecutive workdays on a specific shift, followed by two off-time days, then work on another shift. Thus rotation

Table 7.3.5 Work System with Permanent Hours and Discontinuous Work Weeks Using Three Schedules[a]

M,T,W,R,F,S,K,M,T,W,R,F,S,K,M,T,W,R,F,S,K,M,T,W,R,F,S,K	
1, 1, 1, 1, 1,O,O, 1, 1, 1, 1, 1,O,O, 1, 1, 1, 1, 1,O,O, 1, 1, 1, 1, 1,O,O	S1M1
2, 2, 2, 2, 2,O,O, 2, 2, 2, 2, 2,O,O, 2, 2, 2, 2, 2,O,O, 2, 2, 2, 2, 2,O,O	S2M1
3, 3, 3, 3, 3,O,O, 3, 3, 3, 3, 3,O,O, 3, 3, 3, 3, 3,O,O, 3, 3, 3, 3, 3,O,O	S3M1

[a] Basic sequence = major cycle = 7 days.

occurs once every 7 days. In each case, rotation is from first to second to third shift. This is referred to as *forward* rotation. Again, the basic sequence and the major cycle are of equal length, but in this case they are 21 days.

The work systems in Table 7.3.7 provide an initial example of how rotation rate can vary. This work system also uses discontinuous work weeks and three rotating shift schedules. The table shows the work schedule for 16 consecutive weeks for each of the three schedules. For all these schedules, rotation is from first to second to third shift (forward rotation), but rotation occurs only once every 28 days. Once again the basic sequence and the major cycle are equal, but here the length is 84 days.

Backward rotation is demonstrated by the example in Table 7.3.8. This work system is identical to that in Table 7.3.7, except for the direction of rotation. In the case of Table 7.3.8, rotation is from first to third to second shift. This is referred to as backward rotation. The basic sequence and major cycle for all these schedules is 84 days.

Table 7.3.6 Work System with Rotating Hours and Discontinuous Work Weeks Using Three Schedules and Forward Rotation[a]

M,T,W,R,F,S,K,M,T,W,R,F,S,K,M,T,W,R,F,S,K,M,T,W,R,F,S,K	
1, 1, 1, 1, 1,O,O, 2, 2, 2, 2, 2,O,O, 3, 3, 3, 3, 3,O,O, 1, 1, 1, 1, 1,O,O	S1M1
2, 2, 2, 2, 2,O,O, 3, 3, 3, 3, 3,O,O, 1, 1, 1, 1, 1,O,O, 2, 2, 2, 2, 2,O,O	S2M1
3, 3, 3, 3, 3,O,O, 1, 1, 1, 1, 1,O,O, 2, 2, 2, 2, 2,O,O, 3, 3, 3, 3, 3,O,O	S3M1

[a] Basic sequence = major cycle = 21 days.

Table 7.3.7 Work System with Slowly Rotating Hours and Discontinuous Work Weeks Using Three Schedules and Forward Rotation[a]

M,T,W,R,F,S,K,M,T,W,R,F,S,K,M,T,W,R,F,S,K,M,T,W,R,F,S,K	
1, 1, 1, 1, 1,O,O, 1, 1, 1, 1, 1,O,O, 1, 1, 1, 1, 1,O,O, 1, 1, 1, 1, 1,O,O	S1M1
2, 2, 2, 2, 2,O,O, 2, 2, 2, 2, 2,O,O, 2, 2, 2, 2, 2,O,O, 2, 2, 2, 2, 2,O,O	S1M2
3, 3, 3, 3, 3,O,O, 3, 3, 3, 3, 3,O,O, 3, 3, 3, 3, 3,O,O, 3, 3, 3, 3, 3,O,O	S1M3
1, 1, 1, 1, 1,O,O, 1, 1, 1, 1, 1,O,O, 1, 1, 1, 1, 1,O,O, 1, 1, 1, 1, 1,O,O	S1M4
2, 2, 2, 2, 2,O,O, 2, 2, 2, 2, 2,O,O, 2, 2, 2, 2, 2,O,O, 2, 2, 2, 2, 2,O,O	S2M1
3, 3, 3, 3, 3,O,O, 3, 3, 3, 3, 3,O,O, 3, 3, 3, 3, 3,O,O, 3, 3, 3, 3, 3,O,O	S2M2
1, 1, 1, 1, 1,O,O, 1, 1, 1, 1, 1,O,O, 1, 1, 1, 1, 1,O,O, 1, 1, 1, 1, 1,O,O	S2M3
2, 2, 2, 2, 2,O,O, 2, 2, 2, 2, 2,O,O, 2, 2, 2, 2, 2,O,O, 2, 2, 2, 2, 2,O,O	S2M4
3, 3, 3, 3, 3,O,O, 3, 3, 3, 3, 3,O,O, 3, 3, 3, 3, 3,O,O, 3, 3, 3, 3, 3,O,O	S3M1
1, 1, 1, 1, 1,O,O, 1, 1, 1, 1, 1,O,O, 1, 1, 1, 1, 1,O,O, 1, 1, 1, 1, 1,O,O	S3M2
2, 2, 2, 2, 2,O,O, 2, 2, 2, 2, 2,O,O, 2, 2, 2, 2, 2,O,O, 2, 2, 2, 2, 2,O,O	S3M3
3, 3, 3, 3, 3,O,O, 3, 3, 3, 3, 3,O,O, 3, 3, 3, 3, 3,O,O, 3, 3, 3, 3, 3,O,O	S3M4

[a] Basic sequence = major cycle = 84 days.

Table 7.3.8 **Work System with Slowly Rotating Hours and Discontinuous Work Weeks Using Three Schedules and Backward Rotation**[a]

M,T,W,R,F, S, K,M,T,W,R,F, S, K,M,T,W,R,F, S, K,M,T,W,R,F, S, K	
1, 1, 1, 1, 1,O,O, 1, 1, 1, 1, 1,O,O, 1, 1, 1, 1, 1,O,O, 1, 1, 1, 1, 1,O,O	S1M1
3, 3, 3, 3, 3,O,O, 3, 3, 3, 3, 3,O,O, 3, 3, 3, 3, 3,O,O, 3, 3, 3, 3, 3,O,O	S1M2
2, 2, 2, 2, 2,O,O, 2, 2, 2, 2, 2,O,O, 2, 2, 2, 2, 2,O,O, 2, 2, 2, 2, 2,O,O	S1M3
1, 1, 1, 1, 1,O,O, 1, 1, 1, 1, 1,O,O, 1, 1, 1, 1, 1,O,O, 1, 1, 1, 1, 1,O,O	S1M4
2, 2, 2, 2, 2,O,O, 2, 2, 2, 2, 2,O,O, 2, 2, 2, 2, 2,O,O, 2, 2, 2, 2, 2,O,O	S2M1
1, 1, 1, 1, 1,O,O, 1, 1, 1, 1, 1,O,O, 1, 1, 1, 1, 1,O,O, 1, 1, 1, 1, 1,O,O	S2M2
3, 3, 3, 3, 3,O,O, 3, 3, 3, 3, 3,O,O, 3, 3, 3, 3, 3,O,O, 3, 3, 3, 3, 3,O,O	S2M3
2, 2, 2, 2, 2,O,O, 2, 2, 2, 2, 2,O,O, 2, 2, 2, 2, 2,O,O, 2, 2, 2, 2, 2,O,O	S2M4
3, 3, 3, 3, 3,O,O, 3, 3, 3, 3, 3,O,O, 3, 3, 3, 3, 3,O,O, 3, 3, 3, 3, 3,O,O	S3M1
2, 2, 2, 2, 2,O,O, 2, 2, 2, 2, 2,O,O, 2, 2, 2, 2, 2,O,O, 2, 2, 2, 2, 2,O,O	S3M2
1, 1, 1, 1, 1,O,O, 1, 1, 1, 1, 1,O,O, 1, 1, 1, 1, 1,O,O, 1, 1, 1, 1, 1,O,O	S3M3
3, 3, 3, 3, 3,O,O, 3, 3, 3, 3, 3,O,O, 3, 3, 3, 3, 3,O,O, 3, 3, 3, 3, 3,O,O	S3M4

[a] Basic sequence = major cycle = 84 days.

7.3.2.5 Work System Examples—Continuous Work Weeks

As noted earlier, continuous work weeks involve 7 day workplace operations. Because these work systems involve covering more hours per week, they usually include more than three work schedules. Differences in basic sequence and major cycle length are also more common. In practice, systems with rotating hours appear to be more common for continuous systems. This may be related to the fact that although permanent hours are feasible for continuous work weeks, they usually require more work schedules than a comparable rotating system.

Table 7.3.9 shows a frequently used European work system known as the "continental rota." This continuous work week rotating system is also known as the 2–2–3, a concise way of denoting the shift and off-time sequencing. Four schedules are required for this system, and Table 7.3.9 shows four consecutive weeks for each of these schedules. For each schedule, the basic sequence is 9 days and the major cycle is 28 days.

Another European work system is shown in Table 7.3.10. This system is known as the "metropolitan rota" and can more concisely be referred to as the 2–2–2–2. The table shows eight consecutive work weeks for each of the four schedules required. For each of these schedules the basic sequence is 8 days and the major cycle is 56 days.

Shifts of 12 hr duration are also used with systems involving continuous work hours. These are sometimes called "compressed work weeks" because a work week can involve less than five shift days. Table 7.3.11 exhibits one of these systems, referred to as the 4–4 or 4–4–4–4. The table shows 16 consecurive work weeks for each of the four schedules required. For each of these schedules the basic sequence is 16 days and the major cycle is 112 days.

Table 7.3.12 shows a similar work system, sometimes referred to as the 3–3 or the 3–3–3–3. This system also requires four basic schedules. Twelve consecutive work weeks for each of the schedules are presented in this table. For each of these schedules the basic sequence is 12 days and the major cycle is 84 days.

The work system shown in Table 7.3.13 is known as EOWEO, an acronym for "every other weekend

Table 7.3.9 **Work System with Rotating Hours and Continuous Work Weeks That Uses Four Schedules and Forward Rotation—the "Continental"**[a]

M, T, W, R, F, S, K, M, T, W, R, F, S, K, M, T, W, R, F, S, K, M, T, W, R, F, S, K	
1, 1, 2, 2, 3, 3, 3, O,O, 1, 1, 2, 2, 2, 3, 3, O,O, 1, 1, 1, 2, 2, 3, 3, O,O,O	S1M1
2, 2, 3, 3,O,O,O, 1, 1, 2, 2, 3, 3, 3, O,O, 1, 1, 2, 2, 2, 3, 3, O,O, 1, 1, 1	S2M1
3, 3, O,O, 1, 1, 1, 2, 2, 3, 3,O,O,O, 1, 1, 2, 2, 3, 3, 3, O,O, 1, 1, 2, 2, 2	S3M1
O,O, 1, 1, 2, 2, 2, 3, 3, O,O, 1, 1, 1, 2, 2, 3, 3, 3, O,O,O, 1, 1, 2, 2, 3, 3, 3	S4M1

[a] Basic sequence = major cycle = 28 days.

Table 7.3.10   Work System with Rotating Hours and Continuous Work Weeks That Uses Four Schedules and Forward Rotation—the "Metropolitan"[a]

M, T, W,R, F, S, K,M, T, W,R, F, S, K,M, T, W,R, F, S, K,M, T, W,R, F, S, K	
1, 1, 2, 2, 3, 3, O,O, 1, 1, 2, 2, 3, 3, O,O, 1, 1, 2, 2, 3, 3, O,O, 1, 1, 2, 2	S1M1
3, 3, O,O, 1, 1, 2, 2, 3, 3, O,O, 1, 1, 2, 2, 3, 3, O,O, 1, 1, 2, 2, 3, 3, O,O	S1M2
2, 2, 3, 3, O,O, 1, 1, 2, 2, 3, 3, O,O, 1, 1, 2, 2, 3, 3, O, O, 1, 1, 2, 2, 3, 3	S2M1
O,O, 1, 1, 2, 2, 3, 3, O, O, 1, 1, 2, 2, 3, 3, O,O, 1, 1, 2, 2, 3, 3, O,O, 1, 1	S2M2
3, 3, O,O, 1, 1, 2, 2, 3, 3, O,O, 1, 1, 2, 2, 3, 3, O,O, 1, 1, 2, 2, 3, 3, O,O	S3M1
1, 1, 2, 2, 3, 3, O,O, 1, 1, 2, 2, 3, 3, O,O, 1, 1, 2, 2, 3, 3, O,O, 1, 1, 2, 2	S3M2
O,O, 1, 1, 2, 2, 3, 3, O, O, 1, 1, 2, 2, 3, 3, O,O, 1, 1, 2, 2, 3, 3, O,O, 1, 1	S4M1
2, 2, 3, 3, O,O, 1, 1, 2, 2, 3, 3, O,O, 1, 1, 2, 2, 3, 3, O, O, 1, 1, 2, 2, 3, 3	S3M2

[a] Basic sequence = 8 days; major cycle = 56 days.

off." This system features two 3 day weekend off-time days every month for each work schedule. Again, four basic schedules with rotating work hours are used. Four work weeks for each of these schedules are shown. In this case the basic sequence and the major cycle are both 28 days long.

7.3.2.6   Work System Examples—Mixed Schedules

Obviously, the work systems presented thus far represent only some of the many possibilities. The schedules within each of these work systems are similar in that shift length is the same and either all of the schedules are permanent or all of them are rotated. This need not be true. Mixing schedules within one work system that has shifts of different workday lengths, or mixing permanent hour schedules with rotating hour shifts, further increases the variety of work systems possible.

Table 7.3.14 provides an example of a work system with continuous work weeks that includes both 8 and 12 hr shifts. Five schedules are used, each with permanent hours. Four work weeks are shown for each of these schedules. For all five schedules, the basic sequence and the major cycle are both 7 days. A variation of this schedule has the 12 hr shift workers rotating, a further mixing of schedule dimensions.

Table 7.3.11   Rotating Hours and Continuous Work Weeks—"Compressed 4-4-4-4"[a]

M, T, W,R, F, S, K,M, T, W,R, F, S, K,M, T, W,R, F, S, K,M, T, W,R, F, S, K	
D,D,D,D,O,O,O,O,N,N,N,N,O,O,O,O,D,D,D,D,O,O,O,O,N,N,N,N	S1M1
O,O,O,O,D,D,D,D,O,O,O,O,N,N,N,N,O,O,O,O,D,D,D,D,O,O,O,O	S1M2
N,N,N,N,O,O,O,O,D,D,D,D,O,O,O,O,N,N,N,N,O,O,O,O,D,D,D,D	S1M3
O,O,O,O,N,N,N,N,O,O,O,O,D,D,D,D,O,O,O,O,N,N,N,N,O,O,O,O	S1M4
N,N,N,N,O,O,O,O,D,D,D,D,O,O,O,O,N,N,N,N,O,O,O,O,D,D,D,D	S2M1
O,O,O,O,N,N,N,N,O,O,O,O,D,D,D,D,O,O,O,O,N,N,N,N,O,O,O,O	S2M2
D,D,D,D,O,O,O,O,N,N,N,N,O,O,O,O,D,D,D,D,O,O,O,O,N,N,N,N	S2M3
O,O,O,O,D,D,D,D,O,O,O,O,N,N,N,N,O,O,O,O,D,D,D,D,O,O,O,O	S2M4
O,O,O,O,D,D,D,D,O,O,O,O,N,N,N,N,O,O,O,O,D,D,D,D,O,O,O,O	S3M1
N,N,N,N,O,O,O,O,D,D,D,D,O,O,O,O,N,N,N,N,O,O,O,O,D,D,D,D	S3M2
O,O,O,O,N,N,N,N,O,O,O,O,D,D,D,D,O,O,O,O,N,N,N,N,O,O,O,O	S3M3
D,D,D,D,O,O,O,O,N,N,N,N,O,O,O,O,D,D,D,D,O,O,O,O,N,N,N,N	S3M4
O,Q,O,O,N,N,N,N,O,O,O,O,D,D,D,D,O,O,O,O,N,N,N,N,O,O,O,O	S4M1
D,D,D,D,O,O,O,O,N,N,N,N,O,O,O,O,D,D,D,D,O,O,O,O,N,N,N,N	S4M2
O,O,O,O,D,D,D,D,O,O,O,O,N,N,N,N,O,O,O,O,D,D,D,D,O,O,O,O	S4M3
N,N,N,N,O,O,O,O,D,D,D,D,O,O,O,O,N,N,N,N,O,O,O,O,D,D,D,D	S4M4

[a] Basic sequence = 16 days; major cycle = 112 days.

Table 7.3.12   Rotating Hours and Continuous Work Weeks—"Compressed 3–3–3–3"[a]

M, T, W, R, F, S, K, M, T, W, R, F, S, K, M, T, W, R, F, S, K, M, T, W, R, F, S, K	
D, D, D, O, O, O, N, N, N, O, O, O, D, D, D, O, O, O, N, N, N, O, O, O, D, D, D, O	S1M1
O, O, N, N, N, O, O, O, D, D, D, O, O, O, N, N, N, O, O, O, D, D, D, O, O, O, N, N	S1M2
N, O, O, O, D, D, D, O, O, O, N, N, N, O, O, O, D, D, D, O, O, O, N, N, N, O, O, O	S1M3
O, O, O, D, D, D, O, O, O, N, N, N, O, O, O, D, D, D, O, O, O, N, N, N, O, O, O, D	S2M1
D, D, O, O, O, N, N, N, O, O, O, D, D, D, O, O, O, N, N, N, O, O, O, D, D, D, O, O	S2M2
O, N, N, N, O, O, O, D, D, D, O, O, O, N, N, N, O, O, O, D, D, D, O, O, O, N, N, N	S2M3
N, N, N, O, O, O, D, D, D, O, O, O, N, N, N, O, O, O, D, D, D, O, O, O, N, N, N, O	S3M1
O, O, D, D, D, O, O, O, N, N, N, O, O, O, D, D, D, O, O, O, N, N, N, O, O, O, D, D	S3M2
D, O, O, O, N, N, N, O, O, O, D, D, D, O, O, O, N, N, N, O, O, O, D, D, D, O, O, O	S3M3
O, O, O, N, N, N, O, O, O, D, D, D, O, O, O, N, N, N, O, O, O, D, D, D, O, O, O, N	S4M1
N, N, O, O, O, D, D, D, O, O, O, N, N, N, O, O, O, D, D, D, O, O, O, N, N, N, O, O	S4M2
O, D, D, D, O, O, O, N, N, N, O, O, O, D, D, D, O, O, O, N, N, N, O, O, O, D, D, D	S4M3

[a] Basic sequence = 12 days; major cycle = 84 days.

Table 7.3.13   Rotating Hours and Continuous Work Weeks "EOWEO"[a]

M, T, W, R, F, S, K, M, T, W, R, F, S, K, M, T, W, R, F, S, K, M, T, W, R, F, S, K	
D, D, O, O, N, N, N, O, O, D, D, O, O, O, N, N, O, O, D, D, D, O, O, N, N, O, O, O	S1M1
O, O, D, D, O, O, O, N, N, O, O, D, D, D, O, O, N, N, O, O, O, D, D, O, O, N, N, N	S2M1
N, N, O, O, D, D, D, O, O, N, N, O, O, O, D, D, O, O, N, N, N, O, O, D, D, O, O, O	S3M1
O, O, N, N, O, O, O, D, D, O, O, N, N, N, O, O, D, D, O, O, O, N, N, O, O, D, D, D	S4M1

[a] Basic sequence = major cycle = 28 days.

Table 7.3.14   Permanent Hours and Continuous Work Weeks—Weekend Schedules[a]

M, T, W, R, F, S, K, M, T, W, R, F, S, K, M, T, W, R, F, S, K, M, T, W, R, F, S, K	
1, 1, 1, 1, 1, O, O, 1, 1, 1, 1, 1, O, O, 1, 1, 1, 1, 1, O, O, 1, 1, 1, 1, 1, O, O	S1M1
2, 2, 2, 2, 2, O, O, 2, 2, 2, 2, 2, O, O, 2, 2, 2, 2, 2, O, O, 2, 2, 2, 2, 2, O, O	S2M1
3, 3, 3, 3, 3, O, O, 3, 3, 3, 3, 3, O, O, 3, 3, 3, 3, 3, O, O, 3, 3, 3, 3, 3, O, O	S3M1
O, O, O, O, O, D, D, O, O, O, O, O, D, D, O, O, O, O, O, D, D, O, O, O, O, O, D, D	S4M1
O, O, O, O, O, N, N, O, O, O, O, O, N, N, O, O, O, O, O, N, N, O, O, O, O, O, N, N	S5M1

[a] For all schedules, basic sequence = major cycle = 7 days.

Another example of a mixed schedule is shown in Table 7.3.15. This system is for a continuous work week in which all schedules have 8 hr shifts. Four work weeks are shown for each of the schedules used in this system. Four schedules are used, three with permanent hours and one with rotating hours. The schedule with rotating hours is the "continental rota" previously described in Table 7.3.9. For all four schedules, the basic sequence and the major cycle are both 28 days long.

7.3.2.7   Discussion

The wide variety of work system shift combinations described above bespeaks the complexity and difficulty of good work scheduling. At one extreme, using the "wrong" work system can diminish worker productivity and health, whereas at the other extreme, using the "right" work system may improve them considerably. Workplace experience suggests that the merit of a given work system is only relative. That is, a specific work system may prove to be "right" in one workplace and "wrong" in another. The complexity of variables involved suggests that an absolute and ultimate general ranking of all work systems may actually never be possible.

Table 7.3.15   Permanent and Rotating Hours—Continuous Work Week[a]

M, T, W, R, F, S, K, M, T, W, R, F, S, K, M, T, W, R, F, S, K, M, T, W, R, F, S, K	
1, 1, 1, 1, 1, 1, 1, O, O, 1, 1, 1, 1, 1, 1, 1, O, O, 1, 1, 1, 1, 1, 1, 1, O, O, O	S1M1
O, O, 2, 2, 2, 2, 2, 2, 2, O, O, 2, 2, 2, 2, 2, 2, 2, O, O, O, 2, 2, 2, 2, 2, 2, 2	S2M1
3, 3, O, O, 3, 3, 3, 3, 3, 3, O, O, O, 3, 3, 3, 3, 3, 3, 3, O, O, 3, 3, 3, 3, 3	S3M1
2, 2, 3, 3, O, O, O, 1, 1, 2, 2, 3, 3, 3, O, O, 1, 1, 2, 2, 2, 3, 3, O, O, 1, 1, 1	S4M1

[a] S1, S2, and S3 have permanent hours, and S4 is the "continental rota." For all schedules the basic sequence and the major cycle are both 28 days.

Thus far, operational definitions and a notation method for work systems have been presented. Twelve work systems have been charted to provide examples of how the notation system is used and to introduce the reader to the variety of work systems shift combinations available. The work systems presented should *not* be viewed as either ideal or recommended approaches to the scheduling of workers. In fact, problems associated with the use of some of these systems indicate that they should be avoided in many cases.

The next section of this chapter is a review of the complex human-factors variables that should be considered in the selection or evaluation of a work system. When appropriate, the work systems presented in this section will be used as examples and points of discussion. The final goal is a recommended method for the evaluation and selection of work systems, not the presentation of a list of recommended work systems.

7.3.3   HUMAN-FACTORS VARIABLES

Traditionally, concerns within U. S. industry about work systems have centered around two main questions: How can I schedule workers so that I can meet the demand for my product without increasing *immediate* per-unit labor costs? How can I recruit workers for these schedules *within* these cost limits? These questions focus on short-term economic gain and ignore potential long-term human-factors problems. Night work and extended work hours, for example, are viewed in this orientation simply as worker inconveniences having little relation to profit. Within this narrow viewpoint, acute fatigue is the only perceived human-factors problem.

Research during the past 30 years, mainly originating in Europe, has clearly indicated that this viewpoint is too narrow. Variations in biology and performance as a function of time of day have been firmly established. Chronic effects of night work have been demonstrated. Significant interactions between off-the-job and on-the-job behavior are evident. As a result, a contemporary human-factors consideration of work schedules must be directed toward a variety of additional human-factors problems and requires a total systems approach. This section provides a review of these factors.

7.3.3.1   Circadian Variation

Numerous well-controlled laboratory studies have demonstrated that people have an internal biological timekeeping system. This internal timekeeper, usually referred to as an *endogenous clock* or oscillator, has a natural cycle time of around 25 hr in length (Wever, 1979). Under everyday conditions, this clock is reset on a daily basis so that it has a cycle time of exactly 24 hr, and it is therefore termed a *circadian system* (from the Latin *circa,* about, and *dies,* a day). Physical and social time cues (i.e., *exogenous* "clocks"), called *zeitgebers* (from the German for time-giver), are responsible for these fairly modest and normal resettings of the circadian system. Research indicates that the circadian system is remarkably resistant to sudden large changes in routine and normally quite stable. For example, individuals placed in isolation for weeks on end, without a clock and with total freedom to determine their activity schedule, continue to live on a self-selected routine based on a "circadian day" of about 25 hr long (Wever, 1979). The circadian system produces concomitant changes in most physiological and behavioral variables. Some of the many behavioral and biological variables showing circadian variation are body temperature, cell-division rate, hormone secretion, alertness, and reaction time (Moore-Ede, Sulzman, and Fuller, 1982).

Both common sense and theory suggest that good health requires maintenance of the normal synchrony of the behavior rest/activity cycle, including work, with the cycle of the circadian biological system. Thus work schedules that either maintain this synchrony or promote a shift in the circadian system so that it is in synchrony with work activity are favored. Because research indicates that many circadian biological systems are quite resilient to an abrupt change in routine, synchrony with many work schedules may be quite difficult to attain (Wever, 1979). This is especially true in the case of the third shift worker whose off time repeatedly thrusts him or her into the zeitgebers of a daytime-oriented society. Such zeitgebers are usually considerably stronger than those encouraging a "night orientation" of the circadian system, and a few days off time on a day-oriented routine may undo a week's worth of accumulated night adjustment. From both laboratory and field studies, it is clear that well over 10 days of uninterrupted nighttime work is needed before full circadian adjustment even begins to occur (Knauth and Rutenfranz, 1976; Akerstedt, 1977). Research with experienced shift workers suggests that those workers who show a desynchronization of the circadian system, as determined with temperature measures, show an intolerance to shift work (Reinberg et al., 1984).

7.3.3.2   Performance

Given the assumption that an endogenous circadian system determines how well one can perform and when one requires sleep, it is reasonable to suggest that night work is a problem in the sense that it calls for behavior that is not in keeping with the concomitant state of the biological clock. Figure 7.3.1 provides a good example of the face validity of such an assumption, showing a strong link between the body temperature rhythm and the workers' ratings of how alert they felt. Thus in the early hours of morning, both temperature and alertness were at their lowest ebb. These data are from Monk and Embry (1981) and are compatible with other findings (Folkard and Monk, 1979).

However, there is a danger in assuming that *all* tasks will show the same pattern as alertness. Figure 7.3.2 shows the performance of these workers on two tasks administered during the same study. These are two similar tasks, mainly differing only in that one requires a high level of memory load (6-MAST) and the other a relatively low memory load (2-MAST). Performance on the two tasks differed, with performance on the low memory load task most closely matching what one might expect based on circadian variation, such as shown in the data in Figure 7.3.1. Figure 7.3.2 shows a very different pattern occurring for the high memory load (6-MAST) task. In fact, examination of the actual job performance of these workers (entering complex numerical codes and values into a process control computer) showed a pattern of efficiency that more closely followed that of the *high* memory load task, rather than that of the low memory task. A similar dissociation of performance tasks has been demonstrated under more highly controlled laboratory conditions (Monk et al., 1983). From a theoretical viewpoint, these and other data suggest that the human circadian system is controlled by two or more oscillator systems that are differentially sensitive to zeitgebers and that performance efficiency may not be predictable solely from the temperature rhythm (Folkard, Wever, and Wildgruber, 1983).

From a practical work systems viewpoint, these findings are important. They suggest not only that worker performance varies with time of day, but also that interactions with work schedule may result in different time of day effects for various tasks. The simple generalization that workplace performance is poorest in the early morning hours must now be limited and qualified. Some work schedules, for example, may result in the best performance for one type of task in the early morning hours and at the same time yield the poorest performance for another type of task at the same time of day (Folkard and Monk, 1979). In selecting or evaluating a work system one must consider not only if time of day variation in performance is a work problem or benefit, but also whether the kind of work task performed should be considered.

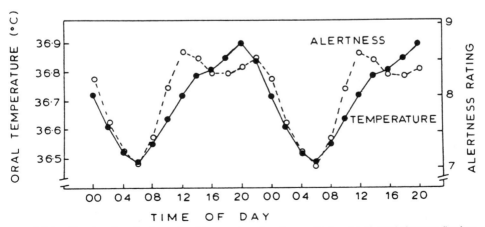

Fig. 7.3.1.   The circadian rhythms of oral temperature (continuous line) and self-rated alertness (broken line) in rotating shift workers. (*Source:* Monk and Embrey, 1981.) Reprinted with permission from A. Reinberg, N. Vieux, and P. Andlauer, Eds., *Night & Shiftwork: Biological & Social Aspects,* Copyright © 1981 by Pergamon Press.

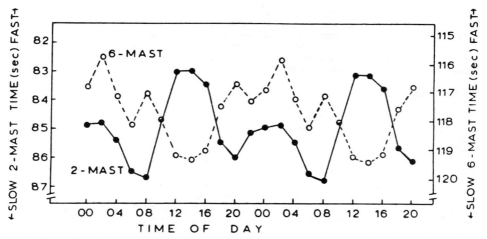

Fig. 7.3.2. The circadian rhythms of low (2-MAST, continuous line) and high (6-MAST, broken line) Memory and Search Task (MAST) performance by rotating shift workers. (*Source:* Monk and Embrey, 1981.) Reprinted with permission from A. Reinberg, N. Vieux, and P. Andlauer, Eds., *Night & Shift Work: Biological & Social Aspects,* Copyright © 1981 by Pergamon Press.

7.3.3.3 Sleep

Sleep is perhaps the most significant variable to show a relation to work schedule. Research studies of workers performing some or all of their work on the third shift have repeatedly found that these workers report a higher incidence of sleep difficulty than permanent first shift workers. Early research with animals, students, and hospital workers suggested that these sleep difficulties were problems in falling asleep, disturbed sleep periods, and spontaneous early awakenings (Tepas, 1982c). A more recent study of employed shift workers sleeping in the laboratory under controlled conditions (Walsh, Tepas, and Moss, 1981) suggests that the primary problem, however, is simply one of obtaining enough sleep time when working on a shift that requires daytime sleep.

Figure 7.3.3 provides a good example of the sleep lengths usually reported by shift workers: second shift workers sleep the most, third shift workers sleep the least, and first shift workers report a sleep length that is somewhere between these two groups. These data are from permanent shift workers (Tepas et al., 1985). For this sample, the main sleep period of third shift workers is almost an hour less per workday than that of second shift workers. The reduction in sleep length when working the third shift is even greater when workers are on rotating shifts (Tepas, Walsh, and Armstrong, 1981), and these differences are consistently reported in a number of studies (Tepas, 1982c).

Reduction in sleep length and sleep deprivation result in significant decrements in performance for many tasks that require vigilant behavior (Johnson, 1982). Thus for some tasks (e.g., quality control, monitoring) performance on the third shift may suffer not only from being at a low ebb of the circadian cycle for that task, but also from sleep deprivation effects. For the rotating shift worker, research suggests that this deterioration in performance is an acute one that may show recovery when the rotation of the shift allows nighttime sleep (Tilley, Wilkinson, Warren, Watson, and Drud, 1982). For the permanent third shift worker, this sleep deprivation appears to be a chronic effect (Tepas, Walsh, Moss, and Armstrong, 1981; Tepas, 1982a).

Animal studies have shown that sleep deprivation can result in death (Rechtschaffen, Gilliland, Bergman, and Winter, 1983), and at least one human epidemiological study suggests that short sleep lengths, such as those reported by permanent third shift workers, may reduce life expectancy (Kripke, Simons, Garfinkel, and Hammond, 1979). However, the question of whether health is impaired by shift work has not yet been fully answered (Rutenfranz, Knauth, and Angersbach, 1981). There are data supporting the hypothesis that many workers leave third shift work for health reasons. However, because some workers leave for reasons other than health or remain on third shift for reasons other than health, the effect of third shift work on health is very difficult to assess accurately (Frese and Okonek, 1984).

A rudimentary assumption is frequently made that the sleep problems of third shift workers are related to the fact that they are trying to sleep at a time out of phase with the dictates of their endogenous circadian oscillators. Indeed, there are good reasons and considerable evidence to link sleep behavior to endogenous circadian systems. For example, a study of research subjects isolated

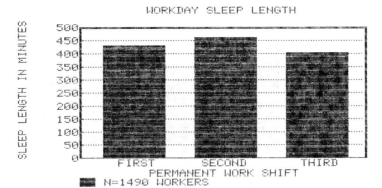

Fig. 7.3.3. Mean workday sleep length in minutes for permanent shift workers. Survey data from a sample of 1490 hourly workers, described by Tepas et al. (1985).

in a time-free environment by Czeisler and colleagues found a high correlation between sleep length and body temperature at sleep onset time (Czeisler, Weitzman, Moore-Ede, Zimmerman, and Knauer, 1980). For these subjects, sleep length was not significantly correlated with the length of the prior wakefulness period. Industrial shift workers when studied during the work week show quite the opposite result. For employed persons, sleep length is frequently not correlated with temperature, and it is significantly correlated with the length of prior wakefulness on workdays (Tepas and Sullivan, 1982). These results should not be viewed as opposing each other, but perhaps rather as a clear demonstration of the degree to which social zeitgebers can override other circadian variables.

7.3.3.4 Fatigue

Development of operational definitions of the various factors commonly referred to as fatigue is an elusive task that is beyond the scope of this chapter. However, fatigue as a variable, regardless of how it is defined, requires discussion in any review of the human-factors dimensions of work schedules and systems. Two dimensions of fatigue would seem to flow logically from the discussion in the preceding sections. First, one should expect fatigue to vary in a circadian manner. The circadian variations in alertness and performance, shown for example in Figure 7.3.1 and 7.3.2, undoubtedly include a fatigue component to some degree. However, systematic efforts to relate some definition of fatigue directly to circadian rhythms are largely lacking.

Second, if some work shifts are associated with off-time sleep deprivation, it should be expected that this will manifest itself in increased fatigue during work hours. The decrements in performance discussed earlier, although gathered in off time, undoubtedly have a fatigue component. However, we are again dealing with an issue that has not been systematically addressed in a worker population. The data presented in Figure 7.3.4 provide a preliminary response to this issue. These data are from the *same* worker sample whose sleep lengths are presented in Figure 7.3.3. These workers were asked how often they felt tired or sleepy at work. When taken together, these two figures suggest that off-time average sleep length is related to the incidence of workplace feelings of fatigue. Obviously it would be good to look at actual production data, but accurate data of this sort is difficult to obtain. Clearly, additional study is needed.

Two additional issues merit brief discussion under the topic of fatigue: shift length and shift breaks. Using a *shift length* longer than the usual 8 hr has both advantages and disadvantages (Tepas, 1985). Common sense and the literature suggest that lengthening the workday for some jobs may increase worker errors and worker accidents. Jobs that are very boring and/or require heavy physical work are probably not good candidates for the extended workday. Following a review of the sparse literature on this topic, Kelly and Schneider (1982) predicted an increase in worker error rate of 80 to 180% for such tasks when going from an 8 to a 12 hr work system.

The issue of commuting time should also be brought up when one considers fatigue and workday lengthening. A 1 hr commute at either end of a 12 hr shift leaves the worker very little time for any other activities if he or she is to get enough sleep. This situation particularly affects female shift workers who have the responsibility for household management and/or child-rearing activities. *Shift breaks* and *split shifts* are ways of breaking up a workday of any length to overcome acute fatigue effects. Abuse of how these breaks are scheduled or used can obviously result in decreases in productivity or significant worker hardship, and commuting time may again be an important factor. In practice, commuting time requirements, shift breaks, and split shifts frequently extend the length of the time

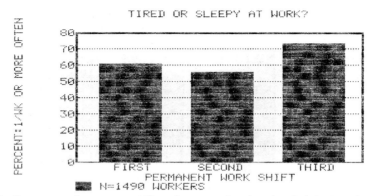

Fig. 7.3.4. Percentage of permanent shift workers reporting that they feel tired or sleepy at work one or more times per week. Survey data from a sample of 1490 hourly workers, described by Tepas et al. (1985).

period dedicated to well beyond 8 hr. There is, on the other hand, modest evidence that appropriate break scheduling will not only prevent decrements in performance but may also *increase* total shift productivity (Bhatia and Murrell, 1969).

7.3.3.5   Off-Time Social Variables

The traditional approach of U. S. industrial employers has been to ignore the off time of hourly workers and regard it as an employee private event with minimal impact on workplace performance. Frequently, when off-time activity interfered with workplace performance, workers were replaced. As employee turnover costs have grown to thousands of dollars, a new concern for worker off-time health and happiness has developed. Shift schedules can have a significant impact on many dimensions of worker off-time activity (Gordon, McGill, and Maltese, 1981), and the resulting off-time behavior can in turn have a significant effect on work satisfaction and performance (Thierry and Jansen, 1981).

In considering off-time behavior, the temptation to simple divide off time into "sleep" and "leisure" must be avoided. As common sense indicates, much of our off time is spent in neither category, but rather in the simple mechanics of living, such as buying, preparing, and eating food, bathing, housekeeping, and commuting to and from work. Thus, for example, most first shift workers have at least 1 or 2 hr of "off time" between waking up and arriving at work, but few would describe this as "leisure time." Activity of this sort must be viewed as a "gray time" that, though not actually work for pay, is not in most cases a recuperative time in any realistic sense.

Sleep is an obvious example of an off-time behavior that may affect work performance. It is important that one recognize that although sleep has a biological and circadian basis, it is also a social behavior. Figure 7.3.5 is a model of when shift workers sleep and work when employed on the first, second, and third shifts. This model holds for the vast majority of both permanent and rotating shift workers. For example, in the sample of permanent shift workers discussed earlier (Tepas et al., 1985), over 92% of the workers followed this model. These workers were job-bound; that is, they must be at work at a specific time and remain there for a given time period. Sleep times, on the other hand, have some flexibility and this model shows what in practice workers choose.

It is reasonable to suggest that these sleep time choices are tempered by personal considerations of fatigue and social factors, in addition to the demands of endogenous circadian rhythms. For one thing, this means that second and third shift work hours begin after a comparatively long period of awake activity, rather than after the period of rest characteristic of first shift workers. Third shift workers need not awake early in order to go to work, yet they show the shortest sleep periods and thus the greatest effects of sleep deprivation. Because it is estimated that better than 85% of these workers use an alarm or some other device to awaken (Tepas, 1982a), it is difficult to argue that these reductions in sleep length are *purely* a function of endogenous circadian variables. It is more reasonable to argue that most third shift workers cut their sleep short to be with family or friends in the late afternoon and early evening, or to meet the mandatory demands of their everyday chores.

A number of studies have clearly demonstrated that the home, family, and social life of workers varies dramatically with work shift schedule. The permanent shift worker sample cited previously (Tepas et al., 1985) provides good examples of this, and is presented in the next three figures. As is shown in Figure 7.3.6, fewer second and third shift workers report that their friends work the same shift schedule as they do. The percentage of workers reporting that they are satisfied with the amount

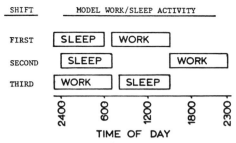

Fig. 7.3.5. Model of work and sleep activity manifested by most permanent and rotating shift workers.

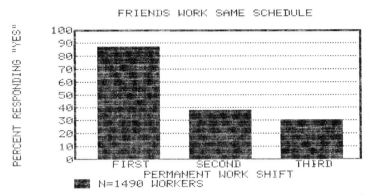

Fig. 7.3.6. Percentage of permanent shift workers reporting that their friends work the same shift schedule as they do. Data from the sample of 1490 hourly workers described by Tepas et al. (1985).

of time they are able to spend with their family and friends is presented in Figure 7.3.7. Second and third shift workers are less satisfied than first shift workers. Figure 7.3.8 shows the percentage of workers in each permanent shift group reporting that they are currently either divorced or separated. The divorce and separation rate of third shift workers is clearly the highest. Many additional examples of interactions between work schedule and off-time social variables have been demonstrated.

It is important to note that the interactions between work schedule and off-time variables are *not* all negative. For example, second and third shift workers in this sample appear to have less difficulty

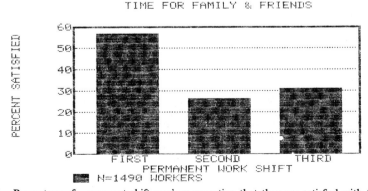

Fig. 7.3.7. Percentage of permanent shift workers reporting that they are satisfied with the amount of time they have to spend with their family and friends. Data from the sample of 1490 hourly workers described by Tepas et al. (1985).

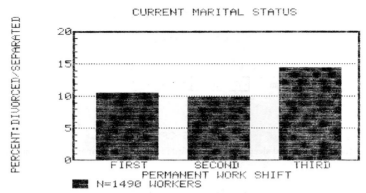

Fig. 7.3.8. Percentage of permanent shift workers reporting that they were divorced or separated at the time they responded. Data from a sample of 1490 hourly workers described by Tepas et al. (1985).

seeing a doctor about their health when not at work. Some divorced second and third shift workers are single parents reporting that their work schedule makes child care easier. It would be premature to conclude that third shift work, or some other work schedule, is stressful or bad in some other respect simply on the basis of selected social variable differences. The important thing is to recognize that there is a great deal of individual variability in the characteristics of the workers on different shift schedules as well as in different plants, and that this variability may warrant special consideration of individual differences if work productivity is to be maintained.

7.3.3.6 Discussion

Many human-factors variables interact with work schedule. These include circadian variation of endogenous biological clock, performance changes, sleep behavior, fatigue problems, and a variety of social variables. Given the complex nature of the interactions among these variables with work schedule, it is obvious that there is no single or universal answer to the simplistic question of "What is the best work schedule?" The more appropriate question is, "What is the best work *system* for a specific type of work, for a given group of workers, and a given demand for product or service?" Above all, the range of human individual differences should not be forgotten. The balance sheet for individuals may sometimes lead to the conclusion that what is good for one worker many not be satisfactory at all for another.

7.3.4 WORK SYSTEM SELECTION ISSUES

In selecting a new work system, and in the evaluation of an existing one, consideration of the human-factors variables discussed in the preceding section should be coupled with consideration of a number of key design and implementation issues. This section provides a brief discussion of some of the more prominent of these issues.

7.3.4.1 Rotating Versus Permanent Hours

It is sometimes argued that permanent hours are to be preferred because they allow the worker to adapt fully to second or third shift work. In practice, there is little to support the notion that most permanent shift workers ever fully adapt to working evening or nighttime hours. They may change their habits over time so that they can cope better with their shift hours, but that is not the same as saying that they live "normal" lives (Tepas, 1982a). Our society is daytime-oriented, and most workers revert to this world during their off-time days. A study of the work systems presented in Tables 7.3.5 and 7.3.6 should make this abundantly clear. Permanent third shift workers will be interested, we expect, just as much as those on rotating shifts, in participation in family and community weekend activities. In a very real sense, permanent shift workers usually practice rotating hours, given their off-time habits.

    If one views nighttime work hours as *the* major hazard to be minimized, then it is appropriate to argue in favor of rotating hours. One can view rotating hours as a method for spreading nighttime work hours across a greater portion of the work force, thereby minimizing exposure to the individual worker. This assumes, of course, that the probable hazards associated with rotating hours are fewer than those associated with nighttime work hours.

A more moderate position argues that changing hours and nighttime work are both hazards with negative factors whose relative strength varies from worker to worker and from job to job. This approach argues that the most appropriate thing to do is to select a work system for a specific job that best matches the existing worker characteristics and job duties.

7.3.4.2 Rotation Rate

Should it be concluded that rotating hours are more appropriate than permanent hours for a given workplace, the question of which rotation rate is the most appropriate remains to be considered. The work systems presented in Tables 7.3.9 and 7.3.10 are usually labeled as swiftly or fast rotating systems, and are practiced at a number of locations in Europe. A work system such as that presented in Table 7.3.7 is labeled a slowly rotating system. Contrary to common U. S. workplace notions, a work system such as that presented in Table 7.3.6 might also be referred to as a slowly rotating system.

Experts are divided as to which is the most appropriate rate. Those who conclude that it is important to minimize the number of consecutive nights worked or argue in favor of maximum maintenance of daytime circadian rhythms may favor swiftly rotating work systems. At the other extreme, if adaptation to a new shift is viewed as a feasible and desirable objective, the slowest rotation rate possible is favored. A direct experimental comparison of the human-factors effects of slowly and swiftly rotating work systems on U. S. industrial workers has never been made. In the United States, it would appear that systems with rotation rates such as that shown in Table 7.3.6 are most common. Although it might be viewed as a good compromise, it may very well represent a hybrid system that incorporates the *worst* features of both rotation rate extremes: It allows the circadian system to start the process of adjustment on a nocturnal routine, without allowing enough time on a particular shift to complete the adjustment process.

7.3.4.3 Direction of Rotation

Many rotating hour work systems allow the option of forward or backward rotation. Tables 7.3.7 and 7.3.8 show forward and backward rotation with essentially the same work system form. Common sense as well as circadian system theory both suggest that forward rotation (from first to second to third shift) would be easier to cope with. Thus most human factors experts recommend it. However, many U. S. plants have for years practiced backward rotation.

No direct experimental comparison of the human-factors effects of forward and backward rotation on industrial workers is available. In one study, Czeisler, Moore-Ede, and Coleman (1982) have reported that when changes to forward rotation *and* a decrease in rotation rate are combined, improvements in worker health, satisfaction, and productivity do occur. Because most experts agree on the issue of perferred shift rotation direction, this would seem to be an easy change for a plant to make (changing from the system shown in Table 7.3.8 to that in Table 7.3.7, for example) in order to test this hypothesis.

7.3.4.4 Basic Sequence and Major Cycle Length

As was pointed out in the work systems presented earlier, shift schedules can vary in their basic sequence and major cycle length. As a general rule, short sequences and cycles are preferred over long ones. Two concepts are relevant here. A short basic sequence makes it easy to *track* a shift schedule. That is, it makes it easy for a worker or a manager to know what shift comes next and when the change is to occur, and makes short-term off-time planning easier. The metropolitan rota shown in Table 7.3.10 is a good example of this. Presumably this decreases absenteeism as well as improves the quality of life of workers.

A major cycle of 7 days, or a multiple of 7 days, makes it easier to *plan ahead*. That is, the worker can easily predict when off time will fall on a particular day of the week, making relatively long-term off-time planning easier. The work system shown in Table 7.3.6 is a good example of such a schedule. The work system shown in Table 7.3.11 is an example of shift schedules with a long major cycle that makes planning ahead difficult. The metropolitan rota in Table 7.3.10, with a short basic sequence and a long major cycle, is a good example of a work system that is said to allow good tracking but poor planning ahead.

Many work systems incorporate work schedules with relatively long sequences and cycles. Table 7.3.12 provides an example of a system of this sort. To overcome the problems inherent in such a system, managers sometimes issue workers a year-long pocket card calendar with off-time days circled to facilitate personal planning. Although the value of such a card should not be questioned for any shift schedule, use of a work system with a shorter sequence and cycle length is often a better solution to the problem. In most cases, it appears that a shift system that makes it difficult for a worker to track and plan ahead also makes it difficult for a manager to schedule adequate personnel on a regular basis.

In conclusion, it must be pointed out that empirical data, in support of the common sense "rules" noted in this subsection, are hard to come by. They may apply more to the problems associated with learning how to use a new system, than to the ability of an individual to cope with a system once it is in practice for a period of time. It can be argued that most shift workers quickly become experts at planning around any work system (Wedderburn, 1981a).

7.3.4.5 Weekend Work

By definition, continuous work weeks require that someone work on weekends. Continuous work systems can approach this in a number of ways. Everyone can be required to work some weekends. Table 7.3.11 is an example of this. Another approach assigns some workers to shift schedules that only work weekends. Table 7.3.14 is an example of this. Other approaches limit the work of some shift schedules to weekdays while assigning in combination with other shift schedules that include both weekday and weekend shifts.

It is very important to recognize that our society is not only daytime-orientated, but also weekend-orientated. Important family, religious, and other social events are almost always associated with week-ends, and for some workers off time on Saturdays may be more important than Sundays (Wedderburn, 1981b). Thus systems that allow all workers some weekend off time are clearly preferable. A system such as the EOWEO, shown in Table 7.3.13, is attractive to some workers because it provides all workers with 3-day weekends twice every 4 weeks on a regular and predictable basis. At the other extreme, a system such as that shown in Table 7.3.14, which requires workers on some shift schedules to work every weekend, may yield some special problems because it requires a commitment to work that is in conflict with community social values.

7.3.4.6 Worker Preferences

Because expert opinion on work system issues is frequently divided and/or good research findings are often nonexistent, worker preferences are a good source of information when evaluating a system, selecting workers for a system, or selecting a new system. The value of the "worker-as-expert" should not be overlooked. Systematic surveys, interviews, record studies, and/or group discussions are all good ways of gathering information from workers as to their shift schedule preferences. They are also good ways of making the work force a part of the evaluation or selection process, as well as a way of testing the accuracy of management and labor perceptions of worker and workplace shift preferences and problems. As has been frequently demonstrated in numerous applications of industrial and organizational psychology, knowledge of how workers *perceive* problems and issues may be just as important as objective facts about the situation.

Demographics and preferences gathered from workers should be viewed as *required* inputs to any decision process related to work systems. Significant care must be taken to assure that the information gathered is reliable, valid, and representative. In most cases, this requires the use of an outside third-party expert to ensure unbiased data collection and comparison of the data collected with appropriate data bases and references. Above all, one must recognize, both in the collection of these data and in the analysis of them, that worker perceptions and preferences may very well be biased or tempered by their own work system experience.

The data presented in Figure 7.3.9 are from the permanent shift worker sample that provided the

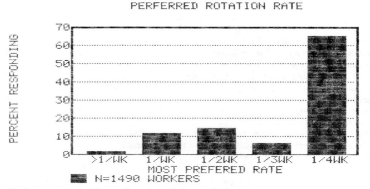

Fig. 7.3.9. Preferred rate of shift rotation reported by a sample of 1490 hourly permanent shift workers. Workers were asked to indicate which one of the following five rates they preferred, if they were required to rotate shifts: more than once a week, every week, every two weeks, every three weeks, every four weeks or longer. Data from the sample described by Tepas et al. (1985).

data presented in previous figures (Tepas et al., 1985). Only 1.1% of these workers have ever worked on rotating shifts. These workers were asked how often they would prefer to change shifts if *required* to work on a rotating shift. As can be seen in Figure 7.3.9, less than 2% of these workers report that they would prefer to change shifts more than once a week. Ninety-two percent reported that they would prefer to work one steady shift rather than rotate. On the surface, this might suggest that swiftly rotating shift schedules, such as the continental or metropolitan rotas, are not appropriate for this worker sample. This conclusion may be quite misleading, however, given the fact that few U.S. plants currently practice these schedules and few of these workers have any rotating shift schedule experience. In this example, the reliability and validity of the worker *perceptions* are *not* questionable, but it must be recognized that these perceptions might change with experience. This clearly highlights the need for a good worker educational program if a work system contrary to worker perceptions is to be considered or successfully implemented.

7.3.4.7 Worker Education

Although expert testimony from workers is an essential input to work system use, this does not mean that one should assume that workers (either managers or hourly workers) are experts in all facets of a given work system and how to cope with it. Historically, workers are inserted into a work system with little preparation. It is not uncommon for workers to change shifts, or for workplaces to change work systems, with little advance planning or notice. Instant shift schedule changes, or work system implementations, invite disaster and probably additional operational costs by promoting absenteeism, worker turnover, and ill health. A proper educational program should minimize these costs and the associated trauma. These educational programs can take a variety of formats: lectures, group discussions, videotapes, and/or literature distribution.

Educational programs should be aimed at two objectives: first, explaining the system to those involved so that they understand it, know what is expected, and understand its rationale; and second, providing workers (and their families) with information that might aid them in coping with the potential problems of the work system in practice. The first of these objectives should ensure that workers know when they are to be at work and minimize absences and complaints. The second of these objectives has the potential of overcoming many of the hazards associated with a particular work system.

Sleep behavior provides a good example of how an educational program might be of help. As we have shown in Figures 7.3.3 and 7.3.5, shift schedule interacts with sleep behavior. The length of the sleep period, the time of day one sleeps, and the relationship of sleep time to work all vary with shift. It is important to remember, however, that these differences are ones selected by workers without the benefit of instruction as to the potential problems of sleep at unusual times of the day and how good sleep hygiene might be practiced. Clearly, there are other strategies that might be practiced (Tepas, 1982b), and workers might develop better coping strategies if this information were made available to them. Most industries cannot and should not control off-time worker behavior, but they can provide helpful educational information that has the potential of improving worker health, happiness, and productivity. Information about sleep, diet, social alternatives, and health are all good examples of helpful information that might be supplied.

7.3.4.8 Health Status

There is evidence indicating that some drug susceptibility and illness symptoms are circadian, with maximum effects during the nighttime (Reinberg, 1974). It would follow then that individuals dependent upon drugs and/or suffering from disease might be more at risk when required to work during the nighttime. In U.S. industry, this possibility appears to be ignored in most cases. Although most employers have physical fitness and health standards that must be met to gain initial employment, few if any have specific medical criteria for nighttime work. Furthermore, many industrial employers do *not* require regular physical examinations once employment occurs, unless a special problem or risk is suggested. It is reasonable to argue that work systems including nighttime shift schedules should be regarded as such a special case, thereby warranting special medical examination for the selection of individuals for nighttime work, as well as periodic health checks for employed nighttime workers (Rutenfranz, 1982). Such health examinations and checks should particularly concentrate on gastrointestinal disorders, sleep problems, and other factors that are likely to result from shift work.

In the United States nighttime work is most frequently assigned to the new employee by default, because the premium pay for nighttime work is small and experienced workers use their seniority to bid out of it. In practice, this means that most (but not all) of those who do nighttime work are the youngest and the most recently hired. Chance, not design, probably means that medical examinations before starting work may keep many with serious illness from nighttime work. However, examinations of this sort have only limited predictive value, and mandatory periodic examinations for nighttime workers should be instituted. A related issue is physician awareness. Physicians giving medical examinations should be informed that the individual being examined would be or is a nighttime worker, and they should be familiar with the work system used as well as the relevant medical literature.

7.3.5   WORK SYSTEM SELECTION METHODOLOGY

Thus far, this chapter has focused on the human factors dimensions of work schedule systems. Table 7.3.16 lists some of the other domains that should be checked in evaluating the net value of a given work system. This is not intended as a comprehensive list, but it does include the major factors that should be considered. It should be recognized that *each* of the items on this list has a human element that interacts with many of the other factors associated with a given work system. Equipment maintenance provides a good example of this.

A good equipment preventive maintenance program is required for the long-term satisfactory operation of a continuous work system. For some work systems, the only possible way to handle the preventive maintenance required for some operations is by purchasing duplicate equipment so that production can continue during maintenance. For other work systems, some downtime is built into the work system so that maintenance can be performed. With this solution maintenance workers usually end up with an odd schedule and some production time is lost, but additional equipment is not required. In some cases it may be possible to change the requirements of preventive maintenance so it can be done during the regular working hours with minimal production loss and little equipment duplication. A good applications effort requires that all reasonable maintenance alternatives be reviewed and evaluated. Because people do the actual maintenance, each of these alternatives has a human element that should be reviewed.

Each of the items listed in Table 7.3.16 should be reviewed in efforts to evaluate an existing work system or in efforts to select a new work system from a number of recommended alternatives. Although *each* of these has a human element associated with it, we shall refer to these as financial and technical factors, as opposed to the human-factors variables discussed earlier in this chapter. Given the range

Table 7.3.16   Checklist of Domains to be Evaluated

Community work system practices	What work systems are practiced by other operations in the immediate community? How well do they seem to work?
Corporate work system practices	What work systems are practiced at other operations within the corporation? How well are they working?
Plant work system history	What work systems have been practiced in this plant? When were they practiced? How well did they work?
Local job market	What is the past, current, and future status of the local job market? What kind of workers are/are not available?
Plant personnel records	Do these records contain demographic, health, and work performance information that might be helpful? Is shift assignment information in the files?
Local utility costs	Do local utility costs vary with time of day or day of the week? How do utility costs vary with volume used?
Legal requirements	What local and national work hour and wage legislation apply to this installation? What is their impact?
Maintenance requirements	How is equipment maintenance handled and does it have any restrictions? How is building maintenance handled?
Supervision and manpower requirements	Does the work system practiced interact with the manpower and supervision requirements of this installation?
Shipping and receiving	Are adequate shipping and receiving services available? Do costs and services vary with time of day or day of week?
Support services	Are the medical, food, transportation, and other installation and community support services adequate?
Productivity indicants	What objective measures of productivity are used at this installation, and do they allow for adequate evaluation of shift schedule differences?
Worker participation	Are adequate mechanisms in place to allow representative and cooperative input and comment from *all* workers?
Demand for product	What are the characteristics of the demand for the goods or services produced by this installation? What is the anticipated future market like? Is it a stable market, or is it subject to cyclic changes?

of items involved, it should be clear that the proper evaluation of a given work system application requires the review of most operations at the organization where it is or might be implemented. When a new work system is implemented, significant organizational change may be required.

As noted earlier, there are a number of ways one can approach the proper assessment of the human-factors variables operating in a given workplace. Table 7.3.17 outlines an evaluation scenario that might be used. This methodology uses a survey for evaluation, and provides a forum for the discussion of significant issues. Other methods and designs may be more appropriate in some situations. In any case, the primary importance of gaining individual worker cooperation in assessing human-factors variables cannot be overstated. Workers must understand what is being considered, agree as to how it is to be done, and be willing to participate and cooperate. For a survey, adequate response rates are necessary. Rates of 90% or higher can be achieved when confidentiality is practiced and workers are full participants (Tepas et al., 1985).

Work system recommendations based primarily on human-factors variables should consider all the issues discussed above. The wide range of variables involved, together with our current lack of knowledge concerning some key issues, makes work system recommendations more an art than science. One should expect that several human-factors-based alternative recommendations may be made. These work system recommendations should consider the full range of alternative work systems available such as flexitime, shared time, and compressed work weeks (Tepas, 1985). Alternative recommendations frequently vary in their financial and technical merit. They may also vary in the degree to which they are acceptable to organized labor. Therefore, a final evaluation of the acceptability of a given work system should be the result of a careful and complete review of all human-factors, financial, technical, and labor considerations.

The importance of this final evaluation cannot be stressed too much. Obviously, errors in the evaluation of the cost of variables such as utility costs, maintenance requirements, and legal requirements can be disastrous. Of equal importance, however, are human-factors errors that may lead, for example, to high turnover rates, a poor safety record, and labor problems (Imberman, 1983). It is easy to overlook these potential human-factors costs, because it is more difficult to attach a dollar value to them than it is when considering financial and technical variables. Thus considerable care must be taken to ensure that these potential human-factors costs are not underrated.

Should these considerations result in the selection of a new work schedule system, much of the planning required for efficient implementation will have been completed if the study and evaluation process described above has been performed. An important remaining effort is that of educating the total work force as to what the new system is, how it operates, *and* how the individual worker might cope with the human-factors variables involved. This may be an appropriate activity even in those instances where no significant changes in the work system are planned. Although many employers provide workers with extensive training as to how their job skills are to be performed, most ignore

Table 7.3.17 An Evaluation Methodology Scenario

Step one	A review of this methodology with all parties, revision of the methodology if needed, and an agreement to participate in the implementation of the agreed-upon methodology
Step two	Development of a survey to collect worker self-reports of their demographics, work system history, perceptions, preferences, and health status. Offering the survey developed to all workers for their voluntary, anonymous, and confidential responses
Step three	Analysis of the survey data and development of work system recommendations based mainly on a human factors/ergonomics basis. A presentation of these recommendations to all parties
Step four	A financial and technical evaluation of all work system recommendations, and the development of a final recommendation. Acceptance by all parties of this work system recommendation and a schedule of implementation
Step five	If a new work system is recommended and agreed to, definition of *criteria* for a future evaluation of the success or failure of the new system. Agreement on a schedule for implementation of the new work system. Selecting a time point at which an evaluation of the new work system will be made
Step six	Development and implementation of an *educational program* for all workers. This is primarily aimed at teaching the workers how to use the new work system as well as how they might best cope with it
Step seven	If a new work system is being installed, implementation of the work system following the agreed schedule
Step eight	Following installation of the new work system, evaluation of that system, as scheduled, and subsequent development of a future plan of action

the fact that learning how to cope with a shift schedule is an equally important task. As noted earlier, this task can be accomplished in a variety of ways. All too often, managers get caught up in a rush to implementation that ignores this important step.

Once a new work system has been implemented, it is important that one track the result. This can, of course, involve the usual personnel and plant record keeping. In most cases, this does not provide a complete picture. An assessment of all the human-factors variables examined in the initial evaluation is in order. Just as it was important to learn how workers perceived the initial work system, it is also important to learn how they perceive the new work system. The workers' view of the changes occurring may differ significantly from those of management or outside observers (Adair, 1984). When viewed from the perspective of the worker, the meaning of some observations may change significantly.

The timing of a follow-up evaluation study is important. Data collected shortly after introduction of the change may be related to *acute* effects, and need not be predictive of any long-term effects. Both positive and negative short-term acute effects are possible. Positive acute effects frequently may make public media reports, but long-term *chronic* effects are more relevant. In most cases, management and labor are interested in permanent changes that require only minimal future tune-ups. Although there are no definitive data available, it is suggested that a new work system should be fully implemented for *at least* 6 months prior to any attempts to measure chronic effects.

7.3.6 CONCLUSION

The range of work schedules and systems available for potential implementation is immense. The shift schedules incorporated in these work systems vary in a number of dimensions, and they interact

Table 7.3.18 Factors Within an Individual That Are Likely to Cause Shift Work Coping Problems

Over 50 years of age

Second job for pay ("moonlighting")

Heavy domestic work load

"Morning type" individuals ("larks")

History of sleep disorders

Neurotic introvert

Psychiatric illness

History of alcohol or drug abuse

History of gastrointestinal complaints

Epilepsy

Diabetes

Heart disease

Table 7.3.19 Factors Associated with Work Systems and Work That are Likely to Cause Shift Work Coping Problems

More than five *third shifts* in a row without *off-time* days

More than four 12 hr *night shifts* in a row

First shift starting times earlier than 0700 hr

Rotating hours that change once per week ("weekly rotation")

Less than 48 hr *off time* after a run of *third shift* work

Excessive regular overtime

Backward *rotating hours* (*first* to *third* to *second shift*)

12 hr *shifts* involving critical monitoring tasks

12 hr *shifts* involving a heavy physical work load

Excessive weekend working

Long commuting times

Split shifts with inappropriate break period lengths

Shifts lacking appropriate *shift breaks*

12 hr *shifts* with exposure to harmful agents and substances

Overly complicated *schedules* that make it difficult to track or plan ahead

with a host of biological, psychological, social, and industrial variables. As a result, it is incorrect to assume that there is any such thing as one ideal work system, or a work system that is the general solution to all industrial applications having the same time demands. Tables 7.3.18 and 7.3.19 represent some of the most obvious factors potentially affecting shift work coping ability, given our current level of understanding. However, these factors should only be considered as *very* preliminary. The best approach to work schedule problems is one that recognizes the complexity involved, the diversity of the human individual differences available, and the need for systematic evaluation. Methodological approaches are available to gather the information needed to assess and evaluate the human-factors dimensions related to work schedules. If a new work schedule system is selected, it is important that prior to implementation workers be educated as to how the system operates and how they might best cope with it. Following implementation, a follow-up evaluation should be conducted to determine the long-term chronic effects of the new work schedules.

REFERENCES

Adair, J. G. (1984). The Hawthorne effect: A reconsideration of the methodological artifact. *Journal of Applied Psychology, 69,* 334–345.

Adler, A., and Roll, Y. (1981). Proposal for a standard shift pattern notation. In A. Reinberg, N. Vieux, and P. Andlauer, Eds., *Night and shift work: Biological and social aspects.* Oxford: Pergamon.

Akerstedt, T. (1977). Inversion of the sleep wakefulness pattern: Effects on circadian variations in psychophysiological àctivation. *Ergonomics, 20,* 459–474.

Bhatia, N., and Murrell, K. F. H. (1969). An industrial experiment in organized rest pauses. *Human Factors, 11,* 167–174.

Czeisler, C. A., Weitzman, E. D., Moore-Ede, M. C., Zimmerman, J. C., and Knauer, R. S. (1980). Human sleep: Its duration and organization depend on its circadian phase. *Science, 210,* 1264–1267.

Czeisler, C. A., Moore-Ede, M. C., and Coleman, R. M. (1982). Rotating shift work schedules that disrupt sleep are improved by applying circadian principles. *Science, 217,* 460–462.

Folkard, S., and Monk, T. H. (1979). Shiftwork and performance. *Human Factors, 21,* 483–492.

Folkard, S., Wever, R. A., and Wildgruber, C. M. (1983). Multi-oscillatory control of circadian rhythms in human performance. *Nature, 305,* 223–226.

Frese, M., and Okonek, K. (1984). Reasons to leave shiftwork and psychological and psychosomatic complaints of former shiftworkers. *Journal of Applied Psychology, 69,* 509–514.

Gordon, G. H., McGill, W. L., and Maltese, J. W., (1981). Home and community life of a sample of shift workers. In L. C. Johnson, D. I. Tepas, W. P. Colquhoun, and M. J. Colligan, Eds., *Biological rhythms, sleep, and shift work,* New York: Spectrum Publications.

Imberman, W. (1983). Who strikes—and why? *Harvard Business Review, 61,* 18–28.

Johnson, L. C. (1982). Sleep deprivation and performance. In W. B. Webb, Ed., *Biological rhythms, sleep and performance.* New York: Wiley.

Kelly, R. J., and Schneider, M. F. (1982). The twelve-hour shift revisited: Recent trends in the electric power industry. *Journal of Human Ergology, 11* (suppl.), 369–384.

Knauth, P., and Rutenfranz, J. (1976). Experimental studies of permanent night and rapidly rotating shift systems. *International Archives of Occupational and Environmental Health, 37,* 125–137.

Knauth, P., Rutenfranz, J., Karvonen, M. J., Undeutsch, K., Klimmer, F., and Ottmann, W. (1983). Analysis of 120 shift systems of the police in the Federal Republic of Germany. *Applied Ergonomics, 14,* 133–137.

Kripke, D. F., Simons, R. N., Garfinkel, L., and Hammond, E. C. (1979). Short and long sleep and sleeping pills. *Archives of General Psychiatry, 36,* 103–116.

Monk, T. H., and Embrey, D. E. (1981). A field study of circadian rhythms in actual and interpolated task performance. In A. Reinberg, N. Vieux, and P. Andlauer, Eds., *Night and shift work: Biological and social aspects.* Oxford: Pergamon.

Monk, T. H., Weitzman, E. D., Fookson, J. E., Moline, M. L., Kronauer, R. E., and Gander, P. H. (1983). Task variables determine which biological clock controls circadian rhythms in human performance. *Nature, 304,* 543–545.

Moore-Ede, M. C., Sulzman, F. M., and Fuller, C. A. (1982). *The clocks that time us.* Cambridge, MA: Harvard University Press.

Presser, H. B., and Cain, V. S. (1983). Shift work among dual-earner couples with children. *Science, 219,* 876–878.

Reinberg, A. (1974). Chronopharmacology in man. In J. Aschoff, F. Ceresa, and F. Halberg, Eds., *Chronobiological aspects of endocrinology.* Stuttgart: F. K. Schattauer Verlag.

Reinberg, A., Andlauer, P., De Prins, J., Malbecq, W., Vieux, N., and Bourdeleau, P. (1984). Desynchronization of the oral temperature circadian rhythm and intolerance to shift work. *Nature, 308,* 272–274.

Rechtschaffen, A., Gilliland, M. A., Bergman, B. M., and Winter, J. B. (1983). Physiological correlates of prolonged sleep deprivation in rats. *Science, 221,* 182–184.

Rutenfranz, J. (1982). Occupational health measures of night- and shiftworkers. *Journal of Human Ergology, 11* (suppl.), 67–86.

Rutenfranz, J., Knauth, P., and Angersbach, D. (1981). Shift work research issues. In L. C. Johnson, D. I. Tepas, W. P. Colquhoun, and M. J. Colligan, Eds. *Biological rhythms, sleep, and shift work.* New York: Spectrum Publications.

Scherrer, J. (1981). Man's work and circadian rhythm through the ages. In A. Reinberg, N. Vieux, and P. Andlauer, Eds., *Night and shift work: Biological and social aspects.* Oxford: Pergamon.

Tasto, D. L., and Colligan, M. J. (1977). *Shift Work Practices in the United States,* Washington, DC: U.S. DHEW Publication No. (NIOSH) 77-148.

Tepas, D. I. (1982a). Adaptation to shiftwork: Fact or fallacy? *Journal of Human Ergology, 11* (suppl.), 1–12.

Tepas, D. I. (1982b). Shiftworker sleep strategies. *Journal of Human Ergology. 11* (suppl.), 325–336.

Tepas, D. I. (1982c). Work/sleep, time schedules and performance. In W. B. Webb, Ed., *Biological rhythms, sleep and performance.* New York: Wiley.

Tepas, D. I. (1985). Flexitime, compressed workweeks, and other alternative work schedules. In S. Folkard and T. H. Monk, Eds., *Hours of work.* New York: Wiley.

Tepas, D. I., and Sullivan, P. J. (1982). Does body temperature predict sleep length, sleepiness and mood in a job-bound population? *Sleep Research, 11,* 42.

Tepas, D. I., Walsh, J. K., and Armstrong, D. (1981). Comprehensive study of the sleep of shift workers. In L. C. Johnson, D. I. Tepas, W. P. Colquhoun, and M. J. Colligan, Eds., *Biological rhythms, sleep, and shift work,* New York: Spectrum Publications.

Tepas, D. I., Walsh, J. K., Moss, P. D. and Armstrong, D. (1981). Polysomnographic correlates of shift worker performance in the laboratory. In A. Reinberg, N. Vieux, and P. Andlauer, Eds., *Night and shift work: Biological and social aspects.* Oxford: Pergamon Press.

Tepas, D. I., Armstrong, D. R., Carlson, M. L., Duchon, J. C., Gersten, A., and Lezotte, D. V. (1985). Changing industry to continuous operations: Different strokes for different plants. *Behavior Research Methods, Instruments, and Computers, 17,* 670–676.

Thierry, H., and Jansen, B. (1981). Potential interventions for compensating shift work inconveniences. In A. Reinberg, N. Vieux, and P. Andlauer, Eds., *Night and shift work: Biological and social aspects.* Oxford: Pergamon Press.

Tilley, A. J., Wilkinson, R. T., Warren, P. S. G., Watson, B., and Drud, M. (1982). The sleep and performance of shift workers. *Human Factors, 24,* 629–641.

Walsh, J. K., Tepas, D. I., and Moss, P. (1981). The EEG sleep of night and rotating shift workers. In L. C. Johnson, D. I. Tepas, W. P. Colquhoun, and M. J. Colligan, Eds., *Biological rhythms, sleep, and shift work.* New York: Spectrum Publications.

Webb, W. B. (1981). Work/rest schedules: Economic, health, and social implications. In L. C. Johnson, D. I. Tepas, W. P. Colquhoun, and M. J. Colligan, Eds., *Biological rhythms, sleep, and shift work.* New York: Spectrum Publications.

Wedderburn, A. A. I. (1981a). How important are the social effects of shiftwork? In L. C. Johnson, D. I. Tepas, W. P. Colquhoun, and M. J. Colligan, Eds., *Biological rhythms, sleep, and shift work.* New York: Spectrum Publications.

Wedderburn, A. A. I. (1981b). Is there a pattern in the value of time off work? In A. Reinberg, N. Vieux, and P. Andlauer, Eds., *Night and shift work: Biological and social aspects.* Oxford: Pergamon Press.

Wever, R. A. (1979). *The circadian system of man,* Berlin: Springer-Verlag.

BIBLIOGRAPHY

Carpentier, J. and P. Cazamian (1977). *Night Work.* Geneva: International Labor Office.

Colquhoun, W. P., Ed. (1971). *Biological rhythms and human performance.* New York: Academic Press.

Colquhoun, W. P., Ed. (1972). *Aspects of human efficiency.* London: English Universities Press.

Colquhoun, W. P., Folkard, S., Knauth, P., and Rutenfranz, J. Eds. (1975). *Experimental studies of shiftwork.* Opladen, West Germany: Westdeutscher Verlag.

Folkard, S., and Monk, T. H., Eds. (1985). *Hours of work: Temporal factors in work scheduling.* New York: Wiley.

Johnson, L. C., Tepas, D. I., Colquhoun, W. P.,and Colligan, M. J., Eds. (1981). *Biological rhythms, sleep and shift work.* New York: Spectrum Publications.

Maurice, M. (1975). *Shift work.* Geneva: International Labor Office.

Melbin, M. (1978). Night as frontier. *American Sociological Review, 43,* 3–22.

Mott, P. E., Mann, F. C., McLoughlin, Q., and Warwick, D. P. (1965). *Shift work.* Ann Arbor: University of Michigan Press.

Palmer, J. D., Ed. (1976). *An introduction to biological rhythms.* New York: Academic Press.

Rentos, P. G., and Shepard, R. D., Eds. (1976). *Shift work and health.* Washington, DC: U.S. DHEW Publication No. (NIOSH) 77-148.

Tasto, D. L., Colligan, M. J., Skjei, E. W., and Polly, S. J. (1978). *Health consequences of shift work.* Washington, DC: U.S. DHEW Publication No. (NIOSH) 78-154.

Thierry, H., and Jansen, B. (1984). Work and working time. In J. D. D. Drenth, H. Thierry, P. J. Willems, and C. J. deWolff, Eds., *Handbook of work and organizational psychology.* New York: Wiley.

Rutenfranz, J., and Colquhoun, W. P. (1979). Circadian rhythms in human performance. *Scandinavian Journal of Work Environment and Health, 5,* 167–177.

Rutenfranz, J., Colquhoun, W. P., Knauth, P., and Ghata, J. N. (1977). Biomedical and psychosocial aspects of shift work. *Scandinavian Journal of Work Environment and Health, 3,* 165–182.

CHAPTER 7.4
OCCUPATIONAL STRESS

MICHAEL J. SMITH

University of Wisconsin, Madison

7.4.1 INTRODUCTION

The purpose of this chapter is to provide the ergonomics practitioner with an understanding of how workplace conditions can produce stress, how to recognize the serious manifestations of stress in a workplace, and how to assess worker stress reactions and various approaches that can be considered for mitigating the effects of stress at the workplace. It is not an exhaustive review of the literature on occupational stress (see Salvendy and Sharit, 1982, and Sharit and Salvendy, 1982 for more comprehensive reviews), nor is it an attempt to reconcile the various theories of stress and ill health. Rather, it is meant to be a compilation of concepts, information, methods, and interventions that can be used by an engineer or psychologist to identify, measure, and control occupational stress in a diversity of workplaces.

Each and every year, as society becomes more complex, as technology proliferates, and as pressure to succeed increases, more and more attention is given to the "stress" involved in daily living. Everyone is touched by stress, from the newborn infant to the aged in nursing homes whiling away their last days. Within the last decade, interest has grown in the role of a person's occupation and aspects of job tasks and work relationships as significant sources of personal stress. That work can contribute to stress is an almost universally accepted fact. However, the extent to which work factors contribute to a person's ill health and disease is less established and much less agreed upon.

In the fall of 1984 The National Institute for Occupational Safety and Health (NIOSH) of the U.S. Government released a manuscript discussing the role of work in generating employee distress and ill health. In this paper it was stated that there are a number of psychological disorders associated with work experiences. These include (1) emotional disturbances such as anxiety and depression, (2) behavioral problems such as excessive alcohol consumption and smoking, and (3) various forms of psychosomatic disorders such as gastrointestinal disturbances.

Such psychological disorders are prevalent in modern society as demonstrated by the President's Commission on Mental Health (1978) estimate that one in four Americans suffer from mild to moderate depression, anxiety, or other emotional disorders at any given time. On a national level, the annual cost of medical treatment for stress disorders in the United States is $14.5 billion, with an additional $10.5 billion for disability payments (Manuso, 1980). These disorders are increasing, as reflected by statistics from California that show threefold increase in occurrence from 1976 to 1980 (Tebb, 1980). Some experts feel that psychological stress is at epidemic proportions and represents the major medical problem of modern society.

A major problem in examining and discussing occupational stress is the lack of a unifying theory about what stress is and how it affects the body adversely. However, this lack of theory does not mean that relationships between workplace conditions and worker reactions to these conditions cannot be studied. What this means is that we do not always know why a relationship exists, only that it does exist. This lack of theory often makes it difficult to develop an intervention to reduce health risk even when a relationship is known. Because stress theory is so diverse, it is reasonable to approach the area of occupational stress by first looking at workplace conditions that produce adverse worker reactions and then going back to the basic body processes that can account for the effects observed. This chapter first examines sources of stress, then their effects, how these effects are measured, and finally some methods that have been proposed for controlling occupational stress.

To help to understand the relationship between working conditions that can produce stress and the health consequences of this stress, Caplan, Cobb, and French (1976) proposed a model that defines these relationships (see Figure 7.4.1). In this model, working conditions such as work load or role conflict can lead to stress responses such as anxiety and/or increased blood pressure. The extent of the stress reaction is mitigated by personal factors such as health status and personality and by the extent of social support received by the individual. If the stress reactions are sufficient and occur over a prolonged time period, then illness occurs in the form of gastrointestinal or cardiovascular or other stress-related diseases.

7.4.2 SOURCES OF OCCUPATIONAL STRESS

Generally, stress in the work environment stems from a mismatch in a combination of the following sources: (1) the environment (social and physical environments at work and at leisure), (2) the task (e.g., mental load, pacing), (3) organizational factors (supervisory style), and (4) the individual (e.g., personality attributes). This is an oversimplification of a complex process of interaction of these sources. For the sake of simplicity, each of these sources is discussed separately.

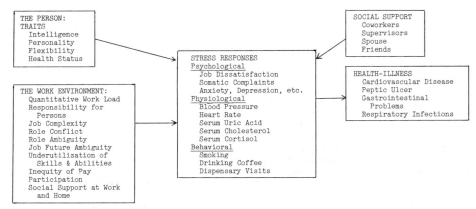

Fig. 7.4.1. A model of occupational stress (adapted from Caplan et al., 1975b).

7.4.2.1 Environmental Sources

Physical working conditions have a role to play in job stress that is typically not a primary, but more of an additive influence. Their main psychosocial impact has been shown to be on lowering worker tolerance to other stressors and on worker motivation. Factory work environments that have excessive noise, poor ventilation, inadequate lighting, and hot or cold shop floors can create both physical and attitudinal problems (Smith, Cohen, Cleveland, and Cohen, 1978). Such problems are not limited to factories; they also occur in modern office buildings. In many offices, workers are packed together with noisy machinery and poor lighting, all of which have a similar effect to those found in factories (Smith, Cohen, Stammerjohn, and Happ, 1981). Essentially, physical working conditions make the individual worker more susceptible to influence by other sources of stress.

7.4.2.2 Organizational Factors

Two organizational factors have been shown to be of special significance for increased job stress and decreased worker health. These are (1) job involvement or participation; and (2) organizational support, as reflected by supervisory style, support from managers, and career development. Lack of participation in work activities has been demonstrated to result in an increase in negative psychological mood (Margolis, Kroes, and Quinn, 1974; Caplan, Cobb, French, Harrison, and Pinneau, 1975b; Smith et al., 1981). In terms of organizational support, it has been shown that close supervision and a supervisory style characterized by constant negative performance feedback are related to high levels of stress and poorer worker health (Caplan et al., 1975b; Smith et al., 1981). Excessive impersonal monitoring of employee performance has been introduced by modern technology and can produce a management style that is stress-producing (Smith et al., 1981). It has been demonstrated that workers' feelings of lack of involvement are related to stress and potentially to health complaints (French and Caplan, 1970).

Career development is another major organizational stressor that has been studied. Concern over chances for promotion has been shown to be a significant stressor for office workers (Smith et al., 1981), whereas being passed over for promotion has been related to increases in both job stress and ill health (Arthur and Gunderson, 1965).

Security is the other side of career development. The threat of job loss is a very potent stressor. It has been tied to serious health disorders such as ulcers, colitis, severe emotional stress, and patchy baldness (Cobb and Kasl, 1977), as well as to increased muscular and emotional complaints (Smith et al., 1981).

7.4.2.3 Job Task Factors

Mental work load factors, such as quantitative underload/overload and work pace have been related to stress (Frankenhaeuser and Gardell, 1976). This relationship is unusual in that a balance is required to avoid negative health consequences. Underload is just as bad for one's health as overload and can affect both psychological well-being and hormone activity. Sedentary jobs create habit patterns that carry over into other aspects of life, thus influencing health risk off the job. Quantitative overload has been shown to be a significant stressor for various occupations including scientists (French and

Caplan, 1970), machine operators (Caplan et al., 1975b), and data-entry clerks (Smith et al., 1981). The impact of work overload varies from psychological disturbances to increased disease risk.

Work pace is a very important work load factor. The speed or rate of work has been shown to be a significant issue in factory-worker ill health. Recent technology, such as computers, that can operate at high speeds on a continuous basis has increased the pacing impact on office workers. Research suggests that pacing produced by computerization may have an even greater effect than factory pacing (Smith et al., 1981; Cakir, Reuter, Von Schmude, and Armbruster, 1978).

Overtime is another work load factor that has been implicated in stress and ill health. Although there are only a few studies concerned with overtime health effects, these indicate a relationship between working 50 or more hours per week and coronary heart disease risk as well as psychological disfunction (Russek and Fohman, 1958; Breslow and Buell, 1960). Overtime also has the potential to affect health in that it takes time away from family and friends and thereby reduces positive social interaction that can buffer stress.

More and more operations are functioning in other than the typical daytime hours of work. This is particularly true of service operations. Shift work, particularly night and rotating shift regimens, has been shown to have a significant impact on health and safety, affecting such diverse areas as industrial injury incidence, worker sleeping and eating patterns, and family and social-life satisfaction (Smith, Colligan, Frockt, and Tasto, 1979; Smith, Colligan, and Tasto, 1982; Rutenfranz, Colquhoun, Knauth, and Ghata, 1977).

Time pressure, such as having to meet deadlines, is a stressor that may interact with both work hours and workplace. Studies have shown increases in stress level as difficult deadlines draw near (Friedman, Rosenman, and Carroll, 1958).

Work role includes a number of job factors such as responsibility for others, job conflict, role ambiguity, accountability, authority, discretionary control, participation, and job status. A number of studies have demonstrated that role ambiguity, job conflict, and responsibility for persons are related to job stress and psychological problems (Margolis, Kroes, and Quinn, 1974; Caplan et al., 1975b). The lack of discretionary control over work activities has also been shown to be related to increased risk of coronary heart disease (Karasek, 1979). Lack of participation in work decisions is yet another source of increased worker stress (French, 1963).

Other work-task factors that have been researched include task variety, task clarity (confusion), challenge, complexity, utilization of skills and abilities, and activity level. All these factors have been related to increased stress and negative psychological states such as boredom, confusion, and frustration, and have also been related to increased risk of health disorders (Cooper, and Marshall 1976).

7.4.2.4 Individual Sources

Three main sources of occupational stress that are related to the individual are (1) health-related factors, (2) the degree of match between the job workers actually perform and their capabilities, likes, and dislikes as they relate to the work environment, and (3) the personality makeup of the individual.

In terms of health considerations, the physical constitution of the individual influences the capacity to respond to the environment. Limitations in the ability to exert control over adverse stimuli owing to low capacity or disease can lead to increased stress reactions. Thus those workers with preexisting health problems of a stress-related origin may be at greater risk to job stress.

With respect to the worker–job mismatch, certain approaches used to define and predict job stress have considered the relationship between stress and the degree of fit between the person and the work environment. Within these so-called person–environment fit models, the match between the characteristics of the person and the work environment determines the degree of stress response (either coping or maladaptive behavior) and subsequent stress-related symptoms. The greater the mismatch, the greater the stress experienced by the worker.

Some personality features have been implicated as significant factors in modifying the individual's response to various work-related situations, such as the degree of extroversion or neuroticism (Eysenck, 1967), and the Type A versus Type B personality (Jenkins, Rosenman, and Friedman, 1967). Assembly-line work has received much attention in terms of the delineation of the kinds of personalities that may either conflict with or successfully adapt to this type of work. Existing evidence implies the following: (1) anxious or autonomically labile individuals (i.e., individuals with very active physiological systems) will find assembly-line work especially frustrating; (2) persons with high ego strength will cope most successfully with paced operations; (3) persons high on sensation seeking will do poorly on the assembly line; and (4) persons high on authoritarianism (individuals resistant to changes in their duties) will adapt well to assembly-line jobs.

7.4.3 THEORIES OF STRESS

The field of occupational stress developed from the concept that factors associated with particular work activities produce worker maladaptation that can lead to ill health. This concept is grounded in many theories that define the relationship between environmental conditions, stress, and disease.

Selye (1956) pioneered the medical model that describes how noxious environmental stimuli can produce a specific syndrome of disease. He called this process the "general adaptation syndrome" to reflect the organism's attempts to adapt to noxious stimuli by mobilizing energy resources to fight or escape. This syndrome was characterized by three stages. The first was "alarm" at the intrusion of the noxious stimuli. This stage produces strong hormonal responses to get the organism ready to respond to the intrusion. The next stage was "adaptation" to the stimuli using the body's energy resources. This stage produces prolonged depletion of energy stores and high hormone levels. The final stage was "exhaustion" in which the organism is no longer able to continue to produce the energy necessary to fight the noxious stimuli. The end results of this syndrome were enlargement of the adrenal cortex, a gland which produces critical hormones for energy mobilization, deep bleeding ulcers of the stomach and intestines, most likely due to overexcretion of gastric juices, and finally, atrophy of the thymus and lymphatic system related to overuse of the organism's immune response mechanism.

This first conceptualization of the medical consequences of stress establishes a linkage between the environment and adverse health effects. It further suggests that the primary mechanism of the maladaptation is through overextension of energy mobilization and disease fighting systems (hormone and immunity response). Thus the organism develops diseases because the mechanisms for fighting diseases are overused to the point of exhaustion. In addition, the mobilization of energy resources produces an internal biological environment that is damaging to certain tissues such as the stomach, intestines, and cardiac muscle. The organism, in an attempt to save itself from the noxious stimuli, kills itself slowly. Although this theory defined some of the biological reactions produced by adverse environmental conditions and some possible mechanisms for the adverse health outcomes, it did not establish the process by which this influence occurred. It lacked the sophistication to define stress effects due to environmental conditions that were only mildly noxious, or that were noxious only to some persons and not others. In essence, it failed to define the perceptual and psychological processes that are of importance to more subtle stress situations.

The relationship between emotional states and biological reactions similar to those defined by Selye (1956) was suggested by Cannon (1914). He proposed that emotional states could have an influence on hormone production and related physiological reactions such as heart rate, and that this influence was reciprocal. Thus increased hormone response due to emotional distress could lead to greater emotional distress. As with Selye's theory, Cannon's established some basic relationships between environmental stresses and bodily reaction, but did not define the psychological and perceptual processes that mediate the stress responses. These processes are still not well understood.

Lazarus (1977) has suggested that adaptation to the environment is mediated by emotions. In his theory cognitive processes determine the quality as well as the intensity of emotional reactions. The cornerstone of this process is cognitive appraisal which is determined by the interaction of individual personality factors and the environmental stimuli encountered. Personality is influenced by psychosocial factors such as beliefs, motives, fears, and desires, which dynamically interact with environmental conditions to shape responses. In this theory emotion is a complex process with three main aspects: (1) subjective affect, (2) physiological changes related to mobilization for action, and (3) behavior having instrumental and expressive features. Somatic disturbance arises from a need for action which influences emotion and reflects the behavioral response. The quality or intensity of the emotion and its resultant behavior depend on cognitive appraisal of the present or anticipated significance of the interaction with the environment in terms of the person's well-being. This appraisal process includes symbolic thought, which is influenced by prior learning. This theory differs from Selye's (1956) in that cognitive appraisal and resultant behavior define the hormone responses to environmental stimuli. Thus hormonal response to stressors is not a general adaptation that is similar for all conditions, but is specific to the cognitive process that defines the significance of the stressors to the well-being of the person.

McGrath (1976) indicates that stress can occur when an environmental situation is perceived by the individual as presenting a demand that threatens to exceed the individual's capabilities and resources for meeting the demand, given that the individual expects a significant difference between the costs in meeting the demand and the rewards from meeting or not meeting it. This complicated formulation identifies significant aspects of the psychological conditions for stress. First, it requires some perception of a demand by the individual. Then there is a cognitive appraisal process to determine the impact of the demand, that is, whether it will be pleasant or unpleasant and what it will cost in terms of resources. This implies that the individual has past experiences to use to evaluate the situation. It also suggests that an intellectual process is necessary to make judgments about demands and to make decisions about actions to take based on a reward structure. This approach illustrates the complexity of the stress process from a psychological perspective and the necessity to take a systems viewpoint of how stress influences an individual.

In this regard, French (1963) and Harrison (1978) have defined a systematic approach for examining stress and strain that focuses on the degree of "fit" between a person and the environment. In this approach, the person consists of two distinct and often mutually exclusive elements, the objective person and the subjective person. The objective person refers to the actual physical, mental, and emotional attributes of the person, and the subjective person represents the individual's perception of himself and his abilities. The environment is also composed of two elements, objective and subjective. The objective environment refers to the physical and social environment as it actually exists, and the subjective environment represents the environment as the individual perceives it. French (1963) indicates that there are four aspects of person environment fit that can produce discrepancies that can cause stress. These are discrepancy between the objective and subjective environment that is the person's "contact with reality." Then there is the discrepancy between the objective and subjective person that is the "accuracy of self-assessment." The other two areas of discrepancy are the degree to which the objective environment meets the wants, needs, and desires of the person, and the abilities of the person to meet the demands of the job in terms of his knowledge, skills, and talents. French (1963) suggests that discrepancies in these four aspects of person/environment fit can lead to job stress. The greater the discrepancy, the higher the probability of stress.

Harrison (1978) provides an example of this concept:

> A secretary who thinks she is able to type 55 words per minute (subjective person) may in actuality only type 40 words per minute (objective person). The secretary's boss may expect her to type at a rate of 70 words per minute (objective environment), while the secretary thinks the boss expects her to type about 60 words per minute (subjective environment). The secretary's objective P–E fit, subjective P–E fit, contact with reality, and accessibility to self can all be determined with respect to typing speed because this same conceptual dimension was used to measure the objective and subjective environment and the objective and subjective person.

Based on an examination of the various theories of stress presented, it can be hypothesized that particular working conditions can influence emotional and biological reactions, as well as the behavior of workers. If these conditions are perceived as unpleasant for a prolonged time, the resultant stress reactions can influence the development of disease. Thus psychological factors, in terms of perceptions and emotions, influence the disease process. In addition, worker behavior plays a significant role in exacerbating or mitigating these emotional and biological reactions, and adding to or reducing the potential for disease and/or accidents.

7.4.4 EFFECTS OF STRESS

The effects of stress are multidimensional including psychophysiological, emotional, and behavioral reactions. As originally hypothesized by Cannon (1914) and Selye (1956), stressors have an immediate and direct influence on hormone excretion. The primary pathway appears to be a response by the pituitary gland that is initiated by the limbic system of the brain. The pituitary excretes a hormone (ACTH) which stimulates other glands in the body to mobilize energy resources for "fight or flight." Among these other glands are the thyroid gland, the pancreas, the liver, and the adrenal glands. The adrenal glands, in particular, are of significance because they excrete hormones that have been associated with adverse stress reactions. These hormones include adrenaline, nonadrenaline, cortisol, and steroids. Some of the hormones, such as adrenaline and noradrenaline, are stimulated directly by nervous impulses from the limbic system in the brain, whereas others are stimulated by ACTH. Evidence is mounting that increases in these "stress hormones" can cause a variety of adverse health effects of a physical and emotional nature.

7.4.4.1 Cardiovascular Effects

Cardiovascular responses and effects are influenced by the hormone reactions and by direct neural stimulation. Certain stress hormones, such as adrenaline, cause the heart muscle to beat faster to increase blood flow for energy mobilization, and other hormones cause peripheral vasoconstriction to minimize blood loss in case of a wound. There are also influences on blood pressure. Owing to the hormones and vasoconstriction, blood pressure rises. Studies on air traffic controllers have demonstrated long-term health consequences of elevated blood pressure. Repeated blood pressure measures of 380 controllers during various job tasks demonstrated that high work load produced increases in blood pressure. During the course of this 5 year study 36 controllers developed hypertension (Hurst, Rose, and Jenkins, 1978).

A number of early studies have examined the relationship between occupational status and the prevalence of coronary heart disease (CHD). Hadley (1939), Ryle and Russell (1949), and Morris, Heady, Raffle, Roberts, and Parks (1953) demonstrated an increased incidence of CHD as occupational status increased and physical job demands decreased. However, Mortenson Stevenson, and Whitney (1959), Stamler, Kjelsberg, and Hall (1960), and Pell and D'Alonzo (1961, 1963) demonstrated the reverse trend. That is, the lower status, higher physically demanding occupations had higher CHD incidence. Rene Marks (1967), in reviewing the relationship between work status and CHD, concluded, "While occupation is probably a risk factor in coronary heart disease, the association is by no means a clear and simple one, but is probably dependent on the specific characteristics of an occupation as well as the social background of the individual who engages in a given type of work."

Although these early studies examining the association between CHD incidence and occupational status did not provide clear-cut evidence of a directional relationship, they did demonstrate that certain types of jobs had an elevated prevalence of CHD. Since these early studies, there have been a number of major epidemiological evaluations of CHD, such as the Framingham study and the Western Collaborative study, which continue to examine a host of risk factors prospectively. The data from these efforts have been applied to the area of occupational stress.

Haynes, Feinleib, Levine, Scotch, and Kannel (1978), Haynes, Feinleib, and Kannel (1980), and Haynes and Feinleib (1980) have evaluated data from the Framingham heart study, which has collected longitudinal information on cardiovascular risk for a number of years. In these evaluative efforts, between 1300 and 1800 individuals (males and females) were examined using questionnaire measures of psychosocial concerns and clinical symptoms which were then related to CHD incidence. Psychosocial variables such as Type A behavior, aging fears, marriage satisfaction, work load, job type, promotion, and emotional states were examined for their relationship with CHD incidence and risk factors. Early results (Haynes et al., 1978) indicated that job factors such as nonsupportive boss or excessive work load did not differentiate CHD incidence, whereas other non-job factors such as Type A behavior, marital satisfaction, and aging fears did relate to CHD incidence. However, having a job was shown to be moderator of these other psychosocial variables, in particular, Type A behavior.

Later work (Haynes et al., 1980) examined differences in psychosocial influences between men and women. Older men (over 55) who developed CHD reported greater work load, more promotions, and higher Type A scores than peers who didn't develop CHD. Under the age of 55, these relationships were not evident. For all men under age 65, white collar workers who scored higher on the Type A scale had an increased risk of CHD, whereas blue collar workers with high Type A scores did not have an increased CHD risk.

For women between the ages of 45 and 64, Type A behavior was linked to CHD incidence, with no differences between employed women and housewives in Type A scores. However, behavior patterns linking CHD risk differed between employed women and housewives. Suppressed hostility predicted CHD incidence for working women, but not for housewives. Emotional factors such as tension, anger, anxiety, and being more easygoing predicted CHD incidence for housewives, but not for the working women. Thus job activity may have influenced the development of behavior patterns that had a negative impact on health (CHD incidence).

In defining the relationship between working women and CHD, Haynes and Feinleib (1980) demonstrated that working women had a slightly higher but statistically nonsignificant CHD incidence (6.4%) than housewives (5.4%), but less than working men (12.50%). However, particular segments of the female working population had a high incidence of CHD, for example, female office/clerical workers (10.6%). When other factors such as being married and having children were taken into account, women clerical workers showed a CHD incidence of 15.4%, which was second only to white collar men in CHD incidence in this study group (white collar men 19.8%). This research demonstrated that the evaluation of CHD is a complex process that requires critical examination of a variety of factors that can potentially influence CHD development. The research results clearly illustrate the significance of work in both directly influencing CHD incidence and indirectly modifying or conditioning behavior patterns that are related to CHD incidence, such as Type A behavior. However, these studies do not demonstrate how work can produce these influences or what it is about work that is most significant in the process.

Karasek (1979) has expanded the examination of work and CHD risk by keying in on specific job characteristics that might potentiate the relationship. Specifically, the amount of control over the job as reflected in decision latitude and freedom to make decisions on how work time will be used has been examined in regard to CHD risk. Karasek (1981) examined 1915 employed males in 1968 and followed up on 1635 of them in 1974. A questionnaire was used to evaluate CHD symptoms and feelings of job control. In addition, a case control study was undertaken to confirm the questionnaire findings. Multiple logistic regression analysis was used to predict CHD risk considering many factors such as (1) age, (2) weight, (3) smoking habits, (4) job demands, (5) intellectual discretion, and (6) personal work schedule freedom. The responses to job demands reported in 1968 were predictive of CHD symptoms in 1974. In particular, measures of job control (intellectual discretion and work schedule freedom) were significant predictors of CHD risk, with lesser control increasing the risk of CHD. The findings were corroborated by the case control study.

In a more extensive study conducted for the National Institute for Occupational Safety and Health (NIOSH), Chadwick, Chesney, Black, Rosenman, and Sevelius (1979) examined 397 white collar males in middle management in the aerospace industry. A wide range of job demands measures as well as CHD risk indictors were gathered in a longitudinal fashion over an 18 month period using questionnaires, medical examinations, Type A interviews, standardized psychological tests, electrocardiograms, and biochemical assays. The results of this study indicated that psychosocial job stress contributed as much to CHD risk, which was measured objectively and by self-report, as any single traditional risk factor such as smoking, lipid levels, diet, or blood pressure. However, there were several distinct patterns in the relationships among psychological variables which defined job stress, the observed behavior patterns of the workers, and physiological variables known or believed to be related to CHD risk. Thus multiple pathways were defined for increasing CHD risk as mediated by job stress. To add confusion, the psychological variables with the strongest relationships with self-reported CHD symptoms did not have the strongest relationships to physiological indicators of CHD risk. Individual worker characteristics (age, personality), the job environment (perceived stress level), and the interaction between these two were equally implicated in CHD risk. Life events stress showed a strong correlation with psychological distress, but only weak correlations with various physiological CHD risk indicators. There was no relationship between Type A behavior measures and CHD risk.

There is no doubt that the literature demonstrates that working makes a contribution to CHD risk. However, an examination of the earlier cardiovascular disease and work relationship literature (see Marks, 1967), and more current efforts by Haynes et al. (1978, 1980), Karasek (1981), and Chadwick et al. (1979) do not provide a clear relationship between working and CHD risk. Certain workers (female clericals with children) have an increased CHD incidence, and particular working conditions (low job control) increase CHD risk. However, as the Chadwick et al. (1979) study demonstrates, various mechanisms seem to influence CHD risk, including job stress, and often in different ways at different times. Thus attempts to show clear-cut relationships between single variables and CHD risk have been limited in their success.

7.4.4.2 Central Nervous System, Psychological, and Behavioral Effects

There are also central nervous system effects such as diminished capability to respond to stimuli, lack of vigilance, changes in brain electrical potential, and emotional disturbances. High stress environments that require mental alertness are particularly vulnerable to these effects. Branton and Oborne (1979) found that anesthesiologists in long surgical operations often have lapses of vigilance due to monotony. As they came out of these lapses they experience extreme stress reactions until they could determine that the patient was alright. These lapses have produced emotional effects defined as "mini-panics."

Psychological and emotional effects have also been frequently observed in relation to job demands and stress. Colligan, Smith and Hurrell (1977) examined 130 distinct occupations for differences in the incidence of mental health disorders such as depression, anxiety, and alcohol abuse. It was shown that a greater than expected incidence of these disorders was noted for workers in the health care field and the helping professions. Other studies have demonstrated that specific occupations report a greater level of emotional and mood disturbances which can be linked to their job requirements. Smith et al. (1981) found that data entry clerks on computerized jobs reported more psychological fatigue and higher psychosomatic complaints than data entry clerks in noncomputerized jobs. Broadbent and Gath (1979) found that assembly-line workers reporting higher levels of job dissatisfaction also reported higher levels of somatic symptoms. Those engaged in paced work tasks reported more anxiety and to a lesser extent more depression.

Stress can have a wide range of effects on worker behavior that can be costly to an organization. Behavioral reactions to stress can be looked on as coping responses (adaptive or maladaptive) that may have some long-term health implications. Smoking, drinking alcohol, and using drugs are behavioral reactions to stress with recognized health outcomes. Other behaviors such as eating habits, sleeping patterns, and exercise are behavioral reactions to stress with potential to produce negative health consequences. In addition, stress may produce changes in worker performance and productivity as well as absenteeism, turnover, and accidents. Although it is difficult to get accurate data on the extent of alcohol and drug use by workers, there is some research tying increased use to occupational stress. Margolis et al. (1974) found that escapist drinking was higher in workers reporting higher levels of job stress in a range of occupations. Smith et al. (1982) found shift workers to have a slightly higher tendency to use alcohol than day workers. In an extensive investigation of alcohol use among working people in Sweden, Kuhlhorn (1971) questioned managers, union representatives, and physicians about causes of drinking in workers. Of eleven possible causes, work-related factors, such as job dissatisfaction, haste, and job stress, were rated the most significant reasons for drinking, after marriage problems.

Roman and Trice (1970) has identified a number of work features that are perceived as contributing factors in problem drinkers. These include (1) lack of visibility, (2) work addiction, (3) occupational obsolescence, (4) new work roles, (5) absence of structured work, (6) jobs that require alcohol consumption, (7) jobs in which deviant drinking benefits others in the organization, and (8) jobs in which deviant drinking is an expected method of tension release.

The use of tobacco has also been related to job stress, tension, and anxiety (Bosse, 1978; Caplan et al., 1975a). Driken (1973) found that the decision to give up smoking was negatively related to the extent of job stress. Caplan (1975a) has demonstrated that increased stress levels are related to increased consumption of cigarettes.

The impact of stress on eating and sleeping habits is best illustrated in shift workers. The influences on these behaviors are both biological, owing to circadian rhythm disturbances, and emotional with stress-mediated hormone effects. Rutenfranz et al. (1977) have reviewed the shift work research literature and defined adverse gastrointestinal effects as one of the most common influences of shift work. Smith et al. (1982) have shown that shift workers have eating habits that differ from day workers and has postulated that this may be responsible for the observed gastrointestinal disorders. Smith et al. (1982) has also demonstrated that shift workers have different sleeping patterns than day workers and that these may lead to sleep disturbances. Tepas, Walsh, and Armstrong (1981) have demonstrated specific sleep disorders in shift workers, some of which may be due to maladaptive sleep patterns.

Based on a number of estimates (WHO, 1984) sickness and absenteeism from work has increased in all industralized nations over the preceding decade. The frequency of absence has increased at a greater pace than the extent of time off. Hoiberg (1982) demonstrated a relationship between sicknesses requiring hospitalization and job stress potential for specific worker groups. Porter and Steers (1973) in a major literature review found that job satisfaction is a key factor in worker withdrawal, absenteeism, and turnover. Gardell (1976) has shown that workers in high stress jobs are absent more often and for longer durations than their less stressed counterparts. Clegg (1983) has shown that there is a tendency for a progression from lateness to absenteeism, and Muchinsky (1977) demonstrated a tendency for a progression from absenteeism to turnover. Thus stress can influence worker sickness absence, which may produce a behavior pattern that leads to turnover.

Work performance and productivity are difficult behaviors to evaluate in regard to stress. Workers often sacrifice their health and well-being to help keep production levels up so as to maintain their job. Thus it is hard to determine the actual productivity influence of stress. Salvendy and Smith have suggested that a number of productivity measures must be assessed to get a true picture of the influence of stress. These include quantity of output, quality, absenteeism, turnover, accidents, arbitrary work breaks, and physiological and psychological costs to the worker.

7.4.5 MEASUREMENT OF STRESS

Because the causes and effects of stress are multidimensional, a multiple factor approach to the measurement of stress must be undertaken. Occupational stress can be assessed by utilizing some or all of the following four categories of measures: (1) physiological, (2) biochemical, (3) psychological, and (4) behavioral.

The task of selecting suitable measure(s) within each of these categories is often difficult. The choice typically is reflected in the specialized expertise of the investigator(s), the availability of equipment, the perceived degree of obtrusiveness of the measurements, and the unique information needed about the task under investigation. In general, it appears most useful to collect both subjective measures of stress, which serve to indicate the degree of stress perceived by the worker, and the more objective physiological, biochemical, and behavioral measures.

7.4.5.1 Physiological Measures

Physiological methods can be used for evaluating both physically and mentally demanding work. With respect to physically demanding work, these methods generally have been employed as a basis for determining rest allowances or for improving work methods.

Recording of physiological measures such as heart rate (HR), heart rate variability (HRV), blood pressure (BP), respiratory rate (RR), and electromyography (EMG) can be made over a continuous time period. These measures are associated with a physiological system that can be considered fast acting in response to internal and external stimulation. They have the potential for conveying information that is of a momentary nature. As a result, these measures frequently make it possible to isolate the effects of job components by evaluating second-to-second changes, or changes at longer intervals (e.g., 10 min time periods).

Advances in physiological recording and processing technology have fueled a recent trend toward the return of on-the-job evaluations of the physiological cost associated with various tasks. These improvements include the use of magnetic cassette tape recorders, which are worn across the worker's belt and allow for up to 24 hr of continuous recording (Smith and O'Brien, 1976); telemetry techniques (radio transmission of signals) (Reischl, Marschall, and Reischl, 1977); and a respirometer for measuring oxygen consumption and ventilation volume relatively free from interference with normal activities (Eley, Goldsmith, Laman, and Wright, 1976). As a result of these innovations, physiological measures can now be evaluated over rest and sleep in addition to the work cycles (Rodahl and Vokac, 1977).

A study by O'Brien, Smith, Goldsmith, Fordham, and Tan (1979) of truck assembly-line workers and medical nurses illustrates this strategy. The use of continuous HR recordings enabled a detailed breakdown of the work activities for these two groups. Results demonstrated that the HR responses for the nurses, in contrast to those for the assemblers, were sensitive to emotional stress. The HR responses for the assemblers were found to be more a function of their physical work activity. These findings emerged despite data on energy expenditure indicating that both occupations could be classified as light industrial work.

Utilizing modern computer and sensor technology, a methodology has been developed for unobtrusively monitoring industrial assembly line workers for HR, RR, and BP (Knight, Geddes, and Salvendy, 1980). Salvendy and Knight (1983) used an automated system for on-line monitoring of blood pressure to demonstrate that industrial workers in machine-paced tasks did not display any greater stress response than in self-paced tasks. The methodology used allowed for on-line data collection simultaneously from multiple workers in an industrial setting. Hennigan and Wortham (1975) in a study using a similar methodology, transmitted HR data from managers throughout the workday. These data were related to the activities experienced by the managers during the day, and illustrate the use of these measures for evaluating white collar work stress.

7.4.5.2 Biochemical Measures

Biochemical measures can be obtained from various fluids of the body, such as urine, blood, and saliva. Analysis of urine has been the most popular choice, largely because of minimum disturbance of collection in the work situation. However, in some cases it is necessary to analyze biochemicals in blood. The most often measured biochemicals include adrenaline, noradrenaline, cortisol, steroids, glucose, uric acid, triglycerides, and lipids.

Justification for the use of these measures in the analysis of occupational stress stems from (1) clinical evidence that has demonstrated the chronic elevation of biochemicals (adrenaline, noradrenaline, triglycerides) due to emotional reactions which can lead to functional disturbances in various organs and organ systems, which in turn may lead to psychosomatic and cardiovascular diseases; (2) natural daily rhythms of these measures which allow for evaluation of the degree to which these measures change from their normal rhythmic pattern when changes in the work requirements are initiated; (3) the availability of numerous automated processes for isolation and subsequent measure of the biochemical constituents of interest; (4) the ability to obtain the needed fluids easily during and after work periods.

For example, the biochemical measures, especially those that display daily rhythmic behavior (i.e., high and low levels of excretion throughout the day), are very appropriate for the examination of various shift work policies. In one study, railroad workers followed 3 weeks of day work with 3 weeks on the night shift (Akerstedt, 1977). On examining the pattern of adrenaline excretion in the workers, it was found that very little adjustment occurred after the first 3 days of night work. What this showed was that adrenaline levels in the workers still remained high during the day, which was not consistent with the normal pattern of adrenaline secretion. Unfortunately, the workers were then expected to sleep during this time and displayed sleep problems, implying that the high adrenaline levels could have been one of the sources of their sleep problems. In fact, the adrenaline excretion patterns for these workers still have not adjusted by the third week of night work.

The ability to obtain indexes of worker stress after work hours is a distinct advantage. At times there are reasons to believe that constraints in a particular work situation may impede the individual's response to stress and produce a delayed reaction that occurs hours after the individual has left the workplace. Also, the worker may be more apt to exhibit certain responses after work hours. The relative ease involved in urine collection could then enable the determination as to whether work-related stress is being manifested after work hours.

In contrast to measures such as HR and RR, the practical restriction on the collection procedures for biochemical measures (and the relatively slow response they exhibit) necessarily dictates that they reflect responses to stress over fairly long periods (e.g., several hours or days). As a result, biochemical measures are relatively insensitive for identifying which components of the task are most stressful. This factor alone is most critical in selecting between these two categories of measures for the purpose of evaluating specific occupational stress producing situations. Practical problems with biochemical measures concern the need to exercise rigid control in the kinds and quantity of food, liquid, and drugs ingested by the worker prior to and during the measurement process. A more complete treatment of the considerations governing the use of biochemical measures is given by Levi (1972).

7.4.5.3 Psychological Measures

Many standardized measurement devices are available that can be used for evaluating stress. Drawing on Selye's notion that stress may be characterized by the intensity of the demand for readjustment or adaptation brought about by some agent or situation, Holmes and Rahe (1967) have developed a Social Readjustment Rating Scale. This scale attempts to quantify the stress potential of various life events that often follow a series of life changes. A relationship has been found between the magnitude of the life crises and the risk of change in health. When applied to studies on occupational stress, the scale offers a method for recognizing and evaluating which types of responses (e.g., psychosomatic complaints) may have actually originated from factors extraneous to the workplace.

A popular method of validating psychological response criteria has been the use of mood checklists. These measurement devices serve to gauge the worker's feelings at any given time and are relatively simple to administer. One such checklist has been found to be capable of differentiating stress from arousal (Mackay, Cox, Burrow, and Lazzerini, 1978). The validity of the checklist has been supported by studies that have indicated a differential sensitivity of these two factors to a variety of environmental and task effects. As an example, after a prolonged and monotonous repetitive task, significant increases in self-reported stress were found together with significant decreases in self-reported arousal. Another mood checklist, the Multiple Affect Adjective Check-list, has proved useful as a measure of immediate or daily anxiety levels and has been found to correlate with work performance over time (Zuckerman, Lubin, Vogel, and Valerius, 1964).

There is another class of measures that evaluate perceptions and feelings about the job situation. These have been referred to as psychosocial aspects of working, such as job satisfaction or perception of work load. There have been a number of research questionnaires developed to ask workers about these factors, most of which examine similar concerns but often have done so in a different manner. Thus there are very few standardized questionnaire instruments for evaluating job factors such as work load, work pace, career opportunities, supervisory style, and organizational environment, all of which relate to stress-producing potential. One of the most often cited and utilized questionnaires was developed by the Institute for Social Research at the University of Michigan and is included in the report *Job Demands and Worker Health* (Caplan et al., 1975b). This instrument has been used to measure various psychosocial aspects of jobs for 23 occupations and has been able to differentiate between occupations on these factors. A second instrument is the Work Environment Scale (Insel and Moos, 1974), which has been used in a number of studies and has also been sensitive to differences across occupations.

7.4.5.4 Behavioral Measures

There is a wide range of behavioral measures that can be used to evaluate work stress. These include observations of behavior to determine deviations from normal patterns, examination of coping behaviors (smoking and alcohol consumption), and evaluation of job performance (productivity and error rate). The field of observing and evaluating behavior has been termed "applied behavioral analysis." In this approach, behaviors that reflect the responses, emotions, or motivations of interest are defined and categorized. Then checklists are made for recording the occurrence of these behaviors. Next, observations of worker behaviors are made for a variety of individuals for varying amounts of time to verify that the behaviors are occurring and can be quantified when they occur. A program of observations is established for randomly observing a sample of employees that represent the group of interest. The times of observation must be random and of sufficient quantity, quality, and duration to be representative of worker behavior over the course of a regular work cycle.

There are a host of electronic techniques for monitoring employee behavior from activity counters to videotape systems. The level of sophistication used is related to the need for accuracy, the amount of data to be collected, and the level of employee knowledge about observation that is desired. These behavior monitoring methods can be used for determining deviations from normal patterns that can indicate stress responses or for determining coping styles. In addition to these techniques are subjective methods such as questionnaires and interviews can be used to ask workers about their behavior patterns and coping styles. In this instance, the necessity for appropriate sampling strategies and item development are needed.

Of special interest in evaluating employee behavior related to stress is the examination of work performance. In utilizing performance measures in studies of occupational stress, the worker's particular task will necessarily govern which performance criteria will be selected. Quantity, quality, and variability of work performance are among the most frequent measures used. Individual attributes often can significantly affect performance and the ensuing inferences regarding the severity of stress associated with the worker's task.

7.4.6 MANAGEMENT OF OCCUPATIONAL STRESS

Ideally, we should not have to consider strategies for the management of occupational stress. With the vast body of literature on industrial-organizational psychology, principles of worker selection and placement, consideration of the human factor in the design of the workplace, and knowledge about task- and job-related stress responses of workers employed in numerous occupations, it would perhaps appear more realistic to set our goals on the prevention of occupational stress. However, the long-term effects of the design of workplaces and organizations are often not sufficiently understood for designers of workplaces to accommodate the interactive complexity of the various workplace features that produce stress. Therefore, individual approaches to stress management must be considered along with organizational approaches for a comprehensive effort for preventing stress. Many job design and organizational interventions have not been completely tested in a systematic way for controlling workplace stress, and therefore the following recommendations are based mainly on theories of job redesign rather than on conclusive research evidence.

7.4.6.1 Ergonomic/Work-Station Redesign Solutions

Ergonomic redesign deals with providing a work area that minimizes the physical demands on the worker. Such physical demands are of significance to emotional stress because they influence fatigue, which is a close relative of stress, and also influence worker attitude and behavior, factors that affect worker response to stressors. There are three major concerns when defining ergonomic stress and associated control strategies. These are the extent of load (exposure) that is placed on (1) the senses (especially visual and auditory), (2) posture, and (3) the manipulative muscles (arms, hands). The ergonomic intervention designs the workplace to minimize the load(s) on each of these bodily systems by providing (1) an appropriate sensory environment, (2) proper work-station design (workbench, shelving, chair), and (3) comfortable environmental conditions.

These considerations will provide the necessary physical environment for the work activity and therefore will not add to the stress load produced by the psychological requirements of the job. Each type of work activity has its own special ergonomic requirements that can be defined by examining the loads imposed by the job tasks, and then applying ergonomic principles to minimize these loads.

7.4.6.2 Job Design Solutions

It is clear from the literature that the greatest job design difficulties occur for jobs that have little inherent job content to begin with. For work activity to provide adequate job satisfaction, there must be meaningful content for the individual to derive a sense of accomplishment and a positive feeling of self-esteem. Many jobs are fragmented and simplified so that the content is so limited that very little meaning or satisfaction can be derived from conducting the work activities. Therefore, boredom and fatigue predominate in these jobs. To enhance job content, meaning has to be built into the job. This can be accomplished by enlarging the use of worker's skills, as opposed to simplifying the work. Work should not be overly repetitive to the extent that the worker uses only simple perceptual motor skills and no social or cognitive skills. In addition, job tasks should be designed to utilize existing skills as much as possible, to enhance worker confidence and performance.

Control of the work process is a significant factor in the occurrence of job stress. Lack of job control has been demonostrated to be one of the primary causes of psychological and physiological dysfunction (Karasek, 1979; Smith et al., 1981). Providing a greater amount of control of work activities, by increasing worker decision-making and use of alternative work procedures, can reduce the stress imposed by machine-controlled work processes. It also enhances the job content aspects of the work activity by giving more individual meaning and satisfaction with individual accomplishments.

Feedback about performance is a significant aspect of worker control of the work process. If the first-level supervisor has continuous information about worker performance, and uses this information to intimidate the employee, then the employee will perceive a lack of control over the work process. Rather than providing performance feedback to the first-level supervisor, it may be better to give this type of information directly to each employee on a frequent basis. This information could be given automatically by the machine on which the worker is operating, or by a written communication from higher management. The information should be presented in a nonthreatening way. There is a large body of literature that indicates that such direct performance feedback to the employee has a positive influence on performance (Smith and Smith, 1966). On the other hand, having the first-level supervisor provide the performance feedback may create tension and stress and thus have a negative influence on performance and social interaction, which is a stress buffer.

Completeness is another aspect of job content and the meaningfulness of work that is often missing in automated work. As work is fragmented through simplification, the relationship of the task activity to the organization and the end product is diminished. Thus workers fail to identify with the work process and the product. They fail to appreciate that a lack of quality in their small component of the product can have a major impact on the completed product and their fellow co-workers' performance. Fragmentation of work must be avoided, so that employees can attain a personal identity with the organization, a product identity, and an organizational pride. If work tasks absolutely have to be simplified and fragmented, then it is imperative that employees understand their contribution to the end product and the organization. They must feel that their contribution is significant and meaningful for a positive feeling of self-esteem. Otherwise, health and productivity will suffer owing to increased job stress. The operator's tasks should be broad enough to provide some closure, and therefore understanding of the significance of the work to the entire organization.

Isolation of individual workers at a fixed work station greatly reduces social interaction and thus should be avoided. As social support is removed by isolation, the positive benefits of socialization in stress control are eliminated. However, with some work processes, it is often not possible to have social interaction during the work task activity. Therefore, social interaction during nontask periods must be enhanced and encouraged. This can be accomplished by providing special work break facilities in close proximity to the work areas, and by allowing groups of workers to go on break together.

A critical job design issue deals with the determination of reasonable work load. Often work load is set by the limits or capacity of the machinery, rather than by the capacity of the operator. Given that the machinery is an expensive investment for any company, it is understandable that production goals are often set based on the cost of the machinery and the need to improve productivity. However understandable this action, it may not be based on sound engineering or psychological principles for determining the proper work load, and thus produce excessive work load and fatigue. To determine the appropriate work load, industrial psychologists and industrial engineers, as well as employee representatives, must be part of the team that designs and implements the work system and sets work load requirements.

7.4.6.3 Organizational Solutions

A major factor that produces worker resistance to new work activities is that changes often appear at the workplace "out of the blue" without worker knowledge of an impeding change. It is very important for the successful implementation of changes in work processes, and subsequent enhancement of worker health, performance, and satisfaction, that organizations have a transition policy that includes worker participation in all stages of the change process. That is, workers should participate from planning, through selection of equipment, to the daily operation of the work system. First, worker representatives should be involved in the planning phase of .automation. This will aid in employee acceptance of the changes in work processes, and ensure that employee concerns are aired. Secondly, the employee representative should be involved in the design of the work system, to ensure that human concerns and capabilities are included, as well as the machinery capabilities. Finally, employees who are affected by automation should be involved in the implementation of the automation. This will provide them with a fuller understanding of the machinery, its capabilities, and their role in the work process.

Training of the operators in job requirements is one of the most neglected aspects of organizations. Of course, most companies provide an introduction to workers as to how their equipment operates and what the various features are. Additionally, the manufacturer typically provides a manual that explains how the machinery works, and how specific functions can be carried out. Often, however, the extent of operator training is to be told to read the manual and to start working. Sometimes on-the-job training also takes place with an experienced operator teaching the new employee. This situation can be stressful, depending on the ability of the experienced operator to serve as a teacher and the social dynamics of the teaching situations. Rather than relying on such informal training, there is a need to have comprehensive training procedures that will develop skills and enhance worker confidence and self-esteem. Training should start with an explanation of the technology, how it works, and its benefits to the company and to the worker. The equipment and computer system should be thoroughly explained, indicating the strengths *and* weaknesses of the system. Then there should be intensive training from the manual that explains how the sytem works, and specific machine functions. Each classroom teaching or individual reading session should be followed up by practice in the functions learned, with a skilled operator available to coach the trainees in a positive way.

After the trainee has successfully passed the training course, he should not be required to work at full speed until becoming accustomed to the work situation. This may take 1 day or 1 month, depending on the complexity of work activities. In addition to initial training, all operators should have periodic retraining to keep skills and confidence at peak levels.

Fear for job security is one of the greatest concerns of workers. This is natural because it is commonly believed that automation produces unemployment. As indicated above, a company that invests time and energy in developing the skills of workers shows a desire to keep a valuable resource and thus reduces worker fears of job loss. However, there are other job security problems related to automation. One is the possibility of being downgraded because the machinery takes over some of the worker's functions, making the job less complex. This is called job deskilling and often reduces labor costs by simplifying work. However, such efforts are almost always doomed to failure, because they produce extensive morale and motivation problems in workers which directly influence productivity. The main purpose of automation is to do work more efficienty and productively, not just to simplify it. Thus companies should establish a procedure that work tasks not be deskilled when automating.

In addition, companies must develop career paths for workers so that advancement can be attained for those who are good performers. Being locked into a highly repetitive job, which has very little content or meaning, with no chance for advancement, is a major source of job stress and a demotivating force for many workers. Companies will have to be innovative in advancing workers through a career path to enhance performance and reduce job stress.

Close employee monitoring creates a dehumanizing work environment in which the worker feels controlled by the machine. When performance monitoring is also used by supervision to control performance, worker perception of work pressure and work load is very high, thus producing stress responses. It also creates an adversarial relationship between the supervisor and the employee. This is especially troublesome, because the supervisory/employee relationship, if positive, can be a stress buffer. For the most effective employee performance and to enhance stress reduction, supervisors should use positive motivational and employee support approaches. This suggests that first-line supervisors should not be involved directly in the employee monitoring and performance feedback. Rather, supervisors should receive training in employee support approaches, thus helping to buffer the effects of other stressful job demands. It must be understood that this approach changes the basic role of the first-line supervisor by establishing this individual as a positive link between employees and other levels of management.

7.4.6.4 Individual Coping Solutions

The development of individual coping strategies is rapidly becoming recognized as an effective means of reducing high levels of worker stress. A number of programs have been developed to apply psychophysiological methods to reduce stress responses, and some have been applied in workplace settings. Psychophysiological methods (biofeedback, relaxation, meditation) have been used successfully to deal with stress related problems since the early 1930s (Jacobson, 1934). Those treatments have been applied mainly on an individual basis as part of a therapy regimen and not as a preventive measure. However, the cost and time requirments of such individual treatment approaches may make them unfeasible for employer acceptance; thus these techniques may have to be modified to decrease their cost and increase their applicability to workplace stress problems.

In the first reported study to reduce worker stress symptoms using a workplace setting, Peters, Benson, and Porter (1977) studied the effectiveness of a relaxation/meditation program in reducing blood pressure for a group of normotensive employees at the Converse Rubber Company in Boston. One hundred and twenty-six workers volunteered for a 12 week relaxation training program using a meditation approach. These workers demonstrated a reduction in blood pressure that was statistically significant but so small to be of questionable clinical significance. In a different study, a multimodal technique using muscle relaxation, breathing exercises, biofeedback, and psychotherapy to assist 30 individuals with headache and general anxiety symptoms demonstrated significant reductions in all measures of distress 3 months after treatment was completed. In addition, significant cost/benefit to the company was demonstrated ($5.50 saved for each $1 invested in the program during the first 3 years of treatment) (Manuso, 1983). Murphy (1980, 1983) assessed the relative effectiveness of progressive muscle relaxation and biofeedback to self-relaxation in a group of nurses in an acute care facility and with 38 blue collar workers. The results indicated positive effects on some of the distress measures for all treatments but only limited effects on other measures. These effects were not sustained after a 3 month follow-up.

From these demonstration studies, it is apparent that psychophysiological methods for reducing stress stymptoms can be successfully implemented at the workplace. There are no special impediments to their application if there is cooperation between managers and workers in the provision of appropriate facilities and in the prudent use of worker time in the program. This means that the programs have to fit into the work schedule, cannot be too time consuming (should probably not exceed 30 min for the employer's convenience), and should be amenable to group application. Although implementation does not seem to pose major problems, the effectiveness of these programs in reducing worker stress symptoms has been mixed and, when taken in total, far from spectacular. Certainly there have been demonstrations of statistically significant reductions in physiological measures of stress. However, the clinical significance of such reductions is questionable.

The treatment regimens have been most successful in bringing about reductions in adverse emotional states and feelings of stress. Such changes are not always reflected in physiological indicators, but they may have significant meaning for long-term health status. Thus the treatments may be extremely effective in improving worker health, but the studies to date have not been designed to evaluate the long-term health benefits, and thus do not adequately assess treatment effectiveness. Manuso (1983) has shown that there are some other secondary benefits from these types of interventions which can include reduced absenteeism, better employee performance, and lower health insurance costs. Murphy (1983) has confirmed the finding of increased employee performance using worker self-reports.

Various approaches that can be used to control job stress have been examined. In most instances, it will be necessary to use some combination of these approaches. The primary stress control approach that should be used first is to eliminate the exposure(s) to stressors. This can be achieved by defining the sources of stress and then applying ergonomic, job design, and/or organizational interventions. Sometimes it is not possible to eliminate a stressor completely. Then the approach should emphasize minimizing the stress load. In such instances, individual coping strategies can be applied to reduce the stress symptoms of workers. Although this does not address the source of the problem, it does reduce the health risk by controlling the stress response. Not all the individual coping techniques are equally effective and each worker will have to experiment with different approaches to find out which will work best for him or her.

REFERENCES

Akerstedt, T. (1977). Inversion of the sleep–wakefulness pattern: Effects on circadian variations in pschophysiological activation. *Ergonomics, 20*, 459–474.

Arthur, R. J., and Gunderson, E. K. (1965). Promotion and mental illness in the Navy, *Journal of Occupational Medicine, 7*, 452–456.

Bosse, R. (1978). Anxiety, extraversion and smoking. *British Journal of Social and Clinical Psychology, 17*, 269–273.

Branton, P., and Oborne, D. (1979). A behavioral study of anaesthetists at work. In D. Oborne, ed., *Psychology and medicine.* London: Academic.

Breslow, L., and Buell, P. (1960). Mortality and coronary heart disease and physical activity on work in California, *Journal of Chronic Disease, 11*, 615–626.

Broadbent, D. E., and Gath, D. (1981). Symptom levels in assembly-line workers. In G. Salvendy and M. J. Smith, (Eds., *Machine pacing and occupational stress* London: Taylor and Francis. 243–252).

Cakir, A., Reuter, H., Von Schmude, L. and Armbruster, A. (1978). Investigations of the accommodations of human psychic and physical functions to data display screens in the workplace. Berlin: *Institut fur Arbeitswissenschaft der Technischen Universitat Berlin.*

Cannon, W. B. (1914). The interrelations of emotions as suggested by recent physiological researchers. *American Journal of Psychology, 25,* 256–282.

Caplan, R. D., Cobb, S. and French, J. R. P., Jr. (1975a). Relationship of cessation of smoking with job stress, personality and social support, *Journal of Applied Psychology, 60,* 211–219.

Caplan, R. D., Cobb, S., French, J. R. P., Harrison, R. V., and Pinneau, S. R. (1975b). *Job demands and worker health.* Washington DC: U.S. Government Printing Office.

Chadwick, J., Chesney, M., Black, G., Rosenman, R. H., and Sevelius, G. (1979). *Psychological job stress and coronary heart disease.* Cincinnati, OH: National Institute for Occupational Safety and Health.

Clegg, C. W. (1983). Psychology of employee lateness, absence, and turnover: A methodological critique and an empirical study. *Journal of Applied Psychology, 68,* 88–101.

Cobb, S. and Kasl, S. V. (1977). Termination: The consequences of job loss. Washington, DC: U.S. Government Printing Office.

Colligan, M. J., Smith, M. J., and Hurrell, J. J. (1977). Occupational incidents rates of mental health disorders, *Journal of Human Stress, 3,* 34–39.

Cooper, C. L., and Marshall, J. (1976). Occupational sources of stress: A review of the literature relating to coronary heart disease and mental ill health, *Journal of Occupational Psychology, 49,* 11–28.

Driken, J. M. (1973). Stress and cardiovascular health: An international cooperative study: II. The male population of a factory in Zurich. *Social Science and Medicine, 7,* 573–584.

Eley, C., Goldsmith, R., Laman, D., and Wright, B. M. (1976). A miniature indicating and sampling respirometer (MISER). *Journal of Physiology, 256,* 59–60.

Eysenck, H. J. *Biological basis of personality,* Springfield, IL: Charles C Thomas.

Frankenhaeuser, M., and Gardell, B. (1976). Underload and overload in working life: Outline of a multidisciplinary approach. *Journal of Human Stress, 2,* 35–46.

French, J. R. P. (1963). The social environment and mental health. *Journal of Social Issues, 19,* 39–56.

French, J. R. P. and Caplan, R. D. (1970). Psychological factors in coronary heart disease. *Industrial Medicine, 39,* 383–397.

Friedman, M., Rosenman, R. H. and Carroll, V. (1958). Changes in the serum cholesterol and blood clotting time in men subjected to cyclic variation of occupational stress. *Circulation, 17,* 852–861.

Gardell, B. (1976). Technology, alienation and mental health. *Acta Sociologica, 19,* 83–94.

Hadley, O. R. (1939). Analysis of 5,116 deaths reported due to acute occlusion in Philadelphia, 1933–1937. *Public Health Reports, 54,* 972–1012.

Harrison, R. V. (1978). Person–environment fit and job stress. in C. Cooper and R. Payne, Eds., *Stress at work.* New York: Wiley. 175–209.

Haynes, S. G. and Feinleib, M. (1980). Women, work and coronary heart disease: Prospective findings from the Framingham heart study. *American Journal of Public Health, 70,* 133–141.

Haynes, S. G., (1978). Feinleib, M., Levine, S., Scotch, N., and Kannel, W. B. (1978). The relationship of psychosocial factors to coronary heart disease in the Framingham study. *American Journal of Epidemiology, 107,* 384–402.

Haynes, S. G., Feinleib, M., and Kannel, W. B. (1980). The relationship of psychosocial factors to coronary heart disease in the Framingham study. *American Journal of Epidemiology, 111,* 37–58.

Hennigan, J. K. and Wortham, A. W. (1975). Analysis of workday stresses and industrial managers using heart rate as a criterion. *Ergonomics, 18,* 674–681.

Hoiberg, A. (1982). Occupational stress and illness incidence. *Journal of Occupational Medicine, 24,* 445–451.

Holmes, T. H., and Rahe, R. H. (1967). The social readjustment rating scale. *Journal of Psychosomatic Research, 11,* 213–218.

Hurst, M. W., Rose, R., and Jenkins, D. (1978). *Air traffic controller health study.* Washington, DC: U.S. Department of Transportation (Report No. FAA-AM-78-39).

Insel, P. and Moos, R. (1974). Work environment scale-form S. Palo Alto, CA: Consulting Psychologist Press.

Jacobson, E. (1934). *You Must Relax.* New York: McGraw-Hill.

Jenkins, C. D., Roseman, R. H., and Friedman, R. (1967). Development of an objective psychological test for the determination of the coronary prone behavior pattern. *Journal of Chronic Diseases, 20,* 371–379.

Karasek, R. A., Jr. (1979). Job demands, job decision latitude and mental strain: Implications for job redesign. *Administrative Science Quarterly, 24,* 285–306.

Karasek, R. A. (1981). "Job decision latitude, job design, and coronary heart disease. In G. Salvendy and M. J. Smith, Eds., *Machine pacing and occupational stress* (pp. 45–56). London: Taylor and Francis.

Knight, J. L., Geddes, L. A., and Salvendy, G. (1980). "Continuous Unobstrusive Performance and Physiological Monitoring of Industrial Workers," *Ergonomics, 23,* 501–506.

Kuhlhorn, E. (1971). Spriten och Jobbet. *Alkoholfragan, 65.* 222–230.

Lazarus, R. S. (1977). "Cognitive and coping processes in emotion. In A. Monat and R. Lazarus, Eds., *Stress and coping.* New York: Columbia University Press.

Levi, L. (1972). Methodological considerations in psychoendocrine research. *Acta Medica Scandinavica, 191 Suppl. 528* 28–54.

Mackay, C., Cox, T., Burrows, G., and Lazzerini, T. (1978). An inventory for the measurement of self-reported stress and arousal. *British Journal of Social and Clinical Psychology, 17,* 283–284.

Manuso, J. S., Jr. (1980). *Stress management training in large organizations.* New York: Equitable Life Assurance Society.

Manuso, J. S., Jr., Ed. (1983). *Occupational clinical psychology.* New York: Praeger.

Margolis, B., Kroes, W. M., and Quinn, R. (1974). "Job stress: An unlisted occupational hazard. *Journal of Occupational Medicine, 16,* 654–661.

Marks, R. (1967). Social stress and cardiovascular disease. *The Millbank Memorial Fund Quarterly, 45,* 51–107.

McGrath, J. E. (1976). Stress and behavior in organizations. in M. Dunnette, Ed., *Handbook of industrial and organizational psychology* (pp. 1353–1395). Chicago: Rand McNally.

Morris, J., Heady, J., Raffle, P., Roberts, C., and Parks, J. (1953). Coronary heart disease in different occupations. *The Lancet, 2,* 1053–1057.

Mortenson, J., Stevenson, T., and Whitney, L. (1959). "Mortality due to coronary heart disease analyzed by broad occupational groups. *American Medical Association Archives of Industrial Health, 19,* 1–4.

Muchinsky, P. M. (1977). Employee absenteeism: A review of the literature. *Journal of Vocational Behavior, 10,* 316–340.

Murphy, L. (1980). *Stress management at the workplace.* Cincinnati, OH: National Institute for Occupational Safety and Health.

Murphy, L. (1983). A comparison of relaxation methods for reducing stress in nursing personnel. *Human Factors, 25,* 431–440.

NIOSH (1984). *Occupational risk factors in psychological disorders: An overview.* Cincinnati, OH: National Institute for Occupational Safety and Health.

O'Brien, C., Smith, W. S., Goldsmith, R., Fordham, M., and Tan, G. L. E. (1979). A study of the strains associated with medical nursing and vehicle assembly. In C. Mackay and T. Cox, Eds., *Response to stress: Occupational aspect* (p. 22). London: IPC Science and Technology Press.

Pell, S., and D'Alonzo, C. (1961). A three year study of myocardial infarction in a large employee population. *Journal of the American Medical Association, 175,* 463–470.

Pell, S. and D'Alonzo, C. (1963). Acute myocardial infarction in a large industrial population. *Journal of the American Medical Association, 185,* 831–833.

Peters, R., Benson, H., and Porter, D. (1977). Daily relaxation breaks in a working population: Effects on self-reported measures of health, performance, and well-being. *American Journal of Public Health, 67,* 946–953.

Porter, L. W., and Steers, R. M. (1973). "Organizational, work, and personal factors in employee turnover and absenteeism. *Psychological Bulletin, 80,* 151–176.

Reischl, U., Marschall, D. M., and Reischl, P. (1977). Radiotelemetry-based study of occupational heat stress in a steel factory. *Biotelemetry, 4,* 115–130.

Rodahl, K., and Vokac, Z. (1977). "Work stress in Norwegian trawler fishermen. *Ergonomics, 6,* 633–642.

Roman, P. M., and Trice, H. M. (1970). The development of deviant drinking behavior. Occupational risk factors. *Archives of Environmental Health, 20,* 424–435.

Russek, H., and Zohman, B. (1958). Relative significance of heredity, diet, and occupational stress in CHD of young adults. *American Journal of Medical Science, 235,* 266–275.

Rutenfranz, J., Colquhoun, W., Knauth, P., and Ghata, J. (1977). Biomedical and psychosocial aspects of shift work. *Scandinavian Journal of Work Environment and Health, 3,* 165–182.

Ryle, J., and Russell, W. (1949). The natural history of coronary disease: A clinical and epidemiological study. *British Heart Journal, 11,* 370–389.

Salvendy, G., and Knight, J. (1983). Circulatory responses to machine-paced and self-paced work: An industrial study. *Ergonomics, 26,* 713–717.

Salvendy, G., and Sharit, J. (1982). Occupational stress. In G. Salvendy, Ed., *Handbook of industrial engineering.* (pp. 1–15) New York: Wiley.

Salvendy, G., Smith, M. J. (1981). *Machine pacing and occupational stress* (p. 366). London: Taylor & Francis.

Selye, H. (1956). *The stress of life.* New York: McGraw-Hill. 1956.

Sharit, J., and Salvendy, G. (1982). Occupational stress: Review and reappraisal. *Human Factors, 24,* 129–162.

Smith, W. S., and O'Brien, C. (1976). A system for rapid analysis of long-term recordings of heart rate and other physiological parameters. *Biomedical Engineering, 11.*

Smith, K. U., and Smith, M. F. (1966). *Cybernetic principles of learning and educational design.* New York: Holt, Rinehart and Winston.

Smith, M. J., Cohen, H. H., Cleveland, R., and Cohen, A. (1978). Characteristics of successful safety programs. *Journal of Safety Research, 10,* 5–15.

Smith, M. J., Colligan, M. J., Frockt, I. J., and Tasto, D. L. (1979). Occupational injury rates among nurses as a function of shift schedule. *Journal of Safety Research, 11,* 181–187.

Smith, M. J., Cohen, B. G. F., Stammerjohn, L. W., Jr., and Happ, A. (1981). An investigation of health complaints and job stress in video display operations. *Human Factors, 23,* 389–400.

Smith, M. J., Colligan, M. J., and Tasto, D. L. (1982). Health and safety consequences of shift work in the food processing industry. *Ergonomics, 25,* 133–144.

Stamler, J., Kjelsberg, M., and Hall, Y. (1960). Epidemiologic studies on cardiovascular-renal disease: I. Analysis of mortality by age–race–sex–occupation. *Journal of Chronic Disease, 12,* 440–455.

Tebb, A. (1980). Dimensions of cumulative injury. In R. Schwartz, Ed, *New developments in occupational stress* (pp. 30–38). Cincinnati, OH: National Institute for Occupational Safety and Health.

Tepas, D., Walsh, J., and Armstrong, D. (1981). Comprehensive study of the sleep of shift workers. In L. Johnson, D. Tepas, W. Colquhoun, and M. Colligan, Eds., *The twenty-four hour workday.* Cincinnati, OH: National Institute for Occupational Safety and Health. (1984).

The President's Commission on Mental Health. 1978. *Report to the President, Vol. 1,* Washington, DC: U.S. Government Printing Office.

WHO (1984). Psychosocial factors and health: Monitoring the psychosocial work environment and workers' health. Geneva: World Health Organization.

Zuckerman, M., Lubin, B., Vogel, L., and Valerius, E. (1964). Measurement of experimentally induced affects. *Journal of Consulting Psychology, 28,* 418–425.

CHAPTER 7.5

THE USE OF SAFETY DEVICES AND SAFETY CONTROLS AT INDUSTRIAL MACHINE WORK STATIONS

JOHN R. ETHERTON

National Institute for Occupational Safety and Health
Morgantown, West Virginia

The author gratefully acknowledges the assistance of Jim McGlothlin, Herb Linn, and Roger Jensen at Division of Safety Research, NIOSH in preparing the manuscript for publication.

7.5.1 INTRODUCTION

This chapter is about human-factors considerations for the use of devices and methods that are designed into or installed on work stations at hazardous industrial machines for the intended purpose of preventing traumatic amputations, crushing injuries, and fatalities to industrial workers. Guidelines given here focus, in particular, on helping persons who are evaluating the relative merits of various machine safeguarding alternatives to understand how the human-factors design aspects of these devices and methods contribute to their effective use. Topics that are considered include human reliability in using machine safety devices and methods to avoid access to industrial machine dangers; inadvertent human entry into machine danger areas and machine controls that respond to these entries in such a way that dangerous machine action will not occur; and possible injuries owing to the *misuse* of machine safety devices or methods.

This chapter includes considerations of human anthropometry, information processing, physiology, and behavior as they relate to the use of safeguarding on industrial machine work stations. Discussion is devoted to the problems of why people (1) fail to use a safeguard that has been provided; (2) fail to adjust and maintain a safeguard that has been provided; and (3) improperly use or erroneously use a safeguard that has been provided.

For overviews of other human-factors aspects of industrial machine work station design (posture, lighting, noise, etc.), see Garg and Kohli (1979) or Percival (1977). The references at the end of the chapter provide more particular information on the construction and operating principles of work station safety devices.

7.5.2 MACHINE SAFETY DEVICES FROM THE HUMAN-FACTORS DESIGNER'S POINT OF VIEW

The industrial machine work station is a focus for ergonomic attention as a place where people's capabilities and limitations must be considered in order to achieve safe and productive industrial output. In the industrial workplace, processing machines (e.g., lathes, saws, spot welders, forming and punching presses, shears) and materials handling machines (e.g., robots, conveyors, coil feeders) are used to work by themselves or to work under the control of people in doing industrial production tasks. They are used because they contribute some machine attribute of speed, power, repetition, or accuracy to the task. People operate and maintain this industrial machinery in order to accomplish desired production goals.

When people work interactively with machines, exchanges occur between the human and the machine. What is to be prevented by the use of safeguarding devices or methods is any brief moment of exchange, as illustrated in Figure 7.5.1, when a person intrudes into a machine location that has become hazardous owing to a machine energy transfer at that location. Machines are hazardous because they accomplish energy transfers (mechanical, electrical, heat) too great for human tolerance. Acute, traumatic injuries, such as amputations, can result.

Injuries at industrial machines remain a persistent and costly problem of the industrial workplace. The U.S. National Institute for Occupational Safety and Health (NIOSH) estimates that in 1982 machine operators suffered 2400 amputation injuries (mostly fingers) and 24,800 fractures. More severe amputations and fatalities occur as well. Workers in other industrial occupations are also potential victims of machine injury. The National Safety Council in the United States estimated for 1983 that 10% of occupational injuries were machine related, and that half of these resulted in permanent partial disability.

Repetitive trauma disorders (Armstrong, 1981) are another injury type associated with machine work stations. These are a group of injuries to the tendons and nerves of the hands, wrists, and shoulders (e.g., tenosynovitis and carpal tunnel syndrome). These injuries are not instantaneous or acute as is the case with traumatic amputations, which are associated with unexpected events (i.e., accidents). Repetitive trauma disorders occur progressively while machine operators are performing the same "normal" task over and over again. These musculoskeletal injuries have, in some cases, been associated with two-hand palm button safeguards used at work stations. It has been found that the tendons or nerves in the wrists of some workers became permanently numb or dysfunctional after working at tasks requiring repetitive exertions with a sharply flexed wrist, such as would be the case for some installation locations for dual push button safeguards. This type of occupational injury is being diagnosed in increasing numbers as repetitive trauma becomes better understood among occupational physicians.

The mutilation and suffering that are consequences of occupational injury can ruin individual lives and are morally deplorable. However, occupational injury can also be costly to business. Worker

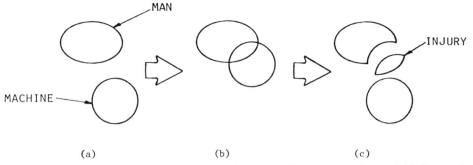

Fig. 7.5.1. The functional states of a dangerous, human–machine energy exchange: (*a*) before work begins; (*b*) a dangerous intrusion into the machine work space; (*c*) injury is sustained.

compensation insurance payments in the United States have increased at a rapid rate. For example, estimated compensation insurance payments in 1978 of $17 billion constituted annual increases of 27 and 20% over 1976 levels. The average cost per $100 of payroll jumped to $1.85 for 1978 compared with $1.73 for 1977. Industrial machine injury can be enormously costly to third parties involved in some aspect of the design of a machine, if a liability lawsuit is pursued by an injured worker as a means of restitution for injury. Awards of several hundred thousand dollars are not uncommon. Finally, consider that safeguarding based on human factors represents an investment not only in the well-being of people who operate and maintain industrial machines, but also in the productive management of the business.

7.5.3   THE PRIMARY REASON FOR USING MACHINE SAFETY EQUIPMENT

When working near or with industrial machines, humans are at risk of acute, traumatic injury from the zone in which the machine tooling does its work (often called the point of operation); workpiece hazards such as chips, sparks, or hot metal; and the components by which power is transmitted to the tooling to perform work (gears, shafts, hydraulic lines, electrical elements). Some of these hazards are illustrated in Figures 7.5.2 and 7.5.3.

7.5.4   HUMAN WORK TASKS AND THE USE OF MACHINE SAFEGUARDING

In the course of doing tasks near machines, workers make various reaching movements that could result in their being in a danger zone (see Figure 7.5.4).

Task design plays a role in the expected hazard exposure. For instance, if one of the tasks to be performed is direct hand placement of work between a machine's tooling or dies, this would involve normal and frequent exposure to a machine hazard. This normal and frequent exposure should lead to the selection of a highly reliable safeguard. In this use of the term "reliable," one would consider the human factors that affect the probability that the safeguard would be in place and functional when the human works at the machine. On the other hand, if the task were designed so that feeding of workpieces does not require a worker's hands to enter the tooling, then a different human-factors

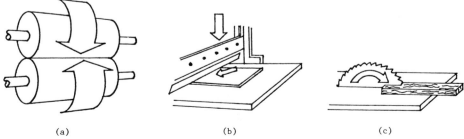

Fig. 7.5.2. Machine dangers due to (*a*) in-running nip point, (*b*) shearing mechanism, and (*c*) cutting mechanism.

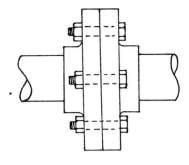

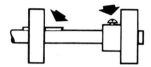

Fig. 7.5.3. Projections that could injure if touched when the shaft rotates at high speed.

evaluation of the work station would be in order. The latter analysis should take into account the unplanned or spontaneous human actions that might lead to an exposure to danger. Some of the tasks associated with industrial machines are discussed in the following sections.

7.5.4.1 Normal Operation Tasks

The machine is processing or handling materials as it was designed to function and the machine operator, feeder, tender, or supervisor is providing input (e.g., manual actuation of switches, manual parts handling, and visual quality surveillance) needed to accomplish the production task.

7.5.4.2 Cleaning and Clearing Tasks

If the machine's operation results in a buildup of scrap or if material becomes caught in the machine, human intervention is necessary to correct the machine feeding process. This is a particularly important area of human–machine interaction where consideration of human task performance can be beneficial.

7.5.4.3 Maintenance Tasks

Achieving safe performance of infrequent or random tasks necessary to keep a machine, process, or system in a state of repair or efficiency presents special human-factors problems. This is because normal

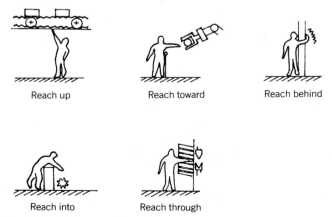

Fig. 7.5.4. Various reaching tasks that might be toward a danger point.

safety methods may not be usable and other safety equipment and procedures are temporarily brought into use. Examples of such tasks would be inspecting, lubricating, repairing, setting up, replacing worn or damaged tooling, or troubleshooting a machine control malfunction that would require close observation of machine action. Motivation, training, and supervision gain increased importance as modes of protection against traumatic injury. Accessibility and postural stability of the human body for the site where maintenance work is to be done should lead to anthropometric evaluation of how these factors could lead to traumatic injury. Stress that leads to risk taking may be associated with efforts to minimize downtime cost and to demonstrate to peers that one is capable of making the repair.

7.5.4.4 Tasks of Other Workers and Bystanders

Persons who do not work with a machine still may be at risk. Access to danger may exist along aisleways on the work floor and at the back or sides of the equipment. Unexpected motion by projecting workpieces can cause injury. This has been the case in injuries at automatic screw machines where passersby have been struck by projecting bar stock which bends out into aisleways while rotating at high speed. Unexpected start-up of automated equipment or of robots could injure passersby who take shortcuts through the machine's working zone.

7.5.5 STANDARDS

7.5.5.1 Regulatory

In the United States, the Occupational Safety and Health Administration (OSHA) has a general, performance-type requirement (General Industry Standard 1910.212) for safeguarding all machines. The standard states in part:

> One or more methods of machine guarding shall be provided to protect the operator and other employees in the machine area from hazards such as those created by point of operation, ingoing nip points, rotating parts, flying chips and sparks. Examples of guarding methods are—barrier guards, two-hand tripping devices, electronic safety devices, etc.

For 1978 to 1981 this standard was the third most often cited violation (10,811 times) in workplace inspection reports by federal safety compliance officers. Other requirements for individual machine categories (mechanical power presses, grinding machines, forging machines, etc.) are also included in Subpart 0 of the General Industry Standards.

7.5.5.2 Voluntary

The consensus, voluntary standards issued by the American National Standards Institute (ANSI) are a source of other suggestions for alternative safeguarding approaches that could be tried on various metalworking, plastics processing, woodworking, and paper processing machines, among others. These voluntary standards may be utilized for specific information and guidance for meeting the performance requirements of the OSHA standards.

Be aware that standards are often only a minimum requirement, and that known risks not covered by standards should be considered. Up-to-date risk information may be found in research articles, patents, technical reports, and industry and trade publications.

7.5.6 VARIOUS SAFEGUARDING METHODS

Comparisons between methods of safeguarding are fruitful only in the context of a particular work station where they are to be used. The most positive means for keeping people away from a danger point (e.g., barriers that do not permit contact with the danger point) is the preferred method of safeguarding. Human factors associated with the use of a particular safeguard to a machine are discussed here in relation to how effective that alternative will be in its primary purpose of preventing injury.

7.5.6.1 Safe Opening

A $\frac{1}{4}$ inch or less opening, as shown in Figure 7.5.5, will not permit any part of a worker's body to contact a danger point. Tooling that closes through a small distance is an example of one way to eliminate a hazard by the design of the process.

7.5.6.2 Positively Fixed Barriers or Enclosures

These guards are designed to prevent contact with hazards by placing a stationary obstruction between workers and hazards.

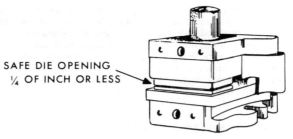

SAFE DIE OPENING
¼ OF INCH OR LESS

Fig. 7.5.5. Safe die opening.

Opening dimensions at different distances from the hazard are given in Figures 7.5.6 and 7.5.7. Fasteners for fixed barriers should be of a type not easily removed, and the material from which the barrier is made should be substantial. Properly maintained barriers are a highly effective safeguard because they do not permit exposure of body parts to dangerous machine actions (Courtney, 1984).

7.5.6.3   Adjustable Barriers

If nonuniformity is present in how the workpiece is presented to a machine, then a variably sized opening adjusts so that the workpiece itself prevents the worker from contacting a danger point. Such barriers may be either manually adjustable or self-adjusting (see Figure 7.5.8).

7.5.6.4   Interlocked Barriers

The goal of an interlocked barrier is to ensure that energy is removed before entry to a hazard area can occur. How frequently the gate is used will indicate how it should function. Two conditions of frequency of use should be considered with interlocked gates. For frequent entry, (Figure 7.5.9), as

Distance of Opening From Point of Operation Hazard, b (Inches)	Maximum Width of Opening, a (Inches)
$\frac{1}{2}- 1\frac{1}{2}$	$\frac{1}{4}$
$1\frac{1}{2}- 2\frac{1}{2}$	$\frac{3}{8}$
$2\frac{1}{2}- 3\frac{1}{2}$	$\frac{1}{2}$
$3\frac{1}{2}- 5\frac{1}{2}$	$\frac{5}{8}$
$5\frac{1}{2}- 6\frac{1}{2}$	$\frac{3}{4}$
$6\frac{1}{2}- 7\frac{1}{2}$	$\frac{7}{8}$
$7\frac{1}{2}-12\frac{1}{2}$	$1\frac{1}{4}$
$12\frac{1}{2}-15\frac{1}{2}$	$1\frac{1}{2}$
$15\frac{1}{2}-17\frac{1}{2}$	$1\frac{7}{8}$
$17\frac{1}{2}-31\frac{1}{2}$	$2\frac{1}{8}$

Fig. 7.5.6. Safe openings between parallel barriers.

Distance of Opening From Nearest Point of Operation Hazard, b (Inches)	Maximum Dimension of Opening, a (Inches)
$< 1\frac{1}{2}$	$\frac{1}{4}$
$1\frac{1}{2}-2\frac{1}{2}$	$\frac{3}{8}$
$2\frac{1}{2}-4$	$\frac{1}{2}$
$4-15$	2

Fig. 7.5.7. Safe square openings.

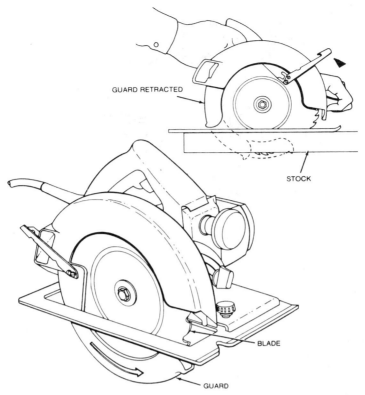

Fig. 7.5.8. An adjustable guard that opens against spring action to accommodate the workpiece.

GUARD RETRACTED

STOCK

BLADE

GUARD

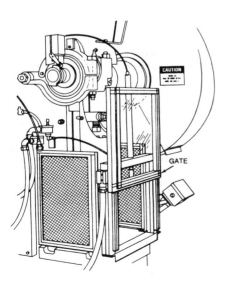

CAUTION

GATE

Fig. 7.5.9. A powered, interlocked gate that must close before dangerous machine motion can begin. The gate is made of clear plastic for viewing the working area.

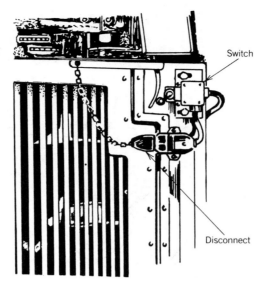

Fig. 7.5.10. An interlocked, sliding door intended for maintenance and servicing tasks.

with manual feeding, gates should either have a latch that holds the gate closed until the dangerous machine action is finished or have a means for stopping machine operation if the gate is opened during machine processing. Precautions should be taken to avoid the use of switching mechanisms and switch configurations that are easily overridden with screwdrivers, pencils, or adhesive tape, or broken so as to make the worker's feeding task more convenient. For infrequent entry, as with cleaning and maintenance tasks (Figure 7.5.10), switch arrangements controlling the machine's power service should be devised to prevent easy override of the switch or closure of the switch by means other than replacement of the barrier. Redundant electrical or other types of additional detection interlock should be considered in high risk situations. Equipment overrun or coasting hazards should also be protected against with interlocks that detect whether such energy sources are completely eliminated before a gate can be opened to permit entry for maintenance and servicing tasks. Visual or auditory warnings if energy remains present with the interlock open (flywheel turning, electrical power on, hydraulic pressure available) are advisable with interlocked barriers. Indications as to the status of the interlock and of machine energy sources when the interlock is open should be given consideration. Easy visual supervision assures the intended use of interlocks.

7.5.6.5 Safe Distance by Presence-Sensing Controls

Noncontact sensing devices can be used to generate an electromagnetic or other form of detection field that is sensitive to hand or body intrusion within a machine hazard zone (Figure 7.5.11). Circuitry connected to the sensor initiates control signals to stop the machine or prevent machine action if such an unsafe intrusion occurs.

In some countries it is also possible to use a presence-sensing device to initiate a machine cycle as well as to act as a safeguard. This mode of operation is governed by very specific regulations. At the date of this chapter, this mode of operation (presence-sensing device initiation) was under study by OSHA, which previously has prohibited its use under any circumstance for mechanical power presses. In a case study comparison, Salvendy, Shodja, Sharit, and Etherton (1983) reported no significant difference in stress measurements in one group of workers between using photo-electronic presence sensing protection and two-hand control device protection.

Other detection principles such as infrared, ultrasonic, and capacitance sensing are available, but care must be taken that the signals processed from the transducer are accurate and precise indicators of human proximity to danger, at any given exposure period and for all persons who may be exposed.

7.5.6.6 Safe Distance by Hand Contact Push Buttons

Keeping both hands away from a point of hazard by using them to depress dual palm buttons in order to start machine action is another method of injury prevention (Figure 7.5.12).

The remote location where the dual palm buttons are placed is determined by both machine and

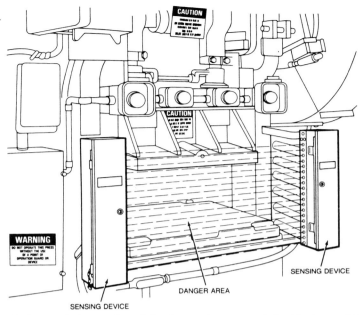

Fig. 7.5.11. A photo-optical sensor intended to issue a stop command should a human hand penetrate the planar sensing field.

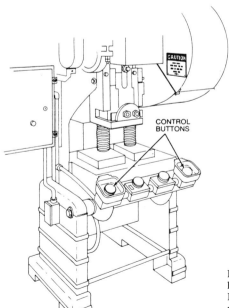

Fig. 7.5.12. Dual palm buttons can be a means of keeping hands at a safe distance from a danger point. Placing the buttons in a high location could cause postures, hand forces, and wrist deviations implicated in repetitive trauma disorders.

human performance factors. Machines do not stop instantaneously, but have a certain brake time. The worker should not be able to reach from the dual palm buttons to a danger point before motion has stopped or the hazard is under control. The human parameter used for average reach speed has generally been either 1.6 or 2.5 m/sec, depending on whether the hazardous reach is considered to have begun with the hand stopped or with the hand already in motion. These values have not always been replicable across variations in reach task imposed by different safety researchers. Musculoskeletal stress caused by location of dual button safety devices should also be considered. The convenience of fixing the buttons above the machine's dies or tooling could create stressful, repetitive reaching tasks. Two-hand control devices have been found to require more muscle effort (4.7 times higher EMG energy) when located above shoulder height than when located in a lower position (Nemeth, 1982).

7.5.6.7 Pullouts and Restraints

These safeguards are attachments to the person's hands, and are either linked to the machine so that the machine pulls the person's hands away from the hazard prior to a dangerous closure or are linked to a fixed point so that movements toward the danger point are stopped short (Figure 7.5.13). Such devices are permitted under current safety regulations in the United States, but their use must be carefully supervised. Human-factors problems may arise owing to inadequate cord adjustment for different size operators using the same device and owing to catch points in the machine tooling that could hold the strap in the tooling when a pullout action begins on the wrist.

7.5.6.8 Awareness Devices

By touch or other human sensory channel, an indication is given that the danger is being approached. These devices include chains hung in front of a hazard that will be brushed against if reaching toward the danger, warning lights that indicate that the power supply to the machine is on, or sirens that signal that a dangerous condition has been created on the machine.

Because they neither stop a person from reaching to a danger point nor initiate a control to deenergize

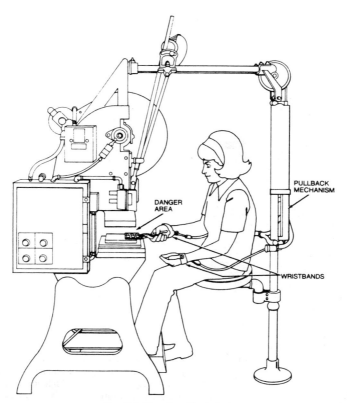

Fig. 7.5.13. A pullout device.

the danger if a danger point is approached, awareness devices are not considered effective primary safeguards. They can, however, provide supplementary information to improve the effectiveness of primary protective devices.

7.5.6.9   Floor Mats

By stepping on a pressure-sensitive floor mat, a person initiates a machine control signal to bring the hazardous machine elements to a safe state.

7.5.6.10   Lockout

To cut off the external power supply to a machine, to block or relieve internal energy sources, and to ensure that these sources remain off are the goals of any lockout method. A study of the various energy sources available is part of this method. Padlocks are one way to achieve external power lockout. One human factor in machine shutdown conditions is that communication between workers may break down. Unless a worker is in control of the restart switch or valve, there is a chance that another worker can accomplish a restart with the co-worker in a danger zone. The method for restart can be made less susceptible to inadvertent human action by requiring special keys and/or use of two or more types of switches.

Special cases within the domain of lockout are situations during machine setup or maintenance when power may be present with a human in or near what would normally be a danger zone. Training, energy reduction, and special supervision are among the considerations that must be carefully reviewed in such cases.

7.5.6.11   Emergency Stop

In case the precautions of separation and responsive control are violated, the elimination of machine power should be readily possible. This can be done with emergency stop buttons or with trip wires (see Figure 7.5.14) or bars near the danger point. Use of these emergency stop devices may lead to a subsequent removal of safeguards. For this reason consideration should be given to a safe restart sequence which assures that safeguards are back in place. When emergency stop buttons are used on control panels with many other control buttons, the color, shape, and location of the emergency stop button should be such as to avoid confusion about use of the button when needed.

7.5.7   HUMAN-FACTORS CONCERNS WITH MACHINE SAFEGUARDS

Lack of human-factors engineering of safety devices and methods plays a part in injuries at machines even when at a first glance might make human factors seem unimportant. In many cases, injuries at machine work stations have not been prevented because although safeguarding was provided, it was not used. When a safeguard is installed it may satisfy the minimum requirement of a safety standard (e.g., an intended purpose to separate the person from the risk), and yet not be effective because it is ill-conceived from a human-factors point of view (e.g., difficult actually to use). An example of this would be a sweep-type device, which is intended to sweep across the opening between a person and a danger point when a machine such as a mechanical power press is closing its dies. The problem with such devices is that they may be a cause of injury if they sweep rapidly. This type of device as a single safeguard on mechanical power presses has not been accepted by OSHA since 1974.

No single, universally satisfactory safeguard has been found that will protect all workers at all times from a machine hazard. Safeguards used in the industrial workplace must be used by people who vary by sex, nationality, size, strength, and age. These safeguards must be used on machines that are old, reworked, and infrequently used, as well as on relatively new machines operating with state-of-the-art control.

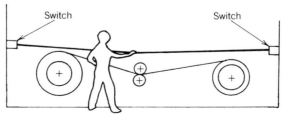

Fig. 7.5.14.   Emergency stop trip wire—it may be possible to pull the wire in such a way that one switch would not detect the emergency condition.

Dissimilar tasks are performed, ranging from repetitive machine loading to occasional clearing of scrap and troubleshooting. Conflicts between the use of safeguards and performance of production tasks in the industrial work environment should be minimized. It would be best if a machine hazard could be eliminated in the design of the process. If this cannot be done, then the human's acceptance and use of a safety device becomes a design criterion. One should avoid safety devices that are satisfactory from an engineering production point of view, but that usually are improperly used by workers.

As machine systems in the workplace evolve and become more sophisticated, the need to provide adequate safeguarding does not diminish. Injuries occur even at machines that have been designed with the intent that human interaction would not normally be needed in their day-to-day operation. Although modern technology permits more machine elements to work autonomously under programmed, microelectronic control, the human presence cannot be discounted. Humans remain proficient and essential in nonroutine machine tasks such as troubleshooting, setup, and adjustment. To do these tasks, workers are required to access machine work zones where an inadvertent startup could be traumatic, such as entry into a robot work envelope to clear a stuck part, or troubleshooting a problem in an automated storage system.

The hazards of stress, fatigue, and cumulative trauma caused by repetitive, lengthy work periods are other elements to be considered in the use of guards and devices for industrial work stations. Motion economy principles, such as the avoidance of excessive reaches, should be used during the study of a safeguard application. It is possible that safety devices will interfere with normal, efficient motion patterns. When such is the case the surcharge imposed by necessary safeguarding must be accounted for in the production standards for the task.

7.5.8 HUMAN-FACTORS ITEMS ASSOCIATED WITH SAFEGUARD USE TRAINING

Untrained and unskilled workers may make dangerous reaching movements that a person familiar with the equipment would not make. However, training is not by itself a total assurance against inadvertent approach to a hazard because mental slips are made by trained workers who momentarily become distracted.

Convenience. Individual preferences, comfort, habit, and job satisfaction can play a role in how a worker uses a safety device provided. The convenience of using a safeguard affects whether people use the safeguard as intended.

No Added Hazard. The misuse of a safety device or an unexpected performance characteristic should be considered. If a gate closes under power, is it creating its own pinching hazard? Can pullout devices catch on a protrusion in the tooling and cause the arm to be jerked back while the hand is caught in the die?

Incentive Production. When work is done on an incentive basis, the worker may be prone to override or remove any safety device that interferes with work or that repeatedly stops work.

Fail-Safe. An ergonomist may not determine the functional control of equipment, but in the process of reviewing how people will use the equipment, it is worthwhile to consider what actions people take if the equipment fails and whether these actions are compatible with fail-safe design. For instance, if a machine can be rather simply restarted after a shutdown due to failure of a safety circuit, it may then be the case that the machine will indeed be restarted and operated with no safety device.

Ease of Maintenance. A safety device that is hard to get parts for, difficult to replace, or otherwise hard to maintain is a candidate for misuse or sabotage.

Verifiability. One technique that has been found useful is painting guards a bright color, such as yellow, making it easy for supervisors to spot guards that are not in place. Pilot lights may also provide indications of safeguard status. When redundant safeguards are used there may be a tendency to assume that all safeguards are operational. Verification that each device is operational is important.

Method of Loading and Unloading. The worker may want to reach into and remove work from a machine before the process is complete. Possible unloading methods to reduce risk include air jet blowouts, mechanical hands, gravity, hand tools, robots, and sliding bolsters. (See Figures 7.5.15 and 7.5.16 for two solutions.)

Error and Inadvertent Use. If an installed safeguard should become ineffective (improper use, failure, inability to operate under unexpected conditions, or other cause), then erroneous or inadvertent use of manual hand or foot controls could lead to an injury.

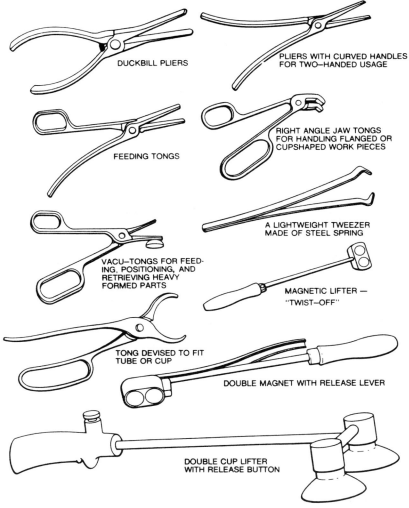

DUCKBILL PLIERS

PLIERS WITH CURVED HANDLES
FOR TWO–HANDED USAGE

FEEDING TONGS

RIGHT ANGLE JAW TONGS
FOR HANDLING FLANGED OR
CUPSHAPED WORK PIECES

A LIGHTWEIGHT TWEEZER
MADE OF STEEL SPRING

VACU–TONGS FOR FEED-
ING, POSITIONING, AND
RETRIEVING HEAVY
FORMED PARTS

MAGNETIC LIFTER —
"TWIST–OFF"

TONG DEVISED TO FIT
TUBE OR CUP

DOUBLE MAGNET WITH RELEASE LEVER

DOUBLE CUP LIFTER
WITH RELEASE BUTTON

Fig. 7.5.15. Hand tools for keeping hands out of danger areas. Caution should be taken to avoid tools that cause excessive fatigue or that create conditions of use associated with repetitive trauma disorders.

Visibility. A safeguard that diminishes visibility into a machine work area, whether by color (reflective vs. nonreflective) or position (vertical vs. horizontal bars), will be used less often and less effectively than a safeguard that does not thus interfere. If work is not visible, a need to reach into a hazardous position may be created.

Posture. If a worker is in an unbalanced seated or standing position, an inadvertent step onto a foot switch or a reach into a danger zone to catch oneself while falling are potential factors in injury causation. Seating height and postural stability are factors to consider when access to a work area is possible. Figure 7.5.17 illustrates an unstable working posture. Knee room and foot rests contribute to postural stability when seated. Standing while using a foot control is not recommended in some cases.

Sabotage. The possibility that a worker will ruin or destroy a safety device that is not accepted can render worthless a well-intended safety expenditure. There persists a human tendency to disregard or underestimate the risk of injury on a machine task. Such attitudes become the rationale for eliminating safety devices that are not adapted to easy, convenient use.

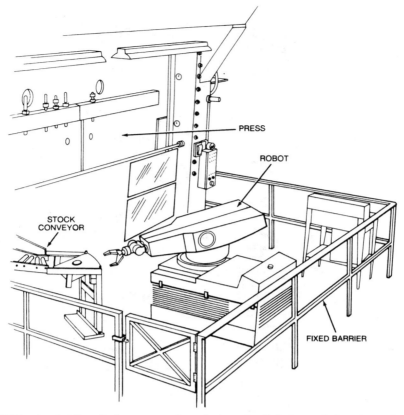

Fig. 7.5.16. A robot for relieving a person from the hazards of placing workpieces into a dangerous machine work area.

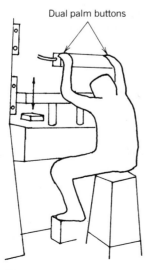

Dual palm buttons

Fig. 7.5.17. Unstable working posture.

7.5.9 CONCLUSION

Giving consideration to human factors in the selection, design, and installation of safety devices will have as its final proof the result that the device is used as intended, that injury is indeed avoided, and that the user of the safeguard has confidence enough in the effectiveness of the safeguard that its use is accepted. The unfortunate side effects of some devices may be that while protecting against a sudden, traumatic injury they may be, over time, causing musculoskeletal injuries. The human-factors specialist must be cognizant of the consequences of both machine safety device use and misuse.

REFERENCES

Armstrong, T. J. (1981). Carpal tunnel syndrome and the female worker. *Transactions of the Forty-third Annual Meeting of the ACGIH,* Portland, Oregon, May 24–29.

Courtney, A. J., and Ng, M. K. (1984). Hong Kong female hand dimensions and machine guarding. *Ergonomics, 27*(2), 187–193.

Garg, A., and Kohli, K. (1979). Human factors in machine design. *Journal of Mechanical Design, 101*(10), 587–593.

Nemeth, S. E. (1982). Ergonomic evaluation of two-hand control location. *Human Factors, 24*(5), 567–571.

Percival, N. (1977). *The role of ergonomics in the design of British machine tools* (pp. 268–275). Geneva: ILO, Occupational Safety and Health Series No. 35.

Salvendy, G., Shodja, S., Sharit, J., and Etherton, J. (1983). A case study of the occupational stress implications of working with two different actuation/safety devices, *Applied Ergonomics, 14*(4), 291–295.

BIBLIOGRAPHY

As an aid to those seeking further information on methods of safeguarding, the following nonexhaustive list of references is offered:

American National Standards Institute *safety standards for the workplace* (B 11). New York.

Blundell, J. K. (1983). *Machine guarding accidents: Trial lawyers guide.* Columbia, MD: Hanrow Press.

British Standards Institution (1975). *Safeguarding of machinery* (BS 5304). London.

Grandjean, E. (1980). *Fitting the task to the man.* London: Taylor and Francis.

McElroy, Frank, Ed. (1974). *Accident prevention manual for industrial operations,* 7th ed. Chicago: National Safety Council.

National Safety Council (1981). *Guards illustrated,* 4th ed. Chicago.

Occupational Safety and Health Administration (1980). *Concepts and techniques of machine safeguarding.* Washington, DC: U.S. Department of Labor.

Occupational Safety and Health Administration (1984). General industry safety standards, subpart "O" 1910.211–1910.219. Washington, DC: U.S. Department of Labor.

VanCott, H. P., and Kinkade, R. G. (1972). *Human engineering guide to equipment design,* revised ed., Washington, DC: U.S. Government Printing Office.

CHAPTER 7.6

PERSONAL PROTECTIVE EQUIPMENT

JOHN B. MORAN

RICHARD M. RONK

National Institute for Occupational Safety and Health
Morgantown, West Virginia

7.6.1 INTRODUCTION

Personal protective equipment, such as respirators or hearing protectors, must be considered as the last line of defense for workers. Every attempt should be made to eliminate the potential of exposure to hazards to the individual worker through administrative and/or engineering control measures such as job rotation or ventilation, respectively. There are, of course, many tasks that are infrequent in nature and often associated with servicing, maintenance, or installation of equipment, where administrative or engineering controls are not practical or feasible. It is in those instances where personal protective equipment can play the key role of protecting the worker from exposures to hazardous substances or physical agents such as noise.

This chapter is devoted to a review of a broad range of personal protective measures available. Its aim is to provide a fundamental understanding of such devices when used in various industrial settings. As such, it is intended to serve as an introduction to the selection, use, and maintenance of personal protective devices. A fundamental assumption is that administrative and/or engineering control means of reducing or eliminating potentially hazardous exposures have been considered and applied where appropriate and feasible.

The reader is cautioned that there are many regulatory standards promulgated by regulatory agencies such as Occupational Safety and Health Administration (OSHA), Mine Safety and Health Administration (MSHA), and Nuclear Regulatory Commission (NRC), which establish specific requirements for the selection and application of personal protective devices. For example, many specific and general OSHA standards require personal protective equipment where forseeable hazards are present. These specific standards define the hazards present and establish the protective equipment required for workers exposed to such hazards. Further, both regulatory standards and consensus standards governing the performance of such devices have been established. These standards and requirements do change as new exposure standards, for instance, are promulgated. Therefore, the reader must ascertain the current applicable standard governing the specific situation. To serve as a reference point, this chapter refers to standards pertinent to specific devices as they presently exist.

7.6.2 PERSPECTIVE

Personal protective equipment includes a very broad range of devices produced and marketed by a large number of diverse suppliers. Such devices range from complex and expensive self-contained breathing apparatus (SCBA) to simple vinyl disposable gloves. Although there is little information available about the size of the protective equipment business, respirators alone were estimated to represent a $285 million market in 1980 and were used by 15% of workers engaged in mining, manufacturing, and construction (approximately 5 million persons) (NIOSH, 1982). Selection of personal protective equipment does not necessarily follow the usual stages of engineering decisions because information vital to making a rational choice is frequently lacking.

Personal protective equipment seldom provides complete protection against a significant stress; rather it serves to attenuate the stress to a level which is acceptable. For example, if protection from a falling pipe wrench is desired for a pipe fitter working 10 ft below his fellow workers, then it would seem desirable to express the stress in engineering terms, dividing by the allowable stress on the head/neck/shoulder system and selecting the hard hat providing the required attenuation. Unfortunately, neither the allowable stress nor the degree of attenuation is known for most applications of personal protective equipment.

Biomechanical research to establish allowable stress is extremely com .ex and costly and there is little, if any, economic incentive for protective equipment manufacture 5 or equipment purchasers to conduct inquiries. Government protective equipment research is also limited, with a majority being conducted for defense purposes, much of this having little transfer to nonmilitary applications.

One researcher in head protection has said "the injury criterion currently used to evaluate industrial safety helmets has no biomechanical justification" (Cook, 1980). Other researchers have called the criteria for respirator certification "technological anachronisms." Many of the criteria for "attenuation" or "performance" of personal protective equipment are based upon design failures that have been frozen into standards for equipment by consensus groups. These groups are chosen largely from manufacturers or distributors of safety equipment who have little, if any, access to research facilities in their organizations and access to only a limited amount of government or academic research. Therefore, it is not surprising that the standards reflect the "art of the profitable" or practical and not necessarily the most effective "state of the art."

Nearly all personal protective devices are manufactured to comply with a "standard." Such performance standards range from federal standards for respirators (30 Code of Federal Regulations Part

11) to consensus standards such as ANSI Z87.1 for eye and face protection devices. Most consensus performance standards for protective devices have, by reference, been adopted by regulatory agencies as establishing acceptable performance criteria. Devices that are stated as complying with such consensus standards are, in essence, self-certified by the individual manufacturer. Studies by the National Institute for Occupational Safety and Health (NIOSH) in the mid-1970s, however, showed that not all such devices in fact met the requirements of the applicable consensus standard. For example, 18 of 21 Class B helmets tested failed to meet the minimum requirements of the ANSI Z89.2 consensus standard (Cook and Groce, 1975). Publication of the results of the NIOSH tests and the recent creation of an independent private certification laboratory (SEI, Safety Equipment Institute) have likely resulted in improved performance standards compliance, however. All devices stated as meeting a consensus standards requirements are so labeled by the manufacturer. Similarly, devices certified by SEI carry the SEI label.

The most common personal protective devices tested and certified by the federal government are respirators. Regulatory agencies require that such approved devices be utilized where respiratory protection is required, and where an appropriate certified device is available. All approved respirators carry a label issued by NIOSH and MSHA which states the limitations of use.

Some protective devices, principally protective clothing such as gloves, splash aprons, and encapsulating suits, are not tested for compliance with standards because none exist to date.

Personal protective devices are largely marketed by safety equipment distributors. Such distributors often represent many manufacturers. Contrary to the fact that personal protective devices are frequently lifesaving equipment such as escape respirators, protective equipment is largely marketed as a commodity item and sold principally on price rather than performance. This is fostered by the fact that the performance standards are largely minimum acceptable performance standards; thus superior performance is not generally marketable for superior price. The selection of personal protective equipment, therefore, should be exercised with care and with preference being given to the supplier and/or manufacturer that can provide information specific to the intended use situation.

The subsequent sections of this chapter are devoted to specific protective devices. The concluding section examines aspects of human factors in personal protective equipment use. Additionally, each section also addresses the negative impact on the user as a consequence of the use of protective equipment. This is an essential consideration, because the use of nearly all protective devices reduces the effectiveness of one or more of the human systems. Respirators, for example, reduce peripheral vision and impede voice communications, increase respiratory and cardiac stress, and may impose psychological stress (Harber, 1984).

7.6.3 HEARING PROTECTION

OSHA standards (29 CFR 1910.95) establish maximum noise levels permitted and establish the requirement for hearing protection programs, including hearing protective devices, when the specified levels are exceeded. The basic philosophy behind the protective equipment component of the standard is that the hearing protection devices should reduce noise levels at the ear to a level at or below the stated standard. This is a common approach to the application of protective devices, which span a wide range of designs including combinations with other protective devices such as hard hats.

There are, however, two basic types of hearing protectors available: those that cover the ear (muffs) and those used in the ear canal, usually referred to as earplugs. The essential feature of relevance, however, is the noise attenuation characteristics.

Prior to 1974, hearing protectors were tested in accordance with ANSI Z24.22-1957. Subsequently, a new method was adopted as ANSI Z3.19-1974, which incorporated the substitution of third-octave bands of noise instead of the discrete tones used in the Z24.22 standard. In addition, reverberant test rooms replaced the earlier anechoic test rooms. Mean attenuation of hearing devices varies with frequency of noise to which exposure is occurring. For example, an earplug may exhibit 25 dB attenuation at a frequency of 250 Hz while exhibiting 35 dB attenuation at 8000 Hz. Earmuffs, on the other hand, may exhibit a 15 dB attenuation at 250 Hz and 37 dB at 8000 Hz.

The present OSHA standards require that no exposure to continuous sound over 115 dB(A) be permitted at any time. Additionally, the standard specifies methods of acceptable noise level measurement that can be achieved with either sound level meters for real time measurements or personal noise dosimeters for integrated work-shift exposure. Noise levels must first be determined before appropriate hearing protection can be selected.

All reputable hearing protector manufacturers provide noise attenuation curves for these protective devices. They should be consulted as an essential step in the process of hearing protector selection. Helpful general guidance with regard to the effectiveness of noise attenuation is available as a result of EPA regulations requiring hearing protector labeling. Key to this labeling is the noise reduction rating (NRR). The NRR rating is based upon noise attenuation tests for continuous noise. In most industrial cases, the attenuated noise exposure to the employee can be reasonably approximated by subtracting the NRR value from the A-weighted noise level. If, however, noise exposure is largely below 500 Hz or is impulse noise, the NRR value approach is less reliable. Typical NRR values for various types of hearing protectors are as follows:

Earmuffs	15 to 25 dB
Premolded and custom molded earplugs	10 to 25 dB
Foam earplugs	29 dB
Disposable earplugs	20 dB

Additional practical factors must be considered in regard to hearing protectors. Earmuff performance often varies depending on whether the headband is worn over the head, under the chin, or behind the head. For example, one such protector offered 26 dB attenuation at 2000 Hz when worn behind the head, and 39 dB attenuation when worn over the head. Earplug performance obviously varies as a function of conformity or fit to the ear canal.

Earmuffs are generally less comfortable than plugs if worn for extended periods such as a whole work shift. Improper fit is most often the result of attempts by the wearer to make this device more comfortable. The attenuation data provided by manufacturers are data obtained in a "perfect" environment, not a working environment. Therefore, care must be taken to provide a comfortable protector to the user, insofar as is possible, because the manufacturer's performance will rarely be realized in the real work setting. Durability, cost, sanitation-hygiene characteristics, and duration of required use both during a shift and during subsequent shifts are additional factors that should be considered in hearing protector selection.

Compensation costs associated with occupationally induced hearing loss are increasing significantly. Noise exposure should be reduced through administrative and/or engineering controls wherever possible. Where hearing protection is required, however, care should be taken to make an informed selection and to ensure an adequate continuing protection program including visitors to "noise" areas. The days of merely providing every employee a set of earplugs are over.

7.6.4 HEAD PROTECTION

Head protection, in the form of the common construction workers' hard hat, is well known. Such protection is often mandated as well by regulating agencies such as OSHA which, under provisions of 29 CFR 1910.135, requires that such head protection be worn where protection to the head from impact and penetration from falling or flying objects or from limited electrical shock and burn is necessary. Such head protective devices must comply with the ANSI Z89.1-1969 consensus standard. Subsequent to the promulgation of the OSHA regulations, an additional improved ANSI standard, ANSI Z89.2-1971 Class B, was issued. This standard relates only to helmets for electrical protection and is now widely used by helmet manufacturers.

There are four categories of industrial head protection devices based on the exposure that may be encountered:

Class A. General industrial head protection devices with limited dielectric protection

Class B. Industrial protective helmets for electrical workers (high-voltage protection)

Class C. Aluminum safety headwear with no dielectric protection

Class D. Fire fighters' helmets

These devices all have two basic elements in common: an outer shell and an inner suspension system. In order to function as designed, the suspension system must be adjusted so that the helmet shell does *not* rest on the head.

Selection must be based on the hazard for which protection is sought. A Class B helmet, for example, meets all the basic requirements of a Class A, yet possesses specific high-voltage dielectric properties to provide protection from electrical contacts. Clearly, a Class C aluminum helmet should never be used where electrical contact is possible. By far the most common helmet in use today is the class B helmet.

Head protection devices are available with a wide range of attachable accessories. Hearing protectors, face shields, welders' face shields, head lamps, and winter liners are the most common. Caution, however, must be exercised in the selection, attachment, and use of these accessories. For example, a winter liner may have a metal zipper that will void the Class B electrical helmet rating and render the helmet unsuitable for an electrical application. Earmuffs should be those specifically adopted for use with the helmet, usually with a swivel mounting on the helmet over each ear.

It is important to note that helmets meeting the noted standards are designed for protection from objects falling directly on the head. Moreover, industrial helmets do not provide protection appropriate for vehicular use. They therefore provide little protection from blows in the horizontal plane. Further, there are many additional helmets available that do not and are not intended to meet these standards. These are often referred to as skullcaps or bump caps. Such devices are intended to provide minimum abrasion or cutting protection to the head. They afford little or no protection from impact.

Helmets often become "personal" to workers and are adorned with names, logos, and so on. Helmets, however, require routine inspection for cracks, breaks, and failure of the suspension, for example. A cracked helmet or a suspension that has one of the four or six suspension attachment points broken offers little protection. Periodic inspection should be required, therefore. When the shell is not damaged, replacement suspensions are available and should be used where the individual has a personal attachment to his/her helmet.

Helmets do not, it must be noted, provide protection against all possible impacts. The ANSI standard requires, for example, that no substantial contact occur between the helmet shell and the suspension when an 8 lb steel ball is dropped on the top of the helmet mounted on a rigid head form from a height of 5 ft. Caution should be exercised with regard to such limitations.

7.6.5 RESPIRATORY PROTECTION

Of all the personal protective devices utilized in the workplace, respiratory protective devices are by far the most complex, diverse, and demanding with regard to proper selection and use. In many instances, such as the fire-fighting environment, these devices are literally life sustaining. In many other applications, however, the misuse or misapplication may not be apparent for several years. An example of this is when exposure to asbestos occurs because of improper respiratory protection and the long-term consequences of the disease manifest themselves 20 to 30 years later. There may also be disagreement among the regulatory agencies, respirator manufacturers, respirator users, and the testers and certifiers of respirators with regard to the degree of respiratory protection required or provided in a specific exposure situation. Sources of advice regarding proper respirator selection for specific exposure situations should be sought from the following:

Respirator manufacturers who market NIOSH/MSHA-approved respirators.

Suppliers of approved respirators if such suppliers have available data or direct supporting guidance from the manufacturers they represent.

NIOSH, through the annual *NIOSH Certified Equipment List,* or by calling the Chief, Certification Branch, Division of Safety Research, NIOSH in Morgantown, West Virginia.

OSHA area, regional, or headquarters offices. For organizations coming under other regulatory agencies, such as MSHA for mining or NRC for nuclear plants, call the appropriate agency office.

The *NIOSH/OSHA Pocket Guide to Chemical Hazards,* DHEW (NIOSH) Publication No. 78-210 or latest revision thereof.

Specific regulatory agency regulations such as OSHA's 29 CFR 1910.1001 standard for asbestos exposures. Substance-specific regulations by OSHA normally contain a respiratory requirements section. The OSHA general industry respirator requirements are contained within 29 CFR 1910.134, which lists the minimum requirements for an acceptable respirator program.

A Guide to Industrial Respiratory Protection, NIOSH publication 76-189, which contains very specific guidelines with regard to establishing an acceptable respirator program.

ANSI standards Z88.2 and Z86.1, available from most respirator manufacturers and suppliers or from American National Standards Institute, 1430 Broadway, New York, NY, 10018.

Respirators are of only two major types: air purifying and atmosphere supplying. *Air-purifying respirators* are those that remove specific contaminants from the atmosphere immediately surrounding the wearer. Examples would be removal of dust in the air by a filter and removal of an organic vapor by a sorbent cartridge. Cartridges may be designed for removal of dusts, mists, fumes, radionuclides, organic vapors, acid gases, alkali gases, or combinations thereof. The wearer, in essence, breathes through a filtering device that does not supply adequate oxygen in an oxygen-deficient atmosphere. *Atmosphere-supplying respirators* are those in which clean, breathing-quality air is delivered to the wearer from an external source. Air may be supplied from compressed air tanks, blowers, or air compressors, or be recirculated through appropriate scrubbers with oxygen recharging (referred to as closed-circuit SCBA). An example would be a fire fighter's SCBA.. Within each of these major categories there are many additional variations. Facepieces may be quarter masks (mouth and nose covered), half masks (nose, mouth, and chin covered), full facepieces (mouth, nose, chin, and eyes covered), helmets with face shields, and hoods, for example.

Respirators range in cost from less than $1 for a disposable dust mask to over $2000 for a long-duration self-contained rebreather. No one respirator is suitable for all possible applications, so selection of a single type does not assure universal application.

Fundamental to the proper selection of a respirator is that the following be known:

The toxic agent to which exposure is expected to occur.

The levels of expected exposure to the toxic agent.

The expected duration of exposure to the toxic agent.

The level of oxygen present in the atmosphere to be entered. Principal interest here is a determination as to whether oxygen depletion *or* enrichment is possible.

The permissible exposure limit to the agent. Sources include OSHA, MSHA, and NRC standards and other knowledgeable sources such as EPA, NIOSH, the ACGIH TLV tables (issued annually), Material Safety Data Sheets (MSDS's) provided by the material supplier, and so on. Where no standard per se exists, it is not safe to assume that the agent is nontoxic. Many new agents have been marketed since the OSHA standards were promulgated in 1971. Do *not* assume an agent is harmless unless a knowledgeable source so states.

The following additional information should also be sought although data sources available may vary with regard to the extent of such information.

If the agent is a gas or vapor, does it exhibit human sensory warning properties at levels at or below the exposure limit? For example, the human can easily smell hydrogen sulfide at very low concentrations. However, at very high concentrations, the olfactory senses are quickly defeated and death can occur.

Does the agent possess an eye irritation potential?

Is the agent absorbed through the skin?

What is the lower flammability limit for the agent?

What is the IDLH concentration of the agent? IDLH refers to the immediately dangerous to life or health concentration. IDLH is defined by NIOSH to include both acute (short-term) and chronic (long-term) consequences of exposure. Thus NIOSH would consider exposure to a known or suspect carcinogen to be an IDLH situation.

Regulatory agency definitions of IDLH may be different and may not include the chronic consequences component.

Is there the possibility of poor respirator cartridge sorbent efficiency at IDLH levels or below? (This information may be difficult to obtain.)

Armed with the above information, one can begin the respirator selection process. The discussion that follows is directed toward development of a fundamental understanding of the selection process and the relevance of the information previously noted. It does not, however, constitute a fully adequate basis for respirator selection, for such would require more space than available here. Knowledgeable manufacturers, suppliers, and other sources previously noted can provide such specific guidance.

The fundamental principle behind the use of any respirator is that the wearer shall not breathe the toxic agent *at levels exceeding* established standards or safe levels. In the simplest of examples, the respirator simply filters toxic dust present in the workers' environment so that the air inhaled by the wearer, inside the respirator facepiece, is at or below the exposure limit for that toxic dust. The ratio of the toxic agent concentration in the atmosphere and the concentration inside the respirator facepiece after filtering is referred to as the respirator protection factor (PF).

Protection factors have been assigned to classes of respirators for use in the workplace.

Various sources give varying protection factors based on laboratory studies of test panels or worker populations. The published values should be considered guidelines because no fully acceptable test method has been developed and therefore no fully valid body of test data exists.

Recent studies by NIOSH suggest that the high protection factor for powered air-purifying respirators demonstrated in the laboratory is not achieved in the field and that much lower values may be more appropriate to the workplace. These lower values are suggested as a guide until performance of these respirators is significantly improved. Escape respirators do not have assigned protection factors, because their use is appropriate only in an emergency situation and exposures are expected to be for very short durations during the egress activity. An example is the belt-worn, chlorine-escape, mouth-bit respirator.

The application of protection factors to the selection process is evident. If the exposure limit to substance X is 10 parts per million and the work atmosphere concentration is 100 parts per million, a suitable respirator with a protection factor of 10× is required.

Based on the substance data collected, the following factors can be included in the respirator selection decision-making process.

The IDLH concentration has as its purpose the establishment of an exposure concentration below which a worker could escape within 30 min without injury or irreversible health consequences in the event the respirator fails. If expected exposure concentrations are at or above the IDLH concentration, only SCBA operating such that a positive pressure is always present inside the facepiece is recommended. Air-purifying or supplied-air respirators are not permitted where exposures are at or above the IDLH concentration.

Oxygen concentration in the exposure environment is clearly of importance where an air-purifying respirator may be considered. These respirators, which merely purify the atmosphere surrounding the wearer, cannot be used where oxygen depletion has occurred or may occur. Only SCBA or supplied air respirators combined with SCBA emergency egress devices may be used in such atmospheres because the source of breathing air is external to the immediate exposure environment.

Warning properties of the agent are important with regard to respirator selection. For example, if a worker is wearing an air-purifying respirator for protection against a gas or vapor with poor warning properties, failure of the respirator or saturation of the sorbent cartridge would not be detectable. Therefore, air-purifying respirators are not appropriate in atmospheres containing toxic substances with warning thresholds above the exposure standard limits. Equally important, as in the hydrogen sulfide case, materials that cause very rapid olfactory fatigue should be approached with caution with regard to the use of air-purifying respirators. If releases of the substance may result in high exposure, air-purifying respirators should not be used.

Eye irritation characteristics are important in respirator selection as well. Clearly, for a substance with eye irritancy potential a respirator with a full facepiece should be used.

Agents that are absorbed through the skin pose potential for serious and often fatal consequences on exposure primarily to liquid substances but, on occasion, to vapor as well. In such instances, air-supplied, fully encapsulating garments provide the maximum protection, although such are not certified as respirators. Material selection is important as well, however, and is addressed in a subsequent section of this chapter.

Exposures to concentrations of a substance above the lower flammable limit are considered IDLH environments. In addition to other protective measures such as nonsparking tools, only positive pressure SCBA or positive pressure supplied-air respirators combined with positive pressure SCBA emergency egress devices are allowed.

Sorbent efficiency is an important consideration in respirator selection. For example, an air-purifying respirator with sorbent cartridge may not be the respirator of choice if sorbent breakthrough may be expected to occur in less than the time required to perform a necessary task at exposure concentrations at or below the IDLH concentration. In such instances, air-purifying respirators are not permitted for escape use either. Sorbent efficiency data is often difficult to obtain. Most reliable respirator manufacturers can provide guidance, however, and their input should be solicited.

Upon selection of the appropriate respirator class and type, air-purifying organic vapor, for example, consideration of additional factors is appropriate and necessary. Under the OSHA regulations, if a respirator is used in the workplace, a respirator program conforming to the requirements specified in 29 CFR 1910.134 is required. Briefly, the requirements are as follows:

1. Written operating procedures governing selection and use shall be established.
2. Selection shall be on the basis of the hazard to which the worker is exposed.
3. The respirator user shall be trained regarding use and limitations of respirators.
4. Respirators shall be regularly cleaned and disinfected.
5. Respirators shall be stored in a convenient, clean, and sanitary location.
6. Respirators shall be inspected and maintained.
7. Work area surveillance is required to assure that the basis for selection remains valid. Where operational or process changes occur, reevaluate the respirator selection logic.
8. Regular inspections and evaluations are required to ascertain program effectiveness.
9. Persons required to use respirators must be medically certified as being capable of doing so.
10. Respirators shall be approved by NIOSH–MSHA.

Many OSHA regulations of specific substances require quantitative respirator fit testing as well. This is a dynamic process where a small room or closed portable envelope is filled with a test aerosol and the actual protection factor of a respirator worn by the individual to whom the respirator will be assigned is determined by comparing chamber concentration to the in-mask concentration. The intent is to permit assignment of the best fitting respirator to an individual.

Qualitative fit testing is required for all respirator programs. This involves exposing a respirator wearer to banana oil or irritant smoke and noting a subjective response in the wearer. Because this test usually does not include a measurement of the challenge concentration and the subjective response threshold is quite variable, the test does not lead to a quantitative determination of fit. Again, selection of the best fitting respirator and recognition of proper fitting adjustments are the objectives.

The single most important testing aspect when using air-purifying respirators is the positive–negative pressure test, which should be performed at every respirator donning. This test involves blocking the exhalation port with the hand and subsequently exhaling into the facepiece, creating a pressure. Missing inhalation valves, ineffective inhalation valves, improperly seated cartridges, and headband fit can all be quickly checked. Next, the cartridges are covered with the hand(s) and the wearer inhales, creating a negative pressure inside the facepiece. In this test, the exhalation valve conditions and facepiece condition including fit can be quickly determined.

Where supplied-air respirators are used, the quality of air delivered to the wearer is important. Requirements are established for both the air quality and the routine monitoring of air compressors which supply the breathing air. These, again, are in the OSHA standards under 29 CFR 1910.134. Manufacturers of air-supplied respirators can also provide comprehensive advice on this issue.

Common problems associated with the use of respirators include costs, discomfort, increased respiratory and cardiac stress, psychological stress, reduction of certain sensory inputs, communication difficulties, proper cartridge or filter change frequencies, corrective eye wear, and beards or long sideburns.

Cost issues should be examined not solely on the basis of the respirator cost. The actual respirator is likely the least expensive part of a respiratory protection *program*. Inspections, cleaning, maintenance, repairs, and so on are normally the major cost factors.

Discomfort and communication difficulties will normally increase the task time for workers wearing respirators. Owing to communication difficulties and 85 dB(A) noise levels in most supplied air hoods and helmets, the wearer is at increased risk of injury from other potential hazards present in the workplace. A related issue, of course, is reduced sensory perception levels, such as reduced peripheral visual fields. These factors should be recognized and measures taken to mitigate the potential consequences.

Cartridge or filter change intervals are difficult to predetermine, because life of these components is related to exposure concentration. If they are used for protection against materials with good warning properties, the wearer is usually aware that the sorbent cartridge is saturated and breakthrough has occurred. High breathing resistance is the change indicator for particulate filters. As a general rule, filters or cartridges should be targeted for change at the end of the normal work shift. More frequent changing of single-use disposable dust masks may be desirable and even required.

Workers who require corrective eye wear and who must use a respirator with a full facepiece can obtain special temples adaptable to the facepiece interior from the respirator manufacturer. A full facepiece respirator may not be worn over standard spectacles because the facepiece fit is compromised. Many authorities also recommend that contact lenses not be worn in full facepiece respirators because they may become difficult to retain. Special attention to workers who wear corrective eye wear is required where use of emergency escape breathing apparatus may be necessary.

For fit compromise reasons, the same logic applies to the use of respirators by individuals with beards or long sideburns as in the case of spectacles. Beards and long sideburns that interfere with respirator fit are not permitted.

Respirators may be but one component of a multiple protection ensemble such as chemical splash goggles, hearing protectors, head protection, gloves, and so forth. In such instances, it is essential that the respirator be carefully evaluated to assure that its purpose is not defeated by another protective device. Of particular concern is the use of a respirator and a welding helmet. Specifically, modified respirators have helped reduce this interface problem. Special combinations that eliminate this potential problem and optimize personal protection do exist in the marketplace. An example is the abrasive blasting hood that provides respiratory protection and protection to the eyes, face, head, and upper torso.

Although respirators may be required for special tasks such as cleaning of pressure vessels or for emergency escape, it should be evident that a comprehensive and adequate respiratory program for routine use of respirators is complex, demanding, and costly and presents a number of disadvantages to the worker that can affect morale and productivity. These issues alone should serve to reinforce the preference for administrative and/or engineering controls as a priority above respirators.

7.6.6 EYE AND FACE PROTECTION

Eye and face protective devices cover a broad range of products including safety glasses, welding goggles, and face shields, all relatively simple devices, to laser goggles, which are extremely complex. Eye and face protection is required where potential for injury can be foreseen by regulatory agencies. OSHA standards are contained within 29 CFR 1910.133, which contains many specific requirements such as being durable, capable of being disinfected, and appropriate for the protection required. Devices manufactured in compliance with the ANSI 87.1 standard are required.

It is of interest to note that current NIOSH studies suggest that over 900,000 serious eye injuries occur annually in the occupational setting and that well over 50% of eye injuries occur to individuals who are not wearing eye protection devices (Cook and Fletcher, 1977). It is evident that a majority of those suffering eye injuries did not perceive that a risk was present. Organizations with mandatory eye protection programs evidence a reduced incidence of such injuries. In addition, new eye wear technology has begun to emerge, polycarbonate safety lenses standing out as a unique example as the impact properties of polycarbonate are vastly superior to glass, yet the spectacles are substantially lighter than glass spectacles.

However, the introduction of this new technology has not been without problems. When first introduced, polycarbonate lenses were available only as plano lenses (no correction). Thus workers who required corrective lenses could not be provided with this superior protection technology. Recent advances have largely resolved this particular problem, however.

Eye and face protection should be chosen based upon the hazard. The ANSI Z87.1 standard provides an excellent guideline for such selection, as do most of the protective eye and face device manufacturers and their suppliers. Proper initial selection requires an analysis of the potential hazards present at the work site and consultation with the manufacturer or supplier. For routine exposures, the process is fairly straightforward, with the major issues of concern being cost and delivery times for prescription lenses. Proper fitting is necessary, of course, as is well known to anyone who wears corrective street eye wear.

There are a number of more complex situations, however. Welders require special protective "welders' lenses and plates." These vary in optical density, depending upon the welding or cutting process. Plate and lens densities are classified by shade number. An excellent guideline is presented in table form in the OSHA standard contained within 29 CFR 1910.252, Welding, Cutting, and Brazing. Welders' plates and lenses, as with spectacles, must also meet other requirements such as impact resistance.

Protective eye wear is available in a wide range of colors ranging from clear to yellow to rose. Each possesses special light transmission characteristics tailored to specific applications. Extreme care should be exercised in selecting the special colored lenses because the visible light transmission spectra and more importantly the IR and UV transmission spectra are not the same as for glass. Glass has a rather high visible light transmittance and low UV and IR transmittance. Thus a special lens that might be used to enhance the yellow spectrum may result in much greater UV exposure to the eye from welding arcs, even at a distance. Reputable eye wear manufacturers can provide transmission spectra and discuss these issues to aid in proper selection.

Polycarbonate has a much greater transmittance of UV and IR than does glass, which it is replacing. UV inhibitors have been developed to improve the UV performance but the IR spectrum remains difficult to resolve. In addition, glass lenses, often referred to as welders' glasses, are commonly used in the welding environment by welders' helpers to reduce eye exposures to indirect or peripheral welding arc emissions. These glass lenses are a distinct green color. Polycarbonate lenses have been marketed in the past in the same green color. Despite the warnings on the packages that they were not suitable for welding arc exposures, many individuals have used them for such.

Protection of the eyes from exposure to laser sources is complex. Laser goggles lenses are designed to be highly absorptive in very narrow frequency bands. Reductions in allowable transmission for very powerful lasers vastly exceed the reductions required in the most intense welding operation. Selection of proper eye protection from lasers requires specific knowledge of the laser type, emission wavelength, and power output, at a minimum. Those manufacturers who provide laser protective eye wear and the laser manufacturer are the only sources that should be relied on to recommend such eye wear. The calculation of the required optical density at one or more frequencies is not simple, nor straightforward. Requests for appropriate protective eye wear requirements should be requested from the laser manufacturer in writing. Recommendations by NIOSH have been developed for more specific spectral exposure selection limits that would assist in these situations. In addition, ANSI Z136.1 provides additional insight.

As is the case with protective devices, eye and face protective devices affect the wearer. Although some limited decrease in the visual field occurs with simple spectacles, such is usually minimal. However, as the degree of protection increases, such as with the addition of side shields or with chippers' goggles, loss of peripheral vision begins to occur and the discomfort index increases. Lens density required for protection against welding flash, lasers, or IR transmittance may also reduce visible light required for task performance or emergency actions to unacceptable levels. The work station layout and job design should consider the imposition of such limiting factors on the worker.

Overall, the selection of suitable and appropriate eye and face protection can be readily accomplished through job analysis surveys and discussion with competent suppliers and manufacturers. Unique protective circumstances, such as eye wear to protect workers from lasers or unusual emission sources, can usually be satisfactorily resolved by the technical experts of the major manufacturers. Most importantly, the majority of industrial eye injuries occurring today are as a consequence of the failure to wear protective eye wear. Review of the noted ANSI standard is strongly encouraged as an important asset in the selection process as well. It will be noted therein, for example, that face shields are to be utilized only over safety spectacles, an important protective measure not commonly recognized.

7.6.7 HAND PROTECTION

There are a vast array of hand protective devices available, most of which are tailored to specify exposure hazards such as chemicals, cold, heat, fire, abrasion, cuts, punctures, and electricity. Standards exist only for electrical linesmen's rubber insulating gloves, which are embodied within ANSI J6.6 covering five voltage classes, designated 0, 1, 2, 3, and 4, with proof test voltages ranging from 5000

V (rms) for the 0 designation to 30,000 V (rms) for 4. Tests conducted by NIOSH of available linesman gloves indicated general compliance with the ANSI standard (Cook and Fletcher, 1977). OSHA requires compliance with this standard for linesmen's electrical insulating gloves. Selection of a designation class suitable to the expected electrical exposure is of course necessary in order to ensure adequate protection to the wearer. Frequent inspection for wear and tear is necessary as well. Such gloves are frequently worn with leather working over-gloves to reduce wear and tear to the insulating glove.

Fire fighters are exposed to a unique range of hand exposures ranging from cold to flame, yet their protective handwear must not unduly encumber the fire fighter in the diverse tasks that are involved in life-threatening environments. The National Fire Protection Association (NFPA) has recently published a standard for gloves for structural fire fighting (NFPA 1973–1983). The recommended standard covers a broad spectrum of exposure situations, many of which may be applicable to other hand protection devices. These include the following:

Resistance to cut
Resistance to puncture
Resistance to conductive heat penetration
Resistance to wet penetration
Flame resistance
Resistance to radiant heat penetration
Heat resistance

In addition, the following human factors were addressed:

Dexterity
Grip

Other types of gloves should also consider the following selection factors:

Comfort
Electrical resistance
Durability
Visibility
Chemical permeation
Chemical degradation
Decontamination
Wrist construction

These criteria encompass many criteria that might be considered in the selection of protective handwear for specific applications and serve as a useful checklist in reviewing the job task and making hand protection selections. This is not to suggest that the specific fire fighters' glove dexterity criterion would be appropriate to the selection of a latex glove used in a semiconductor operation to protect the semiconductor from the worker. The point is, of course, that dexterity must be a consideration in any selection process for protective hand wear for a specific work task.

Fortunately, hand injuries are rarely, if ever, fatal as a consequence of exposures of the hands to hazards associated with cuts, abrasions, cold, punctures, heat, or flame. In applications where such hazards may exist, a wide range of protective gloves is available and the individual worker can usually reach a satisfactory compromise between the type of glove and the work task.

Serious or fatal injury can occur, however, upon exposure to electricity. The recommended electrical criteria in NIOSH fire fighter's glove standard (NIOSH, 1976) and the ANSI J6.6 standard are clearly of major significance where such exposures may occur. Fortunately, based upon earlier NIOSH tests (Cook and Fletcher, 1977), linesmen's electrical insulating gloves appear to comply with the ANSI standard and thus represent a safe choice once the exposure voltage potential has been determined.

Potentially serious injury can also occur upon exposure of the skin to chemicals, Chemical burns, dermatitis, or absorption with subsequent toxic response consequences are possible. Unfortunately, no standards now exist governing the performance of gloves intended to provide protection against chemicals. Most suppliers provide subjective data referred to as chemical degradation characteristics with rating ranges from "excellent" to "not recommended." The degradation test merely subjectively assesses the response of the glove material to the chemical of interest when the glove material is immersed in the chemical. This test provides no information about the ability of the chemical to permeate the material, data that are relevant to the exposure environment. Fortunately, a new test method has been developed to determine the relevant information: ASTM F739-81. This is a permeation test method, which is now being utilized by the major glove and glove material manufacturers. Where exposure to chemicals occurs, the purchaser should also request data on glove performance based upon the ASTM permeation test method rather than relying only upon the degradation performance data.

User problems with hand protection devices are largely obvious and include affects on dexterity, comfort, and so on. The principal comfort problem involves sweating, particularly in chemical barrier gloves. Such gloves are often lined or used with cotton undergloves, several pairs of which may be required for a complete work shift. Recent studies in NIOSH's Division of Safety Research (DSR) have indicated that use of chemical protective gloves can increase errors and assembly time for mechanical assembly work. Time was increased from 15 to 37% over ungloved times, depending upon glove selection. Therefore, the use of gloves may prove to be an economic disincentive for workers on piecework.

7.6.8 FOOT PROTECTION

"Safety shoes" must be worn in the workplace where the potential for foot injury is foreseeable. Such devices must comply with the requirements of ANSI Z41.1, which is a standard established for safety-toe footwear (OSHA regulation 29 CFR 1910.136). The level of protection provided is determined by classifications of 30, 50, and 75, which refer to the weight dropped on the toe from a specified height while deforming the safety cap in the toe less than a specified amount. Tests by NIOSH (Cook, 1976) indicated that at that time only about 57% of the shoes tested met the requirements of the standard and that the principal failures appeared to be associated with overrating, that is, Class 75 select shoes were more likely to perform as Class 50 shoes.

The OSHA regulations do not specify the class required but place the burden on the employer to provide that which is appropriate to protect against the hazards present. In addition, safety-toe shoes principally protect only the toe from impact and compression insults. As in the case of eye injuries, the majority of foot injuries occur to individuals who are not wearing safety footwear. Further, more than half of these injuries result from an object falling on the foot, suggesting that the use of safety-toe shoes could result in a significant reduction in foot injuries while potentially reducing the severity of injuries as well.

In the mid-1970s, when the NIOSH tests were conducted, there were at least 25 manufacturers of safety shoes who marketed at least 703 styles of safety-toe footwear. Clearly, the selection available is large. There is also a large variety of additional foot protective devices, in addition to the basic safety-toe shoe, available for specific applications.

Metatarsal guards are devices that attach over the shoe to provide additional toe protection and, in addition, cover the top of the foot and the lower part of the ankle. ANSI Z41.2 recommends impact and compression standards for these devices. The wearer of metatarsal guards, however, has increased potential for tripping and for falls, particularly when using stairs and ladders. Extensions of the basic metatarsal device include skin-knee-instep, foot-shin guards with side shields and so on.

Flexible safety insoles or special outer sole devices are available as a means of providing protection against sole punctures where such hazards may be encountered. There are presently no recommended standards for these devices, however. Care should be exercised with many of the outer over-sole devices because they often are made of materials, such as wood, that possess very different frictional characteristics from those of the shoe sole.

ANSI Z41.3 recommends standards for a special class of electrical conductive footwear. Two types of conductive footwear are specified (Type I and Type II), each of which is intended for specific electrical hazard applications. Type I is appropriate for use by linesmen working with high voltages. ANSI Z41.4 recommends standards for electrical-hazard safety-toe footwear designed to provide protection against direct open electrical circuit contacts up to 600 V.

Although no specific additional ANSI standards exist beyond the basic ANSI Z41.1, there are additional types of safety footwear tailored to special environments. These include insulated, heat-resistant, nonsparking (for use in explosive atmospheres), water-repellent, chemical-resistant, and foundry/welders' footwear. Foundry/welders' boots serve as an excellent example of tailoring to solve a problem unique to a job task, for they are designed for quick removal should hot metal fall inside the boot top. Additional job-tailoring is evident in the wide range of sole materials and patterns available.

Sole pattern selection presents little selection problem and is largely based upon experience. Sole material, however, is generally categorized by exposure (oil) and/or slip resistance in the anticipated work environment. However, attempts to develop a dynamic floor surface friction measuring device, let alone a means of evaluating or quantifying frictional characteristics between a sole material and a floor, have not met with success. At this time, sole material selection must be based largely upon experience.

As in every case involving selection of personal protective equipment for workers, the job task must be analyzed so that the obvious and perhaps less-than-obvious hazards, such as the need for nonsparking footwear, are identified. As in the case of chemical protective handwear, few data beyond simple degradation data are available to aid in the selection of chemical protective footwear. However, in the footwear instance, the degradation data are perhaps more useful because such devices are usually thick compared to a glove. Nonetheless, permeation data should be sought as an aid in the selection process. Frequent inspection of such devices during use is of course necessary. An additional issue, not addressed as yet by any recognized organization, is decontamination. Chemical boots tend to be used for much longer periods of time than chemical gloves, owing to their cost. Exposure to some chemicals may result in the chemical being retained in the outer boot material, washing the boot after exposure may not remove the chemical, or the washing agent may degrade the boot material. When exposure to toxic chemicals occurs, special care should be exercised in this regard. If off-gassing from a chemical or chemical reaction is occurring, the chemical is clearly adhering to the boot surface, or the washing agent is deteriorating the boot surface, the boot should be properly disposed. Disposable latex booties are frequently worn over chemical resistant boots to allow ease of gross decontamination.

7.6.9   FALL PROTECTION

Fall protection devices include a wide range of products generally categorized as safety belts, lifelines, lanyards, safety nets, and harnesses. No *general* OSHA standard exists for such devices in the General Industry Standards (29 CFR Part 1910). Specific sections of the General Industry Standards, such as Roof Lifelines (29 CFR 1910.28), are addressed, however. The Construction Industry Standards (29 CFR Part 1926) does address certain requirements related to such devices, although details vary with regard to specific application requirements. Therefore, specific applications must be evaluated with regard to regulatory requirements on a case-by-case basis.

The principal consensus standard covering performance requirements of belts, harnesses, and securing lines is ANSI A10-14, which bases performance largely upon weight-drop tests. Safety belts, harnesses, lifelines, nets, and lanyards deserve special attention by those selecting and using such devices, for they are literally, in many cases, the last line of defense for the worker who must rely upon them. It is essential to recognize that such devices are systems composed of a body harness, lanyard, and securing point. *All* elements of the system must function properly and must be mutually compatible. A 100 lb or more lanyard is of little value should a fall occur, regardless of the fact that the harness meets ANSI A10-14 performance requirements.

There are a number of specific use recommendations which must be observed for any effective fall protection program:

Inspection/maintenance
Replacement
Operation

All fall protection devices should be inspected and the warning labels on the devices carefully read to assure that the devices are suitable for the intended application. Each element of protection must be viewed as part of the fall protection *system*. This requires that all elements of the system be appropriately compatible. Prior to every use, the complete system should be inspected. The fall protection system is most often a backup emergency lifesaving device. Its relationship to the work area must be determined. As an example, lifelines are used when workers utilize scaffolds at elevation. Should the scaffold fall, the lifeline is of little value if it gets entangled in the falling scaffold, as has been the case in fatalities investigated by NIOSH. Maintenance per se should not be performed on fall protection devices if it involves repair of a system component. For example, if the snap hook on a securing level requires repair, the line should be replaced or returned to the manufacturer for repair.

The fall protection system—all components—should be replaced regardless of appearance if it has been subjected to impact loads such as occurs in arresting a fall. In addition, most manufacturers require that only their components be used in the fall protection system. This ensures integrated system performance and proper attachment of various elements of the system.

In operation, there are a number of precautions that must be observed. As noted above, the work site should be inspected with regard to potential entanglement of the fall protection system. The wearer must be continually alerted to changes in the work task which might similarly degrade or eliminate the value of the fall protection system. Manufacturers' instructions must be closely adhered to with regard to attachment methods, methods of safe operation, and so on. Fall protection systems are life or death devices. Expert assistance and guidance is necessary in the selection of appropriate systems and to assure compliance with the appropriate regulatory standard that applies. Special and diligent attention to such systems through inspections, maintenance, and operational procedures are required as a defective component can result in a fatality.

7.6.10   BODY PROTECTION

OSHA has established general regulations regarding body protection under 29 CFR 1910.132 which require that protective devices shall be used to protect workers from foreseeable hazards. In addition, many specific OSHA standards exist, such as 29 CFR 1910.261(g)5, Pulp, Paper, and Paperboard Mills, which requires that protective aprons and other appropriate protective equipment must be worn during lime slaking. Such specific requirements are far too numerous to outline here. Specific standards applicable to operations for which protective equipment is required must be reviewed to determine what is mandated by the cognizant regulatory agency.

Body protection exclusive of that covered above under separate headings in the chapter encompasses a wide range of devices including, but not limited to, the following:

Aprons

Disposable garments
   Caps
   Jackets
   Pants
   Shoe booties
   Full coveralls

Fire fighters' turnout gear

Fire-approach suits

Fully encapsulating suits

Life jackets

Splash/rain gear
   Jackets
   Pants
   Hoods

Liquid-cooling garments

Air-cooling devices

Cold weather suits

Knee pads

There are essentially no performance standards with which these types of protective equipment need comply. Experience, recommendations from major suppliers and manufacturers, and a thorough job task hazard analysis are the principal components of proper selection.

There are, however, a number of important issues to consider that may be easily overlooked in the selection process. Aprons, garments, and so on intended to provide protection against metal splashes or chemical splashes must clearly be suitable for protection against those specific hazards. Particular care with regard to chemical splash protection is important. In the latter instance, chemical material degradation data are a suitable frame of reference because continuous contact, which is expected with gloves, is not intended. An apron material that is destroyed by sulfuric acid is, for example, clearly not an appropriate protective device where such protection is required.

Disposable clothing is becoming more widespread in industrial applications. Several fabrics are available and several fabrics that are coated with liquid barrier films are also available. The coated fabrics, however, generally increase the potential for heat stress because they do not "breathe." The purpose served by the disposables is to provide a reasonable degree of protection from dirt, oil, grease, and liquid splashes. They should be discarded daily or whenever tears, punctures, and so forth develop.

Fire fighters' turnout gear continues to improve in quality and performance. Currently, available gear is able to withstand exposures to higher heat levels and direct flame exposures more effectively than older equipment. As a consequence, fire fighters are being exposed to higher temperature extremes that have resulted in heat damage to their other supporting equipment, principally the respirator and helmets. Improved helmets are becoming available and respirator manufacturers are beginning to examine the respirator problem. Additionally, NIOSH and other organizations are working toward the development of high-performance SCBA standards for specific application to the fire service.

Fully encapsulating impermeable suits, and fire-approach suits to a degree, represent a major potential heat stress threat to the wearer. NIOSH has published a personal protective equipment selection guideline for hazardous materials exposures. Encapsulating garments are included in this document. Should work be required while wearing such garments, extreme care must be exercised to avoid excessive heat stress. In addition, appropriate medical facilities should be available to heat stress victims. Most heat stress fatalities result from the failure to apply appropriate medical measures quickly. Additionally, workers often are overcome by heat stress without being aware that their situation is becoming critical. Auxiliary cooling is frequently used with such garments to reduce the heat stress potential. Two approaches are utilized: air cooling and liquid cooling. Air cooling involves passing breathing quality air through a vortex tube and subsequently passing the cold air from the tube into the garment, thus providing both breathing air and some cooling. Liquid cooling is achieved by wearing a vest that contains small tubes throughout. Within the tubes, cold water is circulated by a battery-powered pump after circulation through a heat exchange containing ice or cold blocks. The cool vest, pump, battery, and heat exchange are carried by the individual. For optimal effectiveness, the vest must be directly in contact with the skin and the cold blocks or ice must be changed hourly. Current data suggest that liquid-cooling garments offer little advantage when the individual is working at high metabolic rates because the cooling provided is nearly offset by the added work associated with carrying the device.

Applications involving the use of impermeable encapsulating garments frequently involve exposure to chemicals. In such instances where the chemicals are known, permeation data is a necessity for proper selection. It should be recognized that such data are also necessary with regard to the gloves and boots, which are usually of different materials from the garment. Where the chemicals may be unknown, such as at hazardous waste sites, every effort should be made to determine the probable materials present with subsequent decisions made based upon permeation data. In every instance where such garments must be utilized, training and emergency preparedness are vital.

The use of much of the body protective equipment previously noted places additional physical and sensory burdens on the wearer. While an impermeable encapsulating garment is worn, a simple job task may be difficult and time consuming. A thorough job task analysis is necessary in order to assess the potential for adverse consequences associated with the use of such equipment. For example, mechanical power transmission equipment oilers are not permitted to wear loose-fitting clothing.

7.6.11 COMBINATION PROTECTION

The preceding sections of this chapter have discussed specific body protection keyed to a body part, such as the head. In practical reality, many workers utilize more than one protective device at the same time. Wearing a hard hat, safety glasses, safety-toe boots, and work gloves, for example, would not be uncommon for a construction worker. In this instance, the devices do not interfere with the respective function of each. There are instances where combinations of protective devices do interfere. Earmuffs over safety glasses are one simple example. A vastly more complex example is that of using a SCBA with an impermeable encapsulating garment. Such factors must be considered when utilizing protective equipment. For the simple muff-safety glasses example, earplugs might be substituted where the individual finds the muff uncomfortable. The complex cases, those involving potentially serious consequences, demand much more attention.

Proper selection and utilization of personal protective equipment requires thoughtful job task analysis, knowledge about the limitations of protective equipment, and the commitment of the individual who must use such equipment. Improper selection, improper use, and improper maintenance, for example, will quickly defeat the purpose for which protective equipment is used, that is, to protect the worker when administrative or engineering control measures are not available or feasible.

7.6.12 HUMAN FACTORS IN PERSONAL PROTECTIVE EQUIPMENT USE

Recommendations for medical surveillance of employees using personal protective equipment, particularly respirators, are frequently found in regulations or guidelines. There is no general agreement on what constitutes an adequate surveillance program nor on the criteria desirable in selecting users of personal protective equipment. The American National Standards Institute has developed a consensus standard, "Respirator Use—Physical Qualifications for Personnel," ANSI Z88.6-1984, in an attempt to fill this void.

Harber (1984) has suggested an interim three-level protocol of medical examination and qualification for selection of respirator users based on physical examination and job evaluation. He points out that a basic program component should include a period of observation of the worker using the respirator, preferably at the exercise level required by the job (Harber, 1984).

Suggestions of a single spirometric test as a screening device have been made by Raven, Moss, Garmon, and Skjaggs (1981). They recommend the 15 sec maximum voluntary ventilation ($MVV_{.25}$) without a respirator. The results of this test along with clinical evaluation and a knowledge of respirator resistances and job ventilatory requirements could then serve as the basis for selection. Harber (1984) points out the limitations of simple spirometry:

1. Sensitivity to restrictive lung disease, respiratory control disorders, and endurance is limited.
2. There is extreme variability in $MVV_{.25}$.
3. No direct relationship exists between the $MVV_{.25}$ and ability to breathe effectively for a prolonged period.
4. $MVV_{.25}$ is difficult to perform with a respirator.
5. The respiratory pattern during an $MVV_{.25}$ test differs from that during normal work.

The ANSI standard mentions spirometry as an addition to the clinical examination. ANSI further states that the employer should consider restriction of respirator use by any employee exhibiting an FVC (forced vital capacity) of less than 80% of the predicted volume or an FEV-1 (forced expiratory

Volume, 1 sec) of less than 70% of the predicted volume. Microprocessor-controlled spirometers are available which give instant readouts of such data as well as the predicted capacities; these can be utilized by trained personnel as a quick screening technique.

Similar discussions of the medical implications of other items of personal protective equipment use are virtually nonexistent in the literature.

Physiological studies of the effects of personal protective equipment have been largely restricted to respirators and specialized protective clothing, particularly fully encapsulating chemical suits and items of military use such as body armor.

Atterbom and Mossman (1978), Raven, Dodson, and Davis (1979a, b), Goldman and Breckenridge (undated), Myhre, Holden, Baumgardner, and Tucker (1979), Ronk et al. (1983), and Ronk, White, and Linn (1984) have published useful reviews of the literature on the physiological consequences of using various respirators and clothing combinations.

Performance decrements and imposed stress when using personal protective equipment are severe. The magnitude of the effect varies considerably between individuals and depends on the type of personal protective equipment worn, environmental conditions, and the work rate of the task to be performed.

Morgan (1980) has suggested that as many as 10% of the work force may be psychologically precluded from the effective use of respirators. He has found that 75% of his test subjects who experienced anxiety attacks while wearing respirators had elevated scores on the trait anxiety scale (STAI) of Spielberger, Gorsuch, and Lushene (1969).

This led him to the conclusion that certain "types" of individuals are more likely to experience anxiety attacks while wearing SCBA than others and that these individuals can be identified in advance with acceptable precision. As of yet, no extension of Morgan's work has been seen to other stressful personal protective equipment such as fully encapsulating suits.

Implementation of personal protective equipment as a control mechanism imposes significant risks of increased physiological and psychological stress. The stressor mechanisms are not well identified and predictive instruments are undeveloped. Caution is urged so that the use of personal protective equipment does not create more problems than it can solve.

Sizing of personal protective equipment that is similar to ordinary clothing is usually based upon commercial practice. That is, a size 12D safety-toe shoe is approximately the same size as a 12D dress shoe. More specialized protective equipment, such as respirator facepieces, may be single sized or of multiple sizes designed to fit a wide range of face geometries.

Fitting to size, especially in respirator facepieces, where the face seal is critically important to minimize inboard leakage, has been addressed in several studies.

McConville and Churchill (1972, 1974), using anthropometric data collected on armed forces personnel, concluded that these data could be used for design and sizing of respirators for civilian respirator users with a considerable degree of confidence. They pointed out, however, that both racial and sex characteristics were variables to be considered. Intraracial variation, they felt, was accommodated by the overall size variation within the sample. Sex differences may not necessarily equate small male facial sizes and female facial characteristics.

In applying McConville's work to the selection of respirator face fit test panels, Hack et al. (1974) used panels consisting of 40% Spanish-Americans and found that the mean values were within 2 mm of the base data. The assumption was made that male and female faces within the same two key dimensions (face length and width) were equivalent.

Respirator face-fit testing using the panel concept was conducted at several laboratories but, owing to problems of sample size and selection, the basic premises of racial and sexual equivalency have not been adequately resolved.

Individual fitting using either a subjective response to a challenge aerosol or vapor (qualitative fitting) or the measurement of both in-mask and ambient (challenge) concentrations (quantitative fitting) is frequently practiced to ensure adequate protection. Studies by Myers and Peach (1983), Myers and Peach (1984), and Lenhart and Campbell (1984), indicate that for at least two types of respirators, the powered air-purifying helmet for particulate protection and the half facepiece powered air-purifying respirator for particulate protection, these levels of protection assigned by laboratory studies are too high by a factor of some 10 to 40. This is in line with common practice in nuclear materials handling, whereby the employee is required to pass a quantitative facepiece fit test with a protection factor 10 times higher than that expected to be used in practice. Therefore, a full facepiece respirator would have to provide a protection factor of 500 in the fitting test but only could be used at concentrations of 50 times the permissible exposure (protection factor of 50).

Other protection that varies with sizing or fitting, such as shoes, hard hats, or muff hearing protectors and safety spectacles, has not been addressed in any meaningful manner.

Training in the use of personal protective equipment, which includes practice in nonhostile environments, has proved beneficial in increasing protection afforded by personal protective equipment, enhancing utilization of personal protective equipment, and reducing physiological and psychological stress. The American Occupational Medical Association (1982) has taken the position that training is a "very important" aspect of a hearing conservation program.

OSHA requires training in use of respirators [29 CFR 1910.134(e)(5)] and hearing protectors [29

CFR 1910.95(K)]. Yet few objective studies have been published to document the value of training. McDonagh (1982) has demonstrated decreased air consumption (hence physiological stress) in fire fighters who completed a 40 hr "smoke divers" course in the use of SCBA in comparison to the air consumption of the same students prior to their training.

The design of certain complex personal protective equipment makes safe use and maintenance unnecessarily difficult. Specific examples noted below illustrate these two points.

One fully encapsulating chemical protective suit has a zippered front opening that is extremely difficult to put on. However, after several complaints, and at least one overexertion injury, the zipper was extended to increase the opening length.

Investigation of several deaths involving the use of SCBA indicated that maintenance of the units was difficult and that users performing the maintenance were not fully informed as to critical variables such as screw torques (NIOSH, 1979). The units were redesigned and parts retrofitted in the field.

Use of SCBA under fully encapsulating chemical protective suits renders the essential bypass and main-line valves inaccessible in an emergency caused by the failure of the air regulator. No full solution to this problem has surfaced.

A sophisticated SCBA designed for escape from shipboard fires failed by releasing a highly caustic aerosol into the hood. The unit had to be redesigned before it could be marketed.

Another SCBA designed for emergency escape from underground coal mines was contained in a case that repeatedly could not be opened. The case required redesign before acceptance.

Interference between items of protective equipment such as hard hats and hearing protectors or respirators, hard hats and chemical protective clothing, or spectacles and respirators are common.

No area of a personal protective equipment program is more critical to its effectiveness than the factors affecting whether or not such equipment is actually used. Yet little research has been done on factors that provide incentives or disincentives for wearing personal protective equipment.

These factors may take several forms and can, for our purposes, be categorized as physical, psychological, social, and economic.

Physical factors that may influence the wearing of personal protective equipment include respiratory or cardiovascular disorders that are magnified when using personal protective equipment. Individuals may be sensitive to changes in oxygen, carbon dioxide, inhalation/exhalation resistance, relative humidity or temperature of inhaled air or oxygen, and/or weight of the personal protective equipment. Similar factors such as preexisting skin disorders, hypohydrosis, or previous heat disorders, may also adversely affect the user's ability to wear protective clothing or equipment. Individuals with low work capacities may be particularly reluctant to use personal protective equipment because of the added metabolic work load required with the additional weight and burden of the equipment.

Social factors, or how the work group perceives the wearing of personal protective equipment, may also influence an individual decision as to personal use. Training and educational techniques can be used to enhance peer acceptance of personal protective equipment use. Direct incentives, such as premium pay or tokens redeemable for merchandise, have been shown in some studies to be effective in increasing utilization of personal protective equipment (Zohar, Cohen, and Azar, 1980; Zohar, 1980).

Economic disincentives such as piecework, in which the worker perceives a decrease in his productivity, and hence pay, through the wearing of personal protective equipment, or the requirement that the worker pay all or part of the cost of purchases or maintenance of his personal protective equipment inhibit personal protective equipment use.

Whatever mechanism is used to motivate workers to wear personal protective equipment must ensure a very high degree of utilization if it is to be successful. For example, given a hazardous atmosphere at 100 times the permissible exposure limit (PEL) for an 8 hr workday, an employee using a protective ensemble having a protection factor of 10,000 will be overexposed if utilization of the ensemble is less than 99%, that is, if he or she removes it for 5 min during the 8 hr of exposure. Figure 7.6.1 illustrates how severely protection decreases with minimal periods of nonwear.

Worker input into the personal protective equipment program is vital if it is to gain acceptance. One national union has stated six general concerns regarding respirator programs which offer goal guidance for any personal protection (O'Brien, 1981):

1. *Greater emphasis must be given to the proper selection of respirators (PPE);*

2. *The fitting of respirators (personal protective equipment) must be more controlled than hit or miss, positive and negative fit tests. This problem includes the availability of facepieces of different sizes and fit-testing procedures that are more quantitative than qualitative;*

3. *An element of employee choice in the selection of the type of respirator (personal protective equipment);*

4. *Provisions for limiting time for respirators (personal protective equipment) wear either by designating certain areas as break areas and providing time for using them or by averaging time of use based on the exposure level and protection factor of the respirator;*

5. *Determining whether an employee is able to wear respiratory (personal protective equipment) protection and economic protection for those workers whose health has already been damaged to such an extent that they cannot wear currently available respirators;*

6. *Concern for the actual development and enforcement of a respirator (personal protective equipment) program in the workplace where such questions as discipline and discharge for failure to wear a respirator or the unilateral imposition of a policy on facial hair can strain even the best of labor–management relations.*

Studies by Plummer et al. (1985) and Poppelsdorf and Cramer (1982) have shown significant increase in time required to assemble small parts while wearing protective gloves over the time required bare-handed. This could directly translate into loss of productivity, wages, and social position for affected workers.

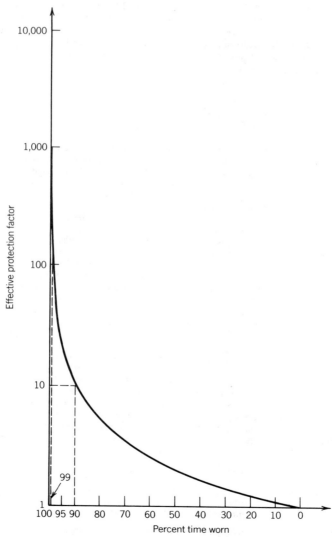

Fig. 7.6.1 Decrease of protection with increasing nonwear of personal protective equipment.

7.6.13 SUMMARY

In summary, personal protective equipment, by establishing an artificial microenvironment for the user, may seriously impact virtually any human–machine–environment interface.

The decision to prescribe the use of personal protective equipment in lieu of, or as a supplement to, effective administrative or engineering controls is not to be made lightly. At best, effective personal protective equipment programs are not inexpensive; at worst, ineffective ones may stifle productivity and increase occupational injuries and disease.

REFERENCES

American Occupational Medical Association. (1982, October). Guidelines for the conduct of an occupational hearing conservation program. *Journal of Occupational Medicine, 24* (10), 772–773.

Atterbom, Hemming A., and Mossman, Paul B. (1978, January). Physiological effects on worker performance of vapor-barrier clothing and full-face respirator. *Journal of Occupational Medicine, 20* (1), 45–52.

Cook, William I., and Groce, Dennis W. (1975, July). Report on tests of class B industrial helmets (NIOSH 76-106). National Institute for Occupational Safety and Health.

Cook, William I. (1976, July). A report on the performance of men's safety-toe footwear (NIOSH 77-113). National Institute for Occupational Safety and Health.

Cook, William I., and Fletcher, William G. (1977, July). A report on the performance of linemen's rubber insulating gloves (NIOSH 77-196). National Institute for Occupational Safety and Health.

Cook, William I. (1980, Fall). Mathematical models of head injury: A ranking experiment. *Journal of Safety Research, 12* (3), 127–137.

Goldman, Ralph F., and Breckenridge, John R. (undated). Current approaches to resolving the physiological heat stress problems imposed by chemical protective clothing systems. U.S. Army Research Institute of Environmental Medicine.

Hack, Alan, Hyatt, Edwin C., Held, Bruce J., Moore, Tom O., Richards, Charles P., and McConville, John T. (1974, March). Selection of respirator test panels representative of U.S. adult facial sizes (LA-5488). Los Alamos National Laboratory.

Haber, Philip. (1984, July). Medical evaluation for respirator use. *Journal of Occupational Medicine, 26* (7), 496–502.

Lenhart, Steven W., and Campbell, Donald L. (1984). Assigned protection factors for two respirator types based upon workplace performance testing. *Annals of Occupational Hygiene, 28* (2), 173–182.

McConville, John T., Churchill, Edmund, and Laubach, L. L. (1972, April). Anthropometry for respirator sizing (NIOSH Contract No. HSM 099-71-11). National Institute for Occupational Safety and Health.

McConville, John T., and Churchill, Edmund (1974, May). Anthropological research related to the sizing and fit of respirators (NIOSH Contract No. HSM 099-73-15). National Institute for Occupational Safety and Health.

McDonagh, Joseph M. (1982). Private communication.

Morgan, William P. (1980, September). Psychological problems associated with the wear of industrial respirators. Presented at the International Respirator Research Workshop, Morgantown, WV, National Institute for Occupational Safety and Health.

Myers, Warren R., and Peach, Michael J., III. (1983). Performance measurements on a powered air-purifying respirator made during actual field use in a silica bagging operation. *Annals of Occupational Hygiene, 27* (3), 251–259.

Myers, Warren R., Peach, Michael J. III, Cutright, K, and Iskander, Wafik H. (1984, October). Workplace protection factor measurements on powered air-purifying respirators at a secondary lead smelter: Results and discussion. *American Industrial Hygiene Association Journal, 45* (10), 681–688.

Myhre, L. G., Holden, R. D., Baumgardner, F. W., and Tucker, D. (1979, June). Physiological limits of firefighters. U.S. Air Force School of Aerospace Medicine.

NIOSH. (1976). The development of criteria for firefighters' gloves. NIOSH (77-134-A), Glove Requirements Vol. 1: NIOSH (77-134-B), Glove Criteria and Test Methods, Vol. 2, National Institute for Occupational Safety and Health.

NIOSH. (1979, June). Tests of self-contained breathing apparatus received from Lubbock, Texas, Fire Department. National Institute for Occupational Safety and Health.

NIOSH. (1982, March 10). Preliminary survey of existing data and economic overview of the respirator industry (NIOSH Contract No. 210-81-1102) (unpublished draft). National Institute for Occupational Safety and Health.

O'Brien, Mary Winn (1980, September). Don't forget your muzzle. Presented at the International Respirator Research Workshop, Morgantown, WV. National Institute for Occupational Safety and Health.

Poppelsdorf, Nina, and Cramer, Caryn. (1982, April). An investigation of the effects of gloves on manual and finger dexterity and fatigue. (Traineeship Grant No. 5T01-01T00161) University of Michigan.

Plummer, Ralph, Stobbe, Terry, Ronk, Richard, Myers, Warren R., Hyunwook, K., and Jaraiedi, M. (1985, September). Manual dexterity evaluation of gloves used in handling hazardous materials. Proceedings of the Human Factors Society 29th Annual Meeting, Baltimore, MD.

Raven, Peter B., Dodson, Ann T., and Davis, Thomas O. (1979, June). The physiological consequences of wearing industrial respirators: A review. *American Industrial Hygiene Association Journal, 40* (6), 517–534.

Raven, Peter B., Dodson, Ann T., and Davis, Thomas, O. (1979, July). Stresses involved in wearing PVC supplied-air suits: A review. *American Industrial Hygiene Association Journal, 40* (7), 592–599.

Raven, Peter B., Moss, Raymond F., Page, Kimberly, Garmon, Robert, and Skaggs, Barbara. (1981, December). Clinical pulmonary function and industrial respirator wear. *American Industrial Hygiene Association Journal, 42* (12), 897–903.

Ronk, Richard, White, Mary Kay, and Linn, Herbert. (1983, October). A selected bibliography on hazardous materials control (NIOSH, unnumbered publication). National Institute for Occupational Safety and Health.

Ronk, Richard, White, Mary Kay, and Linn, Herbert. (1984, October). Personal protective equipment for hazardous materials incidents: A selection guide (NIOSH 84-114). National Institute for Occupational Safety and Health.

Spielberger, Charles D., Gorsuch, R. L., and Lushene, R. E. (1969). *The State-Trait Anxiety Inventory Manual.*

Zohar, Dov, Cohen, Alexander, and Azar, Naomi. (1980, February). Promoting increased use of ear protectors in noise through information feedback. *Human Factors, 22* (1), 69–79.

Zohar, Don. (1980). Promoting the use of personal protective equipment by behavior modification techniques. *Journal of Safety Research, 12* (2), 78–85.

CHAPTER 7.7
HEALTH INDEX

MASAMITSU OSHIMA

**The Medical Information System Development Center
Tokyo, Japan**

This paper reviews two issues in the study of health. First, definitions of health and self-perception of health are explained. Second, methods that are available to calculate quantitative measures of health for both populations and individuals are discussed.

7.7.1 GENERAL ISSUES IN THE STUDY OF HEALTH

There are many possible definitions of health. Some of the common definitions of health are listed in Table 7.7.1. Health is usually considered to be the absence of disease and pain, but it also can be viewed as an optimal adjustment of the organism to its environment. This adjustment to the environment includes mental and social factors in addition to physical ones.

Health is characterized as a combination of three components of well-being: physical, mental, and social. This definition is the one that will be used in subsequent discussions.

Health should not be approached as a simple dichotomy of disease versus the absence of disease. Instead, it should be viewed as a continuum with many levels. Figure 7.7.1 illustrates the health continuum. The implication of this continuum is that an individual's health condition is not permanent; it can be improved or worsened. An improvement generally means an expansion of activity in the routine of life whereas a worsening health condition reduces the amount of potential activity. An individual's position on the health continuum determines the activity level, as can be seen from Figure 7.7.2, which is a diagram of activity levels. People at the unhealthy end of the health continuum can maintain only the basic metabolic activity level, whereas a semihealthy person might be able to work but not enjoy the higher activity level of leisure pursuits.

Self-perception of health depends upon many variables such as appetite, fatigue, and work. Figure 7.7.3 and 7.7.4 show the factors that contribute to an individual's perception of his/her health. The integration of these different factors is known as health feeling.

The elements related to health can be classified as environmental or individual (see Table 7.7.2). Environmental elements such as quality of water, weather, drug prescriptions, and health examinations can influence health. Individual or internal factors include philosophy of life, religion, personality attributes, physical condition, age, and ability to enjoy life.

Health can be checked using any of the methods outlined in Figure 7.7.5. These methods for evaluating health vary from behavioral and medical analyses to analyses of patients' subjective feelings about health. However, there are some drawbacks to these methods, and they are shown in Figure 7.7.6.

7.7.2 HEALTH INDEX

Several quantitative methods have been proposed for measuring health. This measure will be referred to as the health index; it can be calculated for both populations and for individuals. Among the

Table 7.7.1 Definitions of Health

1. Absence of disease
2. Sound condition: freedom from disease, ailment, and pain
3. Optimal fitness
4. Wholeness
5. Flourishing condition
6. Physical, mental, and social well-being, not merely the absence of disease and pain
7. An equilibrium
8. An undisturbed rhythm
9. Perfect adjustment of an organism to its environment
10. Adaptation
11. A balance
12. A general condition
13. A level
14. A process of human–environment interaction
15. A purchasable product (a good or a commodity)

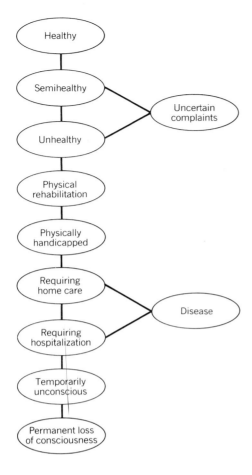

Fig. 7.7.1. Continuum of possible health conditions.

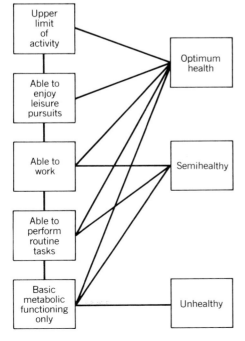

Fig. 7.7.2. Activity levels and their relationships to health.

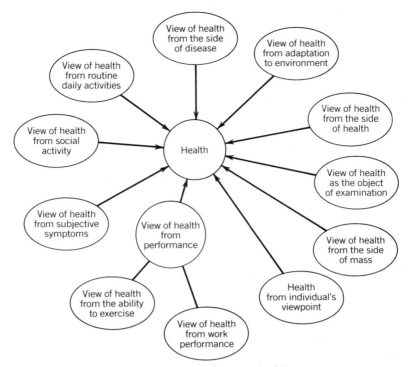

Fig. 7.7.3. Contributors to health.

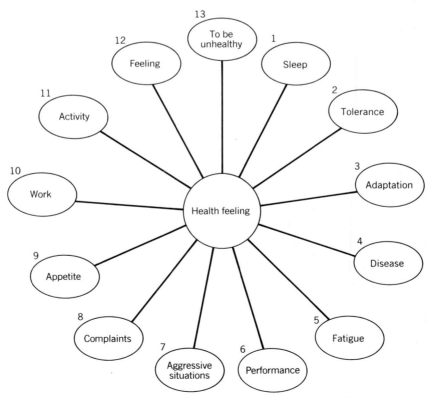

Fig. 7.7.4. Multiphasic conditions of health and health feeling.

898

Table 7.7.2   Individual and Environmental Elements for Health

Individual	Environmental
Physical power	Public health
Health feeling	Safety
Fatigue	Health examination
Growth	Sports
Taste	Drug prescriptions
Aging	Time difference (Jet lag)
Sleep	Character of community
Emotion	Human relations
Sex	Hobbies
Feeling of life	Job, task
Nutrition	Friends
Personality	Leisure
Religion	Welfare
Volunteer spirit	Economy
Ambition	School history
	Water
	Season
	Weather

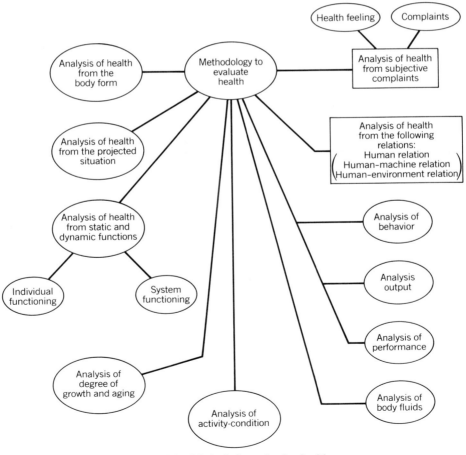

Fig. 7.7.5. Methods for evaluating health.

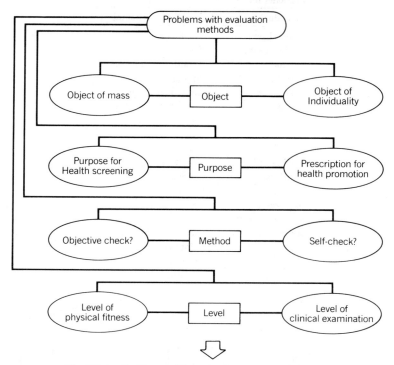

Fig. 7.7.6. Problems with methods of evaluating health.

different kinds of health indexes that are discussed here are measures for a population health index, such as the Proportional Mortality Indicator (PMI) and the Cornell Medical Index (CMI), and measures for an individual health index, such as the Life Checker System.

A health index indicates the degree of health of either a population or an individual. Information regarding health should be collected from three sources: medical examinations, subjective complaints, and evaluations of life-style (see Figure 7.7.7). In calculating a health index, several factors must be evaluated in unison. These factors can be seen in Figure 7.7.8. The arrows in the figure indicate the items that should be evaluated together. A summary of the methodology for indexing health is depicted in Figure 7.7.9. Items labeled one through eight are the different methods of health indexing. Items labeled *a, b,* and *c* represent the considerations that should be made in evaluating an index. Evaluation

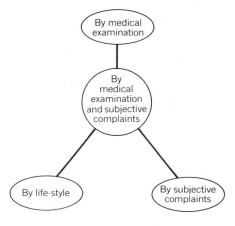

Fig. 7.7.7. Evaluating health to determine the health index.

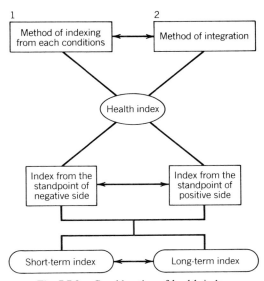

Fig. 7.7.8. Consideration of health index.

of an index must be done considering short-term and long-term events, and positive versus negative standpoints.

7.7.3 POPULATION HEALTH INDEX MEASURES

In this section five measures of population health index are presented. The measures vary in their complexity; the first two are based on general statistics, whereas the last three are based on subjective questionnaires.

The first measure of population health index incorporates the following seven items:

1. Infant mortality
2. General death rate
3. Morbidity
4. Incidence rate or prevalence rate
5. Disease recovery rate
6. Accident recovery rate
7. Life expectancy

Another proposal for evaluating population health index uses two scales, impairment and disability, which would be considered in conjunction with each other. Figure 7.7.10 demonstrates how the two scales work to measure health.

The third population health index is the Proportional Mortality Indicator (PMI), which is a more comprehensive index that was developed by the United Nations Committee of Experts on International Definition and Measurement of Standard and Level of Living have identified twelve components of life levels as follows:

1. Health, including demographic conditions
2. Food and nutrition
3. Education including literacy and skills
4. Conditions of work
5. Employment
6. Aggregate consumption and saving
7. Transportation
8. Housing, including household facilities
9. Clothing

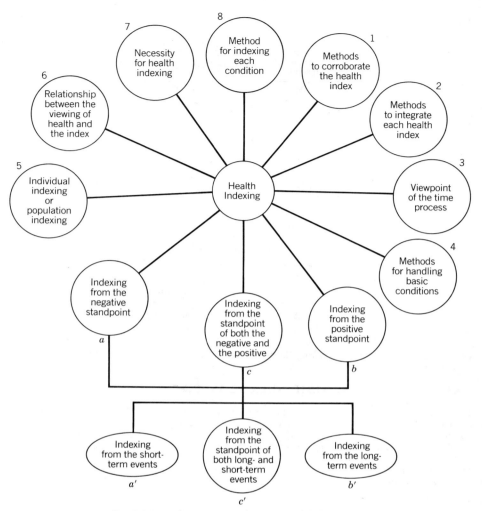

Fig. 7.7.9. Calculating and evaluating health indexes.

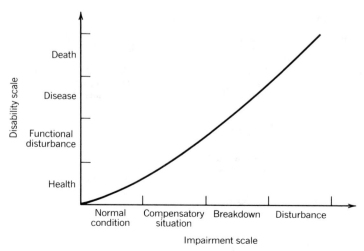

Fig. 7.7.10. Impairment scale and disability scale.

Table 7.7.3 Sections on the CMI

Section	Questions Referring to	Number of Questions
A	Eyes and ears	9
B	Respiratory system	18
C	Cardiovascular system	13
D	Digestive system	23
E	Musculoskeletal system	8
F	Skin	7
G	Nervous system	18
H	Genitourinary system	11
I	Fatiguability	7
J	Frequency of illness	9
K	Miscellaneous diseases	15
L	Habit	6
M	Inadequency	13
N	Depression	5
O	Anxiety	9
P	Sensitivity	6
Q	Anger	9
R	Tension	9
	Total	195

10. Recreation and entertainment
11. Social society
12. Human freedoms

PMI is calculated as the ratio of 50% of deaths in the population over 50 years old to total number of deaths per year.

The fourth health index, the Cornell Medical Index (CMI) utilizes a subjective questionnaire that includes 195 components; 144 are physical components and 50 are mental ones. Table 7.7.3 provides a breakdown of the topic areas covered in the questionnaire.

The fifth measure was established by the Human Research Laboratory and is a system of health index evaluation which contains 46 items pertaining to the physical situation, physical fitness, resistance power, psychological situation, and social adaptability. A sample question follows:

1. What is your physical situation?

enough	almost	normal	want to	always
satisfied	satisfied	satisfied	become healthy	bad

Participants being evaluated with the Human Research Laboratory index complete a comprehensive preliminary questionnaire. The items included in the questionnaire are listed in Table 7.7.4. A study

Table 7.7.4 Items Included in Preliminary Questionnaire from the Human Research Laboratory

1.	Age	12.	Income per year
2.	Weight (fatness)	13.	Time spent commuting
3.	Wife	14.	Duration of sleep
4.	Child	15.	Chronic disease, constitutional disease
5.	Grandchild		
6.	Age of father	16.	Medical history
7.	Age of mother	17.	Blood pressure
8.	School career	18.	Amount of cigarettes
9.	Situation of duty	19.	Frequency of alcohol use
10.	Daily work or shift work	20.	Amount of alcohol use
11.	Task	21.	Adaptation to alcohol

Table 7.7.5   The Subjects (Human Research Laboratory)

Age (years)	Number of Subjects
20	215
30	289
40	258
50	260
60	505
Total	1527

using this index was conducted with 1527 subjects. A breakdown of the subjects by age is contained in Table 7.7.5. The results were as follows. Physical powers decline in parallel with age and social adaptation increases with age. The higher partial correlation coefficient in all age categories is shown in the group of factors: school history, duty situation; contents of work, blood pressure, age and amount of alcohol consumption. The higher correlation with age is seen in the following factors:

Table 7.7.6   Ratios for Calculating Health Index for Individuals

1. Ratio of subjective complaints:
$$\frac{\text{checked number of complaints}}{\text{all items of complaints}} \times 100(\%)$$

2. Ratio of absence:
$$\frac{\text{absence-days}}{\text{days of duty}} \times 100(\%)$$

3. Ratio of unhealthy symptoms:
$$\frac{\text{number of unhealthy symptoms}}{\text{all number of symptoms}} \times 100(\%)$$

4. Ratio of abnormal test items:
$$\frac{\text{number of abnormal test items}}{\text{number of all test items}} \times 100(\%)$$

5. Ratio of use of drugs:
$$\frac{\text{days of use of drugs}}{365 \text{ days}} \times 100(\%)$$

6. Ratio of amounts of work:
$$\frac{\text{practical amounts of work}}{\text{routine amounts of work}} \times 100(\%)$$

7. Ratio of holiday:
$$\frac{\text{practical holiday}}{\text{normal holidays}} \times 100(\%)$$

8. Ratio of adaptation to load:
$$\frac{\text{this performance} - \text{routine performance}}{\text{maximum performance}} \times 100(\%)$$

9. Ratio of evaluation of fatiguability:
$$\frac{\text{this grade of fatigue} - \text{usual grade of fatigue}}{\text{usual grade of fatigue}} \times 100(\%)$$

10. Ratio of sleep, ratio of going to bed:
$$\frac{\text{this sleep time} - \text{usual sleep time}}{\text{usual sleep time}} \times 100(\%)$$

11. Ratio of use of alcohol:
$$\frac{\text{this amount of alcohol} - \text{usual amount of alcohol}}{\text{usual amount of alcohol}} \times 100(\%)$$

12. Ratio of evaluation of health feeling
$$\frac{\text{grade of evaluation of health feeling} - \text{usual grade of evaluation}}{\text{usual grade of evaluation}} \times 100(\%)$$

income (over 60 years of age), sleeping time (over 50 years old), tobacco (over 40 years old), frequency of alcohol consumption (over 60 years old), and adaptation to alcohol. The other items have a lower correlation to the health index.

7.7.4 INDIVIDUAL HEALTH INDEX MEASURES

One approach to calculating an individual health index involves the ratios of 12 items. These ratios cover topics such as absenteeism from work, use of drugs, fatigue, alcohol consumption, symptoms of ill health, and subjective complaints. A precise delineation of these ratios is located in Table 7.7.6.

Another approach is the Life Checker System, which estimates an individual's future health. Some of the items investigated are the patients' normal life, physical activities, mental and emotional situations, environments, and current health status. Information obtained from these five items concerning an individual's life-style is entered and risk factors are calculated. Years of remaining life expectancy can be estimated by relating risk factors to lifestyle information. Life expectancy can be increased when risk factors are removed.

A third method, designed by the Society of Industrial Medicine is a questionnaire that evaluates health from the standpoint of fatigue. Physical complaints, mental symptoms, and neurosensory systems contribute to determining the health index. Table 7.7.7 lists the 30 items surveyed. Figure 7.7.11 demonstrates the relationship of health to the results of the questionnaire.

The fourth method of measuring an individual's health index assumes that health can also be viewed as a system with certain functions. When all functions are operating the health index is 100%. Table 7.7.7 lists the critical system functions involved in determining the health index. The health index can be calculated by taking the number of pertinent marked items from the list for a specific individual and subtracting that number from 10, and dividing the result by 100. For instance, if an individual reports disruption of six functions from Table 7.7.7, then the health index would be 40%.

In summary, there are several approaches to calculating a health index for an individual or a population. Although there are some differences between methods, most evaluate mental and social

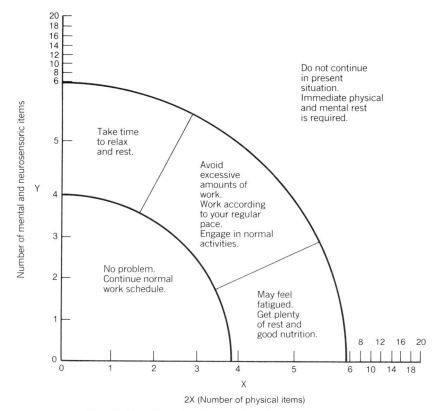

Fig. 7.7.11. Diagnosis of fatigue by subjective complaints.

Table 7.7.7 Subjective Complaints to Fatigue

Physical Complaints

Heavy feeling in head
Headaches
Tired feeling in entire body
Tired feeling in any body part
Pain and/or tension ache in some body part
Stiffness
Choking
Lack of saliva
Yawning
Cold sweat

Mental Symptoms

Lightheaded and disoriented
Inability to collect thoughts
Inability to concentrate
Irritated
Excessive sleepiness
Easily distracted
Disloyal
Failure to remember
Lack of confidence and makes mistakes
Anxiety attacks

Neurosensory Symptoms

Tired eyes and inability to focus
Dry eyes
Lethargic inaccurate movements
Tottering gait
Turning sore
Dizziness
Twitching of eyelids
Ringing in ears and loss of hearing
Arm and leg muscle twitches
Inability to sit up straight

aspects as well as physical ones. The important implication of health indexes is that with a quantitative measure of health, improvements can be implemented more successfully.

BIBLIOGRAPHY

Anderson, K. L., Masironi, R., Rutenfranz, J., and Seliger, V. (1978). *Habitual physical activity and health,* Copenhagen: W.H.O.

Bainbridge, J., and Sapirie, S. (1974). *Health project management,* Geneva: W.H.O.

Charron, K. C. (1975). *Education of the health professions in the context of the health care system: The Ontario experience,* Paris: O.E.C.D.

Drary, C. G., and Fox, J. G., Ed., *Human reliability in quality control,* London: Taylor & Francis.

Dubos, R. (1959). *Mirage of health, utopias, progress and biological change,* New York: Harper & Brothers.

Jazairi, N. T. (1976). *Approaches to the development of health indicators,* Paris: O.E.C.D.

Journal, S. M. (1974). *Healthy personality, an approach from the viewpoint of humanistic psychology,* New York: Macmillan.

Koizumi, Akira, and Tanaka, Tsumeo, Ed., (1983). *Man and health,* Tokyo: Taishukan Shoten.

Kokumion Seikatsu Shingikai Chosa Bukai, Ed., *Social indicators,* Tokyo: Okusasho Insatsukyoku.

Lisitsin, Y. (1972). *Health protection in the USSR,* Moscow: progressive Publishers.

Nihon Tanpa Hoso, (1974). *View of human being in the present time,* Tokyo.

NHK Hoso-Yoron-Chosazyo, Ed., (1981). *The health-feeling of Japanese,* Tokyo: Nihon Hoso, Syreppan Kyokai.

O.E.C.D., (1974). *Subjective elements of well-being,* Paris: O.E.C.D.

Oshima, Masamitsu, (1972). *Introduction to the health,* Tokyo: Ishiyaku Shuppan K.K.

Oshima, Masamitsu, (1981). *The researches of fatigue,* Tokyo: Dobun Shoin.

Oshima, Masamitsu, (1982). *The Human Being—Approach to Its Unknown,* Dobun Shoin, Tokyo, 1982.

Oshima, Masamitsu, et al. (1983). *Industrial stress,* (Translation of proceeding of machine pacing and occupational stress, edited by G. Salvendy and M. J. Smith, 1983), Dobun-Shoin: Tokyo.

Praget, J. (1970). *Introduction—the place of the sciences of man in the system of sciences,* Paris: Unesco.

Tengstom, A. (1975). *Patterns of health care and education in Sweden,* Paris: O.E.C.D.

The Council for National Life, Ed., (1970). *The future image of national life—vision after 20 years,* Tokyo: Okurasho Insatsukyoku.

The Department of Investigation for National Life, Ed., (1976). *The grade of selection for national life,* Tokyo: Okurasho Insatsukyoku.

Wilson, J. M. G., and Jungner, G. (1968). *Principles and practice of screening for disease,* Geneva: W.H.O.

Young, J. Z. (1971). *An introduction to the study of man,* Oxford: Clarendon Press.

differences in obtained data. Methods of studying construct validity include the gamut of experimental and correlational techniques including laboratory experiments, field experiments, and correlational studies. In recent years path analysis, confirmatory factor analysis, and structural equation modeling have become powerful tools for the study of construct validity (Kenny, 1979; Joreskog, 1978). A good example of the use of confirmatory factor analysis and path analysis in investigating construct validity is a study by Hunter, Gerbing, and Boster (1982) identifying four component beliefs that were involved in the construct of Machiavellianism.

Construct validity has not been popular as a strategy for defending the use of predictors, even when used in combination with other validation strategies. One reason for this neglect may have been the emphasis on construct validation as a never-ending process indistinguishable from the validation of any theory of behavior (Cronbach, 1980). Recently there has been more emphasis on combining construct validity with other validation strategies. A good example is an industry-wide validation study involving electric power operators (Dunnette, 1982). This study investigated both predictor and job performance constructs through literature review, job analysis, factor analysis of both predictor and criterion measures, and related methodologies.

8.1.2 CONTENT VALIDITY

Another trend is the increasing attention being paid to the content validity of selection procedures. The extent to which selection procedures actually sample relevant job behaviors is an important consideration in any attempt to develop valid selection instruments. Content validity has become somewhat controversial, but should always be addressed when evaluating the usefulness of predictors for personnel selection. Content validity has also played an important role in methods of establishing the validity of training criteria such as achievement and proficiency tests, and is included as a viable validation procedure in the *Uniform Guidelines on Employee Selection* (1978) and *Principles for the Validation and Use of Personnel Selection Techniques* (APA, 1980).

Content validity refers to the extent to which a test samples a domain of important job behaviors. Content validity is established by showing that (1) the content domain measured by the test consists of important skills or knowledge needed for successful job performance and (2) the test representatively and reliably samples the defined content domain. Content validity is evaluated by examining the adequacy of the operations used to define the content domain and the adequacy of the sampling procedures for selecting the items of the test (Tenopyr, 1977). A third criterion for establishing content validity is to examine the reliability or generalizability of scores on the content sample to the defined domain (Ebel, 1977; Guion, 1977; Guion, 1978).

As has been frequently pointed out, validity refers to the accuracy and appropriateness of inferences made on the basis of test scores (Messick, 1975, 1980; Tenopyr, 1977). Content validity is limited to the extent that test construction operations are relied upon with no analysis of actual test scores as in a generalizability study.

Predictors that measure important job knowledge or skills may be evaluated in terms of their content validity. There are several key procedures involved in establishing the content validity of a selection device. Job analysis is especially important. In addition, careful attention to sampling to assure representativeness of the test to the content domain is important. Finally, study of the generalizability or reliability of the test for measuring the content domain is essential.

8.1.3 CRITERION-RELATED VALIDITY

8.1.3.1 General Features of the Model

The criterion-related validity model dates from the earliest attempts to apply scientific methods to employee selection (Hull, 1928). The model prescribes a specific and straightforward method for the development and validation of employee selection techniques that is best characterized as a series of steps.

1. *Job Analysis.* In criterion-related validity studies there are several purposes of job analysis. One major objective is to discover those abilities and personal qualities that are predictive of successful/unsuccessful job performance. Another important objective is to identify critical job behaviors that should be measured in development of criterion measures. A more generalized objective is to describe the job systematically in order to compare it with similar jobs in other contexts, as is done in validity transportation. Because task analysis is summarized in Chapter 3.4, job analysis is not reviewed here. For a summary of job analysis issues in personnel selection see Sparks (1982).

2. *Criterion Development.* Measurement of job performance has traditionally been the weak link in criterion-related validity studies. Objective measures of job behaviors such as work sample tests have been used successfully, but supervisor evaluations have been the most commonly used criteria.

3. *Selection of Predictors.* Based on job analysis data one or more predictors are selected for trial. A frequent practice is to try out more predictors than will be used for applicant screening.

4. *Data Collection.* In the predictive validity model experimental selection techniques are administered to applicants. In the concurrent validity model selection techniques are administered to job incumbents. Criterion data are collected after sufficient job experience has been obtained. In the predictive validity model data collection can be a lengthy process depending on the hiring rate and the period needed for becoming fully qualified on the job. On the other hand, although the concurrent validity model is usually more efficient, the merits of this model are a matter of some controversy (Barrett, Phillips, and Alexander, 1981; Guion and Cranny, 1982). Use of the concurrent validity model should be accompanied by a careful study of the possible effects of job experience on the predictors, the similarity of the applicant population to the job incumbents in the sample, and the effect of various sources of restriction of range.

5. *Statistical Analysis.* Correlation and analysis of variance are used to investigate the reliability of predictors and criteria. Correlation and regression are used to study the validity of the predictors. Predictors can be combined using ordinary least squares (OLS) procedures, but in practice, equally weighted composites of predictors are often used. When predictors are positively intercorrelated and not too different in validity, there is usually little loss in predictive power and fitting sampling errors is avoided (Wainer, 1976).

6. *Estimation of Population Parameters.* Population validity coefficients and population cross-validity coefficients can be estimated using shrinkage formulas. Validity coefficients typically must be adjusted for the effects of restriction of range of scores on both predictors and criteria (see Section 8.1.3.3). Validity coefficients are frequently adjusted for attenuation due to unreliability of criterion measurements (see Section 8.1.4.2). Both adjusted and unadjusted validity coefficients are typically reported.

8.1.3.2 Problems in Criterion Measurement

As mentioned earlier, criterion measurement has been the weak link in the criterion-related validity model. Studies involving objective criteria such as work sample tests or output measures are most desirable. However, objective criteria are the exception rather than the rule and for many jobs development of objective measures of performance is a formidable and not always feasible task. For these reasons one focus of research in performance measurement has been studies aimed at understanding the subjective rating process and the many sources of error involved.

One thrust of research in subjective methods of performance measurement has been studies of the effectiveness of various types of formats for constructing performance rating scales. Behaviorally anchored rating scale methodology has been the focus of many studies (Smith and Kendall, 1963; Schwab, Heneman, and Decotiis, 1975; Bernardin and Smith, 1981). Another approach has been behavioral observation scale methodology developed by Latham and Wexley (1981). Comparative studies of rating formats generally have shown only small differences or have been inconclusive (Schwab et al., 1975). These results have been summarized by Landy and Farr (1980), who concluded that differences in rating format do not result in marked differences in the psychometric quality of performance ratings.

The question naturally arises, if differences in rating format do not influence the psychometric quality of ratings, then what does? A second area of research in performance measurement has been investigations of halo error and other systematic biases that plague performance ratings (Borman, 1983). This research has been strongly influenced by studies of implicit personality theory, especially work on semantic relationships among traits as they exist in the mind of the beholder. Work on systematic biases has also been influenced by research on interviewing that has uncovered a Pandora's box of systematic biases in interviewer judgments (Schmitt, 1976). Perhaps one of the most persistent distortions in performance ratings is halo error. Research on halo error and ways of minimizing it has been summarized by Cooper (1981). Unfortunately, research has served to underline the pervasiveness of halo error much more so than ways of eliminating it.

A third approach to research on performance measurement is focus on cognitive processes involved in observing, storing, retrieving, and evaluating job performance (Feldman, 1981). These studies point up the complexities involved in observing and rating behavior. Like the investigations of rater biases, results of these studies have so far not been translated very successfully into practical methods for improving subjective ratings of performance. In addition, attempts at training raters to avoid errors in ratings and to improve accuracy have only been partially successful (Ivancevich, 1979; Warmke and Billings, 1979; Bernardin and Pence, 1980; Zedeck and Cascio, 1982; McIntyre, Smith, and Hassett, 1984).

Given the many problems with subjective ratings of performance it is remarkable that validity generalization studies and meta-analyses have shown average adjusted validities in the 0.30 to 0.50 range for predicting rated job proficiency with cognitive ability tests. However, the validities of cognitive tests typically have been higher for predicting training criteria. These results may very well be due to the more frequent use of objective measures such as job knowledge or job proficiency tests as criteria for success in training programs.

8.1.3.3 Adjustments for Restriction of Range

Restriction of range is a serious problem in criterion-related validity studies. Problems arise in comparing the validity of two predictors if each predictor has a different degree of restriction. In cases of severe range restriction the statistical power of a criterion-related validity study may be so low that the feasibility of such a study is severely limited. Range restriction is also a problem when the results of several studies are aggregated as in validity generalization research. If range restriction is not too severe, a validity coefficient can and should be adjusted for range restriction to obtain a more accurate estimate of validity in the applicant population.

Validity coefficients computed on a sample of job incumbents may seriously underestimate the validity of a predictor in the applicant population if the predictor was used in the selection process or if the predictor is correlated with other operational predictors. Because the interest is in estimating the validity of the predictor in the applicant population, it has now become commonplace to adjust validity coefficients for restriction of range. However, when validity coefficients are adjusted for restriction of range, it is important that both the unadjusted and the adjusted coefficients be reported (APA, 1980).

When applicants must exceed a minimum score in order to be selected, the result is restriction of range due to explicit selection on the predictor, and incidental restriction of range on the criterion. When unsuccessful job incumbents are transferred or terminated, the result is restriction of range on the criterion. In addition, restriction of range on the criterion may occur when successful job incumbents are promoted to more responsible positions. Incidental selection on a predictor may occur owing to its correlation with a predictor on which there is explicit selection. The effect of restriction of range is to reduce the validity coefficient in the selected population proportional to the degree of restriction. A predictor with a high validity coefficient in the applicant population may have a validity coefficient close to zero in the selected population.

Restriction of Range Due to Explicit Selection

When there is explicit selection on the predictor and the standard deviation of the predictor in the applicant sample is known, the validity coefficient in the applicant population can be estimated by

$$\hat{\rho}(X, Y) = \frac{r(X, Y)U}{[1 + r^2(X, Y)(U^2 - 1)]^{1/2}} \tag{1}$$

where $\hat{\rho}(X, Y)$ is the estimated validity in the applicant population, $r(X, Y)$ is the sample validity coefficient, and U is the ratio of the standard deviation in the applicant sample to the standard deviation of the selected sample. This equation assumes linearity of regression and homoschedasticity but does not assume bivariate normality (Thorndike, 1949, equation 6, p. 173). A table for estimating the unrestricted validity coefficient as a function of selection ratio and restricted validity coefficient when restriction of range is due to explicit selection has been given by Sands, Alf, and Abrahams (1978).

Monte Carlo studies of the accuracy of the above equation show that the adjusted validity coefficient is a more accurate estimator of the unrestricted validity coefficient than is the unadjusted coefficient. The equation tends to be more accurate for higher values of validity in the applicant population. The equation tends to be robust under violations of homoschedasticity but is sensitive to departures from linearity (Greener and Osburn, 1981). In some concurrent validity designs in which sample members are selected to achieve heterogeneity on the predictor, the standard deviation in the employee group may be larger than that of the applicant group, in which case the adjustment would result in an estimated validity in the applicant population that is lower than the obtained validity (Osburn and Greener, 1978).

The standard error of the adjusted validity coefficient is larger than the standard error of the unadjusted coefficient. Equations for estimating the standard error of correlations adjusted for restriction of range and errors of measurement have been presented by Bobco (1983).

Adjustment for Incidental Selection

There is restriction of range due to incidental selection when a predictor is correlated with a second predictor that is subject to explicit selection. For example, suppose that 50% of applicants are screened out using a minimum score on a cognitive ability test, and the remaining applicants are interviewed. Interviewer evaluations are subject to restriction of range due to incidental selection if interviewer evaluations are correlated with cognitive test performance. When the standard deviation of the variable subject to explicit selection is known for the applicant sample, the validity coefficient in the applicant population for the predictor subject to incidental selection can be estimated by

$$\hat{\rho}(Z, Y) = \frac{r(Z, Y) + r(X, Y)r(X, Z)(U^2 - 1)}{[1 + r^2(X, Y)(U^2 - 1)]^{1/2}[1 + r^2(X, Z)(U^2 - 1)]^{1/2}} \tag{2}$$

where $\hat{\rho}(Z, Y)$ is the estimated validity in the applicant population, $r(X, Y)$ is the correlation between the predictor subject to explicit selection and the criterion, $r(Z, Y)$ is the correlation between the variable subject to incidental selection and the criterion, $r(X, Z)$ is the correlation between the two predictors, and U is the ratio of the standard deviation of X in the applicant sample to the standard deviation of X in the selected sample (Thorndike, 1949, equation 7, p. 174).

When there is incidental selection on a predictor but adjustment for restriction of range is made from Equation (1) (assuming explicit selection), serious errors can result (Linn, 1968). If the predictor subject to incidental selection is moderately correlated with the variable subject to explicit selection the adjustment tends to underestimate the validity in the unselected group. If the two predictors are relatively independent, the adjustment tends to overestimate the validity. However, considering all factors that influence adjustments for restriction of range there is evidence that, in practice, corrections for restriction of range tend to be conservative (Linn, Harnisch, and Dunbar, 1981a). Restriction of range also occurs when there is nonrandom attrition on the criterion owing to such actions as promotions, transfers, or terminations. In practice data are seldom available to adjust for this type of restriction.

Both obtained validity coefficients and coefficients adjusted for restriction of range should be reported in a criterion-related validation study. In interpreting coefficients adjusted for range restriction the following points should be kept in mind. The correction assumes linearity of regression and homoschedasticity. Departures from linearity and homoschedasticity can result in either underestimates or overestimates of the unrestricted validity (Greener and Osburn, 1980). The correction tends to be more accurate for higher levels of unrestricted validities and lower degrees of range restriction. In most cases the adjusted validity coefficient is a more accurate estimate of the unrestricted population validity than is the obtained validity coefficient.

An important use of restriction of range formulas is in comparing the relative validity of two predictors in a situation in which one predictor has been subject to explicit selection and the alternative predictor has been subject to incidental selection. In this situation the predictor with the highest validity in the restricted sample may not have the highest validity in the applicant sample (Thorndike, 1949; Hsu, 1982).

Criterion-related validities are frequently corrected for both errors of measurement and restriction of range. This can be accomplished in two ways: (1) a triple correction in which the obtained validity coefficient is first adjusted for restriction of range, the criterion reliability coefficient is corrected for restriction of range, and then the adjusted validity coefficient is corrected for errors of measurement using the adjusted reliability coefficient; (2) a two-step correction in which the obtained validity coefficient is corrected for errors of measurement using the restricted reliability coefficient, and then the corrected validity coefficient is adjusted for restriction of range. Both procedures give equivalent results (Lee, Miller, and Graham, 1982). However, it is incorrect first to adjust the obtained validity coefficient for restriction of range and then correct the adjusted validity coefficient for errors of measurement using the restricted reliability coefficient.

8.1.3.4   Formula Estimates of Cross-Validated Multiple Correlation

It is customary in criterion-related validity studies to use multiple predictors for screening applicants. Multiple predictors are usually needed to measure all the relevant factors that should be considered in personnel selection. Sometimes scores on multiple predictors are combined judgmentally, that is, by subjective weighting of the data, but probably more often scores are combined into a composite using least squares regression weights (OLS). The advantage of using least squares regression weights is that the sum of squared errors of prediction is minimized and the correlation between the predictor composite and criterion scores is maximized; and each predictor is weighted in relation to its contribution to the overall prediction. A disadvantage is overestimation of the population multiple correlation attributable to the fitting of errors specific to the sample. To overcome this problem investigators have advocated cross-validation of the sample weights on a new sample as the best method of dealing with overfitting. However, cross-validation is a costly and not always feasible procedure involving, as it does, splitting the sample into two parts. Recent Monte Carlo studies have shown that formula estimates provide useful and accurate estimates of the population multiple correlation and the population cross-validated multiple correlation (Schmitt, Coyle, and Rauschenberger, 1977). Cattin (1980) has summarized the various formula estimates.

Estimation of the Population Multiple Correlation.   Wherry's (1931) formula is appropriate:

$$\hat{\rho}^2 = 1 - \frac{(N-1)}{(N-p-1)}(1-R^2) \tag{3}$$

where N is the sample size, p is the number of predictors, and R is the sample multiple correlation. When N is less than 50, Cattin (1980) recommended using an equation developed by Olkin and Pratt (1958, p. 211).

Estimation of the Population Cross-Validated Multiple Correlation Assuming that the Variables are Distributed Multivariate Normal (the Correlation Model). The following formula developed by Browne (1975, p. 82, Formula 2.10) has been shown in simulation studies to be quite accurate (Drasgow, Dorans, and Tucker, 1979).

$$\hat{\rho}^2(C) = \frac{(N - p - 3)\rho^4 + \rho^2}{(N - 2p - 2)\rho^2 + p} \qquad (4)$$

where ρ is estimated by Equation (3).

Estimation of the Population Cross-Validated Multiple Correlation Assuming that the Predictors are Fixed (the Regression Model). For this case Cattin (1980) recommended the following formula:

$$\hat{\rho}^2(C) = \frac{(N - 1)\rho^4 + \rho^2}{(N - p)\rho^2 + p} \qquad (5)$$

where ρ is also estimated using Equation (3).

These formula estimates are appropriate for situations in which the estimation procedure is ordinary least squares (OLS) and the predictor variables have been selected a priori. When these conditions are met, the formula estimates are more accurate and precise than the results of empirical cross-validation procedures (Schmitt et al., 1977; Drasgow et al., 1979).

These estimates of the population multiple correlations are useful in several contexts. For example, in comparing the validity of a composite based on least squares weights with a composite based on some other method of combination such as expert judgment, the relevant comparison is between the estimated cross-validated multiple correlation and the validity of the competing composite. In combining predictors into a composite there is always the possibility of using equal weights rather than OLS weights. The advantage of using equal weights is that equal weights are not subject to sampling error and the problem of overfitting is avoided. In addition, under some conditions encountered in actual practice equal weights provide a similar level of accuracy compared with OLS weights (Wainer, 1976). The disadvantage of equal weights is that some predictors may be overweighted, resulting in lowering the validity of the composite. Formula estimates of the validity of optimally weighted composites facilitate relevant comparisons between competing models.

8.1.3.5 Advantages of the Criterion-Related Validity Model

The criterion-related validity model is an effective technology for applications in which little or nothing is known about the job or the validity of potential selection devices pertaining to the job. In these situations a thorough job analysis is obviously necessary for the identification of potentially valid predictors and for the development of criterion measures. Predictors can be administered to applicants but not used for selection. Data collection can be extended until sufficient sample sizes are obtained to achieve adequate statistical power for detecting practically useful effects.

In addition to the available statistical apparatus mentioned above, the criterion-related validity model also lends itself to the study of the utility of selection decisions that are based upon this model (see Section 8.1.6 for a discussion of utility). If a payoff function can be associated with the criterion measures and the cost of administering the selection procedures estimated, the utility of various selection strategies can be calculated. This is an important advantage of the criterion-related validity model.

The criterion-related validity model is also obviously well suited for the evaluation of newly developed predictors. Even in jobs for which well established predictors are available, study of the incremental validity of experimental selection techniques for increasing the predictability of job performance would be desirable and necessary.

8.1.3.6 Limitations of the Criterion-Related Validity Model

Statistical Power

The statistical power of a criterion-related validity study depends upon sample size, population effect size, attenuation of the population effect size due to criterion unreliability, and restriction of range. A statistical power of approximately .80 to .90 is needed to adequately guard against type II errors (Cohen and Cohen, 1975). Schmidt, Hunter, and Urry (1976) have shown that sample sizes on the order of 75 to 400 (depending on the population validity, the selection ratio, and degree of error variance in the criterion measurements) are required for this level of statistical power assuming explicit selection on the predictor. In many practical situations a criterion-related validity study may not be feasible because sample sizes needed for adequate statistical power are simply not available. However, there has been a trend over the past 30 years toward increasing sample sizes in validation studies published in *Personnel Psychology* (Lent, Aurbach, and Levin, 1971; Monahan and Muchinsky, 1983).

The statistical power of several closely related situations has also been studied. Sample sizes needed for adequate statistical power have been investigated for predictor composites (Cascio, Valenzi, and Silbey, 1978, 1980). Statistical power of a predictor subject to incidental selection has also been investigated, and somewhat smaller sample sizes are needed to attain adequate statistical power under incidental selection (Sackett and Wade, 1983). In addition, tables have been developed for estimating statistical power in testing for differential validity (Trattner and O'Leary, 1980). Drasgow and Kang (1984) investigated statistical power in both differential validity (comparing correlation coefficients) and differential prediction (comparing regression coefficients) and showed that tests of differential prediction were uniformly more powerful than tests of differential validity. Although there are substantial variations in samples sizes required, depending on the above mentioned variables, a conclusion from all these studies is that large sample sizes are needed for adequate statistical power in most conditions encountered in criterion-related validity studies.

Neglect of Cumulative Studies

The criterion-related validity model does not, in and of itself, provide for use of previous research on predictor-job combinations. Thus, although the criterion-related validity model is well suited to situations in which there is little or no information from previous research, this model tends to encourage practitioners to validate and revalidate selection devices for which there is abundant evidence of validity. Recent developments in validity generalization (to be discussed in more detail in Sections 8.1.5 and 8.1.7) have shown the cost ineffectiveness of carrying out a criterion-related validity study on predictor-job combinations for which there is strong evidence of validity in similar contexts. Slavish adherence to the criterion-related validity model and the fear of litigation under civil rights laws can result in failure to use selection devices, especially tests, for which there is strong evidence of validity. Thus applicants, unqualified on the selection device, may be hired in order to garner more conclusive evidence of validity in future studies.

8.1.4 RELIABILITY OF MEASUREMENT

Validity and reliability of measurement are interrelated concepts. If a selection procedure is a valid predictor of job performance, then the greater the reliability of the procedure, the more likely that qualified applicants will be selected and unqualified applicants rejected. High reliability always favors the more qualified and lack of reliability always favors the less qualified. Thus even the most valid selection procedures may be ineffective if reliability is low. In addition, in most situations, reliability of measurement can be improved by increasing the number of measurement observations.

Measurement of human attributes is subject to many sources of error. For example, a reliability coefficient based on the correlation between equivalent forms of a test administered on two different occasions is affected by differences in the content of the two forms, changes in the individuals taking the test during the interval between the administration of the two forms, differences in test administration conditions, and random response variability. A reliability coefficient based on a single administration of a test is not affected by changes in individuals or differences in test administration. On the other hand, such reliability coefficients are highly sensitive to the internal consistency (homogeneity) of the test items. Because different estimates of reliability reflect different sources of error, there are many relevant reliabilities of a measurement, and it is important to note in some detail the type of reliability coefficient that is being reported.

8.1.4.1 Classical True Score Theory

Classical psychometric theory assumes that the observed score on a test can be expressed as the sum of two components: a true score and an error score. Additional assumptions are that (1) the expected error score is zero; (2) the correlation between true and error scores is zero; (3) the correlation between the error score on one measurement and the true score on a second is zero; and (4) the correlation between errors on distinct measurements is zero (Lord and Novick, 1968). These assumptions lead to the definition of parallel measurements as measurements with equal true scores for every individual and equal error variances.

From the aforementioned assumptions and definitions it follows that in the population of persons (1) the expected true score is equal to the expected observed score, (2) the observed score variance is equal to the sum of the true score variance and the error score variance, and (3) the square of the correlation between observed scores and true scores equals the ratio of the true score variance to the observed score variance. The reliability of a test is defined as the squared correlation between observed scores and true scores and is equal to the correlation between parallel measurements. The standard error of measurement is the square root of the error variance, which in turn is equal to the observed score variance multiplied by 1 minus the reliability coefficient.

A number of important consequences can be derived from true score theory including the Spearman-

Brown formula for estimating the reliability of a test that is increased in length and the formula for correcting validity coefficients for attenuation due to errors of measurement. Because corrections for attenuation have become widespread in personnel research, this formula is discussed in some detail in the next section.

8.1.4.2 Correction for Attenuation

An important relation between reliability and validity, the correction for attenuation, can be derived from true score theory,

$$\rho(T_X, T_Y) = \frac{\rho(X, Y)}{[\rho(X, X')\rho(Y, Y')]^{1/2}} \tag{6}$$

where $\rho(X, Y)$ is the correlation between observed scores, $\rho(T_X, T_Y)$ is the correlation between true scores on the X and Y variables, and $\rho(X, X')$ and $\rho(Y, Y')$ are the respective reliability coefficients. Equation (6) adjusts the observed correlation for unreliability in both X and Y. Adjustment can be made for unreliability in X alone by omitting $\rho(Y, Y')$ from Equation (6), or for unreliability in Y alone by omitting $\rho(X, X')$. It is clear from Equation (6) that the correlation between true scores will be overestimated if the reliability of X or Y is underestimated and underestimated if the reliabilities are overestimated. Because one of the frequent applications of reliability estimates is to adjust validity coefficients for attenuation, this is an important issue in personnel selection. The extent to which reliability may be underestimated or overestimated depends in part on the method of estimation, and it can be shown that there is a hierarchy of lower bound estimates of reliability (Guttman, 1945; Jackson and Agunwamba, 1977; Callender and Osburn, 1980).

8.1.4.3 Reliability Estimates

When correcting obtained validity coefficients for errors of measurement, it is important to avoid underestimating the true reliability. If two (or more) parallel forms of a test are available, reliability can be estimated by computing the correlation between equivalent forms. Reliability will be underestimated by this procedure to the extent that (1) the two forms are not exactly parallel owing to variations in factor content, and (2) the two forms are not exactly parallel owing to variations in true scores as a result of events intervening between the administration of the two forms. Thus, in order to avoid underestimating reliability when making corrections for attenuation, equivalent forms should be matched in factor content and administered in reasonably close temporal proximity.

If only one form of a test is available, coefficient alpha can be used to estimate reliability. Coefficient alpha is derived from a persons-by-components two-way analysis of variance with one observation per cell.

$$\alpha = MS(\text{persons})/MS(\text{persons}) - MS(\text{residual}) \tag{7}$$

where α is the estimated reliability of the sum of components and MS(residual) is the interaction mean square. This estimate has been proposed by various investigators (Jackson, 1939; Hoyt, 1941; Cronbach, 1951) and it can be shown that Kuder–Richardson formulas 20 and 21 are special cases.

In large samples coefficient alpha is a lower bound to the reliability coefficient. To the extent that the components are factorially heterogeneous alpha will underestimate reliability. Because it can be shown that coefficient alpha computed on the items of a test is equal to the expected value of coefficient alpha computed on all possible split halves (Cronbach, 1951), it follows that, for heterogeneous tests, there will be an alpha computed on a split half that is larger than coefficient alpha computed on the test items but that, in large samples, is still a lower bound to the true reliability. Thus coefficient alpha, computed on the items of a heterogeneous test, should be avoided when adjusting correlations for attenuation. In most cases a better procedure would be to divide the items into two halves based on similarity of content and compute coefficient alpha on the halves.

A special problem arises when the data are unmatched, that is, not all individuals in the sample are measured on every component (Cronbach, Rajaratnam, and Gleser, 1963). For example, unmatched data frequently occur when the data are supervisor ratings of job performance. If the data are unmatched, mean square within persons instead of mean square interaction must be used to estimate reliability. This results in a reliability estimate that is a lower bound to coefficient alpha. The consequence may be a serious underestimation of reliability if the data are partially matched; for example, some supervisors rate most of the job incumbents and a few supervisors rate relatively few job incumbents.

As mentioned earlier, reliability estimates based upon a single administration of a test do not reflect such sources of error as differences in test administration or changes in individuals taking the test. If these two factors are important to the decisions based on the test scores, coefficient alpha will underestimate the error variance due to these two sources and may overestimate the relevant reliability of the test. Thus we return to the point made in the beginning of this section. There are many reliabilities of a test depending on the sources of error that influence the scores on which decisions or inferences are to be based. The theory of generalizability of tests provides a framework for analyzing components of variance involved in test scores and systematic methods for assigning variance components to measurement error, depending upon the decision parameters involved in the study (Cronbach, Gleser, Nanda, and Rajaratnam, 1972).

Reliability of measurement and corrections for attenuation due to errors of measurement play important roles in research on validity generalization, the subject of the next section.

8.1.5   VALIDITY GENERALIZATION

One of the major innovations in personnel selection over the past 10 years has been the development of the concept and method of validity generalization by Schmidt and Hunter (1977). The central hypothesis is that observed variations in validity coefficients from study to study are at least partially due to statistical artifacts that affect the obtained validity coefficients. Although there are six or seven artifacts that can affect the results, there are four artifacts that can be quantified and used to adjust obtained validities: (1) errors of measurement in the criterion, (2) errors of measurement in the predictor, (3) restriction of range, and (4) sampling errors. Validity generalization represents a significant change in thinking regarding the validity of selection procedures in that it is hypothesized that the validity of tests and other selection techniques is very stable across situations, rather than the opposite hypothesis that validity differs greatly from situation to situation.

Validity generalization involves the accumulation of criterion-related validity studies for certain combinations of predictors and jobs. The procedure consists of correcting the obtained distribution of validity coefficients for variations due to the statistical artifacts mentioned above. Adjustments for variations in restriction of range and criterion reliability are most common. However, when different versions of the same predictor are included in the study, variations in predictor reliability are adjusted to a common mean value. In addition, variance in validity coefficients due to sampling error is subtracted from the variance of the obtained distribution.

The objective of a validity generalization study is to estimate the mean and variance of the distribution of true corrected validities. Several methods for adjusting the obtained validity coefficients for statistical artifacts have been proposed (Schmidt, Hunter, Pearlman, and Shane, 1979; Pearlman, Schmidt, and Hunter, 1980; Callender and Osburn, 1980; Raju and Burke, 1983). Both the Callender and Osburn (1980) procedure and the Schmidt et al. (1979) procedure yield very similar estimates of the mean and variance of the distribution of true validities. In a simulation study it was found that accuracy of the mean and variance estimates were not affected by the shape of the distribution of true validities (Callender, Osburn, Greener, and Ashworth, 1982). Callender and Osburn (1980, 1982) and Raju and Burke (1983) have described the mathematical foundation of validity generalization equations in some detail. For a recent review of validity generalization models see Burke (1984).

One outcome of validity generalization studies is a test of the hypothesis that the true validity variance is zero. Acceptance of this hypothesis suggests that true validity does not vary across situations, jobs, and so on that were involved in the study. If the true validity variance is near zero, the hypothesis is probably tenable. However, the power of tests of this hypothesis tends to be low in cumulating studies involving moderate sample sizes (Osburn, Callender, Greener, and Ashworth, 1983). Of course, it is not necessary that the true validity variance be zero for validity generalization. Validity generalization is said to exist when the confidence interval of the estimated distribution of true validity coefficients exceeds zero or some arbitrarily chosen positive value. Typically this value is set at 0.10. A procedure that has much in common with validity generalization is meta-analysis (Hunter, Schmidt, and Jackson, 1982; Glass, McGaw, and Smith, 1981). Meta-analysis is a method of quantitative analysis of cumulative experimental or correlational studies. When applied to validation studies meta-analysis is similar to validity generalization, but in meta-analysis there is more emphasis on estimating the mean effect size and less emphasis on estimating the standard deviation of the effect sizes.

Validity generalization studies have shown that much of the variation in obtained validity coefficients can be attributed to statistical artifacts. Criterion-related validities, at least for cognitive tests, tend to generalize across similar jobs and there is strong evidence for generalization across jobs with substantial differences in task structure (Pearlman, Schmidt, and Hunter, 1980; Dunnette, 1982). On the other hand, there is evidence that a given cognitive ability test is more valid for some jobs than for others (Schmidt, Hunter, and Pearlman, 1981; Linn, Harnisch, and Dunbar, 1981b). Thus the results of validity generalization studies can be used in two ways: (1) to generalize the validity of cognitive tests to similar jobs and (2) to optimize validity by using validity generalization data to select a predictor with optimal validity for the job in question (Osburn et al., 1983).

One issue in validity generalization is the accuracy of correction for attenuation. As mentioned in the discussion of reliability the correction for attenuation due to errors of measurement in the criterion could either overestimate or underestimate the true validity, depending on the extent to which the true reliability of the criterion is over- or underestimated. When the criterion is supervisor's ratings of job performance, the following factors would tend to result in underestimating reliability: (1) systematic differences among the supervisors in their observation of job performance, (2) the use of the intraclass correlation in estimating reliability, and (3) changes in the job performance of the ratees due to experience, training, changes in jobs, and so on. On the other hand, lack of independence in the ratings by different supervisors or collecting ratings from the same supervisor at different times would tend to result in overestimating the true reliability. In attempting to evaluate the impact of these factors it should be noted that level of supervision is correlated with the kinds of subordinate behavior that can be observed and through necessity first- and second-line supervisors are frequently used. This would tend to result in underestimating reliability, unless the second-line supervisor based his or her judgments on input from the first-line supervisor (lack of independence).

Empirical reliability estimates are often lacking in validity generalization studies. For this reason assumed distributions of reliabilities are frequently used. Schmidt and Hunter (1977) proposed an assumed reliability distribution ranging from 0.30 to 0.90 with a mean of 0.60; this distribution has been used in several studies to adjust validities for errors of measurement in ratings of job proficiency (Pearlman et al., 1980; Schmidt, Gast-Rosenberg and Hunter, 1980). Is 0.60 a reasonable estimate of the average reliability of supervisor ratings? Schmidt and Hunter (1977) have argued that it is probably an underestimate because the appropriate reliability estimate is the correlation across reasonable time intervals between ratings produced by different raters. Because ratings from different supervisors are usually collected at a single point in time, it is argued that reliability is overestimated. However, it also can be argued that using a coefficient of stability in estimating the reliability of supervisor ratings is deficient because coefficients of stability tend to assign real changes in job performance to error variance. The issue is important because using 0.6 as the assumed average reliability results in increasing the average obtained validity by about 30%, whereas using a more conservative estimate, say, 0.8, as the assumed average reliability increases the average obtained validity by only about 12%.

8.1.6 UTILITY OF SELECTION

Criterion-related validity is concerned with the accuracy with which job performance or other criteria are predicted. Utility refers to the gains or losses to an organization as a consequence of using a selection procedure. A predictor may have a high validity coefficient and still have low utility. On the other hand, a predictor may have relatively low validity and still have high utility. Utility is a function of the costs of administering the selection system, the validity of the predictors, the proportion of applicants that are selected, and the magnitude and significance of individual differences in criterion performance. Clearly, utility is the ultimate test of the practical effectiveness of a selection procedure.

Taylor and Russell (1939) proposed a two-valued payoff function for estimating the utility of a predictor. Their tables provide an estimate of the expected proportion of "satisfactory" hires given the selection ratio, base rate, and validity coefficient. Though useful, the Taylor-Russell approach has several limitations. Continuous criterion scores must be dichotomized, resulting in loss of information. Further, choice of the dichotomy is often arbitrary, and this choice may influence the results (Schmidt, Hunter, Mckenzie, and Muldrow, 1979). In addition, the cost of testing is not considered.

Brogden (1946, 1949) presented equations which showed that utility was a continuous, linear function of validity. His equations incorporated the cost of testing and pointed up the importance of the standard deviation of job performance. Brogden (1949) was among the first to propose that the criterion be evaluated in dollars to facilitate the application of utility concepts. Brogden and Taylor (1950) emphasized the application of cost accounting methods to criterion development.

Cronbach and Gleser (1965) in their classic work greatly expanded utility theory to include two-stage and multistage sequential selection, placement decisions, adaptive treatments, and other important concepts. They gave the following equation (equation 2, p. 37) for net gain in utility per applicant evaluated in fixed treatment selection:

$$\Delta U = r(X,Y)s(Y)E(X') - C(x) \tag{8}$$

where ΔU is the net gain in utility per applicant evaluated, $r(X,Y)$ is the validity coefficient of the predictor, $s(Y)$ is the standard deviation of the criterion evaluated in dollars, X' is the cutoff score on the predictor, $E(X')$ is the ordinate of the normal curve corresponding to that score, and $C(x)$ is the cost of collecting predictor information on one applicant. A number of principles can be deduced from the preceding equation.

1. Utility is a linear function of validity.
2. Because $E(X')$ is largest at the median of the normal distribution, net gain in utility per applicant evaluated is greatest when 50% of the applicants are selected from a normal score distribution. However, as shown in the next section net gain in utility per applicant selected is usually greatest when the selection ratio is low.
3. Utility is a function of the magnitude and practical significance of individual differences on the criterion.
4. When the cost of collecting predictor information is considered, the predictor with the highest validity may not have the highest utility.

Dividing the net gain in utility per person evaluated by P, the proportion of applicants selected, yields net gain in utility per person selected (equation 1.10, p. 309):

$$\frac{\Delta U}{\text{Selectee}} = r(X, Y)s(Y)\overline{X} - \frac{C(x)}{P} \tag{9}$$

where scores on the predictor are expressed in standard score units. Equation (9) is important because it shows that when the cost of evaluating applicants is low, large numbers of applicants should be evaluated relative to the number selected. Thus, within limits established by the cost of processing applicants, the lower the selection ratio, the higher the net gain in utility per person selected. Because the number to be selected is usually fixed, gains in utility can be obtained by increasing the number of applicants.

8.1.6.1 Selection Strategies

Using these utility concepts it is possible to compare the utility per person evaluated for various selection strategies. Some possible selection strategies are as follows: *single-screen,* in which selection is based on a single predictor; *battery,* in which applicants are administered all predictors and selected on the basis of their combined score on the battery; *sequential,* in which all applicants are administered the first predictor and on the basis of their score are divided into three groups—select, reject, or administer the second predictor; *pre-reject,* in which some applicants are rejected based on their scores on the first predictor and the remaining applicants are administered the second predictor; *pre-accept,* in which some applicants are accepted on the basis of the scores on the first predictor and the remaining applicants are administered the second predictor.

Depending on the relative size of the validity coefficients of the two predictors and their relative costs of administration any one of the above strategies may be optimal. For example, when the cost of administering the first predictor is considerably less than the cost of administering the second predictor and the selection ratio is low, a pre-reject strategy may yield the greatest utility. This is ordinarily the situation when cognitive tests are combined with an employment interview. It usually pays to screen on the cognitive tests, which tend to have high validity and low costs, using a pre-reject strategy and then screen the remaining applicants on the employment interview, which tends to have low to moderate validity and high cost.

8.1.6.2 Selection with Adaptive Treatments

Utility theory provides a framework for a systems approach to personnel selection in which recruitment, selection, training, and job design are all interrelated parts of a comprehensive personnel system. As mentioned earlier, when each person accepted is assigned the same treatment (fixed treatment selection), utility is a linear function of validity. However, Cronbach and Gleser (1965) have shown that when treatments (job design, training program, etc.) can be adjusted according to the ability of the selectees, the gain in utility (over fixed treatment selection) due to adapting treatment to the average ability of those selected ia a parabolic function of validity. Potentially, considerable gains in utility can be achieved through adaptive treatments. Gains in utility due to selection may be partially offset by modifications of training programs or the design of jobs to fit the characteristics of those selected.

When placement decisions are considered and individuals can be assigned to the treatment that is best for them, the added utility that can occur from optimal assignment depends on differences in the slopes of the payoff functions for the various treatments. For example, there is some evidence that growth need strength (need for autonomy, need for personal growth, etc.) of job incumbents moderates the relationship between the motivating potential of jobs and job satisfaction where motivating potential is assessed by evaluating the job on such variables as skill variety, autonomy, feedback, task significance, and task identity (Hackman and Oldham, 1976).

8.1.6.3 Estimating the Standard Deviation of the Payoff Function

Utility theory has been used effectively in simulation studies to evaluate the consequences of various selection strategies. (Cascio and Silbey, 1979; Hunter, Schmidt, and Rauschenberger, 1977). However, a major obstacle to using utility equations to assess the dollar value of gains from selection has been difficulties in developing a metric for the payoff function. A few attempts have been made to apply cost accounting methods to this problem (Roche, 1965). Recently subjective methods for estimating the standard deviation (S.D.) of the payoff function have been proposed (Schmidt, Hunter, McKenzie, and Muldrow, 1979; Cascio, 1982). In the Schmidt et al. (1979) procedure knowledgeable respondents are required to estimate the dollar value of the mean, as well as the 15th and 85th percentiles of employee contributions. Assuming that the payoff function is normally distributed and that respondents "know" the mean, percentile points, and shape of the distribution, the S.D. can be estimated. This procedure yielded an estimate of $10,413 per year as the S.D. for computer programmers and $11,327 per year for budget analysts (Schmidt et al., 1979). An alternative method is to estimate the S.D. of employee contributions as approximately 40% of average salary or approximately 20 to 30% of average output (Schmidt and Hunter, 1983). When substituted in utility equations, these S.D. estimates result in very large estimates of net gain in utility for situations in which large numbers of applicants are screened with a predictor of substantial validity (Hunter and Schmidt, 1982; Schmidt, Hunter, and Pearlman, 1982; Hunter and Hunter, 1984).

How accurate are the judgmental methods of estimating the standard deviation of employee contributions? Very strong assumptions are involved, but the estimates seem reasonable. One study has compared estimates by supervisors and senior salespersons with actual data on dollar value of volume of sales (Bobco, Karren, and Parkington, 1983). There was fairly good correspondence between the best estimates of the S.D. averaged over judges ($48,800) and actual S.D. of sales volume ($52,300). However, there was considerable variability in the estimates. One respondent estimated $300,000 as the mean annual sales volume, and the lowest estimate was $13,900. A recent study also found wide variability in estimates of average value (Burke and Frederick, 1984). Estimates of the average dollar value of the performance of district sales managers ranged from $25,000 to $500,000. In a modified procedure designed to diminish the effect of such variability it was found that estimated S.D.s were not comparable with either actual sales volume or district manager's salary.

Another problem with using judgmental estimates of the S.D. is that the estimated S.D. of the payoff function is substituted for the actual S.D. of supervisor's rating of overall job performance. This implies that the selection procedure has exactly the same validity in predicting dollar volume of output as it does for predicting supervisor ratings of overall job performance. Either underestimates or overestimates of the utility of the selection procedure could result, depending on the validity of the selection device for predicting output compared with the validity in predicting supervisor's ratings of overall job performance.

Boudreau (1983) has pointed out that current applications of utility models have simply multiplied estimated yearly program benefits by some estimate of average tenure. This procedure does not consider such potential influences as variable costs, the need for discounting costs and benefits, duration and variability of program effects, diminishing returns over time, effects of changes in the size of the work force, and the effects of turnover. For example, the benefits of selection on cognitive ability, rather than remaining level as is usually assumed, may actually diminish over time as new employees learn the job.

8.1.7 PREDICTOR CONSTRUCTS

The impact of civil rights legislation, the development of validity generalization methods for aggregating the results of criterion-related validation studies, and the extraordinary interest in assessment center methodology have all combined to produce a wealth of data on predictor constructs and considerable insight into their relation to job performance. This section discusses research on five important selection procedures: cognitive tests, the selection interview, biographical data, assessment centers, and motor abilities. Research on personality and interest measures has been less extensive and are not reviewed here. Readers interested in this topic are referred to two recent reviews of the use of personality tests in personnel selection (Cornelius, 1983; Hogan, Carpenter, Briggs, and Hanson, 1984). In addition, research on work sample tests used for prediction will not be reviewed here. Interested readers are referred to Asher and Sciarrino (1974).

8.1.7.1 Cognitive Ability

In the discussion that follows the term cognitive ability is used to refer to mental ability that reflects the cumulative influence of nonspecific life experiences (Anastasi, 1968). Cognitive ability is typically measured by paper and pencil tests, and is often used interchangeably with the term aptitude which has a predictive connotation. Cognitive ability is distinguished from achievement, which refers to the outcomes of specific learning experiences as opposed to potential.

There is no universal agreement on a list of cognitive abilities. Thurstone published the primary mental abilities in 1938. Since that period there have been numerous attempts to classify cognitive abilities using factor analytic methods. Guilford (1967) proposed a Structure of Intellect model involving three dimensions: operations, contents, and products. The model implied 120 possible abilities because there were five operations, four contents, and six products all combined independently. However, independent investigations of the Structure of Intellect model have raised serious questions regarding its accuracy (Horn, 1967; Horn and Knapp, 1973). A "kit" of reference tests based upon factor analytic studies was distributed by the Educational Testing Service (French, Ekstrom, and Price, 1963). Twenty-four cognitive ability factors were included with two to five reference tests for measuring each factor. Ekstrom (1973) has reviewed more recent research relevant to these 24 factors. Ekstrom's report has shown that the situation remains unsettled with respect to the identification of cognitive abilities.

This review uses a classification of cognitive abilities that was developed by Pearlman (1979) for use in validity generalization studies. This classification is rather broad and combines some factors that are separable in factor analytic studies but is useful in describing cognitive ability tests involved in validation studies.

General Mental Ability.   The ability to understand verbal and numerical concepts, underlying principles, and the ability to reason or solve problems requiring the perception and understanding of relationships among abstract patterns or symbols. Tests classified as general mental ability contain a combination of any verbal, quantitative, and either reasoning or spatial ability item types. Either a summed score of separate test parts or a spiral omnibus type of test may be used.

Verbal Ability.   The ability to comprehend and use language effectively and correctly. Cognitive tests containing reading comprehension, vocabulary, grammar, spelling, word fluency, sentence completion, and/or any combination of these item types are classified as verbal ability.

Quantitative Ability.   The ability to understand numerical relationships and concepts and to perform routine arithmetic operations quickly and accurately. Cognitive tests containing computations (addition, subtraction, multiplication, division, and/or mixed computations), arithmetic word problems, error location, and graph/table reading or combinations of these item types are classified as quantitative ability.

Reasoning Ability.   The ability to think logically and clearly in factual, symbolic, or figural terms; the ability to understand and apply underlying principles, or to draw correct conclusions or make good decisions from stated conditions or information. Cognitive tests that contain verbal reasoning, abstract reasoning, letter series, number series, judgment, and/or logical order of events type items are classified as reasoning ability.

Perceptual Speed.   The ability to perceive pertinent detail quickly and accurately in verbal, numerical, pictorial, or graphic material; all measures of this ability are speeded. Cognitive tests containing name, number, or name and number comparisons, figure comparisons, cancellations, number filing, coding, name filing, and/or substituting are classified as perceptual speed.

Memory.   The ability to learn, recall, and reproduce or apply visually or orally presented information or associations among verbal, numerical, or figural stimuli. Tests containing memory for oral instructions, classification, coding, substitution, number writing, and immediate memory type items are classified as memory tests.

Spatial/Mechanical Ability.   The ability accurately to perceive or to visualize and manipulate mentally spatial patterns and relationships or the orientation and movement of objects in space; the ability to understand and apply simple physical and mechanical principles. Cognitive tests containing mechanical knowledge, spatial relations, location, mechanical principles, and/or pursuit items are classified as spatial/mechanical.

General Clerical Ability.   The ability to perform tasks of a general clerical nature which require a basic facility with verbal or numerical material and the ability quickly to perceive detail in such material. Clerical aptitude batteries are classified as general clerical ability.

In the review that follows the term "adjusted validity coefficient" is used to describe validity coefficients that have been adjusted for restriction of range and/or criterion unreliability as is commonly done in validity generalization studies or meta-analyses. Adjustments for criterion unreliability have frequently assumed a mean criterion reliability of .60. In addition, the term "estimated mean true validity" is used to designate the mean of the distribution of true validity coefficients as estimated by a formal validity generalization study.

Clerical Occupations

Cognitive ability tests have been widely used for the selection of clerical workers. Ghiselli (1966, 1973) reported average validities of 0.40 for perceptual accuracy (perceptual speed) and 0.33 for spatial/mechanical tests in predicting training criteria, and 0.27 in predicting job proficiency criteria. The pattern of validities was similar for recording and general clerks but spatial/mechanical tests tended to have higher validities for computing clerks.

The very substantial validity of cognitive ability tests for predicting performance in clerical occupations has been amply demonstrated by Pearlman et al. (1980) in a large-scale validity generalization effort summarizing the results of 698 published and unpublished studies. Results showed that for stenographic, typing, filing, and related occupations (DOT Occupational Groups 201–209) estimated mean true validities against criteria of job proficiency ranged from 0.20 for spatial/mechanical tests to 0.50 for general mental ability tests and tests classified as clerical aptitude. In addition, 90% credibility values were all positive. Estimated mean true validities for predicting training criteria ranged from 0.29 for reasoning tests to 0.80 for tests of general mental ability.

For computing and account-recording occupations (DOT Occupational Groups 210–219) estimated mean adjusted validities against proficiency criteria ranged from 0.41 for tests of verbal ability to

0.63 for reasoning tests. Estimated credibility values were 0.10 or larger for all cognitive ability test types. Against training criteria estimated mean true validities ranged from 0.36 for spatial/mechanical test to 0.66 for general mental ability and quantitative ability tests. For production and stock clerks and related occupations (DOT Occupational Groups 221–229) estimated mean true validities against proficiency criteria ranged from 0.31 for tests of reasoning ability to 0.48 for spatial/mechanical tests. Against training criteria estimated mean true validities ranged from 0.26 for reasoning and spatial/mechanical tests to 0.70 for general mental ability tests.

For all clerical occupations included in the study consisting of the above listed occupations (plus information and message distribution occupations, public contact and clerical service occupations, and three categories of unspecified, miscellaneous, or mixed clerical occupations) estimated mean true validity against proficiency criteria ranged from 0.30 for spatial/mechanical tests to 0.52 for general mental ability tests. Against training criteria estimated mean true validities ranged from 0.39 for perceptual speed to 0.71 for general mental ability. These results indicate that cognitive ability tests were valid for clerical occupations, with general mental ability tending to be the most useful and tests of spatial/mechanical ability the least useful. It should be emphasized however, that the pattern of validities for clerical aptitude, verbal, quantitative, and reasoning tests were very similar to the pattern for general mental ability.

Process Occupations

In a review of validity studies in the petrochemical industry Dunnette (1972) reported median validities ranging from 0.19 (quantitative tests) to 0.32 (general ability tests) for operator and processing jobs. For maintenance jobs median validities ranged from 0.20 (general ability tests) to 0.38 (spatial/mechanical tests). For clerical jobs median validities ranged from 0.13 (quantitative tests) to 0.22 (verbal tests). In a recent reanalysis of these data, in which validity coefficients were corrected for errors of measurement, Hunter and Hunter (1984) reported an average adjusted validity of .45 for cognitive tests (excluding perceptual speed).

In addition, two recent validity generalization studies (Schmidt, Hunter, and Caplan, 1981; Callender and Osburn, 1981) using different analyses on the same data set have been reported for operator, maintenance, and laboratory technician jobs in the petroleum industry. Four types of cognitive ability tests were studied: spatial/mechanical (different forms of the Bennett Test of Mechanical Comprehension), general mental ability (primarily RBH Learning Ability), quantitative ability (different forms of RBH Arithmetic Reasoning test), and RBH Chemical Comprehension, a test measuring knowledge of basic chemical concepts. For operator jobs estimated mean true validity ranged from 0.26 for general mental ability and quantitative ability to 0.33 for spatial/mechanical ability. For maintenance jobs mean estimated true validity ranged from 0.15 for quantitative ability to 0.33 for spatial/mechanical. Estimated mean true validities for operator, maintenance, and laboratory jobs combined ranged from 0.20 for quantitative ability to 0.32 for general mental ability against proficiency criteria and 0.47 for chemical comprehension to 0.54 for general mental ability against training criteria. In both studies it was found that the variance of obtained validities was largely accounted for by statistical artifacts.

In a large-scale industry-wide study of electric power plant operators Dunnette (1982) found a mean obtained validity of 0.35 between a predictor composite consisting of spatial, quantitative, general ability, assembly, and table reading tests and a criterion composite measuring job problem-solving ability. When validities were adjusted for errors of measurement and restriction of range, the mean adjusted validity was 0.48 and the variance of the adjusted validities was in accordance with chance expectations, indicating that the validity of the composite generalized across organizations.

Other Occupations

Schmidt, Gast-Rosenberg, and Hunter (1980) reported a study of the validity of the Programmer Aptitude Test (PAT) for predicting job proficiency and training criteria for computer programmers. Results showed a mean estimated true validity for the PAT of 0.73 against proficiency criteria and 0.91 against training criteria. In addition, 90% credibility values were well above zero.

In a review of military studies Vineberg and Joyner (1982) reported a median validity coefficient of 0.12 for mental ability tests in predicting global ratings of job performance, and a median validity of 0.24 for predicting likely suitability for military service. Validities were in the 0.30 to 0.50 range for job knowledge test criteria, but were quite variable. These validities are considerably lower than those reported above for clerical, process, and computer programmer occupations. Two hypotheses were offered to explain these results: (1) differences in job performance become minimal after the first year because selection and training have combined to reduce differences in technical ability; and (2) global ratings were inadequate because of halo, contamination, and lack of opportunity to observe job performance. However, no data were available to assess the accuracy of these hypotheses.

Hunter (1980) cumulated the results of 515 studies of the validity of the General Aptitude Test Battery (GATB) using the job analysis categories of data, people, and things. Three composites of the 9 GATB tests were used: GVN—general ability, verbal aptitude, and numerical aptitude; SPQ—

spatial aptitude, form perception, and clerical perception; and KFM—motor coordination, finger dexterity, and manual dexterity. Average validities of GVN for various job families ranged from 0.13 (feeding/off-bearing) to 0.35 (negotiating/serving). Average validities of SPQ ranged from 0.14 (mentoring) to 0.35 (setup). Average validities of KFM ranged from 0.06 (mentoring) to 0.35 (feeding/off-bearing). Average adjusted validities of 0.54 for GVN, 0.41 for SPQ, and 0.26 for KFM were found against training criteria and 0.45, 0.37 and 0.37 against proficiency criteria.

Job Complexity

Hunter and Hunter (1984), using data from the USES (U.S. Employment Service) study discussed above, found a mean adjusted validity of 0.56 for GVN against performance criteria for jobs of highest complexity compared with a mean adjusted validity of 0.23 for jobs of lowest complexity. Slightly higher and more uniform mean validities were found for training criteria. SPQ showed approximately the same pattern of mean adjusted validities whereas KFM showed higher validities in predicting job performance for jobs of lowest complexity. It was concluded that the validity of cognitive tests is moderated by job complexity. In a study in which position analysis questionnaire (PAQ) dimension scores were correlated (across jobs) with validities of subtests of the general aptitude test battery (GATB) it was found that PAQ dimensions of decision-making and information processing moderated the validity of GATB cognitive ability tests (Gutenberg, Arvey, Osburn, and Jeanneret, 1983). These results also support the hypothesis that job complexity is a moderator of the validity of cognitive tests.

In utilizing the above results on the validity of cognitive tests the following points should be kept in mind. Results from validity generalization studies and meta-analyses provide the most accurate estimates of the true validity of cognitive ability in applicant populations because the obtained validities have been adjusted upward for the effects of statistical artifacts. Summaries of unadjusted, obtained validities are underestimates of the true validity in the applicant population. However, to the extent that applicants are preselected on unmeasured variables that are highly correlated with cognitive ability one can expect lower operational validities. For example, if in hiring professionals only those with a college grade point average of 3.5 or better are considered, the validity of cognitive tests could be expected to be lower for this group. In addition, as shown by the data on job complexity, cognitive tests will have lower validities for jobs with a low level of decision-making and skill variety. Although cognitive ability tests have been shown to be generally valid for broad occupational classifications, the results of validity generalization studies and meta-analyses can and should be used to identify the most valid predictors for the specific application at hand.

8.1.7.2 Employment Interviews

Employment interviewing is a wide-band selection technique in the sense that information obtained from the interview can be used to evaluate many different attributes and behaviors. Because the interview is so useful in gathering pertinent information about applicants, it is far and away the most widely used employee selection procedure. Like other wide-band techniques the interview also tends to have low reliability and validity (Wagner, 1949; Mayfield, 1964; Ulrich and Trumbo, 1965). For this reason the interview has come under sharp criticism (Dunnette and Bass, 1963). More recent investigators have also found low validities for the interview. Dunnette (1972) reported a median validity of 0.30 for clerical jobs, and 0.14 for testing and quality control jobs. Reilly and Chao (1982) reported an average validity of 0.23. Hunter and Hunter (1984) reported an average adjusted validity of 0.14 for predicting supervisor ratings.

There are a number of methodological problems that make it difficult to interpret validity studies on the interview. Sometimes interviewers have access to test and credential information about the applicant that affects their judgments along with information obtained in the interview. Often interviewers make the final hiring decision resulting in severe restriction of range in interviewer ratings. Also, interviewer predictions are to some degree a function of the individual skills of the interviewer, so that generalizations about "the interview" are hazardous.

More recently investigators have sought to improve the reliability and validity of the employment interview by such interventions as training interviewers, developing interview guides, studying the decision-making process, and developing structured interview strategies.

Interventions to Improve the Interview

Training of Interviewers. Several studies have demonstrated the effectiveness of training interviewers (Wexley, Yukl, Kovacs, and Sanders, 1973; Howard and Dailey, 1979). In the program evaluated by Howard and Dailey participants are taught to use a carefully constructed interview guide. Videotaped demonstration interviews are used to model effective behaviors. Practice interviews are audio- and videotape recorded and critiqued by peers and a qualified trainer. Finally, the literature on decision-making in the interview is utilized in training interviewers to avoid common decision-making errors.

Use of Interview Guides.   An interview guide is used to prompt the interviewer in effective ways of opening the interview by establishing rapport and structuring the interaction. Open-ended questioning techniques for obtaining information about work background, education, career goals, and so on are illustrated. Summarizing techniques and effective ways of commenting on information are presented. Finally, practical methods of presenting information about the organization and the job and closing the interview are presented. Interview guides help to standardize the interview and obtain comparable information from all applicants for the same job.

Structuring the Interview.   There is evidence to indicate that structuring the interview can increase both reliability and validity (Wagner, 1949; Mayfield, 1964; Ulrich and Trumbo, 1965). For example, a recent study by Latham, Saari, Pursell, and Campion (1980) developed interview questions based on a critical incidents job analysis. In addition, objective benchmarks for evaluating answers to the structured questions were provided. In two studies interview reliabilities were 0.76 and 0.79, respectively. Two concurrent validity coefficients were 0.46 and 0.30, respectively. Another approach to structuring the interview is to define evaluation dimensions, as is done in assessment centers. Then questions are developed that facilitate obtaining information relevant to these dimensions. This technique facilitates structuring the interview through organizational and job analysis.

Use of Job Information.   There is evidence that providing interviewers with relevant job information can increase interviewer reliability (Schmitt, 1976).

Use of Multiple Interviewers.   The reliability of interviewer judgments can be increased by using multiple interviewers and averaging their judgments or asking them to reach a consensus (Reynolds, 1979).

Interviewer Decision-Making

Edward Webster's *Decision Making in the Employment Interview,* published in 1964, stimulated extensive experimental and field research on factors influencing decision making in the employment interview. This literature has been summarized in several reviews (Ulrich and Trumbo, 1965; Wright, 1969; Schmitt, 1976; Arvey and Campion, 1982). These studies have identified a whole host of biases and response tendencies that can influence interviewer judgments. Some of the more potent biasing factors that have been found to influence interviewer judgments are listed below.

1. *First Impression Bias.*   Information that comes first in the interview tends to have a stronger influence on the final decision than information that comes later.
2. *Negative Information.*   Interviewers tend to weight unfavorable information about applicants more heavily than favorable information.
3. *Stereotyping.*   Interviewers tend to form a stereotype of an ideal applicant, and evaluate the current applicant with respect to that stereotype.
4. *Nonverbal Cues.*   Nonverbal cues (appearance, eye contact, smiling, posture) tend to influence interviewer judgments strongly.
5. *Contrast Effect.*   The qualifications of preceding applicants can influence the evaluation of the current applicant either positively or negatively.

Discrimination in the Interview

Arvey (1979) has summarized research on age, sex, race, and handicapped status. He concluded that females are generally given lower evaluations than males when both have similar qualifications (Dipboye, Arvey, and Terpstra 1977). Evidence also supported an interaction between sex and job type. Females are given lower ratings for "masculine jobs" and vice versa for males. Few studies were found regarding the influence of race on interviewer judgments. Two studies found a strong effect of age on interviewer evaluations and studies of handicapped applicants suggest that handicapped applicants receive lower ratings in terms of hiring but higher evaluations on motivational variables. Litigation over the employment interview has emphasized appropriate areas of inquiry and objective standards of evaluation.

8.1.7.3   Assessment Centers

There has been a steady increase in the use of multiple assessment procedures since the publication of results of the Management Progress Study, a longitudinal study undertaken by AT&T in 1956 (Bray and Grant, 1966). Because this study set the pattern for assessment center applications. It is described briefly in order to outline the general features of the method. Candidates were brought together for $3\frac{1}{2}$ days at an assessment center where they were assessed using a variety of techniques including interviews, paper and pencil tests, projective tests, work samples, and participation in group problem solving and leaderless group discussions. The assessment staff discussed each candidate extensively and rated him or her on 25 dimensions plus an overall rating of management potential. Data on the 422 assessees were held in confidence for research purposes only and had no effect on the careers of the participants. This study reported large and significant differences in management progress between participants who were rated as having high management potential and those rated as having low potential. Correlations between the overall assessment rating and salary progress ranged from 0.46 to 0.57 in various subgroups.

Cohen, Moses, and Byham (1977) define an assessment center program as follows:

(1) Multiple assessment techniques: such as situational exercises, interviews, objective and projective tests, peer ratings, and other performance measures are used; at least some of the techniques require group interaction. The techniques are selected or designed to bring out performance relating to dimensions previously identified as important for job success. . .

(2) From three to six trained assessors staff the program. Usually they are unfamiliar with one another and the assessees, do not directly supervise the assessees, and are two to three supervisory levels above them. Assessment training for the staff from 1 to 3 days to 1 month. A psychologist may serve on the staff or be the director of the program.

(3) From six to twelve participants are assessed at the same time, and there is a low ratio of assessees to assessors (1–1 up to 4–1) which allows systematic and close observation, and multiple evaluations of each assessee by several assessors.

(4) The assessment staff integrates and interprets its observations of each assessee judgmentally and may use psychological tests or other psychometric measures to supplement its judgment. The strengths and weaknesses of an individual on each dimension are evaluated, followed by a global rating which is an overall prediction based on assessment center performance.

(5) The assessment center is conducted away from the job but not necessarily off company premises, usually in a 1 to 3 day period.

Huck (1973) in a review of the external and internal validity of assessment centers reported moderate to high rater agreement on ratings of assessment dimensions. Mean observer reliabilities ranged from about 0.60 to 0.80 in most studies. The test–retest reliability of assessment center ratings was evaluated in two multiple assessment programs in which a group was assessed twice (with a month or more intervening between the two programs). The correlation between overall performance in the two programs was 0.73.

Several reviews of research on assessment centers have reported substantial validity for overall assessment ratings for predicting job potential, management progress, and job performance (Huck, 1973; Cohen et al., 1977; Thornton and Byham, 1982). Cohen et al. (1977), in a summary of 19 studies, reported median validity coefficients of 0.33 for predicting job performance, 0.63 for predicting job potential, and 0.40 for predicting management progress. Unlike the management progress study there was the possibility of criterion contamination influencing the outcomes of these studies. The management progress criterion was especially vulnerable to criterion contamination. However, in a review of five studies in which no results were reported to participants or their managers Thornton and Byham (1982) concluded that all five studies were supportive of the validity of assessment ratings.

Another issue in assessment center research has been the relative validity of assessment center ratings as compared with other assessment techniques such as paper and pencil tests. Thornton and Byham (1982, p. 318) summarize the validity of assessment ratings in comparison with other methods as follows:

Although there are some mixed results in the studies cited, it appears that data generated in assessment centers can yield more accurate predictions of managerial success than paper and pencil tests alone. In most cases the overall assessment rating is more accurate than the typical ability or personality test scores. Situational exercises and dimension ratings by assessors are more valid than single test scores.

Although validity studies of assessment centers have been positive, recent research has raised a number of issues concerning the internal validity of the method. Klimoski and Strickland (1977), noting that assessment centers have consistently shown higher validities in predicting management progress compared with job performance as a manager, have suggested that to some extent assessors may be assessing behaviors that facilitate getting promoted rather than behaviors that reflect true management potential. Sackett and Dreher (1982) have questioned the construct/content validity of assessment centers by showing that correlations between ratings on different dimensions within exercises were substantially higher than correlations between ratings of the same dimension across exercises. Factor analysis of these dimension-exercise ratings produced exercise factors rather than dimension factors.

Assessment center methodology has received favorable rulings in several court cases (Byham, 1982). One issue is the extent to which assessment centers can be validated using content validity. Sackett and Dreher (1982) have argued that the methodology is too complex for validation by the content validity strategy.

8.1.7.4 Biographical Data

Biographical data have been widely used to predict job performance and tenure in a variety of occupations. Reilly and Chao (1982) reported an average validity of 0.38 for biographical data averaged across a wide variety of jobs. Hunter and Hunter (1984), on the basis of their meta-analyses, place biographical data second in general validity to cognitive ability tests with a mean adjusted validity of 0.37 against supervisory ratings of job performance. Brown (1981), in a study of the predictive validity of biographical data for life insurance salesmen, found an average adjusted validity of 0.26, with evidence of some difference in validity across companies. Hunter and Hunter (1984) reported a validity of 0.40 for the Supervisory Profile Record, a biodata instrument for identifying supervisory potential, developed in a consortium study involving a large number of organizations.

Most biographical instruments are empirically keyed; that is, item scores are assigned on the basis of discrimination between known groups. For this reason biographical keys must be cross-validated in order to estimate their validity. One problem with empirical keys is a tendency to lose validity as time passes. Thus such instruments must be rekeyed and revalidated in order to maintain their validity over time. Because large sample sizes are required, biodata predictors are most useful for large organizations or consortium studies.

Recently, a radically different approach to biodata analysis has been proposed by Owens and Schoenfeldt (1979). Biodata dimensions, identified in factor analytic studies, are used to cluster individuals into homogeneous subgroups. When biodata subgroups have been established, the behavior of the subgroups is investigated to identify behavioral patterns that discriminate among the subgroups and that predict the future behavior of members of the subgroup. This approach, though promising, has not been widely used in selection research possibly because the empirical keying method is more practical and efficient, at least in the short range. Eberhardt and Muchinsky (1982a,b) have presented evidence on the factor stability and vocational relevance of factor analytically derived clusters.

8.1.7.5 Motor and Physical Abilities

Motor performance has been studied extensively using factor analytic methods. Fleishman (1962) reported evidence for 11 relatively independent motor abilities.

Control Precision.   Ability to make finely controlled muscular adjustments.

Multilimb Coordination.   The ability to coordinate the movements of a number of limbs simultaneously.

Response Orientation.   The ability to make accurate movements in relation to a stimulus under highly speeded conditions.

Reaction Time.   The speed of response when a stimulus appears.

Speed of Arm Movement.   Speed of gross arm movements where accuracy is not required.

Rate Control.   The ability to make continuous motor adjustments relative to a moving target.

Manual Dexterity.   The ability to make skillful arm and hand movements in handling relatively large objects under speeded conditions.

Finger Dexterity.   The ability to manipulate small objects with the fingers skillfully.

Arm–Hand Steadiness.   The ability to make precise arm–hand positioning movements that do not require strength or speed.

Wrist–Finger Speed.   The ability to make rapid tapping movements with the wrists and fingers.

Aiming.   The ability to place dots in circles as rapidly as possible.

Motor ability tests that have been used for personnel selection have for the most part been limited to tests of finger dexterity, manual dexterity, wrist–finger speed, and aiming. Dunnette (1972) reported an average validity of 0.21 for motor tests used for selection in the petroleum refining industry. These data involved mostly maintenance jobs. Hunter and Hunter (1984) in a reanalysis of Ghiselli's data reported mean adjusted validities for general motor ability ranging from .44 for vehicle operators to 0.17 for salesclerks. Hunter and Hunter also concluded that the validity of motor ability tests is moderated by job complexity. In their analysis of GATB data average validities for motor tests ranged from 0.30 and 0.21 in predicting performance in more complex jobs to 0.48 for predicting performance in jobs of least complexity.

Fleishman (1964) found nine physical ability factors in a factor analytic study of physical fitness measures. These factors were (1) static strength, (2) dynamic strength, (3) explosive strength, (4) trunk strength, (5) extent flexibility, (6) dynamic flexibility, (7) gross body coordination, (8) gross body equilibrium, and (9) stamina. Recently there has been increased interest in measures of physical abilities for screening applicants for physically demanding jobs (Campion, 1983).

Campion (1983) reviewed three different approaches to physical screening. The first is to measure maximum aerobic power and then to select people whose maximum aerobic power is great enough so that they can perform the job without excessive physiological fatigue (Astrand and Rodahl, 1977). A second approach is the analysis of physical strength of workers in relation to the strength requirements of jobs (Chaffin, 1974). A third focus is the work of Fleishman (1964) mentioned above, which involves the measurement of nine physical fitness factors. A modification of the Fleishman approach combining data from factor analytic studies and the physiological literature has been developed by Hogan (1984).

Campion (1983) reported that most studies involving physical strength predictors have found that usually one or two measures could adequately predict the criteria. Criterion-related validity studies have shown high validities when maximum performance physical ability tests were used to predict maximum performance work samples or completion of difficult training programs. However, the sparse data available suggest that correlations tend to be much lower when predicting overall job performance.

8.1.7.6 Comparisons Among Selection Procedures

In a recent meta-analysis Hunter and Hunter (1984) reported that cognitive ability had the highest adjusted validity for entry level jobs (0.53), followed by job tryout (0.44), biodata (0.37), and reference checks (0.26). The interview was sixth (0.14). For predictors used for promotion and certification adjusted validities ranged from 0.54 for work sample tests to .43 for assessment centers. Ability composite again had an adjusted validity of .53.

8.1.8 LEGAL ASPECTS OF PERSONNEL SELECTION

The Civil Rights Act of 1964, related legislation, and executive orders have had a far-reaching impact on personnel selection and the professional practice of personnel psychology. One consequence has been government regulation of personnel functions through a series of guidelines culminating in the *Uniform Guidelines On Employee Selection Procedures* (1978). A second consequence has been extensive litigation in the federal court system, including several important cases that have been decided by the United States Supreme Court (Griggs v. Duke Power Co., 1971; Albemarle Paper Co. v. Moody, 1975; Washington v. Davis, 1976).

The *Uniform Guidelines* were issued jointly by four agencies: Civil Service Commission, Department of Justice, Equal Opportunity Employment Commission, and Department of Labor. The *Guidelines* are divided into five major subdivisions: General Principles, Technical Standards (for Validity Studies), Documentation of Impact and Validity Evidence, Definitions, and Appendix. The stated purpose of the *Guidelines* is to, ". . . . Incorporate a single set of principles which are designed to assist employers, labor organizations, employment agencies, and licensing and certification boards to comply with requirements of Federal law prohibiting employment practices which discriminate on grounds of race, color, religion, sex, and national origin" (1978, Section 1B). The *Guidelines* define discrimination and adverse impact, spell out record-keeping requirements, list detailed technical standards that should be met in conducting criterion-related, content, and construct validity studies, and specify documentation requirements for record-keeping and reporting validity studies. These *Guidelines* (published in the *Federal Register*) should be consulted by anyone planning to use and/or validate a selection procedure.

The *Uniform Guidelines* have been controversial to say the least. For example, it is argued that it is impossible to comply fully with the technical standards required by the *Guidelines*. A case in point is language regarding the consideration of suitable alternative selection procedures: ". . . Accordingly, whenever a validity study is called for by these guidelines, the user should include, as part of the validity study, an investigation of suitable alternative selection procedures and suitable alternative methods of using the selection procedure which have as little adverse impact as possible, to determine the appropriateness of using or validating them in accordance with these guidelines. . ." (1978, Section 3B). A more basic criticism is that the *Guidelines* are so detailed and structured that it is not possible to comply with every provision in any practical validity study. A validity study may meet the highest professional standards and still be in violation of some provisions of the *Guidelines*.

In regard to the way in which the *Uniform Guidelines* were structured one can contrast the *Guidelines* with *Standards for Psychological Tests and Manuals* published by APA (1973) and *Principles for Validation and Use of Personnel Selection Procedures* developed by APA's Division of Industrial/Organizational Psychology (1980). Both the *Standards* and the *Principles* focus on more general principles as compared with the *Guidelines*.

Differential Validity and Test Fairness

Differential Validity. An important issue that has stimulated research and discussion is the extent to which cognitive ability tests may be differentially valid for minority groups. Early investigations found that cognitive ability tests more frequently showed significant validity in majority samples as compared to minority samples (Kirkpatrick, Ewen, Barrett, and Katzell, 1968). However, these studies tended to focus on single group validity, that is, the validity coefficient is statistically significant in one group but not the other (Boehm, 1972). Because sample sizes for majority samples were almost always larger, this artifact alone could have accounted for the discrepancies. A more accurate formulation of the problem was a test of the null hypothesis of no difference between majority and minority samples (differential validity) (Humphreys, 1973; Linn, 1978; Bartlett, Bobco, and Pine, 1977). When studies were conducted with better controls, the results have supported the view that cognitive ability tests are equally valid for majority and minority groups (Bray and Moses, 1972; O'Conner, Wexley, and Alexander, 1975; Schmidt, Berner, and Hunter, 1973; Boehm, 1977; Katzell and Dyer, 1977; Hunter, Schmidt, and Hunter, 1979). Although it can be concluded that cognitive ability tests have comparable validity for minority and majority groups, it should be remembered that the *Guidelines* require a test of differential validity when such a test is feasible in validation studies.

Test Fairness. On the average, minority samples tend to score lower on cognitive ability tests compared with majority samples. Minorities also tend to score lower in job performance. However, the discrepancies between minority and majority job performance tend to be smaller than the discrepancies observed for cognitive ability tests. Analysis of ways to deal with this situation has resulted in sharp controversy regarding the issue of test fairness. Hunter and Schmidt (1976) have summarized statistical definitions of test fairness that have been proposed by various investigators. The Cleary (1968) definition held that a test is unbiased only if the regression lines for blacks and whites are identical. Fairness by the Cleary definition is assessed by a statistical test of equality of regression lines. If the slopes of the two lines are equal, differences in intercepts can be accounted for by using a multiple regression equation with test score and race as predictors. If slopes are unequal, either two separate regression equations must be used or a product term of race times test score must be used in the regression equation. Linn and Wertz (1971) noted that a theoretically perfectly reliable test that is unbiased in the Cleary sense and on which black applicants score lower than white applicants will have unequal regression lines because of unreliability of measurement. However, as Hunter and Schmidt (1976) have pointed out, the effect of unreliability is always to bias the test in favor of less qualified applicants.

Three additional definitions of test fairness have been proposed. Thorndike (1971) suggested that it would be fair to select the same percentage of each group that would have been selected had the selection been made on the criterion. Darlington's definition 3 defined a test as fair only if the correlation between race and test score with the criterion partialed out of both race and test score is zero (Darlington, 1971). Cole (1973) argued that applicants who would be successful if selected should have an equal chance of being selected regardless of race. This will be true only if the regression lines of test on criterion are the same. Because a zero partial correlation implies equality of regression, Darlington's definition 3 (1971) and the Cole (1973) definition are the same.

The Cleary definition is frequently used as a test of fairness in practical situations. Slopes and intercepts of majority and minority regression lines are tested for equality. If, as frequently happens, slopes are equal but the majority intercept is above the minority intercept, then either two regression equations are used or a common regression line is used. In a simulation study using various combinations of validity, minority base rate, and selection ratios Hunter, Schmidt, and Rauschenberger (1977) compared the consequences of using selection strategies implied by each of the three definitions of test fairness plus a strict quota model in which the selection ratio is the same for all subgroups. It was concluded that increases in minority selection ratios across models (Cleary, Thorndike, Darlington, and Quota) were more striking than losses in expected utility, and that the utility of the Darlington and Thorndike models expressed as a percentage of the utility of the Cleary model remained quite high for many commonly occurring selection situations (low minority base rate, moderate validity, and low overall selection ratios).

It is clear, as Hunter and Hunter (1984) have pointed out, that no matter what fairness selection strategy is adopted, using valid selection procedures with relatively low selection ratios results in much higher utility compared with alternative strategies. For example, consider a job for which a cognitive ability test is far and away the most valid predictor. Which of the following two selection strategies will result in lower adverse impact and higher utility? (1) Hire from the top down in both groups based on ability test scores, but use quotas to avoid adverse impact. (2) Set a low cutoff score on the ability test and hire randomly from above that cutoff. As Hunter and Hunter (1984) have shown, option 1, that is, hiring from the top down with quotas results in much higher utility given this situation. Using quotas to minimize adverse impact in selection may be an anathema to some on ethical grounds, but it can be an effective strategy from a utility point of view. A similar point has recently been made in a study of U.S. park rangers (Schmidt, Mack, and Hunter, 1984).

8.1.9 PERSONNEL SELECTION, TRAINING, AND JOB DESIGN

The typical approach to personnel selection is to assume that the job and any associated training programs are fixed and consequently selection procedures and selection criteria must be tailored to fit the abilities and personal characteristics that are required by the job and the associated training programs. This is one reason that there is so much emphasis on job analysis as the initial step in a criterion-related validity study. Although this model has much to recommend it, there are other possibilities that should be considered.

As mentioned in the utility section, gains in utility can be achieved by adapting the job and/or the training program to the average ability of those selected. If differential job placement is feasible, then gains in utility are possible to the extent that the regression of job performance on predictor scores differs from job to job.

One limitation of the job analysis approach is that in many occupations the initial job assignment is only temporary and, as the employee gains experience and expertise, job assignments change, often in such a way that different abilities and skills are required for successful job performance. In addition, jobs change as a result of new technologies such that different abilities and skills are required. Finally, there is an inevitable interaction between definition of the job and the personal characteristics of the incumbent. To a greater or lesser extent most job incumbents mold the job to fit their abilities, skills, and interests. These considerations suggest that selection systems must be sensitive to the long-range needs of the organization and that criteria for personnel selection should be set in the light of the dynamic nature of job change and job mobility.

It can be argued that personnel selection is in some respects more fundamental than training or job design. Training programs can be modified and jobs can be redesigned, but it is more difficult to change the basic abilities and personality characteristics of job incumbents. This may be one reason why selection on cognitive ability has proved so successful in some occupations.

8.1.10 SUMMARY

Although construct and content validity models have received greater attention in recent years, the criterion-related validity model is still dominant in research in personnel selection. This model has been augmented by greater attention to statistical artifacts such as restriction of range, errors of measurement, and sampling error that influence the results. In addition, simulation studies have shown that formula estimates of population multiple correlations and population cross-validated multiple correlations are more accurate and precise than the more cumbersome empirical cross-validation procedures.

There has been increasing attention to utility of selection. The consequences of various selection strategies have been evaluated in simulation studies using utility concepts. Studies involving judgmental estimates of the standard deviation of the payoff function have suggested that the payoff from using valid selection techniques is very substantial.

The number and quality of validity generalization studies have resulted in considerable progress toward scientific generalizations concerning the validity of selection techniques. The evidence is quite strong regarding the generalized validity of cognitive ability tests for clerical and for mechanical and process occupations. One important clue to generalization of cognitive ability is the hypothesis that job complexity moderates the validity of cognitive ability tests (Hunter and Hunter, 1984).

The development of assessment center methodology has been a significant advance in personnel selection. Assessment centers have demonstrated good validity for the prediction of management potential and show promise for a wide variety of selection applications.

Federal regulation of personnel selection has had strong impact on personnel selection practice and research. There has been a marked increase in the quality of validation studies although possibly a decline in the rate of appearance of such studies in journals (Boehm, 1982; Monahan and Muchinsky, 1983). Some issues such as the differential validity of cognitive ability tests for minority groups have been laid to rest for the most part. In addition a consensus seems to be forming regarding the appropriateness and usefulness of the Cleary definition of test fairness.

REFERENCES

Albemarle Paper Co. v. Moody (1975). 422 U.S. 405, 9 EPD #10,230, 10 FEP 1181.

American Psychological Association (1973). *Standards for educational and psychological tests and manuals.* Washington, DC: American Psychological Association.

American Psychological Association (1980). *Principles for the validation and use of personnel selection procedures.* Washington, DC: APA Division of Industrial/Organizational Psychology.

Anastasi, A. (1968). *Psychological testing,* 3rd ed. New York: Macmillan.

Arvey, R. D. (1979). Unfair discrimination in the employment interview: Legal and psychological aspects. *Psychological Bulletin, 86,* 736–765.

Arvey, R. D., and Campion, J. E. (1982). The employment interview: A summary of recent literature. *Personnel Psychology, 35,* 281–322.

Asher, J. J., and Sciarrino, J. A. (1974). Realistic work sample tests: A review. *Personnel Psychology, 27,* 519–533.

Astrand, P. O., and Rodahl, K. (1977). *Textbook of work physiology,* 2nd ed. New York: McGraw-Hill.

Barrett, G. V., Phillips, J. S., and Alexander, R. A. (1981). Concurrent and predictive validity designs. *Journal of Applied Psychology, 66,* 1–6.

Bartlett, C. J., Bobco, P., and Pine, S. M. (1977). Single group validity: Fallacy of the facts. *Journal of Applied Psychology, 62,* 155–157.

Bernardin, H. J., and Pence, E. C. (1980). Effects of rater training: Creating new response sets and decreasing accuracy. *Journal of Applied Psychology, 65,* 60–66.

Bernardin, H. J., and Smith, P. C. (1981). A clarification of some issues regarding the development and use of behaviorally anchored rating scales (BARS). *Journal of Applied Psychology, 66,* 458–463.

Bobco, P. (1983). An analysis of correlations corrected for attenuation and range restriction. *Journal of Applied Psychology, 68,* 584–589.

Bobco, P., Karren, R., and Parkinton, J. J. (1983). Estimates of standard deviations in utility analysis: An empirical test. *Journal of Applied Psychology, 68,* 170–176.

Boehm, V. R. (1972). Negro–white differences in validity of employment and training selection procedures: Summary of research evidence. *Journal of Applied Psychology, 56,* 33–39.

Boehm, V. R. (1977). Differential prediction: A methodological artifact? *Journal of Applied Psychology, 62,* 145–154.

Boehm, V. R. (1982). Are we validating more but publishing less? (The impact of government regulation on published validation research—An exploratory investigation). *Personnel Psychology, 35,* 175–187.

Borman, W. C. (1983). Implications of personality theory and research for the rating of work performance in organizations. In F. Landy, S. Zedeck, and J. Cleveland, eds., *Performance measurement and theory.* Hillsdale, NJ: Erlbaum.

Borman, W. C., Rosse, R. L., and Abrahams, N. M. (1980). An empirical construct validity approach to studying predictor-job performance links. *Journal of Applied Psychology, 65,* 662–671.

Boudreau, J. W. (1983). Effects of employee flows on utility analysis of human resource productivity programs. *Journal of Applied Psychology, 68,* 396–406.

Bray, D. W. and Grant, D. L. (1966). The assessment center in the measurement of potential for business management. *Psychological Monographs, 80* (Whole No. 625).

Bray, D. W., and Moses, J. L. (1972). Personnel selection. *Annual Review of Psychology, 23,* 545–576.

Brogden, H. E. (1946). On the interpretation of the correlative coefficient as a measure of predictive efficiency. *Journal of Educational Psychology, 37,* 65–76.

Brogden, H. E. (1949). When testing pays off. *Personnel Psychology, 2,* 171–183.

Brogden, H. E., and Taylor, E. K. (1950). The dollar criterion—applying the cost accounting concept to criterion construction. *Personnel Psychology, 3,* 133–154.

Brown, S. H. (1981). Validity generalization in the life insurance industry. *Journal of Applied Psychology, 66,* 664–670.

Browne, M. W. (1975). Predictive validity of a linear regression equation. *British Journal of Mathematical and Statistical Psychology. 28,* 79–87.

Burke, M. J. (1984). Validity generalization: A review and critique of the correlation model. *Personnel Psychology, 37,* 93–116.

Burke, M. J., and Frederick, J. T. (1984). Two modified procedures for estimating standard deviations in utility analysis. *Journal of Applied Psychology, 69,* 482–489.

Byham, W. C. (1982). *Review of legal cases and opinions dealing with assessment centers and content validity.* Monograph IV. Pittsburgh, PA: Developmental Dimensions.

Callendar, J. C., and Osburn, H. G. (1980). Development and test of a new model for validity generalization. *Journal of Applied Psychology, 65,* 543–558.

Callender, J. C., and Osburn, H. G. (1981). Testing the constancy of validity with computer-generated sampling distributions of the multiplicative model variance estimate: Results of petroleum industry validation research. *Journal of Applied Psychology, 66,* 274–281.

Callender, J. C., and Osburn, H. G. (1982). Another view of progress in validity generalization: Reply to Schmidt, Hunter and Pearlman. *Journal of Applied Psychology, 67,* 846–852.

Callender, J. C., Osburn, H. G., Greener, J. M., and Ashworth, S. (1982). Multiplicative validity generalization model: Accuracy of estimates as a function of sample size and mean, variance, and shape of the distribution of true validities. *Journal of Applied Psychology, 67,* 859–867.

Campion, M. A. (1983). Personnel selection for physically demanding jobs: Review and recommendations. *Personnel Psychology, 36,* 527–550.

Carroll, J. B. (1979). Measurement of abilities constructs. In *Construct validity and psychological measurement: Proceedings of a colloquium on theory and application in education and measurement.* Princeton, NJ: Educational Testing Service.

Cascio, W. A., (1982). *Costing human resources: The Financial impact of behavior in organizations.* Boston: Kent.

Cascio, W., and Silbey, V. (1979). Utility of the assessment center as a selection device. *Journal of Applied Psychology, 64,* 107–118.

Cascio, W. A., Valenzi, E. R., and Silbey, V. (1978). Validation and statistical power. *Journal of Applied Psychology, 63,* 589–595.

Cascio, W. A., Valenzi, E. R., and Silbey, V. (1980). More on validation and statistical power. *Journal of Applied Psychology, 65,* 135–138.

Cattin, P. (1980). Estimation of the predictive power of a regression model. *Journal of Applied Psychology, 65,* 407–414.

Chaffin, D. B. (1974). Human strength capability and low back pain. *Journal of Occupational Medicine, 16,* 248–254.

Cleary, T. A. (1968). Test bias: Prediction of grades of negro and white students in integrated colleges. *Journal of Educational Measurement, 5,* 115–124.

Cohen, J., and Cohen, P. (1975). *Applied multiple regression/correlation analysis for the behavioral sciences.* New York: Wiley.

Cohen, B. M., Moses, J. L., and Byham, W. C. (1977). *The validity of assessment centers: A literature review.* Pittsburgh, PA: Developmental Dimensions International.

Cole, N. S. (1973). Bias in selection. *Journal of Educational Measurement, 10,* 237–255.

Cooper, W. H. (1981). Ubiquitous halo. *Psychological Bulletin, 90,* 218–244.

Cornelius, III, E. T. (1983). The use of projective techniques in personnel selection. In *Research in personnel and human resources management,* Vol. 1. New York: JAI Press.

Cronbach, L. J. (1951). Coefficient alpha and the internal structure of tests. *Psychometrika, 16,* 297–334.

Cronbach, L. J. (1980). Selection theory for a political world. *Public Personnel Management, 9,* 37–50.

Cronbach, L. J., and Gleser, G. (1965). *Psychological tests and personnel decisions,* (2nd ed.) Urbana: University of Illinois Press.

Cronbach, L. J., and Meehl, P. E. (1955). Construct validity in psychological tests. *Psychological Bulletin, 52,* 281–302.

Cronbach, L. J., Rajaratnam, N., and Gleser, G. C. (1963). Theory of generalizability: A liberalization of reliability theory. *British Journal of Statistical Psychology, 16,* 138–161.

Cronbach, L. J., Gleser, G., Nanda, H., and Rajaratnam, N. (1972). *The dependability of behavioral measurements: Theory of generalizability for scores and profiles.* New York: Wiley.

Darlington, R. B. (1971). Another look at "cultural fairness." *Journal of Educational Measurement, 8,* 71–82.

Dipboye, R. L., Arvey, R. D., and Terpstra, D. E. (1977). Sex and physical attractiveness of raters and applicants as determinants of resume evaluations. *Journal of Applied Psychology, 1977, 62,* 288–294.

Drasgow, F., and Kang, T. (1984). Statistical power of differential validity and differential prediction analysis for detecting measurement nonequivalence. *Journal of Applied Psychology, 69,* 498–508.

Drasgow, F., Dorans, N.J., and Tucker, L. R. (1979). Estimates of the squared multiple correlations coefficient: A Monte Carlo investigation. *Applied Psychological Measurement, 3,* 387–399.

Dunnette, M. D. (1972). *Validity study results for jobs relevant to the petroleum industry.* Washington, DC: American Petroleum Institute.

Dunnette, M. D. (1976). Aptitudes, abilities and skills. In M. D. Dunnette Ed., *Handbook of industrial and organizational psychology.* Chicago: Rand McNally.

Dunnette, M. D. (1982). *Development and validation of an industry-wide electric power plant operator selection system.* Minneapolis, MN: Personnel Decisions Research Institute.

Dunnette, M. D., and Bass, B. M. (1963). Behavioral scientists and personnel management. *Industrial Relations, 2,* 115–130.

Ebel, R. E. (1977). Comments on some problems of employment testing. *Personnel Psychology, 30,* 55–84.

Eberhardt, B. J., and Muchinsky, P. M. (1982a). An empirical investigation of the factor stability of Owen's biographical questionnaire. *Journal of Applied Psychology, 67,* 138–145.

Eberhardt, B. J., and Muchinsky, P. M. (1982b). Biodata determinants of vocational typology: An integration of two paradigms. *Journal of Applied Psychology, 67,* 714–727.

Ekstrom, R. B. (1973). *Cognitive factors: Some recent literature* (Technical Report No. 2, ONR Contract N00014-71-C-0117, NR 150-329). Princeton, NJ: Educational Testing Service.

Farr, J. I., O'Leary, B.S., and Bartlett, C. J. (1971). Ethnic group membership as a moderator of the prediction of job performance. *Personnel Psychology, 24,* 609–636.

Feldman, J. M. (1981). Beyond attribution theory: Cognitive processes in performance appraisal. *Journal of Applied Psychology, 66,* 127–148.

Fleishman, E. A. (1962). The description and prediction of perceptual-motor skill learning. In R. Glaser, Ed., *Training research and education.* Pittsburgh: University of Pittsburgh Press.

Fleishman, E. A. (1964). *The structure and measurement of physical fitness.* Englewood Cliffs, NJ: Prentice-Hall.

French, J. W., Ekstrom, R. B., and Price, L. A., (1963). *Kit of reference tests for cognitive factors.* Princeton, NJ: Educational Testing Service.

Ghiselli, E. E. (1966). *The validity of occupational aptitude tests.* New York: Wiley.

Ghiselli, E. E. (1973). The validity of aptitude tests for personnel selection. *Personnel Psychology, 26,* 461–477.

Glass, G. V., McGaw, B., and Smith, M. L. (1981). *Meta-analysis in social research.* Beverly Hills, CA: Sage.

Greener, J. M., and Osburn, H. G. (1980). Accuracy of corrections for restriction of range due to explicit selection in heteroschedastic and nonlinear distributions. *Educational and Psychological Measurements, 40,* 337–347.

Greener, J. M., and Osburn, H. G. (1981). An empirical study of the accuracy of corrections for restriction in range due to explicit selection. *Applied Psychological Measurement, 3,* 31–41.

Griggs v. Duke Power Co. (1971). 424, 3 EPD #8137, 3 FEP 178.

Guilford, J. P. (1967). *The nature of human intelligence.* New York: McGraw-Hill.

Guion, R. M. (1976). Recruiting, selection and job placement. In M. D. Dunnette, Ed., *Handbook of Industrial and Organizational Psychology.* Chicago: Rand McNally.

Guion, R. N. (1977). Content validity—the source of my discontent. *Applied Psychological Measurement, 1,* 1–10.

Guion, R. N. (1978). "Content validity"—in moderation. *Personnel Psychology, 31,* 205–214.

Guion, R. N., and Cranny, C. J. (1982). A note on concurrent and predictive validity designs: A critical reanalysis. *Journal of Applied Psychology, 67,* 239–244.

Gutenberg, R. L., Arvey, R. D., Osburn, H. G., and Jeanneret, P. R. (1983). Moderating effects of decision-making/information processing job dimensions on test validities. *Journal of Applied Psychology, 64,* 602–608.

Guttman, L. A. (1945). A basis for analyzing test–retest reliability. *Psychometrika, 10,* 255–282.

Hackman, J. R., and Oldham, G. R. (1976). Motivation through the design of work: Test of theory. *Organizational behavior and human performance, 21,* 289–304.

Hogan, J. (1984). *A model of physical performance for occupational tasks.* Paper presented to the annual meeting of the American Psychological Association.

Hogan, R., Carpenter, B. N., Briggs, S. R., and Hanson, O. (1984). Personality assessment and personnel selection. In H. J. Bernardin and D. A. Bownas, Eds., *Personality assessment in organizations.* New York: Praeger.

Horn, J. L. (1967). On subjectivity in factor analysis. *Educational and Psychological Measurement, 27,* 811–820.

Horn, J. L., and Knapp, J. R. (1973). On the subjective character of the empirical base of Guilford's Structure-of-Intellect model. *Psychological Bulletin, 80,* 33–44.

Howard, G. S., and Dailey, P. R. (1979). Response-shift bias: A source of contamination of self-report measures. *Journal of Applied Psychology, 64,* 144–150.

Hoyt, C. (1941). Test reliability by analysis of variance. *Psychometrika, 6,* 153–160.

Hsu, L. M. (1982). Estimation of the relative validity of employee selection tests from information commonly available in the presence of direct and indirect range restriction. *Journal of Applied Psychology, 67,* 509–511.

Huck, J. R. (1973). Assessment centers: A review of the external and internal validities. *Personnel Psychology, 26,* 191–212.

Hull, C. L. (1928). *Aptitude testing.* Yonkers, NY: World Book.

Humphreys, L. G. (1973). Statistical definitions of test validity for minority groups. *Journal of Applied Psychology, 58,* 1–4.

Hunter, J. E. (1980). Test validation for 12,000 jobs: An application of synthetic validity and validity generalization to the General Aptitude Test Battery (GATB). Washington, DC: U.S. Employment Service, U.S. Department of Labor.

Hunter, J. E., and Hunter, R. F. (1984). Validity and utility of alternative predictors of job performance. *Psychological Bulletin, 96,* 72–98.

Hunter, J. E., and Schmidt, F. L. (1976). Critical analysis of the statistical and ethical implications of various definitions of test bias. *Psychological Bulletin, 83,* 1053–1071.

Hunter, J. E., Schmidt, F. L., and Rauschenberger, J. M. (1977). Fairness of psychological tests: Implications of four definitions for selection utility and minority hiring. *Journal of Applied Psychology, 62,* 245–262.

Hunter, J. E., Schmidt, F. L., and Hunter, R. F. (1979). Differential validity of employment tests by race. *Psychological Bulletin, 86,* 721–735.

Hunter, J. E., Gerbing, D. W., and Boster, F. J. (1982). Machiavellian beliefs and personality: Construct invalidity of the Machiavellianism dimension. *Journal of Personality and Social Psychology, 43,* 1293–1305.

Hunter, J. L., and Schmidt, F. L. (1982). The economic benefits of personnel selection using psychological ability tests. *Industrial Relations, 21,* 293–308.

Hunter, J. L., Schmidt, F. L., and Jackson, G. B. (1982). *Advanced meta-analysis: Quantitative methods for accumulating research findings across studies.* Beverly Hills, CA: Sage.

Ivancevich, J. M. (1979). Longitudinal study of the effects of rater training on psychometric error in ratings. *Journal of Applied Psychology, 64,* 502–508.

Jackson, P. H., and Agunwamba, C. C. (1977). Lower bounds for the reliability of the total score on a test composed of non-homogeneous items: Algebraic lower bounds. *Psychometrika, 42,* 567–578.

Jackson, R. W. (1939). Reliability of mental tests. *British Journal of Psychology, 29,* 267–287.

Joreskog, K. G. (1978). Structural analysis of covariance and correlation matrices. *Psychometrika, 43,* 444–447.

Katzell, R. A., and Dyer, F. J. (1977). Differential validity revived. *Journal of Applied Psychology, 62,* 137–145.

Kenny, D. A. (1979). *Correlation and causality.* New York: Wiley.

Kirkpatrick, J. J., Ewen, R. B., Barrett, R. S., and Katzell, R. A. (1968). *Testing and fair employment.* New York: New York University Press.

Klimoski, R. J., and Strickland, W. J. (1977). Assessment centers: Valid or merely prescient. *Personnel Psychology, 30,* 353–363.

Landy, F., and Farr, J. (1980). Performance rating. *Psychological Bulletin, 87,* 72–107.

Latham, G. P., and Wexley, K. N. (1981). *Increasing productivity through performance appraisal.* Menlo Park, CA: Addison-Wesley.

Latham, G. P., Saari, L. M., Pursell, E. D., and Campion, M. A. (1980). The situational interview. *Journal of Applied Psychology, 65,* 422–427.

Lee, R., Miller, K. J., and Graham, W. K. (1982). Corrections for restriction of range and attenuation in criterion-related validity studies. *Journal of Applied Psychology, 67,* 637–639.

Lent, R. H., Aurbach, H. A., and Levin, L. S. (1971). Predictors, criteria, and significant results. *Personnel Psychology, 24,* 519–533.

Linn, R. L. (1968). Range restriction problems in the use of self-selected groups for test validation. *Psychological Bulletin, 69,* 68–73.

Linn, R. L. (1978). Single group validity, differential validity, and differential prediction. *Journal of Applied Psychology, 63,* 507–512.

Linn, R. L., and Wertz, C. E. (1971). Considerations for studies of test bias. *Journal of Educational Measurement, 8,* 1–4.

Linn, R. L., Harnisch, D. L., and Dunbar, S. B. (1981a). Corrections for range restriction: An empirical investigation of conditions resulting in conservative corrections. *Journal of Applied Psychology, 66,* 655–663.

Linn, R. L., Harnisch, D. L., and Dunbar, S. B. (1981b). Validity generalization and situational specificity: An analysis of the predictions of first-year grades in law school. *Applied Psychological Measurement, 5,* 281–289.

Lord, F. M., and Novick, M. R. (1968). *Statistical theories of mental test scores.* Reading, MA: Addison-Wesley.

Mayfield, E. C. (1964). The selection interview: A reevaluation of published research. *Personnel Psychology, 17*, 239–260.

McIntyre, R. M., Smith, D. E., and Hassett, C. E. (1984). Accuracy of performance ratings as affected by rater training and perceived purpose of rating. *Journal of Applied Psychology, 69*, 147–156.

Messick, S. (1975). Meaning and values in measurement and evaluation. *American Psychologist, 30*, 955–966.

Messick, S. (1980). Test validity and the ethics of measurement. *American Psychologist, 35*, 1012–1022.

Monahan, C. J., and Muchinsky. P. M. (1983). Three decades of personnel selection research: A state-of-the-art analysis and evaluation. *Journal of Occupational Psychology, 56*, 215–225.

O'Conner, E. J., Wexley, K. N., and Alexander, R. A. (1975). Single group validity: Fact or fallacy. *Journal of Applied Psychology, 60*, 352–355.

Olkin, I., and Pratt, J. W. (1958). Unbiased estimation of certain correlation coefficients. *Annals of Mathematical Statistics, 29*, 201–211.

Osburn, H. G., and Greener, J. M. (1978). Optimal sampling strategies for validation studies. *Journal of Applied Psychology, 65*, 602–608.

Osburn, H. G., Callender, J. C., Greener, J. M., and Ashworth, S. (1983). Statistical power of tests of the situational specificity hypothesis in validity generalization studies: A cautionary note. *Journal of Applied Psychology, 68*, 115–122.

Owens, W. A., and Schoenfeldt, L. F. (1979). Toward a classification of persons. *Journal of Applied Psychology, 64*, 569–607 (monograph).

Pearlman, K. (1979, August). *The validity of tests used to select clerical personnel: A comprehensive summary and evaluation* (Tech. Study TS-79-1). Washington, DC: U.S. Office of Personnel Management, Personnel Research and Development Center (NTIS No. PB 80-102650).

Pearlman, K., Schmidt, F. L., and Hunter, J. E. (1980). Validity generalization results for tests used to predict job proficiency and training success in clerical occupations. *Journal of Applied Psychology, 65*, 373–406.

Raju, N. S., and Burke, M. J. (1983). Two new procedures for studying validity generalization. *Journal of Applied Psychology, 68*, 382–395.

Reilly, R. R., and Chao, G. T. (1982). Validity and fairness of some alternative employee selection procedures. *Personnel Psychology, 35*, 1–62.

Reynolds, A. H. (1979). The reliability of a scored oral interview for police officers. *Public Personnel Management, 8*, 324–328.

Roche, W. J., Jr. (1965). A dollar criterion in fixed-treatment employee selection. In L. J. Cronbach and G. C. Gleser, *Psychological tests and personnel decisions.* Urbana: University of Illinois Press.

Sackett, P. R., and Dreher, G. F. (1982). Constructs and assessment center dimensions: Some troubling empirical findings. *Journal of Applied Psychology, 67*, 401–410.

Sackett, P. R., and Wade, B. E. (1983). On the feasibility of criterion-related validity: The effects of restriction of range assumptions on needed sample sizes. *Journal of Applied Psychology, 68*, 374–381.

Sands, W. A., Alf, Jr., E. F., and Abrahams, N. W. (1978). Correction of validity coefficients for direct restriction in range occasioned by univariate selection. *Journal of Applied Psychology, 63*, 747–750.

Schmidt, F. L., and Hunter, J. E. (1977). Development of a general solution to the problem of validity generalization. *Journal of Applied Psychology, 62*, 529–540.

Schmidt, F. L., and Hunter, J. E. (1983). Individual differences in productivity: an empirical test of estimates derived from studies of selection procedure utility. *Journal of Applied Psychology, 68*, 407–414.

Schmidt, F. L., and Kaplan, L. B. (1971). Composite vs. multiple criteria: A review and resolution of the controversy. *Personnel Psychology, 24*, 419–434.

Schmidt, F. L., Berner, J. G., and Hunter, J. E. (1973). Racial differences in validity of employment tests: Reality or illusion? *Journal of Applied Psychology, 58*, 5–9.

Schmidt, F. L., Hunter, J. E., and Urry, V. W. (1976). Statistical power in criterion-related validity studies. *Journal of Applied Psychology, 61*, 473–485.

Schmidt, F. L., Hunter, J., McKenzie, R. C., and Muldrow, T. W. (1979). Impact of valid selection procedures on workforce productivity. *Journal of Applied Psychology, 64*, 609–626.

Schmidt, F. L., Hunter, J. E., Pearlman, K., and Shane, G. S. (1979). Further tests of the Schmidt–Hunter Bayesian validity generalization procedure. *Personnel Psychology, 32*, 257–281.

Schmidt, F. L., Gast-Rosenberg, I., and Hunter, J. E. (1980). Validity generalization results for computer programmers. *Journal of Applied Psychology, 65,* 643–661.

Schmidt, F. L., Hunter, J. E., and Caplan, J. R. (1981). Validity generalization results for two job groups in the petroleum industry. *Journal of Applied Psychology, 66,* 261–273.

Schmidt, F. L., Hunter, J. E., and Pearlman, K. (1981). Task differences as moderators of aptitude test validity in selection: A red herring. *Journal of Applied Psychology, 66,* 166–185.

Schmidt, F. L., Hunter, J. E., and Pearlman, K. (1982). Assessing the economic impact of personnel programs on work force productivity. *Personnel Psychology, 32,* 333–347.

Schmidt, F. L., Mack, M. J., and Hunter, J. E. (1984). Selection utility in the occupation of U.S. park ranger for three modes of test use. *Journal of Applied Psychology, 69,* 490–497.

Schmitt, N. (1976). Social and situational determinants of interviewer decisions: Implications for the employment interview. *Personnel Psychology, 26,* 79–101.

Schmitt, N., Coyle, B. W., and Rauschenberger, J. (1977). A Monte Carlo evaluation of three formula estimates of cross-validated multiple correlation. *Psychological Bulletin, 84,* 751–758.

Schwab, D. P., Heneman, M. G., and Decotiis, T. A. (1975). Behaviorally anchored rating scales: A review of the literature. *Personnel Psychology, 28,* 549–562.

Smith, P. C., and Kendall, L. M. (1963). Retranslation of expectations: An approach to the construction of unambiguous anchors for rating scales. *Journal of Applied Psychology, 47,* 149–155.

Sparks, C. P. (1982). Job analysis. In K. M. Rowland and G. R. Ferris, Eds., *Personnel management.* Boston: Allyn and Bacon.

Taylor, H. C., and Russell, J. F. (1939). The relationship of validity coefficients to the practical effectiveness of tests in selection: Discussion and tables. *Journal of Applied Psychology, 23,* 565–578.

Tenopyr, M. L. (1977). Construct-content confusion. *Personnel Psychology, 30,* 47–54.

Thorndike, R. L. (1949). *Personnel selection.* New York: Wiley.

Thorndike, R. L. (1971). Concepts of cultural fairness. *Journal of Educational Measurement, 8,* 63–70.

Thornton, III G. C., and Byham, W. C. (1982). *Assessment centers and managerial performance.* New York: Academic.

Thurstone, L. L. (1938). Primary mental abilities. *Psychometric Monographs,* No. 4.

Trattner, M. H., and O'Leary, B. S. (1980). Sample sizes for specified statistical power in testing for differential validity. *Journal of Applied Psychology, 65,* 127–134.

Ulrich, L., and Trumbo, D. (1965). The selection interview since 1949. *Psychological Bulletin, 63,* 100–116.

Uniform Guidelines on Employee Selection Procedures (1978). *Federal Register 43,* 38295–38309.

Vineberg, R., and Joyner, J. N. (1982). *Prediction of job performance: Review of Military Studies.* San Diego, CA: NPRDC TR 82-37 Naval Personnel Research and Development Center.

Wagner, R. (1949). The employment interview: A critical summary. *Personnel Psychology, 2,* 17–46.

Wainer, H. (1976). Estimating coefficients in linear models: It don't make no nevermind. *Psychological Bulletin, 83,* 213–217.

Warmke, D. L., and Billings, R. S. (1979). Comparison of training methods for improving the psychometric quality of experimental and administrative performance ratings. *Journal of Applied Psychology, 64,* 124–131.

Washington v. Davis (1976). 426 U.S. 229. 12 F.2d 1415.

Webster, E. C. (1964). *Decision making in the employment interview.* Montreal: Eagle.

Wexley, K. N., Sanders, R. E., and Yukl, G. A. (1973). Training interviewers to eliminate contrast effects in employment interviews. *Journal of Applied Psychology, 57,* 233–236.

Wherry, R. J., Sr. (1931). A new formula for predicting the shrinkage of the coefficient of multiple correlation. *Annals of Mathematical Statistics, 2,* 440–457.

Wright, O. R., Jr. (1969). Summary of research on the selection interview since 1964. *Personnel Psychology, 22,* 391–413.

Zedeck, S., and Cascio, W. F. (1982). Performance appraisal decisions as a function of rater training and purpose of the appraisal. *Journal of Applied Psychology, 67,* 752–758.

CHAPTER **8.2**
CONCEPTS OF TRAINING

D. H. HOLDING

Department of Psychology
University of Louisville

The object of systematic training is to ensure that specific skills, jobs, or procedures are learned. The task of the trainer is to ensure that the learned material or behavior is appropriate to the job, to ensure that the learning process is conducted efficiently, and to ensure that what is learned during the training sessions transfers satisfactorily to the job environment. These requirements raise a number of different issues, which must be understood if training is to be successfully undertaken. This chapter attempts to provide guidance for the training process by making use of principles derived mainly from experimental studies in the laboratory; as a consequence, some care is needed in their application to practical situations. In particular, it should be remembered that the content of training is at least as important as the conditions in which training takes place.

The first section, 8.2.1, is concerned with some necessary preliminary issues; it outlines the different types of task for which training may be designed, providing some basic distinctions with regard to skill requirements, and deals with the contribution of motivation to the learning process. Section 8.2.2 considers verbal methods of training, comparing the characteristics of words and actions. These issues are discussed in relation to aging, in the context of rule learning, and as exemplified in programmed learning. Section 8.2.3 is concerned with showing the learner what to do, either by direct demonstration or by guidance methods. Topics include the role of imitation, the use of films, the contribution of various forms of visual cues, and the effectiveness of physical guidance.

Practice variables are covered in Section 8.2.4. This section includes discussions of knowledge of results and feedback, the distribution of practice, and the part–whole issue in learning. Adaptive techniques of training are included, as are team-training issues, and an assessment of mental practice. Section 8.2.5 deals with the transfer of training. The basic principles of transfer are described, the effects of task difficulty are considered, and measures of transfer effectiveness are explained. Finally, Section 8.2.6 presents brief conclusions and recommendations.

8.2.1 BASIC ISSUES

8.2.1.1 Regulating Information

The trainer may be viewed as determining the efficiency of the learning process by regulating the supply of information to the learner (Holding, 1968). For this purpose, information should not be regarded as a merely verbal concept. Rather, following the usage of information theory, the term may be applied to the outcome of any procedure that specifies some out of many alternatives. In this sense, information may be conveyed by words, by the arrangement of signals on a control console, by the cutting pattern of a lathe, or by the configuration of human limb movements. The trainer should arrange to supply the learner with the right amount of information, at the right time, and in the right way.

The requirement to provide the information in the right way tends to resolve itself into a choice between the three broad approaches distinguished in later sections: telling the learner what to do, using verbal methods (Section 8.2.2); showing the learner what to do, by demonstration or guidance (Section 8.2.3); or having the learner practice, under controlled conditions (Section 8.2.4). The final step should be to ensure that the right kind of information has been communicated, by measuring transfer of training (Section 8.2.5). Other issues concerning the regulation of information appear throughout the chapter.

As for determining the right amount of information, for example, it is clear that learning will be inefficient if insufficient information is provided. Distributing practice sessions too sparsely, or covering too little ground in any one session, are obviously wasteful training procedures. However, many tasks and concepts will overload the learner, so that other procedures, like the use of part practice, are often to be recommended. Again, information may be inserted at the right or the wrong point in a sequence of skilled activity, with resulting consequences for efficiency in learning. A guiding word at each of the successive stages of an activity is likely to prove far more effective than a set of elaborate instructions, provided beforehand, which will have to be held in memory throughout the operation.

Some of the central issues can be appreciated in the context of a practical example. In a study comparing several methods for teaching medical suturing (Salvendy and Pilitsis, 1980), analysis of the task requirements led to the development of simulators for wound-stitching and knot-typing. The simulators were used in training one group of medical students, while other groups received the traditional training mix of lectures, visual aids, and stitching practice, or else saw visual demonstrations of experienced surgeons and inexperienced students at work. The trainers measured such factors as the training time, quality of suturing, and instructor ratings, and a variety of physiological indices of stress. In each case, performance was measured on transfer from the training medium to the criterion job task. The results showed that the improved training methods took less than half the time needed for traditional training, yielding faster suturing times, and with better performance on both the training materials and the transfer task.

The study shows the advantage of a systematic approach to training. Note that it compares versions of the three methods that are distinguished in the previous section. One method relied substantially on verbal instruction, in the form of lectures, another used visual training by demonstration, and a third made use of practice on the simulators. The most successful approach for this perceptual–motor task was the practice method, which presumably exposed the trainees to the necessary information in the optimum manner. It is important that a transfer test was conducted, to ensure that the task components embodied in the simulators were appropriate. It also seems important that the design of the simulators was based on a preliminary task analysis.

8.2.1.2   Types of Task

The nature of the task to be learned should obviously determine the content of a training program, even though the general principles that govern the conditions of learning may be largely independent of the task analysis. It is fortunate that the effects of learning variables such as feedback or cuing, or variables affecting the efficiency of practice such as the massing of learning trials, may be viewed as applying to broad classes of human activities without the need for a detailed specification of the component skills. No single classification of human activities has found endorsement, although certain basic distinctions are widely accepted.

Tasks and activities may be broadly classified as either *verbal* or *motor*, or as a composite of both of these. In this context, the verbal category will include not only linguistic activites, but also cognitive processes such as judgment, rule learning, and problem-solving. The motor, or perceptual–motor, category will include all of the other tasks in which bodily action plays a part. These, in turn, are often subdivided into *gross*, whole-body, skills and *fine*, usually manipulative, skills.

Further distinctions within the perceptual–motor category are numerous. One approach to the problem is to consider the structure of human abilities revealed by factor analysis. Research on several hundred airmen, who performed more than 30 different apparatus tasks, showed the existence of a large number of separate ability factors (Fleishman, 1966). Most important were spatial response speed, arm–hand steadiness, limb coordination, and rate control; in addition, there were factors relevant only to specific apparatus tasks. Later work has identified other factors such as manual dexterity, which must be distinguished from finger dexterity and from wrist–finger speed. There are also whole-body factors such as trunk strength and gross body equilibrium. More than 18 separate perceptual–motor factors have now emerged, and there are undoubtedly others still to be identified. A further complication is that the importance of these factors change with practice so that, for example, visual abilities give way to kinesthetic abilities as skill increases.

The situation may be simplified at the cost of some inaccuracy by the device of combining several overlapping distinctions, between tasks rather than abilities. Tasks may be *simple* or *complex*. There is also a distinction between *discrete* tasks, such as throwing a ball, and continuous *tasks*, such as driving a car. The discrete tasks may in turn be either *single*, or else *serial*, as in the repeated procedures used in many industrial tasks. It is often the case that discrete tasks are simpler than continuous tasks, while gross movements are often simpler than fine-adjustment tasks. Hence, for many purposes it is possible to locate simple, gross, and discrete skills at one end of a continuum and assign complex, fine, continuous skills at the other, as shown in Figure 8.2.1.

The other dimension along which skilled tasks may vary is also composite. Some tasks, like process control monitoring, are largely *perceptual;* others, such as hammering railroad spikes, are predominantly *motor*. This dinstinction corresponds very approximately with the difference between *open* skills such as are required in ice hockey, where the appropriate responses are heavily dependent on a rapidly changing, external input, and *closed* skills such as discus throwing, which may be accomplished almost without reference to external signals. The same dimension also reflects the distinction between truly *skilled* activities, and those which are largely *habitual*. Figure 8.2.1 shows how a selection of tasks may be located on the surface provided by the dimensions.

Although the principles of learning are similar in broad outline for most tasks, there are sufficient differences in detail, perhaps, to justify distinguishing different requirements for learning, for different types of performance. Gagné (1967) has presented another taxonomy of human verbal and motor activities, shown in Table 8.2.1, specifying the conditions considered necessary for each type of learning to occur. The taxonomy is assumed to be hierarchical, so that each level of learning is dependent on achievement of the previous levels. Thus, problem-solving requires the use of previously learned rules, which in turn depend on the use of previously acquired concepts, and so on. Note, however, that the approach is based on learning theory, rather than training principles, and that the classification affords little opportunity for distinguishing between different perceptual–motor skills. It is implicit in the scheme that each stage of learning is acquired by repeated practice.

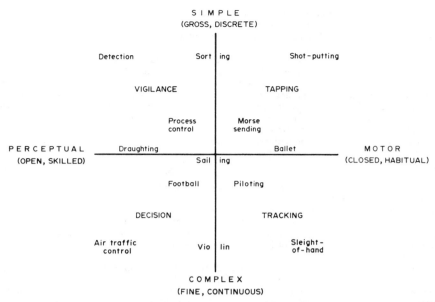

Fig. 8.2.1. A pragmatic classification of skilled tasks along two dimensions [from Holding (1981)].

Table 8.2.1   Summary of Conditions Considered Necessary for Seven Kinds of Learning[a]

Performance Established by Learning	Internal (Learner) Conditions	External Conditions
Specific responding	Certain learned and innate capabilities	Presenting a stimulus under conditions commanding *attention;* occurrence of a response *contiguous* in time; *reinforcement*
Chaining:		
Motor	Previously learned individual connections	Presenting a *sequence* of external cues, effecting a sequence of specific responses *contiguous* in time; repetition to achieve selection of response-produced stimuli
Verbal	Previously learned individual connections, including implicit "coding" connections	Presenting a *sequence* of external verbal cues, effecting a sequence of verbal responses *contiguous* in time
Multiple discrimination	Previously learned chains, motor or verbal	Practicing to provide *contrast of correct and incorrect stimuli*
Classifying	Previously learned multiple discriminations	Reinstating discriminated response chain contiguously with *a variety of stimuli* differing in appearance but belonging to a single class
Rule using	Previously learned concepts	Using external cues (usually verbal), effecting the recall of previously learned concepts contiguously in a suitable sequence; specific applications of the rule
Problem solving	Previously learned rules	Self-arousing and selecting previously learned rules to effect a novel combination

Source: Gagné (1967)

942

8.2.1.3 Motivation

It is unclear to what extent motivation is required for learning to take place, although training schemes often include motivational features. Laboratory studies of animal learning, and analogous studies of human learning, seem to show that learning may occur in the absence of specific motivation. Nevertheless, the amount of motivation that is present will determine the level of performance that is demonstrated. In studies of "latent learning," for example, animals are allowed to roam in mazes with empty goal boxes. At this stage there is no measurable change in their route learning, but, whenever a food reward is made available, a marked improvement in the scores demonstrates that learning has taken place.

With human subjects the issue often takes the form of comparing incidental with intentional learning. Many studies find that the intention to learn makes no difference to what is eventually retained, unless the amount of information processing that the task demands of the learner is too low. One group of subjects may be set to learn a list of words, while another group is asked to carry out either a meaningful or a superficial task using the same words. A subsequent test of memory for the words will show that only those subjects whose processing was superficial will have remembered less than the group that intended to learn the material (Crowder, 1976). Having the intention to learn may affect the amount and kind of rehearsal that subjects perform, but incidental learning will otherwise be equally good.

Similar conclusions have been reached for perceptual–motor tasks. In one recent study (Crocker and Dickinson, 1984), people were given a keyboard task of choice reaction in which a sequence of display lamps prompted the pressing of the appropriate keys. A later test was made for learning of the keying sequence. In normal conditions, a group who knew that later recall would be required did somewhat better than an incidental learning group, although it is noteworthy that both the intentional and the incidental learners showed practice improvements from trial to trial. When all of the subjects were made to carry out a distractor task, such as reading out lists of numbers and answering arithmetic problems during the period before recall, the resulting disruption of rehearsal removed any advantage possessed by the intentional learners.

At the same time, even incidental learners must somehow be induced to practice the task. Although motivation may not affect learning directly, motivation does influence performance during the practice period, which in its turn will affect the learning outcome. There are thus theoretical reasons for expecting that successful training will require some minimum level of motivation. This will not normally constitute a practical problem for groups of willing trainees, whether industrial apprentices, flying personnel, or upwardly mobile managers, although it may be postulated that certain groups, such as inductees to the armed forces, might be less than optimally motivated.

It is normally assumed that extrinsic motivation, as evidenced in pay scales, job security, or penalties of various kinds, is less effective than the intrinsic motivation that may stem from the interest residing in the skill itself. Malone (1981) lists the categories into which intrinsic motivation might be resolved, based on an analysis of the relative interest of a variety of computer games. Self-motivating instruction is thought to depend on challenge, fantasy, and curiosity. Challenge results from such features as the uncertainty of outcome of the task goal, variability of the difficulty level, the presence of multiple goals, hidden information, and unpredictability. The fantasy element may be more important in computer games than in most training situations. The curiosity value of an instructional task depends on both sensory and cognitive components, cognitive curiosity being aroused by making learners doubtful about their own knowledge and by providing surprising and constructive feedback.

Informative feedback, or knowledge of results, has been thought to provide a degree of intrinsic motivation in many situations. Indirect measures such as tardiness and absenteeism, as well as anticipatory performance declines, seems to result from depriving the trainee of knowledge of the results of practice efforts. However, giving knowledge of results implies a comparison between attained performance and an implied or explicit goal, and it can be argued that much of the benefit due to knowledge of results should be ascribed to its goal-setting function (Locke, 1970). Within limits, the higher the goal to which the trainee aspires, the better the level of performance. Knowledge of results may be used to make the learner's goals realistic or to raise the level of aspiration.

Attempting to improve training by introducing additional motivation in other ways has generally been found ineffective. An early industrial study of work at adjusting electrical relays used competition between groups to increase motivation (Williams, 1956). Compared with the noncompetitive groups in the same training scheme, the competing groups worked somewhat faster. However, the increase in speed was accompanied by a reduction in quality of work, so that the number of correctly adjusted relays showed little difference. The effects due to both individual and social motivation therefore appear to be weak.

8.2.2 VERBAL METHODS

The first training method to be considered in detail is the use of verbal instruction. Telling the learner what to do constitutes a natural approach to training objectives. A basic decision at the outset of any training program is the extent to which verbal instruction is to be employed, and what amount of such instruction should precede on-the-job practice. Verbal sequences carry large amounts of information and thus, despite the storage load they place on human memory, are often used in the form of lectures, discussions, notes, and learning programs.

Verbal methods are clearly efficient and economical when used to give initial orientation for a task or job, when used to outline goals, or when used to communicate knowledge of progress to trainees. However, trainees will often be less verbally inclined than their instructors, and the translation from words into the equivalent actions is by no means automatic. Holding's (1965) review of the problem gives a number of reasons for treating the use of verbal instructions with some caution.

8.2.2.1 Words and Actions

There is no doubt that the basic processes of verbal learning and motor learning are essentially similar. Both kinds of process are affected in the same way by practice and reinforcement variables, and the course of forgetting is equivalent when the amounts of prior learning and task difficulty are taken into account. A comparison by Schmidt (1971), for example, shows that verbal and motor tasks are equally susceptible to the interference caused by interpolated activities, and that such interference is reduced in the same way by administering greater amounts of original learning. However, verbal and bodily responses are not always interchangeable. Trainees cannot always translate verbal instructions into perceptual–motor activity, nor report verbally on such activity.

Even in essentially verbal tasks, there may be no direct correspondence between words and performance. Berry and Broadbent (1984) investigated behavior in two widely differing computer tasks (a simulation of sugar production problems in an underdeveloped country and a personal interaction task), in a series of experiments that varied the amounts of practice, the use of verbal instructions, and the use of concurrent verbalization. An increase from 2 to 30 sessions of practice produced a substantial improvement on the performance scores, but had no effect on the trainees' ability to answer related questions after the test. In contrast, verbal instructions explaining the mode of operation of the simulation improved the ability to answer questions, but had no effect on the trainees' control performance. Finally, verbalization in the form of thinking aloud while performing the computer task had no effect either on control performance or on verbal knowledge of the task.

The position is somewhat different when verbal assistance is provided a little at a time. Computer aiding may take this form, as in a study of training for fault diagnosis in logic networks conducted by Brooke, Cook, and Duncan (1983). As trainees made a series of tests to trace faults, the computer might interject "A fault here would not account for the symptoms you see," or "Unit 42 is connected to 43 which you already know has an output of 1." Combined with pretraining on trouble-shooting strategies, this technique was very effective, although more so for testing than diagnosis. Intermittent computer aiding of this type had its greatest effect after the student had already obtained some understanding of the problems during pretraining.

Where verbal instructions are to be given for perceptual–motor performance, it again seems better to break the instructions into separate components, which are given at appropriate intervals in the task (Holding, 1965). It appears unlikely that mastering an entire primary helicopter training manual, for example, with its fund of information concerning asymmetry of lift, antitorque failure, and approach paths, and its welter of precepts like "divide your attention" or "stay aware of wind conditions," will make for successful transfer to the piloting task. Early laboratory work has even shown that overelaborate preliminary instructions, before practice on a tracking task, may not only fail to assist the learner but may obstruct progress over several days of training.

8.2.2.2 Instruction and Aging

Approximately half the work force is over 40 years of age, and thus susceptible to the various changes in behavior that tend to accumulate in the later years of life. The retraining of older workers is quite practicable and often desirable, and many studies have yielded successful results with appropriate methods of training. However, the problem of relating words to actions is seen in a particularly acute form when the learners are older individuals. In many cases, it is recommended that less reliance be placed on verbal methods of instruction.

An early study showed that the words and the actions produced by people of various ages do not necessarily constitute equivalent responses. After people had seen a set of posters depicting road safety principles, their memory for the information was tested in two different ways. The first method was simply to ask for verbal descriptions of the posters. The other method was to assess the use made of the information, by asking each age group to identify the faults depicted in photographs of road traffic scenes. The method of verbal recall showed the younger persons to advantage, as expected. However, although the older groups would have shown a decline on a written test, the information had become more directly incorporated into their behavior, with the result that their use scores tended to be higher.

A further step was taken to ascertain whether action learning might be superior to verbal memorizing

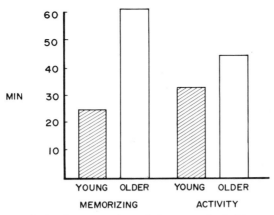

Fig. 8.2.2. The times required to learn a sorting task by two methods. Younger trainees learn faster using memory, but older trainees learn faster through activity [from Belbin (1958)].

as a training technique. In a postal sorting task where numbered cards had to be assigned to differently colored slots, the relation between number and color could be learned by practicing with colored cards or by memorizing a chart of relationships. Sorting was done more quickly and accurately after the activity form of learning, except by the youngest age group, although the time taken to learn by memorizing was shorter. In a similar task, where card numbers had to be correlated with slot positions, the younger and older trainees showed a complete reversal. As illustrated in Figure 8.2.2, the younger group learns faster with memorization than with activity learning, while the older group does better with activity than memorization.

Subsequent work has qualified these conclusions in some respects. For example, highly verbal groups such as university students seem to prefer memorization, and even to impose this upon the activity method. Apparently the value of activity training depends on the way in which it is approached by the learner. Again, in a postal task where village–county associations had to be learned, the activity of posting cards into slots was less effective, particularly for older trainees, than simply turning over the cards while attending to their destinations. Further work on associative learning showed that activity methods only work well if the activities are accompanied by a degree of cognitive effort. In reviewing these developments, Belbin (1970) advocates the use of "discovery" methods, particularly for older trainees. These methods involve the learner in making repeated efforts to discover the correct activity, although trainees may be led progressively from stage to stage by cues supplied by the trainer. The principle is similar to that of behavior shaping (see below), so that discovery learning might be considered an activity form of programmed learning.

8.2.2.3   Rules and Theories

The effectiveness of verbal instructions depends on the form they take, which should be as simple and direct as possible, and on the nature of the task for which the instructions are designed. Verbal methods seem most appropriate for communicating the rules that apply to successful performance, or the theories and concepts that make for an understanding of the task. Even so, the effect on perceptual–motor performance of receiving rules instruction is somewhat ambiguous.

Early work on teaching the principles of the refraction of light, to students learning to throw darts at targets submerged below the surface of the water, found no advantage when the targets were at a constant depth. The principles appeared to transfer well when instructed subjects were made to perform at different depths, although comparison with a later experiment suggests that the effect was due to the extra time spent on theoretical instruction. When an equivalent amount of time was spent in practicing games skills as in learning the simple mechanics of the games, the effects on learning were also equivalent. In a more technical context, it has been shown that giving conceptual instructions, explaining a radar display in terms of the movements of the antenna and the displayed aircraft, has no advantage over practical instructions concerning the use of the controls.

Once learned, general principles seem to be fairly resistant to forgetting. Instrument flying skills, for example, show some loss of the procedural components after four months of inactivity, but little loss of altitude, speed, or bank control understanding. Such losses as are observed for level or heading of flight are soon made good during relearning trials. Similarly, there is little loss over time of the strategic component of trouble-shooting skills acquired in diagnosing faults in an acid-purification plant. However, a knowledge of principles or strategies may be acquired in several different ways.

A systematic study of process control training (Shepherd, Marshall, Turner, and Duncan, 1977) compared three modes of instruction. As theory, one group was given a description of the flow of the product through a chemical plant, together with the functions of the various equipment and the effects of different control loops. Another group of trainees was also given a set of explicit diagnostic rules, such as: "*high* temperature and pressure in column head associated with *low* level in reflux drum indicates overhead condenser failure—*provided* all pumps and valves are working correctly." A further group had no special instruction beyond a description of the instruments on the control panel. After an adaptive form of cumulative-part training (see Section 8.2.4), all trainees were tested on faults that they had previously encountered, and also on a new set of simulated faults.

The rule-instructed group appeared slightly worse on familiar faults, apparently because, unlike the other groups, they had no bias toward repeating earlier diagnoses. However, their versatility was clearly indicated in dealing with the unfamiliar situations, which may be viewed as providing the more stringent test. Both the rules group and the theory group paid a time penalty for their increased sophistication, taking significantly longer than the control group to carry out their diagnoses.

The value of theory depends on the ability of the trainee to apply it to practical situations. In electronics training, knowledge of theory may be high on leaving a school, but fault-finding ability low. As practical skill increases with field experience, knowledge of theory may decline. Even the application of explicit rules may not be automatic. Hence, in the process control situation, simply presenting the diagnostic information may be insufficient (Marshall, Duncan, and Baker, 1981). Withholding some of the information, in order to ensure that trainees have to request the display of appropriate instrument readings, seems to encourage the use of diagnostic rules and thus to improve the results of training.

8.2.2.4 Programmed Learning

When verbal instruction is appropriate, the techniques of programmed learning have much to offer. Although some of the early enthusiasm for programmed methods has dissipated, there remains a residue of useful principles. Some of these principles are also adopted in computer-aided instruction (see Chapter 8.4), which is not specifically considered here. Programmed instruction is systematically designed, and it renders the progress of learning open to inspection. The learner is given new material at an individual rate, with each new step following naturally from the current state of knowledge. Feedback is typically given after each response. Since it is ensured that the new material is within the learner's grasp, and since many programs include a degree of prompting, that feedback is nearly always positive. Modifications may be made at any point in the learning procedure if the results are unsatisfactory.

Learning programs may be broadly classed as either *linear* or *branching*. The linear technique is associated with the name of Skinner (1954), who based the theory on the principles of behavior shaping derived from animal learning. In the well-known example of teaching a dog to touch a door knob with its nose, a response not originally in its repertoire, a secondary reinforcer such as a whistle is first associated with reward. Then, whenever the dog faces the doorway, the reward signal is sounded. As the dog approaches the door nonspecific responses are no longer reinforced, but responses such as raising the head are encouraged instead. As training proceeds, only those responses that include rearing up are given the reward signal. Thus, the animal is gradually led to perform a novel response by using feedback to encourage successive approximation to the desired behavior.

Linear programs, therefore, tend to proceed by small steps, with immediate feedback after each response. The learner typically has to construct an answer to each frame of the program, rather than simply check a multiple-choice alternative, so as to benefit from practicing the target behavior. A single frame might read: "The conduction of heat in liquids is about ten times faster than in gases. In water, heat will flow about . . . faster than in air." Note that the use of heavy prompting tends to ensure that the learner produces the right answer. The learner is then given feedback in the form of the correct answer, perhaps presented by means of a teaching machine. Table 8.2.2 summarizes the main features of these linear programs, contrasting them with the branching form.

Intrinsic, or branching, programs were originated by Crowder (1960), who introduced the "scrambled" textbook. The learner is typically given a page of instruction followed by a multiple-choice question. Learning is thus less searchingly tested, as recognition rather than recall is demanded by the format, and longer periods elapse before each response is required. Such texts often contain motivational phrases, such as "very good," or "right again," which may be omitted without disadvantage. The most important difference is that the branching text makes provision for remedial teaching. If the learner chooses the wrong alternative, as is likely if the alternatives are plausible, the page selected as a choice will correct the error. Programs of this type are sometimes preferred by the users, despite the theoretical disadvantage incurred by their encouragement of errors.

In practice, there is little to choose between the results obtained with the two kinds of program. Both may produce learning as good as or better than classroom learning, often with the use of less training time and fewer instructors. A great many evaluation studies have been conducted using various measures of effectiveness. A survey by Nash, Muczyk, and Vettori (1971) shows 29 studies (out of

Table 8.2.2   Characteristics of Linear and Branching Programs

Feature	Linear	Branching (Intrinsic)
Due to	Skinner	Crowder
Size of step	Small	Large
Degree of prompting	Heavy	Light
Type of response	Constructed	Multiple choice
Memory tested	Recall	Recognition
Presentation mode	Machine (or text)	Scrambled text (or machine)

32) in which programmed instruction gave shorter training times. Immediate tests of learning showed 9 where programmed learning was better, with 20 cases equal to conventional instruction. In 26 studies where later retention of the material was also tested, programmed instruction was superior in 5 and equivalent in 16 cases.

8.2.3   DEMONSTRATING AND GUIDING

Showing the learner what to do might take the form of visual demonstration or else of guiding performance with various degrees of constraint. In its most general form, training by demonstration can be an effective way to teach entire segments of behavior. Moses (1978) describes the success of behavior-modeling techniques for management training. The methods involve the use of films or videotapes, perhaps showing superior–subordinate interactions over problems such as work quality or absenteeism. These are usually followed by the opportunity to rehearse the behaviors demonstrated by the model.

Evaluations of behavior modeling have been carried out in several large corporations. In a study of two telephone companies, 90 trained supervisors were contrasted with 93 others who had not been exosed to behavior modeling but were otherwise comparable. Ratings of their behavior in a set of problem situations showed that 84% of the trained group were considered above average in their performance, against 32% of the untrained supervisors. The success of the method is possibly due to the fact that general goals and procedures are taught, rather than specific actions. Note, too, that practice in the form of direct rehearsal of the behaviors is used to supplement the observed modeling.

8.2.3.1   Imitation

The utility of demonstration methods for training perceptual–motor skills is more dubious, if only because so little research has been conducted on the mechanics of imitation. Industrial observations suggest that imitation by itself has severe limitations, since the opportunity to imitate is all that is offered by many kinds of on-the-job training. The informal, sit-by-me method of instruction is often unfavorable compared with more systematic methods. On the other hand, the early work on laboratory tasks such as the finger maze shows clearly that some learning can occur as a consequence of merely observing other performers at work. It has also been shown that watching others perform can be fatiguing, and that electromyographic measures reveal a degree of vicarious participation in the watchers.

Social variables play some part in imitation learning. In comparison with a control group, several imitation groups do better at balancing on a free-standing ladder (Landers and Landers 1973). It can be seen in Figure 8.2.3 that students given the opportunity to imitate a model all begin their practice at a higher level than a group of untrained students. Only by block 3 of practice does the performance of the control group reach the initial level of performance shown by the imitation group. However, students who have been shown a demonstration perform better when the model is present to observe them. The size of the effect also depends on the status of the model, in conjunction with the skill level of the model. Viewing an unskilled teacher is less effective than viewing an unskilled peer, or than viewing a skilled performer of either kind.

A video monitor can be used to demonstrate the movements of a model, and the use of modeling can be combined with giving visual feedback. In an experiment by Carroll and Bandura (1982), subjects were trained to carry out an action pattern, using hand-held paddles, that they could not normally see in peripheral vision. The action was demonstrated to all of them, but only some were enabled by the video monitor to see their own actions during practice. Subsequent testing showed that those who had received visual feedback, either throughout or in the later stages of training, had learned to make more accurate reproductions of the pattern of action. The visual feedback seemed not to help until the learners had developed an internal model of the action, although their conceptual representation of the pattern seemed to develop as a response to the modeling alone.

Learning by imitation is an indirect process. In order to copy an instructor's actions, the learner

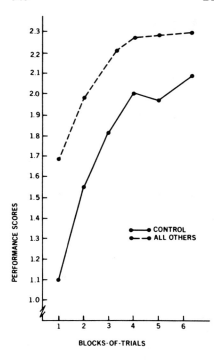

Fig. 8.2.3. Learning a balancing task by imitation. The control group had no training, but the balancing skill was modeled for all others [from Landers and Landers (1973)].

has to try responses that feel as if they will resemble the appearance of the demonstrated responses. Hence, few attempts at training rely on demonstration alone. However, visual demonstrations can draw attention to relevant perceptual cues, and may give the learner a standard against which to match performance. Imitation may be valuable in the earlier stages of learning, since it narrows down the range of alternative actions, but is virtually always supplemented by direct practice.

8.2.3.2 Films

Television and film presentations can incorporate both factual material and demonstration sequences. Material presented through these media is as well remembered as material presented live, and very often the same material could not be presented in a classroom. A potential advantage of the film is that special techniques like the animated diagram or the slow-motion sequence could not be achieved by any other method. Many studies have shown that film and television presentations produce measurable amounts of learning. Films of road scenes, for example, may be used to improve drivers' passing judgments (Lucas, Heimstra, and Spiegel, 1973). However, the demonstration of actual movement may be unimportant for procedural skills. Neither for the dismantling and repair of a sashcord window, for example, nor for understanding the working of a machine gun trigger mechanism, does it matter whether moving films or static filmstrip is used for instruction.

A major problem with film presentation is the inflexibility of its timing. The viewer typically has no control over the rate at which information is presented and, unlike the reader of a text, cannot check back over earlier sections for review. The passivity of the film viewer can be partially overcome by encouraging audience participation. Having the audience answer questions on the content of a film, section by section, gives a clear advantage over passive viewing; this advantage is enhanced when correct answers are also supplied.

The advantages of film presentation (Goldstein, 1974) include (1) known effectiveness when one-way communication is appropriate; (2) replacing otherwise unavailable instructors; (3) unique potential for presenting dynamic events; (4) presenting otherwise unavailable action sequences. However, some disadvantages are evident: (1) due to one-way communication, and to lack of adaptability to variation in the composition of the audience; (2) due to fitting the presented material into a different and perhaps limited learning environment; and (3) due to expense and the need for instructor preparation.

8.2.3.3 Visual Cues

There are many ways in which the learner can be shown what to do in some detail, by presenting supplementary visual cues. A strictly perceptual task, like that of detecting gaps in Landolt rings, can be better assisted by color cuing (such as associating red backgrounds with left-hand gaps, and green backgrounds with right-hand gaps) than by giving right/wrong information during trial-and-error practice. Giving strongly indicative visual cues may be viewed as a form of prompting, which is highly effective in programmed learning (Taylor, 1970) and may be more effective than the use of methods of confirmation in the acquisition of verbal responses.

Visual prompting has been of use in retraining the middle-aged and elderly, as indicated in Section 8.2.2. In training for inspection work, younger learners can benefit from a "lead-in" procedure in which visual markers indicate all the faulty items (Belbin and Shimmin, 1964). However, there is some tendency for learners to rely on the artificial cues, rather than to attend to those inherent in the task. Changing the procedure so that discrimination is stressed, with new faults being introduced one at a time, makes the procedure far more effective for older trainees. Visual magnification, as when textile weaves are shown sufficiently large for their structure to be perceived, also improves discrimination and subsequent learning.

A point of particular practical interest is the function of visual input in learning to type. The traditional practice has been to hide the keyboard from view at an early stage, on the grounds that the trainee must then become familiar with the tactile cues eventually to be used in touch typewriting. However, it has been shown that the proportion of successfully corrected errors is far greater when the keyboard is visible, particularly at the outset of training; with only kinesthetic and tactile feedback, efficiency remains at only 40% for the greater part of training. Long (1976) has separated the effects of viewing the keyboard, the source material, and the typed output. Consulting all three of these is part of normal keying behavior. The speed and accuracy of typing were shown to depend on viewing the keyboard, which allows typists both to locate unfamiliar keys and to coordinate keys with fingers, but having sight of the typed copy is also required for efficiency in the correction of errors.

For motor tasks such as the laboratory maze or the tracking task, giving visual guidance by showing the pathways ahead usually produces greater learning than do trial-and-error procedures in which preview is not available. However, it seems essential that such guidance should not be so restrictive as to obviate knowledge of alternatives. Following a target course in which both the right and the wrong alternatives are clearly marked produces far more improvement in learning than does simply following the correct path. Similar effects have been noted with forced-response forms of guidance (see below), where learning may be facilitated by practice with alternatives to the correct movement.

Visual guidance may be used to provide errorless practice for simple movements, although its effect may be complicated by distracting attention from the learner's internal movement cues. Smyth (1978) administered 30 training trials and 5 test trials, on a slider apparatus with a mechanical stop and a CRT display. The trainees' hand and arm movements were hidden from view. The stop device, which indicates the correct movement length by preventing the trainee from overshooting, was used in half the training conditions. Stop or no stop could be used alone, or combined with visual guidance taking the form of a visible line moving in time with the motion of the slider, or else combined with terminal feedback in which the line became visible only at the end of the movement.

The visual guidance condition did produce learning, relative to the scores of those who had no stop and no feedback. However, the mechanical stop alone produced better learning results, about equivalent to the effects of knowledge of results provided by showing the line after the movement. The learning produced by visual guidance was poorer, whether or not the stop was present, presumably because it tended to preempt the trainees' attention. The visually guided trainees tended to undershoot the target distance, but were less likely to be correct in judging the direction of their errors. Reliance on the visual information may discourage trainees from programming the endpoint before embarking on each movement.

8.2.3.4 Physical Guidance

Teaching movements along a slide by making use of a mechanical stop is an example of guidance by restriction. The trainee powers the movement, but is shown what is required by the prevention of incorrect alternatives. It is also possible for the trainer to take over complete control of the movement, by giving forced-response guidance. These methods are not of great practical importance, although the restriction principle is embodied in such devices as the harness used for practicing gymnastics, and forced-response is used whenever a coach manipulates the leg of a swimmer in order to demonstrate a stroke.

Most of the laboratory work on these methods has been reviewed by Holding (1970). In training for single, discrete movements such as are involved in line-drawing, or for serial responses of the kind required in keyboard tasks, guidance appears to function by providing information about the desired response while also preventing the learning of errors. For these tasks, restriction is most often superior to forced-response. Nevertheless, Decker and Rogers (1973) found substantial improvements in serial reaction times when forced-response training was given. With more complex requirements, as exemplified in the continuous tracking of unpredictable target courses, guidance seems to help by relieving the learner of part of the information overload, a function for which the forced-response method is best suited.

The relative effectiveness of different forms of guidance depends on several variables, including

Table 8.2.3 Comparison of Direct and Reversed Tasks

	Number of Guided Runs	Mean Total Lights Extinguished per Subject	Improvement over Control	Percentage Improvement (%)
Direct	0	106.75		
	1	116.95	10.2	9.6
	9	123.35	16.6	15.6
Reversed	0	81.10		
	1	96.20	15.1	18.6
	9	100.40	19.3	23.8

Source: Macrae and Holding (1965)

the degree of compatibility between the practiced and the tested response. In an experiment on the learning of simple movements, restriction was at least as effective as knowledge of results, but forced-response was somewhat weaker (Holding and Macrae, 1964). Part of the reason is that forcing creates a tendency to pull, which is incompatible with a later pushing movement. Changing the test task to one of releasing, rather than pushing, a slider, reversed the effectiveness of the two techniques. A further experiment showed that a restriction technique was superior to forced-response in a serial tracking task, in which the movements of a lever were used to extinguish a series of lights.

Another factor is the availability of knowledge of alternatives. The trainee receiving forced-response for a single movement length has no access to comparisons with alternative movement lengths. However, this information can be inserted into the training procedure. In an experiment where forced-response was given at a variety of movement lengths, the training value of guidance was substantially increased, and forced-response was as effective as restriction. It should be noted that a number of experiments, for the most part using methods other than physical guidance, have also shown the advisability of using a variety of practice materials and movements. Variety helps the trainee to build up a flexible internal schema for responding in different circumstances. Of course, providing knowledge of alternatives presents no problem for serial or continuous tracking, since these tasks automatically involve the learner in a wide range of movements.

A point to note is that practicing alternative responses is not synonymous with learning to commit errors. The importance of error prevention can be seen in Table 8.2.3, which shows the amounts of learning conferred by guidance in two versions of a serial tracking task. In the direct version, lever movements made to a position corresponding with 1 of 10 signal lamps would extinguish the light. Errors were relatively few, even without guidance, which was accomplished by having the lever move automatically to the correct position. In the reversed version of the task, it was necessary to move the lever to position 10 in order to extinguish lamp 1, and vice versa. Reversing the display–control relationships in this predisposes the learner to make errors, despite the fact that the sequence in which the lamps are illuminated is the same in both cases. The percentage improvement shown in the test trials, after varying amounts of guidance training, is clearly greater when more potential errors are prevented by guidance in the reversed version.

8.2.4 PRACTICE METHODS

Having the learner practice is, or should be, the basic activity in most forms of training. The effectiveness of practice depends on the conditions in which it takes place, including the amount and kind of knowledge of results that is offered. Knowledge of results may be viewed as constituting a large part of the feedback that arises as a consequence of one's own actions.

A person carrying out any kind of skilled activity may need information at three stages (Newell, 1981). Information before action includes knowledge of goals or objectives, and may often be conveyed by lectures or demonstrations. Information during action may include inputs from the task itself, cuing and guidance techniques, and some forms of feedback. Information after action, in the form of feedback resulting from completed actions, seems necessary in order for permanent learning to occur. The cycle of input, output, and feedback is shown in Figure 8.2.4. Note that the comparison between the current state indicated by feedback and the projected state indicated by goal or target information gives rise to knowledge of results.

It is nowadays clear that the complete loop is not always in operation. A person who has acquired proficiency at a skill has built up programs for motor activities, probably organized at a high level of generality in the form of an action schema, that can operate in the absence of feedback. However, feedback is necessary to set the initial conditions for an activity and to adjust the motor program variables during an activity. In any case, some feedback seems essential for acquiring a skill, and many experiments have shown that successful learning is mediated by knowledge of results.

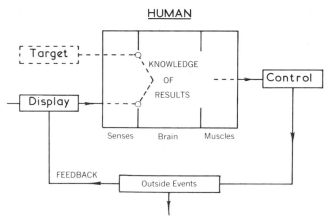

Fig. 8.2.4. The feedback loop in human performance, showing how feedback gives rise to knowledge of results [from Holding (1968)].

8.2.4.1 Knowledge of Results

Manipulating the feedback that a learner receives is one of the devices open to the trainer. Receiving knowledge of results, or feedback in various forms, allows the learner to correct for errors. However, the final learning effect seems to depend on the manner in which the corrections are achieved. Many forms of augmented feedback, such as a buzzer that sounds when a gunlayer is on target, will improve performance temporarily but have no permanent effect on learning as measured when the device is no longer available. The outcome has been said to depend on whether the trainee receives action feedback or learning feedback.

The distinction between action and learning feedback tends to coincide with the distinction that has been made between concurrent and terminal knowledge of results (Holding, 1965). Thus, when trainees who are attempting to achieve the correct pressure on a manual plunger are shown a concurrent visual display representing the effort exerted, in what amounts to visual guidance conditions (see Section 8.2.3), their performance may deteriorate sharply if the supplementary visual cue is removed. A group only shown the result at the end of a movement will perform relatively poorly during training, but this terminal knowledge of results appears to have a more permanent effect on later learning. Instead of being used to amend an ongoing response, as happens with action feedback, the terminal correction is stored for use with the subsequent response and is thus more permanently retained. However, concurrent information in the form of action feedback can also have lasting effect in certain circumstances (Fox and Levy, 1970), since the crucial factor may be whether or not the trainee pays attention to the intrinsic cues in the task.

Unfortunately, many laboratory studies of knowledge of results have failed to carry out tests for the learning that remains after knowledge of results is removed. A major review of the effects of knowledge of results (Salmoni, Schmidt, and Walter, 1984) has analyzed more than 250 studies, distinguishing sharply between temporary performance effects and more permanent learning effects. Table 8.2.4 shows the major conclusions with respect to the frequency, the delay, and the precision of knowledge of results. With respect to frequency, for example, it appears useful to withhold knowledge of results on a proportion of occasions so that trainees are forced to rely on the intrinsic cues.

For similar reasons, delaying knowledge of results is not unfavorable to learning. Of course, delayed action feedback can severely impair performance and thus impede learning, as when delays are inserted between steering wheel movements and a road display (Smith and Kaplan, 1970). However, delays before and after terminal knowledge of results, although difficult to disentangle from variations in the intertrial interval, seem to allow the learner more time for processing the feedback information and planning the forthcoming action. Delays following knowledge of results appear definitely favorable. However, if other activities are interpolated during the delay before knowledge of results is given, the interference will impair learning, as might be expected.

With respect to precision, knowledge of results may vary from simple, qualitative "right/wrong" information to exact measurements of error. The effects of such variation are unclear, as other factors also affect the issue. Tomlinson (1972), for example, found that precise visual feedback for a pedal movement improved performance on an immediate transfer test, and was better during training, but imprecise verbal feedback gave better learning after a 1-day retention interval. An optimum level of

Table 8.2.4   Summary of Various Effects of Knowledge of Results (KR) on Performance and Learning

Increases in Variables	Effects on Motor Performance	Effects on Motor Learning
Absolute frequency of KR	Enhanced	Enhanced
Relative frequency of KR	Enhanced	Degraded
KR delay slightly	None of slightly degraded	None or degraded
Post-KR delay	Enhanced slightly	Enhanced, if KR delay constant
Intertrial interval	Mixed, unclear	Enhanced, if KR delay constant
Interpolated activities in KR delay	Degraded, if demanding	Degraded
Interpolated activities in post-KR delay	Degraded, if demanding	None
Trials delay of KR	Degraded	Enhanced
KR precision	Enhanced	Enhanced

Source: Salmoni et al. (1984)

precision for knob-turning feedback may be present or absent (Rogers, 1974), depending on the length of the delay following knowledge of results. In general, however, increased precision most often gives rise to increased learning.

There are also cases in which the trainer can make available to the learner information that is not normally accessible at all. A dramatic example occurs in the biofeedback work, where the learner can gain control over heartbeat or alpha rhythm, when these are exhibited or amplified. Similar principles can sometimes be used in training perceptual–motor skills, as when the force–time pattern of a runner's foot or the squeeze pattern of a trigger movement are displayed to the learner. Such devices are often successful, although not a great deal of research is available.

8.2.4.2   Timing and Spacing

The efficiency of practice depends not only on its correction through knowledge of results, but also on its distribution over time. However, it seems to make little difference whether trainees concentrate initially on speed or on accuracy. Although it is sometimes argued that the form taken by a response differs when performed at high speed, and thus that training at a slow speed is inappropriate, experimental studies have found little effect of training speed. In a demanding keyboard task, for example, it made little difference to later performance of the skill whether the initial instructions stressed speed or accuracy (Leonard and Newman, 1965). The traditional technique of first learning accuracy and then acquiring speed seems unexceptionable.

Practice is likely to be fatiguing, unless adequate rest pauses are given. Hence, considerable research attention has been paid to the distribution of practice sessions, usually contrasting massed with spaced practice on the basis of the length of rest provided. In this context it is again important to distinguish between temporary and permanent effects on performance and learning. Distribution variables often exert substantial effects on immediate performance, although the residual learning effects may be slight.

Groups receiving spaced practice, with substantial rest pauses, typically show continuing improvement within a practice session. There may be a slight drop between the level reached at the end of one session and the beginning of the next, but progress is relatively steady. In contrast, the progress of groups receiving massed practice may be much more erratic. The overall trend of performance within a practice session may become negative, as fatigue mounts, but trainees will begin the next session at a higher level, when temporary fatigue effects have dissipated. The learning that occurs is thus often hidden by temporary performance deficits. If both the massed and the spaced groups are eventually tested on the same schedule under laboratory conditions, the gains made by both groups during training tend to be very similar.

When massed practice leaves some residual deficit, it may be due to two factors (Holding, 1965). One is the possibility that the spaced trainees can make use of the rest pauses for consolidating and rehearsing the material just learned. The advantage of spacing can disappear when the rest pause is filled with competing activities. The second, and major, factor is that trainees fatigued by massed practice may do measurable less work during the practice sessions and, thus, performing less real practice, they have lesser opportunities for learning.

For practical purposes, there may be significant differences in the efficiency of learning as a function of the distribution of practice, since the amount of effective practice may be uncontrolled. Differences are particularly likely to occur when the length of the work period, rather than the length of the rest period, is the training variable. Baddeley and Longman (1978) evaluated different distributions of practice in training post office employees to type alphanumeric codes, comparing the effects of using one or two sessions per day, each of either 1 or 2 hr. The group receiving 1 × 1 hr per day needed only about 35 hr of actual training to learn the keyboard, whereas the group receiving 2×2 hr needed 50 hr and were still inferior in a retention test nine months later. The 1×2 hr and the 2×1 hr groups were intermediate, and similar, at about 43 hr of training time.

A complication is that the 1×1 hr group, although learning more in each session as measured in both speed and accuracy, take a longer calendar time to train. Based on the figures above (although the actual training time was determined by administrative considerations), training the 1×1 hr group should take 35 days, but training the 2×2 hr group should need only 13 days. The massed practice is inefficient at the level of hours and minutes, but, neglecting any qualitative differences, it is economical in terms of days and weeks. Practical considerations must often, therefore, determine the choice of a schedule of practice.

8.2.4.3 Part Practice

It is often not feasible to learn a skill in its entirety from the beginning, but inefficient to practice too small a part of the skill. The question of part versus whole practice can be resolved by taking into account the size, or complexity, of the task to be learned, and the interdependence, or divisibility, of the aggregation of parts. Other things being equal, the larger the task, the greater the advantage of part practice; however, the more interdependent the parts, the better the whole method becomes. In some cases, such as the recognition of human faces (Woodhead, Baddeley, and Simmonds, 1979), training with isolated features can be completely ineffective.

When the size of a task is varied by the simple accumulation of similar activities, part learning soon becomes superior. In learning mazes of different sizes composed by the combining of identical units, for example, the whole method incurs many times the error score shown when learning the parts separately. In any case, where a reasonable choice is available, part methods are usually recommended for older, less able, or disadvantaged trainees. Where part training is practicable, the order in which the parts are taught seems immaterial (Singer, 1968). Volleyball skill can be subdivided into the components of spiking, setting, digging, and serving, and the various orders in which these subskills are practiced has no effect on learning or retention.

However, certain tasks are less easily divisible into separate parts. Separating a two-dimensional lever movement for individual practice on the left–right and forward–backward dimensions has been shown to be less efficient than practicing the whole task. Similarly, predicting simulated battle formations is best done as a whole when the positions of aircraft carriers and submarines are interdependent (Naylor and Briggs, 1963). Hence, it has been concluded that the value of part training must depend on the tradeoff between task difficulty and the degree of interdependence of the parts. On the other hand, the majority of practical skills can be subdivided in other ways, thus minimizing any violence to the integrity of these skills. The responsibility of the trainer is to make the appropriate subdivisions.

Various forms of cumulative part training have been proposed in order to overcome the potential difficulties in recombining the separately practiced parts. The learner may practice part A (perhaps loading a capstan lathe), then parts A and B (adding the cross-slide operation), and finally parts A, B, and C (now including the turret operation), in the "part–continuous" method. alternatively, in a "progressive-part" scheme, the learner might practice A, B, and C individually until a learning criterion is reached; then separately practice A with B and B with C, before tackling the whole ensemble of A with B and C. However, the evidence suggests that if a task can be meaningfully subdivided, the learner will have little difficulty in reconstituting the task from its components. Such experiments as are available show very little difference between simple part training and cumulative part practice.

8.2.4.4 Adaptive Training

The point of breaking a complex skill into separate parts is to confront the learner with a subtask that is not too difficult to master. An alternative way of achieving the same objective is to begin by simplifying the whole task, increasing the difficulty level in successive steps until the full operational level is reached. If the level of difficulty at any stage is made to depend on the individual learner's level of achievement at that point, the method is adaptive. The principle behind adaptation is not unlike the progression in behavior shaping demonstrated in programmed learning (Section 8.2.2). Adaptive forms of training are potentially superior to fixed difficulty methods, since the learner is always confronted with the right degree of challenge at each level of progress (Kelley, 1969).

A special problem arises if dual tasks are employed. In any practical occupation, there may be conflicting demands on attention. The operator is often expected to carry out time-sharing between two disparate tasks, perhaps making control movements in response to a visual display in a primary task, while processing verbal or numerical information as a secondary task. In these circumstances it has been shown, for example, that training outcomes depend on the task priorities and on the separation of tasks for practice (Gopher and North, 1977). Tracking is more sensitive to changes in emphasis between the primary and secondary tasks, but continues to improve when practiced alone. Digit processing, on the other hand, seems to depend on time-sharing ability, and only improves when practiced in time-sharing conditions.

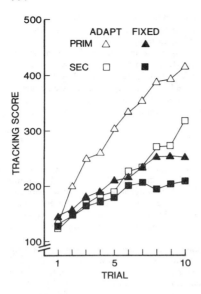

Fig. 8.2.5. Increased learning with adaptive training for a tracking task. The filled symbols represent fixed difficulty training, the open symbols represent adaptive training. [From Johnson and Haygood (1984).]

If dual-task training is to be adaptive, the progressive difficulty level may be based on the performance level reached on either the primary or the secondary task. As Figure 8.2.5 demonstrates, adaptive training of either kind tends to be superior to training at a fixed level of difficulty. In this experiment (Johnson and Haygood, 1984), the primary task was a simulated driving exercise, while the secondary task required a push-button judgment of the position of a subsidiary display located below the driving scene. As might be expected, adapting the difficulty level of training for the primary, driving task on the basis of performance on the secondary, judgment task is less effective than adapting on the basis of the driving task itself.

8.2.4.5 Team Training

Special problems might be expected to arise in team training, where social variables often play a large part in determining behavior. Team training has been shown to be worse than individual training in circumstances where the transfer task, for which the training prepares the learners, is essentially under individual control (Briggs and Naylor, 1965). When the criterion is one of team performance, the results depend on several other variables.

Two-man teams of radar controllers have been trained to carry out aircraft interceptions in a variety of training conditions involving the presence or absence of team coordination, with or without the opportunity for communication (Johnston, 1966). Making use of communication generally interfered with task performance during training, so that the acquisition of team skills adversely affected individual scores. However, when crews were transferred to a coordinated team task at the conclusion of training, the type of training had little effect on the scores. At this stage, the degree of skill acquisition was of primary importance.

The degree of individual skill also seems to determine the effect of replacing team members (Naylor and Briggs, 1965). When a new member is introduced into a three-man crew, team performance may be changed in either direction. However, the positive effect of introducing a highly skilled replacement is smaller than the negative effect resulting from a less-skilled replacement. Adding a greater proportion of untrained personnel to a training team increases the team training load, with some adverse effects.

Synthetic work over long durations on a battery of monitoring and cognitive tasks, including a "code-lock" task only solvable by group cooperation, has been studied by Morgan, Coates, Kirby, and Alluisi (1984). After at least six full days of training for the experienced team members, five-man crews were composed with zero through five untrained personnel in order to allow comparisons between different training loads over a further six days of work. Increasing the training load led to poorer performance over the first two days of work, on both the index of individual efficiency and the measures of team skill, although no differences remained between the differently composed groups by the end of the training period. It is noteworthy that the progress of individual team members was unaffected by working with different proportions of untrained men, while the untrained team members progressed at the same rate, again regardless of the training load in their team.

8.2.4.6   Mental Practice

Apparently, practice need not be overt in order to have some training effect. Having trainees rehearse mentally has been found to produce results that are measurably better than those after no practice, even for perceptual–motor skills such as dart-throwing, bowling, mirror-drawing, and piano playing. Shick (1970) has confirmed by electromyographic methods that mentally practicing the serve at volleyball induces incipient muscle movements in the learner, as if the practice has been partly internalized. It can also be shown that mental practice induces some fatigue, and is thus improved by providing a brief rest period. Such practice is naturally inferior to direct practice at the activity, except in highly cognitive skills, but may be considered as a supplement to normal training.

How mental practice has its effect is not yet fully understood. There may be some motivational benefit, as the mental practice groups normally receive special attention and devote more time to practice, and there might be some advantage from muscle priming. However, the major interest attaches to the way that central processes operate in the organization of skills. An explanation of mental practice based on the possibility of an internal loop of input, incipient output, and resulting feedback, would emphasize the part played by proprioception from the imagined limbs; it would thus predict that the learning effect should be limited to practice with the appropriate limb. In contrast, an outflow type of explanation would emphasize the part played by centrally planned motor programming, and would thus assume that mental practice with a different limb should be almost as effective as imagining the same limb as required in the task.

The outflow, mental programming, type of explanation is favored by the results of transfer tests. Kohl and Roenker (1983) gave trainees either physical or mental practice on a pursuit rotor task, for either the right or left hand and tested the effects of this practice on final performance on only the left hand. Although physical practice was better with the appropriate hand, mental practice was equally good with either hand, and was significantly better than no practice at all. It thus appears that the benefits of mental practice are relatively nonspecific.

8.2.5   TRANSFER OF TRAINING

Carrying over the results of practice with the right hand to tracking with the left hand is an example of transfer of training. To some extent all learning involves transfer of training, since acquiring a new skill is never completely independent of the other activities that precede it. Transfer is a critical issue for most training programs because, except in the case of direct on-the-job training, the value of the training program will depend on the degree to which it transfers to the real task. It has been shown, for example, that only 15 hr of specialized navigation training can produce as much transfer to low-level helicopter flying as more than 2000 hr of general flying experience (Farrell and Fineberg, 1976), a finding which fully justifies the expenditure of training time. Some knowledge of the general principles of transfer is needed in both the development and the evaluation of training programs.

8.2.5.1   Basic Principles

The main issues in transfer of training have been reviewed by Holding (1977). When learning a first task (A) improves the scores obtained in a second task (B), relative to the scores of a control group learning B alone, the transfer from A to B is positive. Task A might consist of pursuit rotor practice with one hand, for example, and task B of practice with the other. It sometimes happens that learning task A hinders the acquisition of task B, in which case the transfer is described as negative. More complex designs may give rise to what are known as retroaction effects, which occur when learning A is followed by learning B and then by retesting on A. If interpolating B in this way has improved performance on task A, there has been retroactive facilitation; if interpolating B has degraded performance on A, there has been retroactive interference (or retroactive inhibition).

The more similar tasks A and B are, the more they interact. Whether the resulting transfer is positive or negative depends on the relationship between the display and control, or stimulus and response, aspects of the two tasks. Osgood's (1949) three-dimensional graph is an attempt to summarize the results of early work on transfer and retroaction relationships. If both the stimulus displays and the required responses are so similar as to be indistinguishable between the two tasks, transfer will obviously be maximal. To all intents and purposes, tasks A and B are versions of the same task, so that practicing A is equivalent to practicing B. Table 8.2.5 lists this effect, together with the other possible combinations of stimulus and response similarity.

The other cases listed in Table 8.2.5 can be appreciated in the context of an example, drawn from shoe manufacture. Let task A consist of producing a seam of shoe stitches (the stimulus component), by applying the correct pressure to a foot treadle (the response component); task B might consist of lighting a row of neon lamps (stimulus) by repeatedly pressing a telegraph key (response). In these circumstances the stimulus and the response aspects of the two tasks are both different, and there is thus no transfer of training (Singleton, 1957). However, if a previously skilled group of shoe seamers is asked to control the row of neon lamps by operating the standard foot pedal, the response is the

Table 8.2.5 Transfer Between Tasks Depends on Both
Perceptual and Response Similarity

	Task Stimuli	Response Required	Transfer
1	Same	Same	High
2	Different	Different	None
3	Different	Same	Positive
4	Same	Different (but similar)	Negative (?)

Source: Holding (1965)

same although the new stimulus display is different. Hence, positive transfer occurs; the experienced seamers do better on this form of task B than do untrained operators.

The final relationship is more elusive, and is inaccurately represented in Osgood's (1949) surface. Requiring a different response to the same stimuli would be equivalent in this case to asking the operators to use the Morse keys to sew shoe seams. This procedure might produce negative transfer. Other things being equal, people tend to do in a new situation what they did in the original task. If the circumstances have changed, but the change is not readily apparent, the old response will occur inappropriately. In the present example, the experienced sewing operatives might occasionally try to press down hard and long on the telegraph keys, instead of giving serial keypresses. However, it is also possible that, despite occasional lapses, the experienced operatives might show better overall performance than the untrained group, because of the general similarities between the two tasks. The result will depend in part on the way in which performance is scored. In any case, it may also happen that initial negative transfer changes to positive transfer as learning on task B proceeds, since lapses will become less frequent.

It is important to prevent the occurrence of negative transfer from a training device, such as a simulator, to the real job, but it is unfortunately not easy to predict when negative transfer will occur. However, an attempt to predict intrusive errors by means of a three-dimensional surface (Holding, 1976), relating stimulus and response similarity to the expected transfer outcome, provides for increasing interference between two tasks as response similarity becomes closer. Whether or not these occasional lapses are important depends on the nature of the task. An occasional error due to negative transfer may be inconsequential in sewing shoe seams, but disastrous while landing an aircraft. Such errors are most likely to occur when the responses in tasks A and B can be easily confused. Trainees are unlikely to confuse riding a bicycle with pouring coffee, even if both are signaled by the same stimulus of a green light. However, such responses as moving a lever up rather than down, or turning a handwheel clockwise instead of counterclockwise, are quite easily confused.

8.2.5.2 Task Difficulty

A separate problem for the prediction of transfer relationships occurs when tasks A and B are at different levels of difficulty. Although the normal educational process assumes progress from easier to more difficult tasks, many experiments have shown greater transfer in the difficult-to-easy direction. In tracking, for example, there may be positive transfer from practice on a difficult, irregular target course to performance on a regular target course, but zero transfer from the regular to the irregular course. As in the case of negative transfer, it is difficult to predict the outcome of asymmetrical transfer relationships.

Better easy-to-difficult transfer has been observed in many perceptual tasks, but also with changes in target speed and amplitude, and some forms of perceptual–motor complexity. However, better difficult-to-easy transfers occurs with other forms of complexity, with changes in various control characteristics, and with differences in display–control relationships. Both directions of effect may occur in the same experimental task, given relatively slight changes in the task variables. There are clearly at least two opposing factors involved, which may be characterized (Holding, 1965) as making for learning more, versus learning better.

Learning more occurs when the difficult version in some sense includes the easier version of a task. It may be that the target motions used in a complex course (A) include those required in a simpler course (B), so that the trainee learning task A has already practiced B, but a trainee who learns task B has no experience of the additional motions in task A. Alternatively, the inclusion may be at the higher level of a strategy for accomplishing the task (Fumoto, 1981). If an intermittent movement strategy is appropriate for task A and viable on an easier task B, but a continuous movement strategy adapted to task B is inappropriate to task A, then better difficult-to-easy transfer will result.

Learning better may occur on the easier version of a task in circumstances where the trainee has the opportunity to acquire higher performance standards, with smaller error tolerances. If these are carried over into performance with the more difficult version, the result will favor transfer from easy to difficult tasks. Again, a change of strategy is sometimes implicated, although this factor may be overshadowed by other effects (Livesey and Laszlo, 1979). Quite large changes in the need for strategies emphasizing speed or accuracy, induced by different rates of target movement, have little effect on transfer. However, learning to stab at large moving dots is a relatively easy task, which transfers better to stabbing at more difficult, smaller dots than does the reverse procedure, despite the fact that both tasks emphasize an accuracy strategy. It will often be found that conducting experimental trials with a training procedure is preferable to attempting to predict the direction of asymmetrical transfer.

8.2.5.3 Measuring Transfer

The amount of transfer to a job tends to increase with the amount of time spent in training. In some cases, the amount of training can be more important than the type of training involved. In a study of automobile driver education (Baron and Williges, 1975), regardless of whether a film program or a driving simulator was used, it appeared that 6 hr of training gave more transfer to road performance than 3 hr of training. However, the amount of training does not increase as a linear function of training time. There are normally diminishing returns from extended training, so that establishing the effectiveness of training will require the continuous measurement of transfer.

The traditional method of measurement is to concentrate on the initial transfer to a new task, calculating the improvement on task B, by those who have previously practiced on task A, relative to the scores obtained by those who have learned task B alone. The difference between the transfer and the control groups (transfer minus control, for accuracy scores; control minus transfer, for speed or error scores), usually on the first trial of task B, is represented as a percentage of the total learning available. A typical formula might show

$$\text{Transfer \%} = \frac{\text{Control score task B, initial} - \text{Transfer (from A) task B, initial}}{\text{Control score task B, initial} \quad \text{Control score task B, final}} \times 100$$

However, these transfer percentages may not remain constant as learning on task B progresses, and more flexible measures may be more appropriate for monitoring training effectiveness.

A sensitive measure of the value returned by different amounts of training is particularly appropriate in using simulators, where the costs of training and of using real devices are often high, but are known and manipulable. Simulators, their degrees of fidelity, and their transfer characteristics are omitted from discussion here (but see Chapter 8.5). However, the aircraft simulator may be taken as a good example of a training task A, whose results are to be transferred to an operational flying task B. A typical problem is to determine how much training should be given on a ground simulator before transferring novice pilots to flight training.

The most useful measures are of the "savings," or "substitution ratio," types. Transfer effectiveness can be calculated in terms of the amount of flight time saved by different amounts of ground training. Roscoe (1971) has outlined incremental and cumulative measures of the ratio. If 10 flying hours are normally needed to attain proficiency, but only 8.6 hr are still needed after the pilot has undergone 1 hr of simulator training on the ground, there has been a saving of 1.4 hr. A second hour may save a little less, perhaps 1.2 hr, so that the cumulative savings after 2 hr on the simulator is 2.6 hr. Dividing this figure by 2 (the number of simulator hours), gives a cumulative transfer effectiveness ratio (CTER) of 1.3, per hour of ground training. The corresponding formula may be rendered as

$$\text{CTER} = \frac{\text{Hours needed control} - \text{Hours needed after training}}{\text{Simulator hours}}$$

As Figure 8.2.6 shows, transfer effectiveness calculated by this index will normally decline in a regular fashion. After 5 hr of simulator training, 5 hr of flight time will still be needed, and the transfer ratio will be reduced to 1.0. Clearly, there is no further gain from continuing ground training beyond this point, in terms of the time required for training. If cost is a factor, as may well be the case, it may be worth continuing the ground training. For example, if 1 hr of flight time costs three times as much as 1 hr on the simulator, it will pay to continue the ground training until the cumulative transfer ratio has fallen to 0.33. Fifteen hours of training on the simulator, together with 5 hr in flight, will only cost the same as 10 hr of flight time. An obvious approach is simply to express the transfer effectiveness ratio in terms of cost, rather than as hours of training. More complex techniques for maximizing cost-effectiveness, based on the methods of calculus, are also available (Carter and Trollip, 1980).

A simpler approach to training effectiveness (Holding, 1977) is also shown in Figure 8.2.6. In the "A + B" method, the training hours (or practice trials, or training costs) for task A are added to

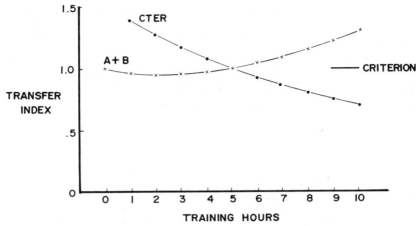

Fig. 8.2.6. Two indices of training effectiveness. The CTER is the cumulative transfer effectiveness ratio (Roscoe, 1971), while A + B is the summed training time (Holding, 1977).

those required on transfer to task B, for each unit of task A practice. Then, whenever the combined value for both tasks exceeds the value for task A alone, it can be concluded that training has become uneconomical. In the graph, this occurs after the point at which 5 hr are needed on both tasks, since 5 hr of simulator time with 5 hr of flight time are clearly no improvement on the standard 10 hr of flight time. The cut-off obviously occurs at the same point located by the transfer effectiveness ratio.

A systematic training program should attempt to maximize transfer to the task for which it is designed. Having achieved good transfer, it should try to optimize the total amount of training time. The use of quantitative transfer measures, like the verbal, visual, and practice techniques reviewed above, should make a contribution toward ensuring training effectiveness.

8.2.6 CONCLUSIONS AND RECOMMENDATIONS

There is very little in the areas of learning and training that is not open to debate. The following statements are given without qualification, in order to provide some practical guidelines. However, in most cases the recommendations should be regarded as representing informed opinion rather than established fact.

8.2.6.1 Basic Issues

The trainer should regulate the supply of information to the learner. Depending on the task involved, a choice must be made between the methods of telling, showing, or doing.

Tasks may be distinguished as verbal or motor. Motor tasks may be simple or complex; gross or fine; discrete or continuous; single or serial. Along another dimension, these tasks may be perceptual or motor; open or closed; skilled or habitual.

Learning at one level, such as rule learning, may be dependent on prior learning at a lower level, such as concept learning.

Motivation does not affect learning directly, but may influence the vigor of the practice needed for learning to occur.

If trainees are sufficiently motivated to undertake learning, additional incentives seem to be unnecessary.

8.2.6.2 Verbal Methods

Words and actions are not equivalent. Verbal instructions may not be utilized in performance, and successful performance may not guarantee the ability to give a verbal report.

Verbal instructions are most effective when spaced throughout the execution of a task, unless used for general orientation and for imparting general principles.

Older trainees may derive less benefit from verbal instructions than younger trainees, and may be less adequately tested by verbal examinations.

Verbal methods can sometimes be employed to convey useful rules, although a knowledge of theory may be ineffective.

Programmed learning provides a desirable model for verbal training, employing systematic ordering of training content, useful prompting, and immediate feedback.

Branching programs use larger steps than linear programs, providing remedial sequences but using weaker (recognition) tests, although both types produce good results.

8.2.6.3 Demonstrating and Guiding

Behavior modeling is a promising technique for higher-level training programs.

Some perceptual–motor learning can be achieved by imitation, although the later stages of skill are best acquired by other means.

Films can present otherwise unavailable lecture or visual material. However, there are disadvantages inherent in the lack of flexibility and audience participation.

Visual cues may be used for prompting perceptual or motor performance, and are important for correcting errors in tasks such as typing.

Visual guidance is effective when it does not obstruct knowledge of alternatives, or direct attention away from cues intrinsic to the task.

Physical guidance can also be effective, when error prevention is important. For most tasks, using restriction techniques will be superior to giving forced-response practice.

8.2.6.4 Practice Methods

Knowledge of results, or learning feedback, is important for learning, although action feedback may also give rise to improvement.

Delaying knowledge of results has little effect, while a delay afterward may be beneficial.

Increased precision of knowledge of results is advantageous up to a point, although there may be an optimum level of precision.

Whether speed or accuracy is stressed during training makes little difference to the eventual outcome.

Massed practice gives poorer temporary performance than distributed practice and may thus indirectly affect learning, but most of the effects of massing are dissipated by rest.

Shorter work sessions may give better results for practical purposes, depending on the choice between time spent actually learning and time spent while training continues.

Part practice becomes preferable to practice of the whole skill as task size or difficulty increases. This effect may be offset in some circumstances by the interdependence of parts.

When a task can be meaningfully subdivided, methods of cumulative part practice confer little benefit.

Adaptive training obviates some part-practice problems by raising the level of difficulty as skill improves, with promising results.

The results of team training depend on the need for coordination and communication. However, individual skill is a major determinant of team performance, and differences due to training load tend to erode as learning progresses.

Covert practice, in the form of mental rehearsal, produces measurable learning in many skills. The effects are normally weaker than those of direct practice, and appear to result from improvement in motor programming.

8.2.6.5 Transfer of Training

Transfer depends on stimulus and response similarity, and the similarity of performance strategies, between two tasks. When the required responses are the same, further task similarities give increasing positive transfer.

Negative transfer may appear in the form of intrusive errors, which are to be avoided in dangerous or expensive contexts. Such errors are likely to occur when similar stimuli indicate different, but confusable, responses.

Differences in task difficulty often produce asymmetrical transfer. Difficult-to-easy transfer is favored when learning a difficult task or strategy includes learning the easy version; the reverse direction of transfer is favored when better performance in the easy version is also appropriate to the difficult version.

Transfer tends to increase, nonlinearly, with increasing training time on a prior task.

Transfer percentages are appropriate for comparing transfer across different tasks, but have limited practical utility.

The optimum length of training before transfer may be decided on the basis of the savings achieved by training, in time or cost, or by adding the pretraining and posttraining time to criterion.

REFERENCES

Baddeley, A. D., and Longman, D. J. A. (1978). The influence of length and frequency of training session on the rate of learning to type. *Ergonomics, 21,* 627–635.

Baron, M. L., and Williges, R. C. (1975). Transfer effectiveness of a driving simulator. *Human Factors, 17,* 71–80.

Belbin, E. (1958). Methods of training older workers. *Ergonomics, 1,* 207–221.

Belbin, E., and Shimmin, S. (1964). Training the middle-aged for inspection work. *Occupational Psychology, 38,* 49–57.

Belbin, R. M. (1970). The discovery method in training older workers. In H. L. Sheppard, Ed. *Towards an industrial gerontology.* Cambridge, MA: Schenkman.

Berry, D. C., and Broadbent, D. E. (1984). On the relationship between task performance and associated verbal knowledge. *Quarterly Journal of Experimental Psychology, 36A,* 209–231.

Briggs, G. E., and Naylor, J. C. (1965). Team versus individual training, training task fidelity, and task organization effects on transfer performance by three-men teams. *Journal of Applied Psychology, 48,* 387–392.

Brooke, J. B., Cook, J. F., and Duncan, K. D. (1983). Effects of computer aiding and pre-training on fault location. *Ergonomics, 26,* 669–686.

Carroll, W. R., and Bandura, A. (1982). The role of visual monitoring in observational learning of action patterns: making the unobservable observable. *Journal of Motor Behavior, 14,* 153–167.

Carter, G., and Trollip, S. (1980). A constrained maximization extension to incremental transfer effectiveness, or, how to mix your training technologies. *Human Factors, 22,* 141–152.

Crocker, P. R. E., and Dickinson, J. (1984). Incidental psychomotor learning: the effects of number of movements, practice and rehearsal. *Journal of Motor Behavior, 16,* 61–75.

Crowder, N. A. (1960). Automatic tutoring by intrinsic programming. In A. A. Lumsdaine and R. Glaser, Eds. *Teaching machines and programmed learning.* Washington, DC: National Education Association.

Crowder, R. G. (1976). *Principles of learning and memory.* Hillsdale, NJ: Erlbaum.

Decker, L. R., and Rogers, C. A. (1973). Forced guidance and distribution of practice in sequential information processing. *Perceptual and Motor Skills, 36,* 415–419.

Farrell, J. P., and Fineberg, M. L. (1976). Specialized training versus experience in helicopter navigation at extremely low altitudes. *Human Factors, 18,* 305–308.

Fleishman, E. A. (1966). Human abilities and the acquisition of skill. In E. A. Bilodeau, Ed. *Acquisition of skill.* New York: Academic.

Fox, P. W., and Levy, C. M. (1970). Learning under conditions of action information feedback. *Journal of Motor Behavior, 2,* 223–228.

Fumoto, N. (1981). Asymmetric transfer in a pursuit tracking task related to change of strategy. *Journal of Motor Behavior, 13,* 197–206.

Gagné, R. M. (1967). Instruction and the conditions of learning. In L. Siegel, Ed. *Instruction: some contemporary viewpoints.* San Francisco: Chandler.

Goldstein, I. L. (1974). *Training: program development and evaluation.* Monterey, CA: Brooks/Cole.

Gopher, D., and North, R. A. (1977). Manipulating the conditions of training in time-sharing performance. *Human Factors, 19,* 583–593.

Holding, D. H. (1965). *Principles of training.* Oxford, England: Pergamon.

Holding, D. H. (1968). Training for skill. In D. Pym, Ed. *Industrial society: social sciences in management.* Harmondsworth, England: Penguin Books.

Holding, D. H. (1970). Learning without errors. In L. E. Smith, Ed. *Psychology of motor learning.* Chicago, IL: Athletic Institute.

Holding, D. H. (1976). An approximate transfer surface. *Journal of Motor Behavior, 8,* 1–9.

Holding, D. H. (1977). Transfer of training. In R. A. Schmidt and E. A. Fleishman, Eds. Section IX: psychomotor learning and performance. In B. B. Wolman, Ed. *International encyclopedia of neurology, psychiatry, psychoanalysis and psychology.* New York: Van Nostrand.

Holding, D. H., Ed. (1981). *Human skills.* New York: Wiley.

Holding, D. H., and Macrae, A. W. (1964). Guidance, restriction and knowledge of results. *Ergonomics, 7,* 289–295.

Johnson, D. F., and Haygood, R. C. (1984). The use of secondary tasks in adaptive training. *Human Factors, 26,* 105–108.

Johnston, W. A. (1966). Transfer of team skills as a function of type of training. *Journal of Applied Psychology, 50,* 102–108.

Kelly, C. R. (1969). What is adaptive training? *Human Factors, 11,* 547–556.

Kohl, R. M., and Roenker, D. L. (1983). Mechanism involvement during skill imagery. *Journal of Motor Behavior, 15,* 179–190.

Landers, D. M., and Landers, D. M. (1973). Teacher versus peer models: effects of model's presence and performance level on motor behavior. *Journal of Motor Behavior, 5,* 129–139.

Leonard, J. A., and Newman, R. C. (1965). On the acquisition and maintenance of high speed and high accuracy in a keyboard. *Ergonomics, 8,* 281–304.

Livesey, J. P., and Laszlo, J. I. (1979). Effect of task similarity on transfer performance. *Journal of Motor Behavior, 11,* 11–21.

Locke, E. A. (1970). Job satisfaction and job performance: a theoretical analysis. *Organizational Behavior and Human Performance, 5,* 484–500.

Long, J. (1976). Visual feedback and skilled keying: differential effects of masking the printed copy and the keyboard. *Ergonomics, 19,* 93–110.

Lucas, R., Heimstra, N., and Spiegel, D. (1973). Part-task simulation training of drivers' passing judgments. *Human Factors, 15,* 269–274.

Macrae, A. W., and Holding, D. H. (1965). Guided practice in direct and reversed serial tracking. *Ergonomics, 8,* 487–492.

Malone, T. A. (1981). Toward a theory of intrinsically motivating instruction. *Cognitive Science, 4,* 333–369.

Marshall, E. C., Duncan, K. D., and Baker, S. M. (1981). The role of withheld information in the training of process plant fault diagnosis. *Ergonomics, 24,* 711–724.

Morgan, B. B., Coates, G. D., Kirby, R. H., and Alluisi, E. A. (1984). Individual and group performances as functions of the team-training load. *Human Factors, 26,* 127–142.

Moses, J. L. (1978). Behavior modeling for managers. *Human Factors, 20,* 225–232.

Nash, A. N., Muczyk, J. P., and Vettori, F. L. (1971). The relative practical effectiveness of programmed instruction. *Personnel Psychology, 24,* 397–418.

Naylor, J. C., and Briggs, G. E. (1965). Team-training effectiveness under various conditions. *Journal of Applied Psychology, 49,* 223–229.

Newell, K. M. (1981). Skill learning. In D. H. Holding, Ed. *Human skills.* New York: Wiley.

Osgood, C. E. (1949). The similarity paradox in human learning: a resolution. *Psychological Review, 56,* 132–143.

Rogers, C. A. (1974). Feedback precision and postfeedback interval duration. *Journal of Experimental Psychology, 102,* 604–608.

Roscoe, S. N. (1971). Incremental transfer effectiveness. *Human Factors, 13,* 561–567.

Salmoni, A. W., Schmidt, R. A., and Walter, C. B. (1984). Knowledge of results and motor learning: a review and critical reappraisal. *Psychological Bulletin, 95,* 355–386.

Salvendy, G., and Pilitsis, J. (1980). The development and validation of an analytical training program for medical suturing. *Human Factors, 22,* 153–170.

Schmidt, R. A. (1971). Retroactive interference and amount of original learning in verbal and motor tasks. *Research Quarterly, 42,* 314–326.

Shepherd, A., Marshall, E. C., Turner, A., and Duncan, K. D. (1977). Diagnosis of plant failures from a control panel: a comparison of three training methods. *Ergonomics, 20,* 347–361.

Shick, J. (1970). Effects of mental practice on selected volleyball skills for college women. *Research Quarterly, 41,* 88–94.

Singer, R. N. (1968). Sequential skill learning and retention effects in volleyball. *Research Quarterly, 39,* 185–194.

Singleton, W. T. (1957). An experimental investigation of sewing-machine skill. *British Journal of Psychology, 48,* 127–132.

Skinner, B. F. (1954). The science of learning and the art of teaching. *Harvard Educational Review, 24,* 86–97.

Smith, K. U., and Kaplan, R. (1970). Effects of visual feedback delay on simulated automobile steering. *Journal of Motor Behavior, 2,* 25–36.

Smyth, M. M. (1978). Attention to visual feedback in learning. *Journal of Motor Behavior, 10,* 185–190.

Taylor, D. S. (1970). The influence of training procedure on multiple choice learning. *Ergonomics, 13,* 193–200.

Tomlinson, R. W. (1972). Control impedance and precision of feedback as parameters in sensorimotor learning. *Ergonomics, 15,* 33–47.

Williams, D. C. S. (1956). Effects of competition between groups in a training situation. *Occupational Psychology, 30,* 85–93.

Woodhead, M. M., Baddeley, A. D., and Simmonds, D. C. V. (1979). On training people to recognize faces. *Ergonomics, 22,* 333–343.

CHAPTER 8.3

THE RELATIONSHIP OF TRAINING GOALS AND TRAINING SYSTEMS

IRWIN L. GOLDSTEIN

**University of Maryland
College Park**

Many of these points were originally made in a presentation at a workshop titled Air Force Learning
Research Laboratory: Proposed Research Agenda as part of a program sponsored by the Manpower
and Personnel Division of the Air Force Human Resources Laboratory at Brooks Air Force Base,
Texas in 1983. Many of the ideas and concepts presented here are also included in a more detailed
version in I. L. Goldstein (1986), *Training in Organizations: Needs Assessment, Development and Evaluation.* Monterey, CA: Brooks/Cole. All such material is reproduced here with the permission of the
author and Books/Cole.

Instructional systems philosophy as developed by Gagné and his colleagues (e.g., Briggs, Campeau, Gagné, and May, 1967; Gagné, 1970) has had a profound impact on the way that such systems are conceptualized. Similar developments stimulated by these instructional theorists have affected the design of training systems in both the public and private sector. These approaches (e.g., Goldstein, 1986, 1978) typically emphasize the specification of instructional objectives based upon needs assessment procedures, precisely controlled learning experiences to achieve these objectives, criteria for performance, and evaluation information. In a training framework, it is possible to examine the components of instructional systems models and ask what the user desires to achieve, that is, what are the goals of the instructional system. This paper examines these goals and the components of instructional systems.

8.3.1 GOALS OF INSTRUCTIONAL PROGRAMS

As presented in Figure 8.3.1, there are four potential goals for training systems. They are as follows:

1. *Training validity.* This particular goal refers only to the validity of the training program. Validity is determined by the performance of trainees on criteria established for the training program.
2. *Performance validity.* This goal refers to the validity of the training program as measured by performance in the transfer of what has been learned to the job setting.
3. *Intraorganizational validity.* This concept refers to the performance of a new group of trainees within the same organization for which the original training program was developed. In this instance, the question is whether generalization will occur to the performance of new trainees based upon the evaluated performance of a previous group.
4. *Interorganizational validity.* In this instance, the goal is to determine whether a training program validated in one organization can be utilized in another organization.

A consideration of these categories indicates that the achievement of validity at each succeeding stage is affected by an increasing number of variables. For example, the establishment of training validity requires the consideration of evaluation procedures including criteria of training success. Trainee performance on the job (performance validity) not only requires the consideration of training criteria of success but also demands an examination of the potential disruptive effects of organizational constraints. As a result, the trainee and training program may be declared inadequate because the training analyst has not considered the relevant variables that determine success or failure at each of the four stages of validity. A consideration of each of these goals and their related instructional processes is presented next.

8.3.2 TRAINING VALIDITY

As noted above, training validity refers to the establishment of treatment effects as a result of the instructional program. In this situation, the pursuit of validity refers only to performance in the instructional program, and does not consider transfer performance. Although training analysts might note that what occurs in the training program does not provide any guarantees about learning transfer to the job, it should be obvious that the vast majority of instructional programs attend only to training performance and not to transfer performance. Of course, many would argue that even when training programs are not designed on the basis of a need assessment of the job, there is still some mental appreciation of the transfer performance around which the instructional program is designed. For example, there are many managerial training programs that have as a goal instructing trainees in the basic theory of management. Evidently, managers are expected to return to their array of jobs and situations with what they have learned and are to determine a way to apply what they have learned. If one questions the developers of such programs, they will readily admit that they have not done a needs assessment and have not measured transfer performance, but somehow or other they have a "feel" for the job setting and thus have designed their program appropriately. Evidence for this viewpoint is nonexistent. It is necessary simply to accept the fact that this type of training program is concerned only with training performance; it is focusing only on the establishment of training validity. As we shall see, that task in itself is not simple.

It is interesting to note that there are some programs that are primarily designed to achieve training validity. Thus instructional programs in academic environments are often designed to teach a particular subject matter primarily because of the value of the content and not necessarily because the learning is expected to transfer to other specific settings. Also, the purpose of the implementation of some

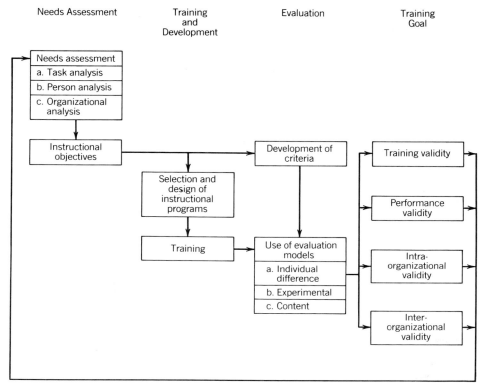

Fig. 8.3.1.   An instructional system. There are many other instructional-system models for military, business, and educational systems. Adapted from I. L. Goldstein, *Training in Organizations: Needs Assessment, Development and Evaluation,* Monterey, CA: Brooks/Cole (1986).

training programs is to provide a research vehicle. For example, an investigator might be concerned with the effects of different types of feedback on the performance of trainees learning a complex motor skill. In this case, the program is concerned primarily with training validity, not necessarily performance validity. It is possible to debate whether or not training programs should be permitted to focus only on training validity without a concern for the transfer setting. However, the value judgment is not the focal point of this discussion. Rather, the point here is that such programs do exist and that the establishment of training validity in itself requires the consideration of many important variables. Some of the most important points to consider include the following.

8.3.2.1   Needs Assessment for Training Validity

First, it is important to realize that some form of needs assessment is necessary. In the case of establishing training validity, the needs assessment would not necessarily focus on the behaviors required in the transfer setting. Rather, the needs assessment would attend to the particular goals and objectives to be achieved as a result of the instructional program. Thus if the program is designed to provide a working knowledge of management philosophy, it is necessary to determine through task analysis and specification of instructional objectives what particular goals the training program is trying to achieve. It is also important to perform a person analysis so that the instructional program is designed to fit the knowledge, skill, and ability characteristics of the trainee population. Indeed, this needs assessment must provide the input for the design of the instructional sequences as well as the suggested criteria by which the training program will be evaluated. It is not reasonable to skip the needs assessment process just because what is being planned is a generalized training program that is not geared to specific behaviors to be performed in a transfer setting. There should still be performance objectives designed for particular trainee populations that are determined by needs assessment procedures. Unfortunately, instead of careful needs assessment the training field often appears dominated by a fads approach. Analysts seldom find out very much about their particular approach before they are off examining another type of program. The fads approach places a heavy emphasis on the development of techniques without consideration of needs assessment followed by a matching of the techniques to the needs. Training analysts still have to be warned about selecting tools and finding something they fit by quotes like the following:

> *If you don't have a gadget called a teaching machine don't get one. Don't buy one don't borrow one; don't steal one. If you have such a gadget get rid of it. Don't give it away, for someone else might use it. This is a most practical rule based on empirical facts from considerable observation. If you begin with a device of any kind you will try to develop the teaching program to fit that device*

GILBERT, 1960, P. 478.

Gilbert is not saying that teaching machines or sensitivity training or CAI or any other technique does not work. He is saying that the design of change programs cannot begin with instructional media. Instead, one must, through needs assessment, determine the objectives of training so that the criteria for evaluation and the choice of instructional program are based on sound decisions. It appears to be unfortunate but true that programs designed to achieve only training validity are also more likely to be designed on the basis of the most popular fad rather than careful needs assessment.

8.3.2.2 A System of Needs Assessment

There are a number of different needs assessment systems that can be used to provide the necessary information for input into the design of training programs. One such system involves the use of both task analysis methodology and the development of the knowledge, skills, and abilities (KSAs) necessary to perform the tasks. With this type of input, the trainer has information regarding what tasks must be performed and what knowledge, skills, and abilities (KSAs) are necessary to perform the tasks. This system also permits the collection of information concerning the tasks and KSAs, such as their importance, where they can best be learned, and so on. These systems are described below. The reader should note that this type of information is required to design a training program even when the major issue of concern is training validity.

Development of Task Statements

The first phase in the task analysis is to specify completely the task performed. The collection of this type of information involves a number of techniques including interviewing job experts, and observing the job being performed. The rules for the specification of tasks have been evolving for a number of years. The following summary is a synthesis of the work of a number of individuals (U.S. Department of Labor, 1972; Ammerman and Pratzner, 1977; Goldstein, Macey, and Prien, 1981; Goldstein, 1986).

1. Use a terse, direct style avoiding long involved sentences that can confuse the organization. The present tense should be used. Words should be avoided that do not give the necessary information.
2. Each sentence should begin with a functional verb that identifies the primary job operation. It is important for the word to describe specifically the work to be accomplished.
3. The statement should describe *what* the work does, *how* the worker does it, to *whom/what*, and *why* the worker does it. The following illustration stems from the job of a secretary:

    What?　　To Whom/What/
    Sorts　correspondence, forms and reports
    　　　　Why?　How?
    in order to facilitate filing them alphabetically.

    The next example comes from the job of a supervisor:

    What　To Whom/What?　Why?
    Inform next shift supervisor　of departmental status
    　　　How?
    through written or verbal reports.
4. The tasks should be stated completely but they should not be so detailed that it becomes a time and motion study. For example, a task could be "slides fingertips over machine edges to detect ragged edges and burrs." However, it would not be useful for the identification of tasks to say that the worker raises his or her hand onto table, places fingers on part, presses fingers on part, moves fingers to the right 6 inches, and so on. Rather, each statement should refer to a whole task that makes sense. Usually, the breaking down of the tasks into a sequence of activities is useful when the task is being taught in the training program. However, that step

doesn't occur until the total task domain is identified and it is determined which tasks should be taught in the training program. Similarly, trivial tasks such as unlocking the file cabinet or turning out the lights when leaving the office should usually be avoided.

Determination of Relevant Task Dimensions

Once the tasks are specified, it is necessary to collect judgments concerning their relevance for the design of training programs. These judgments of tasks are ordinarily collected from groups of subject matter experts such as experienced workers, supervisors, and personnel specialists. The determination of what questions to ask about a task is an important issue. For example, it would not ordinarily be useful to design a training program for tasks that are not important, not performed frequently, and are easily learned on the job. Similarly, a task that is not very important, frequently performed, and easy to learn would not require the attention that should be given to an infrequent but critical task that is difficult to learn. Thus after the tasks are identified, it is necessary to decide which question should be asked about each task. Some examples of the dimensions which can be addressed are presented below as adapted from Ammerman and Pratzner (1977).

A. Importance
   1. Importance in terms of criticality of task for the job
   2. Importance in terms of consequences of error in performing task
B. Time–Frequency
   1. Which tasks were actually performed last year
   2. Frequency that tasks were performed on the job
   3. Most recent time the task was performed on the job
   4. The time spent performing the task on the job
C. Difficulty of learning task
   1. How difficult is it to learn task
   2. How difficult is it to learn the task on the job
   3. How much opportunity is given to learn task on the job
D. Difficulty of performing task
   1. How difficult is it to perform task
   2. Why is it difficult to perform task (complexity of task, lack of training, monotonous work task, etc.)
E. Miscellaneous
   1. Where should task be learned (from training to on the job, etc.)
   2. What level of worker proficiency on task is expected after training

The particular questions that are asked depend on the purpose of the outcomes of the task analysis. If the purpose of the analysis is to determine which tasks should be learned on the job and which tasks should be learned in training, then answers to questions such as how difficult is it to learn this task on the job and where should the task be learned are very useful. Just as it is important to select the questions to meet the purposes of the analysis carefully, it is also critical to design the response dimensions for each question carefully. The data collected are analyzed to determine average responses, variability, degree of agreement between different judges, and so on. Sometimes, in these types of analyses, the researchers discover that different groups of judges such as supervisors and employees don't agree on what tasks are required to perform the job. In those instances, it would be important to resolve the disagreement before training programs are designed. It should be noted that the type of information collected in task analyses is useful for decisions that extend beyond the training program. For example, it is also possible to develop questions related to what tasks individuals should be able to perform before being selected for the job.

Specification of Human Capabilities

There are several different systems for specifying human capabilities. One system advocated by Prien (e.g., Prien, 1977; Goldstein, et al., 1981; Macey, Prien, and Goldstein, 1980) emphasizes the KSAs necessary to perform the tasks developed in the task analysis effectively. Prien defines these categories as follows:

Knowledge (K) is the foundation upon which abilities and skills are built. Knowledge refers to an organized body of knowledge usually of a factual or procedural nature, which, if applied makes adequate job performance possible. It should be noted that possession of knowledge does not insure that it will be used.

Skill (S) refers to the capability to perform job operations with ease and precision. Most often skills refer to psychomotor type activities. The specification of a skill usually implies a performance standard that is usually required for effective job operations.

Ability (A) usually refers to cognitive capabilities necessary to perform a job function. Most often abilities require the application of some knowledge base.

Examples of KSAs obtained from a number of different jobs are as follows:

Knowledge Characteristics

Knowledge of standard accounting principles and procedures

Knowledge of the state and local regulations concerning fire inspection practices

Knowledge of points of wear on the labeling machine

Knowledge of needs assessment procedures for use in the design of training programs

Skills Characteristics

Skill in adjusting cutting blades on labeling machine

Skill in operating a motor vehicle during a high-speed chase

Skill in simultaneously adjusting volume and temperature of water spray

Ability Characteristics

Ability to shift priorities in response to a change in supply conditions

Ability to evaluate the capabilities of subordinates for promotion

Ability to identify causes of employee discrimination complaints

Ability to recognize the usefulness of information supplied by others

Prien recommends the use of interview procedures with job supervisors, personnel specialists, or experienced successful incumbents to develop the KSA information. Often the best procedure is to supply several panels of five to eight knowledgeable persons with a list of the tasks (developed from the task analysis) and ask the following type of questions:

Describe the characteristics of good and poor employees on (*name of task*).

Think of someone you know who is better than anyone else at (*name of task*). What is the reason they do it so well?

What does a person need to know in order to (*name of task*)?

Recall concrete examples of effective or ineffective performance. [Then lead a discussion to explain causes or reasons.]

If you are going to hire a person to perform (*name of task*), what kind of KSAs would you want person to have?

What do you expect person to learn in training that would make them effective at (*name of task*)?

On the basis of this input, the job analyst would obtain the information necessary to write KSA statements. Some of the guidelines for such statements include the following:

1. Maintain a reasonable balance between generality and specificity. Exactly how general or specific the KSAs should be depends on its intended use. When the information is being used to design a training program, it must be specific enough to suggest what must be learned in training.

2. Avoid simply restating a task or duty statement. Such an approach is redundant and usually provides very little new information about the job. It is necessary to ask what knowledge, skills, and abilities are necessary to perform the task. For example, a task might be to "analyze hiring patterns to determine whether company practices are consistent with fair employment practices guidelines." Clearly, one of the knowledge components for this task will involve "knowledge concerning federal, state, and local guidelines on fair employment practices." Another component might involve "ability to use statistical procedures appropriate to perform these analyses." Both the knowledge and ability components would have implications in the design of any training program to teach individuals to perform the required task.

3. Avoid the error of including trivial information when writing KSAs. For example, for a supervisor's job, "knowledge of how to order personal office supplies" might be trivial. Usually, it is possible to avoid many trivial items by emphasizing the development of KSAs only for those tasks that have been identified in the task analysis as important for the performance of job operations. However, because the omission of key KSAs is a serious error, borderline examples should be included. When the KSAs are judged according to their criticality (as described next) unimportant KSAs will be eliminated.

The final step in this procedure is similar to the phase in the analysis of task statements where the items are judged along relevant dimensions. Some of the dimensions which might be used for KSAs are as follows:

Difficulty to learn. Item is judged on how difficult it is to gain competence with reference to the knowledge, skill, or ability for job performance.

Importance. Item is judged on the criticality of the knowledge, skill, or ability for job performance.

Opportunity to Acquire. Judgments are collected pertaining to the opportunity to acquire the requisite knowledge, skill, or ability on the job.

As discussed in the task analysis section, the questions and the response scales must be specifically designed for the purpose of the needs assessment. It is then possible to obtain ratings on the scales for each of the KSAs and use that information for program design. For example, it might be useful to design training content based on a measure that identifies the important KSAs necessary for job performance, with the main emphasis being on KSAs for which there is a minimum opportunity to learn on the job.

The task analysis provides a specification of the required job operations regardless of the individual performing the task. There remains a very critical part of the total process—the human being. At least two populations must be considered; the first consists of those persons who are already performing the job, and the second involves those persons who will be trained. In some instances these are the same individuals, but other cases involve new trainees. Because these new individuals differ from those already performing the task, it is necessary to examine this second population. The key issue is what are the person capabilities necessary to effectively perform the job. After these capabilities are specified, then it is possible to analyze the performance of the target population to determine whether training is necessary.

8.3.2.3 Criteria and Training Validity

In general, evaluation, as described by Goldstein and Buxton (1982), is considered to be an information-gathering process that should not be expected to lead to decisions that declare a program as all good or all poor. One of the purposes of the needs assessment is the determination of the KSAs required for successful job performance. As shown in the model in Figure 8.3.1, that information must provide direct input into the training program to determine the actual content of the instructional material. The same information concerning the KSAs necessary for successful job performance should also provide the input for the establishment of measures of training success. Logically, we should want our training program to consist of the materials necessary to develop the KSAs to perform successfully on the job. Just as logically, we should determine the success of our training program by developing measures (or criteria) that tell the training evaluator how well the training program does in teaching the trainees the same KSAs necessary for job success. Also, these criteria are used at the end of the training program to determine how well our program is doing at that time. And as discussed in the section on transfer validity, these criteria should also be used later, when the trainees are on the job, to determine how much of the KSAs learned in training transferred to the actual job.

8.3.2.4 Evaluation Models and Training Validity

Instructional programs are never complete; instead, they are designed to be revised on the basis of information obtained from evaluations founded on relevant multiple criteria that are free from contamination and reliable. The better experimental procedures control more of the potentially disruptive variables, thereby permitting a greater degree of confidence in specifying program effects. Although the constraints of the training environment may make laboratory type evaluations impossible to achieve, an awareness of the important factors in experimental design makes it possible to conduct useful evaluations even under adverse conditions.

There are several types of evaluation in models employed to assess learning. Perhaps the best-known model is typically titled experimental or quasi-experimental design model. These types of models, which are sometimes employed in evaluating training validity, include considerations related to pre- and posttesting, control groups, random assignment of subjects, and so on. The job of the analyst is to choose the most rigorous design possible and to be aware of its limitations. Unfortunately, there are few studies that employ the most rigorous design within the limitations imposed by the environment. In many instances the authors of research articles apologize in advance for poor designs that obviously could have been strengthened by some initial effort. It appears that most efforts at evaluation are afterthoughts. Appropriate planning and the use of elementary research procedures would significantly increase the amount of valid information gathered.

Recently, however, training researchers have begun to consider other types of evaluation models. One model that is now being used quite frequently is an individual difference model. In this model, learning scores are correlated with later on-the-job performance scores to determine whether the same pattern of individual differences exist. Thus if persons who perform well in training also perform well on the job positive correlations would result. The other model is a content validity model where the analysis examines whether the knowledge, skills, and abilities that are judged critical for job performance are also those emphasized in training. Further discussion of these models is presented in the discussion on transfer performance because they all depend on transfer data as well as training data. They are presented here only because they also depend on training validity data. This also is an appropriate place to hint that issues concerning the interrelationship between these different types of evaluation models pose important questions.

Still another evaluation model in training validity is called a process model. This model is actually a partner to all evaluation models rather than a completely separate way of examining the world of training validity. There is a growing awareness that the use of evaluation designs has not provided the degree of understanding that was originally anticipated. The evaluation designs and specification of outcome criteria have been based upon a product or outcome view of training validity. Thus researchers collected pre- and postmeasures, compared them with control groups, and discovered that they did not understand the results they had obtained. This problem was especially apparent when the collectors of these data were outside consultants who appeared only to collect pre- and postdata but had no conception of the process that had occurred in training between the pre- and postmeasurement. A recent event experienced by this author illustrates this issue. In a study of computer-assisted instruction in a school setting (Rosenberg, 1972), two teachers each agreed to instruct a geometry class by traditional methodology and by computer-assisted instruction (CAI). Thus each teacher taught one traditional and one CAI class. Further, the teachers agreed to work together to design an exam that would cover material that was presented in both classes. At the end of the first testing period the traditional classes taught by each teacher significantly outperformed the CAI groups taught by these same teachers. The same outcome occurred at the second testing. However, at the third testing, one of the CAI groups improved so that it was now equivalent to the two traditional groups. The other CAI group performed significantly worse than the other three instructional groups. One reasonable conclusion for this series of events is that one of the teachers finally learned how to instruct the CAI group so that it was now equivalent to the two traditional groups, whereas the other teacher had not been able to perform that task with his CAI group. Indeed if the investigators had collected only the outcome measures any number of similar erroneous conclusions could probably have been offered as explanations for the data. In this case, the investigators observed the instructional process to provide information about the program. The evaluators learned that the instructor for the CAI groups that eventually improved had become disturbed over the performance of his students. Thus the teacher offered remedial tutoring and essentially turned his CAI class into a traditional group.

Many researchers have become concerned about the lack of understanding about the variables in the training process that affect or determine outcomes. The view is especially well expressed by Cronbach (1973, p. 675):

> *Insofar as possible evaluation should be used to understand how the course produces its effects and what parameters influence its effectiveness. It is important to learn, for example, that the outcome of a programmed instruction depends very much upon the attitude of the teacher; indeed, this may be more important than to learn that on the average such instruction produces slightly better or worse results than conventional programs.*

8.3.3   PERFORMANCE VALIDITY

As noted earlier, the establishment of performance validity has the additional burden of determining whether performance has positively transferred from the training program to the on-the-job environment. For this type of validity, transfer of performance is considered only for the original group being trained in the instructional program. The question of generalizing to new training populations or new organizations is treated in a later section. Considering the points presented in the discussion of training validity it may appear that there are relatively few additional issues to be added to the analysis of performance validity. Unfortunately, that is not true. Now it is necessary to consider conditions that involve the transfer of performance in one environment (training) to another environment (on the job). This perspective has important implications for the needs assessment process because now the objectives and goals of the instructional program must be determined by a needs assessment of the transfer environment.

8.3.3.1   Needs Assessment and Performance Validity

In the preceding section on training validity, it was noted that the needs assessment process was still necessary. However, it was limited to task and person analysis to establish learning objectives that were not necessarily related to transfer performance. In the establishment of performance validity, the concern is with designing training programs based on a needs assessment of on-the-job performance.

There are dozens of studies that can be cited that demonstrate the dangers of designing training programs without this type of needs assessment. Perhaps one of the sadder examples is revealed by an investigation (Miller and Zeller, 1967) of 418 hard-core unemployed trainees in a program to train highway construction machinery operators. The authors were able to obtain information from 270 graduates. Of this group, 61% of the graduates were employed and 39% unemployed at the time of the interview. In addition, more than half of the employed group said they were without jobs more than 60% of the time. Some of the reasons for the unemployment situation were inadequacies in training such as not enough task practice. The details showed that the program was not based on a consideration of job components. One trainee noted that "Contractors laughed when I showed them my training diploma and said come back after you get some schooling, buddy" (p. 32–33). In a familiar lament, the authors of the report wonder how a training program could be designed without a thorough analysis of the tasks and knowledge, skills, and abilities required.

As far as research needs concerning traditional needs assessment, they are all specified in the preceding section on training validity. However, there is a growing realization that training occurs in one organization (training organization) and performance occurs in another organization (the work organization). Thus performance validity forces consideration of the fact that something learned in one environment (training) will be performed in another environment (on-the-job). The trainee will enter a new environment to be affected by all of the interacting components that represent organizations today. Certainly, there are some aspects of the environment that help determine the success or failure of training programs beyond the attributes the trainee must gain as a result of attending the instructional program.

The actual components of an organizational analysis depend on the type of program being instituted and the characteristics of the organization to which transfer will occur. However, the following broad categories should be considered.

The Specification of Goals

When organizational goals are not considered in the implementation of training programs, objectives and criteria that ensue from the needs assessment process are not evaluated. Later, the organizations are not able to specify their achievements because they have not collected the necessary information. Clearly, if the goals are not specified, it is not very likely that they are included in the design of the instructional program.

Organizations might expect training programs to provide trainees with expectations about the job or particular views toward performance requirements. It is not unusual, for example, to discover that police training programs are devoted to skills requirements (e.g., operating a police vehicle or utilizing firearms) or information requirements (knowing the difference between a felony and misdemeanor). Yet organizational analyses often discover that there are organizational expectations concerning interpersonal relationships with the public and a concern for all citizens regardless of race, color, or creed. Obviously, if these organizational philosophies and emphases are not clearly defined and specified during the needs assessment phase, they will not be considered in the design of the training program and at best will receive only passing attention in the instructional sequence. In this case, the problem is further complicated by the fact that it is a lot easier to teach weapons procedure than to instruct in the topic of interpersonal relations. Thus certain topics may be deemphasized and sometimes organizations become aware that they have a problem only when they are facing a series of complaints. At that point, everyone wonders why the organizational objectives were not translated into training and job requirements. In short, many training programs are based on teaching skills requirements rather than the unspecified and unidentified organizational objectives. Yet paradoxically, organizations often judge the value of a program on the basis of their own objectives, which were never specified, never considered in the design of the program, and never utilized in designing the evaluation model. Thus the financial cost of some instructional programs results in their early demise because the instructional system analysis did not identify that variable as an organizational objective.

The Determination of the Organizational Training Climate

As complex as the specification of system-wide organizational objectives appears to be, the determination of the objectives by themselves will not do the job. Unfortunately, many of the situations are also marked by organizational conflicts that are very disruptive. For example, conflicts between government sponsors of a program, employers, and training institutions can completely disrupt programs (Goodman, 1969). Many of these conflicts are based on the different parties to the program having different sets of equally unspecified goals and expectations. A study by Salinger (1973) characterized the negative consequences for a poor climate. In her study, top management in federal agencies had a generally negative view of training and no direct knowledge of its benefits. This resulted in a system that failed to reward managers for effective training efforts. Thus, managers failed to plan or budget for training. It is not difficult to guess that whatever training was provided did not serve much of a purpose for the individual or the organization. Even more likely, there are insidious conflicts that somehow (as with a magic wand) the training program is supposed to resolve. Even assuming that the training program could be appropriately designed in such a situation, it is likely it will succumb to the conflict.

The Identification of Relevant External System Factors

The preceding sections on organizational analysis have identified issues related to the failure to specify organizational goals and the problems of organizational conflicts. The issue examined in this section is the failure to recognize the importance of the interacting constraints acting on an organization and their effects upon training programs. These considerations could be treated as a failure to specify organizational goals or as organizational conflicts. However, external constraints are becoming a very serious problem in the design of training program. Thus they are treated as a seperate section. The design of instructional programs is affected by legal, social, economic, and political factors. Interrelationships of these variables should be carefully specified during organizational analysis. An example of these factors is provided by Salinger's study (1973) which found instances of writing and typing styles taught to government clerical workers that were prohibited on the job. Another example is that federal requirements concerning equal employment opportunity often affect the applicant pool, which in turn affects the design of instructional programs.

Some Organizational Analysis Questions

The issues discussed in this section indicate the importance of organizational analysis procedures as part of a needs assessment. As a result of this type of analysis, it may be possible both to identify organizational goals and to solve potential organizational conflicts. Then, the instructional program is more likely to be relevant and produce transfer to the work organization. It is even possible to suggest that training analysts might take a self-diagnostic test to determine whether training is really ready to begin. Some of the questions that could be asked are listed below. To the extent that the analyst is left with an uneasy feeling of uncertainty in answering these questions, there is probably a lot of work to do before implementing a training program.

Are there unspecified organizational goals that should be translated into training objectives or criteria?

Are the various levels in organization committed to the training objectives?

Have the various levels and/or interacting units in the organization participated in the development of the program beginning at the needs assessment?

Are key personnel ready both to accept the behavior of the trainees as well as to serve as models of the appropriate behavior?

Is training being utilized as a way of overcoming other organization problems or organizational conflicts that require other types of solutions?

Will trainees be rewarded on the job for the appropriate learned behavior?

Is top management willing to commit the necessary resources to maintain the work organization while individuals are being trained?

8.3.3.2 Evaluation and Performance Validity

A reexamination of the evaluation section for training validity should reinforce the point that each type of validity being considered builds on the concepts presented previously. Thus performance validity assumes that the training analyst has already met the issues of criterion development and evaluation design appropriate for the establishment of training validity. Clearly, if the concern is performance validity, there is increasing complexity in the design of proper evaluation methodologies. The additional difficulties are well documented by Blaiwes and Regan's (1970) suggestion that evaluation efforts must consider three criteria: (1) original learning efficiency; (2) transfer of learning to the new task; and (3) retention of original learning. Yet in the area of flight simulation where the greatest effort has been expended, the emphasis has been on the most immediate criteria, original learning, and even those studies have been plagued by serious problems.

In addition, the consideration of performance validity features the discussion of two other evaluation models briefly mentioned in the training validity section. They are *individual difference models* and *content validity models*.

Individual Difference Models

Many psychologists have emphasized the use of training scores as a way to predict the future success of potential employees. There are a number of these types of studies showing meaningful relationships between training performance and on-the-job performance. For example, Kraut (1975) found that peer ratings obtained from managers attending a month-long training course predicted several criteria including future promotion and performance appraisal ratings of job performance. Other investigators have used early training performance to predict performance later on in more advanced training. An example of this approach is offered by Gordon and Cohen (1973), whose study involved a welding program that was part of a larger manpower development project. The program consisted of 14 different tasks that fell into four categories and ranged in difficulty from simple to complex. Advancement from one task to the next depended upon successful completion of all previous tasks. Thus trainees progressed at a rate commensurate with their ability to master the material to be learned. These investigators understood that they were predicting the performance of individuals on a later task (e.g., on the job or later in training) based upon performance in the training program. As a matter of fact, once these relationships have been established in an appropriately designed study, it is possible to select individuals for a job or for later training based upon these training scores. In other words, the training score serves as a validated predictor of future performance. However, an important note of caution should be considered when this technique is used as an evaluation of the training program. The relationship between training performance and on-the-job performance simply means that persons who perform best on the training test also perform well on the job. This does not necessarily mean that the training program is properly designed or that persons have learned enough in training to perform well on the job. It is entirely possible that the training program did not teach anything or that the trainees did not learn anything. In those cases, there could still be individual differences on the training test. Even if the training program did not teach relevant material for the job, there would still be a strong relationship if the person who did well on the test did well on the job and the persons who did poorly on the test did poorly on the job. In other words, the training program had not achieved anything except to maintain the individual differences between trainees that might have existed before they entered training.

Content Validity Models

If the needs assessment is appropriately carried out, and the training program is designed to reflect the KSAs, then the program would be judged as having content validity. In other words, the training program should reflect the domain of KSAs represented on the job that the analyst has determined should be learned in the training program. One way of conceptualizing the content validity of a training program is presented in Figure 8.3.2. In this figure, the horizontal axis across the top of the figure represents the dimension of the importance or criticality of the KSAs as determined by the needs assessment. Although the diagram presents KSAs only as being important or not important, it must be realized that this is an oversimplification of a dimension with many points. The vertical dimension represents the degree of emphasis for the KSAs in training. Again, to simplify the presentation, the dimension is presented as indicating that the KSA is or is not emphasized in the training program. This results in the fourfold table presented in Figure 8.3.2. Using this approach, both boxes A and D provide support for the content validity of the training program. KSAs that fall into box D are judged as being important for the job and emphasized in training. Items in box A are judged as not

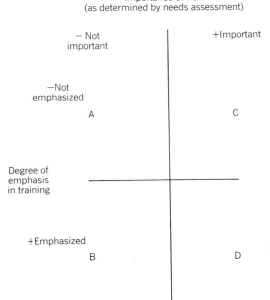

Fig. 8.3.2. A conceptual diagram of content validity of training programs. Adapted from I. L. Goldstein, *Training in Organizations: Needs Assessment, Development and Evaluation,* Monterey, CA: Brooks/ Cole (1986).

important for the job and not emphasized in training. Conceptually, the degree to which KSAs fall into categories A and D make it possible to think about the training program as being content valid. Again, this is an oversimplification. There will be KSAs that are judged as moderately important for the job or KSAs that are moderately emphasized in training. However, it is possible to conceive of this type of relationship and actually measure the degree to which those KSAs judged as important are emphasized (and, it is hoped, learned) in training. Very few persons would be unhappy with a training program that tended to emphasize the objectives associated with KSAs that are judged critical or important for job performance.

Box B represents a potential error and could affect the degree to which a program is judged content valid. KSAs falling into the B category are judged as important for the job but are not emphasized in training. From a systems perspective, these items must be analyzed to determine whether the organization intends for these KSAs to be gained as a result of training. If that is the case, then there is a problem. However, it is also possible that individuals are expected to be selected with that particular KSA or that the individual is expected to learn the material related to that KSA on the job. To that extent, the training program should not be expected to emphasize the item and its content validity would not be questioned. However, it would still be important to determine that KSAs judged as important or critical are covered in another system such as selection. If the item is not covered then the organization must decide whether revision of the training program is necessary or whether some other system has to be redesigned to cover that material.

Box C represents KSAs that are emphasized in training but are judged as not being important for the job. This is often a criticism of training programs. That is, they tend to spend a lot of time emphasizing material that is not job related. Most analysts agree that the use of needs assessment procedures results in a decrease in the amount of training time necessary to complete the program. This usually comes about because of a reduction in the scope of the program based upon an elimination of the type of items present in category C. Interestingly, a systems view of this process might even suggest a reduction of items that would appear in category D. Selections experts might suggest that sometimes KSAs included in category D are unnecessary because they have already been used as a basis for selection. Thus trainees are again subjected to training materials on KSAs that are already in their repertoire.

8.3.4 INTRAORGANIZATIONAL VALIDITY

Intraorganizational validity assumes that the trainer has established training and performance validity and is now concerned with predicting the performance of a new group of trainees. Just as performance validity assumes the consideration of the points established for training validity, intraorganizational validity assumes that the points discussed for training and performance validity have been established. It should be noted that the previously expressed view concerning evaluation stated that evaluation is considered an information-gathering process that provides feedback about the multiple objectives of most training programs. Thus it becomes apparent that evaluation should be a continual process that provides data as a basis for revisions of the program. New data should be collected based upon the performance of a new group of trainees, which provides further understanding about the achievement of objectives or about the variables that affect the achievement of objectives. That does not mean that each effort must start from the very beginning. However, it should be possible to collect further new information about the effects of revisions. Also, it should be possible to collect data that can be checked against previously collected information to make sure that the instructional program is having the same effect.

8.3.5 INTERORGANIZATIONAL VALIDITY

In this instance, the analyst is attempting to determine whether a training program validated in one organization can be utilized in another organization. All the factors discussed in training validity, performance validity, and intraorganizational validity have an effect on this decision. As indicated in the section on intraorganizational validity, when the needs assessment shows differences (i.e., the task, person, or organizational components) or the evaluation is questionable, or the training program has changed, then generalization is dangerous. In this instance, the needs assessment and evaluation have not been performed for the organization that desires to use the training program. Considering the vast number of ways organizations differ, it is dangerous to attempt to generalize the training results in one organization to trainees in another organization.

8.3.6 CONCLUSION

This paper has presented the author's views concerning a hierarchy in the establishment of training, performance, and intraorganizational and interorganizational validity. One way for the training analyst to consider this entire process is to ask, what is the final goal? If the final goal is the establishment of intraorganizational validity, then it is necessary to establish all of the components of validity that precede it (i.e., both training and performance validity).

REFERENCES

Ammerman, H. L., and Pratzner, F. C. (1977). *Performance content for job training,* R. & D Ser. 121–125, Vols. 1–5. Columbus, OH: Center for Vocational Education.

Blaiwes, A. S., and Regan, J. J. (1970). *Integrated approach to the study of learning, retention, and transfer—A key issue in training-device research and development* (NAVTRADEVCEN 1H-178). Orlando, FL: Naval Training Device Center.

Bray, D. W., and Grant, D. L. (1966). The assessment center in the measurement of potential for business management. *Psychological Monographs, 80*(17, Whole No. 625).

Briggs, L. J., Campeau, P. L., Gagné, R. M., and May, M. A. (1967). *Instructional media: A procedure for the design of multi-media instruction, a critical review of research, and suggestions for future research.* Palo Alto, CA: American Institutes for Research.

Catalanello, R. F., and Kirkpatrick, D. L. (1968). Evaluating training programs—the state of the art. *Training and Development Journal, 22,* 2–9.

Cronbach, L. J. (1973). Course improvement through evaluation. *Teachers College Record, 64,* 672–683.

Gagné, R. M. (1965, 1970). *The conditions for learning.* New York: Holt, Rinehart & Winston.

Gilbert, T. F. (1960). On the relevance of laboratory investigation of learning to self-instructional programming. In A. A. Lumsdaine and R. Glaser, Eds., *Teaching machines and programmed instruction.* Washington, DC: (1974). National Education Association.

Goldstein, I. L. (1986). *Training in organizations: Needs assessment, development and evaluation.* Monterey, CA: Brooks/Cole.

Goldstein, I. L. (1978). The pursuit of validity in the evaluation of training programs. In I. L. Goldstein, Ed., Training: methodological considerations and empirical approaches. Special Issue *Human Factors Journal, 20.*

Goldstein, I. L., and Buxton, V. (1982). Training and human performance. In E. A. Fleishman and M. D. Dunnette, Eds., *Stress and performance effectiveness.* Hillsdale, NJ: Erlbaum.

Goldstein, I. L., Macey, W. H., and Prien, E. P. (1981). Needs assessment approaches for training development. In H. Meltzer and W. R. Nord, Eds. *Making organizations humane and productive.* New York: Wiley.

Goodman, P. S. (1969). Hiring and training the hard-core unemployed: A problem of system definition. *Human Organization, 28,* 259–269.

Gordon, M. E., and Cohen, S. L. (1973). Training behavior as a predictor of trainability. *Personnel Psychology, 26,* 261–272.

Kraut, A. I. (1975). Prediction of managerial success by peer and training-staff ratings. *Journal of Applied Psychology, 60,* 14–19.

Macey, W. H., Prien, E. P., and Goldstein, I. L. (1980). Multi-method job analysis: Methodology and applications. Unpublished paper. Memphis, TN: Performance Management Associates.

Miller, R. W., and Zeller, F. H. (1967). *Social psychological factors associated with response to retraining* (Final Report, Office of Research and Development, Appalachian Center, West Virginia University). Washington, DC: U.S. Department of Labor.

Mindak, W. A., and Anderson, R. E. (1971). Can we quantify an act of faith? *Training and Development Journal, 25,* 2–10.

Panell, R. C., and Laabs, G. J. (1979). Construction of a criterion-referenced, diagnostic test of an individual instruction program. *Journal of Applied Psychology, 64,* 255–262.

Prien, E. P. (1977). The function of job analysis in content validation. *Personnel Psychology, 30,* 167–174.

Rosenberg, B. D. (1972). An evaluation of computer-assisted instruction in the Anne Arundel County School System. Master's thesis. University of Maryland.

Salinger, R. D. (1973). *Disincentives to effective employee training and development.* Washington, DC: U.S Civil Service Commission, Bureau of Training.

Swezey, R. (1978). Criterion-referenced measurement in performance. In I. L. Goldstein, Ed., Training: Methodological considerations and empirical approaches. Special Issue *Human Factor Journal, 20.*

U.S. Department of Labor, Manpower Administration (1972). *Handbook for analyzing jobs.* Washington, DC: U.S. Government Printing Office.

Williges, B. H., Roscoe, S. N., and Williges, R. C. (1972). *Synthetic flight training revisited* (Tech. Rep. ARL-72-21/AFOSR-72-10). Savoy, IL: Aviation Research Laboratory.

CHAPTER **8.4**

COMPUTER-ASSISTED AND COMPUTER-MANAGED INSTRUCTION

RAY E. EBERTS

Purdue University
West Lafayette, Indiana

JOHN F. BROCK

Honeywell Systems and Research Center
Minneapolis, Minnesota

Some of the material in this chapter is based on a previous article in the 1984 *The Annual Review of Human Factors* (Eberts and Brock, 1984).

8.4.1 INSTRUCTIONAL TECHNOLOGY

Instructional technology has its roots in the programmed-instruction (PI) movement of the 1940s. Prior to that time, the western world approached instruction with one of two models: the master/ apprentice (or apprentices) model and the classroom model. PI led to individualization. Individualization—on any of several dimensions—first became sacrosanct and then controversial. Instructional technology actually began as a scientific response to a political/philosophical problem: If all persons should be afforded equal opportunity to learn, how could that best be accomplished?

There were two basic types of PI: linear or branching. The former represented the strict behaviorist ideal; it attempted to eliminate the possibility of response error by the student. Branching programs, on the other hand, allowed the student to respond incorrectly, but also used the incorrect responses to teach by correcting the student's misperceptions and misunderstandings.

Soon, instructional technologists were examining all aspects of the instructional process. How long should instructional units be: in time, in number of words, and in number of ideas or tasks to be learned? Did motion in a display instruct better than no motion? How about color? Humor? Graphics? Along how many dimensions do people really differ? These are the questions of instructional technology, many of which will be addressed in this chapter.

The computer has been an effective tool in helping to answer these questions. It can provide many different instructional techniques, and it can collect and process student performance data. As the major part of this discussion will show, the exploitation of instructional technology is possible primarily—but not exclusively—through the use of computer-based instructional systems.

8.4.2 COMPUTER TECHNOLOGY

It was estimated in 1976 that by 1980 there would be 700,000 computers in the United States. In 1981, over 800,000 *personal* computers were sold (Boraiko, 1982). A recent estimate (November 1984) stated that there were 600,000 computers in the elementary and secondary schools alone; one for every 66 students and an increase of 20 times from the 1980 level. We should not be surprised at early underestimates; there is folklore that at one time in the 1940s IBM estimated that 10 computers could meet the data-processing needs of the United States. Computers are a growth industry; whether or not to use a computer in instructional processes may soon cease to be a choice. It will just be done.

There are three aspects of computer detail that we will discuss briefly: size, cost, and operator characteristics. Computers are getting smaller. Most of the readers of this chapter have digital watches that keep time in two time zones, awake the wearer, time a race, beep when a meeting should be over, and glow in the dark. A few readers' watches can balance their checkbooks and play "Happy Birthday" as well. It is axiomatic that computer hardware is becoming smaller and smaller.

Hardware is decreasing in cost, although the bottom may be near. For example, in the last 25 years the price of computer hardware used by the Army for instruction has been reduced by a factor of 5000 (Psotka, 1983). There is probably a limit to how cheaply displays and controls can be made. Software development costs, being labor intensive, are expanding at a tremendous rate. It may be true that if Rolls Royces had followed the path of computers, you could now buy a Silver Cloud for $2.25. But to keep the metaphor sound, you would also be paying $1500.00 for a gallon of gas.

Powerful devices will be available to train most people inexpensively. The instructional and software components of these devices will continue to be expensive to design. Amortization across many systems may be the only cost-effective way to proceed with computer-based devices. As computer accessibility goes up and computer costs go down, the success of computer-based instruction depends on the development of effective methods and instructional designs for implementing them (Montague, Wulfeck, and Ellis, 1980a).

8.4.3 COMPUTER-ASSISTED INSTRUCTION

Computer-assisted instruction (CAI) is a type of computer-based training that is often used in conjunction with other instructional systems. For simulators, CAI can be used to guide the instruction. CAI can also be used as a component of a computer-managed-instruction (CMI) system, which manages and prescribes the mode of instruction to be used; CAI is one of the available modes. Finally, intelligent CAI (ICAI) is a class of instructional programs that is made more intelligent through the use of artificial-intelligence (AI) techniques. This extension of CAI will be discussed in a following section.

The amount of computer control over the complete instructional process varies with the application. CAI can be used as an ancillary part of an instructor's lesson, providing instruction on a small subset of the total material. At the other end of the spectrum, ICAI can be used to control all aspects of the pedagogical interactions with little or no assistance from an instructor.

Actual CAI instructional material, called courseware, can take many forms. Courseware is the special-purpose software that is authored by a subject-matter expert or a computer programmer to provide the instructional interactions. Bunderson (1981) provides a wide definition for courseware: the materials of instruction that constitute the applications programs. Visibly, this can include magnetic tape, a hard disk, or a floppy disk, which contains the courseware code. In addition, according to this definition, the courseware can include off-line adjunctive materials, such as filmstrips or slides on computer-controlled devices, audio tape or disk, videotape or videodisk, a set of microfiche cards, or printed material that can be referenced by the computer.

Bunderson (1981) also makes a distinction between authoring-system courseware and delivery-system software. The authoring-system courseware controls the interaction between an author of instruction and the computer. It provides a system whereby the courseware can be created and later edited (such as changing the material or inserting new text). The authoring system is an important component of a CAI package. The author–computer interaction is made as simple as possible so that subject-matter experts will be able to author the courseware with little software training. The delivery-system software, on the other hand, controls the interaction between the student and the computer. It takes care of providing the text, evaluating student answers, and recording the correctness of the responses.

8.4.3.1 Examples of CAI Systems

Three examples of CAI systems—PLATO, TICCIT, and PLANIT—will be discussed. In discussing them in this order, we will be going from a large-scale system (PLATO), through an intermediate-sized system (TICCIT), to a much smaller authoring language (PLANIT), thus demonstrating what a CAI system can do at each of these different levels. These examples will show the diversity of some of the systems available and how they can be used in a wide variety of settings. They are not by any means the only CAI systems available, but they do represent three major types.

PLATO

PLATO stands for programmed logic for automated teaching operations. The completion of the first design for the PLATO system occurred in June 1960, as a collaborative effort among engineers, physicists, psychologists, and educators at the University of Illinois. During the more than 20 years of funding through various military, industrial, and educational sources, it has gone through four phases. Through the years, about $900 million has been invested in its research and development. As an example of its heavy use, in one 7 year period (July 1974 to September 1981) students and authors recorded roughly 9,600,000 hr of use (Lyman, 1981). In addition, more than 12,000 instructional hours have been authored and are available in more than 150 subject areas (Lyman, 1981). As another example, the University of Delaware's PLATO system has 9000 students taking 112 courses each semester (Hofstetter, 1983). Currently, 17 central PLATO systems are in operation with each capable of handling more than 1000 terminals. Since 1976, the Control Data Corporation (CDC) of Minneapolis, Minnesota, has been marketing the system.

Courseware Authoring. The courseware authoring system for PLATO has been designed by educators and software engineers to be used by educators; those who are most familiar with the content area are the ones who program it. TUTOR IV is the language that is presently used. It has been in use since 1971, developing out of the earlier languages CATO and TUTOR III. Many of the characteristics now included in TUTOR IV have resulted from the suggestions of the users. It is, in fact, a courseware authoring language that has evolved from a long period of interactions between the users (e.g., educators) and the software designers. The implementation of micro TUTOR, an authoring language highly similar to TUTOR but on a smaller scale, allows the PLATO terminal to be used off-line from the central computer terminal.

For PLATO, the emphasis has been placed on creating a courseware language that is easy to use yet highly flexible—objectives that are not always compatible with one another. As an example, in the discussion of TICCIT in the next subsection, we will look at a language that is easy to use but relatively inflexible. PLATO has remained flexible by not being geared to any particular instructional strategy. The strategy used depends on what the individual authors think the most effective teaching technique will be. This aspect of the courseware, however, makes it more difficult to learn how to program. To allow more structure and make programming easier, lesson drivers that are built upon a specific model of student instruction have been developed. Each lesson driver comes with a customized editor. PLATO is flexible enough to do tutorials, drills, and simulation, and to mimic the TICCIT instructional strategy.

Student–Computer Interaction. In the PLATO system, a video display terminal is used to present the courseware material. This material can appear as text, drawings, or animated graphics. The student can input responses through the use of a keyboard or a touch panel on the display surface. Depending on the input, the system can provide feedback on the correctness of the response and can branch the student to the next appropriate point in the lesson.

Hardware.   The current centralized PLATO system at the University of Illinois is run by a CYBER 73–24 with 2 million words of memory. Twenty-four disk drivers are attached to the system so that many thousands of files are available at any one time (Lyman, 1981). This system currently supports approximately 1300 terminals at more than 200 locations (44 are on the University of Illinois campus; the rest are scattered throughout the United States).

PLATO originally used high-resolution plasma panel displays, but CDC has recently gone to CRT displays, which enable greater color capabilities. Besides text and graphics, the following auxiliary equipment has been added to the system (Lyman, 1981): random-access slide projector, random-access audio facility, music synthesizer, and floppy-disk system for off-line use and running of micro TUTOR-based programs. This auxiliary equipment enables highly flexible and varied instructional techniques.

Applications.   CDC has been applying PLATO to industrial training areas both within CDC and outside the company. Some of the industrial training applications include the following:

1. In a project funded by the Department of Labor, nearly 50 Fair Break centers use it to train the chronically unemployed or underemployed for jobs.

2. Control Data Learning Centers offer PLATO courses to individuals and businesses.

3. Twenty-six Control Data Institutes use it to train people for careers in computer programming, operations, and maintenance.

4. Groups of persons with disabilities that prevent them from working outside the home are in a program called HOMEWORK, in which they are trained and employed in jobs using an in-the-home terminal.

5. Prison inmates use it for self-guided instruction to prepare for work after release from prison.

6. United Airlines pilots use it for simulated flight training.

7. Banks use it to train tellers in service skills and banking practices.

8. Farmers use it to learn more efficient farm management.

The PLATO system has also been applied to many military training problems. A discussion of some of the military applications appears later in this chapter under the section on CMI.

PLATO was originally designed for educational instruction at the University of Illinois. Many classroom courses are available and are presently used at the university. Other universities (e.g., California State, Florida State, the University of Delaware, and the University of Brussels) presently have central- ized PLATO systems and use these systems in their classroom instruction. PLATO has also been implemented at schools below the university level. The Chicago City Colleges, Parkland College (a junior college in Champaign, Illinois), and the Urbana and Champaign, Illinois, school districts all use the University of Illinois system (with on-site terminals) for classroom instruction.

TICCIT

TICCIT stands for time-shared interactive computer-controlled information television. This CAI system was developed by the MITRE Corporation in the late 1960s based on research performed at the University of Texas CAI Laboratory and at Brigham Young University. Much of the research and development work on TICCIT has been funded since 1971 by the National Science Foundation. TICCIT and the TICCIT Authoring Language (TAL) are currently being distributed by the Hazeltine Corpora- tion. In discussing TICCIT, we will concentrate on the differences between it and PLATO both to illustrate some of the CAI issues and to better describe the TICCIT system.

Courseware Authoring.   The most fundamental difference between PLATO and TICCIT is that TICCIT is based on and uses a particular instructional strategy: a rule–example–practice model instead of the more flexible tutorial strategies used by PLATO (Reigeluth, 1979). Since TICCIT is so highly structured, it is relatively easy to author. The courseware author is responsible for supplying the content (as in PLATO), but has no control over the instructional strategy.

The underlying philosophies of the two systems are quite different. In TICCIT, there is a sort of courseware versus engineering philosophy. The structuring in TICCIT was created so that courseware could be engineered and produced in an effective, standardized fashion. In PLATO, the philosophy is more courseware-as-art. Recently, in an effort to make TICCIT more flexible, instructional strategies such as drill, practice, games, and simulations, have been added.

Student–Computer Interaction.   Another difference between the two systems is that TICCIT pro- vides a student-centered culture, whereas PLATO uses a teacher-centered environment. For TICCIT, the system, books, and teachers are all used as resources for the student (Bunderson, 1981). Students also have control over the lesson by deciding what lesson to use next, what kind of presentation mode to use, and the sequence of the instructional presentation.

Hardware.   The TICCIT hardware can use smaller minicomputers than the large centralized sys- tems used by PLATO. The Data General Nova 800 and 840 machines are used in some applications.

A maximum of 123 terminals can be linked to a one-computer terminal processor. Another computer acts as the main processor. The TICCIT terminal includes a color CRT display with keyboard.

Applications.   An Educational Testing Service (ETS) evaluation has examined the student completion rates for courses that use TICCIT, comparing it to the completion rates for conventional courses. In the study, only 25% of the junior college mathematics students using TICCIT completed the course, compared with 55% of the students in the conventional course (Bunderson, 1981). Bunderson suggests that part of the problem with the low completion rate might be attributed to the difficulty students had in adapting to a student-centered culture rather than a teacher-centered environment with which they are more familiar. He then cites a study done at Brigham Young University in which the environment was made more conducive to student-centered learning and found a 90% completion rate for TICCIT versus a 76% completion rate for conventional instruction. Although there could be bias problems in comparing a specially constructed environment with a more traditional instructional setting, this study does show that the students must be carefully prepared for TICCIT, and probably most other CAI systems, to make those systems effective.

PLANIT

PLANIT stands for programming language for interactive teaching. It was developed by the Systems Development Corporation (SDC) in the 1960s and is currently being marketing by Charles Frye Software, Inc. Upon the development of PLANIT, SDC's goals were to create courseware that could operate independently of any particular computer system, could be easily transferred to any system, was small enough to use little memory, and was simple enough to learn so that lessons could be quickly authored. PLANIT itself does not necessarily include any hardware; it is modified, if needed, to run on any available systems.

Courseware Authoring.   PLANIT's courseware authoring system is easier to use in a training environment than other software languages, such as BASIC or FORTRAN, because it is specifically structured for instructional training. The text in PLANIT can appear in any of four different types of programming frames—decision (D), programming (P), multiple choice (M), or question (Q). The frames are the basic programming units of PLANIT; they are further divided into individual statements. The M and Q frames are similar to each other in that they both are capable of presenting text (for the M frame, the text is presented in terms of multiple-choice questions), evaluating the correctness of the student's response, and branching to other frames contingent upon the particular response. A D frame is used to branch to other frames contingent upon various sources of information: individual student characteristics, the particular lesson chosen for the student, or patterns of responses. "If–then" statements are the major kind of programming statement used within a D frame. The P frame operates similarly to the way in which a subroutine would function in other software languages. A P frame is called from a statement within one of the frames, the statements within the P frame are executed, and the program execution is then returned to the succeeding statement in the calling frame. In summary, the D and P frames contain characteristics similar to those found in any programming language; the M and Q frames provide the PLANIT courseware with structures that enable it to have instructional and testing capabilities.

Another important feature of the PLANIT courseware is its student answer-matching routines. Three methods can be used to match answers:

1. *Exact Match.*   An answer is evaluated as correct if it is spelled and punctuated correctly.
2. *Key-Word Match.*   The author stipulates the important key words to search for in the answer; it is correct if all of the key words appear.
3. *Numerical Match.*   For mathematical answers, if the student provides an answer containing symbols or variables with particular numeric values, PLANIT can determine the numeric equivalent and evaluate the answer against the author-provided numerical answer.

For each answer evaluation, the author must specify which one of the answer-matching routines to use. In addition, the author can specify individual variants of the answer that can be considered correct and can specify various forms of incorrect answers so that instruction can branch to different parts of the lesson depending on the particular type of incorrect answer.

When writing a lesson in the PLANIT courseware, the author is required to provide the text in the frames, provide possible answers (both correct and incorrect), specify the branching dependent on the student answer, and specify the feedback. A lesson can be simple or complex; the effectiveness of the PLANIT instruction depends on how well the author designs the courseware.

Student–Computer Interaction/Hardware.   These two categories are merged since the transportability of PLANIT results in the student interactions being driven by the hardware in which PLANIT resides. PLANIT and an expanded version called Enhanced PLANIT can exploit all communication facilities in the hardware, including CRTs, electronic line printers, special-purpose keys, illumination of switchcaps, and keyboards. This feature can significantly increase the simulation capabilities of an instructional program. For instance, if the lesson asks, "What button would you press to activate X?," the student can actually press the appropriate button.

PLANIT also has an automatic record-handling capability that allows for the maintenance of student records from session to session. When the student signs on, the program can find out at which point he or she stopped in the last session and where the present session should begin. If sometime during the lesson the system should fail, restart points can be specified during authoring. This capability is included in PLANIT so that the program will not restart at some uninterpretable point within the program, such as the middle of a question, which would make evaluation of the student answer difficult. Both PLATO and TICCIT have similar features. What makes PLANIT unique is that it is machine independent and still manages to perform these functions.

Applications. The Army Research Institute (ARI) and the Honeywell Systems and Research Center (SRC) have completed an automatic testing and teaching program for the Tactical Firing system (TACFIRE) at the U.S. Army Artillery School, Fort Sill, Oklahoma. The operational TACFIRE field system is a custom-designed computer-based unit, which an operator uses to control and coordinate artillery fire. Since this unit was computer based, a PLANIT system was installed on the operational TACFIRE system to allow embedded training and test capabilities. The PLANIT operating system is transported on a cassette tape so that whenever PLANIT is to be run, it is simply placed into the computer's memory; it completely takes over and mimics, with instructional techniques included, the operating system of the TACFIRE software. Thus, instructional techniques can be run on the operational equipment. Presently, TACFIRE operators are tested on the PLANIT-based system at both the Fort Sill school and TACFIRE units in Germany. The main advantage of such a system is that operators can be tested on the same test wherever a TACFIRE unit exists throughout the world. The PLANIT courseware can be transported easily to remote sites through the delivery of the tape cassette.

8.4.3.2 Other Systems

These three systems—PLATO, TICCIT, and PLANIT—have been used to provide examples of CAI systems, from a large-scale system to a relatively small system. There are other courseware packages, but they have not been included here in detail. IBM has marketed a CAI system called COURSE-WRITER, but is not currently doing so. The fastest growing market for courseware and authoring systems is for micro and personal computers. As an example of some of the programs available and the features of these programs, Table 8.4.1 contains a list of some current authoring systems for the IBM Personal Computer (PC).

Table 8.4.1 Examples of Authoring Tools for IBM PC

Software	Vendor	Features	Single User Price
IAS (128K, color	McGraw-Hill New York, NY	Color video (VCR), menu-based	$1500
The Author The Author Plus (64K, monochrome)	Phoenix Performance Systems, Stillwater, MN	Graphics, record keeping, function keys, tutorial	$ 295
PC/PILOT (64K, color (monochrome)	Washington Computer Services, Bellingham, WA	Color, graphics editor	$ 100
TenCORE (128K, color)	Computer Teaching Corp., Champaign, IL	Color, graphics	$2000
Easylearn (128K, color monochrome)	Miracle Computing Lawrence, KS	Create text materials, preprogrammed word, 32,000 screens	$ 250
Teacher Turned Author! (128K, color monochrome)	Raster Technology Longwood, FL	Menu-driven, color graphics, record keeping	$ 399–$999

Source: Kearsley (1984) and *PC World's* 1985 Annual Software Review.

8.4.3.3 Discussion of CAI Features

By comparing three CAI systems the capabilities, applications, features, and issues of some of the courseware packages have been examined. In addition, author–computer and student–computer interactions have been described. In this section, the advantages and disadvantages of CAI systems and the future for CAI will be examined.

Advantages

Measuring the effectiveness of CAI as compared to other instructional techniques is difficult because many examples exist where CAI is better, in terms of material learned, and examples exist where the other techniques are better. Generalizations may not be appropriate because a good CAI lesson is better than a poor conventional lesson and a good conventional lesson is better than a poor CAI; we do not know how to evaluate or define whether a good CAI lesson is better than a good conventional lesson. In later sections, we consider research on what constitutes a good design for CAI. The advantages of CAI can be objectively evaluated in terms of cost per contact hour and a reduction in instructional time due to individualization. The contact-hour estimates are included in Table 8.4.2; Table 8.4.5 in a later section contains the reduction in instruction time. CAI has many inherent advantages; these will also be considered in this section. These advantages may be more important for some systems than others.

Table 8.4.2 contains several estimates of the cost per terminal for PLATO. Many of the estimates are below the estimated 1984 cost of $3 to $7 per student contact hour for an instructor at the college level (Hofstetter, 1983).

Because CAI allows individualized training, training effectiveness can be increased. Performance standards or objectives can be set for each lesson, and individual students are not allowed to continue until those objectives have been met. For conventional training, evaluating individual student performance on several objectives during training is often difficult and time consuming for an instructor; some students are able to advance inappropriately to more complex instruction.

Several other features of CAI systems can be used to increase training effectiveness. CAI can provide:

1. Consistent high-quality instruction on a large scale—instructional strategies of the best instructors can be incorporated and replicated in a CAI lesson.

2. Positive reinforcement and feedback is most effective when it is directly and unambiguously linked to a particular action; the student can quickly see a mistake. Students using CAI systems do not have to wait until someone grades their papers.

3. Privacy—students can succeed or fail in private so that the embarrassment of failure is lessened and students feel free to take more risks.

4. High-quality training at remote sites—the TACFIRE example showed how CAI courseware could be easily transferred to remote sites (e.g., Europe) without having to send students to an instructor or an instructor to students; the student testing can be ensured to be highly similar no matter what the student's location.

5. Hands-on instruction—for TACFIRE it was also shown how CAI courseware could be embedded in a computer-based operational system so that the testing situation is highly similar to the operating environment.

6. Idiosyncratic adaptation—a good CAI system will not only proceed at the student's pace, but it should provide material in a manner compatible with the student's learning characteristics (e.g., abstract versus concrete, verbal versus graphic, concept versus example). This issue will be further considered in later sections.

Table 8.4.2 Cost of PLATO per Terminal

Cost of Terminal per Student-Hour	Method of Estimation	Year of Estimation	Reference
$7.20	Estimate	1976	Kearsley, 1977
$2.47	Use at University of Delaware	1982	Hofstetter, 1983
$1.85	Estimate	1976	Douglas, 1976
$1.15	Using microPLATO at University of Delaware	1982	Hofstetter, 1983
$0.34–0.68	Estimate	1970	Alpert & Bitzer, 1970

CAI applications can also reduce the training costs. Individualized training can reduce the training time, which, in turn, reduces cost; this aspect will be discussed in more detail in the section on CMI. Training costs can be further reduced because CAI systems:

1. Are less costly than expensive or dangerous actual equipment—an RAF study (Computer-Based Military Training Systems, 1980) showed that CAI was a cost-effective alternative if actual equipment costs exceeded $70,000.
2. Can allow a reduction in the number of instructional, administrative, and training support staff.
3. Can reduce travel time—instructional materials can be sent through communications links, thus reducing the need for costly personnel travel.
4. Can provide rapid update of material—it is generally more difficult and expensive to change printed material than to make minor changes to the courseware.

Bunderson (1981) points out that the major breakthrough in instruction that was achieved by printing several centuries ago was that instructional material could be replicated. The breakthrough for CAI, in terms of instructional effectiveness, is that now the instructional interactions (feedback, scoring, branching, and recordkeeping) can also be replicated.

Disadvantages

The major disadvantage of CAI is the length of time it takes to author 1 hr of instruction. Table 8.4.3 contains several examples of the time required to produce 1 hr of instruction. As can be seen from the table, this time is dependent on several factors including the kind of lesson produced and the experience level of the author. The estimates range from over 2000 hr to just a few hours per hour of instruction. Another disadvantage is that the modes of communication between the student and the computer are limited (Hartley, 1980). Presently, the communication is done through a keyboard, touch panel, or limited voice input; natural language, as in conventional classroom instruction, would be ideal but lies far in the future. A third disadvantage is that the responses must be in the range anticipated by the author; CAI courseware cannot handle unexpected or unique answers. A fourth problem is that CAI has limited application in some areas; it is probably best applied where the content to be learned is factual, with specific goals and objectives. CAI is less applicable in areas where the knowledge to be learned is less explicit. It is this limitation of conventional CAI that has led the push for intelligent CAI, discussed in Section 8.4.5.

Future of CAI

In the future, CAI programs will become more intelligent so that some of the problems and disadvantages outlined here can be handled. Research and development laboratories are working on natural-language processing to accept unique student answers and automatic programming, which will decrease authoring time.

In terms of hardware, Fields and Paris (1981) predict that smaller, stand-alone systems will become more common in the future. Several reasons are cited for this: mass production can lower costs; the same system can be used for other purposes, such as entertainment or word processing; the government regulations on communications should limit the growth of large systems; and people seem to prefer things that are personal, controllable, ownable, and physically available.

CAI systems should become more cost-effective. The price of hardware has been reduced significantly and should continue to come down. Courseware authoring costs can be lessened through the use of specialized software. As an example of how authoring time has already been reduced, Table 8.4.4 contains the time to author an hour of instruction after specialized CAI languages were introduced, and after recent advances in "user-friendly" interfaces. It should be pointed out that the 8 hrs for the menu-driven display may be low, having been estimated by the software designer.

In terms of the marketability of CAI systems, CAI applications in military and industrial training have already been achieved. Public education has largely been resistant to the changes in instruction that are implied by the application of CAI techniques. Some countries have accepted CAI techniques better than others. As an example, teachers in France get a year off to learn courseware authoring language so that they can implement these programs in their own schools. All of the companies currently involved with CAI courseware are eagerly awaiting the opening of the home market for instruction on personal computers. The trend toward smaller systems is an example of the convergence of concentration on this market. Finally, Bunderson (1981) suggests that the need to rapidly upgrade skills and knowledge of the general population to reduce unemployment caused by the encroachment of high-technology industries on the established, traditionally high-employment industries will be the impetus for a new courseware industry to open up.

Table 8.4.3 Authoring Time

Hours per Hour of Instruction	Method of Estimation	Description of Project	Experience of Author	Reference
2286	Average of 38 authors	Full range; authored in FORTRAN on PLATO; before July 1967, batch processing	Not given	Avner (1979)
615	Average of 43 authors	Full range; specialized authoring language used; before October 1968, batch processing used	Not given	Avner (1979)
300	Estimate	Sophisticated individualized material; IBM Coursewriter language; first year of new project	Kansas City, experienced school teachers	Avner (1979)
295.1	Controlled experiment	Full range of features on PLATO	New authors	Avner (1979)
292	Average of eight authors	Military application on PLATO IV	New authors	Dare (1975)
250	Estimate	Complete simulation for Air Force automated test station/repair training; used videodisk and touch screen	Not given	Dallman, Pieper, and Richardson (1983)
200–500	Estimate	Moderately sophisticated instructional interaction	Not specified	Fairweather and O'Neal (1984)
200	Estimate	Naval Air Maintenance Training Group; estimate over several applications	Not given	Brackett (1979)
196	Estimate	Full range of capabilities for lessons at University of Delaware	Not given	Hofstetter (1983)
165–610	Controlled experiment	Generating new pedagogical structure on PLATO	Low	Avner (1979)
151	Average of 47 authors	Full range of capabilities on PLATO; authoring in authoring language on-line	Not given	Avner (1979)
100–150	Estimate	Comprehensive instructional system design effort within the Air Force	Not specified	Brackett (1979)
82	Benchmark test	CAI on how to use an oscilloscope; graphics	Best programmer from vendor	Hillelsohn (1984)
80	Estimate	Full range of CAI capabilities, university setting	Highly experienced	Avner (1979)
72	Benchmark test	CAI on how to use an oscilloscope; graphics	Best programmer from vendor	Hillelsohn (1984)
54	Benchmark test	CAI on how to use an oscilloscope; graphics	Best programmer from vendor	Hillelsohn (1984)
49.7	Controlled experiment	Sample course from Chanute Air Force Base	Not given	Brackett (1979)
48.5	Controlled experiment	Sample course from Chanute Air Force Base	Not given	Brackett (1979)
32	Average of 34 authors	Full range of capabilities; authors instructed to design effective instruction	Experienced	Avner (1979)
30–70	Estimate	Instructors in military settings	Not given	Brackett (1979)
27–180	Controlled experiment	Generating new pedagogical structure on PLATO	High experience	Avner (1979)
26.4	Controlled experiment average	Rull range of capabilities on PLATO	Experienced	Avner (1979)

Table 8.4.3 (*Continued*)

Hours per Hour of Instruction	Method of Estimation	Description of Project	Experience of Author	Reference ●
25–50	Estimate	Straight drill and practice lesson	Not given	Avner (1979)
22	Estimate	Language vocabulary drill	Not given	Avner (1979)
21	Estimate	Full range of CAI capabilities	Highly experienced	Avner (1979)
17	Controlled study of 21 authors	Full range of capabilities; author encouraged to produce materials quickly	Experienced	Avner (1979)
13	Controlled experiment	Sample course from Chanute Air Force Base	Not given	Brackett (1979)
8	Estimate	Authors using menu-driven and prompted authoring editor	Not given	Fairweather and O'Neal (1984)
8–63	Controlled experiment	Full range of PLATO capabilities using existing pedagogical structure	Low experience	Avner (1979)
6.9	Controlled experiment	Translating programmed text to computer-based material at Florida State	Not given	Avner (1979)
6–39	Controlled experiment	Full range of PLATO capabilities using existing pedagogical structure	Low experience	Avner (1979)
"Few hours"	Estimate	Drill and practice using TUTOR or APL	Experienced	Avner, Smith, and Tenczar (1984)

8.4.4 COMPUTER-MANAGED INSTRUCTION

Computer-managed instruction (CMI) uses a computer as the housekeeper in an individualized instructional program. Baker (1981) presents an idealized example of the capabilities of a CMI system. We shall look at the characteristics of this system by describing the capabilities from the viewpoint of the student, the instructor, and the administrator. After these viewpoints are presented, the specific functions of the system will be examined.

8.4.4.1 The Student

Upon entering a CMI setting, the student has immediate contact with the computer management of the training process. The student uses a computer form to register and then immediately is given a printed copy of the program of studies and the first lesson assignment. The lessons can be individual study, a field exercise, a CAI lesson, or a lecture. For each learning activity the student is provided with a computer printout of the available resources.

The instructional process approaches complete individualization; as students proceed, they can be distributed throughout the curriculum. An attempt is made to define each learning activity; the student

Table 8.4.4 Reduction in Time to Author

Language	Dates Available	Batch or On-Line	Hours/Hour	Reference
FORTRAN	Before July 1967	Batch	2286	Avner (1979)
Authoring	Before October 1968	Batch	615	Avner (1979)
Authoring	After 1968	On-line	151	Avner (1979)
Authoring with menus and prompts	1984	On-line	8	Fairweather and O'Neal (1984)

should always be aware of the criteria for evaluation. At any time during the process, students can check their progress and the computer can provide an estimated time of completion. As a further evaluation aid, students are scheduled at various times to meet with the training staff to review progress. At the completion of the training process, a computer-generated transcript describes the competencies acquired by the student.

From the students' view, a CMI system provides many benefits. Students are receiving feedback on a needed basis and are always aware of where they stand in the course. The training is self-paced; deficiencies are flagged and progress is clearly noted. Finally, the goals of the training should be clear-cut. The student must know exactly what must be achieved before advancement can be made.

8.4.4.2 The Instructor

Instructors can be assigned to several roles in a CMI program. As a manager, the instructor is required to monitor the students. The computer management can provide instructors with several readily available kinds of assistance. Through interaction with the system, an instructor can determine how each student is performing in each training module and how he or she compares with other students who have already completed the instructional process. The system can also be designed to alert automatically an instructor if a situation arises in the training process that might require immediate attention. Besides monitoring the students, an instructor may be required to record data (e.g., performance ratings or observational data). Because instructors are freed from doing much time-consuming clerical work, they will be able to devote more individual attention to any student problems that might arise during the training process; their instruction will be much more personal and individualized. Also, because so many forms of student evaluation are available, an instructor can enter into the training process well informed about student progress.

As a curriculum developer, the instructor is required to create, implement, evaluate, and optimize training programs. These training programs can take many forms: CAI lessons, field exercises, individual study, or instructional groups. An important function of the curriculum developer would be to evaluate the effectiveness of the training programs that have been produced. A CMI system can provide an instructor with summarized reports such as test-item analyses and attempts-until-mastery that can be used in that evaluation. This information can be used to change the instructional strategy quickly if the evaluation shows a need for a change.

8.4.4.3 The Administrator

From an administrative point of view, a CMI system presents many challenges. Because instruction is individualized, different students can be at different phases in the training process at the same time. The training process is not broken up into discrete time units, such as semesters, so that an administrator can accurately determine when students will finish training. Resources must be managed so that instructional materials will be available to the students when needed. Just as the individualization possible in a CMI system creates some of these problems, a CMI system can be used to solve them as well. The CMI system can keep past records of the students; this information can be used to predict future status and identify the resources that will be needed. The capability of extrapolating future needs from past performance can be used to ensure that student flow is as orderly as possible.

For administrators, CMI can be used to automate many mundane, time-consuming functions. Registration, attendance tallies, scheduling of instructional materials and groups, and resource allocation are all functions that can be performed automatically by the CMI system. This capacity can greatly reduce labor costs and free the administrator's time to deal personally with school or student problems that may arise.

8.4.4.4 Functions of CMI Systems

Not all CMI systems will operate like this described model; they can perform any subset of the functions that were described. Dennis (1979) lists eight activities that can be performed by a CMI system: diagnosing, assigning or prescribing, facilitating study, evaluating, data collecting and manipulation, reporting, resource and space management, and providing an information network. Each of these functions will be discussed in more detail.

Diagnosing

A CMI system should measure student progress toward specified objectives and correctly interpret a student's learning progress. Once the progress has been diagnosed, the system can be used to assign or prescribe remedies.

Assigning or Prescribing

A CMI system should direct the kinds of assignments or prescriptions to be given to the student when needed. Dennis (1979) lists the kinds of non-computer-based instruction that can be assigned or prescribed by the system: individual practice, text assignments, group activity, teacher consultation, laboratory sessions, and alternate media sessions.

Facilitating Study

A CMI system can provide feedback on the correctness of student performance on instructional material; timely help on how to correct errors can also be provided. This type of system is much more efficient than the typical classroom situation, where feedback and prescription are seldom immediate.

Evaluating

A CMI system can provide both the student and the instructor with current information about the student's progress. Furthermore, past history of a student's progress can be used to compare with present data. This information can be used to assign and advise correctly. The CMI system can also evaluate the instructional program so that bad instruction can be identified and CMI procedures can be tuned.

Data Collection and Reporting

This function provides instructors with extensive performance data. Dennis (1979) lists several kinds of data that can be reported: individual students, particular groups of students, individual learning activities, groups of learning activities, and effectiveness of students' decisions. These data can be reported across several levels, such as reports to students, instructors, school administrators, and external agencies. As instruction becomes more learner centered, it is crucial that the student become a primary recipient of these reports.

Resource and Space Management

Dennis (1979) recommends that a CMI system should know what resources are needed for each learning activity, the quantity of these resources in terms of the number of students that may use a resource at one time, and the spaces available in which to use these resources.

Providing an Information Network

In a CMI system it is easy for an instructor to lose contact with the students, who are dispersed throughout the training process. Just as the computer manages the system it can also be used to communicate and provide an information network. This can be done by a computerized mailbox— an individualized depository for information from other sources that can be accessed at will by the owner. The messages from one individual to another are stored until retrieved by the appropriate person. Messages can also be sent to groups of persons. PLATO, for instance, allows students to comment on the instruction they are receiving.

8.4.4.5   Hardware Requirements

As could be seen from the functions discussed above, CMI is basically a management information system that requires a significant amount of data storage. Up until very recently, the essential component of such a system was a large mainframe computer that possessed considerable memory and large amounts of mass storage. This mainframe would be required to manage the system and perform the eight functions listed previously. This section is written from the point of view that a large mainframe is a must. However, as microprocessor technology expands, micros too may have the capabilities to perform the described functions.

The mainframe is required to support a large number of peripheral devices. One of the most important devices would be a high-speed printer that could be used to produce large reports, such as student progress reports or attendance records. VDT displays can be used to retrieve information quickly, send and receive messages, and query the system. VDT displays should be linked with hard-copy devices (Baker, 1981). Besides these two classes of peripherals for particular CMI purposes, Baker suggests that two other kinds of peripherals be included: an optical mark reader and a hand-held terminal. The optical mark reader is a device that can read pencil marks that are used on forms. These forms can be used by the students to enter answers to test questions or for recording attendance data. Many mark readers, ranging from desk-top units to high-capacity machines, are currently on the market. A handheld terminal is a device about the size of a calculator that is used to transmit information to the computer. This device can be used by the instructor to collect observational data and student performance ratings at remote sites. When it is time to enter the data into the computer, the handheld terminal can be attached to a modem that is used to receive and transmit the information to the computer over standard telephone lines.

Instructional systems (e.g., simulators, CAI, or ICAI) can be interfaced to CMI systems. The hardware requirements necessary for these instructional systems are discussed elsewhere in this chapter. Interfacing these systems can greatly increase the efficiency of the total CMI system by automatically recording performance measurements. Even a simple multiple-choice testing capability that could be presented on a CRT screen would be more efficient than having a second party either hand score the test or enter that information into the computer later.

Several CMI systems, including hardware and software, are currently on the market. Control Data is marketing a system called PLATO Learning Management (PLM). PLM is useful for medium- to large-scale individualized instruction and is capable of sequencing activities for individual students, administering placement and diagnostic tests, selecting learning resources, and keeping records of student and courseware performance. The TICCIT system previously discussed has a CMI component. The main features of TICCIT are multimedia (e.g., slides and video) instructional control, interactive communication, extensive data gathering, and computerized mailboxes. Other systems with similar capabilities are also available.

8.4.4.6   An Example of a CMI System: AIS

This example of a large-scale CMI system is used to illustrate how the hardware and software capabilities can be brought together to produce a cost-effective operational system. AIS (Advanced Instruction System) is an operational CMI system at Lowry Air Force Base, Colorado. McDonnell-Douglas Corporation was the prime contractor for the implementation of this system on a four-year $10 million contract that began in May, 1973. The system was designed for a large-scale instructional setting where a variety of training programs are used.

AIS contains many of the features that have been described previously. About 90% of the instruction is off-line. The off-line instruction is capable of using a variety of instructional techniques, such as instruction assisted by audio-visual equipment and traditional printed text. The other 10% of instruction is on-line in the form of CAI and interactive testing. During the instructional process, the computer prescribes assignments for the students based on an evaluation of their progress. The learning materials and other resources are allocated by the computer. Also, by evaluating the student's capabilities, the system can provide a projection of the completion time of each student.

AIS manages training ranging from routine functions to highly technical hands-on equipment and maintenance repairs (Computer-based military training systems, 1980). The courses managed by the system include inventory management, training for weapons mechanics, and operation of precision measuring equipment (Fletcher, 1974).

During the development of this system, Rockway and Yasutake (1974) identified the features that make this system unique in comparison to other systems of this time. The AIS system:

1.   Integrates all of the technology required for individualized instruction on a large scale
2.   Is flexible, in that several approaches for instruction are available (it is not limited to just one instructional approach)
3.   Can incorporate new instructional and management innovations into the system as they become available
4.   Has an inherent capability for continuously evaluating and upgrading its own cost-effectiveness
5.   Is a total system in that it performs all of the functions required to conduct a large-scale training enterprise

The AIS can provide individualized instruction for more than 3500 students per year, and it is estimated that the annual cost savings afforded by this system are about $6 million. Much of this savings is due to decreased training time. When compared with conventional lock-step instruction, AIS students have been found to achieve equal or better performance levels in 30% less time (Rockway and Yasutake, 1974).

8.4.4.7   Evaluation of CMI Systems

CMI systems can be evaluated across two dimensions: cost savings and effectiveness of training.

Cost Savings

We have already seen in the AIS example how installing a CMI system can reduce costs by applying the computer to manage the training. In a CMI system, the computer is doing what it does best—

Table 8.4.5 Reduction in Instruction Time Due to Individualization

Percent Reduction in Time to Complete	Description	Reference
66	Training at United Airlines	Conkwright (1982)
64	Average from evaluation of military systems	Orlansky and String (1977)
63	Technical maintenance training at American Airlines	Conkwright (1982)
50–67	CAI courses in school districts	Morgan (1978)
40	College courses	Avner, Moore, and Smith (1980)
33	Course at Federal Aviation Administration	Buck (1982)
15–27	Navy advanced maintenance course	Montague, Wulfeck, and Ellis (1983b)
15–25	Navy apprentice training	Montague, Wulfeck, and Ellis (1983b)

manipulating, storing, and retrieving large amounts of information. Also, instructors are doing what they do best—giving individualized and group instruction to the students, answering student questions, and personally managing the instructional process. Because much of the clerical work is automated, instructors are free to spend more time interacting with the students. In addition, instruction is not necessarily based on instructional technology that is poorly understood. Live instruction is a time-tested procedure.

Most of the cost-effectiveness comes in being able to individualize training. Table 8.4.5 lists several references that have evaluated the percentage reduction in time to complete a computerized course when compared to a traditional course. For this table, no distinction is made between CMI and CAI lessons. The reduction in instruction time due to individualization ranges from 14% to 67%. In a classic classroom situation, training time for the whole group is driven by the slowest learners. In a CMI system, training time is driven by the individual student's capabilities and motivation. The savings can be particularly high when students as well as instructors must be paid, as occurs in military and industrial environments. We have already seen the cost savings exhibited by the Air Force's AIS program. Another example of cost savings is the U.S. Navy's CMI system at Memphis, Tennessee. In fiscal year 1977, this training center had a throughout of 52,672 graduates. The estimated yearly savings after the CMI system was installed was from $9 million to 10 million.

It is likely that there are other cost savings in CMI as well. If the efficiency of a training program goes up, less material will be wasted, schedules will tend to eliminate duplication of both instructors' and students' efforts, training equipment will be better utilized, and so forth. (Fletcher, 1983).

Training Effectiveness

The emphasis in CMI systems is to manage instruction to increase training effectiveness. Several features of the system should cause this increase. First, instruction is individualized so that students are not forced to spend more or less time in training than needed. Second, instructors have more time to spend with the students because time-consuming clerical work is automated. Third, instructors can better evaluate the students because individual and group summaries of performance are readily available; problem areas can be quickly flagged and remedied. Fourth, students receive detailed feedback immediately upon completion of a training module. Fifth, the goals and objectives are clearly stated to the students. Some of the research that deals with these five issues is considered in Section 8.4.6. In summary, CMI technology is well established and should be used if at all possible. Features such as automated attendance and record-keeping can be implemented in almost any instructional environment. A CMI system also is a good foundation for future growth of automated instruction systems. If a centralized computer-based system is established and running, interfacing automated instruction, such as CAI or ICAI, would be a further logical step to take. The implementation of automated instruction could be done in the future, when those technologies are better established and the bugs have been worked out to the existing CMI system. In other words, a CMI system is a good starting point for making further applications of computers in training.

8.4.5 ICAI

ICAI uses AI techniques to enhance training and instruction for computer-based systems. AI techniques (see Chapter 9.2), using human intelligence as both a model and a source of data, instruct students by interacting with the student naturally, answering questions, and providing a data structure so that facts can be acquired and retrieved efficiently. ICAI programming is generative; it can be run repeatedly by the same student although his or her learning situation is different each time (Camstra, 1977). Thus, ICAI is not completely dependent on programming that attempts to account for every possible interaction future students may require.

ICAI is targeted toward complex learning domains and may offer an alternative to costly high-fidelity simulators or actual equipment. ICAI provides several benefits over the different types of computer-based systems that we have previously discussed. First, the computer instruction can be very flexible. With an intelligent system, the courseware author does not have to anticipate every instructional situation; the program can use its own intelligence to handle an unforeseen situation. Second, training is tailored to the individual student; most ICAI programs try to understand the individual student by deriving a model of what the student does and does not know. Third, the student is allowed to play a more active experimental role in the learning process. Like any good instructor, ICAI programs are designed to direct, not lead. An emphasis is placed on providing situations in which the student can query the computer to try to discover the correct answers. Finally, many of the instructional strategies used in the program are based on modeling expert instructors. If the best instructors can be modeled, then their effective teaching strategies can be be distributed to many students all over the world on an individualized basis. These are the uses, benefits, and goals of ICAI. Next, we will look at particular programs and features of three programs to see how these goals are put to use.

8.4.5.1 Characteristics of ICAI

The ICAI programs that currently exist are the results of intensive research and development efforts. Very few of them have been implemented at this time, so it is difficult to compare them with other types of training. The ICAI programs discussed in this section have been chosen because they use features of ICAI programming that will be important in the development of future systems. In the remainder of this section we will briefly introduce the reader to three programs, abstract the features common to all of them, show in detail how the features are combined in a particular system (STEAMER), and then examine the future of ICAI.

8.4.5.2 Brief Introduction to Three ICAI Systems

SOPHIE is an ICAI system that has been in existence since 1974 and has gone through three phases (Brown and Burton, 1978; Brown, Burton, and Bell, 1974; Brown, Rubinstein, and Burton, 1976; and Burton and Brown, 1979) at Bolt, Beranek and Newman. SOPHIE is a simulated laboratory environment that allows students to experiment with problem solving in electronics troubleshooting. To instruct the student, SOPHIE places a fault somewhere in the system. The student must try to isolate and find the fault. The student interacts with SOPHIE by taking measurements, asking questions about how to interpret the measurements, replacing components in the simulated circuitry, and posing hypotheses about the source of failures.

WUSOR is an ICAI game that was developed by Goldstein and Carr (Goldstein, 1978) at MIT. It is designed to teach students about simple logic and probability while they are playing the game. In the game, the student must avoid dangers while choosing a route through caves in search of the wumpus beast. WUSOR will provide clues to the student on which route to take. For the clues to be effective, students must use the rules of logic and probability. Thus, the subject matter is not taught to the student explicitly, but becomes put to practical use in the course of playing the game.

GUIDON is an ICAI program developed at Stanford by William Clancey and his associates (Clancey, 1979, 1983) in the late 1970s. The goal of this program is to teach medical diagnostic problem solving. During instruction, GUIDON describes the symptoms of a sick patient. The student must make a diagnosis of the illness by interacting with GUIDON. The student actively solves the diagnostic problem by requesting additional information about the sick patient through GUIDON.

8.4.5.3 Components and Features of ICAI Systems

There are three components that are present in most ICAI systems—the representation of subject material, the student model, and the instructional strategy used (including the type of dialogue format used between the computer and the student). Each of these components will be discussed in more detail.

Representation of Subject Material

When representing knowledge in an ICAI system, the information must be organized so that acquisition, retrieval, and reasoning can be done efficiently. For a computer-based system, the machine must be able to acquire knowledge easily in the sense that the author can quickly enter facts into an organized database. An ultimate goal of an ICAI system is for the computer to acquire knowledge automatically by making analogies from old knowledge and applying it to new knowledge. This level of sophistication is still early in its developmental stage. After knowledge is acquired, it must be retrievable by the system. Knowing the information that is relevant to a particular problem and being able to retrieve it is an especially important issue when the problems are complex. Instead of having facts stored for every problem, a feature of an intelligent system would be to have the system infer information and reason from the fixed facts. The form of the data structure often determines how inferences and reasoning can be done. Some simple forms of reasoning are possible and have been implemented in various systems.

In general, two broad categories of representation—declarative and procedural—are used in present ICAI systems. The type of application determines which representation to use; information organized one way can also be organized using the other method. The two differ mainly with regard to which aspects of the database are explicitly represented and which aspects must be inferred by the organization of the data. Declarative representations are explicit about what is represented in the knowledge base. How the knowledge is used is less clear. Procedural representations are explicit about the step-by-step use of knowledge. However, it is difficult to determine what is or is not known from the knowledge that is represented. What is known must be inferred from the behavior of the system.

Semantic networks are a type of procedural representation used by some ICAI systems. A semantic network was originally used by Qullian (1969). It represents knowledge as a linked structure of objects, facts, or events, which are called nodes. Knowledge is acquired by the system through determination by the authors of the relationships between a given fact and the other facts that are already in the system. These relationships can be used to infer knowledge not explicitly represented. As an example used by Collins and Quillian (1969), a system can infer that a bear is an animal by having stored that a bear is a mammal and a mammal is an animal.

Several methods are used for procedural representations such as finite state automation, procedural networks, and augmented transition networks. These procedural representation methods include information such as the constraints of the systems, the state of the system, critical events that will change the state, boundary conditions, and how a system will transition from one state to another. A production system is a particular kind of procedural representation of the form SITUATION–ACTION; the action will occur only if the situation is determined to be true. Anderson (1976), in his computer-simulation system called ACT, demonstrated that a production system could be self-modifying; the situation–action production rules could be changed by the system as the system learns. This type of knowledge representation holds potential promise for future ICAI systems.

Frames (Minsky, 1975) and scripts (Schank and Abelson, 1977) are knowledge representations of knowledge. A frame is used to represent stereotyped situations and is organized as a network of nodes and relations. The top levels of a frame are fixed (they are always true) and the bottom levels have slots that must be filled by particular instances. Once a correct frame is found for a particular stereotyped situation, the computer can infer other knowledge about the situation from information stored with the frame. A script is a similar concept to a frame in that it is used to represent typical or everyday events. A script, however, places even more of an emphasis on expected events. Frames and scripts have prompted much basic research but have seen only limited implementation. GUS (Bobrow, Kaplan, Kay, Norman, Thompson, and Winograd, 1977) is a prototype for an automated airline reservation assistant, which extracts information to fill in particular slots of a frame. It also directs queries for information not yet supplied.

The Student Model

Use of a student model is another feature of ICAI systems that distinguishes them from CAI systems. A student model infers from the student's behavior what the student knows. Thus, by presenting only information that the student does not know already and by correcting the student's misconceptions, instruction can be more individualized than that available on traditional CAI systems. Although not all ICAI systems use a student model, the use of one enables a system simultaneously to be powerful and flexible.

Five general methods are in use to determine what a student knows. The first two methods used, topic marking and the context model, are relatively simple approaches. For the topic-marking approach, the system keeps track of the information with which the student has been presented. For the context model, the extent of the student's knowledge is interpreted in terms of the dialogue and questions that the student asks the system.

Two other methods, the bug and overlay approaches, compare the student's knowledge with the knowledge of an expert. For the bug approach, student knowledge is characterized in terms of the bugs or misconceptions the student has about the subject when compared with an expert's knowledge. Student performance is compared with a variety of bugs until the best match is found. For the overlay approach, student knowledge is characterized as a subset of an expert's knowledge. Once the student demonstrates the correct facts or rules contained in the expert's knowledge, those facts are considered as having been acquired by the student.

The fifth method, generative modeling, is a relatively new approach. Instead of looking at the particular facts acquired, the student's knowledge is characterized in terms of the plans used to solve a particular problem. The computer instruction is then organized according to these perceived plans; factual misconceptions are corrected and feedback is tailored to the particular student's conceptions.

Instructional Strategy

In this section, the pedagogical strategy used by various ICAI systems will be discussed. One of the characteristics of an intelligent system is that it can fulfill many of the roles of traditional instructors. The traditional instructor must interact with the student, format the educational session or manage the dialogue, choose topics, select problems, evaluate answers, provide feedback, and evaluate and modify instructional techniques. Most of these duties have been incorporated into ICAI programs. The following discussion will explain the issues involved in implementing these duties for the systems under discussion.

In a traditional instructional situation, the interaction between the student and the teacher occurs using natural language. ICAI systems can perform some types of elementary natural-language input. Such input capability is important if the task is to focus on the task and not on how to communicate with the machine. Some ICAI systems have the capability to go beyond what it is possible to do in a classroom. In the SOPHIE-III program, a demon faults the system and then the student watches as an expert finds the fault. During the process, the computer queries the student to make sure that he or she can follow the reasoning used by the expert.

The issue in dialogue management is how much control to give to the student and how much control to give to the computer to direct the course of instruction. All programs can be characterized by extensive control by the computer, the student, or both. When control of both, it is a mixed-initiative dialog, which involves a sophisticated set of rules to decide when the student controls and when the computer controls. As an example, GUIDON uses what it knows from the student responses and decides when to program formal instruction. However, ICAI systems have not been able to provide an interactive explanation that is improved using a student's feedback, as can be given by expert teachers (Clancey, Bennett, and Cohen, 1982).

Research in ICAI has done very little to make topic selection an intelligent process. In most ICAI systems, the sequence of topics is determined by program authors. In simulation games, no particular sequence is typically required. There has been some interest among members to organize the topics according to particular principles (Suppes, 1979; Wescourt, Beard, and Gould, 1977).

There are two general approaches for problem selection: select the problem from those stored in memory or direct the program to instruct a new problem. If the problems are prestored in memory, they can be selected from those available, selected according to some preassigned priority, or selected for reasons relating to the individual student. This last approach is the most sophisticated of the three and allows highly individualized instruction. WHY (Stevens, Collins, and Golden, 1978) provides an example of this sophisticated approach by selecting problems that will explain and correct student misconceptions. The other general approach that can be used to select problems—the construction of new problems—promises to be an important contribution to ICAI systems. However, this approach is still being researched (Collins and Stevens, 1980), and has yet to be implemented.

There are several methods, ranging from simple to sophisticated, to evaluate student answers. The simplest methods are to determine whether the student's answer is correct (as in SCHOLAR) or whether it is optimal (as in WUSOR-II). More sophisticated methods can evaluate whether an answer is partially correct and can then determine what part of the answer the student does not know. WUSOR-II, GUIDON, BUGGY (Brown and Burton, 1978), and MACSYMA (Moses, 1975) all use a strategy of understanding the bugs in an answer. The most advanced method for evaluating student answers is the ability to understand novel responses that might be correct but are unanticipated. SCHOLAR uses a hierarchical knowledge structure to show that a response is a superset of the anticipated answer.

Feedback in instruction is provided so that the student can evaluate and modify his or her response. ICAI systems employ a great variety of types of feedback. Socratic tutors, such as WHY, provide counterexamples to incorrect responses. Other systems provide the results of actions, information on whether the answer is correct or partially correct, hints, and advice. Artificial intelligence is used to choose which kind of feedback to give at a particular time. As an example, AI techniques are used in WUSOR-II to determine the appropriate coaching for a particular player's skill level.

The purpose of evaluation and modification of instructional techniques is to build a self-improving program that learns from its experiences with students so that the instructional strategy used can be evaluated and improved. This technique is potentially very useful, but it has not been thoroughly researched or implemented. O'Shea's (1979) work in the domain of teaching quadratic equations is a prime example of an ICAI program that can learn from its experiences with a particular instructional strategy.

Hardware

For ICAI, much of the emphasis is placed on the capabilities of the software or courseware. Many of the hardware issues discussed in the CAI section are also relevant for ICAI systems. ICAI hardware is different from that discussed in the CAI section in that it requires special machines for developmental purposes, larger amounts of memory for representing complex knowledge, and special interactive graphics.

Most ICAI programs are written in list-processing language (LISP). LISP is the language that is used because, unlike other languages that manipulate numbers, it is specialized for the manipulation of symbols, lists, and strings. Computers that run LISP have usually been quite large and memory consumptive. Recently, three smaller computers running LISP—the Xerox Dolphin, the Symbolics LISP Machine, and the LMI LISP Machine—have been marketed. The use of these special-purpose computers should decrease development time for ICAI programs.

ICAI systems are designed so that the student–computer interaction is as natural as possible. Therefore, they tend to use special peripherals that have not been available in older CAI systems. Limited natural language and speech processing are components that are included in many ICAI systems to make student–computer interaction easier. Inexpensive speech recognition systems with vocabularies of about 100 words are available off-the-shelf. More expensive systems with vocabularies of 400 or more words are also available. Another interactive device that is used quite often is the touch panel. The touch panel, as used in STEAMER, will be discussed in the next section.

8.4.5.4 STEAMER—An Example of an ICAI System

In this section, we will show how the instructional and hardware components of an ICAI system can be brought together to produce an intelligent computer-assisted instruction system. STEAMER is a five-year project performed by Bolt, Beranek and Newman (BBN) under sponsorship of the Navy Personnel Research and Development Center (NPRDC) in San Diego, California. It is designed to be an intelligent simulator that will be used to train Navy personnel in propulsion engineering at the Surface Warfare Officer School in Newport, Rhode Island. STEAMER was selected as the example to use in this section because it is a recent project that used advanced technology and addresses a complex training problem. This discussion is taken largely from a paper presented by Hollan, Stevens, and Williams at the Seattle Summer Simulation Conference in 1980. A more complete discussion of the STEAMER project can be found in Hollan (1984) and Williams, Hollan, and Stevens (1981).

Training in propulsion engineering is judged by naval officers to be one of the greatest training problems in the Navy. The purpose of a shipboard steam plant is to convert heat into electricity and to turn the ship's propeller. The collection of operating procedures runs to two or more 4 in. thick volumes. The operational system is physically quite large and contains thousands of complex components. From 10 to 20 watch standers are required to operate the propulsion system. Mistakes in operation can cause expensive damage and even death. In the past, the training of operators has involved the use of a costly high-fidelity simulator located at the Surface Warfare Officer School.

STEAMER, being a simulator itself, has modified the existing simulator by adding an intelligent tutor and making the whole simulation smaller and less costly. The intelligent tutor will do some of the things discussed previously. It will explain the performance of the system, identify and correct student misconceptions, and guide the student in his or her interaction with the system. The simulator itself will consist of a low-cost, transportable, tabletop computer. Many of the complex components of the steam plant will be simulated through the use of color graphics and animation. Students can inspect and manipulate the steam plant through this graphics interface.

Instructional Strategy

STEAMER was designed so that the student could develop a conceptual understanding of the system. The developers of STEAMER felt that the steam plant operation was too complex to learn by rote. This observation was corroborated by looking at how experts think and talk about steam plants (Hollan, Stevens, and Williams, 1980). They found that experts seem to have a mental simulation of how the system works instead of a list of facts or collection of procedures. Therefore, instruction should be geared toward teaching the student conceptual knowledge so that the student can mentally simulate the operation of the system. Hollan, Stevens, and Williams (1980) call this mental simulation the device model.

A device model is not the mathematical model that runs the simulator; rather, it is a conceptual model that has been developed by experts through years of experience. Instead of a mathematical simulation, experts seem to use a qualitative simulation of the system. Experts do not necessarily memorize the operations of the system. They form hypotheses about the system and know where to look to confirm or disconfirm the hypotheses. Expert operators seem to agree that the STEAMER model conception is similar to how they think about the system (Hollan, Stevens, and Williams, 1980).

The intelligent tutor incorporated in STEAMER is used to help the student conceptualize the operation of the system. The tutor answers questions, provides hints, and gives explanations. It analyzes the student's misconceptions and guides the student toward learning the underlying principles of propulsion engineering operations.

Conclusions about STEAMER

STEAMER can be used alone or in conjunction with an existing high-fidelity simulator. In addition to instruction, STEAMER can also be used for testing and evaluating instructional techniques. The tabletop computer used for instruction offers affordability (estimated cost in 1983: $10,000) and transportability (STEAMER can be packed up and taken to other training sites). The simple-to-use student–computer interface means that the student does not have to be a polished computer user to interact with and manipulate STEAMER.

8.4.5.5 Evaluation

ICAI systems are still in their early developmental stages and very few have been implemented. Therefore, it is difficult to compare ICAI against conventional training or CAI. There is no ICAI framework that can be used across several applications. Thus, each ICAI program is a separate entity and must be evaluated in its own right. A much more rigorous evaluation will have to wait until more ICAI systems have been implemented.

Cost and Time Considerations

Software costs are the major expenditure of time and money for intelligent systems. The costs are high now and will probably increase in the future. Perlowski (1980) estimated that in the design of a smart product, if hardware costs are $20,000, the software would cost about $100,000. The gap between the two is probably larger today. However, the rise in software cost could be slowed or reversed by a decline in development time. As future ICAI systems grow from their precedents, it seems reasonable to expect a decrease in development time through automation of the development process.

Efficiency of Training

The same arguments that were advanced for the efficiency of CAI training can also be made for ICAI systems. The inclusion of a student model in intelligent systems allows instruction to be highly individualized and flexible. ICAI can focus on those areas that are misunderstood by the student and, thus, need more instruction. When the instruction is individualized and tailored to the student, as it is in ICAI systems, much training time can be saved when compared with classroom training.

   ICAI has the added advantage that many of the instructional strategies employed are based on well-researched cognitive psychology laboratory studies. ICAI can show how effectively these laboratory studies can be applied to more complex, real-world training situations. The effectiveness in terms of these instructional strategies remains to be seen; the theories, more so than other training computer applications discussed in this chapter, are solidly grounded in laboratory studies. These theories and studies are discussed in detail at the conclusion of this chapter.

8.4.5.6 Future for ICAI Systems

Application Thrusts

The future of ICAI systems depends heavily on the monetary commitment that military, industry, and education managers are willing to place in research and development of these systems. The military and industrial commitments, in addition to the financial aspect of the problem, also signal whether or not important decision makers perceive a need for such systems and whether a potential market is available. These factors are important for the future of ICAI.

   It is difficult to evaluate the industrial commitment to ICAI. Research has been carried out at such heavy research firms as Xerox and Bolt, Beranek and Newman. Researchers in artificial intelligence are forming commercial ventures (e.g., Teknowledge, Inc.) in anticipation of potential markets for AI—including ICAI—products. Firms that are already committed to marketing CAI can be expected to incorporate intelligence into their systems as soon as the technology is established.

Research Thrusts

It is difficult to predict what will be incorporated in ICAI systems of the future. However, it can be expected that the research now being done in cognitive science laboratories will provide both theory and technology for future ICAI systems. Relevant work includes theories of learning, analogical reasoning, and expert/novice differences.

   One of the goals of ICAI systems is to be self-modifying and generative. This type of system would allow ICAI programs to acquire new knowledge from old knowledge and to learn from the interactions with the student. Several lines of research look promising in this direction. Analogical reasoning appears to be an important theme because it underlies how new knowledge is generated from old knowledge; analogies are formed from one domain to apply knowledge to a new domain (see Chapter 9.2 for an example of the uses of analogies in AI). Rumelhart and Norman (1978) have recently formulated a learning theory about how old knowledge structures can be used to generate new knowledge structures. In a knowledge structure composed of production rules, Anderson (1976) has shown that in a limited domain, production rules can be self-modifying and can generate new production rules. Both of these lines of research are creating interest and could result in advancements for ICAI systems in the future.

Another important area of research is expert/novice differences. One goal of training is to provide novices with the foundation they need to become experts as quickly as possible. The information given to the student must be compatible with the type of information that an expert uses. When designing STEAMER, the program authors questioned the types of knowledge that experts have about steam plants and tried to direct instruction so that knowledge structures built up by the student would be compatible with expert knowledge. Similar work on expert/novice differences done in today's laboratories will be applied to ICAI systems.

If efficiency of training is to be increased, the student–computer interaction must be improved. Students must be able to concentrate on the lesson instead of concentrating on how to use the computer. ICAI has greatly improved this interaction. The next large improvement will come when some kind of natural-language processing becomes a reality.

The discussion of the future of ICAI systems has emphasized improvements in the intelligence of the system rather than advancements in hardware. More so than with other computer applications, the future of ICAI depends on advancements in cognitive science theories so that the learning process can be understood and subsequent training can be directed to enhance that learning.

8.4.6 COMPUTER-BASED INSTRUCTION ISSUES AND RESEARCH

As was discussed at the beginning of this chapter, the last two decades have seen a tremendously rapid advance of computer capabilities. We now get much greater computing power at a lot less cost: color, real-time, and animated graphics can be used, as can videodisks and new interface techniques, such as voice. The list is long and growing longer. The range of possible computer applications for training is almost limitless. It is time to ask what the computer should do rather than what it can do. We need theoretical guidance to suggest fruitful approaches to instructional design. The purpose of this section is to survey recent research into the principles of computer-based instruction.

As should be clear from the previous discussions, a poor instructional design that is computerized is not improved; it is still a poor instructional design. In education, much work has been done to specify what constitutes good instructional design. The results of a recent survey of 116 instructional designers by Braden and Sachs (1983) of the most recommended books on instructional design are presented in Table 8.4.6. These books can be consulted in the design of computer-based instructional material.

The next section surveys some of the recent work and many of the issues in computer-based instruction. For this survey, both basic research, which may not have been applied to CAI or CMI, and applied research are considered. The issues are broken down into the following variables, which have an influence on effective instructional design: individual differences, knowledge of results, amount of practice, augmented feedback, part–whole tasks, adaptive instruction, conceptual representations, and motivation. The following is a summary; references are provided for further reading.

Table 8.4.6 Most Recommended Books on Instructional Design

Rank	Title (Short Title)	Author
1	*Principles of Instructional Design*	Gagne and Briggs (1974)
2	*The Conditions of Learning*	Gagne (1977)
3	*Instructional Design*	Briggs (1977)
4	*Instructional Message Design*	Fleming and Levie (1978)
5	*Preparing Instructional Objectives*	Mager (1975)
6	*The Systematic Design of Instruction*	Dick and Casey (1978)
7	*Handbook of Procedures*	Briggs and Wager (1981)
8	*Learning System Design*	Davis, Alexander, and Yelon (1974)
9	*Instructional Design*	Kemp (1977)
10	*A Taxonomy of Educational Objectives*	Bloom (1956)

Source: Braden and Sachs (1983).

8.4.6.1 Individual Differences

Traditionally, individual differences have been studied with the goal of selecting the right personnel for the appropriate job. The premise is that instruction time can be reduced by selecting people who have certain measurable characteristics and training only them. This approach is primarily concerned with matching the person to the job rather than matching training to a student. However, in many cases, matching the person to the job may not be possible. Researchers must learn to be able to choose the correct learning strategy for the person based on a quick evaluation of the students' abilities. Only then will truly individualized instruction be achieved.

Instruction must be neither too difficult nor too easy for the student. If too difficult, the student may give up on the task altogether; if too easy, the student will become bored, losing interest or motivation. It is important to evaluate individual differences and set appropriate task difficulty levels.

Although this has been the goal, this has rarely been achieved (Steinberg, 1977). This lack of success could be due to either choosing inappropriate individual characteristics or choosing inappropriate instructional design based on the individual differences. Table 8.4.7 targets some of the important individual differences and provides examples of how the individual differences have been used to individualize CAI courses.

8.4.6.2 Knowledge of Results

Knowledge of results (KR) is defined as the feedback or reinforcement that is given to students to tell them how well they performed on the task. The form of KR can range from a simple "yes" or "no" to a more detailed "the shot was off-target to the right by 22 mm." KR was one of the earliest variables manipulated in the study of the acquisition of skills and has a long history of research behind it. The most extensive review of KR has been done by Adams (1971). In the development of his closed-loop theory of skill learning, Adams extrapolated several principles from the KR research. For more detail on the effect of KR on performance, Adams' (1971) paper should be consulted. Table 8.4.8 summarizes some of the important basic and applied research findings on feedback and knowledge of results. Much of the work in the table is basic research; the results should be directly applicable to CAI.

8.4.6.3 Amount of Practice

One of the best determinants of the amount of skill students obtain is the amount of practice they have performing the task to be learned. Experts are usually distinguished from novices in that they have had more practice performing a particular task. As an example, it is estimated that expert radiologists have analyzed nearly half a million radiographs (Lesgold, Feltovich, Glaser, and Wang, 1981).

In discussing the amount of practice, several issues will be considered: What causes practice to increase performance? How can practice be made more efficient? How can instruction be achieved in less time? These questions are summarized in Table 8.4.9.

8.4.6.4 Augmented Feedback

Lintern (1980a) defines augmented feedback as experimentally or system-provided perceptual cues that enhance the intrinsic task-related feedback. As examples, in a tracking task, Kinkade (1963) used an augmenting tone to indicate when the subject was on the track; Gordon and Gottlieb (1967) turned on a yellow light that further illuminated the tracking display when subjects were tracking off-target. In a simulated landing task, Lintern (1980b) provided subjects with augmenting visual cues that indicated the correct flight path. Augmented feedback has been used in simulation and for the training of perceptual skills. (Table 8.4.10 summarizes some of the results.)

8.4.6.5 Part–Whole Training Instruction

Part–whole instruction is concerned with the issue of whether it is more beneficial to break a task down and train component parts (part training) or whether it is more beneficial to train the whole task all at once (whole training). This issue has important implications for computer-based instruction because part training is typically less expensive to do on a computer system than is whole training (Wheaton, Rose, Fingerman, Korotkin, and Holding, 1976). Certain parts of the task can be simulated on a small microprocessor-based system, whereas the whole task might required a larger system where the many complex interactions of a particular system are simulated. If part training is more beneficial than whole training, then instruction could be done less expensively. If whole training is more beneficial, the amount of benefit must be determined before the increased cost can be justified. Table 8.4.11 summarizes some of the research results addressing part–whole instruction.

Table 8.4.7 Individual Differences

Individual Difference	Description	Implication	Reference
Time-sharing ability	People vary in their natural ability to time-share, to carry on more than one task at a time	Adjust the task difficulty to take into account this ability	North and Gopher (1976); Ackerman, Wickens, and Schneider (1984)
Holistic versus analytic processors	Evidence indicates that holistic processors process information as a whole and analytic processors process information feature by feature	Present information differently depending on processor type	Cooper (1976)
Learning strategies	Good students in physics analyze problems abstractly and poor students analyze problems literally	Indicate abstractions to the student	Larkin, McDermott, Simon, and Simon (1980a, 1980b); Larkin (1981, 1982)
Field dependent versus field independent	This individual difference was used to structure CAI for foreign language learning; no evaluation of program done	Possible variable to use for other applications	Raschio and Lange (1984)
Measured from pretest	CAI lesson that was structured individually depending on the score of the pretest; no performance differences were found when this group was compared to a control group	The individual differences may not have been important to performance; lesson structuring may not have been optimal	Tatsuoka and Birenbaum (1981)
Student's major	Individualized group was only slightly better than non-individualized group	Student's major may not provide individual difference information	Ross (1984)
General	Potential exists for structuring lessons according to individual differences	Potential, but little or no data to indicate this will work	Cronbach (1967); Farley (1983); Glaser (1982)
General	In reviews of CAI lessons structured according to individual differences, concluded that lesson authors have not been able to acquire and use enough information about the learner to provide ideally individualized instruction	More research needs to be done; do not know how to use individual differences in CAI at this time	Berliner and Cohen (1973); Bracht (1970); Bracht and Glass (1972); Cronbach and Snow (1977); DiVesta (1973); Glaser (1970); Hall (1983); McCann (1983); Tobias (1976)

Table 8.4.8 Knowledge of Results

Issue	Discussion	Reference
Performance improvement in knowledge acquisition depends on KR. The rate of improvement depends on the precision of KR	In a very early experiment on skill training, a group given qualitative KR performed best, a group given quantitative KR (yes or no) performed next best, and groups given no KR or irrelevant KR were worst	Trowbridge and Cason (1932)
Increasing the post-KR interval up to a point will improve the performance level in skill acquisition	In basic skill research, found that students must have time between trials to think about their performance after they have received feedback	Adams (1971)
Feedback should be immediate	Anything over 2 sec and the student may be thinking about something else	Miller (1968)
Delay of feedback may have detrimental effect on retention of material	In a study, delay of 30 sec had detrimental effect on retention of math material	Gaynor (1979)
Feedback on correct responses may not be useful or could be detrimental to performance	In a review of studies, nine studies found no difference or a decrement in performance for feedback given on correct responses	Anderson, Kulhavy, and Andre (1971)
CAI offers good opportunity for giving immediate feedback	Hypothesized that reason CAI group in a music course was better than a control group was because the CAI group knew precisely how they were doing on each drill and the control group only received feedback when the homework was returned	Arenson (1982)
CAI feedback should be used to target misconceptions	In the ICAI program SOPHIE, feedback is used to target misconceptions and to offer alternative ways to conceptualize the problem	Burton and Brown (1979)

Table 8.4.9 Amount of Practice

Issue	Discussion	Reference
Performance during practice seems to follow a power law	Reviewed several areas of research; developed a lawful relationship between performance and practice; performance improves with practice	Newell and Rosenbloom (1981)
Practice can be made more efficient if consistent practice is given	Several experiments look at the acquisition of perceptual skills through computer based training; research shows that the student must be able to see the consistent relationships between the input and the output	Schneider and Fisk (1981); Schneider and Shiffrin (1977); Shiffrin and Schneider (1977)
Reduce computer down time and thus increase amount of practice	Microprocessors can be used for many tasks; they are cheap and portable so that each student can have a computer, backups are available, and the instruction can be given in poorly accessible areas	McDonald and Crawford (1983)
Time compression can be used to present information in a shorter time period	Training on perceptual tasks that look at slowly occurring trajectories, such as radar operation and air traffic control, can be made more efficient by time-compressing the paths	Bloomfield and Little (1985); Scanlan (1975); Schneider (1981)
In a simulation, processes that take a long time with providing little information can be shortened	In STEAMER, long processes like filling up a tank with water can be shortened so that more trials of practice can be given	Hollan, Stevens, and Williams (1980)
Practice can be speeded up by increasing the presentation rate	Presentation rate was increased from 10 characters per second (cps) to 30 cps with no drop in performance	Dennis (1979)
Drill and practice	Important function of drill and practice method is to bring learner to level of "automaticity," on lower-level subskills so that learner can more readily perform some higher-level complex skill	Merrill and Salisbury (1984)

Table 8.4.10 Augmented Feedback

Issue	Discussion	Reference
Augmenting cues can be effective	In a flight simulation, "highway-in-the-sky" augmenting cues were used to increase performance	Lintern (1980b)
Augmenting cues should make the consistencies in the task more salient	In a computer-based tracking task, augmenting cues were only effective if they made the consistencies inherent in the task more salient	Eberts and Schneider (1980)
Students must not form a dependency on the augmenting cues	Augmenting cues should be presented on only some of the trials so that students are forced to do some trials without the augmentation	Wheaton, Rose, Fingerman, Korotkin, and Holding (1976)

Table 8.4.11 Part–Whole Instruction

Issue	Discussion	Reference
Some tasks are so difficult that the student cannot approach the required skill level to perform the task	May have to break the task into components so that the students can build toward this more complex task; finding in basic skill instruction	Gaines (1972)
Part learning is necessary when the amount of learning is large	Conclusion from review of several studies	Wheaton, Rose, Fingerman, Korotkin, and Holding (1976)
Whole learning is slightly better if the sequence is long and complex	Conclusion from review of several studies	Wheaton, Rose, Fingerman, Korotkin, and Holding (1976)
CAI lessons must address one specific objective at a time	The goal must be clearly stated and obtainable; may have to break overall task down into parts to achieve this	Jay (1983)
CAI can be used for part-task simulation rather than for the whole course	Principle used in the Navy beginning in the mid-1970s based on successes and failures in CAI	Montague, Wulfeck, and Ellis (1983a)

8.4.6.6 Adaptive Instruction

Lintern and Roscoe (1980) define adaptive instruction as a "method of individualized instruction in which task characteristics are automatically changed from an easy or simple form to the difficult or complex criterion version." The reasons to use adaptive instruction were discussed for augmented feedback and part–whole instruction. All of these methods are used to decrease or increase the difficulty of the task; the distinctions between them sometimes get hazy.

One of the important issues in CAI is whether the sequencing of instruction should be program or student controlled. Control can be done by the computer depending on the pattern of errors made by the student. This and other issues of adaptive instruction are summarized in Table 8.4.12.

8.4.6.7 Conceptual Representations

Determining how people conceptually represent information is important for instruction because the material to be learned must fit in with how a person eventually comes to conceptualize the problem; the material must fit into the person's memory structures. Ausubel (1968) articulates this concept by stating that the learning of new material depends on its interaction with the cognitive structures held by the student. The cognitive structures of the student provide a kind of "ideational scaffolding" that can be used as an anchor for the new material. The form of the instructional material should be compatible with how students internally represent the information. The following issues should be considered with regard to training and conceptual representations:

1. What special capabilities do humans have for internally representing the outside world?
2. How do experts internally represent complex concepts?
3. How can computers be used to present information that will be compatible with the conceptual representations?

Table 8.4.13 summarizes some of the research that has been done on conceptual representations.

8.4.6.8 Motivation

Much work has been done on motivation and learning. In our presentation of this variable, we limit it to how the special features of computers can be used to increase motivation. Table 8.4.14 summarizes some of this research.

8.4.7 SUMMARY AND CONCLUSION

We have reviewed the technologies that have merged to produce computerized instruction: instruction technology and computer technology. It seems reasonable to suggest that as computers become more available and more friendly, the instructional principles we have elucidated here will become more easily implemented. The danger that seems quite real is that computerization will be increasingly used for instruction without exploiting good instructional principles. It behooves all in the training and human factors professions to guard against this.

In our discussion of CAI we have provided three prototypical examples: PLATO, TICCIT, and PLANIT. All have advantages and disadvantages. If large time-sharing CAI systems remain—a big if—they will probably borrow from all three of these systems. Much of what has been learned from these systems is also applicable to microprocessor instructional systems.

The major advantage of CAI is that it can maximize individualization of instruction for each student. The major disadvantage is the authoring time for CAI, and this will be discussed in our concluding comments.

CMI can introduce a degree of efficiency into a training system that simply is not possible under conventionally (i.e., noncomputer) managed systems. We have described the system from the point of view of the student, the instructor, and the course administrator. We have discussed the eight activities of a CMI system identified by Dennis (1979): diagnosing, assigning or prescribing, facilitating study, evaluating, data collecting, reporting, managing resources, and providing an information network.

Because CMI requires major data manipulations and a variety of peripherals, we concluded that a large mainframe computer is needed. We have described one such CMI system, the Air Force Advanced Instruction System. We conclude that CMI is both cost- and training-effective when student load is large. We have been unable to locate data to indicate how many students will make CMI cost-effective.

ICAI is a logical application of the rapidly expanding technology of artificial intelligence—at least that part of the technology that addresses knowledge-based expert systems. The three ICAI systems discussed—SOPHIE, WUSOR, and GUIDON—represent the state of the art even though all three were developed in the mid-1970s.

Table 8.4.12 Adaptive Instruction

Issue	Discussion	Reference
The variable that is adaptively manipulated must not significantly change the nature of the task	In teaching tasks, variables such as the order of control and the gain of the system were adaptively manipulated; no benefit was found for this manipulation	Crooks (1973); Gopher, Williges, and Damos (1975)
Provide augmenting cues only when students are not performing accurately	Using this method, the difficulty of the task will automatically adapt itself to the level of the student—more help in the beginning and less help once experienced	Gordon (1968); Lintern (1980b)
Analyze student's on-task error patterns to adaptively control CAI material presented	Effective method for adaptively presenting material to students; shown to be better than nonadaptive or learner control	Park and Tennyson (1980); Ross (1984); Tennyson, Tennyson, and Rothen (1980)
Can adaptively manipulate several different variables	Concluded that the following variables can be adaptively manipulated: (1) amount of instruction; (2) sequence of instruction; (3) instructional display time; and (4) control strategy with advisement information	Tennyson, Christenson, and Park (1984)

Table 8.4.13 Conceptual Representation

Issue	Discussion	Reference
The cognitive structures of the student provide an "ideational scaffolding" for new material	The material presented must be given so that it fits into the student's memory structures	Ausubel (1968); Hartley and Davies (1976)
Compatibility must exist for training operators of complex systems	Should be a compatibility between the physical system, the internal representation that the operator has of the system, and the interface (the display) between the two	Wickens (1984)
Humans represent information spatially	Computer graphics can be used to help the student visualize concepts; it has been used especially in engineering, math, and music	DuBoulay and Howe (1982); Forchieri, Lemut, and Molfino (1983); Harding (1983); Lamb (1982); Pohlman and Edwards (1983); Rogers (1981); Stevens and Roberts (1983)
Programming used to structure information	Understand complex subjects, such as math, if the student can program it; helps to specify steps	DuBoulay and Howe (1982)

Table 8.4.14 Motivation

Issue	Discussion	Reference
To make computer learning intrinsically motivating, CAI can utilize challenge, fantasy, and curiosity	In a study of computer games, these features were incorporated; the same features could be used in CAI; much potential exists	Malone (1980)
Can use graphics to express new ideas and thoughts	Children used LOGO to creatively and actively manipulate images	Vaidya (1983)

There are four components in an ICAI system: the representation of subject matter, the instructional strategy, the student model, and the dialogue format between student and computer. If ICAI systems can be made cost-effective, they will be the predominant computer-based instructional method as we enter the 21st century.

In the final discussion section of this chapter, we have attempted to identify those instructional issues that can be exercised by computer-based instructional systems. There is no guarantee that computers will improve instruction; the issues and research discussed should be considered to increase the probability that instructional improvement will occur.

We have addressed a variety of topics in what we hope is a systematic approach to computers in instruction. There is at least one relevant topic not discussed: the design of instruction. The assumption herein has been that appropriate instructional design front-end activity has been performed (e.g., task analysis, learning objective development, criterion test identification). Our concluding comments address this assumption.

It is quite possible to construct a computer-based instructional system that provides easy and friendly access to the instruction and manages the entire program flawlessly while still failing to meet the objectives of the program. Computers have been used to measurably improve the way instruction is conducted; the design of that instruction has been, in large part, ignored.

The individual's idiosyncrasies of learning style and pace can be integrated into the delivery of instruction. What is lacking is the application of the power of the computer to the development of instructional content: the steps that immediately precede software development. An ideal ICAI system could partially compensate for this by providing a complete knowledge base and the strategies for accessing it. This seems practical with only limited knowledge bases, however. A truly intelligent ICAI system should be able to query subject-matter experts to increase that knowledge base.

We therefore conclude with one research recommendation. If the design of instruction can be reasonably automated, then the capabilities of CAI, CMI, and ICAI can be applied to sound instructional objectives. Training research and development dollars should, therefore, be directed toward this process.

REFERENCES

Ackerman, P. H., Wickens, C. D., and Schneider, W. (1984). Deciding the existence of a time-sharing ability: A combined methodological and theoretical approach. *Human Factors, 26,* 71–82.

Adams, J. A. (1971). A closed-loop theory of motor learning. *Journal of Motor Behavior, 3,* 111–149.

Alpert D., and Bitzer, D. L. (1970). Advances in computer-based education. *Science, 167,* 1582–1590.

Anderson, J. R. (1976). *Language, memory, and thought.* Hillsdale, NJ: Erlbaum.

Anderson, R. C., Kulhavy, R. W., and Andre, T. (1971). Feedback procedures in programmed instruction. *Journal of Educational Psychology, 62,* 148–156.

Arenson, M. (1982). The effect of a competency-based computer program of the learning of fundamental skills in a music theory course for non-majors. *Journal of Computer-Based Instruction, 9,* 55–58.

Ausubel, D. P. (1968). *Educational psychology: A cognitive view.* New York: Holt, Rinehart & Winston.

Avner, A. R. (1979). Production of computer-based instructional materials. In H. F. O'Neil, Jr. (Ed.), *Issues in instructional systems development.* New York: Academic.

Avner, A., Moore, C., and Smith, S. (1980). Active external control: A basis for superiority of CBI. *Journal of Computer-Based Instruction, 6,* 115–118.

Avner, A., Smith, S., and Tenczar, P. (1984). CBI authoring tools: Effects on productivity and quality. *Journal of Computer-Based Instruction, 11,* 85–89.

Baker, F. (1981). Computer managed instruction. In H. F. O'Neil, Jr., Ed. *Computer-based instruction: A state-of-the-art assessment.* New York: Academic.

Berliner, D. C., and Cohen, L. S. (1973). Trait-treatment interaction and learning. *Review of Research in Education, 1,* 58–94.

Bloom, B. S. (Ed.) (1956). *A taxonomy of educational objectives, Handbook I: The cognitive domain.* New York: Longman.

Bloomfield, J. R., and Little, R. K. (1985). Operator tracking performance with time-compressed and time integrated moving target indicator (MTI) radar imagery. In R. E. Eberts & C. G. Eberts, Eds. *Trends in Ergonomics/Human Factors, Vol. II.* Amsterdam: North-Holland.

Bobrow, D. G., Kaplan, R. M., Kay, M., Norman, D. A., Thompson, H., and Winograd, T. (1977). GUS, a frame-driven dialog system. *Artificial Intelligence, 8,* 155–173.

Boraiko, A. A. (1982). The chip. *National Geographic, 162*(4), 421–456.

Boulter, L. R. (1964). Evaluation of mechanism in delay of knowledge of results. *Canadian Journal of Psychology, 18,* 281–291.

Bracht, G. H. (1970). Experimental factors related to attribute-treatment interactions. *Review of Educational Research, 40,* 627–645.

Bracht, G. H., and Glass, G. V. (1972). Interaction of personological variables and treatment. In L. Sperry, Ed. *Learning performance and individual differences: Essays and readings.* Glenview, IL: Scott, Foresman.

Brackett, J. W. (1979). TRAIDEX: A proposed system to minimize training duplication. In H. F. O'Neil, Ed. *Issues in instructional systems development* (pp. 101–131). New York: Academic.

Braden, R. A., and Sachs, S. G. (1983). The most recommended books on instructional development. *Educational Technology, 23*(2), 24–28.

Briggs, L. J., Ed. (1977). *Instructional design: Principles and applications.* Englewood Cliffs, NJ: Educational Technology Publications.

Briggs, L. J., and Wager, W. W. (1981). *Handbook of procedures for the design of instruction,* 2nd ed. Englewood Cliffs, NJ: Educational Technology Publications.

Brown, J. S., and Burton, R. R. (1978). Diagnostic models for procedural bugs in mathematical skills. *Cognitive Science, 2,* 155–192.

Brown, J. S., Burton, R. R., and Bell, A. B. (1974). *SOPHIE: A sophisticated instructional environment for teaching electronic troubleshooting (an example of AI in CAI)* (Technical Report No. 2790). Cambridge, MA: Bolt, Beranek, and Newman.

Brown, J. S., Rubinstein, R., and Burton, R. (1976). *Reactive learning environment for computer assisted electronics instruction* (Technical Report No. 3314). Cambridge, MA: Bolt, Beranek, and Newman.

Buck, J. A. (1982). Federal Aviation Administration seventh year computer-based training report. *Journal of Computer-Based Instruction, 8,* 53–55.

Bunderson, C. V. (1981). Courseware. In H. F. O'Neil, Jr., Ed., *Computer-based instruction: A state-of-the-art assessment.* New York: Academic.

Burton, R. R., and Brown, J. S. (1979). Toward a natural-language capability for computer-assisted instruction. In H. F. O'Neil, Ed. *Procedures for instructional system development* (pp. 273–313). New York: Academic.

Camstra, B. (1977). Make CAI smarter. *Computers and Education, 3,* 177–183.

Clancey, W. B. (1979). Dialogue management for rule-based tutorials. In *Proceedings of the Sixth IJCAI,* pp. 155–161. Los Altos, CA: Kaufman.

Clancey, W. B. (1983). Guidon. *Journal of Computer-Based Instruction, 10,* 8–15.

Clancey, W. J., Bennett, J. S., and Cohen, P. R. (1982). Applications oriented AI research: Education. In A. Barr and E. A. Feigenbaum, Eds. *The handbook of artificial intelligence* (Vol. 2). Los Altos, CA: Kaufman.

Collins, A. M., and Quillian, M. R. (1969). Retrieval time from semantic memory. *Journal of Verbal Learning and Verbal Behavior, 8,* 240–247.

Collins, A., and Stevens, A. L. (1980). Goals and strategies of interactive teachers. In R. Glaser, Ed. *Advances in instructional psychology* (Vol. 2). Hillsdale, NJ: Erlbaum.

Computer Based Military Training Systems (1980, July). United Kingdom: Royal Air Force, Technical Cooperation Program Sub-Committee on Non-Atomic Military Research and Development, Subgroup U, Technical Panel UTP-2.

Conkwright, T. D. (1982). PLATO applications in the airline industry. *Journal of Computer-Based Instruction, 8,* 49–52.

Cooper, L. A. (1976). Individual differences in visual comparison processes. *Perception & Psychophysics, 12,* 433–444.

Crooks, W. H. (1973). Varied and fixed error limits in automated adaptive skill training (Technical Report No. IR ARL-73-8, AFOSR 73-4). Savoy, IL: University of Illinois, Aviation Research Laboratory.

Cronbach, L. J. (1967). How can instruction be adapted to individual differences? In R. M. Gagne Ed. *Learning and individual differences* (pp. 23–29). Columbus, OH: Charles E. Merrill.

Cronbach, L., and Snow, R. (1977). *Aptitudes and instructional methods: A handbook for research on interactions.* New York: Irvington Publishers.

Dallman, B. E., Pieper, W. J., and Richardson, J. J. (1983). A graphics simulation system—Task emulation not equipment modeling. *Journal of Computer-Based Instruction, 10,* 70–72.

Dare, F. C. (1975). *Evaluation of the PLATO IV system in a military training environment* (Final Report). Aberdeen Proving Ground: U.S. Army Ordnance Center and School.

Davis, R. H., Alexander, L. T., and Yelon, S. L. (1974). *Learning system design: An approach to the improvement of instruction.* New York: McGraw-Hill.

Dennis, J. R. (1979). Computer managed instruction and individualization (Report No. 1, Illinois Series on Educational Application of Computers). Champaign, IL: University of Illinois.

Dennis, V. E. (1979). The effect of display rate and memory support on correct responses, trials, total instructional time and response latency in a computer-based learning environment. *Journal of Computer-Based Instruction, 6,* 50–54.

Dick, W., and Carey, L. (1978). *The systematic design of instruction.* Glenview, IL: Scott, Foresman.

DiVesta, F. J. (1973). Theory and measures of individual differences in studies of trait by treatment interaction. *Educational Psychologist, 10,* 67–75.

Douglas, J. H. (1976). Learning technology comes of age. *Science News, 110,* 170–174.

DuBoulay, J. B. H., and Howe, J. A. M. (1982). LOGO building blocks: Student teachers using computer-based mathematics apparatus. *Computers & Education, 6,* 93–96.

Eberts, R. E., and Brock, J. B. (1984). Computer applications to instruction. In F. W. Muckler, Ed. *The annual review of human factors* (pp. 239–284). Santa Monica, CA: The Human Factors Society.

Eberts, R. E., and Schneider, W. (1980). Computer assisted displays enabling internalization and reduction of operator workload in higher order systems, or, pushing the barrier of human control beyond second order systems. In B. P. Corrick, E. C. Haseltine, and R. I. Durst, Eds. *Proceedings of the twenty-fourth annual meeting of the human factors society.* Santa Monica, CA: Human Factors Society.

Fairweather, P. G., and O'Neal, A. F. (1984). The impact of advanced authoring systems on CAI productivity. *Journal of Computer-Based Instruction, 11,* 90–94.

Farley, F. H. (1983). Basic process individual differences: A biologically based theory of individualization for cognitive, affective, and creative outcomes. In F. H. Farley and N. J. Gordon, Eds. *Psychology and education: The state of the union* (pp. 9–29). Berkeley, CA: McCutchan.

Fields, C., and Paris, J. (1981). Hardware-software. In H. F. O'Neil, Jr., Ed. *Computer-based instruction: A state-of-the-art assessment.* New York: Academic.

Fleming, M. L., and Levie, W. H. (1978). *Instructional message design.* Englewood Cliffs, NJ: Educational Technology Publications.

Fletcher, J. D. (1974). Computer applications in education and training: Status and trends (Technical Report No. FY75-32). San Diego, CA: U.S. Navy Personnel Research and Development Center.

Fletcher, J. D. (1983). Personal communication. April 26.

Forchieri, P., Lemut, E., and Molfino, M. T. (1983). The GRAF system: An interactive graphic system for teaching mathematics. *Computers & Education, 7,* 177–182.

Gagne, R. M. (1977). *The conditions of learning,* 3rd ed. New York: Holt, Rinehart & Winston.

Gagne, R. M., and Briggs, L. J. (1974). *Principles of instructional design.* New York: Holt, Rinehart & Winston.

Gaines, B. R. (1972). The learning of perceptual-motor skills by man and machines and its relationship to training. *Instructional Science, 1,* 263–312.

Gaynor, P. (1979). The effect of feedback delay on retention of computer-based mathematical material. *Journal of Computer-Based Instruction, 8,* 28–34.

Glaser, R. (1970). Psychological questions in the development of computer-assisted instruction. In W. H. Holtzman, Ed. *Computer-assisted instruction, testing and guidance* (pp. 74–93). New York: Harper and Row.

Glaser, R. (1982). Instructional psychology: Past, present and future. *American Psychologist, 37,* 292–305.

Goldstein, I. (1978). Developing a computational representation for problem solving skills. In *Proceedings of the Carnegie-Mellon Conference on Problem Solving and Education: Issues in Teaching and Research.* Pittsburgh: CMU Press, 1978.

Gopher, D., Williges, R. C., and Damos, D. L. (1975). Manipulating the number and type of adaptive variables in training. *Journal of Motor Behavior, 7,* 159–170.

Gordon, N. B. (1968). Guidance versus augmented feedback and motor skill. *Journal of Experimental Psychology, 77,* 24–30.

Gordon, N. B., and Gottlieb, M. J. (1967). Effect of supplemental visual cues on rotary pursuit. *Journal of Experimental Psychology, 15,* 566–568.

Hall, K. B. (1983). Content structuring and question asking for computer-based education. *Journal of Computer-Based Instruction, 10,* 1–7.

Harding, R. D. (1983). A structured approach to computer graphics for mathematical uses. *Computers & Education, 7,* 1–19.

Hartley, J., and Davies, I. K. (1976). Pre-instructional strategies: The role of pre-test, behavioral objectives, overviews, and advanced organizers. *Review of Educational Research, 46,* 239–265.

Hartley, R. (1980). Computer assisted learning. In H. T. Smith and H. R. G. Green, Eds. *Human interaction with computers.* London: Academic.

Hillelsohn, J. J. (1984). Benchmarking authoring systems. *Journal of Computer-Based Instruction, 11,* 95–97.

Hofstetter, F. T. (1983). The cost of PLATO in a university environment. *Journal of Computer-Based Education, 9,* 248–255.

Hollan, J. (1984). Intelligent object-based graphical interfaces. In G. Salvendy, Ed. *Human computer intraction* (pp. 293–296). Amsterdam: North-Holland.

Hollan, J. Stevens, A., and Williams, N. (1980). STEAMER: An advanced computer-assisted instruction system for propulsion engineering. Paper presented at the Summer Simulation Conference, Seattle.

Jay, T. B. (1983). The cognitive approach to computer courseware design and evaluation. *Educational Technology, 23*(1), 22–26.

Kearsley, G. P. (1977). The costs of CAI: A matter of assumption. *AEDS Journal, 10,* 100–112.

Kearsley, G. (1984). Authoring tools: An introduction. *Journal of Computer-Based Instruction, 11,* 67.

Kemp, J. E. (1977). *Instructional design: A plan for unit and course development,* 2nd ed. Belmont, CA: Fearon.

Kinkade, R. G. (1963). A differential influence of augmented feedback on learning and on performance (Technical Documentation Report 63-12). Wright-Patterson Air Force Base, OH: Aerospace Medical Research Laboratory.

Lamb, M. (1982). An interactive graphical modeling game for teaching musical concepts. *Journal of Computer-Based Instruction, 9,* 59–63.

Larkin, J. H. (1981). Enriching formal knowledge: A model for learning to solve problems in physics. In J. R. Anderson, Ed. *Cognitive skills and their acquisition.* Hillsdale, NJ: Erlbaum.

Larkin, J. H. (1982). Understanding problem representations and skill in physics. In S. Chipman, J. Segal, and R. Glaser, Eds. *Thinking and learning skills: Current research and open questions.* Hillsdale, NJ: Erlbaum.

Larkin, J. H., McDermott, J., Simon, D. P., and Simon, H. A. (1980a). Expert and novice performance in solving physics problems. *Science, 208,* 1334–1342.

Larkin, J. H., McDermott, J., Simon, D. P., and Simon, H. A. (1980b). Models of competence in solving physics problems. *Cognitive Science, 4,* 317–345.

Lesgold, A. M., Feltovich, P. J., Glaser, R., and Wang, Y. (1981). The acquisition of perceptual diagnostic skill in radiology (Office of Naval Research Technical Report No. PDS-1). Pittsburgh, PA: University of Pittsburgh, Learning Research and Development Center.

Lintern, G. (1980a). Augmented feedback for perceptual-motor instruction. Paper presented at the meeting of the American Psychological Association, Montreal.

Lintern, G. (1980b). Transfer of landing skill after training with supplementary visual cues. *Human Factors, 22,* 81–88.

Lintern, G., and Roscoe, S. N. (1980). Visual cue augmentation in contact flight simulation. In S. N. Roscoe, Ed. *Aviation Psychology.* Ames, IA: Iowa State University Press.

Lyman, E. R. (1981). *PLATO highlights.* Urbana, IL: University of Illinois Computer-Based Education Research Laboratory.

Mager, R. F. (1975). *Preparing instructional objectives,* 2nd ed. Palo Alto, CA: Fearon.

Malone, T. W. (1980). What makes things fun to learn? A study of intrinsically motivating computer games (Report CIOS-7 [SSL-80-11]). Palo Alto, CA: Xerox PaRC.

McCann, R. H. (1983). Learning strategies and computer-based instruction. *Computers & Education, 5,* 133–140.

McDonald, B. A., and Crawford, A. M. (1983). Remote site training using microprocessors. *Journal of Computer-Based Instruction, 10,* 83–86.

Merrill, P. F. and Salisbury, D. (1984). Research on drill and practice strategies. *Journal of Computer-Based Instruction, 11,* 19–21.

Miller, R. B. (1968). Response time in man-computer conversational transactions. In *AFIPS Conference Proceeding (Fall Joint Computer Conference).* Washington: Thompson Book Co.

Minsky, M. (1975). A framework for representing knowledge. In P. H. Winston, Ed. *The psychology of computer vision.* New York: McGraw-Hill.

Montague, W. E., Wulfeck, W. H., and Ellis, J. A. (1983a). Computer-based instructional research and development in the Navy: An overview. *Journal of Computer-Based Instruction, 10,* 83.

Montague, W. E., Wulfeck, W. H., and Ellis, J. A. (1983b). Quality CBI depends on quality instructional design and quality implementation. *Journal of Computer-Based Instruction, 10,* 90–93.

Morgan, C. E. (1978). CAI and basic skills information. *Educational Technology, 18*(4), 37–39.

Moses, J. (1975). *A MACSYMA primer* (Mathlab Memo No. 2). Cambridge, MA: Massachusetts Institute of Technology, Computer Science Laboratory.

Newell, A., and Rosenbloom, P. S. (1981). Mechanisms of skill acquisition and the law of practice. In J. R. Anderson, Ed., *Cognitive skills and their acquisition* (pp. 1–55). Hillsdale, NJ: Erlbaum.

North, R. A., and Gopher, D. (1976). Measures of attention as predictors of flight performance. *Human Factors, 18,* 1–14.

Orlansky, J., and String, J. (1977). Cost-effectiveness of computer-based instruction in military training (IDA Paper P-1375, 1977). Arlington, VA: Institute for Defense Analyses (AD A073 400).

O'Shea, T. A. (1979). Self-improving quadratic tutor. *The International Journal of Man-Machine Studies,* 97–124.

Park, O., and Tennyson, R. D. (1980). Adaptive design strategies for selecting number and presentation order of examples in coordinate concept acquisition. *Journal of Educational Psychology, 72,* 362–370.

Perlowski, A. A. (1980). Application of the new technology: The "smart" machine revolution. In T. Forester, Ed. *The microelectronics revolution.* Cambridge, MA: MIT Press.

Pohlman, D. L., and Edwards, B. J., (1983). Desk top trainer: Transfer of training of an aircrew procedural task. *Journal of Computer-Based Instruction, 10,* 62–65.

Psotka, J. (1983). Computer-based instructional research and development in the Army: An overview. *Journal of Computer-Based Instruction, 10,* 73.

Quillian, M. R. (1969). The teachable language comprehender: A simulation program and theory of language. *Communications of the ACM, 12,* 459–476.

Raschio, R., and Lange, D. L. (1984). A discussion of the attributes, role and use of CAI material in foreign languages. *Journal of Computer-Based Instruction, 11,* 22–27.

Reigeluth, C. M. (1979). TICCIT to the future: Advance in instructional theory for CAI. *Journal of Computer-Based Instruction, 6,* 40–46.

Rockway, M. R., and Yasutake, J. Y. (1974). The evolution of the Air Force Advanced Instructional System. *Journal of Educational Technology System, 2,* 217–239.

Rogers, D. F. (1981). Computer graphics at the U.S. Naval Academy. *Computers & Education, 5,* 165–182.

Ross, S. M. (1984). Matching the lesson to the student: Alternative adaptive designs for individualized learning systems. *Journal of Computer-Based Instruction, 11,* 42–48.

Rumelhart, D. E., and Norman, D. A. (1978). Accretion, tuning, and restructuring: Three modes of learning. In J. W. Cotton and R. L. Klatzky, Eds. *Semantic Factors in Cognition.* Hillsdale, NJ: Erlbaum.

Scanlan, L. A. (1975). Visual time compression: Spatial and temporal cues. *Human Factors, 17,* 84–90.

Schank, R. C., and Abelson, R. P. (1977). *Scripts, plans, goals, and understanding.* Hillsdale, NJ: Erlbaum.

Schneider, W. (1981). Automatic control processing concepts and their implications for the training of skills (Technical Report HARL ONRE-8101). Champaign, IL: University of Illinois, Human Attention Research Laboratory.

Schneider, W., and Fisk, A. D. (1981). Degree of consistent training: Improvements in search performance and automatic process development. *Perception & Psychophysics, 31,* 160–168.

Schneider, W., and Shiffrin, R. M. (1977). Controlled and automatic human information processing: I. Detection, search, and attention. *Psychological Review, 84,* 1–66.

Shiffrin, R. M., and Schneider, W. (1977). Controlled and automatic human information processing: II Perceptual learning, automatic attending, and a general theory. *Psychological Review, 84,* 127–190.

Steinberg, E. R. (1977). Review of student control in computer-assisted instruction. *Journal of Computer-Based Instruction, 3,* 84–90.

Stevens, A., and Roberts, B. (1983). Quantitative and qualitative simulation in computer based training. *Journal of Computer-Based Instruction, 10,* 16–19.

Stevens, A. C., Collins, A., and Goldin, S. (1978). *Diagnosing student's misconceptions in causal models* (Report No. 3786). Cambridge, MA: Bolt, Beranek and Newman.

Suppes, P. (1979). Current trends in computer-assisted instruction. In M. C. Yovits, Ed. *Advances in computers.* New York: Academic.

Tatsuoka, K., and Birenbaum, M. (1981). Effects of instructional backgrounds on test performances. *Journal of Computer-Based Instruction, 8,* 1–8.

Tennyson, C. L., Tennyson, R. D., and Rothen, W. (1980). Content structure and management strategy as design variables in concept acquisition. *Journal of Educational Psychology, 72,* 491–505.

Tennyson, R. D., Christenson, D. L., and Park, S. I. (1984). The Minnesota adaptive instructional system: An intelligent CBI system. *Journal of Computer-Based Instruction, 11,* 2–13.

Tobias, S. (1976). Achievement treatment interactions. *Review of Educational Research, 46,* 671–74.

Trowbridge, M. H., and Cason, H. (1932). An experimental study of Thorndike's theory of learning. *Journal of General Psychology, 7,* 245–258.

Vaidya, S. (1983). Using LOGO to stimulate children's fantasy. *Educational Technology, 23*(12), 25–26.

Wescourt, K. T., Beard, M., and Gould, M. (1977). Knowledge-based adaptive curriculum sequences for CAI: Application of a network representation (Technical Report No. 128). Palo Alto, CA: Stanford University, Institute for Mathematical Studies in the Social Sciences.

Wheaton, G. R., Rose, A. M., Fingerman, P. W., Korotkin, A. L., and Holding, D. H. (1976). Evaluation of the effectiveness of training devices: Literature review and preliminary model. Alexandria, VA: U.S. Army Research Institute for the Behavioral and Social Sciences, Research Memorandum 76-6.

Wickens, C. D. (1984). *Engineering psychology and human performance.* Columbus, OH: Charles E. Merrill.

Williams, M., Hollan, J., and Stevens, A. (1981). An overview of STEAMER: An advanced computer-assisted instruction system for propulsion engineering. *Behavior Research Methods and Instrumentation, 13,* 85–90.

CHAPTER 8.5

TRAINING SIMULATORS

RALPH E. FLEXMAN

University of Illinois
Champaign–Urbana, Illinois

EDWARD A. STARK

The Singer Company
Link Flight Simulation Division
Binghamton, New York

8.5.1 INTRODUCTION

This chapter addresses the following questions: What is a training simulator and what functions does it perform? What advantages and benefits can it produce; where and for what purposes are training simulators being used today? How is the general requirement for a training simulator established and its essential capabilities and characteristics determined? How are the design specifications developed and how should its cost and effectiveness be assessed?

The answers to these questions are of value to engineers and to human factors and training specialists considering the use of simulators in new training applications. A list of references is included at the end of the chapter to assist in the resolution of specific problems and to provide a more comprehensive understanding of the subject.

Much of the material in this chapter and much of that referred to in the list of references and readings is oriented toward simulators used in aviation training. This is due in part to the experiences of the authors, but due also to the preponderance of general experience in the development and operation of aviation training systems and devices. Each of the principles relating to the design and application of simulators in aviation skills training is relevant also to the training of skills in other contexts, whether in the operation of ground vehicles, industrial plants, power stations, or communication networks.

Many millions of dollars have already been spent on research relating to the design and use of simulators for training and many millions more will be spent as the emerging technology in microprocessors further expands their potential. Predictions are already lagging actual developments, and the future portends important breakthroughs in training as a result of these developments.

8.5.1.1 Historical Background

The use of simulators for training has a rich and interesting history with roots in both military and civilian programs. The early simulators were devices that attempted to replicate the more important characteristics and environment of operating systems controlled by one or more people. Their function was to provide appropriate cues and response capabilities so that a trainee could practice tasks he or she would use later in the operation of the actual equipment. The simulators were generally used in preference to the operational equipment to save costs and/or to provide a more favorable training environment.

It is difficult to find an anchor point in history where the first simulator was built or used to support a training program. The first reported incident goes back to the years when the U.S. Army was still heavily dependent on the use of horses. General Wood was reported to have invented and successfully used a horse simulator for training cavalry recruits. Following World War I the expanding Army Air Corps was experiencing problems in training their pilots to maintain control of airplanes while flying in clouds. They tried several approaches to solving this problem. One of the more elaborate was the use of an instrument flight simulator called a "Ruggles Orientator" located at Brooks Field, Texas. It was called a simulator because it provided pilot trainees with cues thought to be similar to those they would encounter in actual flight, and a control system like that used in controlling the aircraft about its three body axes, in pitch, roll, and yaw. The device used a cockpitlike enclosure suspended on a gimballed framework that allowed the cockpit to move in all three axes of rotation. It also provided a turn indicator and a magnetic compass.

However, it was not until 1929 when Edwin A. Link built a device that more faithfully replicated an aircraft cockpit, instrumentation, controls, and environment of flight, that simulation for training was clearly established. The purpose of the trainer was to reduce the cost of training civilian pilots in the skills of contact or visual flight. The Link trainer later evolved into an instrument trainer and was used to train a half-million military pilots during World War II on the skills involved in instrument flight and navigation.

The Link trainer of World War II was a "generic simulator" because it did not simulate the performance of a specific make or model airplane. It provided the kind of instruments and controls used in most airplanes of its day and some of the sensations of flight, as well as accurate simulation of the electronic environment used by pilots to navigate and communicate. It was used to save fuel and airplane costs and to accelerate the acquisition of critical flight skills. .

After World War II the technology of simulation experienced spectacular growth and other applications were found for a wide spectrum of vehicles, air- and spacecraft, seagoing vessels, industrial processes, and power plants. The purpose of most of these devices was to facilitate training, but other purposes also evolved, including the support of engineering design, research, and assessment of operator performance. Where they were used specifically for training, the training people pressed the manufacturers of simulators for devices that would duplicate as closely as possible the real-world counterpart of what was being simulated. And because of the lack of any persuasive arguments to the contrary, "realism" in simulation was thought to be the key to designing an effective and efficient training simulator. This presumed relationship between training effectiveness and fidelity of simulation contributed to an escalation in the costs of procuring simulators.

8.5.1.2 Functional Definition

Training simulators have two primary functions. First, to present information like that associated with some real system for which training is required. The simulator stores, processes, and displays information reflecting the functional characteristics of the system, the effects of relevant environmental events, and the effects of control inputs made by the operator.

Second, training simulators incorporate special features that facilitate and enhance their ability to support practice and learning for the express purpose of influencing operator performance in the real system which is simulated.

Many kinds of training devices have characteristics in common with simulators, but go by other designations. While definitions of the various kinds of training devices are somewhat subjective and arbitrary, they are useful in conceiving and organizing the suites or hierarchies of settings typically required in the efficient training of many complex skills. Two of the more common ones are the procedures trainer and the system simulator.

One of the lower-level devices in the spectrum of training equipment is the procedures trainer. It contains enough information to permit fundamental procedural control inputs to be reflected in the displays most directly related to the control. But only enough information is provided to permit the operator to observe and learn the basic procedural steps in system operation or checkout.

Systems trainers are more advanced in the degree of simulation they embody, but they deal only with the tasks associated with a specific subsystem with which critical, difficult, or time-consuming training objectives are associated. For example, a radar systems trainer might be used to support practice in the more difficult perceptual skills required in the effectual use of the gain control. The important nonlinearities and discontinuities in brightness and picture quality resulting from the positioning of this control, compensating for the characteristics of the target or target type being viewed, and from the effects of meteorological and other influences impact on learning how to best use this critical control.

Training simulators have six primary characteristics that differentiate them from other training devices and equipment.

Synthetic

Training simulators are synthetic in the sense that they are constructed only to provide task information rather than supporting real operational mission functions.

In-flight simulators are constrained by the same parameters as those associated with normal aircraft but they have synthetic control systems capable of being adjusted to represent aircraft of a different type than that in which they are installed. They are used in research and training applications that require greater levels of realism than are ordinarily required, but are still simulators because their dynamic characteristics are synthetic.

Other synthetic training simulation is achieved by wiring a real aircraft to a computer, which stores environmental information and systems data that respond to the aircraft controls as though it were flying, even though it is fixed to the ground.

Data Storage and Processing

Simulators store data that represent the dynamics of the system being simulated and the task-relevant portions of the environment in which the system (and the system operator) must perform. Data storage is almost entirely in computers, in the form of mathematical expressions, although a variety of other systems are used in the storage of information used in driving displays of real-world visual and auditory information.

When a control input is made, its effects are displayed only after the computer identifies the parameters effected and computes the effects of that control input in relation to other control and environmental conditions.

System Dynamics

Typically, the response of a system to a control input or to some external influence is a function of many factors whose precise influence is important to the student learning to understand, control, and employ the system in its assigned mission. The simulator's store of system and environmental data is organized and processed to reflect the dynamic response of the real system as it is perceived and used by the student in a learning situation.

Controls and Displays

The primary function of the training simulator is to display information representing system performance in response to control inputs. Displays and controls are only represented in abstract form in many low-level training devices, but the simulator incorporates controls and displays that are as real as is possible, to support learning at both the intellectual and psychomotor levels. In the simulator, all elements of the interface with the operator are represented, to the extent that they are expected to influence the learning of assured perceptions and their use in system operation.

The detail characteristics of the system's controls are crucial in the simulator, where they are used in tasks and under conditions where their recognition by sight, location, feel, and unique response characteristics verify their relevance to the task at hand.

In most cases, even an individual display provides more than one kind of information in the operation of the system in which it appears. An altimeter, for example, tells the pilot or the navigator the height above sea level, and rate information representing the speed and smoothness of movement of the pointer must also be represented as accurately as possible.

Finally, the conditions of viewing must be considered. However the instrument is represented, it must provide as much, and not more nor less information than is available in the viewing conditions expected in the aircraft.

In the simulator, all aspects of the system/student interface relevant to learning must be included, but their relevance must be established before the simulator is designed.

Whole Task Support

Training devices are designed to meet specific, clearly defined, training objectives. These objectives may involve cognitive, perceptual, procedural, psychomotor, interactive, or judgmental objectives. Simulators may be used in individual training or in the training of crews or interactive groups. Their primary function is to provide information needed in developing the specific skills at hand.

Simulators are designed generally to support whole-task rather than part-task training. Their flexibility permits them also to be used in support of difficult or critical task elements, however, where it would be impractical to procure specific part-task trainers. The simulator permits the student(s) to practice in the workload, stresses, and time pressures typical of the job for which they are training. It is also the setting in which the individual skills and skill elements to which prior training has been directed are validated and integrated.

Instructional Control

The training simulator, unlike other kinds of simulators, is characterized by a capability for the control of the information that supports practice, for the purposes of facilitating and enhancing learning. Control is achieved through the intervention of an instructor. The instructor may participate directly in the control of the conditions of training, or he or she may be represented in a computer program incorporating his or her experience and that of the training system in controlling training for maximum learning effectiveness.

In essence, the training simulator contains a mathematical model of the system for which training is required and an instructional model that provides the control necessary for learning to take place in the most efficient manner.

8.5.2 FUNCTIONS OF TRAINING SIMULATORS

Training simulators are designed to be used in training in lieu of the systems they simulate. Use of simulators is more safe, economical, and convenient than training in the real system. Simulators can also support training in a broader range of skills and functions than is possible in the real system. In general, training simulators are designed to replace the practice that would otherwise take place in the system itself plus essential practice, not possible in the real system itself.

Simulators have unique training-relevant capabilities that permit them to perform additional functions which make them superior settings for training, practice, and learning. The simulator performs some functions about as well as they would be performed in the actual environment. It performs some much better than they can be performed in the real-world system, and it performs some that are impossible in the real world, but which facilitate learning. Some of the unique instructional functions of the simulator follow.

8.5.2.1 Briefing and Demonstration

Briefings can be provided in a number of different ways in the simulator, depending on the facilities available and the importance and difficulty of the task being trained. The instructor can provide an oral briefing; the simulator can coach the instructor in maintaining a standardized format and content; or it may conduct the briefing automatically, using a tape recording, a voice synthesis system, or a digitally stored and controlled voice system.

Automated briefings have the advantage of standardizing, while reducing instructor workload. They can also be used to individualize instruction, with careful and detailed planning.

Demonstrations may also be conducted by the instructor, from an instructor's normal position in the system, from special controls at the instructor's station, or from the student's position with the student looking on. Demonstrations may also be automated; in most cases, this means that instrument, aural, visual, and motion cues portray the performance of the task without direct human intervention. Demonstrations may be prerecorded in the simulator computer by an expert operator, or a program can be prepared defining an ideal performance of the task. A major advantage of an automated demonstration is in its ability to show the student how the task is to be performed time after time with no deviation. It can also be prepared to demonstrate the effects of specific conditions and influences, one at a time, and in combination. The instructor may also critique the demonstration, pointing out significant features of the task and task environment, or the student may review any demonstration as many times as he or she wishes, within the constraints of the curriculum.

8.5.2.2 Practice

The simulator supports practice by providing the controls and the information used by the operator in performing his or her assigned tasks. The information provided reflects the status and the condition of the system, the effects of the environment on the system, and the effects of control inputs made during practice.

The simulator's value as a practice setting is due to its ability to represent precisely the conditions needed to practice each task and each task element separately if advantageous. Real-world practice is hard to control because of unpredictable environmental influences and, frequently, by the impracticality of scheduling important task conditions. Especially in the early stages of learning, it is important for the student to have the opportunity to experience and, finally, recognize the effects of each control input on the performance of the system, as the student attempts to get it to perform within the required tolerances. It is also important for the student to be able to experience the effects of each influence on system performance, including variations in system characteristics, modes, and conditions, and those aspects of the outside environment that effect its performance.

The simulator is unique in its ability to present exactly the training setting needed to facilitate learning at each stage of the student's development. Properly designed, it is the most flexible and effective practice setting possible.

8.5.2.3 Performance Analysis

While performance and learning are not the same thing, the rate and quality of learning can, to some extent, be inferred from performance, either in the operator's assigned job or in the training scenario, providing a valid and reliable relationship has been established between performance and learning in that task. Performance on the job is measured, usually, by speed, accuracy, and output. Normally, the observation and measurement of operational performance is difficult to relate to the performance or to the skill of the individual because of influences outside his or her control, and because of the inherent difficulty in achieving the required accuracy of measurement, and in measuring important control events that do not relate directly to specific system outputs.

Since virtually all events represented in the simulator involve a mathematical expression or a measurable input to or output from the simulator computer, they are accessible for measurement. In fact, there is so much information available that the primary problem in performance assessment is not in finding relevant information, but in finding and organizing only that information which reflects important elements and aspects of the performance and learning processes to be assessed.

8.5.2.4 Learning Enhancement

The simulator can enhance learning in a number of ways if its inherent capabilities are properly organized and employed. It can standardize practice settings, systematize student exposure to tasks, subtasks, and task elements, and to conditions relating to the development and exercise of operator skills; it can be operated in fast or slow time; it can be stopped to permit an on-the-spot review of performance; and it can be used to replay the essential features of the student's performance for critique and for problem diagnosis and evaluation. Each of these capabilities can enhance learning efficiency and effectiveness, if it is carefully applied.

Many of the simulator's capabilities for learning enhancement derive from its ability to provide immediate and clearly defined performance feedback to the student. Others relate to its ability to standardize the presentation of task conditions, and others result from the ability of the simulator to provide task conditions of graduated difficulty as the student's skill grows.

Knowledge of Results

The development of complex control skills is, in a sense, the development of information processing skills. The simulator provides information defining the status of the system being simulated, and information reflecting the control inputs made by the student as he or she practices his or her assigned tasks. Information resulting from these control inputs permits the student and the instructor to evaluate the appropriateness of the inputs. In the simulator several methods can be used to support the evaluation, in addition to those resulting from the simulator's representation of the real practice environment. Spears' (1982) discussion of feedback and knowledge of results, the report of Caro, Pohlmann, and Isley, (1979) on the design of instructional features, and Polzella's (1983) evaluation of simulator instructional features are excellent sources of information on the exploitation of the simulator's unique ability to provide effective feedback and knowledge of results in support of skill learning.

Frequently, the simulator's representation of a system ensures the same kind of knowledge of results available in real-world practice. Information provided by the simulator can be more effective in the simulator than in the real world, however, because the student knows that the system's performance can be more closely linked to his or her own inputs than in the real world. If a pilot lands short of the runway, for example, he or she knows that the aircraft's performance was not the result of an inadvertent wind gust, but is the result of errors in his or her own performance unless, of course, the instructor has told him or her to be on the lookout for changes in the wind.

Supplementary Cues

Some operator actions do not provide immediate or readily recognizable performance within real-world operations. Frequently, correct and incorrect performances tend to be differentiated by the nature of events taking place over a long period of time or by comments made by an instructor or expert operator who can verbalize the ultimate consequences of the novice operator's performance. In either case, learning is difficult because the operator lacks a direct, recognizable connection between his or her actions and their effects on the system. In some cases, information may be present, but difficult for the student to interpret at that stage in his or her perception of the system and its performance.

The simulator, because of the manner in which it can store, process, and display information, can be programmed to provide immediate knowledge of results that would not ordinarily be available and to enhance what is there for more rapid interpretation.

An experiment by Lintern (1980) demonstrated that artificial cues can be effective in promoting real-world performance and, also, and equally important, that they could detract from proficiency in the real-world task practiced in the enhanced simulator, if the enhancement were used too extensively. The study is reviewed in some detail, by Roscoe (1980).

It is worth noting that poorly designed simulators can inadvertently provide augmenting cues which can degrade learning and transfer of learning. If the click of a switch always accompanies the instructor's introduction of a simulator failure or if a tank beside a tree is always a hostile tank, the student may learn inappropriate responses that can detract from rather than enhance his or her real-world skills. "Poorly designed" in this context means a failure to identify the information needed for the support of learning and/or a failure to design the simulator to provide only that information.

Cognition

The performance of a complex task requires specific patterns of response to specific patterns of information; but complex operator skills require more than the ability to observe, analyze, interpret, and react to stimuli. It is necessary also for the operator to anticipate the responses he or she must make before they are actually required. In many instances, the time required for analysis and interpretation is simply not available.

Normally the operator of a complex system learns to anticipate its requirements on his or her performance by observing its behavior in response to his or her own inputs and to external events. Eventually, as operator skill develops, responses are anticipated through the recognition of patterns of the events taking place over finite periods of time. In the real system it is difficult to assign a particular system behavior to a specific external event or to a specific control input because the external influences are largely unmeasurable at any given point in time. In the simulator, the value of each parameter influencing the system is precisely known at each point in time. As a result, the simulator gives the novice operator an opportunity to experience the effect of each possible influence, one at a time, in any degree, and in various combinations. As a result, the novice can gain a systematic and objective cognitive understanding of the system and its operating characteristics more readily than is possible through exposure to the real system. This can result in more rapid development of skill in anticipating system performance and system input requirements through the enhancement of cognitive learning.

Automation of Instruction

Another function of high-technology training simulators is the support of automated instruction. Modern simulators provide sufficient instructional features and capabilities to put the training process completely under student control, to the extent that trainees can and will take responsibility for their own learning. Higher motivation, a better opportunity to learn what does not work as well as what will work, less dependency on remembering instructor instructions, and receiving better feedback from both correct and incorrect responses all appear to contribute to the apparent effectiveness of automated training.

8.5.2.5 Performance Assessment

The progress of learning and the judgment that learning has progressed to a point where more complex tasks can be practiced effectively are based on the measurement and assessment of the student's performance. When the student practices in the actual system, performance assessment tends to be subjective, partly because the instructor tends to compare overall performance with the performance of qualified operators and partly because he or she rarely has access to specific relevant performance parameters. In some systems, the instructor is as much concerned with safety as with the student's progress, making performance assessment a secondary responsibility.

Performance measurement and assessment are also difficult in the real system because of the unknown effects of variables that neither the instructor nor the student can control. In the simulator, all relevant parameters are under direct control of either the student or the instructor. The instructor sets the conditions of practice and each input made by the student has a known and measured effect. Every parameter relating to the status of the simulated system and its environment and every input made by the student or the instructor is expressed, stored, and processed in the simulator computer as a set of mathematical relationships. The simulator computer, thus, contains all of the information possible about both overall system and operator performance and about each parameter involved in the operator's responses.

Performance assessment performs three functions in the training program. The first rates the performance of the student with respect to the demands of the tasks and the mission for which his or her training is preparing him or her. This is the mission-oriented function of performance assessment and it is usually accomplished by measuring aspects of performance relating to the requirements of the mission of which the operator's task is a part. Mission performance is frequently defined, in turn, by the demonstrated capabilities of acknowledged experts under a variety of circumstances known to effect performance.

The second function of performance assessment is to determine when the student is ready to progress from one training task to another. This process requires insight into the processes by which complex skills are developed. Most simulator training programs can make use of the simulator's ability to measure many parameters, and to correlate these parameters with performance in various tasks and subtasks. Thus, the simulator can be used to develop a body of relationships defining the criteria on which to judge both the quality of performance and the readiness of the student for practice in more complex tasks and conditions.

The third function of the performance assessment process is in identifying and diagnosing the cause of incorrect performance or performance that appears to reflect substandard progress in learning. Typically, instructors develop insights over a period of time from working with students having varying approaches to learning, which help them to diagnose individual performance problems and to identify the specific behaviors associated with incorrect performance. The simulator provides powerful resources, both for storing these insights and for expanding them through the accumulation of longitudinal data and for organizing those data to reflect significant relationships among various performance parameters and the quality of task performance and learning.

Several studies deal in detail with the use of simulators in performance measurement and assessment. Some of the more relevant are listed at the end of the chapter. Note that there are four primary problems in performance assessment. First, the performance parameters available for measurement must be related to the quality of performance in the mission environment. Perfect performance in the simulator, or in any other training setting, is irrelevant if it does not contribute to the ability of the trainee to perform his or her assigned mission. Many performance parameters can be correlated with mission performance; but in many cases the most important aspects of mission performance are rarely, and, in some cases, never accessible for observation, leaving the identification of mission-relevant parameters and performance criteria to a process of inference from experience in similar systems and missions.

The simulator can be valuable in identifying student performance which relates to mission performance if it includes capabilities for simulating the tasks and conditions under which operators perform in their mission assignments. Since the simulator cannot expose the student to the stresses and hazards involved in some missions, conclusions reached in simulator evaluations must be tempered by the experience of operational and training personnel in the system being simulated and in similar systems.

The second problem in using the simulator in performance assessment is in translating the subjective measures used in the evaluation of performance in the real system, whether it is used in training or only in the performance of its assigned responsibilities, into the objective measures of which the simulator is capable. In addition, of course, additional parameters that are measurable in the simulator but not in the real system must be related to the methods of assessment with which instructors are familiar through experience in the real system itself.

The third problem in maximizing the simulator's potential in performance assessment is in correlating observed and measured performance with learning. Much of performance assessment relates to the comparison of student performance with that of an expert; but expert and novice performances in the same task are, almost by definition, different. The novice observes a task situation, recalls a pattern of information defining the situation as it is supposed to be, makes a comparison between existing and desired conditions, postulates a solution, attempts a control input, observes its effect, iterates the process, and modifies the input accordingly. In contrast, the expert observes the same situation, selects a response that has proved to be effective, makes a control input based on that learning, assesses the results, and modifies his or her performance accordingly. The expert does something else that differentiates him or her from the novice—the expert attempts to use all of the pertinent information available in anticipating requirements on his or her performance in the future so that he or she can prepare to make relevant and effective responses without involving the time or the distractions associated with the real-time analysis of task conditions.

It is important that the relationships among performance, learning, and transfer of learning, as discussed in Spears' (1983) report be considered in the development of learning-relevant performance measures.

Finally, the methods by which performance information is displayed and employed in the simulator and in the training program must be carefully considered. So much information is available that it can be overwhelming to the point of being useless. Some information can be displayed effectively in alphanumeric form, some in graphic form; but regardless of the method, the format must be derived from the requirements of the instructor and the training program as a whole. The instructor will always apply subjective judgment in performance monitoring, diagnosis, and evaluation. The instructor may perform these functions directly or through algorithms stored in the simulator computer. The information available in the simulator must be organized to support these functions by revealing important performance and learning trends.

8.5.2.6 Malfunctions and Failures

Although most man/machine systems are gradually becoming more reliable, failures and malfunctions represent a major consideration in the use of simulators for training. Some training in the recognition and correction of system failures and in the unique skills required to complete or modify a mission in the face of malfunctions can be given in the system itself. But, practical, economic, and safety considerations severely limit this practice.

The increased reliability of complex systems has drastically reduced the rate of failure in some systems. At the same time, the consequences of failures, even though they are rare, have increased even more drastically because of the increased reliance being placed on complex systems. Thus, the skills required in responding to failure are increasingly important, but the possibility of exercising (and learning) them in the real world is becoming less.

The modeling of malfunctions and failures is not easy, because it is not always possible to anticipate the failure modes of a complex system, nor is it always possible to induce a system failure for the purpose of data collection. Because of this, system operators tend to learn as much as they can about the design and performance characteristics of a system or subsystem so that when it does fail, they have a chance of deducing the reasons for the failures, and a chance also of generating responses consistent with its behavior and with the completion of the mission.

8.5.2.7 Adverse Operating Conditions

An operator's initial exposure to a system is usually designed to introduce him or her to the knowledge, procedures, and skills required in operating the system under normal system environmental conditions. In fact, in most exposures to the real system abnormal, unusual, and hazardous conditions are avoided to the greatest extent possible. Occasionally, students are exposed to adverse operating conditions when they can be arranged, but only under exceptional levels of control. As a result, the first exposure of many system operators to the conditions in which high levels of proficiency are really crucial is in actual system operations rather than training.

Increasingly sophisticated technologies are making the simulation of adverse conditions more and more effective in providing the skill levels required of system operators. Simulators are being used in training personnel in the techniques required; but they are also being used in developing the techniques to be trained and employed.

In the past few years wind shear has received a great deal of attention owing in part to a series of airline accidents and incidents attributed to rapid and drastic changes in wind direction and velocity during the final approach and landing. Airplanes in the landing phase encounter wind shear so rarely that most pilots will never experience it in their entire career and yet, when it does occur, it can have profound and far-reaching effects, and it requires special techniques that have been developed through the simulation of wind shear situations. These new techniques have now been incorporated in many training simulators, making it possible for pilots to acquire the special and critical skills required in case they are ever needed, and at, essentially, trivial cost.

8.5.3   ADVANTAGES AND BENEFITS

Simulators have many advantages in the training of both new and experienced operators. For many years they were used primarily as substitutes for the real system in the interest of economy, safety, and convenience. In the past few years they have been used increasingly not only as practical substitutes for the operational task setting, but also as a means to the development of essential skills and skill levels that simply could not be attained in many real systems.

Twenty years ago simulators were able to reduce the cost of training; but, then when they were not available, they could for all intents and purposes be replaced with additional practice in the real system itself. Today, many real systems and their missions are too complex to be used as primary sources of training, and they are generally not configured to provide training with any degree of efficiency. The simulator is deliberately designed to enhance learning and to provide practice in tasks that cannot be scheduled for training.

Electronic warfare training is an interesting example of the unique value of the simulator in supporting tasks that cannot be practiced in the real world. First, of course, many of the system's signals and techniques must be protected for reasons of security, and cannot be used in training. In addition, many systems would interfere seriously with normal television, telephone, and radio communication systems. As a result, the simulator is the only setting available to train the complex and critical skills required in electronic warfare operations.

Simulators are also essential in training personnel who must perform correctly on their first exposure to the prime system as well as to critical or difficult system missions and mission functions. Many space systems, for example, cannot be operated for the purpose of training; astronauts must be fully qualified before the system begins its first operational mission. The success of the Mercury and Gemini orbital flights, the Apollo orbital and lunar missions, and the Skylab and Shuttle missions attest to the value of simulation in achieving virtually 100% of the training requirement without use of the system itself.

8.5.3.1   Advantages

Scheduling

One of the more obvious advantages of simulation is in scheduling training. Simulators can be scheduled at any time because the simulator is not affected by weather or by the need to perform operational missions in addition to training. Subject to its own reliability and the availability of instructors, students, and support personnel, it can be scheduled whenever it is needed in support of a training program.

Cost

In general, the cost of simulator training is significantly lower than the cost of training in the prime system itself. Orlansky and his colleagues in the Institute for Defense Analysis have systematically followed the evolution of the use of training simulators over the past several years, especially in military and commercial flight operations. Orlansky (1982) reports that simulator costs average about 8 to 10% of the cost of the operational system when power, fuel, facilities, and maintenance costs are compared. The cost of using simulators in aviation and weapon system training are even lower, in fact, when the cost of systems damaged or destroyed in training and those lost due to inadequate training are considered. In military training, the cost of ranges, facilities, and ammunition is significant. Someone has estimated that it can cost from $10,000 to $100,000 to fire one modern missile system. The same event can be practiced in the simulator for a few cents. Interestingly enough, the simulated event is likely to have much greater training value than the real weapon launch because of the simulator's superior ability to provide the information needed for learning.

Safety

Simulators are inherently safer than most of the systems they simulate. In fact, simulators have two major implications for safety. First, they reduce the exposure of operators and prime systems to the hazards of the operating environment. Second, they provide training in difficult and critical tasks where proficiency is needed in achieving maximum safety in system operation, but where real-world practice is not feasible.

Control of Training Conditions

Environmental control is essential in effective training because the environment, that is, outside influences on the system, can have many significant effects on the system and on the operator. Many important environmental effects having profound effects on the operator and his or her performance are difficult, if not impossible, to be scheduled for training. Weather effects are classic influences on the performance of aircraft, ground vehicles, and ship crews.

Weather or meteorological effects include fog, rain, snow, haze, clouds, ice, icy surfaces, turbulence, winds, hail, waves, and other influences on the operator and the system. They also include air temperature, pressure, density, and humidity. Water temperatures, gradients and layers, water currents, and runway, road, and terrain surfaces are also important in operator training, and once modeled are available in the simulator computer to be used as required to promote the development of complete skills.

Humans can also be considered a part of the operational environment important in the training of individual, crew, and interactive skills. Training involving other operators can be accomplished in a variety of ways. Two or more students can practice together as members of the same crew or in simulations of systems that interact in real-world operations. Individuals and crews can also be trained using inputs from specially trained training support personnel who act as other members of a crew, operators of communication facilities, enemy personnel operating hostile systems, or other organizations supporting or being supported by the system being simulated.

Prerecorded messages and performances can be employed to represent the human part of the operating environment; but a considerable amount of planning is required to preserve the flexibility associated with human participation. Currently, some programs are capable of providing much of the interactivity required in the simulation of personnel inputs by modeling the responses of performers to reflect different approaches to various tactical problems.

Learning Enhancement

Simulators have the advantage of being able to enhance learning and performance beyond levels achievable in real systems when the tasks and missions are reasonably complex. Operators trained in real systems might achieve high levels of skill in normal operations, in normal circumstances, through proper guidance and coaching; but it is difficult and, in many cases impossible, for the operator to learn to deal effectively with any system malfunctions and with many unusual, difficult, and important environmental conditions and mission requirements in the real world.

It is also rare that training in a prime system can be organized for optimum learning. It is difficult for a student in most cases, to practice an important, difficult element of an overall task in isolation from the rest of the task. He or she must cope with distracting elements of the task while attempting to observe and control crucial relationships in other parts. In the simulator the student can concentrate on difficult portions of the task until he or she has mastered them, at which point the student can integrate each part into the whole task as he or she is able.

Performance Enhancement

One of the major advantages of simulation is in the training of tasks that are mission critical, but which cannot be trained in the real world for one reason or another. Some tasks would require too much terrain, fuel, and ammunition, and too many support facilities to be feasible; while others would involve hazards that are unacceptable in the context of training; still others simply cannot be scheduled for efficient practice. The ability of the simulator to support training in each of these circumstances is a major advantage.

8.5.3.2 Benefits

The use of simulators can produce a number of benefits, both to the training organization and to the organizations responsible for the conduct of the missions toward which training is directed. One of the more noticeable benefits is reduced wear and tear on prime equipment. Equipment wear is reduced because it need not be used as extensively for training, but also because, when it is used, its use is less likely to induce wear because the operators are now more skilled than would have been possible without simulation.

Reduced equipment use avoids costs due to maintenance and equipment support. At first glance, that would seem also to reduce the quality of training for maintenance and support personnel. Frequently, much of the training of maintenance personnel is given on the job as the equipment used in training is maintained. However, good maintenance training requires the same systematic approach required in good operator training. Like operator training, it must be based on mission objectives and on good training practice rather than on extemporaneous practice on problems that result from the day-to-day support of operator training. Maintenance personnel must be proficient in tasks associated with operational performance as well as in those resulting from training operations.

Obvious savings in fuel, spares, ammunition, facilities, and other expendables result from the use of simulators in training. Some training aircraft can cost $2000 or $3000 per hour to fly, with a comparable simulator costing $200 or $300 per hour. Caro (1972) showed that in one helicopter training program 53.5 hours of flight training were saved each year by each student through the use of simulators to replace instrument flight training previously given in the aircraft.

Ammunition is associated with some of the more difficult and critical skills required of military crews. A weapon firing 6000 rounds per minute and weapons firing rounds costing hundreds and even thousands of dollars each can expend a training budget in a very short time and, yet, practice in weapon operation is critical in the development and maintenance of the proficiency levels required for combat readiness.

Another important benefit of simulation is in the release of equipment otherwise needed in training, for the performance of operational missions. Many systems used in training, including airliners, ships, tanks, power plants, automobiles, or fighters can also perform nontraining operational missions such as carrying passengers and cargo, standing watch at a border in a high security area, or generating electricity.

Learning efficiency is another benefit of the use of simulators for training. Simulators can reduce the amount of time spent in the training process, provided the same training objectives are attained with the simulators as would be achieved with the real system. In many cases, the introduction of simulators in the training program reduces total training time because the simulator is insensitive to the scheduling problems associated with the prime system that it replaces. In many cases, however, reductions in total training time are not as great as expected. In most training programs, recognition of the simulator's capabilities for training skills that are not able to be practiced in the real system adds training time which would not otherwise be used. Emergency procedures, operations in degraded modes, and operations in adverse and unusual environmental conditions cannot be practiced in the real system. These training functions are usually added to the training program's objectives when the simulator becomes available.

Learning is not only more complete with simulation than without, it tends to occur more rapidly and efficiently because of the capability for using the simulator to control practice for maximum rates of learning. Parts of tasks can be practiced in isolation; instructors are freer to monitor, guide, and critique; instructors have better information on which to base guidance; feedback can be enhanced to promote more efficient learning; learning progress can be measured more accurately and directly; finally, more time can be spent practicing a given task in a simulator period than can be devoted to a specific learning objective in the real system in the same period of time.

In the real system, part of each training period must be used in setting up the conditions required for practice in that particular practice period. In the simulator practice conditions can be instantaneously selected at the touch of a switch. In flight training, in particular, a significant portion of each flight is spent in taking off, flying to and from the practice area, returning, and landing. In some special cases, each of these phases of the flight has value for a number of secondary training objectives, but more often they result in unnecessary practice at the expense of the training objective for that particular practice session.

8.5.4. REPRESENTATIVE PROGRAMS

Training simulators are used whenever the use of the real system is inefficient, expensive, hazardous, inconvenient, or impossible. Much of the current simulation technology was developed in aviation training, but similar technology has been applied in a wide variety of training needs.

8.5.4.1 Aircraft

Modern simulators were first used in the training of airplane pilots. Modern simulation technology received its initial impetus in aviation training owing, in part, to the cost of training in the aircraft, but owing mostly to the inconvenience of training in actual flight and to the fact that the airplane is a confusing setting for the learning of complex skills.

Figure 8.5.1 illustrates a modern commercial airline simulator, which is used to train crews of

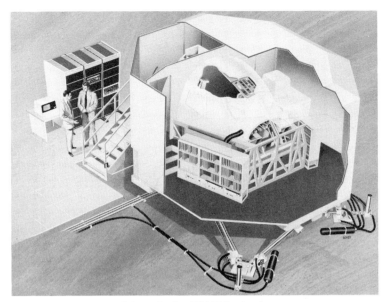

Fig. 8.5.1. Boeing 767 flight simulator.

the Boeing 767 aircraft. Figure 8.5.2 is an artists' concept of another simulator used in military training. In this simulator two crew stations are used to train the pilot and the co-pilot gunner of the Army's AH-1 armed helicopter. This simulator provides training for the pilot and co-pilot gunner independent of each other or as a crew.

Both the 767 and the AH-1 simulators have systems to provide motion information to the crew members, but the AH-1 has seat-vibration systems in addition to the normal 6-degrees-of-freedom cockpit motion system. These provide the unique vibrational cues associated with the helicopter rotor system. Both simulators also provide out-of-the-window visual information to support training in contact flight skills and, in the case of the AH-1, tactical skills.

8.5.4.2 Space

Simulators have been used extensively in the various space programs. They are used somewhat differently in the training of astronauts than in the training of other crews. Pilots, aircrew members, and operators of most systems almost always receive a significant amount of training in the system itself, but the astronaut does not have this luxury: his or her first flight in space is not a training flight, but an operational mission, with virtually no opportunity for further training, and little room for error.

Many of the functions in space missions are automated and computer-controlled, with a great deal of redundancy built into the various systems. The crew's nominal function is to monitor the mission performance of the various systems and to provide manual backup in the event of a system failure. The crews also perform a variety of functions relating to the missions at hand and in several instances have been forced to participate in functions that were intended to be automated.

8.5.4.3 Locomotives

Thousands of locomotive crews are trained each year, and while most of the training is given in locomotives on the job, simulators are also used to systematize and reduce the cost of training. Figure 8.5.3 represents a simulator used in the training of diesel locomotive crews. It contains the instruments and controls used in the locomotive cab, and a visual simulation system displaying the track ahead. The visual system uses a variable-speed, flickerless motion picture projector to provide the image of the track, roadbed and other major features of the engineer's visual scene, as prerecorded on film, on real railroad tracks. The projector can simulate travel at a wide range of speeds, with projector speed varying with the simulated speed of the locomotive.

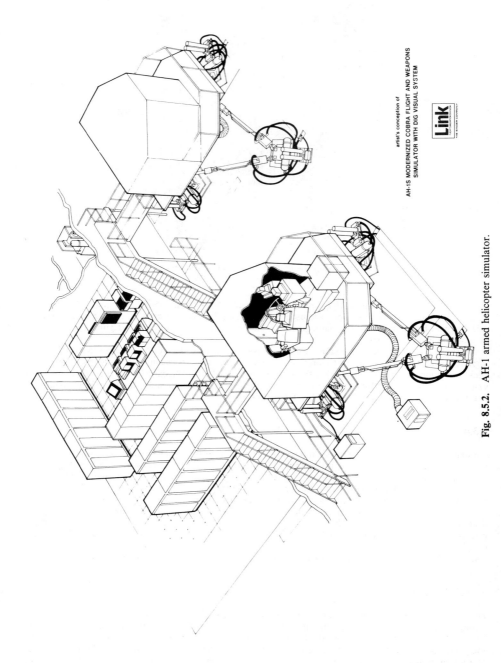

Fig. 8.5.2. AH-1 armed helicopter simulator.

Fig. 8.5.3. Locomotive simulator.

8.5.4.4 Ships

Several kinds of simulators are used to train ships' crews. Some are used primarily for training in steering control, while other specialized simulators are used in training power plant operations, the operation of tactical and weapon systems, and in the control of naval task forces.

Ship steering poses some difficult training and simulation problems. Ships have a great deal of inertia, requiring many miles and many minutes of planned turns. Much of the technology used is effective in drastically reducing operating costs and costs due to damaged ships and lost cargos.

Simulators are also used to train the specialized skills required in submarine and antisubmarine tactics. Much of the technology used in the simulation of the environmental sounds heard and used by system crews was developed to support the training of sonar operators.

In general, environmental sounds are simulated by obtaining under various circumstances recordings of the sounds to be represented, storing the elements of the sounds in a digital computer, and calling them up through the operation of algorithms describing the relationships among sound elements, environmental conditions, and operator control settings.

The submarine periscope, the attack system, and the weapons used by the submarine are simulated to permit crews to practice all of the procedures and techniques involved in detecting, locating, identifying, evading, and attacking submarine and surface targets. Figure 8.5.4 is an artist's concept of a submarine attack team trainer. Antisubmarine training is also given to aircraft and ship crews, using simulators representing the systems and environments involved. The U.S. Navy also uses simulators to train key staff personnel in naval task force operations.

Figure 8.5.5 represents the simulator used in training crews of the P-3C antisubmarine aircraft. The pilot, copilot, and flight engineer are trained in one portion of the device, while the operators of the aircraft's tactical systems are trained in other sections. All crew stations can be interconnected for integrated crew training.

8.5.4.5 Ground Vehicles

Simulators are used in a number of programs for training the operators of ground vehicles. Simple driver trainers have been used for years to teach the procedures, principles, and regulations relating to automobile driving. Some of these expose groups of students to films portraying many different traffic situations. The students sit at replicas of the automobile driver's position making steering, signaling, braking, and acceleration responses as appropriate to the scenarios portrayed in the films. Student inputs do not influence the scenes projected on the screen, but are recorded to indicate the appropriateness of each input and the reaction time required to make it.

More complex interactive driver trainers are also in use, primarily for the training of the crews

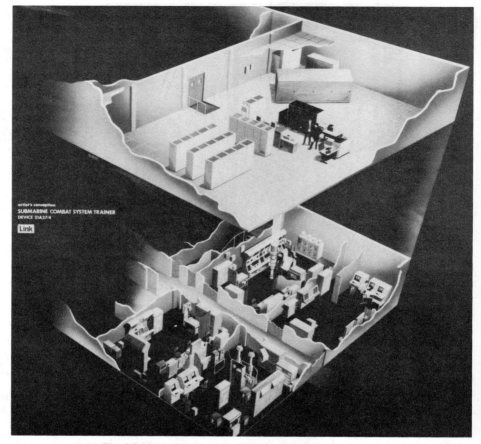

artist's conception
SUBMARINE COMBAT SYSTEM TRAINER
DEVICE 21A37/4

Link

Fig. 8.5.4. Artist's concept of a submarine attack trainer.

of tanks and other military fighting vehicles. The basic driving procedures required in tanks, personnel carriers, and other similar vehicles are relatively simple, but the driving techniques required in combat operations, over a variety of terrain, and in relevant weather conditions are exceptionally complex and difficult.

A number of trainers have been built for training drivers of a variety of armored vehicles. In some armies these simulators are used both for driver training and for driver licensing because of severe limitations on terrain for manuevering by student drivers and because of the superiority of the simulator for training and testing.

Most current tank driver trainers provide visual imagery by means of a camera model system. A lens and a mirror arrangement is held just above the surface of the terrain model on a probe that is in contact with the model surface. The mirror reflects light from the model to the lens, which relays it to the television camera. The camera transmits the model picture to a display system at the simulator crew station. The probe/lens/mirror system is driven around the model by the simulator computer in response to inputs made by the driver's steering, gear shift, accelerator, and brake controls. Changes in the attitude of the probe as it moves over the terrain model produce changes in the attitude of the crew compartment, through its motion system. The tank driver trainer is a good example of a simulator capable of meeting training and mission objectives that cannot be met through practice in the real system itself. Figure 8.5.6 illustrates the major components of a driving simulator for a tank.

8.5.4.6 Industrial Plants

Over the past 30 years more and more simulators have been used in the training of operators of industrial systems. One of the most notable applications has been in the power industry, where training simulators are used in the qualification, testing, and licensing of operators of electrical power generation

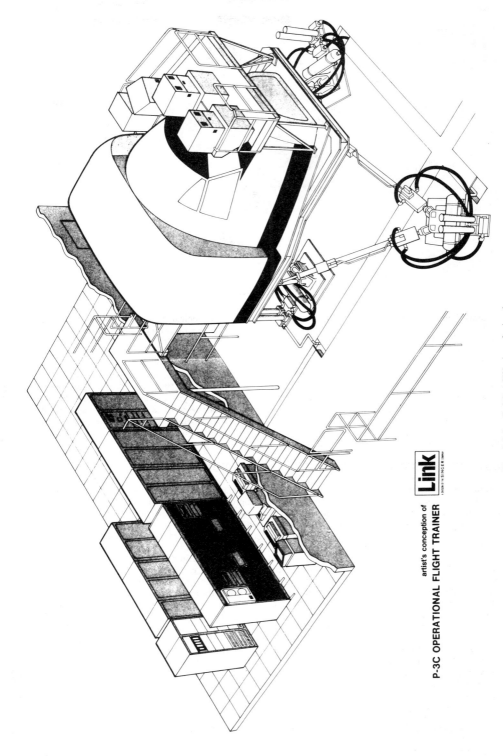

artist's conception of

Link

A DIVISION OF SINGER COMPANY

P-3C OPERATIONAL FLIGHT TRAINER

Fig. 8.5.5.   P-3C ASW aircraft simulator.

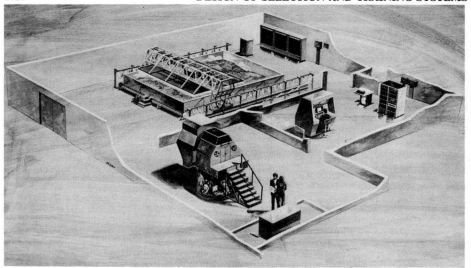

Fig. 8.5.6.   Tank driver simulator.

systems. Simulators have been developed for both fossil-fuel- and nuclear-powered electrical generating systems. These systems have been used in training power plant operators without disrupting normal operations and without inducing any of the abnormal conditions requiring high levels of skill and judgment.

Simulators are also used in the training of operators of industrial processing systems. Simulators of chemical processing and oil refining systems are being used, both to train operators and to evaluate various approaches to the processes being simulated. Rafael and Joaquin (1984) describe the development of a simulator used in the training of operators of a petrochemical processing unit. The simulator is being used to replace on-the-job training in the field because of the rarity with which unusual events occur; it takes about 10 years of field experience to qualify an operator to operate the plant alone. The simulator will reduce the training time required and will expose operators to more situations than they could ever encounter in on-the-job training.

8.5.5   ESTABLISHING THE REQUIREMENTS FOR A TRAINING SIMULATOR

Simulators can contribute significantly to the quality of many kinds of training programs, but, for a variety of reasons, their use is by no means always appropriate. The initial decision on whether or not to use a simulator is usually made on the basis of a general understanding of its potential contribution and its potential cost. Once a tentative decision is made to use a simulator, its specific capabilities and characteristics are established through systematic analysis of the training requirement and of the overall structure of the program itself.

Many training programs are most effective and economical in a mix of real equipment simulators and other less complex training devices and settings are used. In general, the structure of a training program is dependent on the kinds of skills to be developed, the variety of training objectives to be met, the ability of the prime equipment to support important training requirements, and the number of operators to be trained in a given period of time. Each training setting can contribute to training; but each also incurs development, operating, and support costs, which must be balanced against its contribution to training cost and, ultimately, to the effectiveness of the system for which training is to be provided.

The decision to use a simulator in some programs of training is made quite readily. When the operational equipment is not available for training and the required training is very complex, the decision is easily made. In most situations where high-technology simulators are being used today, no complex analyses are needed to establish the simulator requirement. In other cases, the simulator requirement is based partially on schedule problems, student loads, production means, saving of scarce resources, and safety considerations. Here a more complex cost/benefit analysis is required.

8.5.6   DERIVATION OF SIMULATOR CAPABILITIES AND CHARACTERISTICS

The initial approach to determining the need for a simulator is to start with an examination of the operational problems that must be solved through training. However, biases or oversights in describing the operational problems, errors in determining what aspects of the operational problems can be resolved

real scenes are stored on a disk much like earlier systems stored images on motion picture film. The difference is that each of the thousands of frames on the videodisk can be located, accessed, and displayed in a very short time, making it possible to display a sequence of frames like those stored on motion picture film, but in any sequence required rather than only in the sequence in which they were stored.

Digital computers are being used increasingly to store and process visual information. These systems display images on a cathode ray tube and use a combination of caligraphic technique and a raster scan method to display objects having three-dimensional attributes, as well as those seen as light sources.

Currently, the most sophisticated method for simulating real-world visual information, considering scene content, image complexity, gaming area, and overall flexibility, is the computer-generated or digital-image generation system. Expressions representing the location of each point in the visual scene are stored in a data base, along with algorithms and tables describing the conditions under which these points will be displayed, the edges to be drawn among points to create faces, and the colors and shadings to be applied to those faces. Currently, systems are being developed to add texture to the displayed faces and to incorporate images of real scene elements within the computer-generated picture.

The most important consideration in the simulation of any kind of task-related information, whether visual, aural, or olfactory, is in the dynamics of the task to be trained and the relationship between those dynamics and the information needed for learning.

Typically, visual simulation requirements are expressed in terms of resolution, color, and field of view. First, how much detail must be provided to the operator trainee? Second, how much of the trainee's field of view must contain visual information; and, finally, does the scene need to be in color to be effective in training? The answers to at least 10 other questions must also be resolved in designing a visual system for a specific training function. Each relates to the specific training requirement at hand, while profoundly impacting the engineering approach to be employed, and its cost.

1. Does the task to be trained require that the operator sense self-movement through the visual scene or can he or she learn effectively from a single point of view? How fast must he or she move through the scene?

2. Does anything in the scene need to move in relation to the operator or in relation to other objects in the scene?

3. What density of scene elements is required by the task to be learned?

4. If the scene must be colored, how many hues, saturations, and shades are required?

5. What range of scene brightness is needed?

6. Must the sun and/or its shadow effects be included?

7. Must the system represent night or day scenes, or both?

8. Must objects in the scene occlude or hide others?

9. Does the task require the simulation of meteorological or other special effects such as smoke, dust, haze, rain, snow, weapon flashes, or missile flares?

10. How quickly must the scene change in response to operator inputs?

Few, if any, of the imaging sensor systems can be simulated effectively without a definition of the behavioral objectives associated with their use. A detailed analysis of the image elements and relationships needed in performing and learning the required behaviors is essential because of the unique manner in which information is provided, and the unique ways in which it varies in response to operator control inputs and environmental changes.

Motion Simulation. Each other sensory system must also be considered in the simulation of systems and their environment for operator training. Figure 8.5.7 illustrates some of the major sensory systems that appear to be involved in operator performance and skill development. Operators of vehicular systems, and other operators whose tasks involve gross head or body movements, appear to make use of information provided by a variety of sensory systems located in the inner ear, skin, viscera, muscles, and joints. Borah, Young, and Curry (1977), Zacharias (1978), and others have developed sophisticated models describing the interactions of these systems in the support of human performance, but their involvement in learning is not completely clear. An annotated bibliography prepared by Puig, Harris, and Ricard (1978) is an outstanding source of information on research in the significance of motion and its simulation. A number of other relevant reports are also listed at the end of the chapter.

In addition to providing alerting information. The nonvisual systems provide signals that orient the eyes, apparently for more accurate interpretation of the visual scene. Interconnections among visual and nonvisual sensors suggest that most preceptions are the result of the integration of a variety of information, and there is evidence that serious perceptual problems can arise when only incomplete

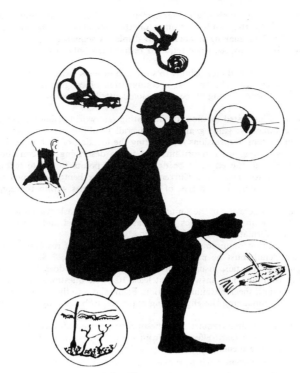

Fig. 8.5.7.   Human sensory systems.

patterns of information are available. McCauley (1984) has edited the proceedings of a workshop on simulator sickness, which is one of the more noticeable results of incomplete simulation of a complex, moving control environment.

Instructional Features

The simulator has many advantages over the real system as a setting for practice, training, and learning. A number of instructional features have been developed to exploit these unique capabilities, based largely on what instructors would like to do to improve real-world training if they were able. While a fairly standard set of instructional features is available, it is crucial that they be designed to support and enhance the instruction being given in the simulator. If they are included in the simulator design simply because they are possible, their overall effect, in addition to adding cost, may range from positive to decidedly negative. Each instructional capability must be designed with a specific goal in mind relating to the enhanced performance of the simulator as an instructional system.

Caro, Pohlmann, and Isley (1979) provide a comprehensive definition of the special features used in simulators. They also provide guidelines in the design and application of 12 commonly employed instructional features.

8.5.6.4   Simulator Support Requirements

An appreciable period of time elapses between the issuance of a simulator specification and the availability of the simulator for training; it frequently takes more time to plan and construct housing for the simulator than to build and deliver the simulator itself. As a result, it is necessary to begin planning for the simulator building as soon as it is apparent that the simulator will be included in the training program.

Simulators make unique demands on their housing and support facilities. Some of these demands are common among various kinds of simulators and training programs while others are specific to specific devices and systems. Virtually all simulation equipment makes use of digital computers, which operate within a narrow band of temperature and humidity. The heat emitted by the simulator and the ancillary equipment to be housed with it must be calculated for worst case conditions. The influence of personnel occupying the building must also be estimated to support planning for heating, air conditioning, air flow, and humidity control.

The consumption of electrical power varies among simulators depending on the types of systems they include, but it is necessary to plan not only for the level of power consumption expected, but plans must also be made for regulating power to avoid fluctuations, which could damage the simulator computer or its programs. Emergency power systems may also be required to retain computer programs during main power failures.

Many vehicle simulators use hydraulic systems to provide appropriate motion of the operator's work station and to provide proper "feel" in some of the simulated control systems, These require power and, in addition, isolation or insulation to shield student, operating, and maintenance personnel from the noise produced by pumps and valves. Pump and valve noise can be distracting and damaging to the ears; it may also provide unwanted information, capable of disrupting learning. Provisions must also be made for the housing of other functions relating to the use, operation, and maintenance of the simulator.

Provisions are required for conducting pre- and posttraining briefings. They may take place at the simulator instructor station, or provisions may be made for using separate briefing rooms, and for including some of the information normally available at the instructor's station in the briefing room.

Crew stations with motion systems impose special requirements for the construction of floors and for the design of normal and emergency access to and from the crew station. Hydraulic motion systems tend to make the crew station inaccessible from the floor, requiring easily deployable access systems. Special egress systems are also needed, since the motion system may malfunction in something other than the normal access position. Simulators tend to be used on a much more compact schedule than real systems when they are used in training, partly because they are relatively free from environmental and operational constraints. As a result, time, facilities, and access for maintenance must be carefully planned. Where possible, provisions should be made to permit simultaneous training and maintenance operations. Not all simulator systems are used in all training sessions, and, with proper planning, they can be checked and repaired without interfering with training.

8.5.7 SIMULATOR DESIGN SPECIFICATIONS

Virtually all simulator designs are governed by one or more formal specifications. Although they vary greatly in detail and content, the function of these specifications is to establish design criteria ensuring that the simulator has characteristics capable of fulfilling its mission as a device for training essential skills. Some specifications relate directly to the fidelity and instructional characteristics of the simulator. Most, however, are concerned with the standardization of design and construction practices and with the assurance of reliability, maintainability, quality, and simulator testing. Both commercial and military simulator designs are governed by written specifications; but, in general, military specifications and standards are more detailed and all-inclusive. As a result, they tend to provide more insight into the specific functions performed by the specifications.

In most cases, the simulator specification is prepared by a customer agency responsible for specification development; but, in some instances, the specification is developed in a contract to a commercial organization. In either case, the basic system engineering and specification standards apply in ensuring the systematic and comprehensive consideration of all parameters capable of influencing the system design.

Dees (1984) has suggested that simulator specifications should define the performance required of the simulator in training, rather than the characteristics assumed to be required in achieving that performance. He has suggested also that the simulator manufacturer be given the responsibility—and the freedom—to design for performance, and that he or she be rewarded or penalized based on the ultimate performance of the simulator in achieving its mission. The measurement of simulator effectiveness is not a simple nor a straightforward process; but the concept of performance-based specifications must be explored. Performance-based specifications have the potential for improving the effectiveness of training systems, while at the same time reducing their cost. Currently, more time, effort, and money is expended in developing, testing, and refining specified design features than in ensuring the contribution of the simulator to training. The use of performance specifications would encourage the objective evaluation of design features by rewarding designs capable of improving training, and deleting those whose value is marginal or indeterminant.

8.5.8 COST AND TRAINING EFFECTIVENESS EVALUATION

The training simulator provides settings in which the skills required of system operators are developed, refined, maintained, and modified. The value of the simulator is in its ability to contribute to these skills and ultimately to the performance of the system mission.

The direct measurement of simulator training effectiveness is relatively straightforward in some applications, but virtually impossible in others. Adams (1978) has reviewed some of the problems involved in evaluating simulator training effectiveness and has suggested some principles which can be applied in evaluating and ensuring the effectiveness of training simulators. Payne (1982) has also developed a set of practical guidelines for conducting the transfer of training studies needed in evaluating training simulators as has Bickley (1980).

Perhaps the most relevant insights are provided by Adams (1978) who suggested that there are two main methods for the evaluation of flight simulators—the transfer of training method and the rating method. He also notes that each is flawed and suggests that five major principles must underlie the design of modern flight simulators. In general, the five principles represent five simulator design criteria that relate to the way in which complex skills are learned and transferred from one setting to another. They directly address the major problem in designing for and evaluating simulator effectiveness—the problem of measuring system performance.

Training simulators can be valuable and effective in three primary ways. First, use of the simulator can reduce the cost and increase the efficiency of training by replacing the real system.

Orlansky (1982) has summarized most of the studies bearing on the cost-effectiveness of simulators in commercial and military training. The identification of all of the elements contributing to the cost of training, whether in simulators or in the real system itself, has been a persistent problem in comparing the cost of programs with and without simulators, particularly in military training programs. Typically, the cost of each element is allocated to a different budget with system procurement costs paid from one account; spares from another; fuel from another; and housing, pay, and allowances from still other accounts. A paper by Knapp and Orlansky (1984) provides a comprehensive structure outlining all of the cost elements relating to military training programs.

The second general contribution of training simulators to training programs and system missions is somewhat more difficult to assess and to predict. Simulators have a direct impact in reducing the hazards associated with training. Some hazards are associated with the utilization of almost all operational systems; but virtually all simulators are safer to operate for training purposes than real systems. A simulator such as the one described by Hard and Guest (1980) not only provides extremely economical training, but it avoids exposure of the crews, the ship, and the adjacent facilities to the hazards of real-world operations.

A third benefit of simulation, which is also difficult to measure and document, is the benefit in skills that are essential to normal system performance, but that cannot be practiced in the system itself. This goes beyond the value of the simulator in training skills needed in dealing with malfunctions and adverse conditions in that it concerns the missions, functions, and tasks assigned to operators that may never be performed in actual real-world operations, but that are a part of the operational mission of the system for which training is given. Again, the systematic measurement and prediction of training value are totally impractical, even though it is obvious with careful analysis that the skills are difficult and critical and, thus, that they must be trained.

Simulators have been used for training for many years. They have succeeded in reducing the costs and hazards associated with training in "real" systems and environments. They have also been able to provide better control over learning than would otherwise be possible.

When a decision has been made to use a simulator in the training program, its ultimate value will depend greatly on its ability to support learning and, only indirectly, on its ability to represent the real system and its environment. The simulator is a method of providing information the student needs in practicing and learning his or her job. The entire process of simulator design, development, and use must thus be oriented around defining, displaying, and controlling the information needed to support these processes.

REFERENCES

Adams, J. A. (1978). On the evaluation of training devices. Convention of the American Psychological Association, Toronto, Canada, 1978.

Bickley, W. R. (1980). Training device effectiveness: Formulation and evaluation of methodology (U.S. Army Research Institute, ARI-RR-1291).

Borah, J., Young, L. R., and Curry, R. E. (1977). Sensory mechanism modeling (Air Force Human Resources Laboratory, TR-77-70).

Caro, P. W. (1972). Transfer of instrument flight training and the synthetic flight training system (Hum RRO pp. 7–72). Human Research Office, George Washington University, Alexandria, VA.

Caro, P. W., Pohlmann, L. D., and Isley, R. N. (1979). Development of simulator instructional feature design guides (Seville Research Corp. Technical Report TR-79-12).

Dees, J. W. (1984). Performance specifications for flight simulator procurement contracting. Proceedings of the Interservice/Industry Training Equipment Conference, Washington, DC.

Hard, D. A., and Guest, F. E. (1980). A review of Marine Safety International's first 3 years of marine simulator training. Summer Computer Simulation Conference, Seattle, Washington.

Knapp, M. I., and Orlansky, J. (1984). A life cycle cost structure for defense training programs. Proceedings of the Interservice/Industry Training Equipment Conference, Washington, DC.

Lintern, G. T. (1980). Transfer of landing skill after training with supplementary visual clues. *Human Factors, 22,* 81–88.

McCauley, M. E., Ed. (1984). Research issues in simulator sickness: Proceedings of a workshop. Washington, DC: National Research Council, Committee on Human Factors.

Orlansky, J. (1982). Cost effectiveness of military training. Alexandria, VA: Institute for Defense Analysis.

Payne, T. A. (1982). Conducting studies of transfer of learning: A practical guide (Air Force Human Resources Laboratory TR-81-25).

Polzella, D. J. (1983). Air crew training devices: Utility and utilization of advanced instructional features (Air Force Human Resources Laboratory Report TR-83-22).

Puig, J. A., Harris, W. T., and Ricard, G. L. (1978). Motion in flight simulation: An annotated bibliography (U.S. Naval Training Equipment Center, IH 298).

Rafael, C. L., and Joaquin, A. G. (1984). Development and implementation of a training simulator for Mexican operators of petrochemical units. *All About Simulators,* Norfolk, VA: The Society for Computer Simulation, pp. 18–20.

Roscoe, S. N. (1980). *Aviation psychology.* Ames, IA: The Iowa State University Press.

Semple, C. A., Cotton, J. C., and Sullivan, D. J. (1981). Air crew training devices: Instructional support features (Air Force Human Resources Laboratory, TR-80-58).

Spears, W. D. (1983). Processes of skill performance: A foundation for the design and use of training equipment (Seville Research Corp. Technical Report TR-82-06).

Zacharias, G. L. (1978). Motion cue models for pilot-vehicle analysis (Air Force Aerospace Medical Research Laboratory, TR-78-2).

Reading List

Berg, T. E., and Oeveraas, P. (1984). Simulator training and applied research at the Maritime Training and Research Center. *All About Simulators.* Norfolk, VA: The Society for Computer Simulation.

Borah, J. and Young, L. R. (1983). Spatial orientation and motion cue environment study in the total in-flight simulator (Air Force Human Resources Laboratory, TR-82-28).

Clutz, D. A. (1976). Flight simulator design—The aero data problem. Proceedings of the Society for Advanced Learning Technology, Washington, DC.

Crampin, T. (1984). Transport delays, texture and scene content as applied to helicopter visual simulation. Conference on Helicopter Simulation, Federal Aviation Administration, Atlanta, GA.

Cross, P., and Olsen, A. (1982). TRIAD—An approach to embedded simulation. Proceedings of the Interservice/Industry Training Equipment Conference, Orlando, FL.

DeMaio, J., Bell, H. H., and Brunderman, J. (1983). Pilot-oriented performance measurement. Proceedings of the Interservice/Industry Training Equipment Conference, Washington, DC

Flexman, R. E., Matheny, W. G., and Brown, E. L. (1950). Evaluation of the school link and special method of instruction in a 10-hour private flight training program. *University of Illinois Bulletin,* Vol. 47, No. 80.

Gamache, R. A. (1982). Simplified control of the simulator instructional system. Proceedings of the Instructional Features Workshop, Naval Training Equipment Center. NTEC-IH-341.

Hagin, W. V., Osborne, S. R., Hockenberger, R. L., and Smith, J. P. (1981). Handbook for operational test and evaluation (OT&E) of the training ability of Air Force air crew training devices. Seville Research Corp. TD 81-09, Vol. I, II, and III.

Holman, G. L. (1979). Training effectiveness of the CH-47 fight simulator (U.S. Army Research Institute, ARI-RR-1209).

Hutchinson, J. (1984). The helicopter simulation data problem? Conference on Helicopter Simulation, Federal Aviation Administration, Atlanta, GA.

Klehr, J. T. (1983). A 4-dimensional thunderstorm model for flight simulators. Proceedings of the Interservice/Industry Training Equipment Conference, Washington, DC.

Kron, G. J. (1975). G-seat development (Technical Report, Air Force Human Resources Laboratory, ASUPT-61).

Mayer, G. B., Jr. (1981). Determining the training effectiveness and cost effectiveness of visual flight simulators for military aircraft. Master's Thesis, Naval Post-Graduate School, Monterrey, CA.

Monroe, E. G., Ed (1977, 1981, 1984). Proceedings of the IMAGE Conferences, Air Force Human Resources Laboratory, Williams AFB, AZ.

Perez, J. M., and Sebastian, M. A. (1984). Simulation of forming processes by computer. *All About Simulators.* Norfolk, VA: The Society for Computer Simulation.

Ricard, G. L., Ed. (1982). Proceedings of the instructional features workshop (Naval Training Equipment Center, NTEC-IH-341).

Ricard, G. L., and Puig, J. A. (1977). Delay of visual feedback in aircraft simulators (U.S. Naval Training Equipment Center, TN-56).

Setty, K. S. L., Epps, R., and Meara, E. (1984). Design and development of user-friendly instructional systems. Proceedings of the Interservice/Industry Training Equipment Conference, Washington, DC.

Stark, E. A., and Wilson J. M., Jr. (1973). Visual and motion simulation in energy maneuvering. AIAA Visual and Motion Simulation Conference, Palo Alto, CA.

Terrass, M. S. (1980). Nuclear power plant simulators after TMI-2. Summer Computer Simulation Conference, Seattle, Washington, DC.

Turner, R. E. (1983). NASA/MSFC FY83 atmospheric research review (NASA CP-2288).

U.S. Department of the Air Force (1973). *Handbook for designers of instructional systems.* Air Force Pamphlet 50-58, Vol. II.

U.S. Department of the Air Force (1978). Handbook for designers of Instructional systems. Air Force Pamphlet 50-58, Vols. I, II, IV, V, and VI.

U.S. Department of the Air Force (1979). *Instructional system development.* Air Force Manual 50-2.

U.S. Department of the Air Force (1983). Cost analysis; USAF cost and planning factors (AFR173-13).

U.S. Department of Defense (1968). Military specification, training devices military; general specification for (MIL-STD-23991C).

U.S. Department of Defense (1968). Military standard, specification practices (MIL-STD-490).

U.S. Department of Defense (1979) Military specification, human engineering requirements for military systems equipment and facilities (MIL-H-46855B).

U.S. Department of Defense (1981). Military standard human engineering design criteria for military systems, equipment and facilities (MIL-STD-1472C).

U.S. Naval Training Equipment Center (1983, 1984). Proceedings Of the Interservice/Industry Training Equipment Conference.

Warrick, M. J. (1955). Effect of transmission-type control lags on tracking accuracy (Air Force Medical Research Lab, TR-5916).

CHAPTER 8.6

DESIGN OF JOB AIDS AND PROCEDURE WRITING

ROBERT W. SWEZEY

Behavioral Sciences Research Center
Science Applications International Corporation
McLean, Virginia

The author wishes to express his gratitude to the following individuals for providing review reports and documents for inclusion in this chapter: G. R. Purifoy and R. P. Joyce of Applied Science Associates, Inc.; T. R. Post of Biotechnology, Inc.; and R. J. Smillie of the U.S. Navy Personnel Research and Development Center.

8.6.1 INTRODUCTION

As systems become more complex, it becomes increasingly apparent that exclusive reliance upon individual skills and memory in order to assemble, operate, and maintain complex systems is futile. As anyone who has ever attempted to assemble even such a simple system as a 10-speed bicycle will readily testify, typical manufacturer-supplied instructions and diagrams are inadequate. The deficiencies exhibited by typical printed instructions provide eloquent testimony that considerable room for improvement exists in the area of operational job aids. In this chapter, recent work in the area of job aids and procedural documentation are discussed, and various techniques and principles for developing job aids are presented. Finally, discussion is devoted to current and future trends in the area of procedural job aid development.

The term "job aid" has been variously defined. Over 20 years ago, Wulff and Berry (1963, p. 273) defined a job aid as "something which guides an individual's performance on the job so as to enable him to do something which he was not previously able to do, without requiring him to undergo complete training for the task." More recently, Rifkin and Everhart (1971, p. 2) defined a job aid as a "device or document which stores information required by a person to perform a particular operation or class of operations and which makes the information available for use on the job." Swezey (1975, p. 45) has referred to job aids as "devices which are designed to increase the human capacity for information storage and retrieval. They reduce not only the amount of decision making necessary to perform a task, but also the need for human retention of procedures and references."

All such definitions have in common the suggestion that job aids are designed to extend human capability to store and process information. Job aids may assume a variety of formats including procedural manuals, color-keyed overlays, microfiche, specialty slide rules, code books, tables, flowcharts, computer printouts, checklists, and audio messages. Such devices have proven beneficial for jobs that involve calculation, stringent memory requirements, speed, accuracy, and difficult decisions or multiple judgments, such as those that are required for troubleshooting. They also appear appropriate for boring and/or repetitive tasks as well as for tasks involving sensory overloading or underloading. In addition to the variety of formating possibilities that are available for job aids, they typically involve at least four additional advantages. First, they are generally based directly on an analysis of what the intended user must do in performing his or her job or task. Second, they contain what-to-do, how-to-do, and when-to-do-it instructions relating to those tasks. Third, they are capable of covering all tasks in a job, such that all decisions can be made using an appropriately designed job aid. Fourth, they tend to present information in small chunks and are designed to alleviate problems involving retention of long procedures in short-term memory (Swezey and Pearlstein, 1974).

Rifkin and Everhart (1971) have suggested that job aids differ in known ways from other systems elements designed to support human performance, such as tools, training aids, and special equipment features. Job aids differ from tools in that tools do not generally store information and make it available on the job. According to Rifkin and Everhart, the primary use of tools is to extend human abilities to (1) manipulate physical objects (e.g., a screwdriver, a wrench, etc.), (2) direct or guide physical objects (e.g., a straight edge), and (3) measure and/or sense environmental characteristics (e.g., yard sticks, voltmeters, etc.). Job aids differ from training aids in that training aids are documents or devices designed to encourage *learning* of particular skills or knowledge, whereas job aids are designed to assist in the *performance* of work activities in the work environment; and job aids differ from special equipment features, such as differently shaped control knobs (intended to allow operators to distinguish one from another by feel), since such equipment features do not, in and of themselves, store and present information.

It has also been suggested (Swezey, 1977; Swezey and Pearlstein, 1974) that the development of job aids has generally focused on tasks which serve a particular role, namely, following procedures. Such a role is, of course, a primary component of many maintenance and/or troubleshooting tasks, and the development of job aids has historically been centered around this context. Job aids may, however, serve in other roles including cueing, as aids to association, as analogs, and as examples. Job aids that serve a procedural role typically provide step-by-step directions for accomplishing an objective, whereas those whose purpose is to provide performance cues are designed to provide signals to act in a specific way without a complete step-by-step set of directions. Checklists, for example, may serve a cue role since in the typical checklist application the user already knows what to do and the brief items on the checklist merely remind him or her of the order in which to do them (Geis, 1984). Sometimes job aids serving a cue role are designed to direct attention to certain characteristics of information, objects, or situations. For example, highlighting, arrows, underlining, circles, and so forth are techniques that tend to focus a user's attention on a specific requirement of importance in a task. Job aids serving an associative role are often designed to enable the user to look up data relating to an existing piece of information. A code book or a graph for converting nautical miles to statute miles are examples of associative aids. Analog aids are those which are designed to provide a means of displaying information that cannot be presented directly. A schematic diagram, for instance, may be considered an analog aid. Aids serving as examples typically are designed to illustrate responses required for completion of a task. A sample form with filled in data, for instance, is a job aid serving an example role.

In the development of the job aid concept, several interesting studies are worthy of mention. One involved the preparation of a job aid for sophisticated radar system (Elliott and Joyce, 1968). The aid was designed for use by low-aptitude novice maintenance technicians. When the development was completed, additional studies were undertaken to determine the time saved by technicians who used the job aid versus those who used the conventional manuals which already existed for the system. In this study, novice technicians using job aids were capable of reducing the time required to isolate and correct malfunctions by as much as 50% over the time required of highly trained technicians using conventional procedures.

A second study compared the performance of high school seniors using job aids to that of highly trained electronics technicians using standard manuals (Elliott and Joyce, 1971). In this study, the high school students were given 12 hr of training on identification of electronic components, use of a volt ohmmeter, and basic soldering procedures. The technicians had considerable formal training in electronics, the majority having three to six years of field experience. Both groups were assigned tasks involving identification, isolation, and correction of malfunctions in complex electronic equipment. Results indicated that the students using job aids outperformed technicians in every phase of the study. One subject in that experiment, who was in the skilled technician group, required approximately 12 hr to troubleshoot a doppler radar system, performing 133 steps and referring to 41 different sections of eight separate manuals. A high school student, using fully proceduralized performance aids, required only five hours, performing 35 steps and referring to only three different sections in three manuals. It appears, thus, that considerable data exist to testify to the potential benefits of well-designed job aids in reducing the need for extensive theoretical training on maintenance tasks. Figure 8.6.1 shows

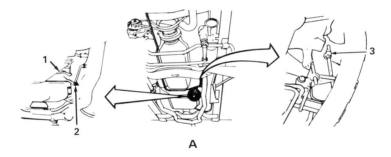

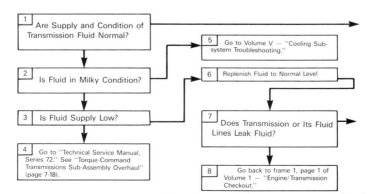

Fig. 8.6.1. Sample dual-level performance aid: troubleshooting procedure for automobile transmissions.

an example of a typical dual-level (i.e., two levels of detail of prescription of aided information) job aid for troubleshooting an automobile transmission.

Folley (1975) has discussed the issue of allocating human performance requirements exclusively to training versus to job aids. Folley hypothesized that if we allocate the entire human performance requirement to training, we may achieve a relatively low level of performance because of the large amount of information to be learned in complex skill areas. Similarly, if we allocate the requirement entirely to job aids, we may also get a relatively low level of performance effectiveness owing to the fact that the performer may become tied to the aid (i.e., if the aid contains an error, the performer gets lost). In any case, in such an aid-dependent situation, a performer must follow the prescription specified by the aid extremely closely; no other way exists for him or her to solve the problem. Folley has suggested that on various gradations between these two extremes we may see an increase in effectiveness up to a maximum point at some optimal mix of this function between training and job aids. Concerning cost, Folley has suggested that if an entire performance requirement is allocated to training, cost may be relatively high because training may be quite long and complicated. Similarly, if complete prescription is included in the job aid, cost may also be relatively high because of the large amount of engineering and analytic effort required to generate and validate the procedures contained in the aid. At various gradations between these two extremes, however, Folley theorized that cost may dip, thus providing appropriate parameters for cost/performance tradeoff considerations. These hypothetical relationships are shown in Fig. 8.6.2.

Smillie (1985), following guidance presented by Joyce, Chenzoff, Mulligan, and Mallory (1973b), has listed several factors as worthy of consideration in determining training versus job aid tradeoffs as follows:

1. *Communication.* Allocate to training tasks that are hard to communicate through printed words (i.e., difficult adjustments); allocate to job aids tasks that would benefit from inclusion of tables, graphs, and charts, and so forth.

2. *Criticality.* Allocate to training tasks in which consequence of errors are serious (such as emergency procedures); allocate to job aids tasks that require verification of readings and tolerances.

3. *Task Complexity.* Allocate to training tasks with difficult procedures that can be achieved only through practice; allocate to job aids tasks that require long and complex behavioral sequences.

4. *Time.* Allocate to training tasks with response rates that do not permit reference to printed instructions, such as a reaction to an emergency; allocate to job aids long tasks or those that require attention to detail.

5. *Frequency.* Allocate to training tasks that are easy to learn via on job experience; allocate to job aids tasks that are performed rarely.

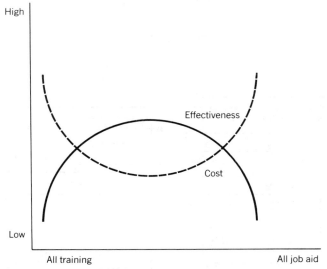

Fig. 8.6.2. Hypothetical cost and effectiveness curves as a function of allocating a performance requirement to job aids versus training [adapted from Folley (1975)].

6. *Psychomotor Aspects of the Task.*   Allocate to training tasks that require extensive practice for acceptable performance (e.g., vehicle operation); allocate to job aids tasks where reference to printed instructions are not disruptive to task performance.

7. *Cognitive Complexity of the Task.*   Allocate to training tasks where evaluation of numerous existing conditions must occur prior to making a decision; allocate to job aids tasks where binary choices can be developed into an aid.

8. *Equipment Complexity and Accessibility.*   Allocate to training tasks where equipment is easily accessed; allocate to job aids tasks that require detailed procedures to properly access the equipment.

9. *Environmental Constraints.*   Allocate to training tasks that require team efforts; allocate to job aids one- or two-person tasks.

10. *Consequences of Improper Task Performance.*   Allocate to training tasks where an occasional error will not cause major damage to equipment; allocate to job aids tasks that require branching logic such as diagnostic decision aids.

Smillie has pointed out that application of various of these ground rules may conflict with one another. He has suggested, therefore, that the ground rules not be applied indiscriminately, but rather, that each be considered in the total context of the task or tasks to be performed.

8.6.2 BEHAVIORAL FOUNDATIONS OF JOB AID TECHNOLOGY

It appears that the historical basis for the technology of job aiding lays in the area of behavioral psychology. Two generic learning theory principles underlie technological development in this area: first, the requirement for a precise description of the specific behaviors necessary to perform a task; and second, immedite feedback or reinforcement for action, whether correct or incorrect.

As early as 1926, S. L. Pressey employed these principles in developing a machine for automated teaching of drill material. This was the foundation of the teaching machine movement, which ultimately led to Skinner's (1954) work on "The Science of Learning and the Art of Teaching" as well as his treatise on teaching machines in 1958. The teaching machine movement subsequently provided a basic frame of reference for the development for proceduralized instruction and later for additional developments in the areas of computer-assisted instruction (CAI) and artificial intelligence (AI). Applications of this technology to instruction have included the hierarchical learning model of Gagné (1965). In this model, eight types of learning are hierarchically arranged from simple to complex, and the presumption is made that each higher-order type depends on mastery of the one below. Gagné's eight types of learning are: signal learning, stimulus-response learning, chaining, verbal association, multiple discrimination, concept learning, principle learning, and problem solving. Gagné's work served as a partial basis for the military's Instructional Systems Development (ISD) movement (Branson, Rayner, Coxx, Furman, King, and Hannum, 1975) and the subsequent development and refinement of procedures for criterion-referenced instruction and measurement (Glaser, 1963; Mager, 1972; Swezey, 1981). While these movements are not discussed here, refer to other chapters of this *Handbook* or to the original sources for such discussions.

An important outcome of this overall behavioral frame of reference, however, was the development of task analysis as a basis for defining both instruction and job aid development. A thorough task analysis is indeed the first step in the process of generating procedural job aids. Task analysis is a systematic technique that enables precise specifications of behaviors and skill levels necessary to accomplish each task within a job, as well as the steps existing within each task that are required in order to achieve adequate overall performance. Again, task analysis is discussed elsewhere in this *Handbook;* however, for purposes of this chapter, let us generally consider task analysis as the development of analytic procedures whose purpose is to obtain the following types of information about a task or a job: (1) the skills necessary to perform a task, (2) the knowledge necessary to perform a task, (3) the facilities and equipment required for task performance, (4) the technical data necessary for task performance, (5) discrete steps in the task performance and their sequence, and (6) critical steps involved in performing the task.

Swezey and Pearlstein (1974) have provided an example of the activity involved in replacing a spark plug as illustrative of such an analysis. After task analysis, for instance, the following types of information might be obtained:

1. *Skills.*   Using a socket wrench, using a torque wrench, and so forth.

2. *Knowledges.*   Direction of turning to unscrew plugs, how to set a spark gap, and so forth.

3. *Facilities and Equipment.*   Sheltered work space, set of sockets, socket wrench, torque wrench, gaping tools, and so forth.

4. *Technical Data.*   Gap to be set on the plug, force with which the plug should be torqued down upon replacement, and so forth.

5. *Discrete Steps.* Identify plug to be replaced, select correct socket size, remove spark plug lead from spark plug, insert socket on plug, attach socket wrench handle, and so forth.

6. *Critical Steps.* Gap tolerance is ±1 mm; failure to meet tolerance will cause misfire in replaced plug; plug should not be torqued down with more than 15 ft/lb of force or threads may be stripped, and so forth.

A number of highly researched task analysis techniques that may be used in job aid development exist. Several of these have been summarized in reviews by Chenzoff (1964) and more recently by Wheaton, Rose, Fingerman, Korotkin, and Holding (1976). As each task is identified, several factors are typically considered by an analyst, including aptitude of the user, type and complexity of the equipment, test equipment to be used, and tradeoff required between information to be presented by the job aid versus that to be stored in the memory of the user. The task analysis may also concern itself with such factors as the content and objectives of existing training programs and the characteristics of current technical documentation. Such considerations result in procedures that are both task-oriented and tailored to the specific skill and knowledge levels of the user population.

Smillie (1985) has listed several facets of the job situation as necessary components for inclusion in a task analysis and for use in the specification of job aids: (1) hardware interface (What are the controls, displays, support equipment, etc., that an individual performing the task will encounter?); (2) criticality (What are the consequences of performing the task incorrectly?); (3) cue (What does the individual see, hear, smell, and feel to initiate the task?); (4) response (What action is required by the task performer when the task is initiated?); (5) feedback (What indication does the task performer have that the task element was completed correctly?); (6) performance criteria (What are the time and accuracy requirements of the task?); and (7) references (What was the source of data used to generate the task elements?).

Major treatises on the bases in formal logic that underlie the job aid approach have been translated from Russian and published by Landa (1974, 1976). When an aid explicitly and/or comprehensively describes either the only procedure or the preferred procedure for attaining a specified result, it may [according to Geis (1984)] be termed an *algorithm.* Landa (1974, p. 11) defines an algorithm as a "precise, generally comprehensible prescription for carrying out a defined (in each particular case) sequence of elementary operations (from some system of such operations) in order to solve any problem belonging to a certain class (or type). Algorithms are defined such that if the steps are precisely followed, the preferred outcome will result; however, not all specified rule systems are algorithms. Landa (1974), for instance, has cited the example of the "Rules of Chess" as a rule system that is not algorithmic according to his definition. The "Rules of Chess" essentially describe how each piece on a chess board may be moved. They do not, however, provide information on how to play a winning game. Although they are rule-based, they are not prescriptive.

Algorithms may be contrasted with *heuristics,* which provide general guidelines, but which often require the user to bring to the task a sophisticated individual repertoire. According to Geis (1984), while heuristics may increase the probability of achieving a desired outcome, they do not guarantee it, whereas algorithms do. Although the attempt to develop heuristics often includes some algorithmic components that are helpful in planning and/or analyzing an event or series of events, a heuristic, even a complex one, cannot guarantee to achieve the preferred outcome. A heuristic, unlike an algorithm, is an appropriate method when the outcome cannot be predicted as a function of its very nature. The intent of job aids and procedural development, of course, is to be algorithmic. Such is definitely the case in developing specific user-oriented documentation. The considerable works of Landa address the formal, logical requirements behind the development of algorithms and their specific use in a wide variety of contexts including, among other things, instructional regulation and control. Landa makes the explicit case that formal logic is the basis for algorithmization and hence for specification for procedures in ways that are optimally designed to achieve desired outcomes. This is necessarily the case with job aids.

8.6.3 DEVELOPMENT OF A JOB AIDING TECHNOLOGY

The primary source of job aid development activity has been in the military. This has occurred within the context of developing maintenance manuals and technical orders for complex military equipment. The basic thrust was developed in the early 1950s in a series of articles by Miller and his coworkers (Miller and Folley, 1952; Miller, Folley, and Smith, 1953a). These involved development of maintenance-oriented job analyses for military equipment and the subsequent development of troubleshooting aids based on these job analyses. A seminal document, entitled "Systematic troubleshooting and the half-split technique" (Miller, Folley, and Smith, 1953b), provided an early logical basis for the development of procedures for anticipating maintenance job requirements. The "half-split" technique is an approach to fault isolation that is designed in an algorithmic fashion such that the fewest troubleshooting steps are required when the chain of troubleshooting activities and each succeeding segment of the chain are split so that the probability is 50% that a malfunction will be on a given side of each checkpoint. Using this technique, Miller and coworkers initiated the development of job aids for various military maintenance tasks. Both the Air Force and Navy have subsequently been active in research and development in advanced type job aids.

A basic Air Force concept has involved the development of so-called "fully proceduralized job performance aids." These have been developed and documented by Joyce, Chenzoff, Mulligan, and Mallory (1973a, 1973b, 1973c). A salient feature of fully proceduralized JPAs is that they are entirely job-oriented and provide illustrated step-by-step instructions for the performance of both linear tasks (that is, tasks whose sequence of steps never vary, such as cleaning, lubrication, inspection, removal, and installation) as well as branching diagnostic routines (such as check-out and troubleshooting activities). Most military specifications for documentation, including those in the Air Force, spell out format requirements in considerable detail. The position, presumably, is that in order to ensure completeness of coverage of hardware as well as completeness of coverage of the tasks performed, a carefully controlled task analysis must be conducted. Thus, a task analytic process that must be completed prior to JPA development is precisely specified. This involves a large matrix in which every possible task performed on each piece of equipment is identified. Furthermore, the training as well as the experience level of a document's user is carefully defined, and each task identified in the matrix is assigned for coverage either via a JPA or via training. Every hardware component whose failure must be identified is listed. In addition to troubleshooting logic, the final product of the Air Force JPA procedure includes step-by-step procedures for access and adjustment tasks. All procedures are necessarily accompanied by illustrations. The Air Force JPA specification is accompanied by two volumes. Volume II (Joyce, Chenzoff, Mulligan, and Mallory, 1973b) is a handbook for JPA developers; Volume III (Joyce, Chenzoff, Mulligan, and Mallory, 1973c) is a handbook intended for use by Air Force personnel who are involved in the process of procuring JPAs, as well as by training specialists who aid in the development of training content that is compatible with the JPAs. Figure 8.6.3 shows an example JPA frame.

The Navy has developed a concept known as Symbolic Integrated Maintenance Manuals (SIMM). SIMM documentation provides easily readable forms of schematics and block diagrams as well as symbolic troubleshooting charts. SIMM manuals further describe an equipment or system at each of four levels. The first is at the level of general information. The second is at the level of blocks of information that pertain to the system as a whole. These include (1) equipment description, (2) functional block diagrams, (3) general operation and installation information, and (4) performance check charts (a check procedure arranged in a columnar form). The third is four-part data packages organized by

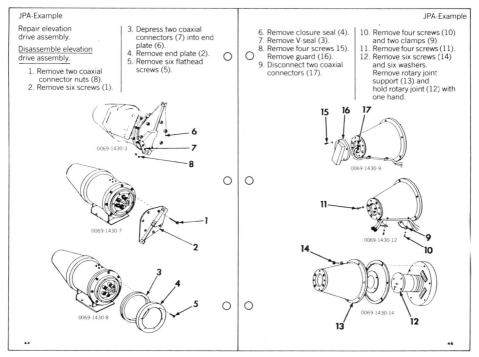

Fig. 8.6.3. Example JPA frame.

equipment function. In this case, each package consists of (1) a maintenance dependency chart (a logic diagram showing in detail the function dependencies among equipment elements), (2) precise access block diagrams (PABD-detailed functional block diagrams each showing one major function of the equipment), (3) precise access block diagram and text that presents a condensed functional theory of operation in juxtaposition with the PABD, and (4) alignment data. Fourth is additional hardware-oriented four-part data packages. These consist of parts data, schematic diagrams, theory of operation, and alignment testing voltage and resistance charts (or similar classes of data for nonelectronic equipment) as well as repair instructions. SIMMs have been found to work well when produced competently and when used by highly qualified, well-trained personnel.

It appears that job aid developers love acronyms, and most such approaches are, in fact, referred to by these short-cut names. Joyce (1974) reviewed the state-of-the-art in U.S. military and industrial documentation and listed numerous maintenance data techniques in use at that time in addition to fully proceduralized JPAs and SIMM. These include:

1. Automated technical order and maintenance sequence (ATOMS), as developed by the Boeing Company. This technique (according to Joyce) uses conventional formats for most information—except troubleshooting data, which are provided in step-by-step form with illustrations included to cover test equipment and test points.

2. Binary fault isolation chart (BFIC). This concept, developed by Westinghouse, presents procedural troubleshooting information in a flowchart format and depends heavily on the use of symbols rather than words to convey information.

3. Block form troubleshooting aids. This technique, developed by Applied Science Associates, uses the same type of fully proceduralized troubleshooting texts and illustrations as those contained in Air Force JPAs; however, the instructions are presented in flowchart format.

4. Condensed servicing data (CONSD). This technique, developed by General Electric, consists of three parts: (a) an alphabetical arrangement of symptom descriptions, (b) schematic diagrams showing all test points, and (c) lists of signal specifications and test points.

5. Design aid for training operation and maintenance (DATOM). These 11 × 17 in. manuals, developed by General Electric, organize troubleshooting information in a manner similar to SIMM.

6. FORECAST. This technique, developed by the Army's Human Resources Research Office, provides theory of operation, block schematics, and functional block diagrams combined with separate tables and signal specifications. Again, the organization is similar to SIMM, but color-coding is omitted.

7. Graphically proceduralized aids for maintenance (GPAM). The Army's publication engineers have developed this highly proceduralized format for presenting instructions on relatively simple tasks. The format is intended for use by lower skill groups and English-deficient personnel and involves cartoon-style illustrations.

8. Maintenance and training in complex equipment (MAINTRAIN). Also developed by the Army's Human Resources Research Office, MAINTRAIN manuals are designed to make maximum use of built-in fault finding circuitry and associated test equipment. These manuals include elaborately color-coded symptom tables and logic diagrams in conjunction with functional schematic diagrams to convey sufficient information.

9. Neostylized manuals. This is another form of cartoon-style manual with which the Army has experimented.

10. Pyramid diagram (PYRAGRAM). Developed by Hughes Aircraft, this concept is organizationally similar to SIMM. It includes text and diagrams shown on facing pages and the troubleshooting information is organized from the system level to the component level.

11. Rapid evaluation system to repair equipment (RESORE). This technique, developed by Martin Marietta, involves printed pages that are smaller in size than the back cover of the manual. A single illustration is printed on the inside back cover in an area not obscured by the pages and is referred to throughout the text.

12. Safeguard maintenance data system (MDS). The MDS was originally developed by the Western Electric Company. It is a complete system of storage, retrieval, and display of task-oriented data. Procedural information is provided in a format similar to the Air Force's JPA technique, and branching troubleshooting procedures are provided in a flowchart form.

13. Transistor radio automatic circuit evaluator (TRACE). The TRACE technique, developed by the Philco Corporation, is designed to simplify troubleshooting in printed circuits. The concept consists of numerous plastic cards containing prepunched holes that are lined directly over designated test points of a printed circuit board. Test-point signal information and other troubleshooting information are printed on the card. The technician positions the plastic card over the printed circuit board, and inserts the probe of a signal projector, meter, or oscilloscope into the various holes in order to make contact with test points.

Joyce concluded from his 1974 review that none of the techniques reviewed were as thoroughly documented as SIMM and none were demonstrated as being as effective.

Recently, the Naval Personnel and Research Development Center has developed a program known as the enlisted personnel individualized career system (EPICS). This technique, as reported by Blanchard, Smillie, and Conner (1984), involves an alternative personnel system concept that is based on simultaneous consideration of training, job aiding, job design, career structure, and personnel resources, coupled with a cost tradeoff model. EPICS features use of job performance aids, deferred formal training, and an individualized career advancement structure. With deferred training, an individual is first sent to sea for an orientation period of 3–12 months and is then returned to shore-based technical training, depending on his or her degree of adaptation to shipboard life and demonstrated level of interest and motivation. During this period, the recruit receives transition training to shipboard life and is made an effective member of the ship's crew through the use of job performance aids. Thus, formal training experiences ashore are distributed throughout an individual's 6-year enlistment, rather than being provided prior to the first shipboard duty assignment.

The EPICS approach has resulted in the development of so-called "enriched" and "hybrid" job performance aids. These techniques, documented by Smillie, Smith, Post, and Sanders (1984), are defined as follows: A *hybrid* performance aid is one that presents troubleshooting information in both directive and deductive formats, whereas an *enriched* aid is one which presents additional job-relevant information to facilitate the transition between directive and deductive formats.

Combination enriched, hybrid performance aids (EHJPAs) have also been developed. These aids are designed for use with inexperienced technicians, and facilitate the use of the more deductive technical manual troubleshooting data. EHJPAs are an integral component of the EPICS program. In EPICS, two types of EHJPAs were developed. One, termed the decision tree functional logic diagram, was developed for relay circuitry, and the other, known as the state table extracted schematic, was developed for digital solid-state circuitry.

Smillie, Smith, Post, and Sanders (1984) evaluated these two types of enriched job aids in a field setting and found that technicians using decision tree functional flow logic diagrams solved all problems in half the time it took technicians who used existing ordnance publications to solve 75% of the problems. Furthermore, the feasibility of state table extracted schematics to facilitate troubleshooting was demonstrated in tabletop troubleshooting scenarios. Thus, both types of EHJPAs were shown to be effective troubleshooting aids. The authors concluded that, by including enrichment information, these aids can support inexperienced technicians during troubleshooting tasks, while, over time, the technicians can acquire the necessary system understanding to use the conventional technical documentation. In addition, it was suggested that EHJPAs can be used as a learning aid during technical training to provide students with a succinct functional description that ties together the theory of operation and functional flow diagrams.

According to Post and Smith (1979), two of the best-known types of early job performance aids were PIMO (presentation of information of maintenance operations) (Goff, Schlesinger, and Parlog, 1969) and AMSAS (advanced manpower concepts for sea-based aviation systems) (Post and Brooks, 1970). In spite of the reported successes of PIMO and AMSAS, however, the military did not adopt the JPA form of presentation of technical information on a large scale until considerably later when Malehorne (1975) suggested that in order to be integrated successfully into a maintenance environment, job aids must be shown capable of working on a continuing basis with existing personnel systems. The Navy's EPICS approach was designed in part to address this need.

Post and Smith (1979) have classified various job aids as applying a directive versus a deductive logical basis. In this context, the term deductive refers to aid forms that both describe a given system (e.g., functional diagrams or electronic schematics) and indicate the behavior required by the troubleshooter (that is, the individual must understand the relationship among symptoms, functions, and signal flow in order to deduce troubleshooting procedures). Table 8.6.1 (Post and Smith, 1979) shows a summary of the contributions of directive and deductive aid forms to various types of job satisfaction criteria, and Figs. 8.6.4 and 8.6.5 show examples of directive and deductive EPICS logic diagrams.

Within the EPICS system, the EHJPA concept provides enrichment strategies according to both enrichment type and message format dimensions. Each of these dimensions contains three elements resulting in a nine-cell matrix as shown in Table 8.6.2. Each cell, thus, represents a different enrichment concept. The three enrichment types are (1) transition (that is, any information designed to help the hybrid user transition from the directive to the deductive aid form); (2) system understanding (that is, information about the operation of the hardware represented in the deductive aid of the hybrid); and (3) theory (that is, information about the operation or principles underlying the class of equipment to which the present hardware belongs.

The three delivery formats in the EHJPA concept are as follows: (1) interpretative (a format which provides the enrichment message as an integral part of the hybrid aid); (2) basic reference cues (a technique whereby either or both of the hybrid elements may contain notes that refer the user to another source document; e.g., another maintenance manual or rate training manual); and (3) extension training reference cues (a technique whereby either or both of the hybrid elements may contain notes that refer the user to a personalized training source available at the local level).

Post and Smith (1979) conducted an evaluation of EHJPAs against various other basic job aids

Table 8.6.1 Summary of Assessing Directive and Deductive Aid Forms Against Selected Job Satisfaction Criteria

Job Satisfaction Criteria	Inexperienced Users		Experienced Users	
	Directive	Deductive	Directive	Deductive
1. Opportunity to learn career-relevant skills	Opportunities are primarily memorization of rote procedures (e.g., test point and part location, turn on–off procedures). Opportunities to learn more career-relevant items (system operation, troubleshooting methods) are limited.	Opportunities to learn system operating principles and methods of troubleshooting are present but of such a complex nature as to require long learning cycles and senior tutors.	Continued use of the directive aid offers only the opportunity to become more proficient in its use (e.g., quicker application).	Each "new" failure represents an opportunity to learn new aspects of the system such as component dependencies, as well as practice in troubleshooting methods.
2. Meaningful work	Good results, at least initially, because the aid permits the technician to do work that would otherwise be beyond his or her capability.	Poor results in that the technician is reduced to observation and menial support chores during his or her long apprenticeship.	Continued use of the directive aid represents rote performance with no true understanding of the logic or reasons behind each sequence.	Good meaning in that the technician is required to develop his or her own strategy and procedures for each use of the aid.
3. Challenge	Some initial challenge that diminishes rapidly because the aid is designed to be mastered in the first usage.	Substantial challenge requiring large investments of time and effort from both the learner and his or her mentor.	No challenge resulting in a boring job that becomes worse with each application of the aid.	Good challenge for a considerable period of time. Eventually, the use of the aid as a "guide" is replaced by its use as a reference and, consequently, the work retains its challenge.

Source: Adapted from Post and Smith (1979).

Procedures

Symptom

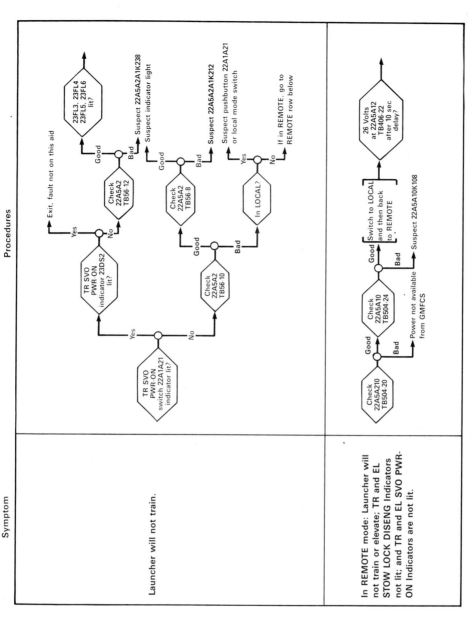

Launcher will not train.

In REMOTE mode: Launcher will not train or elevate; TR and EL STOW LOCK DISENG Indicators not lit; and TR and EL SVO PWR-ON Indicators are not lit.

TR SVO PWR ON switch 22A1A21 indicator lit?

TR SVO PWR ON indicator 23DS2 lit?

Yes

No

Check 22A5A2 TB56·10

Good

Bad

Yes

No

Check 22A5A2 TB56·12

Good

Bad

Check 22A5A2 TB56·8

Good

Bad

In LOCAL?

Yes

No

23FL3, 23FL4 23FL5, 23FL6 lit?

Good

Bad

Exit, fault not on this aid

Suspect 22A5A2A1K238

Suspect indicator light

Suspect 22A5A2A1K212

Suspect pushbutton 22A1A21 or local mode switch

If in REMOTE, go to REMOTE row below

Check 22A5A210 TB504·20

Good

Bad

Check 22A5A10 TB504·24

Good

Bad

Switch to LOCAL and then back to REMOTE

26 Volts at 22A5A12 TB406·22 after 10 sec delay?

Power not available from GMFCS

Suspect 22A5A10K108

Fig. 8.6.4. Directive element of an EPICS logic diagram.

1049

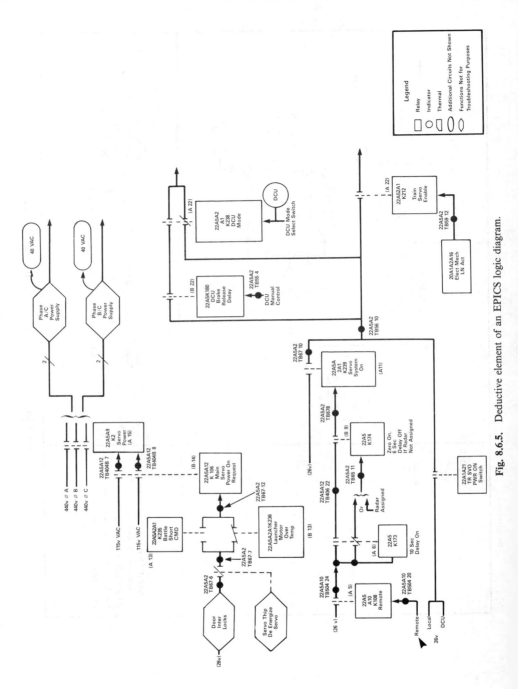

Fig. 8.6.5. Deductive element of an EPICS logic diagram.

Table 8.6.2 EPICS Enrichment versus Format Job Aid Concepts

	Delivery Formats		
Enrichment types	Interpretative	Basic reference cues	Extension training reference cues
Transition	Tr/I	Tr/BRC	Tr/ETRC
System understanding	SE/I	SE/BRC	SE/ETRC
Theory	Th/I	Th/BRC	Th/ETRC

formats. These include (1) fully proceduralized JPAs (Potter and Thomas, 1976), (2) partially proceduralized JPAs (Horn, 1973), (3) maintenance dependency charts (Horn, 1973), (4) functional flow logic diagrams (Atchley and Lehr, 1964), and (5) schematics (Atchley and Lehr, 1964). Post and Smith concluded that the EHJPA concept was feasible in comparison to other formats and recommended as particularly appropriate the following three EHJPA types: transitional/interpretative, system understanding/basic reference cue, and theory/extension reference cue.

8.6.4 TECHNIQUES FOR USE IN JOB AID DESIGN

Over 20 years ago, Folley (1961) provided guidance on the design of job aids. Folley's summary of procedures for designing job aids follows:

1. Identify the task elements for which job aids should be considered: (a) describe the job or position, (b) describe each task involved, (c) analyze each task element. In this process, consideration is given to providing performance aids for each task element where one (or more) of three conditions occurs: (a) it is judged that the anticipated task performer population (including their background as well as the training they are likely to receive) will not be able to achieve the required level of performance without an aid; (b) use of an aid is anticipated to significantly improve performance above the required minimum level; and (c) providing an aid will result in cost savings (in the form of reduced selection and/or training requirements) without a decrement in performance.

2. Determination of what job aid functional characteristics will enhance the performance of each task element. The functional characteristics of a job aid are defined as the operations that an aid must perform in order to accomplish the required improvement. These characteristics are considered to be independent of the physical means of providing the functions; that is, they merely describe what the aid must do with respect to a given task element, not how it must do it.

3. Specification of the physical design characteristics required of the aids to carry out the necessary functions. This activity has three steps: (a) combining the functional characteristics into appropriate groupings, (b) determining the most suitable method of performing each combination of functions, and (c) specifying the physical characteristics of the recommended aid for each combination of functional characteristics. According to this approach, the physical design characteristics of performance aids describe the mechanisms or devices that will perform the required aid function. In turn, they depend primarily on the specific nature of the function (i.e., the ways in which the characteristics can be combined in the situation in which the aids must be used).

4. Evaluation, modification, and updating of the resulting aids. Here, five steps are recommended: (a) review the training versus job aid tradeoff; (b) check that the behavioral requirements involved in using each aid are compatible with the work performance situation; (c) build prototypes of sample job aids and conduct tryouts; (d) modify specifications as indicated by the evaluation; and (e) periodically update the information content of aids in order to keep them current with changes in the system.

This generalized guidance has been updated and expanded in more recent publications by Rifkin and Everhart (1971), and by Pearlstein, Schumacher, Smith, and Rifkin (1973) in response to specific job aid situations. The approaches advocated by these authors, however, differ from the guidance discussed previously only in the specific application requirements for which their recommendations were generated. In a more specific form (that of fully proceduralized job performance aids for the Air Force), Joyce, Chenzoff, Mulligan, and Mallory (1973b) have provided a series of steps for JPA developers' use in constructing performance aids compatible with the Air Force requirement. The steps involved in this process are shown in Table 8.6.3 (adapted from Joyce, Chenzoff, Mulligan, and Mallory, 1973b).

Smillie (1985) has addressed the issue of format in job aid design. According to Smillie, the standard unit for job aid format consideration is the individual frame. Four frame formats are generally employed: (1) complete integration of text and graphics; (2) division of the frame vertically (or horizontally), where text is presented in one portion and graphics in a second; (3) varying sizes of illustrations, where graphics are placed as required in the frame to support the text; and (4) heavy text with graphics

Table 8.6.3	Steps Involved in Job Aid Design as Recommended
by Joyce, Chenzoff, Mulligan and Mallory (1973b)

I.	Maintenance Task Analysis
  Conduct task analysis
  Determine recommended personnel qualifications
  Establish preliminary task identification matrix (TIM)
  Establish JPA/training, tradeoff ground rules
  Annotate TIM for JPA/training tradeoff (ATIM)
  Validate ATIM
  Develop level of detail guide
  Develop test equipment and tool use form
  Construct preliminary information worksheets
  Construct task description and information index
  Construct detailed step description worksheets
  Construct generalized task list
II.	Job Guide Development
  Consider recommended personnel qualifications
  Group tasks into activities
  Assign activities to volumes
  Draft index volume
  Draft frontmatter for each volume
  Format job guide activities
  Draft maintenance support information manual
  Validate job guide
III.	Fully Proceduralized Troubleshooting Aid Development
  Consider recommended personnel qualifications
  Develop list of components and failure modes (LCFM) and
  list of malfunction symptoms
  Develop the checklist procedure
  Develop action trees
  Checkout and action tree validation
IV.	Quality and Accuracy Assurance
  Develop JPA quality assurance plan
  Establish quality and accuracy assurance organization
  Establish quality control responsibilities
  Conduct technical quality control
  Conduct production quality control
  Conduct validation
  Conduct verification

used sparingly only to illustrate areas unfamiliar to the user. Whatever the format, the level of detail of the aid must be designed on the basis of the performance repertoire identified for the users of the aid, that is, on their population abilities and characteristics.

## 8.6.5	EVALUATION CRITERIA

Of necessary concern are criteria against which job aids may be evaluated. Briefly mentioned, several potential criteria for use in evaluating job aids (which are essentially applicable to any information presentation strategy) are: comprehensiveness, degree of prescription, reliability/validity, user opinion, and cost. The comprehensiveness criterion is demonstrated to the extent that each activity required to perform a task or job is included in the aid (i.e., all items included in the aid are critical, and nothing important is omitted). The degree of prescription criterion assesses the extent to which the aid is optimally prescriptive for the user population. In considering this issue, one must consider the degree proceduralization required in order for the user population to achieve the required task performance. The reliability/validity criterion concerns the extent to which various persons using an aid can achieve the same results (reliability) and the extent to which use of the aid enables achievement of the desired performance outcomes (validity). Again, the aptitudes and the individual difference characteristics of the user population must be taken into account; however, a primary purpose of job aids is to reduce such differences. Therefore, a competently developed job aid must be usable across a wide array of task performers who exhibit widely varying individual differences. Swezey (1981) has discussed these individual performance assessment issues in detail from the perspecive of criterion-referenced measurement. Furthermore, user opinion of the aid must be addressed, in terms both of its ease of use and of its utility. Finally, cost considerations are of considerable import in job aid development.

Osga and Smillie (1983) have recently edited a volume that addresses the cost issue in job aid development and production. In that document, Osga (1983) has summarized the following considerations as driving costs in job aid development:

1. *Personnel, Equipment, and Environment Characteristics.* This concerns the extent to which the varying capabilities of personnel must be addressed in the job aid development process, the extent to which the equipment for which a job aid is developed is variously complicated, the extent to which various task types (such as troubleshooting versus remove/replace) have cost ramifications, and the extent to which system parameters, such as the format and/or level of detail of the job aid, must be considered.

2. *Procuring Agency Attributes.* This addresses the extent to which the job aids are institutionalized in a procuring agency versus whether the agency must be educated as to their benefit.

3. *Job Aid Producer Attributes.* This concerns the extent to which the organization developing the job aid can effectively attract, recruit, and retain knowledgeable graphic, analytic, and documentation staffs, and the cost consequences of doing so.

4. *Job Aid Characteristics.* These factors include the physical attributes of the job aid and its materials. The cost of job aids have been compared to that of an iceberg with a small portion visible above the water. The visible portion represents attributes of the aid, such as numbers of pages, density, format, illustrations, and so on, whereas the submerged portion represents other factors such as the input data quality, and the procurement process for obtaining job aids.

8.6.6 CURRENT AND FUTURE DIRECTIONS IN JOB AID TECHNOLOGY DEVELOPMENT

The general trend in the area of job aiding appears to be moving away from hardcopy technical documentation and toward automated aids. Chenzoff and Joyce (1983) have recently listed a variety of automated and electronic job aiding techniques currently in use or in development. These include the following.

MIARS (maintenance information automated automatic retrieval system) used by the Navy, which employs a microfilm cartridge to store and present technical information.

SIMM/FOMM. The acronym SIMM (discussed previously) stands for symbolic integrated maintenance manual, whereas FOMM stands for functionally oriented maintenance manuals. The name was changed from SIMM to FOMM in 1974 by the Navy in order to distinguish it from previous versions of Navy maintenance manuals. FOMM is distinguished from previous Navy job aid techniques by (1) use of color and shading, (2) page size (which was increased), and (3) detailed maintenance dependency charts and logically arranged functional blocked schematics with keyed text.

NTIPS (Navy technical information presentation system) is a major component of the previously discussed EPICS system. NTIPS is currently moving toward a shopbench-type terminal that is capable of presenting a wide range of texts and illustrations on a high-resolution CRT screen. The same device will serve as both a training device and a maintenance performance aid, and will feature voice input and output capability. A single device will be used to requisition parts, to submit technical information deficiency and evaluation reports, and to submit data collection reports. NTIPS is a massive technical information system designed to address a large number of Navy goals and objectives.

NOMAD (Navy on-board maintenance aid device) is a minicomputerized maintenance performance aid developed by the Hazeltine Corporation. NOMAD uses a structured automated diagnostic strategy, which prompts and leads a technician through appropriate procedures in troubleshooting or repair. The NOMAD display console consists of a high-resolution color CRT, a full keyboard, and a light pen, which can be used for rapid selection of menu items and for answering multiple choice questions. Color texts and color graphics and videotaping are also components of the NOMAD approach. NOMAD is basically a modified version of the Hazeltine Corporation's TICCIT (time-shared interactive computer control information television) instructional system, and is designed for use both as a training and as a maintenance performance aid.

PEAM (Personalized electronic aid for maintenance) is a triservice project, with the majority of the work being done by Texas Instruments and the XYZYX Information Corporation; it is essentially a concept for portable electronic job aids featuring the delivery of textual and graphic information to maintenance personnel at the workplace. It will involve a limited vocabulary and voice command interpreted by a voice recognition device. The display involved is an electroluminescent panel, which is portable and can be transported in a large attache case.

ATO (Air Force's automated tech order) program concept is similar in many ways to PEAM. Whereas PEAM is more hardware-oriented, ATO is primarily oriented toward tailoring the presentation of information to the specific needs of the user. The ATO concept incorporates three techniques for technical information delivery: (1) the concept of multiple tracks of procedural information presentation, such that a user can select from among three levels of detail (novice, average, and expert) and can move from one level to another at will; (2) the concept of information "pools" to support the basic procedural information (this is similar in concept to the back-up reference information as employed in enriched Navy aids); and (3) flexible interaction between the device and the user which will enable the user, for example, to stop a presentation at any point in order to access a tool or to move across multiple tracks of level of detail. The ATO program is designed such that the dialogue between the device and the user is modified after natural supervisor technician communication. It is thus designed to be extremely user friendly.

VIMAD (voice interactive maintenance aiding device) is a protable maintenance performance aid that presents a user with all necessary information while allowing freedom for both hands. The user wears a $1\frac{1}{2}$ lb helmet with a 1 in. television tube, microphone, and earpieces, as well as a receiver battery pack with an auxiliary keypad attached to a belt around the waist. This technique, developed originally by Honeywell, and modified by Perceptronics, Inc., may also be used in a configuration where a microprocessor videodisk player and speech recognition device, as well as an audio response unit, can be placed on a roll-around cart that is connected to the belt pack by a cable. A limited 20-word voice vocabulary that has words such as "yes," "no," "continue," "help," and various numerals is included. Auxiliary communication can be achieved through the keypad. Much of the video display material consists of pictures where a maintainer is actually working on the task at hand, with additional guidance provided through overlays and verbal instructions. Such displays may be either still frames or motion pictures. Textual frames such as menus are also available. It is anticipated that future users may also be able to patch into a phone line to access experts if additional assistance is needed. Various voice synthesis techniques have been tried out in the VIMAD approach; however, synthesized speech generally appeared infeasible in noisy environments. Speech recognition, however, was satisfactory. As currently configured, the helmet-mounted TV picture is shown into the operator's right eye. It appears to be about 6×8 in. in size to the user at an apparent viewing distance of about 14 in.

LOGMOD (logic model analysis technique), developed by DETEX Systems, Inc., provides for automated fault isolation through the generation of dependency chains based on a logical model. This model interrelates all dependency chains identified in a systematic fashion in order to facilitate fault isolation through the use of a diagnostic strategy.

Knerr (1977) has provided a description of an Army-generated technique known as ACTS (adaptive computerized training system). ACTS has four components: (1) a task model, which is a model of an electronic circuit in which faults can be simulated; (2) a student model, which is an expected utility model of a student's decisions designed to permit adaptive estimation of utilities for various test measurements (as the student troubleshoots, the model is adjusted to match his or her performance); (3) an expert model, which is also an expected utility decision model of an expert troubleshooter (unlike the student model, the expert model is adjusted only while the expert is developing the model; the model is fixed while the student is being trained); (4) an instructional model, which consists of the adaptive instructions provided to the troubleshooter. Instruction is based on a comparison between student and expert models, focused on the student's decision-making process. It is assumed that the student knows electronic theory and simple procedures like taking test measurements.

Future activity in the development of job aids appears to be headed in the direction of artificial intelligence (AI), and many current and developing systems are already incorporating such approaches. The basic problem (as in the case of all AI research) appears to involve the representation of knowledge. Whereas it is beyond the scope of this chapter to address AI knowledge representation techniques in detail, the reader is encouraged to review them in other sections of this *Handbook*. The general idea must necessarily be to model the user and to codify his or her knowledge data base in a fashion that is designed to enable presentation of correct troubleshooting information on an as-needed basis in a real-time environment. As discussed previously in this chapter, whereas most current job aids employ algorithmic-based logical parameters, the AI-oriented approaches enable movement toward more heuristic perspectives such that job aiding information can be presented adaptively and in a fashion that is compatible with specific task requirements as well as the specific knowledge base of the user. Development of adaptive knowledge-based systems that are powerful, flexible, reliable, and valid are already in progress.

A recent review of the implications of AI for user-designed job aiding systems was prepared by Magliero, McNabb, and Joyce (1983). As discussed by Kern (1985), such work must involve modeling, user's task performance, reading comprehension skills, information seeking behaviors, and various other factors affecting performance at the work site. Kern has identified four cogent features of modeling work site performance that will be of use in the design of adaptive job aids:

1. With repeated performance of the same task, the probability of a user seeking information rapidly diminishes.

2. Most information-seeking behaviors of job aid users represent an effort to resolve a discrepancy between their individual expectations and the observed effects of their actions.

3. Manuals and currently existing electronic job aids are not typically designed to "interrogate users in order to help them identify the problem and the solution."

4. The effectiveness of technical manuals and job aids at the current state-of-the-art is probably more dependent on an individual's prior knowledge than on the goodness of design of the aid itself.

The major problem that must be overcome in the area of job aid technology development is the needs of the workers themselves. Job aids must be designed in a fashion that alleviates feelings on the part of job incumbents that they are merely "trained monkeys" who are qualified only to perform the simplest tasks. The motivational properties of leading individuals by the hand through a series of steps without requiring them to exercise their own judgment and/or logic is typically considered demeaning. This is a particular problem in the job aid domain, since the reliability of job aids is demonstrably higher than is the reliability of typical free-wheeling, seat-of-the-pants maintenance approaches that are presumably based on some unknown level of understanding of underlying theoretical principles. The problem to be resolved is a psychological one. As indicated by Swezey (1984a), for instance, the most easily understandable information presentation techniques are not necessarily those that optimally facilitate performance, recall, and/or retention of alphanumeric material. It is hoped that future endeavors in the area of job aid and proceduralized instructional development will consider, and find ways of dealing with, this thorny problem (Swezey, 1984b).

As indicated by Wulff and Berry (1962), it is necessary to maintain stimulus control over users of job aids; however, this must be done in a user-acceptable fashion. Although the technology of job aiding has made major strides in this period in terms of format, level of detail, degree of proceduralization, and movement toward automation, the points raised by these authors of this cogent article over 20 years ago continue to be of major importance to the field.

REFERENCES

Atchley, W. R., and Lehr, D. J. (1964). Troubleshooting Performance as a Function of Presentation Technique and Equipment Characteristics, *Human Factors,* 257–263.

Blanchard, R. E., Smillie, R. J., and Conner, H. B. (1984). Enlisted personnel individualized career system (EPICS): Design, development and implementation (NPRDC TR 84–15). San Diego, CA: U.S. Navy Personnel Research and Development Center.

Branson, R. K., Rayner, G. T., Coxx, J. L., Furman, J. P., King, F. J., and Hannum W. J. (1975). Interservice procedures for instructional systems development: Executive summary and model (ADAO19486). Fort Benning, GA: U.S. Army Combat Arms Training Board.

Chenzoff, A. P. (1964). A review of the literature on task analysis methods (NAVTRADEVCEN 1218–3). Port Washington, NY: U.S. Naval Training Device Center.

Chenzoff, A. P., and Joyce, R. P. (1983). *Present and future maintenance job aids in the U.S. Navy and other armed services.* Valencia, PA: Applied Science Associates.

Elliott, T. K., and Joyce, R. P. (1968). An experimental comparison of procedural and conventional electronic troubleshooting (AFHRL-TR-68-1). Wright-Patterson Air Force Base, OH: U.S. Air Force Human Resources Laboratory.

Elliott, T. K., and Joyce, R. P. (1971). An experimental evaluation of a method for simplifying electronic maintenance. *Human Factors, 13,* 217–227.

Folley, Jr., J. D. (1961). A preliminary procedure for systematically designing performance aids (ASD Technical Report 61–550). Wright-Patterson Air Force Base, OH.

Folley, Jr., J. D. (1975). Research issues in JPA technology development. Paper presented at the SALT-LMI Conference on Improved Information Aids, Arlington, VA.

Gagné, R. M. (1965). *The conditions of learning.* New York: Holt, Rinehart & Winston.

Geis, G. L. (1984). Checklisting. *Journal of Instructional Development, 7*(1), 2–9.

Glaser, R. B. (1963). Instructional technology and the measurement of learning outcomes: Some questions. *American Psychologist, 18,* 519–521.

Goff, J., Schlesinger, R., and Parlog, J. (1969). Project PIMO final report: Vol. II—PIMO test summary (AFHRL-TR-69-155) (ADAO852102). Wright-Patterson Air Force Base, OH.

Horn, L. (1973). Optimum troubleshooting aids study (Technical Report 73–25). Washington, DC: U.S. Naval Weapons Engineering Support Activity, Washington Navy Yard.

Joyce, R. P. (1974): *Survey of maintenance information needs of U.S. Independent telephone companies and implications of recent technical documentation developments for northern electric practices.* Valencia, PA: Applied Science Associates.

Joyce, R. P., Chenzoff, A. P., Mulligan, J. F., and Mallory, W. J. (1973a). Fully proceduralized job performance aids: Vol. I, Draft specification for organizational and intermediate maintenance (AFHRL-TR-73-43(I)) (AD775-702). Wright-Patterson Air Force Base, OH: U.S. Air Force Human Resources Laboratory.

Joyce, R. P., Chenzoff, A. P., Mulligan, J. F., and Mallory, W. J. (1973b). Fully proceduralized job performance aids: Vol. II, Handbook for JPA developers (AFHRL-TR-73-43(II)) (AD775-705). Wright-Patterson Air Force Base, OH: U.S. Air Force Human Resources Laboratory.

Joyce, R. P., Chenzoff, A. P., Mulligan, J. F., and Mallory, W. J. (1973c). Fully proceduralized job performance aids: Vol. III, JPA manager's handbook (AFHRL-TR-73-43(III)). Wright-Patterson Air Force Base, OH: U.S. Air Force Human Resources Laboratory.

Kern, R. P. (1985). Modeling users and their use of technical manuals. In T. M. Duffy and R. Waller, Eds. *Designing usable text.* New York: Academic.

Knerr, B. W. (1977). Artificial intelligence techniques in electronic troubleshooting training: Development and evaluation. In Training: Technology to policy (Special Publication P-77-6) (ADAO49629). Alexandria, VA: U.S. Army Research Institute.

Landa, L. N. (1974). *Algorithmization in learning and instruction.* Englewood Cliffs, NJ: Educational Technology Publications.

Landa, L. N. (1976). *Instructional regulation and control: Cybernetics, algorithmization and heuristics in education.* Englewood Cliffs, NJ: Educational Technology Publications.

Mager, R. F. (1972). *Goal analysis.* Belmont, CA: Lear Siegler/Fearon.

Magliero, A. J., McNabb, S. D., and Joyce, R. P. (1983). *Implications of artificial intelligence for a user-defined technical information system.* Valencia, PA: Applied Science Associates.

Malehorne, M. K. (1975). Impact of Job Performance Aids on Personal Training. In *Proceedings of Invitational Conference on Improved Information Aids for Technicians.* Society for Applied Learning Technology and Logistics Management Institute.

Miller, R. B., and Folley, Jr., J. D. (1952). *The validity of maintenance job analysis from the prototype of an electronic equipment.* Pittsburgh, PA: American Institute for Research.

Miller, R. B., Folley, Jr., J. D., and Smith, P. R. (1953a). *Troubleshooting in electronics equipment: A proposed method.* Pittsburgh, PA: American Institute for Research.

Miller, R. B., Folley, Jr., J. D., and Smith, P. R. (1953b). Systematic troubleshooting and the half-split technique (Center Technical Report 53–21). Lackland Air Force Base, TX: U.S. Air Force Human Resources Research Center.

Osga, G. A. (1983). Analysis, conclusions and recommendations. In G. A. Osga and R. J. Smillie, Eds. Proceedings of an invitational conference on job performance aid cost factors (Report SR83-39). San Diego, CA: U.S. Navy Personnel Research and Development Center, pp. 79–88.

Osga, G. A., and Smillie, R. J. (Eds.) (1983). Proceedings of an invitational conference on job performance aid cost factors (Report SR-83-39). San Diego, CA: U.S. Navy Personnel Research and Development Center.

Pearlstein, R. B., Schumacher, S. P., Smith, A. P., and Rifkin, K. I. (1973). *Information display design.* Valencia, PA: Applied Science Associates.

Post, T. J., and Brooks, F. A. (1970). *Advanced manpower concepts for sea-based aviation systems (AMSAS).* Washington, DC: U.S. Naval Air Systems Command.

Post, T. J., and Smith, M. G. (1979). Hybrid job performance aid technology definition (Technical Report 79–25). San Diego, CA: U.S. Navy Personnel Research and Development Center.

Potter, N. R., and Thomas, D. L. (1976). Evaluation of three types of technical data for troubleshooting: Results and project summary (AFHRL-TR-76-74(I)). Wright-Patterson Air Force Base, OH.

Pressey, S. L. (1926). A simple apparatus which gives tests and scores—and teaches." *School and Society, 23,* 373–376.

Rifkin, K. I., and Everhart, M. C. (1971). *Position performance aid development.* Valencia, PA: Applied Science Associates.

Skinner, B. F. (1954). The science of learning and the art of teaching. *Harvard Educational Review, 24,* 2.

Skinner, B. F. (1958). Teaching machines. *Science* (October 24).

Smillie, R. J. (1985). Design strategies for job performance aids. In T. M. Duffy and R. Waller (Eds.), *Designing Usable Text.* New York: Academic, pp. 969–977.

Smillie, R. J., Smith, M. G., Post, T. R., and Sanders, J. H. (1984). Enriched hybrid job performance aid development (Technical Report). San Diego, CA: U.S. Navy Personnel Research and Development Center.

Swezey, R. W. (1975). On the use of performance aids in increasing job effectiveness. Proceedings: Systems Engineering Conference, National Institute of Industrial Engineers (pp. 30–33), Las Vegas, NV.

Swezey, R. W. (1977). Performance aids as adjuncts to instruction. *Educational Technology, 3,* 27–32.

Swezey, R. W. (1981). *Individual performance assessment: An approach to criterion-referenced test development.* Reston, VA: Reston.

Swezey, R. W. (1984a). Optimising legibility for recall and retention. In R. Easterby and H. Zwaga, Eds. *Information design* (pp. 145–156). Chichester, England: Wiley.

Swezey, R. W. (1984b). *Cognitive skill learning: A review and assessment.* McLean, VA: Science Applications, Inc.

Swezey, R. W., and Pearlstein, R. B. (1974). Performance aids—Key to faster learning?" *Training: The Magazine of Manpower and Management Development, 8,* 30–33.

Wheaton, G. R., Rose, A. M., Fingerman, P. W., Korotkin, A. L., and Holding, D. H. (1976). Evaluation of the effectiveness of training devices: Literature review and preliminary model (Memorandum 76–16). Alexandria, VA: U.S. Army Research Institute.

Wulff, J. J., and Berry, P. C. (1963). Aids to Performance. In R. M. Gagné, Ed. *Psychological principles in system development* (pp. 273–298). New York: Holt, Rinehart & Winston.

PART 9
PERFORMANCE MODELING

CHAPTER 9.1
DECISION MAKING

WARD EDWARDS

Social Science Research Institute
University of Southern California
Los Angeles, California

This chapter is a highly condensed version of parts of *Decision analysis and behavioral research,* by Detlof von Winterfeldt and Ward Edwards. I am grateful to Cambridge University Press for permission to use many of the same ideas and some of the same words. Preparation of that book, and so of this chapter was supported by the Office of Naval Research under Contract Number 00014-79-C-0529, with cosponsorship by the Army Research Institute. Other supporters included the University of Southern California, The Wood Kalb Foundation, and the Defense Advanced Research Projects Agency. We are grateful to them all, and in particular to Martin A. Tolcott, Robert Sasmor, and Stephen J. Andreole. For years of collaborative effort I am grateful to Detlof von Winterfeldt, who really should be a coauthor of this chapter, but declined. For very helpful criticisms I am grateful to Baruch Fischhoff, Ralph Keeney, Sarah Lichtenstein, Paul Slovic, Martin A. Tolcott, Amos Tversky, and an anonymous reviewer.

9.1.1 INTRODUCTION

Decisions are ubiquitous and so the topic of decision making is important; however, most decisions are trivial. Sometimes triviality results from small stakes. The choice between fish and veal for dinner is unlikely to affect the fate of nations, or even of your digestive system. Triviality may result from obviousness. Decisions about routes from here to there are often obvious; if not, recognition of routes that, although feasible, are not worth considering is always obvious. A formal rule called *the principle of the flat maximum* (see von Winterfeldt and Edwards, 1982, for some mathematics and applications) in effect states that, for discrete sets of decision options, typically it will either be pretty obvious which is best or else choice among them cannot matter much in a percentage sense; this is a manifestation of the relatively insensitive behavior of linear models to which Dawes and Corrigan (1972) and Wainer (1976) have called attention.

Yet the topic of decision making is important both in psychology and in its own right as a technological problem. Some decisions are important and not at all obvious. The stakes may be high; investment decisions are often like that. Formulation of the problem as a decision problem may be difficult. The assumptions leading to flat maxima may not be met. And, in any case, decision making is one of the most common human intellectual activities. The questions of how it is performed, and how it should be performed, are inevitably of interest to psychologists interested in the exercise of intelligence, and are crucial to the design of human–machine systems that have decisions as outputs.

Decision theory is a loose name for the mathematical background of the topic. It is primarily concerned with identifying the best of a set of options. *Behavioral decision theory* is defined as the study of how human beings actually make decisions, a topic within cognitive psychology. *Decision analysis* is a psychologically based technology that provides many tools to facilitate wise decisions. An important auxiliary topic is *inference*, whether statistical or otherwise; this topic too can be treated descriptively or prescriptively.

Descriptive and prescriptive treatments of intellectual tasks are inevitably intertwined. People usually want to infer wisely and decide well, and often manage to do so—sometimes by using the tools of decision analysis and more often not. To the extent that prescriptive ideas become embodied in human thinking, they become descriptive psychological models. The extreme familiar instance is arithmetic; for most educated adults, the formal prescriptions of arithmetic are also descriptions of how arithmetical reasoning is done, given time and tools.

This chapter focuses almost entirely on decision analytic thinking, bringing in behavioral issues only as they illuminate, guide, or challenge decision analytic procedures. The goal of decision analysis could be described as the engineering of thought in certain problem situations; the goal of this chapter is to describe how that can be accomplished. The topic is demanding, and many subtopics within it would be skimpily treated at book length; full treatment at chapter length is impossible. *Decision analysis and behavioral research,* by von Winterfeldt and Edwards (1985) presents at book length all that this chapter covers and much more, from the same perspective. If something said here tickles your curiosity, the chances are good that that book contains a far more extensive treatment.

9.1.2 ELEMENTARY IDEAS

Decisions are made between alternative *courses of action,* or *options,* normally defined so that choosing one precludes choosing all others. An option, selected and executed, is an *act*. Choices are made by the *decision maker*—whom I shall call *you*. For most major decisions, that individual is fictional, since major decisions are typically made by groups or organizations. Such organizations often contain individuals labeled as decision makers. While the roles of these individuals vary widely, they are called that because they take the blame if the decision works out badly. Formal decision theory assumes that the (individual) decision maker has a set of *values,* and chooses acts that, as he or she sees it, will best serve them. This chapter also assumes that decisions serve values, and should serve them well, but does not treat those values as necessarily describing individuals; they may describe groups or organizations instead.

The extent to which the *consequences* of a decision serve your values may depend not only on your act, but also on *events* not under your control. It is convenient to think of your life, as it relates to a decision problem, as a strictly alternating sequence of acts and events. This is a convention, made to fit facts by combining sequences of acts without intervening events into a compound act, and similarly for sequences of events without an intervening act.

Act–event sequences, also known as *decision trees,* extend in principle at least for your life span; indeed, when you make a will you try to extend them further. Patently, any useful representation must be cut short in order not to be a *bushy mess*. The point at which a decision tree is cut is inevitably arbitrary. Two rules govern the choice; a decision tree should not be too complicated (a rule of thumb is that one should be able to draw it on one sheet of $8\frac{1}{2} \times 11\frac{1}{2}$ in. paper, with labels for the nodes) and the tips (or in nonstandard usage that I like, twigs) of the tree, called *outcomes*, should be easier to evaluate than prior points at which the tree could have been pruned. Figure 9.1.1 shows a very simple decision tree concerned with whether to buy for $300 a collision insurance policy with a $200 deductible for a $10,000 car. Since, even in this simple case, the set of possible outcomes is inconveniently large, this tree considers only three possible outcomes of each event or chance node.

In my experience, a useful place at which to prune many decision trees (in fact most of those I have met in practice) is prior to the first chance node. This makes the decision problem one of assessment of values and omits probabilities, but need not omit uncertainty. Recognition of the arbitrariness of pruning decision trees makes irrelevant the traditional distinctions among decisions under certainty, uncertainty, and risk.

The consequences of each act, or of each act–event–act–event . . . sequence, are called *outcomes*. Each outcome is conceived of as having a subjective value or *utility*. Many decision analysts distinguish between values, associated with riskless choices and riskless elicitation procedures, and utilities, associated with risky ones. I find the distinction unnecessary and inconvenient, and will call both utilities; for detailed technical discussion, see von Winterfeldt and Edwards (in press). In most contexts, consequences serve a number of different values to some extent, and so a technical procedure called multiattribute utility technology (or theory, or technique; *MAUT* in any case) is needed to generate a single interval- or ratio-scale utility for each outcome.

If uncertainty is important to a decision, its extent is measured by numbers called *probabilities* attached to events. Fallible inference, the only kind with which this chapter will deal, sometimes uses *Bayes's theorem* to revise probabilities on the basis of new evidence. Simple forms of inference such as are found in statistics can use a simple version of Bayes's theorem, but the mathematics relevant to almost every real-world inference is much more complex than that.

Utilities and probabilities are the normal inputs to a *decision rule*. This chapter considers only three, all trivial and easy to apply. The reason decisions are often seen as difficult is because options may be hard to invent; values may be hard to identify; utilities and probabilities may be hard to assess; and vagueness, self-doubt, and iterative recycling plague every step of every important decision analysis. Once you are comfortable about the inputs, the decision-making part of the process is no problem (Edwards, 1971).

The remainder of this chapter follows the steps of one form of a decision analysis. These are:

1. Structuring the problem as a decision problem.
2. Identifying relevant values.
3. Identifying relevant uncertainties, if needed.
4. Assessing the utilities of possible outcomes on the value dimensions identified at step 2.
5. Making fallible inferences, if evidence is available.
6. Aggregating the outputs of steps 4 and 5 into a tentative decision.
7. Performing sensitivity analyses and recycling.
8. Reaching and implementing a final decision.

Treatment of steps 1 through 6 will inevitably be dry and expository, since real examples are difficult to use until the tools are available. Sensitivity analysis will be presented primarily by means of a real example. Instead of directly discussing step 8, I shall present four real examples from published and not-yet-published literature, selected to display the diversity of decision-analytic problems, methods, and results.

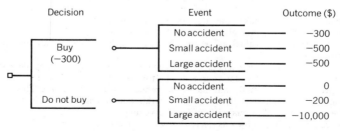

Fig. 9.1.1.   Decision tree for an insurance example.

9.1.3 STRUCTURING THE PROBLEM, IDENTIFYING VALUES, AND IDENTIFYING UNCERTAINTIES

Decision theory starts by presuming that available acts and relevant events are predefined. Real decision analysts are seldom lucky enough to be asked for advice about such problems. A much more typical starting place for an analysis is "This situation is intolerable, for the following reasons. I have thought of actions A, B, C, and D, but all seem to have unavoidable and intolerable consequences. Now what?" Indeed, an analyst is lucky to encounter such a frank admission of perplexity. At least equally often, the situation is as described, but the problem is presented as a choice of A, B, C, or D—all of which are indeed intolerable.

The first task of the analyst is to learn what the function of the organization is, or more broadly put, what the context of the decision is. This includes not only organizational or individual objectives and constraints, but also a broad collection of technical details, folkways, history, organizational structure, and, above all, vocabulary.

In the course of this learning, the analyst must identify the decision maker, or more often, discover what say in the decision exists at each node of an organizational structure. While organization charts are a useful starting place, the informal mores of the organization are at least as important as the formal structure in answering this question. The analyst does not at this point want to be interacting with the boss. The boss's time is limited, and should be saved for tasks that no one else can do, such as weighting values. The early stages of most analyses consist of a lot of interaction with staff, some with line managers, and a lot of reading.

The first step in structuring usually is to get a preliminary idea of what options really exist. This is normally easy, since members of the organization have already thought hard about the question. But the analyst should probe for options that occur to him or her but not to organization members. Do they not make sense? Why not? Often they do not, and the reasons why not reflect either organizational values or organizational folkways and constraints of which the analyst needs to be aware.

Some public decisions involve multiple organizations in conflict; decisions that hinge on the environmental consequences of technological development are examples. In such cases, an analyst must become familiar with the characteristics and values of the various stakeholder groups with voices in the decision. Often, in such cases, the staff of a regulating agency can be the place to start, although the analyst must keep in mind that the regulator is normally a stakeholder too.

Out of this process of exploration, two products should emerge. One is a list of options. The other is a value structure relevant to evaluating the consequences or potential consequences of each. These are closely related and normally are functions of each other. Options suggest values that help to discriminate among them. Values suggest new options that are likely to serve them well. The analyst may be able to be highly innovative in option invention because the organization may not have structured its values explicitly, and so may not have thought hard about innovative options that may serve them. (If so, the analyst also needs dexterity at making the new option occur to and be suggested by an organization member. "Not invented here" is a severe handicap to any option.)

Option invention and value structuring are highly recursive activities. A good rule of thumb for analysts is: if your picture of the options and values relevant to the problem has not changed substantially at least three times, you are not ready for definitive analysis yet. This rule of thumb does not imply that you should not proceed with informal elicitation of numbers and performance of computations. These procedures are often highly stimulating to intuition and can help to produce restructuring. My own experience is that option invention goes more quickly than value structure development. Sometimes, options really do come predefined, at least generically. Values virtually never do. My files suggest that the median number of value structures I come up with for a single problem is about eight, but some of these are fairly minor modifications of their predecessors.

9.1.3.1 Kinds of Structures

Every decision problem that I have encountered requires a list of values to evaluate consequences. The top-down method of elicitation that seems to be most common consists of starting from relatively abstract values and deriving from them more concrete ones, until one gets to values that are clear and specific enough so that each possible consequence of the decision can be scaled on each value. Such scaling may come from measurement, judgment, or a combination of the two. The section on utility measurement returns to the topic.

Figure 9.1.2 presents a top-down value tree that an electric company might use to evaluate alternative technologies for generating more electric power. The twigs (rightmost nodes of the tree) are not sufficiently disaggregated to be measurable. In completed form, such a value tree would extend at least one set of nodes to the right, and perhaps two for some nodes.

Von Winterfeldt and Edwards (in press) propose a seven-item check list for evaluating value trees:

1. Does the value tree repeat subobjectives and attributes? The subordinate values may be redundant and unnecessary.

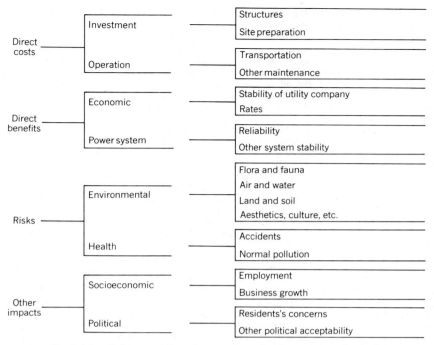

Fig. 9.1.2. Refinement of a value tree for evaluating energy technologies.

2. Can the decision maker think of any value-relevant aspect of the options that has not been captured? If so, the tree may not be complete.

3. Are attributes highly correlated across options? If so, they may be redundant.

4. Can the options be located on each attribute easily? If not, the attribute may be ill-defined.

5. Can the decision maker(s) or experts think of "good" and "bad" scenarios for each objective and value dimension in the hierarchy? If not, the dimension may not be value-relevant.

6. Can the decision maker think of preferences in an attribute independently of the levels in other attributes? If not, the dimension may not be judgmentally independent of other dimensions.

7. Are there some easy rules for aggregating attribute measure (e.g., combining various dollar costs by summation)? If so, the breakdown may be unnecessary.

Test 3 is often violated, and such violations present no special problems. Violation of test 4 is usually a sign of too little thought. Violations of test 6 will inevitably occur, but should be minimized in order to permit additive models and elicitations to be used. Test 7 may be misleading; dollar costs that come from various budgets, for example, may not be suitable for summation. In short, this checklist is intended as a guide for thought rather than as a substitute for it.

Note that only decision trees have been presented. Inference trees, for probability inference, are too specialized for discussion here (see, e.g., von Winterfeldt and Edwards, in press).

9.1.3.2 Research Needs

Structuring is an art form, subject to few rules and little prescription. Perhaps the most interesting ill-understood problem within it is the nature of the creative processes underlying option invention. Options grow out of values, and, later, values grow out of options; but little is known about how such reciprocal invention works. Pearl (1977) and Pearl, Leal, and Saleh (1980) have approached the topic from an artificial intelligence perspective, and Pitz, Sachs, and Heerboth (1980) have published a behavioral study of it. Much more work is needed.

Problems often resemble one another enough to permit generic structures to be developed. Keeney (1980) presents generic structures for siting decisions. Work on generic inference structures is in progress. Computer aids to structuring are also being developed; an interesting example is MAUD (Humphreys, 1980).

9.1.4 SINGLE-DIMENSION UTILITIES

A tentative structure has been chosen. Except for pure inference problems, it requires evaluation of consequences. Usually, a value tree will be available to help specify the dimensions on which consequences must be evaluated. Occasionally, as in Miller, Kaplan, and Edwards (1967, 1969), consequences may be evaluated directly, without disaggregation of the dimensions of evaluation. Either way, it will be necessary to assess the attractiveness of various consequences on one or more dimensions. This is clearly a judgmental task; this section of the chapter deals with how to go about making the judgments.

A key issue underlying disagreements about how to do this has to do with whether or not preferences are different from other judgments. Some utility theorists and decision analysts prefer to base elicitations of all kinds on preference and indifference judgments only. This nonpsychological tradition would make sense if evidence could be found that such judgments are either more stable or more valid than other forms of judgments. I know of no such evidence, and indeed the studies that have looked at the topic seem to imply that skill at managing the details of an elicitation is far more important than any such questions to the merit of the result. Since a broader range of judgments permits considerable simplification of elicitation techniques, this chapter will assume that they are appropriate.

The first question the analyst must consider is whether a given value lies on a natural physical scale, and if so, what that scale is. In choosing where to live, one such scale would be distance from the workplace. While this is physically measurable, it is clearly not a best choice as the basis for evaluation of this aspect of where to live. Identification of a best choice would depend on local circumstances. If it were clear that the decision maker would routinely drive to work, the dimension is the nuisance value of the drive, and a good proxy for it might be mean driving time during commuting hours. If the decision maker will or might use public transportation instead, the convenience and duration of the public transportation might be captured in mean commute time. If the likely mode of transportation will vary from day to day, or the destination is not fixed, it may be appropriate to look for other proxies or perhaps to disaggregate the attribute. In any case, the goal is to find a physical measure with which one would expect preference to be roughly linear. Imagination often helps. The decision maker may well feel that driving is a pain, but care much less how long it takes if it must be done at all. Such a set of values would best be handled by disaggregation, but a nonlinear utility scale over distance from the workplace could also capture it.

Many dimensions do not lend themselves to physical scales. Aesthetic ones are obvious examples. So are such subjective ones as pain, fear, and regret.

Some dimensions lend themselves too easily to physical scales; the most familiar and important are costs and benefits. A decision analyst should always think twice about goals expressed in exclusively financial terms. Dollars are often good proxies for other variables, but few organizations have exclusively financial goals or sharply limited planning horizons. The output of a decision analysis can often be stated in dollars, but this is normally the result of extensive exploration of tradeoffs between dollars and other measures of value.

If a physical scale exists on which to locate the objects of evaluation for a particular dimension, the next step is to establish the relation between that physical scale and judged attractiveness. In doing so, one must begin by establishing a range. Sometimes the range will be naturally specified by the objects of evaluation, extending from the least to the most attractive object on the dimension at hand. Often, however, one may not know the set of objects of evaluation at the time ranges are being selected; the househunter described above is an example, since candidate houses enter and leave the market all the time. Often, even if the set of evaluation objects is fixed, the range should not correspond to what they offer. The important point is that the decision maker must know what the range is, and that it must make sense to him or her. An often-useful device for ensuring this is to use "normal" or "plausible" ranges, defined by the decision maker. For commuting time, for example, the decision maker might take 10 min as best and 45 min as worst. Such a judgment implies that the decision maker does not expect, or perhaps does not wish, to live closer to the office than 10 min away, and would regard a commute of longer than 45 min as unacceptable.

Once range has been established, a question or two can easily identify whether utility within the range is monotonically increasing, monotonically decreasing, or nonmonotonic. Monotonic functions are preferable; nonmonotonic ones invite considering whether the value structure needs more work. But the answer may well be no. A single-peaked preference function with an internal maximum is often acceptable. One example with an internal maximum is racial–ethnic mix in classrooms (Edwards, 1979). The textbook example is the amount of sugar in a cup of coffee. Single-valleyed functions are harder to justify, and functions with multiple peaks or more than one interior valley are forceful critiques of the value structure being used. Of course these issues do not arise for purely judgmental dimensions for which no value-related physical scale exists.

Now scaling can begin. Often a useful way of starting is to try to identify, on the underlying value-relevant scale, a midpoint in value. The analyst often hopes to find that the value-relevant scale is sufficiently linear with the decision-maker's values so that by connecting the end points the utility assessment job can be considered done. Minor deviations from linearity can often be ignored, at least in a first-cut analysis. Major ones may exist and cannot be ignored.

The easiest way of finding a midpoint is to ask for it. "If 10 min of driving time is worth 100 points, and 45 min is worth 0, what amount of driving time is worth 50 points?" Such questions have a hidden danger. They require numerical answers, and a natural tendency is to do arithmetic. An answer like 27.5 min would be a good tip-off that that had occurred; an answer like 25 or 30 min or an indication that the respondent's feelings lie in that range would be preferable.

If the midpoint seems linear, two more bisections should be plenty to satisfy the analyst that the function is sufficiently linear to be so treated. If not, and if the analyst feels confident that the obtained midpoint is genuine, two more bisections of the half-ranges will provide enough points so that a function can be fit to them, either by eye or by choosing some suitable functional form. Figure 9.1.3 suggests some useful functional forms; many others have also been used.

Both to satisfy compulsiveness and to honor the traditions of decision analysis, it is often appropriate to check such psychophysically assessed single-dimensional utilities by means of preference judgments. A typical procedure would be to define a 50–50 gamble between the upper and the lower boundaries of a monotonic function, and to assess the point such that the decision-maker would be indifferent between getting that point for sure and getting the 50–50 gamble. This variable certainty equivalent procedure is appropriate when the underlying continuum is well defined and densely spaced options on it are easy to think about. If the set of options is sparse and/or no underlying value-relevant continuum exists, the analyst will instead need to vary the probability of getting the best or worst option in order to find a value such that the respondent is indifferent between the bet and some intermediate sure thing. This variable probability procedure is a little harder for some respondents to understand.

The value-utility distinction that I rejected earlier says that such a test of psychophysically elicited

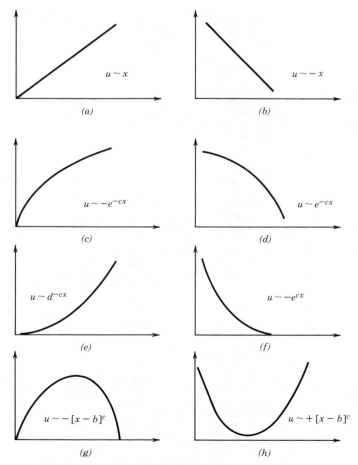

$u \sim x$ (a)

$u \sim -x$ (b)

$u \sim -e^{-cx}$ (c)

$u \sim e^{-cx}$ (d)

$u \sim d^{-cx}$ (e)

$u \sim -e^{cx}$ (f)

$u \sim -[x-b]^c$ (g)

$u \sim +[x-b]^c$ (h)

Fig. 9.1.3. Possible shapes of value functions ($\sim$ means "proportional to").

value scales by means of questions about bets will typically fail because of attitude toward risk. It is certainly true that if a rational decision maker has a utility function shaped like that in Figure 9.1.3c, he or she will reject the gamble consisting of a 50–50 chance of getting the upper or lower ends of the continuum if the alternative is its midpoint. Such a pattern of preferences is frequently though not invariably observed (for gains, not for losses), and is called risk-averse. But it does not follow from observation of risk-averse choices that the decision maker has a curvilinear utility function over the relevant utility dimension. It may instead be the case, as Bell (1982) has suggested, that he or she would prefer not to lose, and, consequently, that phrasing the question in gamble form introduces a new dimension of regret into the utility assessment picture.

Actually, many elicitations based on choices among bets are really direct numerical assessments in disguise. After one or two elicitations in which the variable probability method is implemented by means of a two-sectored green and yellow disk over which a spinner is to be rotated, and the percentage of green is adjusted to find the indifference point, almost any respondent will recognize that the indifference point is the value of the sure thing, and will simply estimate it. Consequently, the distinction between psychophysical and preference-based elicitations is often tenuous. My experience, and that of many analysts, is that the two procedures yield very comparable results. I believe the psychophysical ones to be easier to use.

A nonmonotonic utility function presents a more difficult assessment problem. One should determine with some precision where the peak is for a single-peaked function. (Single-valleyed functions are very rare indeed!) One should also be more careful about boundaries in such cases, since utility may fall off much farther on one side of the peak than on the other. For crude work, a bilinear function from each boundary to the peak may be adequate. For more careful work, several points on each side of the peak will be needed to determine slopes and provide information adequate for curve fitting.

For dimensions that have no underlying value-related continuum, the first step is to establish a preference ordering on the relevant dimension among the objects of evaluation (or, if the list of objects of evaluation is not fixed, among a fairly rich subset). Then, define the best as having a value of 100 and the worst as having a value of 0, and ask for numerical judgments of the objects in between. Once this has been done, check for consistency by asking a few gamble-based questions of the variable probability type. If, later, new objects of evaluation enter the picture, it will often be adequate simply to fit them into the existing preference ordering, and then obtain numerical utilities for them either by direct assessment or by taking the midpoint of the interval in which each falls as its utility.

Inconsistent judgments are valuable at any point in the process of single-dimensional utility assessment. They give the decision analyst a good reason to go back to the decision maker and invite harder thought, possible restructuring, and reconsideration of elicited values. Decision analysis is inherently iterative; any event that encourages and justifies iteration of a structure or numbers about which one still has misgivings is to be sought after and welcomed. On the other hand, numerical precision is not the goal; adequate capturing of considered judgments is. Often, numerical precision is of no use whatsoever. Generally, precision is needed more when the set of options is continuous and infinite than when it is sparse and discrete.

9.1.5   MULTIATTRIBUTE UTILIITY: MODEL FORMS AND WEIGHTS

The preceding discussion was concerned with how to assess numbers that I sometimes call utilities and sometimes call location measures; they indicate where an outcome falls on a dimension of value, in units that are conceived of as having at least interval scale properties. But if analysis is based on a value tree or other value structure (e.g., a list of values), a collection of interval-scale measures of attractiveness of an outcome or an option on a number of different dimensions is not very useful. It may help to eliminate some options by dominance, but it will not help, other than informally, to choose among those still in contention. What is needed is a procedure for aggregating those numbers into a single measure of the attractiveness of each outcome. Any such aggregation procedure must deal with two facts of life. One is that some values are simply more important than others. The other is that how valuable a given location on one dimension is may depend on the location of that same outcome on some other dimension; that is, values may interact.

The notion of importance is very slippery. The word is intuitively meaningful to us all, and indeed that intuition is so strong that people have no trouble making ratio judgments of importance. But the idea is inherently ambiguous. Suppose you are choosing among jobs, and that two major issues are city and salary. Suppose further that cities in which you have offers are Los Angeles, San Francisco, and New York, and that you like Los Angeles best and New York least. Assuming that you prefer more salary to less and that on other value dimensions the three offers are equivalent, you have a decision problem only among pairs of job offers for which the more attractive city is associated with the lower salary. You might regard salary as the more important value if you were offered $30,000 a year in Los Angeles and $60,000 a year in New York, but location as more important if those figures were $30,000 and $30,500.

Given this interaction between importance and range, the options are to use elicitation methods that automatically take ranges into account and do not refer to importance directly, or to exploit the intuitiveness of importance while making sure that the respondent takes range into account in making the necessary weight judgments.

Value interaction is a fact of life. You may feel that Los Angeles has more cultural and recreational advantages than San Francisco, but San Francisco has good public transportation and Los Angeles does not. Consequently, my preference between the two cities, other things being equal, may depend on salary; if I can not afford a car or theater tickets or weekends in Mexico, I may prefer to be in San Francisco so that I can at least ride BART to work, rather than depending on once-an-hour buses or the charity of friends.

Unfortunately, the degree to which it is practical to take value interaction into account in choosing a multiattribute utility aggregation model is very limited; virtually all applications have used either an additive model or a multiplicative one that can incorporate only some very simple interactions. More complex models are easy to define, but the elicitation labor they imply is generally prohibitive. Fortunately, one can often take care of possible interactions in the structuring phase of an analysis. In the Los Angeles versus San Francisco example, I would probably know ahead of time whether the jobs I qualified for would pay enough to make it comfortable to own a car. If I did not, or if the answer were borderline, I might disaggregate city preferences into such subdimensions as ease of use and effectiveness of public transportation, availability of affordable recreational opportunities, and the like. The weight of the public-transportation dimension, for example, would depend on whether or not I expected to use it. The most important point to keep in mind is that the goal of multiattribute utility is not to come up with a fully faithful representation of a respondent's detailed values and preferences, but rather to come up with a tractable one adequate to the purpose at hand. It is both common and appropriate to test for additivity, conclude that it does not apply, and go right ahead and use an additive model anyhow.

Table 9.1.1 summarizes the two common models. Two versions of the additive model are given, depending on whether a nonlinear transformation of the underlying physical continuum is needed or not. The multiplicative model captures all interaction effects into a single parameter w, and reduces to the additive one when $w = 0$. For the multiplicative model $-1 < w < \infty$. The additive model, in any of its versions, assumes that the utility of any object of evaluation on any dimension is independent of its locations on all other dimensions. More formally, additive difference independence (ADI) requires that the relative strength of preference between two objects x and y that have identical fixed levels in some attributes should not change when these fixed levels are changed to some other fixed levels. ADI, if present, justifies the additive model. The multiplicative model assumes multiplicative difference independence (MDI); MDI holds if the only effect of changing a constant attribute from one physical value to another is to shrink or expand the scale of value and to move the elements of the set up or down along it. Simple tests for both assumptions can be formulated; see, for example, Dyer and Sarin (1979). MDI is itself a quite demanding condition; the relaxation from the strength of the ADI assumption is not great.

The preceding paragraph is technically about value rather than utility elicitation methods, since it deals with riskless contexts. The risky version of ADI is more likely to be violated, because of something called multiattribute risk aversion (see von Winterfeldt, 1980). If you would prefer a 50–50 lottery in which heads produced the best value on dimension A and the worst on dimension B, while tails produced the opposite, to a lottery in which heads produced the best on both dimensions and tails produced the worst on both, you are multiattribute risk averse. In thought experiments for

Table 9.1.1   Some Models for Aggregating Single Attribute Utility Functions ($n = 3$)

Model	Formula[a]
Additive with linear value function	$u(x) = w_1 x_1 + w_2 x_2 + w_3 x_3$
Additive	$u(x) = w_1 u_1(x_1) + w_2 u_2(x_2) + w_3 u_3(x_3)$
Multiplicative (extended)	$u(x) = w_1 u_1(x_1) + w_2 u_2(x_2) + w_3 u_3(x_3)$ $+ w w_1 w_2 u_1(x_1) u_2(x_2)$ $+ w w_1 w_3 u_1(x_1) u_3(x_3)$ $+ w w_2 w_3 u_2(x_2) u_3(x_3)$ $+ w^2 w_1 w_2 w_3 u_1(x_1) u_2(x_2) u_3(x_3)$
Multiplicative (compact)	$1 + w u(x) = [1 + w w_1 u_1(x_1)] [1 + w w_2 u_2(x_2)]$ $[1 + w w_3 u_3(x_3)]$

[a] Notation: u : overall value function
x : evaluation object
x_i : measurement (level, degree) of x on attribute i
u_i : single attribute value function
w_i : weight of attribute i

situations involving only gains, most people are. For such reasons, most users of the multiplicative aggregation rule combine it with gamble-based elicitations; users of the additive rule may use gambles or simpler psychophysical procedures. My own advice is to use the additive rule unless some obvious and strong interaction (e.g., between performance and reliability; you won't care much about performance if the system is broken down all the time) makes an interactive rule obviously necessary. The data seem to indicate that differences between risky and riskless elicitations are minor (Fischer, 1977; von Winterfeldt, Barron, and Fischer, 1980). In general, the riskless procedures are easier to use, but it is a good idea, after conducting a riskless elicitation, to try out a few appropriate bets, and in particular, if the context makes it relevant, to explore multiattribute risk independence. If the respondent is not multiattribute risk independent, you should consider whether the violations implicit in using an additive approximation are large enough to worry about. For much more detail and a different point of view, see Keeney and Raiffa (1976). For a more technical exposition of this point of view, see von Winterfeldt and Edwards (in press).

9.1.5.1   Eliciting Swing Weights

If you are using an additive model (with or without linear utilities), you can choose a variety of basically psychophysical weight elicitation techniques. Swing weights is not the simplest, but it does have the technical advantage of taking ranges into account. It begins by defining the utility of the outcome in which all attributes are at the bottom of their ranges as 0, and the utility of the outcome in which all attributes are at the top of their ranges as 1. Then the respondent is asked, attribute by attribute, what increase in utility is produced by increasing that attribute from the bottom to the top of its range, leaving all others at the bottom. The result will be a set of numbers between 0 and 1. These numbers, renormalized to sum to 1, will be the weights of the attributes.

9.1.5.2   Eliciting Ratio Weights

Ratio weights exploit the intuitiveness of the concept of importance. The respondent first arrays the attributes from most to least important, taking ranges into account. Then the least important attribute is assigned a weight of 10, and those above it are assigned larger weights by asking how many items more important each is than the least important attribute. It is crucial that the respondent understand "important" in a swing sense, and have the ranges of the attributes in mind. Consistency checks can be done by using attributes other than the least important one as the base, and getting ratio judgments with respect to them. The final step, as with swing weighting, is to normalize the elicited weights.

The ratio weighting procedure was a part of what Edwards (1971b, 1977) called a simple multiattribute rating technique (SMART). Von Winterfeldt and Edwards (in press) have chosen to use the acronym SMART in a more generic way, as a name for procedures of this type that depend on ratings.

9.1.5.3   Eliciting Weights by Gambling Procedures for the Multiplicative Model

While there is no logical requirement that it be so, in fact a close linkage exists between elicitation methods and choice among model forms in MAUT. Everyone prefers additive to multiplicative models; they are easier to work with, easier to understand and explain, and conform more easily to nonquantitative intuitions like the notion of importance. Perhaps because of that intuitiveness, those who prefer psychophysical to gamble-based elicitation methods also are likely to use additive models, even when the assumptions on which they rest are obviously wrong. Those who are comfortable with the variable probability or variable certainty equivalent elicitation methods use the multiplicative model more often. (No one, so far as I know, uses the multilinear model, which really begins to capture the complexity of the interactions that value dimensions can have.)

The first step in a typical analysis that combines the multiplicative model with gamble-based elicitations is to use one or the other of the gamble-based procedures to construct utility functions, which may not be linear, over each attribute separately. To use the certainty equivalent method, for example, you would start by defining the range of the attribute. Then you would ask the respondent to specify a value of the attribute such that he or she would be indifferent between an option worst on all other attributes and having that value for the attribute in question and a 50–50 gamble in which winning means getting an option worst on all other attributes but best on that one and losing means getting an option worst on all attributes. This specifies the midpoint of the single-dimensional utility function. A further set of such questions, all based on 50–50 bets, but using the previously determined points instead of the best or worst values for the attribute, will provide as many additional points on that function as may be required. Even quite serious nonlinearity will seldom require more than three or four elicitations.

The next step is to elicit the w_i. The key to this is use of the convenient fact that the contribution of each dimension to aggregate utility is 0 if the dimension is at its worst level. Using the variable probability method, determine a value p_i of a probability such that the decision maker is indifferent between receiving for sure an option in which the ith dimension is at its best level and all others are at their worst, and a gamble yielding the best possible option with probability p_i and the worst possible gamble with probability $1 - p_i$. (Such a gamble is called a basic reference lottery ticket, or BRLT. BRLTs are usually used in implementations of the variable probability methods.) The best possible option has utility 1 and the worst has utility 0, by definition. So

$$w_i u_i(x_i^*) = p_i u(x^*)$$

or, from the equation for the multiplicative model in Table 9.1.1,

$$w_i = p_i$$

By solving as many such equations as there are dimensions, one can obtain all the w_i. Note that these are not required to sum to 1 for the multiplicative model, although they would be for the additive model.

The remaining task is to find the interaction constant w. This is done by calculation. If all attributes are set at their best levels, then all of the $u_i(x_i^*)$ are equal to 1. So, from the fundamental equation of the multiplicative model (see Table 9.1.1),

$$1 + w = \prod_{i-1}^{n} (1 + w\, w_i)$$

This equation can be solved for w using iterative procedures presented in Keeney and Sicherman (1976).

9.1.5.4 Multiplying Weights Downward Through a Value Tree

All of the procedures for weighting specified so far depend rather sensitively on the ranges of the attributes, and would be difficult to use if those ranges were not defined. But in any value tree the meanings of the higher-level attributes are defined by the attributes to the right of them. Once a value tree has been used to ascertain what the appropriate values are, its superstructure can be discarded and only its right-hand nodes need be used, since those are the ones for which location measures or single-attribute utilities are obtainable.

Discarding the value tree structure, however, leaves you in two kinds of difficulty. One is that you can deal only with single-dimensional utilities or fully aggregated ones—nothing in between. This may be inconvenient. One reason for using multiattribute procedures is that you feel more confident of the resulting utility assessments, since they depend on more detailed examination of the things being assessed. But another may well be that you are interested in looking at multiple measures of the same objects of evaluation. It is by no means always the case that only full aggregation or full disaggregation is interesting; often a partially aggregated profile describing the objects of evaluation is more informative than either.

The second difficulty is of a more practical kind. If you are dealing with a multiattribute structure with, say, more than 14 attributes, it may well be difficult to make the judgments that trade them off against one another directly. It would be more convenient to make judgments directly only about nodes that derive from a single higher-level node, and then use a computational procedure that derives more general weights from these.

For additive models and psychophysically elicited weights, multiplying through the tree is a convenient device. Figure 9.1.4 illustrates the idea for a hypothetical example. The example is concerned with alternative sites for a treatment center for individuals who have been required to participate in a antidrug-addiction program as a condition of parole. The four top-level dimensions, A–D, have been weighted against one another by judging importance weights directly, using ratio judgments with D as the standard. Similarly, AA–AE, BA–BB, CA–CC, and DA–DB have been compared with one another. Final weights for each twig have been obtained by multiplication. Thus, the weight of AA is $0.17 = 0.43 \times 0.39$. This is one reason for advocacy of importance weights. It would, of course, be possible to work the other way, obtaining the weights of AA through DB directly by some procedure such as swing weighting and then obtaining the weights of the higher-level nodes by addition. I know of no example in which that has been done. Procedures that depend crucially on a clear understanding of ranges, such as swing weights or gamble-based weights, are harder to apply at higher levels of the tree, since the extremes of such higher-level nodes are defined by the conjunction of the extremes of all the nodes that depend from them. While that fact presents no formal difficulty in applying more complex weighting procedures, it does make them more difficult to understand.

9. 1. 6 UNDERSTANDING, ELICITING, AND REVISING PROBABILITIES

In order to understand how probabilities are used in decision analysis, it is first necessary to understand what a probability is. Consider the following two (independent) hypotheses:

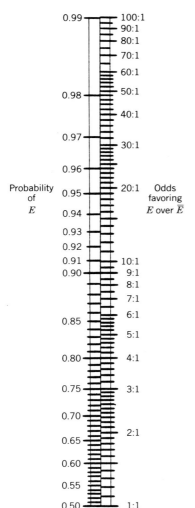

Fig. 9.1.5. Relationship between probabilities, odds, and log odds.

probabilities—and the latter are often very important in applications. Human beings cannot be expected to discriminate judgmentally between 10^{-6} and 10^{-8}, even though the distinction, if it refers to the probability that some industrial or environmental disaster may occur, could be crucial to sensible decision making.

The most common method of assessing very low probabilities is to calculate them from bigger ones. Industrial disasters typically require numerous simultaneous events, not just one. A model called a fault tree can be designed to specify what combinations of events will produce the disaster. Then, with appropriate attention to which independence assumptions are and are not appropriate, one can calculate the probability of the disaster from the probabilities of the component events required for it to happen. Much controversy has arisen from U.S. applications of this technique to calculation of the probability of a core meltdown in a nuclear power plant (U.S. NRC, 1975; McCormick, 1981).

The psychological literature is full of controversy about whether, how well, and under what circumstances people can assess probabilities and perform fallible inferences, and what biases and systematic errors they may be subject to in the process. An adequate review and discussion of the topic would take too long, and lead too far away from the main theme of this chapter. For just such a review and discussion see von Winterfeldt and Edwards (in press) or Edwards and von Winterfeldt (in press); the ideas are the same in both places. The fundamental conclusion to which those discussions lead is that one can indeed hope to elicit probabilities from respondents, especially sophisticated ones, that will be useful guides to decision making—but that the processes appropriate to the task conceal many traps for the unwary.

9.1.7 SINGLE-STAGE BAYESIAN INFERENCE

If probabilities are opinions, then Bayes's theorem specifies how such opinions should be revised as new evidence bearing on them becomes available. Bayes's theorem itself is a simple and uncontroversial consequence of the fact that probabilities sum to 1; controversies about the Bayesian position arise from the assertions that probabilities are opinions and that unique events can have probabilities.

Suppose that you wish to know the probability of some hypothesis H in the light of some datum D. Bayes's theorem says that

$$P(H|D) = P(D|H)\,P(H)/P(D)$$

$$\text{where } P(D) = \sum_{i=1}^{n} P(D|H_i)\,P(H_i)$$

$P(H|D)$, the posterior probability of H after D has been observed, is found by calculating the product of $P(H)$, the probability of H before D was observed, usually called the prior probability, and $P(D|H)$, the probability that datum D would be observed if hypothesis H were true, and the product is divided by the normalizing constant $P(D)$. An especially convenient form of Bayes's theorem is obtained by writing it twice for two different hypotheses, H_1 and H_2, which need not be mutually exclusive nor exhaustive, the probabilities of both of which are revised by the same datum D. One can divide one equation by the other. $P(D)$ cancels out, and the result is

$$\frac{P(H_1|D)}{P(H_2|D)} = \frac{P(D|H_1)}{P(D|H_2)}\frac{P(H_1)}{P(H_2)}$$

$$\text{or } \Omega_1 = L\,\Omega_0$$

This is the odds-likelihood ratio version of Bayes's theorem, and links directly with the previous discussion of odds. Still another way to write Bayes's theorem, of course, is

$$\log \Omega_1 = \log L + \log \Omega_0$$

This formulation is the only one available in which change of opinion is linear with evidence; that is, the amount of change from log prior odds to log posterior odds produced by a given datum is independent of the value of the log prior odds. It can easily be seen from this that $\log L$ is a measure of the ability of a datum to discriminate between H_1 and H_2, and indeed is intimately formally related to the measures of diagnosticity used in information theory.

It is as easy to write and work with Bayes' theorem for continuous cases as for discrete ones; for an exposition of the mathematics involved, see Edwards, Lindman, and Savage (1963) or any recent text in Bayesian statistics (e.g., Phillips, 1973).

The odds-likelihood ratio version of Bayes's theorem suggests an important and very general fact about Bayesian inference: the likelihood principle. Note that, in that formulation, one does not need to know $P(D|H_i)$ and $P(D|H_j)$ in order to perform an inference; it is adequate to know the ratio $P(D|H_i)/P(D|H_j)$. This statement turns out to be true in general for single-stage Bayesian inferences in both discrete and continuous cases. Numbers of the form $P(D|H)$ need be defined only up to an unspecified positive multiplicative constant. This fact, known as the likelihood principle, is directly responsible for most of the controversial aspects of Bayesian statistical inference. It implies, for example, that once data collection has stopped, the stopping rule used is almost never relevant to statistical inference. The natural procedure, taboo under classical rules, of gathering data, stopping every now and then to test to see whether or not you have proved your point, and, if not, gathering some new data to pool with the old, turns out to be not only legitimate but good science. (The procedures a Bayesian statistician would use to "test to see whether or not you have proved your point" are quite different from those of classical statistics, and are not subject to the property inherent in classical procedures that such data gathering and data interpreting would guarantee eventual rejection of any null hypothesis.)

The practical problem of using Bayesian procedures that has captured most attention is that of providing prior probabilities or probability distributions. The difficulties with priors seem to be of two kinds. One is that they seem inescapably subjective, and this is uncomfortable for some. The second is that, even if one accepts their importance and the necessity of producing them, they seem difficult to produce. The first objection is, in my opinion, irrelevant. Posterior probabilities are fully as subjective as prior ones, and indeed every probability is both prior and posterior; posterior to antecedent events and prior to subsequent ones. The second problem is more real, and various devices have been developed to deal with it.

One important one is the principle of stable estimation (Edwards, Lindman, and Savage, 1963). This is simply a set of conditions, often satisfied in the presence of a reasonable number of observations, that justifies one in treating a prior over a continuous quantity as uniform, even though it is not.

When the principle of stable estimation does not apply, either because not enough data are available or because the data, though available, are not sufficiently definitive relative to the prior to swamp out its effects, the next relevant tool becomes that of conjugate distributions. A distribution is conjugate to a data-generating process if the relation between the two is such that data produced by that process will modify only the parameters of the distribution, not its mathematical form. Thus, the beta distribution is conjugate to data produced by a Bernoulli process, normally distributed data are conjugate to the normal distribution, etc. For a compendium of conjugate families and details of the relevant mathematics, see Raiffa and Schlaifer (1961). Note that the relevant question is not whether your actual prior opinions are well fit by the appropriate conjugate prior, but rather whether use of a conjugate prior in combination with the data at hand will produce a posterior distribution that is acceptably close to the one you would obtain by using your real prior and doing Bayes's theorem by brute force. While this question is unrealistic to answer by means of computation, since once you have done the brute-force computation you have no use for the one based on conjugacy, it is nevertheless useful to have in mind; often, situations that do not quite justify the stable-estimation approximation to your prior will fully justify the conjugate-distribution approximation.

If you cannot use either of these approximation techniques, you have no alternative but to assess your prior distribution and calculate your posterior distribution directly. This is often exactly the situation for discrete, unordered lists of hypotheses. The techniques used for them are also the techniques appropriate to brute-force calculations on continuous distributions.

9.1.8 MULTISTAGE INFERENCE

Bayesian procedures are implicit in the mathematics of decision trees, and become explicit when decision trees are used to guide decisions about purchasing information. Otherwise, single-stage Bayesian inference is less often used in decision analyses than one might anticipate from the high correlation between being a decision analyst and subscribing to the Bayesian point of view. The reason is straightforward enough. Single-stage Bayesian inference makes very strong assumptions, which almost never fit real-world inference problems. They are:

1. Datum or data and hypotheses are clearly defined and specifiable.
2. The inference-maker can obtain from somewhere a number, set of numbers, function, or set of functions that link datum to hypotheses in quantitative form. These numbers or functions obey the familiar rules of the probability calculus.
3. Data are conditionally independent of one another; that is, the answer to the question "How likely is datum D if hypothesis H is true?" remains unchanged as the inference-maker learns about data other than D.

In most real inference problems, such as those encountered, for example, in intelligence system contexts, hypotheses are scenarios. More often than not, they are unfolding as the data accumulate, and so are never fully defined. The notion of "a datum" is quite likely to be ill-defined in real contexts. Consider a photoreconnaissance situation in which, on the basis of some aerial photographs, a photointer-preter reports that a tank division is assembling at such-and-such a map location, and, consequently, an attack is imminent. The "data" are photographs. The tanks are blobs on those photographs; the assessment that these blobs are tanks is an inference by the photointerpreter. The ideas that the tanks are assembling and that an attack is imminent are additional inferences based on the first one. Such hierarchical chains of inference characterize not only intelligence analysis, but almost all real-world inference contexts.

The beginnings of a technology of real-world inference exist, thanks to Professor David A. Schum of George Mason University. The context in which the complexity of real inference problems has been most extensively studied is legal inference. Schum discovered the work of a legal scholar named Wigmore (1937), and has considerably amplified it and given it a Bayesian quantitative form. (See Schum (1977, 1980) and Schum and Martin (1982) for progress reports.) So far as I am aware, no real applications of hierarchical inference techniques have yet occurred, although one is currently under development.

The problems of hierarchical inference are formidable. As yet no one has published technological ideas about how to think quantitatively about scenarios that unfold while one watches them. Schum has specified appropriate formal structures for performing numerical inferences in situations in which hypotheses are well-defined but the data are linked to them only through intermediate hypotheses, and therefore are not conditionally independent. Unfortunately, the numerical assessments required to work with such structures are especially difficult to make, in part because of the complexities produced by highly conditional probabilities and in part because in such hierarchical inference structures the likelihood principle fails. The sufficiency principle, so familiar in conventional statistics, depends in Bayesian statistical inference on the likelihood principle; if the likelihood principle does not apply, the sufficiency principle does not apply either. This means that one cannot know *a priori* what aspects of the datum may be relevant, or in how much detail the datum should be specified. Moreover, failure of conditional independence means that old data may affect the diagnostic impact of new data. Consequently, a datum cannot be discarded once its impact on the probability distribution of interest has been determined; it must be retained against the possibility that a later datum may change that impact. Any technology development program that must overcome these intellectual obstacles is guaranteed to be slow, intellectually demanding, and full of pitfalls and requirements for approximate methods. But this seems to be the most promising technology development direction in which decision analysis is now moving, and progress is in fact being made. Ten years from now, it would probably be impossible to write a chapter such as this without including as much how-to-do-it information about complex inference as this one has about how to measure utilities. For a fuller discussion of the present state of this art, but without further details about ongoing efforts toward development of the technology, see Chapter 6 of von Winterfeldt and Edwards (in press).

9.1.9   DECISION RULES

Only three decision rules will be discussed here. Two can be seen as special cases of the third. The literature is well supplied with other rules, which can also be seen as special cases.

9.1.9.1   Dominance

The idea of dominance is perhaps the most obvious and uncontroversial of the decision rules. It can be expressed in the context either of gambles or of multiattribute utilities. In the gamble form: If bet A pays off at least as well as bet B no matter what happens, and is definitely better for at least one outcome, then bet A (ordinally) dominates bet B, and B should be discarded from the choice set. In MAUT form: If act A leads with certainty to an outcome which is at least as good as the outcome of act B on every dimension of evaluation, and definitely better on at least one, then A ordinally dominates B, and B should be discarded from the choice set.

A more subtle version of dominance exists, and is called cardinal dominance. In its probabilistic version, if bet B is not ordinally dominated, but some probability combination of bets can be constructed out of the other bets in the choice set that ordinally dominates B, then B is said to be cardinally dominated, and again should be eliminated from the choice set. This idea becomes clearer on inspection of Figure 9.1.6. Gamble g_1 is ordinally dominated by g_2. Gamble g_3 is not ordinally dominated, but is cardinally dominated by a mixture of gambles g_2 and g_4. Both should be eliminated from the choice set—as, of course, should gamble g_6 also. The choice set, then, consists of the three undominated gambles: g_2, g_4, and g_5. The same kind of thinking applies in multiattribute utility.

It is lucky but rare to encounter a decision problem for which dominance specifies the best option. However, dominance often permits elimination of options from the choice set, and thus reduces the complexity of the problem. A more sophisticated concept, applicable in both decision trees and value trees, is to look for options that can be eliminated on the basis of higher-level dominance—which simply means that the tree is partially aggregated and then the notions of dominance are applied at the partially aggregated level. This notion leads to a form of sensitivity analysis too complex to present here; see von Winterfeldt and Edwards (in press), Chapter 11.

A special case of dominance arises in contexts in which the task is to choose k options out of n contenders, where $k > 1$. It is as appropriate in that case as in any other to delete dominated options before applying any form of analysis, but is much more difficult to determine which options are dominated (at least for the additive case). But the issue is still more complex because, in addition to sure losers,

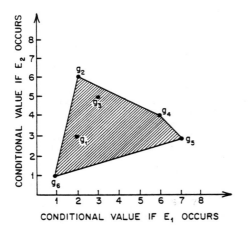

Fig. 9.1.6.   Conditional values of six gambles.

the k-out-of-n case will often have sure winners, options that are certain to be selected by any decision rule. These too should be eliminated from the choice set (and the value of k correspondingly reduced) prior to additional analysis. The effect of reducing the choice set by eliminating sure winners and sure losers is interesting. In many real world applications, such as selection of students for graduate school, the bases for the selection (e.g., examination scores, grade point averages) are likely to be positively correlated with one another. In the presence of such positive correlations, weights applied to these selection criteria are often inappropriate (Dawes and Corrigan, 1974; Wainer, 1976). However, the effect of purging sure winners and sure losers from the choice set is to lower the positive correlations. How far they are lowered depends on the number of dimensions. For two-dimensional cases, the correlations are certain to be negative, and so weights are certain to be important. For larger numbers of dimensions, negative correlations will often appear within the contender set, although obviously it is impossible to have high negative correlations between each predictor and all others.

9.1.9.2 Ruin and Quasiruin

The mathematics of gambler's ruin has been a topic in probability theory for a long time. The idea is simple: A gambler's resources are limited, and so even if the odds are in his or her favor, random fluctuations may make continued play impossible if the opponent has deeper pockets. Thus the principle of accepting favorable bets and rejecting unfavorable ones must be modified if one possible outcome of a favorable bet is ruin for the bettor.

The notion of quasiruin considerably generalizes this idea and makes it applicable to a fairly substantial collection of real-world decisions. The notion is simply that, without removing you from participation in life, some hazards are large enough so that loss would severely impair your life style; such losses are called quasiruinous, and one can easily argue that one should not accept even favorable bets that have quasiruinous outcomes. Thus, for example, it may well be true that you would not be economically ruined if your car were totalled or your house burned down. Yet many people carry, and should carry, collision insurance on their cars, and most people carry fire insurance on their houses, even though in both cases the insurance company prices its insurance to make a profit, and thus buys those bets from you at a cost to you considerably more expensive than their (negative) expected value.

9.1.9.3 Maximization of Subjectively Expected Utility

The expected value (EV) of a gamble is defined by the following equation:

$$EV(g) = \sum_{i=1}^{n} p_i \, v_i$$

where p_i is the probability of obtaining the ith outcome, v_i is the value of doing so, and the summation extends over the exhaustive set of n mutually exclusive events being considered. Exactly the same equation, with SEU substituted for EV and u substituted for v, describes the subjectively expected utility (SEU) of a gamble—keeping in mind that the probabilities are personal. By far the most general decision rule I know of is the one that asserts that you should choose the option, from those available for choice, that maximizes SEU. Indeed, the principles of decision by dominance and by avoidance of quasiruin can easily be seen as special cases of this more general principle.

SEU has a long history, and commands routine acceptance as both normatively appropriate and descriptively correct from many nonexperimentalists. Perhaps the strongest argument for it, at least normatively, is that it has no equally general competitors. A second strong normative argument for it is that it is optimal in a well-defined sense. SEU maximization is demonstrably preferable to any alternative yet proposed in virtually every decision context.

Still, the literature is well supplied with challenges of various kinds to SEU. It has been proposed as a model of how people in fact make decisions in risky situations, and a long literature has challenged that proposal; for a recent instance that contains an alternative proposal see Kahneman and Tversky (1979). The descriptive merit of SEU maximization in specific decision contexts is quite debatable, but I would not defend it as a general descriptive model of decision making.

Other challenges, more important to this chapter, have been directed against SEU as a normative model. They are of two kinds. One asserts that in certain specific kinds of decisions the SEU model leads to prescriptions that no one would willingly accept. One of the older challenges of this kind is the Allais Paradox (see Allais, 1953; Allais and Hagen, 1979). Here is the version of it presented in Kahneman and Tversky (1979).

Problem 1: Choose between	
A. $2500 with probability 0.33	*B.* $2400 with certainty
$2400 with probability 0.66	
$0 with probability 0.01	

Problem 2:   Choose between

C. $2500 with probability 0.33	*D.* $2400 with probability 0.34
$0 with probability 0.67	$0 with probability 0.66

82% of their subjects preferred *B* over *A,* while 83% preferred *C* over *D.* Simple formal arguments show that preference of *B* in Problem 1 and *C* in Problem 2 violates any form of SEU maximization.

Other similar paradoxes exist, and yet others may remain to be discovered. I believe that these paradoxes arise from a combination of two kinds of effects. One is that choices that involve money are virtually always related to one or more kinds of status quo, and such paradoxes leave the status quo ambiguous. Thus, option B changes the status quo with certainty, while option D does not. The other is that transactions of such kinds are assimilated into transaction streams. Both the mechanism and the timing of the assimilation are (1) ill-understood and understudied, and (2) subject to manipulation by a variety of details, including wording and the sizes of the bets. The ideas hinted at previously are spelled out at greater length in Chapter 10 of von Winterfeldt and Edwards (in press); to examine them in detail here would take this chapter far afield. The spirit of these thoughts is in some ways similar to those of Kahneman and Tversky (1979) and Tversky and Kahneman (1981, 1982), but in other respects rather different. In any case, the upshot of the argument is that, so far as I am concerned, the normative status of the SEU model as a tool for decision analysis remains unimpaired by the existence of the paradoxes.

A more challenging kind of attack on SEU as a normative model has been mounted by many authors who have argued, in various ways and on various bases, that the inputs that the model requires are unavailable, inaccurate, or biased. With respect to utilities, this line of thought has become known as the labile values hypothesis, and is well presented by Fischhoff, Slovic, and Lichtenstein (1980). Without examining details here, my view is that some values are labile and others not, and that methods of elicitation make a big difference. See Chapter 10 of von Winterfeldt and Edwards (in press) for a review. The issue is fundamentally empirical, and quite a lot of relevant data exist.

A large literature on biases and systematic errors in probability assessment has accumulated since 1970. I have called such errors "cognitive illusions," and the name seems to have stuck. The issues raised by this literature are numerous and much more complex than those just discussed. This is clearly no place to examine them. Chapter 13 of von Winterfeldt and Edwards (in press) explores them in detail, as does Edwards and von Winterfeldt (in press). The conclusions of both examinations, which have essentially the same content, are that (1) the experimental phenomena are reliable, although rather specific to the detailed experimental conditions; (2) the errors apply not only to probability assessment but to a wide variety of contexts in which people are asked to perform intuitively intellectual tasks that are difficult enough so that one would normally use tools to get them right; (3) that since they are linked to the use of tools, the degree to which they are likely to get in the way of decision analysis depends on how well the analysis uses tools; and (4) that cultural change, by making new intellectual tools available and embodying existing ones in physical tools, changes the amount and kind of cognitive illusions one should expect. While a lot of methodological warnings emerge from this literature, and decision analysts should heed them carefully, nothing that emerges is so overwhelming as to make decision analysis ineffective as a psychotechnology.

A final form of attack on SEU maximization, generated mostly by those interested in organizational decisions, asserts that rationality is an unattainable ideal, and that real decision makers, even when the stakes are high, "satisfice" instead. The basic idea of satisficing is that one selects the first option processed that seems satisfactory, rather than spending the additional effort of processing all options and finding the best. In the absence of criteria for satisfactoriness and of information about the cost of processing, the principle of satisficing seems hard to spell out in detail. In any case, it is clearly intended as descriptive rather than normative in thrust. In contexts in which the cost of processing is higher than the differences among options (and such contexts surely do exist), no conflict exists between satisficing and SEU maximization.

My conclusion from this very sketchy review of SEUs criticisms is that nothing that has emerged from 20 years of such literature should give a decision analyst pause in using SEU as the fundamental normative principle of decision making.

It scarcely needs to be added that, in the special case in which options are conceived of as having determinate outcomes, the principle of maximizing SEU reduces to the principle of choosing the option with the largest utility—typically multiattributed.

9.1.10   SENSITIVITY ANALYSIS

It would be naive to suppose that a decision analyst and a client, working together, can structure or should structure a problem, assess the relevant utilities and weights, assess the relevant probabilities if the structure calls for them, calculate aggregate utilities or SEUs, pick the option with the highest SEU, and then have everyone quit and go home. A lot of anxieties and doubts will have accumulated during the early phases of the analysis. Typically, the analyst will wonder whether the structure of the problem may be oversimplified, or if something that contains its essence may have been inadvertently omitted. Typically, the client(s) will feel uncomfortable about some or all of the assessments. Both should want some basis for belief that the conclusion of the analysis is not accidental; that it is in some sense robust under reasonable amounts of error. In principle, both might like to try the decision out to see what its consequences would be, and then to revise the analysis on the basis of the try-out. Only very rarely does the chance to perform such an empirical test exist. In its absence, analysts have devised a set of procedures collectively called sensitivity analysis that are designed both to give analyst and, to some degree, client more understanding of and security about the analysis and to explore the consequences of alternative representations and alternative sets of numbers. Sensitivity analyses are sometimes disparagingly referred to as "wiggling parameters around." They do in part consist of wiggling both parameters and model forms around, but they are better thought of as "developing expertise about your model."

As an activity, sensitivity analysis is conducted almost entirely by the analyst. It is computationally intensive, iterative, and often yields insight only with difficulty if at all. This fact leads to an important thought. It is usually worth while to distinguish between sensitivity analysis for the analyst and sensitivity analysis for the client. Sensitivity analysis for the analyst is a computer-supported, technically demanding activity. Sensitivity analysis for the client consists of reducing those results of the sensitivity analysis for the analyst that are worth communicating to some simple display form and explaining their meaning and implications to the client. My experience has been that more often than not the sensitivity analyses I did to enhance my own understanding and confidence required massaging and transformation in order to be intelligible to the client, and other practicing analysts report the same.

Two key ideas seem to lie behind the arithmetical part of sensitivity analysis. One is dominance and partial aggregation. Whether working with a value tree or a decision tree, one tends to be more concerned with structural and numerical issues near the beginning (top of the value tree, left-hand side of the decision tree) than with those at the end, simply because the parts of the analysis near the beginning tend to have more leverage on the outcome of the whole analysis than those at the end. Thus, in doing sensitivity analyses of weights, I tend to confine virtually all of my attention to weights at or near the top of the value tree. Often, partial aggregation will cause options to be dominated at a higher level. In many cases, higher-level dominance justifies elimination of the option, at least for the duration of that analysis. Elimination of a subset of options makes it easier and often more illuminating to concentrate on the relative merits of the surviving options in detail.

The other key idea of sensitivity analysis is break-even analysis. That is, one varies interesting parameters of the base-case analysis and checks to find the point at which the conclusion of the analysis changes. Once one has found such a switch-over or break-even point, one can explore the extent to which the numbers that produce the switch are plausible, as compared with those that enter into the base case—or whether the location of the switch-over point is a clue to some even more fundamental problem, such as an initial misstructuring of options.

Rather than go further with abstractions, I next present a real analysis, which focused on a quite sophisticated sensitivity analysis.

9.1.10.1 Sensitivity Analysis in a Medical Decision Problem*

Moroff and Pauker (1983) have recently reported a most sophisticated and elaborate sensitivity analysis in a medical decision-making problem.

The basic analytic tool with which they were working was the declining exponential approximation of life expectancy (the DEALE). This is a technique developed by Pauker and his group (see Beck, Kassirer, and Pauker, 1982; Beck, Pauker, and Gottlieb, 1982) to approximate life expectancy information using the assumption that life expectancy follows a simple declining exponential function—an approximation that seems to work well in many situations. The reason why the DEALE is important is that it provides a clinical decision maker with the utilities needed to make life-or-death decisions. The basic concept is that utility, in life-or-death medical contexts, should be measured by quality-adjusted life years. The adjustments are usually somewhat rudimentary; for example, in this instance they did not take dollar cost into account, but did consider pain and suffering, by subtracting time from life expectancy. Much more challenging to the modeler, of course, is to sort out life expectancies as a function of the wide variety of data bearing on diagnostic procedures and outcomes that the literature has to offer in connection with any medical problem.

In this instance, the patient was a very healthy 95-year-old man who presented with what his primary physician believed to be cancer of the lung. The options were to proceed to surgery, to give radiotherapy without verification of the nature of the lesion (a medically unusual procedure), to administer bronchoscopy and then, if the lesion was diagnosed as cancerous, to proceed with surgery or radiotherapy, or to do nothing. Bronchoscopy was a fairly unattractive option because (a) the prior probability of stage I cancer was 0.90, and (b) even a negative bronchoscopic examination would

* This section is reproduced from Chapter 11 of von Winterfeldt and Edwards, in press, with the permission of Cambridge University Press.

have reduced that probability only to 0.734, mostly because the examination reports cancer when present in only 70% of cases.

The crucial fact is that a 95-year-old healthy man has a life expectancy of 2.9 years. Consequently, the mortality rates associated with diagnostic and therapeutic procedures are of great importance. The fundamental conclusion Moroff and Pauker (1983, pp. 326–327) reached was "that radiotherapy would provide a quality-adjusted life expectancy of our patient of 17.5 months, bronchoscopy 14.1 months, surgery 13.5 months, and no treatment 8.5 months. In a person in whom 'normal' life expectancy is just under 35 months, we regard these differences as meaningful."

Figure 9.1.7 shows the sensitivity analysis for perioperative mortality (i.e., mortality caused by surgery), for which the baseline figure for this patient was 0.35. Clearly, given the other numbers going into the calculation, the decision was not a close call at all between surgery and any other procedure, except bronchoscopy. Bronchoscopy is a close call, and does make a difference, since a negative result would mean that the patient would receive no treatment, given baseline numbers.

A particularly dramatic two-way sensitivity analysis is presented in Figure 9.1.8. It shows the tradeoff between surgery and radiotherapy for various annual mortality rates assumed to go with each. It includes two lines, corresponding to two different perioperative mortality rates for surgery. This figure shows that the conclusion in favor of radiotherapy over surgery in this case is quite insensitive to the specific numbers inferred from the review of the medical literature on which Moroff and Pauker based their baseline figures for mortality rates for the two procedures.

These two displays should whet your appetite for the article itself. The general conclusion to which it leads is that a quite unconventional bit of advice (give radiotherapy without prior bronchoscopy; if the radiotherapist refuses, do the bronchoscopy and give radiotherapy if it comes out positive) is clearly appropriate to this admittedly remarkable case. The numerous other sensitivity analyses contained in the article are something like a textbook example of how sensitivity analyses for decision trees should be done and presented.

It is worth noting that none of the analyses presented took the patient's attitude toward risk into account. If they had done so, the fact that surgery carries a risk of immediate iatrogenic death while radiotherapy does not could only have reinforced the conclusions already reached.

While the computations involved in the DEALE require at least a programmable handheld calculator and fair understanding of the models, the calculations involved in the decision process itself are extremely simple, depending as they do on five probabilities and five utilities. Indeed, for most purposes two of those probabilities and one of the utilities can be ignored, since the option of bronchoscopic examination is never best, and one of the utilities and two of the probabilities are associated with it.

Consequently, a simple display of the two figures should be adequate as sensitivity analysis for the patient. The conclusion is extremely insensitive to the numbers going into it. Inspection of Figure 9.1.7 shows that radiology without bronchoscopy is preferable to any option in which bronchoscopy comes first, for the baseline numbers. That kind of conclusion obviously depends strongly both on the prior probability of cancer and on the two probabilities that describe the bronchoscopic test.

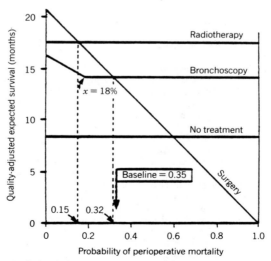

Fig. 9.1.7. Sensitivity analysis of the probability of operative mortality. The heavy arrow marks the baseline assumption of 35%. The dashed vertical lines denote threshold values of 15% and 32%. *X* marks the point above which bronchoscopy, if positive, leads to radiotherapy and below which it leads to surgery. Source: Moroff and Pauker (1983).

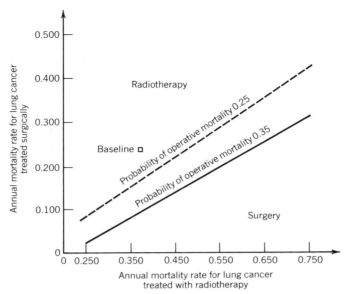

Fig. 9.1.8. Sensitivity analysis of the effect of lung cancer treatment-specific mortality modes on the decision between surgery and radiotherapy. The solid diagonal line represents the baseline assumption of a 35% probability of operative mortality; the dashed line, representing a 25% operative mortality, expands the zone for surgery and shrinks that for radiotherapy. Source: Moroff and Pauker (1983).

A very interesting sensitivity analysis holds the bronchoscopic probabilities constant, but varies the age of the patient. This shows that, for the given input probabilities (adjusted where appropriate for age change) and for ages from 56 on up, bronchoscopy is never the attractive first step, given this high a prior probability of stage I cancer.

A patient might well want to understand better why so common a diagnostic procedure would be inappropriate in all such cases—and dramatically so. The following numbers would help me, and might help the patient. If cancer were present, one would expect a positive diagnosis from bronschoscopy 70% of the time; if absent, one would expect a positive diagnosis (i.e., that cancer was present) 30% of the time. Given the extreme prior probability of cancer, one would expect with probability 0.9(0.7) + 0.1(0.3) = 0.66 to get a positive result from bronchoscopy, followed by radiotherapy. The probability of a negative bronchoscopic examination, and therefore of no treatment, is of course 0.34. But, given a negative bronchoscopic examination, the probability of cancer is still 0.734. So the net result of bronchoscopy would be radiotherapy (needed or not) with probability 0.66, no radiotherapy but cancer with probability 0.34 (0.734) = 0.25, and no radiotherapy and no cancer with probability 0.09. This set of outcomes compares with the strategy of giving radiotherapy with probability 1, in which case it is appropriate with probability 0.9 and inappropriate with probability 0.1. So the decision about bronchoscopy is between a probability of 0.1 of unnecessary radiotherapy, and a probability of 0.25 of failing to give radiotherapy to a patient with cancer, but saving the patient from unnecessary radiotherapy with probability 0.09. The choice does not seem hard to make.

This is a dramatic instance of the effect of extreme priors and dichotomous tests of fairly weak diagnostic power (see Meehl and Rosen, 1955). It is important to note that such effects cannot occur if the initial output of the test can be thought of a continuum of scores, dichotomized by a cut point the location of which reflects prior probabilities and payoffs, as in signal detectability theory examples. So far as I know, bronchoscopy does not fit this description.

9.1.11 APPLICATIONS*

This chapter has so far consisted of a long exposition of methods. But the test of methods is usefulness, and the test of usefulness is use. The remainder of the chapter therefore consists of four real applications, varying in approach, degree of sophistication, degree of success, and kinds of methods used. Other applications are reviewed in Keeney and Raiffa (1976), Howard, Matheson, and Miller (1976), Krischer

---

* This section is condensed from Chapter 12 of von Winterfeldt and Edwards, in press, with the permission of Cambridge University Press.

(1980), and von Winterfeldt and Edwards (in press). Brown (1970) and Ulvila and Brown (1982) review business applications. The journal *Medical Decision Making* is devoted to medical applications, and *Operations Research* and *Management Science* have published special issues on decision analytic applications. Many more applications can be found in technical reports issued by consulting firms like Decisions and Designs, Inc., Decision Science Consortium, the Maxima Corporation, and Woodward-Clyde Consultants.

9.1.11.1 A Suicide Prediction Model

Most people who attempt to commit suicide complain about suicidal thoughts to friends, psychiatrists, or mental health workers first. That is, the probability $p(C/A)$ of a complaint (C) prior to an attempted suicide (A) is much higher than $p(C|\bar{A})$. A good model for predicting suicide attempts from complaint patterns could aid clinicians in their decisions about whether or not to hospitalize a client and what form of treatment to choose.

If c_i is the ith category of a suicide complaint, and if complaint categories and levels within each are conditionally independent, the formal rule for determining the posterior odds of attempt versus nonattempt is

$$\frac{p(A|c_1,\ldots,c_n)}{p(\bar{A}|c_1,\ldots,c_n)} = \frac{p(c_1|A)}{p(c_1|\bar{A})} \frac{p(c_2|A)}{p(c_2|\bar{A})} \cdots \frac{p(c_n|A)}{p(c_n|\bar{A})} \frac{p(A)}{p(\bar{A})}$$

Gustafson, Grest, Stauss, Erdman, and Laughren (1977) developed an elaborate interactive computer program to diagnose suicidal patients and to predict the probability of a suicide attempt using this straightforward Bayesian model.

They first asked a group of experienced psychiatrists and residents in psychiatry to generate a list of possible symptoms of suicide attempts. They generated a list of 32 symptoms having 246 individual symptom levels. Table 9.1.2 shows the symptoms. Next, the researchers asked six clinicians to assess likelihoods for the symptoms and symptom levels by making judgments like the following:

Assume that 100 patients have come in your office with suicidal thoughts, and you know that all of these 100 patients will make a serious attempt within the next three months. Of these 100 patients, how many are:

Single	_____
Married	_____
Divorced	_____
Separated	_____
Widowed	_____

Table 9.1.2 Suicide Prediction Symptoms

Sex	Degree problems can be discussed with most trusted friend
Marital status	
Income	Existence of suicide plan
Effect of patient's suicide on others	Degree of communication of suicide concern to others
Age	
Rage	Living arrangements
Recent loss	Degree of self-like
Expected future loss	Thought disorder
Drug Abuse	Concern about achieving major goal
Religious attendance	Accident proneness
Family history of mental disorder	Previous suicide attempt
Patient's prognosis for next 24 hr	Health concern
Satisfaction with life and daily activities	Religion
Likelihood of solving problems	Frequency of suicide thought in last 24 hr
Feeling of being needed by others	Duration of current suicide thoughts
Degree of impulsivity	Chance of being dead in 1 month
U.S. citizenship adjustment problems	Alcohol abuse

Source: Gustafson, Grest, Stauss, Erdman, and Laughren (1977).

Table 9.1.3 Sample Likelihood Ratios for Selected Symptom Levels[a]

Symptom Category (S_i)	Symptom Level (S_{ij})	Likelihood Ratio	Favoring Attempt?
Marital status	Divorced, second time, less than 1 year	1.8	Yes
Age	13–18	1.63	No
Uses drugs	Not at all	1.08	No
Possibility of solving problems	No chance	3.31	Yes
Frequency of suicide thoughts	One to three times per year	2.34	No
Frequency of suicide thoughts	All the time	5.00	Yes
Suicide plan	Has one and has obtained the means	4.69	Yes
Self image	Moderate	1.13	Yes
Concern about achieving a major goal	Low-moderate	2.60	No
Living arrangements	Alone, do not know neighbors	1.33	Yes

Source: Gustafson, Grest, Stauss, Erdman, and Laughren (1977).

[a] The larger the ratio, the more this symptom level would support the hypothesis in the right-hand column.

Group discussions resolved inconsistent assessments. These assessments generated $p(c_i|A)$. After similarly assessing $p(c_i|\bar{A})$ the analysts could calculate the likelihood ratios, the crucial ingredients in the Bayesian model. For example, a future attemptor was judged to be five times more likely to have a suicide plan and to have obtained whatever was needed to carry it out (handguns, pills, etc.) than a nonattemptor.

An interactive computer program based on the resulting model followed an efficient path in eliciting the information required to calculate the posterior odds of a suicide attempt. A typical result of such an interview is a patient profile like the one shown in Table 9.1.3. The likelihood ratios in it are all expressed as numbers greater than 1. If they are listed as favoring an attempt, they are of the form $p(c_i|A)/p(c_i|\bar{A})$. If not, the form is $p(c_i|\bar{A})/p(c_i|A)$. Given the profile in Table 9.1.3 and a prior probability of a suicide attempt of ½, the model would assign posterior odds of 19.6 or a posterior probability of 0.95 to the hypothesis that this patient would attempt suicide.

The model was validated in an experiment and in field studies, both with positive results. In the experiment, psychiatrists looked at profiles of patients similar to the one shown in Table 9.1.3 and made judgments about the probability of a suicide attempt for each. This assessment was compared with the output of the Bayesian model. Fifty profiles described patients who in fact subsequently attempted to commit suicide; fifty profiles described patients who made no such attempt. Table 9.1.4 shows the percentages of correct predictions for the computer, the residents, and the experienced clinicians. The computer substantially outperformed the clinicians in predicting attempts, and it was only slightly less accurate in predicting nonattempts. The result obviously reflects the fact that clinicians do not assign a prior as high as 0.5 to attemptors, whereas the computer worked with the appropriate prior for the experiment—but the superiority of the computer's performance is too great to explain with that argument alone.

A second dependent measure was the probability assigned to the correct predictions. Obviously one wants the system to assign a very high probability to the prediction that eventually turns out to be correct. Table 9.1.5 shows the average posterior probabilities. On this measure too the computer model outperformed the clinicians substantially, especially by assigning a much higher probability to the correct prediction of a serious attempt. The field studies produced similar results.

This model had a rather sad fate: in spite of its obvious success, and the subsequent media attention it received, it was never implemented in ongoing clinical practice. The reasons for this implementation

Table 9.1.4 Percent Correct Predictions of Suicide Attempts/Nonattempts

	Computer	Residents	Psychiatrists
Attempt	70%	33%	38%
Nonattempt	90%	97%	93%

Source: Gustafson, Grest, Stauss, Erdman, and Laughren (1977).

Table 9.1.5. Comparison of Average Probability Assigned to Correct
Prediction of Attempts and No Attempts by Computer and Clinicians

	Computer	Psychiatrist	Resident
Attempt	0.68	0.40	0.34
Nonattempt	0.94	0.86	0.87

Source: Gustafson, Grest, Stauss, Erdman, and Laughren (1977).

failure are complex. They include the failure of the analysts to turn the program into an aid for the clinician, the complexity and "black box" character of the model, and the clinicians' lack of familiarity with computers.

9.1.11.2 Prioritizing Research Projects for the Construction Engineering Research Laboratory

The Construction Engineering Research Laboratory (CERL) is a major research laboratory of the U.S. Army Corps of Engineers. Its research produces products and systems in response to construction-related Army needs and requirements. Environmental problems, energy systems, and information technologies have received most attention in the past few years. Examples of CERL product/systems (as they call their projects) include systems for energy conservation in Army installations; techniques for noise reduction at training facilities; and computer-aided engineering and architectural design systems. Almost all work is done in-house, not on contract.

As in any other research laboratory, the total amount of money requested to implement all the ideas proposed each year typically exceeds the funds available. Of approximately 200 proposals each year, only about 100 are funded. Consequently, the management of CERL has the yearly task of selecting attractive new research projects, deciding on the desirability of continuing old ones, selecting funding levels for all projects to be funded, and rejecting other projects. This problem is complicated by the diversity of CERL's product/systems and by the fact that several divisions within CERL in effect compete for research funds.

The management of CERL sought to develop more rational procedures for selecting funding levels for product/systems. They wanted a formal prioritization system for research proposals and projects. After several unsuccessful attempts at developing such a system (which included cost-benefit approaches and attempts to assess return on investment for selected product/systems) management decided to try out multiattribute utility measurement as a technique for evaluating and prioritizing projects. What follows is a description of the procedures von Winterfeldt and I used to apply SMART to the task. This time the output was called a multiattribute aid for prioritization (MAP). It seems to be the fate of SMART to be relabeled each time it is used. This summary is adapted from Edwards, von Winterfeldt, and Moody (in press).

The first and most important element of the task was to develop a value tree capturing the values of CERL scientists, managers, and others relevant to product/system priorities. We spent four days at CERL interviewing all key CERL scientists and managers in depth. Of particular interest was the issue of divisional specificity of the value dimensions. CERL has four main divisions: Facility Planning and Design; Construction Management and Technology; Energy and Energy Conservation; and Environmental Quality. The Energy and Environment divisions are budgeted separately, but Facility Planning and Design and Construction Management and Technology essentially compete for the same budget. These four divisions together shared a total budget of over $7 million in fiscal year 1981. In addition, several smaller sets of activities do not fit the divisional boundaries. These include basic research, combat engineering, and mobilization. An early decision was to apply the prioritization system only to the major divisions.

Management wanted to develop a prioritization system that cut across budgetary boundaries. This offered an interesting challenge. Would it be possible to abstract from divisional values and find value dimensions that would apply to all programmatic activities? As we mentioned earlier, the product/systems CERL develops are extremely diverse. If proposals in the environmental division were evaluated using contribution to environmental quality as a criterion, that criterion might be quite inapplicable to new information technologies. Fortunately, we realized very early in our interviews that divisional objectives are in fact quite similar. They have to do with user requirements, extent of Army need for the product/system and of Army perception of that need, and ability to transfer CERL's output to its intended users. We suspect that such abstract values necessarily develop in a laboratory that produces a very broad spectrum of products and services.

Our interviews were relatively informal, but focused on eliciting value structures. We asked questions like "What is a high priority product/system?"; we probed intradivisional values, for example, the need for staff continuity. And, we suggested some values of our own, such as, the value of enhancing professional stature. The result was the Value Tree Model shown in Figure 9.1.9. A copy of this

A. CERL-wide issues
    AA. Relevance to CERL mission
    AB. Conformity to guidance from STDG, QCR, MAD, and similar sources
    AC. Contribution to strategy
        ACA. Conformity to a well-developed and accepted strategy
        ACB. Contribution to or initiation of a developing or new strategy
    AD. Degree of uniqueness
    AE. Contribution to CERL program diversification
B. Division-specific issues
    BA. Prior effort in this product/system needs to be carried to completion
    BB. Appropriate use of available people and/or equipment resources
C. Proponents, users, champions
    CA. Type of external support
        CAA. Rank
        CAB. Organizational location
    CB. "Loudness" of external support
    CC. Importance of the problem addressed by the product system
D. Anticipated degree of success
    DA. Anticipated degree of technical success
    DB. Timeliness
        DBA. Likelihood of completion on schedule
        DBB. Duration of anticipated need for product/system after completion
    DC. Ease of technology transfer
        DCA. Anticipated life-cycle cost of product/system to user(s)
        DCB. Availability of user resources needed for successful transfer
E. Direct cost to CERL
F. Anticipated cost savings to user(s)

Fig. 9.1.9.   Value tree CERL-MOD 1.

value tree was circulated to all division heads and to top-level management for comment. Subsequently, one of us (Edwards) visited CERL again to explore in detail their responses to the proposed tree. The numerous changes that resulted were supposed to put the tree in final form. That final form turned out not quite final, and another visit was necessary to make additional modifications and to establish consensus. The final tree is shown in Figure 9.1.10. It differs from the initial tree not only in wording, but also in the general structure of the tree. The final tree is simple—it only has two levels and it is more precise about the mission-oriented values.

The next task was to obtain weights. Division chiefs, clear that weights were expressions of priorities among values, were quite content to have them provided by the two top managers of CERL. The analyst arranged for an uninterrupted afternoon, away from phones and other distractions. Weight elicitation began with a rather careful explanation by the analyst of what weights were, emphasizing the notion of trade-offs, the fact that ranges are weights, and the counterintuitive nature of range effects. Then each manager, working independently, judged weights, using the SMART procedure with triangular consistency checks. Finally, they compared normalized weights, and revised until they had an agreed-on and consistent additive weight structure with agreed-on and consistent triangular consistency checks. The whole process took about 3½ hr. Table 9.1.6 shows the result. Values related to need (mission-oriented values) and use (anticipated benefits) received the highest weights. The weight for direct cost was relatively low. CERL is a very stable and secure organization; neither its own funding nor that of elements within it is subject to much year-to-year fluctuation.

Once value structure and weights were firmly established, each division chief established a small committee on which he or she sat to rate that division's product/systems on each branch of the tree. The analysts and the program office managers, working together, had prepared detailed worksheets for scoring the product/systems on each twig; each worksheet included careful definition of the scores. Table 9.1.7 presents two examples. These definitions had an interesting history. Initially, the program office managers wanted much more objective definitions, preferably using either more objectively measureable values or proxy variables for the subjective values or both. Attempts to develop such objective definitions were vigorously rejected by top managers and division leaders alike; they wanted values that expressed what they really cared about, subjective or not, but they also wanted guidelines for judgment. In our experience, those two preferences are common in secure organizations. Most managers

A. Mission-oriented issues
    AA. Relevance to Army Mission Areas
    AB. Conformity to validated requirements
    AC. Well-defined problem and solution fits into (mission-related) activities
    AD. Opportunity for technological breakthrough
B. Resources and effort issues
    BA. Future effort required for completion of product/system
    BB. Appropriate use of resources
C. Characteristics of champion and/or champions
    CA. Position of champion
    CB. Emphasis of external support
D. Anticipated degree of success
    DA. Anticipated degree of technical success
    DB. Timeliness
    DC. Duration of anticipated usage for product/system after completion
    DD. Ease of technology transfer
E. Direct remaining R&D cost to CERL
F. Anticipated benefits to users
    FA. Tangible benefits
    FB. Other benefits

Fig. 9.1.10   Value Tree CERL-MOD 4.

acknowledge that evaluations are and should be subjective; they want guidance about how to make a complex subjective task simpler and more orderly, not less subjective. Insecurity, need for organizational self-justification, and knee-jerk belief that "subjective" implies "inferior" has characterized organizations for which use of inherently subjective dimensions has presented me with problems.

Independently of these ratings by the division chiefs, two program office managers rated the 97 product/systems that were evaluated on all attributes. We made arrangements for disagreements to be resolved by top management; they were unnecessary. Disagreements were easily resolved in the discussions between division chiefs and the program office managers.

Table 9.1.6   CERL Weights for the Value Tree (1982)

Node Code		Normalized Weight	Twig Weight
A.	Mission-oriented values	0.39	
	AA.	0.50	0.1950
	AB.	0.13	0.0507
	AC.	0.06	0.0234
	AD.	0.31	0.1209
B.	Resources and effort issues	0.05	
	BA.	0.50	0.0250
	BB.	0.50	0.0250
C.	Characteristics of champion(s)	0.18	
	CA.	0.67	0.1206
	CB.	0.33	0.0594
D.	Anticipated degree of technical success	0.12	
	DA.	0.15	0.0180
	DB.	0.46	0.0552
	DC.	0.08	0.0096
	DD.	0.31	0.0372
E.	Direct remaining cost to CERL	0.02	0.0200
F.	Anticipated benefits to user(s)	0.24	
	FA.	0.67	0.1608
	FB.	0.33	0.0792

Table 9.1.7   Examples of Scoring Instructions for Rating Product/System (P/S)

AA.   Relevance to Army Mission Areas

   100—P/S directly supports five submission areas

   90—P/S directly supports four submission areas

   75—P/S directly supports three submission areas

   60—P/S directly supports two submission areas

   50—P/S directly supports one submission area

   0—P/S supports *no* submission area

   Submission Areas: Base/facility development
   Installation support activities
   Energy conservation and alternate sources
   Environmental quality
   Military Engineering

FA.   Tangible benefits to users

   100—P/S provides for reduction in work efforts and/or improvement in productivity plus reduction in equipment, training, materials, and operating costs

   80—P/S provides reduction in work efforts and/or improvement in productivity plus reduction in equipment and materials

   60—P/S provides reduction in work effort and/or improvement in productivity

   40—P/S provides reduction in work effort and/or improvement in productivity *but* higher costs for materials and equipment

   0—P/S provides no tangible benefits

The overall value of each product/system could now simply be computed using the additive model and the weights provided by top management. Top management and each division leader discussed the results and made decisions about support and funding level or exclusion of that division's existing and proposed product/systems. Surprisingly little disagreement occurred; all participants felt that MAP substantially facilitated these tough decisions.

Uniquely (so far as I know) this application of multiattribute utility measurement involved a large number of objects of evaluation. A look at some of the statistics that describe the judged twig ratings for the 97 product/systems is instructive. Table 9.1.8 shows the means, standard deviations, and ranges of these ratings, together with the twig weights. The most obvious finding was that all mean ratings are greater than 50 except on twig BA. Ratings on the attributes related to the anticipated degree of success are particularly high. This is to be expected; informal processes eliminate obviously unattractive product/systems before they reach formal evaluation. The next finding is that the attribute intercorrela-

Table 9.1.8   CERL Twig Weights and Various Descriptors

Twig	Weight	Mean	Standard Deviation	Range Min	Range Max
AA	0.1950	61	11.0	50	90
AB	0.0507	69	26.5	0	100
AC	0.0234	71	20.1	0	100
AD	0.1209	71	20.7	0	100
BA	0.0250	45	27.0	0	100
BB	0.0250	89	11.1	50	100
CA	0.1206	59	29.1	0	100
CB	0.0594	53	27.6	0	100
DA	0.0180	80	15.7	25	100
DB	0.0552	94	6.5	70	100
DC	0.0096	87	16.7	20	100
DD	0.0372	73	22.4	0	100
E	0.0200	63	21.0	25	100
FA	0.1608	68	15.2	20	100
FB	0.0792	64	20.0	0	100

tions were generally low (see Table 9.1.9). No dramatic halo effects occurred. The five high attribute intercorrelations have obvious explanations. For example, a highly placed champion (CA) is in a position to create validated requirements (AB).

The third finding was that the entire range was used for most attributes. The serious exceptions are AA, BB, and DB. Such instances raise several questions. Were the ranges and end-point definitions initially plausible? In at least the case of DB, I believe the definition of the zero point was not. Do the weight judgments reflect the ranges actually stated, even though they were not realized? It seems likely, and that view is encouraged by some informal ex-post-facto tradeoff judgments made by the decision makers. Should one have expected all of the plausible range to be used for each twig? Clearly not. In instances like this in which the entire evaluation scheme, including weights, must be in place before evaluations are done, even 97 product/systems may not be enough to span the full ranges of 15 variables, especially since they are subject to informal preselection. Far smaller numbers of options characterize most applications; one would expect few if any ranges to be fully covered. So long as the weights are appropriately related to the ranges, that fact makes no difference to order or spacing of the output values. Procedures based on plausible ranges seem best even for contexts in which the single-dimension values for all options are known before the weights are assessed. If procedures based on trade-offs or gambles are to be used for weight elicitation, they can equally well be based on the option locations on scales defined by plausible rather than actual ranges. If you find such procedures harder to use with end points other than 0 and 100, that finding should raise questions about the meaning of intermediate numbers obtained from questions in which the endpoints are 0 and 100. If the trade-off judgments or gambles are expressed in physical units, the issue should not arise, since only by accident (or approximation) will the arbitrary definitions of 0 and 100 coincide with convenient round values of the physical variables.

Both top managers and division chiefs in CERL like MAP very well, and it has now been used unchanged for three years. Top managers and division chiefs have come to realize that MAP, like any useful evaluation procedure, is also a program design tool and a monitoring guide. New internal proposals in CERL are coming to take the MAP format to some extent, and routinely deal with the issues specified by it. In ongoing monitoring of programs in progress, managers make sure that they track the MAP dimensions. If you have an explicit evaluation tool in which you believe, it is natural and entirely appropriate to do whatever you can to optimize the aspects of performance it measures.

A final note. MAP contributed to decision-making about allocations of roughly 10 million dollars in 1982, and even more in 1983. It cost less than $30,000, plus a lot of CERL staff time, to develop and use once. The analysts deliberately gave CERL a bargain, for various reasons not relevant here; but a normal price for this job should not be greater than $75,000, in 1982 currency. CERL actually spent an identifiable 0.3% of the amount to be allocated in order to buy a tool to help apportion the allocation. Either that figure or the 0.75% that we suggest would be a normal price is extremely cheap. A frequently encountered rule of thumb asserts that, for any given expenditure, one should add 10% to be spent in thinking about how to spend the original amount properly. This experience suggests that assistance in such thought processes seldom costs anything like 10%—typically saves much more than that.

9.1.11.3 Evaluation of Pumped-Storage Sites

This section reports Ralph Keeney's evaluation of alternative sites for a pumped-storage electricity generation facility, which used lotteries to assess utility functions. For a full report, see Keeney (1979). An electrical utility that foresees future inability to meet demand can increase its capacity in two ways. It can enhance its ability to supply base loads by building a large generating facility, using oil, coal, or nuclear power. Such facilities are cornerstones of any electrical system. Peak load supply facilities provide additional supplies for short periods of time when demand peaks and threatens to create shortages. A favored method for supplying peak load electricity is to pump water up a hill to a reservoir using the surplus electricity available during off-peak hours. During peak hours, the water moves to a downhill reservoir through turbines that drive generators and produce additional electricity.

While the technology for pumped-storage facilities is relatively straightforward, finding appropriate sites for such facilities is not. The facility has to be reasonably close to the areas it will supply. The site must have a steep gradient to produce a substantial height differential between the uphill and downhill reservoir. The site should be fairly small, and the land should be cheap and accessible. Building and using the facility should not damage the environment. These objectives normally conflict.

A southwest utility company asked Keeney and his colleagues from Woodward-Clyde Consultants to screen potential pumped-storage sites and aid in selecting an appropriate one. They used a multiattribute utility approach, adapted to the sequential nature of the screening, evaluation, and selection process that typifies siting decisions (see also Keeney, 1980).

The first step was to remove regions from the vast area in which pumped-storage facilities that would serve the intended market could be located. Regions with specific scenic, cultural, aesthetic, or archeological significance were eliminated, as were highly populated areas. More technical criteria eliminated other areas. For example, the height differential between reservoirs has to be at least 500

Table 9.1.9 Interattribute Correlation Matrix

	AA	AB	AC	AD	BA	BB	CA	CB	DA	DB	DC	DD	E	FA	FB
A	1.00														
B	0.03	1.00													
C	-0.05	0.25	1.00												
D	-0.13	0.15	0.28	1.00											
BA	-0.16	-0.16	0.05	-0.09	1.00										
BB	-0.10	0.07	0.05	0.02	0.26	1.00									
CA	0.07	0.69	0.32	0.06	-0.11	0.04	1.00								
CB	-0.01	0.54	0.31	0.08	-0.11	-0.06	0.69	1.00							
DA	-0.04	0.01	-0.28	-0.23	0.23	0.29	0.02	-0.06	1.00						
DB	0.00	-0.36	0.03	-0.11	0.24	-0.10	-0.34	-0.32	0.21	1.00					
DC	-0.09	0.26	0.22	-0.14	-0.11	-0.02	0.28	0.22	0.18	-0.25	1.00				
DD	-0.22	0.32	0.55	0.21	0.06	0.09	0.14	0.28	0.30	-0.05	0.29	1.00			
E	-0.24	-0.29	-0.01	-0.16	0.79	0.27	-0.29	-0.27	0.12	0.28	-0.19	0.00	1.00		
FA	-0.04	-0.15	0.20	0.30	0.20	0.02	0.00	-0.07	0.03	0.08	0.05	-0.02	0.14	1.00	
FB	0.20	0.28	0.14	0.37	-0.14	0.03	0.27	0.13	-0.08	-0.09	0.23	-0.03	-0.23	0.10	1.00

Table 9.1.10 A Preliminary List of General Concerns and Considerations

General Concerns	Considerations
Health and safety	Consequences of dam breach
	Impact on water quality due to reservoir development
Environmental effects	Terrestrial ecological impact
	Aquatic ecological impact
	Equalization of species composition between reservoirs
	Ecological impact from disposing of blowdown waters
	Transmission line impacts
Socioeconomic effects	Recreation potential
	Preemption of resources
	Archaeological features
	Land acquisition
	Sociopolitical system effects
Economics	Cost for adequate safety and operability
	System reliability
	Reserve capacity
Public attitudes	Public acceptance

Source: Keeney (1980).

ft, with a relatively steep gradient and large potential reservoirs. This process left about 70 potential candidate areas. Aerial reconnaissance eliminated 50 areas, and visits reduced the candidate set to 10 possible sites.

A detailed multiattribute utility analysis then started. The analysts, working with utility representatives, first generated a list of general concerns and values; see Table 9.1.10. Not all values in Table 9.1.10 differentiated among the sites; not all seemed important. For example, archeological impact seemed irrelevant for all sites and was therefore left out. Table 9.1.11 shows the very simple final set of attributes.

The analysts carefully operationalized each of the four main attributes. Costs were defined as first year capital plus operational costs. Ecological and aesthetic impacts of transmission lines were measured judgmentally on a scale labeled "mile equivalents." One mile equivalent was defined as 1 mile of transmission lines having minimal environmental impacts: not visible from highways, passing through only unpopulated rangeland, not having any harmful effect on endangered species, and not passing through pristine areas. Ten mile equivalents were defined as 1 mile of transmission lines having maximum environmental impact: traversing state or national parks, wildlife refuges, historic monument sites, or habitats that contained unusual or unique animal communities or supported endangered species. The aesthetic and environmental damage produced by transmission lines was assessed by measuring mileage in each area and then multiplying that measurement by the mile equivalent figure judged appropriate to the characteristics of that area. The sum of such products was then treated as a one-dimensional value scale. Note that two points were defined, conforming to the tradition that utilities are defined up to a linear transformation, and thus have two free parameters. If one makes the quite reasonable further assumption, implicit in the method of using mile equivalents, that the (negative) utility of having no transmission line at all is 0, then these definitions amount to the strong assumption that no single mile of transmission line through areas relevant to these sites could be worse than 10 times as bad as the best possible mile, considering only aesthetic and environmental effects. Of course, if the assessors had regarded that assumption as dubious, they would have chosen a larger ratio.

Ecological and environmental impacts at the site were operationalized as acres of pinyon–juniper forest lost and yards of riparian community lost as a result of building the facility. Table 9.1.12 lists the attributes and the plausible ranges chosen for them. The measurements and judgments of the site impacts were performed by financial experts of the utility company for the first attritute, and by

Table 9.1.11 Final Objectives and Attributes

Objective	Attribute
Minimum cost	Capital and operations cost
Minimum transmission line impacts	Aesthetic and environmental damage
Minimum ecological and environmental impacts	Pinyon–Juniper forest destruction; Riparian community

Source: Keeney (1980).

Table 9.1.12   Final Attributes for Pumped-Storage Ranking

Attribute	Measure	Range	
		Best	Worst
$x_1 =$ First-year cost	Millions of 1976 dollars	50	75
$x_2 =$ Transmission line distance	Mile equivalents	0	800
$x_3 =$ Pinyon–juniper forest	Acres	0	800
$x_4 =$ Riparian community	Yards	0	2000

Source: Keeney (1980).

Woodward-Clyde Consultants for the other three. The results for the 10 sites are shown in Table 9.1.13.

The next step was to construct a utility model based on judgments by representatives of the utility company. In tests of the three key assumptions of the additive and multiplicative EU models, preferential independence and multiplicative utility independence were approximately satisfied, and additive utility independence was not. Figures 9.1.11 and 9.1.12 show some of the tests that led to these conclusions. The following model was selected to represent the structure of the utility company's preferences; it is the multiplicative model discussed earlier:

$$1 + wu(x_1, x_2, x_3, x_4) = \prod_{i=1}^{4} [1 + ww_i u_i(x_i)]$$

The lottery procedures outlined earlier in this chapter were used to construct single-attribute utility functions. Figure 9.1.13 shows the results. Note that the scale for utility used in the figure is 0 to 1,

Table 9.1.13   Data for Evaluating UCS Pumped-Storage Candidate Sites

Candidate site	First-year cost (millions $)	Transmission line (mile equivalents)	Pinyon–juniper (acres)	Riparian community (yards)
		Base data		
S_1	56.01	97.8	230	0
S_2	59.18	140.0	150	0
S_3	61.48	163.0	0	0
S_4	59.68	342.3	0	0
S_5	64.47	91.0	270	0
S_6	61.36	152.7	721[a]	2000
S_7	58.23	681.0	0	0
S_8	59.92	704.0	240	0
S_9	49.71	84.2	260	1900[b]
S_{10}	75.42	392.7	419[c]	1600
		Alternative data		
S_9^*	49.71	84.2	260	1900[d]
S_6^*	51.64[e]	152.7	721[a]	2000
S_8^*	52.98[e]	704.0	240	0
S_7^*	65.19[f]	681.0	0	0

Source: Keeney (1980).

[a] Includes addition of 350 acres for damage to arroyo seeps.

[b] Assumes this 1900 yards is a unique riparian community.

[c] Includes addition of 200 acres for impact on raptors.

[d] Assumes this 1900 yards is a "normal" riparian community.

[e] Assumes upper reservoir is not completely lined.

[f] Assumes upper reservoir is completely lined.

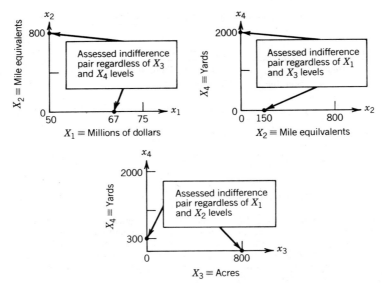

Fig. 9.1.11.   [Source: Keeney (1980)].

not 0 to 100. Additional indifference judgments were needed to obtain the scaling factors w_i and w. The analysts used a version of the BRLTs procedure described earlier. First they fixed x_3 and x_4 at their worst levels. This means that $u_3(x_{3*}) = u_4(x_{4*}) = 0$. This causes all terms of the model that involve u_3 or u_4 or both to drop out. For any such option, the model can therefore be rewritten as

$$u(x_1, x_2, x_{3*}, x_{4*}) = w_1 u_1(x_1) + w_2 u_2(x_2) + w w_1 w_2 u_1(x_1) u_2(x_2)$$

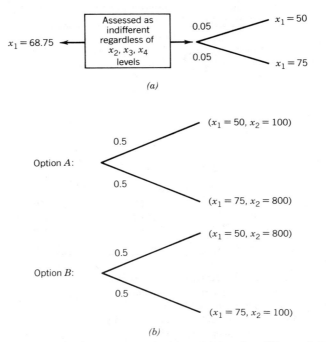

Fig. 9.1.12.   Examining potential independence conditions. (*a*) Invariant difference indicates x_1 utility independent of (x_2, x_3, x_4). (*b*) A preference for option B violates additive independence.

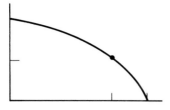
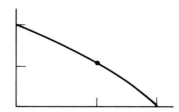

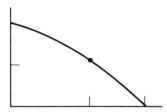
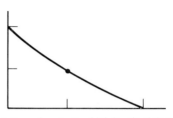

Fig. 9.1.13. Assessed utility functions. ($*x_4$ had a different shape for site 9, which involved destruction of a rare riparian community.) Source: Keeney (1980).

Bear in mind that a BRLT is a lottery between the best and worst options. In this case, since x_3 and x_4 are fixed at their worst levels, the two possible outcomes of the BRLT are $(x_{1*},x_{2*},x_{3*},x_{4*})$ and $(x_{1*},x_{2*},x_{3*},x_{4*})$. By definition, $u_1(x_{1*}) = u_2(x_{2*}) = 0$. Consequently,

$$u(x_{1*},x_{2*},x_{3*},x_{4*}) = w_1 + w_2 + ww_1w_2$$
$$\text{and } u(x_{1*},x_{2*},x_{3*},x_{4*}) = 0$$

Thus a BRLT having probability p of winning $(x_{1*},x_{2*},x_{3*},x_{4*})$ and probability $1 - p$ of winning $(x_{1*},x_{2*},x_{3*},x_{4*})$ has expected utility $\text{EU}(p) = p(w_1 + w_2 + ww_1w_2)$. Next, consider an option in which all dimensions except x_1 are set at their worst levels, and x_1 is set at its best level. Inspection of the basic model, remembering that all u_i's except that for x_1 are zero and that $u_1(x_1)$ is 1, shows that the value of such an option is w_1. This result is general; if the ith attribute is set at its best level and all other attributes are set at their worst levels, the utility of the option thus defined is w_i. So w_2 is the utility of $(x_{1*},x_{2*},x_{3*},x_{4*})$. If p_1 is the BRLT judgment made when x_1 is set at its best level and p_2 is that judgment made when x_2 is set at its best level, then

$$w_1 = p_1(w_1 + w_2 + ww_1w_2)$$
$$\text{and } w_2 = p_2(w_1 + w_2 + ww_1w_2)$$

Keeney found that p_1 and p_2 were 0.75 and 0.40.

We are dealing with a problem having five unknowns, the four w_i's and w. We have two equations; we need three more.

Keeney obtained two more equations from the judgments reported in Figure 9.1.11. The second panel of that figure shows that, regardless of x_1 and x_3 levels, the worst value of x_4, 2000 yards, was judged equivalent to 150 mile equivalents. More precisely, the judgment was

$$u(x_{1*},x_{2*},x_{3*},x_{4*}) = u(x_{1*}, x_2 = 150, x_{3*}, x_{4*})$$

In other words, the respondent could just be compensated for a change in x_2 from its best value, 0 miles, to 150 miles by changing x_4 from its worst value, 2000 yards, to its best value, 0 yards. Moreover, this was true for any fixed values of x_1 and x_3. Therefore,

$$w_2 = w_4 + w_2u_2(150) + ww_4w_2u_2(150)$$

Exactly the same procedure can be applied to the judgment reported in the third panel of Figure 9.1.11. The result is

$$w_4 = w_3 + w_4 u_4(300) + w w_3 w_4 u_4(300)$$

Now we have four equations. But, since there are five unknowns, we need a fifth equation to solve for them. It is easily obtained simply by considering the utility of $(x_{1*}, x_{2*}, x_{3*}, x_{4*})$, the option composed of the best levels on all four attributes. Of course all u_i values for that option are 1. Substitute those 1's into the basic model and you get

$$1 + w = \prod_{i=1}^{4} (1 + w w_i)$$

Now at last we have five equations; solving them for our five unknowns is simply a matter of fairly complex number-crunching. To do that number-crunching, of course, we must know the values of $u_2(150)$ and $u_4(300)$. These come from the equations in panels b and d of Figure 9.1.13.

The final output of this work is

$$w_1 = 0.716$$
$$w_2 = 0.382$$
$$w_3 = 0.014$$
$$w_4 = 0.077$$
$$w = -0.0534$$

The negative value of w means that the decision maker is multiattribute risk averse. Table 9.1.14 shows the resulting utilities for the possible sites. To make interpretation of the utilities easier, the last column indicates equivalent first-year costs. That is, each such number is a dollar amount x_1' such that, according to the model, the decision maker would be indifferent between the vector (x_1, x_2, x_3, x_4) that actually characterized the site and the vector $(x_{1*}, x_{2*}, x_{3*}, x_{4*})$. This procedure can be used to make the results of any (risky or riskless) multiattribute utility procedure that uses money as an attribute easier to interpret; it is often called *pricing out*.

The outcome of this analysis shows the importance of sensitivity analysis and critical thought about models. The final recommendation was that sites S_1, S_6, and S_9 be studied in depth, and the final decision accepted site S_9. Why? S_9 scored poorly primarily because its impact on the riparian community was severe, and that dimension had been given a special heavy weight different from the weight used for other sites. The alternative analysis of it (S_9^*) did not give its riparian effects such heavy weight. Similarly, S_9 was relatively expensive because its cost data assumed a high construction

Table 9.1.14 Evaluation of UCS Pumped-Storage Candidate Sites

Alternative	Rank	Utility	Equivalent First-Year Cost (millions $)
		Base data	
S_1	1	0.931	58.7
S_2	2	0.885	62.0
S_3	3	0.846	64.1
S_4	4	0.820	65.3
S_5	5	0.809	65.8
S_6	6	0.799	66.2
S_7	7	0.732	68.6
S_8	8	0.697	69.7
S_9	9	0.694	69.8
S_{10}	10	0.196	78.7
		Alternative data	
S_9^*		0.941	57.8
S_6^*		0.905	60.7
S_8^*		0.780	66.9
S_7^*		0.596	72.2

Source: Keeney (1980).

cost about which the decision makers had doubts. The final choice of S_9^* represented a choice to give its riparian community effects no greater weight than had been given to riparian community effects at the other contending sites.

9.1.11.4 A Capital Investment Decision

Hax and Wiig (1977) describe an application of decision analysis to a complex capital investment problem faced by a major U.S. mining company. This interesting application dealt mainly with the uncertainty side of the investment problem, had a relatively straightforward objective function (net present cash value and product volume), and involved the highest levels of corporate decision making.

The problem the mining company faced was actually an opportunity: the U.S. government invited bids for two parcels of land (labeled A and B in the subsequent analysis) that had extensive and valuable ore deposits. The company had to decide whether or not to bid, how much to bid, whether to bid alone or with a partner, and how to go about exploration and production should the bid be won. A critical uncertainty was a competing venture that the company was considering. That venture would substantially change the company's overall financial situation, forcing it to carry out exploration and production with a partner whether or not it had bid alone. The project required approximately three years of planning, five years of construction, and 20 years of plant operation. The total capital commitments were $500 million.

The structure of this problem is represented in the decision tree form of figure 9.1.14. The tree is self-explanatory. The key probabilities were:

1. Probability of winning the bid.
2. Probability of success of competing the venture.
3. Probability distribution over amounts of capital investment.
4. Probability distribution over product market prices.
5. Probability distribution over operation costs.

The probability of winning the bid is obviously a function of the amount bid. To construct this relationship, the decision maker assessed fractiles of the cumulative probability distribution of winning as shown in Figure 9.1.15. The tree analyzed only three discrete bids: low, medium, and high, with their associated probabilities of winning and losing. The probability of a successful competing venture was assessed as 0.10. The other three uncertainties were taken into account by using relatively crude ranges of the respective variables and incorporating them in the form of a probabilistic sensitivity analysis.

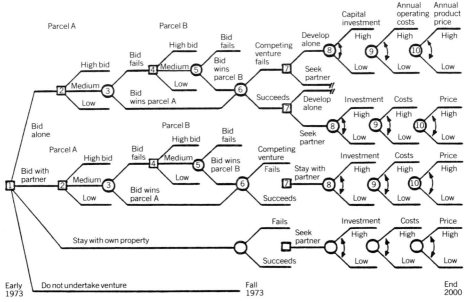

Fig. 9.1.14. Decision tree for capital investment project. Circles represent uncertain events; squares represent decision alternatives. Like nodes and branches are suppressed. Source: Hax and Wiig (1977).

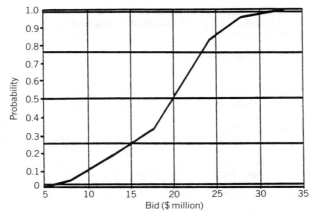

Fig. 9.1.15. Probability of winning bids, encoded by decision maker at 1%, 25%, 50%, 75%, and 99%. Source: Hax and Wiig (1977).

Every path through the tree in Figure 9.1.14 defines an outcome state. From the point of view of the decision maker two variables were important for defining preferences among these states: net present cash value (NPV) and product output (PO). Both were, of course, uncertain.

The analysis considered 26 strategies, a small subset of the total number of paths through the tree specified by acts under the decision maker's control. A probability distribution was constructed for each of the objectives, NPV and PO, for each of the 25 strategies. Strategy 25 appeared stochastically dominant over the other strategies. That is, virtually regardless of the value of p, its cumulative distribution function was to the right of all others, and so should produce a higher NPV.

Next, the analysts assessed utility functions over NPV and PO. They assumed a multiplicative utility function of the familiar form

$$u(\text{NPV},\text{PO}) = w_1 u_1(\text{NPV}) + w_2 u_2(\text{PO})$$
$$+ w w_1 w_2 u_1(\text{NPV}) u_2(\text{PO})$$

The analysts assessed this utility function as

$$u(\text{NPV},\text{PO}) = 0.988[1 - \exp\{-0.005(\text{NPV} + 100)\}] + 0.197[1 - \exp\{-0.03\text{PO}\}]$$
$$+ 0.067[1 - \exp\{-0.005(\text{NPV} + 100)\}] [1 - \exp\{-0.03\text{PO}\}]$$

Figure 9.1.16 is a two-dimensional representation of this two-attribute utility function. Table 9.1.15 shows the expected utilities that will result from the probability distributions over NPV and PO and the two-attribute utility function. The stochastic dominance of strategy 25 disappears in the two-attribute analysis. Instead strategies 1 and 2 appear rather attractive.

The authors summarize the results and impact of the study as follows (Hax and Wiig, 1977, p. 294):

> *The decision maker chose strategy 2 as a result of the analysis outlined in this paper. He had been frustrated by his inability to handle the two objectives and resolve tradeoffs (or conflicts, as he expressed it). With the multiattribute utility analysis he was satisfied that his views and values were properly represented and, hence, he had no hesitation in accepting the optimal strategy.*

9.1.12 WHERE IS DECISION ANALYSIS GOING?

Decision analysis is an up-and-running technology now. Its major virtue in practice seems to be that it does facilitate the task of making difficult decisions, for exactly the reasons that its content would lead one to expect. It has various major flaws.

1. Structuring of decision problems, like structuring of other analytic problems in Operations Research, is an art form, subject to few prescriptions. Research on generic structuring is in progress [see, e.g., Keeney (1980)], but has some distance to go before it produces tools easily used by nonexperts, who need them most.

2. Decision analysis as presently defined requires the participation of a decision analyst. No technol-

Indifferent point.
U(200, 0) indifferent to U(55, 100)
Gamble:

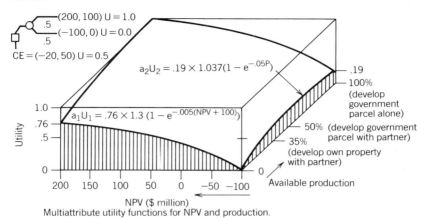

Fig. 9.1.16.   Multidistribute utility functions for NPV and production. Source: Hax and Wiig (1977).

ogy that depends on the availability of expert "high priests" for its effective use can possibly have the general impact to which one would aspire for ideas as general as these.

3.   The field of artificial intelligence has set out to compete with decision analysis in various subject matters—and seems to be winning. Expert systems encode expert knowledge, including expert knowledge about inferences and decisions. Some of the currently exemplary AI systems are built around inferential procedures that no decision analyst would use, since they make very strong and improbable assumptions or even violate rules in ways that guarantee incoherent assessments. Still, these systems are often labeled as AI successes, even though the track record of their successful use seems to be scanty or nonexistent.

4.   The technology of real, as distinct from statistical, inference is severely undeveloped. The fact that real inference problems routinely violate the likelihood and sufficiency principles implies that development of that technology to a numerical form is likely to be very difficult. Projects with exactly that goal are, however, in progress, and seem very promising.

Of the aforementioned flaws, problems 2 and 3 seem central. Fortunately, problem 3 may solve problem 2. As the technology of expert systems develops, it seems very likely that the skills and knowledges of the "high priests" of decision analysis will come to be encoded in computer programs. Research on how to do that is occurring to some extent, especially in connection with work on problem 4. Eventually, the routine practice of decision analysis, like the routine practice of elementary statistics, will be done primarily by computers. The expertise of experts will be used for hard problems, and the funding for research will come from those who face them.

One such hard problem has already been clearly identified. It, along with structuring and inference, are the three key research areas of decision analysis today. The problem is that of decisions made in a social context characterized by multiple stakeholders having conflicting values. Research already complete suggests that the tools of multiattribute utility measurement may have much to contribute

Table 9.1.15   Expected Utilities and Equivalent NPVs of Main Strategies

	1	2	17	25	26
		Bid High	Bid Low	Develop Own	
	Bid High	with	with	Property with	Do
	Alone	Partner	Partner	Partner	Nothing
Expected two-objective utility	0.719	0.722	0.706	0.710	0.386
Certainty equivalent ($ million NPV) (assessed at $P = 50$)	51	52	45	47	—

Source: Hax and Wiig (1977).

to facilitation of such decisions—but as yet no example of a complete application to that kind of problem exists, so far as I know.

Decision analysis can be seen, from one perspective, as a branch of applied cognitive psychology—even though most of those who have developed it are not psychologists and most cognitive psychologists are ill-informed about its central ideas. A task for the immediate future is to remedy the currently poor communication between these appliers and those who, at least in principle, are doing the basic work that underlies the applications.

REFERENCES

Allais, M. (1953). Le comportement de l'homme rationnel devant le risque: Critique des postulats et axiomes de l'ecole americaine. *Econometrica, 21,* 503–546.

Allais, M., and Hagen, J. (Eds.) (1979). *Expected utility hypotheses and the Allais paradox.* Dordrecht, The Netherlands: Reidel.

Alpert, M., & Raiffa, H. (1982). A progress report on the training of probability assessors. In D. Kahneman, P. Slovic, and A. Tversky (Eds.). *Judgment under uncertainty: Heuristics and biases* (pp. 294–305). New York: Cambridge University Press.

Barclay, S., Brown, R. V., Kelly, C. W., III, Peterson, C. R., Phillips, L. D., and Selvidge, J. (1977). *Handbook for decision analysis.* McLean, VA: Decisions and Designs, Inc.

Beck, J. R., Kassirer, J. P., and Pauker, S. G. (1982). A convenient approximation of life expectancy (the 'DEALE'). I. Validation of the method. *American Journal of Medicine, 73,* 883–888.

Beck, J. R., Pauker, S. G., and Gottlieb, J. E. (1982). A convenient approximation of life expectancy (the 'DEALE'). II. Use in medical decision making. *American Journal of Medicine, 73,* 889–897.

Bell, D. E. (1982). Regret in decision making under uncertainty. *Operations Research, 30,* 961–981.

Beyth-Marom, R. (1982). How probable is probable? Numerical translations of verbal probability expressions. *Journal of Forecasting, 1,* 257–269.

Brown, R. V. (1970). Do managers find decision theory useful? *Harvard Business Review, 48,* 78–89.

Dawes, R. M., and Corrigan, B. (1974). Linear models in decision making. *Psychological Bulletin, 81,* 95–106.

deSmet, A. A., Fryback, D., and Thornbury, J. R. (1979). A second look at the utility of radiographic skull examination for trauma. *American Journal of Radiology, 132,* 75–99.

Dyer, J. S., and Sarin, R. A. (1979). Measurable multiattribute value functions. *Operations Research, 22,* 810–822.

Edwards, W. (1971a). Don't waste an executive's time on decision making. In *Decision making in a changing world.* Princeton: Auerbach, pp. 63–75.

Edwards, W. (1971b). Social utilities. *The Engineering Economist,* Summer Symposium Series. 6.

Edwards, W. (1977). How to use multiattribute utility measurement for social decision making. *IEEE Transactions on Systems, Man and Cybernetics, SMC-7,* 326–340.

Edwards, W. (1979). Multiattribute utility measurement: Evaluating desegregation plans in a highly political context. In R. Perloff, Ed. *Evaluator interventions: Pros and cons.* Beverly Hills, CA.: Sage Publications.

Edwards, W., Lindman, H., and Savage, L. J. (1963). Bayesian statistical inference for psychological research. *Psychological Review, 70,* 193–242.

Edwards, W., and von Winterfeldt, D. (in press). Cognitive illusions and their implications for the law. *University of Southern California Law Review.*

Edwards, W., von Winterfeldt, D., and Moody, D. L. (in press). Simplicity in decision analysis: An example and a discussion. In D. Bell (Ed.). *Decision making: Descriptive, normative, and prescriptive interactions.* Cambridge, MA: Harvard Business School.

Fischer, G. W. (1977). Convergent validation of decomposed multiattribute utility assessment procedures for risky and riskless decisions. *Organizational Behavior and Human Performance, 18,* 295–315.

Fischhoff, B., Slovic, P., and Lichtenstein, S. (1980). Knowing what you want: Measuring labile values. In T. Wallsten (Ed.). *Cognitive processes in choice and decision behavior.* Hillsale, NJ: Erlbaum.

Goodman, B. (1972). Action selection and likelihood estimation by individuals and groups. *Organizational Behavior and Human Performance, 7,* 121–141.

Gustafson, D. H., Grest, J. H., Stauss, F. F., Erdman, H., and Laughren, T. A probabilistic system for identifying suicide attemptors. *Computers and Biomedical Research, 10,* 83–89.

Hax, A. C., and Wiig, K. M. (1977). The use of decision analysis in a capital investment problem. In D. Bell, R. L. Keeney, and H. Raiffa, Eds. (pp. 277–297). *Conflicting objectives in decisions.* New York: Wiley.

Howard, R. A., Matheson, J. E., and Miller, R. L., Eds. (1976). *Readings in decision analysis.* Menlo Park, CA: Stanford Research Institute.

Humphreys, P. C. (1980). Decision aids: Aiding decisions. In L. Sjöberg, T. Tyszka, and J. A. Wise, Eds. *Decision analyses and decision processes.* Lund, Sweden: Doxa.

Kahneman, D., and Tversky, A. (1979). Prospect theory: An analysis of decision under risk. *Econometrica, 47,* 263–291.

Keeney, R. L. (1979). Evaluation of proposed pumped storage sites. *Operations Research, 17,* 48–64.

Keeney, R. L. (1980). *Siting energy facilities.* New York: Academic.

Keeney, R. L., and Raiffa, H. (1976). *Decisions with multiple objectives: Preferences and value tradeoffs.* New York: Wiley.

Keeney, R. L., and Sicherman, A. An interactive computer program for assessing and using multiattribute utility functions. *Behavioral Science, 21,* 173–182.

Krischer, J. P. (1980). An annotated bibliography of decision analytic applications to health care. *Operations Research, 20,* 97–113.

Lichtenstein, S., and Fischhoff, B. (1977). Do those who know more also know more about how much they know? The calibration of probability judgments. *Organizational Behavior and Human Performance, 20,* 159–183.

Lichtenstein, S., Fischhoff, B., and Phillips, L. D. (1982). Calibration of probabilities: The state of the art to 1980. In D. Kahneman, P. Slovic, and A Tversky (Eds.). *Judgment under uncertainty: Heuristics and biases.* New York: Cambridge University Press.

Ludke, R. L., Stauss, F. Y., and Gustafson, D. H. (1977). Comparison of methods for estimating subjective probability distributions. *Organizational Behavior and Human Performance, 19,* 162–179.

Lusted, L. B., Roberts, H. V., Edwards, W., Wallace, P. L., Lahiff, M., Loop, J. W., Bell, R. S., Thornbury, J. R., Seale, D. L., Steele, J. P., and Fryback, D. G. (1979). *Efficacy of X-ray procedures.* American College of Radiology.

McCormick, N. J. (1981). *Reliability and risk analysis.* New York: Academic.

Meehl, P. E., and Rosen, A. Antecedent probability and the efficacy of psychometric signs, patterns, or cutting scores. *Psychological Bulletin, 52,* 194–216.

Miller, L. W., Kaplan, R. S., and Edwards, W. (1967). JUDGE: A value-judgment based tactical command system. *Organizational Behavior and Human Performance, 2,* 329–374.

Miller, L. W., Kaplan, R. S., and Edwards, W. (1969). JUDGE: A laboratory evaluation. *Organizational Behavior and Human Performance, 4,* 97–111.

Moroff, S. V., and Pauker, S. G. (1983). What to do when the patient outlives the literature, or DEALE-ing with a full deck. *Medical Decision Making, 3,* 313–338.

Murphy, A. H., and Winkler, R. L. Can weather forecasters formulate reliable forecasts of precipitation and temperature? *National Weather Digest, 2,* 2–9.

Pearl, J. (1977). A framework for processing value judgements. *IEEE Transactions on Systems, Man and Cybernetics,* SMC-7, 349–354.

Pearl, J., Leal, A., and Saleh, J. (1980). GODDESS: A goal directed decision structuring system (UCLA-ENG-8034). Los Angeles, CA: University of California.

Phillips, L. D. (1973). *Bayesian statistics for social scientists.* New York: Thomas Y. Crowell Co.

Pitz, G. F. (1974). Subjective probability distributions for imperfectly known quantities. In L. W. Gregg, Ed. *Knowledge and cognition.* New York: Wiley.

Pitz, G. F., Sachs, N. J., and Heerboth, J. (1980). Procedures for eliciting choices in the analysis of individual decisions. *Organizational Behavior and Human Performance, 26,* 396–408.

Raiffa, H. *Decision analysis.* Reading, MA: Addison-Wesley, 1968.

Raiffa, H., and Schlaifer, R. (1961). *Applied statistical decision theory.* Boston, MA: Harvard University Press.

Savage, L. J. (1954). *The foundations of statistics.* New York: Wiley.

Schum, D. A. (1977). The behavioral richness of cumulative and corroborative testimonial evidence. In N. J. Castellan, Jr., Pisoni, and Potts Eds. *Cognitive theory.* Vol. 2, Hillsdale, NJ: Erlbaum.

Schum, D. A. (1980). Current development in research on cascaded inference. In T. S. Wallsten, Ed. *Cognitive processes in decision and choice behavior.* Hillsdale, NJ: Erlbaum.

Schum, D. A., and Martin, A. W. (1982). Formal and empirical research on cascaded inference in jurisprudence. *Law and Society Review, 17,* 105–157.

Seaver, D. A., von Winterfeldt, D., and Edwards, W. (1978). Eliciting subjective probability distributions on continuous variables. *Organizational Behavior and Human Performance, 21,* 379–391.

Sieber, J. E. (1974). Effects of decision importance on the ability to generate warranted subjective uncertainty. *Journal of Personality and Social Psychology, 30,* 688–694.

Tversky, A., and Kahneman, D. (1981). The framing of decisions and the psychology of choice. *Science, 211,* 453–458.

Tversky, A., and Kahneman, D. (1982). The framing of decisions and the psychology of choice. In R. Hogarth, Ed. *Question framing and response consistency.* San Francisco, CA: Jossey-Bass.

Ulvila, J., and Brown, R. V. (1982). Decision analysis comes of age. *Harvard Business Review, 60,* 130–141.

United States Nuclear Regulatory Commission (1975). Reactor safety study: An assessment of accident risks in U.S. commercial nuclear power plants (NUREG-75/014). Washington, DC: NRC.

von Winterfeldt, D. (1980). Additivity and expected utility in risky multiattribute preferences. *Journal of Mathematical Psychology, 121,* 66–82.

von Winterfeldt, D., Barron, H. F., and Fischer, G. W. (1980). Theoretical and empirical relationships between risky and riskless utility functions (Research Report 80-3). University of Southern California, Los Angeles, CA: Social Science Research Institute.

von Winterfeldt, D., and Edwards, W. (1982). Costs and payoffs in perceptual research. *Psychological Bulletin, 19,* 609–622.

von Winterfeldt, D., and Edwards, W. (in press). *Decision analysis and behavioral research.* New York: Cambridge University Press.

Wainer, H. (1976). Estimating coefficients in linear models: It don't make no nevermind. *Psychological Bulletin, 83,* 713–717.

Wigmore, J. H. (1937). *The science of judicial proof as given by logic, psychology, and general experience, and illustrated in judicial trials,* 3rd ed. Boston: Little, Brown.

Zlotnick, J. A theorem for prediction. *Foreign Service Journal, 45,* 20.

CHAPTER 9.2
ARTIFICIAL INTELLIGENCE

KING-SUN FU

Purdue University
West Lafayette, Indiana

9.2.1 INTRODUCTION

Many human mental activities such as writing computer programs, doing mathematics, recognizing patterns, understanding speech and language, and even driving an automobile are said to demand "intelligence." Over the past few decades, several computer systems have been built or programmed to perform tasks such as these. Specifically, there are computer systems that can diagnose diseases, recognize objects, classify white blood cells, analyze aerial photographs and satellite pictures, inspect manufactured products, analyze electronic circuits, understand limited amounts of human speech and natural language text, or identify fingerprints. We might say that such systems possess some degree of "artificial intelligence." Thus, artificial intelligence (AI) is usually concerned with the use of computers in tasks that are normally considered to require knowledge, perception, reasoning, learning, understanding, and similar problem-solving abilities (Barr, Cohen, and Feigenbaum, 1981–1982; Nilsson, 1971, 1980; Rich, 1983; Winston, 1984).

AI researchers are usually motivated by one of the following two objectives:

1. Understanding the fundamental nature of "intelligence."
2. Making computers more (intelligently) useful.

The first objective helps to explain the close ties between AI, psychology, and education. The second objective leads to a more pragmatic engineering-oriented approach, and is given a more specific name "machine intelligence."

Some typical applications of machine intelligence are listed below (Barr, Cohen, and Feigenbaum, 1981–1982; *Computer,* 1982; Fikes, Hart, and Nilsson, 1972; Fu, 1981–1982, 1982; Hayer-Roth, Waterman, and Lenat, 1983; Ishizuka, Fu, and Yao, 1983; Lee, Gonzalez, and Fu, 1983; Nau, 1983; Weiss and Kulikowski, 1984):

1. Medical diagnosis and consultation.
2. Analysis of signals and images (EEG, ECG, seismic signals, aerial photographs, etc.).
3. Identification of human faces and fingerprints.
4. Automatic inspection and testing.
5. Language and speech understanding.
6. Automated manufacturing.
7. Robotics.
8. Management and decision support.

In this chapter, we briefly review several basic methods in problem solving and artificial intelligence.

9.2.2 STATE-SPACE SEARCH

Perhaps the most straightforward approach to finding a solution to a problem would be to try out various possible solutions until we happen to produce the desired solution. Such an attempt involves essentially a trial-and-error search. To discuss solution methods of this sort it is helpful to introduce the notion of problem states and operators. A problem state, or simply state, is a particular problem situation or configuration. The set of all possible configurations is the space of problem states or the state space. An operator, when applied to a state, will transform the state into another state. A solution to a problem is a sequence of operators that transforms an initial state into a goal state.

It is useful to imagine the space of states reachable from the initial state as a graph containing nodes corresponding to the states. The nodes of the graph are linked together by arcs that correspond to the operators. A solution to a problem could be obtained by a search process that first applies operators to the initial state to produce new states, then applies operators to these, and so on until the goal state is produced. Methods of organizing such a search for the goal state are most conveniently described in terms of a graph representation.

Example 1: "Blocks World"

Consider that a robot's world consists of a table T and three blocks A, B, and C. The initial state of the world is that blocks A and B are on the table, and block C is on top of block A (see Figure 9.2.1). The robot is asked to change the initial state to a goal state in which the three blocks are

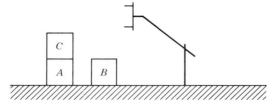

Fig. 9.2.1. A configuration of robot and blocks.

stacked with block A on top, block B in the middle, and block C on the bottom. The only operator that the robot can use is MOVE X from Y to Z, which moves object X from the top of object Y onto object Z. In order to apply the operator, it is required that (a) X, the object to be moved, be a block with nothing on top of it, and (b) if Z is a block, there must be nothing on Z.

We can simply use a graphical description like the one in Figure 9.2.1 as the state representation. The operator MOVE X from Y to Z is represented by MOVE (X, Y, Z). A graph representation of the state-space search is illustrated in Figure 9.2.2. By removing the dotted lines in the graph (that is, the operator is not to be used to generate the same operator more than once), we obtain the state-space search tree. It is easily seen from Figure 9.2.2 that a solution that the robot can obtain consists of the following operator sequence: MOVE(C,A,T), MOVE(B,T,C), MOVE(A,T,B)

Example 2: Monkey-and-Bananas Problem

A monkey is in a room containing a box and a bunch of bananas (Figure 9.2.3). The bananas are hanging from the ceiling out of reach by the monkey. How can the monkey get the bananas?

The four-element list (W, x, Y, z) can be selected as the state representation, where

W = horizontal position of the monkey

x = 1 or 0, depending on whether the monkey is on top of the box or not, respectively

Y = horizontal position of the box

z = 1 or 0, depending on whether the monkey has grasped the bananas or not, respectively

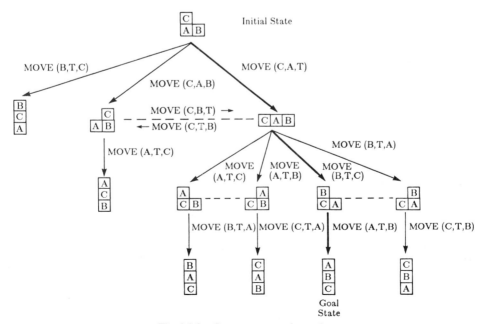

Fig. 9.2.2. State-space search graph.

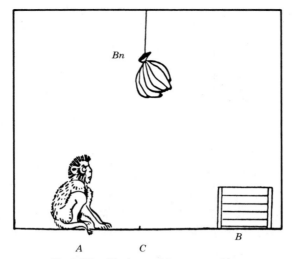

Fig. 9.2.3. Monkey-and-bananas problem.

The operators in this problem are:

1. goto(U)—monkey goes to horizontal position U, or in the form of a production rule,

$$(W, 0, Y, z) \xrightarrow{\text{goto}(U)} (U, 0, Y, z)$$

That is, state $(W, 0, Y, z)$ can be transformed into $(U, 0, Y, z)$ by applying operator goto(U).

2. pushbox(V)—monkey pushes box to horizontal position V, or

$$(W, 0, W, z) \xrightarrow{\text{pushbox}(V)} (V, 0, V, z)$$

It should be noted from the left side of the production rule that, in order to apply the operator pushbox(V), the monkey should be at the same position W as the box, and in the meantime the monkey is not on top of the box. Such a condition imposed on the applicability of an operator is called the precondition of the production rule.

3. climbbox—monkey climbs on top of the box, or

$$(W, 0, W, z) \xrightarrow{\text{climbbox}} (W, 1, W, z)$$

It should be noted that, in order to apply the operator climbbox, the monkey must be at the same position W as the box, and the monkey is not on the top of the box.

4. grasp—monkey grasps the bananas, or

$$(C, 1, C, 0) \xrightarrow{\text{grasp}} (C, 1, C, 1)$$

where C is the location on the floor directly under the bananas. It should be noted that in order to apply the operator grasp, the monkey and the box should both be at position C and the monkey is already on the top of the box.

It is noted that, in this case, both the applicability and the effects of the operators are expressed by the production rules. For example, in rule 2, the operator pushbox(V) is only applicable when its precondition is satisfied, that is, the monkey and the box are both at the same position W and in the meantime the monkey is not on top of the box. The effect of the operator is that the monkey has pushed the box to position V. In this formulation, the set of goal states is described by any list whose last element is 1.

Let the initial state be $(A, 0, B, 0)$. The only operator that is applicable is goto(U), resulting in the next state $(U, 0, B, 0)$. Now three operators are applicable; they are goto(U), pushbox(V), and climbbox (if $U = B$). Continuing to apply all operators applicable at every state, we produce the state space in terms of the graph representation shown in Figure 9.2.4. It can be easily seen that the

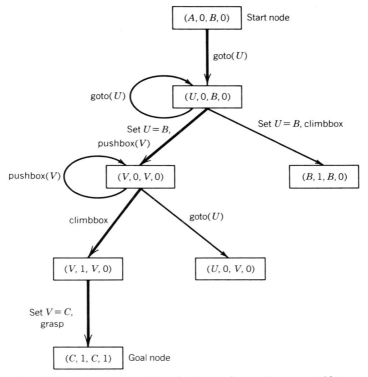

Fig. 9.2.4. Graph representation for the monkey-and-bananas problem.

sequence of operators that transforms the initial state into a goal state consists of goto(B), pushbox(C), climbbox, grasp.

The graph in Figure 9.2.4 is quite small, and a solution path from the initial state to a goal state can be obtained easily. In general, a search process is needed to move through the state (problem) space until a path from an initial state to a goal state is found. One way to describe the search process is to use production systems. A production system consists of:

1. A data base that contains the information relevant to the particular task. Depending on the application, this data base may be as simple as a small matrix of numbers or as complex as a large, relational indexed file structure.
2. A set of rules operating on the data base. Each rule consists of a left side that determines the applicability of the rule or precondition, and a right side that describes the action to be performed if the rule is applied. Application of the rule changes the data base.
3. A control strategy that specifies which applicable rules should be applied and ceases computation when a termination condition on the data base is satisfied.

In terms of production system terminology, the graph in Figure 9.2.4 is generated by the control strategy. The various data bases produced by rule applications are actually represented as nodes in the graph. Thus, a graph-search control strategy can be considered as a means of finding a path in a graph form a (start) node representing the initial data base to one (goal node) representing a data base that satisfies the termination (or goal) condition of the production system.

A general graph-search procedure can be described as follows:

Step 1.   Create a search graph, G, consisting solely of the start node, s. Put s on a list called OPEN.

Step 2.   Create a list called CLOSED that is initially empty.

Step 3.   LOOP: if OPEN is empty, exit with failure.

Step 4.   Select the first node on OPEN, remove it from OPEN, and put it on CLOSED. Call this mode n.

Step 5.   If n is a goal node, exit successfully with the solution obtained by tracing a path along the pointers from n to s in G. (Pointers are established in step 7.)

Step 6.   Expand node n, generating the set M of its successors that are not ancestors of n. Install these members of M as successors of n in G.

Step 7.   Establish a pointer to n from those members of M that were not already in OPEN or CLOSED. Add these numbers of M to OPEN. For each member of M that was already on OPEN or CLOSED, decide whether or not to redirect its pointer to n. For each member of M already on CLOSED, decide for each of its descendants in G whether or not to redirect its pointer.*

Step 8.   Reorder the list OPEN, either according to some arbitrary criterion or according to heuristic merit.

Step 9.   Go LOOP.

If no heuristic information from the problem domain is used in ordering the nodes on OPEN, some arbitrary criterion must be used in step 8. The resulting search procedure is called uninformed or blind. The first type of blind search procedure orders the nodes on OPEN in increasing order of their depth in the search tree.† The search that results from such an ordering is called breadth-first search. It has been shown that breadth-first search is guaranteed to find a shortest-length path to a goal node, providing a path exists. The second type of blind search orders the nodes on OPEN in descending order of their depth in the search tree. The deepest nodes are put first in the list. Nodes of equal depth are ordered arbitrarily. The search that results from such an ordering is called depth-first search. To prevent the search process from running away along some fruitless path forever, a depth bound is set. No node whose depth in the search tree exceeds this bound is ever generated.

The blind search methods described above are exhaustive search methods for finding paths from the start node to a goal node. For many tasks it is possible to use task-dependent information to help reduce search. This class of search procedures is called heuristic or best-first search and the task-dependent information used is called heuristic information In step 8 of the graph search procedure, heuristic information can be used to order the nodes on OPEN so that the search expands along those sectors of the graph thought to be most promising. One important method uses a real-valued evaluation function to compute the "promise" of the nodes. Nodes on OPEN are ordered in increasing order of their values of the evaluation function. Ties among the nodes are ordered arbitrarily, but always in favor of goal nodes. The choice of evaluation function critically determines search results. A useful best-first search algorithm, the A* algorithm, is described below.

Let the evaluation function f at any node n be

$$f(n) = g(n) + h(n)$$

where $g(n)$ is a measure of the cost of getting from the start node to node n, and $h(n)$ is an estimate of the additional cost from node n to a goal node. That is, $f(n)$ represents an estimate of the cost of getting from the start node to a goal node along the path constrained to go through node n.

The A* Algorithm

Step 1.   Start with OPEN containing only the start node. Set that node's g value to 0, its h value to whatever it is, and its f value to $h + 0$, or h. Set CLOSED to the empty list.

Step 2.   Until a goal node is found, repeat the following procedure: If there are no nodes on OPEN, report failure. Otherwise, pick the node on OPEN with the lowest f value. Call it BESTNODE. Remove it from OPEN. Place it on CLOSED. See if BESTNODE is a goal node. If so, exit and report a solution (either BESTNODE if all we want is the node, or the path that has been created between the start node and BESTNODE if we are interested in the path). Otherwise, generate the successors of BESTNODE, but do not set BESTNODE to point to them yet. (First we need to see if any of them have already been generated.) For each such SUCCESSOR, do the following:

    1.   Set SUCCESSOR to point back to BESTNODE. These back links will make it possible to recover the path once a solution is found.

* If the graph being searched is a tree, then none of the successors generated in step 6 has been generated previously. Thus, the members of M are not already on either OPEN or CLOSED. In this case, each member of M is added to OPEN and is installed in the search tree as successors of n. If the graph being searched is not a tree, it is possible that some of the members of M have already been generated, that is, they may already be on OPEN or CLOSED.

† To promote earlier termination, goal nodes should be put at the very beginning of OPEN.

2. Compute g(SUCCESSOR) = g(BESTNODE) + cost of getting from BESTNODE to SUCCESSOR.

3. See if SUCCESSOR is the same as any node on OPEN (i.e., it has already been generated but not processed). If so, call that node OLD. Since this node already exists in the graph, we can throw SUCCESSOR away, and add OLD to the list of BESTNODE's successors. Now we must decide whether OLD's parent link should be reset to point to BESTNODE. It should be if the path we have just found to SUCCESSOR is cheaper than the current best path to OLD (since SUCCESSOR and OLD are really the same node). So see whether it is cheaper to get to OLD via its current parent or to SUCCESSOR via BESTNODE, by comparing their g values. If OLD is cheaper (or just as cheap), then we need do nothing. If SUCCESSOR is cheaper, then reset OLD's parent link to point to BESTNODE, record the new cheaper path in g(OLD), and update f(OLD).

4. If SUCCESSOR was not on OPEN, see if it is on CLOSED. If so, call the node on CLOSED OLD, and add OLD to the list of BESTNODE's successors. Check to see if the new path or the old path is better just as in steps (2) and 3, and set the parent link and g and f values appropriately. If we have just found a better path to OLD, we must propagate the improvement to OLD's successors. This is a bit tricky. OLD points to its successors. Each successor in turn points to its successors, and so forth, until each branch terminates with a node that either is still on OPEN or has no successors. So to propagate the new cost downward, do a depth-first traversal of the tree starting at OLD, changing each node's g value (and thus also its f value), terminating each branch when you reach a node with no successors or a node to which an equivalent or better path has already been found. This condition is easy to check for. Each node's parent link points back to its best known parent. As we propagate down to a node, see if its parent points to the node we are coming from. If so, continue the propagation. If not, then its g value already reflects the better path of which it is part. So the propagation may stop here. But it is possible that with the new value of g being propagated downward, the path we are following may become better than the path through the current parent. So compare the two. If the path through the current parent is still better, stop the propagation. If the path we are propagating through is now better, reset the parent and continue propagation.

5. If SUCCESSOR was not already on either OPEN or CLOSED, then put it on OPEN, and add it to the list of BESTNODE's successors. Compute f(SUCCESSOR) = g(SUCCESSOR) + h(SUCCESSOR).

It is easy to see that the A* algorithm is essentially the graph search algorithm using $f(n)$ as the evaluation function for ordering nodes. Note that because $g(n)$ and $h(n)$ must be added, it is important that $h(n)$ be a measure of the cost of getting from node n to a goal node.

The objective of a search procedure is to discover a path through a problem space from an initial state to a goal state. There are two directions in which such a search could proceed: (1) forward, from the initial states, and (2) backward, from the goal states. The rules in the production system model can be used to reason forward from the initial states and to reason backward from the goal states. To reason forward, the left sides or the preconditions are matched against the current state and the right side (the results) are used to generate new nodes until the goal is reached. To reason backward, the right sides are matched against the current state and the left sides are used to generate new nodes representing new goal states to be achieved. This continues until one of these goal states is matched by an initial state.

By describing a search process as the application of a set of rules, it is easy to describe specific search algorithms without reference to the direction of the search. Of course, another possibility is to work both forward from the initial state and backward from the goal state simultaneously until two paths meet somewhere in between. This strategy is called bidirectional search.

9.2.3 PROBLEM REDUCTION

Another approach to problem solving is "problem reduction." The main idea of this approach is to reason backward from the problem to be solved, establishing subproblems and subsubproblems until, finally, the original problem is reduced to a set of trivial primitive problems, that is, their solutions are obvious. A problem-reduction operator transforms a problem description into a set of reduced or successor problem descriptions. For a given problem description there may be many reduction operators that are applicable. Each of these produces an alternative set of subproblems. Some of the subproblems may not be solvable, however, so we may have to try several operators in order to produce a set whose members are all solvable. Thus it again requires a search process.

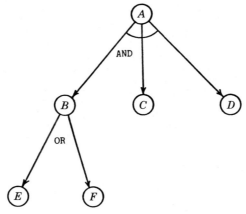

Fig. 9.2.5.   An AND/OR graph.

The reduction of problem to alternative sets of successor problems can be conveniently expressed by a graphlike structure. Suppose problem A can be solved by solving all of its three subproblems B, C, and D; an AND arc will be marked on the incoming arcs of the nodes B, C, and D. The nodes B, C, and D are called AND nodes. On the other hand, if problem B can be solved by solving any one of the subproblems E and F, an OR arc will be used. These relationships can be shown by the AND/OR graph shown in Figure 9.2.5. It is easily seen that the search methods discussed in Section 9.2.2 are for OR graphs through which we want to find a single path from the start node to a goal node.

Example 3

An AND/OR graph for the monkey-and-bananas problem is shown in Figure 9.2.6. Here, the problem configuration is represented by a triple (S, F, G), where S is the set of starting states, F is the set of operators, and G is the set of goal states. Since the operator set F does not change in this problem and the initial state is $(A, 0, B, 0)$, we can suppress the symbol F and denote the problem simply by $(\{(A, 0, B, 0)\}, G)$. One way of selecting problem-reduction operators is through the use of a "difference."*

From Example 1, $F = \{f_1, f_2, f_3, f_4\} = \{goto(U), pushbox(V), climbbox, grasp\}$. First, we calculate the difference for the initial problem. The reason that the list $(A, 0, B, 0)$ fails to satisfy the goal test is that the last element is not 1. The operator relevant to reduce this difference is $f_4 = grasp$. Using f_4 to reduce the initial problem, we obtain the following pair of subproblems: $(\{(A, 0, B, 0)\}, G_{f_4})$ and $(\{f_4(s_1)\}, G)$, where G_{f_4} is the set of state descriptions to which the operator f_4 is applicable and s_1 is that state in G_{f_4} obtained as a consequence of solving($\{(A, 0, B, 0)\}, G_{f_4}$).

To solve the problem $(\{(A, 0, B, 0)\}, G_{f_4})$, we first calculate its difference. The state described by $(A, 0, B, 0)$ is not in G_{f_4} because (1) the box is not at C, (2) the monkey is not at C, and (3) the monkey is not on the box. The operators relevant to reduce these differences are, respectively, $f_2 = pushbox (C)$, $f_1 = goto(C)$, and $f_3 = climbbox$. Applying operator f_2 results in the subproblems $(\{(A, 0, B, 0)\}, G_{f_2})$ and $(\{f_2(s_{11})\}, G_{f_2})$, where $s_{11} \in G_{f_2}$ is obtained as a consequence of solving the first subproblem.

Since $(\{(A, 0, B, 0)\}, G_{f_2})$ must be solved first, we calculate the difference. The difference is that the monkey is not at B, and the relevant operator is $f_1 = goto(B)$. This operator is then used to reduce the problem to a pair of subproblems $(\{(A, 0, B, 0)\}, G_{f_1})$ and $(\{f_1(s_{111})\}, G_{f_2})$. Now the first of these problems is primitive; its difference is zero since $(A, 0, B, 0)$ is in the domain of f_1 and f_1 is applicable to solve this problem. Note that $f_1(s_{111}) = (B, 0, B, 0)$ so the second problem becomes $(\{(B, 0, B, 0)\}, G_{f_2})$. This problem is also primitive since $(B, 0, B, 0)$ is in the domain of f_2 and f_2 is applicable to solve this problem. This process of completing the solution of problems generated earlier is continued until the initial problem is solved.

In an AND/OR graph, one of the nodes, called the start node, corresponds to the original problem description. Those nodes in the graph corresponding to primitive problem descriptions are called terminal

---

* Loosely speaking, difference for (S, F, G) is a partial list of reasons why the goal test defining the set G is failed by the member of S. (IF some member of S is in G, the problem is solved and there is no difference.)

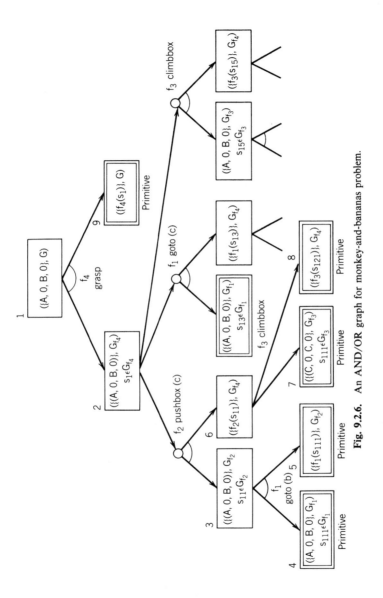

Fig. 9.2.6. An AND/OR graph for monkey-and-bananas problem.

nodes. The objective of the search process carried out on an AND/OR graph is to show that the start node is solved. The definition of a solved node can be given recursively as follows:

1. The terminal nodes are solved nodes since they are associated with primitive problems.
2. If a nonterminal node has OR successors, then it is a solved node if and only if at least one of its successors is solved.
3. If a nonterminal node has AND successors, then it is a solved node if and only if all of its successors are solved.

A solution graph is the subgraph of solved nodes that demonstrates that the start node is solved. The task of the production system or the search process is to find a solution graph from the start node to the terminal nodes. Roughly speaking, a solution graph from node n to a set of nodes N of an AND/OR graph is analogous to a path in an ordinary graph. It can be obtained by starting with node n and selecting exactly one outgoing arc. From each successor node to which this arc is directed, we continue to select one outgoing arc, and so on until eventually every successor thus produced is an element of N.

In order to find solutions in an AND/OR graph, we need an algorithm similar to A*, but with the ability to handle the AND arc appropriately. Such an algorithm for performing heuristic search of an AND/OR graph is called AO* algorithm.

The AO* Algorithm

Step 1. Let G consist only of the node representing the initial state. (Call this node INIT.) Compute h(INIT).

Step 2. Until INIT is labeled SOLVED or until INIT's h value becomes greater than FUTILITY, repeat the following procedure:

1. Trace the marked arcs from INIT and select for expansion one of the as yet unexpanded nodes that occurs on this path. Call the selected node NODE.
2. Generate the successors of NODE. If there are none, then assign FUTILITY as the h value of NODE. This is equivalent to saying that NODE is not solvable. If there are successors, then for each one (called SUCCESSOR) that is not also an ancestor of NODE do the following:
   a. Add SUCCESSOR to the graph G.
   b. If SUCCESSOR is a terminal node, label it SOLVED and assign it an h value of 0.
   c. If SUCCESSOR is not a terminal node, compute its h value.
3. Propagate the newly discovered information up the graph by doing the following: Let S be a set of nodes that have been marked SOLVED or whose h values have been changed and so need to have values propagated back to their parents. Initialize S to NODE. Until S is empty, repeat the following procedure:
   a. Select from S a node none of whose descendants in G occurs in S. (In other words, make sure that for every node we are going to process, we process it before processing any of its ancestors.) Call this node CURRENT, and remove it from S.
   b. Compute the cost of each of the arcs emerging from CURRENT. The cost of each arc is equal to the sum of the h values of each of the nodes at the end of the arc plus whatever the cost of the arc itself is. Assign as CURRENT's new h value the minimum of the costs just computed for the arcs emerging from it.
   c. Mark the best path out of CURRENT by marking the arc that had the minimum cost as computed in the previous step.
   d. Mark CURRENT SOLVED if all of the nodes connected to it through the new marked arc have been labeled SOLVED.
   e. If CURRENT has been marked SOLVED or if the cost of CURRENT was just changed, then its new status must be propagated back up the graph. So add to S all of the ancestors of CURRENT.

It is noted that rather than the two lists, OPEN and CLOSED, that were used in the A* algorithm, the AO* algorithm uses a single structure G, representing the portion of the search graph that has been explicitly generated so far. Each node in the graph points both down to its immediate successors and up to its immediate predecessors. Each node in the graph is associated with an h value, an estimate of the cost of a path from itself to a set of solution nodes. The g value (the cost of getting from the start node to the current node) is not stored as in the A* algorithm, and h serves as the estimate of goodness of a node. A quantity FUTILITY is needed. If the estimated cost of a solution becomes greater than the value of FUTILITY, then the search is abandoned. FUTILITY should be selected to correspond to a threshold such that any solution with a cost above it is too expensive to be practical, even if it could ever be found.

A breadth-first algorithm can be obtained from the AO* algorithm by assigning $h \equiv 0$.

9.2.4 USE OF PREDICATE LOGIC

In many applications, the information to be encoded into the data base of a production system originates from descriptive statements that are difficult or unnatural to represent by simple structures like arrays or sets of numbers. Some problem-solving tasks require the capability for representing, retrieving, and manipulating sets of statements. The language of logic or, more specifically, the first-order predicate calculus, can be used to express a wide variety of statements. The logical formalism is appealing because it immediately suggests a powerful way of deriving new knowledge from old—mathematical deduction. In this formalism, we can conclude that a new statement is true by proving that it follows from the statements that are already known to be true. Thus the idea of a proof, as developed in mathematics as a rigorous way of demonstrating the truth of an already believed proposition, can be extended to include deduction is a way of deriving answers to questions and solutions to problems.

At this point, readers who are unfamiliar with propositional and predicate logic may want to consult a good introductory logic text before reading the rest of this chapter. Readers who want a more complete and formal presentation of the material in this section should consult Chang and Lee (1973).

Let us first explore the use of propositional logic as a way of representing knowledge. Propositional logic is appealing because it is simple to deal with and there exists a decision procedure for it. We can easily represent real-world facts as logical *propositions* written as *well-formed formulas* (wffs) in propositional logic, as shown in the following:

It is raining
   RAINING
It is sunny
   SUNNY
It is foggy
   FOGGY
If it is raining, then it is not sunny
   RAINING $\rightarrow$ ~SUNNY

Using these propositions, we could, for example, deduce that it is not sunny if it is raining. But very quickly we run up against the limitations of propositional logic. Suppose we want to represent the obvious fact stated by the sentence

John is a man.

We could write

JOHNMAN

But if we also wanted to represent

Paul is a man.

we would have to write something such as

PAULMAN

which would be a totally separate assertion, and we would not be able to draw any conclusions about similarities between John and Paul. It would be much better to represent these facts as

MAN(JOHN)
MAN(PAUL)

since now the structure of the representation reflects the structure of the knowledge itself. We are in even more difficulty if we try to represent the sentence

All men are mortal

because now we really need quantification unless we are willing to write separate statements about the mortality of every known man.

So we appear to be forced to move to predicate logic as a way of representing knowledge because it permits representations of things that cannot reasonably be represented in propositional logic. In predicate logic, we can represent real-world facts as *statements* written as wffs. But a major motivation for choosing to use logic at all was that if we used logical statements as a way of representing knowledge, then we had available a good way of reasoning with that knowledge. In this section, we briefly introduce the language and methods of predicate logic.

The elementary components of the predicate logic are predicate symbols, variable symbols, function symbols, and constant symbols. A predicate symbol is used to represent a relation in a domain of discourse. For example, to represent the fact "robot is in room $r1$," we might use the simple atomic formula:

INROOM(ROBOT, $r1$)

In this atomic formula, ROBOT and $r1$ are constant symbols. In general, atomic formulas are composed of predicate symbols and terms. A constant symbol is the simplest kind of term and is used to represent objects or entities in a domain of discourse. Variable symbols are terms also, and they permit us to be indefinite about which entity is being referred to, for example, INROOM(x,y). Function symbols denote functions in the domain of discourse. For example, the function symbol *mother* can be used to denote the mapping between an individual and his or her female parent. We might use the following atomic formula to represent the sentence "John's mother is married to John's father":

MARRIED[*father*(JOHN), *mother*(JOHN)]

An atomic formula has value T (true) just when the corresponding statements about the domain are true and it has the value F (false) just when the corresponding statement is false. Thus INROOM(ROBOT, $r1$) has value T, and INROOM(ROBOT, $r2$) has value F. Atomic formulas are the elementary building blocks of the predicate logic. We can combine atomic formulas to form more complex wffs by using connectives such as $\wedge$ (and), $\vee$ (or), and $\Rightarrow$ (implies). Formulas built by connecting other formulas by $\wedge$'s are called conjunctions. Formulas built by connecting other formulas by $\vee$'s are called disjunctions. The connective "$\Rightarrow$" is used for representing "if–then" statements. For example, the sentence "if the monkey is on the box, then the monkey will grasp the bananas," might be represented by

ON(MONKEY, BOX) $\Rightarrow$ GRASP(MONKEY, BANANAS)

The symbol "~" (not) is used to negate the truth value of a formula; that is, it changes the value of a wff from T to F and vice versa. The (true) sentence "robot is not in room $r2$" might be represented as

~INROOM(ROBOT, $r2$)

Sometimes an atomic formula, $P(x)$, has value T for all possible values of x. This property is represented by adding the universal quantifier ($\forall x$) in front of $P(x)$. If $P(x)$ has value T for at least one value of x, this property is represented by adding the existential quantifier ($\exists x$) in front of $P(x)$. For example, the sentence "all robots are gray" might be represented by

($\forall x$) [ROBOT(x) $\Rightarrow$ COLOR(x, GRAY)]

The sentence "there is an object in room $r1$" might be represented by

($\exists x$) INROOM(x, $r1$)

If P and Q are two wffs, the truth values of composite expressions made up of these wffs are given by the following table:

P	Q	$P \vee Q$	$P \wedge Q$	$P \Rightarrow Q$	$\sim P$
T	T	T	T	T	F
F	T	T	F	T	T
T	F	T	F	F	F
F	F	F	F	T	T

If the truth values of two wffs are the same regardless of their interpretation, these two wffs are said to be equivalent. Using the truth table, we can establish the following equivalences:

$\sim(\sim P)$	is equivalent to	P
$P \lor Q$	is equivalent to	$\sim P \Rightarrow Q$

deMorgan's laws:

$\sim(P \lor Q)$	is equivalent to	$\sim P \land \sim Q$
$\sim(P \land Q)$	is equivalent to	$\sim P \lor \sim Q$

Distributive laws:

$P \land (Q \lor R)$	is equivalent to	$(P \land Q) \lor (P \land R)$
$P \lor (Q \land R)$	is equivalent to	$(P \lor Q) \land (P \lor R)$

Commutative laws:

$P \land Q$	is equivalent to	$Q \land P$
$P \lor Q$	is equivalent to	$Q \lor P$

Associative laws:

$(P \land Q) \lor R$	is equivalent to	$P \land (Q \land R)$
$(P \lor Q) \lor R$	is equivalent to	$P \lor (Q \lor R)$

Contrapositive law:

$P \Rightarrow Q$	is equivalent to	$\sim Q \Rightarrow \sim P$

In addition, we have

$\sim(\exists x)P(x)$	is equivalent to	$(\forall x)[\sim P(x)]$
$\sim(\forall x)P(x)$	is equivalent to	$(\exists x)[\sim P(x)]$

In the predicate logic, there are rules of inference that can be applied to certain wffs and sets of wffs to produce new wffs. One important inference rule is *modus ponens*, that is the operation to produce the wff $W2$ from wffs of the form $W1$ and $W1 \Rightarrow W2$. Another rule of inference, *universal specialization*, produces the wff $W(A)$ from the wff $(\forall x)W(x)$, where A is any constant symbol. Using modus ponens and universal specialization together, for example, produces the wff $W2(A)$ from the wffs $(\forall x)[W1(x) \Rightarrow W2(x)]$ and $W1(A)$.

Inference rules are applied to produce derived wffs from given ones. In the predicate logic, such derived wffs are called *theorems*, and the sequence of inference rule applications used in the derivation constitutes a *proof* of the theorem. In artificial intelligence, some problem-solving tasks can be regarded as the task of finding a proof for a theorem. The sequence of inferences used in the proofs gives a solution to the problem.

Example 4

The state-space representation of the monkey-and-bananas problem can be modified so that the states are described by wffs. We assume that, in this example, there are three operators—grasp, climbbox, and pushbox.

Let the initial state s_0 be described by the following set of wffs:

~ONBOX
AT(BOX, B)
AT(BANANAS, C)
~HB

The predicate ONBOX has value T only when the monkey is on top of the box, and the predicate HB has value T only when the monkey has the bananas.

The effects of the three operators can be described by the following wffs:

1. Grasp
$(\forall s)\{ONBOX(s) \land AT(BOX, C, s) \Rightarrow HB(grasp(s))\}$

meaning "for all s, if the monkey is on the box and the box is at C in state s, then the monkey will have the bananas in the state attained by applying the operator grasp to state s." It is noted that the value of grasp(s) is the new state resulting when the operator is applied to state s.

2. Climbbox
$(\forall s)\{ONBOX(climbbox(s))\}$

meaning "for all s, the monkey will be on the box in the state attained by applying the operator climbbox to state s."

3. *Pushbox*

$(\forall x \forall s)\{\sim\text{ONBOX}(s)\Rightarrow\text{AT}(\text{BOX}, x, \text{pushbox}(x, s))\}$

meaning "for all x and s, if the monkey is not on the box in state s, then the box will be at position x in the state attained by applying the operator pushbox(x) to state s."

The goal wff is

$(\exists s)\text{HB}(s)$

This problem can now be solved by a theorem-proving process to show that the monkey can have the bananas (Green, 1969; Nilsson, 1971; Winston, 1984).

9.2.5 MEANS–ENDS ANALYSIS

So far, we have discussed several search methods that reason either forward or backward, but, for a given problem, one direction or the other must be chosen. Often, however, a mixture of the two directions is appropriate. Such a mixed strategy would make it possible to solve the main parts of a problem first and then go back and solve the small problems that arise in connecting the big pieces together. A technique known as means–ends analysis allows us to do that. The technique centers around the detection of difference between the current state and the goal state. Once such a difference is determined, an operator that can reduce the difference must be found. It is possible that the operator may not be applicable to the current state. So a subproblem of getting to a state in which it can be applied is generated. It is also possible that the operator does not produce exactly the goal state. Then we have a second subproblem of getting from the state it does produce to the goal state. If the difference was determined correctly and if the operator is really effective at reducing the difference, then the two subproblems should be easier to solve than the original problem. The means–ends analysis is applied recursively to the subproblems. From this point of view, the means–ends analysis could be considered as a problem-reduction technique.

In order to focus the system's attention on the big problems first, the differences can be assigned priority levels. Differences of higher priority can then be considered below lower priority ones. The most important data structure used in the means–ends analysis is the "goal." The goal is an encoding of the current problem situation, the desired situation, and a history of the attempts so far to change the current situation into the desired one. Three main types of goals are provided:

Type 1.   *Transform* object A into object B.
Type 2.   *Reduce* a difference between object A and object B by modifying object A.
Type 3.   *Apply* operator Q to object A.

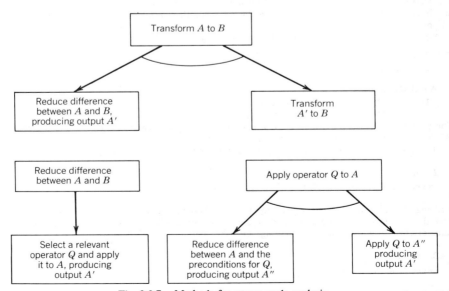

Fig. 9.2.7.   Methods for means–ends analysis.

Associated with the goal types are methods or procedures for achieving them. These methods, shown in a simplified version in Figure 9.2.7, can be interpreted as problem-reduction operators that give rise either to AND nodes, in the case of *transform* or *apply,* or to OR nodes, in the case of a *reduce* goal.

The first program to exploit means–ends analysis was the general problem solver (GPS). Its design was motivated by the observation that people often use this technique when they solve problems. For GPS, the initial task is represented as a *transform* goal, in which A is the initial object or state and B is the desired object or the goal state. The recursion stops if, for a *transform* goal, there is no difference between A and B, or for an *apply* goal the operator Q is immediately applicable. For a *reduce* goal, the recursion may stop, with failure, when all relevant operators have been tried and have failed.

In trying to transform object A into object B, the *transform* method uses a matching process to discover the differences between the two objects. The difference with the highest priority is the one chosen for reduction. A difference-operator table lists the operators relevant to reducing each difference.

Consider a simple robot problem. The available operators are listed as follows:

Preconditions	Operator	Results
1. AT(ROBOT, OBJ) $\land$LARGE(OBJ) $\land$CLEAR(OBJ) $\land$HANDEMPTY	PUSH(OBJ, LOC) $\longrightarrow$	AT(OBJ, LOC) $\land$AT(ROBOT, LOC)
2. AT(ROBOT, OBJ)	CARRY(OBJ, LOC) $\longrightarrow$	AT(OBJ, LOC)
3. $\land$SMALL(OBJ) None	WALK(LOC)	$\land$AT(ROBOT, LOC) AT(ROBOT, LOC)
4. AT(ROBOT, OBJ)	PICKUP(OBJ) $\longrightarrow$	HOLDING(OBJ)
5. HOLDING(OBJ)	PUTDOWN(OBJ) $\longrightarrow$	~HOLDING(OBJ)
6. AT(ROBOT, OBJ2) $\land$HOLDING(OBJ1)	PLACE(OBJ1, OBJ2) $\longrightarrow$	ON(OBJ1, OBJ2)

Figure 9.2.8 shows the difference-operator table that describes when each of the operators is appropriate. Note that sometimes there may be more than one operator that can reduce a given difference, and a given operator may be able to reduce more than one difference.

Suppose that the robot were given the problem of moving a desk with two objects on it from one room to another. The objects on top must also be moved. The main difference between the initial state and the goal state would be the location of the desk. To reduce the difference, either PUSH or CARRY could be chosen. If CARRY is chosen first, its preconditions must be met. This results in two more differences that must be reduced: the location of the robot and the size of the desk. The location of the robot can be handled by applying operator WALK, but there are no operators that can change the size of an object. So the path leads to a deadend. Following the other possibility, operator PUSH will be attempted.

Difference	Operator					
	PUSH	CARRY	WALK	PICKUP	PUTDOWN	PLACE
Move object	✓	✓				
Move robot			✓			
Clear object				✓		
Get object on object						✓
Get hand empty					✓	✓
Be holding object				✓		

Fig. 9.2.8. A difference-operator table.

PUSH has three preconditions, two of which produce differences between the initial state and the goal state. Since the desk is already large, one precondition creates no difference. The robot can be brought to the correct location by using the operator WALK, and the surface of the desk can be cleared by applying operator PICKUP twice. But after one PICKUP, an attempt to apply the second time results in another difference—the hand must be empty. The operator PUTDOWN can be applied to reduce that difference.

Once PUSH is performed, the problem is close to the goal state, but not quite. The objects must be placed back on the desk. The operator PLACE will put them there. But it cannot be applied immediately. Another difference must be eliminated, since the robot must be holding the objects. The operator PICKUP can be applied. In order to apply PICKUP, the robot must be at the location of the objects. This difference can be reduced by applying WALK. Once the robot is at the location of the two objects, it can use PICKUP and CARRY to move the objects to the other room.

The order in which differences are considered can be critical. It is important that significant differences be reduced before less critical ones. The next section describes a robot problem-solving system, STRIPS, which uses the means–ends analysis.

9.2.6   ROBOT PROBLEM SOLVING AND PLANNING

The simplest type of robot problem-solving system is a production system that uses the state description as the data base. State descriptions and goals for robot problems can be constructed from logical statements. As an example, consider the robot hand and configurations of blocks shown in Figure 9.2.1. This situation can be represented by the conjunction of the following statements:

CLEAR(B)	block B has a clear top
CLEAR(C)	block C has a clear top
ON(C, A)	block C is on block A
ONTABLE(A)	block A is on the table
ONTABLE(B)	block B is on the table
HANDEMPTY	the robot hand is empty

The goal is to construct a stack of blocks in which block B is on block C, and block A is on block B. In terms of logical statement, we may describe the goal as ON(B, C) $\wedge$ ON(A, B).

Robot actions change one state, or configuration, of the world into another. One simple and useful technique for representing robot action was used by a robot problem-solving system called STRIPS (Fikes and Nilsson, 1971). A set of rules is used to represent robot actions. Rules in STRIPS consist of three components. The first is the precondition that must be true before the rule can be applied. It is usually expressed by the left side of the rule. The second component is a list of predicates called the delete list. When a rule is applied to a state description, or data base, delete from the data base the assertions in the delete list. The third component is called the add list. When a rule is applied, add the assertions in the add list to the data base. The MOVE action for the block-stacking example is given below:

MOVE(X, Y, Z)	move object X from Y to Z
Precondition	CLEAR(X), CLEAR(Z), ON(X, Y)
Delete list	ON(X, Y), CLEAR(Z)
Add list	ON(X, Z), CLEAR(Y)

If MOVE is the only operator or robot action available, the search graph (or tree) shown in Figure 9.2.2 is generated.

Consider a more concrete example with the initial data base shown in Figure 9.2.1 and the following four robot actions or operations in STRIPS form:

1. PICKUP(X)
   Precondition and delete list: ONTABLE(X), CLEAR(X), HANDEMPTY
   Add list: HOLDING(X)
2. PUTDOWN(X)
   Precondition and delete list: HOLDING(X)
   Add list: ONTABLE(X), CLEAR(X), HANDEMPTY
3. STACK(X, Y)
   Precondition and delete list: HOLDING(X), CLEAR(Y)
   Add list: HANDEMPTY, ON(X, Y), CLEAR(X)

4. UNSTACK(X, Y)
   Precondition and delete list: HANDEMPTY, CLEAR(X), ON(X, Y)
   Add list: HOLDING(X), CLEAR(Y)

Suppose that our goal is ON(B, C) $\wedge$ ON(A, B). Working forward from the initial state description shown in Figure 9.2.1., we obtain the complete state space for this problem as shown in Figure 9.2.9, with a solution path between the initial state and the goal state indicated by dark lines. The solution sequence of actions consists of {UNSTACK(C, A), PUTDOWN(C), PICKUP(B), STACK(B, C), PICKUP(A), STACK(A, B)}. It is called a "plan" for achieving the goal.

If a problem-solving system knows how each operator changes the state of the world or the data base and knows the preconditions for an operator to be executed, it can apply means–ends analysis to solve problems. Briefly, this technique involves looking for a difference between the current state and a goal state and trying to find an operator that will reduce the difference. A relevant operator is one whose add list contains formulas that would remove some part of the difference. This continues recursively until the goal state has been achieved. STRIPS and most other planners use means–ends analysis.

We have just seen how STRIPS computes a specific plan to solve a particular robot problem. The next step is to generalize the specific plan by replacing constants by new parameters. In other words, we wish to elevate the particular plan to a plan schema. The need of a plan generalization is apparent in a learning system. For the purpose of saving plans so that portions of them can be used in a later planning process, the preconditions and effects of any portion of the plan need to be known. To accomplish this, plans are stored in a triangle table with rows and columns corresponding to the operators of the plan. The triangle table reveals the structure of a plan in a fashion that allows parts of the plan to be extracted later in solving related problems.

An example of a *triangle table* is shown in Figure 9.2.10. Let the leftmost column be called the *zeroth column*; then the jth column be headed by the jth operator in the sequence. Let the top row be called the first row. If there are N operators in the plan sequence, then the last row is the $(N + 1)$-th row. The entries in cell (i, j) of the table, for $j > 0$ and $i < N + 1$, are those statements added to the state description by the jth operator that survive as preconditions of the ith operator. The entries in cell $(i, 0)$ for $i < N + 1$, are those statements in the initial-state description that survive as preconditions of the ith operator. The entries in the $(N + 1)$-th row of the table are then those statements in the original-state description, and those added by the various operators, that are components of the goal.

Triangle tables can easily be constructed from the initial-state description, the operators in the sequence, and the goal description. These tables are concise and convenient representations for robot plans. The entries in the row to the left of the ith operator are precisely the preconditions of the operator. The entries in the column below the ith operator are precisely the add formula statements of that operator that are needed by subsequent operators or that are components of the goal.

Let us define the ith *kernel* as the intersection of all rows below, and including, the ith row with all columns to the left of the ith column. The fourth kernel is outlined by double lines in Figure 9.2.10. The entries in the ith kernel are then precisely the conditions that must be matched by a state description in order that the sequence composed of the ith and subsequent operators be applicable and achieve the goal. Thus, the first kernel, that is, the zeroth column, contains those conditions of the initial state needed by subsequent operators and by the goal; the $(N + 1)$-th kernel [i.e., the $(N + 1)$-th row] contains the goal conditions themselves. These properties of triangle tables are very useful for monitoring the actual execution of robot plans.

Since robot plans must ultimately be executed in the real world by a mechanical device, the execution system must acknowledge the possibility that the actions in the plan may not accomplish their intended effects and that mechanical tolerances may introduce errors as the plan is executed. As actions are executed, unplanned effects might either place us unexpectedly close to the goal or throw us off the track. These problems could be dealt with by generating a new plan (based on an updated state description) after each execution step, but obviously, such a strategy would be too costly, so we instead seek a scheme that can intelligently monitor progress as a given plan is being executed.

The kernels of triangle tables contain just the information needed to realize such a plan execution system. At the beginning of a plan execution, we know that the entire plan is applicable and appropriate for achieving the goal because the statements in the first kernel are matched by the initial-state description, which was used when the plan was created. (Here we assume that the world is static; that is, no changes occur in the world except those initiated by the robot itself.) Now suppose the system has just executed the first $i - 1$ actions of a plan sequence. Then, in order for the remaining part of the plan (consisting of the ith and subsequent actions) to be both applicable and appropriate for achieving the goal, the statements in the ith kernel must be matched by the new current-state description. (We assume that a sensory perception system continuously updates the state description as the plan is executed so that this description accurately models the current state of the world.) Actually, we can do better than merely check to see if the expected kernel matches the state description after an

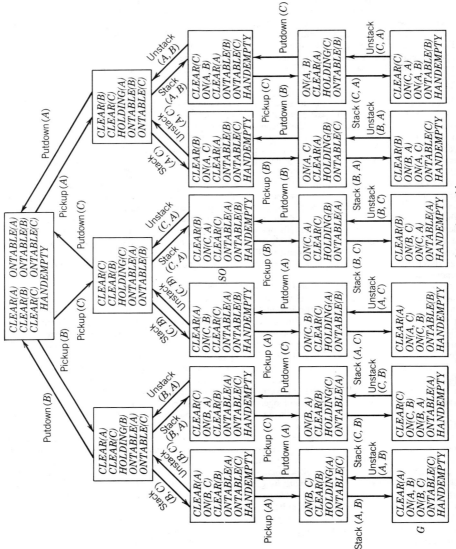

Fig. 9.2.9. The state space for a robot problem.

	0	1	2	3	4	5	6
1	HANDEMPTY CLEAR(C) ON(C, A)	1 unstack(C, A)					
2		HOLDING(C)	2 putdown(C)				
3	ONTABLE(B) CLEAR(B)		HANDEMPTY	3 pickup(B)			
4			CLEAR(C)	HOLDING(B)	4 stack(B, C)		
5	ONTABLE(A)	CLEAR(A)			HANDEMPTY	5 pickup(A)	
6					CLEAR(B)	HOLDING(A)	6 stack(A, B)
7					ON(B, C)		ON(A, B)

Fig. 9.2.10.   A triangle table.

action; we can look for the highest numbered matching kernel. Then, if an unanticipated effect places us closer to the goal, we need only execute the appropriate remaining actions; and if an execution error destroys the results of previous actions, the appropriate actions can be reexecuted.

To find the appropriate matching kernel, we check each one in turn starting with the highest numbered one (which is the last row of the table) and work backward. If the goal kernel (the last row of the table) is matched, execution halts; otherwise, supposing the highest numbered matching kernel is the ith one, then we know that the ith operator is applicable to the current-state description. In this case, the system executes the action corresponding to this ith operator and checks the outcome, as before, by searching again for the highest numbered matching kernel. In an ideal world, this procedure merely executes in order each action in the plan. In a real-world situation, on the other hand, the procedure has the flexibility to omit execution of unnecessary actions or to overcome certain kinds of failures by repeating the execution of appropriate actions. Replanning is initiated when there are no matching kernels.

As an example of how this process might work, let us return to our block-stacking problem and the plan represented by the triangle table in Figure 9.2.10. Suppose the system executes actions corresponding to the first four operators and that the results of these actions are as planned. Now suppose the system attempts to execute the pick-up block-A action, but the execution routine (this time) mistakes block B for block A and picks up block B instead. [Assume again that the perception system accurately updates the state description by adding HOLDING (B) and deleting ON(B, C); in particular, it does not add HOLDING(A).] If there were no execution error, the sixth kernel would now be matched; the result of the error is that the highest numbered matching kernel is now kernel 4. The action corresponding to STACK(B, C) is thus reexecuted, putting the system back on the track.

The fact that the kernels of triangle tables overlap can be used to advantage to scan the table efficiently for the highest numbered matching kernel. Starting in the bottom row, we scan the table from left to right, looking for the first cell that contains a statement that does not match the current-state description. If we scan the whole row without finding such a cell, the goal kernel is matched; otherwise, if we find such a cell in column i, the number of the highest numbered matching kernel cannot be greater than i. In this case, we set a *boundary* at column i and move up to the next-to-bottom row and begin scanning this row from left to right, but not past column. If we find a cell containing an unmatched statement, we reset the column boundary and move up another row to begin scanning that row, and so forth. With the column boundary set to k, the process terminates by finding that the kth kernel is the highest numbered matching kernel when it completes a scan of the kth row (from the bottom) up to the column boundary.

Example 5

Consider the simple task of fetching a box from an adjacent room by a robot vehicle. Let the initial state of robot's world model be as shown in Figure 9.2.11. Assume that there are two operators, GOTHRU and PUSHTHRU:

GOTHRU (d, $r1$, $r2$)   robot goes through door d from room $r1$ into room $r2$

Precondition: INROOM(ROBOT, $r1$) $\wedge$ CONNECTS (d, $r1$, $r2$) the robot is in room $r1$
                    and door d connects room $r1$ to room $r2$

Delete list: INROOM(ROBOT, S) for any value of S

Add list: INROOM(ROBOT, $r2$)

PUSHTHRU(b, d, $r1$, $r2$)   robot push object b through door d from room $r1$ into room $r2$

Precondition: INROOM(b, $r1$) $\wedge$ INROOM(ROBOT, $r1$) $\wedge$ CONNECTS (d, $r1$, $r2$)

Delete list: INROOM(ROBOT, S), INROOM(b, S)

Add list: INROOM(ROBOT, $r2$) INROOM(b, $r2$)

The difference-operator table is shown in Figure 9.2.12.

When STRIPS is given the problem it first attempts to achieve the goal G_0 from the initial state M_0. This problem cannot be solved immediately. However, if the initial data base contains a statement INROOM($B1$, $R1$), the problem-solving process could continue. STRIPS finds the operator PUSHTHRU($B1$, d, $r1$, $R1$) whose effect can provide the desired statement. The precondition G_1 for PUSHTHRU is

G_1: INROOM($B1$, $r1$) $\wedge$ INROOM(ROBOT, $r1$) $\wedge$ CONNECTS(d, $r1$, $R1$)

From the means–ends analysis, this precondition is set up as a subgoal and STRIPS tries to accomplish it from M_0.

Although no immediate solution can be found to solve this problem, STRIPS finds that if $r1=R2$, $d=D1$, and the current data base contains INROOM(ROBOT, $R2$), the process could continue. Again STRIPS finds the operator GOTHRU(d, $r1$, $R2$) whose effect can produce the desired statement. Its precondition is the next subgoal, namely,

G_2: INROOM(ROBOT, $r1$) $\wedge$ CONNECTS(d, $r1$, $R2$)

Using the substitutions $r1=R1$ and $d=D1$, STRIPS is able to accomplish G_2. It therefore applies GOTHRU($D1$, $R1$, $R2$) to M_0 to yield

M_1: INROOM(ROBOT, $R2$) CONNECTS($D1$, $R1$, $R2$)
        CONNECTS($D2$, $R2$, $R3$), BOX($B1$)
        INROOM($B1$, $R2$), . . .
        $(\forall x)(\forall y)(\forall z)$ [CONNECTS(x, y, x) $\Rightarrow$ CONNECTS(x, z, y)]

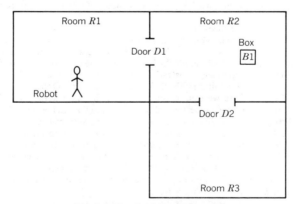

Fig. 9.2.11. Initial world model.

Initial data base M_0:
        INROOM(ROBOT, $R1$)
        CONNECTS($D1$, $R1$, $R2$)
        CONNECTS($D2$, $R2$, $R3$)
        BOX($B1$)
        INROOM($B1$, $R2$)
        $(\forall\ x\ \forall\ y\ \forall\ z)$ [CONNECTS(x, y, z) $\rightarrow$ CONNECTS(x, z, y)]
Goal G_0:
        $(\exists x)$ [BOX(x) $\wedge$ INROOM(x, $R1$)]

Difference	Operator	
	GOTHRU	PUSHTHRU
Location of box		✓
Location of robot	✓	
Location of box and robot		✓

Figure 9.2.12 Difference-operator table for Example 5.

Now STRIPS attempts to achieve the subgoal G_1 from the new data base M_1. It finds the operator PUSHTHRU($B1$, $D1$, $R2$, $R1$) with the substitution $r1=R2$ and $d=D1$. Application of this operator to M_1 yields

M_2: INROOM(ROBOT, $R1$), CONNECTS($D1$, $R1$, $R2$)

CONNECTS($D1$, $R2$, $R3$), BOX($B1$)

INROOM($B1$, $R1$), . . .

$(\forall x \forall y \forall z)[\text{CONNECTS}(x, y, z) \Rightarrow \text{CONNECTS}(x, z, y)]$

Next, STRIPS attempts to accomplish the original goal G_0 from M_2. This attempt is successful and the final operator sequence is

GOTHRU($D1$, $R1$, $R2$), PUSHTHRU($B1$, $D1$, $R2$, $R1$)

We would like to generalize the above plan so that it could be free from the specific constants $D1$, $R1$, $R2$, and $B1$ and could be used in situations involving arbitrary doors, rooms, and boxes. The triangle table for the plan is given in Figure 9.2.13, and the triangle table for the generalized plan is shown in Figure 9.2.14. Hence the plan could be generalized as follows:

GOTHRU($d1$, $r1$, $r2$)

PUSHTHRU(b, $d2$, $r2$, $r3$)

and be used to go from one room to an adjacent second room and push a box to an adjacent third room.

To improve a robot's problem-solving capability to handle more complex tasks and also to speed up the planning process, an approach is to design the system with a learning capability. STRIPS uses a generalization scheme for machine learning (Fikes, Hart, and Nilsson, 1972). Another form of learning would be the updating of the information in the difference-operator table from the system's experience.

Learning by analogy has been considered as a powerful approach and has been applied to robot planning. A robot planning system with learning, called PULP-I, has been proposed (Tangwongsan

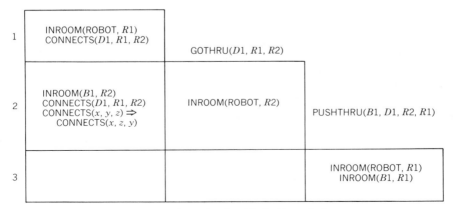

Fig. 9.2.13. Triangle table for Example 5.

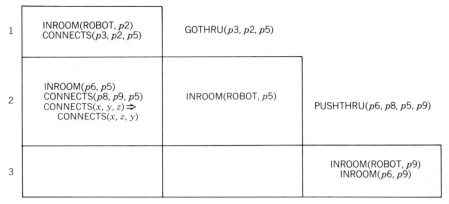

Fig. 9.2.14.   Triangle table for generalized plan.

and Fu, 1979). The system uses an analogy between a current unplanned task and any known similar tasks to reduce the search for a solution. A semantic network, instead of predicate logic, is used as the internal representation of tasks. Initially a set of basic task examples is stored in the system as knowledge based on past experience. The analogy of two task statements is used to express the similarity between them and is determined by a semantic matching procedure. The matching algorithm measures the semantic "closeness," the smaller the value, the closer the meaning. Based on the semantic matching measure, past experience in terms of stored information is retrieved and a candidate plan is formed. Each candidate plan is then checked by its operator's preconditions to ensure its applicability to the current world state. If the plan is not applicable, it is simply dropped out of the candidacy. After the applicability check, several candidate plans might be found. These candidate plans are listed in ascending order according to their evaluation values of semantic matching. The one with the smallest value of semantic matching has the top priority and must be at the beginning of the candidate list. Of course, if no candidate is found, the system terminates with failure.

Computer simulation of PULP-I has shown a significant improvement of planning performance. This improvement is not merely on the planning speed, but also in the capability of forming complex plans from the learned basic task examples.

9.2.7   EXPERT SYSTEMS AND KNOWLEDGE ENGINEERING

In most areas of AI, programs fall far short of the competence of humans or even animals. Computer systems designed to see images, hear sounds, and understand speech can only claim limited success. However, in one area of AI—that of reasoning from knowledge in a limited domain—computer programs cannot only approach human performance, but in some cases they can exceed it (Hayer Roth, Waterman, and Lenat, 1983; Ishizuka, Fu, and Yao, 1983; Nau, 1983; Weiss and Kulikowski, 1984).

These programs use a collection of facts, rules of thumb, and other knowledge about a given field, coupled with methods of applying those rules, to make inferences. They solve problems in such specialized fields as medical diagnosis, mineral exploration, and oil-well log interpretation. They differ substantially from conventional computer programs because their tasks have no algorithmic solutions and because they often must make conclusions based on incomplete or uncertain information.

In building such expert systems, researchers have found that amassing a large amount of knowledge, rather than sophisticated reasoning techniques, is responsible for most of the power of the system. Such high-performance expert systems, previously limited to academic research projects, are beginning to enter the software marketplace, AI programs are turning up for potential use in robots, medical diagnosis, automatic crew-scheduling system for the U.S. Space Shuttle, and flight systems for F-16 military aircraft, among other places.

9.2.7.1   Construction of an Expert System

Not all fields of knowledge are suitable at present for building expert systems. For a task to qualify for "knowledge engineering," the following prerequisites must be met:

1. There must be at least one human expert who is acknowledged to perform the task well.
2. The primary sources of the expert's abilities must be special knowledge, judgment, and experience.

3. The expert must be able to articulate that special knowledge, judgment, and experience, and also explain the methods used to apply it to a particular task.

4. The task must have a well-bounded domain of application.

Sometimes an expert system can be built that does not exactly match these prerequisites; for example, the abilities of several human experts rather than one, might be brought to bear on a problem.

The structure of an expert system is modular: facts and other knowledge about a particular domain can be separated from the inference procedure—or control structure—for applying those facts, while another part of the system—the global data base—is the model of the "world" associated with a specific problem, its status, and its history. It is desirable, though not yet common, to have a natural-language interface to facilitate the use of the system both during development and in the field. In some sophisticated systems, an explanation module is also included, allowing the user to challenge the system's conclusions and to examine the underlying reasoning process that led to them.

An expert system differs from more conventional computer programs in several important respects. In a conventional computer program, knowledge pertinent to the problem and methods for using this knowledge are intertwined, so it is difficult to change the program. In an expert system there is usually a clear separation of general knowledge about the problem (the knowledge base) from information about the current problems (the input data) and methods (the inference machine) for applying the general knowledge to the problem. With this separation the program can be changed by simple modification of the knowledge base. This is particularly true of rule-based systems, where the system can be changed by the simple addition or subtraction of rules in the knowledge base.

9.2.7.2 Rule-Based Systems

The most popular approach to representing the domain knowledge (both facts and heuristics) needed for an expert system is by production rules (also referred to as SITUATION–ACTION rules or IF–THEN rules). A simple example of a production rule is: IF the power supply on the space shuttle fails, AND a backup power supply is available, AND the reason for the first failure no longer exists, THEN switch to the backup power supply. Rule-based systems work by applying rules, noting the results, and applying new rules based on the changed situation. They can also work by directed logical inference, either starting with the initial evidence in a situation and working toward a solution, or starting with hypotheses about possible solutions and working backward to find existing evidence—or a deduction from existing evidence—that supports particular hypothesis.

One of the earliest and most often applied expert systems is Dendral (Barr, Cohen, and Feigenbaum, 1981–1982). It was devised in the late 1960s by Edward A. Feigenbaum and Joshua Lederberg at Stanford University to generate plausible structural representations of organic molecules from mass spectrogram data. The approach called for:

1. Deriving constraints from the data.

2. Generating candidate structures.

3. Predicting mass spectrographs for candidates.

4. Comparing the results with data.

This rule-based system, chaining forward from the data, illustrates the very common AI problem-solving approach of "generation and test." Dendral has been used as a consultant by organic chemists for more than 15 years. It is currently recognized as an expert in mass-spectral analysis.

One of the best-known expert systems is MYCIN (Barr, Cohen, and Feigenbaum, 1981–1982), designed by Edward Shortliffe at Stanford University in the mid-1970s. It is an interactive system that diagnoses bacterial infections and recommends antibiotic therapy. MYCIN represents expert judgmental reasoning as condition-conclusions rules, linking patient data to infection hypotheses, and at the same time provides the expert's "certainty" estimate for each rule. It chains backward from hypothesized diagnoses, using rules to estimate the certainty factors of conclusions based on the certainty factors of their antecedents, to see if the evidence supports a diagnosis. If there is not enough information to narrow the hypotheses, it asks the physician for additional data, exhaustively evaluating all hypotheses. When it has finished, MYCIN matches treatments to all diagnoses that have high certainty values.

Another rule-based system, R1, has been very successful in configuring VAX computer systems from a customer's order of various standard and optional components. The initial version of R1 was developed by John McDermott in 1979 at Carnegie-Mellon University, for Digital Equipment Corp. Because the configuration problem can be solved without backtracking and without undoing previous steps, the system's approach is to break the problem up into the following subtasks and do each of them in order:

1. Correct mistakes in the order.

2. Put components into CPU cabinets.

3. Put boxes in Unibus cabinets and put components in boxes.
4. Put panels in Unibus cabinets.
5. Lay out system floor plan.
6. Do the cabling.

At each point in the configuration development, several rules for what to do next are usually applicable. Of the applicable rules, R1 selects the rule having the most IF clauses for its applicability, on the assumption that that rule is more specialized for the current situation. (R1 is written in OPS 5, a special language for executing production rules.) The system now has about 1200 rules for VAXs, together with information about some 1000 VAX components. The total system has about 3000 rules and knowledge about PDP-11 as well as VAX components.

9.2.7.3 Remarks

Many expert systems are also under development (Hayer-Roth, Waterman, and Lenat, 1983; Ishizuka, Fu and Yao, 1983; Nau, 1983; Weiss and Kulikowski, 1984). Their application areas include medical diagnosis and prescription, medical-knowledge automation, chemical-data interpretation, chemical and biological synthesis, mineral and oil exploration, planning and scheduling, signal interpretation, military threat assessment, tactical targeting, space defense, air-traffic control, circuit analysis, VLSI design, structure damage assessment, equipment fault diagnosis, computer-configuration selection, speech understanding, computer-aided instruction, knowledge-base access and management, manufacturing process planning and scheduling, and expert-system construction.

There appear to be few constraints on the ultimate use of expert systems. However, the nature of their design and construction is changing. The limitations of rule-based systems are becoming apparent: not all knowledge can be structured as empirical associations. Such associations tend to hide causal relationships, and they are also inappropriate for highlighting structure and function. The newer expert systems are adding deep knowledge about causality and structure. These systems promise to be less fragile than current systems and may yield correct answers often enough to be considered for use in autonomous systems, not just as intelligent assistants.

Another change is the increasing trend toward non-rule-based systems. Such systems, using semantic networks, frames, and other knowledge-representation structures, are often better suited for causal modeling. By providing knowledge representations more appropriate to the specific problem, they also tend to simplify the reasoning required. Some new expert systems, using the "blackboard" approach combine rule-based and non-rule-based portions (Barr, Cohen, and Feigenbaum, 1981–1982; Nau, 1983; Winston, 1984), which cooperate to build solutions in an incremental fashion, with each segment of the program contributing its own particular expertise.

The growth of expert systems, coupled with increased computer capability and greater access to computers by the public, promises to give virtually everyone access to expertise. This will lead to profound changes in our society.

9.2.8 CONCLUDING REMARKS

We have briefly reviewed several basic methods in AI. Robot planning is used as an application example to illustrate the methods described. A recent trend indicates that AI methods are actually more effective in solving problems in restricted domains. The resulting systems, often called expert systems, have found very useful applications in several fields, such as medical diagnosis, mineral exploration, and computer-system configuration. Nevertheless, powerful knowledge representation schemes, effective and efficient reasoning mechanisms, and design of systems with learning capabilities are certainly still in high demand.

REFERENCES

Barr, A., Cohen, P., and Feigenbaum, E. A. (1981–1982). *The handbook of artificial intelligence.* Los Altos, CA: William Kaufmann, Vols. 1, 2, and 3.

Chang, C. L., and Lee, R. C. T. (1973). *Symbolic logic and mechanical theorem proving.* New York: Academic.

Fikes, R. E., Hart, P. E., and Nilsson, N. J. (1972). Learning and executing generalized robot plans. *Artificial Intelligence, 3*(4), 251–288.

Fikes, R. E., and Nilsson, N. J. (1971). STRIPS: a new approach to the application of theorem proving to problem solving, *Artificial Intelligence, 2*(3/4), 189–208.

Fu, K. S., Ed. (1981–1982). *Syntactic pattern recognition applications.* New York: Springer-Verlag.

Fu, K. S., Ed. (1982). *Applications of pattern recognition.* Cleveland, OH: CRC Press.

Green, C. (1969). Application of theorem proving to problem solving. *Proc. first international joint conference on artificial intelligence.* Washington, D.C.

Hayer-Roth, F., Waterman, D., and Lenat, D., Eds. (1983). *Building expert systems.* Reading, MA: Addison-Wesley.

Ishizuka, M., Fu, K. S., and Yao, J. T. P. (1983). A rule-based damage assessment system for existing structures. *SM Archives, 8,* 99–118.

Lee, C. S. G., Gonzalez, R. C., and Fu, K. S., Eds. *Tutorial on robotics.* New York: IEEE Computer Society Press.

Nau, D. S. (1983). Expert computer systems. *Computer, 16,* 63–85.

Nilsson, N. J. (1971). *Problem solving methods in artificial intelligence.* New York: McGraw-Hill.

Nilsson, N. J. (1980). *Principles of artificial intelligence.* Palo Alto, CA: Tioga Publ. Co.

Rich, E. (1983). *Artificial intelligence.* New York: McGraw-Hill.

Special Issue on robotics and automation. *Computer, 15*(12).

Tangwongsan, S. and Fu, K. S. (1979). An application of learning to robotic planning. *International Journal of Computer and Information Sciences, 8*(4), 303–333.

Weiss, S. M., and Kulikowski, C. A. (1984). *A practical guide to designing expert system.* Totowa, NJ: Rowman and Allanheld.

Winston, P. H. (1984). *Artificial intelligence,* 2nd ed. Reading, MA: Addison-Wesley.

CHAPTER 9.3
EXPERT SYSTEMS

CHAYA GARG-JANARDAN
RAY E. EBERTS
BERNHARD ZIMOLONG*
SHIMON Y. NOF
GAVRIEL SALVENDY

School of Industrial Engineering
Purdue University

---

* Now at Psychologisches Institut, Rühr-Universität Bochum, Bochum, West Germany.

This chapter discusses the technical and human aspects of expert systems. The former is discussed because an understanding of the technical aspects of expert systems is imperative to the research and study of the human aspects of expert systems. This chapter also provides a broad overview of the areas to which expert systems have been applied.

9.3.1 INTRODUCTION

An expert system has been defined by Weiss and Kulikowski (1984) as one that "handles real world complex problems requiring an expert's interpretation" and one that "solves these problems using a computer model of expert human reasoning, reaching the same conclusions that the human expert would reach if faced with a comparable problem." These systems are characterized by their ability to draw intelligent conclusions by manipulating and exploring their symbolically expressed knowledge bases which are comprised of large bodies of domain-specific knowledge gathered from human experts. This knowledge consists of facts, procedures, and heuristics that human experts have found useful in the solution of difficult problems in the domain. The problems to which expert systems have been applied are those where the need to examine an increasing number of solution possibilities is proportional to problem complexity.

An expert system can also be seen as a computer program, using artificial intelligence (AI) techniques, which is capable of intelligent problem solving of commercial and scientific importance. The basic ideas on which this intelligent problem solving rests are summarized in Table 9.3.1. An assumption behind expert systems is that specific knowledge of the task and general problem solving knowledge will eventuate in expert level analysis. Expert systems derive their capacity for intelligent problem solving by separating the task-specific knowledge from the procedures that manipulate it. This endows the expert system with flexibility since the knowledge base (task-specific knowledge) can be treated as any other data structure.

9.3.1.1 Knowledge

Components of Knowledge

In expert systems, expertise consists of knowledge of domain-specific information as well as skill at solving domain-specific problems. This knowledge can be obtained from both public and private knowledge sources (see Figure 9.3.1), where public sources are those that have been published in some form and private sources are those that are obtained from a subject matter expert. Besides identifying the sources of knowledge, it can also be represented in a dimensional structure. Hayes-Roth (1984a) uses three dimensions to represent knowledge; these dimensions are pictured in Figure 9.3.2 and can be summarized as follows:

1. The scope of knowledge ranging from the general and widely applicable to the specific and more narrowly applicable.
2. The purpose of knowledge ranging from descriptive to prescriptive purposes.
3. The validity of knowledge ranging from the certain to the uncertain. The knowledge base of an expert system employs both certain and probabilistic domain-specific knowledge.

Although the above structure of knowledge is theoretically plausible, in the real world knowledge may not be so easily categorized into the above three dimensions. Instead, humans may encode knowledge in one of the following forms:

1. Empirical associations, which may actually be used as procedures (for example, trouble shooting by maintenance personnel).
2. Heuristics used to solve problems efficiently and also to deal with uncertain or erroneous data.
3. Causal models of the system: operators in process control industries may have an internal (mental) model of the system which they may use to guide their control behavior (Rouse, 1984).

In the context of knowledge bases in expert systems, the descriptions and relationships are separate from the procedures used to manipulate them. The knowledge engineer tries to simulate the human problem solving process as far as possible and gives the expert system the capacity to use heuristics

Table 9.3.1 Basic Ideas of Intelligent Problem Solving

Knowledge = facts + beliefs + heuristics

Success = finding a good enough answer with the resources available

Search efficiency directly affects success

Aids to efficiency:

    Applicable, correct, and discriminating knowledge

    Rapid elimination of "blind alleys"

    Elimination of redundant computation

    Increased speed of computer operation

    Multiple, cooperative sources of knowledge

    Reasoning at varying levels of abstraction

Sources of increased problem difficulty:

    Errorful data or knowledge

    Dynamically changing data

    The number of possibilities to evaluate

    Complex procedures for ruling out possibilities

Source: Hayes-Roth, Waterman, and Lenat (1983).

in problem solving as well as to learn from its interactions with users who require it to solve different problems from day to day.

Types of Engineered Knowledge

At this time, some types of knowledge are easier to "engineer" than other types of knowledge. Table 9.3.2 provides a listing, from Hayes-Roth (1984a), of knowledge that is currently amenable (the first column) and types of knowledge that are not yet available for knowledge engineering (the second column). The types of knowledge that are presented in the first column should be rather self-explanatory. A brief explanation of those types of knowledge represented in the second column follows.

Naive physics refers to a nonverbal understanding of physical operations. As a concrete example of this, when a piece of cloth catches fire, we can mentally act out our response to it and image ourselves trying to extinguish it with water. An attempt to represent this knowledge on a computer has been problematic because we do not understand how humans learn to carry out these procedures. Metarepresentation and metaknowledge refer to the system's ability to know what it knows. Specifically, this is the ability to know what kinds of representations are used and what kinds of knowledge are accessible. As an example, we, as humans, do not have to search through memory to know that we do not know the telephone number of the White House. We have some kind of metaknowledge that

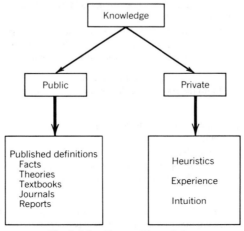

Fig. 9.3.1.   Types of knowledge.

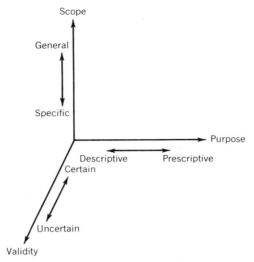

Fig. 9.3.2. The three dimensions of knowledge [Source: Hayes-Roth (1984a)].

this information is not available in memory. Expert systems would be more effective if they were able to reason about their own problem solving methods and knowledge. Expert systems also lack the ability to generalize from problem to problem or to infer that a given problem is a specific instance of a general set of problems. In addition, they lack the ability to decompile their own compiled knowledge unlike human beings who can, for instance, transform high-level programs into lower-level programs. Further research in these areas is needed so that engineered knowledge more closely corresponds to human expert knowledge.

The preceding discussions have helped to point out some of the problem areas in which expert systems can be applied. In particular, the problem area should be amenable to the techniques of applied artificial intelligence using available types of engineered knowledge (refer to Table 9.3.2). In addition, the problem area must be narrow in scope and specific enough for expertise to be encoded (refer to Figure 9.3.2), yet it must be complex enough to warrant an expert approach to solve the problem. An expert system is based on human expert behavior, which implies that the problem area should be solvable by human experts in a reasonable, finite period of time and the solution to the problem should have general agreement among experts. Usually, the problems addressed by an expert system can be solved using a limited subset of the human expert's total knowledge.

9.3.1.2 Characteristics of Expert Systems

An expert system, to perform intelligently and efficiently, should possess certain characteristics that make it unique from other kinds of systems and methods. In this section, the characterisitcs of expert systems will be considered along with a discussion on how expert systems differ from AI, conventional data processing programs, and human experts.

Table 9.3.2 Possible Types of Engineered Knowledge

Currently Practicable	Not Available
Definitions and taxonomies	General problem solving knowledge
Discrete descriptions	Analogs
Simple constraints and invariants	Naive physics
Empirical associations	Metarepresentation knowledge
Perceptual structures	Metaknowledge
Deductive methods	First principles
Simple inductive methods	Compiling and decompiling know-how
Very simple physical models	
Very simple search heuristics	

Source: Hayes-Roth (1984a).

Expert systems have the following characteristic features. Certainly, not every expert system built to date possesses all of the features mentioned. Each feature may be viewed, instead, as constituting a continuum ranging from none to an ideal amount of the feature.

First, expert systems have the capability for arriving at precise and accurate solutions to complex problems. They can reach solutions in a reasonably quick manner; an expert system reaching even the most perfect conclusion after an inordinate delay may render the conclusion invalid. Second, expert systems address problems in narrow specialized domains using specialized techniques. They have the capacity to use heuristics or "rules of thumb" and to avoid blind search; this permits the expert system to reason with judgmental knowledge as well as with formal knowledge of established theories. They also have a capacity to reason by manipulation of symbolic descriptions. (This precludes the assumption that physical systems can be conceptualized and represented by symbols.) Fourth, they have the ability to take a problem stated in some arbitrary initial form and convert it into the form appropriate for processing by expert rules. A good reformulation system is one which knows its limits and does not overzealously force problems into models suited to it. Fifth, expert systems are transparent in that they provide answers as well as explanations about the line of reasoning used to arrive at the answer. This gives the expert system an interactive user interface and the ability to justify its conclusions. Some of the newer expert systems are able to interact with the user in natural language. Sixth, expert systems have an ability to function with erroneous, uncertain, incomplete data and with uncertain judgmental rules. Many times they also have an ability to process two or more competing hypotheses at the same time. Finally, expert systems have an ability to incrementally integrate new knowledge into the existing knowledge base. This allows the expert system to be very flexible.

Expert systems perform difficult tasks at expert levels of performance. Other methods and systems are, however, available to solve problems. Techniques in AI are used to solve some kinds of problems. Conventional data processing programs are used to solve some problems. Human experts also solve problems. A issue that is raised often is the differences between expert systems and these other systems and methods.

Expert systems may be considered to be a subfield of AI with three concepts of AI underlying nearly all the work done in expert systems. The three concepts are symbolic programming, problem solving, and search. In addition, the techniques and applications of AI are more general purpose in nature, whereas expert systems address more narrowly defined domain-specific application problems. Thus, expert systems emphasize domain-specific problem solving strategies over the more general "weak methods" of AI.

Conventional data processing systems automate time-consuming clerical functions by algorithmically amassing and processing large volumes of data. The operations to be performed on the data are specified in the program so that the machine is only required to read and execute the program. Conventional data processing systems organize their knowledge at two levels: data and program. On the other hand, expert systems deal with problems whose solutions require drawing upon a body of knowledge and using heuristics to facilitate quick problem solutions. Expert systems have their own knowledge base and inference engine that they use to draw conclusions with regard to data of a current case given by the user. In addition, most expert systems organize their knowledge on three levels: data, knowledge base, and control.

Despite all their capabilities, expert systems lack a human expert's breadth of knowledge and understanding of fundamental principles. Expert systems are limited in their capacity to examine a single issue from several diverse criteria. They do, however, simulate some aspects of human thought and are able to apply many characteristics of human thought. Expert systems are currently not very good at learning in terms of updating their own knowledge base or system of search. AI learning programs such as Winston's (1984) "Structured Concept Learning Program" and Lenat's (1980) AM program could be applied to expert systems in the near future.

9.3.1.3 Functions Served By Expert Systems

The specific problem areas to which expert systems have been applied may be classified into three main areas (Hayes-Roth, Waterman, and Lenat, 1983): analysis problems, synthesis problems, and problems using a combination of analysis and synthesis. Table 9.3.3 provides a breakdown of these three areas along with examples of applications in each of them.

Analysis problems (refer to Table 9.3.3) are those that explain the current state of a system. The individual problem categories in this class of analysis problem to which expert systems have been and are being applied are interpretation and diagnosis problems. Interpretation refers to the inference of situation descriptions from a systematic, accurate, and rigorous analysis of observed data. This is accomplished by assigning, to the observed data, symbolic meaning descriptions of the situation or system state accounting for the data. Diagnosis systems infer malfunctions or behavioral irregularities in a system based on observed data and relate these inferences to underlying causes. This is achieved either by using a table of associations between behaviors and diagnoses or by comparing and elucidating flaws in system design and implementation based on a knowledge of correct design and implementation.

Synthesis problems are those that construct a solution to satisfy a goal within stated constraints.

Table 9.3.3 Application of Expert Systems

Problem Areas			
Analysis Problems		Synthesis Problems	
Problem Category	Specific Fields	Problem Category	Specific Fields
Interpretation	Surveillance, speech understanding, image analysis, chemical structure elucidation signal interpretation, intelligence analysis	Design	Digital circuits
Diagnosis	Medical, electronic, mechanical, and software diagnosis	Planning	Molecular genetics, automatic programming, military planning, robot, and project and route communications planning

Analysis and Synthesis Problems	
Problem Category	Specific Fields
Prediction	Weather forecasting, demographic predictions, traffic predictions, crop estimation, military forecasting, and economic policy
Debugging and repair	Computer-aided debugging for programs, automotive, network, avionic, and computer maintenance
Instruction	Education and teaching aids
Monitoring and control	Airtraffic control, mission control, process control, disease, regulatory and fiscal management tasks, business management, and battle management

Specific areas of application include the problem categories listed here. Design involves the creation of specifications for object configurations while keeping the constraints of the design problem in mind. Also, the system attempts to minimize cost and other undesirable properties (not specified in the constraints) of the object configuration while laying out the specifications. Planning systems construct a plan or program of actions that achieve goals without consuming excessive resources or violating constraints. A hierarchical ranking of goals is carried out to cope with conflicting goals. The effects of proposed plans are determined by using simulation or modeling.

Finally, expert systems are most often applied to problem areas that are comprised by both analysis and synthesis types of problems, simultaneously. Prediction involves the inference of consequences accruing from different states of a system. Expert systems in this area use dynamic models of the system with relevant parametric values associated with a given desirable state. Any changes in the state are accompanied by changes in the associated parametric values. This change facilitates the prediction of the future based on a given state. Debugging and repair areas perform approximately the same functions and hence are dealt with as a single area here. These systems prescribe remedies for system malfunctioning. They create specifications, make recommendations, and execute plans to remedy a diagnosed problem. The term debugging is used in the area of programming; in other areas, the term repair is used.

Expert systems in instruction construct a hypothetical description of a student's knowledge and then interpret the student's behavior by diagnosing strengths and weaknesses in the student. Based on this interpretation, the system specifies a remedial course of actions that are executed in the form of a tutorial interaction. Monitoring and control entails maintaining the system in a predetermined state. It involves interpreting the existing state of the system to determine if any constraints are currently being violated or if undesirable changes will be carried out in a future state. The underlying causes for the system behavior are determined, the appropriate cautioning alarms are set off, and remedial action is planned and executed.

9.3.2 STEPS IN THE CONSTRUCTION OF AN EXPERT SYSTEM

Expert systems are constructed to solve problems in narrow, specialized domains. In a typical interaction between a user and an expert system, the user inputs information, that describes the problem, to the expert system. The system then uses the problem solving techniques present in its control structure to arrive at solutions and conclusions to the problem in the global data base. It makes decisions and conclusions, thus, on the basis of knowledge and rules stored in its data base. These conclusions are conveyed to the user using its interface facilities. Also, if the expert system is truly an intelligent learning system, it will make changes in its knowledge base dependent on new rules inferred from its present interaction with the user.

Similar to most software development, the construction of an expert system may be divided into the following six stages (Hayes-Roth, Waterman, and Lenat, 1983):

1. Identification stage.
2. Conceptualization stage.
3. Formalization stage.
4. Implementation stage.
5. Testing stage.
6. Prototype revision stage.

During the implementation of these stages, Hayes-Roth, Waterman, and Lenat (1983) have listed a set of heuristics (Table 9.3.4), developed through personal experiences in developing expert systems, which can be used as guidance when building knowledge based systems. Each of the six stages will be described in more detail.

9.3.2.1 Identification Stage

In this stage, the whole project is given a structure and purpose by identifying participants and roles as well as defining the problem, the goals, and the resources needed to achieve the goals.

The participants represent experts from two main streams: (1) the domain for which the expert system is being constructed and (2) the field of knowledge engineering. The number of participants from each area is flexible depending on the complexity and size of the project as well as the availability of resources. The domain experts have to verbalize to try to share with the design experts all the knowledge and heuristics used by them to solve domain-specific problems. The knowledge engineer or design expert records the knowledge as well as tries to direct the expert's thinking so as to draw out as much information as possible. The domain expert and the design experts may have to go through many cycles of discussion, definition, and redefinition before they finally agree on a problem definition. At this stage, the domain expert supplies the clarifications and the design expert raises most of the points listed later in the questions in Table 9.3.5.

Once the problem has been defined, the participants list out the resources that are available and that may eventually constrain the problem definition. Because all participants should be willing to devote the many months that are usually required in the construction of an expert system, time and money become important resources. Other resources that may be important are sources of public knowledge domains such as books and journals. In addition, many expert systems are built from the existing framework of other expert systems so that these software tools, which already exist, could be an important resource.

9.3.2.2 Conceptualization Stage

The key concepts and relations in the now-defined problem are made more explicit in this stage. This involves continuous interaction between the participants. The knowledge engineer may introduce certain general ideas about representation and tools in order to direct the conceptualization. The design expert should avoid choosing certain ideas of representation so as not to influence or restrict the domain expert's analysis of the problems and enumeration of concepts. Below are listed some of the questions that need to be answered in this stage (Hayes-Roth, Waterman, and Lenat, 1983):

What types of data are available?

What is given and what is inferred?

Do the subtasks have names?

Do the strategies have names?

Are there identifiable partial hypothesis that are commonly used? What are they?

How are the objects in the domain related?

Can you diagram a hierarchy and label causal relations, set inclusion, part–whole relations, and so forth? What does it look like?

What processes are involved in problem solution?

What are the constraints on these processes?

What is the information flow?

Can you identify and separate the knowledge needed for solving a problem from the knowledge used to justify a solution?

Table 9.3.4 Heuristics in the Construction of an Expert System

Task Suitability
- Focus on a narrow specialty area that does not involve a lot of common-sense knowledge.
- Select a task that is neither too easy nor too difficult.
- Define the task clearly.
- Commitment from an articulate expert is essential.

Building the Prototype System
- Become familiar with the problem before beginning extensive interaction with the expert.
- Clearly identify and characterize the important aspects of the problem.
- Record a detailed protocol of the expert solving at least one prototypical case.
- Choose a knowledge engineering tool or architecture that minimizes the representional mismatch between subproblems.
- Start building the prototype version of the expert system as soon as the first example is well understood.
- Work intensively with a core set of representative problems.
- Identify and separate the parts that have caused trouble for AI programs in the past.
- Build in mechanisms for indirect reference.
- Separate domain-specific knowledge from general problem solving knowledge.
- Aim for simplicity in the inference engine.
- Do not worry about time and space efficiency in the beginning.
- Find or build computerized tools to assist in the rule-writing process.
- Pay attention to documentation.
- Do not wait until the informal rules are perfect before starting to build the system.
- When testing the system, consider the possibility of errors in input/output characteristics, inference rules, control strategies, and test examples.

Extending the Prototype System
- Build a friendly interface to the system.
- Provide capabilities for examining the knowledge base and line of reasoning.
- Keep a library of cases presented to the system.

Finding and Writing Rules
- Do not just talk with the expert, watch him or her doing examples.
- Use the terms and methods that experts use.
- Look for intermediate-level abstractions.
- If a rule looks big, it is.
- If several rules are very similar, look for an underlying domain concept.

Maintaining the Expert's Interest
- Engage the expert in the challenge of designing a useful tool.
- Give the expert something useful on the way to building a large system.
- Insulate the expert, as well as the user, from technical problems.
- Be careful about feeling expert.

General Advice
- Avoid redundancy.
- Be familiar with the architecture of several expert systems.
- The process of building an expert system is inherently experimental.

Source: Hayes-Roth, Waterman, and Lenat (1983).

9.3.2.3 Formalization Stage

In this stage the design expert takes on a more active role. He or she tries to map the concepts and relations enumerated in the previous stage to more formal representations based on various knowledge-engineering tools or frameworks. The choice of a tool involves the critical examination of the following three factors: (1) the hypothesis space, (2) the underlying model of the process, and (3) the characteristics of the data.

The hypothesis space may be delineated by examining how the concepts link to form hypotheses. The nature of the relations among the concepts should be determined, such as whether causal or spatiotemporal relations exist. The following questions about the hypothesis space should be addressed (Hayes-Roth, Waterman, and Lenat, 1983):

Is the hypothesis space finite?

Does it consist of prespecified classes?

Are the hypotheses ordered hierarchically?

Is there any uncertainty or judgmental reasoning associated with the final and intermediate hypotheses?

Would diverse levels of abstraction be useful?

The design expert should determine (from the concepts and relations given by the domain expert) whether behavioral or mathematical models underlie the problem solving process used in the deomain. The nature of the model will help determine the problem solving techniques that will be applicable in a more formalized representation.

If consideration of the preceding three factors leads to a tool that seems to fit closely, then participants determine the requirements of this tool and proceed to fill it. The choice of a tool also determines the specifications for representing formalized knowledge and for building a prototype.

9.3.2.4   Implementation Stage

In this stage, the prototype system is built by converting the domain knowledge gathered from the domain experts into the chosen knowledge base and control structure using a convenient mode of coding. As this knowledge conversion is completed and made consistent and compatible, a prototype expert system evolves.

9.3.2.5   Testing Stage

In the testing stage, the design experts attempt to elicit from the domain experts potential points of weakness in the expert system; those problems in the domain of expertise where they expect the system to fall. The prototype system is tested with problems that use its entire range of expertise as far as possible with particular attention being paid to the control strategies. Attention is also paid to the interface, learning, and explanation facilities of the system.

9.3.2.6   Prototype Revision Stage

In this final stage, any weaknesses pinpointed in the prior stage are rectified, consistency checks are made, and any additional data available are incorporated. The process of testing and revision is repeated until the participants fail to find any weaknesses in the system.

9.3.3   METHODOLOGICAL ISSUES IN THE CONSTRUCTION OF EXPERT SYSTEMS

At every stage in the construction of an expert system, several methodological issues arise pertaining to:

1. Knowledge acquisition.
2. Knowledge representation.
3. Control structure.
4. Interface facility.
5. Learning capabilities.
6. Tools and languages used.

The complete and satisfactory resolution of these issues results in the ideal expert system.

9.3.3.1   Knowledge Acquisition

An important and often difficult component of expert systems is acquiring the knowledge from the human expert. Anecdotal evidence indicates that experts are often the least able to verbally explain to novices their expertise. From work done on studying experts in process control tasks, Lees (1974) states that "it is well-known, for example, that a pilot may be skilled in flying but may give a quite erroneous account of the control linkages in the aircraft." In other words, the pilot may be an expert

Table 9.3.5 Questions to Be Asked During the Identification Stage

- What class of problems will the expert system be expected to solve?
- How can these problems be characterized or defined?
- What are important subproblems and partitioning of tasks?
- What are the data?
- What are important terms and their interrelations?
- What does a solution look like and what concepts are used in it?
- What aspects of human expertise are essential in solving these problems?
- What is the nature and extent of "relevant knowledge" that underlies the human solutions?
- What situations are likely to impede solutions?
- How will these impediments affect an expert system?

Source: Hayes-Roth, Waterman, and Lenat (1983).

at the task but may not be able to explain how the task is carried out. Another problem is that the task may be done automatically and, thus, not be accessible by conscious processes. In fact, this is a problem that has been studied for some time; Woodworth (1938) cites evidence that conscious content of verbal reports disappeared with extended practice. Despite these problems, methods, ranging from the problem solving techniques of Newell and Simon (1972) to a clinical psychology theory (Kelly, 1955), have been developed to elicit knowledge from experts.

The methods for acquiring knowledge are not always formalized; different approaches may be needed for different subject domains and different experts. Knowledge acquisition is certainly related to the six stages in the construction of an expert system from Hayes-Roth, Waterman, and Lenat (1983), which were discussed in an earlier section. Tables 9.3.5 and 9.3.6 present some of the important questions that should be addressed in the identification and formalization stages, respectively [from Hayes-Roth, Waterman, and Lenat (1983)]. In practice, the process may not always step through these six stages; fine tuning of the experts' knowledge is needed.

To help guide the knowledge acquisition, psychological theories about human behavior are often used. The information processing theory of Newell and Simon (1972) has been used extensively. Recently, techniques from psychotherapy have also been adapted to elicit information from experts. Both of these approaches will be discussed.

Following the influential Newell and Simon (1972) book, the expert is often modeled as an information processor who is presented with a problem that must be solved. The information used for solving a problem is viewed as passing through three successive stages (Newell and Simon, 1972; Rouse, 1984). In the first stage, the human receives information about the problem through the senses; this information is used to interpret the problem and set goals for how to solve the problem. Second, this initial information is used to construct an internal representation of the problem (Newell and Simon, 1972) or a problem space. The problem space is the state of knowledge at a particular point in time. Third, strategies, procedures, and operations are used to transform the knowledge state to match that of the desired goal state. Several strategies may be needed and several subgoals may be formulated to solve the more complex problems.

The Newell and Simon (1972) approach has been used extensively as a framework for acquiring knowledge from experts and is often referred to as the protocol method [for a general review of this method see Ericsson and Simon (1980)]. For the protocol method, experts are given a problem to solve and are asked to verbalize what they are doing as they try to solve the problem. They are

Table 9.3.6 Questions to Be Asked About the Nature of the Data

- Are the data sparse and insufficient or plentiful and redundant?
- Is there uncertainty attached to the data?
- Does the logical interpretation of data depend on their order of occurrence over time?
- What is the cost of data acquisition?
- How are data acquired or elicited? What classes of questions need to be asked to obtain data?
- How can certain data characteristics be recognized when sampled or extracted from a continuous data stream; how can features be extracted from waveforms or pictures, or from parsing natural language input?
- Are the data reliable, accurate, precise (hard); or are they unreliable, inaccurate, or imprecise (soft)?
- Are the data consistent and complete for the problems to be solved?

Source: Hayes-Roth, Waterman, and Lenat (1983).

asked to verbalize their strategies used, the kind of prior knowledge that they are using, and their goals and subgoals as they move through the task. The method requires motivated experts who can easily verbalize what they are doing. The interviewer, or so-called knowledge engineer, also plays a significant role because he or she must know when to prompt the expert if it is felt that steps are being left out. The information processor model can be used as a framework for understanding the protocols and, thus, can be used to determine what kind of information is needed and if any important information is being left out.

The expertise transfer system (ETS) (Boose, 1984) is another technique based on an interviewing technique used in psychotherapy by Kelly (1955) to elicit information from clients. An important part of the technique is to have the client list underlying elements (the knowledge base) and then the interviewer asks a series of "why" questions to determine the superordinate and subordinate relationships. The result of this kind of interview is a hierarchical relationship of the elements and the reasons for using the elements.

ETS has been developed to automatically interview the expert and help construct and analyze an initial knowledge base for the problem. In ETS, knowledge is elicited from the information sources (the experts), placed into an information base, analyzed and organized into knowledge bases, and then combined with other knowledge bases to form a knowledge network. Knowledge elicited for expert systems in this way, then, can be a combination of knowledge from many human experts. ETS tests the knowledge for sufficiency and necessity and will ask the expert to fill in any missing parts. As pointed out by Boose (1984), evaluating the success of ETS and other similar techniques is difficult; inappropriate constructs are easy to weed out, but there is no guarantee that all the important knowledge has been elicited from the expert.

TEIRESIAS is a program similar to ETS that has been used to assist with the extraction of knowledge from an expert to a particular expert system, MYCIN (Davis, 1976, 1977). TEIRESIAS is used mainly by an expert to try to modify an existing knowledge base. The modification process of TEIRESIAS operates in the following manner. The expert can interact with the expert system in its normal manner. When an error is spotted, the expert indicates this to TEIRESIAS and it then works backward through the program to determine the steps that led to the error until the bug is identified. TEIRESIAS then helps the expert fix the bug by adding to or modifying the knowledge base. To modify the knowledge base, TEIRESIAS is capable of keeping track of the interrelationships between the data structures of the expert system. Thus, when new knowledge is incorporated, TEIRESIAS knows what has to be changed in other knowledge structures to incorporate this new knowledge.

Although a knowledge engineer is trained to elicit information from experts, these techniques ultimately rely on the experts being able to verbalize their expertise. Several problems exist with the method of data collection:

Interpretation. The knowledge engineer must interpret the protocols so that his or her biases are not inserted into the data.

Completeness. The expert may leave out important steps in the problem solving task.

Verbalization Assumption. An implicit assumption of these methods is that the procedures and data can be verbalized; some may not be amenable to verbalization.

Problem with Experts. Expert skills are often automatized and, thus, may be difficult to verbalize or to interpret.

The knowledge engineer is usually aware of these problems and tries to work around them. Asking the expert and taking protocols is still the most used knowledge acquisition method.

9.3.3.2 Representation of Knowledge

In the context of expert systems, "representation of knowledge" purports to make a system knowledgeable by using the right combination of data structures and interpretive procedures. The data structures refer to the knowledge base that contains the facts and heuristics, and the interpretive procedures refer to the inference engine or control structure of the system, which consists of processes that use the knowledge base to infer solutions to problems, to perform analyses, and to form hypotheses. Since the knowledge base is separate from the inference engine, the two will be dealt with separately in this chapter too. First, the representation of knowledge in a knowledge base will be considered.

Representation schemes should fulfill several requirements. First, the representation scheme should be such that the knowledge base is extendable without necessitating substantial revision. This implies that the knowledge base should be consistent so that the addition of knowledge should not create unresolvable conflicts between new and old statements or rules. Second, it should allow the control structure to retrieve data efficiently. Using conceptually simple and uniform data structures, keeping the form of knowledge homogeneous, and writing special access functions for nonuniform structures can be used to achieve this goal. Finally, its knowledge base should be syntactically and semantically complete.

The choice of an appropriate representation scheme is particularly important because the power of an expert system resides in its knowledge base. The knowledge to be represented is typically heuristic knowledge and is, therefore, judgmental, experimental, and uncertain. The representation scheme should have some method to cope with this inexactness.

A very controversial and much debated issue is whether representations should be procedural or declarative. In procedural representations, the bulk of the knowledge is represented as procedures for using the knowledge. Procedural representations are advantageous when the knowledge consists mainly of how to do things. Also, procedural representations are usually utilized for probabilistic reasoning and for heuristic knowledge.

A declarative representation scheme is one in which most of the knowledge is represented as a static collection of facts accompanied by a small set of general procedures for manipulating them. The advantage of a declarative scheme is that each fact need be stored only once, regardless of the number of different ways in which it may be used. It is easier to add facts to the system since it need not be accompanied by a change to all the related procedures and subprocedures.

The question of which representation scheme, declarative or procedural, should be used continues to draw an unusual amount of disagreement, which is exemplified in a survey conducted by Brachman and Smith (1980). Each of the representation schemes has distinct advantages associated with it. Declarative representations are flexible, economic, complete, and modifiable. Procedural representations have a direct line of inference and are easy to code, and the reasoning is understandable. It seems unlikely that any one scheme may be used in complete isolation from the other, and the future emphasis will be toward schemes using both forms of knowledge.

An extensive and exhaustive review of knowledge representation techniques is available in Winograd (1975) and Bobrow (1975). The existing representation techniques may be classified into three main kinds: rule-based, logic-based, and structure-based representations. Each of these will be discussed in detail.

Rule-Based Representations: Production Systems

Expert systems whose knowledge bases use rules for the representation of knowledge are called production systems. The following is an overview of the components of a production system, the major areas to which production systems have been applied, the characteristics of production systems, and the advantages and limitations of production systems.

A production system consists of three parts: (a) a rule base composed of a set of production rules; (b) a special, bufferlike data structure called the "content," and (c) an interpreter that controls the system's activity. A production rule is a statement of the form "IF this CONDITION holds, THEN this ACTION is appropriate." To use a concrete example,

IF
      the temperature is below 30°F and you are going out of doors
THEN
      wear your winter jacket

The "IF" part of the rule is called the condition or premise and is a conjunction of one or more clauses. All clauses in this part have to be true before the action in the action part of the rule may be carried out.

The premise of each rule is a boolean combination of one or more clauses each of which is comprised of a predicate function with an associative triple (attribute, object, values) as its argument, with an associated degree of certainty (certainty factor), CF. For example,

PREMISE: ($ AND) (⟨clause 1⟩ · · · ⟨clause n⟩)
⟨clause⟩: (predicate function attribute object value)

Here the $ indicates that the premise is not a logical conjunction, but a plausible conjunction that requires the taking into account of CFs associated with each clause.

The action part of the rule indicates the conclusion made for action to be taken. It implies the addition of a new fact to the data base or the modification of an existing CF. Action is represented as:

ACTION: (CONCLUDE ⟨new fact⟩ ⟨CF⟩)

A sample production rule and its encoded LISP form is given later in Fig. 9.3.10.

Generally, one side of a rule is evaluated with reference to the data base, and if it is true, the action specified by the other side is performed. By evaluation matching, is referred to and by action is implied the addition or replacement of a fact in the data base. The manner in which the rules are organized and accessed has a direct impact on the efficiency of the system. Conflict resolution and other rule evaluation schemes are often employed to enhance the efficiency of the system.

A set of symbols constitute the data base of the simplest production systems. The organization of the data base and the interpretation of the symbols depends on the domain to which the system has been applied. For instance, the data base of systems designed to study how humans process symbols are interpreted as modeling the contents of some memory mechanism (typically the short-term memory) with each symbol representing some chunk of knowledge, with the chunks being constrained to the information processing limitations of the human short-term memory. As another example, the data base of DENDRAL contains complex graph structures that represent molecules and molecular fragments. The data base of a production system, irrespective of the domain to which it is applied, is the only storage mechanism for all state variables of the system. The data base is accessible to every rule in the system.

The primary task of the program is to select rules applicable to the current state of the system and execute them, thereby modifying the data base. This cycle of select–execute is repeated again and again until a decision or solution is arrived at.

The rules in a production system interact with each other only indirectly through the data base. Although this effects a system that is strongly modular, it makes the analysis of the behavior flow of a system very difficult. This occurs because, even for very trivial tasks, the overall behavior of a production system may not at all be evident from a simple review of its rules.

A production system is extremely modular, no two rules interact directly, and its rules may be changed (added, deleted, or replaced) with no unanticipated change to other functional units. Thus, changing a rule may cause the data base to evoke different rules, but the new rule will not directly affect another rule. If rule order is critical to performance of the expert system, then the task of rule insertion should be given due attention.

The visibility of behavior flow of a system indicates the ease with which the overall behavior of a system may be understood by observing it or by reviewing its rule base. Owing to the indirect interaction between the rules of a production system, the visibility of behavior flow for such a system is rather opaque even for relatively simple tasks. The reading of a production system requires scanning the contents of the data base as well as the entire rule set at every cycle.

The format of a production system is very constrained. The constrained format facilitates the dissection and understanding of productions by other parts of the program. A disadvantage of the constrained format is that encoding knowledge that is hard to express in the required form poses difficulties. However, the constrained format of a production rule enables it to not only examine its own rules but to also check for any inconsistencies among its rules.

The question as to whether it is difficult to program in a framework as that required by a production system is addressed by Moran (1973): "Any structure which is added to the system diminishes the explicitness of rule conditions. Thus rules acquire implicit conditions. This makes them (superficially) more concise, but at the price of clarity and precision. Another questionable device in most present production systems (including mine) is the use of tags, markers, and other cute conventions for communicating between rules. Again, this makes for conciseness, but it obscures the meaning of what is intended. The consequence of this in my program is that it is very delicate: one little slip with a tag and it goes off the track. Also, it is very difficult to alter the program; it takes a lot of time to readjust the signals."

The advantages of a production system are that it is modular and rules can be added, deleted, or modified without any danger of these changes having a direct bearing on any other rules. Since all information has to be encoded in the rigid format required by the production rules, a uniform structure is imposed on the knowledge making it easier for the user to understand the reasoning used. In addition, all knowledge can be represented with ease and in a natural manner as though an expert was himself or herself verbalizing.

The disadvantages of a production system are the lack of direct communication among rules, which leads to inefficient problem solving because every action is performed via the match-action cycle, since all information is conveyed by the context data structure. This results in a lack of responsiveness to predetermined sequences of situations which, if possible, would be more efficient. Although it is easy to comprehend which rule or rules lead to the final conclusion, it is hard to determine the flow of control in problem solving. This is because the rules alone are expressed naturally, the algorithmic knowledge is not. Davis and King (1976), in their discussion on production systems, specify domains in which the use of production systems is appropriate (Table 9.3.7). Currently, work is underway to make problem solving using production systems more efficient.

Logic-Based Representation: Predicate Logic

Logic-based representation schemes use predicate logic to encode domain-specific information. Section 9.2 has dealt extensively with knowledge representation using predicate logic and so it will not be repeated here. The chief advantages of using a logic-based representation scheme are that, as stated by McCarthy (1977) and Filman (1979), it seems a natural way to express notions because it is probably

Table 9.3.7 Knowledge Representation Schemes and Domains to which They Are Applicable

Rule Based	Logic Based	Structure Based
When the domain-knowledge is diffuse, consisting of many facts, for example, clinical medicine	When domain-specific knowledge can be represented as simple independent rules	Domain-specific information that is very complex and interrelated
When processes in the domain may be represented as a set of independent actions	When the problem solving technique used in the control structure needs to reformulate the problem	When knowledge can be separated and represented as discrete units
Domains where knowledge can be easily separated from the manner in which it used to be, for example, classificatory taxonomies in biology	Good for domains where special problem solving techniques have to be used, since it is amenable to a wide variety of these techniques	When time-varying states have to be represented
Domains where tasks may be represented as discrete states	Domains where it is difficult to separate facts and heuristics from procedures that manipulate them	When the domain is dynamic

compatible with our intuitive understanding of the domain. In addition, the rules of logic are so precise that there are standard methods for performing many of the functions of expert systems. In particular, determining the meaning of an expression in a formalism is possible, automated versions of theorem-proving techniques are available, and consistency is guaranteed because the set of inferences that may be drawn from a set of logic statements is completely specified by the rules of inference. Finally, logic is flexible. This permits the control structure to use a diverse range of problem solving techniques and allows the program to be easy to update.

The above use of mechanistic theorem provers works fine in relatively small data bases, but in large ones a combinatorial explosion is confronted. Also, pure logic has little provision for representing uncertainty or for representing spatial and temporal problems. Furthermore, pure logic does not allow for default reasoning.

Structural Representations

When domain-specific information is very complex and intertwined, then it is advantageous to represent related properties of objects as units and to then access these units. This forms the basis for structural representations. The three major methods that use structural representations are: (1) semantic networks, (2) frames and scripts, and (3) conceptual dependency theory. Before discussing these methods, we will briefly introduce certain concepts and considerations that are used in all of the three methods mentioned.

Concepts and Considerations in Knowledge Representation. The first concept to be considered is methods to represent relationships between objects. Two types of relationships, ISA and ISPART, are used. ISA relationships describe the relationship between objects in a hierarchical taxonomy. The following example shows how ISA relationships can be used to show hierarchical relationships for a metal:

GOLD	ISA	METAL
METAL	ISA	ELEMENT
ELEMENT	ISA	NONLIVING

ISPART relationships comprise relationships between objects that are made up of a set of components with each of these components in turn being made up of other components. The following example shows how ISPART can be used:

BRICK	ISPART	WALL
WALL	ISPART	HOUSE

Important properties of the above relationships are that they are lowerbound in that the branching factors increase as one descends the hierarchy. Also, hierarchies are transitive. Using the example in Fig. 9.3.3 we know that gold is an element since gold is a metal which is an element.

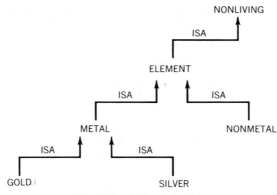

Fig. 9.3.3. ISA relationships.

Another concept which is important in expert systems is that of the level of representation. The level of representation determines the level at which domain-specific information should be represented. The lowest level at which an object may be represented is called a primitive. As an example, the process of hitting may merely be represented as hitting or, when broken down into primitives, hitting could be represented as a process where one individual propels his or her hand toward another individual and achieves physical contact. The major advantage of converting all statements in a representation in terms of primitives is that the rules that are used to draw inferences from that knowledge need only be written in terms of primitives rather than in terms of the many different ways in which the knowledge may originally have appeared. Programs by Schank and Abelson (1977) and Wilks (1972) make wide use of primitives. The main shortcomings in the use of primitives are that the conversion of each high level fact into primitive form requires much programming effort. In addition, the greater the number of objects stored as primitives the greater the memory required. As it turns out, a majority of these primitives are not used very often and, thus, a better idea may be to sort out objects and store only those objects that are used at the primitive level as primitives.

A representation scheme should also provide mechanisms that facilitate the location of relevant structures so that these structures can be accessed by the program. A mechanism incorporated in the representation scheme may allow this by indexing structures directly by the context words, by associating each verb or noun with a structure that describes its meaning. A problem arises when the same word may refer to more than one knowledge structure. The word "red," as an example, could imply color or anger. In this case, it would be necessary to have mechanisms that allow the indexing of the same word to two or more knowledge structures; one structure per context in which the word may appear.

Minsky (1975) outlines several issues which need to be resolved to ensure that the right structure is accessed:

1. How to perform an initial selection of the most appropriate structure.
2. How to fill in appropriate details from current situations.
3. How to find a better structure if the one chosen is inappropriate.
4. What to do if none of the available structures is appropriate.
5. When to create and remember a new structure.

Semantic Networks. Semantic networks use structures comprised of nodes that are connected by links or arcs. In semantic networks, nodes represent objects, concepts, or situations, and links or arcs represent relations between objects and facts. The nodes and arcs usually represent knowledge using the ISA or ISPART relationships discussed earlier. As an example which is used quite often, if the following is represented,

Clyde is a Robin.   Robins are birds.   Birds have wings.   Clyde owns a nest.

the structure in Figure 9.3.4 is used. In this figure, the nodes are represented by rectangles and the links by arrows. The relation between Clyde and Bird is evident by tracing up the hierarchy and using the transitivity property inherent in the structure. The use of this kind of reasoning to determine relations is called property inheritance and the ISA link is referred to as a property inheritance link.

Simmons and Slocum (1972) introduced the concept of case frames and case arcs, which allows nodes to represent changing situations and actions as well as objects. A widely used inference scheme

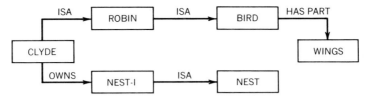

Fig. 9.3.4. Structural representation of knowledge [Source: Barr and Feigenbaum (1982)].

when knowledge is represented in semantic networks is that of network matching. This approach has been adopted in expert systems.

Frames and Scripts

Frames and scripts are methods for organizing the knowledge representation about objects and sequences of events that are so typical to certain situations that they are expected (based on prior experience) to take place. The term frames was coined by Minsky (1975) as a basis for understanding visual perception, natural-language dialogues, and other complex behaviors. Scripts are framelike structures designed by Schank and Abelson (1977) for the representation of a sequence of events. Frames and scripts are used on the premise that when individuals see a context or a broad framework that incompletely represents knowledge about the context, they draw on their past experiences to infer and fill in the incomplete parts. For example, if "table" represents a frame, an individual will draw upon his or her past knowledge and experience and infer that the table is made of wood, has two legs, and so forth. These specifics (table, number of legs) are called slots within the broad framework of "table," and the individual is expected to fill these slots from his or her past learning. A frame sometimes includes default values which are chosen from typical occurrences but which may be changed based on other information. Figure 9.3.5 illustrates some of these concepts for a CHAIR frame. A generic CHAIR frame has certain slots, in the left column, which may be filled in with information in the right column. JOHN'S CHAIR frame shows how the slots are filled in with specific information for that particular chair. Script-based processing has been developed largely by Schank and Abelson (1977) and is now being used to build systems that are capable of constrained natural language processing.

Conceptual Dependency Theory

The most distinctive feature of the conceptual dependency theory is that it endeavors to provide a representation of all actions using a small number of primitives. The theory emphasizes task independence and, when used as the basis of programs that can paraphrase and translate an input text to another language, the theory can draw inferences from its representation scheme. The theory seeks to capture the underlying conceptual structure of the represented knowledge. This implies that the input or representation must be unambiguous even though it has syntactic or semantic ambiguity. Another requirement is that the representation should be unique. The following example shows how distinct sentences with the same conceptual content have the same representation. The representations for the three sentences, when using context dependency theory, would all be the same:

I want a book.
I want to get a book.
I want to have a book.

CHAIR Frame

Specialization-of:	FURNITURE
Number-of-legs:	an integer (DEFAULT=4)
Style-of-back:	straight, cushioned, . . .
Number-of-arms:	0, 1, or 2

JOHN'S-CHAIR Frame

Specialization-of:	CHAIR
Number-of-legs:	4
Style-of-back:	cushioned
Number-of-arms:	0

Fig. 9.3.5. Knowledge representation using frames [Source: Barr and Feigenbaum (1982)].

To facilitate unique representations, Schank's system relies on a set of 11 primitive ACTS and categories (Schank and Abelson, 1977) (Figure 9.3.6).

The manner in which the elements of these categories may be combined into representations is prespecified by fixed rules. There are only two basic kinds of combinations or conceptualizations; one involves an actor (a picture producer—PP) doing a primitive ACT, the other involves an object (again a PP) and a description of its state (a picture aider—PA). Conceptualizations may be tied together by relations of instrumentality or causation.

The representations of text in conceptual dependency are said to be language-free because the primitive elements that occur in conceptualizations are not words but are the concepts which reflect the thought that underlies language. This makes the task of translating from one language to another very easy, consisting in the mere parsing of a language to its conceptual dependency and then generating text in the second language from the conceptual dependency representation.

Schank emphasizes the functional advantage of a language-free representation in terms of the computational ease with which they can be constructed. The parsing of language into its conceptual dependency requires that information implicit in the language be made explicit. Thus, each ACT entails its own set of inferences. The inference drawn from a language thus decides the choice of a primitive as well as defines what the primitive is.

The main criticism leveled against this theory is that it expresses the meaning of an action verb only in terms of its physical realization. Schank and Abelson have been adding several new devices to be able to represent a wider range of human actions such as plans, goals, and themes. Although extensive research work is underway, no scheme exists to date that is capable of an all-encompassing natural language representation.

Physical acts

PROPEL	apply a force to a physical object
MOVE	move a body part
INGEST	take something to the inside of an animate object
EXPEL	force something out from inside an animate object
GRASP	grasp an object physically

Acts characterized by resulting state changes

PTRANS	change the location of a physical object
ATRANS	change an abstract relationship, such as possession or ownership, with respect to an object

Acts used mainly as instruments for other acts

SPEAK	produce a sound
ATTEND	direct a sense organ toward a stimulus

Mental acts

MTRANS	transfer information
MBUILD	construct new information from old information

There are several other categories, or concept types, besides the primitive ACTs in the representational system. They are:

Picture Producers (PPs), which are physical objects. Some special cases among PPs are natural forces like wind and three postulated divisions of human memory: the Conceptual Process (where conscious thought takes place), the Intermediate Memory, and the Long Term Memory.

Picture Aiders (PAs), which are attributes of objects.

Times.

Locations.

Action Aiders (AAs), which are attributes of ACTs.

Fig. 9.3.6. Physical ACTS and concept types used in Schank's conceptual dependency theory [Source: Rich (1983)].

Table 9.3.7 provides a guideline regarding the suitability of a particular form of representation, given on the characteristics of the domain.

9.3.3.3 Reasoning with Uncertainty

Expert systems have been constructed in domains where the available information is either incomplete, uncertain, or both. To contend with this uncertainty, various methods have been developed. In the following discussion, methods to deal with uncertainty are discussed. These methods have varying degrees of success in contending with uncertainty and not all methods are guaranteed to be successful.

One method to contend with uncertainty is the use of monotonic reasoning. Systems that employ monotonic reasoning are those in which the addition of new rules or statements does not necessitate the reformulation or invalidation of an earlier statement. The main advantage of monotonic reasoning is that the addition of a new statement need not be accompanied by a check on old statements in order to find inconsistencies between the new statement and the old statements. Also, whenever a proof is made, it is not necessary to remember the antecedent statements because there is no possibility or danger of statements being changed or removed.

In many ways, the limitations of monotonic logic are the strongholds of nonmonotonic reasoning. Monotonic reasoning cannot be applied when information available is incomplete or when changing and interrelated situations have to be represented. Furthermore, it cannot be used when assumptions have to be generated in the course of complex problem solving.

Nonmonotonic reasoning is more suitable when the system is dealing with a domain that entails reasoning with sets of beliefs for which each belief is supported by some evidence and some personal motivation for maintaining it. The statement of a belief with some level of confidence or certainty as well as the representation of the levels of subjective factors and of heuristic information are also situations in which beliefs play an important role. Quine and Ullian (1978) give a very lucid presentation of the inconsistency that characterizes belief systems.

Default reasoning and circumscription are two classic examples of nonmonotonic reasoning. Default reasoning is illustrated in the following example. Individual A is on his way to visit his friend B, who is ill. A decides to take a bar of chocolate for B since he applies the general rule that most people like chocolates and assumes that B is no exception. This is an instance of default reasoning because the acceptance of one piece of information leads to the preclusion of another piece of information (i.e., that B may not like chocolate).

Circumscription (McCarthy, 1980) occurs when we assume that the only objects that can satisfy some property "P" are those that can be shown to satisfy it. For example, if we can show that A, B, and C satisfy some property P, we assume that only A, B, and C are true and others are false. This assumption that the "rest are not" is default reasoning, and any proof which shows that D also satisfies the property P will entail the revision of beliefs. Reiter (1978, 1980) provides a mathematical treatment of the subject along with a survey of many of the applications of AI where nonmonotonic reasoning comes into play. McDermott and Doyle (1980) also present a formal treatment of nonmonotonic logic.

Several methods can be used to contend with the uncertainty and inconsistency of nonmonotonic reasoning. In the following discussion, truth maintenance systems and probabilistic reasoning will be considered in some detail. In addition, other techniques will be considered.

Truth Maintenance System

The truth maintenance system (TMS) of Doyle (Doyle, 1979a, 1979b) is an implemented system that supports nonmonotonic reasoning. Its chief role is to maintain consistency among statements by accounting for the effects of newer statements and inferences made by reasoning programs. When an inconsistency is detected, TMS evokes its own reasoning mechanism—dependency directed backtracking—to resolve the inconsistency by altering a minimal set of beliefs.

In TMS, each statement or rule is called a node, which is, at any point in time, in one of two states: IN state, believed to be true, or OUT state, not believed to be true because none of the reasons is currently valid or because no reasons currently exist to believe it is true. The determination of the state of a node (IN or OUT) depends on the list of justifications attached to it. For a node to be true, or in the IN state, at least one of the justifications should be valid. Justifications represent how the validity (truth or IN status) of one node can depend on the validity of another node. Justifications are of two kinds:

SUPPORT LIST   (SL (in-nodes) (out-nodes))

CONDITIONAL PROOF (CP ⟨consequent⟩

                       (in-hypothesis) (out-hypothesis))

Support list (SL) justifications are the most common. They are valid if all of the nodes mentioned in the list of in-nodes are currently in and if all the nodes mentioned in the list of out-nodes are currently out. For example,

1. It is summer   (SL ( ) ( )).
2. It is cold   (SL (1) ( )).

The empty IN and OUT list in the SL justification of node 1 indicates that the node does not depend on the belief or lack of belief of any other node. Such nodes are referred to as premises. The IN list of the SL justification of node 2 contains the number 1. This implies that node 2 is true or in the IN state only when node 1 is in the IN state or is true. Nodes like node 2, which have nonempty in and out lists, are called assumptions. If node 1 fails to be in an IN state, node 2 will also become an OUT node.

Conditional proof (CD) justifications represent hypothetical arguments. These justifications are valid if the consequent node is in the IN state, whenever nodes in the IN hypothesis are IN and those in the OUT hypothesis are OUT. CP justifications involve a very detailed discussion and hence the reader is referred to Doyle (1979a, 1979b).

It should be brone in mind that the TMS does not decide the IN/OUT status of premises. The reasoning program which a TMS supports decides that. TMS merely maintains consistency between old and new statements and rules by incorporating a method that in any context determines the status (true/false) of a node. Even if a node is false in a context, it is stored under that status and not discarded since it may change its status to true in another context.

Probabilistic Reasoning

The TMS is applicable when the knowledge being used is such that it is either true or false. However, if knowledge is of a nature that it may be true to an extent or false to an extent, TMS fails. Probabilistic reasoning is a method that can be used to cope with this kind of uncertainty.

Either Bayesian probability theory or certainty theory methods have been used by knowledge-based systems to cope with uncertainty in a probabilistic sense. Bayesian probability theory has been incorporated into the structure of the mineral-locating expert system PROSPECTOR. This theory provides a way of computing the probability of an event based on some given set of observations. In PROSPECTOR, production rules have probability values assigned to them. The rules are then combined using straightforward Bayesian probability theory by computing, as an example, conditional probabilities, or joint probabilities.

In the context of expert systems, Bayesian probability theory has some drawbacks associated with it. In particular, collecting all the prior conditional and joint probabilities is a very time consuming task. If the control structure had to search for all prior probabilities before drawing a conclusion, the search process would be rendered very inefficient. Second, in Bayesian theory the sum of all possible outcomes should equal 1. This would necessitate the modification of existing estimates of rules and outcomes each time a rule was added. Third, the accuracy of the Bayesian formula depends on the availability of a complete set of hypotheses so that at least one of the known hypotheses should be true. In the kinds of problems that expert systems seek to solve, however, this may not be the case. For example, a patient may have a disease that no one has ever diagnosed before. Because of these drawbacks, many expert systems use methods other than Bayesian probability theory to contend with uncertainty.

To illustrate the use of certainty theory in probabilistic reasoning, one particular expert system, MYCIN, will be considered. Each production rule in MYCIN has associated with it a certainty factor (CF) whose value ranges on a scale from $+1$ to -1, where $+1$ indicates absolute certainty of the rule being true and -1 indicates total certainty in the rule being false. The value of the CF is computed from a measure of belief (MB) and a corresponding measure of disbelief (MD). The MB of a hypothesis h, given evidence e, is the proportionate decrease in disbelief in the hypothesis and can be thought of in terms of probabilities as,

$$MB[h, e] = \frac{\max[P(h/e), P(h)] - P(h)}{\max[1, 0] - P(h)}$$

Similarly,

$$MD[h, e] = \frac{\min[P(h/e), P(h)] - P(h)}{\min[1, 0] - P(h)}$$

A particular piece of evidence either increases the probability of h, in which case $MB(h, e) > 0$ and $MD(h, e) = 0$, or it decreases the probability of h, in which case $MD(h, e) > 0$ and $MB(h, e) = 0$. The CF is finally computed using the formula,

$$CF[h, e] = MB[h, e] - MD[h, e]$$

Shortliffe and Buchanan (1975) and Shortliffe (1976) provide a detailed discussion on the use of probabilistic reasoning in a variety of medical diagnosis problems. The use of probabilistic reasoning presents difficulties in the initial coding phase when design experts have to assign numbers to the phrases of human experts as "in most of the cases" or "few, if any." This brings in a slight amount of subjectivity, but, on the whole, the preceding method has been found to function satisfactorily in systems where it has been used.

Other Methods

Other methods have been used for dealing with uncertainty. These other methods will be mentioned, but will not be discussed in any detail. References will be provided for further study.

Fuzzy logic, based on the development of the theory of fuzzy sets, provides a way of representing the fuzzy or continuous properties of objects. Good references on fuzzy theory can be found in Zadeh (1965, 1975, 1978) and Kaufman (1975). Two new methods, the concept of belief spaces and the Dempster–Shafer theory of evidence, have recently been developed to help knowledge-base systems in coping with requirements for nonmonotonic reasoning. Strat (1984) has applied the Dempster–Shafer theories to evidential reasoning.

9.3.3.4   The Control Structure

The control structure, very simply stated, is a computer program that contains the procedures to make decisions about how to use the domain-specific knowledge stored in the knowledge base of the expert system. An expert system derives its power from the separation of domain-specific information from the procedures used to manipulate it. The main problem solving approaches used in the control structure or inference engine of an expert system are briefly mentioned below. Detailed discussions of these approaches are available in Hayes-Roth, Waterman, and Lenat (1983), Winston (1984), Rich (1983) and Nau (1983).

The state-space approach represents the problem in the form of a graph, tree, or array, and searches through this structure by employing rules and operators. Various descriptions of the problem are stored as nodes or states in the graph or tree, and one state is transformed into another by the application of operators. This approach has been discussed in detail in Section 9.2. Two main forms of state-space search exist: blind search and heuristic search. Blind search is predominantly an AI technique. Its utility lies in the fact that it provided the basis for the formulation of heuristic search. Heuristic search avoids combinatorial explosion and, hence, has often been employed in the control structure of expert systems. The A* and the AO* (Nilsson, 1971) algorithms guarantee finding the best solution while incurring a minimum cost in finding the best solution. We will not go into the details of the algorithm; it will, however, be mentioned that heuristic search is often employed in the control structure of an expert system in preference to other blind search methods.

Some of the other techniques that use heuristics to facilitate efficient problem solving are abstracting the solution space and generate-and-test. Abstracting the solution space reduces search time since, depending on the problem, search is carried out at varying levels of specificity. This avoids the need to search the entire problem space at every stage. The generate-and-test technique divides the search process into two parts: a generator of possible solutions and a tester which prunes solutions that fail to meet certain constraints. To increase the efficiency of the technique, heuristics are incorporated in the form of constraints. Thus, only those solutions that do not violate prespecified constraints are chosen. A generator is said to be complete and nonredundant if it generates all possible solutions that are in keeping with the constraints once and only once. The tester applies additional constraints and prunes the solutions further.

Some of the other techniques used in the control structure of an expert system are planning and dependency directed backtracking. These and other reasoning techniques are discussed very extensively in Winston (1984) and Rich (1983).

9.3.3.5   Interface

The interface of an expert system is that mechanism which permits communication between the user and the system. This is an important mechanism because it constitutes the basis for the interaction between the user and the system. The user inputs information pertaining to a domain-specific problem to the global data base of the system via the interface. The system asks for additional information as well as displays its solutions and conclusions via the interface. Also, exchanges of questions–answers and explanations take place via the interface. Many expert system designers feel that the efficiency of the interface can be enhanced by the use of natural language processing and spoken speech input. Each of these techniques will be considered in some detail.

Understanding Natural Language

People communicate with each other in natural languages like English or German. However, machines communicate with people in dialogue modes that are constrained in their syntax and semantics. Human users sometimes experience difficulty in communicating with the computer by using these dialogue modes. This gave rise to an attempt to build computer programs that understood natural language. Natural language processing has been successful when the domain has been limited.

The earliest natural language programs, like Green's BASEBALL, Lindsay's SADSAM, Bobrow's STUDENT, and Weizenbaum's ELIZA, used ad hoc data structures to store facts about a limited domain. The program searched for predeclared key words and patterns when restricted and simple declarative and interrogative forms of sentences were input. These early forms of natural language programs were so restricted that they were able to ignore many of the complexities involved in language.

Another early approach to natural language processing consisted in storing a representation of the text itself in the data base using a variety of clever indexing schemes to retrieve material containing specific words or phrases. PROTOSYNTHEX-I (Simmons, Burger, and Long, 1966) and Semantic Memory (Quillian, 1968) use this approach. This text-based approach restricts systems in that they could only respond with material that had been explicitly prestored. The limited-logic systems, like Raphael's (1968) SIR and Quillian's (1969) TLC, were developed with a view to characterize and use the meaning of sentences. In these systems, a formal notation is used to store information in the data base and mechanisms are provided for performing a semantic analysis by translating input sentences into the internal form. The overall goal of these systems was to make them capable of drawing inferences based on material in their data bases and thereby giving answers (which had not been stored explicitly in the data base) to related questions. The limitations of these systems was that their ability to make deductions was restricted to only that material which was directly or indirectly present in the data base.

Presently, the development of natural language programs is closely intertwined with the representation of knowledge in knowledge-based expert systems. These programs use the domain-specific information present in the knowledge base to understand sentences and to interact with the user in constrained natrual language. Tools like EMYCIN, TEIRESIAS, and KAS have a capability to carry on natural language dialogue with the user.

Understanding Spoken Language

Speech is the natural medium of communication between people. Thus, if the interface mechanism in an expert system has the capability for understanding spoken input, it may be more acceptable to them. Chapter 11.5 on speech recognition presents an extensive review of this topic and interested readers are referred to that chapter.

9.3.3.6 Learning Capabilities in Expert Systems

Learning, in a general sense, refers to the ability to increase one's knowledge and improve one's skills. In the context of expert systems, learning has been defined by Simon (1983) as "any process by which a system improves its performance." Thus, the system could improve its performance by acquiring new knowledge or by improving methods and procedures. Figure 9.3.7 illustrates a simple model of learning systems, developed by Cohen and Feigenbaum (1983), based on Simon's definition. In the figure, the circles denote facts and statements made by the expert and the two sets of rectangles represent procedures used by the system. In this model, information supplied to the learning element from the environment is used to make improvements in an explicit knowledge base, which in turn helps the performance element perform its tasks.

Learning in expert systems certainly does not approach the complexities seen in human learning. However, some low-level kinds of learning can take place. Some of the programs that can learn are considered in this section.

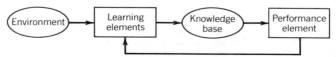

Fig. 9.3.7. A simple model of learning systems [Source: Cohen and Feigenbaum (1983)].

One form of learning is parameter (or coefficient) adjustment and has been used by Samuel (1963) in his checkers program. Checkers, like many other programs, relies on heuristics for efficient problem solving. In this case, the heuristic combines information from several sources (called factors) into a summary statistic. For the checkers game program, the factors used include distance to the goal state, cost, and differences in the attributes of the current state and the goal state. These factors are estimated and weights are assigned to each factor depending on its estimated importance in causing efficient search. Initial weights (of coefficients) are defined by the programmer. The learning capability of the program depends on the degree to which it is able to increase or decrease the weights assigned to each factor depending on the results achieved. For example, if the formula to arrive at the summary statistic is $C_1t_1 + C_2t_2 + \cdots + C_{16}t_{16}$, where t represents each of the 16 factors and C represents their respective weights, the idea is to change the C values (or coefficients). The changing of the values involves the consideration of both the magnitude and the direction of the change. Obviously, factors that accurately predict the final outcome should be increased in value and vice versa. A problem is encountered in the providing of feedback to the program at each step about the state resulting at every step, after every application of an operator with a given summary statistic value. Feedback is provided to a given program by applying a static evaluation function or matching each outcome and determining the benefits derived. The program then reinforces or alternates the influence of the factors.

Another form of learning could be called automated advice taking. In this form, the expert system interacts with a human expert to try to increase its knowledge and knowledge structure. This form of learning is encountered in TEIRESIAS (Davis, 1976) and consists in the system having to interpret and transform general purpose advice or knowledge into a form that can be used readily by the performance element of the system. This transformation is called operationalization and is an active process that may involve activities such as deducing the consequences of what has been told, making assumptions, filling in the details, and asking for more advice. Another system, ETS (Boose, 1984), is similar to TEIRESIAS in that it automatically interacts with a human expert to acquire knowledge and learn from the expert. Since both of these systems deal with knowledge acquisition from experts, they have been considered in more detail in an earlier section.

Other AI learning methods have been applied or have the potential to be applied to expert systems. One such method is concept learning. A concept, in AI, is said to be a predicate that partitions the instance space into positive and negative subsets. Cohen and Feigenbaum (1983) define the single concept learning as follows: Given (1) a representation language for concepts. This implicitly defines the rule space: the space of all concepts representable in the language. (2) A set of positive (and usually negative) training instances. Find: The unique concept in the rule space that best covers all of the positive and none of the negative instances. Most work to date assumes that if enough instances are presented, exactly one concept exists that is consistent with the training instances.

The concept, once learned, can be used for several kinds of performance tasks. The concept can be used in classification when the system is presented with new unknowns and is asked to classify them as positive or negative instances of a concept. It can also be used in prediction when the training instances are successive elements of a sequence. In this case, the system is asked to predict future elements in the sequence. Finally, the concept can be used in data compression when the system is given all possible instances and is asked to find a concept that describes them in a compact form. Programs and algorithms that address the single-concept learning problem are Mitchell's (1977, 1979) Version Space, Langley's (1980) BACON programs, and Dietterich and Michalski's (1981) INDUCE programs for structural learning.

Some expert systems have the capability to learn, or can generalize to multiple concepts. This approach has been used in the learning systems AQ11 (Michalski and Larson, 1978), AM (Lenat, 1977), and META-DENDRAL (Buchanan and Mitchell, 1977). META-DENDRAL is discussed in detail to give the reader an inkling of how this particular learning system functions.

META-DENDRAL was constructed by Buchanan and associates (Buchanan and Mitchell, 1977; Buchanan and Feigenbaum, 1978; Feigenbaum, Buchanan, and Lederberg, 1971; and Lindsay, Buchanan, Feigenbaum, and Lederberg, 1980) at Stanford to acquire and compile knowledge for the heuristic DENDRAL system. META-DENDRAL is used to formulate general rules, concepts, and patterns from examples. META-DENDRAL is provided with a set of descriptions and properties of compounds whose molecular structures have, in the past, been described by human experts. It is the task of the program to infer a small and fairly general set of rules to account for the given molecular structure. The META-DENDRAL program learns by generating production rules using the chemical structures that have been keyed in by the designer. This program is not only able to infer rules but also to add new rules.

A measure of the proficiency of META-DENDRAL is the ability of a DENDRAL program to predict a correct spectra of new rules based on the rules learned by the former. This is accomplished in the following manner. One of the DENDRAL performance programs ranks a list of plausible hypotheses (candidate molecules) according to the similarity of their predictions (predicted spectra) to the observed data. The ranks of the correct hypotheses (i.e., the molecules actually associated with the observed spectrum) provides a quantitative measure of the "discriminatory power" of the rule set. The META-DENDRAL program has not only successfully discovered known published rules of mass spectrometry for two classes of molecules, but has done so for three closely related families of molecules for which rules have not been reported previously.

Two other methods, learning by example and learning by analogy, have been used in AI programs and show some promise for applications to expert systems. In learning by example, the system is taught how to perform a task by presenting it with examples that illustrate how it should behave. The system is then required to make generalizations to higher-level rules that may be used to improve the performance of the system. Pattern-recognition systems employ this form of learning. Learning by analogy has not received much attention by researchers but shows potential for making systems learn. Its premise is that if a system has available to it a knowledge base for a related performance task, it may be able to improve its own performance by recognizing analogies and transferring the relevant knowledge to its knowledge base. Chapter 9.2 describes some of the advances in AI that have been made with this technique.

9.3.4   APPLICATIONS OF EXPERT SYSTEMS

A historical review of expert systems would be incomplete without tracing its origins to its parent field, AI. The methods of AI are an indispensable part of the process of building expert systems. AI itself had its beginnings in mathematical logic and computation (Figure 9.3.8). The logical systems of Frege, Whitehead, and Russell showed that some aspects of reasoning may be formalized in a relatively simply framework. Church and Turing were the first to point out that numbers were not just an integral part of computation but could be used as symbols to represent the internal states of a machine. Turing, often called the father of AI, was the one who first put forward the revolutionary idea that computational machines could behave in a manner perceived as intelligent.

Barr and Feigenbaum (1982, p. 3) define AI as "the part of Computer Science concerned with designing intelligent computer systems, i.e., systems that exhibit the characteristics we associate with intelligence in human behavior—understanding, language, learning, reasoning, ..." The different subfields of AI are depicted in Figure 9.3.8. The subfield of expertise more popularly known as "knowledge engineering" led to the evolution of expert systems. However, expert systems draw upon the other subfields of AI for their construction. As examples, problem solving techniques contribute to the control structure of an expert system and natural language processing can provide the interface in an expert system.

The main application areas for expert systems are indicated in Figure 9.3.8. This is by no means an exhaustive list and many other systems exist. The following sections provide examples of some of the expert systems that have been applied in a variety of areas. Tables 9.3.8 and 9.3.9 provide summaries of these systems along with the other important systems which are not discussed in as much detail in the following sections.

9.3.4.1   Geology Applications

A knowledge-based system by the name of PROSPECTOR was constructed by Duda, Gasching, and Hart, (1979) to assist in the probabilistic interpretation of soil and geological data to locate minerals like molybdenum, copper, and uranium. The global data base of PROSPECTOR is comprised of descriptions of the properties of rock, soil, and other relevant geological data that are input by the user.

The knowledge incorporated in PROSPECTOR was acquired from human experts in the field via design experts who later encoded the knowledge into the expert system. The geological knowledge is represented in its knowledge base using inference networks, which are also used to guide the reasoning of the system. These networks represent both judgmental knowledge (as rules) and static knowledge about objects in the domain. The nodes of the network are comprised of assertions made about known geological knowledge. Because geological data are often inexact, these assertions could be true or false or suspected to be true or false with a degree of probability. The arcs of the inference network define inference rules that specify how the probability of the truth of one assertion affects the probability of the truth of another assertion. To permit certain kinds of logical reasoning by the system, each assertion is represented as a partitioned space in a partitioned semantic network. A typical space asserts the hypothetical existence of physical entities having certain properties and participating in specific reactions. This large taxonomic network is also capable of portraying element subset relations in the present context of deposits.

The control structure uses network matching methods and the backward chaining approach to draw inferences about likely mineral deposits. The inexactness of geological data is handled by using Bayesian probability methods and computing conditional probabilities each time changes are made. The final conclusion has a certainty factor (CF) attached to it ranging on a scale from +5.0 (a situation holds with 100% certainty) to −5.0 (a situation does not hold with 100% certainty). The actual methods for the derivatin of the CF are explained in Duda et al. (1977).

In PROSPECTOR, each rule contains two confidence estimates. For example,

IF: magnetite or pyrite in disseminated form is present

THEN: (2, −4) there is favorable mineralization and texture for the propylitic stage

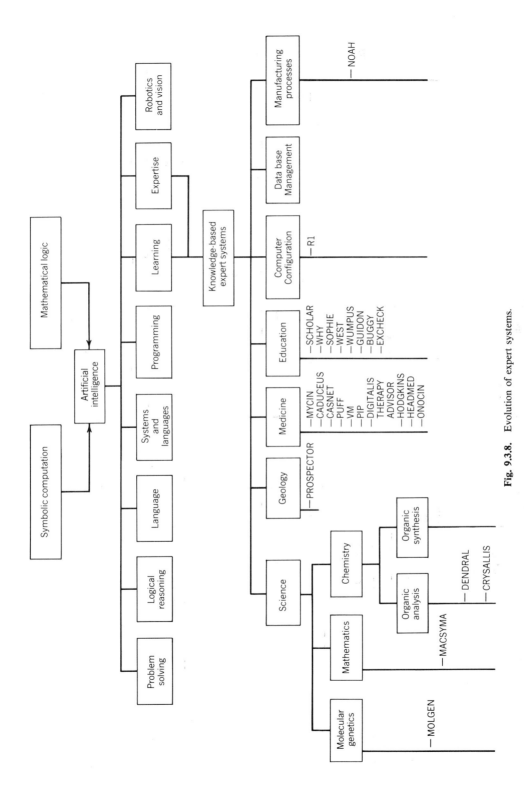

Fig. 9.3.8. Evolution of expert systems.

Table 9.3.8 Historical Overview of Expert Systems

Expert System	References	Purpose	Performance/Comments
MOLGEN	Stefik, Aikins, Blazer, Benoit, Birnbaum, Hayes-Roth, and Sacerdoti (1982)	Design of molecular genetics experiments	Used by researchers in designing experiments
MACSYMA	Martin and Fatemaro (1971); Mathlab group (1977); Moses (1975)	Symbolic problem solving in math	Used by researchers
DENDRAL	Buchanan and Feigenbaum (1978); Feigenbaum (1977); Lindsay, Buchanan, Feigenbaum, and Lederberg (1980)	Chemical structure elucidation using mass spectrometry	Can handle all problems in its domain of expertise
CRYSALIS	Englemore and Nil (1977); Englemore and Terry (1979)	Chemical structure elucidation using crystallography	Not very widely used; being tested by researchers
LHASA	Corey and Wipke (1969)	Organic synthesis	Replaced by SECS and SYNCHEM
SECS	Wipke, Brown, Smith, Choplin, and Sieber (1977)	Organic synthesis	Used widely for organic synthesis
SYNCHEM I, SYNCHEM II	Gelernter et al. (1977)	Organic synthesis	Used widely for organic synthesis
PROSPECTOR	Duda, et al. (1977, 1978); Duda, Gasching, and Hart (1979); Gasching (1980a, 1980b)	Interpretation of geological data for mineral exploration	Helped locate a molybdenum deposit worth $100 million
MYCIN	Shortliffe (1976)	Diagnosis and therapy of infectious blood diseases	Can handle all problems in its domain of expertise
CADUCEUS	Pople (1977); Pople, Myers, and Miller (1975); Pople (1981); Miller, Pople, and Myers (1982)	Diagnosis of internal diseases	Very large knowledge base
CASNET	Weiss, Kulikowski, and Safir (1977, 1978)	Diagnosis and treatment of glaucoma	Used by ophthalmologists in the United States and Japan
PUFF	Feigenbaum (1977); Freiherr (1980)	Pulmonary function tests	Used at a California medical center
CENTAUR	Aikins (1980)	Interpretation of preliminary function tests' results	Used in medical centers
WHEEZE	Smith and Clayton (1984)	Interpretation of preliminary function tests' results	Used in medical centers and clinics
VM	Fagan, Kunz, Feigenbaum, and Osburn (1979); Fagan (1978, 1980)	Interprets a patient's data from a physiological monitoring system	Provides an efficient control structure for coping with time-varying data
PIP	Pauker, Gorry, Kassirer, and Schwartz (1976); Szolovits and Pauker (1976, 1978)	Diagnosis of renal diseases	Undergoing field trials before it is used by doctors
DIGITALIS THERAPY ADVI-	Gory, Silverman, and Pauker (1978); Sil-	Regulation of the administration of the	Has been found to provide an appropri-

System	Purpose	References	Comments
SOR	drug digitalis	verman (1975); Swartout (1977)	ate treatment regimen for the drug
SCHOLAR	To tutor students in South American geography	Carbonell (1970a, 1970b); Collins, Warnock and Passaflume (1974); Collins (1976, 1978a, 1978b)	It is capable of diagnosing a students misunderstandings but continuity between questions is weak
WHY	To tutor students in a complex geophysical process, the causes of rainfall	Stevens and Collins (1977, 1978); Stevens, Collins, and Goldin (1978)	Lack of long-term goals
SOPHIE I, SOPHIE II	Helps the student in trouble-shooting electronic equipment	Brown, Rubenstein, and Burton (1976); Brown and Burton (1978a, 1978b); Brown, Burton, and Bell (1974); Brown, Burton, and Dekleer (1985)	Articulate expert, capability to tutor two or more students simultaneously
WEST	Tutors the student in mathematical expressions	Burton and Brown (1976, 1979a, 1979b); Goldstein (1977)	A satisfactory attempt toward the unobtrusive modeling of the student
WUMPUS	To develop a theory of coaching	Goldstein (1977)	It effected an improved understanding of the learning and teaching processes
GUIDON	To teach the techniques of diagnostic problem solving	Clancey (1979a, 1979b, 1979c)	The framework provided by GUIDON may be used by other related domains too
BUGGY	Helps students develop math skills	Brown and Burton (1978b)	Good at modifying the dialogue to suit a students level of learning
EXCHECK	Teaches a university level course in logic, set theory, and proof theory	Suppes (1981)	Good in carrying out a tutorial dialogue in this relatively difficult to teach domain
RI	Computer configurations for DEC VAX systems	McDermott and Doyle (1980)	Used extensively at DEC
NOAH	Robot planning	Sacerdoti (1977)	The problem solving technique adopted by NOAHCS useful when the domain has interacting subproblems

Table 9.3.9 Characteristics of Expert Systems

Expert System	Knowledge Base	Problem Solving	Interface	Learning Capabilities
MACSYMA	Semantic network	Reformulation, semantic pattern-matching, hill-climbing	Explanation facilities minimal but now ADVISOR acts as an automated consultant	Meagre, improving the system requires extensive reprogramming
DENDRAL	Rules for deriving constraints on molecular structure from experimental data. Procedure for generating candidate structures to satisfy constraints. Rules for predicting spectrographs from structures	Forward chaining, plan, generate, test	Interactive front-end, user may supply constraints, good editing facilities available	By itself not capable of learning but META-DENDRAL capable of inferring rules automatically
LHASA	Procedural representation of information	Backward chaining and problem reduction	Graphics	None
SECS	Production rules	Backward chaining and problem reduction	Graphics	New reactions may be added
SYNCHEM	Production rules	Backward chaining problem reduction and parallel processing of nodes		New reactions may be added in a language called ALCHEM
PROSPECTOR	Semantic network	Network matching and backward chaining probabilistic reasoning based on Bayesian theory	Convenient editing facilities provided by KAS; it is able to give explanations regarding rules used. It alerts the user if he or she has provided incomplete information	Owing to KAS, easy to add knowledge
MYCIN	Production rules	Backward chaining, exhaustive search use of certainty factors	Can give HOW and WHY explanations	TERESIAS provides for very efficient interactive transfer of expertise
CADUCEUS	Disease tree, a dynamic model used for representation	Combination of bottom up and top down approaches	Minimal explanation facilities	No capacity for learning extension of the knowledge base requires reprogramming
CASNET	Causal network, knowledge representation as a dynamic process	Hypothesis-driven data elucidation, question selection, bottom-up approach	Able to give explanations but not able to display the computation of status scores	Updating easier due to modularization of knowledge base; however, the learning capabilities are not as good as MYCINS

The number 2 indicates the extent to which the presence of the evidence described in the condition part of the rule suggests the validity of the rule's conclusion. The second number, −4, shows the extent to which the evidence present in the premise clause is necessary for the validity of the conclusion. This method is different from other expert systems, such as MYCIN, where the premise clauses assert the lack of evidence and the action clauses state the conclusion to be drawn.

The interface facilities of PROSPECTOR provide the user with a set of commands that may be used to request a summary or an explanation about the rules, but not the method, used to compute the conditional probabilities. It alerts the user to different possible interpretations of the data and identifies and asks for additional observations that would be valuable in reaching conclusions. LIFER, a natural language interface facility incorporated in PROSPECTOR, permits the user to input simple statements using constrained English.

The knowledge acquisition system (KAS) in PROSPECTOR was designed to improve its interface and knowledge acquisition capabilities. The primary task confronting a geologist who wants to prepare a new model for PROSPECTOR is the representation of his or her model as an inference network. KAS assists in this process by continually prompting the user until all missing parts of a new structure are filled in. This process is driven by an external grammar that can be changed without difficulty, making it easy to modify KAS as PROSPECTOR evolves. The core of KAS is a network editor that understands various mechanisms in PROSPECTOR and gives the user a limited ability to edit new knowledge in terms of content rather than form.

The developers of PROSPECTOR claim that it is capable of handling any problem from its domain of expertise (Gasching, 1980a, 1980b). Its prediction of a location for a molybdenum deposit was confirmed with a find worth $100 million.

9.3.4.2 Science–Mathematics Applications

A knowledge-based system called MACSYMA (Fig. 9.3.8) was constructed at MIT by Engleman, Martin, and Moses to assist mathematicians, scientists, and engineers in a wide range of algebraic manipulations that rely on symbolic inputs yielding symbolic results. Knowledge was acquired from domain experts and then formalized and encoded in the expert system. This expert system is described in detail in the references given in Table 9.3.8.

The knowledge base (Table 9.3.9) in MACSYMA is represented in the form of a semantic network. This network incorporates hundreds of rules used by math experts with each rule indicating a method to transform an expression into an equivalent simpler one. The solution of a problem requires finding a chain of rules that transforms the original expression into one that is suitably simplified.

Unlike PROSPECTOR, MACSYMA has no capability for reasoning with uncertainty and involves no search. Instead, it uses a reformulation approach to convert the problem into a form so that it is able to recognize the problem and establish an association with a specific solution method. This allows it to apply automatically predefined rules to simplify expressions. Rules to simplify the expressions are chosen by its control structure using either a semantic pattern matching method or a hill-climbing technique where each expression is heuristically broken down into simpler forms.

The interface facilities of MACSYMA were initially incapable of explaining the manner in which the system reached the final expression. Thus, in this kind of situation, users are often wary of employing results that they are unable to explain. Recently, features such as a frame-oriented interactive primer, a program for searching the reference manual, an information network, and an experimental version of an automated consultant for novice users (ADVISOR) have been incorporated in MACSYMA. ADVISOR accepts a description of the difficulty experienced by a user and attempts to reconstruct the user's plan for solving his or her problem. Based on this plan and its knowledge of MACSYMA, the ADVISOR then generates advice tailored to the user's specific need (Genesereth, 1978).

The learning capabilities of MACSYMA are meager and the system can be improved only by extensive reprogramming or by the addition of subroutines.

MACSYMA can perform 600 distinct mathematical operations including differentiation, integration, matrix operations, and vector algebra. It is used by hundreds of engineers and researchers (scientists, plasma physicists) in universities throughout the United States.

9.3.4.3 Education Applications

Knowledge-based systems have made a significant impact in the field of education particularly in the area of computer-assisted instruction (CAI); CAI programs that incorporate AI and expert system techniques are referred to as "intelligent computer-assisted instruction" (ICAI). The focus is on the design of systems that can offer instruction in a manner that is sensitive to each student's unique strengths, weaknesses, and preferred style of learning.

For CAI programs, instructional text was presented (sometimes online, sometimes not) and students were asked questions that required a brief answer. The student would answer the question, and the system would indicate if the answer was right or wrong and possibly branch to a different part of the program based on the correctness of the answer. Although some individualization for a student can occur, most of the interactions with the student are predetermined in a CAI system.

As Carbonell (1970b) states, an ICAI system is one which is not only knowledgeable of the subject matter to be presented but which can also carry on a dialogue with the student and use the mistakes made by the student to diagnose the student's level of understanding. The first step toward the building of intelligent CAI systems resulted in the system's ability to generate problems from a large data base representing the subject taught. Presently, ICAI systems are distinguished by their ability to interact in a truly reactive learning environment (Brown, 1977) in which the student is actively engaged with the instructional system and the student's interests and misunderstandings guide the tutorial dialogue.

An ICAI system has three main components or modules: the problem solving expertise incorporated into the expertise module consisting of the knowledge that the system endeavors to impart to the student, the student-model module that determines what a student does or does not know, and the tutoring module consisting of tutoring strategies that the system uses to present material to the student. The expertise module is comprised of the knowledge specific to the domain in which instruction is given. Knowledge may be represented using any of the standard techniques of representation such as production rules, semantic nets, or procedural representations. This component is called an "articulate expert" if it is able to explain the problem solving decision in terms that correspond to those of a human problem solver. It is called an "opaque expert" in the absence of this explanatory capacity. This module generates questions and evaluates answers given by the student.

The student-model module represents the student's understanding of the material to be taught. Much of the recent work in ICAI has focused on this component. By modeling the student's understanding, the system formulates hypotheses about the students misconceptions and suboptimal performance strategies so that the tutoring module can indicate why they are wrong and suggest corrections. The student model may be constructed by applying simple pattern recognition to the student's history or by using a semantic net to keep track of which nodes (objects and concepts) the student does or does not know. A third method, called the overlay method (Carr and Goldstein, 1977), compares the student's behavior to an expert's and represents the student's knowledge as a subset of the expert's knowledge. Finally, the bug method is another method to determine the deviation of the student's knowledge from that of an expert. This method is different from the overlay method in that it compares and examines reasoning techniques used.

The tutoring module communicates with the student, selects problems for him or her to solve, monitors and criticizes his or her performance, provides assistance upon request, and selects remedial material. The design of this module involves issues such as when should the system give feedback and what should the content of the feedback be.

The ICAI systems SCHOLAR, WHY, SOPHIE, and WEST are discussed in detail in this section. Other ICAI systems, such as WUMPUS, GUIDON, BUGGY, and EXCHECK, are introduced very briefly in Table 9.3.8 where the interested reader is also provided with references for each. Also, a detailed discussion on ICAI systems is included in Chapter 8.4.

SCHOLAR (Table 9.3.10) uses a Socratic style of tutoring in which the student's misconceptions precede and direct the presentation of material in a manner that will force the student to realize and correct mistakes (Collins, Warnock, and Passaflume, 1974). This tutorial was designed by applying human tutorial protocols (Collins, 1976) to determine the strategies adopted by human tutors.

Natural language processing, based on a case grammar system, is used for the interface. The semantic interpretation of a student's response is guided by the knowledge of geography contained in the semantic nets. The capability for natural language processing is important since it allows the student to take the initiative and ask questions that may not have been anticipated in the test. The system's built-in taxonomy of expected question types helps it comprehend questions or words input in natural language. The evaluation of the correctness of a student's response is carried out using plausible (Collins, 1978b) as well as default (Reiter, 1978) reasoning.

WHY is another ICAI system which, in this case, teaches students about the concepts of rainfall. Rainfall is a subject matter that is not factual but where the causal and temporal relations between the concepts in the domain are of greater interest. For example, no single factor can be isolated that is both necessary and sufficient to account for rainfall. WHY was constructed by Stevens and Collins (1977) to incorporate elements of a theory of tutoring into a program, identify the weak elements in the theory, and investigate them further. Collins (1976) argues that the learning of complex processes with interrelated factors, as occurs in rainfall, is best accomplished by dealing with specific cases and trying to generalize from them.

The manner in which the system operates in terms of its components is given in Table 9.3.10. The operation of the modules is directed by 24 heuristics that have been incorporated into the system to control the student–system interaction. For example, one of the heuristic rules is:

IF: The student gives an example of causal dependence one or more factors that are not necessary,

THEN: Select a counter example with the wrong value of the factor and ask the student why his or her causal dependence does not hold in that case.

Table 9.3.10   Characteristics of Intelligent Computer-Assisted Instruction Systems

Name of the System	Main Questions Explored	Expertise Module	Student-Model Module	Tutoring Module	Comments
SCHOLAR	(1) Effective storage of information so that it was quickly and easily retrievable. (2) The general reasoning strategies needed to make appropriate plausible inferences from incomplete data base of the system; this is because the student may ask questions beyond that incorporated in the systems knowledge base. (3) To what extent can the preceding strategies be made independent of the content of the domain (i.e., dependent only on the form of representation).	Knowledge stored as semantic nets, generates questions keeping in mind timing considerations and generates those questions in the diagnosed area that have been ranked as more important. Emphasis on covering a predetermined lesson (chosen by student model) in a fixed period of time. Evaluation of answers to questions made by using default and plausible reasoning.	Makes a model of the student's behavior to that of the computer-based "expert" in the same environment. The modeling component ranks each skill according to whether evidence indicates that the student knows the material or not. It places pointers equivalent to flags in the subject matter semantic net or in the rule base representing areas that the subject has mastered.	Interacts with students in English. Both student and system can take the initiative to ask questions.	Continuity between questions is weak, since it does not plan a series of questions to make a point. It is capable of diagnosing a student's misunderstanding only by following up one question with a related question.
WHY	(1) The goal structure of a Socratic tutor and his or her use of questions, statements, and examples. (2) Diagnosis and classification of student's misconceptions. (3) Abstractions and viewpoints used by tutors to explain physical processes.	Knowledge base represented in scriptlike data structures that encode temporal relations in the rainfall process.	Based on 24 heuristics, it models the students understanding.	The system prompts the student to suggest causes of rainfall, to look for prior or intermediate causes, and finally to suggest a general rule. When a generalization is made, the system gives a counterexample and forces revision of the rule made. This procedure is repeated until a rule of sufficient generality is made and exception to the rule are noted by the student.	It does not have long-term goals for the tutorial dialogue; major emphasis on the form which the student–system interactions should take, not too much emphasis on method of storing data or generating and evaluating questions, emphasis on what and not on how.

Table 9.3.10 (Continued)

Name of the System	Main Questions Explored	Expertise Module	Student-Model Module	Tutoring Module	Comments
SOPHIE I, SOPHIE II	(1) Effective storage and retrieval of information. (2) Effective evaluation of responses given by student. (3) Effective capacity for natural language processing.	Procedural representation of the domain knowledge, that is, procedures simulating the behavior of a circuit. Procedures are represented based on the concept of a performance or semantic grammar (to facilitate, natural language processing). Inference mechanism used is generate and test, that is, hypothesis generation and evaluation. SOPHIE I is an opaque expert. SOPHIE II an articulate expert.	The bug approach used to make a model of the student.	Student has a one-to-one relationship with the system who helps him come up with his or her own ideas, experiment with these ideas, and, when necessary, debug them. Gives the student a convenient way to express his or her ideas in natural language.	Even SOPHIE II, an articulate expert, is unable to provide the student with suggestions or direction which if pursued may lead to success. However, SOPHIE II, in which a competitive environment is provided, with partners sharing and trying to debug, proved more effective and more openings were suggested.
WEST	(1) Diagnostic strategies required to infer a student's misunderstandings from his or her observed behavior. (2) Various explicit tutoring strategies for directing the tutor to say the right thing at the right time.	This consists of knowledge in the form of issues and examples. Issues are concepts that an expert would use at a particular time. Examples show instances of use of the issues. Any move made by the student is evaluated by this model based on the model of issues used by an expert.	Models found by differential modeling. (1) Evaluate the students current move (issue) with respect to the alternate set of moves theorem, which the expert may have used in the same situation. (2) Determine what underlying skills were used to select and compose the move; try to better the student's moves. Issues evaluated by the expertise model indicate a students strengths and weaknesses.	Based on tutoring principles and the students weaknesses the right issues are invoked. Explanations are given using a SPEAKER in the knowledge base. SPEAKER's are procedures responsible for presenting a few lines of the text explaining its issue.	It is often difficult to find or point out the lack of which issue (concept) in the subject prevented him or her from making a noise. Also, what motivated him or her to make the move since the student's mind itself cannot be analyzed. The instructor may make the wrong model and think the subject has weaknesses that he or she does not have.

SOPHIE, another ICAI system, was created to provide a learning environment in which students would be challenged to explore ideas on their own and to come up with conjectures and hypotheses about a problem solving situation. The students receive detailed feedback about the logical validity of their proposed solutions. In cases where the students's ideas have logical flaws, SOPHIE can generate relevant counterexamples. This system combines domain-specific knowledge and powerful domain-independent inferring mechanisms to answer questions that might stymie even human tutors. Figure 9.3.9 represents the structure of SOPHIE.

The student generates hypotheses in natural language and SOPHIE receives these hypotheses in its tutoring module (Table 9.3.10). The expertise module evaluates it using the generate-and-test method of inference. Based on the results of evaluation, it returns an answer in natural language via the tutoring module to the student. SOPHIE I was opaque; however, SOPHIE II is an articulate expert debugger and also contains a trouble shooting game with two teams of students.

WEST, an ICAI system derived from a board game "How the West was Won," teaches students about mathematical expressions. This system endeavors to determine the coaching strategy that an ICAI system should ideally follow in terms of how much advice should be given and how much and how often advice should be given. In this system, the student–model module (Table 9.3.10) makes its diagnoses unobtrusively without asking the student questions. The student model is inferred from a student's behavior by comparing it to the behavior of an expert in a similar situation.

The coaching system in WEST (Table 9.3.10) advocates the use of the "issues and examples" technique to provide instruction. The tutoring principles that determine the timing for the giving of advice are based on the following principles:

1. Coaching philosophy—make a point only when its adoption will increment the student's current level of performance substantially.

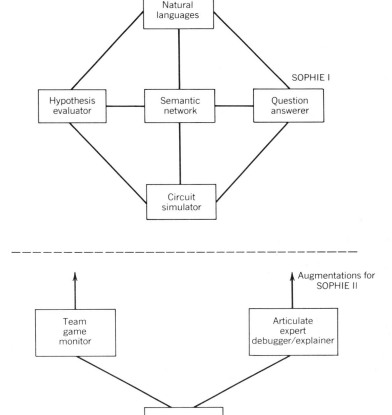

Fig. 9.3.9. SOPHIE I and SOPHIE II [Source: Barr and Feigenbaum (1982)].

2. Maintain the students interest in the session by not interrupting him or her often.

3. Only major weaknesses should be corrected immediately, the optimal method of performance should be demonstrated and an explanation should be given stating why it is optimal.

4. The coaching module should be forgiving.

This system primarily addresses issues related to feedback and to the unobtrusive inference of the student model.

The ICAI systems that have been discussed do not each address all the ICAI issues enumerated; each addresses one or two issues. Nevertheless, these systems have demonstrated potential tutorial skill and have at times shown insights into students's misconceptions. To enable the use of these systems on a large scale, more work needs to be done in this area and systems that address not one but all of the issues pertaining to the representation of expertise, student modeling, and capabilities for carrying on a tutorial dialogue need to be constructed.

9.3.4.4 Science–Chemistry Applications

The application of knowledge-based systems in the two broad areas of organic analysis and organic synthesis. Organic analysis, also called chemical structure elucidation, consists in the determination of the atoms in the molecules and the generation of all the possible ways in which the atoms can be arranged in a three-dimensional structure. The generated arrangements are then examined and discarded or selected depending on whether they meet all the constraints specified. (The constraints are computed based on the component atoms and their properties.) The task of enumerating the component atom is not very difficult for an individual, but the task of generating all possible arrangements of atoms by hand becomes very tedious and error prone owing to the large number of arrangements of atoms that can typically be generated.

The construction of the expert system DENDRAL alleviated greatly the problems entailed in the elucidation of chemical structures using mass spectrometry. This expert system will be discussed in more detail in this section. Another expert system, CRYSALIS, has been constructed that elucidates chemical structures using crystallography. This expert system is summarized in Table 9.3.8 and will not be discussed in detail.

The function of organic synthesis in chemistry consists in the creation of chemical products or in the development of more efficient processes for the manufacturing of old products. In the synthesis of a new product, the desired chemical structure of the new product is specified and then work is done on this required chemical structure in a backward direction by breaking the new product into increasingly simpler chemical structures until chemicals or materials available commercially are arrived at. This process, when carried out manually, is very time-consuming and complex, since a plan for synthesizing a compound, called a synthetic route, may involve dozens of separate reactions. For example, a simple steroid composed of about 20 atoms has over 10^{18} possible direct routes. Expert systems LHASA, SECS, and SYNCHEM have been constructed to perform the function of organic synthesis and will be discussed later in this section.

DENDRAL analyses mass spectrographic, nuclear magnetic resonance, and other data to infer the plausible structures of an unknown compound. Its construction was begun in 1965 at Stanford University. A later version of the system, META-DENDRAL (1970), was developed to automatically infer rules from instances of molecular structures that have been analyzed by human experts in the past so that this data base could be added to that knowledge acquired from experts for DENDRAL.

The knowledge base of DENDRAL consists of rules for deriving constraints on molecular structures from experimental data, procedures for generating structures, and rules for generating spectrographs. The global data base is comprised of mass spectrogram data, constraints, and candidate structures.

DENDRAL uses an efficient variant of the generate-and-test technique in its problem solving strategy to access the data in its knowledge base, which occupies a large but factorable solution space. The control structure of DENDRAL has three parts: plan, generate, and test. In the plan stage, the problem of finding all possible arrangements of atoms in a three-dimensional structure is reformulated into a problem of finding all such combinations that are consistent with the constraints specified. The constraints are stored in the knowledge base of the expert system in the form of IF–THEN rules. In the generate stage, all possible structures, in which constraints are not violated, are generated. In this stage, CONGEN (1976) is a program that is used to satisfy the requirements that all possible structures satisfying the constraints should be generated, no structure satisfying the constraints should be left out, and no structure should be generated more than once. This stage can be truly interactive in that it allows the user to specify constraints. Finally, in the test stage, the generated structures are ranked by simulating their behavior in a mass spectrometer. The structures that correspond most closely to the empirical ones are ranked as highest.

The interface facilities offered by DENDRAL are equipped with an interactive structure editor, EDITSTRUC, a teletype-oriented structure display program, DRAW, and an executive program that ties together the individual subprograms and assists the user in creating, editing, saving, and restoring constraints, substructures, and superatoms (groups of connected atoms). Also, examples of the structure can be produced from the beginning so that the chemist can stop the computation early if the problem has been poorly or incorrectly constrained.

DENDRAL, through the development of META-DENDRAL, has some learning capability. META-DENDRAL is capable of inferring rules from existing molecular structures and, thus, adding more knowledge to DENDRAL's base becomes easier. The molecular structures elucidated by DENDRAL may be used by META-DENDRAL for inferring new rules.

DENDRAL has been found to perform accurately on the structure elucidation task owing to its method of systematic search through the space of all possible molecular structures and its systematic use of what it does know to constrain the list of possibilities. Presently, DENDRAL is used to determine the structures of the following kinds of molecules (Barr and Feigenbaum, 1982): terpenoid natural products from plant and marine animal sources, marine sterols, organic acids in human urine and other body fluids, photochemical rearrangement products, antibiotics, metabolite of microorganisms, and insect hormones and pheromones. It has been applied to published structure elucidation problems in organic chemistry to check the accuracy and completeness of published solutions. In several cases, the program found structures that were plausible alternatives to the published structures (based on problem constraints that appeared in the article). This kind of information served as a valuable check on conclusions drawn from experimental data.

The application of knowledge-based systems in the area of organic synthesis takes the form of the following three expert systems: LHASA (logic and heuristic applied to synthetic analysis), SECS (simulation and evaluation of chemical synthesis) which grew out of LHASA, and SYNCHEM I and II (synthetic chemistry). The knowledge acquisition for all the three systems was carried out by a design expert conversing with and interrogating a domain expert.

In LHASA, knowledge representation takes a procedural format where actual methods or procedures are represented in the knowledge base. In the case of SECS, reverse transforms or reactions (a reverse transform is a reaction written in reverse, e.g., $C + O_2 \rightarrow CO_2$ in reverse is written as $CO_2 \leftarrow C + O_2$) are represented as production rules. In SYNCHEM II, the knowledge-base corresponds to a library of reactions or transforms that can be updated by chemists without reprogramming. Each reaction is compiled automatically into a reverse reaction. The global data base for all the three systems consist of the chemical compound and the desired chemical structure components.

In organic synthesis tasks, search is always carried out in the backward direction, where the path is traced from a given desired compound to its basic chemicals. This results in an AND/OR graph where the tree descends from the root node representing the target module to the terminal nodes at the bottom of the tree representing the chemicals. Branches connecting the nodes represent chemical reactions. The AND links represent methods to combine compounds in the reactions and the OR links represent alternative ways of synthesizing a compound. LHASA and SECS are interactive in the sense that the heuristic used to direct search is supplied by the chemist, whereas in SYNCHEM, it is derived from the program. SYNCHEM's search is particularly efficient because it calculates the cost of reaching a target molecule by taking into account factors such as the efficiency of reaction, monetary cost of materials, and difficulty of synthesizing the precursor nodes from available starting materials.

In terms of the human–computer interface, much human engineering has been involved in all three systems. In particular, graphics have been designed to correspond to the chemist's model of thinking.

None of the systems is capable of automatically learning and reinforcing its knowledge base by inferring rules. In LHASA, the additions of rules by design and domain experts is possible only through very extensive reprogramming due to the procedural presentation of knowledge. In SECS and SYN-CHEM, new transforms or reactions can be added without much effort since the reactions are stored separate from the procedures that manipulate them. The number and complexity of transforms is thus not limited by the size of the core memory. A language called ALCHEM is provided that chemists can use to enter new transforms.

SECS is used to perform organic synthesis by chemists in Europe and North America. SYNCHEM has been used by chemists as well as by AI researchers to explore the AI issues involved in such a search space.

9.3.4.5 Medicine Applications

Expert systems in the field of medicine have been used mainly for the purpose of diagnosis and treatment. These systems must have a general knowledge of diseases including manifestations, causal mechanisms, and diagnostic procedures as well as specific knowledge about the patient including the current medical history and therapies. These expert systems also incorporate methods to contend with the inexact nature of medical knowledge. To carry this out, factors, such as the "strength of association or belief" (CASNET, CADUCEUS) or "certainty factor" (MYCIN), are computed. This also allows the system to weigh different pieces of evidence in order to accept or reject a particular hypothesis. Although differing in specifics, the systems usually compare a numerical value of a hypothesis to a certain preset threshold that is defined by the domain expert. The hypothesis is accepted if the score exceeds the threshold. Another important aspect of these systems is the ability to explain and justify their line of reasoning in order to be accepted as consultation systems by physicians.

Three expert systems in this area—MYCIN, CASNET, and CADUCEUS—are described in more detail.

MYCIN was constructed at Stanford University (Shortliffe, 1976) to provide consultative advice on the diagnosis and therapy for infectious blood diseases and for meningitis infections. The current knowledge base of MYCIN consists of 450 rules that allow it to diagnose and prescribe therapy for diseases caused by blood infections. The knowledge encoded in the rules was acquired from domain experts by design experts and then coded into rules.

The knowledge base of MYCIN consists of production rules. Each rule represents a chunk of domain-specific information indicating an action if a premise holds true. The rules are stored internally in LISP code and from this the English version is generated as shown in Figure 9.3.10. These rules can be displayed or executed when explanations have to be given by the system. The premise of each rule in a Boolean combination of one or more clauses each of which is constructed from a predicate function with an associative triple (attribute, object, and value) as its argument. Thus, each premise clause typically has the following four components:

⟨predicate function⟩ ⟨object⟩ ⟨attribute⟩ ⟨value⟩

Using the example from the figure, the third clause is written in LISP as

(SAME CONTXT PORTAL GI)

In this case, SAME → predicate function, CONTXT → object, PORTAL → attribute, GI → value. A standardized set of 24 domain-independent predicate functions (e.g., SAME, KNOWN, DEFINITE); a range of domain-specific attributes (e.g., IDENTITY, SITE); objects (e.g., ORGANISM, CULTURE), and associated values (e.g., E.COLI, BLOOD) form the vocabulary of conceptual primitives used for constructing rules. A rule premise is always a conjunction of clauses, but it may contain arbitrarily complex conjunctions or disjunctions nested within each clause. The action part of the production rule indicates one or more conclusions that may be drawn if the premises are satisfied. This part also consists of a predicate function and an associative triple.

Since medical diagnosis is judgmental, the MYCIN inference engine combines various computing hypotheses and actions to arrive at a diagnosis. The action has a number assigned to it on a scale ranging from -1.0 to $+1.0$, where -1.0 indicates complete confidence that a proposition is false and $+1.0$ indicates with complete confidence that a proposition is true. These numbers, called certainty factors (CFs), indicate the strength of a rule. Using the example from Figure 9.3.10, the assertion is made with more than mild confidence (0.7).

The global data base in MYCIN is comprised of the symptoms and other routine data of the patient that is keyed in by the user. Routine data consist of information on items such as age, sex, and identification code. Symptoms will include observable symptoms such as heart rate and blood pressure as well as results of laboratory tests like blood and urine analysis reports. From time to time, the system may request that the user supply data about the patient.

The problem solving technique used by the inference engine of MYCIN invokes rules in a simple backward chaining fashion that results in an exhaustive depth first search of an AND/OR goal tree. This backward chaining approach considers all antecedents for all conclusions ignoring only those antecedents where previous data obviate the testing of certain conditions. This backward chaining

RULE 050

PREMISE: (AND (SAME CNTXT INFECT PRIMARY-BACTEREMIA)
             (MEMBF CNTXT SITE STERILESITES)
             (SAME CNTXT PORTAL GI))

ACTION: (CONCLUDE CNTXT IDENT BACTEROIDES TALLY .7)

MYCIN's English translation:

IF   1) the infection is primary-bacteremia, and
       2) the site of the culture is one of the sterile sites, and
       3) the suspected portal of entry of the organism is the gastrointestinal tract,

THEN   there is suggestive evidence (.7) that the identity of the organism is bacteroides.

Fig. 9.3.10.   A MYCIN production rule [Source: Barr and Feigenbaum (1982)].

takes place by regressing from possible conclusions to related antecedent conditions and from the conditions to their required data (recursively, if necessary). The exhaustive search is essential because MYCIN, on account of its domain, must find all possible diseases that might account for a patient's symptoms.

The backward chaining search has been augmented by two other heuristics to make the MYCIN inference engine more efficient. The program, before attempting to retrieve the entire list of rules for a subgoal, tries to first find a sequence of rules that would establish the goal with certainty based on what is known currently. This, in short, is a search for a sequence of rules with CF = +1.0. In addition to efficiency, this process offers the advantage of allowing the program to make commonsense deductions with a minimum of effort. (Rules with CF = +1.0 are largely definitional and typically very few in number; therefore, the search is very brief.) The other heuristic used is that the inference engine looks to see if any one premise in a production rule is bound to be false.

The MYCIN interface accepts inputs from the user in restricted natural language. It is also able to provide explanations to "HOW" and "WHY" questions posed by the user. This is done by tracing the system's inferential process and displaying the English version of the rules used. It also provides literature sources relevant to the rules used.

The learning capabilities of MYCIN have been enhanced with the construction of TEIRESIAS (Davis, 1976). It is now possible for domain experts to examine the line of reasoning used by the MYCIN inference engine and then to add, delete, or modify rules. Details are given in an earlier section.

MYCIN has been found to be very accurate and effective for solving almost any problem in its domain. MYCIN is also used for the purpose of teaching in a medical setting and has contributed to AI methodology. It has led to the development of expert systems with an architecture very similar to its own (PUFF, 1978 and PROSPECTOR, 1980). It has also led to the development of the tools EMYCIN and KS300, which help in the construction of newer expert systems.

The causal association network (CASNET) program was developed by Weiss, Kulikowski, and Safir (1977) at Rutgers University for medical diagnosis and treatment of Glaucoma.

CASNET, unlike MYCIN and DENDRAL which use static rules for the representation of knowledge, models diseases by a causal network. Within this network, the program reasons to determine the effects of a therapy or disease. Knowledge for the system was acquired from domain experts and encoded by design experts.

Knowledge is represented as a network of causally linked pathophysiological states in three planes (Figure 9.3.11). The middle plane is called the plane of pathophysiological states, and the nodes in the plane represent elementary hypotheses about the disease process. Each causal link has a confidence ranging from 1 to 5 (1 corresponds to rarely causes, 5 corresponds to always causes) assigned to it. The bottom plane is the plane of observation and contains information pertinent to the current patient being diagnosed. It has nodes corresponding to the patient's symptoms, signs, and test results. Arcs called associational links connect nodes on this plane and on the middle plane. Each link has a predetermined (defined by the designers of CASNET in consultation with domain experts) confidence attached to it ranging from 1 to 5. Also, one test or symptom may lead to more than one node in the plane of pathophysiological states or two or more nodes may relate to the same symptom sign or test node. Each test has associated with it a cost that reflects both monetary cost and danger to the patient. The third plane contains the classification tables (Table 9.3.11) that contain treatments and define the disease as a set of pathophysiological states. A disease is represented as progressing with time, and states further down a pathway indicate worsening states of the disease. The bottom plane also contains the information for the global data base, which contains the patient's symptoms, signs, and test results are input by the user.

The inference engine or control structure of CASNET performs its tasks by hypothesis-driven data elicitation and question selection. It adopts a bottom-up approach working from observations, symptoms, and test results to hypothesized disease states and finally to disease categories. The values of test nodes and weights associated with causal arcs are used to compute a status or confidence factor for each node in the causal net. The status of a state is affected both by the results of its associated nodes and by the status values of the states around it. An algorithm built into the system computes weights for the states. A state is marked confirmed if its status is greater than a preset threshold and denied if it is lesser than a second threshold; otherwise it is ranked undetermined. After all the status have been computed, classification tables are selected using a strategy where the goal is to find starting states for which causal pathways can be generated that reach the largest number of confirmed states without traversing a denied state. This procedure is repeated until all confirmed states have been covered. Just like a state, a treatment has status values that have been attached to it after computation. The treatment with the highest status value is chosen. At the moment, CASNET has 150 states, 350 tests, and 50 classification tables.

The user interface of CASNET incorporates restricted natural language dialogue formats. It is able to summarize the consultation by displaying the scores of the hypotheses and the status measures of states in the causal network. It is, however, unable to explain the method by which the scores were computed. Extensive references to medical literature are provided in its explanations.

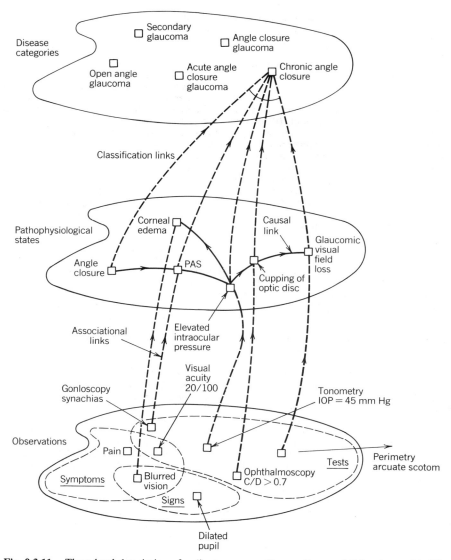

Fig. 9.3.11. Three-level description of a disease process [Source: Barr and Feigenbaum (1982)].

Updating is easy due to the modularization through the use of three planes. The system is improved and extended by the addition of new rules both causal and associated in its knowledge base.

CASNET is used by ophthalmologists in the United States and Japan owing to its speed, efficiency, and accuracy of performance. The program, which was a first attempt to provide a general framework for building expert systems, offers a general framework for modeling diseases instead of specifically modeling glaucoma alone. It led to the development of an expert system building tool, "EXPERT" which has been applied to rheumatology and endocrinology as well as to glaucoma.

CADUCEUS is another consultation system that attempts to make diagnoses, in this case, in the field of internal medicine. It was originally called INTERNIST and was constructed at the University of Pittsburgh. CADUCEUS consists of a large semantic network of diseases and symptoms in internal medicine. It models the manner in which clinicians carry out diagnostic reasoning and, when presented with a list of manifestations of a disease, attempts to distinguish which diseases fit the manifestations most accurately. In 1982, CADUCEUS possessed about 85% of all relevant knowledge in internal medicine and was comprised of 100,000 associations. This knowledge was acquired from domain experts by design experts and encoded into a semantic network.

Table 9.3.11 A Classification Table

STATE	DISEASES	TREATMENTS
ANGLE CLOSURE		
INCR IOP	ANGLE CLOSURE GLAUCOMA	TREATMENT1
CUPPING		
VFL	CHRONIC ANGLE CLOSURE GLAUCOMA	TR1, TR2

Source: Barr and Feigenbaum (1982).

Like CASNET, the knowledge base of CADUCEUS is represented in the form of a disease tree and the program reasons dynamically. The tree is organized such that diseases of any organ form the top level of the tree, classes of diseases are represented on the branches, and individual diseases are represented at the lowest levels of the tree (Figure 9.3.12). Diseases and manifestations are related in that a manifestation can evoke a disease and a disease can manifest certain signs and symptoms. The strength of manifestation is expressed by assigning numbers ranging on a scale from 0 to 5, where 0 implies total lack of association between diseases and manifestations and 5 implies that the manifestation is always associated with the disease. Each disease in the tree is associated with its relevant manifestations. In addition, the causal, temporal, and other associations among trees are highlighted. The global data base consists of signs, symptoms, and test results (all of which are called manifestations here) particular to the patient undergoing diagnosis. These are input to the system by the user.

The control structure or inference engine of CADUCEUS employs special tactics to produce results in a reasonable period of time. The reason this is needed is because the diagnosis problem is very complicated in that a patient may have more than one disease making the number of possible combinations of manifestations and diseases enormous.

For each evoked disease, a model is created which consists of observed manifestations that this disease cannot explain (called the shelf list), observed manifestations that are consistent with the disease, manifestations that should be present if the disease is present but has not yet been observed in the patient, and manifestations observed in the patient but inconsistent with the disease. Each disease model is scored, with each model receiving a positive score for each manifestation it explains and a negative score for each manifestation it does not explain. Both are weighed by IMPORT. It receives a bonus if it is linked causally to a disease that has already been confirmed. The disease models are partitioned into two sets: one called the top-ranked model for diseases that are mutually exclusive to each other (alternatives) and the second set consisting of diseases that are complementary to the top-ranked model. For example, heart diseases, lung diseases, and liver diseases are all complementary to each other but disease within any of these organs are alternatives. When competing diseases exist,

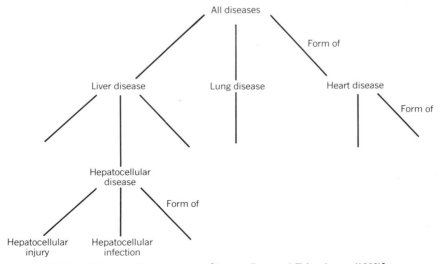

Fig. 9.3.12. INTERNIST's disease tree [Source: Barr and Feigenbaum (1982)].

the diseases that have strongly associated manifestations are accepted and the others are rejected. In conclusion, the reasoning used for CADUCEUS is a combination of bottom-up or data-driven (manifestation-evoking diseases) and top-down or hypothesis-directed reasoning (predictions of manifestations that should be present if a diagnosis is to be confirmed) approaches. It is a purely associational system where the program builds up disease models as it goes up and then dynamically partitions the disease tree into smaller disease areas corresponding to the patients symptoms.

In comparison to the other expert systems discussed, the human–computer interface does not have very many features, although attempts are underway to extend and refine them. Currently, explanation facilities are minimal.

The system is not capable of learning. It is extended by adding disease and symptom entities or by adding inferential links (evocations or manifestations).

CADUCEUS is very accurate within its domain. It encompasses knowledge of 500 diseases, 350 disease manifestations, and 100,000 symptomatic associations. To date, it has the largest knowledge base in comparison to other expert systems. Currently, clinical trials are underway at the National Institute of Health. CADUCEUS has also been used in the teaching of medical students. It has made a substantial contribution to AI methodology with regard to the role hypothesis formulation in problem solving.

Some of the other expert systems built in this area are described in Figure 9.3.8 and Table 9.3.8. Currently, there is no standard along which expert systems across domains may be compared. A first step may be identifying the components of knowledge in the knowledge base and control structure of these expert systems. Tables 9.3.12a and 9.3.12b (from Janardan-Garg and Tsatsoulis, 1985) represent the components of knowledge in the knowledge base and control structure of a few expert systems. Despite the many person-years of work that have been expended in the construction of these systems, PUFF is one of the few systems that is in routine clinical use. The use of expert systems in medicine will be facilitated by a careful choice of the medical problem, the cooperation of interested experts, and the performance of tasks by the expert system which the physician cannot do or which the physician is willing to let the computer do. Expert systems will be used in this area provided they offer substantive assistance to the physician.

9.3.5 HUMAN FACTORS AND EXPERT SYSTEMS

The human factors engineer has many talents that can be incorporated in the design and implementation of expert systems; these talents can be especially important in those areas in which expert systems have been criticized in the past. Hayes-Roth, Waterman, and Lenat (1983) delineated the following factors, which were problematic for expert systems: (1) They have an inability to recognize or deal with problems for which their (expert systems's) own knowledge is inapplicable or insufficient; (2) they have very narrow domains of expertise; (3) they lack the ability to interact with the user in unconstrained natural language; (4) knowledge extraction is a bottleneck; and (5) they have a limited ability to learn and extend their knowledge base. These problems and others can be attacked by the human factors engineer through three different areas that have traditionally been important application areas for human factors: task analysis, human–machine interface, and experimental evaluation of human performance. These three areas will be discussed in more detail.

Table 9.3.12a Components of Knowledge Required in Certain Expert Systems

Components of Knowledge Required in Expert Systems	MY-CIN	PU-FF	CEN-TAUR	WHE-EZE[b]	VM	CAS-NET	INTE-RNIST	R1
Rule grouping	X[a]			X			X	
Data ordering		X	X	X		X		
Contexts	X		X	X	X	X	X	X
Context rules	X		X	X	X			X
Inference rules	X	X	X	X	X	X	X	X
Question guiding rules	X		X	X		X	X	
Certainty strengthening rules	X		X					
Rule antecedent importance			X	X		X	X	
Positive certainty values	X	X	X	X	X	X	X	X
Negative certainty values	X	X	X	X	X		X	

Source: From Janardan-Garg and Tsatsoulis (1985).

[a] An 'X' indicates the presence of a component in an expert system.

[b] In WHEEZE knowledge is represented in the form of frames instead of production rules.

Table 9.3.12b   Components of Knowledge Used by Expert Systems

Components of Knowledge in Expert Systems	Description of the Component of Knowledge
Rule grouping	The grouping of contradicting or redundant rules. This makes control more efficient and accurate.
Data ordering	The ordering of observed facts (input by the user) or of generated possible solutions, based on importance, and frequency, and so on.
Contexts	The rules and knowledge base are divided into (possibly) interrelated contexts. This reduces search time.
Context rules	Also called meta-rules, they guide the system from context to context, activate and deactivate subtasks, and give direction to the search process.
Inference rules	They are generally production and are used to infer new facts from given information.
Question guiding rules	These rules decide which facts are relevant at a certain stage of the inference procedure and ask the appropriate questions.
Certainty strengthening rules	These rules are of the form (IF (A and B) THEN A) and increase (or decrease) the certainty of a known fact.
Rule antecedent importance	Assigned values to the rule indicate the importance of a fact in arriving at a conclusion.
Positive certainty values	Values denoting the certainty with which a conclusion is reached.
Negative certainty values	Values denoting the certainty of the absence of a fact, given a set of observations.

Source: From Janardan-Garg and Tsatsoulis (1985).

9.3.5.1   Task Analysis

A task analysis should be performed before the expert system is implemented to determine if an expert system is needed at all, how the task can be broken down into subtasks, and how the whole system will have to be changed so that the expert system and the users can be efficient and productive. A task analysis is performed by observing and talking to the operators who already perform the task; Hayes-Roth, Waterman, and Lenat (1983) (see Section 9.3.2) delineate a task analysis method as part of their design of an expert system. Human factors engineers often have experience in this area and can be called upon to perform this part of the design process.

The first thing that must be considered in the task analysis is if an expert system is needed at all. The task analysis must try to determine if the tasks being performed by the humans are algorithmic or heuristic is nature; if heuristic (based on rules-of-thumb or experience), an expert system may be appropriate if no algorithmic model can be found to perform the same task. As an example in process control, Eberts, Nof, Zimolong, and Salvendy (1984) present tables of process control operator tasks that are algorithmic and single out those tasks that are heuristic and, therefore, must be performed by an expert system. With the models and computer programs that are available for many tasks, expert systems may not have to be implemented.

The task analysis should also show how the task will change after the expert system is incorporated. This task analysis should try to separate the role of the expert system from that of the human. A possible problem with expert systems and automation in general is that the human operator, often forced to monitor the automation, may not be aware of the limitations in the "knowledge" of the expert or automated system; the operator may think that it is more "intelligent" than it actually is. Wickens (1984) presents two examples of this occurring with automated aviation systems. First, in Detroit in 1975 a DC10 and L-1011 were on a collision course, the air traffic controller saw this but did not act because he knew that if a collision became imminent the system would alert him. About this time, the air traffic controller took a break, the new controller saw the collision course and managed to contact the DC10 in time to just barely avert the crash. In the other example, a crash of an L-1011 in 1972, the pilot put too much faith in the autopilot. When a fault occurred in the plane, the pilot put the plane on autopilot, the autopilot did not hold, and the plane crashed. Automated and expert systems may require some monitoring to make sure that they are properly working in a particular situation.

Similar problems occur for expert systems. Jenkins (1984) ran a simulation where an operator along with an expert system controlled a nuclear power plant. He found that the operators assumed that the expert system "knew" more than it actually did; they thought that the system should have known when a failure in the cooling system occurred because the control panel had conveyed that information to the operator; the operator assumed that the expert system had access to the same information. The operator then assumed that the failure was being attended to by the expert system; in actuality, the expert system did not "know" that a failure had occurred. The solution to these problems is to carefully train operators so that they know the limitations of the expert system; the task analysis should show the limitations of the expert system.

9.3.5.2 The Human–Machine Interface

Human factors specialists have been involved in the design of the interface between the human and the machine. With the incorporation of computers into many different areas, the interface between the human and machine is quite often a problem in human–computer interaction. Since an expert system involves a computer, the interface between the human and the expert system is also a human–computer interaction problem.

Human factors specialists have, in the past, taken four different approaches to solving problems in human–computer interaction [see Eberts (1985) for a review]. One approach, the empirical approach, attempts to determine which of the various methods available is the best to display information, to query the user, or to take responses. The best choice is found by testing humans in an experimental situation with each of the possible methods; an analysis of the data should show which method is best. The ability of human factors engineers to design and run an experiment such as this can be utilized when the empirical approach to human–computer interaction design is incorporated.

The second approach, the anthropomorphic approach, analyzes the human–computer interaction task, determines how that task may be different from a human–human interaction task, and then offers suggestions on how the human–computer interaction can be made more like human–human interaction. Quite often this approach has attempted to make the interactions with computers more natural; one way to do this is to design the task so that some form of natural language interaction can take place. The anthropomorphic approach can be used to try to determine the kinds of features that must be incorporated into the interface to make it more humanlike.

The third approach, the cognitive approach, attempts to apply experimentation and theory from cognitive psychology to the design of the interface. The goal of this approach is to design the interface so that the user receives a clear picture, or mental model, of how the system works. Since the only picture the user has of the system is through the displayed information, this displayed information should be used to convey information that is easily perceived and fits in the user's memory structures. Typically, a human information processing model is utilized to determine how the displayed information will be processed by the user.

The fourth approach, the predictive modeling approach attempts to predict which interface design will be most effective before the interface is designed. The only model available under this approach is the GOMS model of Card, Moran, and Newell (1983). This model is very closely allied with the cognitive approach. In this model, Card, Moran, and Newell identified parameters that are important in the human–computer interaction task. These parameters were then quantified by surveying basic research from cognitive psychology. The quantification of the parameters allows different designs to be evaluated on the time to perform the task using the particular interface; the best design in terms of time can then be predicted. One set of parameters also identifies how the human codes information; this set is used to predict the kinds of errors that will be made. GOMS has been successful at predicting time and error performance.

These four approaches can be used to evaluate and design the human–computer interfaces for expert systems. One particular interface model that can be evaluated is the natural language interface. Hayes-Roth, Waterman, and Lenat (1983) mentioned that a problem with current expert systems is that unconstrained natural language is not possible; work from human factors indicates that natural language may not be the best interface mode. Experimentation has indicated that users commit fewer errors when using a specialized language as opposed to natural language in a data base search problem (Small and Weldon, 1983). Shneiderman (1980) argues that natural language is too inexact, does not apply the appropriate constraints, and can be misleading when compared to the specialized interface languages. He suggests that common interface modes, such as the use of menus, can be more effective at less overhead than natural language interfaces. Natural language may appear to be an ideal interface mode to researchers outside of human factors; work in human factors can be cited to indicate that other possibilities may be better.

9.3.5.3 The Evaluation of Performance

The third role played by the human factors engineer is in the evaluation of performance. Since expert systems try to emulate human experts and take over the roles that these experts have performed, expert behavior can be evaluated, interpreted, and analyzed to determine how to extract knowledge from experts and how to design expert systems. The mismatches between human experts and current expert systems can be especially illuminating to determine where more work needs to be done on expert systems. The following discussion presents some of the information that is currently known about experts.

An expert can be characterized as someone who has been doing the task for a long time. Although difficult to determine when someone becomes an expert, one popular notion is that an expert has practiced the task for at least 5000 hr. As examples, Lesgold, Feltovich, Glaser, and Wang (1981) state that expert radiologists see, on an average, 67 X-rays a day and up to 10,000 a year. The knowledge base accumulated by an expert is enormous. In a study of master chess players, Chase and Simon (1973) estimated that these experts could recognize around 31,000 basic or primitive piece configurations. Brooks (1977) has estimated that expert programmers have available to them between 10,000 and 100,000 rules that can be used to perform programming tasks. Storing and retrieving this vast amount of information is difficult.

All tasks require a certain skill level. The required level is a function of the situational demands and the experience of the operator. Often, skill is conceptualized as occurring along a continuous dimension, but recent theories emphasize that as behavior becomes highly skilled, it is organized differently than the unskilled behavior [see Schneider and Shiffrin (1977) as an example]. Hacker (1978) suggests three skill levels of human performance: (1) sensory–motor, (2) perceptual–conceptual, and (3) intellectual control level. A similar formulation by Rasmussen (1983) also describes three levels of skilled performance: (1) skill (routine) level, (2) rule level, and (3) knowledge-based level. This formulation can be used to not only suggest research directions for expert systems but also to suggest the kinds of information the expert system should convey to the human. Each of these levels, along with the information requirements and the methods of information processing, will be discussed in more detail.

The skill-based behavior represents highly automated sensory–motor and cognitive performances and takes place without conscious control as smooth, highly integrated patterns of behavior. It maps stimuli to responses in a fairly rapid low-level cognition mode. The flexibility of skilled performance is due to the ability to evoke from a large repertoire of automated subroutines the sets suited for the specific situations such as scanning, detection and recognition, information interaction, action selection and sequencing, and action implementation. Information or signals from the environment have no meaning or significance except as cues triggering the appropriate actions (Goodstein, 1981).

At the rule-based level, an action is selected by activating in working memory a hierarchy of rules. After mentally scanning those rules the human will implement the appropriate rule or set of rules. The control knowledge may have been derived empirically during previous occasions, by instruction, or by conscious problem solving and planning. Sequencing of automated subroutines in a familiar work situation is typically controlled by some external schedules, diagrams, or decision rules. At the rule-based level, information or signs are used to select or modify rules controlling the sequencing of skilled subroutines. They cannot be used for functional reasoning to generate new rules or to predict the response of an environment to unfamiliar disturbances. Rules and patterns of behavior are to a certain extent memorizable.

Knowledge-based level behavior is evoked when entirely new, unstructured, or complex problems are encountered. Control knowledge, schedules, or automatic mapping do not exist. Decision making and problem solving can occur at this level. Decision making can involve the identification of options or alternatives, assessing their relative attractiveness, assessing the likelihood of being realized, and integration of considerations to identify what appears to be the best option. In solving problems, the human tries to identify the current state of the process, searches for the target state, and then employs a set of operators or methods to change the current state into the goal state. Information at the knowledge-based level is perceived as symbols. They refer to meaningful concepts tied to functional or physical properties. Owing to the abstraction level of reasoning, they are defined and interpreted as symptoms of the underlying structure.

This characterization of skilled behavior indicates that the kind of information provided to the human from an expert system should take into account the human's skill level. The human's skill level is dependent on individual differences and expertise. As examples, one human may have a rule for solving a particular problem and, thus, would operate at the rule-based level. Another human may have never seen that particular situation before and would have to operate at the knowledge-based level to solve this novel problem. The information should be presented to the individual operator as either a signal, a sign, or a symbol if operating at the skill-, rule-, or knowledge-based levels, respectively.

This characterization of skill performance can also be used in knowledge extraction. For each task, different experts operating at the appropriate skill level may have to be found. Operation of the expert at the skill-based level for a particular task may not be an appropriate candidate for knowledge extraction. As emphasized previously, skill-based behavior operates without conscious control so that introspection by an expert at this level may be difficult or inaccurate. Unintuitively, a less-skilled operator may have to be found. If a compilation of rules is needed for an expert system, then the human expert should be operating at a rule-based level and not a knowledge-based level. If strategies and methods for solving problems or making decisions are needed, then an expert operating at the knowledge-based level for a particular task may have to be found. Several experts may have to be used; this formulation suggests how to interpret the level that an expert is working at.

9.3.6 CONCLUSIONS

The role of human factors in expert systems can be in task analysis, the human–machine interface, and the evaluation of human performance. In surveying these three areas, direct applications to the human–expert system interface and to the extraction of the knowledge from experts were seen. Also, the implementation of an expert system must take place cautiously so that the humans involved in the system do not think that the system is more intelligent than it actually is.

REFERENCES

Aikins, J. S. (1980). Prototypes and production rules: A knowledge representation for computer consultants. (Doctoral Dissertation, Report No. STAN-CS-80-814). Stamford, CA: Stamford University.

Barr, A., and Feigenbaum, E. A. (Eds.). (1982). The handbook of artificial intelligence. Los Altos, CA: William Kaufman, Vols. I and II.

Bobrow, D. G. (1975). Dimensions of representation. In D. G. Bobrow and A. Collins, Eds. *Representation and understanding: Studies in cognitive science.* New York: Academic, pp. 1–34.

Boose, J. (1984). A framework for transferring human expertise. In G. Salvendy, Ed., *Human–computer interaction.* New York: Elsevier, pp. 247–254.

Brachman, R. J., and Smith, B. C. (1980). *SIGART Newsletter 70, Special Issues on Knowledge Representation.*

Brooks, R. E. (1977). Towards a theory of the cognitive process in computer programming. *International Journal of Man-Machine Studies, 9,* 373–451.

Brown, J. S. (1977). Uses of artificial intelligence and advanced computer technology in education. In R. J. Seidel and M. Rubin, Eds. *Computers and communications: Inplications for education.* New York: Academic, pp. 253–270.

Brown, J. S., and Burton, R. R. (1978a). Multiple representations of knowledge for tutorial reasoning. In D. G. Bobrow and A. Collins, Eds. *Representation and understanding: Studies in cognitive science.* New York: Academic, pp. 311–349.

Brown, J. S., and Burton, R. R. (1978b). Diagnostic models for procedural bugs in basic mathematical skills. *Cognitive Science, 2,* 155–191.

Brown, J. S., Burton, R. R., and Bell, A. G. (1974). SOPHIE: a sophisticated instructional environment for teaching electronic troubleshooting (an example of AI in CAI) (BBN Report No. 2790). Cambridge, MA: Bolt, Beranek and Newman.

Brown, J. S., Burton, R. R., and DeKleer, J. (1985). Knowledge engineering and pedagogical techniques in SOPHIE I, II and III. In D. Sleeman and J. S. Brown, Eds. *Intelligent tutoring systems.* London: Academic.

Brown, J. S., Rubinstein, R., and Burton, R. (1976). Reactive learning environment for computer assisted electronics instruction (BBN Report No. 3314). Cambridge, MA: Bolt, Beranek and Newman Inc.

Buchanan, B. G., and Feigenbaum, E. A. (1978). DENDRAL and meta-DENDRAL: Their applications dimension. *Artificial Intelligence, 11,* 5–24.

Buchanan, B. G., and Mitchell, T. M. (1977). Model directed learning of production rules (Report STAN-CS-77-597). Stanford, CA: Computer Science Department, Stanford University.

Burton, R. R., and Brown, J. S. (1976). A tutoring and student modeling paradigm for gaming environments. *SIGCSE Bulletin 8,* 236–246.

Burton, R. R., and Brown, J. S. (1979a). Toward a natural-language capability for computer-assisted instruction. In H. O'Neil, Ed. *Procedures for instructional systems development.* New York: Academic, pp. 273–313.

Burton, R. R., and Brown, J. S. (1979b). An investigation of computer coaching for informal learning activities. *International Journal of Man-Machine Studies, 11,* 5–24.

Carbonell, J. R. (1970a). Mixed-initiative man-computer instructional dialogues (BBN Report No. 1971). Cambridge, MA: Bolt, Beranek and Newman.

Carbonell, J. R. (1970b). AI in CAI: an artificial intelligence approach to computer aided instruction. *IEEE Transactions on Man-Machine Systems, MMS-11* (4), 190–202.

Card, S., Moran, T., and Newell, A. (1983). *The psychology of human computer interaction.* Hillsdale, NJ: Erlbaum.

Carr, B., and Goldstein, I. (1977). Overlays: A theory of modeling for computer aided instruction (AI Memo 406). Cambridge, MA: AI Laboratory, MIT.

Chase, W. G., and Simon, H. A. (1973) Perception in chess. *Cognitive Psychology, 4,* 55–81.

Clancey, W. J. (1979a). Dialogue management for rule-based tutorials. *IJCAI 6,* 155–161.

Clancey, W. J. (1979b). Transfer of rule-based expertise through a tutorial dialogue (Report No. STAN-CS-769). Stanford, CA: Computer Science Department, Stanford University.

Clancey, W. J. (1979c). Tutoring rules for guiding a case method dialogue. *International Journal of Man-Machine Studies 11*, 25–49.

Cohen, P., and Feigenbaum, E. A. (1983). The handbook of artificial intelligence. Los Altos, CA: Kaufman, Vol. III.

Collins, A. (1976). Processes in acquiring knowledge. In R. C. Anderson, R. J. Spito, and W. E. Montague, Eds. *Schooling and the acquisition of knowledge*. Hillsdale, NJ: Erlbaum, pp. 339–363.

Collins, A. (1978a). Reasoning from incomplete knowledge. In D. G. Bobrow and A. Collins, Eds. *Representation and understanding: Studies in cognitive science*. New York: Academic, pp. 383–415.

Collins, A. (1978b). Fragments of a theory of human plausible reasoning. *TINLAP-2*, 194–201.

Collins, A., Warnock, E. H., and Passaflume, J. J. (1974). Analysis and synthesis of tutorial dialogues (BBN Report No. 2789). Cambridge, MA: Bolt, Beranek and Newman.

Corey, E. J., and Wipke, W. T. (1969). Computer assisted design of complex organic synthesis. *Science 166*, 178–191.

Davis, R. (1976). Applications of meta-level knowledge to the construction, maintenance and use of large knowledge-bases (Ph.D. Dissertation, Report STAN-CC-76-564). Stanford, CA: Computer Science Department, Stanford University.

Davis, R. (1977). Interactive transfer of expertise Acquisition of new inference rules. *IJCAI 5*, 321–328.

Davis, R., and King, J. (1976). An overview of productions systems. In E. W. Elcock and D. Mitchie, Eds. *Machine intelligence*, New York: Wiley, Vol. 8, pp. 300–332.

Dietterich, T. G., and Michalski, R. S. (1981). Inductive learning of structural descriptions: Evaluation criteria and comparative review of selected methods. *Artificial Intelligence, 16*, 257–294.

Doyle, J. (1979a). A glimpse of truth maintenance. In *Artificial intelligence: An MIT perspective*. Cambridge, MA: MIT Press, Vol. I.

Doyle, J. (1979b). Truth maintenance system. *Artificial intelligence, 12*,(3), 321–272.

Duda, R. O., Gasching, J., and Hart, P. E. (1979). Model design in the PROSPECTOR consultant system for mineral exploration. In D. Michie, Ed. *Expert systems in the microelectronic age*. Edinburgh University Press, pp. 153–167.

Duda, R. O., Gasching, J., Hart, P. E., Konolige, K., Reboh, R., Barrett, P., and Slocum, J. (1978). Development of the PROSPECTOR consultant system for mineral exploration (Final Report, SRI Projects 5821 and 6415). Menlo Park, CA: SRI International, Inc.

Duda, R. O., Hart, P. E., Nilsson, N. J., Reboh, R., Slocum, J., and Sutherland, G. L. (1977). Development of a computer based consultant for mineral exploration (Annual Report, SRI Projects 5821 and 6415). Menlo Park, CA: SRI International, Inc.

Eberts, R. E. (1985). Human-computer interaction. In P. A. Hancock, Ed. *Human factors psychology*. Amsterdam: Elsevier.

Eberts, R. E., Nof, S. Y., Zimolong, B., and Salvendy, G. (1984). Dynamic process control: Cognitive requirements and expert systems. In G. Salvendy, Ed. *Human-computer interaction*. Amsterdam: Elsevier, pp. 215–228.

Englemore, R. S., and Nii, H. P. (1977). A knowledge based system for the interpretation of protein x-ray crystallographic data (Heuristic Programming Project, Report No. HPP-772). Palo Alto, CA: Computer Science Department, Stanford University.

Englemore, R. E., and Terry, A. (1979). Structure and function of the CRYSALIS system. *IJCAI 6*, 250–256.

Ericsson, K. A., and Simon, H. A. (1980). Verbal reports as data. *Psychological Review, 87*, 215–251.

Fagan, L. (1978). Ventilator manager: a program to guide on-line consultative advice in the intensive care unit (Technical Report HPP-78-16). Stanford, CA: Heuristic Programming Project, Stanford University.

Fagan, L. M. (1980). VM: representing time-dependent relations in a medical setting (Ph.D. Dissertation). Stanford, CA: Computer Science Department, Stanford University.

Fagan, L. M., Kunz, J. C., Feigenbaum, E. A., and Osburn, J. (1979). Representation of dynamic clinical knowledge: Measurement and interpretation in the intensive care unit. *IJCAI 6*, 260–262.

Feigenbaum, E. A. (1977). The art of artificial intelligence: Themes and case studies of knowledge engineering. *IJCAI 5*, 1014–1029.

Feigenbaum, E. A., Buchanan, B. G., and Lederberg, J. (1971). On generality and problem solving: A case study using the DENDRAL program. In B. Meltzer and D. Michie, Eds. *Machine intelligence.* Edinburgh: Edinburgh University Press, Vol. 6, pp. 165–190.

Feigenbaum, E. A., Englemore, R. S., and Johnson, C. K. (1977). A correlation between crystallographic computing and artificial intelligence research. *Acta Crystallographica A33,* 13.

Filman, R. E. (1979). The interaction of observation and inference in a formal representation system. *IJCAI, 6,* pp. 269–276.

Freiherr, G. (1980). The seeds of artificial intelligence (NIH No. 80–2071, SUMEX-AIM). Washington, DC.

Gasching, J. (1980a). An application of the PROSPECTOR system to the DOE's national uranium resource evaluation. *AAAI 1,* 295–297.

Gasching, J. (1980b). Development of uranium exploration models for the PROSPECTOR consultant system (Final Report SRI Project 7856). Menlo Park, CA: SRI International.

Gelernter, H. L., Sanders, A. F., Larsen, D. L., Agarwal, K. K., Boivie, R. H. Spritzer, G. A., and Searleman, J. E. (1977). Empirical explorations of SYNCHEM. *Science 197,* 1041–1049.

Genesereth, M. R. (1978). Automated consultation for complex computer systems (Doctoral Dissertation). Cambridge, MA: Division of Applied Sciences, Harvard University.

Goldstein, I. (1977). The computer as coach: An athletic paradigm for intellectual education (AI MEMO 389). Cambridge, MA: AI Laboratory, MIT.

Goodstein, L. P. (1981). Discriminative display support for process operators. In J. Rasmussen and W. B. Rouse, Eds. *Human detection and diagnosis of systems failures.* New York: Plenum, pp. 433–449.

Gory, G. A., Silverman, H., and Pauker, S. G. (1978). Capturing clinical expertise: A computer program that considers clinical response to digitalis. *American Journal of Medicine 64,* 452–460.

Hacker, W. A. (1978). *Allgemeine arbeits- und ingenieurpshcologie.* Bern: Huber.

Hayes-Roth, F. (1984a). The knowledge-based expert system: A tutorial. *IEEE Computer, 17,* pp. 11–29.

Hayes-Roth, F. (1984b). Knowledge-based expert systems. *IEEE Computer, 17*(10), 263–273.

Hayes-Roth, F., Waterman, D., and Lenat, D., Eds. (1983). *Building expert systems.* Reading, MA: Addison-Wesley.

Janardan-Garg, C., and Tsatsoulis, C. (1985). The organization of expert systems. Unpublished report. Purdue University, School of Industrial Engineering, West Lafayette, Indiana.

Jenkins, J. P. (1984). An application of an expert system to problem solving in process control. In G. Salvendy, Ed. *Human-computer interaction,* Amsterdam: Elsevier, pp. 255–260.

Kaufman, A. (1975). *Introduction to the theory of fuzzy sub-sets.* New York: Academic, Vol. I.

Kelly, G. A. (1955). The psychology of personal constructs. New York: Norton.

Langley, P. W. (1980). Descriptive discovery process: experiments in Baconian science (Report No. CS-80-121). Pittsburgh, PA: Computer Science Department, Carnegie Mellon University (Doctoral Dissertation).

Lees, F. P. (1974). Research on the process operator. In E. Edwards and F. P. Lees, Eds. *The human operator in process control.* New York: Halsted, pp. 386–425.

Lenat, D. B. (1977). On automated scientific theory formation: A case study using the AM program. In J. E. Hayes, D. Michie and L. I. Mikulich, Eds. *Machine intelligence.* New York: Halstead, Vol. 9, pp. 251–286.

Lenat, D. B. (1980). AM: an artificial intelligence approach to discovery in mathematics as heuristic search (Report No. STAN-CS-76-570). Palo Alto, CA: Computer Science Department, Stanford University.

Lesgold, A. M., Feltovich, P. J., Glaser, R., and Wang, Y. (1981). The acquisition of perceptual diagnostic skill in radiology (Tech. Report PDS-1). Pittsburgh, PA: University of Pittsburgh.

Lindsay, R. K., Buchanan, B. G., Feigenbaum, E. A., and Lederberg, J. (1980). Applications of artificial intelligence for organic chemistry: the DENDRAL project. New York: McGraw-Hill.

Martin, W. A., and Fatemaro, R. J. (1971). The MACSYMA system. In *Proceedings of the second symposium of symbolic and algebraic manipulation.* Los Angeles, CA, pp. 59–75.

Mathlab Group. (1977). *MACSYMA reference manual.* Cambridge, MA: Computer Science Laboratory, Massachusetts Institute of Technology.

McCarthy, J. (1977). Epistmological problems of artificial intelligence. *IJCAI, 5,* 1038–1044.

McCarthy, J. (1980). Circumscription—a form of nonmonotonic reasoning. *Artificial Intelligence, 13,* 27–39.

McDermott, D., and Doyle, J. (1980). Non-monotonic logic I. *Artificial Intelligence, 13,* 41–72.

Michalski, R. S., and Larson, J. B. (1978). Selection of most representative training examples and incremental generation of VLI hypotheses: The underlying methodology and the description of programs ESEL and AQ11 (Report No. 867). Urbana, IL: Computer Science Department, University of Illinois.

Miller, R. A., Pople, H. E., and Myers, J. D. (1982). INTERNIST-I, an experimental computer-based diagnostic consultant for general internal medicine. *New England Journal of Medicine*, 468–476.

Minsky, M. (1975). A framework for representing knowledge. In P. Winston, Ed. *The psychology of computer vision*. New York: McGraw-Hill, pp. 211–277.

Mitchell, T. M. (1977). Version spaces: A candidate elimination approach to rule learning. *IJCAI, 5*, 305–310.

Mitchell, T. M. (1979). An analysis of generalization as a search problem. *IJCAI, 6*, 577–582.

Moran, T. P. (1973). *The symbolic imagery hypothesis: A production system model*. Pittsburgh, PA: Computer Science Department, Carnegie Mellon University.

Moses, J. (1975). A MACSYMA primer (Mathlab Memo No. 2). Cambridge, MA: Computer Science Laboratory, Massachusetts Institute of Technology.

Nau, D. (1983). Expert computer systems. *IEEE Computer, 16*, 63–85.

Newell, A., and Simon, H. A. (1972). *Human problem solving*. Englewood Cliffs, NJ: Prentice-Hall.

Nilsson, N. J. (1971). *Problem-solving methods in artificial intelligence*. New York: McGraw-Hill.

Pauker, S., Gorry, G. A., Kassirer, J., and Schwartz, W. (1976). Towards the simulation of clinical cognition—Taking a present illness by computer. *American Journal of Medicine, 60*, 981–996.

Pople, H. E. (1977). The formation of composite hypotheses in diagnostic problem-solving. An exercise in synthetic reasoning. *IJCAI 5*, 1030–1037.

Pople, H. E., Jr. (1981). Heuristic methods for imposing structure on ill-structured problems: the structuring of medical diagnostics. In P. Szolovitz, Ed. *Artificial intelligence in medicine*. Boulder, CO: Westview, pp. 119–185.

Pople, H. E., Jr., Myers, J. D., and Miller, R. A. (1975). DIALOG: A model of diagnostic logic for internal medicine. *IJCAI 4*, 848–855.

Quillian, M. R. (1968). Semantic memory. In M. Minsky, Ed. *Semantic information processing*. Cambridge, MA: MIT Press, pp. 216–270.

Quillian, M. R. (1969). The teachable language comprehender: a simulation program and the theory of language. *CACM 12*, 459–476.

Quine, W. V., and Ullian, J. S. (1978). *The web of belief*. New York: Random House.

Raphael, M. (1968). SIR: a computer program for semantic information retrieval. In M. Minsky, Ed. *Semantic information processing*. Cambridge, MA: MIT Press, pp. 33–145.

Rasmussen, J. (1983). Skills, rules and knowledge; Signals, signs and symbols and other distinctions in human performance models. *IEEE Transactions on Systems, Man, and Cybernetics, SMC-13*, 257–266.

Reiter, R. (1978). On reasoning by default. In *TINLAP2*, pp. 210–218.

Reiter, R. (1980). A logic for default reasoning. *Artificial Intelligence, 13*, 81–132.

Rich, E. (1983). *Artificial intelligence*. New York: McGraw-Hill.

Rouse, W. B. (1984). Design and evaluation of computer-based decision support systems, 1982. In G. Salvendy, Ed. *Advances in human factors/ergonomics I: Human-computer interaction*. New York: Elsevier, pp. 229–247.

Sacerdoti, E. D. (1977). *A structure of plans and behavior*. New York: American Elsevier.

Samuel, A. L. (1963). Some studies in machine learning using the game of checkers. In E. A. Feigenbaum and J. Feldman, Eds. *Computer and thought*. New York: McGraw-Hill, pp. 71–105.

Schank, R. C., and Abelson, R. P. (1977). Scripts, plans, goals and understanding. Hillsdale, NJ: Erlbaum.

Schneider, W., and Shiffrin, R. M. (1977). Controlled and automatic human information processing: I. Detection, search, and attention. *Psychology Review 84*, 1–66.

Schneiderman, B. (1980). *Software psychology*. Cambridge, MA: Winthrop.

Shortliffe, E. H. (1976). Computer-based medical consultations: MYCIN. New York: American Elsevier.

Shortliffe, E. H., and Buchanan, B. G. (1975). A model of inexact reasoning in medicine. *Mathematical BioSciences, 23*, 351–379.

Silverman, H. (1975). A digitalis therapy advisor (Report No. TR-1143 MAC Project). Cambridge, MA: Computer Science Department, Massachusetts Institute of Technology.

Simon, H. A. (1983). Why should machines learn? In R. S. Michalski, J. Carbonell, and T. M. Mitchell, Eds. *Machine learning: an artificial intelligence approach*. Palo Alto, CA: Tioga Press.

Simmons, R. F., Burger, J. F., and Long, R. E. (1966). An approach towards answering English questions from text. AFIPS Conference Proceedings 29. Fall Joint Computer Conference. Washington, DC: Spartan, pp. 357–363.

Simmons, R. F., and Slocum, J. (1972). Generating English discourse from semantic networks. *CACM, 15*(10), 891–905.

Small, D. W., and Weldon, L. J. (1983). An experimental comparison of natural and structured query languages. *Human Factors, 25,* 253–263.

Smith, D. E., and Clayton, J. E. (1984). Another look at frames. In B. G. Buchanan and E. H. Shortcliffe, Eds. *Rule-based expert systems: The MYCIN experiments of the Stamford Heuristic Programming Project.* Reading, MA: Addison-Wesley, pp. 441–450.

Stefik, M., Aikins, J., Blazer, R., Benoit, J., Birnbaum, L., Hayes-Roth, F., and Sacerdoti, E. (1982). The organization of expert systems: a tutorial. *Artificial Intelligence, 18,* 135–173.

Stevens, A. L., and Collins, A. (1977). The goal structure of a socratic tutor (BBN Report No. 3518). Cambridge, MA: Bolt, Beranek and Newman.

Stevens, A. L., and Collins, A. (1978). Multiple conceptual models of a complex system (BBN Report No. 3923). Cambridge, MA: Bolt, Beranek and Newman.

Stevens, A. L., Collins, A., and Goldin, S. (1978). Diagnosing students misconceptions in causal models (BBN Report No. 3786). Cambridge, MA: Bolt, Beranek and Newman.

Strat, T. M. (1984). Continuous belief functions for evidential reasoning. In *Proceedings of the national conference on artificial intelligence.* Austin, TX: University of Texas.

Suppes, P., Ed. (1981). University-level computer assisted instruction at Stanford: 1968–1980. Stanford, CA: Institute for Mathematical Studies in the Social Sciences.

Swartout, W. (1977). A digitalis therapy advisor with explanations (Report No. TR-176 MAC Project). Cambridge, MA: Computer Science Department, Massachusetts Institute of Technology.

Szolovits, P., and Pauker, S. (1976). Research on a medical consultation program for taking the present illness. *Proceedings of the third Illinois conference on medical information systems.*

Szolovits, P., and Pauker, S. G. (1978). Categorical and probabilistic reasoning in medical diagnosis. *AI, 11,* 115–146.

Weiss, S. M., and Kulikowski, C. A. (1984). *A practical guide to designing expert systems.* Location, NJ: Rowman and Allanheld.

Weiss, S., Kulikowski, C., and Safir, A. (1977). A model based consultation system for the long term management of glaucoma. *IJCAI, 5,* 826–832.

Weiss, S. M., Kulikowski, C. A., and Safir, A. (1978). A model-based consultation system for computer-aided medical decision making. *Artificial Intelligence, 11,* 65–172.

Wickens, C. D. (1984). *Human engineering and performance.* Columbus, OH: Merrill.

Wilks, Y. A. (1972). *Grammar meaning and machine analysis of language.* London: Routledge and Kegan Paul.

Winograd, T. (1975). Frame representation and the declarative-procedural controversy. In D. G. Bobrow and A. Collins, Eds. *Representation and understanding: Studies in cognitive science.* New York: Academic, pp. 185–210.

Winston, P. H. (1984). *Artificial intelligence.* Reading, MA: Addison-Wesley.

Wipke, W. T., Brown, H., Smith, G., Choplin, F., and Sieber, W. (1977). SECS-simulation and evaluation of chemical synthesis: Strategy and planning. In W. T. Wipkes and H. J. House, Eds. *Computer-assisted organic synthesis.* Washington, DC: American Chemical Society, pp. 97–127.

Woodworth, R. S. (1938). *Experimental psychology.* New York: Holt.

Zadeh, L. A. (1965). Fuzzy sets. *Information and Control, 8,* 338–353.

Zadeh, L. A. (1975). Fuzzy logic and approximate reasoning. *Synthese, 30,* 407–428.

Zadeh, L. A. (1978). PRUF—a meaningful representation language for natural languages. *International Journal of Man-Machine Studies, 10,* 395–460.

CHAPTER 9.4
STOCHASTIC NETWORK MODELS

RICHARD SCHWEICKERT

Department of Psychological Sciences
Purdue University

DONALD L. FISHER

Department of Industrial Engineering
and Operations Research
University of Massachusetts

The research for this chapter was supported by a grant from the National Institute for Mental Health (1R01 MH38675-01) to Richard Schweickert and by a grant from the National Science Foundation (IST-8309431) to Donald L. Fisher. The authors would like to thank Jerome Busemeyer, Bruce G. Coury, Susan A. Duffy, and Carolyn Jagacinski for their helpful comments and criticisms.

9.4.1 MENTAL PROCESS NETWORKS: ANALYSIS AND SYNTHESIS

Most human factors engineers are familiar with the need for network models in manufacturing. Such models can be used to predict the future behavior of a manufacturing system, to decrease the time it takes to complete all the processes in the system, and to optimize the system, by, for example, optimizing profits. Human mental behavior can also be modeled by a network of processes. More importantly, at least from the standpoint of the human factors specialist, network models of human mental behavior can be put to exactly the same uses as network models of manufacturing.

While mental networks have recently received attention in human factors [e.g., Kantowitz and Sorkin (1983), Rouse (1980), and Wickens (1984)], the models that have received the most attention are serial and parallel models. This chapter will consider these but will also introduce recent advances that have made it possible to synthesize and analyze more complex mental networks.

First, consider the problem of synthesizing or constructing a network of mental processes that mimics the actual network used by subjects. This problem can perhaps best be understood by an analogy. Suppose one could observe the time at which raw materials are delivered to a factory and the time at which finished products are output. The raw materials go through many different processes on their way to becoming finished products. Some of the processes may be in series and some in parallel, so the network may be a complex one. Now, suppose for some reason that the network cannot be observed directly. How could one construct the network of manufacturing processes knowing only the time it takes to produce a finished product from raw materials?

The same problem appears when trying to synthesize a network of mental processes. One can observe the time at which a stimulus arrives and the time at which the response is made. However, one cannot observe the inner workings of the subject, and for this reason the job of writing out the network of mental processes is not simple. A technique for synthesizing the network of mental processes that makes use only of the observed starting and finishing times will be described.

Second, consider a related problem, that of analyzing the properties of a known network of processes. In order to be useful to the human factors specialist, the analysis should lead to predictions of future behavior, to insights into ways to decrease the finishing time, and to methods for optimizing the behavior. For complicated arrangements of mental processes with random durations, these can be difficult problems, and in some cases no analytic solutions are possible. There are many cases, however, for which an analytic solution is possible, and for these we will introduce a technique for computing the mean, variance, and probability distribution of the finishing time. The technique can localize those processes most responsible for long finishing times, and it can be helpful when trying to optimize behavior.

In short, this chapter will present mathematical techniques for both synthesizing and analyzing complex networks of mental processes. Applications of the techniques to problems in human factors and cognitive psychology will be discussed.

It is assumed that the reader is familiar with elementary concepts in probability and in matrix algebra. Concepts from network theory will be introduced as we go along.

9.4.1.1 Mathematical Models in Human Factors

Before beginning a discussion of the various techniques for synthesizing and analyzing complex networks of mental processes, something should be said about the place of the work discussed in this chapter within the broader field of mathematical modeling in human factors. Until quite recently, most of the mathematical modeling within human factors has been in the areas of manual control, information theory, and signal detection [e.g., see Sheridan and Ferrel (1974)]. With few exceptions, work in these areas has not made use of the sort of mental process networks discussed in this chapter, nor does it draw heavily on the more general concepts within cognitive psychology.

Quite recently there has been a rapid change in both the types of mathematical models used within human factors and the relation of these models to work in cognitive psychology. For example, a recent textbook (Rouse, 1980) written largely from an engineering perspective includes chapters on the uses of queueing systems, fuzzy set theory, and production systems as mathematical models of person–machine systems. The connections to work in cognitive psychology are present throughout the text. However, there is little mention in the text of the more detailed mental process networks discussed in this chapter. Quite the opposite situation obtains with a recent textbook (Card, Moran, and Newell, 1983) written from a more psychological perspective. Mental process networks are discussed in detail as models within applied settings. However, there is little systematic discussion of the broad range of mathematical tools one might use to synthesize or analyze such networks. In this chapter an attempt will be made to bridge the gap between mental process networks, on the one hand, and mathematical techniques, on the other hand.

We turn now to a specific example of a network of mental processes, those involved in visual search.

9.4.1.2 Visual Search

In a visual search task, a subject is presented with an array of elements such as digits. A given digit, say 5, is the target on a particular trial. The subject is instructed to press one button if the target is present and another button if it is not. It is commonly found that reaction time increases monotonically, often linearly, with the number of elements in the display, and extra time is typically required for absence responses compared with presence responses.

These results are usually explained by the model in Figure 9.4.1a, see Sternberg (1969). In the encoding process of the model an internal representation of the display is constructed. In the comparison process the target is compared with the display elements to see if there is a match. The linear increase in reaction time with display size is explained by saying that for each element in the display a comparison must be made and each comparison requires the same amount of time. In the response selection process, a presence or absence response is chosen, depending on the outcome of the comparison process, and selecting an absence response requires more time.

In this model the comparison process consists of a set of comparisons, carried out one after the other in series. The time required for one comparison can be found by subtracting the reaction time when there are $n - 1$ elements in the display from the reaction time when there are n elements. This procedure for estimating the duration of a process is called Donders' subtractive method after the Dutch physician who developed it in 1869.

It is also possible that all the comparisons are carried out simultaneously, that is, in parallel, see Figure 9.4.1b. One might think that serial and parallel models would lead to quite different predictions for reaction time. But a surprising and important finding of Townsend (1972) is that the two kinds of models often make identical predictions for reaction time distributions. In particular, observing the way reaction times change as the number of items in the display is increased does not ordinarily give sufficient information to distinguish serial processing from parallel processing. We will have more to say about this.

Another possible arrangement of mental processes is a PERT network like the one in Figure 9.4.2. The network illustrates a situation in which stimulus y_1 is presented followed after an interstimulus interval (ISI) by stimulus y_2. To perform the task, the subject carries out some processes sequentially and some simultaneously. For example, as soon as the subject finishes encoding y_1, he or she immediately begins comparing it with the target, and these two processes are said to be sequential. On the other hand, the subject can encode y_2 at the same time as he or she is comparing y_1 with the target, and these two processes are said to be concurrent or collateral. We are supposing that a subject cannot encode two stimuli at the same time, and he or she cannot compare two stimuli at the same time, perhaps because there is only one processor available for each of these functions. However, the subject can encode one stimulus at the same time as he compares another.

At a given vertex, a process begins execution as soon as all those processes that immediately

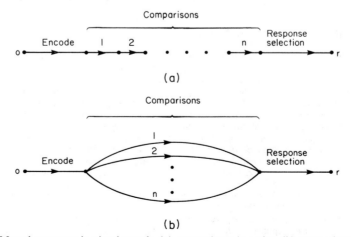

(a)

(b)

Fig. 9.4.1 Mental processes in visual search: (a) comparisons in series; (b) comparisons in parallel.

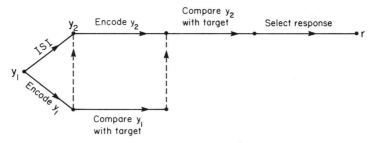

Fig. 9.4.2 A PERT network of mental processes in visual search. Dotted arrows are dummy processes that indicate precedence. In this system, only one stimulus can be encoded at a time and only one can be compared at a time.

precede it are finished. In the situation illustrated, the subject cannot compare y_2 with the target until two processes are finished, namely, encoding y_2 and comparing y_1 with the target. The dotted arrow going from the end of the comparison of y_1 to the start of the comparison of y_2 is called a *dummy process*. Its duration is zero and its only function is to illustrate the precedence order between the two processes. After completing all the processes in the network, the subject makes a response at point r.

This concludes the discussion of several example networks of mental processes in visual search tasks. In Sections 9.4.2 and 9.4.3, we discuss techniques for synthesizing complex networks of mental processes. In Sections 9.4.4–9.4.7, we discuss techniques for analyzing complex networks of mental processes. Finally, in the Section 9.4.8, we will discuss a general problem with network models.

9.4.2 SYNTHESIS OF SERIAL SYSTEMS

If, as Townsend (1972) demonstrated, observing the way reaction times vary with the number of elements to be processed will not usually reveal whether processes are serial or parallel, how can we discover the way processes are arranged? A different approach to the problem, the additive factor method, was introduced by Sternberg in 1969. The key idea is to use experimental factors to prolong processes and thereby increase reaction times. It turns out that experimental factors behave differently depending on whether the processes they prolong are serial or parallel, or more generally, sequential or collateral (Schweickert, 1978; Sternberg, 1969; Townsend and Ashby, 1983; Townsend and Schweickert, 1984).

One of the difficulties in using reaction times to make inferences about processing is that the reaction time in a given condition is not unique, but depends on how willing the subject is to respond faster at the cost of making more errors. This raises a question. Suppose two experimental factors have certain effects on reaction time in a task. Would the effects have been the same if the error rate were different? It turns out that if some plausible assumptions are met, the answer is no. A discussion of errors will be incorporated in the following section on the synthesis of serial systems. See Schweickert (1985) for more information.

Suppose to perform a task a subject executes the mental processes $x_1, \ldots, x_n$ in series. Later we will consider partially ordered processes. Let T_i be a random variable corresponding to the time allocated to process x_i. A particular value taken by T_i will be denoted t_i. Let T denote the mean time required to finish the task. Then

$$T = \Sigma T_i \tag{1}$$

Suppose the more time allocated to process x_i, the greater is its accuracy. When the mean time allocated to process x_i is T_i, let the probability that x_i is executed correctly be denoted P_i. Here, P_i is a function of the random variable T_i. By saying that a process is executed correctly, we mean the following. Suppose process x_i has a set of possible inputs and a set of possible outputs (Bamber and van Santen, 1980). In a particular experimental condition, for a given input, the set of outputs is partitioned into those that are correct and those others that are incorrect. Process x_i is executed correctly if its output is correct, given its input. For example, a process may have as input a visual percept of the letter A, and as a correct output a phonemic representation of A. The input and output sets need not be finite, for example, the output of a process might be one of an infinite number of levels of familiarity. But the relationship between inputs and outputs is categorical in the sense that for a given input, the same output is not correct on one trial and incorrect on another if the experimental conditions have not changed.

We assume that the entire task is performed correctly only if every process is executed correctly.

We assume further that the probability that every process is correct is the product of the probabilities that each process is correct. Let P denote the probability that the task is performed correctly. Then

$$P = \Pi P_i$$

and

$$\log P = \Sigma \log P_i \tag{2}$$

We are not assuming that the errors made by the processes are stochastically independent, by the way. That would impose additional requirements, such as that the probability all processes are in error is the product of the probabilities that each process is in error. The assumption in Equation (2) is slightly weaker.

Sternberg's (1969) additive factor method is based on the idea that if two factors prolong different processes in series, they will have additive effects on reaction time. By comparing Equations (1) and (2), it seems that factors have additive effects on reaction time would have additive effects on log percent correct, and we will now discuss what is required for the latter to hold.

In Equation (1) the expected value of the left-hand side is the sum of the expected values of the terms on the right-hand side. This does not require the quantities on the right-hand side to be stochastically independent, nor does it require any assumptions about their distributions, except that the expected values exist.

Equation (2) is more complicated because it involves a product of functions of random variables. (Note that the value of P_i depends on the value of T_i, which varies randomly.) If the process durations are stochastically independent, then the expected value of the product on the right-hand side is the product of the expected values of its multiplicands, that is,

$$E[P] = \Pi E[P_i]$$
$$\log E[P] = \Sigma \log E(P_i)$$

We can estimate $E[P]$ for a condition by dividing the number of correct trials in that condition by the total number of trials in that condition. When the preceding equation holds, the accuracy for the entire task can be written as the sum of the accuracies of the individual processes. A system with this property is said to satisfy the *separable accuracy model*.

For a given combination of levels of the experimental factors, as the time allocated to process x_j increases, the probability that x_j is executed correctly increases. Suppose experimental factor 1 changes this speed–accuracy function for process x_j (see Figure 9.4.3) but not that of any other process, and suppose experimental factor 2 changes the speed–accuracy function of process x_k alone. Suppose for a given level of factor 1, the point selected on the speed–accuracy curve for process x_j is the same regardless of the level of factor 2. The subject may meet this assumption spontaneously, or he or she may be induced to meet it by a payoff function to be described below.

As the levels of factors 1 and 2 change, the changes in duration and log percent correct of processes x_j and x_k lead to changes in the task completion time and log probability of a correct response. Let $T(1,1)$ be the mean time required to complete the task when factors 1 and 2 are at their baseline

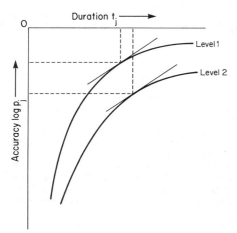

Fig. 9.4.3 There is a speed–accuracy function for every level of the factor affecting process x_j. With the payoff function described in the text, payoff is maximized when the slope of the function is c_1/c_2.

levels. Let $P(1,1)$ be the corresponding probability of a correct response. Let $T(2,1)$ and $P(2,1)$ be the mean task completion time and probability of a correct response, respectively, when x_j is made more difficult but x_k is at its baseline level. Let $T(1,2)$, $P(1,2)$, $T(2,2)$, and $P(2,2)$ be defined analogously.
  Then from Equations (1) and (2), it is easy to see that

$$E[T(2,2)] = E[T(2,1)] + E[T(1,2)] - E[T(1,1)] \tag{3}$$

and if the process durations are independent,

$$\log E[P(2,2)] = \log E[P(2,1)] + \log E[P(1,2)] - \log E[P(1,1)] \tag{4}$$

  Equation (3) states that if the system is serial, the experimental factors will be additive factors for response times. If, further, the process durations are stochastically independent, then Equation (4) states that the factors will be additive factors for log percent correct. Interactive effects of factors on reaction time indicate that the model is wrong in some way, perhaps because the processes are not in series. If the factors have additive effects on reaction time, but not on log percent correct, the processes may be in series, but their durations are not stochastically independent.
  When the experimental conditions change, it is possible for the reaction time to increase while log percent correct decreases; this is a speed–accuracy trade–off. In Figure 9.4.4 changing the level of a factor increases the difficulty of process x_j in the sense that for every value of t_j the log probability of correct execution is smaller at level 2 than at level 1. But at the points selected for operation, changing from level 1 to level 2 makes the process duration increase and the log probability correct increase also, that is, speed and accuracy both increase. Nonetheless, the additive relationship of Equation (4) is still valid.
  Sometimes authors report that factors in an experiment have additive effects on the frequency of errors (Lively, 1972; Schuberth, Spoehr, and Lane, 1981; Shwartz, Pomerantz, and Egeth, 1977). Note that minus the natural log of probability correct is approximately equal to the probability of an error, if the latter is small, since

$$-\log_e P = (1 - P) + (1 - P)^2/2 + (1 - P)^3/3 + \cdots$$

Therefore, under the conditions leading to Equation (4) factors having additive effects on reaction time will have approximately additive effects on the error probabilities.
  The effects of the factors on accuracy can be investigated with a chi-square test, based on a log linear model (Bishop, Feinberg, and Holland, 1975; Bonett and Bentler, 1983; Smith, 1976). Suppose the same number of trials is presented in each condition. Then additivity for the effects of the factors on log percent correct can easily be shown to be equivalent to the statement that in a table giving the expected probability correct in each condition, the cell probabilities should be equal to the product of the marginal probabilities (Schweickert, 1985). To see if this holds for the observations, a chi-square test can be done on the table of correct and incorrect responses. More details on this procedure will be given below.
  There is a problem in applying the chi-square test, however. One of the assumptions underlying this test is that the observations are independent, but in most reaction time experiments the same

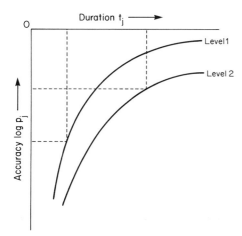

Fig. 9.4.4 A speed–accuracy trade-off. Compared with level 1 of the factor, at level 2 the time allocated to process x_i is greater, but so is the accuracy.

subject is used in several conditions. Methods for analyzing repeated measure designs with the analysis of variance have been developed, but comparable methods are not generally available yet for the chi-square test. One can statistically test for dependence across trials with autocorrelations; but if there is dependence, it is not obvious how to proceed with the chi-square test. Until more suitable tests are devised, one can only use the existing tests for guidance, and interpret them with care.

9.4.2.1 An Application to Choice Reaction Time

Shwartz, Pomerantz, and Egeth (1977) carried out an experiment in which an arrow pointed either left or right, and the subject responded by pressing a button. There were three factors, each with two levels. The arrows were (1) either bright or dim, and (2) either similar or dissimilar. The responses were (3) either compatible or incompatible, depending on whether the subject pressed the button the arrow pointed toward or the button the arrow pointed away from. The mean reaction times in each condition are given in Figure 9.4.5.

An analysis of variance showed that the three factors had additive effects on reaction times, and separate analyses of variance showed that the factors did not have a significant interaction for the variances and higher cumulants of the reaction times (although numerically the variances deviate considerably from additivity). The additive effects on reaction time are explained if each factor prolongs a different process in a series of processes (Sternberg, 1969), and the absence of interactions for the higher cumulants is evidence that the process durations are stochastically independent.

Analysis of accuracy in this experiment also supports these ideas. The observed frequencies of correct and incorrect responses in each condition are in Table 9.4.1. There are three experimental factors—intensity, similarity, and compatibility—and one response factor—correct versus incorrect. The frequencies given in parentheses are those predicted if the factors have additive effects on log percent correct. They are calculated via a log linear model; see Bishop, Feinberg, and Holland (1975); Bonett and Bentler (1983); and Smith (1976). It happens that the model for this experiment is model number 5 in Table 3.4.2 of Bishop, Feinberg, and Holland (1975). Estimates of the expected cell frequencies can be calculated with the iterative procedure available in computer programs such as ECTA for log linear models.

The model fits the data quite well as can be seen by comparing the observed cell frequencies with the predicted ones. The Pearson chi-square is 4.93 with four degrees of freedom, and is not significant. We conclude that the experimental factors have additive effects on log percent correct, supporting the separable accuracy model.

9.4.2.2 A Payoff Inducing Additivity on Accuracy

We assumed above that the subject would locate his or her point of operation on the speed–accuracy curve for a process x_i at the same place regardless of the levels of the factors affecting other processes. The subject might do this spontaneously, but he or she can also be encouraged to do so through the use of the following payoff scheme.

The experimenter can arrange the payoff at the end of a sequence of trials so the subject gains an amount proportional to log P, but loses an amount proportional to T. That is, the payoff is

$$U = -c_1 T + c_2 \log P \qquad (5)$$

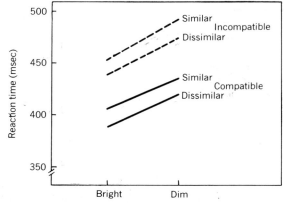

Fig. 9.4.5 Mean reaction times in the experiment of Shwartz, Pomeratz, and Egeth (1977) (by permission).

for positive constants c_1 and c_2 chosen by the experimenter. Equations (9.4.1) and (9.4.2) lead to

$$U = -c_1 \Sigma T_i + c_2 \Sigma \log P_i$$

To find the values t_i that maximize U, the partial derivative of U with respect to t_i is set to 0 for $i = 1, \ldots, n$. The optimal time to allocate to process x_i is given by

$$\frac{d \log P_i}{dt_i} = \frac{c_1}{c_2} \tag{6}$$

The above equation yields a maximum when $d^2 \log P_i/dt_i^2 < 0$, which is likely to hold. An important consequence of Equation (5) is that t_i, the optimal time to allocate to process x_i, does not depend on the optimal time to allocate to any other process.

In some situations, rather than selecting the duration for a process, and letting the duration determine its accuracy, the subject may select the accuracy and let it determine the process duration. As long as the relationship between speed and accuracy is one to one in a given condition, $dt_i/d \log P_i$ is the reciprocal of $d \log p_i/dt_i$, and Equation (6) will hold.

Another useful equation can be derived. Suppose there is only one value of $\log P$ obtained in an experiment for each value of t, and this value of t maximizes the payoff for given values of c_1 and c_2. Then

$$\frac{dU}{dt} = -c_1 + c_2 \frac{d \log P}{dt} = 0$$

or

$$\frac{d \log P}{dt} = \frac{c_1}{c_2}$$

In other words, the slope of the log percent correct curve at any time t is equal to the ratio of the costs. According to this model, it is the ratios of the costs in Equation (5) that determines the subject's strategy, not the individual values.

9.4.3 SYNTHESIS OF PERT NETWORKS

It was Donders' 1869 idea that all the processes in a task are executed in series, but this idea is now known to be too simple. If a person is given two information processing tasks to perform simultaneously, the time to complete both will be shorter than the sum of the times required to complete each one separately. It is easy to notice this phenomenon in daily life; for example, the time needed to make coffee and toast together is less than the time to make coffee plus the time to make toast. Efficiency is achieved by making use of slack, that is, by inserting one activity into the interval supplied when

Table 9.4.1 Frequencies of Correct and Incorrect Responses In the Experiment of Shwartz, Pomerantz, and Egeth (1977)[a]

Intensity	Compatible		Incompatible	
	Dissimilar	Similar	Dissimilar	Similar
		Correct		
Bright	978 (972.5)	958 (956.6)	934 (930.3)	882 (892.6)
Dim	954 (956.3)	927 (931.6)	885 (891.9)	851 (837.2)
		Incorrect		
Bright	22 (27.5)	42 (43.4)	66 (69.7)	118 (107.4)
Dim	46 (43.7)	73 (68.4)	115 (108.1)	149 (162.8)

[a] The empirical frequencies were calculated from Table 1 of Shwartz, Pomerantz, and Egeth (1977). The frequencies predicted if the factors have additive effects on log percent correct are in parentheses.

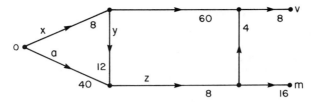

Fig. 9.4.6 A PERT network. The stimulus is presented at o. A manual response is made at m and a verbal response at v. Processes x and a are collateral, while x and z are sequential. Associated with each process is its duration, a nonnegative real number.

the system is waiting for some other activity to be completed. One way humans become efficient at cognitive tasks is by taking advantage of slack in mental processes, and arrangements of processes more complicated than the serial organization will now be considered.

Figures 9.4.2 and 9.4.6 illustrate a common type of process organization, the PERT (program evaluation and review technique) network (Kelley, 1961; Malcolm, Roseboom, Clark, and Fazar, 1959). More general organizations will be discussed in a subsequent section on order-of-processing (OP) networks. In a PERT network, each process in the task is represented by an arrow, and if process x must be completed before process y can start, then the arrow representing x precedes the arrow representing y (see Figure 9.4.6). The network is acyclic, that is, no arrow precedes itself.

Processes in a PERT network are related in one of two ways. Pairs of processes joined by a directed path, for example, x and z in Figure 9.4.6, must be executed in order and are called *sequential*. We assume that no process can begin until all those preceding it are finished. Pairs of processes not joined by a directed path, such as x and a in Figure 9.4.6, are called *collateral*. Collateral processes can be executed simultaneously. Processes in series and in parallel are special cases of sequential and collateral processes, respectively. Two sequential processes a and b are in *series* if whenever another process c is on a path with one of them, it is on a path with the other. Two processes are in *parallel* if they have the same starting point and the same terminating point.

Associated with every process is a nonnegative number—its duration—and the duration of the task—the response time—is the sum of the durations of the processes on the longest path through the network, called the *critical path*. In this section we assume the durations are fixed quantities; later we will consider durations that are random variables. The network in the former case is deterministic; in the latter, it is stochastic. Some authors use the term critical path network, instead of PERT network, when the process durations are fixed quantities, but we will say PERT network throughout.

The problem we will discuss here is that of synthesizing a PERT network that will represent the processes involved in a task and account for the response times. The key to constructing the network is to use the idea from Sternberg's (1969) additive factor method of prolonging processes. The effects of such prolongations are surprisingly informative about the network.

The assumptions here differ from those of the additive factor method in that we allow for the possibility of two processes being executed concurrently. The assumptions are also different from those of McClelland's (1979) cascade model. In his model, a process does not need to be completed before it sends output to the next process in the series; instead, the partially completed results of a process are passed continuously to its immediate successor. In a PERT network, a process begins execution only when all its predecessors are completed.

The assumption made in this section that the process durations are fixed quantities is surely wrong for human information processing, and equations based on it are only approximations to the correct stochastic equations. However, if the prolongations are long and there are a large number of trials, the approximations are not too bad (Schweickert, 1982).

Furthermore, whether the process durations are real numbers or random variables, the procedure of prolonging processes can be used to distinguish sequential and collateral processes (Schweickert, 1978; Townsend and Schweickert, 1984). This information is enough to synthesize a PERT network. An excellent discussion of the procedure for synthesizing a network given this information can be found in Golumbic (1980), and the application of the procedure to psychological tasks is discussed in Schweickert (1983b).

Once a PERT network has been synthesized, its stochastic behavior can be investigated by using the OP diagram procedure to be explained. It gives the exact expressions for the completion time mean and variance as well as for other quantities of interest.

9.4.3.1 Latent Network Theory

Here we will briefly summarize the procedure for constructing a PERT network from the behavior of the processes when prolonged. Let $t(0,0)$ denote the response time when all the processes are at the shortest durations used in the experiment. Let $\Delta t(\Delta x, 0)$ denote the increase in $t(0,0)$ produced by prolonging process x by Δx, leaving all other processes unchanged. Other increases are denoted analogously. If x and y are concurrent, then

$$\Delta t(\Delta x, \Delta y) = \max\{\Delta t(\Delta x, 0), \Delta t(0, \Delta y)\} \tag{7}$$

[All of the equations in this section are derived in Schweickert (1978).]

The situation is more complicated if x and y are sequential. Suppose x precedes y on a directed path. The amount of time by which x can be prolonged without making y start late is called the slack from x to y, written $s(xy)$. A different quantity is the amount of time by which x can be prolonged without delaying the response, r, and thereby increasing the response time. This quantity is called the total slack for x, written $s(xr)$. A process is on a critical path if and only if its total slack is zero. If all the processes are in a sequence, there is only one path, and it is the critical path, so every process has zero total slack.

Slack is important when two sequential processes are prolonged. Suppose x precedes y on a path. If the prolongations Δx and Δy are not too small, then it can be shown that

$$\Delta t(\Delta x, \Delta y) = \Delta t(\Delta x, 0) + \Delta t(0, \Delta y) + k(xy) \tag{8}$$

where $k(xy) = s(xr) - s(xy)$ is called the coupled slack from x to y. By saying that the prolongations are not too small we mean simply that the prolongations are long enough to overcome all of the relevant slacks. The general equations for prolongations of all sizes are derived in Schweickert (1978).

The magnitude of $k(xy)$ does not depend on the magnitudes of x and y. This fact provides a strong test of whether a network analysis applies to a given set of data: All values of Δx and Δy large enough for Equation (8) to hold should yield the same value for $k(xy)$, the interaction term.

If all the processes are sequential, $k(xy) = 0$ for every pair x and y, and Equation (8) becomes the additive relationship of the additive factor method. In general, however, Equations (7) and (8) indicate that when two separate processes in a network are prolonged, their effects will interact.

The Wheatstone Bridge

A negative value of $k(xy)$ is very informative. If x precedes y and $k(xy) < 0$, then the task network must have a subnetwork in the shape illustrated in Figure 9.4.7, called a Wheatstone bridge. Moreover, certain relationships hold among the path durations, although these are beyond the scope of this chapter; see Schweickert (1978) for the details and proof.

A peculiarity of processes x and y arranged in a Wheatstone bridge with $k(xy) < 0$ is that small prolongations of x and y will result in Equation (7) holding, rather than Equation (8). That is, for small prolongations, x and y will behave as collateral processes, even though they are sequential. This situation can only occur with a Wheatstone bridge. This mimicking of collateral processes by sequential ones is the analog in PERT networks of the nonidentifiability Townsend (1972) discovered for serial and parallel processes.

Determining Processing Order

Experiments in which the subject makes two responses on every trial are also very informative. Let the two response times be t_1 and t_2, both measured from the same starting point. This point might be the onset of the warning signal or the point at which the first stimulus is presented. The important thing for our purposes is that the time elapsing for each response is measured with respect to the same initial event.

Each response time considered alone will satisfy Equation (7) or (8) under the appropriate conditions. Furthermore, t_1 and t_2 are related. Suppose x precedes y, which precedes both responses, and the prolongations Δx and Δy are not too small. Then

$$\Delta t_1(\Delta x, 0) + k_1(x,y) = \Delta t_2(\Delta x, 0) + k_2(x,y) \tag{9}$$

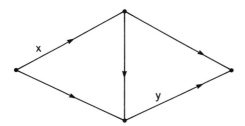

Fig. 9.4.7 When the coupled slack $k(xy)$ is negative, processes x and y are in a Wheatstone bridge.

If y precedes x instead of x preceding y, then $\Delta t_1(0,\Delta y)$ and $\Delta t_2(0,\Delta y)$ are required in Equation (9) in place of $\Delta t_1(\Delta x,0)$ and $\Delta t_2(\Delta x,0)$. The order of x and y is revealed, then, if one version of Equation (9) holds but not the other. If neither version holds, a PERT network model is invalid.

There is another way to find the order of execution of processes. Suppose x precedes y which precedes z. If the prolongations of Δx, Δy, and Δz are not too small, then the combined effect of prolonging all three processes is

$$\Delta t(\Delta x, \Delta y, \Delta z) = \Delta t(\Delta x, 0,0) + k(xy) + \Delta t(0,\Delta y, 0) + k(yz) + \Delta t(0,0,\Delta z) \tag{10}$$

This equation is useful for two reasons. First, since all the parameters in it can be determined by prolonging the processes individually and in pairs, the equation makes a prediction that can be tested. Second, the equation gives information about order: y is executed between x and z. To see this, note that for the three processes there are three coupled slacks, $k(xy)$, $k(yz)$, and $k(xz)$, but $k(xz)$, the one corresponding to the first and last of the three processes, is missing in the above equation. If the order were, say x, then z, then y, the missing term would be $k(xy)$.

It is also possible to learn something about the path durations. As might be expected, the effects of prolonging the processes do not uniquely determine the process durations, but it is possible to find an interval within which each process duration lies. The details are beyond the scope of this chapter, but the following equation illustrates the type of information available. Let the terminal vertex of process x be denoted x''. Suppose there are two responses, one at r_1 and one at r_2. Let the duration of the longest path from x'' to r_1 be denoted $\delta(x'',r_1)$. Then it can be shown that

$$t_1(\Delta x) - t_2(\Delta x) = \delta(x'',r_1) - \delta(x'',r_2) \tag{11}$$

in other words, while we do not learn what the duration of a path is in itself, we do learn what the difference between two path durations is. For more information on path durations see Schweickert (1978, 1983a).

An Example

We will illustrate the technique of latent network theory by synthesizing a critical path network for a set of hypothetical reaction time data. Suppose a task requires two responses, one manual and the other verbal. Suppose there are three experimental factors, for example, (1) the intensity of the stimulus, (2) the stimulus probability, and (3) the memory load. Let x, y, and z, respectively, be the processes prolonged by these three factors.

We let $t_m(0,0,0)$ denote the manual reaction time when all the factors are at their lowest levels, that is, when none of the processes are prolonged. Let $t_v(0,0,0)$ denote the verbal response time in the same condition. Let $t_m(\Delta x,0,0)$ be the manual response time when the intensity of the stimulus is decreased and x is prolonged, and let the corresponding change in the manual response time be denoted

$$\Delta t_m(\Delta x,0,0) = t_m(\Delta x,0,0) - t_m(0,0,0)$$

Other expressions will be defined in the analogous way.

Table 9.4.2 gives the data from this hypothetical experiment. We first consider the effects of prolonging x and y. For the manual response, Equation (8) holds with $k_m(xy) = 20$,

Table 9.4.2   Response Times and Increases in Response Times in a Hypothetical Experiment

	Response Times		Increases in Response Times	
	t_m	t_v	Δt_m	Δt_v
Base	64	76	0	0
Δx	74	106	10	30
Δy	94	90	30	14
$\Delta x\ \Delta y$	124	120	60	44
Δz	104	100	40	24
$\Delta x\quad\Delta z$	114	110	50	34
$\Delta y\ \Delta z$	134	130	70	54
$\Delta x\ \Delta y\ \Delta z$	164	160	100	84

$$\Delta t_m (\Delta x, \Delta y, 0) = \Delta t_m (\Delta x, 0,0) + \Delta t_m (0, \Delta y, 0) + k_m (xy)$$

that is,

$$60 = 10 + 30 + 20$$

For the verbal response, Equation (8) holds with $k_v (xy) = 0$,

$$\Delta t_v (\Delta x, \Delta y, 0) = \Delta t_v (\Delta x, 0,0) + \Delta t_v (0, \Delta y, 0) + k_v (xy)$$

that is,

$$44 = 30 + 14 + 0$$

We conclude that x and y are sequential processes.

What is the order of execution for x and y? If y precedes x, then Equation (9) should hold in the form

$$\Delta t_m (0, \Delta y, 0) + k_m (xy) = \Delta t_v (0, \Delta y, 0) + k_v (xy)$$

However,

$$30 + 20 \neq 14 + 0$$

so we reject the idea that y precedes x. On the other hand, if x precedes y, then Equation (9) should hold in the form

$$\Delta t_m (\Delta x, 0,0) + k_m (xy) = \Delta t_v (\Delta x, 0,0) + k_v (xy)$$

that is,

$$10 + 20 = 30 + 0$$

Since this equation holds while the earlier one does not, we conclude that x precedes y.

Further calculations show the values of the remaining coupled slacks to be

$$k_m (yz) = 0 \quad \text{and} \quad k_v (yz) = 16$$
$$k_m (xz) = 0 \quad \text{and} \quad k_v (xz) = -20$$

The negative coupled slack for x and z is of particular interest, because it indicates that x and z are arranged in a Wheatstone bridge with respect to the verbal response.

The reader can verify that x precedes z using Equation (9). This equation gives no information about the order of y and z. However using the values of the coupled slacks in Equation (10) for the verbal response times reveals that the order of the processes is x, then y, then z. To see this, note that

$$\Delta t_v (\Delta x, \Delta y, \Delta z) - \Delta t_v (\Delta x, 0,0) - \Delta t_v (0, \Delta y, 0) - \Delta t_v (0,0, \Delta z)$$
$$= 84 - 30 - 14 - 24$$
$$= 16$$
$$= k_v (xy) + k_v (yz)$$

The reader can check that the order x, then y, then z is consistent with Equation (10) applied to the manual response times. The manual response times alone cannot be used to determine uniquely the order of the processes x, y, and z because $k_m (yz) = k_m (xz)$, so more than one order of the processes will satisfy Equation (10).

A final piece of information is that m is not on the longest path from the terminal vertex of z to v. Recall that z'' denotes the terminal vertex of z. If m were on the longest path from z'' to v, then this path would consist of two parts, the longest path from z'' to m, and the longest path from m to v. Then

$$\delta(z'', v) = \delta(z'', m) + \delta(m, v)$$

so

$$\delta(z'', v) - \delta(z'', m) \geq 0$$

However, from Equation (11)

$$\delta(z'',v) - \delta(z'',m) = t_v(0,0,\Delta z) - t_m(0,0,\Delta z)$$
$$= 100 - 104 < 0$$

A network incorporating these facts is given in Figure 9.4.6. This network with the given process durations was used to calculate the data in Table 9.4.2, using the prolongations $\Delta x = 30$, $\Delta y = 50$, and $\Delta z = 40$. (Recall that the process durations themselves cannot be determined uniquely from the effects of the prolongations.) The reader is urged to try a few of the calculations in order to develop a feel for how the networks behave.

9.4.3.2   PERT Networks in Information Processing

In 1931 Telford found that if one stimulus is followed closely by another, the time to respond to the second one is longer than if it were presented alone. He called the delay in responding to the second stimulus the psychological refractory period, by analogy with the refractory period for neurons.

One of the leading explanations of the psychological refractory period is the single-channel hypothesis (Broadbent, 1958; Welford 1952, 1959, 1967, 1980). According to this hypothesis, the central processes, or more specifically, the decisions, concerning two separate stimuli cannot overlap in time. If two stimuli arrive close together, one must be held in storage while the central mechanism is occupied by the other.

In the simplest version of the single-channel model, *all* the processing of the first stimulus is completed before processing of the second stimulus starts, that is, processing of the stimuli is sequential. This hypothesis gives a good account of some of the data, but it unequivocally fails to fit certain experiments, notably that of Karlin and Kestenbaum (1968). For details about the failure of this hypothesis, see Kantowitz (1974), Keele (1973), Ollman (1968), and Schweickert and Boggs (1984). One can therefore reject the purely sequential type of model in which the subject completely processes the first stimulus, and only then begins processing the second one.

Evidence for concurrent processing in human cognitive tasks is given by Pashler (1984), for a visual search task. In his experiment 1 a small white bar was presented followed by an array of six letters. With one hand, subjects pressed a button to indicate whether the bar was above or below the fixation point and with the other hand subjects pressed a button to indicate whether the target, L, was present in the display. Two experimental factors were manipulated in the visual search task. The first was the intensity of the display, and the second was whether the target was present or absent. In another condition, the visual search task was done alone without the choice reaction to the bar.

The reaction times for the visual search task are in Figure 9.4.8. Two points are important for our purposes. First, the effect of intensity is significantly smaller in the dual-task condition than in the single-task condition. This is because in the dual-task condition there are processes going on concurrently with encoding of the visual display, and although encoding takes longer when the display is dimmer, part of the increase in the encoding time is needed to overcome the slack, and only the remainder prolongs the search response time; see Figure 9.4.9.

The second point is that in both the single- and dual-task conditions, the effects on response time of intensity and of target presence or absence are additive. This indicates that the processes affected

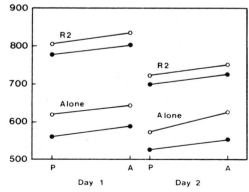

Fig. 9.4.8 Mean reaction times in a visual search task performed alone or as the second of two tasks (R_2). Filled circles, high intensity; unfilled circles, low intensity; P, target present; A, target absent. Experiment 1 of Pashler (1984). (By permission.)

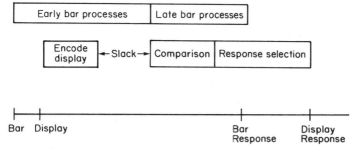

Fig. 9.4.9 When the visual search task is performed in conjunction with a choice reaction to a bar, there is some slack in the encoding process, the process affected by stimulus intensity.

by these factors are sequential (Townsend and Schweickert, 1984). In other words, Pashler's experiment provides evidence for both sequential and concurrent processing.

Dual tasks arise often in human factors. An air traffic controller may search a screen for an aircraft while listening for a signal (Pasmooij, Opmeer, and Hyndman, 1976), or a supervisor in a factory may monitor several displays simultaneously (Curry and Gai, 1976). When a human operator is given an additional task, there is an increase in effort, but often only a small increase in time, because of concurrent mental processing.

At some future time there may be a table in which one could look up a given pair of mental processes to see whether they are collateral or sequential, or, better still, the relationship between a pair of mental processes might be derived from basic cognitive principles. But for now there is little information to go on. Every task must be analyzed separately, and the principles of cognitive operations are glimpsed only dimly. An excellent and thoughtful review of the situation is given in Kerr (1973).

One principle beginning to merge from dual-task experiments is that subjects can execute only one decision at a time. As Kerr explains, according to current versions of the single-channel model there is a central mechanism that is limited in that it can only carry out one process at a time, and this sometimes causes a delay in processing two signals. Processes not needing the central mechanism can be executed concurrently with each other, and with processes that require the central mechanism.

Another source of delays in dual tasks arises somewhere toward the response end of the system. One hypothesis, response interdiction (Keele, 1973), is that only one response can be initiated at a time. Another hypothesis, response conflict, is that while more than one response can be initiated at a time, responses proceed more slowly when concurrent; see Kantowitz (1974). We are unable to say which, if either, of these hypotheses are correct, but it is well established that there is a delay in dual tasks when the responses required in the two tasks conflict.

Both sources of delay—decisions about stimuli and coordination of responses—occur in the Stroop effect, which we will use to illustrate the application of latent network theory to information processing tasks.

Single-Channel Theory and the Stroop Effect

In the Stroop (1935) task, a subject is presented with a color name such as GREEN written in a colored ink such as red. His or her task is to say aloud the color of the ink, or, in the reverse Stroop task, to say aloud the color name. As would be expected, response times are longer when the word and color conflict than when they agree.

In an experiment on the Stroop task by Schweickert (1983b), the subject pressed a button to indicate which word was presented and said aloud the name of the color. Three factors were varied. The first was the number of alternatives for the word, the second was the number of alternatives for the color, and the third was whether the word and color agreed or conflicted. These factors were also manipulated in a reverse Stroop task in which the subject pressed a button to indicate which color was presented and said aloud the color name.

A PERT network representing the processes in the Stroop task is in Figure 9.4.10. The process prolonged by increasing the number of alternative words is denoted W, the process prolonged by increasing the number of colors is denoted H, and the process prolonged when the word and color conflict is denoted C. In accordance with the hypothesis that only one decision can be executed at a time, W and H are sequential. Figure 9.4.10 illustrates the case in which the subject named the color aloud, and in this case W preceded H. Process C follows the two decisions. In this experiment the interval between the manual response and the verbal response did not change when the experimental factors were varied. To represent this fact in the figure, the three processes prolonged by the factors all precede the manual response, m, which in turn precedes the verbal response, v.

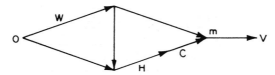

Fig. 9.4.10 A PERT network for the Stroop task. The word and color are presented at o. Processes W and H are the decision about the word and the decision about the color, respectively. The process prolonged when the word and the color conflict is C. The manual response to the word is made at m, and the verbal response to the color is made at v. From Schweickert (1983b). (By permission.)

In the reverse Stroop task in which the subject said the color name aloud, two findings are of interest for our purposes. First, the effect of prolonging both decisions was not significantly longer than the effect of prolonging the color decision alone, so that Equation (7) holds. This is another example of the idea that if a process is not on the critical path, it can be prolonged without increasing the response time. Evidently, when both decisions were prolonged, the word decision was not critical, or was only critical on a small number of trials, so prolonging it had no effect.

The second finding of interest is that in the reverse Stroop task the order of the decisions was interchanged so that H preceded W. If each decision requires the use of the central mechanism, which can serve only one decision at a time, then the decisions must be sequential, but there is no logical constraint on their order. The subject can schedule them in the way that is most convenient.

If two processes are always sequential, but sometimes one is first and sometimes the other, then clearly the output of one is not required as input to the other, and the most likely situation is that the two processes require the exclusive use of some resource. We now turn to the question of the optimal ordering of processes in this situation.

9.4.3.3   The Optimal Order of Execution of Processes

Suppose two processes, x and y, must be executed sequentially, but there is no logical constraint on which comes first. We say such processes commute (Bernstein, 1966). Given that the order of execution of all other processes has been settled, which process, x or y, should come first if the response time is to be minimized? The answer depends on the lengths of several paths in the network in the following way (Schweickert, 1983b).

Let N be the network in which all sequential processes are ordered, but x and y are so far unordered; see Figure 9.4.11. A network N_1 in which x precedes y is made by inserting a path P from the terminal vertex x'' of x to the starting vertex y' or y. Another new network N_2 in which y precedes x is made by inserting the path P from the terminal vertex y'' of y to the starting vertex x' of x. Let t_1 and t_2 be the response times for networks N_1 and N_2, respectively. Clearly, if $t_1 = t_2$, then it does not matter which process comes first, so suppose t_1 and t_2 are not equal. Then

$$t_1 < t_2 \text{ iff } \delta(ox') + \delta(y''r) < \delta(oy') + \delta(x''r)$$

In other words, if certain paths are short, then x should come before y, while if certain other paths are short, then y should come before x. The relevant paths are indicated in Figure (11).

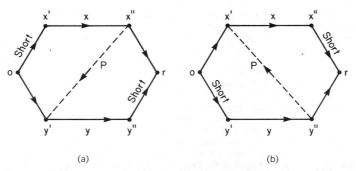

(a)   (b)

Fig. 9.4.11 Suppose all the processes have been ordered except for x and y, which are required to be sequential. If the paths indicated in (a) are short, then it is optimal for x to precede y. If the paths indicated in (b) are short, then y should precede x.

If the inserted path P has a different duration in N_1 and N_2, then x should precede y iff

$$k_1(\text{xy}) < k_2(\text{xy})$$

where $k_1(\text{xy})$ and $k_2(\text{xy})$ are the coupled slacks between x and y in N_1 and N_2, respectively; the proof is in Schweickert (1983b).

9.4.3.4 Speed and Accuracy in PERT Networks

So far our analysis of PERT networks has ignored the fact that subjects can decrease their reaction times at the cost of making more errors. Berman (1964) developed an approach for dealing with time–cost trade-offs in PERT networks that can be applied to cognitive tasks if the experimenter uses an appropriate payoff function.

The time required to complete all the processes in a PERT network is the sum of the durations of the processes on the path with longest duration, the critical path. Several paths from the source to the sink of the network may have the same duration and be longer than the other paths; if so, then all are critical paths. As before, we will assume that the probability that the response is correct is the product of the probabilities that each individual process is correct, so $\log p = \Sigma \log p_i$.

Suppose u is a vertex in a critical path network. Let E be the set of processes terminating at u, and let F be the set of processes starting at u. Then, applying Berman's (1964) result to our situation, to optimize the payoff according to Equation (5) the subject should allocate the execution times t_i so that

$$\sum_{x_i \, \epsilon \, E} \frac{d \log p_i}{dt_i} = \sum_{x_i \, \epsilon \, F} \frac{d \log p_i}{dt_i}$$

This can be thought of as a network flow equation. The sum of the rates of change of log percent correct for processes entering a vertex is equal to the corresponding sum for processes leaving the vertex. Clark (1961) and Elmaghraby (1968) give another way to conceive of the same result. If U and V are sets of vertices, the set of all arrows going from some $u \, \epsilon \, U$ to some $v \, \epsilon \, V$ will be denoted (U, V). A *cut* separating o and r is a minimal set of arcs (U, V) whose removal disconnects the source from the sink of the network. For example, in Figure 9.4.6, $\{x, a\}$ is a cut. To optimize the payoff, the sum of the derivatives of the cost versus time functions for the process in a cut separating o and r is the same for all such cuts. In our application, this means that when the subject is performing optimally, for any cut C separating o and r the sum $\Sigma_{x_i \, \epsilon C} \, d \log p_i / dt_i$ is constant, and equal to c_1/c_2.

This relationship does not ordinarily lead to the conclusion that factors affecting different processes in a critical path network will have additive effects on log percent correct. There is one important situation in which additivity will occur, however. Suppose only one path is critical, so that processes not on this path have ample time to be completed accurately. Then for a noncritical process, $\log p_i$ is close to asymptote, so $d \log p_i / dt_i = 0$. Moreover, every cut separating o and r will contain one of the critical processes, so for a critical process, $d \log p_i / dt_i = c_1/c_2$. But this is the relationship in Equation (9.4.6), so critical processes will behave like processes in series, and additivity for log percent correct should hold.

So far we have presented a way to synthesize a network representing the mental processes in a cognitive task. We now consider how to analyze the behavior of such a network once it has been derived.

9.4.4   CONCURRENT PROCESSING: STOCHASTIC PERT NETWORKS

Many of the systems that human factors engineers hope to model can be represented as complex stochastic networks. Unfortunately, there is space to discuss only a limited number of such networks. Consideration will be given to those four types of networks with the greatest potential for applications in human factors: PERT networks (this section), transient queueing networks (Section 9.4.5), OP networks (Section 9.4.6), and some simple steady-state queueing networks (Section 9.4.7). Until recently, most of the above networks could be used as models of person–machine systems only by individuals with some reasonably sophisticated quantitative skills. However, such is no longer the case. The problems that now stand in the way of a particular appliction are largely conceptual and experimental (i.e., the structure of the processes in the networks must still be determined), not quantitative.

9.4.4.1 Stochastic PERT Networks: Modeling

Deterministic PERT networks have been discussed at length in Section 9.4.3. The focus in this section will be on stochastic PERT networks. As in Section 9.4.3, the processes will be labeled $x_1, x_2, \ldots$, x_n. The structure of the processes in stochastic PERT networks remain the same from trial to trial.

However, unlike the durations of the processes in deterministic PERT networks, the durations of the processes in stochastic PERT networks will vary from one trial to the next. Thus, the duration of process x_i will now be identified as T_i (a random variable) rather than t_i (a constant).

It is assumed throughout the remainder of this section that the durations of the processes are pairwise independent, exponentially distributed random variables. A random variable T has an *exponential distribution* if its probability density function is

$$f(t) = e^{-\lambda t} \text{ for } t \geq 0$$
$$= 0 \text{ otherwise}$$

The mean of the random variable T is $1/\lambda$, and its variance is $1/\lambda^2$. There simply is not space here to consider other distributions, although the methods presented in this section can be extended to other distributions; see Fisher and Goldstein (1983a) and Goldstein and Fisher (1983).

Before concluding this section, it may be helpful to introduce a simple numerical example. This example can help make clear the differences between deterministic and stochastic PERT networks. Suppose that two processes x_1 and x_2 are executed in parallel, that both processes begin at the same time, and that both processes must complete before a task is finished. Furthermore, suppose process x_1 takes on average 800 msec to complete and process x_2 takes on average 600 msec to complete.

To begin, consider computation of the time it takes on average to complete both processes x_1 and x_2. If the network is deterministic, this means that process x_1 has duration $t_1 = 800$ msec and process x_2 has duration $t_2 = 600$ msec. By assumption, the time t that it takes to complete the task is equal to the maximum of the times t_1 and t_2, that is, $t = \max(t_1, t_2)$. Thus, in the example $t = 800$ msec since process x_1 always takes 200 msec longer to complete than process x_2.

Now assume that the network is stochastic. This means that the expected duration $E[T_1]$ of process x_1 equals 800 msec and the expected duration $E[T_2]$ or process x_2 equals 600 msec. However, on any given trial process x_2 could finish before process x_1 since the durations of the processes vary from trial to trial. For purposes of this example, assume that the durations of both processes x_1 and x_2 are independent, exponentially distributed random variables. Then the expected time $E[T]$ to complete the task is equal to the expectation of the maximum of the durations T_1 and T_2 of the processes x_1 and x_2, that is, $E[T] = E[\max(T_1, T_2)]$. In this case, it is easy to show that $E[T] = 1028.57$ msec. The expected value $E[\max(T_1, T_2)] = 1028.57$ msec of the maximum of T_1 and T_2 is larger than the maximum, $\max(E[T_1], E[T_2]) = 800$ msec, of the expected values of T_1 and T_2 just because of the variability in the process durations.

Differences exist on other measures of performance as well. Most notably, the variance of a deterministic network is identically zero, whereas the variance of a stochastic PERT network is always positive.

Computational formulas for the various performance measures of stochastic PERT networks (e.g., the expectation, variance, and distribution of the response time; the probability that a particular path is critical) are described in the sections which follow immediately below.

Order-of-Processing Diagrams

In order to obtain performance measures, it is useful first to translate a PERT network into an order-of-processing (OP) diagram (Fisher and Goldstein, 1983a). To begin, note that at any particular moment in time during the execution of the processes in a PERT network it is possible to partition the processes into three sets: the set A of processes that have not yet begun, the set B of processes that have completed, and the set C of processes that are currently executing. The contents of sets A, B, and C will change over time. It will be useful to assign a single, identifying index to the sets in each unique partition (e.g., A_i, B_i, C_i). During the execution of processes in the network, if the processes are partitioned into sets A_i, B_i, and C_i, the network (or system) will be said to be in *state* s_i.

While computer programs exist that can be used to generate the sets A_i, B_i, and C_i associated with each state s_i [see Fisher, Saisi, and Goldstein (1985)], in many cases inspection of the PERT network is sufficient. For example, consider the network displayed in Figure 9.4.12a. Inspection of the network indicates that there are a total of five different states defined by the following unique partitions (the contents of set A_i are not listed since they can be determined from the contents of sets B_i and C_i):

$s_1 \ (B_1 = \emptyset, \ C_1 = \{x_1, x_2\})$
$s_2 \ (B_2 = \{x_1\}, \ C_2 = \{x_2\})$
$s_3 \ (B_3 = \{x_2\}, \ C_3 = \{x_1\})$
$s_4 \ (B_4 = \{x_1, \ x_2\}, \ C_4 = \{x_3\})$
$s_5 \ (B_5 = \{x_1, x_2, x_3\}, \ C_5 = \emptyset)$

This state information is more conveniently represented in the OP diagram in Figure 9.4.12b. The processes in the current set C_i are listed in the upper half of state s_i; the processes in the completed set B_i are listed in the bottom half of state s_i.

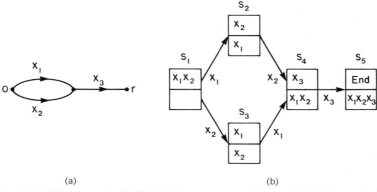

(a)   (b)

Fig. 9.4.12 (a) A PERT network. (b) The OP diagram generated from the PERT network. Processes in the current set of state s_i are listed in the upper half of s_i, processes in the completed set are listed in the bottom half.

A transition can be made in one step from state s_i to state s_j if the completion of exactly one process in the current set of state s_i leads immediately to the partition of processes into the sets that define state s_j. For example, a transition can be made from state s_1 to state s_2 since inspection of the PERT network in Figure 9.4.12a indicates that process x_1 can complete before process x_2 and that no new processes can begin executing until both processes x_1 and x_2 complete. Similarly, a transition can be made from state s_1 to state s_3 since process x_2 can complete before process x_1. Transitions between states are also represented in the OP diagram. For example, the above two transitions are represented in the OP diagram by single directed arcs leading, respectively, from state s_1 to state s_2 and from state s_1 to state s_3. To make the OP diagram as clear as possible, each arc between a pair of states is labeled with the process that completes when a transition is made from one state to the next. Note that since it is assumed that the durations of the processes are mutually independent, transitions where more than one process in the current set completes need not be considered. Any such transitions would have to occur with probability zero.

Since the PERT network is stochastic, the complete sequence of states that are entered in a given trial will vary. For example, note that there are exactly two sequences of states generated by the PERT network in Figure 9.4.12b. If the duration of process x_1 is shorter than the duration of process x_2, then the sequence of states entered will be (s_1, s_2, s_4, s_5): processes x_1 and x_2 begin executing (state s_1); process x_1 then completes before process x_2, since by assumption the duration of process x_1 is shorter than the duration of process x_2 (state s_2); process x_2 completes and process x_3 begins executing (state x_4); finally, process x_3 completes (state x_5). In contrast, if the duration of process x_1 is longer than the duration of process x_2, then the sequence of states entered will be (s_1, s_3, s_4, s_5).

The sequence of states are also represented in an OP diagram. Define state s_j as an *immediate successor* of state s_i if an arc in the OP diagram goes from s_i to s_j. In this case we also say that s_i is an *immediate predecessor* of s_j. Define the start state as the state with no predecessors and label this state s_1. Define the terminating state as the state with no successors and label this state s_{n_0}, where n_0 is the number of states in the OP diagram. Define a path o_i in an OP diagram as a sequence $(s_{i_1}, s_{i_2}, \ldots, s_{i_k})$ of states such that s_{i_1} is the start state (i.e., $s_{i_1} = s_1$), such that state s_{i_j} is an immediate predecessor of state $s_{i_{j+1}}$ and such that state s_{i_k} is the terminating state (i.e., $s_{i_k} = s_{n_0}$). A path so defined represents what has been referred to above as a complete sequence of states. For example, in the OP diagram of Figure 9.4.12b there are two paths represented by the two sequences of states $o_1 = (s_1, s_2, s_4, s_5)$ and $o_2 = (s_1, s_3, s_4, s_5)$. Note that the labeling of paths is arbitrary.

Finally, let α be a sequence of indices of states along a path o_i in an OP diagram. For example, $\alpha = (1, 2, 4, 5)$ along path o_1. Furthermore, state $s_{\alpha_1} = s_1$, $s_{\alpha_2} = s_2$, $s_{\alpha_3} = s_4$, and $s_{\alpha_4} = s_5$. Similar remarks apply to path o_2. It will be convenient to refer to α as a path, even though it is a sequence of indices of states (and not a sequence of states proper).

The Expectation

Once the OP diagram has been constructed, it is a simple matter to compute the expected response time $E[T]$. Define λ_i as the rate parameter of the exponentially distributed duration T_i of process x_i, that is, $P(T_i > t_i) = e^{-\lambda_i t_i}$. (Recall that the reciprocal of the rate parameter is equal to the expected duration of the process. For example, if a comparison process takes 60 msec on average, then the rate parameter is equal to $1/60$.) Set c_i equal to the sum of the rate parameters of the processes current in the upper half of state s_i. For example, in the OP diagram of Fig. 9.4.12b, two processes x_1 and x_2 are in the current set of state s_1. Thus, $c_1 = \lambda_1 + \lambda_2$.

Define the transition probability p_{ij} between each pair of states s_i and s_j as follows. Set p_{ij} equal to zero if state s_j is not an immediate successor of state s_i. Otherwise, set p_{ij} equal to the quantity λ_k/c_i, where λ_k is the rate parameter of the process that gets completed when a transition is made from state s_i to state s_j. For example, in the OP diagram of Figure 9.4.12b, $p_{12} = \lambda_1/c_1 = \lambda_1/(\lambda_1 + \lambda_2)$ since the completion of process x_1 in state s_1 leads to the transition from state s_1 to state s_2. Define a square n_o by n_o transition matrix P, which consists of components p_{ij}, for $i,j = 1, \ldots, n_o$.

The expected response time $E[E]$ can now be computed from the above quantities. Let b_i be the number of processes in the completed set of state s_i. Let e_i be a 1 by n_o column vector with a 1 in the ith row and 0's everywhere else. Then,

Theorem 1

$$E[T] = \sum_{i=1}^{n_0-1} c_i^{-1} e_1' P^{b_i} e_i$$

In words, the expected response time $E[T]$ is equal to the sum of products of the form $c_i^{-1} e_1' P^{b_i} e_i$, where c_i^{-1} is the conditional expected duration of state s_i, given that state s_i is reached, where $e_1' P^{b_i} e_i$ is the probability that state s_i is reached. The proof of this theorem is presented in detail in Fisher and Goldstein (1983a) and will not be reproduced here. Note that the function of the vectors e_1' (a row vector) and e_i (a column vector) is simply to select out the entry in row 1 and column i from the matrix P raised to the power b_i.

An example can make the application of Theorem 1 clear. Consider again the OP diagram in Figure 9.4.12b. Suppose that $\lambda_1 = \lambda_2 = 1$ and $\lambda_3 = 4$. Then,

$$c_1 = \lambda_1 + \lambda_2 = 2 \qquad c_2 = \lambda_2 = 1$$
$$c_3 = \lambda_1 = 1 \qquad c_4 = \lambda_3 = 4$$

Note that since $n_o = 5$ and since the sum in Theorem 1 is taken from 1 to $n_o - 1 = 4$, the quantity c_5 is not computed.

There are a total of five positive transition probabilities in the current example:

$$p_{12} = \lambda_1/c_1 = \tfrac{1}{2} \qquad p_{13} = \lambda_2/c_1 = \tfrac{1}{2}$$
$$p_{24} = \lambda_2/c_2 = 1 \qquad p_{34} = \lambda_1/c_3 = 1$$
$$p_{45} = \lambda_3/c_4 = 1$$

All other transition probabilities are set equal to zero. Thus, the transition probability matrix p can be written as:

$$P = \begin{matrix} 0 & \tfrac{1}{2} & \tfrac{1}{2} & 0 & 0 \\ 0 & 0 & 0 & 1 & 0 \\ 0 & 0 & 0 & 1 & 0 \\ 0 & 0 & 0 & 0 & 1 \\ 0 & 0 & 0 & 0 & 0 \end{matrix}$$

The index i increases down the rows; the index j increases across the columns.

Returning to Theorem 1, the expression for the expected response time can now be evaluated:

$$E[T] = \sum_{i=1}^{n_0-1} c_1^{-1} e_1' P^{b_i} e_i$$
$$= c_1^{-1} e_1' P^{b_1} e_1 + c_2^{-1} e_1' P^{b_2} e_2$$
$$\quad + c_3^{-1} e_1' P^{b_3} e_3 + c_4^{-1} e_1' P^{b_4} e_4$$
$$= (\tfrac{1}{2})(e_1' P^0 e_1) + (1)(e_1' P^1 e_2)$$
$$\quad + (1)(e_1' P^1 e_3) + (\tfrac{1}{4})(e_1' P^2 e_4)$$
$$= (\tfrac{1}{2})(1) + (1)(\tfrac{1}{2}) + (1)(\tfrac{1}{2}) + (\tfrac{1}{4})(1) = 1.75$$

Note that the matrix P raised to the zero power (i.e., P^0) is set equal to the identity matrix I, and thus the entry in the first row and first column of P^0 (i.e., $e_1' P^0 e_1$) is set equal to 1. Other performance measures such as the variance and distribution of the response time T (Fisher and Goldstein, 1983a) and the probability that a particular path is critical (Fisher, Saisi, and Goldstein, 1985) can be computed from the OP diagram. Unfortunately, there is not space enough to do so here.

The Path Duration

It will be useful to have a closed-form expression for the distribution of the compound event that path α is taken and that each state duration is less than or equal to t_i. Let S_{α_i} be the duration of state s_{α_i}. Let $S_\alpha = (S_{\alpha_1}, S_{\alpha_2}, \ldots, S_{\alpha_n})$ be a random vector. Let P_α be an indicator random variable, which is set equal to 1 if path α is followed and is set equal to 0 otherwise. Let $\lambda_{\alpha j}$ be the rate parameter of the process $x_{\alpha j}$ which completes in state s_{α_j}. Then the methods of Fisher and Goldstein (1983a) can be used to derive the density function of the above compound event:

$$f_{S_\alpha, P_\alpha = 1}(t_1, \ldots, t_n) = \prod_{j=1}^{n} \lambda_{\alpha j} \exp[-c_{\alpha j} t_j] \tag{12}$$

9.4.4.2 Stochastic PERT Networks: Applications

The preceding quantitative techniques have only very recently been developed, and so far the primary applications have occurred in the area of visual search (Fisher, 1982, 1984; Fisher and Papazian, 1983). Two such applications are described below. In the first application it is shown how stochastic PERT networks can be used to model the early stages of processing in a visual search task. In the second application it is shown how the model of visual search can be used to optimize the number of words to appear in highlighting on a CRT (video monitor).

Visual Search: Channel Limits

Much of the recent work in visual search has focused on the capacity limitations in the early stages of processing (Egeth, Jonides, and Wall, 1972; Jonides and Gleitman, 1972; Schneider and Shiffrin, 1977; Triesman and Gelade, 1980). Briefly, most theories of visual search assume that in some situations processing is capacity dependent, while in other situations processing is capacity independent. In capacity-dependent situations, subjects appear to search the display one stimulus at a time for the target. In capacity-independent situations, subjects appear to search the display in parallel. In a completely parallel search, the target can be compared simultaneously with each of the stimuli. Each comparison of target and stimulus can be thought of as occurring on a separate visual channel. Thus, in a completely parallel search there are as many channels as stimuli in the visual display. Recently, Fisher (1982, 1984) has argued that there is a limit on the number of channels (i.e., the number of comparison operations that can be performed at any one time). In particular, in tasks which had previously been thought to be capacity independent, he found that on average subjects could make only four comparisons of a target in memory with stimuli in the visual display.

Stochastic PERT networks can be used to model visual search behavior if the number of stimuli in the display is less than or equal to the maximum number of comparisons that can be executed simultaneously (Fisher, 1982). [More general OP networks are needed when the number of stimuli is greater than the maximum number of simultaneous comparisons (see Section 9.4.6).] For example, assume that there are only two stimuli in the display, that encoding must occur in series, that the comparisons of the target in memory with the stimulus in the visual display can occur in parallel, and that the maximum number of simultaneous comparisons is equal to four. Assume that the target is present on some trials, absent on others. Subjects are asked to indicate as quickly as possible whether the target is present or absent.

A PERT network that represents the preceding constraints in target-absent trials is displayed in Figure 9.4.13a. The encoding processes are represented as "e_1" and "e_2," the comparison processes as "c_1" and "c_2," and the response process as "r_1." It is then a simple matter to construct the OP diagram (Fig. 9.4.13b) from the PERT network using the methods of Fisher, Saisi, and Goldstein (1985).

If the rate parameters of the encoding, comparison, and response processes are known, then the expected response time $E[T_{a_i}]$ to search a display with i stimuli when the target is absent can be computed directly from Theorem 1 and the OP diagram. For example, if there are two stimuli in the display, then Theorem 1 is applied directly to the OP diagram in Figure 9.4.13b. Consider the case where the search is *self-terminating*, that is, if the target is present, the subject stops as soon as the target is found. Then, the expected response time $E[T_{p_j}]$ to search a display with j stimuli when the target is present is simply a probability mixture of the target-absent expected response times $E[T_{a_i}]$. In particular, suppose that the probability that the target is the ith stimulus to be encoded does not vary with i. Then, it is easy to show that for $i \leq j$,

$$E[T_{p_j}] = \sum_{i=1}^{j} (1/j) E[T_{a_i}]$$

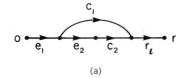

(a)

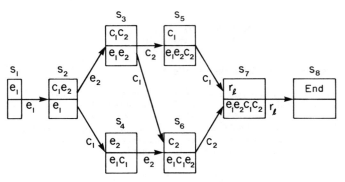

Fig. 9.4.13 (a) A PERT network model of visual search; (b) the OP diagram.

Note that it is assumed that the response time does not vary with the type of response (present or absent). This assumption is not always satisfied, in which case slight modifications to the model would be required.

If the rate parameters are not known, then values for these parameters must be estimated from experimental data. The exact method used to estimate these values will depend greatly on the particular circumstances of the experiment. For purposes of simplicity, suppose that reaction times were collected for display sizes of one, two, three, and four stimuli and for target-present and target-absent trials. This yields a total of eight different conditions. Assume that at least four comparisons can be carried out simultaneously. Assume that the encoding (λ_e) and comparison (λ_c) rates do not vary with the stimulus or the position of the stimulus in the input stream. There is only one response rate parameter (λ_{r_1}). Let t_{a_i} and t_{p_i} be the observed values of the response times for the target-absent and target-present conditions, when the display contains i elements. Then it makes sense to select values of λ_e, and λ_c, and λ_{r_1} that minimize the sum of the squared deviations of the predicted values from the observed values:

$$\sum_{i=1}^{4} \{(E[T_{a_i}] - t_{a_i})^2 + (E[T_{p_i}] - t_{p_i})^2\}$$

One can then compare this sum to the variance of the observations to obtain a measure of the fit of the model to the data (Fisher, 1982).

Visual Search: Highlighting on CRTs

The implications for practice of the model described in the above section are extensive, since visual search is itself a major component of most interactions between individuals and machines. As an example of how this model can be used, consider an area of research of interest to computer manufacturers. Video screens are becoming increasingly cluttered. Words (or whatever) can be presented in a number of different types of highlighting (e.g., boldface, reverse video, blinking). The original, primary purpose of highlighting was to reduce the average search time for target information. This purpose may have been compromised by the excessive use of highlighting. Software companies need to know at what point the use of highlighting actually slows (rather than speeds) the search for target information.

A model of the visual search process such as that discussed above can help determine the optimal number of words to appear in highlighting for a given menu, help screen, or other screen type. The procedure is as follows. Define H as the set of words in highlighting for a particular menu m. Define M as the set of words in menu m. Define S as the set of words searched for when menu m is examined. Thus, set S is the set of potential targets for menu m. Assume for the sake of simplicity that only one word in set S is searched for when menu m is examined. Define $P(w_i)$ as the probability that word w_i is the target when the menu is m. It is assumed that this quantity is known. Define

$E[T_i|w_i]$ as the conditional expectation of the search time T, given word w_i is the target. An expression can now be obtained for the expected search time $E[T]$:

$$E[T] = \sum_i E[T|w_i] P(w_i), \qquad w_i \in S$$

Since $P(w_i)$ is given ($w_i \in S$), the only quantity that needs to be computed is the conditional expected search time $E[T|w_i]$.

This conditional expected search time $E[T|w_i]$ will depend on the subset of words that is highlighted, the set M of words in the menu, the set S of targets, and the model of visual search. For example, suppose that word w_1 is highlighted [i.e., $H = \{w_1\}$], words w_1 through w_4 appear in menu m ($M = \{w_1, w_2, w_3, w_4\}$), and words w_1, w_2, and w_9 are potential targets ($S = \{w_1, w_2, w_9\}$). Furthermore, suppose that subjects search the highlighted words first and suppose that the search is a serial self-terminating one where it takes e units of time to encode and compare each word in the menu with the target and where it takes r units of time to respond. Finally, suppose $P(w_1) = 0.45$, $P(w_2) = 0.05$, and $P(w_9) = 0.50$.

Consider now computation of the expected search time $E[T]$ for the above example. To begin, it seems reasonable to assume that subjects search the highlighted portion of the screen before searching the unhighlighted portions. Therefore, if the subject is searching for word w_1 and word w_1 is highlighted, the conditional expected search time $E[T|w_1] = e + r$ since the subject finds the target (w_1) immediately. If the subject is searching for word w_2, then the target (w_2) is encountered as the second, third, or fourth word in the menu. Since the search is a serial self-terminating one, this means that $E[T|w_2] = 3e + r$. Finally, if the subject is searching for word w_9, then the conditional expected search time $E[T|w_9] = 4e + r$ since the target word w_9 does not appear in the menu set M. Substituting for the quantities $E[T|w_i]$ and $P(w_i)$ in the expression for the expected search time $E[T]$ yields:

$$E[T] = (e + r)(0.45) + (3e + r)(0.05) + (4e + r)(0.50)$$

In order to find the optimum number of words in highlighting, it would now be necessary to compute the expected search time $E[T]$ for the 16 potential sets of H of words in highlighting ($H = \{w_1\}$, $H = \{w_2\}$, $H = \{w_3\}$, $H = \{w_4\}$, $H = \{w_1, w_2\}$, $H = \{w_1, w_3\}$, etc.). This is easy enough to do with a computer.

If subjects's search of a menu were always a serial self-terminating one, there would be no need for the more complex stochastic networks introduced in the previous sections. However, there is good reason to believe that the model of search is considerably more complex. In particular, evidence indicates that subjects can analyze two words in parallel in each eye fixation and that subjects do not make a new fixation until all the information in the current fixation has been analyzed. If this is the case, then the model of processing is simply a series of repetitions of the PERT network represented in Figure 9.4.13a, where "r_1" now represents the movement of the eyes from one fixation to the next. The duration of this movement will depend on the distance between fixations. The conditional expected search time $E[T|w_i]$ must be recomputed using the formulas introduced earlier. However, most important from the current standpoint is the fact that the optimal number of words in highlighting will vary with the model of search. Thus, without the ability to test more complex models than simple serial and parallel ones, it is difficult if not impossible to find the optimum.

9.4.5. CONCURRENT PROCESSING: TRANSIENT QUEUEING NETWORKS

Cognitive systems and person–machine systems that cannot be modeled as stochastic PERT networks can frequently be modeled as queueing networks. Most readers are presumably familiar with queueing networks and their use in industrial settings. However, some readers may not have seen queueing networks used in a psychological setting.

Briefly, queueing networks typically consist of four elements: customers, nodes, servers, and queues. In a psychological context, these elements can be interpreted as follows. Customers are individual stimuli (e.g., a letter in the display of a visual search task) that require processing. Nodes are stages in the cognitive processing of a stimulus (e.g., the encoding and comparison stages). A server is the cognitive equivalent of machinery that actually does the transformation on a stimulus at a particular stage or node (e.g., a server at the encoding stage is the cognitive equivalent of a "machine" which transforms the stimulus from a raw, unprocessed state to an encoded state). The number of servers present at a given node determines the maximum number of stimuli that can be executed in parallel at a given stage. Finally, a queue is equivalent to a memory (e.g., short-term memory). Just as different organizations of memory are possible, so too are different organizations of the queue.

The discussions in this section will be confined to finite source, acyclic Jackson networks. The mathematics of such networks are discussed first (Section 9.4.5.1). This is followed by a discussion of the use of these networks to model short-term memory and traffic sign behavior (Section 9.4.5.2).

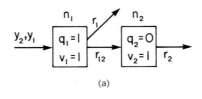

(a)

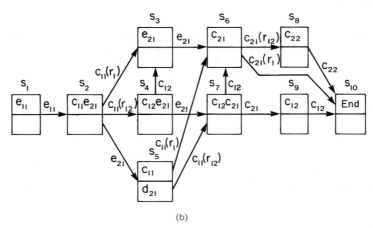

(b)

Fig. 9.4.14 (*a*) A finite source, acyclic Jackson network; (*b*) the corresponding OP diagram.

9.4.5.1 Finite Source, Jackson Networks: Modeling

To begin, consider the example of a finite source, acyclic Jackson network presented in Figure 9.4.14*a*. There are two nodes in the network (labeled n_1 and n_2). Unlike the previous networks, the arcs represent the route that stimuli take, not the processes in the network. The actual processing goes on at the nodes. Two stimuli (y_1, y_2) are waiting to enter the system at node n_1. One server is available for processing at node n_1 ($v_1 = 1$) and one server at node n_2 ($v_2 = 1$). The number of stimuli that can wait in the queue is equal to one at node n_1 ($q_1 = 1$) and is equal to zero at node n_2 ($q_2 = 0$). If a stimulus completes service at node n_1, it is routed with probability r_1 out of the system at node n_1 and is routed with probability r_{12} to node n_2. If a stimulus completes service at node n_2, it is routed with probability r_2 out of the system at node n_2 ($r_1 = 1 - r_{12}; r_2 = 1$). Finally, a stimulus is lost from the system with probability 1 if the stimulus arrives at a node and there is no room in the queue or on the servers. Since this can occur at every node if the appropriate conditions are met, there is no need to represent this loss with an additional arc. In the current context the response time T is defined as the time that elapses between the onset of the arrival of the first stimulus (y_1) from outside of the system and the exit of all stimuli from the system. It should be reemphasized at this point that the arcs do not represent processes: y_1 and y_2 are stimuli; r_1, r_{12}, and r_2 are probabilities.

Finite source, acyclic Jackson queueing networks are defined by the following assumptions: (1) a customer never visits the same node twice; (2) each node n_g in the queueing network has v_g servers where the service times at each server are mutually independent, exponentially distributed random variables each with rate parameter μ_g; (3) the interarrival times of stimuli entering from outside the system at node n_g are exponentially distributed each with rate parameter v_g; (4) the number of stimuli m_g that can enter node n_g from outside the system is finite; (5) the queue size at each node n_g is finite; and (6) once a stimulus receives service at node n_g, the stimulus is sent immediately to node n_h ($h = 1, 2, \ldots, n_q$) with probability r_{gh}, or leaves the network with probability r_g.

As with stochastic PERT networks, it will be necessary to translate the Jackson network into an OP diagram in order to compute the performance characteristics of the network. The actual computations of the performance characteristics (e.g., the expected response time) then follows in a straightforward fashion.

The OP Diagram

In order to construct the OP diagram, it will be necessary to introduce some additional notation. At any moment in time, the stimuli can be partitioned into five sets: the set A of stimuli that have not yet entered the system, the set B of stimuli that have exited from the system, the set C of stimuli that are currently being processed, the set D of stimuli that are delayed and waiting in queue for service, and the set E of stimuli that are currently in the process of entering (arriving) from outside the system. The contents of the sets A, B, C, D, and E will change over time. A single, identifying index will be assigned to each unique partition of the sets C, D, and E. During the processing of the stimuli in the network, the network (or system) will be said to be in state s_i if the stimuli are partitioned into sets C_i, D_i, E_i since only the members of the sets C_i, D_i, and E_i will be of interest in the current context.

Label the stimuli $y_1, y_2, \ldots, y_{n_q}$. The members of set C_i will be designated c_{kg}, where the presence of c_{kg} in set C_i indicates that stimulus y_k is being processed on node n_g. Likewise, the presence of d_{kg} in D_i indicates that stimulus y_k is in the queue at n_g, and the presence of e_{kg} in E_i indicates that stimulus y_k is arriving from outside the system at node n_g. A transition can be made in one step from state s_i to s_j if the completion of exactly one process in state s_i leads to the partition of stimuli represented by state s_j.

An OP diagram can be used to represent the different states and paths (sequences of states) associated with a finite source, acyclic Jackson network (Fisher and Goldstein, 1983c). Stimuli that are either arriving or being processed are placed in the upper half of the state. Again, the process that completes when a transition is made between a pair of states is used to label the arc connecting the two states. In addition, the probabilities r_g and r_{gh} are identified with the appropriate arcs. For example, consider again the Jackson network in Figure 9.4.14a. The associated OP diagram is presented in Figure 9.4.14b. In state s_1 of the OP diagram stimulus y_1 is arriving at node n_1 ($C_1 = \emptyset$, $D_1 = \emptyset$, $E_1 = \{e_{11}\}$). In state s_2, stimulus y_1 is currently executing at node n_1 and stimulus y_2 is arriving at node n_1 ($C_2 = \{c_{11}\}$, $D_2 = \emptyset$, $E_2 = \{e_{21}\}$). If stimulus y_1 is processed before stimulus y_2 arrives (i.e., c_{11} completes before e_{21}), then a transition is made to state s_3 with probability r_1 where stimulus y_1 exits from the system. A transition is made to state s_4 with probability r_{12} where stimulus y_1 begins processing at node n_2. However, if stimulus y_2 arrives before stimulus y_1 is processed (i.e., e_{21} completes before c_{11}), then a transition is made to state s_5 where stimulus y_2 is placed in the queue at node n_1 (since there is only one server at node n_1 and this server is occupied).

Assume that a stimulus is moved along to a new node (say n_g) only if there is room in queue q_g. Otherwise, the stimulus is dropped from the system. For example, note that in state s_7 stimulus y_1 is on the single server at node n_2 and stimulus y_2 is on the single server at node n_1 (i.e., $C_7 = \{c_{12}, c_{21}\}$, $D_7 = \emptyset$, $E_7 = \emptyset$). If the next event to occur is the completion of the processing of stimulus y_2 at node n_1, then stimulus y_2 is lost from the system even if it is routed from node n_1 to node n_2 (since there is no room at node n_2) and the new state (state s_9) to which a transition is made indicates this (i.e., $C_9 = \{c_{12}\}$, $D_9 = \emptyset$, $E_9 = \emptyset$).

Once the OP diagram has been generated, the computation of performance measures can proceed smoothly. Closed-form expressions for two performance measures will be presented: the expected response time and the probability $P(o_i)$ that path o_i is taken.

The Expectation

Before computing the expected response time $E[T]$, it will be necessary to define an n_o by n_o matrix Q similar in all but one respect to the matrix P previously defined. In particular, set q_{ij} equal to 0 if state s_j is not an immediate successor of state s_i. For example, in Figure 9.4.14b $q_{14} = 0$. Otherwise, take one of three actions. If when making a transition from state s_i to state s_j a stimulus y_k does not move either from one node to the next or exit from the system, let $q_{ij} = p_{ij}$. For example, $q_{25} = p_{25} = \mu_1/(\mu_1 + \nu_1)$. If a stimulus y_k does move from node n_g to node n_h when a transition is made from state s_i to state s_j, then let $q_{ij} = p_{ij} r_{gh}$. For example, $q_{24} = p_{24} r_{12} = [\mu_1/(\mu_1 + \nu_1)] r_{12}$. Finally, if a stimulus y_k exits from node n_g when a transition is made from state s_i to state s_j, then let $q_{ij} = p_{ij} r_g$. For example, $q_{23} = p_{23} r_1 = [\mu_1/(\mu_1 + \nu_1)] r_1$.

Computation of the expected response time $E[T]$ can now proceed in a straightforward fashion (Fisher and Goldstein, 1983c):

Theorem 2

$$E[T] = \sum_{i=1}^{n_o-1} \left\{ c_i^{-1} \left[\sum_{k=0}^{n_o-1} e_1' Q^k e_i \right] \right\}$$

The work required to compute the expected response time from Theorem 2 is almost identical to the work required to compute the expected response time for Theorem 1 and thus, given the limited space, will not be repeated here.

The Quantity $P(o_i)$

It will be useful to know how to compute the probability $P(o_i)$ that path o_i is taken. The probability that path o_i is taken is simply the product of the transition probabilities between states along path

o_i. Recall that α represents the sequence of indices of states on path o_i. Set n_α equal to the number of states along path α, then

Theorem 3

$$P(o_i) = \sum_{j=1}^{n_\alpha - 1} q_{\alpha_j \alpha_{j+1}}$$

Other performance measures such as the variance and distribution of the response time can be derived, but will not be presented here [see Fisher and Goldstein (1983c)].

9.4.5.2   Finite Source, Jackson Networks: Applications

As was true of the quantitative techniques described in Section 9.4.4, the quantitative techniques described in this section have only very recently been developed. Two applications are discussed subsequently that should help to illustrate the potential of these techniques. In the first application, it is shown how a finite source, acyclic Jackson network can be used to model behavior in free-recall tasks. In the second application, it is shown how the above model can be used to optimize the presentation rate of information in the new flat-panel, dot-matrix highway traffic signs.

Short-Term Memory: Free Recall

The models of short-term memory discussed by Waugh and Norman (1965) and Atkinson and Shiffrin (1968) are familiar to most engineering psychologists. These models can be used to explain behavior in a number of different tasks. The task of most interest in the current context is a standard free-recall task. In this task subjects are presented with a list of w words and are asked after the presentation of the last word to repeat the words in the list (Murdock, 1961). In almost all situations where such a task is given to subjects, it is found that accuracy is best for words at the beginning and end of the list and is worst for words in the middle of the list. A graph of accuracy versus serial position of the word is U-shaped and frequently referred to as a serial position curve.

   For purposes of discussion, the Jackson network in Figure 9.4.14a can be used to illustrate the two most important characteristics (decay and bump out) of models of short-term memory. Words arrive one at a time and are encoded serially (say at node n_1 in Figure 9.4.14a). The arrival of word i at node n_1 is represented as e_{i1} in the OP diagram; the encoding of word i at node n_1 is represented as c_{i1}. It is assumed that some constant proportion of the words will *decay* because they are not placed in short-term memory before their trace in iconic memory (Sperling, 1960) fades beyond all recognition. This is represented by the exit of a word out of the system from node n_1 with probability r_1. Finally, those words that make it through the encoding stage are then passed on to the next stage (node n_2). The placing of word i at node n_2 into short-term memory is noted as c_{i2}. If the queue at node n_2 is full, the word being routed from the encoding stage is lost from the system (i.e., is *bumped out*).

   Assume as in the model in Figure 9.4.14a that only two words (i.e., two stimuli) are presented. Using Theorem 3, the probability that zero, one, or two words are remembered can then be computed. In particular, the probability $p(i|2)$ that i of the two words are remembered, $i = 1,2$, is the sum of the probabilities of all paths in which i of the stimuli (y_1 or y_2) are codes at node n_2. For example, there are exactly two paths through the OP diagram in which no words are coded: $o_1 = (s_1, s_2, s_3, s_6, s_{10})$ and $o_2 = (s_1, s_2, s_5, s_6, s_{10})$. Thus, $p(0|2) = P(o_1) + P(o_2)$ where the probability $P(o_i)$ is computed from theorem 3. These probabilities can then be used to generate a theoretical serial position curve. Finally, the overall percent correct recall $p(2)$ for a message of two words can easily be computed:

$$p(2) = 0 \times p(0|2) + 1 \times p(1|2) + 2 \times p(2|2)$$

Short-Term Memory: Traffic Signs

Psychologists have long known that if eye movements are made unnecessary, then subjects can read and understand a single sentence much faster than is normally done (Forster, 1970; Potter, 1984). In order to obviate eye movements, words are presented successively in the same spatial location (frequently under the control of a computer). This technique is known as rapid serial visual presentation (RSVP).

   In principle, there is no reason that RSVP could not be used to present information in highway traffic signs. The technology is currently available (flat-panel, dot-matrix displays). However, there is an important difference between the driver's situation on the road and a subject's situation in the laboratory. The driver on the road may glance at a sign at any point in the presentation of the message. The subject in laboratory is always presented with the first word in the message or sentence. Even so, the results from recent experiments indicate that more is remembered in a given unit of time from a sign presented in an RSVP format than from a sign presented in standard format (Fisher, Tengs, and Colen, 1984).

Ideally, one would like to optimize the presentation of highway traffic sign information. In the current context, the cost of too rapid a presentation is a decrease in recall, whereas the cost of too slow a presentation is an increase in the time a driver spends paying attention to the sign (and not to the road). Suppose a message is w words in length. Let d equal the gaze duration (the time it takes to present the message) and let $p(w,d)$ equal the percent recall. Then, using a very simple approach, the benefits or utility $u(w,d)$ of a message of w words presented for a duration of d units can be written as the sum $u(w,d) = w_1 p(w,d) - w_2 d$ where $w_1, w_2 > 0$ are weights to be determined by traffic engineers most familiar with the tradeoffs between accuracy and gaze duration.

The model of processing discussed in the previous section must be modified slightly in order to reflect the special characteristics of the RSVP situation. First, in the RSVP situation words can be presented so fast that they effectively mask one another, making encoding impossible. Thus, an additional exit arc should be attached to node n_1. The probability $r_3(d)$ of exiting from this arc should decrease as d increases. Second, the time to place a word in short-term memory should be made sensitive to which word in a message a subject sees first. In particular, the rate parameter of the placement process (c_{i2}) should decrease as the position of the first presented word in a sentence increases (Fisher, Tengs, and Colen, 1984).

In order to optimize $u(w,d)$ it is necessary to make two additional assumptions. First, assume that a driver is just as likely to glance up at a sign during the presentation of the first word in the sign as the driver is to glance up at a sign during the presentation of any other word in the sign. Second, assume that the driver sees exactly one complete presentation of the message in the sign (this second assumption can easily be changed). Theorem 3 can be then used to compute the probability $p(w,d)$ of recalling a message of w words presented for a duration of d units (see discussion at end of the section on short-term memory and free recall). The value of d which maximizes $u(w,d)$ is then the value which should be used in highway applications.

A numerical example may help make the above discussion more clear. Assume that a message of two words is to be presented ($w = 2$). Assume that prior work establishes the following values for the probabilities r_3, r_2, and r_{12} as a function of the gaze duration d: $r_3 = 1/d$ (d in milliseconds, $d \geq 1$), $r_1 = 0.1(1 - 1/d)$, $r_{12} = 0.9(1 - 1/d)$. Assume that $v_{11} = v_{21} = v = w/d = 2/d$ (recall that v_{11} and v_{21} are, respectively, the rate parameters of the arrival processes e_{11} and e_{21}), that $\mu_{11} = \mu_{21} = \mu_1 = 1/50$ (μ_{11} and μ_{21} are, respectively, the rate parameters of the encoding processes c_{11} and c_{21}), and that $\mu_{12} = \mu_{22} = \mu_2 = 1/100$ (μ_{12} and μ_{22} are, respectively, the rate parameters of the placement processes c_{12} and c_{22}).

Percent recall $p(2,d)$ can now be computed for a given gaze duration d and a sign of two words:

$$p(2,d) = 0 \times p(0|2,d) + 1 \times p(1|2,d) + 2 \times p(2|2,d)$$

For example, consider computation of the conditional probability $p(2|2,d)$ that two words are recalled, given two words are presented for a total of d units of time. Note that there are three paths through the OP network (Figure 9.4.14b) in which both words are recalled:

$$o_3 = (s_1, s_2, s_4, s_3, s_6, s_8, s_{10})$$
$$o_4 = (s_1, s_2, s_4, s_7, s_6, s_8, s_{10})$$
$$o_5 = (s_1, s_2, s_5, s_7, s_6, s_8, s_{10}) \cdot \cdot \cdot$$

From Theorem 3 the probability $p(o_3)$ that path o_3 is taken is equal to the product:

$$p(o_3) = q_{12} q_{24} q_{43} q_{36} q_{68} q_{8,10}$$
$$= [1][r_{12} \mu_1 / (\mu_1 + v)][1][1][1]$$

Substituting for r_{12}, μ_1, and v in the preceding expression, one obtains

$$p(o_3) = [0.9(1 - 1/d)][1/50][(1/50) + (2/d)]$$

The probabilities $p(o_4)$ and $p(o_5)$ that paths o_4 and o_5 are taken can be computed similarly. One can then obtain an expression for the conditional probability $p(2|2,d)$ that two words are recalled, given that two words are presented for a total of d units of time. In much the same fashion, one can obtain expressions for the conditional probabilities $p(0|2,d)$ and $p(1|2,d)$, from which percent recall $p(2,d)$ can be computed.

Finally, in order to obtain the optimal gaze duration d it is necessary to select values for the weights w_1 and w_2, say, $w_1 = 1000$ and $w_2 = 1$. That value of the gaze duration d is chosen that maximizes the quantity $1000p(2,d) - d$.

9.4.6 CONCURRENT PROCESSING: OP NETWORKS

OP diagrams were used to represent information about the order of processing in PERT networks and finite source, acyclic Jackson queueing networks. However, there is no particular reason that an OP diagram has to be generated from the preceding networks. In fact, in the most general case constraints on the order in which processes are executed can be formulated directly in terms of states and transitions between states. This leads to what will be referred to as an OP network. Note that the OP diagrams generated from PERT networks and from finite source, acyclic Jackson networks are special cases of the OP networks defined below.

9.4.6.1 OP Networks: Modeling

Briefly, an OP network will be defined as a directed, acyclic graph with a single start state s_1 and a single terminating state s_{n_0}. Each state in the network represents a unique partition into sets F_1, F_2, ..., F_{n_0} of the set F of stimulus processes z_{ij}. (A stimulus process z_{ij} is defined as a stimulus y_i undergoing process x_j.) As previously, let B_i be the completed set of stimulus processes in state s_i. Let C_i be the current set of stimulus processes in state s_i. And let D_i be the queued set of stimulus processes in state s_i. (Sets B_i, C_i, and D_i are the union of one or more sets F_j.) A transition can be made in one step from state s_i to state s_j if the following two conditions are met. First, the completed set B_j of state s_j must be identical to the completed set B_i of state s_i except for the addition of one stimulus process, say, z_{kl}, from the current set C_i of state s_i (i.e., $B_j = B_i + \{z_{kl}\}$). Second, the current set C_j of state s_j must contain all the members of the current set C_i of state s_i except for stimulus process z_{kl}. Note that the current set C_j can also contain any number of stimulus processes not in the current set C_i. A more general formulation of an OP network can be found in Fisher (1985) and Goldstein and Fisher (1984).

The expected response time $E[T]$ is computed as follows. Let the duration S_i of state s_i be set equal to the time that elapses between the onset of state s_i and the completion of the first process in state s_i. Define the response time T as the sum: $\sum_{i=1}^{n_0-1} S_i$. Suppose that stimulus process z_{kl} completes when a transition is made from state s_i to state s_j. Let p_{ij} be the probability that this stimulus process completes first. Let g_{ij} be the probability that a transition is made from state s_i to state s_j, given that stimulus process z_{kl} completes first. Set the probability h_{ij} that a transition is made from state s_i to an immediate successor state s_j equal to the product $p_{ij}g_{ij}$. If states $s_{j_1}, s_{j_2}, \ldots, s_{j_k}$ are immediate successors of state s_i, then it must be the case that $\sum_{l=1}^{k} g_{ij_l} = 1$. Let H be the matrix of transition probabilities h_{ij}. If c_i is set equal to the sum of the rate parameters of the processes in the current set of state s_i, then it can be shown (Fisher, 1985):

Theorem 4

$$\sum_{i=1}^{n_0-1} c_i^{-1} e_1' H^i e_i$$

9.4.6.2 OP Networks: Applications

In this section two applications are discussed. The model of visual search described in the section on channel limits in visual search is made more general with the help of OP networks. In addition, OP networks are shown to be useful as models of associative memory and paired-associate learning. OP networks have also been useful in obtaining closed-form expressions for the expected response times in GERT networks (Fisher and Goldstein, 1982) and in stochastic shortest route networks (Fisher and Goldstein, 1983b). Unfortunately, there is not space enough to consider these applications here.

Visual Search

In the section on channel limits in visual search, it was noted that PERT networks could be used to model visual search only in cases where the number of stimuli in the display is less than or equal to the number of comparison channels (i.e., the number of simultaneous comparisons of the target and display stimuli). In this section it is shown how OP networks can be used to remove the above constraint.

Consider a very simple example. In particular, assume that there are two stages, an encoding stage (node n_1) and a comparison stage (node n_2). Assume that the number of stimuli in the display is equal to two, that only one comparison operation can occur at a time, that the queue before the encoding and comparison stages is large enough to hold all arriving stimuli, and that a response cannot begin until all comparisons have been completed. Let y_k be the kth stimulus to be encoded. The time y_k spends waiting in queue at a node is called the *delay* for y_k at that node. Let stimulus process e_{k1} (c_{k2}, d_{k2}) represent the encoding (comparison, waiting in the queue or delay) of stimulus y_k at stage n_1 (n_2, n_2). Let r_l represent the response process.

An OP network consistent with the above model of processing is displayed in Figure 9.4.15. Note

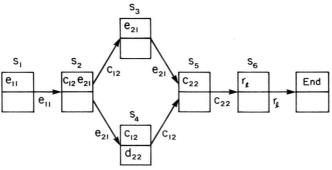

Fig. 9.4.15 An OP network representation of visual search (the number of channels, one, is less than the number of stimuli, two, in the display).

the response process r_1 does not begin until both comparison processes (c_{12} and c_{22}) have completed. And note that two comparison processes are never current. A potential exists for just such a situation if in state s_2 the encoding of stimulus y_2 at node n_1 (e_{21}) completes before the comparison of stimulus y_1 at node n_2 (c_{12}). In order to avoid the simultaneous presence of two or more comparison operations in the current set, the newly encoded stimulus y_2 is placed in the queued set D_4 of state s_4 (i.e., $D_4 = \{d_{22}\}$).

Using Theorem 4, the OP network can be fit to the data from laboratory experiments (see the section on channel limits in visual search). The resulting model (assuming it fits the data reasonably well) can then be used (among other things) to optimize the display of highlighted words on a CRT (see the preceding section).

Associative Networks: Long Term Memory

Associative networks have long been used to represent the structure of what is learned in paired-associate tasks [e.g., Kintsch (1970)] and, more generally, the structure of world knowledge in long-term memory [e.g., Anderson and Bower (1973)]. Unfortunately, computation of the relevant response time indices does not follow in a straightforward fashion from the associative network representation. OP networks can be used to advantage in this context.

First, consider the simple associative network in Figure 9.4.16a. This network can be thought of as a diagram of the sentence "Geraldine likes Ron." Let a_{ij} represent the spreading of activation from node n_i to an immediately adjacent node n_j (formally, a_{ij} is identical to the stimulus-processes previously defined). Suppose that once a node is activated, the node activates in turn all immediately adjacent nodes (with one exception to be mentioned shortly). For example, if node n_2 were the *first* node in the network to be activated, then activation would spread from node n_2 to nodes n_1, n_4, and n_5. Suppose that if a_{ij} activates node n_j, then activation a_{ji} does not spread immediately from node n_j back to node n_i (the aforementioned exception). (This restriction is one way to take into

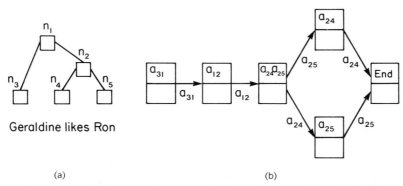

| (a) | (b) |

Fig. 9.4.16 (*a*) An associative network representation of the sentence, "Geraldine likes Ron"; (*b*) the OP network.

account the refractory period.) Finally, assume that a fact (e.g., "Geraldine likes Ron") is retrieved as soon as all terminal nodes (n_3, n_4, n_5) have been activated.

An OP network which is consistent with the above constraints is displayed in Figure 9.4.16b. In this network it is assumed that node n_3 is the first node to be activated. Then activation spreads from node n_3 to node n_1 (i.e., $C_1 = \{a_{31}\}$). When node n_1 is activated, activation spreads to node n_2 (and not back to node n_1 since this would require the addition of a_{13} immediately after a_{31} had completed). The expected time to retrieve the sentence (given that node n_3 is the first to be activated) can then be computed directly from Theorem 4. A different OP network would need to be constructed if some other terminal node were activated first.

9.4.7 STEADY-STATE QUEUEING NETWORKS

The discussion of steady-state queueing networks as models of behavior will be relatively brief. This is primarily because there has been much recent discussion of their application to problems in engineering psychology (Fisher, 1982, 1984) and human factors engineering (Rouse, 1980). Performance indices for one very useful steady-state queueing network will be presented (Section 9.4.7.1). It will then be shown that this network can be used to optimize behavior in dynamic visual inspection tasks (Section 9.4.7.2).

9.4.7.1 Steady-State Queueing Networks: Modeling

Steady-state, acyclic Jackson queueing networks are identical to the finite-state, acyclic Jackson networks discussed in Section 9.4.5 except for the following steady-state requirement. Assume that the supply of stimuli (customers) at each node is infinite. Let $P_n(t)$ be the probability that n stimuli at time t are in the system. (The system includes both the stimuli in service and the stimuli on a queue.) A single-node system is said to be in steady state if as t goes to infinity, the probability $P_n(t)$ ceases to be a function of t [i.e., $dP_n(t)/dt = 0$ as t goes to infinity]. Similar conditions define steady state in a network with several nodes.

There are many different queueing networks one could consider [e.g., see White, Schmidt, and Bennett, (1975)]. In the current section, there is space to consider only one model. To begin, assume that there is a single node n_1 with a finite number k of servers. Assume that the arrivals to the system are Poisson distributed with rate parameter λ. Assume that the service times are independent and identically distributed (the exact shape of the distribution does not matter) with mean $1/\mu$. Finally, assume that the queue size at node n_1 is equal to zero. Given the above assumptions, it can be shown that in steady state the probability p_k that a stimulus is lost from the system can be computed quite simply. Let $u = \lambda/\mu$. Then $p_k = [u^k/k!][\sum_{i=0}^{k} u^i/i!]$. This formula was originally derived by Erlang [see White, Schmidt, and Bennett (1975) for a derivation].

9.4.7.2 Steady-State Queueing Networks: Applications

Many different models of behavior in dynamic visual inspection tasks have been proposed [e.g., Smith and Barany (1970) and Wentworth and Buck (1982)]. Recently, the steady-state queueing network previously described has been used as a model of behavior in such tasks (Fisher, Moroze, and Morrison, 1983). This model will be described in very brief detail. To begin, consider the specifics of the dynamic visual inspection task. An inspector stands in front of a conveyor belt. Items are moving on a conveyor belt one at a time past the inspector. Some percentage of the items are defective. The inspector has to remove the defective items from the conveyor belt.

In order to use a steady-state queueing network as a model of behavior in the above task, some assumptions must be made about the inspector's behavior. First, for the purpose of this discussion assume that there is only one information processing stage of importance, the comparison stage. Second, assume that the arrival rate λ of items at the comparison stage is determined by the speed of the conveyor belt (say t_c) and the interitem spacing (say t_d). In particular, assume that $\lambda = t_c/t_d$. Third, assume that the inspector compares each arriving item with a mental template for good items at a rate equal to μ. Fourth, assume that the interarrival and comparison times are mutually independent, exponentially distributed random variables. Fifth, assume that items cannot be queued at the comparison stage for later analysis. And, finally, assume that the number k of simultaneous comparisons (channels or servers) is finite. If these seemingly reasonable assumptions are made, then Erlang's loss formula can be used to predict the probability p_k that an inspector will miss an item.

More importantly, the model can also be used for purposes of optimizing profits. It can easily be shown that the probability an inspector misses an item will vary with the arrival rate λ. There are costs associated with too rapid a rate of movement of items (e.g., bad items will be missed) and there are costs associated with too slow a rate of movement (e.g., too few good items may be inspected per unit of time to make the operation profitable). Erlang's loss formula together with optimization techniques similar to those described at the end of the section on short-term memory and traffic signs can be used to find the rate that maximizes profits [a detailed account is given in Fisher, Moroze, and Morrison, (1983)].

9.4.8   STRUCTURES THAT MIMIC OTHER STRUCTURES

In this chapter we have discussed several types of networks that can serve as models of human information processing. We will end the chapter with a knotty problem. Sometimes processes arranged in one configuration will behave in the same way as processes arranged in a different configuration. For example, we saw previously that two sequential processes in a Wheatstone bridge will behave like two collateral processes if they are prolonged by small amounts. And in the introduction we mentioned that Townsend (1972) has shown that processes in series and processes in parallel are indistinguishable in terms of their reaction time distributions in many situations. Here will discuss this result of Townsend in more detail.

To demonstrate the serial–parallel equivalence, we will work out the probability density functions for the finishing times of systems consisting of a set of processes in series or in parallel. The reader desiring more information about such systems is referred to Townsend and Ashby (1983). In a parallel system, see Figure 9.4.1b, processing of all the elements starts at the same time, but need not end at the same time. If process times are stochastic, the order in which the elements are completed will vary from trial to trial. On the other hand, in a serial system, see Figure 9.4.1a, only one element is processed at a time, but the order in which the elements are processed might vary from trial to trial. In either type of model, the interval between the completion of one element and the completion of the next will be called a state. In a serial system, in any state there is always one process executing; in a parallel system, in any state several processes are executing, one for each element remaining to be completed.

To simplify the calculations, we will assume within-state independence for the parallel system, that is, during any single state the process durations of all unfinished elements are stochastically independent. We will consider systems with two processes; the generalization to n processes is straightforward, [see Townsend and Ashby (1983)]. Let the two elements to be processed be a and b. In some situations it might be more convenient to let a and b represent the positions of the elements in the display rather than the elements themselves.

We will now work out the probability density functions describing the system. In a serial system, a will be processed first some of the time, and b will be first the other times. Let p be the probability that a is processed first. Let $f_{a\,1}(t)$ be the probability density function for the completion time of a when a is processed first, and let $f_{a\,2}(t)$ be the corresponding probability density function when a is processed second. We define $f_{b\,1}(t)$ and $f_{b\,2}(t)$ analogously. We will sometimes call a probability density function simply a density.

We will represent the event that the processing of a precedes the processing of b as $\langle a,b \rangle$. The density of the joint event that a precedes b, that the processing time for a is $t_{a\,1}$, and that the processing time for b is $t_{b\,2}$ will be denoted $f_{a\,1,b\,2}(t_{a\,1}, t_{b\,2}; \langle a,b \rangle)$. This density can be written as the product of (1) the probability p that a is completed first times (2) the density of a when it is completed first, $f_{a\,1}(t_{a\,1})$, times (3) the conditional density for b given that a required time $t_{a\,1}$. That is,

$$f_{a\,1,b\,2}(t_{a\,1}, t_{b\,2}; \langle a,b \rangle) = p\, f_{a\,1}(t_{a\,1})\, f_{b\,2}(t_{b\,2} | t_{a\,1}) \tag{13}$$

Likewise,

$$f_{b\,1,a\,2}(t_{b\,1}, t_{a\,2}; \langle b,a \rangle) = (1-p)\, f_{b\,1}(t_{b\,1})\, f_{a\,2}(t_{a\,2} | t_{b\,1}) \tag{14}$$

In a parallel system, we let $g_{a\,1}(t)$ denote the probability density for the time at which a is completed, given that a is completed first, and we let $g_{b\,1}(t)$ denote the corresponding density for b when it is completed first. If a is completed first, the density of the second state, during which b is completed, is denoted $g_{b\,2}(t)$, while if b is completed first, the density of the second state, during which a is completed, is $g_{a\,2}(t)$. Note that the latter two densities are not the densities of the total processing time of a or b, they are the densities for the time remaining for one element after the other element is completed.

The distribution function corresponding to a density function will be denoted by a capital letter, for example, $G_{a\,1}(t)$ is the distribution function for the time at which a is completed when a is completed first. Finally, for a given distribution function, say $G_{a\,1}(t)$, the survivor function is $\overline{G}_{a\,1}(t) = 1 - G_{a\,1}(t)$.

We are now ready to write expressions for the density functions for the parallel system. Let $g_{a\,1,b\,2}(t_{a\,1}, t_{b\,2}; \langle a,b \rangle)$ be the joint density for the event that a is completed before b, that a takes time $t_{a\,1}$, and that after a is completed, the remaining time for b is $t_{b\,2}$. Because of the within-state independence assumption, this density can be written as the product of (1) the density for a being completed at time $t_{a\,1}$ times (2) the probability that b is not completed before $t_{a\,1}$, times (3) the density for b being completed at time $t_{b\,2}$ given that a was completed at time $t_{a\,1}$. That is,

$$g_{a\,1,b\,2}(t_{a\,1},t_{b\,2};\langle a,b\rangle) = g_{a\,1}(t_{a\,1})\,\overline{G}_{b\,1}(t_{a\,1})\,g_{b\,2}(t_{b\,2}|t_{a\,1}) \tag{15}$$

Likewise,

$$g_{b\,1,a\,2}(t_{b\,1},t_{a\,2};\langle b,a\rangle) = g_{b\,1}(t_{b\,1})\,\overline{G}_{a\,1}(t_{b\,1})\,g_{a\,2}(t_{a\,2}|t_{b\,1}) \tag{16}$$

A system is said to have across-state independence if the intervals between completions of successive elements are stochastically independent. For a serial system this requires, for example, that $f_{b\,2}(t_{b\,2}|t_{a\,1}) = f_{b\,2}(t_{b\,2})$. A case of particular interest to us is when the intervals between completions of successive elements, that is, the state durations, are exponentially distributed. For a serial system having processes with exponentially distributed durations, and with across-state independence the density functions in Equations (13) and (14) will have the form

$$f_{a\,1,b\,2}(t_{a\,1},t_{b\,2};\langle a,b\rangle) = pu_{a\,1}\,e^{(-u_{a1}t_{a1})}u_{b2}\,e^{(-u_{b2}t_{b2})} \tag{17}$$
$$f_{b\,1,a\,2}(t_{b\,1},t_{a\,2};\langle b,a\rangle) = (1-p)u_{b\,1}\,e^{(-u_{b1}t_{b1})}u_{a2}\,e^{(-u_{a2}t_{a2})} \tag{18}$$

Here, $u_{a\,1}$ is the rate parameter for the completion time of a when a is completed first; the other rate parameters are denoted analogously.

We now consider a parallel system with exponentially distributed intercompletion times. To compute the density in Equation (15), note that for the expressions on the right-hand side,

$$g_{a\,1}(t_{a\,1})\,\overline{G}_{b\,1}(t_{a\,1}) = v_{a\,1}\,e^{(-v_{a1}t_{a1})}\,e^{(-v_{b1}t_{a1})} = v_{a\,1}\,e^{-(v_{a1}+v_{b1})t_{a1}}$$

For the remaining expression on the right-hand side, with the assumption of across-state independence, $g_{b\,2}(t_{b\,2}|t_{a\,1}) = g_{b\,2}(t_{b\,2}) = v_{b\,2}\,e^{(-v_{b2}\,t_{b2})}$. Then

$$g_{a\,1,b\,2}(t_{a\,1},t_{b\,2};\langle a,b\rangle) = v_{a\,1}\,e^{-(v_{a1}+v_{b2})t_{a1}}\,v_{b\,2}\,e^{(-vb2tb2)} \tag{19}$$
$$g_{b\,1,a\,2}(t_{b\,1},t_{a\,2};\langle a,b\rangle) = v_{b\,1}\,e^{-(v_{b1}+v_{a1})t_{b1}}\,v_{a\,2}\,e^{(-va2ta2)} \tag{20}$$

9.4.8.1 Equivalent Serial and Parallel Systems

We will now show what is required to make the densities for the serial system equivalent to those for the parallel system. A comparison of the equations for the two types of system reveals that the expressions for the last state have the same form in the two systems. This reflects the fact that in the last state there is only one element being processed, whether the system is serial or parallel. To make the second states equivalent, then, requires only that $u_{a\,2} = v_{a\,2}$ and $u_{b\,2} = v_{b\,2}$.

There is a difference between the serial and parallel systems in the first state. Note that for the serial system, the duration of the first state depends on which element is completed first. In other words, the exponent of the first e is different in Equations (17) and (18). This does not occur in the parallel system where the duration of the first state does not depend on which element is completed first. In other words, the exponent of the first e in Equation (19) is the same as that in Equation (20), namely, $-(v_{a\,1} + v_{b\,1})$. This shows that if $u_{a\,1} \neq u_{b\,1}$, there is no parallel exponential system equivalent to the serial system described in Equations (17) and (18).

On the other hand, if $u_{a\,1} = u_{a\,2}$, then we can construct a parallel system equivalent to the serial system by setting

$$v_{a\,1} = pu_{a\,1}$$
$$v_{b\,1} = (1-p)u_{a\,1}$$

and $v_{a\,2} = u_{a\,2}$, $v_{b\,2} = u_{b\,2}$. Given a parallel system, we can always construct a serial system equivalent to it by setting

$$u_{a\,1} = u_{b\,1} = v_{a\,1} + v_{a\,2}$$
$$p = v_{a\,1}/(v_{a\,1} + v_{b\,1})$$

and

$$u_{a\,2} = v_{a\,2}, \qquad u_{b\,2} = v_{b\,2}$$

The section on path durations gives another way to think of the density functions used above. For a parallel system, Equation (12) leads directly to Equations (19) and (20). For a serial system, p

is the probability that a is processed first, and Equation (17) follows by using Equation (12) to write the density for the path consisting of the processing of a followed by the processing of b. Similar reasoning leads to Equation (18).

We have made several particular assumptions here, that the process durations are exponentially distributed, that there is across-state independence, and so on. These assumptions are not necessary, and a set of relatively simple and very general conditions under which serial and parallel models will mimic each other is given in Townsend (1972) and in Townsend and Ashby (1983). The latter reference works out the details to show that a parallel model of visual search can produce reaction times linear in the number of elements, with equal slopes for presence and absence trials.

9.4.9 CONCLUSION

Relatively simple networks of mental processes have played an important role in cognitive psychology and human factors engineering. However, more complex networks have largely been ignored. Two problems stood in the way of the use of such networks. First, it was difficult to determine the structure of complex networks of mental processes. Second, even when the structure of the network was known, it was difficult to determine how the network would behave, especially when the durations of the processes in the network varied from trial to trial. Recently, progress has been made on both problems. This chapter focused on these recent advances. Emphasis was placed on the implications of these advances for work in human factors engineering.

The ability to both synthesize and analyze complex networks of mental processes holds much promise. If the information processing metaphor of cognitive behavior remains intact for some time to come, the technical developments described in this chapter should help advance both theory and practice.

REFERENCES

Anderson, J. R., and Bower, G. H. (1973). *Human associative memory.* Washington, DC: Winston.

Atkinson, R. C., and Shiffrin, R. M. (1968). ·Human memory: a proposed system and its central processes. In K. W. Spence and J. T. Spence, Eds. *Advances in the psychology of learning and motivation research and theory.* New York: Academic, Vol. 2.

Bamber, D., and Van Santen, J. P. H. (1980). *Testing discrete state models using conditional probability matrices.* Paper presented at the Mathematical Psychology Meeting, Madison, WI.

Berman, E. B. (1964). Resource allocation in a PERT network under continuous activity time-cost functions. *Management Science, 10,* 734–745.

Bernstein, A. J. (1966). Analysis of programs for parallel processing. *IEE Transactions on Electronic Computers, EC-15,* 757–763.

Bishop, Y. M. M., Feinberg, S. E., and Holland, P. W. (1975). *Discrete multivariate analysis: theory and practice.* Cambridge, MA: MIT Press.

Bonett, D. G., and Bentler, P. M. (1983): Goodness-of-fit procedures for the evaluation and selection of log-linear models. *Psychological Bulletin, 93,* 149–166.

Broadbent, D. E. (1958). *Perception and communication.* New York: Pergamon.

Card, S. K., Moran, T. P., and Newell, A. (1983). *The psychology of human-computer interaction.* Hillsdale, NJ: Erlbaum.

Clark, C. E. (1961). The optimum allocation of resources among the activities of a network. *Journal of Industrial Engineering, 12,* 11–17.

Curry, R. E., and Gai, E. G. (1976). Detection of random process failures by human monitors. In T. B. Sheridan and G. Johannsen, Eds. *Monitoring behavior and supervisory control.* New York: Plenum.

Donders, F. C. (1969). "On the speed of mental processes." In W. G. Koster, Ed. and Trans. *Attention and performance II.* Amsterdam: North-Holland. (Reprinted from *Acta Psychologica, 30,* 1969.)

Egeth, H., Jonides, J. and Wall, S. (1972). Parallel processing of multielement displays. *Cognitive Psychology, 3,* 674–698.

Elmaghraby, S. E. (1968). The determination of optimal activity duration in project scheduling. *Journal of Industrial Engineering, 12,* 11–17.

Fisher, D. L. (1982). Limited channel models of automatic detection: capacity and scanning in visual search. *Psychological Review, 89,* 662–692.

Fisher, D. L. (1984). Central capacity limits in consistent mapping visual search tasks: four channels or more? *Cognitive Psychology, 16,* 449–484.

Fisher, D. L. (1985). Network models of reaction time: The generalized OP diagram. In N. G. D'Ydewalle, Ed., *Cognition, information processing and motivation* (Vol. 3, pp. 229–254). Amsterdam: North-Holland.

Fisher, D. L., and Goldstein, W. M. (1982). *Probabilistic PERT: a continuous time Markov process.* Paper presented at the meetings of the Operations Research Society, Detroit.

Fisher, D. L., and Goldstein, W. M. (1983a). Stochastic PERT networks as models of cognition: derivation of the mean, variance and distribution of reaction time using order-of-processing (OP) diagrams. *Journal of Mathematical Psychology, 27,* 121–151.

Fisher, D. L., and Goldstein, W. M. (1983b). *Closed form expressions for path optimality indices in stochastic shortest route problems.* Paper presented at the meetings of the Operations Research Society of America, Orlando.

Fisher, D. L., and Goldstein, W. M. (1983c). *The sojourn time distribution of a three queue Jackson network.* Paper presented at the meetings of the Operations Research Society, Chicago.

Fisher, D. L., Moroze, M., and Morrison, R. (1983). Dynamic visual inspection: queueing models. *Proceedings of the 27th Annual Meetings of the Human Factors Society.* Santa Monica, CA: The Human Factors Society.

Fisher, D. L. and Papazian, B. (1983). Response times and safety research: application of the critical path and the OP methods. *Proceedings of the 27th Annual Meetings of the Human Factors Society,* Santa Monica: The Human Factors Society.

Fisher, D. L., Saisi, D., and Goldstein, W. M. (1985). Stochastic PERT networks: OP diagrams, critical paths and the expected completion time. *Computers and Operations Research, 12,* 471–482.

Fisher, D. L., Tengs, T., and Colen, S. (1984). *Traffic signs: rapid visual serial presentation* (IEOR Technical Report 84–01). Amherst, MA: Department of Industrial Engineering and Operations Research, University of Massachusetts.

Forster, K. I. (1970). Visual perception of rapidly presented word sequences of varying complexity. *Perception & Psychophysics, 8,* 215–221.

Goldstein, W. M., and Fisher, D. L. (1983). *The completion time distribution of stochastic PERT networks.* Paper presented at the meetings of the Operations Research Society, Chicago.

Goldstein, W. M., and Fisher, D. L. (1984). *Order-of-processing diagrams as models of response time.* Paper presented at the meetings of the Society for Mathematical Psychology, Chicago.

Golumbic, M. C. (1980). *Algorithmic graph theory and perfect graphs.* New York: Academic.

Jonides, J., and Gleitman, H. (1972). A conceptual categorization effect in visual search: O as a letter or digit. *Perception & Psychophysics, 12,* 457–460.

Kantowitz, B. H. (1974). Double stimulation. In B. H. Kantowitz, Ed., *Human information processing: Tutorials in performance and cognition.* New York: Wiley.

Kantowitz, B. H., and Sorkin, R. D. (1983). *Human factors: Understanding people-system relationships.* New York: Wiley.

Karlin, L., and Kestenbaum, R. (1968). Effects of number of alternatives on the psychological refractory period. *Quarterly Journal of Experimental Psychology, 20,* 167–178.

Keele, S. W. (1973). *Attention and human performance.* Pacific Palisades: Goodyear.

Kelley, J. E., Jr. (1961). Critical path planning and scheduling, mathematical basis. *Operations Research, 9,* 296–320.

Kerr, B. (1973). Processing demands during mental operations. *Memory and Cognition, 1,* 401–412.

Kintsch, W. (1970). *Learning, memory and conceptual processes.* New York: Wiley.

Lively, B. L. (1972). Speed/accuracy trade-off and practice as determinants of stage durations in a memory search task. *Journal of Experimental Psychology, 96,* 97–103.

Malcolm, D. G., Roseboom, J. H., Clark, C. E., and Fazar, W. (1959). Applications of a technique for research and development program evaluation. *Operations Research, 7,* 646–669.

McClelland, J. L. (1979). On the time relations of mental processes: An examination of processes in cascade. *Psychological Review, 86,* 287–330.

Murdock, B. B., Jr. (1961). The retention of individual items. *Journal of Experimental Psychology, 62,* 618–625.

Ollman, R. T. (1968). Central refractoriness in simple reaction time: The deferred processing model. *Journal of Mathematical Psychology, 5,* 49–60.

Pashler, H. (1984). Processing stages in overlapping tasks: Evidence for a central bottleneck. *Journal of Experimental Psychology: Human Perception and Performance, 10,* 358–377.

Pasmooij, C. K., Opmeer, C. H. J. M., and Hyndman, B. W. (1976). Workload in air traffic control. In T. B. Sheridan and G. Johannsen, Eds., *Monitoring behavior and supervisory control.* New York: Plenum.

Potter, M. C. (1984). Rapid serial visual presentation (RSVP): a method for studying language processing. In D. Kieras and M. Just, Eds., *New methods in reading comprehension research.* Hillsdale, NJ: Erlbaum.

Rouse, W. B. (1980). *Systems engineering models of human-machine interaction.* New York: North Holland.

Schneider, W., and Shiffrin, R. M. (1977). Controlled and automatic human information processing: I. Detection, search and attention. *Psychological Review, 84,* 1–66.

Schuberth, R. E., Spoehr, K. T., and Lane, D. M. (1981). Effects of stimulus and contextual information on the lexical decision process. *Memory and Cognition, 9,* 68–77.

Schweickert, R. (1978). A critical path generalization of the additive factor method: Analysis of a Stroop task. *Journal of Mathematical Psychology, 18,* 105–139.

Schweickert, R. (1982). The bias of an estimate of coupled slack in stochastic PERT networks. *Journal of Mathematical Psychology, 26,* 1–12.

Schweickert, R. (1983a). Synthesizing partial orders given comparability information: Partitive sets and slack in critical path networks. *Journal of Mathematical Psychology, 27,* 261–276.

Schweickert, R. (1983b). Latent network theory: Scheduling of processes in sentence verification and the Stroop effect. *Journal of Experimental Psychology: Learning, Memory and Cognition, 9,* 353–383.

Schweickert, R. (1985). Separable effects of factors on speed and accuracy: Memory scanning, lexical decision and choice tasks. *Psychological Bulletin, 97,* 530–546.

Schweickert, R., and Boggs, G. J. (1984). Models of central capacity and concurrency. *Journal of Mathematical Psychology, 3,* 223–281.

Sheridan, T. B., and Ferrel, W. R. (1974). *Man-machine systems: Information, control, and decision models of human performance.* Cambridge: MIT.

Shwartz, S. P., Pomerantz, J. R., and Egeth, H. (1977). State and process limitations in information processing: An additive factor analysis. *Journal of Experimental Psychology: Human Perception and Performance, 3,* 402–410.

Smith, J. E. K. (1976). Analysis of qualitative data. *Annual Review of Psychology, 27,* 487–494.

Smith, A. L., and Barany, J. H. (1970). An elementary model of human performance on paced visual inspection. *AIEE Transactions, 2,* 298–308.

Sperling, G. (1960). The information available in brief visual presentations. *Psychological Monographs, 74* (Whole Report No. 11).

Sternberg, S. (1969). The discovery of processing stages: extensions of Donders' method. In W. G. Koster, Ed., *Attention and Performance II.* Amsterdam: North-Holland.

Stroop, J. R. (1935). Studies of interference in serial verbal reactions. *Journal of Experimental Psychology, 18,* 643–662.

Telford, C. H. (1931). Refractory phase of voluntary and associative responses. *Journal of Experimental Psychology, 14,* 1–35.

Townsend, J. T. (1972). Some results concerning the identifiability of parallel and serial processes. *British Journal of Mathematical and Statistical Psychology, 25,* 168–199.

Townsend, J. T., and Ashby, F. G. (1983). *The stochastic modeling of elementary processes.* New York: Cambridge University Press.

Townsend, J. T., and Schweickert, R. (1984). Interactive effects of factors prolonging processes in latent mental networks. *Proceedings of the XXIII International Congress of Psychology,* Acapulco, Mexico. Amsterdam: North-Holland.

Triesman, A. M., and Gelade, G. (1980). A feature-integration theory of attention. *Cognitive Psychology, 12,* 97–136.

Waugh, N. C. and Norman, D. A. (1965). Primary memory. *Psychological Review, 72,* 89–104.

Welford, A. T. (1952). The "psychological refractory period" and the timing of high-speed performance—A review and a theory. *British Journal of Psychology, 43,* 2–19.

Welford, A. T. (1959). Evidence of a single-channel decision mechanism limiting performance in a serial reaction task. *Quarterly Journal of Experimental Psychology, 11,* 193–209.

Welford, A. T. (1967). Single-channel operation in the brain. *Acta Psychologica, 27,* 5–22.

Welford, A. T. (1980). The single channel hypothesis. In A. T. Welford, Ed., *Reaction times.* New York: Academic, pp. 239–252.

Wentworth, R. N., and Buck, J. R. (1982). Presentation effects and eye-motion behaviors in dynamic visual inspection. *Human Factors, 24,* 642–658.

White, J. A., Schmidt, J. W., and Bennett, G. K. (1975). *Analysis of queueing systems.* New York: Wiley.

Wickens, C. D. (1984). *Engineering psychology and human performance.* Columbus, OH: Merrill.

CHAPTER 9.5
FEEDBACK CONTROL MODELS

RONALD A. HESS

Department of Mechanical Engineering
University of California, Davis

9.5.1 INTRODUCTION

There are many tasks in everyday life that require almost continuous human control for their successful and safe completion. Driving an automobile, riding a bicycle, and flying an aircraft are three examples among many. Each of these tasks involve the human being acting as a feedback element in a closed-loop control system. The importance of such human feedback activity in the operation of many engineering systems has led to the development of a separate discipline called manual feedback control, or more simply, manual control. As an engineering discipline, manual control is approaching its 40th year of existence. Probably all of the modern manual control research had its genesis in the work of feedback control engineers during, and immediately after, World War II. Pioneering studies such as those of Tustin (1947) compared the control behavior of the human to that of inanimate automatic feedback devices. This work was dictated by the development of weapons, such as antiaircraft guns, bombsights, etc. which could only function in concert with human operators. Fortunately, the existing mathematical tools for feedback analysis were mature enough to be applied to the problem, and the "servomechanism" model of the human operator was born. Indeed, the control theory paradigm which has evolved in the intervening years has been so useful in quantifying control-related human behavior that it has become a fundamental mode of thinking on the part of most manual control practitioners (McRuer, 1980) and may well have laid the groundwork for the entire science of cybernetics (Miller, 1982).

The control theory description of the human operator will be the basis of the modeling work discussed in this chapter. It will be assumed that the reader has had some background in elementary control theory, equivalent to an introductory undergraduate course in the subject. If this is not the case, it is recommended that a suitable introductory text be selected for reference, for example, Ogata (1970). As a means of establishing a common ground of communication, consider Figure 9.5.1, which illustrates an automobile driving task. Figure 9.5.2 is a simplified block diagram illustration of the probable manual control loop closures used in this task. "Closure" refers to sensing and utilizing vehicle response to form a feedback control loop. In Figure 9.6.2, both vehicle heading error and lane deviation error are sensed by the driver and used as feedback signals in what is termed a multiloop compensatory control system. The term "compensatory" refers to the fact that only error information is used to generate control steering inputs in the driver/vehicle system modeled by Figure 9.5.2. In reality, of course, other information such as the future course of the vehicle would be used by the driver in such tasks. Such "preview" control models will be discussed in Section 9.5.4. However, the compensatory model of Figure 9.5.2 is of primary importance from the standpoint of predicting human/vehicle stability and estimating performance. The steering disturbance is assumed to be a continuous and random (or random-appearing) function of time.

There are, of course, alternate ways of drawing Figure 9.5.2. Consider Figure 9.5.3, where the same task has been represented by a slightly different structure. The feedback diagram of Figure 9.5.3 is often referred to as multiple-loop (as opposed to multiloop) feedback structure. The primary difference lies in the fact that Figure 9.5.2 has distinct inner and outer loops not apparent in Figure 9.5.3. Such diagrams are sometimes referred to as serial (Figure 9.5.2) and parallel (Figure 9.5.3) control structures. We will have reason to return to this difference in discussing human operator modeling approaches in Section 9.5.2.

Now the majority of work in manual control theory has dealt with single-loop compensatory systems typified by the inner-loop closure of Figure 9.5.2, shown in a more general form in Figure 9.5.4. Rather than the steering disturbance of Figures 9.5.2 and 9.5.3, the system of Figure 9.5.4 utilizes a random command input which, in the absence of any cues derived from vehicle motion, is equivalent to the disturbance in terms of human operator behavior and human/vehicle performance.

It was the hopes of the early manual control researchers that the human's dynamic characteristics in tasks such as that implied by Figure 9.5.4 could be described by the same types of mathematical equations used in describing linear servomechanisms, that is, sets of linear, constant coefficient differential equations. In his pioneering report, Tustin stated:

> The object of the series of tests described in the present report was to investigate the nature of the layer's (gunner's) response in a number of particular cases and to attempt to find the laws of relationships of movement to error. In particular it was hoped that this relationship might be found . . . to be approximately linear and so permit the well-developed theory of "linear servomechanisms" to be applied to manual control in the same way as it is applied to automatic following.

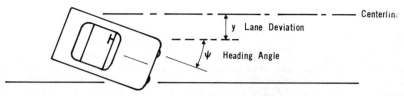

Fig. 9.5.1. An automobile driving task.

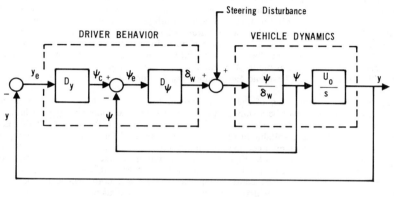

$$D_y, D_\psi = \text{Driver Dynamics}$$

$$\delta_w = \text{Steering Wheel Angle}$$

$$\psi / \delta_w \cong K/s$$

$$U_0 = \text{Automobile Velocity}$$

Fig. 9.5.2. A block diagram representation of manual loop closures in the task of Figure 9.5.1.

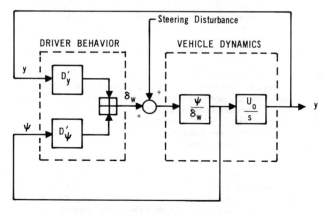

$$D_y', D_\psi' = \text{Driver Dynamics}$$

Fig. 9.5.3. An alternate block diagram from the task of Figure 9.5.1.

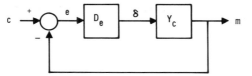

Fig. 9.5.4. A single-loop compensatory manual control system.

In general, Tustin's hopes were well founded. For example, the measurements reported by McRuer, Graham, Krendel, and Reisener (1965) demonstrated that quasilinear describing functions could be used to describe human operator dynamics in single-loop compensatory systems with a variety of controlled elements. This work represented a summation and culmination of much of the manual control research in the intervening years. Figure 9.5.5 shows this describing-function representation in more detail. The signal shown injected with the error e is referred to as "remnant" and when multiplied by the transfer function Y_p, represents that portion of the human's output that is not linearly correlated with the input $c(t)$ (Graham and McRuer, 1971). Remnant can be thought of as representing actual noise injection by the human and/or modeling errors, that is, errors implicit in assuming a linear, time-invariant representation of human operator dynamics. Thus, the human operator describing function consists of a linear transfer function and a random "noise" injected with the displayed or sensed error signal. This noise is assumed to be stationary and is described by its power spectral density defined as (Lee, 1960)

$$\phi_{nn_e}(\omega) = \int_{\infty}^{\infty} \phi_{nn_e}(\omega)e^{-j\omega\tau}\, d\tau \qquad (1)$$

where $\phi_{nn_e}(\tau)$ is called the autocorrelation function and is given by

$$\phi_{nn_e}(\tau) = \lim_{T \to \infty} \frac{1}{2T} \int_{T}^{T} n_e(t)n_e(t+\tau)\, d\tau \qquad (2)$$

Figure 9.5.5 also indicates how $Y_p(j\omega)$ and $\phi_{nn_e}(\omega)$ can be determined from time histories taken from tracking experiments.

The utility of feedback models of the human operator stems from the fact that the relative magnitude of the error-injected remnant signal tends to be rather small for most controlled elements and tasks. Thus the transfer function part of the describing function, D_e, describes most of the human operator behavior. Finally, it should be noted that the spectral relations given in Figure 9.5.5 constitute but one way to estimate human operator dynamics. Other methods, emphasizing time-series analysis (Box and Jenkins, 1976) have been used quite successfully (Shinners, 1974). The spectral approach given in Fig. 9.5.5 is adopted here because it was the method utilized by McRuer, Graham, Krendel, and

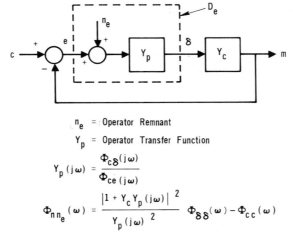

n_e = Operator Remnant

Y_p = Operator Transfer Function

$$Y_p(j\omega) = \frac{\Phi_{c\delta}(j\omega)}{\Phi_{ce}(j\omega)}$$

$$\Phi_{n n_e}(\omega) = \frac{|1 + Y_c Y_p(j\omega)|^2}{Y_p(j\omega)^2}\, \Phi_{\delta\delta}(\omega) - \Phi_{cc}(\omega)$$

Fig. 9.5.5. A human operator describing function.

Reisener (1965) in deriving an extremely important representation of human operator dynamics referred to as the crossover model.

9.5.2   MODELS FOR SINGLE-LOOP COMPENSATORY CONTROL

9.5.2.1   The Crossover Model

The Human Operator Describing Function

The measurement of human operator describing functions undertaken by McRuer, Graham, Krendel, and Reisener (1965) were obtained from an experimental matrix encompassing a variety of controlled-element dynamics and random-appearing input commands with different bandwidths. The results of that study confirmed the applicability of a previous model of the human operator (McRuer and Krendel, 1957) and refined the model parameter adjustment rules. The essence of these results will now be presented. For further details the reader is referred to the report in question and McRuer and Krendel (1974).

In almost all cases of engineering interest, the transfer function

$$Y_p = \frac{K_p e^{-\tau_e s}(T_L s + 1)}{(T_I s + 1)(T_N s + 1)} \tag{3}$$

can be suitably adjusted to provide a satisfactory description of the linear portion of the human operator describing function D_e in a single-loop compensatory systems in a frequency range around crossover. Here, T_L represents a lead time constant, T_I is a lag time constant, and T_N is a second lag time constant associated with a rudimentary model of the human neuromuscular system. In Equation (3), $T_N \ll T_I$. Ignoring T_N for purposes of exposition, the relative magnitudes of T_L and T_I determine whether the model of Equation (3) is using error-rate information or error-integral information in producing an output. If $T_L \gg T_I$, the model's output is essentially proportional to error rate for frequencies $1/T_L < \omega < 1/T_I$. If, on the other hand, $T_I \gg T_L$, the model's output is essentially proportional to the error integral for frequencies $1/T_I < \omega < 1/T_L$. For frequencies above or below the limiting values just given, model output is proportional to error. The crossover frequency is that frequency where the amplitude portion of the Bode plot of the open-loop transfer function obtained by multiplying Y_p by Y_c has a magnitude of unity (or zero dB). A Bode plot of a transfer function $G(s)$ is a graphical representation of the magnitude and phase angle of the complex number $G(j\omega)$ as ω varies in logarithmic fashion over a specified frequency range. For manual control studies, this is typically $0.1 < \omega < 20$ rad/sec. The magnitude of the complex number is expressed in decibels (dB) defined as $20 \times \log_{10} |G(j\omega)|$. The Bode plot is sometimes referred to as a "frequency-response diagram" and represents the most common method for graphically portraying measurements of human operator transfer functions such as given by Equation (3). For example, Figure 9.5.6 shows the Bode plot for a case in which the controlled element was

$$Y_c = \frac{K}{s} \tag{4}$$

and the input bandwidth, ω_{BW_c}, was 1.5 rad/sec. The solid line represents a hand-faired line through the data, which were obtained at discrete frequencies. From this figure one can see that the crossover frequency (denoted ω_c) is approximately 4 rad/sec.

There is extensive evidence that the model of Equation (9.5.3) when combined with the controlled element reduces to a very simple form for a variety of input commands and controlled elements. This form is

$$Y_p Y_c \simeq \frac{\omega_c e^{-\tau_e s}}{s} \tag{5}$$

where

$$\omega_c \simeq \omega_{c_0} + 0.18\omega_{BW_c}$$
$$\tau_e \simeq \tau_0 - 0.08\omega_{BW_c}$$
$$\tau_0 \simeq \pi/2\omega_{c_0}$$

The parameter ω_{c_0} represents the value of open-loop crossover frequency adopted by the human operator as the bandwidth of the command input approaches zero. Table 9.5.1 shows representative values of τ_0 and ω_{c_0} for three different limiting forms of controlled-element dynamics from McRuer, Graham, Krendel, and Reisener (1965). These forms represent (1) amplification or attenuation of human control

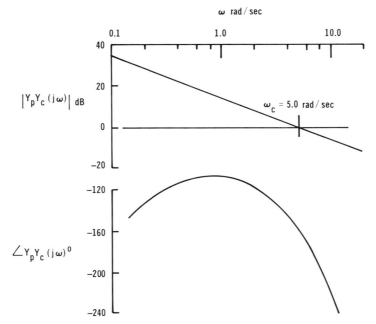

Fig. 9.5.6. A Bode plot of experimental open-loop transfer function for $Y_c = K/s$.

output ($Y_c = K$), (2) amplification or attenuation of the integral of human control output ($Y_c = K/s$), and (3) amplification or attenuation of the second integral of human control output ($Y_c = K/s^2$). A great many of the controlled elements of practical importance in manual control systems can be closely approximated by these limiting forms in the important region of open-loop crossover. The model of Equation (5) is generally referred to as the simplified crossover model of the human operator, or more simply, the crossover model. Knowledge of the form of the controlled-element dynamics near crossover obviously allows one to determine Y_p in the crossover region. Figure 9.5.7 compares the Bode plot of Equation (5) with the "data" curves of Figure 9.5.4. Note the general excellent agreement between model and data except for a discrepancy in the low-frequency phase match. This low-frequency phase effect in the data, often called "phase droop," can be modeled by extensions to the simple crossover model of Equation (5), however the discrepancy in Figure 9.5.7 has almost no effect on the quality of predicted human/vehicle stability or performance.

The remaining part of the human operator describing function is the remnant that is characterized by its power spectral density $\Phi_{nn_e}(\omega)$. Again, extensive experimental evidence suggests that this remnant power spectral density scales with the variance of the error signal to which it is added and can be represented by the following equation:

$$\Phi_{nn_e}(\omega) = \frac{R\bar{e}^2}{\omega^2 + \omega_R^2} \tag{6}$$

Table 9.5.2 gives approximate values for R and ω_R as a function of the limiting controlled-element forms shown in Table 9.5.1.

Table 9.5.1 Approximate Parameter Values for Crossover Model of Equation (4)

Y_c	τ_0 (sec)	ω_{c_0} (rad/sec)
K	0.30	5.0
K/s	0.35	4.5
K/s^2	0.50	3.0

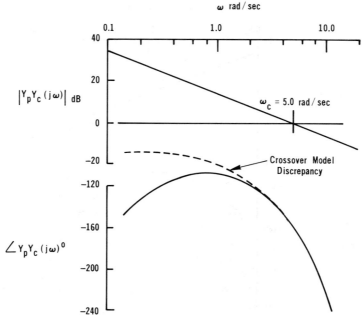

Fig. 9.5.7. A Bode plot comparing experimental open-loop transfer function of Figure 9.5.6 with crossover model of Equation (4).

Simplified Performance Calculations

Block diagram algebra, for example, Ogata (1970, Chap. 4), coupled with fundamental spectral analysis techniques, for example, Lee (1960), yield the following expression for the mean square error $\bar{e}^2$ in the single-loop compensatory manual control system of Figure 9.5.5 assuming a stationary random input with power spectral density $\Phi_{cc}(\omega)$ and bandwidth ω_{BW_c}:

$$\bar{e}^2 = \frac{\bar{c}^2 \int_{-\infty}^{\infty} \dfrac{\Phi_{cc}(\omega)}{|\,1 + Y_p Y_c\,(j\omega)\,|^2}\, d\omega}{1 - F} \tag{7}$$

where

$$F = \int_{-\infty}^{\infty} \Phi_{nn_e}(\omega) \left| \frac{Y_p Y_c(j\omega)}{1 + Y_p Y_c(j\omega)} \right|^2 d\omega \tag{8}$$

The function F can be rewritten as

$$F = \frac{R}{\omega_R^2 \tau_e} I_1 \tag{9}$$

Table 9.5.2 Approximate
Parameter Values for Remnant
Model of Equation (6)

Y_c	R	ω_R (rad/sec)
K	0.1–0.5	3.0
K/s	0.1–0.5	3.0
K/s^2	0.1–0.5	1.0

For $\omega_{BW_c} < \omega_c$, which is normally the case, the numerator term in Equation (7) can be approximated by the "$\frac{1}{3}$rd power law" (McRuer and Krendel, 1974) as

$$\bar{c}^2 \int_{-\infty}^{\infty} \frac{\Phi_{cc}(\omega)}{|1 + Y_p Y_c(j\omega)|^2} d\omega \simeq \frac{1}{3} \bar{c}^2 \left(\frac{\omega_{BW_c}}{\omega_c}\right)^2 \qquad (10)$$

Thus Equation (9.5.7) can be rewritten as

$$\bar{e}^2 = \frac{\frac{1}{3}\bar{c}^2 \left(\dfrac{\omega_{BW_c}}{\omega_c}\right)^2}{1 - \dfrac{R}{\omega_R^2 \tau_e} I_1} \qquad (11)$$

Figure 9.5.8 allows one to determine I_1, given the crossover frequency, ω_c, the appropriate remnant model of Equation (6) and the human operator effective time delay, τ_e.

If ω_{BW_c} approaches the ω_c calculated in Equation (9.5.5) in magnitude, an important phenomenon known as "crossover regression" occurs in which the human sharply reduces the open-loop crossover frequency ω_c by reducing his or her gain. In normal circumstances, with $\omega_{BW_c} \leq \omega_c$, such a regression would result in a substantial increase in mean square tracking error; however, when $\omega_{BW_c} \rightarrow \omega_c$, a decrease in ω_c to a value well below ω_{BW_c} results in a decrease in mean square tracking error as compared to the value that would be obtained if no ω_c regression had occurred. The reason for this is demonstrated in Figure 9.5.9 where it has been assumed for convenience that the input spectrum is rectangular in form, that is, there is no power in the input beyond ω_{BW_c}. Now as shown in the figure, the mean square tracking error, $\bar{e}^2$, can be given as

$$\bar{e}^2 = \int_{-\infty}^{\infty} \Phi_{ee}(\omega) \, d\omega \qquad (12)$$

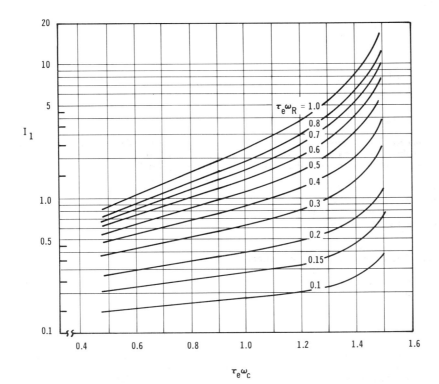

Fig. 9.5.8. Curves for determining integral I_1 in performance calcualtions.

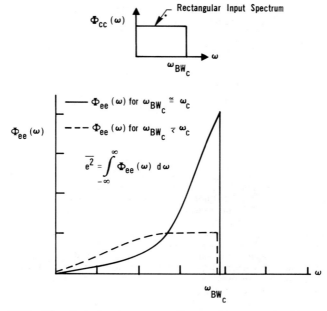

Fig. 9.5.9. The effect of crossover regression on mean square tracking error.

The solid and dashed curves in Figure 9.5.9 represent the error power spectral density for $\omega_{BW_c} \simeq \omega_c$ and $\omega_{BW_c} \leq \omega_c$, respectively. The area under each of these curves represents the mean square tracking error that would result for these two conditions. The performance improvement associated with crossover regression is evident.

For purposes of predicting when crossover regression is likely to occur it is useful to define an effective input bandwidth $\omega_{BW_{ce}}$ as

$$\omega_{BW_{ce}} = \frac{\left[\int_0^\infty \Phi_{cc}(\omega)\, d\omega\right]^2}{\int_0^\infty [\Phi_{cc}(\omega)]^2\, d\omega} \tag{13}$$

then, whenever $\omega_{BW_{ce}}$ becomes greater than $0.8\omega_{c0}$ for nearly rectangular spectra or when $\omega_{BW_{ce}}/\omega_c$ is greater than 1 for more realistic low-pass spectra, then the crossover frequency would be expected to regress to values much lower than ω_{c0} and ω_c, respectively (McRuer and Krendel, 1974). Given that crossover regression is likely, an analysis of mean square tracking performance using Equation (7) would indicate that the "optimum" regressed crossover frequency should be zero, indicating no manual loop closure at all. However, in all cases of practical interest, the human is interested in maintaining some minimum level of control over the system at hand, so a minimum crossover frequency or operator gain is maintained. It is not generally possible to predict this value, however.

9.5.2.2 The Precision Crossover Model

A more detailed human operator model can be presented for cases in which one wishes to match the linear portion of the human operator describing function with precision. This model can be given as

$$Y_p = K_p e^{-s\tau}\left(\frac{T_L s + 1}{T_I s + 1}\right)\left[\frac{T_k s + 1}{T_K' s + 1}\right]\left\{\frac{1}{(T_{N_1} s + 1)[(s/\omega_n)^2 + (2\zeta_n/\omega_n)s + 1]}\right\} \tag{14}$$

Figure 9.5.10 demonstrates the ability of the precision model to match measured describing-function data. It should be noted that the controlled-element dynamics were unstable in this case. As might be expected, the precision model is used almost exclusively for obtaining such matches rather than being a predictive model of human operator behavior.

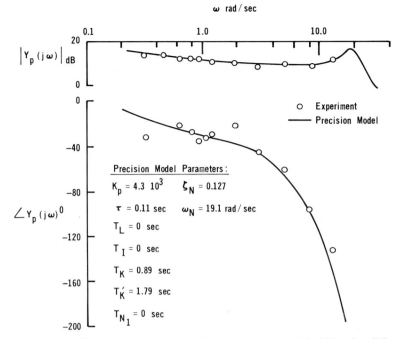

Fig. 9.5.10. Fitting experimental data with the precision model of Equation (14).

9.5.2.3 The Optimal Control Model

The equalization derived from the crossover model is very similar to that which would be prescribed by an experienced control system designer given the same controlled element and feedback variable, that is, system error. This fact has led to the application of modern control system design techniques to human operator modeling. One such technique is based on optimal control and estimation for the so-called stochastic linear quadratic Guassian problem (Athans, 1971). This problem and its application to manual control is best explained by means of a simple example.

Figure 9.5.11 represents the single-axis compensatory tracking task of Figure 9.5.4, modified slightly

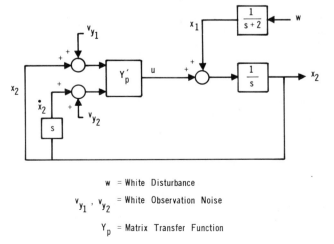

w = White Disturbance

v_{y_1}, v_{y_2} = White Observation Noise

Y_p = Matrix Transfer Function

Fig. 9.5.11. An optimal estimation and control formulation of a compensatory manual control system.

to simplify the discussion. In the figure, the controlled element is an integrator. A disturbance added to the human operator's output is utilized rather than a command input. The disturbance is modeled as white noise passed through a first-order filter with a break frequency of 2.0 rad/sec. The human is hypothesized to sense the controlled-element output and its time rate of change. Each of these variables is assumed to be corrupted by white noise referred to as "observation noise." The stochastic linear quadratic Guassian control problem then is one of finding the linear system Y_p' (here a matrix transfer function) that will result in a scalar index of performance being rendered a relative minimum. This index of performance can be given by

$$J = E \left\{ \lim_{T \to \infty} \frac{1}{T} \int_0^T [qx_1^2(t) + r\dot{u}^2(t)] \, dt \right\} \tag{15}$$

where $\{E\}$ represents the expected value of the functional in brackets, which is a random variable because of the random disturbance exciting the system. Here q and r are called "weighting coefficients" and weight the respective contributions of $x^2(t)$ and $\dot{u}^2(t)$ to the value of the integrand at each time instant.

The details of precisely what occurs in the element Y_p' of Figure 9.5.11 is of only secondary interest to the modeling efforts of this chapter. However, for the sake of completeness and to familiarize the reader with some of the terminology associated with the optimal control model of the human operator, a brief discussion is in order here. The linear system denoted by Y_p' in Figure 9.5.11 is an optimal linear regulator in combination with an optimal linear estimator, or Kalman filter. The Kalman filter produces optimal estimates of the values of the system state variables [here $x_1(t)$ and $x_2(t)$] at each instant of time given noisy measurements of linear combinations of these state variables. In Figure 9.5.11, these linear combinations take the form of the state variables themselves, but this need not always be the case. In fact, any combination is acceptable, as long as the resulting system possesses a property called "observability." The linear regulator multiplies the optimal estimates of the state variables by optimal regulator gains to produce the control output $u(t)$. The regulator gains can be found if the system possesses a property called "controllability." The properties of observability and controllability are almost always met in manual control systems so no further discussion on these topics will be pursued here. The interested reader is referred to any standard text on the subject, for example, Bryson and Ho (1975).

The equations describing the Kalman filter and optimal linear regulator can be given, respectively, by

$$u = -l_{11}x_1 - l_{12}x_2 \tag{16}$$

and

$$\begin{aligned} \dot{\hat{x}}_1 &= -2\hat{x}_1 + k_{11}(x_2 + v_{y_1} - \hat{x}_2) + k_{12}(\dot{x}_2 + v_{y_2} - \dot{\hat{x}}_2) \\ \dot{\hat{x}}_2 &= \hat{x}_2 + u + k_{21}(x_2 + v_{y_1} - \hat{x}_2) + k_{22}(\dot{x}_2 + v_{y_2} - \dot{\hat{x}}_2) \end{aligned} \tag{17}$$

The Kalman filter gains K_{ij} and optimal estimator gains l_{ij} are found from the solution of matrix Riccati equations (Bryson and Ho, 1975) and depend on the covariances of the observation noises and disturbance (for the filter Riccati equation) and on the coefficients q and r in the index of performance defined in Equation (9.5.15) (for the regulator Riccati equation).

One of the obvious differences between the optimal control model formulation and the crossover model formulation is that the former assumes a vector input to the operator, even for "single-loop" tasks like that of Figure 9.5.11. It has been shown (Levison, Kleinman, and Baron, 1969) that this vector representation leads to an equivalent single-loop remnant model similar to that given in Equation (9.5.6). Of course, the optimal control model formulation is not limited to single-loop tasks. Indeed, one of the advantages of the multivariable, state-space approach is the relative ease with which multiloop tasks are modeled. This subject will be treated in greater detail in Section 9.5.3.

The model discussed in Equations (15) to (17) and shown in Figure 9.5.11 is a more simplified version of the optimal control model of the human operator than is usually employed in human/machine studies, for example, Curry, Kleinman, and Hoffman (1977) and Baron and Levison (1977). Figure 9.5.12 shows the more complete model emphasizing the state-space representation of the system. Note that the operator time delay is now explicitly modeled and the Kalman estimator is followed by a predictor that finds an optimal estimate of the state vector at time t given an optimal estimate at time $t - \tau$. The block labeled "neuromotor dynamics" is actually part of the optimal regulator and is a direct consequence of defining the index of performance so as to include control rate. The noise term $v_u(t)$ is referred to as "motor noise" and is included to provide more realistic performance predictions. The index of performance defined by Equation (15) can be rewritten in matrix form as

$$J = E \left\{ \lim_{T \to \infty} \frac{1}{T} \int_0^T [y^T(t)Qy(t) + \dot{u}(t)R\dot{u}(t)] \, dt \right\} \tag{18}$$

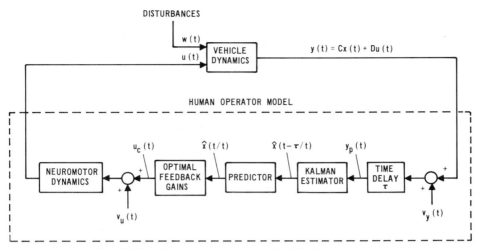

Fig. 9.5.12. The optimal control model of the human operator.

Details of the complete optimal control model (OCM) of the human operator are discussed by Kleinman, Levison, and Baron (1970).

OCM Parameter Selection

The OCM is essentially specified by (1) the weighting matrices Q and R in the index of performance, (2) the covariances of the observation and motor noises, denoted V_y and V_u, and (3) the magnitude of the operator time delay τ. As in the case of the precision crossover model, these parameters can be selected to yield OCM transfer functions, which closely match those obtained in experiment. In addition, the OCM formulation also provides root-mean-square performance predictions and remnant spectral densities. For example, consider the controlled element and disturbance given in Figure 9.5.11. The equations of state that describes this system can be given by

$$
\begin{aligned}
\dot{x}_1 &= -2x_1 + w \\
\dot{x}_2 &= x_1 + u \\
y_1 &= x_2 + v_{y_1} \\
y_2 &= x_1 + u + v_{y_2}
\end{aligned}
\tag{19}
$$

Using the following OCM parameter values

$$
J = E\left\{ \lim_{T \to \infty} \frac{1}{T} \int_0^T [x_2^2(t) + 0.0017\, \dot{u}^2(t)]\, dt \right\}
$$
$$
V_{y_1} = 0.01\pi E[y_1^2] \qquad V_u = 0.003\pi E[u^2]
$$
$$
V_{y_2} = 0.01\pi E[y_2^2] \qquad \tau = 0.15 \text{ sec}
\tag{20}
$$

where the 0.01 and 0.003 factors are called "noise-ratios," the experimental and model generated describing functions (operator transfer function and remnant) of Figure 9.5.13 result (Kleinman, Levison, and Baron, 1970). The weighting coefficient selection of $q = 1.0$ and $r = 0.0017$ will be justified in what follows. It should be noted that only the ratio of these values is of consequence in the OCM. Multiplying the integrand of Equation (15) by a constant in no way influences the optimization strategy. The present formulation of the OCM does not capture the low-frequency phase droop evident in the data of Figure 9.5.13. However, a modification of the model to include "pseudomotor noise" can alleviate this problem (Levision, Baron, and Junker, 1976). A point which cannot be overemphasized in the discussion of single-loop applications of the OCM is that the optimization procedure inherent in the model tends to produce human operator transfer functions very similar in form to those predicted by the crossover model, that is, in the region of open-loop crossover, the product of the operator and controlled-element transfer functions appears as shown in Equation (5).

The effect of OCM parameter variations on the resulting model describing funcion is not as transparent as in the case of the crossover model of the human operator. This is owing to the fact that the OCM parameters are essentially inputs to an optimization scheme that entails the solution of sets of nonlinear algebraic equations (the steady-state Riccati equations). Perhaps the most important parame-

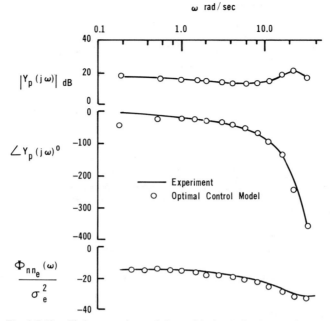

Fig. 9.5.13. Fitting experimental data with the optimal control model.

ters in the OCM formulation are the weighting matrices Q and R, which appear in the index of performance definition. In single-loop applications these matrices are often scalars, and the index of performance takes the form shown in Equation (15). In this case, an approximate but very useful relationship exists between the scalars q and r, the controlled-element dynamics, and the closed-loop system bandwidth. It can be shown (Hess, 1984; Kwakernaak and Sivan, 1972) that

$$\omega_{BW} \simeq [K(q/r)^{1/2}]^{1/(n-m+1)} \tag{21}$$

The parameters K, m, and n are obtained from the controlled-element dynamics when expressed as

$$Y_c = \frac{K[s^m + a_{m-1}s^{m-1} + \cdots + a_1 s + a_0]}{s^n + b_{n-1}s^{n-1} + \cdots + b_1 s + b_0} \tag{22}$$

The precise definition of ω_{BW} as given above is the magnitude of that closed-loop pole closest to the frequency where the amplitude of the closed-loop transfer function is 6 dB below its zero-frequency value. Now, the following relation allows one to approximate the open-loop crossover frequency given the closed-loop bandwidth calculated by Equation (21):

$$\omega_c \simeq 0.56\omega_{BW} \tag{23}$$

Equation (5) and the relation for ω_c is, of course, still useful for OCM applications. The time delay in the OCM is more representative of an actual, as opposed to an effective, delay and so a nominal value of 0.2 sec is employed. This leaves the covariances of the observation and motor noises to be specified. There is fairly extensive evidence available to suggest the following values for single-loop tracking tasks using ideal displays and manipulators:

$$\begin{aligned} V_{y_1} &= 0.01\pi E[y_1^2] \text{ (displacement noise)} \\ V_{y_2} &= 0.01\pi E[y_2^2] \text{ (rate noise)} \\ V_u &= 0.001 \rightarrow 0.005\pi E[u^2] \text{ (motor noise)} \end{aligned} \tag{24}$$

9.5.2.4 An Example

At this juncture it is useful to consider an example exercising both the crossover model and OCM in calculating the performance and dynamic characteristics of a well-trained well-motivated human

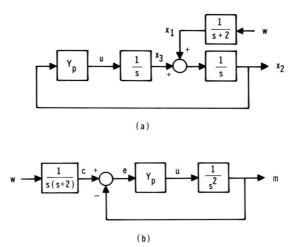

(a)

(b)

Fig. 9.5.14. (a) A tracking task with a velocity disturbance. (b) An equivalent tracking task with a command input.

operator in a simple single-loop tracking task. The task and controlled-element dynamics are shown in Figure 9.5.14a and are based on experimental tasks conducted by Kleinman, Levison, and Baron (1970) in validating the OCM for single-loop applications. The frequency-domain and state-variable representations for the controlled element and disturbance can be given as

$$\dot{x}_1 = -2x_1 + w \qquad w/x_1 = 1/(s+2)$$
$$\dot{x}_2 = x_3 + x_1 \qquad x_2 = (x_1 + x_3)/s$$
$$\dot{x}_3 = u \qquad x_3 = u/s$$
$$\bar{x}_1^2 = 0.054$$

(25)

The OCM computer program utilized in this study is documented by Curry, Hoffman, and Young (1976) and Doyle and Hoffman (1976).

In applying either the crossover model or the OCM, the bandwidth of the input is of obvious importance. Figure 9.5.14a utilizes a disturbance from which an equivalent input can be defined using simple block diagram algebra. This is shown in Figure 9.5.14b. The change in sign in the "error" signal in going from part a to b is of no consequence here. Now the bandwidth of the equivalent input needs to be determined. The bandwidth discussed in connection with the crossover model possesses a rectangular spectrum, while that of the equivalent input in Figure 9.5.14b does not. In addition, Equation (13) is not appropriate for input spectra in which the power spectral density contains free values of ω in the denominator. In this case, Equation (13) can be modified as

$$\omega_{BW_{ce}} = \frac{\left[\int_0^\infty \omega^p \, \Phi_{cc}(\omega) \, d\omega\right]^2}{\int_0^\infty [\omega^p \Phi_{cc}(\omega)]^2 \, d\omega}$$

(26)

where p is the power of the free ω in the denominator of the input power spectral density. Here, $p = 2$. Evaluating Equation (26) for the equivalent input in question yields $\omega_{BW_{ce}} = 2.0$ rad/sec. The equivalent input in Figure 9.5.14b also has a mean square value that is undefined because of the free ω^2 in the denominator of its power spectral density. For purposes of using the relations derived for the crossover model, the mean square value of the equivalent input will be defined

$$\bar{c}^2 = \int_{-\infty}^\infty \omega^p \, \Phi_{cc}(\omega) \, d\omega = 0.054$$

(27)

The Crossover Model

For the dynamics at hand, Equations (5) and (26) together with Tables 9.5.1 and 9.5.2 suggest the following crossover model parameter values:

$$\omega_c = 3.0 - 0.18(2.0) = 3.36 \text{ rad/sec}$$
$$\tau_e = 0.5 - 0.08(2.0) = 0.34 \text{ sec}$$
$$\omega_R = 1.0 \text{ rad/sec} \tag{28}$$
$$R = 0.1$$

Also note that $\omega_{BW_{ce}}/\omega_c$ is less than unity, so crossover frequency regression will not be a problem. Since the data to be utilized were generated from a laboratory tracking task with an ideal display and manipulator, the lowest value of R was used in the remnant model. Now Figure 9.5.8 can be used to find I_1 for use in Equation (11) to estimate mean square tracking error. The required abscissa value in Figure 9.5.8 is $\tau_e \omega_c = 1.14$ and the required curve is $\tau_e \omega_R = 0.34$. This yields $I_1 = 0.9$. Now Equation (11) can be evaluated as follows:

$$\bar{e}^2 = \frac{\frac{1}{3}(0.054)\left(\dfrac{2}{3.36}\right)^2}{1 - \dfrac{0.1}{(1)(0.34)}(0.9)} = 0.0087 \tag{29}$$

Now the actual form of the transfer function portion of the human operator describing function (Y_p) is not uniquely specified in application of the crossover model. All that is known in this case is that the operator must generate lead in order for $Y_p Y_c$ to resemble ω_c/s around crossover. This means,

$$Y_p \simeq K_p(T_L s + 1)e^{-\tau_e s} \tag{30}$$

It should be noted at this juncture that the requirement for lead equalization correlates strongly with human operator estimates of task difficulty, that is, the greater the lead time constant T_L the more difficult the task appears to the operator.

The Optimal Control Model

The index of performance for the example can be given by

$$J = E\left\{\lim_{T \to \infty} 1\frac{1}{T}\int_0^T [qx_1^2(t) + r\ddot{u}^2(t)]\,dt\right\} \tag{31}$$

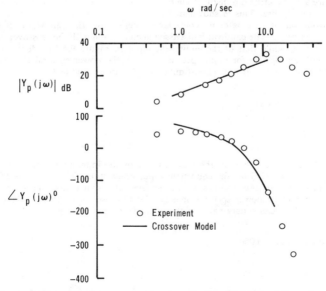

Fig. 9.5.15. A comparison of the human operator transfer function derived from the crossover model with experiment.

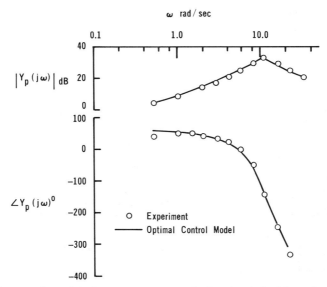

Fig. 9.5.16. A comparison of the human operator transfer function derived from the optimal control model with experiment.

Selecting $q/r = 5.0 \times 10^5$ would, according to the approximate relations of Equations (21) and (23), yield a crossover frequency of 3.4 rad/sec, quite close to the value suggested by Equations (28) (which are also valid for the OCM). The remaining OCM parameters were identical to those given in Equation (24), with the motor noise ratio set to 0.001. The operator time delay was set to the nominal value of 0.2 sec.

Results

Figure 9.5.15 shows the operator dynamics implied by the crossover model around open-loop crossover. The experimental data are taken from the study reported by Kleinman, Levison, and Baron (1970). Figure 9.5.16 shows the same experimental data, this time with the operator transfer function generated by the OCM. In both figures, the modeling results are quite acceptable, particularly around the important crossover region. Table 9.5.3 compares model-generated and experimental crossover frequencies and tracking performance. The performance comparison could be improved as both the crossover model and OCM yield mean square error scores which are too small. This is to be expected since the remnant models associated with each human operator model have parameter values that can vary somewhat depending on the experiment and operator. For example, by increasing the R value in the crossover model remnant to 0.29, the resulting square error can be made to agree with the experimental 0.014 value. Likewise, increasing the motor noise ratio in the OCM from 0.001 to 0.047 will bring the OCM mean square error to 0.14. This latter change has a minimal effect on the OCM transfer function shown in Figure 9.5.15.

Table 9.5.3   Model Parameters and Performance Compared with Experiment

	ω_c (rad/sec)	$\bar{e}^2$
Experiment	3.5	0.014
Crossover model	3.4	0.0087
Optimal control model	3.4	0.0100

9.5.2.5   More General Controlled Element Dynamics

The Crossover Model

The controlled-element dynamics discussed thus far have really been stereotypes representing the characteristics of more realistic dynamics in the region of crossover. For example, consider the following controlled-element dynamics:

$$Y_c = \frac{\omega_n^2}{s^2 + 2\zeta_n \omega_n s + \omega_n^2} \tag{32}$$

For the purposes of argument let $\zeta_n < 1.0$. Now, considering Table 9.5.1 and Equation (5) if $\omega_n << \omega_c$, then the dynamics of Equation (32) closely resemble K/s^2 over a broad frequency range including probable crossover. One can, with a fair degree of confidence, model the human operator using the results of Table 9.5.1 and Equation (5) assuming K/s^2 dynamics. Likewise, if $\omega_n >> \omega_c$, then the dynamics of Equation (32) closely resemble K over a broad frequency range including probable crossover. Again, one can, with a fair degree of confidence, use the results of Table 9.5.1 and Equation (5) assuming K dynamics. Now, if $\omega_n \simeq \omega_c$, the issue is not as clear cut. Obviously the operator will have to generate some lead equalization to force $Y_p Y_c$ into the ω_c/s form around crossover. However, selection of ω_c itself and the effective time delay is not straightforward in this case since the controlled-element dynamics do not fall into either the K or K/s^2 class.

   In the case just described, the following procedure is recommended (McRuer and Krendel, 1974). First make a reasonable estimate of ω_c. Then use Figure 9.5.17 to yield τ_0. Next use

$$\Delta\tau_e \simeq 0.08\omega_{BW_{ce}} \tag{33}$$

to determine $\Delta\tau_e$, the decrement in time delay due to forcing function bandwidth. Now the nominal crossover frequency can be determined using

$$Y_p \simeq K_p(T_L s + 1)e^{-(\tau_0 - \Delta\tau_e)s} \tag{34}$$

and adjusting K_p to minimize mean square error. This will yield a nominal value for ω_c. If this value differs significantly from the estimated value, repeat the process. The remnant model associated

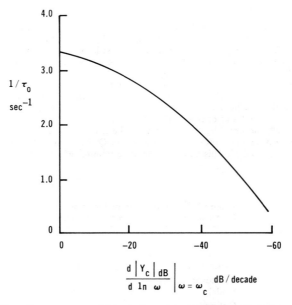

Fig. 9.5.17.   Variation of crossover model basic time delay with controlled-element amplitude slope at crossover.

with K/s^2 dynamics (i.e., a controlled element requiring lead equalization) in Table 9.5.2 should be utilized.

The Optimal Control Model

An analogous procedure to that just discussed can be utilized with the OCM. First, an estimate of ω_c is made. Then Equations (21) to (23) are used to obtain an estimate of q/r. Next, using the OCM parameters suggested by Equation (24) (with the motor noise ratio ≥ 0.003) the ratio of q/r is varied until the mean square tracking error is minimized. It should be pointed out that this process is sensitive to the magnitude of the motor noise ratio. The value suggested is representative of that obtained by matching OCM characteristics with those of the well-trained, well-motivated human operator in single-loop laboratory tracking tasks.

9.5.3 MODELS FOR MULTILOOP COMPENSATORY CONTROL

9.5.3.1 The Classical Approach Using the Crossover Model

The automobile driving task that began this chapter is an example of a multiloop, single-point control problem. The descriptor "single point" refers to the fact that one control variable, the steering wheel angle, is utilized in controlling both vehicle heading and lane position. Figure 9.5.2 is referred to as a "series" configuration, while Figure 9.5.3 is called a "parallel" configuration. Obviously, both descriptions can be considered valid representations of the control problem. There are a number of reasons for the existence of multiloop control systems, one of which has just been mentioned, that is, the desire to control more than one vehicle output quantity with a single control variable. Other reasons include the desire to use auxiliary output quantities in controlling a single primary output variable. An example of this would be the use of lateral acceleration by the driver of an automobile to aid in the control of vehicle heading. Finally, multiloop systems naturally evolve when one desires to achieve control coupling or decoupling.

The block diagrams of Figures 9.5.2 and 9.5.3 represent a classical approach to the control system design problem. They emphasize a very important fact, namely, that, in multiloop control, one constructs feedback loops about an "effective" controlled element, which may itself involve the original controlled element modified by other loop closures. This approach is often missing from the modern multivariable control system design approaches as exemplified by the linear quadratic Guassian method discussed in connection with the OCM. Given a manual control problem like the automobile driving example, modeling the human operator or driver is equivalent to determining (1) the transfer functions D_ψ and D_y in Figure 9.5.2 or D'_ψ and D'_y in Figure 9.5.3, and (2) determining operator remnant that may exist in the loop closures.

The crossover and optimal control models of the human operator for single-loop tasks have successfully been developed largely due to the existence of a high-quality data base for such tasks. Unfortunately, a similar data base does not exist for multiloop tasks. Part of the problem lies in the fact that the classical spectral measurement techniques that have been used to determine human operator describing functions for single-loop tasks are somewhat restricted in their applicability to multiloop tasks. It can be shown (Stapleford, McRuer and Magdeleno, 1967) that the number of measurable operator transfer functions in a multiloop task is equal to the number of uncorrelated inputs times the number of operator outputs or controls. Thus, for example, in the block diagram of Figure 9.5.2, only a linear combination of the D_ψ and D_y could be measured with the single disturbance shown.

Despite the relative scarcity of multiloop data, it has been demonstrated (Stapleford, Craig, and Tennant, 1969; Weir and McRuer, 1972) that many multiloop tasks of engineering interest can be modeled by series loop structures. In addition, in cases where display scanning is minimal, it appears that the major contributor to operator remnant is in inner-loop operation. As in the case of single-loop tasks, many of the measured human operator characteristics in multiloop tasks are coincident with those that would be selected by an experienced control system engineer in designing inanimate compensators to accomplish the same task.

A convenient stereotype of a series loop structure is shown in Figure 9.5.18. Here, three successive loop closures characterized by the crossover frequencies ω_{c_1}, ω_{c_2}, and ω_{c_3} are shown, with

$$\omega_{c_1} > \omega_{c_2} > \omega_{c_3} \tag{35}$$

The structure of this diagram typifies almost any feedback system whether under automatic or manual control. As an example, Figure 9.5.18 might represent the flight control system of a hovering vertical takeoff and landing (VTOL) aircraft. Here, the innermost loop represents vehicle attitude control, the next loop represents vehicle velocity control, and the last loop represents vehicle position control. The crossover frequency (and hence bandwidth) separation implied by the inequality Equation (35) allows some simplification in the analysis of a system like that of Figure 9.5.18. For example, with

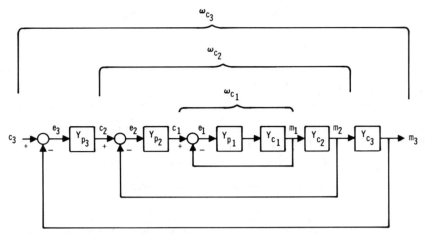

Fig. 9.5.18. Block diagram for a single-point, multiloop manual control problem.

the inner loop closed, the effective open-loop dynamics for the second closure, defining m_2/c_2, takes the approximate form

$$\frac{m_2}{e_2} \simeq Y_{p_2} Y_{c_2} \bigg|_{|s|=\omega_{c_2}} \tag{36}$$

Likewise, with the first and second loops closed, the effective open-loop dynamics for the final loop closure, defining m_3/c_3, takes the approximate form

$$\frac{m_3}{e_3} \simeq Y_{p_3} Y_{c_3} \bigg|_{|s|=\omega_{c_3}} \tag{37}$$

Equations (36) and (37) are useful, albeit approximate, relations for analysis and design. Their utility is attributable to the fact that closed-loop dynamic characteristics are essentially determined by the characteristics of the open-loop system in the region of crossover. Of course, this is precisely the reason that the crossover model of the human operator is so useful in analyzing single-loop tasks.

The following general guidelines can be used to formulate a single-point multiloop manual control problem, using the crossover model and given the structure of Figure 9.5.18:

1. The operator's dynamic characteristics are distributed in serial fashion as Y_{p_1}, Y_{p_2}, and Y_{p_3}. Y_{p_1} can be assumed to be identical in form to the single-loop models that have been discussed. As has been stated, in the absence of significant scanning, operator remnant can be considered to be injected only in the inner loop. Here "significant" scanning can be interpreted as that requiring distinct shifts in the operator's eye point-of-regard throughout the task.

2. The operator dynamics represented by Y_{p_2} and Y_{p_3} ultimately create the input signal c_1, which drives the innermost loop. It is within this loop that the actual human/machine interface occurs through the action of the control manipulator. Note that the signals c_1 and c_2 are internally generated by the operator as opposed to being explicitly displayed or perceived from the environment.

3. The operator's effective time delay is usually placed in the innermost loop. This is desirable since a considerable part of the effective delay is attributable to neuromuscular effects that occur only at the human/manipulator interface.

4. The operator dynamics are estimated by repetitive application of the crossover model starting from the innermost loop and working out. The operator dynamics in outer loops should consist of simple gains, that is, the effective open-loop dynamics given by Equations (36) and (37) should appear as integrators in the appropriate crossover regions. If this requirement is not met, it is likely that the assumed loop closures will not be those adopted by the human.

5. Given simple gain equalization in outer loop closures, the crossover frequencies should be separated by a factor of approximately 2–3. This is a rule of thumb suggested by limited experimental data, that is, Hess and Beckman (1984) and Ringland, Stapleford, and Magdaleno (1971), and by sound control system design principles.

6. Estimation of loop crossover frequencies is not a simple task, even using the separation factor suggested in item 5. Initially, at least, one should begin at the outermost loop and estimate the bandwidth

of the likely command or disturbance existing there (c_3 in Figure 9.5.18). The crossover frequency of the outer loop can then be estimated based on this bandwidth. Estimates of remaining crossover frequencies follow using the separation factor if 2–3. If the crossover frequency of the innermost loop exceeds 0.55 of the magnitude calculated by Equation (5) and Table 9.5.1, suitable reductions of all crossover frequencies should be made. The 0.55 factor is again suggested by the limited experimental data mentioned in item 5. Moderate adjustments to these estimates can be made by dynamically simulating the human/machine system (including as many realistic disturbances as possible) and varying the crossover frequencies as a group until outer loop performance is optimized.

Very few systems of engineering interest are single-variable systems, that is, possess a single output m_1 in response to a control input. In the block diagram of Figure 9.5.18, for example, the dynamics of the controlled element, like those of the operator, have been distributed over three loops. In reality, the controlled element is multivariable in nature as shown in Figure 9.5.19. In order to analyze the system as indicated in Figure 9.5.18, one must be able to interpret elements like Y_{c_1}, Y_{c_2}, and Y_{c_3} in terms of the multivariable dynamic system indicated in Figure 9.5.19. This can be done as follows:

$$
\begin{aligned}
Y_{c_1} &= \frac{m_1}{\delta} \\[2mm]
Y_{c_2} &= \frac{m_2}{m_1} = \frac{m_2}{\delta Y_{c_1}} = \frac{m_2}{\delta}\frac{1}{Y_{c_1}} \\[2mm]
Y_{c_3} &= \frac{m_3}{m_2} = \frac{m_3}{\delta Y_{c_1} Y_{c_2}} = \frac{m_3}{\delta}\frac{1}{Y_{c_1}}\frac{1}{Y_{c_2}}
\end{aligned}
\tag{38}
$$

9.5.3.2 The Modern Approach Using the Optimal Control Model

Being essentially a multivariable design technique, the OCM appears well suited to the multiloop control problem. No changes in Figure 9.5.12, nor in the computational technique used to generate the OCM, is needed in order to address the single-point, multiloop problem. Of course, with the OCM, no assumptions regarding loop structure are made. For each observed variable, the operator is also assumed to perceive its first time derivative, and white observation noise is added to all such variables. As in the single-loop problem, the observation noise is assumed to scale with the mean square value of the signal to which it is added. Again, motor noise and an operator pure time delay are utilized. Finally, the index of performance of Equation (18) defines the optimization problem.

Appropriate selection of the weighting matrices in the index of performance is obviously of some importance in the problem formulation. The following general guidelines can be used to formulate a single-point multiloop manual control problem using the OCM, given the controlled element of Figure 9.5.19.

1. The operator's effective time delay is set to a nominal value of 0.2 sec.
2. The covariances of the observation and motor noises are set to values derived from single-loop applications, that is, $V_y = n(0.01)\pi E[y^2]$ and $V_u \geq 0.003\pi E[u^2]$. Here,

$$
\begin{array}{lll}
y_1 = m_1 & y_3 = m_2 & y_5 = m_3 \\
y_2 = \dot{m}_1 & y_4 = \dot{m}_2 & y_6 = \dot{m}_3
\end{array}
\tag{39}
$$

and n is the number of explicitly displayed variables. In this case $n = 3$. Including the factor n is an approximate means of handling the effects of attention sharing on the part of the operator (Baron and Levison, 1973).

3. Variables to be included in the index of performance are determined from the task definition. Unfortunately, *a priori* selection of index of performance weighting coefficients is something of an art. The following procedure is recommended for preliminary analysis: The weighting coefficients of each variable appearing in the index of performance, with the exception of the control rate term, is selected as the reciprocal of the estimated "maximum allowable deviation" of that variable. This method

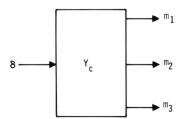

Fig. 9.5.19. A multivariable controlled element.

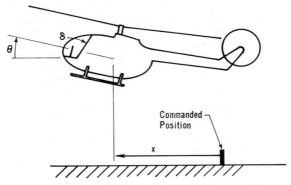

Fig. 9.5.20. A helicopter longitudinal hover task.

has been offered for optimal control formulations, in general (Bryson and Ho, 1975), and has shown promise in OCM applications (Baron, 1982). Finally, the weighting coefficient on control rate is selected to yield an "inner-loop" crossover frequency calculated in a manner identical to that just described for the classical approach using the crossover model. Strictly speaking, of course, no "inner-loop" exists for the OCM; however, the pertinent operator transfer functions can be calculated between the control variable and the appropriate "inner-loop" feedback variable to allow determination of a crossover frequency. Equations (21) and (22) can still be used to give an approximate relationship between the weighting coefficient on the inner-loop feedback variable (q) and the control rate (r).

9.5.3.3 An Example

A very simple example exercising the crossover model and optimal control model in a multiloop task can now be discussed. The task consists of the longitudinal control of a hovering helicopter, specifically, maintaining the vehicle over a specified position on the ground in the presence of atmospheric disturbances. Figure 9.5.20 shows the task. The simplified vehicle equations of motion can be given by

$$\dot{x} = u$$
$$\dot{u} = -g\theta + X_u u$$
$$\dot{\theta} = \delta + d_g \tag{40}$$
$$\dot{d}_g = -d_g + w$$

Here, $g = 9.8$ m/sec, and $X_u = -0.1$/sec. The variable d_g represents a pitch rate disturbance due to atmospheric turbulence. It is represented as white noise through a first-order filter with a break frequency of 1.0 rad/sec. The turbulence is assumed to have a root mean square value of 5.73 deg/sec (0.10 rad/sec).

The Crossover Model

Figure 9.5.21 shows the block diagram of the single-point multiloop manual control system. It is quite similar to Figure 9.5.18. Using the guidelines offered in the preceding discussion on the use of the crossover model for multiloop systems, one can proceed as follows: Crossover frequency selection begins with the inner rather than the outer loop. This is due to the fact that the outer-loop command is identically zero and the primary disturbance is the inner loop d_g. Equation (26) yields

$$\omega_{BW_{ce}} = 1.0 \text{ rad/sec} \tag{41}$$

Now the vehicle equations of motion, Equation (40) give

$$\frac{\theta}{\delta} = \frac{1}{s} \tag{42}$$

Equation (5) and Tables 9.5.1–3 indicate that

$$\omega_{c_\theta} = 0.55[\omega_{c_o} + 0.18\omega_{BW_{ce}}] = 2.56 \text{ rad/sec}$$
$$\tau_e = \tau_0 - 0.08\omega_{BW_{ce}} = 0.27 \text{ sec}$$
$$R = 0.5$$
$$\omega R = 3.0 \text{ rad/sec}$$

(43)

Note the 0.55 factor included in the ω_{c_θ} calculation and the fact that crossover regression is not suggested by these results. The largest R value in Table 9.5.2 was used since the operator is actually sensing three feedback variables. Now the crossover model indicates that the approximate form of the pilot equalization in the inner loop of Figure 9.5.21 is

$$Y_{p_\theta} \simeq \omega_{c_\theta}e^{-\tau_e s} = 2.56e^{-0.27s}$$

(44)

Applying a crossover frequency separation factor of 3 indicates that

$$\omega_{c_u} \simeq 0.85 \text{ rad/sec}$$
$$\omega_{c_x} \simeq 0.28 \text{ rad/sec}$$

(45)

The second of Equation (38) indicates

$$Y_{c_u} = \frac{-g}{s - X_u} \simeq \frac{-g}{s}\Bigg|_{|s| \simeq \omega_{c_u}}$$

(46)

Now again applying the crossover model to the effective open-loop system for the second-loop closure (deleting the time delay) yields

$$Y_{p_u} \simeq \frac{-\omega_{c_u}}{g}$$

(47)

The last of Equation (38) indicates

$$Y_{c_x} = \frac{1}{s}$$

(48)

Finally, applying the crossover model to the effective open-loop system for the third and final loop closure (deleting the time delay) yields

$$Y_{p_x} \simeq \omega_{c_x} = 0.28$$

(49)

A digital simulation of the human/machine system just defined was implemented for the purposes of calculating performance scores. These will be presented and compared with results obtained using the OCM.

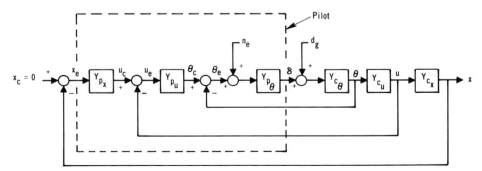

Fig. 9.5.21. The single-point, multiloop, serial representation of the manual control task of Figure 9.5.20 (crossover model applications).

The Optimal Control Model

Figure 9.5.22 shows the block diagram of single-point, multiloop parallel representation of the manual control task of Figure 9.5.20. The preceding discussion on the modern approach using the OCM was followed here. The motor noise ratio was set to 0.003. The index of performance was selected as

$$J = E\left\{ \lim_{T \to \infty} \frac{1}{T} \int_0^T \left[\frac{x^2(t)}{x_{max}^2} + \frac{u^2(t)}{u_{max}^2} + \frac{\theta^2(t)}{\theta_{max}^2} + r\delta^2(t) \right] dt \right\} \tag{50}$$

In Equation (50) the maximum allowable deviations were chosen as

$$\theta_{max} = 0.088 \text{ rad } (5°) \quad x_{max} = 7.8 \text{ m}$$
$$u_{max} = 2.57 \text{ m/sec} \quad \delta_{max} = 3.33 \text{ rad/sec}^2 \tag{51}$$

The maximum allowable deviations were chosen in the following way. First, θ_{max} is chosen. Its magnitude is immaterial. Equations (46) and (48) indicate simple integral relationships between θ and u, and u and x. The remaining maximum allowable deviations were chosen as

$$u_{max} = g \int_0^T \theta_{max} dt \tag{52}$$

$$x_{max} = \int_0^T u_{max} dt$$

The upper limits on the integrals were chosen as 3 sec simply on the basis of experience. In words, Equation (52) states that the maximum allowable deviation on u is the value that would be reached if θ were held at its maximum allowable value over some specified time period. Likewise, the maximum allowable deviation on x is the value that would be reached if u were held at its maximum allowable value over a specified time period. There is no analytical justification for this procedure. It is utilized simply because it exercises the maximum allowable deviation method often used in specifying weighting coefficients in linear quadratic regulator problems. Equations (21) to (23) were used to obtain an approximation for q/r, where "q" in this case is $1/(\theta_{max})^2$. The first of Equations (38) was used for Y_c in Equation (22). The resulting value of r was 0.3.

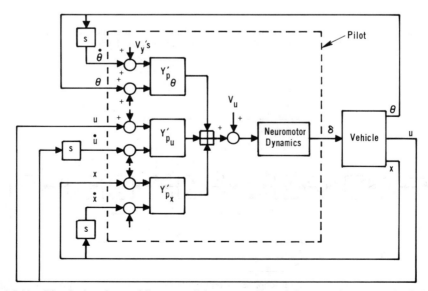

Fig. 9.5.22. The single-point, multiloop, parallel representation of the manual control task of Figure 9.5.20 (OCM applications).

Optimal Control Model (OCM). A model of the human operator consisting of a combination of a linear optimal estimator or Kalman filter and a linear regulator. The dynamic and information processing limitations of the human operator are incorporated in the OCM through the use of observation noise, time delays, and index-of-performance weighting coefficients.

Power. The power in a continuous function of time $f(t)$ can be defined as

$$P = \lim_{T \to \infty} \frac{1}{T} \int_0^\infty f^2(t)\,dt$$

and is equivalent to the mean square value of $f(t)$.

Power Spectral Density. A function of frequency ω that when integrated over all frequencies gives the total amount of power in a continuous function of time.

Preview Control System. A system in which information about the system error and the input is displayed to the operator.

Remnant. That portion of the human operator's output which is not linearly correlated with the input. Remnant can consist of actual noise injection by the operator or can represent modeling errors.

Root Mean Square Value. The square root of the mean square value.

Servomechanism. A device that measure the difference between an actual state (e.g., a position) and a desired state and uses the difference to drive the actual state toward the desired state.

Single-Loop System. A feedback system involving only a single feedback closure.

Single-Point System. A feedback system involving a single control variable that may have more than one feedback loop.

Single-Variable System. A system whose mathematical description involves a single differential equation.

State Variables. Those variables whose values describe all the information about a system at each instant of time. Values of the state variables at some instant of time along with the system model and future input behavior allow one to completely describe the future behavior of the system.

White Noise. A fictitious continuous function of time that posssesses a power spectral density which is constant over all frequencies.

REFERENCES

Allen, R. W., and McRuer, D. T. (1979). The man/machine control interface—Pursuit control. *Automatica, 15,* 683–686.

Athans, M. (1971). The role and use of the stochastic linear-quadratic-Guassian problem in control system design. *IEEE Transactions on Automatic Control, AC-16*(6), 529–534.

Baron, S. (1982). An optimal control model analysis of data from a simulated hover task. *Proceedings of the eighteenth annual conference on manual control,* pp. 186–206.

Baron, S., and Levison, W. H. (1973). A display evaluation methodology applied to vertical situation displays. *Proceedings of the ninth annual conference on manual control,* pp. 121–132.

Baron, S., and Levison, W. H. (1977). Display analysis with the optimal control model of the human operator. *Human Factors, 19*(5), 437–457.

Box, G. E. P., and Jenkins, G. M. (1976). *Time series analysis: forecasting and control.* San Francisco: Holden-Day.

Bryson, A. E., and Ho, Y. C. (1975). *Applied optimal control.* New York: Wiley.

Curry, R. E., Hoffman, W. C., and Young, L. R. (1976). Pilot modeling for manned simulation, vol. 1 (Air Force Fight Dynamics Laboratory, AFFDL-TR-76-124).

Curry, R. E., Kleinman, D. L., and Hoffman, W. C. (1977). A design procedure for control/display systems. *Human Factors, 19*(5), 421–436.

Doyle, K. M., and Hoffman, W. C. (1976). Pilot modeling for manned simulation, vol. 2 (Air Force Flight Dynamics Laboratory, AFFDL-TR-76-124).

Graham, D., and McRuer, D. T. (1971). *Analysis of nonlinear control systems.* New York: Dover, Chap. 10.

Hess, R. A. (1981). Pursuit tracking and higher levels of skill development in the human pilot. *IEEE Transactions on Systems, Man, and Cybernetics, SMC-11*(4), 262–273.

Hess, R. A. (1984). Analysis of aircraft attitude control systems prone to pilot induced oscillations. *Journal of Guidance, Control, and Dynamics, 7*(1), 106–112.

Hess. R. A. (1985). A model based theory for analyzing human control behavior. In W. B. Rouse, Ed. *Advances in Man-Machine Systems Research.* New York: North Holland, Vol. 2, 129–175.

Hess, R. A., and Beckman, A. (1984), An engineering approach to determining visual information requirements for flight control tasks. *IEEE Transactions on Systems, Man and Cybernetics, SMC-14*(2), 286–298.

Kleinman, D. L., Levison, W. H., and Baron, S. (1970). An optimal control model of human response, part I: Theory and validation. *Automatica, 6*(3), 357–369.

Kwakernaak, H., and Sivan, R. (1972). *Linear optimal control systems.* New York: Wiley-Interscience.

Lee, Y. W. (1960). *Statistical theory of communication.* New York: Wiley, Chap. 13.

Levison, W. H., and Junker, A. M. (1977). A model for the pilot's use of motion cues in roll-axis tracking tasks (Aerospace Medical Research Laboratory, AMRL-TR-77-40).

Levison, W. H., Kleinman, D. L., and Baron, S (1969). A model for human controller remnant. *IEEE Transactions on Man-Machine Systems, MMS-10*(4), 101–108.

Levison, W. H., Baron, S., and Junker, A. M. (1976). Modeling the effects of environmental factors on human control and information processing (Aerospace Medical Research Laboratory, AMRL-TR-76-74).

McRuer, D. T. (1980). Human dynamics in man-machine systems. *Automatica, 16*(3), 237–253.

McRuer, D. T., and Krendel, E. (1957). Dynamic response of human operators (Wright Air Development Center, WADC TR 56-524).

McRuer, D. T., and Krendel, E. (1974). Mathematical models of human pilot behavior (AGARDograph No. 188).

McRuer, D. T., Graham, D., Krendel, E., and Reisener, W., Jr. (1965). Human pilot dynamics in compensatory systems (Air Force Flight Dynamics Laboratory: AFFDL-TR-65-15).

McRuer, D. T., Hofmann, L. G., Jex, H. R., Moore, G. P., Phatak, A. V., Weir, D. H., and Wolkovitch, J. (1968). New approaches to human-pilot/vehicle dynamic analysis (Air Force Flight Dynamics Laboratory, AFFDL-TR-67-150).

Miller, J. (1982). *The body in question.* New York: Random House, Chap. 8.

Ogata, K. (1970). *Modern control engineering.* Englewood Cliffs, NJ: Prentice-Hall.

Reid, L. D., and Drewell, N. H. (1972). A pilot model for tracking with preview. *Proceedings of the eighth annual conference on manual control,* pp. 191–204.

Ringland, R. F., Stapleford, R. L., and Magdaleno, R. E. (1971). Motion effects on an IFR hovering task (National Aeronautics and Space Administration, NASA CR-1933).

Shinners, S. M. (1974). "Modeling of human operator performance utilizing time series analysis. *IEEE Transactions on Systems, Man, and Cybernetics, SMC-4*(5), 446–458.

Stapleford, R. L., Craig, S. J., and Tennant, J. A. (1969). Measurement of pilot describing functions in single-controller multiloop tasks. (National Aeronautics and Space Administration, NASA CR-1238).

Stapleford, R. L., McRuer, D. T., and Madgaleno, R. E. (1967). Pilot describing function measurement in a multiloop task. *IEEE Transactions on Human Factors in Electronics, HFE-8*(2), 113–125.

Tustin, A. (1947). The nature of the operator's response in manual control and its implication for controller design. *Journal of the IEE, 94* (Part IIA, No. 2).

Wier, D. H., and McRuer, D. T. (1972). Pilot dynamics for instrument approach tasks: Full panel multiloop and flight director operations (National Aeronautics and Space Administration, NASA CR-2019).

CHAPTER 9.6
SUPERVISORY CONTROL

THOMAS B. SHERIDAN

Massachusetts Institute of Technology
Cambridge, Massachusetts

This chapter is a tutorial on "supervisory control," drawing heavily on experiments done at MIT. It is not a comprehensive or even-handed review of the literature in human–computer interaction, monitoring, diagnosis of failures, human error, mental workload, or other closely related topics. Recent reviews by Rouse (in press) and Moray (in press) cover these aspects more fully.

9.6.1 WHAT IS SUPERVISORY CONTROL?

The term *supervisory control* is derived from the close analogy between the characteristics of a supervisor's interaction with subordinate human staff members and a person's interaction with "intelligent" automated subsystems. A supervisor of people gives general directives that are understood and translated into detailed actions by staff members. In turn, staff members aggregate and transform detailed information about process results into summary form for the supervisor. The degree of intelligence of staff members determines the level of involvement of their supervisor in the process. Automated subsystems permit the same sort of interaction to occur between a human supervisor and the process (Ferrell and Sheridan, 1967; Sheridan, Fischhoff, Posner, and Pew, 1983). Supervisory control behavior is interpreted to apply broadly to vehicle control (aircraft and spacecraft, ships, and undersea vehicles), continuous process control (oil, chemicals, power generation), and robots and discrete tasks (manufacturing, space, undersea, mining).

In the strictest sense, the term supervisory control indicates that one or more human operators are setting initial conditions for, intermittently adjusting, and receiving information from a computer that itself closes a control loop (i.e., interconnects) through external sensors, effectors, and the task environment. In a broader sense, supervisory control is involved with a computer makes complex transformation of data to produce integrated (chunked) displays, or retransforms operator commands to generate detailed control actions. Figure 9.6.1 compares supervisory control with direct manual control and full automatic control. Figures 9.6.1c and 9.6.1d characterize supervisory control in the strict formal sense; Figure 9.6.1b characterizes supervisory control in the latter (broader) sense.

The essential difference between these two characterizations of supervisory control is that in the first and stricter definition the computer can act on new information independently of and with only blanket authorization and adjustment from the supervisor; that is, the computer implements discrete sets of instructions by itself, closing the loop through the environment. In the second definition the computer's detailed implementations are "open loop," that is, feedback from the task has no effect on computer control of the task except through the human operator. The two situations may appear similar to the supervisor, since he or she always sees and acts through the computer (analogous to a staff) and therefore may not know whether it is acting open loop or closed loop in its fine behavior. In either case the computer may function principally on the efferent or motor side to implement the supervisor's commands (e.g., do some part of the task entirely and leave other parts to the human, or provide some control compensation to ease all of the task for the human). Alternatively the computer may function principally on the display side (e.g., to integrate and interpret incoming information from below, or to give advice to the supervisor as to what to do next as an "expert system"). Or it may do both.

9.6.2 THE EMERGENCE OF SUPERVISORY CONTROL IN TECHNOLOGICAL SYSTEMS

Supervisory control is emerging rapidly in many industrial, military, medical, and other contexts, although this form of human interaction with technology is relatively little recognized or understood in a formal way.

From the pyramid building pharaohs of Egypt through all of the history of technology there surely has been concern about how best to extend the capabilities of human workers. Early in the present century, against the backdrop of the newly mechanized production line, Taylor's "scientific management" (Taylor, 1911) catalyzed a formal intellectual consciousness about the human factors involved. Taylor intended a new interest in the sensori-motor aspects of human performance. What he did not intend was the subsequent criticism of his essentially mechanistic approach that it was dehumanizing.

The 1940s and 1950s saw "human factors" ("ergonomics" in Europe) emerge, first in essentially empirical "knobs and dials" form, concentrating on the human–machine interface itself. This was supported over the next decade by the theoretical underpinnings of "man–machine systems" (Sheridan

The writing of this chapter and much of the research discussed herein were supported under the Office of Naval Research Contract No. N00014-83-K-0193.

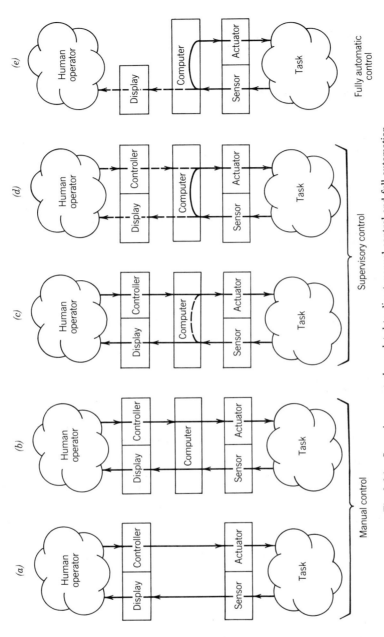

Fig. 9.6.1. Supervisory control as related to direct manual control and full automation.

and Ferrell, 1974). Such theories included control, information, signal detection and decision theories originally developed for application to physical systems but now explicitly applied to the human operator. As contrasted with human factors engineering at the interface, man–machine systems analysis considers characteristics of the entire causal "loop" of decision, communication, control, and feedback—through the operator's physical environment and back again to the human.

From the late 1950s the computer began to intervene in the causal loop: electronic compensation and stability augmentation for control of aircraft and similar systems, electronic filtering of signal patterns in noise, electronic generation of simple displays. It was obvious that if vehicular or industrial systems were equipped with sensors that could be read by computers, and by motors that could be driven by computers, then—even though the overall system was still very much human controlled—control loops between those sensors and motors could be closed automatically. Thus the chemical plant operator was relieved of keeping the tank at a given level or the temperature at a reference—he or she needed only to set in that desired level or temperature signal from time to time. So, too, after the autopilot was developed for the aircraft the human pilot needed only to set in the desired altitude or heading; an automatic system would strive to achieve this reference, with the pilot monitoring to ensure that the aircraft did in fact go where desired. The automatic building elevator, of course, has been in place for many years, and is certainly one of the first implementations of supervisory control. Recently, developers of new systems for word processing and handling of business information (i.e., without the need to control any mechanical processes) have begun thinking along similar lines [e.g., see Card (1984)].

The full generality of the idea of supervisory control came to the author and his colleagues (Ferrell and Sheridan, 1967; Sheridan, 1960) as part of research on how people on earth might control vehicles on the moon through 3 sec round-trip time delays (imposed by the speed of light). Under such constraint remote control of lunar roving vehicles or manipulators was shown to be possible only by performing in "move-and-wait" fashion. This means the operator can commit only to a small incremental movement "open loop," that is, without feedback (which actually is as large a movement as is reasonable without risking collision or other error), then stopping and waiting one delay period for feedback to "catch up," then repeating the process in steps until the task is completed.*

It was shown that if, instead of the human operator remaining within the control loop, he or she communicates a goal state relative to the remote environment, and if the remote system incorporates the capability to measure proximity to this goal state, then the achievement of this goal state can be turned over to a remote subordinate control system for implementation. In this case there is no delay in the control loop implementing the task and thus there is no instability.

There necessarily remains, of course, a delay in the supervisory loop. This delay in the supervisor's confirmation of desired results is acceptable so long as (1) the subgoal is a sufficiently large "bite" of the task, (2) the unpredictable aspects of the remote environment are not changing too rapidly (i.e., disturbance bandwidth is low), and (3) the subordinate automatic system is trustworthy. More will be said of each of these points.

If these conditions obtain, and as computers gradually become more capable both in hardware and software (and as "machine intelligence" finally makes its real if modest appearance), it is evident that telemetry transmission delay is in no way a prerequisite to the usefulness of supervisory control. The incremental goal specified by the human operator need not be simply a new steady-state reference for a servomechanism (as in resetting a thermostat) in one or even several dimensions (e.g., resetting both temperature and humidity, or commanding a manipulator endpoint to move to a new position including three translations and three rotations relative to its initial position). Each new goal statement can be the specification of a whole trajectory of movements (as the performance of a dance or a symphony) together with programmed branching conditions (what to do in case of a fall or a broken violin string, or how to respond contingent upon audience applause).

In other words the incremental goal statement is a program of instructions in the full sense of a computer program, which makes the human supervisor an intermittent real-time computer programmer, acting relative to the subordinate computer much the same as a teacher or parent or boss behaves relative to a student or child or subordinate worker. The size and complexity of each new program is necessarily a function of how much the computer can (be trusted to) cope with at once, which in turn depends on the computer's own sophistication (knowledge base) and the complexity (uncertainty) of the task.

Supervisory control is emerging in various forms in various industries—usually without being called

---

* Attempts to drive or manipulate continuously only produce instability, as simple control theory predicts (i.e., where loop gain exceeds unity at a frequency such that the loop time delay is one half cycle, instead of errors being nulled out they are only reinforced).
Performing remote manipulation with delayed force feedback was shown by Ferrell (1966) to be essentially impossible since forces at unexpected times act as significant disturbances to produce instability. At least the visual feedback can be ignored by the operator.

that. (More likely, each developer or vendor has its own cute acronym emphasizing how "smart" and easy it is to use the new product.) Aircraft autopilots are now "layered," meaning the pilot can select among various forms and levels. At the lowest level he or she can set in a new heading or rate of climb. Or he or she can program a sequence of heading changes at various way-points, or a sequence of climb rates initiated at various altitudes. Or he or she can program his or her inertial guidance system to take him or her to (within a fraction of a mile of) a distant city. Given the existence of certain ground-based equipment, he or she can program an automatic landing on a given runway, and so on. Wiener and Curry (1980) provide a good review of how such automation is creeping into the aircraft flight deck. Modern chemical plants can similarly be programmed to perform heating, mixing, and various other processes according to a time line, but including various sensor-based conditions for shutting down or otherwise aborting the operation.

More and more a multiplicity of computers are used in a supervisory control system, as shown in Figure 9.6.2. One typically large computer is in the control room to generate displays and interpret commands. We call this a "human-interactive computer" (HIC), part of a "human-interactive system" (HIS). It in turn forwards that command to various microprocessors that actually close individual control loops through their own associated sensors and effectors. We call these "task-interactive computers" (TICs), each part of its own "task-interactive system" (TIS).

The examples cited above characterize the first or stricter definition of supervisory control previously given (Figure 9.6.1c and 9.6.1d), where the computer, once programmed, makes use of its own artificial sensors to sense completion of the assigned task. Many familiar systems such as automatic washing machines, dryers, dishwashers, or stoves, once programmed, perform their operations "open loop," that is, there is no measurement or knowledge of results. If the task can be performed in such open-loop fashion, and if the human supervisor can anticipate the task conditions and is good at selecting the right open-loop program, there is no reason not to employ this approach. To the human supervisor whether the lower-level implementation is open loop or closed loop is often opaque and/or of no concern; his or her only concern is whether the goal is achieved satisfactorily. For example, a programmable microwave oven without the temperature sensor in place operates open loop, while the same oven with the temperature sensor in place operates closed loop. To the human supervisor/programmer they both look the same.

A very important aspect of supervisory control is the ability of the computer to "package" information for visual display to the human operator, including data from many sources, from past, present, or even predicted future, and presented in words, graphs, symbols, pictures or some combination. The ubiquitous examples of such integrated displays in aircraft and air traffic control, chemical and power plants, and various other industrial or military settings are too numerous to review here.

General interest in supervisory control became evident in the mid-1970s (Edwards and Lees, 1974; Sheridan and Johannsen, 1976), and continues to grow. A recent report by the National Research Council (Sheridan and Hennessy, 1984) outlines current problems of supervisory control, especially with regard to experimental research (which is particularly difficult because of the inherent complexity and capital cost of real supervisory control systems, inhibiting both simulation and experimental control) and system design.

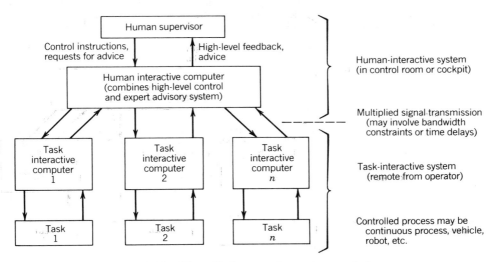

Fig. 9.6.2. Hierarchical nature of supervisory control.

9.6.3 SUPERVISORY ROLES, LOCI, AND LEVELS OF HUMAN AND COMPUTER

The human supervisor's *roles* are: (1) *planning* off-line what task to do and how to do it; (2) *teaching* (or programming) the computer what was planned; (3) *monitoring* the automatic action on-line to make sure all is going as planned and to detect failures; (4) *intervening*, which means the supervisor takes over control after the desired goal state has been reached satisfactorily, or he or she interrupts the automatic control in emergencies to specify a new goal state and reprogram a new procedure; and (5) *learning* from experience so as to do better in the future. These are usually time-sequential steps in task performance.

We may view these steps as being within three nested loops, as shown in Figure 9.6.3. The innermost loop, monitoring, closes on iteself, that is, evidence of something interesting or completion of one part or cycle of a monitoring strategy leads to more investigation and monitoring. We might include minor on-line tuning of the process as part of monitoring. The middle loop closes from intervening back to teaching, that is, human intervention usually leads to programming of a new goal state to the process. The outer loop closes from learning back to planning; intelligent planning for the next subtask is usually not possible without learning from the last one.

The three supervisory loops operate at different time scales relative to one another. Revisions in fine-scale monitoring behavior take place at brief intervals. New programs are generated at somewhat longer intervals. Revisions in significant task planning occur only at still longer intervals. These differences in time scale further justify Figure 9.6.3.

For each of the five roles or stages of the supervisory process there are three *loci* of function in a physiological sense: *sensory* functions (accessing displays, observing, perceiving); *cognitive* functions internal to the supervisor (evaluating the situation, accessing memory, making decisions); and *response* functions. S, C, R are the classic designators to differentiate these functional elements of causation through the operator.

Finally, we may appeal to the *levels* of behavior introduced by Rasmussen (1976); *skill-based* behavior (continuous, typically well-learned, sensory-motor behavior analogous to what can be expected from a servomechanism); *rule-based* behavior (what an "artificially intelligent" computer can do in recognizing a pattern of stimuli, then triggering an "if–then" algorithm to execute an appropriate response); and finally *knowledge-based* behavior ("high-level" situation assessment and evaluation, consideration of alternative actions in light of various goals, decision and scheduling of implementation—a form of behavior machines are not now good at).

Considering the above three metacharacteristics of supervision, namely, *role, loci,* and *level,* as three independent dimensions of such behavior, we may then represent any behavioral element to be within one cell of a three-dimensional array, as shown at the top of Figure 9.6.4. Immediately below this array of supervision behaviors in Figure 9.6.4 lies the human-interactive computer (HIC). This is conceived to be a large enough computer to communicate in a human-friendly way, using near-natural language, good graphics, and so on. This includes being able to accept and interpret commands and to give the supervisor useful feedback. The HIC should be able to recognize patterns in data sent up to it from below and decide on appropriate algorithms for response which it sends down as instructions. Eventually the HIC should be able to run "what would happen if . . ." simulations and be able to give useful advice from a knowledge base, that is, include an "expert system."

The HIC, located near the supervisor in a control room or cockpit, may communicate across a "barrier" of time or space with a multiplicity of task-interactive computers (TICs), which probably are microprocessors distributed throughout the plant or vehicle. The latter are usually coupled intimately with artificial sensors and actuators, in order to deal in low-level language and to close relatively tight control loops with objects and events in the physical world.

The human supervisor can be expected to communicate with the HIC intermittently in information "chunks" (alphanumeric sentences, video pages, etc.) while the task communicates with the TIC continuously in computer words at the highest possible bit rates. The availability of these computer aids means that the human supervisor, while retraining the knowledge-based behavior for him- or herself, is likely to "down-load" some of the rule-based and almost all of the skill-based programs into the HIC. The HIC, in turn, should down-load a few of the rule-based programs, and most of the skill-based programs, to the appropriate TICs.

In the sections which follow the various supervisory roles are discussed in more detail, bringing in examples of research problems and prototype systems to aid the supervisor in these roles.

9.6.4 PLANNING AND LEARNING: COMPUTER REPRESENTATIONS OF RELEVANT KNOWLEDGE

The first and fifth supervisory roles previously described—planning and learning—may be considered together since they are similar activities in many ways. Essentially, in planning the supervisor asks "what would happen if . . . ?" questions of the accumulated knowledge base and considers what the

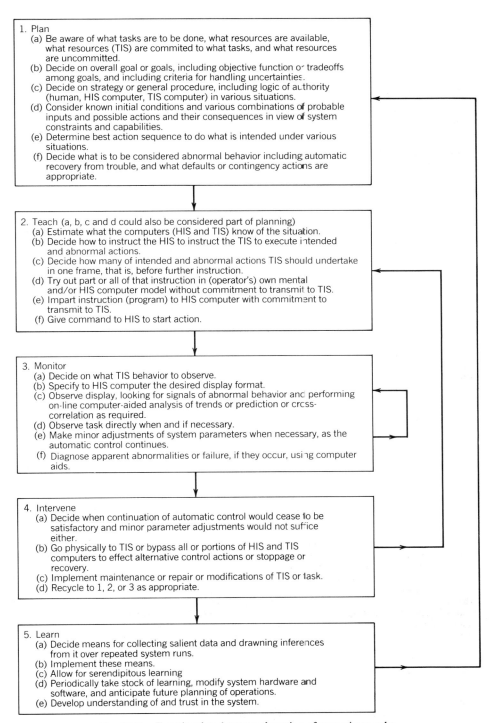

1. Plan
   (a) Be aware of what tasks are to be done, what resources are available, what resources (TIS) are commited to what tasks, and what resources are uncommitted.
   (b) Decide on overall goal or goals, including objective function or tradeoffs among goals, and including criteria for handling uncertainties.
   (c) Decide on strategy or general procedure, including logic of authority (human, HIS computer, TIS computer) in various situations.
   (d) Consider known initial conditions and various combinations of probable inputs and possible actions and their consequences in view of system constraints and capabilities.
   (e) Determine best action sequence to do what is intended under various situations.
   (f) Decide what is to be considered abnormal behavior including automatic recovery from trouble, and what defaults or contingency actions are appropriate.

2. Teach (a, b, c and d could also be considered part of planning)
   (a) Estimate what the computers (HIS and TIS) know of the situation.
   (b) Decide how to instruct the HIS to instruct the TIS to execute intended and abnormal actions.
   (c) Decide how many of intended and abnormal actions TIS should undertake in one frame, that is, before further instruction.
   (d) Try out part or all of that instruction in (operator's) own mental and/or HIS computer model without commitment to transmit to TIS.
   (e) Impart instruction (program) to HIS computer with commitment to transmit to TIS.
   (f) Give command to HIS to start action.

3. Monitor
   (a) Decide on what TIS behavior to observe.
   (b) Specify to HIS computer the desired display format.
   (c) Observe display, looking for signals of abnormal behavior and performing on-line computer-aided analysis of trends or prediction or cross-correlation as required.
   (d) Observe task directly when and if necessary.
   (e) Make minor adjustments of system parameters when necessary, as the automatic control continues.
   (f) Diagnose apparent abnormalities or failure, if they occur, using computer aids.

4. Intervene
   (a) Decide when continuation of automatic control would cease to be satisfactory and minor parameter adjustments would not suffice either.
   (b) Go physically to TIS or bypass all or portions of HIS and TIS computers to effect alternative control actions or stoppage or recovery.
   (c) Implement maintenance or repair or modifications of TIS or task.
   (d) Recycle to 1, 2, or 3 as appropriate.

5. Learn
   (a) Decide means for collecting salient data and drawning inferences from it over repeated system runs.
   (b) Implement these means.
   (c) Allow for serendipitous learning
   (d) Periodically take stock of learning, modify system hardware and software, and anticipate future planning of operations.
   (e) Develop understanding of and trust in the system.

Fig. 9.6.3. Functional and temporal nesting of supervisory roles.

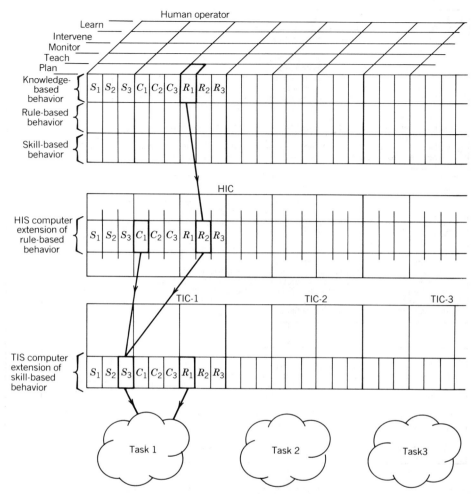

Fig. 9.6.4. Attributes and offloading among the human, HIS, and TIS.

implications are for hypothetical control decisions. In learning the supervisor asks "what did happen?" questions of the data base for the more recent subtasks, and considers whether initial assumptions and final control decisions were appropriate.

The designer of an automatic control system or a manual control system (see Chapter 2.7 on manual control) must ask him- or herself "what variables do I wish to make do what, subject to what constraints and what criteria"? The planning role in supervisory control requires that the same kinds of questions be answered, because, in a sense, the supervisor is redesigning an automatic control system each time he or she programs a new task and goal state. Absolute constraints on time, tools, and other resources available need to be clear, as do the criteria of tradeoff between time, dollars and resources spent, accuracy, and risk of failure.

Just as computer simulation figures into planning so too it figures into supervisory control—the difference being that such stimulation may be more likely subject to time stress in supervisory control. Simulation requires acquiring some idea of how the process or system to be controlled works, that is, a set of equations relating the various controllable variables, the various uncontrollable but measurable variables (disturbances), and the degree of unpredictability (noise) on measured system response variables. This is a common representation of knowledge. Given measured inputs and outputs there are well-established means to infer the equations—if the processes are approximately linear and differentiable.

Once such a model is in place the supervisor can posit hypothetical inputs and observe what the outputs would be. Also, one may use such a process model as an "observer" (in the sense of modern control theory). Namely, when control signals are put into both the model and actual processes, and

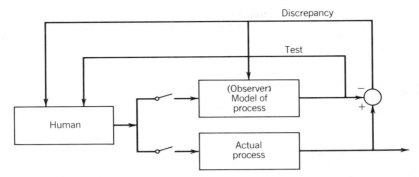

Fig. 9.6.5. Use of computer-based observer as an aid to supervisor.

the model parameters are then trimmed to force certain model outputs to conform to corresponding actual process outputs that can be measured (Figure 9.6.5), then other process outputs that are inconvenient to measure may be estimated ("observed") from the model. Just as this is a theoretical prerequisite to optimal automatic control of physical systems, so it is likely to be useful practice to aid humans in supervisory control (Sheridan, 1984a).

A different type of knowledge representation is that used by the artificial-intelligence (AI) community. Here knowledge is usually couched in the form of "if–then" logical statements called "production rules," semantic association networks, and similar forms, and is usually programmed in LISP. The input to a simulation program usually represents in cardinal numbers a hypothetical physical input to a simulated physical system. In contrast, the input to the AI knowledge base can be a question about relationships for given data or a question about data for given relationships. This can be in less restrictive ordinal form (e.g., networks of diadic relations) or in nominal form (e.g., lists).

Currently there is a great interest in how best to transfer expertise from the human brain (knowledge representation, mental model) into the corresponding representation or model within the computer, how best to transfer it back, and when to depend on each of those sources of information. This research on mental models has a lively life of its own (Falzon, 1982; Gentner and Stevens, 1983; Rouse and Morris, 1984; Sheridan, 1984a) quite independent of supervisory control.

According to Card (1984) the "third law of interactive application systems" (there apparently being no first or second laws) states that every interactive application system has to contain knowledge about two domains: (1) the subject area of application and (2) how to interact with the user. Card illustrates the point by describing a game called WEST (Burton and Brown, 1982) designed to motivate children to practice simple arithmetic. At each move the player is given three random digits and must combine them using addition, subtraction, multiplication, and/or division with at least two digits in brackets, the idea being to determine the correct number of steps to land on a desirable reward. The computer keeps track of the player's decisions for each move, forms a running statistical model of his or her tendencies and oversights, and after each move gives advice as to what he or she might have done or could do in the future. Such a smart system, of course, raises the question of the degree to which the computer should play "coach" as contrasted to that of a delegated decision maker that is smarter than its boss.

9.6.5 TEACHING THE COMPUTER

Teaching or programming a task, including a goal state and a procedure for achieving it, and including constraints and criteria, can be formidable or quite easy, depending on the command hardware and software. By command hardware is meant the way in which human response—hand, foot, or voice— is converted to physical signals to the computer. Command hardware can be either *analogic* or *symbolic*. Analogic means that there is a spatial or temporal isomorphism among human response, semantic meaning, and/or feedback display. For example, moving a control up rapidly to increase the magnitude of a variable quickly, which causes a display indicator to move up quickly, would be a proper analogic correspondence.

Symbolic command, by contrast, is accomplished by depressing one or a series of keys (as in typing words on a typewriter), or uttering one or a series of sounds (as in speaking a sentence), each of which has a distinguishable meaning. For symbolic commands a particular series or concatenation of such responses has a different meaning from other concatenations. Spatial or temporal correspondence to the meaning or desired result is not a requisite. Sometimes analogic and symbolic can be combined, for example, where up–down keys are both labeled and positioned accordingly.

It is natural for people to intermix analogic and symbolic commands or even to use them simultaneously. Typical industrial robots are taught by a combination of grabbing hold and leading the end point of the manipulator around in space relative to the workpiece, at the same time using a switch box on a wire (a "teach pendant") to key in codes for start, stop, speed, etc., between various reference positions. This happens, for example, when a person talks and points at the same time, or plays the piano and conducts a choir with head or free hand.

Supervisory command systems have been developed for mechanical manipulators which utilize both analogic and symbolic interfaces with the supervisor and which enable teaching to be both rapid and available in terms of high-level language. Brooks (1979) developed such a system he called SUPER-MAN, which allows the supervisor to use a master arm to identify objects and demonstrate elemental motions. He showed that even without time delay for certain commands, which refer to predefined locations, supervisory control—including both teaching and execution—took less time and had fewer errors than manual control.

Yoerger (1982) developed a more extensive and robust supervisory command system that enables a variety of arm–hand motions to be demonstrated, defined, called on, and combined under other commands. In one set of experiments Yoerger compared three different procedures for teaching a robot arm to perform a continuous seam weld along a complex curved workpiece. The end effector (welding tool) had to keep 1 in. away and retain an orientation perpendicular to the curved surface to be welded and move at constant speed.

Yoerger tested his subjects in three command (teaching) modes. The first mode was for the human teacher to first move the master (with slave following in master–slave correspondence) relative to the workpiece in the desired trajectory. The computer would memorize the trajectory, and then cause the slave end effector to repeat the trajectory exactly. The second mode was for the human teacher to move the master (and slave) to each of a series of positions, pressing a key to identify each. The human would then key in additional information specifying the parameters of a curve to be fit through these points and the speed at which it was to be executed, and the computer would then be called upon for execution. The third mode was to use the master–slave manipulator to contact and trace along the workpiece itself, to provide the computer with knowledge of the location and orientation of the surfaces to be welded. Then, using the typewriter keyboard, the human teacher would specify the positions and orientations of the end effector *relative to* the workpiece. The computer could then execute the task instructions relative to the geometric references given.

Figure 9.6.6 shows the average results for three experimental subjects, based on running measures of both position error and orientation error in system performance following teaching in each of the three modes. Identifying the geometry of the workpiece analogically, and then giving symbolic instructions relative to it, proved the constant winner. The reasons for this advantage apparently are the same as for Brooks's results previously described, provided of course the time spent in the teaching loop (8 of Figure 9.6.11) is sufficiently short.

Along with the advance of computer science in natural language understanding, it will be important to learn how to cope with the "fuzziness" (Zadeh, 1984) inherent in the way people think about, and therefore communicate about, their tasks. That is, both memorized "rules" and typed or spoken messages would by nature be sentences consisting of fuzzy terms. As an example, a fuzzy rule for driving a car might be: "If your car is going *fast,* and if the car ahead is *very* close or *moderately* close and going *slow,* brake." The italicized terms are fuzzy sets, which may be defined with varying degrees of "membership" over a range of numerical values of speed and distance. Given a number of statements like the one above, and given membership functions for each fuzzy term over the physical variables, the "relative truth" of each of several control actions (e.g., brake, accelerate, coast) can be determined. Buharali and Sheridan (1982) demonstrated that a computer could be taught to drive a car by repeatedly giving rules, where the computer thereby would come to "know what it didn't know" with regard to various combinations of conditions and could ask the supervisor–teacher for additional rules to cover its "domains of ignorance."

9.6.6 MONITORING OF DISPLAYS AND DETECTION OF FAILURES

The human supervisor monitors the automated execution of the task in order to ensure proper control. This includes intermittent adjustment or trimming if the process performance remains within satisfactory limits, to detect if and when it goes outside limits, and to diagnose failures or other abnormalities. The subject of failure detection in human–machine systems has received considerable attention lately (Rasmussen and Rouse, 1981). Moray (in press) regards such failure detection and diagnosis as the most important human supervisory role. I prefer the view that all five supervisory roles are essential and no one can be placed above the others.

The supervisory controller tends to be removed from full and immediate knowledge about the controlled process. The physical processes he or she must monitor tend to be large in number and tend to be distributed widely in space (e.g., around a ship or plant). The physical variables may not be immediately sensible by him or her (e.g., steam flow and pressure) and may be computed from remote measurements on other variables. Sitting in the control room or cockpit the supervisor is dependent on various artificial displays to give him or her feedback of results as well as knowledge

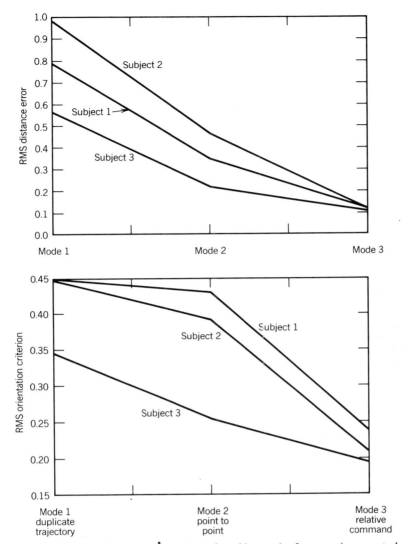

Fig. 9.6.6. Results of Yoerger's comparison of teaching modes for supervisory control.

of new reference inputs or disturbances. These factors greatly affect how he or she detects and diagnoses abnormalities in the process, but whether removal from active participation in the control loop makes it harder (Ephrath and Young, 1981) or easier (Curry and Ephrath, 1977) remains an open question. Curry and Nagel (1974), Niemala and Krendel (1974), Rouse (1974), Gai and Curry (1978), and Wickens and Kessel (1979, 1981) have studied various psychophysical aspects of this problem.

In traditional control rooms and cockpits the tendency has been to provide the human supervisor with an individual and independent display of each and every variable, and for a large fraction of these to provide a separate additional alarm display that lights up when the corresponding variable reaches or exceeds some value. Thus modern aircraft may easily have over 1000 displays and modern chemical or power plants 5000 displays. In the writer's experience in one nuclear plant training simulator during the first minute of a "loss of coolant accident" 500 displays were shown to have changed in a significant way, and in the second minute 800 more.

Clearly no human being can cope with so much information coming simultaneously from so many seemingly disconnected sources. Just as clearly such signals in any real operating system actually are highly correlated. In real life situations in which we move among people, animals, plants, or buildings our eyes, ears, and other senses easily take in and comprehend vast amounts of information—just as much as in the power plant. Our genetic makeup and experience enable us to integrate the bits of information from different parts of the retina and from different senses from one instant to the next—presumably because the information is correlated. We say we "perceive patterns," but do not pretend to understand how. In any case the challenge is to designs displays in technological systems to somehow integrate the information to enable the human operator to perceive patterns in time and space and across the senses. As with teaching (command), the forms of display may be either analogic (e.g., diagrams, plots) or symbolic (e.g., alphanumerics) or some combination.

In the nuclear power industry the "safety parameter display system" (SPDS) is now required of all plants in some form. The idea of the SPDS is to select a small number (e.g., 6–10) of variables that tell the most about plant safety status, and to display them in "integrated" fashion, such that by a glance the human operator can see whether something is abnormal, if so what, and to what relative degree. Figure 9.6.7 shows an example of an SPDS. Figure 9.6.7a gives the "high-level" or overview display (a single computer "page"). If the operator wishes more detailed information about one variable or subsystem he or she can "page down" (select lower levels), such as Figure 9.6.7b. These can be diagrams having lines or symbols that change color or flash to indicate changed status, and alphanumerics to give quantitative or more detailed status. These can also be bar graphs or cross plots, or "integrated" in other forms. One novel technique is the "Chernoff face" (Figure 9.6.7c) in which the shapes of eyes, ears, nose, and mouth systematically differ to indicate different values of variables, the idea being that facial patterns are easily perceived. The Nuclear Regulatory Commission, fearful that some enterprising designer might employ this technique before it was proved, formally forbade it as an acceptable SPDS.

Since detection and diagnosis of system failure is a critical task for the supervisor, computer aiding by the HIC in comparing, computing, and displaying has great potential. Various techniques have been proposed for doing this. One such technique (Tsach, Sheridan and Buharali, 1983) continuously compares key measurements from the plant to corresponding variables of an on-line computer model; then a computer-graphic display focuses the operator's attention on the discrepancies which indicate abnormality. Figure 9.6.8 shows one type of iconic display developed for this system—a polygon whose vertices indicate the degree to which each variable (of one subsystem in this case) is below or above a normal range (torus). The display therefore "points" to corresponding discrepancies between the measured and model variables as they evolve in time.

As previously noted (Figure 9.6.5) an important potential of the HIC is for modeling the controlled process. Such a model may then be used to generate a display of "observed" state variables that cannot be seen or measured directly. Another use is to run the model in fast time to predict the future, given of course that the model is calibrated to reality at the beginning of each such predictive run. A third use, now being developed for application to remote control of manipulators and vehicles in space, helps the human operator cope with telemetry time delays (as discussed in Section 9.6.2, wherein video feedback is necessarily delayed by at least several seconds). By sending control signals to a computer model in addition to the actual manipulator or vehicle (Figure 9.6.9), then using the model as a basis for superposing the corresponding graphic model on the video, the graphic model will "lead" the video picture and indicate what the video will do several seconds hence. This has been shown to speed up the execution of simple manipulation tasks by 70–80% (Noyes and Sheridan, 1984).

A final aspect of supervisory monitoring and display concerns format adaptivity—the ability to change both the format and the logic of the display as a function of the situation. Displays in aerospace and industrial systems now have fixed formats (e.g., the labels, scales, ranges are designed into the display). Alarms have fixed set points. However, future computer-generated displays even for the same variables may be different at various mission stages or in various conditions. Thus formats may differ for aircraft take-off, landing, and on-route travel, and be different for plant startup, full capacity operation, and emergency shutdown. Some alarms have no meaning, or may be expected to go off when certain equipment is being tested or taken out of service. In such a case adaptive formatted alarms may be suppressed, or the setpoints changed automatically to correspond to the operating mode. Future displays and alarms could also be formatted or adjusted to the personal desires of the supervisor, to provide any time scale, or degree of resolution, etc., necessary at the time. Ideally some future displays could adapt based on a running model of how the human supervisor's perception was being enhanced. There are hazards, of course, in allowing emergency displays to be too flexible, to the point where they cause errors rather than prevent them.

9.6.7 INTERVENING AND HUMAN RELIABILITY

The supervisor decides to intervene when the computer has completed its task and must be retaught, when the computer has run into difficulty and requests of the supervisor a decision as to which way to go, or when the supervisor decides to stop the automatic action because he or she judges that system performance is not satisfactory.

It is at this stage that human error most reveals itself. Errors in learning from past experience, planning, teaching, and monitoring will surely exist. Many of these are likely to be corrected as the

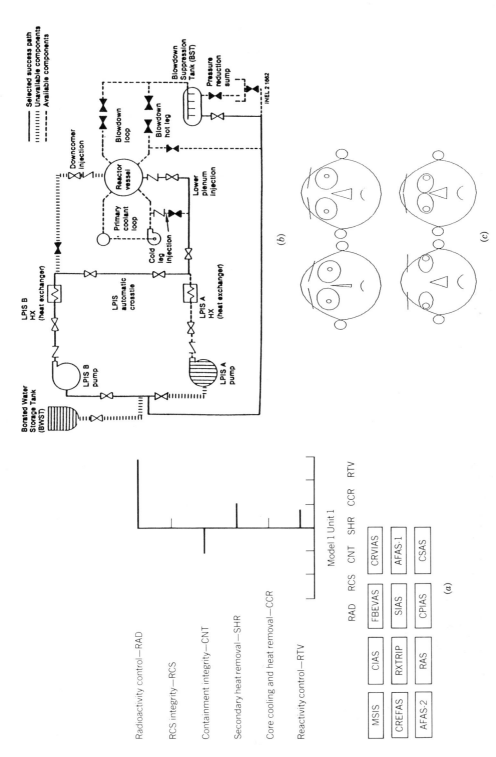

Fig. 9.6.7. Safety parameter display system for a nuclear power plant.

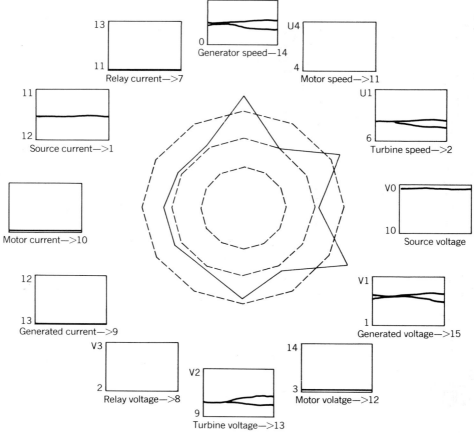

Fig. 9.6.8. Computer display of Tsach, Sheridan, and Buharali system for failure detection and location in process control.

supervisor notes them. In operational systems such errors are relatively unlikely to be noted or counted. It is after the automatic system is functioning that those human errors make a difference and where it is therefore critical that the human supervisor intervene in time and take appropriate action. Thus the intervention stage is where human error is most manifest.

If human error is not caught by the supervisor, it is perpetuated slavishly by the computer, much as happened to the Sorcerer's Apprentice. For this reason supervisory control may be said to be especially sensitive to human error.

There are several factors affecting the supervisor's decision to intervene and/or his or her success in doing so.

1. *Tradeoff between Collecting More Data and Taking Action in Time.* The more data collected from the more sources, the more reliable is the decision of what, if anything, is wrong, and what to do about it. Weighted against this is that if the supervisor waits too long, the situation will likely get worse, and corrective action may be too late. Formally the optimization of this decision is called the "optional stopping problem."

2. *Risk Taking.* The supervisor may operate from either risk-averse criteria such as minimax (minimize the worst outcome that could happen) or more risk-neutral criteria such as expected value (maximize the subjectively expected gain). Depending on the criterion the design of a supervisory control system may be very different in complexity and cost.

3. *Mental Workload.* This problem is aggravated by supervisory control. When a supervisory control system is operating well in the automatic mode, the supervisor may have little concern. When there is a failure and sudden intervention is required, the mental workload may be considerably higher than in direct manual control, where in the latter case the operator is already actively participating

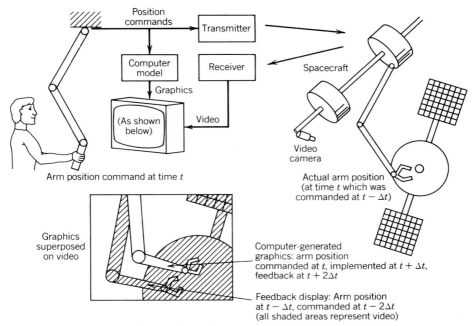

Fig. 9.6.9. Predictor display for delayed telemanipulation.

in the control loop. In the former case the supervisor may have to undergo a sudden change from initial inattention, moving physically and mentally to acquire information and learn what is going on, then making a decision on how to cope. Quite likely this will be a rapid transient from very little to very high mental workload.

Mental workload can be at issue in any human operation. The topic is reviewed by Moray (1979, 1982), Williges and Wierwille (1979), and Hart and Sheridan (1984). Ruffle-Smith (1979) studied pilot errors and fault detection under heavy cognitive workload in a realistic flight simulator and found that crews made approximately one error every 5 min.

Although the subject of human error is currently of great interest, there is no consensus on either a taxonomy or a theory of causality of errors. One common error taxonomy relates to locus of behavior: sensory, memory, decision, or motor. Another useful distinction is between errors of omission and those of commission. A third is between slips (correct intentions that inadvertently are not executed) and mistakes (intentions that are executed but that lead to failure).

In supervisory control there are several problems of human error worth particular mention. One is the type of slip called "capture." This occurs when the supervisor intends to do several steps of, but then deviates from, a well-rehearsed (behaviorally) and well-programmed (in the computer) procedure. Somehow habit, augmented by other cues from the computer, seems to "capture" behavior and drive it on to the next (unintended) step in the well-rehearsed and computer-reinforced routine.

A second supervisory error, important in both planning and failure diagnosis, results from the human tendency to seek confirmatory evidence for a single hypothesis currently being entertained (Gaines, 1976). It would be better if the supervisor could keep in mind a number of alternative hypotheses and let both positive and negative evidence contribute symmetrically in accordance with the theory of Bayesian updating (Sheridan and Ferrell, 1974). Norman (1981), Reason and Mycielska (1982), Rasmussen (1982), and Rouse and Rouse (1983) provide reviews of human error research from their different perspectives.

Theoretically anything that can be specified in an algorithm can be given over to the computer, so that the reason the human supervisor is present is to add novelty and creativity—precisely those ingredients that cannot be prespecified. This means, in effect, that the best or most correct human behavior cannot be prespecified, and that variation from precise procedure must not always be viewed as errant noise. The human supervisor, by the nature of his or her function, must be allowed room by the system design for what may be called "trial and error" (Sheridan, 1983).

What training should the human supervisory controller receive in order to do a good job at detecting failures and intervening to avoid errors? As the supervisor's task becomes more cognitive, is the answer to provide training in theory and general principles? Curiously the literature seems to provide a negative answer (Duncan, 1981). In fact Moray (in press) in his review concludes that "There seems to be no case in the literature where training in the theory underlying a complex system has produced a dramatic change in fault detection or diagnosis." Rouse (1985) similarly concludes that the evidence [e.g., Morris and Rouse (1985)] does not support a conclusion ". . . that diagnosis of the unfamiliar requires theory and understanding of system principles." Apparently frequent hands-on experience in a simulator (i.e., with simulated failures) is the best way to enable a supervisor to retain an accurate mental model of a process.

9.6.8 MODELING SUPERVISORY CONTROL

Modeling supervisory control is a challenge. For 15 years various models of supervisory control have been proposed. Mostly these have been models of particular aspects of supervisory control, not apparently claiming to model all or even very many aspects of it.

The simplest model of supervisory control might be that of nested control loops (Figure 9.6.10) where one or more inner loops are automatic and the outer one is manual. In aerospace vehicles the innermost of four nested loops is typically called "control," the next "guidance," and the next "navigation," each having a set point determined by the next outer loop. Hess and McNally (in press) have shown how conventional manual control models can be extended to such multiloop situations. The outer loop in this generic aerospace vehicle includes the human operator, who, given mission goals, programs in the destination. In driving a car the functions of navigation, guidance, and control are all done by a person, and can be seen to correspond roughly to knowledge-based, skill-based, and rule-based behavior. (See Chapter 2.7.)

Figure 9.6.11 (Sheridan, 1984b) is a qualitative functional model of supervisory control, showing the various cause–effect loops or relationships among elements of the system, and emphasizing the symmetry of the system as viewed from top and bottom (human, task) of the hierarchy.

Figure 9.6.12 is an abbreviated version of Rasmussen's qualitative model referred to above in Section 9.6.3, showing in particular the nesting of skill-based, rule-based, and knowledge-based behavioral loops. In Figure 9.6.13 I extend Rasmussen's model to show various interactions with computer aids having comparable levels of intelligence.

One problem of interest to the supervisor is how often he or she should sample the input, how often he or she should update a control setting, or both, particularly if there is a cost incurred each time he or she does so. Given assumptions on the magnitude distribution and autocorrelation of inputs, a utility function for value of performance resulting from a particular input and particular control action in combination, and a discrete cost of sampling, Sheridan (1970a) showed how an optimal sampling strategy could be derived to maximize expected gain. Sheridan (1976) suggested a framework for how a supervisor equipped with a variety of sensing options and a variety of motor options could try various combinations of these in "thought experiments" or simulations with an "internal model" of the controlled process and utility function. Using Bayes theorem it is shown how expected utility can be maximized.

One problem the supervisor faces is allocating his or her own attention between different tasks, where each time he or she switches tasks there is a time penalty in transfer—typically different for different tasks, and possibly involving uses of different software procedures, different equipment, and even bodily transportation of him- or herself to different locations. Given relative worths for time spent attending to various tasks, it has been shown (Sheridan, 1970b) that dynamic programming enables the optimal allocation strategy to be established. Moray et al. (1982) applied this model to deciding whether human or computer should control various variables at each succeeding moment. For simpler experimental conditions the model fit the experimental data (subjects acted like utility maximizers), but as task conditions became complex, apparently it did not. Wood and Sheridan (1982) did a similar study where supervisors could select among alternative machines (differing in both rental cost and productivity) to do assigned tasks or do the tasks themselves. Results showed the supervisors to be suboptimal, paying too much attention to costs and too little to productivity, and in some cases using machines when they could have done the tasks more efficiently by themselves. Govindaraj

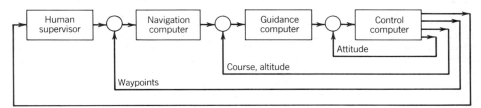

Fig. 9.6.10. Nested control loops of aerospace vehicle.

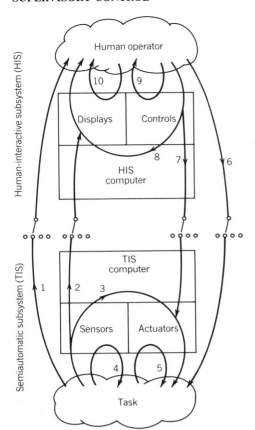

1. Task is observed directly by human operator's own senses.

2. Task is observed indirectly through artificial sensors, computers and displays. This TIS feedback interacts with that from within HIS and is filtered or modified.

3. Task is controlled within TIS automatic mode.

4. Task is affected by the process of being sensed.

5. Task affects actuators and in turn is affected.

6. Human operator directly affects task by manipulation.

7. Human operator affects task indirectly through a controls interface, HIS/TIS computers, and actuators. This control interacts with that from within TIS and is filtered or modified.

8. Human operator gets feedback from within HIS, in editing a program, running a planning model, etc.

9. Human operator orients him- or herself relative to control or adjusts control parameters.

10. Human operator orients him- or herself relative to display or adjusts display parameters.

Fig. 9.6.11. Multiloop model of supervisory control.

and Rouse (1981) modeled the supervisor's decisions to turn away from a continuous task in order to perform or monitor a discrete task.

Rouse (1977) utilized a queueing theory approach to model whether from moment to moment a task should be assigned to a computer or to the operator him- or herself. The allocation criterion was to minimize service time under cost constraints. Results suggested that human–computer "misunderstandings" of one another degraded efficiency more than limited computer speed. In a related flight simulation study Chu and Rouse (1979) had a computer perform those tasks that had waited in the queue beyond a certain time. Chu, Steeb, and Freedy (1980) extended this idea to have the computer learn the pilot's priorities and later make suggestions when the pilot is under stress.

Tulga and Sheridan (1980) and later Pattipatti, Kleinman and Ephrath (1983) utilized a model of allocation of attention among multiple task demands, a task displayed on the computer screen to the subject as is represented in Figure 9.6.14. Instead of being stationary, these demands appear at random times (not being known until they appear), exist for given periods of time, then disappear at the end of that time with no more opportunity to gain anything by attending to them. While available they take differing amounts of time to complete, and have differing rewards for completion, which information may be available after they appear and before they are "worked on." The human decision-maker in this task need not allocate his or her attention in the same temporal order in which the task demands became known, nor in the same order in which their deadlines will occur. Instead he or she may attend first to that task which has the highest payoff or takes the least time, and/or he or she may try to plan ahead a few moves so as to maximize his or her gains. The Tulga–Sheridan experimental results suggest that subjects approach optimal behavior which, when heavily loaded (i.e., there are more opportunities than he or she can possibly cope with), amounts to simply selecting the task with highest payoff regardless of time to deadline. These subjects also reported that their sense of subjective workload was greatest when by arduous planning they could barely keep up with all tasks presented. When still more tasks came at them and they were free to select which they should do, subjective workload decreased.

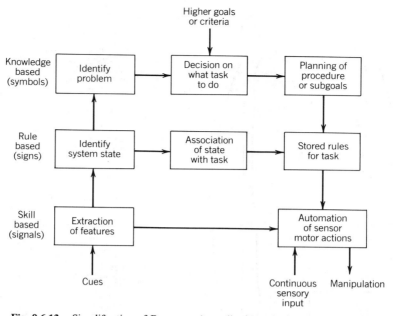

Fig. 9.6.12.   Simplification of Rasmussen's qualitative model of human behavior.

The samples of research cited above have tended to focus on probabilistic allocation of resources. Coming at supervisory control from a somewhat more continuous and deterministic perspective, Kok and Van Wijk (1978) extended the well-known Baron and Kleinman (1969) model of the human operator as an optimal controller. They considered processes that are slowly responding. Their model of the supervisor (Figure 9.6.15) consists of an observer on both system input and output to provide a best estimate of system state and error variance, an optimal controller to determine a best control

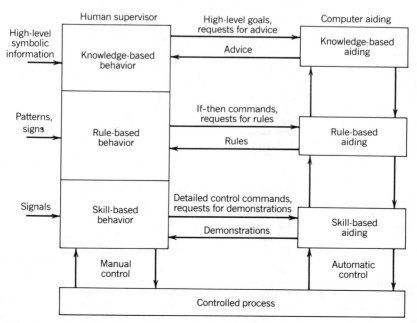

Fig. 9.6.13.   Supervisor interactions with computer decision aids at knowledge, rule, and skill levels.

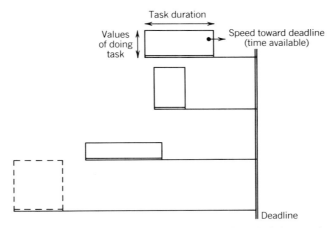

Fig. 9.6.14. Multitask computer display used in Tulga–Sheridan experiment.

law for the given conditions, and a decision-making element to decide when best to sample, observe, and make control set-point adjustments. These authors fit their model to data on the helmsman of a large ship. White (1981) fit the same model to the control of a process plant.

Using as a "front end" attention allocation mechanisms similar to those of the Tulga–Sheridan and Pattipatti–Kleinman–Ephrath models, Baron, Zacharias, Muraldiharan, and Lancraft (1980) extended the Baron and Kleinman optimal control model and called it PROCRU. This is diagrammed in Figure 9.6.16. It was originally built to model crew selection and implementation of procedures in aircraft approach and landing. Optimum decision and control algorithms maximize expected gain for given nominal procedure requests from the ground, aerodynamic disturbances, vehicle dynamics, and objective function.

There are a number of questions that researchers and designers of supervisory control systems must cope with. Among these are (1) how much autonomy is appropriate for the TIC, (2) how much should the TIC tell the HIC and the HIC tell the human supervisor, and (3) how should responsibilities be allocated among the TIC, HIC and the supervisor (Johannsen, 1981)?

Rouse (in press) concludes his discussion of models of supervisory control with the interesting comment that "perhaps the most important result of the emergence and clashing of models over the past decade has been a shift away from monolithic, computationally overwhelming models to frameworks or categorizations of models, each of which may be quite simple, involving elementary control laws, a few heuristics, or pattern recognition rules. Thus, as knowledge and understanding of supervisory control has grown, researchers have come to realize that neither they nor operators need to approach tasks in such a global, brute-force manner."

The most difficult, and it might even be said impossible, aspect of supervisory control to model is that of setting in goals, conditions, and values. Even though overall goals may be given to an actual system (or given in an experiment) how those are translated into subgoals and conditional statements remains elusive. The same is true for communicating values (criteria, coefficients of utility, etc.). Although

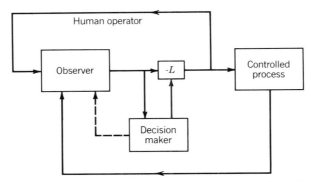

Fig. 9.6.15. Kok and Von Wijk model of supervisory control. (Dotted line indicates corrections.)

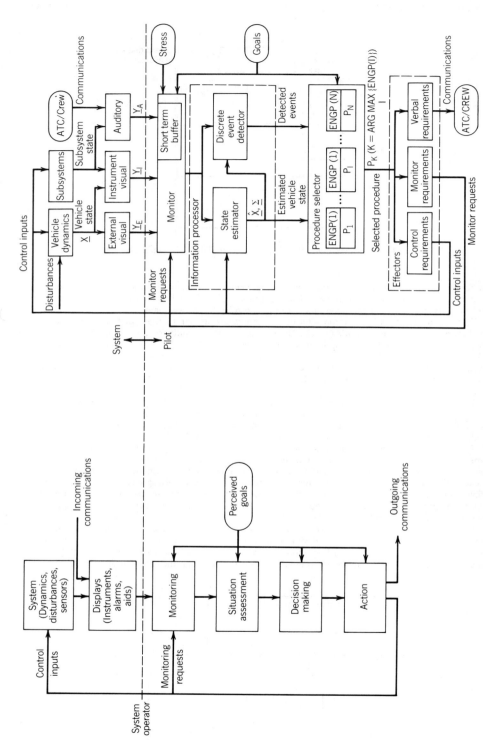

Fig. 9.6.16. PROCRU model of Baron et al. (1980).

this act of evaluation remains the sine qua non of why human participation in system control must remain, there is little prospect for mathematical modeling of this aspect in the near future.

9.6.9   SOCIAL IMPLICATIONS AND THE FUTURE OF SUPERVISORY CONTROL

No one can predict the future with certainty, but it is ethically mandatory that we predict it as best we can.

One near certainty is that as technology of computers, sensors, and displays improves, supervisory control will become more prevalent. This should occur in two ways: (1) a greater number of semiauto-mated tasks will be controlled by a single supervisor (a greater number of TICs will be connected to a single HIC), and (2) the sophistication of cognitive aids, including expert systems for planning, teaching, monitoring, failure detection, and learning, will increase, and include more of what we now call "knowledge-based behavior" in the HIC.

Concurrently the understanding by the layman (including those of both corporate and government bureaucracies) should come to understand the potential of supervisory control much better. At the present time the layman tends to see automation as "all-or-none"—a system is controlled either manually or automatically, with nothing in between. In robotized factories the media tend to focus on the robots, with little mention of design, installation, programming, monitoring, fault detection and diagno-sis, maintenance, and various learning functions that are performed by people. In the space program the same is true—options are seen to be either "automated," "astronaut in EVA," or "astronaut or ground controlling tele-manipulator"—without much appreciation for the potential of supervisory con-trol.

In considering the future of supervisory control relative to various degrees of automation, and to the complexity or unpredictability of task situations to be dealt with, a representation such as Figure 9.6.17 comes to mind. The meaning of the four extremes of this rectangle are quite identifiable. Supervi-sory control may be considered to be a frontier (line) advancing gradually toward the upper right-hand corner.

The tendency, for obvious reasons, has been to automate what is easiest and to leave the rest to the human. This has sometimes been called the "technological imperative." From one perspective this dignifies the human contribution, from another it may lead to a hodge podge of partial automation, making the remaining human tasks less coherent and more complex than need be, resulting in overall degradation of system performance (Bainbridge, 1982).

As previously discussed, supervisory control may involve varying degrees of computer aiding on the afferent or incoming side, as well as on the efferent or control execution side. Table 9.6.1 suggests a scale of "degrees of automation" that separates the afferent (sensing) from the efferent (taking action)

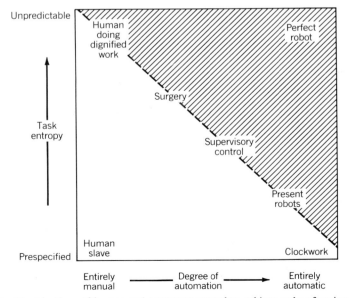

Fig. 9.6.17.   Combinations of human and computer control to achieve tasks of various difficulty.

Table 9.6.1 Degrees of Automation

	Degree of Computer Aiding in Acquiring and Analyzing Data			
	Past Data	Current Data	Situation Assessment	Action Alternatives
0. Human does it all (within capability)	Measures and stores relevant past data	Observes relevant current data	Assesses current system state or situation relative to goals	Conjures up various action alternatives within resource constraints
1. Computer acquires all relevant data	Stores all relevant past data	Measures all relevant current data	Estimates current state or situation	Determines all relevant action alternatives
2. Computer displays all relevant data to human	Displays relevant past data, trends, and so on.	Displays relevant current data, relates it to past	Displays current state or situation	Displays all relevant action alternatives
3. Computer selects and displays narrow set of data to human	Selects and displays narrow set of past data	Selects and displays narrow set of current data	Assesses and displays current state or situation relative to assumed goals	Determines and displays a few most salient action alternatives
4. Computer selects and displays a single recommendation with justification	Selects and displays only that past data sufficient to support recommendation	Selects and displays only that current data sufficient to support recommendation	Assesses and displays only limited information sufficient to specify current state relative to assumed goals	Determines and displays a single recommendation for action

	Degree of Computer Aiding in Implementation	
	Selection of Action Alternative	Conditions Accompanying Implementation
0. Human does it all (within capability)	(a) Decides on action independent of computer (b) Selects an action proposed by computer	Implements action if, when, and how he or she decides
1. Computer implements, with human say over if, when, or how	(a) What human selects (b) What computer recommended	(a) Only if and when human approves (b) By a deadline if human does not stop it in time
2. Computer implements, independent of human	What computer recommended	(a) And necessarily tells human after the fact what it did (b) And tells human after the fact what it did only if human asks (c) And tells human after the fact what it did if it thinks he or she should be told

and breaks the afferent down into components dealing with (1) experience, (2) sensing present data, (3) interpreting present data, and (4) formulating action alternatives.

I have written elsewhere about the long-term social implications of supervisory control (Sheridan, 1980; Sheridan, Vamos, and Aida, 1983). My concerns might be reviewed here very briefly:

1. *Unemployment.* This is the factor most often considered. More supervisory control means more efficiency, less direct control, fewer jobs.

2. *Desocialization.* Although cockpits and control rooms now require two to three person teams, the trend is toward fewer people per team, and eventually one person will be adequate in most installations. Thus cognitive interaction with computers will replace that with other people. As supervisory control systems are interconnected, the computer will mediate more and more interpersonal contact.

3. *Remoteness from the Product.* Supervisory control removes people from hands-on interaction with the workpiece or other product. They become not only separated in space but also desynchronized in time. Their functions or actions no longer correspond to how the product itself is being handled or processed mechanically.

4. *Deskilling.* Skilled workers "promoted" to supervisory controller may resent the transition because of fear that when and if called on to take over and do the job manually, they may not be able to. They may also feel loss of professional identity built up over an entire working life.

5. *Intimidation due to Higher Stakes.* Supervisory control will encourage larger aggregations of equipment, higher speeds, greater complexity, higher cost of capital, and probably greater economic risk if something goes wrong and the supervisor does not take the appropriate corrective action.

6. *Discomfort in the Assumption of Power.* The human supervisor will be forced to assume more and more ultimate responsibility. Depending on one's personality, this could lead to insensitivity to detail, anxiety about being up to the job requirements, or arrogance.

7. *Technological Illiteracy.* Supervisory controllers may lack the technological understanding of how the computer does what it does. They may come to resent this and resent the elite class who do understand.

8. *Mystification.* Human supervisors of computer-based systems could become mystified, superstitious about the power of the computer, even seeing it as a kind of magic or "big brother" authority figure.

9. *Sense of Not Being Productive.* Although the efficiency and mechanical productivity of a new supervisory control system may far exceed that of an earlier manually controlled system which a given person has experienced, that person may come to feel no longer productive as a human being.

10. *Eventual Abandonment of Responsibility.* As a result of the factors previously described, supervisors may eventually feel they are no longer responsible for what happens; the computers are.

These 10 potential negatives may be summarized with a single word—alienation. In short, if human supervisors of the new breed of computer-based systems are not given sufficient social contacts, sufficient familiarization with and feedback from the task, sufficient sense of retaining their old skills, or ways of finding identity in new ones, they may well come to feel alienated. They must be trained to feel comfortable with their new responsibility, must come to understand what the computer does and not be mystified, and must realize that they are ultimately in charge of setting the goals and criteria by which the system operates. If these principles of human factors are incorporated into the design, selection, training, and management, supervisory control has a positive future.

9.6.10   CONCLUSION

Computer technology, both hard and soft, is driving the human operator to become a supervisor (planner, teacher, monitor, and learner) of automation and an intervener within the automated control loop only for abnormal situations. A number of definitions, models, and problems have been discussed. There is little or no present consensus that any one of these models characterizes in a satisfactory way all or even very much of supervisory control with sufficient predictive capability to entrust to the designer of such systems. It seems that for the immediate future we are destined to run breathless behind the lead of technology, trying our best to catch up.

REFERENCES

Bainbridge, L. (1982). Ironies of automation. *Proceedings of the IFAC/IFIP/IFORS/IEA conference on analysis, design and evaluation of man-machine systems.* Baden-Baden, FRG, pp. 151–157.

Baron, S. and Kleinman, D. L. (1969). The human as an optimal controller and information processor. *IEEE Transaction on Man-Machine Systems, MMS-10,* 9–17.

Baron, S., Zacharias, G., Muraldiharan, R., and Lancraft, R. (1980). PROCRU; a model for analyzing flight crew procedures in approach to landing. *Proceedings of 16th annual conference on manual control.* Cambridge, MA: MIT, pp. 488–520.

Brooks, T. L. (1979). SUPERMAN: a system for supervisory manipulation and the study of human computer interactions (SM Thesis). Cambridge, MA: MIT.

Buharali, A. and Sheridan, T. B. (1982). Fuzzy set aids for telling a computer how to decide. *Proceedings of IEEE international conference on cybernetics and society.* Seattle, WA, pp. 643–647.

Burton, R. R. and Brown, J. S. (1982). An investigation of computer coaching for informal learning activities. In. D. Sleeman, and J. S. Brown (Eds.). *Intelligent tutoring systems.* New York: Academic.

Card, S. K. (1984). Human factors and the intelligent interface. Paper presented at Combining human factors and artificial intelligence: a new frontier for human factors. Metropolitan Chapter, Human Factors Society, New York, Nov. 15.

Chu, Y. Y. and Rouse, W. B. (1979). Adaptive allocation of decision making responsibility between human and computer in multi-task situations. *IEEE Tansaction in Systems, Man and Cybernetics, SMC-9,* 769–778.

Chu, Y. Y., Steeb, R., and Freedy, A. (1980). Analysis and modeling of information handling tasks in supervisory control of advanced aircraft (PATR-1080-80-6). Woodland Hills, CA: Perceptronics.

Curry, R. E. and Ephrath, A. R. (1977). Monitoring and control of unreliable systems. In T. B. Sheridan and G. Johannsen (Eds.). *Monitoring behavior and supervisory control.* New York: Plenum, pp. 193–203.

Curry, R. E. and Nagel, D. C. (1974). Decision behavior with changing signal strengths. *Journal of Mathematical Psychology, 14,* 1–24.

Duncan, K. D. (1981). Training for fault diagnosis in industrial process plants. In J. Rassmussen and W. B. Rouse (Eds.). *Human detection and diagnosis of system failures.* New York: Plenum, pp. 553–524.

Edwards, E. and Lees, F. (1974). *The human operator in process control.* London: Taylor and Francis.

Ephrath, A. R. and Young, L. R. (1981). Monitoring vs. man-in-the-loop detection of aircraft control failures. In J. Rasmussen and W. B. Rouse (Eds.). *Human detection and diagnosis of system failures.* New York: Plenum, pp. 143–154.

Falzon, P. (1982). Display structures: compatability with the operator's mental representation and reasoning processes. *Proceedings of the second annual conference on human decision making and manual control,* pp. 297–305.

Ferrell, W. R. (1966). Delayed force feedback. *Human Factors,* October, 449–455.

Ferrell, W. R., and Sheridan, T. B. (1967). Supervisory control of remote manipulation. *IEEE Spectrum, 4(10),* 81–88.

Gai, E. G. and Curry, R. E. (1978). Perseveration effects in detection tasks with correlated decision intervals. *IEEE Transactions on Systems, Man and Cybernetics, SMC-8,* 93–110.

Gaines, B. R. (1976). On the complexity of causal models. *IEEE Transactions on Systems, Man and Cybernetics, SMC-6,* 56–59.

Gentner, D. and Stevens, A. L. (Eds.) (1983). *Mental Models.* Hillsdale: NJ: Erlbaum Press.

Govindaraj, T. and Rouse, W. B. (1981). Modeling the human controller in environments that include continuous and discrete tasks. *IEEE Transactions on Systems, Man and Cybernetics, SMC-11,* 411–417.

Hart, S. G. and Sheridan, T. B. (1984). Pilot workload, performance, and aircraft control automation. *Proceedings of AGARD symposium on human factors considerations in high performance aircraft.*

Hess, R. A. and McNally, B. D. (in press). Automation effects in a multi-loop manual control system. *IEEE Transactions on Systems, Man and Cybernetics,* New York: IEEE.

Johannsen, G. (1981). Fault management and supervisory control of decentralized systems. In J. Rasmussen and W. B. Rouse (Eds.). *Human detection and diagnosis of system failures.* New York: Plenum, pp. 353–360.

Kok, J. J. and Van Wijk, R. A. (1978). Evaluation of models describing human operator control of slowly responding complex systems. Delft: DUT, Report of Laboratory for Measurement and Control.

Moray, N. (Ed.) (1979). *Mental workload.* New York: Plenum.

Moray, N. (1982). Subjective mental workload. *Human Factors, 24,* 25–40.

Moray, N. (in press). Monitoring behavior and supervisory control. In K. Boff (Ed.). *Handbook of perception.* New York: Wiley.

Moray, N., Sanderson, P., Shiff, B., Jackson, R., Kennedy, S., and Ting, L. (1982). A model and experiment for the allocation of Man and Computer in supervisory control. *Proceedings of IEEE international conference on cybernetics and society.* Seattle, WA, pp. 354–358.

Morris, N. M. and Rouse, W. B. (1985). The effects of type of knowledge upon human problem solving in a process control task. *IEEE Transactions on Systems, Man and Cybernetics, SMC-15. 6,* 698–707.

Niemala, R. and Krendel, E. S. (1974). Detection of a change in plant dynamics in a man-machine system. *Proceedings of 10th annual conference on manual control*, pp. 97–112.

Norman, D. A. (1981). Categorization of action slips. *Psychological Review, 88*, 1–15.

Noyes, M. and Sheridan, T. B. (1984). A novel predictor for telemanipulation through a time delay. *Proceedings of 1984 annual conference on manual control*. Moffett Field, CA: NASA Ames Research Center.

Pattipatti, K. R., Kleinman, D. L., and Ephrath, A. R. (1983). A dynamic decision model of human task selection performance. *IEEE Transactions on Systems, Man and Cybernetics, SMC-13*, 145–166.

Rasmussen, J. (1976). Outlines of a hybrid model of the process plant operator. In T. B. Sheridan and G. Johannsen (Eds.). *Monitoring behavior and supervisory control*. New York: Plenum, pp. 371–383.

Rasmussen, J. (1982). Human errors: a taxonomy for describing human malfunction in industrial installations. *Journal of Occupational Accidents, 4*, 311–333.

Rasmussen, J. and Rouse, W. B. (Eds.) (1981). *Human detection and diagnosis of system failures.* New York: Plenum.

Reason, J. and Mycielska, K. (1982). *Absent minded: the psychology of mental lapses and everyday errors.* Englewood Cliffs, NJ: Prentice-Hall.

Rouse, W. B. (1974). The effect of display format on the human perception of statistics. *Proceedings of 10th annual conference on manual control.*

Rouse, W. B. (1977). Human-computer interaction in multi-task situations. *IEEE Transactions in Systems, Man and Cybernetics, SMC-7*, 384–392.

Rouse, W. B. and Morris, N. M. (1984). *On looking into the black box: prospects and limits in the search for mental models.* Norcross, GA: Search Technology, Inc.

Rouse, W. B. and Rouse, S. H. (1983). Analysis and classification of human error. *IEEE Transactions on Systems, Man and Cybernetics, SMC-13*, 539–549.

Rouse, W. B. (1985). Supervisory control and display systems. In J. Zeidner (Ed.). *Human productivity enhancement.* New York: Praeger.

Ruffell Smith, H. P. (1979). A simulator study of the interaction of pilot workload with errors, vigilance and decisions. (NASA TM-78482). Moffett Field, CA: NASA Ames Research Center.

Sheridan, T. B. (1960). Human metacontrol. Paper presented at Conference on Manual Control, Wright Patterson AFB, OH.

Sheridan, T. B. (1970a). On how often the supervisor should sample. *IEEE Transactions on System Science and Cybernetics, SSC-6*, 140–145.

Sheridan, T. B. (1970b). Optimum allocation of personal presence. *IEEE Transactions on Human Factors in Electronics, HFE-10*, 242–244.

Sheridan, T. B. (1976). Toward a general model of supervisory control. In T. B. Sheridan and J. Johannsen (Eds.). *Monitoring behavior and supervisory control*. New York: Plenum, pp. 271–282.

Sheridan, T. B. (1980). Computer control and human alienation. *Technology Review*, October, 61–73.

Sheridan, T. B. (1983). Measuring, modeling and augmenting reliability of man-machine systems. *Automatica, 19(6)*, 637–646.

Sheridan, T. B. (1984a). Interaction of human cognitive models and computer-based models in supervisory control (report). Cambridge, MA: MIT Man-Machine Systems.

Sheridan, T. B. (1984b). Supervisory control of remote manipulators, vehicles and dynamic processes. In W. B. Rouse (Ed.). *Advances in man-machine systems research*. New York: JAI Press, vol. 1, pp. 49–137.

Sheridan, T. B., and Ferrell, W. R. (1974). *Man-machine systems.* Cambridge, MA: MIT Press.

Sheridan, T. B., Fischoff, B., Posner, M., and Pew, R. W. (1983). Supervisory control systems. In *Research needs in human factors*. Washington, DC: National Academy Press.

Sheridan, T. B. and Hennessy, R. T. (Eds.) (1984). *Research and modeling of supervisory control behavior: report of a workshop.* Washington, DC: National Academy Press.

Sheridan, T. B. and Johannsen, G. (Eds.) (1976). *Monitoring behavior and supervisory control.* New York: Plenum.

Sheridan, T. B., Vamos, T., and Aida, S. (1983). Adapting automation to man, culture and society. *Automatica, 19(6)*, 605–612.

Taylor, F. W. (1911). *Principles of scientific management.*

Tsach, U., Sheridan, T. B. and Buharali, A. (1983). Failure detection and location in process control—integrating a new model-based technique with other methods. *Proceedings of 1983 American control conference,* San Francisco, CA.

Tulga, M. K. and Sheridan, T. B. (1980). Dynamic decisions and workload in multi-task supervisory control. *IEEE Transactions on Systems, Man and Cybernetics, SMC-10,* 217–232.

White, T. N. (1981). Modeling the human operator's supervisory behavior. *Proceedings of the first European annual conference on human decision making and manual control.* Delft: DUT, pp. 203–217.

Wickins, C. and Kessell, C. (1979). The effects of participatory model and task workload on the detection of dynamic system failures. *IEEE Transactions on Systems, Man and Cybernetics, SMC-9,* 24–34.

Wickins, C. and Kessell, C. (1981). Failure detection in dynamic systems. In J. Rasmussen and W. B. Rouse (Eds.). *Human detection and diagnosis of system failures.* New York: Plenum, pp. 155–169.

Wiener, E. L., and Curry, R. E. (1980). Flight deck automation: promises and problems. *Ergonomics, 23,* 995–1011.

Williges, R. C. and Wierwille, W. W. (1979). Behavioral measures of aircrew mental workload. *Human Factors, 21,* 549–574.

Wood, W. and Sheridan, T. B. (1982). The use of machine aids in dynamic multi-task environments: a comparison of an optimal model to human behavior. *Proceedings of IEEE international conference on cybernetics and society.* Seattle, WA, pp. 668–672.

Yoerger, D. (1982). Supervisory control of underwater telemanipulators: design and experiment (Ph.D. Thesis). Cambridge, MA: MIT.

Zadeh, L. A. (1984). Making computers think like people. *IEEE Spectrum, 21,* No. 8, 26–32.

PART **10**
SYSTEM EVALUATION

CHAPTER **10.1**

SYSTEM EFFECTIVENESS TESTING

DAVID MEISTER

U.S. Navy Personnel Research and Development Center
San Diego, California

10.1.1 INTRODUCTION AND OVERVIEW

System effectiveness testing is human factors (HF) test and evaluation (T&E). We shall not discuss personnel performance testing accomplished on civilian (nonmilitary) systems, because we lack specifications and data concerning it. In consequence, this chapter is written almost entirely from the standpoint of testing as accomplished in the development and operation of American military systems. As was pointed out, however, in Chapter 1.2 with regard to system development as a whole, much of what is said in this chapter should be applicable to the testing of civilian systems, since testing requirements stem from common needs to learn certain things about the system and to make design decisions. Military system testing is, however, likely to be more formal and detailed than that of civilian systems.

The requirement for HF T&E on military systems is called out by MIL-H-46855B (Department of Defense, 1979). Section 3.2.3 of that document (Human Engineering or Test and Evaluation) is divided into three subsections: planning, implementation, and failure analysis. The purpose of HF T&E is to (1) ensure fulfillment of requirements; (2) demonstrate conformance of system, equipment, and facility design to human engineering design criteria; (3) confirm compliance with performance requirements where a person is a performance determinant; (4) require quantitative measures of system performance that are a function of human–machine interaction; and (5) determine whether undesirable design or procedural features have been introduced into design.

The planning subsection emphasizes the need for early, timely conduct of testing. Without the proper scheduling of HF tests, they may be of little use to the system design. It is important to demonstrate that test results will be available at a time when they can properly impact that design. Unfortunately, HF T&E is often performed merely as documentation that a contractual requirement has been satisfied (Geer, 1977).

The implementation section of 3.2.3 specifies that HF tests shall include, where applicable, a simulation or actual conduct of mission or work cycle; tests of tasks in which human participation is critical with regard to such factors as speed, accuracy, reliability, or cost; a representative sample of scheduled and unscheduled maintenance tasks; proposed job aids, training equipment, and so forth; utilization of personnel representative of the ultimate user population; collection of task performance data; identification of discrepancies between required and obtained task data; and criteria for the acceptable performance of the task. The failure analysis subsection requires analysis of equipment failures occurring during the test to determine those failures resulting from human error.

This chapter discusses testing under four conditions: (1) the use of *mockups* for performance testing; (2) the testing of *prototype systems* under conditions that approximate their operational use; (3) the testing of *operational systems* under operational conditions (items 2 and 3 are termed "operational testing" or OT); and (4) performance measurement in equipment and system *simulators.*

Mockups were briefly discussed in Chapter 1.2. Performance measurement in simulators is a variation (albeit an important one) on OT, but with special characteristics. In this chapter we shall be discussing mostly OT because the mockup, being very limited, offers only a limited scope for testing; and much of what we have to say about OT applies as well to simulator testing.

Since system effectiveness testing is inextricably bound up with the system development process, the reader should refer to Chapter 1.2 for context, if he or she has not already done so. Briefly, the use of mockups (particularly those of a static, nonfunctioning nature) as a performance measurement instrument is characteristic of early system development; OT is performed with a prototype system just before the system goes into full-scale production; and performance measurement in simulators is characteristic of a mature system or a research program, because simulators, in view of their high cost, are not built until the system configuration is fixed. System operation, the routine life of the functioning system, is considered simply as the prolonged end phase of system development. Figure 10.1.1 is a diagram of the kinds of system effectiveness tests as a function of the developmental cycle.

The reader will note that there has been no mention of developmental tests conducted in the engineering facility. As was pointed out in Chapter 1.2 (for which see the discussion on this topic), these tests are primarily hardware-oriented tests (e.g., nondestructive reliability tests) and offer the HF engineer (HFE) little of a behavioral performance value.

Personnel performance testing usually takes place in four contexts: (1) the engineering development facility for mockups; (2) the test site for the prototype system, (3) the training simulator [most simulators are developed for training purposes, although a few have been built specially for research, e.g., the advanced simulator for pilot training (ASUPT) at Williams Air Force Base, Arizona]; and (4) the functioning operational system for performance testing in the real world of systems. Of these test environments the HFE probably works more with mockups and the prototype system than he or she does with the others, but much depends on the particular job he or she has.

We will have occasion to use the term "system" quite frequently because, outside of the human

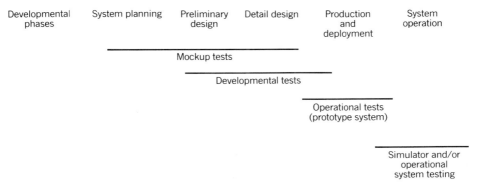

Figure 10.1.1. Sequence of system effectiveness tests. (See Chapter 1.2 for definitions.)

operator/maintainer, the system is the major player in the performance testing game. The system is an assembly of men and women who operate and maintain equipment and perform tasks to accomplish some goal required by the system as a whole. Chapter 1.2 describes the system orientation in greater detail.

10.1.2 THE FOUR TEST CONDITIONS

Taking the operational system as the criterion, because that system is the goal of all preceding developmental processes, it is possible to place the mockup, the prototype system, and the simulator on several continua. In terms of similarity to the operational system the mockup is least similar, the simulator, most, the prototype system being much closer to the simulator than to the mockup. In terms of the proportion of system units each includes, the mockup usually represents only a single equipment, although it *can* be made to represent the entire system; the prototype system represents the entire system, as does the simulator (for the simulator, at least those equipment that represent the human–machine interface for the operator). In terms of how many personnel actions or personnel-related system aspects one can measure, one can test least with mockups, most with simulators and the prototype system, but not as much of course as one can test with the actual operational system.

Most mockups are far less complex from an electronic component standpoint than simulators but as one adds microcomputers, software, functioning controls and displays, and so forth, to the mockup, it eventually metamorphoses into a simulator.

Mockup tests are performed as the HFE feels they are necessary, because they are not ordinarily part of routine, scheduled engineering development. OT of the prototype system is mandated by Department of Defense regulations (Department of Defense, 1977). Like mockup tests, performance measurement in the simulator and in the operational system depends on special circumstances such as the need to test a system modification; occasionally these measurements are made for research purposes (e.g., to verify that a training simulator does in fact train personnel).

The simulator is a physical device that replicates the operational equipment (in terms of its human–machine interfaces). It must be distinguished from a computer simulation, which is essentially a conceptual system.

Beyond evaluating the training simulator to see if it trains, the HFE would use a simulator in preference to the actual operational system when any of the following conditions obtain: (1) the operational system is not available for measurement; (2) it is necessary to vary system parameters to make experimental comparisons and the operational system cannot be controlled for this purpose; (3) the scenarios to be measured involve emergency conditions that are too dangerous to elicit in real life.

There are, moreover, definite advantages in simulator measurement: the ability to start and stop the simulated system as desired; the ability to "freeze" a behavioral segment and examine it more closely; the ability to vary modes of operation as desired; automatic data recording instrumentation may be available. On the other hand, there are disadvantages: the full range of system capability may not be available in the simulator (and certainly this relates to maintenance functions); the fidelity of the simulation may occasionally be less than desired; the personnel assigned as system operators (test subjects) are usually trainees and hence may be less skilled than fully operational personnel; the scenarios one may have to accept in exercising the simulated system are training scenarios and therefore may not be completely representative of operational conditions; the length of time the HFE can get the simulator exclusively for measurement purposes may be limited.

The other test situations have their own disadvantages. The operational system is embedded in an environment in which extrasystem influences sometimes penetrate the system boundary and contaminate the study, for example, interruptions from higher command. This is what causes much of the lack of control for which the operational environment is known. The prerequisites for OT of the prototype system may not be completely fulfilled: for example, stable system configuration (both equipment and procedures) and test subjects representative of the users; because of continuing development problems, equipment may fail more often than it would operationally. The mockup is so limited in the stimuli it can present and the personnel responses it can induce that a good deal of extrapolation is required when analyzing test results [which is not to say, however, that it lacks usefulness; see Bohn (1983)].

The general purpose of a system effectiveness test may be any or all of the following: (1) to verify that personnel can perform their assigned tasks and that this can be accomplished without excessive error or workload; (2) to verify that operating procedures and other equipment/system characteristics present no insuperable obstacles to effective personnel performance and that these characteristics satisfy desired human engineering attributes; (3) to determine the effect of any contrasting variables inherent in the equipment/system mission (e.g., day/night operations) on personnel performance; (4) to determine the adequacy from a behavioral standpoint of any modification in the job, equipment/system, or its mission and of the adequacy of any design solution to a problem; (5) to determine by measuring the performance of personnel in operational systems whether their continuing performance is adequate; and, if not, what factors are causing performance deficiencies.

The more complete the system under test is, and the more it functions operationally (i.e., all its subsystems are exercised fully under all mission conditions exactly as they would be performed operationally), the more these test purposes can be accomplished. That is why only some of these test purposes can be accomplished and then only partially with a static mockup (the most common type of mockup); the more similar the test situation is to the operational system, the more these test purposes can be accomplished. This implies that it is possible to get partial data to satisfy a test purpose partially. For example, assuming that one has a static mockup of a fighter cockpit, it is possible to verify that controls can be reached and displays seen, as required for effective aircraft operation. That is part— but only from an anthropometric standpoint—of the verification that personnel can perform their tasks. It requires a simulator at least to determine that pilots will be able to respond quickly enough to changing events.

The test purposes described above apply as much to operational functions as to system development. As the system matures, more is learned about it and modifications in equipment and procedures may be proposed; testing is needed to determine just how effective these modifications are. Or the system must be tested repeatedly over its life because in normal functioning it cannot display how well it can achieve its goals; this is particularly characteristic of military systems, which are not programmed to function in a combat mode until war breaks out. Testing of combat systems can be accomplished in war games; for example, two brigades of infantry with supporting tanks, artillery, and aircraft may be set against each other. If a problem arises in system operations (unexplained and consistent performance degradation), it may be necessary to test in order to explore the parameters of the problem. Or the HFE seeks to determine the effect of potentially important inherent variables (like varying work shifts) on performance. And sometimes OT and simulator testing are performed simply to collect research data.

System effectiveness testing (OT) has certain characteristics (Johnson and Baker, 1974) that show little relationship to traditional controlled experimental (e.g., laboratory) research. This type of testing addresses real but messy problems; is time and resource limited; measures in macro units (minutes) rather than micro units (seconds); evaluates both personnel and equipment; employs a system approach; has high face validity; has somewhat less control over the conduct of the test; has multiple purposes and multiple criteria (intermediate as well as ultimate); and includes many levels of entry into the test and/or into the system.

System effectiveness testing implies that the HFE is measuring personnel *performance*. Attribute testing (which is the making of subjective judgments about system/equipment/job characteristics as distinguished from the performance of personnel interacting with these characteristics) is not really performance measurement. Judging the adequacy of the human engineering characteristics of an equipment is the most common type of attribute testing. Geer (1977) makes a distinction between informal and formal T&E, the former referring to attribute testing (the evaluation of equipment design characteristics either in the form of drawings or of a nonfunctioning mockup or equipment), the latter referring to performance measurement. Attribute evaluation of equipment sometimes forms part of performance testing and will therefore also be considered here. This applies also to such subjective measurement tools as interviews, questionnaires, and ratings that are used to get information *about* performance, although they may not measure performance directly.

10.1.3 THE PERFORMANCE MEASUREMENT TEST PLAN

This section is organized around the development of a performance measurement test plan. Such a plan is required in every test situation but obviously is most required when the test situation is complex. The categories listed in Table 10.1.1 apply even if the measurement is performed in a laboratory.

Table 10.1.1 Outline of the Personnel Performance Section of a System Effectiveness Test Plan

1.0 Purpose of the Test

    1.1 *General.* Example: Verify that system personnel can perform required tasks.

    1.2 *Specific.* Example: Determine the type and magnitude of errors made by personnel; determine the effect of low-temperature (arctic) conditions on personnel ability to maintain a tank.

2.0 Description of System Being Evaluated

    2.1 *List of equipment* to be operated/maintained by personnel and for which personnel performance data are to be collected.

    2.2 *List of equipment tests* during which personnel performance data are to be collected. Example: Installation and checkout of the *XYZ* fire direction console.

    2.3 *List of tasks* for which personnel performance data will be collected. Example: Alignment of the M-113 theodolite.

    2.4 *Applicable technical manuals* or other procedures.

3.0 Experimental Comparisons (as required) Example: Comparison of accuracy of low-level navigation during daytime and nighttime reconnaissance flights.

4.0 Standards, Criteria, and Measures

    4.1 *Personnel performance standards.* Example: Receive, code, and transmit between 12 and 15 messages per hour.

    4.2 *Personnel performance criteria.* Example: Detection range; number of messages sent.

    4.3 *Personnel performance measures.* Example: Time taken to load handheld missile; officer evaluation of squad reconnaissance performance.

5.0 Data Collection Methods

    5.1 Data collectors

        5.1.1 *Number.* Example: 4

        5.1.2 *Tasks to be performed.* Example: Record start/stop time in operation of laser tracking set.

        5.1.3 *Training* (if required). Example: All data collectors will receive 3 hr of instruction in gathering data on the F42 machine gun (see training schedule appended).

    5.2 *Data collection forms.* Example: Data sheet for retractable machine gun (appended); postmission debriefing questionnaire for truck driver (appended).

    5.3 *Data collection procedures.* Example: See Appendix A to test plan.

    5.4 *Instrumentation* (only if required). Example: Videotape machine; two tape recorders with three rolls of tape per data collector.

6.0 Subject Characteristics

    6.1 *Number.* Example: three squads.

    6.2 *Required characteristics.* Example: All subjects will have 20/20 vision (corrected) and will have been qualified in operation of the retractable machine gun.

7.0 Constraints

8.0 Data Analysis

    Example: Determine mean number (and standard deviation) of messages transmitted between forward observers and batteries; develop equation relating gun loading speed and operator errors.

9.0 Testing Schedule

    Example: Concurrent with other tests.

Source: Meister (1978)

However, the test plan categories have a different meaning when the test is system-oriented. For example, as the reader will see, criteria and measures are necessary in every laboratory test, but standards have a particular significance in system effectiveness testing

In the course of expanding on these categories, the many variables affecting system performance measurement will be considered with special reference to the several settings that are the subject of this chapter. It is of course possible to measure performance without first developing a test plan, but the results are likely to be less than desired. The utility of developing a test plan is obvious. It requires the HFE to specify precisely what must be done in the test; the effort of writing a planning statement requires and may even encourage clarity of thought. The test plan also communicates information to others who are concerned in both test planning and implementation.

In the course of testing, the HFE will make use of at least one of the techniques listed in Table 10.1.2; more than one, depending in part on how comprehensive the test is (e.g., whether it is a test of an equipment or of an entire system). One of the fascinating things about system effectiveness testing is the variety of methods from which the HFE can select.

On the other hand, lack of sophistication in the test situation will limit the number of methods that can be applied. For example, obviously in a static mockup one cannot take advantage of automated data collection, but this is possible in an operational system or in a simulator. Some methods are more generalizable than others. For example, observation can be used as a measurement tool in every data collection context, but automation cannot.

One of the interesting aspects of performance testing is that the tester attempts to measure under conditions of maximum fidelity to the operational system and the operational environment, regardless of how poor that fidelity may be, as in the static mockup. The effort to achieve maximum operational fidelity drives much test planning and data collection. In this respect system effectiveness testing differs fundamentally from laboratory psychological studies that are only minimally concerned, if at all, with fidelity to any particular setting.

10.1.3.1 Purpose of the Test

Section 1.0, Table 10.1.1, describes the purpose of the performance measurement test. Although that purpose may seem self-evident, it is often not. This is especially so in testing the prototype system to determine compliance with HF requirements, since these are usually not called out explicitly. The requirement for evaluating a system is usually phrased in terms of the need to verify or examine gross constructs, such as operability, maintainability, and reliability, but these must be broken down into specifics because as abstractions they offer few guidelines for selection of appropriate methods or test strategy. The goals of system effectiveness testing listed in the previous section are quite general. The reason for specifying the test purpose(s) in very precise detail is because, as will be seen, these purposes suggest methods to be employed. Only by breaking down the general purpose into specifics (e.g., to verify that personnel can perform certain tasks satisfactorily; that training and technical documentation are adequate) can one deduce how to measure properly. A specific purpose of personnel performance measurement is simply a question about that performance which the HFE wishes to answer. After the HFE lists every question he or she wishes the personnel performance data to answer, each question becomes a test purpose. Consequently, because there are many unknowns in new systems, personnel performance measurement usually has more than one specific purpose. The evaluator must tailor his or her test methodology to secure data relevant to those goals.

If there is a possibility that personnel performance will be affected by other system elements (the equipment configuration, its procedures, environment, or logistics), then a question must be answered by testing. For example, if it is possible that heat/humidity within a tank may degrade tank personnel performance, then the HFE wants to accept or reject this hypothesis. Therefore, he or she must arrange to measure personnel performance in relation to heat and humidity. A specific purpose of the test plan might therefore be phrased as follows. Example: To determine the capability of tank personnel to operate tank controls while "buttoned up" under high heat/humidity conditions.

This specific purpose tells the HFE that he or she must arrange to collect data on operation of tank controls under high heat/humidity conditions. This means that he or she may have to (1) install a thermometer and hygrometer inside the tank to determine internal temperature/humidity; (2) require tank personnel to perform and measure that performance while recording temperature/humidity conditions; (3) compare recorded temperature with heat/humidity standards specified in MIL-STD 1472C (Department of Defense, 1981); (4) determine whether performance varies with temperature/humidity changes; (5) gather subjective opinions about the degree of comfort/discomfort within the tank. None of these would have been suggested by the general purpose of determining whether personnel can perform their tasks effectively.

In Chapter 1.2 it was pointed out that a fundamental question that underlies all personnel performance testing in a system framework is how personnel subsystem performance influences total system output. These specific, discrete test purposes do not deal with such general questions; the HFE must infer the relationship from more detailed tests rather than test it directly.

10.1.3.2 Description of the System/Selection of Tasks to Be Evaluated

Section 2.0, Table 10.1.1, of the test plan pertains only to complete system evaluations, such as those with a prototype system and simulators, in which multiple subsystems must be exercised and tasks performed. If testing is to be performed on only selected tasks and equipment, these must be listed. If the test is to be performed on a static mockup, if only one or two tasks must be performed, it is hardly likely that these tasks need be described in planning the test. On the other hand, if those system operations for which data are to be collected are not described, data collectors may not have a clear idea of how to accomplish the measurement. The larger the system, the more necessary this section is.

Table 10.1.2 Elements of System Effectiveness Performance Measurement

Data Collection Methods	Data Collection Contexts	Data Reduction Methods
Automated data collection	Attribute evaluation (checklist type)	*Descriptive statistics*
Checklists	Case studies	Frequency distributions
Document review	Controlled experimental studies	Histograms
Interviews (standardized/nonstandardized; single/group)	Operational system testing	Mean, median, range, standard deviations, and so forth
Narrative recording	Operational prototype system testing	*Comparative statistics*
Observation (participant–observer, overt/covert; non-participant–observer, overt/covert)	Problem investigation	Parametric/nonparametric
Performance testing (standardized)	Simulator measurement	Measures of correlation (e.g., product-moment)
Questionnaire	Surveys	Significance tests (e.g., analysis of variance)
Ranking	*Instrumentation*	*Reporting Methods*
Rating scales	Automated mechanical/electrical instruments	Test planning documents
Self-report (diary)	Communication devices, for example, radio/telephone	Data collection documents (e.g., data recording forms)
Semiautomated data collection (time/event recorders)	Computerized measurement	Test reports (published/unpublished)
Survey techniques	Environmental instruments, for example, light meter, sound level meter, thermometer	
Walkthrough	Paper and pencil forms	
	Physiological instruments	
	Tape recorder	
	Timers	
	Videotape/film	

Describing the system to be evaluated may appear unnecessary to those who are familiar with it. However, in most systems of any size, only *some* of the system equipment and only *some* of the tasks involved in operating that equipment will be of interest for performance measurement. If the system under development has had a predecessor system and certain equipment/tasks that have been carried over to the new system have been tested previously, one would not repeat those tests. If the system is large and complex, test time is short, or data collectors are few, it may not be possible to record data on all tasks performed in all situations. For routine tasks and equipment it may not be cost effective to do so. Under these circumstances it will be necessary to select those tasks/equipment for which personnel performance will be measured. The bases for selection of tasks will be (1) task criticality to mission accomplishments; (2) task complexity; and (3) likelihood of significant error.

Altman (1969) has addressed the considerations involved in the selection of tasks for measurement. To make the selection, the HFE should identify: (1) tasks with critical time limits (these are likely to be tasks with higher error potential); (2) tasks in which personnel interact frequently and in which the task is a team operation; (3) tasks involving ambiguous stimuli and/or that require decision-making; (4) tasks whose failure or inadequate performance could have significant negative effects on system output; (5) mission scenarios that impose heavy demands on system personnel; (6) tasks with high error potential and that possess critical system design features. Information describing those tasks should be available from task, workload and error analyses.

Where systems have changing functions and missions as a consequence of contingent occurrences, one must decide which functions/missions will be exercised during the test. The Navy (Department of the Navy, 1979) makes a distinction between *scenario-oriented* and *operation-oriented* testing. Scenario-oriented testing is for systems whose mode of operation or functions change according to a changing operational situation. For example, a shipboard-antiair war (AAW) fire control system that is mostly in the search mode until an attack occurs is a prime candidate for scenario-oriented testing. In such testing the system being evaluated is introduced into a realistic simulation of a developing operational situation, that is, a scenario. For the AAW system a raid consisting of strike aircraft could be programmed at a designated but unannounced time. Multipurpose systems may require several scenarios to exercise all their capabilities. The significance of scenario testing for the HFE is that he or she may be asked to help develop appropriate scenarios and he or she must anticipate what the effect will be on his or her analysis and measurement activities, for example, a requirement for more observers or special measures of effectiveness.

Operation-oriented testing is used for equipment or subsystems whose mode of operation remains constant throughout an exercise. The tasks associated with such equipment will require less intensive behavioral measurement.

It is unnecessary for this section of the test plan to provide highly detailed task and equipment descriptions; these are probably already described in procedures and technical documentation; however, the procedures should be identified, as also the equipment associated with these procedures.

10.1.3.3   Experimental Comparisons

If the mockup, equipment, or system is to be exercised in varying operational modes (e.g., reconnaissance and attack, preventive and corrective maintenance, day or night) or if there are variables inherent in system operations that may influence or reflect differences in personnel performance, these variables must be studied (Section 3.0, Table 10.1.1). If they are not studied, salient qualities of the system may remain obscure.

Each of the factors to be studied should be listed. Sometimes the factors represent different system missions (e.g., search and attack); at other times they represent variations in the normal functioning of the system (e.g., the individual shifts that a crew works); or they may be differences in training and experience, as, for example, when auxiliary personnel such as clerks may be called on to function in jobs requiring combat experience. Alternative missions and personnel differences should always be tested; those inherent in routine system operations *may be,* if it is possible that they make a difference to performance. Consequently, each factor involving a contrasting condition establishes the requirement for an experimental comparison. Most systems of any complexity have a number of missions and factors that should be compared.

Where behavioral conditions to be compared are inherent in a system mission (e.g., night/day operations), they do not require the setting up of special test situations (other than the development of the necessary scenarios). Test managers prefer minimal disruption of routine system operations.

In these comparisons it is possible to apply experimental designs of the type ordinarily used in more highly controlled research studies (e.g., analysis of variance). However, where system operations are dependent on uncontrolled contingencies, such as the weather or the action of a simulated enemy, it may be difficult to use a highly controlled design. One study (Stinson, 1979) involved testing a prototype vehicle for transporting Marines from amphibious ships to shore at high speed under various sea states. The question was whether personnel performance on shore (firing rifles, assaulting barriers, etc.) would be hampered by the previous 2 hr at sea. Ideally, data would be collected following sea state 1, sea state 2, and so forth. In the event, certain sea state conditions could not be realized because the ocean was uncooperative. This effectively eliminated the possibility of analyzing the data according to the original analysis of variance design.

Within limits, therefore, it is possible to apply standard experimental designs to system tests. However, the experimental design comparisons described in the data analysis section (8.0) may be in part *post facto* and contingent on the kind of data ultimately gathered.

10.1.3.4　Standards, Criteria, and Measures

Overview

MIL-H-46855B (Department of Defense, 1979) requires that HF tests be based on criteria that describe adequate performance of tasks. A criterion is, however, relatively abstract, for example, speed, response time. What MIL-H-46855B means is that measures must be developed and that *standards* of acceptable task performance should be available so that actual personnel performance can be compared with that standard (Section 4.0, Table 10.1.1).

A measure is simply a description of personnel performance or system output. A standard is much more important. It is some output that personnel are *required* by the system to produce or to achieve. Criteria, measures, and standards organize the ways in which performance (personnel and system) can be described; they are closely related and may therefore confuse the HFE in his or her planning of the measurement situation.

The HFE wishes to develop measures and standards, but he or she must begin with the criterion because it is the most molar and the developmental sequence is (1) identify the criterion: (2) identify the measure that makes the criterion specific and precise; and (3) establish the standard, which is coordinated with the measure.

In this discussion we reverse the process, starting with the standard, because it is what we ultimately hope to derive and without standards true evaluation is impossible; we describe criteria next, because measures are developed out of criteria; and finish with a discussion of measures.

Standards

If the reader will review again the general purposes for which testing is performed (previous section), he or she will see that performance standards are implied in most of them. One cannot verify that personnel perform effectively unless a performance standard exists (e.g., X performance is adequate, $X + n$ is "super," $X - n$, unsatisfactory). Without a standard any variation in performance is inexplicable. Even when comparisons are made between conditions A and B, one wishes to know whether the less effective performance of B is still within standard. The nonavailability of a standard renders all measurement values difficult to interpret.

The primary goal of system effectiveness testing as it applies to personnel is to verify that personnel can perform to standards imposed by the system and its mission. The need for standards is peculiar to evaluation. One does not have this problem in general behavioral research because verification of a particular condition or status is not a research objective.

Standards set by system requirements in formal documents are almost always physical. In evaluating a new fighter, for example, the aircraft must meet altitude, range, fuel consumption, and so on, requirements; it is *assumed* that whoever flies it is qualified. Normative performance standards are behavioral because they were developed specifically to describe personnel performance. The reason why system requirements are almost always physical parameters is that mission requirements describe the terminal outputs of the system (what the system is supposed to accomplish), and these outputs are always physical.

Behavioral requirements are almost always inferred, for example, the pilot of an aircraft should have sufficient external visibility to maneuver safely. Because behavioral requirements are only inferred, they usually do not have quantitative standards associated with them. Such standards may only be implicit in the system operation. It is possible to derive these quantitative standards in two ways: by asking subject-matter experts, for example, pilots, to describe those implicit standards; or by empirical data describing what personnel typically do in a system like the one under development (normative data).

Theoretically, the human performance standard should be implicit in the task description, but many task descriptions lack explicit performance values, although these can be derived by the process known in instructional system development (Chapter 7.3) as development of job performance measurement standards. In addition, many proceduralized tasks have binary standards in which the range of performance variability—which is what we are interested in as far as standards are concerned—is highly restricted. For discrete subtasks and tasks, like throwing a switch or performing a sequence of switchlike actions, the operator either performs the action or he or she does not. No question of performance *quality* enters in nor could it because the switch can be thrown in only one way. When the operator does not throw the switch, we simply record an error.

When the performance standard is binary, it is easy to determine whether or not the operator has satisfied the standard. On the other hand, with such highly proceduralized tasks the amount of error is usually very small, since unless the operator is unskilled or under pacing pressure, he or she rarely makes a mistake. Consequently, unless the HFE can measure the operator's performance many times (e.g., 100–200 trials), the HFE may observe no errors at all in the relatively few trials performed in the average test program.

With a nonproceduralized task that can be performed in a variety of ways and therefore provides more variety in the data, the performance standard becomes more important but is more difficult to derive. The index of acceptable performance is more difficult to derive than is the binary standard, may be recordable only in qualitative terms, and is often judgmental. If performance quality is measured qualitatively, for example, the quality of a singer's voice, only a few highly trained experts may be able to recognize that quality.

Theoretically, the standard for any task can be derived by the deductive logic of function/task analysis from initial system/mission requirements. However, the thread relating an individual task standard to a system requirement may be too long and convoluted to permit this.

If human performance standards are lacking, it is impossible for testing to verify human performance except in a limited way, that is, if the system performs its missions adequately, and no outstanding operator performance deficiencies are noted, this *suggests* (but does not prove) that personnel can do their jobs effectively. Geer (1977), p. 29) recommends the experienced evaluator's judgment as the basis for "making such determinations as test precision, personnel skill, degree of personnel training, etc." He notes that it would be "difficult to measure these qualities in any other way during a test."

This is a very unsatisfactory situation for the HFE, even though it will satisfy most test managers. Where objective standards are lacking, performance testing is essentially only a context in which to discover human performance inadequacies, although from a purely developmental standpoint the uncovering of such deficiencies and their subsequent remediation is important.

The specification of precise quantitative standards presents a number of difficulties. Since a standard is ultimately a subjective value judgment, it is reasonable to utilize subject-matter experts to attempt to derive a specific quantitative standard for a task. This can be done informally, by interview, or more formally by some sort of psychometric judgment.

However, even where there is a predecessor system it is often difficult to persuade skilled system personnel to provide precise standards of their performance. A frequent response is, "it all depends," which may reflect the feeling that the performance depends on so many interactive and contingent factors that it cannot be specified.

The absence of performance standards does not mean that the HFE cannot utilize test data to determine how well personnel are performing; he or she must do so, however, through substitute secondary methods not involving the comparison of empirical data with an objective standard. For example, supervisor and peer ratings of personnel performance can be substituted.

Moreover, even if the meaning of human performance data is ambiguous, it is possible to do useful things with the information. For example, performance vastly different from that which the HFE would anticipate on the basis of previous tests will suggest that something is wrong and must be investigated. The particular kinds of difficulties experienced by test subjects will suggest areas for further examination.

Criteria

Before one can develop standards one must have criteria that are system attributes or characteristics in the abstract. The difference between the two is a matter of specificity; a major criterion in operator detection of aircraft, for example, is detection range; a standard might be 3.5 miles. Obviously, the more molar criterion, detection range, precedes the standard, which specifies and quantifies the criterion.

There are three distinct types of performance criteria: those describing the functioning of the system; those describing how missions are performed; and those describing how personnel respond. System-descriptive criteria include such attributes as reliability, maintainability, vulnerability, and cost of operation. Mission-descriptive criteria include effectiveness in mission accomplishment, output quality and accuracy, reaction time, performance duration, queues, and delays.

Each of these includes personnel elements that must be differentiated from nonpersonnel elements. On the other hand, personnel performance criteria describing operator and crew responses (reaction time, accuracy, number of responses, response consistency, speed, etc.) lack meaning unless they are considered in relation to system and mission criteria.

Performance criteria may act as independent or dependent variables. As independent variables (e.g., the requirement to produce N units) they impose on the operator a requirement that serves as a forcing function for his or her performance (he or she must produce N units). As dependent variables they describe his or her performance (i.e., the operator has provided N or $N - 1$ or $N + 1$ units) and can be used to measure that performance. The latter is what we are concerned with.

In evaluative performance measurement the criterion requires a standard of performance that permits accomplishment of the system mission. In measurement research, in general, criteria are also necessary (as dependent variables) but they do not imply or require standards. In evaluative measurement a criterion (e.g., efficiency) is meaningless without a standard because it does not provide a means of determining whether personnel are performing well or poorly. Thus, for example, the number of requests for information handled is a criterion that can be used in research on telephone operator performance but the evaluation of that performance makes it necessary to specify in advance of measurement the minimum number of requests that *must* be handled in a given time period.

The criterion must be precise and quantitative. A criterion such as one that is occasionally found in system procurement descriptions, "the system shall be so designed that personnel perform their duties with minimum difficulty," is meaningless because it is undefined in quantitative terms, which means that it can be understood in any way one wishes. With undefined criteria one must rely on the evaluator's ability to translate the criterion into concrete terms; and unless those terms are specified in writing, it is almost impossible to communicate their meaning to others.

Not all criteria are equally relevant and valuable for performance measurement. The level of adrenaline in the blood of subjects performing a visual vigilance task may be related to target detection, but adrenaline level is not the most desirable criterion one can find to measure sonar detection because it is only *indirectly* output-related. The HFE should examine the criteria he or she has available and select those that seem most directly related to the performance at issue. The relevance and importance of a potential criterion can be determined by asking how seriously failure to achieve this criterion would affect system performance. If the relationship of the criterion to system output is weak, the potential criterion is not a very satisfactory candidate.

Objective criteria are preferable, because the performance described by such criteria can be observed and recorded simply and directly without the HFE's interpretation. Unfortunately, many criteria fail this test. Some performances (primarily perceptual and cognitive) are inherently subjective. For example, it is not presently possible to measure the quality of decision making in a combat situation with instrumentation. The cues needed to describe quality may be so tenuous that only a specialist can perceive them. For qualitative criteria we must call on the expert because only he or she has the requisite experience to recognize the performance involved. However, use of expert judgment may have disadvantages. For example, the specialist's unrecognized biases may produce error. Moreover, we can accept conclusions based on his or her judgments but with a somewhat lesser level of confidence, because, not being experts ourselves, we can never be quite sure just how expert he or she is.

Complex systems may also have multiple criteria because personnel must perform a variety of functions. If so, one must use all the criteria (assuming they are all substantially related to system output); the HFE should not pick and choose among criteria (especially not *post facto*) even though it is sometimes embarrassing when he or she secures one set of results with one criterion (e.g., personnel perform effectively) and another set with another criterion (e.g., personnel perform with only minimal efficiency).

Criteria interact with the organizational structure of the system. As the focus of performance shifts from individual operator to team or from subsystem to system, criteria may change. In measuring team performance, for example, one must consider member interactions, a factor that is obviously irrelevant to single operator performance.

Measures

Lacking criteria, in a system of any reasonable size the number of performance outputs that *could* be measured might bewilder the HFE because he or she literally has an embarrassment of riches with no basis for selection. The variety of performance measures available is shown in Table 10.1.3.

Table 10.1.3   Classification of Generic Performance Measures

Time

1. Reaction time, that is, time to
   a. perceive event
   b. initiate movement
   c. initiate correction
   d. initiate activity following completion of prior activity
   e. detect trend of multiple related events
2. Time to complete an activity already in process; that is, time to
   a. identify stimulus (discrimination time)
   b. complete message, decision, control adjustment
   c. reach criterion value

3. Overall (duration) time
   a. time spent in activity
   b. percent time on target
4. Time sharing among events

Accuracy

1. Correctness of observation; that is, accuracy in
   a. identifying stimuli internal to system
   b. identifying stimuli external to system
   c. estimating distance, direction, speed, time
   d. detection of stimulus change over time

Table 10.1.3 (*Continued*)

e. detection of trend based on multiple related events	d. measures of achieved maintainability
f. recognition: signal in noise	e. equipment failure rate (mean time between failure)
g. recognition: out-of-tolerance condition	f. cumulative response output
2. Response-output correctness; that is, accuracy in	g. proficiency test scores (written)
a. control positioning or tool usage	2. Magnitude achieved
b. reading displays	a. terminal or steady-state value (e.g., temperature high point)

e. detection of trend based on multiple related events

f. recognition: signal in noise

g. recognition: out-of-tolerance condition

2. Response-output correctness; that is, accuracy in
   a. control positioning or tool usage
   b. reading displays
   c. symbol usage, decision making, and computing
   d. response selection among alternatives
   e. serial response
   f. tracking
   g. communicating

3. Error characteristics
   a. amplitude measures
   b. frequency measures
   c. content analysis
   d. change over time

Frequency of Occurrence

1. Number of responses per unit, activity, or interval
   a. control and manipulation responses
   b. communications
   c. personnel interactions
   d. diagnostic checks

2. Number of performance consequences per activity, unit, or interval
   a. number of errors
   b. number of out-of-tolerance conditions

3. Number of observing or data gathering responses
   a. observations
   b. verbal or written reports
   c. requests for information

Amount Achieved or Accomplished

1. Response magnitude or quantity achieved.
   a. degree of success
   b. percentage of activities accomplished
   c. measures of achieved reliability (numerical reliability estimates)

d. measures of achieved maintainability

e. equipment failure rate (mean time between failure)

f. cumulative response output

g. proficiency test scores (written)

2. Magnitude achieved
   a. terminal or steady-state value (e.g., temperature high point)
   b. changing value or rate (e.g., degrees change per hour)

Consumption or Quantity Used

1. Resources consumed per activity
   a. fuel/energy conservation
   b. units consumed in activity accomplishment

2. Resources consumed by time
   a. rate of consumption

Physiological and Behavioral State

1. Operator/crew condition
   a. physiological
   b. behavioral

Behavior Categorization by Observers

1. Judgment of performance
   a. rating of operator/crew performance adequacy
   b. rating of task or mission segment performance adequacy
   c. estimation of amount (degree) of behavior displayed
   d. analysis of operator/crew behavior characteristics
   e. determination of behavior relevancy
      i. omission of relevant behavior
      ii. occurrence of nonrelevant behavior
   f. causal description of out-of-tolerance condition

2. Subjective reports
   a. interview content analysis
   b. self-report of experiences ("debriefing")
   c. peer, self-, or supervisor ratings

Source: Smode, Gruber, and Ely (1962)

The criterion allows the HFE to select a subset of all possible performance outputs (measures). The situation is not so bad at the individual equipment level, but if one is dealing with a subsystem or the system as a whole, the number of operations that can be measured may be excessive. For example, Vreuls, Obermayer, Goldstein, and Lauber (1973) generated over 800 measures for a simple captive helicopter performing common maneuvers.

A single criterion, for example, effectiveness of corrective maintenance (CM), may have a number of measures, one of which is downtime: the time it takes the technician to restore malfunctioning equipment to operating condition. However, other measures are possible. For example, the *number* of malfunctions correctly diagnosed, the *number* of tests made, or the *speed* of malfunction diagnosis (not the same as remedying the fault).

All measures like all criteria are not equally useful in describing performance because some of them may be only *indirectly* related to performance. *Within the limits of the HFE's resources* and to the extent that there is reason to suspect a substantial relationship between the measure and the system output, he or she should record all indirectly as well as directly relevant measures.

In translating the criterion into the measure, the HFE examines what the operator does. For example, if a forklift operator in a warehouse has to move boxes from the loading dock to an assigned storage area, the number of boxes moved per hour is an excellent measure of his or her effectiveness.

Measures may describe terminal performance (the output of the operator action) or intermediate performance (an operator behavior leading to an output). Intermediate measures are more likely to be more detailed than terminal ones. To measure rifle-firing proficiency, for example, one could record the tremor of the finger in squeezing the trigger (an intermediate measure) or one could record the number of hits on target (a terminal measure). Terminal measures are more valuable than intermediate ones, because the evaluator is interested in outputs, although the intermediate measures are useful for diagnosing operator inadequacies. If, for example, the system output one is concerned with is the accuracy of rifle fire, then highly molecular measures like trigger pressure would be less useful than number of hits, since squeeze pressure is less directly related to firing accuracy than number of hits. On the other hand, the reason why a novice fails to hit the target may be because he or she is not squeezing properly.

The particular measures selected will determine in part *how* the data are collected. For example, instrumentation would probably be required if one wished to measure trigger pressure but not if one measured error in hitting the target.

The difficulty in developing meaningful performance measures increases when the system has complex task interrelationships. Measures of secondary or dependent tasks may appear less relevant to the terminal system output. It is necessary to work through the dependency relationships among tasks before deciding to accept or reject a measure.

Another difficulty is that the most common objective measures—time and errors—may not be hard to measure but harder to make sense of them. Unless response/reaction time is *critical* for a task, time as a measure has little meaning. Only if time is unduly prolonged will it affect system output; small variations in performance time do not reveal very much. Errors may be indicative of performance quality but only if the errors have a significant effect on system performance. In well-trained personnel significant errors may be so few that data collectors have difficulty observing them.

If the tasks being measured are cognitive (more and more likely these days) or if significant performance dimensions are so tenuous that only a specialist can pick them up, the HFE is in measurement trouble, because they may not be observable (or, if observable, only by a specialist).

The fact that common response measures (e.g., time and errors) do not make much difference to many tasks may make it necessary to select critical tasks (in which variations in response time and errors do make a difference) for measurement.

If it is necessary to select among measures, the HFE should, all other things being equal, select those that are (1) highly related to the output or product of the performance being measured; (2) objective; (3) quantitative; (4) unobtrusive; (5) easy to collect; (6) require no specialized data collection techniques; (7) are not excessively molecular and therefore require no specialized instrumentation; (8) cost as little as possible monetarily and in terms of data collection effort. Few measures satisfy all these criteria.

The reason for emphasizing criteria, measures, and standards is that unless these are very carefully considered in the test planning process, much useful data will not be gathered and much of the data secured in the test will be uninterpretable.

The process of selecting criteria, measures, and standards is more complex than many HFEs realize. Often HFEs accept the obvious ones without working through their relationships to system outputs and, consequently, their real meanings.

10.1.3.5 Data Collection

Section 5.0, Table 10.1.1, of the plan deals with *who* will collect data and more importantly, *how* they will collect it. Section 8.0, Table 10.1.1 (Data Analysis) describes how the data will be analyzed. Naturally, data analysis considerations will in part determine what data will be collected and consequently how these will be collected.

To ensure any possibility of success the HFE must develop for his or her performance measurement plan a scenario describing in as much detail as possible the data he or she is going to collect and how he or she intends to collect them. He or she must do this even though it is extremely likely that last minute changes will be required because of changes in overall test circumstances.

The specification of data collection methodology is necessary for a number of reasons: to communicate information to other test planners and to the data collection team; to enable the HFE to have firmly in mind what he or she intends his or her data collectors to do; to expose to examination any difficulties or inadequacies that may exist in the HFE's plans.

The number of personnel who are serving as data collectors (Section 5.1.1, Table 10.1.1) should be noted as a matter of record. Where more than one data collector is required, how they are to be scheduled becomes important.

The specific data-collection tasks which the collectors will perform (Section 5.1.2, Table 10.1.1) should be indicated. These are *not* the same as system operating tasks for which data are to be collected. Rather, they are the activities involved in gathering information about the task performance being measured. For example: At the conclusion of each truck driving cycle, data collectors will administer a questionnaire concerning ride quality.

Unless data-collection tasks are complex, they need be described only as general functions, for example, administer questionnaires. On the other hand, if data collection may pose special problems, the tasks involved should be described in step-by-step fashion. If there is much to say about them, Section 5.1.2 (Table 10.1.1) can be an appendix to the test plan.

If data collectors must receive special training to enable them to perform their duties effectively, that training should be described (Section 5.1.3, Table 10.1.1). In OT as performed by the military, data collectors are often military personnel who have no particular training or expertise as data collectors; under these circumstances intensive training may be necessary. Even if the HFE uses specialists in the system being tested, for example, infantry sergeants to evaluate squad performance, they should receive training and practice in methods of recording data.

If data collectors are not familiar with the system under test, they will require special training to give them this familiarity; a data collector cannot function effectively without knowledge. Data-collection training should be oriented toward enabling collectors to recognize the events they have to record and how these should be recorded (as well as any instrumentation to be used). Several data-collection "dry runs" should be held to habituate data collectors to their tasks.

All data-recording forms should be noted in Section 5.2, Table 10.1.1, with the actual forms appended to the test plan. This permits everyone concerned to examine them.

The rationale for the data-collection methods that will be employed should be described, as also any demands on the time of test personnel acting as subjects (e.g., interview time, the completion of questionnaires, etc.). Instrumentation used to record data is described in a Section 5.4, Table 10.1.1.

The specific data-collection methods to be used will depend in part on the criteria/measures that have been developed previously, in part on convenience, where several methods can provide roughly the same information. It would however be a capital mistake to select one's data-collection methods before determining criteria and measures.

Military test managers prefer manual methods, in large part because of their lesser cost, to some degree also because laymen understand interviews, questionnaires, and rating scales more than they do instrumentation. They are after all familiar with and make great use of opinions in their routine work.

Where testing is performed in the field (e.g., test site, aboard ship), it is fair to say that behavioral data-collection methods cannot be overly sophisticated because of the constraints that affect testing efficiency in the field environment (see Section 7.0, Table 10.1.1). OT, for example, is a relatively crude measurement environment. Greater sophistication is possible in the simulator because the simulator is a more controlled environment; on the other hand, unless the mockup is a functioning one, data-collection methods in mockup tests will be limited to observation and recording of performance time and errors.

Most performance-measurement tools, particularly those that involve subjective judgment, like ratings or interviews, are not standardized or developed using psychometric methods; HFEs tend to make up their own tools. They should therefore be tried out before the test under conditions that approximate the ones in which they will be used and should be modified where deficiencies are found.

Unless the data collection procedure is very simple, it should be described (Section 5.3, Table 10.1.1), including the following: (1) The hours data collectors will work, or the sequence of operations (their beginning/completion) that will determine the data-collection period. (2) How data collectors should process their data, for example, do they pass the data on to the evaluation personnel immediately or hold on to them? Do they partially analyze data while they are being recorded? (3) What data collectors should do if an emergency occurs (e.g., if an exercise is suddenly stopped before it is completed), or if something not covered by operating procedures occurs. (4) The level of detail to which they should record data (principally relevant when reporting qualitative observations). (5) The extent to which the data collector is permitted to interact with the personnel whose performance is being evaluated (e.g., the distance they must remain away from participants in the operation). (6) Any equipment data collectors will be required to operate.

Data-collection instrumentation sometimes used in field operations includes small magnetic tape recorders for recording observer notes and communications, and handheld video tape recorders or motion picture cameras for recording events visually. To assess the environmental conditions under which performance occurs, if these are relevant, light meters and accelerometers (for vibration effects) may be employed or sound level meters may be used to record noise levels.

Automated data collection becomes increasingly feasible as the system becomes more computerized. When system control is exercised by means of commands to a computer, it is a relatively simple matter to arrange for subroutines to record all operator inputs and their timing in relation to equipment processes [see as an example Cohill and Ehrich (1983)]. Even so, the automated data-collection mechanisms pick up only overt responses, meaning that perceptual and cognitive activities will be ignored.

10.1.3.6 Subject Characteristics

Section 6.0, Table 10.1.1, of the test plan describes test personnel; that is, those personnel who are selected to operate and maintain the system during its testing and whose performance is used to evaluate personnel effectiveness.

Subject characteristics will vary as a function of the type of testing performed. In mockup testing that is conducted during early and middle system development, subjects are likely to be the HFE's engineering colleagues. At this stage the HFE is likely to be somewhat less concerned about the similarity of test subjects to the anticipated user personnel because the questions he or she asks at this time deal with more fundamental parameters such as anthropometric suitability, for example, can controls be reached and displays read? The system configuration is still fluid, so that the relationship between subject characteristics and system design is largely conjectural.

If initial OT (see Chapter 1.2 for the variations in OT mandated by government regulation) is conducted by a contractor, subject personnel may be either engineers or user personnel recruited from the customer's ranks. In final OT (conducted by the customer) subjects will be operational personnel because the customer will use his or her own personnel for the test. Moreover, if the purpose of final OT is to accept the system and determine what defects still exist, the system must be exercised by operational personnel, since engineers are likely to overcome emergent problems that might not be so easy for operational personnel, who are less skilled than engineers.

If performance is measured in a simulator, it is highly probable that subjects will be operational personnel, since almost all simulators are built after the system configuration is fixed and the primary system has been turned over to the customer.

If OT is to be valid, therefore, it must be performed with personnel who are representative of those who will eventually operate and maintain the system. [This requires that one know as much as possible about the characteristics of the ultimate system operators as well as of those personnel who are intended to represent them. Geddie (1976) has pointed out that test personnel may be the major source of variance in test data.] If test personnel are much more trained or experienced than eventual system users, evaluation results will fail to describe correctly the performance of the latter.

Characteristics to be considered in system users and test personnel include (as relevant to the task): (1) Physical—for example, vision, hearing, height, weight, and strength. (2) Aptitude—general intelligence; special aptitudes, for example, mechanical. (3) Training (as required by the special characteristics of the system). (4) Experience—number of years operating a system of the type being tested.

Attitudinal and motivational factors are important also but cannot be precisely specified. Personnel used as subjects should be willing volunteers or at least not resistant to the notion of serving as test personnel.

In the military, when OT is performed as part of operational exercises, there is no question of personnel volunteering for the exercises. When the test is something above and beyond normal military duties, subjects must be volunteers because governmental regulations require that the potential subject have the opportunity to reject the privilege. This situation arises more in connection with research than system development. The question of motivation arises only when military subjects are impelled to volunteer when they would rather not. In engineering development, test subjects are most often engineers and technicians who exercise prototype systems at test sites as part of their normal duties. The problem of volunteering does not arise; but any special HF tests other than those mandated as part of engineering development do require volunteers.

Whether volunteers or not, it is important to explain to subjects why they are being asked to participate in the test, the importance of their participation, and the fact that there will be no negative consequences for them. It would be incorrect to think of test personnel simply as bodies to be manipulated.

Most frequently, subject requirements will involve training and experience. For example, it is obvious that, to evaluate a new type of bus, for example, the subjects must be qualified bus drivers and, if driving the new bus requires new skills, driver subjects must receive special training.

Section 6.1, Table 10.1.1, notes the number of subjects or the organizational units in which they function (e.g., the squad, the sonar team, the production unit). This item is for information only.

Section 6.2, Table 10.1.1, describes required subject characteristics. These are characteristics that, in the HFE's judgment, will significantly affect the accuracy of the data if not possessed by personnel acting as subjects.

A special case exists where both the ultimate users and the test personnel must be highly skilled, for example, pilots, navigators, and electronic technicians. These personnel possess a continuum of capability ranging from mediocre through average to exceptionally highly qualified. The question is whether one selects test personnel who are at about the 50th percentile in skill or those who are exceptionally well qualified? The military services have always selected specially trained and qualified personnel to test their aircraft (automobile companies also employ highly skilled test drivers for their new cars); and this selection is justified because these systems are always very dangerous and require the greatest skill to test. Where the new system is *not* dangerous, however, although it may be complex and demanding, the question remains: average or exceptional personnel? The use of more skilled personnel will undoubtedly make the new system look better in the eyes of potential critics, because these personnel will be able to compensate for any deficiencies the new system may have; this solution has a political value. On the other hand, the new system will by definition ultimately be operated by average personnel and system managers will not discover the problems that are likely to disturb average personnel if exceptional people have managed to overcome them. (It is of course possible for sensitive test subjects to recognize factors that would present problems for personnel less skilled than they are.) The use of the most highly skilled personnel makes the OT somewhat less operational. Whom one selects as test subjects is a value judgment, but if one of the OT goals is to discover areas of "weakness" in the system, personnel of average skill are more likely to do so.

One undesirable aspect of OT is that in some tests operators are military personnel, but to ensure that undesirable consequences do not occur (because an equipment may fail or an operator make an error) they are backed up by contractors (engineers) who step in and take over control of the tasks being performed whenever a difficulty arises. From the standpoint of operational fidelity such a procedure completely invalidates the test, even if it makes the new system appear more effective, because the system in routine operations will probably not have access to engineers.

10.1.3.7   Constraints

A major goal in system effectiveness testing is to simulate the operational system and its operating environment as closely as possible. When an evaluation is performed on an actual operational system performing its assigned missions in its assigned manner, the simulation is complete (or almost complete; an operational system can, after all, be exercised in a nonoperational manner). The simulation cannot be complete where the system is only a prototype, probably incompletely debugged and often functioning in a special test facility (e.g., a test station like Vandenberg AFB); even a simulator is incomplete because even if its equipment functions operationally, the *context* in which it is exercised is not completely operational.

In the effort to simulate the operational system, there is an inverse tradeoff between the amount of control the HFE can exercise and the system's fidelity: the more fidelity, the less control. That is because the system, when performing operationally, is subject to many extrasystem effects over which the HFE lacks control.

If the system under test is to be operated with maximum fidelity, once operations have been initiated, it must function without external interference (except for safety reasons) until its mission or goal is naturally accomplished or aborted. In consequence, (1) the system must not be stopped or influenced in its operation in any way to allow the HFE or anyone else special time or facilities to interrogate system personnel, to gather special data, or to require test personnel to perform other than test duties; (2) data collectors cannot interfere with the normal exercise of the system to gather data and data collection activities cannot be too obtrusive.

There are many constraints (Section 7.0, Table 10.1.1) on testing in the OT and simulator settings (the major mockup constraint is the relative nonflexibility of the static mockup). The physical configuration of the system may constrain data collection. For example, the lack of a necessary power source in the system may prevent the use of particular instrumentation; the space in which system personnel work may be so small that it precludes the use of an observer.

Another constraint is motivation. Where operational personnel serve as test subjects they may feel that they are being asked to exert themselves unduly without receiving any particular reward or compensation for playing the role of "guinea pigs." To ensure cooperation the evaluator may have to offer them an inducement. For example, to install and evaluate the effectiveness of nontactical computers aboard warships to perform certain administrative functions, it may be necessary to provide ship personnel with a special computer program that presents recreational video "game" programs.

Lacking an inducement, system personnel may become indifferent to the test or noncommunicative or may "act out" latent hostility by deliberately providing incorrect or partial information when asked or by concealing their task activities from an observer. It is good practice to enlist managers and system personnel as integral participants and planners of the test, if this is possible.

Apart from all that, if the test is performed on an operational system that has contingent demands placed on it, as many military systems do, the system may change its mission abruptly, for example, a warship switching immediately from a training mission to one of search for an unknown intruder. This may disrupt carefully laid test plans.

Where constraints are anticipated that may affect the efficiency of data collection and the resultant data, for example, small size of personnel sample, the HFE should describe these in the test plan so that possible remedies can be sought.

10.1.3.8 Data Analysis

The test plan must include a procedure for statistical analysis of the data (Section 8.0, Table 10.1.1). If the HFE waits until after data are collected before developing his or her statistical plans, the chances are excellent that too much or too little data will be collected and—much more serious—much of his or her data may be unusable because they will not fit statistical requirements. (On the other hand, because of uncontrollable events, data analysis methods may have to be changed somewhat at the last moment to accommodate those events.)

Table 10.1.4 suggests the kind of data that will be available for analysis and the analyses that can be performed. Table 10.1.5 expands on Table 10.1.4 in terms of steps to follow in the data analysis process.

Table 10.1.4 Generic Categories of System Test Data Analysis

Analyses by Mission

1. Frequency and percentage of mission accomplishment in terms of
   a. overall mission goals
   b. comparison of segments and phases
   c. goal-relevant criteria (e.g., miss distance)
   d. time to accomplish the mission
   e. expenditure of system resources (fuel, etc.)
2. Measures of system reliability in terms of ratio of mission success to failures
   a. achieved reliability measurement
   b. comparison of achieved with predicted or required reliability
3. Analysis of frequency, type, and severity of discrepancies occurring during the mission
   a. equipment malfunctions
   b. personnel error and/or difficulties
   c. time discrepancies

Analyses by Equipment

1. Determination of equipment subsystem reliability.
   a. mean time between failure for major equipment components
   b. comparison of minimum acceptable reliability for each component with its achieved reliability
2. Determination of the nature, frequency, and impact of equipment failures

Analyses by Behavioral Elements

1. Frequency and percentage of tasks accomplished by personnel
   a. percentage of time operator/crew track target correctly
   b. percentage of time operator/crew detects signal and identifies stimulus correctly
   c. ratio of tasks accomplished correctly to tasks attempted (achieved reliability)
   d. probability of task accomplishment (human reliability index)
      1. in terms of time to accomplish task

2. in terms of time to react to initiating stimulus
2. Frequency, percentage, magnitude, and classification of human errors analyzed
   a. by task and mission phase
   b. by operator position
   c. by impact of error on mission or task
   d. in emergency conditions
   e. as human-initiated failures
   f. by probability of error occurrence
   g. by equipment operated
3. Comparison of mission phases in terms of
   a. frequency and type of human error
   b. error effect
4. Comparison of tasks
   a. required versus actual duration
   b. required versus actual reaction time
   c. other criteria requirements versus actual accomplishments

Analyses by System Characteristics

1. Frequency and classification of system discrepancies
   a. communications errors
   b. nonavailability of required system elements
      1. prime equipment, test equipment, tools, spares
      2. personnel
      3. technical data
   c. logistics inadequacies
   d. technical data inadequacies
2. Measures of operability (see behavioral analyses)
3. Measures of system maintainability (as a whole and by mission segments) in terms of
   a. equipment downtime
   b. number and duration of holds and delays in mission performance for other than maintenance reasons
   c. amount of preventive maintenance

Source: Meister and Rabideau (1965)

Table 10.1.5 Steps in the Evaluation of Behavioral Data

Analysis	Criteria Methods
1. a. Determine frequency and percentage of tasks successfully completed. b. Establish probability of successful completion in future.	1. a. Examine tasks in terms of their terminal outputs b. Establish failure causes; eliminate tasks failing for equipment reasons; apply probability theory
2. Determine effects of task noncompletion on performance of a. subsequent tasks b. other system elements c. overall mission, segments, and phases	2. a. Examine related task pairs; establish relevant dependent relationships b. Determine which tasks present serious problems
3. a. Determine task duration b. Determine which tasks were significantly delayed	3. a. Determine minimum performance time criteria b. Compare with actual performance time
4. Determine whether task reaction time requirements were met	4. a. Determine task reaction time requirements b. Compare with actual reaction times c. Assess human factors causes of reaction time failures
5. Determine effects of task duration and reaction time delay on a. subsequent tasks b. other system elements c. overall mission, segments, and phases	5. Identify tasks causing major portion of system delays
6. Determine frequency and types of errors a. in types of tasks and functions b. on successive mission trials	6. a. Establish human factors causes of errors b. Estimate probability of error recurrence
7. Determine impact of errors on a. task in which error occurred b. on subsequent tasks c. on other system elements d. on overall mission, segments, and phases	7. Establish significance of specific errors to task completion

Source: Meister and Rabideau (1965)

The types of data available (not every test will produce every data type) are: (1) Descriptive data: narrative of what took place to compare with the mission scenario. (2) Error data: nature of error; by whom performed; part of which procedure; effect of the error. (3) Start/stop times for task performance; reaction time (where appropriate); duration of runs; excessive delays in performance. (4) Tasks completed successfully or unsuccessfully or aborted; effects of task failure or noncompletion; equipment indications following performance of procedures (e.g., boiler values, course headings). (5) Communications recordings. (6) Logistics data (documentation): for example, fuel or ammunition expended; sorties flown; number of shells or boxes loaded; trucks dispatched. (7) Self-report data (interviews, questionnaires, ratings): difficulties encountered and reasons for these; attitudes toward system/job design characteristics; ratings of fatigue or workload; checklist evaluations of equipment design.

Table 10.1.5 suggests three categories of data analysis:

1. *Verification Analysis.* Comparison of personnel performance with a system requirement or performance standard; this verifies that personnel can perform required tasks *effectively* (general purpose of the test, Section 1.1, Table 10.1.1). Comparatively little of the data analysis will involve this type of comparison because few if any performance requirements are specified in advance of the test; HFEs must make a deliberate effort to develop these requirements and performance standards and one suspects that few do.

As was pointed out in Section 4.0 (Table 10.1.1) of the test plan, many personnel requirements are inferred only; for example, although no document explicitly says so, the operator should be able to perform without excessive workload. For such inferred requirements the basis for evaluation is likely to be the HFE's own expert opinion, since objective standards are not available.

The inferred personnel requirement may also be one of task accomplishment, that is, errorless performance or at least performance with no *uncorrected* errors; where all tasks are necessary for mission accomplishment, one can infer that an error will fail the task and the failure of any task will fail the mission.

However, not all tasks may be necessary to mission accomplishment and here the performance analysis will involve two things: (1) determination of the error probability, that is, frequency of observed error as a function of error *opportunities;* and (2) analysis of those errors that are particularly critical to system success. Errors in critical tasks will indeed fail the system, and the probability of critical error would provide some indication of operator effectiveness. The causes of such errors may also dictate revisions to the job. Without a substantial number of trials, however, it is impossible to secure enough data to determine the operator's error likelihood and probability of task accomplishment: few operational tests involving a prototype system run for hundreds of trials. This is where mathematical models of the Siegel/Wolf type (Siegel and Wolf, 1969) are useful because the usual operating sequence in this type of model involves 100–200 trials.

There may also be an inferred reaction/duration time requirement. If the system is exercised over a number of trials, a subject matter expert can, for example, examine the durations of each trial and indicate whether any of these times is grossly disproportionate and can constitute a mission failure. Again, given that the system has been exercised a sufficient number of times, it is possible to develop a measure of time adequacy; say 1 out of 10 system trials was grossly overextended; one could say then that the probability of accomplishing the mission in sufficient time is 0.90.

The reader will not have failed to note the importance of the subject matter expert in determining the criticality of behavioral data to system outputs and mission success. In contrast to controlled experiments in which the meaning–significance of data has been preestablished by the experimental variables, much of the data in OT may have equivocal meanings that can only be resolved by "expert" judgment.

2. *Special Conditions Analysis.* The second comparison is between any special conditions that were tested, for example, operator performance under daytime versus nighttime conditions [Section 3.0 (Table 10.1.1) of test plan]. Standard statistical techniques can be applied here, assuming the number of trials per subject are large enough (which must be determined on a case by case basis). If, for example, the system being tested is exercised 30 times under each of three mission phases and the mean frequency of error for 12 subjects is 2.8, 3.6, and 1.2 for each phase, respectively, conventional analysis of variance techniques make it possible to determine whether the error frequency differences are statistically significant.

3. *Qualitative Analysis.* The third data analysis method method, which may involve the bulk of the data, is descriptive only. Any qualitative data must of course be descriptive only because qualitative data cannot be compared either with a requirement or with contrasting system conditions. Data gathered from interviews and questionnaires (e.g., explanation of why errors were made or courses of action taken) would be of this nature. Subjective data that can be quantified, for example, rating scale values, can be statistically compared, but the opportunity to do so does not often arise.

Yet much of the most meaningful behavioral test data is of this type. It has been pointed out a number of times that often objective measures of time, task accomplishment or error frequency are difficult to understand unless they are related to a system requirement or an effect on the system. To make such data comprehensible, it may be necessary to gather subjective qualitative (hence descriptive) data. Qualitative data may include (1) test subject explanations of why behaviors and performance occurred (Why did you wander off course in navigating the helicopter? I was confused by the lack of landmarks in this area); (2) analysis of error or failure causes and effects; (3) descriptions of human engineering inadequacies; (4) attitudes of test personnel to system characteristics, the test situation, job design, their training, and so forth.

The data analysis section of the test plan should therefore not confine itself solely to quantitative comparisons but should also describe the descriptive analyses to be performed and the rationale for them.

10.1.3.9 Testing Schedule

Section 9.0 (Table 10.1.1) of the test plan describes the data collection schedule. If the system test is very complex, and only some of the system operations will be used as occasions for the gathering of personnel performance data, then a daily, weekly, or monthly schedule of data collection activities should be appended to the test plan. For example: In a flight navigation testing program for helicopter pilots conducted by the author, pilots of varying levels of experience were to be tested over a year's time. Since pilot navigation performance was hypothesized to depend in part on the appearance of the terrain, it was necessary to arrange the subject schedule so that pilots with different experience levels could be tested during both summer (heavy foliage) and winter (bare trees, snow) conditions.

If the personnel performance data are to be collected during all test events, a detailed subject schedule is unnecessary because the overall test schedule will determine when data will be collected.

10.1.4   METHODS OF PERFORMANCE MEASUREMENT

10.1.4.1   Introduction and Overview

Section 10.1.3 described how the system test was to be performed. We have said little, however, about the methods to be used in conducting that test. In this section we discuss the following methods: (1) observation; (2) self-report; (3) objective measurement; (4) ratings; (5) interview/questionnaire; (6) automated performance measurement; (7) checklists. It will of course be impossible because of space restrictions to describe each method in all the detail it deserves; the interested reader is referred to Meister (1985).

10.1.4.2   Observation

Observation (either by the HFE or by the test subject) is involved in all the measurement methods with the possible exception of automated measurement; in fact, it has been suggested that observation is the generic basis of all behavioral measurement (Meister, 1985). Because of this, observation can take a number of forms:

1.   Observation of task performance, the most common form of which is to have the operator perform a task with the data collector checking to see that the operator has performed correctly or, if not, what errors have been made. This is most easily done when the task is highly proceduralized (step by step) and is in written form, and the observer can check off each step in the procedure as the operator completes it. The written procedure serves as a "template," as it were, of correct performance. The activity to be observed must be clearly evident to the observer, must not involve too many activities occurring concurrently and rapidly, and should not require complex judgments of the observer.

2.   Observation is involved also in the *walkthrough,* which is a form of design review (Chapter 1.2) making use of a static mockup and simulation as the test situation. In the walkthrough the operator conceptually simulates his or her performance (because he or she cannot actually perform on the mockup). The purpose of the walkthrough is to check design adequacy, to verify operating procedures, and to discover problems requiring modification.

In the walkthrough the operator who acts as test subject points to or touches the mockup control or display which he or she would activate or monitor in the actual situation. As he or she does so, he or she explains (self-report) what he or she is doing, in which sequence and why and indicates whether he or she would experience any difficulties in doing so. All of this is observed and recorded by the HFE in a descriptive (narrative) manner or as answers to a series of questions. Since the walkthrough activities are not or at least not initially performed in real time, there is an opportunity for interchange between the HFE and the test subject as the former questions the latter to secure more information. It should be noted that the walkthrough is applied during development, and, since it is a design review, the test subject will not be an operational user but a skilled engineer.

The author once participated in a nuclear power control room design assessment in which the control room was mocked up with full-scale two-dimensional cardboard/paper representations of the control panels arranged in accordance with initial design. Test personnel were two highly experienced engineer–operators who followed standard procedures.

At the start of the procedure initial plant conditions were defined. The operator indicated where these conditions could be monitored, that is, which control room displays provided the information. The first step of the procedure was then read aloud. The operator proceeded to whatever control panel locations were necessary, explaining what controls and displays were required to implement the task, what positions or readings he expected, what he expected the system response time to be, and any other pertinent data. The engineer observing interpreted the accuracy or intent of a step, if doubt existed, and simulated parameter readings, where necessary. Once the evaluators had gathered all the data they needed, the next step was read. The initial walkthrough frequently took as long as several hours, slowly stepping through the first time and answering all the questions. Subsequent walkthroughs with the same equipment but with different operating modes took less time.

For those procedures that were time-critical a second walkthrough was conducted with no interruptions for questions. The operator performed actions as close to real time as possible, with a minimum of verbalization. A separate data sheet was prepared for every step of the procedure. The information gathered for each step included: where the operator had to go (the panel, e.g., PCC Left Section); which controls and displays were used, identified by number; the cues used to identify when to start and stop the action; the feedback provided; required communications; possible backup modes of operation or other redundant sources of information; and any obvious potential errors and likely consequences.

At the conclusion of the walkthroughs, the participants reviewed all data sheets and marks made on the panels during the walkthroughs and generated the list of comments, deficiencies, and recommendations defining the nature of the problem, its possible adverse effect on operations, and suggested improvements or recommendations.

3. In the walkthrough the operating procedure was preestablished and relatively invariable. A form of observation that may be used when the operator performs a number of varied activities and the HFE wishes to learn what is performed, how frequently, and when, is called *activity analysis*. The jobs for which one would use activity analysis are those in which there is no set pattern in which things must be done: the order in which they are performed and the frequency of their performance may be dependent on contingent events. For example, what a maintenance technician does during troubleshooting (what he or she observes, the tests he or she makes, the information he or she examines, etc.) depends in large part on the nature of the equipment failure and the clarity of the failure symptoms. It is likely that the job being analyzed has already been developed, but because of its contingent characteristics it has not yet been documented as a procedure.

Activity sampling is the systematic observation of a system through some sort of sampling procedure to describe accurately what the operator does. [See, for example, Christensen and Mills (1967).] The HFE might use activity sampling if he or she were going to analyze the job, to develop a training curriculum for it, or to determine where improvements in job design might be made.

The unit of observation may be either the operator and everything he or she does in a particular time period or everything that happens during a particular stage of a job or mission.

The basic technique is simple enough. The observer, making use of a timing mechanism, records what the operator/worker is doing at that moment at predetermined times. The recording is usually done on a specially prepared report form, but more recently fairly sophisticated instrumentation for recording data in an observational situation has been developed. After the data are recorded, the investigator can then estimate (1) the percentage of the operator's total time spent in various activities, (2) the average length of time spent in each activity; and (3) the sequence in which the worker performs various parts of the job. A more detailed description of activity analysis is provided in Chapanis (1959) and Meister (1985).

Observation can be useful in almost all test situations. It can be used to answer the following questions: (1) What happened; what did the operator do? (2) How frequently did an event occur? (3) When in the sequence of events did an event occur? (4) Was the task accomplished satisfactorily? (5) What errors, if any, occurred and when, in relation to what equipment and procedural step? (6) What team member interactions occurred? (7) What was the subject's appearance and manner, for example, alert, confused, calm? (8) What information sources did the operator consult, and so forth?

10.1.4.3 Self-Report

Self-report is the reporting of actual or anticipated events and performance by the operator who serves as the test subject. The report can be made during actual task performance, as we saw it in the walkthrough, or while the operator is actually performing his or her job, as in troubleshooting a failed equipment ("I am doing this and the reason for it is . . ."), or the report may be made subsequent to task performance. Variations of self-report are the *diary* in which the operator records items of desired information on paper at intervals during his or her performance. One can even think of the completion of a standardized report form (e.g., failure, problem incident, or accident report) as a species of self-report.

The self-report is an artificial device to elicit subjective information because it is not an activity the respondent or test subject would perform on his or her own. It is designed to elicit information that cannot be acquired by observation and is most valuable in connection with covert (e.g., cognitive, perceptual) functions that are nonobservable by other than the test subject.

In self-report the test subject selects that material that he or she wishes to communicate, and, therefore, unless very explicit instructions are given as to the categories of information to be reported, the possibility of subject bias and report distortion (by choosing to reveal certain types of information while concealing others) must always be considered.

If the self-report is made during task performance, there is ordinarily no way for the HFE to interact with the operator during his or her report without influencing task performance; the HFE cannot probe or prod. This distinguishes the self-report from the more common interview in which interaction with the interviewee is possible.

Data can be secured about: (1) What does the operator intend to do in a particular situation? (2) What does he or she see; what are his or her thought processes? (3) How does he or she feel about certain aspects of the task, the job, the system? The self-report is particularly useful for exploring the mental model the operator has in analyzing diagnostic situations in which there is a high degree of uncertainty (e.g., malfunction situations or process control systems drifting away from equilibrium).

One would not ordinarily utilize a self-report technique if the desired information could be gathered more easily by interview or observation.

10.1.4.4 Objective Measurement

The scientific ideal is of course to measure objectively. In the test setting this reduces largely to time and error measurement (leaving aside any physiological measures that, except for workload considerations, are generally irrelevant to task performance). As has been pointed out, no objective measurement means very much unless it relates to a personnel performance requirement; consequently, time and error data may be uninterpretable without such a requirement, which is usually not specified. Lacking such a firm requirement, the HFE is often reduced to asking a subject matter expert whether a particular task duration or reaction time and a specific error or compilation of errors represent good, poor, or average performance and what its effect, if any, on system output or mission accomplishment is. Unless the time and error measures are critical to personnel performance and the system output, these measures are simply descriptive. Which is not of course to say that such data should be ignored if it is possible to secure them without excessive effort. In the mockup and OT situations, unless instrumentation is available (usually it is not), only a few relatively gross time/error measures are collectable. If instrumentation is available in the simulator situation, obviously a great many fairly fine time/error measurements can be made; in this situation it is wise to consider the utility of each measure collected, lest the HFE become overwhelmed by the amount of data provided to him or her. In the operating situation (i.e., aboard an operationally functioning system like a ship) instrumentation is unlikely to be available unless a specific automated performance measurement system (which will be discussed) has been installed. Any data secured must be interpreted, and it can be embarrassing when purely descriptive data are uninterpretable.

10.1.4.5 Ratings

Although the scientific goal is completely objective data, the experienced tester is aware of the value of ratings, or quantified opinion. The rating permits estimation of degrees of quantity in subjective responses, which one cannot derive from interview or questionnaire formats. Ratings can be secured either from an observer acting as data collector or evaluator, or from the test participant. Almost anything can be turned into a rating scale (and is) because HFEs usually develop their own. Obviously these scales should be developed using psychometric methods and should be standardized and validated; but this would require much more time and effort than the HFE working on an individual T&E project has. If one ignores the fact that almost all rating scales used in tests are *ad hoc* instruments and never validated, such scales may provide answers to the following questions: (1) quantitative evaluation by an observer or a peer of the test subject's performance, for example, highly competent, average, inadequate; (2) how the test subject him- or herself evaluates his or her own performance; (3) evaluation by the test subject of adequacy of individual aspects of the job, for example, equipment, procedures, technical data, and prior training; (4) how effective was the team (ratings by all team members). Ratings can be applied in all test settings and can be used to study any question that can be reduced to an opinion. Whether quantification of responses is sufficiently important to outweigh the interview as a means of data gathering depends on the individual test situation. Quantification has little value unless the numbers produced are meaningful in terms of some test purpose.

10.1.4.6 Interview/Questionnaire

No test situation should be "closed out" without interviewing test participants (subjects, data collectors, and evaluators). It is important to learn how the subjects viewed the test situation because their reaction to the situation may differ (significantly sometimes) from what the HFE intended; and this may affect the data he or she collected. For example, the HFE may have intended the test situation to be only mildly stressful, but subjects may report that they were excessively stressed. The HFE may have intended that all necessary information be available to them, but they may report that some information they needed was not available. Or subjects may have misunderstood instructions so that their responses were essentially distorted or random. This kind of information is necessary to interpret test responses correctly.

The HFE interviews following the test to determine: (1) Did subjects view the test situation as the HFE had anticipated they would? (2) Were there any aspects of the test that surprised subjects, appeared incongruous or unrealistic to them, or stressed them excessively (even only temporarily)? What did they do as a consequence? (3) Was their training adequate for the tasks assigned them? (4) Were the technical data they used as reference material complete and understandable? (5) If they made errors, why did they make these? (6) Were there any aspects of the system that they thought was poorly human engineered or gave them particular difficulty?

A great deal has been written about the interview, but each interview must be developed for the individual test situation because the topics to be discussed and the questions asked will differ depending on the particular mission scenario the test followed. In developing the interview the following principles should be followed: (1) Determine the topics the HFE wishes to explore in the interview. (2) Pretest the interview with a few test subjects and then revise it, if necessary. (3) Conduct the interview in a quiet place away from system operations but not so far that the HFE has to travel to it. (4) Explain the reason for the interview, that all information is confidential, and that subjects are not being evaluated in terms of their performance. (5) Begin the interview with general questions and become progressively more focused and detailed. (6) Allow the interviewee to control what he or she wishes to talk about as long as he or she does not engage in irrelevancies; an interview is not a question and answer session. Where necessary, draw the subject back to the topic at hand. (7) Probe for greater detail when the respondent makes statements that are particularly interesting and provocative. (8) Complete the interview after 30 min at most, since most respondents (certainly the interviewer as well) become fatigued after only 20 min. (9) If an interview is conducted after each of several trials, do not ask the same questions in precisely the same way after each trial. The HFE should be prepared to compress the interview if he or she gets repetitive answers (ones received in previous interviews). The HFE should be aware that his or her respondent can become bored with the interview. (10) Tape record the interviews unless the respondent indicates that recording disturbs him or her. It is permissable to slow the interview down if necessary to record responses manually. (11) Where team responses are being probed, it is permissable to conduct a group interview (all team members present).

The one advantage the questionnaire has over the interview is that it can be administered to large groups, whereas it is difficult to use even a group interview for a very large team. There are many detailed instructions about how to develop a questionnaire (Meister, 1985), which means that one is less likely to go wrong with the questionnaire. Except for these qualifications the interview is that preferred method of eliciting subjective test data, primarily because it is more flexible than the questionnaire (which has a fixed format) and it enables the interviewer to probe for the data he or she wants. The "guidance" the interviewer can provide may lead to more consistent subject responses. Since most systems do not make use of extremely large groups (let us say, arbitrarily, 10 or more personnel), the interview is much more frequently used than the questionnaire.

10.1.4.7 Automated Performance Measurement

To date only a few operational systems have been developed to record operator responses automatically, although as systems become increasingly computerized one can look forward to this development. The Naval Ocean Systems Center developed an automatic performance measurement system (APM) called OPREDS (operator performance recording and evaluation data system) (Osga, 1981) to be used on a command/control system aboard major ships, but they found that OPREDS was inadequate to the requirements for data and the system was discarded without useful data being collected. The performance measurement system developed by General Physics Corporation has been used to collect error data in nuclear control room simulators (Beare and Dorris, 1983). The ASUPT flight simulator has an elaborate APM (Fuller, Waag, and Martin, 1980; Waag, Eddowes, Fuller, and Fuller, 1976).

Increasingly, training simulators have been programmed to record student responses automatically; this is the essence of what has been termed "computer-managed instruction." However, unless the system under development is heavily computerized, it is unlikely that the prototype being tested in OT will have an APM installed, and certainly a mockup will not have one either. This leaves the simulator as the only test situation in which APM is reasonable.

It is tempting to visualize automatic data collection, but this methodology has several disadvantages: (1) Development of such a measurement system is very costly and time consuming; hence, such systems have been used primarily for research rather than evaluation. (2) The system must be developed specifically to record data on a particular system; general purpose automatic data recording systems are probably beyond the present state of the art. (3) At present such measurement systems can record only overt operator responses to physical stimuli. This means that they cannot record the operator's perceptual and cognitive responses; these must be inferred from the operator's control responses. This last severely curtails the utility of APM. (4) Many test managers prefer to rely on simple, noninstrumental data collection devices.

Technological advances may make the automated measurement system more common but presently it is a rather rare phenomenon.

10.1.4.8 Checklists

Checklists for measurement purposes are used primarily for what we have termed "attribute testing," that is, evaluational judgments that a human–machine interface contains or fails to contain certain desired characteristics. Checklists are used primarily in the evaluation of equipment drawings, to a lesser extent with static mockups and occasionally in OT; they are essentially mnemonic devices to stimulate the HFE to make certain judgments. Although a human engineering checklist can be used in every test situation, it is less likely to be used when performance can be measured, because measurement is a more powerful tool than the HFE's judgment. For example, it is possible by analyzing the nature of the operator's errors to deduce the human engineering inadequacies of the interface.

Chapter 1.2 indicated some of the deficiencies of the checklist as a method: its items are binary when many of the dimensions of the human-machine interface are continuous; the relative value of each item/dimension in the checklist is unknown; there is no way of combining checklist judgments to derive a single quantitative value for the human engineering adequacy of the interface; there are no standardized, psychometrically developed human engineering checklists (although some checklists have been published) and most HFEs develop their own. The checklist is therefore primarily a technique to be used in early design when performance measurement is not feasible.

10.1.5   HOW EFFECTIVE IS EFFECTIVENESS TESTING?

As compared with attribute testing or design analysis, which have significant weaknesses, HF test methodology is reasonably adequate (which does not mean of course that it could not be improved). This is as it should be, since, as was pointed out in Chapter 1.2, most HFEs, university-trained in behavioral science, have specialized to a greater or lesser extent in measurement.

The judgment of adequacy must be tempered however. First of all, what has been described in this chapter is how effectiveness testing should be conducted; in practice the constraints of measurement in a field environment produce test weaknesses. Although HF test methodology is adequate for picking up gross behavioral deficiencies in design, it is much less effective in supplying data to predict the eventual performance of system personnel. On the other hand, test managers are not very demanding in what they want and expect of HF testing; they are interested primarily in gross answers and do not expect the kind of precision researchers expect of their experimental results.

Another qualification to the evaluation of testing adequacy is that it refers only to operator performance. Rarely has any attempt been made to evaluate the effectiveness of maintenance functions, primarily because the behaviors involved are cognitive and hence covert and because performance in troubleshooting is highly contingent on the nature of the equipment failure. Although there are troubleshooting standards in terms of anticipated mean-time-to-repair, these are (at least initially) pious hopes only and no one realistically expects the technician to satisfy these requirements. Moreover, in OT the failures that are experienced are only a small subset of those that will eventually occur over the system's life, so that any conclusions the HFE might reach about them cannot be generalized to later maintenance technician performance.

In addition, relating performance at a lower system level to one higher in the system is extremely difficult, which means that the HFE cannot say anything quantitative about the contribution of personnel performance to system output. He or she can, for example, say that the sonar team's performance was 95% effective, but he or she cannot say how much that 95% contributed to the success or failure of an attack on a submarine. This makes it very difficult for him or her to demonstrate the importance of personnel performance to system functioning, a difficulty that has two effects: (1) It is impossible to induce designers and developers to give HF the attention it deserves in system development; (2) it is difficult for the HFE to take steps to improve personnel performance because he or she does not know how weak that performance is.

The difficulty in relating personnel performance to system outputs is a function of the multiplicity of variables inherent in any system situation. In any real-world system, even a single seat fighter, the number of factors influencing operator performance is numbing. If the experimental researcher had to deal in his or her experiment with one or two dozen variables that varied concurrently during his or her study, he or she, too, would find it difficult to test his or her hypotheses. The experiment reduces the number of variables to a manageable few, although in the process the experiment becomes somewhat artificial, since in the real world all variables function concurrently. Unfortunately, in dealing with real systems (although fortunately not their simulators) we cannot arbitrarily reduce the number of covariant variables.

Another restriction on test effectiveness is one that has been emphasized previously: the lack of either required or normative standards with which to compare actual personnel performance. Without such standards, even if only inferred, it is difficult to make sense out of performance data.

Some authors might suggest that, if it were possible to install instrumentation to record operator performance automatically, these problems would disappear; but that is the triumph of optimism over reality. Even with instrumentation, one must have standards to determine whether performance is adequate or inadequate. As systems become increasingly computerized, there will undoubtedly be a greater reliance on automated performance measurement (since it is so easy—comparatively—to program the computer to pick off certain measures), but the difficulties will persist.

System effectiveness testing is more rough and ready, more circumscribed by limitations of logistics, time, money, personnel, and so forth, than most experimental situations. The meaning of the data resulting from an experiment is inherent in the way in which the researcher sets up his or her experiment, but this clarity does not exist in a test situation.

How effectively is the personnel subsystem tested in system effectiveness testing? We do not really know because each system test is an individual test; there are few efforts to compile the "corporate" experience of system tests [although see General Accounting Office (1984)]. It is well known, for example, that in system testing of the war game variety, such as those involving naval units, little usable personnel performance data are collected and performance results are routinely "fudged" to enable commanding officers to stand in well with their superiors.

We have then a set of unanswered questions: (1) How well do our test methods work in actual testing? (2) Which methods are in fact utilized by the HFE and how? (3) What factors enhance or inhibit the use of these test methods? Here we need feedback from the actual test situation to answer these questions [see Meister (1983)].

From what little we know or can surmise measurement methodology has a number of applied research priorities. Someone should develop psychometrically sound and standardized human engineering checklists and rating scales that can be used with slight modifications in any test situation. Feasible methods of relating performance at the various system levels (individual, team, subsystem, system) should be developed. We need accurate historical accounts of what goes on in system effectiveness testing.

Despite all these caveats, the HFE can make a very definite contribution to the evaluation of system performance.

10.1.6 SUPPLEMENTAL READING

There is a rather extensive literature on personnel performance testing that the interested reader may wish to consult. For a history of early efforts, up until 1969, see Snyder (1969). Askren and Newton (1969) reviewed and analyzed 95 documents related to personnel subsystem test and evaluation; their abstracts of these reports supply the unique flavor of OT. Keenan, Parker, and Lenzycki (1965) published an inventory of assessment practices in the Air Force. Kinkaid and Potempa (1969) edited the proceedings of a human factors testing conference, which contains some excellent papers on varied aspects of test and evaluation. The author's own book (Meister and Rabideau, 1965) is a good primary source, as well as certain sections of his later one (Meister, 1971). His most recent work on methodology (Meister, 1985) contains material on all the topics referred to in this chapter. The special issue on field testing in the *Human Factors* journal (1974) is also worth consulting.

An excellent general description of OT is Geer (1977). Reports by Holshouser (1975, 1977) also summarize policy and procedures for performing OT; the latter is particularly useful in preparing a HF test plan.

Myers, Carter, and Hosteller's (1966) work was developed as an aid in the assessment of HF effects on system performance. Rabideau (1964) is a good review of problems associated with field testing and subjectively oriented data collection methods.

To get the real flavor of OT it is useful to read not only the test plans but also the final reports of completed tests. It would therefore be useful to read the following: Lathrop, Grave, and Lahey (1960); Martin Company (1963); Department of the Air Force (1965). One of the few design check lists published is Philco-Ford (1964). Peters and Hall (1963) provide a compilation of the kinds of human engineering discrepancies one finds in an OT.

Attempts have been made to develop new T&E methods; these have been only slightly successful, but it is illuminating to read Potempa (1969); Crites (1969a, 1969b, 1969c); and Askren, Bower, Schmid, and Schwartz (1969).

A number of T&E manuals that are oriented around human engineering characteristics are in use in both the Army and Navy. HEDGE (Department of the Army, 1974) was followed by HFTEMAN (Department of the Navy, 1975). The author also wrote a T&E guide that was more than human engineering oriented (Meister, 1978). The intent behind the development of these manuals was to enable nonspecialists to perform HFE responsibilities (since many evaluators are laymen), but the documents are useful to specialists also. These documents do not solve the problems we have discussed in this chapter, but they have value in showing how far we have been able to push the state of the art (not far enough).

The reader cannot have failed to note that many of the references cited so far are from the 1960s and early 1970s. This does not mean that there has been a cessation of T&E activity, but rather that scholarly interest in the problems of system effectiveness measurement is less obvious than in somewhat "hotter' HF areas such as workload. System effectiveness evaluations continue to be performed, as the following suggest: Bohn (1983); Edwards et al. (1981); Jones (1983); Malone, Micocci, and Bradley (1974); Malone et al. (1980); Test Design Division (1976); and Waag, Eddowes, Fuller, and Fuller (1976). However, the problems noted previously still remain; one hopes that they will be addressed more continuously and systematically in the future.

REFERENCES

Altman, J. W. (1969). Choosing human performance tests and measures. In J. P. Kinkaid and K. W. Potempa, Eds., Proceedings of the human factors testing conference 1–2 October 1968 (Report AFHRL-TR-69-6). Wright-Patterson AFB, OH: Air Force Human Resources Laboratory.

Askren, W. B., and Newton, R. R. (1969). Review and analysis of personnel subsystem test and evaluation literature (Report AFHRL-TR-68-7). Wright-Patterson AFB, OH: Air Force Human Resources Laboratory.

Askren, W. B., Bower, S. M. Schmid, M. D. and Schwartz, N. F. (1969). A voice-radio method for collecting human factors data (Report AFHRL-TR-68-10). Wright-Patterson AFB, OH: Air Force Human Resources Laboratory.

Beare, A. N., and Dorris, R. E. (1983). A simulator-based study of human errors in nuclear power plant control room tasks. *Proceedings, Human Factors Society Annual Meeting,* pp. 170–174.

Bohn, C. A. (1983). Use of low cost, low fidelity mockups for preproduction testing. *Proceedings, Human Factors Society Annual Meeting,* pp. 589–591.

Chapanis, A. (1959) *Research techniques in human engineering.* Baltimore: Johns Hopkins Press.

Christensen, J. M., and Mills, R. G. (1967). What does the operator do in complex systems. *Human Factors, 9,* 329–340.

Cohill, A. M., and Ehrich, R. W. (1983). Automated tools for the study of human/computer interaction. *Proceedings, Human Factors Society Annual Meeting,* pp. 897–900.

Crites, C. D. (1969a). Video tape recording as a technique for personnel subsystem test and evaluation (Report AFHRL-69-18). Wright-Patterson AFB, OH: Air Force System Command.

Crites, C. D. (1969b). Press camera with Polaroid back technique for personnel subsystem test and evaluation (Report AFHRL-69-17). Wright-Patterson AFB, OH: Air Force System Command.

Crites, C. D. (1969c). Miniature event recording as a technique for personnel subsystem test and evaluation (Report AFHRL-69-16). Wright-Patterson AFB, OH: Air Force System Command.

Department of the Air Force (1965). Category II system development test and evaluation of the 482 L emergency mission support system, personnel subsystem test and evaluation (Report APGC-TR-65-61, Vol. III). Eglin AFB, FL: Deputy for Test Operations, Air Proving Ground Center (AD 476 172).

Department of the Army (1974). *Human factors engineering data guide for evaluation (HEDGE).* Aberdeen Proving Ground, MD: U.S. Army Test and Evaluation Command.

Department of Defense (1977), Test and Evaluation (DoD Directive 5000.3), Washington, DC.

Department of Defense (1979). Human engineering requirements for military systems (MIL-H-46855B). Washington, DC.

Department of Defense (1981). Human engineering design criteria for military systems, equipment and facilities (MIL-STD 1472C). Washington, DC.

Department of the Navy (1975). Human factors test and evaluation manual (HFTEMAN), Vol. 1 Data Guide, Vol. II Support Data, Vol. III Methods and Procedures (Report TP-76-11A, B, C). Point Mugu, CA: Pacific Missile Test Center.

Department of the Navy (1979). Operational test director guide, COMOPTEVFOR Instruction 3960.1B. Norfolk, VA: Operational Test and Evaluation Force.

Edwards, J. M., Bloom, R. F., Oates, J. R., Sipitowski, S., Brainin, P. A., Eckenrode, R. J., and Zeidler, P. C. (1981). *An annotated bibliography of the manned systems measurement literature.* Darien, CT: Dunlap and Associates.

Fuller, J. H., Waag, W. L. and Martin, E. L. (1980). Advanced simulator for pilot training: design of automated performance measurement system (Report AFHRL-TR-79-57). Williams Air Force Base, AZ: Air Force Human Resources Laboratory.

Geddie, J. C. (1976). Profiling the characteristics of the developmental test participant (Technical Memorandum 31–76). Aberdeen Proving Ground, MD: U.S. Army Human Engineering Laboratory.

Geer, C. W. (1977). User's Guide for the Test and Evaluation Sections of MIL-H-46855 (Report D194-10006-1). Seattle, WA: Boeing Aerospace Company (AD A045 097).

General Accounting Office (1984). The Army needs more comprehensive evaluations to make effective use of its weapon system testing (Report GAO/NSIAD-84-40). Washington, DC.

Holshouser, E. L. (1975). Human factors engineering policy and procedures for test and evaluation of Navy systems (Report TP-75-15). Point Mugu, CA: Pacific Missile Test Center (AD-B0006035L).

Holshouser, E. L. (1977). Guide to human factors engineering general purpose test planning (GPTP) (Airtask A3400000/054C/7W0542-001) (Technical Publication TP-77-14). Point Mugu, CA: Pacific Missile Test Center.

Johnson, E. M., and Baker., J. D. (1974). Field testing: the delicate compromise. *Human Factors, 16,* 203–214.

Jones, D. T. (1983). Human factors in support of the coast guard advanced marine vessels test and evaluation effort. *Proceedings, Human Factors Society Annual Meeting,* pp. 584–588.

Keenan, J. J., Parker, T. C., and Lenzycki, H. P. (1965). Concepts and practices in the assessment of human performance in Air Force systems (Report AMRL-TR-65-168). Wright-Patterson AFB, OH: Aerospace Medical Research Laboratories.

Kincaid, J. P., and Potempa, K. W., Eds. (1969). Proceedings of the human factors testing conference

1-2 October, 1968 (Report AFHRL-TR-69-6). Wright-Patterson AFB, OH: Air Force Human Resources Laboratory.

Lathrop, R. C., Grave, C., and Lahey, S. G. (1960) Evaluation of the human factors aspects of the ground aircraft missile (GAM-77) (Hound Dog) (Report APGC-TN-60-19). Eglin AFB, FL: Human Factors Office, Air Proving Ground Center (AD 236 953).

Malone, T. B., Kirkpatrick, M., Mallory, D., Elke, D., Johnson, J. G. and Walker, R. W. (1980). Human factors evaluation of control room design and operator performance at Three Mile Island-2 (Report NUREG/CR-1270, Vol. I). Alexandria, VA: Essex Corporation. (Available from National Technical Information Service, Springfield, VA.)

Malone, T. B., Micocci, A. J., and Bradley, J. G. (1974). Man-machine evaluation of the M60A2 tank system (Research Problem Review 74-4). Ft. Hood, TX: Army Research Institute (AD A077 756).

Martin Company (1963). Personnel subsystem test and evaluation, test cycle report on missile SM68-11 (Report CR-63-43). Denver, CO (AD 405 382).

Meister, D. (1971). *Human factors: theory and practice.* New York: Wiley.

Meister, D. (1978). Human factors in operational system testing: a manual of procedures (Report SR 78-8). San Diego, CA: Navy Personnel Research and Development Center.

Meister, D. (1983). Are our methods any good? A way to find out. *Proceedings,* Human Factors Society Annual Meeting, pp. 75-79.

Meister, D. (1985). *Behavioral analysis and measurement methods.* New York: Wiley.

Meister, D., and Rabideau, G. F. (1965). *Human factors evaluation in system development.* New York: Wiley.

Myers, L. B., Carter, R. G., and Hosteller, R. S. (1966). Guidebook for the collection of human factors data (Report PTB 66-3). State College, PA: HRB-Singer, Inc. (AD 631 023).

Osga, G. (1981). *Guidelines for development, use and validation of a human performance data bank for NTDS combat operations.* San Diego, CA: Systems Exploration, Inc.

Peters, G. A., and Hall, F. S. (1963). Missile system safety: an evaluation of system test data (Atlas MA-3 engine system) (Report ROM 3181-1001, R-5135). Canoga Park, CA: Reliability Operations, Human Factors, Rocketdyne Engineering. March 1963.

Philco-Ford (1964). Human engineering design check list (Report WDL-TR-1968A), Palo Alto, CA: (AD 829 426).

Potempa, K. W. (1969). A catalog of human factors techniques for testing new systems (Report AFHRL-TR-68-15). Wright-Patterson AFB, OH: Air Force Human Resources Laboratory.

Rabideau, G. F. (1964). Field measurement of human performance in man-machine systems. *Human Factors, 6,* 663-672.

Siegel, A. I., and Wolf, J. J. (1969). *Man-machine simulation models: psychosocial and performance interaction.* New York: Wiley.

Smode, A. F., Gruber, A., and Ely, J. H. (1962). The measurement of advanced flight vehicle crew proficiency in synthetic ground environments (Technical Documentary Report No. MRL-TDR-62-2). Wright-Patterson AFB, OH: Behavioral Sciences Laboratory.

Snyder, M. T. (1969). Historical development and current trends of human factors testing on Air Force systems. In J. P. Kinkaid and K. W. Potempa, Eds. Proceedings of the human factors testing conference 1-2 October 1968 (Report AFHRL-TR-69-6). Wright-Patterson AFB, OH: Air Force Human Resources Laboratory, pp. 7-31.

Stinson, W. J. (1979). Evaluation of LVA full-scale hydrodynamic vehicle motion effects on personnel performance (Report NPRDC TR-79-16). San Diego, CA: Navy Personnel Research and Development Center.

Test Design Division (1976). Test design plan, new Army battle tank, XM1, operational test 1 (Report TDP-OT-931). Falls Church, VA: U.S. Army Operational Test and Evaluation Agency (AD A021 726).

Vreuls, D., Obermayer, R. W., Goldstein, I., and Lauber, J. W. (1973). Measurement of trainee performance in a captive rotary-wing device (Report NAVTRAEQUIPCEN 71-C-0194-1). Orlando, FL: Naval Training Equipment Center.

Waag, W. L., Eddowes, E. E., Fuller, Jr., J. H., and Fuller, R. R. (1976). ASUPT automated objective performance measurement system (Report AFHRL-TR-75-3). Williams AFB, AZ: Air Force Human Resources Laboratory (AD A014 799).

CHAPTER 10.2

SIMULATING MANNED SYSTEMS

GERALD P. CHUBB

SofTech, Inc.
Fairborn, Ohio

K. RONALD LAUGHERY, JR.

Micro Analysis and Design
Boulder, Colorado

A. ALAN B. PRITSKER

Pritsker & Associates, Inc.
West Lafayette, Indiana

10.2.1 INTRODUCTION

The design of systems that include human operators is a complex process because the assessment of total system performance should be based on the interacting behavior of humans, hardware, and software. The nature of the activities performed by humans in systems has evolved from one of manual control to the management of a set of automated or semiautomated functions. The introduction of automation has the potential to improve system performance. However, automation often requires a shift in the allocation of the operator's effort from a relatively static set of predefined tasks to the dynamic supervision of the automated system. Hence, predictable direct tasks may decrease in number, and the need to act depends on the results from and quality of supervisory task performance. The pattern of activities behaves probabilistically rather than deterministically. Fully evaluating system performance during system design is therefore becoming more difficult with the set of static analysis tools available to human factors engineers, such as job and task analysis or operational sequence diagrams (Geer, 1981). What is needed are more sophisticated analysis tools that permit careful evaluation of human operator/ hardware/software dynamics.

The current approach to evaluating these complex human operator/system dynamics is largely through the development of prototypes and experimentation with human subjects. Often, by the time a prototype is available, the "hard" decisions about system design have been made.

Therefore, if human factors engineering is to impact system design, tools must be available to analyze alternative human operator/system designs. Human factors engineers need to produce quantitative predictions of human behavior in response to realistic operating conditions for the various system design alternatives. While there are certainly other ways of quantitatively predicting human performance, computer simulation of the human operator in a system/mission context represents a method that is internally consistent and compatible with other contemporary system engineering evaluative techniques. Simulations usually resolve many design and development problems in an effective manner earlier than experiments with human subjects can. That does not eliminate the need for empirical tests, but, properly exploited, such simulations can also help to focus test time on appropriate problems.

10.2.1.1 What Is Computer Simulation?

Computer simulation is the process of designing a mathematical/logical model of a proposed or existing human–machine system and experimenting with this model on a computer (Pritsker, 1974; Shannon, 1975). The use of computer simulation encompasses a model-building process, as well as the design and implementation of an appropriate experiment involving that model.

Simulation is a substitute for some real (or imagined) system. Why use a substitute? Well, perhaps the real system is not available for use or what we want to do to it (or with it) would not be permitted. Or perhaps what we wish to do with (or to) the real system is so expensive or risky that a more affordable approach is desired. The question then is how to create this suitable substitute, study its behavior, and generalize the results to the system of interest.

Most engineering disciplines have been using computer simulation of systems during design for many years. Today, it is rare that an integrated circuit, suspension bridge, or production line is designed without some type of computer simulation study to evaluate designs. Recently, simulation tools have been developed specifically for modeling the human operator of a system. The analyst interested in evaluating human performance can use these tools to construct a computer model of the human in much the same way that a bridge designer can build a computer model of a bridge. Then, the computer model can be used to test various concepts of human operator/system design in a reasonable manner long before the system actually exists.

10.2.1.2 Limitations of Computer Simulation

Simulation inherently requires that the model developer selectively include variables of interest. Selection means exclusion, and that means something has been left out of the model (just as laboratory experiments will intentionally and unintentionally leave out elements that operate in the so-called "real world"). Techniques will be discussed in this chapter for ensuring that the appropriate variables and structures have been included in a model. However, one is never quite sure what impact this has unless there is a serious attempt to compare predicted behavior with actual behavior. This is true whether the model is embedded in a computer simulation exercise or in the equations used to perform the ANOVA on data from a laboratory study. The process of validation is similar in each case. However, a valid interpretation of simulation model results allows conclusions to be drawn about the real system's behavior that often would otherwise be impossible to obtain.

How good is "good enough" for any substitute depends on one's purpose, one's constraints, and one's talents for creating appropriate models. The "goodness" criterion is not something one person should prescribe for someone else. The modeler faces many of the same (or at least similar) tough decisions and tradeoffs that any empirical scientist faces in designing paradigms for laboratory experiments.

At the fundamental level, models are simply descriptions of systems. In an abstract sense, our notions about the world are based on models. Sensation and perception provide an internal representation of objective reality. The concepts we form from perception become the mental models from which we draw our expectations about our social and physical environment. In the fields of science and engineering, these models are formalized with statements in mathematics, logic, and semantics. In order to perform computer experiments with a model, it is generally necessary to have a formal and precise statement of the model. Developing such a model is easier if (1) physical laws are known that govern the system; (2) pictorial or graphic representation can be made of the system; and (3) the variability of system inputs, elements, and outputs is manageable (Pritsker, 1984).

The modeling of complex, human–machine systems is often more difficult than the modeling of physical systems because (1) there are few fundamental laws or "first principles" in behavioral science; (2) relevant procedural elements are often difficult to describe and represent; (3) strategies and policies often guide or constrain behavior and their impact is hard to quantify; (4) random components may be significant elements in many aspects of behavior; and (5) human decision making and problem solving are often integral parts of such systems. A simulation approach allows the modeler to deal directly with these issues whenever, and however, it seems appropriate to do so.

Since a model is a description of a system, it is also an abstraction of a system. Abstraction is typically achieved by aggregating (recombining, chunking, or lumping things together that are really separable) or by filtering. The abstraction process should, itself, be guided by the purpose for which the model is being built. Reference to this purpose should be made when deciding if an element of a system is significant and hence should be modeled. Success depends on how well one picks significant elements and defines the relationship among these elements.

Simulation provides the flexibility to build either very aggregate or very detailed models. Differing levels of detail of various system components may appear in the same model. Simulation also supports the concept of interactive model building: allowing simple models at first, then embellishing and refining them through a series of direct additions.

10.2.1.3 Types of Human Operator Models

Different experts in the field have identified different types of models. For example, Rouse (1980) identified five types of models including estimation-theory models, control-theory models, queueing-theory models, fuzzy-set models, and production-system models. Many psychologists have identified detailed models of cognitive behavior [e.g., Newell and Simon (1972)]. There is no clear consensus as to what the "appropriate" model categories are. For the purpose of discussion, let us consider three types of human performance models that are extensively in use today:

1. Cognitive models.
2. Control-theory models.
3. Task-network models.

Cognitive models represent high-level intellective processes such as active memory, storage and retrieval, comparative judgment and evaluation, decision making, and problem solving. Such models may portray some presumed logic or exploit artificial-intelligence concepts to mimic human cognitive behavior. In the context of system-design evaluation, a good cognitive simulation model should correctly predict some specific aspect of operator problem-solving and decision-making behavior. That does not necessarily mean the model must make the best decisions, nor solve the problems fastest.

In modeling the human operator in systems, cognitive modeling has been used to describe various types of behavior. There are clearly many aspects of operator/system interaction that involve high-level cognitive behavior. Cognitive science can and will provide an increasingly significant contribution to modeling system performance in the future, as already evidenced by Card, Moran, and Newell (1983). At this time, however, cognitive models tend to be rather cumbersome and limited in scope. Consequently, some of the cognitive model development techniques are still impractical for use by human engineers in the context of system design. Cognitive behavior is also strongly influenced by emotive reactions (Chubb, 1983). Cognitive modeling will not be discussed further, although such models should be considered for inclusion whenever the system design problems become focused on highly cognitive operator tasks.

Control theory models are models of fine motor control. While they have been applied to other types of modeling problems, control models are best suited to modeling skilled psychomotor performance for trained operators. Control models treat the human as an "optimal controller" with an associated set of parameters that define input lags, response leads, and other parameters related to human capabilities and limitations [e.g., Baron, Kleinman, and Levison (1970)]. Although they need empirical performance data to set the values of the parameters, once the parameters are established, these models are very good at predicting performance under different conditions.

Task-network modeling is a technique that is an outgrowth of some of the task-description technologies developed in the 1970s such as function flow diagrams and operational sequence diagrams. Computer languages exist that are specifically designed for the study of task network models. These languages facilitate the study of dynamic events and their effects on human performance for a given system architecture (as defined by a static system description such as an operational sequence diagram).

In a task-network model, human performance is separated into a series of subtasks where relationships among those subtasks are represented by a network that connects them. Each node of the network represents a discrete subtask performed by the human. The network structure defines the order in which the human performs the subtasks. This can include branching pathways to portray decisions or action alternatives. Loops may represent repetitions, the consequences of error, or the impact of frustrating environmental conditions. Therefore, the subtask that the human is performing at any point in the simulation is represented by which node of the network is currently being executed. Associated with each node are the following types of information:

1. Time to complete the subtask.
2. If the subtask is always to be followed by the same subtask, the identity of that subsequent subtask.
3. If the subtask can be followed by several tasks, the list of possible subtasks that may be next performed and the rule that will be used to select one (and only one) member of that set of subtasks.
4. Optionally, a user-defined relationship that specifies the effect the system has on human performance, the effect human performance has on the system, and how either may affect variables used in the rule for selecting which subtask is performed next.
5. Subtask name to identify this node in the network.

Once these attributes are determined for each subtask, a computer model of the task network can be readily developed via any of several computer modeling or simulation languages designed for network analysis. Once the model has been implemented in the chosen simulation language, experiments can be conducted by varying the subtask attributes or the structure of the network itself. For example, if we want to explore the overall effect on the system of slower performance on one or several of the subtasks, we could change the "time to perform" attribute of those subtasks and then run the simulation model. Since performance times may be probabilistic (i.e., they are randomly sampled from some specified distribution), we can obtain estimates of the probability distribution representing the total time to perform the overall task network.

To provide an example of a task network, consider dialing a telephone. Figure 10.2.1 presents a visual depiction of our proposed network of a human dialing a telephone. Examining the network, one can note the human picks up the receiver, determines whether the call is long distance, dials the appropriate numbers, and then redials if the number is busy or a mistake is made. This model was used to study the time savings that would be realized if a touch-tone system were installed to replace a dial system as a function of the number of long distance calls.

While task-network models appear best suited for human tasks that are largely procedural in nature, the modeling languages being used for studying task networks permit inclusion of cognitive and control models as well. For example, subtasks of a task-network model that involve decision-making behavior could be modeled via cognitive models; those tasks that involve psychomotor skills could be modeled via control-theory models. Because many systems require combinations of control tasks, procedures, and tasks that involve some level of cognitive behavior (e.g., decision making), it is proposed that available task-network simulation languages are ideally suited for modeling humans in many kinds of systems, not just those where tasks are simply procedural.

Additionally, since network models are structured around procedural operator tasks, which are defined during normal task analysis, they represent a natural extension of current human engineering practices.

In this chapter, the focus is on the development of computer simulations of manned systems via task network models. As one develops a simulation, however, they should keep in mind the other types of operator models that can be incorporated into task-network simulations.

10.2.1.4 Constructing and Using a Simulation of Manned Systems

Successful development of any simulation model usually begins with a simple model that is enriched in an evolutionary fashion until problem-solving requirements are met. Within this process, the following stages of development may occur:

1. *Problem Formulation.* The definition of the problem-solving objective. What design uncertainties need to be resolved? How will system performance be evaluated so alternatives can be compared?

2. *Model Building.* The abstraction of the human/machine system into mathematical/logical relationships in accordance with the problem formulation.

3. *Data Acquisition.* The identification, specification, and collection of data. For example, estimates of subtask durations, the frequency/probability of taking optional pathways, and so forth.

4. *Model Translation.* Preparing the model for computer processing as prescribed by whatever simulation language is selected.

5. *Verification.* The process of establishing that the computer program executes as intended.

6. *Validation.* The process of establishing that some desired accuracy or correspondence exists between the simulation model and the real system.

7. *Strategic and Tactical Planning.* The process of establishing the experimental conditions for using the model, including statistical experimental design.

8. *Experimentation.* The execution of the simulation model to obtain output values replicating the run conditions to achieve the desired statistical precision in estimating results.

9. *Analysis of Results.* The process of analyzing the simulation outputs to draw inferences, fit regression equations to run results, and make recommendations for problem resolution.

10. *Utilization.* The process of implementating decisions resulting from the analysis of simulation results.

11. *Documentation.* The detailed description of the model, its implementation, the results, and their utilization.

These idealized stages of simulation development are rarely performed exactly in this structured sequence. A simulation project may involve false starts, erroneous assumptions that must later be abandoned, reformulation of the problem objectives, and repeated evaluation and redesign of the model. If properly done, however, this interactive process should result in a simulation model that properly assesses alternatives and enhances the decision-making process during design and perhaps beyond (Pritsker, 1984).

10.2.1.5 Chapter Organization

This chapter is grouped into three remaining subsections. In the next subsection, simulation approaches are discussed. While we have briefly discussed the general approach to simulating human/machine systems using task-network modeling, a more detailed discussion of the concepts and steps taken to prepare for computer simulation modeling will be provided. This discussion provides the reader with an understanding of the elements behind a simulation model. In the subsection following that discussion, several currently available simulation languages are described. One of the factors that has greatly contributed to the increased use of computer simulation is the availability of these languages which are invaluable tools to the modeler. Lastly, a discussion of the statistical issues when running computer simulation experiments is presented. Because of the nature and low cost of most simulations, the statistical issues for computer modeling differ from those in actual human experimentation.

This chapter does not contain a significant discussion of two major components of simulation model development: (1) developing descriptions of the human operator(s) and (2) determining which aspects of the rest of the system to include in the simulation model (e.g., the hardware, the software, the external environment). With respect to developing descriptions of human activity, the reader should refer to other chapters where detailed discussions are provided on how an operator's performance can be decomposed into a series of discrete actions that can be used to "feed" the development of a task-network computer model. The type of simulation emphasized in this chapter is primarily an extension of task analysis/system decomposition techniques. Computer simulation allows one to explore the dynamic behavior of the human/machine system rather than the simple static viewpoints obtained from the examination of a task analysis.

With respect to determining the aspects of the system that must be modeled, the reader should look beyond this volume. When modeling complex, closed-loop human/machine systems, one usually represents more than the simple abstraction of a human performing a series of tasks. At the simplest level, there are at least three major aspects of a human/machine system that may need to be modeled, as in depicted in Figure 10.2.2. The operator is the "system component" with which human engineers are most familiar and which will be described by task analytic techniques. However, the *system* also includes the equipment (i.e., hardware/software) with which the human is interacting, and the environment in which these activities will be occurring. The model should be an abstraction of the entire

ENTER THE NETWORK

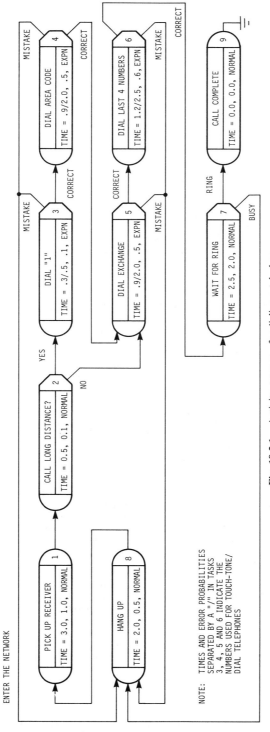

Fig. 10.2.1. Activity sequence for dialing a telephone.

NOTE: TIMES AND ERROR PROBABILITIES SEPARATED BY A "/" IN TASKS 3, 4, 5 AND 6 INDICATE THE NUMBERS USED FOR TOUCH-TONE/DIAL TELEPHONES

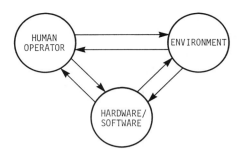

Fig. 10.2.2. Major aspects of the human/machine model.

system that selectively includes only *some* aspects of the human operator, the equipment, and the environment. One is typically caught between the desire to include everything in a model, and the requirement to produce a parsimonious model quickly and cheaply. The question is "which aspects are important?" Unfortunately, there are no simple ways to answer the abstraction question. Alternatively, one should ask, "what can be left out?" How these questions get answered determines the validity of the simulation, with respect to its ability to answer the questions for which it was designed.

It is suggested that the reader consider top-down or structured system decomposition techniques for help in model definition and development. One such technique found useful in human/machine system modeling is $IDEF_0$ (Bachert, Evers, and Rolek, 1982, 1983). Other equally useful techniques are available, and an integration of various techniques is presently under development (Kornfeld, 1984).

10.2.2 SIMULATION APPROACHES AND FUNDAMENTAL CONCEPTS

There are many possible approaches one might take in developing a model of manned system behavior. One could begin by identifying an analog (people are servomechanisms, people are servers of an event queue, people are multiprocessors, etc.) and then develop that description. Alternately, one might postulate a set of primitives or axioms and attempt to deduce the validity of propositions about manned system behavior. Alternately, one might begin somewhat abstractly with a holistic view of manned system behavior and by reductionism or partitioning attempt to identify contributors to the behavior. The last approach is particularly attractive and its elaboration is treated by Wulff, Jepson, Alden, and Leonard (1968). It is the essence of task analysis and task-network modeling. Also, as was previously discussed, it does not preclude adopting one of the other approaches later in the formulation and mechanization of an explicit model of a specific system. Thus any partitioning can be viewed as a global structure that guides the modeler. It elaborates the perspective of the whole by identifying more elemental contributors and their relationships.

In developing a simulation model, an analyst can benefit from a conceptual framework to guide model definition.

A conceptual framework identifies what might be considered without dictating what must be included. An example can be found in Chubb (1982). Table 10.2.1 summarizes some of the possible considerations. One also needs a simulation framework or perspective within which the model's functional description and relationships are captured and exercised in computer code. The remainder of this section summarizes the alternative perspectives or "world views" for simulation modeling.

10.2.2.1 Discrete Simulation

Discrete event simulation occurs when the dependent system variables change by fixed amounts at specified points in simulated time, referred to as "event times." In other words, the human/machine system is only modified by the occurrence of an "event" (normally the start or completion of an operator task in human operator modeling). This type of simulation is appropriate when continuously changing system elements are not of interest to the modeler (e.g., position of one's automobile on the road). In discrete simulation, updating of the simulation "clock" (the variable which represents time in the simulation) is normally advanced by the occurrence of an event.

As an example of discrete simulation, consider an operator who has 10 tasks that may need to be performed (Table 10.2.2) as part of an assembly operation. Perhaps not all parts that are to be assembled require all 10 operations (i.e., some parts may only require operations 1 through 3, 4 through 10, 4 and 6, or all 10). This model might be represented by the task network in Figure 10.2.3. Components for assembly are presented to the operator as completely unassembled (possibly requiring all 10 operator tasks to be performed), or partially assembled (therefore requiring only a subset of the operator tasks to be performed). If the operator is busy, then they are put in the "queue" and wait until the operator is available and the first appropriate operator task can be initiated. When the operator is no longer

Table 10.2.1 Inputs/Outputs of a Conceptual Framework for Manned Systems

	Inputs	
	Constant(s)	Variable(s)
System drivers	Givens (plans)	Unplanned (events)
Machine properties affecting human work requirements	Hardness	Complexity
	Yield	Difficulty
	Quality	Accuracy
	Latency	Speed
Human attributes impacting work	Capability	Preferability
	Training	Deferrability
	Drill	Thresholds
	Values	Tolerances
	Essentiality	Sensitivities
	Blockage(s)	Fatigue
	Negativity	

	Outputs	
	Short Term	Long Term
	System performance	Mission effectiveness
	Operator fatigue	Operator proficiency

Table 10.2.2 Discrete Simulation of an Operator

Task Name	Resources	Preceding Task	Average Time[a] To Perform Task	Task Time[a] Standard Deviation	Next Task Number	Selection Rule
Lift component	I	Start	3.8	1.5	2	N/A
Rotate component	I	1	3.5	1.7	3	N/A
Attach screws	I, II	2	4.2	1.9	4 or 7	A
Tighten screws	I, III	3 or Start	3.1	1.5	5	N/A
Torque screws	I, IV	4	2.4	1.1	6	N/A
Rotate component	I	5	4.3	2.0	7 or end	B
Attach meter	I, V	6	2.7	1.0	8	N/A
Read meter	I, V	7 or 9	2.1	0.9	9 or 10	C
Adjust	I, III	8	2.5	0.9	8	N/A
Final inspection	I	8	3.4	0.9	End	N/A

Resource List	Selection Rules
I. Operator	N/A = not applicable (no choice)
II. Screws	A = probabilistic: 40% of the assemblies do not have to meet torque specification
III. Screwdriver	B = conditional: If assembly is Type X, there is no adjustment; all other models must be checked: adjust as required
IV. Torque gauge	C = Conditional: If meter reading is at or above specified value, no adjustment is required: repeat tasks 8 and 9 as necessary.
V. Meter	

[a] These data are fictitious and serve only as an illustration of the kind of values one needs to implement the discrete event simulation of the task network.

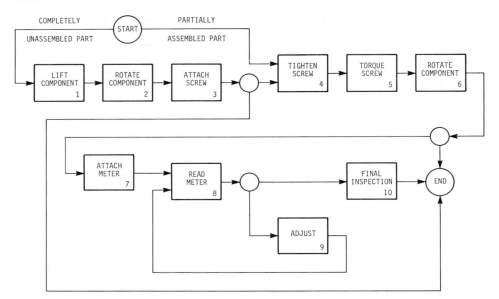

Fig. 10.2.3. Task network for a hypothetical assembly operation.

busy, the task is then initiated, and once all servicing is completed, the part then leaves the system. Since different processing pathways are portrayed, a model for determining how part routing is to be determined and controlled can be developed. To build such a model, we define the variables that portray the system's status and then identify those events that can change system status. The system status variables to be represented depend on the purpose of the model. If the purpose is to study the effectiveness of the operator and workstation, then we may need to represent variables such as the operator's orientation (e.g., facing the window or facing the workspace); the location and state of the operator's tools; likelihood that adjustments will need to be made; and perhaps even the operator's posture. On the other hand, if the model is intended to study only the efficiency of the operator in performing tasks, then maybe the representation of the operator could be aggregated and simply treated in terms of the time it takes to process tasks and whether the operator is busy or not.

For this exposition, the latter purpose is presumed, and the state of the system can be completely specified by the status of the operator and by the number of tasks waiting for processing. The state of the system is changed by (1) the arrival of a part for assembly and (2) the completion of task processing by the operator. To simulate this system for the case where a part goes through all 10 tasks, we need to generate a stream of unassembled part arrivals and corresponding task service times—perhaps by sampling from appropriate input probability distributions or reading an input file. We also need to determine whether processing will be strictly sequential (as implied by the diagram in Figure 10.2.3) or whether some overlap might occur. For example, as one hand rotates the component, the other may reach for the screw to be attached. In that case, task 3 "arrival" may occur before task 2 ends, permitting task 3 to start. If we allow overlapped task execution, Table 10.2.3 summarizes results one might obtain for a single sample of 10 simulated tasks.

In this table the events are listed in chronological order. A graphic portrayal of the status variables over time is shown in Figure 10.2.4. The model is a discrete model because the states of the system status variables (number of ongoing tasks in the system and operator utilization) change only at certain epochs or event times. Between these event times, the state of the system does not change. Moreover, when the state of the system does change, it changes in nonzero "jumps." From the data generated from the simulation, performance measures for the system can be calculated. For example, one can determine that the operator is idle 20% of the time, the average number of tasks waiting and being performed is 1.45, the average waiting time for a task is 2.5 min, the maximum wait was 6.9 min, and the minimum waiting time was zero. This simple example provides a basis for the general discussion of discrete simulation that follows.

The objects within the boundaries of the discrete system, such as people, equipment, and raw materials, are called "entities." There may be many types of entities, and each may have its own characteristics or attributes.

The aim of a discrete simulation model is to reproduce the activities that the entities engage in and thereby learn something about the behavior and performance potential of the system. This is

Table 10.2.3 Event-Oriented Description of an Operator Simulation

Event Time	Task Name	Event Type	Number in Queue	Number in System[a]	Operator Status	Operator Idle Time
0.0	—	Start	0	0	Idle	—
3.2	Lift component	Arrival	0	1	Busy	3.2
7.0	Lift component	Departure	0	0	Idle	
10.9	Rotate component	Arrival	0	1	Busy	3.9
13.2	Attach screws	Arrival	1	2	Busy	
14.4	Rotate component	Departure	0	1	Busy	
14.8	Tighten screws	Arrival	1	2	Busy	
17.7	Torque screws	Arrival	2	3	Busy	
18.6	Attach screws	Departure	1	2	Busy	
19.8	Rotate component	Arrival	2	3	Busy	
21.5	Attach meter	Arrival	3	4	Busy	
21.7	Tighten screws	Departure	2	3	Busy	
24.1	Torque screws	Departure	1	2	Busy	
26.3	Read meter	Arrival	2	3	Busy	
28.4	Rotate component	Departure	1	2	Busy	
31.1	Attach meter	Departure	0	1	Busy	
32.1	Adjust	Arrival	1	2	Busy	
33.2	Read meter	Departure	0	1	Busy	
35.7	Adjust	Departure	0	0	Idle	
36.6	Final inspection	Arrival	0	1	Busy	
40.0	—	Departure	0	0	Idle	

[a] That is, the number of ongoing or active tasks.

done by defining the states of the system and constructing activities that move the system from state to state. The state of a system is defined in terms of the values assigned to the attributes of the entities. In human/machine simulation, the most common type of activity is the operator task. Entities could include components for assembly (as in the previous example), aircraft approaching an airport, or even the human operators themselves. As the human operator performs his or her activities, the state-of-the-entities change (e.g., the components are assembled, the aircraft are assigned to flight paths, or the operator brushes his or her teeth).

In discrete simulation, the state of the system can change only at event times. Since status remains constant between event times, a complete dynamic portrayal of the state of the system can be obtained by advancing simulated time from one event to the next. This timing mechanism is referred to as the "next event approach" and is used in most discrete simulation languages and task-network models.

A discrete simulation model can be formulated by (1) defining the changes in state that occur at each event time, (2) describing the activities in which the entities in the system engage, or (3) describing the processes through which the entities in the system flow. The relationship among the concepts of an event, an activity, and a process is depicted in Figure 10.2.5. An event takes place at a point in time at which decisions are made to start or end activities. A process is any time-ordered sequence of events and may encompass several more elemental activities. These concepts lead, naturally, to three alternative world views for discrete simulation modeling. These are commonly referred to as the "event," "process," and "activity scanning" world views.

In the event-oriented world view, a system is modeled by defining the changes that occur at event times (Kiviat, Villanueva, and Markowitz, 1969; Pritsker and Kiviat, 1969). The modeler must determine the events that can change the state of the system and then develop the logic associated with each event type. A simulation of the system is produced by executing the logic associated with each event in a time-ordered sequence.

In task-network models, event times might be part arrival, task completion, or inspection results. Part arrival might trigger the task "rotate component" or an internal model process like determining whether the operator is busy now. In this manner, simple studies of the range of times required to accomplish a set of operator tasks can be examined via a series of discrete events that collectively constitute a task-network simulation. In more complicated systems, operator tasks can include changes

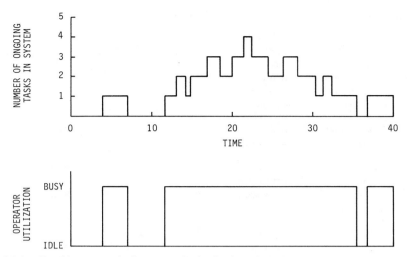

Fig. 10.2.4. Graphic portrayal of status variables for hypothetical simulation of operator assembly task network.

in some state of the system variables such as moving a steering wheel, setting a switch, communicating a message, or changing the operator's position/orientation.

To create a simulation of an operator using the event orientation, we would maintain a calendar of events and cause operator task execution to occur at the proper points in simulated time. The event calendar would initially contain event notices corresponding to the first arrival event (e.g., commencement of the first operator task). As the simulation proceeds, additional arrival and end of service events would be scheduled onto the calendar as prescribed by the logic associated with the operator's job. Each task would then be executed by the simulated operator in a time-ordered sequence, with simulated time being advanced from one event to the next as task execution continues to job completion.

In the process-oriented world view, entities arrive and then wait for processing by an operator if certain conditions cannot be satisfied immediately (i.e., on arrival). The logic associated with such a sequence of conditional activities can be generalized and defined by a single statement. For example, a statement can identify where arrivals are stored while waiting for service to start. An activity statement can represent some type of service provided. A simulation language could then translate statements into the appropriate sequence of operator activities that is automatically executed by the simulation language as the entities move through each process (Franta, 1977).

The process orientation provides a description of the flow of entities through servicing resources. Its simplicity is derived from the fact that the event logic associated with the statements is contained within the model description. Commonly used process-oriented simulation languages include GPSS (Schriber, 1974) and Q-GERT (Pritsker, 1979; Pritsker and Sigal, 1983).

In the activity-scanning-oriented world view, the modeler describes the tasks the operator engages in and prescribes the conditions that cause a task to start or end. The events that start or end the activity are not scheduled by the modeler, but are initiated from the conditions specified for the task. As simulated time is advanced, the conditions for either starting or ending a task are scanned. If the

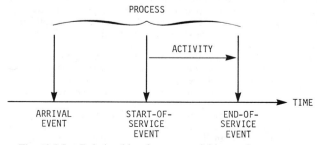

Fig. 10.2.5. Relationship of events, activities, and processes.

prescribed conditions are satisfied, then the appropriate action for the task is taken. To ensure that each activity is accounted for, it is necessary to scan the entire set of tasks or conditions at each time advance.

For certain types of problems, the activity-scanning approach can provide a concise modeling framework. The approach is particularly well suited for situations where an activity duration is indefinite and is determined by the state of the system satisfying a prescribed condition. Because of the need to scan each activity at each time advance, the approach is relatively inefficient when compared to the discrete event orientation. As a result, the activity-scanning orientation has not been widely adopted as a simulation framework. However, this approach seems well suited to implementing models based on artificial-intelligence techniques (e.g., production systems), so advances in this area can be expected and may evolve quite rapidly.

10.2.2.2   Continuous Simulation

In a continuous-simulation model, the state of the system is represented by dependent variables that change continuously over time. To distinguish continuous change variables from discrete change variables, the former are referred to as "state variables." A continuous-simulation model is constructed by defining equations for a set of state variables whose dynamic behavior simulates the real system.

Many types of operator activities involve continuous monitoring and/or control and are, therefore, most appropriately modeled by continuous simulation. Examples are vehicle control, tracking a target, and maintaining process variables within close tolerances over periods of time. Even if the operator's tasks do not involve continuous activities, the modeler may need to represent system elements that do change continuously over time (e.g., vehicle position, target location, process values). Consequently, many task-network simulations may involve some components of continuous simulation.

Models of continuous systems are frequently written in terms of differential equations. The reason for this is that it is often easier to construct a relationship for the rate of change (e.g., acceleration) of the state variable (e.g., velocity) than to devise a relationship for the state variable directly. For example, our modeling effort might produce the following differential equation describing the behavior of the state variables (e.g., speeds) as a function of time t together with an initial condition at time 0:

$$\frac{ds(t)}{dt} = s^2(t) - t^2$$
$$s(0) = k \tag{1}$$

Here, $ds(t)/dt$ is acceleration or change in speed that occurs with a change in time. The simulation analysts's objective is to determine the response of the variable s over a specified time period.

In some cases it is possible to determine an analytical expression for the state variable s given an equation for ds/dt. However, in many cases of practical importance, an analytical solution for s cannot be easily obtained. As a result, we must obtain the response s by integrating ds/dt over time. This integration could be performed by either an analog or digital computer.

An analog computer represents the state variables in the model by electrical charges. The dynamic structure of the system is modeled using circuit components such as variable resistors, capacitors, and amplifiers. The principal shortcoming of an analog computer is its limited accuracy. In addition, the analog computer lacks the logical control functions and data-storage capability of the digital computer.

To overcome these difficulties, a number of continuous-simulation languages have been developed for use on digital computers. A digital computer performs the common mathematical operations, such as addition, multiplication, and logical testing, with great speed and accuracy, and it uses numerical methods to perform the integration operation required in continuous simulation. These methods divide the independent variable (normally time) into small slices, referred to as "steps." The values for the state variables requiring integration are obtained by employing an approximation (normally a Taylor series) to the derivative of the state variable over time. In this situation, there is a tradeoff between accuracy of state variable calculations and computer run time. A description of the various numerical integration algorithms can be found in introductory texts on numerical analysis [e.g., Carnahan, Luther, and Wilkes (1969)].

Sometimes a continuous system is modeled using difference equations. In these models the time axis is decomposed into time periods of fixed length h. The dynamics of the state variables are described by specifying an equation that calculates the values of the state variable at the next period $k + 1$ from the value of the state variable at the present period k. For example, the following difference equation could be employed to describe the dynamics of the state variable s:

$$s_{k+1} = s_k + r \cdot h \tag{2}$$

where r is the rate of change of s. When using difference equations, the essential structure of a continuous-simulation model is reflected in the relationship between the rate r projected over some time (of duration h) to period $k + 1$ and the value s_k of the state variable at period k.

10.2.2.3 Combined Discrete–Continuous Models

In combined discrete–continuous models, the variables in the model may change both discretely and continuously. The behavior of the system model is simulated by recomputing the values of the variables that are of a continuous nature, at small time steps, and recomputing values of variables that only change with the occurrence of events at event time (Pritsker, 1984).

There are two types of events that can occur in combined simulations: First, "time events" are those commonly thought of in terms of discrete-simulation models (e.g., operator tasks). Second, "state events" are unscheduled and occur when some system variable reaches a particular state (value or level). For example, as illustrated in Figure 10.2.6, a state event could be specified to occur whenever state variable X crosses state variable Y in the positive direction. Note that the notion of a state event is similar to that of activity scanning in that the event is not scheduled, but is initiated by the state of the system. The possible occurrence of the state event must be tested at each time advance in the simulation.

Combined discrete–continuous modeling is particularly attractive for simulations of human–machine systems and is the fundamental approach presented in most task network modeling languages. Systems in which operators control machines or processes whose performance is continuous are the most common instances. Two examples are: (1) pilots operate aircraft that move continuously in space and (2) nuclear power reactor operators monitor both processes and displays whose states change continuously in time. For such cases, operator actions and control inputs (both continuous and discrete controls can be represented) are usually modeled as discrete events, and the system responses (including continuous displays) are represented as state variables. State events usually trigger operator actions. For example, the displayed value on an altimeter dropping below the required altitude will trigger a control input by a pilot.

10.2.3 SIMULATION LANGUAGES

The widespread use of simulation as an analysis tool has led to the development of a number of languages specifically designed for simulation. Shannon (1975) has identified the following advantages of using such a special-purpose language when performing a simulation study:

1. Reduction of the programming task.
2. Guidance in concept articulation and model formulation.
3. Aid in communication and documentation of the study.
4. Flexibility in embellishment or revision of the model.
5. Provision of the common support functions required in any simulation.

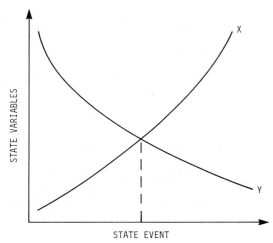

Fig. 10.2.6. Example of state event occurrence.

Emshoff and Sisson (1970) list the following support functions as requisites for any simulation language:

1. Generation of random variates.
2. Management of the simulated clock.
3. Collection and recording of output data.
4. Summarization and statistical analysis of output data.
5. Detection and reporting of error conditions.
6. Generation of standard output reports.

This section compares the major characteristics of the most widely used simulation languages based on each modeling approach. The intent is not to provide detailed descriptions of each language, but to highlight the similarities and differences. A simulation model can sometimes be implemented in more than one language. For example, the FORTRAN based two-person model by Siegel and Wolf (1967) has also been captured in GPSS (Kochhar and Willis, 1971) and in SAINT (Pritsker, Wortman, Seum, Chubb, and Seifert, 1974).

10.2.3.1 Discrete Languages

The major process-oriented languages discussed here are GPSS and Q-GERT. SIMSCRIPT is described as an event-oriented language, and GASP IV and SLAM are discussed as combined discrete–continuous languages.

GPSS

General purpose simulation system (GPSS) is a process-oriented simulation language for modeling discrete variable representations of systems. It exists in a number of dialects, with GPSS/360 and GPSS/H the most widely circulated and used versions. GPSS V is a superset of GPSS/360, and, therefore, programs written for GPSS/360 are compatible with GPSS V (Gordon, 1975). Schriber (1974) is an excellent text for learning GPSS.

The principal appeal of GPSS is its modeling simplicity. A GPSS model is constructed by combining a set of standard blocks into a block diagram that defines the logical structure of the system. Entities are represented in GPSS as transactions that move sequentially from block to block as the simulation proceeds. Learning to write a GPSS program consists of learning the functional operation of GPSS blocks and how to combine the blocks logically to represent a system of interest.

The GPSS processor interprets and executes the block diagram description of a system. The language is limited in computing power and lacks a capability for floating point or real arithmetic. Since it is most widely used on IBM computers, suitable compilers re not always available for some other vendors' machines.

The GPSS language provides almost all of the basic simulation functions listed in the beginning of this section, including extensive data collection and summarization capabilities. On the other hand, GPSS models usually execute more slowly (hence are more expensive to run), and GPSS has a limited capability for generating random variates. However, in GPSS/H execution speeds have been substantially improved, and a wider variety of random variate generators has also been provided.

Q-GERT

Developed by Pritsker (1979), Q-GERT is a network-oriented simulation language. The letters "GERT" represent an acronym for graphical evaluation and review technique; the "Q" is appended to indicate that queueing systems can be modeled in graphic form.

The language employs an activity-on-branch network philosophy in which a branch represents an activity that models a processing time or delay. Nodes are used to separate branches and to model milestones, decision points, and queues. A Q-GERT network consists of nodes and branches. Flowing through the network are entities, referred to as "transactions." Different types of nodes are included in Q-GERT to allow for the modeling of complex queueing situations and project management systems. Attributes are used to distinguish transactions (entities) flowing through a Q-GERT model. Attribute values can be assigned at any node. Activity durations are prescribed by a distribution type and a parameter set number.

The procedures for constructing a model in Q-GERT are similar to those used in GPSS. The modeler combines the Q-GERT network elements into a network model that pictorially represents the system of interest. This network model is then transcribed into a set of input records for interpretation and processing by the Q-GERT analysis program.

Although Q-GERT and GPSS are similar in some respects, their differences should be noted. Both languages provide for automatic collection of statistics on many standard system entities over a single simulation run, however, Q-GERT also collects such statistics over a set of independent runs.

This capability facilitates the analysis of model outputs. Because of its small number of node types, Q-GERT is also quite easy to learn. Unlike GPSS, Q-GERT has a real-valued clock and provides functions to generate all of the commonly used random variates.

Simscript

The simulation language SIMSCRIPT and SIMSCRIPT II was developed at the RAND Corporation (Kiviat, Villanueva, and Markowitz, 1969; Markowitz, Karr, and Hausner, 1963). SIMSCRIPT II is divided into five levels:

Level 1. A simple teaching language designed to introduce programming concepts to nonprogrammers.

Level 2. Statement types that are comparable in power to FORTRAN.

Level 3. Statement types that are comparable in power to ALGOL or PL/1.

Level 4. Statement types that allow modeling using the concepts of entity, attribute, and set.

Level 5. Statement types for time advance, event processing, generation of samples, and accumulation and analysis of simulation-generated data.

One of the principal appeals of SIMSCRIPT as a programming and simulation language is its Englishlike and free-form syntax. Programs written in SIMSCRIPT tend to be self-documenting. It also provides dynamic allocation of computer storage so the modeler does not have to reserve memory space in advance of running the model.

The discrete simulation modeling framework of SIMSCRIPT is primarily event oriented. In SIMSCRIPT, the state of the system is defined by entities, their associated attributes, and logical groupings of entities, referred to as "sets." The dynamic structure of the system is described by defining the changes that occur at event times.

In SIMSCRIPT, the attributes of entities are separately named, not numbered, thereby enhancing model description. For example, we could define a temporary entity named CUSTOMER that has an attribute named MARK.TIME. Sets are also named as opposed to being numbered, further enhancing the model description. For example, a set containing customers waiting for service could be named QUEUE.

The language provides all of the standard support facilities previously outlined. Particularly notable are its flexible statement types for creating output reports. Functions are provided for the usual statistical computations as well as random variate generation. Since the coding of the event routines is left to the user, the debugging aids in SIMSCRIPT are not as extensive as in GPSS and Q-GERT. However, the Englishlike structure of SIMSCRIPT facilitates communication and documentation. Recent versions of SIMSCRIPT II.5 contain process-oriented capabilities (Russell, 1983). NETWORK II.5 is also available to SIMSCRIPT users for analyzing network models. Like GPSS, SIMSCRIPT requires its own special compiler and may not always be available at every computer installation.

10.2.3.2 Continuous Languages

Although a wide variety of special purpose continuous-system simulation languages (CSSLs) have been developed since the 1950s, the structure and functions of most CSSLs have largely standardized in recent years. Whereas, early CSSLs were block oriented, so that a continuous model was constructed using a block diagram similar to that for an analog computer, currently most CSSLs are equation oriented and have FORTRAN-like syntax. Among the CSSLs of this second type is CSMP III for IBM 360 and 370 computers (Graybeal and Pooch, 1980).

A program written in a continuous language is composed of three types of statements:

1. Data statements establish initial conditions for state variables and assign numerical values to constants and to parameters varied over multiple runs.

2. Structural statements specify the way in which solutions to the model equations are calculated. A number of standard functions are available for use in structural statements. For example, the integrator function computes the integral over time of a state variable subject to a specified initial condition. In addition, most standard FORTRAN mathematical function subprograms are available.

3. Control statements specify options for program execution and input/output formatting. For example, the TIMER statement specifies the duration of the run, the integration step width, and the data recording interval. The PRINT statement is used to obtain standard printed output.

Two recently developed CSSLs with extended features are ACSL (advanced continuous simulation language) (Mitchell and Gauthier, 1976) and DARE-P (differential analyses replacement evaluation)

(Korn and Wait, 1978). Both languages are written in ANSI FORTRAN IV and are, therefore, machine independent. Both are also available in versions that will run on microcomputers.

Systems dynamics, as developed by Forrester (1971), is a problem-solving approach to complex problems that emphasizes the structural aspects of models of systems. State variables, called "levels," are defined in difference equation form. The DYNAMO programming language (Pugh, 1973) was developed to provide a vehicle for analyzing systems dynamics models. The language uses a fixed step size to evaluate the level variables over time.

A set of numbered prototype equations are defined in DYNAMO, and the user must structure his or her model to conform to these equation forms. To employ a particular form, the user codes the equation number and the combination of variables or functions required in that form. In addition to standard mathematical functions, such as the exponential and logarithmic functions, DYNAMO provides for operations involving step functions, table functions, clipping, smoothing, and delays. Printing and plotting features are also provided. Perhaps the most serious drawback of DYNAMO is its use of a fixed step size; if the user selects an interval that is too large, a serious loss of accuracy can occur. Selecting small step sizes increases run time and, therefore, cost.

10.2.3.3 Combined Discrete–Continuous Languages

Although much research has been devoted to the development of combined discrete–continuous simulation languages, the general activity simulation program (GASP IV) is the only such language that has achieved widespread use. The more recently developed simulation language for alternative modeling (SLAM) is based on the GASP IV design for discrete–continuous simulation, and it adds a process-oriented view. It also incorporates the network-oriented view used in SAINT (Seifert and Chubb, 1978), and provides new interface capabilities.

GASP IV

This language provides an organization structure that allows system descriptions to be written in terms of discrete event models, continuous models, or a combination of the two (Pritsker, 1974). This structure specifies procedures for writing differential or difference equations as well as methods for defining the logical conditions that affect system status variables. Within this framework, the GASP executive can perform the time advancement functions required by a simulation model and can call specific user-written routines to obtain system status updates. The details of operation for GASP IV are embodied within SLAM and are not given here.

A recently developed language, GASP V (Cellier and Blitz, 1976), expands the continuous capabilities of GASP IV. The new features in GASP V involve the inclusion of different integration algorithms, which can be user selected; procedures for handling partial differential equations; and logic, memory, and generator functions. Examples of logic functions are input switches, flip-flops, and gates. Memory functions include hysteresis and delays, and generator functions for step, ramp, impulse, and dead spaces are available.

SLAM

The SLAM II language (Pritsker, 1984) incorporates the process-oriented features of Q-GERT and the combined discrete–continuous features of GASP IV. In SLAM, the alternate modeling world views are combined to provide a unified modeling framework. A discrete change system can be modeled within an event orientation, a process orientation, or both. Continuous change systems can be modeled using either differential or difference equations. Combined discrete–continuous change systems can be modeled by combining the event and/or process orientation with the continuous orientation. In addition, SLAM incorporates a number of features that correspond to the activity scanning orientation.

The process orientation of SLAM employs a network structure composed of nodes and branches. These symbols model elements in process, such as queues, servers, and decision points. The modeling task consists of combining these symbols into a network model that pictorially represents the system of interest. The entities in the system flow through the network model. The pictorial representation of the system is transcribed by the modeler into an equivalent statement model for input to the SLAM processor.

For event-oriented simulation models, the user is required merely to code the processing logic corresponding to each event type in separate support routines. In this case the SLAM executive always advances time to the next event occurrence in order to update system status. This is accomplished by including a file or calendar of events and processing the next event whenever the processing of a current event is completed.

An important aspect of SLAM is that alternate world views can be combined within the same simulation model. There are six specific interactions that can take place among the network, discrete event, and continuous world views of SLAM:

1. Entities in the network model can initiate the occurrence of discrete events.
2. Events can alter the flow of entities in the network model.
3. Entities in the network model can cause instantaneous changes in values of the state variables.
4. State variables reaching prescribed threshold values can initiate entities in the network model.
5. Events can cause instantaneous changes to the values of state variables.
6. State variables reaching prescribed threshold values can initiate events.

10.2.3.4 Specialized Simulation Capabilities

The discussion thus far has focused on general purpose modeling languages that are not specifically designed for human/machine system simulation. Since modeling languages that address operator performance in a system context are relatively rare, these aforementioned languages have been used to simulate human/machine system with varying degrees of success. Reviews of other operator models can be found in Pew, Feeher, and Miller (1977); Strieb, Glenn, Fisher, and Fitts (1977); Greening (1976); and Baron (1984). Most of these models represent only a specific portion of an operator's duties such as tracking, reading displays, or control, and, therefore, are not modeling languages per se. In fact, there are few modeling languages that include cognitive or decision-making activity by the operator or attempt to evaluate operator mission performance in the total system setting. A few techniques have been developed recently that have focused on mission performance, for example, TLA (timeline analysis program) (Anderson and Miller, 1977), WAM (workload assessment model) (Edwards, Curnow, and Ostrand, 1977), and others (Asiala, 1975; Klein and Cassidy, 1972). Each of these methods requires the user to estimate the performance time for each task, and none provides a complete description of operator behavior in the system.

The approaches previously described tend to concentrate on the microactions of operators and do not generally address issues of strategic and tactical decision making that the operator must accomplish to perform or execute a mission. Thus, they tend to be highly detailed but lack an overall mission orientation. Models that concentrate on system performance have generally been of the optimal control type (e.g., Kleinman, Baron, and Levison, 1970) that aggregate the operator's performance to a degree which precludes analysis of many operator effectiveness issues.

The human/machine system modeling languages—HOS, SAINT, and Micro-SAINT—each have the potential to close the gap between the task-analytic and optimal control orientations. Each has capabilities that permit a mission scenario to be simulated while preserving a detailed operator representation. Cognitive behavior of the operators can also be modeled with both HOS and SAINT.

The Human Operator Simulator (HOS)

The HOS approach (Lane, Strieb, Glenn, and Sherry, 1981) is based on the following four assumptions:

1. Human behavior is predictable and goal oriented, particularly for trained operators.
2. Human behavior can be described as a sequence of microevents that, in the aggregate, explain task performance.
3. Humans are single-channel information processors, but can be executing more than one procedure concurrently.
4. Trained operators rarely forget procedures or make procedural errors.

The first assumption provides a framework for modeling cognitive behavior of operators; the second assumption permits micromodels of behavior to be built into the language. The third assumption fixes the internal simulation methodology, and the fourth assumption permits HOS to focus on the effects of system configuration on the "average" operator. It avoids the difficult and poorly understood error processes that are independent of the system, that is, solely attributable to the operator.

Unlike most other operator models, the modeler does not specify activity times for the actions of the operator. HOS internally constructs the activity times using micromodels of behavior. The micromodels included in HOS are:

1. Information absorption.
2. Recall of information.
3. Mental calculation.
4. Decision making.
5. Anatomy movement.
6. Control manipulation.
7. Relaxation.

Each operator action is a combination of the preceding activities. HOS internally selects the combination of functions for any action, the sequence of functions, and the total time required to perform all functions. The micromodels are based on the human factors literature, empirical evidence, and judgment. A micromodel may be changed readily by reprogramming.

HOS contains an input processor to read and interpret statements describing the mission, operating procedures, hardware functions, and so on. These are converted to table entries used by a second program that performs the simulation and accumulates output files. A third program examines the output data and creates reports tailored to the user's specifications. In simulation, all activity times are deterministic. Random components are included in the micromodels primarily to determine thresholds. For example, the success of information recall is determined by a random draw against a calculated probability of recall.

HOS maintains a record of control settings and the position of operator body parts (eyes, left hand, right foot, etc.). The micromodels use these body part positions in computing activity times and a detailed event trace of body part activities can be printed.

The task-sequencing and decision-making activity is represented in HOS by computing a "criticality" or priority for each task, and the procedure with the highest criticality is selected. This procedure is internal to HOS, and its calculations are performed with parameters and limits specified in the input statements.

HOS provides the modeler with an extensive set of built-in capabilities and allows a flexible definition of procedures and variables specific to the system under study. The following are some limitations of HOS:

1. The HOS operator is stationary.
2. Only one operator may be modeled in detail.
3. HOS has no explicit model of communication.
4. The level of detail is not easily controlled by the modeler because the internal computation of activity times requires operator actions to be specified at a relatively low level.

Conceptually, HOS models the operator as the analog of a real-time, event drive computer. While not intentionally designed to do so, it mimics the operating system software that controls such computers. HOS is particularly well suited to performing analyses in which the operator's workstation layout is the main concern and where performance degradation results primarily from work congestion in reading displays and manipulating controls. Unfortunately, the level of detail required by HOS and the restrictions on multiple operators and communications may make HOS inappropriate for large systems involving complex interactions among more than one operator.

System Analysis of Integrated Networks of Tasks (SAINT)

SAINT is a simulation methodology (Chubb, 1981; and Wortman, Duket, Seifert, Hann, and Chubb, 1978) that was designed specifically for task network simulation. Operator tasks, or rule-based activities, are modeled by SAINT as a network of nodes and branches. Nodes represent task elements at which time delays may occur. Branches represent the ordering among task elements. When more than one branch emanates from a node, it typically indicates a decision by the operator. The activities of the operator can be visualized by tracing through the network graph, pausing for the appropriate time period when a node is reached, and branching along the prescribed output path.

To provide another example of a SAINT network (Figure 10.2.1 presents one for dialing a telephone) let us use an example of the tasks involved in catching a fish. Figure 10.2.7 represents a visual depiction of our proposed network of a human fishing. As one follows through the network, you can see the human cast, wait for a nibble, and attempt to catch a fish if a nibble occurs. If, for example, we wanted to explore the effect on the time required to catch a fish or reeling in the line more frequently to check the bait, we would simply change the attribute of subtask number 4 from 120 sec to whatever number is desired. This, of course, would confound with the probability of a nibble occurring in subtask 3. A parametric experiment testing the effects of these two variables could easily be conducted using SAINT. This example is intended to provide the reader with an example of SAINT modeling using a familiar task. Far more practical applications have been found for SAINT, including the modeling of Air Defense System operation (Laughery and Polito, 1983), visual search (Kraiss and Knaeuper, 1982), and aircraft piloting (Muralidharan, Baron, and Feehrer, 1979). Other applications are cited in Seifert and Chubb (1978) and in Chubb (1986).

Operator tasks that involve decision making can be connected to a decision-making submodel that sequences the operator tasks to satisfy the operator's mission. To perform simulations, a data description is prepared for each node and branch in the network. Also, there may be user-written FORTRAN programs for each node, which implement special features of the system under study. These may be used to reset selected attributes or to modify the duration of this particular activity. Since the output branching can be conditional on attribute values, this feature allows SAINT to portray

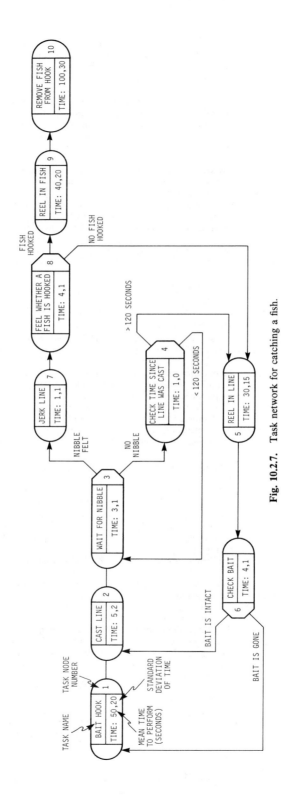

Fig. 10.2.7. Task network for catching a fish.

situations that are very dynamic and complicated. The SAINT simulation software (linked with the user's FORTRAN) reads the data descriptions. This creates an internal representation of the network that allows simulations to be performed.

SAINT is a combined discrete–continuous language, and, therefore, contains automatic features for representing the dynamics of system performance and/or continuous controls. SAINT can also solve differential equations, difference equations, and/or algebraic equations that describe system parameters and performance, as part of or in parallel with, the operator simulation. Furthermore, the operator and continuous models can interact through state and time events that affect parameters in both models. This permits, for example, representing the dynamic change in an aircraft's position, speed, and heading as well as the readout on the altimeter, which may lag actual altitude during rapid maneuvers (Seifert, 1979).

SAINT differs fundamentally from HOS in that it has few built-in features that dictate an approach or require a particular level of detail. This leaves much to be done by the modeler, but it also allows a modeler to aggregate or disaggregate selectively various aspects of the systems model according to the requirements of the present problem. In this way, a project can concentrate on the important issues under study. It also means that there are no limitations on such things as number of operators, how they communicate, and whether or not they can move around. For example, one operator might be represented in great detail while other operators in the system who are not of central concern can be represented in lesser degrees of detail. Furthermore, external modules can be provided for use by SAINT that emulate built-in paradigms of behavior modeling (Laughery and Gawron, 1984).

SAINT provides the simulation processing capability for a complete problem-solving system MO-PADS (models of operator performance in air defense systems) (Polito, 1983; Polito and Laughery, 1983; Walker and Polito, 1982). MOPADS provides a friendly user interface to SAINT, and data base capabilities to describe, store, and process the following data elements:

1. Scenarios to be simulated including the location and characteristics of air defense units, the command and control system, and the coordinate reference system.
2. Characteristics of the operators of air defense systems and their environment.
3. Dynamic relationships of operator tasks.
4. Statistics collected during simulations.

MOPADS combines modern data base and computer technology with SAINT and advanced human performance modeling.

Micro SAINT

Micro SAINT is, in essence, a microcomputer version of the SAINT modeling language (Laughery, 1984, 1985). It captures many of the features of the full version of SAINT as discussed in the previous section with some differences. It is primarily designed to represent operator task networks in a straightforward manner. Additionally, Micro SAINT has continuous simulation capabilities but lacks the ability to solve differential equations.

The reason for the development of Micro SAINT was that SAINT, while a powerful language for modeling human/machine systems, lacked model development tools. In other words, setting up a SAINT task-network simulation involves learning a relatively complicated, often cumbersome model description language. Micro SAINT, on the other hand, allows all aspects of the model to be developed via an interactive model construction language. During model development, the modeler first specifies what he or she would like to work on (e.g., the task network, continuous simulation variables, the simulation scenario). Then, the model development language provides the modeler with a set of menus and questions that facilitate model development. The need to develop actual computer code has been eliminated through the use of a parser. The need to generate "entry codes" describing other aspects of the simulation has also been eliminated in Micro SAINT via the model construction software. The goal of Micro SAINT is ease of use by modelers. Anyone who can draw the task network (i.e., conduct a task analysis) and describe other relevant aspects of the system being modeled (e.g., variable changes as a function of task performance) should be able to build a Micro SAINT simulation.

Summary

The issue of how much "help" should be provided in the form of built-in features is a classic concern of those who develop specialized simulation languages. As more specialized features are included, the scope of the language is reduced, and more modeling decisions are made *a priori* for the modeler. This usually gives the language a powerful capability within its restricted scope and may lead to less expensive application because the data collection and modeling for the built-in features has already been done.

In this section we have given two examples with different types of help features. HOS provides highly detailed representations of a restricted class of single-operator systems with a reasonable amount of effort. Such languages, however, become cumbersome when systems are addressed that do not fit easily into the problem class for which the language was developed. Moreover, not everyone agrees with a particular modeler's choice of representation, so built-in features may prove unattractive.

SAINT and Micro SAINT, on the other hand, have few built-in system restrictions and make no assumptions about behavioral paradigms. As a consequence, they can be used for an extremely wide range of application areas and for systems of arbitrary complexity. The cost of this flexibility is, of course, that the modeler must develop representations of the behavioral aspects, as well as the system dynamics. That puts a larger burden on the prospective user. Selection of a simulation methodology must be made based on which tool is most appropriate for the problem to be solved.

10.2.3.5 Choosing a Simulation Language

The selection of a simulation language is frequently based on knowledge and availability as opposed to a formal comparison of language features. However, if the frequent use of simulation is anticipated, then a comprehensive evaluation of the available languages and anticipated modeling needs is warranted. Shannon (1975) provides a review and diagrams a procedure for making such an evaluation. Table 10.2.4 is a summary of important factors to consider in comparing simulation languages.

10.2.4 ISSUES IN SIMULATION MODEL DEVELOPMENT AND USE

Developing and running a simulation model on a computer is, in essence, a complex sampling experiment. Thus, the procedures for designing and analyzing simulation runs are similar to the techniques used in other scientific experiments; the main difference is that the simulation analyst has greater control over the experimental conditions. An appropriate statistical analysis is a necessary part of a simulation study in order to (1) use simulation-generated data efficiently in the estimation of system performance measures, and (2) reveal the scope and limitations of the conclusions based on the data. The first consideration involves the selection of a method for data collection and estimation, as well as the resolution of tactical questions concerning how to start, execute, and stop the simulation. The second consideration reflects both the validity of the simulation model and the suitability of the overall layout specifying the runs to be performed. This section surveys the major statistical issues facing the simulation analyst in the conduct of a simulation study.

To clarify the subsequent discussion, the following classification of simulation is made with respect to output analysis.

1. *Steady-State (Nonterminating) Simulations.* In this type of simulation study, we assume that, after a sufficiently long period of operation, the probability law governing the behavior of the real system will stabilize to an asymptotically steady level. For such a system we seek to estimate the steady-state or long-run average measures of system performance. Most of the statistical procedures discussed here apply to this situation.

Table 10.2.4   Features on which to Evaluate a Simulation Language

Feature	Consideration
Training required	Ease of learning the language
	Ease of conceptualizing simulation problems
Coding considerations	Ease of coding, including random sampling and numerical integration
	Degree to which code is self-documenting
Portability	Language availability on other or new computers
Flexibility	Degree to which language supports different modeling concepts
Processing considerations	Built-in statistics gathering capabilities
	List processing capabilities
	Ability to allocate core
	Ease of producing standard reports
	Ease of producing user-tailored reports
Debugging and reliability	Ease of debugging
	Reliability of compilers, support systems, and documentation
Run-time considerations	Compilation speed
	Execution speed

2. *Transient (Terminating) Simulations.* When it is of interest to analyze the behavior of a system over a fixed period during which the underlying probability law changes, then the system is simulated only over the specified period. Such a terminating simulation is appropriate (1) when the corresponding real system shuts down at regular intervals or (2) when we want to study the short-term system response to certain "shocks." As is discussed subsequently, replication analysis is the appropriate technique for the investigation of terminating systems.

10.2.4.1 Choosing Input Probability Distributions

In the formulation of a simulation model, it is frequently necessary to characterize the random elements of a system by particular probability distributions. To select an appropriate distribution for an input process, the analyst must understand some of the basic properties of the common distributions and the circumstances in which those distributions arise. Pritsker (1984) provides a good introduction to these topics. Hastings and Peacock (1975) summarize the properties for a large number of distribution types.

In the data-acquisition stage of a simulation study, empirical frequency distributions should be collected from the real system for each of the input processes to be modeled. This enables the analyst to apply goodness-of-fit tests to his or her hypothesized input distributions. A monograph by Phillips (1972) describes the most popular goodness-of-fit tests and describes a FORTRAN program to evaluate an empirical frequency distribution against the most commonly used theoretical distributions. Programs are available for employing both graphical and statistical goodness-of-fit tests to sample records (Law and Vincent, 1984; Musselman, Penick, and Grant 1981).

10.2.4.2 Model Validation

Validation consists of determining that the simulation model is a reasonable representation of the real system (Fishman and Kiviat, 1967; Sargent, 1984; Van Horn, 1971). Validation is normally performed in levels. The authors recommend that a validation be performed on data inputs, model elements, subsystems, and interface points. Validation of simulation models, although difficult, is a significantly easier task than validating other types of models, such as a linear programming formulation. In simulation models there should be a correspondence between the model elements and system elements. Hence, testing for reasonableness involves a comparison of model and system structure and comparisons of the number of times elemental decisions or subsystem tasks are performed.

Specific types of validation involve evaluating reasonableness using specific values in the simulation model or assessing sensitivity of outputs for some systematic variation of data inputs. For example, how sensitive is model behavior to a specific operator decision-making strategy? In conducting validation studies, the comparison yardstick should be both past system outputs and experiential knowledge of system performance behavior. A point to remember is that past system outputs are but one sample record of what could have happened.

Although validation of a simulation model is not solely a statistical issue, there are a number of statistical techniques that can aid in this stage of a simulation study. Shannon (1975) discusses a validation procedure that is based on a comparison of input–output transformations. This involves comparing the outputs of the real system with those of the model, using inputs as nearly identical as possible. By employing an appropriate two-sample test, it is possible to evaluate the hypothesis that the two sets of responses came from the same (or nearly the same) population. A variety of tests are presented, including the nonparametric Mann–Whitney test (Hollander and Wolfe, 1973), goodness-of-fit tests, spectral analysis tests, and the usual tests based on normality assumptions. In addition, Shannon discusses some appropriate two-sample tests when the analyst has multiple performance measures with respect to which model validation is required. It should be pointed out, however, that he emphasizes the superiority of the professional judgment of operating personnel in assessing the validity of a model of an existing system. This is in contrast to the test and measurements viewpoint where empirical tests of predictive validity are emphasized.

10.2.4.3 Estimation Methods

When a simulation model incorporates random input processes, its output performance measures are also subject to random variation, just as actual human experiments are. It is necessary to take this into account when making inferences about the corresponding real system. In particular, techniques are required that specify how to calculate a good estimator. These techniques should also provide a meaningful assessment of the reliability associated with that estimator. One measure of reliability for a simulation-based estimator is the probability that the inherent estimation error falls within some acceptable limits. These are the upper and lower confidence bounds on the estimator, as discussed in elementary statistical texts (Friedman, Pisani, and Purvis, 1978; Spence, Cotton, Underwood, and Duncan, 1983).

Replication Analysis

To estimate the mean waiting time μ_w for tasks an operator is to process (see the example in Section 10.2.2), one approach is to execute, say, k independent runs of the simulation model and record the waiting times of the first b tasks of each run. The independence of successive runs can be ensured by independently selecting a new set of starting seeds for all random number streams at the beginning of each run. If w_{ij} is the waiting time of the jth customer on the ith run, then the average waiting time computed over the ith run is given by

$$\bar{w}_i = \frac{1}{b} \sum_{j=1}^{b} w_{ij} \tag{3}$$

and over the set of runs each of the observations ($\bar{w}_i$) constitutes a random sample of size k. The same mean $\bar{w}_i$ is an unbiased estimator of the mean waiting time μ_w, and by the central limit theorem, the distribution of $\bar{w}_i$ is asymptotically normal. This will be true no matter what the distribution is for the w_{ij} themselves. These properties form the basis for replication analysis. The grand mean

$$\bar{\bar{w}} = \frac{1}{k} \sum_{i=1}^{k} \bar{w}_i \tag{4}$$

is taken as the point estimator of μ_w, and the overall sample variance

$$s^2 = \frac{1}{k-1} \sum_{i=1}^{k} (\bar{w}_i - \bar{\bar{w}})^2 \tag{5}$$

is used to construct the following $100(1 - \alpha)\%$ confidence interval for μ_w:

$$\left[\bar{\bar{w}} - (t_{\alpha/2, k-1}) \frac{S}{\sqrt{k}}, \bar{\bar{w}} + t_{\alpha/2, k-1} \cdot \frac{S}{\sqrt{k}} \right] \tag{6}$$

Note $t_{\alpha/2, k-1}$ is the critical value cutting off a tail of size $\alpha/2$ for Student's t distribution, with $k-1$ degrees of freedom.

A disadvantage of replication analysis is that it restarts system operation anew at the beginning of each run; thus there can be a warm-up period during which atypical behavior is observed. In the operator-task simulation example (in Section 10.2.2), small waiting times observed for the early tasks introduce a bias or systematic error when estimating steady-state values such as the average waiting time $\bar{w}_i$ for each run. (See subsequent section on the design of experiments for further discussion of the start-up problem.)

When applying replication analysis with a fixed total sample size of $n = k \cdot b$ observations, the analyst must evaluate the tradeoff between the run length b and the replication count k. Increasing the run length reduces the effects of initial bias, but it also reduces the number of degrees of freedom in the overall sample variance given by Equation (5), so that the confidence interval becomes wider. Reducing the run length will simultaneously shrink this confidence interval while shifting it away from its target point μ_w. Law (1977) discusses these characteristics of replication analysis and provides guidelines for the application of this technique.

Note that in terminating simulations the run length b is fixed by the problem, and we simply seek to study the transient behavior of the system over the first b observations by averaging across a sufficiently large number of independent replications. In this case classical statistical techniques are entirely appropriate.

Subinterval Sampling

An alternative approach to the estimation of steady-state waiting time is simply to execute a single run of $n = k \cdot b$ simulated customers (i.e., tasks) and then to group the observations into k batches of b customers each. In effect, this experiment is equivalent to repeating a run of length b a total of k times, where the final system state for one run constitutes the initial condition for the next run. In this situation we let w_{ij} denote the waiting time of the jth customer in the ith batch, and the ith batch mean $\bar{w}_i$ is calculated according to Equation (3). As in the case of replication analysis, the grand mean $\bar{\bar{w}}$ over all the batches provides a point estimator of μ_w.

Subinterval sampling eliminates the problem of initial bias, at least beyond the first batch. However, successive batch means $\bar{w}_i$ are no longer independent since observations inside each batch are correlated with observations inside other batches. Correlation between the batch means influences the variance estimator for $\bar{\bar{w}}$ based on Equation (5). The remedy for this problem is to make the batch size b large enough so that, for practical purposes, the batch means are independent. Schmeiser (1982) has shown that it is typically unnecessary to have more than 30 batches, and in most cases, 10 are sufficient.

Regeneration Analysis

The operator-task system possesses the following important property: every time an arriving task finds the operator idle, the operation of the system "starts over" probabilistically. Such a point in time is called a "regeneration epoch," and beyond that point the future evolution of the system is independent of its past history. The portion of the process describing system behavior that occurs between two consecutive regeneration epochs is called a "cycle" or "tour." We may accumulate the following quantities over each tour:

Y_i = sum of the customer (task) waiting times during the ith tour.

X_i = number of customers (tasks) served during the ith tour.

The regenerative property of the system ensures that the pairs (Y_i, X_i) are independent and identically distributed and that the steady-state average waiting time μ_w is given by the ratio of the expected values of the variates:

$$\mu_w = \frac{E(Y_i)}{E(X_i)}$$

To estimate μ_w using regenerative analysis, we simulate the operation of the system until k tours have been completed and compute the sample means

$$\bar{Y} = \frac{1}{k} \sum_{i=1}^{k} Y_i$$

$$\bar{X} = \frac{1}{k} \sum_{i=1}^{k} X_i$$

Then we obtain the following ratio estimator of μ_w:

$$\hat{r} = \frac{\bar{Y}}{\bar{X}} \tag{7}$$

Next we compute the variance by the equation:

$$S^2 = \frac{1}{k-1} \sum_{i=1}^{k} (Y_i - \hat{r} X_i)^2$$

The variance of the regenerative estimator $\hat{r}$ is approximated by $S^2/\bar{X}^2 k$. An approximate $100(1 - \alpha)\%$ confidence interval for the steady-state mean waiting time μ_w is therefore given by

$$\left[\hat{r} - Z_{\alpha/2} \cdot \frac{S}{(\bar{X}\sqrt{k})}, \hat{r} + Z_{\alpha/2} \cdot \frac{S}{(\bar{X}\sqrt{k})} \right] \tag{8}$$

where $Z_{\alpha/2}$ is the critical value of the standard normal distribution that cuts off a tail of size $\alpha/2$.

In general, a regenerative simulation is characterized by a tour-defining state to which the system returns periodically so that successive cycles are independent replicates of steady-state system behavior. This allows the application of classical statistical techniques to measurements accumulated over each cycle; in addition, the problem of initial condition bias is completely avoided. The major disadvantage of this method is revealed in systems with a large number of possible states: If the tour-defining state occurs infrequently, then it may be necessary to run the simulation for a very long time in order to obtain an adequate number of cycles. It should be noted that the confidence interval given in Equation (8) is valid only for large samples. Crane and Lemoine (1977) provide an excellent introduction of this method of simulation analysis.

Design of Experiments

A simulation run is an experiment in which an assessment of the performance of a system is estimated for a prescribed set of conditions. In the jargon of design of experiments, the conditions are referred to as "factors" and "treatments," where a treatment is a specific level of a factor. The literature in the field of design of experiments for simulation includes Kleijnen (1974, 1985) and Naylor (1969). Although the basic principles of classical experimental design also apply to simulation studies, several features of simulated experimentation distinguish it from more traditional applications. Among these distinguishing characteristics of simulation experiments are the following: (1) the simulation modeler knows in detail the structure of the process that produces the response variables; (2) additional observations of these variables are usually easy to obtain; and (3) the variance of these variables can sometimes be controlled.

The major problem involved in simulation experiments is associated with the definition of the inference space associated with the simulation model. Making *a priori* assessments of how widely the results obtained from the simulation model are to be applied and developing a thorough understanding of the inferences that can be made seem to be the most neglected aspects of the design of experiments associated with simulation studies. Kleijnen (1985) provides an extensive survey of these issues.

In general, the objectives of simulation experiments are:

1. To obtain knowledge of the effects of controllable factors on experimental outputs.
2. To estimate system parameters of interest.
3. To make a selection from among a set of alternatives.
4. To determine the treatment levels for all factors that produce an optimum response.

When multiple factors are involved, the approach to the first two objectives listed is to select one of the many possible experimental designs, and to hypothesize a model for the analysis of variance for the experimental design selected. The experimental design specifies the combination of treatment levels along with the number of replications for each combination for which the simulation model must be exercised. Using the data obtained from the experiment, the parameters of the hypothesized model are determined along with the estimation of the error terms. The significance of each factor is then judged, based on the derived model, and from this, estimates of system parameters of interest can be calculated.

In the problem of making choices among alternatives, the statistical procedures of ranking and selection are used. Kleijnen (1985) and Dudewicz (1979) present state-of-the-art reviews that summarize past research in this area and how it can be used in simulation analysis. Many procedures have been developed for specifying the sample size required in order to select the alternative whose population mean is greater than the next best population mean by a prescribed value with a given probability. The test procedures typically involve the computation of the sample mean based on the sample size specified and the selection of the largest sample mean observed. However, other descriptive statistics could also be used and ranked (Gibbons, Olkin, and Sobel, 1977).

A final topic relating to the design of experiments is the optimization of a simulation model. This problem differs from those previously described, in that, values for the controllable variables are sought that either maximize or minimize an objective function.

There have been two basic approaches to optimization, using simulation models. The first approach involves a direct evaluation of the independent variables, using the simulation model. Farrell (1977) divides these techniques into three categories: mathematically naive techniques, such as heuristic search, complete enumeration, and random search; methods appropriate to unimodal objective functions, such as coordinate search and pattern search; and methods for multimodal objective functions.

The second approach uses the response surface methodology (Myers, 1971). In this method, one fits a surface to experimental observations, using a factorial design in the vicinity of an initial search point. Then an optimization algorithm is applied, such as the gradient method, to determine the optimum values of the controllable variables relative to the fitted equation. The optimum values for the fitted surface are then used to define the next search point.

10.2.4.4 Tactical Planning

The objective of tactical planning is to make the most efficient use of the simulation model in executing the runs specified by the overall experimental design. This topic includes three major issues: (1) determining a policy for starting up the model that avoids initial bias, (2) controlling the execution of a model in order to reduce the variance of the outputs, and (3) determining a stopping point in the experimentation when a sufficient amount of data has been collected in order to ensure acceptable reliability in the final results.

Start-Up Policies

Start-up policies, typically prescribe a method for setting the initial conditions for a simulation model and for selecting a truncation point beyond which sample observations are to be recorded. This is necessary when the system behaves differently when it reaches a steady state as opposed to when it is commencing operation. Initial conditions are sought that will minimize the duration of the model's warm-up period, and a truncation procedure is used to try to identify the end of that warm-up period. In practice, these decisions are usually made by inspecting the results of some pilot runs of the model and by applying an intuitive test to detect the onset of steady-state behavior. Several authors have attempted to formalize this procedure in order to provide automatic start-up policies that can be incorporated into the operation of a simulation model. A survey of these policies is presented in Wilson and Pritsker (1978a,b).

It should be noted that theoretical studies of start-up policies have been restricted to small, well-behaved models. For such models the variability associated with sample values during start-up is not too different from the steady-state variability. Thus, the theoretical research tends to indicate that no truncation point needs to be identified and, therefore, start-up is unnecessary. Practical applications, however, indicate that this is not the case and that truncation is a reasonable policy to follow. This is especially true when many sequential operations must be performed before the system becomes heavily loaded.

Variance Reduction Techniques

To improve the precision of simulation-based performance estimators, a number of variance reduction techniques have been developed for simulation experiments. Some of these techniques are designed to exploit special information about the structure of the model, whereas, other techniques actually distort the structure or operation of the model in some way. Only two of these methods are widely used in practice: common random number streams and antithetic variates.

The technique of common random numbers enables the experimenter to sharpen the comparison between alternative system configurations by using exactly the same input sequence of random numbers to drive the simulation of each alternative. If the simulator can arrange for proper synchronization in the use of successive random numbers across all alternatives, then this technique ensures that all alternatives are compared under identical experimental conditions (Kleijnen, 1974). In the language of experimental design, the use of common random numbers creates a random block effect, and standard analysis procedures for block designs are appropriate when applying this variance reduction technique.

The basic idea behind the method of antithetic variates is to make complementary pairs of runs on the same simulation model so that the corresponding responses X, X' tend to fall on opposite sides of the mean μ_x. If this can be arranged, then the average of two antithetic responses X, X'

$$Y = \frac{X + X'}{2}$$

will have a smaller variance than the average of two independent replications. The usual method for implementing this technique is to supply complementary random number input sequences to the two runs: Where the random number r is sampled on the first run, the corresponding sampled value on the second run is $1 - r$. If K independent pairs of antithetic runs $[(X_i, X'_i)]$ are executed, then point and confidence interval estimators of the expected value μ_x of the performance measure X can be obtained by applying replication analysis to the k observations of Y.

It should be noted that the effectiveness of both common streams and antithetic sampling depends largely on the behavior of the simulation model to which they are applied; neither technique is guaranteed to improve the precision of simulation-based performance estimators. Various other variance reduction techniques have been based on concepts of regression analysis, stratified sampling, and importance sampling. Kleijnen (1974) thoroughly discusses the advantages and pitfalls of each of these techniques.

Stopping Rules

If the analyst specifies beforehand that a particular level of reliability is required in the final estimator of system performance, this requirement is usually translated into a maximum half-length H for the corresponding $100(1 - \alpha)\%$ confidence interval estimator. To ensure that an adequate sample size is accumulated to satisfy this reliability requirement, a sequential stopping procedure should be used. For example, a simple stopping rule for replication analysis is to continue increasing the number of replications k and recomputing the sample standard deviation S given by Equation (5) until the condition

$$t_{\alpha/2,\, k-1} \cdot \frac{S}{\sqrt{k}} < H \tag{9}$$

is finally satisfied. Law and Carson (1979) have developed a stopping rule based on subinterval sampling, and Fishman (1977) has proposed a sequential stopping procedure for use with regenerative analysis.

In addition to determining the sample size to meet desired confidence interval specifications, there are practical issues associated with the stopping of a simulation run. Such questions involve the consideration of what to do about entities in the model at the end of a run. The answers to such questions are problem-specific. If such entities are representative of the other entities on which statistics were collected, the further processing of them should not matter. However, if they are atypical and are of direct interest in the corresponding real system, then their processing should be considered and treated separately.

10.2.5   SUMMARY

Simulation provides a way to experiment with a model of a system. It complements and supplements both empirical testing and analytical modeling. It may be preferred to empirical testing when use of the real system is impossible, impractical, or imprudent. The first step in the simulation process is to build an appropriate model and check its validity. This may simply involve the adaption of some previously used model or it may require a full-scale development of some new representation of the system to be studied. In either case, once the model has been outlined, its implementation in a simulation language needs to be considered. The prudent modeler will usually have in mind some simulation framework as the system model is being formulated. This facilitates the modeling process and simplifies the translation into computer analyzable form.

Once a valid model is in-hand, the procedures for using it to study behavior are the same in principle, to the experimental design techniques used in laboratory research. It is important to remember that generalizations for a simulation model are analogous to extrapolations from laboratory research. Their range of validity is determined by how well the experimenter's paradigm (model or laboratory study) really represents the system and its operating environment. In many cases, some of these issues must be left unexplored because one cannot afford or is not permitted, the luxury of a full and complete validation test. For example, one can simulate the human performance implications of exposure to lethal doses of ionizing radiation, but one cannot be sure the predictions are correct. The validating experiment would be unethical. But then the use of models in such contexts is often very useful, even if basically speculative. It is predicated on the presumption that some analysis is better than no thought at all.

Simulation may then be viewed as a substitute for empirical testing. Given the advances in the technology of human/machine systems modeling, it is now a technology that should be considered by human factors engineers, particularly as a tool for early system design. In military system acquisition, it can be used and progressively refined throughout the development cycle. Once operational, system modeling can be used to identify and explain performance deficiencies and to evaluate proposed engineering changes. Eventually, this would include exploration of concepts for specifying performance requirements for the next system. If pursued in this fashion, the simulation models developed could be progressively refined, extended, and validated.

To date, applications have fallen short of such sustained, continuing efforts. They occur more commonly in other engineering disciplines. At a minimum, the human factors engineer should become more knowledgeable of simulation in order to participate in and affect these system-oriented simulation efforts of his or her peers and colleagues. Ideally, human factors engineers will more often adopt and use these techniques extensively to achieve predictive capabilities comparable to other engineering disciplines, illustrating how human factors impact system performance.

REFERENCES

Anderson, A., and Miller, K. (1977). Timeline analysis program (TLA-1) user's manual (Report D180-20247-3). Seattle, WA: Boeing Aerospace.

Asiala, C. (1975). *Advanced man/machine evaluation techniques.* Huntsville, AL; American Defense Preparedness Association.

Bachert, R. F., Evers, K. H. and Rolek, E. P. (1982). Static and dynamic model integration for a SAM C³ simulation via IDEF/SAINT, *Proceedings of the IEEE international conference on cybernetics and society.*

Bachert, R. F., Evers, K. H., and Rolek, E. P. (1983). IDEF/SAINT SAM simulation: hardware/human submodels. *Proceedings of the IEEE national aerospace and electronics conference.*

Baron, S. (1984). A control theoretic approach to modeling human supervisory control of dynamic systems. *Man-Machine Systems Research, I,* pp. 1–47.

Baron, S., Kleinman, D. L., and Levison, W. H. (1970). An optimal control model of human response. *Automatica, 6.*

Card, S. K., Moran, T. P., And Newell, A. (1983). *The psychology of human computer interactions.* Hillsdale, NJ: Lawrence Erlbaum Associates.

Carnahan, B., Luther, H., and Wilkes, J. O. (1969). *Applied numerical methods,* New York: Wiley.

Cellier, F., and Blitz, A. E. (1976). GASP V: a universal simulation package. *Proceedings, IFAC Conference.*

Chubb. G. P. (1981). SAINT, a digital simulation language for the study of manned systems. In J. Moraal and K. F. Kraiss, Eds. *Manned systems design: methods, equipment and applications. New York: Plenum.*

Chubb, G. P. (1982). A comparison of anxiety and frustration impacts on performance in manned systems AFAMRL-TR-81-129. Wright-Patterson Air Force Base, OH: Air Force Aerospace Medical Research Laboratories.

Chubb, G. P. (1983). Emotive disruptions: performance implications. In R. S. Jensen Ed. *Proceedings of the second symposium on aviation psychology.* Columbus, OH: Ohio State University, Aviation Psychology Laboratory.

Chubb, G. P. (1986). Factors in systems. *Proceedings of the international topical meeting on advances in human factors in nuclear power systems.*

Crane, M. A., and Lemoine, A. J. (1977). *Introduction to the regenerative method for simulation analysis.* New York: Springer-Verlag.

Dudewicz, E. J. (1979). *Introduction to statistics and probability.* Columbus, OH: American Sciences Press.

Edwards, R., Curnow, R., and Ostrand, R. (1977). Workload assessment model: user's manual (Report D180-202-47-3). Seattle, WA: Boeing Aerospace.

Emshoff, J. P., and Sisson, R. L. (1970). *Design and use of computer simulation models.* London: MacMillan.

Farrell, W. (1977). Literature review and bibliography of simulation optimization. *Proceedings, 1977 Winter Simulation Conference,* pp. 116–124.

Fishman, G. S. (1977). Achieving specific accuracy in simulation output analysis, *Communications of the ACM, 20,* 310–315.

Fishman, G. S., and Kiviat, P. J. (1967). Analysis of simulation generated time series. *Management Science, 13,* 525–557.

Forrester, J. W. (1971). *Principles of systems.* Cambridge, MA: Wright-Allen Press. 1971.

Franta, W. R. (1977). *The process view of simulation.* New York: Elsevier North-Holland.

Friedman, D., Pisani, R., and Purvis, R. (1978), *Statistics.* New York: Norton.

Geer, C. W. (1981). *Human engineering procedures guide* (AFAMRL-TR-81-35). Wright-Patterson Air Force Base, OH: Air Force Aerospace Medical Research Laboratory.

Gibbons, J. D., Olkin, D., and Sobel, M. (1977). *Selecting and ordering populations.* New York: Wiley.

Gordon, G. (1975). *The application of GPSS V to discrete systems simulation.* Englewood Cliffs, NJ: Prentice-Hall.

Graybeal, W. and Pooch, U. W. (1980). *Simulation: principles and methods.* Cambridge, MA: Winthrop.

Greening, C. P. (1976). Mathematical modeling of air to ground target acquisitions. *Human Factors, 18*(2), 111–148.

Hastings, N. A. J., and Peacock, J. B. (1975). *Statistical distributions.* London: Butterworth.

Hollander, N. A. J., and Wolfe, D. A. (1973). *Nonparametric statistical methods.* New York: Wiley.

Kiviat, P. J., Villanueva, R., and Markowitz, H. (1969). *The SIMSCRIPT II programming language.* Englewood Cliffs, NJ: Prentice-Hall.

Kleijnen, J. P. C. (1974). *Statistical techniques in simulation, Part 1.* New York: Dekker.

Kleijnen, J. P. C. (1985). *Statistical tools for simulation practitioners,* New York: Dekker.

Klein, T., and Cassidy, W. (1972). Relating operator capabilities to system demands. *Proceedings of the Human Factors Society,* 16th Annual Meeting.

Kleinman, D. L., Baron, S., and Levison, W. H. (1970). An optimal control model of human response, part I: theory and validation. *Automatica,* 357–369.

Kochhar, D. S., and Willis, B. L. (1971). Simulation of a two-man interaction system. *Proceedings of the 5th winter simulation conference.* New York, pp. 56–62.

Korn, G. A., and Wait, J. V. (1978). *Digital continuous-system simulation.* Englewood Cliffs, NJ: Prentice-Hall.

Kornfeld, J. R. (1984). Specification and preliminary validation of IAT methods: executive summary (TR-223). Burlington, MA: ALPHATECH, Inc.

Kraiss, K. F., and Knaeuper, A. (1982). Using visual lobe area measurements to predict visual search performance. *Human Factors, 24*(6).

Lane, N. E., Strieb, M. I., Glenn, F. A., and Sherry, R. A. (1981). The human operator simulator: an overview. In J. Moraal and K. F. Kraiss, Eds. *Manned system design: methods, equipment. and applications.* New York: Plenum.

Laughery, K. R. (1984).Instructions for the use of Micro SAINT (Internal Report). Boulder, CO: Micro Analysis and Design.

Laughery, K. R. (1985). Network modeling on microcomputers. *Simulation,* January, 10–16.

Laughery, K. R., and Polito, J. (1983). MOPADS year 1 progress report. Ft. Bliss, TX: Army Research Institute Field Unit.

Laughery, K. R., and Gawron, V. (1984). Human factors moderator functions (MOPADS Volume 5.10). Fort Bliss, TX: U.S. Army Research Institute Field Unit.

Law, A. M. (1977). Confidence intervals in discrete event simulation: a comparison of replication and batch means. *Naval Research Logistics Quarterly, 24*(4).

Law, A. M., and Carson, J. S. (1979). A sequential procedure for determining the length of a steady-state simulation. *Operations Research, 27*(5), 1011–1025.

Law, A. M., and Vincent, S. G. (1984). UNFIT: an interactive computer package for fitting probability distributions to observed data. *Proceedings of the 1984 winter simulation conference.* Dallas, TX., p. 21.

Markowitz, H. M., Karr, H. W., and Hausner, B. (1963). *SIMSCRIPT: a simulation programming language.* Englewood Cliffs, NJ: Prentice-Hall.

Mitchell, E. L., and Gauthier, J. S. (1976). Advanced continuous simulation language (ACSL). *Simulation, 25,,* 72–78.

Muralidharan, R., Baron, S., and Feehrer, C. (1979). A decision, monitoring and control model of the human operator applied to an RPV control problem (Air Force Office of Scientific Research Report, AFOSR-TR-79-0675). Bolling AFB, DC.

Musselman, K. J., Penick, W. R., and Grant, M. E. (1981). *AID: fitting distributions to observations.* West Lafayette, IN: Pritsker and Associates.

Myers, R. H. (1971). *Response surface methodology.* Boston, MA: Allyn & Bacon, 1971.

Naylor, T. H. (1969). *The design of computer simulation experiments.* Durham, NC: Duke University Press.

Newell, A., and Simon, H. (1972). *Human problem solving.* Englewood Cliffs, NJ: Prentice-Hall.

Pew, R. B., Feehrer, C. E., and Miller, D. C. (1977). Critical review of performance models applicable to man-machine systems evaluation (Report No. 3446). Cambridge, MA: Bolt, Beranek, and Newman.

Phillips, D. T. (1972). Applied goodness of fit testing (AIIE-OR-72-1). Norcross, GA.

Polito, J. (1983). *The MOPADS data base control system (MOPADS Vol. 5.13).* West Lafayette, IN: Pritsker & Associates.

Polito, J. and Laughery, K. R. (1983). *MOPADS final report.* West Lafayette, IN: Pritsker and Associates.

Pritsker, A. A. B. (1974). *The GASP IV simulation language.* New York: Wiley.

Pritsker, A. A. B. (1979). *Modeling and analysis using Q-GERT networks.* New York: Halstead.

Pritsker, A. A. B. (1984). *Introduction to simulation and SLAM II.* New York: Wiley, and West Lafayette, IN: Systems Publishing Corporation.

Pritsker, A. A. B., and Kiviat, P. J. (1969). *Simulation with GASP II: A FORTRAN based simulation language.* Englewood Cliffs, NJ: Prentice-Hall.

Pritsker, A. A. B., Wortman, D. B., Seum, C. S., Chubb, G. P., and Seifert, D. J. (1974). SAINT: volume I, systems analysis of integrated networks of tasks (AMRL-TR-73-126). Wright-Patterson Air Force Base, OH: Aerospace Medical Research Laboratories.

Pugh, A. L. (1973). *Dynamo II user's manual.* Cambridge, MA: MIT Press.

Rouse, W. (1980). *Systems engineering models of human-machine interaction,* New York: North Holland.

Russell, E. C. (1983). *Building simulation models with SIMSCRIPT II.5.* Los Angeles, CA: CACI.

Sargent, R. G. (1984). Simulation model validation. T. Oren et al., Eds. *Simulation and model based methodologies.* New York: Springer Verlag.

Schriber, T. J. (1974). *Simulation using GPSS.* New York, Wiley.

Schmeiser, B. W. (1982). Batch size effects in the analysis of simulation outputs. *Operations Research, 30,* 556–568.

Seifert, D. J. (1979). Combined discrete network-continuous control modeling of man-machine systems AMRL-TR-79-34. Wright-Patterson Air Force Base, OH: Aerospace Medical Research Laboratory.

Seifert, D. J., and Chubb, G. P. (1978). SAINT: a combined simulation language for modeling large complex systems (AMRL-TR-78-48). Wright-Patterson Air Force Base, OH: Aerospace Medical Research Laboratory.

Shannon, R. E. (1975). *System simulation: the art and science.* Englewood Cliffs, NJ: Prentice-Hall.

Siegel, A. I., and Wolf, J. J. (1967). *Man-machine simulation models: performance and psychosocial interaction.* New York: Wiley.

Spence, J. T., Cotton, J. W., Underwood, B. J., and Duncan, C. P. (1983). *Elementary statistics,* 4th ed. Englewood Cliffs, NJ: Prentice-Hall.

Strieb, M. I., Glenn, F. A., Fisher, C., and Fitts, L. (1977). Computer aids for station design (Technical Report 1285-A). Willow Grove, PA: Analytics.

Van Horn, R. L. (1971). Validation of simulation results. *Management Science, 17,* 247–258.

Walker, J. L., and Polito, J. (1982). *Development methodology for MOPADS air defense system modules.* West Lafayette, IN: Pritsker & Associates.

Wilson, J. R., and Pritsker, A. A. B. (1978a). A survey of research on the simulation startup problem. *Simulation, 31,* 55–58.

Wilson, J. R., and Pritsker, A. A. B. (1978b). A procedure for evaluating startup policies in simulation experiments. *Simulation, 31,* 79–89.

Wortman, D. B., Duket, S. D., Seifert, D. J., Hann, R. L., and Chubb, G. P. (1978). Simulation using SAINT: a user-oriented instruction manual (AMRL-TR-77-61). Wright-Patterson Air Force Base, OH: Aerospace Medical Research Laboratory.

Wulff, J., Jepson, P., Alden, F., and Leonard, J. N. (1968). A descriptive model for determining optimal human performance in systems (NASA-CR-876). Washington DC: National Aeronautics and Space Administration, Vol. I.

CHAPTER 10.3

MAINTAINABILITY

NICHOLAS A. BOND, JR.

U.S. Office of Naval Research (Tokyo Branch)

10.3.1 BASIC CONSIDERATIONS

First comes word that "the system is down"; and very soon after that, "When is that thing going to be fixed?" Both sentences are heard wherever complex devices are found, and both have their overtones of irritability and disappointment. Surely, the simple "when" question does define the essential "maintainability" ($\overline{M}$) problem of designing and implementing systems so that restoration can be achieved with acceptable ease and expediency.

10.3.1.1 The Persistent Maintenance Crisis

System failures can have catastrophic consequences in domains such as red-phone diplomacy, early warning radar, weapons testing, air control, aerospace missions, nuclear power, chemical process control, medicine, and communications. Even without catastrophe the bill can be high. United Airlines has estimated that when its computerized reservation system goes down, a lost-sales-opportunity cost of $36,000 a minute is incurred (Hightower, 1982). Banks and insurance companies pay heavily for duplex or fallback facilities in their 24-hr data centers; they do this because even a few hours delay in billing or electronic money transfer can be expensive. In nuclear power the "expense" of troubles may far exceed simple money costs, as unexpected fault conditions may lead legislative and executive authorities to close down whole operations, regardless of what the "real" hazards may be.

In the military, some of the numbers are disturbing. At any given moment, only about one-half of the combat aircraft on a U.S. Navy carrier are able to fly off the ship with all systems in an "up" condition; for each hour it flies, a plane spends about 30 hr in the shop (Halff, 1984). From military maintenance reports, it appeared that from 4% to 43% of all "faulty" components sent back from the field for repair were really good; these actions accounted for from 9% to 32% of maintenance manhours. In about 10% of the corrective actions undertaken, the defect was never found, or a good component was damaged by the maintenance crew (Orlansky and String, 1981).

As one concrete illustration of present state-of-the-art in big-system corrective maintenance, consider the case of the E-3A radar, a major electronics element in AWACS surveillance aircraft. This radar is a well-seasoned system, with sophisticated built-in test (BIT) capabilities. Figure 10.3.1 tabulates the results of some 12,000 trouble indications. Right away, one sees that 85% of the detected "troubles" quickly proved to be false alarms. Thus only 8% or 1039 of the incidents could be counted as real troubles. However, 25% of them could not be confirmed at the first check. That left some 761 confirmed troubles, and the BIT "detected" 746 or nearly 98% of them. Further attempts were then initiated to *isolate* these 746 detected failures; automatic test equipment (ATE) found about one-third of them, 15% tested good and so once again were not real troubles, and manual techniques finally had to be used to isolate the remaining 51% of the failures (Coppola, 1984). The illustration is quite typical: many false alarms, high detection rates, but much of the troubleshooting still done by people.

From the maintenance personnel angle, there are also plenty of disquieting signs. Service costs for commercial computer systems recently were rising at the rate of 25% a year, and it always takes a long time to produce an effective technician. Not everybody can do the work. Right now, the median U.S. Navy recruit reads at about a 10th grade level, while the average Navy tech manual "readability" is at the 10th grade level; this means that a typical manual may not be fully understandable by half the Navy recruits. The prime U.S. recruit population will shrink from 4.25 million in 1978 to 3.25 million in 1990; while that is happening, analysts also expect that the incessant shortage of good technicians will mean that the better ones will have to stay "on the line."

10.3.1.2 Response to the Crisis

The main response options to the maintenance crisis are quite evident. One could try to improve the numbers of personnel in the recruit sample, to raise the input quality by setting high "cutoff" scores, and to improve radically the training of the people involved. Such approaches are heavily constrained by the realities of the selection pool and the limited training time. One may contemplate the formation of a few elite crews of specialists, who could be transported to "tough trouble" sites; something like that has been implemented on a few occasions, but is not an acceptable long-range solution. One could refuse to accept equipment until most of the maintenance bugs have been identified and removed; but that would require years of burn-in and thus would deprive the user of new capabilities and possibilities. One could press for ever more sophisticated and effective automatic test equipment and computerized fault location. However, as the data in Figure 10.3.1 indicate, even the best of such devices cannot yet find all the troubles, they require additional resources, and they may themselves

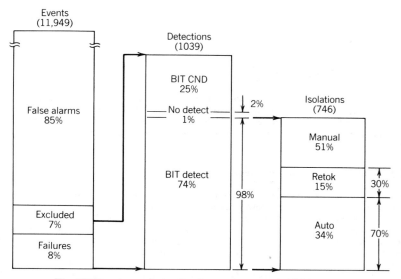

Fig. 10.3.1. An analysis of maintenance events in E-3A radar.

cause new complexities and maintenance events. Although he was undoubtedly exaggerating for effect, we can cite De Kleer's statement: "You write 100,000 lines in Pascal code for your million dollar piece of diagnostic equipment which finds 30% of the faults in the device" (De Kleer, 1984).

Then there are some insidious things about these maintenance problems and unfindable troubles. Since some failures are so difficult to locate, heroic hardware/software efforts have been initiated. The engineering design community has made great progress in designing equipment so that failure is masked when it does occur. For critical signal flows, several redundant or parallel paths may be provided, with the data flow proceeding according to a "voting" criterion among members of redundant set. So a fault in a single element will not prevent the system from working. Perhaps no one would ever know that a fault happened; in fact, a system could be "full of troubles" and still continue to operate for a long time. Another design technique is to arrange things so that the system reconfigures itself when certain failures occur, perhaps by operating at a lower level of performance. All this is very clever engineering and certainly admirable for some applications. However, all troubles cannot be anticipated or masked; and when a big system starts to reconfigure itself, it may get itself in such a state that literally nobody can figure out where it is or what it is doing (Rouse, 1984).

Among M professionals, there is much talk these days about "5% troubles." These are the tough ones, the bridging faults, the interactives, and the intermittents—the ones that elude the automatic test routines, the fault-tolerant engineering, and the massive parts-substitution schemes. Even at present, the "5% troubles" may account for nearly half the total maintenance effort (Maxion, 1984).

10.3.1.3 A Human Factors Approach

Given these critical and persistent maintenance problems, an old question can be raised: Do we really know how to make complex but maintainable systems? The standpoint of this chapter is that we do have, right now, many of the requisite tools and knowledges; the formulas, principles, and practices are well known. However, continuing pressure for expanded capabilities pushes equipment designers into "edge-of-the-art" technologies. It may take most of the energies of the design team just to get a new system to work, so that aspects like testing, maintainability, and logistics support are not fully validated at the time of delivery.

There are naturally many reasons why a complex system could prove to be a maintenance nightmare, and some of these have nothing to do with human factors. When a system proves to be a continuing maintenance problem, however, it is often possible to trace this undesirable condition to definite human factors aspects that either were ignored or were recognized too late. We assume here that instead of "getting the human out of the system," a preferred approach is to get the human effectively *into* the system. This chapter therefore undertakes a rapid survey of the main ideas, types of information, and resources available to a human-factors-oriented maintainability analyst. Some of this information is quite explicit and can be stated in cookbook terms; examples are the design of a meter face or the dimensions for an accessibility opening. But for some of the most important problems, one must rely more on a psychological, empirical, and even skeptical stance toward the utilization of the human. The present equipment trends are moving toward more computerization, miniaturization, and abstraction, so the human roles will surely follow those trends. Fully realized and applied, we believe that utilization of the information now at hand will increase the likelihood that a system will have troubles that are findable and fixable. To accomplish this requires the integration of technical work in several domains and a strong commitment by management, but it is entirely feasible, and in fact is being achieved every day.

10.3.2 MODELING

Of the many definitions, the Department of Defense version is probably the best known: "Maintainability is the probability that an item will be retained in or restored to a specified condition within a given period of time, when the maintenance is performed in accordance with prescribed procedures and resources" (U.S. Department of Defense, 1966). As noted by many analysts, this definition is quite imprecise. "Prescribed" procedures and resources, for example, always embrace some definite operating environments and support arrangements, but these are implicit and often unique to particular users. A "given period of time" sounds firm but is statistically evaluated in real projects, with certain risk levels on both the contractor and user ends. Even the idea of "probability" itself is not as clear-cut in the $\bar{M}$ domain as one might think. And there are various trade practices and subtleties, such as the very important "prime maintenance" assumption: when something like an ailing mainframe is restored to full service within, say, 23 min, the "defective" modules taken out will enter some kind of a test and repair chain that could extend over many miles, days, and dollars; but the prime maintenance time of 23 min is still taken as the key $\bar{M}$ parameter.

Fortunately all the ambiguities do not constrain a maintainability effort very much, since everybody agrees on the basic goal of rapid and reliable system restoration, using ordinary trained people and reasonable support facilities. As one comment on $\bar{M}$ definitions, it may be noted that while "maintainability" and "maintenance" sound very much alike, maintainability is concerned with the achievement of some specific capability by means of design and other variables; after the design has been accomplished and is in the field, "maintenance" embraces all the technical activities that are devoted to keeping the equipment up.

Because $\bar{M}$ target parameters appear in contracts as sacred numbers to be achieved, the human factors analyst may encounter several kinds of "maintainability models" in any fair-sized project. Often there are two main efforts: "availability" modeling and "life-cycle" costing. For either type, there can be interesting problems of estimation and application, and these problems may extend from the proposal stage all the way to project completion.

10.3.2.1 Availability

Availability starts from the idea that a key feature of any design is the proportion of operating time that is "downtime." Thus for a customer who buys a system and puts it into the field, "achieved availability" (A_A) or "operational availability" (A_O) would be the bottom line of eventual usefulness, and would be expressed as follows:

$$A_A = A_O = 1 - \frac{\text{downtime}}{\text{up time}}$$

Clearly, high availability numbers are to be desired, and it is evident that when "downtime" exceeds a certain fraction, a system might be considered to be useless. Although the preceding equation is simple and appealing, the general "downtime" term is very inclusive. It includes all the administrative delays, encumbrances, unpredictables, and shortages of the field environment. Since these factors are not under the control of a contractor, a better index of a contractor's work is "intrinsic" or "inherent" Availability (A_I), expressed as

$$A_I = \frac{\text{MTBF}}{\text{MTBF} + \text{MTTR}}$$

where

MTBF = mean time between failures

MTTR = mean time to repair

The A_I index has separate index numbers for reliability and maintainability, and it specifies the close interaction and trade off possibilities of those numbers. A_I measurement is done under near-ideal support conditions, perhaps during a demonstration at a factory or exemplary field site. Thus, there will be adequate tools, test equipment, service personnel, spare parts, and documentation on hand for an A_I demonstration. While all parties know that these ideal arrangements will not exist in the field, an A_I estimate may give an appropriate benchmark for evaluating "the best the system can do," from the standpoint of utilization. Nearly always, such items as preventive maintenance, supply lags, and training problems are treated later as important off-line factors, but they are left out of formulas for general maintainability goals.

With so much emphasis on time-to-repair, the system modeler naturally uses some statistical model of repair times, so that strict probabilistic inferences can be carried out when evaluating $\overline{M}$ in a real item or system. Although various distributions have been proposed from time to time, the log-normal is probably the most widely used model, because many actual samples of repair times can be fitted well to the log-normal, and also because there is now much experience with estimating parameters and setting risk levels, so that the procedures are well known to both customer and contractors. Figure 10.3.2, taken from Cunningham and Cox (1972), shows one large set ($N = 567$) of repair times in large electronics equipment of the early 1970s era. Clearly visible there is the "hump" of very short repair times, with a long tail extending out some hours to the right. In such log-normal distributions, the mean is always greater than the median, and the mode is smaller than either mean or median.

Once a statistical distribution is proposed, actual (or assumed) repair times can be fitted to the model and the fit can be subjected to standard goodness-of-fit tests. If the fit is satisfactory, the analyst can go ahead and obtain estimates of the probability of achieving a given $\overline{M}$ level, under a given set of "risk" assumptions. The logic is exactly the same as in quality control theory. Suppose that a reasonable number of demonstration troubles, say, some 50 or 100, will be inserted into a system when it is evaluated for acceptance. The contractor wants to be "reasonably sure" that his or her system will meet some stated $\overline{M}$ time goal such as an MTTR. Suppose that "reasonably" means 80% sure, or in other words "a contractor's risk" of 0.2 is adopted. If the statistical assumptions hold, then about 20% of the time the equipment is basically good enough to pass, but *does not* pass because of random factors that produce failure during a particular evaluation. On the other side of the evaluation table, the customer takes a risk too. His or her risk is that an equipment is "really" unacceptable, but passes the demonstration because of random factors that produced, on a particular occasion, a false-positive score. Standard statistical procedures exist for setting "target" values that

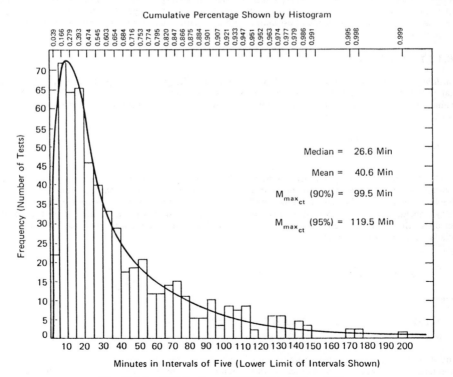

Fig. 10.3.2. Distribution of MTTR times ($N = 567$).

minimize the other risk, given one fixed risk. The results of these procedures match intuition quite well. Cunningham and Cox (1972), in a worked out example with realistic parameters, show that if an MTTR of 30 min is specified in a contract, then a "customer's risk" cutoff point of observed MTTR of about 27 min would be a good value; furthermore, after calculating that point, the contractor might shoot for an even lower MTTR value (23.8 min, in the example). Both contractor and user risks, of course, will be reduced if more demonstration troubles are inserted; but demonstrations are expensive and the variance reduces as the square root of the number of troubles, which means diminishing returns after the first few dozen test problems. By the way, most demonstrations use less then 100 troubles, if by "trouble" some real hardware malfunction is meant. (For certain automatic test routines, hundreds or thousands of slight stimulus excursions and parameter variations can occur in a self-test mode.)

Preventive maintenance modeling is not always specified in $\overline{M}$ contracts, but it can be a most useful addition if good maintenance cost data are available for both preventive and corrective maintenance over time. One form of analysis postulates certain fixed costs for each preventive maintenance procedure, and also a fixed cost for each corrective maintenance action. (A corrective action usually costs several times as much as a preventive action.) An optimum maintenance policy may then be sought that will minimize the total maintenance cost, given the cost data and the effects of preventive maintenance on reducing the need for corrective maintenance, and with everything proceeding under the constraint that the item or system target reliability goal is always met. The mathematical details of the optimization are well known (Rau, 1970), and commercial software packages are available for application of the model. Lamarre (1980) gives an instructive worked-out example with preventive and corrective maintenance cost estimates, and realistic values for failure, estimated parts replaced, reliability targets, and so forth. He also notes that models exist for deciding among alternative configurations; it may, for instance, be cheaper to provide several alternative or parallel systems, rather than to mount major and expensive attempts to improve the availability of one equipment. Occasionally, the results of such comparisons are counterintuitive, or what may be worse, contrary to some stated policy. A purchasing authority may insist on providing just one "adequate" system, whatever the cost, rather than accepting a two-or-three-system configuration that would be better on nearly all performance/cost/maintainability calculations.

10.3.2.2 Life-Cycle Modeling

In many procurement agencies, life-cycle-cost (LCC) analysis is another requirement which a system producer has to meet. The basic idea is simple. Each product can be considered to go through several phases, for example, preliminary concept, design, manufacture, installation and burn-in, and operate–maintain. Furthermore, each of these phases has its own costs, and so alternative proposals to meet some functional requirement should be evaluated on total costs and returns over the expected life of the systems. Maintenance costs are often most important in the LCC estimates. The design stages can be expected to take about 15% of the expense, with 35% for production and 50% for operation and support. Such proportions, incidentally, are often assumed in U.S. Department of Defense procurements.

While nearly all interested people will agree that a LCC concept is meaningful and useful, there is often some controversy over such matters as the exact factors to put into the analysis, how uncertainty is to be modeled, and how a customer can evaluate the credibility of an elaborate LCC submittal. Summaries of LCC factors and formulas are given by Brannan (1976) and the U.S. Department of Defense manual (1970). Usually the "locked-in" or nonrecurring costs like R&D design, manufacture, and documentation are segregated from the recurring costs of operating, sparing, and maintaining the system. A maintenance-oriented analyst may develop separate tabulations of costs that depend on a failure and those that do not depend on a failure. Specific curves and relations can be postulated between various cost factors and the reliability and longevity of a system. Thus, maintenance should drop when a high MTBF is actually realized, but original procurement expense increases with MTBF. Logical curves for these relations are often postulated, but it is difficult to find validations of them. For LCC evaluation, many trade offs and summations must be done before deciding which of several competing proposals is the preferred one. The human factors community has written computer programs as aids in effecting these trade offs, but much of the experience is proprietary and there are few published studies on the utility of such aiding. Often it is sufficient to gather as much *information* about the likely trade off curves as can be found, and then to convene an informal trade off meeting between interested parties. In all such trading and estimating, the data-processing limitations of people should be remembered. The Philco approach to trade off management is a good general guide (Cunningham and Cox, 1972).

Many ingenious proposals for LCC performance indexes have been made, often by defining a criterion ratio score. The U.S. Air Force evaluated aircraft tires by calculating the "tire" cost of each landing. Reportedly, an LCC-based review of actual landing records produced a sharp improvement in landings per dollar. Such calculations can be done for maintenance matters too. In one LCC support study, a portable UHF radio proposal was analyzed. For this radio, the "number of maintainers needed" was obtained by starting from the MTBF/MTTR data. Thus, MVH was then divided by the expected working hours/man/year to get the number of technicians needed. Field troubleshooting time requirements are often twice or three times as big as those recorded in "ideal" demonstrations (Cunningham and Cox, 1972). Such calculations of "maintenance manhours per year" may well be serious underestimates.

10.3.2.3 Special Human Factors Problems and Hazards: Modeling

Deceptive precision. Statistical analyses such as the log-normal procedure yield "hard numbers," with an "exact science" look. But sometimes the data base is only reliable enough for very gross inferences. Ideally, time and other scores should meet the usual psychometric standards for measurement error (Nunnally, 1978). A large tabulation of empirical maintenance times, along with predicted times obtained from MIL-STD methods, has been published by the U.S. Air Force. The compilation identifies equipments generically, not by specific model or manufacturer (U.S. Air Force, undated).

Management impatience with experimentation. If there are severe time and cost pressures, neither engineering management nor the customer may be willing to sponsor realistic experimentation just to get good MTTR or other estimates. In such a case, sometimes even a few "walk through" runs will be enough to show where the information needs are most crucial, and may alert the human factors specialist to the areas where predictions and projections are soft. In well-conceived maintainability projects, the key modeling parameters should become firmer and more credible as the project moves toward the end of the development phase.

Belief that radical engineering changes automatically produce radical human factors changes. The hardware changes from vacuum tube to transistor to circuit board to complex chips were profound. However, as seen in Figure 10.3.3, the relative proportions of time that humans spend on such functions as "localize" and "isolate" has remained rather constant, with fault isolation taking the most time. (Class 1 is the most advanced; class III is the old vacuum tube technology.) Perhaps a revised statement would be true: a radical engineering change produces a maintainability impact *only if new diagnostic features are built into the new design, in such a fashion as to be satisfactorily utilized by maintenance people.*

10.3.3   THE PSYCHOLOGY OF TROUBLESHOOTING

Since people have to find and fix at least some of the troubles, the effectiveness of their search behavior is often the crucial element of a maintenance event. In this section we touch on some of the main findings and design implications from 30 years of troubleshooting research.

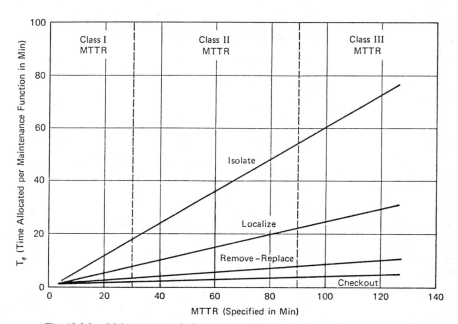

Fig. 10.3.3.   Maintenance task time versus MTTR, modular-designed equipment.

10.3.3.1   Early Studies

Because real troubleshooting situations on Navy ships or Air Force flightlines were rather turbulent places to do controlled research, most early investigations used special simulator formats. Thus, schematic diagrams of real circuits were used but the behavior variance associated with setting up test equipment and parts was eliminated. Subjects could take a "voltage" by lifting a paper tab on a board (Glaser and Phillips, 1954) or by opening a window in a special machine; nowadays, of course, a computerized simulator would be used. Performances were scored for such features as whether the trouble was found or not, overall time-to-success, and "efficiency" as defined by the number of checks required or parts removed. These scoring practices have persisted for some decades now. Present maintainability demonstrations usually call for several overall time scores, along with subscores for access, correction, special test procedures, and administrative lags (Cunningham and Cox, 1972).

A first result is that much of the flavor or "logical essence" of troubleshooting can indeed be captured in *simulator and paper–pencil formats.* Subjects become engaged in a well-simulated troubleshooting task, and at least for relatively simple radio and radar circuits, there is a substantial correlation between simulator scores and "hardware scores" on the same circuits. This is still a useful observation: If technicians cannot troubleshoot a clean simulator diagram of a system efficiently, they probably cannot perform fault isolation on the real system either. Also, local hypotheses about fault-locating performance and aiding can be checked in a simulator format, and arguments about the logical efficacy of aiding schemes can be quickly evaluated (Duncan and Shepherd, 1975). A thorough review of the concepts and controversies surrounding simulator fidelity appears in Su (1984).

Individual differences in trouble-shooting skills are usually marked, even in groups with similar training and experience. Suppose that a dozen inserted troubles are administered to advanced Navy trainees or working technicians; most likely, a few subjects will find one or two troubles only, while others might locate nearly all of them within a time limit of 30 min or so per problem. One practical implication is that it is difficult simply to "produce" good troubleshooters via regular training programs and field exposure. Some performance variance should always be expected, and a sizable fraction of people may never become very proficient.

Troubleshooting performance shows a moderate relationship with other indicators of *technical knowledge,* such as ability to solve textbook problems using Ohm's and Kirchhoff's laws. This result might seem to be the discovery of the obvious, since much troubleshooting is done by means of if–then relations as assumed or remembered by the troubleshooter, and these relations themselves are derived from physical laws and principles. To cite just one experiment, nearly 60% of the individual variance in troubleshooting scores on an oscilloscope could be accounted for by "technical knowledge" of that instrument, and the correlation with "electronics fundamentals" was also moderately high (Highland, Newman, and Waller, 1956). In contrast, for decades investigators have found that correlations between troubleshooting and *aptitude* test scores were quite low. In Air Force research, only 7% of the troubleshooting variance was attributable to tested "reasoning ability" (Highland, Newman, and Waller, 1956). Perhaps the low correlations between troubleshooting and ability are due to restriction of ability range in the test sample. U.S. Navy technical schools, for instance, limit admission to those with upper-percentile scores on verbal and numerical tests, and this practice could reduce variance and hence correlation (Henneman and Rouse, 1984).

Fault-locating behavior is *seldom optimal.* Even in clean simulator setups where the subject is not bothered by the noise, discomfort, accessibility, supply problems, and other inconveniences of the real world, an actual performance has redundancies, mistakes, and missed logical opportunities. A technician may do all the checks logically necessary to accomplish fault isolation, but often he or she will not appreciate fully the implications of his or her test sequences, and so he or she will never locate the trouble. This "inefficiency" can be due to plain forgetting, to imperfect "strategy" or search logic, or to ignorance of the system relations. Forgetting surely takes place: after all, many readings, tests, and removals can be made in a few minutes, and it is often easy to repeat a check if there is any doubt about it. Technicians do not often keep a written record of "where they are" in a search attempt, either, and this tendency would lead to repeated (redundant) testing and checking. Technicians may also realize that there will usually be several critical and correlated test cues, and that the troubleshooting problem will be "over determined," in the sense that several test sequences can be effective in bracketing the trouble. And they may even realize that the correlation among cues will alleviate the effects of forgetting or omitting some of the cues (Einhorn and Hogarth, 1982). So "redundant troubleshooting" can be quite rational. Sometimes it is possible to score a subject separately for "optimality of search" and "optimality of data base." Suppose that a subject has in mind a given symptom–malfunction matrix for an equipment, and that this matrix can be elicited from the subject. (The matrix shows the relation of every fault to every test symptom.) An "optimal" troubleshooter presumably would take the information in the matrix and act about like a Bayesian statistician would. (A Bayesian always selects that check which best reduces ambiguity, given the present information state.) In fact, trained and experienced technicians "act Bayesian" only half the time or less. And their unaided subjective beliefs about symptom–fault relations are apt to be quite inaccurate, so that even a good Bayesian often could not logically find the trouble using their "defective" matrices (Bond and Rigney,

1966). Of course, a real troubleshooter may be acting under conditions and constraints that a Bayesian processor could not know about. Again the practical implications are obvious: do not assume that technicians know the basic cue–fault relations, or that they will process the information they do have in anything like an efficient manner. Explicit aids to prevent forgetting, to suggest a good search strategy, and to provide an effective system "data base" of relations should be provided.

10.3.3.2 Recent Research

It has gradually become clear that *knowledge representation* is more important than the quality of an individual's processing capabilities; in fact, the representation is perhaps the key thing that separates experts from novices. This finding has cropped up in troubleshooting and in many other technical domains, and is now one of the best established results of cognitive psychology. When Master chess players and weaker players look briefly at a chessboard and try to memorize the position of the pieces, Masters remember *meaningful* positions almost perfectly, whereas ordinary players show a very fragmentary recall. But when pieces are put on the board *randomly,* the Masters are no better at recall than the novices. Clearly, the Masters have a large store of meaningful patterns or configurations, and so they can code a meaningful position in terms of the set of stored elements (Chase and Simon, 1973; DeGroot, 1965). Similar "expert recall" results come from the game of GO, and from the memory performance of military experts who look at toy battlefield displays. When experts and novices solve problems in physics, the experts tend to orient their efforts around the laws of physics involving energy and momentum, while the novice is more likely to focus on the literal "objects" and surface features of a given problem. Differences are even found in the way that experts and novices *categorize* problems which are presented to them. The expert's "principles" and "deeper" structures are also more apt to be tightly interrelated (Chi and Glaser 1983; Larkin, McDermott, Simon, and Simon, 1980).

In a similar way within the electronics circuit domain, experts have a "richer" cognitive map, and they remember diagrams not so much from the pictorial layout, but by recalling the functional units themselves, and the way that the units are electrically related. And most crucially, experts are more influenced by the *logical* features of the circuit than they are by the surface physical details; this is reflected in the fact that they do very much better when they have to "reorganize" a circuit layout (Wickens, Geiselman, Sanet, and Yelvington, 1982).

The "knowledge focus" suggested by all these cognitive studies has strong implications for maintainability projects and practitioners. Perhaps the first task for the analyst is to determine exactly what an effective knowledge representation *is* for a given system. Surprisingly, the basic system charts suitable for technicians often have to be developed by human factors staff people. A good representation will probably be more pictorial, hierarchial, and "chunkable" than the usual fixed information found in schematics and in automatic testers. Attention then may be directed to how the aided or unaided memories of humans can express, encode, and operate within this knowledge base, and to how easily the base can be interrogated and modified. A knowledge representation viewpoint strongly suggests evaluation of mental representation systems during the early design stages of a project. There is a marvelous opportunity here for interdisciplinary cooperation. There is no intrinsic reason why a physical system cannot be described in terms that an ordinary technician can understand.

At some level, *complexity* of the prime equipment is clearly a most powerful factor affecting human performance. Some systems, like the flight control on the F-18, are so complicated that there is literally no person on earth who "understands it all." At a less exotic level, complexity effects have been demonstrated experimentally. Thus, putting more components into a flow-chart simulator leads to more errors, and a definite "complexity effect" is also obtained with actual circuit boards (Brooke and Duncan, 1981; McDonald, Waldrop, and White, 1983).

Complexity has still not been well defined, at least for the behavioral scientist. Parts, stages, subunits, data paths, and connectors can be counted, of course, and in some aggregate sense the numbers are related to complexity. But the complexity of real interest is probably something like a perceived or weighted "intellectual load" associated with the explication of test data and information flows through a configuration, as this load is levied on the average technician when the system fails. These "explications" are themselves determined by technological practice and the designer's attention to such matters as modularization, signal "traceability," and the ease of sequencing critical tests. So complexity often gets to be extremely equipment specific or technology specific. Perhaps a working measure of "effective complexity" could be developed by selecting a probability sample of the troubles to be found, tracing the psychological "inference constituents" required to bracket each trouble, and then assessing the availability of these constituents in the average technician, either in his or her own memory or in an aid. Given such a set of required "constituents" and information about the psychological difficulties associated with them, training and aiding programs might be oriented rather differently from those existing today.

Towne (1984) has a scheme that predicts quite well the maintenance sequences of real technicians on a computer system of moderate complexity. Surprisingly few items of information are necessary to do this: the reliabilities of replaceable elements, times to accomplish a test, a rather simple adjustment of parts-failure likelihoods as successive tests are made, test and unit "cost" information, and that is about all. A very compact symptom–malfunction matrix seems to be sufficient for driving an optimizer inference engine. If different failure modes in a suspect unit affect an indicator in different ways, then the indicator-unit relation is inserted in the model as the rather "fuzzy" statement of "MIXED ABNORMALS." Towne's results suggest that such gross test information may be just what the real technician uses in a troubleshooting attempt. The design implication is that such information must be available, if assembly schemes of the Towne variety are to be used as predictors. Wohl (1982) also has a method for estimating "effective complexity" by counting the number of significant electrical *relations* that exist in a configuration. Rouse (1984) calculates an entropy-type complexity over a set of rather abstract components, and finds that human search time corresponds rather well to the computed entropy. Until better cognitive indexes of complexity are validated, the Towne, Rouse, and Wohl schemes may be worthwhile as indicators of perceived complexity, especially when systems are assembled from standard off-the-shelf components.

One might well wonder whatever happened to the "maintainability prediction" movement. In the 1960s and 1970s, firms like ARINC, RCA, and TRW came up with definite techniques for forecasting maintenance times. Back then, many analysts thought that these proposals would undergo a "shake out." Then, it was hoped, a standard method would emerge, which would generalize across hardware categories such as radar, ECM, and digital communications sets.

Little is heard now about any universally applicable method. What apparently has happened, though, is that the big electronics companies have adopted their own slightly different versions of a reliability-based model. That is, a prime equipment is broken down into subsystems or units at such a level that the failure rates are known for each unit; then the conventional stages of troubleshooting like localizing, isolation, and repair are given "standard times," either from in-company experience tables or from official sources like the Air Force RADC tabulation (Fuqua, undated). The weighted time–cost summations then proceed routinely, and the final estimated times are reported statistically. There seem to be no radically new ideas in this approach.

A *"positive"* indication of abnormality or failure occurs when an expected signal or quantity is absent, or when some parameter like a voltage, pressure, or resistance is observed to be widely variant from the prescribed value. In contrast, a "negative information" test is one that shows a given unit or data chain to be "good" (operating normally). Theoretically, of course, the establishment of some units as "good" conveys critical information in just the same way as an out-of-tolerance observation; and automatic fault-location programs take full advantage of data about the set of "known good" units, as the search for malfunctioning units is carried out. Human subjects, however, have a tendency to *seek out "positive"* indications (out-of-tolerance readings and outputs), and to *ignore the "negative"* or good unit information. Somehow the "positive" (bad) indications are psychologically more obtrusive and are most sought after (Mackie, 1974; Rouse, 1978).

Again the implication for the system designer is clear: since troubleshooters will be looking for positive "clearly bad" test indications, one should make it easy for them to find and use such indications. And if positive or "already proved good" information must be employed in a typical troubleshooting attempt, this facet of the search process should be aided, or otherwise presented so as to be naturally utilized by a positive-test-seeking technician. There is a training implication, too: "aiding the negatives" (identifying trouble candidates that have been logically eliminated) can assist positive transfer from an aided to an unaided situation, at least for advanced students (Rouse, 1979).

Troubleshooters often have *incomplete hypothesis sets.* Studies show that even experienced auto mechanics may have rather "bare" sets of possibilities that they are considering; and at the same time they can have great confidence in the completeness of their sets of candidates (Mehle, 1980). More generally, people seem not to be especially good at generating hypotheses, even in relatively familiar domains (Gettys, Manning, Mehler, and Fisher, 1980). To practical troubleshooters, a few obvious possibilities that often have been culprits in the past may loom large and thus may choke off any further hypothesis-generating effort.

A simple-minded solution would be to furnish a complete list of admissibles to a troubleshooter; he or she then could use the list as a device for eliminating candidates; entered in a computer, a program could help him or her to keep track of them. Among the possible difficulties are that such a list may not be available, that some troubles (e.g., bridge or "new wire" faults) may not be amenable to description in a list, or that the system may have reconfigured itself so that the "old" documentation no longer holds. However the problem is faced, the maintainability analyst may safely predict that working technicians will not automatically generate a complete hypothesis set. Even a mild "be system-atic" attempt can help; when subjects were required to "preplan" a simulated troubleshooting effort, by listing all possible causes and then outlining critical checks for each cause, the search proceeded more efficiently (Moore, Saltz, and Hoehn, 1955). In a computerized environment when much of the preplanning could be done for the technician, a key design question is how much of the planning should be *required* of the human, and how much should be *provided* to him or her via some gadget or aid. At present, the best practice probably is to provide much of the search planning.

Cognitive style is sometimes a correlate of troubleshooting performance, in those situations that are relatively "free." Style indexes such as field-dependence–independence and reflectivity–impulsivity are related to fault-finding achievement, and such scores may also have plausible interpretations in training. For instance, "impulsives" can be expected to make more errors and to benefit less from practice. "Reflectives," on the other hand, tend to be more efficient and to be better interpreters of test information (Morris and Rouse, 1984). Not too much should be expected from such variables; special practice in finding hidden figures and in spatial scanning did not produce positive transfer to a simulated troubleshooting situation (Levine, Bahkk, Eisner, and Fleishman, 1979; Levine, Schulman, Brahlek, and Fleishman, 1980). Also, there are measurement and conceptual problems associated with cognitive style, and there is good reason to believe that the effects of style can be "overwhelmed" by technological aiding or by any number of practical conditions. Perhaps the style idea may be of most practical use as one component in the *selection* of people to be troubleshooters in very demanding and not-fully-structured environments; "reflectives" with good skills in logical reasoning and numerical manipulation should be good prospects for technical training (Federico, 1982).

Much real troubleshooting is *opportunistic*. An actual technician uses whatever is at hand: his or her knowledge of the functioning of the equipment; his or her skill at turning it on and tweaking it up; his or her beliefs about failure likelihoods of different subunits; his or her schematic diagrams, published test sequences, manuals, aids, and so forth. It is this opportunistic feature that allows people without much theoretical knowledge to troubleshoot and repair very complex radars and sonars. A typical performance begins with "operator-type" connections and adjustments of the equipment; as noted by Kieras (1984), experienced people become quite smooth in this work of connecting test gear and adjusting equipment. When a front-panel adjustment effort fails and a real trouble state is confirmed, equipment manuals are usually consulted, and signal generation or signal tracing techniques are employed in some kind of bracketing strategy. Test convenience, rather than test optimality, is often a major determinant of what to do next, as is the technician's knowledge that some subsystems simply cannot be easily accessed or interpreted. Once *some* isolation or bracketing is achieved, then the technician may soon start to replace elements he or she perceives as failure-prone. As testing proceeds, rather inelegant "jerry-rigged" test setups often may be seen; one technician will shout out readings while a second compares these with a table of desired values. If an equipment manual is known to have a pretty good troubleshooting sequence, then that routine might be followed more or less according to the book, although the sequence will seldom be completed just as published. Some quite sticky normal–abnormal judgments are often required in waveform analysis; it is often difficult to get a waveform on a test so that it can be compared with an expected waveshape in the prime equipment manual. At any time in a real troubleshooting attempt, maintenance histories, replacement parts summaries, and statements from other people can contribute information too.

One of the most human and most positive features of troubleshooting behavior, then, is the technician's ability to use different resources and methods. Unlike some inflexible automatic test routines, the human tends to keep going regardless of glitches and "holds" in a test sequence. This flexibility often produces some surprisingly effective performances. There are informal indications, too, that people enjoy looking at a troubleshooting effort from several standpoints, and that "jumping around" from one aspect to another therefore is not really as random as it looks to the casual observer.

10.3.3.3 Caution and Hazards: Psychology of Troubleshooting

Performance criteria. Although "user" authorities always call for quick restoration and precise MTTR numbers, technicians themselves do not necessarily try to do a troubleshooting job in the quickest way. Perhaps that is just as well, because under severe time pressure subjects tend to show less optimal behavior (Rouse, 1978). And in many situations, the manager would gladly accept *nearly certain troubleshooting success within a reasonable time* as being a good criterion. Therefore, designers might do well to focus on the *logical effectiveness* of a maintenance procedure, and not allow time to drive the whole effort.

Impervious nature of troubleshooting behavior. It has proved to be very difficult to cause marked changes in fault-location performance parameters, when really complex systems are involved. A cliché is that it took an average of an hour to find a trouble in a major equipment in 1955; whereas in 1985, after 30 years of progress, it takes an hour to find a trouble in a major equipment. Many bright ideas, training programs, aids, and computerized gadgets have been proposed, and though they are often most useful, nothing has eliminated the troubleshooting problem. One key reason is that the recognition, preparation, and administrative time segments associated with a fault are often practically irreducible in the field, and perhaps half the time is spent waiting (Kane, 1981). Another reason is that a new aid or training scheme may not enjoy instant acceptance by the target population. In any event, skepticism is urged toward any claim that some new device or procedure will effect radical changes in human performance.

Common sense beliefs about technical expertise are often either *difficult to validate, or downright wrong*. For many decades, electronics schools have started their students out by teaching "basic electronics theory," on the grounds that the teaching and learning of it is necessary to effective on-the-job maintenance work. Actually, the data show that training in "fundamentals" or principles is of little use, and that most of the material is quickly forgotten (Foley, 1977; Morris & Rouse, 1984; Williams

& Whitmore, 1959). Kieras (1984) also reported a rather counterintuitive result from an experiment with two different equipment-operating instruction formats. One type of instruction was a plodding step-by-step procedure, with each little action listed at the lowest descriptive level. An alternative "menu" format gave the same step-by-step list at the lowest level, but also organized the steps into two or three higher levels or clusters of procedures. The common-sense expectation was that the "menu" presentation would lead to better performance, because it had more hierarchial structure, and it also permitted the user to select the level of discourse appropriate to his or her knowledge and preference. In fact, the menu format was *not* any better. Subjects appeared to have overestimated their knowledge, and they started doing things after only a quick perusal of the menu; later they would discover their mistakes, and perhaps then they would have to plod through the step-by-step sequence anyway. Thus, even the most "quick-and-dirty" empirical checks on behavioral assumptions may reveal surprises.

Is this all we know about troubleshooting behavior? While any rapid survey is incomplete, we probably have mentioned most of the main research results in the preceding paragraphs, and more exhaustive reviews are available (Morris and Rouse, 1984). Somehow the state of our knowledge seems unsatisfactory. Although much good advice can be found, nobody has "wrapped up" troubleshooting into a neat package that can serve human factors and other practitioners. The variables listed above, for instance, are only loosely related, and their effects can be weak, compensatory, and situation bound. It may be hard to know which factor is most important for a given project.

Part of the difficulty may be due to a wrong formulation of the basic human factors problem. Most researchers were psychologists, and they naturally emulated the general psychology of problem solving; under that model, a troubleshooter is viewed as having to put together bits of information into a "new" synthesis, perhaps similar to the solutions of the classical Maier, Duncker, and Wertheimer problems (Johnson, 1972). A problem-solving research is successful if it identifies one or more factors affecting solution. For experimental psychology, the problems used as vehicles often were specially contrived to require novelty, and they had little real world significance.

Practically, it is often much better to consider successful troubleshooting *not* as some "creative" new realization, but rather as *highly skilled utilization of specific knowledges and materials.* On this view, a successful troubleshooting research demonstrates marked improvement in achieving efficient search performance. This might be done by devising special procedures, aiding devices, equipment arrangements, practice, and so forth. Novel thinking and originality are not essential. The good trouble-shooter would have to be careful in monitoring and interpreting equipment indications, but the ordinary trouble search would not require new discovery. Good surgery, accounting, and dentistry do not require creativity in every application, either, but they do depend on "heads-up" and competent application of known techniques.

This formulation should not lure us into the old trap of "trained-ape" maintenance, where the human is considered to be merely a dumb sensor, meter reader, or wire connector, while some master computer program or check list contains all the essential intelligence. The goal should be the provision of a situation where the human has about the right amount of intellectual and perceptual challenge to keep him or her interested, but the task is still feasible for the ordinary trained person. Fortunately, this can be realized, as the mainframe computer manufacturers have shown. Failures in their big machines are usually repaired smoothly and rapidly, and without the necessity for creative problem solving in every trouble. And there is still plenty of room for the exercise of individuality, high competence, and job satisfaction. We shall address this point again at the end of the chapter.

10.3.4 DESIGN FOR MAINTAINABILITY

10.3.4.1 Design Specifications

Many of the features that enhance maintainability are quite obvious, as anybody knows who has tried to reach an inaccessible bolt in an automobile engine compartment. Human factors people started collecting and categorizing design ideas some years ago, and most of the early recommendations still hold good (Folley and Altman, 1956; Munger and Willis, 1959; Rigney, Cramer, and Towne, 1965). If a maintainability contract is large, and especially if it includes several subcontractors who have disparate practices for labeling and documenting, then a special guide often will be standardized for that program alone.

Qualitative design specifications are often stated at two or three levels of generality. Cunningham and Cox (1972) give a good example of a fairly high level of discourse:

> *Plugs and connectors: All plugs and connectors shall be of the quick-disconnect type. All plugs and connectors shall be coded by varying size, shape, pin arrangement, or any other method to prevent accidental incorrect connection.*

Of course, a far more detailed treatment of connectors can be called out. Among the possible factors related to connectors are such things as the requirement for removal with ordinary hand tools only, U-lugs instead of O-lugs, visibility and adjacency, clearance for hands and tools, keyways to facilitate pin lineup, matching stripe or arrow coding, switch interlocking, quick release clamps, and so forth. A good way to keep track of compliance is to follow the Philco procedure of setting up a single checklist sheet for each design topic, with a rating system for each design element. One version has four spaces on each line for scores on:

1. *Applicability* or significance of the item to the present design (0, 1, 2, or 3).
2. *Adequacy* of the item (0, 1, 2, or 3).
3. *Product* of the numbers in the first two columns; this furnishes a rough index of overall importance.
4. *Correctability* of the inadequacy, or its likelihood of being corrected (0, 1, 2, or 3); this number can be multiplied by the "product" number above, if desired.

CONNECTORS

_/ _/ _/ _/

1. Connectors are provided wherever equipment separation is likely. _ _ _ _
2. Quick-disconnect or plug-in connectors are used where feasible.   _ _ _ _
3. Connectors are visible, reachable, operable without disassembly. _ _ _ _

4. Connectors are operable by hand, replaceable with common tools. _ _ _ _
5. Adequate workspace & tool clearance surrounds each connector.   _ _ _ _
6. Connecting auxiliary & maintenance equipment requires no tools. _ _ _ _

7. Each connector can be removed without disturbing others.   _ _ _ _
8. Rear of plug is accessible for testing where practicable.   _ _ _ _
9. Adaptors are provided if needed for test/auxiliary equipment. _ _ _ _

10. Connectors are designed, placed, coded to prevent misconnection. _ _ _ _
11. Delicate parts are protected & overtightening is prevented.   _ _ _ _
12. LRUs are never soldered in & plugs are not safety-wired.   _ _ _ _

13. Connector mounting points are supported against breakage. _ _ _ _
14. An open connector is obvious, but design prevents shorting. _ _ _ _
15. All connectors are replaceable; leads allow six replacements. _ _ _ _

16. All receptacles, terminal boards, etc. are readily replaceable.   _ _ _ _
17. Extra connectors, pins, receptacles are provided as appropriate. _ _ _ _
18. Dust caps are supplied & protect connectors when not in use.   _ _ _ _

19. Plugs & receptacles are clearly identified by color, tags, etc. _ _ _ _
20. Connector labels/codes correlate with function, jack, diagrams. _ _ _ _
21. Stripes, arrows, etc. indicate position for proper insertion.   _ _ _ _

22. Plugs/receptacles are provided with aligning pins or devices.   _ _ _ _
23. Aligning pins in uniform position, extend beyond electric pins. _ _ _ _
24. Pins are clearly coded and are arranged in standard fashion.   _ _ _ _
25. Symmetrical pin arrangements are keyed to prevent misconnection. _ _ _

(0 = None, 1 = Slight, 2 = Moderate, 3 = Extreme) Subtotals: _ _ _ _

Name Equipment Evaluated   Name of Evaluator & Date   TOTAL

CONNECTORS

Fig. 10.3.4.   Checklist of design items for connectors.

Figure 10.3.4 shows a sample sheet for the connector category. For any real item the Philco system would have at least two dozen other sheets. One list of topic sheets is shown in the left column of Table 10.3.1. The Human Engineering Guide (Crawford and Altman, 1972) also shows a design feature breakdown; it has 12 topics in the list. Usually there is no controversy over the positive value of meeting the specifications from standard design guides; the difficulties arise when a specific "compliance" is expensive, when it is in conflict with some other desirable factor, and when even partial compliance will mean that a compromise must be negotiated. Sometimes the field conditions can impose nearly insuperable maintainability problems, as when structural constraints cause different units of a large Navy system to be spread "all over the ship." Even if inescapable, such situations may still benefit from a very thorough design-checklist review. For example, the fact that prime-equipment cabinets are (necessarily) separated might lead to the provision of special test-point arrangements, extenders,

Table 10.3.1   Checklist of Maintainability Design Factors

Factors	N/A	SAT	COND	UNSAT	Remarks–identify item number
Accessibility (relative to maintenance items)					
1. Openings, panels, and doors	( )	( )	( )	( )	
2. Covers, plates, and caps	( )	( )	( )	( )	
3. Drawers, frames, and slides	( )	( )	( )	( )	
4. Internal work space/volume	( )	( )	( )	( )	
5. Internal lighting, paint	( )	( )	( )	( )	
6. Location, Arrangement	( )	( )	( )	( )	
7. Other	( )	( )	( )	( )	
Maintenance Items					
8. Test and service points	( )	( )	( )	( )	
9. Cases, shields, and guides	( )	( )	( )	( )	
10. Interlocks, overrides, and stops	( )	( )	( )	( )	
11. Lines, cables, and connectors	( )	( )	( )	( )	
12. Disconnects, latches, catches	( )	( )	( )	( )	
13. Fasteners, pins, safety wiring	( )	( )	( )	( )	
14. Mounting and packaging	( )	( )	( )	( )	
15. Controls and displays	( )	( )	( )	( )	
16. Coding, labeling, and pathways	( )	( )	( )	( )	
17. Parts, assemblies, modules	( )	( )	( )	( )	
18. Fuses, circuit breakers	( )	( )	( )	( )	
19. Other	( )	( )	( )	( )	
Maintenance Steps and Processes					
20. Detection	( )	( )	( )	( )	
21. Localization	( )	( )	( )	( )	
22. Isolation	( )	( )	( )	( )	
23. Correction and repair	( )	( )	( )	( )	
24. Testing and verification	( )	( )	( )	( )	
25. Removal and replacement	( )	( )	( )	( )	
26. Inspection, servicing, cleaning	( )	( )	( )	( )	
27. Adjustment, calibration	( )	( )	( )	( )	
28. Other	( )	( )	( )	( )	
Support Provisions					
29. Handling (powered) (manual)	( )	( )	( )	( )	
30. Mobility and transportability	( )	( )	( )	( )	
31. Stands, ladders, and rests	( )	( )	( )	( )	
32. Test and service equipment	( )	( )	( )	( )	
33. Tools and accessories	( )	( )	( )	( )	
34. Spares, parts, and material	( )	( )	( )	( )	
35. Other	( )	( )	( )	( )	

Table 10.3.2   Correlation of Design Factors with Maintainability[a]

Factors		All equipment $(N = 45)$	Electronic $(N = 25)$	Nonelectronic $(N = 20)$
I	Maintenance safety	−0.185	−0.249	−0.408
II	Maintenance information	−*0.357	−0.298	−0.401
IV	Fasteners and tools	−*0.310	−0.353	−0.324
V	Alignment and keying	−*0.356	−0.134	−**0.536
VI	Manual control layout	−**0.464	−*0.455	−*0.481
VII	Workspace configuration	−0.190	−0.080	−0.289
VIII	Accessibility	−0.190	−0.186	−0.061
	R	*0.523	*0.609	*0.682
	R^2	0.274	0.371	0.465

[a] * = 0.05 and ** = 0.01, significance values based on Wallace and Snedecor tables (1931). R = multiple correlation; R^2 = percent of total criterion variance accounted for by R.

cables, displays, and test routines that can compensate for the fact that the subunits are not installed in adjacent spaces.

The "salience" of the different design elements for maintainability can be estimated by relating the score for each feature to some overall index, such as general ease of maintenance, or mean time to repair. A comprehensive study from the late vacuum-tube era used a 112-item design checklist, which was arranged into eight categories. This checklist was applied to 25 electronic and 20 nonelectronic equipments, and then correlated with reported maintenance times on those 45 systems, with the results shown in Table 10.3.2 (Topmiller, 1964). Although the coefficients jump around across categories and equipments, many regression weight numbers are rather high, certainly enough so to encourage the idea of following the design guidelines. The argument becomes even stronger when one considers that a large proportion of human error is attributable to poor human engineering (Shapero, Cooper, Rappaport, Schaeffer, and Bates, 1960).

10.3.4.2   Aiding the Troubleshooter

A technician's "aid" can be anything from a plastic card with a list of checks on it, to a large configuration of computers, programs, and automatic test equipments. Although aiding is obviously dependent on many equipment and situational factors, our emphasis here is primarily on the "cognitive" aspects of aiding.

Partitioning of the Prime Equipment

One of the best things an equipment designer can do for the technician is to separate *logically* the different units of the equipment. If this is done, a relatively simple check sequence can decide whether an output is "good" or not. It is desirable that the inputs and outputs in a network be "clean," that is, it should be feasible for an ordinary competent human to make tests at different points and to determine whether the quantities obtained at the ends of units or subunits are in or out of tolerance. To suggest the power of this clean partitioning, imagine an extremely large system, composed of some hundreds of related elements, but with the discrete "good" or "bad" test readings available at an input or output connection on each element. It still would take only a few tests to isolate a trouble; even nonoptimal sequences probably would not need more than a few dozen checks, and performance planning for the next check would be quite simple to make. Rasmussen (1980) reviews much of the work that has been done on the "decomposition" of complex functions for human use and understanding.

A designer's reaction to this demand for "clean" separation among units and stages may be that the relationships existing are intrinsically complex, and cannot be expressed simply as a series of "in" or "out" readings. The answer is then to design the circuit so that aspects of the circuit action can be *specially transformed* into the simpler GO–NO–GO check, *even if this is done just for test purposes.* The same kind of analysis is demanded by other analytic tools, such as fault-tree analysis. If the response to that suggestion is that it is now too late in the design cycle for such special signal transforms and "cleaning up" of unit interactions to be accomplished, then the maintainability analyst must make do with special test equipment or procedural approaches which still might work.

In the field of fault-tree analysis, an interesting example of such simplification is the GETREE system for depicting AND/OR logic trees. When a very complicated and confusing "spaghetti" of nodes is being explored, the display can be commanded to show only a particular subtree and its "children." A commitment to "clean partitioning," though, is a very effective aid all by itself. And as all digital computer people know, very rapid isolation of a defective card in a large series of similar cards often is quite easily achieved. One might even say that a clean partitioning makes strategy

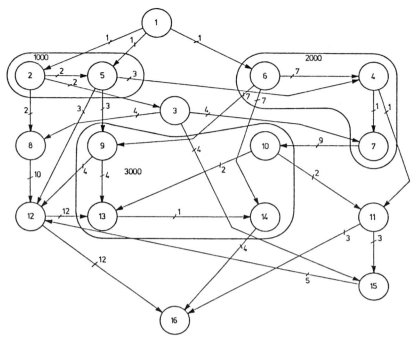

Fig. 10.3.5. Small digital system, with minimum set of test points.

rather irrelevant; while half-split search is usually optimal, optimality does not matter much if either back-to-front, bracketing, or front-to-back search patterns find the trouble very quickly anyway.

There are two techniques that are not well known in the human factors community, but which are plainly related to the partitioning problem. One of them is the application of graph theory to failure diagnosis. Under the reasonable assumption of a single, nonintermittent fault, a graph-theory model can provide accurate diagnosis, with a minimum set of test points. The essential steps in the procedure are:

1. Model the system by means of a single-entry single exit (SEC) graph.
2. Translate the graph into a connectivity matrix C.
3. Compute a "reachability" matrix R.
4. Find a covering matrix Z.
5. Solve Z as a minimal covering problem.

Figure 10.3.5, taken from Arsenault, Des Marais, and Williams (1980), shows an exemplary configuration. This particular system requires exactly 32 test points, and the possible faulty vertexes can be identified in a 16 × 12 fault signature table. The entire diagnostic layout can be computed from the original SEC graph. We can expect more use of such abstract models. Projects will almost certainly benefit from attempting to apply them.

Another technique is "simulation for diagnosis." Here the goal is to assemble an effective representation that, when run, will locate a great majority of the faults. "Testability" is the key idea, and to have a testable system you must arrange things so that you can *initialize* or set each element for its specific response, so that you can *observe* the response of each element, and so that you can *control* by sending a specified test signal to each element. Methods for modifying a circuit to achieve these three ends are well known; for instance, buffering of complex sequential circuits, provision of separate "full clock control" for tests, and the systematic breaking of feedback paths and long counter chains (Des Marais, 1980).

Computerized Aids

A computer is not necessary to a good aid. In fact, a well-designed human-oriented booklet can be extraordinarily useful. For some years, though, a popular idea has been to house, in a portable package about the size of a briefcase, a "complete" troubleshooting-aiding system. Such a system would have a microcomputer and a large test data base for one or more systems, along with associated interfaces for prescribing optimal test sequences, assessing the meaning of readings and checks taken, keeping a record of the whole troubleshooting performance, and printing a hard copy report of the fault-locating attempt. The portable aid might also be able to communicate with a distant master computer or consultant group, and it could convey the latest field changes, manual revisions, and failure probability information (Smillie, 1984). Packages that achieve some or all of these goals have been built and demonstrated, although it is difficult to find a convincing documented proof of their field effectiveness. Some of the same conditions that plague all complex automatic testers prevail, of course: incomplete sets of admissible troubles, long linear test sequences, an unfriendly human interface (the user often cannot "jump in" to the same place in the test sequence, but must wait as it runs through a prescribed routine), results are ambiguous because the test directed by the aid are themselves difficult to make or are wrongly documented, and so forth.

However, since portable computerized aids could conceivably be so convenient, there remains a strong motivation to produce them and to make them more general and more powerful. Much of the data-base information in such devices could be obtained from properly designed manuals, so that the unique capability of a computerized aid to *interact* with the technician should be emphasized. Among the many obvious ways to enhance such interaction is the provision of several levels of generality as the aiding device analyses a performance. A technician may need to "stand back" from the details of an attempt and look at it afresh, from a "big-picture" standpoint. There might also be benefit from responding to "individual style" in aid preferences; maybe some technicians do not necessarily want to have the aid suggest the "next best" check, but would prefer it to propose several checks, either one of which advances the search somewhat, if not optimally. The computer-based maintenance aids system (CMAS) concept would allow the user to specify his or her own kind of aid, organized around a basic "kernel" of system information (Thomas, 1982). But we pointed out before that presenting a "menu" of information may not necessarily lead to superior troubleshooting.

In the largest computers, testing may be semiautomatic. A technician compares obtained nodal states according to a computerized directive; the human makes the judgment as to whether the signal corresponds to the expected value. A variant is to have the automatic tester narrow the trouble to a few components, then for the humans to do the final isolation (Des Marais, 1980).

Artificial Intelligence and Troubleshooting

There is a gray area between the most complex automatic test equipment (ATE) and artificial intelligence (AI), with AI generally being credited with some kind of abstraction capabilities not found in a fixed automatic test routine. In the AI domain, terms like "expert system," "theorem prover," and "natural language processing" sound so good that it is easy to think that the applications to practical search problems are fairly direct, although in fact the realizations may be quite difficult. For example, conceivably one could ask a theorem prover to prove that all components in an equipment were faulty; maybe only one or a few would survive the proof test, and these would then be designated as the culprits (Coppola, 1984). Or one could say that expert troubleshooters exist, and a significant part of their knowledge might be coded into rules, which could then be stored in an "expert system." In application, the system would ask the user for data, and could effect dialog and control via an interaction with the human user. Thus the expert knowledge is preserved, and can be transferred indefinitely to others besides the original expert. Among the other AI possibilities are robotics, machine "perceivers" and recognizers, combinatorial mathematics (for preventive maintenance and fault-tolerant design), automatic programming (converting a statement in a high-level language to a specific set of codes), and language processing. For general reviews, Nilsson (1980) is a standard general textbook on AI, and Coppola (1984) summarizes the findings of a Department of Defense panel that was charged to discover what AI might accomplish in maintenance applications. The review was fairly optimistic, but had the usual "no panacea" disclaimer.

Expert systems may be the best near-term AI bet for maintenance. Whatever the exact representation chosen, there is always a data or knowledge base and some inference system to process a prevailing state of knowledge. A human expert may offer much knowledge from memory; these are usually declarative statements. Starting from a fairly fixed initial set of queries, some atomic facts are obtained by automatic or human measurements, then other facts are deduced by a logical inference processor. There are many rules, usually of an if–then type: if a state X exists, then do action Y. As information is entered, the logic processor searches for rules whose "if" part has been confirmed. When the "then" state is fired, the "action" may be a request to the user for more information, or it may be a prescription or interpretation. Until the system is "satisfied" that it has reached a goal (such as one or a few components left in the admissible set of faults), the interaction continues.

An expert system of medium complexity is the General Electric DELTA configuration for troubleshooting diesel electric locomotives. One experimental version now running has over 500 rules, which were elicited from an experienced engineer. The rules are organized hierarchically so that one kind of rule tries to find out whether some new knowledge is required, while another tries to figure out what can be inferred, given a just-received "new fact." It is planned to increase the rule set up to about

1200. One encouraging feature is that a relatively simple language is employed, with only nine types of predicates and eight verbs (Bonissone and Johnson, 1984). Such work is complex; it took GE about 12 months to produce a 50-rule feasibility test model, and another year for the 500-rule DELTA to reach the field-test stage. Davison (1984) estimates that another 2 years, and a megabyte or so of memory, will be necessary before the 1200-rule "ultimate" version runs. These facts illustrate how expert systems have a way of requiring large resources in time, money, and computing power.

The Lockheed Expert System (LES) is now being evaluated on a large signal-switching network containing 3000 circuit boards and 1000 cables. LES encapsulates much of the essence of the network by identifying the places where a given signal flow path exists; when a relatively simple test for signal presence does not work because there are not enough test points, a human expert provided rules that the system can use for bracketing (Laffey, Perkins, and Firschein, 1984). Neither DELTA nor LES seems to have well-developed plans for interaction with humans.

Perhaps the most exciting AI area to the maintainability analyst is the new emphasis on "deep structure" models of physical systems. Instead of the hundreds of empirical rules found in MYCIN or DELTA, it may be that relatively few but "deeper" relations can be programmed. Since electronic devices under consideration are physical systems and are well understood both physically and functionally, it is sometimes possible to encode knowledge about a system in very compact ways. DeKleer (1984) claims that a fairly complex regulated power supply with more than 30 components could be well represented by about 25 "causal" rules, whereas it might take 100 or more empirical "if–then" rules to do as much reasoning as this reduced set of 25. Sixteen of the "deep" rules are presented in DeKleer (1984); they might be rather unintelligible to a naive person. Davis (1984) designed a hierarchy of assumption-rule sets; by "layering the models," the program works along at a certain "layer" of inference until some contradiction is encountered. Such a contradiction occurs, for example, when one branch of the model says that the fault must be either $R6$ or $C17$, whereas another branch of the model asserts that *neither $R6$ nor $C17$* can be faulted. It must be that one of the assumptions in the layer is erroneous; therefore, a slightly less restrictive model is tried next. For some intrinsically difficult faults, like "bridges" in a digital computer module (unwanted "wires" or connectors across components, perhaps caused by solder spattering), the Davis approach is effective in isolating the trouble, whereas the usual symptom–malfunction matrix or automatic testing approach fails totally in bridging troubles. An emergent by-product of the Davis "deep structure" approach was more insight into the "locality" aspects of a physical representation. Bridging faults are local in a physical sense, but may fan out through a device in the functional sense (Davis, 1984). Since many of the ultimate circuit interactions are physically local, the search interpretation may be greatly facilitated by an awareness of physically significant locality *possibilities* in a circuit.

An immediate challenge is to determine how the "deep structure" of circuits can be appreciated and utilized by humans. A look at DeKleer's rules shows that they are rather abstract, and perhaps also rather impenetrable in their meaning at first reading. Can ordinary technicians learn them, or perhaps even originate them? Will such rules prove to be general across circuits and equipment configurations? An optimist might predict a convergence of automatic test programs, cognitive psychology, and deep structure models of physical devices, in the aiding devices of the future. And among the several ironies, it could turn out that the old "theory first" concept is basically correct, but that "theory" now must be structured and learned in very special ways indeed, in order to facilitate use of "deep knowledge" search algorithms.

10.3.4.3 Software Maintainability

The Apollo lunar program software cost $600 million; it was developed by some of the best programmers in the world, with two teams checking each others work. Yet most of the faults in the Apollo were blamed on software errors. Again the big problem seems to be management of complexity; and though the analogies cannot be pushed too far, it seems that the same logical approaches that work to make a piece of *hardware* reliable and maintainable also apply to software. For instance, development of reliable hardware requires disciplined monitoring and continuous refinement of failure rates for individual parts. Assembly of that hardware into a maintainable unit requires attention to partitioning, modularization, and the clear GO–NO-GO checking of performance at informative test points.

The software domain also has its techniques for dealing with complexity and testability. One powerful idea is "top-down" design, which emphasizes the decomposition of a system into a set of small modules that will support an upstream logical requirement. The top-down approach can also be expected to increase the independence among modules. Structured programming, which is based on a half dozen or so basic software structures, is often proposed as an aid to control programming effort, and, theoretically, it should improve the making of maintainability changes because of the small program modules and standard structures employed (Wirth, 1973). Sometimes the nature of a software reliability problem can be drastically changed by structured programming. Ronback (1975) gives an illustration, which is shown in Figure 10.3.6. The program has eight decision blocks and two nested loops. Yet a brute-force search would take over 500 years to test every path. The realistic way to verify the module is to consider it as made up of 13 blocks, with each block having a single input and single output. If it

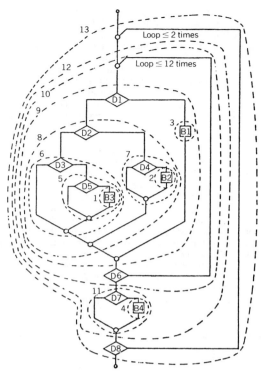

Fig. 10.3.6. Software module. There are 1.6×10^{19} possible paths. Testing one path every 10^{-9} sec would require 500 years.

is known that for each block, the correct output is produced for all possible inputs, the program can be considered checked. Networking programs are available to assist in organizing these block checks.

There are formulas for predicting the number of bugs in a program, just as there are formulas for assembling parts-failure likelihoods into an overall availability estimate. To give just one example, "vocabulary" can be defined as the sum of the number of distinct operators (n_1) and operands (n_2) appearing in a program; "length" N is the total number of *usages* (N_1 and N_2, respectively) of the terms in the vocabulary. Then, the error content (E) can be formulated:

$$E = \frac{N \, \log_2(n_1 + n_2)}{2n_2/n_1N_2}$$

B or estimated number of bugs is predicted as follows:

$$B = \frac{E^{2/3}}{3000}$$

There seems to be little validation of such formulas. Sukert (1977) gives a critical review of software reliability modeling.

"Sneak circuit" analysis is a set of techniques for discovering *unwanted* but *designed in* circuits in electronics equipment. "Sneaks" also plague software, and there are systematic methods for discovering software bugs. Surprisingly, there seem to be only a handful of topologically basic sneak patterns: five in electrical circuits and four in software. It has also been established that there are relatively few "clues" that identify sneak paths and circuits. Topological network searches for the "clues" are carried out by computer, using pattern-recognition techniques. The sneak technology is still developing, but it holds promise for maintenance people who must produce and debug their own software. Among the types of software faults identified are improper branching, unnecessary instructions, and the like. In the future, technicians will have to be fluent in the use of such software analysis aids.

10.3.5   IMPLEMENTATION

Suppose that the analysts and engineers have done their work according to a good project plan, and that the technical aspects of a maintainability project are well in hand. There is still the task of *implementing* the plan and the system, of making it all happen in the desired environment. This can be one of the most difficult stages of the work; doing the *development* and doing the *introduction* are quite different things. Nobody knows how many well-conceived training plans, aiding devices, preventive schedules, and utilization schemes have never been realized after the maintainability "acceptance" demonstration; but there certainly have been many "technical successes" and "practical failures." In this section, we touch on some of the factors involved in realizing the full potential of a maintainability effort.

10.3.5.1   General Factors Affecting Implementation

Know the scene and the setting where your system is going! Banal advice, surely. Yet it has within it several issues that must be faced, if a successful application is to result. When a new program is to be installed somewhere, there is always a *user constituency;* this constituency is made up primarily of technicians, but will also include those people at lower and middle field supervisory levels, along with staff from ancillary supply and support units. Nothing much will happen without the cooperation and enthusiasm of these people; hence, they deserve study just as the hardware configuration does. There are other constituencies, of course, perhaps in Washington, in the contractor's plant, and so forth. One primary consideration is the extent to which the "working" part of the constituency perceives a need for the maintainability policies, devices, and programs. An example comes from experience with a troubleshooting aid for a military radio transceiver set. This equipment had been around for a few years, and there were real down-time problems associated with it. Laboratory studies had confirmed beyond any reasonable doubt that a systematic search based on a symptom–malfunction matrix could quickly bracket most troubles in the prime equipment. All told, some man-months went into the development and validation of a fault locator based on the research. But it was seldom or never used in the field. Follow-up interviews strongly suggested that while Washington and other command authorities had felt a need for such an aid, the working technician did not perceive this same need. So it went on the shelf, along with all the other expensive and seldom-used compilations of information. Because field people ignored the presumably good procedures that research had supplied, the case was an implementation failure. It was not unusual. One field review attributes 40% of maintenance errors to this simple failure-to-follow the outlined procedures (Young, 1980).

Any new program or system must eventually operate within an *existing social structure*. This structure may have its own vested policies and ways of doing things. And it can stall, change, or stultify what it perceives as an intrusion into its domain. Human factors people often see this in the training area. For decades "basic electricity/electronics" courses have been taught in about the same way; even so mild an innovation as computerized objective testing was actively resisted for some time in one large training command. And here again, to the researchers and planners who worked to develop a working system, the resistance seemed almost incredible: the new testing scheme should free instructors from the onerous chore of making and scoring tests, it would automate student individuation and tracking, and it would do much of the course bookkeeping off-line, automatically, and efficiently. However, some senior instructors saw the idea not as an opportunity, but rather as a reduction in their autonomy as expert teachers and a partial takeover by "a gang of programmers." Even a small probability sample of interviews might have revealed such feelings, prevented much of the delays and passive resistance, and led to a more insightful implementation plan.

When the information about a new concept is announced and distributed, there may be *skepticism because of previous experience with a similar approach.* Experienced technicians on Navy ships or Army bases have heard glowing promises from manufacturers before, they have been briefed about the wonders of a new test procedure or computer program, and they know that other items in this category have seldom lived up to the claims. Any particular new project, then, may be seen as "just another one" of those previous (and unsatisfactory) schemes. This kind of skepticism is almost always encountered when complex new test equipment is accompanied to the field with a team of engineers who give a quick briefing and then depart.

A related negative factor can be the presence of *a small number of obvious bugs* in the new devices or procedures; these obvious mistakes are then used as an excuse for rejecting or diluting a whole program. In a satellite antenna installation, although the information items, manuals, and procedures were quite well conceived and accurate, there were a few major errors that had slipped through the review process. These were quickly noted by the field crews. An evaluation team visiting the site found that much of the essential maintenance procedural instruction was being ignored. When asked for an interpretation, some senior crew members pointed to the offending errors, and said "this program is all wrong; just look at these entries here." As far as anyone could tell, only a very few wrong values were in the materials, and they could easily have been corrected. Yet the whole package was rejected.

The matter of *incentives* for specific actions may often be worth separate attention. An outstanding example in military electronics maintenance is the filling out of "maintenance incident" forms. This task is universally disliked by the field people, and their position is understandable enough; some of the forms are indeed hard to complete, they may ask for much information or for data that are difficult or impossible to get; and anyway, why should a busy field technician do a careful job in completing the forms? If asked whether he or she wants to improve maintenance records on a system, any technician would say yes. But he or she knows that any possible gains to the system from his or her form-completion activities will be in the distant future, long after he or she has left the scene. There can even be disincentives to fill out failure forms; failure might make the unit look bad, or reflect on the supply system if certain repairs cannot be made because of inadequate spares, and so forth. Although evidence on the point is not strong, it may be more cost effective to remove technician disincentives than to attempt to add new incentives (Young, 1980).

Because many technicians have to face the turbulence of the field, which nobody else really faces, they may come to regard themselves as the *only experts*. Thus, the people who formulate aiding or on-the-job training programs are seen merely as distant experts. And since these distant experts are thousands of miles away, they cannot appreciate the practical difficulties such as making delicate electronic tests on a heaving ship or working under the extreme distractions at flightline. This perception of oneself as an expert, then, may furnish a pseudorationale for local changes in key elements of a maintenance plan. It seems to be quite rare for local technicians actually to improve on a well-designed package of procedures and manuals, but they may believe that they can do so.

One of the best techniques for promoting identification with a new program is *"bottom-up" participation and break-in*. Mainframe computer manufacturers often exploit this idea by sending their maintenance technicians back to the factory when a new fault-locator program is being developed. The home office of the company may assert (truthfully) to the technician that he or she is being sent there so that the company can receive the benefit of his or her input and field experience. Another and equally powerful benefit, though, is the technician's strong identification with and enthusiasm for the new program. When it finally reaches the technician in the field, it won't be something dropped on him or her from on high; the technician will already know a lot about it, and will trust it more because of his or her association with the development back at the factory. There is even reason to expect this trust to extend generally toward the host institution (the company in this example).

Any or all of the factors mentioned in the preceding paragraphs could be important in a particular effort. It often seems that, the more unusual or "disruptive" a maintainability program appears to be, the most apt it is to cause resistance in the field application.

10.3.5.2 Technician Effectiveness and Morale

As noted throughout this Handbook, an extremely large number of psychological, sociological, and organizational variables serves to constrain or to facilitate human performance. In the maintainability domain, the working organizational unit is often a maintenance crew or at most a maintenance squadron; such groups are usually small enough to have fairly delimited environment and work roles, and fairly local goal-setting procedures. Here we mention some of the "nontechnical" variables that certainly are worth the attention of the maintainability analyst, although they do not appear in contract requirements.

Maintenance personnel can be analyzed into two large categories: (1) technician experience, skill, and knowledge factors; and (2) technician productivity and morale factors. Table 10.3.3 gives one breakdown used by an Air Force team. Every variable listed has reportedly been shown to have an impact on either technical performance or job satisfaction.

With over 30 variables like these listed and measured, an effectiveness model of a maintenance effort could be formulated simply as a multivariate prediction system, with each factor being weighted for its contribution to some overall criterion measure. In fact, a few empirical trials have been conducted within this framework. Meister's team showed that such elements as the diagnostic strategy used and the number of times assistance was required could be scored. *Technical information* was the best single predictor (Meister, Finley, and Thompson, 1971). Askren et al. (1976) had a sample of 120 technicians who performed maintenance on a missile-handling task; among the findings was that "work motivation," as measured by survey questionnaires, contributed about 25% of the variance on a "performance time" criterion. Another SRAM missile-mechanic study had several dozen opinion survey variables in the data set. Again, some technician motivation variables came out as being rather good predictors of a "performance speed" criterion. (Some of the launcher system checkout tasks took several hours to perform; the "land launcher to aircraft" segment of the work required about an hour.) Regression weights and correlation coefficients against a "performance quality" criterion were even higher; "job curiosity," "persistence trait," and "self-starter trait" were among the best predictors (Sauer, Campbell, and Potter, 1977).

Young's (1980) questionnaire covered 137 items and some 18 factors. This instrument was given

Table 10.3.3 A Taxonomy of Human Resources in Maintenance

Technician Experience, Skill, and Knowledge Factors	Technician Productivity and Morale Factors
Experience	Organizational Climate and Group Morale
AFSC level (3, 5, 7, etc.)	Competency of supervision
Rank	Supervisory conditioning of tasks
Months in career field	Structure/warmth/standards/identity/risk
Skills (supervisor/observer ratings)	Group satisfaction of individual motives
For components worked on	Satisfaction with interpersonal relationships
For test equipment used	Team cohesiveness
For use of test equipment in general	Personal Traits and Motivators
For equipment repair	Job curiosity trait
Knowledge (supervisor/observer ratings)	Persistence trait
Of maintenance procedures	Emotional stability trait
Of equipment-handling procedures	Fatigue trait
Of use of equipment when operational	Responsibility trait
Of equipment maintenance procedures, in general	Self-starter trait
	Ascendency trait
	Organizational identification
	Professional identification
	Operational and Environmental Conditioners
	Pay and benefits, as perceived
	Assignment locality and climate
	Airmen/civilian relationships
	Participation in interest/service clubs
	Social status of occupation, as perceived
	Lighting/noise/workplace size/clothing

to 180 technicians at two Air Force bases. Immediate supervisors filled out a technical information and performance-ranking form for each technician who answered the questionnaire. Thus, each of the many opinion items was studied for its correlation with the performance rankings. In an official Air Force report, all the questionnaire items, score distributions, and regression weights are published, so that the report is a good source of material for those contemplating opinion measurement in maintenance crews (Young, 1980). One interesting result is that performance *quality* ratings tended to have more significant correlations with the questionnaire measures than did performance *time*. A reasonable hypothesis is that performance time is most related to the technical information and aiding provided, whereas work quality encapsulated the softer human resource variables to a greater extent.

Based on informal weights obtained from maintenance executives (Air Force officers), Young (1980) gave "speed of work" a weight of 0.4, while "quality of work" received an 0.6 weight." Predictions from the survey were accurate enough for Young to recommend that an opinion survey might actually substitute for supervisor ratings. Since there are, as expected, plenty of similarities and differences across squadrons, statistical validation on several samples and generations of crews or squadrons will be required before the particular variables that are most useful will be known. The important point is that questionnaire indexes, though seldom included in maintainability documents and procedures, can reflect important segments of field performance.

One can go farther with the criterion problem than the usual MTTR time indexes. Criteria of maintenance effectiveness can be cast into a utility-function framework. A well-documented example has been reported by White (1980), who obtained specific effectiveness attributes from Air Force maintenance squadron commanders. A dozen or more factors were elicited from each judge, followed by the usual steps of identifying a utility function for each factor, checking for independence, additivity, and so forth. Measures of each factor were also defined, along with current status numbers. Table 10.3.4 shows one attribute array. Given such material, a utility evaluation model for a judge can be described algebraically. In principle, the function could be used for evaluating a unit, and for tracking changes over time or as a result of some intervention. Among the practical questions are how well do competent judges agree on the key attributes, and how similar are their final utility functions and trees; and of course an ultimate question concerns the usefulness of the utility approach in a field setting.

Table 10.3.4 Attributes and Utility Functions in Squadron Maintenance

Name	Attribute Number	Units	Range	K_i	Current Status	Utility Function
Availability	X_{111}	AC here/assigned	$92\text{-}100_{852}$ $80\text{-}100_{135}$	0.06	93	$U(X_{111}) = 0.0625[X_{852} - 92] + 0.025[X_{135} - 80]$
Sortie proficiency	X_{112}	% flow/scheduled	$92\text{-}100$	0.1	95	$U(X_{112}) = 1, X_{112} = 100$ $U(X_{112}) = 0, X_{112} < 100$
Parts timely	X_{121}	% deferred	$0\text{-}10$	0.08	5	$U(X_{121}) = 0.1[10 - X_{121}]$
Parts quality	X_{122}	% failed	$0\text{-}10$	0.08	6	$U(X_{122}) = -0.1 X_{122} + 1$
Training	X_{21}	% unqualified MSEP score	$0\text{-}20$	0.15	15	$U(X_{21}) = 0.05[20 - X_{21}]$
Morale	X_{221}	—	$0\text{-}10$	0.015	8	$U(X_{221}) = 0.1 X_{221}$
Working conditions	X_{222}	% of time	$0\text{-}100$	0.015	57	$U(X_{222}) = 0.01 X_{222}$
Manning	X_{23}	% hr on job/hr avail.	$50\text{-}70$	0.2	60	$U(X_{23}) = 0.05[X_{23} - 50]$
Safety	X_{24}	Man-days lost	$0\text{-}3$	0.075	0	$U(X_{24}) = 1 - 0.3333 X_{24}$
Funded/Requested	X_{31}	% received/requested	$85\text{-}100$	0.04	85	$U(X_{31}) = -(1/225)(100 - X_{31})^2 + 1$
Cost/aircraft hour	X_{32}	Cost per	$\$200\text{-}400$	0.02	227	$U(X_{32}) = -(1/40{,}000)(X_{32} - 200)^2 + 1$

As White's data showed, there was enough agreement on key attributes and weights to suggest considerable transferability of utility functions across people. The technique seems not to have been fully tried yet. It seems logical that a maintenance manager would be highly interested in tracking the effectiveness of his or her organization by means of a utility function derived from his or her own preferences. Since standard and friendly computer programs are available for systematic elicitation and refinement of all the values, perhaps more managers will try and use the approach. The attribute lists already published by White (1980) may be suggestive to those who plan to apply the technique to field maintenance managers.

10.3.5.3 Maintainability Demonstrations

Government contracts for big equipment items have very detailed acceptance requirements, and these are spelled out in such documents as MIL-STD-470 (1966) and MIL-STD-471. Specialists have also accumulated much experience in handling the various technical and social problems associated with any acceptance-test occasion. It is widely accepted that the tests should be performed at the eventual use site if possible, that training and user personnel should select the set of troubles, that the faults should be inserted by putting in failed parts and not by actually shorting them in the equipment, that the "starting time" for a trouble-locating effort is when it is detected (not when it is inserted), and the best time to terminate a given problem is when the technician seems to be stymied. As automatic test equipment becomes more central in big-system diagnostics, decisions will have to be made on how to assess its performance, as "fault isolation time" could depend on how the automatic tests are arranged and queued. The Cunningham and Cox (1972) chapter on demonstrations is still the best source; it even gives a comparison of "demonstrated" versus "field-achieved" maintainability for four systems (field MTTRs are always higher).

Two methodological advances in demonstration testing have been proposed. Instead of inserting a probability sample of failures and taking the actual repair times for each level of maintenance, the faults are still a probability sample, but the measure is the difference between *predicted time and obtained time.* The statistical quality-control scheme is the same; but in the second case the hypothesis being tested is that the predicted estimates are correct. The distribution of "deviation-from-predicted" times would be expected to have a much smaller variance, and hence greater statistical power than the distribution of "raw" times on the same set of troubles. A graph of the summed difference between predicted and observed times could then be applied to a Wald-type sequential acceptance scheme.

A second innovation is to include probability information about failure *modes,* and to include this in the sampling process. If the level of troubleshooting is to a card, then "card failure modes" would be in the sampling plan; if isolation to the part is required, then the conceivable ways that the part could fail would be taken into account, and sampled proportionally to their likelihoods of occurrence. Since the different kinds of failure can have different effects all through a system, a formal failure modes and effects analysis (FMEA) is often very useful in predicting failures. It shows all possible (single) failures in the fault signature table. However, except for small and medium-size systems, a full FMEA listing often would be prohibitively large.

10.3.5.4 Postscript and Prospects

The maintainability domain is full of interesting people and there is discernible progress; but the human factors part of the field moves slowly enough for us to project things at least a short distance into the future. There are several positive indications. One of the most important of these is the realization, now becoming fairly widespread, that the military services have been expecting the impossible from their technicians. It is simply unreasonable to hope that high school graduates with a year or so of technical school can regularly do effective troubleshooting on really complex systems, using present manuals and test gear. Many good performances and exceptional individuals can be observed, but in many situations there are not enough of these to count on. Extremely powerful aiding, extremely effective training, extremely thorough human engineering, and honest evaluation of effectiveness will all be necessary. These "supplementary" requirements cannot be hastily added on as a design nears the field, but must be seen as essential components from the beginning.

The concept of the "exemplar" or ideal performer is also receiving more attention (Gilbert, 1978). Some of these effective people are well known, and their effectiveness can be investigated, elucidated, and emulated. One good exemplar is the contract field technician of the big computer companies. Although surprisingly few published reports are available on these people, the reasons for their success are evident enough: strong management commitment to strict performance and monitoring; provision for aids and training at the right level; a status, pay, and career situation that encourages stable groups; "specialization but not boredom" in job assignments; moderate selection requirements; supervision and promotion strictly from within the technically skilled personnel pool so that managers know the game; reasonable but not excessive room for discretionary time and individual development. There is little mystery in these conditions and requirements. Another positive sign from the engineering side is that *some* automatic fault-locators are now quite effective, and will surely get better as terminals and search software are improved.

Perhaps the key negative cloud on the maintainability horizon is the *degree of commitment* by the military and by other government services. Only in exceptional cases are nonmaintainable equipments torn out of ships and combat aircraft, and returned to the manufacturers (although there *are* cases of this sort in the submarine and space communities). In an early paragraph, we mentioned some data on a fault-detector system in an Air Force radar; as far as could be determined at time of writing, this system has not been significantly altered, although it obviously needs improvement. (Actually, as Coppola (1984) reported, this particular fault-detector is rather good, compared to many others.) Meetings of $\overline{M}$ specialists are held all the time with much consensus, but there often is no really effective voice, with enough power and "teeth" to cause *radical* change. Without a very strong office or agency, much of the R & D effort in maintainability is done in rather small individual projects, where company A builds a fault locator for equipment B, or contractor X shows that a state-space diagram is a better display than a regular schematic. These circumstances have led to a lot of rediscovery and wasted resources.

A couple of urgent research needs are quite clear. One of these is to establish one place where a critical mass of high-quality developmental and evaluation work can be done. In such a location, instead of just putting together another automatic test routine for airsearch radar Z, there would be several major prime equipments undergoing the most intensive maintainability analyses. To take just one example of the work to be done, "problem" equipments could be brought in for *intensive* $\overline{M}$ rework and technician aiding; after a few such attempts, we would have a much better idea of what can be expected from *the best possible revision and aiding.* Within a few years, and within a few million dollars, there should be a substantial bank of experience and accumulated wisdom about the limits of improvement. This is not just a lame request for still another "center" to be run by one of the present government agencies; perhaps such an activity best could be operated under contract by a leading university, technical institute, or manufacturer. In any case, the establishment should be able to realize the "best possible $\overline{M}$," and to do this over a wide enough range of hardware and software to produce effective practices in design and analysis.

At a more basic research level, probably two or three separate programs in cognitively oriented $\overline{M}$ research should be funded, say for 5 years at a time, and at several times the present support levels. There are many good ideas. The best current illustrations are probably in areas like "deep structure" and "representation"; but as truly intelligent interactive terminals get closer, the interaction processes will be more complex and they may offer substantial payoffs in performance as well as in the intrinsic motivation of troubleshooters and trainees. As in many other areas of maintainability, the *commitment and implementation* problems may be more difficult to solve than the technical ones.

As a final research suggestion, something really should be done about the tantalizing social and "soft" variables in maintenance work. The hints from the questionnaire studies cited earlier are just hints until they are pursued vigorously. It may be that, as many people believe, the "soft" factors largely take care of themselves when motivated people are doing effective, interesting, and valuable work. But it is more likely that we will need "soft" research, along with the more strictly technical contributions, if we really intend to have maintainable complex systems.

REFERENCES

Arsenault, J. E., Des Marais, P. J., and Williams, S. D. G. 1980. System design. In J. E. Arsenault and J. A. Roberts, eds. *Reliability and maintainability of electronic systems* Rockville, MD: Computer Science Press, pp. 414–434.

Askren, W. B., Campbell, W. B., Seifert, D. J., Hall, T. J., Johnson, R. C., and Sulzen, R. H. 1976. Feasibility of a computer simulation method for evaluating human effects on nuclear systems safety (Technical Report AFHRL-TR-76-18). Wright-Patterson Air Force Base, OH.

Bond, N. A., Jr., and Rigney, J. W. 1966. Bayesian aspects of troubleshooting behavior. *Human Factors, 8,* 377–383.

Bonissone, P., and Johnson, H. E. 1984. DELTA: An expert system for diesel electric locomotive repair. In Denver Research Institute, *Artificial intelligence in maintenance* (pp. 397–414). Brooks AFB, TX: Air Force Human Resources Laboratory.

Brannan, R. C. 1976. Army life cycle cost model (AD-A021-900). Alexandria, VA: Defense Technical Information Center.

Brooke, J. B., Duncan, K. D., and Marshall, E. C. 1978. Interactive instruction in solving fault-finding problems. *International Journal of Man-Machine Studies, 10,* 603–611.

Chase, W. G., and Simon, H. A. 1973. The mind's eye in chess. In W. G. Chase, Ed. *Visual information processing.* New York: Academic.

Chi, M. T. H., Feltovich, P. J., and Glaser, R. 1981. Categorization and representation of physics problems by experts and novices. *Cognitive Science, 5*(2), 121–152.

Chi, M. T. H., and Glaser, R. 1983. Problem solving abilities (Technical Report No. 8). Pittsburgh, PA: University of Pittsburgh, Learning Research and Development Center.

Coppola, A. 1984. Artificial intelligence applications to maintenance. In Denver Research Institute, Ed. *Artificial intelligence in maintenance.* (pp. 23–44). Brooks AFB, TX: Air Force Human Resources Laboratory.

Crawford, B. M., and Altman, J. W. (1972). Design for maintainability. In H. P. Vancott and R. G. Kinkade, Eds. *Human engineering guide to equipment design* Washington, DC: Superintendent of Documents, U.S. Printing Office, pp. 586–631.

Cunningham, E. E., and Cox, W. 1972. *Applied maintainability engineering.* New York: Wiley.

Davis, R. (1984). Diagnosis via causal resoning: paths of interaction and the locality principle. In Denver Research Institute, Ed. *Artificial intelligence in maintenance.* (pp. 109–115). Brooks AFB, TX: Air Force Human Resources Laboratory.

Davison, J. 1984. Expert systems in maintenance diagnostics for self-repair of digital flight control systems. In Denver Research Institute, Ed. *Artificial intelligence in maintenance.* (pp. 293–304). Brooks AFB, TX: Air Force Resources Laboratory.

DeGroot, A. 1966. Perception and memory versus thought: some old ideas and recent findings. In B. Kleinmuntz, Ed. *Problem solving.* New York: Wiley.

DeKleer, J. 1984. AI approaches to troubleshooting. In Denver Research Institute, Ed. *Artificial intelligence in maintenance.* (pp. 83–96). Brooks AFB, TX: Air Force Human Resources Laboratory.

Des Marais, P. J. (1980). Assembly design. In J. E. Arsenault & J. A. Roberts, Eds. *Reliability and maintainability of electronic systems* (pp. 451–469). Rockville, MD: Computer Science Press.

Duncan, K. D., and Shepherd, A. 1975. A simulator and training technique for diagnosing plant failures from control panels. *Ergonomics, 18,* 627–641.

Einhorn, H. J., and Hogarth, R. M. 1982. Prediction, diagnosis, and causal thinking in forecasting. *Journal of Forecasting, 1,* 1–14.

Federico, P. A. 1982. Individual differences in cognitive characteristics and computer-managed mastery learning. *Journal of Computer-Based Instruction, 9,* 10–18.

Foley, J. P., Jr. 1977. Performance measurement of maintenance (Technical Report AFHRL-TR-77-76). Wright-Patterson AFB, OH: Air Force Human Resources Laboratory.

Folley, J. D., Jr., and Altman, J. W. 1956. Guide to design of electronic equipment for maintainability (Report WADC-TR-56-218). Wright-Patterson AFB, OH.

Fuqua, N. B. (undated). Electronic equipment maintainability data (EEMD-1). Griffiss AFB, NY: Rome Air Development Center.

Gettys, C. F., Manning, C., Mehle, T., and Fisher, S. (1980). Hypothesis generation: a final report of three years of research (Technical Report 15-10-80). Norman, OK: Decision Processes Laboratory, University of Oklahoma.

Gilbert, T. F. 1978. *Human competence.* New York: McGraw-Hill.

Glaser, R., and Phillips, J. C. 1954. An analysis of proficiency for guided missile personnel: patterns of troubleshooting behavior (Technical Bulletin 55-16). Washington: American Institute for Research.

Halff, H. 1984. Overview of training and aiding. In Denver Research Institute, Ed., *Artificial intelligence in maintenance.* (pp. 67–80). Brooks AFB, TX: Air Force Human Resources Laboratory.

Henneman, R. L., and Rouse, W. B. 1984. Measures of human problem solving performance in fault diagnosis tasks. *IEEE Transactions on Systems, Man, and Cybernetics, SMC-14,* 99–112.

Highland, R. W., Newman, S. E., and Waller, H. S. 1956. A descriptive study of electronic trouble shooting. In Air Force human engineering, personnel, and training research (Technical Report 56-8). Baltimore, MD: U.S. Air Force Research and Development Command.

Hightower, K. 1982. Experience with UAL's Apollo system. Paper presented at the Lake Arrowhead workshop on practical approaches to high availability of computer systems, Lake Arrowhead, CA.

Johnson, D. M. 1972. *Systematic introduction to the psychology of thinking.* New York: Harper.

Kane, W. D., Jr. 1981. Task accomplishment in an Air Force maintenance environment (AFOSR-80-0146). Washington, DC: United States Air Force.

Kieras, D. E. 1984. The psychology of technical devices and technical discourse. In Denver Research Institute, Ed. *Artificial intelligence in maintenance* (pp. 227–254). Brooks AFB, TX: Air Force Human Resources Laboratory.

Laffey, T. J., Perkins, W. A., and Firschein, O. (1984). LES: a model-based expert system for electronic maintenance. In Denver Research Institute, Ed. *Artificial intelligence in maintenance.* (pp. 429–450). Brooks AFB, TX: Air Force Human Resources Laboratory.

Lamarre, B. G. 1980. Mathematical modeling. In J. E. Arsenault and J. A. Roberts, Eds. *Reliability and maintainability of electronic systems* (pp. 351–379). Rockville, MD: Computer Science Press.

Larkin, J. H., McDermott, J., Simon, D. P., and Simon, H. A. 1980. Models of competence in solving physics problems. *Cognitive Science 4,* 317–345.

Levine, J. M., Brahlek, R. E., Eisner, E. J., and Fleishman, E. A. (1979). Trainability of abilities: training and transfer of abilities related to electronics fault-finding (Technical Report ARRO-3010-TR2). Washington, DC: Advanced Research Resources Organization.

Levine, J. M., Schulman, D., Brahlek, R. E., and Fleishman, E. A. (1980). Trainability of abilities: training and transfer of spatial visualization (Technical Report ARRO-3010-TR3). Washington, DC: Advanced Research Resources Organization.

Mackie, J. L. 1974. *The cement of the universe: a study of causation.* Oxford: Clarendon Press.

Maxion, R. 1984. Artificial intelligence approaches to monitoring system integrity. In Denver Research Institute, Ed. *Artificial intelligence in maintenance.* (pp. 257–273). Brooks AFB, TX: Air Force Human Resources Laboratory.

McDonald, L. B., Waldrop, G. P., and White, V. T. 1983. Analysis of fidelity requirements for electronic equipment maintenance (Technical Report 81-C-0065-1). Orlando, FL: Naval Training Equipment Center.

Mehle, T. 1980. Hypothesis generation in an automobile malfunction inference task. (Technical Report TR 25-2-80). Nerman, OK: Decision Processes Laboratory, University of Oklahoma.

Meister, D., Finley, D. L., and Thompson, E. A. 1971. Relationship between system design, technician training, and maintenance job performance on two autopilot subsystems (Bunker-Ramo Report AFHRL-TR-70-20). Wright-Patterson AFB, OH.

Moore, J. V., Saltz, E., and Hoehn, A. J. 1955. Improving equipment maintenance by means of a preplanning technique (Technical Report AFPTRC-TN-55-26). Lackland AFB, TX: Air Force Personnel Training Research Center.

Morris, N. M., and Rouse, W. B. 1984. Review and evaluation of empirical research in troubleshooting (Report 8402-1). Norcross, GA: Search Technology Inc.

Munger, M. R., and Willis, M. P. 1959. *Development of an index of electronic maintainability* [Report No. AIR-B75-59-FR-207(I)]. Pittsburgh, PA: American Institutes for Research.

Nilsson, N. 1980. *Principles of artificial intelligence.* Palo Alto, CA: Tioga Press.

Nunnally, J. C. 1978. *Psychometric theory.* New York: McGraw-Hill.

Orlansky, J., & String, J., 1981. Cost-effectiveness of maintenance simulators for military training (Paper 1568). Arlington, VA: Institute for Defense Analyses.

Rasmussen, J. 1980. The human as a systems component. In H. T. Smith & R. G. Green, Eds. *Human interaction with computers.* Orlando, Florida: Academic Press.

Rau, J. G. 1970. *Optimization and probability in systems engineering.* New York: Van Nostrand-Reinhold.

Rigney, J. W., Cramer, R. H., and Towne, D. M. (1965). Design of electronic equipment for ease of maintenance: current engineering design practices (Technical Report 41). Los Angeles, CA: University of Southern California, Behavioral Technology Laboratories.

Ronback, J. A. 1975. Software reliability—how it affects system reliability. In *Proceedings, 1975 Canadian reliability symposium.* New York: Pergamon Press.

Rouse, W. B. 1978. A model of human decision making in a fault diagnosis task. *IEEE Transactions on Systems, Man, and Cybernetics, SMC-8,* 357–361.

Rouse, W. B. 1979. Problem solving performance of first semester maintenance trainees in two fault diagnosis tasks. *Human Factors, 21,* 611–618.

Rouse, W. B. 1984. Models of natural intelligence in fault diagnosis tasks: implications for training and aiding of maintenance personnel. In Denver Research Institute, Ed. *Artificial intelligence in maintenance.* (pp. 193–214). Brooks AFB, TX: Air Force Human Resources Laboratory.

Sauer, D. W., Campbell, W. B., and Potter, N. R. 1977. Human resource factors and performance relationships in nuclear missile handling tasks (Technical Report AFHTL-TR-76-85). Wright-Patterson AFB, OH.

Shapero, A., Cooper, J. I., Rappaport, M., Schaeffer, K. H., and Bates, C., Jr., (1960). Human engineering testing and malfunction data collection in weapon system test programs (WADD Technical Report 60-36). Wright-Patterson AFB, OH.

Smillie, R. J. 1984. Implications of artificial intelligence for a user defined technical information system. In Denver Research Institute, Ed. *Artificial intelligence in maintenance.* Brooks AFB, TX: Air Force Human Resources Laboratory, pp. 354–358.

Smillie, R. J., and Porta, M. M. 1981. Comparison of state tables and ordance publications for troubleshooting digital equipment (Technical Report NPRDC TR 82-7). Fullerton, CA: Hughes Aircraft Corp.

Su, Y.-L. D. 1984. A review of the literature on training simulators: transfer of training and simulator fidelity (Report No. 84-1). Atlanta, GA: Georgia Institute of Technology, School of Industrial and Systems Engineering.

Sukert, A. N. 1977. An investigation of software reliability models. *IEEE 1977 Annual Reliability and Maintainability Symposium* (pp. 478–488), New York: IEEE.

Thomas, D. 1982. Computer-based maintenance aids system: preliminary development and evaluation of a prototype (Report AFHRL-TP-82-84). Wright-Patterson AFB, OH: Air Force Human Resources Laboratory.

Topmiller, D. A. 1964. A factor analytic approach to human engineering analysis and prediction of system maintainability (Report AMRL-TR-64-115). Wright-Patterson AFB, OH: Aerospace Medical Research Laboratories.

Towne, D. M. (1984). *A generalized model of fault isolation performance.* In Denver Research Institute, Artificial Intelligence in Maintenance, Brooks AFB, TX: AFHRL-TR-8425.

U.S. Department of Defense. 1966. Maintainability program requirements (MIL-STD-470). Washington, DC: U.S. Government Printing Office.

U.S. Department of Defense. 1970. Life cycle costing procurement guide (Report LCC1-LCC3). Washington, DC: U.S. Government Printing Office.

White, C. R. 1980. Maintenance productivity (Report AFOSR-79-0016). Auburn, AL: Auburn University, Department of Industrial Engineering.

Wickens, T. D., Geiselman, R. E., Samet, M. G., and Yelvington, C. L. 1982. Mental representation of circuit diagrams: individual differences in structural knowledge (Report TR-PATR-1109-82-6). Woodland Hills, CA: Perceptronics.

Williams, W. L., Jr., and Whitmore, P. G., Jr. 1959. The development and use of a performance test as a basis for comparing technicians with and without field experience (Technical Report 52). Washington, DC: George Washington University, Human Resources Research Office.

Wirth, N. 1973. *Systematic programming—an introduction.* Englewood Cliffs, NJ: Prentice-Hall.

Wohl, J. G. 1982. Maintainability prediction revisited: diagnostic behavior, system complexity, and repair time. *IEEE Transactions on Systems, Man, and Cybernetics, SMC-12,* 241–250.

Young, H. H. 1980. Effectiveness planning and evaluation model for Air Force maintenance organizations (AFOSR Final Report 79-0111). Washington DC: U.S. Air Force.

PART 11
HUMAN FACTORS IN THE DESIGN AND USE OF COMPUTING SYSTEMS

CHAPTER 11.1

DESIGN OF VDT WORKSTATIONS

ETIENNE GRANDJEAN

Former Director of the Department of Ergonomics
Swiss Federal Institute of Technology
Zurich, Switzerland

11.1.1 WHY DO WE NEED AN ERGONOMIC DESIGN OF VDT WORKSTATIONS?*

At present VDTs are invading offices and specific departments in industry, that is to say, a world where machines have not been used until quite recently. The result is a significant change of working conditions.

At the traditional office desk an employee carries out a number of physical activities and has much space for various postures and movements: he or she might look for documents, take notes, use the telephone, read a text, exchange information with colleagues, type for a while, and perform many other tasks during the course of his or her working day. A desk that is too low or too high or has other ergonomic shortcomings is not likely to cause annoyance or physical discomfort to such a person.

The situation is, however, entirely different for an operator working with a VDT for several hours without interruption or even for a whole day. *Such a VDT operator is tied to a human–machine system.* His or her movements are restricted, his or her attention is directed toward the screen, and his or her hands are fixed to the keyboard. Operators are more exposed to ergonomic shortcomings, to inadequate lighting conditions, and to uncomfortable furniture. They are more sensitive to visual strain and to unsuitable desk levels that cause constrained postures. Such circumstances call for Ergonomics and that is how this science has found its way into the office world.

As long as engineers and other highly motivated experts operated VDTs, nobody complained about adverse effects. However, the situation changed drastically with the expansion of VDTs to workplaces where traditional working methods had formerly been applied. Now complaints from VDT operators about visual strain and physical discomfort in the neck–shoulder area became more and more frequent. They provoked different reactions: some believed that the complaints were highly exaggerated and mainly a pretext for social and political claims; others considered the health hazard to be a serious problem requiring immediate measures to protect operators from injuries to their health. *Ergonomists stand between the opposite poles; it is their duty to analyze the situation objectively and to deduce guidelines for the appropriate design of VDT workstations.* This is also the aim of this chapter.

11.1.2 FIELD STUDIES ON COMPLAINTS OF VDT OPERATORS

The frequent complaints of VDT operators induced field studies in Sweden [Oestberg (1975), Gunnarsson and Söderberg (1983) as well as Johansson and Aronsson (1980)], in Germany [Cakir, Reuter, von Schmude, and Armbruster (1978)], in Switzerland [Hünting, Läubli, and Grandjean (1981), Läubli, Hünting, and Grandjean (1981), and Meyer, Gramoni, Koral, and Rey (1979)], in the United States [Smith, Cohen, Stammerjohn, and Happ (1981), Dainoff, Happ, and Crane (1981), Stare, Thompson, and Shute (1982), and Sauter (1984)], in France [Elias and Cail (1983)], and in Japan [Nishiyama, Uehata, and Nakaseko (1984)]. A few other studies are discussed by the U.S. Panel on Impact of Video Viewing on Vision of Workers (1984). Another survey of several field studies was published by Dainoff (1982).

Most authors record an increase of complaints about visual discomfort or musculoskeletal troubles. But the results of these field studies are in most cases not comparable, since the methods (questionnaires) as well as the professional groups studied differ greatly from each other. Some surveys failed to include control groups; others used non-VDT groups as control, but these again differed from the VDT group not only in the use of VDTs but also in many other respects. The Panel of the National Academy of Sciences concludes: "Thus it is not possible to determine from existing studies to what extent complaints reported by VDT operators have resulted from the VDT itself as opposed to such factors as workstation or job design."

The intricate problem of control groups was analyzed by Läubli and Grandjean (1984). They emphasize that several studies reveal the incidence of discomfort in the VDT groups to be of the same order as in the control groups. This applies to the following control groups: full-time typists (Hünting et al., 1981; Läubli et al. 1981), telephone-directory-assistance operators searching for required telephone numbers in printed paper books (Starr et al., 1982), and an experimental group engaged in proofreading from hardcopy (Gould and Grischkowsky, 1984). In these studies VDT operating is compared to other strenuous office jobs or tasks. But different results are obtained when the VDT group is compared with groups performing traditional office jobs with a lower productivity and a great variety of activities. These control groups reveal a much lower incidence of visual and musculoskele-

---

* Portions of this chapter are reprinted with permission of Burroughs Corporation.

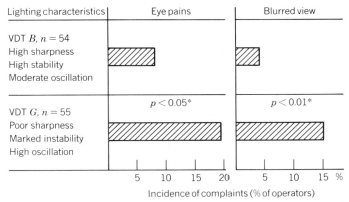

Fig. 11.1.1. The incidence of complaints of eye troubles in two groups of VDT operators using two different makes of VDTs. The two groups have the same job and are comparable in relation to sex and age. (n = number of operators per group; * = statistical significance of differences of complaints.)

tal discomfort than the VDT groups (Hünting et al., 1981; Läubli et al., 1981). It must be concluded that control groups in field studies are very often problematic. To put it cynically, with field studies every hypothesis can be proved; it all depends on the choice of an appropriate control group.

More conclusive are those studies that reveal significant relationships between physical characteristics of VDT workstations on the one hand and subjective or objective effects on operators on the other hand (Elias and Cail, 1983; Grandjean, 1982; Hünting et al., 1981; Läubli et al., 1981). The most interesting finding is the significant difference of adverse effects between an operator group using a VDT of good lighting qualities and a comparable group with a VDT of poor lighting qualities (Grandjean, 1982). These results are presented in Figure 11.1.1. They disclose that the operators of group B using a VDT with high sharpness, good stability, and moderate oscillation degree of characters report a lower incidence of eye pains and blurred vision than the VDT group G with poor lighting characteristics of the display.

Other results of Läubli et al. (1981) and of W. Hünting et al. (1981) confirm the importance of a proper ergonomic design: the frequency of eye troubles increases at workplaces with great brightness contrasts between screen and surrounding (source documents, windows, bright walls and dark screen) (Läubli et al., 1981). The same applies to the incidence of physical troubles in the neck–shoulder–arm area (Hünting et al., 1981).

11.1.3 BEHAVIORAL STUDIES AT VDT WORKSTATIONS

From the point of view of Ergonomics it is important to know for how long an operator is looking at his or her source material, watching the display, and operating the keyboard in turn. Among several studies the very systematic investigation of Elias and Cail (1983) has been chosen here for brief discussion. These authors used the NAC eye recorder equipment on operators engaged in a data-entry task in banking and on operators doing a conversational task in a storage control. The results are given in Figure 11.1.2. In the data-entry job the operators cast a short look at the screen from time to time, while their eyes are chiefly directed toward the source documents. In the conversational job the operators rarely change the direction of sight; their eyes are focused on the display for much longer periods.

In a job like data enquiry, looking at the display lasts a mean duration of 80% of the working time with glances up to 135 sec.

van der Heiden, Braeuninger, and Grandjean (1984) carried out a work-sampling study on computer-aided-design (CAD) workstations to determine the use of keyboard, digitizer tablet, and screen; 38 workstations with different engineering tasks (mechanical design, printed-wiring-board design, and electrical schematics) were involved. The results are reported in Table 11.1.1.

The operator's glance was frequently directed to the screen, namely, between 46% and 68% of the working time at the CAD terminal. This means that an operator observed the graphic display between $2\frac{1}{2}$ and $3\frac{1}{2}$ hr per day on average. It follows that CAD operating takes about as much time for screen viewing as in a conversational VDT job.

Operating the keyboard required 14% to 22% of the working time. These figures are lower than the ones found at other terminals. The reason for this is that frequently used command strings were integrated in the tablet menu and activated by the digitizing pen. The authors emphasize that the use of two input mediums, keyboard and tablet, gives rise to interference problems. Since the tablet was the primary input medium, the keyboard was usually placed next to it, either to the left or to

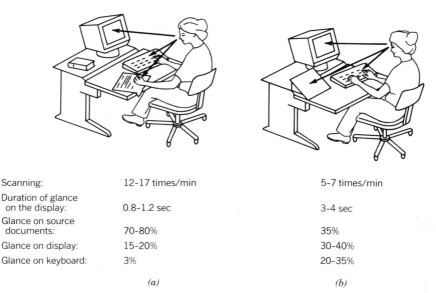

Scanning:	12-17 times/min	5-7 times/min
Duration of glance on the display:	0.8-1.2 sec	3-4 sec
Glance on source documents:	70-80%	35%
Glance on display:	15-20%	30-40%
Glance on keyboard:	3%	20-35%
	(a)	*(b)*

Fig. 11.1.2. Frequency of scanning and duration of the glance fixed on the screen in (*a*) data entry task and in (*b*) conversational task. The percentages refer to the approximate duration related to the working time according to Elias and Cail (1983).

the right of the tablet. The normally practiced two-handed keying induced a twisted position of the trunk. In case of extensive use of the keyboard many operators placed it on top of the tablet. The result was an unnatural position of the hands and wrists while keying. It was concluded that a flat keyboard that can easily be put on and taken off the tablet would be advisable.

11.1.4 VISUAL FUNCTIONS AND LIGHTING CONDITIONS

The main visual functions involved at VDT workstations are *accommodation, convergence,* and *adaptation.*

Accommodation and convergence are the ability of the eye to bring objects at varying distances from infinity to the nearest point of distinct vision—the so-called "near point of accommodation"— into *sharp focus.* Focusing on near objects is achieved by adapting the curvature of the lens through contraction of the ciliary muscle (muscles of accommodation). The ciliary muscle changes the curvature of the lens so that it bulges just the right amount to throw the sharp image back onto the plane of

Table 11.1.1 Results of a Work-Sampling Study on 38 CAD Workstations

Activity	Percentage of Working Time Spent Interactively on CAD Workstation		
	Mechanical Design (%)	pwb[a] Design (%)	Electrical Schematics (%)
Watching the screen	52	68	46
Operating keyboard	14	14	22
Operating tablet	48	43	26
Hardcopy manipulation	<1	<1	1
Document manipulation	14	9	15
Number of observations	1681	1670	1177

Source: Van der Heiden (1984).

[a] pwb = printed wiring board.

the retina. The movements of each eyeball are controlled by six external muscles. Convergence is regulated by eye movements and achieved through focusing. The optical axis of both eyes is brought to the observed object, thus permitting both eyes to converge on an object. If convergence is not well regulated, double images occur.

The level of illumination and the sharpness of contrasts between characters and background are essential for accommodation and convergence and therefore legibility. If lighting is poor and contrasts are low, both speed and precision of accommodation and convergence are reduced.

Adaptation is the capacity of the eye to adapt its sensitivity to the incident light flow. In fact, the aperture of the pupil, as well as the sensitivity of the retina, are continuously adapted to the prevailing light conditions. Both aperture of pupil and retinal adaptation prevent over- or underlighting of the retina. The adjustment of the pupil aperture takes a measurable time between a few tenths and 1 sec. The adaptation of the sensitivity of the retina takes a comparatively long time of about 30 min.

At VDT workstations the visual field of an operator often contains dark and bright areas. In such situations only a partial adaptation takes place: the bright area projected on the retina reduces the retinal sensitivity, whereas the dark area increases it. This form of disturbance is called *relative glare;* it reduces the general visual capacities such as visual acuity and visual sensitivity to contrasts. Adaptation conflicts are a frequent phenomenon at VDT workstations where the dark background of the screen lies near bright source documents or other bright surfaces. An important conclusion can be drawn from these considerations: *all surfaces within the visual field of an operator should be of a similar order of brightness.* Surfaces in the middle of the visual field should not have a relative brightness contrast of more than 3:1. Relative contrasts between the middle and the periphery of the visual field should not exceed the ratio 10:1. (See also Chapters 7.7 and 7.8.)

The temporal uniformity of the surface luminance is as important as the static spatial uniformity. Rhythmically fluctuating surface luminances in the visual field are very annoying and reduce visual performances. Such unfavorable conditions prevail if the work requires the operator to glance alternatingly at a bright and a dark surface or if the light source generates an oscillating light. As mentioned previously, pupil and retina can cope with changes in brightness only after a certain delay, so that oscillating brightness leaves the eyes either under- or overexposed for most of the time.

Since fluorescent tubes operate from alternating current, the light is generated with 100 light oscillations per second in Europe and with 120 per second in the United States. Below a certain level, oscillation is perceived as a flickering light; above this level, oscillation is not seen. *The threshold of perception is called the critical fusion frequency* (CFF). The 100 and 120 Hz oscillations of fluorescent tubes are usually not perceived as flicker. It can, however, become noticeable as a stroboscopic effect on moving reflective objects.

When the tubes wear out or are defective, they develop a slow, easily perceptible flicker, especially at the periphery. Earlier studies have revealed that many tubes show a small 50-Hz oscillation superimposed on the main 100-Hz one. This 50-Hz oscillation apparently comes from asymmetrical emissions of the electrodes; it is perceptible and likely to cause visual discomfort. *Visible flicker has adverse effects on the eye mainly because of the repetitive overexposure of the retina. Flickering light is extremely annoying and causes visual discomfort.*

When fluorescent lighting was first introduced on a large scale in European offices, several complaints of irritated eyes and eye strain were reported. On the assumption that the oscillating character of fluorescent light was the cause of visual discomfort, the lighting technology developed phase-shifting equipments that produced an almost constant light. Complaints seem to have stopped in offices where phase-shifted fluorescent tubes were installed. But the question of which degree of oscillation can be tolerated and which is likely to produce visual discomfort remains open. In Europe, then, it was generally concluded that offices should never be lighted with single fluorescent tubes but always with two or more phase-shifted tubes inside one luminary.

Another important problem arises from *light as a source of glare.* Avoiding glare inside an office is one of the essential ergonomic considerations in the design of workplaces. Some principles can be summarized as follows:

No unscreened light should appear in the visual field of any working employee.

The line from eye to light must show an angle of more than 30° with the horizontal plane. If a smaller angle cannot be avoided, for example, in large offices, the lamps must be effectively shaded.

To avoid glare from the reflection of the desk, the lights should be arranged on either side of the workplace in order to avoid a coincidence of the line of sight with the line of the reflected light.

11.1.5 THE GENERATION OF CHARACTERS ON DISPLAYS

Most VDTs in use today are based on the cathode ray tube (CRT) technology. On the inner surface of the screen there is a phosphor layer which, stimulated by the electron beam, generates a light emission.

The greater number of VDTs use luminous characters on a rather dark screen background. In Europe this is called *negative presentation;* in the United States, it is called *positive contrast.* However, there is an increasing tendency on the market to offer VDTs with dark characters on a bright background, which is called *positive presentation* in Europe and *negative contrast* in the United States.

The most common techniques of generating characters on the screen are the dot matrix method or, less frequently, the continuous stroke method. The dot matrix system has been proved superior in legibility and is generally preferred by operators. With this technique characters are generated on an ideal grid of dots covering the entire surface of the screen. The CRT draws horizontal lines (scanlines) on the screen. The electron beam is turned on or off as required to produce line segments of symbols and characters. The dot spacing depends on the size of the scanning spot and on the raster pitch. Raster pitch is caused by the fact that the horizontal scan lines setting up the raster are not quite horizontal but slightly curved. If the scan line spacing is equal to the spot size of the scanning beam, then the spots composing the characters will partially overlap, producing almost strokelike characters. The more scan lines are used to form a character, the better the legibility will be.

A 525-line raster display presents visible spaces between raster lines, which cause dot visibility. A well-designed display consisting of 729 or 1029 lines is likely to have raster lines which are barely visible.

A visible raster structure should be avoided. It is detrimental to legibility, as studies by Beamon and Snyder (1975) have shown. More important still is the visibility of the matrix structure of the individual characters, caused by dot spacing greater than dot diameter. Snyder and Maddox (1978) could show that an increase of the spaces between dots leads to a prolonged reading time. The more a dot matrix character resembles a stroke character, the more readable is the text. The same authors demonstrated that character font and matrix size can have a significant effect on legibility and readability. A 5×7 dot matrix font is less legible than a 7×9 dot matrix font, which, in turn, is less legible than a 9×11 font [see also Snyder (1984)].

11.1.6 COLORS

The eye is more sensitive to the central part of the visible spectrum, which appears as a yellow–green color and seems to be brighter than other colors. Many VDT makes have characters of this yellow–green color. Some operators prefer green colors, although they cannot give rational arguments for this choice. It is possible that green characters are more easily distinguished if disturbing reflections appear on the glass surface of the screen.

A few VDT makes have a screen background of an amber color and characters of a shining yellow phosphor; operators like this combination, too. There is no scientific reason for recommending one color rather than another. The color of characters is mainly a matter of personal preference.

Several displays today use more than one color. Different colors can emphasize certain parts of the text; they function as codes and facilitate identification processes.

Colors may cause problems for the accommodation mechanism through chromatic aberration. In fact, red colors are focused behind the retina, blue colors are focused in front of it, and yellow–green is focused right on the retina. However, Krueger and Mader (1982) could show that the colors used in VDTs are not associated with noticeable chromatic aberration.

11.1.7 DISPLAY CHARACTERISTICS

The first question that arises when visual comfort at VDTs is at stake is: What may be the difference between reading a printed text and reading a text displayed on a VDT? Compared with a printed text, the main and often observed differences are as follows:

The characters are luminous on a dark background.

The face may be flickering; its luminance is of an oscillating kind.

The text often discloses low sharpness.

The contrasts between characters and background may be low.

The characters may be moving and unstable.

The text sometimes shows an insufficient geometric design of characters and face.

It must be assumed that sharpness, contrasts, and poor design of characters influence the speed and precision of accommodation and that contrasts of surface luminances as well as a flickering face may disturb the adaptation of the retina. *One important consequence of disturbed accommodation and adaptation might be lower legibility and occasional visual fatigue.* Some of the lighting characteristics determining visual comfort will be discussed in the following chapters.

11.1.7.1 Techniques of Display Measurement

Measuring display characteristics involves two photometric parameters: luminance and illumination. Luminance measures the brightness in candela/m² units (cd/m²). An older unit is the footlambert (ftL); 1 ftL = 3.426 cd/m². Illumination measures the stream of light falling onto a surface in lux units (lx). An older unit of illumination is the footcandle (ftc); 1 ftc = 10.76 lx. Since luminance is a function of the light that is emitted or reflected from the surface of walls, furniture, and other objects, it is greatly affected by the reflective power of the respective surface. If the luminances of various surfaces are compared, they can also be expressed as reflectance (percentage of the reflected luminous flux).

The luminance in cd/m² and the illumination in lx are related as follows:

$$\text{reflectance } (\%) = \frac{0.32 \text{ cd/m}^2}{\text{lx}}$$

Fellmann Bräuninger, Gierer, and Grandjean (1982) and Bräuninger et al. (1982, 1983, 1984) designed the following equipment to measure the different lighting characteristics of displays: a microscope picks up the luminance of a small dot of 0.1 mm inside a bar of a character, leads it to a photomultiplier that amplifies the signals and transfers them to an oscilloscope, a dc voltmeter, an ac voltmeter, a Fourier analyzer, and a linearcorder. The bandwidth of the system was 1 MHz. In order to establish the degree of oscillation they measured the luminance of a 5 × 7-cm display surface by means of a camera. The luminances of the various surfaces at the VDT workstations were measured with a Tektronix instrument (Types J 16 and J 6523). All measurements were carried out under standardized lighting conditions: indirect constant light with 400 lx vertical and 160 lx horizontal. Many of the measurements were conducted with an adjusted luminance of the characters, the so-called "preferred" luminance. These preferred figures were between 20 and 50 cd/m² and were assessed by the experimenters. In practice, operators adjust luminances between 9 and 77 cd/m², the mean value being 33 cd/m² (Läubli et al., 1981). With the equipment previously described the following measurements were conducted: brightness contrasts of surfaces, oscillation degrees of characters, sharpness, contrast ratios and stability of characters, reflections, and dimensions of characters and face.

Another procedure to measure image quality is the *modulation transfer function* (MTF). Snyder (1980) applied the MTF to quantify the quality of the displayed image of VDTs. This measurement is based on a Fourier transformation of a luminance contrast. Snyder explains the MTF "as the contrast (modulation) expressed as a function of the size of the bars on a sine-wave grating, with increasing spatial frequency (e.g., cycles per unit visual angle) denoting decreasing bar width. More modulation per unit spatial frequency indicates greater contrast and perceived sharpness to the displayed image" (Snyder, 1984). With the MTF, sharpness of characters, reflected glare and character contrast were determined.

11.1.7.2 Oscillating Luminances of Characters

The light of characters, generated by the stimulation of the phosphor through the electron beam, is composed of light flashes with the frequency of the refreshing rate of the CRT. For that reason the light of the characters is not constant but oscillating.

The *critical fusion frequency* (CFF), the threshold of perceived flicker, depends not only on the refreshing rate but also on the phosphor persistence which determines how long the phosphor remains illuminated after the electron beam has excited it. Figure 11.1.3 shows the oscillating luminances of two VDT makes, one with a slow and the other with a fast phosphor.

If the phosphor persistence is too slow, a "smearing" of the image may occur. This is also called a "ghost image"; it appears when scrolling procedures are used. If the phosphor persistence is too fast, characters appear as flickering. The VDTs on the market today use very different phosphors; the figures listed in Table 11.1.2 are taken from the IBM brochure *Human factors of workstations with visual displays* (IBM, 1984).

To illustrate the importance of the phosphor persistence the oscillograms of a slow and a fast phosphor are shown in Figure 11.1.3. The phosphors of the two makes disclose different decay times. VDT *A* has a slow decay time and the luminance does not come to zero between two stimulations of the electron beam. VDT *G* has a very short decay time and the luminance falls to zero after a few milliseconds. The result is obvious: In order to get a mean luminance of characters of 40 cd/m² with VDT *A*, the flash needs a peak value of about 75 cd/m²; in order to get the same mean luminance of characters with VDT *G*, a peak value of more than 700 cd/m² is necessary. This is the reason why the oscillation degree *a* is much higher for *G* than for *A*.

There are important individual differences in the sensitivity to see flicker. Furthermore, the size and brightness of the target as well as the waveform influence the threshold of perceived flicker. As mentioned in Section 11.1.4 *visible flicker is extremely annoying and causes strong visual discomfort*.

The majority of VDT makes today have refresh rates of 50 or 60 Hz. These rates seem to be in the critical range where some operators might already begin to perceive flicker. Gyr, Nishiyama, Läubli, and Grandjean (1984) used simulated VDT equipment with bright characters on a dark background and measured CFF levels between 45 and 55 Hz on 28 subjects. Bauer (1984) measured the CFF of a bright screen background (80 cd/m²) with a rather fast phosphor on 30 subjects and observed a

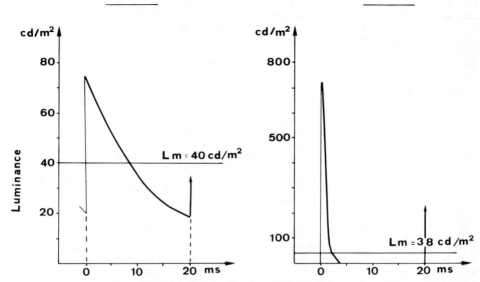

Fig. 11.1.3. Oscillograms of two VDTs with phosphors of different decay times. VDT *A* has a slow phosphor; VDT *G* has a fast phosphor decay time. (Oscillograms of dots of 0.1 mm in a dark room with approximately the same mean luminance of characters, 40 and 38 cd/m², respectively.)

range between 73 and 93 Hz. At present there exists no scientific basis to assess the tolerance limit for the oscillation degree of character luminances. Fellmann et al. (1982) recommend an oscillation degree which does not exceed that of a phase-shifted fluorescent light.

The oscillation degree of a light source can be determined by recording the amplitude of the oscillation over the mean luminance. Bräuninger et al. (1982, 1983, 1984) adopted a procedure based on the following formula:

$$a = \frac{1}{Lm} \sqrt{\sum_{n=1}^{20} A_{n\,\text{eff}}^2}$$

where

a = oscillation degree

Lm = mean luminance

$A_{n\,\text{eff}}$ = amplitude of the groundwave and of 20 first harmonics of a Fourier transformation

Table 11.1.2   Characteristics of Commonly Available Phosphors for VDTS

Phosphor Type	Color	Decay-Times (msec)	
		to 10% Brightness	to 1% Brightness
P 1	Green	24	50
P 4	White	0.15	0.9
P 20	Yellow	6.5	16
P 22 G	Green	6	16
P 22 B	Blue	4.8	17
P 22 R	Red	1.5	23
P 31	Green	7	19
P 38	Orange	1100	6000
P 39	Green	400	1300
P 45	White	1.5	5.2

Source: IBM (1984).

Table 11.1.3   Oscillation Degree a of 33 VDT Models According to the Formula Given in Text[a]

Number of Models	Oscillation Degree a	Evaluation
11	0.02–0.19	Low oscillation, no flicker visible
10	0.2–0.39	High oscillation, like unshifted fluorescent lamps
12	0.4–1.0	Unacceptable, very strong flicker

Source: Bräuninger et al. (1982, 1983, 1984)

[a] Measures of screen surface 5 × 7 cm, lighting conditions standardized (see Chapter 7.1).

These authors determined the oscillation degree a of 33 VDT models (of 20 different makes from Europe and the United States).

The results and a tentative evaluation are given in Table 11.1.3. The results reveal that only one-third of the studied models have a low oscillation degree and a display that is free of flicker. Phase-shifted fluorescent tubes have figures for a below 0.1. Oscillation degrees between 0.2 and 0.4 might be associated with visible flicker, whereas figures exceeding 0.4 exhibit strong flicker; such VDTs must be considered unacceptable.

Preference is to be given to CRTs with a degree of oscillation of character luminances comparable to figures shown by phase-shifted fluorescent tubes; a should be lower than 0.2. Refreshing rates of 80–100 Hz with phosphor decay times of about 10 msec for the 10% luminance level would be suitable.

11.1.7.3   Sharpness of Characters

An important characteristic of image quality is the sharpness of characters or image resolution.

It is generally accepted that *characters with sharp outlines guarantee a comfortable legibility, whereas characters with blurred edges offer lower visual comfort.* Gomer and Bish (1978) studied the effects of image resolution on evoked potentials of the brain. An image of higher resolution produced a stronger and more clearly defined evoked potential than an image of lower resolution. Rupp, McVey, and Taylor (1984), using five subjects, investigated the effects of a sharp and a blurred text on a display terminal. The focused display had a blurred border zone of about 0.25 mm, whereas the defocused charactes had one of approximately 0.35 mm. The mean values for accommodation and stability of accommodation were of the same order for both conditions. These results do not explain why operators prefer characters with sharp edges.

Printed texts of good quality have sharp outlines, whereas VDT characters display a relatively blurred border area. The extent of the blurred border area determines the sharpness of characters. Snyder (1980) and Snyder and Maddox (1978) used the MTF procedure to assess the degree of sharpness of displayed characters. Bräuninger et al. (1982, 1983, 1984) adopted the following method: the microscope of this equipment was moved across the capital letters "U" with a speed of 5 mm/min. The procedure of measuring the blurred border area of these characters is illustrated in Figure 11.1.4.

The tangent of the slope of the increasing luminance in the border zone of the letter "U" is determined, the distance r in millimeters characterizes the blurred border zone. With this procedure Bräuninger et al. (1982, 1983, 1984) determined the blurred border area of characters of 33 different VDT models (of 20 VDT makes). The results are reported in Table 11.1.4.

The evaluation is to some extent arbitrary and chiefly based on the observation that a border zone r of less than 0.3 mm is not perceived, whereas higher values for r reveal visible blurred edges. From Table 11.1.4 it is evident that only 5 models out of 33 show good sharpness with a blurred border zone of less than 0.3 mm; 19 models have blurred border areas of more than 0.4 mm, and they must be considered to have poor sharpness.

The reason for poor character sharpness is often found in an insufficient focusing device of the CRT. In some cases antireflective equipment, such as Micromesh filters, substantially reduces the sharpness of characters.

11.1.7.4   Character Contrasts

Various parameters have been proposed to describe the luminance difference between the image and its background. Among them are contrast ratio, the MTF, and several formulas to define contrast and percent contrast. Contrast ratio is certainly the easiest figure to be assessed. Although the MTF is a more appropriate contrast specification, contrast ratio shall be used here because this parameter is easy to understand and can be imagined to some extent.

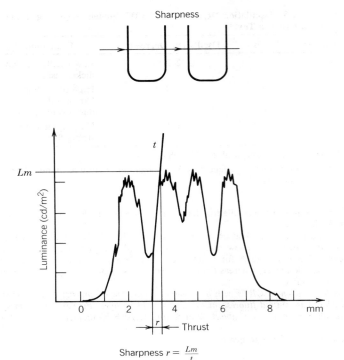

Fig. 11.1.4. Sharpness of characters is expressed in millimeters according to the formula given in the text.

Printed texts of good quality usually have high contrast ratios of $1:20$ and more; ratios of $1:10$ are considered to be suitable.

Shurtleff (1980), Snyder and Maddox (1978), and other authors have studied the influence of character contrasts on the legibility of display symbols. Shurtleff concludes that the minimum contrast ratio acceptable for general display conditions is within the range of $10:1$ to $18:1$. Field studies revealed (Läubli et al., 1981) that operators in conversational VDT jobs adjust the character luminances in such a way as to keep character contrasts in the range between $2:1$ and $31:1$, the mean contrast ratio being $9:1$. *The ratio of $10:1$ has become a generally accepted industrial standard for display design.*

This recommendation is valid for character sizes between 16 and 25 arcmin. The smaller the size of characters, the higher the contrast ratio should be.

There is also an important reciprocal relationship between sharpness and contrasts of characters: poor sharpness requires higher character contrasts if good legibility is to be maintained (Snyder and

Table 11.1.4   Sharpness of Characters of 33 VDT Models Expressed as the Blurred Border Area r (mm)[a]

Number of Models	Border Area r (mm)	Evaluation of Sharpness
5	0.1–0.29	Good
9	0.3–0.39	Unsatisfactory
12	0.4–0.5	Poor
7	>0.5	Unacceptable

Source: Bräuninger et al. (1982, 1983, 1984).

[a] The figures are mean values of nine measurements taken from nine locations on the screen at preferred luminances between 20 and 50 cd/m².

Table 11.1.5   Character Contrasts of 33 VDT Models[a]

Number of Models	Rest Luminance (%)	Approximate Contrast Ratios	Evaluation
8	<17	More than 6:1	Good
10	17–25	6:1 to 4:1	Acceptable
6	26–33	4:1 to 3:1	Insufficient
9	>33	Less than 3:1	Very poor contrast

[a] Expressed as rest luminance in the space between the "U"'s measured with preferred character luminances between 20 and 50 cd/m².

Maddox, 1980). This applies in particular if the contrast ratio between bar luminance and interspace luminance (space between two characters) is considered. In fact, CRTs with an insufficiently focused electron beam disclose an increased luminance of the space between characters and the corresponding contrast ratios can be as low as 1:2. The luminance in the space between two characters is called rest luminance and is expressed as percentage of the luminance of the bars. It is recommended that *the rest luminance of the space between two characters should not exceed 15%, which corresponds to a contrast ratio of about 1:6.*

Snyder and Maddox (1980) conclude that any symbol luminance above roughly 65 cd/m² is adequate as long as a sufficient contrast is maintained. Field studies (Läubli et al., 1981) revealed that operators engaged in conversational jobs prefer character luminances in the range of 9–77 cd/m² with a median value of 33 cd/m².

The luminance of the screen background depends on the luminous flux in the room and on the reflection characteristics of the display screen. It is therefore hardly possible to recommend a precise luminance of the screen background. In the previously cited field study (Läubli et al., 1981) the measured background luminances range between 1 and 11 cd/m² with a median figure of 4 cd/m². This gives a mean contrast ratio for characters of 9:1. A Swedish study at a telephone information center showed screen background luminances between 0.2 and 5.6 cd/m² with a mean luminance of 1.3 cd/m² during the day and of 2.0 cd/m² during the night shift (Shahnavaz, (1982).

Some authors and organizations (Cakir, Hart, and Stewart, 1979) recommend a rather high screen background luminance of between 15 and 20 cd/m². Such background luminances will mislead operators into adjusting high character luminances of more than 100 cd/m² with the risk of obtaining poor sharpness and visible flicker. *Thus, it is advisable to keep the screen background luminances below 8 cd/m².*

Bräuninger et al. (1982, 1983, 1984) studied the character contrasts of 33 VDT models under standardized conditions. They measured the luminance of the space between two "U"'s and expressed it as percentage of the preferred luminance measured inside the bars of "U" (rest luminance). The measured rest luminances and contrast ratios are reported in Table 11.1.5.

The results reveal that 18 models out of 33 had good or acceptable character contrasts with rest luminances of less than 25%. On the other hand, nine models showed character contrasts of less than 3:1, which must be considered a very poor contrast.

The previously mentioned relationship between sharpness and contrast of characters led the authors to conclude that a sharpness r of less than 0.3 mm combined with a character contrast of 1:5 should guarantee fairly good legibility. If one of these parameters does not meet the recommended level, then the other one at least should show an optimum figure. The best of the studied models revealed a high sharpness of $r = 0.19$ mm and a rest luminance of 8% (contrast ratio 1:12).

11.1.7.5   Stability of Characters

If the electron beam is well regulated, the face appears stable. If the regulation is insufficient, the characters show a poor stability. This phenomenon occurs as drift, jitter, or disturbances of linearity. Drift is a change in the position of a symbol and can cause a merging of characters, which is illustrated in the lower cutting of Figure 11.1.5. These movements are rather slow. Jitter is a brief, small, abrupt, and repetitive change in the position of a symbol. Disturbances of linearity refer to bends in the displayed lines. Such electronic interferences may produce annoying sensations.

There may be two main reasons for jitter: first, the noise in the electronic line and image-deflection circuits may cause irregular displacements of the single dot. Second, jitter can be produced by ac fields of external sources that may be superimposed on the deflection field. The operators report irregular movements or additional blurring of characters. Bauer (1984) studied the phenomenon of jitter under the conditions of reversed presentation (dark characters on bright background). For 10 subjects the mean threshold value for jitter movements at 10 Hz was 25.4 μm, which corresponded to a visual angle of 17.5″ of arc. The author concluded that for VDT operators using 80 cd/m² bright screens, the physical jitter at 10 Hz must be less than 15″ of arc. At frequencies above 30 Hz, movements

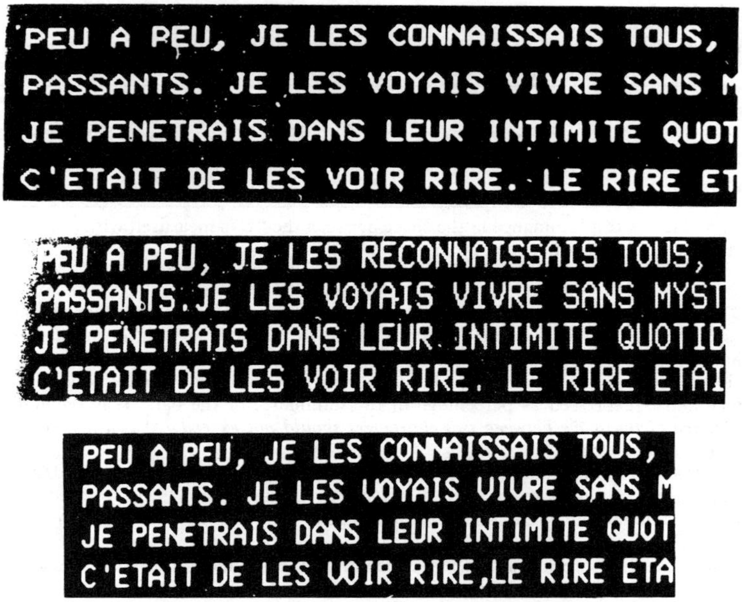

Fig. 11.1.5. Three cuttings of faces of three VDTs. Upper cutting: bars of letters are sharp and well contrasted to immediate surrounding. The letters have a good proportion of height to width, the space between letters and lines is large. *Middle cutting:* the proportion of height to width is not appropriate. The space between letters and lines is too small. *Lower cutting:* The space between letters is not constant, the bars are often merged. This phenomenon is, to some extent, due to the instability of characters (drift of symbols). The whole face is poor.

became blurred. Even if the jitter-induced movements are eliminated, other flickerlike interferences may occur on the bright background.

Fellmann et al. (1982) as well as Bräuninger et al. (1982, 1983, 1984) assessed the stability of characters in a simple way: The luminance of a dot of 0.1 mm in the bar of a character was continuously recorded with a microscope. CRTs with high stability produced a straight line, whereas those with a low stability revealed characteristic changes of the luminance. Figure 11.1.6 shows such recordings from different VDT makes.

In order to compare and to evaluate the stability of characters, the recorded variations were determined and expressed as variance in percentage of the maximal luminance. It was observed that a variance of less than 5% is not perceived; this figure served as basis for the evaluation of 34 different VDT models, reported in Table 11.1.6.

The results reveal that 8 models out of 34 disclose a high instability of characters, which shows in letters merging into each other.

11.1.7.6 Reflections on Screen Surfaces

The surface of the screen is made of glass that reflects about 4% of the incident light; this suffices to reflect clear images of the office surroundings such as lights, the keyboard, or the operator. The luminance of the reflections decreases character contrasts and dusturbs legibility; it can be so strong that it produces a glare.

Image reflections are annoying, especially since they also interfere with focusing mechanisms; the eye is induced into focusing alternately the text and the rflected image. Thus, reflections are also a source of distraction.

Stammerjohn Smith, and Cohen (1981) as well as Elias and Cail (1983) observed that bright reflections on the screen are often the principal complaint of operators. The reflected luminances reached values between 3 and 50 cd/m².

The most important preventive measures are the correct positioning of the screen with respect to lights, windows, and other objects, and the adjustment of the screen angle; a more-detailed discussion of this problem will follow in Section 11.1.8.

It is certainly not always possible to completely avoid disturbing bright reflections by these measures.

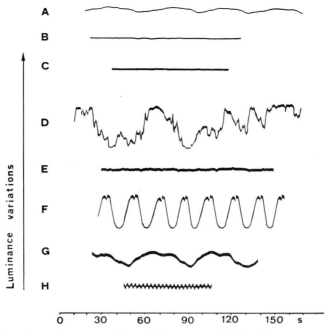

Fig. 11.1.6. Recordings of image stability of eight makes of VDTs. The luminance of a dot of 0.1 mm, located in the middle of a bar of a character, is continuously recorded. Poor stability generates characteristic luminance variations.

That is why many manufacturers have developed antireflective technologies and devices. Some of these systems are described and evaluated in the following paragraphs.

1. *Micromesh filters* are fine fabrics placed directly onto the surface or in front of the display screen. They give the screen a black appearance and are certainly efficient in reducing reflections. They have one main drawback: the sharpness of characters and their luminance are reduced since part of the emitted light is absorbed and diffused. These effects induce operators to increase the character luminance, which causes even poorer sharpness and restluminance between the characters.

2. *Etching or roughening the screen glass* is an often applied procedure. This is a chemical or mechanical treatment of the outer surface of the front glass in order to produce an optically irregular surface. It does not reduce much the total amount of reflected light but it breaks up the reflected image and makes it more diffuse. The more the surface is roughened, the more the reflection is broken up. The reflected image becomes softer because it is dispersed over a large area. The light rays generated by the excited phosphor are also dispersed when they pass through the front glass. This produces a blurring of the display characters. The roughening procedure is moderately efficient in reducing reflections, and the effects on character sharpness can be kept slight if the surface is not roughened too much. Bräuninger et al. (1982, 1983, 1984) tested several VDT models with roughened front glass, finding reduced reflections but, nevertheless, good character sharpness.

3. The *coating* of the screen with a thin *antireflective film* is a very efficient means of reducing

Table 11.1.6   The Stability of Characters[a]

Number of Models	Deviation (%)	Evaluation of Stability
8	<5	Good
15	5–20	Acceptable
3	21–40	Insufficient
8	>40	Unacceptable, letters merge

[a] Expressed as luminance deviations in percentage of the mean luminance of a bar. The figures are mean values of nine measurements on nine locations on the screen with preferred luminances.

Table 11.1.7 Reflected Luminances of a Light Source of 100 W on Screen Surfaces[a]

Tested Antireflective Device	Number of Models	Reflected Luminance (cd/m²)	Drawbacks
Makes without antireflective devices	5	525–2450	High reflections
Lambda/4 layer	1	26	Easily soiled
Etching or roughening treatment	18	143–235	Reduce sharpness to some extent
Micromesh filters	7	33–72	Poor sharpness
Polarization filters	2	160–1480	Low efficiency, double images
Colored glass	2	205–470	Poor sharpness

[a] The light source is positioned at a distance of 168 cm from the screen, and the luminance is measured at a reflection angle of 52°.

reflections. This film layer, usually a Lambda/4 layer, has the thickness of a quarter of the wavelength of light and does not diminish the sharpness of characters. The only drawbacks are the sharp outlines of the remaining image reflections and the fact that the surface is easily soiled by fingerprints.

4. *Polarization filters* polarize the incident light and partially reduce reflections. The main drawback of this device is the occurrence of double images. Moreover, the outer surface of the polarization filter is easily soiled. Bräuninger et al. (1982, 1983, 1984) have compared the reflected luminances of 30 VDT models, 5 of them having no protective device. A standardized procedure was adopted: A light source of 100 W was placed at a certain distance and angle to the screen and the reflected luminances were measured. The results and comments on observed drawbacks are reported in Table 11.1.7. All antireflective technologies have serious drawbacks. If efficiency is weighed against drawbacks, the Lambda/4 coatings and the etching-roughening procedures are preferable, whereas micromesh and polarization filters cannot be recommended. The correct placement of lights and an appropriate positioning of the screen with respect to windows remain the most efficient preventive measures.

11.1.7.7 Luminances of Surfaces

The surfaces in the visual field of a VDT operator are the screen, the frame, the surfaces of the VDT set, the desk, the keyboard, the source documents, and some elements of the surroundings, such as walls and windows. An example of excessive luminance is given in Figure 11.1.7.

If the ergonomic recommendations concerning brightness contrast ratios in the visual field are applied to a VDT workstation, the luminance contrasts between the dark screen and the neighboring surfaces (including source documents, parts of the VDT set, and keyboard) should not exceed a ratio of 1:3. Some ergonomists object to this and claim that the eyes looking at the screen are focusing the bright characters only and not the dark background. These experts recommend surface contrasts between screen and source documents that do not exceed the ratio 1:10. Conversely, if the eyes are directed toward the source document, they adapt to the brightness of that surface and the screen is in the periphery of the visual field; this contrast should not exceed 10:1.

Although nowadays not all problems of spatial and temporal differences of luminance in the visual field of VDT operators are solved, it is reasonable and realistic to make the following propositions: *The luminance contrast between dark screen and source document should not exceed the ratio 1:10. All other surfaces in the visual field should have a luminance of an average value between that of screen and source document.*

These recommendations are illustrated in Figure 11.1.8, where the luminances are expressed as reflectances (percentage of reflected light with respect to luminous flux). The instructions for the designer of a VDT workstation can therefore be summed up as follows: *Select colors of similar brightness for the different surfaces, replace eye-catching effects with black and white contrasts, avoid reflecting materials, and give preference to dim colors.*

11.1.7.8 Design of Characters

Ergonomics of Reading

There is a distinction between *reading*, as a taking in of information, and *search*, where the needed information must be located. For both functions the eyes move in quick jumps, called saccades, along a line. Between the jumps the eyes are steady and fix a certain small surface, which is projected on

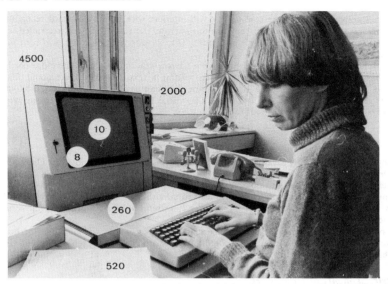

Fig. 11.1.7. Excessive brightness contrasts in the visual field of an operator. The figures in the circle indicate the measured luminance expressed in cd/m². According to Läubli et al. (1981).

the fovea. Only in the fovea and in the adjacent area, the parafovea, is the detail vision sufficiently accurate for the recognition of normal print. Two reading saccades are of importance here: the rightward reading saccades and the leftward line saccades. The reading saccades cover about 8 ± 4 letters. The line saccades start before the end of a line is reached and jump to the beginning of the next line. Correct line saccades require sufficiently large distances between lines; line distance should increase with line length. The eye pauses between the saccades last mostly between 120 and 300 msec (Bouma, 1980). During the eye pauses characters are recognized in foveal and parafoveal vision. Recognition requires that the characters are acceptable and identified. Acceptability is the degree to which characters on the screen correspond to the internal model readers have of them. This is the fundamental process of reading.

During the eye pause the fovea picks up visual information from a small surface and the adjacent area, the visual reading field. As to code numbers without much redundancy, only a few symbols can be picked up in a single glance. As to words, merely part of the information needs to be seen to ensure correct recognition.

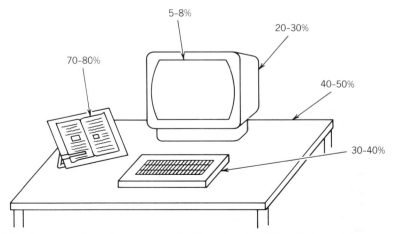

Fig. 11.1.8. Recommended reflectances for a VDT workstation. The reflectance is the percentage of reflected light related to the luminous flux falling onto the surface concerned. Reprinted with permission of Burroughs Corporation.

The lines above and below the reading line will interfere with parafoveal word recognition, unless line distances are sufficiently wide. If they are too narrow, the reading field is restricted, so that less information can be picked up in a single eye pause. *Thus, a wide visual reading field calls for sufficient interline distance.*

Nonredundant code numbers, however, have a narrow reading field and therefore need less space between the lines than a continuous text. When reading a text, the eyes make about four fixations per second. In well-printed texts the visual reading field can easily be as wide as 20 letters, about 8 to the left of the fixation and 12 to the right. The visual fields overlap, that is to say words within the visual reading field may appear at least twice.

Contrast and Colors

According to Timmers (1978) parafoveal word recognition is critically dependent on character contrast. *The lower the contrast, the narrower is the visual reading field and the lower, therefore, is the readability.* Similar effects were observed with colored letters. Engel (1980) could show that colored letters and digits can only be read when quite close to the fixation, although color itself may well be discernible far away from the fixation. This indicates that *color is a useful aid for visual search but actual reading takes place in a restricted visual field.*

If a reader is familiar with the significance of colors, then colors will help him or her to locate the required information quickly, but the recognition of a word or symbol itself depends on the legibility of the characters and not on the color.

Geometric Elements of Characters

The main geometric factors of characters are

Height (character size)
Width
Stroke width
Height-to-width ratio
Font
Horizontal spacing
Vertical spacing

These elements of characters and space are of great importance for easy identification and visual comfort.

The requirements for printed texts are well known and are described in all ergonomic text books. Since the resolution of CRT displays is usually poorer than that of printed material, character size and character contrast call for special attention. Shurtleff (1980), Snyder and Maddox (1978, 1980, 1984), Vartabedian (1971), and Stewart (1979) conducted the most detailed studies in this field, the main parameters being identification accuracy (legibility) and in some cases readability.

In general, character size is defined as the physical height of upper case letters. The classical measurement of a character is from top to bottom edge. If the edge is not sharp, it is recommended to use the 50% luminance criterion. In this case the edge is defined by the mean value of the bright and dark areas.

There is a general agreement among the authors as to the size of characters; provided that suitable contrast ratios and character luminances are given (see Section 11.1.7.4) *the range for appropriate character sizes is 16–25 arcminutes. Thus a suitable character height is 3 mm at a viewing distance of*

Table 11.1.8 Recommended Geometric Elements of Characters

Faces for Viewing Distances of 50–70 cm	Recommended Figures
Height of capital letters (mm)	3–4.3
Width of capital letters (percentage of height)	75
Stroke width (percentage of height)	20
Distance between characters (percentage of height) (two dots are recommended)	25
Space between lines (mm)	4–6

Source: Reprinted with permission of Burroughs Corporation.

Table 11.1.9 Dimensions of Characters and Face of 26 VDT Models

	Recommended Figures	Mean of 26 Models	Range
Height of upper case letters (mm)	3–4.3	3.4	2.5–4.4
Width (percentage of height)	75	55	31–81
Space between letters (mm)	0.7–1.0	1.0	0.7–1.7
Space between lines (mm)	4–6	3.3	2.1–5

Source: Reprinted with permission of Burroughs Corporation.

50 cm and 4.3 mm at a viewing distance of 70 cm (corresponding to 20 arcminutes). Larger sizes cause dot dissociation, rendering legibility more difficult. Furthermore, too big and too distant characters are associated with a lower number of letters inside the visual reading field, causing lower readability. All recommended geometric sizes of VDT characters are reported in Table 11.1.8. In the relevant literature no references are made to recommended sizes of small letters. With 25 different VDT makes the mean height of small letters was 2.4 mm, and the mean width was 1.6 mm.

All experts agree that upper case characters increase readability and accuracy. On the other hand, texts with both upper and lower case characters are generally easier to read than texts with upper case characters only. Both alternatives should be available.

Bräuninger et al. (1983, 1984) studied the character sizes and faces of 26 different VDT models. Table 11.1.9 shows the results and compares them with generally accepted recommendations. As can be seen above many models present too narrow letters and too small spaces between lines. Some of these insufficient characteristics are illustrated in the middle cutting of Figure 11.1.5.

On CRT displays different types of font are used. Shurtleff (1980) explored several of them and arrived at the conclusion that the *Lincoln–Mitre* font can be recommended, since experiments had minimized intrasymbol confusions. Snyder (1984) and Maddox (1978) as well as Abramson, Mason, and Synder (1983) observed significant effects of fonts on legibility and readability. These studies disclosed the superiority of the *Huddleston* font, particularly for flat panel displays that are subject to failures of single picture elements (pixels) or entire lines of addressable pixels. It could be demonstrated that the Huddleston font produces shorter text reading times for all levels of percent pixel failure and for all types of display failure (Abramson et al., 1983). Moreover, the interaction of font and matrix size had a substantial effect on legibility and readability; on a matrix of the size 5 × 7, the legibility of the Huddleston font was superior to that of three other fonts. However, the Huddleston and Lincoln–Mitre fonts were equally legible on 7 × 9 or 9 × 11 matrix sizes (Snyder, 1984; Snyder and Maddox, 1978). The two fonts are represented in Figure 11.1.9.

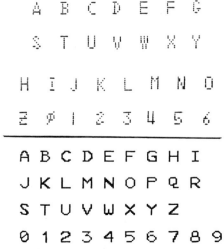

Fig. 11.1.9. The Lincoln–Mitre font (top) (7 × 5 matrix) and the Huddleston font (7 × 9 matrix). Reprinted with permission of Burroughs Corporation.

11.1.8. LIGHTING AND GENERAL ARRANGEMENTS IN VDT OFFICES

In traditional offices illumination levels between 500 and 700 lx are generally considered suitable. This is not valid for a VDT operator who is alternately looking at the dark screen and a bright source document. Since the contrast ratio between these two elements should not exceed 1:10, the illumination on the source document should be low. On the other hand, however, the reading task requires that the source document is well illuminated. This conflicting situation calls for a compromise. It is not surprising, therefore, that the definition of the optimum level is a controversial matter.

Walking through offices with VDT workstations one can often note that single fluorescent tubes have been removed or switches off by operators. Upon enquiry they cannot give plausible reasons for this but claim that a lower illumination suits them better. An interesting study on preferred lighting conditions was carried out by Shahnavaz (1982) in a Swedish telephone information center. The operators could adjust the level of illumination on the working desk. The preferred mean illumination levels on the telephone catalogue was 322 lx during the day and 241 lx for night shifts with similar levels on the desk and on the keyboard.

van der Heiden et al. (1984) observed at CAD workstations that many operators had switched off some of the lights, or drawn the curtains in front of the windows or let down the blinds. The measured illumination levels at 38 CAD workstations are reported in Table 11.1.10.

The authors concluded that the mean illumination levels were rather low under such conditions and that figures below 100 lx were insufficient for reading documents. Indeed 22% of the operators thought that there was not enough light for reading purposes.

Snyder (1984) says that ambient office illumination can be reduced to a level compatible with that of the VDT. Such levels are in the order of 200 lx and generally cause the office to appear dimly illuminated. This light is inadequate for reading hard-copy documents. To meet this insufficiency Snyder recommends local lighting fixtures designed and arranged so as to avoid reflected or direct glare for other workers.

General experience and field studies (Läubli et al., 1981; Shahnavaz, 1982) lead to the following compromise and recommendations: *For well-printed source documents an illumination level of 300 lx can be recommended. For source documents with reduced readability an illumination level of up to 500 lx may be necessary. Data entry tasks require a level of 500 to 700 lx.*

In Section 11.1.7.6 the view is expressed that an adequate arrangement of light sources is a more efficient means of preventing image or glare reflections than antireflective devices and procedures applied to the screen surface. Indeed, the reflected glare on VDTs is often related to the position and the design of lighting fixtures. If the light source is behind the VDT operator, it can easily be reflected on the screen and cause reflected glare. If it is in front of the operator, it can cause direct glare. These conditions are shown in Figure 11.1.10. Thus it is preferable to install the light source parallel to and on either side of the operator–screen axis.

In an office windows play a similar role as lights or lamps: a window in front of an operator disturbs through direct glare, behind the operator it produces reflected glare. *Therefore, the VDT workstation must be placed at a right angle to the window front.*

Glare elimination of lights is another important measure to reduce glare reflections on the screen. The most common types of glare shields are the prismatic pattern shield, the grid or louvre pattern shield, and the smoked glass shield. The former two types of glare elimination are preferred today because they effect a more favorable light distribution. The luminous flux angle should not exceed 45° with a vertical line. The glare elimination of a ceiling lamp is illustrated in Figure 11.1.11. Light sources that are recessed in the ceiling effect a lower luminance of the ceiling than lamps that are suspended below the ceiling. Thus, suspended luminaries are less suitable for operating VDTs.

Some lighting engineers propose indirect lighting for offices with a VDT. Such systems throw 90% or more of the light onto the ceiling and walls, which reflect it back into the room. This may produce a pleasant aesthetic effect, which is, however, undesirable because of the bright reflections of surrounding objects on the screen.

Table 11.1.10   Illumination Levels at 38 CAD Workstations

Workplace Element	Illumination Levels at the Workplace (lx)	
	Median	90% Range
Tablet	125	15–440
Keyboard	125	15–505
Reference table	118	15–500

Source: Van der Heiden et al. (1984).

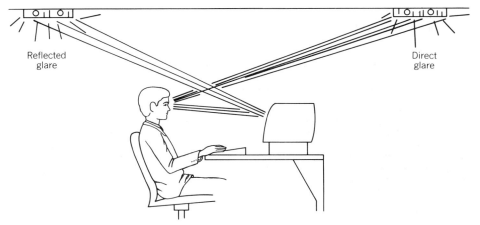

Fig. 11.1.10. Light sources behind the operator constitute a risk of reflected glare; light sources in front of the operator cause direct glare. Reprinted with permission of Burroughs Corporation.

11.1.9 ERGONOMIC DESIGN OF VDT WORKSTATIONS

11.1.9.1 Medical Aspects

As already mentioned in Section 11.1.1 the introduction of VDTs leads to an integration of employees in a man-machine system. The relationship between the operator and his machine is reciprocal, it is a kind of closed system. One of the consequences of this is a restriction of space for physical activities and constrained postures, together with long lasting static contractions of muscles. Static effort reduces blood irrigation of the muscles and causes local fatigue. The symptoms are tiredness, pains and even cramps.

Postural efforts do not only reduce performance and productivity, in the long run they also affect well-being and health. In fact, if postural efforts are repeated daily over a long period, more or less permanent aches will affect the limbs concerned, and may involve not only muscles but also other tissues. Thus long-lasting postural efforts can lead to a deterioration of joints, ligaments and tendons. Several field studies as well as general experience show that postural efforts are associated with a higher risk of

Inflammation of the joints
Inflammation of the tendon-sheaths
Inflammation of the attachment-points of tendons

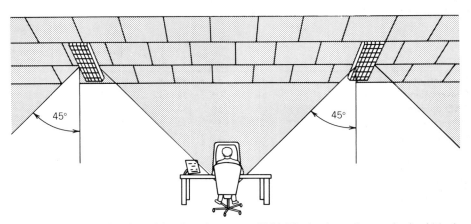

Fig. 11.1.11. Ceiling lighting with prismatic pattern shield. The luminous flow angle should be less than 45°. Reprinted with permission of Burroughs Corporation.

Symptoms of chronic degeneration of the joints in the form of chronic arthroses
Painful induration of the muscles
Disc troubles

Persistent pains in the overstrained tissues occur particularly among elderly operatives.

11.1.9.2   Field Studies on Postural Discomfort

In the last decade many studies have shown that constrained postures are often observed in office jobs involving regular work with machines, such as full-time typing, operating accounting or punching machines, as well as working with VDTs; these studies include Hünting et al. (1980); Laville (1980); Maeda, Hünting, and Grandjean (1980); Duncan and Ferguson (1974); Komoike and Horiguchi (1971); Elias and Cail (1983); Smith et al. (1981); Dainoff et al. (1981); Starr et al. (1982); Sauter (1984), Nishiyama et al. (1984); Cantoni et al. (1984); G. van der Heiden et al. (1984); Ch. N. Ong (1984); and Läubli and Hünting (1985).

A review of several field studies on VDT jobs was published by M. J. Dainoff (1982) and by the National Academy, Washington (1984). All authors reported complaints about physical discomfort, often located in the neck–shoulder–arm–hand area, but also in the back. The studies on VDT operators have the same omissions as those discussed in Section 11.1.2: the control groups in most cases are not comparable. In fact, a VDT job cannot be compared to a "non-VDT job" without reservations; there are always differences apart from the use or nonuse of a VDT.

The results of four field studies by Hünting et al. (1981), Maeda et al. (1980), van der Heiden et al. (1984) and Läubli and Hünting (1985) reveal the crucial importance of the control groups. In each survey the same questionnaire was used. Some characteristics of the eight groups are presented in Table 11.1.11. The jobs of the eight groups can be described as follows:

1. *Data-Entry Terminal.* Full-time numeric data entry with the right hand; 12,000–17,000 strokes/hr. Gaze mainly at documents.
2. *Conversational Terminal.* Payment transactions in two banks. Both hands operate keyboard. Gaze ~50% of the time on screen and ~50% on source document. Low stroke speed.
3. *CAD Operating.* Mechanical design, and printed circuit board and electrical schematics design. Tablets are used for ~40% and keyboard for ~20% of the working time. Gaze ~60% on the screen. Great diversity of body movements.
4. *Mechanical and Wiring Board Design.* The same job as CAD operating, but without VDT.
5. *Space Control in Airline.* Conversational type of job. No source documents; all information is given on the screen; 70% of the machines have fixed keyboard.
6. *Accounting Machine Operating.* Full-time numeric data entry, taken from coupons and typed with the right hand only. Gaze mainly on coupons; 8000–12,000 strokes/hr.

Table 11.1.11   The Eight Groups of the Surveys of Maeda et al. (1980), Hünting et al. (1981), G. van der Heiden et al. (1984), and Läubli and Hünting (1985)

	Survey	n	Age $\bar{x} \pm SD$	Women (%)	>6 hr/day at Keyboard or Terminal (%)
Data-entry terminal	a	53	30 ± 8	94	81
Conversational terminal	a	109	34 ± 12	50	73
CAD operators	b	69	33 ± 8	23	20
Mechanical and wiring board design	b	52	34 ± 12	31	–
Space control in airline	c	45	29 ± 7	58	100
Accounting operators	d	119	21 ± 3	100	~80
Full-time typists	a	78	34 ± 13	95	65
Traditional office work	a	55	28 ± 11	60	30

a Hünting et al. (1981).
b Van der Heiden et al. (1984).
c Läubli and Hünting (1985).
d Maeda et al. (1980).

Table 11.1.12   Incidence of "Almost Daily" Pains Reported from Four Surveys

Groups	Survey	Subjects with Almost Daily Pains in			
		Neck (%)	Shoulder (%)	Right Arm (%)	Right Hand (%)
Data-entry terminal	a	11	15	15	6
Conversational terminal	a	4	5	7	11
CAD operators	b	3	3	3	3
Mechanical and wiring board design	b	2	1	2	0
Space control in airline	c	7	11	4	6
Accounting operators	d	3	4	8	8
Full-time typists	a	5	5	4	5
Traditional office work	a	1	1	1	0

[a] Hünting et al. (1981).
[b] Van der Heiden (1984).
[c] Läubli and Hünting (1985).
[d] Maeda et al. (1980).

7. *Full-Time Typing.*   Partly copying documents, partly using dictating machine. Gaze mainly on documents. High typing speed.

8. *Traditional Office Work.*   Payment transactions in a branch office of a bank (without VDT); keyboards are used only occasionally; great diversity of body movements.

Some of the results of the questionnaires are reported in Table 11.1.12. The results reveal that serious impairments in the neck–shoulder–arm–hand area were observed in each group. The highest figures, though, are found in the group operating data-entry terminals, whereas the lowest figures are reported by CAD operators and their control group (draftsmen for mechanical and other design) as well as by the group representing traditional office work.

It is striking that the three groups characterized by a great diversity of body movements clearly present the lowest incidence of physical discomfort. On the other hand, it seems that jobs involving repetitive work on machines are likely to cause physical discomfort in the neck–shoulder–arm–hand area. Some of the VDT jobs certainly belong to this type of work, but not all of them. *The determining causal factor, therefore, is not the question of "use" or "nonuse" of VDTs.*

Läubli and Grandjean (1984) compared the incidence of neck and arm pains of four groups using VDTs with four groups not using a VDT. Furthermore, the approximate typing speed was assessed for each group. Six groups were identical with those listed in Tables 11.1.11 and 11.1.12, two control groups were added. Figure 11.1.12 discloses that the difference between use and nonuse of VDTs among the eight groups is much less important than the differences concerning the daily keystrokes.

These results confirm what was said previously: A VDT as such does not cause physical discomfort, it is the way it is used that is responsible for constrained postures and corresponding troubles.

Läubli examined all subjects of the field study (Hünting et al., 1981; Läubli; and Grandjean (1984) with methods used in clinical investigations of musculoskeletal troubles. Some of these results are shown in Table 11.1.13 and Figure 11.1.13.

Table 11.1.13   Incidence of Medical Findings in the Neck–Shoulder–Arm Area[a]

Medical Findings	Data-entry Terminal ($n = 53$) (%)	Conversational Terminal ($n = 109$) (%)	Typists ($n = 78$) (%)	Tradition Office Job ($n = 54$) (%)
Tendomyotic pressure pains in shoulders and neck	38	28	35	11
Painfully limited head movability	30	26	37	10
Pains in isometric contractions of forearm	32	15	23	6

[a] n = number of subjects is equal to 100% in each group.

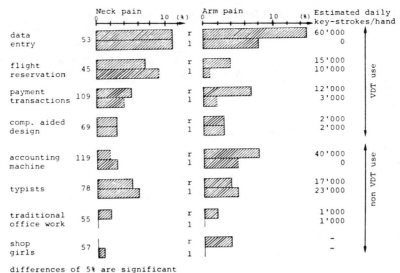

Fig. 11.1.12. Incidence of "almost daily pains" in neck and arm in eight different occupational groups. Abscissa: Percentage of subjects. Differences of more than 5% are significant. According to Läubli and Grandjean (1984).

The results of Table 11.1.13 confirm the complaints reported in Table 11.1.12: Medical findings, indicating musculoskeletal troubles in muscles, tendons, and joints are frequent in the groups using data-entry terminals and among full-time typists, whereas the group with traditional office work consisting of a great variety of activities and movements show the lowest figures. The palpation findings in shoulders, listed in Figure 11.1.13, disclose a similar distribution of symptoms.

The complaints as well as the medical findings must be taken seriously, especially since 13–27% of the examined employees had consulted a doctor for this reason.

11.1.9.3 Physical Discomfort Related to Workstation Elements and Postures

In the field studies of Hünting et al. (1981) and Läubli and Hünting (1985) several significant relationships were discovered between the design of workstations or postures, on one hand, and the incidence of complaints or medical findings, on the other hand. These results can be summarized as follows: Physical discomfort and/or the number of medical findings in the neck–shoulder–arm–hand area are likely to increase when

The keyboard level above the floor is too low.

Forearms and wrists cannot rest on an adequate support.

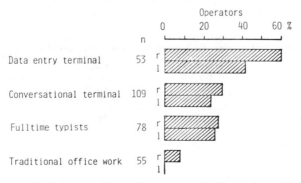

Fig. 11.1.13. Palpation findings in shoulders of four groups of office workers. Painful pressure points at tendons, joints and muscles: r = right; l = left; n = number of examined operators.

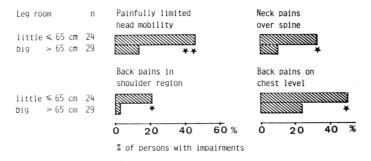

Fig. 11.1.14. Leg room and physical discomfort in 53 VDT operators at conversational terminals; n = number of operators. According to Läubli and Grandjean (1984).

The keyboard level above the desk is too high.

Operators have a marked head inclination.

Operators adopt a slanting position of the thighs under the table due to insufficient space for the legs. This is illustrated in Figure 11.1.14.

Operators disclose a marked sideward twisting (*ulnar abduction*) of hands while operating the keyboard; this is shown in Figure 11.1.15, where the subjects are divided into two subgroups: one with an angle of ulnar abduction of the hand of 20° and more, and the other with an abduction of less than 20°.

The authors discovered the same relationship when plotting the incidence of medical findings in the forearm muscles against the observed ulnar abduction of the hands.

Cantoni et al. (1984) carried out an interesting investigation on 300 operators (50% males) working in the switchbard control room of the Italian National Telephone Company in Milan where a gradual transformation of workstations from the traditional electromechanical switchboard to a VDT-operated switchboard was in progress. The old switchboard has a fixed working surface and a vertical panel

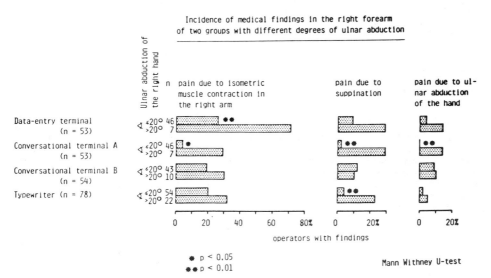

Fig. 11.1.15. Incidence of medical findings in the right forearm of two groups presenting different degrees of ulnar abduction of the right hand. Conversational terminal *A:* large space to rest wrists and hands; keyboards movable. Conversational terminal *B:* a narrow rim of about 3 cm to rest the balls of the thumbs; keyboards recessed in the table surface. p = significance between the two angle groups.

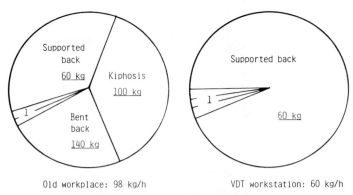

Old workplace: 98 kg/h   VDT workstation: 60 kg/h

Fig. 11.1.16. Average time of performance in four different positions of four subjects on an electrome-chanical telephone switchboard (on the left) and on a VDT-operated switchboard (on the right). For each condition the mean load on the lumbar disc L3 per hour is given. Reprinted with permission of Burroughs Corporation.

in which the operators insert the jacks; the chairs have fixed backrests and the legroom is too small. The new workstation has a screen that is adjustable in a vertical, horizontal, and sagittal plane. The keyboard is recessed in the table; a large space of 25 cm is left to rest wrists and forearms. The chairs are provided with a flexible backrest, which is 28 cm high. The survey revealed a high incidence of troubles in the cervical spine area (~60% over 36 years old) and in the lumbar region (~55% older than 36). In order to study the influence of the new workstation on postural behavior a limited group of four subjects, two males and two females aged between 37 and 44 years with body heights between 159 and 170 cm were selected. With a work-sampling procedure the characteristic trunk postures were rated and the duration of each posture determined. For each postural variant the load on the intervertebral disc of L3 was calculated according to a procedure by Molteni et al. (1983) and an unpublished modification of it for supported upper limbs by Occhipinti et al. In Figure 11.1.16 the observed duration of each posture and the mean load on L3 is shown for the old and the new workstation.

Two results should be pointed out: the change from the old to the new workstation caused a decrease of the calculated load on the third lumbar disc from 97.5 to 60.2 kg/hr. This is a significant improvement and a substantial relief of the lumbar region. Another important result are the prolonged periods of the back leaning on the backrest (from about 30% to nearly 100% of the time). The positions "bent back" and "kyphosis" nearly disappeared. On the other hand, a marked decrease of the diversity of postures was observed at the VDT workstation. The authors consider this increased postural fixation to be a disadvantage that could be met by a 15 min break every 2 hr.

11.1.9.4 Preferred Settings of Adjustable VDT Workstations in Laboratory Experiments

At workplaces for traditional office jobs the risk of constrained postures is low, since workers perform a variety of activities (see Section 11.1.1). Given such working conditions no employee will mind or complain if the design of the workplace is not optimal. However, unsuitable settings will be of crucial importance for people who adopt a constrained posture when working with VDTs or other office machines. Every inadequacy of design or dimension will, in the long run, generate static efforts associated with muscle fatigue, stiffness, and pains in the neck–shoulder–arm–hand area. This is the reason why adjustable VDT workstations appeared on the market in the last few years, mainly with the argument that a workstation should be adaptable to the different anthropometric data of employees.

Several experiments were carried out under laboratory conditions with the aim to assess the preferred settings. Since there was no agreement about the results, it is necessary to give a brief description of the different designs and conditions of the experiments.

1. Miller and Suther (1981) used 22 male and 15 female subjects; 10 operators belonged to the 5th percentile (very short people), 11 to the 50th, and about 16 to the 95th percentile (very tall people) of physical stature. The average body height was 171 cm. For the assessment of preferred settings the subjects were set the task of typing one page. It was not necessary to look at the screen. Nothing is said about the distance between keyboard and table edge or the possibility of resting forearms and wrists.

2. Brown and Schaum (1980) used 100 subjects: 40 females (body height $\bar{x} = 164$ cm) and 60 males (body height $\bar{x} = 179$ cm). The subjects played a word guessing game. After each of the six

words they had guessed, the subjects could adjust the VDT components. There was no typing involved and there was no obligation to look at the screen. The ranges of adjustability were insufficient, since 17 subjects set the screen at the upper limit of 81 cm above the floor and 9 subjects set the keyboard (home row) at the lower limit of 71 cm above the floor.

     3. Grandjean, Nishiyama, Hünting, and Piderman (1982) used 30 trained female typists. The group had a normal distribution of body statures, exceeding the mean body height of European women by 5 cm (= 166 cm); 13 subjects wore glasses. Only two subjects reported pains in neck and shoulders during the last few weeks. The home row of the keyboard was 8 cm above desk level. A support for forearms and wrists was provided (see Figure 11.1.19). the chair had a high backrest with an adjustable inclination. The subjects typed a text of five lines on the screen and afterward recopied the same text for 10 min. The preferred dimensions were assessed before, during, and after the 10-min typing test. After that the subjects had to repeat the typing tasks with imposed settings.

     4. Cushman (1984) tested 20 experienced female VDT operators who entered text from paper copy for 50 min. Their average stature was 164 cm with a standard deviation of ± 8 cm. The subjects performed the task for 10 min for each of five test heights (from 70 to 86 cm above floor). Keying rate and error data as well as subjective judgments were obtained for all five test conditions. The keyboard was 7 cm high and movable. An adjustable chair with a fixed backrest inclination was provided. There was no hand rest in front of the keyboard.

     5. Rubin and Marshall (1982) tested 25 men and 25 women aged between 17 and 73 with three different positions of a VDT workstation. Five groups were formed, each consisting of five males and five females, corresponding to the 5th, 25th, 50th, 75th, and 95th percentile of the British civilian population. All subjects were naive users of keyboards and VDTs, so they had to glance frequently from the screen to the keyboard to ensure correct key selection. The three positions are defined as follows: a "standard position," corresponding to dimensions that might be found in a typical office; a "user-preferred position," taking into account the preferred settings of the subjects; an "ergonomist determined position," which meets the recognized human factors recommendations. Each experiment lasted 10–15 min.

     6. Weber, Sancin, and Grandjean (1984) recorded the EMG of the m. trapezius at the preferred keyboard height as well as 5 cm above and below it. Each condition was tested with and without forearm–wrist support. Furthermore, the pressure load of forearms, wrists, and hands on the support and the keyboard was recorded. Twenty trained subjects had to imitate a VDT job by operating the keyboard and looking alternately at source document and screen. Each experiment lasted 10 min.

     The preferred settings obtained in these six laboratory studies are assembled in Table 11.1.14.

     As was previously mentioned, the experimental conditions of the six studies differed greatly from each other (simulated VDT work versus other test activities, different choice of stature distribution, "with" versus "without" wrist support, trained versus naive subjects). The disagreement about the

Table 11.1.14 Preferred Settings of Adjustable VDT Workstations of Six Laboratory Experiments[a]

		Miller and Suther (1981)	Brown and Schaum (1980)	Grand-jean et al. (1982)	Cushman (1984)	Rubin and Marshall (1982)	Weber et al. (1984)
Keyboard height	$\bar{x}$ (cm)	71	74	77	74–78[f]	70.5	78[g]
	Range	64–80	72–84	71–84	—	—	74–84
Screen height[c]	$\bar{x}$ (cm)	92	100	109	—	86.7	97
	Range	78–106	88–108	94–118	—	—	85–108
Screen angles[d]	Degrees	3°	10°	0°	—	—	11°
	Range	0–7°	3–17°	0–16°	—	—	0–21°
Screen distance	$\bar{x}$ (cm)	—	52	65	—	—	71
	Range	—	44–66	47–94	—	—	60–96
Seat height	$\bar{x}$ (cm)	41	50	47	—	41.8	47
	Range	32–49	44–52	43–51	—	—	43–55

[a] $\bar{x}$ = mean values.

[b] Home row height above floor.

[c] Center of the screen above floor.

[d] Upward tilted screens related to a vertical line.

[e] Screen center to table edge.

[f] Settings with best subjective ratings, highest keying performances, and lowest error rates.

[g] With wrist support.

results of Table 11.1.14 is therefore not surprising. Nevertheless, a few tendencies emerge from these results.

In all studies the range of preferred settings is rather wide. Taking into account all extreme figures the following ranges can be observed:

1. Keyboard heights: 64–84 cm.
2. Screen heights: 78–118 cm.
3. Screen distances from table edge: 44–96 cm.
4. Screen angles: 0–21°.
5. Seat heights: 32–55 cm.

The heights of the keyboard are not in accordance with the usual ergonomic recommendations, which are mainly based on anthropometric considerations. The ranges reveal that a large number of operators prefer keyboard heights above the recommended levels, between 72 and 75 cm. Cushman (1984) obtained preferred heights that were 5–10 cm above elbow level, as opposed to ergonomic text books, which recommend elbow height.

The preferred values for the screen level are in general also higher than recommended. About 50% of the subjects fixed the screen center at levels exceeding 95 cm above floor. That means that many operators prefer a nearly horizontal line of sight when looking at the screen or a slightly downward visual angle.

All six laboratory studies have one important drawback in common: The experiments were carried out only over a short period of 10 min or even less. It is very doubtful whether subjects engaged in a short-term experiment will have the same postures or prefer the same settings as those working at a VDT workstation for months or years.

In spite of these shortcomings the laboratory studies disclosed other interesting results that shall be discussed here briefly. The experiments with preferred and imposed settings (Grandjean et al., 1982) revealed an increase of physical discomfort in the neck–shoulder–arm area under the conditions of imposed keyboard heights and screen distances. These results, compared with those of Cushman (1984), lead to the conclusion that the subjects are guided by a feeling of relaxation when assessing the preferred workstation settings, which are, in their turn, associated with high keying performance and low error rates.

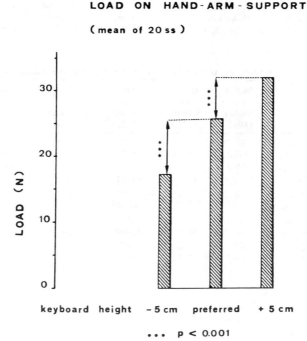

Fig. 11.1.17. Mean pressure load on wrist support and keyboard during each experimental condition with support. Means of 10-min periods for 20 trained subjects. According to Weber et al. (1984).

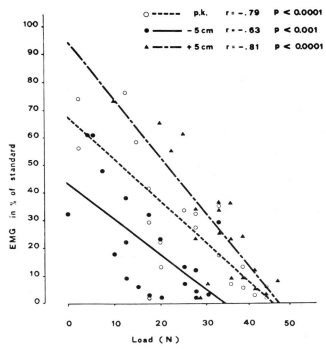

Fig. 11.1.18. Relationship between EMG activity and exerted pressure load for each experimental condition with wrist supports. r = Pearson correlation coefficient; p.k. = preferred keyboard height; ± 5 cm = below and above the p.k. Means of 10-min working periods for each of the 20 subjects. According to Weber et al. (1984).

Twenty out of 30 subjects preferred a keyboard with wrist support and 24 subjects claim that the wrist support does not impede typing activities (Grandjean et al., 1982). Weber et al. (1984) examined the effects of wrist support in a more systematic way. The pressure load exerted on the support remained surprisingly constant over the 10-min typing periods. Without wrist support the pressure load on the keys was nearly zero; when working with support, it ranged between 15 and 35 N (~1.5–3.5 kPa) on average and increased significantly with higher keyboards (see Figure 11.1.17). In each working condition with wrist support there was a significant negative correlation between EMG figures and exerted pressure load. This implies that the more the arms and hands rest on the support, the lower is the electrical activity in the m. trapezius. A comparison of the experiments with and without wrist support showed that in the former the EMG activity of the m. trapezius was always lower, independent of keyboard height. At the end of the experiment 12 out of 20 subjects preferred a keyboard with wrist support.

The results of the previously cited studies were not confirmed by the other experimental approach to keyboard operation carried out by Life and Pheasant (1984). A preceding survey of 14 typists had shown that they worked with high keyboard levels (~80 cm) with a mean elbow–keyboard distance of 8.7 cm. The authors assumed that the reported complaints of discomfort were related to the dimensions of the workplace. For the ensuing experiments the authors engaged 12 skilled female typists who performed a typing task first at elbow level, and then 5, 10, and 20 cm above elbow height. (No information is given about the existence or the use of a wrist support!) Six subjects read the copy script from a stand, whereas for the other six the copy was laid flat on the desk. The subjects did not look at the screen during the task. The performance was virtually unaffected when the keyboard was raised from elbow level to 20 cm above elbow level. But there was a consistent increase of the torque in the shoulder as the working height was increased. This means that a higher keyboard level requires more static activity by the shoulder muscles to support the weight of the upper limb. The torque at the neck (C7 articulation) was slightly decreased when the keyboard level was raised. The six subjects with script in stand complained about a significant increase of discomfort as the keyboard level was put up. This effect was less pronounced with the subjects working with the laid out script. The strongest discomfort was felt in the forearms, arms, and shoulders. The authors conclude that the home row of the keyboard should be approximately at the elbow height of the operator to reduce shoulder load. This conclusion might be valid for typists who adopt an upright trunk posture and do not rest the wrists on a support. The study of Life and Pheasant (1984) demonstrates how delicate laboratory experiments can be and how dangerous it is to draw general conclusions. Who knows what the latter experiments would have revealed if the subjects had been given chairs with proper backrests, suitable supports for the wrists, and a task requiring visual contact with the screen? These objections might well be the reason for the disagreement with all the previously cited studies.

Another study, carried out by Launis (1984), shows that a few jobs with VDTs require very peculiar postures associated with unusual dimensions of workstations. With 30 female operators engaged in computer-aided selling of railway tickets, the optimum level and acceptable range of seat and table heights were assessed. The test was carried out at a workstation, the subjects simulated their job of selling tickets with computer-aided equipment. The operators were constantly focused on the customers so that they were able to "read from the lips." Under noisy conditions they used to lean forward and stretch themselves as far as possible. The preferred levels of the seats were 44 cm (range: 36–52 cm) and the optimum table heights were between 17 and 26 cm above seat level. The latter was on average 1.4 cm lower than elbow height. This particular working situation, with the readiness to serve customers, decisively determined these results.

11.1.9.5   Preferred Settings of VDT Workstations Observed in Field Studies

Preferred Settings and Postures at VDT Workstations in Offices

An extensive study on postures and preferred settings of adjustable VDT workstations during subjects' usual working activities was carried out by Grandjean, Hünting, and Piderman (1983). The experiments were conducted on 68 operators (48 females and 20 males aged 28 years on average) in four companies: 45 subjects had a conversational job in an airline company, 17 subjects had primarily data entry activities in two banks, and 6 subjects were engaged in word-processing operations. Each subject used the adjustable workstation shown in Figure 11.1.19 for one week. The keyboard height was 8 cm above desk level. A chair was provided with a high backrest and an adjustable inclination. For the first two days a forearm–wrist support was used; on the following two days the subject operated the keyboard without support; and on the last day the subjects were given the option to use it or not. Document holders were provided as an optional supporting device for each subject. The preferred settings and postures were assessed and determined every day.

The analysis of the results of preferred settings disclosed no noticeable differences among the five days. In other words, the mean values remained practically the same for the whole week, independent of the use of wrist support. Thus the data obtained during the week could be put together for evaluation.

The frequency distribution of all preferred keyboard heights is reported in Figure 11.1.20. The

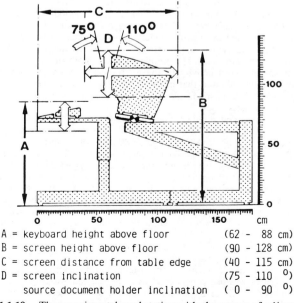

A = keyboard height above floor   (62 – 88 cm)
B = screen height above floor   (90 – 128 cm)
C = screen distance from table edge   (40 – 115 cm)
D = screen inclination   (75 – 110 °)
    source document holder inclination   (0 – 90 °)

Fig. 11.1.19.   The experimental workstation with the ranges of adjustability.

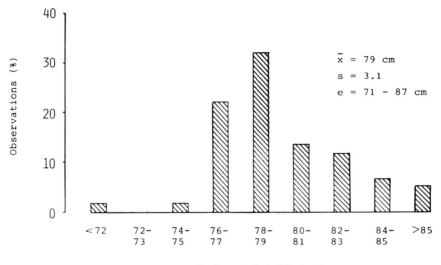

Keyboard height (cm)

Fig. 11.1.20. Preferred keyboard heights (236 observations) of 59 operators while performing their usual daily jobs (home row above floor). (s = standard deviation; e = range; $\bar{x}$ = mean value.)

95% confidence interval lies between 73 and 85 cm. A desk level between 63 and 79 cm suits a keyboard of 8 cm; a level between 68 and 84 cm suits a keyboard of 3 cm. Assuming the 95% confidence interval *the range for the adjustability of desk levels lies between 65 and 82 cm.* This seems to be a reasonable recommendation for workstation manufacturers.

The results obtained in this field study reveal slightly higher levels of keyboards than those obtained in comparable laboratory studies (cited in Table 11.1.14). It is assumed that in short-term experiments subjects are less relaxed, sit more upright and try to keep the elbows low and the forearms in a horizontal plane, thus giving preference to a slightly lower keyboard height.

All the results of preferred settings are assembled in Table 11.1.15. The preferred screen heights and screen inclinations are in some cases influenced by the attempt of operators to reduce reflections. In fact, many operators reported less annoyance by reflections if they could adjust the screen.

The capital letters on the screen were 3.4 mm high; this corresponds to a comfortable visual distance of 68 cm. At the adjustable VDT workstation the operators tended to choose greater viewing distances; 75% of them had visual distances between 71 and 93 cm. No explanation can be found for this behavior.

The visual down angles do not correspond to the comfortable viewing angles of 38° for reading subjects in a sitting posture (Lehmann and Stier, 1961). VDT operators obviously prefer slightly declined visual down angles of 0–15° (= 90% confidence).

The calculation of Pearson correlation coefficients between anthropometric data and preferred settings

Table 11.1.15   Preferred VDT Workstation Settings and Eye Levels during Habitual Working Activities[a]

Adjustable Dimensions	n_1	n_2	Mean	Range
Seat height (cm)	58	232	48	43–57
Keyboard height above floor (cm)	59	236	79	71–87
Screen height above floor (cm)	59	236	103	92–116
Visual down angles (eye to screen center) (deg)	56	224	−9°	+2° to −26°
Visual distances (eye to screen center, cm)	59	236	76	61–93
Screen upward inclination (deg)	59	236	94°	88–103°
Eye levels above floor (cm)	65	65	115	107–127

[a] n_1 = number of subjects; n_2 = number of observations; Visual down angles and screen inclination are related to a horizontal plane.

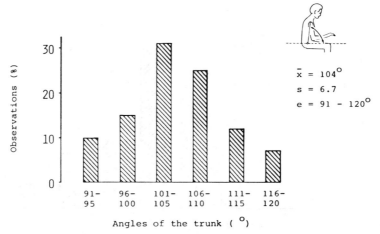

Fig. 11.1.21. Distribution of trunk postures (236 observations of 59 operators while performing their usual daily jobs. ($\bar{x}$ = mean value; s = standard deviation; e = range.)

revealed only poor relationships: between eye levels and screen heights $r = 0.25$ ($p = 0.03$) and between body length and keyboard heights $r = 0.13$ (not significant). Some of the laboratory studies revealed similar results of poor or no relationships. It can be concluded, therefore, that *the preferred settings of VDT workstations are hardly influenced by anthropometric factors*.

The most striking result of this field study concerns the postures associated with the preferred settings. The operators moved very little and did not noticeably change the main postural elements, which are obviously determined by the position of keyboard and screen. Figure 11.1.21 shows the distribution of determined trunk postures expressed as angles of a line "shoulder articulation to trochanter" related to a horizontal plane.

The trunk inclinations approximate a normal distribution. The majority of subjects prefer trunk inclinations between 100° and 110°. Only 10% demonstrate an upright trunk posture. Figure 11.1.22 illustrates the mean and the range of observed trunk postures. It is obvious that the majority of operators lean backward. This is the basis for all other adopted postural elements: the upper arms are kept higher and the elbow angles slightly opened. The mean figures for preferred trunk–arm positions are listed in Figure 11.1.23.

It must be pointed out here that about 80% of the subjects do rest their forearms or wrists if a proper support is available. If no special support is provided, about 50% of the subjects rest forearms and wrists on the desk surface in front of the 8-cm-high keyboard.

The observed postures are not due to the experimental workstation since the measurements carried out at the previous workstations had already revealed nearly the same trunk and arm inclinations.

This study confirms a general impression one gets when observing the sitting posture of many VDT operators in offices: most of them lean backward and often stretch out their legs. They seem to put up with having to bend the head forward and lift their arms. In fact, *many VDT operators in offices have postures very similar to those of car drivers*. This is understandable: Who would like to adopt an upright trunk posture when driving a car for hours?

The preferred posture does not at all correspond to the commonly published and recommended

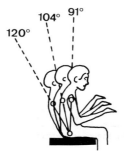

Fig. 11.1.22. Mean and range of observed trunk postures of 59 operators.

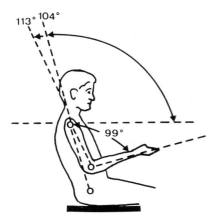

Fig. 11.1.23. The "average posture" with the workstation settings preferred by operators.

values for postures (Bell Labs, 1984; Calir et al., 1979; Din Norm 4549, 1981; Verwaltungs-Berufsgenossenschaft, 1981). Figure 11.1.24 illustrates the great gap between "wishful thinking" (recommendations) and actual postures.

An important question suggests itself here. Is the upright posture healthy and therefore recommendable or is the relaxed position with the backward-leaning trunk to be preferred? Interesting experiments by the Swedish surgeons Nachemson and Elfstrom (1970) and Andersson and Orten Green (1974) offer an answer to this question. These authors measured the pressure inside the intervertebral discs as well as the electrical activity of the back muscles in relation to different sitting postures. *When the backrest angle of the seat was increased from 90 to 120°, subjects exhibited an important decrease of the intervertebral disc pressure and of the electromyographic activity of the back.* Since heightened pressure inside intervertebral discs means that they are stressed and will wear out more quickly, the conclusion is obvious that a sitting posture with reduced disc pressure is healthy and desirable. The results of the Swedish studies indicate that by leaning the back against an inclined backrest some of the weight of the upper part of the body is transferred to the backrest. This considerably reduces the physical load on the intervertebral discs and the static strain of the muscles. Thus, VDT operators instinctively do the right thing when they prefer a backward-leaning posture and ignore the recommended upright trunk position.

Most ergonomic standards for VDT workstations are based on traditional views on healthy sitting postures. Mandal (1982) reports that the "correct seated position" goes back to 1884 when a German surgeon called Staffel recommended the well-known upright position. In 1951 Ellis (1951) was able

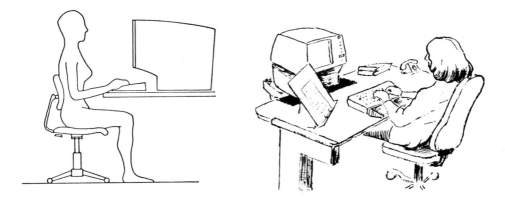

Wishful thinking   Preferred body posture

Fig. 11.1.24. Recommended and actual postures at office VDT workstations. (*a*) wishful thinking: The upright trunk posture, keeping the elbows down and the forearms nearly horizontal, demanded in many brochures and standard works. (*b*) preferred body posture: The actual posture most commonly observed in offices at VDT workstations resembles the posture of a car driver.

to confirm the old empirical rule: the maximum speed of operation for manual jobs carried out in front of the body was achieved by keeping the elbows down at the sides and the arms bent at right angles. In another investigation 12 female packers of foodstuffs were studied by Tichauer (1976), to find out how the position of the upper arms affected performance and energy expenditure. The best performance and lowest energy expenditure were not observed with elbows kept vertically down but with the arms abducted from the body at an angle of 8–23° with the vertical.

Mandal (1982) stated with exaggerated subtlety: "But no normal person has ever been able to sit in this peculiar position [upright trunk, inward curve of the spine in the lumbar region and thighs in a right angle to the trunk] for more than 1–2 min and one can hardly do any work as the axis of vision is horizontal. Staffel [the previously mentioned German surgeon] never gave any real explanation why this particular posture should be better than any other posture. Nevertheless, this posture has been accepted ever since quite uncritically by all experts all over the world as the only correct one."

It is indeed a fact: the sitting posture of students in the lecture hall or of any other audience is very seldom a "correct upright position of the trunk"; on the contrary, the majority lean backward (even with unsuitable chairs!) or in some cases lean forward with elbows resting on the desk. It is most probable that these two preferred trunk positions are associated with a substantial decrease of intervertebral disc pressure as well as lessened tension of muscles and other tissues in the lumbar and thoracic spine (Nachemson and Elfstrom, 1970; Andersson and Ortengreen, 1974).

One restriction suggests itself here: Some special work situations (such as manual work requiring freedom of movement or physical effort) might call for an upright trunk position with elbows down and forearms in a horizontal plane. Presumably the old mechanical typewriters requiring key forces of several hundred grams were more easily operated in such a posture. But the advances in electronic keyboard technology today permit very rapid keying with low key forces of 40–80 g and key displacements of 3–5 mm. The new keyboard is mainly operated by finger movements with hardly any assistance of forearms. These conditions might to some extent explain why VDT operators in offices prefer to lean backward, keep the upper arms slightly forward with the wrists on a support (which can be the desk itself), and adjust the keyboard at a rather high level.

Preferred Settings at CAD Workstations

van der Heiden and Krueger (1984) examined the use and acceptance of an adjustable workstation for CAD operations. Height and tilt of work surface as well a as height, tilt, rotation, and distance of the monitor were adjustable with motorized devices. To study the use of the adjustment fixtures a continuous registration of settings was carried out during one week. The majority of operators had more than 6 weeks' experience in using the adjustable workstation. In the test week eight women and three men were studied during their normal CAD work consisting of mechanical design. A total of 67 CAD work sessions was registrated. Questionnaires and preferred settings were obtained from 11 female and 4 male operators. Of a total of 166 registrated adjustments 142 (86%) were preadjustments and 24 (14%) readjustments. Small operators used the adjustment device more frequently than tall operators. Furthermore, operators who had not received specific instructions adjusted less frequently than others who had been given such instructions. The preferred settings are presented in Table 11.1.16.

The mean seat height of 54 cm is quite unusual, but might be related to the group of rather tall operators with a body length between 158 and 185 cm. Another striking result is the forward tilting of the monitor with a preferred mean angle of $-8°$. The operators claimed that by setting reflections resulting from windows behind them could be avoided. The forward tilting of the monitors reduced their height; for that reason most operators preferred a relatively high setting of the monitor. All other preferred dimensions are similar to those of the VDT operators as shown in Table 11.1.15.

Table 11.1.16 Preferred Settings of 15 Operators of an Adjustable CAD Workstation

	$\bar{x}$ [a]	e [a]
Seat height (cm)	54	50–57
Work surface height (cm)	73	70–80
Monitor centre above floor (cm)	113	107–115
Monitor visual distance	70	59–78
Work surface tilt (degree)	8.6°	2°–13°
Monitor tilt (degree) [b]	−7.7°	−15°–+1°

Source: Reprinted with permission of Burroughs Corporation.

[a] $\bar{x}$ = mean values; e = range.

[b] Negative tilt = a forward monitor inclination (top of the screen toward the operator).

11.1.10   ERGONOMIC DESIGN OF KEYBOARDS

The keyboard for typing letters was invented in 1868. It was a mechanical device that required a design of four parallel rows of keys. To operate these keys quickly the typist must keep the hands parallel to the rows. This requires an unnatural posture of wrists and hands, defined by an inward turning of forearms and wrists and an ulnar abduction of hands. These constrained postures often cause physical discomfort and in some cases even inflammations of tendons in the forearms of keyboard operators. Figure 11.1.25 illustrates such a constrained posture of wrists and hands at a keyboard. With the development of electronics the mechanical typewriter was replaced by the electric one. The mechanical resistance of keys was much reduced and the operation of the keyboard was made easier, but the unnatural position of wrists and hands remained.

At VDT workstations the typing activity is similar to the traditional operation of typewriters. There are some slight differences, though: first, the number of keys increased with specially arranged numerical keys and several functional keys for operating the computer. Second, in conversational jobs operators must frequently wait for the response of the computer. This response time can last from one to several seconds. According to the Swedish study of Johansson and Aronsson (1980) response times of more than 5 sec were experienced as annoying and stressing. During these unwanted pauses operators like to rest forearms and wrists on suitable supports. This induced some VDT designers to develop flat keyboards that allow operators to rest forearms and wrists on the desk. For this reason many ergonomists now recommend the use of *a flat keyboard with a home row not higher than 3 cm above the desk and the possibility for the operator to shift the keyboard on the desk according to his or her needs.*

The next step should be an ergonomic design of the keyboards in order to avoid the constrained posture of hands by reducing the inward turning and ulnar abduction of hands and providing a large support for resting forearms and wrists. Such an ergonomic design of a keyboard is shown in Figure 11.1.26. There is a general agreement among experts about other dimensions of keyboards which can be summarized as follows:

Keyboard height above desk (middle row): 30 mm

Keyboard height (front side): < 20 mm

Inclination: 5–15°

Distance between key tops: 17–19 mm

Resistance of keys: 400–800 mN

Key displacement: 3–5 mm

Fig. 11.1.25.   Position of wrists and hands operating a traditional keyboard. The parallel position to the rows requires an inward turning of forearms and wrists and a sideward twisting (ulnar *abduction*) of hands.

Fig. 11.1.26. A keyboard design in accordance with ergonomic principles. Two keyboard halves show an opening angle in order to avoid ulnar abductions of hands and have lateral slopes in order to lessen the extent of inward turning of forearms and wrists. According to Nakaseko, Grandjean, Hünting, and Gierer (1984).

The operator should feel when the stroke has been accepted; this is called the tactile feedback. The best feedback quality is achieved when the point of acceptance and pressure is located about halfway down the key displacement. Recommendations for colors and reflectances were given in Section 11.1.7.7.

11.1.11 THE CHAIR AT THE VDT WORKSTATION

For thousands of years designing chairs was mainly a question of form. Even in the early 20th century chairs tended to be status symbols rather than useful. Only in the last decades have sitting posture and seats become topics for scientific research, especially of ergonomics and orthopedics. Studies revealed that the sitting position reduces static muscular efforts in legs and hips, but increases the physical load on the intervertebral discs in the lumbar region of the spine. It is not surprising that the frequency of lumbar back complaints is high in all sedentary jobs. From these findings the most relevant recommendations for office chairs and others can be deduced: *the profile of the backrest must have a "lumbar cushion" to support this area of the back.* A properly designed backrest gives a good lumbar support in the forward sitting position as well as in the backward position. The studies on VDT workstations disclosed that operators sit like car drivers and adjust the angles of the backrest between 90° and 120°. These observations lead to the conclusion that *a VDT chair needs a backrest with a height of 50 cm above seat level and a range of adjustable inclination from 90° to 120°. It should be possible to fix the inclination at any position desired.* Figure 11.1.27 presents a chair with a proper backrest. Traditional office chairs with a relatively small back support are not suitable for VDT jobs, since they do not allow relaxation of the whole back. Furthermore, it is obvious that a VDT chair must meet all requirements of a modern office chair: adjustable height, swivel, rounded front edge of the seat surface, castors or glides, five-arm base, and user-friendly control devices.

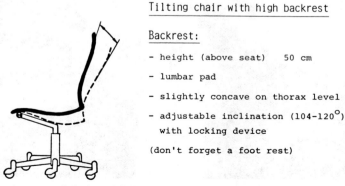

```
Tilting chair with high backrest

Backrest:

- height (above seat)    50 cm

- lumbar pad

- slightly concave on thorax level

- adjustable inclination (104-120°)
  with locking device

(don't forget a foot rest)
```

Fig. 11.1.27. A proper chair for a VDT workstation must have a high backrest of about 50 cm with an adjustable inclination that can be fixed at any position desired. Reprinted with permission of Burroughs Corporation.

11.1.12   SUMMARY OF RECOMMENDATIONS

11.1.12.1   Rules of Thumb for VDT Users

Positioning of VDTs

In order to reduce the risk of reflections on the screen surface the front of the VDT must be arranged in a right angle to the window front. Windows should be neither behind nor in front of the operator. If necessary, an intermediate screen can be placed against windows or bright surfaces.

Lighting

Light sources behind the operator can produce bright reflections on the screen; lights should be placed on either side of and parallel to the line of vision. Fluorescent light sources should be provided with prismatic pattern shields or with grid shields; the radiation angle should not exceed 45° (to a vertical line).

For a good legibility of source documents high illumination is required, but, to avoid excessive luminance contrasts, a low illumination is recommended. Operators prefer figures between 200 and 300 lx. Illumination levels should be adapted to the test quality of source documents; figures from 300 to 500 lx are suitable.

It is important to use only light sources with phase shifted fluorescent tubes.

11.1.12.2   Recommendations Addressed to Manufacturers of VDTs

Luminances and Contrasts

The background of the screen should be neither too dark nor too bright. A luminance contrast of $1:6$ between background and characters guarantees good legibility. Given a character luminance of 40 to 50 cd/m² a background luminance of 6 to 8 cd/m² would be appropriate.

The luminances of the VDT set and the keyboard should lie between that of the screen background and that of the source documents.

Oscillation Degree of Character Luminances

As a general rule it is recommended to lower the degree of oscillation of characters to a level comparable to figures shown by phase-shifted fluorescent tubes. Refreshing rates up to 80 or 100 Hz with a phosphor decay time of approximately 10 msec for the 10% luminance level are recommended.

Sharpness of Characters

The blurred border area of characters should not exceed 0.3 mm.

Character Contrasts

The contrasts of luminances in the spaces between characters and the bars of characters should have a ratio of $1:6$. That means that the brightness between characters should not exceed 15% of the brightness of characters.

Stability of Characters

The electronic control of the electron beam must secure a good stability of characters. A light dot recorded from the middle of a character bar should not exceed a variance of luminance of 20%.

Reflections on the Screen

Reflected glare on the screen should be reduced by 5–10 times, but the chosen technique should not decrease the sharpness of characters nor darken the screen background too much.

Size of Characters and Face

Height of capital letters: 3–4.2 mm
Width of characters: 75% of height
Distance between characters: 25% of height
Space between lines: 100–150% of height

Keyboard

The design of the keyboard must be flat and movable on the desk. A support surface of 15 cm depth is recommended to rest wrists and forearms.

11.1.12.3   Recommendations Addressed to Designers of Office Furniture

Adjustable Workstations

To reduce constrained postures and physical discomfort the furniture should, in principle, be conceived as flexible as possible. A proper VDT workstation should be adjustable in the following ranges:

> Keyboard height (middle row to floor): 70–85 cm
>
> Screen center above floor: 90–115 cm
>
> Screen backward inclination to horizontal plane: 88–105°
>
> Keyboard (middle row) to table edge: 10–26 cm
>
> Screen distance to table edge: 50–75 cm

A VDT workstation without adjustable keyboard height and without adjustable height and distance of the screen is not suitable for continuous work with a VDT. The controls for adjusting the dimensions of a workstation should be easy to handle, particularly at workstations for rotating shift work.

It is nearly impossible for an operator to adjust the workstation dimensions by him- or herself. Another person should be in charge of handling the controls while the operator is working at the VDT workstation.

Space for the Legs

Insufficient space for the legs causes unnatural postures. The space at the level of the knees should be at least 60 cm from table edge and at least 80 cm at the level of the feet.

The Chair for a VDT Workstation

A backward-leaning posture is justified since it allows a relaxation of the back muscles and decreases the load on the intervertebral discs. The traditional office chairs with relatively small backrests are not suitable for a VDT workstation.

The chair should have a 50-cm-long backrest (above the seat surface) and an adjustable inclination. The backrest should have a lumbar support (10–20 cm above seat level) and a slightly concave form on the thoracic level. It should be possible to fix the backrest at any position desired.

REFERENCES

Abrammson, S. R., Mason, H. L., and Snyder, H. L. (1983). The effects of display errors and font styles upon operator performance with a plasma panel. *Proceedings of the Human Factors Society,* 28–32.

Andersson, B. J. G., and Ortengreen, R. (1974). Lumbar disc pressure and myoelectric back muscle activity. *Scandanavian, Journal of Rehabilitation Medicine, 3,* 115–121.

Bauer, D., (1984). What causes flicker in bright-background VDUs and how to cure it. In E. Grandjean, Ed., *Ergonomic and health aspects in modern offices.* London: Taylor and Francis.

Beamon, W. S. and Snyder, H. L. (1975). An experimental evaluation of the spot wobble method of suppressing raster structure visibility (Technical Report AMRL-TR-75-63). Ohio: Wright-Patterson Air Force Base.

Bell Labs. (1984). Video Display Terminals. Preliminary Guidelines for Selection, Installation and Use. Holmdel, NJ.

Bouma, H., (1980). Visual reading processes and the quality of text displays. In E. Grandjean and E. Vigliani, Eds., *Ergonomic aspects of VDTs.* London: Taylor and Francis.

Bräuninger, U., (1983). Lichttechnische Eigenschaften der Bildschirmgeräte aus ergonomischer Sicht (Thesis). Zürich: Swiss Federal Institute of Technology, Department of Ergonomics and Hygiene, E.T.H.

Bräuninger, U., Grandjean, E., Fellmann, T. and Gierer, R. (1982). Lighting characteristics of VDTs. Proceedings of the Zurich seminar on digital communication.

Bräuninger, U., Grandjean, E., van der Heiden, G., Nishiyama, K., and Gierer, R. (1984). Lighting characteristics of VDTs from an ergonomic point of view. In E. Grandjean, Ed., *Ergonomic and health aspects in modern offices*. London: Taylor and Francis.

Brown, C. R., and Schaum, D. L. (1980). User adjusted VDT parameters. In E. Grandjean and E. Vigliani, Eds., *Ergonomic aspects of visual display terminals*. London: Taylor and Francis.

Cakir, A., Hart, D. J., and Stewart, T. F. M. (1979). *The VDT manual*. Darmstadt: Inca-Fiej Research Ass.

Cakir, A., Reuter, H. J., von Schmude, L., and Armbruster, A. (1978). Anpassung von Bildschirmarbeitsplätzen an die physische und psychische Funktionsweise des Menschen. Das Bundesministerium für Arbeit und Sozialordnung, Referat Presse, Postfach, 5300 Bonn.

Cantoni, S., Colombini, D., Occhipinti, E., Grieco, A., Frigo C., and Pedotti, A. (1984). Postural analysis and evaluation at the old and new workplace in a telephone company. In E. Grandjean, Ed., *Ergonomic and health aspects in modern offices*. London: Taylor and Francis.

Cushman, W. H. (1984). Data-entry performance and operator preferences for various keyboard heights. In E. Grandjean, Ed., *Ergonomic and health aspects in modern offices*, London: Taylor and Francis.

Dainoff, M. J. (1982). Occupational stress factors in VDT's operation: a review of empirical research. *Behaviour and Information Technology, 1*, 141–176.

Dainoff, M. J. Happ, A. and Crane, P. (1981). Visual fatigue and occupational stress in VDT operators. *Human Factors, 23*, 421–437.

DIN Norm 4549. (1981). Schreibtische, Büromaschinentische und Bildschirmarbeitstische. Berlin: Beuth Verlag.

Duncan, J., and Ferguson, D. (1974). Keyboard operating posture and symptoms in operating. *Ergonomics, 17*, 651–662.

Elias R., and Cail, F. (1983). Contraintes et astreintes devant les terminaux à écran cathodique. Institut National de Recherche et de Sécurité, Rapport No. 1109/Re Juin, 1982; and in *Travail Humain, 46*, 81–92.

Ellis, D. S. (1951). Speed of manipulative performance as a function of work surface height. *Journal Applied Psychology, 35*, 289–296.

Engel, F. L. (1980). Information selection from visual display units. In E. Grandjean and E. Vigliani, Eds., *Ergonomic aspects of VDTs*. London: Taylor and Francis.

Fellmann, T., Bräuninger, U., Gierer, R., and Grandjean, E. (1982). An ergonomic evaluation of VDTs. *Behavior and Information Technology, 1*, 69–80.

Gomer, F. E., and Bish, K. G. (1978). Evoked potential correlates of display image quality. *Human Factors, 20*, 589–596.

Gould, J. D., and Grischkowsky, N. (1984). Doing the same work with paper and with cathode ray tube displays. In E. Grandjean, Ed., *Ergonomic and health aspects in modern offices*. London: Taylor and Francis. 1984.

Grandjean, E. (1982). Ergonomics related to the VDT work workstation. Proceedings of the international Zurich seminar on digital communication.

Grandjean, E., Hünting, W., and Pidermann, M. (1983). VDT workstation design: preferred settings and their effects. *Human Factors, 25*, 161–175.

Grandjean, E., Nishiyama, K., Hünting, W., and Piderman, M. (1982). A laboratory study on preferred and imposed settings of a VDT workstation. *Behaviour and Information Technology, 1*, 289–304.

Gunnarsson, E. and Söderberg, I. (1983). Eye strain resulting from VDT work at the Swedish Telecommunications Administration. *Applied Ergonomics, 14*, 61–69.

Gyr, St. Nishiyama, K., Läubli T., and Grandjean, E., (1984). The effects of various refresh rates in positive and negative displays. In E. Grandjean, Ed., *Ergonomic and health aspects in modern offices*. London: Taylor and Francis.

Hünting, W., Läubli, Th., and Grandjean, E. (1981). Postural and visual loads at VDT workplaces. Part 1: Constrained Postures. *Ergonomics, 24*, 917–931.

IBM. (1984). Human factors of workstations with visual displays Report G 320-6102-2).

Johansson, G. and Aronsson, G. (1980). Stress reactions in computerized administrative work. Reports from the Department of Psychology, University of Stockholm, Suppl. 50.

Komike, Y., and Horiguchi, S. (1971). Fatigue assessment on key punch operators, typists and others. *Ergonomics, 14*, 101–109.

Krueger, H., and Mader, R. (1982). Der Einfluss der Farbsättigung auf den chromatischen Fehler der Akkomodation des menschlichen Auges. *Fortschritte der Ophthalmologie, 79*, 171–173.

Läubli, Th., and Grandjean, E. (1984). The magic of control groups in VDT field studies. In E. Grandjean, Ed., *Ergonomic and health aspects in modern offices*. London: Taylor and Francis.

Läubli, Th., and Hünting, W. (1985). Gesundheitsprobleme bei ganztägiger Bildschirmarbeit am Beispiel der Flugreservationskontrolle. *Zeitschrift für Arbeitswissenschaft.*

Läubli, Th. Hünting, W., and Grandjean, E. (1981). Postural and visual loads at VDT workplaces. Part 2: Lighting conditions and visual impairments. *Ergonomics, 24,* 933–944.

Laville, A. (1980). Postural reactions related to activities on VDU. In E. Grandjean and E. Vigliani, Eds., *Ergonomic aspects of VDTs.* London: Taylor and Francis.

Launis, M. (1984). Design of a VDT workstation for customer service. In E. Grandjean, Ed., *Ergonomic and health aspects in modern offices.* London: Taylor and Francis.

Lehmann, G., and Stier, F. (1961). Mensch und Gerät. In *Handbuch der gesamten Arbeitsmedizin,* Band 1. Berlin: Urban und Schwarzenberg.

Life, M. A., and Pheasant, S. T. (1984). An integrated approach to the study of posture in keyboard operation. *Applied Ergonomics, 15,* 83–90.

Maeda, K., Hünting, W., and Grandjean, E. (1980). Localized fatigue in accounting-machine operators. *Journal of Occupational Medicine, 22,* 810–816.

Mandal, A. C. (1982). The seated man: theories and realities. *Proceedings of the Human Factors Society,* pp. 520–524.

Meyer, J. J. Gramoni, R., Korol, S., and Rey, P. (1979). Quelques aspects de la charge visuelle aux postes de travail impliquant un écran de visualisation. *Le Travail Humain, 42,* 275–301.

Miller, I, and Suther, T. W. Preferred height and angle setting of CRT and keyboard for a display station input task. *Proceedings Human Factors Society 25th Annual Meeting.* Santa Monica, CA: Human Factors Society.

Molteni, G., Grieco, A., Colombini, D., Occhipinti, E., Pedotti, A., Boccardi, S., Frigo, G., and Menoni, O., (1983). Analisi delle posture. *C.E.E., 6,* 6.

Nachemson, A., and Elfstrom, G. (1970). Intravital dynamic pressure measurements in lubar discs. *Scandanavian Journal of Rehabilitation Medicine,* Suppl. 1.

Nakaseko, M., Grandjean, E., Hünting, W., and Gierer, R. (1984). Studies on ergonomically designed alphanumeric keyboards. *Human Factors.*

Nishiyama, K., Uehata, T., and Nakaseko, M., (1984). Health aspects of VDT operators in the newspaper industry. In E. Grandjean, Ed., *Ergonomic and health aspects in modern offices.* London: Taylor and Francis.

Oestberg, O. (1975). Health problems for operators working with CRT displays. *International Journal of Occupational Health and Safety,* Nov./Dec., 24–52.

Ong, C. N. (1984). VDT workplace design and physical fatigue. In E. Grandjean, Ed., *Ergonomic and health aspects in modern offices.* London: Taylor and Francis.

Panel on Impact of Video Viewing on Vision of Workers. (1984). *Video displays, work and vision.* Washington DC; National Academy Press.

Rubin, T., and Marshall, C. J. (1982). Adjustable VDT workstations: can naive users achieve a human factors solution?" International conference on man-machine systems, Manchester, IEE Conference Publication No. 212.

Rupp, B. A., McVey, B. W., and Taylor, S. E. (1984). Image quality and the accommodation response. In E. Grandjean, Ed., *Ergonomic and health aspects in modern offices.* London: Taylor and Francis.

Sauter St. L. (1984). Predictors of strain in VDT users and traditional office workers. In E. Grandjean, Ed., *Ergonomic and health aspects in modern offices.* London: Taylor and Francis.

Shahnavaz. H. (1982). Lighting conditions and workplace dimensions of VDU operators. *Ergonomics, 25,* 1165–1173.

Shurtleff, D. A. (1980). *How to make displays legible.* La Mirada, CA: Human Interface Design.

Smith, M. J., Cohen, B. C. F., Stammerjohn, L. W., and Happ, A. (1981). An investigation of health complaints and job stress in video display operations. *Human Factors, 23,* 387–399.

Snyder, H. L. (1980). Human visual performance and flat panel display image quality (Technical Report HFL-80-1). Blacksburg, VA: Virginia Polytechnic Institute and State University.

Snyder, H. L. (1984). Lighting characteristics, legibility and visual comfort at VDTs. In E. Grandjean, Ed., *Ergonomic and health aspects in modern offices.* London: Taylor and Francis.

Snyder, H. L., and Maddox, M. E. (1978). Information transfer from computer-generated dot matrix displays (Technical Report HFL-78-3). Blacksburg, VA: Virginia Polytechnic Institute and State University.

Snyder, H. L., and Maddox, M. E. (1980). On the image quality of dot-matrix displays. *Proceedings of the SID, 21,* 3–7.

Stammerjohn, L. W., Smith, M. J., and Cohen, B. G. F. (1981). Evaluation of workstation design factors in VDT operations. *Human Factors, 23,* 401–412.

Starr, St. J., Thompson, C. R., and Shute, St. J. (1982). Effects of video display terminals on telephone operators, *Human Factors, 24,* 699–711.

Stewart, T. F. M. (1979). Eye strain and visual display units: a review. *Displays* (April).

Tichauer, E. R. (1976). Biomechanics sustains occupational safety and health. *Industrial Engineering, 27,* 46–56.

Timmers, H. (1978). An effect of contrast on legibility of printed text. *I.P.O. Annual Progress Report,* No. 13, 64–67.

van der Heiden, G. H., Braeuninger, U., and Grandjean, E., (1984). Ergonomic studies on computer aided design. In E. Grandjean, Ed., *Ergonomic and health aspects in modern offices.* London: Taylor and Francis.

van der Heiden, G., and Krueger, H. (1984). Evaluation of ergonomic features of the Computer Vision Instaview Graphics Terminal. (Report). Zurich: Department of Ergonomics of the Swiss Federal Institute of Technology.

Vartabedian, A. G. (1971). Legibility of symbols on CRT displays. *Applied Ergonomics, 2,* 130–132.

Verwaltungs-Berufsgenossenschaft. (1981). Sicherheitsregeln für Bildschirm-Arbeitsplätze im Bürobereich. Hamburg: Überseering.

Weber, A., Sancin, E., and Grandjean, E. (1984). The effects of various keyboard heights on EMG and physical discomfort. In E. Grandjean, Ed., *Ergonomic and health aspects in modern offices.* London: Taylor and Francis.

CHAPTER 11.2

HUMAN FACTORS OF COMPUTER PROGRAMMING

MARK WEISER
BEN SHNEIDERMAN

University of Maryland
College Park, Maryland

Portions of this work were supported by the NASA Goddard Human Factors Committee and by the IBM Federal Systems Division. John Gannon, Elliot Solway, S. Iyengar, and Vic Basili made useful comments on previous drafts.

Human factors of software design and development is a subset of the science of "software psychology," the study of human performance in using computer and information systems. We focus in this chapter on human factors issues of the *process* of software design and development, and do not discuss the human factors of the software product itself as seen by its end-users. Some aspects of human factors of the software product are covered instead in Chapter 11.7. Other reviews of software psychology were done by Curtis (1984), Sheil (1981), Moher and Schneider (1981), and Shneiderman (1980).

The human factor is pervasive in software creation. Unlike cars, which are stamped out in factories by the millions, software is individually handcrafted. The cost of a software product is mostly related to the intellectual effort of researchers, designers, implementers, and maintainers, with very little manufacturing cost.

Of all the steps in creating software, the one called programming is the best understood. Other steps—specification, design, testing, debugging and maintenance—are less well understood. This chapter contains much information about the programming step, and less about the other steps. They are not less important, just less studied.

Applying human factors techniques to software development in large part means measuring what is happening. We do not have a good grasp on all the relevant human factors that contribute to successful software development. Some of the factors that are known are subsequently described, but there is no guarantee that these are more important than other factors that have not been studied. Also lacking are strong predictive or even explanatory theories, although there are useful fragments and promising attempts. (See Section 11.2.4.)

All the steps of software creation require a high level of knowledge of the problem domain and computer-related concepts and a high level of skill in problem solving. Software creation is highly innovative and therefore unpredictable. Because of these factors there is extreme variability in individual performance. One person may take 10 or 20 weeks to accomplish what someone else does in 1 week (Curtis, 1981). Because of this high variability, within subjects (repeated measures) experimental designs are preferred.

The impact of experience and knowledge is difficult to assess. Months of programming experience have not been shown to correlate well with performance measures. Stronger correlations occur with diversity of skills. (See Section 11.2.6.)

11.2.1 METHODOLOGY

When analyzing the human factors of the programming process, the key problem is gathering reliable information about what the programmers are doing. Several methods are outlined below.

11.2.1.1 Introspection and Protocol Analysis

The simplest form of gathering information about the programming process is *introspection,* in which the experimenters or the subjects simply reflect on how they write, study, and debug programs or how they use the software product. Unfortunately, introspection is done differently by each individual, and the conclusions that one person reaches may not be shared by others. Folklore about the diverse idiosyncrasies of programmers should be enough to convince most people that introspection may not produce results that would be applicable to a wide range of users. On the other hand, introspection is the way that most new ideas are discovered—by an individual working alone in his or her office and thinking quietly. Introspective judgments based on experience in using, designing, and teaching systems play an essential role in generating novel ideas.

Introspection experiments might be conducted by asking a group of subjects to evaluate their use of indentation, commenting techniques, mnemonic variable names, flowcharts, modularity, or debugging tools. Such forced thinking may compel subjects to understand their usage patterns for long or short variable names, for subscripts or statement labels, and for module or system names. As soon as one person makes his or her style explicit, it becomes possible to verify the utility of that rule and teach it to others. If the rule is widely accepted, "old-timers" will claim that they have been following it implicitly for decades, but, without an explicit rule, discussion is uncommon and teaching difficult.

A variant of introspection is *protocol analysis* in which the experimenter or the subject keeps a written or taped record of his or her perceived thought processes. This permanent record or transcript can be reviewed at leisure and analyzed for frequency counts of certain words, first or last occurrence of a word or behavior, or clusters of behavioral patterns. Lewis (1982) discusses how to use this method, calling it "thinking aloud."

Standish (1973) produced some interesting protocols of his work on a few popular problems, such as the eight queens problem, with summary observations and hypothesis. Brooks (1977) has performed

Table 11.2.1   Introspective Methods

Video or audio tape recording
Thinking aloud, with or without recording
"Pencil movement" studies
Keystroke recording
Command recording

extensive protocol analyses of program composition tasks using computer string processing facilities to help develop a model of the cognitive processes in program development. Adelson and Soloway (1984) used protocol analysis to follow the design process with expert and novice programmers. Grantham and Shneiderman (1984) used thinking aloud to study the cognitive processes in programming.

Introspection is worthwhile when the subject is a capable and sensitive programmer, since important insights may be obtained. But, there is no guarantee that other programmers will behave in the same manner or even that the subject will repeat the same process tomorrow. Analyzing protocols for substantial numbers of individuals is difficult, time-consuming, and expensive.

11.2.1.2   Case Studies and Field Studies

Case or field studies involve careful study of programming practices or computer usage at one or more sites. This approach has been used to compare management techniques, programming languages, and error patterns and is effective in discovering how people actually use computer systems. Many researchers performing case or field studies collect voluminous amounts of data in the hope that "something interesting" will emerge. Worthwhile insights may be gained, but the lack of experimental controls means that there is no guarantee that results are replicable or generalizable. The same study conducted at a different time, by a different researcher, or at a different site may not give the same results. In spite of these problems, case and field studies are popular since they provide worthwhile data to compare performance against and can reveal unexpected usage patterns.

Knuth's famous "Empirical study of FORTRAN programs" (1972) showed heavy use of the simplest forms of FORTRAN statements: 86% of the assignment statements involved no more than one arithmetic operator, 95% of DO loops used the default increment of 1, and 87% of the variables had no more than one subscript. Similar studies were done for PL/1 (Elshoff, 1976a) and APL (1977). These studies were conducted by capturing samples of programs from program libraries. A similar set of studies using programs submitted for execution during normal production focus on errors that programmers make during program development (Boies and Gould, 1974; Gannon, 1975; Gould, 1975; Young, 1974). Studies of terminal usage provide data about use of interactive systems (Boles, 1974) and programming productivity (Thadhani, 1984).

One of the most famous field studies was the IBM/New York Times Information Bank Project in which new programming technologies such as chief programmer team and structured coding were tested (Baker, 1972). This study showed dramatic improvements in productivity with reduced error rates when the new techniques were used, but it has been criticized for lack of experimental controls and exaggerated reporting. The project's high visibility and the dedicated work of expert programmers may have been as important as the new techniques. A field study by IBM in England (Lambert, 1984) found improved productivity when a team had shorter response time and individual terminals. A recent study at the University of Maryland showed that programmers do quite well without running their programs at all (Selby and Basili, 1984).

Even if no initial hypothesis is advanced and no new technique is being "tested," data collecting case or field studies are useful in developing an image of actual computer usage and programmer performance. Often the statistical analysis, coupled with informal interviews of participants and experimenters, can suggest insights that are immediately useful or provide the basis for a controlled experiment.

11.2.1.3   Controlled Experimentation

Controlled experimentation is the fundamental paradigm of scientific research. By limiting the number of independent variables, controlling for external bias, carefully measuring dependent variables, and performing statistical tests, it is possible to verify hypotheses within stated confidence levels. Controlled experimentation depends on a reductionist approach, which limits the scope of the experiment, but yields a clear convincing result. Critics complain that controlled experimentation concentrates on minor issues, but supporters argue that each small result is like a tile in a mosaic: a small fragment with clearly discernible color and shape that contributes to the overall image of programming behavior. We summarize the shape of many of the software design and development tiles in this section.

11.2.1.4 Experimental Design

It is not our intent to include a summary of experimental design in this subsection. The interested reader can consult one of the many good textbooks on the subject (Kirk, 1968; Runyon and Haber, 1984). However, some aspects of experimental design for programming are specific to programming. In particular the choice of dependent and independent variables. The choice of independent variables depends heavily on what is being studied, and in fact the entire contents of Chapter 11.2 could be considered as a long treatment of independent variables for programming studies.

Dependent variables are less governed by the topic of study and more by issues of expediency and relevance. Table 11.2.2 summarizes the major types of dependent variables that have been used in programming studies. Of these, error counts are the most realistic, but also the most difficult to perform. Timing measurements are easier to gather, but slightly more suspect because they result from a blending of times for many different tasks. Both error rate and time have immediate on-the-job consequences.

Realistic performance tasks, such as modifying or simulating a program, are suitable for a controlled environment, but are also close to tasks a professional programmer might actually perform. They thus form the most widely used group of dependent measures.

Slightly more removed from professional practice are the unrealistic performance tasks. These are easy to control and measure, but do not obviously relate to any actual programmer activity. Memorization is probably the best measure here, since it has been shown to correlate with programming performance as well as performance in a number of other skilled tasks such as chess and physics.

Program-product-based measures are the most suspect for independent measures, in part because it is not clear what these measures have to do with anything. A small object code size is important for a program on a limited memory computer, but it says nothing about that program's quality. Code metrics are even worse, so much so that they are themselves a primary dependent variable for a number of experiments. (See Section 11.2.5.) Automatic measures of programs have great future potential but are not yet to be relied on.

11.2.1.5 Experimental Ethics

Experiment designers should make every effort to protect the integrity of experimental subjects. Since programming experiments rarely place subjects at physical risk, the more serious problems of medical experimentation are avoided. Emotional threats such as heightened anxiety, increased fear of failure, or decreased confidence; and personal threats such as the misuse of experimental data or invasion of

Table 11.2.2 Some Dependent Variables for Programming Language Experiments

Type of Measurement[a]	Specific Measure
Error counts	Compile-time errors
	Run-time errors
	Design errors
	Numbers of runs
More realistic	Comprehension tests
	Program modification
	Hand simulation
Less realistic	Memorization
	Cloze
	Quality judgement
	Exam scores
Times	Time to debug
	Time to program
	Time to compile
Code metrics	Source code size
	Object code size
	Metrics[b]

[a] See text.
[b] See Section 11.2.5.

privacy must be carefully reviewed and minimized. Experimental results should not influence a student's course grade or a professional's career, and participation should be voluntary. Subject names or other identifying information should not be collected, unless absolutely necessary. Anonymous forms protect the subjects from invasion of privacy and protect experimenters from accusations that privacy has been invaded. Subjects should be informed that they are participating in an experiment and give their informed consent.

In the following sections we review, in brief, experimental results in software design and development. Generally, only statistically significant results and results that form a pattern of failure to find significance are reported. While this means some well-known studies are not represented, it permits the human factors practitioner to better rely on our tables. Many of the original papers referenced below have been collected by Curtis (1982).

11.2.2 PROGRAMMING STYLE

Programming style describes nonalgorithmic variations among programmers in the use of a programming language. Style affects comments, variable names, indenting, choice of modules, all of which have little or no bearing on the algorithm being computed, but which may have a very great bearing on how that program is understood by its current and future programmers. Of these four stylistic features "choice of module" is the one least studied, perhaps because of the difficulty of studying the very large software projects for which modularity is predicted to make a difference. This is in area of high current interest, however, and some studies in progress may soon remedy the lack of concrete results (Britton, Parker, and Parnas, 1981).

11.2.2.1 Commenting

A comment in a program is a section of text that is ignored by the programming language. Comments may be just a few words or many pages (see Figure 11.2.1). They are usually in a natural language (e.g., English) but may include diagrams or pseudocode. Commenting facilities exist in all programming languages, but there is controversy over their benefits and much discussion about the best kind of comments. Most introductory programming texts encourage students to comment [comments "serve a valuable documentary purpose" (Cress, Dirksen, and Graham, 1970)]. But some critics claim that they obscure the code, interfere with debugging by misleading the programmer, and are dangerous if not updated when the program is changed. In an early book Weinberg wrote "the population of programmers seems hopelessly split on the desirability of using comments in programs. Some see them as a distraction which is likely to draw attention and energy from more fruitful documentation efforts. Others, equally skilled and conscientious, advocate the liberal use of comments, sometimes to the extent of explaining every statement in the program" (Weinberg, 1971). For further information about commenting techniques see Chapter 11.7.

The experimental results for comments are mixed, but it is low-level comments that fare the worse. In practice, comments will be more useful for large programs than for small, and even then are likely to be helpful only if applied to large segments of code (no smaller than a subroutine or module), and harmful otherwise.

11.2.2.2 Variable Names

Except for length limitations, choice of variable name is usually unrestricted by any rules of a specific programming language. For the human reader, poorly chosen names can obscure or even disguise the meaning of a program. There is no excuse for names such as "X" or "I," when "MAX" and "NEXT" convey so much more, in a program whose useful life is more than a few days.

(a)   X := 1827 ;/* Comment: LVB RIP */

(b)   ```
/*
FACTOR finds the prime factors of NUM and returns them in array FACT.
When FACTOR is called, N should contain the maximum length of FACT.
On return, N is set to the number of factors found or 0 in case of error.
*/
SUBROUTINE FACTOR (NUM , FACT , N)
. . .
```

**Fig. 11.2.1.**  Examples of comments: (a) short comment in poor style; (b) longer comment in better style.
```

Table 11.2.3 Comments

Result	Reference
On a hand interpretation task, no comments were better than comments, and comments were better than incorrect comments	Okimoto, 1970
On a hand interpretation task, comments aided speed of completion at the expense of errors	Weissman, 1974
On recall and modification tasks, a single explanatory (high-level) comment per routine was better than many detailed (low-level) comments	Shneiderman, 1977
Comments were of limited help for short programs	Sheppard, Curtis, Milliman, and Love, 1979a

Table 11.2.4 Variable Names

Result	Reference
Programmers feel they understand programs using mnemonic names better, although comprehension tests don't bear them out	Weissman, 1974
On a comprehension task, non-mnemonic names were better than mnemonic names when all names were defined	Newsted, 1975
On a comprehension task, mnemonic names helped more on hard programs than on easy programs	Newsted, 1975
Mnemonic variable names were of little help for short programs	Sheppard et al., 1979a
On a comprehension task for uncommented programs, mnemonic names were better than nonmnemonic	Shneiderman, 1980, p. 70

Meaningful mnemonic names can help the reader grasp the semantic structure and therefore aid comprehension and retention. As for comments, mnemonic names are more useful for larger, more difficult programs, but are of little importance for short programs.

11.2.2.3 Indentation

Blank space (or "blanks") in a program text is usually not significant for the algorithm being computed, leaving the programmer free to use blanks for stylistic purpose. The horizontal positioning of lines of code is called *indentation*. [Vertical layout has received little attention, except for some weak results regarding the difficulty of reading program functions printed on more than one sheet of paper (Weissman, 1974). This result has led to the widely used, but not strongly justified, rule-of-thumb limiting functions to no more than 60 lines of code—one printed page.]

Indenting has often been advocated as a means of illustrating the control dependencies of a program. However, there are as many experimental results against it as in favor of it. Indentation may be bad because it can interfere with rapid scanning up and down a listing, lead to additional line breaks,

Table 11.2.5 Indentation

Result	Reference
Indentation had no significant effect in a hand simulation task	Weissman, 1974
In a hand simulation task, comments and indenting together are worse then either alone	Weissman, 1974
Indentation had no significant effect in a modification task	Shneiderman and McKay, 1976
Indentation had no significant effect in a comprehension task	Love, 1977
Reconstruction of FORTRAN programs was not aided by indentation	Norcio & Kerst, 1978
Filling in missing lines of a program was facilitated by the presence of indentation	Norcio, 1982
Indenting, any style, can aid program comprehension; 2–4 spaces were better than 0 or 6 spaces	Miara, Musselman, Mavarro, and Shneiderman, 1983

(a)   if b > **max**
    then b := max
    else if c > max
    then c := max
    else a := max

(b)   if b > **max**
    then b :=
    max
    else if c > max
    then c :=
    max
    else a :=
    max

Fig. 11.2.2. Examples of indenting: *(a)* an unindented program; *(b)* an indented program, showing the broken line (exaggerated) that can hinder program reading.

and emphasize the syntactic structure while clouding the high-level semantic structure (see Figure 11.2.2).

Indentation is sometimes called "prettyprinting," and there are many different styles. Most prettyprinting articles have appeared in *Sigplan Notices*, the newsletter of the Special Interest Group on Programming Languages of the Association for Computing Machinery. No consensus has yet emerged on which styles are better.

11.2.3   PROGRAMMING LANGUAGE FEATURES AND TOOLS

11.2.3.1   Conditional Statements

Programming language features are what distinguishes one language from another. Of the many ways in which languages differ, only a few have been studied, primarily those having to do with flow-of-control. A great many studies have compared "structured" control features, such as WHILE–DO and IF–THEN–ELSE, with the once ubiquitous GOTO statement. Apparently, GOTO *is* to be avoided [but see the results of Smith and Dunsmore (1982)], and structured constructs are good.

Sime, Green, and Guest (1977) and Embley (1978) each studied a novel feature which no real programming language has yet incorporated but which proved better than the standard features. Apparently there is room for improvement in most programming languages.

Table 11.2.6   Structured versus Unstructured Constructs

Result	Reference
On a program-correctness measure, GOTO's after IF's were worse than IF–THEN–ELSE	Sime et al., 1973
Students preferred using structured constructs	Weissman, 1974
Programmers untrained in structured concepts take more computer runs to code with GOTO's	Lucas and Kaplan, 1976
Avoiding GOTO's decreased programming time, compile time, and object code size	
Structured FORTRAN programs were easier to remember than unstructured programs	
On program memorization and modification tasks performed by professional programmers, structured FORTRAN code was better than unstructured	
IF–THEN is better than GOTO for comprehension by novices of FORTRAN	Smith and Dunsmore, 1982
GOTO is better than IF–THEN/WHILE–DO for comprehension by novices of FORTRAN	
A study of BEGIN–END versus IF–ENDIF styles of scope delimiters suggests that IF–ENDIF is better	Sykes, Tillman, and Shneiderman, 1983

Table 11.2.7   Other Control Constructs

Result	Reference
On a comprehension task, FORTRAN arithmetic-IF's were harder for novice programmers than logical-IF's, but not for intermediate programmers	Shneiderman, 1976
On a program correctness measure, IF–condition–THEN–ELSE–not-condition was better than normal IF–condition–THEN–ELSE	Sime et al., 1977
On comprehension and self-evaluation measures, combined CASE and ITERATION construct did better than either IF–THEN–ELSE or CASE	Embley, 1978

Gannon's unique study (Gannon, 1977) demonstrates that typed languages (such as Pascal, Ada, Mesa, and Modula-2) are better than untyped languages (such as Lisp and Fortran) for at least some tasks. Most new languages designed since Gannon's paper have been typed, with the notable exceptions of Forth and Prolog.

11.2.3.2   Flowcharts

The use of detailed and system flowcharts has been popular since the earliest days of programming. In recent years, critics have reflected the increasing anger that many programmers have when they are required to produce detailed flowcharts. Brooks (1975) called flowcharts "a curse." Ledgard and Chmura (1976), using more moderate language, argue that "program flowcharts can easily suppress much useful information in favor of highlighting sequential control flow, something which distracts the programmer from the important functional relationship in the overall design." Advocates of flowcharting, such as Bohl (1971) claim that the flowchart is "an essential tool in problem solving."

If the flow charts are much longer than the program, then they may be more difficult to study and may distract attention. Compact higher-level flowcharts can reveal relationships among program modules that are difficult to recognize from studying the code. For further information about flowcharts see Chapter 11.7.

The practitioner should consider high-level language [program design language (PDL)] descriptions of algorithms instead of flowcharts, and in any case low-level flowcharts should be avoided. Documentation of program algorithms should be supplemented by complementary documentation, such as descriptions of data structures.

11.2.3.3   Debugging

Debugging is the process of finding and correcting errors in programs. Studies of debugging have looked at types of bugs, methods of debugging, and programming language constructs that influence debugging.

The debugging results present no clear picture to the practitioner. It appears that debugging aids are not used, but that does not mean their use would not be beneficial. Perhaps the strongest results are those showing the influence of language features on bugs. To have the fewest errors use a strongly typed language without arrays.

Table 11.2.8   Flowcharts

Result	Reference
Prose procedures are easier to remember than flowcharts	Wright and Reid, 1973
Fewer errors were made following a flowchart than following a written (English) procedure	Kammann, 1975 Wright and Reid, 1973
On a set of diverse tasks (composition, comprehension, debugging, and modification), flowcharts failed to be significantly useful	Shneiderman, Mayer, McKay, Heller, 1977
Flowcharts are useful for tracing program control flow, but do not help identify faults in the program	Brooke and Duncan, 1980a, 1980b
Data structure documentation was better than either flowchart or PDL documentation for program comprehension	Shneiderman, 1982
Programmers wrote higher-quality PDL designs that flowchart designs PDLs were no better and no worse than flowcharts for comprehension or subsequent coding	Ramsey, Atwood, and Van Doren, 1983

Table 11.2.9   Types of Bugs in Student Programs

Types of Bugs	Percentage of Bugs
None	31
Execution	33
I/O	21
Declaration	10
Arithmetic	2
Compile	36
Assignment	9
Format	8
Identifiers	5
DO	3
Other	11

Source: Moulton and Muller, 1967.

Table 11.2.10   Types of Bugs in Professional Programs

Language	Percentage of Bugs
PL/1	17%
FORTRAN	16%
Assembler	12%

Source: Boies and Gould, 1974.

Table 11.2.11   Debugging Results

Result	Reference
Experienced and novice programmers had error types in approximately the same percentages	Youngs, 1974
Novices and professionals have the same number of bugs on their first run, but professionals eliminate bugs faster	Youngs, 1974
Experienced programmers did not use an interactive debugging tool.	Gould, 1975
Telling programmers the line number of bug reduced median time from 6.5 to 3 min	Gould, 1975
Automatic debugging aids generally did not help debugging times.	Gould, 1975
Bugs in assignment statements were harder to identify than iteration or array bugs.	Gould, 1975
Graduate student programmers using a statically typed language made fewer errors than those using a typeless language.	Gannon, 1977
Programmers made more errors using a typeless language even if they had already solved the same problem in a statically typed language.	Gannon, 1977
When debugging, programmers focus on coherent program subsets called "slices"	Weiser, 1982

11.2.4   COGNITION IN PROGRAM DESIGN

What goes on in programmers' minds when they are programming? What makes the difference between an experienced programmer and a beginning programmer? Answering these questions is important to understanding human factors of programming, but work is just beginning. Some examples are the work of Shertz and Weiser (1981); Soloway, Bonar, and Ehrlich (1983a); Soloway, Ehrlich, Bonar, and Greenspan (1983b); and Ehrlich and Soloway (1984). This research studies the "tacit plan knowledge" that programmers have of programs. Plan knowledge enables programmers to quickly make sense of programs they have never seen before by categorizing programs into a relatively few different types. This new area of research has great potential payoff in terms of educating new programmers, testing for programming knowledge, and developing better automatic aids to programming. [For instance, see Rogers (1984).]

11.2.4.1 Cognitive Models

Individual experiments can provide guidance on specific issues in programming language design, programming style, or design strategies, and at the same time contribute to the development of a theory of programmer behavior. Each experimental result is small but clear in its implications. Collections of experiments and replications of results will lead to deeper theories about programming.

Useful theories not only explain previous results, but also suggest predictions in novel situations. Unfortunately, there are few theories about how programmers do their work. Brooks (1977) attempted to model programmer behavior by designing a computer program. Anderson (1983) designed a computer program which modeled the learning process for LISP. Shneiderman and Mayer (1979) proposed a cognitive model of programming knowledge and processes. We expand below on the Shneiderman and Mayer model (also called the "syntactic/semantic model"), distinguishing two kinds of semantic knowledge.

A successful model of programmer behavior should help analyze performance during multiple stages of the programming process:

Design of a new program.

Composition of a program based on an explicit design.

Comprehension of a program.

Debugging of a given program.

Modification of a given program to accommodate new demands.

Learning new programming techniques.

Novice learning of programming.

A successful model should describe both the knowledge structures and the cognitive processes that people use. Knowledge structures are stored in human long-term memory that can be studied by asking subjects to describe their knowledge, by observing programmers doing work, or by arranging specific tasks that reveal the knowledge structure. Cognitive processes involve short-term memory as people study programs, and also working memory when they problem-solve to design, compose, debug, or modify programs. Long-term knowledge influences the way inputs from perception are organized and facilitates problem-solving in working memory (Figure 11.2.3).

11.2.4.2 Knowledge Structures in Long-Term Memory

In the syntactic/semantic model, long-term knowledge is of three types.

1. *Semantic Knowledge of Application Domain.* This is the programmers's knowledge of payroll practices, orbital mechanics, liver disease treatment, or chess strategies. It may take a lifetime to acquire this knowledge, and it is independent of the computer implementation.

Semantic knowledge of the application domain is level-structured (Figure 11.2.4). In chess there are low-level details about how each chess piece moves, middle-level plans about developing an attack or defense, and high-level strategies about controlling the center of the board or pursuing the end-

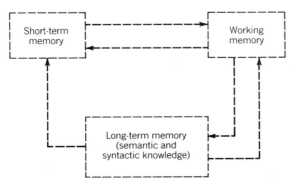

Fig. 11.2.3. Components of memory in problem solving.

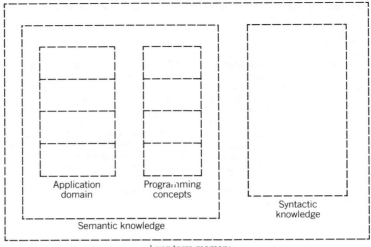

Fig. 11.2.4.   Long-term knowledge in the syntactic/semantic model.

game. Novices learn the low-level details and then spend a life-time acquiring more and more experience with patterns of attack or strategies for dominating play. A similar decomposition could be made about orbital mechanics or other application domains. Hierarchical and level-structured patterns are a vital approach for the human organization of complex knowledge. Experts have a clear sense of when they are pursuing details and when they are debating high-level plans; novices sometimes pursue details and forget high-level plans.

2. *Semantic Knowledge of Programming Concepts.*   This is the programmers's knowledge of programming practices, algorithms, file structures, data structures, programming language features, operating systems, text editors, or disk libraries. This knowledge may also take a long time to acquire, and it is independent of the application domain of a particular program.

Semantic knowledge is also language independent, that is, a programming concept such as finding the largest element of an array may be learned in one programming language, but the concept is easily applied in another programming language.

Semantic knowledge is acquired through "meaningful learning," which connects the concept to the programmer's large base of knowledge. Analogies, specific examples, or general theories can be used to convey semantic knowledge. Once acquired, semantic knowledge is relatively stable in long-term memory.

Semantic knowledge of programming concepts is level-structured, ranging from low-level semantic items to high-level concepts. Low-level items may be the effect of an assignment statement, a conditional, or a multiplication operator. These items are clustered into knowledge about patterns of use, such as three assignment statements to form a swap operation, or nested conditionals, or arithmetic expressions. These midlevel patterns are woven together to form higher-level program plans or templates, such as performing a sort, implementing a decision table, or carrying out numerical integration.

3. *Syntactic Knowledge.*   These are the details about how to express a semantic knowledge concept in a programming language. The syntax of similar concepts may be very different across different programming languages. The strings of characters are often arbitrary and, therefore, difficult to maintain in memory without frequent rehearsal. For example, SQR is the square root function in BASIC, but it squares a number in PASCAL, and SQRT is the square root function in FORTRAN. WRITE, PUT, PRINT, and OUTPUT are the keywords for printing a string or variables in different programming languages. Special characters, such as the assignment operator, may vary from an equals sign, to a colon and an equals sign, to a back arrow. Syntax for looping strategies also varies widely across programming languages.

Syntactic knowledge is language-dependent, is arbitrary, must be rote memorized, and is subject to forgetting unless frequently rehearsed. Programmers who stay away from a programming language for a few months may have difficulty recalling the syntax for composing programs, but when they read a program, the syntax provides the cues for the semantics of the language. By imitating the syntax of a given program the programmer's memory is refreshed.

The boundaries between semantic knowledge of the application domain, semantic knowledge of computer programming concepts, and syntactic knowledge are not always as sharp as in Figure 11.2.4, but this separation can be useful. The syntactic/semantic model suggests that when teaching programming, novices can be best motivated by examples from the application domain, that the algorithms should then be described with computer programming concepts, and that the syntactic details should be presented last. By using examples from familiar application domains, the students can more quickly associate the programming concepts, than if nonsense programs are used as examples.

For programmers who are learning a new programming language, there is no need to teach the computer concepts again, assuming that the languages have similar semantic structures. A simple translation of the syntax is all that is necessary to show the linkage between the known and the novel language. Of course, where there are semantic differences, extra care must be taken to highlight these differences.

For programmers learning a new application domain, there may be a substantial investment necessary before they become productive. Expert programmers recognize that they are novices in new application domains and take the time to learn the low-level details and the high-level goals before beginning to write programs.

11.2.4.3 Cognitive Processes in the Syntactic/Semantic Model

Comprehending a given program is a major intellectual effort that requires all three types of long-term knowledge. First, the programmer must recognize the syntax and separate the comments from the program, isolate the names of procedures and variables, and discriminate among statement types. Then the low-level semantic structure of the program is organized by recognizing the function of individual statements, module boundaries in the code, and distinctions between procedural and declarative information. Next higher-level computer programming concepts are spotted such as initializations of arrays, familiar algorithms, output formats, and error conditions. Finally, the reader of a program grasps the relationship between a line of code and some intention in the application domain: finding the shortest route for a milk delivery truck, computing the payroll, or determining the preferred dosage for a patient.

Again, this model is an idealization of reality. Sometimes knowledge of the application domain will guide the insecure programmer in searching for the line of code that determines whether the delivery route is optimal. A person who is knowledgeable in an application domain might learn something about programming by studying programs in his or her domain. Similarly, programmers can learn about an application domain by reading relevant programs. Programs are a means of communication and education.

In summary, readers of programs rapidly associate syntactic details with low-level semantics, organize this knowledge into higher-level units, and further convert this knowledge into compact language-independent concepts in the application domain. After reading a 300-line program, a knowledgeable programmer can convey its contents briefly, for example, "this program does Dijkstra's spanning tree algorithm for the northeastern pipeline system." Another programmer, familiar with the application domain and the programming concepts, could take this brief description and reconstruct a semantically equivalent program.

The capacity of expert programmers to encode a program into some internal semantic structure was demonstrated in a program memorization and reconstruction task (Shneiderman, 1980). After 2 min of study, expert programmers were able to reconstruct the 20-line program successfully, although they made orderly syntactic errors. They would replace the FORTRAN statement numbers consistently, after variable names consistently, or change the order of execution when it did not affect the semantics of the program.

The presence of a language-independent representation of programs was demonstrated in a recent informal experiment. Advanced undergraduates were given a 22-line Pascal program to study for 2 min. Then half the students were asked to reconstruct the program from memory. The other half of the students were asked to rewrite the program in any other programming language in which they were fluent. Both groups succeeded in reconstructing the program successfully. Participants could easily describe the program's semantics without using the syntax or keywords of any programming language.

This informal experiment and others contribute to the clarification of this model, but much work remains to be done to formulate a predictive model.

During program composition expert programmers begin with the problem domain and can construct a top-down, level-structured design for the program. They can then iteratively refine the programming concepts and finally produce the syntactic details for a specific language. Novices tend to emphasize syntactic details and have erratic patterns of moving from problem domain to programming concepts and from high-level to low-level details (Adelson and Soloway, 1984).

11.2.5 SOFTWARE QUALITY EVALUATION

Software quality measurement is an infant discipline. As this speciality matures, the ability to measure will develop, but in its youthful phase, there are conflicting opinions as to what and how software characteristics should be measured. With time, reliable and useful standard measuring concepts will emerge.

Boehm, Brown, and Lipow (1977) identify key issues such as the definition of criteria for software quality that are measurable, nonoverlapping, and evaluated automatically. They conclude that an automated tool for software quality evaluation should not merely produce a variety of metrics but should identify where and how a product is deficient. This is an ambitious goal, since we do not even agree on which metrics are useful or on satisfactory values for metrics. The simple quantitative formulas for quality have counterexamples in which programs having high ratings are of low quality. Software design methodology is evolving so rapidly that it is difficult to establish useful metrics and to "write metrics in stone" could reinforce practices that might later prove to be undesirable.

Through the years a great many metrics have been proposed. [Harrison (1984) offers a guide to further reading.] Most metrics we report on below have been subjected to experimental varification.

11.2.5.1 "Software Science" Metrics

The software science metrics (Halstead, 1977) are defined in Table 11.2.12. The efficacy of the software science metrics are in dispute: some relevant papers are Gordon and Halstead (1976); Fitzsimmons and Love (1978); Elshoff (1976b, 1978); Curtis (1980); Love and Bowman (1976). A critical review is by Shen, Conte, and Dunsmore (1983).

It appears that simply counting lines of code is as good as computing complicated metrics for predicting programmer effort and number of errors.

11.2.5.2 Productivity Metrics

Measuring productivity in programming is an extremely elusive goal because of the difficulties of measuring program quality. Most productivity results in Table 11.2.13 use a measure such as lines-of-code per programmer per month, and ignore the difference between low-quality and high-quality lines of code. It is even unknown whether high quality and lines-of-code rate are related. Section 11.2.6 has more suggestions and results for managing and measuring quality.

Perhaps the most important productive result is Weinberg and Schulman's (1974), which states that programmers will be productive on whatever dimension they are told is important. Careful instructions to programmers are essential.

11.2.5.3 Complexity Metrics

Program complexity can be logical, structural, or psychological. Logical complexity involves program characteristics that make proofs of correctness difficult, long, or not possible. For example, an increase in the complexity of expressing a program's function in first-order logic will increase its logical complexity. Structural complexity is determined by the organization of the program such as the number of distinct program paths. Psychological complexity (better called "comprehensibility") refers to characteristics that make it difficult for humans to understand software. Psychological complexity can be influenced by both logical and structural complexity, and also by other things such as quality of commenting or external documentation.

Structural complexity is the easiest to measure but the least interesting by itself. Therefore, most studies attempt to relate structural complexity metrics (usually just called "program metrics") to some other measure of psychological complexity such as time to debug or time to understand.

Table 11.2.12   Some Software Science Equations

n_1 = number of unique operands

n_2 = number of unique operators

n = vocabulary = $n_1 + n_2$

N_1 = total number of operands

N_2 = total number of operators

N = implementation length = $N_1 + N_2$

n_1^* = minimum number of unique operands

n_2^* = minimum number of unique operators

Volume: $V = N \log_2 n$

Potential Volume: $V^* = (2 + n_2^*)\log_2(2 + n_2^*)$

Program level: $L = V^*/V$

Programming effort: $E = V/L$

Table 11.2.13 Some Productivity Results

Result	Reference
Programming teams, when given different objectives (minimum core, output clarity, program clarity, minimum source statements, minimum hours), beat all other teams on their selected objective	Weinberg and Schulman, 1977
$R = L/ST$ where R is lines of source code per person-month, L is lines of code in finished product, S is staffing level, T is scheduled calendar time in months	Zak, 1977
A least squares fit to productivity data on 60 projects produced the equation: person-months $= 5.2 \times$ thousands-of-delivered-source-lines Productivity is highest when source code is *not* reused. Productivity decreases when development is spread over more than one location	Walston and Felix, 1977
Programmer productivity can be predicted by nine program characteristics and five programmer characteristics	Chrysler, 1978
Disciplined programming teams worked faster with fewer errors than undisciplined teams or individuals	Basili and Reiter, 1979
For student programmers, a log-normal random variable provides a good model of cpu and number-of-runs resource measures during programming	McNicholl and Magel, 1984

The most comprehensive empirical study of program metrics was done by Basili and Reiter (1979), who compared more than 100 metrics across different programming methods. Sheppard, Borst, and Love (1978) showed program comprehension was correlated with program length, McCabe's metric (McCabe, 1976), and program structuring, but not with Halstead's effort metric E. Dunsmore and Gannon (1979) showed that metrics based on number of data references and number of live variables could predict programming and maintenance effort.

11.2.6 PROGRAMMING MANAGEMENT

The first two decades of programming history produced the image of the introverted, isolated programmer surrounded by stacks of output. Fortunately, this image is becoming only a wild caricature of

Table 11.2.14 Some Results for Management

Result	Reference
Professional programmers vary in performance by 20:1 or more.	Sackman, Erickson, and Grant (1968) Curtis, 1981
"Egoless programming" is a state of mind in which programmers separate themselves from their products, thus aiding team effort.	Weinberg, 1971
"Chief programmer teams" surround a superior programmer with many assistants, but crucial coding is carried out by the chief programmer.	Baker, 1972
Program inspections can save one programming month per 1000 noncomment source statements.	Fagan, 1974
Programmers rating other programmers tend to agree among themselves on program quality.	Anderson and Schneiderman, 1977
Three person teams doing debugging were only modestly more effective than individuals, but consumed much more time.	Myers, 1978
Student programmers required to meet with colleagues during class time and make written critiques of each others work did better on the final exam than students who had no group debugging experience.	Lemos, 1979
For less experienced programmers, diversity of experience counts for more than number of years experience.	Sheppard et al., 1979a
Type of programming task can determine the proper choice among Chief Programmer or Egoless programming methods.	Mantei, 1981

reality. The lonely days of the programming frontier are giving way to community, interdependency, and stability.

Personality studies of programmers (Couger and Zawacki, 1978) still show their social need for interaction is significantly lower than for many other professionals. Some other important results for managing programmers are summarized below.

11.2.7 CONCLUSION

Researchers in software design and development have a golden opportunity to improve programmer's productivity and quality, as well as to gain fundamental insights into human cognition. We see opportunities to refine contemporary programming languages, coding style guidelines, design strategies, quality and productivity measures, management techniques, and software tools. Human factors are crucial to better software design and development, and objective experimentation and measurement are crucial to better human factors.

REFERENCES

Adelson, B., and Soloway, E. (1984). Designing software: novice/expert differences. In G. Salvendy, Ed., *Proceedings of the first USA/Japan on human/computer interfaces.* Amsterdam: Elsevier.

Anderson, J. R. (1983). Learning to Program. *Proceedings of the 8th International Joint Conference on Artificial Intelligence.* (pp. 57–62) Los Altos, CA: Morgan, Kaufmann.

Anderson, N., and Shneiderman, B. (1977). Use of peer ratings in evaluating computer program quality. *Proceedings of the 15th Annual Conference of the ACM Special Interest Group on Computer Personnel Research.* New York: ACM.

Baker, F. T. (1972). System quality through structured programming. *Proceedings of the fall joint computer conference,* (pp. 339–343) Montvale, NJ: AFIPS Press.

Basili, V. R., and Reiter, R. W. (1979). An investigation of human factors in software development. *Computer, 12*(12), 21–38.

Boehm, B. W., Brown, J. R., and Lipow, M. (1977). Quantitative evaluation of software quality. *Software phenomenology: working papers of the software lifecycle management workshop.* (pp. 81–94) New York: IEEE.

Bohl, M. (1971). *Flowcharting Techniques.* Chicago, IL: Science Research Associates.

Boles, S. J. (1974). User behavior on an interactive computer system. *IBM Systems Journal, 13*(1), 1–18.

Boles, S. J., and Gould, J. D. (1974). Syntactic errors in computer programming. *Human Factors, 16,* 253–257.

Britton, K. H., Parker, R. A., Parnas, D. L. (1981). A procedure for designing abstract interfaces for device interface modules. *Proceedings of the fifth international conference on software engineering.* (pp. 195–204) New York: IEEE.

Brooke, J. B., and Duncan, K. D. (1980a). Experimental studies of flowchart use at different stages of program debugging. *Ergonomics, 23*(11), 1057–1091.

Brooke, J. B., and Duncan, K. D. (1980b). An experimental study of flowcharts as an aid to identification of procedural faults. *Ergonomics, 23*(4), 387–399.

Brooks, F. P., Jr. (1975). *The Mythical Man-Month.* Reading, MA: Addison-Wesley.

Brooks, R. (1977). Towards a theory of the cognitive processes in computer programming. *International Journal of Man-Machine Studies 9,* 737–751.

Chrysler, E. (1978). Some basic determinants of computer programming productivity. *Communications of the ACM, 21*(6), 472–483.

Couger, J. D., and Zawacki, R. A. (1978). What motivates DP professionals. *Datamation 24*(9), 116–123.

Cress, P., Dirksen, P., and Graham, J. W. (1970). *FORTRAN IV with WATFOR and WATFIV.* Englewood Cliffs, NJ: Prentice-Hall.

Curtis, B. (1980). Measurement and experimentation in software engineering. *Proceedings of the IEEE, 68*(9), 1144–1157.

Curtis, B. (1981). Substantiating programmer variability. *Proceedings of the IEEE, 69*(7), 533.

Curtis, B. (1982). *Human factors in software development.* Piscataway, NJ: IEEE Computer Society.

Curtis, B. (1984). Fifteen years of psychology in software engineering: individual differences and cognitive science. *Proceedings of the 7th international conference on software engineering.* (pp. 97–106) Orlando, FL.

Dunsmore, H. E., and Gannon, J. D. (1979). Data referencing: an empirical investigation. *IEEE Computer, 12,* 50–59.

Ehrlich, K., and Soloway, E. (1984). An empirical investigation of the tacit plan knowledge in programming. In M. L. Schneider, Ed., *Human Factors in Computer Systems*. (pp. 113–134) Norwood, NJ: Able.

Elshoff, J. L. (1976a). An analysis of some commercial PL/1 programs. *IEEE Transactions on Software Engineering, SE-2,* 113–121.

Elshoff, J. L. (1976b). Measuring commercial PL/I programs using Halstead's criteria. *ACM SIGPLAN Notices, 7*(5), 38–46.

Elshoff, J. L. (1978). An investigation into the effects of the counting method used on software science measurements. *ACM SIGPLAN Notices, 13*(2), 30–45.

Embley, D. W. (1978). Empirical and formal language design applied to a unified control structure. *International Journal of Man-Machine Studies, 10,* 197–216.

Fagan, M. (1974). Design and code inspections, and process control in the development of programs (IBM Technical Report 21.572).

Fitzsimmons, A., and Love, T. (1978). A review and evaluation of software metrics. *Computing Surveys, 10*(1), 3–18.

Gannon, J. D. (1975). Language design to enhance programming reliability. (Ph.D. Thesis). Toronto: University of Toronto.

Gannon, J. D. (1977). An experimental evaluation of data type conventions. *Communications of the ACM, 20*(8), 584–595.

Gordon, R. D., and Halstead, M. H. (1976). An experiment comparing FORTRAN programming times with the software physics hypothesis. *Proceedings of the National Computer Conference 45,* 935–937, Montvale, NJ: AFIPS Press.

Gould, J. D. (1975). Some psychological evidence on how people debug computer programs. *International Journal of Man-Machine Studies, 7,* 151–182.

Grantham, C., and Shneiderman, B. (1984). Programmer behavior and cognitive activity: an observational study. *Proceedings ACM Washington DC chapter annual technical symposium.* Washington, D.C.: ACM.

Halstead, M. (1977). *Elements of Software Science.* New York: Elsevier Computer Science Library.

Harrison, W. (1984). Software complexity metrics: a bibliography and category index. *ACM Sigplan Notices, 19*(2), 17–27.

Kammann, R. (1975). The comprehensibility of printed instructions and flowchart alternative. *Human Factors, 17,* 183–191.

Kirk, R. E. (1968). *Experimental design: procedures for the behavior sciences.* Monterey, CA: Brooks-Cole.

Knuth, D. E. (1972). An empirical study of FORTRAN programs. *Software—Practice and Experience, 1,* 105–133.

Lambert, G. N. (1984). A comparative study of system response time on program developer productivity. *IBM Systems Journal, 23*(1), 36–43.

Ledgard, H., and Chmura, L. (1976). *COBOL with Style.* Rochelle Park, NJ: Hayden.

Lemos, R. S. (1979). An implementation of structured walkthroughs in teaching COBOL programming. *Communications of the ACM, 22*(6), 335–340.

Lewis, C. (1982). Using the "thinking-aloud" method in cognitive interface design (RC 9265). Yorktown Heights, NY: IBM.

Love, T. (1977). Relating individual differences in computer programming performance to human information processing abilities (Ph.D. Dissertation). University of Washington.

Love, T., and Bowman, A. B. (1976). An independent test of the theory of software physics. *ACM SIGPLAN Notices, 7*(10), 42–49.

Lucas, H. C., and Kaplan, R. B. (1976). A structured programming experiment. *The Computer Journal, 19*(2), 136–138.

Mantel, M. (1981). The effect of programming team structures on programming tasks. *Communications of the ACM, 24*(3), 106–113.

McCabe, T. J. (1976). A complexity measure. *IEEE Transactions on Software Engineering, SE-2*(4), 308–320.

McNicholl, D. G., and Magel, K. (1984). Stochastic modeling of individual resource consumption during the programming phase of software development. In M. L. Schneider, Ed., *Human Factors in Computer Systems.* (pp. 79–112) Norwood, NJ: Ablex.

Miara, R. J., Musselman, J. A., Navarro, J. A., and Shneiderman, B. (1983). Program indentation and comprehensibility. *Communications of the ACM 26*(11), 861–867.

Moher, T., and Schneider, G. M. (1981). Methods for improving controlled experimentation in software engineering. *Proceedings of the fifth international conference on software engineering,* (pp. 224–233) New York: IEEE.

Moulton, P. G., and Muller, M. E. (1967). DITRAN—a compiler emphasizing diagnostics. *Communications of the ACM, 10,* 45–52.

Myers, G. J. (1978). A controlled experiment in program testing and code walkthroughs/inspections. *Communications of the ACM, 21*(9), 760–768.

Newsted, P. R. (1975). Grade and ability prediction in an introductory programming course. *ACM SIGCSE Bulletin, 7*(2), 87–91.

Norcio, A. F. (1982). Indentation, documentation and programmer comprehension. *Proceedings of human factors in computer systems,* Washington, DC: ACM. pp. 118–120.

Norcio, A. F., and Kerst, S. M. (1978). *Human memory organization for computer programs.* Washington, DC: Catholic University of America.

Okimoto, G. H. (1970). The effectiveness of comments: a pilot study (IBM SDD Technical Report TR 01.1347). IBM.

Ramsey, H. R., Atwood, M. E., and Van Doren, J. R. (1983). Flowcharts versus program design languages: an experimental comparison. *Communications of the ACM, 26*(6), 445–449.

Rogers, J. B. (1984). Inferring cognitive focus from students' programs. *Proceedings of the Computer Science Education Conference.* New York: ACM.

Runyon, R. P., and Haber, A. (1984). *Fundamentals of behavioral statistics,* 5th ed. Reading, MA: Addison-Wesley.

Saal, H. J., and Weiss, Z. (1977). An empirical study of APL programs. *Computer Languages, 2*(3), 47–60.

Sackman, H., Erickson, W. J., and Grant, E. E. (1968). Exploratory experimental studies comparing online and offline programming performance. *Communications of the ACM, 11*(1), 3–11.

Selby, R. W., Jr., and Basili, V. R. (1984). CLEANROOM software development: an empirical evaluation (Tech. Rep. TR-1415). College Park, MD: Dept. Com. Sci. Univ. Maryland.

Shell, B. A. (1981). The psychological study of programming. *Computing Surveys, 13,* 101–120.

Shen, V. Y., Conte, S. D., and Dunsmore, H. E. (1983). Software science revisited: a critical analysis of the theory and its empirical support. *IEEE Transactions on Software Engineering, SE-9*(2), 155–165.

Sheppard, S. B., Borst, M. A., and Love, L. T. (1978). Predicting software comprehensibility (Technical Report TR 77-388100-1). Arlington, VA: General Electric Information Systems Programs.

Sheppard, S. B., Curtis, B., Milliman, P., and Love, T. (1979a). Modern coding practices and programmer performance. *IEEE Computer,* December, 41–49.

Sheppard, S. B., Milliman, P., and Curtis, B. (1979b). Experimental evaluation of on-line program construction. Arlington, VA: General Electric Information Systems Programs.

Shertz, J., and Weiser, M. (1983). Programming problem representation in novice and expert programmers. *International Journal of Man-Machine Studies,* Dec.

Shneiderman, B. (1976). Exploratory experiments in programmer behavior. *International Journal of Computer and Information Sciences, 5*(2), 123–143.

Shneiderman, B. (1977). Measuring computer program quality and comprehension. *International Journal of Man-Machine Studies, 9.*

Shneiderman, B. (1982). Control flow and data structure documentation: two experiments. *Communications of the ACM, 25*(1), 55–63.

Shneiderman, B. (1980). *Software psychology: human factors in computer and information systems.* Boston: Little, Brown.

Shneiderman, B., and Mayer, R. (1979). Syntactic/semantic interactions in programmer behavior: a model and experimental results. *International Journal of Computer and Information Sciences, 7,* 219–239.

Shneiderman, B., Mayer, R., McKay, D., and Heller, P. (1977). Experimental investigations of the utility of detailed flowcharts in programming. *Communications of the ACM, 20,* 373–381.

Shneiderman, B., and McKay, D. (1976). Experimental investigations of computer program debugging and modification. *Proceedings of the 6th international congress of the International Ergonomics Association.* International.

Sime, M. E., Green, T. R. G., and Guest, D. H. (1973). Scope marking in computer conditionals—a psychological evaluation. *International Journal of Man-Machine Studies, 5,* 105–113.

Sime, M. E., Green, T. R. G., Guest, D. H. (1977). Scope marking in computer conditionals—a psychological evaluation. *International Journal of Man-Machine Studies, 9,* 107–118.

Smith, C. H., and Dunsmore, H. E. (1982). On the relative comprehensibility of various control structures by novice Fortran programmers. *International Journal of Man-Machine Studies, 17,* 165–171.

Soloway, E., Bonar, J., and Ehrlich, K. (1983). Cognitive strategies and looping constructs: an empirical study. *Communications of the ACM, 26*(11), 853–860.

Soloway, E., Ehrlich, K., Bonar, J., and Greenspan, J. (1983b). What do novices know about programming? In A. Badre, Ed., *Directions in Human-Computer Interactions.* Norwood, NJ: Ablex.

Standish, T. A. (1973). Observations and hypotheses about program synthesis mechanisms (Automatic Programming Memo 9, Report No. 2780). Cambridge, MA: Bolt Beranek and Newman.

Sykes, F., Tillman, R. T., and Shneiderman, B. (1983). The effect of scope delimiters on program comprehension. *Software—Practice and Experience, 13,* 817–824.

Thadhani, A. J. (1984). Factors affecting programmer productivity during application development. *IBM Systems Journal, 23*(1), 19–35.

Walston, C. E., and Felix, C. P. (1977). A method of programming measurement and estimation. *IBM Systems Journal, 16*(1), 54–73.

Weinberg, G. M. (1971). *The psychology of computer programming.* New York: Van Nostrand.

Weinberg, G. M., and Schulman, E. L. (1974). Goals and performance in computing programming. *Human Factors, 16*(1), 70–77.

Weiser, M. (1982). Programmers Use Slices When Debugging. *Communications of the ACM, 25*(7), 446–452.

Weissman, L. (1974). A methodology for studying the psychological complexity of computer programs (Ph.D. Thesis). Toronto, University of Toronto.

Wright, P., and Reid, F. (1973). Written information: some alternatives to prose for expressing the outcomes of complex contingencies. *Journal of Applied Psychology, 57,* 160–166.

Youngs, E. A. (1974). Human errors in programming. *International Journal of Man-Machine Studies, 6,* 361–376.

Zak, D. (1977). Initial experiences in programming productivity realized using implementation language. *Fifth annual conference principles of software development.* Control Data Corporation, pp, 191–219.

CHAPTER 11.3

SOFTWARE INTERFACE DESIGN

ROBERT C. WILLIGES

BEVERLY H. WILLIGES

JAY ELKERTON

Virginia Polytechnic Institute and State University
Blacksburg, Virginia

Support for this effort was provided in part by the Army Research Institute under contract number MDA903-84-C-0217. Dr. Milton S. Katz serves as technical monitor for this contract. The views, opinions, and findings contained in this section are those of the authors and should not be construed as an official Department of the Army position, policy, or decision, unless so designated by other official documentation. The first author is indebted to discussions during his participation in a National Research Council workshop on Software Human Factors and his subsequent discussions with Dr. John A. Whiteside of Digital Equipment Corporation to form the basis of the iterative software design process presented in this section. The authors also thank Ms. Kathy J. Atkins who efficiently managed the word processing of the various versions of this section.

11.3.1 INTRODUCTION

With the proliferation of interactive computer systems and the growing number of computer-unsophisticated users, it is clear that the key to optimizing the human–computer interface is the appropriate design and management of dialogue. Traditionally, human factors design of the human–computer interface has been restricted primarily to hardware and workplace layout considerations. User considerations in the design of computer hardware include such topics as keyboard layout, system response delays, and quality assessment of the visual display screens. Workplace design, on the other hand, incorporates information related to anthropometrics of the user and human factors in the working environment of the computer user such as lighting, ventilation, and equipment layout.

The information interface between the human and the computer, particularly with computer-unsophisticated users, may be even more important than hardware and workplace layout considerations. Transmitting and receiving data through this information interface can be characterized as a communication or dialogue problem. Management of human–computer dialogue is essential for the enhancement of the information processing and decision-making capabilities of computer users working in real-time, demanding tasks. The purpose of this section is to review the human factors design process for creating these software interfaces and to discuss the potential for developing more-sophisticated adaptive interfaces appropriate for inexperienced users.

11.3.1.1 Interface Design Philosophy

Recently, Gould and Lewis (1983) discussed four critical factors that need to be considered in developing and evaluating a computer system for end users. These factors include early focus on users, interactive design, empirical measurement, and iterative design.

Early focus on users means the design team should have direct contact with potential users prior to overall system design, and the user interface should be the first component developed in the system design. Interactive design involves the inclusion of typical users who become members of the design team at least for some short period of time at the beginning of the design cycle. Empirical measurement includes both evaluating the learnability and usability of a software interface as well as conducting empirical and experimental studies throughout the development process. Finally, iterative design involves the incorporation of the results of behavioral testing into the next version of the system.

Although these four design features seem rather straightforward to human factors specialists, results of six surveys conducted by Gould and Lewis (1983) of 447 system designers showed little recognition of these principles and even less agreement among them as to the intention of these four factors. Subsequently, Gould and Boies (1984) described the evolution of a highly usable interface for the 1984 Olympic Message System through the application of these design features.

11.3.1.2 Stages of Human–Computer Software Interface Design

The design of quality human–computer software interfaces is not a rigid, static procedure. The content and context of each interface varies according to the specific application, and the design team often involves the collaborative efforts of both human factors specialists and computer specialists. Additionally, design of software interfaces is usually iterative in nature and requires several evaluations. For example, Rubinstein and Hersh (1984) describe a top-down design approach to software interfaces, which includes collecting information, producing a design, building a prototype, evaluating the system, and delivering the system. Their structured approach with 93 design guidelines, however, still includes a revision cycle during design production and prototype building.

Figure 11.3.1 depicts a generalized flow diagram of the iterative design process for developing human–computer software interfaces. Three general stages are shown in Figure 11.3.1. Stage 1 is the initial design stage in which the software interface is specified. Stage 2 is an evaluation stage during which the software interface evolves through iterative design as shown by the dashed feedback loop in Figure 11.3.1. The resulting operational software interface developed through this iterative design process is evaluated in Stage 3 and should provide new insights that form the basis for additional human factors software design guidelines as shown in the feedforward loop in Figure 11.3.1.

The two stages of evaluation shown in Figure 11.3.1 are not unlike parallel design procedures used in other human factors application areas. Specifically, the process often used in instructional technology is analogous to the human–computer software interface design process. Generally, evaluation falls into two stages during design (Wright, 1983). This first stage is called formative evaluation and deals with techniques for obtaining user feedback to aid the designer in making decisions for design

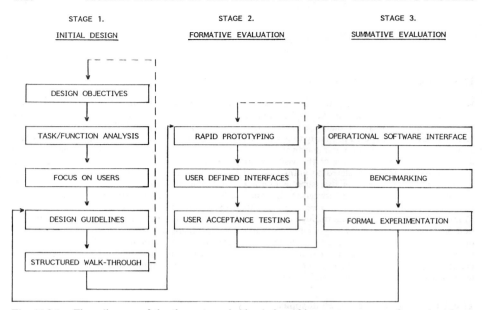

STAGE 1.   STAGE 2.   STAGE 3.

INITIAL DESIGN   FORMATIVE EVALUATION   SUMMATIVE EVALUATION

Fig. 11.3.1.   Flow diagram of the three stages in the design of human–computer software interfaces.

revisions during the iterative design process. The second stage is called summative evaluation and is used to test the final design configuration to ensure that it is functioning properly.

Although the iterative design process itself is typical of accepted human factors design procedures, the specific steps and techniques used to accomplish these tasks are often unique to the software interface domain. In many instances, new tools and procedures need to be developed to use this design procedure effectively. Details concerning specific steps that may be appropriate within the three stages of this iterative design process follow. The emphasis is to summarize a variety of design methods and tools that can be used to develop quality human–computer software interfaces.

11.3.2 STAGE 1. INITIAL DESIGN

The first stage in iterative design is to develop an initial design configuration. If the initial design is close to the configuration specified by the design goals, then fewer iterations of redesign will be needed. Figure 11.3.2 provides a detailed breakdown of the various alternatives that need to be considered during the initial design stage. All of these procedures deal with initial planning, specification, and basic configuration of the human–computer software interface. The results of this stage of design should provide the necessary information to construct the initial version of the software interface, which is then iteratively improved during formative evaluation.

11.3.2.1 Design Objectives

Before a software interface can be designed, clear objectives or design goals must be specified. Without carefully stating and designing to these objectives, the resulting software interface can easily be either under- or over-designed. If the interface is under-designed, it will not meet the design objectives. If the interface is over-designed, it will exceed the design objectives, thereby expending unnecessary time and effort.

Stating design objectives also implies the development of design criteria that can be used to evaluate the software interface. These criteria should be in the form of user-oriented metrics. Usability metrics similar to those discussed in Chapter 11.2 are appropriate to consider. The extent to which the design objectives and criteria can be specified in quantitative terms, the more precise the resulting evaluations will be. These evaluations can be both analytical and empirical. Rouse (1984), for example, proposes both a top-down analytical evaluation of design objectives as well as a bottom-up empirical evaluation of the resulting design configuration.

Basic human factors principles fundamental to good system design in other applications are equally important in the design of computer-based systems. Often the guidelines proposed for the design of human–computer interfaces are a restatement of these basic human factors principles in terms that

STAGE 1.

INITIAL DESIGN

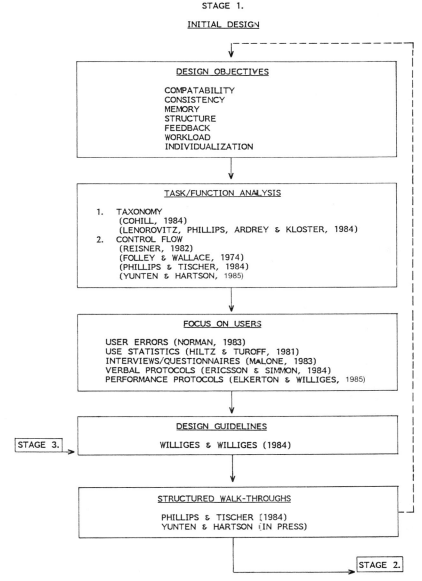

Fig. 11.3.2. Summary of Stage 1: initial design procedures.

relate to specific aspects of a human–computer interface. These general principles provide a basis for developing design objectives and a comprehensive set of metrics to evaluate the design. Table 11.3.1 summarizes the principles of compatibility, consistency, memory, structure, feedback, workload, and individualization.

Compatibility

In its most general form, the principle of compatibility predicts that high information transfer will occur when the amount of information recoding required of the user is minimal. Translated to the design of human–computer interfaces, this would suggest that the interface must be compatible with human perception, memory, problem solving, action, and communication (Barnard, Hammond, Morton, and Long, 1981.)

Table 11.3.1   Fundamental Principles of Human–Computer Dialogue Design

Compatibility Principle:	Minimize the amount of information recoding that will be necessary
Consistency Principle:	Minimize the difference in dialogue both within and across various human–computer interfaces
Memory Principle:	Mimimize the amount of information that the user must maintain in short-term memory
Structure Principle:	Assist the users in developing a conceptual representation of the structure of the system so that they can navigate through the interface
Feedback Principle:	Provide the user with feedback and error-correction capabilities
Workload Principle:	Keep user mental workload within acceptable limits
Individualization Principle:	Accommodate individual differences among users through automatic adaption or user tailoring of the interface

Gaines and Facey (1975) emphasized the importance of adhering to the user's organization of the information as well as vocabulary and language for dealing with the information. The choice of terminology, format, and system action should be consistent with user population stereotypes. Furnas and coworkers (1982, 1984) discussed some of the issues that make language selection for human–computer dialogues a very difficult task. These include the diversity of language use and the imprecision of its application by humans. According to Payne and Green (1983), traditional computer science tools, such as Backus Normal Form (BNF) notation, fail to reveal semantic groupings of grammatical objects, and these groupings are critical to the performance of end users. They advocate the use of set-grammar in language development to organize semantic clusters.

Clarity or understandability of the information presented is critical. The input required of the user should not be ambiguous, and the output from the computer should be clear and, therefore, useful. The wording for commands, menus, error messages, and HELP displays should be selected with care. In iconic systems the function represented by each graphic icon should be clear to the user and should conform to any population stereotypes. The need to translate, transpose, interpret, or refer to documentation should be minimized.

Spatial and movement compatibility are also important in dialogue design. Fitts and Seeger (1953) demonstrated the advantages of using compatible spatial arrangements of displays and controls. This principle is generally referred to as stimulus–response (S-R) compatibility. For example, movement of objects displayed on the screen should be consistent with the direction of the input movement of the user. Barnard et al. (1981) demonstrated a sizable advantage for a compatible left-to-right display of command elements in terms of the order in which the items had to be entered versus the incompatible right-to-left arrangement: a 12% reduction in viewing time, a 19% reduction in instruction requests, a 56% reduction in reversed argument errors, and a 37% reduction in total argument errors.

The importance of selecting compatible stimulus and response modalities for effective human–computer interfaces has been examined by Wickens and his colleagues (Wickens, 1984). The results of their research suggest that auditory input and speech response are compatible with verbal tasks, whereas visual input and manual response are more compatible for spatial tasks. These results are summarized in a principle that extends the principle of S-R compatibility. Stimulus–central processing–response (S-C-R) compatibility predicts that performance will be optimal only when the association among display, input device, and short-term memory code are compatible.

Consistency

Nickerson (1981) suggests that one of the primary reasons people do not like to use computer systems is the lack of consistency and integration. Interfaces differ both internally and across systems resulting in the need for users to remember several different techniques to accomplish the same thing. Nickerson points out that the problem arises from the fact that in large application programs various pieces of software are written by different individuals, and the user is forced to communicate with several pieces of software.

However, problem solution occurs most readily in a consistent environment. Ideally, system consistency should be derived from the user's natural task solution, not from a logical formalism or system model that the user must learn (Hammond, Jorgensen, MacLean, Barnard, and Long, 1983). To ensure consistency it may be necessary at times to require operations that appear to decrease system throughput, such as requesting information from the user that is already known by the system in order that similar procedures require identical user actions. On the other hand, permitting users to jump between branches of a menu structure may be necessary even at the expense of clarity of the software architecture.

Both the input required of the user and the output of the system should be consistent across the display, module, program, and information system. For example, the computer action associated with a set of special function keys should not vary across menus or tasks nor should the syntax requirements of the command language. The system should perform in a generally predictable manner, without exception.

The costs for failing to provide a consistent interface can be high. Teitlebaum and Granda (1983) found that positional inconsistency in menu items resulted in a 73% increase in search time. Unfortunately, a recent survey of system designers indicated that command languages currently are developed piecemeal with little regard for consistency (Jorgensen, Barnard, Hammond, and Clark, 1983).

Various techniques have been proposed to improve consistency in dialogue design. Reisner (1981) demonstrated the impact of inconsistencies or multiple rules on the number of user errors made with an interactive graphics system. Reisner (1982) proposed the use of formal grammar to predict user performance with various interfaces as a function of consistency. On the other hand, Nickerson (1981) suggested the development of an intermediary program between the user and the applications software to provide consistency. This intermediary program would serve as a translator to prevent inconsistencies within and across software systems and would relieve the user of the need to learn the details of operation of many different systems. Waterman (1978) used a similar concept of a user agent to interpret human–computer dialogue in the context of flexibility in dialogue. The fundamental goal in designing for consistency is to aid the user in developing a conceptual model or internal representation of the structure of the system.

The consistency principle strongly suggests that previous experience with similar computer systems should lead to ease of use and not difficulty in learning the new system. However, one caution is necessary. Only ensuring that the dialogue design is consistent with previous products may not yield a good interface. Sometimes the result will be nothing more than a justification for a previous poor design.

Memory

In the design of human–computer dialogues it is important to minimize the amount of information the user must maintain in short-term memory, particularly if the other information processing is required simultaneously. Theories of short-term memory propose the existence of some upper limit in information that can be recalled soon after it has been presented. Miller (1956) suggested that the upper limit of short-term memory is five to nine items. However, the number of items one can remember is also related to the complexity of the items (Simon, 1974), the sequence of presentation (Badre, 1982), the length of time the items must be remembered (Melton, 1963), and the amount of competing information processing (Murdock, 1965).

Consider the implications of this limitation in human information processing for the design of selection menus. If the user must consider all options simultaneously to determine the optimal choice, the menu will be more effective if the number of options does not exceed human memory limitations (Wickens, 1984). The problem is even more difficult when the menu will be presented by synthesized speech for telephone access to a database system. Kidd (1982) outlined the following characteristics of auditory menus that conflict with human information processing capabilities: (1) user generally cannot control the rate at which information is presented, (2) user may not know the number of menu items that will be presented, (3) user cannot scan the list but must listen to each item individually before making a selection, and (4) user has no visual cues to indicate the structure of the menu.

Whenever large amounts of information must be conveyed to the user, meaningful units should be grouped together in chunks to reduce the load on short-term memory. To increase the number of bits of information processed in one input sequence, larger chunks, each containing more information, should be built.

Maintaining information in short-term memory requires rehearsal. But, simultaneous information processing or decision-making tasks reduce the human's ability to rehearse information in short-term memory. Conversely, when users must remember large amounts of information or complex codes, their ability to solve problems and make decisions is restricted. Obviously, the dialogue designer should strive to reduce the short-term memory load on the user.

However, brevity is not always appropriate in system messages. Simpson, Coler, and Huff (1982) reported that synthesized voice warnings were understood more quickly and accurately in a background of competing human speech when the synthetic messages were short sentences rather than keywords. Because listeners were using top-down processing including syntactic, semantic, and pragmatic information to perceive the message, failing to provide this higher-order information in the keyword messages reduced their effectiveness.

Structure

One fundamental aspect of human behavior is that humans seek structure or organization in their environment, even in cases where none exists. The purpose of the structure is to unify conflicting data gathered through discovery or insight. For example, when asked to recall unstructured lists of items, humans will attempt to order the lists (Tulving, 1962). In addition, when semantically anomalous sentences are used in speech intelligibility testing and the speech is not completely intelligible, listeners will try to determine the words in the sentence by using the syntactical structure and ascribing meaning to the sentence.

Likewise, users of computer-based systems seek to determine the structure of the dialogue and control systems. This internal representation forms the basis of the user's understanding of the system and determines user decisions and actions. If the user fails to understand the structure, errors will occur, such as typing a command in input mode on a text editor. Norman (1981) provides several excellent examples of these structure errors with the complex and powerful UNIX operating system.

Providing an interface where the structure is both compatible with user expectations and internally consistent will assist the user in learning the structure by reducing misrepresentations in the user's internal representation. User experience, accompanied by appropriate system feedback, helps the user develop an accurate representation of the system. Interfaces that involve direct manipulation, such as a word processor where the formatted document is displayed, may make the structure easier to discern.

Various aids have been proposed for teaching the system structure. Human–computer interfaces are often designed around another familiar interface, such as a typewriter for a word processing interface. These analogies can be used to teach the user the structure of the system, but success depends to some degree on the learners being able to distinguish the limits of the analogy (Douglas and Moran, 1983). Another approach uses spatial representations so that the structure becomes explicit. An example might be the use of a graphical aid to depict the hierarchial organization of a database. These spatial representations serve as maps by which the user can "navigate" through the system. In fact, Billingsley (1982) has found that maps inprove user's retrieval performance in unfamiliar sections of a menu database.

Feedback

A human–computer system should be closed loop with information feedback to the users about the quality of their performance, the condition of the system, and the steps necessary to cause some desired outcome. Generally, feedback should occur in close temporal proximity to the related event, such as a user's request for information, computer detection of a user error or missing information, or a change in the status of the system based on a user input. When the response to a user's request will be delayed, the user should be given some indication that the request is being processed. Otherwise users may perceive that no action has been taken by the computer and may reenter the request or attempt another approach to the problem. Timing is also important in handling user errors. Maguire (1982) points out that some consideration should be given to the user's sensitivity to interruption, that is, there may be circumstances where error feedback should be provided after the user has completed an entry rather than immediately upon computer detection of an error. Research is needed to evaluate various factors that influence the timing of feedback.

In computer-based systems users should at all times be aware of where they are, what they have done, and whether or not it was successful. System feedback provides the user assurance that the system is available and information concerning whether user queries are being processed or delayed. Subtle feedback, such as a change in the shape of a cursor to indicate a mode change or an error message displayed in the border areas of the display, may not be noticed by the user.

The importance of feedback is underscored by the recent interest in systems that incorporate direct manipulation. Shneiderman (1983) summarized the central ideas: objects of interest are directly visible or clearly represented, results of an action are displayed immediately, actions can be easily reversed, and actions occur by direct manipulation of physical objects, such as special function keys or a mouse.

Recent research (Carroll and Carrithers, 1984) indicates that no feedback or excessive negative consequences when errors occur during training can interfere with users' learning the system. A simplified presentation of functions to users in order to prevent common errors and mitigate the consequences of other errors is proposed. This learning principle is not new. More than 25 years ago behavioral psychologists proposed that avoidance of errors and subsequent negative feedback would enhance training, and this principle was fundamental to the design of linear programmed instruction. Using a simplified word processing system, Carroll and Carrithers (1984) reduced training time by 21%.

Nonetheless, errors do occur in human–computer dialogues, and feedback is required so that the user can determine if corrective action is needed and what form it should take. Error messages should be specific and written from the point-of-view of the user, not in some formal notation used by the programmer to represent various classes of errors. The user should be given every opportunity to correct errors. This is particularly important in systems where hardware errors may compound user errors (e.g., automatic speech recognizers). In one study where speech recognition was used for data entry (Schurick, Williges, and Maynard, 1985) accuracy of data fields was increased from 70% to 97% through feedback and user-error correction.

Feedback need not be limited to the detection of user error, replies to user requests for information, or status messages. McCoy (1983) proposed that human–computer interfaces should include a means for the computer to correct user misconceptions. The absence of these corrections actually may serve to confirm the misconceptions. The approach taken by McCoy to correct user misconceptions involves a complete system knowledge base, a "model" of the user's knowledge base that is continually updated during the discourse, and a set of domain-independent heuristics to determine what is wrong with the user's information and what faulty reasoning led to user acceptance of the incorrect piece of information. Another form of intelligent feedback provides the computer with an ability to make suggestions in the absence of a request from the user. These topics are covered later in this section.

Workload

An assessment of the potential effects of the interface on user mental workload is an essential step in the design of human–computer dialogues. Because the probability of user error or failure to act increases in an overload or underload situation, the overall goal should be to keep the workload of the user within acceptable limits when defining the operator's task and dialogue requirements. For a review of various behavioral workload assessment measures, see Williges and Wierwille (1979).

The density of the information displayed or presented to the user is an important factor in workload. Displayed output should be organized to minimize the scanning required of the user, and only information essential to the user's current needs should be displayed. In some applications, slow rates of information display may be preferrable even though the hardware is capable of faster display rates. It may be necessary to allow operator control of the rate of presentation of auditory displays. Relevant data should be presented in a usable form to reduce the processing requirements of the user (Mitchell and Miller, 1983). It may be necessary for the system to integrate information from several sources into one coherent display (Goldsmith and Schvaneveldt, 1984). Additional workload considerations include allocating tasks between the user and computer, determining the modalities to be used for dialogue, and providing for redundant information sources.

Individualization

Human behavior is characterized by individual differences. In human–computer systems these differences may stem from factors such as language usage, problem-solving style, or level of expertise with computer systems. The diversity of language usage becomes obvious when computer users are free to select command names, and a variety of names are used for each system function (Furnas, et al., 1984). See Good, Whiteside, Wixon, and Jones (1984) for a discussion of user-defined language. This diversity in language usage is a major problem addressed in the design of natural language (Kelley, 1983) and user-defined interfaces.

Differences in interaction style may be necessary for different classes of users. For example, novice users may prefer a computer-guided dialogue, whereas expert users may prefer a user-guided dialogue. A further complication is the fact that users change as they gain experience with computers and a particular system. These changes may occur slowly or quite rapidly. The interface must be designed to accommodate various types of individual differences among users. Individualization can often be achieved by providing a continuum of ways of accomplishing the same function (e.g., a system that provides menu choices for the novice user and then as the user becomes more experienced allows commands to be entered in place of one or more menu choices).

At least two approaches to individualization are possible: a flexible interface and an adaptive interface. A flexible interface allows the user to tailor the interface to his or her own needs or permits various types of interaction. Examples include allowing for command synonyms, alternate syntax, and user-defined commands. Adaptive interfaces accommodate the individual user automatically and may change over time. Benbasat and Wand (1984) proposed the concept of a dialogue generator to adapt the interface to the evolving needs of users. Adaptive interfaces are discussed in more detail later in this chapter.

11.3.2.2 Task/Function Analysis

Human–computer software interfaces are designed to perform specific tasks and functions. An analysis of the requirements of these tasks and functions must precede detailed design. In general, the task/function analysis of software interfaces requires specification of three major components. First, a specification of the interface input/output representations must be provided. Second, a specification of the sequence of dialogue must be stated. And, third, the control structure for interfaces, dialogues, and software computations must be specified.

Some attempts have been made to develop taxonomies of generic functions and actions performed through human–computer interfaces. For example, Cohill (1984) listed 10 functions representative of common communication interfaces. These functions were separated into four categories as follows: system driven, peripheral functions (news and menus); user driver, main functions (update, temporary storage, and inquiry); user driven, peripheral functions (mail and help); and system driven, monitoring functions (error/message handling and statistics). The final function in her taxonomy is a linkage function to interconnect the four main categories.

Perhaps the most comprehensive effort in developing a generic taxonomy of user–system interface actions is provided by Lenorovitz et al. (1984). As shown in Table 11.3.2, the user action subtaxonomies

Table 11.3.2   User Action Subtaxonomies for User–System Interface Actions

User-Input Taxonomy

			Name
			Group
	Associate Introduce		Insert
Create	Assemble		Aggregate
			Overlay
	Replicate		Copy
			Instance
Indicate	Select (position/object) Reference		

			Cut
	Remove		Delete
			Suspend
	Stop		Terminate
Eliminate			Rename
	Disassociate		Ungroup
			Segregate
	Disassemble		Filter
			Suppress
			Set-aside

Manipulate — Transform (change attribute)

Activate — Execute (____ function)

User-Internal Taxonomy

			Detect
			Search
	Acquire		Scan
Perceive			Extract
			Cross-reference
	Identify		Discriminate
			Recognize

			Categorize
	Analyze		Calculate
			Itemize
			Tabulate
			Estimate
			Interpolate
Mediate	Synthesize		Translate
			Integrate
			Formulate
			Project/extrapolate
	Access		Compare
	Decide		Evaluate

			Call
			Acknowledge
			Respond
	Transmit		Suggest
Communicate			Direct
			Inform
			Instruct
	Receive		Request

Source: Lenorovitz et al. (1984).

are described in terms of either a user-input taxonomy or a user-internal taxonomy. These generic actions and functions form the basic building blocks for specifying human–computer tasks.

Formal procedures have been used to represent the control flow of information and procedures in human–computer tasks. Representative procedures include formal grammars (Reisner, 1982); lexical, syntatic, and semantic interaction analysis (Foley and Wallace, 1974); composition graphs, task description language, and dialogue description language (Phillips and Tischer, 1984); and supervised flow diagrams (Yunten and Hartson, 1985). The advantage of these formal procedures is that they can provide task analysis and descriptions that can readily be translated into operational code to build the software interface. Unfortunately, most of these procedures, have only limited use in the design of operational human–computer interfaces. Kloster and Tischer (1984), however, attempted to build a formal software interface design process for requirements specification. A case study using this approach in the design of the software interface for an air traffic control system is presented in Chapter 12.6

11.3.2.3 Focus on Users

The overall philosophy of iterative design is to focus the design process on the end user. Consequently, the initial design phase should incorporate user inputs to the greatest extent possible. Gould and Lewis (1983) stress the requirement of early focus on users and recommend that users participate interactively with the design team during early stages of design. Inputs from end users can be obtained in a variety of ways including analysis of user errors (Norman, 1983; Shneiderman, 1984), statistics on both user characteristics and usage rates (Hiltz and Turoff, 1981); interviews/questionnaires with the representative user population (Malone, 1983); verbal protocol analysis (Ericsson and Simon, 1984); and performance protocol analysis (Elkerton and Williges, 1985).

11.3.2.4 Dialogue Design Guidelines

In the past several years, a number of technical reports, journal articles, and books have offered collections of specific guidelines for the design of computer-based dialogues. Although these design guidelines might be useful for developing a requirements definition for the interface (Smith, 1982), they are used more frequently to improve the quality of the initial interface design. For example, Tullis (1983) recently summarized guidelines focusing on the formatting of alphanumeric displays. The factors of overall information density, local information density, grouping, and layout complexity were addressed, and objective measurement techniques for these factors were offered. These measurements, derived from empirically based dialogue guidelines, permit the dialogue designer to evaluate the format of an alphanumeric display proposed for a human–computer interface prior to collection of performance on the entire interface.

Williges and Williges (1984) compiled over 500 software design guidelines dealing directly with the human–computer dialogue from 16 sources as listed in Table 11.3.3. Information related to the

Table 11.3.3 Reports and Books Reviewed by Williges and Williges (1984)[a]

Barmack and Sinaiko, 1966

Brown, Burkleo, Mangelsdorf, Olsen, and Williams, 1981

Ehrenreich, 1981

Engel and Granda, 1975

Foley, Wallace, and Chan, 1981

Galitz, 1981

Gebhardt and Stellmacher, 1978

Hiltz and Turoff, 1978

Martin, 1973

Miller and Thomas, 1976

Newman and Sproull, 1979

Parrish, Gates, Munger, and Sidorsky, 1981

Pew and Rollins, 1975

Ramsey and Atwood, 1979

Shneiderman, 1980

Smith and Aucella, 1983

[a] From *Human Factors Review,* 1984, p. 168. Copyright © 1984, by The Human Factors Society, Inc. and reproduced by permission.

design of computer hardware and workspace design of computer-based systems was not included. The purpose was simply to compile dialogue design guidelines from a variety of sources into one document to aid in structuring behavioral research directed toward empirical evaluation of guidelines. Because this compilation was not developed to serve as a handbook for designers of human–computer interfaces, no evaluation of these guidelines was given nor was any indexing or cross-referencing provided.

Obviously, the relevant user considerations vary depending on the task environment and end user population, and each dialogue designer must determine which are appropriate for a particular human–computer interface. In cases where conflicting dialogue guidelines exist, the designer must determine which to adopt until behavioral research can resolve these conflicts and/or establish the appropriate context area for each.

Classification Scheme

Because of the overlapping nature of design guidelines, various classification schemes have been proposed. The Williges and Williges (1984) compilation was organized into six major sections including data organization, dialogue modes, user input devices, feedback and error management, security and disaster prevention, and multiple users. The topical outline shown in Table 11.3.4 is included to suggest the variety of information one must consider to develop quality human–computer dialogues and to demonstrate the need to provide information retrieval aids for the dialogue author who must implement these guidelines.

Data organization deals with aspects of structuring information on the visual display in an interactive environment. The design of auditory information displays is not covered in the Williges and Williges (1984) compilation although interest in the application of speech displays is increasing. The major topics of consideration include methods of coding visual information, control of the amount or density of information displayed, the use of labeling to organize the information displayed, various techniques for formatting the display, and considerations for the overall layout of information fields on the display screen. For additional information on coding schemes applicable to human–computer systems, see Parrish et al. (1981) and Barmack and Sinaiko (1966). Foley and Wallace (1974), Foley et al. (1981), Martin (1973), Prince (1971), and Newman and Sproull (1979) provide some additional information on the use of graphics in human–computer communication. Simpson (1983) and Simpson and Navarro (1984) review issues in the design of synthesized speech displays.

Design considerations dealing with dialogue mode are organized into six general types of interaction in the Williges and Williges (1984) review as shown in Table 11.3.4 including form-filling, computer inquiry, menu selection, command languages, formal query languages, and restricted natural language. (Natural language is discussed later in this chapter.) The first three dialogue modes represent primarily computer-initiated dialogues, whereas the last three are primarily user-initiated dialogues. As yet, mixed-initiative dialogues have seen little application, and few data exist concerning their optimal design (Ehrenreich, 1981).

Form-filling is a structured dialogue mode in which the user provides information in designated fields on the interactive display. Considerations in form-filling involve the selection of default values for data fields, feedback to the user, screen layout, data-entry procedures, and cursor movement to designated fields on the form. Computer inquiry dialogue is a computer-initiated query mode generally used when the data items are known and their order is constrained, computer response is fast, and novice end users are involved. Menu selection is a type of structured dialogue in which the user must select among a variety of options. Dialogue design considerations include the order of menu options, the selection codes for the options, the display layout of the menu, the content of the menu, and the control sequence of the menu dialogue. Command language dialogues allow the user to communicate with the computer by providing specific commands that specify various functions to be performed. Topics discussed include command organization and nomenclature, default values used in commands, editor orientation, user control of command processing, command operation, system lockout of user input, and special purpose commands. Formal query languages involve a set of syntactical and lexical rules with which the user can question the computer in order to retrieve information from a database. Restricted natural language is the most unstructured dialogue and is used as a flexible method to query a database. Although sentencelike commands are used, vocabulary size and/or syntax may be restricted.

Various user considerations must be made to select the appropriate device by which the user makes a dialogue entry to the computer and to structure the use of the device. The guidelines presented in this section of the Williges and Williges (1984) compilation are concerned with the selection of an input device, keyboard considerations for special-function keys and cursor control, the use of pointing controls such as light pens and touch panels, the use of continuous controls such as trackballs and joysticks, the choice of graphic tablets for graphical data entry, and considerations for voice input. These topics are discussed in detail in Chapter 11.5.

Feedback and error management guidelines deal primarily with communications from the computer to the user including feedback, error recovery procedures, user control of the transaction sequence, help/documentation, and computer aiding. Feedback information provided by the computer includes

Table 11.3.4 Classification Scheme for Dialogue Considerations (from Williges and Williges, 1984)[a]

1. Data Organization
   1.1 Information coding
      1.1.1 Color codes
      1.1.2 Shape codes
      1.1.3 Blinking codes
      1.1.4 Brightness codes
      1.1.5 Alphanumeric codes
   1.2 Information density
   1.3 Labeling
   1.4 Format
      1.4.1 Prompts
      1.4.2 Tabular data
      1.4.3 Graphics
      1.4.4 Textual data
      1.4.5 Numeric data
      1.4.6 Alphanumeric data
   1.5 Screen layout
2. Dialogue Modes
   2.0 Choice of dialogue mode
   2.1 Form-filling
      2.1.1 Default values
      2.1.2 Feedback
      2.1.3 Screen layout
      2.1.4 Data entry procedures
      2.1.5 Cursor movement
   2.2 Computer inquiry
   2.3 Menu selection
      2.3.2 Selection codes
         2.3.2.1 Letter codes
         2.3.2.2 Number codes
         2.3.2.3 Graphic symbols
         2.3.2.4 Mnemonic codes
      2.3.3 Menu layout
      2.3.4 Menu content
      2.3.5 Control sequencing
   2.4 Command languages
      2.4.1 Command organization
      2.4.2 Command nomenclature
         2.4.2.1 Abbreviations
         2.4.2.2 Argument formats
         2.4.2.3 Separators/terminators
      2.4.3 Defaults
      2.4.4 Editor orientation
      2.4.5 User control
         2.4.5.1 Command stacking
         2.4.5.2 Macros

         2.4.5.3 Immediate commands
      2.4.6 Command operation
      2.4.7 System lockout
      2.4 8 Special operations
   2.5 Formal query languages
   2.6 Restricted natural language
3. User Input Devices
   3.0 Data entry procedures
   3.1 Selection of input device
   3.2 Keyboards
      3.2.1 Special function keys
      3.2.2 Cursor control
   3.3 Direct pointing controls
   3.4 Continuous controls
   3.5 Graphics tablets
   3.6 Voice analyzers
4. Feedback and Error Management
   4.1 Feedback
      4.1.1 Status messages
      4.1.2 Error messages
      4.1.3 Hard copy output
   4.2 Error recovery
      4.2.1 Immediate user correction
      4.2.2 User correction procedures
      4.2.3 Metering and automatic error checks
      4.2.4 Automatic correction
      4.2.5 Stacked commands
   4.3 User control
   4.4 Help and documentation
      4.4.1 Offline documentation
      4.4.2 Online documentation
   4.5 Computer aids
      4.5.1 Debugging aids
      4.5.2 Decision aids
      4.5.3 Graphical input aids
5. Security and Disaster Prevention
   5.1 Command cancellation
   5.2 Verification of ambiguous or destructive actions
   5.3 Sequence control
   5.4 System failures
6. Multiple Users
   6.1 Separating messages/inputs
   6.2 Separating work areas
   6.3 Communications records

[a] From *Human Factors Review*, 1984, p. 172. Copyright © 1984 by The Human Factors Society, Inc. and reproduced by permission.

status messages, error messages, and hard copy output. User actions required to correct an error involve user correction procedures, computer metering of transactions, automatic error correction, and stacking of multiple commands. Guidelines are considered for the level, amount, and type of user control of feedback and error messages. User considerations for on-line and off-line documentation, as well as help information to enhance feedback and error management, are presented. (See Chapter 11.7 for a complete discussion of documentation for software systems.) Finally, guidelines for computerized aids for program debugging, decision making, and graphical input are listed. Approaches to computer assistance in the human–computer interface are considered later in this chapter.

Dialogue guidelines for security and disaster prevention deal with catastrophic circumstances, such as inadvertent deletion of files or the premature termination of a computer session. Topics included are methods for the cancellation of data entry and command sequences, the requirement to confirm ambiguous or destructive actions, the control of destructive actions, and the handling of system crashes.

The final section of the Williges and Williges (1984) compilation deals with interfaces where there are multiple users sharing the same system or database. Although most human–computer dialogue considerations are concerned with the dialogue between a single user and the computer, a few guidelines have been offered for the multiple-user environment. Major topics include the separation of messages and inputs of multiple users, the use of cursors in multiuser displays, and computerized record keeping of inter-user messages.

Limitations of Guidelines

Although the dialogue design considerations summarized by Williges and Williges (1984) and others deal primarily with human–computer dialogues in the form of alphanumeric information, some consideration has been given to dialogues dealing with graphic information. Few guidelines deal with the design of "intelligent" systems incorporating rule-based or other artificial intelligence techniques. A need clearly exists to extend the dialogue guideline database in these areas.

Obviously, a fairly sizable database of dialogue design considerations currently exists, and certainly it will expand rapidly with the growing interest in human factors issues in the design of human–computer interfaces. Unfortunately, many of these dialogue guidelines are not presented within the limited context in which they were researched. Maguire (1982) suggests that the apparent contradictory nature of various recommendations may actually represent alternative strategies to be followed under certain specific circumstances.

A recent paper by Meister (1984) proposed a list of contents for human–computer interface design guidelines. They include: (1) a description of the hardware or software requirement to be incorporated into the design; (2) alternate methods for incorporating the requirement into the design; (3) limits within which the design feature will function; (4) expected performance of an operator using the design feature; (5) expected variability in operator performance; (6) advantages of incorporating the design feature; (7) cost of failing to incorporate the feature; and (8) empirical evidence to support the cost-benefit tradeoffs. Unfortunately, most guidelines available today provide only a description of the requirement, and as Meister (1984) suggests the designer may not know how to incorporate the human factors requirement into the design.

Even if a completely comprehensive and nonconflicting database of design guidelines was available, it is critical to determine how this information can be best conveyed to the designer of dialogue software. The usual approach for delivering dialogue principles is merely to compile these considerations into a handbook with no retrieval assistance beyond a table of contents and/or index. Smith and Mosier (1984) are currently evaluating, through a user survey, the usefulness of a handbook of dialogue principles. Due to the complexity and overlapping nature of any comprehensive dialogue guideline database, the structural organization of handbooks and the subsequent search for procedures to locate relevant guidelines may become unmanageable. Consequently, various forms of computer aiding should be considered. The computer aiding may be no more than tree searching procedures for data retrieval or may incorporate sophisticated rule-based procedures to aid in decision making.

A computer-aided implementation of information for the dialogue author should include three basic stages, as shown in Figure 11.3.3. First, a comprehensive set of empirically derived dialogue guidelines must be available in a computerized database. Obviously, this database will need to be updated continually as additional research is completed. Second, an on-line acquisition procedure should be developed to retrieve relevant dialogue principles for a specific application. Third, providing some form of decision aiding to select the appropriate set of dialogue guidelines for a particular application environment may be helpful. Rule-based systems can be developed to check whether design rules have been violated in a specific application, to select the appropriate set of rules, or to design the specific human–computer interface based on the rules provided. Tools for implementing dialogue guidelines are covered under iterative design and rapid prototyping.

Merely applying dialogue design guidelines to a particular system design does not guarantee a good human–computer interface because various factors in a specific design may interact and the applicability of each guideline to a specific design situation varies. Evaluation of any design with the designated end users is essential. However, use of dialogue design guidelines should provide an improved initial design or prototype requiring fewer design iterations.

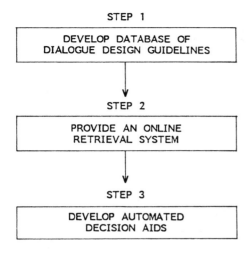

STEP 1

DEVELOP DATABASE OF DIALOGUE DESIGN GUIDELINES

STEP 2

PROVIDE AN ONLINE RETRIEVAL SYSTEM

STEP 3

DEVELOP AUTOMATED DECISION AIDS

Fig. 11.3.3. Stages in the development of a comprehensive and usable dialogue design guideline data-base.

11.3.2.5   Structured Walk-Throughs

A final step in the initial design stage is to combine all the information collected during initial design into a form that represents the control structure of the software interface. Users and designers must then exercise this representation to evaluate its adequacy. This can be completed quite informally as a paper–pencil exercise, or it can be conducted more formally by having users and designers evaluate task description languages and dialogue description languages (Phillips and Tischer, 1984) as discussed in Chapter 11.6.

One of the most comprehensive approaches is presented by Yunten and Hartson (1985) in which they suggest a holistic approach to the management of dialogue design in the form of a supervisory methodology and notation (SUPERMAN) system. A basic premise of their approach is to separate the applications software from the dialogue software to facilitate iterative design and modification. They developed a graphical, supervised flow diagram whereby specification of the communication transactions and computational sequences can be prototyped from a high-level supervisory structure to an expanded interface structure of pure computation and pure dialogue components. Implementation becomes a straightforward mapping of final interface structural representation to source code. Although the Yunten and Hartson (1985) approach is a potentially powerful design tool, its applicability for the design of large-scale systems still needs to be tested.

11.3.2.6   Initial Design Modifications

Often the results of structured walk-throughs require some modification in the design objectives and specifications (Carroll and Rosson, 1985). The feedback loop shown in Figure 11.3.2 represents the iteration phase that continues until adequate results are obtained in the structured walk-through analysis.

11.3.3   STAGE 2. FORMATIVE EVALUATION

Formative evaluation is the crux of the iterative design process. During this stage of human–computer software development, the interface specified in Stage 1 is implemented, evaluated, and redesigned in an iterative fashion until the desired design objectives are reached. A summary of the major approaches to iterative design using formative evaluation is shown in Figure 11.3.4. Various versions of the software interface can be implemented efficiently by procedures described in the rapid prototyping and user-defined interface sections. User acceptance testing includes the set of alternatives to consider for collecting user evaluations of candidate interface configurations. The feedback loop shown in Figure 11.3.4 depicts the redesign phase in which the interface is modified according to the acceptance testing results. This iterative cycle of design, testing, and redesign allows the software interface to evolve in an efficient manner.

In addition, various forms of documentation need to be considered early in the design of software interfaces. These alternatives include hard copy user manuals, embedded tutorials, on-line manuals, and user help. The material presented in Chapter 11.7 provides an overview of guidelines to consider in the design of hard copy manuals. On-line assistance systems are discussed later in this chapter in the context of adaptive systems.

Regardless of the documentation presentation format, there is still a need to test the efficacy of the end user documentation during formative evaluation. Tradeoffs need to be made between the software

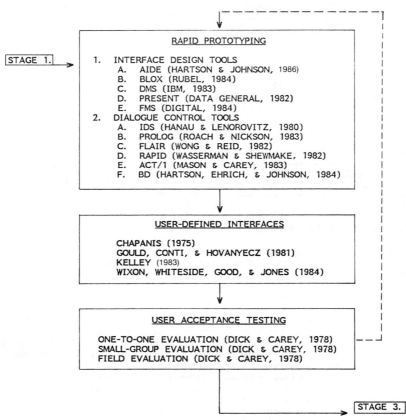

Fig. 11.3.4.   Summary of Stage 2: formative evaluation procedures.

interface design configuration and the form of documentation during this iterative design process. Sullivan and Chapanis (1983), in fact, recommend the use of an iterative design approach for the development of the documentation material.

11.3.3.1   Rapid Prototyping

Once the designer begins to construct the software interface and documentation, tools are needed to provide rapid development of a candidate interface configuration that can be easily modified after user evaluation. Hartson and Johnson (1986) reviewed a variety of approaches to rapid prototyping. Some of these approaches are incorporated into more general software design approaches described as User Interface Management Systems (UIMS). For example, Hartson, Johnson, and Ehrich (1984) describe a Dialogue Management System in which iterative design is facilitated by separating the dialogue from the computational components of the software. Rapid prototyping of the dialogue is facilitated by various software tools in this system. Wasserman and Shewmake (1982) describe a User Software Engineering (USE) information system design methodology that uses state transition diagrams as a basis for rapid prototype implementation.

Most rapid prototyping techniques have been developed as specific tools to aid the designer in generating the software interface configuration in an efficient manner so that it can be easily modified in the iterative design process. Two considerations are important in developing rapid prototyping tools. First, the tool must result in software that is easy to modify in order to facilitate iterative redesign. Often this software provides only a simulation of the interface, but it can be readily translated into operational code once the iterative design process is complete. Second, the rapid prototyping tool must be compatible with the level of computer sophistication of the dialogue designer. Some rapid prototyping procedures require in-depth programming skills and necessitate professional system analysts to use them properly. Other tools are primarily software shells using macros or menu-driven displays that permit the construction of software interfaces by individuals who are not sophisticated computer programmers.

Although the development of rapid prototyping tools is continuing at an accelerated pace, application of these techniques to large-scale software interface design projects is still needed in order to test the applicability of these tools and to detail areas needing additional development. Alavi (1984) described the application of rapid prototyping and iterative redesign to 12 information system development projects varying in size from 3 to 89 person-month efforts. Interviews of the project managers and system analysts noted several positive as well as negative features of using a rapid prototyping approach to software design. Some potential benefits of prototyping include the presentation of real interfaces for users to evaluate, a common baseline for users and designers to identify problems, and a practical way to elicit user participation. Some of the potential shortcomings of the prototyping procedures evaluated included the maintenance of user enthusiasm through several design iterations and the potential limits of prototyping procedures to represent all features of the interface, the management and control of the iteration process, and the ability to prototype large information systems.

Even though techniques are only emerging and have not been evaluated on many large-scale system applications, several rapid prototyping tools are available for consideration. Essentially, these tools can be considered either as interface design tools or as dialogue control tools as summarized in Figure 11.3.4.

Interface Design Tools

Interface design tools are most useful in aiding the human factors designer to select among the various alternatives available for screen layout and device selection so as to modify the human–computer interface without having to rewrite source code. See Chapters 5.1 and 11.4 for a description of the various input/output alternatives. These interface tools, along with dialogue control flow procedures, facilitate development of candidate interface designs for evaluation by end users and subsequent revision through iterative design.

Johnson and Hartson (1982) describe their concept of authoring tools in terms of an Author's Interactive Dialogue Design Environment (AIDE), which is part of the Dialogue Management System. The AIDE system allows the designer to view directly the changes and modifications in the dialogue through the use of various dialogue-editing tools including dialogue language, graphics formatters, menu formatters, keypad formatters, voice input/output management, language implementers, and design expertise. Once editing of the dialogue is completed, source code for implementation is generated. Several of the authoring tools of AIDE are currently operational, but large-scale applications to the complex software interfaces still need to be evaluated.

Several commercial rapid prototyping packages are available to aid in human–computer interface design. Most of these packages are of more limited scope and were developed for specific design purposes. Examples of these rapid prototyping systems include BLOX (Rubel, 1984) for window design of displays, Development Management System (IBM, 1983) for data entry screen management, PRESENT (Data General, 1982) for formatting data into reports and graphics, and the Forms Management System (Digital, 1984) for menu driven form design.

Dialogue Control Tools

Since most human factors design of software interfaces centers around the human–computer communication interface, rapid prototyping tools allow the human factors specialist a means of readily modifying the dialogue in an iterative fashion without having to rewrite source code. Various general approaches have been taken to develop such tools. For example, Hanau and Lenorovitz (1980) developed an interactive dialogue synthesizer that incorporates a dialogue specification language as input to a simulation mechanism for presenting the appearance of the interface for a variety of interactive devices. More recently, Roach and Nickson (1983) recommended the use of a rule-based programming language PROLOG as a dialogue specification technique which allows rapid prototyping of alternative human–computer dialogue configurations.

In addition, several specific dialogue control tools have been developed. These include a Behavioral Demonstrator (BD) developed by Hartson, Johnson, and Ehrich (1984) to prototype dialogues in their Dialogue Management System; Rapid Prototypes of Interactive Dialogues (RAPID), a dialogue prototyping tool using state-transition diagrams, as part of USE developed by Wasserman and Shewmake (1982); a Functional Language Articulated Interactive Resources (FLAIR) developed by Wong and Reid (1982) to prototype dialogues specified by their Dialogue Design Language specification using a series of menu screens to direct designers; and ACT/1 developed by Mason and Carey (1983) for prototyping user interface scenarios using a screen-oriented system.

Since most rapid prototyping tools are still in the embryonic stage of development, software implementation is currently being performed by highly talented programmers. These programmers, in fact, embody the rapid prototyping capability in most software interface development projects. Whether the implementation is performed by programmers or rapid prototyping tools, the purpose of this design phase is still to facilitate interface development that can be easily redesigned through iterative procedures.

11.3.3.2 User-Defined Interfaces

A novel approach to design of software interfaces is to employ a facade for design purposes. This approach is quite often useful in simulating new interface configurations where the technology may not be currently available. Chapanis (1975) used simulation to describe human–computer communication as a method for determining design features for human-computer communication interfaces. Gould et al. (1981) used a facade to evaluate the potential value of speech recognition as a means of composing documents. Currently, speech recognition technology is not advanced to the point of recognizing speaker independent, continuous, unlimited vocabularies. By interposing an expert typist who was conversant with the interactive text editor between the user verbally composing text and the computer display, Gould et al. (1981) were able to provide a convincing simulation in which to evaluate parameters of speech recognition.

Kelley (1983) developed a user-defined natural language interface for calendar management applications. He used a six-step iterative design approach consisting of task analysis, deep structure program development, facade of the natural language interface, first-approximation language processor, facade of interface with limited experimenter intervention, and cross-validation testing. The final calendar program recognized approximately 86% of the natural language commands generated by users.

Simulation was also used as a means of building a user-defined interface by Good et al. (1984). The electronic mail system that was implemented employed a facade for iterative development. Computer novices used a command-line interface without the benefit of documentation or assistance. A hidden operator intercepted the commands and created the illusion of a true interactive session. Based on the user's commands the software interface was iteratively redesigned, and user command recognition improved from 7% to 76% recognition. Based on the convincing results of this study, user-defined interfaces developed through facades appear to be an important technique to consider during formative evaluation.

11.3.3.3 User-Acceptance Testing

As shown in Figure 11.3.4 the final procedure in formative evaluation involves some form of user-acceptance testing that is the culmination of the iterative design process. If the stated design goals and objectives are not reached during the user-acceptance testing, then the iterative design process and formative evaluation process continues as designated by the feedback loop shown in Figure 11.3.4. Once the design goals are reached, then the iterative design process involving user-acceptance testing terminates and the final stage or summative evaluation begins. It must be emphasized that any user-acceptance testing must be conducted on the appropriate population of end user in order for the results to be valid.

Williges (1984) noted that the evaluation techniques used during formative evaluation in instructional technology seem quite appropriate to consider during user-acceptance testing of human–computer software interface design. Dick and Carey (1978) describe three major stages in formative evaluation of instructional systems. These stages include one-to-one evaluation, small-group evaluation, and field evaluation.

One-to-One Evaluation

The first stage of formative evaluation described by Dick and Carey (1978) is an extremely informal evaluation in which a student sits with an instructor and attempts to complete the instruction. At this stage of evaluation, there is more discussion with the instructor as to difficulties being encountered than there is actual instruction. Likewise, the first form of evaluation in software interface design could include a potential user sitting with the designer and describing the difficulties encountered in attempting to use the system.

Small-Group Evaluation

Once a potentially usable instructional system evolves through one-to-one evaluation, Dick and Carey (1978) recommend the use of small-group evaluation. During this stage of evaluation, a group of students completes the instruction with a minimum amount of intervention from the instructor. Errors and difficulties are noted, and the instructional system is redesigned in an iterative fashion to eliminate these bottlenecks. The cycle of small-group testing and redesign is continued as shown by the feedback loop in Figure 11.3.4 until the desired level of acceptance is reached. Often specific criteria for an acceptable level of training performance are used as a means of determining when to terminate the iterative redesign cycle of small-group evaluation (Vasek, 1983).

In an analogous way, small-group evaluation could be used in software interface design by having potential users evaluate a prototype interface. Attitude questionnaires, structured interviews, and on-line performance assessment each could be used as a means of isolating problem areas of the interface that need to be redesigned.

Field Evaluation

The final stage of the Dick and Carey (1978) formative evaluation process involves a test of the revised instructional system in a situation that closely resembles the actual training environment. This form of evaluation is the borderline between formative and summative evaluation. Beta testing, or site testing of software, is one example of the field evaluation stage as it relates to new software interfaces. However, behavioral data must be collected during beta testing so that the human–computer interface, as well as software execution, can be evaluated.

11.3.3.4   Iterative Redesign

The final component of Stage 2. Formative Evaluation shown in Figure 11.3.4 is the feedback loop from user acceptance testing to rapid prototyping. This loop is the crux of the iterative design approach in which the results of user acceptance testing define the redesign issues to be addressed in the next design iteration. This redesign is implemented by rapid prototyping and user-defined interface procedures, and continues until the desired design objectives are reached. At this point the formative evaluation process is completed.

11.3.4   STAGE 3. SUMMATIVE EVALUATION

Following the completion of the formative evaluation process, a summative evaluation should be conducted to test the final design configuration. This evaluation can be a comparison of the final design with a competing design alternative, or it could consist of a comparison to a previous version of the software interface if the emphasis is a redesign project. Summative evaluation can occur in both laboratory and field research environments. Obviously, the laboratory provides more control over extraneous variables, whereas the field environment provides a more realistic evaluation configuration. Figure 11.3.5 summarizes various alternatives the interface designer should consider during summative evaluation.

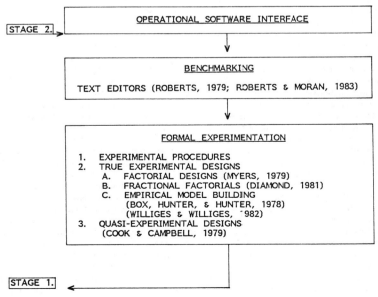

STAGE 3.

SUMMATIVE EVALUATION

Fig. 11.3.5.   Summary of Stage 3: summative evaluation procedures.

11.3.4.1 Operational Software Interface

Since summative evaluation involves the final design configuration, the production software and complete documentation for the resulting human–computer interface must be completed before this phase of evaluation can begin. In effect, the summative evaluation compares this completed software interface with competing designs or earlier versions of the design in the case of a redesign effort.

11.3.4.2 Benchmarking

Standard tests and tasks should be used in conducting summative evaluation. These so-called benchmark tests would be quite useful in helping to standardize summative evaluation. Unfortunately, few benchmark tests are available for evaluation of human–computer software interfaces. One notable exception is the benchmark test for text editors (Roberts, 1979; Roberts and Moran, 1983). This benchmark task evaluates user performance on a set of core editing tasks along four dimensions: the time it takes experts to perform basic editing tasks, the error cost for experts, the learning of basic editing tasks by novices, and the functionality of the text editor over a wide range of editing tasks. Their task was based on a taxonomy of 212 editing tasks that potentially can be performed using a text editor. Additional tests such as this one need to be developed for a variety of software interface tasks in order to improve the summative evaluation process.

11.3.4.3 Formal Experimentation

Most summative evaluation involves some type of formal experimentation. In the design of these summative evaluation experiments, care must be given to the experimental procedures and the choice of various true experimental designs and quasi-experimental designs.

Experimental Procedures

As Williges (1984) discussed, human–computer interface studies require that careful attention be given to the selection of users, the control of the experiment, the research environment, and the procedures used for data collection and analysis.

The correct choice of user population is critical to the evaluation. If the software interface is being designed for first-time or novice users, then evaluations on expert users would be inappropriate. Determining level of expertise is often difficult, because the user interface varies across several dimensions. Often a user may be an expert in one area, such as knowledge of a software language, but novice in terms of a particular interactive editor. Defining the characteristics of the appropriate user population and the level of expertise is central to a successful evaluation.

Closely related to user selection is the issue of the amount of training provided to the users. Often it is possible to achieve a more homogeneous group of users by providing them with a certain amount of training before beginning formal evaluation. In this way, each user is performing at a relative stable level before evaluation begins. The amount and type of training to be provided is situation specific and needs to be determined through pretesting.

One characteristic of human–computer software interface evaluations is that it is usually automated. In other words, the computer system itself is used in the presentation and collection of data. Consequently, the person performing the evaluation is often in a monitoring role using repeat terminals or closed-circuit video. It is important, nonetheless, that this monitoring be carefully performed to ensure that aberrant user behavior is not due to equipment malfunctions or misunderstandings. Even in completely automated evaluations, an evaluator should always be present in a monitoring role. Additionally, careful attention should be given to the development and pretesting of instructions to ensure that the user has a complete understanding of the tasks to be performed.

Various characteristics of the research environment also need to be considered. If the summative evaluation is being conducted in a controlled laboratory environment, characteristics of the operational environment need to be analyzed so that the key factors are present in the laboratory. Workplace clutter, lighting, layout, and types of interruptions are all related variables that can affect the evaluation of an office automation interface and need to be considered for inclusion in the laboratory evaluation. Depending on the type of task, the nature of the evaluation, and the skill level of the user, either actual task scenarios or task facades need to be developed for the evaluation.

Both user performance and user subjective ratings of the interface are used in summative evaluation. Although affective measures of user preferences are usually made during summative evaluation, these measuring devices are not well developed. Often, overall measures of user satisfaction may be in variance with actual performance measures. This can represent either a true difference in satisfaction as compared to usability of the interface, or it can represent an insensitivity of the subjective measures. More research is needed to develop sensitive affective measures, and, currently, rating scales should be used in conjunction with actual performance measures during evaluation. Techniques using magnitude evaluation (Cordes, 1984) and bipolar adjectives (Coleman, Williges, and Wixon, 1985) hold promise as subjective evaluation tools.

One advantage of software interface evaluation is the possibility of embedding the performance assessment into the software so that the evaluation is not apparent to the user. Cohill and Ehrich (1983) describe the use of both metering files to collect user performance on-line and data collapsing routines to format the data for subsequent data analysis off-line. Neal and Simons (1983) incorporate a playback procedure in their metering package whereby user keystroke sequences can be reviewed selectively after data collection at the character and function level. In developing the performance metering systems, tradeoffs need to be made in terms of level of detail of assessment. For example, metering at the keystroke level will certainly provide data to collapse in a variety of ways, but this level of detail could require tremendous storage requirements and off-line processing for data analysis. Preplanning of major data analysis strategies can avoid subsequent frustrations.

True Experimental Designs

A variety of techniques are currently available that can be considered for formal experimentation on human–computer software evaluation. Most of these procedures fall under the rubric of experimental design and are well documented in experimental design textbooks (Box et al. 1978; Diamond, 1981; Keppel, 1982; Myers, 1979). All true experimental designs assume random assignment of subjects to evaluation conditions in order to control for systematic error. Often some form of an analysis of variance factorial design is used.

The complexity of software interfaces often requires the evaluation of several factors simultaneously in order to provide a generalizable evaluation. Since factorial designs completely cross all the levels of one factor with all the levels of other factors, the resulting treatment combinations increase in a multiplicative fashion. A solution to the generalizability/cost dilemma is to use more economical data collection designs.

Williges (1982) suggested three economical data-reduction alternatives including single observation factorial designs, hierarchial designs, and fractional factorial designs. One efficient alternative to reduce the number of factors considered in summative evaluation is to use a two- or three-level factorial design with only one observation per cell as opposed to multiple observations. Higher-order interactions can be used as error terms for lower-order effects.

Hierarchial designs can be used to reduce the number of resulting treatment combinations, because the levels of one factor appear only at one level of another factor. Factors that occur hierarchically cannot interact, so care must be taken in defining hierarchical relationships when these designs are used in summative evaluation.

Probably the most widely used alternative to complete factorial designs is the fractional factorial design. These designs are economical because only subsets of complete factorial designs are used. Diamond (1981) provides a complete treatment of the use of fractional factorials as efficient procedures for conducting large, multifactor evaluations.

Often the primary aim of the evaluation is to determine a quantitative relationship among user performance and various parameters of the design configuration. These functional relationships, in turn, represent empirical models of the human–computer software interface. Box et al. (1978) describe the use of polynomial regression as a procedure for specifying these empirical models. Data collection is conducted in a sequential manner according to response surface methodology.

A design often used in the development of empirical models is a central-composite design (Box et al. 1978; Williges, 1981), because it minimizes the data collection required to solve polynomial regression models. Williges and Williges (1982) demonstrated the use of this procedure in evaluating the effects of four system timing parameters (system delay, display rate, keyboard echo rate, and keyboard buffer length) on user performance, satisfaction, and work sampling during computer-based data entry. They also discussed the use of these modeling procedures as a means of including evaluation data into a readily retrieval database to aid in the fundamental understanding of human–computer interface configurations that can be used in future design implementations.

Quasi-experimental Designs

When summative evaluation is not conducted in laboratory environments, it is often difficult to use completely controlled experimental designs. Cook and Campbell (1979) recommend the use of both nonequivalent control groups and time-series quasi-experimental designs in field testing environments where random assignment of users to evaluation configuration is not possible.

11.3.4.4   Feedforward Results

The results of summative evaluation are important for two reasons. First, these results complete the iterative design process by providing empirical evidence as to the efficacy of the software interface. Without this evidence, it is difficult to document both the advances made by this interface and the future improvements that are still needed. Second, the results of summative evaluations can be fed forward, as shown in Figure 11.3.1, to augment the growing database of software design guidelines so as to improve and perhaps shorten the iterative development process of future human–computer software interfaces.

11.3.5   ADAPTIVE INTERFACES FOR INEXPERIENCED USERS

Human–computer interfaces for inexperienced users can be optimized according to the iterative design process depicted in Figure 11.3.1. Generally, the resulting design is a static interface to be used by all users of the computer system. Human–computer dialogues, however, are beginning to evolve into more dynamic and adaptive interfaces. Inexperienced users demand and deserve an interface that fits their needs, experience, and preference. Since providing an adaptive interface is a means toward these ends, some of the design considerations for developing such interfaces are reviewed.

The need for adaptive human–computer systems is predicated on human, machine, and task differences. Human variability in computer systems is well established for interindividual differences (Ehrlich and Soloway, 1984; Elkerton and Williges, 1984a; Reisner, 1977; Rosson, 1983; Scapin, 1981) and is beginning to be explored for intraindividual differences (Badre, 1984; Gilfoil, 1982). On the other side of human–computer dialogue, machine and task differences are highly visible variations that compound user difficulties with interfaces. Users can perform a myriad of tasks from word processing to database management on a variety of computer systems. Designing dialogues for a human–computer system with these differences is a difficult problem.

Two general design approaches can address these problems. The first is a flexible-interface design philosophy; the second is an adaptive-interface design philosophy. The flexible-interface method is the more traditional approach toward designing human–computer interfaces. The mismatch between user and computer is addressed through careful design that typically evaluates a continuum of procedures in the hopes of providing at least one method that is "natural" for the user. The user-defined interface (Good et al., 1984) is a good example of flexible design and illustrates its explicit objective to accommodate a wide range of users.

In contrast, the adaptive-interface design philosophy recognizes the inherent mismatch between human and computer in most real-world applications. An adaptive interface communicates with the user and the application task to assist, instruct, or present alternative dialogues to the user. This additional communication may be direct or indirect, and aids the human in the use of a complex computer system. In general, the adaptive philosophy also stresses the development of interfaces that can change dynamically with respect to user, machine, and task differences (Innocent, 1982; Thomas, 1981). The adaptive-interface philosophy further assumes that the user desires to communicate with a computer as a cooperative partner rather than strictly as a tool (Licklider, 1960). Since adaptive interfaces are beginning to emerge as viable human–computer interfaces for future systems, the remainder of this section will discuss design alternatives that appear to be appropriate for this class of software interface.

To build these "intelligent" human–computer interfaces requires adaptive communication techniques and strategies for knowledge representation. Adaptive communication techniques are important due to the active and dynamic dialogue inherent in adaptive interfaces, while strategies for knowledge representation must be considered, since models of the user, machine, and/or task are needed to reference user styles and skills with computer-based procedures. However, representation and communication techniques cannot be addressed in isolation. Both are required for the design and evaluation of adaptive interfaces, since knowledge stored in the interface must be transmitted effectively to the inexperienced user.

11.3.5.1   Strategies for Adaptive Communication

Software and techniques for adapting communication to the user, machine, and task are presented in Table 11.3.5. Although Table 11.3.5 is by no means complete, the list illustrates the large number of alternatives available to the designer of adaptive software. As shown in Table 11.3.5, adaptive software for human–computer communication can be broadly categorized as adaptive dialogue systems, adaptive assistance systems, and adaptive instructional systems. Each of these alternative adaptive systems attempts to aid the user; however, different aspects of human–computer interaction are emphasized.

Adaptive Dialogue Systems

The alternative dialogue techniques listed in Table 11.3.5 include input/output technologies such as natural language, voice interaction, and graphical displays. In providing these different communication channels an adaptive dialogue system attempts to speak the "language" of the inexperienced user. For example, with electronic mail an executive vice president would not have to contend with a keyboard command interface when an adaptive dialogue system also provides a natural language, voice-driven interface. Clearly, implementing several alternative dialogues will be required for a truly adaptive interface (Anderson and Sibley, 1972; Hayes, Ball, and Reddy, 1983).

Table 11.3.5 **Design Alternatives for Adaptive Human–Computer Communication**

Adaptive Dialogue Systems

1. Natural language
   a. Clarification dialogues
      (Cohen, Perrault, and Allen, 1981)
   b. Task-constrained interfaces
      (Weizenbaum, 1966; Wilensky, Arens, and Chin, 1984; Woods, 1973)
   c. Menu-based interfaces
      (Tennant, Ross, and Thompson, 1983)
2. Speech recognition
   a. Adaptive enrollment
      (Holden and Haley, 1982; North, Young, and Graffunder, 1984)
   b. Context-specific error correction
      (Biermann, et al., 1983, 1984; Brown, 1984; Spine, Maynard, and Williges, 1983)
3. Graphical displays
   a. Pictorial mental models
      (Hollan, Hutchins, and Weitzman, 1984)
   b. Windows
      (Card, Pavel, and Ferrell, 1984)

Adaptive Assistance Systems

1. Active, passive, and creative user agents
   (Waterman, 1978)
2. Technical expert assistants
   (Buchanan and Feigenbaum, 1978; Shortliffe, 1976)
3. Decision aids
   (Adelman, Donnell, Phelps, and Patterson, 1982; Andriole, 1982)
4. Implicit communication
   (Revesman and Greenstein, 1983)
5. "Do What I Mean"
   (Teitelman and Masinter, 1981)
6. Command summary, step-by-step procedures and error explanation
   (Cullingford, Krueger, Selfridge, and Bienkowski, 1982)
7. Context-sensitive help
   (Fenchel and Estrin, 1982)
8. Core command aids
   (Markus, 1982)
9. Concrete examples
   (Grignetti, Hausmann, and Gould, 1975)
10. Performance critics
    (Gingrich, 1983; Macdonald, 1983)
11. Excursion tours
    (Darlington, Dzida, and Herda, 1983)

Adaptive Instructional Systems

1. Knowledge-of-results displays
   (Kelley, 1969)
2. Socratic dialogues
   (Stevens and Collins, 1978)
3. Case-method dialogues
   (Clancey, 1982)
4. Guided discovery learning
   (Burton and Brown, 1982)
5. Diagnosis and remediation
   (Elkerton and Williges, 1984b)

Development of robust natural language interfaces has been a goal of several researchers. However, the problems with unrestricted natural language stem from the difficulties people have in forming precise natural expressions for complex procedures (Miller, 1981; Shneiderman, 1980). This would suggest that natural language programs require the ability to deal with user inputs that lack precision. Clarification dialogues as suggested by Cohen et al. (1981) may be necessary. That is, if the natural language program detects ambiguities in the user's input, it could query the user with a yes/no question, ask the user to select an interpretation from a set of alternatives, or request the user to reformulate the input.

Natural language software may also restrict the communication domain to improve dialogue interface. The epitome of this approach is embodied in ELIZA (Weizenbaum, 1966). ELIZA gave the illusion of intelligence in a doctor–patient interaction with only keyword matching capabilities. Only very limited understanding of a user's input was accomplished by ELIZA since the program simply decomposed and reassembled text in terms of keywords. Nevertheless, a keyword approach may be more than adequate for some adaptive interfaces.

More realistic applications of task-constrained interfaces for natural language dialogue exist in database retrieval (Woods, 1973) and user help systems (Wilensky et al. 1984). Both of these efforts demonstrated that natural language could be a successful dialogue strategy if semantic and pragmatic information is extracted from the task domain for language understanding. LUNAR (Woods, 1973) exploited specific information on lunar geology, while the UNIX Consultant (Wilensky et al. 1984) used the explicit context of file management in UNIX.

The third natural language alternative in Table 11.3.5 even further restricts the communication problem. Tennant et al. (1983) have demonstrated the usefulness of constrained natural language with a menu-based, window-oriented interface. The menu-based interface builds a natural language command for an application through selection of keywords displayed in windows. The user of the menu-based interface is guided through constructing a natural language command, and imprecise or invalid requests cannot be formed. Furthermore, users do not seem to have navigational problems with the menu dialogue since the constructed sentence provides a map of past selections. Support for this design alternative can be seen in the behavioral research indicating that people naturally restrict vocabularies or perform adequately with a restricted subset of words (Kelly and Chapanis, 1977; Miller, 1981). That is to say, approximations to natural language may be possible for adaptive dialogues.

The voice-interactive dialogues listed in Table 11.3.5 focus explicitly on the use of adaptive techniques with speech recognition devices to improve recognition accuracy. For example, current commercial speech recognizers are usually speaker dependent, meaning that speech samples are required of each user to provide a template for each vocabulary item. This process is termed enrollment. Speech recognition during task completion is often impaired because of various factors that affect the speech signal which are not reflected in the original template, such as changes in speech due to stress, high mental workload, or noise in the environment. Adaptive procedures for automatically updating speech templates may provide substantial improvement in recognition performance over single-time enrollment (Holden and Haley, 1982; North et al., 1984).

Intelligence can also be instilled into speech recognition interfaces by providing context-specific automatic error correction procedures. In syntactically constrained environments, the interface can be designed to detect an erroneous input sequence, change it to the most likely legal sequence, and then, if desired, present the correction to the user for verification. Brown (1984) incorporated features such as command-field backtracking, default word values, and second-choice word checking in a parser developed for speech applications. Using these techniques, Spine et al. (1983) estimated that recognition rates could be raised from 76% to 97% accuracy.

More sophisticated approaches to error handling with speech recognition have been developed including the use of personalized synophone sets, the union of illegal utterances to predict the intended legal utterance, and the application of pragmatic information, such as the prior communication of users in similar dialogues (Biermann, et al., 1983). Tests of a natural language problem-solving environment incorporating speech recognition and pragmatic information for error correction resulted in 77% of the 6000 sentences spoken correctly processed (Biermann et al., 1984).

The final set of alternative dialogues for adaptive interfaces shown in Table 11.3.5 pertains to graphical displays. The use of graphics is becoming increasingly popular in conventional computing environments and may be useful in adaptive interfaces to summarize large amount of information. The effectiveness of graphics as a dialogue technique in an adaptive sysem is exemplified by STEAMER (Hollan et al., 1984). A graphical interface in STEAMER was implemented to evaluate its usefulness in presenting qualitative models of a complex simulated propulsion system to students. The graphical models could be manipulated by the students facilitating the development of mental schemas that may be helpful in understanding the quantitative relationships of several variables in the propulsion system. This demonstrates the complex interactions between representation and communication in adaptive human–computer systems. Quantitative "black box" models are typically difficult to understand. However, as STEAMER has illustrated, graphical data presentation may alleviate these communication problems.

The evolution of graphical displays has also introduced high-resolution bit-mapped displays in command languages and word processing applications. Such a capability allows the user several "windows" of the task environment, thereby permitting cooperative parallel processing between human and computer (Card et al., 1984). Thus, an inexperienced user could communicate with an adaptive interface through several communication windows on a display.

Adaptive Assistance Systems

In addition to speaking the language of the inexperienced user, an adaptive interface could assist the user in understanding the language of the computer. For example, the first alternatives for adaptive assistance in Table 11.3.5 illustrates that a user agent could be interposed between the task and the user to monitor and provide advice on automated procedures. Several types of alternative assistants could be designed and have been suggested by Waterman (1978). For example, the agent could be active, passive, or even creative. An active agent may communicate with the user and translate this communication for the computer, while the passive agent lets the user communicate directly with the computer. A more creative agent may personalize the computer system by implementing other user agents to perform mundane or repetitious tasks.

Providing technical assistance has also been the focus in the development of expert systems. (See Chapter 9.3 for a more comprehensive discussion of expert systems.) Briefly, these artificial intelligence programs attempt to solve difficult real-world problems in specific domains. Examples of expert systems as technical consultants for inexperienced users include DENDRAL (Buchanan and Feigenbaum, 1978) for structural elucidation problems in organic chemistry and MYCIN (Shortliffe, 1976) for infectious blood diseases in medicine. Note that the inexperienced users in these expert assistance systems are a different class of inexperienced computer users. Highly skilled technicians and professionals expect a sophisticated interface that can explain its reasoning and also accept advice. Therefore, the transfer of the task-specific knowledge captured in an expert system is critical to its success.

A third adaptive assistance system for inexperienced users listed in Table 11.3.5 is concerned with aiding the human in complex decision processes. The need for automated assistance in decision making is based on the wide variety of suboptimal human decision strategies (Benbasat and Taylor, 1982; Wickens, 1984). Therefore, it has been proposed that decision aids should attempt to structure decision making without intruding on the human's style or preferences (Adelman et al., 1982; Andriole, 1982). Indeed, closely allied with these decision aids is the development of human–computer dialogues with implicit communication (Revesman and Greenstein, 1983). Implicit communication permits a computer to work around the inexperienced user, aiding but not interfering with the computer-based task.

The next adaptive assistance strategy listed in Table 11.3.5 is a "do what I mean" (DWIM) capability that was developed in the INTERLISP programming environment (Teitelman and Masinter, 1981). The DWIM facility was basically a spelling corrector that could be invoked from several subsystems in INTERLISP and as a result was capable of detecting and correcting errors in commands. This capability freed the programmer from tedious changes in code and also gave the INTERLISP limited autonomy to act based on contextual information. Needless to say, the designer of adaptive interfaces should be extremely careful when implementing DWIM capabilities. For instance, a DWIM interface might be disastrous if the user attempts to load disk files with the command "DEPRINT *", but mistypes it as "DEPENT *". Depending on the correction algorithm, the DWIM interface might erase all the user's files with a "DELETE *" command. Previously mentioned confirmation dialogues and explicit constraints need to be designed and evaluated with DWIM interfaces to prevent the execution of unwanted actions.

Unburdening the user was also the goal in a computer-aided-design assistant reported by Cullingford et al. (1982). These researchers implemented an explanation system (CADHELP) that was tailored to a user's needs through three levels of description: command summary and purpose, normal explanation with step-by-step procedures, and summary information with suspected error explanations. This action explanation in CADHELP attempted to provide the user with knowledge of the CAD system so that more time could be spent on the primary task of circuit design. An inexperienced user expects and rightfully deserves to be able to create and to design freely on these computer systems rather than having to learn complicated procedures.

In a similar design approach, user help systems are also capable of taking a more active role in assisting the inexperienced user. Providing context-specific help for computing systems has been reported by Fenchel and Estrin (1982). These investigators have developed a method for integrating assistance information with the system software so that syntactic guidance is available at every step in a user's command. Likewise, Markus (1982) has illustrated that inexperienced users of bibliographic retrieval systems could retrieve a substantial amount of information when provided an assistant that standardized retrieval for a core set of commands across several databases. Lastly, Grignetti et al. (1975) have demonstrated that concrete examples using the inexperienced user's workspace is feasible in a text-editing domain.

The last two communication strategies listed in Table 11.3.5 for assistance systems center on establishing an advisory role for the adaptive interface. For example, the Writer's Workbench (Macdonald, 1983) provides stylistic advice to writers on the UNIX operating system. In the behavioral evaluation of the Writer's Workbench, Gingrich (1983) found that the writing aid increased the number of errors that were discovered and decreased the time required for editing. In general, subjects liked the objective and private advice from a computer-based assistant.

In a similar application, Darlington et al. (1983) have used an excursion tour methodology to prevent interactive deadlock between the inexperienced user and computer. These researchers developed a system to detect whether the user could accomplish a task with a specific plan. If the user could not finish the task, then interactive deadlock exists and a context-sensitive tutorial, called an excursion tour, is provided to the user.

Adaptive Instructional Systems

This final class of adaptive communication techniques in Table 11.3.5 recognizes the need for training and instructing inexperienced users (Paxton and Turner, 1984). If a user cannot be provided an alternative language or cannot be assisted in the task, possibly the only remaining option is to train the user on the automated procedures. With this option, an individualized training environment is desirable due to the heterogeneous instructional needs of inexperienced users. An individualized approach requires the development of diagnostic student models and the implementation of instructional strategies. Both adaptive training and intelligent, computer-assisted instruction provide methods for attaining these objectives.

Behavioral research on adaptive training systems is extensive and suggests that adaptive instructional systems should adapt on feedback variables to improve performance by providing information that guides the trainee through the task. Indeed, Kelley (1969) stressed that effective adaptive training environments may need knowledge-of-results displays since adaptive training deprives students of performance feedback. As a result of this research, the possibilities for adapting human–computer communication seem worthwhile and deserve further research.

With intelligent, computer-assisted instruction, adaptive communication techniques typically address how the computerized tutor should educate the inexperienced user. Stevens and Collins (1978), for instance, have adopted a Socratic method for educating users. This system actively questions the user and attempts to isolate misconceptions by entrapment. That is, the Socratic tutor would generate questions that will test explicitly for user misconceptions. An analogous instructional dialogue is the case-method approach introduced by Clancey (1982). Clancey used the case-method dialogue to experiment with tutoring users on the diagnostic rules in MYCIN. The tutor would explain the nuances of MYCIN's reasoning through specific cases. Therefore, a case-method dialogue focuses directly on the computer-based task and seems applicable to human–computer instruction.

In contrast to these two instructor-driven approaches, a more user-driven dialogue has been suggested by Burton and Brown (1982) with guided-discovery learning. The user of an instructional system in this dialogue is given freedom to explore and discover concepts and relationships of the task with the subtle guidance of a computer-based coach. This coach interrupts the inexperienced user only when specific instruction can be given on user misconceptions.

From these three instructional dialogue techniques it is obvious that diagnosis and remediation are a primary concern in adaptive instructional systems. The instructional interface should monitor the user and provide training when appropriate.

In demonstrating the importance of diagnostic and remedial variables, Elkerton and Williges (1984b) illustrated that inexperienced users can be instructed on expert file-search strategies. The results of their investigation revealed that suggestive advice with a strict diagnostic model aided inexperienced users to acquire expert search strategies when compared to forceful advice and more lenient diagnostic models. However, as Elkerton and Williges (1984b) stated, further design and evaluation is required owing to the intrusiveness of the instructional approach to novice file-search performance.

11.3.5.2 Strategies for Knowledge Representation

To implement adaptive interfaces and the communication strategies that were just discussed, knowledge of the user, task, or machine must be represented explicitly. In fact, knowledge representation is of even more importance when developing an adaptive interface for an inexperienced user, since the user should have easy access to the information interface. Fortunately, a myriad of representational styles exist and have been implemented. Table 11.3.6 provides a partial list of representation methods that seem to be applicable to adaptive dialogue, assistance, and instructional systems. In general, the sources of these models are artificial intelligence (Barr and Feigenbaum, 1982) and the systems engineering disciplines (Rouse, 1980). (The interested reader is referred to Chapter 9.2 for a more detailed description of methods in artificial intelligence.)

Subsets of the methods in Table 11.3.6 share common characteristics. Analysis of the representational strategies yields five dimensions on which to define models for adaptive interfaces. Each of these dimensions is presented in Table 11.3.7 as a set of five dichotomies. These model dimensions in Table 11.3.7 are discussed with respect to the actual models in Table 11.3.6 and the adaptive communication strategies

Table 11.3.6 Representational Strategies for Adaptive Human–Computer Systems

Blackboards
(Erman and Lesser, 1980)
Rules
(Clancey, 1982; Markus, 1982; Shortliffe, 1978)
Semantic networks
(Grignetti, Hausmann, and Gould, 1975)
Scripts
(Cullingford, Krueger, Selfridge, and Bienkowski, 1982; Schank and Abelson, 1977)
Plans
(Darlington, Dzida, and Herda, 1983; Wilensky, Arens, and Chin, 1984)
Differential models
(Burton and Brown, 1982)
Discriminant analysis
(Revesman and Greenstein, 1983)
Performance profiles
(Elkerton and Williges, 1984b, 1985; Gingrich, 1983; Macdonald, 1983)

in Table 11.3.5. It is hoped that this discussion will give designers an appreciation for the decisions that have to be made in selecting a representational strategy.

Symbolic and Quantitative Models

The distinction between symbolic and quantitative models is sometimes a fuzzy one. Surely a quantitative model is also a symbolic one. The distinction made here is that a model should be considered quantitative if it is based primarily on numbers (e.g., frequencies, statistics, and mathematical constructs). A symbolic model, in contrast, will represent basic concepts in higher-order symbols (e.g., textual, spatial, and verbal constructs). Using this operational definition, the symbolic models in Table 11.3.6 would be rules, semantic, networks, scripts, and plans; and the quantitative models would be discriminant analysis and performance profiles. Blackboards and differential models can be either or can contain both symbolic and quantitative information.

To explicate this distinction fully, the semantic nets of Grignetti et al. (1975) can be taken as an example of a symbolic model. The semantic nets were implemented to provide help and concrete examples to users of the NLS-SCHOLAR text-editing system. The information contained in these models included concept definitions, descriptions of commands, and command sequence information. This is to be compared to the performance profiles of Elkerton and Williges (1984b, 1985) where only command frequencies were stored in order to provide inexperienced users search strategy advice. Obviously, there are strengths and weaknesses with both approaches. Explanation capabilities are inherent in symbolic models as demonstrated by NLS-SCHOLAR as well as by CADHELP (Cullingford et al., 1982). With quantitative models, however, the strength is with predictive capability. For example, development of the implicit dialogue by Revesman and Greenstein (1983) was predicated upon predicting the operator in the process control task.

Therefore, a distinction that frequently can be made between symbolic and quantitative representation strategies rests on model expertise. A transparent "glass box" expert is usually associated with symbolic models, whereas an opaque "black box" expert is often associated with quantitative models. Glass box experts attempt to use problem-solving techniques that are similar to human methods, while the

Table 11.3.7 Modeling Dimensions for Representational Strategies

Symbolic	↔	Quantitative
Performance	↔	Cognitive
Skilled	↔	Unskilled
Static	↔	Dynamic
Singular	↔	Multiple

more efficient black box experts may not be constrained to these procedures (Burton and Brown, 1982). Thus, black box, quantitative models may require elaborate dialogue strategies, such as the graphical techniques used in STEAMER (Hollan et al., 1984), to describe the expertise to an inexperienced user. In short, there are tradeoffs in the development of adaptive symbolic and quantitative models with respect to their expository power and predictive efficiency.

Performance and Cognitive Models

The second modeling dimension in Table 11.3.7 illustrates that the source of knowledge is also important for the construction of adaptive interfaces. Two different sources of information can be used. One is based on actual user performance, and the other is based on detailed interviews with users.

The performance-sampling approach has been used by Macdonald (1983) and Elkerton and Williges (1984b, 1985) with their performance profile models. These investigations have revealed that performance-based models may be more than adequate for providing specific information to inexperienced users. Moreover, performance-based models can be quickly and efficiently constructed. The performance-sampling model only requires user monitoring and statistical summary. Therefore, performance-based models can be built while the user performs a computer-based task and may not, necessarily, require sophisticated system experts.

Nevertheless, complex tasks and behaviors may be difficult to capture and describe with performance-based models. Both Gingrich (1983) and Elkerton and Williges (1984b) have reported that inexperienced users have difficulty using the advice from a performance profile perhaps due to its quantitative nature. Consequently, the development of adaptive interfaces for inexperienced users may require intricate explanations of the cognitive processes in a task that may not be apparent with performance-based approaches. A collaborative effort between the adaptive system designers and task experts is often required to explicate cognitive processes and detailed task procedures.

Several examples of the models in Table 11.3.6 reveal the cognitive approach to modeling. The use of scripts by Cullingford et al. (1983) in a CAD application with several levels of active explanation illustrates the detailed knowledge that must be encoded into the adaptive interface. Scripts are ideal representation procedures since these knowledge structures contain detailed procedures for accomplishing goals (Schank and Abelson, 1977). Other examples include rule-based expert systems like MYCIN (Shortliffe, 1976) and intelligent, computer-assisted instruction interfaces as described by Clancey (1982). In an expert system, expert processes and procedures are mimicked at some level of abstraction to encourage participation and acceptance by the technical user. Likewise, with intelligent, computer-assisted instruction the inexperienced user must be able to understand the expert procedures in order for intelligent, computer-assisted instructions to be effective.

There are problems, however, in developing cognitive models. First, the construction of these models frequently relies on verbal protocols (Ericsson and Simon, 1984). Analyses of these data are difficult, time intensive, and perhaps even unreliable. Second, the adaptive interfaces developed with cognitive models are not infallible since they can do no better than the human expert.

Skilled and Unskilled Models

Models of task performance by skilled users are often used in building adaptive interfaces. The work by Clancey (1983) with rule-based tutors, Darlington et al. (1983) with planning models for active user help, and even the performance profile approach of Macdonald (1983) and Elkerton and Williges (1984b, 1985) all assume that inexperienced users can be diagnosed through skilled user models. In essence, these are differential models as described by Burton and Brown (1978) that index the capabilities of inexperienced users with respect to skilled-based models.

Differential models can easily describe skill presence or absence, but may ignore erroneous or unskilled behavior that is a significant component of human–computer systems. Models of unskilled performance occur in CADHELP's error explanations (Cullingford et al., 1982) and have been suggested as extensions to the plan-based models in the UNIX Consultant (Wilensky et al., 1984).

An unskilled modeling approach can be valuable to adaptive interfaces for inexperienced users, since user problems can be identified quickly without inferring a skill absence. For example, a catalog of alternative or unskilled behavioral processes could be constructed from inexperienced users. If a user is observed using one of these unskilled approaches, then the computer model would have direct evidence for this alternative user process, and, if necessary, could act immediately. However, it may be difficult and time consuming to represent all alternative or inappropriate skills in such an adaptive interface.

Static and Dynamic Models

Many of the models in Table 11.3.6 are static. Rules, semantic networks, scripts, differential models, and discriminant analysis are all primarily static models of behavior. Interestingly, adaptivity can be achieved with static representations through a large and comprehensive knowledge base. Expert systems

(Shortliffe, 1976), for example, appear adaptive due to the large amount of information available at the interface. Indeed, for many systems a large information reserve may be exploited for static adaptation.

However, the closed-world assumption that all behavior can be defined in a static model is an extreme limitation. The development of dynamic models in adaptive interfaces has been motivated by this limitation as well as by the evolutionary nature of inexperienced user behavior. Not surprising, both quantitative and symbolic approaches have been considered as potential learning paradigms. Both Elkerton and Williges (1984b, 1985) with performance profiles and Darlington et al. (1983) with plans have hinted at the possibility of dynamic models. Unfortunately, the problems in developing learning programs are difficult and will require extensive research. Specifically, the development of a dynamic program will require a model to converge rapidly to a sufficient representation of the user. A dynamic model that is slow or fluctuates widely may be unusable.

Singular and Multiple Models

All of the models in Table 11.3.6 are singular representations of the user, task, or machine with the exception of blackboards (Erman and Lesser, 1980). Blackboards have not been described previously, but have developed out of the needs in natural language and voice recognition for representing several levels of knowledge (e.g., acoustic, phonemic, morphemic, prosodic, syntactic, semantic, and pragmatic sources).

The blackboard representation in the HEARSAY-II voice understanding system was a framework for handling several independent sources of information in a system. Conceptually, hypotheses about the speech signal in HEARSAY-II were posted on a blackboard. Other knowledge sources could then use this information to derive additional information. Therefore, the blackboard data structure allowed the system to analyze a speech signal at several levels concurrently.

The success of a blackboard method for speech understanding was limited. However, the representational strategy seems appropriate to adaptive human–computer interfaces where user performance is typically multidimensional. More specifically, a blackboard of hypotheses might aid in forming a cohesive user model in an adaptive interface. That is, conclusions from one information source could be verified or updated by another. Further research is necessary and encouraged on this multiple representation technique in speech understanding and other adaptive dialogues.

11.3.5.3 Future Implications

Most of the alternatives for adaptive interface design are exploratory in nature, and few behavioral studies have been conducted to evaluate the efficiency and effectiveness of these approaches. These alternatives, however, represent future trends in the design of human–computer software interfaces. The various design alternatives for adaptive human–computer communication and knowledge representation, if substantiated by behavioral research support, will provide methods for building adaptive software interfaces through iterative design procedures.

11.3.6 CONCLUSION

The design of software interfaces for inexperienced users is a complex process. Central to this process is the concept of iterative design to reach specific design goals by incorporating knowledge about the end-user population, methods of optimizing human–computer communications, and user acceptance testing. Although the iterative design process is currently used as an efficient means of optimizing static software interface configurations, the design process itself should also be amenable to the development and testing of adaptive human–computer interfaces. As the methods of adaptive human–computer communication and knowledge representation are further developed and tested, adaptive interfaces should become viable alternatives to the conventional software interfaces used in most current systems.

REFERENCES

Adelman, L., Donnell, M. L., Phelps, R. H., and Patterson, J. F. (1982). An interative Bayesian decision aid: toward improving the user-aid and user-organization interfaces. *IEEE Transactions on Systems, Man, and Cybernetics, SMC-12,* 733–743.

Alavi, M., (1984). An assessment of the protyping approach to information systems development. *Communications of the ACM, 27*(6), 556–563.

Anderson, R. H., and Sibley, W. L. (1972). A new approach to man-machine interfaces (Report No. R-876-ARPA). Santa Monica, CA: Rand Corporation.

Andriole, S. J. (1982). The design of microcomputer-based personal decision-aiding systems. *IEEE Transactions on Systems, Man, and Cybernetics, SMC-12,* 463–469.

Badre, A. N. (1982). Designing chunks for sequentially displayed information. In A. Badre and B. Shneiderman, Eds., *Directions in human-computer interaction.* Norwood, NJ: Ablex Publishing.

Badre, A. N. (1984). Designing transitionality into the user-computer interface. In G. Salvendy, Ed., *Human-computer interaction* (pp. 27–34). Amsterdam, Elsevier.

Barmack, J. E., and Sinaiko, H. W. (1966). Human factors problems in computer-generated graphic displays (AD-63617C). Washington, DC: Institute for Defense Analysis

Barnard, P. J., Hammond, N. V., Morton, J., and Long, J. B. (1981). Consistency and compatibility in human-computer dialogue. *International Journal of Man-Machine Studies, 15,* 87–134.

Barr, A., and Feigenbaum E. A., Eds. (1982). *The handbook of artificial intelligence.* Los Altos, CA: William Kaufmann, Vols. I and II.

Benbasat, I., and Taylor, R. N. (1982). Behavioral aspects of information processing for the design of management information systems. *IEEE Transactions on Systems, Man, and Cybernetics, SMC-12,* 439–450.

Benbasat, I., and Wand, Y. (1984). A structured approach to designing human-computer dialogues. *International Journal of Man-Machine Studies, 21,* 105–126.

Biermann, A., Rodman, R., Ballard, B., Betancourt, J., Bilbro, G., Deas, H., Fineman, L., Fink, P., Gilbert, K., Gregory, D., and Heidlage, F. (1983). Interactive nautral language problem solving: a pragmatic approach. *Proceedings of the conference on applied natural language processing,* Santa Monica, CA, pp. 180–191.

Biermann, A., Rodman, R., Rubin, D., and Heidlage, F. (1984). Natural language with discrete speech as a mode for human to machine communication (Technical Report). Durham, NC: Duke University, Department of Computer Science.

Billingsley, P. A. (1982). Navigation through hierarchical menu structures: does it help to have a map? *Proceedings of the Human Factors Society 26th annual meeting.* Santa Monica, CA: The Human Factors Society, pp. 103–107.

Box, G. E. P., Hunger, W. G., and Hunter, J. S. (1978). *Statistics for experimenters: an introduction to design, data analysis, and model building.* New York: Wiley.

Brown, D. M., Burkleo, H. V., Mangelsdorf, J. E., Olsen, R. A., and Williams, A. R., Jr. (1981). *Human factors engineering criteria for information processing systems.* Sunnyvale, CA: Lockheed.

Brown, R. L. (1984). Automatic error correction for a voice data entry task (unpublished master's project). Blacksburg, VA; Virginia Polytechnic Institute and State University.

Buchanan, B. G., and Feigenbaum, E. A. (1978). Dendral and Meta-Dendral: their applications dimension. *Artificial Intelligence, 11,* 5–24.

Burton, R. R., and Brown, J. S. (1982). An investigation of computer coaching for informal learning activities. In D. Sleeman and J. S. Brown, Eds., *Intelligent tutoring systems.* New York: Academic.

Card, S. K., Pavel, M., and Farrell, J. E. (1984). Window-based computer dialogues. *Proceedings of INTERACT'84 first IFIP conference on human-computer interaction.* London, pp. 355–359.

Carroll, J. M., and Carrithers, C. (1984). Training wheels in a user interface. *Communications of the ACM, 27*(8), 800–806.

Carroll, J. M., and Rosson, M. B. (1985). Usability specifications as a tool in iterative development. In H. R. Hartson, Ed., *Advances in human-computer interaction* (pp. 1–28). Norwood, NJ: Ablex.

Chapanis, A. (1975). Interactive human communications. *Scientific America, 232*(3), 36–42.

Clancey, W. J. (1982). Tutoring rules for guiding a case method dialogue. In D. Sleeman and J. S. Brown, Eds., *Intelligent tutoring systems.* New York: Academic.

Cohen, P., Perrault, C., and Allen, J. (1981). Beyond question-answering (Report No. 4644). Cambridge, MA: Bolt Beranek and Newman.

Cohill, A. M., and Ehrich, R. W. (1983). Automated tools for the study of human/computer interfaces. *Proceedings of the Human Factors Society 27th annual meeting.* Santa Monica, CA: The Human Factors Society, pp. 897–900.

Cohill, L. F. (1984). A taxonomy of user-computer interface functions. In G. Salvendy, Ed., *Human-computer interaction* (pp. 125–128). Amsterdam: Elsevier.

Coleman, W. D., Williges, R. C., and Wixon, D. R. (1985). Collecting detailed user evaluations of software interfaces. *Proceedings of the Human Factors Society 29th Annual Meeting.* (pp. 240–244). Santa Montica, CA: The Human Factors Society.

Cook, T. D., and Campbell, D. T. (1979). *Quasi-experimentation.* Chicago: Rand-McNally.

Cordes, R. E. (1984). Application of a magnitude estimation for evaluating software ease-of-use. In G. Salvendy, Ed., *Human-computer interaction.* Amsterdam: Elsevier, pp. 199–202.

Cullingford, R. E., Krueger, M. W., Selfridge, M., and Bienkowski, M. A. (1982). Automated explanations as a component of computer-aided design system. *IEEE Transactions on Systems, Man, and Cybernetics, SMC-12,* 168–181.

Darlington, J., Dzida, W., and Herda, S. (1983). The role of excursions in interactive systems. *International Journal of Man-Machine Studies, 18,* 101–112.

Data General (1982). Data General's PRESENT information presentation facility user's manual (Document 093-000168). Data General Corporation.

Diamond, W. J. (1981). *Practical experimental designs for engineers and scientists.* Belmont: Lifetime Learning.

Dick, W., and Carey, L. (1978). *The systematic design of instruction.* Glenview, IL: Scott, Foresman.

Digital Equipment Corporation (1984). DEC's VAX11 form management system, (Document SPD AE-R440C-TE). Burlington, MA: Digital Equipment Corporation.

Douglas S. A., and Moran, T. P. (1983). Learning text editor semantics by analogy, *Proceedings of human factors in computing systems* (pp. 207–211). Boston, ACM.

Ehrenreich, S. L. (1981). Query languages: design recommendations derived from the human factors literature. *Human Factors, 23,* 709–725.

Ehrlich, K., and Soloway, E. (1984). An empirical investigation of the tacit plan knowledge in programming. In J. C. Thomas and M. L. Schneider, Eds., *Human factors in computer systems* (pp. 113–133. Norwood, NJ: Ablex.

Elkerton, J., and Williges, R. C. (1984a). Information retrieval strategies in a file-search environment. *Human Factors, 26,* 171–184.

Elkerton, J., and Williges, R. C. (1984b). The effectiveness of a performance-based assistant in an information-retrieval environment. *Proceedings of the Human Factors Society 28th annual meeting* (pp. 634–638). Santa Monica, CA: Human Factors Society.

Elkerton, J., and Williges, R. C. (1985). A performance profile methodology for implementing assistance and instruction in computer-based tasks. *International Journal of Man-Machine Studies, 23,* 135–151.

Engel, S. E., and Granda, R. E. (1975). Guidelines for man/display interfaces (Technical Report 00.2720). Poughkeepsie, NY: IBM.

Ericsson, K. A., and Simon, H. A. (1984). *Protocol analysis verbal reports as data.* Cambridge: MIT Press.

Erman, L. D., and Lesser, V. R. (1980). The Hearsay-II speech understanding system: a tutorial. In W. A. Lea, Ed., *Trends in speech recognition* (pp. 316–339). Englewood Cliffs, NJ: Prentice Hall.

Fenchel, R. S., and Estrin, G. (1982). Self-describing systems using integral help. *IEEE Transactions on Systems, Man, and Cybernetics, SMC-12,* 162–167.

Fitts, P. M., and Seeger, C. M. (1953). S-R compatability: spatial characteristics of stimulus and response codes. *Journal of Experimental Psychology, 46,* 199–210.

Foley, J. D., and Wallace, V. L. (1974). The art of natural graphic man-machine conversation. *Proceedings of IEEE, 63,* 462–471.

Foley, J. D., Wallace, V. L., and Chan, P. (1981). The human factors of graphic interaction tasks and techniques (Technical Report No. GWU-11ST-81-3). Washington, DC: George Washington University.

Furnas, G. W., Gomez, L. M., Landauer, T. K., and Dumais, S. T. (1982). Statistical semantics: how can a computer guess what people mean when they name things? *Proceedings of Human Factors in Computer Systems* (pp. 251–253). New York: Association for Computing Machinery.

Furnas, G. W., Landauer, T. K., Gomez, L. M., and Dumais, S. T. (1984). Statistical semantics: analysis of the potential performance of keyword information systems. In J. C. Thomas and M. L. Schneider, Eds., *Human factors in computing systems* (pp. 187–242). Norwood, NJ: Ablex.

Gaines, B. R., and Facey, P. V. (1975). Some experience in interactive system development and application. *Proceedings of the IEEE, 63,* 155–169.

Galitz, W. O. (1981). *Handbook of screen format design.* Wellesley, MA: Q.E.D. Information Science.

Gebhardt, F., and Stellmacher, I. (1978). Design criteria for documentation retrieval languages. *Journal of the American Society for Information Science, 29,* 191–199.

Gilfoil, D. M. (1982). Warming up to computers: a study of cognitive and effective interaction over time. *Proceedings of human factors in computing systems* (pp. 171–175). New York: Association for Computing Machinery.

Gingrich, P. S. (1983). The UNIX writer's workbench software: results of a field study. *The Bell System Technical Journal, 62,* 1909–1921.

Goldsmith, T. E., and Schvaneveldt, R. W. (1984). Facilitating multiple-cue judgments with integral information displays. In J. C. Thomas and M. L. Schneider, Eds., *Human factors in computer systems* (pp. 243–270). Norwood, NJ: Ablex.

Good, M. D., Whiteside, J. A., Wixon, D. R., and Jones, S. J. (1984). Building a User-Derived Interface. *Communications of the ACM, 27*(10), 1032–1043.

Gould, J. D., and Boies, S. J. (1984). Human factors of the 1984 Olympic message system. *Proceedings of the Human Factors Society 28th annual meeting* (pp. 547–551). Santa Monica, CA: The Human Factors Society.

Gould, J. D., Conti, J., and Hovanyecz, T. (1981). Composing letters with a simulated listening typewriter. *Proceedings of the human factors 25th annual meeting* (pp. 505–508). Santa Monica, CA: The Human Factors Society.

Gould, J. D., and Lewis, C. (1983). Designing for usability—key principles and what designers think. *Human factors in computing systems* (pp. 50–53). New York: Association for computing Machinery.

Grignetti, M. G., Hausmann, C., and Gould, L. (1975). An intelligent on-line assistant for tutor—NLS—SCHOLAR. *Proceedings of the National Computer Conference, 44,* 775–781.

Hammond, N., Jorgensen, A., MacLean, A., Barnard, P., and Long, J. (1983). Design practice and interface usability: evidence from interviews with designers. *Proceedings of human factors computing systems,* Boston, ACM. (pp. 40–44).

Hanau, P. R., and Lenorovitz, D. R. (1980). Prototyping and simulation tools for user/computer dialogue design. *Proceedings of computer graphics SIGGRAPH 1980 conference.* New York: Association for Computing Machinery, Vol. 14.

Hartson, H. R., and Johnson, D. H. (1986). Human-computer interface development: Concepts and systems for its management. (Technical Report TR-86-07). Blacksburg VA: Virginia Polytechnic Institute and State University.

Hartson, H. R., Johnson, D. H., and Ehrich, R. W. (1984). A computer dialogue management system. *Proceedings of INTERACT'84 first IFIP conference on human-computer interaction.* London, pp. 57–61.

Hayes, P., Ball, E., and Reddy, R. (1983). Breaking the man-machine communication barrier. *Computer, 14* (March), 19–30.

Hiltz, S. R.; and Turoff, M. (1978). *The network nation.* Reading, MA: Addison-Wesley.

Hiltz, S. R., and Turoff, M. (1981). The evolution of user behavior in a computerized conferencing system. *Communications of the ACM, 24,* 739–751.

Holden, A. D. C., and Haley, P. V. (1982). Adaptive methods in speech understanding systems. *IEEE proceedings of the international conference on cybernetics and society.* Seattle, WA, pp. 161–165.

Hollan, J. D., Hutchins, E. L., and Weitzman, L. (1984). STEAMER: an interactive inspectable simulation-based training system. *AI Magazine,* 15–27 (Summer).

IBM (1983). IBM system productivity facility for MVS, general information (Document GC34-2039-0). White Plains, NY: IBM.

Innocent, P. R. (1982). Towards self-adaptive interface systems. *International Journal of Man-Machine Studies, 10,* 287–289.

Johnson, D. H., and Hartson, H. R. (1982). The role and tools of a dialogue author in creating human-computer interfaces (Technical Report CSIE-82-8). Blacksburg, VA: Virginia Polytechnic Institute and State University.

Jorgensen, A. H., Barnard, P., Hammond, N., and Clark, I. (1983). Naming commands: an analysis of designer's naming behavior. In T. R. G. Green, S. J. Payne, and G. C. Van Der Veer, Eds., *The psychology of computer use.* London: Academic.

Kelley, C. R. (1969). What is adaptive training?" *Human Factors, 11,* 547–556.

Kelley, J. F. (1983). An empirical methodology for writing user-friendly natural language computer applications. *Proceedings of human factors in computing systems* (pp. 193–196). New York: Association for Computing Machinery.

Kelly, M. J., and Chapanis, A. (1977). Limited natural language dialogue. *International Journal of Man-Machine Studies, 9,* 479–501.

Keppel, G. (1982). *Design and analysis: a researcher's handbook.* New York: Wiley.

Kidd, A. L. (1982). Problems in man-machine dialogue design. *Proceedings of the Telecommunications Conference.* Zurich.

Kloster, G. V., and Tischer, K. (1984). Man-machine interface design process. *Proceedings of INTERACT'84 first IFIP conference on human-computer interaction.* London, pp. 236–241.

Lenorovitz, D. R., Phillips, M. D., Ardrey, R. S., and Kloster, G. V. (1984). A taxonomic approach to characterizing human-computer interfaces. In G. Salvendy, Ed., *Human-computer interaction* (pp. 111–116). Amsterdam: Elsevier.

Licklider, J. C. R. (1960). Man-machine symbiosis. *IRE Transactions on Human Factors in Electronics.* 4–11.

Macdonald, N. H. (1983). The UNIX writer's workbench software: rationale and design. *The Bell System Technical Journal, 62,* 1891–1908.

Maguire, M. (1982). An evaluation of published recommendations on the design of man-computer dialogues. *International Journal on Man-Machine Studies, 16,* 237–261.

Malone, T. W. (1983). How do people organize their desks? Implications for the design of office information systems. *ACM Transactions on Office Information Systems, 1,* 99–112.

Markus, R. S. (1982). User assistance in bibliographic retrieval networks through a computer intermediary. *IEEE Transactions on Systems, Man, and Cybernetics, SMC-12,* 116–133.

Martin, J. (1973). *Design of man-computer dialogues.* Englewood Cliffs, NJ: Prentice-Hall.

Mason, R. E. A., and Carey, T. T. (1983). Prototyping interactive information systems. *Communications of the ACM, 26,* 347–352.

McCoy, K. F. (1983). Correcting misconceptions: what to say when the user is mistaken. *Proceedings of human factors in computing systems* (pp. 197–201). New York: Association for Computing Machinery.

Meister, D. (1984). New opportunities in the human factors engineering of computerized systems. In G. Salvendy, Ed., *Human-computer interaction* (pp. 43–54). Amsterdam: Elsevier.

Melton, A. W. (1963). Implications of short-term memory for a general theory of memory. *Journal of Verbal and Verbal Behavior, 2,* 1–21.

Miller, G. A. (1956). The Magical number seven, plus or minus two: some limits on our capacity for processing information. *Psychological Review, 63,* 81–97.

Miller, L. A. (1981). Natural language programming: styles, strategies, and contrasts. *IBM Systems Journal, 20,* 184–215.

Miller, L. A., and Thomas, Jr., J. C. (1976). Behavioral issues in the use of interactive systems (RC6326). Yorktown Heights, NY: IBM.

Mitchell, C. M., and Miller, R. A. (1983). Design strategies for computer-based information displays in real-time control systems. *Human Factors, 25,* 353–369.

Murdock, B. B. (1965). Effects of a subsidiary task on short-term memory. *British Journal of Psychology, 56,* 413–419.

Myers, J. L. (1979). *Fundamentals of experimental design.* Boston: Allyn and Bacon.

Neal, A. S., and Simons, R. M. (1983). Playback: a method for evaluating the usability of software and its documentation. *Proceedings of human factors in computing systems* (pp. 78–82). New York: Association for Computing Machinery.

Newman, W. M., and Sproull, R. F. (1979). *Principles of interactive computer graphics.* New York: McGraw-Hill.

Nickerson, R. S. (1981). Why interactive computer systems are sometimes not used by people who might benefit from them. *International Journal of Man-Machine Studies, 15,* 469–483.

Norman, D. A. (1981). The trouble with UNIX. *Datamation, 27*(12), 139–150.

Norman, D. A. (1983). Steps toward cognitive engineering: design rules based on analysis of human error. *Proceedings of Human Factors in Computing Systems* (pp. 378–382). New York: Association for Computing Machinery.

North, R. A., Young, M., and Graffunder, K. (1984). Dynamic training approaches for an airborne speech recognition system. Minneapolis, MN: Honeywell.

Parrish, R. N., Gates, J. L., Munger, S. F., and Sidorsky, R. C. (1981). Development of design guidelines and criteria for user/operator transactions with battlefield automated systems, volume IV: provisional guidelines and criteria for the design of user/operator transactions (Draft Report, Phase I). Alexandria, VA: U.S. Army Research Institute.

Paxton, A. L., and Turner, E. J. (1984). The application of human factors to the needs of the novice computer user. *International Journal of Man-Machine Studies, 20,* 137–156.

Payne, S. J., and Green, T. R. G. (1983). The user's perception of the interaction language: a two-level model. *Proceedings of Human Factors in Computing Systems* (pp. 40–44). New York: Association for Computing Machinery.

Pew, R. W., and Rollins, A. M. (1975). Dialog specification procedures (Report No. 3129). Cambridge, MA: Bolt Beranek and Newman.

Phillips, M. D., and Tischer, K. (1984). Operations concept formulation for next generation air traffic control systems. *Proceedings of INTERACT'84 first IFIP conference on human-computer interaction.* London, pp. 242–247.

Prince, M. D. (1971). *Interactive graphics for computer-aided design.* Reading, MA: Addison-Wesley.

Ramsey, H. R., and Atwood, M. E. (1979). Human factors in computer systems: a review of the literature (Report No. SAI-79-111-DEN). Englewood, CO: Science Applications.

Reisner, P. (1977). Use of psychological experimentation as an aid to development of a query language. *IEEE Transactions on Software Engineering, SE-3,* 218–229.

Reisner, P. (1981). Formal grammar and human factors design of an interactive graphics system. *IEEE Transactions on Software Engineering, SE-7,* 229–240.

Reisner, P. (1982). Further developments toward using formal grammar as a design tool. *Proceedings of human factors in computer systems* (pp. 304–308). New York: Association for Computing Machinery.

Revesman, M. E., and Greenstein, J. S. (1983). Application of a model of human decision making for human/computer communication. *Proceedings of human factors in computing systems* (pp. 107–111). New York: Association for Computing Machinery.

Roach, J. W., and Nickson, M. (1983). Formal specifications for modeling and developing human/computer interfaces. *Proceedings of Human Factors in Computing Systems* (pp. 35–39). New York: Association for Computing Machinery.

Roberts, T. L. (1979). Evaluation of computer text editors (Report SSL-79-9). Palo Alto, CA: Xerox, System Science Laboratory, Palo Alto Research Center.

Roberts, T. L., and Moran, T. P. (1983). The evaluation of text editors: methodology and empirical results. *Communications of the ACM, 26.*

Rosson, M. B. (1983). Patterns of experience in text editing. *Proceedings of human factors in computing systems* (pp. 171–175). New York: Association for Computing Machinery.

Rouse, W. B. (1980). *System engineering models of human-machine interaction.* New York: North Holland.

Rouse, W. B. (1984). Design and evaluation of computer-based decision support systems. In G. Salvendy, Ed., *Human-computer interaction.* Amsterdam: Elsevier.

Rubel, A. (1984). Graphic based applications—tools to fill the software gap. *Digital Design.*

Rubenstein, R., and Hersh, H. M. (1984). *The human factor: designing computer systems for people.* Burlington, MA: Digital Press.

Scapin, D. (1981). Computer commands in restricted natural language: some aspects of memory and experience. *Human Factors, 23,* 365–375.

Schank, R. C., and Abelson, R. P. (1977). *Scripts, plans, goals, and understanding.* Hillsdale, NJ: Erlbaum.

Schurick, J. M., Williges, B. H., and Maynard J. F. (1985). User feedback requirements with automatic speech recognition. *Ergonomics, 28*(11), 1543–1555.

Shneiderman, B. (1980). *Software psychology: human factors in computer and information systems.* Cambridge, MA: Winthrop.

Shneiderman, B. (1983). Direct manipulation: a step beyond programming languages. *Computer, 16*(8), 57–69.

Shneiderman, B. (1984). Correct, complete operations and other principles of interaction. In G. Salvendy, Ed., *Human-computer interaction* (pp. 135–146). Amsterdam: Elsevier.

Shortliffe, E. H. (1976). *Computer-based medical consultations: MYCIN.* New York: American Elsevier.

Simon, H. A. (1974). "How big is a chunk?" *Science, 183,* 482–488.

Simpson, C. A. (1983). Integrated voice controls and speech displays for rotorcraft mission management. *Proceedings of the SAE aerospace congress and exposition.* Long Beach, CA.

Simpson, C. A., Coler, C. R., and Huff, E. M. (1982). Human factors of voice I/O for aircraft cockpit controls and displays. *Proceedings of workshop on standardization for speech I/O technology* (pp. 159–166). Gaithersburg, MD: National Bureau of Standards.

Simpson, C. A., and Navarro, T. (1984). Intelligibility of computer generated speech as a function of multiple factors. *Proceedings of the IEEE national aerospace and electronics conference.* Dayton, OH.

Smith, S. L. (1982). User-system interface design for computer-based information systems (Technical Report ESD-TR-82-132). Bedford, MA: Mitre Corporation.

Smith, S. L., and Aucella, A. F. (1983). Design guidelines for the user interface to computer-based information systems (ESD-TR-83-122). Bedford, MA: Mitre.

Smith, S. L., and Mosier, J. N. (1984). The user interface to computer-based information systems: a survey of current software design practice. *Proceedings of the IFIP INTERACT '84 conference on human-computer interaction.* London.

Spine, T. M., Maynard, J. F., and Williges, B. H. (1983). Error correction strategies for voice recognition. *Proceedings of the voice data entry systems applications conference.* Chicago, IL.

Stevens, A., and Collins, A. (1978). The goal structure of a Socratic tutor. *Proceedings of the ACM annual conference.* (pp. 256–263). New York: Association for Computing Machinery.

Sullivan, M. A., and Chapanis, A. (1983). Human factoring a text editor manual. *Behavior and Information Technology, 2,* 113–125.

Teitlebaum, R. C., and Granda, R. E. (1983). The effects of positional constancy on searching menus for information. *Proceedings of Human Factors in Computing Systems.* (pp. 40–44). New York: Association for Computing Machinery.

Teitelman, W., and Masinter, L. (1981). The Interlisp programming environment. *Computer, 14* (April), 25–32.

Tennant, H. R., Ross, K. M., and Thompson, C. W. (1983). Usable natural language through menu-based natural language understanding. *Proceedings of human factors in computing systems* (pp. 154–160). New York: Association for Computing Machinery.

Thomas, R. C. (1981). The design of the adaptable terminal. In M. J. Coombs and J. L. Alty, Eds., *Computing skills and the user interface* (pp. 427–463). New York: Academic.

Tullis, T. S. (1983). The formatting of alphanumeric displays: a review and analysis. *Human Factors, 25,* 657–682.

Tulving, E. (1962). Subjective organization in free recall of unrelated words. *Psychological Review, 69,* 344–354.

Vasek, J. R. (1983). Using formative evaluation to develop a microcomputer assisted instruction program to teach simple microcomputer language acquisition (unpublished doctoral dissertation). Blacksburg, VA: Virginia Polytechnic Institute and State University.

Wasserman, A. I., and Shewmake, D. T. (1982). Rapid prototyping of interactive systems. *ACM SIG-SOFT Software Engineering Notes,* 1–18.

Waterman, D. A. (1978). Exemplary programming in RITA. In D. A. Waterman and F. Hayes-Roth, Eds., *Pattern directed inference systems.* New York: Academic, pp. 261–279.

Weizenbaum, J. (1966). ELIZA—a computer program for the study of natural language communication between man and machine. *Communications of the ACM, 9,* 36–45.

Wickens, C. D. (1984). *Engineering psychology and human performance.* Columbus, OH: Charles E. Merrill.

Wilensky, R., Arens, Y., and Chin, D. (1984). Talking to UNIX in English: an overview of UC. *Communications of the ACM, 27,* 574–593.

Williges, R. C. (1981). Development and use of research methodologies for complex system/simulation experimentation. In J. Moraal and K. F. Kraiss, Eds., *Manual system design methods, equipment, and applications* (pp. 59–87). New York: Plenum.

Williges, R. C. (1984). Evaluating human-computer software interfaces. *Proceedings of the 1984 international conference on occupational ergonomics.* Toronto, pp. 81–87.

Williges, R. C., and Wierwille, W. W. (1979). Behavioral measures of aircrew mental workload. *Human Factors, 21,* 549–574.

Williges, R. C., and Williges, B. H. (1982). Modeling the human operator in computer-based data entry. *Human Factors, 24,* 285–299.

Williges, B. H., and Williges, R. C. (1984). Dialogue design considerations for interactive computer systems. In F. A. Muckler, Ed., *Human factors review 1984.* (pp. 167–208). Santa Monica, CA: The Human Factors Society.

Wong, P. C. S., and Reid, E. R. (1982). FLAIR—user interface dialog design tool. *Computer Graphics, 16,* 87–98.

Woods, W. A. (1973). Progress in natural language understanding: an application, lunar geology. *Proceedings of the National Computer Conference, 42,* 441–450.

Wright, P. (1983). Manual dexterity: a user-oriented approach to creating computer documentation. *Proceedings of human factors in computing systems.* (pp. 11–18) New York: Association for Computing Machinery.

Yunten, T., and Hartson, H. R. (1985). A SUPERvisory methodology and notation (SUPERMAN) for human-computer system development. In H. R. Hartson, Ed., *Advances in Human-Computer Interaction* (pp. 243–281). Norwood, NJ: Ablex.

CHAPTER 11.4
HUMAN FACTORS ASPECTS OF MANUAL COMPUTER INPUT DEVICES

JOEL S. GREENSTEIN

Clemson University
Clemson, South Carolina

LYNN Y. ARNAUT

Virginia Polytechnic Institute and State University
Blacksburg, Virginia

11.4.1 INTRODUCTION

A list of the more common computer input devices would include keyboards, touch screen devices, graphic tablets, mice, trackballs, joysticks, and light pens. Additionally, although the ability of computers to recognize speech is limited at this time, speech recognition systems are now being employed effectively in certain applications, particularly those in which it is not possible or desirable to dedicate the hands to computer data entry. This chapter presents recommendations for the design and selection of manual computer input devices. In Section 11.4.2, the human factors considerations affecting the design of each type of input device are considered in turn. Section 11.4.3 presents comparison data to aid in the selection of an appropriate device type for a given application. Chapter 11.5 considers the use of voice as an alternative or additional medium for the input of information to a computer.

11.4.2 DESIGN CONSIDERATIONS

11.4.2.1 Keyboards

Of the various devices available for transmitting information from the human to the computer, the keyboard is certainly the most generally applied. It is clearly the device of choice for applications involving significant amounts of textual input. It is also the computer input technology most affected by historical precedent. The data entry keyboard achieved some degree of design standardization long before its linkage to the computer. While this linkage has had a substantial effect on the design of current keyboards, the functionality, layout, dimensions, and mechanics of these devices are still in many ways a reflection of nineteenth-century technology.

Fixed-Function and Variable-Function Keyboards

In many applications, an initial consideration in the selection of a keyboard concerns whether the keys shall have fixed or variable functions. Examples of fixed-function-keyboard applications include point-of-sale terminals and basic hand-held calculators. General advantages of fixed-function keyboards include:

1. Simplicity of operation—typically, only one key at a time is pressed and the same function is always performed by the same key.
2. All available functions can be determined directly by scanning the keys.
3. Relatively little software support is necessary.
4. Keys can be arranged in logical groups.

Disadvantages of fixed-function keyboards include:

1. A large number of required functions requires a large number of keys.
2. Frequent visual search and arm/hand movement may be required over a large area.
3. Changes require hardware modification.
4. It may be difficult to group keys logically for all operating procedures.

In general, the selection of a fixed-function keyboard appears to be appropriate when one set of functions is frequently employed, when functions must be executed quickly, and when correct selection of the functions implemented by the keyboard is critical to satisfactory operation of the system.

Key labels of variable-function keyboards are generally varied in one of three ways: shift keys may be used, permitting the user to shift a key's function among several fixed alternatives; labeled overlays appropriate to a given mode of operation may be positioned above the keys; or the functions of the keys may be placed under software control with the user informed of the key–function relationships via an associated video display unit. Applications representative of the use of variable-function keyboards include upper case/lower case video display terminal keyboards (shifted keys), keyboards associated with video game systems (different overlays for different games), and menu items offered on a video display (software control). General advantages of variable-function keyboards include:

1. Fewer keys are needed relative to a fixed-function keyboard of equivalent power; this difference may range to a ratio of 1 to 1000 for video-display-unit labeling versus a customized-format fixed-function process-control keyboard.

2. Less visual search and arm/hand movement are required.

Software-controlled key–function relationships have several additional advantages:

1. The operating procedure and sequence can be guided by programmed instructions displayed on the associated video display.
2. Changes require software rather than hardware modification.
3. Labels can be logically grouped for each operating procedure.
4. Keys for functions inappropriate to the current situation may be made inactive.

There are also certain disadvantages associated with the different types of variable-function keyboards:

1. With shift keys, the user must press more than one key to execute any shifted function. This additional complexity increases entry time and number of errors. It also becomes difficult to label clearly the keys with their multiple functions and to group labels logically for all procedures.
2. With overlays, the user must select and attach the appropriate overlay.
3. With software-controlled labeling, the user must select the currently appropriate label-to-key assignment. It is therefore likely that he or she will require more training than would the user of a fixed-function keyboard. The training problem is complicated further when the user employs a variety of software packages. There can easily be a lack of equivalence in function-to-label and function-to-key assignments among the different software packages. Thus, a given function may go by different labels, a given label may represent different functions, and a given function may be associated with different keys across the software packages.

In general, the selection of variable-function keyboards appears to be appropriate when there are frequently used subsets of functions, when the pacing of entries is not forced, and when relatively sophisticated prompting and feedback are available. Additionally, software-controlled labeling seems particularly appropriate to applications that experience continual modification.

The relative advantages and disadvantages of fixed- and variable-function keyboards are summarized in Table 11.4.1. The circumstances in which each of these types of keyboards tends to be more useful are indicated in Table 11.4.2.

Keyboard Layout

The arrangement of components within a workspace is guided by the importance, frequency-of-use, function, and sequence-of-use of the individual components. These general considerations apply to the layout of keys on a keyboard as well. Hanes (1975) elaborates upon these fundamental notions to provide a more detailed set of guidelines for keyboard layout:

1. Determine the characters and numbers of keys required.
2. Arrange the keys according to their frequency of use and according to user characteristics.
3. Follow historical precedent.
4. Follow established standards.

Table 11.4.1   Advantages and Disadvantages of Fixed- and Variable-Function Keyboards

	Fixed-Function Keyboards	Variable-Function Keyboards
Advantages	Simplicity of operation	Fewer keys
	Function is evident from key	Less visual search
	Minimal software support	Less arm/hand movement
	Logical key grouping	Can be modified by software changes
Disadvantages	Numerous functions require numerous keys	Increased function selection time
	Frequent visual search	Decreased clarity of key labeling
	Frequent arm/hand movement	Increased prompting and feedback requirements
	Changes require hardware modification	Increased training requirements

Table 11.4.2   When to Use Fixed- and Variable-Function Keyboards

Use Fixed-Function Keyboards When	Use Variable-Function Keyboards When
One set of functions is frequently employed	Several subsets of functions are frequently used
Functions must be executed quickly	Pacing of entries is not forced
Correct function selection is critical	Sophisticated prompting and feedback are available

5. Group frequently used keys under the resting position of the hand where the user can determine their locations by touch.

6. Group related functions together

7. Group logically and according to sequence of use.

8. Locate according to importance.

9. Code the keys.

10. Consider all factors, including the intended applications, the costs, and the manufacturing requirements.

As Hanes notes, these rules can conflict. The QWERTY keyboard layout (named after the leftmost six keys in the top row of letters) and its variants have become standard for alphanumeric data entry in many countries. The appropriateness of the QWERTY layout has been the subject of debate for more than 50 years. Its continued use illustrates an adherence to guideline 3 in preference to guideline 2.

Keyboards for computer input typically include alphanumeric, function, auxiliary numeric, and cursor control key groups (see Figure 11.4.1). Most of the attention to date has focused on the layout of the alphanumeric group, composed of the upper and lower case alphanumeric characters, 10 numerals, punctuation marks, and special symbols such as the ampersand and asterisk. The QWERTY layout is generally used in English-speaking countries and specifies the location of the alphabetic characters, as well as most of the numerals and symbols. Minor variations from this layout have been adopted elsewhere, the AWERTY layout being used in France, and the QWERTZ layout in German-speaking countries.

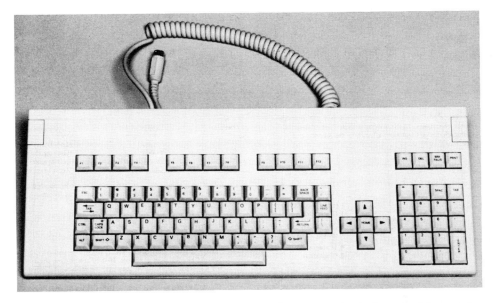

Fig. 11.4.1. A general purpose alphanumeric keyboard for computer input. Separate alphanumeric, cursor control, and auxiliary numeric key groups are positioned from left to right. Above the alphanumeric keyset, a row of 12 programmable function keys is arranged in three groups of four keys each. Four fixed-function keys are located above the auxiliary numeric keypad. (Courtesy of Texas Instruments, Inc.).

The function keys provide such rudimentary functions as mode changes (shift key) and communication (enter key). Such frequently used functions are typically included within the periphery of the alphanumeric keyset. Other specialized functions, such as for text editing (insert, delete), may be offered by additional function keysets on the keyboard. These keysets may also be programmable, enabling their functions to vary with the application in which the keyboard is employed.

An auxiliary numeric keypad provides an efficient adjunct to the alphanumeric keyset in applications requiring numeric data entry. A cursor control keyset provides a key-oriented means to control current position on an associated video display.

The QWERTY Layout.   The QWERTY keyboard layout has been adopted as the basis for a standard alphanumeric keyboard arrangement. Considerations that led to the adoption of this arrangement included the many QWERTY keyboards in use, as well as the time and money already invested in the training and texts for this arrangement (Ancona, Garland, and Tropsa, 1971; Lohse, 1968). The standard arrangement is depicted in Figure 11.4.2. Keys E00, D12, C12, and B00 allow for the selection and placement of characters appropriate to a specific keyboard application area. Recommendations for assigning characters to these keys, as well as recommendations for character replacements to accommodate other characters, are included in the notes of Figure 11.4.2. For keyboards used for information interchange, the character assignments and substitutions in Table 11.4.3 should be used with the keyboard arrangement of Figure 11.4.2 to provide the appropriate character set for the intended application. Notes 2 and 3 of Figure 11.4.2 can also be applied to keyboards for information interchange, but may be restricted to specific applications in order to maintain keyboard code compatibility (American National Standards Institute, 1982).

The widespread acceptance of the QWERTY keyboard is reason enough to caution against deviation from this layout. Although the considerations which led to the creation of this layout are unclear (Noyes, 1983), recent data on the typing of English text indicate that the QWERTY arrangement offers other advantages as well. Kinkead (1975) timed and analyzed 115,000 keystrokes from 22 touch

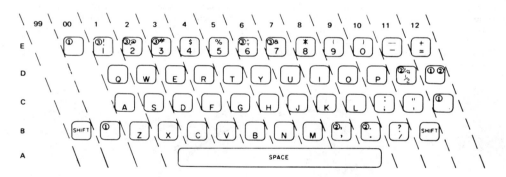

① These positions are reserved for primary use unassigned graphic characters. The following primary use unassigned graphic characters and their allocation are recommended for these blank keys.

Key Position	Graphic
E00 u/c	° (Degree)
E00 l/c	± (Plus/minus)
D12 u/c	[(Left bracket)
D12 l/c	] (Right bracket)
C12 u/c	³ (Superscript 3)
C12 l/c	² (Superscript 2)
B00 u/c	¶ (Paragraph)
B00 l/c	§ (Section)

② These assigned graphic characters should be allocated as shown. If a specific application allows for graphic replacement to accommodate other primary use unassigned graphic characters, the following is recommended.

Key Position	Assigned	Replacement
B08 u/c	, (Comma)	< (Less than)
B09 u/c	. (Period)	> (Greater than)
D11 u/c	¼ (One-quarter)	[(Left bracket)
D11 l/c	½ (One-half)	] (Right bracket)

The following graphic character assignment is recommended for key position D12 replacing the square brackets recommended in Note 1 above.

Key Position	Graphic
D12 u/c	μ (Mu)
D12 l/c	· (Product dot)

③ These assigned graphic characters (@, ¢, #, ! and &) should be allocated as shown. If they are not required for a specific application, they may be replaced by other primary use unassigned graphic characters.

④ Physical characteristics and locations (that is, size, shape, skew, etc) of the space bar or of the keys are not to be inferred.

Fig. 11.4.2.   The QWERTY keyboard layout. This material is reproduced with permission from American National Standard (X4.23-1982), copyright 1982 by the American National Standards Institute. Copies of this standard may be purchased from the American National Standards Institute at 1430 Broadway, New York, N.Y. 10018.

Table 11.4.3   Character Assignment for Keyboards for Information Interchange

Key Position	Word Processing	American Standard Code for Information Interchange (ASCII)	Optical Character Recognition (Style A) (OCR-A)*	Optical Character Recognition (Style B) (OCR-B)*
E00 u/c	°	~	∫	NA†
E00 l/c	±	`	ꓩ	NA†
D12 u/c	[	[	[	[
D12 l/c	]	]	]	]
C12 u/c	³	{	{	{ ‡
C12 l/c	²	}	}	} ‡
B00 u/c	¶	>	>	>
B00 l/c	§	<	<	<
D11 u/c	¼	\	\	\
D11 l/c	½	|	|	|
E06 u/c	¢	^	^	^
E11 u/c	_	_	Ψ	_
B08 u/c	´	´	CE†	CE†
B09 u/c	.	.	GE†	GE†

*Extended repertoire of OCR graphics should be allocated as presented in the OCR standards (ANSI X3.17-1981 and ANSI X3.49-1975).

†NA = not applicable; CE = character erase; GE = group erase.

‡Proposed.

Source: This material is reproduced with permission from American National Standard (X4.23-1982), copyright 1982 by the American National Standards Institute. Copies of this standard may be purchased from the American National Standards Institute at 1430 Broadway, New York, N.Y. 10018.

typists to determine the differences in keying time between hands, rows, columns, and individual keys. Alternate-hand keying was found to be 24% faster than same-hand keying. Keystrokes to the bottom row were slower than those to the top and middle rows. Keying by the index and middle fingers was more rapid than keying by the ring and little fingers. Successive keystrokes by the same finger resulted in the slowest keystrokes. These data suggest, then, that an efficient keyboard would use the frequency of character digrams (sequences of two letters) to ensure that most keystrokes alternate from hand to hand; would assign the least frequent letters to the bottom row, and to the ring and little fingers; and would minimize successive keystrokes by the same finger. Kinkead's analyses indicate that the QWERTY layout conforms to these constraints remarkably well.

The QWERTY layout is not without its disadvantages, however, and these disadvantages have motivated the development of many alternative arrangements. Noyes (1983) provides a historical review of the origins of the QWERTY keyboard and other keyboard layouts put forth since to improve upon it. Some of the criticisms directed at the QWERTY layout over the past 50 years include (Noyes, 1983):

1. It overloads the left hand—57% of typing is carried out by the nonpreferred hand for the majority of the population.
2. It overloads certain fingers. (The differential strength of fingers, however, is perhaps less an issue with today's electronic keyboards than it was with earlier manual typewriters.)
3. Too little typing is carried out on the home row of keys (32%). Too much typing (52%) is carried out on the top row. [Most critics of the QWERTY layout have assumed that home-row keying is the fastest. Kinkead (1975) noted, however, that while this may have been so on manual typewriters, top-row keying appears to be fastest for skilled typists on electric typewriters.]
4. Excessive row hopping is required in frequently used sequences, often from the bottom row to the top row and down to the bottom again.
5. Many common words are typed by the left hand alone.

6. Forty-eight percent of all motions to reposition the fingers laterally between consecutive strokes are one-handed rather than easier two-handed motions.

The Dvorak Layout. Of the many efforts to improve upon the QWERTY layout, the Dvorak layout has proven the most enduring. A variant of this layout has, in fact, been accepted by the American National Standards Institute as an alternative standard (American National Standards Institute, 1983). The alternative standard arrangement is depicted in Figure 11.4.3. As with the standard arrangement based on the QWERTY layout, certain character assignments and substitutions should be used with this arrangement to provide the appropriate character set for an information interchange application.

The Dvorak Simplified Keyboard was arranged on the basis of the frequencies with which letters and letter sequences occur in English text. It was designed to the following criteria (Noyes, 1983):

1. The right hand was given more work (56%) than the left hand (44%).
2. The amount of typing assigned to different fingers was proportional to their skill and strength.
3. Seventy percent of typing was carried out on the home row—the most frequently used letters were arranged on this row. Only 22% and 8% of typing was carried out on the top and bottom rows, respectively.
4. Letters often occurring together were assigned positions so that alternate hands could strike them.
5. Finger motions from row to row and difficult, awkward reaches from the home row were minimized.
6. Thirty-five percent of the words typically used were typed exclusively on the home row.

Seibel (1972) notes that a subject's stage of learning on a data entry device has a large effect on performance. In the study of data entry performance, effective elimination of stage of learning as a confounding variable requires extended periods of practice. (Seibel suggests at least six months, and often a year or more.) As a result, the conduct of a valid experimental comparison of skilled typing with the QWERTY and Dvorak layouts is an extremely difficult and expensive proposition. And, in fact, no studies reported in the literature have been accepted generally as valid.

Yamada (1980) reviews the development and testing of the Dvorak Simplified Keyboard and provides an extensive set of references. While the testing of the Dvorak layout has generally been criticized for lack of careful experimental control, it is worthwhile to note some of the claims that have been made on the basis of these tests. Succinctly, it has been claimed that, relative to the QWERTY layout, the Dvorak layout is easier to learn, is less fatiguing to use, and permits faster data entry with fewer errors (Yamada, 1980).

Kinkead (1975) used keying time data collected from skilled typing on a QWERTY keyboard to predict the speed increase possible with a Dvorak layout. He predicted a keying speed advantage of only 2.6% for the Dvorak layout, this advantage largely the result of the larger proportion of alternate-hand keying possible with the Dvorak layout. Kinkead's assumption that key-to-key time data from QWERTY keying are also valid for Dvorak keying is suspect, however. Seibel (1972) notes that average times for low motor difficulty entries tend to be faster in isolation than when mixed with higher motor difficulty entries. Because the Dvorak layout was consciously designed to minimize the awkward finger movements that commonly occur in QWERTY typing, it is quite possible that it would achieve reduced key-to-key times relative to those achieved with the QWERTY layout.

Thus, Kinkead's prediction of a 2.6% speed advantage for the Dvorak layout is probably pessimistic.

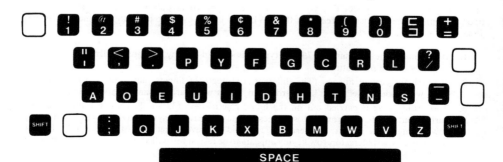

Fig. 11.4.3. The Dvorak keyboard layout. (Copyright, Dvorak International Federation, 1983. Printed with permission of Virginia de Ganahl Russell, Dvorak International Federation.)

Seibel (1972) offers a guess that the upper limit of the daily production advantage for such an arrangement would be about 10%, noting that daily production rate in keypunching tasks is typically about half that of speed-test performance. There are no data available, however, to verify that such an adjustment for estimation of daily production would hold for data entry with a Dvorak keyboard.

The advantages of the QWERTY and Dvorak keyboard layouts are summarized in Table 11.4.4.

The Alphabetical Layout. It seems reasonable to assume that the familiar order of an alphabetically arranged keyboard would enhance both the speed and accuracy of occasional users employing a "hunt-and-peck" approach to data entry. In contrast to the arbitrary appearing structure that the QWERTY and Dvorak layouts present to the inexperienced user, an alphabetical layout should provide an easily understood structure that aids in the search for desired keys. Several studies of alphabetical layouts have been reported in the literature and their results are quite consistent. The alphabetical layout does not appear to offer any practical performance advantages relative to the QWERTY layout.

Hirsch (1970) reported the results of a study that sought to determine whether the performance of unskilled typists might be improved through use of an alphabetically ordered keyboard. Subjects identified themselves as nontypists and achieved pretest typing rates below 24 words per minute. After 7 hr of practice, subjects using the alphabetical keyboard were still unable to type more quickly with this layout than they had typed in the pretest, without practice, on the QWERTY layout. A second group of subjects with 7 hr of practice on the QWERTY layout showed a significant improvement in typing rate over their QWERTY pretest scores. Hirsch concluded that the alphabetical layout is certainly not better than, and perhaps not as good as, the QWERTY layout for relatively low-skilled typists. He offered two explanations for the superiority of the QWERTY layout. First, many of the most frequently used letters happen to be clustered in the center of the QWERTY layout, permitting the hunt for a letter to focus upon a small visual area. Second, while hunt-and-peck use of the QWERTY layout involves a purely visual search that can often be focused upon a small area, use of the alphabetical layout may first involve a memory search to locate the letter's position in the alphabet, followed by a visual search for the key on the board. The focused visual search with the QWERTY layout may be more efficient than the combination of memory and visual search required with the alphabetical layout.

Michaels (1971) conducted a second experiment comparing performance on the QWERTY and alphabetical keyboard layouts. Subjects ranging in typing skill from almost none to secretarial level operated the keyboards for 10 half-hour sessions. Half of the subjects started on the alphabetical layout and half on the QWERTY layout, with the two groups switching keyboards after five sessions. There were no significant differences in the performance of low-skill subjects across the two keyboard arrangements. For the medium- and high-skill subjects, keying speed and overall work output were significantly greater on the QWERTY keyboard. This last result is to be expected given the prior experience these subjects had with the QWERTY layout.

Norman and Fisher (1982) noted that in testing the potential superiority of alphabetical keyboards for novice users, it may be inappropriate to use the QWERTY layout as the basis of comparison. Because there are few people who have not had some exposure to the QWERTY layout, the results of such a comparison tend to be biased in favor of the QWERTY layout. Knowledge of the QWERTY layout would enhance performance with it and perhaps interfere with use of the alphabetical layout as well. They suggested that the proper control for a test of the alphabetical layout would be a randomly organized keyboard, since it would be completely unfamiliar to the novice typist. Accordingly, they tested two alphabetically arranged keyboards and a randomly structured keyboard with subjects who classified themselves as nontypists. Potential subjects were first given a 10 min pretest on the QWERTY keyboard. Those who typed less than 25 words per minute were then tested for 10 min on each of the other three keyboards in counterbalanced order. The results revealed a small (10%) but significant increment in typing speed for the two alphabetical keyboards over the random layout, but no difference

Table 11.4.4 Advantages of QWERTY and Dvorak Keyboard Layouts

Advantages of the QWERTY Keyboard	Advantages of the Dvorak Keyboard
Widespread acceptance and use	Increased efficiency
Accepted as an American National Standard	Accepted as an American National Standard
Most keystrokes alternate between hands	Increased use of alternate hand keying
Bottom row contains least frequent letters	Increased use of home row
Ring and little fingers key least frequent letters	Amount of keying assigned to fingers is proportional to finger strength and skill
Relatively few successive keystrokes by same finger	Minimal awkward finger movement
	Increased use of right hand

between the alphabetical keyboards. A QWERTY layout was not tested within the experimental design and, thus, was not included in the analysis. However, the speed achieved on the QWERTY keyboard pretest was 66% greater than that later reached with the alphabetical keyboards and 82% greater than that reached with the random keyboard. Norman and Fisher conclude that the potential assistance an alphabetical structure might provide the user in keying is negated by the mental computation required to make use of the structure. They also suggest that many people may not know the alphabet well enough to take full advantage of the structure.

Thus, an alphabetically ordered key layout does appear to enhance the performance of unskilled users when an equally unfamiliar layout, such as a random arrangement, is used as the basis of comparison. But the increment in performance obtained is modest. When the basis of comparison is the familiar QWERTY layout, performance with the alphabetical arrangement is at best no better, and may not be as good. Finally, as would be expected, users who have already developed some degree of skill with the QWERTY layout perform significantly better with the QWERTY layout than with the alphabetical layout. Table 11.4.5 Summarizes the results of the three studies cited for the performance of unskilled typists.

QWERTY versus Alternative Alphanumeric Layouts. The QWERTY layout is the standard arrangement for alphanumeric data entry. Whatever its origins, it is a reasonably efficient layout as well. More efficient arrangements have been derived and the Dvorak layout has been accepted as an alternative standard. But there are no reliable data to indicate the increase in system throughput that would be achieved through adoption of a more efficient keyboard arrangement. Thus, the cost of adoption cannot be traded against a known increase in productivity.

It is clear, however, that the cost of adopting an alternative layout today is lower than ever before. New keyboard technologies have substantially lowered manufacturing costs. Removable key tops and programmable machine logic permit easy switching of key arrangements within the constraints of the traditional four-row layout. The capability to produce a low-cost switchable keyboard for both the QWERTY and an alternative arrangement removes two barriers to the adoption of alternative layouts. Those already skilled in the use of the QWERTY layout are assured its continued availability, and those considering the investment of time and effort to acquire skill on a more efficient layout can be assured of the alternative layout's general availability.

Of the alternative layouts that have been proposed, the Dvorak layout has the advantage of an enduring following as well as an accepted standard legitimizing its use. The Dvorak layout represents a more efficient arrangement than the QWERTY standard, although the increment in efficiency to be expected is still unclear. It may have ease-of-learning and ease-of-use advantages as well. Designed in the 1930s for use with mechanical typewriters and English-text entry, the Dvorak layout is probably not the optimal arrangement for today's input technologies and programming tasks. One might wish for a more modern alternative layout based on analysis of current usage and validated through controlled testing. But history has shown that those willing to invest great effort into keyboard redesign are likely to find their efforts ignored.

Kinkead (1975) argues that reassignment of keys within the conventional layout will achieve only small improvements in raw typing speed. If new arrangements are to represent real improvements, then they must demonstrate advantages in other areas as well, such as training time, error rates, and fatigue. Kinkead notes, however, that opportunities for increasing speed of data entry remain. He reports speed improvements of 7–8% using automatic carriage returns, for example. A word processing system monitored keystrokes and automatically inserted "enter" codes when the space bar was hit toward the end of a line.

Table 11.4.5 Performance of Unskilled Typists Using the QWERTY and Alphabetical Layouts

	Practice Time		
	7 hr[a]	2.5 hr[b]	10 min[c]
QWERTY	Significant increase in typing rate over the QWERTY pretest rate	No significant difference between the two layouts	QWERTY pretest rate 6% higher than rates attained on two alphabetical layouts
Alphabetical	No significant increase in typing rate over the QWERTY pretest rate		Rates 10% higher than rate attained on randomly arranged keyboard

[a] Hirsch (1970).

[b] Michaels (1971).

[c] Norman and Fisher (1982).

Rather than invest great effort in keyboard rearrangement for questionable return, emphasis might be placed on the availability of features that permit rapid and easy error correction. The DWIM ("Do what I mean") error correction facility available within the INTERLISP programming environment provides an example of such an approach. This facility corrects various classes of input errors, in some cases automatically without interruption of the user's input process.

Numeric Keypad Layout. The generally accepted keypad for numeric data entry consists of 10 keys, one for each of the 10 digits, arranged in a three-by-three matrix, with the zero key either above or below the matrix. Two key arrangements are commonly encountered on the three-by-three matrix. The first, commonly found on hand calculators, assigns the digits 1, 2, 3 from left-to-right on the bottom row, with the digits 7, 8, 9 from left-to-right on the top row. The second, used on touch telephones, assigns the digits 1, 2, 3 from left-to-right on the top row, while the digits 7, 8, 9 are assigned from left-to-right on the bottom row. These layouts, with the zero key placed below the bottom row, are illustrated in Figure 11.4.4.

The highly practiced user can perform about equally well with either the calculator or the telephone arrangement (Seibel, 1972). The selection of a particular arrangement becomes an issue in situations where less dedicated use is made of the keypad or when the user must alternate between the two arrangements. Lutz and Chapanis (1955) sought to determine whether a population stereotype existed for the arrangement of digits on each of six different 10-key keyset configurations. They found that subjects generally expected to find numbers arranged in left-to-right order in horizontal rows starting with the top row. For the three-by-three plus one-key-below arrangement, the telephone layout was expected by 55 of the 100 subjects. The calculator layout was a distant second choice, selected by 8 of the 100 subjects.

Conrad and Hull (1968) conducted an experiment to compare speed and accuracy of data entry with the calculator and telephone layouts. One group of subjects worked exclusively with the calculator layout, a second group worked with the telephone layout, and a third group alternated frequently between the two layouts. Subjects claimed no previous experience with either of the layouts. There was no significant difference in speed of entry between the calculator and the telephone conditions. The telephone layout did, however, achieve significantly greater accuracy than the calculator layout. Subjects working exclusively with the telephone layout were also significantly faster and more accurate than subjects alternating between the telephone and calculator layouts.

For the occasional user, then, the telephone layout offers both conceptual compatibility and performance advantages over the calculator layout. It is also undesirable, given the results of the Conrad and Hull study, to require users to alternate between the two arrangements. The considerations guiding the selection of a numeric keypad layout are summarized in Table 11.4.6. With the telephone layout accepted internationally as the standard for pushbutton telephones, and with computer input devices now serving as input devices for telecommunications as well, the telephone layout appears to be the layout of choice for many applications.

Cursor-Control Key Layout. Cursor-control keys provide a key-oriented means to control current position on an associated visual display. Typically, cursor-control keysets direct the cursor to the left, right, up, down, and, perhaps, to a "home" position on the display screen. The human factors literature provides little specific guidance regarding layout of cursor control keys. More attention has been focused on determining the applications for which keys represent an appropriate technology for cursor control. This issue will be addressed in Section 11.4.3.

Figure 11.4.5 illustrates two reasonable cursor control key configurations. Both appear to offer a clear relationship between key location and key function. The inverted "T" layout might be particularly appropriate for applications in which the user generally tends to move downward through displayed

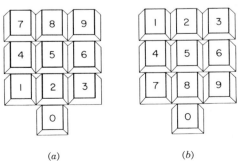

(a)   (b)

Fig. 11.4.4. Numeric keypad layouts: (a) calculator layout; (b) telephone layout.

Table 11.4.6 Numeric Keypad Layout Considerations

The telephone and calculator layouts are generally accepted.

The telephone layout is consistent with many users' expectations.

Highly practiced users perform well with either keypad layout.

Occasional users achieve increased accuracy with the telephone layout.

Alternation between layouts degrades performance.

---

data. In such a situation, this layout positions the three most frequently used keys on one compact and easily touch operated home row.

Function Key Layout. The location of function keys on a keyboard is guided by the general considerations of importance, frequency-of-use, function, and sequence-of-use. Beyond these general considerations, guidelines for function key layout tend to be highly application-specific. A study reported by Hollingsworth and Dray (1981) does, however, indicate a means for limiting the complexity of selecting the appropriate key from among many on a function keyboard. This study investigated the effect of backlighting those keys corresponding to permissible response options at different stages in the use of a function keyboard. The keyboard consisted of two horizontal rows of eight keys. All 16 keys were uniformly backlit until a stimulus was presented. Upon stimulus presentation (a function key label presented on a video display), the backlighting of 4, 8, or all 16 of the keys switched to a discriminably higher intensity level. The key corresponding to the label on the display was always one of the high intensity keys. The subject's task was to press the key indicated on the video display. Reaction time was substantially shorter when the target key was one of only four high intensity keys than when the lighting pattern carried no information (all 16 keys at high intensity). Hollingsworth and Dray suggest that even greater reductions in response time might be achieved if the lighting pattern could be made available before stimulus presentation. Their work indicates that complex function keyboards should be designed to provide response cueing.

An additional consideration in the design and placement of some function keys may be the effect of inadvertent operation. Measures taken to prevent inadvertent entries invariably slow rate of entry as well. However, where unintended invocation of a particular function is highly undesirable, the following approaches can be taken:

1. The key corresponding to the function may be located such that inadvertent operation is unlikely.
2. The key corresponding to the function may be designed to require a larger activation force than other keys.
3. Invocation of the function may require the simultaneous activation of two or more nonadjacent keys. (This approach, however, may make such a function inaccessible to handicapped users.)

General considerations guiding the layout of function keys are summarized in Table 11.4.7.

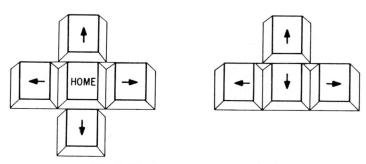

Fig. 11.4.5. Cursor-control key layouts.

Table 11.4.7   Function Key Layout Considerations

Locate according to importance, frequency-of-use, function and sequence-of-use.

Indicate the current subset of appropriate function keys.

Guard against inadvertent activation of critical functions by remote key location, large activation force, or simultaneous key depression.

Key Dimensions, Mechanics, and Feedback

Alden, Daniels, and Kanarick (1972) conducted an extensive review of the literature on keyboard design and operation. They conclude that a good deal of research has been conducted, but there are few definitive findings on which to base design standards. The sponsorship and motivation of much of this keyboard research has resulted in work primarily concerned with the evaluation of specific products. While the work may determine one keyboard design to be superior to another, it often does not isolate the effects of individual keyboard parameters on user performance. Keyboard design research also frequently involves proprietary material, and thus is not published in the open literature.

Key Dimensions and Mechanics.   The size and spacing of keys on general purpose alphanumeric keyboards are largely based on design conventions, rather than empirical data. Key diameters of 0.5 in. (13 mm) with center-to-center spacings of 0.75 in. (19 mm) are typical (Alden et al., 1972). The keytop is typically square (offering greater surface area than an equivalently sized circular keytop) with a slightly concave surface to assist proper finger placement. Key activation forces from 0.9 to 5.3 ozf (0.25 to 1.47 N) and total key displacements between 0.05 and 0.25 in. (1 to 6 mm) appear to be preferred by operators, although, within certain limits, force and displacement appear to have little effect on the keying performance of experienced users (Alden et al., 1972). These general specifications are summarized in Table 11.4.8. Typical production keyboards for computer input have key activation forces ranging from 0.4 to 1.2 N with key displacements between 3 and 5 mm.

There are few studies in the open literature dealing with the relationship between keying force and resulting key displacement. Most keyboards appear to exhibit a rapid buildup of force as the key is depressed, with a reduction in the required force in the region of activation, followed by a second increase in force thereafter. Brunner and Richardson (1984) conducted a study involving three keyboards with the different force–displacement characteristics shown in Figure 11.4.6. The first keyboard utilized a snap-spring mechanism to produce a buildup of resistance on the downstroke with a sharp dropoff at the point of activation. This keyboard also exhibited the most pronounced hysteresis (the tendency of a key switch to remain closed even after partial reduction of applied force; hysteresis is used to reduce the possibility of inadvertent multiple entries). The second keyboard had an elastomer key action that produced a double-peaked force–displacement curve (the second peak followed switch closure, but preceded the point at which the key bottomed out). This keyboard operated without mechanical hysteresis, instead using an electronic polling mechanism to control for unintended switch contacts. The third keyboard utilized a linear spring mechanism, exhibiting light linearly increasing resistance to switch closure, rapid doubling of resistance at switch closure, and low hysteresis. Performance data were obtained for both occasional and expert typists. Somewhat lower error rates and greater throughput speed were achieved with the elastomer key-action keyboard than with the snap- and linear-spring key-action boards. This facilitation was greater for occasional typists than for experts. Subjects also committed more inadvertent entries with the light resistance, low hysteresis linear-spring keyboard. This effect was again more pronounced for occasional typists.

Table 11.4.8   Key Dimensions and Mechanics—General Recommendations

Diameter: 0.5 in. (13 mm)

Spacing: 0.75 in. (19 mm) center-to-center

Shape: square with slightly concave surface

Activation force: 0.9–5.3 ozf (0.25–1.47 N)

Total displacement: 0.05–0.25 in. (1–6 mm)

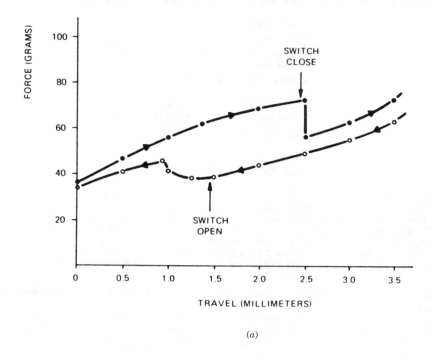

(a)

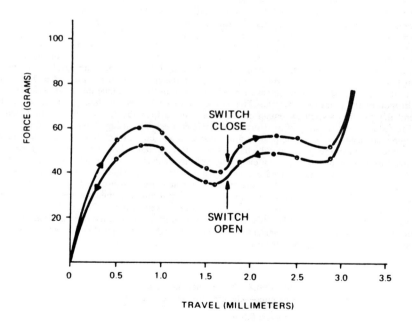

(b)

Fig. 11.4.6. Force-displacement functions of keyboards used in a study by Brunner and Richardson (1984): (a) snap spring; (b) elastomer spring; (c) linear spring. [From *Proceedings of the Human Factors Society 28th annual meeting,* 1984, Vol. 1, p. 271. Copyright (1984), by The Human Factors Society, Inc. and reproduced by permission.]

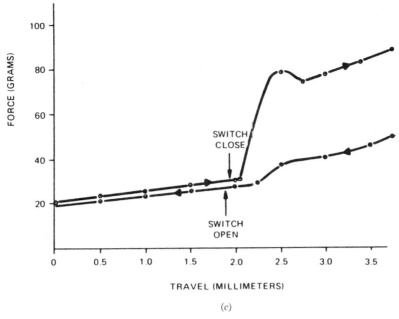

(c)

Fig. 11.4.6.   (*Continued*)

Feedback.   The major source of feedback for the highly skilled user of a full-travel keyboard appears to be the kinesthetic–proprioceptive–tactual feedback that the user receives by actually making the movement and striking the key (Seibel, 1972). Visual feedback does appear to be important during training, however (Alden et al., 1972; Klemmer, 1971), and for the correction of errors. Rosinski, Chiesi, and Debons (1980) investigated the effect of displaying typed input on the input performance of novice, semiskilled, and professional typists. They found that regardless of skill level, the presence or absence of visual feedback did not affect input speed or number of input errors. There was, however, a significant effect of visual feedback on error correction. Greater amounts of visual feedback permitted the subjects to review their performance and correct their errors. The authors conclude that there is no advantage gained by providing visual feedback if the main interest is in the initial speed and accuracy of input. But where errors must be monitored and corrected, or when editing is necessary, visual feedback is advantageous.

The effect of supplemental auditory feedback upon keying performance with a full-travel keyboard is less clear. Pollard and Cooper (1979) investigated the effect of auditory feedback on numeric data entry performance with a touch telephone keypad. The four feedback conditions included multifrequency tones, single tone, click on depression and release, and a baseline condition in which no supplemental feedback was provided. They found no significant differences in keying speed, error rate, or user preference across the four conditions. They conclude that the naturally occurring sounds of operation, as well as the tactual and kinesthetic sensations experienced during keying, provide adequate feedback and eliminate the need for additional, electronically generated feedback.

Monty, Snyder, and Birdwell (1983) investigated the effect of auditory feedback on text entry performance with several alphanumeric keyboards. They reported a small (2%) but significant improvement in text entry time when electronically generated click options were employed. Subjects also indicated a substantial preference for the supplemental auditory feedback. Brunner and Richardson (1984) also studied the effect of auditory feedback on text entry performance with an alphanumeric keyboard. They utilized an elastomer key-action keyboard that was noiseless in operation. The keyboard provided for auditory feedback with an option for electronic click at the instant of switch closure. The addition of auditory feedback resulted in slightly higher throughput, with the facilitation greater for expert typists than for occasional typists. Upon initial exposure, subjects indicated a preference for the keyboard with auditory feedback. After an hour of use, however, preference differences were nonsignificant.

The effects of the different forms of feedback possible with full-travel keyboards are summarized in Table 11.4.9.

Table 11.4.9   Sources of Feedback with Full-Travel Keyboards

Kinesthetic–proprioceptive–tactual: major source of feedback, especially for skilled-users

Visual: important during training, and for error correction and editing tasks; does not affect input accuracy and speed

Supplemental auditory: may provide modest enhancement, but inherent feedback is typically sufficient

Membrane Keypads.   While the effect of supplemental feedback (whether kinesthetic, visual, or auditory) on input performance with full-travel keyboards appears to be rather small, there are keyboard technologies for which supplemental feedback may achieve more substantial results. The membrane keypad, consisting of mechanical contacts on two layers of material separated by a spacer layer less than 1 mm thick, is one such technology (see Figure 11.4.7). Switch activation occurs when the user depresses the flexible upper layer through holes in the spacer layer. When pressure is removed from the membrane, it breaks contact with the shorting pad and returns to its original position.

Membrane keypads offer several engineering advantages. They are inexpensive to produce, their thin profile offers considerable design flexibility, and switches can be protected from hostile environments, dust, and spills. Their design lacks much of the feedback inherent in full-travel keyboards, however. Key travel is negligible (0.15–0.20 mm), they are noiseless in operation, and while key locations are depicted graphically on the membrane, the smooth surface offers no tactual cues for key location. Pollard and Cooper (1979) compared keying performance with full-travel and membrane telephone keysets. Subjects entered 10-digit sequences on each keyset. An auditory tone as well as visual feedback were provided upon digit entry with the membrane keyset. While 3.17% of the sequences entered on the conventional keypad contained errors, there were errors in 11.55% of the sequences entered with the membrane keypad. The authors note, however, that preliminary work with the membrane keypad without provision of any supplemental feedback resulted in error rates as high as 20%. Thus, supplemental feedback appears to enhance performance with membrane keypads.

This conclusion is reinforced by research recently reported by Roe, Muto, and Blake (1984). Their work addressed the additional key discriminability provided by embossed key edges as well as the supplemental feedback provided by metal domes and auditory tones on membrane keypads. The metallic domes increased key travel (from 0.25 mm without domes to 0.41 mm with domes) and provided some snap-action and auditory feedback. The study focused on applications where only occasional entries are made. The results of this study indicated that the pairing of auditory tones with embossed key edges and/or metal domes best enhanced performance and preference scores with membrane keys.

Membrane keypads are typically used in applications involving only occasional data entry. Cohen Loeb (1983) suggests that with extended practice and additional design optimization, membrane key-

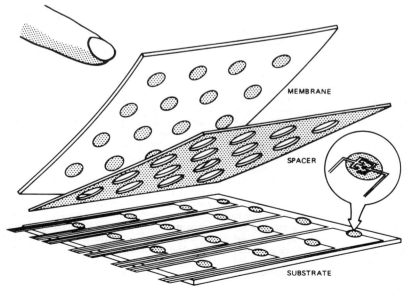

Fig. 11.4.7.   Membrane switch technology. [Reprinted with permission from the *Bell System Technical Journal.* Copyright 1983, AT&T.

Table 11.4.10   Advantages and Disadvantages of Membrane Keypads

Advantages	Disadvantages
Inexpensive	Negligible inherent feedback
Thin profile	Supplemental kinesthetic–tactual feedback necessary
Protective switch enclosure	Supplemental auditory feedback necessary
Ease of cleaning	Input rates equivalent to full-travel keypads not yet demonstrated

boards may prove effective in more intensive data entry applications as well. She compared data entry performance using a membrane keyboard with that using a full-travel keyboard for subjects with different levels of typing skill. Auditory feedback was provided with both keyboards. Touch typists performed considerably better on the full-travel keyboard than on the membrane keyboard. (The excellent touch typists, for example, typed 26% fewer words per minute with the membrane keyboard.) The advantage of the full-travel keyboard was considerably smaller for nontouch typists, who typed 6.5% fewer words per minute with the membrane keyboard. Although subjects were provided only 3 hr of exposure to the membrane keyboard, rapid learning effects were apparent. Cohen Loeb concludes that the cost advantages and design flexibility offered by membrane keyboards certainly warrant additional research and development.

The advantages and disadvantages of membrane keypads are summarized in Table 11.4.10.

11.4.2.2   Touch-Screen Devices

A touch-screen device produces an input signal in response to a touch or movement of the finger at a place on the display. There are two basic principles by which touch-screen devices work. One method uses an overlay that responds to pressure. The other method is activated when the finger interrupts a signal. Five representative types of touch-screen technologies will be discussed. The characteristics of the different touch-screen technologies are summarized in Table 11.4.11.

Types of Touch-Screen Devices

Touch Wires.   Johnson (1967) reported one of the earliest types of touch-screen devices. A set of copper "touch wires" is set in rows above a transparent mask placed over a CRT display surface. When an operator touches the wire with his or her finger, a connection is made with another wire on the mask. The operator's body provides a path to earth, unbalancing a bridge circuit (Hopkin, 1971; Ritchie and Turner, 1975). A signal results, which in turn is sent to the computer.

Table 11.4.11   Characteristics of Touch-Screen Technologies

Characteristic	Touch Screens[a]				
	TW	IR	AC	PS	CO
May obscure display	X				
Unreliable detection	X				
Limited resolution		X			
Parallax		X	X	X	
Sensitive to ambient lighting		X			
Inadvertent activation		X	X		
Does not use whole screen			X		
Awkward drawing device	X	X	X		X
Extra training required				X	
Reduced light from screen					X
Relatively easily damaged					X

[a] TW: touch wire; IR: infrared; AC: acoustic; PS: pressure sensitive; CO: conductive.

A more recent version of the touch-wire device uses both horizontal and vertical wires set in transparent sheets placed on the display (Schulze and Snyder, 1983). In this *cross-wire* device, a current is applied to either the vertical or horizontal wires and a signal is produced when the wires are touched at an intersection.

Pfauth and Priest (1981) note two disadvantages of the touch-wire technology. First, touches are not always reliably detected. Second, the wires may obscure parts of the display.

Infrared Beams.   Another type of touch-screen device uses infrared or light-emitting diode (LED) beams (Pfauth and Priest, 1981; Schulze and Snyder, 1983). The LEDs are paired with light detectors and placed along all sides of the display, as shown in Figure 11.4.8. Thus, no overlays are used with this device and consequently no display obstruction or decrease in image quality is experienced. When the operator touches his or her finger to the screen, two light beams are interrupted and the resulting X and Y coordinates for that position are calculated.

There are several disadvantages associated with the use of light beams. First, the resolution is limited to the number of light beams included. Second, because light beams travel in a straight line, the LEDs must be removed from the curved CRT screen, and thus parallax is a problem. Third, the ambient lighting may affect the response of the device. However, the illumination in the workplace may be sampled and stored in memory to be used as a correction factor (Schulze and Snyder, 1983). Fourth, because this device is activated when the light beams are interrupted and not when the screen is actually touched, inadvertent activation may be a problem (Mims, 1984).

Acoustic Devices.   In this type of touch-screen device, a glass plate is placed over the CRT screen and ultrasonic waves are generated on the glass by transducers placed along both the X and Y axes (de Bruyne, 1980; Hlady, 1969). When a waveform is interrupted, the horizontal and vertical waves are reflected back and detected by the transducer, and X and Y coordinates are calculated based on the time between wave transmission and detection.

Acoustic touch-screen devices have two advantages: they do not obscure the display and they allow higher resolution than do infrared devices. However, the device may be activated by dirt or scratches on the glass. In addition, owing to the fact that there is a short period of time after a transducer emits a wave before that transducer may act as a detector, there is a $1\frac{1}{2}$–2-in. strip on the outside of the glass that does not respond to touch input (Pfauth and Priest, 1981).

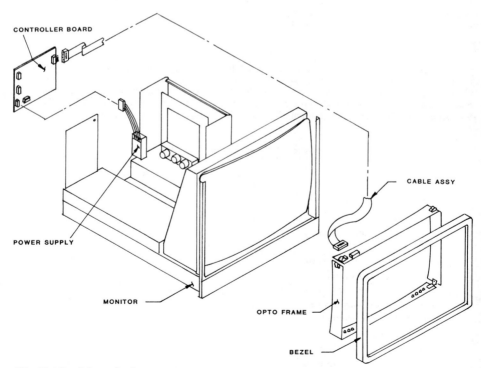

Fig. 11.4.8.   Schematic diagram of the infrared touch-screen device. (Courtesy of Carroll Touch.)

Pressure-Sensitive Devices.   One problem with the acoustic touch-screen device is that if the user moves his or her finger across the glass, friction is created, preventing smooth movements. This problem limits the usefulness of such a device to pointing gestures. However, one potential use of a touch-screen device may be to input graphic information through drawing gestures.

In an attempt to overcome this problem, Herot and Weinzapfel (1978) developed a pressure-sensitive device that employs strain gauges placed between the display and a glass overlay. Two strain gauges are mounted on each side of the display; one measures forces perpendicular to the glass, such as occurs in a pointing motion, and one measures forces parallel to the glass, as might occur when drawing. Force and torque values for X, Y, and Z axes are generated based on the output from the strain gauges, and these values are used to calculate X and Y coordinates and cursor acceleration (Schulze and Snyder, 1983).

Because this touch-screen device uses an overlay, parallax may be a problem. Additional training may be required to use such a device owing to the addition of the force vectors. The advantages of such a device, however, are that the entire surface may be used (in contrast to the acoustic device), and that the glass overlay does not contain any wires that would obscure the screen. In addition, drawing is made easier through the use of the strain gauges.

Conductive Devices.   There are several types of touch-screen devices that use a transparent conductive material placed over the display. One method uses two conductive layers, each with an electrode grid in both X and Y directions, which are placed over the display. When pressure is applied, the two surfaces touch, an electrode pair shorts, and a circuit is completed. The two voltage levels from the X and Y axes are encoded into X and Y coordinates (Pfauth and Priest, 1981; Schulze and Snyder, 1983).

A second method, called a capacitive touch-screen device, operates through the use of a conductive film, which is deposited on the back of a glass overlay. Such a touch screen is depicted in Figure 11.4.9. As with the touch wires, the body's capacitance causes an electrical signal to be generated when an individual touches the overlay (Ritchie and Turner, 1975).

Although there are no wires to obscure the display in these methods, the overlay may reduce the amount of light that is transmitted through the screen. In addition, the first conductive method uses plastic sheets that may be easily damaged (Pfauth and Priest, 1981).

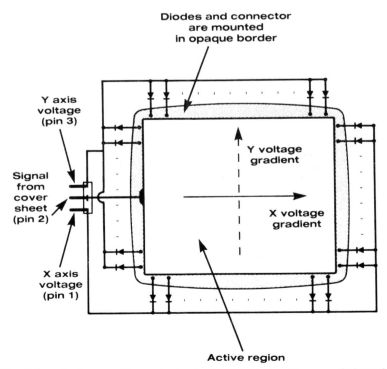

Fig. 11.4.9.   Schematic diagram of the conductive touch-screen device. (Courtesy of Elographics, Inc.)

Comparison of Touch-Screen Devices

Schulze and Snyder (1983) performed a comparison of the following five touch-screen technologies: acoustic, capacitive, conductive film, cross wires, and infrared. They found that the infrared device provided the highest resolution, owing to the fact that no overlay was placed over the screen. The capacitive device resulted in the least display noise. In a second experiment, subjects were required to perform three tasks: searching for and touching a specified alphanumeric character, assigning seats on an aircraft, and hierarchical menu selection. On the basis of errors and total time to complete these three tasks, the authors reported that the infrared and cross-wire devices provided the best performance. Overall, the cross-wire device received the highest subjective rankings from the participants.

Advantages and Disadvantages of Touch-Screen Devices

One of the most obvious advantages of touch-screen devices is that the input device is also the output device. That is, there is direct eye–hand coordination, and, consequently, there is a direct relationship between the user's input and the displayed output. A second advantage is that possible inputs are limited by what is displayed on the screen; thus, no memorization of commands is required, and input errors are minimized. In addition, the possible inputs can change as the display changes, so that the operator may be led through an appropriate sequence of inputs.

With the exception of Herot and Weinzapfel's (1978) pressure-sensitive device, which can be used for drawing, the only movement required by a touch-screen device is a natural pointing gesture. Thus, training is minimized for a touch-screen device and so is the need for operator selection procedures. Individuals can become quite skilled at fast target selection in a relatively short period of time. Consequently, there may be high user acceptance of such a device.

There are at least five disadvantages related to the use of touch-screen devices. First, the user must continually lift his or her hand to the display, a movement which may lead to arm fatigue. A second disadvantage is that there is limited resolution possible owing to the size of the operator's finger in relation to the screen. A related problem is that, owing to the low resolution, touch-screen devices are inappropriate for selection of small items such as a single character. Fourth, because only one finger may be used at a time, data entry will also be slower than it will be with a keyboard. A fifth disadvantage is that the finger or arm may block the screen. For further discussion of these advantages and disadvantages, see Hopkin (1971), Pfauth and Priest (1981), and Ritchie and Turner (1975).

For those touch-screen devices that are not located on the same plane as the screen, that is, those that use a separate overlay, parallax may be a problem. For those devices using an interrupt system, inadvertent activation is a potential problem. For all devices, there is a problem with dirt and smudges present on the screen or the overlay owing to the fingertip touching the display. In the case of an acoustically activated device, this may lead to inadvertent activation; for other touch-screen devices these smudges may obscure parts of the display, or at least be somewhat annoying for the operator. The advantages and disadvantages of touch screen devices are summarized in Table 11.4.12.

Types of Tasks

Owing to the nature of the touch-screen device, it is best suited to certain task types. Pfauth and Priest (1981) state that these devices are best used when working with data that are already displayed on the screen, while inputting new graphic information is best performed with other input devices. Touch-screen devices are quite useful in menu selection tasks. In addition, it has already been stated that selection or entry of single characters is slow and may be beyond the resolution capabilities of the touch-screen device.

Table 11.4.12   Advantages and Disadvantages of Touch Screen Devices

Advantages	Disadvantages
Direct eye–hand coordination	Arm fatigue
No command memorization needed	Limited resolution
Operator may be led through correct command sequence	Hard to select small items
Minimal training needed	Slow data entry
High user acceptance	Finger/arm may obscure screen
	Overlays may lead to parallax
	Inadvertent activation

Fig. 11.4.10. A touch-screen device for the use of bank customers. (Courtesy of Carroll Touch.)

Touch-screen devices are useful in applications where it is time consuming or perhaps even dangerous to divert attention from the display. There is evidence, for example, which indicates that touch-entry devices may work well in air-traffic-control tasks (Gaertner and Holzhausen, 1980; Stammers and Bird, 1980). These devices are also useful in reducing workload in situations where the possible types of inputs are limited and well defined. For instance, Beringer (1979) states that touch-screen devices may decrease workload if used in plane cockpits for navigation purposes. Similarly, touch-screen devices are potentially beneficial in high-stress environments owing to the limited number of possible inputs (Pfauth and Priest, 1981). Finally, if many potential users are unfamiliar with the system, touch panels may prove helpful; for example, touch-screen devices have been used successfully with information displays in shopping malls, banks, and hotels, as shown in Figure 11.4.10.

11.4.2.3　Graphic Tablets

Graphic or data tablets consist of a flat panel that is placed on a table in front of the display. The surface of the tablet represents the display. There are two major categories of graphic tablets. One category, typically called digitizing tablets or digitizers, works through the use of a special stylus or puck, which is attached to the tablet by a cable. This stylus produces signals indicating coordinate values for cursor positioning. The other type of graphic tablet is typically called a touch tablet or touch-sensitive tablet. This device is activated without the use of a special stylus; rather, it responds to a touch by a finger or pen, and it uses information from the tablet instead of a stylus to calculate cursor position. Figure 11.4.11 illustrates a digitizer.

As was the case with touch-screen devices, there are several principles by which graphic tablets operate. Five of these techniques will be discussed. The characteristics of these technologies are briefly summarized in Table 11.4.13.

Types of Graphic Tablets

Matrix-Encoded Tablets. One method by which digitizers work is through the use of electrical or magnetic fields. As the special stylus or puck is passed over the tablet surface it detects signals produced by horizontal and vertical conductors or wires in the tablet, as shown in Figure 11.4.12. Each wire carries a unique code. These signals are sent to a control unit where they are used to determine X and Y coordinates for the cursor (Ohlson, 1978; Scott, 1982). One advantage of such a system is that it allows for high resolution of cursor control.

Fig. 11.4.11. A touch tablet. (Courtesy of Elographics, Inc.)

Table 11.4.13. Characteristics of Graphic Tablet Technologies

Characteristic	Graphic Tablet[a]					
	ME	VG	AD	EA	AT	ML
High resolution	X	X				
Three-dimensional capability			X			
Requires stylus	X	X	X	X		
Produces noise			X			
Sensitive to ambient noise			X			
Inadvertent activation					X	X

[a] ME: matrix-encoded; VG: voltage-gradient; AD: acoustic digitizing; EA: electroacoustic; AT: acoustic touch; ML: multilayer.

Voltage-Gradient Tablets. With a voltage-gradient tablet, a conductive sheet forms the surface of the tablet. A potential is applied to the point of the stylus, and a decrease in potential on the plate at the stylus position is measured. Using the distance from the sides of the tablets as a reference, X and Y coordinate values are calculated (Ohlson, 1978; Ritchie and Turner, 1975).

Ohlson (1978) states that this method has two disadvantages. First, the plate must be uniformly conductive for the coordinate values to be generated accurately. Second, the stylus must be in direct contact with the tablet surface. Thus, no paper may be placed on the tablet for such operations as tracing or digitizing.

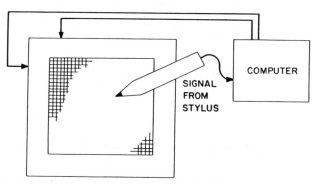

Fig. 11.4.12. Schematic diagram of a matrix-encoded tablet [adapted from Ohlson (1978)].

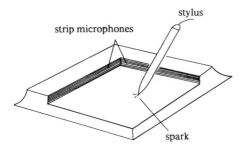

Fig. 11.4.13. An acoustic digitizing tablet [reprinted, by permission, from W. M. Newman and R. F. Sproull, 1979. *Principles of Interactive Computer Graphics,* New York: McGraw Hill. Copyright © 1979. Used with permission.

Acoustic Digitizing Tablets.   Acoustic digitizing tablets operate through the use of a stylus that generates a spark at its tip (see Figure 11.4.13). The sound impulses that result are detected by two strip microphones mounted on adjacent sides of the tablet. By determining the delay between the time the sound originated and its reception, X and Y coordinate values may be calculated. A third microphone placed perpendicular to the tablet surface may be added to provide coordinates for the production of three-dimensional drawings. Note that this method does not require that the stylus touch the tablet surface, and in fact allows the stylus to be used in the air above the tablet to produce three-dimensional drawings (Newman and Sproull, 1979; Scott, 1982). Ohlson (1978) states, however, that the acoustic tablet does not provide high resolution.

An alternative acoustic technique, called the electroacoustic method, requires that electric pulses be generated on the tablet. These pulses are detected by the stylus and the delay between pulse generation and reception is used to calculate the cursor position. Electroacoustic tablets are quieter than acoustic tablets and are less sensitive to noise in the environment (Newman and Sproull, 1979).

Acoustic Touch Tablets.   One technique that may be used without a special stylus is the acoustic-touch tablet. High-frequency waves are produced on a glass surface. When these waves are interrupted by a finger or pen, they are reflected back to the tablet edge. Acoustic couplers detect the reflection, and, as with the acoustic digitizing tablet, the delay between the wave generation and reception is used to calculate X and Y coordinates. This device is actually the same as that used in acoustic touch screens (Hlady, 1969). However, by placing the glass on a flat surface, it may be used for drawing and tracing. One disadvantage of this technique is that if the arm and/or hand touch the tablet surface, inadvertent activation will result.

Multilayer Tablets.   There are several tablets available that use two or three conducting sheets. When these sheets are pressed together by a passive stylus or finger, an electrical potential is generated, and X and Y coordinate values are calculated based on the origin of the signal (Ritchie and Turner, 1975). As with the acoustic touch tablets, the user's arm may not be in contact with the tablet, else inadvertent operation may result.

There are other types of touch tablets that may be used without concern for inadvertent activation caused by arm pressure. For example, Scott (1982) discusses a touch-sensitive digitizer that responds only to a point of pressure such as that from a pen. For a more detailed discussion of these and other technologies, see Mims (1984) or Ritchie and Turner (1975).

Parameters of Graphic Tablets

Although the fact that graphic tablets are separate from the display means that there is an indirect relationship between the output and input devices, it also means that they are more flexible than touch-screen devices. The size of the tablet is free to vary from one that can fit on or next to a keyboard to an entire digitizing table. Additionally, other parameters of the graphic tablet can be modified. For example, the amount of movement of the cursor relative to a movement on the tablet, or what is referred to as *control-display gain,* may be changed. [See Chapanis and Kinkade (1972) for a discussion of control-display relationships.] Arnaut and Greenstein (1984) found that for a touch-sensitive tablet, a low gain of between 0.875 and 1.0 resulted in superior performance on a target-selection task than did higher gains. This result is probably due to the fact that with high gains, while gross movements are faster, fine positioning movements are more difficult.

Another feature of the tablet that may be changed is the method of cursor movement. For example, when an individual places his or her finger on the tablet, the display cursor may move from its current position and appear at a position that corresponds to the location of the finger on the tablet. Subsequent movement of the finger on the tablet will then produce cursor movement such that the cursor location is continually referenced to the location of the finger on the tablet. This method may be referred to as an *absolute* mode of cursor control. A second possibility is that when the finger is placed anywhere on the tablet, the display cursor remains in its current position. Movement of the finger in this case leads to a corresponding cursor movement relative to this cursor location; consequently, this method is referred to as a *relative* mode of cursor control. Arnaut and Greenstein (1984) found that an absolute mode of target selection resulted in faster target selection rates than did a relative mode.

Owing to the indirect nature of the graphic tablet, Swezey and Davis (1983) suggest that it is important to include some sort of feedback and/or confirmation mechanism, especially where incorrect entries may be detrimental. Requiring that an operator press a confirmation area or button on the tablet before data entry is finalized may aid in decreasing inadvertent inputs. An audible click from a button on the stylus or tablet may also be helpful. See Swezey and Davis (1983) for additional guidelines on tablet usage.

Advantages and Disadvantages of Graphic Tablets

Graphic tablets have several advantages. First, the movement required and the control-display relationship are natural to many users. Similarly, Swezey and Davis (1983) suggest that graphic tablets may improve productivity because the user is not required to "translate" a command or movement into a series of keypresses.

In comparison to touch-screen devices, in which the user points directly at the screen to input data, Whitfield, Ball, and Bird (1983) suggest that the graphic tablet provides four distinct advantages. First, both the display and the tablet may be positioned separately according to user preference. Second, the user's hand does not cover any parts of the display. Third, there are no problems with parallax due to the viewing angle of the user. Fourth, drift in the display will not affect the input. In addition, the user of a graphic tablet is not likely to experience fatigue associated with continually lifting his or her hand to the screen, as is typical with a lightpen or a touch screen.

Graphic tablets have several disadvantages. Foley and Wallace (1974) indicate that touch-sensitive tablets may not provide high positioning accuracy. In comparison with a touch screen, graphic tablets do not allow for direct eye–hand coordination, since they are somewhat removed from the display. Large graphic tablets take up space on the work surface, although as depicted in Figure 11.4.14, small tablets may be inserted in a keyboard for cursor control. Finally, for those tablets that require

Fig. 11.4.14. A touch tablet inserted into a keyboard may be used for cursor control. (Courtesy of Elographics, Inc.)

Table 11.4.14   Advantages and Disadvantages of Graphic Tablets

Advantages	Disadvantages
Natural control-display relationship	May have low positioning accuracy
Hand does not obscure display	Indirect eye–hand coordination
No parallax	Requires space
No arm fatigue	May lose/break digitizer stylus
Can modify control-display gain	Slow character data entry
Can use absolute or relative mode	

a stylus, there may be a problem with loss or breakage of the stylus (Rouse, 1975). See Table 11.4.14 for a summary of these advantages and disadvantages.

Types of Tasks

Graphic tablets may be used effectively for several types of tasks. A graphic tablet is virtually the only input device that may be used for drafting or hardcopy data entry (Ohlson, 1978; Rouse, 1975), freehand sketching (Ellis and Sibley, 1967; Hornbuckle, 1967), or producing a three-dimensional picture (Sutherland, 1974). Parrish, Gates, Munger, Grimma, and Smith (1982), in an attempt to standardize military usage of input devices, recommend that graphic tablets be used for all drawing purposes. Touch tablets are also appropriate in situations in which the user is required to select or point to an item from an array or menu. Because of their inherent graphic nature and the fact that all fingers may not be used at one time, graphic tablets are slow when they are used for data entry.

A digitizer has several advantages over a touch-sensitive tablet (White, 1983). First, it is especially suited for digitizing drawings such as maps, graphs, and the like. Thus, it is well suited for many CAD/CAM applications. Digitizers typically have a higher resolution than touch-sensitive tablets, in part because of the small tip on the stylus. In addition, digitizers do not respond to pressure from the arm as some touch-sensitive tablets do. On the other hand, a touch-sensitive tablet has the advantage of being used without a special stylus. Thus, loss of or damage to the stylus is not a problem. Finger-activated tablets also eliminate the extra movement required to pick up the stylus. These tablets are useful for cursor movement applications when digitizing is not required.

11.4.2.4   Mice

A mouse is a hand-held input device, which is typically attached to the computer by a wire. The traditional mouse is a small plastic box, which can fit under the palm or fingertips. Movement of the mouse on a flat surface is used to calculate cursor position. Mice usually have from one to three buttons on their top and/or sides that may be pressed to perform such functions as changing menus, drawing lines, or confirming inputs. Two mouse technologies will be discussed.

Types of Mice

Mechanical Mice.   One method by which a mouse can operate is through the use of two wheels mounted on the bottom of the device, oriented at right angles to each other. Movement of the mouse, and consequently, movement of the wheels, leads to output from potentiometers, which is then used to calculate X and Y coordinates (Ohlson, 1978; Ritchie and Turner, 1975). Figure 11.4.15 shows a mechanical mouse.

Another mechanical technique involves mounting a small ball in the bottom of the mouse; essentially this device is then an upside-down trackball. Movement of the mouse again leads to voltage output from potentiometers, which is used to determine orientation information.

One problem with mechanical mice is that they may produce noise during movement. Additionally, Mims (1984) states that debris from the table surface may become lodged inside the mouse. However, mechanical mice have the advantage of working on any surface (compare to the optical mouse).

Optical Mice.   Another mouse technology uses optical sensors, which emit pulses as the mouse is moved across a special grid. The number of lines in the grid are counted as the mouse crosses the grid, and X and Y coordinates are calculated accordingly. An optical mouse is shown in Figure 11.4.16. The control-display gain of the mouse may be changed by changing the spacing of the lines on the grid (Somerson, 1983). These mice make no noise and require no moving parts; thus, they will not pick up debris from the surface. However, optical mice require that the special grid be used,

Fig. 11.4.15.   A mechanical mouse. (Photos courtesy of Measurement Systems, Inc. Norwalk, Connecticut.)

else they will not operate. Resolution may also be lower than that obtained with mechanical mice (Somerson, 1983).

Parameters of the Mouse

As indicated previously, mice typically have one or more buttons that may be pressed for various functions. Price and Cordova (1983) compared two methods of button depression: multiple depressions on one button and depression of multiple buttons. The authors report that for a task in which one item was repeatedly selected, performance was faster for clicks of single rather than multiple buttons. However, for a task that involved several actions (in this case, cursor movement and an indication

Fig. 11.4.16.   An optical mouse. (Courtesy of Mouse Systems.)

of the correctness of a mathematical sum), performance using multiple buttons was better than when multiple clicks of the same button were used.

As was the case with the graphic tablet, the gain of the mouse may be changed. Thus, movement of the mouse may result in either more, less, or the same amount of movement of the cursor on the screen. However, unlike the graphic tablet, the mouse will only work in relative mode.

Advantages and Disadvantages of the Mouse

Mice have become quite popular as computer peripheral devices for several reasons. First, they can work in small spaces because the mouse can be picked up and repositioned. Second, as mentioned above, the control-display gain for a mouse can be modified. However, most mechanical mice do not give the user control over gain selection. (As indicated previously, the optical mouse's gain may be changed by a change in the grid.) A third advantage of mice is that they are inexpensive in comparison to other devices, such as graphic tablets. Fourth, the operator can usually locate and move the mouse while still looking at the screen. For further discussion of these advantages, see Rubinstein and Hersh (1984), Somerson (1983), and Warfield (1983).

Disadvantages of the mouse include the following features (McGeever, 1984; Mims, 1984; Scott, 1982). First, while it is true that the mouse may only require a small surface, it does require some space in addition to that allotted to a keyboard. Thus, it is not compatible with many portable or lap computers. Second, as stated previously, a mouse can only be operated in relative mode, a feature that may limit its usefulness for drawing tasks. Other features that limit the use of a mouse for graphic applications include an inability to trace a drawing or to handprint characters. Drawing with a mouse is not as natural as is drawing with a pen or pencil, thus some experience may be necessary to use the mouse effectively for graphic tasks.

In comparison with digitizers, mice typically have lower resolution capabilities. They also do not transmit information as quickly as do most digitizers; the difference may be on the order of a 10-fold decrease from digitizer to mouse (McGeever, 1984; Newman and Sproull, 1979). For a brief review of the advantages and disadvantages of the mouse, see Table 11.4.15.

Types of Tasks

Owing to the limitations of the mouse, it appears to be best suited for pointing and selection tasks, while graphics tablets are more suited to drawing and design tasks. In addition, as with the touch-screen devices, the mouse is not well suited for single character data entry. Thus, mice are typically used as peripheral devices only and not as the sole input device.

11.4.2.5 Trackballs

A trackball is composed of a fixed housing holding a ball that can be moved freely in any direction by the fingertips, as shown in Figure 11.4.17. This input device is similar to the mouse in operation. Two types of trackballs will be discussed: mechanical trackballs that make use of potentiometers and trackballs that use optics. For a more detailed description of these trackballs, see Ritchie and Turner (1975) and Scott (1982).

Types of Trackballs

Mechanical Trackballs. Movement of this type of trackball leads to movement of two shaft encoders. This in turn causes output to be generated from internal potentiometers. The output pulses from the potentiometers correspond to changes in X and Y directions, and the display cursor is moved accordingly.

Table 11.4.15. Advantages and Disadvantages of Mice

Advantages	Disadvantages
Work in small spaces	Requires space beside keyboard
Can modify control-display gain	May have low resolution and information transmission rates
Inexpensive	Unnatural drawing movements
Can keep eyes on screen	Relative mode only
Mechanical mice use any surface	Optical mice require grid
Optical mice are noiseless	Mechanical mice produce noise and pick up debris

Fig. 11.4.17.   A trackball. (Courtesy of Measurement Systems, Inc. Norwalk, Connecticut.)

Optical Trackballs.   In an optical trackball, optical encoders generate signals or pulses. These pulses are used to determine increments in rotation in each of four directions: up and down on both the X and Y axes. Both cursor distance (from the number of rotations) and cursor direction (from the direction of ball rotation) are calculated.

Parameters of the Trackball

There are several features of the trackball that may be adjusted (Ohlson, 1978; Scott, 1982). First, the frictional forces present during rotation may be increased or decreased. In addition, the cursor movement can be made a nonlinear function of the ball's rotational velocity; this permits the use of a low control-display gain at low rotational velocities coupled with progressively higher gains as rotational velocity increases. The trackball's gain function may then be adjusted so that it is optimal for both gross movement and fine positioning; that is, rapid movements of the ball result in large changes in cursor position per unit or rotation, while slower movements result in smaller changes in cursor position. Thus, the trackball may be fairly flexible, permitting rapid movements and accurate positioning.

Advantages and Disadvantages of Trackballs

The trackball has several advantages that have contributed to its selection in many military and radar applications (Ritchie and Turner, 1975; Rubinstein and Hersh, 1984). It is comfortable to use for an extended period of time if the user is allowed to rest his or her forearm. The trackball provides direct tactile feedback from the ball's rotations and speed. In addition, the trackball provides high resolution of movement. A fourth advantage is that a trackball requires only a small fixed amount of space, and it can be installed in a keyboard. Finally, owing to these advantages and the flexibility of its features described in the previous section, a trackball allows for very rapid cursor movement.

The disadvantages of the trackball include its expense, which is typically greater than other devices such as mice and joysticks. Additionally, a trackball cannot be used for tracing, input of hand-drawn characters, or three-dimensional drawings. The advantages and disadvantages of trackballs are summarized in Table 11.4.16.

Table 11.4.16   Advantages and Disadvantages of the Trackball

Advantages	Disadvantages
Direct tactile feedback	Expensive
High resolution	Not well-suited for drawing
Requires little space	No three-dimensional input
Allows rapid cursor positioning	
Can modify control-display gain	

Types of Tasks

Owing to the limitations of the trackball just described, it is best suited for tasks requiring rapid cursor positioning. While trackballs may be used to draw lines and sketch, Parrish et al. (1982) suggest that they should only be used when requirements for drawing speed and accuracy are not very stringent. These authors also suggest that trackballs are excellent for moving and indicating symbols on a display, especially when high speed and accuracy are necessary.

11.4.2.6 Joysticks

A joystick consists of a lever approximately 1–4 in (25–100 mm) long, which is mounted vertically in a fixed base. The joystick works using one of three basic operating mechanisms: displacement, force operation, or digital switches. These three joystick technologies will be discussed briefly; refer to Mims (1984), Ohlson (1978), or Scott (1982) for further information on these devices.

Types of Joysticks

Displacement Joysticks. With a displacement joystick the user moves the joystick in any direction and the displayed cursor moves proportionally. Movements of the lever are detected by potentiometers. Both on- and off-axis movements may be made. When the user takes his or her hand off the joystick, it will either remain where it is or a set of springs will return it to center. These joysticks typically operate in absolute mode; that is, the cursor position corresponds to the joystick position. When the joystick returns to center, the cursor will return to the center of the display. Figure 11.4.18 is an example of a displacement joystick.

Force-Operated Joysticks. A force-operated or *isometric* joystick is a rigid lever that does not move noticeably in any direction (see Figure 11.4.19). Rather, strain gauges measure pressure applied to the joystick. Thus, cursor position is dependent on the force applied to the joystick, not to the position of the device. As with the displacement joystick, the force-operated joystick will respond to pressure in any direction. Both direction of cursor movement (from direction of pressure) and cursor speed (from amount of pressure) are calculated. When pressure is released, the cursor stops moving.

Switch-Activated Joysticks. A switch-activated or *digital* joystick is like the displacement joystick in that the stick itself can be moved; however, only movement in eight directions will be detected.

Fig. 11.4.18. A displacement joystick. (Courtesy of Measurement Systems, Inc. Norwalk, Connecticut.)

Fig. 11.4.19. A force-operated joystick. (Courtesy of Measurement Systems, Inc. Norwalk, Connecticut.)

Movement of the joystick generates output by closing one or more switches that are connected to the base of the stick. When the joystick is released, springs return the lever to center and the switches are no longer activated.

Digital joysticks are typically operated such that the cursor moves with a constant velocity in the same direction as the stick displacement but does not return to center when the joystick is released. These joysticks are limited by the fact that they may only be moved in eight directions; while this feature provides a relatively inexpensive joystick, it does not lend itself to graphic applications.

Parameters of the Joystick

There are several features of joysticks that may be modified. As noted previously, displacement joysticks may either return to center or remain in the position to which they are moved. Some joysticks provide the capability to switch between these modes. In addition, the friction force and force–displacement relationship in many displacement joysticks may be adjusted.

As with touch tablets, mice, and trackballs the gain of the joystick may be changed. Jenkins and Karr (1954) conducted a study in which they found that the optimal gain for movement of a pointer was approximately 0.4. However, joystick gain is typically greater than one due to the difference between the size of the display and the joystick displacement possible. Foley and van Dam (1982) state that it is difficult to use a joystick in absolute mode because, owing to this gain, movements of the hand are magnified on the display. The authors suggest that the joystick is better utilized when movements of the joystick control the velocity or rate of cursor displacement in addition to direction. That is, with rate control, a constant joystick displacement results in a constant rate of cursor movement.

Finally, it is possible to include a three-dimensional capability on the joystick by allowing the stick to be twisted in a clockwise or counterclockwise direction (Foley and van Dam, 1982). Alternatively, a rotatable knob can be placed on the top of the joystick (Scott, 1982).

Advantages and Disadvantages of the Joystick

The advantages of the joystick include the fact that it requires only a small fixed amount of desk space, and it can be made small enough to fit into a keyboard. If a palm or hand rest is provided, the joystick may be used for extended periods of time with little fatigue (Ritchie and Turner, 1975). Because the joystick has been used in varied applications, both military and otherwise, there are many models available. These range from inexpensive, typically switch-activated joysticks, which come with home computers, to more expensive joysticks that allow the user to modify many of the features discussed above.

Disadvantages of the joystick include low accuracy and low resolution (Rubinstein and Hersh, 1984; Scott, 1982). In addition, they cannot be used to trace or digitize drawings, to input hand-drawn characters, or to input single characters. Table 11.4.17 summarizes the advantages and disadvantages of joysticks.

Table 11.4.17   Advantages and Disadvantages of the Joystick

Advantages	Disadvantages
Requires little space	Low accuracy
Can be used without fatigue	Low resolution
Many models available	Difficult to use for drawing
Can modify control-display gain	

Types of Tasks

Joysticks tend to be most suited to tracking tasks or to pointing tasks that do not require a great deal of precision (Mims, 1984). Parrish et al. (1982) suggest that a joystick in absolute mode may be used for line drawing if high accuracy and speed are not required. The authors also suggest that joysticks can be useful in placing and moving symbols, and in menu selection if rate control is used.

11.4.2.7   Light Pens

The light pen, depicted in Figure 11.4.20, is a stylus that generates information when it is pointed at the screen. Unlike the other input devices that have been discussed, all light pens use the same operating principles, although the actual implementation of this principle may vary (Newman and Sproull, 1979; Scott, 1982).

When the light pen is pointed at a cathode-ray-tube (CRT) display, a lens or optical system focuses any light that is emitted from the screen onto a light detector or photocell. When the electron beam in the CRT passes over and refreshes the phosphor at the spot where the light pen is pointing, the increase in brightness causes an electrical signal to be sent to the computer. Based on the timing of this signal, the coordinates of the spot on the display are calculated. Light pens are equipped with either a shutter or finger-operated mechanical switch, which, when pressed, allows light to reach the light detector. In this manner, inadvertent activation is avoided.

Because the light pen is activated by the increase in brightness of the CRT phosphor, it may typically only be used with CRT displays. However, Ritchie and Turner (1975) state that with a complex interface, a storage display or plasma panel may be used. The method of operation also requires that the hardware and software be extremely precise, so that the exact spot on the screen

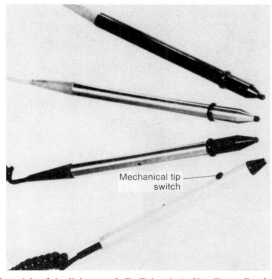

Mechanical tip switch

Fig. 11.4.20.   Several models of the light pen. J. D. Foley & A. Van Dam, *Fundamentals of Interactive Computer Graphics,* © 1982, Addison-Wesley, Reading, Massachusetts, Pg. 193, Fig. 5.11. Reprinted with permission.

may be calculated before the screen is refreshed again. Typically, a refresh interrupt occurs so that the phosphor refresh is halted for a brief imperceptible period of time during the calculation.

Modes of Light Pen Operation

There are two modes in which the light pen may be activated: pick or pointing mode, and tracking mode. In *pointing* mode, a character or figure may be selected using the method previously described in which the operator selects a spot on the display and then enables the light pen. In *tracking* mode, the light pen is used to position a cross hair or cursor present on the display. The operator aims the light pen at the cross hair and then moves the pen. As long as the image remains in the light pen's field of view, a line will be traced where the pen has been moved. It is necessary to move the light pen at a steady rate or the cross hair will be lost and the tracking will be interrupted.

Parameters of the Light Pen

There are two basic features of the light pen that are subject to modification. These are the field of view of the pen and the type of switch that is used to enable/disable the pen (Foley and van Dam, 1982). Because the light pen interacts directly with the display as does the touch screen, concepts such as control-display gain are meaningless for this input device.

Advantages and Disadvantages of the Light Pen

The light pen is the only input device besides the touch-sensitive screen that uses the output display as the input interface. It therefore allows natural pointing and/or drawing gestures to be used to input data (Mims, 1984; Scott, 1982). Thus, this interface provides a direct relationship between output and input. The light pen may also be mounted in a device that is worn on the head, providing an input device for handicapped users or for those operators who must use their hands for other tasks (see Figure 11.4.21). Additional advantages include the availability of inexpensive models, and the fact that the light pen does not require extra desk space.

The light pen has several disadvantages (Mims, 1984; Scott, 1982). The light pen must be held up to the screen; this will be fatiguing over long work periods unless the screen is horizontal. In addition, the light pen and the operator's arm will obscure parts of the display. The light pen lacks high-resolution capabilities, and Foley and van Dam (1982) indicate that it may be activated by what they refer to as false targets, such as adjacent characters or ambient lighting. Light pens are highly dependent on the hardware and software with which they are used, and any problems or inaccuracies in either of these components may lead to inaccuracies in light-pen performance. Finally, the light pen requires the added gesture of picking up and putting down an input device when selection or drawing is required. See Table 11.4.18 for a summary of the light pen's advantages and disadvantages.

Fig. 11.4.21. A light pen may be used as a hands-free input device. (Courtesy of Inkwell Systems.)

Table 11.4.18   Advantages and Disadvantages of Light Pens

Advantages	Disadvantages
Direct eye–hand coordination	Arm fatigue
Inexpensive models available	Arm and light pen obscure display
No extra space required	Dependent on hard- and software
	Inadvertent activation

Types of Tasks

Because the light pen provides a natural pointing gesture, this input device is well suited to menu-selection tasks (Rubinstein and Hersh, 1984). While a light pen may be used for drawing, Scott (1982) suggests that it is not accurate for precise sketching needs, especially since a constant rate of movement is necessary in tracking mode. The light pen is not capable of tracing from hard copy since it must be in contact with the screen. Parrish et al. (1982) suggest that light pens are most useful in placing and moving symbols on a display, regardless of required speed.

11.4.2.8   Miscellaneous Input Techniques

In addition to the input devices that have already been discussed, there are many alternative techniques available. Several of these will be briefly described.

Thumb Wheels

A thumb wheel consists of a serrated knob or dial that is placed on the keyboard and is moved by the operator's thumb. Movement of the thumb wheel causes output from a potentiometer leading to a change in cursor position. Typically, two knobs are positioned at right angles to each other, with one knob leading to cursor movement along the X axis and the other providing cursor movement along the Y axis. Thumb wheels are not as accurate as are joysticks or trackballs for diagonal or curved motions (Scott, 1982).

Puck Pointer

In an attempt to combine the features of a mouse and a joystick, the Puck pointer was developed (Barney, 1984). As can be seen in Figure 11.4.22, this device fits into a space of approximately 4 in.[2]

Fig. 11.4.22.   The Puck pointer installed in a keyboard. (Courtesy of KA Design Group.)

(2500 mm²) on a keyboard. Using a light-emitting diode, the Puck generates X and Y coordinates through an optical sensor when a small handle placed in the center of the space is moved. The handle is approximately the size of a standard key on a keyboard. The advantages of a Puck pointer are that it is extremely inexpensive and that it does not require the operator to lift his or her hand off the keyboard. In addition, it may be used for three-dimensional input by pressing the handle down.

Eye-Controlled Switching

Calhoun, Arbak, and Boff (1984) have begun to investigate the possibility of using eye movement as a method of input to computers in aircraft cockpits. The pilot wears a helmet fitted with an oculometer, which tracks the position of the eyes (see Figure 11.4.23). Helmet and thus head position is also monitored. Using a cockpit simulator, the authors performed a preliminary investigation into the use of eye control as a method of activating switches. They report that eye-controlled switching appears to be a possible alternative to manual switching, with the total switch operation requiring approximately $1\frac{1}{2}$ sec for two subjects tested. An alternative eye-control technique requires that the operator wear eyeglasses, which track the eye through the use of LEDs. It is also possible to track the eye from a remote point using a television camera that detects an infrared light source. Infrared light shines into the eye, the camera detects the reflection, eye position is calculated, and the display coordinates are determined (Bolt, 1984).

The obvious advantages of this potential input method are speed, accuracy, and the freeing of the hands for other tasks. The disadvantage is that this method may interfere with visual tracking and surveillance tasks. In addition, the first two technologies require that the operator wear special equipment, thus they may be obtrusive.

Gesture-Based Input

Another unique technology utilizes operator gestures as a basis for input. Just as touch-screen devices may be used for small screens, a gesture-based system may be used to point with a large-screen display. Bolt (1984) describes the development of such a technique. This method works as follows: both a small "transmitter" cube and a large "sensor" cube are surrounded by magnetic fields. The sensor is placed several feet away from the transmitter. Movement of the transmitter by the user is sensed by the large cube based on changes in the magnetic fields. This information is used to calculate X, Y,

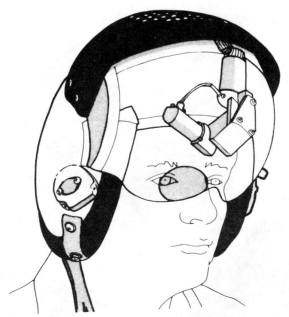

Fig. 11.4.23. The helmet-mounted oculometer. [From *Proceedings of the Human Factors Society 28th annual meeting*, 1984, Vol. 1, p. 259. Copyright (1984) by The Human Factors Society, Inc. and reproduced by permission.]

and Z coordinates. While such a technology is not available for use on a large-scale basis, it illustrates the potential input techniques that may be developed for use in the future.

11.4.3 EMPIRICAL COMPARISONS

As shown in the preceding discussion, there are many computer input devices from which to choose. Although an attempt has been made to suggest the types of tasks for which each device is best suited, it is obvious that there is some degree of overlap in the potential use of these devices. In the following section, the results of empirical studies aimed at comparing several input devices will be presented.

11.4.3.1 Target Selection and Tracking Tasks

Most of the experimental comparisons which have been performed have required that the subjects perform target selection tasks; that is, a stationary target is displayed and the subject is required to position the display cursor at or inside the target position. As soon as the subject confirms the position, a new target is typically presented. Other tasks require that a moving target be tracked continuously over some period of time.

In one of the earliest input device comparisons reported, English, Engelbart, and Berman (1967) compared a light pen, a knee control, a Grafacon arm, a joystick in both rate and absolute mode, and a mouse in a target-selection task. Subjects were required to move a cursor to select a target. For experienced subjects, the mouse resulted in the shortest selection times, while the longest times occurred with the use of the absolute joystick. Inexperienced subjects performed best in terms of selection time with the knee control and the light pen, and worst with the two joysticks. All subjects made the fewest errors in target selection with the mouse. The light pen and the absolute joystick resulted in the highest error rates for experienced subjects, and the light pen and rate joystick resulted in the highest error rates for the inexperienced subjects.

Card, English, and Burr (1978) compared four input devices for a text selection task. The input devices were a mouse, an isometric joystick, step keys (that is, keys that move the cursor either up/down or right/left), and text keys (function keys that position the cursor at the previous or next character, word, line or paragraph). The subjects were required to position the cursor at a target in text displayed on a CRT. Total response time was divided into homing time (time in seconds from subject's initiation of the new task to cursor movement) and positioning time (time in seconds from initial cursor movement to selection).

The results were similar to those reported by English et al. (1967); the mouse was superior to the other devices in terms of total response time, positioning time, and target-selection error rate. This result occurred regardless of either the distance of the target from the current cursor position, or the target size. The authors attribute the superiority of the mouse to the continuous nature of the movement allowed by the mouse, and they suggest that less "cognitive load" is needed to translate desired into actual cursor movement.

Karat, McDonald, and Anderson (1984) compared a touch screen, a mouse, and a keyboard for target selection, menu selection, and selection with typing tasks. Contrary to card et al.'s results, these authors found that the touch screen was superior in terms of selection rate and task completion time across the task types; the keyboard was second, and the mouse was the worst in terms of selection rate and task completion times. In addition, subjects preferred the keyboard or touch screen to the mouse for all tasks.

Whitfield et al. (1983) compared touch screens and touch-sensitive tablets for three target-selection tasks. The first two tasks involved selecting a target from a menu with low- and medium-resolution targets, respectively. The third task also involved the selection of a target from a menu, but several levels of target resolution were included in this one task. Total response times were a combination of both target selection and confirmation time.

Across all three tasks the touch screen resulted in shorter total response times than the touch tablet. The longer response time with the touch tablet was attributed to the longer confirmation time required owing to a need to reverse finger pressure to confirm an entry. The touch tablet was especially slow with the high-resolution targets. The authors attribute this result to the direct eye–hand coordination present with the touch screen and not with the touch tablet. With respect to errors, the two input devices were comparable. An additional comparison with a trackball was made in the third study. The trackball resulted in somewhat slower response times than either the touch screen or the touch tablet. However, the trackball resulted in a lower error rate than the two touch input devices at all levels of resolution. The authors suggest that touch input devices in general should not be used with high-resolution targets or when the task is paced. However, they feel that the touch screen and the touch tablet provide comparable performance levels.

Albert (1982) compared the performance of seven input devices in a cursor positioning task. The devices were touch screen, light pen, data tablet with puck, trackball, displacement joystick, force-operated joystick, and keyboard. The light pen and the touch screen were used both with and without a footswitch for confirmation. The subject's task involved positioning the cursor within a target and then confirming that position. The trackball, graphic tablet, and force joystick resulted in the three most accurate performances, while the touch screen without a footswitch led to the most inaccurate performance. The touch screen and light pen (both without footswitches) resulted in the fastest positioning speed, while the keyboard and both joysticks were slowest. Albert attributes this second result to the direct eye–hand coordination involved with the touch screen and the light pen, as did Whitfield et al. (1983). Preference ratings indicated that subjects preferred the touch screen, light pen, and graphic tablet to the other four devices.

In a comparison of a trackball, joystick, and mouse, Swierenga and Struckman-Johnson (1984) had subjects perform a compensatory tracking task (that is, one in which only the direction and distance of the error between the target and the tracking cursor is displayed). They found that tracking error was lower for the trackball and joystick than for the mouse. The trackball and joystick did not differ significantly. Subjective responses indicated that the subjects preferred the joystick to either the trackball or the mouse for this task.

Gomez, Wolfe, Davenport, and Calder (1982) report a comparison of a trackball with a touch-sensitive tablet for a task in which the subjects were required to superimpose a cursor over a target. Half of the subjects had previous experience with a trackball, and half did not. Response times were not significantly different for the two devices. However, the trackball resulted in significantly less error than did the touch tablet. The authors attribute this latter result to the higher precision characteristics of the trackball, in particular to the fact that the hand was stabilized with the trackball and not with the touch tablet. The response times for the two devices did not differ as a result of prior subject training. However, both the experienced and the unexperienced subjects had lower error rates when using the trackball than when using the touch tablet.

In a comparison of a light pen and light gun with a keyboard, Goodwin (1975) had subjects perform tasks in which they were required to position a cursor in both arbitrary and sequential positions on a display, and also to read a passage and indicate the position of randomly generated incorrect letters. She reported that for cursor positioning, both a light pen and a light gun were faster than a keyboard for all three tasks. Errors were not considered.

Mehr and Mehr (1972) tested the following input devices: (1) a center-return displacement joystick in rate mode; (2) a remain-in-position displacement joystick in absolute mode; (3) a rate-mode thumb-operated isometric joystick mounted in a hand-held grip; (4) a finger-operated isometric joystick in rate mode; and (5) a trackball. Subjects used each of these devices to position a target within a stationary circle presented on a CRT. The trackball resulted in the shortest positioning times, followed closely by the finger-operated isometric joystick while the two displacement joysticks resulted in the longest positioning times. The trackball and the isometric finger-operated joystick provided the most accurate performance.

11.4.3.2   Text Entering and Editing Tasks

In an attempt to provide more realistic tasks for their subjects, several researchers have used tasks in which the subjects are required to actually enter or make changes in text on a display. In an early look at text entry, Earl and Goff (1965) found that a point-in data entry method which was similar to a light pen was superior with respect to input time and errors to a type-in method (i.e., a keyboard) in entering alphabetic material. This surprising result is probably due to the nature of the task. The subjects did not use the point-in method to enter one character at a time, but rather to mark or point to three words. In the type-in task, subjects were required to type three words. These results would argue for the use of a menu-selection method of data entry for some tasks, but not for the use of a point-in method to input individual letters.

Haller, Mutschler, and Voss (1984) compared a light pen, digitizing tablet, mouse, trackball, cursor keys, and voice recognizer for a text correction task. This task required the location of 18 separate characters and then replacement of each with the correct character. Location was performed using one of the six input devices, and correction was performed through the use of either keyboard or voice. The positioning time for the task, excluding correction, was lowest for the light pen and longest for voice input. The average times for the graphic tablet, mouse, trackball, and cursor keys fell between the light pen and voice, and these devices did not differ significantly from each other. The average positioning errors were as follows: light pen and cursor keys, 0%; voice input, 0.9%; graphic tablet, 1.4%; mouse, 4.2%; and trackball, 5.6%. Both of the correction methods—the keyboard and the voice input—resulted in equivalent performance. The authors recommend that the light pen is well suited for cursor positioning, as are the mouse and trackball if their gain is sufficiently low.

In another study, Struckman-Johnson, Swierenga, and Shieh (1984) compared a keyboard, a joystick, a trackball, and a light pen for a text-editing task. In terms of time to complete the editing task, significant differences were found among all four devices. The light pen was the fastest, followed by the trackball, joystick, and keyboard, respectively. With respect to errors, the keyboard and the trackball resulted in the best performance, with the joystick worst. (A measure of errors could not be obtained for the light pen.) The trackball was the most preferred device, followed by the light pen, keyboard, and joystick, respectively.

Other researchers have used tasks besides text editing which are currently being performed in certain jobs. Beringer (1979) reported that the use of a touch screen on a map display led to better performance on a plotting task in terms of errors and time to task completion than did keyboard-based systems. On other tasks, including continuous flight control and navigation information updating, the touch- and keyboard-based systems provided similar results.

11.4.4 CONCLUSIONS

Table 11.4.19 summarizes the results of the studies just discussed. In general, trackballs tend to result in high accuracy for cursor positioning tasks. Touch screens and light pens lead to fast cursor positioning and selection.

Several devices have shown conflicting results in different studies. For example, Card et al. (1978) and English et al. (1967) found that the mouse was the best device for text and target selection tasks, respectively. On the other hand, Swierenga and Struckman-Johnson (1984) reported that the mouse was the worst device for a tracking task, and Karat et al. (1984) report the same results for both target- and menu-selection tasks. Dissimilar results have also been found for the light pen and the joystick. In addition, not only have there been conflicting results across studies, there have also been disagreements within the same studies; thus, for instance, Albert (1982) found that the touch screen without a footswitch was the fastest target-selection device, yet it also resulted in the most errors.

These conflicting results illustrate what Gruenenfelder and Whitten (1984) have discussed as one problem with generic human factors research, or research aimed at discovering general human factors design principles. That is, the results of these studies may not apply to different settings and uses of the design. Thus, while a particular mouse may provide faster and more accurate positioning for one task, that mouse may work poorly given another task and another user population. Conversely, two different mice may lead to entirely different performance on the same task with the same users. Future comparisons should be performed with devices that have first been optimized for the task and users at hand. In addition, Gruenenfelder and Whitten suggest that the limitations of a study's results should be investigated so that the generalizability of results may be determined.

In summary, the choice of an input device should involve the following considerations (see Table 11.4.20). First, the characteristics of the task, users, and working environment should be determined. Not only present, but foreseeable future demands should be considered. Next, the potential input

Table 11.4.19   Summary of Experimental Comparisons of Input Devices

Study	Task	Most Accurate	Shortest Positioning	Most Preferred
Albert 1982	Target Selection	Trackball/graphic tablet	Touch screen/ light pen	Touch screen/ light pen
Card et al. (1978)	Text selection	Mouse	Mouse	—
Earl and Goff (1965)	Data entry	Light pen	Light pen	—
English et al. (1967)	Target selection	Mouse	Mouse	—
Gomez et al. (1982)	Target selection	Trackball	Trackball/ graphic tablet	—
Goodwin (1975)	Cursor positioning	—	Light pen	—
Haller et al. (1984)	Text correction	Light pen/cursor keys	Light pen	Light pen
Karat et al. (1984)	Target/menu selection	Touch screen	—	Keyboard/ touch screen
Mehr and Mehr (1972)	Target selection	Trackball/isometric fingertip joystick	Trackball	—
Struckman-Johnson et al. (1984)	Text editing	Keyboard/trackball	Light pen	Trackball
Swierenga et al. (1984)	Compensatory tracking	Trackball/joystick	—	Joystick
Whitfield et al. (1983)	Target selection	Trackball	Touch screen	Touch screen

Table 11.4.20   Guidelines for Selection of Input Devices

Consider present and future characteristics of users, tasks, working environments.

Match input device characteristics to demand requirements.

Consider previous research and user preference.

Test input device in working environment.

Optimize modifiable device characteristics.

devices should be compared with the requirements of the working environment to narrow down the possible choices. Previous research concerning the input devices under consideration should be reviewed at this point. It is also important to consider user preferences in making the selection; as has been shown, subjective preferences do not always correspond to the device that will provide the best performance. Yet it is important to provide the operators with a tool that will be used.

Once a choice has been made (or preferably, before the choice has been made), the input device should be tested in the working environment with the user population. Within these constraints, the device should be optimized if such optimization is possible given the features of the input device. In this manner, the device chosen may be matched to the use environment. Given a systematic approach to the selection of an input device, it is possible to provide a tool that will be accepted by the users and that will be matched to the tasks and environment.

REFERENCES

Albert, A. E. (1982). The effect of graphic input devices on performance in a cursor positioning task. In *Proceedings of the Human Factors Society 26th annual meeting* (pp. 54–58). Santa Monica, CA: Human Factors Society.

Alden, D. G., Daniels, R. W., and Kanarick, A. F. (1972). Keyboard design and operation: a review of the major issues. *Human Factors, 14,* 275–293.

American National Standards Institute. (1982). *American national standard for office machines and supplies—alphanumeric machines—keyboard arrangement. ANSI X4.23-1982.* New York: ANSI.

American National Standards Institute. (1983). *American national standard for office machines and supplies—alphanumeric machines—alternative keyboard arrangement. ANSI X4.22-1983.* New York: ANSI.

Ancona, J. P., Garland, S. M., and Tropsa, J. J. (1971). At last: standards for keyboards. *Datamation, 17*(5), 32–36.

Arnaut, L. Y., and Greenstein, J. S. (1984). Digitizer tablets in command and control applications: The effects of control-display gain and method of cursor control (Technical Report). Blacksburg, VA: Virginia Polytechnic Institute and State University, Department of Industrial Engineering and Operations Research.

Barney, C. (1984). "Puck pointer" combines functions of mouse and joystick in number-pad-sized package. *Electronics Week,* July 23, 26.

Beringer, D. B. (1979). The design and evaluation of complex systems: application to a man-machine interface for aerial navigation. In *Proceedings of the Human Factors Society 23rd annual meeting* (pp. 75–79). Santa Monica, CA: Human Factors Society.

Bolt, R. A. (1984). *The human interface: where people and computers meet.* Belmont CA: Lifetime Learning Publications.

Brunner, H., and Richardson, R. M. (1984). Effects of keyboard design and typing skill on user keyboard preferences and throughput performance. In *Proceedings of the Human Factors Society 28th annual meeting* (pp. 267–271). Santa Monica, CA: Human Factors Society.

Calhoun, G. C., Arbak, C. L., and Boff, K. R. (1984). Eye-controlled switching for crew station design. In *Proceedings of the Human Factors Society 28th annual meeting (pp. 258–262). Santa Monica, CA: Human Factors Society.*

Card, S. K., English, W. K., and Burr, B. J. (1978). Evaluation of mouse, rate-controlled isometric joystick, step keys, and text keys for text selection on a CRT. *Ergonomics, 21,* 601–613.

Chapanis, A., and Kinkade, R. G. (1972). Design of controls. In H. P. Van Cott and R. G. Kinkade, eds., *Human engineering guide to equipment design* (rev. ed.). (pp. 345–379). Washington, DC: U.S. Government Printing Office.

Cohen Loeb, K. M. (1983). Membrane keyboards and human performance. *Bell System Technical Journal, 62,* 1733–1749.

Conrad, R., and Hull, A. J. (1968). The preferred layout for numeral data-entry keysets. *Ergonomics, 11,* 165–173.

de Bruyne, P. (1980). Acoustic radar graphic input device. *Computer Graphics, 14*(3), 25–31.

Earl, W. K., and Goff, J. D. (1965). Comparison of two data entry methods. *Perceptual and Motor Skills, 20,* 369–384.

Ellis, T. O., and Sibley, W. L. (1967). On the development of equitable graphic I/O. *IEEE Transactions on Human Factors in Electronics, HFE-8,* 15–17.

English, W. K., Engelbart, D. C., and Berman, M. L. (1967). Display-selection techniques for text manipulation. *IEEE Transactions on Human Factors in Electronics, HFE-8,* 5–15.

Foley, J. D., and van Dam, A. (1982). *Fundamentals of interactive computer graphics.* Reading, MA: Addison-Wesley.

Foley, J. D., and Wallace, V. L. (1974). The art of natural graphic man-machine conversation. *Proceedings of the IEEE, 62,* 462–471.

Gaertner, K. P., and Holzhausen, K. P. (1980). Controlling air traffic with a touch sensitive screen. *Applied Ergonomics, 11,* 17–22.

Gomez, A. D., Wolfe, S. W., Davenport, E. W., and Calder, B. D. (1982). *LMDS:* lightweight modular display system (NOSC Technical Report 767). San Diego, CA: Naval Ocean Systems Center.

Goodwin, N. C. (1975). Cursor positioning on an electronic display using lightpen, lightgun, or keyboard for three basic tasks. *Human Factors, 17,* 289–295.

Gruenenfelder, T. M., and Whitten, W. B. (1984). Augmenting generic research with prototype evaluation experience in applying generic research to specific products. In *Proceedings of the Interact '84 conference, first IFIP conference on "human-computer interaction"* (Vol. 2, pp. 315–319).

Haller, R., Mutschler, H., and Voss, M. (1984). Comparison of input devices for correction of typing errors in office systems. In *Proceedings of the Interact '84 conference, first IFIP conference on "human-computer interaction"* (Vol. 2, pp. 218–223).

Hanes, L. F. (1975). Human factors in international keyboard arrangement. In A. Chapanis, Ed., *Ethnic variables in human factors engineering* (pp. 189–206). Baltimore: Johns Hopkins University Press.

Herot, C. S., and Weinzapfel, G. (1978). One-point touch input of vector information for computer displays. *Computer Graphics, 12*(3), 210–216.

Hirsch, R. S. (1970). Effects of standard versus alphabetical keyboard formats on typing performance. *Journal of Applied Psychology, 54,* 484–490.

Hlady, A. M. (1969). A touch sensitive X-Y position encoder for computer input. *AFIPS Conference Proceedings, 35,* 545–551.

Hollingsworth, S. R., and Dray, S. M. (1981). Implications of post-stimulus cueing of response options for the design of function keyboards. In *Proceedings of the Human Factors Society 25th annual meeting* (pp. 263–265). Santa Monica, CA: Human Factors Society.

Hopkin, V. D. (1971). The evaluation of touch displays for air traffic control tasks. *IEE Conference on Displays* (Conf. Publ. #80), 83–90.

Hornbuckle, G. D. (1967). The computer graphics user/machine interface. *IEEE Transactions on Human Factors in Electronics, HFE-8,* 17–20.

Jenkins, W. L., and Karr, A. C. (1954). The use of a joystick in making settings on a simulated scope face. *Journal of Applied Psychology, 38,* 457–461.

Johnson, E. A. (1967). Touch displays: a programmed man-machine interface. *Ergonomics, 10,* 271–277.

Karat, J., McDonald, J. E., and Anderson, M. (1984). A comparison of selection techniques: touch panel, mouse, and keyboard. In *Proceedings of the Interact '84 conference, first IFIP conference on "human-computer interaction"* (Vol. 2, pp. 149–153).

Kinkead, R. (1975). Typing speed, keying rates, and optimal keyboard layouts. In *Proceedings of the Human Factors Society 19th annual meeting.* (pp. 159–161). Santa Monica, CA: Human Factors Society.

Klemmer, E. T. (1971). Keyboard entry. *Applied Ergonomics, 2,* 2–6.

Lohse, E., Ed. (1968). Proposed USA standard—general purpose alphanumeric keyboard arrangement for information interchange. *Communications of the ACM, 11,* 126–129.

Lutz, M. C., and Chapanis, A. (1955). Expected locations of digits and letters on ten-button keysets. *Journal of Applied Psychology, 39,* 314–317.

McGeever, C. (1984). Graphics and digitizers. *InfoWorld, September,* 46–48.

Mehr, M. H., and Mehr, E. (1972). Manual digital positioning in 2 axes: a comparison of joystick and trackball controls. In *Proceedings of the Human Factors Society 16th annual meeting* (pp. 110–116). Santa Monica, CA: Human Factors Society.

Michaels, S. E. (1971). QWERTY versus alphabetic keyboards as a function of typing skill. *Human Factors, 13*, 419–426.

Mims, F. M., III (1984). A few quick pointers. *Computers and Electronics, May,* 64–117.

Monty, R. W., Snyder, H. L., and Birdwell, G. G. (1983). Keyboard design: an investigation of user preference and performance. In *Proceedings of the Human Factors Society 27th annual meeting* (pp. 201–205). Santa Monica, CA: Human Factors Society.

Newman, W. M., and Sproull, R. F. (1979). *Principles of interactive computer graphics,* New York: McGraw-Hill.

Norman, D. A., and Fisher, D. (1982). Why alphabetic keyboards are not easy to use: keyboard layout doesn't much matter. *Human Factors, 24*, 509–519.

Noyes, J. (1983). The QWERTY keyboard: a review. *International Journal of Man-Machine Studies, 18*, 265–281.

Ohlson, M. (1978). System design considerations for graphics input devices. *Computer, 11*, 9–18.

Parrish, R. N., Gates, J. L., Munger, S. J., Grimma, P. R., and Smith, L. T. (1982). Development of design guidelines and criteria for user/operator transactions with battlefield automated systems. Phase II final report: volume II. Prototype handbook for combat and material developers (Technical Report). Synectics Corp., U.S. Army Research Institute for the Behavioral and Social Sciences.

Pfauth, M., and Priest, J. (1981). Person-computer interface using touch screen devices. In *Proceedings of the Human Factors Society 25th annual meeting* (pp. 500–504). Santa Monica, CA: Human Factors Society.

Pollard, D., and Cooper, M. B. (1979). The effects of feedback on keying performance. *Applied Ergonomics, 10*, 194–200.

Price, L. A., and Cordova, C. A. (1983). Use of mouse buttons. In *Proceedings of the CHI '83 Conference on Human Factors in Computing Systems* (pp. 262–266). New York: ACM.

Ritchie, G. J., and Turner, J. A. (1975). Input devices for interactive graphics. *International Journal of Man-Machine Studies, 7*, 639–660.

Roe, C. J., Muto, W. H., and Blake, T. (1984). Feedback and key discrimination on membrane keypads. In *Proceedings of the Human Factors Society 28th annual meeting.* (pp. 277–281). Santa Monica, CA: Human Factors Society.

Rosinski, R. R., Chiesi, H., and Debons, A. (1980). Effects of amount of visual feedback on typing performance. In *Proceedings of the Human Factors Society 24th annual meeting* (pp. 195–199). Santa Monica, CA: Human Factors Society.

Rouse, W. B. (1975). Design of man-computer interfaces for on-line interactive systems. *Proceedings of the IEEE, 63*, 847–857.

Rubinstein, R., and Hersh, H. M. (1984). *The human factor: designing computer systems for people.* Burlington, MA: Digital Press.

Schulze, L. J. H., and Snyder, H. L. (1983). A comparative evaluation of five touch entry devices (Tech. Report No. HFL-83-6). Blacksburg, VA: Virginia Polytechnic Institute and State University, Department of Industrial Engineering and Operations Research.

Scott, J. E. (1982). *Introduction to interactive computer graphics.* New York: Wiley.

Seibel, R. (1972). Data entry devices and procedures. In H. P. Van Cott and R. G. Kinkade, Eds., *Human engineering guide to equipment design* (rev. ed.). (pp. 311–344). Washington, DC: U.S. Government Printing Office.

Somerson, P. (1983). The tale of the mouse. *PC Magazine, 1*(10), 66–71.

Stammers, R. C., and Bird, J. M. (1980). Controller evaluation of a touch input air traffic data system: an "indelicate" experiment. *Human Factors, 22, 581–589.*

Struckman-Johnson, D. L., Swierenga, S. J., and Shieh, K. (1984). Alternative cursor control devices: an empirical comparison using a text editing task (Final Report: Task II.2). Vermillion, SD: University of South Dakota, Human Factors Laboratory.

Sutherland, I. E. (1974). Three-dimensional data input by tablet. *Proceedings of the IEEE, 62*, 453–461.

Swezey, R. W., and Davis, E. G. (1983). A case study of human factors guidelines in computer graphics. *IEEE Computer Graphics and Applications,* 21–30.

Swierenga, S. J., and Struckman-Johnson, D. L. (1984). Alternative cursor control devices: an empirical comparison using a tracking task (Final Report: Task II.3). Vermillion, SD: University of South Dakota, Human Factors Laboratory.

Warfield, R. W. (1983). The new interface technology: an introduction to windows and mice. *Byte,* *8(12)*, 218–230.

White, G. M. (1983). Video pointing devices: enter the touch tablet. *Byte, 8*(12), 218–219.

Whitfield, D., Ball, R. G., and Bird, J. M. (1983). Some comparisons of on-display and off-display touch input devices for interaction with computer generated displays. *Ergonomics, 26,* 1033–1053.

Yamada, H. (1980). A historical study of typewriters and typing methods: from the position of planning Japanese parallels. *Journal of Information Processing, 2,* 175–202.

CHAPTER 11.5
SPEECH CONTROLS AND DISPLAYS

CAROL A. SIMPSON

Psycho-Linguistic Research Associates
Menlo Park, California

MICHAEL E. McCAULEY

Monterey Technologies, Inc.
Carmel, California

ELLEN F. ROLAND

Rolands and Associates Corporation
Monterey, California

JOHN C. RUTH

McDonnell Douglas Electronics Co.
St. Louis, Missouri

BEVERLY H. WILLIGES

Virginia Polytechnic Institute and State University
Blacksburg, Virginia

Substantial portions of this chapter are from *Human Factors*, (1985), Vol. 26, pp. 115–141. Copyright (1985), by the Human Factors Society, Inc. and reproduced by permission.

The speech recognition sections of this chapter were based largely upon discussions among the authors and material generated while they served on the Committee for Computerized Speech Recognition Technologies, Commission on Engineering and Technical Systems, of the National Research Council (NRC). The committee was sponsored by members of the Voice SubTechnical Advisory Group (Sub-TAG) of the Department of Defense. The committee's report, entitled "Automatic Speech Recognition in Severe Environments," was published by the National Research Council in October, 1984. James L. Flanagan, AT&T Bell Laboratories chaired the Committee. From the National Research Council, Dennis F. Miller served as Study Director, and Howard Clark was the Staff Officer. The authors gratefully acknowledge the support of the Voice SubTag member agencies which funded the NRC Committee and subsequently encouraged the preparation of this article. The knowledge gained from our fellow committee members is gratefully acknowledged. However, the authors take full responsibility for any errors and for the opinions expressed in this article. A shorter and prior version of this chapter appeared in substantially the same form as an article in *Human Factors*. Permission from the Human Factors Society for reprinting this material is gratefully acknowledged.

11.5.1 INTRODUCTION

Language is one of the outstanding capabilities of humans. Using this complex, multilevel code, we transmit and receive extremely detailed information amongst ourselves. We have incorporated written language into human–machine systems (e.g., alphanumeric displays and keyboards), but spoken language has been used only for interpersonal communications. Automatic speech recognition and speech generation by machine now offer the promise of person–system transactions via spoken language. Speech input/output (I/O) systems accept spoken input (speech controls) or "display" information to the user by means of the spoken word (speech displays).

No single discipline can claim a monopoly in research on human communication via language. Philosophy, linguistics, psychology, psycholinguistics, logic, and computational linguistics have all contributed to our understanding of human language communication. Traditionally, human factors investigations have concentrated on the written form in the context of training materials, reading comprehension, and alphanumeric display design. Aside from work on the measurement of speech intelligibility, spoken language production and perception have received minimal interest from human factors and ergonomics compared to the attention bestowed upon them by researchers in psycholinguistics, on the one hand, and by speech acoustics engineers and computer scientists, on the other hand. Meanwhile, speech engineers have applied basic research in speech acoustics and articulation to the building of speaking and listening machines. Paradoxically, this technological leap now makes it imperative for human factors psychologists and engineers to better understand human speech perception and production in the context of human–machine interfaces.

Better understanding of human speech communication can contribute to better design of human–machine speech interfaces in two ways. To some extent it is desirable for automatic systems to model or mimic humans just because humans are very good at speech production and perception. We cannot model what we have not yet described. And, a better understanding of the way humans produce and process speech and of the constraints on their capabilities will lead to better design principles for systems that speak to or listen to humans.

Because speech communication until recently has been neglected by human factors and ergonomics researchers, this chapter will differ from many others in this book in that few hard guidelines and specifications will be given for the design of systems using automatic speech technology. We will review the state of the human factors art of applying this technology. At the end of major sections, general rules of thumb, to be consumed with many grains of proverbial salt, will be offered.

Automatic speech technology is of interest within the human factors community because it is a tool to help the human operator perform certain tasks. Its potential lies in reducing or reallocating operator workload by providing an alternative I/O channel to the normally overloaded visual–manual channel. But it is only a tool, not a panacea, for the overloaded operator.

The challenge to the human factors field is to determine when, where, and how automated speech technology should be used in person–system transactions. This challenge is formidable because the technology is evolving, and guidelines for its application will depend on many variables. These variables include the characteristics of the users, the physical environment, the communications environment, the operator's workload, the constraints imposed by the task, and the stress on the operator.

In general, the strategy for the human factors contribution to the field of speech interactive systems is three-pronged: (1) to provide methodologies for identifying appropriate applications of speech technology, (2) to select appropriate speech recognition or generation algorithms and system characteristics, and (3) to integrate speech subsystems within the context of the user's task. The current state of the art provides some but not all of the procedures with which to implement this strategy. By their absence, the missing pieces suggest directions for future research.

11.5.2 TERMS AND DEFINITIONS

Figure 11.5.1 illustrates a human–machine controls and displays system (Chapanis, 1976). Its components are a human, a machine, one or more controls, one or more displays, and an environment.

A *voice-interactive system* is a system that includes speech recognition as one form of user control or information input, and speech generation as one form of information display. There will also be systems that use one but not both of these technologies.

A superficial examination of the terminology in speech recognition and in speech generation obscures the commonality of concepts in these two technologies, which are, in many respects, mirror images of each other.

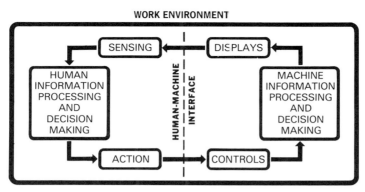

Fig. 11.5.1. Components of a voice interactive system [from National Research Council (1984)], after Chapanis (1976).

11.5.2.1 Speech Recognition Terms

From a human factors perspective, a *speech recognition system* is composed of a human speaker, a recognition algorithm, and a device that responds appropriately to the recognized speech. The algorithm recognizes different human speech utterances and translates them into symbol strings. Those utterances could be words, phrases, or, at a lower level, syllables or *phonemes*—the vowel and consonant sounds of the language. The device assigns meaning to the symbol strings in the context of the human's task.

The term *voice* recognition is sometimes used interchangeably with *speech* recognition, leading to confusion with the related technology of speaker identification or voice identification. Speaker identification is the automatic identification of a given human speaker. To avoid such confusion, this paper will use the term *speech* recognition exclusively.

11.5.2.1.1 Methods of Recognition

Figure 11.5.2 shows the general processes that are employed by speech recognition algorithms, in the laboratory and in commerically available products. A human utterance is converted from its analog form to a digital representation by means of various techniques including filter banks or various time series analyses such as Fourier analysis or linear predictive coding. These techniques compress the data to reduce the computer memory required for storage. The resulting digital data are normalized to control for different speech rates, amplitudes, and physical contexts of the input utterance (e.g., due to background noise). Acoustic features that are known to discriminate different utterances may be extracted from the digital representation. One such feature is the presence of periodicity in the data. Certain speech sounds, those that are voiced due to vibration of the vocal folds during pronunciation, are periodic in nature. The digital representation of the user's utterance together with its associated features provides a data pattern which can then be compared to templates of possible utterances that are stored in the recognition system.

Current commercial systems perform the comparison between whole utterances (see left branch of Figure 11.5.2). They make use of a dictionary of whole utterance templates (words and short phrases) that have been previously stored. Some laboratory systems attempt to analyze the digital representation of the user's utterance into acoustic segments. The resulting string of segments is compared for a match against a dictionary of stored word templates, also represented as segment strings. Rules that specify permissible sentence syntax or compare the possible interpretations for meaningfulness against a stored data base of information may help limit the number of alternative interpretations the recognition algorithm must compare. Ultimately, for either recognition method, the system decides to select a match for the user's utterance or to reject it as being unrecognizable. The selection decision is based on the value of a distance metric computed on the differences between the user's utterance and the closest matching template.

Speaker Dependence

Speech recognition systems vary with respect to several parameters. Speaker dependence refers to the extent to which the system must have data about the voice characteristics of the particular human speaker(s) using it. *Speaker-dependent* recognition systems can recognize the speech of a particular human speaker only if examples of that person's speech have been provided. The vast majority of

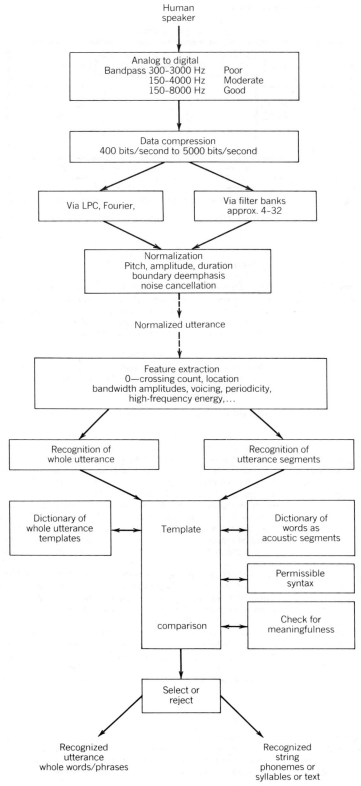

Fig. 11.5.2. Methods of speech recognition [after Simpson, in Moore, Moore, and Ruth (1984).]

current speech-recognition systems are speaker-dependent. *Speaker-independent* systems theoretically can recognize speech spoken by any human in a particular language. Speaker-independent speech recognition is available for small vocabulary sets of 10–20 utterances. In practice, recognition accuracy depends on the similarity of the speech characteristics of the group of users that use the system. So-called "speaker-independent" systems could be said to be *group-dependent.* The less variability among speakers in the user group, the better will be the average recognition accuracy for the group using the system. Speech spoken with a foreign accent, for example, is less reliably recognized than speech spoken with the accent for which the system has been developed. Also, in practice, it is difficult for a speaker-independent system to recognize both male and female speech. For example, a speaker-independent system designed to recognize only male or only female speakers will tend to have greater recognition accuracy than a system designed to recognize both male and female speakers (Rollins, 1984).

Speech Variability

Linguists recognize at least five levels of variability in spoken language, including language families, individual languages, dialects, idiolects, and variations in the speech of individual speakers over time. These levels, with examples, are shown in Table 11.5.1. At the highest level are families of languages. For example, French, Italian, Spanish, and Portuguese are similar in grammatical structure, vocabulary, and phoneme inventory. They all belong to the Romance language family, which is a member of the larger Indo-European family that includes the Germanic languages, including English. The next level is that of individual languages. Then, within languages, there are dialects. For example, British English dialects differ substantially in vocabulary and pronunciation from American English and Australian English. The categorization of dialects themselves is multidimensional and can be made geographically, by social class, by age, and even by neighborhood [for an introduction to dialectology, see Allen and Underwood (1971)]. An *idiolect* is the speech of a single individual. An idiolect varies over time as a function of physiological, psychological, and sociological factors. Similar idiolects can be grouped according to various dimensions, for example, sex, accent, or dialect.

The current practice that distinguishes between speaker-dependent and speaker-independent systems grossly simplifies the range of speaker variability. Even speaker-dependent systems will recognize people who have not enrolled in the system, but the recognition accuracy will be poor. The distinction between speaker-dependent and speaker-independent systems is based largely on engineering strategy for establishing templates. It belies the range of variability in speech and the factors responsible for it, such as regional accent, sex, stress/workload, fear, and so forth. Human speech variability and large vocabulary size are two major challenges for speech recognition systems. Advances in these areas will depend on fundamental research in linguistics at all levels of language structure (Fujimura, 1984).

Speaking Mode

Another parameter of recognition systems is the speaking mode, the manner in which utterances are spoken to the system (National Research Council [NRC], 1984). *Isolated word* systems are most prevalent. With isolated word systems, the user must pause briefly (approximately 100 msec) between vocabulary items when speaking to the system. *Connected* word systems are able to recognize words within utterances spoken without artificial pauses between words. However, the individual words are spoken with the same intonation pattern that would be used if they were read from a list. The term *continuous* speech recognition has often been used to refer to what is here called *connected* word recognition. In this article, *continuous* speech recognition is reserved for recognition of utterances spoken with natural speech rhythm and intonation (prosodics). The final term, continuous speech *understanding,* adds another dimension to the recognition task. It has been used to refer to systems that attempt to correctly accomplish tasks using continuous speech input (NRC, 1984). The measure of successful performance of such systems is thus neither word recognition accuracy nor message recognition accuracy, but

Table 11.5.1 Levels of Linguistic Variability

Linguistic Level	Example
Language family	Romance
Individual language	French
Dialect	Parisian French
Idiolect	Individual female person
Idiolect variation	Individual speaking November 2, 1984, while complaining angrily about a billing error

rather response accuracy. At a primitive level, such systems can be said to "understand" what is spoken to them and to assign meaning to the messages they receive.

Vocabulary Size

A third parameter is vocabulary size. Speech-recognition systems with *fixed vocabulary* must be provided with samples of each word or phrase they are to recognize. They perform acoustic pattern matching at the word and phrase level and typically handle vocabularies of 100–200 utterances (Kersteen and Damos, 1983). Algorithms are under development for *unlimited vocabulary* systems that analyze the speech into phonetic segments, determine the words spoken, and perhaps generate correctly spelled text.

Enrollment

A fourth parameter is the type of *enrollment*. Enrollment is the process of providing templates to the recognition system for the different vocabulary items. These templates are derived from human speech examples of each item. Speaker-dependent systems must be enrolled separately for each speaker who will use them if good recognition accuracy is to be obtained. Typically, the user speaks each item one or more times into a microphone connected to the recognition system. The recognition system converts these human speech waveforms into digital templates. Most systems provide a procedure for *user-enrollment*. Some systems are more flexible than others in the permissible procedures. Speaker-independent systems, in contrast, may be designed for *vendor-enrollment*. This means that the vendor develops the templates that the vendor believes will result in the best speaker-independent recognition accuracy. Some researchers have turned speaker-dependent systems into group-dependent or quasi-speaker-independent systems by means of creative enrollment procedures [e.g., Poock, Schwalm, Martin, and Roland (1982)].

11.5.2.2 Speech Generation Terms

A speech generation system is the mirror image of a recognition system. It is composed of a device that generates messages in the form of symbol strings; a speech generation algorithm, which converts the symbol strings to an acoustic imitation of speech; and a human listener. A speech generation system operates in the context of the user's task environment.

Method of Generation

Speech generation systems, like recognition systems, vary with respect to several parameters. One is the method of speech generation. The two primary methods are synthesized and digitized speech. *Synthesized* speech refers to speech generated entirely by rule, without the aid of an original human recording. The term *digitized* speech applies to human speech that was originally recorded digitally, and then (usually) transformed into a more compressed data format. The most common compression techniques include, but are not limited to, Fourier transform, linear predictive coding (LPC), and waveform parameter encoding. Another pair of terms used to describe these methods are *synthesis-by-rule* for speech synthesis and *synthesis-by-analysis* for digitized speech generation that uses a data compression technique (Flanagan, 1972).

   Figure 11.5.3 illustrates the two general approaches to speech generation. Synthesis-by-analysis, resulting in digitized speech, is shown on the left. Synthesis-by-rule (on the right) uses a set of rules to convert text or some sort of phonetic transcription into the necessary acoustic or vocal tract parameters to generate synthesized speech.

Vocabulary Size

Another parameter is vocabulary size. Speech-generation systems can have a *fixed vocabulary* or an *unlimited vocabulary*. Fixed-vocabulary systems contain a set of words or phrases that can be combined to produce messages. Unlimited-vocabulary systems can produce an unlimited number of messages from normally spelled text, from phonemes, or from phonetic segments (Simpson, 1981a, 1981c). Digitized speech systems are limited to fixed vocabularies. Synthesized speech systems can have either a fixed or unlimited vocabulary. Fixed-vocabulary systems are *user-programmable* only if the user can change the vocabulary items. They are *vendor-programmable* if the user must rely on the manufacturer or some other third party for new vocabulary.

Voice Type

Digitized speech systems can have an unlimited variety of different voices, since they depend on human speakers for their vocabulary. However, once a particular speaker has been selected for an application, the digitized speech system, in order to sound consistent, is dependent on that same human speaker

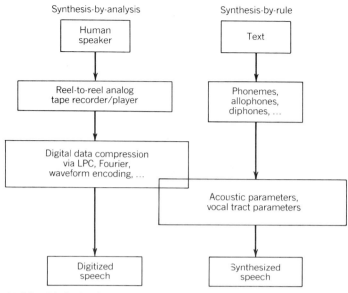

Fig. 11.5.3. Methods of speech generation [from Simpson, in Moore et al. (1984)].

for new vocabulary. Synthesized speech systems do not depend on a human speaker for new vocabulary. But the number of different voice types that can be obtained from a given system is limited and varies currently from one to about six. Most synthesized voices can be varied under program control with respect to fundamental frequency (perceived as voice pitch) and speaking rate. Most commercially available synthesizers produce male-sounding speech, although a few also produce female-sounding speech. With software control of the pronunciation of individual phonemes, some variation in dialect or accent can be obtained. For reviews of the commercially available speech generation devices see Butler, Manaker, and Obert-Thorn (1981), Sherwood (1979), Simpson (1981a), and Smith (1984).

Data Rate, Intelligibility, and Naturalness

Three parameters often used to evaluate speech generation systems are *data rate, intelligibility,* and *naturalness.* Data rate terms are often confusing in the speech product literature because they can refer either to the amount of storage needed to store speech data or to the rate at which speech data are transmitted to the speech device or to the rate at which the resulting speech is actually spoken (Simpson, 1983b). The terms naturalness and intelligibility often are confused in today's product literature and, unfortunately, the scientific literature as well. The term *intelligibility* has a very precise meaning—the percentage of speech units correctly recognized by a human listener out of a set of such units. The units may be words, sentences, individual speech sounds (called phonemes), or even the perceptual acoustic features of those phonemes. Kryter (1972) provides a comprehensive guide to intelligibility testing. *Naturalness* refers to a listener's judgment on a scale of the degree to which the speech sounds as though it were spoken by a human. Intelligibility and naturalness can be measured independently, although there are no standardized tests of naturalness (Simpson, 1983b). Furthermore, naturalness and intelligibility are not necessarily correlated (Thomas, Rosson, and Chodorow, 1984). For example, a radio announcer may sound natural in a background of static noise, but the speech may have low intelligibility. Conversely, synthesized speech warning messages tha are well known to a pilot may sound mechanical, yet pilots have rated such messages more intelligible than human voice messages transmitted via aircraft radio (Simpson, 1983b; Simpson and Williams, 1980).

11.5.2.3 Measures of Algorithm Performance

Recognition accuracy is the most commonly used measure of performance for speech-recognition algorithms. Its counterpart is speech intelligibility, which is the most commonly used performance measure for speech-generation algorithms. Both measures are simply the percentage of speech utterances correctly recognized by the "listener" out of a set of such utterances presented under a particular set of listening conditions. When measuring recognition accuracy, the "listener" is the recognition algorithm. Conversely, human listeners are used to measure the intelligibility of speech generated by algorithm.

The classes of errors that occur when speech is presented to either humans or machines are the same, but human and machine performance may differ substantially. Errors fall into one of four mutually exclusive categories: (1) substitution errors (one utterance from the vocabulary is mistaken for another), (2) insertion errors (an utterance is reported that was not spoken), (3) deletion errors (an utterance that was spoken was not reported), and (4) rejection errors (an utterance that is a legal item in the vocabulary is detected but not recognized). Rejection errors are frequently reported in the speech-recognition literature. They are not reported as such in the speech intelligibility literature, but occur when the subject responds with "don't know."

Both machines and humans also can make correct rejections. For machines, that is, recognition algorithms, a correct rejection is made when the algorithm refuses to process an utterance that is not in the legal vocabulary, for example, if the user coughed or said something to another human in the work environment. Similarly, human listeners will correctly reject utterances spoken in an unfamiliar foreign language. And, under conditions of poor signal-to-noise ratio, they will also correctly reject nonsense words in their own language and often will then substitute a word that makes sense [Garnes and Bond (1977) as reported in Bond and Garnes (1980)]. This human capability (to substitute a word that makes sense) requires knowledge of syntax, semantics, and pragmatics that is well beyond that found in commercially available speech recognition systems (Fujimura, 1984; NRC, 1984).

11.5.2.4 Models of Speech Perception

Models of human speech perception can provide a useful framework for assessing human performance with speech-generation algorithms. They can also suggest methods for improving the performance of speech-recognition algorithms. Most current models of human speech perception are based on the concept of analysis-by-synthesis (Stevens, 1972). This model postulates that humans decode speech by mentally synthesizing alternative interpretations and comparing them to the incoming speech stimulus for goodness of fit. The level of linguistic structure over which analysis-by-synthesis takes place is the subject of considerable debate. Some would limit its domain of operability to distinctive features of individual phonemes (so called because they distinguish one phoneme from another by their presence or absence or by their value for one phoneme compared to another). At the other extreme, examples from natural language have been introduced which suggest that analysis-by-synthesis could operate at the discourse level over multiple clauses.

The basic idea that humans use their knowledge of the linguistic structure of their language when decoding speech is well accepted. Table 11.5.2 illustrates these different linguistic structure levels and the units that compose them.

At each linguistic level, there are constraints on possible combinations and ordering of units. Semantic constraints at the discourse and sentence levels restrict legal ideas and words to those which make sense in the context of the other ideas and words. Usually pragmatic constraints apply also to limit the interpretations of a sentence to what the listener knows to be possible in the real world, given his or her knowledge about the current state of the real world. Syntactic constraints limit possible word orders within sentences. In English, adjectives precede the nouns they modify. In French, they usually follow the noun. Morphemes are the smallest meaningful units in a language. They include roots and affixes (prefixes and suffixes). The English word *faithfully* has three morphemes: the root *faith,* the suffix *-ful* meaning "having the attribute of," and the suffix *-ly* indicating that this word is a modifier of a verb. Phonological constraints limit the ordering and position of phonemes within words. English has words that end in the phoneme that is at the end of the word *sing,* written with the International Phonetic Alphabet (International Phonetic Association, 1949) symbol /ŋ/. English also has words ending in the phoneme /n/, as in *sin.* However, while /n/ can also come at the

Table 11.5.2 Levels of Linguistic Structure

Level	Units
Dialog or discourse	Ideas, information, intentions
Encoded as sentences or clauses	Ordered words
Encoded as morphemes	Ordered roots, prefixes, affixes
Encoded as phonemes	Distinctive features
Encoded as context-sensitive, co-occurring acoustic patterns	Frequency components, amplitudes, duration, periodicity, noise, silence, rate of change of acoustic patterns

beginning of English words, /ŋ/ cannot. Phonetic constraints limit the acoustic or articulatory details of a particular phoneme in a particular phonetic environment with the result that a given phoneme is not acoustically invariate, despite our invariate conscious perception of it. For example, the phoneme /k/ sounds different in each of the following words: *keep, kill, kelp, cat, cool, coal, call.* Yet an English language speaker will report that all seven of these words begin with the "kay sound." It is postulated that a speaker of English uses knowledge about such rule-governed variation to mentally synthesize acoustically detailed representations of possible words which can then be compared to the input speech signal. [For a more detailed introduction to theories and models of human speech perception see the chapters by several authors in Cole (1980).]

In large part because speech recognition algorithms do not utilize knowledge of linguistic structure to the extent that humans do, the performance of speech recognition algorithms "listening" to human speakers is quantitatively and qualitatively different from the performance of speech generation algorithms speaking to human listeners. Accordingly, the remainder of this chapter treats separately research on speech recognition, on speech generation, and on integration of recognition and generation into voice-interactive systems.

11.5.3 SPEECH RECOGNITION RESEARCH

In the Section 11.5.1, a three-pronged strategy for human factors effort in automatic speech technology was stated. Two of the three levels, namely, (1) identification of speech recognition applications, and (2) selection of appropriate recognition system characteristics, will be the subject of this section on speech recognition research. The third, integration of speech subsystems within the user's task, will be discussed in the section on system integration.

11.5.3.1 Applications

Selection of test-beds for speech technology has not been systematic. Nevertheless, the application research has proven useful in helping to identify characteristics of appropriate applications, potential human–machine interaction problems, required system capabilities, and the need for an integrated systems approach to incorporating speech recognition systems. The following application examples, extracted from NRC (1984), are provided as a cross section of attempted applications of speech recognition technology and lessons learned.

Case Study: Integrated Information Display

Each major Navy Fleet Command Center can access current information on the location of friendly or known enemy ships in an operating area. Most of this information is updated automatically from inputs sent directly from U.S. ships concerning their present capability, position, and planned movements, or from other automated intelligence sensors. This information can be accessed, updated manually, and displayed tabularly or graphically using the integrated information display system.

Speech recognition equipment has undergone an initial testing phase in the integrated information display system (Poock and Roland, 1984). The Navy is considering its use as a tool to reduce training time for new or infrequent users. This will permit the more expert users to undertake more complicated tasks such as correlating tracks and ensuring that the information in the system is correct.

During the test phase, both users and managers were enthusiastic about the recognition system but seem to have lost interest over time. This is attributed to inadequate vocabulary size and hardware interface problems that interfere with a full implementation of terminal operations using voice commands. The equipment tested was limited to approximately 250 utterances, and it was estimated that well over 1000 utterances would be required for full implementation.

Case Study: Advanced Fighter Technology Integrator (AFTI) F-16

The Air Force, Navy, and National Aeronautics and Space Administration (NASA) sponsored a program to flight test interactive voice systems specifically designed for the harsh environment of modern fighter aircraft. The program consisted of extensive stand-alone and integration tests of the voice recognition systems prior to flight tests. Considerable improvements in the recognition accuracy rate were noted during these phases of the program. A speaker-dependent, isolated-word recognition scheme with a small (35 word) vocabulary was defined for the flight test, and a 10-word subset of that vocabulary was tested in "active" mode, to actually control selected aircraft systems.

Lessons learned in this program are being used in the operational utility evaluation phase presently being conducted. Pilot attitudes concerning voice interactive systems were found to affect system performance. Improvements in the recognition accuracy helped to secure pilot acceptance during the flight test (Moore et al., 1984).

Case Study: Precision Approach Radar Training System (PARTS) and Air Controller Exerciser (ACE)

The Naval Training Equipment Center sponsored the development of two prototype training systems, one for precision approach radar controllers and the other for air intercept controllers (Breaux, 1977; Grady, 1982). Both prototype systems demonstrated the feasibility of eliminating the need for a person to act as a pseudopilot and for automated instruction and performance measurement of verbal tasks, thus reducing instructor workload. Air controller tasks are amenable to speech recognition because they involve highly structured speech as the primary output of the trainee.

These training systems were among the first to be designed from the outset around speech technology rather than retrofitted with a speech capability. The recognition accuracy of both (PARTS and ACE) fell short of providing graceful interaction between trainees and the system. PARTS used a discrete utterance recognizer, while ACE used a connected speech recognizer. Both were insufficiently tolerant of trainees' speech variability under the stress of simulated operational conditions. Trainees were particularly at a disadvantage in the use of speech-recognition systems because they generally did not have extensive prior experience with voice communications and because they were required to learn both the new job and voice control techniques.

McCauley and Semple (1980) reported that it was perhaps unfortunate that the system development program was so ambitious, including automated instruction, automated performance measurement, adaptive syllabus control, and modeling of pilot behavior and environmental variables. All of these subsystems were directly interactive with the speech recognition subsystems and, therefore, any errors in speech recognition were amplified by subsequent system functions (McCauley and Semple, 1980; McCauley, Root, and Muckler, 1982).

Nevertheless, these prototype training systems demonstrated the potential for speech interaction in real-time simulation and in powerful combination with several automated technologies. The system evaluations indicated that slight increases in recognition accuracy would lead to considerably more effective training.

Observations From Case Studies

A review of these and other case studies leads to some observations about speech-recognition technology and its integration into operational systems, condensed here from the NRC (1984) report:

1. Recognition accuracy was one of the main limitations.
2. The variability in human speech under stressful conditions contributed to unacceptable performance.
3. The success of voice-interactive systems in most applications arose from its integration with other procedures or automation features.
4. Projects designed from inception to incorporate a voice-interactive system had a greater probability of success than when the capability was added to an existing system.
5. Highly connected systems that depend on accurate speech recognition input tended to amplify the effects of recognition errors.
6. A staged process of voice-interactive system development, including regular checks and tests by users, was more likely to lead to successful systems.
7. Speaker enrollment (in a speaker-dependent system) was usually more effective when conducted in the context of the operational task.
8. Other voice communication functions in the task environment sometimes interfered with the speech recognition task.
9. For externally paced tasks, the timing of the task sequence was disrupted by either long recognition time or recognition errors.
10. The lack of an appropriate recognition feedback mechanism tended to confuse operators regarding the status of the system.

The types of applications that might benefit from speech recognition can be classified into major functional categories. Three such categories that were used to classify military applications are (1) data base management systems, (2) command and control of weapons systems, and (3) training systems (NRC, 1984). The required speech recognition system characteristics for these three categories differ, as shown in Table 11.5.3 [adapted from NRC (1984)]. These functional categories were derived in the context of military applications. However, they apply to a wider range of applications. For example, the characteristics of the data base management and the training functional categories also apply to industrial and medical applications for speech recognition. And, the characteristics of the command and control of weapons systems functional category apply to the design of speech controls for assistive devices for the physically handicapped. For example, speech controls for a wheelchair must be highly

Table 11.5.3 Typical Voice-Interactive System (VIS) Characteristics by Functional Application

	Data Base Management Systems	Command and Control of Systems	Interactive Training Systems
Vocabulary size in words	Large, 1000–5000	Small, less than 100	Medium, 50–500
Recognizer type	Connected utterance	Discrete with connected digit capability	Discrete with connected digit capability
Speaker dependence	Speaker independent	Speaker dependent	Speaker dependent
Enrollment time	Not applicable	Varies	Less than 5% of total training time
Typical noise	Quiet	Moderate to high	Quiet
Typical operator stress	Low to moderate	Moderate to severe	Moderate to high
Operational requirements[a]	Less than 5% error	Less than 1% error	Less than 3% error
VIS response time	Less than 5 sec	Less than 1 sec	Less than 2 sec
System integration requirements	Moderate	Critical	Important
Physical constraints (size weight, power, cooling)	Minimal	Severe	Minimal

[a] There are certain safety and survivability conditions that mandate minimal error tolerance for portions of the vocabulary.

Source: NRC (1984).

reliable, must provide rapid VIS response, must operate under moderate to severe levels of operator stress, and must satisfy severe physical constraints.

11.5.3.2 Speech Recognition Task Selection

One compelling reason to incorporate speech recognition into complex systems is the potential for reducing the visual–manual task load. However, the decision to use speech for a particular task requires a matching of speech mode features with task characteristics (Simpson, 1984) and analysis of the advantages and constraints of the manual mode versus the speech mode in the context of the tasks to be performed (North and Lea, 1982).

The research on task selection has been conducted on two major fronts. Some researchers have aimed to develop and apply methodologies for selecting appropriate tasks for speech (North and Lea, 1982) and user-preference questionnaires for application of voice recognition and speech generation (Brown, Bertone, and Obermeyer, 1968; Cotton, McCauley, North, and Strieb, 1983; Kersteen and Damos, 1983; Williams and Simpson, 1976). Others have investigated human speech data entry performance when simultaneous verbal and manual tasks are required (Wickens, Sandry, and Vidulich, 1983).

One study determined that speech is useful primarily for complex tasks requiring cognitive and/or visual effort, while simple tasks involving the copying of numeric data were accomplished more quickly and accurately with keyboard entry compared to voice entry (Welch, 1977). A series of dual task tracking and data entry studies (Coler, Plummer, Huff, and Hitchcock, 1977; Simpson, Coler, and Huff, 1982) conducted in the presence of helicopter noise and helicopter motion, respectively, found that, across all noise and motion conditions, tracking performance was less degraded with data entry by speech recognition than it was with data entry by keyboard. Recognition accuracy and keyboard accuracy for the no-noise and no-motion conditions were 99%. Recognition accuracy, however, declined slightly in noise and motion, while keyboard accuracy did not. Another study found that in the presence of a simultaneous verbal task, voice data entry resulted in less decrement in tracking performance than keyed data entry (Harris, North, and Owens, 1977). However, it has also been found that recognition error rates can increase by as much as 39% with concurrent tracking, suggesting that task stress has a sufficiently large effect on human speech production to degrade recognition accuracy (Armstrong, 1980).

Research comparing speed and accuracy of voice versus manual keyboard input has produced conflicting results, depending on the unit of input (alphanumerics or functions) and other task-specific variables. For example, one study on the use of voice input to a computerized war game concluded that the manual method of entry was faster than voice input (McSorley, 1981). Another study, conducted in the same laboratory, to assess speech recognition for operation of a distributed network system, showed speech input to be superior to manual entry in both speed and accuracy (Poock, 1981a). Task requirements were cited as the primary reason for these different results, since the majority of other factors, user group composition, training, equipment, and environment were constant. The results of these studies suggest that the benefit to be derived from voice input and output is highly dependent on the specific task and environment.

In summary, the selection of potential tasks for speech recognition should be based on specific task requirements. Speech is not a useful substitute for manual data entry when such tasks already are being performed successfully (Welch, 1977). Speech input is likely to improve system throughput only in complex tasks that involve high cognitive, visual, and manual loading. Such limits on improvements to system throughput using the speech mode are likely to exist irrespective of any future improvements in the technology simply because of the characteristics of speech itself. These characteristics and their implications will be discussed in detail in the section on system integration. Clearly, more research is needed to better understand the complex interaction of the speech mode, voice recognition technology, the user, and the task being accomplished.

Methods and guidelines are needed for identifying tasks that are amenable for speech recognition applications. Interview and questionnaire techniques are helpful but they are limited in predictive power because the potential user community is familiar with their job but naive with respect to the capabilities and limitations of speech technology.

Finally, no analytic procedure for selecting speech tasks is likely to be sufficiently accurate to enable detailed specification of the speech system requirements. Further work is needed on simulation techniques to establish the speech system requirements early in the system design process. This will be discussed in the next section on the second level of the research strategy, the selection of recognition system characteristics.

11.5.3.3   Selecting Speech Recognition System Characteristics

Given an appropriate task for the speech mode and given a set of performance requirements for the application system that will incorporate that speech task, the characteristics of the speech recognition system must be carefully selected. The algorithm, the human operator, and the interface that links them are all components of the speech recognition system for which the proper characteristics are to be chosen. Therefore, research on algorithm performance and on human–system performance is essential to recognition system selection. Research to date has documented a variety of factors that affect recognition system performance.

Speech Recognition Algorithm Performance

Many factors influence recognition accuracy. They can be viewed in terms of the characteristics of the physical speech signal itself and the context in which it is spoken. Today's commercially available recognition algorithms, however, do not take advantage of the pragmatic and linguistic context of an utterance. Some applications software has used context by adding simple syntactic and pragmatic constraints downstream of the recognizer output. Syntactically illegal messages are rejected by the application system, and the user is asked to repeat the message. Another technique, which has been called "syntax subsetting," limits the set of templates that are compared to the user's utterance to just that set of vocabulary items that are legal given the previously recognized item. These techniques can improve recognition accuracy, but such pre- or postrecognition application of contextual constraints severely limits their power. If contextual constraints were incorporated into the recognition algorithm itself, they could act at the level of the recognition decision to influence its outcome. As of this writing, the only algorithms that make use of context are still in the laboratory and do not operate in real time.

The performance of current recognition algorithms is far more fragile than humans' listening performance with respect to degradations or changes in the speech signal or the physical context in which it is spoken. Currently, speaker-dependent, isolated word systems can perform in the laboratory with vocabularies up to 100 words with an error rate of less than 1%. However, recognizer performance demonstrated favorably in the laboratory often degrades dramatically under the effects of noise, user stress, and operational demands on the user (NRC, 1984).

User Characteristics

User characteristics can affect speech recognition system performance. Successful applications of speech recognition to date usually involve a small number of carefully selected talkers who have been trained to speak distinctly and use the equipment correctly. Doddington and Schalk (1981) reported that three-fourths of the talkers they tested had better-than-average recognition scores, indicating that a few people had a majority of the problems.

Enrollment

Enrollment is another critical element in speaker-dependent speech recognition systems. Enrollment techniques that avoid any systematic bias in the speech samples seem to be most successful. For example, recognition accuracy is better when the several tokens of each vocabulary item are sampled randomly instead of collecting all tokens of a vocabulary item in sequence (Poock, 1981b). Recognition performance is also enhanced when enrollment occurs in an acoustic or motion environment similar to that of operational conditions (Simpson et al., 1982). In a subsequent study, with a different recognition system, it was found that enrollment in a quiet environment resulted in no adverse effects on recognition accuracy in cockpit noise levels up to 100 dB SPL (Coler, 1982). In general, performance of different commercial systems varies considerably as a function of enrollment environment (NRC, 1984). For enrollment prompts to the user, the visual mode is usually used. The use of synthesized speech to prompt enrollment has been questioned, because some speakers tend to mimic the prompt (McCauley, 1984).

Adaptive Recognition Algorithms

Adaptive recognition algorithms for speaker-dependent systems are one method for dealing with speech variability over time for a given speaker. The algorithm alters its reference template to reflect slow changes in the user's pronunciation over time. To do this, it needs feedback on the accuracy of each recognition attempt. One study (Coler and Plummer, 1974) reported an improvement from 95% to 99.9% recognition accuracy using an adaptive algorithm.

System Feedback

Feedback by the system to the user may enhance performance, either by altering the user's speech or by allowing for user error correction. Poock, Martin, and Roland (1983) found no conclusive evidence that different levels and types of feedback contributed to changes in speaking patterns nor improved recognition accuracy. However, it was shown that feedback in general affects recognition performance. Recognition performance with subjects not accustomed to feedback improved when some type of feedback was presented, and, conversely, was degraded if the feedback to which a user was accustomed was reduced.

In the absence of feedback, a user may assume, incorrectly, that a sequence of voice commands was executed properly by the system. For example, one study (Schurick, Williges, and Maynard, 1985) demonstrated that accuracy in a data base entry task using speech could be increased from 70% to 97% correct with feedback and user error correction. Although there is general agreement about the need for feedback, an important issue for human factors integration is how to best provide feedback to avoid interfering with the operator's primary task and to maximize throughput.

Error Correction

Speech recognition system performance can be improved with two types of error correction. The system can be designed to detect illegal input sequences automatically and correct them to the most likely legal sequence. It can then optionally present the correction to the user for verification. For example, with syntactically constrained dialogues it has been suggested that the recognizer could select both the first- and second-choice vocabulary items by using standard parsing techniques (Spine, Maynard, and Williges, 1983). Another suggestion is the use of subject-specific confusion matrices as well as the logical "anding" of utterances when users are asked by the system to repeat (Bierman, Rodman, Rubin, and Heidlage, 1984). Schmandt and his colleagues point out that redundant information sources such as gestures detected by electronic pointing devices may be useful in clarifying speech commands (Schmandt, 1982). While such techniques can improve recognition accuracy, they do not ultimately guarantee a semantically correct message or command. Thus, the human ought to remain in the loop, at least for critical entries.

In addition to automatic error correction, provision should be made for error correction by the user. Three documented types of user errors include failure to remember the vocabulary set, failure to follow the speech cadence restrictions, and conversing with co-workers with an active microphone. Vocabulary errors involve speaking other words outside the vocabulary, including synonyms. Cadence errors include using connected speech with discrete word recognizers. Other types of user errors are to be expected. Lack of a rapid error-correction capability can be frustrating to the user who is engaged in a dynamic, time-critical task, as seen in the case study on the Precision Approach Radar Training System (McCauley and Semple, 1980) and can drastically increase the time to achieve a desired system goal via speech recognition.

Environmental Factors

The task environment comprises a number of factors that must be studied for their effect on human performance and therefore on speech task design. Physical, physiological, emotional, and workload factors can be expected to partially determine the success or failure of a particular speech system design. Only after the effects of these factors are known can speech systems de designed in ways that will enhance rather than hinder human performance and thus systems performance.

The major environmental factor that has been studied is the effect of background noise on recognition accuracy, but little is known of its effect on human performance while using speech-recognition devices. There is qualitative information available on the effects of environmental stress on human speaking performance, but little quantitative data. Relationships between psychophysiological state and voice parameters have been investigated, including changes in laryngeal tension, rise in the fundamental frequency, pitch perturbations, and breath noises [see NRC (1948), for references]. One study manipulated task-induced stress to determine the effects of speaker stress on speech (Hecker, Stevens, von Bismark, and Williams, 1968). This study documented the variety of differences between speech spoken with and without stressful conditions as well as differences in the effects of task-induced stress on the speech of individual speakers. Because stress-related speech changes can take many forms and are neither consistent among people nor tasks, speech recognition performance may vary dramatically as a function of the work environment. This may be why most successful applications of speech recognition do not involve severe time constraints nor life-threatening situations.

11.5.3.4 Human–System Performance Measurement

Although general methods for performance measurement are available for different levels of human/ system performance, the measurement of speech recognition performance in a complex control and display task is more difficult and less well understood. In addition to speed and accuracy of operator performance of complex tasks, it is necessary to measure variables such as operator workload and operator ability to deal with novel situations. Also, conflicts between other controls and displays and speech controls and displays have to be assessed.

11.5.3.5 Simulating Recognition Systems

By simulating speech recognition hardware, various levels of speech recognition capability can be controlled and evaluated experimentally. Research using simulations of speech controls and displays originated with studies of how people communicate to solve problems (Chapanis, 1975). Problem solution occurred most rapidly whenever the voice link was available. Other modes were typing and handwriting. Since this study did not restrict the speech channel in vocabulary, syntax, or permissible speaking cadence, its relevance to current speech recognition capabilities is limited, but it illustrates the power of voice communication for problem solving, emphasizes the importance of further development of speech recognition technology, and demonstrates simulation of speech recognition as a research methodology.

Since those first studies, several attempts have been made to study system performance and acceptability when the speech channel is restricted in various ways to simulate the use of speech recognition hardware. One study simulated a listening typewriter where speech was constrained either in terms of vocabulary size or speech pause requirements (Gould, Conti, and Hovanyecz, 1983). Shortcomings of the simulation included slow response time, failure to simulate misrecognition errors as well as nonrecognitions, and inconsistent restriction of discrete data entry when the spelling mode was used to enter words not in the vocabulary. However, the simulation contributes to development of techniques to simulate speech recognition for human factors research.

Another study demonstrates the difficulty of designing a good simulation of speech recognition. The study attempted to evaluate user acceptance of various levels of recognition accuracy (Poock and Roland, 1982). Because subjects read words in a prescribed order, it was difficult to control appropriate feedback when the subject spoke the wrong word or made a detectable noise. Also, since the subjects had no real task to accomplish, they often failed to read the visual feedback provided, and were unaware of errors. As a result, all levels of speech-recognition accuracy tested in the simulation were judged acceptable by the subjects, probably indicating simply that they liked the concept of voice input. Avoidance of these and related problems in future simulation designs will be no trivial task.

A recent study (Zoltan-Ford, 1984) demonstrated a simulation technique that was quite believable for subjects and provides encouraging data on successful methods for constraining users' syntax and vocabulary when they speak to a recognition system. Subjects conversing with a computer were not constrained to use any particular syntax or vocabulary. However, the computer, simulated by the experimenter, "responded" with a constrained vocabulary and syntax. The subjects imitated the "computer" and gradually adopted its vocabulary and syntax over the course of the experiment.

11.5.3.6   Future Speech Recognition Research

Simulation

Simulation techniques are needed to provide controlled variation along dimensions such as speed of recognition and feedback, recognition accuracy level, and types of recognition errors. In addition, system performance measures must be developed that integrate recognizer performance, human performance, task workload, system utility, and user acceptance.

Important issues to be addressed include the following:

1. Speed and accuracy requirements for various applications.
2. Criticality of errors, by type.
3. Appropriate forms of error correction.
4. The need for speaker independence.
5. The need for connected or continuous speech.
6. The effects of large vocabulary size.
7. Human ability to constrain speech in terms of vocabulary, syntax, and speaking patterns (NRC, 1984).

Data from these simulations can be used to determine candidate tasks for speech, and the speech-recognition performance required for successful use of a speech data base. The simulations can provide samples of speech produced under various task conditions, such as noise, mental workload, stress, and various levels of recognition error rate. Finally, simulation would provide a research environment for developing general guidelines on how speech data entry should be integrated into different task environments.

Enrollment Methods and User Training

Better enrollment methods are needed for speaker-dependent systems. These methods should permit enrollment of the speech recognition system in a benign environment when it is to be used in a more hostile environment. This will reduce enrollment costs in terms of equipment operation as well as operator time, stress, and fatigue. Also, if techniques can be perfected for automatic updating of speech samples while the system is in use, recognition systems will be better able to handle slow changes in the user's voice over time due to fatigue, for example.

Better methods are needed for predicting the speech recognition performance on the basis of user characteristics. For example, the user's dialect may influence recognition performance. Research is needed on techniques for predicting low performance users and on potential remediating methods. Training users to modify their speech patterns will be difficult because speech is a highly overlearned behavior that is difficult to modify. The extent to which training can reliably alter speaking habits, particularly under stressful conditions, has yet to be determined. This is an important research issue especially for the types of applications envisioned by the military sector.

Performance Measurement

Improved performance measurement is essential for providing data for decisions about system design and effectiveness. More detailed analysis of recognition algorithm errors will permit a better understanding of the effects of different user characteristics, environmental factors, and task-related factors on recognition accuracy. Errors should be displayed in a confusion matrix format at the task, utterance, and phoneme levels. High-fidelity quality audio recordings of subjects' utterances spoken to speech recognizers under known, controlled experimental conditions ought to be routinely made and analyzed to discover speech variability factors that affect recognition performance.

Speech recognition performance should be measured within a realistic task scenario, both within the laboratory and in the actual operational setting, including worst case conditions. Laboratory benchmark tests using standard vocabularies, experienced users, and controlled environments are useful for comparing recognizers, but they are not sufficient for predicting actual performance in operational systems. Adequate methods for the measurement of both human and recognizer performance under realistic conditions remain to be developed. Importance of speed versus accuracy will vary with the application. Speed of command entry will not always be the primary measure of effectiveness when the user is engaged in simultaneous manual tasks. For example, performance on a primary manual task may be facilitated with the use of voice on a secondary task even though that secondary task is then accomplished at a slower but still acceptable rate. Generic measures need to be developed that can be applied to task- or mission-specific events.

Operator workload is an important measure because it can be used to compare system design alternatives. Currently, there is no single, reliable method for assessing human workload in a variety of tasks (Wierwille and Connor, 1983). Although some research is ongoing in this area, an emphasis on this topic would be valuable, not only for speech-recognition applications, but for many other issues in human–system interface design. Chapter 3.5 discusses workload in detail.

Because the state of the art of knowledge regarding the human factors/ergonomics of speech-recognition system design is at present qualitative in nature, this chapter can offer few quantitative guidelines. Table 11.5.4 offers a list of design considerations for implementing speech recognition into user interfaces. These design considerations have been derived from the research reviewed in this section on speech-recognition research.

Table 11.5.4   Speech Recognition System Design Considerations

Task Selection

Expect the greatest payoffs in task performance speed and accuracy using speech recognition compared to manual input for complex information entry tasks that must be performed concurrently with other visual or manual tasks.

Base selection of speech recognition mode on an analysis of specific application task requirements.

Message Design

Conduct tradeoff studies of vocabulary size, recognition processing time (which generally increases with vocabulary size), and recognition vocabulary item unit (e.g., letters, words, phrases).

Avoid acoustically similar vocabulary items for the recognition vocabulary.

Try to use terminology for the recognition vocabulary that is familiar to the users, or, for speaker-dependent systems, let individuals choose their own words for the system functions to be controlled by speech.

Incorporate syntax subsetting into the recognition system to limit the number of alternative recognition templates that are compared to a user-spoken utterance to those that are permissible, given the utterance or utterances that have been previously recognized.

Performance Measurement

To predict operational effectiveness, measure recognition performance under actual or simulated operational conditions.

Analyze recognition errors by type (substitution, rejection, insertion, deletion) to better evaluate application vocabularies.

To estimate message recognition accuracy from word recognition accuracy, take the word error rate and raise it to the power that is equal to the number of words in a typical message. The estimated message recognition accuracy will be 1 minus the message error rate. Note that this formula does not take into account syntax branching factors due to permissible message syntax.

Measure performance of the human–machine system in terms of operationally relevant measures, system response time and accuracy, and user acceptance.

User Training

Provide training for speech recognition system users to improve the consistency of their pronunciation and microphone usage. Expect poor speech consistency from about 25% of the population at large.

Provide practice time with the system as part of user training.

Recognizer Enrollment

For speaker-dependent systems that are to be used in noise levels above 85 dB SPL or under high acceleration levels, make provision for speaker enrollment and/or adaptive recognition in the operational environment.

Present vocabulary items for enrollment in random order for initial enrollment and updating templates.

Use the visual mode rather than the speech mode for presenting prompts to the operator during enrollment.

System Design

Include voice system design considerations in system design from the beginning.

Because the severity of consequences of recognition errors varies, depending on the application, consider the relative importance of different types of errors for a particular application.

Where feasible, incorporate adaptive recognition algorithms into the operation of the system. In any case, provide for updating of templates during operation of the system.

Table 11.5.4 (*Continued*)

Consider potential conflicts between speech messages intended for the recognizer and other speech communications in the task environment.

Consider also potential conflicts between requirements for the human operator to speak to the recognition system and to other humans and to listen to other humans and other, possibly machine-generated speech messages.

For systems that do not use syntactic and pragmatic constraints, minimize message length, when possible, to improve message recognition rate. (See estimation of message recognition accuracy under *Performance Measurement*).

Provide feedback to the operator regarding recognition system status and what has been recognized. In general, within the constraints of the overall task, the more immediate this feedback, the less confusion will be created for the human operator.

Make feedback presentation compatible with task demands by selection of appropriate feedback modality and verbosity.

Provide for user correction of recognition errors.

Provide system checks for illegal utterances.

11.5.4 SPEECH GENERATION RESEARCH

Properly designed speech displays can potentially unload a user's visual system when performing visually demanding tasks. Examples of such tasks are reading technical maintenance or operations manuals while operating or repairing a system, looking through a microscope or other visual system to position one's work, reading flight charts while flying in busy airspace, checking multiple visual readouts while operating a nuclear power station, simultaneously controlling a robotic arm and multiple cameras onboard a space station or underwater, monitoring multiple vital signs displays during surgery, and editing text on a visual display. In such situations, not only is the user engaged in a visual task, but efficiency of task performance also depends on the user being able to maintain eye point of regard. Spoken messages, delivered by speech displays, carrying certain information might be more effective and result in more efficient overall task performance than if the same information were displayed visually.

The strategy for effective use of speech generation, like that for speech recognition, is threefold. Methodologies are needed for (1) identifying applications for speech generation, (2) selection of appropriate algorithms and system characteristics, and (3) integration of speech generation into the design of voice-interactive systems.

11.5.4.1 Applications

The most common approach for identifying applications for speech generation has been to select a particular human–machine system as a candidate for speech messages and to simulate a version of the system that uses speech messages. Usually, an existing problem such as high visual workload or poor performance is the basis for investigating the speech mode in place of the visual mode. Typically, an experiment is performed, using the current system as a control condition, and various measures of task performance are used to determine the relative merits of visual and speech output for the task in question. The results of such studies support decisions regarding the utility of speech displays for that particular application, but are difficult to generalize to other applications. However, they may suggest areas for more generic research and provide valuable insight into human factors issues regarding speech display design. Three speech display case studies will be reviewed briefly.

Case Study: SYNCALL

A flight simulation study, sponsored jointly by NASA Ames Research Center and American Airlines Flight Academy, evaluated the concept of a synthesized voice approach callout (SYNCALL) system for airline operations (Simpson, 1981b). This study was conducted in a training simulator using line-qualified crews. Half the approaches were flown using the current pilot-not-flying approach callouts. These are altitude annunciation and flight-path-deviation callouts made by the pilot who is not actively flying. Approaches varied in degree of difficulty. For the less difficult types of approaches, there were no differences in flight performance attributable to the SYNCALL system. But, for the most visually, manually, and cognitively demanding approaches, performance with the synthesized voice system was better than when the normal procedure of pilot-not-flying callouts was used. Numerous other observations were made. For the one approach for which SYNCALL consistently (by experimental design) made false callouts, flight performance was significantly degraded (more variable across pilots with adverse effects on the performance of some) compared to performance on the same approach flown with pilot-not-flying callouts that were correct. Overall, pilots rated the SYNCALL system as less informative, less helpful, less coordinating, but more dependable than the pilot-not-flying callouts. SYNCALL was rated more informative, more helpful, more coordinating, and more dependable than a system currently in use in airline cockpits for aiding altitude awareness, the Ground Proximity Warning System (GPWS). The study demonstrated the concept of an automatic approach callout system using synthesized speech and provided a large list of pilot-recommended refinements to the system. The importance of eliminating false information callouts was underscored. Integration of GPWS callouts and SYNCALL callouts to avoid multiple voice messes was recommended.

Case Study: I/O Modes for Aircraft Flight Data Link

A series of experiments was conducted in Link model GAT-1 and GAT-2 flight simulators to evaluate different modes of cockpit input and output devices (Hilborn, 1975). The airline pilot subjects flew the simulator with information provided by visual display, speech display, and printed paper display. They preferred visual displays for all but warning information. For warnings, they preferred synthesized speech messages. A large-letter LED display was preferred as a recall instrument for currently assigned heading, altitude, and airspeed information. In-cockpit printout was preferred for less time-critical information, which also must be remembered or referred to over a period of time after receipt. Such information included air traffic control clearances and weather.

Case Study: Voice Interactive Electronic Warning System (VIEWS)

A helicopter nap-of-the-earth (NOE) functional flight simulation included an experimental radar threat detection system that made use of synthesized speech and visual symbols to indicate the lethality and direction of various types of radar threats (Voorhees, Bucher, Huff, Simpson, and Williams, 1983). One of the important features of the system design was the voice message/visual display integration logic, which determined communication priorities in the event of multiple threats. All seven helicopter pilot subjects in the study rated the experimental system as better for threat detection and avoidance and easier to use than the current system. The current system uses tone codes and an analog display of detected threat signals. All the pilots judged the speech and visual systems to be well integrated.

A main variable in this study was voice type (Simpson, Marchionda-Frost, and Navarro, 1984). Male digitized, female digitized, and a digitized version of synthesizer-generated speech were compared. While there were no flight performance differences associated with voice type, pilots expressed a preference for a distinctive, slightly mechanical sounding voice. They reported extreme dissatisfaction with the slow speaking rate of all three digitized voices, caused by the artificial pauses that were introduced by the word concatenation method used to generate the messages. Direct synthesized speech with more natural prosodics was judged by these same pilots as preferable to both the digitized synthesized and the digitized human female speech used in the VIEWS study. The study concluded that direct synthesized speech should be used for a VIEWS type of system. The VIEWS voice and visual displays and integration logic were subsequently incorporated into an operational prototype system for field testing.

General Guidelines for Use of Speech Displays

Despite the limited generalizability of results from such application-specific research, there are some general guidelines for selecting functions for speech. These are based to a small degree on experimental data, but mostly on a combination of deductive and inductive reasoning. For example, Deathridge (1972) lists general guidelines for deciding, first, between audio and visual displays and, then, for deciding between speech and nonspeech audio displays. Reasons for using auditory (speech or nonspeech) rather than visual display include:

1. To give warning signals, because the auditory sense is omnidirectional.
2. When there are too many visual displays.
3. When information must be presented independently of head movement or body position.
4. When darkness limits or precludes vision.
5. For conditions of anoxia, because of the greater resistance of auditory sensitivity to anoxia compared to visual sensitivity.

Reasons given by Deathridge for using speech rather than nonspeech are:

1. Flexibility.
2. Identification of a message source.
3. For listeners without special training in coded signals.

4. Rapid, two-way information exchanges are required.
5. The message deals with a future time, requiring preparation.
6. Situations of stress, which might cause the operator to forget the meaning of coded signals.

The state of the art in selecting voice functions has not really progressed beyond this philosophical stage.

Simpson (1983a) and Williges and Williges (1982) independently added the same two items to the inventory:

1. Spoken information should be highly reliable;
2. Spoken information should be intended for use in the immediate future, owing to its poor retention in short-term memory.

To these we add a corollary to Deathridge's third reason (above) for using speech:

3. Use speech rather than nonspeech to minimize information processing requirements of the listener by eliminating the need for decoding nonspeech signals.

11.5.4.2 Selection of Functions for Speech Displays

It is important to select the best functions for speech displays. These functions can be classified according to the speech acts (Searle, 1969) they represent. Simpson (1985) lists five basic types of information (i.e., speech acts) for which speech displays may be useful. These basic information types transcend specific applications. They are warnings, advisories, responses to user queries, feedback from control inputs, and commands. A sixth class, not listed, is spoken prompts from the system to the user to elicit user action such as data entry. It is unlikely that any particular type of speech act will be amenable to speech output in all situations. Rather, the combination of task and user characteristics associated with a particular application will dictate the applicability of speech displays.

Warnings

Of the six types of speech acts, warnings have received the majority of attention in speech display research. Results from a series of studies [summarized in Simpson and Navarro (1984)] suggest that voice warnings should be worded as short phrases containing a minimum of four or five syllables to minimize listener attention needed for what they call "perceptual copying" and to ensure high message intelligibility for unexpected messages in the presence of competing noise and speech.

The voice used for cockpit displays needs to be distinctive (Brown et al., 1968; Simpson and Williams, 1980; U.S. Dept. of Defense, 1981) in order to stand out against other human speech. Female voice for environments where male voices prevail has frequently been suggested for warnings because of its unique voice quality [e.g., Brown et al. (1968)], but there are few such environments today. There is also an accumulation of reports from pilots who have served in speech display flight simulation studies that the voice ought not to sound too human (Cotton et al., 1983; Simpson, 1981b; Voorhees et al., 1983) lest it be confused with human speech, such as radio or intercom communications. The underlying concept here is that a machine should have a machine voice as a cue to its identity when it speaks.

As discussed in the case study on VIEWS (Simpson et al., 1984; Voorhees, et al., 1983) proper speaking rate and prosodics are important features for voice warnings, regardless of voice type. A study of helicopter pilot preferences for synthesized speech speaking rate for warnings presented warnings to pilots at three rates (123, 156, and 178 words per minute) while they were engaged in an attention-demanding "flying" task (Simpson and Marchionda-Frost, 1984). While neither flying task performance nor message comprehension was affected by speaking rate, pilots preferred the 156 word per minute rate.

A series of studies has addressed system response time (the interval from onset of signal to user's first response) for synthesized voice warnings with and without preceding alerting tones or words. First (Simpson and Williams, 1980), it was found that an alerting tone preceding a synthesized voice warning increased system response time, but that lengthening message wording with an extra word to add semantic context did not increase system response time. A subsequent study (Hakkinen and Williges, 1984) replicated these results but also found that when a synthesized voice was used for multiple functions with the alerting tone as a variable only for warning messages, then an alerting tone used exclusively before warnings improved detection of urgent messages without increasing system response time to these urgent messages. Studies of voice warning prefixes [Bucher, Karl, Voorhees, and Werner (1984); Bucher et al. reported in Simpson and Navarro (1984)] found no difference in system response time as a function of prefix type, despite differences in actual length of the different prefixes (tone, neutral word, one of three semantic cue words). These studies support the possibility that synthesized speech is somehow distinctive, compared to human speech, and can perform the alerting function concurrently with the information transfer function. The physical correlates of this distinctiveness remain to be determined experimentally.

Prompts

Prompts by the system to the user have been studied by Mountford and her colleagues (Mountford, North, Metz, and Graffunder, 1982). They studied different levels of verbosity for voice messages used as feedback and prompts to users of a simulated voice data entry system for flight planning and navigation. They found that short dialogues with little prompting and terse feedback provided the best data entry performance. Future research may well find that the tradeoff between verbosity and time spent to complete voice transactions depends on the criticality of an error. The more catastrophic the effects of an error, the more willing users may be to invest the time required for more verbose prompts and feedback. More work is needed in this area.

Prompts can also be used in spoken menu-driven systems. Spoken prompts were used in the design of a voice mail system for the 1984 Olympics (Gould and Boies, 1984). User feedback during testing with preliminary versions of the system indicated that the number of spoken prompts for any menu should be limited to three, probably due to limits on human short-term memory for spoken items. These results apply to computer-naive users, but might extend to sophisticated users as well, to be determined by further research.

Feedback

Feedback in response to discrete user control inputs is frequently mentioned as a function for speech displays. Relevant research was discussed above in the section on speech recognition. In passing, it should be noted that feedback can be provided by prompts [cf. the Mountford et al. (1982) study just discussed]. That is, if the system prompts the user for a reasonable next data entry or control input, the user will assume that the system correctly received the previous input. The real-world conditions under which the user can safely make such assumptions, however, need to be understood. The type of feedback employed, in terms of modality, information conveyed, and verbosity of linguistic encoding, will depend on the time criticality of the control input and the severity of the consequences of a speech-recognition error.

Responses to User Queries

User queries were studied in a computer-graphic simulation of NOE helicopter flight (Voorhees, Marchionda, and Atchison, 1982). Subjects could ask the helicopter to state airspeed, torque, and altitude as they attempted to fly their simulated craft through a maze on a visual display. Maze flying performance was better when subjects used voice queries and received synthesized voice responses than when they had to obtain this information from either a head up display or conventional dial gauges.

Advisories

The utility of speech to provide advisories may depend on the other functions for which speech is being employed in a particular application. When advisories were given in conjunction with voice warnings, it was seen previously (Hakkinen and Williges, 1984) that warning detection suffered, unless an alerting cue was also used. Studies of civilian and military pilot preferences for warning system design (Brown et al., 1968; Cooper, 1977; Cotton et al., 1983; Williams and Simpson, 1976) have found repeatedly that pilots wanted speech reserved for only the most critical, that is, warning, information. They prefer to receive advisories and less critical warnings visually. This preference may stem from the fact that pilots can decide when to attend to a visual display. A spoken advisory message, on the other hand, intrudes on the pilot's ongoing cognitive processing because the pilot must attend to it while it is speaking. Pilots' preference for the visually presented advisories may reflect a reluctance to be interrupted during periods of high workload for anything but urgent information. For example, a survey of pilot opinion on airline flight deck warning system design found that pilots wanted noncritical alerts inhibited during high workload periods, such as takeoff and landing (Veitengruber, 1978).

For applications in environments other than aircraft cockpits and in situations that do not include speech messages for the warning function, spoken advisories may be useful. Also, pilots or other operators who are sensitive to unwanted spoken interruptions, might nonetheless find some utility in a system that announced advisory information only on request by the user. The utility and user acceptance of spoken advisories clearly require further research.

Commands

There is some research and discussion in the literature on the advisability of giving commands by automatic speech generation. Simpson and Williams (1975) argue that great caution should be exercised in the use of commands, at least in the aircraft cockpit environment, because pilots are reluctant to follow a command without knowing the reason for it. In partial support of this argument, a study by DuBord (1982) reported by Palmer and Ellis (1983) found that giving pilots a visual display of traffic situation information reduced their response time to a visual collision avoidance command, compared to when they were given the command without benefit of the traffic situation display. A similar effect is likely for spoken commands. For users other than pilots, speech commands issued as instructions, in non-time-critical situations or in conjunction with advisories, could perhaps be useful in a variety of applications.

Alternative Warning System Design Philosophies

The cockpit voice warning system design philosophy presented here is characterized by the following principles: (1) use speech only for the most time-critical warnings; (2) use a distinctive voice quality for spoken warnings; (3) do not include any nonspeech alerting signal before a spoken warning; (4) do not repeat a spoken warning until a period of time has passed during which the pilot has not corrected the problem; and (5) use a minimum of four syllables for any voice warning message.

Recommendations in Chapter 5.2 by Sorkin [after Patterson (1982)] were derived from the same data base of research findings but, taken individually, contradict several of the principles just listed. Patterson recommends (1) a keyword format for time-critical warnings (with no minimum number of syllables specified); (2) a short sentence or phrase format for less critical warnings; (3) an aural alerting signal before and after all voice warnings; and (4) amplitude reduction but not silencing of the aural signal and speech message after the initial presentation. The purpose of the amplitude reduction is to reduce pilot annoyance and cognitive disruption. The major differences between the two philosophies can be attributed to differences in basic design features. The present authors recommend using a distinctive, highly intelligible but machine-quality voice for cockpit warnings and limiting the use of voice to time-critical warnings only. Given these features, an alerting prefix signal is contraindicated. Patterson does not specify a distinctive voice, nor does he limit speech displays to the annunciation of time-critical warnings only. Given these two features, an alerting prefix signal may be required.

The differences between recommended message formats can be explained in terms of different assumptions about the duration of a speech message. Patterson bases his use of relatively shorter (4.8 sec) situation-specific non speech sounds to initiate a warning signal on his assumption that a speech message requires an unacceptably long period of 6–7 sec for presentation. The voice message for Patterson is a backup to the aural signal and may additionally provide advisory information. The present authors assume that four to eight syllable warning messages can be spoken in 1.5–3 sec. This shorter message duration permits more rapid information transfer to the pilot via speech than is assumed by Patterson and causes less interference with other cockpit communications.

When these two design philosophies are thus analyzed, their differences are minimized. Each one taken as a whole is consistent with its basic assumptions and with the experimental literature. Finally, the Patterson system can be characterized as treating the nonspeech aural signal as primary and the voice signal as a backup. The system proposed by the present authors treats speech as primary and makes use of nonspeech signals in a secondary role. For different types of flight operations, one or the other system may be more effective than the other.

Simulation of Human Communications

The speech acts discussed above would be performed by machines speaking qua machines to human operators. Another important speech display application is the simulation of human speech communications, for example, for training purposes to eliminate the need for human speakers playing a role. For example, speech generation has been proposed and evaluated for training systems for precision approach radar controllers and for air intercept controllers (Breaux, 1977; Grady, 1982).

Comparative Display Modes

Comparative speech and visual display research has addressed user preferences, response time, accuracy, and task accomplishment for speech versus visual displays and for speech combined with visual displays, for various speech acts.

Early voice warning research using taped voice messages found that pilot response time to voice warnings is faster than to visual warning displays (Lilleboe, 1963) and that a voice-warning-augmented visual display results in faster responses to emergencies than does a tone-augmented display (Kemmerling, Geiselhart, Thornburn, and Cronburg, 1969).

Another difference between the visual and speech mode may be users' tolerance for and ability to perform their task when presented with false information. A study of airline pilots' preferences for design of cockpit warning systems found that pilots expressed less tolerance for false speech messages than for false visual messages (Williams and Simpson, 1976). And, flight performance was degraded in the presence of false voice warnings (Simpson, 1981b) compared to when the voice warnings gave accurate warning information.

Pilot preference data also suggest that, in the cockpit, the speech mode ought to be reserved for urgent, time-critical messages. As previously noted (Hilborn, 1975) airline pilots preferred visual displays for all but warning information. For warnings, they preferred speech messages. Similarly, other researchers (Brown, et al., 1968; Cooper, 1977; Cotton, et al, 1983; Williams and Simpson, 1976), report that pilots want the speech mode used for warnings but not advisories. Further research is needed to determine if this preference pattern holds for other user populations, for example, automobile drivers, power plant operators, computer users.

In some situations, users may object to the speech mode for certain types of information. For example, a recent study (Stern, 1984) compared speech and visual displays for prompting and giving error messages to users of an automated teller machine. While there were no performance differences between text and speech displays, subjects did not like spoken error messages because other customers could hear them.

For sensory-handicapped users, the selection of visual or speech displays will depend on the handicap. For blind users, the challenge is to design the speech interface in ways that facilitate performance in the speech mode of the functions that are normally better accomplished using visual displays.

11.5.4.3   Selecting Speech Generation System Characteristics

Research on appropriate speech generation systems for particular applications has been done at two levels. The performance of speech generation algorithms has been assessed as a function of multiple factors that influence intelligibility. Also, human–system performance in simulations has been assessed to determine what benefits may derive from using speech displays.

Speech Generation-Algorithm Performance

Intelligibility is influenced by the physical characteristics of the speech signal and by the context in which the speech is spoken. In addition to intelligibility, comprehension and human information retention and retrieval in the speech mode must be measured.

Operational Intelligibility

A recent review of research on intelligibility of computer-generated speech (Simpson and Navarro, 1984) defines three types of context that interact with the speech signal to produce what the authors call the "operational intelligibility" of speech. The operational intelligibility of a particular algorithm is the intelligibility of its speech in a particular set of physical, pragmatic, and linguistic contexts, and it can differ considerably from basic phoneme intelligibility. Figure 11.5.4 depicts the four major factors, that is, the physical signal and the three types of context, that contribute to operational intelligibility. The physical speech signal can vary with respect to sex and voice characteristics of the speaker, speaking rate, fundamental frequency, amplitude, accuracy of pronunciation and prosodics, accent, and dialect, among other parameters. The physical context includes aspects of the physical environment such as noise, other audio signals, vibration, and acceleration forces. The pragmatic context is essentially the real-world situation in which the message is spoken. It includes the current events, the ongoing

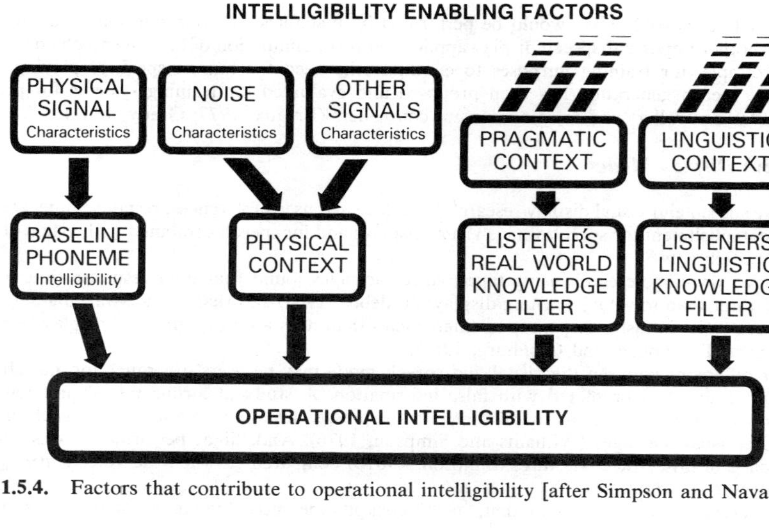

INTELLIGIBILITY ENABLING FACTORS

Fig. 11.5.4.   Factors that contribute to operational intelligibility [after Simpson and Navarro (1984)].

task, the time and place, past events, and logically possible future events. The effect of the pragmatic context will be filtered by the listener's knowledge of that pragmatic context. The linguistic context of a speech signal influences intelligibility by providing cues to the listener that limit the possible interpretations of the incoming speech signal. This limit on possible interpretations is a complex type of closed response set. It has long been known that as size of response set decreases, intelligibility of human speech heard in noise increases, all other factors held constant (Miller, Heise, and Lichten, 1951). Linguistic context limits the size of the response set in more complex ways than simply limiting message set size. This is due to interactions among the constraints provided by the different levels of linguistic encoding. Simpson and Williams (1975) list these levels and provide references to the literature on the effects of various types of linguistic context on human speech intelligibility.

The effects of linguistic context are filtered by the listener's linguistic knowledge of the language being spoken. For example, someone just learning the Swedish language will not be able to utilize linguistic context to perceive Swedish speech as effectively as a native speaker of Swedish. At the lexical level, familiarity with a particular vocabulary and phraseology can facilitate speech perception.

Often the effect of factors that enable intelligibility is stronger for synthesized than for human speech (Nye and Gaitenby, 1974; Simpson, 1975). Simpson and Navarro (1984) report that with sufficient assistance from intelligibility-enabling factors, synthesized speech from commercially available devices has been found to be 100% intelligible; without such assistance it has been measured as low as 19%. Factors included as physical characteristics of the speech signal are fundamental frequency, speech rate, prosodics, intonation, learnability of the speech accent, voice type, and phonetic accuracy of the generated speech. Under physical context, the authors review research on effects of background noise and competing speech. Pragmatic context factors include listener familiarity with the speech accent, with the phraseology and vocabulary, and with the real-world situation in which the messages will be spoken. Linguistic context factors include semantic and syntactic context and number of syllables.

The relative importance of the three types of context varies. Under ideal listening conditions, characterized by high signal-to-noise ratio, no competing speech or other audio signals, and listeners familiar with the accent of the machine speech, sentence intelligibility of synthesized speech is 99–100%. Reducing the peak signal-to-noise ratio to levels in the range of −10 to −23 dB and leaving the other factors constant has shown little or no detriment in operational intelligibility, at least for aircraft cockpit messages (Simpson, 1984). High intelligibility (99–100%) has been obtained for short, familiar phrases, heard in simulated cockpit noise, using both LPC-encoded digitized speech and synthesis-by-rule speech.

Table 11.5.5 summarizes the intelligibility results of studies reviewed in Simpson and Navarro (1984). The reader is cautioned, however, to avoid direct comparisons between studies since in most cases they differ by more than one and often by several variables.

Intelligibility of digitized speech varies as a function of speaker sex. LPC-encoded and adaptive-predictive-coding (APC) encoded female speech is more susceptible to bit errors that might be expected during transmission than male speech encoded using the same algorithms. The difference is consistent across a wide range of bit error rates (Smith, 1983). Similar comparisons are needed between female-synthesized and male-synthesized speech.

When linguistic context or pragmatic context, rather than signal-to-noise ratio, is reduced, substantial degradation of intelligibility occurs (Simpson and Navarro, 1984). For this reason, Simpson and Navarro recommend a minimum of four syllables for warnings or other unexpected speech messages.

The excellent intelligibility reported by Simpson and Navarro (1984) was achieved at the expense of phonetic hand editing by experts in speech acoustics. Intelligibility of speech generated by text-to-speech algorithms can be poorer and depends on the particular algorithm. For example, intelligibility of Harvard Psycho-Acoustic Laboratory (PAL) sentences (Egan, 1948) was 93.2% when spoken by one system (Pisoni and Hunnicutt, 1980) compared to 87% for PAL sentences spoken by another system (Nusbaum and Schwab, 1983).

The main deficiencies of text-to-speech algorithms are (1) phonetic errors of pronunciation for words that are exceptions to English spelling-to-sound correspondences and (2) inadequate rules for generating correct word stress and sentence intonation for some syntactic, semantic, and discourse structures. Until these deficiencies are corrected, the need for hand editing of individual speech messages (noted previously) will remain.

Comprehension

Although synthesized speech can be 100% intelligible to listeners familiar with its accent, the phraseology, and the pragmatically possible messages, further research is needed on comprehension of synthesized speech messages compared to human speech. Luce, Feustel, and Pisoni (1983) found deficiencies in speech-processing capacity for speech synthesized by one text-to-speech system compared to human speech, when they loaded their subjects with additional short-term-memory recall tasks. They interpret these results to mean that synthesized speech places increased demands on encoding and/or rehearsal processes in short-term memory and argue that synthesized speech ought not to be used for cockpit displays. However, their subjects were unfamiliar with the accent of the synthesizer prior to the experi-

Table 11.5.5 Results of Some Operational Intelligibility Studies

Intelligibility Enabling Factors	Experimental Conditions				
Physical signal	Synthesized speech from phonetic segments	Synthesized from phonetic segments versus LPC-encoded version of synthesized	Synthesized from phonetic segments	Synthesized speech from phonetic segments versus high-quality human speech recording	Synthesized from phonetic segments
Physical context	Helicopter noise at S/N ratio of −23 dB	Helicopter noise at S/N ratio −23 dB	Widebody jet noise at S/N ratio −10 dB	No noise	Competing human speech recorded from radio weather broadcast; signal-to-competing speech ratio +8 dB
Pragmatic context	Flight-relevant messages heard while "flying" (threat warnings)	None	Flight-relevant messages (altitude callouts)	Flight-relevant messages (warnings and air traffic control clearances) versus everyday messages	Flight relevant messages (cockpit warnings)
				Subjects: unfamiliar with messages; pilots versus nonpilots	Subjects: pilots unfamiliar with messages
Linguistic context	Phrases of four to eight words in length	Minimal: messages were real words (PB words)	Considerable: messages were sentences; altitudes in flight phraseology	Considerable: messages were sentences	Two keywords: monosyllabic versus polysyllabic; versus two keywords in sentence format: monosyllabic versus polysyllabic

Intelligibility	98.7–99.8%	Synthetic: 44% LPC-encoded: 19%	99.7%	Pilots
				  flight message
				    synthetic: 96%
				    human: >99%
				  everyday message
				    synthetic: 93%
				    human: >99%
				Nonpilots
				  flight message
				    synthetic: 86%
				    human: 96%
				  everyday message
				    synthetic: 93%
				    human: >99%
					Polysyllabic
					  keywords: 94%
					  sentences: 93%
					Monosyllabic
					  keywords: 78%
					  sentences: 96%

Reference	Simpson and Marchionda-Frost (1984)	Simpson and Navarro (1984)	Simpson (1980)	Simpson (1975)	Simpson (1976)

Source: Simpson and Navarro (1984).

ment. The applicability of their findings to pilots' comprehension of familiar messages encoded with sentence-level linguistic context and spoken in a familiar pragmatic context remains to be determined. There is experimental evidence that pilots can store information presented by synthesized speech and later retrieve that information while flying a flight simulator under high workload (Simpson and Marchionda-Frost, 1984). Little is known about how efficiently information in synthesized speech messages can be recalled and under what circumstances listeners will become overloaded in the speech channel. Statements about synthesized speech system performance in general must be made cautiously with attention to the particular conditions under which the results were obtained.

Voice Characteristics

Desirable voice characteristics are application dependent. When an inanimate system is speaking qua machine to the user, a machine-voice quality is preferred by some user populations (e.g., pilots), as previously discussed. On the other hand, when a system simulates human communications, as in an ATC training system, a natural-sounding voice, using digitized human speech, is preferred (Cotton and McCauley, 1983). As with machine-sounding speech, it is important to incorporate natural prosodics into the generation process because the temporal characteristics of the speech may influence the user's comprehension of the messages and performance of his or her task.

Voices can also be varied with respect to pitch and apparent sex of voice. Voice pitch has been suggested for indicating the urgency of a message with higher pitch signaling greater urgency (Simpson and Marchionda-Frost, 1984). Recent research (Brokx and Nooteboom, 1982) also suggests that differences in voice pitch can help listeners track one or the other of two concurrent messages. The extent to which users can deal with multiple messages needs to be studied and may be a function of the degree of difference among voice types heard on the job.

11.5.4.4   Human–System Performance Measurement

Operational Relevance

The human factors of speech generation system performance extend far beyond effects on speech intelligibility and recognition accuracy. Message comprehension, human storage and retrieval of information presented in the voice mode, and interactions between speech comprehension and human performance of other concurrent tasks are equally important. Such measures should be "operationally relevant" to the task for which a voice display is used (Simpson, 1981b). An operationally relevant measure of system performance is one that provides users and designers with information about how the system will perform in terms that are meaningful to the operator. A measure of the effectiveness of a navigation computer with voice controls and displays might be the time it took a pilot flying in turbulence in a busy ATC environment to change a waypoint by means of voice and by means of manual keys and a visual display.

For example, the SYNCALL case study previously described measured flight performance in terms of percentage time out of airline operational tolerance for flight parameters. Measures are also needed that will predict the costs and benefits of using speech technology in terms of time saved, more efficient utilization of personnel and equipment, and safer operations. Such measures may follow a generic format, but will be application specific in content.

System Response Time

Another operationally relevant measure is system response time. The fact that a speech message takes time to be delivered gives particular importance to what Simpson and Williams (1980) have called "system response time." The authors defined system response time for voice warnings as the time interval starting with the onset of a warning signal and continuing until the listener has decided upon and initiated his or her first action. System response time thus includes detection, perceptual copying, comprehension, storage, retrieval, and decision making. System response time, rather than simple reaction time or human response latency, is a critical variable for voice warning display systems.

System response time alone cannot always be the determining factor. Response accuracy must also be considered. The particular response-time–accuracy tradeoffs that are made will depend on the size of the time window available for the user to respond and the consequences of an inaccurate response.

Other examples of operationally relevant measures are those used in the VIEWS helicopter simulation study (Voorhees et al., 1983). They included survivability, that is, the percentage of craft shot down while flying the threat-evasion mission, the number of times pilots' craft were "painted" (detected) by enemy radar, and number of tree strikes, among others. Such "event-counting" measures will be specific to the particular application being assessed. Selection of the types of events to be counted for a particular application should be guided by the operational relevance of those events to the users' goals.

Another important measure is user acceptance. Different populations may vary in their acceptance of speech for particular functions. For example, acceptance by the general public of talking automobile displays may differ from blind users' acceptance of speech displays for computer terminals.

11.5.4.5 Future Speech-Generation Research

Research directed toward speech displays in general will support effective design for voice-interactive controls and displays. Specific issues relevant to the design of integrated voice I/O systems are selection of voice type (human or machine-sounding; male, or female); message wording and syntax as a function of speech act; assignment of priorities to functionally different speech display messages; and methods for integrating voice and visual messages when they present the same information. Research on speech display aspects of dialog design must also deal with the issue of how to handle concurrent speech messages. Two cases must be handled: (1) user speaking to speech recognition device while speech display, radio, or intercom is enunciating a message, and (2) triggering more than one speech display message at a time. General design principles are not yet known for assigning speech message priorities and the logic that will decide whether to interrupt and ongoing message and that will select the next message to be spoken.

Speech displays also require improvements to text-to-speech generation algorithms to eliminate the need for hand editing of speech data. The relative importance of various types of phonetic and prosodic accuracy for synthesized speech intelligibility, learnability, and comprehension is another area that requires further investigation. The degradation of operational intelligibility due to inaccurate vowels, consonants, phoneme transitions, word stress, and prosodics has not been measured systematically. Because of missing perceptual cues in synthesized speech and, to some extent, in human speech digitized at low bit rates, the fidelity of audio transmission systems may be more critical than it is for human speech. Just how much redundancy and what type is optimal (syntactic, semantic, phonetic) has not been determined experimentally for all types of speech acts. Also, for listener populations with possible high-frequency hearing loss, computer speech perception may present special problems just because it does not contain all the perceptual cues of human speech. Specifications for intelligibility of speech to be used by such groups or speech to be used in high noise environments must take this into account.

As for speech recognition, the type of design guidance that can be derived from human factors research to date is mostly qualitative in nature with a few, highly qualified, quantitative items. Table 11.5.6 lists design considerations for implementing speech-generation systems and serves as a summary of the research findings reviewed in this section on speech generation.

11.5.5 SYSTEM INTEGRATION

System integration, the third level of the research strategy, must consider the research requirements for speech recognition and generation as synergistic technologies. The human visual and manual modalities are commonly associated with perceptual and motor (input/output) characteristics. Similarly, the speech modality has identifiable human speech perception and production characteristics that will become the basis (either unwittingly or by design) for the interface characteristics of speech I/O systems.

The critical issues in human factors integration are task design specifically for the speech modality and human–system dialog design.

Task Design

Applications using the speech modality must be designed around the characteristics of speech. Certain unique features of the speech mode preclude a one-to-one mapping of individual manual controls to speech controls and of visual display elements to speech display messages (Cotton and McCauley, 1983; Simpson, 1984; Williges and Williges, 1982).

Speech is a discrete, single-channel, omnidirectional, well-known semantically sophisticated encoding system for the transmission of information. It commands the user's attention and should not be allowed to deliver false information. Used for control of systems, speech can, if properly implemented, reduce the need for the user to learn computer-programming like languages and can give an alternative to manual input systems. Speech messages require time to be spoken and may be misunderstood by human or machine "listeners" in the presence of other, competing voice messages, aural signals, or noise. These constraints imposed by the speech mode (time and single channel) must be considered in any implementation of speech displays and controls. Furthermore, certain features of speech constrain the way in which it can be used in human–machine systems. Speech may not always provide the most rapid means of interacting with the system. The time required for an operator to execute a speech command is strongly influenced by variables such as vocabulary selection, syntax design, and especially, dialog design.

The receiver of a speech message, whether human or machine, has great difficulty processing more than one message at a time, with the result that speech is a single-channel code in two senses: neither

Table 11.5.6   Speech Generation System Design Guidelines

Task selection

For aircraft cockpit applications, limit the use of unexpected spoken messages to warnings for only the most time-critical situations. Limit any other functions for speech displays to information that is requested by the pilot. Note that the pilot could *request* advisories for particular operations or for particular flight or system parameters.

Consider use of speech displays when the user is visually occupied in a manner that would place unacceptable additional demands on visual workload capacity. Note that the classification of additional demands as unacceptable is task-dependent.

Use the speech mode to announce discrete events, not to present readings of continuously changing analog parameters.

Do not use the speech mode to present long, detailed messages or to present information that must be remembered by the user without benefit of a concurrent visual display.

User training

Provide user training with speech accent.

Speech characteristics

Select voice type according to the source of the speech messages: for machine messages to the operator, use machine-sounding voice quality; when simulating human speech, use human-sounding voice quality.

For the simulation of some human speech, it may be desirable to simulate regional accents, sex, and age of speaker.

Regardless of voice type, use the best possible approximation to natural prosodics.

For warning messages, use a speaking rate of approximately 150 words per minute. A slower rate may be desirable for training listeners unfamiliar with the speech accent or for other speech functions. Pending further research, the best rate for a given application will have to be determined experimentally.

When machine-quality voice is used exclusively for warnings, do not put any alerting nonspeech sound before the speech warning message.

When machine-quality voice is used for warnings and for other functions, for example, advisories, responses to user queries, and so forth, incorporate an alerting characteristic into the voice warnings. Possible alerting features may include higher voice pitch, alerting speech or nonspeech prefixes, or other features that make the warning message distinctive and can be shown to increase detectability without also increasing human–system response time.

Message design

For warning messages use a minimum of four syllables to provide sufficient linguistic context for warning comprehension after first ennunciation of the message.

Make message content appropriate for the task, and use terminology that is familiar to the users.

Tailor the verbosity of spoken prompts and feedback to the time and error criticality of the task.

Performance measurement

Measure intelligibility in the operational environment or in a simulation of it.

Measure performance of the human–machine system in terms of operationally relevant measures, system response time and accuracy, and user acceptance.

System design

Do not present unreliable information in the voice mode.

Consider conflicts between multiple voice messages and between listening to voice messages and speaking voice messages.

When delivering time-critical information by voice, as in warnings, incorporate a priority system to order concurrently triggered voice messages in order to present the most critical first.

For warning messages, provide a user cancel capability once the message has been presented.

For warning messages, repeat the message after an appropriate time interval (see next guideline) only if the condition that triggered the warning message is still true.

For warning messages, the length of time to wait before presenting the same message again should depend on the severity of the consequences of the user not correcting the problem.

For spoken menus, without concurrent visual display, limit number of menu items to three.

humans nor current machines can talk and listen accurately at the same time, and both have great difficulty processing more than one speech message at a time. One implication of this constraint is that speech commands cannot be allowed to interfere with, or suffer interference from, other speech messages within the system.

Speech messages have a transitory existence, unless they are recorded for later playback. Human memory limits may make it difficult for the operators to remember their location in the command structure of a recognition system without the aid of feedback messages and prompts. The task conditions under which this holds true require further investigation, building on previous findings (Mountford et al., 1982).

In general, current speech-recognition technology requires a vocabulary that consists of acoustically distinct words. Vocabularies and syntax also must be constrained to be compatible with current recognition technology. These limitations of the technology may be reduced by degrees in the future, but cannot be expected to disappear without major advances in fundamental understanding of human speech variability and incorporation of this knowledge into recognition algorithms.

Irrespective of future advances in recognition technology, human performance limitations will dictate vocabulary and syntax constraints. To minimize human cognitive load and the time required to issue speech commands, the number of words in each command should be small. More information is needed on human memory for constrained verbal material and on the effects of such constraints on system performance. Information about the effects of harsh environments and stress on verbal versus motor memory and performance would be particularly relevant. Research in this area would lead to guidelines for establishing vocabularies that are flexible, are easy to remember, reduce acoustic confusion, and minimize awkward speech stylization. Similar guidelines are needed for developing formal grammatical rules that facilitate recognition without placing undue constraints on the user.

Human–System Dialog Design

Careful design of all the interchanges between the human and the system, not just the speech interchanges, will have major effects on the overall system performance. There are at least two subsets of dialog design—the dialog between the user and the speech system and the dialog between the user and all the subsystems under his or her control.

The human–machine interchanges, that is, dialog, must be designed with regard to the total set of control and display options for all subsystems. Mission and task scenarios will have to be analyzed for speech and other audio load, and the likelihood of concurrent interfering speech messages. The properties of potential functions to be controlled by speech must be assessed, along with the priorities of all speech messages within the system. Voice commands and displays will have to be applied in ways that complement rather than conflict with other controls and displays. Future research and development efforts should address these issues.

To improve system throughput with speech, it is essential to design a speech system dialog that facilitates rapid information transfer between human and machine. The dialog design also should minimize the potential for error and the subsequent time required for error correction by the user. Not only the speech commands, but also dialog elements such as prompts, system feedback, and query responses must be carefully designed and a timeline of the total dialog evaluated. The desired type and amount of linguistic redundancy for a particular application should be determined experimentally. Syntax design should be viewed as an integral part of speech system design, rather than simply a technique for improving the performance of a marginal recognition or generation system.

Possibly the error rates obtained with current systems can be reduced if system designers provide aids to the user such as tonal prompts for cadence, menus of acceptable entries, consistent feedback, and convenient error correction commands. The best format for these dialog elements should be determined by further research.

New techniques are needed to capitalize on syntactic and semantic constraints in the dialog. These techniques would improve automatic error detection and correction, thereby increasing recognition accuracy and reducing the user's burden of detecting and correcting errors.

At a higher level of dialog design, speech controls and displays need to be carefully integrated into the total control and display system in order to preclude overloading the speech channel (Simpson, 1984). Certain types of information may be better processed if presented via speech; others may require pictorial, textual, spatial, or other forms of information representation. Some basic research has addressed the issue of task/modality compatibility (Wickens et al., 1983) and provides evidence that speech is a better communications mode for some types of tasks, compared to manual input and visual output. When subjects performed two tasks simultaneously—one spatial and the other verbal—spatial task performance was better when the verbal task was accomplished using speech recognition and generation compared to the condition in which both tasks had to compete for the manual and visual channels. This basic research, however, has not involved voice-interactive dialogs. More work is needed on these compatibility issues, using realistic tasks, to support decisions about selecting appropriate tasks for speech-interactive systems with the objective of reducing operator workload. Successful speech system performance for a particular task will not guarantee successful performance of the total application system. Basic limitations of human memory and information processing must be accounted for in the design of any human–machine interface, and especially for those using speech.

Many of the problems of today's complex control and display systems, from the operator's viewpoint, may be solvable by better design at the overall system level. Speech controls and displays may play a role in those solutions, but this can be determined only after considerable analysis or simulation research to compare speech to alternative modes of control and display.

11.5.6 OVERALL DIRECTIONS FOR FUTURE RESEARCH

Human factors research on the design of integrated voice systems is limited and reports are spread among conference proceedings, government technical reports, and journal articles [see, for example, Cotton and McCauley (1983); McCauley (1984); Pallett (1982); and Simpson (1984)]. While several new design guidelines have been suggested in these reviews, the standard references used by system design engineers such as U.S. Department of Defense (1981), Van Cott and Kinkade (1972), and Woodson (1981) do not incorporate the new knowledge summarized by these reviews.

If the benefits of speech technology are to be realized, a major effort in human factors research will be needed at many levels of integrated system design: task selection; determination of task-specific recognition and generation system performance requirements; human factors integration to incorporate the speech modality; speech controls and display design; task environment effects; and system performance assessment.

A substantial effort in human factors research is needed to develop procedures for selecting appropriate tasks for the voice mode and for integrating voice interaction into the total system design. There is no single area that can be chosen for particular emphasis. However, all of the recommended directions for research should emphasize the total context, that is, the integrated system in which the speech I/O is to be used.

11.5.7 CONCLUSIONS

While the human factors literature includes research that supports certain principles of speech system design, this knowledge has not yet been formulated as design guidelines. Human factors methodology *is* sufficiently developed to permit comparison of task-specific speech systems experimentally but does not yet have the tools required for the generation of generic speech system design guidelines. For the near term, simulation of speech system capabilities in conjunction with the development of improved system performance measures should be a productive methodology for accomplishing this work.

Speech generation algorithms may seem to be more advanced than speech-recognition algorithms. Reasonably intelligible text-to-speech from standard English spelling is available commercially. The recognition counterpart, speech-to-text, that is, machine conversion of human speech to correctly spelled and punctuated text, will not be available commercially in the foreseeable future and is limited to highly constrained laboratory systems that operate on utterances composed from a finite vocabulary with a constrained syntax and a small set of known speakers. Another interpretation of this discrepancy is that it merely illustrates the human's great superiority over machine-processed algorithms when it comes to dealing with variability in the speech signal. Humans quickly learn the strange accent of computer-generated speech, thereby compensating for the deficiencies of the algorithm that generates it. On the other hand, when humans speak to current machines, they must eliminate as much as possible of the normal variability in their speech in order to provide the recognition algorithm with as little variability as possible in the input signal. Figure 11.5.5 illustrates what could be called "operational recognition." The figure obviously is derived from Figure 11.5.4 on operational intelligibility.

Currently, operational recognition makes only rudimentary use of pragmatic and linguistic context. As knowledge and understanding increase of the systematic linguistic and pragmatic variability in speech, and as this knowledge is incorporated into computer-speech-system algorithms, then recognition algorithms can be expected to perform well over a wider range of speaking contexts, and generation algorithms can be expected to provide speech that contains additional cues for the human listener. Such technology advances will enlarge the overlap between tasks appropriate for the speech mode and speech systems with characteristics that match the task requirements. However, the basic design issues discussed here will apply regardless of the state of speech technology.

For the near term, the current recognition algorithms appear adequate for use in environments characterized by low to moderate noise (up to 85 dB SPL) for applications that require small vocabularies and that do not place the user under severe task-induced stress. Advances in techniques for dealing with background noise are reducing the impact of this source of variability in the utterances to be recognized. However, in the case of stress-induced variability, the obtainable recognition accuracy and the limits of acceptable recognition algorithm performance when the user is under stress are not known. Therefore, great caution must be exercised with current technology for stress-inducing applications.

Speech generation algorithms, on the other hand, have demonstrated acceptable performance under

RECOGNITION ENABLING FACTORS

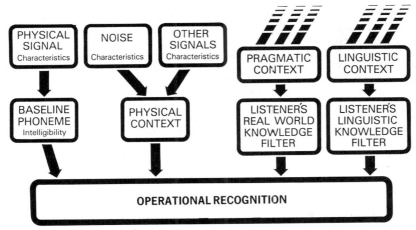

Fig. 11.5.5. Factors that contribute to operational recognition [from Simpson, in Moore, Ruth, and Simpson (1984)].

conditions of severe noise and high workload. This technology is ready to be applied appropriately, with careful attention to the human factors integration issues discussed here.

Together, these two technologies offer near-term potential for selected applications. The critical issues for near-term application of voice technology are primarily in the human factors domain. For the longer term, substantial efforts in both algorithm development and human factors will be required in order to extend the range of speech variability that can be accommodated by speech recognition and generation technology and hence the possible applications of speech technology.

REFERENCES

Allen, H. B., and Underwood, G. N., Eds. (1971). *Readings in American dialectology*. New York: Appleton-Century-Crofts.

Armstrong, J. W. (1980). The effects of concurrent motor tasking on performance of a voice recognition system (Masters thesis). Monterey, CA: Naval Postgraduate School.

Bierman, A., Rodman, R., Rubin, D., and Heidlage, F. (1984). *Natural language with discrete speech as a mode for human to machine communication*. Durham, NC: Duke University, Computer Science Department.

Bond, Z. S., and Garnes, S. Misperceptions of fluent speech. In R. A. Cole, Ed., *Perception and production of fluent speech*. Hillsdale NJ: Erlbaum.

Breaux, R. (1977). Laboratory demonstration of computer speech recognition in training. In R. Breaux, M. Curran, and E. M. Huff, Eds., *Voice technology of interactive real-time command/control systems applications*. Moffett Field, CA: NASA-Ames Research Center.

Brokx, J. P. L., and Nooteboom, S. G. (1982). Intonation and the perception of simultaneous voices. *Journal of Phonetics, 10.*

Brown, J. E., Bertone, C. M., and Obermeyer, R. W. (1968). Army aircraft warning system study (U.S. Army Technical Memorandum 6–68). Aberdeen Proving Ground, MD: U.S. Army Engineering Laboratories.

Butler, F., Manaker, E., and Obert-Thorn, W. (1981). Investigation of a voice synthesis system for the F-14 aircraft: final report (Report No. ACT 81–001). Bethpage, NY: Grumman Aerospace Corporation.

Bucher N. M., Karl, R., Voorhees, J., and Werner, E. (1984). Alerting prefixes for speech warning messages. *Proceedings of the national aerospace and electronics conference (NAECON)*. New York: IEEE.

Chapanis, A. (1975). Interactive human communication. *Scientific American, 232*(3).

Chapanis, A. (1976). Engineering psychology. In M. D. Dunnette, Ed., *Handbook of industrial and organizational psychology*. Chicago: Rand McNally.

Cole, R. A., Ed. (1980). *Perception and production of fluent speech*. Hillsdale, NJ: Lawrence Erlbaum.

Coler, C. (1982). Helicopter speech-command systems: recent noise tests are encouraging. *Speech Technology, 1*(3).

Coler, C. R., and Plummer, R. P. (1974). Development of a computer speech recognition system for flight systems applications. *Preprints of the 45th annual scientific meeting*. Washington, DC: Aerospace Medical Association.

Coler, C., Plummer, R. Huff, E., and Hitchcock, M. (1977). Automatic speech recognition research at NASA-Ames Research Center. *Proceedings of the voice-interactive real-time command/control systems application conference*. Moffett Field, CA: NASA Ames Research Center.

Cooper, G. E. (1977). A survey of the status of and philosophies relating to cockpit warning systems (NASA CR-152071). Moffett Field, CA: NASA Ames Research Center.

Cotton, J. C., and McCauley, M. E. (1983). Voice technology design guides for Navy training systems: final report for the period 23 April, 1980–2 January, 1982 (Report No. NAVTRAEQUIPCEN 80-C-0057-1). Orlando, FL: Naval Training Equipment Center.

Cotton, J. C., McCauley, M. E., North, R. A., and Strieb, M. (1983). Development of speech input/output interfaces for tactical aircraft (AFWAL-TR-83-3073). Dayton OH: Flight Dynamics Laboratory, Wright-Patterson AFB.

Deathridge, B. H. (1972). Auditory and other sensory forms of information presentation. In H. P. Van Cott and R. G. Kinkade, Eds., *Human engineering guide to equipment design*. Washington, DC: U.S. Government Printing Office.

Doddington, G., and Schalk, T. (1981). Speech recognition: turning theory to practice. *IEEE Spectrum, 18*.

Dubord, M. J. (1982). An investigation of response time to collision avoidance commands with a cockpit display of traffic information (unpublished Masters Thesis). San Jose, CA: San Jose State University.

Egan, J. P. (1948). Articulation testing methods. *Laryngoscope, 58*.

Flanagan, J. L. (1972). *Speech analysis, synthesis, and perception,* 2nd ed. New York: Springer Verlag.

Fujimura, O. (1984). The role of linguistics for future speech technology. *LSA Bulletin,* No. 104. Baltimore, MD: Linguistic Society of America.

Garnes, S., and Bond, Z. S. (1977). The influence of semantics on speech perception. Paper presented at the 93rd Meeting of the Acoustical Society of America, University Park, PA.

Gould, J. D., Conti, J., and Hovanyecz, J. (1983). Composing letters with a simulated listening typewriter. *Communications of the ACM. 26*(4), pp. 295–308.

Gould, J. D., and Boies, S. J. (1984). Human factors of the 1984 Olympic message system. *Proceedings of the Human Factors Society 28th annual meeting*. Vol. 2, Santa Monica, CA: Human Factors Society.

Grady, M. W. (1982). Air intercept controller prototype training system (NAVTRAEQUIPCEN 78-C-0182-14). Orlando, FL: Naval Training Equipment Center.

Hakkinen, M. T., and Williges, B. H. (1984). Synthesized voice warning messages: effects of alerting cue in single- and multiple-function voice synthesis systems. *Human Factors, 26*(2).

Harris, S. D., North, R. A., and Owens J. M. (1977). A system for the assessment of human performance in concurrent verbal and manual control tasks. Paper presented at the 7th annual meeting of the national conference on the use of on-line computers in psychology. Washington, DC.

Hecker, M. H., Stevens, K. N., von Bismark, G., and Williams, C. E. (1968). Manifestations of task-induced stress in the acoustical speech signal, *Journal of the Acoustical Society of America, 44*.

Hilborn, E. H. (1975). Human factors experiments for data link: final report (FAA-RD-75-170). Cambridge, MA: Department of Transportation Systems Center.

International Phonetic Association (IPA). (1949). *The principles of the International Phonetic Association*. London: International Phonetic Association. (Available from the Secretary of the International Phonetic Association, Department of Phonetics, University College, London, W.C. 1, England.)

Kemmerling, P., Geiselhart, R., Thornburn, D. E., and Cronburg, J. G. (1969). A comparison of voice and tone warning systems as a function of task loading (Technical Report ASD-TR-69-104). Dayton, OH: Wright Patterson Air Force Base.

Kersteen, Z., and Damos, D., (1983). Human factors issues associated with the use of speech technology in the cockpit (Final Technical Report, U.S. Army Grant No. NAG2-217). Available from J. Voorhees, U.S. Army Aeromechanics Laboratory, NASA-Ames Research Center, Moffett Field, CA 94035.

Kryter, K. (1972). Speech communication. In H. P. Van Cott and R. G. Kinkade, Eds., *Human engineering guide to equipment design*. Washington, DC: U.S. Government Printing Office.

Lilleboe, M. L. (1963). Final report: evaluation of Astropower, Inc. auditory information display installed in the VA-3B airplane (Technical Report ST 31-22R-63) (AD-831823). Patuxent River, MD: U.S. Naval Air Station, Naval Air Test Center.

Luce, P. A., Feustal, T. C. and Pisoni, D. B. (1983). Capacity demands in short-term memory for synthetic and natural speech. *Human Factors, 25.*

McCauley, M. E. (1984). Human factors in voice technology. In F. A. Muckler, Ed., *Human factors review.* Santa Monica, CA: Human Factors Society.

McCauley, M. E., Root, R. W., and Muckler, F. A. (1982). Training evaluation of an automated air intercept controller training system (NAVTRAEQUIPCEN 81-C-0055-1). Orlando, FL: Naval Training Equipment Center.

McCauley, M. E., and Semple, C. A. (1980). Precision approach radar training system (PARTS) (NAVTRAEQUIPCEN 79-C-0042-1). Orlando, FL: Naval Training Equipment Center.

McSorley, W. J. (1981). Using voice recognition equipment to run the warfare environmental simulator (WES) (Masters thesis). Monterey, CA: Naval Postgraduate School.

Miller, G. A., Heise, G. A. and Lichten W. (1951). The intelligibility of speech as a function of the context of the test materials. *Journal of Experimental Psychology, 41.*

Moore, C. A., Moore, D. R., and Ruth, J. C. (1984). Applications of voice interactive systems— military flight test and the future. *Proceedings of the sixth digital avionics systems conference.* New York: IEEE.

Moore, C. A., Ruth, J. C., and Simpson, C. A. (1984). Voice interactive systems applications and implementation (Tutorial presented at the 6th Digital Avionics Systems Conference). Baltimore, MD: IEEE/AIAA.

Mountford, S. J., North, R. A., Metz, S. V., and Graffunder, K. (1982). Methodology for identifying voice functions for airborne voice-interactive control systems (Contract No. N62269-81-R-0344). Minneapolis, MN: Honeywell Systems Research Center.

National Research Council, Committee on Computerized Speech Recognition Technologies. (1984). *Automatic speech recognition in severe environments.* Washington, D.C.: National Research Council, Commission on Engineering and Technical Systems.

North, R. A., and Lea, W. (1982). Application of advanced speech technology in manned penetration bombers (AFWAL-TR-82-3004). Wright Patterson AFB, OH: Flight Dynamics Laboratory.

Nusbaum, H. C., and Schwab, E. C. (1983). The effects of training on intelligibility of synthetic speech: II. the learning curve for synthetic speech (paper presented at the 105th meeting of the Acoustical Society of America).

Nye, P. W., and Gaitenby, J. (1974). The intelligibility of synthetic monosyllabic words in short, syntactically normal sentences. Status report on speech research (SR-37/38). New Haven, CT: Haskins Laboratories.

Pallett, D., Ed., (1982). *Proceedings of the workshop on standardization for speech I/O technology.* Gaithersburg, MD: National Bureau of Standards.

Palmer, E., and Ellis, S. R. (1983). Potential interactions of collision avoidance advisories and cockpit displays of traffic information (SAE Technical Paper Series 831544). In *Proceedings of the second aerospace behavioral engineering technology conference, aerospace congress and exposition.* Warrendale, PA: SAE.

Patterson, R. D. (1982). Guidelines for auditory warning systems (CAA Paper 82017). London: Civil Aviation Authority.

Pisoni, D. B., and Hunnicutt, S. (1980). Perceptual evaluation of MITalk: the MIT unrestricted text-to-speech system. *IEEE international conference record on acoustics, speech, and signal processing.* New York: IEEE.

Poock, G. K. (1981a). A longitudinal study of computer voice recognition performance and vocabulary size (NPS-55-81-013). Monterey, CA: Naval Postgraduate School.

Poock, G. K. (1981b). To train randomly or all at once—that is the question. Proceedings of the voice data entry systems applications conference. Sunnyvale, CA: Lockheed Missiles & Space Co.

Poock, G. K., Martin, B. J. and Roland, E. F. (1983). The effect of feedback to users of voice recognition equipment (NPS Technical Report NPS-55-83-003). Monterey, CA: Naval Postgraduate School.

Poock, G. K., and Roland, E. F. (1982). Voice recognition accuracy: what is acceptable? (NPS Technical Report, NPS55-82-030). Monterey, CA: Naval Postgraduate School.

Poock, G. K., and Roland, E. F. (1984). A feasibility study for integrated voice recognition input into the integrated information display system (IID) (NPS Technical Report NPS-55-84-008). Monterey, CA: Naval Postgraduate School.

Poock, G. K., Schwalm, N. D., Martin, B. J., and Roland, E. F. (1982). Trying for speaker independence in the use of speaker dependent voice recognition equipment (NPS Technical Report NPS-55-82-032). Monterey, CA: Naval Postgraduate School.

Rollins, A. M. (1984). "Composite" templates for speech recognition for small groups. *Proceedings of the Human Factors Society 28th annual meeting.* Vol. 2 Santa Monica, CA: Human Factors Society.

Schmandt, C. (1982). Voice interaction: putting intelligence into the interface. *Proceedings of the IEEE interaction conference on cybernetics and society,* Seattle, WA.

Schurick, J. M., Williges, B. H., and Maynard, J. F. (1985). User feedback requirements with automatic speech recognition. *Ergonomics.* (11) pp. 1543–1555.

Searle, J. (1969). *Speech acts.* London: Cambridge University Press.

Sherwood, B. A. (1979). The computer speaks. *IEEE Spectrum,* August.

Simpson, C. A. (1975). Occupational experience with a specific phraseology: group differences in intelligibility for synthesized and human speech. *Journal of the Acoustical Society of America, 58* (Suppl. 1).

Simpson, C. A. (1976). Effects of linguistic redundancy on pilots' comprehension of synthesized speech. *Proceedings of the twelfth annual conference on manual control* (NASA TMX-73170). Moffett Field, CA: NASA Ames Research Center.

Simpson, C. A. (1980). Synthesized voice approach callouts for air transport operations (NASA Contractor Report NASA CR-3300). Moffett Field, CA: NASA Ames Research Center.

Simpson, C. A. (1981a). Access of speech synthesis and its applications. In J. C. Warren, Ed., *The best of the computer faires. Volume VI: conference proceedings of the sixth west coast computer faire.* Woodside, CA: West Coast Computer Faire.

Simpson, C. A. (1981b). Evaluation of synthesized voice approach callouts (SYNCALL). In J. Moraal and K. F. Kraiss, Eds., *Manned systems design: methods, equipment, and applications.* New York: Plenum.

Simpson, C. A. (1981c). Programming "phoneme" voice synthesizers phonetically. In J. C. Warren, Ed., *The best of the computer faires. Volume VI: conference proceedings of the sixth west coast computer faire.* Woodside, CA: West Coast Computer Faire.

Simpson, C. A. (1983a). Advanced technology—new fixes or new problems? Verbal communications in the aviation system. Paper presented at *Beyond pilot error: a symposium of scientific focus.* Sponsored by the Air Line Pilots Association, Washington, DC. In C. A. Simpson, Ed., *Third aerospace behavioral engineering technology conference proceedings "automation workload technology: friend or foe?.* Warrendale, PA: SAE.

Simpson, C. A. (1983b). Evaluating computer speech devices for your application. In J. C. Warren, Ed., *Proceedings of the seventh west coast computer faire.* Woodside, CA: West Coast Computer Faire.

Simpson, C. A. (1984). Integrated voice controls and speech displays for rotorcraft mission management (SAE Technical Paper Series 831523). In *SAE 1983 transactions.* Vol. 92, Section 4. Warrendale, PA: SAE.

Simpson, C. A. (1985). Voice displays for single pilot IFR (Report prepared for NASA Langley Research Center under subcontract to Honeywell, Inc.), NASA Contract Report CR-172422). Hampton, VA: NASA Langley Research Center.

Simpson, C. A., Coler, C. R., and Huff, E. M. (1982). Human factors of voice I/O for aircraft cockpit controls and displays. In Pallett, D., Ed., *Proceedings of the workshop on standardization for speech I/O technology.* Gaithersburg, MD: National Bureau of Standards.

Simpson, C. A., and Marchionda-Frost, K. (1984). Synthesized speech rate and pitch effects on intelligibility of warning messages for pilots. *Human Factors, 26*(5). 1984.

Simpson, C. A., Marchionda-Frost, K., and Navarro, T. N. (1984). Comparison of voice types for helicopter voice warning systems (SAE Technical Paper Series 841611). *Proceedings of the third aerospace behavioral engineering technical conference, 1984 SAE aerospace congress and exposition.* Warrendale, PA: SAE.

Simpson, C. A., and Navarro, T. N. (1984). Intelligibility of computer generated speech as a function of multiple factors. *Proceedings of the National Aerospace and Electronics Conference* (84CH1984-7 NAECON). New York: IEEE.

Simpson, C. A., and Williams, D. H. Human factors research problems in electronic voice warning system design. *Proceedings of the 11th annual conference on manual control* (NASA TMX-62,464). Moffett Field, CA: NASA Ames Research Center.

Simpson, C. A., and Williams, D. H. (1980). Response time effects of alerting tone and semantic context for synthesized voice cockpit warnings. *Human Factors, 22*(3).

Smith, C. (1983). Relating the performance of speech processors to the bit error rate. *Speech Technology, 2*(1).

Smith, G. (1984). Five voice synthesizers. *Byte, 9*(10).

Spine, T. M., Maynard, J. F., and Williges, B. H. (1983). Error correction strategies for voice recognition. Proceedings of the voice data entry systems application conference. Chicago, IL.

Stern, K. R. (1984). An evaluation of written, graphics, and voice messages in proceduralized instructions. *Proceedings of the Human Factors Society 28th annual meeting.* Vol 1. Santa Monica, CA: Human Factors Society.

Stevens, K. N. (1972). Segments, features, and analysis-by-synthesis. In J. F. Cavanaugh and I. G. Mattingly, Eds., *Language by eye and by ear.* Cambridge, MA: MIT Press.

Thomas, J. C., Rosson, M. B., and Chodorow, M. (1984). Human factors and synthetic speech. *Proceedings of the Human Factors Society 28th annual meeting.* Santa Monica, CA: Human Factors Society, vol. 2.

U.S. Department of Defense. (1981). Human engineering design criteria for military systems, equipment, and facilities (MIL-STD-1472C). Washington, DC: U.S. Department of Defense.

Van Cott, H. P., and Kinkade, R. G., Eds. (1972). *Human engineering guide to equipment design* (rev. ed). Washington, DC: U.S. Government Printing Office.

Veitengruber, J. E. (1978). Design criteria for aircraft warning, caution, and advisory alerting systems, *Journal of Aircraft, 5*(9).

Voorhees, J. W., Bucher, N. M., Huff, E. M., Simpson, C. A., and Williams, D. H. (1983). Voice interactive electronic warning system (VIEWS). *Proceedings of the IEEE/AIAA 5th digital avionics systems conference* (83CH1839-0). New York: IEEE.

Voorhees, J. W., Marchionda, K., and Atchison, V. (1982). Auditory display of helicopter cockpit information. *Proceedings of workshop on standardization for speech I/O technology.* Gaithersburg MD: National Bureau of Standards.

Welch, J. R. (1977). Automated data entry analysis (RADC TR-77-306). Rome, NY: Rome Air Development Center, Griffiss AFB.

Wickens, C. D., Sandry, D. L., and Vidulich, M. (1983). Compatibility and resource competition between modalities of input, central processing, and output. *Human Factors, 25.*

Wierwille, W. W., and Connor, S. A. (1983). Evaluation of twenty pilot workload measures using a psychomotor task in a moving-base aircraft simulator. *Human Factors. 25.*

Williams, D. H., and Simpson, C. A. (1976). A systematic approach to advanced cockpit warning systems for air transport operations: line pilot preferences. In Proceedings of the aircraft safety and operating problems conference (NASA SP-416). Norfolk, VA: NASA Langley Research Center.

Williges, B. H., and Williges, R. C. (1982). Structuring human/computer dialogue using speech technology. *Proceedings of the workshop on standardization for speech I/O technology.* Gaithersburg, MD: National Bureau of Standards.

Woodson, W. E. (1981). *Human factors design handbook.* New York: McGraw-Hill.

Zoltan-Ford, E. (1984). Reducing variability in natural language interactions with computers. *Proceedings of the Human Factors Society 28th annual meeting.* Vol 2., Santa Monica, CA: Human Factors Society.

CHAPTER 11.6

TEXT EDITORS

TERESA L. ROBERTS

Xerox Office Systems Division
Palo Alto, California

The author would like to thank the following people for suggestions and information generously given to improve this chapter: Bob Ayers, Stu Card, Jeff Johnson, Bill Verplank, and Bruce Whittaker.

The text editor is one of the oldest and best understood of interactive computer applications; it is also one of the most widely used and studied. Thus text editing merits its own chapter in this Handbook. Much of the knowledge that may be applied to text editor design is the same as that which applies to the design of any other interactive application. Since interactive applications in general are covered in Chapter 11.7, we refer the reader to that chapter for general information. This chapter emphasizes those aspects of text editor design that are specific to this particular application. Even more specifically, this chapter concentrates on providing guidelines for the practitioner, someone who is interested in either designing or evaluating text editors. An excellent survey of research, including more theoretical work, has been written by Embley and Nagy (1981); we do not try to duplicate all of it here.

In the text that follows, we break the editing process into its component pieces and examine how each piece can be provided optimally. First, Section 11.6.1 briefly covers the variety of text editors in existence, which meet the requirements of different tasks and which are constrained by different hardware facilities. In Section 11.6.2, we explore fundamental issues about the process of text editing and about command languages in general. Section 11.6.3 discusses the portion of text editing having to do with getting the *content* of the document correct. Similarly, Section 11.6.4 discusses operations by which the user specifies the *appearance* of the document, an area that is becoming a requisite for all but the most bare-bones of editors. Section 11.6.5 explores more advanced global issues.

We must keep in mind, however, that the true quality of a text editor lies in how well it gets the user's whole job done. Quality in this area is more than just the sum of the quality of the constituent parts; it also has to do with how well the parts fit together and match the task. Section 11.6.6 returns to this holistic view and presents evaluation methodologies that examine a text editor in terms of end-user performance.

11.6.1 VARIETIES OF TEXT EDITORS

Although the basics of text editing—insertion and correction of letters and punctuation in a document—have been around for long enough that many of the alternatives have been explored, there is still new development. The new developments have been in response to two basic forces: new users with their new tasks and new technologies that make computer aid possible in ways the old technologies did not support.

11.6.1.1 Users and Their Tasks

From the beginning of interactive computing, programmers developed text editors with which they wrote and modified their programs. The programmers then used the same tool for writing their documentation and for writing other natural-language documents. This usage spawned the offshoot of text editors meant for the office, now known as word processors. Similarly, the publishing domain discovered that text editors could make writers' work more timely and could also automate the typesetting process, so an important branch of text-oriented computers formed to specialize in publishing needs (Seybold, 1982). These three areas reinforce each other, with office-oriented priorities contributing to ease-of-use of all text editing systems, and publishing functionality drifting down to become available to less specialized users of text editors.

Because of the wide variety of tasks being performed with text editors, editors with diverse capabilities and emphases are needed.

11.6.1.2 Effect of Hardware

Concurrent with the evolution of text editors caused by their application to different environments, there has been an evolution due to a change in the hardware on which the text editors run. The original line-at-a-time hardcopy terminals were useful for providing a history of the interaction, but they were unable to give the user a constantly updated view of the document. Thus, they resulted in editors that had sophisticated search capabilities and that moved words from line to line only at the user's explicit request. Now there are CRT terminals that run the same editors as these hardcopy terminals. These often run faster than hardcopy terminals, but they do not offer greater capability (and in fact generally prevent the user from viewing more than 24 lines of output at once, a limitation hardcopy machines do not have). Such terminals are colloquially called "glass teletypes"; further discussion in this section categorizes them with hardcopy terminals.

CRT terminals that allow the user to control the cursor position have made it possible for the user to indicate the position of a change merely by pointing to the appropriate place on the screen, eliminating the need to describe that position in a command language. In addition, the continuous display of the text on the screen has been taken advantage of in most document-oriented systems by allowing automatic word-wrap (filling lines with as many words as will fit and then placing subsequent words on the next line). Experimental evidence suggests that display-oriented text editors are generally superior to line-oriented editors for both manuscript editing (Roberts and Moran, 1983) and online composition (Card, Robert, and Keenan, 1984). Still, on character-mapped terminals the text generally appears only as fixed-width characters in one type style (at most, boldface and underlining can be presented with reasonable accuracy).

The current hardware advance is the bit-mapped display. This high-resolution screen supports placement of characters anywhere and allows use of different type faces and sizes; the user no longer needs to wait until the document is printed to see its appearance. Bit-mapped technology also allows interactive creation of graphic illustrations and display of the illustrations along with the text. This capability is summed up in the phrase "what you see is what you get." (Of course, to take advantage of a system with such screen capabilities, one also needs a printer of like ability, such as a laser printer or a phototypesetter).

11.6.2   FUNDAMENTAL ISSUES

This section covers some background issues in text editing: the text editor's relationship with other applications running on the same machine, the process of text editing, and issues about command languages for an interactive application such as a text editor.

11.6.2.1   Relationship with Other Applications

First of all, in studying text editing we must recognize that text editing does not occur in isolation from all other operations. The user frequently needs to interact with other areas of a computer system in the process of editing a document. The user may want to find out the name of a file on this workstation or on a public file server; the user may want to send off a piece of electronic mail asking for some information, and then read the response; the user may want to perform a quick calculation or a detailed statistical analysis, and incorporate the results directly into the document. The user would like to do all of these things without interrupting the flow of editing, either by having to close down the editing of the current document or by having to wait a long time while another process gets started.

Traditional computer systems in which the user can run only one job at a time usually prevent easy switching between text editing and other programs. There may be certain conveniences built into the text editor to allow the most common of interruptions, such as looking at the names of the user's other files or receiving some electronic mail. But in most cases there's no general facility for running an arbitrary program on the side and for the user controlling that program along with the text editor. Such a capability is well handled by the newer "windowing" computer systems, in which the screen can be partitioned into separate areas or "windows" for each application. The user need only point into a new application's window to be able to give it instructions; there is no need to lose context in the editing or other processes that may also be on the screen.

A windowing system works best if it has the following characteristics:

1.   The user interfaces of the different applications are similar and certainly not contradictory, so switching between applications does not confuse the user.
2.   Data can be interchanged among the applications. At the minimum, the user must be able to copy plain text between applications. Even better is for the user to be able to copy whole tabular structures at once.
3.   The system must be able to switch between applications in a matter of seconds so the user does not lose his or her train of thought.

11.6.2.2   The Process of Editing

The process of text editing consists of a sequence of individual changes to a document; even initial composition, while it may contain long periods of typing new text, is punctuated by alterations to previously typed text. Card, Moran, and Newell (1983) studied text editing as one of the primary domains in their work on the psychology of human–computer interaction. They describe the manuscript-editing process as being broken into unit tasks; these, in turn, are divided into the steps:

Acquire unit task (find out from the manuscript what to do next).
Locate position of the change in the online version.
Modify text.
Verify edit.

Manuscript editing (making changes that are marked on a hardcopy in the computer's version of a document) is only one aspect of text editing, but the steps in making each change to the document are much the same even when the user is inventing the tasks on the fly.

These steps require that the system provide methods of displaying the text either to find potential places to change or to ensure that previous changes were performed correctly, methods of saying where in the document a change is to be made, and a command language by which the user indicates changes to be made. The first two requirements are described in the following paragraphs; command languages, a large topic in themselves, are covered in Section 11.6.2.3.

Display

Since display of the text has two functions, one of letting the user read the text to decide on further edits and the other of letting the user verify the correctness of a just-completed edit, a system must allow for both kinds of use. The first requires that large amounts of text be displayed on request; the second that specific areas be displayed at the appropriate moment.

The display function is one in which CRT and hardcopy terminals differ the most, but both provide the capability to respond to both display functions. Editors running on hardcopy terminals typically allow the user to request the display of a certain range of text, which the user can then read at leisure. After performing a modification, some such editors automatically redisplay any changed lines, so the user can verify the correctness of the change. This feature answers the second display need. Such an automatic display should, however, be optional, since an expert may be happy to edit "blind" in exchange for not having to wait for potentially slow feedback.

CRT-based editors that display a range of text continuously make proofreading and verification more straightforward. Many studies have been done to find the optimal screen size and appearance for various tasks. For instance, Neal and Darnell (1984) showed that for manuscript editing a partial-line display is almost as good as a full-page display. Duchnicky and Kolers (1983), in studying reading comprehension, showed that text displayed in 80-character lines is read faster than text in 40-character lines, but did not show much improvement between a 4-line screen and a 20-line screen. In addition, Beldie, Pastoor, and Schwarz (1983) showed that text displayed with variable-width characters is faster to read on a screen than text with fixed-width characters. None of these studies, however, have looked at general text-editing tasks in which proofreading and editing are combined; one might anticipate that in these tasks a larger screen size becomes more useful.

Sometimes it is even necessary for the user to view two widely separated positions in the document at once. The capability to do this is called a *split screen* or split window, in which the different areas of the screen show different parts of the document at once. Not all editors have this capability; if it is missing, the user must flip back and forth between the needed parts of the document as quickly as possible, remembering the information that needed to be shared. While the common 24-line screens are reasonably adequate for document display, a heavy user often wishes for more space, especially when the screen is split.

The other function of display is verification of a change. Here, CRTs do best if they leave the caret or cursor at the position of the modification and if that mark is noticeable enough to draw the user's attention easily. Automatic word-wrap can move text around for the user and make the focus of attention hard to find without such aids.

Finally, both hardcopy and CRT displays do well if they allow the user to look at more than one document at once. For a hardcopy terminal, this usually means just allowing printing of selected parts of a new document while the document being edited is still available. For CRTs, this usually means, again, splitting the screen. Allowing editing of both documents, and copying text between them, is even better.

Addressing

Text to be deleted, the point where an insertion is to go, and the source and destination of a move operation all require specification of place(s) in text: a range of text, a point between existing characters, or both. If the user can see where the operation is to be performed on a hardcopy version of the document, the user can specify the location in any manner convenient to the editor. Some such ways include pointing at the screen, giving the number of the line the text is on, or searching for the string according to its content. Each of these is employed by successful systems as their primary tool for addressing; pointing is obviously the most straightforward mechanism, but the others are quite adequate.

There are other occasions, however, when the user wants to perform some operation on all pieces of text that have a certain characteristic. At such times it is necessary for the system to support addressing according to that characteristic. The most common such characteristic, again, is the content of the text, and most systems have a method of finding all occurrences of a certain string. A generalization of this capability is to relax the criteria for the sought string. For instance, characters could match whether they are in upper or lower case; any letter, digit, or punctuation mark might match certain elements in the string; or "wild cards" might allow matching anything at all. Further search capability could be based on the format rather than the content of the text, or its position in the document's structure. Such functionality is not required in a text editor, but occasionally is just what the expert needs.

11.6.2.3 Command Language

Any computer program, and a text editor is no exception, needs some kind of a command language by which the user tells the system what to do next; it should be as well suited as possible to the functionality of the editing process. It could be a traditional verbal language or it could be language of pointing and gesturing, but it still must be capable of stating all of the actions that will be required. Many different kinds of languages have worked successfully, so there is no one type that we can recommend above others; even controlled studies have failed to find the large differences among systems when all that is varied is the command language style (Hauptmann and Green, 1983).

There is suggestive evidence that command languages that are simpler in syntax not only make the *average* learning time faster for novices, but in addition reduce the *variability* among the novices by being disproportionately easier for users who normally have more trouble (Gomez, Egan, Wheeler, Sharma, and Gruchacz, 1983). Such simplified command languages are often associated with screen editors, since the process of addressing may be done by pointing rather than by describing a location verbally.

Most commands given to a computer, like commands in natural language, consist of a verb indicating the action to take and a noun indicating the object of the action. (In some cases more than one noun may be required, and/or there may be adverbs giving details of how the action is to be performed.) Because of this parallel with natural language, some people have suggested that the syntax of computer languages be made as much like natural language as possible, to aid learning. In particular, since English commands put the verb before the object, it has been hypothesized that a verb–object order is preferable for computer languages used by English-speaking people. In her recent dissertation Cherry (1983) disproved this hypothesis, finding no significant difference in performance between groups using a verb–object order and groups using an object–verb order. This result applied to both novice and experienced computer users performing text editing tasks.

The only firm criteria for forming the command language are that it be simple and consistent. The order of the elements should be well defined, or else the function of the different elements should be clear (to avoid unintentional results caused by putting operands in the wrong order). The command must end with some indication to the system that the command is complete and that the system is to execute it. If the last piece of the command is a clearly recognizable, fixed-length action (such as pressing a function key), there is often no need for a separate "do it" operation; this can make the editor slightly faster to use by omitting the "do it" action.

Finally, the language should include STOP and UNDO commands, so that the user can recover easily from errors, and can explore new areas of the editor without fear of irreparable damage.

Verb

There are a variety of ways in which the user might be able to give the verb of the command. Many of the common ways are listed in Table 11.6.1, along with their advantages and disadvantages. A given text editor can pick one way of expressing verbs, or it may use more than one. It could offer the user alternative ways of saying the same thing (for instance, a "canonical" method and a faster method that requires more learning). Or the editor could have one style of language for common commands and a different style for the less common commands (for instance, a small number of function keys augmented by control keys or perhaps pop-up menus). More than two ways of invoking commands, however, are likely to overwhelm the user.

A common problem is what to do with large numbers of commands, without either overflowing the user's memory or making the process of invoking the commands too awkward. Studies of command naming have shown that the best words to use for command names are words which are infrequent and discriminating (Black and Moran, 1982), semantically specific (Barnard, Hammond, MacLean, and Morton, 1982; Grundin and Barnard, 1984), and of consistent parts of speech (Carroll, 1982). For typing larger numbers of command names, single-letter abbreviations are no longer feasible. The common solution when using command abbreviation is to require the user to type at least enough of the command name to make the word unique (Hirsh-Pasek, Nudelman, and Schneider, 1982); this even allows for future additions of commands if the system does not assume that the user will type any fixed number of letters in the command name. For menu systems, large numbers of commands are often grouped onto separate menus that the user must learn to navigate among. This can be made more palatable if shortcuts to the appropriate menu (or directly to the desired command) are provided. In this way, the experienced user can "chunk" several related actions into one, and the system does not need to spend time in repeated menu painting. [More considerations on the use of menus may be found in Card (1982) and Shneiderman (1983).]

Table 11.6.1   Mechanisms for Specifying Verb in Command

Mechanism	Advantages	Disadvantages
Keyboard, abbreviated word	Mnemonic No special hardware required Can leave hands on home position	Multiple keystrokes Prevents "modeless" insertion
Keyboard, single character	Very quick No special hardware required Can leave hands on home position	Not all commands very mnemonic Limit of ~26 commands Prevents "modeless" insertion
Control keys [hold down CTRL, press letter(s)]	Hardware common Can leave hands on home position Can serve as "do it" signal	Awkward to type very much
Function keys, labeled	Very clear Can operate with other hand on pointing device Can serve as "do it" signal	Requires special hardware Keys *always* present Limited number
Function keys, unlabeled	Hardware fairly common Can operate with other hand on pointing device Can serve as "do it" signal	Non-mnemonic, unless meanings displayed some other way Limited number
Menu on screen, user presses corresponding key	Continuous reminder Can give whole name	Takes up screen space Limited number
Menu on screen, user clicks it	Continuous reminder Fast if already using pointing device Can give whole name	Takes up screen space Limited number May require pointing away from target object
Pop-up menu	Can hold more commands than permanent menu Can give whole name	Commands hidden Several mouse actions May require pointing away from target object

Object (especially, Addressing)

The most common object of a command is a place in the text of the document. The most intuitive method for indicating an address is by pointing to it on the screen. The mouse has been shown to be an optimal pointing device (once learned, using the mouse is as fast as using a finger) (Card, English, and Burr, 1978); the other pointing device in common use is the step keys (or "arrow" keys). Hardcopy terminals, which do not allow pointing, have editors that generally use the line number as the normal method for addressing. Line numbers can either be absolute, in which case the number of a given line does not change until the user explicitly orders the lines to be renumbered, or relative, in which case lines are just counted from the beginning of the document and the number of any line changes if lines are inserted or deleted before it. In the former scheme, users generally give explicit line numbers in their commands. In the latter, the system remembers a position in the file and the user issues commands to move the pointer from that position forward or backward; the user typically moves the pointer in separate operations from issuing a command to make a change.

The preceding operations typically only indicate one point in the document. If the user wants to specify a whole range, some systems allow extending the range of affected text directly (by giving starting and ending line numbers, for instance, or by moving the mouse and clicking a second mouse button). Other systems, however, make "marking" the beginning of the desired text and moving on to indicate the end of the desired text into separate operations. Such a process adds unnecessarily to the number of steps required to perform a command.

There are times when object(s) or other modifier(s) of a command are not places in the document's text. Such modifiers include, for instance, the number of times an operation is to be repeated, and whether uppercase and lowercase letters are to be treated as equivalent in a search. How these are handled depends largely on how verbs are specified. For systems with textual commands, it is most common for the other modifiers of a command to be expressed as additional textual arguments, given before the signal to execute the command. Or the system can lead the user through a dialogue in which the system queries the user for each of the necessary pieces of information. Such a dialogue, however, can be tedious for an expert, and it often prevents easy correction of an erroneous response even if it is discovered before the command is invoked. Menu-based systems can allow hierarchies of menus to hold all of the command's details. Screen-oriented systems can display a little form (sometimes called an "option sheet") in which the user can set nondefault options for a command. This is an improvement over the dialogues previously described since the user, rather than the system, directs the interaction; but it is still too high-profile for constant use.

11.6.3 CONTENT

The primary purpose of text editors is to put the right text, that is, content, into the document. This section discusses alternatives for the way the text might be presented to the user, both in the user's mental model and on the text editor's display, and discusses operations on the text.

11.6.3.1 Model of Data

The most obvious model that many users bring to text editors is the model of typing on a typewriter. In this model, characters (letters, numbers, and punctuation) are placed in certain positions on a piece of blank paper. The typewriter model makes some operations easy, such as "typing over" an erroneous character; but it does not allow for other operations that are straightforward for a computer, such as inserting a word in the middle of a line. Many studies have demonstrated that users do approach text editors in this way, by inference from common errors (Galambos, Wikler, Black, and Sebrechts, 1983; Moran, 1983). How much the designer of a new system wants to take advantage of this model or to divorce the new system from it depends on how thoroughly the data and operations of the system can be made to conform to the model. The "typewriter" model has been used as the basis of the "quarter-plane" model of Irons (Meyrowitz and Van Dam, 1982).

Traditional text editors regard text as a stream of characters: letters, numbers, punctuation, and space characters. Some, in addition to this aspect of the model, also rely heavily on the *line* concept. These editors consider the break between lines to be very important. Such editors tend to be those that run on hardcopy terminals, since they use lines as the user's primary mechanism for finding a given place in the document; words would become very hard to find if they could flow onto new lines. Other such editors are those meant for editing programs, which typically do not make use of line wraparound. Line-oriented editors sometimes even have different commands for editing between lines from the ones that edit within a line, since there are different ways of addressing those places (typically line numbers versus string match).

Screen editors that are not particularly aimed at line-oriented tasks regard the entire document as one stream of characters. Within paragraphs, lines simply consist of those words that currently fit within the page margins; as the user types more words than will fit on a line, the new words move down to the next line. This feature, already mentioned several times previously, is called *word-wrap* or *wraparound*. Some editors perform word-wrap only on initial type-in and when explicitly requested by the user; others will ripple the effect of any insertion or deletion through a paragraph interactively.

A specific difference between the typewriter model and the stream-of-characters model is the treatment of the blank space at the end of each line of text: in the typewriter model it is full of blank space characters, whereas typically in the stream model there is nothing there at all. The operational difference appears when the user tries to point there and type: this can be done with the typewriter model but with the stream model the user must select the last real character on the line and then space over to the desired position before being able to type text. The typewriter model seems to work well in a line-oriented application such as program code; the stream model is better able to deal with proportional spacing and interactive word-wrap.

In stream-model systems there are times when fixed line breaks are needed. Carriage-return characters embedded in the text serve this purpose; they are simultaneously ordinary characters that can be inserted and deleted like any others, and special characters that have the function of causing the next character to begin a new line. Paragraphs are variously broken by the same carriage return character, two carriage-returns in a row, or a different special character.

Many systems include special treatment for certain groupings of text. Characters, words, sentences, and paragraphs are the units in which text is manipulated in document-oriented systems. Program-oriented systems deal with characters, words, and lines.

Some systems also deal with a hierarchical organization of text, program editors using the normal structure of the language they support, and document editors the chapter/section/subsection structure of many documents. This development for document editors is a descendent of Engelbart's NLS system (Engelbart and English, 1968), developed to "augment the human intellect." These systems typically allow users to choose to view the document only at the highest levels, so they can skim the overview.

Then the users can either modify the organization by moving whole sections around if they are authors, or delve into the sections that seem most interesting to them if they are readers.

A final element of the user's model of the document's content is the treatment of unusual characters. These include publishing symbols such as the dagger †, mathematical symbols such as the less-than-or-equal sign $\leq$, Greek letters $\alpha\beta\gamma$, and so forth. There are two primary ways of handling these. The traditional way is to type these with normal characters, but indicate that they are to be printed in a different type face called a *pi font*. Another approach is to recognize that these characters have identities of their own, and augment the system's character set to include more than plain ASCII or whatever. This is a current topic of standards activity, both by international standards organizations and by vendors of publishing equipment (International Organization for Standardization, 1973a, 1973b, 1982; Xerox Corp., 1984).

11.6.3.2 Display of Text

The display of text is, as we have discussed above, dependent on the terminal on which the editor is running. With a hardcopy terminal, the best that can be done is to display selected areas of the text at the user's request. With CRTs, however, the text can be displayed continuously.

The particular area of text displayed on a CRT is determined by a process called *scrolling*. The document is envisioned as being printed on a long, continuous scroll of paper. Only a small segment of that paper can be seen through the window on the CRT at once. The user issues commands to change what part of the scroll is displayed through the window. This scrolling can be continuous, one or a few lines at a time, window-full at a time, or it may be to distant places in the document.

There is an interesting dichotomy in the way the relationship between the "scroll" and the "window" may be perceived. If one is looking at the middle of the document and wants to look at the next segment toward the end of the document, does one scroll "up" or "down"? If one envisions the window moving over the paper, one scrolls down. If one envisions the paper moving behind the window, one scrolls up. (In fact the characters do move up on the screen.) A study by Bury, Boyle, Evey, and Neal (1982) has shown a preference for the window-moving ("down") model, but there are successful systems that use each model.

Another way in which text editors for CRT terminals vary is in how closely they try to approximate the printed output in the length of line they choose to display. Most systems allow the user to specify the line length of the output. If this is shorter than or the same as the width of the screen, then the system can display the line just as it will be printed. But if the hardcopy line length is longer than the screen width, the system has the choice of breaking the line at the edge of the window, making the screen view different from the printed view, or letting the line extend beyond the edge of the window, requiring the user to scroll horizontally in order to read the whole line. Each approach has its uses depending on the user's current task, so it is probably the user who should have the option of how this situation is treated at any time. If that is impossible, the most generally useful solution is probably the wrapping choice, as long as the application of the system emphasizes the content over the appearance.

11.6.3.3 Operations

The most common operations on text are inserting, deleting, and replacing a few characters or words (Allen, 1982). Other basic operations are insertion of large amounts of text, moving, copying, and transposing (which could be performed with two move operations). Of these operations, a few deserve special notice.

Insertion

As was mentioned in discussing the model of a document, the obvious analogy that people bring to text editors is that of typewriters. The ability of new text to "squeeze" between old text is magical and counterintuitive to people with this model. Systems that try to take advantage of the model and mimic a typewriter make the default function of typing be character-for-character replacement of the existing text. Such systems require that the user give a special command that opens up a blank area in the midst of the text, if the new text is to be inserted among the existing characters. At the end of the insertion, the user types a special key to close up the blank area. This model seems to be intuitive, but it is dangerous since the easiest mode of operation is destructive. It is safer to have the default be that as the user types the text is inserted without destroying any old text.

Systems that use normal typing keys for issuing commands (without requiring that the user hold down a Control or Code key simultaneously) have a problem since those same keys are also used for typing text. The system only knows what is meant by a given key-press by what "mode" it is in. Typically, the system is normally in "command mode," in which keystrokes are interpreted as commands; there is a command that puts the system into "insertion mode," in which keystrokes are entered into the document's text. As one might expect, errors are common due to the user typing commands or text when the system is in the "wrong" mode. Because of this problem, many systems are consciously "modeless," that is, normal keystrokes always enter text into the document, and never have another meaning. Experimental studies have not shown the expected clear benefit of modeless editing (Poller and Garter, 1984), possibly because of other problems introduced by the alternative command languages. Nevertheless, if all else is equal, the system may as well aid the user by preventing mode errors.

Move and Copy

A move or copy operation involves two text locations, the text to be moved or copied and the destination location. The user may be paying attention to either of those locations at the point when the decision is made that the operation needs to take place. If the insertion point (or "caret") is already in the destination location because the user has been inserting text, it is convenient if the user can just specify the text to be moved/copied, without losing the place where the caret is. Likewise, if the text to be moved/copied happens already to be selected, it is most convenient if the command can be issued with that text as the first argument. These two orderings of the arguments imply two different versions of these commands to satisfy the two different task requirements. A system that is able to work both of these in will meet more needs than a system with either of the single move/copy styles.

The simple fact of having more than one argument may cause some command languages difficulty. This problem may be avoided for these particular operations in some systems by dividing the move or copy into two operations: cut and paste. The user first selects the text to be moved (copied) and invokes the "cut" (or "duplicate") command. At this point the selected text goes into a buffer from which it may be recalled at any time until something new is put into the buffer. The user then goes to the destination spot and issues the "paste" command to copy the text from the buffer. This method of performing a move/copy can simplify the command language. But it is best implemented in a way such that the user can see the contents of the buffer, rather than having to work blind and trust that text is really there.

Global Replace

Many edits to a document are the one-at-a-time kinds of changes described above. But occasionally it is necessary to change the wording of something throughout a document; at such a time a global replace feature is invaluable. This feature is related to the search capability discussed in Section 11.6.2.2 under "Addressing." The user specifies a search string, and then also supplies a string with which it's to be replaced. The system makes this substitution throughout the entire document, or over whatever range the user requests. An additional feature is to allow the user to confirm or veto each change before it is made.

The flexibility in specifying wild cards that was useful for searches carries over to replacing. In addition, replacement facilities make use of variables, particularly for leaving a portion of the string as it was originally. For instance, the search string might be "$a\langle 1\rangle b$," which means that it will match any string beginning with "a" and ending with "b." If the replacement string is "$c\langle 1\rangle d$," the operation will change the "a" to "c" and the "b" to "d" without affecting the text in between. Or replacing "$/\langle 1\rangle:\langle 2\rangle/$" with "$/\langle 2\rangle:\langle 1\rangle/$" reverses the strings on each side of the colon. Such a capability approaches the ability to program, without necessarily requiring a large amount of learning overhead.

Writing Assistance

At an even higher level are tools that aid the writer with the meaning of the text. Research is taking place on tools that aid in picking an appropriate overall organization and in sorting ideas into that organization. Existing tools check spelling, perform hyphenation, check grammar (make sure that subjects and verbs are both singular or both plural, for instance), and find out the difficulty level of the text as a function of the length of sentences and words. The most comprehensive package of writers' tools is the Writer's Workbench from Bell Laboratories (Alexander, 1984; Cherry and Mac-Donald, 1983).

11.6.4   APPEARANCE

It is no longer sufficient for a document-oriented text editor to deal only with the content of the document. At the low end, text editors generally allow simple formatting that most printers can handle: boldface (obtained by typing a character twice at approximately the same place), underlining, and justifying text by adding extra spaces between words. Editors that have a more capable printing device can go much further: allowing italic text, different typefaces, continuous control over line spacing, kerning, and more. Besides character-oriented formatting, we also see variations in page layout, ranging from control over page margins to multiple columns, variable headers and footers, footnotes, and widow and orphan control.

11.6.4.1 Model of Appearance Information

Placement of Information

There is a choice of user models for the information about the appearance of the text: Traditionally, such information has been encoded directly inline along with the content of the document. It is typically set apart from the true content by beginning with an unusual character. A characteristic that is turned on with one code remains in effect until another inline code turns it off or supersedes it. Such embedded formatting controls, not surprisingly, make the content of the document hard to read and to work with. They also tend to involve very brief, cryptic encoding to minimize the disruption of the text, but such an encoding leads in the process to errors and to learning difficulties.

An alternative model for including formatting information with the document's text is to associate *properties* with the normal objects (characters, paragraphs, pages, and so forth) of the document. Thus, a character might have its typeface, size, weight, position relative to the baseline, and so forth, as its properties. A paragraph might have information on whether it is to be justified or its right margin is to remain ragged, and how much raggedness is allowed. In such a model, the user makes formatting changes by selecting affected text and changing the properties on a form which lists all applicable properties; such a form is called a *property sheet.* (Special-purpose commands may offer a shortcut for changing common properties.)

There is a model for formatting information that falls midway between the preceding two. In such a case, the user inserts a nonprinting character into the text and associates the formatting information with it. Such a character does not get in the way of reading the text very much (and may even be able to be suppressed for the display). And it may hold properties that are not appropriate for normal characters to carry (such as page-formatting information). More than one special character may be used for holding different kinds of formatting information, with the appearance of the character giving the user a hint about its purpose. This makes it easier for the user to compartmentalize different types of formatting and to locate the place in the document that is causing a certain effect.

Which model is chosen for a given situation depends largely on hardware capability. Inline coding is necessary if there is no way to insert special-purpose characters and no good way to implement properties and display their effects.

Higher-Level Models

Additional aspects of formatting cover how formatting options are combined. One method of combining formatting controls is to allow the user to group several settings together under one name and then to invoke the whole effect at once. An extension of this concept is for the system to remember the name of the grouping and not just its effect each time it is referred to. Then the system could allow the user to change the effect associated with that name, and all places in the text that used that name would automatically change their formatting. Such a mechanism is sometimes called *styles.* For instance, the definition of "A-Level Head" could be "centered, bold, 12-point" for one publication and "left-justified, bold, 10-point" for another; with a styles mechanism, the user need only associate new style definitions with a document to format it for one publication or another.

A second higher-level aspect of formatting involves allowing one formatting command to have several sequential effects. Thus if the user were, say, using a business-letter style, a single command might cause the first paragraph to be on the right side of the page (the date), the next paragraph to be on the left (for the recipient's address), and the third paragraph (the salutation) not to be indented like a normal body paragraph.

These higher-level forms of formatting add to both the power and the complexity of a system. They are probably not necessary for simple applications. When they are used, a typist must be able to invoke them easily, without having to know the details of how the styles are defined. The style definition process may be somewhat more difficult to use, in accordance with the complexity of what it is trying to accomplish, since style definition is performed much less regularly than style invocation. But since the definer is likely to be a graphic artist with minimal interest in arcane computer skills, ease of use is still important. One of the primary formatting systems using these concepts is Scribe (Reid, 1980; Reid and Walker, 1980).

Type of Language

A final general issue in the formatting of text is whether the language used is descriptive or procedural. Many results may be specified in either way, as in "this range of text is bold" as opposed to "turn bold on here" and "turn bold off here." The language types differ more as specifications get more and more complex. For instance, to determine how to avoid widows (the first line of a paragraph alone at the bottom of a page), a descriptive system might specify the required bottom page margin and the minimum amount of text permissible alone on a page, and leave it up to the system to figure out whether a given paragraph should be started on the current page or should be moved to the next page. A procedural specification, however, might inquire how close to the bottom of the page the current text is about to be placed, test that value against some threshold, and then conditionally force the system to begin a new page.

Descriptive languages generally take the burden of programming off the user and give it to the system (the user describes a desired result and it's up to the system to figure out how that might be achieved). However, the user is at the mercy of the capability that the system's designers have foreseen and allowed for. A procedural language, while it may force the user to know implementation details irrelevant to the task at hand, may provide more flexibility in the long run. Experimental evidence does not clearly favor either type of language, but there is some evidence that even though programming is difficult for many people, in complicated situations people do better with a procedural language than with a descriptive language in which they must specify a comparable amount of detail (Reisner, 1981a; Welty and Stemple, 1981).

11.6.4.2 Display of Formatting

The most effective display of formatting information is, in general, a simulation of the effects of the formatting. It allows a maximum amount of work to be performed accurately at the workstation, without the need for a delay while a proof is printed. This is the "what you see is what you get," sometimes abbreviated as WYSIWYG, principle. There are limits to the effectiveness of a WYSIWYG philosophy, however, both in the subtlety of what must be displayed and in the capabilities of our display screens, so other approaches must sometimes be used.

Starting at the lowest end of display technology, there are some machines that are very limited in what they can show with any fidelity. So their primary way of displaying format changes is to show marks in the text where the change takes place; typically this is the very inline encoding that the user inserted to cause the formatting to take place. The alternative is to show nothing at all, giving a clean picture of the text's content.

At the next level is the attempt to display whatever formatting the hardware can handle, such as underlining and brightening (for boldface text). An extension of this notion is to represent a formatting change in a way that makes it noticeable, but which does not necessarily represent the type of change accurately at all. Such an appearance substitution might be the use of underscore to represent italics, or reverse video to represent a typeface change.

An additional level up is a system which theoretically could display formatting, but which for reasons of economy limits the precision of such a display. An example of this is the use of "generic typefaces": typically one serif-type family and one sans-serif family are available on the workstation, and font changes are represented with these. The character widths of the true printer fonts are used so that line breaks are accurately displayed, but the character shapes may not be quite correct. In general, it is possible for there to be any number of screen fonts; but if a printer font is ever needed for which the screen font is unavailable, font substitution similar to the use of generic fonts must be performed.

A similar practical problem arises with display of page layout. The screen may be capable of approximating the true layout, but performance problems may make its continuous display prohibitive. The solution is generally to display a "manuscript" form that contains only that appearance information which is readily computed. Layout is performed as a batch process, and a "soft proof" displayed on the CRT; only limited amounts of editing might be possible at this stage.

At the highest level of capability come displays having full flexibility, power, and speed to show whatever can be printed. But they are still quite likely to have a lower resolution than the eventual printer (60–120 dots per inch versus 300–2000 dots per inch), so the representation is still inexact.

But beyond these hardware considerations, there's another major aspect of formatting information for which WYSIWYG is completely inadequate: since WYSIWYG displays the *effect* of the formatting, it does not always include the user's *intent*. For instance, if the user notices that the system did not hyphenate any words in a whole chapter, there's no way of telling from looking at the document as rendered whether that was a coincidence or whether the formatting of that chapter forbade hyphenation. Similarly, one does not know whether the distinctive appearance of a chapter title is caused by explicit formatting specifications or by the invocation of a style. Such information may be displayed by any of the aforementioned techniques. An additional mechanism available on high-quality screens is to annotate the text at the edge of the screen or of the window to mark format changes.

11.6.4.3 Operations

The operations associated with formatting are, generally, invoking a certain effect and turning that effect off. How this is done depends primarily on the user model of the formatting information. If that information is encoded as plain text inserted into the document's content, the commands for affecting formatting are identical to those for altering the content. If, however, the notion of properties is used, new commands are required. Such commands must fit gracefully into the overall command language of the system (using control keys, function keys, menus, or whatever).

There are a few formatting operations that do not fit precisely into the category of invoking or removing an attribute, but that require special mention since they are quite common operations. These include making the text plain (removing all special formatting and setting all parameters to their default values), making one sequence of text look just like another (copying formatting information), and changing the case of the text to upper case or lower case (case may or may not be regarded as a formatting attribute in a given system). These operations must be easy to invoke.

Finally, there's the process of specifying the page layout. This may be done parametrically as with other formatting information (two columns, each 20 picas wide, with $\frac{1}{2}$ in. between them). But since page layout is meant to result in two-dimensional representation that is visually pleasing, a layout designer is likely to prefer creating the specification with graphical tools. For this purpose it will be convenient if the formatting package is compatible with a graphics application, or, better yet, if a graphics capability is incorporated into the editor itself.

11.6.5 ADVANCED FEATURES

11.6.5.1 Special applications

Few documents contain text only. Most contain tables and illustrations in addition to the text, and some documents contain specialized types of information such as mathematical formulas [see Knuth (1984), which describes TEX, a high quality, batch-oriented formatting system that specializes in mathematics.] If a document editor is to make such documents possible, it must make some provision for nontextual objects, both in entering and editing their content and in laying them out appropriately on the hardcopy page.

The least an editor can do to allow for a special object is to just leave white space for it; the idea is that the user will manually insert the appropriate figure after the text is completed.

The next choice is for the system to provide another application program with which the user creates the figure. Then the user performs some kind of a copy operation to move the figure from the application that created it into the document. Typically, the user cannot then subsequently edit the figure in the document; that must be done by returning to the original application and editing the copy of the figure that (hopefully) was left there.

The third and preferred mechanism is to integrate other applications with the document editor itself. In such a system, the document text may contain *frames:* boxes that implement different kinds of applications. The user inserts a frame into the document, makes it the right size, and specifies how it is to fall in the document layout; all this uses the normal document editing procedures. But then when the user starts working within the frame, the user is no longer interacting with the text editor, but with an application that implements whatever type of figure is needed: tables, line drawings, photographic enhancement, mathematics, music, or whatever. In such a system, the nontextual material is always present right where it belongs in the document, and it can be edited there at any time.

Tables

Tables are a very common type of special application, and since they consist primarily of text, they are often implemented in at least a rudimentary way by most text editors. The minimum capability is that of the typewriter: the user can set tab stops. Then in the process of typing a line, when the user presses the tab key, the type-in point moves over to the nearest tab stop and starts inserting text left-justified at that position. In such a model there is nothing that prevents the typing from continuing through the next tab stop.

Enhancements to this basic form of a table come in two primary categories: alignment and ruling. Computer systems make it possible to align the text in other ways at the tab stop, besides left-alignment. Text can be centered over the tab (or, on some systems, centered between boundary positions that the user has set up), right justified, or decimal aligned for numeric data.

Printed tables also often have horizontal and/or vertical lines (or *rules*) between rows and columns. This is sometimes provided in text editors by supplying the user with vertical and horizontal line characters and letting the user piece them together into the appropriate ruling. Better, though, is for the system to draw lines of the appropriate length automatically. If this is to be done without the system drawing its rules under every line of text (not desirable if table entries can span multiple lines), the systems must have some way of indicating how much text is to be considered as one whole entry. Once such information is available, further enhancements are possible, such as providing automatic line-wrap within a table cell and justifying the text in the cell.

11.6.5.2 Programming

The ability for an end-user to program within any computer application can lead to tremendous efficiency on that system. Anything that needs to be done repetitively is best automated. Places where the appropriate thing to do can vary depending on what its context turns out to be (such as where it falls on a page during layout) can be handled with conditionals in a programming language.

A simple form of "programming" is *abbreviation expansion.* The user types an abbreviation, presses a special key (or perhaps the user just presses the particular key that represents a single expansion), and the system automatically replaces the abbreviation with an arbitrary amount of boilerplate text. Abbreviation expansion can cause actions to take place as well as inserting content in systems with inline formatting codes, of course. But beyond that, some systems even allow normal editing commands to be part of an expansion, so that the invocation not only inserts text but also may perform arbitrary edits.

Another form of programming is the global replace function covered earlier. A third involves macros inserted into the text of the document, which can affect the content and/or appearance. These typically occur in systems with procedural, rather than descriptive, formatting languages, but are possible with either.

Finally, there is a kind of programming in which the program is stored in some other place than in abbreviation expansions or within the document. Such programs can potentially have arbitrarily complex effects on the document. These programs may be written in any programming language, but they generally include operations that are the same as the text editor's commands. Since programming is not easy for many people to do, one method of making the programming process more straightforward is for the user to "write" the program by simply demonstrating to the system what is to be done. This technique is called *programming by example* (Halbert, 1984).

11.6.6   EVALUATION

The foregoing has been primarily an analysis of the issues that must be considered in the design of a whole text editor. It has discussed many alternatives for various aspects of the user interface, giving considerations for when each might be appropriate, but generally avoiding prescriptions. This is because the research does not support any one type of user interface above all others, because the various types of terminals in common use place different requirement on the systems, because different users and uses put different requirements on the system, and because the variety of successful systems provides an existence proof that different approaches can be quite adequate. Indeed, the quality of a system may well stem from how well it integrates all of its requisite functionality within a parsimonious system model, rather than from any set of aspects of that model itself.

Formal methods have been proposed to analyze the consistency of a user model and command language, but they remain unwieldy for systems of realistic magnitude (Moran, 1981; Reisner, 1981b). The most effective way of evaluating a system for a class of users and their tasks remains trying it out on them.

A good paper-and-pencil means for quickly evaluating any interactive computer system is the Keystroke-Level Model (Card, Moran, and Newell, 1983). This predicts the amount of time it will take experts to perform tasks, as a function of the amount of typing, pointing, thinking, waiting, and so forth, that the method for performing the task requires. Specifically, the evaluator divides the total job to be done into individual *unit tasks* and lists precisely how the user would perform each unit task. Then the evaluator counts the number of occurrences of each type of operation (typing, pointing, and so forth) that appear, multiplies by a time constant, and adds the components together. A simplified version of the components and their multipliers is shown in Table 11.6.2. The rules for placing "mental" operators into the sequence of actions are given in Table 11.6.3.

More specific to text editing, and giving experimental results in addition to analytical ones, is a benchmark battery of texts for evaluating a text editor [discussion in Roberts and Moran (1983), materials in Roberts (1980)]. This scheme contains experiments testing expert performance on common content-oriented tasks and testing novice learning of these tasks. Its primary strength, besides having

Table 11.6.2   Components of the Keystroke-Level Model

Operator Code	Description	Time (sec)
Acquire	Find out what the next task is (usually by looking at a marked-up manuscript).	1.8 (non-CRT) 4.0 (CRT)
K	Keystroke or button press.	0.1–1.0
P	Pointing to a target with a mouse.	1.10
H	Homing the hand(s) on the keyboard or other device.	0.40
M	Mentally preparing for physical actions.	1.35
R(*t*)	Response time by the system. (Only if it causes the user to wait.)	*t*

Source: Card et al. (1983, adapted from Figure 8.1)

Table 11.6.3 Heuristic Rules for Placing the M Operators

	Begin with a method encoding that includes all physical operations and response operations. Use Rule 0 to place candidate **M**s, and then cycle through Rules 1 to 4 for each **M** to see whether it should be deleted.
Rule 0.	Insert **M**s in front of all **K**s that are not part of argument strings proper (e.g., text or numbers). Place **M**s in front of all **P**s that select commands (not arguments).
Rule 1.	If an operator following an **M** is *fully anticipated* in an operator just previous to **M**, then delete the **M** (e.g., **PMK → PK**).
Rule 2.	If a string of **MK**s *belongs to a cognitive unit* (e.g., the name of a command), then delete all **M**s but the first.
Rule 3.	If **K** is a *redundant terminator* (e.g., the terminator of a command immediately following the terminator of its argument), then delete the **M** in front of it.
Rule 4.	If a **K** *terminates a constant string* (e.g., a command name), then delete the **M** in front of it; but if the **K** terminates a variable string (e.g., an argument string), then keep the **M** in front of it.

Source: Card et al. (1983, Figure 8.2)

some validity in the domain of text editing, is that it allows quantitative comparison of a new system with nine existing text editors. It is weak, however, in evaluating the usability of formatting and layout facilities. It also does not point out particular places where the user interface might be improved.

In the end, the most trustworthy means of validating the usability of a text editing system is the same as with any human-machine system: with standard human factors techniques. The designer can start with *gedanken* experiments on tasks that the system is targeted to perform frequently. When prototype implementations become available, "tough" users (tough in the sense of not minding system crashes, and also in the sense of not being tolerant of a poor user-interface design) can try it; they should perform both tasks supplied by the testers and their own tasks that they would like to be able to perform on the system. Finally, as the implementation becomes more robust, testing can expand to a broader class of users. But even then, one must still concentrate on getting the depth of knowledge attainable by watching the experiences of a few people, in addition to getting the breadth attainable by observing many users.

11.6.7 CONCLUSION

In developing a new text editor, there is no substitute for experience to give the designer a repertoire of ideas. There are enough text editors on the market these days that obtaining such experience is merely a matter of going to the nearest computer store and becoming familiar with several demonstrations. Such an exercise can also serve to broaden the perspective of someone who has had years of experience with editors, but who may have working knowledge of only a few. In addition, many publications (ACM, 1981; Furuta, Scofield, and Shaw, 1982; Meyrowitz and Van Dam, 1982; Seybold Reports) describe enough editors to show the range of possibilities and also include analyses of those editors.

As was mentioned in the preceding section, there is no one *form* of text editor that has a consensus as being the best, even though individual editors have been shown to vary greatly in quality. Thus, the editor developer does best by designing with the following ideas in mind:

1. Follow general cognitive psychology principles (Newman and Sproull, 1979).
   a. Keep the user model as parsimonious as is reasonable.
   b. Give the user feedback about the state of the process (whether the last command completed correctly, what's selected) and the state of the document.
   c. Make sure that system response time is appropriate for each task being done. It may be necessary to trade off functionality for a system that is fast enough to be usable.

2. Include functionality required by the user population, and not a lot more. This will prevent the user interface and the response time from being bogged down by unused options.

3. Implement the riskier parts of the user interface first, test them on realistic users, and be prepared to make alterations.

REFERENCES

Alexander, G. A., (1984). Computer aids for authors and editors. *The Seybold Report on Publishing Systems, 13*(10).

Allen, R. B., (1982). Patterns of manuscript revisions. *Behaviour and Information Technology, 1*(2), 177–184.

ACM Special Interest Group on Office Automation (1981). *Proceedings of the ACM SIGPLAN SIGOA symposium on text manipulation.* Portland, Oregon.

Barnard, P. J., Hammond, N. V., MacLean, A., and Morton, J. (1982). Learning and remembering interactive commands in a text-editing task. *Behaviour and Information Technology, 1*(4), 347–358.

Beldie, I. P., Pastoor, S., and Schwarz, E. (1983). Fixed versus variable letter width for televised text. *Human Factors, 25*(3), 273–277.

Black, J. B., and Moran, T. P. (1982). Learning and remembering command names. *Proceedings of the human factors in computer systems conference* (pp. 8–11). Gaithersburg, MD.

Bury, K. F., Boyle, J. M., Evey, R. J., and Neal, A. S. (1982). Windowing vs scrolling on a visual display terminal. *Proceedings of the human factors in computer systems conference* (pp. 41–44). Gaithersburg, MD.

Card, S. K. (1982). User perceptual mechanisms in the search of computer command menus. *Proceedings of the human factors in computer systems conference* pp. 190–196. Gaithersburg, MD.

Card, S. K., English, W. K., and Burr, B. J. (1978). Evaluation of mouse, rate-controlled isometric joystick, step keys, and text keys for text selection on a CRT. *Ergonomics, 21*(8), 601–613.

Card, S. K., Moran, T. P., Newell, A. (1983). *The psychology of human-computer interaction.* Hillsdale, NJ: Erlbaum.

Card, S. K., Robert, J. M., Keenan, L. N. (1984). Online composition of text. *Proceedings of Interact '84.* (pp. 231–236). London: Elsevier.

Carroll, J. M. (1982). Learning, using and designing filenames and command paradigms. *Behaviour and Information Technology, 1*(4), 327–346.

Cherry, J. M. (1983). Command languages: effects of word order on user performance (Ph.D. dissertation). University of Pittsburgh.

Cherry, L. L., and MacDonald, N. H. (1983). The Unix Writer's WorkBench software. *Byte, 8*(10), 241–248.

Duchnicky, R. L., and Kolers, P. A. (1983). Readability of text scrolled on visual display terminals as a function of window size. *Human Factors, 25*(6), 683–692.

Embley, D. W., and Nagy, G. (1981). Behavioral aspects of text editors. *ACM Computing Surveys, 13*(1), 33–70.

Engelbart, D. C., and English, W. K. (1968). A research center for augmenting human intellect. *Proceedings of the 1968 fall joint computer conference* (pp. 395–410). Montvale, NJ: AFIPS.

Furuta, R., Scofield, J., and Shaw, A. (1982). Document formatting systems: survey, concepts, and issues. *ACM Computing Surveys, 14*(3), 417–472. Also in Nievergelt, J., Coray, G., Nicoud, J.-D., and Shaw, A. C. Eds. (1982). *Document preparation systems.* New York: North Holland.

Galambos, J. A., Wikler, E. S., Black, J. B., and Sebrechts, M. M. (1983). How you tell your computer what you mean: ostension in interactive systems. *Proceedings of the CHI'83 human factors in computing systems conference* (pp. 182–185). Boston, MA: ACM.

Gomez, L. M., Egan, D. E., Wheeler, E. A., Sharma, D. K., and Gruchacz, A. M. (1983). How interface design determines who has difficulty learning to use a text editor. *Proceedings of CHI'83 human factors in computing systems conference* (pp. 176–181). Boston, MA: ACM.

Grudin, J., and Barnard, P. (1984). The cognitive demands of learning and representing command names for text editing. *Human Factors, 26*(4), 407–422.

Halbert, D. C. (1984). Programming by example (Ph.D. dissertation), Berkeley, CA: University of California at Berkeley.

Hauptmann, A. G., and Green, B. F. (1983). A comparison of command, menu-selection and natural-language computer programs. *Behaviour and Information Technology, 2*(2), 163–178.

Hirsh-Pasek, K., Nudelman, S., and Schneider, M. L. (1982). An experimental evaluation of abbreviation schemes in limited lexicons. *Behaviour and Information Technology, 1*(4), 359–369.

International Organization for Standardization. (1973a). *7-bit coded character set for information processing exchange* [ISO 646–1973(e)].

International Organization for Standardization. (1973b). *Code extension techniques for use with the ISO 7-bit coded character set* [ISO 2022–1973(E)].

International Organization for Standardization. (1982). *Information processing-ISO 7-bit and 8-bit coded character sets—code extension techniques.* (Submitted on 02–04–1982, ISO/DIS 2022.2, ISO/TC 97).

Knuth, D. E. (1984). *The TₑXbook.* Reading, MA: Addison-Wesley.

Meyrowitz, N., and Van Dam, A. (1982). Interactive editing systems, parts I and II. *ACM Computing Surveys, 14*(3), 321–415. Also in Nievergelt, J., Coray, G., Nicoud, J.-D., and Shaw, A. C. Eds. (1982). *Document preparation systems,* New York: North Holland.

Moran, T. P. (1981). The command language grammar: a representation for the user interface of interactive computer systems. *International Journal of Man-Machine Studies, 15*(1), 3–50.

Moran, T. P. (1983). Getting into a system: external-internal task mapping analysis. *Proceedings of the CHI'83 human factors in computing systems conference* (pp. 45–49). Boston, MA: ACM.

Neal, A. S., and Darnell, M. J. (1984). Text-editing performance with partial-line, partial-page, and full-page displays. *Human Factors, 26*(4), 431–441.

Newman, W. M., and Sproull, R. F. (1979). *Principles of interactive computer graphics,* 2nd ed. New York: McGraw-Hill.

Poller, M. F., and Garter, S. K. (1984). The effects of modes on text editing by experienced editor users. *Human Factors, 26*(4), 449–462.

Reid, B. K. (1980). Scribe: a document specification language and its compiler (Ph.D. dissertation), Pittsburgh, PA: Carnegie-Mellon University.

Reid, B. K., and Walker, J. H. (1980). *Scribe user's manual,* 3rd ed. Pittsburgh, PA: Unilogic Ltd.

Reisner, P. (1981a). Human factors studies of database query languages: a survey and assessment. *ACM Computing Surveys, 13*(1), 13–31.

Reisner, P. (1981b). Using a formal grammar in human factors design of an interactive graphics system. *IEEE Transactions on Software Engineering, SE-7*(2), 229–240.

Roberts, T. L. (1980). Evaluation of computer text editors (Ph.D. dissertation). Stanford, CA: Stanford University. Available as Report AAD 80–11699 from University Microfilms, Ann Arbor, MI.

Roberts, T. L., and Moran, T. P. (1983). The evaluation of text editors: methodology and empirical results. *Communications of ACM, 26*(4), 265–283.

Seybold, J. W. (1982). Document preparation systems and commercial typesetting. In Nievergelt, J., Coray, G., Nicoud, J. D., and Shaw, A. C. Eds., *Document preparation systems* (pp. 243–264). New York: North Holland.

The Seybold report on professional computing; The Seybold report on office systems; The Seybold report on publishing systems, Media, PA.

Shneiderman, B. (1983). Design issues and experimental results for menu selection systems (Technical Report CS-TR-1303). University of Maryland Computer Science Center.

Welty, C., and Stemple, D. W. (1981). Human factors comparison of a procedural and a nonprocedural query language. *ACM Transactions on Database Systems, 6*(4), 626–649.

Xerox Corporation. (1984). Xerox system integration standard: Xerox character code standard (XSIS 058404). Stamford,CT: Xerox Corporation.

CHAPTER 11.7

DOCUMENTATION FOR SOFTWARE SYSTEMS

SYLVIA B. SHEPPARD

Computer Technology Associates
Lanham, Maryland

The author would like to express her appreciation to Dr. John J. O'Hare and Dr. Deborah A. Boehm-Davis for their comments on an early draft of this chapter.

11.7.1 THE NEED FOR DOCUMENTATION

Appropriate documentation is necessary in order to provide for the efficient and cost-effective development, operation, and maintenance of a software system. Documentation supports the systematic exchange of information among managers, system developers, and users throughout the stages of the software life cycle. A variety of different documents fulfills the following four functions (Department of Commerce, 1976):

1. Provides a description of the capabilities of the system, which enables users to determine whether the system serves (or will serve) their needs.
2. Provides a record of design decisions made and implemented, thus permitting later development, use, or modification of the software.
3. Provides technical materials for review at planned milestones during the development and acquisition of the system.
4. Provides procedural information about operation or maintenance of the software.

11.7.2 THE COMPOSITION OF DOCUMENTS

Writing documentation for computer systems is like writing prose: manuals and other documents should be clear, concise, accurate, and complete. The normal rules of language composition apply, but additional rules are needed for style, format, and content because of the technical nature of the text.

Before beginning a manual there are several questions the writer should answer in order to focus its style, format, and content for maximum effectiveness (Adams and Halasz, 1983; Hendricks, Kilduff, Brooks, Marshak, and Doyle, 1983; Thomas and Carroll, 1981):

1. What information is to be documented? Are there some areas of information that should be emphasized more than other areas? It is important to focus on the objectives of the document. Concern with presenting minute details may result in failure to provide the materials in a coherent, understandable format.
2. What is the nature of the audience? How abstract should the presentation be? Potential users of the manuals should be identified, and the contents should be tailored to their skill levels and requirements.
3. What kinds of tasks does the user need to know how to do? Instructions need to be clear and complete. Errors should be anticipated, and procedures should be provided for actions to amend errors.
4. How will the manual be used? Training workbooks will have different contents and be organized differently from reference books. The contents of the manual should be tailored to its intended use.
5. Will reference materials be readily available? If not, the manual must be self-contained.
6. In what environment will the manual be used? Desks and workstations vary in available surface area, and the size and shape of documents should be varied to accommodate the environment in which the documentation will be used.
7. What is the budget for producing the documentation? How many people can be supported and at what level? What is the timeframe for completing the documentation? Planning ahead will aid in meeting time and budget constraints.
8. Who will evaluate the manual for accuracy, completeness, and clarity of presentation? A reviewer who understands the content area and is familiar with the intended users should examine the documentation for inaccuracies, omissions, and words and phrases that might be confusing.
9. How will the documentation be revised and maintained? It is important to provide some mechanism for periodic evaluation and for making changes to the documentation after it has been tested and used. A revised edition of a manual should clearly mark paragraphs that differ from the previous version. Inserting brackets in the margins is one method for annotating changes.

11.7.2.1 Style

A number of suggestions for good writing style follow.

1. Write at a reading level appropriate for the audience. Most documents are designed to be read at about the eighth-grade educational level. To determine the average grade-level, use the Gunning Readability Formula:
    a. Randomly select a sample segment of 100 words from the document.
    b. Count the number of sentences and divide by 100 to get an average sentence length.
    c. Count the words in the sample with three or more syllables.
    d. Add that number to the average sentence length.
    e. Multiply the sum by 0.4. This figure is an estimate of the grade level for the documentation.
    f. When the grade-level estimate exceeds eight, edit the document until the estimate falls within that limit (Adams and Halasz, 1983).
2. Present actions in logical sequences and describe steps in the order in which they are to be performed. For example:
    Poor: "Before answering the questions, read the instructions."
    Better: "Read the instructions before you answer the questions."
3. Avoid negatives. For example:
    Poor: "Do not write in column 7."
    Better: "Leave column 7 blank."
4. Use active voice instead of passive voice. For example:
    Poor: "Your name should be entered on line 5."
    Better: "Enter your name on line 5."
5. Use simple, grammatical structures.
6. Use short paragraphs that have a single topic.
7. Place the main topic of a sentence near its beginning.
8. Use prose decision trees to guide users through difficult tasks; for example, "If the answer to Question 6 is 'Yes,' go to Section 3. If the answer is 'No,' go to Section 4."
9. Sequence the content of the material from simple to complex or more general to more specific.
10. Use jargon and technical terms sparingly. Define each term as it appears and give its acronym in parentheses, if appropriate. Reference the definition if you use the term in a later section of the text.
11. Use terms that the user will understand. Give complete, explicit definitions to explain their connotations, particularly if the terms are used in a different context than found in normal speech.
12. Allow a term to have only one meaning. Terms should be used consistently throughout the text.
13. Use abbreviations consistently. Be sure to define them. Make them distinct from the other abbreviations that are used.
14. Abbreviate only if an abbreviation is considerably shorter or more meaningful than the original term.
15. Define meta-symbols carefully; for example, for the command " COPY file-1, file-2 " explain that file-1 and file-2 are filenames to be supplied by the user. Use examples to accompany textual descriptions (e.g., " COPY SALARY-83, SALARY-ARCHIVE ").
16. Use a reference system that is appropriate for the user; for example, a 12-hr clock is familiar to civilians, but a 24-hr clock is appropriate for military users (Hendricks et al., 1983).

Instructional Materials

A considerable portion of documentation is used as instructional material. Adams and Halasz* offer practical suggestions for the design of materials that help people to learn:

First, *gain the users' attention.* Entice them into wanting to know more.

Second, *inform users of what they are going to learn.* Summarize the most important content through concise objectives.

Third, *stimulate recall of prior learning.* Help software users tie what they are going to learn to what they already know.

---

* From Kay A. Adams & Ida M. Halasz, *25 Ways to Improve Your Software User Manuals,* 1983. Used with permission.

Fourth, *present the heart of the content.* Explain rules and concepts.

Fifth, *guide learning through prompts, hints, and examples.* The more time spent on a particular task, the greater the achievement level.

Sixth, *ask users to perform.* Provide real problems that users must solve. People learn by doing.

Seventh, *provide feedback.* User success is the most important stimulus for continued learning.

Eighth, *enhance retention and transfer of what is learned.* Include varied examples and applications. Tie concepts to a larger meaningful context.

Ninth, *summarize the essentials of what has been taught.* Repetition facilitates learning.

Tenth, *assess overall performance.* Once feedback has been given, allow the user to demonstrate mastery of all the learned knowledge.

11.7.2.2 Formatting

Careful formatting makes text more appealing and encourages reader interest. The following list has suggestions for formatting prose:

1. Arrange the material to provide enough "white" space. Unbroken strings of text may discourage the reader.

2. Use visual markers to break text into meaningful sections. Use dashed or solid lines, boxes, underlining, lists, bullets, dashes, and variations in printing size and type font to vary the visual appearance of the text.

3. Include tables, drawings, charts, photographs, or other illustrative materials, where appropriate, to add visual interest and better comprehensibility. Illustrations can eliminate pages of text and enhance the reader's understanding of different and complex information. "A picture *is* worth a thousand words."

4. Identify instructional objectives and use the concept of advance organizers to highlight those objectives. An advance organizer is a short explanation of the purpose of a body of material and is presented prior to the material. It does not contain material from the text but provides general concepts and ideas the reader can use to put what is about to be read in context. Organizers also help the reader to find the way from one section of a document to another. Ausubel (1968) contends that learning of technical material is enhanced by advance organizers. Figure 11.7.1 (Hendricks et al., 1983, p. 9–1), which illustrates the primary purposes of a chapter on screens and printers, is a good example of the use of an advance organizer.

Alternatives to Prose for Expressing Complex Information

Up to this point the discussion has focused on the production of documentation in natural language. Normal prose, however, is not the only medium for representing system and program specifications and design decisions. A variety of other media are used successfully to augment or replace prose. A representative set of these types of documentation is presented briefly.

Program Design Language. Program design language (PDL), sometimes called structured English, is a constrained form of natural language useful for representing descriptions of program processes. As formulated by Caine, Farber, and Gordon, PDL was based on IF···THEN and DO ···ENDDO control structures (Caine and Gordon, 1975). However, a myriad of variations now exist because the concept is easily adaptable to the structures of many programming languages.

CHAPTER 9
SCREENS AND PRINTERS

The purposes of this chapter are to:

- Identify various physical screen characteristics of the display that affect legibility of the display.
- Identify printer and paper qualities that produce readable, usable output.

Section I. SCREENS

LEGIBILITY

One of the main requirements of a visual display is that the displayed information must background and the characters.

- The level of luminance recommend

Fig. 11.7.1. An example of the effective use of advance organizers. [From Hendricks et al. (1983)].

PDL is useful because it can represent either high-level design or detailed design. It lends itself well to structured decomposition and stepwise refinement techniques. See Figure 11.7.2, taken directly from DeMarco (1978).

POLICY FOR INVOICE PROCESSING

If the amount of the invoice exceeds $500,

    If the account has any invoice more than 60 days overdue, hold the confirmation pending resolution of the debt.

    Else (account is in good standing), issue confirmation and invoice.

Else (invoice $500 or less),

    If the account has any invoice more than 60 days overdue, issue confirmation, invoice and write message on the credit action report.

    Else (account is in good standing), issue confirmation and invoice.

Fig. 11.7.2. A problem expressed in a program design language. Reprinted by permission. From *Structured Analysis and System Specification,* by Tom DeMarco. Copyright © 1978, 1979 by Yourdon Inc., (New York; Yourdon Press, 1979), p. 43.

Flowchart. A flowchart is a diagrammatic technique used to express the control structures of design. A standard set of symbols is used to encode information about processes, decisions, hardware, and entry and termination points. The symbols are connected by arrows showing the order of the

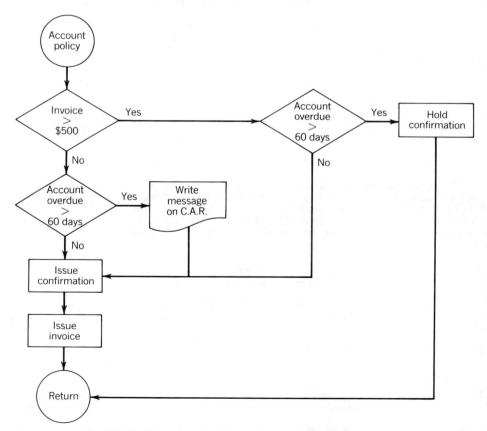

Fig. 11.7.3. A flowchart for the problem expressed in Figure 11.7.2.

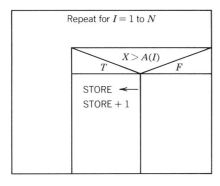

Fig. 11.7.4. A Nassi-Shneiderman iteration.

operations, which are performed in sequence beginning at the top of the page. Figure 11.7.3 shows a flowchart derived from the example in Figure 11.7.2.

Nassi–Shneiderman Chart. Nassi–Shneiderman charts (Nassi and Shneiderman, 1973) provide an alternative to flowcharts. A single process is completely contained within a rectangular box. "Insideness" or containment is used to represent the hierarchical relationship between program blocks. There are three basic elements of a Nassi–Shneiderman chart:

1. A rectangle is used to represent the assignment function, for example, $I = I + 1$.
2. A half-diamond shape represents a decision, for example, Yes or No, True or False.
3. A box enclosed within a larger box and located in the lower right-hand corner indicates repetition, that is, iteration. See Figure 11.7.4.

Figure 11.7.5 shows the example in Figure 11.7.2 illustrated in a Nassi-Shneiderman chart.

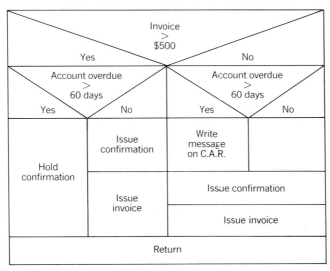

Fig. 11.7.5. A Nassi-Shneiderman chart for the problem expressed in Figure 11.7.2.

Decision Table. A decision table is a special purpose tool for representing complex logical relationships (Montalbano, 1974; Pollack, Hick, and Harrison, 1971). A list of initial conditions (tabulated in a left-hand column) and the system responses, actions, or rules (column headings) form the framework of the table. Individual cells in the table indicate the presence or absence of the action for each condition. See Figure 11.7.6 from DeMarco (1978).

Rules				
Conditions	1	2	3	4
1. Invoice > $500	Y	N	Y	N
2. Account overdue by 60+ days	Y	Y	N	N
Actions				
1. Issue Confirmation	N	Y	Y	Y
2. Issue Invoice	N	Y	Y	Y
3. Message to C.A.R.	N	Y	N	N

Fig. 11.7.6. A decision table. Reprinted by permission. From *Structured Analysis and System Specification* by Tom DeMarco. Copyright © 1978, 1979 by Yourdon Inc., (New York: Yourdon Press, 1979), page 43.

Decision tables can be used to represent complex problem specifications, but they are not useful for illustrating design. The basic premise of the decision table is that the responses or actions are predetermined for every condition (or set of conditions) that can occur. In program design it is necessary to account for any condition that is changed during the execution of the algorithm; decision tables cannot be used to show those changes (Fitter and Green, 1979).

Decision Tree. A decision tree is an alternative way to represent the information contained in a decision table. See Figure 11.7.7, modified from DeMarco (1978).

Data Flow Diagrams. A data flow diagram (DFD) is a graphic aid for representing the flow of data in a system. It is one of the tools of the structured-analysis approach recommended by DeMarco (1978). There are four basic elements of the DFD:

1. Circles, or "bubbles," which represent processes.
2. Vectors, or directed lines, which represent the flow of data.
3. Straight lines, which indicate files.
4. Boxes, which show either external sources or sinks for the data. A source is an originator of data; a sink is a receiver. See Figure 11.7.8 from DeMarco (1978).

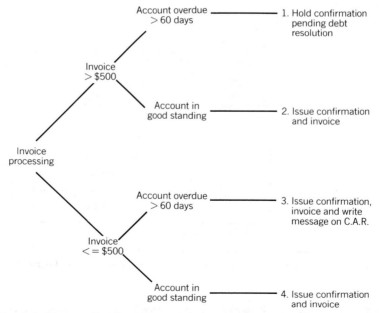

Fig. 11.7.7. A decision tree. Reprinted by permission. From *Structured Analysis and System Specification* by Tom DeMarco. Copyright © 1978, 1979 by Yourdon Inc. (New York: Yourdon Press, 1979), page 44.

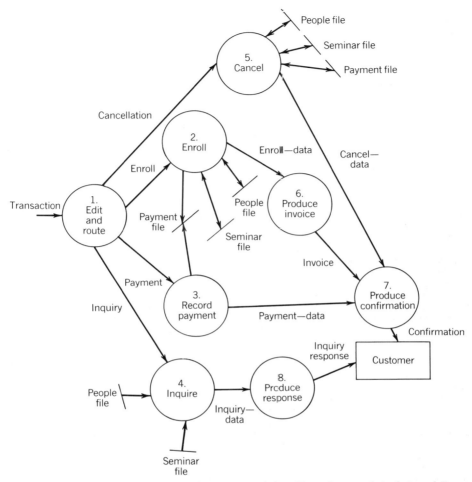

Fig. 11.7.8. A data flow diagram. Reprinted by permission. From *Structured Analysis and System Specification* by Tom DeMarco. Copyright © 1978, 1979 by Yourdon Inc. (New York: Yourdon Press, 1979), page 44.

DFDs differ from flow charts in that information about the initiation of a process and the sequence of control occurring within the process does not appear. DFDs are, however, useful for validating the correctness of information presented, because material presented in this manner more closely approximates a user's view of the system (DeMarco, 1978).

Hierarchy Plus Input, Process, Output (HIPO). HIPO is a graphical way to represent the relationships among the processes of a system and to describe the inputs and outputs of those processes (Stay, 1976). Each process is a named, numbered box in the visual table of contents. See Figure 11.7.9. Each individual process is further described with a HIPO chart. See Figure 11.7.10.

Related Research

There is evidence to suggest that media other than natural language promote better comprehension of complex procedural instructions. Wright and Reid (1973) asked nontechnical participants to use narrative, short sentences, decision tables, and tree charts to solve travel problems with time, cost, and distance tradeoffs. Narrative proved to be the least effective representation for both easy and difficult problems. The other formats were equally useful for easy problems, but tree charts were most effective for difficult problems.

In a similar study, Blaiwes (1974) studied flowcharts and short sentences as alternate ways to

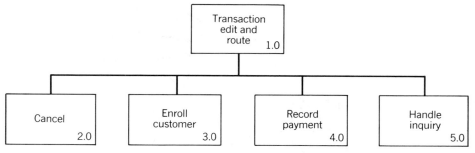

Fig. 11.7.9. A visual table of contents.

provide instructions for a control console. For easy problems, participants in the experiment performed equally well with either form of instruction; but, for difficult problems, flowcharts were associated with fewer errors.

Kamman (1975) compared the effectivness of prose with two forms of flowcharts as presentation media for telephone dialing instructions. Two groups of participants, housewives and engineers, found flowchart formats to be more effective than prose in following the instructions.

Miller (1981) asked computer-naive students to write procedures for file-searching problems. The resulting prose instructions were unclear because they contained a high degree of contextual referencing and because the use of pronouns made it difficult to determine what entity was being referenced. Miller concluded that it is difficult to deal with complex procedural instructions in a medium that is not suited to structuring the problem and the implementation process. He suggested that a constrained subset of natural language would be more useful than normal prose for program instructions.

Research has also been conducted with various documentation formats during programming tasks. Ramsey, Atwood, and Van Doren (1978) asked graduate computer science students to express a design in the form of either a flowchart or a PDL. While they found no difference in code implementation, the designs expressed in PDL were judged of better quality because they contained more procedural detail than those expressed in flowcharts.

Sheppard, Bailey, and Kruesi-Bailey (1984) and Sheppard, Kruesi, and Curtis (1980) conducted a series of four experiments with a variety of programming tasks (comprehension, coding, debugging, and modification). They combined four types of symbology (natural language, abbreviated natural language, PDL, and flowcharts) with three types of spatial arrangement (sequential or vertical flow, branching or flowchartlike, and hierarchical or treelike) to produce 12 types of specification formats. The differences among the types of symbology were clear. The natural language format was associated with significantly poorer performance; the PDL evoked the best performance. The effect of spatial arrangement was less pronounced; however, the branching arrangement was mildly superior to the others, particularly in reducing the number of errors related to control flow. The sequential and branching PDL versions were clearly associated with good performance; they both represented succinct formats that clearly described the structure of the algorithms.

In summary, natural language is not always the best medium for expressing complex, procedural instructions. Other formats appear to be associated with better performance. For program design, a PDL appears quite workable. As always, the user must select his or her medium of presentation based on the nature of the task and the environment in which the task is to be performed.

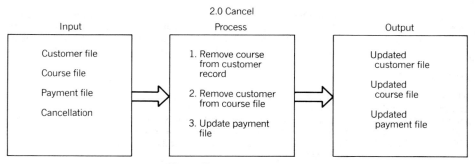

Fig. 11.7.10. A HIPO chart.

11.7.2.3 Content

The following organizational aids are useful for organizing the content of computer system manuals and guides. The major aspects are presented in order of their inclusion in a document.

1. *Title Page.* Name of document, author(s), date published, and name of sponsor or sponsoring organization, where appropriate. Security information may also be described here as needed.
2. *Preface, Foreword, or Abstract.* A short description of the document and its contents. It identifies the audience for which the document is intended and gives a brief description of any general instructions or caveats for its use.
3. *Table of Contents.* Topic headings or topic sentences and the page number upon which each topic begins. It divides the text into chapters, units, sections, and subsections.
4. *List of Figures, Exhibits, Tables and/or Illustrations.* The captions of all nontextual, illustrative inserts and the page numbers upon which they appear.
5. *Chapters or Major Sections.* The main part of the document. Divisions are made easier to locate when separated by index tabs or heavy divider sheets
6. *Summary.* A short synopsis of key concepts that helps the reader to remember what has just been read. There may be a summary following each chapter or one following the combined chapters.
7. *References.* A full citation of the works that have been cited in the text. Where many references are cited for each chapter or major section, the chapters or major sections may contain individual reference lists. Otherwise, one reference list is sufficient.
8. *Bibliography.* A citation of works that were consulted but are not specifically cited in the text and listed in the references.
9. *Chapter Checklists, Quizzes, or Exercises.* Optional material to illustrate how concepts are used in various situations and to allow the user to practice what has just been presented. Feedback is helpful to someone who is using a piece of documentation as a learning device; material that is not understood thoroughly can be reviewed before the reader proceeds to new material.
10. *Glossary of Terms and Acronyms.* An alphabetical listing of technical terms and acronyms employed in the text.
11. *Appendices.* Reserve for materials that are very long, detailed, or peripheral to the subject matter.
12. *Index.* Page numbers for key concepts, subjects, techniques, and names that appear in the text.
13. *Blank Pages.* Sometimes included for insertion of the reader's reference notes or memoranda.

11.7.2.4 Documentation Throughout the Software Life Cycle

The software life cycle can be divided into three phases—the initiation phase, the development phase, and the operations phase. See Figure 11.7.11.

Initiation Phase

During the initiation phase, the need for a new or revised system is recognized. The problems, opportunities and goals are defined. Decisions include whether or not the project should be pursued and, if so,

Initiation Phase	Development Phase				Operation Phase
	Requirements definition stage	Design stage	Programming stage	Testing stage	
	Functional requirements document	System/ subsystem specification			
		Program specification	Operations manual		
	Data requirements document	Database specification	Program maintenance manual		
		User's manual			
		Test Plan		Test analysis report	

Fig. 11.7.11. Documentation within the software life cycle. [Adapted from Department of Commerce (1976)].

what priority it should be given. Activities during this phase might include a feasibility study and/or a cost-benefit analysis.

Development Phase

The development phase of a system can be subdivided into four stages: requirements definition, design, programming, and testing. The stages provide planned intervals at which completed work can be reviewed. There are opportunities for appropriate technical approvals at the end of each interval, and there are also opportunities for cost and schedule reviews. A project may have a limited commitment, and continuation to the next stage may be contingent on the progress achieved in the preceding stage.

Operations Phase

This is the period of the life cycle during which the software is used. Remaining errors in the software are identified and corrected, and changes are made to accommodate new or additional requirements. A post-installation evaluation may be conducted to compare the benefits and cost savings realized by the system to the benefits and savings that had been projected (Department of Commerce, 1976).

System Development Documents

A variety of system documents may be produced during the development phase of the software life cycle. Some projects will use all of these documents; other projects may omit some of them, depending on the size, complexity, and inherent nature of the particular application. Furthermore, several documents may be able to be combined into one report (e.g., functional and data requirements are often combined).

Each stage of the development phase is described below, and typical types of documentation are listed. The purpose of each type of documentation, its intended audience, and a suggested format for its construction are included. The reader will note some redundancy in the various documents; this is deliberate. Introductory information and descriptions of inputs, outputs, and the environment are repeated for two reasons: to ensure the self-contained nature of each document and to provide a common format or pattern of organization. Patterns help to increase understanding and ease the preparation of documents and the search for information. Standard formats help to tie a series of documents together to provide perception of an integrated whole. The formats given in the following paragraphs are merely suggestions; they are taken from Department of Commerce (1976).

Requirements Stage.   During the requirements stage the user's problem or need is defined in detail. Activities include an analysis of the deficiencies of the present system or method being used; a definition of the basic requirements for the hardware, software, and documentation of the target system; and an analysis of the benefits expected from the target system. In addition, considerable effort may be expended upon staging the product. Staging provides for the early release of a minimally adequate system and includes plans for a graceful evolution to full capacity. A functional requirements document and a data requirements document are produced during the requirements stage of the development phase.

1. *Functional Requirements Document.*   This defines the needs of the proposed system and provides a basis for users and designers to agree on an initial description of a method for satisfying those needs. The emphasis here is on what is to be done and not on how to do it. Descriptions of the existing system or manual procedure and its inherent deficiencies are often included. Software and hardware requirements are described independently where possible. See Figure 11.7.12 for specific information that might be included.

2. *Data Requirements Document.*   This provides a description of the data element, its sources, input and output media, initial values, and measurement scales. The data collection requirements are also included. This information is sometimes included as part of the functional requirements document instead of being defined in a separate document. Designers need to be sure that users understand and agree with the data requirements. See Figure 11.7.13.

Design Stage.   During the design stage, the specific functions to be performed and the constraints on those functions are analyzed, and alternate solutions are considered. In the requirements stage decisions are made about what is to be done. In this stage decisions are made about how it is to be done. Documentation in this stage includes system and subsystem specifications. On some projects user's manuals and test plans are also completed during the design stage. Traditionally, these two forms of documentation were completed during later stages of the development phase. However, it has become increasingly obvious that production of these documents early in the life cycle aids in improving the quality of the system. For example, if the user's guide is developed before the system design is completed, the designer can include features in the design that are deemed desirable from a human factors perspective. An awareness of these features comes from the processes involved in writing the user's guide. Conversely, if the user's guide is not documented until after the code has been developed, the user interface may have undesirable features that are difficult to correct at that point in the development cycle.

The test plan is another type of documentation that is more useful when completed during the design stage. Early knowledge of the test plan, against which code will be validated, can stimulate more thorough and complete design and coding processes.

1. *System/Subsystem Specification.* This is sometimes called the external specification or high-level design. It is a description of the system logic for meeting the needs described in the functional

---

Functional Requirements Document

1. GENERAL INFORMATION

    1.1. Summary. Summarize the general nature of the software to be developed.

    1.2. Environment. Identify the project sponsor, developer, user, and computer center or network where the software is to be implemented.

    1.3. References. List applicable references, such as:

        a. Project request (authorizations).
        b. Previously published documents on the project.
        c. Documentation concerning related projects.
        d. Other reference documents.

2. OVERVIEW

    2.1. Background. Present the purpose and scope of the software, and any background information that would orient the reader. Explain relationships with other software.

    2.2. Objectives. State the major performance objectives of the software, including examples. Identify anticipated operational changes that will affect the software and its use.

    2.3. Existing Methods and Procedures. Describe the current methods and procedures that satisfy the existing objectives. Include information on:

        a. Organizational and personnel responsibilities.
        b. Equipment available and required.
        c. Volume and frequency of inputs and outputs.
        d. Deficiencies and limitations.
        e. Pertinent cost considerations.

    Illustrate the existing data flow from data acquisition through its processing and eventual output. Explain the sequence in which operational functions are performed by the user.

    2.4. Proposed Methods and Procedures. Describe the proposed software and its capabilities. Identify techniques and procedures from other software that will be used or that will become part of the proposed software. Identify the requirements that will be satisfied by the proposed software. Include information on:

        a. Organizational and personnel responsibilities.
        b. Equipment available and required.
        c. Volume and frequency of inputs and outputs.
        d. Deficiencies and limitations.
        e. Pertinent cost considerations (developmental as well as operational).

    Illustrate the proposed data flow to present an overall view of the planned capabilities. Describe any capabilities in the existing software that may be changed by the proposed software. State the reasons for these changes. Explain the sequence in which operational functions are to be performed by the user.

Fig. 11.7.12. Contents of a functional requirements document [from Department of Commerce (1976, pp. 16–19)].

2.5. **Summary of Improvements.** Itemize improvements to be obtained from the proposed software, such as:

a. New capabilities.
b. Upgraded existing capabilities.
c. Elimination of existing deficiencies.
d. Improved timeliness, e.g., decreased response time or processing time.
e. Elimination or reduction of existing capabilities that are no longer needed.

2.6. **Summary of Impacts.** Summarize the anticipated impacts of the proposed software on the present system, in the following categories:

2.6.1. Equipment Impacts. Summarize changes to currently available equipment, as well as new equipment requirements and building modifications.

2.6.2. Software Impacts. Summarize any additions or modifications needed to existing applications and support software in order to adapt them to the proposed software.

2.6.3. Organizational Impacts. Summarize organizational impacts, such as:

a. Functional reorganization.
b. Increase/decrease in staff level.
c. Upgrade/downgrade of staff skills.

2.6.4. Operational Impacts. Summarize operational impacts, such as modifications to:

a. Staff and operational procedures.
b. Relationships between the operating center and the users.
c. Procedures of the operating center.
d. Data (sources, volume, medium, timeliness).
e. Data retention and retrieval procedures.
f. Reporting methods.
g. System failure consequences and recovery procedures.
h. Data input procedures.
i. Computer processing-time requirements.

2.6.5. Developmental Impacts. Summarize developmental impacts, such as:

a. Specific activities to be performed by the user in support of development of the proposed software.
b. Resources required to develop the data base.
c. Computer processing resources required to develop and test the new software.

2.7. **Cost Considerations.** Describe resource and cost factors that may influence the development, design, and continued operation of the proposed software. Discuss other factors which may determine requirements, such as interfaces with other automated systems and telecommunication facilities.

2.8. **Alternative Proposals.** If alternative software has been proposed to satisfy the requirements, describe each alternative. Compare and contrast the alternatives. Explain the selection reasoning.

Fig. 11.7.12. (*Continued*)

Functional Requirements Document

---

3. REQUIREMENTS

3.1. Functions. State the functions required of the software in quantitative and qualitative terms, and how these functions will satisfy the performance objectives.

3.2. Performance. Specify the performance requirements.

    3.2.1. Accuracy. Describe the data-accuracy requirements imposed on the software, such as:

        a. Mathematical.
        b. Logical.
        c. Legal.
        d. Transmission.

    3.2.2. Validation. Describe the data-validation requirements imposed on the software.

    3.2.3. Timing. Describe the timing requirements imposed on the software, such as, under varying conditions:

        a. Response time.
        b. Update processing time.
        c. Data transfer and transmission time.
        d. Throughput time.

    3.2.4. Flexibility. Describe the capability for adapting to changes in requirements, such as:

        a. Changes in modes of operation.
        b. Operating environment.
        c. Interfaces with other software.
        d. Accuracy and validation timing.
        e. Planned changes or improvements.

    Identify the software components which are specifically designed to provide this flexibility.

3.3. Inputs-Outputs. Explain and show examples of the various data inputs. Specify the medium (disk, cards, magnetic tape), format, range of values, accuracy, etc. Provide examples and explanation of the data outputs required of the software, and any quality control outputs that have been identified. Include descriptions or examples of hard copy reports (routine, situational and exception) as well as graphic or display reports.

3.4. Data Characteristics. Describe individual and composite data elements by name, their related coded representations, as well as relevant dictionaries, tables, and reference files. Estimate total storage requirements for the data and related components based on expected growth.

3.5. Failure Contingencies. Specify the possible failures of the hardware or software, the consequences (in terms of performance), and the alternative courses of action that may be taken to satisfy the information requirements. Include:

Fig. 11.7.12. (*Continued*)

a. Back-up. Specify back-up techniques, i.e., the redundancy available in the event the primary system element goes down. For example, a back-up technique for a disk medium would be to record periodically the contents of the disk to a tape.

b. Fallback. Explain the fallback techniques, i.e., the use of another system or other means to accomplish some portion of requirements. For example, the fallback technique for an automated system might be manual manipulation and recording of data.

c. Recovery and Restart. Discuss the recovery and restart techniques, i.e., the capability to resume execution of software from a point in the software subsequent to which a hardware or software problem occurred, or the re-running of the software from the beginning.

4. OPERATING ENVIRONMENT

4.1. Equipment. Identify the equipment required for the operation of the software. Identify any new equipment required and relate it to specific functions and requirements to be supported. Include information such as:

a. Processor and size of internal storage.
b. Storage, online and offline, media, form, and devices.
c. Input/output devices, online and offline.
d. Data transmission devices.

4.2. Support Software. Identify the support software and describe any test software. If the operation of the software depends on changes to support software, identify the nature and planned date of these changes.

4.3. Interfaces. Describe the interfaces with other software.

4.4. Security and Privacy. Describe the overall security and privacy requirements imposed on the software. If no specific requirements are imposed, state this fact.

4.5. Controls. Describe the operational controls imposed on the software. Identify the sources of these controls.

5. DEVELOPMENT PLAN

Discuss in this section the overall management approach to the development and implementation of the proposed software. Include a list of the documentation to be produced, time frames and milestones for the development of the software, and necessary participation by other organizations to assure successful development.

Fig. 11.7.12. (*Continued*)

Data Requirements Document

1. **GENERAL INFORMATION**

    1.1. **Summary.** Summarize the general nature of the software for which these data requirements are being defined.

    1.2. **Environment.** Identify the project sponsor, developer, user organization, and computer center where the software is to be installed. Show the relationships of these data requirements and those of other software.

    1.3. **References.** List applicable references, such as:

    a. Project request (authorization).
    b. Previously published documents on the project.
    c. Documentation concerning related projects.
    d. Other reference documents.

    1.4. **Modification of Data Requirements.** Describe or reference procedures for implementing and documenting changes to these data requirements.

2. **DATA DESCRIPTION**

    Separate the data description into two categories, static data and dynamic data. Static data are defined as that data which are used mainly for reference during operation and are usually generated or updated in widely separated time-frames independent of normal runs. Dynamic data include all data which are intended to be updated and which are input during a normal run or are output. Arrange the data elements in each category in logical groupings, such as functions, subjects, or other groupings which are most relevant to their use.

    2.1. **Static Data.** List the static data elements used for either control or reference purposes.

    2.2. **Dynamic Input Data.** List the dynamic input data elements which constitute the data intended to be changed by a normal run or during online operation.

    2.3. **Dynamic Output Data.** List the dynamic output data elements which constitute the data intended to be changed by a normal run or during online operation.

    2.4. **Internally Generated Data.** List the internally generated data of informational value to the user or developer.

    2.5. **Data Constraints.** State the constraints on the data requirements. Indicate the limits of the data requirements with regard to further expansion or utilization, such as the maximum size and number of files, records, and data elements. Emphasize the constraints that could prove critical during design and development.

Fig. 11.7.13. Contents of a data requirements document [from Department of Commerce (1976, pp. 22–23)].

3. DATA COLLECTION

   3.1. **Requirements and Scope.** Describe the type of information required to document the characteristics of each data element. Specify information to be collected by the user and that to be collected by the developer. It should be logically grouped and presented. Include:

      a. Source of Input. Identify the source from which the data will be entered, e.g., an operator, station, organizational unit, or its component group.
      b. Input Medium and Device. Identify the medium and hardware device intended for entering the data into the system. In those cases where only certain special stations are to be legitimate entry points, they should be specified.
      c. Recipients. Identify the intended recipients of the output data.
      d. Output Medium and Device. Identify the medium and hardware device intended for presenting output data to the recipient. Specify whether the recipient is to receive the data as part of a hard copy printout, a symbol in a CRT display, a line on a drawing, a colored light, an alarm bell, etc. If the output is to be passed to some other automated system, the medium should be described, such as magnetic tape, punched cards, or an electronic signal to a solenoid switch.
      e. Critical Value(s). One value from a range of values of data may have particular significance to a recipient.
      f. Scales of Measurement. Specify for numeric scales, units of measurement, increments, scale zero-point, and range of values. For non-numeric scales, any relationships indicated by the legal values should be stated.
      g. Conversion Factors. Specify the conversion factors of measured quantities that must go through analog or digital conversion processes.
      h. Frequency of Update and Processing. Specify the expected frequency of data change and the expected frequency of processing input data. If the input arrives in a random or in an "as occurred" manner, both the average frequency and some measure of the variance must be specified.

   3.2. **Input Responsibilities.** Provide recommendations as to responsibilities for preparing specific data inputs. Include any recommendations regarding the establishment of a data input group. Specify by source those data inputs dependent on interfacing software or unrelated organizations.

   3.3. **Procedures.** Provide specific instructions for data collection procedures. Include detailed formats where applicable, and identify expected data communications media and timing of inputs.

   3.4. **Impacts.** Describe the impacts of these data requirements on equipment, software and the user and developer organizations.

Fig. 11.7.13. (*Continued*)

requirements document. The specification includes descriptions of the major functions to be performed by the software; the contents of all files, output reports, and displays; system security and control features; and performance requirements. This document is targeted for the user, who must agree to the form and function of the design before work is continued. See Figure 11.7.14.

    2. *Program Specification.* This is sometimes called the internal specification or the software design definition. It describes the logical flow of a program; the detailed specifications for input, output, or display formats; record and file structures; and storage and performance requirements. Unlike previous documents, this document is very technical in nature. It is intended for the programmers who will do the coding. See Figure 11.7.15.

System/Subsystem Specification

1. GENERAL INFORMATION

    1.1. Summary. Summarize the specifications and functions of the system/subsystem to be developed.

    1.2. Environment. Identify the project sponsor, developer, user, and computer center or network on which the system is to be implemented.

    1.3. References. List applicable references, such as:

        a. Project request (authorizations).
        b. Previously published documents on the subject.
        c. Documentation concerning related projects.
        d. Other reference documents.

2. REQUIREMENTS

    2.1. Description. Provide a general description of the system/subsystem to establish a frame of reference for the remainder of the document. Include a summary of functional requirements to be satisfied by this system/subsystem. Show the general interrelationship of the system/subsystem components.

    2.2. Functions. Specify the system/subsystem functions in quantitative and qualitative terms and how the functions will satisfy the functional requirements.

    2.3. Performance. Specify the performance requirements.

        2.3.1. Accuracy. Describe the data accuracy requirements imposed on the system or subsystem, such as:

            a. Mathematical.
            b. Logical.
            c. Legal.
            d. Transmission.

        2.3.2. Validation. Describe the data validation requirements imposed on the system/subsystem.

        2.3.3. Timing. Describe the timing requirements imposed on the software, such as, under varying conditions:

            a. Response time.
            b. Update processing time.
            c. Data transfer and transmission time.
            d. Throughput time.

        2.3.4. Flexibility. Describe the capability for adapting the program to changes in requirements, such as:

Fig. 11.7.14. Contents of a system/subsystem specification (from Department of Commerce (1976, pp. 26–27)].

System/Subsystem Specification

a. Changes in modes of operation.
b. Operating environment.
c. Interfaces with other software.
d. Accuracy and validation and timing.
e. Planned changes or improvements.

Identify the system/subsystem components which are specifically designed to provide this flexibility.

3. OPERATING ENVIRONMENT

3.1. Equipment. Identify the equipment required for the operation of the system/subsystem. Identify any new equipment required and relate it to specific functional requirements to be supported. Include information, such as:

a. Processor and size of internal storage.
b. Storage, online and offline, media, form, and devices.
c. Input/output devices, online and offline.
d. Data transmission devices.

3.2. Support Software. Identify the support software and describe any test software. If the operation of the system/subsystems depends on changes to support software, identify the nature and planned date of these changes.

3.3. Interfaces. Describe the interfaces with other software.

3.4. Security and Privacy. Describe the overall security and privacy requirements imposed on the system/subsystem. If no specific requirements are imposed, state this fact.

3.5. Controls. Describe the operational controls imposed on the system/subsystem. Identify the sources of these controls.

4. DESIGN CHARACTERISTICS

4.1. Operations. Describe the operating characteristics of the user and computer centers where the software will be operational.

4.2. System/Subsystem Logic. Describe the logic flow of the entire system/subsystem in the form of a flowchart. The flow should provide an integrated presentation of the system/subsystem dynamics, of entrances and exits, computer programs, support software, controls, and data flow.

5. PROGRAM SPECIFICATIONS

5.1. Program (Identify) Specification. Specify the system/subsystem functions to be satisfied by the computer program.

a. Describe the program requirements.
b. Describe the operating environment.
c. Describe the design characteristics of the program including inputs, program logic, outputs, and data base.

5.N. Program (Identify) Specification. Describe the remaining computer programs in a manner similar to the paragraph above.

Fig. 11.7.14. (*Continued*)

Program Specification

1. **GENERAL INFORMATION**

    1.1. **Summary.** Summarize the specifications and functions of the computer program to be developed.

    1.2. **Environment.** Identify the project sponsor. developer, user, and computer center where the computer .program is to be run.

    1.3. **References.** List applicable references, such as:

    a. Project request (authorization).
    b. Previously published documents on the subject.
    c. Documentation concerning related projects.
    d. Other reference documents.

2. **REQUIREMENTS**

    2.1. **Program Description.** Provide a general description of the program to establish a frame of reference for the remainder of the document. Include a summary description of the system/subsystem functions to be satisfied by this program.

    2.2. **Functions.** Specify the functions of the program to be developed. If the program in itself does not fully satisfy a system/subsystem function, show the relationship to other programs which in aggregate satisfy that function.

    2.3. **Performance.** Specify the performance requirements.

    2.3.1. Accuracy. Describe data-accuracy requirements imposed on the program, such as:

    a. Mathematical.
    b. Logical.
    c. Legal.
    d. Transmission.

    2.3.2. Validation. Describe the data–validation requirements imposed on the program.

    2.3.3. Timing. Describe the timing requirements imposed on the program, such as, under varying conditions:

    a. Response time.
    b. Update processing time.
    c. Data transfer and transmission time.
    d. Throughput and internal processing time.

    2.3.4. Flexibility. Describe the capability for adapting the program to changes in requirements, such as:

Fig. 11.7.15. Contents of a program specification [from Department of Commerce (1976, pp. 30–32)].

a. Modes of operation.
b. Operating environment.
c. Interfaces with other programs.
d. Accuracy, validation, and timing.
e. Planned changes or improvements.

Identify the components of the program which are designed to provide this flexibility.

3. **OPERATING ENVIRONMENT**

    3.1. **Equipment.** Identify the equipment required for the operation of the program. Include information on equipment required, such as:

        a. Processor and size of internal storage.
        b. Storage, online and offline, media, form, and devices.
        c. Input/Output devices, online and offline, and capacities.
        d. Data transmission devices.

    3.2. **Support Software.** Identify the support software and describe any test programs. If the operation of the program depends on changes to support software, identify the nature and planned date of these changes.

    3.3. **Interfaces.** Describe all interactions with the operator. Describe all interactions with other software, including sequence or procedure relationships and data interfaces.

    3.4. **Storage.** Specify the storage requirements and any constraints and conditions.

        a. Internal. Describe and illustrate the use of internal storage areas, including indexing and working areas. Briefly state the equipment constraints and design considerations that affect the use of internal storage.
        b. Device. List by device type all peripheral storage required. Briefly state any constraints imposed on storage requirements by each storage device. State requirements for permanent and temporary storage, including overlays.
        c. Offline. Describe the form, media and storage requirements of all offline storage.

    3.5. **Security and Privacy.** Describe the security and privacy requirements imposed on the program, the inputs, the outputs, and the data bases. If no specific requirements are imposed, state this fact.

    3.6. **Controls.** Describe the program controls such as record counts, accumulated counts, and batch controls. Identify the sources of these controls.

4. **DESIGN CHARACTERISTICS**

    4.1. **Operating Procedures.** Describe the operating procedures and any special program functions or requirements necessary for its implementation. Describe the load, start, stop, recovery, and restart procedures. Describe all other interactions of the program with the operator.

Fig. 11.7.15. (*Continued*)

Program Specification

4.2. Inputs. Provide information about the characteristics of each input to the program, such as:

a. Title and tag.
b. Format and type of data, such as a record layout.
c. Validation criteria.
d. Volume and frequency.
e. Means of entry.
f. Source document and its disposition, or specific interface source.
g. Security and privacy conditions.

4.3. Program Logic. Describe the program logic. The logical flow should be presented in graphic form (flowcharts, decision-logic tables) supplemented by narrative explanations.

4.4. Outputs. Provide information about the characteristics of each output from the program, such as:

a. Title and tag.
b. Format specifications, such as a report format.
c. Selection criteria for display, output, or transfer.
d. Volume and frequency.
e. Output media.
f. Description of graphic displays and symbols.
g. Security and privacy conditions.
h. Disposition of products.
i. Description of sequence of displays, display contents, fixed and variable formats, and display of error conditions.

4.5. Data Base. Describe the logical and physical characteristics of any data base used by the program.

4.5.1. Logical Characteristics. Describe for each unique set, file, record, element, or item of data, its identification, definition, and relationships.

4.5.2. Physical Characteristics. Describe in terms of this data base, the storage requirements for program data, specific access method, and physical relationships of access (index, device, area), design considerations, and access security-mechanisms.

Fig. 11.7.15. (*Continued*)

3. *Data Base Specification.* This describes the logical and physical characteristics of the data base. It also provides instructions and design considerations for using the data base. This document is technical and is intended for designers and programmers. See Figure 11.7.16.

4. *User's Manual.* This is written in nontechnical terms for the users of the system. Its purpose is to explain everything that the user needs to know in order to use the system effectively. Depending on need, the user's manual may be either a tutorial or a reference manual. However, in either case, it includes detailed descriptions of how to input data and program parameters, interpret results, and correct for errors. Where complex user–computer interaction takes place, pictures or illustrations of the contents of the CRT screen should be provided, and keyboard actions should be described in conjunction with the pictures or illustrations.

Data Base Specification

1. **GENERAL INFORMATION**

   1.1. **Summary.** Summarize the purpose of the data base and general functions of the software that will use it.

   1.2. **Environment.** Identify the project sponsor, developer, user organization, and computer center where the software and data base are to be installed.

   1.3. **References.** List applicable references, such as:

   a. Project request (authorization).
   b. Previously published documents on the project.
   c. Documentation concerning related projects.
   d. Other reference documents.

2. **DESCRIPTION**

   2.1. **Identification.** Specify the code name, tag, or label by which the data base is to be identified. If the data base is to be experimental, test, or temporary, specify this characteristic and effective dates or interval. Any additional identification information should also be given.

   2.2. **Using Software.** Identify all software intended to use or access this data base. Identify for each: the software name, code name, and any release or version number.

   2.3. **Conventions.** Describe all labeling or tagging conventions essential for a programmer or analyst to use this data base specification.

   2.4. **Special Instructions.** Provide any special instructions to personnel who will contribute to the generation of the data base, or who may use it for testing or operational purposes. Such instructions include criteria, procedures, and formats for:

   a. Submitting data for entry into the data base and identification of a data control organization.
   b. Entering data into the data base.

   Where these instructions are extensive, reference appropriate sections of other documents.

   2.5. **Support Software.** Describe briefly all support software directly related to the data base. Descriptions should include name, function, major operating characteristics, and machine-run instructions for using the support software. Cite the support-software documentation by title, number, and appropriate sections.

   Examples of support software are:

   a. Data base management systems.
   b. Storage allocation software.
   c. Data base loading software programs.
   d. File processing programs.
   e. Other generating, modifying, or updating software.

Fig. 11.7.16. Contents of a data base specification [from Department of Commerce (1976, pp. 34–35)].

Data Base Specification

3. **LOGICAL CHARACTERISTICS**

A data base is a logical arrangement of data. Sets (aggregates), files, records, elements, and items of data may vary in their logical arrangement and relationships. The organization of the content of this section should provide a meaningful presentation of the logical organization of the data base.

Define each unique set (aggregate), file, record, element, or item of data providing information, such as:

  a. Identification. Name and tag, or label.
  b. Definition. Standard or unique; purpose in data base; using software; media; form; format and size; update criteria and conditions; security and privacy restrictions, limitations, or conditions (update or access); integrity and validity characteristics; controlling data-elements or items; and graphic representation.
  c. Relationships. Superior and inferior relationships; update and access relationships.

4. **PHYSICAL CHARACTERISTICS**

  4.1. Storage. Specify the storage requirements for the data base and any constraints and conditions.

    a. Internal. Describe and illustrate the use of internal storage areas set aside for data including indexing and working areas. Briefly state the equipment constraints and design considerations that affect the use of internal storage.
    b. Device. List by device type all peripheral storage required for the data base. Briefly state any constraints imposed on storage requirements by each storage device. State requirements for permanent data storage and temporary data storage, including overlays.
    c. Offline. Describe the form, media and storage requirements of all offline data storage.

  4.2. Access. Describe the access method and specify the physical relationships of access (index, device, area). Describe all physical-access security mechanisms.

  4.3. Design Considerations. State the design considerations for the handling of this data base, such as blocking factors. Emphasize those physical relationships important to the efficient utilization of the data base.

Fig. 11.7.16. (*Continued*)

A user interface specification written early in the design stage can substitute for early development of the user's manual. Such a specification aids design decisions and forms a basis for the development of the user's manual. In any case, if systems are to be easy to learn and use, the human–computer interface must be planned early in the life cycle, not included as an afterthought. See Figure 11.7.17.

5. *Test Plan.*   This provides the programmer with detailed descriptions of all tests to be performed on the software. Test data and evaluation criteria are included. See Figure 11.7.18.

User's Manual

1. GENERAL INFORMATION

    1.1. Summary.   Summarize the application and general functions of the software.

    1.2. Environment.   Identify the user organization and computer center where the software is installed.

    1.3. References.   List applicable references, such as:

        a. Project request (authorization).
        b. Previously published documents on the project.
        c. Documentation concerning related projects and software.
        d. Other reference documents.

2. APPLICATION

    2.1. Description.   Describe when and how the software is used and the unique support provided to the user organization. The description should include:

        a. Purpose of the software.
        b. Capabilities and operating improvements provided.
        c. Functions performed.

    2.2. Operation.   Show the operating relationships of the functions performed to the organization that provides input to and receives output from the software. Describe security and privacy considerations. Include general charts and a description of the inputs and outputs shown on the charts.

    2.3. Equipment.   Describe the equipment on which the software can be run.

    2.4. Structure.   Show the structure of the software and describe the role of each component in the operation of the software.

    2.5. Performance.   Describe the performance capabilities of the software including where appropriate:

        a. Quantitative information on inputs, outputs, response time, processing times, and error rates.
        b. Qualitative information about flexibility and reliability.

    2.6. Data Base.   Describe all data files in the data base that are referenced, supported, or kept current by the software. The description should include the purpose for which each data file is maintained.

    2.7. Inputs, Processing, and Outputs.   Describe the inputs, the flow of data through the processing cycle, and the resultant outputs. Include any applicable relationships among inputs or outputs.

Fig. 11.7.17.   Contents of a user's manual [from Department of Commerce (1976, pp. 38–40)].

3. PROCEDURES AND REQUIREMENTS

This section should provide information about initiation procedures, and preparation of data and parameter inputs for the software. The scope, quality, and logical arrangement of the information should enable the user to prepare required inputs and should explain in detail the characteristics and meaning of the outputs. It should also describe error, recovery, and file-query procedures and requirements.

3.1. Initiation. Describe step-by-step procedures required to initiate processing.

3.2. Input. Define the requirements of preparing input data and parameters. Typical considerations are:

a. Conditions—e.g., personnel transfer, out of stock.
b. Frequency—e.g., periodically, randomly, as a function of an operational situation.
c. Origin—e.g., Personnel Section, Inventory Control.
d. Medium—e.g., keyboard, punched card, magnetic or paper tape.
e. Restrictions—e.g., priority and security handling, limitations on which files may be accessed by this type of transaction.
f. Quality control—e.g., instructions for checking reasonableness of input data, action to be taken when data appears to be in error, documentation of errors.
g. Disposition—e.g., instructions necessary for retention or release of all data files received, other recipients of the inputs.

3.2.1. Input Formats. Provide the layout forms used in the initial preparation of program data and parameter inputs. Explain each entry, and reference it to the sample form. Include a description of the grammatical rules and conventions used to prepare input, such as:

a. Length—e.g., characters/line, characters/item.
b. Format—e.g., left justified.
c. Labels—e.g., tags or identifiers.
d. Sequence—e.g., the order and placement of items in the input.
e. Punctuation—e.g., spacing and use of symbols (virgule, asterisk, character combinations, etc.) to denote start and end of input, of lines, of data groups, etc.
f. Combination—e.g., rules forbidding use of groups of particular characters, or combinations of parameters in an input.
g. Vocabulary—e.g., an appendix which lists the allowable character combinations or codes that must be used to identify or compose input items.
h. Omissions and Repeats—e.g., indicate those elements of input that that are optional or may be repeated.
i. Controls—e.g., header or trailer control data.

3.2.2. Sample Inputs. Provide specimens of each complete input form. Include:

a. Control or header—e.g., entries that denote the input class or type, date/time, origin, and instruction codes to the software.
b. Text—e.g., subsections of the input representing data for operational files, request parameters for an information retrieval program.

Fig. 11.7.17. (*Continued*)

    c. Trailer—e.g., control data denoting the end of input and any additional control data.

    d. Omissions—e.g., indicate those classes or types of input that may be omitted or are optional.

    e. Repeats—e.g., indicate those positions of the input that may be repeated.

3.3. **Output.** Describe the requirements relevant to each output. Typical considerations are:

    a. Use—e.g., by whom and for what.
    b. Frequency—e.g., weekly, periodically, or on demand.
    c. Variations—e.g., modifications that are available to the basic output.
    d. Destination—e.g., computer area, remote terminal.
    e. Medium—e.g., printout, CRT, tape, cards.
    f. Quality control—e.g., instructions for identification, reasonableness checks, editing and error correction.
    g. Disposition—e.g., instructions necessary for retention or release, distribution, transmission, priority, and security handling.

    3.3.1. Output Formats. Provide a layout of each output. Explanations should be keyed to particular parts of the format illustrated. Include:

        a. Header—e.g., title, identification, date, number of output parts.

        b. Body—e.g., information that appears in the body or text of the output, columnar headings in tabular displays, and record layouts in machine readable ouputs. Note which items may be omitted or repeated.

        c. Trailer—e.g., summary totals, trailer labels.

    3.3.2. Sample Outputs. Provide a sample of each type of output. For each item on a sample, include:

        a. Definition—e.g., the meaning and use of each information variable.

        b. Source—e.g., the item extracted from a specific input, from a data base file, or calculated by software.

        c. Characteristics—e.g., the presence or absence of the item under certain conditions of the output generation, range of values, unit of measure.

3.4. **Error and Recovery.** List error codes or conditions generated by the software and corrective action to be taken by the user. Indicate procedures to be followed by the user to ensure that any restart and recovery capability can be used.

3.5. **File Query.** Prepare this paragraph for software with a file query retrieval capability. Include detailed instructions necessary for initiation, preparation, and processing of a query applicable to the data base. Describe the query capabilities, forms, commands used, and control instructions required.

If the software is queried through a terminal, provide instructions for terminal operators. Describe terminal setup or connect procedures, data or parameter input procedures, and control instructions. Reference related materials describing query capabilities, languages, installation conventions and procedures, program aids, etc.

Fig. 11.7.17. (*Continued*)

Test Plan

1. GENERAL INFORMATION

1.1. Summary. Summarize the functions of the software and the tests to be performed.

1.2. Environment and Pretest Background. Summarize the history of the project. Identify the user organization and computer center where the testing will be performed. Describe any prior testing and note results that may affect this testing.

1.3. References. List applicable references, such as:

a. Project request (authorization).
b. Previously published documents on the project.
c. Documentation concerning related projects.
d. Other reference documents.

2. PLAN

2.1. Software Description. Provide a chart and briefly describe the inputs, outputs, and functions of the software being tested as a frame of reference for the test descriptions.

2.2. Milestones. List the locations, milestones events, and dates for the testing.

2.3. Testing (Identify Location). Identify the participating organizations and the location where the software will be tested.

2.3.1. Schedule. Show the detailed schedule of dates and events for the testing at this location. Such events may include familiarization, training, data, as well as the volume and frequency of the input.

2.3.2. Requirements. State the resource requirements, including:

a. Equipment. Show the expected period of use, types, and quantities of the equipment needed.
b. Software. List other software that will be needed to support the testing that is not part of the software to be tested.
c. Personnel. List the numbers and skill types of personnel that are expected to be available during the test from both the user and development groups. Include any special requirements such as multishift operation or key personnel.

2.3.3. Testing Materials. List the materials needed for the test, such as:

a. Documentation.
b. Software to be tested and its medium.
c. Test inputs and sample outputs.
d. Test control software and worksheets.

2.3.4. Test Training. Describe or reference the plan for providing training in the use of the software being tested. Specify the types of training, personnel to be trained, and the training staff.

Fig. 11.7.18. Contents of a test plan (from Department of Commerce (1976, pp. 50–52)].

2.4. Testing (Identify Location). Describe the plan for the second and subsequent locations where the software will be tested in a manner similar to paragraph 2.3.

3. **SPECIFICATIONS AND EVALUATION**

   3.1. **Specifications.**

      3.1.1. Requirements. List the functional requirements established by earlier documentation.

      3.1.2. Software Functions. List the detailed software functions to be exercised during the overall test.

      3.1.3. Test/Function Relationships. List the tests to be performed on the software and relate them to the functions in paragraph 3.1.2.

      3.1.4. Test Progression. Describe the manner in which progression is made from one test to another so that the entire test cycle is completed.

   3.2. **Methods and Constraints.**

      3.2.1. Methodology. Describe the general method or strategy of the testing.

      3.2.2. Conditions. Specify the type of input to be used, such as live or test data, as well as the volume and frequency of the input.

      3.2.3. Extent. Indicate the extent of the testing, such as total or partial. Include any rationale for partial testing.

      3.2.4. Data Recording. Discuss the method to be used for recording the test results and other information about the testing.

      3.2.5. Constraints. Indicate anticipated limitations on the test due to test conditions, such as interfaces, equipment, personnel, data bases.

   3.3. **Evaluation.**

      3.3.1. Criteria. Describe the rules to be used to evaluate test results, such as range of data values used, combinations of input types used, maximum number of allowable interrupts or halts.

      3.3.2. Data Reduction. Describe the techniques to be used for manipulating the test data into a form suitable for evaluation, such as manual or automated methods, to allow comparison of the results that should be produced to those that are produced.

4. **TEST DESCRIPTIONS**

   4.1. **Test (Identify).** Describe the test to be performed.

      4.1.1. Control. Describe the test control, such as manual, semi-automatic, or automatic insertion of inputs, sequencing of operations, and recording of results.

Fig. 11.7.18. (*Continued*)

4.1.2. Inputs. Describe the input data and input commands used during the test.

4.1.3. Outputs. Describe the output data expected as a result of the test and any intermediate messages that may be produced.

4.1.4. Procedures. Specify the step-by-step procedures to accomplish the test. Include test setup, initialization, steps, and termination.

4.2. **Test (Identify).** Describe the second and subsequent tests in a manner similar to that used in paragraph 4.1.

Fig. 11.7.18. (*Continued*)

Programming Stage. The program code is the main product of the programming stage. Program maintenance and operations manuals are produced; the code itself may be included in the program maintenance manual.

1. *Operations Manual.* This provides all pertinent information that is necessary to operate the system. It is intended for computer operations personnel. The operations manual includes an overview of the software organization, procedures for operating the system and ensuring the integrity of the data, and specific schedule requirements. See Figure 11.7.19.

Operations Manual

1. **GENERAL INFORMATION**

    1.1. **Summary.** Summarize the general functions of the software.

    1.2. **Environments.** Identify the software sponsor, developer, user organization, and the computer center where the software is to be installed.

    1.3. **References.** List applicable references, such as:

        a. Project request (authorization).
        b. Previously published documents on the project.
        c. Documentation concerning related projects.
        d. Other reference documents.

2. **OVERVIEW**

    2.1. **Software Organization.** Provide a diagram showing the inputs, outputs, data files, and sequence of operations of the software. Runs may be grouped by periods of time cycles, by organizational level where they will be performed, or by other groupings.

    2.2. **Program Inventory.** Identify each program by title, number, and mnemonic reference.

    2.3. **File Inventory.** Identify each permanent file that is referenced, created, or updated by the system. Include the title, mnemonic reference, storage medium, and required storage.

3. **DESCRIPTION OF RUNS**

    3.1. **Run Inventory.** List the various runs possible and summarize the purpose each run. Show the programs that are executed during each run.

    3.2. **Run Progression.** Describe the manner in which progression advances from one run to another so that the entire run cycle is completed.

    3.3. **Run Description (Identify).** Organize the information on each run into the most useful presentation for the operating center and operations personnel involved.

        3.3.1. Control Inputs. List the run–stream control statements needed for the run.

        3.3.2. Operating Information. Provide information for the operating center personnel and management, such as:

            a. Run identification.
            b. Operating requirements.
            c. Initiation method, such as on request, at predetermined time, etc.
            d. Estimated run time and turnaround time.
            e. Operator commands and messages.
            f. Contact personnel for problems with the run.

Fig. 11.7.19. Contents of an operations manual [from Department of Commerce (1976, pp. 42–43)].

3.3.3. Input-Output Files. Provide information for files created or updated by the run, such as:

    a. File name or label.
    b. Recording medium.
    c. Retention schedule.
    d. Disposition of file.

3.3.4. Output Reports. For each output report or type of report, provide information such as:

    a. Report identification.
    b. Medium.
    c. Volume of report.
    d. Number of copies.
    e. Distribution.

3.3.5. Reproduced Output Reports. For those reports that are computer-generated and then reproduced by other means, provide information such as:

    a. Report identification.
    b. Reproduction technique.
    c. Dimensions of paper or other medium.
    d. Binding method.
    e. Distribution.

3.3.6. Restart/Recovery Procedures. Describe procedures to restart the run or recover from a failure.

3.4 **Run Description (Identify).** Present information about the subsequent runs in a manner similar to that used in paragraph 3.3.

4. NON-ROUTINE PROCEDURES

Provide any information necessary concerning emergency or non-routine operations, such as:

    a. Switchover to a back-up system.
    b. Procedures for turnover to maintenance programmers.

5. REMOTE OPERATIONS

Describe the procedures for running the programs through remote terminals.

Fig. 11.7.19. (*Continued*)

2. *Program Maintenance Manual.*   This provides the maintenance programmer with all the information needed to modify the program. It includes descriptions of existing programs, utilities, and runtime services; programming conventions; error correction procedures; and verification of the changes. Design documentation and code may also be included. See Figure 11.7.20.

Testing Stage.   During the testing stage, the software is validated to ensure its correctness and its adherence to the specifications developed during the design stage. The documentation completed during this stage is the test analysis report.

Program Maintenance Manual

1. GENERAL INFORMATION

    1.1. Summary.   Summarize the general nature of the software to be maintained.

    1.2. Environment.   Identify the project sponsor, developer, user and computer center or network where the software is implemented.

    1.3. References.   List applicable references, such as:

        a. Project request (authorizations).
        b. Previously published documents on the project.
        c. Documentation concerning related projects.
        d. Other reference documents.

2. PROGRAM DESCRIPTIONS

Describe the program and programs in the system/subsystem for the maintenance programmer. If a complex system is being described, provide a general description of that system identifying each program and its functions.

    2.1. Program (Identify) Description.   Identify the program by title, tag or label, and programming language.

        2.1.1.   Problem and Solution Method.   Describe the problem to be solved or the program function and the solution method used.

        2.1.2.   Input.   Describe the input to the program and provide a layout. Identify the medium used. Include information, such as codes, units of measurement, format, range of values, or reference a data-element directory.

        2.1.3.   Processing. Describe processing features and purposes important to the maintenance programmer, such as:

            a. Processing logic.
            b. Linkages.
            c. Variables and constants.
            d. Formulas.
            e. Error handling provisions.
            f. Restrictions and limitations.
            g. Locations, settings, internal switches and flags.
            h. Shared storage.

        2.1.4.   Output.   Describe the output of the program and provide a layout. Identify the medium used.

        2.1.5.   Interfaces.   Describe the interfaces with other software, such as data formats, messages, parameters, conversion requirements, interface procedures, and media.

        2.1.6.   Tables.   Identify each table and its items. Describe the location, structure, and purpose of each.

Fig. 11.7.20.   Contents of a program maintenance manual [from Department of Commerce (1976, pp. 46–47)].

2.1.7. Run Description. Describe or reference the operating procedures to run the program, including loading, operating, terminating, and error handling.

2.2. **Program (Identify) Description.** Describe the second through nth computer program in a manner similar to that used in paragraph 2.1.

3. OPERATING ENVIRONMENT

3.1. **Hardware.** Identify the equipment required for the operation of the system. Describe any unusual features used. Relate the hardware to each program. Include information such as:

a. Processor and size of internal storage.
b. Storage online or offline, media, form, and devices.
c. Input/output devices, online and offline.
d. Data transmission devices.

3.2. **Support Software.** Identify the support software needed for each computer program.

3.2.1. Operating System. Identify and describe the operating system including the version or release number and any unusual features used.

3.2.2. Compiler/Assembler. Identify and describe the compiler or assembler including the version or release number and any special features used.

3.2.3. Other Software. Identify and describe any other software used including data management systems, report generators, etc.

3.3. **Data Base.** Describe or reference documentation on the data base used. Include information such as codes, units of measurement, format, range of values, or reference a data-element directory.

4. MAINTENANCE PROCEDURES

4.1. **Programming Conventions.** Identify and describe the programming conventions used.

4.2. **Verification Procedures.** Describe the verification procedures to check the performance of the programs, either general or following modifications. Include a reference to test data and testing procedures.

4.3. **Error Correction Procedures.** Describe all error conditions, their sources, and procedures for their correction.

4.4. **Special Maintenance Procedures.** Describe any special procedures required for the maintenance of the programs. Include information such as periodic purges of the data base, temporary modifications needed for leap years or century changes, etc.

4.5. **Listings and Flowcharts.** Reference, append, or describe the method for obtaining copies of listings of the programs and flowcharts.

Fig. 11.7.20. (*Continued*)

1. *Test Analysis Report.*   This describes the testing that was done and compares the results to the expected results. Capabilities and deficiencies are reported, and estimates are given of the readiness of the software for release. See Figure 11.7.21.

11.7.3 THE COMPOSITION OF CODE

Thus far we have discussed the importance of producing clear, readable manuals for computer systems. The writing of clear, readable code is extremely important as well, not only for the software development team but also for those who will maintain the code at a later date. Maintenance of code is particularly

---

Test Analysis Report

1. **GENERAL INFORMATION**

    1.1. **Summary.** Summarize both the general functions of the software tested and the test analysis performed.

    1.2. **Environment.** Identify the software sponsor, developer, user organization, and the computer center where the software is to be installed. Assess the manner in which the test environment may be different from the operational environment and the effects of this difference on the tests.

    1.3. **References.** List applicable references, such as:

        a. Project request (authorization).
        b. Previously published documents on the project.
        c. Documentation concerning related projects.
        d. Other reference documents.

2. **TEST RESULTS AND FINDINGS**

    Identify and present the results and findings of each test separately in paragraphs 2.1 through 2.N.

    2.1. **Test (Identify).**

        2.1.1. Dynamic Data Performance. Compare the dynamic data input and output results, including the output of internally generated data, of this test with the dynamic data input and output requirements. State the findings.

        2.1.2. Static Data Performance. Compare the static data input and output results, including the output of internally generated data, of this test with the static data input and output requirements. State the findings.

    2.N. **Test (Identify).** Present the results and findings of the second and succeeding tests in a manner similar to that of paragraph 2.1.

3. **SOFTWARE FUNCTION FINDINGS**

    Identify and describe the findings on each function separately in paragraphs 3.1 through 3.N.

    3.1. **Function (Identify).**

        3.1.1. Performance. Describe briefly the function. Describe the software capabilities that were designed to satisfy this function. State the findings as to the demonstrated capabilities from one or more tests.

        3.1.2. Limits. Describe the range of data values tested, including both dynamic and static data. Identify the deficiencies, limitations, and constraints detected in the software during the testing with respect to this function.

Fig. 11.7.21.   Contents of a test analysis report [from Department of Commerce (1976, pp. 54–55)].

3.N. **Function (Identify).** Present the findings on the second and succeeding functions in a manner similar to that of paragraph 3.1.

4. **ANALYSIS SUMMARY**

4.1. **Capabilities.** Describe the capabilities of the software as demonstrated by the tests. Where tests were to demonstrate fulfillment of one or more specific performance requirements, prepare findings showing the comparison of the results with these requirements. Assess the effects any differences in the test enviroment as compared to the operational environment may have had on this test demonstration of capabilities.

4.2. **Deficiencies.** Describe the deficiencies of the software as demonstrated by the tests. Describe the impact of each deficiency on the performance of the software. Describe the cumulative or overall impact on performance of all detected deficiencies.

4.3. **Recommendations and Estimates.** For each deficiency provide any estimates of time and effort required for its correction and any recommendations as to:

a. The urgency of each correction.
b. Parties responsible for corrections.
c. How the corrections should be made.

State the readiness for implementation of the software.

Fig. 11.7.21. (*Continued*)

critical from the standpoint of software life-cycle costs. Boehm (1981) has shown that maintenance is now responsible for the major portion of software expenditures, and he predicts that this trend will continue. As much as 75% of programming work involves modifying or enhancing old software (Shneider 1980), and maintenance has been found to be much more error-prone than development coding (Mair, Wood, and Davis, 1976). Programmers who originally create code seldom maintain it throughout its life cycle; they change jobs or take on more challenging tasks, leaving maintenance to junior programmers or to those who are new in an organization (Lientz and Swanson, 1979). Thus it is essential to produce complete, readable, self-documenting modules of code that can be understood by those other than the original developers.

Program code, like prose in natural language, should be constructed according to a set of structural and stylistic guidelines. In some cases, experimentation has helped to define a methodology for writing code. In other cases, the evidence for selecting one methodology over another is not clear and more research is needed. In such cases, the choice should be whatever seems most natural for a given application.

11.7.3.1 Style

Structured Coding

Structured coding became a real issue in 1968 when Dijkstra (1968) pointed out that the unconditional jump statement (i.e., GO TO) caused havoc with the structure of a computer program. The structured programming movement was born, and although it came to represent a variety of changes in software engineering practices, one concept emerged that is of interest here. Programs that are developed with a limited number of well-defined control flow structures can be constructed and modified more quickly and with fewer errors than those that have a "spaghetti" structure (Clifton, 1978; Green, Sime, and Fitter, 1980; Lucas and Kaplan, 1974; Sime, Arblaster, and Green, 1977). Figure 11.7.22 taken from Sheppard, Curtis, Milliman, and Love (1979) shows the superiority of two versions of structured code to "spaghetti," or convoluted, code in a comprehension task. Numerous control flow structures are provided by the various programming languages, but the salient point shown by the data in this figure is that it is the general discipline of a top-down flow of control that is helpful in communicating information about the contents of the code.

Commenting

A comment is a natural language statement that is used to describe what is being done in a given part of a program, for example, in a loop. It is important to ensure that comment statements accurately describe the functions being performed in the code. Otherwise, a programmer can be misled during debugging, resulting in a longer time to locate an error. Furthermore, the comments must be updated

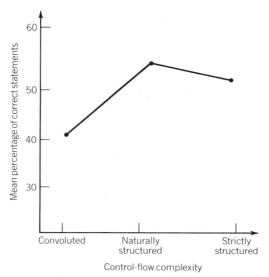

Fig. 11.7.22. Mean percentage of statements correctly recalled for three types of program structure [from Sheppard et al. (1979, p. 44)]. Reprinted with permission.

with the code. If the code is altered, but the comment is not changed in conjunction with the code, the comment may mislead someone at a later time.

The following guidelines are suggested for the construction of comments.

1. A good rule-of-thumb is that there should be one comment per 10–30 lines of code. Assume that the reader understands the purpose of simple program statements. Too many comments or comments at too low a level can detract from an understanding of the code.

2. The number of comments should be dependent on the structure of the program. For example, provide a comment as an explanation for a large iteration, but not necessarily one for a five-line iteration.

3. Comments should be provided to describe data structures that are complex or difficult to understand.

4. Comments should be set off by dashes or otherwise isolated from the code. (See Figure 11.7.23).

5. Comments should be descriptive of real-world situations. Comments are most effective when they relate to the abstract nature of a problem as opposed to a description of what is being done. Poor: "Find the largest value in the table."
Better: "Print the name of the student with the highest term average." (Shneiderman, 1980, p. 69).

Several programming experiments have cast doubt on the usefulness of comments inserted in the code (Newsted, no date; Okimoto, 1970; Sheppard et al., 1979). The experiments suggested that comments contributed little to the comprehension of code. However, the programs used in the experiments were very small, and it is possible that comments become of value only after a given program size has been reached. Furthermore, low clarity and poor formatting of the comments may have detracted from their usefulness in some cases.

A more recent experiment, conducted by Woodfield, Dunsmore, and Shen (1981), investigated the effects of comments on programmers' comprehension of the units of a larger program. Four different types of modularization of the same program were each tested with and without comments. Code with comments was associated with significantly better performance than code without comments, regardless of the type of modularization. Furthermore, the comments made more of a difference in performance for three types of modularization where there were multiple comments per section of code than for the modularization that had only one comment per section. These results suggest that comments supply useful information not readily available from the code itself. ". . ."

Variable Names

Variable names should be chosen to enhance program readability and comprehensibility. The following criteria for selection seem reasonable:

1. Choose common, well-understood terms that relate to the application being coded.
2. Use names that are easily differentiated from each other.

```
--Ask user for name of desired Mba Coverage File. Append the suffix ".MCF" to the name, open
--the file, and return the data.
```

```
LOOP
    Prompt (Control, Prompt-String);     --Ask user for 9-character Mba Coverage File name.
    Coverage-Name := Get-Input (Control) (1 . . 9);
    BEGIN
        Open (Coverage-Name & ".MCF");

                                . . .

        Close (Coverage-Name);
        EXIT;
    EXCEPTION
        WHEN OTHERS =>     --Error in file access; tell user file is busy or not correct.
            Put-Line (Control);
    END;
END LOOP;
```

```
--The Beam Contribution Matrix is calculated for use in producing the desired plots.
```

```
Beam-Matrix := Calculate-Beam-Contributions (Control, Coverage-File, Beam-Weights);
```

Fig. 11.7.23. Comments are inserted in code in two ways: (a) set off by horizontal dashed lines or (b) included to the right of a line of code.

3. Use names that are long enough to be meaningful but not so long that they cause extensive wraparound of program lines.

Research on the effectiveness of using mnemonic variable names has produced contradictory results. Shneiderman and McKay (1976) performed an experiment in which four programs were presented to novice programmers in mnemonic or nonmnemonic forms. The mnemonic forms were associated with significantly better performance on a comprehension test. (See Figure 11.7.24.) This result, however, differs from other experiments that showed no particular benefit from the use of mnemonic names (Newsted, no date; Sheppard et al., 1979). One explanation for these differing results is that names that are truly mnemomic (i.e., that aid memory and understanding) are helpful. However, it is undoubtedly easier to aid one's own understanding than to select names to aid another person's understanding. Newsted expressed the problem succintly: "One programmer's mnemonic is another's gobbledegook" (Newsted, no date, p. 21). The aim should be to use understandable variable names, but improvements in performance are not assured because of the names selected.

11.7.3.2 Formatting

Identation

Overton et al. (1973) studied problems of software maintenance and concluded that to modify a program efficiently a programmer must be able to recognize the structure to be modified and to identify the conceptual blocks within it. Indentation is one method for highlighting structure. The general rule is that each statement belonging to a control statement is indented to the right an equal number of spaces. Two to four spaces seem to be adequate. Languages with an end-of-construct feature are easier to indent than languages with constructs that are less rigidly structured; that is, it is easier to indent PASCAL with its IF $\cdots$ THEN $\cdots$ END IF structure than the IF $\cdots$ GOTO of BASIC (Leinbaugh, 1980).

Indentation is currently presented as a standard by leading computer scientists. See Figure 11.7.25 from Mills et al. (1986). However, research exploring the usefulness of indentation has produced contradictory results. Experiments by Love (1977) and Norcio and Kerst (1979) showed no differences between indented listings and unindented listings in the ability of programmers to reconstruct short programs from memory. Shneiderman and McKay (1976) found a similiar result in a debugging experiment: there were no significant differences in performance between programs that were or were not indented. An experiment by Weissman (1974) suggested that two-space indentation, combined with in-line comments, produced programs that were difficult to hand-simulate. Furthermore, Krall and Harris (1980) demonstrated that indentation, combined with blank lines (vertical spacing), was associated with poorer comprehension of a COBOL program than indentation alone.

One interesting experiment that supports the use of indentation was conducted by Miara, Musselman, Navarro, and Shneiderman (1983). They tested four levels of indentation: none, two-space, four-space, and six-space, and they found that two-space or four-space indentation was associated with higher

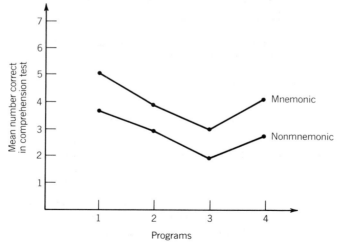

Fig. 11.7.24. Comprehension with mnemonic and nonmnemonic variable names [from Newsted (no date, p. 71)].

```
PROGRAM OldNew(INPUT,OUTPUT);
VAR
   Old, New: CHAR;
BEGIN {OldNew}
   READ(Old);
   WHILE Old <> '0' DO
      IF Old = '1'
         THEN
            BEGIN
               New := Old;
               Old := '0'
         ELSE
               Old := New;
   WRITELN(Old)
END.  {OldNew}
```

Fig. 11.7.25. Recommended styles of indentation and capitalization (from Mills et al. (1986, p. 7–7). Reprinted with permission.

program comprehension than no indentation or six-space indentation. A total lack of indentation restricted the user's view of the control structure of the program. Six-space indentation caused exaggerated spacing, made control structures harder to locate, and caused wraparound of program lines to occur more frequently; thus comprehension was affected in a negative manner. It appears that two- to four-space indentation aids in comprehension as long as other factors do not intervene to interrupt the visual scanning of the program.

Capitalization

A final guideline pertains to the practice of selecting uppercase letters for the reserved words in a programming language and lowercase letters with an initial capital for all others. See Figure 11.7.25. This procedure requires more effort on the part of the programmer but is recommended by leading computer scientists (Mills et al., 1986). It is another method for demonstrating the structure of the program, both on paper and on a terminal, and the variation in type appears to increase visual readability on-line.

11.7.3.3 Content of a Typical Module or Unit of Code

In order to be self-documenting, program code should include a variety of information other than the code itself. Depending on the system and the method of development of the design and code, units of code may be whole programs or individual modules. Each separate unit should contain the following information in addition to the code:

1. File identification: system access name.
2. Author(s): the name(s) of the programmer(s).
3. Creation date: the date on which the code was originally entered into the program library or put under configuration management.
4. Modification history: dates and explanations for each change to the file. If change requests are formalized and filed in a program library, the program change-request number is sufficient.
5. Abstract: an English description of the purpose of the code and the general approach used. Relationships to other units of code are described if they are relevant to an understanding of the unit.
6. Data structures: descriptions of all data internal to the unit and all interfaces with other units of code, including inputs; outputs; and procedure, package, and function calls.
7. Comments: English statements interspersed among the lines of code to describe the functions being performed in the code.

Figures 11.7.23 and 11.7.26 illustrate these features.

11.7.4 SUMMARY

There are multiple ways to document the development of a computer system. No one method is "the right way." The numbers and types of documents produced and the issues of formatting, style, and content must depend on the nature of the project, the development environment, the end users,

```
--FILE NAME: [ADAPGM.PLOT]MBAPLOT.PDL
--
--AUTHOR: A. DAVID
--
--CREATION DATE: 6/25/84
--
--MODIFICATION HISTORY:
--
--DATE          CHANGE REQUEST NUMBERS
--6/16/84       2
--6/24/84       15
--6/25/84       23
--
-------------------------------------------------------------------

TASK BODY Mba-Plotting IS

--This task controls the execution of the Mba_Plotting function.
--It is initiated by the Executive when plots are requested from
--the high-level menu.  It receives access rights to a control
--terminal and a graphics terminal from the Executive.  If Mba-
--Plotting is initiated from a graphics terminal, the graphics
--terminal also serves as the control terminal; otherwise, another
--terminal is the control terminal.  After start-up, the beam
--contributions are calculated once and used until the task
--terminates.  The user is prompted for the desired plot, which
--is then drawn using the calculated 37 X 37 matrix of beam
--contributions.  When the task terminates, a message is sent to
--the Executive, and the access rights to the terminal(s) are
--cancelled.

-------------------------------------------------------------------

--INPUTS:  User input to Plotting_Menu
--         Calculated Beam_Contribution_Matrix
--         Terminal pointers for console and graphics terminals
--
--OUTPUTS: Pointers for console and graphics terminals
--         to subroutines and to rendevous with Executive
--         Beam_Contribution_Matrix to subroutines
--         Plotting_Menu on console terminal
--         Any applicable error messages on console terminal

-------------------------------------------------------------------

  BEGIN
```

Fig. 11.7.26.   Contents of a typical unit of code.

and the individual preferences of the developers. However, in each case, the overall goal is to produce a set of clear, concise, accurate and complete documents.

REFERENCES

Adams, K. A., and Halasz, I. M. (1983). *25 ways to improve your software user manuals.* Worthington, OH: Technology Training Systems.

Ausubel, D. P. (1968). *Educational psychology: a cognitive view.* New York: Holt, Rinehart & Winston.

Blaiwes, A. S. (1974). Formats for presenting procedural instructions. *Journal of Applied Psychology, 59,* 683–686.

Boehm, B. (1981). *Software engineering economics.* Englewood Cliffs, NJ: Prentice-Hall.

Caine, S. H., and Gordon, E. K. (1975). PDL—a tool for software design. In *Proceedings of the 1975 national computer conference,* pp. 168–173. American Federation of Information Processing Societies.

Clifton, M. H. (1978). A technique for making structured programs more readable. *ACM SIGPLAN Notices, 13,* 58–63.

DeMarco, T. (1978). *Structured analysis and system specification.* New York: Yourdon.

Department of Commerce (1976). Guidelines for documentation of computer programs and automated data systems (Federal Information Processing Standards Publication 38). Washington, DC: U.S. Department of Commerce.

Dikstra, E. W. (1968). GO TO statement considered harmful. *Communications of the ACM, 11,* 147–148.

Fitter, M., and Green, T. R. G. (1979). When do diagrams make good computer languages? *International Journal of Man-Machine Studies, 11,* 235–261.

Green, T. R. G., Sime, M. E., and Fitter, M. J. (1980). The problem the programmer faces. *Ergonomics, 23* (9), 893–907.

Hendricks, D. E., Kilduff, P. W., Brooks, P., Marshak, R., and Doyle, B. (1983). *Human engineering guidelines for management information systems.* Aberdeen Proving Ground, MD: U.S. Army Human Engineering Laboratory.

Kamman, R. (1975). The comprehensibility of printed instructions and the flowchart alternative. *Human Factors, 17,* 183–191.

Krall, A., and Harris W. (1980). An investigation of program style on the readability/understandability of a simple COBOL program: the effects of indentation and vertical spacing (Unpublished manuscript report). College Park, MD: University of Maryland.

Leinbaugh, D. W. (1980). Indenting for the computer. *ACM SIGPLAN Notices, 15* (5), 41–48.

Lientz, B. P., and Swanson, E. G. (1979). The use of productivity aids in systems development and maintenance (Tech. Rep 79–1). Los Angeles, CA: Graduate School of Management, UCLA.

Love, T. (1977). Relating individual differences in computer programming performance to human information procession abilities (Unpublished doctoral dissertation). Seattle, WA: University of Washington.

Lucas, H. C., and Kaplan, R. B. (1974). A structured programming experiment. *The Computer Journal, 19* (2), 136–138.

Mair, W. C., Wood, D. R., and Davis, K. W. (1976). *Computer control and audit.* Altamonte Springs, FL: The Institute of Internal Auditors, Inc.

Miara, R. J., Musselman, J. A., Navarro, J. A., and Shneiderman, B. (1983). Program indentation and comprehensibility, *Communications of the ACM, 26* (11), 861–867.

Miller, L. A. (1981). Natural language programming: styles, strategies, and contrasts. *IBM Systems Journal, 20* (2), 184–215.

Mills, H., Basili, V., Gannon, J., and Hamlet, R. (1986). *Principles of computer programming: A mathematical approach.* Boston, MA: Allyn & Bacon.

Montalbano, M. (1974). *Decision tables.* Chicago, IL: Science Research Associates, Inc.

Nassi, I., and Shneiderman, B. (1973). Flowchart techniques for structured programming. *SIGPLAN Notices, 8,* 12–26.

Newsted, P. R. (no date). FORTRAN program comprehension as a function of documentation. (Unpublished manuscript). Milwaukee, WI: School of Business Administration, University of Wisconsin.

Norcio, A. F., and Kerst, S. M. (1979). Documentation and indentation effects on memory organization of computer programs. Presented at the Annual meeting of the American Psychological Association. New York.

Okimoto, G. H. (1970). The effectiveness of comments: a pilot study (SDD Tech. Rep. No. TR-01-0147). New York: IBM Corp, Systems Development Division.

Overton, R. K., Colen, P., Freeman, P., Wersan, S. J., Viegel, M. L., and Steelman, R. (1973). Research toward ways of improving software maintenance: Ricasm final report (Tech. Rep. No. ESC-TR-73-125) (NTIS No. AD 760819). Claremont, CA: Corporation for Information Systems Research and Development. January 1973.

Pollack, S. L., Hick, Jr., H. T., and Harrison, W. F. (1971). *Decision tables, theory and practice.* New York: Wiley-Interscience.

Ramsey, H. R., Atwood, M. E., and Van Doren, J. R. (1978). Flowcharts vs. program design languages: an experimental comparison. In *Proceedings of the 22nd annual meeting of the Human Factors Society,* pp. 709–713. Santa Monica, CA: The Human Factors Society.

Sheppard, S. B., Bailey, J. W., and Kruesi-Bailey, E. (1984). An empirical evaluation of software documentation formats. In J. Thomas and M. Schneider, Eds., *Human factors in computer systems* (pp. 135–164). Norwood, NJ: Ablex.

Sheppard, S. B., Curtis, B., Milliman, P., and Love, T. (1979). Modern coding practices and programmer performance. *Computer, 12* (12), 41–49.

Sheppard, S. B., Kruesi, E., and Curtis, B. (1980). The effects of symbology and spatial arrangement on the comprehension of software specifications (Tech. Rep. No. TR-80-388-200-2). Arlington, VA: General Electric.

Shneiderman, B. (1980). *Software psychology.* Cambridge, MA: Winthrop Publishers, Inc.

Shneiderman B., and McKay, D. (1976). Experimental investigations of computer program debugging and modification. In *Proceedings of the 6th Congress of the International Ergonomics Association.* Santa Monica, CA: The Human Factors Society.

Sime, M. E., Arblaster, A. T., and Green, T. R. G. (1977). Structuring the programmer's task. *Journal of Occupational Psychology, 50,* 205–216.

Smith, S. L. User-system interface design for computer based information systems (Tech. Rep. No. ESD-TR-82-132), Bedford, MA: Mitre Corp.

Stay, J. F. (1976). HIPO and integrated program design. *IBM Systems Journal, 2,* 143–154.

Thomas, J. C., and Carroll, J. M. (1981). Human factors in communications. *IBM Systems Journal, 20* (2), 237–263.

Weissman, L. (1974). A methodology for studying the psychological complexity of computer programs (Unpublished doctoral dissertation). Toronto: University of Toronto.

Woodfield, S., Dunsmore, H. E., and Shen, V. Y. (1981). The effect of modularization and comments on program comprehension. In *Proceedings of the fifth international conference on software engineering* (pp. 215–223). New York: IEEE Computer Society.

Wright, P., and Reid, F. (1973). Written information: some alternatives to prose for expressing the outcomes of complex contingencies. *Journal of Applied Psychology, 57,* 160–166.

PART 12
SELECTED APPLICATIONS OF HUMAN FACTORS IN COMPUTER SYSTEMS

CHAPTER 12.1

HUMAN FACTORS IN OFFICE AUTOMATION

SARA J. CZAJA

Department of Industrial Engineering
State University of New York at Buffalo

12.1.1 INTRODUCTION

Information-handling activities are beginning to dominate the economy. Currently about 50–60% of the total labor force is engaged in some way in the handling and processing of information. White collar work represents the single largest and fastest growing sector of the economy. In 1900, white collar workers accounted for only 17% of the labor force; today about 50% of the workers in the United States are white collar workers (Figure 12.1.1).

A distinction should be made between white collar workers and office workers, since these are similar but not identical occupational categories. White collar workers include office workers as well as other types of workers who are not formally classified office workers, such as retail clerks, cashiers, and bank tellers (Panko, 1984). This distinction may explain some of the discrepancies that appear in the literature regarding the number of office workers. It has been suggested that "information work" would be a more useful classification, and using this term, 54% of the labor force can be labeled as "information workers" (Panko, 1984).

Within the white collar work force, the largest group of workers is clerical workers and the second largest group is professional technical workers (Figure 12.1.2). These trends are also linked to the current demand for information. With respect to the future, it is predicted that the explosive growth in the number of white collar workers will level off (Table 12.1.1). However, despite this leveling off, white collar work and the demand for information will dominate in most industries.

In addition to changes in the composition of the work force, changes are also occurring in the way work is performed. Computer and information technologies are increasingly being used to automate routine tasks and procedures. Also, an increasing number of workers are now relying on computer-based information systems to carry out their job activities. It is estimated that by 1990 between 40% and 50% of all American workers will use some type of electronic terminal equipment on a daily basis (Giuliano, 1982).

During the next decade, manufacturing organizations are planning to invest heavily in robotics, computer-aided design and computer-aided manufacturing, automated materials handling devices, and distributed process control. It has been suggested that the "factory of the future" will be characterized by three major elements: flexible workstations with embedded intelligence, large-scale information networks, and simulation and modeling packages to aid decision making and planning (Messina, 1984). These projections for the factory of the future are paralleled in the office.

The diffusion of computer and information technologies into the office environment has narrowed

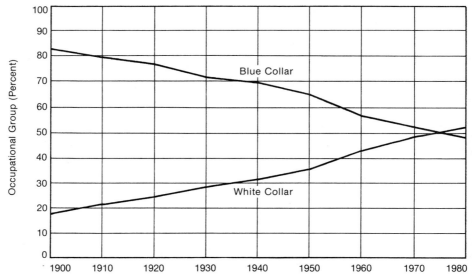

Fig. 12.1.1. Trends in employment in the United States. [From: Guliano, Vincent E. Copyright © (1982) by Scientific American, Inc. All rights reserved.]

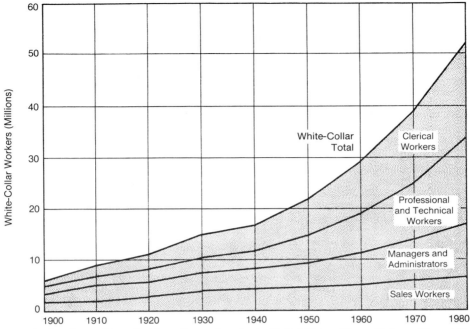

Fig. 12.1.2. Composition of white collar workers in the United States. [From: Guliano, Vincent E. Copyright © (1982) by Scientific American, Inc. All rights reserved.]

the gap between white collar and blue collar work. Many office tasks are becoming increasingly specialized with greater speed and accuracy requirements, production deadlines, and shift work (Hoos, 1983). In the near future, the worlds of the factory and the office may merge in the form of an integrated operation, where organization of a manufacturing corporation will be structured as an information heirarchy utilizing a communications network. In this type of arrangement, office and factory functions are integrated, and as a result the classic distinctions between the office and factory are blurred.

Initially computer technologies had a fairly limited application in offices (e.g., payroll), and their access was restricted to a handful of specialists. However, as microelectronics become more powerful and less expensive, they are increasingly being used for gathering, storing, manipulating, and communicating information. Currently, telecommunication costs are declining about 10% per year and computer hardware costs are declining about 25–40% per year (Helander, 1983). Total spending on office technology in 1980 was $120 billion; see Table 12.1.2.

There have also been significant increases in computing power. A few years ago a personal computer had 48K of working memory and an eight-bit microprocessor running at a rate of 1 MHz. Machines now have 512K of memory, a 16-bit microprocessor, and run at 4 MHz or more. Storage capacity and processing power will continue to increase as integrated-circuit manufacturing firms are currently experimenting with 4M-byte devices.

These advances in technology, as well as growing concerns about lagging productivity are largely responsible for the rapid growth of microcomputers for business applications. Most organizations are experiencing a growth in their use of word processing, electronic mail, personal computing, and so forth. The application of computer technology to office work not only promises to improve office productivity by allowing for more effective information handling but also to improve the quality of work life and job satisfaction of office employees. It is argued that the use of the new technologies will rid jobs of tasks that are tedious and boring and allow people more time for tasks that are creative and challenging. This will ultimately result in work that is more rewarding, productive, and satisfying.

While these are the promises of office automation (OA), there is currently some doubt about the extent to which they will be realized. Several analysts (Hoos, 1983; Klein and Hirschman, 1983) feel that office automation may negatively affect quality of work life by deskilling and fragmenting jobs. Also, developments in office automation fall short of previous predictions. The demand for OA has developed more slowly than expected, and, as Uttal (1982) points out, we are still far from witnessing the "paperless" office. This is due in part to a limited understanding of the impact of office automation on people and on organizations.

Table 12.1.1   Projected Growth Rates for White Collar Work

	Percentage of Work Force			
Category	1900	1940	1980	1990[a]
White collar				
Managers	5.8	7.2	10.9	10.7
Professional/technical	4.3	7.4	15.8	14.8
Clerical	3.0	9.6	18.3	19.0
Sales	4.5	6.7	6.2	6.7
White collar total	17.6	31.1	51.1	51.2
Other	82.4	68.9	48.8	48.8
Total	100	100	100	100

Source: Panko, R. R. (1984). Office work. *Office Technology and People,* vol. 2, pp. 205–238. Reprinted with permission of North Holland Publishing Company, Amsterdam.

[a] Estimated.

Table 12.1.2   Estimates of U.S. Spending on Office Technology, 1980

Category	Spending ($ billions)
Telephone	36
Office systems support labor[a]	30
Office data processing[b]	27
Text processing	
Reproduction	8.3
Postal delivery	6
Paper, not otherwise classified	5
Word processing	1.5
Typewriters	1.3
Electronic mail	1
Micrographics[c]	0.9
Calculators	0.6
Mailroom equipment	0.6
Pens and mechanical pencils	0.6
Miscellaneous[d]	1
Text processing total	26.8
Office furniture	2
Total (rounded)	120
Per office worker (rounded)[e] $3000	

Source: Panko, R. R. (1984). Office work. *Office Technology and People,* vol. 2, pp. 205–238. Reprinted with permission of North Holland Publishing Company, Amsterdam.

[a] Includes the labor needed to plan, operate, and maintain office systems. Data processing center labor would fall into this category. Does *not* include end user labor.

[b] Includes all data processing done to handle what would normally be considered office function.

[c] An unknown portion of this micrographics spending really belongs under office data processing.

[d] Includes answering systems, dictation machines and many other items.

[e] Assume 38 million office workers.

The influx of information technology changes organizational structure, work roles, job content, decision making, interaction patterns, and the physical environment. Yet only a limited amount of effort has been directed toward assessing the impact of these changes on end users and organizations. Before successful OA systems can be developed and implemented, we need to have an understanding of how office work is performed and also of the needs of workers and organizations so that we can consider the potential impact of automation on office activities and people.

This chapter will present an overview of OA and discuss the implications for office work and office workers including issues surrounding the implementation and evaluation of OA such as needs assessment, training of office workers, and resistance to change. Finally, some perspectives about the future of OA will be presented. Since the domain of OA is so vast and the field is constantly evolving, discussion of specific technologies and their application will be limited. The emphasis will be on examining automated office systems from a human factors perspective.

12.1.2 OVERVIEW OF OFFICE AUTOMATION

Although the term office automation is widely used, the introduction of computer technologies into offices does not constitute automation in a strict sense. Automation usually implys a process of substitution; substituting some device or machine for human activity (Parsons, 1983). In office environments, the implementation of computer technologies serves instead to enhance and augment worker tasks. Therefore, this chapter will be discussing the use of computer technology with respect to the facilitation of office functions. However, this will be done under the heading of office automation, since this term is typically used to describe this augmentation process.

Specifically, OA refers to the application of computer technologies to the creation, manipulation, storage, reproduction, retrieval, and dissemination of information. While this is a generally agreed upon definition of OA, it is far from adequate as it tends to focus on individual office tasks and not on the integration of these tasks or the functions of these tasks within the larger organization system. A working definition of OA will be presented later in the chapter.

Computers were first introduced into the office in the early 1960s. These machines were expensive and limited in application. However, by the late 1960s, most large corporations used a computer to facilitate functions such as payroll, inventory, and billing. With advances in microelectronics, more powerful and remote terminals consisting of a keyboard and visual display began to appear. These remote terminals were linked to large mainframe computers. In the early 1970s, minicomputers, public branch exchanges (PBX), and word processors made their way into the marketplace, and, by the end of the 1970s, the microcomputer paved the way for automated offices. In fact, word processing is often cited as a hallmark of office automation. Table 12.1.3 presents some of the technological developments that have been milestones in the growth of OA.

There has been some controversy regarding what falls within the domain of office automation. This has largely resulted from a division between office automation and data processing. Proponents of office automation argue that data processing does not fall under the heading of OA since it has limited applications in offices, is not performed by office workers, and is a product of the 1960s. However, similar claims can also be made for word processing, since its applications are limited and it is typically used by a small segment of office workers. Unfortunately, in many instances word processing is used synonymously with office automation. This is a bit ironic since clerical salaries account for only a small percentage of office costs and, therefore, do not represent the biggest potential for savings.

The antagonism between office automation and data processing is no longer very relevant. In recent years the scope of office automation has broadened and the focus has shifted away from clerical workers and toward professional and technical workers. OA is now aimed at providing tools that augment tasks such as information management and decision making. OA has come to imply an integrated

Table 12.1.3 Some Milestones in the Development of Office Automation

1968	Integrated circuits used in hardware
1970	Word processing introduced
1972	Electronic typewriters introduced
1974	VDTs commonly used for word processing
1978	Intelligent fiber-optics printers
1980	Merging of word processing and data processing
	Local area networks
1981	Xerox Star executive workstation introduced
1982	Personal computers common
	Systems integration with voice, data, and video
1983	16-bit portable, personal computer with complete word processing

Source: Helander (1983) Reprinted with permission.

office system (IOS) or office information system (OIS) where all components necessary to support office personnel are integrated. An IOS is designed to aid office personnel in a wide range of activities, which include document preparation, information management, and decision making.

In today's office, large mainframe computers are rapidly being replaced or augmented by personal computers. These personal computers, often referred to as "intelligent workstations," allow a person to perform a variety of functions such as word processing, data analysis and storage, graphics production, and electronic filing from a single workstation. They also enable a worker to access internal as well as external data bases.

For example, the Xerox 8010 Star is a personal workstation that was designed for business and professional workers. The Star is a distributed processing system that can act as a microcomputer and word processor, and also as an information channel when it is linked to data bases and other workstations within a network. The Star was designed for computer novices and does not require users to have any special computer skills. The system makes use of icons; office objects such as documents, folders, file drawers are represented on the screen by small pictures. Data icons such as documents are printed, mailed, or filed by moving them to icons such as printers, outbaskets, or files by manipulating a cursor on the screen. Thus command names are not needed for these types of operations. Other sophisticated systems include Prime's Automated Office Systems, IBM's Profs, and Wang's Office. These systems are similar to the Star and offer generic tools such as electronic messaging, text editing, programming, and filing.

There are several advantages associated with personal computers. They are designed for nonspecialists and thus are intended to be easy and simple to use. They are designed for a single user and therefore private data files can be created and stored. They are independent of the mainframe computer; thus users are not affected by the response time of the mainframe system and can continue to work if the mainframe fails (Goldstein, Heller, Moss, and Wladawsky-Berger, 1984). Local area networks (LAN) represent one of the major advances in office technology. A LAN is a system where information can be transferred among office system terminals and peripherals. It allows workstations within an office and offices within an organization to communicate with one another. It also provides the capability of integrating functions such as word processing, electronic mail, and video conferencing. Large amounts of information may be transferred at a rate of several millions of bits per second.

Ethernet, developed by Xerox, operates at data rates of 1 million to 10 million bits per second. These rates allow file transfer at a rate similar to that between the disk secondary storage of a computer and its central processor (Newell and Sproull, 1982). Microcomputers within this system communicate with one another through the use of a coaxial cable. The system is controlled by "workstations" and "servers," which have direct control over the data network. VM Pass-Through is a networking facility developed by IBM that supports terminals in the IBM 3270 family. It provides remote terminal connections for time sharing and data query applications and allows a single VM/SP terminal to access many different computers using a wide variety of software systems. One claimed advantage of Pass-Through is that it is easy to install and operate; it can usually be installed and tested in a few hours (Mendelsohn, Lenehan, and Anzick, 1983).

One important distinction that characterizes a network is whether the network is considered proprietary or open. Proprietary networks can only be used with a vendor's product line, for example, Xerox's Ethernet and IBM's Pass-Through, or designed to fit other vendor's product lines, for example, an operating system such as MS-DOS. Open networks are designed to accommodate any vendor's system. The emphasis is on the functions and features of the system as opposed to workstations that can be connected to it. Current open networks include 3M, Sytek, and Corvus. The obvious advantage of open networks is their flexibility to combine several systems. However, since they are not vendor specific, management and maintenance of these networks is more difficult than for proprietary networks.

LANs are increasingly used for simultaneous processing of voice, video, and data in digital form. We can expect routine use of speech recognition and voice store and forward systems. Long-distance communications will also become available. For example, Teletex is a service for worldwide telecommunications. It allows users to exchange electronic documents among office text machines such as electronic typewriters and word processors. It has been designed as an international standard and has the ability to satisfy almost all users of the Latin-based alphabet. Teletex offers higher speeds, improved print quality, and lower transmission costs than Telex, and it is anticipated that it will be widely used in the near future (Moore, 1983).

It is anticipated that LANs will move more toward open systems and provide a means of integrating equipment from different vendors. Presently, lack of standardization of hardware and software is one of the major roadblocks facing office automation as devices and systems are incompatible, limiting information transfer.

Recent trends also include the development of integrated software packages such as Framework or Symphony. Symphony (Lotus Development Corp.) incorporates several functions including word processing, communications (between workstations and/or mainframe data bases), spreadsheet, a forms-oriented data base manager, and graphics. Other developments in OA will include low-cost color graphics, micrographics, and optical desk storage systems. Optical memories with 10,000–100,000 megabyte capacities will be commonly used.

However, despite these advances in technology, the future of OA is dependent on the extent to which these systems can be used effectively and are accepted by the user. For example, software design represents one of the major challenges facing office automation. Current software is typically intricate and complex and often designed for sophisticated computer specialists. Many software packages are designed without consideration of users, who, in turn, become frustrated and less willing to use the system. Unless software is designed with a focus on users, it will continue to hinder future progress. The emphasis in the design of office systems must be on the identification of useful functions and the development of easy-to-use human interfaces (Gould and Boies, 1983).

12.1.3 WHY AUTOMATE?

The most commonly cited reasons for office automation include:

Declines in productivity.
Increase in number of white collar workers.
Increase in office costs.
Increase in demand for information.

Office productivity has received a great deal of attention in the last decade. A commonly held perception is that white collar productivity has been stagnant, increasing only 4% in the last 10 years. At the same time, the cost of doing business has been rising steadily. For example, between 1952 and 1981 the cost of producing a single-page business letter more than quadrupled, increasing from $1.15 in 1952 to $6.62 in 1981 (Barcomb, 1981). In 1979, total office costs were $800 billion, 73% was spent on managers and professionals and 27% was spent on clerical workers (Helander, 1983). Additionally, the demand for information is increasing; each year in the United States more than 100 billion telephone calls are initiated and 70 billion documents are created (Lieberman, Selig, and Walsh, 1982). These widely quoted figures have generated a lot of concern given that office work is so extensive and has a direct impact on our position in the competitive market.

OA has been considered as the panacea for the malaise of white collar work. This is partly based on the premise that limited investment in office equipment has thwarted increases in productivity. It is often cited that the average white collar worker is backed by only $2000 in capital investment. However, this statistic is not based on empirical data, and a recent study (Panko, 1984) indicated that the annual spending on office systems, including maintenance and capital investment, is approximately $3000 per office worker. Given the correlation between equipment investment and performance in other industries, it is assumed that if spending on office technology is increased, the performance and productivity of office workers will also increase. In fact, Booz, Allen and Hamilton (1980) maintain that properly implemented office automation can result in a 15% improvement in productivity through time savings generated by use of office equipment. It is also estimated that OA will reduce office costs. For example, word processing can reduce secretarial costs from $7 per letter to less than $2, and electronic mail can further reduce the costs of sending a message to 30 cents or less.

Despite these claims, the results of automating office operations are often disappointing. One reason is that the statistics indicating a decline in growth in office productivity are misleading. Panko (1984) indicated that the 4% growth estimate is erroneous. It was derived by dividing the Gross National Product by the total number of office workers, disregarding improvements in the quality of work. In fact, Panko (1984) surveyed federal office productivity and found that from 1977 to 1981 the productivity of federal office workers increased 14.8%, corresponding to a compounded annual rate of 2.8%. However, these statistics must also be viewed with some skepticism. The real issue surrounding office productivity is that it is difficult to define and subsequently difficult to measure.

Productivity is usually defined as the ratio of input to output. This definition works well in factory environments where input is measured in person hours and capital, and output refers to tangible products. However, it is not applicable in office environments. Offices do not always produce a tangible product that can be used as a measure of productivity.

Some analysts focus on objective measures such as number of pages typed or time required to type x number of pages when evaluating the productivity of offices. Such estimates ignore both the quality and value of outcomes. It may well be that additional pages can be produced through use of a word processor, but these pages may have little value to the organization. Increasing the amount of information produced or the rate at which information is produced does not necessarily contribute to the quality of the organization. For example, the use of copying machines has both positive and negative consequences. It allows documents to be reproduced quickly and easily, but, at the same time, contributes to the problem of paper overload.

Measurement of productivity becomes even more difficult when attempting to quantify the output of unstructured office activities associated with managers and professionals. How can activities such as decision making or meeting and communicating be quantified? Sometimes, office automation is justified through estimates of time savings associated with using the new technologies. However, this is only a partial measure of productivity, since it does not take into account lost time associated with using OA technologies and it is based on the assumption that the time saved will be spent productively. What is needed is a different approach to studying the impact of office automation. Hammer (1982) suggests that the real objective of office automation is to increase the effectiveness of business operations, rather than to increase the efficiency of information handling.

The distinction between *efficiency* and *effectiveness* of performance was specified by Drucker (1967) and it is useful when studying performance in offices. Efficiency of performance refers to producing output at the least possible cost with the fewest possible labor hours and the least number of errors. Effectiveness of performance is concerned with "doing the right things," by selecting the correct goals and then pursuing them efficiently (Panko, 1984). This type of measure may be more appropriate for assessing managers and professionals.

Generally measures of the "success" of office automation are based on estimates of efficiency with little consideration of effectiveness. This is based on the assumption that the objective of office automation is to improve the efficiency of office tasks such as typing, filing, communicating, and so forth. This approach emphasizes the routine tasks in the office and how performance of these tasks can be improved with the use of office technologies.

There are a number of problems with this approach. First, it emphasizes using new tools for conventional tasks under the assumption that these tasks and the current method of doing business is optimal. It does not consider benefits of office automation obtained by restructuring office activities. Second, it assumes that the performance of these tasks is an end in itself and does not consider how they relate to the goals and needs of an organization. Office tasks are not goals in and of themselves but exist to enable an office to accomplish its business function. Finally, the major focus of this approach is on clerical workers, and it overlooks managers and professional workers. This latter group of workers account for the largest percentage of office costs and enhancing their performance represents the biggest potential for benefits to the organization (Hammer and Sirbu, 1980).

Effectiveness should be the basis for office automation. This is not to imply that efficiency should be ignored, but it should not be viewed as an end in itself. The assessment must be based on the extent to which it improves the ability of an organization to accomplish its business goals. This type of assessment focuses on the content of office work as opposed to the form of office work. It is based on a functional definition of OA where OA is defined as the use of technology to realize business functions (Hammer and Sirbu, 1980). This definition views OA as a means of supporting office functions where tasks and technologies are evaluated in relation to the office and its goals.

Thus an effective measure of the success of OA depends on an understanding of the functions of the organization. OA can be said to be successful or contribute to an increase in productivity if improvements in these functions are realized. These improvements may take several forms, including cost reductions, achievement of more desirable goals, added capabilities, or quality enhancement.

The traditional cost/benefit analysis is not appropriate for office automation. The costs associated with office automation are difficult to calculate. They are not restricted to hardware and software maintenance, but also include organizational stress, work disruption, training, and adaptation. Similarly, the benefits derived from OA cannot always be expressed in hard dollar terms but may include things like improved quality, enhanced capabilities, increased morale, or improved customer relations.

In identifying the measures that are to be improved through OA, it is necessary to understand the nature of the organization. Obviously performance measures will vary from office to office, however generic measures of performance that may cut across organizations have been identified (Hammer, 1982, Helander, 1983). They include:

1. Financial measures—cost reduction, improved cash flow, cost avoidance
2. Quality measures—speed, accuracy, flexibility, reduced turnaround time
3. Competitive measures—size, growth rate, number of customers
4. Organizational health—absenteeism, turnover, morale indices
5. Enhanced capabilities—access to more current information, simulation providing "what if" capabilities
6. Reduction in unproductive tasks—misdialed numbers, busy signals, travel

The extent to which these measures are applicable depends on the organization. However, any OA effort must be based on a system approach where the goals of automation are viewed in terms of improving the overall operation of the office system. A systems approach to OA will be discussed later in the chapter. The next section will focus on the impact of automation on people and organizations as this is closely related to the topic of productivity.

12.1.4   THE IMPACT OF OFFICE AUTOMATION ON PEOPLE AND ORGANIZATIONS

The evolution of changes in the organization of offices has been characterized by Giuliano (1982) according to three stages: preindustrial office, industrial office, and information-age office. The differences in style among these three types of offices are depicted in Figure 12.1.3. The preindustrial office is characteristic of many small businesses. It depends on the performance of individuals without much reliance on systematic work organization or modern information technologies. This type of organization works well for low-volume operations, but it is inadequate for offices that handle large volumes or engage in complex procedures that require a variety of data bases. The industrial office is organized according to the principles of work specialization and simplification and time and motion efficiency. It closely resembles a production line and has been considered to be well suited for offices that handle a large volume of transactions. This type of office has several postural disadvantages: the flow of information may be slow, mistakes are difficult to identify, and much of the work is boring and routine. The information-age office represents the automated "office of the future," and it developed in response to the limitations of the industrial office. It attempts to combine terminal-based workstations, data bases, and communication networks. Although the information-age office is considered to represent the highest level on the evolutionary scale, Giuliano (1982) maintains that many large organizations are still in the industrial stage. Despite the rapid developments in information technology, many organizations are slow in adopting and taking advantage OA. In fact, only 15,000 LANs currently exist within the United States.

There are several reasons why the information-age office is so slow in coming. One reason is that the benefits are difficult to assess. Many organizations are unwilling to make initial investments in technology since they cannot be discussed in terms of traditional measures of ROI. Another reason, often overlooked, is that technology has a pronounced impact on people and on organizational structure. The nature of this impact must be understood before any implementation will be successful.

It is widely recognized that one reason for the limited success is a lack of sensitivity to human issues. Martin (1981) studied the impact of office automation in a number of organizations and found several barriers that limit the success of office automation (Table 12.1.4). He concluded that the goal of OA should be the integration of the three components of office systems: the people, the organization, and the technology. Unfortunately, in many instances the major impetus for automation has been technological capabilities rather than user requirements.

OA is a means of performing traditional office work in new ways. It presents opportunities to eliminate boring and repetitive tasks, introduces greater variety and skill requirements into jobs, and allows workers the chance to perform jobs that require more responsibility and are more challenging. However, while these are potential consequences of OA, they are not often realized and thus increases in efficiency and effectiveness of performance do not occur. As Hoos (1983) points out, automation is not an unmitigated blessing for all people. It often results in adverse affects such as labor displacement, skill obsolescence, stress, and worker dissatisfaction. For example, the implementation of word processing has had mixed reviews. In many instances it was received with negative reactions. Clerical workers expressed concern that their jobs were becoming depersonalized and fragmented, and they no longer had control over their work. These concerns were often expressed by lowered morale or decrements in performance (Rubin, 1983). This does not imply negative consequences are always necessary. However, it does point out the need to address human and organizational issues. Otherwise there is a risk of interjecting a significant source of stress and frustration into the work environment (Cohen, 1984).

One critical variable is user acceptance of the system. The degree of user acceptance is determined to a large degree by the way in which the technology is introduced into the organization. There is a wide body of literature that has addressed the issue of how the introduction of change impacts on the adoption of change. Although a majority of these studies have focused on social change as opposed to technological change, the results are still generalizable to situations where technology is being introduced. The findings clearly indicate that the way change is introduced has a significant effect on the degree to which it will be adopted within an organization (Grummon, 1982).

The introduction of office technology not only alters the objective work situation but also the worker's perception of the situation, which in turn has a major impact on behavioral response to that environment (Craik and Lockhart, 1972; Kahneman and Tversky, 1973; Nisbett and Ross, 1980). To date, insufficient attention has been given to the attitudes, perceptions, and expectations of the users of office systems. Yet, these are major factors in determining their reactions toward innovation in the workplace (Park and Freedman, 1984).

Zuboff (1981) maintains that information technology alters the relationships between the person, the task, and the organization. According to Stout (1981), there are eight major job related and personnel changes brought about by office automation:

1. Work groups: people are often regrouped to make the best use of equipment and long-term relationships may be disrupted.
2. Social relationships: number and type of social contacts may change.
3. Job content: the use of technology may eliminate some tasks while creating new job tasks. Thus skills may become obsolete and retraining is required.
4. Career paths: new job opportunities may be created and old ones may no longer exist.
5. Responsibility: certain tasks may be allocated to a machine that were formerly the responsibility of a person.

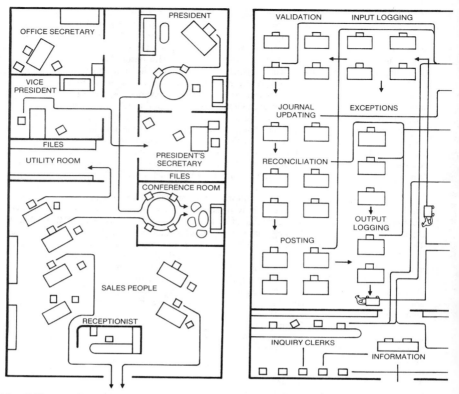

Fig. 12.1.3. Differences in organizational structure among preindustrial, industrial, and information-age offices. [From: Guliano, Vincent E. Copyright © (1982) by Scientific American, Inc. All rights reserved.]

6. Job autonomy: freedom of choice in the performance of tasks may be altered. Workers may have to change their work habits to conform with machines.

7. Work environment: spatial arrangements, furnishings, and the ambient environment must be altered to accommodate the new equipment.

8. Status: changes in work roles may create changes in status, which in turn can create changes in a person's self-image.

The extent of these changes depends of course on the organization, the amount and type of technology, and the implementation strategy. However, they will occur to some degree in all organizations where technology is introduced.

Parsons (1983) considers the aspects of automation or changes brought about by automation as independent variables and the reactions of people to automation as dependent variables. Reactions to or effect of automation on people can be manifested in attendance, locomotion/transfer, social behavior, perceptions, feeling, and health and safety. The extent to which worker's reactions to automation are positive or negative depends on the outcomes of automation. Also, there will be individual differences in reactions. For example, some individuals react positively to the need for retraining since it presents challenge and the opportunity to acquire new skills. Others may react negatively because of perceived threats to security and fear of failure. Parson's distinction (Table 12.1.5) provides a good conceptual framework for studying the impact of automation on people. The next section will discuss some of the changes brought about by automation and potential problems that may emerge as a result of these changes. The solution advocated in this chapter to these potential problems is a systems approach to implementation, which will be discussed in a later section.

12.1.4.1 Job Design

One important aspect of automation is the way in which it affects the structure and content of jobs. If consideration is not given to people, the use of computer technologies can lead to routinization,

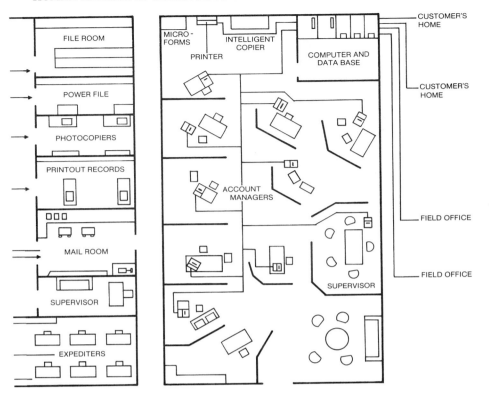

Table 12.1.4   Barriers to Successful Office Automation

Organizational
    Dominated by short-term goals/earnings.
    Failure to understand the potential for improved operations.
    Failure to understand productivity/cost benefits.
    Wants benefits without costs.
People
    Resist change.
    Want control.
    Do not understand what OA requires them to do.
Implementors
    Unsure of user requirements.
    Seeking "ultimate" solutions.
    Lack planning expertise.
    Do not understand the anxiety of employees toward change.
Technological
    Number of choices.
    Incompatibility of technology.
    Incompatibility of media.
    Too few communication standards to permit standards integration.
    Questionable security.
    Changing user skills.
    Poor planning of physical space in the office.

Source: Martin, J. (1981) Successful Office Automation, Part I. *Computer Decisions,* vol. 13, pp. 108–122, 161–162. Copyright © 1981. Reprinted with permission of Hayden Publishing Co., Rochelle Park, N.J.

Table 12.1.5   Framework for Analyzing the Impact of Office Automation

A.   Aspects of OA Impacting Workers
     Employment opportunities
     Job/task design
     Equipment design
     Environmental design
     Organizational factors
     Social interactions
     Health and safety factors
     Motivational factors (e.g., incentives)
B.   Sources of Reactions of Workers to the Outcomes of OA
     Job attendance
     Job performance
     Social behavior
     Health and safety
     Perceptions
     Attitudes/morale
     Learning

Source: Adapted from Parsons (1983).

simplification, and fragmentation of jobs. Thus skills are underutilized and opportunities for challenge, achievement, and self-fulfillment become limited. Jobs may become boring and repetitive instead of challenging and interesting. This has often been the case for clerical and secretarial workers. Studies have indicated that despite the introduction of automation, the amount of low-level clerical jobs has remained the same or even increased (Working Women, 1980). This is a bit ironic given that clerical workers represent the largest sector of the white collar work force.

Automation can also reduce the freedom of choice people have in performing their tasks, making jobs restrictive and depersonalized. For example, workers may experience some loss of creativity in problem-solving tasks because of available software packages (Office of Technology Assessment, 1983).

Control over workload is another important job design factor. The use of computers may result in some tasks becoming machine paced rather than operator paced. Smith, Cohen, Stammerjohn, and Happ (1981) found that the workload of VDT operators is often determined by the capabilities of the machine without consideration for the limits of the operator. They also found that loss of control over work is a contributor to job stress. Because computer technologies are designed to make work faster and easier, work overload may result. For example, word processing enables documents to be changed quickly and easily; thus people may request excessive changes, which creates additional work (Cohen, 1984).

Finally, changes in job content may make current skills obsolete and require workers to learn new skills. Zuboff (1981) maintained that computer work typically demands more conceptual skills and is more abstract. Skill obsolescence is a common outcome of automation in both factories and offices. This can create problems with job displacement and cause workers to experience a loss of confidence and self-esteem and a fear of failure.

The types of changes to job content previously discussed that may result from office automation can create negative reactions on the part of workers. People will most likely experience feelings of dissatisfaction and stress and strain. It is well documented that job design factors are major contributors to stress in the workplace. For example, work overload is a major source of job stress (French and Caplan, 1973; Sharit and Salvendy, 1982). Hall (1984) indicated that key sources of stress for clerical and secretarial workers include conflicting instructions, rapid pacing and machine-paced work, monotonous work, lack of respect and recognition, few opportunities for self-expression, and lack of input into decision making. In general, jobs that are characterized by low status, low pay, repetitive work, and low amounts of physical activity are linked to higher rates of stress and job dissatisfaction. Stress and dissatisfaction not only impact on the individual but also on the organization. These outcomes are linked to low morale, high absenteeism, and high turnover rates (Cohen, 1984). Figure 12.1.4 depicts the relationships between core job dimensions and job outcomes. For a more complete discussion of these issues, refer to Chapters 4.1 and 7.4.

Office automation does not necessarily have to have a negative impact on job design. It has the potential for eliminating routine repetitive tasks and increasing skill requirements. As Taylor (1982) points out:

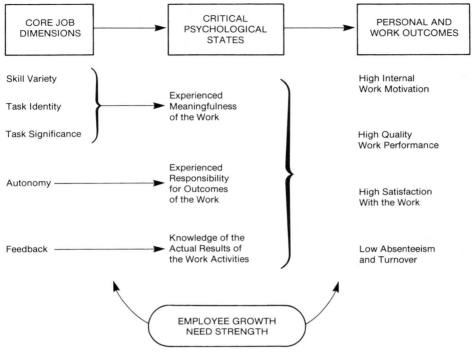

Fig. 12.1.4. Relationships among core job dimensions, critical psychological states, and on-the-job outcomes. © 1975 by the Regents of the University of California. Reprinted/Adapted/condensed from *California Management Review,* volume xxii, no. 4, p. 58 by permission of the Regents.

Computerization does not require fragmented jobs of the assembly-line approach. The technology itself is largely neutral. Whether jobs are made more challenging or congenial or more fragmented or dreary depends on integrative organizational design. The difference lies in whether people are regarded as extensions of the machines or the machine is designed as an extension of people.

12.1.4.2 Organizational Design

The introduction of office automation has an impact on organizational structure and working relationships. It makes information needed for decisions more readily available. It also provides access to a wider variety of data bases and allows more people access to information. This may have an impact on the structure of an organization.

Ideally, the organization should become more decentralized. Many of the intermediary levels in the decision making process should disappear and involvement in decision making may cut across more levels of the organization. This should create more cooperation between managers and their subordinates (Sell, 1984). However, many times organizations remain tied to a top-down hierarchical structure. This can create problems with role ambiguity and role conflict, both of which can contribute to feelings of dissatisfaction.

Workers become uncertain as to what constitutes their job duties and what other people expect of them in terms of responsibility. For example, when tools such as electronic mail and calendar are introduced, questions arise between managers and secretaries regarding who is responsible for maintaining the calendar and screening the mail. Managers may prefer to perform these tasks and thus traditional roles break down. Eason and Gower (1984) found that introducing information technology into an organization altered people's perceptions of what can be done and who should do it. For example, managers perceived that word processing enabled secretaries to get more work done in less amount of time. Thus they concluded that secretaries had more spare time for other activities such as administrative duties.

Changes in job duties can also create problems with role conflict. Workers may feel that they are being forced to perform tasks that are outside of their job description or beneath their status. Managers, for example, often balk at the idea of using a keyboard, likening keyboard usage to secretarial work.

Adversary relationships may also arise between managers and their subordinates because the computer allows managers or supervisors direct access to the employee's work. Smith (1984) points out that employee monitoring by computers creates negative reactions, because the employee feels controlled by the machine. It can also reduce the amount of interaction between the employee and the supervisor.

Changes in status are also linked to computerization. When job duties and responsibilities change, position in the organizational hierarchy may also change. Even if an objective change in status does not occur, workers may perceive a change in status because of changes in job demands. One common fear among workers is that their job has been downgraded because the computer has taken over their responsibilities. This is especially common for secretarial workers who find that many of their tasks are performed by a machine or by other people in the organization. Not only do they experience perceptions of lowered status, but they also have fears about job security. Workers may also find that former career paths are no longer open to them. They may become uncertain as to their position in the organizational hierarchy and future opportunities for advancement.

12.1.4.3   Communication Interaction

The introduction of technology may modify the social environment and create changes in working relationships and communication/interaction patterns. In an automated office, the computer becomes a person's primary focus of interaction. One consequence is that the amount of face-to-face interaction between coworkers may be reduced. People may come to rely on communication tools such as electronic mail and teleconferencing as opposed to face-to-face meetings. They may also find it easier to access a data base when they need information instead of asking a coworker. Also, informal interactions may be reduced as people will no longer need to move around to access or file information. All of this will be possible from a terminal at their workstation.

This reduction in face-to-face interactions is not entirely negative, since managers and professional workers currently spend a large portion of their time in meetings; however, problems with social isolation may arise. Workers may perceive a loss of support from their coworkers and feel that they are no longer a part of a work group. For example, Smith et al. (1981), in their study of VDT operators, found that clerical workers who used VDTs reported lower levels of coworker support and less group cohesion than did workers who did not use VDTs. Salvendy (1984) suggests that social isolation and worker constraint are the most significant changes associated with office automation.

12.1.4.4   Physical Environment

The advent of office automation has vast implications for office design. Conventional arrangements and traditional office furnishings are not compatible with the new technologies. There is an abundance of research which indicates that operators of computer equipment experience both physical and psychological problems if the design of the physical environment is inadequate. Since the design of the physical environment is covered extensively in other chapters of the handbook (6.1–6.7, 11.1 and 11.4), only the most important issues will be discussed in this section.

There are two major effects on buildings and office spaces that result from the introduction of information technology (Ellis, 1984). One is the direct effect that results from the physical equipment. The proliferation of terminals in offices generates new types of requirements for amount of space, layout, furnishings, and the ambient environment. Second, there are indirect effects that stem from changes in sizes of and configuration of working groups and differences in interaction patterns. More space for social interaction and group meetings will be required in automated offices. As Ellis (1984) indicates, social areas, both formal and informal, are more important in the automated office than in the conventional office for three reasons:

1. To compensate for long periods of time spent interacting with a machine in isolation.
2. To accommodate group projects that are characteristic of decentralized organization.
3. To allow for and promote informal communication in order to balance the formalization resulting from using electronic media.

Additionally, communication centers that accommodate "electronic meetings" will be required. Also, since computer activities require new skills and generate a need for retraining, more space may be necessary for meeting and training (Rubin, 1983).

The most pertinent design concern is *flexibility*. Environments must be planned so that they are capable of accommodating rapidly occurring advances in technology and organizational structure. Flexibility is needed in location and configuration of workstations, electrical outlets, lighting, and office partitions (Rubin, 1983).

Office automation generates a need for a different approach to design. Design must become an interactive process between the user groups and the design team (Gould and Lewis, 1983). User involvement is a key element in the design process. The emphasis can no longer be on developing plans and configurations that are standardized across the entire organization but rather in determining the specific requirements for user subgroups and tasks (Lodahl, 1982). Also, the design team must be aware of and plan for future changes that are likely to occur within the organization. Kaplan (1982) observed that many office layouts and plans for office automation do not succeed because of inadequate architectural programming and planning. In order for programming to be effective, it must encompass organizational, motivational, communication, functional, and technical factors. The extent to which the physical environment provides support for individuals and their activities and equipment will have a major impact on the success of the automation effort.

The preceding sections highlight some of the issues surrounding office automation. The discussion should clearly indicate the need for human factors engineering in the design and implementation of office systems. In summary, human factors support is needed in the following areas (Parsons, 1983):

Task allocation—human machine division of labor.

Hardware design.

Software design.

Furniture design and selection.

Environmental design—layout and ambient.

Safety.

Job design.

Determination of skill requirements for selection.

Development of training techniques.

Design of evaluation methods.

The next section will discuss a strategy for the design and implementation of office automation systems.

12.1.5   A SYSTEMS APPROACH TO THE DESIGN OF OFFICE AUTOMATION SYSTEMS

The design of an automated office system is a complex process for a variety of reasons. Office work is ill-defined; procedures are often defined by convention. Offices vary greatly in their business functions and thus use different technical systems and have different social systems. Finally, the domain of office automation is vast and constantly evolving.

The systems approach has been used quite successfully to solve complex military and space design problems and recently has been used for other applications (e.g., software design). The intent of this approach is to develop an optimal integration of person, machine, and environment. There are two main features: the objectives of the system are clearly defined and stated in performance terms and the emphasis in design is on the interrelationships among the components of the system. The application of the systems approach to the design of office environments is increasingly important. This section will describe a systems methodology that can be used for the design and implementation of office automation systems. It should be noted that this represents a generic methodology, and there may be differences in emphasis or application according to the needs of a specific situation.

The methodology that is outlined is based on a sociotechnical view of office systems. An important element of any methodology is the conceptual framework that is adopted to guide the analysis. The framework that is chosen specifies which aspects of the system are emphasized in the system analysis and thus ultimately determines what type of system will be designed. The sociotechnical view focuses on the technical and social systems of work and the interaction between them. The technical system encompasses tools and work procedures and the social system focuses on how work is divided and coordinated and the relationships between individuals and their work. This framework has proven to be successful for the analysis of office work (Ranney and Carder, 1984).

For the planning of office automation, six steps are suggested:

1. Formation of a design team.
2. Systems definition.
3. Needs assessment.
4. System selection.
5. Implementation.
6. Evaluation

As shown in Figure 12.1.5, this is a closed-loop process where the results of the evaluation are fed back into the design cycle. The methodology chosen is based on four principles that have been proven to be critical to the design of user acceptable systems (Gould and Lewis, 1983). These principles are:

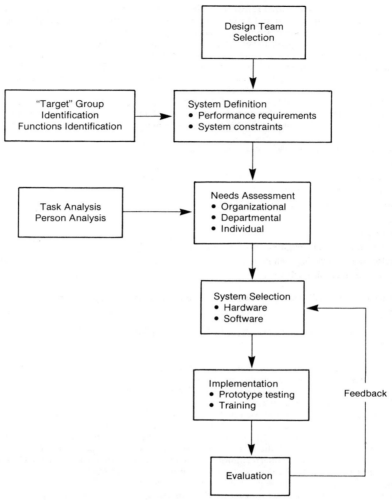

Fig. 12.1.5. A systems approach to the design of office automation systems. [From: Guliano, Vincent E. Copyright © (1982) by Scientific American, Inc. All rights reserved.]

1. There must be an early focus on users.
2. Design must be interactive—users should be involved in the design process.
3. Simulations and prototype designs should be tested by users and performance should be measured.
4. Design must be an interative process, characterized by design, test, measure, and redesign.

12.1.5.1 Formation of a Design Team

Prior to the design of any OA system, a design team should be formed. This team should have the responsibility of designing, implementing, and evaluating the office system. Generally, the design team should include managers, supervisors, technical experts, human factors personnel, and a representative sample of users (e.g., workers from all work groups who will be affected by the technology). If an organization lacks the technical expertise, it may be desirable to have outside consultants as team members. The number of people on the team depends on the type and scope of the system that is being designed.

When the team is formed, it must be provided with direction, and responsibilities within the team must be clearly identified. If the team is large, this may involve the formation of subcommittees.

However, providing a team with a mission does not guarantee its success (Grummon, 1983). The team must be given support by management. For example, workers should not be penalized for attending team meetings. If feasible, they should be relieved of some of their normal responsibilities so they have sufficient time.

12.1.5.2 Systems Definition

Before any system can be designed, its objectives must be clearly identified. The definition of system objectives depends on:

1. Identification of the work group who is going to be the target of the system.
2. Identification of the function and mission of this work group.

The size and nature of the work group that is going to be automated has an impact on the type of system which is going to be designed. It is first necessary to identify who is going to be the target of the automation effort. For example, is the whole organization going to be automated or will the system be restricted to a certain department or certain group of workers? Additionally, what job types are represented in the target group? Does the group consist mainly of clerical workers or does it include managers and professional workers? Different groups and levels within an organization have different needs with respect to system design.

The objective of any automation system is to improve the realization of office functions. After the target group has been identified, the functions and goals of the group must be described. These functions must not be considered in isolation but must be viewed in terms of the larger organizational system. If an entire organization (or office unit) is going to be automated, the function of the various groups within the organization must be described, and the relationships between these functions must be delineated.

The task of describing group and organizational functions is not always clear cut. The functions of a work group are not always evident. Also different individuals may have different perceptions regarding the nature of the group's or organization's objectives. Therefore, it is important to gain exposure to a variety of viewpoints and try to obtain some type of consensus.

Once the objectives of the system are defined, the performance specifications of the system must be stated. This typically includes specifying both the performance requirements and the system constraints. The performance requirements specify what the system must be able to do. They are the operational characteristics that detail the goals or objectives of the system. For example, if one of the objectives of the system is to enhance communication throughout an office, a performance requirement might be the development of a LAN that links all departments within the office and also links the office to other offices within the organization. Additional requirements might be that the network include terminals for all employees and transmit information at a rate of 100,000 bits per second.

System constraints are the limits within which the performance requirements must be developed. They include environmental, resource, cost, and time constraints. For example, it may not be possible to give a terminal to each employee because of budgetary constraints or because of insufficient space to accommodate the equipment. It is important to specify existing constraints before the system is developed. Specifying constraints before system design reduces problems with false expectations and subsequent disappointments. For example, if each employee was initially informed that he or she would be receiving a terminal and then did not, negative reactions toward the system might develop.

12.1.5.3 Needs Assessment

The object of this phase is to specify user requirements. It is necessary to analyze needs at all levels of the organization. A useful taxonomy to guide this analysis is to examine office needs at three levels: organizational, departmental, and individual (Panko, 1984).

The organizational level refers to the relationships among departments and divisions. The focus is on the development of integrated management information and text-processing systems that cut across the entire organization.

The departmental level is an important area for OA because many office tasks such as document creation and information filing take place at department levels. Panko (1984) suggests that it is useful to distinguish between two types of departments, Type I and Type II, as they have different needs. Type I departments handle the routine information processing tasks and have in the past been a focus of office automation. For example, a word processing center is a Type I department. Type II departments handle the nonroutine information processing tasks such as legal departments or engineering departments. Type II offices are going to be the focus of the future. Currently, tools to support activities in these types of offices are limited. Type II departments typicaly need both generic and specialized tools that handle both text and data processing.

The individual level refers to the person and his or her workstation. There are a variety of office automation tools available that support the individual level. Needs at this level are very diverse and needs differ according to job type. For example, the needs of managers are becoming increasingly specialized as organizations become more complex. Many managers perform jobs such as financial planning, marketing, and human resource management. These types of jobs generate a need for individualized systems. Professional workers also perform diverse and specialized tasks. Current efforts in office automation are aimed at aiding managers and professionals by developing decision support systems. To support needs at the individual level, OA systems must be equipped with generic tools such as word processing or electronic mail and individualized tools such as CAD or a financial planning program.

Defining User Needs

The specification of user needs depends on two types of analyses:

1. An analysis of the activities that are performed within an office.
2. An analysis of the potential user population.

Understanding what people do in offices is not an easy task, since office work is so diverse and often ill-defined. A number of investigations have studied office tasks and classified how different types of office workers spend their time. Panko (1984) summarized the results of most of these studies (Table 12.1.6). As shown in Table 12.1.6, managers and professionals spend a majority of their time communicating. Klemmer and Snyder (1972) found that the principal focus of activity for most office workers is communication with other people. Managers and professional workers also spend a large portion of their time performing noncommunication tasks such as data analysis and reading and writing. Clerical workers also perform a variety of tasks. According to a study performed by AT&T (Knopf, 1982), clerical workers spend about 36% of their time typing and 20% talking on the telephone. The remainder of their time is spent on activities such as filing, copying, record keeping, and tabulating.

The results of these studies are useful in providing an initial understanding of office activities. They also can be used when trying to determine what types of generic tools are needed by office workers. For example, electronic mail and teleconferencing would support communication activities, and word processing would aid document preparation. However, this type of information cannot be the sole source of input for the specification of user needs. The type of activities performed in offices and the people who perform these activities vary across situations and task data must be collected, so that the system may be tailored to the needs of the users.

Task analysis is a tool that can be used to collect information on tasks. A task analysis describes job procedures and also delineates skills that are needed to perform task activities. Types of information that need to be collected include task procedures, equipment, sources of needed job information, types of needed job information, and interaction patterns. There are several methods available for collecting task analysis data including observation, interview, video taping, diary, and job participation. Each method yields different types of information and has specific advantages and disadvantages. As a rule of thumb, several methods should be combined to collect task information. A more complete description of task analysis is given in Chapter 3.4.

After the task information is collected, it should be organized according to some framework so that it will be easier to analyze with respect to system requirements. Helander (1985) developed a taxonomy of office tasks (Figure 12.1.6) in an attempt to provide a conceptual framework for classifying different types of office tasks and human–machine interactions. Using Helander's taxonomy, office tasks are ordered according to job type. For example, inputting and transcribing correspond to clerical tasks, and persuading and negotiating correspond to management tasks. Human skills that are used in performing these tasks are also listed.

Taxonomies of this sort provide a useful framework for organizing information about office activities. They may be used to initially identify which automation features are needed by different job types. However, there are some limitations with this approach. The boundaries between office tasks and job types are not always clear cut and thus the needs of some workers may be overlooked. Also the information on human skills is oversimplified.

It is also necessary to classify tasks according to the extent to which they are related to the attainment of organizational or office goals. Tasks that are not related to goal attainment can be classified according to two categories: tasks that are needed for social and organizational reasons and tasks that may need to be reconsidered. An analysis of office activities may identify tasks that no longer contribute to the attainment of office goals but are only performed out of habit. Some of these tasks are needed as they may fulfill social functions, but others need to be eliminated or restructured before the new system is implemented (Bracchi and Pernici, 1984).

Information about the potential user population is also important. This includes information regarding the skill level(s) of the user group, experience with office automation, attitudes toward office automation, needs, and preferences. This information is important for specifying user requirements as well

Table 12.1.6 Managers' and Professionals' Use of Time

Source	Number of Respondents	Percentage of Working Day				
		Face to Face	Phone	FTF & Phone	Reading, Writing	Total Communication
Managers only						
Teger	a	47	9	56	29	85
Stogdill and Shartle	470	60	—	—	—	—
Engel et al.	199	27	13	40	34	74
Stewart	160	54	6	60	28	88
Hinrichs	136	45	8	53	25	78
Burns	76	—	—	52	24	76
Horne and Lupton	66	54	9	63	24	87
Doktor	8	75	3	78	—	—
Dubin and Spray	8	55	6	61	5	66
Croston and Goulding	6	56	7	63	18	81
Ives and Olson	6	68	10	78	—	—
Mintzberg	5	64	6	70	—	—
Kurke and Aldrich	4	62	8	70	—	—
Kelley	4	61	—	—	—	—
Choran	3	36	17	53	—	—
Palmer and Beishon	1	54	6	60	15	75
Notting	1	—	—	59	17	76
Professionals only						
Case Institute	15000	—	—	—	—	53
Teger	a	23	17	40	42	82
Engle et al.	130	10	11	21	30	51
Hinrichs	96	25	5	30	27	57
Managers and professionals						
Xerox	17,000	18	5	23	47	70
Booz-Allen	b	—	—	46	21	67
Managers, professionals, and secretaries/clerks						
Klemmer and Snyder	c	35	7	42	26	68
Blair and Nelson	175	26	14	40	21	62

Source: Panko, R. R. (1984). Office work. *Office Technology and People*, vol. 2, pp. 205–238. Reprinted with permission of North Holland Publishing Company, Amsterdam.

a 5300 questionnaires and personal interviews for managers and professionals combined.

b 90,000 time samples for 299 managers and professionals.

c 3132 time samples

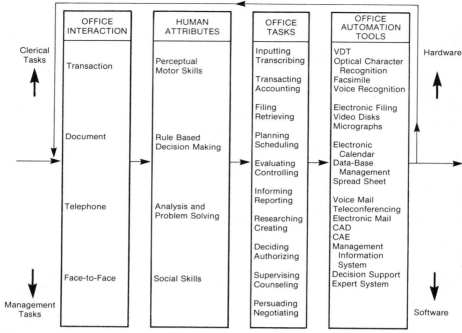

Fig. 12.1.6. Taxonomy of office interactions, human attributes, office tasks, and office automation tools. From Helander, M. G. (1984). *Human Factors,* 1984, vol. 27, pp. 3–20. Copyright 1985, by the Human Factors Society, Inc. and reproduced by permission.

as for the design of training programs. Users of office systems may be very diverse and the system must be able to support a range of skill levels including those who are inexperienced and unsophisticated technologically to those who are experienced and computer sophisticated.

As previously indicated, user input is critical at this stage. However, this may be problematic as it may be difficult for users to articulate their needs and preferences especially if they lack technical expertise and are unfamiliar with office automation features. One technique that can be used to overcome this problem is to conduct a user needs workshop. Essentially, a workshop consists of educational segments, group discussions, brainstorming sessions, and questionnaire modules (Johansen and Baker, 1984). Users are first given information about the features of the automation system. They are then asked to identify problems with current procedures and desired benefits from future systems.

After task and user abilities, attitudes, needs, and preferences are analyzed, the information can be used to allocate office functions or tasks among automation devices. In general, some tasks are routine and highly structured and are good candidates for automation. In these cases, the computer serves to replace human labor. Other tasks are nonroutine and semi- or unstructured and cannot easily be performed by a computer but are better suited to a human. In these cases, the computer can serve to augment human labor (Hammer and Zisman, 1979). As indicated, most office technologies serve to augment office tasks. Criteria that can be used as guidelines when deciding which tasks are likely candidates for automation are presented in Table 12.1.7. The key to successful allocation is to design a system where all components are integrated within the overall framework of the organization.

Table 12.1.7   Criteria for Determining Candidates for Automation

The following criteria can be applied to office tasks to determine whether they are good candidates for a computer application. A "yes" answer to any of these questions indicates a likely candidate.

Is the task a simple, repetitive sequence affecting data, and is this sequence executed often?

Is there a sizable and/or volatile body of information that must be maintained or that is frequently referred to by a variety of individuals?

Is the function too time-consuming to do at all without the computer?

Is the function one that does not yet exist, that is, one that is not possible without the computer?

Is the task one that can be enhanced by the use of an interactive computing environment?

Source: Gruhn and Hohl (1979).

AUTOMATION FEATURES	OFFICE TASKS						
	Document Preparation	Meeting and Conferring	Filing and Retrieving	Planning and Scheduling	Communicating	Decision Making	Transacting
Word Processing	●		●				●
Electronic Mail					●		
Electronic Filing			●				
Voice Mail					●		
Data-Base Management				●		●	
Teleconferencing					●		
Electronic Calendar				●			
Spread Sheet				●		●	●
Decision Support System						●	●

Fig. 12.1.7. Example of a task-by-feature matrix.

At this stage, it might be useful to develop matrices to identify which automation features and hardware components are needed and also how these components should be allocated within an organization. Three matrices are suggested:

1. Task by feature matrix.
2. Job type by feature matrix.
3. Feature by hardware matrix.

A task by feature matrix identifies which tasks can be supported by automation features and vice versa. It also can be used to identify which automation features support the most tasks or most critical tasks; this can be useful if there are situational constraints which mandate that only a limited number (type) of features may be selected and thus features must be prioritized. An example of this type of matrix is given in Figure 12.1.7.

A job type by features matrix identifies which personnel should get which features. This matrix may be derived from the task analysis that specifies which job types perform which tasks and also from the task by feature matrix that specifies which tasks are supported by which features.

The hardware by features matrix is developed after necessary features are selected. It identifies which hardware components are needed to support which features and subsequently who needs what types of equipment.

12.1.5.4 System Selection

The objective of this stage is to select hardware and software features that meet system requirements and fulfill user needs. Generally criteria that can be used to guide hardware and software selection include ease of use, cost, estimated life span, flexibility, convenience, comfort, compatibility, safety, training, and maintenance requirements. It is beyond the scope of this chapter to evaluate existing technologies in terms of these criteria. However, some generic automation features will be discussed and organized according to the following office activities: communicating, document manipulation, decision making, and meeting and conferring. For a more complete discussion of these issues, see also Helander (1983), Rubin (1983) and Chapters 9.1, 11.5, and 11.7.

Communicating

The enhancement of communication is a primary goal of OA since office workers spend a majority of their time communicating. Therefore, a number of technological innovations have been developed to aid communication tasks. The majority of these tools are generic and can be used by a variety of office workers.

Local Area Networks.   LANs supply the systems context for the organization of office automation. They allow information to be transferred among individual workstations, terminals and peripherals, clustered controllers, or host networks. As indicated previously, one characteristic that distinguishes LANs is whether the network is open or proprietary. Open networks are more flexible since they are not vendor specific but they require more technical expertise for installation and maintenance.

There are three common topologies for LANs: the star, the ring, and the bus. In the star network, several terminals are connected to a central controlling device. This configuration is not very economical as a data link between different workstations. It is mostly used with mainframes and PABX systems. In the ring configuration, all workstations are arranged in a circular network. It consists of a series of nodes (any machine that attaches to and uses the LAN) (Dixon, Strole, and Markov, 1983) that are connected by a unidirectional transmission node to form a single closed path. Information passes from node to node and is regenerated as it passes through each node. This type of network has greater reliability than the star network; however, it is difficult to expand and there are problems with data security. The bus network has a bidirectional transmission facility to which all nodes are attached. The transmission facility may consist of multiple cables and information may be transmitted from the originating node in both directions to the terminated ends of the bus. This topology can easily be expanded, it does not require a centralized controller, and it can handle short messages economically. However, it may have security problems, and if it breaks down, it may cripple all downline communication.

When designing or selecting LAN, several requirements must be satisfied. The network must be flexible to allow users to intermix different types of equipment and technologies over the common network and also to accommodate new developments in technology. It should allow for communications between geographically remote locations, and fault detection and isolation requirements are essential to ensure that the network is available and reliable. Finally, the network should be relatively easy to use (Dixon, Strole, and Markov, 1983).

Voice Mailbox.   The analysis of office tasks performed by AT&T showed that use of the telephone was one of the most important office activities. However, in a majority of cases, the telephone calls made were unsuccessful because the receiver was out of the office or in a meeting, or the line was busy. This creates problems with time and information losses. Voice mail allows a caller to leave a message regardless of the availability of the office party.

A voice mailbox stores a message for delivery at a later time. Most voice mailbox systems digitize the message and store in on a disk or random access device. The system can stand alone or be connected to a CRT terminal. There are a number of features that may be incorporated into the system, such as ability to send a message at a predetermined time and sending the same message to a group of people.

There are several advantages associated with voice mailbox systems. Use of voice mailbox represents a potential time saver; the length of an average telephone message is about 6 min, whereas the average length of a voice mailbox message is about 1 min (Helander, 1983). Additionally, it allows people the opportunity to send and receive messages at any time of the day, decreasing interruption of other activities. However, some users have difficulties because of the impersonal nature of the interaction. Other user problems include lack of feedback confirming that the message was received, messages played out at nonoptimal rates, either too fast or too slow, and inadequate help systems.

Electronic Mail.   Electronic mail is offered as a substitute for telephone calls and business letters. Messages are transmitted on private lines or through telephone systems.

The use of electronic mail within organizations has been limited. Most of this has been due to insufficient planning (McQuillan, 1983). Issues associated with the use of electronic mail include problems with keying, privacy, disruption of communication patterns, etiquette, and lack of hard documentation of messages. Also, it can be difficult to prioritize messages, and questions may arise regarding who has the responsibility for sorting the mail. This is traditionally a clerical task. However, managers may decide to screen their own messages, and problems with role ambiguity may arise.

In summary, there are several criteria that must be met if a communication system is going to be successful including ubiquity, simplicity, flexibility, reliability, privacy, and authenticity (Sirbu, 1978).

Document Manipulation

Several automation devices are available for document preparation. These include word processing, electronic filing, optical character recognition, electronic typewriters, and graphic systems. This discussion will be limited to word processing and electronic filing.

Word Processing.   Word processing is currently the most commonly used automation feature. It is used for typing and editing text and can also be used for creating data files. The task of word processing involves an individual at a computer terminal, which has a keyboard for input and a video display for text viewing. Text is entered into the system and manipulated by the use of command functions.

The use of word processing offers several advantages over traditional typing. For example, editing functions such as moving, deleting, or inserting text can be performed quickly and easily and do not necessitate retyping of the entire document. Also, since information is stored on a disk, it can be transmitted to other people who are attached to a network. Additionally, computer programs such as a sorting procedure can be applied to the information.

However, there are also some problems with existing word processing systems. The difficulty of word processing systems varies with the structure of the commands, the number of steps required to perform a function, the operating system of the computer, the amount of text to be entered or edited, the keyboard design, and the user's familiarity with word processing or computer activities in general (Card, Moran and Newell, 1983). Some of these issues are addressed below. Some packages have command functions that are difficult to learn and execute. Roberts (1979) evaluated four word procesing packages and found significant differences in performance and time to learn. It was suggested that the source of the difference may have been related to differences in cognitive demands required by the packages. Also, in most word processing systems only a portion of text is available on the screen at any one point in time. This implies that users must remember what information is hidden. Many users find this difficult as they like the ability to view a whole page of text, especially when composing or editing. Some packages make use of windowing to alleviate these problems. For example, the program Word (Microsoft Inc.) provides windows to allow users to simultaneously view different documents or different parts of the same document.

The help and menu functions in packages are inadequate. For example, Wordstar (Micropro Inc.) has several menus designed to eliminate the need for users to memorize command functions. Each menu has a different name (e.g., Block) and contains different commands (e.g., ^g = delete). However, in order to access this information, the user has to know which menu contains which command or skim through all of the menus. A recent study (Czaja, Hammond, and Blascovich, 1984) found this task to be especially difficult for older learners. This may be due in part to declines in Memory functions that occur with advanced age.

Electronic Filing

The purpose of electronic filing is to provide a means for rapid storage and retrieval of information. The file system can either be closed loop, in which the location and the information are electronically encoded, or open loop, in which the location is electronically encoded but the information is not (Barcomb, 1981).

The use of electronic file systems can offer several advantages: faster access to information, reductions in misfiling, storage efficiency through shared access, and increases in amount of usable floor space (Barcomb, 1981). Space for file cabinets is expensive, \$8.00–\$25.00 per square foot per month. The estimated compression factor for electronic digital storage is 300 to 1. However, there are some user issues that need to be considered when designing these systems. One important issue is how to organize information so that it can be easily retrieved. This requires an easy-to-use indexing system. Design of such a system may draw from knowledge of how people typically organize information. Several studies have been undertaken that attempt to model human filing behavior by observing how individuals retrieve and file information. The intent is to create filing systems that mimic human filing habits (Helander, 1983). However, this is a difficult task since there are a variety of ways in which information can be filed and many methods are idiosyncratic. Malone (1983) studied how people typically organize their desks and found that people have dificulties deciding how to classify information, and this often prevents them from filing that information. Malone also found that an important function of desk organization is reminding. Much of the information that is visible on top of desks is there to remind the user to do something. Electronic file systems should therefore have a reminding function. To design this type of function requires an understanding of how people schedule work and also of forgetting and procrastination habits. In general, electronic filing systems are still in the early stages of development. Many current systems are cumbersome and difficult to use. Also many users are hesitant to use existing systems because they have fears of permanently losing information if the system fails. Most of these fears stem from a lack of understanding of the system and are indicative of inadequate training.

Decision Making

The recent focus of OA is aimed toward the development of support tools for managers and professional workers. The intent of these tools is to increase the effectiveness of task performance. This is a major challenge to system designers since management and professional work is often specialized and unstructured. Most of the automation features that are being developed are decision support tools. Decision makers need quick and easy access to a wide variety of continuously updated data bases. Therefore, one of the most important attributes of an automated office is the organization of a data base so that workers are able to access the information that they need (Rubin, 1983).

Data Base Management Systems. In most offices, a variety of data bases are needed to perform both routine and nonroutine tasks. Basically, these data bases fall into one of two categories: external, which may be accessed over a communications network, and internal, which exist within an office. One problem surrounding the use of data bases is that they generally have been developed independently to meet specific needs. The data is often fragmented and incompatible. As a result, it is difficult for workers to access information, and files and records are often duplicated, which is inefficient and costly (Rubin, 1983).

Data base management systems (DBMS) have been developed in response to these problems. Generally, a DBMS is characterized by (Rubin, 1983):

User requesters of data, for example, read–date, modify–data, add–delete data.

Software to retrieve, write, or modify data.

An applications program, which writes codes to process user requests.

The control of information resource by generating data descriptions and selecting user access policies.

Several DBMSs, are currently available for personal computers. These include Selector V, dBase II, and QSORT. One important issue related to the use of these systems is design of the software that is used to manipulate the data. The key criterion is usability. This is of primary importance, since a variety of users need access to data bases and the users vary in their level of technical sophistication. This is especially true for managers who often have limited time to learn new skills; use of the system should require a minimum amount of training.

Decision Support Systems. Decision support systems are important elements in OA. They are designed to support the decision maker in the performance of unstructured tasks such as strategic planning or risk analysis. Decision support systems do not automate decision making but are used to support nonroutine problem-solving tasks, to examine a wide range of alternatives, and to observe the effects of manipulating certain variables on outcomes ("what if"). Initially these systems were restricted to mainframe computers, but now they can be used on personal computers. "Generic" support systems such as MYCIN, DAISY, and the Interactive Financial Planning Package (Execum) are available. The Execum package can be used for a variety of tasks including planning, model building, model analysis, risk analysis, simulation, and forecasting.

However, in most offices, tasks performed by managers and professional workers are specialized and require different types of decisions and access to different information banks. Therefore, in order to be effective, decision support systems must be individualized. They must be designed to support specific user groups in specific decision situations. This necessitates that the systems are flexible, easy to build, and relatively inexpensive. It also necessitates a clear understanding of the decision making process, factors that affect this process, and the nature of decision problems. System designers must address issues regarding what type of information should be presented to the decision maker and when and how to present this information. For example, the order in which information is presented, has implications for decision making. The sequence of information can establish an order for eliminating the options available to decision makers. This is a factor that is commonly controlled by the computer system, but its effects often go unrecognized (Didner and Butler, 1982).

How much information to present is another important issue. Too much information can result in overload, and too little information can constrain decisions prematurely. Other considerations include identifying what types of decisions can be augmented. To resolve these issues, practical models of the decision making process and of the universe of decision problems must be developed.

Zachary and Hopson (1981) have developed a methodology for designing decision augmentation systems. The methodology is composed of two parts. The first part outlines the steps that need to be followed in the development cycle, and the second part is a set of guidelines for completion of these steps. The steps identified in their methodology include:

1. Definition and analysis of the decision problem.
2. Determination of the way in which the problem is amenable to augmentation.
3. Identification of the appropriate level of decision automation.
4. Allocation of decision making functions between humans and the computer.
5. Specification of the optimal mix of techniques to accomplish the functions allocated in whole or in part to the computer.
6. Identification of the type and amount of information that must be transmitted between the human and the computer.
7. Development of ways to communicate effectively this information between the computer and the human and the human and the computer.

Zachary and Hopson (1981) indicate that there is some arbitrariness in the sequencing of these steps. However, this methodology provides a framework for designing consistent and well-engineered systems. For a more complete discussion of decision support systems, refer to Chapter 8.4.

Meeting and Conferring

A large portion of a manager's or a professional's time is spent in meetings. However, an analysis of meeting activities has shown that the purposes of meetings are often ill-defined and the results of meetings are often questionable. Meetings typically lead to other meetings (Brecht, 1978). Providing computer support for meeting activities seems to represent a potential for gains in productivity and time savings. For example, it offers the potential for reducing travel times and costs. However, meetings are both formal and informal and some (e.g., counseling) are difficult to automate. Also, it is doubtful whether the majority of office workers will find the various types of technologically aided conferencing preferable to face-to-face meetings.

Teleconferencing. Teleconferencing is a meeting that involves two or more persons at different locations with discussions being transmitted through audio (audio conferencing), digital (computer conferencing), or video (video conferencing) techniques. In general, teleconferencing is not a substitute for face-to-face interactions. However, it is an effective substitute for many kinds of meetings including educational conferences, announcements, task force reviews, and emergencies. Investigations of attitudes toward teleconferencing have shown that teleconferencing is preferred by persons who know each other and when the discussion is neutral. People often have difficulties using these systems if they need to meet with strangers or if the topic of the meeting is of a personal or confidential nature (Sheridan et al., 1981).

Audio conferencing can be either public or private. For private conferences, rooms are equipped with microphones and speakers and the telephone is used in conjunction with a switching center to interconnect the participants. Current audioconferencing systems are characterized by several problems—difficulties identifying speakers and establishing speaking order and inadequate acoustics.

Video conferencing offers great potential for substituting meetings because of the availability of powerful technologies. Although it is increasingly used by large corporations, to date its use has been limited. This is mainly because there are large costs involved. However, users of these systems indicate that in many cases they would rather "attend" a video conference than have to travel to a meeting (Duncanson and Williams, 1973).

In computer conferencing, participants use terminals to communicate with one another. Computer conferencing can be distinguished from VDT conferencing. In VDT conferences there is no use of computer power terminals to transmit messages. In computer conferences software and computer memory are available for storing and sorting information. Examples of commercially available software packages for computer conferencing include EIES, COM, HUB, and AUGMENT.

This type of conferencing is claimed to be especially useful for researchers and policy makers because it permits communications to be rich and interactive and allows real-time exchange of information (Barcomb, 1981). It also accommodates more participants than video or audio conferencing and allows asynchronous exchange of information, and the conferences can be augmented by graphics.

The preceding section presents a limited discussion of available automation features. The intent was to highlight some of the human factor issues that need to be considered when selecting system components.

12.1.5.5 Implementation

The manner in which an automated system is introduced has a direct influence on its degree of success within an organization. All too often systems are purchased and put into offices without any strategic planning. Two key factors that are critical to successful implementation are user participation and training.

A major factor that impacts on acceptance of innovation in the workplace is management's willingness to let workers participate in the decision making process. Schmidt and Libby (1984) interviewed employees at two large organizations where OA had recently been implemented. They found that over 30% of all managers and 80% of all clerical personnel who would be using the new equipment were not involved in the design or implementation of the new system. Lack of participation in the change process produces resistance to change. People may perceive a loss of freedom of choice and that they no longer have control over their own destiny.

To overcome some of the organizational problems previously outlined, management should involve workers in the design and implementation of OA systems. If possible, workers who will be affected by the automation should be included in all phases of decision making, including decisions about whether to automate, hardware and software selection, hardware and software distribution, and responsibilities for system operation and maintenance. They should also be involved in the implementation process, that is, a select group of workers may become part of the implementation team. Worker involvement in decision making lowers resistance to change and increases commitment to change. Margulies (1979) found that worker participation in the introduction of automation resulted in higher morale, more positive attitudes toward automation, and less absenteeism and turnover.

Insufficient training is one of the most common barriers to office automation. Park and Freedman (1984) studied factors that influenced the acceptance of electronic systems and created negative feelings among the workers. The majority of workers indicated that the training programs were inadequate and that they experienced stress when attempting to learn to use the system. Most training programs do not extend beyond explaining what features are available and how to operate the equipment. This results in a number of problems. Individuals may become frustrated, suffer a loss of confidence in their abilities, and develop negative reactions toward the system. A recent study (Czaja et al., 1984), which examined the effectiveness of various training strategies to teach word processing, found that the attitudes of the trainees toward computers was related to learning performance. Subjects who exhibited poor learning performance also developed negative attitudes toward computer technologies.

Insufficient training can also lead to problems with system underutilization. If individuals are forced to learn a system without adequate training or by a process of self-learning, they may not become aware of all of the capabilities of the system or how to use the system to its fullest advantage. For example, insufficient training is one reason why multifunction workstations are underutilized. Although ease-of-use features have been incorporated into these workstations, the user still has to understand how to apply and integrate the variety of applications offered by the system to their particular set of tasks. The effort required to acquire these new skills often overwhelms managers and professionals (Smith, 1984).

Training programs must go beyond equipment operation and demonstrate how to use the equipment as tools to accomplish desired work processes. Users need to learn new ways of accomplishing job tasks and be able to assimilate the applications offered by the new technologies into their work procedures. Smith (1984) refers to these types of skills as "augmentation" skills and maintains that augmentation level skills are necessary before multifunction workstations can be used in an integrated manner. Augmentation skills generally do not develop on an ad hoc basis; they must be carefully defined, introduced, and supported.

Training programs must be systematically designed and tailored to meet the needs of the trainees. For example, unsophisticated users should not receive the same type of training as experienced users who may only need information regarding specific operating procedures. Often individuals with different skill levels are placed into the same training program, and the needs of both low-and high-skill-level users are not effectively addressed. Training programs need to be flexible, giving detailed instruction to novices but responding to the needs of the expert with abbreviated information. For example, when computerized office aids were introduced at the IBM Thomas J. Watson Research Center, training procedures included individual instruction and formal classes. Newcomers to the system received 5–10 hr of individualized instruction, and when they were able to do simple work on the system, they attended regularly scheduled classes. In addition, minicourses and seminars were available. This approach was found to be very successful in the areas of text processing and administrative computer usage (Gruhn and Hohl, 1979). In addition to formal training, workers should also be given time to experiment with the system and acquire certain skills on their own or in collaboration with other colleagues. Availability of easy-to-use reference materials and help systems is also important.

The actual implementation of the OA system should occur in stages and involve prototype testing. Basic system features can be introduced to a representative group of workers and used on a trial basis. These systems can be evaluated and user problems can be identified. The result of this evaluation should then be used to modify the features accordingly. After the basic system is operating smoothly, additional features can gradually be introduced. For example, prior to introducing their OA system, Citibank installed 16 prototype systems for a trial period. The systems were evaluated and several problems were identified including difficult-to-use operating procedures, the need to type in commands as opposed to using function keys, and the inability to view a full page of text on the screen. The system was changed to alleviate these problems and actual implementation was successful (McNurlin, 1978).

Implementing OA systems in incremental stages not only allows system problems to be identified but also provides workers greater opportunities to absorb new knowledge and cope with new work patterns.

12.1.5.6   Evaluation

The objective of this phase is to assess the effectiveness of the system and identify user problems. Evaluation should be an ongoing process and the results from evaluation efforts should be incorporated into system design. The future of office automation is dependent to some degree on rigorous evaluation of current systems. Future developments in OA must be based on knowledge of strengths and weaknesses of current designs.

To some extent, the topic of evaluation was discussed in the section addressing worker productivity

(Section 12.1.3). To reiterate, the evaluation of OA systems must be based on measures of both efficiency and effectiveness of performance. Effectiveness is especially important as systems are more geared toward managers and professionals. To this end three classes of measures can be identified:

Measurement of user satisfaction.

Measurement of system utilization.

Measurement of task performance (efficiency and effectiveness).

The way in which these dimensions are specifically assessed will vary from organization to organization. However, some generic measures of system effectiveness might include:

User satisfaction survey (specifically designed for system).

Job satisfaction [e.g., Job Description Index (Smith, Kendall, and Hulsh, 1969)].

Job stress questionnaire [e.g., NIOSH (Smith, Cohen, Stammerjohn and Happ, 1981)].

Amount and quality of output.

Time savings.

Improved quality of information (access to sources of information not previously available).

Ease of use.

Timeliness of retrieving and sending information, meeting deadlines.

System availability.

System reliability.

Improvement in communications.

A repeated measures design should be used for the evaluation. Data about the aforesaid issues should be gathered using a pretest, posttest design where measures are taken at least at two points in time—prior to OA implementation and after OA implementation. This type of design is more adequate than simple studies, where measures are taken at some point in time after the automation has been introduced. A pretest, posttest design provides baseline data that can be used for comparative purposes and thus allow stronger causal inferences to be made regarding the impact of office automation. The after measures should be taken at several points in time following the change (e.g., 3 months, 6 months, 12 months). Also, if feasible, two groups should be included in the evaluation study—one experimental group and one control group. The experimental group consists of workers (office) within an organization who have received the OA system, and the control group consists of workers (office) who have not received the new technologies. This technique can control for the many possible extraneous causes of change that may invalidate the single group results (Helander, Billingsley and Schurick, 1984; Rittenhouse, 1983). The groups (offices) selected for study should be as similar in size, structure, and business function as possible.

The payoffs of OA can be realized in a variety of ways including reductions in costs, improvements in performance, time savings, easier access to information, added capabilities, and improvements in employee morale. Many of these results cannot be expressed in monetary terms; however, they still have an impact on the realization of office functions.

12.1.6 CONCLUDING REMARKS

The introduction of information technologies into the office offers the potential for increasing the efficiency and effectiveness of office work. The primary obstacle limiting the success of office systems is lack of attention to human issues. To date, most systems have been technology driven without consideration of the impact of the new technologies on end users and organizations. More effort needs to be devoted to assessing the impact on job design, organizational structure, communication/interaction patterns, and the physical environment.

The future of office automation is dependent on the extent to which the new technologies can be effectively used and are accepted by office workers. Technologically well-conceived systems do not necessarily lead to implementation success. The intent of this chapter has been to highlight some of the issues surrounding office automation. There is a great need for further research in human factors engineering for both the design and implementation of office systems.

REFERENCES

Barcomb, D. (1981). *Office automation, a survey of tools and technology.* Bedford, MA: Digital Press.

Booz, Allen and Hamilton, Inc. (1980). *Why automate?* Cambridge, MA: Booz, Allen and Hamilton, Inc.

Bracchi, G., and Pernici, B. (1984). The design requirements of office systems. *ACM Transactions on Office Information Systems, 2,* 151–170.

Brecht, (1978). A meeting is not a meeting is not a meeting: implications for teleconferencing. *Proceedings of the 22nd annual meeting of the Human Factors Society* (pp. 335–338). Santa Monica, CA: The Human Factors Society.

Card, S. K., Moran, T. P., and Newell, A. (1983). *The psychology of human-computer interaction.* Hillsdale, NJ: Erlbaum.

Cohen, B. G. F. (1984). Organizational factors affecting stress in the clerical worker. In B. F. Cohen, Ed. *Human aspects in office automation* (pp. 33–42). Amsterdam: Elsevier.

Craik, F. I. M., and Lockhart, R. S. (1972). Levels of processing: a framework for memory research. *Journal of Verbal Learning and Verbal Behavior, 11,* 671–684.

Czaja, S. J., Hammond, K., and Blascovich, J. (1984). Age-related differences in learning to use a computerized text editor as a function of training techniques (Unpublished manuscript). Buffalo, NY: State University of New York at Buffalo, Department of Industrial Engineering.

Didner, R. S., and Butler, K. A. (1982). Information requirements for user decision support: designing systems from back to front (Paper 0360-8913/81/000-8415). *Proceedings of the IEEE conference* (pp. 415–419). New York, IEEE.

Dixon, R. C., Strole, N.C., Markov, J. D (1984). A token-ring network for local data communications. *IBM Systems Journal, 22,* 47–62.

Drucker, P. F. (1967). *The effective executive,* New York: Harper and Row.

Duncanson, J. P., and Williams, A. D. (1973). Video conferencing: reactions of users. *Human Factors, 15,* 471–485.

Eason, K. D., and Gower, J. C. (1984). Implications of new technology on work organisation: a case study. In H. W. Hendrick and O. Brown, Jr., Eds. *Human factors in organizational design and management* (pp. 361–366). Amsterdam: North-Holland.

Ellis, P. (1984). Office planning and design: the impact of organizational change due to advanced information technology. *Behavior and Information Technology, 3,* 221–234.

French, J. R., and Caplan, R. D. (1973). Organizational stress and individual strain. In A. J. Morrow, Ed. *The failure of success* (pp. 30–66. New York: AMACOM.

Giuliano, V. E. (1982). The mechanization of office work. *Scientific American, 247,* 148–165.

Goldstein, B. C., Heller, A. R., Moss, F. H., and Wladawsky-Berger, I. (1984). Directions in cooperative processing between workstations and hosts. *IBM Systems Journal, 23,* 224–235.

Gould, J. D., and Boies, S. J. (1983). Human factors challenges in creating a principal support system—the speech filing system approach. *ACM Transactions on Office Information Systems, 1,* 273–298.

Gould, J. D., and Lewis, C. (1983). Designing for usability—key principles and what designers think. *Proceedings CHI '83, conference, human factors in computing systems* (pp. 50–53). Boston, MA.

Gruhn, A. M., and Hohl, A. C. (1979). A research perspective on computer-assisted office work. *IBM Systems Journal, 18,* 432–456.

Grummon, P. T. H. (1983). Managing the introduction of new technology: methods from organizational development and practice (Technical Paper MM83-472). Dearborn, MI: Society of Manufacturing Engineers.

Hackman, J. R., Oldman, G., Janson, R., and Purdy, K. (1975). A new strategy for job enrichment. *California Management Review, XVII,* 57–71.

Hall, J. H. (1984). Mental health of secretarial and clerical personnel. In B. F. Cohen, Ed. *Human aspects in office automation* (pp. 167–176). Amsterdam: Elsevier.

Hammer, M. H. (1982). Improving business performance: the real objective of office automation. *Proceedings, office automation conference, AFIPS* (pp. 247–254). San Francisco, CA.

Hammer M. M., and Sirbu, M. (1980). What is office automation? *Proceedings of the national computer conference on office automation* (pp. 37–49). Arlington, VA: AFIPS.

Hammer, M. M., and Zisman, M. (1979). Design and implementation of office information systems. *Proceedings of the NYU symposium on automated office systems.* pp. 13–23.

Helander, M. G. (1983). The automated office: a description and some human factors design consideration (Technical Report). Farsta, Sweden: Teldok, Swedish Telecommunications.

Helander, M. G. (1985). Emerging office automation systems. *Human Factors, 27,* 3–20.

Helander, M. G., Billingsley, P. A., and Schurick, J. S. (1984). An evaluation of human factors research on visual display terminals in the workplace. In F. Muckler, Ed. *Human Factors Review, 1,* 55–129.

Hoos, I. R. (1983). When the computer takes over the office. *Office Technology and People, 2,* 57–68.

Johansen, R., and Baker, E. (1984). User needs workshops: a new approach to anticipating user needs for advanced office systems. *Office Technology and People, 2,* 103–120.

Kahneman, D. E., and Tversky, A. (1973). On the psychology of prediction. *Psychological Review, 80,* 237–251.

Kaplan, A. (1982). The ergonomics of office automation. *Modern Office Procedures,* May, 51–64.

Klein, H. K., and Hirschman, R. (1983). Issues and approaches to appraising technological change in the office: a consequentialist perspective. *Office Technology and People, 2,* 15–42.

Klemmer, E. T., and Snyder, F. W. (1972). Measurement of time spent communicating. *The Journal of Communication, 22,* 142–158.

Knopf, C. (1982). Presentation at voice processing seminar sponsored by Probe Research, Inc., New York.

Landau, J. Blair, J., and J. Siegman, Eds. *Emerging office systems,* Northwood, NJ: Ablex.

Lieberman, M. A., Selig, G. J., and Walsh, J. J. (1982). *Office automation, a manager's guide for improving productivity.* New York: Wiley.

Lodahl, T. (1982). Designing the automated office. In R. Landau, J. Blair, J. Siegman, Eds. *Emerging office systems.* Northwood, NJ: Ablex.

Malone, T. W. (1983). How do people organize their desks? Implications for the design of office information systems. *ACM Transactions on Office Information Systems, 1,* 99–112.

Margulies, N. (1979). A literature review to investigate the empirical research previously accomplished on the relationship of human factors to technological processes (Technical Report, ICAM). Northrop Corporation.

Martin, J. (1981). Successful office automation. *Computer Decisions, 13,* 108–122, 161–162.

McNurlin, B. (1978). The automated office: part I. *EDP Analyzer, 16.*

McQuillan, J. M. (1983). Electronic mail: The planning gap. *Proceedings of conference on office automation.* Arlington, VA: AFIPS Press.

Mendelsohn, N., Linehan, M.H., and Anzick, W. J. (1983). Reflections on VM/Pass-Through: a facility for interactive networking. *IBM Systems Journal, 22,* 63–80.

Messina, A. (1984). Automated factory, automated office. *Computer World Office Automation,* June, 71–75.

Moore, D. J. (1983). Teletex–a worldwide link among office systems for electronic document exchange. *IBM Systems Journal, 22,* 30–46.

Newell, A., and Sproull, R. F. (1982). Computer networks: prospects for scientists. *Science, 215* (12), 843–852.

Nisbett, R. E., and Ross, L. (1980). *Human inference: strategies and shortcomings of social judgement.* Englewood Cliffs, NJ: Prentice-Hall.

Office of Technology Assessment (1983). Automation and the workplace: selected labor, education and training issues (Technical Report). Washington, DC: U.S. Congress.

Panko, R. R. (1984). Office work. *Office Technology and People, 2,* 205–238.

Park, N. W., and Freedman, J. L. (1984). Factors influencing acceptance of electronic systems (Technical Paper MM84-626). Dearborn, MI: Society of Manufacturing Engineers.

Parsons, M. (1984). People-related parallels between the automated factory and the automated office (Technical Paper MM83-480). Dearborn, MI: Society of Manufacturing Engineers.

Ranney, J. M., and Carder, C. E. (1984). Socio-technical design methods in office settings: two cases. *Office Technology and People, 2,* 169–186.

Rittenhouse, R. G. (1983). Productivity assessment issues on office automation. *Proceedings of sixteenth Hawaii international conference on system sciences.* pp. 576–580.

Roberts, T. L. (1979). Evaluation of computer text-editors (Technical Report SSL-79-9). Palo Alto, CA: Xerox Palo Alto Research Center.

Rubin, A. I. (1983). The automated office—an environment for productive work, or an information factory? A report on the state of the art (Technical Report NBSIR 83-2784-1). Washington, DC: General Service Administration.

Salvendy, G. (1984). Research issues in the ergonomics, behavioral, organizational and management aspects of office automation. In B. F. Cohen, Ed. *Human aspects in office automation* (pp. 115–126). Amsterdam: Elsevier.

Schmidt, M. L., and Libby, W. L., Jr. (1984). Successful implementation of office technology: a cross-national impact analysis (Technical Paper MM84-639). Dearborn, MI: Society of Manufacturing Engineers.

Sell, R. G. (1984). New technology and the effects on jobs. In H. W. Hendrick and O. Brown, Jr., Eds. *Human factors in organizational design and management.* Amsterdam: North-Holland.

Sharit, J., and Salvendy, G. (1982). Occupational stress: review and reappraisal. *Human Factors, 24,* 129–162.

Sheridan, T., Senders, J., Moray, M., Stoklosa, J., Guillaume, J., and Makepeace, D. (1981). *Experimentation with a multi-disciplinary teleconference and electronic journal on mental workload.* Cambridge, MA: Massachusetts Institute of Technology.

Sirbu, M. (1978). Innovation strategies in the electronic mail marketplace. *Telecommunication Policy, 2.*

Smith, J. (1984). Beyond User friendly—towards the assimilation of multifunction-workstation capabilities. *Journal of Behavior and Information Technology, 3,* 205–220.

Smith, M. J. (1984). Ergonomic aspects of health problems in VDT operators. In B. F. Cohen, Ed. *Human aspects of office automation* (pp. 97–114). Amsterdam: Elsevier.

Smith, M. J., Cohen, B. G. F., Stammerjohn, L. W., Jr., and Happ, A. (1981). An investigation of health complaints and job stress in video display operation. *Human Factors, 23,* 387–400.

Smith, P. C., Kendall, L., and Hulsh, C. (1969). *The measurement of satisfaction in work and retirement: a strategy for the study of attitudes.* Chicago, IL: Rand McNally.

Stout, E. (1981). The human factor in productivity: the next frontier in the office. *Journal of Micrographics,* April.

Taylor, J. (1982). Integrating computer systems in organizational design. *National Productivity Review.*

Uttal, B. (1982). What's detaining the office of the future? *Fortune,* May, 176–196.

Working Women. (1980). Race against time: Automation of the office (Technical Report). Cleveland, OH: Working Women, National Association of Office Workers.

Zachary, W., and Hopson, J. (1981). A methodology for decision augmentation system design. *Proceedings of the American Institute of Aeronautics and Astronautics conference on computers in aerospace III.* San Diego, CA.

Zuboff, S. (1981). Psychological and organizational implications of computer-mediated work (CISR #71). Cambridge, MA: MIT.

CHAPTER 12.2

TECHNICAL AND HUMAN ASPECTS OF COMPUTER-AIDED DESIGN (CAD)

WOODROW BARFIELD
TIEN-CHIEN CHANG
ANN MAJCHRZAK
RAY EBERTS
GAVRIEL SALVENDY

Purdue University
West Lafayette, Indiana

This chapter reviews the hardware and software technologies of computer-aided design (CAD), and discusses the human aspects in the design and use of CAD systems. The allocation of functions between the designer and CAD system components is discussed, while the cognitive, perceptual, and workspace design issues in the design and use of CAD systems are reviewed. Lastly, job design issues for CAD operators are reviewed and discussed.

12.2.1 USE OF CAD SYSTEMS

The term "computer-aided design" (CAD) describes any system that uses a computer to assist in the creation or modification of a design. A typical CAD system is shown in Figure 12.2.1. In general, a CAD workstation consists of a color graphics display terminal, a digitizing tablet, a keyboard, a printer/plotter, and a local graphics processor (under the table). A user can design a part, build an assembly, and conduct an engineering analysis directly and interactively on the CAD system.

Although the use of CAD systems can save up to about 70% of the total manual design time, the real benefits of using CAD are a reduction in product lead time, better product quality and design documentation, and increases in the creativity and productivity of the designer.

The design process when performed manually involves many trials and errors and is a very time-consuming task. Since the manual design process is very tedious, it is amenable to human error. Each time a modification of the design is performed, an engineering analysis, such as building a finite-element model (from which stress and heat relationships may be derived), must be conducted. When performed manually, it may take several weeks for the designer to receive feedback from the results of the design analysis.

In contrast, a CAD system integrates the functions of drafting and analysis by storing a design in the computer electronically. The results of the engineering analysis can be shown graphically by displaying the design on the CAD terminal, thus, providing immediate feedback to the designer. Also, use of CAD allows the designer to go through many design iterations in a reasonably short time. The quality of the final product is usually much higher and the lead time much shorter when using CAD.

Another advantage of using CAD systems is the ability to interface CAD with computer-aided manufacturing (CAM). When CAD is interfaced with CAM, the total design–production cycle is shortened. CAD becomes not only a tool to save costs, but, more importantly, a means to survive in the high competition marketplace of manufacturing.

Today there are more than 10,000 CAD workstations delivered for use each year. With the price of these systems decreasing and the performance improving, we will see increasing numbers of CAD workstations being used by industry.

12.2.1.1 History of CAD Developments

The development of CAD can be dated back to the late 1950s when researchers at the Massachusetts Institute of Technology (MIT) started work on the SAGE system (CRT display). In the early 1960s, systems began using APT-like (see Chapter 12.4) commands to control a plotter to draw 2-D (two-dimensional) engineering diagrams. The seminal work of Sutherland (1963) at MIT lead to the development of interactive computer graphics, which allowed the CRT to be used as an electronic drafting board.

The capability to produce a 3-D (three-dimensional) wireframe model was introduced into the CAD system in the 1960s. In addition, advances in semiconductor technology made the minicomputer available for CAD purposes. In the 1970s, software for 3-D solid modeling, shaded color graphics, and so forth, were developed, while advances in seminconductor RAM memory for microprocessors made computer and graphics terminals more powerful and more affordable. Software for engineering analysis applications was also integrated into CAD systems. At this time CAD was moving from computer-aided drafting to computer-aided design.

The 1980s marked the real boom in CAD usage. CAD systems are now available on all three levels of computers—mainframe, mini, and micro.

At the time of this writing, CAD has grown into nearly a $2 billion dollar industry. Although a mainframe based system may cost as much as $100,000 per workstation, a PC-based system may cost as little as $5000 (including peripheral devices). However, the backbone of CAD systems are the 'mega-mini' computer-based workstations. These systems have a 32-bit processor, 2–8 Mbyte of RAM memory, and are able to handle all the computing needs of the designer. In the future, one can expect to see more software power packed into a workstation. Realistic image synthesis, expert design software, and integrated design/manufacturing software are all becoming available on CAD systems.

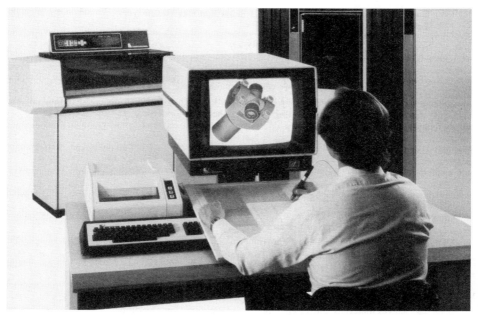

Fig. 12.2.1. A CAD workstation. Photograph Courtesy of Computervision Corporation, Bedford, Massachusetts.

12.2.2 RATIONALE FOR CAD USAGE

There are many human factors issues involved in the design and use of CAD systems. The primary goal of designing the human–CAD interface is to create a system that augments the designer's creativity and decision making abilities. In order to accomplish this objective the hardware and software components of the CAD system should be made compatible with the designer's cognitive processes and physical requirements for interacting with the system. Such a system should result in an increase in design creativity and productivity, an increase in the quality of design decision making, and greater job satisfaction (Barfield and Salvendy, 1984).

The human–CAD system is a hybrid intelligent system in that the creative and decision making abilities of the human are coupled with the information processing capabilities of the computer. The resulting system exceeds the capabilities of either system component when the interaction between the two is synergistic.

Groover and Zimmers (1984) and Pao (1984) discussed several reasons for implementing a CAD system. These include:

1. CAD may result in significant productivity increases. The graphic capabilities of a CAD system allow the designer to visualize the product and to synthesize, analyze, and document the design interactively. Productivity improvements for CAD, in comparison to manual design, have been reported to range from 3:1 (Gold, 1983) and 4:1 (Krouse, 1982; Sheldon, 1983) to 10:1 (Rosenbaum, 1983). Productivity improvements largely depend on the complexity of the drawing, level of design detail required, and repetitiveness of the designed parts (Groover and Zimmers, 1984; Krouse, 1981). In a survey of 33 current CAD users by Datapro (1984), 78% of the respondents reported increased productivity, 76% reported shortened cycle time, and 70% reported reduced costs. In addition, 78% of the respondents indicated that the CAD system did what they expected it to do, while 75% would recommend the particular system they were using to a colleague.

2. The use of CAD may lead to an improvement in design quality and accuracy. The survey by Datapro (1984) indicated that 84% of the designers reported an improvement in the accuracy of their drawings.

3. CAD may improve the communication among designers mainly due to the use of a common data base, standardization of drawings, usage of common graphic symbols, and greater legibility of drawings (Groover and Zimmers, 1984).

4. CAD may aid the manufacturing process. For example, a CAD user can define a part shape, analyze stresses and deflections for the part, and check its mechanical action. Furthermore, the geometric

Table 12.2.1   Checklist of Potential Benefits that may Result from Implementing CAD

1. Improved engineering productivity
2. Shorter lead times
3. Reduced engineering personnel requirements
4. Customer modifications are easier to make
5. Faster response to requests for quotations
6. Avoidance of subcontracting to meet schedules
7. Minimized transcription errors
8. Improved accuracy of design
9. In analysis, easier recognition of component interactions
10. Provides better functional analysis to reduce prototype testing
11. Assistance in preparation of documentation
12. Designs have more standardization
13. Better designs provided
14. Improved productivity in tool design
15. Better knowledge of costs provided
16. Reduced training time for routine drafting tasks and NC part programming
17. Fewer errors in NC part programming
18. Provides the potential for using more existing parts and tooling
19. Helps ensure designs are appropriate to existing manufacturing techniques
20. Saves materials and machining time by optimization algorithms
21. Provides operational results on the status of work in progress
22. Makes the management of design personnel on projects more effective
23. Assistance in inspection of complicated parts
24. Better communication interfaces and greater understanding among engineers, designers, drafters, management, and different project groups.

Source: Mikell P. Groover, Emory W. Zimmers, Jr., *CAD/CAM* Computer-Aided Design and Manufacturing, © 1984, pp. 59, 67. Reprinted by permission of Prentice-Hall, Inc., Englewood Cliffs, N.J.

description of the part provided by the CAD user can be used to generate NC tapes, instruct robots, and manage plant operations (Krouse, 1980).

Table 12.2.1 adopted from Groover and Zimmers (1984) lists the advantages and benefits which may accrue from the implementation of CAD in an integrated CAD/CAM system.

As improvements are made in the areas of software, hardware, and the human–CAD interface, further benefits can be expected to occur from the use of CAD (Barfield and Salvendy, 1984).

This chapter first discusses various design issues that relate to CAD followed by a discussion on the technology of CAD. Although the hardware and software components of CAD are conceptually integrated, each is discussed separately for convenience of presentation.

12.2.3   DESIGN PROCESS

The traditional design process is illustrated in Figure 12.2.2 (Groover and Zimmers, 1984).

The design process as illustrated by Figure 12.2.2 includes the major activities of the designer and CAD system for each major stage of design. The recognition of a need to develop a product is a human-initiated activity. The definition of the problem also involves human action and includes a specification of the item to be designed.

The process of design is most often iterative. When decisions on product characteristics are made based on certain heuristic rules, we refer to this stage of design as synthesis. During the synthesis stage a designer will construct a graphical image of the product on the CRT screen by inputting data, such as points, lines, and arcs, and the commands that join these elements together.

Geometric modeling is concerned with the computer-compatible mathematical description of the geometry of the object appearing on the CRT screen (Groover and Zimmers, 1984). The mathematical description allows the image of the object to be displayed and manipulated to a graphics terminal through signals generated by the CPU of the CAD system. The software that provides geometric modeling capabilities must be designed for efficient use both by the computer and the designer.

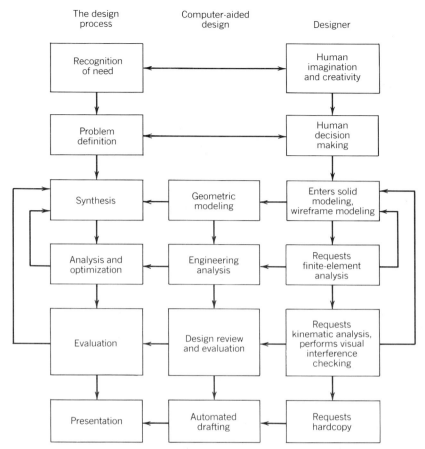

Fig. 12.2.2. Major tasks of the designer and CAD system for the design process. Mikell P. Groover, Emory W. Zimmers, Jr., *CAD/CAM Computer-Aided Design and Manufacturing,* © 1984, pp. 59, 67. Reprinted by permission of Prentice-Hall, Inc., Englewood Cliffs, N.J.

Design analysis and optimization refer to the process where a subsystem of the design is subjected to analysis, improved through analysis, and then redesigned. This process is repeated until the design has been optimized within the constraints imposed upon the designer. CAD affects this process by allowing the designer the capability to perform a finite-element analysis. A finite-element analysis allows the designer to analyze the product for stress–strain and for heat-transfer relationships.

Checking the accuracy of the design can be accomplished by evaluating the figure on the CRT screen through several CAD options. Semiautomatic dimensioning and tolerancing routines, which assign size specifications to surfaces indicated by the users, help to reduce the possibility of dimensioning errors. The designer can zoom in on part details and magnify the image on the graphics screen for close scrutiny. The final stage in the design process is the presentation of the design. This includes design documentation by means of drawings, material specifications, assembly lists, and so on (Groover and Zimmers, 1984).

The computer can be used in four of the six design steps indicated in Figure 12.2.2—synthesis, analysis and optimization, evaluation, and image presentation. Since graphics is essential in both design synthesis and presentation, and is useful for presenting analytical results, computer graphics has become an essential part of CAD. Although computer graphics is not CAD, without it, CAD is not complete. The main functions of a CAD system are shown in Figure 12.2.3.

The extent to which a CAD system can aid a designer depends upon human factors (Thomas, 1983). The various computer tools to aid design are successful to the extent that they are based on what we know about design and designers. Recent studies by van der Heiden, Brauninger, and Grandjcan, (1984) and Encarnacao and Schlechtendahl (1983) point out that a designer can spend up to 50% of

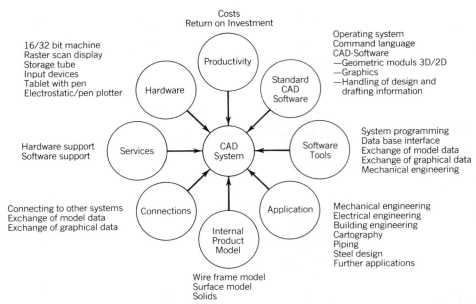

Fig. 12.2.3. Characteristics of CAD-systems [after Grabowski and Anderl, (1983)]. Reprinted with permission.

his or her time away from the workplace requesting and verifying information while only 25% of the designer's time is spent actually drawing the product or part being designed.

The use of a CAD system has changed the range of skills required to perform design. The most obvious change is that a certain amount of knowledge about computers and how to interact with them is required, although a human-factored system should minimize the need to know about computers.

The introduction of CAD into the workplace may also affect the skill requirements for performing design. Computer-based systems are well suited for performing analytical work when algorithmic solutions are necessary; but are, at present, ill-suited for creative decision making. For example, CAD may aid designers whose work requires geometric modeling and an engineering analysis. However, these designers may be subject to more stress as a result of increased levels of computer-paced decision making (Cooley, 1981). On the other hand, designers performing tasks that require routine drawing skills, rather than engineering skills, may be needed less often as CAD systems proliferate into the design workplace (Arnold and Senker, 1982).

12.2.4   CAD HARDWARE

A CAD system consists of a computer, graphics I/O devices, and various software packages. CAD systems can be classified in several ways. For example, a categorization by CAD applications might include mechanical design, electrical/electronics design, architectural design, civil/structural engineering, and mapping. The classification of CAD systems by hardware includes mainframe-based, minicomputer-based, workstation-based, and microcomputer-based systems. A classification by modeling methods includes 2-D, 3-D drafting, wireframe modeling, and 3-D solid modeling.

The center of a CAD system is the workstation (see Figure 12.2.1). A typical workstation consists of a local graphics processor, a CRT display, a keyboard, a tablet or a lightpen, and a designer. The workstation also supports local image transformations and limited manipulations. Functions such as editing the CAD data base, evaluating the geometric model, generating a display list, and performing an engineering analysis are handled by the CAD system computer.

12.2.4.1   Graphics Terminals

The CRT is used to display both graphics and alphanumeric information. There are three major types of CRT displays (Neuman and Sproul, 1979):

1.   The direct view storage tube stores an image on the tube, similar to writing on a sheet of paper. With this technique an image of any complexity can be drawn. This method provides a very-

high-resolution display at a low cost, but does not allow the user the capability to partially erase the image. Any required change on the image is done by erasing the entire image and then completely redrawing it. This process may require a substantial amount of CPU and operator time.

2. The refresh vector drawing display is another type of CRT display. Refresh means that the image is redrawn at a fixed cycle, for example, 30 times per second. The vectorized image is stored in a display list memory. A dedicated circuit redraws the image on the CRT 30 times per second. The ability to partially alter the image is made possible by changing the display list. However, when an image becomes very complex, the display may start to flicker if the image is not redrawn or refreshed quickly enough.

3. The refresh raster scan display eliminates the problems associated with the storage tube and vector drawing display. Pictures are drawn on the display from computer memory (memory frames). Each picture element or pixel (the smallest addressable dot on the display) has a corresponding location in the memory frame that stores its intensity and color. On a fixed cycle each pixel is refreshed at the same rate. What one sees on the screen corresponds directly to what is located in the memory frame. The computer therefore draws on the memory frame rather than on the CRT. Since the refresh raster scan display has a fixed refresh cycle and fixed frame, the complexity of the displayed picture no longer poses a problem. Also partial erasure can easily be accomplished with refresh raster scan displays. However, a large amount of frame memory is required for this capability, for example, a display with $1,000 \times 1,000$ pixels and 16 colors requires a minimum of 500K of memory. However, with the drastic reduction of RAM memory costs, this problem is no longer a limiting factor. In addition, a raster scan display can produce a shaded solid object providing realism to the design. This technique has become more and more popular in recent years.

In some CAD workstations two terminals are used, one for alphanumeric data I/O and another for graphics I/O. (The workstation shown in Figure 12.2.1 has two displays.) Commands can be entered from the keyboard or through other input devices, such as a tablet, light pen, or mouse.

12.2.4.2 Input Devices

Input devices are used for entering commands and for digitizing data. Digitizing refers to the processing of inputting data such as the X, Y coordinates of a point into the CAD system. Input devices, for general use, have been discussed in detail in Chapter 11.4. Commands can be entered from the keyboard or selected from a menu by a pointing device such as a joystick, light pen, table, or mouse. Design data are normally entered through a digitizing table or a pointing device used with the menu. When entering an existing drawing, a digitizing table is normally used, whereas for interactive design a pointing device is preferred when a menu is used.

CAD I/O peripherals are illustrated in Figure 12.2.4. A digitizing table consists of a large flat surface (3 ft $\times$ 4 ft or 4 ft $\times$ 5 ft) and a movable crosshair. Whenever the "enter" button is pressed, the X-Y position of the crosshair on the table is relayed to the computer. With this technique, a drawing or map can be digitized quickly and with high accuracy. The resolution of these devices is normally about 0.001 in.

A tablet is the most commonly used pointing device for CAD systems. It consists of a flat surface approximately 1–3 ft^2 in size. The tablet also includes a stylus. The tablet is used either to point to a command from a menu overlaid on the tablet surface or to control the movement of a crosshair cursor on the graphics display. The crosshair cursor moves on the CRT display in relation to the movement of the stylus over the tablet surface.

A light pen can be used to locate a line or to point on the CRT screen. Since the operator has to hold the pen in an elevated position at the CRT surface, this input process can be very tiring after an extended period of time. Another disadvantage of using the light pen is that the accuracy of positioning it on the CRT may be a problem. Although a few systems still use the light pen for input, most CAD systems do not.

The mouse is a device that can be moved on any flat surface. It controls the relative position of the cursor on the CRT display. A few (from one to three) push buttons may be found on the mouse, which allow a user to select commands. The mouse is popular for CAD applications when the relative position of the cursor needs to be located, for example, when selecting an item from the screen menu.

The joystick is primarily used to locate and manipulate the screen cursor. It is not used frequently for CAD applications because of the difficulty in fine-tuning its movement to an accurate position.

12.2.4.3 Menu Design for Digitizing Tablet

A digitizing tablet and alphanumeric keyboard are the two primary devices that are used to assist the designer in creating, storing, and manipulating the image displayed by the CAD system. Figure 12.2.1 shows a CAD system that allows the designer both options for inputting design information.

The keyboard is an interactive input device that results in immediate image generation as commands are typed in. Some disadvantages of the keyboard are that the user may have to remember cryptic

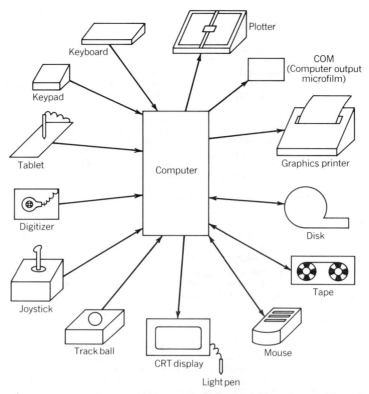

Fig. 12.2.4.   Input and output devices for a CAD system. Tien-Chien Chang, Richard A. Wysk, *An Introduction to Automated Process Planning Systems,* © 1985, pp. 52, 72, 73. Reprinted by permission of Prentice-Hall, Inc., Englewood cliffs, N.J.

commands and also data coordinates, which may or may not exceed certain valid ranges. Human factors can assist this problem by recommending that these commands are short, easily remembered abbreviations such as L for line, C for circle, and S for stop.

The digitizing tablet is used to submit commands as well as to control the cursor on the screen, both of which can be accomplished using a "menu" (Artwick, 1984). An area of the digitizing table can be set aside as a menu area, and rectangles with graphics function names can be printed on a thin plastic card that goes over this area (Price, 1982). Figure 12.2.5 shows an example of such an overlay. When the user wants to enter a line in the design, he or she moves the stylus to the proper box, presses the stylus button, then moves the stylus to the start and end points of the line on the screen to submit it.

Menu cards may contain hundreds of graphic functions. The menu is either located on the screen or fastened onto the digitizing tablet.

When the menu is placed on-screen, hierarchical command menus can be used (1) when certain entered commands are used to predict the following commands and (2) when a small, changing menu is preferable to a large menu that includes the entire command set (Price, 1982).

Menu-driven systems have several advantages over keyed-entry systems. Function names do not need to be memorized because they appear on the menu. Furthermore, the user does not have to keep alternating between the keyboard and graphics tablet to submit graphics information. However, the user must still divide his or her attention between the tablet and the screen. This problem can be eliminated by placing the menu information on the screen and allowing the user to move the cursor in order to make a particular command selection (Artwick, 1984).

Menus also have some disadvantages. For example, van der Heiden et al. (1984) in a survey of 69 CAD operators reported that 35% of the users felt a need for improvement of the table-menus and 26% stated that the menu contained a number of items they never actually used. In addition, manually digitizing a large complex drawing can be laborious because of the difficulty in maintaining the required hand–eye coordination over a long period of time. In fact, the mechanical precision of

Fig. 12.2.5. The degree of reduction necessary to reproduce this menu card for a large CAD systems makes many of the entries unreadable, but it is easy to see the very broad specification of functions that can be stipulated with such a menu. An alternative menu arrangement is to display the data on the screen so elements can be specified with a screen cursor. Bruce A. Artwick, *Applied Concepts in Microcomputer Graphics*, © 1984, pp. 38, 167. Reprinted by permission of Prentice-Hall, Inc., Englewood Cliffs, N.J.

the digitizer may exceed the manipulative capabilities of the designer. In order to alleviate the problem of fatigue, many CAD systems have a function menu on which frequently used commands can be issued merely by pressing a single key.

12.2.4.4   Output Devices—Hardcopy

A CAD system terminal displays an image temporarily. If a permanent copy of a drawing is needed, a hardcopy is required. For example, hardcopies are usually required when checking the design for errors, for production or subcontracting, or for a permanent file. The hardcopy can be generated by a printer, a plotter, or a film recorder.

The most popular hardcopy device is the X-Y plotter. Presently, there are pen plotters, photo plotters, electrostatic plotters, and ink jet plotters. Descriptions of these plotters are given below:

1.   Pen plotters are the most commonly used of all the plotters. Essentially, a pen plotter is a vector plotting device: given any two points, a pen plotter can draw a line between them. Some plotters have built-in routines to plot circles, arcs, letters, and so forth. However, these printed forms are all approximated by line segments. Pen plotters may use multiple pens, so that lines with different widths and colors can be plotted. Pen plotters are slow but relatively accurate, that is, they may display 400 dots per inch.

2.   Photo plotters operate in a similar manner to a pen plotter except that light sources replace the pen and film replaces the paper. For example, light beams of different sizes can be selected like selecting pens. The film records the trace of the light beam. Photo plotters are used mainly for IC (integrated circuit) layout or for printed circuit board (PCB) applications, such as preparing a photo mask for a PCB etching.

3.   Electrostatic plotters are dot matrix or raster plotters. A series of mechanical styli (400/in) deposit an electrical charge on specially treated paper, which is then developed. Each stylus can either be on or off, thereby leaving a black or white dot on the paper. Electrostatic plotters are much faster than a pen plotter. A pen plotter is similar to a vector drawing display (either storage tube or refresh screen) and an electrostatic plotter is similar to a raster scan display. The major problem associated with electrostatic plotters is the conversion of CAD data, which are in vector form, into the raster image before plotting. This process may require substantial amounts of CPU time. However, after the data are converted, the plotter can produce a drawing in a very short time.

A few other types of hardcopy devies, (i.e., ink jet, impact printer, and film recorder) can also be found in CAD systems; however, they are mostly considered secondary hardcopy devices.

12.2.5.   SOFTWARE TECHNOLOGY OF CAD

12.2.5.1   Nonintelligent Software Issues

The heart of a CAD system is its software. Programming and testing are essential parts of software design. The cost of testing programs can reach between 20% and 40% of the cost of the software development (Daly, 1977).

A fundamental goal of software development for CAD is to avoid user errors. Altman and Weber (1977) list the following common types of errors:

Execution errors: constructional errors, specification errors, design errors, and errors in requesting resources.

Logic errors: omitted control flow segments, wrong conditions, wrong or omitted actions.

Encarnacao and Schlechtendahl (1983) report that the following variables should be considered when designing software:

The required user knowledge about the system.

Design of help functions and error messages that are adaptable to diverse user backgrounds.

The documentation and training needed to support the CAD user on- and off-line.

Software affects the quality of a CAD system more than the hardware does. CAD software includes four main parts (see Figure 12.2.6): system control, communication, methods, and a data base. The system control routines control and coordinate the operation of the entire system. The communication routines handle the interface between the human operator and the system. Typical interfaces such as menus, on-line help facilities, cursor selection, and so forth are controlled by the communication routines. This is the most critical aspect from a human factors point of view.

Method routines are those mathematical formulas required for geometry computation, dimensioning and tolerancing routines, algorithms for design rule checks, automatic dimensioning, hidden line or

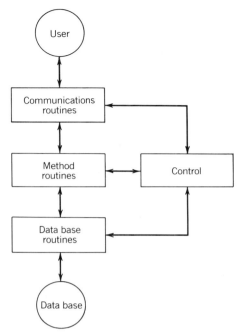

Fig. 12.2.6. The software structure between the CAD user and CAD data base.

surface removal, and so forth. All of the geometric and technological information on products is stored in a data base. Method routines use information stored in the data base to perform calculations. Data base routines represent the interface between the method routines and the actual data base. Data base routines are used to manipulate (insert, delete, modify, retrieve, and save) the design information in the data base. Different method routines can gain access to the data base through the same set of data base routines.

The way information is stored in the data base depends on the CAD system's internal model (Figure 12.2.7) or internal representation of the system. The internal representation affects the way products are modeled and also how communication routines are built. There are several internal representations, such as the 2-D model, 3-D wireframe model, and the 3-D solid model.

In the following section, figure representations, along with their applications and limitations, will be discussed. Representations will be classified as 2-D drawings, 3-D wireframe drawings, and 3-D solid modeling drawings.

12.2.5.2   Two-Dimensional Drawing

The 2-D line drawing is sufficient for applications such as IC design and layout, PCB design, pattern nesting (garment and sheet metal), and so forth. It is also used as a substitute for drafting. Since traditional engineering designs of any complexity are represented in 2-D drafting, 3-D objects are represented in 2-D using drafting rules. For example, drafting uses three views (top, front, and side) to represent an 3-D object; the invisible shapes are drawn with dashed lines. Engineers are trained to read 2-D drawings and to interpret them in 3-D in their mind. Two-dimensional CAD drafting systems provide drafting functions as a substitute for drafting instruments and paper. Since the drafting is done on an electronic sheet, it is much easier to create and modify.

IC design and layout can be accomplished on a grid 2 μin. wide. On a 0.25 in. $\times$ 0.25 in. chip, one can put 125,000 $\times$ 125,000 grids to "layout" the design. The layout is performed with special routines that interactively spot design logic errors. (Such checks are almost impossible to perform manually.) CAD can also output the design to a photoplotter in order to generate a pattern mask for IC chip manufacturing.

Two-dimensional drafting applications (see Figure 12.2.8) are a substitute for conventional drafting methods. The CAD system can perform the functions of a graphics editor in this mode. Basic geometry such as points, lines, circles, arcs, conics, and polygon and spline curves can be defined interactively. Other drafting symbols such as tolerancing, crosshatching, dimensioning lines, standard threads, cham-

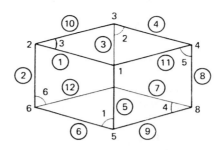

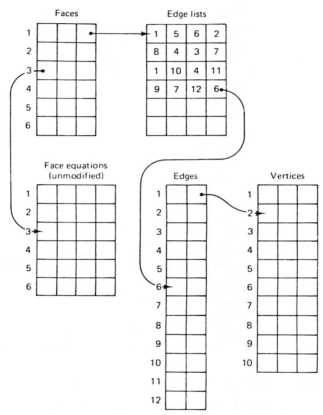

Fig. 12.2.7. The BUILD data structure for geometric modeling [after Baer, Eastman, and Henrion (1979)].

fers, and so on are also available to the operator. Since its applications are simple to produce, 2-D drafting systems are available on mini- as well as microcomputers. In mechanical design (except pure punch) drill and turn parts can be fully represented, although most other parts cannot be completely represented. Also, functions such as NC part programming cannot be performed with 2-D figures. Lastly, engineering analysis functions cannot be performed directly on the system when using 2-D images.

12.2.5.3   Three-Dimensional Wireframe Model

Three-dimensional wireframe applications include mechanical and architecture design, NC part programming, piping, mapping, and so on. Figure 12.2.9 shows an example of a 3-D wireframe model. With 3-D wireframe modeling, in addition to the geometry included in 2-D drafting, ruled and sculptured

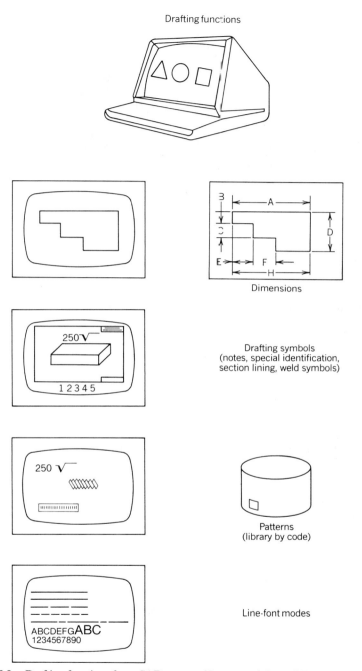

Fig. 12.2.8. Drafting functions for a CAD system. (Courtesy of Control Data Corporation.)

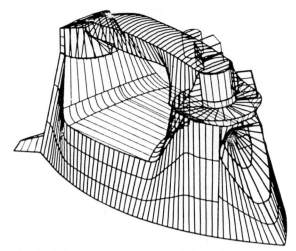

Fig. 12.2.9. Example of wireframe model with hidden lines. (Reprinted by Permission of Calma Company.)

surfaces can be used to model the product. Building a 3-D wireframe model is similar to building a 2-D drawing: lines and surfaces are located on the screen and then connected.

Three-dimensional wireframe systems represent a product by its edges and outlines of curves. There is no way to distinguish whether a point is inside, on the surface, or outside the product. Therefore, wireframe models are not sufficient for engineering analysis applications that require the calculation of mass properties (volumes, weights, centers of gravity, moments of inertia, and surface areas). Wireframe models may also be ambiguous to interpret (Figure 12.2.10).

12.2.5.4   Three-Dimensional Solid Model

Three-dimensional solid modeling is used mainly in mechanical design. In general, there are six different types of solid modeling schemes (Figure 12.2.11).

A solid model can be used not only for display and drafting applications, but also for a whole array of engineering analysis applications. With solid modeling, functions such as simulation of the products functionality can be performed and a realistic result displayed. Since the entire solid is defined, a hidden surface can be removed automatically. In addition, color shading and ruled surfaces can be added to the model.

When different types of solid modeling schemes are used, a CAD user can input data in a variety of ways. For example, in Constructive Solid Geometry (CSG), products are constructed from primitives. Each primitive has a set of attributes representing its dimension, location, and some additional informa-

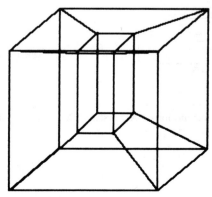

Fig. 12.2.10.   Ambiguity of wireframe model.

tion such as color. Boolean operations (union, difference, and intersection) are applied to these primitives in order to construct a product. For example, the following Boolean operation represents two boxes, $B1 = Box (0,0,0,2,2,2)$ and $B2 = BOX(1,0,1,1,1,1)$. The first three Boolean attributes are the x,y,z coordinates of the box, and the following three are the length, width, and height of the box. After a difference operation, $B3 = B1.dif.B2$; that is, $B3$ is a box missing one corner (Figure 12.2.12). Complex shapes can be constructed step-by-step from primitives in this manner.

There are times when it is very difficult to select the right primitives and operations to construct the designed product. Since the internal model consists of primitives and the display is the final result of the model, there is no one-to-one correspondence between the display surface and the primitives. In order to modify a model, one must go back to the level of primitives.

The Boundary model is another popular solid modeling scheme. In this scheme, a product is modeled by its boundary surfaces (in contrast to a wireframe model which uses bounding edges.) Faces on a model are defined by their bounding edges and a direction (to point either inside or outside of the product). Euler's law is applied to the model to make sure that the model is valid. Otherwise, it is possible to generate geometrically meaningless models.

Boundary models are easy to display and interact with from a user's point of view. NC applications can easily use boundary models to generate part programs. Sweeping is normally available as part of boundary modeling system.

12.2.5.5 Graphics Data Exchange

Currently, many different vendors supply CAD equipment. These systems often differ with respect to the data format used to represent CAD models. Even systems that use the same representation may have different data formats. This information is not transferable from one system to another. When a company uses several different CAD systems or in exchanges between manufacturers and suppliers, CAD data may have to be remodeled several times. The Initial Graphics Exchange Standard (IGES), an ANSI-approved standard, has defined an intermediate graphics data format to help exchange information between two systems. Each CAD system need only to translate its internal data format to and from the IGES format, then the IGES file can be exchanged with any CAD system. A postprocessor for each system is designed to translate the internal data format to the IGES format. A preprocessor translates the IGES format into the CAD-acceptable format. As of this writing, IGES is only designed for 2-D drafting and 3-D wireframe systems. Most CAD installations have IGES pre- and postprocessors operating with the system. Thus, IGES allows an interface between different CAD systems.

12.2.5.6 Intelligent Software Issues

An expert system is a computer system that uses the experience of human experts in some problem domain and applies their problem-solving expertise to make useful inferences for the user of the system (Hayes-Roth, Waterman, and Lenat, 1983). (See Chapter 9.2 for a discussion of artificial intelligence and Chapter 9.3 for a discussion on expert systems.) An expert system is characterized as being:

1. *Heuristic.*   It employs judgmental as well as formal reasoning in solving problems.
2. *Transparent.*   It has the ability to explain and justify its line of reasoning to users.
3. *Flexible.*   Domain-specific knowledge is generally separate from domain-independent inference procedures.

One of the main steps in developing an expert system for CAD is the extraction of knowledge from an expert designer. The skills possessed by an expert designer include pattern recognition, knowledge of design rules and relationships, and complex problem solving and decision making.

There have been some attempts to create "intelligent software" to aid the designer. For example, Lafue (1978) developed a program (ORTHO) that retrieves the 3-D structure implied by two or three orthographic projections drawn on a digitizing tablet. A major concern in the design of ORTHO was to free the user from input constraints usually imposed to avoid ambiguities. The program assists the designer by analyzing the 3-D structure of the design for geometric arrangements that are veridical. This task frees the designer from visual pattern recognition for complex figures.

Begg (1984) states that an intelligent CAD system will include knowledge of design attributes and relationships so that relevant information is deducible from the arrangement of lines and geometric forms.

An expert CAD system can assist the designer in several ways. Intelligent software can decrease the amount of perceptual decision making required by the designer by using several different scene analysis rules. For example, the software can ensure that fixed solid objects do not overlap in space or that lines that should connect do connect.

There are a number of issues that developers of intelligent CAD software should consider in order to create user-friendly systems:

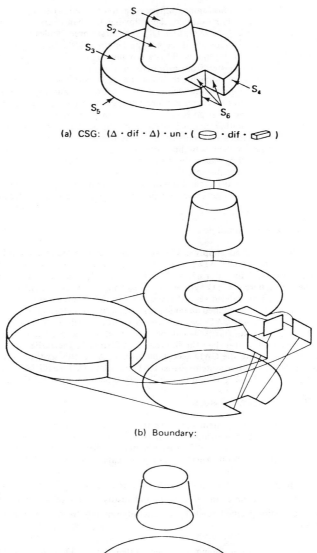

(a) CSG: $(\Delta \cdot \text{dif} \cdot \Delta) \cdot \text{un} \cdot (\ominus \cdot \text{dif} \cdot \square)$

(b) Boundary:

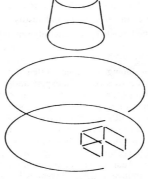

(c) Wire frame

Fig. 12.2.11. Three-dimensional representation schemes. Tien-Chien Chang, Richard A. Wysk, *An Introduction to Automated Process Planning Systems,* © 1985, pp. 52, 72, 73. Reprinted by permission of Prentice-Hall, Inc., Englewood Cliffs, N.J.

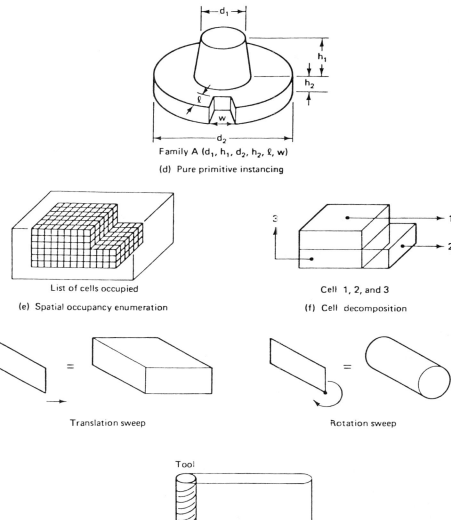

Family A $(d_1, h_1, d_2, h_2, \ell, w)$

(d) Pure primitive instancing

List of cells occupied

(e) Spatial occupancy enumeration

Cell 1, 2, and 3

(f) Cell decomposition

Translation sweep

Rotation sweep

Tool

Translation sweep of a tool

(g) Sweeping

Fig. 12.2.11. (*Continued*)

1. Information retrieval, verification, and simulation should be faster than current available methods (the noncreative elements of the system) (Begg, 1984).

2. The user should be allowed to query the system and receive an explanation for any decision that is made.

3. The user's model of how the CAD system functions should be incorporated into the decision making algorithms.

Although some cognitive skills can be automated by intelligent software systems, human creativity is still the main ingredient for any successful CAD system.

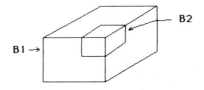

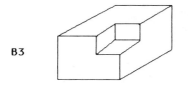

Fig. 12.2.12.   Boolean difference between two boxes.

12.2.6   HUMAN INFORMATION PROCESSING FOR CAD SYSTEMS

In this section, theories in perception, cognitive science, and cognitive psychology are applied to CAD tasks to learn about the operator's cognitive and visual processes when interacting with the CAD system. If the designers' information processing methods can be determined, then clues as to how the information should be displayed on the screen can be identified. Cognitive factors such as human problem solving, creativity, and information processing are becoming increasingly important for the design of complex systems like CAD. Greater information-processing and decision-making skills are required of humans who use CAD than of those who design manually because a CAD system allows the designer to ask "what if" questions and receive an immediate answer. A CAD operator can perform kinematic analysis, engineering analysis, and finite element analysis at the touch of the fingertips. Of course, the information derived from these analyses requires that the designer process the information and make decisions regarding the different designs.

Information processing underlies all aspects of perception and cognition. An extensive analysis of human information processing is presented in Chapter 2.2 of this handbook and will not be discussed in great detail in this chapter. Generally, however, the human is viewed as a channel through which information flows. In most models three information-processing stages or subsystems are considered: perception, cognition (spatial reasoning, decision making, and problem solving), and response control. Three memory systems for storing information are usually considered: sensory, short-term, and long-term memories. Human performance, including performance when using a CAD system, depends on, and is limited by, the information processing capabilities of these three major stages and by the storage characteristics of the memory systems. In addition, the time needed to perform a task on a CAD system is dependent on the time the information needs to pass through each of the stages. Card, Moran, and Newell (1983) present an extensive quantification of the parameters present in the information processing model.

The information processing model has many important implications for the design of CAD systems. Some specific research on CAD systems as they relate to human information processing has been performed. The next sections review this work in the context of the information processing model. The information processing model is usually represented as occurring in stages for convenience rather than accuracy. The stages, processes, and memories are all characterized by extensive interactions with each other. The stages, however, provide a convenient method for discussing and categorizing the diverse work in the area.

12.2.6.1   Perception of CAD Displays

Although graphics on CAD systems operate on basic geometric entities such as surfaces (faces), lines (edges), and points (vertices), the CAD operator works at a higher conceptual level dealing with shapes and features (Staley, Henderson, and Anderson, 1983). For example, in three-dimensional mechanical design, holes, slots, and pockets consist of basic entities such as surfaces, lines, and points. Perception involves the process of sensing the light patterns on the screen, combining the patterns into basic features, and then organizing the features into objects. An obvious problem is that 3-D objects must be represented on a 2-D screen. Another problem is that physical objects have hidden areas; but these areas must be viewed by the CAD operator and this can cause clutter and confusion on how the lines fit together. A not-so-obvious problem is that objects on the screen appear to float in space with little reference to up and down or right and left. For example, Rock (1973) and Rock, Halper, and Clayton (1972) found that changes in orientation about a horizontal axis perpendicular to the plane of the figure can produce alternations of the up–down and left–right sides of a figure. These transformations produced marked changes in the perception, recognition, and identification of the figure. These findings were explained by proposing that perceivers implicitly assign the directions top, bottom, left, and right to a figure on the basis of other information such as gravity or the visual frame of reference. These directions are missing on most CAD systems.

Although these potential problems exist for CAD systems, research on human perception can suggest methods for solving some of them. The following discussion examines two issues, perception in three dimensions and color displays, which can have a bearing on the design of CAD systems and the features which should be included in an existing system.

Many of the objects designed using a CAD system are made to appear three-dimensional. The main difficulty in displaying these items effectively is that the 3-D objects must be displayed on a 2-D surface. This can be accomplished by providing depth cues to the operators. In visual perception many depth cues have been identified. Some of these have been applied to CAD systems.

The perceptual depth cues can be divided into monocular and binocular cues. The monocular cues are those that can be viewed by one eye and thus could be represented on a 2-D surface. The binocular cues are those that need to be viewed by both eyes and thus are dependent on the existence of the third dimension when viewing. Murch (1973) identifies the following monocular depth perception cues. First, size indicates depth in that if two objects are considered to be roughly the same size then the smaller one appearing on a 2-D surface will be perceived as further away than the larger one. Second, interposition is a cue in that if one object partially covers a second object the blocked object is perceived as behind and beyond the blocking object. Third, the linear perspective cue of depth depends on the fact that as distance increases, parallel objects appear to converge. Fourth, contrast, clarity, and brightness together provide depth cues in that the nearer objects appear to be sharper and more distinct and the further objects appear to be duller. Fifth, the shadow cast by an object provides some cues to the relative position of objects. Sixth, in texture gradient, the texture of a detailed surface becomes finer as the distance increases. Finally, for motion parallax, objects in motion appear to be moving at different velocities depending on their distance. Objects that are far away appear to move more slowly than objects that are closer.

The binocular cues provide stereopsis, the depth perception commonly associated with 3-D viewing. The binocular cue of depth is dependent on the separation of the eyes. For close objects, the visual pattern on the two eyes will be offset from each other and slightly different. This effect is usually referred to as the disparity of the object and accounts for the stereoscopic perception of depth. When objects are further away, the disparity between the objects on the two eyes will not be as large.

Several techniques for CAD have been developed to take advantage of the monocular depth cues so that objects displayed on a 2-D surface can appear to be 3-D (see Table 12.2.2). One such technique is the elimination of hidden lines or hidden surfaces. In this case, lines or portions of lines and surfaces that normally would be behind objects are not shown. This technique uses the depth cue of interposition to indicate three dimensions. A second technique is linear perspective; parallel lines are displayed to converge at their vanishing points. A third technique, kinetic depth effect, is related to the depth cue of motion parallax in that it relies on the motion of the object relative to the viewer's position to indicate a sense of depth. When a single object is rotated about a vertical axis the lines close to the viewer move more slowly than those far away. A fourth technique, the use of intensity cues, relies on the depth cues of contrast, clarity, and brightness. The intensity of lines are modulated so that distant lines are less intense than close lines. Foley and Van Dam (1984) point out that as the complexity of the displayed image increases, the effectiveness of this depth technique decreases. Finally, artificial texturing uses the depth cue of texture to convey three dimensions. The density of objects can be changed depending on the distance of the object.

Table 12.2.2   Perceptual Three-Dimensional Depth Cues for CAD

Perceptual Depth Cue	Corresponding CAD Techniques
Size of objects	
Interposition	Elimination of hidden lines or surfaces
Linear perspective	Parallel lines converge at vanishing point
Contrast, clarity, and brightness	Intensity cues, density, artificial texturing
Shadows	Limited availability of cue
Motion parallax	Kinetic depth effect
Stereopsis	Limited availability of cue
—	Depth clipping

Other means are available to create realistic 3-D images not necessarily related to the monocular cues of depth listed above. In particular, depth clipping is used in some CAD systems by cutting a plane through the object being displayed. By varying this back-clipping plane, the operator can dynamically display depth information. In general, ease of visualization for a CAD drawing relates directly to the projection and type of graphic feature used (Groover and Zimmers, 1984). Orthographic views are less comprehensible than isometric views, especially for depth relationships. In addition, perspective views that include color or animation result in the best visualization capability.

These perceptual issues have a bearing on the ease of identification of wireframe models and solid modeling. Groover and Zimmers (1984) discuss some of the limitations of wireframe modeling that are most relevant for 3-D objects. One problem is hidden line removal on complex images when the lines that define the edges of an object remain visible in the image, resulting in an image that may have multiple interpretations. Another problem is that wireframe models do not convey information about the surfaces of objects and the designer cannot differentiate between the inside and the outside of objects. For this reason, wireframe models are again ambiguous and subject to different interpretations.

The ambiguities inherent in wireframe modeling are not as prevalent in solid modeling. Using solid modeling, unambiguous objects can be designed if the following criteria from Doo and Sabin (1978) are adhered to:

1. *Validity.* There must exist a real 3-D object corresponding to any given representation of the object.

2. *Uniqueness.* There can be only one 3-D object for any given representation.

3. *Conciseness.* The representation of the solids should not contain redundant information.

4. *Ease of Creation and Modification.* The internal representation of the object should be as close as possible to the designer's model of the object he or she is creating.

The second perceptual issue to be considered is the usefulness of color displays. The use of color on the terminal could increase the cost of the total system, decrease the resolution of the display, decrease the data transmission rate, and increase the complexity of programming and interaction with the system. Color can be useful to sell the system to the CAD operator, to highlight information on the display, and to make the drawings easier to see. If color systems are chosen, the number of colors for a system is another decision to be made. Differences in the perceptibility and speed of response to colors exist. From the fastest color to identify to the slowest color, the search speed is: red, blue, yellow, green, black, white. This order persists across all background colors (Ramsey and Atwood, 1979). Color identification decreases with decrements in illumination although some colors appear to be more resistant to performance decrements than others. Red is superior for distance viewing but will shift markedly in brightness with low illumination. Yellow, on the other hand, has been shown to be the most accurately identified color (Davis and Swezey, 1983).

Experimental results indicate that color is a salient cue to detect highlighted areas quickly. Laboratory experiments on human perception of colors on computer terminals have been performed to determine the effectiveness of color coding. A typical experiment records the time to detect a target in a field of distractors when the target can be distinguished by color differences or by shape differences. Results indicate that color differences are detected faster than shape differences (Schneider and Eberts, 1980; Zwaga and Duijnhouwer, 1984). In the Schneider and Eberts results, color discrimination could occur 2.25 times faster than shape discrimination. The subjects did not have to scan the display to perceive that a color occurred; processing of the display occurred in parallel for all items on the screen.

Although these results indicate that color can be useful if speed is stressed, other experimental results show little advantage for color over shape coding. For example, Tullis (1981) tested information display methods to diagnose problems over telephone lines using both color and shape coding. The time and errors to make decisions were tested. Tullis found no difference in terms of speed or errors in decision making when these two conditions were compared. When the operators rank ordered their preferences, however, the color display was preferred over the black and white display.

A final issue is the choice of colors for information highlighting. People can discriminate about 50 colors (see Helander, Chapter 5.1), therefore, purchasing CAD systems with more colors than this would not be useful. Investigators suggest that color should be used conservatively to avoid an appearance of clutter on the screen (Brown et al., 1980). The maximum range of colors that should be used in an information display is between 6 and 11 (Barmack and Sinaiko, 1966; Cropper and Evans, 1968; Engel and Granda, 1975). Table 12.2.3 compares color coding to other methods in terms of the number of codes which can be incorporated for each method. To limit the number of colors used on CAD systems, Frome (1983) suggested that black and white be used for menus, screen divisions, and information messages. Color should be reserved to highlight important messages for images that violate design rules. One consideration with color coding is to design the display so that operators with color weaknesses or color blindness can still perceive the important information. Approximately 8% of males have a problem with color blindness. The largest percentage of this group has trouble distinguishing red from green.

Table 12.2.3 Various Coding Methods for Encoding Information

Coding Method	Maximum Number of Codes for Nearly Error-Free Recognition
Color	6
Geometric shapes	10
Line width	2
Line type	5
Intensity	2

Source: Foley and Van Dan *Fundamentals of Interactive Computer Graphics,* © 1982, Addison-Wesley, Reading, Massachusetts. Pgs. 3–40 (text only). Reprinted with permission.

12.2.6.2 Cognitive Aspects of CAD

After the visual information is sensed and the perception of the objects continues, the information is manipulated based on the requirements of the task. The cognitive stage of human information processing includes reasoning, decision making, and problem solving. Two important aspects of cognition involving CAD systems are spatial reasoning and problem solving. Each of these issues will be discussed (see Table 12.2.4 for a summary).

The ability to internally visualize the 3-D structure of a part of a figure is one of the most important mental characteristics of a good designer. From a cognitive point of view, an important question for the CAD designer is the form of that internal or mental representation of an object. The internal form of the representation has implications for the kinds of mental transformations that can be carried out on it. To differentiate between theories on the form of the representation, a mental rotation task was studied in several psychological experiments. In the mental rotation task, the experiment participant is required to mentally rotate an object to compare it with a reference object. The time and accuracy of participants who perform this task are measured and studied when manipulating several variables such as the complexity of the figure or the amount of rotation of the object.

Two theories have been postulated about the form of the underlying representation of an object. The imagery or analog model theory states that people can represent visual information in memory through images. These images bear a one-to-one correspondence to the physical representation of the object. The propositional theory states that visual information in memory is transformed into proposi-

Table 12.2.4 Summary of Cognitive Research Related to CAD

Research Topic	Findings	References
Spatial reasoning	Time to rotate a 3-D object was a linear function of the amount of angular separation between the figures; supports the analog model	Shepard and Metzler (1971)
	Time to perform mental rotation tasks was statistically independent from the complexity of the figure; supports the analog model	Cooper and Podgorny (1976)
	Experimental participants mentally rotated simple figures and for more complex figures they seemed to use a feature-by-feature comparison of parts of the figures; supports a hybrid model	Barfield (1986)
Problem solving	CAD task was described in terms of a hierarchical goal structure; the structure is composed of an overall goal, phases representing the subgoals, unit tasks, and events representing the commands to satisfy the unit tasks	Card, Moran, and Newell (1983)

tions. A proposition is the smallest unit of knowledge that can stand as a separate assertion. Although propositions have been hypothesized to occur in many forms, Kintsch (1974) represents each proposition as a list containing a relation (i.e., a verb) followed by an ordered list of arguments (i.e., the relations between the verbs).

These two theories make different predictions about how visual information is internally manipulated. Since the analog theory states that mental rotation should correspond to external manipulation of objects, the complexity of the figure being rotated should have no effect on the time to rotate, however, the amount to rotate should have an effect on the comparison times. On the other hand, the propositional theory states that each part of the figure must be transformed separately for a mental rotation to occur. The complexity of the figure, therefore, would have an effect on the time to mentally rotate it but the amount of rotation should have little effect on the comparison time. The mental rotation task can differentiate between these two theories.

In a mental rotation task, the experimental participant is provided with a figure in either two or three dimensions. When the experiment begins, the participant is provided with another figure that is either a rotated version of the first figure or a rotated mirror image of the first. The subject is asked to determine whether the new figure is a rotation or a mirror image; time and accuracy are measured.

Manipulations on the amount of rotation and the figure complexity help differentiate between the theories. Shepard and Metzler (1971) used pairs of three dimensional block figures in their mental rotation experiment in which the angular separation between the two figures (i.e., the comparison figure and the to-be-rotated figure) was experimentally manipulated. They found that the time to mentally rotate was a linear function of the amount of angular separation between the two figures. Cooper and Podgorny (1976) investigated the effects of figure complexity to the time to mentally transform rotated or mirror-image figures. Two-dimensional figures, varying the number of angles or points as a measure of complexity, were used in the experiment. They found that the time to perform mental rotation tasks was statistically independent from the complexity of the figure. Both experiments provided support for the analog theory over the propositional theory.

Barfield (Barfield and Salvendy, 1984; Barfield, 1986) performed a mental rotation experiment using wireframe modeling on a high-resolution CAD system. The experimental participants memorized the graphics figures and then were asked to answer questions concerning a rotated version of the same figure. They were asked to indicate if the displayed figure was rotated either in depth or in the picture plane, whether the figure was rotated in the x, y, or z axes, and to indicate the angle of rotation. The complexity of the figures had effects on axes of rotation and the angle of rotation but not on the depth–picture plane question. For all three questions, the rotation needed had statistically significant effects on the time to perform the task.

The results from the Barfield experiment indicate that performance does not fit neatly into one of the theoretical frameworks to the exclusion of the other. The effect of angle disparity on the time needed to answer all three questions and the lack of an effect of complexity on the depth–plane question would seem to indicate the analog theory was appropriate to explain the data. The effects of complexity on the other two questions would seem to indicate that the propositional theory is appropriate to account for the data. A hybrid model (e.g., Yuille and Steiger, 1982) for the mental rotation task appears to be the most appropriate to explain the data. For simple levels of figure complexity, people mentally rotate the figure in an analog manner into congruence with the stored version of the memorized test figure. As figure complexity increases, people seem to perform the task by a feature-by-feature analysis. In verbal reports to the experimenter, some of the experimental participants indicated that they used this strategy. For example, one of the participants said, "as figures become more complex, I keyed on specific parts and rotated them separately."

Another interesting result from the Barfield experiments was a quantification of the amount of information transmitted by the CAD operator as a function of the complexity of the figure. If the operator is viewed as an information channel, then percent of information transfer refers to the amount of information in the original stimulus that is retained in the response after passing through the human information processor. The results indicate that moderately complex figures had the lowest percentage of information transferred when compared to simple and highly complex figures. These results again fit in with the hybrid model. For the simple figures, the experimental participants were probably using the analog model by mentally rotating the figure with relative ease. For the highly complex figures, the experimental participants were probably performing feature comparisons that would be relatively easy because the complex figures have more features to choose from. For the moderately complex figures, neither method was optimal for the task, and performance suffered.

The results from the experiments on mental rotation of figures have implications for CAD tasks. In particular, the results predict when CAD designers can perform transformations mentally and when the CAD system would be needed to display these transformations. If the figures are relatively simple, CAD operators would not need to physically transform the objects on the screen to "look" at them in different locations or orientations. These transformations could be done mentally on visual images similar to the predictions from the analog theory. If the figures are relatively complex, however, CAD operators could perform some mental transformations on individual features of the object and not on the whole object. This could have important implications about the kinds of errors that are made. Working on individual features in memory means that the integrity of the object is destroyed and the features may be placed back together incorrectly. The operator's memory could not be relied on in this case, and the objects should be physically rotated on the screen. For simple figures, the integrity of the object could be maintained and the features would remain in the same relationships after mental transformations.

The second aspect of cognition in CAD use is that of problem solving. The design process for CAD entails many problems that must be solved. Modeling problem solving is helpful in understanding this process. A recent model of problem solving for computer tasks is the GOMS model, developed by Card, Moran, and Newell (1983) and based on the earlier work by Newell and Simon (1972) on human problem solving. GOMS stands for goals, operations, methods, and selection rules, which is descriptive of how the model works. GOMS has been applied extensively to the use of text editors but is also applicable to interaction with a CAD system. The purpose of the GOMS model is to predict performance based on the particular CAD system and to analyze the task according to both the input to the system and the plans and actions for carrying out the task or solving problems presented by the task. It has the capability to provide accurate time estimates of tasks for different systems and to assist in the understanding of the operator's problem solving behavior and expertise.

Under this analysis, a CAD task can be considered as an overall problem that must be solved by the operator. Solving the problem involves breaking the task into subproblems and then solving each of these to obtain the overall solution. The methods and the corresponding operations must be combined by using selection rules to solve the subproblems. An overall map of the solution can be constructed by dividing the task into goals and subgoals. By analyzing the behavior of the operator through the actions, videotape recordings, and verbal protocols (i.e., having the operator tell the experimenter his or her mental operations at each point in the problem solution), the goal structure can be analyzed. This goal structure, according to GOMS, can be divided into four different levels. Starting from the highest conceptual level and moving to the lowest levels, they are: the overall goal, phases which represent the division of the goal into subgoals, unit tasks which is a further division of the subgoals, and events which correspond to the system commands to satisfy the unit tasks. Each phase, unit task, and event can be determined and labeled by analyzing the time record of the CAD task especially considering the pauses which occur. These pauses indicate that the operator may be moving on to another event, unit task, or phase.

Card et al. (1983) used this analysis technique on a CAD Very Large Scale Integration (VLSI) circuit design task. The experimenters tested one expert CAD operator who had a year's experience and around 300 hours practice on the system. The particular task tested was the modification of an existing design; this task took about 40 min. In analyzing the CAD task, Card et al. (1983) assumed that all keystrokes that occurred within 0.3 sec of each other could be considered as part of one command. The pauses longer than 0.3 sec but less than 5 sec were hypothesized to be pauses between unit tasks. Any pauses longer than this, up to 80 sec in the task analyzed, were considered to be pauses between phases. The experimenters analyzed the data by computer to categorize these goal levels, and made some corrections as a result of the verbal protocols of the CAD operator.

In analyzing all the sources of data, Card et al. (1983) identified the goal structures of the CAD task by identifying the unit tasks, phases, and the times associated with each. The unit tasks were composed of four different types and the phases could be divided into three kinds. The draw unit task was used to create new circuit elements to the layout. The alter unit task was used to move circuit elements or change their configurations. The dimension unit task was used to measure the dimensions of substructures, the distances between circuit elements, or the alignment of elements. The check unit task was used to check for connectivity or look for VLSI design rule violations. These four unit tasks were combined into three phase structures. The transcription phase lasted for about 14 min and was used by the operator to transcribe the layout of the circuit from the hand sketch into the CAD system. The vertical compression phase lasted about 7 min and was used to compress the circuit vertically by moving substructures around to make them fit more closely. The horizontal compression phase lasted about 15 min and was used to compress the circuit horizontally.

By dividing the CAD task into basic components, GOMS is able to predict the times for those components when they are recombined for other tasks. Card et al. (1983) found that GOMS could predict the times to execute all commands in the experimental session within 16%. The GOMS model predicted 1192 seconds as the time to execute all the commands but the CAD operator actually executed the commands in 1028 sec. The discrepancy between the two times was hypothesized to be due to the expertise of the CAD operator, who required less time for the mental operation than expected.

The analysis of the problem-solving behavior of the CAD operator and the corresponding goal structures through the GOMS model has several possible applications. Testing operators and determining the average times for performing basic tasks that could be combined into other CAD tasks would certainly provide useful data. Overall goal structures could be obtained for other design problems and time predictions could be determined for finishing them. Also, alternative CAD workstation designs could be analyzed to determine which CAD system would be faster and easier to use. This analysis could also be used in training CAD operators. Expert CAD operators could be compared to novice

CAD operators in terms of the goal structures for each. If the novice is not using the same kinds of structures, methods, or operations, the comparison would show this clearly and suggest how help could be given to allow the novice to act more like the expert. The novice may not be aware that a shorter sequence of operations could satisfy the same command. Finally, this method has the potential to identify levels of expertise. Card et al. (1983) found that the expert CAD operator they tested devoted 96% of the time to physical operations and only 4% to the mental operations. Novices would have to think about what they were doing and, thus, would devote more time to mental operations.

12.2.6.3 Response Execution

The last stage in the human information processing model is response. Several kinds of responses can occur. The stylus can be moved to a different position of the digitizing tablet. The CAD operator may look at a different part of the display. Help may be requested from someone else if a problem cannot be solved. An important consideration from a response point of view are the kinds of input devices to be used (see Section 12.2.4.2). Also, since the CAD task shares many characteristics with other computerized tasks, a general discussion and comparison of the possible computer input devices is useful (see Greenstein and Arnaut, Chapter 11.4). An important consideration in CAD, not covered elsewhere in this volume, is the quantification of the discrete movement of the hand or stylus to another position. This movement time is represented by the following equation which has been termed Fitts' Law (1954):

$$\tau_{\text{pos}} = I_M \ln \left[\frac{2A}{W} \right]$$

where

τ_{pos} is the total movement time to be calculated,

I_M corresponds to the time to correct a movement,

A is the amplitude of the movement, and

W is the width of the target.

I_M has been estimated to be 63 msec/bit.

Fitts' Law shows the tradeoff that occurs between speed and accuracy. Accuracy is represented by the width of the target (W). The speed of movement can be fast if the accuracy does not have to be high. On the other hand, speed would have to be reduced if the accuracy is increased.

As an example of how Fitts' Law can be applied to CAD tasks, consider the design of a menu overlay on the digitizing tablet. The times to move the stylus from one menu item to another can be predicted by Fitts' Law. Having these time predictions would then allow for the consideration of optimal menu layout designs. An important consideration is the amplitude of the movement from one item to another. Another important consideration is the size of the area for a menu item which can be touched by the stylus. Table 12.2.5 provides the predicted movement times for Fitts' Law at several widths and amplitudes. The values from the table show clearly the tradeoffs between speed and size of the space for the menu item. It may be optimal to have the frequently used items large and the lesser used items small. But having many large spaces for menu items would increase the needed amplitude for movements from one item to another.

12.2.7 ALLOCATION OF FUNCTIONS AMONG CAD SYSTEM COMPONENTS

The foregoing discussion describes how humans process the information presented on the CAD system. In turn, the CAD system processes the information provided to it by the CAD user. Since both the human and the CAD system process information, one issue is how should the information processing be divided between the human and the computer. Another issue is the kind of capabilities with which the CAD system should provide the user so that he or she can efficiently process the information. Both of these issues are considered in the following section.

12.2.7.1 Dividing Functions between Human and Machine

Bailey (1985) distinguished the tasks performed best by the human and CAD system components when design is performed in a logical sequence. The results are shown in Table 12.2.6. Lists such as these may assist in structuring the tasks performed in the CAD system and in the development of the optimal allocation of functions between humans and CAD equipment.

In general, information processing and decision making capabilities of a CAD system are limited in the following ways:

1. Computer systems are limited in their ability to recognize novel situations.

Table 12.2.5   Predicted Movement Times for Various Amplitudes and Widths from Fitts' Law

Amplitude (A) (cm)	Width (W) (cm)	Movement Time (τ_{pos}) (msec)
1	1	44
10	1	189
20	1	261
30	1	276
10	10	44
20	10	87
30	10	113
40	10	131
20	20	44
30	20	69
40	20	87
20	30	18
30	30	44
40	30	62
30	40	26
40	40	44

2. The CAD system cannot use heuristic rules but must instead rely on algorithmic solutions to solve problems.
3. The CAD system is not efficient in handling model changes in the synthesis stage of design as compared with its excellent capabilities for analysis.

Basically, three skills are needed by the designer when using a CAD system:

1. *Creativity.*   This is the most critical aspect of CAD. The features of a CAD system are successful only to the extent that they enhance the creativity of the designer.
2. *Decision Making and Problem Solving.*   Most features of CAD such as figure rotation, scaling, translation, and engineering analysis assist the designer in decision making and problem solving. Types of decisions include aesthetic analysis of scenes, tolerancing problems, visual checks, and so on.
3. *Psychomotor Skills.*   These include skills such as typing, light pen, or mouse manipulation.

The allocation of CAD tasks between the human and computer have been most clearly defined for psychomotor skills, somewhat defined for decision making and problem solving, and not defined for creativity.

12.2.7.2   Analysis of CAD Options for Design

Options available on a CAD system should assist the designer in decision making and problem solving. The following list describes some CAD options that are useful for creating, editing, and displaying an image for design purposes:

1. *Rotation.*   This option allows the CAD user to display a different orientation of the object. Some systems allow smooth rotations of a whole image. For example, a whole screen might be rotated about its center but off-center axes of rotation may be beyond the capabilities of the system. Truly versatile rotations can be performed by CAD software using matrix multiplications of trigonometric functions.
2. *Scaling and Zoom.*   The ability to see detail or to shrink an image to put it in proper perspective is often necessary. In addition, one display dimension (the "X" screen direction for instance) may need to be stretched or compressed to compensate for nonsquareness (Artwick, 1984). Scaling is used to perform such operations. When scaling is performed to all elements in the "X" and "Y" screen directions, the result is "zoom."

Table 12.2.6 Proposed Allocation of Functions and Responsibilities among Humans and Machines in CAD systems*

Activity	Goal	Human Does	Machines Does	Implication
Creativity	The designer needs to create various concepts for the design.	What he/she does best: thought, creativity, decision making.	What the designer does not do well: manage 3-D information, solve differential equations, and so on.	The designer has a partner—one who fills in where the designer is deficient. Designer has more freedom to create and experiment.
Geometric modeling	Create a mathematical model of the design in the computer.	Enters solid part model through use of geometric primitives, extrusions, sweeps, and Boolean operations.	Maintains geometric and topological data bases. Performs mathematics of Boolean operations.	Designer deals with geometry in solid terms, rather than wireframe edges. Enough information is in the data base to perform significant analyses.
Design parametrization	The mathematical geometric model will need to be edited as part of design iterations.	Recognizes the dependent relationships within the design. Uses symbols wherever possible, instead of numbers.	Records key dimensions as symbols. Can regenerate entire geometric model if a different number is assigned to one of those symbols.	A family of design possibilities is actually created instead of just a single design.
Geometric editing	The mathematical geometric model will need to be edited as part of design iterations.	Graphically interacts by picking and pointing to geometric entities on the screen.	Recognizes picks; recognizes and performs geometric editing.	Designer can edit (thus, iterate) quickly through the use of graphical interaction.
Visualization: wireframe	Create dynamic graphical rendering with edge information only.	Can interactively enter rotation, translations, and scaling information.	Special transformational hardware performs wireframe transformations in real time.	Dynamic transformations enhance visualization and understanding.
Visualization: solid	Create graphical rendering of the solid object, possibly dynamically.	Supplies light source position and surface rendering properties. Supplies transformation information.	Uses lighting model to generate solid rendering. Some new systems have specialized hardware to do this dynamically.	Visualization and understanding enhanced. Also, this is a good way to display the results of the various analyses.

1642

Analysis: mass properties	Determine mass, center-of-mass, and moments of inertia of the design.	Specifies density of the materials in the design. May also need to input an "accuracy factor."	Computes mass properties based on geometry and desired accuaracy.	Physical properties are determined easily and can become part of design iterations.
Analysis: mechanical dynamics	Determine the kinematic/dynamic response of a proposed design.	Enters kinematic connection (joint) information. Enters driving force functions.	Determines mass properties first, if necessary. Solves differential equations of motion. Animates results. Saves position, velocity, acceleration, and reaction force information for future use.	Dynamic response of a design can be easily determined and thus iterated upon.
Analysis: stress	Determine structural response of the design to force input.	Supplies elastic property constants for the material of the design. Enters boundary and forcing conditions. Approves finite-element mesh.	Generates finite-element mesh. Determines any force conditions it does not already have, possibly by running a dynamic analysis.	Strength/deformation of a design can be easily determined, and thus can be iterated upon.
Interfaces to manufacturing	Bridge the CAD/CAM gap.	Supplies tolerances, surfaces finishes, and so on.	Feature recognition, process planning, N/C path generation, N/C verification.	A design can be manufactured with little additional intervention. CAD system can constantly examine design during creation/editing for impending manufacturing difficulties.
Production of drawings	Produce drafted, annotated drawings.	Asks for it. Possibly needs to indicate where dimensions will be placed.	Projects just the bounding edges of the designed solid. Dimensions drawing where appropriate.	Drawings are no longer the primary goal, but simply a side effect of the entire system.

Overall Implication: The designer creates a "virtual prototype," with enough information to visualize, analyze, and manufacture a design.

* The authors wish to thank Dr. Michael J. Bailey, formerly of the School of Mechanical Engineering at Purdue University, for developing this table. Dr. Bailey is now Director of Advanced Development for Megatek Corporation.

3. *Viewports.*   The area of the screen used for graphic displays is called a viewport. Viewports can be defined to be square, rectangular, circular, or any desired shape. In addition, more than one viewport may be placed on a screen.

4. *Windows.*   Windows occur once a viewport has been specified, either by default or by the user, as available for display. A conversion or scaling procedure must be carried out before data can be displayed in the chosen viewport. Some displays allow up to 64 windows to appear on the screen.

5. *Translation.*   This option allows the CAD user to move an object from one location to another on the CRT screen.

An example to show how these features work together for a specific application is shown in Figure 12.2.13. An area of the map of the United States (window) is rotated and translated, scaled to new coordinate positions, mapped onto the viewport coordinator, and displayed on the user's screen (viewport) with screen coordinates.

Another task where the aforementioned features could be useful is in interference checking. Interference checking involves removing hidden lines and uncluttering an image when the CAD user is checking the image for errors. This process involves visually searching an image (such as a chemical plant with complicated piping structures) to make sure no components occupy the same space. Rotating the image, zooming in on selected portions, and removing hidden lines are all useful techniques for assisting the designer in interference checking.

12.2.8   CAD WORKSTATION DESIGN

The proper selection and design of a CAD workstation can have a profound effect on the productivity, creativity, and mental and physical health of the designer. For example, Foley and Van Dam (1984) state that two different graphics drafting systems, designed for the same tasks, may show differences of up to 100% in the overall time to complete a given task. In van der Heiden et al.'s (1984) survey of 69 CAD users, 41% of the respondents reported difficulty with the placement of the keyboard and placed it on top of the tablet for tasks that required substantial amounts of keying. Encarnacao and Schlechtendahl (1983) have developed a list of pertinent factors for CAD workstation design. The following recommendations are offered for the design of CAD workstations (see Chapter 11.1 for a more detailed discussion of computer workstation design requirements):

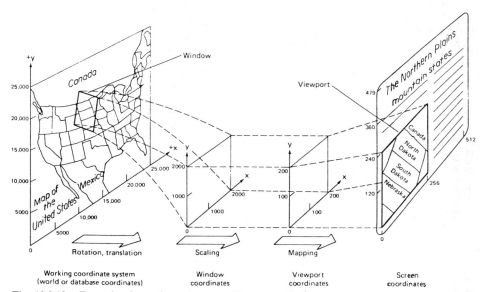

Fig. 12.2.13. Example of transformation capabilities of a CAD system. Bruce A. Artwick, *Applied Concepts in Microcomputer Graphics,* © 1984, pp. 38, 167. Reprinted by permission of Prentice-Hall, Inc., Englewood Cliffs, N.J.

1. Physical layout of the graphics terminal:
   a. Keyboard—The keyboard should be detachable to allow maximum user flexibility in keying tasks.
   b. Workstation table—The CAD table should be height adjustable and should allow for adequate workspace so that tasks can be done without excessive twisting, turning, and stretching, thus minimizing muscular strain.
   c. Chair—The CAD workstation chair should provide adequate back support and allow for freedom of postural movement. The chair should be height adjustable.
   d. CRT—For a random-scan screen the refresh rate is generally performed at about a 60–70 Hz to minimize flicker. In addition, the minimum resolution of a raster display should be about 70 dots/in. (Stahlin, 1980). The optimum luminance depends on the ambient illumination, which is usually in the range of 20–70 foot-lamberts (Davis and Swezey, 1983). Optimum luminance is then approximately 50 mL.

2. Environmental workstation factors: It is common practice for a ventilation system to be installed in computer rooms where the most sensitive hardware is located (Constantinou, Rathmill, and Leonardt, 1982).
   a. Humidity should be between 50% and 55% relative humidity.
   b. Noise should not exceed 50 dB if the work requires extreme mental concentration (Aronson, 1976).
   c. The level of lighting necessary for CAD varies among individuals and according to the task being performed. Van der Heiden et al. (1984) suggested that the illumination at CAD workstations be kept low to improve CAD image visibility and also avoid reflections. In addition, luminance contrasts of more than 10:1 should be avoided.

Please note that while the above are generally good design criteria they have not been specifically researched in the context of CAD workstations.

In addition to the general hardware requirements discussed above for CAD workstation design, the following factors should also be considered to exploit the full potential of a CAD system (Constantinou et al., 1982):

1. Designers requirements:
   a. A personal work area that permits concentration, that screens distraction, discourages interruptions, and has adequate space to store drawings should be provided.
   b. The work area should have proximity to common facilities such as the printer and/or computer room.
   c. The work environment should contain safety features such as sensors, alarms, and fire control.

It should be pointed out that contradictions do occur to the above requirements. For example, the desire for an individual customized workstation contrasts with the need for maximum CAD-user interactions when design is performed by a team.

12.2.9 JOB DESIGN ISSUES

Loth (1983) has estimated that in 1985, 36,000 CAD workstations will be in use, each requiring one operator per shift of operation. The selection and training of these operators and the design of the job they will perform is of particular interest to human factors specialists.

Changes to jobs with CAD may take several different forms. Routine activities could be eliminated from jobs as CAD absorbs these tasks; traditional communication processes may change as CAD allows designs to be more easily shared, and work between jobs may flow in a more reciprocal than sequential fashion as CAD allows for a more reiterative design process.

Since there is only limited research on CAD effects, conclusions about job changes with CAD must rely in part on research done with computerized information technologies such as electronic mail, decision support systems, and management information systems. Summarized in Table 12.2.7*, literature on information technologies suggest that information systems affect communication patterns in several ways:

1. They increase the choice of when to communicate as well as the speed and explicitness of communication.
2. They discount the importance of distance as a barrier to communication.
3. They stimulate the use of other communication media such as face-to-face interactions.
4. They are less effective when opinion change due to peer or nonverbal pressure is needed.

CAD research corroborates these findings, at least in terms of a general increased need for communication.

Table 12.2.7   Summary of Literature on Changes to Job with CAD

Job Dimension	Author	Sample	Finding
	Information Technology		
Routineness	Bjorn-Andersen, Hedberg, Mercer, Mumford, and Sole (1979)	Clerks	More routine, more tasks
	Gutek (1982)	Secty., Mgr.	No change
	Mann and Williams (1960)	Clerks	Fewer routine and minor decision-making tasks
Workflow	Whisler (1970)	Clerks	More interdependence
	Mann and Williams (1960)	Clerks	More need for coordination
Communication	Whisler (1970)	Clerks	Less interpersonal communication
	Kerr and Hiltz (1982)	Review	More options for communication
	Edwards (1978)	Professionals	Increased degree and options for communication
	Hiltz (1978)	Engineers & Scientists	More communication
	Rice and Case (1983)	Professionals	More communication
	Short, Williams, and Christie (1976) Rice (1984)	Review	Less persuasiveness
	CAD		
Routineness	Norton (1981)	Operators; managers	Less routine for designers; more routine for operators
	Newton (1984)	Operators	No change in autonomy
	Wingert, Rader, and Riehm (1981)	Managers	Less routine
	Majchrzak, Collins, and Mandeville (1985)	Engineers; drafters	No change in predictability of job
Workflow	Majchrzak et al. (1985)	Engineers; drafters	No change in degree of reciprocal interdependence
Communication	AIDD (1983)	Operators	More need for communication skills
	Majchrzak et al. (1985)	Engineers; drafters	More use of coordination models

* This literature and others in the following sections are described in more detail in Majchrzak, et al. (1986).

These information technology studies also indicate that no single pattern emerges about the impact of information systems on the content and routinity of jobs. The CAD literature, as small as it is, suggests that the lack of a clear pattern for job changes may be due in part to different changes occurring to engineers versus drafters. Engineers appear to spend less time performing nonroutine tasks with CAD. For example, Cooley (1977) reports that the use of a CAD system for design may increase the designer's rate of decision making by 1900%, and decrease the designer's efficiency by 30 to 40% in the first hour, to 70 to 80% in the second hour (see also Bernholz, 1973). In contrast to engineers, drafters appear to spend more time with routine activities. As a result of the difference between these two groups, the sum change is zero.

Finally, workflow interdependence does not appear to change substantially with CAD. While some increased need for coordination is experienced with CAD, translating this experience into feelings of reciprocal or shared interdependence seems to be less a function of the technology than of other (e.g., organizational) factors.

In addition to a consideration of communication, workflow, and content of jobs, job designs with CAD may change because of the high obsolescence rate of CAD systems. Sophisticated CAD systems can become obsolete in three years (Cooley, 1977). This high obsolescence rate for CAD equipment necessitates the introduction of shift work for CAD operators (Cooley, 1981). The problems associated with shift work are numerous: ulcer rates are eight times higher and divorce rates are 60% higher among shift workers than control groups.

12.2.10. SELECTION OF PERSONNEL FOR CAD SYSTEMS

While job descriptions may not change across all jobs with CAD, personnel to fill CAD jobs will apparently need a whole new range of skills, as Table 12.2.7 indicates. These skills include conceptualizing, problem-solving, human relations, communications, broad-based knowledge, and creativity. Moreover, an AIDD survey (1983) suggests that an additional primary selection criteria for new CAD personnel should be their expressed interest in learning CAD. Note that the new skills are not necessitated by the CAD system per se (as programming skills would be). Rather, these new skills are needed in order to optimally benefit from the design changes brought about by CAD.

As part of this issue of selection is whether personal background characteristics of individuals should be considered in the selection of CAD personnel. Numerous studies have attempted to test the validity of the popular notion that older workers are more resistant to new technology such as CAD. As summarized in Table 12.2.8, the literature indicates that age, in and of itself, is not an important predictor of resistance to CAD. Rather, Tull (1984) suggests that initially older workers tend to be harder to motivate because their extensive experience base necessitates that the new system be substantially better than their current way of working. However, Tull also found that once trained, older workers tended to bring to the CAD system a work experience not found in younger workers, and that this work experience helped them to recognize CAD/CAM's full potential. Thus, the data do not support using age as a selection criteria for CAD personnel.

12.2.11 TRAINING OF CAD PERSONNEL

Numerous studies have found training to be an important human factor issue for CAD (e.g., Staehle, 1982). Moreover, Lazear (1984) estimates that there is a need to train 100,000 drafters and engineers on CAD over the next few years at an average cost of $10,000 per trainee. This cost includes the direct and indirect expense for instructor and student labor, supplies, use of a CAD terminal and the cost of less productive output. "This means the United States faces a staggering expenditure of $1 billion in training costs alone" (p. 49).

Despite the importance of training, however, several studies suggest that it has not been given sufficient attention by organizations installing CAD. In a survey of 850 engineering CAD users, only 55% responded that they had received formal training in CAD (Design Engineering, 1983). A recent Harbor Research Group (Boston, MA) survey of CAD implementors found that the biggest single problem mentioned by CAD users was education and training (Wagner, 1985). In a recent study by

Table 12.2.8 Summary of Literature on Importance of Age as Selection Criteria for CAD Personnel

Author	Sample	Finding
Information Technology		
Kaufman (1979)	Engineers	Age unrelated to self-reported job obsolescence
Lucas (1975)	Salesmen	No relationship between age and system use
Kerr and Hiltz (1982)	Review	No consensus on relationship among studies
Hiltz (1984)	Professionals	No relationship between age and system use
CAD		
Majchrzak et al. (1985)	Engineers; drafters	Age relatively unimportant predictor of nonusers' fears of CAD. Age was positively related to users' adjustment to CAD.
Newton (1983)	Operators	Age unrelated to satisfaction with CAD

Jacobs (1985), 202 Michigan auto suppliers responded to a mailed questionnaire on training for CAD. Jacobs found that there was generally a lack of an organized approach to training. Of the firms with CAD 72% performed training only through informal on-the-job-training; neither apprenticeships nor more formal training programs were provided. In addition, a full 40% of the firms had no training coordinator, with virtually all firms using vendor training and private training companies as the most popular source of training.

Other studies attest to the lack of CAD training. Majchrzak, Nieva, and Newman (1985) surveyed a national probability sample of manufacturing firms and found that for those firms using CAD, few (19%) offered CAD training. Hamilton and Sheehan (1982) also interviewed CAD users (81 users at Rockwell) and found that one of the implementation problems most commonly mentioned by users was the lack of adequate training. Moreover, users complained that management failed to support any training or learning inefficiencies inherent in adjusting to a new system.

One reason for users' reports of inadequate training may be the form in which this training typically occurs, that is, informal training with heavy reliance on peers and with little customization of vendor training for specific organizational work processes (Jacobs, 1985; AIDD, 1983; Majchrzak, Nieva, and Newman, 1985). Tapscott (1982) in reviewing failures of implementation in office automation systems found that "vendors install standard products in customer environments. The large body of failure literature indicates that such an approach has not and will not enable the successful implementation of integrated office systems" (p. 31). He offers several reasons for a need not to rely on vendors: (1) the vendor cannot understand all the different needs of the various subgroups; (2) a training program appropriate at one point in time may not be appropriate later given the dynamic nature of organizations; (3) vendors tend to ignore human problems in the training and change process; and (4) vendors tend to be poor at adapting to user requirements because they start with a solution in search of a problem. Tull (1984) extends Tapscott's conclusions to the CAD environment. In his experience implementing training programs for CAD users, Tull observes that vendor training only provides a minimal level of system indoctrination; without customizing training to address the company's specific operating procedures, the major benefits of CAD are difficult to achieve.

While vendors' training packages should not be used to substitute for inhouse initial training programs, Ettlie and Rubenstein (1980) have found a special purpose for vendors. In their longitudinal study of automation projects, they found that successful implementation efforts were characterized by vendors modeling for organizations with appropriate behavior for technology use and trouble-shooting. From this finding, the authors recommend that the interaction between vendors and user personnel be exploited to facilitate the acquisition of new behaviors among users.

In addition to a new approach for using vendors, initial training can benefit from a recognition of the type of training actually needed during the initial implementation period. Johnson et al. (1985) found that more sophisticated use of new systems was achieved when training focused on general education in concepts (e.g., work process) rather than in specific machine functions. The researchers also found that training which focused on learning new ways to use a system rather than simply more proficiently using specific functions yielded a more sophisticated use of new systems by the organizations.

Tull (1984) also recommends a move away from the traditional training focused on specific machine operation. He points out that a shortcoming of most CAD training programs is that they teach operators how to draw on a VDT in the same way as with manual drafting. Instead, operators must be trained to produce computer models that provide all end-product information required by downstream users.

Training programs of the future should not only teach a broader range of skills than those previous, but should be directed toward a broader audience. Design managers themselves need to be trained in the basic use of CAD. Hamilton and Sheehan (1982) found that such training of design managers decreased resistance to change not only among the managers but among those working for them. Tull (1984) also suggests a familiarization course for managers. Such a course would explain how CAD will change managers' jobs, such as the changes to monitoring subordinates' work given new output cycles, and the changes to work scheduling and budgeting with CAD. Such a course can help managers to understand the frustrations of users in learning and adapting to the new system by exposing them, for example, to anxieties associated with slow response time. Tull suggests that such a course need not be longer than ten hours.

A final recommendation for improvements to initial training programs is offered by Hiltz's (1984) research. In her longitudinal study of the implementation of an electronic information exchange system, Hiltz found that the first sessions on the terminal were characterized by users expressing the need for feedback from others. One colorful example of an expression from one user was: "This is the first message of a wandering wordsmith caught in a time warp" (p. 95). This observation suggests that initial training programs need to incorporate extensive feedback as users become adapted to the new system. This need for feedback is corroborated by Holding (1986) in his recent review of the training literature. In his words: "Some feedback seems essential for acquiring a skill and many experiments have shown that successful learning is mediated by knowledge of results" (p. 32).

These studies lead to the following conclusions about characteristics important for future programs focused on helping users through their initial training period:

1. Customize training to organizations, decrease reliance on vendors to provide the training, and use real organizational examples during training sessions.
2. To facilitate customization of training not only within the organization but across individual users, conduct pretraining screening assessments.
3. Formalize training plans.
4. Help management to understand the learning curve.
5. Broaden training to include general concepts and system capabilities.
6. Broaden the audience of training to include those who are only indirectly involved in the new system.
7. Recognize that training will create anxiety and frustration; prepare management accordingly such that users do not feel unduly pressured to perform too early.
8. To reduce reliance on other users during initial learning stages and thus enhance time spent on system and not helping others, establish a user consultant with primary respnsibilities to offer help to new users.

12.2.12   IMPACT OF CAD ON ORGANIZATION

The implementation of CAD appears to have primary effects on two aspects of the organization: formalization of procedures and structural arrangements. As depicted in Table 12.2.9, CAD creates a need to formalize a variety of work rules; a need so strong that implementation failure may result if not carried through. For example, Palitto (1984), in his case study of CAD implementation at Firestone, found that a mistake they made in the implementation process was transferring the workload immediately from manual to CAD without allotting sufficient time to formulate procedures and policies for CAD use. Hall (1984) also describes a case study of a Fortune 500 electronic company that complained of inadequate productivity gains from CAD. Hall found that the design work process has not been changed to be aligned with the new technology. Specifically, he found that a logical, formalized approach to drawing creation was not being used by drafters, there were no formalized procedures for using existing drawings; design engineers were giving schematics to drafters without any procedures for ensuring readability, accuracy, and quick interpretation; and symbols used on CAD were personally trailored so that different users could not easily converse with one another.

In addition to its effect on organizational procedures, CAD appears to affect the structure of the organization. Structure could be affected in three ways. The manner and degree to which departments or organizational units are differentiated could change with CAD, such as the addition of a unit specifically for CAD or the reorganization of drafting and engineering departments under one manager. CAD may also affect the way differentiated departments are integrated such as by introducing matrix organizations or coordinating committees. Finally, CAD may affect the degree to which important decisions are decentralized, such as allowing CAD operators or user departments to select locations for installing new CAD terminals.

Table 12.2.10 presents results of quantitative empirical studies examining the effect of system use

Table 12.2.9   Summary of Literature on Impact of CAD on Formalization of Procedures

Author	Sample	Finding
Information Technology		
Alter (1976)	Managers	Formalized mechanics of communication process that provided common conceptual basis across departments
Stewart (1971)	Managers	More use of formal procedures to complete work
Bjorn-Andersen et al. (1979)	Clerks	More formalization of work procedures
Mann and Williams (1960)	Clerks	More formalization of decision rules
CAD		
Majchrzak et al. (1985)	Engineers; drafters	More formalization in job duties
AIDD (1983)	Operators	More formal procedures for releases, signatures on drawings, and log books

Table 12.2.10   Summary of Literature on Impact of CAD on Organizational Structure

Author	Sample	Finding
Information Technology		
Bjorn-Andersen et al. (1979)	Banks	More need for integrative mechanisms; more centralization of data processing decisions
Whisler (1970)	Life insurance companies	More structural changes among highly interdependent, problem-solving, and time pressure departments
Mann and Williams (1960)	Utility company	More meetings and centralized control over work procedures
Sanders (1968)	Service	Decentralization of decision-making to lower-level managers
Pfeffer and Leblevici (1977)	Variety	More differentiated departments and decentralization
CAD		
Schaffitzel and Kersten (1985)	German companies	More use of structural interfaces between design and manufacturing
Majchrzak et al. (1985)	Engineers and drafters' perceptions	No relationship between CAD use and perceived decentralization

on organizational structure. In addition, several case studies of CAD have been done and are reviewed in Majchrzak, et al. (1986). From this research, it appears that CAD's primary effect on organizational structure is through its effect on an organization's integrative mechanisms. That is, implementation of CAD leads to an increased use of integrative mechanisms ranging from matrix structures suggested by Benn (1983) to committees, as well as the movement of engineering and manufacturing into the same facility.

The precise effect of CAD on the degree of differentiation among organizational units and centralization of decision-making authority is not clear as yet. A likely conclusion from the research yet to be done, however, is probably one suggested by Ellis (1984): that technology does not have a determining effect on centralization or decentralization; rather it increases options for organizations to centralize or decentralize, including the option to do both. Thus, organizations can, with CAD, control information at the same time they decentralize their operating responsibilities.

12.2.13   IMPLEMENTATION ISSUES

Two issues to consider in the implementation of any CAD system is the degree of involvement by top management and users.

12.2.13.1   Top Management Involvement

The need for top management support and involvement in implementing organizational change is such a well-accepted doctrine that it is almost folk-wisdom (e.g., Beer, 1980). However, it is essential to ascertain how top management should become involved in the process. Several studies provide suggestions.

Ettlie (1973), analyzing organizational data, found that organizations were more successful in their implementation when their top management was committed to the concept rather than the hardware of the new technology. That is, the involvement of top management in the general objectives and uses of the equipment rather than specific details of the hardware were necessary.

Three studies have found that while top management should be more focused on the concept than the hardware, they need to understand the new technology well enough to know how it fits into the work process (Johnson et al, 1985; Collins and Moores, 1983; and Ackoff, 1967). That is, a lack of understanding of the system results in incorrect judgments about equipment expenditures and optimizing system utilization.

Not only should top management be involved in both conceptualizing the technology and exploring how it fits into the organization, but numerous studies suggest that top management should be involved in the consensus-building process concerning system objectives (e.g., Mann and Williams, 1960; Kling, 1980). For example, Mann and Williams found that a lack of consensus between managers of different departments regarding the understanding, intended use, and priorities of the CBIS resulted in cross-departmental conflicts that slowed implementation.

This literature review of management involvement suggests several specific ways in which top management should be involved in the implementation of CAD. First, top management must understand enough of the technology and the organization's business and workflow process to make reasonable judgments on how the technology fits in with the company's needs. In the words of one author:

Wittgenstein: Whereoft one does not understand, one should not speak.

Hammer: Whereoft one does not understand, one should not automate.

HAMMER (1982); CITED IN TAPSCOTT (1982)

Second, top management needs to be involved in the implementation process primarily from a top-down perspective. That is, their focus needs to be on strategic concepts rather than the specific details of either the implementation or the technology. Too much involvement in operational decisions will remove the discretion needed by lower-level systems developers as they adjust to the concerns of lower-level users of the system.

Third, top management must recognize a need to become involved not only in specifying objectives and uses, but in resolving the conflict likely to occur as objectives, priorities, and uses are defined across departments and users. Interests of different departments and users are likely to be in conflict as strategic decisions are made about whether CAD should serve the needs of all designs, whether its output should feed directly to manufacturing tools, how CAD designs are approved, at what point in the design process modifications to CAD designs are halted, and so on. Thus, an important role for top management is to resolve these conflicts as they arise—and preferably prior to full-scale implementation.

Finally, this research suggests that, upon agreement as to system strategy, priorities, and use, this agreed-upon strategic direction should be communicated to employees affected by CAD. This is particularly important because peoples' perceptions of top management's role influence their optimism; thus, a strategic direction not communicated cannot combat employee pessimism. Moreover, technological change tends to disrupt peoples' lives in ways over which they have no control. Thus, as Mann and Williams (1960) found:

During much of the transition to using a CBIS, the employees operated on faith: faith that the demands placed upon them would be rewarded in the future and faith in management's ability to effect the new change. (p. 235).

Any communication that can help employees to maintain their faith is needed.

12.2.13.2   User Involvement

Not only does top management support affect the success of the implementation process, but user involvement in the implementation process also appears to be important (c.f., studies by Clausen, 1979; Alter, 1976; Johnson et al, 1985; Swanson, 1974; Collins and Moores, 1983; Lucas, 1975; AIDD, 1983). However, despite the clear consensus in the literature on the importance of user involvement, studies indicate that users are rarely involved in system development or implementation (Bjorn-Andersen, Hedberg, Mercer, Munford, and Sole, 1979; Danziger, Dutton, Kling, and Kraemer, 1982; BNA, 1984).

Danziger et al. (1982), in their study of government users of CBISs, suggest that a reason why user participation is so poor may be the way in which the gap between developers and users is bridged. Developers who find it difficult to communicate to users or solicit information in a way that leaves users feeling inadequately involved discourage user participation.

Bjorn-Andersen et al. (1979) suggest that the lack of user involvement is not due to a lack of management desire. They quote a French computer manager as saying: "It is not a good thing for the computer staff to work out problems alone. There should be more contact between the computer staff and the users" (p. 330). Given that the desire and perceived need for user involvement is there, the authors suggest that the reasons for the lack of user involvement seems to be "due to the ignorance on the part of the designers on how best to involve users, and to an absence of pressure from users to be involved" (p. 330).

Bostrom and Heinen (1977) reviewed the literature on MIS implementation failures and found that the lack of user involvement may be in part due to a choice made by the users themselves. If participation is perceived by users as undertaken by management for the express purpose of gaining acceptance rather than collaborative problem-solving, users will balk at such an involvement. This perception tends to be fueled by system developers who rarely allow users to actually accept responsibility for the success of the system; responsibility and thus credit and blame are typically in the hands of the developer. The authors suggest that users may also choose not to participate because of the time needed to make reasonable judgments. For example, in a survey of 2000 managers, the authors found that 80% wanted to influence the design of the MIS but were unwilling to invest the time.

In her longitudinal survey of 150 users of an electronic information exchange system, Hiltz (1984) also found users to be partly to blame for the lack of user involvement. She found that the more hours users spent on the system, the more features of the system they used. From this finding, she concludes that users may not be involved because they cannot tell the system developer what they need prior to using the technology.

Finally, Johnson and associates (1985) examined user involvement in their study of organizations implementing word processing. They found the lack of participation to be attributable to both managers and users. Managers were found to underestimate the capability of low-level staff to generate new ideas and applications of equipment and thus, involved them only minimally in system development. Moreover, operators were found to not necessarily want participation in all types of decisions. Rather, they preferred involvement in decisions about training, but cared little about involvement in equipment choice.

The aforementioned studies about user involvement suggest several reasons why users are not as involved in the implementation process as the literature would recommend. These reasons are attributable equally to the users themselves, to managers of the users, and to the system developers. First, users are not more involved because they don't want the time investment to learn the system. Thus, they want involvement only if their participation is truly problem-solving and only on those issues that matter to them and they can understand (such as training). Second, users are not more involved because system developers find it difficult to work with users. This difficulty is due in part to language barriers and in part to differences in focus. Finally, users are not more involved because managers fail to recognize how much users can offer to the implementation and system development process. Thus,the level and type of encouragement necessary from the management staff if users are to become more involved is missing.

These findings paint a rather bleak picture of the current state and future for user involvement. Yet, as the literature described at the beginning of this section indicates, user involvement is essential for successful implementation. Thus, three recommendations are suggested for enhancing the level of user involvement.

The first recommendation for enhancing user involvement (by Hedberg, cited in Bostrom and Heinen, 1977) suggests that design decisions can be broken into two types: strategic (which estimates limits and objectives for the system) and technical (which works within the strategic guidelines to develop specific software and hardware options). Given the level of technical knowledge necessary to become involved in technical decisions, and given users' adversity to spending such time, Hedberg recommends that users be involved in the strategic decisions only, leaving technical decisions for the system developers to solve. Thus, users can feel involved in setting guidelines on issues about which they personally have concerns and in a way in which their involvement is productive.

A second recommendation for enhancing user involvement stems from Hiltz's (1984) finding that users' needs change as they become familiar with a new system. Due to this finding, Hiltz suggests that the typical approach to involving user i.e., one-shot interviews conducted prior to system development and installment, is wrong. Rather, "an evolutionary design approach" in which continuous problem-solving discussion with users on a range of issues from software development to training would provide both a more knowledgeable involvement of users and the sense of continued improvement necessary to maintain a lengthy and changing implementation process. In support of this notion, the AIDD survey of CAD users found that a suggestion offered by several users for organizing a CAD department was providing time for people to work out system development problems on a continuous basis.

A final recommendation for enhancing user involvement revolves arround the management of system developers. Clearly, something is wrong when system developers cannot speak the users' language or leave users with the impression that developers are concerned less with system use than computer hardware. Thus efforts must be made to identify ways to appropriately manage the system developers to help overcome the chasm between developers and users.

REFERENCES

Ackoff, R. L. (1967). Management misinformation systems. *Management Science, 14*(4), B147–B156.

AIDD CADD Committee (1983). *Survey of CAD system users.* Unpublished manuscript. American Institute for Design and Drafting, Gretna, LA.

Alter, S. L. (1976, November–December). How effective managers use information systems. *Harvard Business Review.*

Altman, A., and Weber, G. (1977). Programmiermethodik, Arbeitsberichte des IMMD, Erlangen, 10, Nr. 3, 1977 (in German).

Arnold, E., and Senker, P. (1982). The effects of computer-aided design on manpower and skills in the U.K. engineering industry. *International Conference on Man/Machine Systems,* 209–212.

Aronson, B. (1976). Germany's war on noise. *Machine Design, 48*(9), 152.

Artwick, B. A. (1984). *Applied concepts in microcomputer graphics.* Englewood Cliffs, NJ: Prentice-Hall.

Baer, A., Eastman, C., and Henrion, M. (1979). Geometric modelling: a survey. *Computer-Aided Design, 11*(5).

Barfield, W. (1986). Cognitive and perceptual aspects of three-dimensional rotations for computer aided design. Unpublished Ph.D. dissertation, School of Industrial Engineering, Purdue University.

Barfield, W., and Salvendy, G. (1984). Computer aided design: human factors considerations. *Proceedings of the Human Factors Society 28th Annual Meeting* (pp. 654–658). San Antonio, Texas.

Barmack, J. E., and Sinaiko, H. W. (1966). Human factors problems in computer–generated graphic displays. (Contract SD-50, ARPA Assignment 15). Institute For Defense Analysis Research and Engineering Support Division.

Beer, M. (1980). *Organization changes and development: A systems view.* Santa Monica, CA: Goodyear.

Begg, V. (1984). *Developing expert CAD systems.* New York: Unipub.

Benn, J. A. (1983). CADCAM organizational charts and productivity. *Effective CADCAM.* Institute of Mechanical Engineering Conference Publications, 1983–7, London.

Bernholz, A. (1973). Computer-aided design—an extension of man. In J. Vlietstra and R. F. Wielinga, Eds., *Proceedings of the CAD Conference, International and Federation of Information Processing.* Eindhoven, The Netherlands.

Bjorn-Andersen, N., Hedberg, V., Mercer, D., Mumford, E., and Sole, A., eds. (1979). *The impact of systems change in organization.* Netherlands: Sijthoff & Noordhoff.

Bostrom, R. P., and Heinen, J. S. (1977). MIS problems and failures, Part I. *MIS Quarterly,* 17–32.

Brown, C. M., Burkleo, H. V., Mangelsdorf, J. E., Olsen, R. A., and Williams, Jr., A. R. (1980). *Human factors engineering criteria for information processing systems.* Sunnyvale, CA: Lockheed.

Card, S. K., Moran, T. P., and Newell, A. (1983). *The psychology of human–computer interaction.* Hillsdale, NJ: Erlbaum.

Chang, T. C., and Wysk, R. A. (1985). *An introduction to automatic process planning.* Englewood Cliffs, NJ: Prentice-Hall.

Clausen, H. (1979). Concepts and experiments with participative design approaches. In E. Grochla, and N. Szyperski, Eds., *Design and Implementation of Computer-Based Information Systems.* Holland: Sijthoff and Noordhoff.

Collins, F., and Moores, T. (1983). Microprocessors in the office: A study of resistance to change. *Journal of Systems Management,* 17–21.

Constantinou, S., Rathmill, K., and Leonardt, R. (1982). Ergonomic aspects of installing graphics CAD systems. *Computer-Aided Design, 14*(3), 161–165.

Cooley, M. J. E. (1977). Impact of CAD on the designer and the design function. *Computer-Aided Design,* (4), 238–242.

Cooley, M. J. E. (1981). Some social aspects of CAD. *Computers in Industry, 2,* 209–215.

Cooper, L. A., and Podgorny, P. (1976). Mental transformations and visual comparison processes: Effects of complexity and similarity. *Journal of Experimental Psychology: Human Perception and Performance, 2,* 503–514.

Cropper, A. G., and Evans, S. J. W. (1968). Ergonomics and computer display design. *The Computer Bulletin, 20,* 369–384.

Daly, E. B. (1977). Management of software development. *IEEE Transactions, SE 31,* 229–242.

Danziger, J. N., Dutton, W. H., Kling, R., and Kraemer, K. (1982). *Computers and politics.* New York: Columbia University Press.

Datapro Research Corporation. (1984). CAD/CAM user survey evaluation. *IEEE Computer Graphics and Application, 4*(2), 21–24.

Davis, E., and Swezey, R. W. (1983). Human factors guidelines in computer graphics: A case study. *Human Factors, 26,* 113–133.

Design engineering, (1983). *Surveying the Profession* (p.57).

Doo, D., and Sabin, M. (1978). Behavior of recursive division surfaces near extraordinary points. *Computer Aided Design, 10,* 356–360.

Edwards, G. C. (1978). Organizational impacts of office automation. *Telecommunication Policy,* 128–136.

Ellis, P. (1984). Office planning and design: The impact of organizational change due to advanced information technology. *Behavior and Information Technology, 3*(3), 221–233.

Encarnacao, J., and Schlechtendahl, E. G. (1983). *Computer aided design.* Berlin: Springer-Verlag.

Engel, S. E., and Granda, R. E. (1975). *Guidelines for man/display interfaces* (Technical Report 00.2720). Poughkeepsie, NY: IBM.

Ettlie, J. E. (1973). Technology transfer—From innovators to users. *Industrial Engineering, 15,* 16–23.

Ettlie, J. E. and Rubenstein, A. H. (1980). Social learning theory and the implementation of production innovations. *Decision Sciences, 11,* 648–668.

Fitts, P. M. (1954). The information capacity of the human motor system in controlling the amplitude of movement. *Journal of Experimental Psychology, 47,* 381–391.

Foley, J. D., and Van Dam, A. (1984). *Fundamentals of interactive computer graphics.* Menlo park, CA: Addison-Wesley.

Frome, F. S. (1983). Incorporating the human factor in color CAD systems. *IEEE Proceedings of the 20th Design Automation Conference* (pp. 189–195). New York: IEEE.

Gold, R. (1983). CAD/CAM—The key to higher productivity. *Precision Metal,* 7–12.

Grabowski, H., and Anderi, R. (1983). CAD-systems and their interface with CAM. In U. Rembold, Ed., *Lecture Notes of the Advanced Course on Computer Integrated Manufacturing (CIM '83).* University of Karlsruhe, Federal Republic of Germany.

Groover, M. P., and Zimmers, E. W. (1984). *CAD/CAM: computer-aided design and manufacturing.* Englewood Cliffs, NJ: Prentice-Hall.

Gutek, B. A. (1982). Effects of office of the future technology on users: Results of a longitudinal field study. In G. Mensch and J. Niehaus, Eds., *Work, Organizations, and Technological Change.* New York: Plenum.

Hall, L. (1984). Policies and procedures for efficient CAD system use. In R. M. Dunn, and B. Herzog, Eds., *CAD/CAM Management Strategies.* Pennsauken, NJ: Auerbach.

Hamilton, H. E., and Sheehan, D. F. (1982). *Human Factors in CAD/CAM.* Paper presented at the Meeting of the Society of Manufacturing Engineers, Los Angeles, CA.

Hayes-Roth, F., Waterman, D. A., and Lenat, D. B., Eds. (1983). *Building expert systems.* London: Addison-Wesley.

Hiltz, S. R. (1984). *Online communities.* Norwood, NJ: Ablex.

Huncke, D. J., and Kent, D. P. (1977). Ergonomics of a large interactive graphics operation. *Computer-Aided Design, 9*(4), 262–266.

Jacobs, J. (1985). *The training needs of Michigan auto suppliers: Interim report.* Unpublished manuscript. Industrial Technology Institute, Ann Arbor, MI.

Johnson, B. M. and Associates (1985). *Innovation in office systems implementation* (NSF Report No. 8110791). University of Oklahoma.

Kaufman, H. G. (1969). Technical obsolescence: Work and organizations are the key. *Engineering Education, 69*(8), 826–830.

Kerr, E. B., and Hiltz, S. R. (1982). *Computer-mediated communcication systems: Status and evaluation.* New York: Academic.

Kintsch, W. (1974). *The representation of meaning in memory.* Hillsdale, NJ: Erlbaum.

Kling, R. (1980). Social analyses of computing: Theoretical perspectives in recent empirical research. *Computing Surveys, 12*(1), 61–113.

Krouse, J. K. (1980, June 12). CAD/CAM—Bridging the gap from design to production. *Machine Design,* 117–125.

Krouse, J. K. (1981, May 21). Automated drafting: The first step to CAD/CAM. *Machine Design,* 50–55.

Krouse, J. K. (1982). *What every engineer should know about computer-aided design and computer-aided manufacturing.* New York: Marcel Dekker.

Lafue, G. (1978). A theorem for recognizing 2-D representations of 3-D objects. In J. Latombe, Ed., *Artificial Intelligence and Pattern Recognition in Computer Aided Design.* New York: North-Holland.

Lazear, T. (1984, January). The challenge of CAD training. *Computer Graphics World,* 49–50.

Loth, R. W. (1983). Searching for CAD/CAM operations? *CAD/CAM Technology, 2*(3), 19–21.

Lucas, H. C. (1975). *Why information systems fail.* New York: Columbia.

Majchrzak, A., Chang, T-C., Barfield, W., Eberts, R., and Salvendy, G. (1986). *Human aspects of computer-aided design.* London: Taylor Francis.

Majchrzak, A., Collins, P., and Mandeville, D. (1985, November). *A quantitative assessment of change in work activities resulting from CAD.* Paper presented at the meeting of the American Institute of Decision Sciences, Las Vegas, Nevada.

Majchrzak, A., Nieva, V. F., and Newman, P. D. (1985). *Computer-automated technological innovation in three manufacturing industries.* (Final report submitted to NSF). Rockville, MD: Westat.

Mann, F. C. and Williams, L. K. (1960). Observations on the dynamics of a change to electronic data-processing equipment. *Administrative Science Quarterly, 5,* 217–256.

Murch, G. M. (1973). *Visual and auditory perception.* Indianapolis, IN: Bobbs-Merrill.

Neuman, W. M., and Sproul, R. F. (1979). *Principles of interactive computer graphics.* New York: McGraw-Hill.

Newell, A., and Simon, H. A. (1972). *Human problem solving.* Englewood Cliffs, NJ: Prentice-Hall.

Newton, R. D. (1984). *Job satisfaction and somatic complaints among computer aided design drafters.* Ph.D. Dissertation, Claremont Graduate School, Claremont, CA.

Norton, F. J. (1981). Interactive graphics and personnel selection. In J. Mermet, Ed., *CAD in Medium-Sized and Small Industries.* Amsterdam: North-Holland.

Palitto, R. C. (1984). Policies for system implementation and use: The Firestone Tire & Rubber Company case study. In R. M. Dunn and B. Herzog, Eds., *CAD/CAM Management Strategies.* Pennsauken, NJ: Auerbach.

Pao, Y. C. (1984). *Elements of computer-aided design and manufacturing.* New York: Wiley.

Pfeffer, J., and Leblebici, H. (1977). Information technology and organization structure. *Pacific Sociological Review, 20*(2), 241–261.

Price, L. A. (1982). Design of command menus for CAD systems. *IEEE 19th Design Conference* (pp. 453–459). New York: IEEE.

Ramsey, H. R., and Atwood, M.E., (1979). Human factors in computer systems: a review of the literature (Report No. SAI-111-DEN). Englewood, CO: Science Applications.

Rice, R. E., and Associates (1984). *The new media: Communication, research, and technology.* Beverly Hills, CA: Sage.

Rice, R. E., and Case, D. (1983). Electronic message systems in the universe: Description of use and utility. *Journal of Communication, 33*(1), 131–152.

Rock, I. (1973). *Perception and orientation.* New York: Academic Press.

Rock, I., Halper, F., and Clayton, T. (1972). The perception and recognition of complex figures. *Cognitive Psychology, 3,* 655–673.

Rosenbaum, J. D. (1983). A propitious marriage: CAD and manufacturing. *IEEE Spectrum,* 49–52.

Ruch, W. A. (1980). *Productivity measurement.* In Second Annual International Computervision User Conference. Atlanta, GA.

Sanders, D. H. (1968). *Computers in business.* New York: McGraw-Hill.

Schaffitzel, W., and Kersten, U. (1985). Introducing CAD systems: Problems and the role of user-developer communication in their solution. *Behavior and Information Technology, 4*(1), 47–61.

Schneider, W., and Eberts, R. E. (1980). *Automatic processing and the unitization of two features* (Technical Report No. 8008). Champaign, IL: University of Illinois Human Attention Research Laboratory.

Sheldon, D. F. (1983). The present state of the art on computer-aided draughting and design. *IEE Proceedings, 130,* 173–179.

Shepard, R. N., and Metzler, J. (1971). Mental rotation of three-dimensional objects, *Science, 171,* 701–703.

Short, J., Williams, E., and Christie, B. (1976). *The social psychology of telecommunications.* New York: Wiley.

Staehle, W. H. (1982). Technology and organizational change in office work: The case of the VDU. In G. Mensch and R. J. Niehaus, Eds., *Work, Organizations, and Technological Change.* New York: Plenum.

Stahlin, R. (1980). Optimizing computer graphics. *Computer Graphics World,* 33–40.

Staley, S. M., Henderson, M. R., and Anderson, D. C. (1983). Using syntactic pattern recognition to extract feature information from a solid geometric data base. *Computers in Mechanical Engineering,* 61–66.

Stewart, R. (1971). *How computers affect management.* London: Macmillan.

Sutherland, I. E. (1963). Sketchpap: *A man-machine graphical communication system* (Technical Report No. 296). Cambridge, MA: Lincoln Laboratory, Massachusetts Institute of Technology.

Swanson, E. B. (1974). MIS's: Appreciation and involvement. *Management Science, 21,* 178–188.

Tapscott, D. (1982). *Office automation: A user-driven method.* New York: Plenum.

Thomas, J. (1983). *Human factors in computer-aided design* (unpublished report). IBM Corporation.

Tull, K. (1984). Planning an in-house CAD/CAM training program. In R. M. Dunn, and B. Herzog, Eds., *CAD/CAM Management Strategies.* Pennsauken, NJ: Auerbach.

Tullis, T. S. (1981). An evaluation of alphanumeric, graphics, and color information displays. *Human Factors, 23,* 541–550.

van der Heiden, G. H., Brauninger, U., and Grandjean, E. (1984). Ergonomic studies on computer aided design. In E. Grandjean, Ed., *Ergonomics and Health in Modern Offices.* (pp. 119–128). London: Taylor & Francis.

Wagner, P. (1985). New directions in mechanical engineering. *Computer Graphics World, 34.*

Whisler, T. L. (1981). *Information Technology and Organizational Change.* Belmont, CA: Wadsworth.

Wingert, B., Rader, M., and Riehm, U. (1981). Changes in working skills in the fields of design caused by use of computers. In J. Mermet, Ed., *CAD in Medium Sized and Small Industries.* Amsterdam: North-Holland.

Yuille, J. C., and Steiger, J. H. (1982). Nonholistic processing in mental rotation: some suggestive evidence. *Perception and Psychophysics, 31,* 201–209.

Zwaga, H. J. G., and Duijnhouwer, F. (1984). The influence of a fixed background in a VDU display on the efficiency of colour and shape coding. In *Proceedings of the Human Factors Society—28th Annual Meeting* (pp. 331–335). Santa Monica, CA: Human Factors Society.

CHAPTER 12.3

HUMAN ASPECTS OF ROBOTIC SYSTEMS

HANS-JÖRG BULLINGER
VOLKER KORNDÖRFER

Institute for Industrial Engineering
Stuttgart, West Germany

GAVRIEL SALVENDY

School of Industrial Engineering
Purdue University

12.3.1 OVERVIEW

This chapter reviews the technology of industrial robots, lists the allocation of functions between humans and robots, and discusses the impact of the utilization of industrial robots on job and organizational design, supervisory control, qualification requirements, safety, and social issues.

12.3.2 INTRODUCTION

An industrial robot is an automatic machine that is designed and often programmed to perform a variety of material handling, positioning, and processing tasks via tools connected to it. Industrial robots can be characterized and classified according to their physical structure, motion control, and intelligence capabilities.

Physical structure characteristics include the number of limbs and their arrangement, the number of degrees of freedom of motion, physical dimensions and operating space, and load capacity. For instance, there are robot models that can move only back and forth without any vertical or angular motion, as well as robot models that can move in all directions. Recent developments permit robot wrist motions of 360°, whereas earlier models did not have a wrist at all. Certain robot models extend to a height of several meters, while other models are only a few centimeters tall.

Motion control in robots is characterized by the programmability of motion sequence, coordinate systems, velocity, and position accuracy. For instance, while certain robot models are limited to a fixed sequence of motions, other robot models can vary the motion sequence easily. Similar to numerical control machines, some models move from point to point, while others can follow a continuous path.

Intelligence characteristics represent the richest frontier in future robot developments, since they determine the degree of task sophistication that robots will perform. Included here are the methods of teaching and communicating with the robot, sensing and decision making abilities, and learning and adapting abilities. For instance, robots without vision have a limited use for inspection purposes, while a robot with no sensing ability is sufficient to perform a predetermined, repetitive task.

A fundamental fact about robot-oriented job and skills analysis is that robots can be designed with the necessary set of specifications. The designed specifications can be controlled to include selected abilities and exclude others, which are considered unnecessary. If certain abilities are specified but not available based on the prevailing technology, new research may be stimulated to advance robot technology in the required direction. In contrast, human-oriented job and skill analysis has relatively much less control over the operator specification. It may lead to personnel selection and specialized training, but people typically have a basic, given set of limbs, characteristics, and abilities.

The tremendous spurt in the rate of use of robots has necessitated the application of the principles of human factors in order to effect optimal productivity in industrial settings where both humans and robots are used.

Table 12.3.1 lists the world robot population at the end of 1982. Figures are given for each robot-producing country in actual numbers and as a percentage of world robot population. Japan with 51% emerges a clear leader in this category followed by the United States (18%). It has also been forecasted that by 1990 (Table 12.3.2) Japan will employ approximately 130,000 robots; also, other countries will register substantial increases in the use of industrial robots. Thus, aspects of human factors planning for industries using robots becomes imperative.

12.3.3 TASK ALLOCATION: ROLE OF THE HUMAN IN ROBOTIC SYSTEMS

A first step in the allocation of functions between human operators and robots is a detailed examination of the capabilities and shortcomings of both robots and humans. Therefore, the advantages and disadvantages of employing robots and humans in various industrial settings have been examined below.

As illustrated in Table 12.3.3, robots have a wide range of capabilities based on their configurations. The three main functions which robots are capable of performing may be listed as:

1. Initiate, perform, and terminate motions.
2. Store and sequence the exact position data in its memory.
3. Interface with other machines and humans.

Robots can perform many complex tasks, but the complexity of the controller determines the robot's capability. Of course, the more complex the robot, the more expensive it is.

Robots can generally be classified as nonservo, servo-controlled, point-to-point, and continuous.

Table 12.3.1   World Robot Population, End of 1982

	Number	Percentage (%)[b]
Japan	18,000	51%
United States	6,200	18
Western Europe:		
West Germany	2,800	8
Sweden	1,600	5
United Kingdom	800	2
France	700	2
Italy	500	1
Norway	400	1
Other	400	1
Total	7,200 [a]	20
U.S.S.R.	3,000	9
Eastern Europe	600	2
Total world population	35,000	100

Source: Tech Tran (1983)

[a] 7200 is the total robot population in Western Europe.

[b] Percentages reflect percentage of total world robot population.

A nonservo is a pick-and-place machine with a limited capability of only two positions—the start and the end. However, a servo-controlled robot has more applications and can be commanded to stop or operate within its specified limit rather than just at the end, like a nonservo. A point-to-point robot is taught to operate an initial program, which is modified later on during program execution. Finally, a continuous robot is programmed on a time basis rather than on the basis of determined points in space.

Factors considered in determining the advantages and disadvantages of robots (Table 12.3.3) were (1) cost, (2) reliability, (3) speed, (4) flexibility, (5) working range, and (6) memory capacity. The advantages and disadvantages of using a human operator are listed in Table 12.3.4.

In addition to the presented information in Table 12.3.4, the behavioral aspects of human work performance should be considered. These aspects include the psychological processes involved in human performance. Performance may be defined as the product of skill and motivation in the operator, since he or she must have the willingness to perform the task as well as the required skills.

Another major factor that needs to be considered in resolving task allocation issues is that machines and robots perform consistently while there is variation between humans with regard to knowledge, personality, capability, and so forth.

Many factors may account for the variance in performance of a human operator. One major factor is pacing, that is, whether a task is machine-paced and the operator has to keep up with the machine,

Table 12.3.2   Forecasted Growth of Annual Robot Production, Worldwide

	1980			1985			1990	
Units		Value	Units		Value	Units		Value
World 7500–8500		$660 mil	52,000–56,000		$3.4–3.5 bil	130,000–140,000		$9.5–10.0 bil
Japan			31,000		$2150 mil	57,500		$4450 mil
United States			7,700		$ 445 mil	31,300		$2100 mil
West Germany			5,000		$ 360 mil	12,000		$ 950 mil
United Kingdom			3,000		$ 200 mil	21,500[a]		$1420 mil
Sweden			2,300		$ 90 mil	5,000		$ 180 mil
Italy			1,250		$ 75 mil	3,500		$ 225 mil
Norway			1,000		$ 50 mil	2,000		$ 103 mil
France			1,000		$ 50 mil	2,800		$ 150 mil

Source: U.S. Department of Commerce (1983).

[a] May be exaggerated but is based on substantial current government support in the United Kingdom.

Table 12.3.3 Advantages and Disadvantages of Different Types of Robots

Robots	Advantages	Disadvantages
Nonservo	High reliability (all) Programmable (all) Repeatability to 0.01 in. Relatively high speed	Needs safeguard (all) Limited flexibility of movements Can stop at end points
Servo-controlled	Move and stop anywhere in their travel Accelerate or decelerate any axis Large memory capacity Very accurate Smooth motions are possible Can execute more than one program	More expensive than nonservo Less reliable than nonservo
Point-to-point	Provides great flexibility Can carry highload capacity Has a high working range	Only can do "record–playback" jobs Path in which the manipulator moves is not programmed (various parts) Can be tedious to program
Continuous	Smooth continuous motion High storage capacity Speed of manipulator can be varied Small size and light weight	

Source: Kamali, Moodie, and Salvendy (1981)

or is self-paced and the operator works with no specific time constraints. Another factor is the degree to which a job is enlarged and the opportunity provided for psychological growth, for example, by job enrichment; thereby, giving the operators a sense of responsibility and autonomy. If the task is meaningless and the operator cannot exercise judgment, then he or she may feel demotivated. Communication in the workplace also influences the behavior of the employees. The degree to which an operator is isolated or is in a work group is crucial. Obviously, physical and social factors like parking facilities, lighting noise, workplace layout, lunch groups, interest groups, outings, and so on, play a role in the attitude of the operator.

Guidance in the allocation of function between human and machine has traditionally been provided by tables listing their respective strengths and weaknesses. Such human versus machine tables are

Table 12.3.4 Advantages and Disadvantages of Different Human Capabilities and Behavior

Advantages	Disadvantages
Can use judgment	Not reliable
Able to see and feel for correct items	High cost training/wages
Can react to an accident	Leaves for sickness, and so on
Can change activity without reprogramming	Gets fatigued
Can do different things and is not restricted to a specific job	Cannot handle heavy objects
Has control over all machines in case something goes wrong (only within operator's range of capability)	Cannot handle dangerous chemicals, and so on.
Does not need maintenance	Has limited mental and physical capabilities
Typically, gradual rather than sudden degradation in performance	Dissatisfaction with monotonous jobs
Low investment of capital	Typically, gradual rather than sudden degradation in performance
Frequently available	High variability in both quantity and quality of work output
	Not always available

Source: Kamali et al. (1982)

usually variants of the one proposed by Fitts et al. (1951). The essence of Fitts et al.'s table as distilled by Jordan (1963) is that "men are flexible but cannot be depended upon to perform in a consistent manner whereas machines can be depended on to perform consistently but they have no flexibility, whatsoever." Tabulations such as Fitts et al.'s were developed before the advent of robot technology and, therefore, here the more current and relevant table presented by Nof, Knight, and Salvendy (1980), based on a detailed task analysis, is recommended. Table 12.3.5 (Nof 1985) compares the relative abilities of humans and industrial robots in the following categories: (1) action and manipulation, (2) brain and control, (3) energy and utility, (4) interface, and (5) miscellaneous factors.

This table differs from the traditional human–machine table in two principal ways. First, the brain and control system section is presented in far more detail for both robots and humans, since decision making abilities are becoming an increasingly significant aspect of robot performance, and distinctions between robot and human decision making abilities are not always immediately obvious. Second, the robot portion of the charts contains considerably more specific detail than the human portion, since an essential difference between robot- and human-oriented job analysis is that the range of available robot configurations must be considered.

In addition to Table 12.3.5 the factors that may generally influence the allocation of function between humans and robots are:

1. Is the organization centralized or decentralized in its system of authority?
2. Are the assignment of duties static or demand-dependent (changing)?
3. Is the emphasis on division of labor by category or by quantity?
4. What are the jobs and processes where automation is being proposed?
5. How responsive is the labor?
6. How susceptible is the equipment or process to faults?

Specifically, the following "rules of thumb," which are evident from a close scrutiny of Table 12.3.5, are used to guide the allocation of tasks between humans and robots.

1. The greater the similarity between two jobs, the easier it is to transfer either robot or human from one job to the other. For humans, such transfer is almost entirely a question of learning or retraining. For robots, however, as job similarity decreases, robot reconfiguration, as well as reprogramming (retraining), become necessary for economical interjob transfer.

2. Humans possess a set of basic skills and experience accumulated over the years, and therefore may require less detail in their job description. Robots, on the other hand, perform each new task essentially from scratch and require a high degree of detail for every micromotion and microactivity.

3. Robots do not have significant individual differences within a given model. Thus an optimized job method may have more generality with robot operators than with human operators.

4. Robot sensing, manipulative, and decision making abilities can be designed for a given specialized task to a much greater degree than can a human's abilities. Of course, this specialization may entail the cost of decreased transferability from one task to another.

5. Robots are unaffected by social and psychological effects (such as boredom) that often impose constraints upon the engineer attempting to design a job for a human operator.

To sum up, it is concluded that:

1. *A human operator must perform the job* because the task is too complex to be performed by any available robot.

2. *A robot must perform the job* because of safety reasons, space limitation, or special accuracy requirements.

3. *A robot can replace a human operator* on an existing job, and the shift to robot operation could result in improvements such as higher consistency, better quality, and increased productivity. Labor shortages in certain types of jobs may also result in robot assignments.

12.3.4 THE USE OF ROBOTS: EFFECTS ON HUMAN OPERATORS

The use of robots in industry has been recommended as strongly as it has been condemned. This is in lieu of the fact that although robots may free the operator from accident-prone work situations, they also threaten replacing the human operator; thereby, the robots are viewed as potential causes of unemployment.

A direct consequence of the use of robots is the reduction in the work force. However, the more important consequence is the increase in criticality of the functions performed by the operator due to the use of robots (Figure 12.3.1).

Robots have an all-pervading effect on the human operator. All the plausible effects are depicted

Table 12.3.5 Comparison of Robots vs. Human Skills and Characteristics

Characteristics	Robot	Human
a. Comparison of Robot and Human Physical Skills and Characteristics		

1. Manipulation
   A. Body

   a. One of four types:
   Uni- or multiprismatic
   Uni- or multirevolute
   Combined revolute/prismatic
   Mobile

   a. A mobile carrier (feet) combined with 3 DF wristlike (roll, pitch, yaw) capability at waist.

   b. Typical maximum movement and velocity capabilities:
   Right-left traverse
   5–18 m at
   500–1200 mm/sec
   Out-in traverse
   3–15 m
   500–1200 mm/sec

   b. Examples[a] of waist movement:
   Role: $\simeq 180°$
   Pitch: $\simeq 150°$
   Yaw: $\simeq 90°$

   B. Arm

   a. One of four primary types:
   Rectangular
   Cylindrical
   Spherical
   Articulated

   a. Articulated arm comprised of shoulder and elbow revolute joints.

   b. One or more arms, with incremental usefulness per each additional arm.

   b. Two arms, cannot operate independently (at least not totally).

   c. Typical maximum movement and velocity capabilities:
   Out-in traverse
   300–3000 mm
   100–4500 mm/sec
   Right-left traverse
   100–6000 mm
   100–1500 mm/sec
   Up-down traverse
   50–4800 mm
   50–5000 mm/sec
   Right-left rotation
   50–380°[b]
   5–240°/sec
   Up-down rotation
   25–330°
   10–170°/sec

   c. Examples of typical movement and velocity parameters:
   Maximum velocity: 1500 mm/sec in linear movement.
   Average standing lateral reach: 625 mm
   Right-left traverse range: 432–876 mm
   Up-down traverse range: 1016–1828 mm
   Right-left rotation (horizontal arm) range: 165–225°
   Average up-down rotation: 249°

   C. Wrist

   a. One of three types:
   Prismatic
   Revolute
   Combined prismatic/revolute
   Commonly, wrists have 1–3 rotational DF: roll, pitch, yaw; however, an example of right-left and up-down traverse was observed.

   a. Consists of three rotational degrees of freedom: roll, pitch, yaw.

   b. Typical maximum movement and velocity capabilities:
   Roll
   100–575°[c]
   35–600°/sec
   Pitch
   40–360°
   30–320°/sec

   b. Examples of movement capabilities:
   Roll: $\simeq 180°$
   Pitch: $\simeq 180°$
   Yaw: $\simeq 90°$

Table 12.3.5 (*Continued*)

Characteristics	Robot	Human
	Yaw 100–530° 30–300°/sec Right-left traverse (uncommon) 1000 mm 4800 mm/sec Up-down traverse (uncommon) 150 mm 400 mm/sec	

D. End effector	a. The robot is affixed with either a hand or a tool at the end of the wrist. The end effector can be complex enough to be considered a small manipulator in itself.
	b. Can be designed to various dimensions.
		a. Consists of essentially 4 DF in an articulated configuration. Five fingers per arm each have three pitch revolute and one yaw revolute joints.
		b. Typical hand dimensions:
		Length: 163–208 mm
		Breadth: 68–97 mm
		         (at thumb)
		Depth: 20–33 mm
		         (at metacarpal)

2. Body dimensions	a. Main body: Height: 0.10–2.0 m Length (arm): 0.2–2.0 m Width: 0.1–1.5 m Weight: 5–8000 kg b. Floor area required: from none for ceiling-mounted models to several square meters for large models.	a. Main body (typical adult): Height: 1.5–1.9 m Length (arm): 754–947 mm Width: 478–579 mm Weight: 45–100 kg b. Typically about 1 m² working radius:

3. Strength and power	a. 0.1–1000 kg of useful load during operation at normal speed; reduced at above normal speeds.
	b. Power relative to useful load.
		a. Maximum arm load: < 30 kg; varies drastically with type of movement, direction of load, etc.
		b. Power: 2 hp ≈ 10 sec
		         0.5 hp ≈ 120 sec
		         0.2 hp ≈ continuous
		               5 kc/min
		subject to fatigue; may differ between static and dynamic conditions.

4. Consistency	Absolute consistency if no malfunctions.	a. Low b. May improve with practice and redundant knowledge of results. c. Subject to fatigue: physiological and psychological. d. May require external monitoring of performance.
5. Overload/underload performance	a. Constant performance up to a designed limit, and then a drastic failure. b. No underload effects on performance.	a. Performance declines smoothly under a failure. b. Boredom under local effects is significant.
6. Environmental constraints	a. Ambient temperature from −10°C to 60°C. b. Relative humidity up to 90%.	a. Ambient temperature range 15–30°C. b. Humidity effects are weak.

Table 12.3.5 (*Continued*)

Characteristics	Robot	Human
	c. Can be fitted to hostile environments.	c. Sensitive to various noxious stimuli and toxins, altitude, and airflow.

b.   Comparison of Robot and Human Mental and Communicative Skills

Characteristics	Robot	Human
1. Computational capability	a. Fast, e.g., up to 10 Kbits/ sec for a small minicomputer control.	a. Slow—5 bits/sec.
	b. Not affected by meaning and connotation of signals.	b. Affected by meaning and connotation of signals.
	c. No evaluation of quality of information unless provided by program.	c. Evaluates reliability of information.
	d. Error detection depends on program.	d. Good error detection correction at cost of redundancy.
	e. Very good computational and algorithmic capability by computer.	e. Heuristic rather than algorithmic.
	f. Negligible time lag.	f. Time lags increased, 1–3 sec.
	g. Ability to accept information is very high, limited only by the channel rate.	g. Limited ability to accept information (10–20 bits/sec).
	h. Good ability to select and execute responses.	h. Very limited response selection/execution (1/sec); responses may be "grouped" with practice.
	i. No compatability limitations.	i. Subject to various compatibility effects.
	j. If programmable—not difficult to reprogram.	j. Difficult to program.
	k. Random program selection can be provided.	k. Various sequence/transfer effects.
	l. Command repertoire limited by computer compiler or control scheme.	l. Command repertoire limited to experience and training.
2. Memory	a. Memory capability from 20 commands to 2000 commands, and can be extended by secondary memory such as cassettes.	a. No indication of capacity limitations.
	b. Memory partitioning can be used to improve efficiency.	b. Not applicable.
	c. Can forget completely but only on command.	c. Directed forgetting very limited.
	d. "Skills" must be specified in programs.	d. Memory contains basic skills accumulated by experience.
		e. Slow storage access/retrieval.
		f. Very limited working register: $\simeq$ 5 items.
3. Intelligence	a. No judgment ability of unanticipated events.	a. Can use judgment to deal with unpredicted problems.
	b. Decision making limited by computer program.	b. Can anticipate problems.
4. Reasoning	a. Good deductive capability, poor inductive capability.	a. Inductive.
	b. Limited to the programming ability of the human programmer.	b. Not applicable.

Table 12.3.5 (*Continued*)

Characteristics	Robot	Human
5. Signal processing	a. Up to 24 input/output channels, and can be increased, multitasking can be provided.	a. Single channel, can switch between tasks.
	b. Limited by refractory period (recovery from signal interrupt).	b. Refractory period up to 0.3 sec.
6. Brain-muscle combination	a. Combinations of large, medium, and small "muscles" with various size memory, velocity and path control, and computer control can be designed.	a. Fixed arrangement.
7. Training	a. Requires training through teaching and programming by an experienced human.	a. Requires human teacher or materials developed by humans.
	b. Training doesn't have to be individualized.	b. Usually individualized is best.
	c. No need to retrain once the program taught is correct.	c. Retraining often needed owing to forgetting.
	d. Immediate transfer of skills ("zeroing") can be provided.	d. Zeroing usually not possible.
8. Social and psychological needs	a. None.	a. Emotional sensitivity to task structure—simplified/enriched; whole/part.
		b. Social value effects.
9. Sensing	a. Limited range can be optimized over the relevant needs.	a. Very wide range of operation (10^{12} units).
	b. Can be designed to be relatively constant over the designed range.	b. Logarithmic: vision: 1. visual angle threshold—0.7 min 2. brightness threshold—4.1 $\mu\mu l$ 3. response rate for successive stimuli $\simeq$ 0.1 sec audition: 1. threshold—0.002 dynes/m² tactile: 1. Threshold—3 g/mm²
	c. The set of sensed characteristics can be selected. Main senses are vision and tactile (touch).	c. Limited set of characteristics can be sensed.
	d. Signal interference ("noise") may create a problem.	d. Good noise immunity (built-in filters).
	e. Very good absolute judgment can be applied.	e. Very poor absolute judgment (5–10 items).
	f. Comparative judgment limited by program.	f. Very good comparative judgment.
10. Interoperator communication	Very efficient and fast intermachine communication can be provided.	Sensitive to many problems, e.g., misunderstanding.
11. Reaction speed	Ranges from long to negligible delay from receipt of signal to start of movement.	Reaction speed ¼–⅓ sec.

Table 12.3.5 (*Continued*)

Characteristics	Robot	Human
12. Self-diagnosis	Self-diagnosis for adjustment and maintenance can be provided.	Self-diagnosis may know when efficiency is low.
13. Individual differences	Only if designed to be different.	100–150% variation may be expected.

c. Comparison of Robot and Human Energy Considerations

	Robot	Human
1. Power requirements	Power source 220/440 V, 3 phase, 50/60 Hz, 0.5–30 KVA. Limited portability.	Power (energy) source is food.
2. Utilities	Hydraulic pressure: 30–200 kg/cm² Compressed air: 4–6 kg/cm²	Air: Oxygen consumption 2–9 liters/min.
3. Fatigue, downtime, and life expectancy	a. No fatigue during periods between maintenance. b. Preventive maintenance required periodically. c. Expected usefulness of 40,000 hr (about 20 one-shift years). d. No personal requirements.	a. Within power ratings, primarily cognitive fatigue (20% in first 2 hr; logarithmic decline). b. Needs daily rest, vacation. c. Requires work breaks. d. Various personal problems (absenteeism, injuries, health).
4. Energy efficiency	a. Relatively high, e.g. (120–135 kg)/(2.5–30 KVA). b. Relatively constant regardless of workload.	a. Relatively low, 10–25%. b. Improves if work is distributed rather than massed.

[a] Where possible, fifth and ninety-fifth percentile figures from Woodson[1981] are used to present min. and max. values. Otherwise, a general average value is given.
[b] A continuous right-left rotation is available.
[c] A continuous roll movement is available.

Source: Shimon Nof, (1985). "Robot Ergonomics: Optimizing Robot Work," *Handbook of Industrial Robotics.* New York: Wiley. Copyright © 1985. Reprinted with permission.

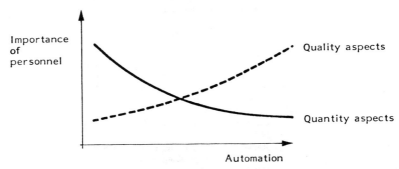

Fig. 12.3.1. Quantitative and qualitative importance of the personnel with increasing automation (schematic).

in Figure 12.3.2. A survey conducted in one-third of the metalworking factories in the Federal Republic of Germany revealed that the greater the use of robots in factories, the greater the trend toward (1) decreasing amounts of physical strain and (2) increasing amounts of psychological stress (Figure 12.3.3).

A survey of studies revealed the following effects of the use of robots on the human work force (Table 12.3.6): (1) automation leads a polarization of qualifications, that is, there is a decrease in the qualification requirements for the operation of equipment, whereas there is an increase in qualification requirements for maintenance and programming personnel; (2) as stated previously, there is an increase in the psychological stress experienced, but a decrease in the amount of physical strain; (3) instead of engaging in direct production activities, operators will be required to engage more in indirect production activities such as monitoring, maintenance, repair, and reprogramming; (4) safety issues; and (5) the work situations available to personnel will be of the following kinds:

1. The work situation will be more intensely determined by the tools than by the subject of work.
2. The personnel will be required to perform direct production supporting activities, for example, material placing or commissioning will be created. These activities will make little qualification demands on the human operator and will be repetitive in nature.
3. The work situation will create simple function-retaining activities (e.g., correction of faults in parts supply) that will require the performance of tasks as vigilance of uniform events, but will make little demand on the qualification requirements of operators.
4. The work situation may require the coupling or decoupling of manual and automated tasks and hence the synchronization of clocktimes.

12.3.5 ORGANIZATION AND JOB-DESIGN ISSUES

The introduction of industrial robots to the workplace changes the requirements for the design of new organizational structures that link computational hierarchy through behavioral hierarchy to organizational hierarchy (Figures 12.3.4 and 12.3.5).

The command and control structure for successful organizations of great complexity is invariably hierarchical, wherein goals, or tasks, selected at the highest level are decomposed into sequences of subtasks that are passed to one or more operational units at the next lower level in the hierarchy. Each of these lower level units decomposes its input command in the context of feedback information obtained from other units at the same or lower levels, or from the external environment, and issues sequences of subtasks to a set of subordinates. This same procedure is repeated at each successive hierarchical level until at the bottom of the hierarchy there is generated a set of sequences of primitive actions that drive individual actuators such as motors, servo values, hydraulic pistons, or individual muscles. This basic scheme can be seen in the organizational hierarchy on the left of Figure 12.3.5.

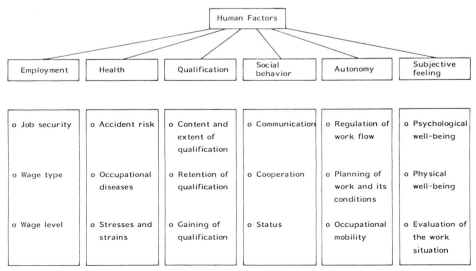

Fig. 12.3.2. Schematic evaluation of factors for humanization.

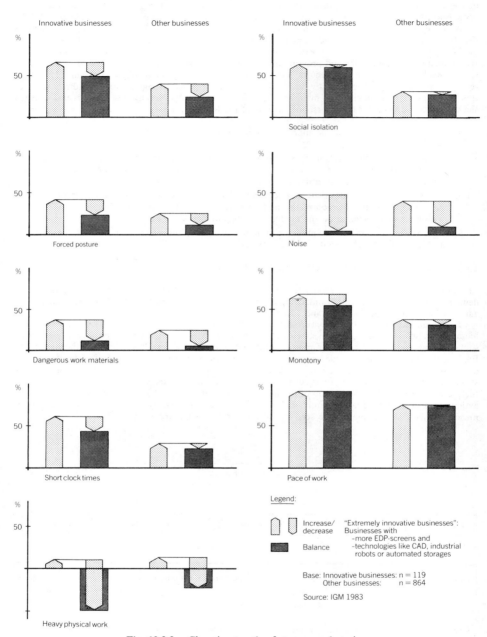

Fig. 12.3.3. Changing trends of stresses and strains.

A single chain of command through the organizational hierarchy on the left is shown as the computational hierarchy in the center of Figure 12.3.5. This computational hierarchy consists of three parallel hierarchies: a task-decomposition hierarchy, a sensory-processing hierarchy, and a world-model hierarchy. The sensory-processing hierarchy consists of a series of computational units, each of which extracts the particular features and information patterns needed by the task-decomposition unit at that level. Feedback from the sensory-processing hierarchy enters each level of the task-decomposition hierarchy.

Table 12.3.6 Summary of the Chances and Risks in the Application of Industrial Robots for those Directly Affected

Consequences Regarding	Manipulation System Application Possibilities	
	Improvement	Deterioration
Accident hazards	Greater distance from danger spot reduces the risk of accident (industrial robot performs dangerous activities).	For repair workers, there may be a greater risk of danger under certain circumstances, if safety devices have to be put out of operation during repair work.
Negative environmental influences	Greater distance from negative environmental influences will reduce their effect (especially in heat, dirt, dust); little change in noise to be expected, since the noise normally propagates (except in the case of an enclosure).	A new source of noise may be created by hydraulic equipment; new source of stress and strain in the absence of shielding.
Physical strain	Reduction of heavy and/or one-sided physical work.	One-sided muscular strain develops in residual activities (e.g., loading).
Psychological stress	It is possible to reduce the dependence on the machine by decoupling.	Danger of increased monotony in residual activities.
	More scope of action and diversion conceivable by job enrichment (machine monitoring, setting-up, programming).	Growing inspection activities requiring increased concentration.
		More intensive working due to increased operating speed of machine.
		Tied more closely to the clock cycle (being in the hands of the machine).
Employment	In a three-shift operation, the third shift could be run automatically (matter of making materials available, magazines, etc.).	Redundancies can be expected (with the associated danger of unemployment).
	Workplace safety by ensuring competitiveness.	The conditions do not normally improve for redundant workers.
		Danger of longer shifts.
Wages	Wage increases for activities of higher qualification (e.g., in job enrichment due industrial-robot-related activities).	Wage reduction to be expected, particularly in the case of analytical job rating.
Job content, autonomy, possibility of personal development, scope of action	The use of short-term production control enhances possibilities of planning and decision-making.	Reduction (particularly marked in tool manipulation).
	Increased scope of action due to job enrichment.	Residual (manual) activities that are difficult to automate will remain.
Qualification	Higher qualification in repair and maintenance staff.	Reduction (particularly in tool manipulation).
	Increase in qualification in production workers as a result of work structuring measures.	Further reduction in training periods and periods of vocational adjustment in workpiece manipulation, if only residual activities remain.
Social behavior	Possibility of improved communication in the case of decoupling.	Reduction of cooperation and communication due to greater isolation.

Source: Arge, HHS (1979)

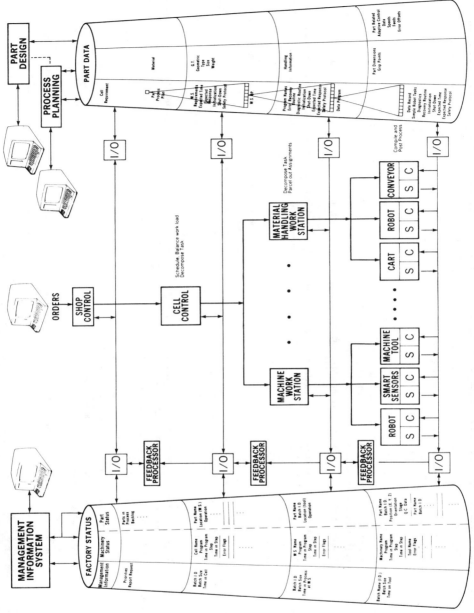

Fig. 12.3.4. A generic system that can be applied to a wide variety of automatic manufacturing facilities and can be extended to much larger applications [reproduced from Albus (1982)].

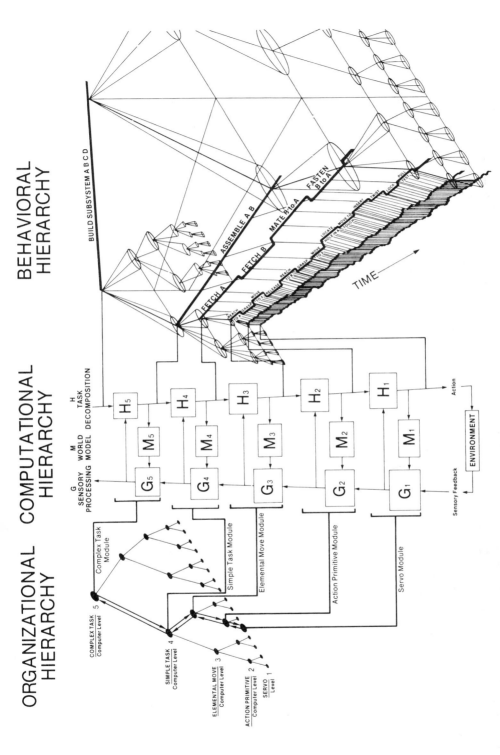

ORGANIZATIONAL HIERARCHY   COMPUTATIONAL HIERARCHY   BEHAVIORAL HIERARCHY

Fig. 12.3.5. The nature and interrelationship of organizational, computational, and behavioral hierarchies [reproduced from Albus (1982)].

This feedback information comes from the same or lower levels of the hierarchy or from the external environment. It is used by the modules in the task-decomposition hierarchy to sequence their outputs and to modify their decomposition function to accomplish the higher-level goal in spite of perturbations and unexpected events in the environment.

The world-model hierarchy consists of a set of knowledge bases that generate expectations against which the sensory-processing modules can compare the observed sensory data stream. Expectations are based on stored information, which is accessed by the task being executed at any particular time. The sensory-processing units can use this information to select the particular processing algorithms that are appropriate to the expected sensory data and can inform the task decomposition units of whatever differences, or errors, exist between the observed and expected data. The task-decomposition unit can then respond, either by altering the action to bring the observed sensory data into correspondence with the expectation or by altering the input to the world model to bring the expectation into correspondence with the observation.

Each computational unit in the task-decomposition, sensory-processing, and world-modeling hierarchies can be represented as a finite-state machine. At each time increment each unit reads its input and, based on its present internal state, computes an output with a very short time delay.

If the output of each unit in the task-decomposition hierarchy is described as a vector, and plotted versus time in a vector space, a behavioral hierarchy such as is shown on the right side in Figure 12.3.5 results. In this illustration a high-level goal or task (BUILD SUBSYSTEM ABCD) is input to the highest level of subtasks, of which (ASSEMBLE AB) is the first. This "complex" subtask command is then sent to the H4 task-decomposition unit. H4 decomposes this "complex" subtask into a sequence of "simple" subtasks (FETCH A), (FETCH B), (MATE B to A), (FASTEN B TO A). The H3 unit subsequently decomposes each of the "simple" subtasks into a string of "elemental moves" of the form (REACH TO A), (GRASP), (MOVE to X), (RELEASE), and so on. The H2 decomposition unit then computes a string of trajectory segments in a coordinate system fixed in the work space, or in the robot gripper, or in the workpiece itself. These trajectory segments may include acceleration, velocity, and deceleration profiles for the robot motion. In H1 each of these trajectory segments is transformed into joint angle movements, and the joint actuators are served to execute the commanded motions.

At each level, the G units select the appropriate feedback information needed by the H modules in the task-decomposition hierarchy. The M units generate predictions, or expected values, of the sensory data based on the stored knowledge about the environment in the context of the task being executed.

The operational effects of this hierarchical control are illustrated in Figure 12.3.5. This shows the information flow from the robots in a computerized flexible manufacturing system. Such organizational structures create a supervisory control in which the production processes and productivity are controlled by the operator by computer terminals.

The computer architecture shown in Figure 12.3.4 is intended as a generic system that can be applied to a wide variety of automatic manufacturing facilities and can be extended to much larger applications. The basic structure is hierarchical, with the computational load distributed evenly over various computational units at various levels of the hierarchy. At the lowest level in the hierarchy are individual robots, N/C machining centers, smart sensors, robot carts, conveyors, and automatic storage systems, each of which may have its own internal hierarchical control system. These machines are organized into workstations under the control of a workstation control unit. Several cell control units may be organized under and receive input commands from a shop control unit, and so on. This hierarchical structure can be extended to as many levels with as many modules per level as are necessary.

On the right-hand side of Figure 12.3.4 is shown a data base that contains the part programs for the machine tools; the part-handling programs for the robots; the materials requirements, dimensions, and tolerances derived from the part design data base; and the algorithms and process plans required for routing, scheduling, tooling, and fixturing. These data are generated by a computer-aided-design (CAD) system and a computer-aided-process-planning (CAPP) system. This data base is hierarchically structured so that the information required at the different hierarchical levels is readily available when needed.

On the left-hand side of Figure 12.3.4 is a second data base, which contains the current status of the factory. Each part in process in the factory has a file in this data base, which contains information as to the part's position and orientation, its stage of completion, the batch of parts that it is with, and quality control information. This data base is also hierarchically structured. At the lowest level, the position of each part is referenced to a particular tray or table top. At the next higher level—the workstation—the position of each part refers to which tray the part is in. At the cell level, position refers to which workstation holds the part. The feedback processors on the left scan each level of the data base and extract the information of interest to the next higher level. A management information system makes it possible to query this data base at any level and determine the status of any part or job in the shop. It can also set or alter priorities on various jobs.

This resulting organizational design raises a number of critical questions such as:

What is the optimal allocation of functions between human supervisory controller and the computer?

What is the relationship between the number of machines controlled by one supervisor and the productivity of the overall system?

What is the optimal number of machines that a supervisor should control?

What is the impact of work isolation of the supervisor in a computer-controlled work environment on the quality of working life and mental health of the operator?

The preceding strategy offered for organizational redesign toward the introduction of industrial robots ignores the human element. An adequate and complete strategy for organization and job redesign should deal with all the factors listed in Table 12.3.7 and incorporate any valid changes in this redesign phase. The final conclusions reached regarding redesign would depend on the individual organization's policy regarding the factors listed in Table 12.3.8.

12.3.5.1 Approaches to Planning Automation

Two main approaches to planning automation with regard to organizational and job redesign have been suggested: (1) equipment-oriented approach and (2) problem-oriented approach. Both approaches have been outlined in Figure 12.3.6. The problem-oriented approach is discussed in detail here. It facilitates the paying of attention to human factors considerations and the making of decisions selectively and consciously instead of implicitly. This approach emphasizes that use of robots should be preceded by organizational and job redesign based on factors that may or may not be observed directly. This proposed work redesign should encompass all aspects of organization, job, and work design, owing to their interrelatedness (Figure 12.3.7). Thus factors like equipment and materials used, scheduling, product planning as well as manual work, worker needs, and safety requirements should be considered. This approach emphasizes the human aspect with regard to the design of the workplace tools and job shop scheduling.

Figure 12.3.8 highlights how the work situation of an operator may influence his or her performance negatively and therefore, particular care should be taken to provide an optimal work and job situation. As automation increases, the jobs performed by a human operator become more critical as well as more difficult to monitor directly. This further necessitates that working conditions be conducive to optimal performance.

The authors assert that the delineation of specific areas of redesign is not possible since the size of organizations and the purpose toward which they are applied differs from organization to organization. Figure 12.3.9 depicts a subset of the range of technical organizational characteristics to which robotic systems have been applied. In Figure 12.3.9: (1) Type a represents the use of an individual device for a relatively simple and well defined task, for example, a welding robot with a rotary table; (2) type b represents the use of several robots working together, for example, several welding robots; (3) type c represents the integrated use of several devices; and (4) type d illustrates a mass application for spot welding, for example, in an automotive industry for spot welding.

The requirements for work design vary considerably from one type to the next, for instance, higher specialization among workers will be needed for type d than for type a.

12.3.6 SAFETY ISSUES

Safety may be effected in an industrial robot system by designing it so that the operators are kept physically away from the robot.

In West Germany, robot manufacturers have spent one-third of the total robot programming time on programming for safety. But, even in these carefully designed situations, accidents and injury to the operator do occur. Hence, it is safest and most effective for the human to exercise supervisory control through computer-based information networks; a subject that is discussed later in this chapter. However, potential injury to the set-up and service personnel continues to exist during maintenance and repair work. Potential accidents may be reduced (though not entirely eliminated) when careful consideration is given to safe job design for the maintenance personnel.

Based on safety studies of industrial robots in Japan (Sugimoto and Kuwaguchi, 1983), the percentage distribution of near accidents caused by industrial robots is illustrated in Table 12.3.9.

From this summary it can be seen that about 28% of near accidents were human related, 61% were equipment related, and 11% were not classified. The 61% that were equipment related owing to the low reliability of robot systems are illustrated in Table 12.3.10.

12.3.7 SUPERVISORY CONTROL OF ROBOTIC SYSTEMS

The concepts, models, and applications of supervisory control are presented in Chapter 9.6. Humans can supervise industrial robots in one of the following two ways:

1. Humans may work adjacent to industrial robots. However, this type of supervision is strongly discouraged owing to the greater chance of accidents and the potentially hazardous psychological

Table 12.3.7   Parameters to be Considered in Organizational and Job Design

Division of labor
  Vertical
  Horizontal
Local assignment of labor
  Single-station/multistation job
  Degree of being tied to the job
Social form
  Individual work
  Room structure (e.g., several isolated workplaces in one room)
  Successive structure (e.g., belt conveyor)
  Integrative structure (e.g., workshop manufacture)
  Group work
Dependencies
  Functional
  Temporal/local
Time structure
  Working hours (longer shifts?)
  Breaks, regulations on taking turns
  Priority of time/tied to time
  Volume of work to be handled (above all for repairmen)
Feedback system
  Regarding quality, production progress, and so on
  Range and speed of the feedback loop
Pay
  Job rating system
  Wage type
  Transitional arrangement in the event of shifts in the requirements of work
Manpower planning
  Redundancy
  Redeployment
  Regrouping
  Qualification
  Promotion possibilities
Deployment of labor
  Required minimum
  Rotation
  Teaming
Management of personnel
  Changed demands on managers
    Compatibility of structure and flow organization in systems using industrial robots with that in conventional areas
Strategy of introduction
  Information policy
  Workers involvement
  Participation of affected workers

Source: Reprinted by permission of the Society of Manufacturing Engineers from the *Proceedings of the 13th International Symposium on Industrial Robotics/Robots 7,* copyright 1983.

Table 12.3.8 Job Content and Work Organization

Remaining tasks	PRODUCTION-SUPPORTING ACTIVITIES				
	Loading/re-moval	Product control	Rework	Supply of auxiliary materials	Waste removal

	PRODUCTION-PREPARING ACTIVITIES		
	Programming new/re-programming/ change	Conversion IR Peripherals	

	PRODUCTION-MAINTAINING ACTIVITIES		
	Supervising with/without possible intervention	Maintenance	Preventive maintenance

	RESTORING ACTIVITIES		
	Fault diagnosis	Fault correction	

Time structure	Duration per occurrence	0 . . . sec . . . min . . . hours . . . days . . . weeks . . . months	
Time volume (gross, including redundancy)	Frequency	0 . . . sec . . . min . . . hours . . . days . . . weeks . . . months	
	Sequence -cyclical	0 . . . sec . . . min . . . hours . . . days . . . weeks . . . months	
	-intermittent	foreseeable	abruptly
	Precedence of reaction	0 . . . sec . . . min . . . hours . . . days . . . weeks . . . months	

Place of action	USER'S FACTORY			MANUFACTURER
	Workshop	Auxiliary facilities	Work planning/ Planning	

Competent depart-ment/ person	USER'S FACTORY						MANUFACTURER	
	Opera-tor	Inspec-tor	Set-up man/ Fore-man	Master	Repair-man	Work-planner/ Planner	Service	Devel-opment

Form of cooperation	Individual work	Room structure	Successive structure	Integrative structure	Group work

Planning and decision making	HIERARCHY LEVEL
	Operator . . . master . . . works manager

Source: Reprinted by permission of the Society of Manufacturing Engineers from the *Proceedings of the 13th International Symposium on Industrial Robotics/Robots 7,* copyright 1983.

and mental health implications. The only partial justification for this work arrangement may possibly exist for low-reliability robots that require a high degree of operator attention.

2. Robots may form a part of a flexible manufacturing system (FMS) (Figure 12.3.10). The entire system would be jointly supervised by a computer and a human. The human supervisor sits in front of a computer terminal and has an impact on certain parts of the total systems functioning.

This second mode of supervisory control is recommended for current and future modes of supervision of industrial robots.

Problem-oriented approach
to planning automation

Equipment-oriented
approach to automation

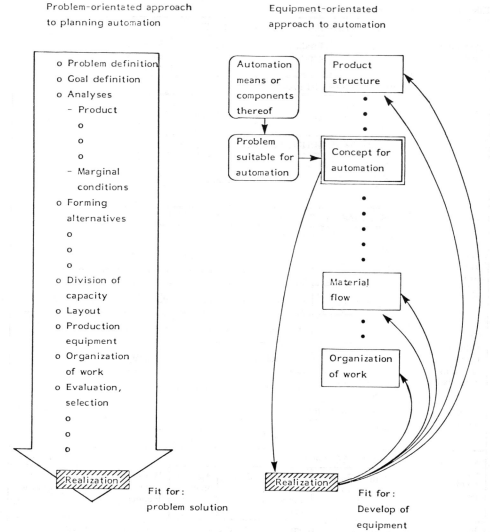

Fig. 12.3.6. Polarized comparison of different approaches to planning a solution to automation [reproduced from Korndörfer (1982)].

12.3.7.1   Concept of Supervisory Control

A valid reason for advocating the role of the human in supervising FMS accrues from reliability data. Figure 12.3.11 compares the reliability of completely automatic systems at different levels of redundancy, against a single-redundant system in which one of the components is a well-trained human.

The human should feel in control of the plant and the computer software at his or her disposal. He or she should have the option to override the computer if he or she feels that it is necessary. This is because the human is able to cope with novel situations more effectively than the computer can. An important point is that the human's role should be coherent. This coherence of the human's role must be ensured in the initial stages of the design process when system tasks are allocated between man and computer (Figure 12.3.12). As Figure 12.3.12 indicates, an inverted "U" relationship exists between arousal level, productivity, and job satisfaction.

If the operator supervising a FMS through the computer is given too little to do, boredom results, which leads to degraded performance and less productivity. On the other hand, if the operator is

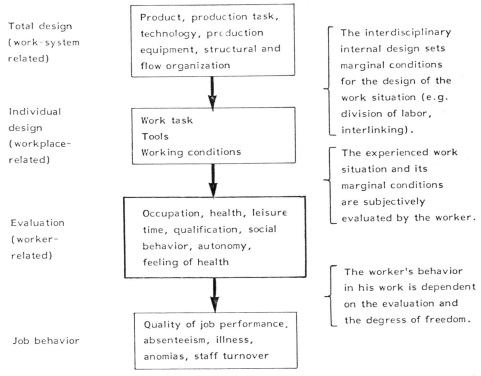

Total design
(work-system
related)

Product, production task,
technology, production
equipment, structural and
flow organization

The interdisciplinary
internal design sets
marginal conditions
for the design of the
work situation (e.g.
division of labor,
interlinking).

Individual
design
(workplace-
related)

Work task
Tools
Working conditions

The experienced work
situation and its
marginal conditions
are subjectively
evaluated by the worker.

Evaluation
(worker-
related)

Occupation, health, leisure
time, qualification, social
behavior, autonomy,
feeling of health

The worker's behavior
in his work is dependent
on the evaluation and
the degress of freedom.

Job behavior

Quality of job performance,
absenteeism, illness,
anomias, staff turnover

Fig. 12.3.7. Schematic representation showing the interrelations of technical planning, perception of the work situation, and job behavior.

given too much to do, mental overload occurs, which also leads to decreased performance and less productivity.

Thus some level between the two extremes in Figure 12.3.12 will result in maximum performance. Salvendy (1978) has hypothesized that enriched jobs result in lower fatigue and psychological stress than simplified jobs. The rationale is tht simplified jobs involve less decision making than enriched jobs.

In assigning functions to humans in a FMS, many decision making responsibilities are associated with an enriched job, whereas low arousal levels and minimum decision making for the operator ensue from a simplified job. Thus, in allocating responsibilties between humans and computers in a FMS, one must be cognizant of levels of arousal for the human as well as the degree of decision making involved in various tasks.

Figure 12.3.13 shows that allocation of functions takes place after tasks that the human performs and tasks that the computer performs are separated. Also, some overlap may be allowed between tasks performed by the human and by the computer to increase reliability and efficiency in the FMS. Tables that show primarily what the human does and what the computer does are useful in assigning tasks to computer or to humans. Such human/computer comparison tables to help decide which tasks go primarily to the computer or primarily to the human in the FMS have been developed by Hwang, Barfield, Chang, and Salvendy, 1984. At the initial stages of allocation, one must assign tasks to the operator that are meaningful.

12.3.7.2 Models of Supervisory Control

The operator in a FMS may shift his or her attention among many machines, rendering to each in turn as much attention as is necessary to service it properly or to keep it under control. The human tends to have more responsibility for multiple and diverse tasks. It is appropriate to view the human as a time-shared computer with various distributions of processing times and a priority structure that allows preemption of tasks. This can be done by using a queueing theory formulation (Chu and Rouse, 1979; Rouse, 1977; Walden and Rouse, 1978).

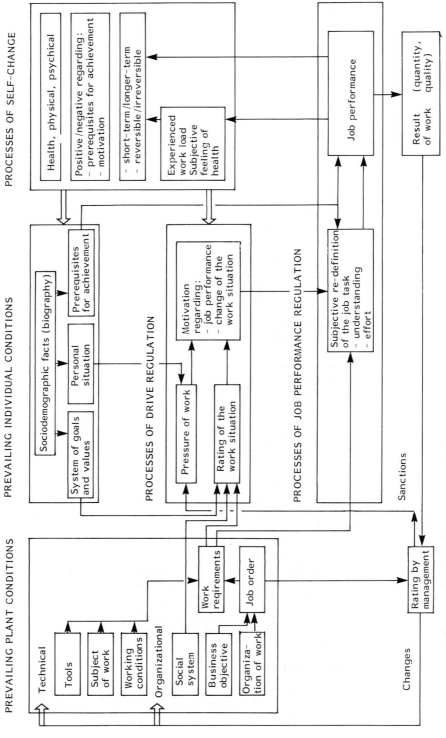

Fig. 12.3.8. Schematic representation of the individual regulation of the work activity, [developed on the ideas of **Hacker** (1980)].

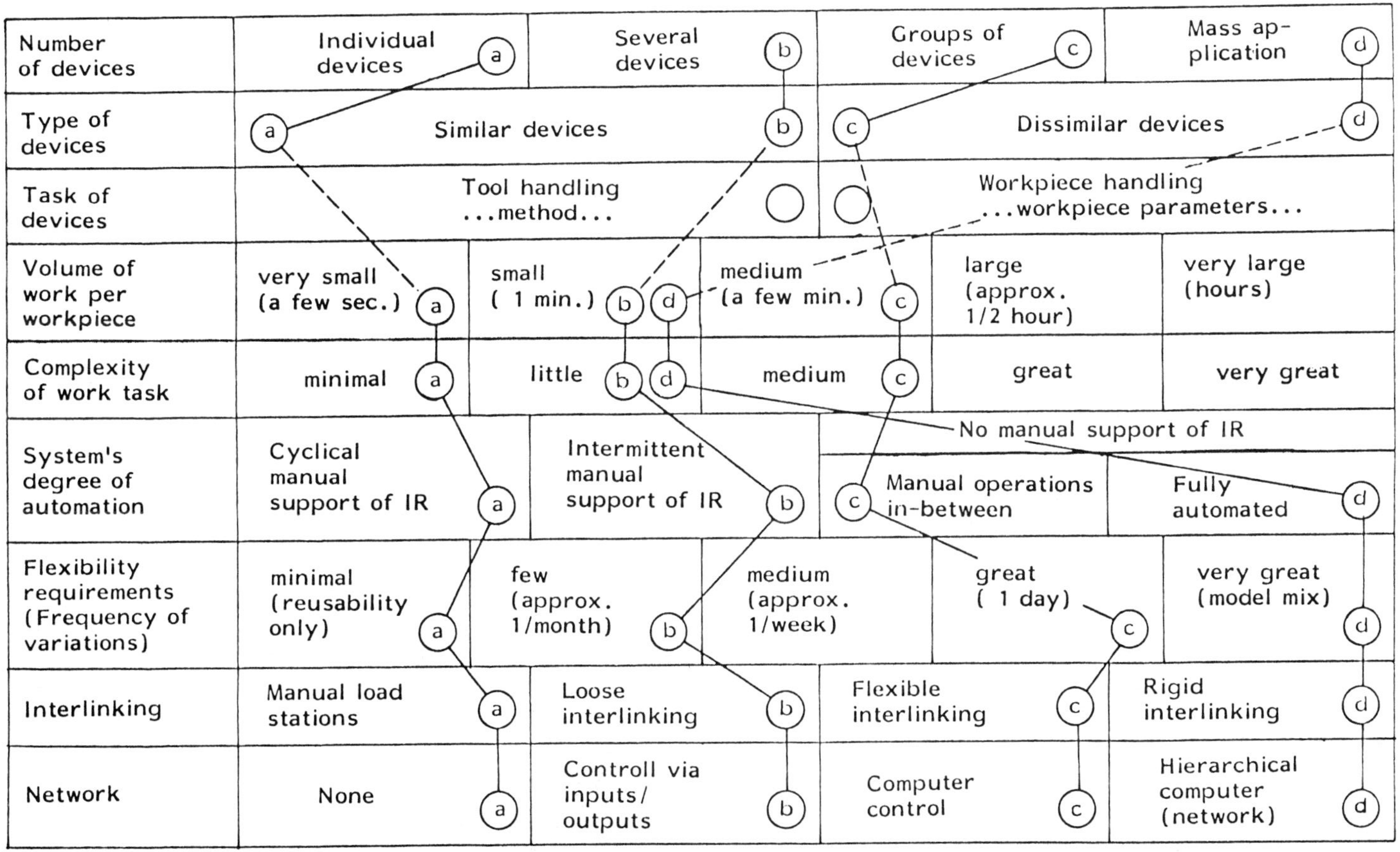

Criterion				
Number of devices	Individual devices (a)	Several devices (b)	Groups of devices (c)	Mass application (d)
Type of devices	Similar devices		Dissimilar devices	
Task of devices	Tool handling ...method...		Workpiece handling ...workpiece parameters...	
Volume of work per workpiece	very small (a few sec.)	small (1 min.) — medium (a few min.)	large (approx. 1/2 hour)	very large (hours)
Complexity of work task	minimal	little — medium	great	very great
System's degree of automation	Cyclical manual support of IR	Intermittent manual support of IR	Manual operations in-between — No manual support of IR	Fully automated
Flexibility requirements (Frequency of variations)	minimal (reusability only)	few (approx. 1/month)	medium (approx. 1/week) — medium (approx. 1 day)	very great (model mix)
Interlinking	Manual load stations	Loose interlinking	Flexible interlinking	Rigid interlinking
Network	None	Controll via inputs/outputs	Computer control	Hierarchical computer (network)

Fig. 12.3.9. Typologies of applications of industrial robots.

Table 12.3.9   Near-Accidents Caused by Industrial Robots

Trouble in Robots (%)		Mean Time between Failure of Robots	
Faults of control system	66.9	Under 100 hr	28.7%
Faults of robot body	23.5	100–250 hr	12.2
Faults of welding gun and tooling parts	18.5	250–500 hr	19.5
Runaway	11.1	500–1000 hr	14.7
Programming and other operational errors	19.9	1000–1500 hr	10.4
Precision deficiency, deterioration	16.1	1500–2000 hr	4.9
Incompatibility of jigs and other tools	45.5	2000–2500 hr	1.2
Other	2.5	Over 2500 hr	8.5

Source: Sugimoto and Kuwaguch: (1983)

Some investigators have developed models of human decision making in multitask situations. Sanders (McCormick and Sanders, 1982) has modeled instrument-monitoring behavior of humans. They assumed that the human used his or her limited input capacity to sequentially observe a number of instruments in a random order. The fraction of time spent observing a particular instrument served as a measure of work load.

Tulga and Sheridan (1980) developed a multitask dynamic design paradigm with such parameters as interarrival rate of task, time before tasks hit the deadline, task duration, the productivity of the human for performing tasks, and task value densities. In this experiment, a number of task-sharing finite completion times and different payoffs appear on a screen. The subject must decide which task to perform at various times to maximize the payoff. When the human performs a variety of tasks with the help of a computer, allocation of responsibility for different tasks is important for optimum performance.

Several investigators (Chu and Rouse, 1979; Rouse, 1977) have studied multitask decision making where the human is required to allocate his or her attention between control tasks and discrete tasks. In a queueing theory formulation, the human "server" serviced various tasks, arriving at exponentially distributed interarrival times. The growing model predicted the mean waiting time for each task as well as the mean fraction of attention devoted to each task.

For optimal performance of different tasks, Govindaraj and Rouse (1981) developed a model with a number of parameters: (1) ratio of weights on control to weights on later error, (2) ratio of nominal weights on control to weights on control over discrete task intervals, and (3) threshold on changes in control. The model could be useful in evaluating displays where the future reference is known for

Table 12.3.10   Reliability of Robots

Cause	Percentage (%)
Erroneous action of robot in normal operation	5.6
Erroneous action of peripheral equipment in normal operation	5.6
Careless approach to robot by human	11.2
Erroneous action of robot in teaching and test operation	16.6
Erroneous action of peripheral equipment during teaching and test operation	16.6
Erroneous action during manual operation	16.6
Erroneous action during checking, regulation, and repair	16.6
Other	11.2

Source: Sugimoto and Kuwaguch: (1983)

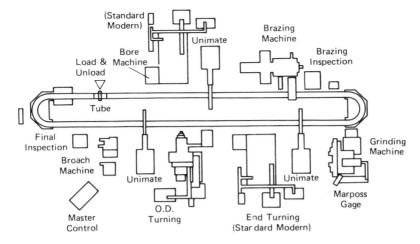

Xerox Robotized Line

Fig. 12.3.10. Examples of robotized flexible conveyor lines where parts must go through each of the machines and where the entire system is controlled by a combined decision support of computer and supervisor who interacts with the computer terminal. Reprinted by permission of the Society of Manufacturing Engineers. Copyright 1982, from the *Journal of Manufacturing Systems*, volume 1/ No. 1, 1982.

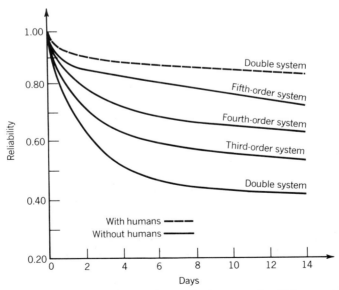

Fig. 12.3.11. The reliability of a double redundant navigation system in which one redundant component is a human (dashed lines) compared with the reliability of systems with various orders of redundancy in which all components are machines (solid lines) From Grodsky (1962). Risk and reliability, *Aerospace Engineering* (pp. 28–33). Copyright American Institute of Aeronautics and Astronautics and reprinted with permission.

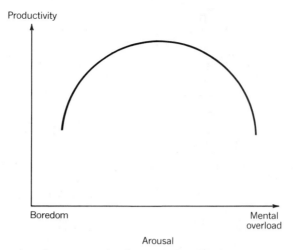

Fig. 12.3.12. When task performance requires low arousal level, it then results in low human attention and increased job productivity. High arousal level results in mental overload and decreased productivity. For each job and each individual an optimal level of arousal exists that results in maximum productivity. Reprinted by permission of the Society of Manufacturing Engineers from the *Proceedings of the 13th International Symposium on Industrial Robotics/Robots 7,* copyright 1983.

a certain distance. If the discrete task characteristics are known, the amount of time required for the discrete tasks and when they should be performed can be determined. An experiment was conducted where subjects controlled an airplane symbol over a map, shown a fixed distance into the future; results revealed that the model compared favorably with experimental data.

12.3.8 QUALIFICATION AND TRAINING ISSUES

The qualifications and training requirements of personnel depends on the kind of robot, the function served by the robot, and the particular production methods employed (Table 12.3.11).

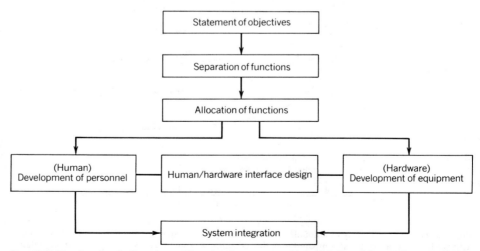

Fig. 12.3.13. Levels of allocation of function in a system From McCormick and Sanders (1982), *Human Factors in Engineering, and Design.* New York: McGraw-Hill. Copyright 1982. Reprinted with permission.

Table 12.3.11   Differentiation of the Requirements Made on the Qualification of Industrial Robots

Equipment-Related Differentiation	Process-Related Differentiation	Organization-Related Differentiation
According to type of programming	In workpiece handling knowledge of peripherals	Organizational allocation of the work tasks
key programming	In tool handling	degree of specialization, division of labor, centralization
key programming using the menu technique	knowledge of the process	
simple text programming		range of shifting tasks from the workshop
According to control type		
PTP		
CP		
According to type of drive		
pneumatic		
hydraulic		
electric		

12.3.8.1   Differentiation by Techniques of Programming

The manner in which a robot is programmed determines to a large extent the qualification requirements of personnel. The main kinds of programming are (1) pure key programming, (2) key programming using the menu technique, and (3) simple text programming.

Pure Key Programming

All commands for the control of the industrial robot are entered by means of keys (or switches). The keys used are directional keys and function keys; in classical key programming, each key corresponded to each function served by the industrial robot. The motion path of the industrial robot across a number of points, the teaching of the points, and their integration in a program are exclusively programmed on a manual operating device (manual programming device). The special features of the qualification requirements in this method are:

1. Theoretical knowledge (additive acquisition structure):
   a. Knowledge of meaning and function of all keys/commands.
   b. Knowledge of the underlying syntax (not all key sequences/combinations are permissible).
   c. Knowledge of the (coded) error messages.
2. Practical knowledge:
   a. Motion of the industrial robot by operating axis-related keys (faculty of 3-D imagination and use of coordinate systems).
   b. Techniques of program preparation without having a synoptic program overview.

This operating technique makes great demands on the operator's memory in memorizing the (often historically developed and not mnemonically designed) keys, their descriptions and functions, and the syntax rules, so that constant exercise will be necessary if safe operation is to be maintained.

A qualification concept will primarily have to try to create supersigns to clarify the structural characteristics of the command language and the operating procedure and to assist in acquiring a lasting routine (and at the same time retaining flexibility).

Key Programming Using the Menu Technique

In this technique, only a few central function keys are permanently retained on the control cabinet or on the programming device; the majority of the commands and functions are assigned: some unoccupied keys are assigned meaning by a changing text appearing on the display. As the operator chooses one function from those displayed to him or her, a set of possible parameters is presented to him or her. Thus prompting is effected. This has the advantage that possible errors can be reduced (by reducing the possible choices) and that the operator can concentrate much more on passive recognition than on active recalling of commands. This permits a higher degree of safe operation, even if the operator has not had much practical training.

The special features of the qualification requirements of this operating technique are:

1. Theoretical knowledge:
   a. Active knowledge of only a few function keys.
   b. Passive knowledge of the majority of functions (usually more than in pure key programming).
   c. Almost no active knowledge of the syntax (prompting).
2. Practical knowledge:
   a. Use of prompting.
   b. Techniques of program preparation without having a synoptic program overview.

This operating technique is easier to learn than pure key programming, since the operator is not required to exactly and actively memorize a large number of keys and functions. Instead the operator's task has been largely reduced to a passive function of the memory with a good menu technique displaying most of the required information.

A qualification concept will also have to make clear the structural features of the language; in this method, however, the emphasis will probably lie on making clear the different language levels (function keys, editor functions, selection of functions, selection of parameters), so that a safe orientation of the state of the system is possible at any time.

If the menu technique and the manual programming device is additionally combined with a control for the motion of the industrial robot by a joystick instead of by keys, some of the complex sensory–motor response requirements will be further dispensed with.

Text Programming

In simple systems using text programming, the points are programmed by means of the manual programming device, which, for example, has motion and/or directional keys, function keys for speed and analog output signals, a display for segment indication, and switches to set the motion type (point-to-point or continuous path). The points input by the manual programming device are integrated in the executive program on the keyboard of the operator's terminal and are indicated on the display. The entry is made in the editor mode; this permits the program to be corrected and supplemented by means of editor functions (inserting, deleting, replacing lines and characters, etc.). It is possible to test the input points in a test run. Points can also be selected directly through a command that is entered on the keyboard of the operator's terminal.

In this technique there is no prompting (operator guidance) through the menu. Nevertheless, the entered program structures can be conveniently read on the display, which means an improvement in clarity.

Safe operation that does not only cover the motion of the industrial robot, but also the preparation of the executive program on the terminal, will only be possible after longer practical training, since in addition to the prescribed programming sequence, a good command of all the formal requirements is required and the formal data must also be entered as text.

The special features of the qualification requirements of this operating technique are:

1. Theoretical knowledge:
   a. Active knowledge of only a few function keys.
   b. Knowledge of the majority of functions and parameters.
   c. Exact, active knowledge of the syntax of the programming language.
2. Practical knowledge:
   a. Techniques of program preparation.
   b. Exhaustive use of the possible combinations of manual programming device/terminal.

Compared to the large number of function keys to be memorized, this operating technique is easier to learn than key programming. The active preparation of the program on the keyboard of the operator's terminal is imperative, and this calls for a detailed knowledge of the syntax as well as of the semantics and the usually coded indications on the terminal's display. Knowledge of the English commands is often a problem for semiskilled and skilled workers. Nevertheless, there are some systems for which off-line program preparation is possible (e.g., in the work planning department).

The advanced systems for text programming based on high-level programming languages (e.g., PASRO, VAL, AML, ROBEX) can be used for additional programming techniques and possibilities, for example, the free formation of variables or dependent jump and branch functions, so that a very differentiated executive program will be implemented.

In such systems, the operator/machine interface can largely be determined by the programmer, as he or she defines by parameters and variables the extent to which and how the operator can or must intervene in the program or production flow.

Since such systems have not yet found wide use in industrial practice at the present, but since they constitute a realistic trend in a certain number of applications, a differentiated analysis of the subject of interfaces from the aspect of the effects and the capability of technical design will be necessary, in the long run.

12.3.8.2 Strategies to Train Personnel

The simplest, but most inflexible and probably the most costly, solution is to have tasks that require additional training to be handled externally, that is, by the industrial robot manufacturer or by an engineer's office.

An alternative would be to hire the right people with the right qualifications, provided the company can afford it.

Any training program first answers questions pertaining to who will attend the program and what are the estimated expenses. The manufacturers usually offer three- to five-day courses for programming; for modern key programming using the menu technique and joystick, this may be adequate as a first step. For text programming, this will certainly not lead to the command of the language and the industrial robot. (The case of a welder who needed a three-month "trial period" after such a course is more or less typical.)

The following trends can be observed currently in the selection of participants in such courses:

1. Highly qualified personnel (engineers) are sent to a manufacturer's course; these engineers will then train the operators in in-house courses. This is suitable for large manufacturers.

2. Qualified skilled workers with a knowledge of hydraulics, electronics, or pneumatics are sent to such courses, above all for maintenance work.

3. "Clever" people of all qualification levels.

It is recommended that such a course should precede the purchase of the industrial robot, so that the future maintenance staff can be actively involved in the installation and, thus, be able to learn.

Practically all manufacturers offer operating, programming, and maintenance courses. However, it needs to be said that the courses are generally not always tailored to the specific requirements of the learner. Throughout they are oriented toward technical factual logic (as the specialist sees things) and not at the learning logic of newcomer (who just does not have the overview and the background of the specialist). This and the de facto separation of theory and practice (". . . now let's do it the way the practitioner does it") have a learning-inhibiting effect.

It is, however, recommended that such a course be bought together with the equipment, but it should not be expected that the operator will then be able to immediately handle all tasks.

12.3.8.3 Content and Methods for Training on Industrial Robots

The content of a course offering training in the operation and programming of industrial robots should include device-related and process-related knowledge and skills. An example is shown in Table 12.3.12. The structure of the general goals of a training program are summarized in Table 12.3.13.

The following broad guidelines are given to direct the structure of training programs:

1. In the presentation of the theoretical subjects, the learner must get an immediate insight into interrelated actions. This does not imply the teaching of everything from the atom to Ohm's law to the working point of the welding current source, but in setting the working point and using Ohm's law as an aid to explain the setting strategy.

2. Dispense with passive teaching techniques in favor of active teaching techniques.

3. Proceed from the concrete to the abstract, when teaching semiskilled and unskilled workers, that is, start with the concrete problem and end with an explanation of the interrelations.

4. Verbal training is very helpful, especially during learning on programmable production equipment, that is, the teacher should explain at every juncture "what, why, and how."

5. The course should be organized in the following steps:
   a. Orientation on the subject matter to be taught.
   b. Setup of a practical problem.
   c. Presentation of the necessary knowledge and skills.
   d. Problem treatment by the teacher.
   e. Final reflection on the interrelations between theory and practice.

12.3.9 SOCIAL ISSUES*

There are at least two major social consequences of widespread use of industrial robots: worker displacement and worker retraining.

---

* This section has been reproduced from Salvendy, G. (1983): "Review and reappraisal of human aspects in planning robotics systems," *Behaviour and Information Technology, 2,* 263–287, and Salvendy, G. (1985): Human aspects in planning and utilization of industrial robots. In S. Y. Nof, Ed. *Handbook of industrial robotics.* New York: Wiley. Reprinted with permission.

Table 12.3.12   Course in the Operation and Programming of Industrial Robots—Content

Practical and theoretical examination	

VII. *Maintenance and repair*
Troubleshooting
Simple maintenance work

VI. *Welding with industrial robots*
Product-related selection of the welding parameters
Ensuring of the product quality (judgment of weld seams; detection of defects; program correction)
Welding on parts for training purposes

V. *Programming II*
Preparation of complete welding programs
Special welding commands (weaving motion, weld seam locating, etc.)
Programming of the electric power source
Selection of peripherals

IV. *Equipment for arc welding with industrial robots*

Characteristics of industrial robots	Instruction in equipment
Programmable electric power sources	
Positioning and clamping devices	Worker safety
Aids, sensors	

⟶ Continuation course, approx. 70 hr

III. *Programming I*
Programming of industrial robots
(key programming, text programming)

II. *Operation of industrial robots*
Motion of axes
Instruction in equipment
Worker safety

I. *Fundamentals*
Fundamentals of EDP
Design of industrial robots
Fundamentals of welding

Basic course, approx. 50 hr

Table 12.3.13   Training Goals

Goal of qualification:
  Autonomy of action
    Avoidance of wrong actions
    Avoidance of misdirected stresses and strains
    Quick reaction and flexibility in the defined spectrum of duties
  Handicraft safety and precision
  Analytical thought
  Sense of responsibility
Cognitive educational goals:
  Functional understanding/functional knowledge of
    Equipment
    Production process
    Business organization
  Precise knowledge of the alternatives of action, their interrelations and consequences
  Knowledge of the methodology of work activities
Affective educational goals:
  Intentional planning of work performance (in graduated details)
  Overcoming of fears/euphorias

12.3.9.1 Worker Displacement

The extent of worker displacement due to automation is difficult to ascertain. Historical data in relation to the application of automation in manufacturing are not particularly reliable indices of future trends. Technological change does not necessarily create jobs or avoid job displacements. Senker (1979) states that there are two phases in major technological revolutions. In the initial phase new technology primarily generates employment. The latter phase, or mature phase, tends to displace labor. Senker asserts that the mature phase has been reached in the "electronics technological revolution." The extension of this, as it applies to industrial robots, is that the low cost and high reliability of microprocessors aid in decreasing robotics costs and concurrently enlarges their range of applicability. The result is the expansion of production without a proportionate increase in employment (Senker, 1979). In the past, increased product demand has caused an increase in workforce demand. The impact of automation has therefore been masked.

To demonstrate a possible net decrease in work as a result of robotization of manual production operations, a simple material-handling operation is presented. If these manual material-handling systems are required to feed three numerical control (NC) machines, a great deal of direct and indirect costs are incurred owing to manual labor. A worker may incur a total first-shift cost of nearly $30.00 per hour. Tote bins typically cost $125–150 (Mangold, 1981). Other expenses may include forklift operators to move pallets of tote bins, and so on.

Supose these NC machines were arranged in a manufacturing cell; owing to the electronics and software capabilities currently available, the same robot may be capable of tending each machine even though each may perform a different operation. It must be realized that some type of materials-feeding system must be designed to suit the robot. It may be possible to eliminate the tote bins, which may consume production space. Forklifts and their operators may be modified or eliminated, and the material-handling personnel may also be displaced. Over an extended time period, humans can work only one shift per day. Since a robot is capable of working more than one shift, it may replace more than one worker (Mangold, 1981). In this particular situation, it can be seen that direct and indirect cost savings may be quite significant and that robots can contribute to disproportionate displacement per job.

This obviously results in fewer hours of work, fewer jobs, and lesser job security for unskilled and semiskilled workers. This has spurred organizations such as the United Auto Workers (UAW) to develop positions on integrated automation, chiefly concerning industrial robots. Precarious as it is to permit one specific organization to speak for all production employees, the UAW does encompass a large proportion of employees in an industry that utilizes the greatest number of industrial robots. The UAW does not place a specific emphasis on robots, but instead includes them as another technological advancement that it must consider (Weekley, 1979). The union also recognizes that enhanced productivity is necessary for long-term economic viability (Mangold, 1981; Weekley, 1979). It is, however, aware of possible detrimental impacts upon its membership primarily due to job insecurity. The union believes that technological advancement is acceptable and is encouraged as long as the current workforce retains job security (Weekley, 1979). The UAW is well aware of the Japanese workers "life-time employment" status (Weekley, 1979), and it's (the UAW's) response to job security and robotics follows along these lines:

Management provides advance notice of new technology to enable discussion.

Introduction of new technology should displace as few workers as possible by using normal workforce attrition.

When increases in productivity outpace attrition rate, the protection of workers against displacement is an appropriate first charge against productivity.

Bargaining-unit integrity must be maintained; bargaining-unit work must not be transferred to out-of-unit employees.

In-unit employees must be given adequate training to perform jobs introduced by new technology.

Work time must be reduced to afford adequate job opportunities to all who want to work (Mangold, 1981; Weekley, 1979).

Douglass Frasier asserts that to achieve these objectives we need to reduce the number of work hours per job (Mangold, 1981). The UAW asserts that a work-time decrease is an alternative to bargaining for higher wages (Mangold, 1981; Sugarman, 1980; Weekley, 1979). Furthermore, the union is adamant that there must explicitly be pay for lost work hours (Mangold, 1981). The extent to which these issues affect the entire workforce is impossible to determine.

Quality of work life (QWL) is also affected by the introduction of robotics. In reference to job security, it is possible that robots may ultimately improve job security. Robot adaptability enables the robot to be assigned and tooled to many production tasks; the degree to which those tasks are similar to tasks performed by humans increases the likelihood that a human and robot are interchangeable in task performance. In the case of consumer items where market fluctuations may be quite drastic, robots can offer a distinct advantage in production assemblies (Sugarman, 1980). In QWL terms, a company may initiate "robot layoffs" due to downward fluctuation in demand and temporarily assign humans to the assembly line, thus minimizing the displacement effects of a market downturn.

It is often argued that technological change (i.e., industrial robots) will create jobs. There is a wide range of estimates on the extent to which displaced jobs will be compensated by newly developed jobs due to increases in complex robotic utilization (Allbus, 1982). This may well be the case, but it is naive and myopic to compare only quantities. To discuss adequately job displacement and job creation as one counteracting the other, the type of job created must be compared to the type of job eliminated. The literature evidences a discrepancy in skill level between those jobs that will become available and those that will be eliminated. There is and probably will be a significant demand for highly trained personnel in computer programming, mechanical engineering, electronic design, and so forth, all highly skilled positions to implement, utilize, and/or maintain the industrial robot (Allbus, 1982). In all probability, those to be displaced will be workers in unskilled and semiskilled jobs. A large void in skill level exists between those jobs eliminated and those jobs created. Companies in the United States who currently utilize robots do not appear to attempt any major effort at retraining displaced employees (Aron, 1982). The United Auto Workers observes that benefits derived from automation (robotics) are applied to a smaller hourly workforce (Weekley, 1979).

12.3.9.2 Worker Retraining

One of the most acute problems associated with introducing and utilizing flexible manufacturing and industrial robots is that the skill requirements in these new technologies do not capitalize and build on the skills, perception, and knowledge accumulated by the industrial worker (Rosenbrock, 1982). This implies that acquired industrial skills, which were widely utilized in the premicroelectronics-automation era (Salvendy and Seymour, 1973), are completely lost and have become redundant for the industrial robot revolution era.

This has two major implications. First, workers who can be retrained for the new skills must be assessed. This can be achieved by analyzing skills and knowledge requirements for robotics jobs. From this analysis, either work samples or tests that simulate the job can be developed. After assessing the reliability and validity of these tests, the samples can be administered to displaced workers to assess the likelihood of their success in mastering new skills (Borman and Peterson, 1982). Based on this evaluation and on the nature of human abilities, it may be estimated that more than one-half of these displaced workers will not possess employable abilities for the new robot-oriented and computer-based manufacturing work environment. If we do not provide careful manpower planning, we may end up with more than 20 million unemployed Americans by the year 2000, an intolerable social and economic situation. To eliminate or reduce this situation, the industrial robotics systems must be designed, developed, and operated so as to capitalize on (as far as possible) acquired and used human skills.

12.3.9.3 Major Issues to be Considered for the Effective Integration of Human Factors Principles in Planning Robotic Systems

According to the U.S. Congress Office of Technology Assessment (1982), there are a number of institutional and organizational barriers to the use of information technology that also have a bearing on the use of industrial robots. These include high initial cost, the lack of high-quality programming, and the dearth of local personnel with adequate training. In this connection Rosenbrock (1982) addresses two vital behavioral and social issues, namely, (1) the skills that robots call for will usually be new, yet there is no reason why they should not be based on older skills and developed from them and (2) industrial robots will facilitate the process of dividing jobs into pieces and enable the use of machines to perform some of these pieces, leaving others to be performed by humans. This has broad implications for the training and retraining of personnel and for the design of the psychological contents of jobs. Many of these implications associated with the introduction of industrial robots have been managed effectively in Japan (Hasegawa, 1982). In this regard, seven positive and seven negative aspects of the social impact of robots are summarized in Table 12.3.14.

The most significant study yet conducted on the impact of robotics on supervision, management, and organization was published by the Fuji Corporation (1983) both in Japanese and English. This 300-page report summarizes a large-scale study undertaken by the Japan Management Association. In this study, more than 20 "robotized" Japanese companies were selected to be used as case studies to determine just what kind of impact the introduction of robots to the workplace has had on management. For instance, what kind of problems did these companies encounter during the introduction process and how did they go about overcoming them? What have they experienced following the robotization of their respective manufacturing processes? How have they coped with day-to-day worker-related problems? How has robotization of their various operations affected them overall? What type of utilization is envisioned for the future? These and other key points were taken up and discussed in detail in this study on the effects of the use of robots. The various case studies and other data

Table 12.3.14 Social Impacts of Robot Diffusion

Positive Impacts	Negative Impacts
1. Promotion of worker's welfare	1. Unemployment problems
2. Improvement of productivity	2. Elimination of pride in old skills
3. Increase in safety of workers	3. Shortage of engineers and newly trained skilled workers
4. Release of workers from time restrictions	4. Production capacity nonproportional to the size of the labor force
5. Ease in maintaining quality standards	5. Decrease in flow of labor force from underdeveloped to developed countries
6. Ease of production scheduling	6. Safety and psychological problems of robot interaction with human
7. Creation of new high-level jobs	7. Great movement of labor population from the second to the third sector of industry

Source: Hasegawa (1982)

contained in this report should prove an effective tool for any manufacturer considering the robotization of its operations. Hence the study objectives, methods and results are presented below.

Study Objectives

The 1980s are expected to witness more widespread and advanced use of industrial robots; thereby, the impact this technology and its extensive applications will have is certain to be far-reaching. Accelerating this trend toward industrial robotization are the following several factors:

1. Improvement of productivity.
2. Prevention of labor accidents and occupational hazards.
3. Conservation of materials and energy.
4. Improvement of production control.
5. Improvement of working environment.
6. Coping with the shortage of skilled labor.

These are some of the advantages of introducing robots into the manufacturing process. At the same time, however, a company that intends to robotize its manufacturing process cannot avoid coming to grips with some very serious problems in terms of management, such as labor–management relationships, worker displacement, surplus manpower, retraining, optimum investment levels, and assessment of the effects of robitization.

This study is an in-depth analysis and assessment of the technology, applications, labor–management issues, and demand trends of robotics. As such, it provides significant insight into the issues surrounding technological innovations and management, thus proving an effective tool for corporate managers, labor union leaders, government-related agencies, and research institutes worldwide that are seriously considering introducing robot technology.

Study Method

Interviews were conducted with top-level managers from more than 20 companies at varying stages of robotization from a variety of industries. This interview process was supported by extensive and in-depth independent research to generate a complete and comprehensive picture of the present and future aspects of robot utilization in Japan, with special emphasis on management. The companies interviewed were chosen from among the general machinery, transportation machinery, electric equipment, and precision instrument industries, with equal numbers selected from among different corporate sizes (large, medium, and small).

Study Results

The study summarizes the social and economic factors contributing to the spread of robots (Table 12.3.15) and the results derived from robot utilization. A close look at the situation in Japanese industry shows that industrial robots are being utilized primarily because (1) there is a shortage of skilled labor at worksites that have been called "hazardous or otherwise undesirable working environments"—

Table 12.3.15   Social and Economic Factors Contributing to the Spread of Robots

Source: Fuji Corporation (1983)

the few skilled workers that are present are mostly older workers; (2) competition between companies is intensifying as productivity increases and quality improves, market needs have become more advanced; and (3) regulations for the prevention of labor accidents are being strengthened as per Labor Ministry requirements. The development of industrial robots in Japan is keeping pace with the needs of the manufacturing industry.

Robot utilization is fulfilling the previously listed requirements and can be said to be achieving (1) increased productivity and improved adaptability vis-à-vis the product changes that occur in multi-product, small-batch production; (2) improved and consistent quality; (3) more efficient equipment investments (reduced time until returns on investments are realized); (4) worker protection and prevention of labor accidents; (5) labor savings; (6) more efficient production planning resulting in a more stable quality implies that the amount of stock on hand can be reduced, thus lowering stock costs and making product planning easier; and (7) elimination of problems related to skilled labor shortages.

There are also quite a few companies that have attained results not anticipated prior to the introduction of industrial robots. For instance, (1) improved worker attitudes, for example, workers, stimulated by the new technology, have been motivated to take a more active part in the improvement of work processes, submitting suggestions concerning how best to utilize that technology; (2) those work processes

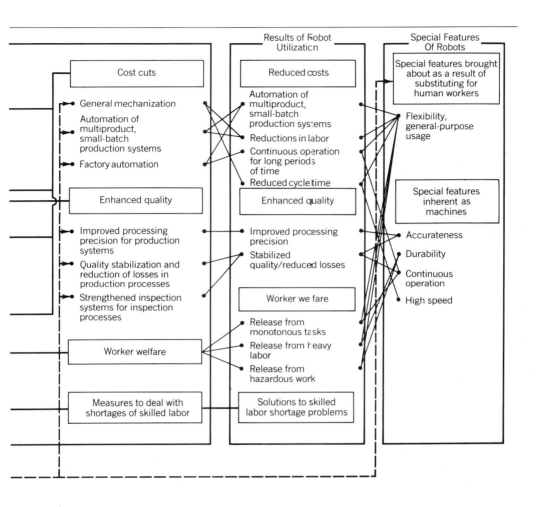

that are preformed before and after the processes using industrial robots have resulted from improved peripheral equipment and overall production technology introduced to keep pace with the robots; (3) the realization of total production systems, that is, the installation of robots made it possible to encode the job knowhow of skilled laborers, thus approaching a total production system for the entire assembly lines; (4) the stabilization of production output, that is, output may be stabilized regardless of the number of workers who show up for work each day or their degree of skill; and (5) more reliable technological capabilities have led to increased product orders.

The strategic implications of industrial robots for those enterprises utilizing them in their manufacturing processes can, in broad terms, be summarized as follows:

1. By enhancing productivity and improving product quality, companies are capable of increasing their market shares and thus improving their market positions.
2. Companies utilizing industrial robots find it possible to open new lines of business (including the development of new products) and enter into new markets.
3. Industrial robots ensure companies of a stable labor force and provide labor itself with improved benefits.

Based on results such as these, it seems safe to conclude that robot utilization not only provides the user company with a competitive edge in the market, but also plays a major role in stabilizing its labor situation.

Nevertheless, in order to effect optimal performance, the managers of companies that use robots have a number of areas to which they must give proper consideration. Principal among these are:

1. The reeducation and retraining of employees in line with the introduction of robots.
2. The carrying out of robot engineering for the purpose of installing and operating industrial robots.
3. The implementation of measures aimed at improving the worksite and ensuring worker safety.

There still exist a number of jobs being performed by humans that are classified as dangerous or heavy labor both within and outside the manufacturing industry. The installation and utilization of industrial robots to perform these kinds of jobs is seen as a possibility in the near future.

12.3.10 SUMMARY

The scientific discussion and the experience gained in companies in the application of robots clearly show that the use of industrial robots increases the qualitative importance of labor. Therefore, a careful weighing of the technical system design and the organization of work with regard to human factors (stresses and strains, qualification) will become more and more important in the application of industrial robots. Although there can be no general rules due to the wide variety of conceivable applications of industrial robots, it can be expected that the pressure for flexibility and the high degree of availability of the devices will make Taylor-type simplified work design solutions less valid and important. The adoption of a particular robotic technology in a factory will determine the qualification requirements of the workforce. It is particularly important to realize the close interlinking between the theoretical knowledge and the functional understanding of the applications of industrial robots.

REFERENCES

Albus, J. (1982). Industrial robot technology and productivity improvement. In Exploration Workshop on the Social Impact of Robotics (No. 90-240 0-82-2). Washington, DC: Office of Technology Assessment, U.S. Congress.

Arge, HHS (1979). Arbeitsgemeinschaft Handhabungssysteme: Forschungs- und Entwicklungsvorhaben HHS, Neue Handhabungssysteme, Erschliessung neuer Anwendungsgebiete und Ausbau der Technologie, Jahresbericht 1979, Teil 4, S. 92 ff. (unveröffentlicht).

Aron, P. (1982). Robot revisited: one year later. Exploratory workshop on the social impact of robotics (U.S. Congress Number 90-240 0-82-2). Washington, DC: Office of Technology Assessment.

Borman, W. C., and Peterson, N. G. (1982). Selecting and training of personnel. In G. Salvendy, Ed. *Handbook of industrial engineering* (Chap. 5.2). New York: Wiley.

Chu, Y. Y., and Rouse, W. B. (1979). Adaptive allocation of decision making responsibility between human and computer in multitask situations. *IEEE Transactions on Systems, Man, and Cybernetics, SMC-9*(12), 769–778.

Congress of the United States, Office of Technology Assessment (OTA). (1983). Automation and the workplace—selected labor, education and training issues. Washington DC: OTA.

DuPont-Gateland, C. (1982). A survey of flexible manufacturing systems. *Journal of Manufacturing Systems, 1*(1), 1–15.

Fuji Corporation. (1983). Robotics and the manager. Business Building 5-29-7 Jungu-mae, Shibuya-ku, Tokyo 150 Japan.

Fitts, P. M., et al. (1951). *Human engineering for an effective air navigation and traffic control system.* Washington: National Research Council.

Govindaraj, T., and Rouse, W. B. (1981). Modeling the human controller in environments that include continuous and discrete tasks. *IEEE Transactions on Systems, Man, and Cybernetics, SMC-11* (6), 410–417.

Grodsky, M. A. (1962). Risk and reliability. *Aerospace Engineering,* January, 28–33.

Hacker, W. (1980). Allgemeine Arbeits- und Ingenieurpsychologie. Psychische Struktur der Regulation von Arbeitstätigkeiten. Berlin: VEB Verlag der Wissenschaften.

Hasegawa, Y. (1982). How robots have been introduced into the Japanese society. Presented at the *Microelectronics international symposium.* Osaka, Japan.

Hwang, S. L., Barfield, W., Jr., Chang, T. C., and Salvendy, G. (1984). Integration of human and computers in the operation and control of flexible manufacturing systems. *International Journal of Production Research, 22,* (5), 841–856.

Jordan, J. (1963). Allocation of functions between man and machines in automated systems. *Journal of Applied Psychology, 47*(3), 161.

Kamali, J. K., Moodie, C. L., and Salvendy, G. (1983). A framework for integrated assembly systems: humans, automation and robots. *International Journal of Production Research, 20*(4), 431–448.

Mangold, V. (1980). The industrial robot as transfer device. *Robotics Age,* 20–26.

McCormick, E. J., and Sanders, M. S. (1982). *Human factors in engineering and design.* New York: McGraw-Hill.

Nof, S., Knight, J., and Salvendy, G. (1980). Effective utilization of industrial robots: a job and skills analysis approach. *AIIE Transactions, 12*(3), 216.

Office of Technology Report. (1982). Information technology and its impact on American education. (GPO Stock No. 052-003-00888-2). Washington, DC.

Qualifizierung an Industrie-Robotern (QIR)-Teilvorhaben: Qualifizierung beim Einsatz von Schweiss-Robotern; Gefördert vom Bundesminister für Forschung and Technologie (BMFT) (FKZ. 01VC 163 A 5) (Projektleiter: V. Korndörfer).

Rouse, W. B. (1977). Human-computer interaction in multitask situations. *IEEE Transactions on Systems, Man, and Cybernetics, SMC-7*(5), 384–392.

Rosenbrock, H. H. (1982). Robots and people. *Measurement and Control, 15,* 105–112.

Salvendy, G. (1978). An industrial dilemma: simplified versus enlarged jobs. In R. Murumatsu and N. A. Dudley, Eds. *Production and industrial systems.* London: Taylor and Francis.

Salvendy, G. (1983). Review and reappraisal of human aspects in planning robotic systems. *Behavior and Information Technology, 2*(2), 263–287.

Salvendy, G., and Seymour, W. D. (1973). *Prediction and development of industrial work performance* (pp. 105–125). New York: Wiley.

Senker, P. (1979). Social implications of automation. *The industrial robot* (Vol. 6, No. 2, pp. 59–61). Oxford, England: Lolswold Press.

Sugarman, R. (1980). The blue collar robot. *IEEE Spectrum, 17*(9), 52–57.

Sugimoto, N. and Kawaguchi, K. (1983). *Proceedings 13th international symposium on industrial robots.* Chicago, IL.

Tech Tran. (1983). Industrial robots—a summary and a forecast (Tech Rept.). Naperville, IL: Tech Tran.

Tulga, M. K., and Sheridan, T. B. (1980). Dynamic decisions and work load in multitask supervisory control. *IEEE Transactions on Systems, Man, and Cybernetics, SMC-10*(5), 217–232.

U.S. Department of Commerce. (1983). The robotics industry (Technical Report). Washington, DC: U.S. Department of Commerce.

Walden, R. S., and Rouse, W. B. (1978). A queueing model of pilot decision making in a multitask flight management task. *IEEE Transactions on Systems, Man, and Cybernetics, SMC-8*(12), 867–875.

Weekley, T. L. (1979). A view of the United Automobile Aerospace and Agricultural Implement Workers of America (UAW) stand on industrial robots (SME Technical Paper MS79-776). Dearborn, MI, SME.

Woodson, W. E. (1981). *Human factors design handbook.* New York: McGraw-Hill.

BIBLIOGRAPHY

Ayres, R. V. and Miller, S. M. (1983). *Robotics applications and social implications,* Cambridge, MA: MIT.

Nof, S. Y., Ed. (1985). *Handbook of industrial robotics.* New York: Wiley.

Office Technology Assessment, Exploratory Workshop on the Social Impact of Robotics, U.S. Superintendent of Document Catalogues, No. 90-240-0-82-2, Washington, D.C., 1982.

Salvendy, G., Ed. *Handbook of industrial engineering.* New York: Wiley.

Salvendy, G., Ed. *Human-computer interaction.* Amsterdam, Netherlands: Elsevier.

Warnecke, H. J., and Schraft, R. D. (1982). *Industrial robots: application experience.* Bedford, England: I.F..S. Publication Ltd.

TECHNICAL AND HUMAN ASPECTS OF COMPUTER-AIDED MANUFACTURING

JOSEPH SHARIT

Department of Industrial Engineering
State University of New York at Buffalo
Buffalo, New York

TIEN-CHIEN CHANG
GAVRIEL SALVENDY

School of Industrial Engineering
Purdue University
West Lafayette, Indiana

Computer-aided manufacturing (CAM) is a sociotechnical system that, for effective system design, requires both a knowledge of the technical and the human component of the system. Thus in order to design the human side of the system, we first need to have an understanding of the technical aspects within which the human works. This chapter briefly presents the technologies of CAM and discusses the effective role of humans in the design and operation of CAM systems. The importance of effective linkage between computer-aided design (CAD) and CAM is emphasized. The technologies of numerical control machines, which are the core of CAM systems, are reviewed as are the technologies of flexible manufacturing systems (FMS). The allocation of functions between humans and computers in CAM systems is given consideration, and job design methodologies and effective supervisory control systems are discussed. The role of decision aids in CAM supervisory control are evaluated. Reference is made to the impact of CAM on job boredom, job stress, changes in skill levels, and operator training.

12.4.1. TECHNOLOGY OF COMPUTER-AIDED MANUFACTURING

12.4.1.1 Introduction

CAM is broadly defined as "the effective utilization of computers in the management, control, and operation of the manufacturing facility" and has, in recent years, become the most important technology in the manufacturing industry. Although not implemented by every shop, a growing number of manufacturing facilities are beginning to adopt this new technology, and CAM has become accepted by the manufacturing industry as necessary for surviving the formidable competition expected in the 21st century.

The basic element of CAM is numerical control (NC) machines. At the time of this writing there are about 100,000 NC machines in operation in the United States alone. Another important development in CAM was the flexible manufacturing system (FMS), which has grown worldwide from one installation in 1971 (completion date) to more than 200 in 1984. In 1981, the United States had approximately 25 FMSs, and several more systems have been built since (Kochan, 1984).

CAM encompasses both direct and indirect applications of computers in manufacturing. Direct applications include NC, computer numerical control (CNC), direct numerical control (DNC), robotics, automated guided vehicles, conveyor control, process control, and system monitoring. Indirect applications include NC part programming, process planning, scheduling, inventory control, production control, and factory management.

In a conventional manufacturing system, machines are operated by human operators. Workpieces, tools, fixtures, and so on, are transported between machines by human or human-operated devices such as carts or forklifts. The first step in automation utilizes mechanical control devices such as cams, timers, and counters. Since hardware devices are used to control the operation sequence, this approach is referred to as hard automation. Although such machines are relatively inflexible, when large quantities of identical parts are produced, hard automation has proven very economical, especially when several of these types of machines are integrated into systems known as transfer lines.

In order to allow flexibility in manufacturing automation, programmable control devices need to be integrated. NC machine tools, robots, automated guided vehicles, and so on use computer-based controllers to enable operation sequences to be changed easily. A manufacturing cell can be thought of as a group of machines that can produce one or several families of parts. (A family is defined as parts that possess similar production features.) A conventional manufacturing system requires human operators; however, with CAM, a cell can run without a human operator. Figure 12.4.1 depicts a small manufacturing cell for turning parts, with a robot tending two turning centers. An expanded computer-controlled manufacturing cell is referred to as an FMS, and can produce up to several hundred different parts.

CAM allows human labor to be reduced or eliminated, although this is not the sole justification for its implementation. Productivity and quality improvement, shorter production lead time, and elimination of routine and hazardous jobs are other benefits resulting from the utilization of CAM.

Design and manufacturing are intimately related activities; whatever is manufactured had to be designed. Full industrial automation can be achieved only after design and manufacturing are integrated through CAD/CAM. Figure 12.4.2 shows a computer-integrated manufacturing (CIM) concept. Centering around the CIM kernel are four major activities: engineering design, factory automation, manufacturing planning, and manufacturing control. Subtasks for each activity are also defined. In this chapter we will discuss only the CAM functions; CAD is discussed in the previous chapter.

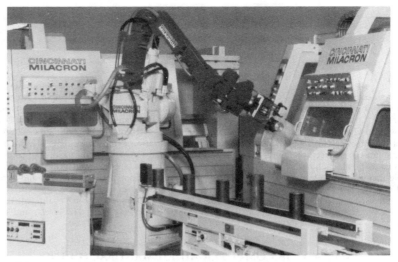

Fig. 12.4.1. A manufacturing cell. (Courtesy Cincinnati Milacron, Cincinnati, Ohio 45209.)

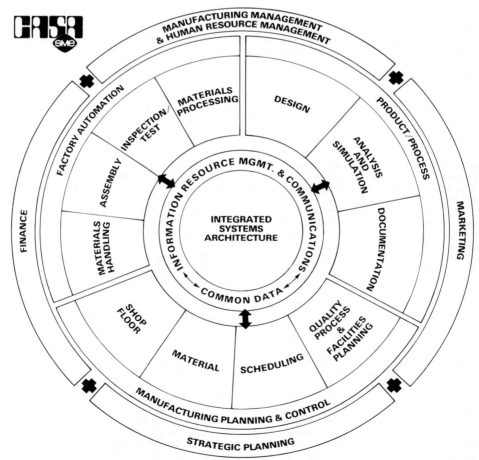

Fig. 12.4.2. CIM concept. CIM "wheel" developed by the Technical Council of the Computer Automated Systems Association of SME (CASA/SME)—Copyright 1985 CASA/SME. Second Edition, Revised November 5, 1985.

12.4.1.2   Development of CAM

The development of CAM can be traced back to the late 1940s when the concept of NC machine tools was conceived. The basic idea was to control the complex motions of a milling machine with numerical data in order to cut sophisticated aircraft components. Under the support of the U.S. Air Force, the first NC milling machine was successfully demonstrated by MIT in 1951. However, it was soon discovered that the preparation of the input numerical data (part program) can be tedious and difficult. The development of APT (automated programming tool) began in the mid-1950s (IITRI, 1967), and was followed by several other part programming languages such as ADAPT (Kelley, 1969), COMPACT II (MSDI, 1984), and SPLIT, AUTOSPOT, and EXAPT (State, 1969). A part programming language allows the user to define the geometry and cutter motion in a high-level language. Computation of final cutter path is handled by the computer, significantly simplifying the part programming task.

The CAM concept was formally introduced in the late 1960s and early 1970s. The 1970s marked the introduction of the FMS. In a FMS, NC machines, robots, material handling devices, flow control software, and system communication functions are integrated into one single manufacturing cell. The system is essentially controlled by computers, with humans assuming responsibility for supervision, monitoring, emergency handling, and loading parts into and out of the system.

Since the early 1970s, tremendous progress has been made in the area of system control. Owing to the advances in microelectronics, localized computer control has been made possible. Microprocessor systems are now used to substitute for complex analog or electromechanical control devices, providing faster and more reliable control. The development of programmable controllers, process control computers, and local area networks has made system control more efficient.

Another notable improvement has been in the interface between CAD and CAM. By integrating NC part programming into the design phase, the same design data base can support both design functions and NC part programming, thereby eliminating the need to redefine part geometry. With this type of system, often referred to as a turnkey CAD/CAM system, a cutter path can be verified directly on a CAD CRT screen.

12.4.1.3   Current Technology of CAM

The basic elements in CAM are machine tools, inspection devices, and material-handling devices. They are controlled by computers or computerlike controllers. Machines are linked by material-handling devices and computers to form a manufacturing cell (Figure 12.4.1). A hierarchy of computers coordinate and control the operation of the entire manufacturing system. In this section, we will discuss each element and how it interrelates within the overall system.

Numerical Control

Numerical control is commonly used to refer to those machine tools that are controlled by numerical data. In a conventional machine shop, a human operator uses the information on a blueprint and the instructions from a process plan (or operation sheet) to control the machine tool. Dials, knobs, and tools are set up manually, and by turning two handwheels simultaneously, the person can cut the material as specified on the drawing. In order to produce a high-quality product, experienced workers with good eye–hand coordination are needed.

Numerical control removes the human from control of the machine. The operator sets up the workpiece and tools on the machine and then loads a part program through a paper tape reader. From this point on, the entire operation is performed automatically. A typical NC machine is depicted in Figure 12.4.3.

NC Structure

An NC system consists of three major components: machine tool, controller, and part program (Figure 12.4.3). The machine tool conducts the actual machining operation, the controller directs the machine tool, and the part program contains the part-specific operation sequence.

Machine Tool.   The different types of machine tools include milling machines, lathes, drill presses, punch presses, and saws. As long as there is one or more moving axes and/or a spindle down feed mechanism, NC can be used effectively. Although a conventional machine tool can be retrofitted with motors, they are usually built specifically for NC, and typically contain steel-structured frames that are more rigid than conventional machine tools.

The axial movement of machine tools is driven by motors (usually DC or AC servomotors). Through a reduction gear box the speed is reduced, and the rotational movement is translated into linear table movement. The speed of the servomotor can be precisely controlled, and a feedback device, usually a shaft encoder or resolver, provides location feedback to ensure positioning accuracy. Machining is

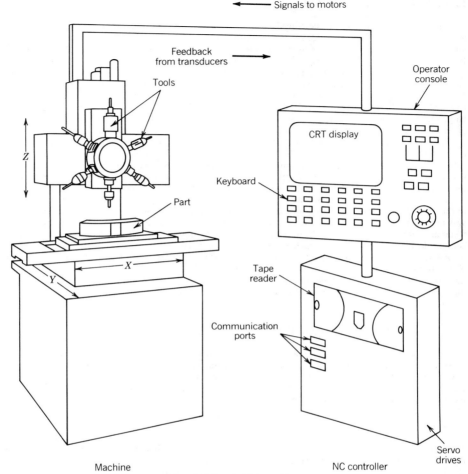

Fig. 12.4.3. A NC machine.

controlled by varying the cut (displacement), feed (table speed), and speed (spindle speed). An NC machine tool can easily achieve an accuracy of 0.001 in., satisfying most machining requirements.

Some of the new machine tools (Figure 12.4.4) also include a tool changer and pallet changer in the system. Referring to the figure, the left side of the machine contains a tool magazine with 32 tools. Tools can be automatically removed from the magazine and inserted in the machine spindle (center of the picture) under part program control. A pallet changer in the front end consists of two machine tables whose positions can be swapped. Several machining operations can be performed with one setup. A workpiece is mounted on a pallet, which is then slid onto one of the machine tables. Since the operator can set up a second workpiece while the machine is working on the first, machine utilization increases. These machines are known as machining centers. In recent years, additional devices such as a robotic part loader have been added. Some recent designs even use a robotlike structure to provide more degrees of freedom.

Controller.   A conventional NC controller is a hardwired electronic device. Modern NC controllers are computer-based devices referred to as CNC, and are capable of communicating with other computers and control devices.

The control panel on a typical CNC is shown in Figure 12.4.5. It consists of a CRT display, a keyboard, some special functional keys, and a few knobs and dials. The CRT displays the current program being executed, the machine status, and different control menus. The keyboard is used to either enter instructions or modify programs. Special functional keys allow the operator to assume manual control of the machine.

Fig. 12.4.4. A machining center with tool changer and pallet changer. (Courtesy Cincinnati, Milacron, Cincinnati, Ohio 45209.)

The program is normally fed into the controller through a paper tape reader. Several programs can be saved in the controller's memory, which is usually of a nonvolatile type such as bubble memory. Basically, a controller first reads in a user program (part program) and then sends out control signals to servomotor drives. [There is an international standard on the part program format (EIA, RS-273-A and RS-274-B); however, minor differences can usually be found between machines.]

In the past, most CNC controllers displayed alphanumeric data on the CRT. Some modern machine tool controllers provide an interactive color graphics interface (Figure 12.4.6). Design information can be keyed in interactively, and the controller will calculate the cutter path and display the results in color, making programming as well as debugging of the program easier. However, this type of system costs more and may actually interfere with production since the machine must be idle when the operator is using the controller for programming. Although it may not be economical to implement this type of controller, an alternative is to use a stand-alone graphics programming system for programming and verification.

There are also NCs that can be controlled by computer-speech recognition using a microphone instead of a keyboard as the input device. The full instruction set can be entered through voice command. Two problems with this system, however, have prevented its widespread adoption:

1. The high noise in the shop may make the voice input unrecognizable.

Fig. 12.4.5. A CNC controller operator console. (Courtesy Industrial Computer Groups, Allen-Bradley Company, 747 Alpha Drive, Highland Heights, Ohio 44143.)

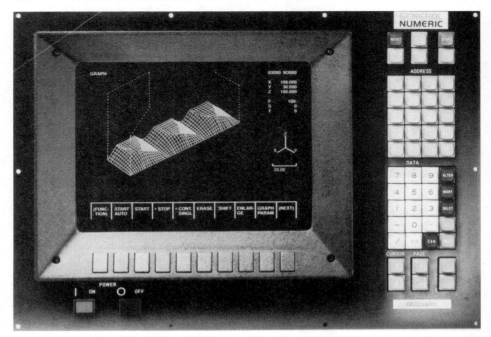

Fig. 12.4.6. A CNC controller with graphics display.

2. Every machine operator would have to perform the tedious task of teaching the controller in order to use it.

Part Programs. A part program provides the instructions to the controller to move the machine. It consists of blocks of instructions, with each block containing a sequence number, XYZ coordinates, ABC rotational axis movements, feed, spindle speed, tool selection, preparatory commands, and miscellaneous commands (Figure 12.4.7). The most commonly used media to store part programs is paper tape, where each character is represented by one row of punched holes. The two standard codes (hole patterns) used are EIA and ASCII (also ISO). To distinguish between these two standards, the EIA code always has an odd number of holes punched (odd parity) while the ASCII code has an even number of holes (even parity). Machine controllers generally accept either code.

12.4.1.4   Part Programming

Part programming, the task of preparing machine control instructions, is quite involved. The programmer must be able to interpret engineering drawings, understand geometry and trigonometry, and translate instructions (including those concerning machine movements) into part program format. The cutter must be normal to the part surface, and calculating cutter location can be an especially tedious task when curved surfaces are involved and line segments are used to interpolate the surface. The part programmer must also take tolerance limits into consideration. Shorter interpolation steps provide better tolerance, but at the expense of longer programs.

There are two approaches to preparing part programs: manual and computer-assisted. In manual part programming, a programmer calculates the final cutter path (based on part geometry). With computer-assisted part programming, a programmer uses a high-level language to define the part geometry and cutter path; calculations and formatting are performed by the computer program.

Manual Part Programming

Manual part programming can be illustrated by the following example. From an engineering drawing, the part's dimensions are specified. The part programmer then uses the process plan to obtain information on the cutter, feed, and cutting speed. For example, the cutter could be a four-flute milling cutter with a diameter of 0.5 in. The feed is normally given in inches per revolution per tooth, and the cutting speed in surface feet per minute. The controller accepts feed in inches per minute and speed in RPM. The programmer must therefore convert these variables into the appropriate units. The next

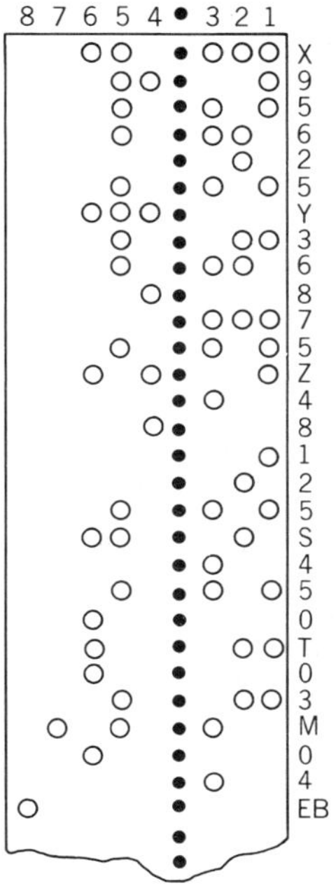

PROGRAMMER'S PROCESS SHEET						
X AXIS	Y AXIS	Z AXIS	FEED	SPEED	TOOL	AUX.
9.5625	3.6875	4.8125		45	03	04
	13.7250		615	489		

Fig. 12.4.7. A part program segment.

step is to calculate the cutter center offset (cutter path). Once all the cutter center points have been found, they can be translated into part program format.

The first line in the part program has a rewind stop symbol ("%") indicating the start of the program. When the tape reader rewinds the tape, it will stop at this symbol. Each of the program lines (blocks) are lead by a N word, followed by a line number, for example N/O. The letter G indicates the preparatory code, preparing the machine or controller for a particular operation. For example, G90 signifies the use of an absolute coordinate system. The letter M denotes a miscellaneous code. For example, M03 turns on the machine spindle clockwise and M08 turns on the coolant. In the block "N30G01X10000Y1500F55200," G01 informs the controller to perform linear interpolation, X,Y provides the initial cutter location, and F and S represent the feed and speed codes, respectively. A block "N40G02X1100Y2000I500J400" would cut a circular arc, where the I and J codes provide the circle center. A last block "N50X0000Y0000M30" moves the cutter back to the origin. M30 turns off the machine and rewinds the tape.

Computer-Assisted Part Programming

Computer-assisted part programming languages were developed to simplify the programming task. Currently there are many part programming languages available such as APT (IITRI, 1967), COMPACT II (MDSI, 1984), EXAPT (State, 1969), and ADAPT (Kelley, 1969). A program written in a part programming language consists of the following major sections:

Environment (or cutting condition) statements.
Geometry statements.
Motion statements.
Postprocessor statements.

Environment definition defines the cutting condition, geometry definition defines part geometry, motion statements guide the cutter, and postprocessor statements prepare machine-specific codes.

A part programming language is composed of a processor and postprocessors, and essentially allows the same program to be used for different machine tools. The only change required is in the postprocessor statements, since each machine has its own part program format and special code requirement. The processor performs all the interpretation and mathematical calculations. The processor's output is in an intermediate form called CL-data (cutter location data). A postprocessor translates this CL-data into the format acceptable by a specific machine.

Although a program written in a part programming language appears longer than the manual part program, it is normally much shorter for a sophisticated part requiring curve interpolation. Some advantages in using a part programming language are:

1. Elimination of manual calculation.
2. Ease in understanding and debugging.
3. Ease in making changes.
4. Provision of additional capabilities such as graphics interface.

Through plotting of both the part geometry and the cutter path, a programmer can verify the program.

Graphics Part Programming

Interactive graphics further facilitates the part programming task. A system with interactive graphics enables the programmer to work with pictures instead of language statements and numbers, further reducing the difficulty involved in defining part geometry. An interactive graphics part programming system is similar to a CAD/CAM system, but usually does not provide functions such as dimensioning, hidden line removal, and rotation of drawings.

Turnkey CAD/CAM systems also support interactive part programming capability. On a CAD/CAM system, the geometry of the part has already been entered by the designer. The part must then be designed to full scale; however, hidden lines should not be removed and dimensioning lines should not be added. A majority of CAD/CAM systems also support graphics verification of cutter path.

The most efficient way to perform part programming is to use the CAD/CAM system for design as well as part programming. When the design and manufacturing departments use different CAD/CAM systems, IGES (Chapter 12.2) pre- and postprocessors must be installed on both systems. Otherwise, the part geometry has to be entered manually.

12.4.1.5 Direct Numerical Control (DNC)

In the previous sections we discussed individual machine controls. The basic concept underlying DNC is to use a computer to control several machines. The concept emerged in the late 1960s prior to the widespread implementation of microcomputers and when NC controllers were still hardwired. The initial concept was to use a large computer to control several machine tools simultaneously and thereby eliminate the NC controllers. However, it failed due to difficulties in software development and in the long signal transmission links between the computer and the motor drives.

The concept later evolved into using direct data links between the computer and NC machines, which substituted for the paper tape reader. Part programs that were developed and stored in the DNC computer could be downloaded directly into the individual NC controllers. Since the paper tape and the tape reader are the most unreliable components of a NC system, DNC reduces errors and increases machine uptime. The DNC computer can also perform other functions such as monitoring the status of machines, keeping track of in-process inventory, and preparing management reports, enabling it to serve as the information manager in the shop.

In recent years, the local area network (LAN) concept has been applied to DNC. A coaxial cable can link as many as 256 computers/controllers on the shop floor. Data travels in the cable at a speed of several million bits per second, and information can be sent back and forth between several devices at the same time. Flexibility and ease of integration are built into these networks, and there is a trend toward standardizing them [e.g., the GM MAP (GM, 1984)] so that computers and control devices from different vendors can be linked into the same network and DNC achieved on a plant-wide basis.

12.4.1.6 Flexible Manufacturing System (FMS)

When automated material-handling and transportation devices are linked to DNC, the system is called a FMS (Figure 12.4.8). The advantages of a FMS are not only in reducing labor costs, but also in:

1. Reducing manufacturing lead time.
2. Reducing inventory.
3. Increasing machine utilization.
4. Improving product quality.
5. Increasing productivity.

The high capital investment cost is currently the primary factor responsible for hindering the implementation of FMSs in industry.

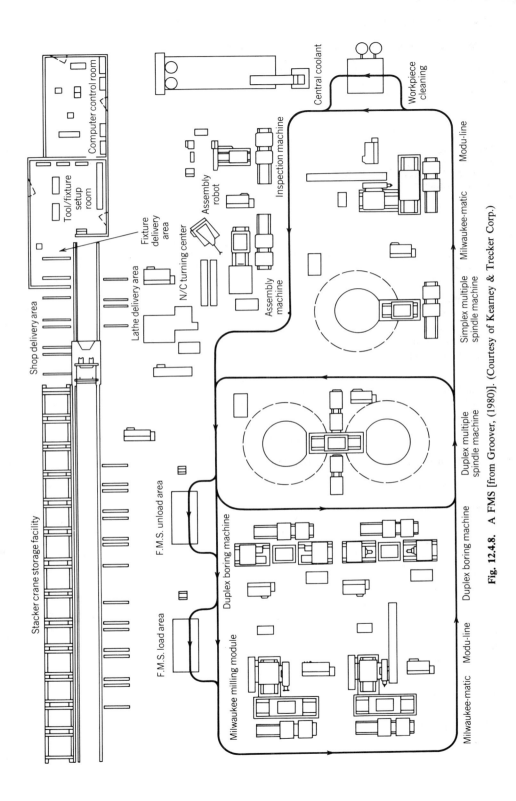

Fig. 12.4.8. A FMS [from Groover, (1980)]. (Courtesy of Kearney & Trecker Corp.)

Central coolant

Workpiece cleaning

Inspection machine

Assembly robot

Assembly machine

N/C turning center

Fixture delivery area

Lathe delivery area

Tool/fixture setup room

Computer control room

Shop delivery area

Stacker crane storage facility

F.M.S. unload area

F.M.S. load area

Duplex boring machine

Milwaukee milling module

Milwaukee-matic   Modu-line

Duplex boring machine

Duplex multiple spindle machine

Simplex multiple spindle machine

Milwaukee-matic   Modu-line

Machines used in FMSs include NC machine tools and cleaning and inspection devices, and all are computer controlled. The machine tools used in FMSs are mostly horizontal machining centers and turning centers, both of which provide greater flexibility for machining parts.

Material-handling devices used in FMSs include robots and push bars. Carousels and stackers may be used as buffer storage devices. Transportation devices include automated guided vehicles (AGVs), powered roller conveyors, belt conveyors, tow lines, and shuttle carts. These devices transport both tools and workpieces. Normally, a workpiece is fixtured on a standard pallet. The operator mounts the workpiece on the pallet and then launches it into the system.

A hierarchy of computers/controllers is typically employed to control and monitor the system. The top-level computers perform DNC and material flow control functions, and enable workpieces and tools to be dispatched into the system to the desired machines at the appropriate times. Since all the manual operations in an FMS have been eliminated, the operator's primary role is as monitor and troubleshooter.

12.4.1.7   Automated Factory

The FMS can be considered as a subsystem of the automated factory, the ultimate goal of CAM. In such a facility, all equipment is automated and integrated, as are production planning and control functions such as MRP, capacity planning, process planning, and inventory control. Ideally, an automated factory can operate without the presence of humans (Figure 12.4.9), and its major functions can be identified as:

1. Production planning and control.
2. Material storage and retrieval.
3. Material processing.
4. Material handling.
5. System and equipment monitoring and control.

All the above functions are under computer control, with LAN established to facilitate the communication. Figure 12.4.2 depicts the concept of CASA (Computer and Automated Systems Association) of SME (Society of Manufacturing Engineers), and it is only after we integrate all the functions shown in that figure that CAM, in a true sense, can be achieved.

12.4.2.   JOB DESIGN IN AUTOMATED MANUFACTURING SYSTEMS

12.4.2.1   Automation and Job Fragmentation

The underlying issues human factor specialists are most likely to confront when predicting or analyzing the effects of computerized manufacturing automation (CMA) are those associated with job design. Fundamental changes in job design as a function of mechanization and automation were noted during the industrial revolution, and provide a compelling argument for reconsidering the current course automation in manufacturing appears to be heading (Rosenbrock, 1983a). Quite often, tasks would first become fragmented to the extent where the human's skill was no longer the essence of the task, leading to a reversal in the primary and secondary roles originally defined in the human–machine system. The jobs remaining for the human were typically repetitive and trivial, and ultimately became totally automated if technologically feasible. Macek (1982, p. 199) views this situation leading to worker dissatisfaction due to a breakdown in job cohesiveness: "When automation is introduced in a factory and everything that can be automated, a number of tasks will be left over. It is tempting to take a jumble of these left over tasks and present them as if they were a single job. This is clearly related to the traditional approach of having the worker carry out the unmechanized parts of a process. This approach would be undesirable. When workers have jobs in which they have a number of unrelated tasks, mistakes are more likely, and they will find the job unsatisfactory." It appears then that the process of automating jobs defines the worker's functions by default as those task components remaining due to infeasibility in their automation.

Job Fragmentation: The Robotics Example

The effects of introducing robots into the manufacturing environment exemplifies, to some degree, this systematic degradation of the worker's role. A case in point is continuous arc welding where a robot is now capable of putting in the weld. The human's task essentially consists of removing the workpart from the jig and mounting a fresh workpart. Not only has the human's real skill in this task, undoubtedly the welding, been taken over, but the leftover tasks that the current generation of robots cannot yet perform have become machine-paced. Edwards (1984) has noted that the integration of humans and robots in arc welding and spray painting operations led to arc welders asking to be

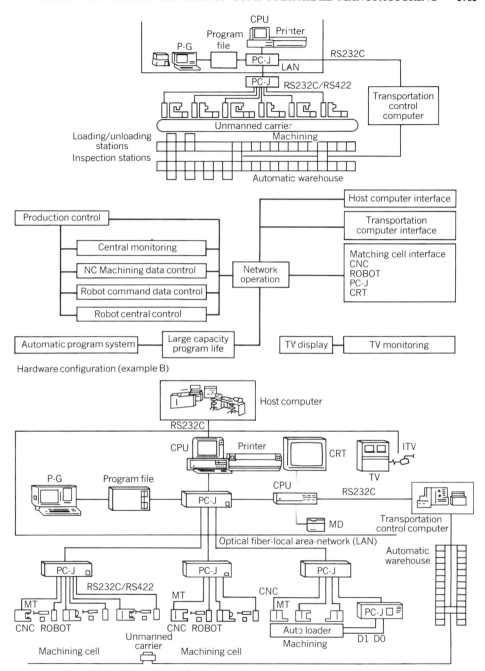

Fig. 12.4.9. A structure of automated factory. (Courtesy of General Numeric.)

transferred back to manual welding and the spray painters remarking that their jobs were less satisfying than manual spraying.

Considering the greater consistency and accuracy associated with robot operations relative to the human operator, economic realities clearly argue in favor of their utilization. Their application to work considered strenuous, monotonous, and hazardous to humans has, in many instances, also enabled the human's work environment to become improved. However, the human factors problem of residual tasks cannot be ignored. These tasks typically fall into four categories (Edwards, 1984): loading, monitoring, manual intervention in the event of robot malfunction, and manual operation of the robot's task. In addition to problems in worker satisfaction and underutilization of human resources, removing the human from the overall control of the task increases the probability of human error due to the operator's lack of familiarity with the manual task. For instance, many material-handling tasks designed for robots present a significant risk to the human when it becomes necessary, as in the event of robot breakdown, to assume the robot's task. Many other types of problems associated with the human–robot interface, for example, the design of information concerning the robot's state, are endemic to the general class of human–machine design problems, and certainly deserve attention. Various fundamental human factors techniques could be applied toward making more effective use of the human–robotics system. Along these lines, Parsons and Kearsley (1982) provide a useful framework for analyzing joint human–machine functions in a robotics system for the purposes of designing synergistic systems whereby the human performs functions the robot is less capable of in order that their combined effort surpasses their individual efforts.

Job Fragmentation: The Flexible Manufacturing System Example

In discrete-parts manufacturing operations, fragmentation of worker skills occurred dramatically with the arrival of NC machines. Under the guise of DNC, the first generation of CIM systems appeared, representing the true predecessors of FMS. These systems potentially contain DNC, robots, automated material handling, and advanced computer capabilities in both control and organizational functions. The FMS therefore affords an excellent opportunity for investigating and solving the dilemma deriving from automated systems that increase the efficiency of and, in this case, production at the batch manufacturing level while fragmenting and trivializing the human role. A characteristic effect of automation in batch manufacturing has been the stepwise removal of the human worker from the physical process as manufacturing operations become more computer integrated (Kegg and Carter, 1982). Although a precise delineation of the human role in future batch manufacturing depends heavily on the evolution of computing power and software, the idealized view of the factory of the future expects very few people on the shop floor. Instead, their primary tasks would be as monitors or supervisors who will spend most of their time at computer stations managing a number of machines or systems (Groover, 1980).

Whether this view of the human role is consistent with the trivialization of human skills discussed earlier is debatable. In particular, FMSs appear to exhibit many of the same drawbacks with respect to job fragmentation and trivialization as its component robot and NC systems have been known to do. Parts still have to be loaded and unloaded from the FMS since robots cannot yet perform this type of task. However, it is only a question of time before this task becomes automated out. Manual replacement of broken tools is also still required in many FMSs. Other examples of human intervention that can be considered as tasks performed by default include (1) clearing parts that may have jammed the transportation pathway; (2) rerunning a NC part program from the point where a tool required replacement; and (3) responding to a signal from a NC machine to change or sharpen a cutting tool.

An Alternative View of the Human Role in Automated Manufacturing

There are viewpoints that seem to imply that the effects of CIM on the jobs of workers will be to the contrary of what the foregoing discussion conveyed. A forecast on the future of CAM (Hatvany, Merchant, Rathmill, and Yoshikawa, 1982) implied that manufacturing industries will be compelled to offer more intellectually taxing jobs. The basis for this belief was the anticipation that human workers will need to become significantly more computer literate and more knowledgeable in general in future CIM systems. New skills will obviously have to be learned that would enable industrial robots and other relatively sophisticated hardware to be properly maintained. If one views CIM in its broadest terms whereby management, engineering, and production functions become integrated to allow effective response to the entire manufacturing operation, various human skills are not only still required but can take on added significance (Adams, 1983). These include the human functions of flexibility, reasoning (especially of the inductive type found in fault diagnosis), sensory functions for dealing with subtle detection problems (e.g., in part orientation), and highly developed manual dexterity for functions related to equipment repair. Although computer algorithms dictate many CIM functions, the complexity of the information associated with the product flows demands synchronization of pattern-recognition and reasoning skills that require the presence of the human. The true demand for human intellectual skills, however, would appear to stem from concerns related to operations management and organizational design. As Gunn (1982, p. 15) states: "The opportunities for mechanization in the factory are widely misunderstood. The emphasis has been almost exclusively on the production process itself, and complete mechanization has come to be symbolized by the industrial robot, a machine designed to replace the production worker. Actually the direct work of making or assembling a product is not where mechanization is now likely to have the greatest effect. . . . The major challenge now, and the major opportunity for improved productivity, is in organizing, scheduling, and managing the total manufacturing enterprise, from product design to fabrication, distribution and field service. . . . For this reason, the most important contribution to the productivity of the factory offered by new data-processing technology is its capacity to link design, management, and manufacturing into a network of commonly available information. The social outcome of the linkage may be to alter far more white-collar jobs than blue-collar ones." In order for this linkage to become a reality, the assimilation of computer technology not only needs to be substantially increased, but it must also be applied to all phases of manufacturing operations.

12.4.2.2 Levels of Human Decision Making in Automated Manufacturing

Based on the foregoing, three levels of human decision making functioning in CIM can be broadly identified (see Table 12.4.1). At the highest level, planning and organizational skills are involved which, with increased assimilation of CIM, should receive more attention. The human's overall goals are generally known. The primary task is to develop subgoals that become partitioned into lesser and lesser subgoals. This conceptualization of planning can be thought of as an iterative process within which lower-level subgoals can be temporarily ignored in favor of the more consciously planned higher-level goals (Johannsen and Rouse, 1979). Eventually, the human must deal with a lower-level subgoal. At that point, a more clearly defined decision making model of the human is needed to understand and hence predict human performance. While a concise decision making model of the human is also necessary to appreciate human planning of high-level goals, the goals at this level are inherently less clearly defined. Equivalently, we could say that the human's internal model, that is, the human's perceptions of how system inputs translate into system outputs, becomes less well defined as one moves up the hierarchy of goals. It is unlikely that the human is attempting to optimize planning activity at the higher levels in the hierarchy but rather, as Rouse (1980) suggests, attempts to "satisfice," that is, attempts to discover a satisfactory solution.

In order to draw an analogy between levels of decision making (Table 12.4.1) and an idealized automated manufacturing facility, Williams (1980) proposed hierarchical structure for an automated steelmaking process can be utilized (Figure 12.4.10). The different areas could represent systems such as FMSs, automated warehouses, automated transport facilities, robotic work groups, and so forth, engaged in coordinated, redundant, or independent operations. In attempting to "satisfice" the integration of these operations, the human would need to generate a plan that is dependent on information from all these sources. The large and varied amount of dynamic information would suggest the need for computerized decision support so that the demand on the human's decision making capabilities are maintained within limits consistent with the human's information processing capacities (see Chapter 2.2). The need for decision support, however, leads to considerations in both allocation of responsibility and the form of communication between human and computer (Table 12.4.2). The choice of interactive design strategy would depend on, among other factors, task characteristics such as the speed at which system events are occurring, and the values and costs associated with making correct and incorrect decisions.

The assembly and maintenance skills depicted at the lowest level (Table 12.4.1) represent the skills that industry appears content to automate, but, owing to present limitations in technology, become

Table 12.4.1 Levels of Human Skills in Automated Manufacturing

Planning and organizational skills

Decide which products should be manufactured and how much

Decide on product priorities

Allocate human and material resources

Where to automate

Coordinate production operations between facilities

Coordinate information from supervisory controllers

Supervisory control and programming skills

Computer-aided process planning/design

Part programmers

Human supervisory controllers of FMSs, robotic systems, automated material handling and other subsystems

Assembly, machine operation, and maintenance skills

Maintenance operators

Human NC operators

Human–robot tasks

Human assembly operators

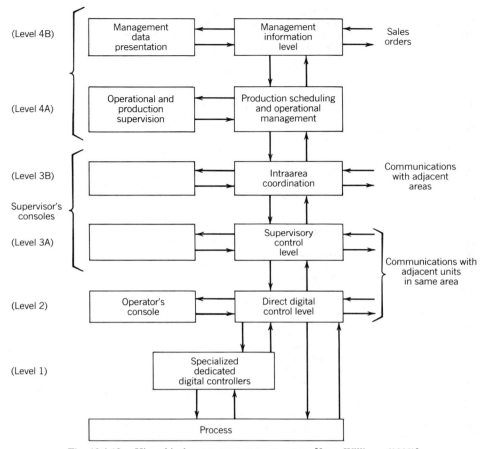

Fig. 12.4.10. Hiearchical computer system structure [from Williams (1980)].

Table 12.4.2 Assignment of Decision Making Responsibility between Human and Computer

1. Human considers decision alternatives, makes and implements a decision.
2. Computer suggests sets of decision alternatives, human may ignore them in making and implementing decision.
3. Computer offers restricted set of decisions, human decides on one of these and implements it.
4. Computer offers restricted set of decision alternatives and suggests one. Human decides on any one alternative and implements it.
5. Computer offers restricted set of decision alternatives and suggests one which it, the computer, will implement if human approves.
6. Computer makes decision and necessarily informs human in time to stop its implementation.
7. Computer makes and implements decision, necessarily tells human after the fact what it did.
8. Computer makes and implements decision, tells human after the fact what it did only if human asks.
9. Computer makes and implements decision, tells human after the fact what it did only if it, the computer, thinks the human should be told.
10. Computer makes and implements decision if it thinks it should, tells human after the fact if it thinks the human should be told.

Source: Sheridan (1980)

fragmented and trivialized. In order to determine the basic task components of a manufacturing operation with CIM technology available, a task analysis (see Chapter 3.4) would need to be performed. Information concerning the abilities and limitations of humans, robots, and automation would need to be integrated in a manner that enables the design of the assembly (or other) workplace system to be optimized. Guidelines for this type of process can be found in Kamali, Moodie, and Salvendy (1982). Listing advantages and disadvantages of humans and robots is a necessary step that needs to be done early in the process (see Chapter 12.3). Ultimately, the relative capabilities between humans, robots, and automation need to be established, and Kamali et al. (1982) provide an example of such an attempt for the case of an assembly workplace system.

The intermediate level (Table 12.4.1) is closely related to automated systems or subsystems whose output information is integral for performance at the level above. Given the relatively large number of automated components present, system reliability should decrease (Melster, 1966), arguing for the presence of at least one human in order to increase system reliability (Figures 12.4.11 and 12.4.12). It is at this level, however, where the greatest uncertainty exists with respect to defining the nature of the human role.

Assuming the system in question is an FMS, the human's role as a system controller would likely be supervisory (see Chapter 9.6). A task allocation scheme for the human and computer (actually the FMS's computer control system) has been developed by Hwang, Barfield, Chang, and Salvendy (1984) for a particular FMS. For this system, the relative advantages and disadvantages of the human versus computer are compared on the basis of functions such as tool replacement, scheduling of work-parts, machine tool control, traffic control, monitoring of tool life and system status, job priority determination, emergency handling, and repair and maintenance (Table 12.4.3).

Although acknowledging that implementation of FMSs results in only certain residual tasks remaining for workers, Kohler and Shultz-Wild (1983) strongly oppose the more frequently adopted job organization policy that consists of a hierarchical and highly specialized job structure. Instead, they suggest that a homogeneous (i.e., nondivision of labor) and enlarged job structure at high-skill levels be integrated into FMSs to counter the fragmentation of skills systems such as FMSs have induced. Long-term predictions on the viability of this approach depends, as these authors point out, on many political factors. Perhaps even more critical are the technological advances that could change the thinking altogether with respect to the utilization of human skills in batch manufacturing environments. In any case, their advocacy of a homogeneous job-structure is consistent with more optimal human performance in that it utilizes human abilities at flexibility and adaptability. At the same time, this

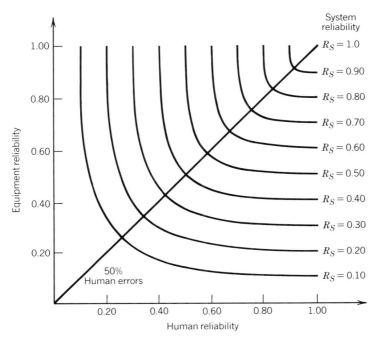

Fig. 12.4.11. Effect of human and equipment reliabilities on system reliability: $R_e \times R_h = R_x$ [from Meister and Rabideau (1965)].

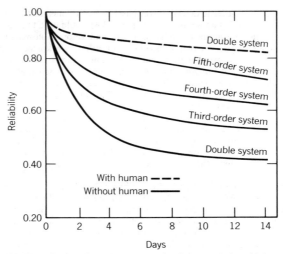

Fig. 12.4.12. The reliability of a double redundant navigation system in which one redundant component is a human (dashed lines) as compared with the reliability of systems with various orders of redundancy in which all components are machines (solid lines) [from Grodsky (1962)].

approach can avoid the problem of monotony that often arises with job structures characterized by a considerable division of labor.

12.4.2.3   Human Supervisory Control in Automated Manufacturing

Consideration of Cognitive Abilities

Consistent with the concept espoused by Sheridan (1976) of a *coherent* role as applied to human supervisory control (HSC), supervision of an FMS should demand a significant portion of the human's attention and abilities. The supervisor's role should be clearly distinguished from that of an operator's by calling for the properties of displayed information to be studied as opposed to being only scanned for problems. The characteristics of an FMS suggests that the human would be well suited to active involvement in system control rather than to a passive role where actions are taken only when something goes wrong. The latter reduces the task to monitoring, a function humans generally perform poorly. Cognitive capabilities that could give the human supervisor a distinct advantage (Table 12.4.4) over the FMS's computer control system include (1) *flexibility,* particularly in handling unexpected events; (2) *pattern recognition,* to categorize system events, abstract similarities from these events, disregard information that appears conflicting, and utilize graphic information; (3) the *ability to teach,* through intervention with software such as the rule-based structure in expert systems (see Chapter 9.3) and with robots, where the human would supervise the motions being carried out through interfacing of the robot computer with the FMS computer control system; (4) *inductive logic,* which would allow the human to develop general rules from events or patterns of activity based on alphanumeric or graphic information of system status; (5) *making adjustments* to changes in priorities; (6) *diagnosis* of system faults based on information supplied from the computer; and (7) *making decisions* dictating the type and extent of (automated) inspection. Many of these human decision making functions in HSC of a FMS reduce to abilities in making difficult tradeoff compromises. Other CIM systems, for example, automated (conveyor) systems that bring products into and out of the plant and are primarily responsible for sorting and routing (Stallard, 1980), would likely require some subset of the abilities outlined above.

Issues in Designing the Human Supervisory Role

HSC of FMSs can be considered as part of the larger concern of human–computer interaction (Figures 12.4.13, 12.4.14, and 12.4.15). The *physical* human–computer interface issues (Greenstein and Wysk, 1980) such as operating system characteristics, dialogue style, and input and display devices all deserve consideration. However, the issues those authors focus on that concern the nature of the human role are more immediately critical to design. Although there is no consensus among manufacturers, some of the views are contrary to the concept of a coherent role discussed earlier. As a report made available by the Kearney and Trecker Corporation (1976) indicates: "Control of the (flexible manufacturing) system is complete. Once loaded with all operational and production parameters, the system will

Table 12.4.3 Capabilities of Human and Computer in the Operation and Control of FMSs

FMS	Computerized Devices	Human
Physical operation		
System loading and unloading	Can direct robot to handle up to 900kg Gives part type, part number and quantity Shows a diagram on-line of how parts are to be placed Difficult to determine workpiece orientation	Maximum arm load: less than 30 kg; varies drastically with type of movement, direction of load Power: 2 hp $\simeq$ 10 sec 0.5 hp $\simeq$ 120 sec 0.2 hp $\simeq$ continuous 5 kc/min subject to fatigue; may differ between static and dynamic conditions Checks part information, notifies computer that part has been loaded
Workpiece setup on machine	Very difficult to set up if the workpiece is not prefixtured on a pallet Machine vision is required Uses locking device when the workpiece is fixtured on a pallet	Slow, but good pattern recognition skills for complex parts
Tool setup (off line)	High precision on complex geometry is not available	Human reliably performs this task Positioning accuracy: relative (with feedback)—good: 0.1 mm; absolute (without feedback)—very poor: 3–8 cm Fitt's law governs movement
Tool replacement	Automatically changes tool magazine: 5 min required Single tool changed in 2 sec Computer knows when a tool magazine is due and instructs a robot crane to perform this job	Inserts tool manually Checks tools when signal is received from computer
Shop scheduling	Computer heuristics can provide "good" schedule Combinatorial problem, no guarantee on optimal solution Computer simulation evaluates scheduling Plots the number of pallets required during the scheduling period	Human intuition helps to determine algorithm Human and computer help determine optimal schedule Human inputs workload data and uses computer simulation to see the effects of scheduling
Equipment control		
Machine tool control	CNC controls axle and spindle motors, tool changer, pallet changer, etc. DNC computer downloads part program to CNC Faster and more accurate than human Preprograms cutter motion	Human interacts with computer to modify on-line program Human supervises machine tool control, primarily monitoring machine status
Robot control	Robot controller controls the motion of robot Sensing devices, for example, vision camera, texture sensor, strain gauge, and so on, provide feedback data	Human trains robot via computer, which requires human intelligence transfer to robot Human good at pattern recognition

1711

Table 12.4.3 (*Continued*)

FMS	Computerized Devices	Human
	System control computer downloads robot control program and operation instructions	Amenable to learning and flexible adaptation to the shop environment
	Very limited learning or environment adaptation	
Inspection device control	Computer can control automatic measurement and gauging devices	Human sets criteria for inspection device
		Good absolute judgement
	Presents inspection results used for quality control analysis	Smallest detectable threshold 10^{-6} ml
	Limited range which can be optimized under relevant needs	
	Can handle position tolerance below ± 0.03 mm	
Flow control		
Workpiece input rate control, 0.5–10 pieces/hour	Ability to accept information is very high, limited only by channel rate	Limited ability to accept information (10–20 bits/sec)
		Human monitors and identifies unusual flow rates
	Computer uses production schedule and capacity information to determine input rate	Human determines the rate limit, maximum and minimum
	Physical workpiece input is accomplished by workpiece loading function	
Traffic control	Computer can optimize the transportation of workpieces, fixtures and tools in the system	Human has good comparative judgment to correct traffic control problems
	The correctness of traffic control is dependent on sensory feedback	Human "programs" program controller to transport items in the system
	When feedback information is not correct, computer may make mistake	Human good at detecting incorrect traffic control
	AVG, powered conveyor, tow line, overhead trolley, and so on, are used to transport items in the system	Can identify unusual flow rates, mean response latency 113–528 msec
System monitoring		
Tool life	Computer records the time a tool has been used. When the cumulative time exceeds the rated life (30–60 min), computer informs the control system to change tool	Human helps supervise tool life by entering tool transactions into data base
		Has control over tool life, that is, can alter program if catastrophe occurs
	Sensing devices monitor the tool wear and catastrophic tool breakage. Not easy to sense the minute change of tool wear	
Quality control	Computer analyses product quality change, detects machine wear and other condition changes	Human makes decisions for product quality control
		Can adopt and identify new plans for quality control
	Reports to human for decision making	
	Computer compares the finished part geometry with the desired geometry and calculates the adjustments needed to correct	

Table 12.4.3 (*Continued*)

FMS	Computerized Devices	Human
System status	Checks and updates the status of the entire system at least once every minute	Human monitors overall system status with flexible decision-making abilities
	Equipment failure report, computer can perform diagnostic test rapidly and accurately	Human queries computer when system is down
	Reports workpiece jam, or redirects workpiece if possible	Computational capability slow—5 bits/sec
Other information handling functions		
Machine control program storage	Computer can store vast amount of part programs and robot programs	Human reviews computer output of program storage
	Program can be downloaded at a speed of 10 byte/sec to several million byte/sec	Human decides if new programs or variables need to be added to program
	Reliability can be increased by using redundant storage device	Human increases reliability of program storage as back-up decision maker
Tool and workpiece inventory records	Maintain a large data base	Human updates data base for tool and inventory, adds tool to inventory
	Updated data can be automated or manually handled	
Job priority determination	Computer can apply a predetermined rule to set priority	Human writes software which determines job priority
Emergency handling	Computer can shut off the system based on a preprogrammed sequence	Reaction time 0.25–0.33 sec
		Good error detection/correction at cost of redundancy
Maintenance	Can keep a maintenance schedule	Human determines maintenance standards
	Updates status data base at each time increment	Human performs periodical maintenance by visual spot checks
Repair	Can diagnose source of problem in priority order	Human deals with repair problems manually
	The most serious problem causes an immediate shutdown, less serious one alerts operator	System is dependent on human involvement for repair. Human initiates a hold on system and corrects the problem

Source: Hwang, S. L., Barfield, W., Chang, T. C., and Salvendy, G. (1984) Integration of humans and computes in the operation and control of flexible manufacturing systems. *International Journal of Production Research,* 22 (5): 36. Reprinted with permission of Taylor & Francis.

manage and operate the entire manufacturing network continuously without any intervention. Only in extraordinary situations will the control system require the manager to input decisions." At the FMS of the Fuji factory in Japan (Inaba, 1981), possibly the largest FMS currently in operation, the supervisor's role consists of monitoring, for the sole purpose of detecting abnormal operating conditions, an array of TV screens.

To have the human assume less and less of a role in the actual production process clearly represents a sound objective on the part of manufacturers who are considering or are in the process of automating. However, we must at least recognize that this approach may also have its shortcomings. First, by not being actively involved in system control, the human's response efficiency to critical system events is likely to be reduced or, even worse, their presence could go entirely unnoticed. Second, it fails to utilize information processing skills (previously discussed) that humans possess but are lacking in the computer control systems of FMSs. System states could therefore develop that might be difficult for the computer to control given its predetermined logic structure, but perhaps be relatively easy for the human to manage, especially with computerized decision support. Third, from an economic standpoint, this type of design is very inefficient since it leads to severe underutilization of human resources. If supervisory controllers continue to assume responsibilities for specific systems or subsystems, underuti-

Table 12.4.4   Human–Machine Function Allocation

Function	Human	Machine/Computer
Flexibility	Able to handle unexpected events	Limited flexibility
Ability to generalize	Able to recognize events as belonging to a given class, to abstract similarities, and to disregard conflicting characteristics	Limited capability
Sensitivity	Simultaneously sensitive to a wide range of kinds of inputs	Threshold is higher and can sense energies beyond human spectrum, but usually sensitive to only one kind of input
Monitoring capacity	Poor	Good
Variability in performance efficiency	High	Low
Reaction time	Relatively slow (200 msec)	Usually as fast as relay operating time
Physical force	Relatively weak	Practically limitless in power
Fatigue	Prone to acute and chronic fatigue	No factor except for physical limitations (heat, corrosion, etc.)
Environmental requirements	Only a very narrow band of environmental conditions is possible	Restricted only by design specifications
Channel capacity	Limited	Limited only by design
Overload operation	Graceful degradation; can tolerate temporary overloads without complete disruption	For a fixed information handling capacity, overload leads to system disruption
Survival	Dangerous situations produce stress and performance degradation	No factor
Computational ability	Poor	Excellent
Deductive logic capability	Right premises sometimes lead to wrong conclusions	Excellent
Inductive logic capability	Can formulate general rules or laws from specific cases	Poor
Distraction	Easily distracted by competing stimuli	Cannot be distracted

Source: Meister (1966)

lization can become a serious problem as more aspects of the overall batch manufacturing process become automated and interrelated. Lastly, a design strategy negating the human's decision making role has associated with it the more underlying concern for the quality of working life. In a preliminary study by Ekkers, Pasmooij, Brouwers, and Janosch (1979) in process control, high coherence of process information, high process complexity, and high process controllability were all associated with low levels of stress and good health. High process controllability, good interface ergonomics, and a varied and challenging array of activity were all associated with feelings of high achievement. These findings suggest a positive relationship between an active supervisory role and quality of working life. Commenting on worker satisfaction during the implementation of automation, Macek (1982, p. 198) notes: "If there is any power that is more forceful in today's world of work than automation, it is the power that the present day work force exerts through its very high expectations and demands for what jobs will be like."

Blumberg and Alber (1982) provide evidence from a case study of an FMS of benefits such as increased system utilization (ratio of hours the system was used to theoretical capacity) that are obtained through consideration of employee satisfaction, motivation, incentive programs, and work group design. These authors noted that despite the intention of removing the human element from the manufacturing process, the implementation of the FMS (they investigated) required movement away from specialized job classifications and toward a strong organizational network among the employees interacting with the FMS. Blumberg and Alber (1982, p. 51) admit that "the major effect humans have on (flexible manufacturing) system performance appears to stem from the manner in which they respond to the system's requirements for raw materials, maintenance, and the clearing of faults brought about by operating anomalies. Until equipment can be brought on-line which is totally automatic and completely

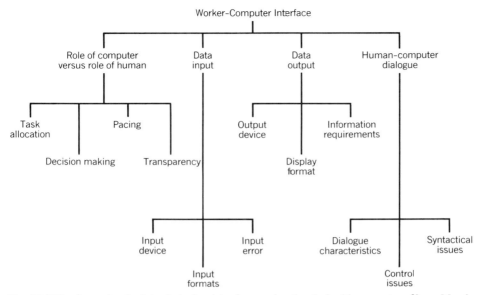

Fig. 12.4.13. Issues involved in designing jobs for people who deal with computers [from Macek (1982)].

reliable at an acceptable cost and can procure its own raw materials and dispose of its finished products, the behavior of human beings will continue to be important."

It is precisely toward achieving these latter ends that many manufacturers dedicated to CIM are directing their efforts. The issue concerning design of the human supervisory role resurfaces since it is reasonable to assume that there will be a need for such a role. Perhaps those manufacturers who are reluctant to allow the human to intervene with a relatively complex system such as an FMS are justified. Although HSC has been extensively studied in continuous process and control environments (Edwards and Lees, 1974), very little is known about HSC capabilities in the batch manufacturing environment. Characteristic differences in HSC between these two environments (Table 12.4.5) raise doubts about human capabilities in controlling systems such as FMSs. The indirect nature of human control and the absence of direct relationships between system inputs and outputs undermine much of the human's ability to act as a closed-loop system. Furthermore, the structure of systems such as FMSs make them relatively unpredictable with respect to changes in input or to various perturbations in the system (e.g., a machine breakdown), precluding the human from effectively utilizing anticipatory skills. However, as discussed earlier, these systems do favor the application of human pattern-recognition, inductive-logic, and other problem-solving skills.

An understanding of human limitations and capabilities at supervisory control in the discrete-parts manufacturing environment is fundamental to effective design of the human role. The foregoing discussion indicates that a reasonable approach to this problem would be to complement the human with computerized decision support to compensate for human limitations that are a function of structural characteristics of the system (Table 12.4.5).

Experimental Approaches to Design of the Human Supervisory Role in Flexible Manufacturing Systems

Human Abilities and Limitations at System Control. Various approaches have been taken in attempting to evaluate the human supervisory role in FMSs. One of these (Sharit, 1984) was based on the characteristic differences in HSC referred to previously (Table 12.4.5), and focused on the abilities and limitations of humans at *active* control of such systems. An interactive computer model of an FMS was developed and simulated in real time on a microcomputer. Various experimental conditions were employed representing hypothetically distinct dimensions of the mental workload construct (see Chapter 3.5). Through the availability of a graphic display of the FMS, the human's pattern-recognition skills (the utilization of which was considered implicit to the type of control environment) were enhanced. Results from this study suggested that humans were better at controlling the FMS under task conditions characterized by increased activity within the FMS (e.g., more parts to process and an increased likelihood of machine breakdowns) as compared to conditions where the human's

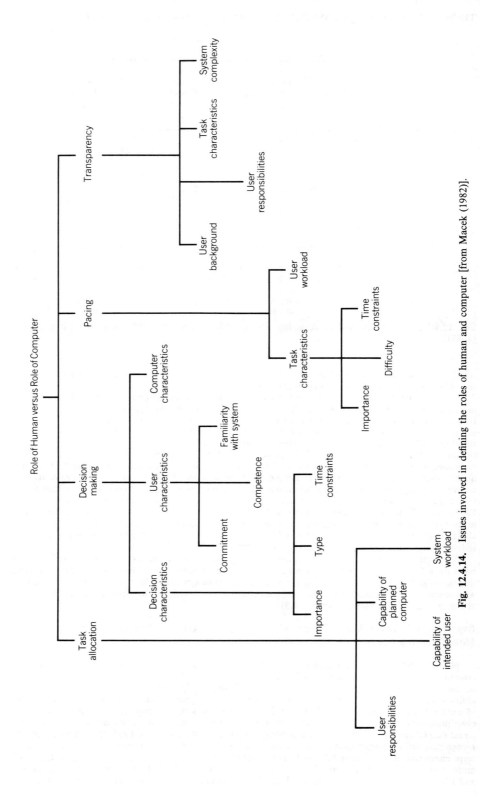

Fig. 12.4.14. Issues involved in defining the roles of human and computer [from Macek (1982)].

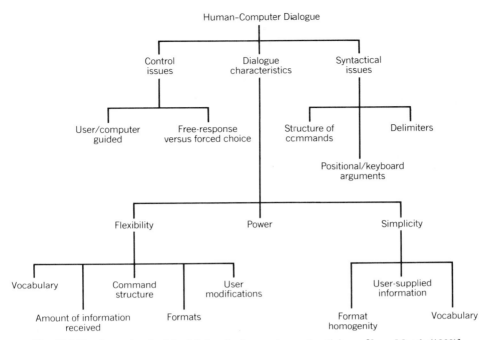

Fig. 12.4.15. Issues involved in defining the human/computer dialogue [from Macek (1982)].

Table 12.4.5 Differences in Human Supervisory Control between Computerized Discrete Part and Process Manufacturing

Variables Considered	Discrete-Parts Manufacturing	Process Industries
Direct versus indirect control	Almost exclusive reliance on indirect control	Control is more direct, and there is a greater reliance on manual control
Qualitative assessment versus quantitative assessment	Assessment of system is more qualitative; the emphasis during perception is on pattern recognition	Assessment of system status is often based on magnitudes and rates of change; the emphasis during perception is on detection
Predictability	Predicting the effects of control actions and anticipating time lags between these actions and system response are difficult	The more direct relationship between control inputs and system outputs allows for more accurate and reliable prediction
Effect of plant size	Larger and/or more complex systems can totally alter the control strategies utilizes	The effect of larger and/or more complex systems on control strategies is usually more limited in scope
Effect of output	Severe disturbances to the system do not necessarily jeopardize the overall functioning of the system	Severe disturbances often have serious effects on the entire process
Control strategy	In formulating ongoing control strategies, system output is used more heuristically	System output is generally used to update the parameters associated with the relationship between inputs and outputs

Source: Hwang, Sharit, and Salvendy (1983).

workload was more "internally" induced. The latter was brought about by requiring the human to acknowledge an additional goal during task performance. More surprisingly, graphics decision support that provided real-time displays of the relative distribution of workparts in the system did not prove very effective. Overall, there were indications that features unique to the FMS environment complicated supervisory control. However, it also seemed plausible that carefully designed decision support and training could enable the human to assume an important role in control of the FMS.

The methodology employed in this study provided both (1) a useful conceptual framework for development of expert systems (see Chapter 9.3) of the human supervisory controller of a FMS; and (2) the foundation for development of a generalized training simulator for FMSs. A "completely" modular interactive computer simulation model would easily permit alternative system designs to be evaluated from the standpoint of HSC. Redesign features might include the number and types of machines and their relative locations, the type of workpart handling system(s), presence of robots, and so on. Such a device would allow manufacturers to evaluate HSC capabilities for configurations that match their own without having to risk the economic consequences of training workers on a real system. Changes in the human–FMS interface would allow evaluation of various decision support devices, dialogue styles, and alternative sets of control actions in order to produce an interface design that is optimal from the standpoint of both FMS and human performance.

Allocation of Tasks Between Human and Computer.   In a somewhat different approach to design of the HSC role in FMSs, Hwang and Salvendy (1984) focused on the more broadbased design issues, specifically (1) the number of machines the human can effectively monitor and (2) the problem of optimal allocation of tasks between human and computer. Two types of FMSs, differing only in degree of automation, were simulated. In the less automated FMS, the human was required to check whether machine queues were overloaded and whether the appropriate tools for workparts were available. In the more automated system, part and tool status were automatically controlled, representing a higher level of task allocation to the computer. Measures of performance and stress were found to be more sensitive to level of task allocation than to number of machines, suggesting that (1) the workload imposed by the number of machines should not be considered a critical factor (during the FMS design stages) and (2) individual stress level can serve as a useful indicator of the level of decision making responsibility that should be allocated between human and computer.

A Sociotechnical Approach.   Rosenbrock's (1982) approach to design of the human element in FMSs was to avoid giving initial consideration to the technology and instead consider both the human and technical aspects simultaneously. This approach circumvents many of the problems that arise when the human is forced to adapt to the technology. Rosenbrock and his colleagues are currently pursuing a research program involving the development of an FMS with sociotechnical (Pasmore and Sherwood 1978) as well as economic objectives, and which would (1) allow for satisfactory system performance; (2) be economically viable; and (3) act as an aid in enhancing the skills of those workers interacting with it. The FMS is relatively small, consisting of an NC lathe, NC milling machine, Unimate robot, and PDP 11–34 computer.

One concept being pursued by these investigators represents a sharp break from current practice, and has the human within the FMS making the first part of a batch by programming at or near the NC machines. Information about the part to be made could be obtained from a display interfacing the FMS to the CAD system, and in a form independent of the machine tool in question. The worker on the shop floor can therefore utilize his or her judgment skills in deciding on which of a number of candidate NC machines should be utilized, and also use immediate feedback on the programming effort in order to determine whether further editing and/or dynamic simulation of the tool path is needed. The current approach has the programming done by specialist part programmers who cannot make effective use of shop floor knowledge in the overall design and processing of the part. Lower machine utilization results due to the segmentation of human skills that ultimately occurs when systems are designed in ways that do not consider the integration of the human's knowledge and skills. The methodological program prescribed in Rosenbrock (1983b) also advocates that programming of the robot be done by these workers partly through a terminal (to the same local computer used for part programming and verification) and partly by physical lead-through. The human would likely also be given production and scheduling responsibilities. The investigators involved in this research program hope that through the FMS example, the groundwork can be laid for a general methodology whereby emerging technologies not only integrate existing human skills but allow these skills to develop into new types in relation to the new technology.

Aiding the Human in Supervisory Control

Dealing With Unexpected Events.   Bainbridge (1983) has noted that in the case of process control, the introduction of automation has in fact made the role of the human operator more critical. With respect to the effects of automation, some useful analogies to HSC in discrete-part manufacturing systems can be drawn from the process control operator. For example, unexpected events in process control require a highly skilled operator. In having previously been monitoring an automated process and suddenly needing to assume manual control, various factors work in favor of causing ineffective control on the part of the operator (Bainbridge, 1983, p. 775): ". . . a formerly experienced operator who has been monitoring an automated process may now be an inexperienced one. If he takes over, he may set the process into oscillation. He may have to wait for feedback, rather than controlling by open-loop, and it will be difficult for him to interpret whether the feedback shows that there is something wrong with the system or more simply that he has misjudged his control action. He will need to make actions to counteract his ineffective control, which will add to his work load." Other factors influencing the human's ability to "take over" involve retrieval of knowledge from long-term memory and utilization of both working memory, which contains the ongoing results of the predictions and decisions made about the process, and short-term memory, containing information concerning the current state of the system. For the human controller to deal effectively with unexpected events, a very skilled supervisor in somewhat continuous interaction with the system would appear necessary. This premise seems to contradict objectives of automation that seek to remove the operator from the system.

Computerized Decision Support: Issues in Training. Dealing with unexpected or problem events also leads to the issue of training supervisory controllers. In an FMS, it is unlikely that the human can be prepared for specific events, since they are typically context dependent, and the number of these contexts is infinite. Training must therefore focus on the development of general strategies rather than specific control actions for particular states, and should consider the potential impact of computerized decision support. Two aspects can be identified: (1) the types of displays/information that need to be available and (2) the allocation of decision making responsibility between human and computer (Figure 12.4.14). With respect to the former, an intuitive dichotomy is the contrast between menu-based displays that provide quantitative information (about the various components of the system) and graphic displays that provide symbolic-type information for the purpose of accentuating the spatial relationships between all components of the system. In addition to information content, another dimension deserving serious consideration is that of time, where displays could be made available providing both current and preview information (predicted information based on current system status or based on some other method).

All these devices have problems of their own that could lead to counterproductive decision support. In contrasting symbolic or spatial versus menu-based information, dependencies on one type of information over the other can develop despite the ongoing status within the system dictating that this strategy not be taken (Sharit, 1984). An additional complication arises when one considers that the order in which spatial and menu-based information is sampled or processed could be critical, again providing opportunities for decision support to prove counterproductive Similarly, various difficulties result in the use of predictor displays. Smith and Crabtree (1975) explored the use of a predictor aid in controlling production output in a simulated job shop. Their aid enabled the human to observe the consequences of any (allowable) action on future system status and thereby ask "what if" questions. There were no significant differences in performance between a group using the predictive aid as compared to a group operating without it. This finding was attributed to the fact that the aid was primarily used for short-term error correction as opposed to deep search (perhaps reflecting limitations in processing resources imposed by the relatively complex interactive scheduling task).

In the predictive or, more accurately, preview display employed by Sharit (1984), the spatial relationships between components of the system could be previewed either 3 or 8 min earlier at any point during task performance. These displays presumably facilitated development of the human's internal model by associating current system states with future states. Counterproductivity in decision support is, however, again a factor. How much earlier in time should the system be previewed? The preview obviously becomes less accurate as the time interval increases. Determining how much earlier in time the system should be previewed (or whether not to preview at all) depends on the internal dynamics of the system as well as on the context implied by the current state of the system. Until a clearer understanding of these types of decision support devices is obtained, we have to accept the possibility that our attempts to reduce the human's workload to levels that lead to optimal system response may actually increase the human's workload, and with it the likelihood that optimal control strategies will not be undertaken.

Computerized Decision Support: Allocation of Decision Making Responsibility. The second aspect to computerized decision support in automated manufacturing relates to strategies in allocation of decision making responsibility between human and computer (Table 12.4.2). This area has evolved from (1) the purely static allocation policies such as "Fitts list" approaches to automation that assign to the human and computer those tasks or functions that each is best at (Table 12.4.4 offers a basis to this approach), through (2) Licklider's (1960) suggestion that the human and computer develop a cooperative association that allows the computer to expand its role in order to facilitate the human in formulative thinking, up to (3) Rouse's (1976, 1981) concept of adaptive allocation of decision making responsibility between human and computer that implies a purely dynamic allocation policy.

This latter concept represents quite sophisticated decision support that is not unidimensional but rather very knowledge-dependent. For example, the computer might need to know what the human is doing or even planning on doing, either through implicit or explicit communication. Advantages in a dynamic approach to task allocation include (1) the ability for the human to access knowledge related to the overall system, which is likely to lead to a more fault-tolerant system in the event of computer failure; (2) a more optimal human workload due to the computer's ability to adapt to ongoing task demands; and (3) a more effective utilization of system resources (Rieger and Greenstein, 1982). From a practical standpoint, decision support methodologies based on dynamic allocation policies raise so many issues to which the only solution appears to be further advances in computer hardware and software. Additionally, these allocation schemes appear more appropriate in process or continuous control tasks, where the notion of dynamic allocation of decision making responsibility is intuitively appealing. It is at least conceivable that the computer could determine what the human is doing utilizing artificial intelligence techniques based on statistical pattern recognition (Greenstein and Rouse, 1982). However, automated manufacturing systems would necessarily require more advanced artificial intelligence techniques where knowledge representation schemes (Barr and Feigenbaum, 1981) would need to be carefully considered. The lack of predictability inherent to these systems (Table 12.4.5) essentially dictates that the potentially useful artificial intelligence methodologies be knowledge rather than statistically based, which leads to issues concerning the development and implementation of expert systems. These issues, in turn, make it well worth considering whether efforts should be directed toward developing adaptive computer aiding or for replacing the human altogether as the primary supervisory controller of the system.

12.4.2.4   The Office of Technology Assessment's Report

The Four Case Studies

In a major study undertaken by the Office of Technology Assessment (OTA) on the impact of computerized manufacturing automation (CMA) on employment education and the workplace (OTA, 1984), four case studies were conducted in companies considered leading users of CAD/CAM technology. Three of these were large companies representing, respectively, the automobile, aircraft, and agricultural implements industries. The fourth case study was based on a group of seven small metalworking shops. The technologies studied were NC, FMSs, management information systems, automated materials-handling robots, and CAD. The overall themes emerging from the four case studies are summarized in Table 12.4.6. Further elaboration on several of these case study themes is given in the following paragraphs.

Changing Skill Levels

CMA has given rise to increased needs for conceptual skills as might be associated with programming and for workers to monitor and maintain systems. At the same time, there is a lesser need for motor and decision making skills, the latter often being directly programmed into the technology. The impact of these changing levels is, however, dependent on the technology. In an FMS, human judgment, although perhaps only occasionally required, can still be a critical factor in efficient system operations. Relative to these complex systems, NC machines have the potential to significantly lower skill requirements.

Training

The OTA study found that many operators perceived themselves as lacking training in the capabilities of their machines. Company-sponsored courses that were designed to train operators about the new computerized technologies were criticized for not addressing the specific capabilities of their machines. The increased sophistication in electromechanical equipment brought about by these technologies have resulted in similar perceptions being voiced by maintenance workers, many of who felt they were inadequately prepared for their tasks. Their perceptions of unpreparedness were further compounded by the rate of technological change. These problems are closely related to issues discussed in Sharit and Salvendy (1982) where it was suggested that uncertainty, especially from workers perceiving having increasingly less knowledge about the systems they are interacting with, would be a major source of stress in future work environments.

Boredom and Stress

Many of the components of boredom and stress that arise from CMA technologies are based on the fact that these technologies are designed to require minimal (and typically routine) operator intervention, yet their smooth operation depends on the operator's alertness and intelligent response to unexpected difficulties that develop. NC operators typically reported boredom when parts required long running

Table 12.4.6   Summary of Overall Themes Emerging From OTA's Case Studies of Companies Considered Leading Users of CAD/CAM Technology

Changes in skill requirements and occupational structure—There was a tendency to embody skill in machines or to move skill to an earlier point in the design and manufacturing process. In occupations such as highly skilled machining this meant that fewer skills would be required on the job. Maintenance work, however, tended to require more skills.

Training—Some operators and maintenance workers expressed a strong desire for more training that would allow them more effectively to run or to repair the machines to which they were assigned.

Increased interdependence—The introduction of Programmable Automation (PA) brought about a greater interdependence among production workers, greater collaboration among maintenance workers, and the necessity for increased cooperation between production and maintenance workers.

Decreased autonomy—Computer-based automation is used in ways that result in decreased autonomy for workers, stemming from the removal of production decisions from the shop floor, the electronic monitoring of some work areas, and the attempt on the part of management to establish an even flow of parts through the plant and of information through the company.

Boredom—One of the consequences of systems intended to minimize operator intervention is that machines may run for longer, although not indefinite, periods of time without active intervention by the operator. For some machine operators, boredom on the job has become a widespread complaint. Some maintenance tasks, however, have become more challenging.

System downtime—Because of the complexity of programmable systems and their high level of integration, the effects of problems with any unreliable element of the system tend to spread, affecting the work pace of production workers and putting great pressure on those involved in the maintenance of the system. Downtime may decrease with better system design and more reliable components.

Stress—Two major sources of automation-related stress were identified: (1) working on very complicated, very expensive, and highly integrated systems and (2) the lack of autonomy at work, extending in some cases to computerized monitoring by management.

Safety—Some applications of PA make the workplace safer, either by eliminating hazardous jobs altogether or by allowing the operator to stand farther from the machine during operation. Other applications introduce hazards of their own, such as automated carriers, clamps, and fixtures that move and close without direct human initiation and sometimes without warning. The net effect of PA, however, is a reduction in traditional physical hazards.

Cleaner and lighter physical work for operators—Some forms of PA have reduced or eliminated heavy or dirty work. In some cases, new jobs requiring physical labor are created in the place of the old, heavier jobs.

Job security—The combination of substantial layoffs at all the large companies and the widespread perception among workers that the introduction of computerized automation caused significant displacement raised strong apprehensions among workers.

Source: OTA (1984)

times. In these instances, the time between new setups could be hours or even days. Large lot sizes, which demand the same part be made repeatedly, also contributed to boredom. The assignment of programming functions to specialist part programmers has most likely contributed to these problems since, in effect, this fragmentation has resulted in shifting the most interesting part of NC machining work from the NC operator. Boredom appeared to be less of a problem with FMS operators, apparently due to the larger number and variety of unexpected events that occurred. However, during the nonproblematic periods, boredom was a factor, since it was difficult to remain alert when the task consisted primarily of monitoring. The lack of active intervention and application of decision making skills served as the catalyst for this effect (although, as the experimental work on HSC in FMSs suggested, the human–FMS interface could conceivably be designed in a manner that allows for active intervention). This alternation between periods of relative inactivity and stress stemming from unanticipated problems appears to be a characteristic of complex and costly computerized work environments and reflects a nonoptimal utilization of human resources.

Concluding Remarks

In its summary on the effects of CMA on the work environment, the study by the OTA notes that one of the beneficial aspects of computerized manufacturing technology is its ability to provide a wide range of choices for systems design and implementation, not only in hardware but in the organizational management aspects of production as well. Yet many problems of a fundamental human factors nature exist, suggesting that the potential for optimization of computerized automation in the work environment has not nearly been realized. Identifying the factors and contexts that govern the impact of CAD/CAM technologies in the work environment is the logical first step. However, the true challenge emerges from the issues that follow. Ample opportunities for solving these issues will derive not only from existing human factors methodologies, but from new human factors methodologies, whose development and discovery will be motivated by these new and challenging problems.

REFERENCES

Adams, T. L. (1983). Computer integrated manufacturing: opportunities and barriers (Technical Paper MM83-477). Dearborn, MI: Society of Manufacturing Engineers.

Bainbridge, L. (1983). Ironies of automation. *Automatica, 19,* 775–779.

Barr, A. and Feigenbaum, E. (1981). *The handbook of artificial intelligence* (vol. 1). Stanford, CA: Heuristech Press.

Blumberg, M., and Alber, A. (1982). The human element: its impact on the productivity of advanced batch manufacturing systems. *Journal of Manufacturing Systems, 1,* 43–52.

Edwards, M. (1984). Robots in industry: an overview. *Applied Ergonomics, 15*(1), 45–53.

Edwards, E., and Lees, F. P. (1974). *The human operator in process control.* London: Taylor & Francis.

Ekkers, C. L., Pasmooij, C. K., Brouwers, A. A. F. and Janosch, A. J. (1979). Human control tasks: a comparative study in different man-machine systems. In J. E. Rijusdorp, Ed. *Case studies in automation related to humanization of work.* Oxford: Pergamon.

GM (1984). *Manufacturing automation protocol (MAP) Specification.* Warren, MI: APMES A/MD, GM Technical Center.

Greenstein, J. S., and Wysk, R. A. (1980). Human-computer interaction in automated manufacturing systems. (Unpublished manuscript). Blacksburg, VA: Virginia Polytechnic Institute and State University, Department of Industrial Engineering and Operations Research.

Greenstein, J. S., and Rouse, W. B. (1982). A model of human decision making in multiple process monitoring situations. *IEEE Transactions on Systems, Man, and Cybernetics, SMC-12,* 182–193.

Grodsky, M. A. (1962). Risk and reliability. *Aerospace Engineering, 21,* 28.

Groover, M. P. (1980). *Automation, production systems and computer-aided manufacturing.* Englewood Cliffs, NJ: Prentice-Hall.

Gunn, T. G. (1982, September). The mechanization of design and manufacturing. *Scientific American, 247,* 114.

Hatvany, J., Merchant, M. E., Rathmill, K. and Yoshikawa, H. (1982). Results of a world survey of computer-aided manufacture. In H. Akashi, Ed. *Control science and technology for the progress of society* (pp. 2169–2176). Oxford: Pergamon Press.

Hwang, S. L., Sharit, J., and Salvendy, G. (1983). Management strategies for the design, control and operation of flexible manufacturing systems. *Proceedings of the Human Factors Society, 27th annual meeting* (pp. 297–301). Norfolk, VA.

Hwang, S. L., Barfield, W., Chang, T. C. and Salvendy, G. (1984). Integration of humans and computers in the operation and control of flexible manufacturing systems. *International Journal of Production Research, 22,* 841–856.

Hwang, S. L., and Salvendy, G. (1984). Human supervisory performance in flexible manufacturing systems. *Proceedings of the 28th annual meeting of the Human Factors Society.* (pp. 664–669). San Antonio, TX.

Inaba, S. (1981). An experience and effect of FMS in machine factory. *Preprints of the IFAC VIII triennial world congress.* Kyoto, Japan.

IITRI (1967). *APT part programming.* New York: McGraw-Hill.

Johannsen, G. and Rouse, W. B. (1979) Mathematical concepts for modeling human behavior in complex man-machine systems. *Human Factors, 21,* 733–747.

Kamali, J., Moodie, C. L., and Salvendy, G. (1982). A framework for integrated assembly systems: humans, automation and robots. *International Journal of Production Research, 20,* 431–448.

Kearney and Trecker Corp. (1976). *Understanding manufacturing systems* (Vol. I). Milwaukee, WI: Author.

Kegg, R. L., and Carter, C. F., Jr. (1982). The batch manufacturing factory of the future (Technical Paper, No. MS82-952). Dearborn, MI: Society of Manufacturing Engineers.

Kelley, R. S. (1969, June 22). The production man's guide to APT-ADAPT. *American Machinist, 97.*

Kochan, A. (1984, July). FMS: an international overview of application. *The FMS Magazine,* 153–156.

Kohler, C., and Schultz-Wild, R. (1983). Flexible manufacturing systems—Manpower problems and policies (Technical Paper, No. MM83-470). Dearborn, MI: Society of Manufacturing Engineers.

Licklider, J. C. R. (1960). Man-computer symbiosis. *IRE Transactions on Human Factors in Electronics,* 4–11.

Macek, A. J. (1982). Human factors facilitating the implementation of automation. *Journal of Manufacturing Systems, 1,* 195–205.

MDSI. (1984). *COMPACT II Programming manual.* Ann Arbor, MI: MDSI.

Meister, D. (1966). Human factors in reliability. In W. G. Ireson, Ed. *Reliability Handbook.* New York: McGraw-Hill.

Meister, D., and Rabideau, G. F. (1965). *Human factors evaluation in system development* (p. 16). New York: Wiley.

Office of Technology Assessment. (1984). Computerized manufacturing automation: employment, education, and the workplace. (OTA-CIT-235). Washington, DC: U.S. Government Printing Office.

Parsons, H. M., and Kearsley, G. P. (1982). Robotics and human factors: Current status and future prospects. *Human Factors, 24,* 535–552.

Pasmore, W. A., and Sherwood, J. J. (1978). *Sociotechnical systems: a source book.* La Jolla, CA: University Associates Inc.

Rieger, C. A., and Greenstein, J. S. (1982). The allocation of tasks between the human and computer in automated systems. *Proceedings of the IEEE conference on systems, man and cybernetics,* (pp. 204–208). New York: IEEE.

Rosenbrock, H. H. (1982). A flexible manufacturing system in which operators are not subordinate to machines (Unpublished Paper). The University of Manchester Institute of Science and Technology.

Rosenbrock, H. H. (1983a). Designing automated systems—need skill be lost? *Science and Public Policy, 10,* 274–277.

Rosenbrock, H. H. (1983b). Draft outline of future software research programme. (Unpublished paper). The University of Manchester Institute of Science and Technology, England.

Rouse, W. B. (1976). Adaptive allocation of decision making responsibility between supervisor and computer. In T. B. Sheridan and G. Johannsen, Eds. *Monitoring behavior and supervisory control* (pp. 295–306). New York: Plenum.

Rouse, W. B. (1980). *Systems engineering models of human-machine interaction.* New York: North Holland.

Rouse, W. B. (1981). Human-computer interaction in the control of dynamic systems. *Computing Surveys, 13,* 71.

Sharit, J. (1984). Supervisory control of a flexible manufacturing system: an exploratory investigation (Ph.D. Dissertation). West Lafayette, IN: Purdue University.

Sharit, J., and Salvendy, G. (1982). Occupational stress; review and reappraisal. *Human Factors, 24,* 129–162.

Sheridan, T. B. (1976). Toward a general model of supervisory control. In T. B. Sheridan and G. Johannsen, Eds. *Monitoring behavior and supervisory control.* London: Plenum.

Sheridan, T. B. (1980). Theory of man-machine interaction as related to computerized automation. In E. J. Kompass and T. J. Williams, Eds. *Man-machine interfaces for industrial control* (pp. 35–50). Barrington, IL: Control Engineering.

Smith, H. T., and Crabtree, R. G. (1975). Interactive planning: a study of computer aiding in the execution of a simulated scheduling task. *International Journal of Man-Machine Studies, 7,* 213–231.

Stallard, D. W. (1980). Computer modules provide for material handling system expansion. *Industrial Engineering, 12,* 44.

State, G. (1969). EXAPT, Möglichkeiten und Anwendung der Automatisierten Programmierung für NC Maschinen. Carl Hanser, Munich.

Williams, T. J. (1980). Hierarchical computer control systems—the medium for industrial control in the future. West Lafayette, IN: Purdue Laboratory for Applied Industrial Control Report.

Further Reading

Groover, M. P., and Zimmers, E. W. (1984). *CAD/CAM,* Englewood Cliffs, NJ: Prentice-Hall.

Hitomi, K. (1979). *Manufacturing systems engineering.* London: Taylor & Francis.

Koren, Y. (1983). *Computer control of manufacturing systems.* New York: McGraw-Hill.

Panky, P. (1983). *The design and operation of FMS.* IFS Publications.

Pressman, R. S., and Williams, J. E. (1977). *Numerical control and computer-aided manufacturing.* New York: Wiley.

SME. (1983). *Numerical control,* Society of Manufacturing Engineers.

CHAPTER 12.5

HUMAN FACTORS CHALLENGES IN PROCESS CONTROL: THE CASE OF NUCLEAR POWER PLANTS

DAVID D. WOODS

Westinghouse Research & Development Center

JOHN F. O'BRIEN

Electric Power Research Institute

LEWIS F. HANES

Westinghouse Research & Development Center

12.5.1 INTRODUCTION

The domain of process industries is a particularly important application for the human factors field because of the demands placed on the human element in these systems. In process environments (regardless of the particular process—continuous or batch, energy or chemical related), there is a dynamic real-time element, systems are complex and often highly automated, and there can be high safety and economic costs associated with failures in human performance.

This chapter does not present an exhaustive review of all human factors activities in all process industries; the industries are too diverse and the activities too extensive to describe in one chapter. Instead the chapter samples a range of recent human factors work primarily in the context of nuclear power plants and the U.S. nuclear industry, in particular [see Edwards and Lees (1974) for a classic work on a broad range of process industries]. This process industry was chosen because it exemplifies three of the challenges and trends that are important, in general, for human factors in complex systems.

First, the importance of human performance in the safe and economic operation of these processes requires identification and correction of human–machine interface deficiencies in existing control panels. One part of this chapter provides an overview and examples of the identification and correction of human–machine problems in conventional control rooms for operating nuclear power plants. These cases illustrate how human–machine performance can be improved through the application of proven human factors techniques and guidelines in order to produce control panels compatible with the operator's sensory and physical limits.

Second, the human's role in increasingly complex and highly automated environments is primarily one of information processor and decision maker—the cognitive factor. Ensuring adequate human performance in these environments demands techniques and knowledge to identify and support operator decision making. Another part of this chapter describes samples of the work that have begun to analyze, model, and aid operator decision making, again primarily in the context of the nuclear power control room application.

Finally, advances in technology have increased the potential range and power of advanced human–machine interfaces. However, the significant cost and the large range of possible kinds of advanced systems that could be developed and implemented requires sophisticated evaluation techniques to identify the benefits that can be achieved. A third part of the chapter addresses the characteristics of good, that is, informative, evaluation studies, particularly studies to evaluate aids to operator decision making.

12.5.1.1 Process Control

One of the earliest processes under human control was the making and tending of a fire. Those responsible for a fire had to add chunks of wood of an acceptable size and condition, at the correct time and in the proper amount, to maintain the fire so that heating and cooking could take place. Control of this process was an art, dependent on the "operator's" ability to sense process conditions directly and to perform appropriate control actions as they were required.

As civilization and industrialization developed, processes became larger and more complicated. It became important that the products of a process meet some minimum consistency standards, be of reasonable quality, and be produced at a reasonable cost. It was no longer possible for the operator simply to sense process state directly and to base control decisions solely on experience.

Sensing and measuring devices were installed to provide more accurate and consistent data to all operators responsible for controlling a process. Standardized procedures were introduced. But operators still controlled the process directly, albeit with varying degrees of success depending on their abilities and the complexity of the process.

The next important event in process control was the introduction of regulators, or feedback controllers. It was no longer necessary for a human operator to take every action to control a process. Appropriate process levels, pressures, flows, temperatures, and so on could be selected and automatically maintained close to setpoints.

The number of instruments provided to aid process operators grew rapidly as more and more sensors and regulators were added to handle the increasingly more complex processes. Operators could no longer handle all of the data from the large number of independent instruments. Centralized control boards with graphic displays were introduced to help organize the data in a more understandable fashion.

Digital control computers were added to the process control systems. This was necessary due to the time lags from control inputs to process responses, interactions between variables being controlled, concern about safety limits on process conditions, time available for certain control actions, and so on.*

Thus, humans have progressed from direct sensing and control of the process (the fire) to the situation typical today: indirect knowledge of the process through instruments fed by sensors and computed measurements and computer control of most elements of the process.

The process industry involves systems where material and energy flows are made to interact and to transform each other. Examples include the generation of electricity in conventional fuel and nuclear power plants; the separation of petroleum by fractional distillation in refineries into gas, gasoline, oil, and residue; hot-strip rolling in steel production; chemical pulping in the production of paper; pasteurization of milk; and high-pressure synthesis of ammonia.

Many of the production systems in the process industry require human operators to perform various monitoring, decision making, and control tasks. Human factors knowledge and criteria have been applied in varying degrees within the process industry. Many modern display and control devices used in control systems have been designed based on human factors data. Size of alphanumeric characters, directions of display and control element movements, and so on, often satisfy handbook recommendations.

It is not uncommon for such human factors panel layout recommendations as frequency of use, sequence of use, and importance of use to have been considered during panel design.

Creation of some of the more complex process control systems, however, has not involved consideration of human factors throughout the development process. One result is that function and task allocations between automatic and human system components may not have been optimized. Also, the range of possible process conditions or states may not have been considered during design.

Improper consideration of the human operator during design may lead to human error during operation. The outcome may be that the process and production facility are damaged or placed in jeopardy, or production may be lost for some period of time, or the quality of the product may become unsatisfactory.

12.5.1.2 The Nuclear Power Plant Application: History of Human Factors in the U.S. Nuclear Industry

This chapter deals primarily with one segment of the process industry: the generation of electricity in a nuclear power plant. This application has been selected primarily because a large amount of recent human factors effort has been applied to nuclear power plants. Also, knowledge about human cognitive functioning is being developed and applied extensively in nuclear power plant operations. Selection of this example permits important cognitive research and application results to be presented.

Many people believe that human factors knowledge and criteria were first applied to nuclear power plants following an incident in 1979 at the Three Mile Island (TMI) Unit 2 plant. During this incident, the control room operators spent almost 2 hr attempting to control an accident condition that they did not understand, although they were sure it was not leading to any threats to safety. Unfortunately, the problem was a small break, an incident known as a loss of coolant accident. The crew interpreted the control board instruments as best they could, but the state of the plant and process was not assessed correctly. The water level kept dropping until the nuclear core was partially uncovered and severaly damaged (Ahearne, 1982).

In fact, human factors knowledge had been applied by several organizations in several countries many years prior to TMI. For example in 1975 the U.S. Nuclear Regulatory Commission (NRC) published a reactor safety study known as WASH-1400 (U.S. Nuclear Regulatory Commission, 1975) that documented the increased risk attributable to discrepencies in the design of control rooms, panels, and instruments. This document stimulated interest in and studies of human reliability while operators performed control room tasks. The Electric Power Research Institute (EPRI) began a program in the 1975 time frame to review representative nuclear power plants, and documented many basic and important violations of human factors principles and criteria (Seminara, Gonzalez, and Parson, 1977; Seminara et al., 1979).

There were several investigations conducted following the TMI incident to identify causes and to propose safety improvements. The reports of the Kemeny (President's) Commission, the NRC Special Inquiry Group (Rogovin Committee), and several other review groups all concluded that a major factor in the TMI accident was the failure to consider adequately the human element. Malone et al. (1980), based on a human factors evaluation of the incident, concluded that the human errors experienced were not due to operator deficiencies but rather to inadequacies in equipment design, information presentation, emergency procedures, and training.

Interest in the human factors field and in application of its knowledge and criteria to nuclear power plants grew rapidly following the TMI incident. The NRC, the electrical utilities that operate

---

* This description of the development of process control is based in part on material presented by Savas (1965).

nuclear plants, architect engineers that design plants, vendors that design and supply equipment for plants, and the relevant professional societies all enlarged their human factors activities. [A good description of human factors programs following TMI to about 1982 is contained in Volume 2 of Hopkins et al. (1982).]

At the time of the TMI accident, there were no activities devoted solely to human factors within the NRC. (The NRC is the government agency responsible for overseeing and regulating the nuclear industry.) Within NRC there was human reliability research being conducted as part of an overall risk assessment program. Following the TMI episode, the Division of Human Factors Safety was established in the Office of Nuclear Reactor Regulation in 1980. About a year later, a Human Factors Branch was established within a Division of the Office of Research.

The NRC has had a major influence on the application of human factors. The Division of Human Factors Safety has been involved in utility nuclear power plant (NPP) licensing reviews. Attention has been given to the following human factors areas (U.S. Nuclear Regulatory Commission, 1983a, pp. 1–2):

Review of NPP staffing to ensure that the numbers, functions, and qualifications of personnel are adequate for safe operation.

Review of training programs for both licensed and nonlicensed NPP staff to ensure that personnel are able to meet existing job performance requirements.

Review of procedures and startup testing programs to ensure their adequacy and effectiveness.

Review of NPP control rooms and remote shutdown panels to ensure that they are designed to facilitate the human–machine interface.

Review of utility management and organization to ensure its adequacy to support safe NPP operation.

In addition, the NRC imposed requirements on operating nuclear power plants regarding human–machine interfaces, procedures, and training.

These NRC reviews and requirements have stimulated the U.S. nuclear industry to consider and utilize human factors knowledge and criteria in designing new control rooms, evaluating and retrofitting existing control rooms, developing new human–machine interface products for control rooms, developing operating procedures, and developing simulator and other training mechanisms. Much of the resulting human factors effort has involved application of existing human factors methods and knowledge [e.g., Van Cott and Kinkade (1972)]. But a major exception to this is in the area of operator decision making and cognitive functioning where research in the nuclear industry is in the forefront of the human factors field.

The electrical utility industry has responded to the NRC requirements and reviews. Some utilities have added human factors professionals to their organizations; others have utilized consultants. The Electric Power Research Institute (EPRI), supported by most of the U.S. utilities with nuclear power plants, has funded extensive human factors programs [cf., EPRI (1984)], many of which have resulted in human factors guidelines for use by the nuclear industry [e.g., Pine et al. (1982) and Kinkade and Anderson (1984)].

The Institute of Nuclear Power Operations (INPO) was established in 1979 by the electric utility industry. Unlike EPRI which sponsors research, INPO has emphasized establishing industry wide benchmarks, conducting independent evaluations, collecting and sharing operating experiences, accrediting training programs, and conducting workshops.

Professional societies became concerned about human factors and nuclear power. The Institute of Electrical and Electronic Engineers (IEEE) has sponsored three workshops on the topic (Hall, Fragola, and Luckas, 1982; Parris, 1984; Schmall, 1980), and established a Subcommittee on Human Factors and Control Facilities (Hanes, O'Brien, and DiSalvo, 1982). This Subcommittee is developing guidelines for human performance evaluation; for the application of human factors engineering to systems, equipment, and facilities of nuclear power plants; and for human reliability assessment. The American Nuclear Society (ANS) has established a Technical Group for Human Factors Systems that has organized sessions at ANS meetings. In addition, an ANS Subcommittee has developed standards related to qualification and training of NPP personnel. The Human Factors Society has performed a long-range human factors plan for nuclear reactor regulation (Hopkins et al., 1982), and included sessions concerned with NPP and human factors at its annual meetings. Finally, the Instrument Society of America has a Control Center Committee that has drafted human engineering recommended practices for control centers.

The chapter is split into three sections. The first examines how established human factors knowledge has been applied to identify and correct human–machine interface problems in existing control rooms. The second describes portions of the emerging body of work on cognitive factors in process control including empirical results and models of operator decision making and developments in systems to aid operator decisions. The third section contains some of the important issues in designing effective and informative evaluations of human–machine system improvements, particularly decision aids.

12.5.2 IMPROVEMENTS TO CONTROL ROOMS

This section illustrates how human factors knowledge and techniques have been applied to identify and correct human–machine problems in existing conventional control rooms in the nuclear industry. The problems are primarily due to failures to match control room characteristics to the operator's sensory and physical limits; experience has shown that mismatches in this area can lead to human performance problems, for example, slips of action (Norman, 1981a; Reason and Mycielska, 1982). The focus is restricted to examples of human factors as applied to control room operations; equally impressive examples of good human engineering are also available in the area of power plant maintenance [e.g., Seminara and Parsons (1981); Seminara, Gonzalez, and Parsons (1984); Siegel et al. (1984)]. Moreover, the control room problems discussed in this section are constrained by the types of solutions that can be realized through modifications to control room hardware in place in operating plants. Human factors problems that are best resolved by developing advanced, computer-based information and control technology are discussed in the next section.

Over the last five years, operating and near operating nuclear power stations in the United States have been subjected to an intense analysis to identify and correct human–machine performance deficiencies that affect plant operational safety. This section (1) describes the methods that have been used to identify problems in operational plants, (2) characterizes the major types of problems found, and (3) discusses some of the solutions under development to correct these problems. A brief characterization of control rooms in typical operating stations is included as background to the discussion.

The quality of written procedures provided plant personnel to aid them in carrying out assigned tasks and the adequacy of the training they receive are unquestionably key elements in the human–machine systems of nuclear power plants. Emphasis in this section, however, is placed on equipment design issues since this aspect of the total human–machine interface has been most intensely investigated and more progress can be reported. [Other sources that report research underway or completed on improving written procedures and personnel training include Haas, Selby, Hanley, and Mercer (1983); Jones et al. (1980); and Nelson et al. (1981).]

12.5.2.1 Control Rooms in Operating Plants

Since nuclear power stations in the United States are procured and operated by many independent utility companies, industry design practices have not led to a high degree of control room standardization. Even so, an attempt will be made to characterize important features common to most, if not all, plant control rooms.

Like most modern process plants, nuclear power plants are remotely controlled by human operators from a central control room located at the site (Figure 12.5.1). The control philosophy followed in most plants represents a mix of automation and manual control. During plant startup, full power operation, and planned shutdown, most major plant components are under automatic control. The human operator's role is to monitor and make adjustments to these automatic controllers to ensure efficient plant output.

Plant shutdown (trip) in the face of abnormal occurrences is also automated. Detection of predetermined conditions requiring plant trip and the initiation of shutdown are under automatic control. The crew of control room operators, however, plays a significant role in achieving a safe and efficient shutdown of the plant. Manual actions must be performed at the appropriate time in support of and in parallel to the actions initiated of automatic signals.

Control room operators, then, must be capable of analyzing the current status of the plant and selecting the appropriate manual actions to bring the plant to a safe state. Even though written procedures are provided to aid the control room crew, successful performance of these tasks requires that operators possess comprehensive and up-to-date knowledge of plant systems and a high degree of cognitive skill.

Information about plant state and manual control access are provided to the control room operators by hardwired displays (meters, binary indicators, chart recorders, etc.) and controls (switches, push buttons, etc.). The control-display philosophy in use at most plants has been aptly characterized as "one measurement–one indicator" (Goodstein, 1981), which means that what the operator sees on a meter or dial in the control room (e.g., value for pressure or temperature) is the output of a single sensor out in the plant. Current generation plants are not characterized by machine processing of the raw sensor data to provide higher-level and more intelligent information for plant control. Instead, operators are provided information at the lowest level of detail (Rasmussen, 1983). (As will be discussed later, attempts are currently under way to provide more intelligent process information.)

Given that nuclear plants are large and complex and contain many components that must be monitored and controlled, the use of hardwired display and control technology has resulted in control rooms that are physically large (some control boards for a single unit are 28 m long) and that display numerous signals (upward of 3000) that must be monitored by human operators (see Figure 12.5.1).

Since operators cannot possibly monitor all of this information, process alarm systems have been provided to aid control room personnel in carrying out their monitoring responsibilities. For each

Fig. 12.5.1. Conventional nuclear power plant control room.

process state judged important for plant operation, circuitry is provided that alerts the control room crew both aurally and visually when a process variable deviates from a predetermined limit (setpoint). The alarm messages, which are displayed via annunciator tiles (Figure 12.5.2), are arrayed at the top of most control boards (Figure 12.5.3).

12.5.2.2 Methods to Identify Control Room Problems

A considerable range of human engineering methods and techniques have been applied to identify problems with the control room human–machine interface. Review methodologies have played a dominant role in the identification of human factors deficiencies including (as indicated previously, the NRC has required that the control room in each operating plant be reviewed to identify deficiencies in the man–machine interface that have safety implications):

Checklist Guided Observations. Detailed human engineering criteria are applied to specific equipment features and the level of compliance is recorded by a trained observer. Checklist of criteria relevant to power plant equipment are adapted from standard human factors references [see U.S. Nuclear Regulatory Commission (1981)].

Structured Interviews. Power plant personnel with operational or maintenance experience are interviewed by a member of a human factors review team. Prepared questions designed to elicit human–machine strengths and weaknesses are used [see Seminara et al. (1977)].

Direct Observations of Personnel Behavior. A member of the human factors review team observes personnel performing assigned tasks. Errors or difficulties in performing tasks are recorded.

Task Analyses and Procedure Evaluation. Operator tasks and plant procedures are reviewed using standard function and task analysis techniques. The information and control requirements needed to complete functions and tasks are described and compared to displays and controls available in the control room.

Analysis of Operating Data. Records of reported operational problems are reviewed to identify human errors and the factors that contributed to these errors.

Critical Incident Analysis. Plant personnel are asked to describe accidents or near accidents in which they were involved. Factors contributing to these accidents are recorded.

Analysis of Workplace Dimensions. Measures of key workplace dimensions are recorded for subsequent analysis of the adequacy of control panel and control room configuration.

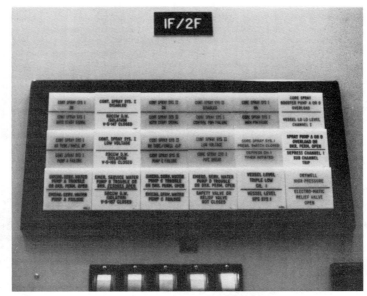

Fig. 12.5.2. Conventional control room annunciator tiles. Copyright © 1984, Electric Power Research Institute. ERPI NP-3448, *A Procedure for Reviewing and Improving Power Plant Alarm Systems Vol. 1. Reprinted with permission.*

Typically, the application of these techniques produces a sizable body of data that is analyzed by a review team comprised of individuals with plant experience as well as human factors specialists. The end product of this analysis is a list of equipment features that are judged to adversely affect operational safety or effectiveness. These design deficiencies or "human engineering deficiencies" typically represent equipment features not in compliance with human engineering criteria and confirmed to be problematic in interviews with operational personnel or as reported in plant operating records or critical incidents.

While there are many salient methodological issues that could be raised regarding the use of these

Fig. 12.5.3. Conventional control room alarm system: the horizontal band of message files above the control board.

techniques, two in particular are worth noting. On the positive side, these techniques are not costly to implement and do yield usable results quickly. The one exception is task analysis, which, if used comprehensively for all or a significant portion of personnel tasks, would be very costly. However, this technique has been applied primarily to the more critical control room tasks.

On the negative side, interpretation of the data generated by application of these review techniques can be highly judgmental. This is not a problem when the identified design deficiencies represent significant departures from human engineering criteria, are independently confirmed to be problems by experienced personnel, and have significant safety impact. But in other cases, where the effect of deficiencies on personnel performance are less clear and the deficiencies can be corrected only at considerable costs, review team judgments are more suspect and must be supplemented with more objective data. Empirical techniques include the collection and analysis of operator performance data from full-scope, control room simulators (Crowe, Beare, Kozinsky, and Haas, 1983; Woods, Wise, and Hanes, 1982), and the statistical analysis of human errors reported in plant accident records (Luckas, Lettieri, and Hall, 1982; Speaker, Thompson, and Luckas, 1982). Analytical techniques focus primarily on the use of probabilistic risk assessment techniques to define opportunities for safety related human errors (Hannaman & Spurgin, 1984; Swain & Gutman, 1983).

12.5.2.3　Problems and Solutions

Application of these techniques has resulted in the identification of numerous design deficiencies in the control room and elsewhere in the plant. Some of these problems are currently being addressed through modifications to installed equipment, changes in procedures, or improved training. No attempt will be made to review all of these problems; they are far too numerous and complex to be treated in a brief presentation. Instead examples are presented that reflect the range of problems identified and the types of solutions that are feasible.

It is not uncommon for operational reviews to uncover in excess of 100 design deficiencies in a typical control room of the type shown before in Figure 12.5.1 (Hanes et al., 1982). These deficiencies represent a broad spectrum of human–machine issues and concerns. In particular, equipment features have been identified that are incompatible with the anthropometric, sensory, perceptual, and cognitive capabilities of the personnel who must interact with control room equipment. For purposes of discussion these deficiencies have been grouped in terms of the specific operator behaviors they affect. Operator activities not fully supported by installed equipment include:

Reading indications.

Reaching controls.

Transforming information.

Activating controls.

Interpreting coding.

Locating individual displays and controls.

Responding to alarms.

While all of these problems may not be present in each control room, they are typical of those found in most operating plants. Each problem is discussed and illustrated with examples taken from operating plants.

Reading Indications

The ability of the human eye to resolve visual targets is a function of the visual angle subtended by the target, the viewing angle, and the illumination present in the viewing area. Obviously the viewer must also have an unobstructed view of the target. Human factors standards are available for each of the preceding sources of variation in visual acuity (McCormick, 1976). Several violations of these standards, however, were observed in control rooms.

Figure 12.5.4 shows an example of one of these violations. Indications located above the black line in Figure 12.5.4 are more than 71 cm from the nearest possible eye position of the average (50th percentile) operator. Seventy-one centimeters is the viewing distance used to establish the size of scale markings and letter and number heights for meters. Viewing distances greater than this require markings larger than those found on standard meters; hence 71 cm is used as the recommended maximum viewing distance given letter sizes on normal meters (McCormick, 1976). Moreover, operators interviewed during the evaluation for the plant shown in Figure 12.5.4 did report difficulty reading the top row of meters on the vertical panel.

In addition to their distance from the viewer, the visibility problem is further compounded by the design features of these meters. They are approximately 20 cm in length, convex in shape, have a fixed scale with a moving pointer, and are mounted with a transparent plastic cover. Since they are convex, the top half of the scale bends away from the viewer; thus this part of the scale must be

Fig. 12.5.4. Displays located in an area of poor visibility.

viewed at an angle greater than 60°. The optimum angle is 90° and the acceptable range is between 60–90° (Thompson, 1972, p. 429). This problem is even worse for other indications on this panel since the pointer position for normal operation is in the top half of the scale. The plastic cover also creates glare problems. Similar problems were observed in secondary panels where indications were located so high on vertical panels that a short operator needed to stand on a chair to have adequate vision.

Visibility problems of the type previously discussed are difficult to correct given the cost associated reconfiguring control panels. Some utilities, however, are replacing meter faces to improve readability and guidance has been provided for utility designers to use in correcting other visibility problems in operating plants (Pine et al., 1982) and to avoid such problems in the future (Kinkade & Anderson, 1984).

Reaching Controls

A properly designed console locates controls within the functional reach of the smallest operator who will use the equipment. Since there are no NRC licensing restrictions on the height and weight of operators, console designs used in power plants should seek to accommodate the fifth percentile (smallest) operator. Figure 12.5.5 shows the extent to which consoles used in the one plant meet this objective. (The dimensions of the console shown may not be exact since they were obtained from direct measures of the console.) Any controls located on the vertical panel or the back edge of the bench panel are outside the functional reach [85 cm including shoulder extension (Damon, Stoudt, and McFarland, 1971, p. 97)] of the smallest operator. Exceeding the functional reach of the smallest operator does not mean that he or she cannot reach controls on the vertical panel, but to reach them he or she must assume an undesirable working posture by extending the upper half of his or her body.

One operator complained that controls on the vertical panel shown in Figure 12.5.5 were difficult to reach. Similarly, another operator reported that when leaning forward to operate feedwater regulating valve controllers located at the bottom of the vertical panel in the steam generator section, he bumped against and accidentally activated the feedwater pump speed controller located on the front edge of the bench panel. Thus, in addition to operator discomfort, an undesirable working posture can have adverse effects on plant operation.

Again, such problems are difficult to correct in operating plants. Some utilities, however, are considering guard rails along the front edge of the console to prevent accidental activation. Also, industry specific anthropometric data (Parris and McConville, 1981) and guidance have been developed for use in preventing these problems in the future (Kinkade and Anderson, 1984).

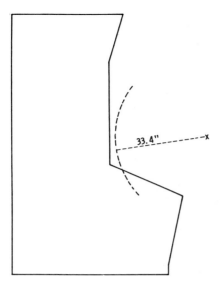

Fig. 12.5.5. Configuration of a control panel from one power plant.

Transforming Data

In the design of data display systems every effort should be made to present data to the operator in the exact form in which he or she intends to use it. Any transformation of the measured data should be done by the control system prior to presentation to the operator. Operators at one station, however, complained about having to make such transformations. In particular, they singled out indications for feedwater header pressure and steam header pressure. Reading from these two meters, which are located side by side, must be compared; yet they have different scales. Similar complaints were made about letdown and charging flow indications.

In addition to delaying operator response and contributing to transformation errors, these scale differences can lead to another type of operator error. Pointer position, which operators often use as data instead of actual readings, could be misleading in a case where scales differ. As shown in Figure 12.5.6 an operator utilizing pin position might conclude that pressure is lower in component A than B when in reality it is higher in A.

Recalibration of meters and modifications of meter faces are being investigated as a means of correcting information transformation problems. Another possible solution is to make use of parameters that must be compared on video display terminals (VDTs) that are now being installed in most plants. In such applications the computer can be used for raw data transformation and presentation in the correct format for each comparison.

Activating Controls

Individual controls should be designed so that the direction in which the control is moved to achieve a desired setting or result is clear to the operator and does not violate any expectations he or she may have developed as a result of past experience. In particular, nomenclature for control position settings should communicate clearly the result of a given control movement, and the direction in which controls are moved to achieve a given result should be consistent throughout those parts of the control system used by the same operator. For example, turning equipment "off" should always

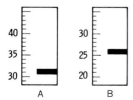

Fig. 12.5.6. Adjacent meters with different scales.

Fig. 12.5.7. Inconsistent direction-of-movement relationships.

involve movement of the control in the same direction, preferably to the left since this direction of movement for off conforms to operator expectation (Chapanis, 1972, p. 350).

For example, Figure 12.5.7 depicts inconsistent direction-of-movement relationships. To raise the valve, the valve control must be moved to the right; yet to raise the speed of the feedwater pump its control must be moved to the left. Since a control movement to the right to achieve an increase in some measure conforms with operator expectation (Chapanis, 1972, p. 150), the direction-of-movement relationship for the pump speed control is incorrect; it should be made consistent with its neighboring valve control.

Shown just above the speed control in Figure 12.5.7 is its corresponding display; the two are related in the following manner. A control movement to the left produces a needle deflection to the right in the display. Thus the control-display direction of movement relationship is also inconsistent. This inconsistency would also be corrected by the change in direction of movement previously prescribed.

Interpreting Coding

Color and illumination coding are used on control panels to communicate information to the operator. When used properly, coding can be a very effective form of communication and can greatly reduce the difficulty in performing certain tasks. The time required for locating a given control or display, for example, can be reduced by the use of color coding. Successful use of codes, though, is contingent upon the ease with which the operator can learn the correct color/control function associations; naturally these associations must be free of inconsistencies.

The color coding schemes used for the controls shown in Figure 12.5.8 were not chosen with the above rule in mind. The back area of the turbine panel trip control is color coded red; this is a good color selection since the association between red and emergency is well learned by most individuals. The back area of the control located on the reactor panel, however, is coded black; yet it is identical in function and criticality to the turbine panel control. An association learned while using the turbine panel control could lead to poor performance when applied to the reactor panel. It might take the

(a)   *(b)*

Fig. 12.5.8. Inconsistent labeling and coding on critical controls (the control on the left is coded with a black background; the one on the right with a red background). (*a*) Control located in reactor control area, (*b*) control located on turbine panel.

operator longer to locate the reactor trip control if he or she searched for a control color red than if he or she did not use color as a search aid.

In addition to its use to signify emergency or alarm conditions such as on reactor trip controls and first out annunciators, the color red is also used to indicate the status of valves and pumps at the plant where the controls in Figure 12.5.8 were observed. When a circuit breaker to a pump is closed (pump operating), a red light on the pump control is illuminated. Similarly, when a valve is open, a red light on the valve control is illuminated. Each of these red lights signifies activity (flow), and during normal operation many red lights on the panel are illuminated. The extensive use of the color red as a status indication could diminish its attention getting value in emergency situations. As with valve and pump nomenclatures, the use of red and green to indicate valve and pump status is a standard industry practice and should not be changed; however, such color philosophies should be avoided as additional functions are color coded.

While problems associated with control activation and use of coding techniques are technologically feasible to correct, solutions must be approached with a great deal of caution. Clearly, the direction of movement relationships and coding schemes used in some cases are far from optimum. These relationships and schemes, however, are well learned and to change them may cause more errors than are prevented. Attention, in most cases, has focused on elimination of inconsistencies that operational experience has shown to be problematic.

Locating Components

Searching for and correctly locating individual components on the board was reported as a major problem by personnel interviewed in many plants. Given the large number of controls and displays found on most control boards, many of which are similar in appearance, this finding is not surprising. Even experienced operators report that they catch themselves reaching for the wrong control and that occasionally such errors are not detected until after the control has been activated. Correct identification of an individual control or display, then, can be time consuming and can delay operator responses at a time when speed of response may be important.

Many design features contribute to the problem of locating control board components. Figure 12.5.9 shows a section of a control panel which manifests some of these problems. In particular, functional relationships among individual controls and displays are not readily apparent and individual components can be identified only by laboriously reading the detailed labels for each component. Moreover, nomenclature used on labels (not visible in Figure 12.5.9) is inconsistent in some cases.

Figure 12.5.10 shows proposed modifications to this same panel. A detailed task analysis was performed to discern functional relationships between components and these are indicated with functional demarcation (Seminara et al., 1979). Generic labels and color coding [not visible, but see Hanes et al. (1982)] have also been applied to aid in component location. Such simple enhancements clearly communicate functional relationships and make it easier to locate individual control board components.

Responding to Alarms

Several significant deficiencies with process alarm systems have also been identified in operational reviews. The most significant of these—the avalanche of alarms that can occur during major plant transients and the problem of standing alarms—can be attributed to the philosophy of single variable

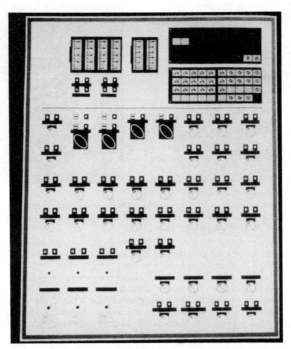

Fig. 12.5.9. Control panel before enhancements. From Seminara et al. (1979); Copyright © 1979, Electric Power Research Institute. EPRI NP-1118 volume 1, *Human Factors Method for Nuclear Power Control Room Design.* Reprinted with permission.

alarming used in most plants. This philosophy is followed because most alarm system hardware is hardwired and provides little logic capability for multivariable conditioning. Consequently, the control room crew is subjected to an avalanche of incoming alarms following a major transient when many process parameters shift above or below limits (Control rooms may contain in excess of 2000 individual alarm points and as many as 200–300 of these may be active in the first few minutes of a transient.) Likewise, shifts in plant operating mode (e.g., from full power operation to hot standby) typically produce numerous standing alarms (active alarms that do not require operator action) since alarm setpoints are usually established for only one mode of plant operation (full power operation). These problems cannot be easily overcome via modifications to installed hardware given the inherent limitations of hardwired logic. Computer-based alarm handling approaches discussed later in the chapter, however, do offer promise.

Apart from the problems attributable to single variable alarming, difficulties related to the visual aspects of alarm presentation can be reduced through enhancements of hardwired annunciators. Figure 12.5.2 shows an annunciator panel observed in one plant (Fink, 1984). Readability is poor due to letter size on the engraved annunciator tiles, and functional relationships are not apparent. Figure 12.5.11 illustrates an enhancement of this panel. Demarcation, generic labels, and regrouping of tiles have been employed to convey functional relationships. Readability is improved through engraving the tiles with larger lettering.

Only a few of the observed problems related to alarm presentation have been discussed; several sources are available which provide more in depth coverage of these problems (Banks & Boone, 1981; Fink, 1984). Also, guidance has been developed to aid utilities in conducting detailed reviews and enhancement of alarm systems (Fink, 1984).

To a large measure, the kinds of human–machine problems that have been identified in operating nuclear power plants can be attributed to the failure of early contol room designs to follow sound human engineering practice, and the inability of control philosophies and technologies now in use to meet the control room operator's demand for intelligent process information. To date, the application of proven human factors techniques and knowledge has been successful in correcting many of the earlier designer failures. For the most parts, these enhancements are producing control boards that are more compatible with the operator's sensory and physical capabilities. If, however, the operator's cognitive needs are to be better met, the inherent limitations of the hardwired technology typically

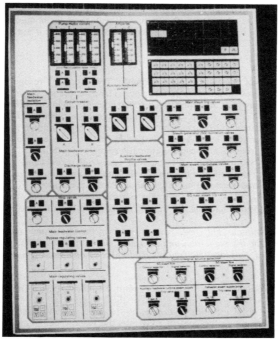

Fig. 12.5.10. Same control panel as Figure 12.5.9, after enhancements. From Seminara et al. (1979); Copright © 1979, Electric Power Research Institute. EPRI NF-1118 Volume 1, *Human Factors Method for Nuclear Power Control Room Design.* Reprinted with permission.

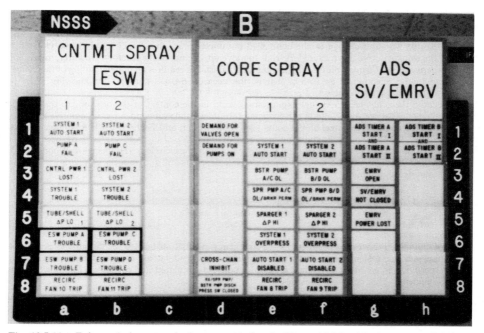

Fig. 12.5.11. Enhanced alarm panel: the same panel as in Figure 12.5.2 with alarms organized and labeled in well-defined groups. Copyright © 1984, Electric Power Research Institute. EPRI NP-3448, *A Procedure for Reviewing and Improving Power Plant Alarm Systems Vol. 1. Reprinted with permission.*

used today must be overcome by the effective use of process computers to provide more intelligent process information. In addition, given the significant cost of backfitting computers to operating control rooms, greater emphasis will be placed on the use of evaluation techniques to clearly demonstrate the benefits attributable to more advanced interface systems and operator aids.

12.5.3   COGNITIVE FACTORS

Assessments of the role of the human element in the TMI accident recognized the importance of "cognitive" factors in successful operator performance. These factors are becoming increasingly important in all process domains (e.g., flightdecks, air traffic control, chemical processes, energy processes, data network control) as increased applications of computers transform the human's role in complex systems from manual controller to supervisor. As a supervisor, the human monitors and manages a partially self-controlling process, handles the unexpected, and provides backup control when automatic systems fail or when disturbances are beyond automatic response capabilities. This means that operators function primarily in a cognitive mode (set/monitor goals, solve problems, make choices) and only secondarily as a simple sensor or effector mechanism (Rasmussen and Rouse, 1981; Sheridan & Hennesey, 1984; Sheridan, Jenkins, and Kisner, 1982; Sheridan and Johannsen, 1976). For example, the following quote (Mertes and Jenny, 1974) about air traffic control systems captures the shift in the human's role in complex systems (note that this is a description of a "fully automatic" system):

> Level V—Full Automation: This level represents a hypothetical system in which man has no direct responsibility for the regulation and control of air traffic. Man's role has become that of a system monitor and manager. He controls a complex of automated resources which, in turn, control aircraft.

The distinction between manual and supervisory control is reflected in a distinction in the mechanisms underlying human error between "slips," errors in the *execution* of an intention, and "mistakes," errors in the *formation* of an intention (Norman, 1981a; Reason and Mycielska, 1982). If a nuclear plant operator decides to move the control rods into the reactor but grabs the control device and begins to move them out, then a failure to execute an intention occurred (action not as planned). On the other hand, if the above operator drove the rods in when circumstances did not demand this response, then a failure in the formation of an intention occurred. However, from the operator's point of view, the action may have been exactly what he intended, *given his or her assessment of process state.*

The traditional ergonomic guidelines discussed in the previous section can address slip prevention (support of *how* to take actions). In the above example, the slip might have occurred due to a violation of the ergonomic guideline on control movement compatibility: the controlled device movement should be compatible with the change in the controlled parameter (e.g., this principle would be violated if pulling the control device down caused the rods to move up, i.e., out of the reactor). However, improving human performance through preventing or minimizing mistakes requires an understanding of the factors that lead to the selection of an action, that is, the decision process (support for knowing *when* to choose one response given the set of possible responses).

Recently, there has been a virtual explosion of work, in the context of the operator in the control room of a nuclear power plant and in other domains, on how to identify and support cognitive activities in work environments (Hollnagel and Woods, 1983; Pew, Miller, and Feehrer, 1981; Rasmussen, 1983, 1984; Woods, 1984b). The techniques, designs, and data developed in these studies have been one of the major sources for the development of an applied cognitive psychology (Hollnagel & Woods, 1983; Norman, 1981b). This section samples some empirical results on and models of operator decision making and decision aiding.

12.5.3.1.   Operator Decision Making

Empirical Results

Understanding how operators perform during emergency events is critical to the design and evaluation of human–machine systems in process control applications. One data base on operator decision making in emergencies has emerged through a series of studies of simulated and actual nuclear power plant critical incidents. This data base includes:

   1.   Woods et al. (1982). This study examined the decision making of experienced operators in simulated emergency events including multiple failure events. Events were run both with and without a computerized operator aid (39 scenarios from 8 crews in 7 events). Figure 12.5.12 contains an example of the decision flowchart format used in this study to summarize performance protocols.
   2.   Pew et al. (1981). This study was a retrospective analysis of operator decision making in four actual plant critical incidents. Figure 12.5.13 is a sample of the decision protocols produced in this

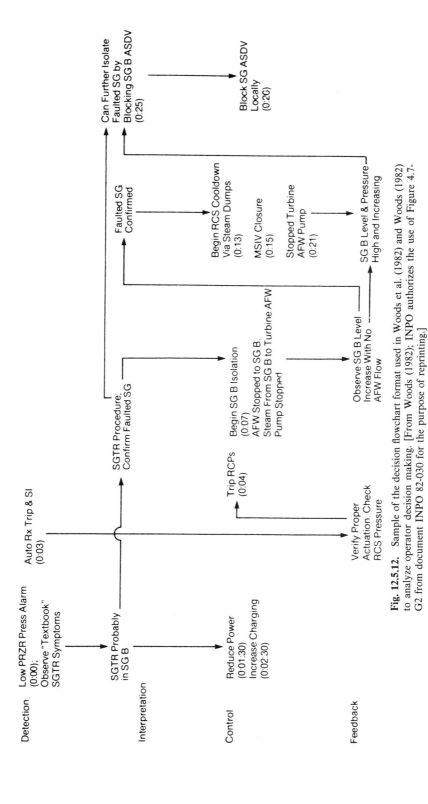

Fig. 12.5.12. Sample of the decision flowchart format used in Woods et al. (1982) and Woods (1982) to analyze operator decision making. [From Woods (1982); INPO authorizes the use of Figure 4.7-G2 from document INPO 82-030 for the purpose of reprinting.]

OPERATOR DECISION/ACTION SUMMARY

Location: OC-BWR Event: LET Page 5 of 12

Date: 5/2/79

Time	Avail. Info. or Stimulus (Info./Loc.)	Event Signaled	Knowledge and/or Belief State Components	Intention	Expectation	Decision/ Action	Source for D/A	Immediate Feedback	Comments
1351: 36	Low Water Level Alarm Clears	Water Level Above Core Greater Than 11' 5"	Water Has Entered Annular Region From Isolation Condenser as Intended. Water Level Within Shroud is Correlated With Water Level in Annular Region	Continue Restoration of Level	Level and Pressure Within Shroud Will Increase. Temperature Will Decrease Because of Input of Cold Water to Annular Region	Monitor Level	SOP	"Yarway" Water Level Instrumentation.	Although Water Level in Annular Region is Increasing, Level Within Shroud Continued to Decrease Because of Closure of B and C Discharge Values and Continued Boiling Within Core.
1352: 52	Triple Lo-Level Alarm ♦ "Yarway" and G/MAC Water Level Indicators	Water Within Shroud 4' 8" Over Top of Active Fuel	Level in Annular Region Equal to or Greater Than 11' 5". Water Level Within Shroud Must Be Higher Than Lo-Level Alarm Suggests Because of Design Correlation Between Annulus and Shroud Water Levels	Rely on "Yarway" and G/MAC Indicators ♦ Continue to Attempt Resolution of Discrepancy	Water Level Within Shroud Satisfactory	Continuing Monitoring, Water Level Indications From "Yarway" and G/MAC	Training and SOP Emphasize Use of the 2 "Yarway" and 2 G/MAC Instruments for Gauging Water Level	No Information Pertinent to the Disparity Between Annulus and Shroud Levels Available	Control Room Contains No Instrumentation Related to Level Within Shroud. Only Available Information on Status Comes From Alarm Annunciator Panel. Level Within Shroud Can Be Determined Exactly by Reading Gauge Located in Reactor Building.
1353: 06	Recirculation Loop Discharge Valve Indicators ♦ Recirculation Loop Temperature Indicators ♦ Reactor Pressure Indicators	All Valves Closed ♦ A, B, C, Temps Still Increasing ♦ Reactor Pressure Continuing to Decrease	B & C Discharge Valves Open as Required by Procedure ♦ Plant Cooling Rate in Excess of Maximally Desirable Rate (100°/Hr.)	Cooldown Rate Must be Reduced	With Reduced Cooldown, Pressure Will Stabilize	Prepare to Close Isolation Condenser Valve	SOP		Status of A, B, C, and E Discharge Valves Could Have Been Determined by Examining Panel Indicators, But Operators Were Under Impression That Required Procedure Had Been Followed Earlier and Did Not Check.

Fig. 12.5.13. Portion of an operator decision protocol from Pew et al. (1981). Copyright © 1981, Electric Power Research Institute. EPRI NP-1982, *Evaluation of Proposed Control Room Improvements.* Reprinted with permission.

analysis. The analysis also used "Murphy diagrams" (Figure 12.5.14) to trace the relationship between the decision making process and performance problems.

3. Woods and Roth (1982). This study examined the decision making of experienced operators in simulated emergency events (15 crews in 4 events) via a more efficient version of the Woods et al. (1982) methodology (Figure 12.5.15). Quality of performance was measured through instructor ratings of crew performance (Figure 12.5.16).

4. Woods (1982). This study was a retrospective analysis of operator decision making in one actual power plant emergency event using the Woods et al. (1982) decision flowchart representation. Of particular interest in this study was the fact that an additional facility was active to support control room emergency operations, the Technical Support Center.

This data base exists because all of these studies used a process tracing methodology and a common perspective on decision making (Rasmussen, 1983) to analyze operator performance. Operator behavior was analyzed with respect to categories of monitoring plant data, interpretation of these data (knowledge or belief state, intention, expectation, recovery strategy), the resulting control actions, and feedback on the results of recovery actions. These concepts were used to chart, not just what actions the crews took, but also the decision process and context that led to the actions. In each study, the critical decisions identified in these protocols were subjected to a more detailed analysis based on the particular objectives of that study.

In general, the operators in these studies performed well. Recovery actions were prompt, there were relatively few operational problems, and those that did occur were usually of little consequence to the final outcome. With respect to the operational problems that did occur, the crews rarely had any difficulties with the initial state identification following the onset of the emergency. In general, when operational problems did occur, it was because operator state identification and control activities gradually became decoupled from actual process state as a function of execution problems or the unexpected.

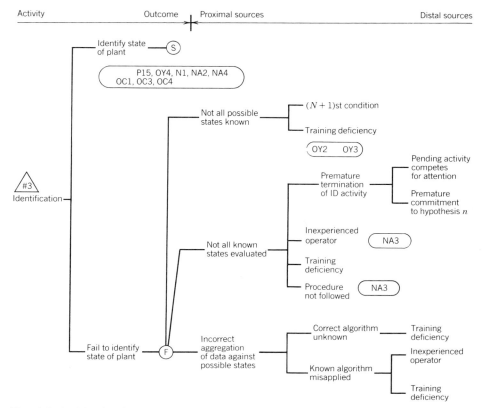

Fig. 12.5.14. Murphy diagram for the decision stage—Identification of system state; this diagram is a technique to analyze operator decision making. From Pew et al. (1981); Copyright © 1981, Electric Power Research Institute. EPRI NP-1982, *Evaluation of Proposed Control Room Improvements . . .* Reprinted with permission.

Decision Flowsheet—Steam Generator Tube Rupture (SGTR)

Decision	Observe/monitor	Discuss	Intent/expect	Action	Feedback/followup
Diagnosis process**	Time:				
—Symptoms considered					
Pzr level					
Pzr press — 0:20					
SG level					
Feed flow					
SJAE monitor — 1:10					
B/D monitor					
Containment					
Other:					
—Intermediate hypothesis —	Sprays shut				
—Confirmation process —	PORV shut				
—Who (SRO, whole crew; etc) —	SRO looking 1:57—Looking for leakrate				
—Diagnosis complete at	1 : 16 min : sec				
SG Identification**	Time: 4:17				
—Symptoms considered		SRO telling BOP to look for different feed flows		SRO 5:33 "Don't want SI" unless you have to!	
SG level — D 5:34				6:10 "Lower Power before trip less of a transient on SI"	
Feed flow — D 5:30					
B/D sample — Asked for one 8:35					
Steamline radiation					
Other:					
—Intermediate hypothesis —	Look like SG"D"				
—Confirmation process —	7:44 SRO "Are we sure its D?"	BOP—YES			
—Identification complete at	7 : 45 Min : Sec	8:44 checked again			
Procedure selection	Time:				
SRO opens correct procedure at	1 : 45 Min : Sec				

Fig. 12.5.15. Sample of the decision protocol format developed in Woods and Roth (1982) to increase the efficiency of protocol generation. The decision protocol was customized for each event in the study (this example is a portion of the form for a steam generator tube rupture) and completed by a domain expert (training instructor) during each simulated event [cf. also, Hollnagel (1982); Woods and Hollnagel (1982)].

MODIFIED COOPER-HARPER SCALE FOR
OBSERVER EVALUATION OF CREW DECISIONS

Directions: Circle one in each column

Scenario: *BY24b*   Date: *5-8-82*   Event: *SGTR*   Decision: *Cooldown*

DECISION ACCEPTABLE?	RANK	COMMENTS
	No improvement possible	
Acceptable (no improvement necessary)	No improvement required	
	No practical improvement required	
	Improvement desirable	*Cooled down at high speed past target*
Reluctantly acceptable (some improvement necessary)	Improvement needed	
	Major improvement needed	
	Considered wrong factors	
Unacceptable (improvement required)	Considered many wrong factors	
	Considered only improper factors	

Fig. 12.5.16. Sample of the form used by domain experts (training instructors) to rate the performance of a crew of operators in simulated emergencies [from Woods and Roth (1982)].

The data in Tables 12.5.1, 12.5.2 and 12.5.3 show what happened following the occurrence of some operational problem. [For a broader synthesis of this data base see Woods (1984b)]. When a crew misidentified plant state or had action execution difficulties, they generally failed to correct their understanding of plant state or to identify execution problems within the duration of the test events. (The execution problems can be considered identification problems in that the concern is not the initial source of the execution difficulty, but rather, the operator's failure or success in assessing the effect of the action or nonaction on process state and goals.) Often, when a problem was corrected, a relatively long time period had elapsed (about 7 min on average in the simulated emergencies). None of these delayed detections resulted from a systematic search or problem solving factors; rather incidental factors dominated [e.g., while moving around the control board for other reasons, an operator would happen to note misaligned valves (Pew et al., 1981, pp. B-17–B-34)]. In addition, some operational problems were corrected through the intervention of an external agent—instructor hints in the simulator exercises or, in actual incidents, when a fresh viewpoint entered the situation. These data indicate that operational problems are often due to deficiencies in the identification of process state that are related to poor feedback about the effect of control actions on process state and on goal achievement.

Analysis of the data base also revealed that when events or maneuvers required a relatively static sequence of component actions with little or highly stereotypical feedback from the process, few problems occurred. But when events demanded a relatively variable sequence of component actions and extensive feedback from the environment in order to adapt behavior to unpredictable constraints or disturbances, two kinds of problems began to emerge:

1. Type A problems where rote rule following persisted in the face of changing circumstances that demanded adaptable responses;
2. Type B problems where adaptation to unanticipated conditions was attempted without the

Table 12.5.1   Operator Performance: Error Correction in Simulated Emergencies[a]

	Detection Failure	Detection via External Agent	Delayed Detection	Immediate Detection
Problem in process state identification	14	5	0	0
Problem in execution	10	0	6	4
			7 min average	
Total	24	5	6	4

[a] Data on operator performance in simulated emergencies from Woods et al. (1982) and Woods and Roth (1982). These data account for 70% of all of the operational problems in these studies; the remaining problems were responses judged as too slow or that were noncorrectable. Given the number of test scenarios (99 using 23 experienced crews in 8 different events), there were very few operational problems, and those that did occur were of little consequence to the final outcome.

Table 12.5.2   Operator Performance: Error Correction in Five Actual Incidents

	Detection Failure	Detection via External Agent	Delayed Detection	Immediate Detection
Problem in process state identification	2	1	0	0
Problem in execution	2	0	1	2
			33 min average	
Total	4	1	1	2

Source: Pew et al. (1981) and Woods (1982).

Table 12.5.3   Brief Description of Error Correction Problems in Actual Incidents (Data from Table 12.5.2)

Failure to detect execution problem: shift supervisor ordered Main Steam Isolation Valve closure, but operator was interrupted before action was taken; shift supervisor was unaware action was not taken (Pew et al., 1981, pp. B-35, B-41).

Failure to detect execution problem: execution error during realignment of normal charging and letdown; 19 min were available to detect and correct error (Pew et al., 1981, pp. B-37, B-45–B-46).

Delayed detection of execution problem: serendipitous detection and correction of misaligned valves by external agent 33 min after error occurred (Pew et al., 1981, p. B-17).

Failure to correct problem in state identification: owing to preceding execution error, there were conflicting measures of reactor water level and no natural circulation; crew was unable to explain conflicting measures, unable to accurately assess inventory status, and erroneously assumed natural circulation was occurring until serendipitous discovery of misaligned valves by external agent after 31 min had elapsed (Pew et al., 1981, pp. B-18–B-19, B-22, B-28–B-34).

Failure to correct problem in state identification: crew and technical support center failed to reevaluate an earlier action in light of later changes in plant state; as a result, a potential safety function threat was not detected (Woods, 1982, p. 3-5–3-6).

Failure to correct problem in state identification: failure to detect and correct conditions (19 min available) leading to pressurizer relief tank rupture into containment (Woods, 1982, pp. 3-9, 3-11–3-13).

complete knowledge or guidance needed to manage resources successfully to meet recovery goals.

There are several factors that demand or create problems for adapting responses to changing circumstances during emergencies. One is the mismatch among procedures (especially procedures embodied in a paper medium) as a static, sequential list of activities and the dynamic nature of actual operations (events occur at indeterminate times, operations can occur in parallel, items may need continuous or semicontinuous monitoring). For example, in continuous control tasks* there is a component to skilled performance based on manipulating inputs contingent on output-goal relations; tuning inputs to achieve output goals is a skill that defies complete expression as a fixed sequence of discrete actions. As a result, when a disturbance in the normal sequence of actions occurred; problems arose when operators either failed to adapt the standard sequence (Type A error) or attempted to adapt component responses with incomplete knowledge (e.g., constraints, interactions) of how the component actions work to achieve goals (Type B error). In one case from simulated emergencies (Woods and Roth, 1982), a crew violated an outcome measure of performance (maintain greater than 50° subcooling) in one continuous control task (primary system cooldown in a steam generator tube rupture event). The crew rotely followed and correctly executed the sequence of procedure steps relevant to this task. However, the outcome measure violation occurred because the crew failed to monitor the goal of the task (subcooling), so that when the event later deviated from expected course and the crew was further along in the rule sequence, the crew failed to detect and correct the deviation over a period of 14 min.

A second kind of situation that arose to challenge operator performance is how does an operator assess the relevance of the currently or nominally active rule set to actual plant conditions, for example, how does a crew decide if a procedure or subprocedure is still relevant after a second failure occurs? In two cases in actual incidents (Meyer, personal communication; Woods, 1982). The crews involved thought that a second failure invalidated the procedure they had been using. In another actual event (Brown and Wyrick, 1982, pp. 3-21-3-52), a crew correctly began to follow one procedure; but later, when the leak increased in size, they failed to activate either the newly relevant procedure or all of the relevant recovery goals. One technique that is often used to deal with this problem is to annotate procedures with error checks. Operational problems such as those described in the previous example are often treated in procedure space by placing instructions to perform error checks downstream in the rule sequence. While there are specific places in action sequences that suggest or demand such checks (e.g., permissives such as do not begin primary system cooldown until the faulted steam generator is isolated), the data base shows that the dynamic aspects of emergency events allow for the possibility that the symptom that activates the check may not occur until after the crew has passed the particular step containing the rule (Meyer, personal communication).

Another factor that produces problems in adapting procedural guidance to actual circumstances is what happens after an error occurs in the execution of a procedure. There were examples in the data base where rote procedure execution broke down once an operator error occurred. In one case (Woods and Roth, 1982), the original error invalidated subsequent steps; the crew's attempts to continue to follow the procedure led to further erroneous actions.

The operational problems that can occur when emergency recovery requires adaptable responses to changing circumstances show that rotely followed procedures are inherently "brittle"; that is, they are not readily able to handle novel situations, to adapt routines to special conditions or contexts, to adapt to underspecified instructions (e.g., "stop all unnecessary safety injection pumps"), and to recover from errors [cf. Suchman (1982) and Brown, Moran and Williams (1982) for more discussion of the semantics of procedures and similar results on instruction following from other work domains]. Good operations require more than rote rule following; it requires the operator to "understand how the various steps of a procedure work together to produce intended effects" (Brown et al., 1982, p. 2). Furthermore, the results from this data base highlight, as have other studies of operator performance (Bainbridge, 1974; Umbers, 1976), that the criterion for skilled operator performance is the ability to adapt responses to changing circumstances in the pursuit of goals.

Models

The operator decision making results show that, when there is some mismatch between actual and perceived process state, it is very difficult for those involved to modify their view given today's control rooms. Misperceptions of process state are often corrected only when a fresh viewpoint enters the situation. This pattern can be described as a type of fixation or perseveration effect, in that, an assessment of plant state tends to persist independent of supporting data. The crew's problems in the TMI accident could also be described as the result of a fixation effect, and some of the recommendations/requirements to prevent reoccurrence of such accidents can be seen as classic methods to prevent or reduce this type of decision problem: the requirement for a Safety Parameter Display System to provide a concise statement of plant safety status can be seen as a way to provide improved feedback about goal achievement (cf., the section on evaluation of decision aids); the requirement for a Shift Technical Advisor can be seen as a way to institutionalize a fresh viewpoint.

---

* In these tasks a continuously variable output parameter is to follow some target trajectory or avoid some dynamic limit, even if the controls available to the operator (the inputs) are discrete and are adjusted intermittently (e.g., cooldown and depressurization tasks in steam generator tube rupture events). In discrete control, the operator places a component or system into one of several discrete states (e.g., align a series of valves to provide a flow path from a source to a destination).

The decision making results are consistent with the Bartlett (1943) multilevel view of cognitive processing (cf., Broadbent (1977), Rasmussen (1983), Rouse (1983) for particular examples of this class of models). The key elements of this view are:

1. Human processing of information is organized at different levels, which operate simultaneously over different time spans.
2. Some (presumably "higher") levels modify or control the operations of others.
3. Lower levels are capable of independent function.

The result is that there is a multilevel (not necessarily hierarchical) architecture of cognitive processes. Lower-level processes, which might be characterized as more ballistic or parameter-driven behavior modes, bring efficiencies to behavior (e.g., stereotypical jumps in Rasmussen's step ladder model) but operate only over a limited range of situations. This tradeoff is balanced by other executive layers of processing that monitor and modify the application of the lower-level routines with respect to error correction (are skilled routines proceeding correctly), goal achievement/setting (are skilled routines achieving the desired effects), and responses to novel situations (which includes the ability to distinguish novel from stereotypical situations). If only the upper levels are available, performance would be slow, awkward, and demanding as when we learn a new skill. On the other hand, if only lower levels were available, performance would be effective only when the novel or unexpected do not occur.

Norman (1981a) and Reason and Mycielska (1982) have characterized the conditions for and classes of failures in lower-level skilled routines (actions not as planned). Some of these categories can be seen in the slips during action execution that occurred in this and in other operator performance data bases. However, the assessment of operator decision making (especially after a problem has arisen or a slip occurred) reveals information about the kinds of breakdowns that can occur in executive processes (Woods, 1983), such as:

Coordination failures.

Errors of omission—failing to correct/adjust lower level activities.

Errors of the third kind—solving the wrong problem.

Attention failures—inability to find, integrate, and interpret all of the "right" data at the "right" time.

The multilevel class of models emphasizes that all human behavior is partially knowledge (or concept) driven; what varies with layer of processing is the type of knowledge that drives behavior: knowledge embedded in a motor program, knowledge built into automatic mental skills (e.g., the heuristics of figure/ground organization and other forms of tacit knowledge), knowledge in the form of explicit rules (derived, learned, or written), knowledge in the form of conscious reasoning. Figure 12.5.17 contains an analysis of one simulated emergency developed to highlight which internal knowledge structures are activated and how they guide observation and action. This point is particularly important with respect to information search behavior that will be discussed further in the context of advanced decision aids.

Countermeasures

There is a mismatch between the model of the operator implicit in past process control interface systems and the model of operator performance presented above. The result, as the data base review reveals, is that two general kinds of operational problems occur. In one case, lower-level routines are misapplied or run on in the absence of higher-level control based on goal monitoring. In the second case, operators attempt to adapt component skills or maneuvers to achieve perceived goals, but problems occur due to incomplete knowledge of the relevant goals and constraints.

There are various techniques that have been applied to reduce this mismatch. For example, foldout sheets appended to paper procedures can be used to specify important operator activities that could be triggered at indeterminate times during some procedure and items that need continuous or periodic monitoring during that segment of operation.

Computerized procedures can also aid continuous monitoring and control tasks by showing items that should be monitored at any stage in the recovery process, through the ability to show actual values as well as criterion values, and through the ability to alert crew members when boundaries are approached or crossed. Aids can be developed to support periodic monitoring of recovery goals such as Safety Parameter Display Systems and safety function status monitors. These and other improvements can be seen as part of a process to improve human–machine system performance (e.g., a more error corrective system) by modeling the control room functional architecture on the multilevel structure of internal cognitive processes. This establishes a criterion of cognitive diversity, that is, a philosophy of defense in-depth with respect to cognitive behavior. A large number and variety of new process

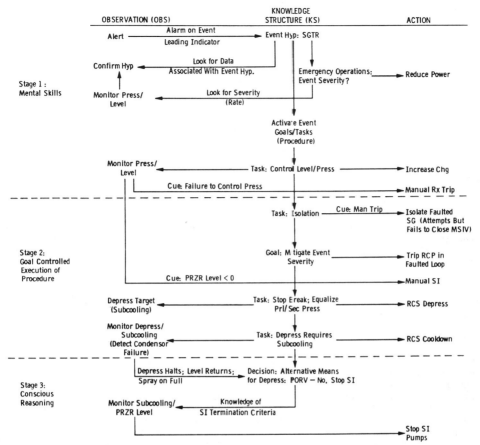

Fig. 12.5.17. Decision flowchart of crew response in a simulated emergency [steam generator tube rupture from Woods et al. (1982)]. The format was developed to map what knowledge structures are activated and how they control observation and action. From Woods (1984b), reprinted with permission from Institution of Chemical Engineering Symposium Series 90, *Ergonomic Problems in Process Operations,* copyright © 1984, Pergamon Press.

control interfaces and decision aids are being developed, tested, and implemented that begin to accomplish this goal.

12.5.3.2 Decision Aiding

The results from studies of operator decision making in emergencies clearly show the role of information problems when operational problems occur: the critical information is not detected among the ambient data load, or is not assembled from data distributed over time or space, or is not looked for because of misunderstandings or erroneous assumptions. However, most human factors guidelines on computer display system design attempt to ensure that human sensory limits are not strained or that users can potentially access data. If the design process stops at this level, there is an implicit assumption that if the user can potentially see/read the data, then he or she will and should find, integrate, and interpret all of the "right" data at the "right" time. But the cognitive performance problems identified in the operator decision making data base [as well as in other studies of user cognitive performance, e.g., Moray (1981) and Woods (1984a)] illustrate that the *potential* to see, read, or access data does not guarantee successful user information extraction. As a result, aiding human cognitive performance requires description of application-specific cognitive activities [Figure 12.5.18 and Hollnagel and Woods (1983), Woods and Hollnagel (1986)] through empirical studies (e.g., the operator decision making data base), through models of cognitive activities in work domains [e.g., Sorkin and Woods (1985)],

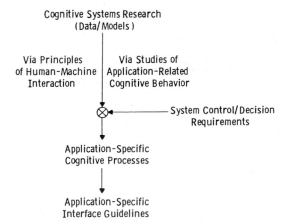

Fig. 12.5.18.   Knowledge from cognitive psychology applied to produce principle-driven interface system design. From Woods (1984a). Reproduced with permission from the *International Journal of Man-Machine Studies,* vol. 21, no. 3, 1984, pp. 235 and 244. Copyright by Academic Press Inc. (London) Ltd.

and through identification of the cognitive activities demanded by the process being controlled and by the structure of the control system design [e.g., Rasmussen (1984)].

Diagnostic Search

One example of the potential of this cognitive task analysis (i.e., the description of cognitive activities in work environments) is Rasmussen's (1981, 1984) work identifying diagnostic search strategies [cf. also, Rouse (1983)]. Rasmussen identifies two classes of diagnostic strategies. In *symptomatic search,* stored data (in the operator's head, in procedures, or in decision aids such as disturbance analysis systems) on the set of observations produced by each abnormality are matched to observations on the actual state of the process in order to identify disturbances (Figure 12.5.19). Examples of types of symptomatic search are pattern recognition where the link between referent symptom patterns and actual observations is automatized; decision table look up where there is an explicit, conscious comparison between stored symptom patterns and actual observations; and hypothesis and test where the diagnostician constructs the symptom patterns for hypothesized abnormalities.

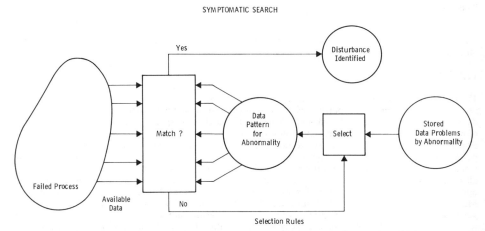

Fig. 12.5.19.   Information flow map for diagnosis through symptomatic search, which operates by finding a match between stored data on the sets of observations produced by an abnormality (symptoms) and actual observations of process state [adapted from Rasmussen (1981)].

TOPOGRAPHIC SEARCH

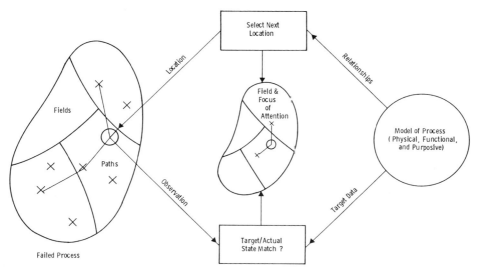

Fig. 12.5.20. Information flow map for diagnosis based on topographic search; it operates by using knowledge of the physical–functional–goal interconnections between locations in a model of the process topology and the results of actual-state–target-state comparisons to progressively limit the field of attention onto important disturbances for current operating context [adapted from Rasmussen (1981)].

In *topographic search,* process state is assessed relative to normal or reference state, for example, is primary system inventory within the proper target band for the current operating context. The results of this target/actual state check are combined with knowledge of the relationships between locations in the process topology (e.g., physical, casual–consequential, goal–means, among many kinds of links) to define and progressively limit the field of attention onto the important disturbances for the current context (Figure 12.5.20). It is the critical role played by the model of the structure of the process and control systems in guiding the search process that gives topographic search its name. This model can represent the topology of the process at multiple levels: in terms of physical geography, functional interconnections, and purpose. One advantage to topographic search is that it only depends on knowledge of where the process should be in different operating contexts with respect to goals. This is relevant to the empirical finding that experienced operators perform better in emergency events than less experienced operators even though there may be no significant difference in their experience *in actual or simulated emergencies.* The extra experience in normal operations may support better or more fluent models of the structure of the process and/or better feel for target states.

To illustrate the role cognitive task analysis can play in the development of decision aids [cf., Woods and Hollnagel (1986)], consider the match between the requirements for effective topographic search and the characteristics of conventional nuclear power plant control rooms. Topographic search requires a model of the structure of the process in terms of physical, functional, and goal interconnections, data on target states, and, as a result, an emphasis on relationships among data points rather than just the magnitude of variables. Past conventional control rooms fail to support these needs since (a) there is only one level of representation of plant state, the operator must construct other levels mentally; (b) there are few indications of normal states, particularly under dynamic conditions (e.g., what is a normal reactor trip), the operator must rely on his or her memory of reference states; and (c) the one-measurement–one-indicator display philosophy does not show relationships among data, the operator must integrate data mentally. The result is a mismatch between the demand for effective diagnosis and the characteristics of the human–machine system, and an increase of the operator's mental workload with a concomitant increase in the possibilities for errors.

Many efforts at developing decision aids are addressing this mismatch. For example, Gallagher (1982) and Lind (1981) have worked on mapping the goal and functional interconnections for power plant processes, and Goodstein (1983), Johnson (1984), and Woods (1986) have worked on building display systems to explicitly represent goal and functional interconnections.

Cognitive Task Allocation

One new challenge for human factors created by the physical to cognitive shift in the human's role in complex systems is the problem of allocating cognitive activities between the human and machine portions of the total system. Figure 12.5.21 describes in cognitive terms the allocation of tasks for the automatic shutdown of a nuclear reactor. This example uses Rasmussen's (1983) step ladder model of operator decision making to describe the cognitive activities performed by the designer and allocated to automatic systems and to the operator. The operator's activities, given the allocation decision, include:

1. Monitor goal achievement—is shutdown necessary?
2. Monitor automatic system performance—has automatic shutdown been carried out correctly?

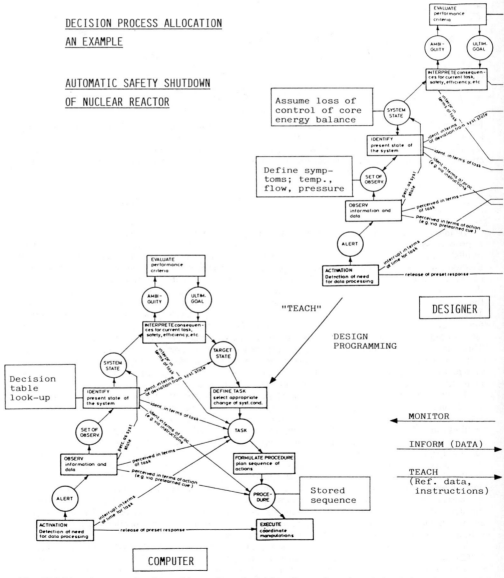

Fig. 12.5.21. An example of cognitive task analysis that shows the roles of the designer, operator, copyright 1984 JAI Press, and reproduced by permission.]

3. Manual control—several methods for manual shutdown either independent of or as backup to the automatic system [e.g., in one nuclear power plant critical incident, operators correctly and quickly detected the abnormal conditions requiring shutdown independent of the automatic system (U.S. Nuclear Regulatory Commission, 1983b)].

In addition, operator cognitive activities that depend on the dynamics of real situations can be described. For example, after automatic shutdown, the operator must decide if it was needed (i.e., emergency conditions exist) or if it is a spurious shutdown (i.e., a "normal" reactor trip due perhaps to an error during a test of the automatic shutdown logic); after a needed automatic shutdown, the operator must monitor that the relevant goal of the task (the reactor off) continues to be met during the recovery (e.g., avoid return to criticality conditions).

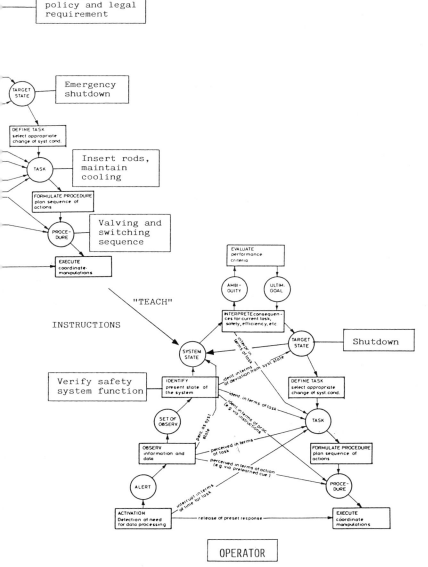

and computer with respect to the task of automatic reactor shutdown. [From Rasmussen (1984);

The allocation of tasks for automatic shutdown illustrates an important characteristic of allocation decisions [cf., Woods and Hollnagel (1986)]: there are varying cognitive consequences to allocation decisions that affect user performance, interface requirements, and, therefore, the allocation decision. The allocation of some task to the machine *creates* new tasks for the human element, which must be uncovered and analyzed. In a process analogous to proving theorems by contradiction, if undesirable consequences come from an allocation decision [such as tasks with high potential for human error, e.g., Norman (1983), an error-intolerant joint system architecture, or additional interface system requirements to be met], then the initial decision for structuring the human–machine system must be reexamined and alternatives explored. Otherwise, the initial decision can be accepted as a workable allocation pending the evaluation of the consequences of other allocation decisions.

Parallel versus Serial Data Presentation

Building interface systems to support operator cognitive activities involves matching the operator's mental model of the system to actual system characteristics and demands, that is, the earlier examples of cognitive activities that are imposed on the operator by the nature of the system. In addition, there must be a match between the system's image of the user and actual user characteristics at a cognitive level (Hollnagel and Woods, 1983).

One example concerns the issue of serial versus parallel presentation of data. Existing display construction guidelines primarily address the characteristics of individual displays. But failure to consider the requirements for how the user integrates data across successive displays can produce devastating effects on user performance [cf., Badre (1982) and Woods (1984a)]. When the requirements of across-display information processing are ignored, computer-based display systems become a serial data presentation medium. The narrow "keyhole" the visual display unit (VDU) when provided can degrade user information extraction compared with so-called "parallel" presentation modes where all of the data are displayed simultaneously (Pope, 1978), for example, on conventional control boards.

However, the "keyhole" phenomenon is not an inevitable consequence of using computer-based displays; nor does it represent human limitations (for example, short-term memory) that must be compensated for through memory aids or walls of VDUs. Across-display processing difficulties are the result of a failure to match the system's image of the user's processing mechanisms to the actual characteristics of human cognitive function.

In contrast to the problems associated with serial data presentation, "parallel" (that is, simultaneous) presentation of data is claimed to be superior because, as Pople (1978, p. 4) states, "the human is used to having his total information system displayed and being able to sample and timeshare from his system *by a movement of the eyes and his interpretive skills*" (emphasis added). This suggests that the advantage attributed to parallel over serial data presentation is based on the characteristics of human perception and attention, rather than the mode of data presentation. Even when the entire data base is simultaneously available, the narrow field of view (2°) of the fovea (the high-resolution portion of the retina) constrains the amount of data a viewer can acquire in any single glance. This is no limitation when viewing real-world scenes because there are psychological mechanisms that convert a "serial" input through a succession of eye fixations into what we commonly think of and experience as "parallel" data acquisition. Woods (1984a) has shown how user information extraction across displays can be improved if knowledge from cognitive psychology about the above perceptual and attentional mechanisms is applied to display system design.

"Visual search is an active interrogation of the visual world during which people systematically detect and use meaningful patterns of relationships to decide where to look first and in what sequence to seek further information" (Rabbitt, 1984, p. 273). The degree to which a display system supports the attentional mechanisms that underlie this ability [or, after Woods (1984), the amount of *visual momentum*] determines whether data presentation is effectively parallel or serial.

Low visual momentum degrades the user's ability to extract information from a display system. For example, the following passage, from a study of process control operator performance (Hollnagel, 1981, p. 133, emphasis added), illustrates how the characteristics of a computer display system produced low visual momentum in the control performance of one operator.

> The most interesting thing about the S's model of the system was the lack of correspondence between the Hot Well and the Feedwater Tank. As mentioned in the discussion of the S's performance, it was not until rather late that he realized, that the water which was pumped from the Hot Wells was ending up in the Feedwater Tank. He afterwards commented that he found the pictures of the Feedwater System quite difficult to use, among other things because the subsystems (Condensate and Feedwater) looked different from the Feedwater System (cf. the pictures). He complained that it was difficult for him to see on which side of the Feedwater Tank he was, when he only looked at a picture of a subsystem. Although this may be the case, it is nevertheless not sufficient to explain his curious forgetting of the Hot Well/Feedwater Tank relation since that should be evident from a purely functional analysis of the system. That means that he was heavily influenced by the way the system was represented in the pictures, and that

this surface representation therefore limited the S's ability to reason about the deep structure (*i.e.*, *the functional structure*) *of the system.*

This operator's performance shows how the perceptual characteristics of displays can affect human problem-solving behavior. Many studies in both applied and basic research contexts have shown that the form of problem representation can greatly influence problem-solving performance. Brooke and Duncan (1981) found, in studies of fault-finding performance, that display format affects "the ability of the diagnostician to *perceive* what is relevant and what is not" (Brooke & Duncan, 1981, p. 188, emphasis added).

The breakdown in the viewer's attentional processes represented by low visual momentum can also be seen in what has been described as "disintegration of the visual field" (Bartlett, 1943) or "cognitive tunnel vision" (Moray, 1981). Cognitive tunnel vision occurs when the user's attention is locked on a subset of variables to the exclusion of others. This decrease in the size of the field of attention (or deviation from optimum sampling patterns) can lead to monitoring failures, especially when the unexpected occurs or when correct state identification is a function of integrating data from several sources. The relationship between these consequences of low visual momentum and the problems in operator assessment of process state can be observed in the decision making data/base.

Based on studies of how people integrate data across successive views, there are a series of techniques available to increase the visual momentum a display system supports [cf., Figure 12.5.22 and Woods (1984)]. The key element in all of these concepts is to provide the viewer with data about the location of one view with respect to another or, more generally, with data about the relationships across display frames (e.g., what are the physical or functional relationships between successive views?). The goal is to use the perceptual context to help the user construct and maintain a cognitive map or schema of the data structure. It is this internal model that results in the simultaneous representation of information.

A *long shot* provides an overview of the display structure as well as summary status data. It is a map of the relationships among data that can be seen in more detailed displays and acts to funnel the viewer's attention to the "important" details. For a summary display to provide an effective world view, the display system structure must explicitly incorporate a set of interdisplay relationships that are important to the user's tasks to be portrayed in the long shot. Merely summarizing data is insufficient for effective across-display information extraction. For example, user performance suffered in the example from Hollnagel (1981) because the summary display did not portray the important functional relationships in the lower-level displays.

Another technique to join together successive views is to provide *perceptual landmarks* across displays. Clear landmarks help the viewer to integrate successive displays by providing an easily discernible feature that anchors the transition and that provides a relative frame of reference to establish relationships across displays. When some feature or object is immediately recognizable in a scene, knowledge can be quickly activated to guide subsequent looking behavior. "Once an object is identified in a scene, we may quickly know the kind of company it keeps" (Biederman, 1981, p. 239).

Another type of "glue" to enhance comprehension across-display transitions is the use of *display overlap*. Physically overlapping displays is a standard cartographic technique to increase viewer comprehension. *Functional overlap* is a technique to present pictorially the functional relationships that cut across display frame boundaries. A display frame presents data with respect to a single topic; functional

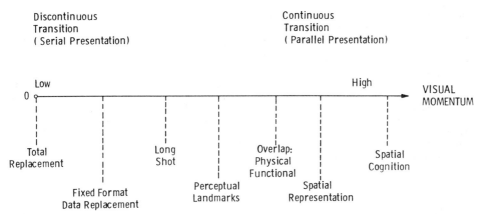

Fig. 12.5.22. Techniques to enhance the visual momentum in a display system. [From Woods (1984a). Reproduced with permission from the International Journal of Man-Machine Studies, Vol. 21, no. 3, 1984, pp. 235 and 244. Copyright by Academic Press Inc. (London) Ltd.

overlap occurs when each frame also contains data or pointers to data on semantically related topics such as goals or functional siblings (i.e., alternative means to achieve a goal). Displays of functional, rather than physical, form have long been used to portray electronic circuits. In process control applications, if a system is designed to transport material to maintain inventory in some reservoir, the display should show data about system operation (for example, is there flow?) and data about goal achievement (for example, is inventory at target levels?). It is important to note that functional overlap can be implemented only if the functional relationships among data points are specified (i.e., how data relate to the user's task). By identifying relationships between data and user tasks and by paralleling those relationships in the structure of the display system, the user can more easily locate "important" and "informative" data.

The long-shot, landmark and overlap techniques all increase visual momentum by providing data about the location of one view with respect to another, that is, a *spatial representation*. Spatial organization of data (spatial coding) is a potent aid to human cognitive processing [e.g., Haber (1981)]. The priority of space as an organizing principle is so compelling that nonspatial data are often given a spatial representation to improve user comprehension, for example, taxonomic trees in biology or computer program flowcharts. Spatial organization translates the normative user internal model into a perceptual map. The user sees, rather than remembers, the organization of data in the system and can move within the system just as he moves in an actual spatial layout.

The above set of techniques for increasing the visual momentum within a display system provides designers with data on how to improve across-display continuity. These techniques lead to the advantages associated with parallel data presentation through support for the user's perceptual and attentional skills at locating "informative" data.

These skills can be supported through the construction of a spatial framework that reflects meaningful relationships among data elements, that is, by constructing a conceptual or virtual space to represent data, particularly data that are not directly or necessarily spatial in character.

Integral or Object Displays

Data presentation in industrial control rooms even with the introduction of VDTs is dominated by a one-measurement–one-indication display philosophy. This approach to data presentation has unfortunate consequences for operator information processing [e.g., Goodstein (1981) and Moray (1981)]. The one-measurement–one-indication tradition induces a sequential, piecemeal form of data gathering; the burden is on the operator to find, integrate, and interpret all of the relevant data with respect to some issue. Separable displays are no problem when operator decisions are predicated on the value of a single variable, but they can obscure important relationships among data points when multiple variables must be integrated to assess the status of higher-order concerns (e.g., is inventory adequate? what is the safety status of the process?), and to diagnose faults. The result has been a large amount of work to develop integral or object displays where groups of data related to an issue are organized and represented in a single perceptual object such as a geometric pattern or a face. Goldsmith and Schvaneveldt (1984) and Wickens (1984) contain overviews of the psychological basis for integrated displays. Integral displays have been used in flight applications [e.g., contact-analog displays (Roscoe and Eisele, 1976)], in statistics to represent the results of multivariate analyses [e.g., Chernoff, (1973) and Kleiner and Hartigan (1981)], in medicine to aid monitoring of patient status [e.g., Siegel, Goldwyn, and Friedman (1971)], and have often been recommended to aid monitoring of system status in process control [e.g., Geiger and Schumacher (1976) and Goodstein (1981)].

Figure 12.5.23 is an example of one kind of object display [originally proposed by Coekin (1969)] that has been developed and applied to depict the overall safety status of a nuclear reactor (Little and Woods, 1981; Woods, Wise, and Hanes, 1981). The different spokes are dynamically scaled so that a regular polygon always represents normal conditions, while distortions in the figure represent a developing abnormality. Displays like this have many advantages for improved operator information extraction (Goldsmith and Schvaneveldt, 1984; Jacob, Egeth, and Bevan, 1976; Smith, 1976; Wickens, Boles, Tsang, and Carswell, 1984). For example, Wickens et al. (1984) examined the role of integral displays by examining fault detection performance in a simple dynamic system. The data available were related through the dynamics of the system and needed to be integrated to determine if the system was working normally. The study found a significant performance advantage for fault diagnosis with an object display similar to the one shown in Figure 12.3.23 over separated display (individual bars) of the data. In general, integral displays can:

Enhance multiparameter decision making.

Take advantage of human pattern recognition abilities.

Enhance discriminability of changes in system state because of the visual system's sensitivity to changes in patterns.

Aid operator identification of system state since operators tend to rely on familiar and stereotypical data points to assess system state and can thus be trapped into making an incomplete and perhaps incorrect assessment when their familiar signs are insufficient.

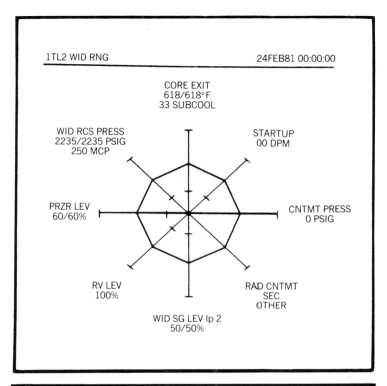

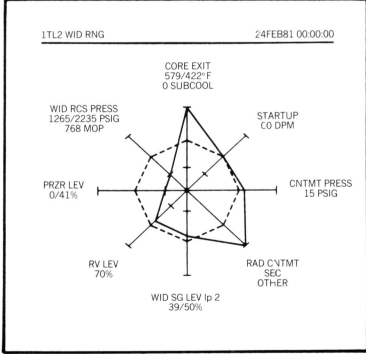

Fig. 12.5.23. Example of one kind of object display [originally proposed by Coekin (1969)] that has been developed and applied to depict the overall safety status of a nuclear reactor. The different spokes are dynamically scaled so that a regular polygon always represents normal conditions (panel *A*), while distortions in the polygon represent a developing abnormality (panel *B*). (Copyright 1980 Westinghouse Electric Corporation, and reproduced by permission.)

Reduce operator's memory load by bringing relevant information to a single location and by providing a mnemonic device (by remembering a single chunk—the pattern—the user can reconstruct all parameter values).

Communicate more data more effectively, for example, the object display in Figure 12.5.23 integrates over 100 sensor values into a representation of process health.

In other words, properly designed integral displays enhance the operator's ability to extract information across the control room interface because they produce a better match between human cognitive characteristics and the cognitive demands of the tasks to be performed.

While there are potential advantages to integral displays, there is also the problem of *how to construct a specific, effective integral display* for a specific process control application [e.g., Frey et al. (1984)]. First, most work on integral displays uses them in a static context [e.g., Kleiner and Hartigan (1981)] rather than for real-time applications. For example, there is no single pattern associated with a particular fault; rather during the incident the pattern changes in relatively characteristic ways that depend on the severity, operating context, other failures, and operator actions. Second, there is the problem of selecting the multiple parameters to be integrated in the display. Here the critical question is what information is the designer trying to communicate to the operator and what data set is needed to accomplish this objective. Third is the problem of scaling the chosen data so that the observer can see the desired patterns. This includes problems of making disparate data comparable (e.g., pressure and temperature or continuous versus discrete dimensions), setting scale resolution, and, most critically, identifying the reference normal and boundary conditions as a function of operating context. For example, the display in Figure 12.5.23 can be seen as a dynamic alarm that shows continuous variations between target and boundary conditions with respect to the health of the reactor (Sorkin and Woods, 1985). Finally, there is the problem of assigning data to the characteristics of the object format selected (e.g., to facial features or to the spokes of the geometric figure). Careful assignment of data is necessary because there are inherent relationships in the graphic format that must be related to (or at least not conflict with) the data relationships to be portrayed. For example, in Figure 12.5.23, the pressure and temperature scales together define subcooling (temperature at high limit market and pressure at low limit marker); placing these data on adjacent scales emphasizes this underlying relationship.

These problems of selecting, scaling, and assigning data to the graphic form are critical for constructing an effective example of the class of integral displays. For example, Kleiner and Hartigan (1981, and the accompanying commentary by Wainer), contains three different solutions to these problems for a single integral format (faces) that visibly change the salience of a single underlying data pattern. But, when these requirements for good integral displays are met during the design process, integral displays are a particularly effective tool to support enhanced operator information extraction.

Fault Management

Fault management, the identification of and response to abnormal conditions, is a major component of the human's role in process control as well as in other complex technological environments such as flightdecks and air traffic control. Alarms in traditional systems consist of messages about simple set-point or state (e.g., open–closed) violations displayed as one dedicated message per set-point (Figure 12.5.2). However, experience in the form of operational history (Kemeny et al., 1979), design reviews (Banks and Boone, 1981; Cooper, 1977), and evaluation studies (Kortlandt and Kragt, 1980; Kragt and Bonten, 1983) have shown that operators frequently have problems identifying, prioritizing, and responding to abnormal conditions with this type of alarm system.

Investigators have noted that one source of these difficulties is "alarm inflation," that is, the large number of messages that can be active simultaneously. Alarm inflation is associated (1) with cases where operators fail to detect a highly important alarm given the large number of simultaneously active but less important messages (high ambient data noise), and (2) with cases where operators fail to integrate disparate low-level alarm messages into accurate higher-level assessments of process abnormalities.

Another closely related source of difficulties is "nuisance alarms," that is, "alarm" messages that are not abnormal, at least for current plant conditions (Seminara et al., 1977):

Operators at all plants complained about the high number of nuisance alarms. The reasons for their occurrence varied. At one plant there was a blank, supposedly nonfunctional, annunciator window that would occasionally alarm. The maintenance and operational people had been unable to determine its cause, but an acknowledgement, silence, and reset were required on each occasion. In many cases alarm set-points were known by operators to be too sensitive to normal transients. As a consequence slight deviations or transients, thought of as normal, would set the alarm off even though no further operational action was required. Maintenance or calibration operations often caused recurring alarms that were a nuisance. The net results of the many false alarms is a "cry-wolf" syndrome which leads to a lack of faith in the system and a casual attitude towards the constant presence of certain alarms. On many occasions operators were observed to casually silence and acknowledge alarms without further concern or surveillance of plant status; the alarms had become "old friends."

The relationship of a given alarm message to different operating contexts is one major contributor to nuisance alarms, that is, a message indicates abnormal conditions in one context but the same state is normal for other situations.

A third source of alarm inflation is "alarms" that are really equipment or system status messages, for example, a message on the status (on–off) of a pump. Often status messages are associated with automatic changes in the state of a system. These messages do provide very useful data to operators (Kragt and Bonten, 1983), for example, because they can provide a wide field of view or summary of plant status and because the operator needs to be informed of properly occurring automatic changes in system state. However, these status messages are not alarms, unless the status is abnormal in the current operating context (e.g., pump A is off but it should be running following a safety injection signal).

The above is the classical description of the "alarm problem" (Lees, 1983). However, the alarm problem can also be seen as a specific example of the significance-of-data problem (Woods, 1986), that is, the problem is rarely the lack of messages, but rather their overabundance.

The significance-of-data problem represents an inability to find, integrate, or interpret the "right" data at the "right" time (e.g., critical information is not detected among the ambient data load, or not assembled from data distributed over time or space, or not looked for due to misunderstandings or erroneous assumptions). This problem occurs in situations where large amounts of potentially relevant data must be sifted in order to find the significant subset for the current context. In other words, most information-handling problems are not due to a lack of data but rather to an overabundance of data.

Operational staff members in dynamic process environments must detect, evaluate, and respond to abnormal conditions. Operators must sift through large numbers of "alarm": messages in order to find and identify abnormal conditions. This task is complicated because (1) the meaning of a particular alarm message depends on context (e.g., plant mode, message history such as the leaky pressurizer power operated relief valve at TMI-2, and the status of other messages) and (2) the individual alarm messages must be selected and integrated to assess process status since each message is only a partial and indirect indicator of an abnormality (e.g., pressurizer level less than 17% is only one datum relevant to abnormal primary system inventory). This means that a single disturbance will often initiate a large number of "alarms" and that a given alarm will be associated with many disturbances. The important point for alarm system design is the critical distinction between the available data and the meaning or information that a person extracts from that data (Smith, 1963, pp. 296–297).

When we examine the process of man-computer communication from the human point of view, it is useful to make explicit a distinction which might be described as contrasting "information" with "data." Used in this sense, information can be regarded as the answer to a question, whereas data are the new materials from which information is extracted. A man's questions may be vague, such as "What's going on here?" or "What should I do now?" Or they may be much more specific. But if the data presented to him are not relevant to some explicit or implicit question, they will be meaningless.

What the computer can actually provide the man are displays of data. What information he is able to extract from those displays is indicated by his responses. How effectively the data are processed, organized, and arranged prior to presentation will extract the information he requires from his display. Too frequently these two terms data and information are confused, and the statement, "I need more information," is assumed to mean, "I want more symbols (data)." The reason for the statement, usually, is that the required information is not being extracted from the data. Unless the confusion between data and information is removed, attempts to increase information in a display are directed at obtaining more data, and the trouble is exaggerated rather than relieved.

One example of the troubles caused by confusion between data and information is that computerization of alarm systems, when it leads to a proliferation of types and degrees of "alarm" messages without consideration of the information needed to perform fault management tasks, has failed to improve or even exacerbated system deficiencies (Pope, 1978).

The goal of fault management is to aid operator identification of and response to abnormal conditions by calling his or her attention to conditions that may require corrective action. This objective means that a good alarm system should (1) alert the user that a particular disturbance or abnormality may exist, (2) provide a means to evaluate the abnormality, and (3) provide a means to determine corrective action. The first criterion means that there should be a close link between the alerting signal and the particular abnormal condition it signals. Thus, the alerting value of a given message depends on a variety of factors beyond ergonomic considerations of auditory signals and acknowledgement systems such as:

The context of simultaneously or recently active signals.

The proportion of signals that do not indicate an abnormal condition.

The ratio of active signals to underlying abnormal conditions.

The other criteria are concerned with how the operator goes from an alert to some corrective action in process control. There is some initiating disturbance that leads to other disturbances through interconnections among plant processes or challenges to and through failures of other parts of the plant, automatic control, or human responses. This means disturbances can be responded to through (1) adjustment or replacement of the disturbed process; (2) responses to the consequences of the disturbance, for example, if a break occurs it produces consequences such as water/energy/radiation in abnormal locations which demand responses; (3) tracing the causal chain to identify correctable causes (diagnosis), for example, if water/energy/radiation is in an abnormal location, determining how it got there can lead to a response that breaks the disturbance chain. The evaluation of and response to a given disturbance can include any or all of these elements at various times or in various combinations during an emergency. The interaction among these categories is further complicated since multiple disturbance chains resulting from independent initiators can coexist and can interact. Indeed, a major component of fault management is shifting among these types of response under various conditions (e.g., severity, disturbance propagation, interactions among disturbance chains) and at various times during emergencies.

Efforts to alleviate the alarm problem all begin with an improved definition of abnormal process conditions. Conventional alarm messages (regardless of the form of implementation—annunciator tiles or VDT) focus on parameter set-point violations and component failures; the operator must assess the meaning of that violation for the current operating context. More sophisticated alarm messages integrate parameter and component level data to describe abnormal conditions in terms of higher-order units such as systems that are not or could not perform properly, functions that are not satisfied, or the underlying causes of disturbances. For example, Goodstein (1983) and Johnson (1984) contain descriptions of disturbances in terms of mass and energy flows and balances, Gallagher (1982) and Johnson (1984) in terms of the effect on critical safety functions, Bastl and Felkel (1981) and Gimmy and Nomm (1982) in terms of the underlying failures.

In complex processes, there is a topology of relationships between systems, functions, and objectives (Rasmussen and Lind, 1981). For example, there is a many-to-many mapping between plant systems and functions so that a given component failure can effect several functions, and problems in a given function can be caused or countered through several means (Figure 12.5.24). This topology of entities and relationships forms one method to produce high-level descriptions of process status that can aid operator fault management, for example, Gallagher (1982), Goodstein (1983), Lind (1981), Nelson (1984), and Woods (1986). Improved understanding of what is abnormal in the process has been used in two ways to mitigate the alarm problem. Alarm-handling systems attempt to better organize data on abnormal conditions to aid operator detection, interpretation, and response to abnormal conditions (Goodstein, 1981; Thompson, 1981; Visuri, 1982). Alarm-handling approaches include separating alarm and status messages, better definition of what constitutes an alarm, and improved organization and display of abnormal data such as integral graphic formats [Figure 12.5.23; Johnson (1984); Visuri (1982); and see Hanes et al. (1982) for some color examples] and displays of goal–function system topologies (Goodstein, 1983; Woods, 1986). Disturbance analysis systems, on the other hand, attempt to inform the operator directly about faults and appropriate corrective strategies. Techniques that have been applied to build disturbance analysis systems include decision tables (Gimmy and Nomm, 1982), cause-consequence trees (Bastl and Felkel, 1981), alarm trees, and rule-based expert systems (Andow, 1984; Nelson, 1984).

12.5.4 EVALUATION OF DECISION AIDS

The development of new interface concepts and operator aids has also sparked a considerable amount of work on the evaluation of the performance of a human–machine system. Evaluation studies are needed to identify weak points where a human–machine system can be improved, to provide design feedback during the development of new concepts, and to determine the ultimate effectiveness of new or modified interface systems. These needs, in an industry undergoing a rapid pace of development of potential user aids, have advanced the methodologies available to the human factors practitioner, especially in the area of evaluations of decision aids, that is, systems that aid the operator's action selection as well as his or her action execution.

This section uses the experience gained from several of the evaluation studies of human–machine system performance in the nuclear industry to describe some of the characteristics of good evaluation studies (Pew et al., 1981; Rouse, 1984; Rouse et al., 1984; Woods et al., 1982). Examples from actual

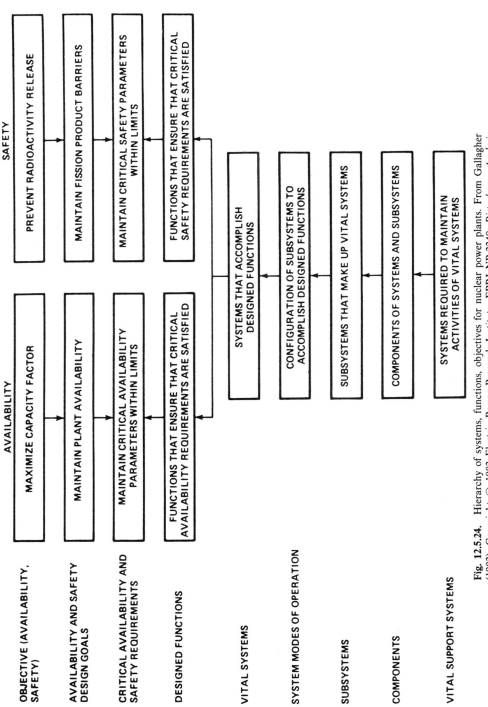

Fig. 12.5.24. Hierarchy of systems, functions, objectives for nuclear power plants. From Gallagher (1982). Copyright © 1982 Electric Power Research Institute. EPRI NP-2240, *Disturbance Analysis and Surveillance System.* . . . Reprinted with permission.

evaluations are used to illustrate these characteristics. The techniques discussed are of special relevance to measuring the performance of human–machine systems on cognitive tasks such as monitoring process and goal status, fault management, and planning. The characteristics of good evaluation studies presented here include:

The role of a conceptual framework or model of relevant decision problems.

The importance of measures of the user's decision process as well as the outcome of that process.

An emphasis on mixed fidelity simulation for cost effective testing.

12.5.4.1   Methodological Issues

Process and Outcome Measures of Performance

It is not enough to record only the outcome of a given scenario; how the operator reaches the outcome provides crucial data on where and why a particular interface helps or fails to help user performance (Duncan, 1981; Woods et al., 1981). Data on the background and context of a user problem can reveal the factors that contribute to successful or unsuccessful human performance and can therefore help identify potential improvements or additions to interface systems.

Tracing the decision process is also necessary because interface systems are multidimensional (Rouse, 1984). Outcome measure alone are not powerful enough to determine which of the multiple potential factors active in any evaluation of interface systems contributed to a specific outcome result (Woods et al., 1981). For example, if a new display concept includes flow path coding (e.g., energy flow, material flow, or data flow in some system), are outcome results due to the form of implementation (is the display legible? are the specific coding techniques good at communicating flow path status?) or to the value of the display concept in some task context (do displays that communicate flow network status support improved user performance at some task?).

Table 12.5.4 contains an example of outcome and process performance criteria for one operational task in one process control event (simulated nuclear power plant emergency). Data from a study that included this task (Woods and Roth, 1982) illustrate the importance of process measures of performance. In some cases there were no outcome measure violations, but performance was rated only reluctantly acceptable (see Figure 12.5.16 for an acutual rating form) because some process measure was violated (5 out of 15 cases). In one case (out of 15), there was an outcome measure violation (less than 50° subcooling). Here, the process measures provide insight as to why the outcome failure occurred. In this case, the crew failed to monitor the goal of the task (subcooling) so that, when the event deviated from expected course, the crew failed to detect and correct the deviation for 14 min.

Tracing the decision process requires trained observers (both domain experts and human performance experts; see Figure 12.5.25) to efficiently produce protocols that describe user decision activities (Hollnagel, 1982; Woods and Roth, 1982). Inputs to decision protocols can include information search measures (e.g., eye-movement records), operator actions and context, plant behavior, and verbal reports. The key is that these types of data are analyzed, *as a set,* in terms of the classification of decision situations of interest to the particular interface evaluation in order to produce a description of operator *performance* (e.g., Figure 12.5.12, 12.5.13, 12.5.15, and 12.5.17 are samples of decision protocol formats).

Concept-Specificity/Context-Independence Tradeoff

Performance is first analyzed or described in the language of the profession: this user, in this domain, on this simulator, in this event, closed valve *x* at time *y*. Analysis at this level (what can be called a *description of actual performance*) is relatively concept free but highly context dependent, in other

Table 12.5.4   Outcome and Process Performance Criteria (Cooldown Task During Steam Generator Tube Rupture)

Outcome Measures[a]	Process Measures (Examples):
Greater than 50° subcooling	Timely initiation of cooldown
Greater than 0° subcooling	Select correct target
	Cooldown rate (rapid, slow)
	Monitor subcooling
	Stop cooldown at target

[a] Note that, from the point of view of the entire event, task outcome measures become event process measures; event outcome measures focus on more global properties such as, for the case of nuclear power plant safety, radiation release and core damage.

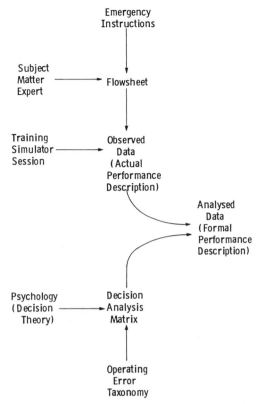

Fig. 12.5.25. Procedure for efficient generation and analysis of decision protocols (Woods and Hollnagel, 1982).

words, it is difficult to generalize the results to other users, tasks, events. It is only by using concepts or models of user performance that generalizable results can be obtained. Just as operator data search is an iterative cycle between data-driven and concept-driven behavior [i.e., observation is a process of looking for, where the object of the search becomes more refined with successive iterations; cf. Woods (1984a)], the evaluation process is also a cycle of data- and concept-driven activity (Figure 12.5.26). Since concepts or models form the basis for this second level of analysis, it can be called a *formal performance description* [cf., Hollnagel, Pederson, and Rasmussen (1981) for more discussion of the role of these levels of analysis and description in the evaluation process]. The formal performance description is critical because it is concepts, not the data itself, that are transportable from one specific situation to another (except in special cases with highly developed theories).

To illustrate the difference between these two levels of description, consider the relationship between avian flight and human attempts to fly. One approach is to copy or mimic how the bird behaves. This surface description leads to attempts to fly via flapping imitation wings. Real understanding only comes when it is discovered that both avian and jet flight are based on the concept of lift provided by an airfoil. The concept of an airfoil provides guidance about what are the critical variables, or the

Fig. 12.5.26. Concept-driven and data-driven processes in evaluation.

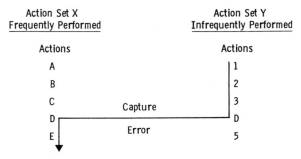

Fig. 12.5.27. Illustration of one human error category, capture error, where a highly practiced behavioral routine inadvertently takes control from a less established sequence.

effective stimuli, to be measured or controlled, that is, what are the aspects of the situation that really matter with respect to the behavior of interest.

To take an example from process control applications, imagine a user executing action set Y (an infrequently performed task) who erroneously substitutes actions from set X, a frequently performed and closely related task (Figure 12.5.27). The actual performance description would state that the user committed an error in maneuver Y, closing valve E rather than valve 4. The formal performance description would state that a "capture" error occurred, where capture error is a category in one model of the internal mechanisms that produce human errors (Norman, 1983).

In this example, a model of human error has been used to analyze user performance; as a result, the data can be combined with and generalized to other users in other events. In addition, measures to mitigate or prevent this general type of human error (if the domain data show it to be a frequent or important error category) can be formulated and tested. Finally, the traditional problems of identifying evaluation criteria and selecting test scenarios are mitigated because the proper formulation of the question to be tested contains a measure of successful performance and the essential conditions that must be produced in any test scenario. For the capture-error example, a test of the effectiveness of interface changes to combat this type of error requires some measure of the occurrence of capture errors and some test situation that is likely to produce capture errors. (Figure 12.5.28 summarizes the concept- and data-driven spiral for this example.)

Question Formulation

Many evaluation studies simply ask whether Interface System A is better than Interface System B. This is a poorly formulated test objective, and subsequent test results are bound to be disappointing, that is, not very informative relative to costs. Since interfaces are used to communicate data to a user to aid in the performance of tasks, at a minimum the evaluation question should be of the form— is Interface System A better than Interface System B at some task or over some range of tasks. The original question also misses any analysis of the relationship between Interface Systems A and B: Why should they lead to different user performance? What problem are they trying to address? Do they represent different interface concepts or different ways to implement one concept?

Similarly, interface evaluation studies are often unable to separate questions related to interface system goals (what is to be achieved) and those related to means (how can it be achieved). Evaluation studies must distinguish between these issues (is concept x valuable in task context g versus what is a good way to communicate x given the context of other items to be communicated), and they must be able to attribute outcomes to either source.

This difference between higher-order questions (Level 2 issues) on what an interface should communicate to the user (e.g., do displays that communicate flow network status support improved user performance at some task?) and lower-order questions (Level 1 issues) on how to implement a given interface

Fig. 12.5.28. Example of the spiral of concept-driven and data-driven processes in evaluation.

goal (e.g., are specific coding techniques x, y, z good at communicating flow network status in the context of the total set of items to be coded for this application) is the crux of the central problem in interface evaluation.

If results on what should be communicated across the interface (Level 2) are available only after a system is designed and iteratively tested at Level 1, it is difficult for those results to have any major impact on the system due to the costs of changes at that point in the design process. These costs can be due to hardware changes, they will certainly be due to software changes, but they will also be due to Level 1 redesign since Level 1 issues are conditional on Level 2 choices. This is simply a restatement of top-down design: how to communicate something depends on what is to be communicated.* On the other hand, if one tries to test Level 2 choices early in the design process, the problem is that Level 1 adequacy is a necessary condition for Level 2 success. As a result, there is a need to distinguish, control, or eliminate Level 1 effects in order to provide meaningful results on Level 2 choices [cf. also, Rouse (1984)].

Thus, the central problem in evaluation is to discriminate results due to Level 1 issues (can the user potentially see, read, access data) from those due to Level 2 issues (does the user find the "right" data at the "right" time). This decoupling can be accomplished in many ways. First, the evaluation method should be able to provide insight as to the source of observed results, that is, process as well as outcome measures are needed. Second, advances in technology make it possible to quickly prototype human–machine system designs (Hollan, 1984). The key is to provide a mechanism to substantiate Level 2 options and at the same time fulfill the necessary condition for adequate Level 1 design. It is important to note that the prototype's Level 1 characteristics are not necessarily (nor even likely) to be the same as the production system's (1) because of the consequences of the Level 2 evaluation results on Level 1 design requirements and (2) because of hardware and software differences between the prototyping system and the production system. Thus, the criteria for the Level 1 characteristics of the prototyped version of the human–machine system is not that they duplicate the production system, but rather that they are adequate to allow the user and the experimenter to see past Level 1 issues in order to examine the underlying Level 2 concepts.

Of course, the preceding discussion means that the first step in the evaluation process, regardless of whether the evaluation is pursued in parallel with or subsequent to design, is to determine the Level 2 issues the design is attempting to address and to separate these questions on what the interface should communicate from Level 1 questions about how to communicate. A top-down design process will greatly enhance quality and efficiency of this step.

Event Selection/Design

Testing user performance with new aids should not be based on selecting a set of domain events per se; rather, it should be based on selecting or designing a set of decision making tasks that are relevant to the problems addressed by the new display or aid. Particular events are chosen in order to create the decision situations of interest. For example, if the goal of an interface design is to provide a more error-corrective human–machine system, then events can be developed to measure/test for this characteristic. In process control, this goal might be studied by simulating operator errors as well as plant malfunctions, perhaps by timing instrumentation or component malfunctions to occur just after a relevant operator action or by instructing a confederate on the crew to commit an execution error. The test would then measure how the supervisor detects and corrects the simulated execution error.

One advantage of this approach to test event selection is that the classification of problems provides a mechanism to generalize results across events. This is because the decision situations in the test can occur in many particular events, including events that are not part of the test bed. In other words, this approach mitigates against the completeness problem in event selection.

To summarize, a poor test evaluates whether performance on problems $1, 2, \ldots, N$ is improved; while a good test evaluates whether performance on a class of problems represented (sampled) by problems $1, 2, \ldots, N$ is improved.

Mixed Fidelity/Focused Tasks

The mixed fidelity or focused task approach to evaluation (Duncan, 1981; Rouse, 1982) is based on the recognition that realistic simulation of all aspects of human performance in complex systems is impractical if not impossible (Sheridan and Hennessy, 1984). Instead, the critical features of the work environment with respect to evaluation goals must be identified. It is these critical features, as defined by the concepts or problems guiding the evaluation, that require high-fidelity simulation; low fidelity can suffice for other aspects. Furthermore, some low-fidelity elements may be deliberately introduced in order to create the problems of interest. For example, to study the effect of stress on user performance,

---

* Note also that this is simply a restatement of the basic result of Shannon's communication theory, namely, that information transfer in a communication channel depends, not on the item to-be-communicated alone, but rather on the set of items that *could* be communicated.

low-fidelity features (e.g., time pressure) may be introduced in order to produce (i.e., simulate) the feature of interest in the actual work environment (stress).

When evaluations are based on full-scale simulation, the evaluator replicates as much as possible of the actual work environment, asks users to behave as if they are in the real system, and then waits for the operational problems of interest to occur. While this can be appropriate for some research goals, such naturalistic observation is an inefficient (i.e., expensive, time consuming, and capable of only weak tests of evaluation objectives) means to *evaluate* new or modified interfaces. Using mixed-fidelity simulation, the evaluator can create the situations of interest by how he or she selects and constrains the task. This approach produces meaningful results on interface strengths and weaknesses and increases the efficiency of the evaluation process. These focused task results can then be confirmed through converging measures of performance derived from a small number of full-scale scenarios or from retrospective analyses of performance in actual incidents (Woods, 1984b).*

12.5.4.2   Case Study in Evaluation of Decision Aids: Safety Parameter Display Concepts Evaluation (Woods et al., 1982)

The goal in this study was to explore the kinds of situations where a computer-based operator aid, a safety parameter display system, might improve operator assessment of nuclear power plant safety status. Therefore, test events were chosen to represent various concepts about where this aid might support improved operator performance (categories such as diagnosis problems, strategy problems, multiple failures, and baseline events that served as a reference point both within this study and to other studies). The crew's decision process in each event was analyzed by a decision flowchart based on categories of detection, interpretation, control, and feedback. (Figure 12.5.12 contains an example of the flowchart format used in this study.) This allowed the experimenters to identify where operational problems occurred in the decision process and where in the decision process prototypes of the new concept were used successfully.

The results without the computer-based aid (Table 12.5.1) showed that errors of decision making tend to persist [see Woods (1984b) for a wider survey and assessment of these data]. As previously discussed, when operators misidentified plant state or had execution difficulties (which was infrequently), they generally failed to correct their understanding of plant state or to identify and correct execution problems within the duration of the test. Operators rarely had problems with the initial state identification following onset of the emergency. Rather, their state identification or recovery actions gradually became decoupled from actual process state as a function of execution problems or the unexpected. This result indicates that operation problems are often related to poor feedback about the effect of control actions on process state and on goal achievement.

Usage results for the prototype display systems suggest that computer-based aids can reduce this tendency by providing improved feedback especially to senior operators acting in an emergency or process management role. Table 12.5.5 contains examples where feedback was obtained from a computer-based aid to correct operational problems. The prototype operator aids were also used successfully in other decision making contexts; however, operators also performed well on these decisions when the aids were not available (e.g., initial event diagnosis).

12.5.5   FUTURE DIRECTIONS

Because the human element is an important contribution to the safe and economic operation of process plants, these industries will continue to be concerned with how to improve human performance. Some of the important areas to watch for further application and development of human factors are:

Maintainability.

Computer-based decision aids.

Advanced control and display interfaces.

Automation of control.

Training.

There are areas in process plants other than control rooms that can benefit from human factors knowledge, especially in the area of maintainability (Seminara and Parsons, 1981). For example in the U.S. nuclear industry, many plant outages are due to less than optimum performance by maintenance personnel. These deficiencies may be caused by such factors as difficulty in accessing equipment to be maintained; inadequate coding and labeling; communications problems; overcrowded workshops;

---

* Methodological considerations are not addressed here in detail for retrospective analyses of decision making. See Pew et al. (1981), Reason and Mycielska (1982), and Woods (1982) for explicit discussions of techniques and case studies on retrospective analyses.

Table 12.5.5   Examples where the Feedback Used to Correct Operational Problems Was Obtained from a Safety Parameter Display Prototype

In event TR3-H the reactor operator detected that the faulted steam generator level was within the wide range instrumentation from Safety Panel B plant status and wide range iconic displays. The balance of plant operator had reported earlier that the faulted steam generator was empty by misreading narrow for wide range level from the control board.

The senior reactor operator (TR2-H) detected low steam generator levels in two unaffected steam generators from Safety Panel B plant status display. (The balance of plant operator had been slow in reestablishing auxiliary feedwater flow to the unaffected steam generators after stopping all auxiliary feedwater to aid in the steam generator tube rupture diagnosis.) The senior reactor operator then directed the balance of plant operator to increase auxiliary feedwater flow to the unaffected steam generators.

Safety Panel B plant status display helped the senior reactor operator detect that auxiliary feedwater had not been isolated completely from the faulted steam generator (TR2-H). The faulted steam generator had been isolated, but the balance of plant operator turned on the turbine-driven auxiliary feedwater pump to increase unaffected steam generator levels. However, auxiliary feedwater flow also began to the faulted steam generator.

In event TR2-H the senior reactor operator detected from the Safety Panel B plant status display that only two of the three unaffected steam generators were being used to cool the primary system. He directed the balance of plant operator to open the third steam generator power operated relief valve.

In another event (FW3-G), a crew detected that pressurizer level was low and decreasing. The crew isolated letdown and then consulted Safety Panel B wide range iconic for feedback. The iconic display showed the crew that the pressurizer level decrease halted.

Source: Woods et al. (1982)

heat, radiation, and other environmental stresses; ineffectual coordination of work tasks; inadequate procedures and job performance aids; and inadequate training. In response to some of these problems, the EPRI is supporting a major program to develop design guidelines for maintainability (Pack, 1984). Other human factors efforts will be mounted to address these problems because of the large potential payoff in increased plant availability.

Computer-based aids will continue to be developed and introduced to support operator decision making. Procedural information can be stored electronically, and will be available through display terminals in the control room or other parts of the plant. Dynamic plant operating data and actual plant configuration descriptions can then be integrated with procedural guidance. A variety of alarm-handling systems will be developed, tested, and introduced to improve operator identification and response to abnormal conditions (Visuri, 1982). There will be continued research into the development of disturbance analysis systems to directly inform operators about failures and about how to best correct those failures. Work is just beginning on the development of expert systems and other artificial intelligence systems to support problem solving and decision making in process control (Andow, 1984; Nelson, 1984). Such systems may have applications in both control room operations and plant maintenance.

Advanced control rooms are being developed that use computers and CRT displays to process and present plant data (Hanes et al., 1982). These advanced systems may include speech output and voice input, and touch-sensitive panels and tracker mechanisms (e.g., track ball, mouse, joystick) are already being introduced. One of the most exciting opportunities is the development of innovative display presentations to support operator cognitive functioning and decision making.

The tasks humans perform will change as more automation is introduced into more process industries. For example, the Japanese are developing a system that will involve automated nuclear power plant start-up, shutdown, load following, and system surveillance (Hanes et al., 1982). This increase automation of control has important implications for operator monitoring, decision making, and information. As a result, there will be more emphasis and research on the human as supervisory controller.

The emphasis on high levels of worker performance will place even greater emphasis on training. The qualitative training evaluation methods common today may be supplemented with more quantitative performance measuring systems. Computer-based instruction and the use of artificial-intelligence-based learning methods may become important in the training process.

The domain of process control will continue to be an exciting one for human factors work. Challenges to human factors include the design of human–computer interfaces to support decision tasks, the development of new kinds of alarm systems and other decision aids, the design of tools to support human supervisory control, the effect on the operator of the introduction of "intelligent" interfaces, and evaluation techniques to determine which of many possible new systems will result in effective improvements in human–machine system performance.

REFERENCES

Ahearne, J. (1982). Keynote address. In R. Hall, J. Fragola, and W. Luckas, Eds. *Conference record for 1981 IEEE standards workshop on human factors and nuclear safety.* New York: IEEE.

Andow, P. K. (1984). Expert systems in process plant fault diagnosis. In D. Whitfield, Ed., *Ergonomic problems in process operations.* London: 1984. Institution of Chemical Engineers, Symposium Series 90.

Badre, A. N. (1982). Designing chunks for sequentially displayed information. In A. Badre and B. Shneiderman, Eds. *Directions in human/computer interaction.* Norwood, NJ: Ablex Publishing.

Bainbridge, L. (1974). Analysis of verbal protocols from a process control task. In E. Edwards and F. P. Lees, Eds. *The human operator in process control.* London: Taylor and Francis.

Banks, W. W., and Boone, M. P. (1981). *Nuclear control room annunciators: problems and recommendations* (NUREG/CR-2147). Springfield, VA: National Technical Information Service.

Bartlett, F. C. (1943). Fatigue following highly skilled work. *Royal Society of London, Proceedings, B-131,* 247–257.

Bastl, W., and Felkel, L. (1981). Disturbance analysis systems. In J. Rasmussen and W. B. Rouse, Eds. *Human detection and diagnosis of system failures.* New York: Plenum.

Biederman, I. (1981). On the semantics of a glance at a scene. In M. Kubovy and J. R. Pomerantz, Eds. *Perceptual organization.* Hillsdale, NJ: Erlbaum.

Broadbent, D. E. (1977). Levels, hierarchies, and the locus of control. *Quarterly Journal of Experimental Psychology, 29,* 181–201.

Brooke, J. B., and Duncan, K. D. (1981). Effects of system display format on performance in a fault location task. *Ergonomics, 24,* 175–189.

Brown, J. S., Moran, T. P., and Williams, M. D. (1982). The semantics of procedures (Tech. Rep.). Palo Alto, CA: Xerox Palo Alto Research Center.

Brown, W., & Wyrick, R., Eds. (1982). *Analysis of steam generator tube rupture events at Oconee and Ginna* (82-030). Institute of Nuclear Power Operations.

Chapanis, A. (1972). Design of controls. In H. P. Van Cott, and R. G. Kincade, Eds. *Human engineering guide to equipment design.* New York: McGraw-Hill.

Chernoff, H. (1973). The use of faces to represent points in k-dimensional space graphically. *Journal of the American Statistical Association, 68,* 361–368.

Coekin, J. A. (1969). A versatile presentation of parameters for rapid recognition of total state. In *International Symposium on Man-Machine Systems* (58-MMS 4). New York: IEEE Conference Record 69.

Cooper, G. E. (1977). A survey of the status and philosophies relating to cockpit warning systems (Tech. Rep. NASA-CR-152071). NASA Ames Research Center.

Crowe, D., Beare, A., Kozinsky, E., and Haas, P. (1983). *Criteria for safety-related nuclear power plant operator actions: 1982 pressurized water reactor simulator exercises* (NUREG/CR-3123). Springfield, VA: National Technical Information Service.

Damon, A., Stoudt, H., and McFarland, R. (1971). *The human body in equipment design.* Cambridge, MA: Harvard University Press.

Duncan, K. D. (1981). Training for fault diagnosis in industrial process plants. In J. Rasmussen and W. B. Rouse, Eds. *Human detection and diagnosis of system failures.* New York: Plenum.

Edwards, E., and Lees, F. P., Eds. (1974). *The human operator in process control.* London: Taylor and Francis.

Electric Power Research Institute. (1984). *The EPRI guide.* Palo Alto, CA: Electric Power Research Institute.

Fink, R. T. (1984). *A procedure for reviewing and improving power plant alarm systems* (NP-3448). Palo Alto, CA: Electric Power Research Institute.

Frey, P. R., et al. (1984). *Computer-generated display system guidelines. Vol. 1: Display design* (NP-3701). Palo Alto, CA: Electric Power Research Institute.

Gallagher, J. (1982). *Disturbance analysis and surveillance system scoping and feasibility study* (NP-2240). Palo Alto, CA: Electric Power Research Institute.

Geiger, G., and Schumacher, W. (1976). Parallel vs. serial instrumentation for multivariable manual control in control rooms. In T. B. Sheridan and G. Johannsen, Eds. *Monitoring behavior and supervisory control.* New York: Plenum.

Gimmy, K. L., and Nomm, E. (1982). Automatic diagnosis of multiple alarms for reactor control rooms. *Transactions of the American Nuclear Society, 41,* 520.

Goldsmith, T. E., and Schvaneveldt, R. W. (1984). Facilitating multiple-cue judgements with integral information displays. In J. C. Thomas and M. L. Schneider, Eds. *Human factors in computer systems.* Norwood, NJ: Ablex Publishing.

Goodstein, L. (1981). Discriminative display support for process operators. In J. Rasmussen and W. B. Rouse, Eds. *Human detection and diagnosis of system failures.* New York: Plenum.

Goodstein, L. (1983). An integrated display set for process operators. In G. Johannsen and J. E. Rijnsdrop Eds. *Analysis, design and evaluation of man-machine systems.* New York: Pergamon.

Haas, P., Selby, D., Hanley, M., and Mercer, R. *Evaluation of training programs and entry level qualifications for nuclear power plant control room personnel based on the systems approach to training* (NUREG/CR-3414). Springfield, VA: National Technical Information Service.

Haber, R. N. (1981). The power of visual perceiving. *Journal of Mental Imagery, 5,* 1–40.

Hall, R., Fragola, J., and Luckas, W., Eds. *Conference record for 1981 IEEE standards workshop on human factors and nuclear safety.* New York: IEEE.

Hanes, L. F., O'Brien, J. F., and DiSalvo, R. (1982). Control room design: lessons from TMI. *IEEE Spectrum, 19*(6), 46–52.

Hannaman, G. W., and Spurgin, A. J. *Systematic human action reliability procedure* (NP-3583). Palo Alto, CA: Electric Power Research Institute.

Hollan, J. D. (1984). Intelligent object-based graphical interfaces. In G. Salvendy, Ed. *Human-computer interaction.* Amsterdam: Elsevier.

Hollnagel, E., (1981). Report from the third NKA/KRU experiment: the performance of control engineers in the surveillance of a complex process (Tech. Rep. N-14-81). Roskilde, Denmark: Risø National Laboratory.

Hollnagel, E. (1982). Training simulator analysis method (Tech. Rep. N-1-82). Roskilde, Denmark: Risø National Laboratory.

Hollnagel, E., and Woods, D. D. (1983). Cognitive systems engineering: new wine in new bottles. *International Journal of Man-Machine Studies, 18,* 583–600.

Hollnagel, E., Pederson, O. M., and Rasmussen, J. Notes on human performance analysis (Tech. Rep. Risø-M-2285). Roskilde, Denmark: Risø National Laboratory.

Hopkins, C. O., Snyder, H. L., Hornick, R. J., Mackie, R., Smillie, R., and Sugarman, R. C. (1982). *Critical human factors issues in nuclear power regulation and a recommended comprehensive human factors long-range plan* (NUREG/CR-2833). Springfield, VA: National Technical Information Service.

Jacob, R., Egeth, H., and Bevan, W. (1976). The face as a data display. *Human Factors, 18,* 189–200.

Johnson, S. E. (1984). *DASS: a decision aid integrating the safety parameter display system and emergency functional recovery procedures* (NP-3595). Palo Alto, CA: Electric Power Research Institute.

Jones, D. W., et al. (1980). *Nuclear power plant simulators: their use in operator training and requalification* (NUREG/CR-1482). Springfield, VA: National Technical Information Service.

Kemeny, J. G., et al. (1979): *The President's Commission on the accident at Three Mile Island.* Springfield, VA: National Technical Information Service.

Kincade, R. G., and Anderson, J., Eds. (1984). *Human factors guide for nuclear power plant control room development* (NP-3659). Palo Alto, CA: Electric Power Research Institute.

Kleiner, B., and Hartigan, J. A. (1981). Representing points in many dimensions by trees and castles. *Journal of the American Statistical Association, 76,* 260–269.

Kortlandt, D., and Kragt, H. (1980). Process alarm systems as a monitoring tool for the operator. In *Proceedings third international symposium on loss prevention and safety promotion in process industries.* Basel, Switzerland.

Kragt, H., and Bonten, J. (1983). Evaluation of a conventional process-alarm system in a fertilizer plant. *IEEE Transactions on Systems, Man, and Cybernetics, SMC-13,* 586–600.

Lees, F. P. (1983). Process computer alarm and disturbance analysis: review of the state of the art. *Computers and Chemical Engineering, 7,* 669–694.

Lind, M. (1981). The use of flow models for automated plant diagnosis. In J. Rasmussen and W. B. Rouse, Eds. *Human detection and diagnosis of system failures.* New York: Plenum.

Little, J. L., and Woods, D. D. (1981). A design methodology for the man-machine interface in nuclear power plant emergency response facilities (Tech. Rep. 81-1C57-CONRM-P1). Westinghouse Research and Development Center, 1981.

Luckas, W. J., Lettieri, V., and Hall, R. E. (1982). *Initial quantification of human error associated with specific instrumentation and control system components in licensed nuclear power plants* (NUREG/CR-2416). Springfield, VA: National Technical Information Service.

Malone, T. B., Kirkpatrick, M., Mallory, K. M., Eike, D., Johnson, J. H., and Walker, R. W. (1980).

Human factors evaluation of control room design and operator performance at Three Mile Island (NUREG/CR-1270). Springfield, VA: National Technical Information Service.

McCormick, E. J. (1976). *Human factors in engineering and design.* New York: McGraw-Hill.

Mertes, F., & Jenny, L. (1974). Automation applications in an advanced air traffic management system (Tech. Rep. DOT-TSC-OST-74-14). TRW, Vol. III.

Meyer, O. (Personal communcation). Idaho National Engineering Laboratory.

Moray, N. (1981). The role of attention in the detection of errors and the diagnosis of failures in man-machine systems. In J. Rasmussen and W. B. Rouse, Eds. *Human detection and diagnosis of system failures.* New York: Plenum.

Nelson, W. R. (1984). *Response trees and expert systems for nuclear reactor operations* (NUREG/CR-36317. Springfield, VA: National Technical Information Service.

Nelson, W. R., et al. (1981). *Applications of functional analysis to nuclear reactor operations* (NUREG/CR-1995). Springfield, VA: National Technical Information Service.

Norman, D. A. (1981a). Categorization of action slips. *Psychological Review, 88,* 1–15.

Norman, D. A. (1981b). Steps towards a cognitive engineering (Tech. Rep.). University of California at San Diego, Program in Cognitive Science.

Norman, D. A. (1983). Design rules based on analyses of human error. *Communications of the ACM, 26,* 254–258.

Pack, R. (1984). (Personal communication).

Parris, H. L. (1984). (Personal communication). December.

Parris, H. L., and McConville, J. T. (1981). *Anthropometric data base for power plant design* (NP-1918-SR). Palo Alto, CA: Electric Power Research Institute.

Pew, R. W., Miller, D. C., and Feehrer, C. E. (1981). *Evaluation of proposed control room improvements through analysis of critical operator decisions* (NP-1982). Palo Alto, CA: Electric Power Research Institute.

Pine, S. M., et al. (1982). *Human engineering guide for enhancing nuclear control rooms* (NP-2411). Palo Alto, CA: Electric Power Research Institute.

Pope, R. H. (1978). Power station control room and desk design: alarm system and experience in the use of CRT displays. In *International symposium on nuclear power plant control and instrumentation.* Cannes, France.

Rabbitt, P. (1984). The control of attention in visual search. In R. Parasuraman and D. R. Davies, Eds. *Varieties of attention.* New York: Academic.

Rasmussen, J. (1981). Models of mental strategies in process plant diagnosis. In J. Rasmussen and W. B. Rouse, Eds. *Human detection and diagnosis of system failures.* New York: Plenum.

Rasmussen, J. (1983). Skills, rules, and knowledge, signals, signs, and symbols, and other distinctions in human performance models. *IEEE Transactions on Systems, Man, and Cybernetics, SMC-12,* 257–266.

Rasmussen, J. (1984). Strategies for state identification and diagnosis in supervisory control tasks, and design of computer-based support systems. In W. B. Rouse, Ed. *Man-Machine Research,* Greenwich, CT: JAI, Vol. I.

Rasmussen, J., and Lind, M. (1981). Coping with complexity. In H. G. Stassen, Ed. *First European annual conference on human decision making and manual control.* New York: Plenum.

Rasmussen, J., and Rouse, W. B. (1981). *Human detection and diagnosis of system failures.* New York: Plenum.

Reason, J. and Mycielska, K. (1982). *Absent minded? The psychology of mental lapses and everday errors.* Englewood Cliffs, NJ: Prentice-Hall.

Roscoe, S. N., and Eisele, J. E. (1976). Integrated computer-generated cockpit displays. In T. B. Sheridan and G. Johannsen, Eds. *Monitoring behavior and supervisory control.* New York: Plenum.

Rouse, W. B. (1982). A mixed fidelity approach to technical training. *Journal of Educational Technology Systems, 11,* 103–115.

Rouse, W. B. (1983). Models of human problem solving. In G. Johannsen and J. E. Rijnsdorp, Eds. *Analysis, design and evaluation of man-machine systems.* New York: Pergamon.

Rouse, W. B. (1984). *Computer-generated display system guidelines. Vol. 2: Developing an evaluation plan.* (NP-3701). Palo Alto, CA: Electric Power Research Institute.

Rouse, W. B., et al. (1984). *A method for analytical evaluation of computer-based decision aids* (NUREG/CR-3655). Springfield, VA: National Technical Information Service.

Savas, E. S. (1965). *Computer control of industrial processes.* New York: McGraw-Hill.

Schmall, T. M., Ed. (1980). *Conference record for 1979 IEEE standards workshop on human factors and nuclear safety.* New York: IEEE.

Seminara, J. L., and Parsons, S. O. (1981). *Human factors review of power plant maintainability* (NP-1567). Palo Alto, CA: Electric Power Research Institute.

Seminara, J. L., et al. (1979). *Human factors methods for nuclear control room design*. Vol. 1: *Human factors enhancements of existing nuclear control rooms* (NP-1118). Palo Alto, CA: Electric Power Research Institute.

Seminara, J. L., Gonzalez, W. R., and Parsons, S. O. (1977). *Human factors review of nuclear power plant control room design* (NP-309). Palo Alto, CA: Electric Power Research Institute.

Seminara, J. L., Gonzalez, W. R., and Parsons, S. O. *Maintainability assessment methods and enhancement strategies for nuclear and fossil fuel power plants* (NP-3588). Palo Alto, CA: Electric Power Research Institute.

Sheridan, T. B., and Hennessy, R., Eds. (1984). *Research and modeling of supervisory control behavior.* Washington DC: National Academy Press.

Sheridan, T. B., and Johannsen, G., Eds. (1976). *Monitoring behavior and supervisory control.* New York: Plenum.

Sheridan, T. B., Jenkins, J. P., and Kisner, R. A., Eds. (1982). *Proceedings of workshop on cognitive modeling of nuclear plant control room operators* (NUREG/CR-3114). Springfield, VA: National Technical Information Service.

Siegel, A., et al. (1984). *Maintenance personnel performance simulation model* (NUREG/CR-3626). Springfield, VA: National Technical Information Service.

Siegel, J., Goldwyn, R., and Friedman, H. (1971). Pattern and process of the evolution of human septic shock. *Surgery, 70,* 232–245.

Smith, H. T. (1976). Perceptual organization and the design of the man-computer interface in process control. In T. B. Sheridan and G. Johannsen, Eds. *Monitoring behavior and supervisory control.* New York: Plenum.

Smith, S. L. (1963). Man-computer information transfer. In J. H. Howard, Ed. *Electronic information display systems.* Washington DC: Spartan.

Sorkin, R. D., and Woods, D. D. (1985). Systems with human monitors: a signal detection analysis. *Human-Computer Interaction, 1,* 49–75.

Speaker, D. M., Thompson, S. R., and Luckas, W. J. (1982). *Identification and analysis of human errors underlying pump and valve related events reported by nuclear power plant licensees* (NUREG/CR-2417). Springfield, VA: National Technical Information Service.

Suchman, L. A. (1982). Towards a sociology of human-machine interaction: pragmatics of instruction following (Tech. Rep.). Palo Alto, CA: Xerox Palo Alto Research Center.

Swain, A. D., and Guttman, H. E. (1983). *Handbook of human reliability analysis with emphasis on nuclear power plant applications* (NUREG/CR-1278). Springfield, VA: National Technical Information Service.

Thompson, D. (1981). Commercial aircrew detection of system failures: state of the art and future trends. In J. Rasmussen and W. B. Rouse, Eds. *Human detection and diagnosis of system failures.* New York: Plenum.

Thompson, R. M. (1972). Design of multi-man-machine work areas. In H. P. Van Cott and R. G. Kincade, Eds. *Human engineering guide to equipment design.* New York: McGraw-Hill.

Umbers, I. G. (1976). A study of cognitive skills in complex systems (Unpublished doctoral dissertation). University of Aston in Birmingham.

U.S. Nuclear Regulatory Commission. (1975). Reactor safety study: an assessment of accident risk in U.S. commercial nuclear power plants (WASH-1400; NUREG-75/014). Springfield, VA: National Technical Information Service.

U.S. Nuclear Regulatory Commission. (1981). *Guidelines for control room design reviews* (NUREG-0700). Springfield, VA: National Technical Information Service.

U.S. Nuclear Regulatory Commission. (1983a). *U.S. Nuclear Regulatory Commission human factors program plan* (NUREG-0985). Springfield, VA: National Technical Information Service.

U.S. Nuclear Regulatory Commission. (1983b). *NRC fact-finding task force report on the ATWS event at Salem Nuclear Generating Station, Unit 1, on February 25, 1983* (NUREG-0977). Springfield, VA: National Technical Information Service.

Van Cott, H. P., and Kincade, R. G. (1972). *Human engineering guide to equipment design.* New York: McGraw-Hill.

Visuri, P. J. (1982). Multi-variate alarm handling and display. In *Proceedings of the international meeting on thermal nuclear reactor safety* (NUREG/CP-0027). Springfield, VA: National Technical Information Service.

Wickens, C. D. (1984). *Engineering psychology and human performance.* Columbus, OH: Charles E. Merrill.

Wickens, C. D., Boles, D., Tsang, P., and Carswell, M. (1984). *The limits of multiple resource theory in display formatting: Effects of task integration* (AD-P003 321). Springfield, VA: National Technical Information Service.

Woods, D. D. (1982). Operator decision making behavior during the steam generator tube rupture at the Ginna nuclear power station. In W. Brown and R. Wyrick, Eds. *Analysis of steam generator tube rupture events at Oconee and Ginna.* Institute of Nuclear Power Operations. (Also Westinghouse Research and Development Center Report 82-1C57-CONRM-R2.)

Woods, D.D. (1983). Human error and decision making. Position paper for NATO workshop on the "Theory and nature of human error."

Woods, D. D. (1984a). Visual momentum: a concept to improve the cognitive coupling of person and computer. *International Journal of Man-Machine Studies, 21,* 229–244.

Woods, D. D. (1984b). Some results on operator performance in emergency events. In D. Whitfield, Ed., *Ergonomic problems in process operations.* London: Institution of Chemical Engineering, Symposium Series 90.

Woods, D. D. (1986). On the significance of data: the display of data in context. (Manuscript in preparation.)

Woods, D. D., and Hollnagel, E. (1982). A technique to analyze human performance in training simulators. In *Proceedings of the Human Factors Society, 26th annual meeting.*

Woods, D. D., and Hollnagel, E. (1986). Cognitive task analysis. (Manuscript in preparation.)

Woods, D. D., and Roth, E. (1982). Operator performance in simulated process control emergencies (Unpublished study).

Woods, D. D., Wise, J. A., and Hanes, L. F. (1981). An evaluation of nuclear power plant safety parameter display systems. In *Proceedings of the Human Factors Society, 25th annual meeting.*

Woods, D. D., Wise, J. A., and Hanes, L. F. (1982). *Evaluation of safety parameter display concepts* (NP-2239). Palo Alto, CA: Electric Power Research Institute.

CHAPTER 12.6

HUMAN FACTORS REQUIREMENTS ENGINEERING FOR AIR TRAFFIC CONTROL SYSTEMS

DAVID R. LENOROVITZ
MARK D. PHILLIPS

Computer Technology Associates, Inc.

Englewood, Colorado

The authors thank Valerio R. Hunt, Director of the FAA's Advanced Automation Program and Dr. Andres Zellweger, Systems Engineering Division Manager, for their support and vision in allowing CTA, Inc. to apply this methodology. Efforts by Larry Fortier, Rodman Bourne, and Delbert Weathers of the FAA to facilitate and guide this process are also greatly appreciated. These methods were created by a dedicated CTA team including G. V. Kloster, H. A. Ammerman, R. S. Ardrey, L. J. Bergen, K. Bruce, M. C. Fligg, G. W. Jones, and K. Tischer.

This chapter describes the process of systematically identifying, analyzing, defining, and documenting the human factors (HF) related system requirements for a particular type of application area, namely that of air traffic control.

12.6.1 OVERVIEW

Like other "applications" treatments in this *Handbook,* this chapter attempts to amplify and integrate some of the methodological and procedural HF information presented in the front portion of this book by presenting it within the context of a significant "real-world" problem domain. When one makes the transition from the theoretical to practical domain, one is often treading on untested ground and is not always afforded the opportunity of prior empirical verification and validation. Nevertheless, systems are developed—albeit imperfectly and with imprecise and/or insufficient information. The systematic methodology for analyzing, developing, and documenting the HF requirements—that is, *engineering* those requirements—for a user–computer interactive system that is presented in this chapter has never been comprehensively tested or empirically evaluated per se. However, the approach is based on the collective engineering judgment and experience of professionals who have spent a considerable number of years observing and participating in the development of such systems. Much of what has been learned through these prior attempts has been incorporated. This approach has been critically reviewed, analyzed, and enhanced by several of our colleagues, including Dr. James Foley, Mr. Harold (Smoke) Price, Dr. Sidney Seidenstein, Dr. Robert Williges, and Dr. Harold Van Cott (Phillips, 1984). Although little is presented herein that is either particularly innovative or unique in and of itself, it is the thoroughness and comprehensiveness of this integrated approach that we believe has value. Specific attempts have been made to productively marry the top-down and bottoms-up approaches of both functional decomposition and analysis, as well as operational task analysis. The underlying theme of this effort has consistently been to provide a traceable rationale for the specification of requirements— rather than to compile an unsubstantiated "wish list" of seemingly desirable features or characteristics for the system under consideration.

The application problem being dealt with here is that of providing for the safe, orderly, and expeditious control of airspace traffic. The selection of air traffic control as a model application area within which to explore the practice of HF is particularly appropriate. This is so for a number of reasons. First, the air traffic system embodies a complex human–machine system in which the human operator's (controller's) role is crucial to overall system success. Also, because controllers can only indirectly receive information about, and execute "control" of, air traffic via sophisticated display, control, and communication equipment, they are heavily dependent on the appropriate design of said equipment. Furthermore, controllers must be able to interface successfully not only with this equipment, but also with an extensive network of personnel—for example, pilots, other controllers, meteorologists, and supervisors—in order to properly perform their task. In addition, the performance of the air traffic control system is so critical to human safety, and so pervasive in its impact upon the functioning of modern society in general, that the level of individual responsibility and commitment borne by each and every controller becomes a significant consideration in overall system design. Finally, the fact that air traffic control tasks are dynamic, continuous, attention-demanding, and time-critical, combine to make air traffic control a complex, important, effective, and highly visible area within which to properly apply HF principles and procedures.

The complexity and importance of HF in air traffic control is obviously such that entire volumes have been devoted to the subject (Hopkin, 1982; Wiener, 1980a, 1980b). However, our intent here is more simply (1) to provide sufficient background information to allow an appreciation of the air traffic control process and (2) to convey an understanding of the engineering requirements analysis approach that has been taken in defining and developing the HF-associated requirements for the coming upgrade of the U.S. Air Traffic Control System.

A final point to be noted at the outset of this chapter—one which is key to the fidelity and validity of the entire approach—is the crucial role that the user population can (and must) play in the entire process of requirements definition and design. As has been previously noted (Miller and Pew, 1981; Ramsey and Atwood, 1979), noncritical acceptance of user inputs may often constitute the first step on the road to design disaster—but failure to adequately incorporate user data and concerns into the design process will most assuredly guarantee one's early arrival at that undesirable destination.

The approach described in this chapter did not merely "consider" the users, ask them to be passive participants, or make feeble attempts at identifying requirements that would be "likely" to result in a "user-oriented" design. Instead, the FAA and our Human Factors Requirements Engineering Team made a joint *commitment to ensure the significant and sustained involvement* of a representative set of actively working air traffic controllers, line supervisors, and area managers throughout the entire requirements analysis and definition process (and to continue said involvement on through the design competition; design evaluation, selection, and award; implementation; and testing phases of the air traffic control system upgrade). Toward this end, the FAA established a Sector Suite (the name given to the controller's workstation) Requirements Validation Team, or SSRVT. This SSRVT was formed so as to represent both veteran and less experienced controllers, en-route and tower-type controllers, line controllers and management, and the various geographic regions of the U.S. Air Traffic Control System. The SSRVT, with its assigned cadre of Human Factors Requirements Engineers, has convened numerous times over the past several years to symbiotically develop the Sector Suite's HF requirements according to the process herein described. None of the analyses described, nor the events, functions, tasks, requirements, or design guidelines that were developed were realized without the technical and methodological guidance and support of the HF Requirements Engineering Team, and the knowledgeable critique and practical experience of the SSRVT members.

12.6.2   AIR TRAFFIC CONTROL IN THE UNITED STATES

For purposes of this chapter we will focus exclusively on the National Airspace System (NAS) of the United States. The NAS encompasses the common network of U.S. airspace; air navigation facilities, equipment, and services; airports or landing areas; aeronautical charts, information, and services; rules, regulations, and procedures; technical information, personnel, and material. Included in the NAS are system components shared jointly with the military.

Within the NAS is a service that promotes the safe, orderly, and expeditious flow of air traffic (including airport, approach, and en-route air traffic control). This organized system of service is called Air Traffic Control (ATC). The U.S. ATC system incorporates a combination of control equipment, techniques, procedures, and skills that have evolved over 40 years. This evolution has produced a mixture of equipment of many ages, technologies, and types. In its current configuration, it is the safest, most efficient ATC system in the world but it is at the same time very expensive to operate and maintain, and, because its expansion capability is limited, it is quite difficult to add new ATC functions.

Figure 12.6.1 provides an overview description of the operating elements of today's ATC system. As shown in the figure, the primary manned operational elements include the Air Route Traffic Control Center (ARTCC), the Airport Traffic Control Tower (ATCT), Terminal Radar Approach Control (TRACON), and Terminal Radar Approach Control in the Tower Cab (TRACAB). Not shown in this figure, but also in existence at a number of smaller installations, are the various nonradar approach control facilities.

The surveillance systems are composed of airport surveillance and long-range radars and common digitizers that provide digital input into the air traffic computer systems. The en route and terminal computer systems provide automated tools to the controllers in the form of radar data displays, printed flight progress strips, aircraft identification information, tracking, and associated data display. Facilities are interconnected either by the National Communications (NATCOM) facility or by local telecommunications interfaces. Pilots and controllers communicate through air/ground radio frequencies.

12.6.2.1   History of U.S. Air Traffic Control

The earliest form of ATC took place via Airway Radio Stations or Airway Communications Stations. These early stations were principally used to communicate weather observations. The radio was shared by other government agencies such as the Department of Agriculture, which reported on crop marketing. The stations were later called (INSACs) for Interstate Airway Communications, and they eventually developed into today's Flight Service Station (FSS) facilities.

The initial en route ARTCCs were established by the airlines, but they were later integrated into a network of 20–50 continental and offshore ARTCCs by the Bureau of Air Commerce. By the late 1930s, at least 20 cities had established airport traffic control towers equipped with light guns for visual signalling and low powered radios for verbal communication. By 1941 the Civil Aeronautics Authority started taking over the towers, and shortly thereafter all three facility types fell under the jurisdiction of the Civil Aeronautics Administration.

For many years air traffic controllers provided separation of en route air traffic by communicating with the pilot through a third party, either an airline company or an agency communication station. There was heavy dependence on visual separation or achieving separation through assignment to low-frequency airways.

A significant growth in the number of controllers, use of remote air/ground communications, and use of radar began after the collision of two airliners over the Grand Canyon in the late 1950s. Soon controllers were working with long-range radar, direct pilot contact, and Very High Frequency Omnidirectional Range (VOR) navigational aids. These improvements in the system, along with increased controller productivity, and the introduction of air carrier jet aircraft led to the establishment of the

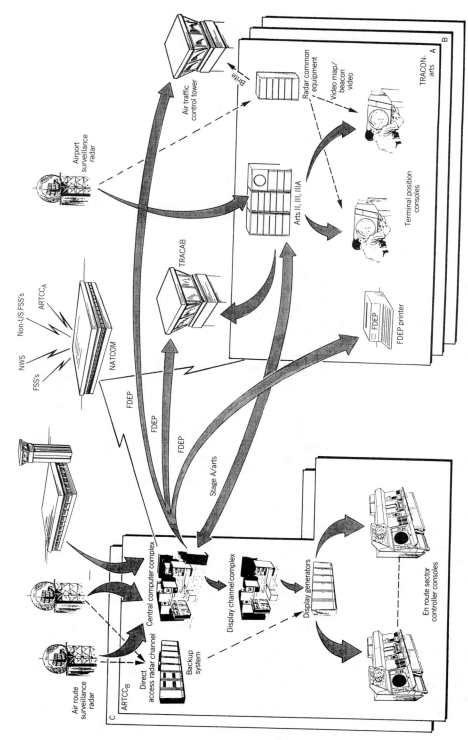

Fig. 12.6.1. Operational elements of current ATC system.

concept of positive control. The positive control concept is now implemented to exercise control and ensure separation of all air traffic at or above 18,000 ft (flight level 180).

By the early 1960s the FAA began efforts to apply automation techniques to its flight data processing systems. Computer equipment was installed in six centers in the northeast and, later, at Atlanta Tower, to improve safety and increase the productivity of controllers. The computers used in the centers produced flight progress strips that required support by flight data personnel. The initial systems did not have any updating capability to reflect dynamic changes in flight plans due to weather, heavy air traffic, and so on. These early computer systems featured a radar-tracking system that was very dependent on controller keyboard entries.

In 1968 a plan to automate many of the functions of the en route air traffic control system was approved. The first version of this system, called NAS Stage A, was installed and tested at the Jacksonville ARTCC. This system was an important development, but was not implemented nationally. A National Air Traffic Automation Coordinating Committee, comprised of air traffic facility data systems specialists (who were former journeyman controllers) and air traffic headquarters specialists, was formed to assist a contractor in defining air traffic requirements and specifications. This led to the successful merging of the flight data processing and radar data processing in subsequent models of NAS Stage A. By 1973 all centers had implemented flight data processing programs and by 1975 the centers were operational with the radar data processing computer program.

Automation in the terminal facilities was accomplished by implementing various versions of a system called ARTS in some 150 medium to large facilities during the 1970s. Collectively, these systems provided the capability for automated radar functioning, tracking aircraft, and predicting particular aircraft positions.

In short, between 1940 and 1980 the ATC system underwent a series of evolutionary changes that took it from a series of separate, fairly simple, primarily manually operated implementations up through today's integrated computer-based network of ATC nodes. These changes were stimulated by increased demand for air travel, and were made possible by technological advances in sensory, computational, and communicational equipment.

12.6.2.2. Current Developments in the ATC Environment

As is apparent from the above description, air traffic control has been and is currently undergoing a number of changes. First, we are entering what has been described as an era of advanced automation. There has been an ever increasing level of automation appearing in the cockpits of today's commercial aircraft. Not only are more capable, "intelligent" navigational and control systems being installed in planes, but enhanced capabilities are also being provided to automatically and continuously communicate with ground stations over bidirectional data links (i.e., not just voice links). Among other things, this kind of capability can support more complete, accurate, and recent positional, speed, and heading information being entered into the ATC data base for a given flight. Correspondingly, these better data, coupled with faster, more efficient computer hardware, and more accurate and comprehensive computational algorithms, can lead to better representations of current status, earlier and more accurate projections of impending situations that might affect the way in which air traffic is being controlled (e.g., weather delays in one location which might have a cumulative impact on traffic coming from other locations), as well as more effective analysis and evaluation of proposed solution alternatives to pending air traffic problems.

However, along with the potential for improvement offered by these technological advances, have come problems associated with concomitant changes in the level and mix of air traffic—many of which have served to complicate and exacerbate the task of the air traffic controller. The level of general aviation traffic has increased tremendously in recent years, and current projections indicate that this trend will continue. In dealing with general aviation traffic, controllers must interface with a much more heterogeneous population of pilots (in terms of skill, experience, aviation knowledge, and geographic familiarity), as well as deal with a wider range of aircraft performance characteristics and on-board navigational and communications equipment configurations. Since air traffic controllers have the responsibility of preventing confrontations between the aircraft that they are "controlling" and other general aviation aircraft (some of which may have voluntarily requested ATC support, and others of which may not be under direct ATC "control") this increase in "noncontrolled" air traffic significantly contributes to the load being handled by today's ATC system.

Thus, lack of flexibility and cost of maintenance of the current ATC system, availability of new, more capable technologies, and changes in the level and type of air traffic, have jointly contributed to the need for upgrading today's ATC system. The initial phase of this upgrade entails developing and implementing what has been termed the Advanced Automation System (AAS). Future upgrades will take place via staged implementation of a program called Automated En Route Air Traffic Control (AERA). Although there has been considerable discussion about the features and characteristics that might be incorporated into the various stages of AERA, detailed analyses of such capabilities are at this time speculative in nature. This section instead focuses on systematically analyzing and identifying the needs and requirements of the controller in the more immediate AAS ATC System. The AAS design will provide the baseline from which all future stages of AERA will evolve. The functions, tasks, and controller requirements identified within the AAS analyses will form the core of that baseline, and the methodology described herein will provide a mechanism for the evolution of any new AERA requirements.

In planning for such a system it must be remembered that the controller will still continue playing a crucial, human-in-the-loop type of role. No current or anticipated technological advances are foreseen that could entirely usurp the human's role as creative problem solver, integrator of seemingly disparate pieces of information, final decision maker, and party of ultimate responsibility for the lives and safety of air travelers.

12.6.2.3. Human Factors Challenge of Advanced Automation in ATC

In trying to gain a better understanding of some of the significant HF issues to be addressed in planning for the ATC upgrade, one must look carefully at both the job which controller's perform and at the characteristics of the current (and anticipated) population of controllers who perform it. Some of the key characteristics of the job itself are:

1. It carries a tremendous amount of personal responsibility.
2. It demands sustained optimum performance.
3. It consists of highly visible, closely monitored tasks—everything is recorded and archived.
4. It requires a tremendous amount of coordination (almost entirely verbal in nature) among colleagues, supervisors, and pilots.
5. There is a tremendous amount of complexity involved in recognizing and properly responding to particular combinations of conditions—for example, weather, traffic conditions, aircraft performance characteristics, airport facility status, and so on.
6. It consists of a set of highly proceduralized subtasks, but creative problem solving and decision making skills are needed to determine which procedures to apply and when to apply them.
7. The information being processed is very dynamic in nature, and there is a critical need to respond to those changes in status in a timely fashion.
8. There is a crucial dependency on the equipment suite that supports the controller. He or she must develop a certain type of rapport with the equipment, and must be able to have complete confidence in its performance and reliability.

These and other ATC environmental factors point toward the characterization of the air traffic controller as an event-sensitive, interruptible information processor. In this role, the controller must respond to simultaneous events, so constant prioritization of responses, pattern recognition, short-term memory, rapid decision making, coordination, and command execution are integral attributes of the controller's job. Because many of these attributes are cognitive in nature, a task analytically based methodology must go beyond observable behaviors in order to accurately represent the ATC task composition.

Finally, it must be noted that an effective ATC system must not only work well, but it must do so continuously. Reliability and maintainability are therefore critical issues of concern. Fault-tolerant system designs must preclude interruptions of service due to breakdowns of operational components (e.g., via redundancy, spare capacity, reconfigurability schemes, etc.). However, equally important concerns, such as accommodation of scheduled/preventive maintenance, must also be taken into account. These issues also have impact within the HF domain, owing to the fact equipment component specifications have to accommodate the accessibility and testability requirements of those personnel responsible for system maintenance and repair.

12.6.3. HUMAN FACTORS REQUIREMENTS ENGINEERING METHODOLOGY

Given the preceding orientation to some of the relevant HF issues, problems, and challenges associated with the ATC application area, we now describe an approach that was taken in helping the FAA to systematically identify, analyze, and develop a set of HF requirements, guidelines, and specifications to be presented to a pair of competitive vendors, each of which was charged with developing an integrated design for the controller–system interface within the ATC system upgrade.

12.6.3.1. Human Factors in Systems Development

The development of large-scale, complex systems, such as the AAS, typically follows a life cycle demarcated by several well-defined milestones. These range from mission definition, through initial operational acceptance testing, to full deployment. The methods used and outputs required of HF change commensurate with these evolutionary states. While fundamentally an iterative processes, these methods usually follow a course of requirements definition, design concept development, prototyping/testing, and performance evaluation. The methods described here focus on the "front-end" of systems development—the definition of user–system interface requirements. Accordingly, we will be describing these HF requirements engineering methods in the context of developing the controller–system interface requirements for the FAA's AAS.

12.6.3.2. Controller–System Interface Operations Concept Definition

The primary objective of our operations concept definition (Phillips and Tischer, 1984) was to decompose controller tasks to the level of detail such that the controller's job could be described in terms of:

1. Sequences of tasks that the controller performs when responding to a given ATC event.
2. A high-level outline of the dialogue between the controller and his or her workstation.
3. Interactions with other controllers, pilots, and supervisory personnel.
4. Information needed by the controller to successfully execute tasks accurately and in a timely fashion.

Meeting this primary objective enables system designers to understand the controller's job and to use operations concept task descriptions as a basis for workstation hardware and software design.
   A secondary objective was to characterize controller tasks in terms of:

1. Human capacity and workload.
2. Machine aids required to enhance controller performance.
3. Required controller training experience and skill development.

Meeting this secondary objective enables human resource management personnel to identify and develop necessary controller training and skills acquisition programs.

12.6.3.3. Controller Information Processing Task Analysis

In order to adequately define an overall operations concept for ATC we first needed a comprehensive identification and decomposition of the set of controller tasks. This was approached via the definition of a hierarchical structure of events to which the ATC system must respond, activities that are initiated in response to those events, subactivities that jointly constitute a main activity, tasks that must be accomplished in order to complete a given subactivity, and the individual task elements that make up a task (see Figure 12.6.2). An event is defined as a distinct occurrence that the controller perceives and responds to in some specific way. To identify tasks, we modeled the controller as an event-sensitive multitasking information processor. The advantage of using this type of model as an analysis tool is that if a comprehensive list of events that the controller observes is documented along with a similarly extensive list of tasks, one may achieve a degree of accuracy in describing the controller's work. A disadvantage in modeling the controller as an information processor—one whose sole purpose is to perform tasks in response to events—is that much of the dynamic operational complexity of the controller's job is not captured by such a simplistic model. One event may trigger another event before an appropriate controller response occurs. In addition, in some cases it may not be appropriate for the controller to take action immediately in response to an event. Furthermore, many of the appropriate responses are mental in nature, and a mental response is difficult to capture and reflect unless an overt action also occurs. Therefore, in viewing the controller as an individual who solely processes information in response to discrete events, we do not obtain an entirely accurate picture of the controller's overall job. However, we feel this does provide a sufficiently accurate and useful model to obtain an initial understanding of user–system interface (USI) requirements.
   Having identified the set of ATC events, and a top-level set of controller activities, we performed a logical decomposition into subactivities and, finally, tasks. The decomposition was intended to preserve consistency, completeness, and transitivity of event stimuli and controller output responses. Composition graphs (Alford, Smith, and Smith, 1979) were used to show the multiprocessing nature of the controller's job. Figure 12.6.3 illustrates composition graph symbols that define sequential, concurrent, iterative, and decision making (path selection) flows of subactivities.
   Tasks were defined as basic statements of *what* must be accomplished by the controller. Low-level procedures or precise steps that would detail *how* a task is to be performed on a given set of equipment were not defined. Rather, the intent was to reflect what is done without unnecessarily implying a particular type of dialogue design (e.g., object-oriented versus command-oriented) or display equipment selection.
   The resultant graphical task decomposition was then augmented with a tailored version of structured English called Task Description Language (TDL). The TDL helped to ensure logical consistency in the graphical task decomposition and communicate the task structure both to controller and engineering personnel. The TDL provides essentially the same information as the composition graphs, but presents

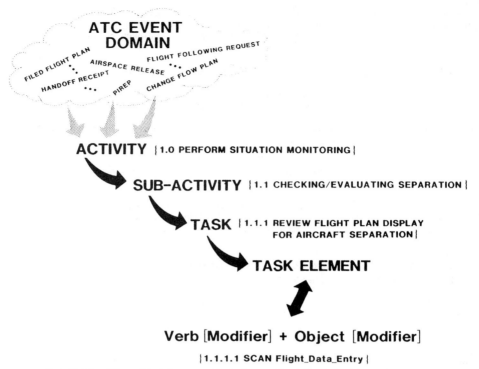

Fig. 12.6.2. Hierarchical decomposition of controller information processing tasks.

it in a way that forces an analysis of the logical connections among the tasks. This was done by first standardizing the task statement in terms of a set of well-defined, mutually exclusive verbs, objects, and qualifiers to ensure consistent use of terminology. A set of logical constructs was then applied to organize the tasks. The use of TDL in conjunction with the graphs serves as a validation tool which ensures that the task analysis is both internally consistent and complete. Figure 12.6.4 shows the relationship between composition graphs and TDL.

12.6.3.4. Characterization of Controller Information Processing Tasks

The levels of detail represented in the composition graphs and TDL allowed for characterizations of tasks based on information inputs and controller output requirements. Tasks were characterized in terms of both ATC complexity factors and sector type (e.g., low altitude arrival sectors, high altitude en route sectors, etc.). ATC complexity factors included level of coordination required, traffic density, traffic orientation, traffic separation, sequencing, and time responsiveness.

These initial characterizations were later used as a basis to form engineering judgments about operator workload levels and to assess the crew/team organization model. Cognitive and perceptual task attributes were inferred from these characterizations, and they were then used as a means of deriving machine-aiding requirements (e.g., the need for display highlighting of key information items, or the need for alarm indicators). Another series of task characterizations provided the basis for estimating the controller skill level requirements and developmental training requirements associated with each task.

12.6.3.5. Controller Dialogue Definition

The translation of controller tasks to USI requirements began with the Dialogue Description Language (DDL) analysis. The DDL required a multistep development process for each controller information processing task which had been identified. The DDL aids in the conceptualization of the model of user interaction, the identification of display information requirements, and the development of logical interaction techniques at the task level. The first step in developing the DDL was to analyze each of the tasks with respect to the following components:

1. Task type:
   a. Entry.
   b. Receipt.
   c. Analytical.
   d. Verbal coordination.
2. Logical display requirement: Aggregates of related data that were identified in base-level documentation as being needed for task performance—for example, the "Situation Display" which was called out in the System Level Specification.
3. Characteristic action type: Application and device-independent methods of controller input to the workstation—for example, select, position, text entry (Foley, Wallace and Chan, 1981).
4. Display content: Information directly viewed or manipulated in the course of task accomplishment—for example, runway list.

By characterizing each task in terms of these four components, recommended information presentation, coding, and interaction techniques could be added to the task statement.

The task statements were then enhanced to add semantic meaning to the tasks and serve as unequivocal requirements statements. The final step in DDL development involved the documentation of inferences the controller would make as a result of task performance. These inferences served as a means of validating the individual DDL statements and can be used as inputs for training program development.

The construction of the DDL established the link between the event-sensitive controller information processing tasks (shown in composition graph and TDL form) and the input and display requirements of the controller–system interface. The operations concept definition, therefore, ensured that ATC controller–system interface requirements were in all cases directly derived from and traceable to controller task requirements. An example of the DDL is presented in Figure 12.6.5.

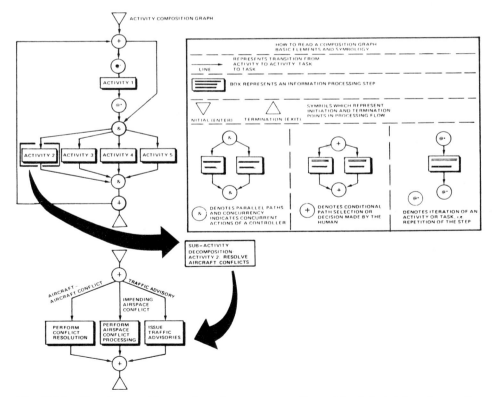

Fig. 12.6.3. Use of composition graph to decompose controller activity [adapted from Lenorovitz, Phillips, Ardrey, and Kloster (1984, Fig. 2), by permission].

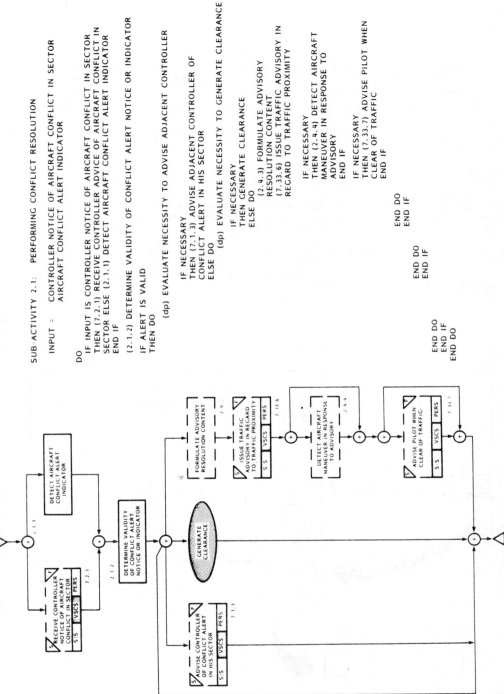

SUB-ACTIVITY 2.1:   PERFORMING CONFLICT RESOLUTION

INPUT =   CONTROLLER NOTICE OF AIRCRAFT CONFLICT IN SECTOR
           AIRCRAFT CONFLICT ALERT INDICATOR

DO
    IF INPUT IS CONTROLLER NOTICE OF AIRCRAFT CONFLICT IN SECTOR
    THEN (7.2.1) RECEIVE CONTROLLER ADVICE OF AIRCRAFT CONFLICT IN
    SECTOR ELSE (2.1.1) DETECT AIRCRAFT CONFLICT ALERT INDICATOR
    END IF

    (2.1.2) DETERMINE VALIDITY OF CONFLICT ALERT NOTICE OR INDICATOR

    IF ALERT IS VALID
    THEN DO

        (dp) EVALUATE NECESSITY TO ADVISE ADJACENT CONTROLLER

            IF NECESSARY
            THEN (7.1.3) ADVISE ADJACENT CONTROLLER OF
                 CONFLICT ALERT IN HIS SECTOR
            ELSE DO
                (dp) EVALUATE NECESSITY TO GENERATE CLEARANCE

                    IF NECESSARY
                    THEN GENERATE CLEARANCE
                    ELSE DO

                        (2.4.3) FORMULATE ADVISORY
                                RESOLUTION CONTENT
                        (7.33.6) ISSUE TRAFFIC ADVISORY IN
                                 REGARD TO TRAFFIC PROXIMITY

                            IF NECESSARY
                            THEN (2.4.4) DETECT AIRCRAFT
                                 MANEUVER IN RESPONSE TO
                                 ADVISORY
                            END IF

                            IF NECESSARY
                            THEN (7.33.7) ADVISE PILOT WHEN
                                 CLEAR OF TRAFFIC
                            END IF

                        END DO
                    END IF

            END DO
            END IF

        END DO
        END IF
END DO

Fig. 12.6.4. Relationship between composition graph and TDL [from Lenorovitz et al. (1984, Fig. 3), by permission].

Task Statement	Task Type	Logical Display	Characteristic Action Type	Display Content	Enhanced Task Statement	Controller Inference
Observe range/bearing between aircraft.	Entry/ Receipt.	Situation.	Select & text entry.	Range/ bearing list.	Observe range/bearing between selected aircraft on the Situation Display obtained by identifying/ selecting the desired aircraft and invoking the range/bearing function.	Utilize range/bearing information to determine multiple aircraft relationships.
Observe air-space intrusion by a non-controlled object.	Receipt	Situation.	N/A	Target(s).	Observe appearance on Situation Display of any target reflecting intrusion into controlled airspace by non-controlled object(s).	Detect/monitor non-controlled objects which could become hazard to controlled aircraft.
Compose/ enter reminder note of airspace intrusion.	Entry.	Situation.	Text entry & select.	Note field, limited data block, or a full data block.	Compose/enter reminder note and/ or track i.d. associated with target on Situation Display to annotate non-controlled intruding object.	N/A
Flight-follow an observed non-controlled object.	Receipt/ Analytical.	Situation.	Text entry & select.	Note field, limited data block, or a full data block.	Flight-follow (continually monitor the movements/behavior of) a non-controlled object on the Situation Display to determine possibility of a hazardous situation.	Monitor behavior of unpredictable, non-controlled object for possible impact on controlled aircraft.

Fig. 12.6.5. Example of controller DDL [from Lenorovitz et al. (1984, Fig. 4), by permission].

12.6.3.6. Controller–System Requirements Specification

Having defined the controller task profile and the DDL, the HF work turned its focus to deriving and specifying the implicit controller–system interface requirements. Figure 12.6.6 depicts a seven-step paradigm employed to specify these requirements. The activities included are as follows:

1. Perform task element analysis.
2. Analyze and decompose controller input messages and display outputs.
3. Develop a conceptual model of interaction.
4. Decompose functional capabilities.
5. Allocate functional capabilities to system components.
6. Define requirements specifications for each component.
7. Define data base requirements.

In this section, we focus on the first three analytic activities because they parallel closely other HF methods and outputs described in this text. The later four activities are briefly described in order to show how the controller task profile (established in the operations concept definition) can be transformed into engineering requirements for the controller–system interface.

Activity 1: Perform Task Element Analysis

As shown in Figure 12.6.6, Activities 1 and 2 occurred in parallel. Each DDL statement (derived from the operations concept definition) reduced to a series of atomic task element statements (TES). Together with the analysis of controller input messages and displayed images, these statements formed the basis for defining the user interface language.

Each TES was in the form of a simple sentence structure composed of a verb and an object, accompanied by optional object modifiers. Objects were defined to encompass input messages originating from the controller, displayed images output by the workstation, and conceptual "images" that exist in the mind of the controller. This syntax was as follows:

VERB [OBJECT MODIFIER] OBJECT [OBJECT MODIFIER]

for example:

SCAN Situation—Display for emphasized Data—Block

In the preceding example, the objects are Situation—Display (which is also termed a "logical display") and Data—Block. The verb is SCAN. Each TES was uniquely categorized as either an act of perception, mediation (cognition), interpersonal communication, or user initiation (input). An example of the reduction from a DDL statement to a series of TES is shown below:

DDL Statement
Delete Flight—Data—Entry and associated notations from a selected display

TESs:
DETERMINE need to suppress Flight—Data—Entry
SELECT Flight—Identifier

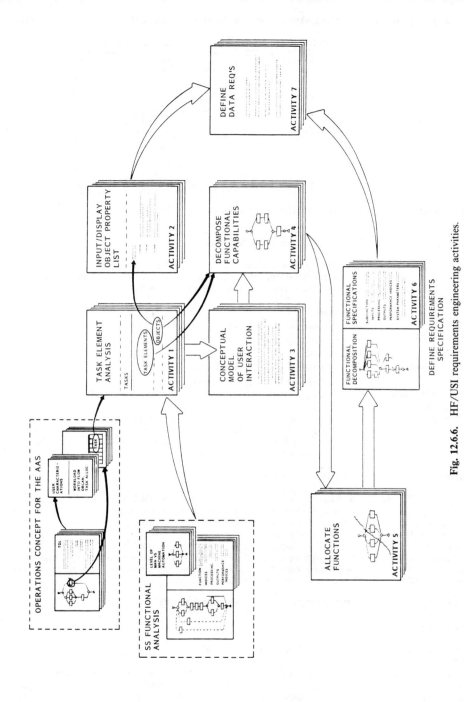

Fig. 12.6.6. HF/USI requirements engineering activities.

EXECUTE Flight—Data—Entry—Suppress—Function
DETECT Flight—Data—Entry—Suppression

Relative frequency of occurrence, necessary system response time, importance of performing the task element, and type and quantity of objects dealt with were also identified for each task element.

The verbs used in the Task Element Analysis were chosen from taxonomies developed by Lenorovitz, Phillips, Ardrey, and Kloster (1984). These taxonomies were developed to help more consistently and unambiguously describe or characterize various aspects of a given user–computer interface. The taxonomies were built on the earlier work of Berliner, Angell, and Shearer (1964) and directly incorporate many of their terms. However, an important addition is that we have explicitly defined each and every term we have used (see Table 12.6.1). This difference is key because one of the major benefits of a taxonomy should be its ability to support clear and consistent communication between its users. The taxonomy's precision of meaning and consistency of application across analysts are crucial to ensuring traceability and completeness when analyzing and engineering a complex USI. Our experience indicates that these taxonomies should be useful to anyone faced with the task of trying to describe and capture the essence of a USI.

Table 12.6.1   Dictionary of USI Action Taxonomy Terms

Term	Definition
Detect	Discover or note an occurrence (usually unsolicited).
Search	Purposeful exploration or looking for specified item(s).
Scan	Glance over quickly, usually looking for overall patterns or anomalous occurrences (not details).
Extract	Directed, attentive reading, observing, or listening with the purpose of gleaning the meaning or contents thereof.
Cross-reference	Accessing or looking up related information usually by means of an indexing or organized structuring scheme set up for that purpose.
Discriminate	Roughly classify or differentiate an entity in terms of a gross level grouping or set membership—frequently on the basis of only a limited number of attributes.
Recognize	Specific, positive identification of an entity.
Categorize	Classify or sort one or more entities into specific sets or groupings, usually on the basis of a well-defined classification scheme.
Calculate	Reckon, mentally compute, or computationally determine.
Itemize	List or specify the various components of a grouping.
Tabulate	Tally or enumerate the frequencies or values of the members of an itemized list or table.
Estimate	Mentally gauge, judge, or approximate, often on the basis of incomplete data.
Interpolate	Assign an approximate value to an interim point based on knowledge of values of two or more bracketing reference points.
Extrapolate/project	Assign an approximate value to a future point based on the value(s) of preceding point(s).
Translate	Convert or change from one form or representational system to another according to some consistent "mapping" scheme.
Formulate	Generate and put together a set of ideas so as to produce an integrated concept or plan.
Integrate	Pull together, and mentally organize a variety of data elements so as to extract the information contained therein.

Table 12.6.1 (*Continued*)

Term	Definition
Compare	Consider two or more entities in parallel so as to note relative similarities and differences.
Evaluate	Determine the value, amount, or worth of an entity, often on the basis of a standard rating scale or metric.
Decide	Arrive at an answer, choice, or conclusion.
Call	Signal to a specific recipient or set of recipients that a message is forthcoming.
Acknowledge	Confirm that a call or message has been received.
Respond	Answer or reply in reaction to an input.
Suggest	Offer for consideration.
Direct	Provide explicitly authoritative instructions.
Inform	Pass on or relay new knowledge or data.
Instruct	Teach, educate, or provide remedial data.
Request	Solicit, query, or ask for.
Receive	Get, obtain, or acquire an incoming message.
Name	Give title to or attach label to for purposes of identification/reference.
Group	Link together or associate for purposes of identification.
Introduce	Originate or enter new data into the system (e.g., type in a free-form message).
Insert	Make space for and place an entity at a selected location within the bounds of another such that the latter wholly encompasses the former, and the former becomes an integral component of the latter.
Aggregate	Combine two or more components so as to form a new composite entity.
Overlay	Superimpose one entity on top of another so as to affect a composite appearance while still retaining the separability of each component layer.
Copy	Reproduce one or more duplicates of an entity (no links to "master").
Instance	Reproduce an original ("master") entity in such a way as to retain a definitional link to the master— that is, such that any subsequent changes or modifications made to the master will automatically be reflected in each and every "instance" created therefrom.
Select	Opt for or choose an entity (e.g., a position or an object) by "pointing" to it.
Reference	Opt for or choose an entity by invoking its name.
Delete	Remove and (irrevocably) destroy a designated portion of an entity.
Cut	Remove a designated portion of an entity and place in in a special purpose buffer (residual components of the original entity usually close in around "hole" left by "cut-out" portion).
Set-aside	Remove entire contents of current (active) work area, and store in a readily accessible buffer (for future recall).
Suspend	Stop a process and temporarily hold in abeyance for future restoration.
Terminate	Conclude a process such that it cannot be restarted from the point of interruption, only by complete re-initiation.

Table 12.6.1 (*Continued*)

Term	Definition
Suppress	Conceal or keep back certain aspects or products of a process without affecting the process itself (i.e., affects appearance only).
Rename	Change an entity's title or label without changing the entity itself.
Un-group	Eliminate the common bond or reference linkage of a group of entities.
Segregate	Partition and separate an entity into one or more component parts such that the structure and identity of the original is lost.
Filter	Selectively eliminate one or more layers of an over-layed composite.
Transform	Manipulate or change one or more of an entity's attributes (e.g., color, line type, character font, size, orientation) without changing the essential content of the entity itself.
Execute	Initiate or activate any of a set of predefined utility or special purpose functions (e.g., sort, merge, calculate, update, extract, search, replace).

Source: Hopkin (1982, Table 1, by permission)

One of the taxonomies, the User-Internal Taxonomy, is depicted in Figure 12.6.7. It deals with actions central to the user, but transparent to the computer system. Its hierarchical structure incorporates three major classes of behaviors that users exhibit independently of the computer system—"perception," "mediation," and (interpersonal) "communication." Perception deals with the process of getting information into the (human) system, as well as some initial level of recognition/classification/identification of that information. Mediation (or, alternatively, cognition) is concerned with the human information processing activities that users perform on the information once it has been perceived. Finally, communications between people (via either some indirect telephonic medium or direct personal interchange) were included here because such communication links are critical components of many real-world human–computer interactive systems; they frequently impact a system's USI design (especially when source data capture is one of the system's principal functions); and finally, as an activity class, they share the characteristic of being transparent to the computer.

In a similar manner, Figure 12.6.7 and the corresponding definitional entries in Table 12.6.1, also present the User-Input Taxonomy. Again, in the interest of generalizability, the "logical" level of input activity description is represented, rather than a more "physically oriented" scheme (i.e., one which would have been more dependent on particular hardware/software implementation technologies). This facilitates this taxonomy's use in requirements specification—the process to which Task Element Analysis is central.

Activity 2: Analyze and Decompose Controller Input Messages and Displayed Outputs

This analysis was done in concert with the Task Element Analysis, such that data objects were identified and defined as they surfaced in the reduction of the DDL statements. A hierarchical schema was defined such that logical displays were defined as the top-level constructs that specified the outputs to the controller workstation. Each logical display could then be reduced to a set of composite objects, composite objects reduced to a set of objects, and objects reduced to sets of primitive data elements.

In the example that follows, "Situation—Display" is the logical display; Target/Track—Descriptor, Weather—Descriptor, Background—Descriptor, Conflict—Resolution—Options, and Weather—Products are composite objects; Data—Block is an object; and Leader—Line is a primitive data element. A structured English format was used to express the relationships between entities.

```
Situation_Display =
              Target/Track_Descriptor
      and     Weather_Descriptor
      and     Background_Descriptor
      and     Conflict_Resolution_Options
      and     Weather_Products
```

Target/Track_Descriptor =
                    [Data_Block]
      and   [Target_Status]
      and   [Track_Status]
      and   [Controlled_Status]
      and   [Route_Display]
      and   [Track_Vector]
Data_Block =
                    [Leader_Line]
      and   (Full_Data_Block
      or   Limited_Data_Block)

By defining the USI data domain in such an unequivocal manner, information density, coding presentation, and input requirements could be associated at both the task and data item level.

Activity 3: Develop a Conceptual Model of Interaction

The conceptual model analysis results in a set of heuristics that guide design decisions. These heuristics are required to ensure consistency in the overall controller–system interface design and to translate operational characteristics into a set of HF dialogue development principles.

Our conceptual model of interaction was developed by examining the nature and frequency of

USI ACTION TAXONOMY

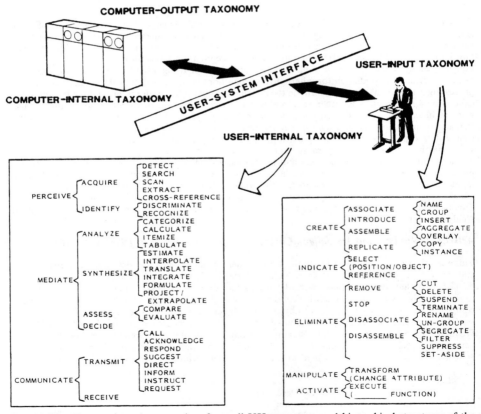

Fig. 12.6.7. Four major subtaxonomies of overall USI taxonomy, and hierarchical structures of the user-internal and user-input taxonomies [from: D. R. Lenorovitz, et al., G. Salvendy (Ed.)], A taxonomic approach to characterizing Human-Computer Interfaces. In: *Human-Computer Interaction* (Elsevier, Amsterdam, 1984), pp. 113.

operator actions (from the Task Element Analysis) and then integrating HF design principles and guidelines (Engle and Granda, 1975; Pew and Rollins, 1975; Smith, 1981) with the actions required for the controller to achieve his or her goals. The conceptual model encompassed strategies for both the input of information by the controller and for the output of information to the controller.

Some of the application-oriented factors that contributed to the selection of an ATC interaction strategy included:

1. Accommodation of task variability from position to position, sector to sector, facility to facility.
2. Meeting specified response time requirements.
3. Minimization of operator input errors.
4. Accommodation of multiple operators with varying degrees of experience.

Some of the HF guidelines applied in selecting the interaction strategy included providing for:

1. Continuity within and between tasks.
2. Consistency across all display coding, dialogue, interaction techniques, symbology, and display structure techniques.
3. Informative, timely feedback.
4. Minimization of keystrokes.
5. Ease of learning.

The Task Element Analysis was central to the derivation of the conceptual model of interaction. Figure 12.6.8 shows the mean frequency/priority rankings of the itemized controller task element verbs (rankings were assigned by 15 experienced controllers using a modified Delphi technique). Overall, controllers tended to assign much higher ratings to task element priority than they did to task element frequency. Given the nature of air traffic control and the potential catastrophic consequences of task performance faults, one can understand the reason behind such high priority ratings. This fact dictates the need for careful USI design in order to support such critical ATC events in an efficient, error-free manner.

The data presented in Figure 12.6.8 also reveal that an important aspect of the controller's job is cognitive in nature. Some of the verbs rated particularly high in priority were "assess," "synthesize," "analyze," and "decide." These ratings indicate that special cognitive processing is required on the part of the controller. It is also apparent that there is substantial information handling that takes place. "Perceive," defined as getting information into the human (controller) system and identifying or recognizing that information, appears in many of the controller tasks. "Indicate" and "execute" describe the actions taken by the controller to get information into, or cause actions to take place within the ATC computer system. The highly rated frequencies of these verbs point to the need for optimal coding conventions and display formatting to accommodate visual search and identification tasks, and the need for efficient input techniques to obviate time-consuming, error-prone entries. The item-by-item relationships of these task element verbs with their respective data objects (as embodied in the Task Element Analysis) are also crucial considerations to be factored into the controller–system interface design.

Activity 4: Decompose MMI Functional Capabilities

Controller–system interface functions were derived by analyzing the input message requirements, application processing requirements, display presentation requirements, and expected response time requirements specified in the previous Task Element Analysis and data decomposition. The conceptual model of interaction was examined to derive additional USI functions. Requirements were defined in terms of the inputs, outputs, transforms, completion criteria, global parameters, and performance indices associated with each identified USI function.

Activity 5: Allocate MMI Functional Capabilities to System Components

This activity involved the partitioning and explicit allocation of functions to the elements comprising the USI subsystem. During this allocation process, additional layers of functions were derived to deal with coordination, fault handling, and resource management. This step enabled analysis of the effects of failures of system functional components allocated to the workstation.

Activity 6: Define Requirements Specification for each Component

This entailed the refinement of the functional and performance requirements defined in earlier activities. Performance requirements for accuracy, response time, and capacity were mapped onto each function, based on information gathered during the above analytic activities.

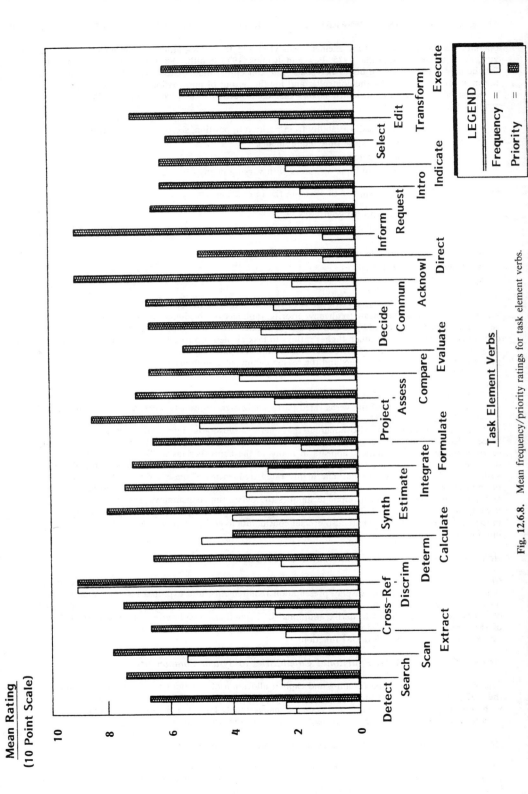

Mean Rating
(10 Point Scale)

Task Element Verbs

Fig. 12.6.8. Mean frequency/priority ratings for task element verbs.

Activity 7: Define MMI Data Base Requirements

The requirements associated with the USI data base were also explicitly addressed. These requirements included the conceptual data base organization (schema), and the control of integrity, privacy, and security of the data. The schema identifies the types of data that are used, and specifies the relations between them. It is the framework into which the values of the data items can be fitted. The schema, in turn, is derived from/dependent on the user's view of the data base (subschema). Data security refers to protection of data against accidental or intentional disclosure to unauthorized persons, or unauthorized modifications or destruction. Privacy refers to the rights of individuals and organizations to determine for themselves when, how, and to what extent information about them is to be accessible to others. Integrity refers to the characteristics and constraints that ensure data correctness.

12.6.4. CONCLUSION

This chapter has described a comprehensive approach to performing HF requirements engineering and analysis. The approach incorporates a set of rigorous analytic methodologies that have been applied in deriving the controller–machine interface requirements for the next generation ATC system. The formalisms applied relate to the foundations of HF described throughout this *Handbook* (e.g., task analysis, workload modeling, conceptual model development). The paradigm employed was designed to synergistically utilize the results of each successive analytic activity to enable the derivation of a complete, consistent, and verifiable set of USI requirements for an extremely complex system. In doing so, we have demonstrated that a structured analytic discipline can be imposed on traditional HF techniques to yield valid systems engineering requirements, and that user tasks are the appropriate starting point for system design.

By following this top-down methodology, appropriately geared to the system development life cycle, HF activities became fully integrated into the overall AAS engineering effort. This permitted optimal application of HF methods and techniques, and should result in a more appropriately designed controller–system interface for the next generation ATC system.

REFERENCES

Alford, M. W., Smith, T. C., and Smith, D. L. (1979). Formal decomposition applied to axiomatic requirements engineering. Final report prepared for Ballistic Missile Defense Advanced Technology Center, Contract No. DASG60-78-C0158. TRW DSSG.

Berliner, D. C., Angell, D., and Shearer, J. W. (1964). Behaviors, measures, and instruments for performance evaluation in simulated environments. *Proceedings of the symposium and workshop on the quantification of human performance.* University of New Mexico.

Engle, S. E., and Granda, R. E. (1975). Guidelines for man/display interfaces (Technical Report 00.2720). Poughkeepsie, NY: IBM Poughkeepsie Labratory.

Foley, J. D., Wallace, V. L., and Chan, P. (1981). The human factors of graphic interaction tasks and techniques (Report Number GWU-11st-81-3). Washington, DC: The George Washington University. January, 1981.

Hopkin, V. D. (1982). Human factors in air traffic control (NATO Tech. Rep. No. AGARD-AG-275). NATO.

Lenorovitz, D. R., Phillips, M. D., Ardrey, R. S., and Kloster, G. V. (1984). A taxonomic approach to characterizing human-computer interfaces. In G. Salvendy, Ed. *Human-computer interaction* (pp. 111–116). Amsterdam: Elsevier.

Miller, D. C., and Pew, R. W. (1981). Exploiting user involvement in interactive system development, *Proceedings of the Human Factors Society.* pp. 401–405.

Pew, R. W., and Rollins, A. M. (1975). Dialogue specification procedures (Revised Edition, Technical Report 3129). Cambridge, MA: Bolt, Baranek, and Newman, Inc.

Phillips, M. D. (1984). Proceedings of the AAS sector suite blue ribbon review team (Contract DTF-A01-83-Y-10554). Englewood, CO: Computer Technology Associates, Inc.

Phillips, M. D., and Tischer, K., (1985). Operations concept formulation for next generation air traffic control systems. In B. Shackel, Ed., *Human-Computer Interaction—INTERACT '84* (pp. 895–898). Amsterdam: Elsevier.

Ramsey, H. R., and Atwood, M. E. (1979). Human factors in computer systems: a review of the literature (Technical Report SAI-79-111-DEN). Englewood, CO: Science Applications, Inc.

Smith, S. L. (1981). Man-machine interface (MMI) requirements definition and design guidelines (Technical Report No. ESD-TR-81-113). Bedford, MA: The Mitre Corporation.

Wiener, E. L., Ed. (1980a). Air traffic control I. (Special Issue) *Human Factors, 22*(5), 517–639.

Wiener, E. L., Ed. (1980b). Air traffic control II. (Special Issue) *Human Factors, 22*(6), 645–691.

CHAPTER 12.7

HUMANS, COMPUTERS, AND COMMUNICATIONS

KOJI KOBAYASHI

NEC Corporation
Tokyo, Japan

12.7.1 WHY I HAVE ADVOCATED "MAN AND 'C&C'" CONCEPT

12.7.1.1 The Coming of the "C&C" Era

Telecommunications, a century after the development of the telephone, is now on the verge of a new era. This new era can be called "C&C," the integration of computers and communications.

Progress in telecommunications over the past century has gone hand in hand with the development of automatic switching systems, multiplex transmission systems, microwave communications systems, and other remarkable technological accomplishments. However, what deserves attention among the various trends in technological progress, is the transition from analog to digital coding of the electric signals used by these various systems. Telecommunications technology is becoming congruent with digital-based computer technology in a trend that signals the intersection and integration of communications and computers.

Meanwhile, computers have also gone through many changes since the construction of ENIAC in 1946, moving from vacuum tubes to transistors and integrated circuits. Now they have come to the age of LSIs and even VLSIs. One additional trend is manifest in the architecture of each new generation of computer, a movement away from the concentration of computing power toward diverse distributed processing.

Distributed processing means the formation of a total system linking each computer unit and subsystem via networks of communications lines. This is where the technological intersection and integration of computers and communications occur.

The use of the term "C&C" is an attempt to express the trend toward mutual contact and integration that can be seen emerging in these two leading areas of electronics. The implications of the new transition away from analog are still being explored. Figure 12.7.1 is an attempt to represent graphically an overview of "C&C." One aspect that has not been shown in Figure 12.7.1, however, is broadcasting systems.

12.7.1.2 Technological Factors in the Advent of "C&C"

Why "C&C" appeared on the scene, and what are the factors that have made it possible? As can be seen from Figure 12.7.1, the responsibility lies with the development of semiconductor circuit elements. The progress began with transistors, continued through ICs to LSIs, and now VLSIs accompany the development of microelectronics. These circuit elements are indispensable to the structure of both communications and computers.

LSIs and VLSIs are coming into use in fields other than communications and computers. The effects of this development can be interpreted as the addition of intelligence. The progress and growth of VLSIs, may have future economic effects which should be kept under consideration.

Critical new technologies that will be indispensable to the development of "C&C" in the years ahead should be discussed. The first of these technologies is the practical application of optical fibers. The diffusion of optronics will result not only in the transformation of communication transmission circuits, but must also be seen as a potent force behind future technological innovation.

The second technological factor will be the wider use of satellites. In addition to intercontinental communications, satellites will be used on a regional or national scale. Other fields, such as broadcasting, maritime navigation, weather forecasting, and exploration for resources are awaiting future satellite applications. One of the most pressing issues in the use of satellites is the overcoming of conventional concepts of national borders.

Many other new technologies will be incorporated into the practical development of "C&C." These include voice recognition and synthesis technologies, sensor technology, and robotics. These new technologies comprise a cluster of high technologies with a promising future.

12.7.1.3 Human Factors in the Advent of "C&C"

Human factors must be considered along with technological factors. Human beings are the main actors who will be using and commanding data through the "C&C" medium. At the same time, they are, in every sense of the word, the principal developers of "C&C." Therefore, the integration of computers and communications around humans will be referred to as "Man and 'C&C.'" In other words, all technological development should be directed toward making both hardware and software easier for humans to use.

Figure 12.7.2 elaborates upon Figure 12.7.1 with the addition of a human axis to create a three-

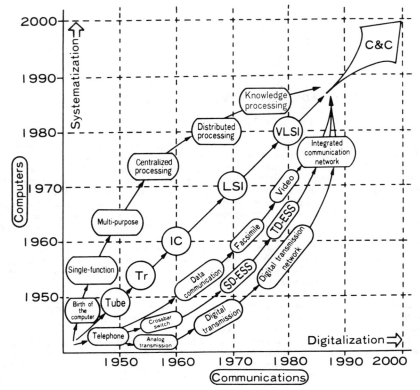

Fig. 12.7.1. Perspective of "C&C."

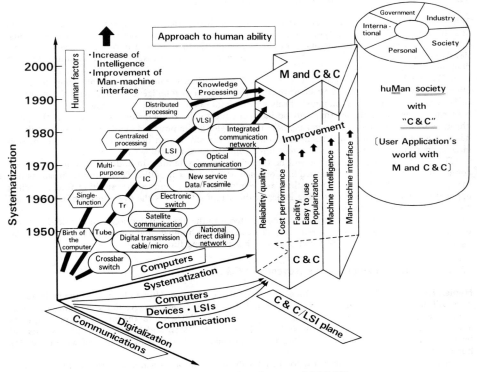

Fig. 12.7.2. Perspective of man and "C&C."

dimensional conceptual image. In using computers and communications, it has conventionally been necessary for people to learn the machine system in order to use it effectively. While the amount of labor required has now been reduced, it is still necessary to put considerable effort into studying and using sophisticated systems. The ideal at NEC is to arrive at a more fulfilling social and cultural life through the use of information systems that anyone can use quickly and easily.

The field that requires the most labor is software which is presently burdened with serious problems, the so-called "software crisis." It is becoming difficult to keep up with the burgeoning need for software. The rational development of the means of software production is crucial if "C&C" is to find appropriate use for mankind's benefit.

Figure 12.7.3 adds the interface between humans and machines to the earlier "C&C" construct. This chart indicates the reduced effort required of humans to interface with the machines as a result of the increasing sophistication of technology. In other words, the upgrading of the human–machine interface is following the slope rising toward the right-hand side of the figure.

As Figure 12.7.4 shows, the reduction in labor has been made possible by the enrichment of software, beginning with more sophisticated machine languages, and by the advances in terminal design developing from the addition of intelligence through the use of LSIs and VLSIs. In other words, it can be considered as a measure of the advance in machine and system intelligence, which has been made possible by the progress in semiconductor circuit elements. By incorporating software within these semiconductor circuit elements, the elements themselves have become intelligent.

Figures 12.7.3 and 12.7.4 are further elaborations on the human axis depicted in Figure 12.7.2 with Figure 12.7.4 providing even greater detail.

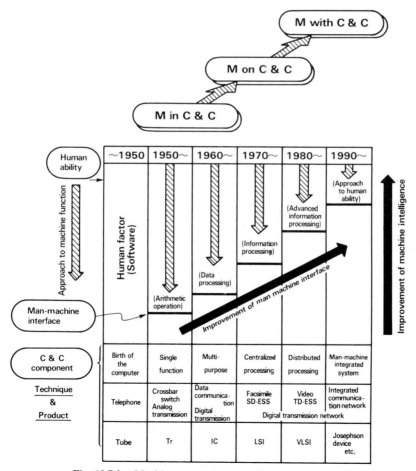

Fig. 12.7.3. Machine approaches man (Kobayashi, 1977).

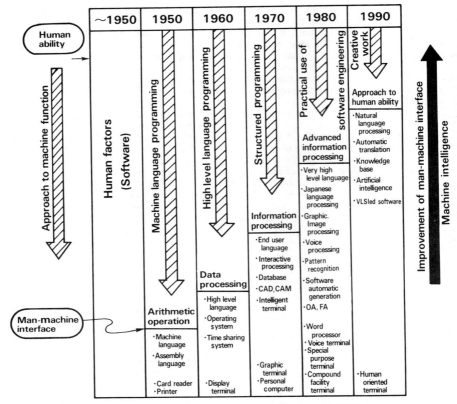

Fig. 12.7.4. Machine approaches man (Kobayashi, 1978).

12.7.2   THE INTERACTION BETWEEN MAN AND "C&C"

12.7.2.1   The Development of "Man and 'C&C' "

In the three decades since the first computers came into use, computer technology has progressed just as spectacularly as communications and semiconductor technologies.

During this time, computer users and those developing software and computer systems have been pursuing two goals. One is to close the wide gap between the standard level of human concepts, thought and behavior, and the level of corresponding computer intelligence. The other is to make the optimum use of computers and related hardware available at any stage of the development process. These efforts will be continued in the future as well.

The interaction between Man and "C&C" can be divided into three stages, as follows:

Era	Feature
1. M in "C&C," 1950–1960	Humans had to use machine level language to utilize systems
2. M on "C&C," 1960–1980	Machines became more user friendly, offering limited assistance to humans
3. M with "C&C" since 1980	Machines provide high-level assistance to humans

Figure 12.7.3 charts these three stages in terms of decades. Three indices of the human factor, M have been chosen:

1. Improvement of the human–machine interface.
2. Increases in the intelligence of machines and systems.
3. Human efforts to more closely approach machines and systems.

Up to now, the closer humans have approached "C&C" systems, the greater the relative amount of human labor and operations the interface has required. In fact, more time is spent in preparing to use the systems than in the actual use of them.

Since the early 1970s, methodology for the human side of this interaction has been developed and applied. For example, high-level languages, TSS on-line systems, and structured programming methods have been used. Thus the systems have become easier both to use and to construct.

At the same time, "C&C" systems have gained in intelligence owing to the development of the facsimile, digital transmission networks, and TD-ESS; the formation of an integrated communications network; and the development of LSIs as component devices.

The 1980s: Toward "C&C" Systems That Interact With Humans

In the 1980s, there should be further progress in the establishment of a useful methodology for the human side of the interaction. As information processing technology advances, these efforts will allow "C&C" systems to be used more effectively. The advances are likely to include research on application technologies, or the use of computers and communications in particular industries or operations. The objectives are to widen the area of "C&C" applications and to standardize the common features of diversified "C&C" usages. Efforts will focus on software engineering, especially in the development and maintenance of software products, and in requirement analysis, so that software production and quality control can proceed rationally and efficiently in spite of software's growing diversity, volume, and scale.

This progress will need to be matched by research in the human–machine interfaces—for example, in the development of technologies for verbal input and output, image processing, diagram processing, word processing, very-high-level languages, and pattern recognition. Rapid development will occur in high-level languages, specialized machines for data base handling, and ultra-high-speed machines, called supercomputers. One aspect of terminal-level development will include intelligent terminals with more sophisticated functions, compound terminals incorporating facsimiles, and industrial-use terminals with optimized application designs—as well as information network systems to link them. Another aspect will be the enhancement of the functions beyond simple terminals to create workstations that provide a much higher level of assistance to humans. Office automation combining these features will be researched, developed, and put into widespread use.

The 1990s: Toward A Human-Machine Integrated System

In the 1990s, "C&C" should approach the level of human faculties, and human beings should be able to devote their energies to more creative, intellectual undertakings. The decade will see the practical use of knowledge bases, artificial intelligence, and natural speech processing as basic functions, as well as the development of software incorporating into VLSIs in each applicable area. In terms of the human–machine interface, terminals will probably have been created to approximate human functions.

At the same time, there should be rapid development in communications technologies in PABX, commercial satellite communications, optical communication systems, and image transmission.

These predicted advances will accelerate the trend away from centralized processing, which was emphasized so heavily in the 1970s, toward decentralized processing adapted to the age of human–machine integrated systems—a trend backed up by the progress in "M and 'C&C'" technology.

The progress of developing the software to fill the gap between human and machine is illustrated in Fig. 12.7.4.

12.7.2.2 The Interrelationship Between Man and "C&C"

People have four relationships with "C&C" systems, as (1) owners, (2) users, (3) managers, and (4) those indirectly affected (see Fig. 12.7.5).

Full account needs to be taken of the benefits and effects that "C&C" systems may have on society or individuals. This requires assessment of the impact of "C&C" systems or technology.

As an example, the introduction of the on-line banking system using teller machines and cash dispensers has enabled banks to extend their hours of service and establish minibranches. These machines have made the banks highly convenient for depositors, as banking hours are increased and funds can be transferred to or from any part of the country. However, any system breakdown has a very large impact. It not only affects bank customers, but also renders it impossible to cash checks from third parties who have business dealings with those customers.

Moreover, when society at large changes through the introduction of various "C&C" equipment and systems, a new "C&C"-based culture is likely to arise. For example, a new information society will appear in which all organizations and individuals will be able to fully utilize the information infrastructure to build such a society. This will no doubt have a gradual but definite impact on our lives.

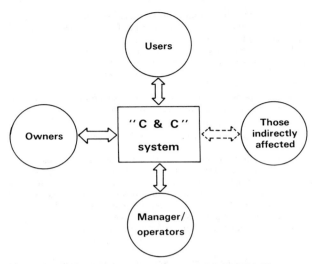

Fig. 12.7.5. Mutual relations between Man and "C&C" system.

12.7.3 BASIC ANALYSIS OF "M AND 'C&C' " SPACE

It is simple to talk of "the human factor," but humans are complex beings. Thus, the requirements are diverse when using "C&C" technology and products, and it would be difficult to represent them all by a single index.

In other words, the human factor incorporates a wide variety of aspects, including philosophy, ideology, emotions, individual behavioral criteria apparent in human character, group culture and traditions, and biological considerations. However, this chapter deals only with those human factors related to the use of "C&C" technology and/or products.

12.7.3.1 Formal Expression of the Human Factor M

If the component factors that make up the human factor M are expressed in terms of m_1, m_2, . . . m_n (where n is an arbitrary number), then M itself becomes a set of m_1, m_2, . . . , m_n.

Letting the space whose Z axis is the human factor M be V_M = "M and 'C&C,' " and the space whose Z axis is the single component factor m_i be V_{m_i} = "m_i and C&C," V_M can be visualized as the sum of all the spaces V_{m_1}, V_{m_2}, . . . , V_{m_n} superimposed on one another. Stated in equation form:

$$V_M = V_{m_1} + V_{m_2} + \cdots + V_{m_n}$$
$$\text{"M and 'C\&C' "} = (m_1 + m_2 + \cdots + m_n) \text{ and C\&C}$$

If $n = 3$, for example, the space shown in Figure 12.7.6 will be obtained.

12.7.3.2 Component Factors of M on the Z Axis

The components of the human factor M and the indices of measures which apply to them should be discussed in detail. The following three factors and indices have been selected as examples, since they are relatively comprehensive in meaning:

First, m_1 gives the trend shown in Fig. 12.7.7a, that is, those who use "C&C" systems are gradually applying their thought processes (or information processing) to contents of a higher level; they are moving from *data* (objective facts) to *information* (data as understood by the human mind) to *knowledge* (the results of reasoning, experimentation, etc.) In other words, the "C&C" level of service is approaching the level of human thought:

Data → Information → Knowledge

Next, the movement of m_2 over time is shown in Figure 12.7.7b. The relationship between hardware and software progresses as follows:

Independent → Parallel → Unified

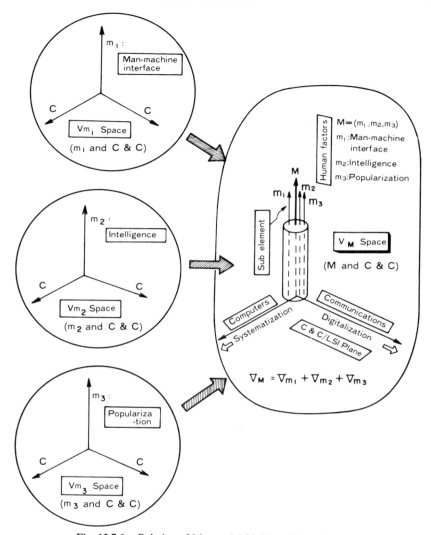

Fig. 12.7.6. Relation of Man and "C&C" and its subspaces.

The development of LSIs and VLSIs has enabled hardware components to have "intelligence."

Put another way, "intelligent products" have appeared with software incorporated into VLSIs themselves. In this way, an age in which software is treated as intellectual merchandise will be realized. It will be an age in which the true fusion of computers and communications will be seen.

Already software is available on cassette tapes and floppy disks and is being widely distributed and sold for the microcomputer and personal computer markets. Greater emphasis will be given to raising productivity and quality in the so-called "knowledge-intensive industries," which will be responsible for the software throughout its life cycle, including both development and maintenance.

The third component factor, m_3, is shown in Figure 12.7.7c. As the human–machine interface and "intelligence" become more advanced, the social distribution of "C&C" use will continue to change. A movement can be traced away from the era in which it will serve broader groups—business, corporations, society as a whole—to an era in which "C&C" will work for the benefit and convenience of individuals.

Initially, computers were used jointly by universities, research institutes, and large organizations. Then individual laboratories and divisions within these organizations gradually came to have their own minicomputers. Now personal computers are being used by private individuals. Microcomputers

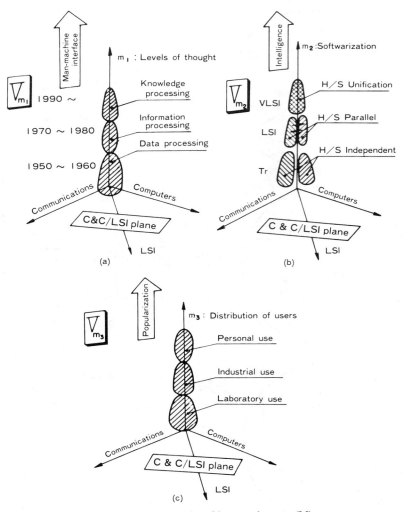

Fig. 12.7.7.   Examples of human elements (M).

have become adopted as a core element in a variety of home electronics devices, starting with air conditioning and heating equipment.

12.7.3.3   "C&C" Space with Humans as an Axis

The course of "C&C" and LSI development has been expressed on a plane diagram using X and Y axes (see Figure 12.7.1). The improvement of the intelligence content and the human–machine interface (i.e., the degree to which the technology approximates human faculties) can be represented as a component of the human factor on the Z axis. The course of these developments can be expressed as a three-dimensional spatial diagram (see Fig. 12.7.2). By projecting "M and 'C&C'" directly on to the X–Y plane along the Z axis, the same "C&C/LSI" plane diagram as in Figure 12.7.1 can be obtained.

Contributing to Society: the Essential Goal of "C&C"

Figure 12.7.2 reflects an important fact about "M and 'C&C'": its aim is to advance "C&C" technology, and to contribute to social development by refining and applying systems to meet the individual user's needs, through the use of increasingly advanced "C&C" technologies and products.

The ultimate aim of "M and 'C&C'" is to create a society in which humans and "C&C" interact—

an "M and 'C&C' " society—as represented by the cylinder on the right in Figure 12.7.2. This is the great challenge of the future.

In meeting this challenge, special consideration must be given to many components of the human factor, including the improvement of the human–machine interface and intelligence content, popularization, and cost reduction. Ultimately, by making "C&C" approach human faculties, its contribution to our lives can be enhanced.

The design of "C&C" systems in the future must satisfy these basic needs, while adequate consideration must also be given to their cost efficiency and quality as systems. The importance of the role played by software and its volumetric expansion has already been pointed out. Software production methods, quality improvement policies, and many other aspects of "C&C" require further study. One of the goals of the Man and "C&C" is to develop the expert system which incorporates expertise of every field into the software. These systems will provide the highest level of assistance to humans, making humans and machines true coworkers.

12.7.4 WORLD TRENDS AND GLOBAL ROLES OF "C&C"

While there may be some differences in the labels used, people in electronics and opinion leaders are in agreement on the concepts of "C&C" and "Man and 'C&C' " outlined in this chapter. The major trend in electronics during the last 15 years of the twentieth century is likely to be the initiation of the "C&C" age. "C&C" will follow a steady path of development as the year 2000 is approached.

This is not merely speculation, but a real trend in industrial society. For instance, this trend can be seen in the fact that the American telecommunications organization, AT&T, is entering the computer field and the computer company, IBM, is moving into communications. This is an inevitable consequence of technological progress and it is based on the necessity for industrial civilizations to progress through full utilization of information and knowledge resulting from advances in "C&C."

The world industrial map is being redrawn around a new axis of information, knowledge, and the equipment required. At NEC, telecommunications has been pursued in the almost 90 years since the company's founding. The company has also been involved in computers and semiconductor products for more than 30 years. NEC should continue to orient its future corporate concerns along the "C&C" growth path, and work to realize the actual development of this concept.

Working along these three business axes, NEC is already actively developing new systems and products in anticipation of the "C&C" era. For instance, the following have been developed: optical-fiber electronic switching networks, satellite earth stations for use in teleconferences, the NEAX61—digital electronic switching system, the ACOS computer series based on distributed information processing, network architecture (DINA), the PC-8800, PC-9800, and APC III series and other lines of personal computers, and Japanese language word processors. These products have sold well in Japan and overseas. In addition, NEC has an outstanding track record in the broadcasting field, which is an important mass medium. Their broadcast program automation systems, frame synchronizers, and other equipment are among the keys to "C&C" in broadcasting.

Recently, NEC succeeded in the practical development of VLSIs with memories in the 4 Mbit range, and is working to create other high-level functional elements. This steady progress can be seen as a kind of warranty for the future development of "C&C."

The advanced technology that came up for discussion in the 1982 Versailles Summit in France as a tool for reactivating the world economy is basically the same as the "C&C" advocated in electronics. Herman Kahn has praised the growing "informationalization" seen in Japan, and compares "C&C" to his C^4I^2 theory—in other words, command, communication, computing, and control, plus information and intelligence. So "C&C," besides being a corporate goal of NEC, is recognized as an important global trend.

"C&C" has won recognition as a development that promises to reconstruct our civilization around knowledge and information. This is true for Japan, which lacks adequate natural resources and raw materials, and for the entire modern world, where constraints on economic growth are increasing day by day. "C&C" can contribute significantly to conserving energy, reducing demands on raw materials, and enhancing the efficiency of every kind of system in use.

12.7.4.1 Actual Examples: The User's Perspective

In the beginning of this chapter, "C&C" was introduced and placed in perspective from the developer and supplier's viewpoints. However, there is also a great and growing interest in the effects that "C&C" will have on the common citizen and society. Therefore, a general user's perspective is required. There are a number of institutions trying to develope outlooks for the future, in Japan and other countries. Several difficult factors need to be considered during development and explanation of outlooks based on "C&C."

First, these areas can be identified for "C&C" applications based on market characteristics: the public sector, the business sector, and the home sector. The business sector can be divided into the office and the factory. Both the public and the business sectors can be considered from an international level as well as at a national or regional level.

The other important consideration is chronological. The three sectors can be divided further based on what is expected to come about in the next few years, and on what may not be available until the year 2000. Taken together, these market and chronological divisions form a matrix framework within which the many variations of "C&C" will develop. Figures 12.7.8 and 12.7.9 are graphic representations of the totality of "C&C" applications.

As the practical development of "C&C" continues, there will be a significant change in the way these systems are operated. Instead of having a single enterprise responsible for running these systems, there will be a rise of a multidimensional structure involving government, public organizations, private industry—even hospitals and libraries—as the public and market dimensions of "C&C" make themselves felt. This will include the multilayered development of broadcasting business on par with telecommunications enterprises. Therefore, it is necessary to study and to consider this multilayered structure of "C&C" from an operational perspective.

12.7.4.2 Communications Networks

Some specific predictions about "C&C" can be made. First communications networks are either wired or wireless. In wired systems, fiber optics will take over from traditional cable circuits. Moreover, voice transmission will be supplemented by the transmission of data and character, graphic, and video information, all by digital coding. The conversion of this infrastructure will require from 10 to 20 years.

In wireless systems, satellites will take center stage alongside conventional microwave circuits, and will be used to establish communications and broadcasting networks. The question of whether satellite reception will be handled on a community or an individual basis is really an economical one rather than a technological one.

The intersections of these networks will rely on an electronic switching system, which in addition to switching will provide answering services such as when the recipient is out or the line is busy. Also, the appearance of new terminals boasting a full range of new functions to cope with the new networks can be predicted.

Meanwhile, a host of dedicated satellites will be used to track the weather, aid maritime navigation, and fulfill other specialized tasks. Mass media broadcasting will also see increased interplay with its audiences as interactive systems make their appearance.

12.7.4.3 Trend to More User-Friendly Computers

The second example is in computers, where radical transformations in both form and function mean that the current image of the mainframe computer will be changed entirely.

Accurate predictions cannot be made as to when commercialization of compound semiconductor elements and Josephson junctions will occur. But megabit VLSIs will become dominant in the future. Old computers based on arithmetic calculations will evolve into intelligent, information-processing machines capable of making their own inferences. This development will permit machines to stand in for many functions of the human brain—the kind of machines the general public envisioned when computers first appeared. The biggest advance in computer functions will be that computers will cease to be something that can only be mastered by trained specialists. Instead, anyone will be able to use computers by following relatively simple sets of procedures.

As equipment becomes more sophisticated, it will also become more compact, and there will be greater awareness of terminal configurations. The new terminals should reflect current advances in ergonomics.

12.7.4.4 The Compounding of Personal Computer Function

The third example of progress in "C&C" will concern personal computers. Japan has begun to see the rapid spread of personal computers and Japanese language word processors. Since Japan is not part of the Roman alphabet cultural sphere, ideographs cannot be dispensed with. Given the number of characters required, a Japanese language word processor is 100 times more complex than its English counterpart. It seems likely that instead of waiting for improvements to be made in the performance of dedicated Japanese language word processors, Japanese language word processing will be popularized as an additional use for personal computers.

As this example shows, a steady cycle of improvements in the functions and performance of personal computers will lead them to share, or in some cases share out, the functions of mainframe computers. Moreover, personal computers with multiple interfaces will come to be numbered among the vital components of any "C&C" system.

Personal computers with word processing functions, along with facsimile terminals, will form the terminals of a future electronic mailing system.

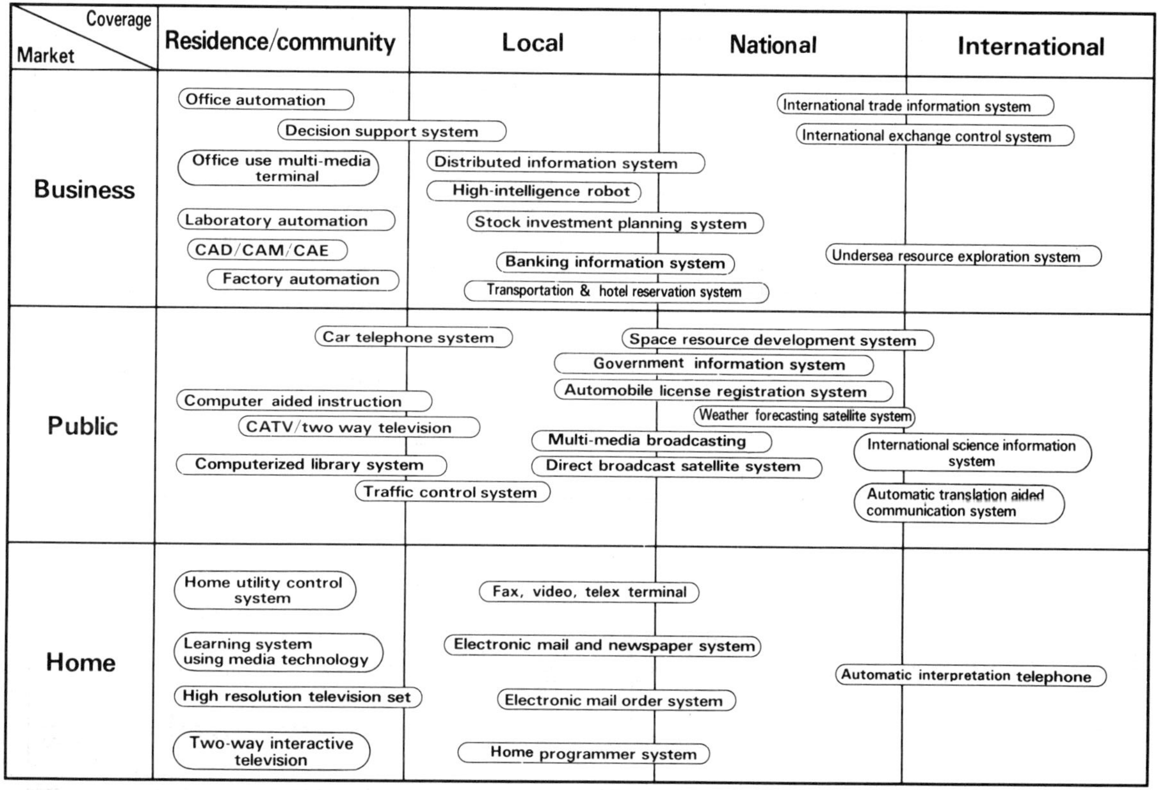

Fig. 12.7.8. Expectations of "C&C" applications (Kobayashi, 1977).

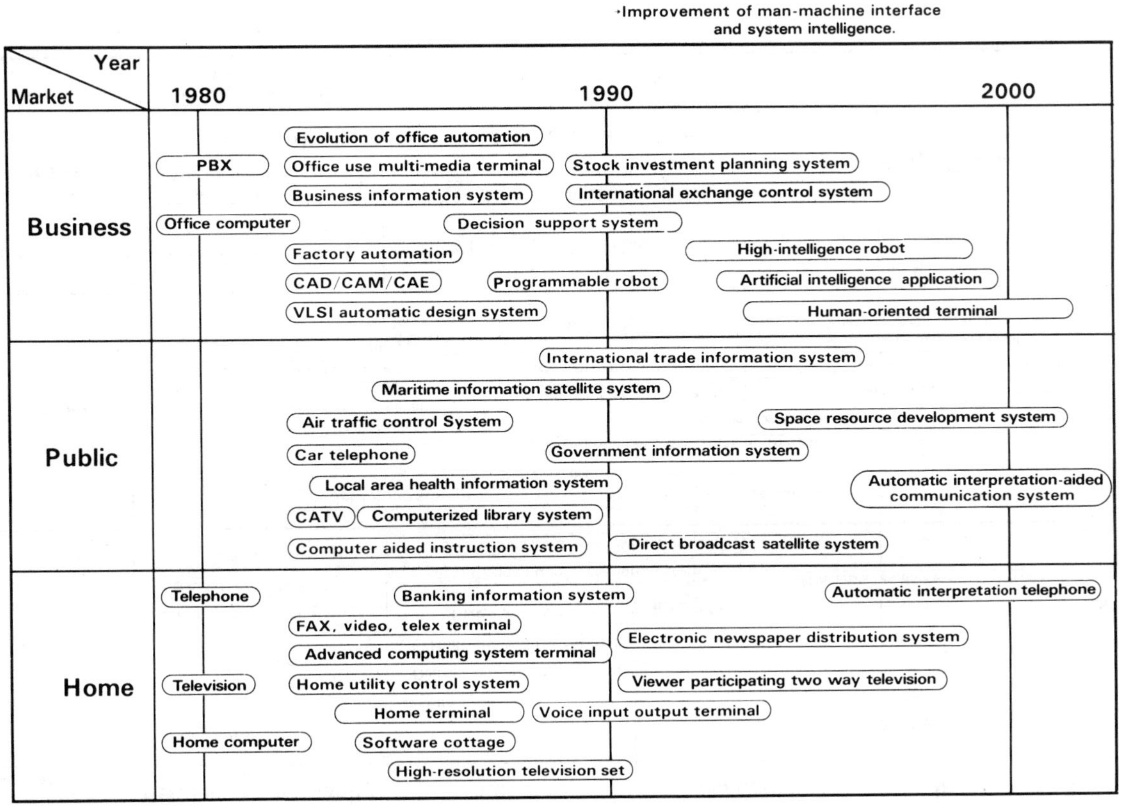

Fig. 12.7.9. Expectations of "C&C" applications (Kobayashi, 1978).

Undoubtedly, personal computers will be linked into the communications networks to create the third most important household electronics center, after the telephone and the television receiver. When that happens, it will be easy to equip the home with all the equipment required to create the so-called "electronic cottage."

A similar situation is developing faster in offices, and on a larger scale, in the form of office automation.

In the field of satellite offices, NEC has established an experimental "C&C" remote office in the suburbs of Tokyo, which is connected to the head office by communication lines. This office is equipped with various kinds of "C&C" equipment that process work otherwise done at the head office.

12.7.4.5 The Role of Robots

The fourth and final example is the progress in the development and use of intelligent robots. The robots dreamed of by the Czechoslovakian playwright, Karel Capek, in 1921 are taking their first steps 60 years later, thanks to the progress and cost reduction in microelectronics. Since these new machines have the direct effect of replacing human workers, concern has arisen as to whether they will aggravate unemployment. The battle between human and robot wits should be considered essentially as "no contest." When humans learn to use these devices wisely, the benefits of robots will far outweigh the disadvantages.

Robots should be employed in tasks that are dangerous or harmful to human health, or in jobs requiring extremely precise assembly. Human workers should be shifted to higher-level jobs more suited to human qualities. There is a tremendous need to devise ways in which robots can be used to their fullest effect such as in the development of earth's resources discovered by satellites or in ocean floor development.

There is no end to examples of what can be expected in the coming "C&C" age. I have pointed out that the distinctive characteristics of "C&C" do not manifest themselves in the individual functions of individual products, but rather, in the integrated use of all these different functions. I have also stressed that healthy market needs should be the standard for determining what effects "C&C" technologies will have and for setting the direction in which system and product development should be guided.

12.7.5 CREATING A "MAN AND 'C&C' " SOCIETY

The goal of "M and 'C&C' " is to benefit society by supplying technology and products that will promote, in the future, a new world of "C&C" uses and applications. To achieve this goal, some thought should be given to the future environment for the use of "C&C." This can be summed up as follows.

12.7.5.1 More Diversified, Broader Fields of Application

As "C&C" products and systems gradually extend to all areas of society, they will be used in a wide variety of forms. Office automation, for example, will come about through the combination of office machines and computers. In 1982, some 120,000 mainframe computers were in use in Japan, a figure that is expected to exceed 200,000 by 1990. Personal computers are spreading rapidly and there will be 2–3 million units in use in 1985.

In these circumstances, it will become important to study the fields of application of "C&C." It will also be important to standardize methods in different industries and businesses. Such measures will help increase the efficiency of software development and production.

Moreover, as the number and range of users increase, there will be a drive to make "C&C" systems easier to use. This need will have to be met by further improvements in the human–machine interface and the intelligence of the systems.

12.7.5.2 Progress in Systems for Public and Personal Use

In the past, efforts to increase the utilization of computers and communications have tended to be concentrated on huge systems such as those used by governments, institutions, and companies. However, in the future, there will be increased emphasis on individual-based systems.

There are two spheres of human life: private and public. There is a growing interest in the potential application of computers to leisure activities and study for individuals in their private capacity as opposed to members of a company or society. But, on the other hand, no one can remain detached from what is happening in society at large. Thus, computers are finding applications as "home utilities" for information and knowledge in the home, and for the pooling and sharing of knowledge among individuals.

What about people within organizations? There is a growing awareness that human beings should be freed to do the work most suited to human capabilities. Most of the routine office work involving the preparation of documents or the transmission of information should be left to computers.

To make this possible, there is an urgent need to develop simpler, cheaper word processors and terminals and to provide computers and communications equipment featuring high quality, reliability, and low maintenance.

12.7.5.3   Greater Quantity and Diversification of Data and Information

The trends discussed above give rise to a major problem: how to cope with the growing volume and diversity of data and information.

As information resources expand, better hardware will be needed. At the same time, the changing nature of the data being processed has resulted in a new demand for smaller-sized memory devices and other components.

As companies expand, their organizations become decentralized. This trend has accelerated the development of distributed processing systems that collect and process data where it is to be used, instead of concentrating all information in one place.

In the future, mutual development of local intelligence that is provided by a personal computer and host computers will be an important factor in computer operation.

12.7.5.4   Increasing Costs of Social Resources

Since the 1973 oil crisis, problems such as food and energy shortages, destruction of the environment, and the population explosion have taken on extreme urgency and seriousness. Our planet faces numerous problems that must be solved over the next few decades from both national and international perspectives.

"C&C" can be expected to play an effective role in speeding up the flow of information and assisting in accurate decision making.

Already the use of computers has achieved many positive results in terms of higher productivity and rationalization of labor, and is helping to counteract spiraling labor costs and the effects of the energy shortage. For these reasons, the role of computers in the modernization of industry will continue to grow.

While the developing countries are facing a population explosion, industrialized countries are experiencing a rapid shift of their population structures toward the older age group. This shift causes serious problems in welfare and health care. Therefore a strong demand is created for the development and introduction of information systems and hospital computer systems to assist medical research and provide diverse medical services.

Furthermore, new types of educational systems and "refresher" systems for individual skills will be needed for use in "lifetime education," that is, to facilitate the absorption of new knowledge and encourage constructive use of leisure time.

12.7.5.5   Components of an "M with 'C&C' " Society

It can be seen from Figure 12.7.10 that an "M with 'C&C' " society (a society in which human beings and "C&C" interact) consists of an environment, the fields of "C&C" application, and various "M with C&C" system groups. The fields of application may be divided into personal, social, industrial, national, and international. These fields interact with such factors as politics, economics, industry, distribution, trade, transportation, natural resources, environmental pollution, food, education, medical treatment, welfare, technology, administration, and international relations, to create the overall environment.

"Man and 'C&C' " technology creates multifaceted, diverse, and easy-to-use systems that support all fields of human activity and enable humans to lead richer lives. "Man and 'C&C' " system groups are set up at each level of human activity, be it decision making, management, and operations in factory and office work. The set of systems depicted in Figure 12.7.10 is a fraction of the total; it is clear that even more systems and products will be developed in the future.

12.7.5.6   Construction of an "M with 'C&C' " Society

Next, the formation of user systems that constitute an "M with 'C&C' " society will be considered (see Figure 12.7.11). The basic components of "C&C" systems include: computer systems, terminals, communication systems, office automation system products, network architecture products, and application programs satisfying the user requirements. These components are called "C&C" products.

Technologies are needed for constructing a variety of easy-to-use and efficient "C&C" products that possess high-level artificial intelligence and human–machine interfaces. These technologies will be referred to as "C&C" technologies.

LSIs, VLSIs, and optical-fiber communications are the fundamental driving forces behind the rapid development of "C&C" products and technologies. These technologies are considered to be basic "C&C" technologies.

Both software and system product engineering will assume central roles in combining these "C&C" products and technologies into a total technology for constructing systems that meet the needs of society. This concerns the whole system life cycle, ranging from the analysis of system requirements

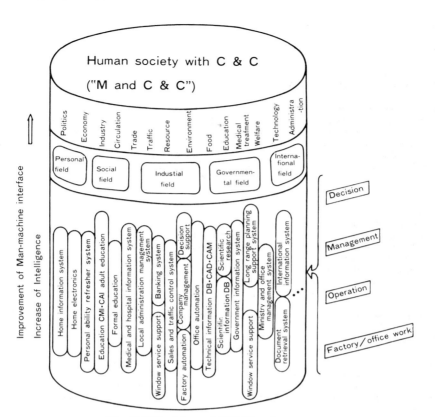

Fig. 12.7.10. Man and "C&C" society.

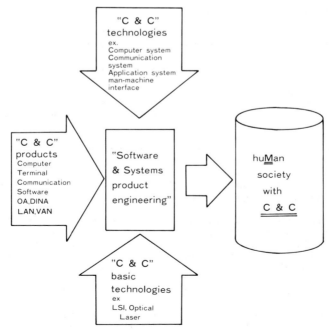

Fig. 12.7.11. Realization of human society with "C&C."

to its development, manufacture, and maintenance. Considerable research and development is required for future incorporation of these systems, particularly those where human factors are concerned.

12.7.6   IMPROVING SOFTWARE QUALITY AND PRODUCTIVITY

Software plays an important role in the creation of a "C&C" society. The tremendous increase in the application of microcomputers is causing a rapid expansion in the software sector and its workforce, in the field of computer applications for information processing and in virtually every industry.

If software development and maintenance technologies remain at their present levels, within a few years, software development will be unable to keep pace with demand. At that point, it is predicted, the "software crisis" that has been felt in some areas for several years will spread throughout the industrial sector. Therefore it is vital to establish modern and practical technologies for the production, management, and quality control of software.

Standardization in the production of software and distribution in modules or packages will improve productivity and quality.

Between 60% and 70% of the processes in the lifetime of software are thought to involve updating and maintenance. Maintenance needs to be regarded more positively in the future, and improving the productivity and quality of maintenance will be important tasks.

Software production can be thought of as a manual industry. Software is a product of mental effort applied manually, which is why each software production process has had a distinctly manual flavor. This has been a primary contribution factor to the incorporation of software bugs. Moreover, it is known that software productivity and quality levels vary widely according to the individual developer.

For these reasons, it is necessary to develop and educate software engineers in the use of effective design, production, and testing methodologies in conjunction with operation standardization, provisional tools to aid development and maintenance, and modernization of the working environment. Knowledge of human engineering will have to be applied to these areas.

The physical working environment has an important bearing on software. It will be important to make comprehensive improvements in this area to create an environment suitable for software work. This entails attention to such details as the size and shape of desks, the area and structure of the room, the siting of files and terminals, and the size and location of conference, programming, and computer rooms.

In addition to software work by individuals, teamwork is also important. Therefore, it will be important to study and optimize the makeup of development teams.

New Ideas Needed by Software Developers

NEC is moving to modernize its software production by introducing a large number of terminals in its software plants, with the aim of enhancing programming and the testing phase by utilizing interactive systems.

NEC has a substantial history, expertise, and methodological tools in hardware production but software production processes have been less visible and have not made as much progress. Although not all of the production management tools and approaches for hardware can be applied to software, those that can should be identified and used. A positive approach is important. New ideas from software developers will have the largest effect in modernizing software development and maintenance work so that software can become an industrial product.

The Importance of Recognizing Software as an Industrial Product with Commercial Value

Human factors will increase in the "Man and 'C&C'" society. Software is an extremely essential medium between humans and computers. In fact, software is accounting for a rapidly increasing proportion of systems.

Two aspects of software must be considered. One aspect is the rationalization of software production and the measures taken to ensure that there are enough personnel skilled in software as an industrial product. The second aspect is the economic value placed on software. This latter aspect has been taken up by computer manufacturers, but users in Japan have shown understanding of software's proper economic value. Thus there has been progress in this regard.

The proper economic evaluation of software is not a problem only of computer manufacturers; it is a problem for users, too, especially in view of the trend toward increasing the number of personnel employed for software work.

At present about 400,000 people are engaged in software work in Japan. This figure is predicted to increase to more than 1,000,000 in five years. The majority of these people will be employed by government offices, public bodies, universities, research institutes, and industry. Attaching economic value to software will lead to proper recognition of the worth of their work.

Higher Software Production Efficiency—Like Climbing Mt. Fuji

The production of software may be compared to climbing Mt. Fuji's beautiful volcanic cone, as visualized in Figure 12.7.12. Mountaineering, by definition, means relying on one's legs. No one needs to have a mountaineer's legs to reach Mt. Fuji's fifth station, which is halfway up the mountain, since the slope is gentle and easy to climb. It is even approachable by car. The real ascent of Mt. Fuji starts at the steep slope stretching from the fifth station to the summit. Here, we have to rely on the strength of our own legs, and the going gets very hard.

The ratio of the cubic volume of the mountain above and below the fifth station might be about 1:10. This applies to software, too. Software corresponding to the part of Mt. Fuji below the fifth station is overwhelmingly greater than that above it. Therefore, the ordinary methods of quality and production control that have so far been exercised over hardware can be applied to this level of software as well. Greater efforts should be made to develop a rational form of software production that does not rely entirely on individual ability. Software standardization would make it possible to have full-scale packaging and modularization for joint use. Computer-aided design (CAD) and computer-aided manufacturing (CAM) will be used in the future wherever possible, just as in other industries.

At any rate, software must become a complete industrial product. This will heighten the morale of software developers. Accordingly, both manufacturers and users must recognize the economic value of software. Manufacturers and users alike should continue their efforts to lower software costs.

12.7.7 SOFTWARE QUALITY CONTROL (SWQC) (FIG. 12.7.13)

The improvement of software quality will be the most important task of the 1980s, and NEC is engaged in software quality control. The easiest area to understand and implement is the reduction of "bugs." Nothing is gained by spending money making software containing bugs and then spending more money taking them out.

Many activities are involved in the overall life cycle of software: those of systems engineers and designers, who define users' needs and perform specification requirement analysis to achieve optimum systems; those of programmers; evaluation tests; and the various software-related activities involved in application. All these are targets of NEC's efforts to improve quality control.

The first goal is to obtain the active involvement and participation of the people working with software. Groups of between 4 and 10 people are organized depending on the work load. These groups seek out the immediate causes of bugs, but more importantly, they trace them back to their origins. Attempts are made to eliminate the sources of the bugs. The groups also seek to make suggestions for improvements—often covering aspects of the environment. NEC expects a quantum leap in software quality, and ultimately a great increase in software productivity, as a result of these quality control activities.

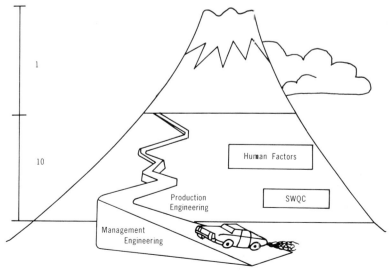

Fig. 12.7.12. Software production and climbing Mt. Fuji.

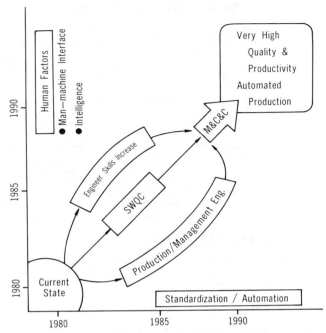

Fig. 12.7.13. The future of software production.

12.7.8 CONCLUSION

Computers and communications will be integrated into one system toward the end of this century in step with technological advances and in response to the needs of mankind. This chapter has expressed the process of computer and communications development in a plane, as shown in Figure 12.7.1, and forecasted the direction of this development and the problems involved. But this development is not simple, in fact, it is complicated by the human factors involved. Therefore, a Z axis for the human factor was added to the two-dimensional expression of this development through the X axis for communications and the Y axis for computers, thus creating a three-dimensional expression. In this way the essence of the problems can be grasped and clarified easily.

This idea was first described in my presidential address at the National Convention of the Information Processing Society of Japan (IPSJ) in May 1980, and later in my keynote speech at the International Conference of Computer Use in Radiography in Tokyo in October 1980. I have also discussed the concept in my opening remarks at the 8th World Computer Conference held in Tokyo by the International Federation for Information Processing (IFIP) in October 1980.

In the years ahead I would like to continue to work toward building a new society for the era beyond the 1980s—a society where all can realize a wider range of human activities, both as individuals and as members of organizations, nations, and the world. I will base my efforts on the concept of "Man and 'C&C' " as I have outlined it here, for I believe it will lead to greater happiness and prosperity for mankind.

However, I feel the world still lacks the mutual understanding among peoples of different nations that is necessary to fully realize this concept. We in Japan feel that linguistic differences are a major barrier, yet I believe that all people should not have to learn one common language. The languages of different peoples are something that must be respected.

In this regard, I believe that in parallel with understanding of "C&C," the development of automatic interpretation telephone systems will be one of the indicators for the realization of the "C&C" concept. Fortunately, NEC has sophisticated voice recognition and synthesis technologies that have been developed over a period of more than 20 years. We hope that by wedding these technologies to techniques for sentence analysis we will be able to achieve the dream of automatic interpretation. If we do, then I could be spoken to in English, it would reach me in Japanese, and my own thoughts would be transmitted to the other party translated into that person's language (see Figure 12.7.14).

For the past 20 years, I have witnessed how ideas for new technologies, be they pulse code modulation, geostationary satellites, or even optical fibers, were brought into practical use. Therefore, I am confident

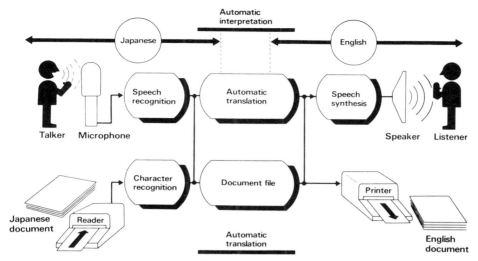

Fig. 12.7.14. An image of automatic interpretation telephone system.

that automatic interpretation systems will be realized before the year 2000. I have made it my life's goal to be able to confirm for myself, the coming of that day. I find encouragement in my efforts from the thought that this marvelous technology will be the greatest gift that "C&C" can give to mankind.

REFERENCES

Kobayashi, K. (1977). Shaping a communications industry to meet the ever-changing needs of society. Presented at Intelecom 77, Atlanta, Georgia.

Kobayashi, K. (1978). The Japanese computer industry, its roots and development. *Third U.S.A.-Japan computer conference.* San Francisco, CA.

Kobayashi, K. (1980). Computers, communications and man—the integration of computers and communications with man as the axis. Presented at The Information Processing Society of Japan.

Kobayashi, K. (1981). Computers, communications and man—the integration of computers and communications with man as an axis—the role of software. Presented at the 23rd IEEE Computer Society International Conference, Washington, DC.

Kobayashi, K. (1982). Man and "C&C" concept and perspectives. Presented at The International Institute of Communications Annual Conference, Helsinki, Finland.

Kobayashi, K. (1982). "Future role of 'C&C' in the home. Presented at The 1982 General Meeting of MIT Alumni Association.

AUTHOR INDEX

SUBJECT INDEX